BASIC RULES OF DIFFERENTIATION

1. $\dfrac{d}{dx}(c) = 0, \quad c$ a constant

2. $\dfrac{d}{dx}(u^n) = nu^{n-1}\dfrac{du}{dx}$

3. $\dfrac{d}{dx}(u \pm v) = \dfrac{du}{dx} \pm \dfrac{dv}{dx}$

4. $\dfrac{d}{dx}(cu) = c\dfrac{du}{dx}, \quad c$ a constant

D1312249

5. $\dfrac{d}{dx}(uv) = u\dfrac{dv}{dx} + v\dfrac{du}{dx}$

6. $\dfrac{d}{dx}\left(\dfrac{u}{v}\right) = \dfrac{v\dfrac{du}{dx} - u\dfrac{dv}{dx}}{v^2}$

7. $\dfrac{d}{dx}(e^u) = e^u\dfrac{du}{dx}$

8. $\dfrac{d}{dx}(\ln u) = \dfrac{1}{u}\cdot\dfrac{du}{dx}$

BASIC RULES OF INTEGRATION

1. $\displaystyle\int du = u + C$

2. $\displaystyle\int kf(u)\,du = k\int f(u)\,du, \quad k$ a constant

3. $\displaystyle\int [f(u) \pm g(u)]\,du = \int f(u)\,du \pm \int g(u)\,du$

4. $\displaystyle\int u^n\,du = \dfrac{u^{n+1}}{n+1} + C, \quad n \neq -1$

5. $\displaystyle\int e^u\,du = e^u + C$

6. $\displaystyle\int \dfrac{du}{u} = \ln|u| + C$

Resources on the Web

Students and instructors will now have access to these additional materials at the book companion Web site: www.cengage.com/math

- Review material and practice chapter quizzes and tests
- Group projects and extended problems for each chapter
- Instructions, including keystrokes, for the procedures referenced in the text for specific calculators (TI-82, TI-83, TI-85, TI-86, and other popular models)
- Coverage of additional topics such as Indeterminate Forms and L'Hopital's Rule

LIST OF APPLICATIONS

BUSINESS AND ECONOMICS

(continued)

List of Applications (*continued*)

List of Applications (*continued*)

SOCIAL SCIENCES

(*continued*)

List of Applications (*continued*)

List of Applications (*continued*)

College Mathematics

for the Managerial, Life, and Social Sciences

Seventh Edition

S. T. TAN

STONEHILL COLLEGE

BROOKS/COLE
CENGAGE Learning™

Australia • Brazil • Japan • Korea • Mexico • Singapore • Spain • United Kingdom • United States

BROOKS/COLE
CENGAGE Learning™

College Mathematics for the Managerial, Life, and Social Sciences, Seventh Edition
S. T. Tan

Acquisitions Editor: Carolyn Crocket

Development Editor: Danielle Derbenti

Assistant Editor: Beth Gershman

Editorial Assistant: Ashley Summers

Technology Project Manager: Donna Kelley

Marketing Manager: Joe Rogove

Marketing Assistant: Jennifer Liang

Marketing Communications Manager:
   Jessica Perry

Project Manager, Editorial Production:
   Janet Hill

Creative Director: Rob Hugel

Print Buyer: Becky Cross

Permissions Editor: Bob Kauser

Production Service:
   Newgen–Austin—Jamie Armstrong

Text Designer: Diane Beasley

Photo Researcher: Kathleen Olson

Cover Designer: Irene Morris

Cover Image: Portrait of Jonathan D. Farley
   by Peter Kiar

Compositor: Newgen

For product information and technology assistance, contact us at
Cengage Learning Customer & Sales Support, 1-800-354-9706
For permission to use material from this text or product,
submit all requests online at **www.cengage.com/permissions**
Further permissions questions can be e-mailed to
permissionrequest@cengage.com

Library of Congress Control Number: 2006935209

Student Edition:
ISBN-13: 978-0-495-01583-3
ISBN-10: 0-495-01583-0

Brooks/Cole Cengage Learning
20 Davis Drive
Belmont, CA 94002-3098
USA

Cengage Learning is a leading provider of customized learning solutions with office locations around the globe, including Singapore, the United Kingdom, Australia, Mexico, Brazil, and Japan. Locate your local office at **www.cengage.com/global**

Cengage Learning products are represented in Canada by Nelson Education, Ltd.

To learn more about Brooks/Cole, visit **www.cengage.com/brookscole**

Purchase any of our products at your local college store or at our preferred online store **www.cengagebrain.com**

Printed in Canada
2 3 4 5 14 13 12 11

TO PAT, BILL, AND MICHAEL

Contents

Note: Sections marked with an asterisk are not prerequisites for later material.

v

CHAPTER 3

Linear Programming: A Geometric Approach   167

CHAPTER 4

Linear Programming: An Algebraic Approach   215

CHAPTER 5

Mathematics of Finance   277

CHAPTER 13

Applications of the Derivative   771

CHAPTER 14

Exponential and Logarithmic Functions   855

Preface

Math is an integral part of our daily life. *College Mathematics for the Managerial, Life, and Social Sciences,* Seventh Edition, covers the standard topics in mathematics and calculus that are usually covered in a two-semester course for students in the managerial, life, and social sciences. The only prerequisite for understanding this book is a year of high school algebra. Our objective for this Seventh Edition is twofold: (1) to write an applied text that motivates students and (2) to make the book a useful tool for instructors. We hope that with this present edition we have come closer to realizing our goal.

General Approach

- **Coverage of Topics** Since the book contains more than enough material for the usual two-semester or three-quarter course, the instructor may be flexible in choosing the topics most suitable for his or her course. The following chart on chapter dependency is provided to help the instructor design a course that is most suitable for the intended audience.

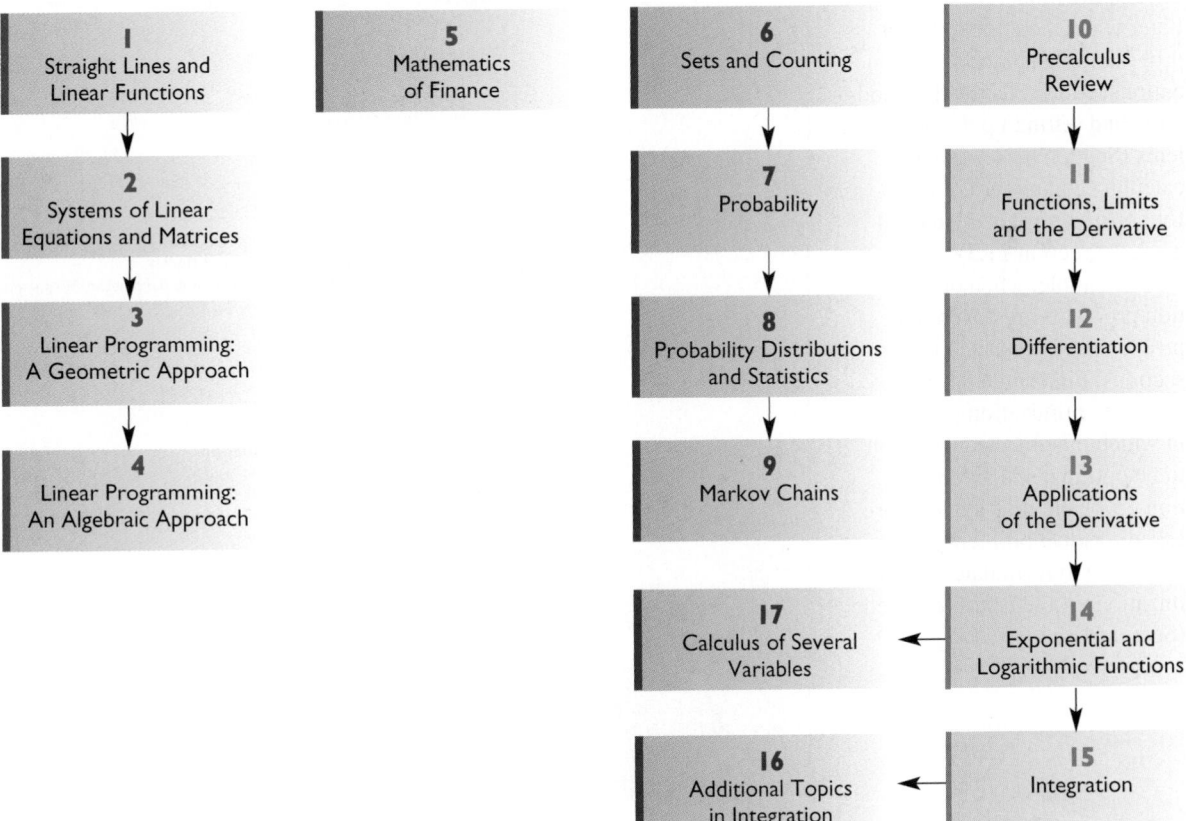

■ **Custom Publishing**   Due to the flexible nature of the topic coverage, instructors can easily design a custom text containing only those topics that are most suitable for their course. Please see your sales representative for more information on Brooks/Cole's custom publishing options.

■ **Level of Presentation**   Our approach is intuitive, and we state the results informally. However, we have taken special care to ensure that this approach does not compromise the mathematical content and accuracy.

■ **Approach**   A problem-solving approach is stressed throughout the book. Numerous examples and applications are used to illustrate each new concept and result in order to help the students comprehend the material presented. An emphasis is placed on helping the students formulate, solve, and interpret the results of the problems involving applications. Very early on in the text, students are given practice in setting up word problems (Section 1.3) and developing modeling skills. Later when the topic of linear programming is introduced, one entire section is devoted to modeling and setting up the problems (Section 3.2). Also, in calculus, guidelines are given for constructing mathematical models (Section 11.3). As another example, when optimization problems are covered the problems are presented in two sections. First students are asked to solve optimization problems in which the objective function to be optimized is given (Section 13.4) and then students are asked to solve problems where they have to formulate the optimization problems to be solved (Section 13.5).

■ **A Maximization Problem**

As an example of a linear programming problem in which the objective function is to be maximized, let's consider the following simplified version of a production problem involving two variables.

APPLIED EXAMPLE 1   A Production Problem   Ace Novelty wishes to produce two types of souvenirs: type A and type B. Each type-A souvenir will result in a profit of $1, and each type-B souvenir will result in a profit of $1.20. To manufacture a type-A souvenir requires 2 minutes on machine I and 1 minute on machine II. A type-B souvenir requires 1 minute on machine I and 3 minutes on machine II. There are 3 hours available on machine I and 5 hours available on machine II for processing the order. How many souvenirs of each type should Ace make in order to maximize its profit?

Solution   As a first step toward the mathematical formulation of this problem, we tabulate the given information, as shown in Table 1.

TABLE 1

	Type A	Type B	Time Available
Machine I	2 min	1 min	180 min
Machine II	1 min	3 min	300 min
Profit/Unit	$1	$1.20	

Let x be the number of type-A souvenirs and y be the number of type-B souvenirs to be made. Then, the total profit P (in dollars) is given by

$$P = x + 1.2y$$

Guidelines for Constructing Mathematical Models

1. Assign a letter to each variable mentioned in the problem. If appropriate, draw and label a figure.
2. Find an expression for the quantity sought.
3. Use the conditions given in the problem to write the quantity sought as a function f of one variable. Note any restrictions to be placed on the domain of f from physical considerations of the problem.

APPLIED EXAMPLE 4   Enclosing an Area   The owner of the Rancho Los Feliz has 3000 yards of fencing with which to enclose a rectangular piece of grazing land along the straight portion of a river. Fencing is not required along the river. Letting x denote the width of the rectangle, find a function f in the variable x giving the area of the grazing land if she uses all of the fencing (Figure 57).

Solution

1. This information was given.

2. The area of the rectangular grazing land is $A = xy$. Next, observe that the amount of fencing is $2x + y$ and this must be equal to 3000 since all the fencing is used; that is,

$$2x + y = 3000$$

3. From the equation we see that $y = 3000 - 2x$. Substituting this value of y into the expression for A gives

$$A = xy = x(3000 - 2x) = 3000x - 2x^2$$

Finally, observe that both x and y must be nonnegative since they represent the width and length of a rectangle, respectively. Thus, $x \geq 0$ and $y \geq 0$. But the latter is equivalent to $3000 - 2x \geq 0$, or $x \leq 1500$. So the required function is $f(x) = 3000x - 2x^2$ with domain $0 \leq x \leq 1500$. ■

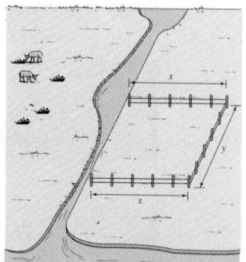

FIGURE 20
The rectangular grazing land has width x and length y.

■ **Intuitive Introduction to Concepts** Mathematical concepts are introduced with concrete real-life examples, wherever appropriate. Our goal here is to capture students' interest and show the relevance of mathematics to their everyday life. For example, curve-sketching (Section 13.3) is introduced in the manner shown here.

Consider, for example, the graph of the function giving the Dow-Jones Industrial Average (DJIA) on Black Monday, October 19, 1987 (Figure 47). Here, $t = 0$ corresponds to 8:30 a.m., when the market was open for business, and $t = 7.5$ corresponds to 4 p.m., the closing time. The following information may be gleaned from studying the graph.

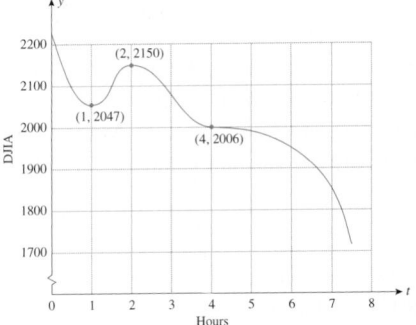

FIGURE 45
The Dow-Jones Industrial Average on Black Monday

Source: Wall Street Journal

The graph is *decreasing* rapidly from $t = 0$ to $t = 1$, reflecting the sharp drop in the index in the first hour of trading. The point $(1, 2047)$ is a *relative minimum* point of the function, and this turning point coincides with the start of an aborted recovery. The short-lived rally, represented by the portion of the graph that is *increasing* on the interval $(1, 2)$, quickly fizzled out at $t = 2$ (10:30 a.m.). The *relative maximum* point $(2, 2150)$ marks the highest point of the recovery. The function is decreasing in the rest of the interval. The point $(4, 2006)$ is an *inflection point* of the function; it shows that there was a temporary respite at $t = 4$ (12:30 p.m.). However, selling pressure continued unabated, and the DJIA continued to fall until the closing bell. Finally, the graph also shows that the index opened at the high of the day [$f(0) = 2247$ is the *absolute maximum* of the function] and closed at the low of the day [$f\left(\frac{15}{2}\right) = 1739$ is the *absolute minimum* of the function], a drop of 508 points!*

Motivation

Illustrating the practical value of mathematics in applied areas is an important objective of our approach. What follows are examples of how we have implemented this relevant approach throughout the text.

■ **Real-life Applications** Current and relevant examples and exercises are drawn from the fields of business, economics, social and behavioral sciences, life sciences, physical sciences, and other fields of interest. In the examples, these are highlighted with new icons that illustrate the various applications.

APPLIED EXAMPLE 4   Financing a Car   After making a down payment of $4000 for an automobile, Murphy paid $400 per month for 36 months with interest charged at 12% per year compounded monthly on the unpaid balance. What was the original cost of the car? What portion of Murphys total car payments went toward interest charges?

Solution   The loan taken up by Murphy is given by the present value of the annuity

$$P = \frac{400[1 - (1.01)^{-36}]}{0.01} = 400 a_{\overline{36}|0.01}$$

■ **Developing Modeling Skills** We believe that one of the most important skills a student can acquire is the ability to translate a real problem into a model that can provide insight into the problem. Many of the applications are based on mathematical models (functions) that the author has constructed using data drawn from various sources, including current newspapers, magazines, and data obtained through the Internet. Sources are given in the text for these applied problems. In Sections 1.3 and 11.3, the modeling process is discussed and students are asked to use models (functions) constructed from real-life data to answer questions about the Market for Cholesterol-Reducing Drugs, HMO Membership, and the Driving Costs for a Ford Taurus.

15. BLACKBERRY SUBSCRIBERS According to a study conducted in 2004, the number of subscribers of BlackBerry, the handheld e-mail devices manufactured by Research in Motion Ltd., is expected to be

$$N(t) = -0.0675t^4 + 0.5083t^3 - 0.893t^2 + 0.66t + 0.32$$
$$(0 \le t \le 4)$$

where $N(t)$ is measured in millions and t in years, with $t = 0$ corresponding to the beginning of 2002.

a. How many BlackBerry subscribers were there at the beginning of 2002?

b. What is the projected number of BlackBerry subscribers at the beginning of 2006?

Source: ThinkEquity Partners

■ **Connections** One example (the maglev example) is used as a common thread throughout the development of calculus—from limits through integration. The goal here is to show students the connection between the concepts presented—limits, continuity, rates of change, the derivative, the definite integral, and so on.

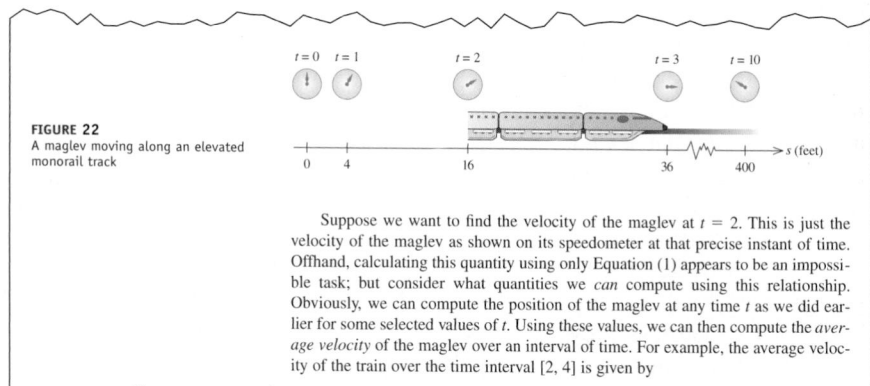

FIGURE 22
A maglev moving along an elevated monorail track

Suppose we want to find the velocity of the maglev at $t = 2$. This is just the velocity of the maglev as shown on its speedometer at that precise instant of time. Offhand, calculating this quantity using only Equation (1) appears to be an impossible task; but consider what quantities we *can* compute using this relationship. Obviously, we can compute the position of the maglev at any time t as we did earlier for some selected values of t. Using these values, we can then compute the *average velocity* of the maglev over an interval of time. For example, the average velocity of the train over the time interval [2, 4] is given by

Utilizing Tools Students Use

■ **Technology** Technology is used to explore mathematical ideas and as a tool to solve problems throughout the text.

■ **Exploring with Technology Questions** Here technology is used to explore mathematical concepts and to shed further light on examples in the text. These optional questions appear throughout the main body of the text and serve to enhance the student's understanding of the concepts and theory presented. Often the solution of an example in the text is augmented with a graphical or numerical solution. Complete solutions to these exercises are given in the *Instructor's Solution Manual*.

EXPLORING WITH TECHNOLOGY

Investments that are allowed to grow over time can increase in value surprisingly fast. Consider the potential growth of $10,000 if earnings are reinvested. More specifically, suppose $A_1(t), A_2(t), A_3(t), A_4(t)$, and $A_5(t)$ denote the accumulated values of an investment of $10,000 over a term of t years and earning interest at the rate of 4%, 6%, 8%, 10%, and 12% per year compounded annually.

1. Find expressions for $A_1(t), A_2(t), \ldots, A_5(t)$.

2. Use a graphing utility to plot the graphs of $A_1, A_2, \ldots, A_5$ on the same set of axes, using the viewing window [0, 20] × [0, 100,000].

3. Use TRACE to find $A_1(20), A_2(20), \ldots, A_5(20)$ and then interpret your results.

■ **Using Technology** These are optional subsections that appear after the exercises. They can be used in the classroom if desired or as material for self-study by the student. Here the graphing calculator and Excel spreadsheets are used as a tool to solve problems. The subsections are written in the traditional example-exercise format with answers given at the back of the book. Illustrations showing graphing calculator screens are extensively used. In keeping with the theme of motivation through real-life examples, many sourced applications are again included. Students can construct their own models using real-life data in Using Technology Section 11.3. These include models for the growth of the Indian gaming industry, the population growth in the fastest growing metropolitan area in the U.S., and the growth in online spending, among others. In Using Technology Section 13.3, students are asked to predict when the assets of the Social Security "trust fund" (unless changes are made) will be exhausted.

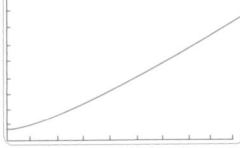

FIGURE T3
The graph of f in the viewing window $[0, 8] \times [0, 10]$

APPLIED EXAMPLE 3 Indian Gaming Industry The following data gives the estimated gross revenues (in billions of dollars) from the Indian gaming industries from 1990 ($t = 0$) to 1997 ($t = 7$).

Year	0	1	2	3	4	5	6	7
Revenue	0.5	0.7	1.6	2.6	3.4	4.8	5.6	6.8

a. Use a graphing utility to find a polynomial function f of degree 4 that models the data.
b. Plot the graph of the function f, using the viewing window $[0, 8] \times [0, 10]$.
c. Use the function evaluation capability of the graphing utility to compute $f(0)$, $f(1), \ldots, f(7)$ and compare these values with the original data.
Source: Christiansen/Cummings Associates

Solution

a. Choosing P4REG (fourth-order polynomial regression) from the STAT CALC (statistical calculations) menu of a graphing utility, we find

$$f(t) = 0.00379t^4 - 0.06616t^3 + 0.41667t^2 - 0.07291t + 0.48333$$

b. The graph of f is shown in Figure T3.
c. The required values, which compare favorably with the given data, follow:

t	0	1	2	3	4	5	6	7
$f(t)$	0.5	0.8	1.5	2.5	3.6	4.6	5.7	6.8

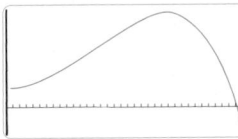

FIGURE T5
The graph of $f(t)$

APPLIED EXAMPLE 3 Solvency of Social Security Fund Unless payroll taxes are increased significantly and/or benefits are scaled back drastically, it is a matter of time before the current Social Security system goes broke. Data show that the assets of the system—the Social Security "trust fund"—may be approximated by

$$f(t) = -0.0129t^4 + 0.3087t^3 + 2.1760t^2 + 62.8466t + 506.2955 \quad (0 \le t \le 35)$$

where $f(t)$ is measured in millions of dollars and t is measured in years, with $t = 0$ corresponding to 1995.

a. Use a graphing calculator to sketch the graph of f.
b. Based on this model, when can the Social Security system be expected to go broke?
Source: Social Security Administration

Solution

a. The graph of f in the window $[0, 35] \times [-1000, 3500]$ is shown in Figure T5.
b. Using the function for finding the roots on a graphing utility, we find that $y = 0$ when $t \approx 34.1$, and this tells us that the system is expected to go broke around 2029.

Exercise Sets The exercise sets are designed to help students understand and apply the concepts developed in each section. Three types of exercises are included in these sets.

■ **Self-Check Exercises** offer students immediate feedback on key concepts with worked-out solutions following the section exercises.

■ New **Concept Questions** are designed to test students' understanding of the basic concepts discussed in the section and at the same time encourage students to explain these concepts in their own words.

■ **Exercises** provide an ample set of problems of a routine computational nature followed by an extensive set of application-oriented problems.

11.6 Self-Check Exercises

1. Let $f(x) = -x^2 - 2x + 3$.
   a. Find the derivative f' of f, using the definition of the derivative.
   b. Find the slope of the tangent line to the graph of f at the point $(0, 3)$.
   c. Find the rate of change of f when $x = 0$.
   d. Find an equation of the tangent line to the graph of f at the point $(0, 3)$.
   e. Sketch the graph of f and the tangent line to the curve at the point $(0, 3)$.

2. The losses (in millions of dollars) due to bad loans extended chiefly in agriculture, real estate, shipping, and energy by the Franklin Bank are estimated to be

$$A = f(t) = -t^2 + 10t + 30 \qquad (0 \le t \le 10)$$

   where t is the time in years ($t = 0$ corresponds to the beginning of 1994). How fast were the losses mounting at the beginning of 1997? At the beginning of 1999? At the beginning of 2001?

   Solutions to Self-Check Exercises 11.6 can be found on page 676.

11.6 Concept Questions

1. Let $P(2, f(2))$ and $Q(2 + h, f(2 + h))$ be points on the graph of a function f.
   a. Find an expression for the slope of the secant line passing through P and Q.
   b. Find an expression for the slope of the tangent line passing through P.

2. Refer to Question 1.
   a. Find an expression for the average rate of change of f over the interval $[2, 2 + h]$.
   b. Find an expression for the instantaneous rate of change of f at 2.

   c. Compare your answers for part (a) and (b) with those of Exercise 1.
3. a. Give a geometric and a physical interpretation of the expression
   $$\frac{f(x + h) - f(x)}{h}$$
   b. Give a geometric and a physical interpretation of the expression
   $$\lim_{h \to 0} \frac{f(x + h) - f(x)}{h}$$
4. Under what conditions does a function fail to have a derivative at a number? Illustrate your answer with sketches.

11.6 Exercises

1. AVERAGE WEIGHT OF AN INFANT The following graph shows the weight measurements of the average infant from the time of birth ($t = 0$) through age 2 ($t = 24$). By computing the slopes of the respective tangent lines, estimate the rate of change of the average infant's weight when $t = 3$ and when $t = 18$. What is the average rate of change in the average infant's weight over the first year of life?

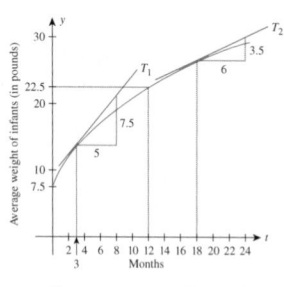

2. FORESTRY The following graph shows the volume of wood produced in a single-species forest. Here $f(t)$ is measured in cubic meters/hectare and t is measured in years. By computing the slopes of the respective tangent lines, estimate the rate at which the wood grown is changing at the beginning of year 10 and at the beginning of year 30.
   Source: The Random House Encyclopedia

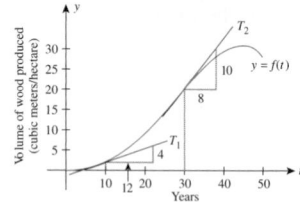

Review Sections   These sections are designed to help students review the material in each section and assess their understanding of basic concepts as well as problem-solving skills.

■ **Summary of Principal Formulas and Terms** highlights important equations and terms with page numbers given for quick review.

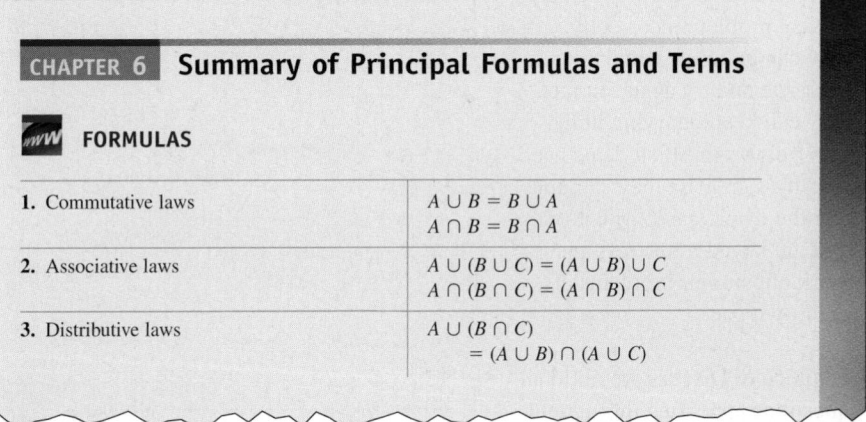

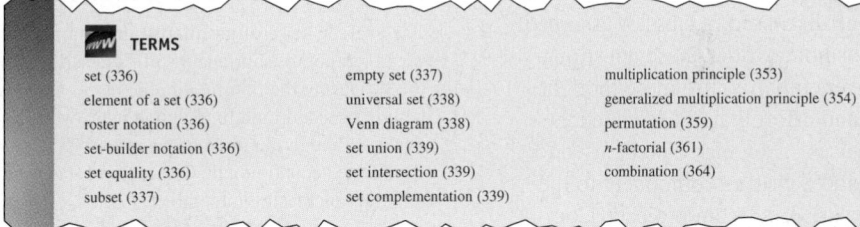

■ **New Concept Review Questions** give students a chance to check their knowledge of the basic definitions and concepts given in each chapter.

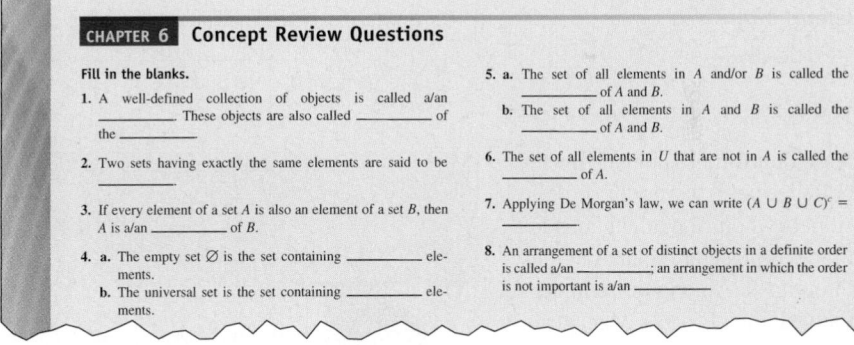

■ **Review Exercises** offer routine computational exercises followed by applied problems.

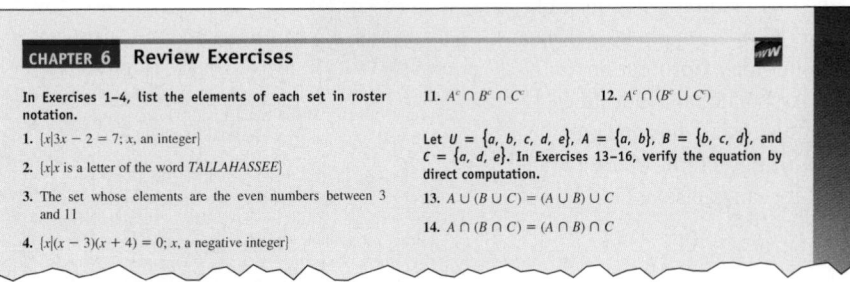

■ **New Before Moving On . . . Exercises** give students a chance to see if they have mastered the basic computational skills developed in each chapter. If they solve a problem incorrectly, they can go to the companion Web site and try again. In fact, they can keep on trying until they get it right. If students need step-by-step help, they can utilize the *CengageNOW* Tutorials that are keyed to the text and work out similar problems at their own pace.

CHAPTER 6   **Before Moving On . . .**

1. Let $U = \{a, b, c, d, e, f, g\}$, $A = \{a, d, f, g\}$, $B = \{d, f, g\}$, and $C = \{b, c, e, f\}$. Find
   a. $A \cap (B \cup C)$
   b. $(A \cap C) \cup (B \cup C)$
   c. A^c
2. Let A, B, and C be subsets of a universal set U and suppose $n(U) = 120$, $n(A) = 20$, $n(A \cap B) = 10$, $n(A \cap C) = 11$, $n(B \cap C) = 9$, and $n(A \cap B \cap C) = 4$. Find $n[A \cap (B \cup C)^c]$.

3. In how many ways can four compact discs be selected from six different compact discs?
4. From a standard 52-card deck, how many 5-card poker hands can be dealt consisting of 3 deuces and 2 face cards?
5. There are six seniors and five juniors in the Chess Club at Madison High School. In how many ways can a team consisting of three seniors and two juniors be selected from the members of the Chess Club?

■ **Explore & Discuss** are optional questions appearing throughout the main body of the text that can be discussed in class or assigned as homework. These questions generally require more thought and effort than the usual exercises. They may also be used to add a writing component to the class or as team projects. Complete solutions to these exercises are given in the *Instructor's Solutions Manual*.

EXPLORE & DISCUSS

The average price of gasoline at the pump over a 3-month period, during which there was a temporary shortage of oil, is described by the function f defined on the interval $[0, 3]$. During the first month, the price was increasing at an increasing rate. Starting with the second month, the good news was that the rate of increase was slowing down, although the price of gas was still increasing. This pattern continued until the end of the second month. The price of gas peaked at the end of $t = 2$ and began to fall at an increasing rate until $t = 3$.

1. Describe the signs of $f'(t)$ and $f''(t)$ over each of the intervals $(0, 1)$, $(1, 2)$, and $(2, 3)$.
2. Make a sketch showing a plausible graph of f over $[0, 3]$.

■ **New Portfolios** The real-life experiences of a variety of professionals who use mathematics in the workplace are related in these interviews. Among those interviewed are a Process Manager who uses differential equations and exponential functions in his work (Robert Derbenti at Linear Technology Corporation) and an Associate on Wall Street who uses statistics and calculus in writing options (Gary Li at JPMorgan Chase & Co.).

PORTFOLIO   Gary Li

TITLE   Associate
INSTITUTION   JPMorgan Chase

As one of the leading financial institutions in the world, JPMorgan Chase & Co. depends on a wide range of mathematical disciplines from statistics to linear programming to calculus. Whether assessing the credit worthiness of a borrower, recommending portfolio investments or pricing an exotic derivative, quantitative understanding is a critical tool in serving the financial needs of clients.

I work in the Fixed-Income Derivatives Strategy group. A derivative in finance is an instrument whose value depends on the price of some other underlying instrument. A simple type of derivative is the forward contract, where two parties agree to a future trade at a specified price. In agriculture, for instance, farmers will often pledge their crops for sale to buyers at an agreed price before even planting the harvest. Depending on the weather, demand and other factors, the actual price may turn out higher or lower. Either the buyer or seller of the

with interest rates. With trillions of dollars in this form, especially government bonds and mortgages, fixed-income derivatives are vital to the economy. As a strategy group, our job is to track and anticipate key drivers and developments in the market using, in significant part, quantitative analysis. Some of the derivatives we look at are of the forward kind, such as interest-rate swaps, where over time you receive fixed-rate payments in exchange for paying a floating-rate or vice-versa. A whole other class of derivatives where statistics and calculus are especially relevant are options.

Whereas forward contracts bind both parties to a future trade, options give the holder the right but not the obligation to trade at a specified time and price. Similar to an insurance policy, the holder of the option pays an upfront premium in exchange for potential gain. Solving this pricing problem requires statistics, stochastic calculus and enough insight to win a Nobel prize. Fortunately for us, this was taken care of by Fischer Black, Myron

■ **New Skillbuilder Videos**, available through *CengageNOW* and Enhanced WebAssign, offer hours of video instruction from award-winning teacher Deborah Upton of Stonehill College. Watch as she walks students through key examples from the text, step by step—giving them a foundation in the skills that they need to know. Each example available online is identified by the video icon located in the margin.

 APPLIED EXAMPLE 7   Oxygen-Restoration Rate in a Pond   When organic waste is dumped into a pond, the oxidation process that takes place reduces the pond's oxygen content. However, given time, nature will restore the oxygen content to its natural level. Suppose the oxygen content t days after organic waste has been dumped into the pond is given by

$$f(t) = 100 \left[\frac{t^2 + 10t + 100}{t^2 + 20t + 100} \right] \qquad (0 < t < \infty)$$

percent of its normal level.

Other Changes in the Seventh Edition

■ **New Applications**   More than 150 new real-life applications have been introduced. Among these applications are Sales of GPS Equipment, Broadband Internet Households, Switching Internet Service Providers, Digital vs. Film Cameras, Online Sales of Used Autos, Financing College Expenses, Balloon Payment Mortgages, Nurses Salaries, Revenue Growth of a Home Theater Business, Same-Sex Marriage, Rollover Deaths, Switching Jobs, Downloading Music, Americans without Health Insurance, Access to Capital, Volkswagen's Revenue, Cancer Survivors, Spam Messages, Global Supply of Plutonium, Testosterone Use, BlackBerry Subscribers, Outsourcing of Jobs, Spending on Medicare, Obesity in America, U.S. Nursing Shortage, Effects of Smoking Bans, Google's Revenue, Computer Security, Yahoo! in Europe, Satellite Radio Subscriptions, Gastric Bypass Surgeries, and the Surface Area of the New York Central Park Reservoir.

■ **Two New Sections on Linear Programming** Sensitivity Analysis is now covered in Section 3.4 and The Simplex Method: Nonstandard Problems is now covered in Section 4.3.

■ **Expanded Coverage of Markov Chains** Markov Chains is now covered in Chapter 9 in three sections—9.1 Markov Chains, 9.2 Regular Markov Chains, and 9.3 Absorbing Markov Chains.

■ **Using Technology** subsections have been updated for Office 2003 and new dialog boxes are now shown.

■ **A Revised and Expanded Student Solutions Manual**   Problem-solving strategies and additional algebra steps and review for selected problems have been added to this supplement.

■ **Other Changes**   In Functions and Mathematical Models (Section 11.3), a new model describing the membership of HMOs is now discussed by using a scatter plot of the real-life data and the graph of a function that describes that data. Another model describing the driving costs of a Ford Taurus is also presented in this same fashion. A discussion of the median and the mode has been added to Section 8.6.

In Section 13.2 an example calling for the interpretation of the first and second derivatives to help sketch the graph of a function has been added. In Section 15.4, the definite integral as a measure of net change is now discussed along with a new example giving the Population Growth in Clark County. Also, Section 16.2, Integration Using Tables of Integrals has been added.

Teaching Aids

- **Instructor's Solutions Manual** includes complete solutions for all exercises in the text, as well as answers to the *Exploring with Technology* and *Explore & Discuss* questions.
- **Instructor's Suite CD** includes the **Instructor's Solutions Manual** and **Test Bank** in formats compatible with Microsoft Office®.
- **Printed Test Bank,** which includes test questions (including multiple-choice) and sample tests for each chapter of the text, is available to adopters of the text.
- **Enhanced WebAssign** offers an easy way for instructors to deliver, collect, grade, and record assignments via the web. Within **WebAssign** you will find:
  - 1500 problems that match the text's end-of-section exercises.
  - Active examples integrated into each problem that allow students to work step-by-step at their own pace.
  - Links to Skillbuilder Videos that provide further instruction on each problem.
  - Portable PDFs of the textbook that match the assigned section.
- **ExamView® Computerized Testing** allows instructors to create, deliver, and customize tests and study guides (both print and online) in minutes with this easy-to-use assessment and tutorial system, which contains all questions from the **Test Bank** in electronic format.
- **JoinIn™ on Turning Point®** offers instructors text-specific JoinIn content for electronic response systems. Instructors can transform their classroom and assess students' progress with instant in-class quizzes and polls. Turning Point software lets instructors pose book-specific questions and display students' answers seamlessly within Microsoft PowerPoint lecture slides, in conjunction with a choice of "clicker" hardware. Enhance how your students interact with you, your lecture, and one another.

Learning Aids

- **Student Solutions Manual** contains complete solutions for all odd-numbered exercises in the text, plus problem-solving strategies and additional algebra steps and review for selected problems. 0-495-11974-1
- **CengageNOW for Tan's** *College Mathematics for the Managerial, Life, and Social Sciences,* **Seventh Edition**, designed for self-study, offers text-specific tutorials that require no setup or involvement by instructors. (If desired, instructors can assign the tutorials online and track students' progress via an instructor gradebook.) Students can explore active examples from the text as well as Skillbuilder Videos that provide additional reinforcement. Along the way, they can check their comprehension by taking quizzes and receiving immediate feedback.
- **vMentor™** allows students to talk (using their own computer microphones) to tutors who will skillfully guide them through a problem using an interactive whiteboard for illustration. Up to 40 hours of live tutoring a week is available and

can be accessed through www.cengage.com via the access card shrink-wrapped to every new book.

Acknowledgments

I wish to express my personal appreciation to each of the following reviewers, whose many suggestions have helped make a much improved book.

Faiz Al-Rubaee
University of North Florida

James V. Balch
Middle Tennessee State University

Albert Bronstein
Purdue University

Kimberly Jordan Burch
Montclair State University

Michael Button
San Diego City College

Peter Casazza
University of Missouri–Columbia

Matthew P. Coleman
Fairfield University

William Coppage
Wright State University

Lisa Cox
Texas A&M University

Frank Deutsch
Penn State University

Carl Droms
James Madison University

Bruce Edwards
University of Florida at Gainesville

Janice Epstein
Texas A&M University

Gary J. Etgen
University of Houston

Charles S. Frady
Georgia State University

Howard Frisinger
Colorado State University

Larry Gerstein
University of California at Santa Barbara

Matthew Gould
Vanderbilt University

Harvey Greenwald
California Polytechnic State University–San Luis Obispo

John Haverhals
Bradley University

Yvette Hester
Texas A&M University

Frank Jenkins
John Carroll University

David E. Joyce
Clark University

Mohammed Kazemi
University of North Carolina–Charlotte

James H. Liu
James Madison University

Norman R. Martin
Northern Arizona University

Sandra Wray McAfee
University of Michigan

Maurice Monahan
South Dakota State University

Dean Moore
Florida Community College at Jacksonville

Ralph J. Neuhaus
University of Idaho

Gertrude Okhuysen
Mississippi State University

James Olsen
North Dakota State University

Lloyd Olson
North Dakota State University

Wesley Orser
Clark College

Jane Smith
University of Florida

Richard Porter
Northeastern University

Devki Talwar
Indiana University of Pennsylvania

Virginia Puckett
Miami Dade College

Larry Taylor
North Dakota State University

Richard Quindley
Bridgewater State College

Giovanni Viglino
Ramapo College of New Jersey

Mary E. Rerick
University of North Dakota

Hiroko K. Warshauer
Texas State University–San Marcos

Thomas N. Roe
South Dakota State University

Lawrence V. Welch
Western Illinois University

Donald R. Sherbert
University of Illinois

Jennifer Whitfield
Texas A&M University

Anne Siswanto
East Los Angeles College

I also wish to thank my colleague, Deborah Upton, who did a great job preparing the videos that now accompany the text and who helped with the accuracy check of the text. I also wish to thank Kevin Charlwood and Tao Guo for their many helpful suggestions for improving the text. My thanks also go to the editorial, production, and marketing staffs of Brooks/Cole: Carolyn Crockett, Danielle Derbenti, Beth Gershman, Janet Hill, Joe Rogove, Becky Cross, Donna Kelley, Jennifer Liang, and Ashley Summers for all of their help and support during the development and production of this edition. I also thank Jamie Armstrong and the staff at Newgen for their production services. Finally, a special thanks to the mathematicians—Chris Shannon and Mark van der Lann at Berkeley, Peter Blair Henry at Stanford, Jonathan D. Farley at MIT, and Navin Khaneja at Harvard for taking time off from their busy schedules to describe how mathematics is used in their research. Their pictures appear on the covers of my applied mathematics series.

S. T. Tan

About the Author

SOO T. TAN received his S.B. degree from Massachusetts Institute of Technology, his M.S. degree from the University of Wisconsin–Madison, and his Ph.D. from the University of California at Los Angeles. He has published numerous papers in Optimal Control Theory, Numerical Analysis, and Mathematics of Finance. He is currently a Professor of Mathematics at Stonehill College.

"By the time I started writing the first of what turned out to be a series of textbooks in mathematics for students in the managerial, life, and social sciences, I had quite a few years of experience teaching mathematics to non-mathematics majors. One of the most important lessons I learned from my early experience teaching these courses is that many of the students come into these courses with some degree of apprehension. This awareness led to the intuitive approach I have adopted in all of my texts. As you will see, I try to introduce each abstract mathematical concept through an example drawn from a common, real-life experience. Once the idea has been conveyed, I then proceed to make it precise, thereby assuring that no mathematical rigor is lost in this intuitive treatment of the subject. Another lesson I learned from my students is that they have a much greater appreciation of the material if the applications are drawn from their fields of interest and from situations that occur in the real world. This is one reason you will see so many exercises in my texts that are modeled on data gathered from newspapers, magazines, journals, and other media. Whether it be the market for cholesterol-reducing drugs, financing a home, bidding for cable rights, broadband Internet households, or Starbucks' annual sales, I weave topics of current interest into my examples and exercises to keep the book relevant to all of my readers."

1 Straight Lines and Linear Functions

© Jim Arbogast/PhotoDisc

Which process should the company use? Robertson Controls Company must decide between two manufacturing processes for its Model C electronic thermostats. In Example 4, page 44, you will see how to determine which process will be more profitable.

THIS CHAPTER INTRODUCES the Cartesian coordinate system, a system that allows us to represent points in the plane in terms of ordered pairs of real numbers. This in turn enables us to compute the distance between two points algebraically. We also study straight lines. *Linear functions*, whose graphs are straight lines, can be used to describe many relationships between two quantities. These relationships can be found in fields of study as diverse as business, economics, the social sciences, physics, and medicine. In addition, we see how some practical problems can be solved by finding the point(s) of intersection of two straight lines. Finally, we learn how to find an algebraic representation of the straight line that "best" fits a set of data points that are scattered about a straight line.

1.1 The Cartesian Coordinate System

The Cartesian Coordinate System

The real number system is made up of the set of real numbers together with the usual operations of addition, subtraction, multiplication, and division. We assume that you are familiar with the rules governing these algebraic operations (see Appendix A).

Real numbers may be represented geometrically by points on a line. This line is called the **real number**, or **coordinate**, **line**. We can construct the real number line as follows: Arbitrarily select a point on a straight line to represent the number 0. This point is called the **origin**. If the line is horizontal, then choose a point at a convenient distance to the right of the origin to represent the number 1. This determines the scale for the number line. Each positive real number x lies x units to the right of 0, and each negative real number x lies $-x$ units to the left of 0.

In this manner, a one-to-one correspondence is set up between the set of real numbers and the set of points on the number line, with all the positive numbers lying to the right of the origin and all the negative numbers lying to the left of the origin (Figure 1).

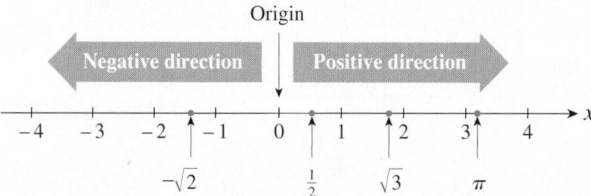

FIGURE 1
The real number line

In a similar manner, we can represent points in a plane (a two-dimensional space) by using the Cartesian coordinate system, which we construct as follows: Take two perpendicular lines, one of which is normally chosen to be horizontal. These lines intersect at a point O, called the **origin** (Figure 2). The horizontal line is called the **x-axis**, and the vertical line is called the **y-axis**. A number scale is set up along the x-axis, with the positive numbers lying to the right of the origin and the negative numbers lying to the left of it. Similarly, a number scale is set up along the y-axis, with the positive numbers lying above the origin and the negative numbers lying below it.

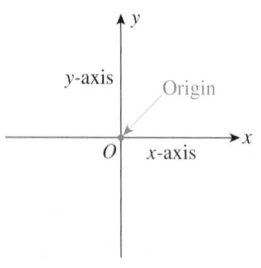

FIGURE 2
The Cartesian coordinate system

Note The number scales on the two axes need not be the same. Indeed, in many applications different quantities are represented by x and y. For example, x may represent the number of cell phones sold and y the total revenue resulting from the sales. In such cases it is often desirable to choose different number scales to represent the different quantities. Note, however, that the zeros of both number scales coincide at the origin of the two-dimensional coordinate system. ■

We can represent a point in the plane uniquely in this coordinate system by an ordered pair of numbers—that is, a pair (x, y), where x is the first number and y the second. To see this, let P be any point in the plane (Figure 3). Draw perpendiculars from P to the x-axis and y-axis, respectively. Then the number x is precisely the number that corresponds to the point on the x-axis at which the perpendicular through P hits the x-axis. Similarly, y is the number that corresponds to the point on the y-axis at which the perpendicular through P crosses the y-axis.

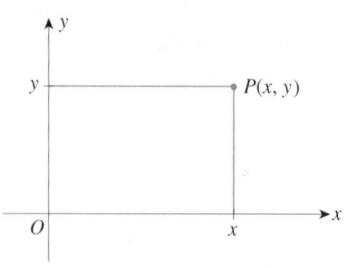

FIGURE 3
An ordered pair in the coordinate plane

Conversely, given an ordered pair (x, y) with x as the first number and y the second, a point P in the plane is uniquely determined as follows: Locate the point on the x-axis represented by the number x and draw a line through that point parallel to the y-axis. Next, locate the point on the y-axis represented by the number y and draw a line through that point parallel to the x-axis. The point of intersection of these two lines is the point P (Figure 3).

In the ordered pair (x, y), x is called the **abscissa**, or **x-coordinate**, y is called the **ordinate**, or **y-coordinate**, and x and y together are referred to as the **coordinates** of the point P. The point P with x-coordinate equal to a and y-coordinate equal to b is often written $P(a, b)$.

The points $A(2, 3)$, $B(-2, 3)$, $C(-2, -3)$, $D(2, -3)$, $E(3, 2)$, $F(4, 0)$, and $G(0, -5)$ are plotted in Figure 4.

Note In general, $(x, y) \neq (y, x)$. This is illustrated by the points A and E in Figure 4.

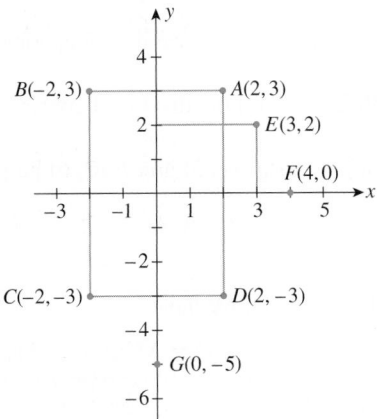

FIGURE 4
Several points in the coordinate plane

The axes divide the plane into four quadrants. Quadrant I consists of the points P with coordinates x and y, denoted by $P(x, y)$, satisfying $x > 0$ and $y > 0$; Quadrant II, the points $P(x, y)$ where $x < 0$ and $y > 0$; Quadrant III, the points $P(x, y)$ where $x < 0$ and $y < 0$; and Quadrant IV, the points $P(x, y)$ where $x > 0$ and $y < 0$ (Figure 5).

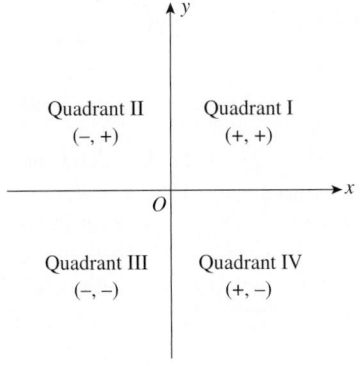

FIGURE 5
The four quadrants in the coordinate plane

FIGURE 6
The distance between two points in the coordinate plane

The Distance Formula

One immediate benefit that arises from using the Cartesian coordinate system is that the distance between any two points in the plane may be expressed solely in terms of the coordinates of the points. Suppose, for example, (x_1, y_1) and (x_2, y_2) are any two points in the plane (Figure 6). Then the distance d between these two points is, by the Pythagorean theorem,

$$d = \sqrt{(x_2 - x_1)^2 + (y_2 - y_1)^2}$$

For a proof of this result, see Exercise 45, page 9.

Distance Formula

The distance d between two points $P_1(x_1, y_1)$ and $P_2(x_2, y_2)$ in the plane is given by

$$d = \sqrt{(x_2 - x_1)^2 + (y_2 - y_1)^2} \tag{1}$$

In what follows, we give several applications of the distance formula.

EXPLORE & DISCUSS

Refer to Example 1. Suppose we label the point $(2, 6)$ as P_1 and the point $(-4, 3)$ as P_2.
(1) Show that the distance d between the two points is the same as that obtained earlier.
(2) Prove that, in general, the distance d in Formula (1) is independent of the way we label the two points.

EXAMPLE 1   Find the distance between the points $(-4, 3)$ and $(2, 6)$.

Solution   Let $P_1(-4, 3)$ and $P_2(2, 6)$ be points in the plane. Then we have

$$x_1 = -4 \quad \text{and} \quad y_1 = 3$$
$$x_2 = 2 \qquad\qquad y_2 = 6$$

Using Formula (1), we have

$$d = \sqrt{[2 - (-4)]^2 + (6 - 3)^2}$$
$$= \sqrt{6^2 + 3^2}$$
$$= \sqrt{45}$$
$$= 3\sqrt{5}$$

APPLIED EXAMPLE 2   The Cost of Laying Cable   In Figure 7, S represents the position of a power relay station located on a straight coastal highway, and M shows the location of a marine biology experimental station on a nearby island. A cable is to be laid connecting the relay station with the experimental station. If the cost of running the cable on land is $1.50 per running foot and the cost of running the cable underwater is $2.50 per running foot, find the total cost for laying the cable.

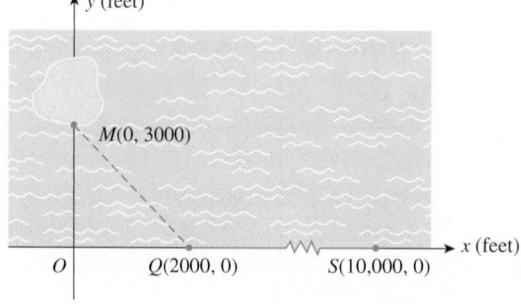

FIGURE 7
The cable will connect the relay station S to the experimental station M.

Solution The length of cable required on land is given by the distance from S to Q. This distance is $(10{,}000 - 2000)$, or 8000 feet. Next, we see that the length of cable required underwater is given by the distance from Q to M. This distance is

$$\sqrt{(0 - 2000)^2 + (3000 - 0)^2} = \sqrt{2000^2 + 3000^2}$$
$$= \sqrt{13{,}000{,}000}$$
$$\approx 3605.55$$

or approximately 3605.55 feet. Therefore, the total cost for laying the cable is

$$1.5(8000) + 2.5(3605.55) = 21{,}013.875$$

or approximately \$21,014. ∎

EXAMPLE 3 Let $P(x, y)$ denote a point lying on the circle with radius r and center $C(h, k)$ (Figure 8). Find a relationship between x and y.

Solution By the definition of a circle, the distance between $C(h, k)$ and $P(x, y)$ is r. Using Formula (1), we have

$$\sqrt{(x - h)^2 + (y - k)^2} = r$$

which, upon squaring both sides, gives an equation

$$(x - h)^2 + (y - k)^2 = r^2$$

that must be satisfied by the variables x and y. ∎

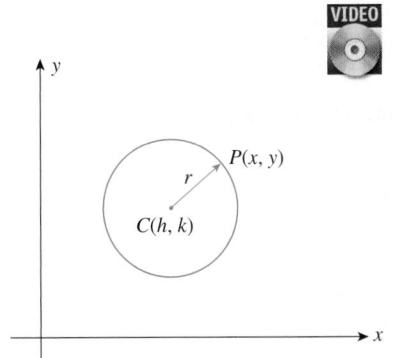

FIGURE 8
A circle with radius r and center $C(h, k)$

A summary of the result obtained in Example 3 follows.

Equation of a Circle
An equation of the circle with center $C(h, k)$ and radius r is given by

$$(x - h)^2 + (y - k)^2 = r^2 \tag{2}$$

EXAMPLE 4 Find an equation of the circle with (a) radius 2 and center $(-1, 3)$ and (b) radius 3 and center located at the origin.

Solution

a. We use Formula (2) with $r = 2$, $h = -1$, and $k = 3$, obtaining
$$[x - (-1)]^2 + (y - 3)^2 = 2^2$$
$$(x + 1)^2 + (y - 3)^2 = 4$$

(Figure 9a).

b. Using Formula (2) with $r = 3$ and $h = k = 0$, we obtain
$$x^2 + y^2 = 3^2$$
$$x^2 + y^2 = 9$$

(Figure 9b).

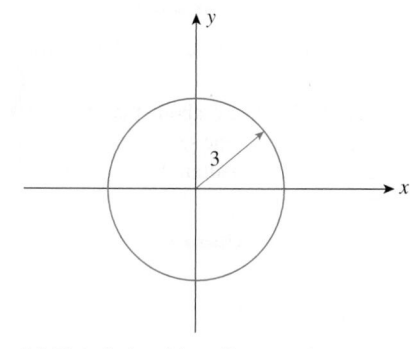

(a) The circle with radius 2 and center (−1, 3)

(b) The circle with radius 3 and center (0, 0)

FIGURE 9

EXPLORE & DISCUSS

1. Use the distance formula to help you describe the set of points in the xy-plane satisfying each of the following inequalities.

   a. $(x - h)^2 + (y - k)^2 \leq r^2$ **c.** $(x - h)^2 + (y - k)^2 \geq r^2$

   b. $(x - h)^2 + (y - k)^2 < r^2$ **d.** $(x - h)^2 + (y - k)^2 > r^2$

2. Consider the equation $x^2 + y^2 = 4$.

   a. Show that $y = \pm\sqrt{4 - x^2}$.

   b. Describe the set of points (x, y) in the xy-plane satisfying the equation

           (i) $y = \sqrt{4 - x^2}$ (ii) $y = -\sqrt{4 - x^2}$

1.1 Self-Check Exercises

1. a. Plot the points $A(4, -2)$, $B(2, 3)$, and $C(-3, 1)$.

    b. Find the distance between the points A and B, between B and C, and between A and C.

    c. Use the Pythagorean theorem to show that the triangle with vertices A, B, and C is a right triangle.

2. The accompanying figure shows the location of cities A, B, and C. Suppose a pilot wishes to fly from city A to city C but must make a mandatory stopover in city B. If the single-engine light plane has a range of 650 mi, can the pilot make the trip without refueling in city B?

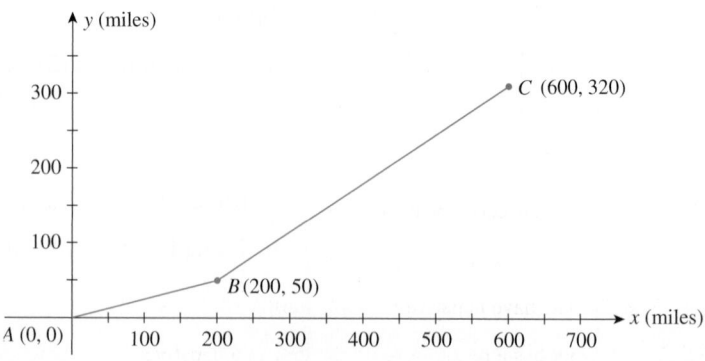

Solutions to Self-Check Exercises 1.1 can be found on page 9.

1.1 Concept Questions

1. What can you say about the signs of a and b if the point $P(a, b)$ lies in (a) the second quadrant? (b) The third quadrant? (c) The fourth quadrant?

2. **a.** What is the distance between $P_1(x_1, y_1)$ and $P_2(x_2, y_2)$?
   b. When you use the distance formula, does it matter which point is labeled P_1 and which point is labeled P_2? Explain.

1.1 Exercises

In Exercises 1–6, refer to the accompanying figure and determine the coordinates of the point and the quadrant in which it is located.

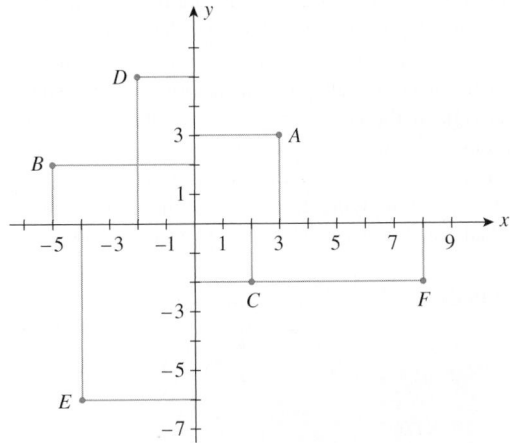

1. A 2. B 3. C
4. D 5. E 6. F

In Exercises 7–12, refer to the accompanying figure.

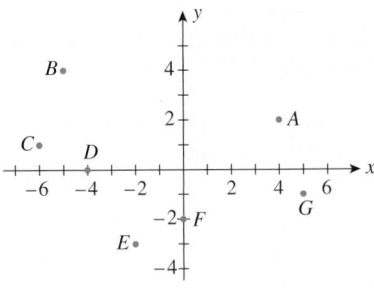

7. Which point has coordinates (4, 2)?

8. What are the coordinates of point B?

9. Which points have negative y-coordinates?

10. Which point has a negative x-coordinate and a negative y-coordinate?

11. Which point has an x-coordinate that is equal to zero?

12. Which point has a y-coordinate that is equal to zero?

In Exercises 13–20, sketch a set of coordinate axes and then plot the point.

13. $(-2, 5)$ 14. $(1, 3)$

15. $(3, -1)$ 16. $(3, -4)$

17. $(8, -7/2)$ 18. $(-5/2, 3/2)$

19. $(4.5, -4.5)$ 20. $(1.2, -3.4)$

In Exercises 21–24, find the distance between the points.

21. $(1, 3)$ and $(4, 7)$ 22. $(1, 0)$ and $(4, 4)$

23. $(-1, 3)$ and $(4, 9)$

24. $(-2, 1)$ and $(10, 6)$

25. Find the coordinates of the points that are 10 units away from the origin and have a y-coordinate equal to -6.

26. Find the coordinates of the points that are 5 units away from the origin and have an x-coordinate equal to 3.

27. Show that the points $(3, 4)$, $(-3, 7)$, $(-6, 1)$, and $(0, -2)$ form the vertices of a square.

28. Show that the triangle with vertices $(-5, 2)$, $(-2, 5)$, and $(5, -2)$ is a right triangle.

In Exercises 29–34, find an equation of the circle that satisfies the given conditions.

29. Radius 5 and center $(2, -3)$

30. Radius 3 and center $(-2, -4)$

31. Radius 5 and center at the origin

32. Center at the origin and passes through $(2, 3)$

33. Center $(2, -3)$ and passes through $(5, 2)$

34. Center $(-a, a)$ and radius $2a$

35. DISTANCE TRAVELED   A grand tour of four cities begins at city A and makes successive stops at cities B, C, and D before returning to city A. If the cities are located as shown in the accompanying figure, find the total distance covered on the tour.

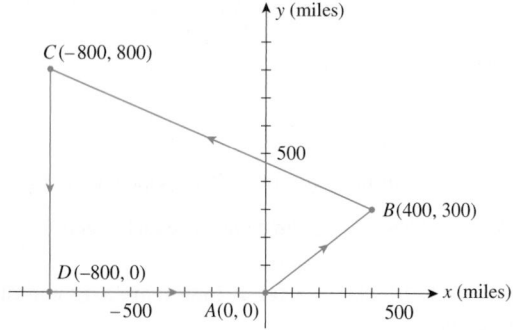

36. DELIVERY CHARGES   A furniture store offers free setup and delivery services to all points within a 25-mi radius of its warehouse distribution center. If you live 20 mi east and 14 mi south of the warehouse, will you incur a delivery charge? Justify your answer.

37. OPTIMIZING TRAVEL TIME   Towns A, B, C, and D are located as shown in the accompanying figure. Two highways link town A to town D. Route 1 runs from town A to town D via town B, and Route 2 runs from town A to town D via town C. If a salesman wishes to drive from town A to town D and traffic conditions are such that he could expect to average the same speed on either route, which highway should he take in order to arrive in the shortest time?

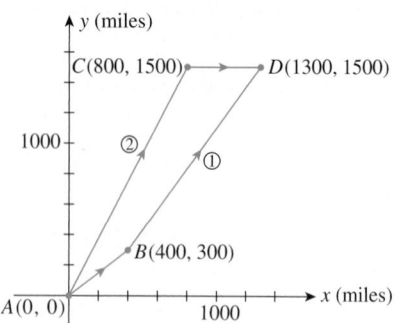

38. MINIMIZING SHIPPING COSTS   Refer to the figure for Exercise 37. Suppose a fleet of 100 automobiles are to be shipped from an assembly plant in town A to town D. They may be shipped either by freight train along Route 1 at a cost of 44¢/mile/automobile or by truck along Route 2 at a cost of 42¢/mile/automobile. Which means of transportation minimizes the shipping cost? What is the net savings?

39. CONSUMER DECISIONS   Will Barclay wishes to determine which antenna he should purchase for his home. The TV store has supplied him with the following information:

Range in miles			
VHF	UHF	Model	Price
30	20	A	$40
45	35	B	50
60	40	C	60
75	55	D	70

Will wishes to receive Channel 17 (VHF), which is located 25 mi east and 35 mi north of his home, and Channel 38 (UHF), which is located 20 mi south and 32 mi west of his home. Which model will allow him to receive both channels at the least cost? (Assume that the terrain between Will's home and both broadcasting stations is flat.)

40. COST OF LAYING CABLE   In the accompanying diagram, S represents the position of a power relay station located on a straight coastal highway, and M shows the location of a marine biology experimental station on a nearby island. A cable is to be laid connecting the relay station with the experimental station. If the cost of running the cable on land is $1.50/running foot and the cost of running cable underwater is $2.50/running foot, find an expression in terms of x that gives the total cost of laying the cable. Use this expression to find the total cost when $x = 1500$ and when $x = 2500$.

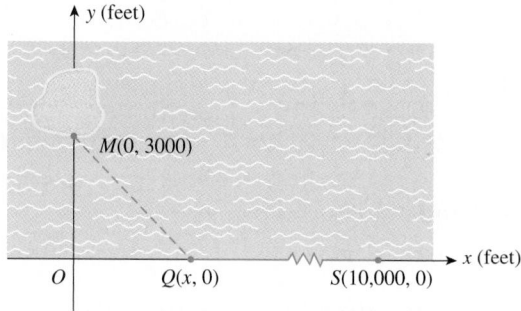

41. Two ships leave port at the same time. Ship A sails north at a speed of 20 mph while ship B sails east at a speed of 30 mph.
   a. Find an expression in terms of the time t (in hours) giving the distance between the two ships.
   b. Using the expression obtained in part (a), find the distance between the two ships 2 hr after leaving port.

42. Sailing north at a speed of 25 mph, ship A leaves a port. A half hour later, ship B leaves the same port, sailing east at a speed of 20 mph. Let t (in hours) denote the time ship B has been at sea.
   a. Find an expression in terms of t that gives the distance between the two ships.
   b. Use the expression obtained in part (a) to find the distance between the two ships 2 hr after ship A has left the port.

In Exercises 43 and 44, determine whether the statement is true or false. If it is true, explain why it is true. If it is false, give an example to show why it is false.

43. If the distance between the points $P_1(a, b)$ and $P_2(c, d)$ is D, then the distance between the points $P_1(a, b)$ and $P_3(kc, kd)$ ($k \neq 0$) is given by $|k|D$.

44. The circle with equation $kx^2 + ky^2 = a^2$ lies inside the circle with equation $x^2 + y^2 = a^2$, provided $k > 1$.

45. Let (x_1, y_1) and (x_2, y_2) be two points lying in the xy-plane. Show that the distance between the two points is given by

$$d = \sqrt{(x_2 - x_1)^2 + (y_2 - y_1)^2}$$

Hint: Refer to the accompanying figure and use the Pythagorean theorem.

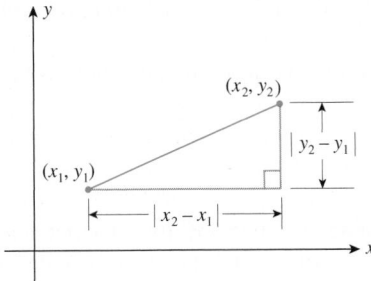

46. In the Cartesian coordinate system, the two axes are perpendicular to each other. Consider a coordinate system in which the x-axis and y-axis are noncollinear (that is, the axes do not lie along a straight line) and are not perpendicular to each other (see the accompanying figure).
  a. Describe how a point is represented in this coordinate system by an ordered pair (x, y) of real numbers. Conversely, show how an ordered pair (x, y) of real numbers uniquely determines a point in the plane.
  b. Suppose you want to find a formula for the distance between two points, $P_1(x_1, y_1)$ and $P_2(x_2, y_2)$, in the plane. What advantage does the Cartesian coordinate system have over the coordinate system under consideration? Comment on your answer.

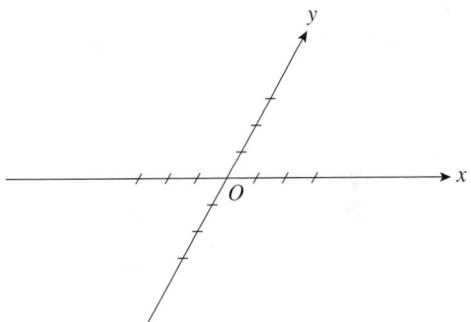

1.1 Solutions to Self-Check Exercises

1. a. The points are plotted in the accompanying figure.

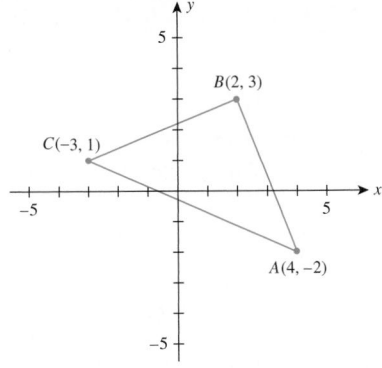

b. The distance between A and B is

$$d(A, B) = \sqrt{(2 - 4)^2 + [3 - (-2)]^2}$$
$$= \sqrt{(-2)^2 + 5^2} = \sqrt{4 + 25} = \sqrt{29}$$

The distance between B and C is

$$d(B, C) = \sqrt{(-3 - 2)^2 + (1 - 3)^2}$$
$$= \sqrt{(-5)^2 + (-2)^2} = \sqrt{25 + 4} = \sqrt{29}$$

The distance between A and C is

$$d(A, C) = \sqrt{(-3 - 4)^2 + [1 - (-2)]^2}$$
$$= \sqrt{(-7)^2 + 3^2} = \sqrt{49 + 9} = \sqrt{58}$$

c. We will show that

$$[d(A, C)]^2 = [d(A, B)]^2 + [d(B, C)]^2$$

From part (b), we see that $[d(A, B)]^2 = 29$, $[d(B, C)]^2 = 29$, and $[d(A, C)]^2 = 58$, and the desired result follows.

2. The distance between city A and city B is

$$d(A, B) = \sqrt{200^2 + 50^2} \approx 206$$

or 206 mi. The distance between city B and city C is

$$d(B, C) = \sqrt{(600 - 200)^2 + (320 - 50)^2}$$
$$= \sqrt{400^2 + 270^2} \approx 483$$

or 483 mi. Therefore, the total distance the pilot would have to cover is 689 mi, so she must refuel in city B.

1.2   Straight Lines

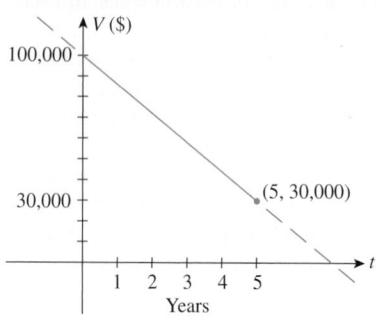

FIGURE 10
Linear depreciation of an asset

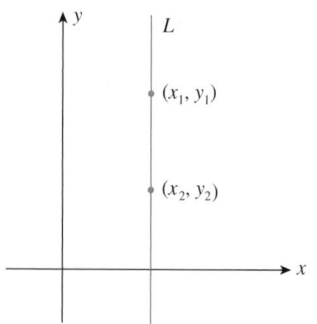

FIGURE 11
The slope is undefined.

In computing income tax, business firms are allowed by law to depreciate certain assets such as buildings, machines, furniture, automobiles, and so on over a period of time. *Linear depreciation*, or the *straight-line method*, is often used for this purpose. The graph of the straight line shown in Figure 10 describes the book value V of a computer that has an initial value of $100,000 and that is being depreciated linearly over 5 years with a scrap value of $30,000. Note that only the solid portion of the straight line is of interest here.

The book value of the computer at the end of year t, where t lies between 0 and 5, can be read directly from the graph. But there is one shortcoming in this approach: The result depends on how accurately you draw and read the graph. A better and more accurate method is based on finding an *algebraic* representation of the depreciation line. (We will continue our discussion of the linear depreciation problem in Section 1.3.)

To see how a straight line in the xy-plane may be described algebraically, we need to first recall certain properties of straight lines.

■ Slope of a Line

Let L denote the unique straight line that passes through the two distinct points (x_1, y_1) and (x_2, y_2). If $x_1 = x_2$, then L is a vertical line, and the slope is undefined (Figure 11). If $x_1 \neq x_2$, we define the slope of L as follows.

Slope of a Nonvertical Line
If (x_1, y_1) and (x_2, y_2) are any two distinct points on a nonvertical line L, then the slope m of L is given by

$$m = \frac{\Delta y}{\Delta x} = \frac{y_2 - y_1}{x_2 - x_1} \qquad (3)$$

(Figure 12).

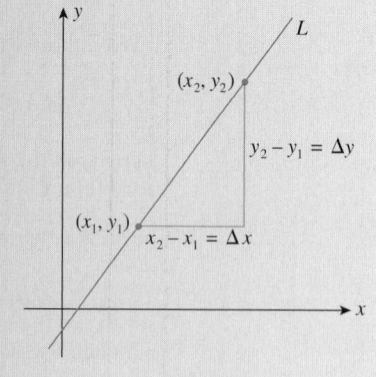

FIGURE 12

Thus, the slope of a straight line is a constant whenever it is defined.

The number $\Delta y = y_2 - y_1$ (Δy is read "delta y") is a measure of the vertical change in y, and $\Delta x = x_2 - x_1$ is a measure of the horizontal change in x as shown in Figure 12. From this figure we can see that the slope m of a straight line L is a

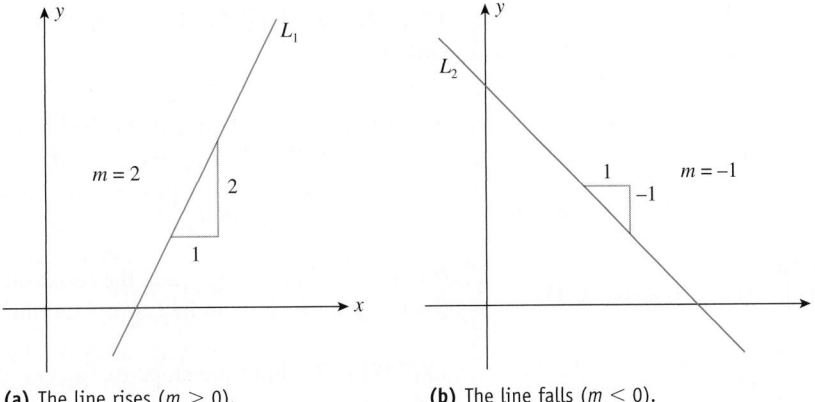

FIGURE 13

(a) The line rises ($m > 0$).

(b) The line falls ($m < 0$).

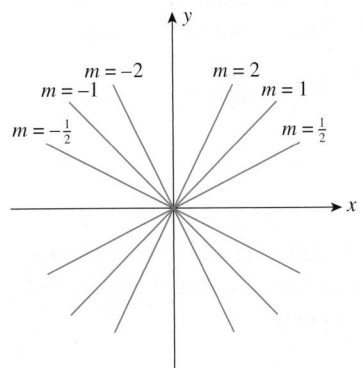

FIGURE 14
A family of straight lines

measure of the *rate of change of y with respect to x.* Furthermore, the slope of a nonvertical straight line is constant, and this tells us that this rate of change is constant.

Figure 13a shows a straight line L_1 with slope 2. Observe that L_1 has the property that a 1-unit increase in x results in a 2-unit increase in y. To see this, let $\Delta x = 1$ in Equation (3) so that $m = \Delta y$. Since $m = 2$, we conclude that $\Delta y = 2$. Similarly, Figure 13b shows a line L_2 with slope -1. Observe that a straight line with positive slope slants upward from left to right (y increases as x increases), whereas a line with negative slope slants downward from left to right (y decreases as x increases). Finally, Figure 14 shows a family of straight lines passing through the origin with indicated slopes.

EXPLORE & DISCUSS

Show that the slope of a nonvertical line is independent of the two distinct points used to compute it.

Hint: Suppose we pick two other distinct points, $P_3(x_3, y_3)$ and $P_4(x_4, y_4)$ lying on L. Draw a picture and use similar triangles to demonstrate that using P_3 and P_4 gives the same value as that obtained using P_1 and P_2.

EXAMPLE 1   Sketch the straight line that passes through the point $(-2, 5)$ and has slope $-\frac{4}{3}$.

Solution   First, plot the point $(-2, 5)$ (Figure 15). Next, recall that a slope of $-\frac{4}{3}$ indicates that an increase of 1 unit in the x-direction produces a *decrease* of $\frac{4}{3}$ units in the y-direction, or equivalently, a 3-unit increase in the x-direction produces a $3(\frac{4}{3})$, or 4-unit, decrease in the y-direction. Using this information, we plot the point $(1, 1)$ and draw the line through the two points.

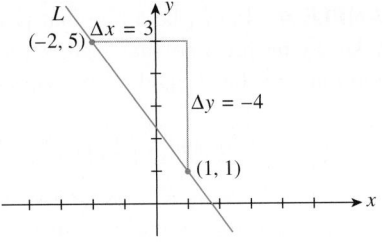

FIGURE 15
L has slope $-\frac{4}{3}$ and passes through $(-2, 5)$.

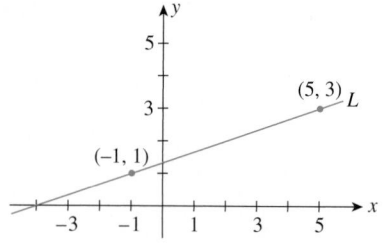

FIGURE 16
L passes through (5, 3) and (−1, 1).

EXAMPLE 2   Find the slope m of the line that passes through the points $(-1, 1)$ and $(5, 3)$.

Solution   Choose (x_1, y_1) to be the point $(-1, 1)$ and (x_2, y_2) to be the point $(5, 3)$. Then, with $x_1 = -1$, $y_1 = 1$, $x_2 = 5$, and $y_2 = 3$, we find, using Equation (3),

$$m = \frac{y_2 - y_1}{x_2 - x_1} = \frac{3 - 1}{5 - (-1)} = \frac{2}{6} = \frac{1}{3}$$

(Figure 16). Try to verify that the result obtained would be the same had we chosen the point $(-1, 1)$ to be (x_2, y_2) and the point $(5, 3)$ to be (x_1, y_1). ∎

EXAMPLE 3   Find the slope of the line that passes through the points $(-2, 5)$ and $(3, 5)$.

Solution   The slope of the required line is given by

$$m = \frac{5 - 5}{3 - (-2)} = \frac{0}{5} = 0$$

(Figure 17).

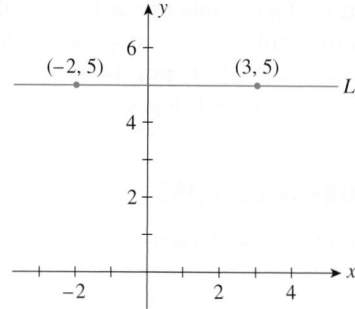

FIGURE 17
The slope of the horizontal line L is zero.

Note   In general, the slope of a horizontal line is zero. ∎

We can use the slope of a straight line to determine whether a line is parallel to another line.

> **Parallel Lines**
> Two distinct lines are **parallel** if and only if their slopes are equal or their slopes are undefined.

EXAMPLE 4   Let L_1 be a line that passes through the points $(-2, 9)$ and $(1, 3)$ and let L_2 be the line that passes through the points $(-4, 10)$ and $(3, -4)$. Determine whether L_1 and L_2 are parallel.

Solution   The slope m_1 of L_1 is given by

$$m_1 = \frac{3 - 9}{1 - (-2)} = -2$$

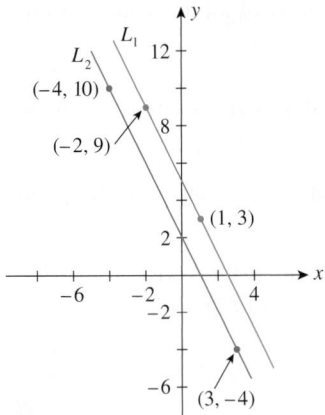

FIGURE 18
L_1 and L_2 have the same slope and hence are parallel.

The slope m_2 of L_2 is given by

$$m_2 = \frac{-4 - 10}{3 - (-4)} = -2$$

Since $m_1 = m_2$, the lines L_1 and L_2 are in fact parallel (Figure 18).

■ Equations of Lines

We now show that every straight line lying in the xy-plane may be represented by an equation involving the variables x and y. One immediate benefit of this is that problems involving straight lines may be solved algebraically.

Let L be a straight line parallel to the y-axis (perpendicular to the x-axis) (Figure 19). Then L crosses the x-axis at some point $(a, 0)$ with the x-coordinate given by $x = a$, where a is some real number. Any other point on L has the form $(a, \bar{y})$, where $\bar{y}$ is an appropriate number. Therefore, the vertical line L is described by the sole condition

$$x = a$$

and this is accordingly an equation of L. For example, the equation $x = -2$ represents a vertical line 2 units to the left of the y-axis, and the equation $x = 3$ represents a vertical line 3 units to the right of the y-axis (Figure 20).

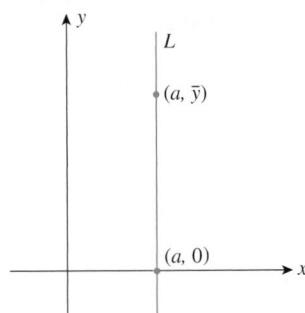

FIGURE 19
The vertical line $x = a$

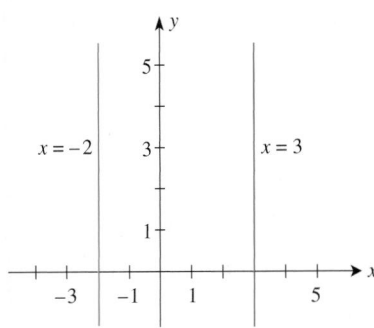

FIGURE 20
The vertical lines $x = -2$ and $x = 3$

Next, suppose L is a nonvertical line so that it has a well-defined slope m. Suppose (x_1, y_1) is a fixed point lying on L and (x, y) is a variable point on L distinct from (x_1, y_1) (Figure 21). Using Equation (3) with the point $(x_2, y_2) = (x, y)$, we find that the slope of L is given by

$$m = \frac{y - y_1}{x - x_1}$$

Upon multiplying both sides of the equation by $x - x_1$, we obtain Equation (4).

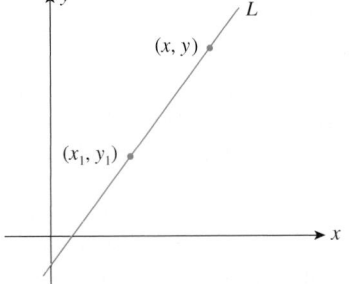

FIGURE 21
L passes through (x_1, y_1) and has slope m.

Point-Slope Form
An equation of the line that has slope m and passes through the point (x_1, y_1) is given by

$$y - y_1 = m(x - x_1) \tag{4}$$

Equation (4) is called the *point-slope form* of the equation of a line because it utilizes a given point (x_1, y_1) on a line and the slope m of the line.

EXAMPLE 5 Find an equation of the line that passes through the point $(1, 3)$ and has slope 2.

Solution Using the point-slope form of the equation of a line with the point $(1, 3)$ and $m = 2$, we obtain

$$y - 3 = 2(x - 1) \qquad y - y_1 = m(x - x_1)$$

which, when simplified, becomes

$$2x - y + 1 = 0$$

(Figure 22).

EXAMPLE 6 Find an equation of the line that passes through the points $(-3, 2)$ and $(4, -1)$.

Solution The slope of the line is given by

$$m = \frac{-1 - 2}{4 - (-3)} = -\frac{3}{7}$$

Using the point-slope form of the equation of a line with the point $(4, -1)$ and the slope $m = -\frac{3}{7}$, we have

$$y + 1 = -\frac{3}{7}(x - 4) \qquad y - y_1 = m(x - x_1)$$
$$7y + 7 = -3x + 12$$
$$3x + 7y - 5 = 0$$

(Figure 23).

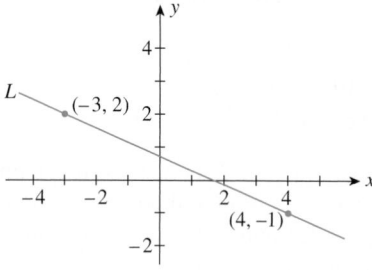

FIGURE 23
L passes through $(-3, 2)$ and $(4, -1)$.

We can use the slope of a straight line to determine whether a line is perpendicular to another line.

Perpendicular Lines

If L_1 and L_2 are two distinct nonvertical lines that have slopes m_1 and m_2, respectively, then L_1 is **perpendicular** to L_2 (written $L_1 \perp L_2$) if and only if

$$m_1 = -\frac{1}{m_2}$$

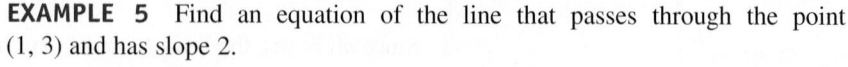

FIGURE 22
L passes through $(1, 3)$ and has slope 2.

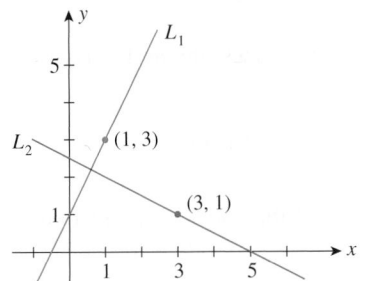

FIGURE 24
L_2 is perpendicular to L_1 and passes through (3, 1).

If the line L_1 is vertical (so that its slope is undefined), then L_1 is perpendicular to another line, L_2, if and only if L_2 is horizontal (so that its slope is zero). For a proof of these results, see Exercise 90, page 23.

EXAMPLE 7   Find an equation of the line that passes through the point (3, 1) and is perpendicular to the line of Example 5.

Solution   Since the slope of the line in Example 5 is 2, it follows that the slope of the required line is given by $m = -\frac{1}{2}$, the negative reciprocal of 2. Using the point-slope form of the equation of a line, we obtain

$$y - 1 = -\frac{1}{2}(x - 3) \qquad {\scriptstyle y - y_1 = m(x - x_1)}$$

$$2y - 2 = -x + 3$$

$$x + 2y - 5 = 0$$

(Figure 24). ∎

EXPLORING WITH TECHNOLOGY

1. Use a graphing utility to plot the straight lines L_1 and L_2 with equations $2x + y - 5 = 0$ and $41x + 20y - 11 = 0$ on the same set of axes, using the standard viewing window.
   a. Can you tell if the lines L_1 and L_2 are parallel to each other?
   b. Verify your observations by computing the slopes of L_1 and L_2 algebraically.

2. Use a graphing utility to plot the straight lines L_1 and L_2 with equations $x + 2y - 5 = 0$ and $5x - y + 5 = 0$ on the same set of axes, using the standard viewing window.
   a. Can you tell if the lines L_1 and L_2 are perpendicular to each other?
   b. Verify your observation by computing the slopes of L_1 and L_2 algebraically.

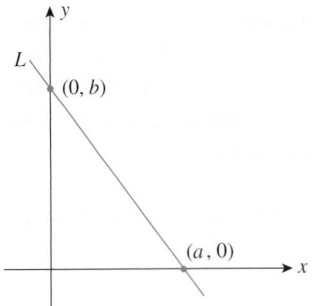

FIGURE 25
The line L has x-intercept a and y-intercept b.

A straight line L that is neither horizontal nor vertical cuts the x-axis and the y-axis at, say, points $(a, 0)$ and $(0, b)$, respectively (Figure 25). The numbers a and b are called the **x-intercept** and **y-intercept**, respectively, of L.

Now, let L be a line with slope m and y-intercept b. Using Equation (4), the point-slope form of the equation of a line, with the point given by $(0, b)$ and slope m, we have

$$y - b = m(x - 0)$$

$$y = mx + b$$

This is called the *slope-intercept form* of an equation of a line.

Slope-Intercept Form
The equation of the line that has slope m and intersects the y-axis at the point $(0, b)$ is given by

$$y = mx + b \qquad (5)$$

EXAMPLE 8   Find an equation of the line that has slope 3 and y-intercept -4.

Solution   Using Equation (5) with $m = 3$ and $b = -4$, we obtain the required equation:

$$y = 3x - 4$$

EXAMPLE 9   Determine the slope and y-intercept of the line whose equation is $3x - 4y = 8$.

Solution   Rewrite the given equation in the slope-intercept form and obtain

$$y = \frac{3}{4}x - 2$$

Comparing this result with Equation (5), we find $m = \frac{3}{4}$ and $b = -2$, and we conclude that the slope and y-intercept of the given line are $\frac{3}{4}$ and -2, respectively.

EXPLORING WITH TECHNOLOGY

1. Use a graphing utility to plot the straight lines with equations $y = -2x + 3$, $y = -x + 3$, $y = x + 3$, and $y = 2.5x + 3$ on the same set of axes, using the standard viewing window. What effect does changing the coefficient m of x in the equation $y = mx + b$ have on its graph?

2. Use a graphing utility to plot the straight lines with equations $y = 2x - 2$, $y = 2x - 1$, $y = 2x$, $y = 2x + 1$, and $y = 2x + 4$ on the same set of axes, using the standard viewing window. What effect does changing the constant b in the equation $y = mx + b$ have on its graph?

3. Describe in words the effect of changing both m and b in the equation $y = mx + b$.

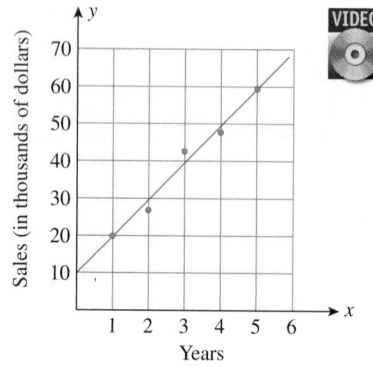

FIGURE 26
Sales of a sporting goods store

APPLIED EXAMPLE 10   Predicting Sales Figures   The sales manager of a local sporting goods store plotted sales versus time for the last 5 years and found the points to lie approximately along a straight line (Figure 26). By using the points corresponding to the first and fifth years, find an equation of the *trend line*. What sales figure can be predicted for the sixth year?

Solution   Using Equation (3) with the points $(1, 20)$ and $(5, 60)$, we find that the slope of the required line is given by

$$m = \frac{60 - 20}{5 - 1} = 10$$

Next, using the point-slope form of the equation of a line with the point $(1, 20)$ and $m = 10$, we obtain

$$y - 20 = 10(x - 1)$$
$$y = 10x + 10$$

as the required equation.

The sales figure for the sixth year is obtained by letting $x = 6$ in the last equation, giving

$$y = 10(6) + 10 = 70$$

or $70,000.   ∎

APPLIED EXAMPLE 11   Predicting the Value of Art   Suppose an art object purchased for $50,000 is expected to appreciate in value at a constant rate of $5000 per year for the next 5 years. Use Equation (5) to write an equation predicting the value of the art object in the next several years. What will be its value 3 years from the purchase date?

Solution   Let x denote the time (in years) that has elapsed since the purchase date and let y denote the object's value (in dollars). Then, $y = 50,000$ when $x = 0$. Furthermore, the slope of the required equation is given by $m = 5000$ since each unit increase in x (1 year) implies an increase of 5000 units (dollars) in y. Using Equation (5) with $m = 5000$ and $b = 50,000$, we obtain

$$y = 5000x + 50,000$$

Three years from the purchase date, the value of the object will be given by

$$y = 5000(3) + 50,000$$

or $65,000.   ∎

EXPLORE & DISCUSS

Refer to Example 11. Can the equation predicting the value of the art object be used to predict long-term growth?

General Form of an Equation of a Line

We have considered several forms of the equation of a straight line in the plane. These different forms of the equation are equivalent to each other. In fact, each is a special case of the following equation.

> **General Form of a Linear Equation**
> The equation
>
> $$Ax + By + C = 0 \tag{6}$$
>
> where A, B, and C are constants and A and B are not both zero, is called the general form of a linear equation in the variables x and y.

We now state (without proof) an important result concerning the algebraic representation of straight lines in the plane.

> An equation of a straight line is a linear equation; conversely, every linear equation represents a straight line.

This result justifies the use of the adjective *linear* in describing Equation (6).

EXAMPLE 12   Sketch the straight line represented by the equation

$$3x - 4y - 12 = 0$$

Solution Since every straight line is uniquely determined by two distinct points, we need find only two such points through which the line passes in order to sketch it. For convenience, let's compute the points at which the line crosses the x- and y-axes. Setting $y = 0$, we find $x = 4$, so the line crosses the x-axis at the point $(4, 0)$. Setting $x = 0$ gives $y = -3$, so the line crosses the y-axis at the point $(0, -3)$. A sketch of the line appears in Figure 27.

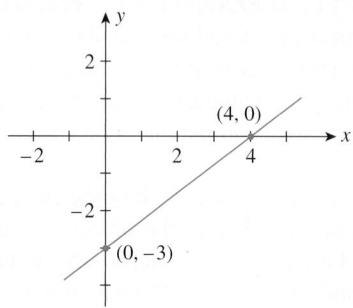

FIGURE 27
The straight line $3x - 4y = 12$

Here is a summary of the common forms of the equations of straight lines discussed in this section.

Equations of Straight Lines

Vertical line: $x = a$

Horizontal line: $y = b$

Point-slope form: $y - y_1 = m(x - x_1)$

Slope-intercept form: $y = mx + b$

General form: $Ax + By + C = 0$

1.2 Self-Check Exercises

1. Determine the number a such that the line passing through the points $(a, 2)$ and $(3, 6)$ is parallel to a line with slope 4.

2. Find an equation of the line that passes through the point $(3, -1)$ and is perpendicular to a line with slope $-\frac{1}{2}$.

3. Does the point $(3, -3)$ lie on the line with equation $2x - 3y - 12 = 0$? Sketch the graph of the line.

4. The percentage of people over age 65 who have high-school diplomas is summarized in the following table:

Year, x	1960	1965	1970	1975	1980	1985	1990
Percent with Diplomas, y	20	25	30	36	42	47	52

Source: U.S. Department of Commerce

a. Plot the percentage of people over age 65 who have high-school diplomas (y) versus the year (x).

b. Draw the straight line L through the points (1960, 20) and (1990, 52).

c. Find an equation of the line L.

d. Assume the trend continued and estimate the percentage of people over age 65 who had high-school diplomas by the year 1995.

Solutions to Self-Check Exercises 1.2 can be found on page 23.

1.2 Concept Questions

1. What is the slope of a nonvertical line? What can you say about the slope of a vertical line?

2. Give (a) the point-slope form, (b) the slope-intercept form, and (c) the general form of an equation of a line.

3. Let L_1 have slope m_1 and let L_2 have slope m_2. State the conditions on m_1 and m_2 if (a) L_1 is parallel to L_2 and (b) L_1 is perpendicular to L_2.

1.2 Exercises

In Exercises 1–4, find the slope of the line shown in each figure.

1.

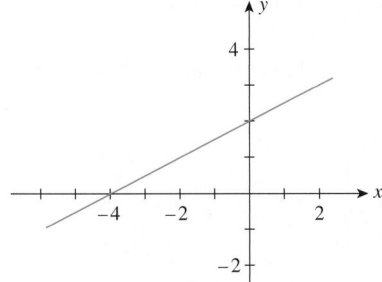

2.

3.

4.

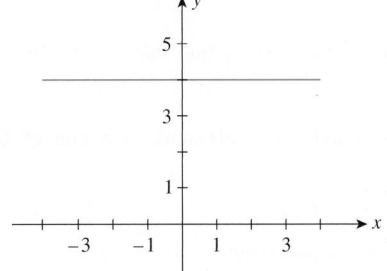

In Exercises 5–10, find the slope of the line that passes through the given pair of points.

5. (4, 3) and (5, 8)

6. (4, 5) and (3, 8)

7. (−2, 3) and (4, 8)

8. (−2, −2) and (4, −4)

9. (a, b) and (c, d)

10. (−a + 1, b − 1) and (a + 1, −b)

11. Given the equation $y = 4x − 3$, answer the following questions.
   a. If x increases by 1 unit, what is the corresponding change in y?
   b. If x decreases by 2 units, what is the corresponding change in y?

12. Given the equation $2x + 3y = 4$, answer the following questions.
   a. Is the slope of the line described by this equation positive or negative?
   b. As x increases in value, does y increase or decrease?
   c. If x decreases by 2 units, what is the corresponding change in y?

In Exercises 13 and 14, determine whether the lines through the pairs of points are parallel.

13. $A(1, −2)$, $B(−3, −10)$ and $C(1, 5)$, $D(−1, 1)$

14. $A(2, 3)$, $B(2, −2)$ and $C(−2, 4)$, $D(−2, 5)$

In Exercises 15 and 16, determine whether the lines through the pairs of points are perpendicular.

15. $A(-2, 5), B(4, 2)$ and $C(-1, -2), D(3, 6)$

16. $A(2, 0), B(1, -2)$ and $C(4, 2), D(-8, 4)$

17. If the line passing through the points $(1, a)$ and $(4, -2)$ is parallel to the line passing through the points $(2, 8)$ and $(-7, a + 4)$, what is the value of a?

18. If the line passing through the points $(a, 1)$ and $(5, 8)$ is parallel to the line passing through the points $(4, 9)$ and $(a + 2, 1)$, what is the value of a?

19. Find an equation of the horizontal line that passes through $(-4, -3)$.

20. Find an equation of the vertical line that passes through $(0, 5)$.

In Exercises 21–26, match the statement with one of the graphs a–f.

21. The slope of the line is zero.

22. The slope of the line is undefined.

23. The slope of the line is positive, and its y-intercept is positive.

24. The slope of the line is positive, and its y-intercept is negative.

25. The slope of the line is negative, and its x-intercept is negative.

26. The slope of the line is negative, and its x-intercept is positive.

a.

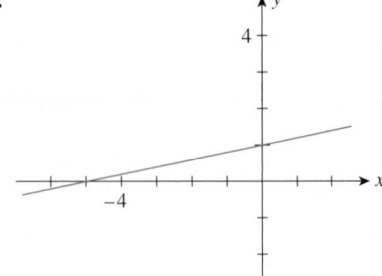

b.

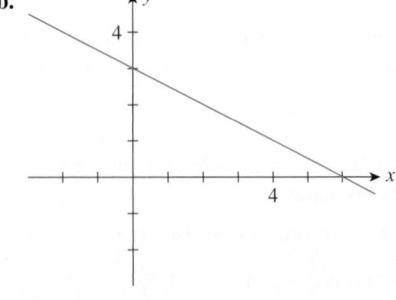

c.

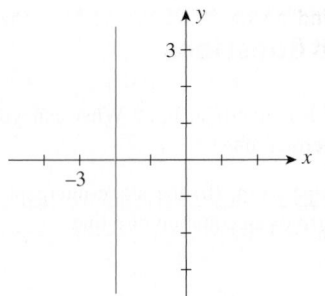

d.

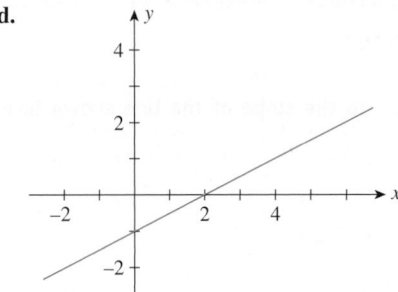

e.

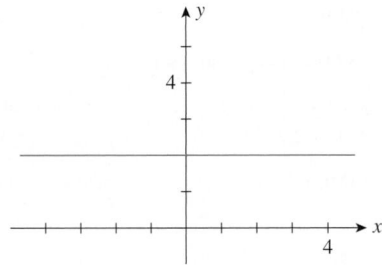

f.

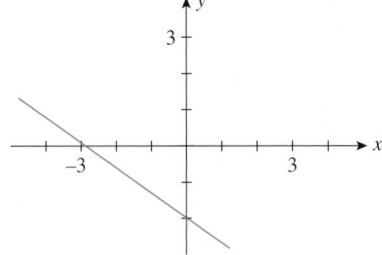

In Exercises 27–30, find an equation of the line that passes through the point and has the indicated slope m.

27. $(3, -4); m = 2$ **28.** $(2, 4); m = -1$

29. $(-3, 2); m = 0$ **30.** $(1, 2); m = -\dfrac{1}{2}$

In Exercises 31–34, find an equation of the line that passes through the given points.

31. $(2, 4)$ and $(3, 7)$ **32.** $(2, 1)$ and $(2, 5)$

33. $(1, 2)$ and $(-3, -2)$ **34.** $(-1, -2)$ and $(3, -4)$

In Exercises 35–38, find an equation of the line that has slope *m* and *y*-intercept *b*.

35. $m = 3; b = 4$

36. $m = -2; b = -1$

37. $m = 0; b = 5$

38. $m = -\dfrac{1}{2}; b = \dfrac{3}{4}$

In Exercises 39–44, write the equation in the slope-intercept form and then find the slope and *y*-intercept of the corresponding line.

39. $x - 2y = 0$

40. $y - 2 = 0$

41. $2x - 3y - 9 = 0$

42. $3x - 4y + 8 = 0$

43. $2x + 4y = 14$

44. $5x + 8y - 24 = 0$

45. Find an equation of the line that passes through the point $(-2, 2)$ and is parallel to the line $2x - 4y - 8 = 0$.

46. Find an equation of the line that passes through the point $(2, 4)$ and is perpendicular to the line $3x + 4y - 22 = 0$.

In Exercises 47–52, find an equation of the line that satisfies the given condition.

47. The line parallel to the *x*-axis and 6 units below it

48. The line passing through the origin and parallel to the line passing through the points $(2, 4)$ and $(4, 7)$

49. The line passing through the point (a, b) with slope equal to zero

50. The line passing through $(-3, 4)$ and parallel to the *x*-axis

51. The line passing through $(-5, -4)$ and parallel to the line passing through $(-3, 2)$ and $(6, 8)$

52. The line passing through (a, b) with undefined slope

53. Given that the point $P(-3, 5)$ lies on the line $kx + 3y + 9 = 0$, find k.

54. Given that the point $P(2, -3)$ lies on the line $-2x + ky + 10 = 0$, find k.

In Exercises 55–60, sketch the straight line defined by the linear equation by finding the *x*- and *y*-intercepts.

Hint: See Example 12.

55. $3x - 2y + 6 = 0$

56. $2x - 5y + 10 = 0$

57. $x + 2y - 4 = 0$

58. $2x + 3y - 15 = 0$

59. $y + 5 = 0$

60. $-2x - 8y + 24 = 0$

61. Show that an equation of a line through the points $(a, 0)$ and $(0, b)$ with $a \neq 0$ and $b \neq 0$ can be written in the form

$$\frac{x}{a} + \frac{y}{b} = 1$$

(Recall that the numbers a and b are the *x*- and *y*-intercepts, respectively, of the line. This form of an equation of a line is called the **intercept form**.)

In Exercises 62–65, use the results of Exercise 61 to find an equation of a line with the *x*- and *y*-intercepts.

62. *x*-intercept 3; *y*-intercept 4

63. *x*-intercept -2; *y*-intercept -4

64. *x*-intercept $-\dfrac{1}{2}$; *y*-intercept $\dfrac{3}{4}$

65. *x*-intercept 4; *y*-intercept $-\dfrac{1}{2}$

In Exercises 66 and 67, determine whether the points lie on a straight line.

66. $A(-1, 7)$, $B(2, -2)$, and $C(5, -9)$

67. $A(-2, 1)$, $B(1, 7)$, and $C(4, 13)$

68. TEMPERATURE CONVERSION The relationship between the temperature in degrees Fahrenheit (°F) and the temperature in degrees Celsius (°C) is

$$F = \frac{9}{5}C + 32$$

a. Sketch the line with the given equation.

b. What is the slope of the line? What does it represent?

c. What is the *F*-intercept of the line? What does it represent?

69. NUCLEAR PLANT UTILIZATION The United States is not building many nuclear plants, but the ones it has are running at nearly full capacity. The output (as a percent of total capacity) of nuclear plants is described by the equation

$$y = 1.9467t + 70.082$$

where *t* is measured in years, with $t = 0$ corresponding to the beginning of 1990.

a. Sketch the line with the given equation.

b. What are the slope and the *y*-intercept of the line found in part (a)?

c. Give an interpretation of the slope and the *y*-intercept of the line found in part (a).

d. If the utilization of nuclear power continues to grow at the same rate and the total capacity of nuclear plants in the United States remains constant, by what year can the plants be expected to be generating at maximum capacity?

Source: Nuclear Energy Institute

70. SOCIAL SECURITY CONTRIBUTIONS For wages less than the maximum taxable wage base, Social Security contributions by employees are 7.65% of the employee's wages.

a. Find an equation that expresses the relationship between the wages earned (x) and the Social Security taxes paid (y) by an employee who earns less than the maximum taxable wage base.

b. For each additional dollar that an employee earns, by how much is his or her Social Security contribution increased? (Assume that the employee's wages are less than the maximum taxable wage base.)

c. What Social Security contributions will an employee who earns $65,000 (which is less than the maximum taxable wage base) be required to make?

Source: Social Security Administration

71. COLLEGE ADMISSIONS   Using data compiled by the Admissions Office at Faber University, college admissions officers estimate that 55% of the students who are offered admission to the freshman class at the university will actually enroll.

a. Find an equation that expresses the relationship between the number of students who actually enroll (y) and the number of students who are offered admission to the university (x).

b. If the desired freshman class size for the upcoming academic year is 1100 students, how many students should be admitted?

72. WEIGHT OF WHALES   The equation $W = 3.51L - 192$, expressing the relationship between the length L (in feet) and the expected weight W (in British tons) of adult blue whales, was adopted in the late 1960s by the International Whaling Commission.

a. What is the expected weight of an 80-ft blue whale?

b. Sketch the straight line that represents the equation.

73. THE NARROWING GENDER GAP   Since the founding of the Equal Employment Opportunity Commission and the passage of equal-pay laws, the gulf between men's and women's earnings has continued to close gradually. At the beginning of 1990 ($t = 0$), women's wages were 68% of men's wages, and by the beginning of 2000 ($t = 10$), women's wages were 80% of men's wages. If this gap between women's and men's wages continued to narrow *linearly*, then women's wages were what percentage of men's wages at the beginning of 2004?

Source: Journal of Economic Perspectives

74. SALES OF NAVIGATION SYSTEMS   The projected number of navigation systems (in millions) installed in vehicles in North America, Europe, and Japan from 2002 through 2006 follow. Here, $x = 0$ corresponds to 2002.

Year x	0	1	2	3	4
Systems Installed, y	3.9	4.7	5.8	6.8	7.8

a. Plot the annual sales (y) versus the year (x).

b. Draw a straight line L through the points corresponding to 2002 and 2006.

c. Derive an equation of the line L.

d. Use the equation found in part (c) to estimate the number of navigation systems installed in 2005. Compare this figure with the sales for that year.

Source: ABI Research

75. SALES OF GPS EQUIPMENT   The annual sales (in billions of dollars) of global positioning systems (GPS) equipment from 2000 through 2006 follow. Here, $x = 0$ corresponds to 2000.

Year x	0	1	2	3	4	5	6
Annual Sales, y	7.9	9.6	11.5	13.3	15.2	17	18.8

a. Plot the annual sales (y) versus the year (x).

b. Draw a straight line L through the points corresponding to 2000 and 2006.

c. Derive an equation of the line L.

d. Use the equation found in part (c) to estimate the annual sales of GPS equipment in 2005. Compare this figure with the projected sales for that year.

Source: ABI Research

76. IDEAL HEIGHTS AND WEIGHTS FOR WOMEN   The Venus Health Club for Women provides its members with the following table, which gives the average desirable weight (in pounds) for women of a given height (in inches):

Height, x	60	63	66	69	72
Weight, y	108	118	129	140	152

a. Plot the weight (y) versus the height (x).

b. Draw a straight line L through the points corresponding to heights of 5 ft and 6 ft.

c. Derive an equation of the line L.

d. Using the equation of part (c), estimate the average desirable weight for a woman who is 5 ft, 5 in. tall.

77. COST OF A COMMODITY   A manufacturer obtained the following data relating the cost y (in dollars) to the number of units (x) of a commodity produced:

Units Produced, x	0	20	40	60	80	100
Cost in Dollars, y	200	208	222	230	242	250

a. Plot the cost (y) versus the quantity produced (x).

b. Draw a straight line through the points (0, 200) and (100, 250).

c. Derive an equation of the straight line of part (b).

d. Taking this equation to be an approximation of the relationship between the cost and the level of production, estimate the cost of producing 54 units of the commodity.

78. DIGITAL TV SERVICES   The percentage of homes with digital TV services stood at 5% at the beginning of 1999 ($t = 0$) and was projected to grow linearly so that, at the beginning of 2003 ($t = 4$), the percentage of such homes was 25%.

a. Derive an equation of the line passing through the points $A(0, 5)$ and $B(4, 25)$.

b. Plot the line with the equation found in part (a).

c. Using the equation found in part (a), find the percentage of homes with digital TV services at the beginning of 2001.

Source: Paul Kagan Associates

79. SALES GROWTH   Metro Department Store's annual sales (in millions of dollars) during the past 5 yr were

Annual Sales, y	5.8	6.2	7.2	8.4	9.0
Year, x	1	2	3	4	5

a. Plot the annual sales (y) versus the year (x).
b. Draw a straight line L through the points corresponding to the first and fifth years.
c. Derive an equation of the line L.
d. Using the equation found in part (c), estimate Metro's annual sales 4 yr from now ($x = 9$).

80. Is there a difference between the statements "The slope of a straight line is zero" and "The slope of a straight line does not exist (is not defined)"? Explain your answer.

81. Consider the slope-intercept form of a straight line $y = mx + b$. Describe the family of straight lines obtained by keeping
a. The value of m fixed and allowing the value of b to vary.
b. The value of b fixed and allowing the value of m to vary.

In Exercises 82–88, determine whether the statement is true or false. If it is true, explain why it is true. If it is false, give an example to show why it is false.

82. Suppose the slope of a line L is $-\frac{1}{2}$ and P is a given point on L. If Q is the point on L lying 4 units to the left of P, then Q is situated 2 units above P.

83. The point $(-1, 1)$ lies on the line with equation $3x + 7y = 5$.

84. The point $(1, k)$ lies on the line with equation $3x + 4y = 12$ if and only if $k = \frac{9}{4}$.

85. The line with equation $Ax + By + C = 0$ ($B \neq 0$) and the line with equation $ax + by + c = 0$ ($b \neq 0$) are parallel if $Ab - aB = 0$.

86. If the slope of the line L_1 is positive, then the slope of a line L_2 perpendicular to L_1 may be positive or negative.

87. The lines with equations $ax + by + c_1 = 0$ and $bx - ay + c_2 = 0$, where $a \neq 0$ and $b \neq 0$, are perpendicular to each other.

88. If L is the line with equation $Ax + By + C = 0$, where $A \neq 0$, then L crosses the x-axis at the point $(-C/A, 0)$.

89. Show that two distinct lines with equations $a_1x + b_1y + c_1 = 0$ and $a_2x + b_2y + c_2 = 0$, respectively, are parallel if and only if $a_1b_2 - b_1a_2 = 0$.
   Hint: Write each equation in the slope-intercept form and compare.

90. Prove that if a line L_1 with slope m_1 is perpendicular to a line L_2 with slope m_2, then $m_1m_2 = -1$.
   Hint: Refer to the accompanying figure. Show that $m_1 = b$ and $m_2 = c$. Next, apply the Pythagorean theorem and the distance formula to the triangles OAC, OCB, and OBA to show that $1 = -bc$.

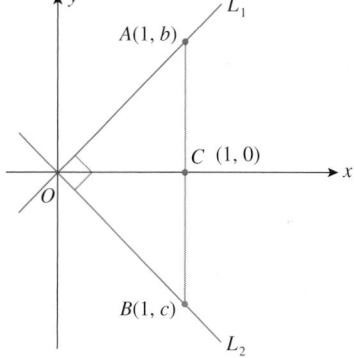

1.2 Solutions to Self-Check Exercises

1. The slope of the line that passes through the points $(a, 2)$ and $(3, 6)$ is

$$m = \frac{6 - 2}{3 - a} = \frac{4}{3 - a}$$

Since this line is parallel to a line with slope 4, m must be equal to 4; that is,

$$\frac{4}{3 - a} = 4$$

or, upon multiplying both sides of the equation by $3 - a$,

$$4 = 4(3 - a)$$
$$4 = 12 - 4a$$
$$4a = 8$$
$$a = 2$$

2. Since the required line L is perpendicular to a line with slope $-\frac{1}{2}$, the slope of L is

$$m = -\frac{1}{-\frac{1}{2}} = 2$$

Next, using the point-slope form of the equation of a line, we have

$$y - (-1) = 2(x - 3)$$
$$y + 1 = 2x - 6$$
$$y = 2x - 7$$

3. Substituting $x = 3$ and $y = -3$ into the left-hand side of the given equation, we find

$$2(3) - 3(-3) - 12 = 3$$

which is not equal to zero (the right-hand side). Therefore, $(3, -3)$ does not lie on the line with equation $2x - 3y - 12 = 0$.

Setting $x = 0$, we find $y = -4$, the y-intercept. Next, setting $y = 0$ gives $x = 6$, the x-intercept. We now draw the line passing through the points $(0, -4)$ and $(6, 0)$, as shown in the accompanying figure.

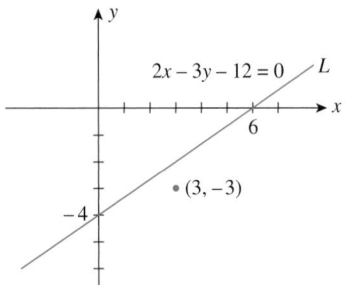

4. a and b. See the accompanying figure.

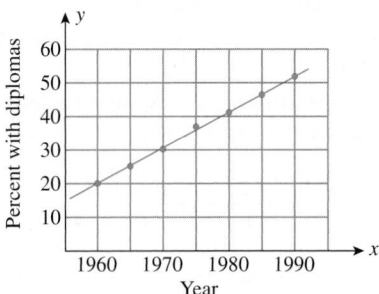

c. The slope of L is

$$m = \frac{52 - 20}{1990 - 1960} = \frac{32}{30} = \frac{16}{15}$$

Using the point-slope form of the equation of a line with the point $(1960, 20)$, we find

$$y - 20 = \frac{16}{15}(x - 1960) = \frac{16}{15}x - \frac{(16)(1960)}{15}$$

$$y = \frac{16}{15}x - \frac{6272}{3} + 20$$

$$= \frac{16}{15}x - \frac{6212}{3}$$

d. To estimate the percentage of people over age 65 who had high-school diplomas by the year 1995, let $x = 1995$ in the equation obtained in part (c). Thus, the required estimate is

$$y = \frac{16}{15}(1995) - \frac{6212}{3} \approx 57.33$$

or approximately 57%.

USING TECHNOLOGY

■ Graphing a Straight Line

Graphing Utility

The first step in plotting a straight line with a graphing utility is to select a suitable viewing window. We usually do this by experimenting. For example, you might first plot the straight line using the **standard viewing window** $[-10, 10] \times [-10, 10]$. If necessary, you then might adjust the viewing window by enlarging it or reducing it to obtain a sufficiently complete view of the line or at least the portion of the line that is of interest.

EXAMPLE 1 Plot the straight line $2x + 3y - 6 = 0$ in the standard viewing window.

Solution The straight line in the standard viewing window is shown in Figure T1.

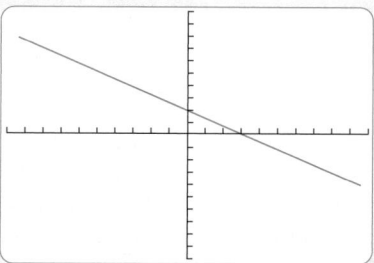

FIGURE T1
The straight line $2x + 3y - 6 = 0$ in the standard viewing window

EXAMPLE 2 Plot the straight line $2x + 3y - 30 = 0$ in (a) the standard viewing window and (b) the viewing window $[-5, 20] \times [-5, 20]$.

Solution

a. The straight line in the standard viewing window is shown in Figure T2a.
b. The straight line in the viewing window $[-5, 20] \times [-5, 20]$ is shown in Figure T2b. This figure certainly gives a more complete view of the straight line.

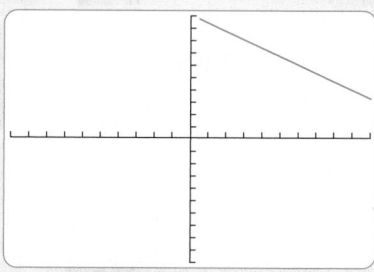 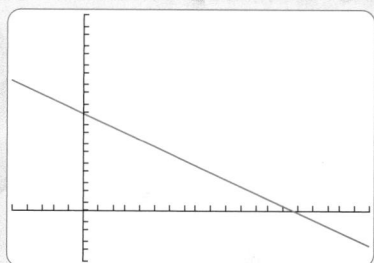

FIGURE T2

(a) The graph of $2x + 3y - 30 = 0$ in the standard viewing window

(b) The graph of $2x + 3y - 30 = 0$ in the viewing window $[-5, 20] \times [-5, 20]$

Excel

In the examples and exercises that follow, we assume that you are familiar with the basic features of Microsoft Excel. Please consult your Excel manual or use Excel's Help features to answer questions regarding the standard commands and operating instructions for Excel.

EXAMPLE 3 Plot the graph of the straight line $2x + 3y - 6 = 0$ over the interval $[-10, 10]$.

Solution

1. *Write the equation in the slope-intercept form:*

$$y = -\frac{2}{3}x + 2$$

2. *Create a table of values.* First, enter the input values: Enter the values of the endpoints of the interval over which you are graphing the straight line. (Recall that

(continued)

we need only two distinct data points to draw the graph of a straight line. In general, we select the endpoints of the interval over which the straight line is to be drawn as our data points.) In this case, we enter −10 in cell A2 and 10 in cell A3.

Second, enter the formula for computing the y-values: Here we enter

$$= -(2/3)*A2+2$$

in cell B2 and then press ⌜Enter⌝.

Third, evaluate the function at the other input value: To extend the formula to cell B3, move the pointer to the small black box at the lower right corner of cell B2 (the cell containing the formula). Observe that the pointer now appears as a black + (plus sign). Drag this pointer through cell B3 and then release it. The y-value, − 4.66667, corresponding to the x-value in cell A3(10) will appear in cell B3 (Figure T3).

	A	B
1	x	y
2	−10	8.666667
3	10	−4.66667

FIGURE T3
Table of values for x and y

3. *Graph the straight line determined by these points.* First, highlight the numerical values in the table. Here we highlight cells A2:A3 and B2:B3. Next, click the ⌜**Chart Wizard**⌝ button on the toolbar.

Step 1   In the Chart Type dialog box that appears, select ⌜**XY(Scatter)**⌝ . Next, select the second chart in the first column under Chart sub-type. Then click ⌜**Next**⌝ at the bottom of the dialog box.

Step 2   Click ⌜**Columns**⌝ next to Series in: Then click ⌜**Next**⌝ at the bottom of the dialog box.

Step 3   Click the ⌜**Titles**⌝ tab. In the Chart title: box, enter y = −(2/3)x + 2. In the Value (X) axis: box, type x. In the Value (Y) axis: box, type y. Click the ⌜**Legend**⌝ tab. Next, click the ⌜**Show Legend**⌝ box to remove the check mark. Click ⌜**Finish**⌝ at the bottom of the dialog box.

The graph shown in Figure T4 will appear.

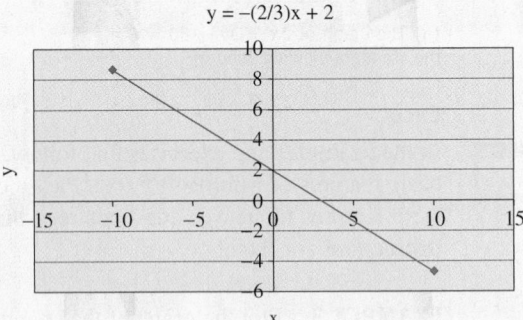

FIGURE T4
The graph of $y = -\frac{2}{3}x + 2$ over the interval $[-10, 10]$

If the interval over which the straight line is to be plotted is not specified, then you may have to experiment to find an appropriate interval for the x-values in your graph. For example, you might first plot the straight line over the interval $[-10, 10]$.

Note: Boldfaced words/characters enclosed in a box (for example, ⌜Enter⌝) indicate that an action (click, select, or press) is required. Words/characters printed in blue (for example, Chart Type) indicate words/characters that appear on the screen. Words/characters printed in a typewriter font (for example, =(−2/3)*A2+2) indicate words/characters that need to be typed and entered.

If necessary you then might adjust the interval by enlarging it or reducing it to obtain a sufficiently complete view of the line or at least the portion of the line that is of interest.

EXAMPLE 4 Plot the straight line $2x + 3y - 30 = 0$ over the intervals (a) $[-10, 10]$ and (b) $[-5, 20]$.

Solution   a and b. We first cast the equation in the slope-intercept form, obtaining $y = -\frac{2}{3}x + 10$. Following the procedure given in Example 3, we obtain the graphs shown in Figure T5.

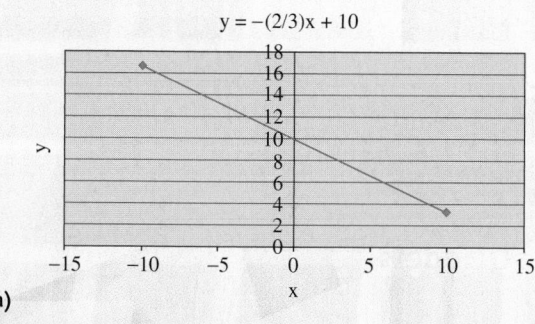

(a)

(b)

FIGURE T5
The graph of $y = -\frac{2}{3}x + 10$ over the intervals (a) $[-10, 10]$ and (b) $[-5, 20]$

Observe that the graph in Figure T5b includes the x- and y-intercepts. This figure certainly gives a more complete view of the straight line.

TECHNOLOGY EXERCISES

Graphing Utility

In Exercises 1–6, plot the straight line with the equation in the standard viewing window.

1. $3.2x + 2.1y - 6.72 = 0$ **2.** $2.3x - 4.1y - 9.43 = 0$

3. $1.6x + 5.1y = 8.16$ **4.** $-3.2x + 2.1y = 6.72$

5. $2.8x = -1.6y + 4.48$ **6.** $3.3y = 4.2x - 13.86$

In Exercises 7–10, plot the straight line with the equation in (a) the standard viewing window and (b) the indicated viewing window.

7. $12.1x + 4.1y - 49.61 = 0$; $[-10, 10] \times [-10, 20]$

8. $4.1x - 15.2y - 62.32 = 0$; $[-10, 20] \times [-10, 10]$

9. $20x + 16y = 300$; $[-10, 20] \times [-10, 30]$

10. $32.2x + 21y = 676.2$; $[-10, 30] \times [-10, 40]$

In Exercises 11–18, plot the straight line with the equation in an appropriate viewing window. (*Note:* The answer is *not* unique.)

11. $20x + 30y = 600$

12. $30x - 20y = 600$

13. $22.4x + 16.1y - 352 = 0$

14. $18.2x - 15.1y = 274.8$

15. $1.2x + 20y = 24$

16. $30x - 2.1y = 63$

17. $-4x + 12y = 50$

18. $20x - 12.2y = 240$

(*continued*)

Excel

In Exercises 1–6, plot the straight line with the equation over the interval [−10, 10].

1. $3.2x + 2.1y − 6.72 = 0$ **2.** $2.3x − 4.1y − 9.43 = 0$

3. $1.6x + 5.1y = 8.16$ **4.** $−3.2x + 2.1y = 6.72$

5. $2.8x = −1.6y + 4.48$ **6.** $3.3y = 4.2x − 13.86$

In Exercises 7–10, plot the straight line with the equation over the given interval.

7. $12.1x + 4.1y − 49.61 = 0; [−10, 10]$

8. $4.1x − 15.2y − 62.32 = 0; [−10, 20]$

9. $20x + 16y = 300; [−10, 20]$

10. $32.2x + 21y = 676.2; [−10, 30]$

In Exercises 11–18, plot the straight line with the equation. (*Note:* The answer is *not* unique.)

11. $20x + 30y = 600$ **12.** $30x − 20y = 600$

13. $22.4x + 16.1y − 352 = 0$

14. $18.2x − 15.1y = 274.8$ **15.** $1.2x + 20y = 24$

16. $30x − 2.1y = 63$ **17.** $−4x + 12y = 50$

18. $20x − 12.2y = 240$

1.3   Linear Functions and Mathematical Models

Mathematical Models

Regardless of the field from which a real-world problem is drawn, the problem is solved by analyzing it through a process called **mathematical modeling**. The four steps in this process, as illustrated in Figure 28, follow.

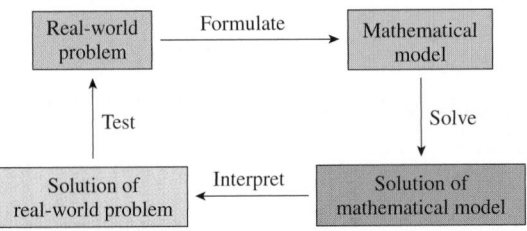

FIGURE 28

1. **Formulate** Given a real-world problem, our first task is to formulate the problem using the language of mathematics. The many techniques used in constructing mathematical models range from theoretical consideration of the problem on the one extreme to an interpretation of data associated with the problem on the other. For example, the mathematical model giving the accumulated amount at any time when a certain sum of money is deposited in the bank can be derived theoretically (see Chapter 5). On the other hand, many of the mathematical models in this book are constructed by studying the data associated with the problem. In Section 1.5, we will see how linear equations (models) can be constructed from a given set of data points. Also, in the ensuing chapters we will see how other mathematical models, including statistical and probability models, are used to describe and analyze real-world situations.
2. **Solve** Once a mathematical model has been constructed, we can use the appropriate mathematical techniques, which we will develop throughout the book, to solve the problem.

3. **Interpret** Bearing in mind that the solution obtained in step 2 is just the solution of the mathematical model, we need to interpret these results in the context of the original real-world problem.

4. **Test** Some mathematical models of real-world applications describe the situations with complete accuracy. For example, the model describing a deposit in a bank account gives the exact accumulated amount in the account at any time. But other mathematical models give, at best, an approximate description of the real-world problem. In this case we need to test the accuracy of the model by observing how well it describes the original real-world problem and how well it predicts past and/or future behavior. If the results are unsatisfactory, then we may have to reconsider the assumptions made in the construction of the model or, in the worst case, return to step 1.

We now look at an important way of describing the relationship between two quantities using the notion of a function. As you will see subsequently, many mathematical models are represented by functions.

▬ Functions

A manufacturer would like to know how his company's profit is related to its production level; a biologist would like to know how the population of a certain culture of bacteria will change with time; a psychologist would like to know the relationship between the learning time of an individual and the length of a vocabulary list; and a chemist would like to know how the initial speed of a chemical reaction is related to the amount of substrate used. In each instance, we are concerned with the same question: How does one quantity depend on another? The relationship between two quantities is conveniently described in mathematics by using the concept of a function.

> **Function**
>
> A *function* f is a rule that assigns to each value of x one and only one value of y.

The number y is normally denoted by $f(x)$, read "f of x," emphasizing the dependency of y on x.

An example of a function may be drawn from the familiar relationship between the area of a circle and its radius. Let x and y denote the radius and area of a circle, respectively. From elementary geometry we have

$$y = \pi x^2$$

This equation defines y as a function of x, since for each admissible value of x (a nonnegative number representing the radius of a certain circle) there corresponds precisely one number $y = \pi x^2$ giving the area of the circle. This *area function* may be written as

$$f(x) = \pi x^2 \tag{7}$$

For example, to compute the area of a circle with a radius of 5 inches, we simply replace x in Equation (7) by the number 5. Thus, the area of the circle is

$$f(5) = \pi 5^2 = 25\pi$$

or 25π square inches.

Suppose we are given the function $y = f(x)$.* The variable x is referred to as the **independent variable**, and the variable y is called the **dependent variable**. The set of all values that may be assumed by x is called the **domain** of the function f, and the set comprising all the values assumed by $y = f(x)$ as x takes on all possible values in its domain is called the **range** of the function f. For the area function (7), the domain of f is the set of all nonnegative numbers x, and the range of f is the set of all nonnegative numbers y.

We now focus our attention on an important class of functions known as linear functions. Recall that a linear equation in x and y has the form $Ax + By + C = 0$, where A, B, and C are constants and A and B are not both zero. If $B \neq 0$, the equation can always be solved for y in terms of x; in fact, as we saw in Section 1.2, the equation may be cast in the slope-intercept form:

$$y = mx + b \qquad (m, b \text{ constants}) \tag{8}$$

Equation (8) defines y as a function of x. The domain and range of this function is the set of all real numbers. Furthermore, the graph of this function, as we saw in Section 1.2, is a straight line in the plane. For this reason, the function $f(x) = mx + b$ is called a linear function.

Linear Function

The function f defined by

$$f(x) = mx + b$$

where m and b are constants, is called a **linear function**.

Linear functions play an important role in the quantitative analysis of business and economic problems. First, many problems arising in these and other fields are linear in nature or are linear in the intervals of interest and thus can be formulated in terms of linear functions. Second, because linear functions are relatively easy to work with, assumptions involving linearity are often made in the formulation of problems. In many cases these assumptions are justified, and acceptable mathematical models are obtained that approximate real-life situations.

APPLIED EXAMPLE 1   Market for Cholesterol-Reducing Drugs In a study conducted in early 2000, experts projected a rise in the market for cholesterol-reducing drugs. The U.S. market (in billions of dollars) for such drugs from 1999 through 2004 is given in the following table:

Year	1999	2000	2001	2002	2003	2004
Market	12.07	14.07	16.21	18.28	20	21.72

A mathematical model giving the approximate U.S. market over the period in question is given by

$$M(t) = 1.95t + 12.19$$

where t is measured in years, with $t = 0$ corresponding to 1999.

---

*It is customary to refer to a function f as $f(x)$.

a. Sketch the graph of the function M and the given data on the same set of axes.

b. Assuming that the projection held and the trend continued, what was the market for cholesterol-reducing drugs in 2005 ($t = 6$)?

c. What was the rate of increase of the market for cholesterol-reducing drugs over the period in question?

Source: S. G. Cowen

Solution

a. The graph of M is shown in Figure 29.

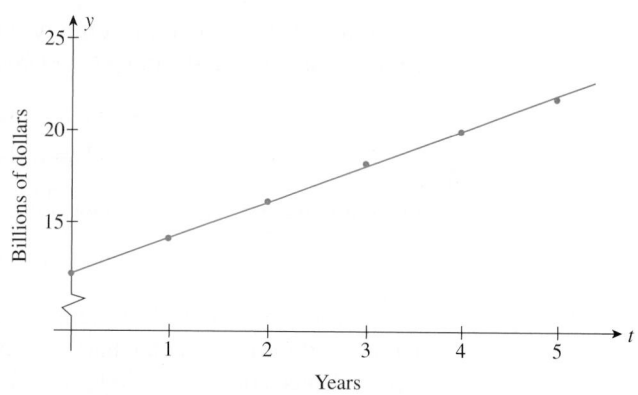
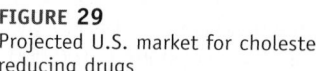

FIGURE 29
Projected U.S. market for cholesterol-reducing drugs

b. The projected market in 2005 for cholesterol-reducing drugs was approximately

$$M(6) = 1.95(6) + 12.19$$
$$= 23.89$$

or $23.89 billion.

c. The function M is linear; hence we see that the rate of increase of the market for cholesterol-reducing drugs is given by the slope of the straight line represented by M, which is approximately $1.95 billion per year. ∎

In Section 1.5, we will show how the function in Example 1 is actually constructed using the method of least squares.

In the rest of this section, we look at several applications that can be modeled using linear functions.

■ Simple Depreciation

We first discussed linear depreciation in Section 1.2 as a real-world application of straight lines. The following example illustrates how to derive an equation describing the book value of an asset that is being depreciated linearly.

APPLIED EXAMPLE 2   Linear Depreciation   A printing machine has an original value of $100,000 and is to be depreciated linearly over 5 years with a $30,000 scrap value. Find an expression giving the book value at the end of year t. What will be the book value of the machine at the end of the second year? What is the rate of depreciation of the printing machine?

Solution Let V denote the printing machine's book value at the end of the tth year. Since the depreciation is linear, V is a linear function of t. Equivalently, the graph of the function is a straight line. Now, to find an equation of the straight line, observe that $V = 100{,}000$ when $t = 0$; this tells us that the line passes through the point $(0, 100{,}000)$. Similarly, the condition that $V = 30{,}000$ when $t = 5$ says that the line also passes through the point $(5, 30{,}000)$. The slope of the line is given by

$$m = \frac{100{,}000 - 30{,}000}{0 - 5} = -\frac{70{,}000}{5} = -14{,}000$$

Using the point-slope form of the equation of a line with the point $(0, 100{,}000)$ and the slope $m = -14{,}000$, we have

$$V - 100{,}000 = -14{,}000(t - 0)$$
$$V = -14{,}000t + 100{,}000$$

the required expression. The book value at the end of the second year is given by

$$V = -14{,}000(2) + 100{,}000 = 72{,}000$$

or \$72,000. The rate of depreciation of the machine is given by the negative of the slope of the depreciation line. Since the slope of the line is $m = -14{,}000$, the rate of depreciation is \$14,000 per year. The graph of $V = -14{,}000t + 100{,}000$ is sketched in Figure 30. ∎

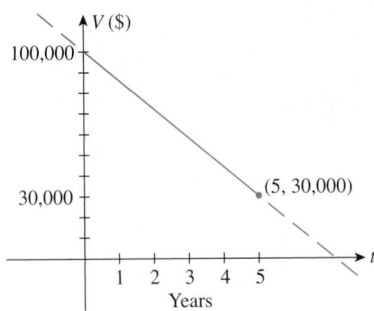

FIGURE 30
Linear depreciation of an asset

▬ Linear Cost, Revenue, and Profit Functions

Whether a business is a sole proprietorship or a large corporation, the owner or chief executive must constantly keep track of operating costs, revenue resulting from the sale of products or services, and, perhaps most important, the profits realized. Three functions provide management with a measure of these quantities: the total cost function, the revenue function, and the profit function.

Cost, Revenue, and Profit Functions

Let x denote the number of units of a product manufactured or sold. Then, the **total cost function** is

$C(x) = $ Total cost of manufacturing x units of the product

The **revenue function** is

$R(x) = $ Total revenue realized from the sale of x units of the product

The **profit function** is

$P(x) = $ Total profit realized from manufacturing and selling x units of the product

---

Generally speaking, the total cost, revenue, and profit functions associated with a company will probably be nonlinear (these functions are best studied using the tools of calculus). But *linear* cost, revenue, and profit functions do arise in practice, and we will consider such functions in this section. Before deriving explicit forms of these functions, we need to recall some common terminology.

The costs incurred in operating a business are usually classified into two categories. Costs that remain more or less constant regardless of the firm's activity level are called **fixed costs**. Examples of fixed costs are rental fees and executive salaries. Costs that vary with production or sales are called **variable costs**. Examples of variable costs are wages and costs for raw materials.

Suppose a firm has a fixed cost of F dollars, a production cost of c dollars per unit, and a selling price of s dollars per unit. Then the *cost function* $C(x)$, the *revenue function* $R(x)$, and the *profit function* $P(x)$ for the firm are given by

$$C(x) = cx + F$$
$$R(x) = sx$$
$$P(x) = R(x) - C(x) \qquad \text{Revenue} - \text{cost}$$
$$= (s - c)x - F$$

where x denotes the number of units of the commodity produced and sold. The functions C, R, and P are linear functions of x.

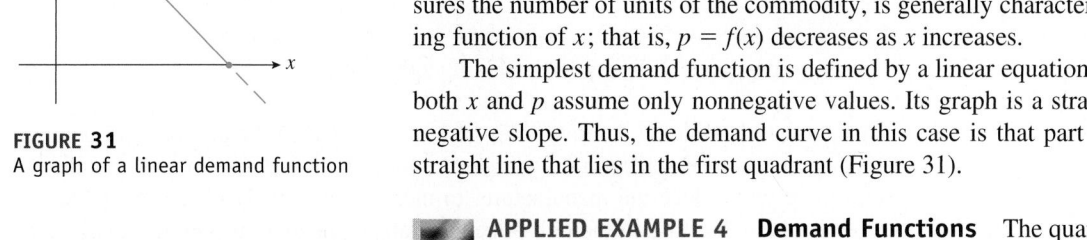

APPLIED EXAMPLE 3   Profit Functions   Puritron, a manufacturer of water filters, has a monthly fixed cost of $20,000, a production cost of $20 per unit, and a selling price of $30 per unit. Find the cost function, the revenue function, and the profit function for Puritron.

Solution   Let x denote the number of units produced and sold. Then

$$C(x) = 20x + 20{,}000$$
$$R(x) = 30x$$
$$P(x) = R(x) - C(x)$$
$$= 30x - (20x + 20{,}000)$$
$$= 10x - 20{,}000$$

Linear Demand and Supply Curves

In a free-market economy, consumer demand for a particular commodity depends on the commodity's unit price. A **demand equation** expresses this relationship between the unit price and the quantity demanded. The corresponding graph of the demand equation is called a **demand curve**. In general, the quantity demanded of a commodity decreases as its unit price increases, and vice versa. Accordingly, a demand function defined by $p = f(x)$, where p measures the unit price and x measures the number of units of the commodity, is generally characterized as a decreasing function of x; that is, $p = f(x)$ decreases as x increases.

The simplest demand function is defined by a linear equation in x and p, where both x and p assume only nonnegative values. Its graph is a straight line having a negative slope. Thus, the demand curve in this case is that part of the graph of a straight line that lies in the first quadrant (Figure 31).

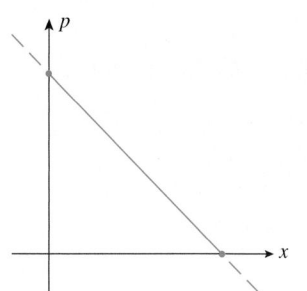

FIGURE 31
A graph of a linear demand function

APPLIED EXAMPLE 4   Demand Functions   The quantity demanded of the Sentinel alarm clock is 48,000 units when the unit price is $8. At $12 per unit, the quantity demanded drops to 32,000 units. Find the demand equation, assuming that it is linear. What is the unit price corresponding to a quantity demanded of 40,000 units? What is the quantity demanded if the unit price is $14?

Solution   Let p denote the unit price of an alarm clock (in dollars) and let x (in units of 1000) denote the quantity demanded when the unit price of the clocks is $\$p$. If $p = 8$ then $x = 48$ and the point $(48, 8)$ lies on the demand curve. Similarly, if $p = 12$ then $x = 32$ and the point $(32, 12)$ also lies on the demand curve. Since the demand equation is linear, its graph is a straight line. The slope of the required line is given by

$$m = \frac{12 - 8}{32 - 48} = \frac{4}{-16} = -\frac{1}{4}$$

So, using the point-slope form of an equation of a line with the point $(48, 8)$, we find that

$$p - 8 = -\frac{1}{4}(x - 48)$$

$$p = -\frac{1}{4}x + 20$$

is the required equation. The demand curve is shown in Figure 32. If the quantity demanded is 40,000 units ($x = 40$), the demand equation yields

$$y = -\frac{1}{4}(40) + 20 = 10$$

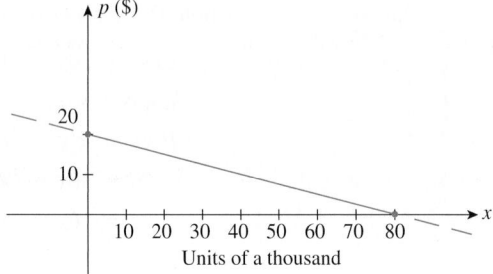

FIGURE 32
The graph of the demand equation
$p = -\frac{1}{4}x + 20$

and we see that the corresponding unit price is $10. Next, if the unit price is $14 ($p = 14$), the demand equation yields

$$14 = -\frac{1}{4}x + 20$$

$$\frac{1}{4}x = 6$$

$$x = 24$$

and so the quantity demanded will be 24,000 units.

In a competitive market, a relationship also exists between the unit price of a commodity and its availability in the market. In general, an increase in a commodity's unit price will induce the manufacturer to increase the supply of that commodity. Conversely, a decrease in the unit price generally leads to a drop in the supply. An equation that expresses the relationship between the unit price and the quantity supplied is called a **supply equation**, and the corresponding graph is called a **supply curve**. A supply function, defined by $p = f(x)$, is generally characterized by an increasing function of x; that is, $p = f(x)$ increases as x increases.

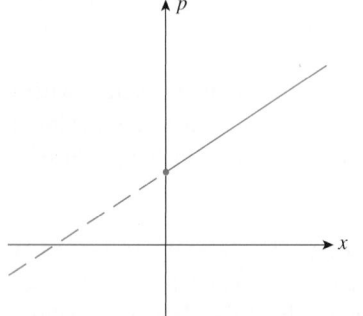

FIGURE 33
A graph of a linear supply function

As in the case of a demand equation, the simplest supply equation is a linear equation in p and x, where p and x have the same meaning as before but the line has a positive slope. The supply curve corresponding to a linear supply function is that part of the straight line that lies in the first quadrant (Figure 33).

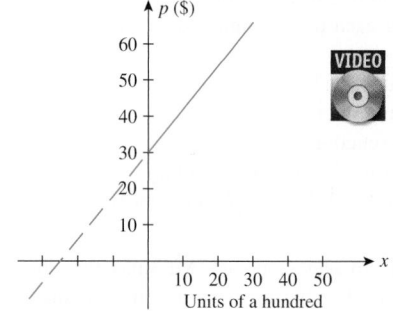

FIGURE 34
The graph of the supply equation
$4p - 5x = 120$

APPLIED EXAMPLE 5   Supply Functions   The supply equation for a commodity is given by $4p - 5x = 120$, where p is measured in dollars and x is measured in units of 100.

a. Sketch the corresponding curve.
b. How many units will be marketed when the unit price is $55?

Solution

a. Setting $x = 0$, we find the p-intercept to be 30. Next, setting $p = 0$, we find the x-intercept to be -24. The supply curve is sketched in Figure 34.
b. Substituting $p = 55$ in the supply equation, we have $4(55) - 5x = 120$ or $x = 20$, so the amount marketed will be 2000 units.

1.3   Self-Check Exercises

1. A manufacturer has a monthly fixed cost of $60,000 and a production cost of $10 for each unit produced. The product sells for $15/unit.
   a. What is the cost function?
   b. What is the revenue function?
   c. What is the profit function?
   d. Compute the profit (loss) corresponding to production levels of 10,000 and 14,000 units.

2. The quantity demanded for a certain make of 30-in. × 52-in. area rug is 500 when the unit price is $100. For each $20 decrease in the unit price, the quantity demanded increases by 500 units. Find the demand equation and sketch its graph.

Solutions to Self-Check Exercises 1.3 can be found on page 38.

1.3   Concept Questions

1. **a.** What is a *function*? Give an example.
   b. What is a *linear function*? Give an example.
   c. What is the domain of a linear function? The range?
   d. What is the graph of a linear function?

2. What is the general form of a linear cost function? A linear revenue function? A linear profit function?

3. Is the slope of a linear demand curve positive or negative? The slope of a linear supply curve?

1.3   Exercises

In Exercises 1–10, determine whether the equation defines y as a linear function of x. If so, write it in the form y = mx + b.

1. $2x + 3y = 6$

2. $-2x + 4y = 7$

3. $x = 2y - 4$

4. $2x = 3y + 8$

5. $2x - 4y + 9 = 0$

6. $3x - 6y + 7 = 0$

7. $2x^2 - 8y + 4 = 0$

8. $3\sqrt{x} + 4y = 0$

9. $2x - 3y^2 + 8 = 0$ 10. $2x + \sqrt{y} - 4 = 0$

11. A manufacturer has a monthly fixed cost of $40,000 and a production cost of $8 for each unit produced. The product sells for $12/unit.
    a. What is the cost function?
    b. What is the revenue function?
    c. What is the profit function?
    d. Compute the profit (loss) corresponding to production levels of 8000 and 12,000 units.

12. A manufacturer has a monthly fixed cost of $100,000 and a production cost of $14 for each unit produced. The product sells for $20/unit.
    a. What is the cost function?
    b. What is the revenue function?
    c. What is the profit function?
    d. Compute the profit (loss) corresponding to production levels of 12,000 and 20,000 units.

13. Find the constants m and b in the linear function $f(x) = mx + b$ such that $f(0) = 2$ and $f(3) = -1$.

14. Find the constants m and b in the linear function $f(x) = mx + b$ such that $f(2) = 4$ and the straight line represented by f has slope -1.

15. **LINEAR DEPRECIATION** An office building worth $1 million when completed in 2005 is being depreciated linearly over 50 yr. What will be the book value of the building in 2010? In 2015? (Assume the scrap value is $0.)

16. **LINEAR DEPRECIATION** An automobile purchased for use by the manager of a firm at a price of $24,000 is to be depreciated using the straight-line method over 5 yr. What will be the book value of the automobile at the end of 3 yr? (Assume the scrap value is $0.)

17. **CONSUMPTION FUNCTIONS** A certain economy's consumption function is given by the equation

$$C(x) = 0.75x + 6$$

where $C(x)$ is the personal consumption expenditure in billions of dollars and x is the disposable personal income in billions of dollars. Find $C(0)$, $C(50)$, and $C(100)$.

18. **SALES TAX** In a certain state, the sales tax T on the amount of taxable goods is 6% of the value of the goods purchased (x), where both T and x are measured in dollars.
    a. Express T as a function of x.
    b. Find $T(200)$ and $T(5.60)$.

19. **SOCIAL SECURITY BENEFITS** Social Security recipients receive an automatic cost-of-living adjustment (COLA) once each year. Their monthly benefit is increased by the same percentage that consumer prices have increased during the preceding year. Suppose consumer prices have increased by 5.3% during the preceding year.
    a. Express the adjusted monthly benefit of a Social Security recipient as a function of his or her current monthly benefit.

    b. If Carlos Garcia's monthly Social Security benefit is now $1,020, what will be his adjusted monthly benefit?

20. **PROFIT FUNCTIONS** AutoTime, a manufacturer of 24-hr variable timers, has a monthly fixed cost of $48,000 and a production cost of $8 for each timer manufactured. The timers sell for $14 each.
    a. What is the cost function?
    b. What is the revenue function?
    c. What is the profit function?
    d. Compute the profit (loss) corresponding to production levels of 4000, 6000, and 10,000 timers, respectively.

21. **PROFIT FUNCTIONS** The management of TMI finds that the monthly fixed costs attributable to the production of their 100-watt light bulbs is $12,100.00. If the cost of producing each twin-pack of light bulbs is $.60 and each twin-pack sells for $1.15, find the company's cost function, revenue function, and profit function.

22. **LINEAR DEPRECIATION** In 2005, National Textile installed a new machine in one of its factories at a cost of $250,000. The machine is depreciated linearly over 10 yr with a scrap value of $10,000.
    a. Find an expression for the machine's book value in the tth year of use $(0 \le t \le 10)$.
    b. Sketch the graph of the function of part (a).
    c. Find the machine's book value in 2009.
    d. Find the rate at which the machine is being depreciated.

23. **LINEAR DEPRECIATION** A server purchased at a cost of $60,000 in 2006 has a scrap value of $12,000 at the end of 4 yr. If the straight-line method of depreciation is used,
    a. Find the rate of depreciation.
    b. Find the linear equation expressing the server's book value at the end of t yr.
    c. Sketch the graph of the function of part (b).
    d. Find the server's book value at the end of the third year.

24. **LINEAR DEPRECIATION** Suppose an asset has an original value of $$C$ and is depreciated linearly over N yr with a scrap value of $$S$. Show that the asset's book value at the end of the tth year is described by the function

$$V(t) = C - \left(\frac{C - S}{N}\right)t$$

Hint: Find an equation of the straight line passing through the points $(0, C)$ and (N, S). (Why?)

25. Rework Exercise 15 using the formula derived in Exercise 24.

26. Rework Exercise 16 using the formula derived in Exercise 24.

27. **DRUG DOSAGES** A method sometimes used by pediatricians to calculate the dosage of medicine for children is based on the child's surface area. If a denotes the adult dosage (in mil-

ligrams) and if S is the child's surface area (in square meters), then the child's dosage is given by

$$D(S) = \frac{Sa}{1.7}$$

a. Show that D is a linear function of S.
Hint: Think of D as having the form $D(S) = mS + b$. What is the slope m and the y-intercept b?
b. If the adult dose of a drug is 500 mg, how much should a child whose surface area is 0.4 m² receive?

28. **DRUG DOSAGES** Cowling's rule is a method for calculating pediatric drug dosages. If a denotes the adult dosage (in milligrams) and if t is the child's age (in years), then the child's dosage is given by

$$D(t) = \left(\frac{t + 1}{24}\right)a$$

a. Show that D is a linear function of t.
Hint: Think of $D(t)$ as having the form $D(t) = mt + b$. What is the slope m and the y-intercept b?
b. If the adult dose of a drug is 500 mg, how much should a 4-yr-old child receive?

29. **BROADBAND INTERNET HOUSEHOLDS** The number of U.S. broadband Internet households stood at 20 million at the beginning of 2002 and is projected to grow at the rate of 7.5 million households per year for the next 6 years.
a. Find a linear function $f(t)$ giving the projected U.S. broadband Internet households (in millions) in year t, where $t = 0$ corresponds to the beginning of 2002.
b. What is the projected size of U.S. broadband Internet households at the beginning of 2008?

Source: Strategy Analytics Inc.

30. **DIAL-UP INTERNET HOUSEHOLDS** The number of U.S. dial-up Internet households stood at 42.5 million at the beginning of 2004 and is projected to decline at the rate of 3.9 million households per year for the next 4 yr.
a. Find a linear function f giving the projected U.S. dial-up Internet households (in millions) in year t, where $t = 0$ corresponds to the beginning of 2004.
b. What is the projected number of U.S. dial-up Internet households at the beginning of 2008?

Source: Strategy Analytics Inc.

31. **CELSIUS AND FAHRENHEIT TEMPERATURES** The relationship between temperature measured in the Celsius scale and the Fahrenheit scale is linear. The freezing point is 0°C and 32°F, and the boiling point is 100°C and 212°F.
a. Find an equation giving the relationship between the temperature F measured in the Fahrenheit scale and the temperature C measured in the Celsius scale.
b. Find F as a function of C and use this formula to determine the temperature in Fahrenheit corresponding to a temperature of 20°C.
c. Find C as a function of F and use this formula to determine the temperature in Celsius corresponding to a temperature of 70°F.

32. **CRICKET CHIRPING AND TEMPERATURE** Entomologists have discovered that a linear relationship exists between the rate of chirping of crickets of a certain species and the air temperature. When the temperature is 70°F, the crickets chirp at the rate of 120 chirps/min, and when the temperature is 80°F, they chirp at the rate of 160 chirps/min.
a. Find an equation giving the relationship between the air temperature T and the number of chirps/min N of the crickets.
b. Find N as a function of T and use this formula to determine the rate at which the crickets chirp when the temperature is 102°F.

33. **DEMAND FOR CLOCK RADIOS** In the accompanying figure, L_1 is the demand curve for the model A clock radio manufactured by Ace Radio, and L_2 is the demand curve for the model B clock radio. Which line has the greater slope? Interpret your results.

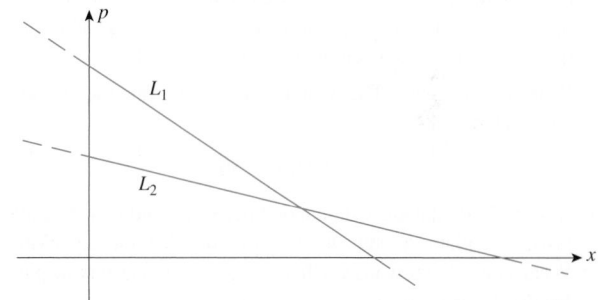

34. **SUPPLY OF CLOCK RADIOS** In the accompanying figure, L_1 is the supply curve for the model A clock radio manufactured by Ace Radio, and L_2 is the supply curve for the model B clock radio. Which line has the greater slope? Interpret your results.

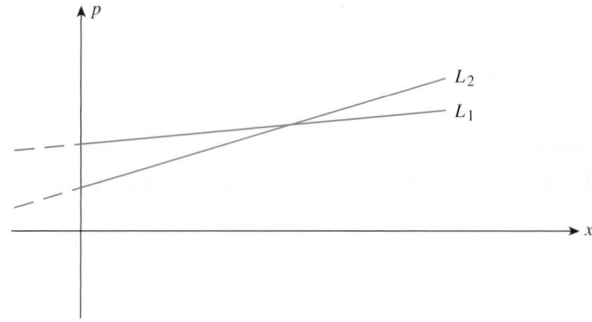

For each demand equation in Exercises 35–38, where x represents the quantity demanded in units of 1000 and p is the unit price in dollars, (a) sketch the demand curve and (b) determine the quantity demanded corresponding to the given unit price p.

35. $2x + 3p - 18 = 0; p = 4$

36. $5p + 4x - 80 = 0; p = 10$

37. $p = -3x + 60; p = 30$ **38.** $p = -0.4x + 120; p = 80$

39. DEMAND FUNCTIONS At a unit price of $55, the quantity demanded of a certain commodity is 1000 units. At a unit price of $85, the demand drops to 600 units. Given that it is linear, find the demand equation. Above what price will there be no demand? What quantity would be demanded if the commodity were free?

40. DEMAND FUNCTIONS The quantity demanded for a certain brand of portable CD players is 200 units when the unit price is set at $90. The quantity demanded is 1200 units when the unit price is $40. Find the demand equation and sketch its graph.

41. DEMAND FUNCTIONS Assume that a certain commodity's demand equation has the form $p = ax + b$, where x is the quantity demanded and p is the unit price in dollars. Suppose the quantity demanded is 1000 units when the unit price is $9.00 and 6000 when the unit price is $4.00. What is the quantity demanded when the unit price is $7.50?

42. DEMAND FUNCTIONS The demand equation for the Sicard wristwatch is

$$p = -0.025x + 50$$

where x is the quantity demanded per week and p is the unit price in dollars. Sketch the graph of the demand equation. What is the highest price (theoretically) anyone would pay for the watch?

For each supply equation in Exercises 43–46, where x is the quantity supplied in units of 1000 and p is the unit price in dollars, (a) sketch the supply curve and (b) determine the number of units of the commodity the supplier will make available in the market at the given unit price.

43. $3x - 4p + 24 = 0; p = 8$

44. $\frac{1}{2}x - \frac{2}{3}p + 12 = 0; p = 24$

45. $p = 2x + 10; p = 14$ **46.** $p = \frac{1}{2}x + 20; p = 28$

47. SUPPLY FUNCTIONS Suppliers of a certain brand of digital voice recorders will make 10,000 available in the market if the unit price is $45. At a unit price of $50, 20,000 units will be made available. Assuming that the relationship between the unit price and the quantity supplied is linear, derive the supply equation. Sketch the supply curve and determine the quantity suppliers will make available when the unit price is $70.

48. SUPPLY FUNCTIONS Producers will make 2000 refrigerators available when the unit price is $330. At a unit price of $390, 6000 refrigerators will be marketed. Find the equation relating the unit price of a refrigerator to the quantity supplied if the equation is known to be linear. How many refrigerators will be marketed when the unit price is $450? What is the lowest price at which a refrigerator will be marketed?

In Exercises 49 and 50, determine whether the statement is true or false. If it is true, explain why it is true. If it is false, give an example to show why it is false.

49. Suppose $C(x) = cx + F$ and $R(x) = sx$ are the cost and revenue functions of a certain firm. Then, the firm is making a profit if its level of production is less than $F/(s - c)$.

50. If $p = mx + b$ is a linear demand curve, then it is generally true that $m < 0$.

1.3 Solutions to Self-Check Exercises

1. Let x denote the number of units produced and sold. Then

  a. $C(x) = 10x + 60,000$

  b. $R(x) = 15x$

  c. $P(x) = R(x) - C(x) = 15x - (10x + 60,000)$
  $$= 5x - 60,000$$

  d. $P(10,000) = 5(10,000) - 60,000$
  $$= -10,000$$
  or a loss of $10,000 per month.
  $P(14,000) = 5(14,000) - 60,000$
  $$= 10,000$$
  or a profit of $10,000 per month.

2. Let p denote the price of a rug (in dollars) and let x denote the quantity of rugs demanded when the unit price is p. The condition that the quantity demanded is 500 when the unit price is $100 tells us that the demand curve passes through the point $(500, 100)$. Next, the condition that for each $20 decrease in the unit price the quantity demanded increases by 500 tells us that the demand curve is linear and that its slope is given by $-\frac{20}{500}$, or $-\frac{1}{25}$. Therefore, letting $m = -\frac{1}{25}$ in the demand equation

$$p = mx + b$$

we find

$$p = -\frac{1}{25}x + b$$

To determine b, use the fact that the straight line passes through (500, 100) to obtain

$$100 = -\frac{1}{25}(500) + b$$

or $b = 120$. Therefore, the required equation is

$$p = -\frac{1}{25}x + 120$$

The graph of the demand curve $p = -\frac{1}{25}x + 120$ is sketched in the accompanying figure.

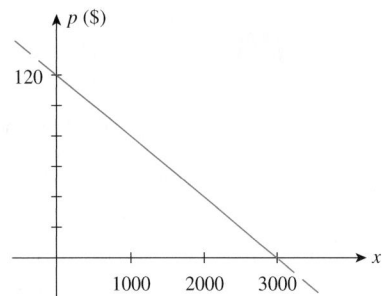

USING TECHNOLOGY

▬ Evaluating a Function

Graphing Utility

A graphing utility can be used to find the value of a function f at a given point with minimal effort. However, to find the value of y for a given value of x in a linear equation such as $Ax + By + C = 0$, the equation must first be cast in the slope-intercept form $y = mx + b$, thus revealing the desired rule $f(x) = mx + b$ for y as a function of x.

EXAMPLE 1   Consider the equation $2x + 5y = 7$.

a. Plot the straight line with the given equation in the standard viewing window.
b. Find the value of y when $x = 2$ and verify your result by direct computation.
c. Find the value of y when $x = 1.732$.

Solution

a. The straight line with equation $2x + 5y = 7$ or, equivalently, $y = -\frac{2}{5}x + \frac{7}{5}$ in the standard viewing window, is shown in Figure T1.
b. Using the evaluation function of the graphing utility and the value of 2 for x, we find $y = 0.6$. This result is verified by computing

$$y = -\frac{2}{5}(2) + \frac{7}{5} = -\frac{4}{5} + \frac{7}{5} = \frac{3}{5} = 0.6$$

when $x = 2$.

c. Once again using the evaluation function of the graphing utility, this time with the value 1.732 for x, we find $y = 0.7072$. The efficacy of the graphing utility is clearly demonstrated here!

FIGURE T1
The straight line $2x + 5y = 7$ in the standard viewing window

When evaluating $f(x)$ at $x = a$, remember that the number a must lie between xMin and xMax.

(continued)

APPLIED EXAMPLE 2   Demand for Electricity   According to Pacific Gas and Electric, the nation's largest utility company, the demand for electricity (in megawatts) in year t is approximately

$$D(t) = 295t + 328 \qquad (0 \le t \le 10)$$

with $t = 0$ corresponding to 1990.

a. Plot the graph of D in the viewing window $[0, 10] \times [0, 3500]$.
b. What was the demand for electricity in 1999?
Source: Pacific Gas and Electric

Solution

a. The graph of D is shown in Figure T2.

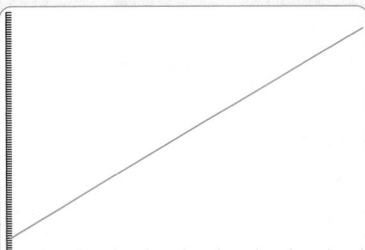

FIGURE T2
The graph of D in the viewing window
$[0, 10] \times [0, 3500]$

b. Evaluating the function at $t = 9$, we find $y = 2983$. Therefore, the demand for electricity in 1999 was approximately 2983 megawatts.   ∎

Excel

Excel can be used to find the value of a function at a given value with minimal effort. However, to find the value of y for a given value of x in a linear equation such as $Ax + By + C = 0$, the equation must first be cast in the slope-intercept form $y = mx + b$, thus revealing the desired rule $f(x) = mx + b$ for y as a function of x.

EXAMPLE 3   Consider the equation $2x + 5y = 7$.

a. Find the value of y for $x = 0$, 5, and 10.
b. Plot the straight line with the given equation over the interval $[0, 10]$.

Solution

a. Since this is a linear equation, we first cast the equation in slope-intercept form:

$$y = -\frac{2}{5}x + \frac{7}{5}$$

Next, we create a table of values (Figure T3), following the same procedure outlined in Example 3, pages 25–26. In this case we use the formula `=(-2/5)*A2+7/5` for the y-values.

	A	B
1	x	y
2	0	1.4
3	5	-0.6
4	10	-2.6

FIGURE T3
Table of values for *x* and *y*

Note: Words/characters printed in a typewriter font (for example, `=(-2/3)*A2+2`) indicate words/characters that need to be typed and entered.

b. Following the procedure outlined in Example 3, page 25, we obtain the graph shown in Figure T4.

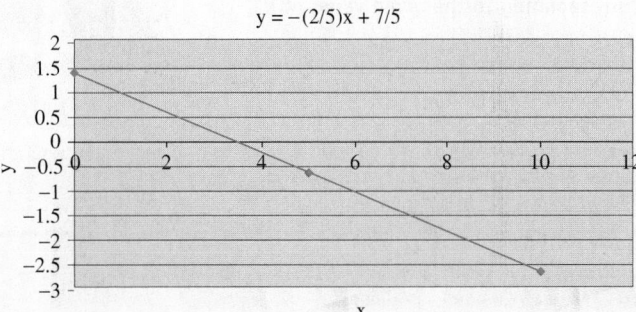

$$y = -(2/5)x + 7/5$$

FIGURE T4
The graph of $y = -\frac{2}{5}x + \frac{7}{5}$ over the interval [0, 10]

APPLIED EXAMPLE 4 Demand for Electricity According to Pacific Gas and Electric, the nation's largest utility company, the demand for electricity (in megawatts) in year t is approximately

$$D(t) = 295t + 328 \qquad (0 \le t \le 10)$$

with $t = 0$ corresponding to 1990.

a. Plot the graph of D over the interval [0, 10].
b. What was the demand for electricity in 1999?
Source: Pacific Gas and Electric

Solution

a. Following the instructions given in Example 3, page 25, we obtain the spreadsheet and graph shown in Figure T5. [*Note*: We have added the value 9 to the table because we are asked to compute the value of $D(t)$ when $t = 9$ in part (b). Also, we have made the appropriate entries for the title and x- and y-axis labels.]

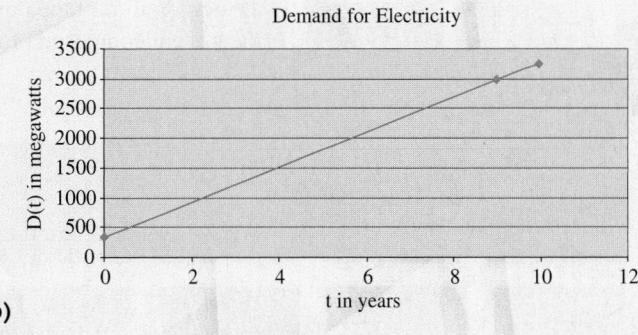

	A	B
1	t	D(t)
2	0	328
3	9	2983
4	10	3278

(a)   **(b)**

FIGURE T5
(a) The table of values for t and $D(t)$ and (b) the graph showing demand for electricity

b. From the table of values, we see that

$$D(9) = 2983$$

or 2983 megawatts.

(continued)

TECHNOLOGY EXERCISES

Find the value of y corresponding to the given value of x.

1. $3.1x + 2.4y - 12 = 0; x = 2.1$

2. $1.2x - 3.2y + 8.2 = 0; x = 1.2$

3. $2.8x + 4.2y = 16.3; x = 1.5$

4. $-1.8x + 3.2y - 6.3 = 0; x = -2.1$

5. $22.1x + 18.2y - 400 = 0; x = 12.1$

6. $17.1x - 24.31y - 512 = 0; x = -8.2$

7. $2.8x = 1.41y - 2.64; x = 0.3$

8. $0.8x = 3.2y - 4.3; x = -0.4$

1.4   Intersection of Straight Lines

Finding the Point of Intersection

The solution of certain practical problems involves finding the point of intersection of two straight lines. To see how such a problem may be solved algebraically, suppose we are given two straight lines L_1 and L_2 with equations

$$y = m_1 x + b_1 \quad \text{and} \quad y = m_2 x + b_2$$

(where m_1, b_1, m_2, and b_2 are constants) that intersect at the point $P(x_0, y_0)$ (Figure 35).

The point $P(x_0, y_0)$ lies on the line L_1 and so satisfies the equation $y = m_1 x + b_1$. It also lies on the line L_2 and so satisfies the equation $y = m_2 x + b_2$. Therefore, to find the point of intersection $P(x_0, y_0)$ of the lines L_1 and L_2, we solve the system composed of the two equations

$$y = m_1 x + b_1 \quad \text{and} \quad y = m_2 x + b_2$$

for x and y.

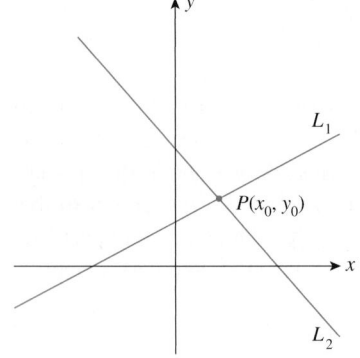

FIGURE 35
L_1 and L_2 intersect at the point $P(x_0, y_0)$.

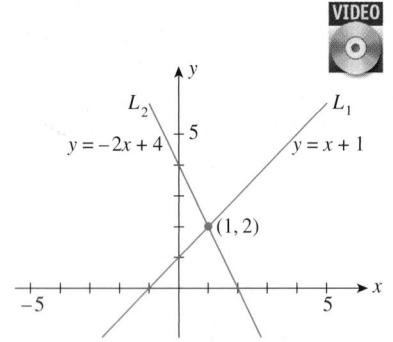

FIGURE 36
The point of intersection of L_1 and L_2 is $(1, 2)$.

EXAMPLE 1   Find the point of intersection of the straight lines that have equations $y = x + 1$ and $y = -2x + 4$.

Solution   We solve the given simultaneous equations. Substituting the value y as given in the first equation into the second, we obtain

$$x + 1 = -2x + 4$$
$$3x = 3$$
$$x = 1$$

Substituting this value of x into either one of the given equations yields $y = 2$. Therefore, the required point of intersection is $(1, 2)$ (Figure 36).

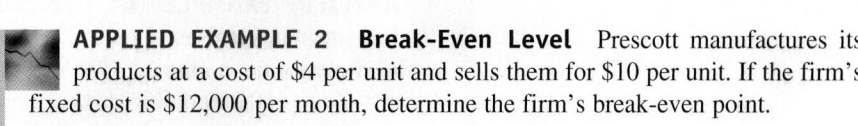

EXPLORING WITH TECHNOLOGY

1. Use a graphing utility to plot the straight lines L_1 and L_2 with equations $y = 3x - 2$ and $y = -2x + 3$, respectively, on the same set of axes in the standard viewing window. Then use TRACE and ZOOM to find the point of intersection of L_1 and L_2. Repeat using the "intersection" function of your graphing utility.

2. Find the point of intersection of L_1 and L_2 algebraically.

3. Comment on the effectiveness of each method.

We now turn to some applications involving the intersections of pairs of straight lines.

Break-Even Analysis

Consider a firm with (linear) cost function $C(x)$, revenue function $R(x)$, and profit function $P(x)$ given by

$$C(x) = cx + F$$
$$R(x) = sx$$
$$P(x) = R(x) - C(x) = (s - c)x - F$$

where c denotes the unit cost of production, s the selling price per unit, F the fixed cost incurred by the firm, and x the level of production and sales. The level of production at which the firm neither makes a profit nor sustains a loss is called the **break-even level of operation** and may be determined by solving the equations $p = C(x)$ and $p = R(x)$ simultaneously. At the level of production x_0, the profit is zero and so

$$P(x_0) = R(x_0) - C(x_0) = 0$$
$$R(x_0) = C(x_0)$$

The point $P_0(x_0, p_0)$, the solution of the simultaneous equations $p = R(x)$ and $p = C(x)$, is referred to as the **break-even point**; the number x_0 and the number p_0 are called the **break-even quantity** and the **break-even revenue**, respectively.

Geometrically, the break-even point $P_0(x_0, p_0)$ is just the point of intersection of the straight lines representing the cost and revenue functions, respectively. This follows because $P_0(x_0, p_0)$, being the solution of the simultaneous equations $p = R(x)$ and $p = C(x)$, must lie on both these lines simultaneously (Figure 37).

Note that if $x < x_0$, then $R(x) < C(x)$ so that $P(x) = R(x) - C(x) < 0$, and thus the firm sustains a loss at this level of production. On the other hand, if $x > x_0$, then $P(x) > 0$ and the firm operates at a profitable level.

APPLIED EXAMPLE 2   Break-Even Level   Prescott manufactures its products at a cost of $4 per unit and sells them for $10 per unit. If the firm's fixed cost is $12,000 per month, determine the firm's break-even point.

Solution   The cost function C and the revenue function R are given by $C(x) = 4x + 12,000$ and $R(x) = 10x$, respectively (Figure 38).

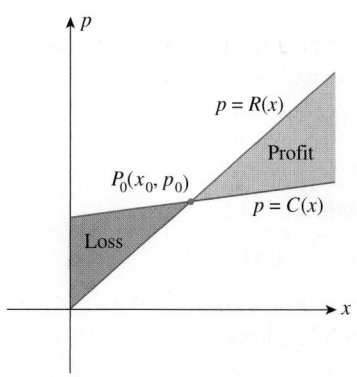

FIGURE 37
P_0 is the break-even point.

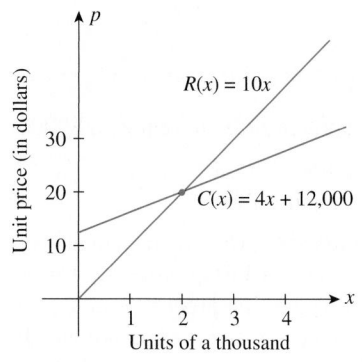

FIGURE 38
The point at which $R(x) = C(x)$ is the break-even point.

Setting $R(x) = C(x)$, we obtain

$$10x = 4x + 12,000$$
$$6x = 12,000$$
$$x = 2000$$

Substituting this value of x into $R(x) = 10x$ gives

$$R(2000) = (10)(2000) = 20,000$$

So, for a break-even operation, the firm should manufacture 2000 units of its product, resulting in a break-even revenue of $20,000 per month. ∎

 APPLIED EXAMPLE 3   Break-Even Analysis   Using the data given in Example 2, answer the following questions:

a. What is the loss sustained by the firm if only 1500 units are produced and sold each month?
b. What is the profit if 3000 units are produced and sold each month?
c. How many units should the firm produce in order to realize a minimum monthly profit of $9000?

Solution   The profit function P is given by the rule

$$P(x) = R(x) - C(x)$$
$$= 10x - (4x + 12,000)$$
$$= 6x - 12,000$$

a. If 1500 units are produced and sold each month, we have

$$P(1500) = 6(1500) - 12,000 = -3000$$

so the firm will sustain a loss of $3000 per month.
b. If 3000 units are produced and sold each month, we have

$$P(3000) = 6(3000) - 12,000 = 6000$$

or a monthly profit of $6000.
c. Substituting 9000 for $P(x)$ in the equation $P(x) = 6x - 12,000$, we obtain

$$9000 = 6x - 12,000$$
$$6x = 21,000$$
$$x = 3500$$

Thus, the firm should produce at least 3500 units in order to realize a $9000 minimum monthly profit. ∎

APPLIED EXAMPLE 4   Decision Analysis The management of Robertson Controls must decide between two manufacturing processes for its model C electronic thermostat. The monthly cost of the first process is given by $C_1(x) = 20x + 10,000$ dollars, where x is the number of thermostats produced; the monthly cost of the second process is given by $C_2(x) = 10x + 30,000$ dollars. If the projected monthly sales are 800 thermostats at a unit price of $40, which process should management choose in order to maximize the company's profit?

Rev = Cost

Solution The break-even level of operation using the first process is obtained by solving the equation

$$40x = 20x + 10,000$$
$$20x = 10,000$$
$$x = 500$$

giving an output of 500 units. Next, we solve the equation

$$40x = 10x + 30,000$$
$$30x = 30,000$$
$$x = 1000$$

giving an output of 1000 units for a break-even operation using the second process. Since the projected sales are 800 units, we conclude that management should choose the first process, which will give the firm a profit. ■

 APPLIED EXAMPLE 5 Decision Analysis Referring to Example 4, decide which process Robertson's management should choose if the projected monthly sales are (a) 1500 units and (b) 3000 units.

Solution In both cases, the production is past the break-even level. Since the revenue is the same regardless of which process is employed, the decision will be based on how much each process costs.

a. If $x = 1500$, then

Cost   Production Cost   Fixed Cost

$$C_1(x) = (20)(1500) + 10,000 = 40,000$$
$$C_2(x) = (10)(1500) + 30,000 = 45,000$$

Hence, management should choose the first process.
b. If $x = 3000$, then

$$C_1(x) = (20)(3000) + 10,000 = 70,000$$
$$C_2(x) = (10)(3000) + 30,000 = 60,000$$

In this case, management should choose the second process. ■

 EXPLORING WITH TECHNOLOGY

1. Use a graphing utility to plot the straight lines L_1 and L_2 with equations $y = 2x - 1$ and $y = 2.1x + 3$, respectively, on the same set of axes, using the standard viewing window. Do the lines appear to intersect?

2. Plot the straight lines L_1 and L_2, using the viewing window $[-100, 100] \times [-100, 100]$. Do the lines appear to intersect? Can you find the point of intersection using TRACE and ZOOM? Using the "intersection" function of your graphing utility?

3. Find the point of intersection of L_1 and L_2 algebraically.

4. Comment on the effectiveness of the solution methods in parts 2 and 3.

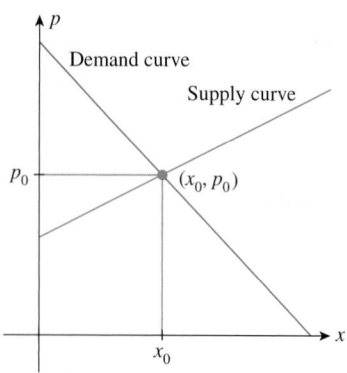

FIGURE 39
Market equilibrium is represented by the point (x_0, p_0).

Market Equilibrium

Under pure competition, the price of a commodity eventually settles at a level dictated by the condition that the supply of the commodity be equal to the demand for it. If the price is too high, consumers will be more reluctant to buy, and if the price is too low, the supplier will be more reluctant to make the product available in the marketplace. Market equilibrium is said to prevail when the quantity produced is equal to the quantity demanded. The quantity produced at market equilibrium is called the equilibrium quantity, and the corresponding price is called the equilibrium price.

From a geometric point of view, market equilibrium corresponds to the point at which the demand curve and the supply curve intersect. In Figure 39, x_0 represents the equilibrium quantity and p_0 the equilibrium price. The point (x_0, p_0) lies on the supply curve and therefore satisfies the supply equation. At the same time, it also lies on the demand curve and therefore satisfies the demand equation. Thus, to find the point (x_0, p_0), and hence the equilibrium quantity and price, we solve the demand and supply equations simultaneously for x and p. For meaningful solutions, x and p must both be positive.

APPLIED EXAMPLE 6   Market Equilibrium   The management of ThermoMaster, which manufactures an indoor–outdoor thermometer at its Mexico subsidiary, has determined that the demand equation for its product is

$$5x + 3p - 30 = 0$$

where p is the price of a thermometer in dollars and x is the quantity demanded in units of a thousand. The supply equation for these thermometers is

$$52x - 30p + 45 = 0$$

where x (measured in thousands) is the quantity that ThermoMaster will make available in the market at p dollars each. Find the equilibrium quantity and price.

Solution   We need to solve the system of equations

$$5x + \ 3p - 30 = 0$$
$$52x - 30p + 45 = 0$$

for x and p. Let us use the *method of substitution* to solve it. As the name suggests, this method calls for choosing one of the equations in the system, solving for one variable in terms of the other, and then substituting the resulting expression into the other equation. This gives an equation in one variable that can then be solved in the usual manner.

Let's solve the first equation for p in terms of x. Thus,

$$3p = -5x + 30$$
$$p = -\frac{5}{3}x + 10$$

Next, we substitute this value of p into the second equation, obtaining

$$52x - 30\left(-\frac{5}{3}x + 10\right) + 45 = 0$$

$$52x + 50x - 300 + 45 = 0$$

$$102x - 255 = 0$$

$$x = \frac{255}{102} = \frac{5}{2}$$

The corresponding value of p is found by substituting this value of x into the equation for p obtained earlier. Thus,

$$p = -\frac{5}{3}\left(\frac{5}{2}\right) + 10 = -\frac{25}{6} + 10$$

$$= \frac{35}{6} \approx 5.83$$

We conclude that the equilibrium quantity is 2500 units (remember that x is measured in units of a thousand) and the equilibrium price is $5.83 per thermometer.

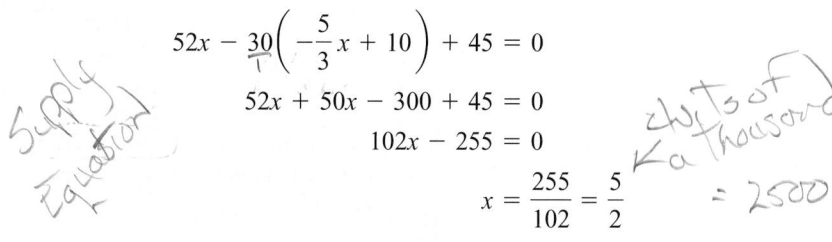

APPLIED EXAMPLE 7 Market Equilibrium The quantity demanded of a certain model of DVD player is 8000 units when the unit price is $260. At a unit price of $200, the quantity demanded increases to 10,000 units. The manufacturer will not market any players if the price is $100 or lower. However, for each $50 increase in the unit price above $100, the manufacturer will market an additional 1000 units. Both the demand and the supply equations are known to be linear.

a. Find the demand equation.
b. Find the supply equation.
c. Find the equilibrium quantity and price.

Solution Let p denote the unit price in hundreds of dollars and let x denote the number of units of players in thousands.

a. Since the demand function is linear, the demand curve is a straight line passing through the points $(8, 2.6)$ and $(10, 2)$. Its slope is

$$m = \frac{2 - 2.6}{10 - 8} = -0.3$$

Using the point $(10, 2)$ and the slope $m = -0.3$ in the point-slope form of the equation of a line, we see that the required demand equation is

$$p - 2 = -0.3(x - 10)$$
$$p = -0.3x + 5 \qquad \text{Figure 40}$$

b. The supply curve is the straight line passing through the points $(0, 1)$ and $(1, 1.5)$. Its slope is

$$m = \frac{1.5 - 1}{1 - 0} = 0.5$$

Using the point $(0, 1)$ and the slope $m = 0.5$ in the point-slope form of the equation of a line, we see that the required supply equation is

$$p - 1 = 0.5(x - 0)$$
$$p = 0.5x + 1 \qquad \text{Figure 40}$$

c. To find the market equilibrium, we solve simultaneously the system comprising the demand and supply equations obtained in parts (a) and (b)—that is, the system

$$p = -0.3x + 5$$
$$p = 0.5x + 1$$

Subtracting the first equation from the second gives

$$0.8x - 4 = 0$$

and $x = 5$. Substituting this value of x in the second equation gives $p = 3.5$. Thus, the equilibrium quantity is 5000 units and the equilibrium price is $350 (Figure 40).

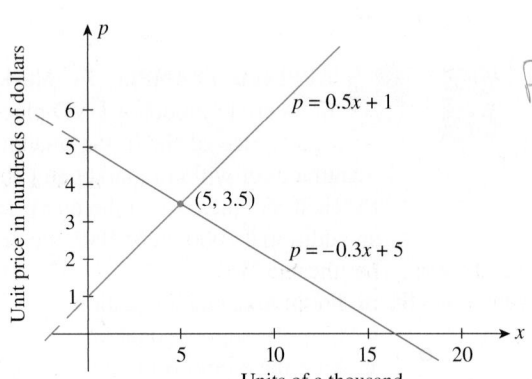

FIGURE 40
Market equilibrium occurs at the point $(5, 3.5)$.

1.4   Self-Check Exercises

1. Find the point of intersection of the straight lines with equations $2x + 3y = 6$ and $x - 3y = 4$.

2. There is no demand for a certain model of a disposable camera when the unit price is $12. However, when the unit price is $8, the quantity demanded is 8000/wk. The suppliers will not market any cameras if the unit price is $2 or lower. At $4/camera, however, the manufacturer will market 5000 cameras/wk.

Both the demand and supply functions are known to be linear.
   a. Find the demand equation.
   b. Find the supply equation.
   c. Find the equilibrium quantity and price.

Solutions to Self-Check Exercises 1.4 can be found on page 51.

1.4 Concept Questions

1. Explain why the intersection of a linear demand curve and a linear supply curve must lie in the first quadrant.

2. Explain the meaning of each term:
   a. Break-even point
   b. Break-even quantity
   c. Break-even revenue

3. Explain the meaning of each term:
   a. Market equilibrium
   b. Equilibrium quantity
   c. Equilibrium price

1.4 Exercises

In Exercises 1–6, find the point of intersection of each pair of straight lines.

1. $y = 3x + 4$
   $y = -2x + 14$

2. $y = -4x - 7$
   $-y = 5x + 10$

3. $2x - 3y = 6$
   $3x + 6y = 16$

4. $2x + 4y = 11$
   $-5x + 3y = 5$

5. $y = \dfrac{1}{4}x - 5$
   $2x - \dfrac{3}{2}y = 1$

6. $y = \dfrac{2}{3}x - 4$
   $x + 3y + 3 = 0$

In Exercises 7–10, find the break-even point for the firm whose cost function C and revenue function R are given.

7. $C(x) = 5x + 10{,}000; R(x) = 15x$

8. $C(x) = 15x + 12{,}000; R(x) = 21x$

9. $C(x) = 0.2x + 120; R(x) = 0.4x$

10. $C(x) = 150x + 20{,}000; R(x) = 270x$

11. BREAK-EVEN ANALYSIS AutoTime, a manufacturer of 24-hr variable timers, has a monthly fixed cost of $48,000 and a production cost of $8 for each timer manufactured. The units sell for $14 each.
    a. Sketch the graphs of the cost function and the revenue function and thereby find the break-even point graphically.
    b. Find the break-even point algebraically.
    c. Sketch the graph of the profit function.
    d. At what point does the graph of the profit function cross the x-axis? Interpret your result.

12. BREAK-EVEN ANALYSIS A division of Carter Enterprises produces "Personal Income Tax" diaries. Each diary sells for $8. The monthly fixed costs incurred by the division are $25,000, and the variable cost of producing each diary is $3.

    a. Find the break-even point for the division.
    b. What should be the level of sales in order for the division to realize a 15% profit over the cost of making the diaries?

13. BREAK-EVEN ANALYSIS A division of the Gibson Corporation manufactures bicycle pumps. Each pump sells for $9, and the variable cost of producing each unit is 40% of the selling price. The monthly fixed costs incurred by the division are $50,000. What is the break-even point for the division?

14. LEASING Ace Truck Leasing Company leases a certain size truck for $30/day and $.15/mi, whereas Acme Truck Leasing Company leases the same size truck for $25/day and $.20/mi.
    a. Find the functions describing the daily cost of leasing from each company.
    b. Sketch the graphs of the two functions on the same set of axes.
    c. If a customer plans to drive at most 70 mi, from which company should he rent a truck for a single day?

15. DECISION ANALYSIS A product may be made using machine I or machine II. The manufacturer estimates that the monthly fixed costs of using machine I are $18,000, whereas the monthly fixed costs of using machine II are $15,000. The variable costs of manufacturing 1 unit of the product using machine I and machine II are $15 and $20, respectively. The product sells for $50 each.
    a. Find the cost functions associated with using each machine.
    b. Sketch the graphs of the cost functions of part (a) and the revenue functions on the same set of axes.
    c. Which machine should management choose in order to maximize their profit if the projected sales are 450 units? 550 units? 650 units?
    d. What is the profit for each case in part (c)?

16. ANNUAL SALES   The annual sales of Crimson Drug Store are expected to be given by $S = 2.3 + 0.4t$ million dollars t yr from now, whereas the annual sales of Cambridge Drug Store are expected to be given by $S = 1.2 + 0.6t$ million dollars t yr from now. When will Cambridge's annual sales first surpass Crimson's annual sales?

17. LCDs VERSUS CRTs   The global shipments of traditional cathode-ray tube monitors (CRTs) is approximated by the equation

$$y = -12t + 88 \qquad (0 \le t \le 3)$$

where y is measured in millions and t in years, with $t = 0$ corresponding to 2001. The equation

$$y = 18t + 13.4 \qquad (0 \le t \le 3)$$

gives the approximate number (in millions) of liquid crystal displays (LCDs) over the same period. When did the global shipments of LCDs first overtake the global shipments of CRTs?

Source: IDC

18. DIGITAL VERSUS FILM CAMERAS   The sales of digital cameras (in millions of units) in year t is given by the function

$$f(t) = 3.05t + 6.85 \qquad (0 \le t \le 3)$$

where $t = 0$ corresponds to 2001. Over that same period, the sales of film cameras (in millions of units) is given by

$$g(t) = -1.85t + 16.58 \qquad (0 \le t \le 3)$$

a. Show that more film cameras than digital cameras were sold in 2001.
b. When did the sales of digital cameras first exceed those of film cameras?

Source: Popular Science

19. U.S. FINANCIAL TRANSACTIONS   The percent of U.S. transactions by check between 2001 ($t = 0$) and 2010 ($t = 9$) is projected to be

$$f(t) = -\frac{11}{9}t + 43 \qquad (0 \le t \le 9)$$

whereas the percent of transactions done electronically during the same period is projected to be

$$g(t) = \frac{11}{3}t + 23 \qquad (0 \le t \le 9)$$

a. Sketch the graphs of f and g on the same set of axes.
b. Find the time when transactions done electronically first exceeded those done by check.

Source: Foreign Policy

20. BROADBAND VERSUS DIAL-UP   The number of U.S. broadband Internet households (in millions) between 2004 ($t = 0$) and 2008 ($t = 4$) was estimated to be

$$f(t) = 7.5t + 35 \qquad (0 \le t \le 4)$$

Over the same period, the number of U.S. dial-up Internet households (in millions) was estimated to be

$$g(t) = -3.9t + 42.5 \qquad (0 \le t \le 4)$$

a. Sketch the graphs of f and g on the same set of axes.
b. Solve the equation $f(t) = g(t)$ and interpret your result.

Source: Strategic Analytics Inc.

For each pair of supply-and-demand equations in Exercises 21–24, where x represents the quantity demanded in units of 1000 and p is the unit price in dollars, find the equilibrium quantity and the equilibrium price.

21. $4x + 3p - 59 = 0$ and $5x - 6p + 14 = 0$

22. $2x + 7p - 56 = 0$ and $3x - 11p + 45 = 0$

23. $p = -2x + 22$ and $p = 3x + 12$

24. $p = -0.3x + 6$ and $p = 0.15x + 1.5$

25. EQUILIBRIUM QUANTITY AND PRICE   The quantity demanded of a certain brand of DVD player is 3000/wk when the unit price is \$485. For each decrease in unit price of \$20 below \$485, the quantity demanded increases by 250 units. The suppliers will not market any DVD players if the unit price is \$300 or lower. But at a unit price of \$525, they are willing to make available 2500 units in the market. The supply equation is also known to be linear.
a. Find the demand equation.
b. Find the supply equation.
c. Find the equilibrium quantity and price.

26. EQUILIBRIUM QUANTITY AND PRICE   The demand equation for the Drake GPS Navigator is $x + 4p - 800 = 0$, where x is the quantity demanded per week and p is the wholesale unit price in dollars. The supply equation is $x - 20p + 1000 = 0$, where x is the quantity the supplier will make available in the market when the wholesale price is p dollars each. Find the equilibrium quantity and the equilibrium price for the GPS Navigators.

27. EQUILIBRIUM QUANTITY AND PRICE   The demand equation for the Schmidt-3000 fax machine is $3x + p - 1500 = 0$, where x is the quantity demanded per week and p is the unit price in dollars. The supply equation is $2x - 3p + 1200 = 0$, where x is the quantity the supplier will make available in the market when the unit price is p dollars. Find the equilibrium quantity and the equilibrium price for the fax machines.

28. EQUILIBRIUM QUANTITY AND PRICE   The quantity demanded each month of Russo Espresso Makers is 250 when the unit price is \$140; the quantity demanded each month is 1000 when the unit price is \$110. The suppliers will market 750 espresso makers if the unit price is \$60 or lower. At a unit price of \$80, they are willing to market 2250 units. Both the demand and supply equations are known to be linear.

a. Find the demand equation.
b. Find the supply equation.
c. Find the equilibrium quantity and the equilibrium price.

29. Suppose the demand-and-supply equations for a certain commodity are given by $p = ax + b$ and $p = cx + d$, respectively, where $a < 0$, $c > 0$, and $b > d > 0$ (see the accompanying figure).

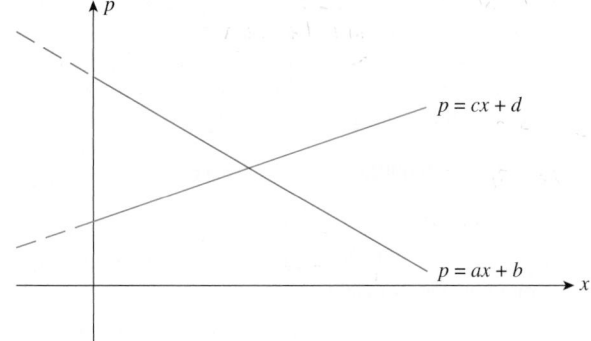

a. Find the equilibrium quantity and equilibrium price in terms of a, b, c, and d.
b. Use part (a) to determine what happens to the market equilibrium if c is increased while a, b, and d remain fixed. Interpret your answer in economic terms.
c. Use part (a) to determine what happens to the market equilibrium if b is decreased while a, c, and d remain fixed. Interpret your answer in economic terms.

30. Suppose the cost function associated with a product is $C(x) = cx + F$ dollars and the revenue function is $R(x) = sx$, where c denotes the unit cost of production, s the unit selling

price, F the fixed cost incurred by the firm, and x the level of production and sales. Find the break-even quantity and the break-even revenue in terms of the constants c, s, and F, and interpret your results in economic terms.

In Exercises 31 and 32, determine whether the statement is true or false. If it is true, explain why it is true. If it is false, give an example to show why it is false.

31. Suppose $C(x) = cx + F$ and $R(x) = sx$ are the cost and revenue functions of a certain firm. Then, the firm is operating at a break-even level of production if its level of production is $F/(s - c)$.

32. If both the demand equation and the supply equation for a certain commodity are linear, then there must be at least one equilibrium point.

33. Let L_1 and L_2 be two nonvertical straight lines in the plane with equations $y = m_1x + b_1$ and $y = m_2x + b_2$, respectively. Find conditions on m_1, m_2, b_1, and b_2 such that (a) L_1 and L_2 do not intersect, (b) L_1 and L_2 intersect at one and only one point, and (c) L_1 and L_2 intersect at infinitely many points.

34. Find conditions on a_1, a_2, b_1, b_2, c_1, and c_2 such that the system of linear equations

$$a_1x + b_1y = c_1$$
$$a_2x + b_2y = c_2$$

has (a) no solution, (b) a unique solution, and (c) infinitely many solutions.
Hint: Use the results of Exercise 33.

1.4 Solutions to Self-Check Exercises

1. The point of intersection of the two straight lines is found by solving the system of linear equations

$$2x + 3y = 6$$
$$x - 3y = 4$$

Solving the first equation for y in terms of x, we obtain

$$y = -\frac{2}{3}x + 2$$

Substituting this expression for y into the second equation, we obtain

$$x - 3\left(-\frac{2}{3}x + 2\right) = 4$$
$$x + 2x - 6 = 4$$
$$3x = 10$$

or $x = \frac{10}{3}$. Substituting this value of x into the expression for y obtained earlier, we find

$$y = -\frac{2}{3}\left(\frac{10}{3}\right) + 2 = -\frac{2}{9}$$

Therefore, the point of intersection is $\left(\frac{10}{3}, -\frac{2}{9}\right)$.

2. **a.** Let p denote the price per camera and x the quantity demanded. The given conditions imply that $x = 0$ when $p = 12$ and $x = 8000$ when $p = 8$. Since the demand equation is linear, it has the form

$$p = mx + b$$

Now, the first condition implies that

$$12 = m(0) + b \quad \text{or} \quad b = 12$$

Therefore,

$$p = mx + 12$$

Using the second condition, we find

$$8 = 8000m + 12$$

$$m = -\frac{4}{8000} = -0.0005$$

Hence, the required demand equation is

$$p = -0.0005x + 12$$

b. Let p denote the price per camera and x the quantity made available at that price. Then, since the supply equation is linear, it also has the form

$$p = mx + b$$

The first condition implies that $x = 0$ when $p = 2$, so we have

$$2 = m(0) + b \quad \text{or} \quad b = 2$$

Therefore,

$$p = mx + 2$$

Next, using the second condition, $x = 5000$ when $p = 4$, we find

$$4 = 5000m + 2$$

giving $m = 0.0004$. So the required supply equation is

$$p = 0.0004x + 2$$

c. The equilibrium quantity and price are found by solving the system of linear equations

$$p = -0.0005x + 12$$
$$p = 0.0004x + 2$$

Equating the two expressions yields

$$-0.0005x + 12 = 0.0004x + 2$$
$$0.0009x = 10$$

or $x \approx 11{,}111$. Substituting this value of x into either equation in the system yields

$$p \approx 6.44$$

Therefore, the equilibrium quantity is 11,111, and the equilibrium price is $6.44.

USING TECHNOLOGY

■ Finding the Point(s) of Intersection of Two Graphs

Graphing Utility

A graphing utility can be used to find the point(s) of intersection of the graphs of two functions. Once again, it is important to remember that if the graphs are straight lines, the linear equations defining these lines must first be recast in the slope-intercept form.

EXAMPLE 1   Find the points of intersection of the straight lines with equations $2x + 3y = 6$ and $3x - 4y - 5 = 0$.

Solution   Solving each equation for y in terms of x, we obtain

$$y = -\frac{2}{3}x + 2 \quad \text{and} \quad y = \frac{3}{4}x - \frac{5}{4}$$

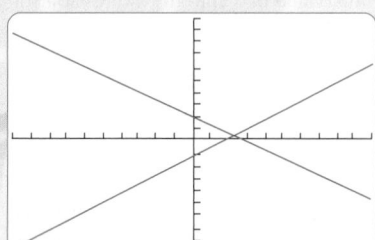

FIGURE T1
The straight lines $2x + 3y = 6$ and $3x - 4y - 5 = 0$

as the respective equations in the slope-intercept form. The graphs of the two straight lines in the standard viewing window are shown in Figure T1.

Then, using TRACE and ZOOM or the function for finding the point of intersection of two graphs, we find that the point of intersection, accurate to four decimal places, is (2.2941, 0.4706). ■

TECHNOLOGY EXERCISES

In Exercises 1–6, find the point of intersection of the pair of straight lines with the given equations. Round your answers to four decimal places.

1. $y = 2x + 5$ and $y = -3x + 8$

2. $y = 1.2x + 6.2$ and $y = -4.3x + 9.1$

3. $2x - 5y = 7$ and $3x + 4y = 12$

4. $1.4x - 6.2y = 8.4$ and $4.1x + 7.3y = 14.4$

5. $2.1x = 5.1y + 71$ and $3.2x = 8.4y + 16.8$

6. $8.3x = 6.2y + 9.3$ and $12.4x = 12.3y + 24.6$

7. BREAK-EVEN ANALYSIS PhotoMax makes disposable cameras that sell for $7.89 each and cost $3.24 each to produce. The weekly fixed cost for the company is $16,500.
   a. Plot the graphs of the cost function and the revenue function in the viewing window $[0, 6000] \times [0, 60{,}000]$.
   b. Find the break-even point by using the viewing window $[0, 6000] \times [-20{,}000, 20{,}000]$.
   c. Plot the graph of the profit function and verify the result of part (b) by finding the x-intercept.

8. BREAK-EVEN ANALYSIS The Monde Company makes a wine cooler with a capacity of 24 bottles. Each wine cooler sells for $245. The monthly fixed costs incurred by the company are $385,000, and the variable cost of producing each wine cooler is $90.50.
   a. Find the break-even point for the company.
   b. Find the level of sales needed to ensure that the company will realize a profit of 21% over the cost of producing the wine coolers.

9. LEASING Ace Truck Leasing Company leases a certain size truck for $34/day and $.18/mi, whereas Acme Truck Leasing Company leases the same size truck for $28/day and $.22/mi.
   a. Find the functions describing the daily cost of leasing from each company.
   b. Plot the graphs of the two functions using the same viewing window.
   c. Find the point of intersection of the graphs of part (b).
   d. Use the result of part (c) to find a criterion that a customer can use to help her decide which company to rent the truck from if she knows the maximum distance that she will drive on the day of rental.

10. BANK DEPOSITS The total deposits with a branch of Randolph Bank currently stand at $20.384 million and are projected to grow at the rate of $1.019 million/yr for the next 5 yr. The total deposits with a branch of Madison Bank, in the same shopping complex as the Randolph Bank, currently stand at $18.521 million and are expected to grow at the rate of $1.482 million/yr for the next 5 yr.
   a. Find the function describing the total deposits with each bank for the next 5 yr.
   b. Plot the graphs of the two functions found in part (a) using the same viewing window.
   c. Do the total deposits of Madison catch up to those of Randolph over the period in question? If so, at what time?

11. EQUILIBRIUM QUANTITY AND PRICE The quantity demanded of a certain brand of a cellular phone is 2000/wk when the unit price is $84. For each decrease in unit price of $5 below $84, the quantity demanded increases by 50 units. The supplier will not market any of the cellular phones if the unit price is $60 or less, but the supplier will market 1800/wk if the unit price is $90. The supply equation is also known to be linear.
   a. Find the demand and supply equations.
   b. Plot the graphs of the supply and demand equations and find their point of intersection.
   c. Find the equilibrium quantity and price.

12. EQUILIBRIUM QUANTITY AND PRICE The demand equation for the Miramar Heat Machine, a ceramic heater, is $1.1x + 3.8p - 901 = 0$, where x is the quantity demanded each week and p is the wholesale unit price in dollars. The corresponding supply equation is $0.9x - 20.4p + 1038 = 0$, where x is the quantity the supplier will make available in the market when the wholesale price is p dollars each.
   a. Plot the graphs of the demand and supply equations using the same viewing window.
   b. Find the equilibrium quantity and the equilibrium price for the Miramar heaters.

1.5   The Method of Least Squares (Optional)

■ The Method of Least Squares

In Example 10, Section 1.2, we saw how a linear equation may be used to approximate the sales trend for a local sporting goods store. The *trend line*, as we saw, may be used to predict the store's future sales. Recall that we obtained the trend line in Example 10 by requiring that the line pass through two data points, the rationale being that such a line seems to *fit* the data reasonably well.

In this section we describe a general method known as the **method of least squares** for determining a straight line that, in some sense, best fits a set of data points when the points are scattered about a straight line. To illustrate the principle behind the method of least squares, suppose, for simplicity, that we are given five data points,

$$P_1(x_1, y_1), P_2(x_2, y_2), P_3(x_3, y_3), P_4(x_4, y_4), P_5(x_5, y_5)$$

describing the relationship between the two variables x and y. By plotting these data points, we obtain a graph called a **scatter diagram** (Figure 41).

If we try to *fit* a straight line to these data points, the line will miss the first, second, third, fourth, and fifth data points by the amounts $d_1, d_2, d_3, d_4,$ and d_5, respectively (Figure 42). We can think of the amounts $d_1, d_2, \ldots, d_5$ as the errors made when the values $y_1, y_2, \ldots, y_5$ are approximated by the corresponding values of y lying on the straight line L.

The **principle of least squares** states that the straight line L that fits the data points *best* is the one chosen by requiring that the sum of the squares of $d_1, d_2, \ldots, d_5$, that is,

$$d_1^2 + d_2^2 + d_3^2 + d_4^2 + d_5^2$$

be made as small as possible. In other words, the least-squares criterion calls for minimizing the sum of the squares of the errors. The line L obtained in this manner is called the **least-squares line**, or *regression line*.

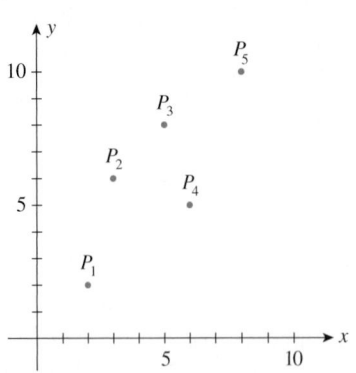

FIGURE 41
A scatter diagram

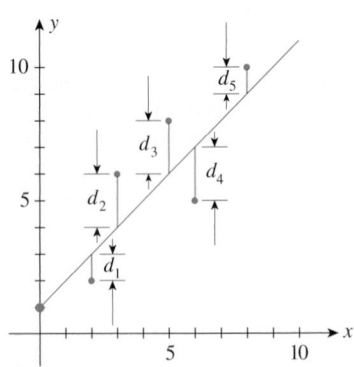

FIGURE 42
d_i is the vertical distance between the straight line and a given data point.

The method for computing the least-squares lines that best fits a set of data points is contained in the following result, which we state without proof.

The Method of Least Squares

Suppose we are given n data points

$$P_1(x_1, y_1), P_2(x_2, y_2), P_3(x_3, y_3), \dots, P_n(x_n, y_n)$$

Then the least-squares (regression) line for the data is given by the linear equation (function)

$$y = f(x) = mx + b$$

where the constants m and b satisfy the **normal equations**

$$nb + (x_1 + x_2 + \cdots + x_n)m = y_1 + y_2 + \cdots + y_n \qquad \textbf{(9)}$$

$$(x_1 + x_2 + \cdots + x_n)b + (x_1^2 + x_2^2 + \cdots + x_n^2)m$$
$$= x_1 y_1 + x_2 y_2 + \cdots + x_n y_n \qquad \textbf{(10)}$$

simultaneously.

EXAMPLE 1 Find the least-squares line for the data

$$P_1(1, 1), P_2(2, 3), P_3(3, 4), P_4(4, 3), P_5(5, 6)$$

Solution Here we have $n = 5$ and

$$x_1 = 1 \qquad x_2 = 2 \qquad x_3 = 3 \qquad x_4 = 4 \qquad x_5 = 5$$
$$y_1 = 1 \qquad y_2 = 3 \qquad y_3 = 4 \qquad y_4 = 3 \qquad y_5 = 6$$

Before using Equations (9) and (10), it is convenient to summarize this data in the form of a table:

	x	y	x^2	xy
	1	1	1	1
	2	3	4	6
	3	4	9	12
	4	3	16	12
	5	6	25	30
Sum	15	17	55	61

Using this table and (9) and (10), we obtain the normal equations

$$5b + 15m = 17 \qquad \textbf{(11)}$$

$$15b + 55m = 61 \qquad \textbf{(12)}$$

Solving Equation (11) for b gives

$$b = -3m + \frac{17}{5} \qquad \textbf{(13)}$$

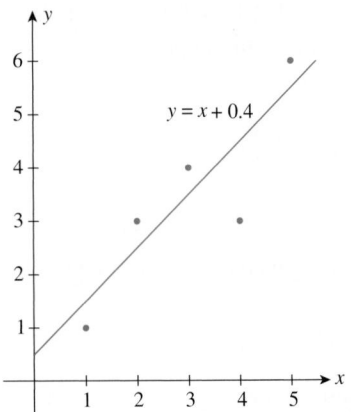

FIGURE 43
The least-squares line $y = x + 0.4$ and the given data points

which upon substitution into Equation (12) gives

$$15\left(-3m + \frac{17}{5}\right) + 55m = 61$$

$$-45m + 51 + 55m = 61$$

$$10m = 10$$

$$m = 1$$

Substituting this value of m into Equation (13) gives

$$b = -3 + \frac{17}{5} = \frac{2}{5} = 0.4$$

Therefore, the required least-squares line is

$$y = x + 0.4$$

The scatter diagram and the least-squares line are shown in Figure 43. ∎

APPLIED EXAMPLE 2   Advertising and Profit   The proprietor of Leisure Travel Service compiled the following data relating the annual profit of the firm to its annual advertising expenditure (both measured in thousands of dollars):

Annual Advertising Expenditure, x	12	14	17	21	26	30
Annual Profit, y	60	70	90	100	100	120

a. Determine the equation of the least-squares line for these data.
b. Draw a scatter diagram and the least-squares line for these data.
c. Use the result obtained in part (a) to predict Leisure Travel's annual profit if the annual advertising budget is $20,000.

Solution

a. The calculations required for obtaining the normal equations are summarized in the following table:

	x	y	x^2	xy
	12	60	144	720
	14	70	196	980
	17	90	289	1,530
	21	100	441	2,100
	26	100	676	2,600
	30	120	900	3,600
Sum	120	540	2646	11,530

The normal equations are

$$6b +\ 120m = 540 \tag{14}$$

$$120b + 2646m = 11{,}530 \tag{15}$$

Solving Equation (14) for b gives

$$b = -20m + 90 \qquad \text{(16)}$$

which upon substitution into Equation (15) gives

$$120(-20m + 90) + 2646m = 11{,}530$$
$$-2400m + 10{,}800 + 2646m = 11{,}530$$
$$246m = 730$$
$$m \approx 2.97$$

Substituting this value of m into Equation (16) gives

$$b = -20(2.97) + 90$$
$$= 30.6$$

Therefore, the required least-squares line is given by

$$y = f(x) = 2.97x + 30.6$$

b. The scatter diagram and the least-squares line are shown in Figure 44.

c. Leisure Travel's predicted annual profit corresponding to an annual budget of $20,000 is given by

$$f(20) = 2.97(20) + 30.6 = 90$$

or $90,000. ∎

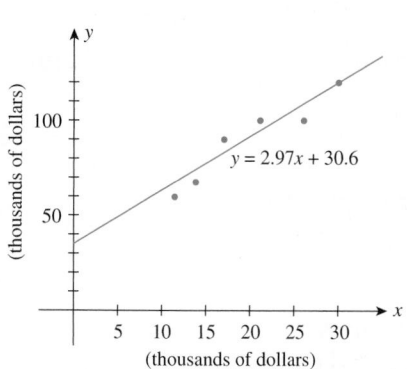

FIGURE 44
Profit versus advertising expenditure

APPLIED EXAMPLE 3   Market for Cholesterol-Reducing Drugs
Refer to Example 1 of Section 1.3. In a study conducted in early 2000, experts projected a rise in the market for cholesterol-reducing drugs. The U.S. market (in billions of dollars) for such drugs from 1999 through 2004 is given in the following table:

Year	1999	2000	2001	2002	2003	2004
Market	12.07	14.07	16.21	18.28	20	21.72

Find a function giving the U.S. market for cholesterol-reducing drugs between 1999 and 2004, using the least-squares technique. (Here, $t = 0$ corresponds to 1999.)

Solution The calculations required for obtaining the normal equations are summarized in the following table:

	t	y	t^2	ty
	0	12.07	0	0
	1	14.07	1	14.07
	2	16.21	4	32.42
	3	18.28	9	54.84
	4	20.0	16	80
	5	21.72	25	108.6
Sum	15	102.35	55	289.93

The normal equations are

$$6b + 15m = 102.35 \tag{17}$$
$$15b + 55m = 289.93 \tag{18}$$

Solving Equation (17) for b gives

$$b = -2.5m + 17.058 \tag{19}$$

which upon substitution into Equation (18) gives

$$15(-2.5m + 17.058) + 55m = 289.93$$
$$-37.5m + 255.87 + 55m = 289.93$$
$$17.5m = 34.06$$
$$m \approx 1.946$$

Substituting this value of m into Equation (19) gives

$$b = -2.5(1.946) + 17.058 = 12.193$$

Therefore, the required function is

$$M(t) = 1.95t + 12.19$$

as obtained before.

1.5 Self-Check Exercises

1. Find an equation of the least-squares line for the data

x	1	3	4	5	7
y	4	10	11	12	16

2. In a market research study for Century Communications, the following data were provided based on the projected monthly sales x (in thousands) of a DVD version of a box-office-hit adventure movie with a proposed wholesale unit price of p dollars.

x	2.2	5.4	7.0	11.5	14.6
p	38	36	34.5	30	28.5

Find the demand equation if the demand curve is the least-squares line for these data.

Solutions to Self-Check Exercises 1.5 can be found on page 62.

1.5 Concept Questions

1. Explain the terms (a) scatter diagram and (b) least-squares line.

2. State the principle of least squares in your own words.

1.5 Exercises

In Exercises 1–6, (a) find the equation of the least-squares line for the data and (b) draw a scatter diagram for the data and graph the least-squares line.

1.

x	1	2	3	4
y	4	6	8	11

2.

x	1	3	5	7	9
y	9	8	6	3	2

3.

x	1	2	3	4	4	6
y	4.5	5	3	2	3.5	1

4.

x	1	1	2	3	4	4	5
y	2	3	3	3.5	3.5	4	5

5. $P_1(1, 3), P_2(2, 5), P_3(3, 5), P_4(4, 7), P_5(5, 8)$

6. $P_1(1, 8), P_2(2, 6), P_3(5, 6), P_4(7, 4), P_5(10, 1)$

7. COLLEGE ADMISSIONS The accompanying data were compiled by the admissions office at Faber College during the past 5 yr. The data relate the number of college brochures and follow-up letters (x) sent to a preselected list of high school juniors who had taken the PSAT and the number of completed applications (y) received from these students (both measured in units of a thousand).

x	4	4.5	5	5.5	6
y	0.5	0.6	0.8	0.9	1.2

a. Determine the equation of the least-squares line for these data.

b. Draw a scatter diagram and the least-squares line for these data.

c. Use the result obtained in part (a) to predict the number of completed applications expected if 6400 brochures and follow-up letters are sent out during the next year.

8. STARBUCKS STORE COUNT According to *Company Reports*, the number of Starbucks stores worldwide between 1999 and 2003 are as follows:

Year, x	0	1	2	3	4
Stores, y	2135	3501	4709	5886	7225

(Here, $x = 0$ corresponds to 1999.)

a. Find an equation of the least-squares line for these data.

b. Use the result of part (a) to estimate the rate at which new stores were opened annually in North America for the period in question.

Source: Company Reports

9. SAT VERBAL SCORES The accompanying data were compiled by the superintendent of schools in a large metropolitan area. The table shows the average SAT verbal scores of high-school seniors during the 5 yr since the district implemented its "back to basics" program.

Year, x	1	2	3	4	5
Average Score, y	436	438	428	430	426

a. Determine the equation of the least-squares line for these data.

b. Draw a scatter diagram and the least-squares line for these data.

c. Use the result obtained in part (a) to predict the average SAT verbal score of high-school seniors 2 yr from now ($x = 7$).

10. NET SALES The management of Kaldor, a manufacturer of electric motors, submitted the accompanying data in its annual stockholders report. The table shows the net sales (in millions of dollars) during the 5 yr that have elapsed since the new management team took over.

Year, x	1	2	3	4	5
Net Sales, y	426	437	460	473	477

(The first year the firm operated under the new management corresponds to the time period $x = 1$, and the four subsequent years correspond to $x = 2, 3, 4,$ and 5.)

a. Determine the equation of the least-squares line for these data.

b. Draw a scatter diagram and the least-squares line for these data.

c. Use the result obtained in part (a) to predict the net sales for the upcoming year.

11. MASS-TRANSIT SUBSIDIES The accompanying table gives the projected state subsidies (in millions of dollars) to the Massachusetts Bay Transit Authority (MBTA) over a 5-yr period.

Year, x	1	2	3	4	5
Subsidy, y	20	24	26	28	32

a. Find an equation of the least-squares line for these data.

b. Use the result of part (a) to estimate the state subsidy to the MBTA for the eighth year ($x = 8$).

Source: Massachusetts Bay Transit Authority

12. INFORMATION SECURITY SOFTWARE SALES As online attacks persist, spending on information security software continues to rise. The following table gives the forecast for the worldwide sales (in billions of dollars) of information security software through 2007:

Year, t	0	1	2	3	4	5
Spending, y	6.8	8.3	9.8	11.3	12.8	14.9

(Here, $t = 0$ corresponds to 2002.)

a. Find an equation of the least-squares line for these data.

b. Use the result of part (a) to forecast the spending on information security software in 2008, assuming that the trend continues.

Source: International Data Corp.

13. U.S. DRUG SALES The following table gives the total sales of drugs (in billions of dollars) in the United States from 1999 ($t = 0$) through 2003:

Year, t	0	1	2	3	4
Sales, y	126	144	171	191	216

a. Find an equation of the least-squares line for these data.

b. Use the result of part (a) to predict the total sales of drugs in 2005, assuming that the trend continued.

Source: IMS Health

14. NURSES SALARIES The average hourly salary of hospital nurses in metropolitan Boston (in dollars) from 2000 through 2004 is given in the following table:

Year	2000	2001	2002	2003	2004
Hourly Salary, y	27	29	31	32	35

a. Find an equation of the least-squares line for these data. (Let $x = 0$ represent 2000.)

b. If the trend continued, what was the average hourly salary of nurses in 2006?

Source: American Association of Colleges of Nursing

15. NET-CONNECTED COMPUTERS IN EUROPE The projected number of computers (in millions) connected to the Internet in Europe from 1998 through 2002 is summarized in the accompanying table:

Year, x	0	1	2	3	4
Net-Connected Computers, y	21.7	32.1	45.0	58.3	69.6

(Here, $x = 0$ corresponds to the beginning of 1998.)

a. Find an equation of the least-squares line for these data.

b. Use the result of part (a) to estimate the projected number of computers connected to the Internet in Europe at the beginning of 2005, assuming the trend continued.

Source: Dataquest Inc.

16. MALE LIFE EXPECTANCY AT 65 The Census Bureau projections of male life expectancy at age 65 in the United States are summarized in the following table:

Year, x	0	10	20	30	40	50
Years beyond 65, y	15.9	16.8	17.6	18.5	19.3	20.3

(Here, $x = 0$ corresponds to 2000.)

a. Find an equation of the least-squares line for these data.

b. Use the result of part (a) to estimate the life expectancy at 65 of a male in 2040. How does this result compare with the given data for that year?

c. Use the result of part (a) to estimate the life expectancy at 65 of a male in 2030.

Source: U.S. Census Bureau

17. FEMALE LIFE EXPECTANCY AT 65 The Census Bureau projections of female life expectancy at age 65 in the United States are summarized in the following table:

Year, x	0	10	20	30	40	50
Years beyond 65, y	19.5	20.0	20.6	21.2	21.8	22.4

(Here, $x = 0$ corresponds to 2000.)

a. Find an equation of the least-squares line for these data.

b. Use the result of part (a) to estimate the life expectancy at 65 of a female in 2040. How does this result compare with the given data for that year?

c. Use the result of part (a) to estimate the life expectancy at 65 of a female in 2030.

Source: U.S. Census Bureau

18. U.S. ONLINE BANKING HOUSEHOLDS The following table gives the projected U.S. online banking households as a percentage of all U.S. banking households from 2001 ($x = 1$) through 2007 ($x = 7$):

Year, x	1	2	3	4	5	6	7
Percent of Households, y	21.2	26.7	32.2	37.7	43.2	48.7	54.2

a. Find an equation of the least-squares line for these data.

b. Use the result of part (a) to estimate the projected percentage of U.S. online banking households in 2008.

Source: Jupiter Research

19. SALES OF GPS EQUIPMENT The annual sales (in billions of dollars) of global positioning system (GPS) equipment from the year 2000 through 2006 follow:

Year, x	0	1	2	3	4	5	6
Annual Sales, y	7.9	9.6	11.5	13.3	15.2	16	18.8

(Here, $x = 0$ corresponds to the year 2000.)

a. Find an equation of the least-squares line for these data.

b. Use the equation found in part (a) to estimate the annual sales of GPS equipment for 2008, assuming the trend continued.

Source: ABI Research

20. ONLINE SALES OF USED AUTOS   The amount (in millions of dollars) of used autos sold online in the United States is expected to grow in accordance with the figures given in the following table:

Year, x	0	1	2	3	4	5	6	7
Sales, y	1	1.4	2.2	2.8	3.6	4.2	5.0	5.8

(Here, $x = 0$ corresponds to 2000.)
a. Find an equation of the least-squares line for these data.
b. Use the result of part (a) to estimate the sales of used autos online in 2008, assuming that the predicted trend continues through that year.

Source: comScore Networks Inc.

21. WIRELESS SUBSCRIBERS   The projected number of wireless subscribers y (in millions) from 2000 through 2006 is summarized in the accompanying table:

Year, x	0	1	2	3
Subscribers, y	90.4	100.0	110.4	120.4

Year, x	4	5	6
Subscribers, y	130.8	140.4	150.0

(Here, $x = 0$ corresponds to the beginning of 2000.)
a. Find an equation of the least-squares line for these data.
b. Use the result of part (a) to estimate the projected number of wireless subscribers at the beginning of 2006. How does this result compare with the given data for that year?

Source: BancAmerica Robertson Stephens

22. AUTHENTICATION TECHNOLOGY   With computer security always a hot-button issue, demand is growing for technology that authenticates and authorizes computer users. The following table gives the authentication software sales (in billions of dollars) from 1999 through 2004:

Year, x	0	1	2	3	4	5
Sales, y	2.4	2.9	3.7	4.5	5.2	6.1

(Here, $x = 0$ represents 1999.)
a. Find an equation of the least-squares line for these data.
b. Use the result of part (a) to estimate the sales for 2007, assuming the trend continues.

Source: International Data Corporation

23. ONLINE SPENDING   The convenience of shopping on the web combined with high-speed broadband access services are spurring online spending. The projected online spending per buyer (in dollars) from 2002 ($x = 0$) through 2008 ($x = 6$) is given in the following table:

Year, x	0	1	2	3	4	5	6
Spending, y	501	540	585	631	680	728	779

a. Find an equation of the least-squares line for these data.
b. Use the result of part (a) to estimate the rate of change of spending per buyer between 2002 and 2008.

Source: Commerce Dept.

24. CALLING CARDS   The market for prepaid calling cards is projected to grow steadily through 2008. The following table gives the projected sales of prepaid phone card sales (in billions of dollars) from 2002 through 2008:

Year, x	0	1	2	3	4	5	6
Sales, y	3.7	4.0	4.4	4.8	5.2	5.8	6.3

a. Find an equation of the least-squares line for these data.
b. Use the result of part (a) to estimate the rate at which the sales of prepaid phone cards will grow over the period in question.

Source: Atlantic-ACM

25. SOCIAL SECURITY WAGE BASE   The Social Security (FICA) wage base (in thousands of dollars) from 2001 to 2006 is given in the accompanying table:

Year	2001	2002	2003
Wage Base, y	80.4	84.9	87

Year	2004	2005	2006
Wage Base, y	87.9	90	94.2

a. Find an equation of the least-squares line for these data. (Let $x = 1$ represent 2001.)
b. Use the result of part (a) to estimate the FICA wage base in 2010.

Source: The World Almanac

26. OUTPATIENT VISITS   With an aging population, the demand for health care, as measured by outpatient visits, is steadily growing. The number of outpatient visits (in millions) from 1991 through 2001 is recorded in the following table:

Year, x	0	1	2	3	4	5
Number of Visits, y	320	340	362	380	416	440

Year, x	6	7	8	9	10
Number of Visits, y	444	470	495	520	530

(Here, $x = 0$ corresponds to 1991.)
a. Find an equation of the least-squares line for these data.
b. Use the result of part (a) to estimate the number of outpatient visits in 2004, assuming that the trend continued.

Source: PriceWaterhouse Coopers

In Exercises 27–30, determine whether the statement is true or false. If it is true, explain why it is true. If it is false, give an example to show why it is false.

27. The least-squares line must pass through at least one data point.

28. The error incurred in approximating n data points using the least-squares linear function is zero if and only if the n data points lie on a straight line.

29. If the data consist of two distinct points, then the least-squares line is just the line that passes through the two points.

30. A data point lies on the least-squares line if and only if the vertical distance between the point and the line is equal to zero.

1.5 Solutions to Self-Check Exercises

1. The calculations required for obtaining the normal equations may be summarized as follows:

	x	y	x^2	xy
	1	4	1	4
	3	10	9	30
	4	11	16	44
	5	12	25	60
	7	16	49	112
Sum	20	53	100	250

The normal equations are

$$5b + 20m = 53$$
$$20b + 100m = 250$$

Solving the first equation for b gives

$$b = -4m + \frac{53}{5}$$

which, upon substitution into the second equation, yields

$$20\left(-4m + \frac{53}{5}\right) + 100m = 250$$
$$-80m + 212 + 100m = 250$$
$$20m = 38$$
$$m = 1.9$$

Substituting this value of m into the expression for b found earlier, we find

$$b = -4(1.9) + \frac{53}{5} = 3$$

Therefore, an equation of the least-squares line is

$$y = 1.9x + 3$$

2. The calculations required for obtaining the normal equations may be summarized as follows:

	x	p	x^2	xp
	2.2	38.0	4.84	83.6
	5.4	36.0	29.16	194.4
	7.0	34.5	49.00	241.5
	11.5	30.0	132.25	345.0
	14.6	28.5	213.16	416.1
Sum	40.7	167.0	428.41	1280.6

The normal equations are

$$5b + 40.7m = 167$$
$$40.7b + 428.41m = 1280.6$$

Solving this system of linear equations simultaneously, we find that

$$m \approx -0.81 \quad \text{and} \quad b \approx 39.99$$

Therefore, an equation of the least-squares line is given by

$$p = f(x) = -0.81x + 39.99$$

which is the required demand equation provided

$$0 \le x \le 49.37$$

USING TECHNOLOGY

■ Finding an Equation of a Least-Squares Line

Graphing Utility

A graphing utility is especially useful in calculating an equation of the least-squares line for a set of data. We simply enter the given data in the form of lists into the calculator and then use the linear regression function to obtain the coefficients of the required equation.

EXAMPLE 1 Find an equation of the least-squares line for the data

x	1.1	2.3	3.2	4.6	5.8	6.7	8
y	−5.8	−5.1	−4.8	−4.4	−3.7	−3.2	−2.5

Plot the scatter diagram and the least-squares line for this data.

Solution First, we enter the data as

$$x_1 = 1.1 \qquad y_1 = -5.8 \qquad x_2 = 2.3 \qquad y_2 = -5.1 \qquad x_3 = 3.2$$
$$y_3 = -4.8 \qquad x_4 = 4.6 \qquad y_4 = -4.4 \qquad x_5 = 5.8 \qquad y_5 = -3.7$$
$$x_6 = 6.7 \qquad y_6 = -3.2 \qquad x_7 = 8 \qquad y_7 = -2.5$$

Then, using the linear regression function from the statistics menu, we obtain the output shown in Figure T1a. Therefore, an equation of the least-squares line $(y = a + bx)$ is

$$y = -6.3 + 0.46x$$

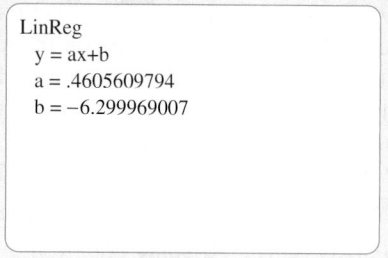

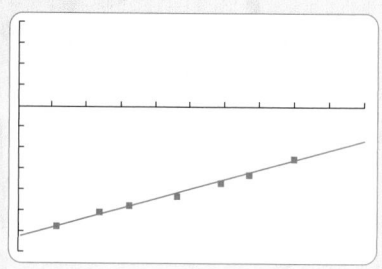

FIGURE T1

(a) The TI-83 linear regression screen

(b) The scatter diagram and least-squares line for the data

The graph of the least-squares equation and the scatter diagram for the data are shown in Figure T1b. ■

APPLIED EXAMPLE 2 Demand for Electricity According to Pacific Gas and Electric, the nation's largest utility company, the demand for electricity from 1990 through the year 2000 is summarized in the following table:

(continued)

t	0	2	4	6	8	10
y	333	917	1500	2117	2667	3292

Here $t = 0$ corresponds to 1990, and y gives the amount of electricity demanded in the year t, measured in megawatts. Find an equation of the least-squares line for these data.
Source: Pacific Gas and Electric

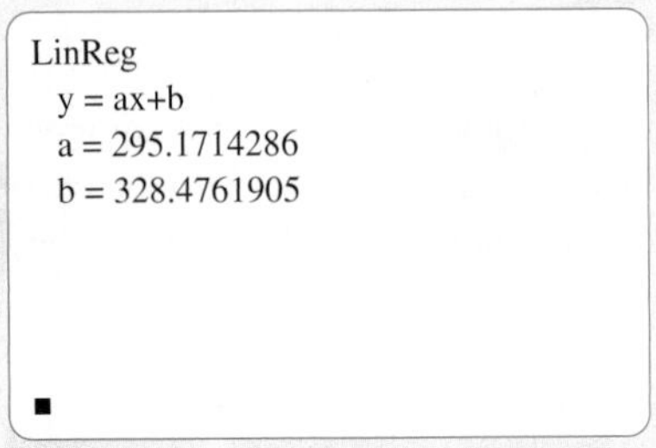

Solution   First, we enter the data as

$$x_1 = 0 \qquad y_1 = 333 \qquad x_2 = 2 \qquad y_2 = 917 \qquad x_3 = 4 \qquad y_3 = 1500$$
$$x_4 = 6 \qquad y_4 = 2117 \qquad x_5 = 8 \qquad y_5 = 2667 \qquad x_6 = 10 \qquad y_6 = 3292$$

Then, using the linear regression function from the statistics menu, we obtain the output shown in Figure T2. Therefore, an equation of the least-squares line is

$$y = 328 + 295t$$

LinReg
  y = ax+b
  a = 295.1714286
  b = 328.4761905

FIGURE T2
The TI-83 linear regression screen

Excel

Excel can be used to find an equation of the least-squares line for a set of data and to plot a scatter diagram and the least-squares line for the data.

EXAMPLE 3   Find an equation of the least-squares line for the data given in the following table:

x	1.1	2.3	3.2	4.6	5.8	6.7	8
y	−5.8	−5.1	−4.8	−4.4	−3.7	−3.2	−2.5

Plot the scatter diagram and the least-squares line for this data.

Solution

1. *Set up a table of values on a spreadsheet* (Figure T3).
2. *Plot the scatter diagram.* Highlight the numerical values in the table of values. Click the $\boxed{\textbf{Chart Wizard}}$ button on the toolbar. Follow the procedure given in Example 3, page 25, with these exceptions: In step 1, select the first chart in the first column under Chart sub-type:; in step 3, in the Chart title: box, type `Scatter diagram and least-squares line`. The scatter diagram will appear.
3. *Insert the least-squares line.* First, click $\boxed{\textbf{Chart}}$ on the menu bar and then click $\boxed{\textbf{Add Trendline}}$. Next, select the $\boxed{\textbf{Linear}}$ graph and then click the $\boxed{\textbf{Options}}$ tab. Finally select $\boxed{\textbf{Display equation on chart}}$ and then click $\boxed{\textbf{OK}}$. The equation

$$y = 0.4606x - 6.3$$

will appear on the chart (Figure T4).

	A	B
1	x	y
2	1.1	−5.8
3	2.3	−5.1
4	3.2	−4.8
5	4.6	−4.4
6	5.8	−3.7
7	6.7	−3.2
8	8	−2.5

FIGURE T3
Table of values for *x* and *y*

Note: Boldfaced words/characters enclosed in a box (for example, $\boxed{\textbf{Enter}}$) indicate that an action (click, select, or press) is required. Words/characters printed blue (for example, Chart sub-type:) indicate words/characters that appear on the screen. Words/characters printed in a typewriter font (for example, `=(−2/3)*A2+2`) indicate words/characters that need to be typed and entered.

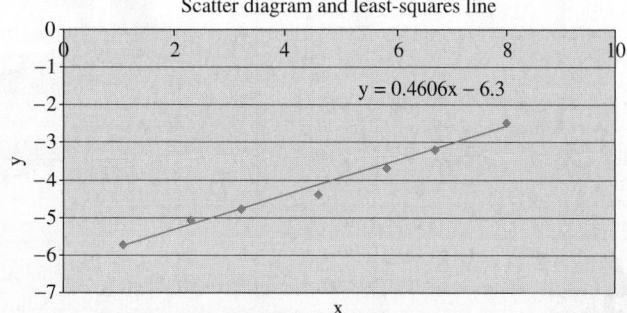

Scatter diagram and least-squares line

$y = 0.4606x - 6.3$

FIGURE T4
Scatter diagram and least-squares line
for the given data

APPLIED EXAMPLE 4   Demand for Electricity According to Pacific Gas and Electric, the nation's largest utility company, the demand for electricity from 1990 through 2000 is summarized in the following table:

t	0	2	4	6	8	10
y	333	917	1500	2117	2667	3292

	A	B
1	x	y
2	0	333
3	2	917
4	4	1500
5	6	2117
6	8	2667
7	10	3292

FIGURE T5
Table of values for *x* and *y*

	Coefficients
Intercept	328.4762
X Variable 1	295.1714

FIGURE T6
Entries in the SUMMARY OUTPUT box

Here $t = 0$ corresponds to 1990, and y gives the amount of electricity in year t, measured in megawatts. Find an equation of the least-squares line for these data.

Solution

1. *Set up a table of values on a spreadsheet* (Figure T5).
2. *Find the equation of the least-squares line for this data.* Click **Tools** on the menu bar and then click **Data Analysis**. From the Data Analysis dialog box that appears, select **Regression** and then click **OK**. In the Regression dialog box that appears, select the **Input Y Range:** box and then enter the y-values by highlighting cells B2:B7. Next select the **Input X Range:** box and enter the x-values by highlighting cells A2:A7. Click **OK** and a SUMMARY OUTPUT box will appear. In the third table in this box, you will see the entries shown in Figure T6. These entries give the value of the y-intercept and the coefficient of x in the equation $y = mx + b$. In our example, we are using the variable t instead of x, so the required equation is

$$y = 328 + 295t$$

TECHNOLOGY EXERCISES

In Exercises 1–4, find an equation of the least-squares line for the given data.

1.

x	2.1	3.4	4.7	5.6	6.8	7.2
y	8.8	12.1	14.8	16.9	19.8	21.1

2.

x	1.1	2.4	3.2	4.7	5.6	7.2
y	−0.5	1.2	2.4	4.4	5.7	8.1

3.

x	−2.1	−1.1	0.1	1.4	2.5	4.2	5.1
y	6.2	4.7	3.5	1.9	0.4	−1.4	−2.5

4.

x	-1.12	0.1	1.24	2.76	4.21	6.82
y	7.61	4.9	2.74	-0.47	-3.51	-8.94

5. STARBUCKS ANNUAL SALES According to company reports, Starbucks annual sales (in billions of dollars) for 1998 through 2003 are as follows:

Year, x	0	1	2	3	4	5
Sales, y	1.28	1.73	2.18	2.65	3.29	4.08

(Here, $x = 0$ corresponds to 1998.)
a. Find an equation of the least-squares line for these data.
b. Use the result to estimate Starbucks sales for 2006, assuming that the trend continued.
Source: Company reports

6. SALES OF DRUGS Sales of drugs called analeptics, which are used to treat attention deficit disorders, were rising even before some companies began advertising them to parents. The following table gives the sales of analeptics (in millions of dollars) from 1995 through 2000:

Year, x	0	1	2	3	4	5
Sales, y	382	455	536	618	664	758

(Here, $x = 0$ represents 1995.)
a. Find an equation of the least-squares line for these data.
b. Use the result of part (a) to estimate the sales for 2002, assuming that the trend continued.
Source: IMS Health

7. WASTE GENERATION According to data from the Council on Environmental Quality, the amount of waste (in millions of tons per year) generated in the United States from 1960 to 1990 was:

Year	1960	1965	1970	1975
Amount, y	81	100	120	124

Year	1980	1985	1990
Amount, y	140	152	164

(Let x be in units of 5 and let $x = 1$ represent 1960.)
a. Find an equation of the least-squares line for these data.
b. Use the result of part (a) to estimate the amount of waste generated in the year 2000, assuming that the trend continued.
Source: Council on Environmental Quality

8. ONLINE TRAVEL More and more travelers are purchasing their tickets online. According to industry projections, the U.S. online travel revenue (in billions of dollars) from 2001 through 2005 is given in the following table:

Year, t	0	1	2	3	4
Revenue	16.3	21.0	25.0	28.8	32.7

(Here, $t = 0$ corresponds to 2001.)
a. Find an equation of the least-squares line for these data.
b. Use the result of part (a) to estimate the U.S. online travel revenue for 2006.
Source: Forrester Research Inc.

9. WORLD ENERGY CONSUMPTION According to the U.S. Department of Energy, the consumption of energy by countries of the industrialized world is projected to rise over the next 25 years. The consumption (in trillions of cubic feet) from 2000 through 2025 is summarized in the following table.

Year	2000	2005	2010	2015	2020	2025
Consumption, y	214	225	240	255	270	285

(Let x be measured in 5-year intervals and let $x = 0$ represent 2000.)
a. Find an equation of the least-squares line for these data.
b. Use the result of part (a) to estimate the energy consumption in the industrialized world in 2012.
Source: U.S. Department of Energy

10. ANNUAL COLLEGE COSTS The annual U.S. college costs (including tuition, room, and board) from 1991/1992 through 2001/2002 are given in the accompanying table:

Academic Year, x	0	1	2	3	4
Cost ($), y	7077	7452	7931	8306	8800

Academic Year, x	5	6	7	8	9	10
Cost ($), y	9206	9588	10,076	10,444	10,818	11,454

(Here, $x = 0$ corresponds to academic year 1991/1992.)
a. Find an equation of the least-squares line for these data.
b. Use the result of part (a) to plot the least-squares line.
c. Use the result of part (a) to determine the approximate average rate of increase of college costs per year for the period in question.
Source: National Center for Education Statistics

CHAPTER 1   Summary of Principal Formulas and Terms

FORMULAS

1. Distance between two points	$d = \sqrt{(x_2 - x_1)^2 + (y_2 - y_1)^2}$
2. Equation of a circle	$(x - h)^2 + (y - k)^2 = r^2$
3. Slope of a nonvertical line	$m = \dfrac{y_2 - y_1}{x_2 - x_1}$
4. Equation of a vertical line	$x = a$
5. Equation of a horizontal line	$y = b$
6. Point-slope form of the equation of a line	$y - y_1 = m(x - x_1)$
7. Slope-intercept form of the equation of a line	$y = mx + b$
8. General equation of a line	$Ax + By + C = 0$

TERMS

Cartesian coordinate system (2)   dependent variable (30)   demand function (33)

ordered pair (2)   domain (30)   supply function (34)

coordinates (3)   range (30)   break-even point (43)

parallel lines (12)   linear function (30)   market equilibrium (46)

perpendicular lines (14)   total cost function (32)   equilibrium quantity (46)

function (29)   revenue function (32)   equilibrium price (46)

independent variable (30)   profit function (32)

CHAPTER 1   Concept Review Questions

Fill in the blanks.

1. A point in the plane can be represented uniquely by a/an _____ pair of numbers. The first number of the pair is called the _____, and the second number of the pair is called the _____.

2. **a.** The point $P(a, 0)$ lies on the _____ axis, and the point $P(0, b)$ lies on the _____ axis.
   b. If the point $P(a, b)$ lies in the fourth quadrant, then the point $P(-a, b)$ lies in the _____ quadrant.

3. The distance between two points $P(a, b)$ and $P(c, d)$ is _____.

4. An equation of a circle with center $C(a, b)$ and radius r is given by _____.

5. **a.** If $P_1(x_1, y_1)$ and $P_2(x_2, y_2)$ are any two distinct points on a nonvertical line L, then the slope of L is $m =$ _____.

b. The slope of a vertical line is _____.
c. The slope of a horizontal line is _____.
d. The slope of a line that slants upward is _____.

6. If L_1 and L_2 are nonvertical lines with slopes m_1 and m_2, respectively, then: L_1 is parallel to L_2 if and only if _____; and L_1 is perpendicular to L_2 if and only if _____.

7. **a.** An equation of the line passing through the point $P(x_1, y_1)$ and having slope m is _____. It is called the _____ form of an equation of a line.
   b. An equation of the line that has slope m and y-intercept b is _____. It is called the _____ form of an equation of a line.

8. **a.** The general form of an equation of a line is _____.
   b. If a line has equation $ax + by + c = 0$ ($b \neq 0$), then its slope is _____.

9. A linear function is a function of the form $f(x) =$ _____.

10. **a.** A demand function expresses the relationship between the unit _____ and the quantity _____ of a commodity. The graph of the demand function is called the _____ curve.

   b. A supply function expresses the relationship between the unit _____ and the quantity _____ of a commodity. The graph of the supply function is called the _____ curve.

11. If $R(x)$ and $C(x)$ denote the total revenue and the total cost incurred in manufacturing x units of a commodity, then the solution of the simultaneous equations $p = C(x)$ and $p = R(x)$ gives the _____ point.

12. The equilibrium quantity and the equilibrium price are found by solving the system composed of the _____ equation and the _____ equation.

CHAPTER 1   Review Exercises

In Exercises 1–4, find the distance between the two points.

1. $(2, 1)$ and $(6, 4)$

2. $(9, 6)$ and $(6, 2)$

3. $(-2, -3)$ and $(1, -7)$

4. $\left(\dfrac{1}{2}, \sqrt{3} \right)$ and $\left(-\dfrac{1}{2}, 2\sqrt{3} \right)$

In Exercises 5–10, find an equation of the line L that passes through the point $(-2, 4)$ and satisfies the given condition.

5. L is a vertical line.

6. L is a horizontal line.

7. L passes through the point $\left(3, \dfrac{7}{2} \right)$.

8. The x-intercept of L is 3.

9. L is parallel to the line $5x - 2y = 6$.

10. L is perpendicular to the line $4x + 3y = 6$.

11. Find an equation of the line with slope $-\frac{1}{2}$ and y-intercept -3.

12. Find the slope and y-intercept of the line with equation $3x - 5y = 6$.

13. Find an equation of the line passing through the point $(2, 3)$ and parallel to the line with equation $3x + 4y - 8 = 0$.

14. Find an equation of the line passing through the point $(-1, 3)$ and parallel to the line joining the points $(-3, 4)$ and $(2, 1)$.

15. Find an equation of the line passing through the point $(-2, -4)$ that is perpendicular to the line with equation $2x - 3y - 24 = 0$.

In Exercises 16 and 17, sketch the graph of the equation.

16. $3x - 4y = 24$

17. $-2x + 5y = 15$

18. **SALES OF MP3 PLAYERS** Sales of a certain brand of MP3 players are approximated by the relationship

$$S(x) = 6000x + 30{,}000 \qquad (0 \le x \le 5)$$

where $S(x)$ denotes the number of MP3 players sold in year x ($x = 0$ corresponds to the year 2003). Find the number of MP3 players expected to be sold in 2008.

19. **COMPANY SALES** A company's total sales (in millions of dollars) are approximately linear as a function of time (in years). Sales in 2001 were \$2.4 million, whereas sales in 2006 amounted to \$7.4 million.

   a. Find an equation giving the company's sales as a function of time.

   b. What were the sales in 2004?

20. Show that the triangle with vertices $A(1, 1)$, $B(5, 3)$, and $C(4, 5)$ is a right triangle.

21. **CLARK'S RULE** Clark's rule is a method for calculating pediatric drug dosages based on a child's weight. If a denotes the adult dosage (in milligrams) and if w is the child's weight (in pounds), then the child's dosage is given by

$$D(w) = \frac{aw}{150}$$

   a. Show that D is a linear function of w.

   b. If the adult dose of a substance is 500 mg, how much should a 35-lb child receive?

22. **LINEAR DEPRECIATION** An office building worth \$6 million when it was completed in 2000 is being depreciated linearly over 30 years.

   a. What is the rate of depreciation?

   b. What will be the book value of the building in 2010?

23. **LINEAR DEPRECIATION** In 2000 a manufacturer installed a new machine in her factory at a cost of \$300,000. The machine is depreciated linearly over 12 yr with a scrap value of \$30,000.

   a. What is the rate of depreciation of the machine per year?

   b. Find an expression for the book value of the machine in year t ($0 \le t \le 12$).

24. **PROFIT FUNCTIONS** A company has a fixed cost of \$30,000 and a production cost of \$6 for each disposable camera it manufactures. Each camera sells for \$10.

a. What is the cost function?

b. What is the revenue function?

c. What is the profit function?

d. Compute the profit (loss) corresponding to production levels of 6000, 8000, and 12,000 units, respectively.

25. **Demand Equations** There is no demand for a certain commodity when the unit price is $200 or more, but the demand increases by 200 units for each $10 decrease in price below $200. Find the demand equation and sketch its graph.

26. **Supply Equations** Bicycle suppliers will make 200 bicycles available in the market per month when the unit price is $50 and 2000 bicycles available per month when the unit price is $100. Find the supply equation if it is known to be linear.

In Exercises 27 and 28, find the point of intersection of the lines with the given equations.

27. $3x + 4y = -6$ and $2x + 5y = -11$

28. $y = \dfrac{3}{4}x + 6$ and $3x - 2y + 3 = 0$

29. The cost function and the revenue function for a certain firm are given by $C(x) = 12x + 20{,}000$ and $R(x) = 20x$, respectively. Find the break-even point for the company.

30. Given the demand equation $3x + p - 40 = 0$ and the supply equation $2x - p + 10 = 0$, where p is the unit price in dollars and x represents the quantity demanded in units of a thousand, determine the equilibrium quantity and the equilibrium price.

31. **College Admissions** The accompanying data were compiled by the Admissions Office of Carter College during the past 5 yr. The data relate the number of college brochures and follow-up letters (x) sent to a preselected list of high-school juniors who took the PSAT and the number of completed applications (y) received from these students (both measured in thousands).

Brochures Sent, x	1.8	2	3.2
Applications Completed, y	0.4	0.5	0.7

Brochures Sent, x	4	4.8
Applications Completed, y	1	1.3

a. Derive an equation of the straight line L that passes through the points $(2, 0.5)$ and $(4, 1)$.

b. Use this equation to predict the number of completed applications that might be expected if 6400 brochures and follow-up letters are sent out during the next year.

32. **Equilibrium Quantity and Price** The demand equation for the Edmund compact refrigerator is $2x + 7p - 1760 = 0$, where x is the quantity demanded each week and p is the unit price in dollars. The supply equation for these refrigerators is $3x - 56p + 2680 = 0$, where x is the quantity the supplier will make available in the market when the wholesale price is p dollars each. Find the equilibrium quantity and the equilibrium price for the Edmund compact refrigerators.

The problem-solving skills that you learn in each chapter are building blocks for the rest of the course. Therefore, it is a good idea to make sure that you have mastered these skills before moving on to the next chapter. The Before Moving On exercises that follow are designed for that purpose. After taking this test, you can see where your weaknesses, if any, are. Then you can go to http://series .brookscole.com/tan/ where you will find a link to our Companion Website. Here, you can click on the Before Moving On button, which will lead you to other versions of these tests. There you can re-test yourself on those exercises that you solved incorrectly. (You can also test yourself on these basic skills before taking your course quizzes and exams.)

If you feel that you need additional help with these exercises, you can use the *CengageNOW Tutorials*, as well as *vMentor*™ for live online help from a tutor.

CHAPTER 1 # Before Moving On . . .

1. Plot the points $A(-2, 1)$ and $B(3, 4)$ on the same set of axes and find the distance between A and B.

2. Find an equation of the line passing through the point $(3, 1)$ and parallel to the line $3x - y - 4 = 0$.

3. Let L be the line passing through the points $(1, 2)$ and $(3, 5)$. Is L perpendicular to the line $2x + 3y = 10$?

4. The monthly total revenue function and total cost function for a company are $R(x) = 15x$ and $C(x) = 18x + 22{,}000$, respectively, where x is the number of units produced and both $R(x)$ and $C(x)$ are measured in dollars.

a. What is the unit cost for producing the product?

b. What is the monthly fixed cost for the company?

c. What is the selling price for each unit of the product?

5. Find the point of intersection of the lines $2x - 3y = -2$ and $9x + 12y = 25$.

6. The annual sales of Best Furniture Store are expected to be given by $S_1 = 4.2 + 0.4t$ million dollars t yr from now, whereas the annual sales of Lowe's Furniture Store are expected to be given by $S_2 = 2.2 + 0.8t$ million dollars t yr from now. When will Lowe's annual sales first surpass Best's annual sales?

2

Systems of Linear Equations and Matrices

How fast is the traffic moving? The flow of downtown traffic is controlled by traffic lights installed at the intersections. One of the roads is to be resurfaced. In Example 5, page 101, you will see how the flow patterns must be altered in order to ensure a smooth flow of traffic even during rush hour.

© Spike Mafford/PhotoDisc

THE LINEAR EQUATIONS in two variables studied in Chapter 1 are readily extended to the case involving more than two variables. For example, a linear equation in three variables represents a plane in three-dimensional space. In this chapter, we see how some real-world problems can be formulated in terms of systems of linear equations, and we also develop two methods for solving these equations.

In addition, we see how *matrices* (ordered rectangular arrays of numbers) can be used to write systems of linear equations in compact form. We then go on to consider some real-life applications of matrices. Finally, we show how matrices can be used to describe the Leontief input–output model, an important tool used by economists. For his work in formulating this model, Wassily Leontief was awarded the Nobel Prize in 1973.

2.1   Systems of Linear Equations: An Introduction

■ Systems of Equations

Recall that in Section 1.4 we had to solve two simultaneous linear equations in order to find the *break-even point* and the *equilibrium point*. These are two examples of real-world problems that call for the solution of a system of linear equations in two or more variables. In this chapter we take up a more systematic study of such systems.

We begin by considering a system of two linear equations in two variables. Recall that such a system may be written in the general form

$$ax + by = h$$
$$cx + dy = k \tag{1}$$

where a, b, c, d, h, and k are real constants and neither a and b nor c and d are both zero.

Now let's study the nature of the solution of a system of linear equations in more detail. Recall that the graph of each equation in System (1) is a straight line in the plane, so that geometrically the solution to the system is the point(s) of intersection of the two straight lines L_1 and L_2, represented by the first and second equations of the system.

Given two lines L_1 and L_2, *one and only one* of the following may occur:

a. L_1 and L_2 intersect at exactly one point.
b. L_1 and L_2 are parallel and coincident.
c. L_1 and L_2 are parallel and distinct.

(See Figure 1.) In the first case, the system has a unique solution corresponding to the single point of intersection of the two lines. In the second case, the system has infinitely many solutions corresponding to the points lying on the same line. Finally, in the third case, the system has no solution because the two lines do not intersect.

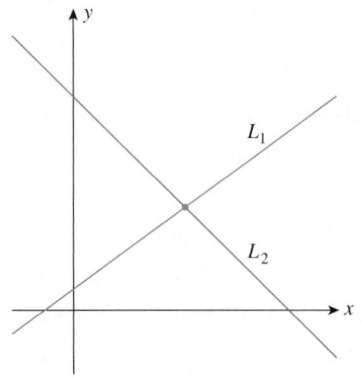

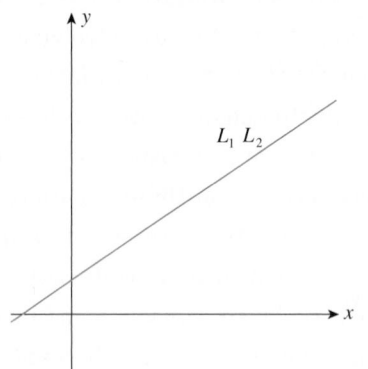

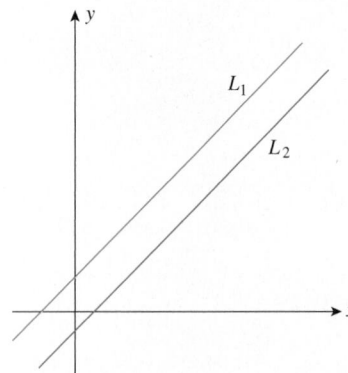

FIGURE 1
(a) Unique solution   **(b)** Infinitely many solutions   **(c)** No solution

---

EXPLORE & DISCUSS

Generalize the discussion on this page to the case where there are three straight lines in the plane defined by three linear equations. What if there are n lines defined by n equations?

Let's illustrate each of these possibilities by considering some specific examples.

1. A system of equations with exactly one solution Consider the system

$$2x - y = 1$$
$$3x + 2y = 12$$

Solving the first equation for y in terms of x, we obtain the equation

$$y = 2x - 1$$

Substituting this expression for y into the second equation yields

$$3x + 2(2x - 1) = 12$$
$$3x + 4x - 2 = 12$$
$$7x = 14$$
$$x = 2$$

Finally, substituting this value of x into the expression for y obtained earlier gives

$$y = 2(2) - 1 = 3$$

Therefore, the unique solution of the system is given by $x = 2$ and $y = 3$. Geometrically, the two lines represented by the two linear equations that make up the system intersect at the point $(2, 3)$ (Figure 2).

Note We can check our result by substituting the values $x = 2$ and $y = 3$ into the equations. Thus,

$$2(2) - (3) = 1 \quad \checkmark$$
$$3(2) + 2(3) = 12 \quad \checkmark$$

From the geometric point of view, we have just verified that the point $(2, 3)$ lies on both lines.   ∎

2. A system of equations with infinitely many solutions Consider the system

$$2x - y = 1$$
$$6x - 3y = 3$$

Solving the first equation for y in terms of x, we obtain the equation

$$y = 2x - 1$$

Substituting this expression for y into the second equation gives

$$6x - 3(2x - 1) = 3$$
$$6x - 6x + 3 = 3$$
$$0 = 0$$

which is a true statement. This result follows from the fact that the second equation is equivalent to the first. (To see this, just multiply both sides of the first equation by 3.) Our computations have revealed that the system of two equations is equivalent to the single equation $2x - y = 1$. Thus, any ordered pair of numbers (x, y) satisfying the equation $2x - y = 1$ (or $y = 2x - 1$) constitutes a solution to the system.

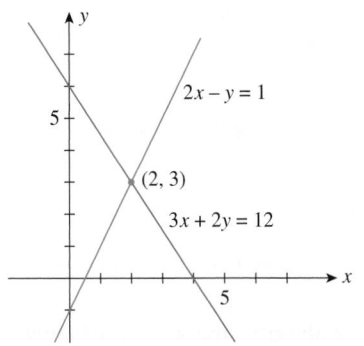

FIGURE 2
A system of equations with one solution

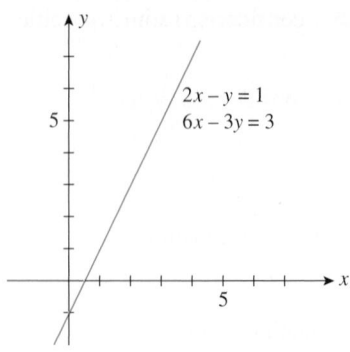

FIGURE 3
A system of equations with infinitely many solutions; each point on the line is a solution.

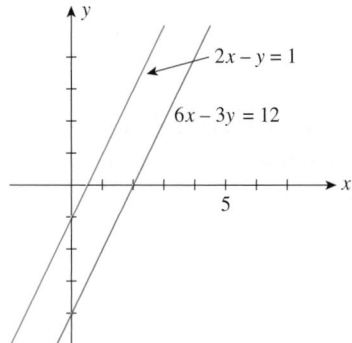

FIGURE 4
A system of equations with no solution

In particular, by assigning the value t to x, where t is any real number, we find that $y = 2t - 1$ and so the ordered pair $(t, 2t - 1)$ is a solution of the system. The variable t is called a **parameter**. For example, setting $t = 0$ gives the point $(0, -1)$ as a solution of the system, and setting $t = 1$ gives the point $(1, 1)$ as another solution. Since t represents any real number, there are infinitely many solutions of the system. Geometrically, the two equations in the system represent the same line, and all solutions of the system are points lying on the line (Figure 3). Such a system is said to be **dependent**.

3. A system of equations that has no solution Consider the system

$$2x - y = 1$$
$$6x - 3y = 12$$

The first equation is equivalent to $y = 2x - 1$. Substituting this expression for y into the second equation gives

$$6x - 3(2x - 1) = 12$$
$$6x - 6x + 3 = 12$$
$$0 = 9$$

which is clearly impossible. Thus, there is no solution to the system of equations. To interpret this situation geometrically, cast both equations in the slope-intercept form, obtaining

$$y = 2x - 1$$
$$y = 2x - 4$$

We see at once that the lines represented by these equations are parallel (each has slope 2) and distinct, since the first has y-intercept -1 and the second has y-intercept -4 (Figure 4). Systems with no solutions, such as this one, are said to be **inconsistent**.

EXPLORE & DISCUSS

1. Consider a system composed of two linear equations in two variables. Can the system have exactly two solutions? Exactly three solutions? Exactly a finite number of solutions?

2. Suppose at least one of the equations in a system composed of two equations in two variables is nonlinear. Can the system have no solution? Exactly one solution? Exactly two solutions? Exactly a finite number of solutions? Infinitely many solutions? Illustrate each answer with a sketch.

Note   We have used the method of substitution in solving each of these systems. If you are familiar with the method of elimination, you might want to re-solve each of these systems using this method. We will study the method of elimination in detail in Section 2.2.   ■

In Section 1.4, we presented some real-world applications of systems involving two linear equations in two variables. Here is an example involving a system of three linear equations in three variables.

APPLIED EXAMPLE 1   Manufacturing: Production Scheduling Ace Novelty wishes to produce three types of souvenirs: types A, B, and C. To manufacture a type-A souvenir requires 2 minutes on machine I, 1 minute on machine II, and 2 minutes on machine III. A type-B souvenir requires 1 minute on machine I, 3 minutes on machine II, and 1 minute on machine III. A type-C souvenir requires 1 minute on machine I and 2 minutes each on machines II and III. There are 3 hours available on machine I, 5 hours available on machine II, and 4 hours available on machine III for processing the order. How many souvenirs of each type should Ace Novelty make in order to use all of the available time? Formulate but do not solve the problem. (We will solve this problem in Example 7, Section 2.2.)

Solution   The given information may be tabulated as follows:

	Type A	Type B	Type C	Time Available (min)
Machine I	2	1	1	180
Machine II	1	3	2	300
Machine III	2	1	2	240

We have to determine the number of each of *three* types of souvenirs to be made. So, let x, y, and z denote the respective numbers of type-A, type-B, and type-C souvenirs to be made. The total amount of time that machine I is used is given by $2x + y + z$ minutes and must equal 180 minutes. This leads to the equation

$$2x + y + z = 180 \qquad \text{Time spent on machine I}$$

Similar considerations on the use of machines II and III lead to the following equations:

$$x + 3y + 2z = 300 \qquad \text{Time spent on machine II}$$
$$2x + y + 2z = 240 \qquad \text{Time spent on machine III}$$

Since the variables x, y, and z must satisfy simultaneously the three conditions represented by the three equations, the solution to the problem is found by solving the following system of linear equations:

$$
\begin{aligned}
2x + y + z &= 180 \\
x + 3y + 2z &= 300 \\
2x + y + 2z &= 240
\end{aligned}
$$

Solutions of Systems of Equations

We will complete the solution of the problem posed in Example 1 later on (page 89). For the moment, let's look at the geometric interpretation of a system of linear equations, such as the system in Example 1, in order to gain some insight into the nature of the solution.

A linear system composed of three linear equations in three variables x, y, and z has the general form

$$
\begin{aligned}
a_1x + b_1y + c_1z &= d_1 \\
a_2x + b_2y + c_2z &= d_2 \\
a_3x + b_3y + c_3z &= d_3
\end{aligned}
\tag{2}
$$

Just as a linear equation in two variables represents a straight line in the plane, it can be shown that a linear equation $ax + by + cz = d$ (a, b, and c not simultaneously equal to zero) in three variables represents a plane in three-dimensional space. Thus, each equation in System (2) represents a *plane* in three-dimensional space, and the *solution(s) of the system* is precisely the point(s) of intersection of the three planes defined by the three linear equations that make up the system. As before, the system has one and only one solution, infinitely many solutions, or no solution, depending on whether and how the planes intersect one another. Figure 5 illustrates each of these possibilities.

In Figure 5a, the three planes intersect at a point corresponding to the situation in which System (2) has a unique solution. Figure 5b depicts the situation in which there are infinitely many solutions to the system. Here, the three planes intersect along a line, and the solutions are represented by the infinitely many points lying on this line. In Figure 5c, the three planes are parallel and distinct, so there is no point in common to all three planes; System (2) has no solution in this case.

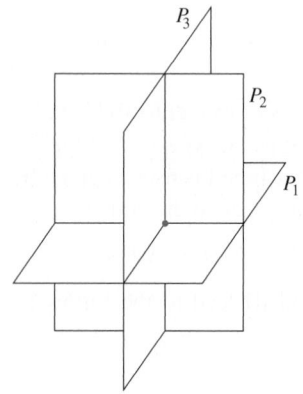

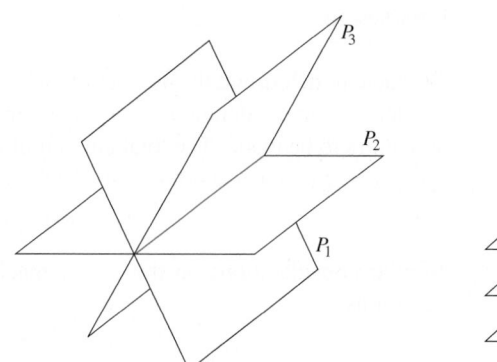

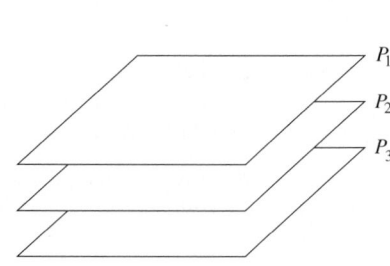

FIGURE 5
(a) A unique solution   **(b)** Infinitely many solutions   **(c)** No solution

Note The situations depicted in Figure 5 are by no means exhaustive. You may consider various other orientations of the three planes that would illustrate the three possible outcomes in solving a system of linear equations involving three variables.

◼

Linear Equations in *n* Variables
A linear equation in n variables, $x_1, x_2, \ldots, x_n$ is one of the form

$$a_1x_1 + a_2x_2 + \cdots + a_nx_n = c$$

where $a_1, a_2, \ldots, a_n$ (not all zero) and c are constants.

For example, the equation

$$3x_1 + 2x_2 - 4x_3 + 6x_4 = 8$$

is a linear equation in the four variables, x_1, x_2, x_3, and x_4.

EXPLORE & DISCUSS

Refer to the Note on page 76.

Using the orientation of three planes, illustrate the outcomes in solving a system of three linear equations in three variables that result in no solution or infinitely many solutions.

When the number of variables involved in a linear equation exceeds three, we no longer have the geometric interpretation we had for the lower-dimensional spaces. Nevertheless, the algebraic concepts of the lower-dimensional spaces generalize to higher dimensions. For this reason, a linear equation in n variables, $a_1x_1 + a_2x_2 + \cdots + a_nx_n = c$, where $a_1, a_2, \ldots, a_n$ are not all zero, is referred to as an *n-dimensional hyperplane*. We may interpret the solution(s) to a system comprising a finite number of such linear equations to be the *point(s) of intersection* of the hyperplanes defined by the equations that make up the system. As in the case of systems involving two or three variables, it can be shown that only three possibilities exist regarding the nature of the solution of such a system: (1) a unique solution, (2) infinitely many solutions, or (3) no solution.

2.1 Self-Check Exercises

1. Determine whether the system of linear equations

$$2x - 3y = 12$$
$$x + 2y = 6$$

has (a) a unique solution, (b) infinitely many solutions, or (c) no solution. Find all solutions whenever they exist. Make a sketch of the set of lines described by the system.

2. A farmer has 200 acres of land suitable for cultivating crops A, B, and C. The cost per acre of cultivating crops A, B, and C is $40, $60, and $80, respectively. The farmer has $12,600 available for cultivation. Each acre of crop A requires 20 labor-hours, each acre of crop B requires 25 labor-hours, and each acre of crop C requires 40 labor-hours. The farmer has a maximum of 5950 labor-hours available. If she wishes to use all of her cultivatable land, the entire budget, and all the labor available, how many acres of each crop should she plant? Formulate but do not solve the problem.

Solutions to Self-Check Exercises 2.1 can be found on page 79.

2.1 Concept Questions

1. Suppose you are given a system of two linear equations in two variables.
   a. What can you say about the solution(s) of the system of equations?
   b. Give a geometric interpretation of your answers to the question in part (a).

2. Suppose you are given a system of two linear equations in two variables.
   a. Explain what it means for the system to be dependent.
   b. Explain what it means for the system to be inconsistent.

2.1 Exercises

In Exercises 1–12, determine whether each system of linear equations has (a) one and only one solution, (b) infinitely many solutions, or (c) no solution. Find all solutions whenever they exist.

1. $x - 3y = -1$
   $4x + 3y = 11$

2. $2x - 4y = 5$
   $3x + 2y = 6$

3. $x + 4y = 7$
   $\frac{1}{2}x + 2y = 5$

4. $3x - 4y = 7$
   $9x - 12y = 14$

5. $x + 2y = 7$
   $2x - y = 4$

6. $\frac{3}{2}x - 2y = 4$
   $x + \frac{1}{3}y = 2$

7. $2x - 5y = 10$
   $6x - 15y = 30$

8. $5x - 6y = 8$
   $10x - 12y = 16$

9. $4x - 5y = 14$
   $2x + 3y = -4$

10. $\dfrac{5}{4}x - \dfrac{2}{3}y = 3$
   $\dfrac{1}{4}x + \dfrac{5}{3}y = 6$

11. $2x - 3y = 6$
   $6x - 9y = 12$

12. $\dfrac{2}{3}x + y = 5$
   $\dfrac{1}{2}x + \dfrac{3}{4}y = \dfrac{15}{4}$

13. Determine the value of k for which the system of linear equations

$$2x - y = 3$$
$$4x + ky = 4$$

has no solution.

14. Determine the value of k for which the system of linear equations

$$3x + 4y = 12$$
$$x + ky = 4$$

has infinitely many solutions. Then find all solutions corresponding to this value of k.

In Exercises 15–27, formulate but do not solve the problem. You will be asked to solve these problems in the next section.

15. AGRICULTURE   The Johnson Farm has 500 acres of land allotted for cultivating corn and wheat. The cost of cultivating corn and wheat (including seeds and labor) is $42 and $30 per acre, respectively. Jacob Johnson has $18,600 available for cultivating these crops. If he wishes to use all the allotted land and his entire budget for cultivating these two crops, how many acres of each crop should he plant?

16. INVESTMENTS   Michael Perez has a total of $2000 on deposit with two savings institutions. One pays interest at the rate of 6%/year, whereas the other pays interest at the rate of 8%/year. If Michael earned a total of $144 in interest during a single year, how much does he have on deposit in each institution?

17. MIXTURES   The Coffee Shoppe sells a coffee blend made from two coffees, one costing $5/lb and the other costing $6/lb. If the blended coffee sells for $5.60/lb, find how much of each coffee is used to obtain the desired blend. (Assume the weight of the blended coffee is 100 lb.)

18. INVESTMENTS   Kelly Fisher has a total of $30,000 invested in two municipal bonds that have yields of 8% and 10% interest per year, respectively. If the interest Kelly receives from the bonds in a year is $2640, how much does she have invested in each bond?

19. RIDERSHIP   The total number of passengers riding a certain city bus during the morning shift is 1000. If the child's fare is $.50, the adult fare is $1.50, and the total revenue from the fares in the morning shift is $1300, how many children and how many adults rode the bus during the morning shift?

20. REAL ESTATE   Cantwell Associates, a real estate developer, is planning to build a new apartment complex consisting of one-bedroom units and two- and three-bedroom townhouses. A total of 192 units is planned, and the number of family units (two- and three-bedroom townhouses) will equal the number of one-bedroom units. If the number of one-bedroom units will be 3 times the number of three-bedroom units, find how many units of each type will be in the complex.

21. INVESTMENT PLANNING   The annual interest on Sid Carrington's three investments amounted to $21,600: 6% on a savings account, 8% on mutual funds, and 12% on bonds. The amount of Sid's investment in bonds was twice the amount of his investment in the savings account, and the interest earned from his investment in bonds was equal to the dividends he received from his investment in mutual funds. Find how much money he placed in each type of investment.

22. INVESTMENT CLUB   A private investment club has $200,000 earmarked for investment in stocks. To arrive at an acceptable overall level of risk, the stocks that management is considering have been classified into three categories: high-risk, medium-risk, and low-risk. Management estimates that high-risk stocks will have a rate of return of 15%/year; medium-risk stocks, 10%/year; and low-risk stocks, 6%/year. The members have decided that the investment in low-risk stocks should be equal to the sum of the investments in the stocks of the other two categories. Determine how much the club should invest in each type of stock if the investment goal is to have a return of $20,000/year on the total investment. (Assume that all the money available for investment is invested.)

23. MIXTURE PROBLEM—FERTILIZER   Lawnco produces three grades of commercial fertilizers. A 100-lb bag of grade-A fertilizer contains 18 lb of nitrogen, 4 lb of phosphate, and 5 lb of potassium. A 100-lb bag of grade-B fertilizer contains 20 lb of nitrogen and 4 lb each of phosphate and potassium. A 100-lb bag of grade-C fertilizer contains 24 lb of nitrogen, 3 lb of phosphate, and 6 lb of potassium. How many 100-lb bags of each of the three grades of fertilizers should Lawnco produce if 26,400 lb of nitrogen, 4900 lb of phosphate, and 6200 lb of potassium are available and all the nutrients are used?

24. BOX-OFFICE RECEIPTS   A theater has a seating capacity of 900 and charges $4 for children, $6 for students, and $8 for adults. At a certain screening with full attendance, there were half as many adults as children and students combined. The receipts totaled $5600. How many children attended the show?

25. MANAGEMENT DECISIONS   The management of Hartman Rent-A-Car has allocated $1.5 million to buy a fleet of new automobiles consisting of compact, intermediate, and full-size cars. Compacts cost $12,000 each, intermediate-size cars cost $18,000 each, and full-size cars cost $24,000 each. If Hartman purchases twice as many compacts as intermediate-size cars and the total number of cars to be purchased is 100, determine how many cars of each type will be purchased. (Assume that the entire budget will be used.)

26. INVESTMENT CLUBS   The management of a private investment club has a fund of $200,000 earmarked for investment in stocks. To arrive at an acceptable overall level of risk, the stocks that management is considering have been classified into three categories: high-risk, medium-risk, and low-risk. Management estimates that high-risk stocks will have a rate of return of 15%/year; medium-risk stocks, 10%/year; and low-risk stocks, 6%/year. The investment in low-risk stocks is to be twice the sum of the investments in stocks of the other two categories. If the investment goal is to have an average rate of return of 9%/year on the total investment, determine how much the club should invest in each type of stock. (Assume that all the money available for investment is invested.)

27. DIET PLANNING   A dietitian wishes to plan a meal around three foods. The percentage of the daily requirements of proteins, carbohydrates, and iron contained in each ounce of the three foods is summarized in the accompanying table:

	Food I	Food II	Food III
Proteins (%)	10	6	8
Carbohydrates (%)	10	12	6
Iron (%)	5	4	12

Determine how many ounces of each food the dietitian should include in the meal to meet exactly the daily requirement of proteins, carbohydrates, and iron (100% of each).

In Exercises 28–30, determine whether the statement is true or false. If it is true, explain why it is true. If it is false, give an example to show why it is false.

28. A system composed of two linear equations must have at least one solution if the straight lines represented by these equations are nonparallel.

29. Suppose the straight lines represented by a system of three linear equations in two variables are parallel to each other. Then the system has no solution or it has infinitely many solutions.

30. If at least two of the three lines represented by a system composed of three linear equations in two variables are parallel, then the system has no solution.

2.1   Solutions to Self-Check Exercises

1. Solving the first equation for y in terms of x, we obtain

$$y = \frac{2}{3}x - 4$$

Next, substituting this result into the second equation of the system, we find

$$x + 2\left(\frac{2}{3}x - 4\right) = 6$$

$$x + \frac{4}{3}x - 8 = 6$$

$$\frac{7}{3}x = 14$$

$$x = 6$$

Substituting this value of x into the expression for y obtained earlier, we have

$$y = \frac{2}{3}(6) - 4 = 0$$

Therefore, the system has the unique solution $x = 6$ and $y = 0$. Both lines are shown in the accompanying figure.

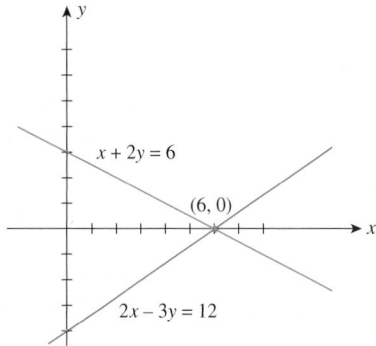

2. Let x, y, and z denote the number of acres of crop A, crop B, and crop C, respectively, to be cultivated. Then, the condition that all the cultivatable land be used translates into the equation

$$x + y + z = 200$$

Next, the total cost incurred in cultivating all three crops is $40x + 60y + 80z$ dollars, and since the entire budget is to be expended, we have

$$40x + 60y + 80z = 12{,}600$$

Finally, the amount of labor required to cultivate all three crops is $20x + 25y + 40z$ hr, and since all the available labor is to be used, we have

$$20x + 25y + 40z = 5950$$

Thus, the solution is found by solving the following system of linear equations:

$$
\begin{aligned}
x + \quad y + \quad z &= \quad\; 200 \\
40x + 60y + 80z &= 12{,}600 \\
20x + 25y + 40z &= \;\; 5{,}950
\end{aligned}
$$

2.2 Systems of Linear Equations: Unique Solutions

■ The Gauss–Jordan Method

The method of substitution used in Section 2.1 is well suited to solving a system of linear equations when the number of linear equations and variables is small. But for large systems, the steps involved in the procedure become difficult to manage.

The **Gauss–Jordan elimination method** is a suitable technique for solving systems of linear equations of any size. One advantage of this technique is its adaptability to the computer. This method involves a sequence of operations on a system of linear equations to obtain at each stage an **equivalent system**—that is, a system having the same solution as the original system. The reduction is complete when the original system has been transformed so that it is in a certain standard form from which the solution can be easily read.

The operations of the Gauss–Jordan elimination method are:

1. Interchange any two equations.
2. Replace an equation by a nonzero constant multiple of itself.
3. Replace an equation by the sum of that equation and a constant multiple of any other equation.

To illustrate the Gauss–Jordan elimination method for solving systems of linear equations, let's apply it to the solution of the following system:

$$
\begin{aligned}
2x + 4y &= 8 \\
3x - 2y &= 4
\end{aligned}
$$

We begin by working with the first, or x, column. First, we transform the system into an equivalent system in which the coefficient of x in the first equation is 1:

$$
\begin{aligned}
2x + 4y &= 8 \\
3x - 2y &= 4
\end{aligned}
\tag{3a}
$$

$$
\begin{aligned}
x + 2y &= 4 \\
3x - 2y &= 4
\end{aligned}
\tag{3b}
$$

Multiply the first equation in (3a) by $\frac{1}{2}$ (operation 2).

Next, we eliminate x from the second equation:

$$x + 2y = 4$$
$$-8y = -8$$

Replace the second equation in (3b) by the sum of $-3 \times$ the first equation + the second equation (operation 3): **(3c)**

$$-3x - 6y = -12$$
$$\underline{3x - 2y = 4}$$
$$-8y = -8$$

Then, we obtain the following equivalent system in which the coefficient of y in the second equation is 1:

$$x + 2y = 4$$
$$y = 1$$

Multiply the second equation in (3c) by $-\frac{1}{8}$ (operation 2). **(3d)**

Next, we eliminate y in the first equation:

$$x = 2$$
$$y = 1$$

Replace the first equation in (3d) by the sum of $-2 \times$ the second equation + the first equation (operation 3):

$$x + 2y = 4$$
$$\underline{- 2y = -2}$$
$$x = 2$$

This system is now in standard form, and we can read off the solution to (3a) as $x = 2$ and $y = 1$. We can also express this solution as (2, 1) and interpret it geometrically as the point of intersection of the two lines represented by the two linear equations that make up the given system of equations.

Let's consider another example involving a system of three linear equations and three variables.

EXAMPLE 1   Solve the following system of equations:

$$2x + 4y + 6z = 22$$
$$3x + 8y + 5z = 27$$
$$-x + y + 2z = 2$$

Solution   First, we transform this system into an equivalent system in which the coefficient of x in the first equation is 1:

$$2x + 4y + 6z = 22$$
$$3x + 8y + 5z = 27$$
$$-x + y + 2z = 2$$ **(4a)**

$$x + 2y + 3z = 11$$
$$3x + 8y + 5z = 27$$
$$-x + y + 2z = 2$$

Multiply the first equation in (4a) by $\frac{1}{2}$. **(4b)**

Next, we eliminate the variable x from all equations except the first:

$$x + 2y + 3z = 11$$
$$2y - 4z = -6$$
$$-x + y + 2z = 2$$

Replace the second equation in (4b) by the sum of $-3 \times$ the first equation + the second equation: **(4c)**

$$-3x - 6y - 9z = -33$$
$$\underline{3x + 8y + 5z = 27}$$
$$2y - 4z = -6$$

$$x + 2y + 3z = 11$$
$$2y - 4z = -6$$
$$3y + 5z = 13$$

Replace the third equation in (4c) by the sum of the first equation + the third equation:

$$x + 2y + 3z = 11$$
$$\underline{-x + y + 2z = 2}$$
$$3y + 5z = 13$$

(4d)

Then we transform System (4d) into yet another equivalent system, in which the coefficient of y in the second equation is 1:

$$x + 2y + 3z = 11$$
$$y - 2z = -3$$
$$3y + 5z = 13$$

Multiply the second equation in (4d) by $\frac{1}{2}$.

(4e)

We now eliminate y from all equations except the second, using operation 3 of the elimination method:

$$x + 7z = 17$$
$$y - 2z = -3$$
$$3y + 5z = 13$$

Replace the first equation in (4e) by the sum of the first equation + $(-2) \times$ the second equation:

$$x + 2y + 3z = 11$$
$$\underline{ - 2y + 4z = 6}$$
$$x + 7z = 17$$

(4f)

$$x + 7z = 17$$
$$y - 2z = -3$$
$$11z = 22$$

Replace the third equation in (4f) by the sum of $(-3) \times$ the second equation + the third equation:

$$-3y + 6z = 9$$
$$\underline{3y + 5z = 13}$$
$$11z = 22$$

(4g)

Finally, multiplying the third equation by $\frac{1}{11}$ in (4g) leads to the system

$$x + 7z = 17$$
$$y - 2z = -3$$
$$z = 2$$

Eliminating z from all equations except the third (try it!) then leads to the system

$$x = 3$$
$$y = 1$$
$$z = 2$$

(4h)

In its final form, the solution to the given system of equations can be easily read off! We have $x = 3$, $y = 1$, and $z = 2$. Geometrically, the point $(3, 1, 2)$ lies in the intersection of the three planes described by the three equations comprising the given system. ◾

▬ Augmented Matrices

Observe from the preceding example that the variables x, y, and z play no significant role in each step of the reduction process, except as a reminder of the position of each coefficient in the system. With the aid of **matrices**, which are rectangular arrays of numbers, we can eliminate writing the variables at each step of the reduction and thus save ourselves a great deal of work. For example, the system

$$2x + 4y + 6z = 22$$
$$3x + 8y + 5z = 27 \tag{5}$$
$$-x + y + 2z = 2$$

may be represented by the matrix

$$\begin{bmatrix} 2 & 4 & 6 & | & 22 \\ 3 & 8 & 5 & | & 27 \\ -1 & 1 & 2 & | & 2 \end{bmatrix} \tag{6}$$

The augmented matrix representing System (5)

The submatrix, consisting of the first three columns of Matrix (6), is called the **coefficient matrix** of System (5). The matrix itself, (6), is referred to as the **augmented matrix** of System (5) since it is obtained by joining the matrix of coefficients to the column (matrix) of constants. The vertical line separates the column of constants from the matrix of coefficients.

The next example shows how much work you can save by using matrices instead of the standard representation of the systems of linear equations.

EXAMPLE 2   Write the augmented matrix corresponding to each equivalent system given in (4a) through (4h).

Solution   The required sequence of augmented matrices follows.

Equivalent System	**Augmented Matrix**

a. $2x + 4y + 6z = 22$
$3x + 8y + 5z = 27$
$-x + y + 2z = 2$
$$\begin{bmatrix} 2 & 4 & 6 & | & 22 \\ 3 & 8 & 5 & | & 27 \\ -1 & 1 & 2 & | & 2 \end{bmatrix} \tag{7a}$$

b. $x + 2y + 3z = 11$
$3x + 8y + 5z = 27$
$-x + y + 2z = 2$
$$\begin{bmatrix} 1 & 2 & 3 & | & 11 \\ 3 & 8 & 5 & | & 27 \\ -1 & 1 & 2 & | & 2 \end{bmatrix} \tag{7b}$$

c. $x + 2y + 3z = 11$
$2y - 4z = -6$
$-x + y + 2z = 2$
$$\begin{bmatrix} 1 & 2 & 3 & | & 11 \\ 0 & 2 & -4 & | & -6 \\ -1 & 1 & 2 & | & 2 \end{bmatrix} \tag{7c}$$

d. $x + 2y + 3z = 11$
$2y - 4z = -6$
$3y + 5z = 13$
$$\begin{bmatrix} 1 & 2 & 3 & | & 11 \\ 0 & 2 & -4 & | & -6 \\ 0 & 3 & 5 & | & 13 \end{bmatrix} \tag{7d}$$

e. $x + 2y + 3z = 11$
$y - 2z = -3$
$3y + 5z = 13$
$$\begin{bmatrix} 1 & 2 & 3 & | & 11 \\ 0 & 1 & -2 & | & -3 \\ 0 & 3 & 5 & | & 13 \end{bmatrix} \tag{7e}$$

f. $x + 7z = 17$
$y - 2z = -3$
$3y + 5z = 13$
$$\begin{bmatrix} 1 & 0 & 7 & | & 17 \\ 0 & 1 & -2 & | & -3 \\ 0 & 3 & 5 & | & 13 \end{bmatrix} \tag{7f}$$

g.
$$\begin{aligned} x \phantom{{}+{}} + \phantom{{}} 7z &= 17 \\ y - 2z &= -3 \\ 11z &= 22 \end{aligned}$$
$$\left[\begin{array}{ccc|c} 1 & 0 & 7 & 17 \\ 0 & 1 & -2 & -3 \\ 0 & 0 & 11 & 22 \end{array}\right] \qquad \textbf{(7g)}$$

h.
$$\begin{aligned} x \phantom{{}={}} &= 3 \\ y \phantom{{}={}} &= 1 \\ z &= 2 \end{aligned}$$
$$\left[\begin{array}{ccc|c} 1 & 0 & 0 & 3 \\ 0 & 1 & 0 & 1 \\ 0 & 0 & 1 & 2 \end{array}\right] \qquad \textbf{(7h)} \qquad \blacksquare$$

The augmented matrix in (7h) is an example of a matrix in row-reduced form. In general, an augmented matrix with m rows and n columns (called an $m \times n$ matrix) is in **row-reduced form** if it satisfies the following conditions.

Row-Reduced Form of a Matrix
1. Each row consisting entirely of zeros lies below any other row having nonzero entries.
2. The first nonzero entry in each row is 1 (called a **leading 1**).
3. In any two successive (nonzero) rows, the leading 1 in the lower row lies to the right of the leading 1 in the upper row.
4. If a column contains a leading 1, then the other entries in that column are zeros.

EXAMPLE 3   Determine which of the following matrices are in row-reduced form. If a matrix is not in row-reduced form, state the condition that is violated.

a. $\left[\begin{array}{ccc|c} 1 & 0 & 0 & 0 \\ 0 & 1 & 0 & 0 \\ 0 & 0 & 1 & 3 \end{array}\right]$ **b.** $\left[\begin{array}{ccc|c} 1 & 0 & 0 & 4 \\ 0 & 1 & 0 & 3 \\ 0 & 0 & 0 & 0 \end{array}\right]$ **c.** $\left[\begin{array}{ccc|c} 1 & 2 & 0 & 0 \\ 0 & 0 & 1 & 0 \\ 0 & 0 & 0 & 1 \end{array}\right]$

d. $\left[\begin{array}{ccc|c} 0 & 1 & 2 & -2 \\ 1 & 0 & 0 & 3 \\ 0 & 0 & 1 & 2 \end{array}\right]$ **e.** $\left[\begin{array}{ccc|c} 1 & 2 & 0 & 0 \\ 0 & 0 & 1 & 3 \\ 0 & 0 & 2 & 1 \end{array}\right]$ **f.** $\left[\begin{array}{cc|c} 1 & 0 & 4 \\ 0 & 3 & 0 \\ 0 & 0 & 0 \end{array}\right]$

g. $\left[\begin{array}{ccc|c} 0 & 0 & 0 & 0 \\ 1 & 0 & 0 & 3 \\ 0 & 1 & 0 & 2 \end{array}\right]$

Solution   The matrices in parts (a)–(c) are in row-reduced form.

d. This matrix is not in row-reduced form. Conditions 3 and 4 are violated: The leading 1 in row 2 lies to the left of the leading 1 in row 1. Also, column 3 contains a leading 1 in row 3 and a nonzero element above it.

e. This matrix is not in row-reduced form. Conditions 2 and 4 are violated: The first nonzero entry in row 3 is a 2, not a 1. Also, column 3 contains a leading 1 and has a nonzero entry below it.

f. This matrix is not in row-reduced form. Condition 2 is violated: The first non-zero entry in row 2 is not a leading 1.

g. This matrix is not in row-reduced form. Condition 1 is violated: Row 1 consists of all zeros and does not lie below the other nonzero rows. $\blacksquare$

The foregoing discussion suggests the following adaptation of the Gauss–Jordan elimination method in solving systems of linear equations using matrices. First, the three operations on the equations of a system (see page 80) translate into the following **row operations** on the corresponding augmented matrices.

Row Operations
1. Interchange any two rows.
2. Replace any row by a nonzero constant multiple of itself.
3. Replace any row by the sum of that row and a constant multiple of any other row.

We obtained the augmented matrices in Example 2 by using the same operations that we used on the equivalent system of equations in Example 1.

To help us describe the Gauss–Jordan elimination method using matrices, let's introduce some terminology. We begin by defining what is meant by a **unit column**.

Unit Column
A column in a coefficient matrix is in unit form if one of the entries in the column is a 1 and the other entries are zeros.

For example, in the coefficient matrix of (7d), only the first column is in unit form; in the coefficient matrix of (7h), all three columns are in unit form. Now, the sequence of row operations that transforms the augmented matrix (7a) into the equivalent matrix (7d) in which the first column

$$2$$
$$3$$
$$-1$$

of (7a) is transformed into the unit column

$$1$$
$$0$$
$$0$$

is called **pivoting** the matrix about the element (number) 2. Similarly, we have pivoted about the element 2 in the second column of (7d), shown circled,

$$2$$
$$\textcircled{2}$$
$$3$$

in order to obtain the augmented matrix (7g). Finally, pivoting about the element 11 in column 3 of (7g)

$$7$$
$$-2$$
$$\textcircled{11}$$

leads to the augmented matrix (7h), in which all columns to the left of the vertical line are in unit form. The element about which a matrix is pivoted is called the *pivot element*.

Before looking at the next example, let's introduce the following notation for the three types of row operations.

Notation for Row Operations

Letting R_i denote the *i*th row of a matrix, we write:

Operation 1 $R_i \leftrightarrow R_j$ to mean: Interchange row *i* with row *j*.

Operation 2 cR_i to mean: Replace row *i* with *c* times row *i*.

Operation 3 $R_i + aR_j$ to mean: Replace row *i* with the sum of row *i* and *a* times row *j*.

EXAMPLE 4   Pivot the matrix about the circled element.

$$\begin{bmatrix} ③ & 5 & | & 9 \\ 2 & 3 & | & 5 \end{bmatrix}$$

Solution   Using the notation just introduced, we obtain

$$\begin{bmatrix} 3 & 5 & | & 9 \\ 2 & 3 & | & 5 \end{bmatrix} \xrightarrow{\frac{1}{3}R_1} \begin{bmatrix} 1 & \frac{5}{3} & | & 3 \\ 2 & 3 & | & 5 \end{bmatrix} \xrightarrow{R_2 - 2R_1} \begin{bmatrix} 1 & \frac{5}{3} & | & 3 \\ 0 & -\frac{1}{3} & | & -1 \end{bmatrix}$$

The first column, which originally contained the entry 3, is now in unit form, with a 1 where the pivot element used to be, and we are done.

Alternate Solution   In the first solution, we used Operation 2 to obtain a 1 where the pivot element was originally. Alternatively, we can use Operation 3 as follows:

$$\begin{bmatrix} 3 & 5 & | & 9 \\ 2 & 3 & | & 5 \end{bmatrix} \xrightarrow{R_1 - R_2} \begin{bmatrix} 1 & 2 & | & 4 \\ 2 & 3 & | & 5 \end{bmatrix} \xrightarrow{R_2 - 2R_1} \begin{bmatrix} 1 & 2 & | & 4 \\ 0 & -1 & | & -3 \end{bmatrix}$$ ■

Note   In Example 4, the two matrices

$$\begin{bmatrix} 1 & \frac{5}{3} & | & 3 \\ 0 & -\frac{1}{3} & | & -1 \end{bmatrix} \quad \text{and} \quad \begin{bmatrix} 1 & 2 & | & 4 \\ 0 & -1 & | & -3 \end{bmatrix}$$

look quite different, but they are in fact equivalent. You can verify this by observing that they represent the systems of equations

$$x + \frac{5}{3}y = 3 \quad \text{and} \quad x + 2y = 4$$
$$-\frac{1}{3}y = -1 \quad\quad\quad -y = -3$$

respectively, and both have the same solution: $x = -2$ and $y = 3$. Example 4 also shows that we can sometimes avoid working with fractions by using the appropriate row operation.   ■

A summary of the Gauss–Jordan method follows.

The Gauss–Jordan Elimination Method

1. Write the augmented matrix corresponding to the linear system.
2. Interchange rows (operation 1), if necessary, to obtain an augmented matrix in which the first entry in the first row is nonzero. Then pivot the matrix about this entry.
3. Interchange the second row with any row below it, if necessary, to obtain an augmented matrix in which the second entry in the second row is nonzero. Pivot the matrix about this entry.
4. Continue until the final matrix is in row-reduced form.

Before writing the augmented matrix, be sure to write all equations with the variables on the left and constant terms on the right of the equal sign. Also, make sure that the variables are in the same order in all equations.

EXAMPLE 5   Solve the system of linear equations given by

$$3x - 2y + 8z = 9$$
$$-2x + 2y + z = 3 \qquad \textbf{(8)}$$
$$x + 2y - 3z = 8$$

Solution   Using the Gauss–Jordan elimination method, we obtain the following sequence of equivalent augmented matrices:

$$\begin{bmatrix} ③ & -2 & 8 & | & 9 \\ -2 & 2 & 1 & | & 3 \\ 1 & 2 & -3 & | & 8 \end{bmatrix} \xrightarrow{R_1 + R_2} \begin{bmatrix} 1 & 0 & 9 & | & 12 \\ -2 & 2 & 1 & | & 3 \\ 1 & 2 & -3 & | & 8 \end{bmatrix}$$

$$\xrightarrow[R_3 - R_1]{R_2 + 2R_1} \begin{bmatrix} 1 & 0 & 9 & | & 12 \\ 0 & 2 & 19 & | & 27 \\ 0 & 2 & -12 & | & -4 \end{bmatrix}$$

$$\xrightarrow{R_2 \leftrightarrow R_3} \begin{bmatrix} 1 & 0 & 9 & | & 12 \\ 0 & ② & -12 & | & -4 \\ 0 & 2 & 19 & | & 27 \end{bmatrix}$$

$$\xrightarrow{\frac{1}{2}R_2} \begin{bmatrix} 1 & 0 & 9 & | & 12 \\ 0 & 1 & -6 & | & -2 \\ 0 & 2 & 19 & | & 27 \end{bmatrix}$$

$$\xrightarrow{R_3 - 2R_2} \begin{bmatrix} 1 & 0 & 9 & | & 12 \\ 0 & 1 & -6 & | & -2 \\ 0 & 0 & ㉛ & | & 31 \end{bmatrix}$$

$$\xrightarrow{\frac{1}{31}R_3}\begin{bmatrix} 1 & 0 & 9 & | & 12 \\ 0 & 1 & -6 & | & -2 \\ 0 & 0 & 1 & | & 1 \end{bmatrix}$$

$$\xrightarrow[R_2 + 6R_3]{R_1 - 9R_3}\begin{bmatrix} 1 & 0 & 0 & | & 3 \\ 0 & 1 & 0 & | & 4 \\ 0 & 0 & 1 & | & 1 \end{bmatrix}$$

The solution to System (8) is given by $x = 3$, $y = 4$, and $z = 1$. This may be verified by substitution into System (8) as follows:

$$3(3) - 2(4) + 8(1) = 9 \qquad \checkmark$$
$$-2(3) + 2(4) + 1 \quad = 3 \qquad \checkmark$$
$$3 \;+ 2(4) - 3(1) = 8 \qquad \checkmark \qquad \blacksquare$$

 When searching for an element to serve as a pivot, it is important to keep in mind that you may work only with the row containing the potential pivot or any row *below* it. To see what can go wrong if this caution is not heeded, consider the following augmented matrix for some linear system:

$$\begin{bmatrix} 1 & 1 & 2 & | & 3 \\ 0 & 0 & 3 & | & 1 \\ 0 & 2 & 1 & | & -2 \end{bmatrix}$$

Observe that column 1 is in unit form. The next step in the Gauss–Jordan elimination procedure calls for obtaining a nonzero element in the second position of row 2. If you use row 1 (which is *above* the row under consideration) to help you obtain the pivot, you might proceed as follows:

$$\begin{bmatrix} 1 & 1 & 2 & | & 3 \\ 0 & 0 & 3 & | & 1 \\ 0 & 2 & 1 & | & -2 \end{bmatrix} \xrightarrow{R_2 \leftrightarrow R_1} \begin{bmatrix} 0 & 0 & 3 & | & 1 \\ 1 & 1 & 2 & | & 3 \\ 0 & 2 & 1 & | & -2 \end{bmatrix}$$

As you can see, not only have we obtained a nonzero element to serve as the next pivot, but it is already a 1, thus obviating the next step. This seems like a good move. But beware, we have undone some of our earlier work: Column 1 is no longer in the unit form where a 1 appears first. The correct move in this case is to interchange row 2 with row 3.

The next example illustrates how to handle a situation in which the entry in row 1 of the augmented matrix is zero.

EXPLORE & DISCUSS

1. Can the phrase "a nonzero constant multiple of itself" in a type-2 row operation be replaced by "any constant multiple of itself"? Explain.

2. Can a row of an augmented matrix be replaced by a row obtained by adding a constant to every element in that row without changing the solution of the system of linear equations? Explain.

EXAMPLE 6   Solve the system of linear equations given by

$$2y + 3z = 7$$
$$3x + 6y - 12z = -3$$
$$5x - 2y + 2z = -7$$

Solution   Using the Gauss–Jordan elimination method, we obtain the following sequence of equivalent augmented matrices:

$$\begin{bmatrix} 0 & 2 & 3 & | & 7 \\ 3 & 6 & -12 & | & -3 \\ 5 & -2 & 2 & | & -7 \end{bmatrix} \xrightarrow{R_1 \leftrightarrow R_2} \begin{bmatrix} ③ & 6 & -12 & | & -3 \\ 0 & 2 & 3 & | & 7 \\ 5 & -2 & 2 & | & -7 \end{bmatrix} \xrightarrow{\frac{1}{3}R_1} \begin{bmatrix} 1 & 2 & -4 & | & -1 \\ 0 & 2 & 3 & | & 7 \\ 5 & -2 & 2 & | & -7 \end{bmatrix} \xrightarrow{R_3 - 5R_1}$$

$$\begin{bmatrix} 1 & 2 & -4 & | & -1 \\ 0 & ② & 3 & | & 7 \\ 0 & -12 & 22 & | & -2 \end{bmatrix} \xrightarrow{\frac{1}{2}R_2} \begin{bmatrix} 1 & 2 & -4 & | & -1 \\ 0 & 1 & \frac{3}{2} & | & \frac{7}{2} \\ 0 & -12 & 22 & | & -2 \end{bmatrix} \xrightarrow[R_3 + 12R_2]{R_1 - 2R_2} \begin{bmatrix} 1 & 0 & -7 & | & -8 \\ 0 & 1 & \frac{3}{2} & | & \frac{7}{2} \\ 0 & 0 & ㊵ & | & 40 \end{bmatrix} \xrightarrow{\frac{1}{40}R_3}$$

$$\begin{bmatrix} 1 & 0 & -7 & | & -8 \\ 0 & 1 & \frac{3}{2} & | & \frac{7}{2} \\ 0 & 0 & 1 & | & 1 \end{bmatrix} \xrightarrow[R_2 - \frac{3}{2}R_3]{R_1 + 7R_3} \begin{bmatrix} 1 & 0 & 0 & | & -1 \\ 0 & 1 & 0 & | & 2 \\ 0 & 0 & 1 & | & 1 \end{bmatrix}$$

The solution to the system is given by $x = -1$, $y = 2$, and $z = 1$; this may be verified by substitution into the system.   ■

APPLIED EXAMPLE 7   Manufacturing: Production Scheduling Complete the solution to Example 1 in Section 2.1, page 75.

Solution   To complete the solution of the problem posed in Example 1, recall that the mathematical formulation of the problem led to the following system of linear equations:

$$2x + y + z = 180$$
$$x + 3y + 2z = 300$$
$$2x + y + 2z = 240$$

where x, y, and z denote the respective numbers of type-A, type-B, and type-C souvenirs to be made.

   Solving the foregoing system of linear equations by the Gauss–Jordan elimination method, we obtain the following sequence of equivalent augmented matrices:

$$\begin{bmatrix} ② & 1 & 1 & | & 180 \\ 1 & 3 & 2 & | & 300 \\ 2 & 1 & 2 & | & 240 \end{bmatrix} \xrightarrow{R_1 \leftrightarrow R_2} \begin{bmatrix} 1 & 3 & 2 & | & 300 \\ 2 & 1 & 1 & | & 180 \\ 2 & 1 & 2 & | & 240 \end{bmatrix}$$

$$\xrightarrow[R_3 - 2R_1]{R_2 - 2R_1} \begin{bmatrix} 1 & 3 & 2 & | & 300 \\ 0 & ⑤ & -3 & | & -420 \\ 0 & -5 & -2 & | & -360 \end{bmatrix}$$

$$\xrightarrow{-\frac{1}{5}R_2} \begin{bmatrix} 1 & 3 & 2 & | & 300 \\ 0 & 1 & \frac{3}{5} & | & 84 \\ 0 & -5 & -2 & | & -360 \end{bmatrix}$$

$$\xrightarrow[R_3 + 5R_2]{R_1 - 3R_2} \begin{bmatrix} 1 & 0 & \frac{1}{5} & | & 48 \\ 0 & 1 & \frac{3}{5} & | & 84 \\ 0 & 0 & \textcircled{1} & | & 60 \end{bmatrix}$$

$$\xrightarrow[R_2 - \frac{3}{5}R_3]{R_1 - \frac{1}{5}R_3} \begin{bmatrix} 1 & 0 & 0 & | & 36 \\ 0 & 1 & 0 & | & 48 \\ 0 & 0 & 1 & | & 60 \end{bmatrix}$$

Thus, $x = 36$, $y = 48$, and $z = 60$; that is, Ace Novelty should make 36 type-A souvenirs, 48 type-B souvenirs, and 60 type-C souvenirs in order to use all available machine time. ∎

2.2 Self-Check Exercises

1. Solve the system of linear equations

$$2x + 3y + z = 6$$
$$x - 2y + 3z = -3$$
$$3x + 2y - 4z = 12$$

using the Gauss–Jordan elimination method.

2. A farmer has 200 acres of land suitable for cultivating crops A, B, and C. The cost per acre of cultivating crop A, crop B, and crop C is $40, $60, and $80, respectively. The farmer has $12,600 available for land cultivation. Each acre of crop A requires 20 labor-hours, each acre of crop B requires 25 labor-hours, and each acre of crop C requires 40 labor-hours. The farmer has a maximum of 5950 labor-hours available. If he wishes to use all of his cultivatable land, the entire budget, and all the labor available, how many acres of each crop should he plant?

Solutions to Self-Check Exercises 2.2 can be found on page 94.

2.2 Concept Questions

1. **a.** Explain what it means for two systems of linear equations to be equivalent to each other.
   b. Give the meaning of the following notation used for row operations in the Gauss–Jordan elimination method:

   a. $R_i \leftrightarrow R_j$ **b.** cR_i **c.** $R_i + aR_j$

2. **a.** What is an augmented matrix? A coefficient matrix? A unit column?
   b. Explain what is meant by a pivot operation.

3. Suppose that a matrix is in row-reduced form.
   a. What is the position of a row consisting entirely of zeros relative to the nonzero rows?
   b. What is the first nonzero entry in each row?
   c. What is the position of the leading 1s in successive nonzero rows?
   d. If a column contains a leading 1, then what is the value of the other entries in that column?

2.2 Exercises

In Exercises 1–4, write the augmented matrix corresponding to each system of equations.

1. $2x - 3y = 7$
$3x + y = 4$

2. $3x + 7y - 8z = 5$
$x + 3z = -2$
$4x - 3y = 7$

3. $-y + 2z = 6$
$2x + 2y - 8z = 7$
$3y + 4z = 0$

4. $3x_1 + 2x_2 = 0$
$x_1 - x_2 + 2x_3 = 4$
$2x_2 - 3x_3 = 5$

In Exercises 5–8, write the system of equations corresponding to each augmented matrix.

5. $\begin{bmatrix} 3 & 2 & | & -4 \\ 1 & -1 & | & 5 \end{bmatrix}$

6. $\begin{bmatrix} 0 & 3 & 2 & | & 4 \\ 1 & -1 & -2 & | & -3 \\ 4 & 0 & 3 & | & 2 \end{bmatrix}$

7. $\begin{bmatrix} 1 & 3 & 2 & | & 4 \\ 2 & 0 & 0 & | & 5 \\ 3 & -3 & 2 & | & 6 \end{bmatrix}$

8. $\begin{bmatrix} 2 & 3 & 1 & | & 6 \\ 4 & 3 & 2 & | & 5 \\ 0 & 0 & 0 & | & 0 \end{bmatrix}$

In Exercises 9–18, indicate whether the matrix is in row-reduced form.

9. $\begin{bmatrix} 1 & 0 & | & 3 \\ 0 & 1 & | & -2 \end{bmatrix}$

10. $\begin{bmatrix} 1 & 1 & | & 3 \\ 0 & 0 & | & 0 \end{bmatrix}$

11. $\begin{bmatrix} 0 & 1 & | & 3 \\ 1 & 0 & | & 5 \end{bmatrix}$

12. $\begin{bmatrix} 0 & 1 & | & 3 \\ 0 & 0 & | & 5 \end{bmatrix}$

13. $\begin{bmatrix} 1 & 0 & 0 & | & 3 \\ 0 & 1 & 0 & | & 4 \\ 0 & 0 & 1 & | & 5 \end{bmatrix}$

14. $\begin{bmatrix} 1 & 0 & 0 & | & -1 \\ 0 & 1 & 0 & | & -2 \\ 0 & 0 & 2 & | & -3 \end{bmatrix}$

15. $\begin{bmatrix} 1 & 0 & 1 & | & 3 \\ 0 & 1 & 0 & | & 4 \\ 0 & 0 & -1 & | & 6 \end{bmatrix}$

16. $\begin{bmatrix} 1 & 0 & | & -10 \\ 0 & 1 & | & 2 \\ 0 & 0 & | & 0 \end{bmatrix}$

17. $\begin{bmatrix} 0 & 0 & 0 & | & 0 \\ 0 & 1 & 2 & | & 4 \\ 0 & 0 & 0 & | & 0 \end{bmatrix}$

18. $\begin{bmatrix} 1 & 0 & 0 & | & 3 \\ 0 & 1 & 0 & | & 6 \\ 0 & 0 & 0 & | & 4 \\ 0 & 0 & 1 & | & 5 \end{bmatrix}$

In Exercises 19–26, pivot the system about the circled element.

19. $\begin{bmatrix} ②　 & 4 & | & 8 \\ 3 & 1 & | & 2 \end{bmatrix}$

20. $\begin{bmatrix} 3 & 2 & | & 6 \\ ④ & 2 & | & 5 \end{bmatrix}$

21. $\begin{bmatrix} ⊖1 & 2 & | & 3 \\ 6 & 4 & | & 2 \end{bmatrix}$

22. $\begin{bmatrix} ① & 3 & | & 4 \\ 2 & 4 & | & 6 \end{bmatrix}$

23. $\begin{bmatrix} ② & 4 & 6 & | & 12 \\ 2 & 3 & 1 & | & 5 \\ 3 & -1 & 2 & | & 4 \end{bmatrix}$

24. $\begin{bmatrix} 1 & 3 & 2 & | & 4 \\ ② & 4 & 8 & | & 6 \\ -1 & 2 & 3 & | & 4 \end{bmatrix}$

25. $\begin{bmatrix} 0 & 1 & 3 & | & 4 \\ 2 & 4 & ① & | & 3 \\ 5 & 6 & 2 & | & -4 \end{bmatrix}$

26. $\begin{bmatrix} 1 & 2 & 3 & | & 5 \\ 0 & ⊖3 & 3 & | & 2 \\ 0 & 4 & -1 & | & 3 \end{bmatrix}$

In Exercises 27–30, fill in the missing entries by performing the indicated row operations to obtain the row-reduced matrices.

27. $\begin{bmatrix} 3 & 9 & | & 6 \\ 2 & 1 & | & 4 \end{bmatrix} \xrightarrow{\frac{1}{3}R_1} \begin{bmatrix} \cdot & \cdot & | & \cdot \\ 2 & 1 & | & 4 \end{bmatrix} \xrightarrow{R_2 - 2R_1}$

$\begin{bmatrix} 1 & 3 & | & 2 \\ \cdot & \cdot & | & \cdot \end{bmatrix} \xrightarrow{-\frac{1}{5}R_2} \begin{bmatrix} 1 & 3 & | & 2 \\ \cdot & \cdot & | & \cdot \end{bmatrix} \xrightarrow{R_1 - 3R_2} \begin{bmatrix} 1 & 0 & | & 2 \\ 0 & 1 & | & 0 \end{bmatrix}$

28. $\begin{bmatrix} 1 & 2 & | & 1 \\ 2 & 3 & | & -1 \end{bmatrix} \xrightarrow{R_2 - 2R_1} \begin{bmatrix} 1 & 2 & | & 1 \\ \cdot & \cdot & | & \cdot \end{bmatrix} \xrightarrow{-R_2}$

$\begin{bmatrix} 1 & 2 & | & 1 \\ \cdot & \cdot & | & \cdot \end{bmatrix} \xrightarrow{R_1 - 2R_2} \begin{bmatrix} 1 & 0 & | & -5 \\ 0 & 1 & | & 3 \end{bmatrix}$

29. $\begin{bmatrix} 1 & 3 & 1 & | & 3 \\ 3 & 8 & 3 & | & 7 \\ 2 & -3 & 1 & | & -10 \end{bmatrix} \xrightarrow[R_3 - 2R_1]{R_2 - 3R_1} \begin{bmatrix} 1 & 3 & 1 & | & 3 \\ \cdot & \cdot & \cdot & | & \cdot \\ \cdot & \cdot & \cdot & | & \cdot \end{bmatrix} \xrightarrow{-R_2}$

$\begin{bmatrix} 1 & 3 & 1 & | & 3 \\ \cdot & \cdot & \cdot & | & \cdot \\ 0 & -9 & -1 & | & -16 \end{bmatrix} \xrightarrow[R_3 + 9R_2]{R_1 - 3R_2}$

$\begin{bmatrix} \cdot & \cdot & \cdot & | & \cdot \\ 0 & 1 & 0 & | & 2 \\ \cdot & \cdot & \cdot & | & \cdot \end{bmatrix} \xrightarrow[-R_3]{R_1 + R_3} \begin{bmatrix} 1 & 0 & 0 & | & -1 \\ 0 & 1 & 0 & | & 2 \\ 0 & 0 & 1 & | & -2 \end{bmatrix}$

30. $\begin{bmatrix} 0 & 1 & 3 & | & -4 \\ 1 & 2 & 1 & | & 7 \\ 1 & -2 & 0 & | & 1 \end{bmatrix} \xrightarrow{R_1 \leftrightarrow R_2} \begin{bmatrix} \cdot & \cdot & \cdot & | & \cdot \\ \cdot & \cdot & \cdot & | & \cdot \\ 1 & -2 & 0 & | & 1 \end{bmatrix}$

$\xrightarrow{R_3 - R_1} \begin{bmatrix} 1 & 2 & 1 & | & 7 \\ 0 & 1 & 3 & | & -4 \\ \cdot & \cdot & \cdot & | & \cdot \end{bmatrix} \xrightarrow[R_3 + 4R_2]{R_1 + \frac{1}{2}R_3} \begin{bmatrix} \cdot & \cdot & \cdot & | & \cdot \\ 0 & 1 & 3 & | & -4 \\ \cdot & \cdot & \cdot & | & \cdot \end{bmatrix}$

$\xrightarrow{\frac{1}{11}R_3} \begin{bmatrix} 1 & 0 & \frac{1}{2} & | & 4 \\ 0 & 1 & 3 & | & -4 \\ \cdot & \cdot & \cdot & | & \cdot \end{bmatrix} \xrightarrow[R_2 - 3R_3]{R_1 - \frac{1}{2}R_3} \begin{bmatrix} 1 & 0 & 0 & | & 5 \\ 0 & 1 & 0 & | & 2 \\ 0 & 0 & 1 & | & -2 \end{bmatrix}$

31. Write a system of linear equations for the augmented matrix of Exercise 27. Using the results of Exercise 27, determine the solution of the system.

32. Repeat Exercise 31 for the augmented matrix of Exercise 28.

33. Repeat Exercise 31 for the augmented matrix of Exercise 29.

34. Repeat Exercise 31 for the augmented matrix of Exercise 30.

In Exercises 35–50, solve the system of linear equations using the Gauss–Jordan elimination method.

35. $x - 2y = 8$
$3x + 4y = 4$

36. $3x + y = 1$
$-7x - 2y = -1$

37. $2x - 3y = -8$
$4x + y = -2$

38. $5x + 3y = 9$
$-2x + y = -8$

39. $x + y + z = 0$
$2x - y + z = 1$
$x + y - 2z = 2$

40. $2x + y - 2z = 4$
$x + 3y - z = -3$
$3x + 4y - z = 7$

41. $2x + 2y + z = 9$
$x + z = 4$
$4y - 3z = 17$

42. $2x + 3y - 2z = 10$
$3x - 2y + 2z = 0$
$4x - y + 3z = -1$

43. $ - x_2 + x_3 = 2$
$4x_1 - 3x_2 + 2x_3 = 16$
$3x_1 + 2x_2 + x_3 = 11$

44. $2x + 4y - 6z = 38$
$x + 2y + 3z = 7$
$3x - 4y + 4z = -19$

45. $x_1 - 2x_2 + x_3 = 6$
$2x_1 + x_2 - 3x_3 = -3$
$x_1 - 3x_2 + 3x_3 = 10$

46. $2x + 3y - 6z = -11$
$x - 2y + 3z = 9$
$3x + y = 7$

47. $2x + 3z = -1$
$3x - 2y + z = 9$
$x + y + 4z = 4$

48. $2x_1 - x_2 + 3x_3 = -4$
$x_1 - 2x_2 + x_3 = -1$
$x_1 - 5x_2 + 2x_3 = -3$

49. $x_1 - x_2 + 3x_3 = 14$
$x_1 + x_2 + x_3 = 6$
$-2x_1 - x_2 + x_3 = -4$

50. $2x_1 - x_2 - x_3 = 0$
$3x_1 + 2x_2 + x_3 = 7$
$x_1 + 2x_2 + 2x_3 = 5$

The problems in Exercises 51–63 correspond to those in Exercises 15–27, Section 2.1. Use the results of your previous work to help you solve these problems.

51. AGRICULTURE The Johnson Farm has 500 acres of land allotted for cultivating corn and wheat. The cost of cultivating corn and wheat (including seeds and labor) is $42 and $30 per acre, respectively. Jacob Johnson has $18,600 available for cultivating these crops. If he wishes to use all the allotted land and his entire budget for cultivating these two crops, how many acres of each crop should he plant?

52. INVESTMENTS Michael Perez has a total of $2000 on deposit with two savings institutions. One pays interest at the rate of 6%/year, whereas the other pays interest at the rate of 8%/year. If Michael earned a total of $144 in interest during a single year, how much does he have on deposit in each institution?

53. MIXTURES The Coffee Shoppe sells a coffee blend made from two coffees, one costing $5/lb and the other costing $6/lb. If the blended coffee sells for $5.60/lb, find how much of each coffee is used to obtain the desired blend. (Assume the weight of the blended coffee is 100 lb.)

54. INVESTMENTS Kelly Fisher has a total of $30,000 invested in two municipal bonds that have yields of 8% and 10% interest per year, respectively. If the interest Kelly receives from the bonds in a year is $2640, how much does she have invested in each bond?

55. RIDERSHIP The total number of passengers riding a certain city bus during the morning shift is 1000. If the child's fare is $.50, the adult fare is $1.50, and the total revenue from the fares in the morning shift is $1300, how many children and how many adults rode the bus during the morning shift?

56. REAL ESTATE Cantwell Associates, a real estate developer, is planning to build a new apartment complex consisting of one-bedroom units and two- and three-bedroom townhouses. A total of 192 units is planned, and the number of family units (two- and three-bedroom townhouses) will equal the number of one-bedroom units. If the number of one-bedroom units will be 3 times the number of three-bedroom units, find how many units of each type will be in the complex.

57. INVESTMENT PLANNING The annual interest on Sid Carrington's three investments amounted to $21,600: 6% on a savings account, 8% on mutual funds, and 12% on bonds. The amount of Sid's investment in bonds was twice the amount of his investment in the savings account, and the interest earned from his investment in bonds was equal to the dividends he received from his investment in mutual funds. Find how much money he placed in each type of investment.

58. INVESTMENT CLUB A private investment club has $200,000 earmarked for investment in stocks. To arrive at an acceptable overall level of risk, the stocks that management is considering have been classified into three categories: high-risk, medium-risk, and low-risk. Management estimates that high-risk stocks will have a rate of return of 15%/year; medium-risk stocks, 10%/year; and low-risk stocks, 6%/year. The members have decided that the investment in low-risk stocks should be equal to the sum of the investments in the stocks of the other two categories. Determine how much the club should invest in each type of stock if the investment goal is to have a return of $20,000/year on the total investment. (Assume that all the money available for investment is invested.)

59. MIXTURE PROBLEM—FERTILIZER Lawnco produces three grades of commercial fertilizers. A 100-lb bag of grade-A fertilizer contains 18 lb of nitrogen, 4 lb of phosphate, and 5 lb of potassium. A 100-lb bag of grade-B fertilizer contains 20 lb of nitrogen and 4 lb each of phosphate and potassium. A 100-lb bag of grade-C fertilizer contains 24 lb of nitrogen, 3 lb of phosphate, and 6 lb of potassium. How many 100-lb bags of each of the three grades of fertilizers should Lawnco produce if 26,400 lb of nitrogen, 4900 lb of phosphate, and 6200 lb of potassium are available and all the nutrients are used?

60. **Box-Office Receipts**   A theater has a seating capacity of 900 and charges $4 for children, $6 for students, and $8 for adults. At a certain screening with full attendance, there were half as many adults as children and students combined. The receipts totaled $5600. How many children attended the show?

61. **Management Decisions**   The management of Hartman Rent-A-Car has allocated $1.5 million to buy a fleet of new automobiles consisting of compact, intermediate, and full-size cars. Compacts cost $12,000 each, intermediate-size cars cost $18,000 each, and full-size cars cost $24,000 each. If Hartman purchases twice as many compacts as intermediate-size cars and the total number of cars to be purchased is 100, determine how many cars of each type will be purchased. (Assume that the entire budget will be used.)

62. **Investment Clubs**   The management of a private investment club has a fund of $200,000 earmarked for investment in stocks. To arrive at an acceptable overall level of risk, the stocks that management is considering have been classified into three categories: high-risk, medium-risk, and low-risk. Management estimates that high-risk stocks will have a rate of return of 15%/year; medium-risk stocks, 10%/year; and low-risk stocks, 6%/year. The investment in low-risk stocks is to be twice the sum of the investments in stocks of the other two categories. If the investment goal is to have an average rate of return of 9%/year on the total investment, determine how much the club should invest in each type of stock. (Assume all of the money available for investment is invested.)

63. **Diet Planning**   A dietitian wishes to plan a meal around three foods. The percent of the daily requirements of proteins, carbohydrates, and iron contained in each ounce of the three foods is summarized in the accompanying table:

	Food I	Food II	Food III
Proteins (%)	10	6	8
Carbohydrates (%)	10	12	6
Iron (%)	5	4	12

Determine how many ounces of each food the dietitian should include in the meal to meet exactly the daily requirement of proteins, carbohydrates, and iron (100% of each).

64. **Investments**   Mr. and Mrs. Garcia have a total of $100,000 to be invested in stocks, bonds, and a money market account.

The stocks have a rate of return of 12%/year, while the bonds and the money market account pay 8%/year and 4%/year, respectively. The Garcias have stipulated that the amount invested in the money market account should be equal to the sum of 20% of the amount invested in stocks and 10% of the amount invested in bonds. How should the Garcias allocate their resources if they require an annual income of $10,000 from their investments?

65. **Box-Office Receipts**   For the opening night at the Opera House, a total of 1000 tickets were sold. Front orchestra seats cost $80 apiece, rear orchestra seats cost $60 apiece, and front balcony seats cost $50 apiece. The combined number of tickets sold for the front orchestra and rear orchestra exceeded twice the number of front balcony tickets sold by 400. The total receipts for the performance were $62,800. Determine how many tickets of each type were sold.

66. **Production Scheduling**   A manufacturer of women's blouses makes three types of blouses: sleeveless, short-sleeve, and long-sleeve. The time (in minutes) required by each department to produce a dozen blouses of each type is shown in the accompanying table:

	Sleeveless	Short-Sleeve	Long-Sleeve
Cutting	9	12	15
Sewing	22	24	28
Packaging	6	8	8

The cutting, sewing, and packaging departments have available a maximum of 80, 160, and 48 labor-hours, respectively, per day. How many dozens of each type of blouse can be produced each day if the plant is operated at full capacity?

67. **Business Travel Expenses**   An executive of Trident Communications recently traveled to London, Paris, and Rome. He paid $180, $230, and $160 per night for lodging in London, Paris, and Rome, respectively, and his hotel bills totaled $2660. He spent $110, $120, and $90 per day for his meals in London, Paris, and Rome, respectively, and his expenses for meals totaled $1520. If he spent as many days in London as he did in Paris and Rome combined, how many days did he stay in each city?

68. **Vacation Costs**   Joan and Dick spent 2 wk (14 nights) touring four cities on the East Coast—Boston, New York, Philadelphia, and Washington, D.C. They paid $120, $200, $80, and $100 per night for lodging in each city, respectively, and their total hotel bill came to $2020. The number of days they spent in New York was the same as the total number of days they spent in Boston and Washington, D.C., and the couple spent 3 times as many days in New York as they did in Philadelphia. How many days did Joan and Dick stay in each city?

In Exercises 69 and 70, determine whether the statement is true or false. If it is true, explain why it is true. If it is false, give an example to show why it is false.

69. An equivalent system of linear equations can be obtained from a system of equations by replacing one of its equations by any constant multiple of itself.

70. If the augmented matrix corresponding to a system of three linear equations in three variables has a row of the form $[0 \ \ 0 \ \ 0 \ | \ a]$, where a is a nonzero number, then the system has no solution.

2.2 Solutions to Self-Check Exercises

1. We obtain the following sequence of equivalent augmented matrices:

$$\begin{bmatrix} 2 & 3 & 1 & 6 \\ 1 & -2 & 3 & -3 \\ 3 & 2 & -4 & 12 \end{bmatrix} \xrightarrow{R_1 \leftrightarrow R_2} \begin{bmatrix} \textcircled{1} & -2 & 3 & -3 \\ 2 & 3 & 1 & 6 \\ 3 & 2 & -4 & 12 \end{bmatrix}$$

$$\xrightarrow[R_3 - 3R_1]{R_2 - 2R_1} \begin{bmatrix} 1 & -2 & 3 & -3 \\ 0 & 7 & -5 & 12 \\ 0 & 8 & -13 & 21 \end{bmatrix} \xrightarrow{R_2 \leftrightarrow R_3}$$

$$\begin{bmatrix} 1 & -2 & 3 & -3 \\ 0 & \textcircled{8} & -13 & 21 \\ 0 & 7 & -5 & 12 \end{bmatrix} \xrightarrow{R_2 - R_3} \begin{bmatrix} 1 & -2 & 3 & -3 \\ 0 & 1 & -8 & 9 \\ 0 & 7 & -5 & 12 \end{bmatrix}$$

$$\xrightarrow[R_3 - 7R_2]{R_1 + 2R_2} \begin{bmatrix} 1 & 0 & -13 & 15 \\ 0 & 1 & -8 & 9 \\ 0 & 0 & 51 & -51 \end{bmatrix} \xrightarrow{\frac{1}{51}R_3} \begin{bmatrix} 1 & 0 & -13 & 15 \\ 0 & 1 & -8 & 9 \\ 0 & 0 & \textcircled{1} & -1 \end{bmatrix}$$

$$\xrightarrow[R_2 + 8R_3]{R_1 + 13R_3} \begin{bmatrix} 1 & 0 & 0 & 2 \\ 0 & 1 & 0 & 1 \\ 0 & 0 & 1 & -1 \end{bmatrix}$$

The solution to the system is $x = 2$, $y = 1$, and $z = -1$.

2. Referring to the solution of Exercise 2, Self-Check Exercises 2.1, we see that the problem reduces to solving the following system of linear equations:

$$\begin{aligned} x + \quad y + \quad z &= \quad 200 \\ 40x + 60y + 80z &= 12{,}600 \\ 20x + 25y + 40z &= \quad 5{,}950 \end{aligned}$$

Using the Gauss–Jordan elimination method, we have

$$\begin{bmatrix} 1 & 1 & 1 & 200 \\ 40 & 60 & 80 & 12{,}600 \\ 20 & 25 & 40 & 5{,}950 \end{bmatrix} \xrightarrow[R_3 - 20R_1]{R_2 - 40R_1} \begin{bmatrix} 1 & 1 & 1 & 200 \\ 0 & 20 & 40 & 4600 \\ 0 & 5 & 20 & 1950 \end{bmatrix}$$

$$\xrightarrow{\frac{1}{20}R_2} \begin{bmatrix} 1 & 1 & 1 & 200 \\ 0 & 1 & 2 & 230 \\ 0 & 5 & 20 & 1950 \end{bmatrix} \xrightarrow[R_3 - 5R_2]{R_1 - R_2} \begin{bmatrix} 1 & 0 & -1 & -30 \\ 0 & 1 & 2 & 230 \\ 0 & 0 & 10 & 800 \end{bmatrix}$$

$$\xrightarrow{\frac{1}{10}R_3} \begin{bmatrix} 1 & 0 & -1 & -30 \\ 0 & 1 & 2 & 230 \\ 0 & 0 & 1 & 80 \end{bmatrix} \xrightarrow[R_2 - 2R_3]{R_1 + R_3} \begin{bmatrix} 1 & 0 & 0 & 50 \\ 0 & 1 & 0 & 70 \\ 0 & 0 & 1 & 80 \end{bmatrix}$$

From the last augmented matrix in reduced form, we see that $x = 50$, $y = 70$, and $z = 80$. Therefore, the farmer should plant 50 acres of crop A, 70 acres of crop B, and 80 acres of crop C.

USING TECHNOLOGY

■ Systems of Linear Equations: Unique Solutions

Solving a System of Linear Equations Using the Gauss–Jordan Method

The three matrix operations can be performed on a matrix using a graphing utility. The commands are summarized in the following table.

	Calculator Function		
Operation	**TI-83**	**TI-86**	
$R_i \leftrightarrow R_j$	**rowSwap**([A], i, j)	**rSwap**(A, i, j)	or equivalent
cR_i	***row**(c, [A], i)	**multR**(c, A, i)	or equivalent
$R_i + aR_j$	***row+**(a, [A], j, i)	**mRAdd**(a, A, j, i)	or equivalent

When a row operation is performed on a matrix, the result is stored as an answer in the calculator. If another operation is performed on this matrix, then the matrix is erased. Should a mistake be made in the operation, the previous matrix is lost. For this reason, you should store the results of each operation. We do this by pressing STO, followed by the name of a matrix, and then **ENTER**. We use this process in the following example.

EXAMPLE 1   Use a graphing utility to solve the following system of linear equations by the Gauss–Jordan method (see Example 5 in Section 2.2):

$$
\begin{aligned}
3x - 2y + 8z &= 9 \\
-2x + 2y + z &= 3 \\
x + 2y - 3z &= 8
\end{aligned}
$$

Solution   Using the Gauss–Jordan method, we obtain the following sequence of equivalent matrices.

$$
\begin{bmatrix}
3 & -2 & 8 & | & 9 \\
-2 & 2 & 1 & | & 3 \\
1 & 2 & -3 & | & 8
\end{bmatrix}
\xrightarrow{\ *\text{row} + (1, [A], 2, 1) \, \blacktriangleright \, B\ }
$$

$$
\begin{bmatrix}
1 & 0 & 9 & | & 12 \\
-2 & 2 & 1 & | & 3 \\
1 & 2 & -3 & | & 8
\end{bmatrix}
\xrightarrow{\ *\text{row} + (2, [B], 1, 2) \, \blacktriangleright \, C\ }
$$

$$
\begin{bmatrix}
1 & 0 & 9 & | & 12 \\
0 & 2 & 19 & | & 27 \\
1 & 2 & -3 & | & 8
\end{bmatrix}
\xrightarrow{\ *\text{row} + (-1, [C], 1, 3) \, \blacktriangleright \, B\ }
$$

$$
\begin{bmatrix}
1 & 0 & 9 & | & 12 \\
0 & 2 & 19 & | & 27 \\
0 & 2 & -12 & | & -4
\end{bmatrix}
\xrightarrow{\ *\text{row}(\frac{1}{2}, [B], 2) \, \blacktriangleright \, C\ }
$$

$$
\begin{bmatrix}
1 & 0 & 9 & | & 12 \\
0 & 1 & 9.5 & | & 13.5 \\
0 & 2 & -12 & | & -4
\end{bmatrix}
\xrightarrow{\ *\text{row} + (-2, [C], 2, 3) \, \blacktriangleright \, B\ }
$$

$$
\begin{bmatrix}
1 & 0 & 9 & | & 12 \\
0 & 1 & 9.5 & | & 13.5 \\
0 & 0 & -31 & | & -31
\end{bmatrix}
\xrightarrow{\ *\text{row}(-\frac{1}{31}, [B], 3) \, \blacktriangleright \, C\ }
$$

(continued)

$$\begin{bmatrix} 1 & 0 & 9 & 12 \\ 0 & 1 & 9.5 & 13.5 \\ 0 & 0 & 1 & 1 \end{bmatrix} \xrightarrow{\text{*row} + (-9, [C], 3, 1) \blacktriangleright B}$$

$$\begin{bmatrix} 1 & 0 & 0 & 3 \\ 0 & 1 & 9.5 & 13.5 \\ 0 & 0 & 1 & 1 \end{bmatrix} \xrightarrow{\text{*row} + (-9.5, [B], 3, 2) \blacktriangleright C} \begin{bmatrix} 1 & 0 & 0 & 3 \\ 0 & 1 & 0 & 4 \\ 0 & 0 & 1 & 1 \end{bmatrix}$$

The last matrix is in row-reduced form, and we see that the solution of the system is $x = 3$, $y = 4$, and $z = 1$.   ∎

Using rref (TI-83 and TI-86) to Solve a System of Linear Equations

The operation **rref** (or equivalent function in your utility, if there is one) will transform an augmented matrix into one that is in row-reduced form. For example, using **rref**, we find

$$\begin{bmatrix} 3 & -2 & 8 & 9 \\ -2 & 2 & 1 & 3 \\ 1 & 2 & -3 & 8 \end{bmatrix} \xrightarrow{\text{rref}} \begin{bmatrix} 1 & 0 & 0 & 3 \\ 0 & 1 & 0 & 4 \\ 0 & 0 & 1 & 1 \end{bmatrix}$$

as obtained earlier!

Using SIMULT (TI-86) to Solve a System of Equations

The operation **SIMULT** (or equivalent operation on your utility, if there is one) of a graphing utility can be used to solve a system of n linear equations in n variables, where n is an integer between 2 and 30.

EXAMPLE 2   Use the SIMULT operation to solve the system of Example 1.

Solution   Call for the SIMULT operation. Since the system under consideration has three equations in three variables, enter $n = 3$. Next, enter a1, 1 = 3, a1, 2 = −2, a1, 3 = 8, . . . , b1 = 9, a2, 1 = −2, . . . , b3 = 8. Select <SOLVE> and the display

$$x1 = 3$$
$$x2 = 4$$
$$x3 = 1$$

appears on the screen, giving $x = 3$, $y = 4$, and $z = 1$ as the required solution.   ∎

TECHNOLOGY EXERCISES

Use a graphing utility to solve the system of equations (a) by the Gauss–Jordan method, (b) using the rref operation, and (c) using SIMULT.

1. $x_1 - 2x_2 + 2x_3 - 3x_4 = -7$
$3x_1 + 2x_2 - x_3 + 5x_4 = 22$
$2x_1 - 3x_2 + 4x_3 - x_4 = -3$
$3x_1 - 2x_2 - x_3 + 2x_4 = 12$

2. $\begin{aligned} 2x_1 - x_2 + 3x_3 - 2x_4 &= -2 \\ x_1 - 2x_2 + x_3 - 3x_4 &= 2 \\ x_1 - 5x_2 + 2x_3 + 3x_4 &= -6 \\ -3x_1 + 3x_2 - 4x_3 - 4x_4 &= 9 \end{aligned}$

3. $\begin{aligned} 2x_1 + x_2 + 3x_3 - x_4 &= 9 \\ -x_1 - 2x_2 \quad\quad - 3x_4 &= -1 \\ x_1 \quad\quad - 3x_3 + x_4 &= 10 \\ x_1 - x_2 - x_3 - x_4 &= 8 \end{aligned}$

4. $\begin{aligned} x_1 - 2x_2 - 2x_3 + x_4 &= 1 \\ 2x_1 - x_2 + 2x_3 + 3x_4 &= -2 \\ -x_1 - 5x_2 + 7x_3 - 2x_4 &= 3 \\ 3x_1 - 4x_2 + 3x_3 + 4x_4 &= -4 \end{aligned}$

5. $\begin{aligned} 2x_1 - 2x_2 + 3x_3 - x_4 + 2x_5 &= 16 \\ 3x_1 + x_2 - 2x_3 + x_4 - 3x_5 &= -11 \\ x_1 + 3x_2 - 4x_3 + 3x_4 - x_5 &= -13 \\ 2x_1 - x_2 + 3x_3 - 2x_4 + 2x_5 &= 15 \\ 3x_1 + 4x_2 - 3x_3 + 5x_4 - x_5 &= -10 \end{aligned}$

6. $\begin{aligned} 2.1x_1 - 3.2x_2 + 6.4x_3 + 7x_4 - 3.2x_5 &= 54.3 \\ 4.1x_1 + 2.2x_2 - 3.1x_3 - 4.2x_4 + 3.3x_5 &= -20.81 \\ 3.4x_1 - 6.2x_2 + 4.7x_3 + 2.1x_4 - 5.3x_5 &= 24.7 \\ 4.1x_1 + 7.3x_2 + 5.2x_3 + 6.1x_4 - 8.2x_5 &= 29.25 \\ 2.8x_1 + 5.2x_2 + 3.1x_3 + 5.4x_4 + 3.8x_5 &= 43.72 \end{aligned}$

2.3 Systems of Linear Equations: Underdetermined and Overdetermined Systems

In this section, we continue our study of systems of linear equations. More specifically, we look at systems that have infinitely many solutions and those that have no solution. We also study systems of linear equations in which the number of variables is not equal to the number of equations in the system.

■ Solution(s) of Linear Equations

Our first example illustrates the situation in which a system of linear equations has infinitely many solutions.

EXAMPLE 1   A System of Equations with an Infinite Number of Solutions   Solve the system of linear equations given by

$$\begin{aligned} x + 2y - 3z &= -2 \\ 3x - y - 2z &= 1 \\ 2x + 3y - 5z &= -3 \end{aligned} \tag{9}$$

Solution   Using the Gauss–Jordan elimination method, we obtain the following sequence of equivalent augmented matrices:

$$\begin{bmatrix} \boxed{1} & 2 & -3 & -2 \\ 3 & -1 & -2 & 1 \\ 2 & 3 & -5 & -3 \end{bmatrix} \xrightarrow[R_3 - 2R_1]{R_2 - 3R_1} \begin{bmatrix} 1 & 2 & -3 & -2 \\ 0 & \boxed{-7} & 7 & 7 \\ 0 & -1 & 1 & 1 \end{bmatrix} \xrightarrow{-\frac{1}{7}R_2}$$

$$\begin{bmatrix} 1 & 2 & -3 & -2 \\ 0 & 1 & -1 & -1 \\ 0 & -1 & 1 & 1 \end{bmatrix} \xrightarrow[R_3 + R_2]{R_1 - 2R_2} \begin{bmatrix} 1 & 0 & -1 & 0 \\ 0 & 1 & -1 & -1 \\ 0 & 0 & 0 & 0 \end{bmatrix}$$

The last augmented matrix is in row-reduced form. Interpreting it as a system of linear equations gives

$$x - z = 0$$
$$y - z = -1$$

a system of two equations in the three variables x, y, and z.

Let's now single out one variable—say, z—and solve for x and y in terms of it. We obtain

$$x = z$$
$$y = z - 1$$

If we assign a particular value to z—say, $z = 0$—we obtain $x = 0$ and $y = -1$, giving the solution $(0, -1, 0)$ to System (9). By setting $z = 1$, we obtain the solution $(1, 0, 1)$. In general, if we set $z = t$, where t represents some real number (called a parameter), we obtain a solution given by $(t, t - 1, t)$. Since the parameter t may be any real number, we see that System (9) has infinitely many solutions. Geometrically, the solutions of System (9) lie on the straight line in three-dimensional space given by the intersection of the three planes determined by the three equations in the system.   ■

Note   In Example 1 we chose the parameter to be z because it is more convenient to solve for x and y (both the x- and y-columns are in unit form) in terms of z.   ■

The next example shows what happens in the elimination procedure when the system does not have a solution.

EXAMPLE 2   A System of Equations That Has No Solution   Solve the system of linear equations given by

$$x + y + z = 1$$
$$3x - y - z = 4 \tag{10}$$
$$x + 5y + 5z = -1$$

Solution   Using the Gauss–Jordan elimination method, we obtain the following sequence of equivalent augmented matrices:

$$\begin{bmatrix} \boxed{1} & 1 & 1 & | & 1 \\ 3 & -1 & -1 & | & 4 \\ 1 & 5 & 5 & | & -1 \end{bmatrix} \xrightarrow[R_3 - R_1]{R_2 - 3R_1} \begin{bmatrix} 1 & 1 & 1 & | & 1 \\ 0 & -4 & -4 & | & 1 \\ 0 & 4 & 4 & | & -2 \end{bmatrix}$$

$$\xrightarrow{R_3 + R_2} \begin{bmatrix} 1 & 1 & 1 & | & 1 \\ 0 & -4 & -4 & | & 1 \\ 0 & 0 & 0 & | & -1 \end{bmatrix}$$

Observe that row 3 in the last matrix reads $0x + 0y + 0z = -1$—that is, $0 = -1$! We therefore conclude that System (10) is inconsistent and has no solution. Geometrically, we have a situation in which two of the planes intersect in a straight line but the third plane is parallel to this line of intersection of the two planes and does not intersect it. Consequently, there is no point of intersection of the three planes.

Example 2 illustrates the following more general result of using the Gauss–Jordan elimination procedure.

Systems with No Solution
If there is a row in the augmented matrix containing all zeros to the left of the vertical line and a nonzero entry to the right of the line, then the system of equations has no solution.

It may have dawned on you that in all the previous examples we have dealt only with systems involving exactly the same number of linear equations as there are variables. However, systems in which the number of equations is different from the number of variables also occur in practice. Indeed, we will consider such systems in Examples 3 and 4.

The following theorem provides us with some preliminary information on a system of linear equations.

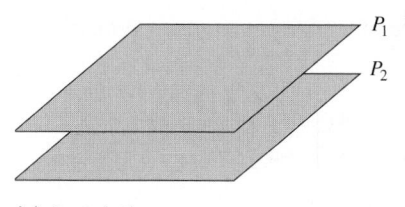

P_1
P_2

(a) No solution

THEOREM 1

a. If the number of equations is greater than or equal to the number of variables in a linear system, then one of the following is true:
  i. The system has no solution.
  ii. The system has exactly one solution.
  iii. The system has infinitely many solutions.
b. If there are fewer equations than variables in a linear system, then the system either has no solution or it has infinitely many solutions.

Note   Theorem 1 may be used to tell us, before we even begin to solve a problem, what the nature of the solution may be.   ■

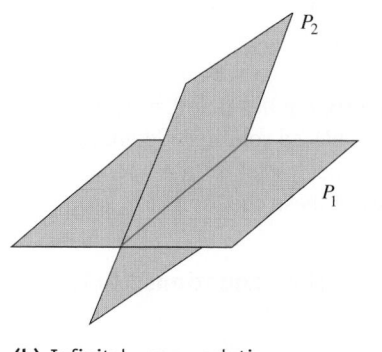

P_2
P_1

(b) Infinitely many solutions

Although we will not prove this theorem, you should recall that we have illustrated geometrically part (a) for the case in which there are exactly as many equations (three) as there are variables. To show the validity of part (b), let us once again consider the case in which a system has three variables. Now, if there is only one equation in the system, then it is clear that there are infinitely many solutions corresponding geometrically to all the points lying on the plane represented by the equation.

Next, if there are two equations in the system, then *only* the following possibilities exist:

1. The two planes are parallel and distinct.
2. The two planes intersect in a straight line.
3. The two planes are coincident (the two equations define the same plane) (Figure 6).

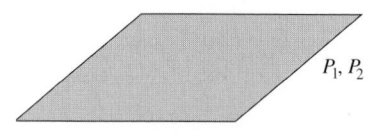

P_1, P_2

(c) Infinitely many solutions
FIGURE 6

EXPLORE & DISCUSS

Give a geometric interpretation of Theorem 1 for a linear system composed of equations involving two variables. Specifically, illustrate what can happen if there are three linear equations in the system (the case involving two linear equations has already been discussed in Section 2.1). What if there are four linear equations? What if there is only one linear equation in the system?

Thus, either there is no solution or there are infinitely many solutions corresponding to the points lying on a line of intersection of the two planes or on a single plane determined by the two equations. In the case where two planes intersect in a straight line, the solutions will involve one parameter, and in the case where the two planes are coincident, the solutions will involve two parameters.

EXAMPLE 3   A System with More Equations Than Variables   Solve the following system of linear equations:

$$\begin{aligned} x + 2y &= 4 \\ x - 2y &= 0 \\ 4x + 3y &= 12 \end{aligned}$$

Solution   We obtain the following sequence of equivalent augmented matrices:

$$\begin{bmatrix} ① & 2 & | & 4 \\ 1 & -2 & | & 0 \\ 4 & 3 & | & 12 \end{bmatrix} \xrightarrow[R_3 - 4R_1]{R_2 - R_1} \begin{bmatrix} 1 & 2 & | & 4 \\ 0 & ㊀4 & | & -4 \\ 0 & -5 & | & -4 \end{bmatrix} \xrightarrow{-\frac{1}{4}R_2}$$

$$\begin{bmatrix} 1 & 2 & | & 4 \\ 0 & 1 & | & 1 \\ 0 & -5 & | & -4 \end{bmatrix} \xrightarrow[R_3 + 5R_2]{R_1 - 2R_2} \begin{bmatrix} 1 & 0 & | & 2 \\ 0 & 1 & | & 1 \\ 0 & 0 & | & 1 \end{bmatrix}$$

The last row of the row-reduced augmented matrix implies that $0 = 1$, which is impossible, so we conclude that the given system has no solution. Geometrically, the three lines defined by the three equations in the system do not intersect at a point. (To see this for yourself, draw the graphs of these equations.)   ∎

EXAMPLE 4   A System with More Variables Than Equations   Solve the following system of linear equations:

$$\begin{aligned} x + 2y - 3z + w &= -2 \\ 3x - y - 2z - 4w &= 1 \\ 2x + 3y - 5z + w &= -3 \end{aligned}$$

Solution   First, observe that the given system consists of three equations in four variables and so, by Theorem 1b, either the system has no solution or it has infinitely many solutions. To solve it we use the Gauss–Jordan method and obtain the following sequence of equivalent augmented matrices:

$$\begin{bmatrix} \textcircled{1} & 2 & -3 & 1 & | & -2 \\ 3 & -1 & -2 & -4 & | & 1 \\ 2 & 3 & -5 & 1 & | & -3 \end{bmatrix} \xrightarrow[R_3 - 2R_1]{R_2 - 3R_1} \begin{bmatrix} 1 & 2 & -3 & 1 & | & -2 \\ 0 & \textcircled{-7} & 7 & -7 & | & 7 \\ 0 & -1 & 1 & -1 & | & 1 \end{bmatrix} \xrightarrow{-\frac{1}{7}R_2}$$

$$\begin{bmatrix} 1 & 2 & -3 & 1 & | & -2 \\ 0 & 1 & -1 & 1 & | & -1 \\ 0 & -1 & 1 & -1 & | & 1 \end{bmatrix} \xrightarrow[R_3 + R_2]{R_1 - 2R_2} \begin{bmatrix} 1 & 0 & -1 & -1 & | & 0 \\ 0 & 1 & -1 & 1 & | & -1 \\ 0 & 0 & 0 & 0 & | & 0 \end{bmatrix}$$

The last augmented matrix is in row-reduced form. Observe that the given system is equivalent to the system

$$x - z - w = 0$$
$$y - z + w = -1$$

of two equations in four variables. Thus, we may solve for two of the variables in terms of the other two. Letting $z = s$ and $w = t$ (s, t, parameters), we find that

$$x = s + t$$
$$y = s - t - 1$$
$$z = s$$
$$w = t$$

The solutions may be written in the form $(s + t, s - t - 1, s, t)$, where s and t are any real numbers. Geometrically, the three equations in the system represent three hyperplanes in four-dimensional space (since there are four variables) and their "points" of intersection lie in a two-dimensional subspace of four-space (since there are two parameters).   ■

Note   In Example 4, we assigned parameters to z and w rather than to x and y because x and y are readily solved in terms of z and w.   ■

The following example illustrates a situation in which a system of linear equations has infinitely many solutions.

APPLIED EXAMPLE 5   Traffic Control   Figure 7 shows the flow of downtown traffic in a certain city during the rush hours on a typical weekday. The arrows indicate the direction of traffic flow on each one-way road, and the average number of vehicles per hour entering and leaving each intersection appears beside each road. 5th Avenue and 6th Avenue can each handle up to 2000 vehicles per hour without causing congestion, whereas the maximum capacity of both 4th Street and 5th Street is 1000 vehicles per hour. The flow of traffic is controlled by traffic lights installed at each of the four intersections.

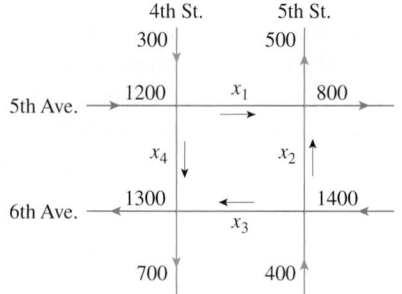

FIGURE 7

a. Write a general expression involving the rates of flow—x_1, x_2, x_3, x_4—and suggest two possible flow patterns that will ensure no traffic congestion.

b. Suppose the part of 4th Street between 5th Avenue and 6th Avenue is to be resurfaced and that traffic flow between the two junctions must therefore be reduced to at most 300 vehicles per hour. Find two possible flow patterns that will result in a smooth flow of traffic.

Solution

a. To avoid congestion, all traffic entering an intersection must also leave that intersection. Applying this condition to each of the four intersections in a clockwise direction beginning with the 5th Avenue and 4th Street intersection, we obtain the following equations:

$$1500 = x_1 + x_4$$
$$1300 = x_1 + x_2$$
$$1800 = x_2 + x_3$$
$$2000 = x_3 + x_4$$

This system of four linear equations in the four variables x_1, x_2, x_3, x_4 may be rewritten in the more standard form

$$
\begin{aligned}
x_1 \qquad\qquad + x_4 &= 1500 \\
x_1 + x_2 \qquad\qquad &= 1300 \\
x_2 + x_3 \qquad &= 1800 \\
x_3 + x_4 &= 2000
\end{aligned}
$$

Using the Gauss–Jordan elimination method to solve the system, we obtain

$$
\left[\begin{array}{cccc|c}
1 & 0 & 0 & 1 & 1500 \\
1 & 1 & 0 & 0 & 1300 \\
0 & 1 & 1 & 0 & 1800 \\
0 & 0 & 1 & 1 & 2000
\end{array}\right]
\xrightarrow{R_2 - R_1}
\left[\begin{array}{cccc|c}
1 & 0 & 0 & 1 & 1500 \\
0 & 1 & 0 & -1 & -200 \\
0 & 1 & 1 & 0 & 1800 \\
0 & 0 & 1 & 1 & 2000
\end{array}\right]
$$

$$
\xrightarrow{R_3 - R_2}
\left[\begin{array}{cccc|c}
1 & 0 & 0 & 1 & 1500 \\
0 & 1 & 0 & -1 & -200 \\
0 & 0 & 1 & 1 & 2000 \\
0 & 0 & 1 & 1 & 2000
\end{array}\right]
$$

$$
\xrightarrow{R_4 - R_3}
\left[\begin{array}{cccc|c}
1 & 0 & 0 & 1 & 1500 \\
0 & 1 & 0 & -1 & -200 \\
0 & 0 & 1 & 1 & 2000 \\
0 & 0 & 0 & 0 & 0
\end{array}\right]
$$

The last augmented matrix is in row-reduced form and is equivalent to a system of three linear equations in the four variables x_1, x_2, x_3, x_4. Thus, we may express three of the variables—say, x_1, x_2, x_3—in terms of x_4. Setting $x_4 = t$ (t a parameter), we may write the infinitely many solutions of the system as

$$
\begin{aligned}
x_1 &= 1500 - t \\
x_2 &= -200 + t \\
x_3 &= 2000 - t \\
x_4 &= t
\end{aligned}
$$

Observe that for a meaningful solution we must have $200 \le t \le 1000$, since x_1, x_2, x_3, and x_4 must all be nonnegative and the maximum capacity of a street is 1000. For example, picking $t = 300$ gives the flow pattern

$$x_1 = 1200 \qquad x_2 = 100 \qquad x_3 = 1700 \qquad x_4 = 300$$

Selecting $t = 500$ gives the flow pattern

$$x_1 = 1000 \qquad x_2 = 300 \qquad x_3 = 1500 \qquad x_4 = 500$$

b. In this case, x_4 must not exceed 300. Again, using the results of part (a), we find, upon setting $x_4 = t = 300$, the flow pattern

$$x_1 = 1200 \qquad x_2 = 100 \qquad x_3 = 1700 \qquad x_4 = 300$$

obtained earlier. Picking $t = 250$ gives the flow pattern

$$x_1 = 1250 \qquad x_2 = 50 \qquad x_3 = 1750 \qquad x_4 = 250$$

2.3 Self-Check Exercises

1. The following augmented matrix in row-reduced form is equivalent to the augmented matrix of a certain system of linear equations. Use this result to solve the system of equations.

$$\begin{bmatrix} 1 & 0 & -1 & \bigm| & 3 \\ 0 & 1 & 5 & \bigm| & -2 \\ 0 & 0 & 0 & \bigm| & 0 \end{bmatrix}$$

2. Solve the system of linear equations

$$\begin{aligned} 2x - 3y + z &= 6 \\ x + 2y + 4z &= -4 \\ x - 5y - 3z &= 10 \end{aligned}$$

using the Gauss–Jordan elimination method.

3. Solve the system of linear equations

$$\begin{aligned} x - 2y + 3z &= 9 \\ 2x + 3y - z &= 4 \\ x + 5y - 4z &= 2 \end{aligned}$$

using the Gauss–Jordan elimination method.

Solutions to Self-Check Exercises 2.3 can be found on page 106.

2.3 Concept Questions

1. **a.** If a system of linear equations has the same number of equations or more equations than variables, what can you say about the nature of its solution(s)?
   b. If a system of linear equations has fewer equations than variables, what can you say about the nature of its solution(s)?

2. A system consists of three linear equations in four variables. Can the system have a unique solution?

2.3   Exercises

In Exercises 1–12, given that the augmented matrix in row-reduced form is equivalent to the augmented matrix of a system of linear equations, (a) determine whether the system has a solution and (b) find the solution or solutions to the system, if they exist.

1. $\begin{bmatrix} 1 & 0 & 0 & | & 3 \\ 0 & 1 & 0 & | & -1 \\ 0 & 0 & 1 & | & 2 \end{bmatrix}$ 2. $\begin{bmatrix} 1 & 0 & 0 & | & 3 \\ 0 & 1 & 0 & | & -2 \\ 0 & 0 & 1 & | & 1 \end{bmatrix}$

3. $\begin{bmatrix} 1 & 0 & | & 2 \\ 0 & 1 & | & 4 \\ 0 & 0 & | & 0 \end{bmatrix}$ 4. $\begin{bmatrix} 1 & 0 & 0 & | & 3 \\ 0 & 1 & 0 & | & 1 \\ 0 & 0 & 0 & | & 0 \end{bmatrix}$

5. $\begin{bmatrix} 1 & 0 & 1 & | & 4 \\ 0 & 1 & 0 & | & -2 \end{bmatrix}$ 6. $\begin{bmatrix} 1 & 0 & 0 & 0 & | & 3 \\ 0 & 1 & 1 & 0 & | & -1 \\ 0 & 0 & 0 & 1 & | & 2 \end{bmatrix}$

7. $\begin{bmatrix} 1 & 0 & 0 & 0 & | & 2 \\ 0 & 1 & 0 & 0 & | & 1 \\ 0 & 0 & 1 & 0 & | & 3 \\ 0 & 0 & 0 & 0 & | & 1 \end{bmatrix}$ 8. $\begin{bmatrix} 1 & 0 & 0 & | & 4 \\ 0 & 1 & 0 & | & -1 \\ 0 & 0 & 1 & | & 3 \\ 0 & 0 & 0 & | & 1 \end{bmatrix}$

9. $\begin{bmatrix} 1 & 0 & 0 & 0 & | & 2 \\ 0 & 1 & 0 & 0 & | & -1 \\ 0 & 0 & 1 & 1 & | & 2 \\ 0 & 0 & 0 & 0 & | & 0 \end{bmatrix}$ 10. $\begin{bmatrix} 0 & 1 & 0 & 1 & | & 3 \\ 0 & 0 & 1 & -2 & | & 4 \\ 0 & 0 & 0 & 0 & | & 0 \\ 0 & 0 & 0 & 0 & | & 0 \end{bmatrix}$

11. $\begin{bmatrix} 1 & 0 & 3 & 0 & | & 2 \\ 0 & 1 & -1 & 0 & | & 1 \\ 0 & 0 & 0 & 0 & | & 0 \\ 0 & 0 & 0 & 0 & | & 0 \end{bmatrix}$ 12. $\begin{bmatrix} 1 & 0 & 3 & -1 & | & 4 \\ 0 & 1 & -2 & 3 & | & 2 \\ 0 & 0 & 0 & 0 & | & 0 \\ 0 & 0 & 0 & 0 & | & 0 \end{bmatrix}$

In Exercises 13–32, solve the system of linear equations using the Gauss–Jordan elimination method.

13. $\begin{aligned} 2x - y &= 3 \\ x + 2y &= 4 \\ 2x + 3y &= 7 \end{aligned}$ 14. $\begin{aligned} x + 2y &= 3 \\ 2x - 3y &= -8 \\ x - 4y &= -9 \end{aligned}$

15. $\begin{aligned} 3x - 2y &= -3 \\ 2x + y &= 3 \\ x - 2y &= -5 \end{aligned}$ 16. $\begin{aligned} 2x + 3y &= 2 \\ x + 3y &= -2 \\ x - y &= 3 \end{aligned}$

17. $\begin{aligned} 3x - 2y &= 5 \\ -x + 3y &= -4 \\ 2x - 4y &= 6 \end{aligned}$ 18. $\begin{aligned} 4x + 6y &= 8 \\ 3x - 2y &= -7 \\ x + 3y &= 5 \end{aligned}$

19. $\begin{aligned} x - 2y &= 2 \\ 7x - 14y &= 14 \\ 3x - 6y &= 6 \end{aligned}$ 20. $\begin{aligned} x + 2y + z &= -2 \\ -2x - 3y - z &= 1 \\ 2x + 4y + 2z &= -4 \end{aligned}$

21. $\begin{aligned} 3x + 2y &= 4 \\ -\tfrac{3}{2}x - y &= -2 \\ 6x + 4y &= 8 \end{aligned}$ 22. $\begin{aligned} 3y + 2z &= 4 \\ 2x - y - 3z &= 3 \\ 2x + 2y - z &= 7 \end{aligned}$

23. $\begin{aligned} 2x_1 - x_2 + x_3 &= -4 \\ 3x_1 - \tfrac{3}{2}x_2 + \tfrac{3}{2}x_3 &= -6 \\ -6x_1 + 3x_2 - 3x_3 &= 12 \end{aligned}$

24. $\begin{aligned} x + y - 2z &= -3 \\ 2x - y + 3z &= 7 \\ x - 2y + 5z &= 0 \end{aligned}$ 25. $\begin{aligned} x - 2y + 3z &= 4 \\ 2x + 3y - z &= 2 \\ x + 2y - 3z &= -6 \end{aligned}$

26. $\begin{aligned} x_1 - 2x_2 + x_3 &= -3 \\ 2x_1 + x_2 - 2x_3 &= 2 \\ x_1 + 3x_2 - 3x_3 &= 5 \end{aligned}$

27. $\begin{aligned} 4x + y - z &= 4 \\ 8x + 2y - 2z &= 8 \end{aligned}$ 28. $\begin{aligned} x_1 + 2x_2 + 4x_3 &= 2 \\ x_1 + x_2 + 2x_3 &= 1 \end{aligned}$

29. $\begin{aligned} 2x + y - 3z &= 1 \\ x - y + 2z &= 1 \\ 5x - 2y + 3z &= 6 \end{aligned}$ 30. $\begin{aligned} 3x - 9y + 6z &= -12 \\ x - 3y + 2z &= -4 \\ 2x - 6y + 4z &= 8 \end{aligned}$

31. $\begin{aligned} x + 2y - z &= -4 \\ 2x + y + z &= 7 \\ x + 3y + 2z &= 7 \\ x - 3y + z &= 9 \end{aligned}$ 32. $\begin{aligned} 3x - 2y + z &= 4 \\ x + 3y - 4z &= -3 \\ 2x - 3y + 5z &= 7 \\ x - 8y + 9z &= 10 \end{aligned}$

33. **MANAGEMENT DECISIONS** The management of Hartman Rent-A-Car has allocated $1,008,000 to purchase 60 new automobiles to add to their existing fleet of rental cars. The company will choose from compact, mid-sized, and full-sized cars costing $12,000, $19,200, and $26,400 each, respectively. Find formulas giving the options available to the company. Give two specific options. (*Note*: Your answers will *not* be unique.)

34. **NUTRITION** A dietitian wishes to plan a meal around three foods. The meal is to include 8800 units of vitamin A, 3380 units of vitamin C, and 1020 units of calcium. The number of units of the vitamins and calcium in each ounce of the foods is summarized in the accompanying table:

	Food I	Food II	Food III
Vitamin A	400	1200	800
Vitamin C	110	570	340
Calcium	90	30	60

Determine the amount of each food the dietitian should include in the meal in order to meet the vitamin and calcium requirements.

35. **NUTRITION** Refer to Exercise 34. In planning for another meal, the dietitian changes the requirement of vitamin C from 3380 units to 2160 units. All other requirements remain the same. Show that such a meal cannot be planned around the same foods.

36. INVESTMENTS   Mr. and Mrs. Garcia have a total of $100,000 to be invested in stocks, bonds, and a money market account. The stocks have a rate of return of 12%/year, while the bonds and the money market account pay 8%/year and 4%/year, respectively. The Garcias have stipulated that the amount invested in stocks should be equal to the sum of the amount invested in bonds and 3 times the amount invested in the money market account. How should the Garcias allocate their resources if they require an annual income of $10,000 from their investments?

37. TRAFFIC CONTROL   The accompanying figure shows the flow of traffic near a city's Civic Center during the rush hours on a typical weekday. Each road can handle a maximum of 1000 cars/hour without causing congestion. The flow of traffic is controlled by traffic lights at each of the five intersections.

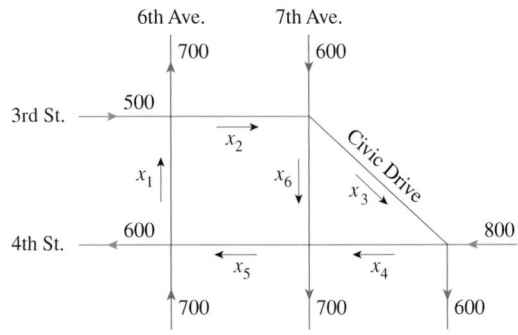

a. Set up a system of linear equations describing the traffic flow.
b. Solve the system devised in part (a) and suggest two possible traffic-flow patterns that will ensure no traffic congestion.
c. Suppose 7th Avenue between 3rd and 4th Streets is soon to be closed for road repairs. Find one possible flow pattern that will result in a smooth flow of traffic.

38. TRAFFIC CONTROL   The accompanying figure shows the flow of downtown traffic during the rush hours on a typical weekday. Each avenue can handle up to 1500 vehicles/hour without causing congestion, whereas the maximum capacity of each street is 1000 vehicles/hour. The flow of traffic is controlled by traffic lights at each of the six intersections.

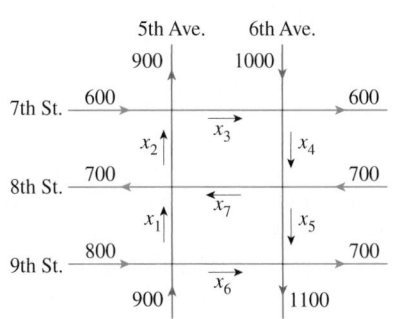

a. Set up a system of linear equations describing the traffic flow.
b. Solve the system devised in part (a) and suggest two possible traffic-flow patterns that will ensure no traffic congestion.
c. Suppose the traffic flow along 9th Street between 5th and 6th Avenues, x_6, is restricted because of sewer construction. What is the minimum permissible traffic flow along this road that will not result in traffic congestion?

39. Determine the value of k such that the following system of linear equations has a solution, and then find the solution:

$$2x + 3y = 2$$
$$x + 4y = 6$$
$$5x + ky = 2$$

40. Determine the value of k such that the following system of linear equations has infinitely many solutions, and then find the solutions:

$$3x - 2y + 4z = 12$$
$$-9x + 6y - 12z = k$$

In Exercises 41 and 42, determine whether the statement is true or false. If it is true, explain why it is true. If it is false, give an example to show why it is false.

41. A system of linear equations having fewer equations than variables has no solution, a unique solution, or infinitely many solutions.

42. A system of linear equations having more equations than variables has no solution, a unique solution, or infinitely many solutions.

2.3   Solutions to Self-Check Exercises

1. Let x, y, and z denote the variables. Then, the given row-reduced augmented matrix tells us that the system of linear equations is equivalent to the two equations

$$\begin{aligned} x \quad - z &= 3 \\ y + 5z &= -2 \end{aligned}$$

Letting $z = t$, where t is a parameter, we find the infinitely many solutions given by

$$\begin{aligned} x &= t + 3 \\ y &= -5t - 2 \\ z &= t \end{aligned}$$

The last augmented matrix, which is in row-reduced form, tells us that the given system of linear equations is equivalent to the following system of two equations:

$$\begin{aligned} x \quad + 2z &= 0 \\ y + \quad z &= -2 \end{aligned}$$

Letting $z = t$, where t is a parameter, we see that the infinitely many solutions are given by

$$\begin{aligned} x &= -2t \\ y &= -t - 2 \\ z &= t \end{aligned}$$

2. We obtain the following sequence of equivalent augmented matrices:

$$\begin{bmatrix} 2 & -3 & 1 & 6 \\ 1 & 2 & 4 & -4 \\ 1 & -5 & -3 & 10 \end{bmatrix} \xrightarrow{R_1 \leftrightarrow R_2}$$

$$\begin{bmatrix} 1 & 2 & 4 & -4 \\ 2 & -3 & 1 & 6 \\ 1 & -5 & -3 & 10 \end{bmatrix} \xrightarrow[R_3 - R_1]{R_2 - 2R_1}$$

$$\begin{bmatrix} 1 & 2 & 4 & -4 \\ 0 & -7 & -7 & 14 \\ 0 & -7 & -7 & 14 \end{bmatrix} \xrightarrow{-\frac{1}{7}R_2}$$

$$\begin{bmatrix} 1 & 2 & 4 & -4 \\ 0 & 1 & 1 & -2 \\ 0 & -7 & -7 & 14 \end{bmatrix} \xrightarrow[R_3 + 7R_2]{R_1 - 2R_2} \begin{bmatrix} 1 & 0 & 2 & 0 \\ 0 & 1 & 1 & -2 \\ 0 & 0 & 0 & 0 \end{bmatrix}$$

3. We obtain the following sequence of equivalent augmented matrices:

$$\begin{bmatrix} \boxed{1} & -2 & 3 & 9 \\ 2 & 3 & -1 & 4 \\ 1 & 5 & -4 & 2 \end{bmatrix} \xrightarrow[R_3 - R_1]{R_2 - 2R_1}$$

$$\begin{bmatrix} 1 & -2 & 3 & 9 \\ 0 & 7 & -7 & -14 \\ 0 & 7 & -7 & -7 \end{bmatrix} \xrightarrow{R_3 - R_2} \begin{bmatrix} 1 & -2 & 3 & 9 \\ 0 & 7 & -7 & -14 \\ 0 & 0 & 0 & 7 \end{bmatrix}$$

Since the last row of the final augmented matrix is equivalent to the equation $0 = 7$, a contradiction, we conclude that the given system has no solution.

USING TECHNOLOGY

■ Systems of Linear Equations: Underdetermined and Overdetermined Systems

We can use the row operations of a graphing utility to solve a system of m linear equations in n unknowns by the Gauss–Jordan method, as we did in the previous technology section. We can also use the **rref** or equivalent operation to obtain the row-reduced form without going through all the steps of the Gauss–Jordan method. The **SIMULT** function, however, cannot be used to solve a system where the number of equations and the number of variables are not the same.

EXAMPLE 1   Solve the system

$$x_1 - 2x_2 + 4x_3 = 2$$
$$2x_1 + x_2 - 2x_3 = -1$$
$$3x_1 - x_2 + 2x_3 = 1$$
$$2x_1 + 6x_2 - 12x_3 = -6$$

Solution   First, we enter the augmented matrix A into the calculator as

$$A = \begin{bmatrix} 1 & -2 & 4 & 2 \\ 2 & 1 & -2 & -1 \\ 3 & -1 & 2 & 1 \\ 2 & 6 & -12 & -6 \end{bmatrix}$$

Then using the **rref** or equivalent operation, we obtain the equivalent matrix

$$\begin{bmatrix} 1 & 0 & 0 & 0 \\ 0 & 1 & -2 & -1 \\ 0 & 0 & 0 & 0 \\ 0 & 0 & 0 & 0 \end{bmatrix}$$

in reduced form. Thus, the given system is equivalent to

$$x_1 = 0$$
$$x_2 - 2x_3 = -1$$

If we let $x_3 = t$, where t is a parameter, then we find that the solutions are $(0, 2t - 1, t)$.   ■

TECHNOLOGY EXERCISES

Use a graphing utility to solve the system of equations using the rref or equivalent operation.

1. $2x_1 - x_2 - x_3 = 0$
$3x_1 - 2x_2 - x_3 = -1$
$-x_1 + 2x_2 - x_3 = 3$
$2x_2 - 2x_3 = 4$

2. $3x_1 + x_2 - 4x_3 = 5$
$2x_1 - 3x_2 + 2x_3 = -4$
$-x_1 - 2x_2 + 4x_3 = 6$
$4x_1 + 3x_2 - 5x_3 = 9$

3. $2x_1 + 3x_2 + 2x_3 + x_4 = -1$
$x_1 - x_2 + x_3 - 2x_4 = -8$
$5x_1 + 6x_2 - 2x_3 + 2x_4 = 11$
$x_1 + 3x_2 + 8x_3 + x_4 = -14$

4. $x_1 - x_2 + 3x_3 - 6x_4 = 2$
$x_1 + x_2 + x_3 - 2x_4 = 2$
$-2x_1 - x_2 + x_3 + 2x_4 = 0$

5. $x_1 + x_2 - x_3 - x_4 = -1$
$x_1 - x_2 + x_3 + 4x_4 = -6$
$3x_1 + x_2 - x_3 + 2x_4 = -4$
$5x_1 + x_2 - 3x_3 + x_4 = -9$

6. $1.2x_1 - 2.3x_2 + 4.2x_3 + 5.4x_4 - 1.6x_5 = 4.2$
$2.3x_1 + 1.4x_2 - 3.1x_3 + 3.3x_4 - 2.4x_5 = 6.3$
$1.7x_1 + 2.6x_2 - 4.3x_3 + 7.2x_4 - 1.8x_5 = 7.8$
$2.6x_1 - 4.2x_2 + 8.3x_3 - 1.6x_4 + 2.5x_5 = 6.4$

2.4   Matrices

▬ Using Matrices to Represent Data

Many practical problems are solved by using arithmetic operations on the data associated with the problems. By properly organizing the data into *blocks* of numbers, we can then carry out these arithmetic operations in an orderly and efficient manner. In particular, this systematic approach enables us to use the computer to full advantage.

Let's begin by considering how the monthly output data of a manufacturer may be organized. The Acrosonic Company manufactures four different loudspeaker systems at three separate locations. The company's May output is described in Table 1.

TABLE 1				
	Model A	**Model B**	**Model C**	**Model D**
Location I	320	280	460	280
Location II	480	360	580	0
Location III	540	420	200	880

Now, if we agree to preserve the relative location of each entry in Table 1, we can summarize the set of data as follows:

$$\begin{bmatrix} 320 & 280 & 460 & 280 \\ 480 & 360 & 580 & 0 \\ 540 & 420 & 200 & 880 \end{bmatrix}$$

A matrix summarizing the data in Table 1

The array of numbers displayed here is an example of a matrix. Observe that the numbers in row 1 give the output of models A, B, C, and D of Acrosonic loudspeaker systems manufactured at location I; similarly, the numbers in rows 2 and 3 give the respective outputs of these loudspeaker systems at locations II and III. The numbers in each column of the matrix give the outputs of a particular model of loudspeaker system manufactured at each of the company's three manufacturing locations.

More generally, a matrix is an ordered rectangular array of real numbers. For example, each of the following arrays is a matrix:

$$A = \begin{bmatrix} 3 & 0 & -1 \\ 2 & 1 & 4 \end{bmatrix} \quad B = \begin{bmatrix} 3 & 2 \\ 0 & 1 \\ -1 & 4 \end{bmatrix} \quad C = \begin{bmatrix} 1 \\ 2 \\ 4 \\ 0 \end{bmatrix} \quad D = \begin{bmatrix} 1 & 3 & 0 & 1 \end{bmatrix}$$

The real numbers that make up the array are called the **entries**, or *elements*, of the matrix. The entries in a row in the array are referred to as a **row** of the matrix, whereas the entries in a column in the array are referred to as a **column** of the matrix. Matrix A, for example, has two rows and three columns, which may be identified as follows:

$$\begin{array}{ccc} \text{Column 1} & \text{Column 2} & \text{Column 3} \end{array}$$

$$\begin{array}{c} \text{Row 1} \\ \text{Row 2} \end{array} \begin{bmatrix} 3 & 0 & -1 \\ 2 & 1 & 4 \end{bmatrix}$$

A 2×3 matrix

The **size**, or *dimension*, **of a matrix** is described in terms of the number of rows and columns of the matrix. For example, matrix A has two rows and three columns and is said to have size 2 by 3, denoted 2×3. In general, a matrix having m rows and n columns is said to have size $m \times n$.

Matrix

A **matrix** is an ordered rectangular array of numbers. A matrix with m rows and n columns has size $m \times n$. The entry in the ith row and jth column is denoted by a_{ij}.

A matrix of size $1 \times n$—a matrix having one row and n columns—is referred to as a **row matrix**, or *row vector*, of dimension n. For example, the matrix D is a row vector of dimension 4. Similarly, a matrix having m rows and one column is referred to as a **column matrix**, or *column vector*, of dimension m. The matrix C is a column vector of dimension 4. Finally, an $n \times n$ matrix—that is, a matrix having the same number of rows as columns—is called a **square matrix**. For example, the matrix

$$\begin{bmatrix} -3 & 8 & 6 \\ 2 & \frac{1}{4} & 4 \\ 1 & 3 & 2 \end{bmatrix}$$

A 3×3 square matrix

is a square matrix of size 3×3, or simply of size 3.

APPLIED EXAMPLE 1 Organizing Production Data Consider the matrix

$$P = \begin{bmatrix} 320 & 280 & 460 & 280 \\ 480 & 360 & 580 & 0 \\ 540 & 420 & 200 & 880 \end{bmatrix}$$

representing the output of loudspeaker systems of the Acrosonic Company discussed earlier (see Table 1).

a. What is the size of the matrix P?
b. Find a_{24} (the entry in row 2 and column 4 of the matrix P) and give an interpretation of this number.
c. Find the sum of the entries that make up row 1 of P and interpret the result.
d. Find the sum of the entries that make up column 4 of P and interpret the result.

Solution

a. The matrix P has three rows and four columns and hence has size 3×4.
b. The required entry lies in row 2 and column 4 and is the number 0. This means that no model D loudspeaker system was manufactured at location II in May.

c. The required sum is given by

$$320 + 280 + 460 + 280 = 1340$$

which gives the total number of loudspeaker systems manufactured at location I in May as 1340 units.

d. The required sum is given by

$$280 + 0 + 880 = 1160$$

giving the output of model D loudspeaker systems at all locations of the company in May as 1160 units.   ∎

▬ Equality of Matrices

Two matrices are said to be *equal* if they have the same size and their corresponding entries are equal. For example,

$$\begin{bmatrix} 2 & 3 & 1 \\ 4 & 6 & 2 \end{bmatrix} = \begin{bmatrix} (3-1) & 3 & 1 \\ 4 & (4+2) & 2 \end{bmatrix}$$

Also,

$$\begin{bmatrix} 1 & 3 & 5 \\ 2 & 4 & 3 \end{bmatrix} \neq \begin{bmatrix} 1 & 2 \\ 3 & 4 \\ 5 & 3 \end{bmatrix}$$

since the matrix on the left has size 2×3 whereas the matrix on the right has size 3×2, and

$$\begin{bmatrix} 2 & 3 \\ 4 & 6 \end{bmatrix} \neq \begin{bmatrix} 2 & 3 \\ 4 & 7 \end{bmatrix}$$

since the corresponding elements in row 2 and column 2 of the two matrices are not equal.

Equality of Matrices
Two matrices are equal if they have the same size and their corresponding entries are equal.

EXAMPLE 2   Solve the following matrix equation for x, y, and z:

$$\begin{bmatrix} 1 & x & 3 \\ 2 & y-1 & 2 \end{bmatrix} = \begin{bmatrix} 1 & 4 & z \\ 2 & 1 & 2 \end{bmatrix}$$

Solution   Since the corresponding elements of the two matrices must be equal, we find that $x = 4$, $z = 3$, and $y - 1 = 1$, or $y = 2$.   ∎

Addition and Subtraction

Two matrices A and B of the *same size* can be added or subtracted to produce a matrix of the same size. This is done by adding or subtracting the corresponding entries in the two matrices. For example,

$$\begin{bmatrix} 1 & 3 & 4 \\ -1 & 2 & 0 \end{bmatrix} + \begin{bmatrix} 1 & 4 & 3 \\ 6 & 1 & -2 \end{bmatrix} = \begin{bmatrix} 1+1 & 3+4 & 4+3 \\ -1+6 & 2+1 & 0+(-2) \end{bmatrix} = \begin{bmatrix} 2 & 7 & 7 \\ 5 & 3 & -2 \end{bmatrix}$$

Adding two matrices of the same size

and
$$\begin{bmatrix} 1 & 2 \\ -1 & 3 \\ 4 & 0 \end{bmatrix} - \begin{bmatrix} 2 & -1 \\ 3 & 2 \\ -1 & 0 \end{bmatrix} = \begin{bmatrix} 1-2 & 2-(-1) \\ -1-3 & 3-2 \\ 4-(-1) & 0-0 \end{bmatrix} = \begin{bmatrix} -1 & 3 \\ -4 & 1 \\ 5 & 0 \end{bmatrix}$$

Subtracting two matrices of the same size

Addition and Subtraction of Matrices

If A and B are two matrices of the same size, then:

1. The *sum $A + B$* is the matrix obtained by adding the corresponding entries in the two matrices.
2. The *difference $A - B$* is the matrix obtained by subtracting the corresponding entries in B from A.

APPLIED EXAMPLE 3   Organizing Production Data   The total output of Acrosonic for June is shown in Table 2.

TABLE 2

	Model A	Model B	Model C	Model D
Location I	210	180	330	180
Location II	400	300	450	40
Location III	420	280	180	740

The output for May was given earlier in Table 1. Find the total output of the company for May and June.

Solution   As we saw earlier, the production matrix for Acrosonic in May is given by

$$A = \begin{bmatrix} 320 & 280 & 460 & 280 \\ 480 & 360 & 580 & 0 \\ 540 & 420 & 200 & 880 \end{bmatrix}$$

Next, from Table 2, we see that the production matrix for June is given by

$$B = \begin{bmatrix} 210 & 180 & 330 & 180 \\ 400 & 300 & 450 & 40 \\ 420 & 280 & 180 & 740 \end{bmatrix}$$

Finally, the total output of Acrosonic for May and June is given by the matrix

$$A + B = \begin{bmatrix} 320 & 280 & 460 & 280 \\ 480 & 360 & 580 & 0 \\ 540 & 420 & 200 & 880 \end{bmatrix} + \begin{bmatrix} 210 & 180 & 330 & 180 \\ 400 & 300 & 450 & 40 \\ 420 & 280 & 180 & 740 \end{bmatrix}$$

$$= \begin{bmatrix} 530 & 460 & 790 & 460 \\ 880 & 660 & 1030 & 40 \\ 960 & 700 & 380 & 1620 \end{bmatrix}$$

The following laws hold for matrix addition.

Laws for Matrix Addition

If A, B, and C are matrices of the same size, then:

1. $A + B = B + A$ Commutative law
2. $(A + B) + C = A + (B + C)$ Associative law

The *commutative law* for matrix addition states that the order in which matrix addition is performed is immaterial. The *associative law* states that, when adding three matrices together, we may first add A and B and then add the resulting sum to C. Equivalently, we can add A to the sum of B and C.

A *zero matrix* is one in which all entries are zero. The zero matrix O has the property that

$$A + O = O + A = A$$

for any matrix A having the same size as that of O. For example, the zero matrix of size 3×2 is

$$O = \begin{bmatrix} 0 & 0 \\ 0 & 0 \\ 0 & 0 \end{bmatrix}$$

If A is any 3×2 matrix, then

$$A + O = \begin{bmatrix} a_{11} & a_{12} \\ a_{21} & a_{22} \\ a_{31} & a_{32} \end{bmatrix} + \begin{bmatrix} 0 & 0 \\ 0 & 0 \\ 0 & 0 \end{bmatrix} = \begin{bmatrix} a_{11} & a_{12} \\ a_{21} & a_{22} \\ a_{31} & a_{32} \end{bmatrix} = A$$

where a_{ij} denotes the entry in the ith row and jth column of the matrix A.

The matrix obtained by interchanging the rows and columns of a given matrix A is called the *transpose* of A and is denoted A^T. For example, if

$$A = \begin{bmatrix} 1 & 2 & 3 \\ 4 & 5 & 6 \\ 7 & 8 & 9 \end{bmatrix}$$

then

$$A^T = \begin{bmatrix} 1 & 4 & 7 \\ 2 & 5 & 8 \\ 3 & 6 & 9 \end{bmatrix}$$

Transpose of a Matrix

If A is an $m \times n$ matrix with elements a_{ij}, then the **transpose** of A is the $n \times m$ matrix A^T with elements a_{ji}.

Scalar Multiplication

A matrix A may be multiplied by a real number, called a **scalar** in the context of matrix algebra. The scalar product, denoted by cA, is a matrix obtained by multiplying each entry of A by c. For example, the scalar product of the matrix

$$A = \begin{bmatrix} 3 & -1 & 2 \\ 0 & 1 & 4 \end{bmatrix}$$

and the scalar 3 is the matrix

$$3A = 3\begin{bmatrix} 3 & -1 & 2 \\ 0 & 1 & 4 \end{bmatrix} = \begin{bmatrix} 9 & -3 & 6 \\ 0 & 3 & 12 \end{bmatrix}$$

Scalar Product

If A is a matrix and c is a real number, then the **scalar product** cA is the matrix obtained by multiplying each entry of A by c.

EXAMPLE 4 Given

$$A = \begin{bmatrix} 3 & 4 \\ -1 & 2 \end{bmatrix} \quad \text{and} \quad B = \begin{bmatrix} 3 & 2 \\ -1 & 2 \end{bmatrix}$$

find the matrix X satisfying the *matrix equation* $2X + B = 3A$.

Solution From the given equation $2X + B = 3A$, we find that

$$2X = 3A - B$$
$$= 3\begin{bmatrix} 3 & 4 \\ -1 & 2 \end{bmatrix} - \begin{bmatrix} 3 & 2 \\ -1 & 2 \end{bmatrix}$$
$$= \begin{bmatrix} 9 & 12 \\ -3 & 6 \end{bmatrix} - \begin{bmatrix} 3 & 2 \\ -1 & 2 \end{bmatrix} = \begin{bmatrix} 6 & 10 \\ -2 & 4 \end{bmatrix}$$
$$X = \frac{1}{2}\begin{bmatrix} 6 & 10 \\ -2 & 4 \end{bmatrix} = \begin{bmatrix} 3 & 5 \\ -1 & 2 \end{bmatrix}$$

APPLIED EXAMPLE 5 Production Planning The management of Acrosonic has decided to increase its July production of loudspeaker systems by 10% (over its June output). Find a matrix giving the targeted production for July.

Solution From the results of Example 3, we see that Acrosonic's total output for June may be represented by the matrix

$$B = \begin{bmatrix} 210 & 180 & 330 & 180 \\ 400 & 300 & 450 & 40 \\ 420 & 280 & 180 & 740 \end{bmatrix}$$

The required matrix is given by

$$(1.1)B = 1.1 \begin{bmatrix} 210 & 180 & 330 & 180 \\ 400 & 300 & 450 & 40 \\ 420 & 280 & 180 & 740 \end{bmatrix}$$

$$= \begin{bmatrix} 231 & 198 & 363 & 198 \\ 440 & 330 & 495 & 44 \\ 462 & 308 & 198 & 814 \end{bmatrix}$$

and is interpreted in the usual manner.

2.4 Self-Check Exercises

1. Perform the indicated operations:

$$\begin{bmatrix} 1 & 3 & 2 \\ -1 & 4 & 7 \end{bmatrix} - 3 \begin{bmatrix} 2 & 1 & 0 \\ 1 & 3 & 4 \end{bmatrix}$$

2. Solve the following matrix equation for x, y, and z:

$$\begin{bmatrix} x & 3 \\ z & 2 \end{bmatrix} + \begin{bmatrix} 2-y & z \\ 2-z & -x \end{bmatrix} = \begin{bmatrix} 3 & 7 \\ 2 & 0 \end{bmatrix}$$

3. Jack owns two gas stations, one downtown and the other in the Wilshire district. Over 2 consecutive days his gas stations recorded gasoline sales represented by the following matrices:

	Regular	Regular plus	Premium
Downtown	1200	750	650
Wilshire	1100	850	600

$A =$ (above)

and

	Regular	Regular plus	Premium
Downtown	1250	825	550
Wilshire	1150	750	750

$B =$ (above)

Find a matrix representing the total sales of the two gas stations over the 2-day period.

Solutions to Self-Check Exercises 2.4 can be found on page 117.

2.4 Concept Questions

1. Define (a) a matrix, (b) the size of a matrix, (c) a row matrix, (d) a column matrix, and (e) a square matrix.

2. When are two matrices equal? Give an example of two matrices that are equal.

3. Construct a 3×3 matrix A having the property that $A = A^T$. What special characteristic does A have?

2.4 Exercises

In Exercises 1–6, refer to the following matrices:

$$A = \begin{bmatrix} 2 & -3 & 9 & -4 \\ -11 & 2 & 6 & 7 \\ 6 & 0 & 2 & 9 \\ 5 & 1 & 5 & -8 \end{bmatrix}$$

$$B = \begin{bmatrix} 3 & -1 & 2 \\ 0 & 1 & 4 \\ 3 & 2 & 1 \\ -1 & 0 & 8 \end{bmatrix}$$

$$C = \begin{bmatrix} 1 & 0 & 3 & 4 & 5 \end{bmatrix}$$

$$D = \begin{bmatrix} 1 \\ 3 \\ -2 \\ 0 \end{bmatrix}$$

1. What is the size of A? Of B? Of C? Of D?

2. Find a_{14}, a_{21}, a_{31}, and a_{43}. **3.** Find b_{13}, b_{31}, and b_{43}.

4. Identify the row matrix. What is its transpose?

5. Identify the column matrix. What is its transpose?

6. Identify the square matrix. What is its transpose?

In Exercises 7–12, refer to the following matrices:

$$A = \begin{bmatrix} -1 & 2 \\ 3 & -2 \\ 4 & 0 \end{bmatrix} \quad B = \begin{bmatrix} 2 & 4 \\ 3 & 1 \\ -2 & 2 \end{bmatrix}$$

$$C = \begin{bmatrix} 3 & -1 & 0 \\ 2 & -2 & 3 \\ 4 & 6 & 2 \end{bmatrix} \quad D = \begin{bmatrix} 2 & -2 & 4 \\ 3 & 6 & 2 \\ -2 & 3 & 1 \end{bmatrix}$$

7. What is the size of A? Of B? Of C? Of D?

8. Explain why the matrix $A + C$ does *not* exist.

9. Compute $A + B$.   **10.** Compute $2A - 3B$.

11. Compute $C - D$.   **12.** Compute $4D - 2C$.

In Exercises 13–20, perform the indicated operations.

13. $\begin{bmatrix} 6 & 3 & 8 \\ 4 & 5 & 6 \end{bmatrix} - \begin{bmatrix} 3 & -2 & -1 \\ 0 & -5 & -7 \end{bmatrix}$

14. $\begin{bmatrix} 2 & -3 & 4 & -1 \\ 3 & 1 & 0 & 0 \end{bmatrix} + \begin{bmatrix} 4 & 3 & -2 & -4 \\ 6 & 2 & 0 & -3 \end{bmatrix}$

15. $\begin{bmatrix} 1 & 4 & -5 \\ 3 & -8 & 6 \end{bmatrix} + \begin{bmatrix} 4 & 0 & -2 \\ 3 & 6 & 5 \end{bmatrix} - \begin{bmatrix} 2 & 8 & 9 \\ -11 & 2 & -5 \end{bmatrix}$

16. $3\begin{bmatrix} 1 & 1 & -3 \\ 3 & 2 & 3 \\ 7 & -1 & 6 \end{bmatrix} + 4\begin{bmatrix} -2 & -1 & 8 \\ 4 & 2 & 2 \\ 3 & 6 & 3 \end{bmatrix}$

17. $\begin{bmatrix} 1.2 & 4.5 & -4.2 \\ 8.2 & 6.3 & -3.2 \end{bmatrix} - \begin{bmatrix} 3.1 & 1.5 & -3.6 \\ 2.2 & -3.3 & -4.4 \end{bmatrix}$

18. $\begin{bmatrix} 0.06 & 0.12 \\ 0.43 & 1.11 \\ 1.55 & -0.43 \end{bmatrix} - \begin{bmatrix} 0.77 & -0.75 \\ 0.22 & -0.65 \\ 1.09 & -0.57 \end{bmatrix}$

19. $\frac{1}{2}\begin{bmatrix} 1 & 0 & 0 & -4 \\ 3 & 0 & -1 & 6 \\ -2 & 1 & -4 & 2 \end{bmatrix} + \frac{4}{3}\begin{bmatrix} 3 & 0 & -1 & 4 \\ -2 & 1 & -6 & 2 \\ 8 & 2 & 0 & -2 \end{bmatrix}$

$- \frac{1}{3}\begin{bmatrix} 3 & -9 & -1 & 0 \\ 6 & 2 & 0 & -6 \\ 0 & 1 & -3 & 1 \end{bmatrix}$

20. $0.5\begin{bmatrix} 1 & 3 & 5 \\ 5 & 2 & -1 \\ -2 & 0 & 1 \end{bmatrix} - 0.2\begin{bmatrix} 2 & 3 & 4 \\ -1 & 1 & -4 \\ 3 & 5 & -5 \end{bmatrix}$

$+ 0.6\begin{bmatrix} 3 & 4 & -1 \\ 4 & 5 & 1 \\ 1 & 0 & 0 \end{bmatrix}$

In Exercises 21–24, solve for u, x, y, and z in the given matrix equation.

21. $\begin{bmatrix} 2x - 2 & 3 & 2 \\ 2 & 4 & y - 2 \\ 2z & -3 & 2 \end{bmatrix} = \begin{bmatrix} 3 & u & 2 \\ 2 & 4 & 5 \\ 4 & -3 & 2 \end{bmatrix}$

22. $\begin{bmatrix} x & -2 \\ 3 & y \end{bmatrix} + \begin{bmatrix} -2 & z \\ -1 & 2 \end{bmatrix} = \begin{bmatrix} 4 & -2 \\ 2u & 4 \end{bmatrix}$

23. $\begin{bmatrix} 1 & x \\ 2y & -3 \end{bmatrix} - 4\begin{bmatrix} 2 & -2 \\ 0 & 3 \end{bmatrix} = \begin{bmatrix} 3z & 10 \\ 4 & -u \end{bmatrix}$

24. $\begin{bmatrix} 1 & 2 \\ 3 & 4 \\ x & -1 \end{bmatrix} - 3\begin{bmatrix} y - 1 & 2 \\ 1 & 2 \\ 4 & 2z + 1 \end{bmatrix} = 2\begin{bmatrix} -4 & -u \\ 0 & -1 \\ 4 & 4 \end{bmatrix}$

In Exercises 25 and 26, let

$$A = \begin{bmatrix} 2 & -4 & 3 \\ 4 & 2 & 1 \end{bmatrix} \quad B = \begin{bmatrix} 4 & -3 & 2 \\ 1 & 0 & 4 \end{bmatrix}$$

$$C = \begin{bmatrix} 1 & 0 & 2 \\ 3 & -2 & 1 \end{bmatrix}$$

25. Verify by direct computation the validity of the commutative law for matrix addition.

26. Verify by direct computation the validity of the associative law for matrix addition.

In Exercises 27–30, let

$$A = \begin{bmatrix} 3 & 1 \\ 2 & 4 \\ -4 & 0 \end{bmatrix} \quad \text{and} \quad B = \begin{bmatrix} 1 & 2 \\ -1 & 0 \\ 3 & 2 \end{bmatrix}$$

Verify each equation by direct computation.

27. $(3 + 5)A = 3A + 5A$ **28.** $2(4A) = (2 \cdot 4)A = 8A$

29. $4(A + B) = 4A + 4B$ **30.** $2(A - 3B) = 2A - 6B$

In Exercises 31–34, find the transpose of each matrix.

31. $\begin{bmatrix} 3 & 2 & -1 & 5 \end{bmatrix}$ **32.** $\begin{bmatrix} 4 & 2 & 0 & -1 \\ 3 & 4 & -1 & 5 \end{bmatrix}$

33. $\begin{bmatrix} 1 & -1 & 2 \\ 3 & 4 & 2 \\ 0 & 1 & 0 \end{bmatrix}$ **34.** $\begin{bmatrix} 1 & 2 & 6 & 4 \\ 2 & 3 & 2 & 5 \\ 6 & 2 & 3 & 0 \\ 4 & 5 & 0 & 2 \end{bmatrix}$

35. CHOLESTEROL LEVELS Mr. Cross, Mr. Jones, and Mr. Smith each suffer from coronary heart disease. As part of their treatment, they were put on special low-cholesterol diets: Cross on diet I, Jones on diet II, and Smith on diet III. Progressive records of each patient's cholesterol level were kept. At the beginning of the first, second, third, and fourth months, the cholesterol levels of the three patients were:

> Cross: 220, 215, 210, and 205
> Jones: 220, 210, 200, and 195
> Smith: 215, 205, 195, and 190

Represent this information in a 3 × 4 matrix.

36. INVESTMENT PORTFOLIOS The following table gives the number of shares of certain corporations held by Leslie and Tom in their respective IRA accounts at the beginning of the year:

	IBM	GE	Ford	Wal-Mart
Leslie	500	350	200	400
Tom	400	450	300	200

Over the year, they added more shares to their accounts, as shown in the following table:

	IBM	GE	Ford	Wal-Mart
Leslie	50	50	0	100
Tom	0	80	100	50

a. Write a matrix A giving the holdings of Leslie and Tom at the beginning of the year and a matrix B giving the shares they have added to their portfolios.

b. Find a matrix C giving their total holdings at the end of the year.

37. BANKING The numbers of three types of bank accounts on January 1 at the Central Bank and its branches are represented by matrix A:

	Checking accounts	Savings accounts	Fixed-deposit accounts
Main office	2820	1470	1120
A = Westside branch	1030	520	480
Eastside branch	1170	540	460

The number and types of accounts opened during the first quarter are represented by matrix B, and the number and types of accounts closed during the same period are represented by matrix C. Thus,

$$B = \begin{bmatrix} 260 & 120 & 110 \\ 140 & 60 & 50 \\ 120 & 70 & 50 \end{bmatrix} \quad \text{and} \quad C = \begin{bmatrix} 120 & 80 & 80 \\ 70 & 30 & 40 \\ 60 & 20 & 40 \end{bmatrix}$$

a. Find matrix D, which represents the number of each type of account at the end of the first quarter at each location.

b. Because a new manufacturing plant is opening in the immediate area, it is anticipated that there will be a 10% increase in the number of accounts at each location during the second quarter. Write a matrix E to reflect this anticipated increase.

38. BOOKSTORE INVENTORIES The Campus Bookstore's inventory of books is:

Hardcover: textbooks, 5280; fiction, 1680; nonfiction, 2320; reference, 1890

Paperback: fiction, 2810; nonfiction, 1490; reference, 2070; textbooks, 1940

The College Bookstore's inventory of books is:

Hardcover: textbooks, 6340; fiction, 2220; nonfiction, 1790; reference, 1980

Paperback: fiction, 3100; nonfiction, 1720; reference, 2710; textbooks, 2050

a. Represent Campus's inventory as a matrix A.
b. Represent College's inventory as a matrix B.
c. The two companies decide to merge, so now write a matrix C that represents the total inventory of the newly amalgamated company.

39. **INSURANCE CLAIMS** The property damage–claim frequencies per 100 cars in Massachusetts in the years 2000, 2001, and 2002 are 6.88, 7.05, and 7.18, respectively. The corresponding claim frequencies in the United States are 4.13, 4.09, and 4.06, respectively. Express this information using a 2×3 matrix.

Sources: Registry of Motor Vehicles; Federal Highway Administration

40. **MORTALITY RATES** Mortality actuarial tables in the United States were revised in 2001, the fourth time since 1858. Based on the new life insurance mortality rates, 1% of 60-year-old men, 2.6% of 70-year-old men, 7% of 80-year-old men, 18.8% of 90-year-old men, and 36.3% of 100-year-old men would die within a year. The corresponding rates for women are 0.8%, 1.8%, 4.4%, 12.2%, and 27.6%, respectively. Express this information using a 2×5 matrix.

Source: Society of Actuaries

41. **LIFE EXPECTANCY** Figures for life expectancy at birth of Massachusetts residents in 2002 are 81, 76.1, and 82.2 years for white, black, and Hispanic women, respectively, and 76, 69.9, and 75.9 years for white, black, and Hispanic men,

respectively. Express this information using a 2×3 matrix and a 3×2 matrix.

Source: Massachusetts Department of Public Health

42. **MARKET SHARE OF MOTORCYCLES** The market share of motorcycles in the United States in 2001 follows: Honda 27.9%, Harley-Davidson 21.9%, Yamaha 19.2%, Suzuki 11%, Kawasaki 9.1%, and others 10.9%. The corresponding figures for 2002 are 27.6%, 23.3%, 18.2%, 10.5%, 8.8%, and 11.6%, respectively. Express this information using a 2×6 matrix. What is the sum of all the elements in the first row? In the second row? Is this expected? Which company gained the most market share between 2001 and 2002?

Source: Motorcycle Industry Council

In Exercises 43–46, determine whether the statement is true or false. If it is true, explain why it is true. If it is false, give an example to show why it is false.

43. If A and B are matrices of the same order and c is a scalar, then $c(A + B) = cA + cB$.

44. If A and B are matrices of the same order, then $A - B = A + (-1)B$.

45. If A is a matrix and c is a nonzero scalar, then $(cA)^T = (1/c)A^T$.

46. If A is a matrix, then $(A^T)^T = A$.

2.4 Solutions to Self-Check Exercises

1.
$$\begin{bmatrix} 1 & 3 & 2 \\ -1 & 4 & 7 \end{bmatrix} - 3\begin{bmatrix} 2 & 1 & 0 \\ 1 & 3 & 4 \end{bmatrix} = \begin{bmatrix} 1 & 3 & 2 \\ -1 & 4 & 7 \end{bmatrix} - \begin{bmatrix} 6 & 3 & 0 \\ 3 & 9 & 12 \end{bmatrix}$$
$$= \begin{bmatrix} -5 & 0 & 2 \\ -4 & -5 & -5 \end{bmatrix}$$

2. We are given
$$\begin{bmatrix} x & 3 \\ z & 2 \end{bmatrix} + \begin{bmatrix} 2 - y & z \\ 2 - z & -x \end{bmatrix} = \begin{bmatrix} 3 & 7 \\ 2 & 0 \end{bmatrix}$$

Performing the indicated operation on the left-hand side, we obtain
$$\begin{bmatrix} 2 + x - y & 3 + z \\ 2 & 2 - x \end{bmatrix} = \begin{bmatrix} 3 & 7 \\ 2 & 0 \end{bmatrix}$$

By the equality of matrices, we have
$$2 + x - y = 3$$
$$3 + z = 7$$
$$2 - x = 0$$

from which we deduce that $x = 2$, $y = 1$, and $z = 4$.

3. The required matrix is
$$A + B = \begin{bmatrix} 1200 & 750 & 650 \\ 1100 & 850 & 600 \end{bmatrix} + \begin{bmatrix} 1250 & 825 & 550 \\ 1150 & 750 & 750 \end{bmatrix}$$
$$= \begin{bmatrix} 2450 & 1575 & 1200 \\ 2250 & 1600 & 1350 \end{bmatrix}$$

USING TECHNOLOGY

■ Matrix Operations

Graphing Utility

A graphing utility can be used to perform matrix addition, matrix subtraction, and scalar multiplication. It can also be used to find the transpose of a matrix.

EXAMPLE 1 Let

$$A = \begin{bmatrix} 1.2 & 3.1 \\ -2.1 & 4.2 \\ 3.1 & 4.8 \end{bmatrix} \quad \text{and} \quad B = \begin{bmatrix} 4.1 & 3.2 \\ 1.3 & 6.4 \\ 1.7 & 0.8 \end{bmatrix}$$

Find (a) $A + B$, (b) $2.1A - 3.2B$, and (c) $(2.1A + 3.2B)^T$.

Solution We first enter the matrices A and B into the calculator.

a. Using matrix operations, we enter the expression $A + B$ and obtain

$$A + B = \begin{bmatrix} 5.3 & 6.3 \\ -0.8 & 10.6 \\ 4.8 & 5.6 \end{bmatrix}$$

b. Using matrix operations, we enter the expression $2.1A - 3.2B$ and obtain

$$2.1A - 3.2B = \begin{bmatrix} -10.6 & -3.73 \\ -8.57 & -11.66 \\ 1.07 & 7.52 \end{bmatrix}$$

c. Using matrix operations, we enter the expression $(2.1A + 3.2B)^T$ and obtain

$$(2.1A + 3.2B)^T = \begin{bmatrix} 15.64 & -.25 & 11.95 \\ 16.75 & 29.3 & 12.64 \end{bmatrix}$$

APPLIED EXAMPLE 2 John operates three gas stations at three locations, I, II, and III. Over 2 consecutive days, his gas stations recorded the following fuel sales (in gallons):

	Day 1			
	Regular	Regular Plus	Premium	Diesel
Location I	1400	1200	1100	200
Location II	1600	900	1200	300
Location III	1200	1500	800	500

	Day 2			
	Regular	Regular Plus	Premium	Diesel
Location I	1000	900	800	150
Location II	1800	1200	1100	250
Location III	800	1000	700	400

Find a matrix representing the total fuel sales at John's gas stations.

Solution   The fuel sales can be represented by the matrix A (day 1) and matrix B (day 2):

$$A = \begin{bmatrix} 1400 & 1200 & 1100 & 200 \\ 1600 & 900 & 1200 & 300 \\ 1200 & 1500 & 800 & 500 \end{bmatrix} \quad \text{and} \quad B = \begin{bmatrix} 1000 & 900 & 800 & 150 \\ 1800 & 1200 & 1100 & 250 \\ 800 & 1000 & 700 & 400 \end{bmatrix}$$

Next, we enter the matrices A and B into the calculator. Using matrix operations, we enter the expression $A + B$ and obtain

$$A + B = \begin{bmatrix} 2400 & 2100 & 1900 & 350 \\ 3400 & 2100 & 2300 & 550 \\ 2000 & 2500 & 1500 & 900 \end{bmatrix}$$

Excel

First, we show how basic operations on matrices can be carried out using Excel.

EXAMPLE 3   Given the following matrices,

$$A = \begin{bmatrix} 1.2 & 3.1 \\ -2.1 & 4.2 \\ 3.1 & 4.8 \end{bmatrix} \quad \text{and} \quad B = \begin{bmatrix} 4.1 & 3.2 \\ 1.3 & 6.4 \\ 1.7 & 0.8 \end{bmatrix}$$

a. Compute $A + B$.   **b.** Compute $2.1A - 3.2B$.

Solution

a. *First, represent the matrices A and B in a spreadsheet.* Enter the elements of each matrix in a block of cells as shown in Figure T1.

	A	B	C	D	E
1		A			B
2	1.2	3.1		4.1	3.2
3	-2.1	4.2		1.3	6.4
4	3.1	4.8		1.7	0.8

FIGURE T1
The elements of matrix A and matrix B in the spreadsheet

Second, compute the sum of matrix A and matrix B. Highlight the cells that will contain matrix $A + B$, type =, highlight the cells in matrix A, type +, highlight the cells in matrix B, and press $\boxed{\textbf{Ctrl-Shift-Enter}}$. The resulting matrix $A + B$ is shown in Figure T2.

	A	B
8		A + B
9	5.3	6.3
10	-0.8	10.6
11	4.8	5.6

FIGURE T2
The matrix $A + B$

Note: Boldfaced words/characters enclosed in a box (for example, $\boxed{\textbf{Enter}}$) indicate that an action (click, select, or press) is required. Words/characters printed blue (for example, Chart sub-type:) indicate words/characters that appear on the screen. Words/characters printed in a typewriter font (for example, =(-2/3)*A2+2) indicate words/characters that need to be typed and entered.

(continued)

b. *Highlight the cells that will contain matrix* (2.1A − 3.2B). Type = 2.1*, highlight matrix A, type −3.2*, highlight the cells in matrix B, and press Ctrl-Shift-Enter . The resulting matrix (2.1A − 3.2B) is shown in Figure T3.

	A	B
13		2.1A - 3.2B
14	−10.6	−3.73
15	−8.57	−11.66
16	1.07	7.52

FIGURE T3
The matrix (2.1A − 3.2B)

APPLIED EXAMPLE 4   John operates three gas stations at three locations I, II, and III. Over 2 consecutive days, his gas stations recorded the following fuel sales (in gallons):

| | Day 1 | | | |
	Regular	Regular Plus	Premium	Diesel
Location I	1400	1200	1100	200
Location II	1600	900	1200	300
Location III	1200	1500	800	500

| | Day 2 | | | |
	Regular	Regular Plus	Premium	Diesel
Location I	1000	900	800	150
Location II	1800	1200	1100	250
Location III	800	1000	700	400

Find a matrix representing the total fuel sales at John's gas stations.

Solution   The fuel sales can be represented by the matrices A (day 1) and B (day 2):

$$A = \begin{bmatrix} 1400 & 1200 & 1100 & 200 \\ 1600 & 900 & 1200 & 300 \\ 1200 & 1500 & 800 & 500 \end{bmatrix} \quad \text{and} \quad B = \begin{bmatrix} 1000 & 900 & 800 & 150 \\ 1800 & 1200 & 1100 & 250 \\ 800 & 1000 & 700 & 400 \end{bmatrix}$$

We first enter the elements of the matrices A and B onto a spreadsheet. Next, we highlight the cells that will contain the matrix $A + B$, type =, highlight A, type +, highlight B, and then press Ctrl-Shift-Enter . The resulting matrix $A + B$ is shown in Figure T4.

	A	B	C	D
23		A + B		
24	2400	2100	1900	350
25	3400	2100	2300	550
26	2000	2500	1500	900

FIGURE T4
The matrix $A + B$

TECHNOLOGY EXERCISES

Refer to the following matrices and perform the indicated operations.

$$A = \begin{bmatrix} 1.2 & 3.1 & -5.4 & 2.7 \\ 4.1 & 3.2 & 4.2 & -3.1 \\ 1.7 & 2.8 & -5.2 & 8.4 \end{bmatrix}$$

$$B = \begin{bmatrix} 6.2 & -3.2 & 1.4 & -1.2 \\ 3.1 & 2.7 & -1.2 & 1.7 \\ 1.2 & -1.4 & -1.7 & 2.8 \end{bmatrix}$$

1. $12.5A$
2. $-8.4B$
3. $A - B$
4. $B - A$
5. $1.3A + 2.4B$
6. $2.1A - 1.7B$
7. $3(A + B)$
8. $1.3(4.1A - 2.3B)$

2.5 Multiplication of Matrices

▬ Matrix Product

In Section 2.4, we saw how matrices of the same size may be added or subtracted and how a matrix may be multiplied by a scalar (real number), an operation referred to as scalar multiplication. In this section we see how, with certain restrictions, one matrix may be multiplied by another matrix.

To define matrix multiplication, let's consider the following problem. On a certain day, Al's Service Station sold 1600 gallons of regular, 1000 gallons of regular plus, and 800 gallons of premium gasoline. If the price of gasoline on this day was $2.69 for regular, $2.79 for regular plus, and $2.89 for premium gasoline, find the total revenue realized by Al's for that day.

The day's sale of gasoline may be represented by the matrix

$$A = \begin{bmatrix} 1600 & 1000 & 800 \end{bmatrix} \qquad \text{Row matrix } (1 \times 3)$$

Next, we let the unit selling price of regular, regular plus, and premium gasoline be the entries in the matrix

$$B = \begin{bmatrix} 2.69 \\ 2.79 \\ 2.89 \end{bmatrix} \qquad \text{Column matrix } (3 \times 1)$$

The first entry in matrix A gives the number of gallons of regular gasoline sold, and the first entry in matrix B gives the selling price for each gallon of regular gasoline, so their product $(1600)(2.69)$ gives the revenue realized from the sale of regular gasoline for the day. A similar interpretation of the second and third entries in the two matrices suggests that we multiply the corresponding entries to obtain the respective revenues realized from the sale of regular, regular plus, and premium gasoline. Finally, the total revenue realized by Al's from the sale of gasoline is given by adding these products to obtain

$$(1600)(2.69) + (1000)(2.79) + (800)(2.89) = 9406$$

or $9406.

This example suggests that if we have a row matrix of size $1 \times n$,

$$A = \begin{bmatrix} a_1 & a_2 & a_3 & \cdots & a_n \end{bmatrix}$$

and a column matrix of size $n \times 1$,

$$B = \begin{bmatrix} b_1 \\ b_2 \\ b_3 \\ \vdots \\ b_n \end{bmatrix}$$

then we may define the **matrix product** of A and B, written AB, by

$$AB = \begin{bmatrix} a_1 & a_2 & a_3 & \cdots & a_n \end{bmatrix} \begin{bmatrix} b_1 \\ b_2 \\ b_3 \\ \vdots \\ b_n \end{bmatrix} = a_1b_1 + a_2b_2 + a_3b_3 + \cdots + a_nb_n \quad \textbf{(11)}$$

EXAMPLE 1 Let

$$A = \begin{bmatrix} 1 & -2 & 3 & 5 \end{bmatrix} \quad \text{and} \quad B = \begin{bmatrix} 2 \\ 3 \\ 0 \\ -1 \end{bmatrix}$$

Then

$$AB = \begin{bmatrix} 1 & -2 & 3 & 5 \end{bmatrix} \begin{bmatrix} 2 \\ 3 \\ 0 \\ -1 \end{bmatrix} = (1)(2) + (-2)(3) + (3)(0) + (5)(-1) = -9$$

APPLIED EXAMPLE 2 Stock Transactions Judy's stock holdings are given by the matrix

$$\begin{matrix} \text{GM} & \text{IBM} & \text{BAC} \end{matrix}$$
$$A = \begin{bmatrix} 700 & 400 & 200 \end{bmatrix}$$

At the close of trading on a certain day, the prices (in dollars per share) of these stocks are

$$B = \begin{bmatrix} 50 \\ 120 \\ 42 \end{bmatrix} \begin{matrix} \text{GM} \\ \text{IBM} \\ \text{BAC} \end{matrix}$$

What is the total value of Judy's holdings as of that day?

Solution Judy's holdings are worth

$$AB = \begin{bmatrix} 700 & 400 & 200 \end{bmatrix} \begin{bmatrix} 50 \\ 120 \\ 42 \end{bmatrix} = (700)(50) + (400)(120) + (200)(42)$$

or $91,400.

Returning once again to the matrix product AB in Equation (11), observe that the number of columns of the row matrix A is *equal* to the number of rows of the column matrix B. Observe further that the product matrix AB has size 1×1 (a real number may be thought of as a 1×1 matrix). Schematically,

Size of A Size of B

$(1 \times n)$ $(n \times 1)$

(1×1)

Size of AB

More generally, if A is a matrix of size $m \times n$ and B is a matrix of size $n \times p$ (the number of columns of A equals the numbers of rows of B), then the *matrix product* of A and B, AB, is defined and is a matrix of size $m \times p$. Schematically,

Size of A Size of B

$(m \times n)$ $(n \times p)$

$(m \times p)$

Size of AB

Next, let's illustrate the mechanics of matrix multiplication by computing the product of a 2×3 matrix A and a 3×4 matrix B. Suppose

$$A = \begin{bmatrix} a_{11} & a_{12} & a_{13} \\ a_{21} & a_{22} & a_{23} \end{bmatrix}$$

$$B = \begin{bmatrix} b_{11} & b_{12} & b_{13} & b_{14} \\ b_{21} & b_{22} & b_{23} & b_{24} \\ b_{31} & b_{32} & b_{33} & b_{34} \end{bmatrix}$$

From the schematic

Same

Size of A (2×3) (3×4) Size of B

(2×4)

Size of AB

we see that the matrix product $C = AB$ is defined (since the number of columns of A equals the number of rows of B) and has size 2×4. Thus,

$$C = \begin{bmatrix} c_{11} & c_{12} & c_{13} & c_{14} \\ c_{21} & c_{22} & c_{23} & c_{24} \end{bmatrix}$$

The entries of C are computed as follows: The entry c_{11} (the entry in the *first* row, *first* column of C) is the product of the row matrix composed of the entries from the *first* row of A and the column matrix composed of the *first* column of B. Thus,

$$c_{11} = [a_{11} \quad a_{12} \quad a_{13}] \begin{bmatrix} b_{11} \\ b_{21} \\ b_{31} \end{bmatrix} = a_{11}b_{11} + a_{12}b_{21} + a_{13}b_{31}$$

The entry c_{12} (the entry in the *first* row, *second* column of C) is the product of the row matrix composed of the *first* row of A and the column matrix composed of the *second* column of B. Thus,

$$c_{12} = [a_{11} \quad a_{12} \quad a_{13}] \begin{bmatrix} b_{12} \\ b_{22} \\ b_{32} \end{bmatrix} = a_{11}b_{12} + a_{12}b_{22} + a_{13}b_{32}$$

The other entries in C are computed in a similar manner.

EXAMPLE 3   Let

$$A = \begin{bmatrix} 3 & 1 & 4 \\ -1 & 2 & 3 \end{bmatrix} \quad \text{and} \quad B = \begin{bmatrix} 1 & 3 & -3 \\ 4 & -1 & 2 \\ 2 & 4 & 1 \end{bmatrix}$$

Compute AB.

Solution   The size of matrix A is 2×3, and the size of matrix B is 3×3. Since the number of columns of matrix A is equal to the number of rows of matrix B, the matrix product $C = AB$ is defined. Furthermore, the size of matrix C is 2×3. Thus,

$$\begin{bmatrix} 3 & 1 & 4 \\ -1 & 2 & 3 \end{bmatrix} \begin{bmatrix} 1 & 3 & -3 \\ 4 & -1 & 2 \\ 2 & 4 & 1 \end{bmatrix} = \begin{bmatrix} c_{11} & c_{12} & c_{13} \\ c_{21} & c_{22} & c_{23} \end{bmatrix}$$

It remains now to determine the entries $c_{11}, c_{12}, c_{13}, c_{21}, c_{22}$, and c_{23}. We have

$$c_{11} = [3 \quad 1 \quad 4] \begin{bmatrix} 1 \\ 4 \\ 2 \end{bmatrix} = (3)(1) + (1)(4) + (4)(2) = 15$$

$$c_{12} = [3 \quad 1 \quad 4] \begin{bmatrix} 3 \\ -1 \\ 4 \end{bmatrix} = (3)(3) + (1)(-1) + (4)(4) = 24$$

$$c_{13} = [3 \quad 1 \quad 4] \begin{bmatrix} -3 \\ 2 \\ 1 \end{bmatrix} = (3)(-3) + (1)(2) + (4)(1) = -3$$

$$c_{21} = [-1 \quad 2 \quad 3] \begin{bmatrix} 1 \\ 4 \\ 2 \end{bmatrix} = (-1)(1) + (2)(4) + (3)(2) = 13$$

$$c_{22} = \begin{bmatrix} -1 & 2 & 3 \end{bmatrix} \begin{bmatrix} 3 \\ -1 \\ 4 \end{bmatrix} = (-1)(3) + (2)(-1) + (3)(4) = 7$$

$$c_{23} = \begin{bmatrix} -1 & 2 & 3 \end{bmatrix} \begin{bmatrix} -3 \\ 2 \\ 1 \end{bmatrix} = (-1)(-3) + (2)(2) + (3)(1) = 10$$

so the required product AB is given by

$$AB = \begin{bmatrix} 15 & 24 & -3 \\ 13 & 7 & 10 \end{bmatrix}$$

EXAMPLE 4   Let

$$A = \begin{bmatrix} 3 & 2 & 1 \\ -1 & 2 & 3 \\ 3 & 1 & 4 \end{bmatrix} \quad \text{and} \quad B = \begin{bmatrix} 1 & 3 & 4 \\ 2 & 4 & 1 \\ -1 & 2 & 3 \end{bmatrix}$$

Then

$$AB = \begin{bmatrix} 3 \cdot 1 & + 2 \cdot 2 + 1 \cdot (-1) & 3 \cdot 3 & + 2 \cdot 4 + 1 \cdot 2 & 3 \cdot 4 & + 2 \cdot 1 + 1 \cdot 3 \\ (-1) \cdot 1 + 2 \cdot 2 + 3 \cdot (-1) & (-1) \cdot 3 + 2 \cdot 4 + 3 \cdot 2 & (-1) \cdot 4 + 2 \cdot 1 + 3 \cdot 3 \\ 3 \cdot 1 & + 1 \cdot 2 + 4 \cdot (-1) & 3 \cdot 3 & + 1 \cdot 4 + 4 \cdot 2 & 3 \cdot 4 & + 1 \cdot 1 + 4 \cdot 3 \end{bmatrix}$$

$$= \begin{bmatrix} 6 & 19 & 17 \\ 0 & 11 & 7 \\ 1 & 21 & 25 \end{bmatrix}$$

$$BA = \begin{bmatrix} 1 \cdot 3 & + 3 \cdot (-1) + 4 \cdot 3 & 1 \cdot 2 & + 3 \cdot 2 + 4 \cdot 1 & 1 \cdot 1 & + 3 \cdot 3 + 4 \cdot 4 \\ 2 \cdot 3 & + 4 \cdot (-1) + 1 \cdot 3 & 2 \cdot 2 & + 4 \cdot 2 + 1 \cdot 1 & 2 \cdot 1 & + 4 \cdot 3 + 1 \cdot 4 \\ (-1) \cdot 3 + 2 \cdot (-1) + 3 \cdot 3 & (-1) \cdot 2 + 2 \cdot 2 + 3 \cdot 1 & (-1) \cdot 1 + 2 \cdot 3 + 3 \cdot 4 \end{bmatrix}$$

$$= \begin{bmatrix} 12 & 12 & 26 \\ 5 & 13 & 18 \\ 4 & 5 & 17 \end{bmatrix}$$

The preceding example shows that, in general, $AB \neq BA$ for any two square matrices A and B. However, the following laws are valid for matrix multiplication.

Laws for Matrix Multiplication

If the products and sums are defined for the matrices A, B, and C, then

1. $(AB)C = A(BC)$ Associative law
2. $A(B + C) = AB + AC$ Distributive law

The square matrix of size n having 1s along the main diagonal and 0s elsewhere is called the identity matrix of size n.

Identity Matrix
The identity matrix of size n is given by

$$I_n = \begin{bmatrix} 1 & 0 & \cdot & \cdot & \cdot & 0 \\ 0 & 1 & \cdot & \cdot & \cdot & 0 \\ \cdot & \cdot & \cdot & & \cdot & \cdot \\ \cdot & \cdot & & \cdot & & \cdot \\ \cdot & \cdot & & & \cdot & \cdot \\ 0 & 0 & \cdot & \cdot & \cdot & 1 \end{bmatrix} \quad n \text{ rows}$$

n columns

The identity matrix has the properties that $I_n A = A$ for any $n \times r$ matrix A and $B I_n = B$ for any $s \times n$ matrix B. In particular, if A is a square matrix of size n, then

$$I_n A = A I_n = A$$

EXAMPLE 5   Let

$$A = \begin{bmatrix} 1 & 3 & 1 \\ -4 & 3 & 2 \\ 1 & 0 & 1 \end{bmatrix}$$

Then

$$I_3 A = \begin{bmatrix} 1 & 0 & 0 \\ 0 & 1 & 0 \\ 0 & 0 & 1 \end{bmatrix} \begin{bmatrix} 1 & 3 & 1 \\ -4 & 3 & 2 \\ 1 & 0 & 1 \end{bmatrix} = \begin{bmatrix} 1 & 3 & 1 \\ -4 & 3 & 2 \\ 1 & 0 & 1 \end{bmatrix} = A$$

$$A I_3 = \begin{bmatrix} 1 & 3 & 1 \\ -4 & 3 & 2 \\ 1 & 0 & 1 \end{bmatrix} \begin{bmatrix} 1 & 0 & 0 \\ 0 & 1 & 0 \\ 0 & 0 & 1 \end{bmatrix} = \begin{bmatrix} 1 & 3 & 1 \\ -4 & 3 & 2 \\ 1 & 0 & 1 \end{bmatrix} = A$$

so $I_3 A = A I_3$, confirming our result for this special case.   ■

APPLIED EXAMPLE 6   Production Planning   Ace Novelty received an order from Magic World Amusement Park for 900 "Giant Pandas," 1200 "Saint Bernards," and 2000 "Big Birds." Ace's management decided that 500 Giant Pandas, 800 Saint Bernards, and 1300 Big Birds could be manufactured in their Los Angeles plant, and the balance of the order could be filled by their Seattle plant. Each Panda requires 1.5 square yards of plush, 30 cubic feet of stuffing, and 5 pieces of trim; each Saint Bernard requires 2 square yards of plush, 35 cubic feet of stuffing, and 8 pieces of trim; and each Big Bird requires 2.5 square yards of plush, 25 cubic feet of stuffing, and 15 pieces of trim. The plush costs $4.50 per square yard, the stuffing costs 10 cents per cubic foot, and the trim costs 25 cents per unit.

a. Find how much of each type of material must be purchased for each plant.
b. What is the total cost of materials incurred by each plant and the total cost of materials incurred by Ace Novelty in filling the order?

Solution The quantities of each type of stuffed animal to be produced at each plant location may be expressed as a 2×3 *production matrix P*. Thus,

$$P = \begin{array}{c} \text{L.A.} \\ \text{Seattle} \end{array} \begin{array}{ccc} \text{Pandas} & \text{St. Bernards} & \text{Birds} \\ \begin{bmatrix} 500 & 800 & 1300 \\ 400 & 400 & 700 \end{bmatrix} \end{array}$$

Similarly, we may represent the amount and type of material required to manufacture each type of animal by a 3×3 *activity matrix A*. Thus,

$$A = \begin{array}{c} \text{Pandas} \\ \text{St. Bernards} \\ \text{Birds} \end{array} \begin{array}{ccc} \text{Plush} & \text{Stuffing} & \text{Trim} \\ \begin{bmatrix} 1.5 & 30 & 5 \\ 2 & 35 & 8 \\ 2.5 & 25 & 15 \end{bmatrix} \end{array}$$

Finally, the unit cost for each type of material may be represented by the 3×1 *cost matrix C.*

$$C = \begin{array}{c} \text{Plush} \\ \text{Stuffing} \\ \text{Trim} \end{array} \begin{bmatrix} 4.50 \\ 0.10 \\ 0.25 \end{bmatrix}$$

a. The amount of each type of material required for each plant is given by the matrix *PA*. Thus,

$$PA = \begin{bmatrix} 500 & 800 & 1300 \\ 400 & 400 & 700 \end{bmatrix} \begin{bmatrix} 1.5 & 30 & 5 \\ 2 & 35 & 8 \\ 2.5 & 25 & 15 \end{bmatrix}$$

$$= \begin{array}{c} \text{L.A.} \\ \text{Seattle} \end{array} \begin{array}{ccc} \text{Plush} & \text{Stuffing} & \text{Trim} \\ \begin{bmatrix} 5600 & 75,500 & 28,400 \\ 3150 & 43,500 & 15,700 \end{bmatrix} \end{array}$$

b. The total cost of materials for each plant is given by the matrix *PAC*:

$$PAC = \begin{bmatrix} 5600 & 75,500 & 28,400 \\ 3150 & 43,500 & 15,700 \end{bmatrix} \begin{bmatrix} 4.50 \\ 0.10 \\ 0.25 \end{bmatrix}$$

$$= \begin{array}{c} \text{L.A.} \\ \text{Seattle} \end{array} \begin{bmatrix} 39,850 \\ 22,450 \end{bmatrix}$$

or \$39,850 for the L.A. plant and \$22,450 for the Seattle plant. Thus, the total cost of materials incurred by Ace Novelty is \$62,300. ◼

■ Matrix Representation

Example 7 shows how a system of linear equations may be written in a compact form with the help of matrices. (We will use this matrix equation representation in Section 2.6.)

EXAMPLE 7   Write the following system of linear equations in matrix form.

$$\begin{aligned} 2x - 4y + z &= 6 \\ -3x + 6y - 5z &= -1 \\ x - 3y + 7z &= 0 \end{aligned}$$

Solution   Let's write

$$A = \begin{bmatrix} 2 & -4 & 1 \\ -3 & 6 & -5 \\ 1 & -3 & 7 \end{bmatrix} \qquad X = \begin{bmatrix} x \\ y \\ z \end{bmatrix} \qquad B = \begin{bmatrix} 6 \\ -1 \\ 0 \end{bmatrix}$$

Note that A is just the 3×3 matrix of coefficients of the system, X is the 3×1 column matrix of unknowns (variables), and B is the 3×1 column matrix of constants. We now show that the required matrix representation of the system of linear equations is

$$AX = B$$

To see this, observe that

$$AX = \begin{bmatrix} 2 & -4 & 1 \\ -3 & 6 & -5 \\ 1 & -3 & 7 \end{bmatrix} \begin{bmatrix} x \\ y \\ z \end{bmatrix} = \begin{bmatrix} 2x - 4y + z \\ -3x + 6y - 5z \\ x - 3y + 7z \end{bmatrix}$$

Equating this 3×1 matrix with matrix B now gives

$$\begin{bmatrix} 2x - 4y + z \\ -3x + 6y - 5z \\ x - 3y + 7z \end{bmatrix} = \begin{bmatrix} 6 \\ -1 \\ 0 \end{bmatrix}$$

which, by matrix equality, is easily seen to be equivalent to the given system of linear equations.   ▪

2.5   Self-Check Exercises

1. Compute

$$\begin{bmatrix} 1 & 3 & 0 \\ 2 & 4 & -1 \end{bmatrix} \begin{bmatrix} 3 & 1 & 4 \\ 2 & 0 & 3 \\ 1 & 2 & -1 \end{bmatrix}$$

2. Write the following system of linear equations in matrix form:

$$\begin{aligned} y - 2z &= 1 \\ 2x - y + 3z &= 0 \\ x + 4z &= 7 \end{aligned}$$

3. On June 1, the stock holdings of Ash and Joan Robinson were given by the matrix

$$A = \begin{matrix} & \overset{\text{AT\&T}}{} & \overset{\text{TWX}}{} & \overset{\text{IBM}}{} & \overset{\text{GM}}{} \\ \begin{matrix} \text{Ash} \\ \text{Joan} \end{matrix} & \begin{bmatrix} 2000 & 1000 & 500 & 5000 \\ 1000 & 2500 & 2000 & 0 \end{bmatrix} \end{matrix}$$

and the closing prices of AT&T, TWX, IBM, and GM were $54, $113, $112, and $70 per share, respectively. Use matrix multiplication to determine the separate values of Ash's and Joan's stock holdings as of that date.

Solutions to Self-Check Exercises 2.5 can be found on page 133.

2.5 Concept Questions

1. What is the difference between scalar multiplication and matrix multiplication? Give examples of each operation.

2 a. Suppose A and B are matrices whose products AB and BA are both defined. What can you say about the sizes of A and B?

b. If A, B, and C are matrices such that $A(B + C)$ is defined, what can you say about the relationship between the number of columns of A and the number of rows of C? Explain.

2.5 Exercises

In Exercises 1–4, the sizes of matrices A and B are given. Find the size of AB and BA whenever they are defined.

1. A is of size 2×3, and B is of size 3×5.

2. A is of size 3×4, and B is of size 4×3.

3. A is of size 1×7, and B is of size 7×1.

4. A is of size 4×4, and B is of size 4×4.

5. Let A be a matrix of size $m \times n$ and B be a matrix of size $s \times t$. Find conditions on m, n, s, and t such that both matrix products AB and BA are defined.

6. Find condition(s) on the size of a matrix A such that A^2 (that is, AA) is defined.

In Exercises 7–24, compute the indicated products.

7. $\begin{bmatrix} 1 & 2 \\ 3 & 0 \end{bmatrix}\begin{bmatrix} 1 \\ -1 \end{bmatrix}$

8. $\begin{bmatrix} -1 & 3 \\ 5 & 0 \end{bmatrix}\begin{bmatrix} 7 \\ 2 \end{bmatrix}$

9. $\begin{bmatrix} 3 & 1 & 2 \\ -1 & 2 & 4 \end{bmatrix}\begin{bmatrix} 4 \\ 1 \\ -2 \end{bmatrix}$

10. $\begin{bmatrix} 3 & 2 & -1 \\ 4 & -1 & 0 \\ -5 & 2 & 1 \end{bmatrix}\begin{bmatrix} 3 \\ -2 \\ 0 \end{bmatrix}$

11. $\begin{bmatrix} -1 & 2 \\ 3 & 1 \end{bmatrix}\begin{bmatrix} 2 & 4 \\ 3 & 1 \end{bmatrix}$

12. $\begin{bmatrix} 1 & 3 \\ -1 & 2 \end{bmatrix}\begin{bmatrix} 1 & 3 & 0 \\ 3 & 0 & 2 \end{bmatrix}$

13. $\begin{bmatrix} 2 & 1 & 2 \\ 3 & 2 & 4 \end{bmatrix}\begin{bmatrix} -1 & 2 \\ 4 & 3 \\ 0 & 1 \end{bmatrix}$

14. $\begin{bmatrix} -1 & 2 \\ 4 & 3 \\ 0 & 1 \end{bmatrix}\begin{bmatrix} 2 & 1 & 2 \\ 3 & 2 & 4 \end{bmatrix}$

15. $\begin{bmatrix} 0.1 & 0.9 \\ 0.2 & 0.8 \end{bmatrix}\begin{bmatrix} 1.2 & 0.4 \\ 0.5 & 2.1 \end{bmatrix}$

16. $\begin{bmatrix} 1.2 & 0.3 \\ 0.4 & 0.5 \end{bmatrix}\begin{bmatrix} 0.2 & 0.6 \\ 0.4 & -0.5 \end{bmatrix}$

17. $\begin{bmatrix} 6 & -3 & 0 \\ -2 & 1 & -8 \\ 4 & -4 & 9 \end{bmatrix}\begin{bmatrix} 1 & 0 & 0 \\ 0 & 1 & 0 \\ 0 & 0 & 1 \end{bmatrix}$

18. $\begin{bmatrix} 2 & 4 \\ -1 & -5 \\ 3 & -1 \end{bmatrix}\begin{bmatrix} 2 & -2 & 4 \\ 1 & 3 & -1 \end{bmatrix}$

19. $\begin{bmatrix} 3 & 0 & -2 & 1 \\ 1 & 2 & 0 & -1 \end{bmatrix}\begin{bmatrix} 2 & 1 & -1 \\ -1 & 2 & 0 \\ 0 & 0 & 1 \\ -1 & -2 & 2 \end{bmatrix}$

20. $\begin{bmatrix} 2 & 1 & -3 & 0 \\ 4 & -2 & -1 & 1 \\ -1 & 2 & 0 & 1 \end{bmatrix}\begin{bmatrix} 2 & -1 \\ 1 & 4 \\ 3 & -3 \\ 0 & -5 \end{bmatrix}$

21. $4\begin{bmatrix} 1 & -2 & 0 \\ 2 & -1 & 1 \\ 3 & 0 & -1 \end{bmatrix}\begin{bmatrix} 1 & 3 & 1 \\ 1 & 4 & 0 \\ 0 & 1 & -2 \end{bmatrix}$

22. $3\begin{bmatrix} 2 & -1 & 0 \\ 2 & 1 & 2 \\ 1 & 0 & -1 \end{bmatrix}\begin{bmatrix} 2 & 3 & 1 \\ 3 & -3 & 0 \\ 0 & 1 & -1 \end{bmatrix}$

23. $\begin{bmatrix} 1 & 0 \\ 0 & 1 \end{bmatrix}\begin{bmatrix} 4 & -3 & 2 \\ 7 & 1 & -5 \end{bmatrix}\begin{bmatrix} 1 & 0 & 0 \\ 0 & 1 & 0 \\ 0 & 0 & 1 \end{bmatrix}$

24. $2\begin{bmatrix} 3 & 2 & -1 \\ 0 & 1 & 3 \\ 2 & 0 & 3 \end{bmatrix}\begin{bmatrix} 1 & 0 & 0 \\ 0 & 1 & 0 \\ 0 & 0 & 1 \end{bmatrix}\begin{bmatrix} 1 & 2 & 0 \\ 0 & -1 & -2 \\ 1 & 3 & 1 \end{bmatrix}$

In Exercises 25 and 26, let

$$A = \begin{bmatrix} 1 & 0 & -2 \\ 1 & -3 & 2 \\ -2 & 1 & 1 \end{bmatrix} \quad B = \begin{bmatrix} 3 & 1 & 0 \\ 2 & 2 & 0 \\ 1 & -3 & -1 \end{bmatrix}$$

$$C = \begin{bmatrix} 2 & -1 & 0 \\ 1 & -1 & 2 \\ 3 & -2 & 1 \end{bmatrix}$$

25. Verify the validity of the associative law for matrix multiplication.

26. Verify the validity of the distributive law for matrix multiplication.

27. Let

$$A = \begin{bmatrix} 1 & 2 \\ 3 & 4 \end{bmatrix} \quad \text{and} \quad B = \begin{bmatrix} 2 & 1 \\ 4 & 3 \end{bmatrix}$$

Compute AB and BA and hence deduce that matrix multiplication is, in general, not commutative.

28. Let

$$A = \begin{bmatrix} 0 & 3 & 0 \\ 1 & 0 & 1 \\ 0 & 2 & 0 \end{bmatrix} \quad B = \begin{bmatrix} 2 & 4 & 5 \\ 3 & -1 & -6 \\ 4 & 3 & 4 \end{bmatrix}$$

$$C = \begin{bmatrix} 4 & 5 & 6 \\ 3 & -1 & -6 \\ 2 & 2 & 3 \end{bmatrix}$$

  a. Compute AB.
  b. Compute AC.
  c. Using the results of parts (a) and (b), conclude that $AB = AC$ does *not* imply that $B = C$.

29. Let

$$A = \begin{bmatrix} 3 & 0 \\ 8 & 0 \end{bmatrix} \quad \text{and} \quad B = \begin{bmatrix} 0 & 0 \\ 4 & 5 \end{bmatrix}$$

Show that $AB = 0$, thereby demonstrating that for matrix multiplication the equation $AB = 0$ does not imply that one or both of the matrices A and B must be the zero matrix.

30. Let

$$A = \begin{bmatrix} 2 & 2 \\ -2 & -2 \end{bmatrix}$$

Show that $A^2 = 0$. Compare this with the equation $a^2 = 0$, where a is a real number.

31. Find the matrix A such that

$$A \begin{bmatrix} 1 & 0 \\ -1 & 3 \end{bmatrix} = \begin{bmatrix} -1 & -3 \\ 3 & 6 \end{bmatrix}$$

Hint: Let $A = \begin{bmatrix} a & b \\ c & d \end{bmatrix}$

32. Let

$$A = \begin{bmatrix} 3 & 1 \\ 0 & 2 \end{bmatrix} \quad \text{and} \quad B = \begin{bmatrix} 4 & -2 \\ 2 & 1 \end{bmatrix}$$

  a. Compute $(A + B)^2$.
  b. Compute $A^2 + 2AB + B^2$.
  c. From the results of parts (a) and (b), show that in general $(A + B)^2 \neq A^2 + 2AB + B^2$.

33. Let

$$A = \begin{bmatrix} 2 & 4 \\ 5 & -6 \end{bmatrix} \quad \text{and} \quad B = \begin{bmatrix} 4 & 8 \\ -7 & 3 \end{bmatrix}$$

  a. Find A^T and show that $(A^T)^T = A$.
  b. Show that $(A + B)^T = A^T + B^T$.
  c. Show that $(AB)^T = B^T A^T$.

34. Let

$$A = \begin{bmatrix} 1 & 3 \\ -2 & -1 \end{bmatrix} \quad \text{and} \quad B = \begin{bmatrix} 3 & -4 \\ 2 & -2 \end{bmatrix}$$

  a. Find A^T and show that $(A^T)^T = A$.
  b. Show that $(A + B)^T = A^T + B^T$.
  c. Show that $(AB)^T = B^T A^T$.

In Exercises 35–40, write the given system of linear equations in matrix form.

35. $2x - 3y = 7$
$3x - 4y = 8$

36. $2x \quad\quad = 7$
$3x - 2y = 12$

37. $2x - 3y + 4z = 6$
$2y - 3z = 7$
$x - y + 2z = 4$

38. $x - 2y + 3z = -1$
$3x + 4y - 2z = 1$
$2x - 3y + 7z = 6$

39. $-x_1 + x_2 + x_3 = 0$
$2x_1 - x_2 - x_3 = 2$
$-3x_1 + 2x_2 + 4x_3 = 4$

40. $3x_1 - 5x_2 + 4x_3 = 10$
$4x_1 + 2x_2 - 3x_3 = -12$
$-x_1 \quad\quad + x_3 = -2$

41. INVESTMENTS William's and Michael's stock holdings are given by the matrix

$$A = \begin{array}{c} \\ \text{William} \\ \text{Michael} \end{array} \begin{array}{cccc} \text{BAC} & \text{GM} & \text{IBM} & \text{TRW} \\ \begin{bmatrix} 200 & 300 & 100 & 200 \\ 100 & 200 & 400 & 0 \end{bmatrix} \end{array}$$

At the close of trading on a certain day, the prices (in dollars per share) of the stocks are given by the matrix

$$B = \begin{array}{c} \text{BAC} \\ \text{GM} \\ \text{IBM} \\ \text{TRW} \end{array} \begin{bmatrix} 54 \\ 48 \\ 98 \\ 82 \end{bmatrix}$$

a. Find AB.

b. Explain the meaning of the entries in the matrix AB.

42. **Foreign Exchange** Kaitlyn just returned to London from a European trip and wishes to exchange the various currencies she has accumulated for euros. She finds that she has 80 Austrian schillings, 26 French francs, 18 Dutch guilders, and 20 German marks. Suppose the foreign exchange rates are €0.0727 for one Austrian schilling, €0.1524 for one French franc, €0.4538 for one Dutch guilder, and €0.5113 for one German mark.

a. Write a row matrix A giving the number of units of each currency that Kaitlyn holds.

b. Write a column matrix B giving the exchange rates for the various currencies.

c. If Kaitlyn exchanges all her foreign currencies for euros, how much will she have?

43. **Real Estate** Bond Brothers, a real estate developer, builds houses in three states. The projected number of units of each model to be built in each state is given by the matrix

$$A = \begin{array}{c} \\ \text{N.Y.} \\ \text{Conn.} \\ \text{Mass.} \end{array} \begin{array}{c} \text{Model} \\ \begin{array}{cccc} \text{I} & \text{II} & \text{III} & \text{IV} \end{array} \\ \begin{bmatrix} 60 & 80 & 120 & 40 \\ 20 & 30 & 60 & 10 \\ 10 & 15 & 30 & 5 \end{bmatrix} \end{array}$$

The profits to be realized are $20,000, $22,000, $25,000, and $30,000, respectively, for each model I, II, III, and IV house sold.

a. Write a column matrix B representing the profit for each type of house.

b. Find the total profit Bond Brothers expects to earn in each state if all the houses are sold.

44. **Box-Office Receipts** The Cinema Center consists of four theaters: cinemas I, II, III, and IV. The admission price for one feature at the Center is $4 for children, $6 for students, and $8 for adults. The attendance for the Sunday matinee is given by the matrix

$$A = \begin{array}{c} \text{Cinema I} \\ \text{Cinema II} \\ \text{Cinema III} \\ \text{Cinema IV} \end{array} \begin{array}{c} \begin{array}{ccc} \text{Children} & \text{Students} & \text{Adults} \end{array} \\ \begin{bmatrix} 225 & 110 & 50 \\ 75 & 180 & 225 \\ 280 & 85 & 110 \\ 0 & 250 & 225 \end{bmatrix} \end{array}$$

Write a column vector B representing the admission prices. Then compute AB, the column vector showing the gross

receipts for each theater. Finally, find the total revenue collected at the Cinema Center for admission that Sunday afternoon.

45. **Politics: Voter Affiliation** Matrix A gives the percentage of eligible voters in the city of Newton, classified according to party affiliation and age group.

$$A = \begin{array}{c} \\ \text{Under 30} \\ \text{30 to 50} \\ \text{Over 50} \end{array} \begin{array}{c} \begin{array}{ccc} \text{Dem.} & \text{Rep.} & \text{Ind.} \end{array} \\ \begin{bmatrix} 0.50 & 0.30 & 0.20 \\ 0.45 & 0.40 & 0.15 \\ 0.40 & 0.50 & 0.10 \end{bmatrix} \end{array}$$

The population of eligible voters in the city by age group is given by the matrix B:

$$B = \begin{array}{c} \begin{array}{ccc} \text{Under 30} & \text{30 to 50} & \text{Over 50} \end{array} \\ \begin{bmatrix} 30{,}000 & 40{,}000 & 20{,}000 \end{bmatrix} \end{array}$$

Find a matrix giving the total number of eligible voters in the city who will vote Democratic, Republican, and Independent.

46. **401(k) Retirement Plans** Three network consultants, Alan, Maria, and Steven, each received a year-end bonus of $10,000, which they decided to invest in a 401(k) retirement plan sponsored by their employer. Under this plan, each employee is allowed to place their investments in three funds: an equity index fund (I), a growth fund (II), and a global equity fund (III). The allocations of the investments (in dollars) of the three employees at the beginning of the year are summarized in the matrix

$$A = \begin{array}{c} \\ \text{Alan} \\ \text{Maria} \\ \text{Steven} \end{array} \begin{array}{c} \begin{array}{ccc} \text{I} & \text{II} & \text{III} \end{array} \\ \begin{bmatrix} 4000 & 3000 & 3000 \\ 2000 & 5000 & 3000 \\ 2000 & 3000 & 5000 \end{bmatrix} \end{array}$$

The returns of the three funds after 1 yr are given in the matrix

$$B = \begin{array}{c} \text{I} \\ \text{II} \\ \text{III} \end{array} \begin{bmatrix} 0.18 \\ 0.24 \\ 0.12 \end{bmatrix}$$

Which employee realized the best returns on his or her investment for the year in question? The worst return?

47. **College Admissions** A university admissions committee anticipates an enrollment of 8000 students in its freshman class next year. To satisfy admission quotas, incoming students have been categorized according to their sex and place of residence. The number of students in each category is given by the matrix

	Male	Female
In-state	2700	3000
$A = $ Out-of-state	800	700
Foreign	500	300

By using data accumulated in previous years, the admissions committee has determined that these students will elect to enter the College of Letters and Science, the College of Fine Arts, the School of Business Administration, and the School of Engineering according to the percentages that appear in the following matrix:

	L. & S.	Fine Arts	Bus. Ad.	Eng.
$B = $ Male	0.25	0.20	0.30	0.25
Female	0.30	0.35	0.25	0.10

Find the matrix AB that shows the number of in-state, out-of-state, and foreign students expected to enter each discipline.

48. **PRODUCTION PLANNING** Refer to Example 6 in this section. Suppose Ace Novelty received an order from another amusement park for 1200 Pink Panthers, 1800 Giant Pandas, and 1400 Big Birds. The quantity of each type of stuffed animal to be produced at each plant is shown in the following production matrix:

	Panthers	Pandas	Birds
$P = $ L.A.	700	1000	800
Seattle	500	800	600

Each Panther requires 1.3 yd^2 of plush, 20 ft^3 of stuffing, and 12 pieces of trim. Assume the materials required to produce the other two stuffed animals and the unit cost for each type of material are as given in Example 6.
  a. How much of each type of material must be purchased for each plant?
  b. What is the total cost of materials that will be incurred at each plant?
  c. What is the total cost of materials incurred by Ace Novelty in filling the order?

49. **COMPUTING PHONE BILLS** Cindy regularly makes long-distance phone calls to three foreign cities—London, Tokyo, and Hong Kong. The matrices A and B give the lengths (in minutes) of her calls during peak and nonpeak hours, respectively, to each of these three cities during the month of June.

	London	Tokyo	Hong Kong
$A = [$	80	60	40	$]$

and

	London	Tokyo	Hong Kong
$B = [$	300	150	250	$]$

The costs for the calls (in dollars per minute) for the peak and nonpeak periods in the month in question are given, respectively, by the matrices

London		.34
$C = $ Tokyo		.42
Hong Kong		.48

and

London		.24
$D = $ Tokyo		.31
Hong Kong		.35

Compute the matrix $AC + BD$ and explain what it represents.

50. **PRODUCTION PLANNING** The total output of loudspeaker systems of the Acrosonic Company at their three production facilities for May and June is given by the matrices A and B, respectively, where

	Model A	Model B	Model C	Model D
Location I	320	280	460	280
$A = $ Location II	480	360	580	0
Location III	540	420	200	880

	Model A	Model B	Model C	Model D
Location I	210	180	330	180
$B = $ Location II	400	300	450	40
Location III	420	280	180	740

The unit production costs and selling prices for these loudspeakers are given by matrices C and D, respectively, where

Model A	120
$C = $ Model B	180
Model C	260
Model D	500

and

Model A	160
$D = $ Model B	250
Model C	350
Model D	700

Compute the following matrices and explain the meaning of the entries in each matrix.
  a. AC **b.** AD **c.** BC **d.** BD **e.** $(A + B)C$
  f. $(A + B)D$ **g.** $A(D - C)$
  h. $B(D - C)$ **i.** $(A + B)(D - C)$

51. **DIET PLANNING** A dietitian plans a meal around three foods. The number of units of vitamin A, vitamin C, and calcium in each ounce of these foods is represented by the matrix M, where

	Food I	Food II	Food III
Vitamin A	400	1200	800
$M = $ Vitamin C	110	570	340
Calcium	90	30	60

The matrices A and B represent the amount of each food (in ounces) consumed by a girl at two different meals, where

	Food I	Food II	Food III
$A = $	[7	1	6]

	Food I	Food II	Food III
$B = $	[9	3	2]

Calculate the following matrices and explain the meaning of the entries in each matrix.
  a. MA^T **b.** MB^T **c.** $M(A + B)^T$

52. PRODUCTION PLANNING Hartman Lumber Company has two branches in the city. The sales of four of its products for the last year (in thousands of dollars) are represented by the matrix

$$B = \begin{array}{c} \text{Branch I} \\ \text{Branch II} \end{array} \begin{array}{cccc} \text{A} & \text{B} & \text{C} & \text{D} \\ \end{array} \\ \begin{bmatrix} 5 & 2 & 8 & 10 \\ 3 & 4 & 6 & 8 \end{bmatrix}$$

For the present year, management has projected that the sales of the four products in branch I will be 10% over the corresponding sales for last year and the sales of the four products in branch II will be 15% over the corresponding sales for last year.

a. Show that the sales of the four products in the two branches for the current year are given by the matrix AB, where

$$A = \begin{bmatrix} 1.1 & 0 \\ 0 & 1.15 \end{bmatrix}$$

Compute AB.

b. Hartman has m branches nationwide, and the sales of n of its products (in thousands of dollars) last year are represented by the matrix

$$B = \begin{array}{c} \text{Branch 1} \\ \text{Branch 2} \\ \vdots \\ \text{Branch } m \end{array} \begin{bmatrix} a_{11} & a_{12} & a_{13} & \cdots & a_{1n} \\ a_{21} & a_{22} & a_{23} & \cdots & a_{2n} \\ \vdots & & & & \\ a_{m1} & a_{m2} & a_{m3} & \cdots & a_{mn} \end{bmatrix}$$

Also, management has projected that the sales of the n products in branch 1, branch 2, ..., branch m will be $r_1\%$, $r_2\%$, ..., $r_m\%$, respectively, over the corresponding sales for last year. Write the matrix A such that AB gives the sales of the n products in the m branches for the current year.

In Exercises 53–56, determine whether the statement is true or false. If it is true, explain why it is true. If it is false, give an example to show why it is false.

53. If A and B are matrices such that AB and BA are both defined, then A and B must be square matrices of the same order.

54. If A and B are matrices such that AB is defined and if c is a scalar, then $(cA)B = A(cB) = cAB$.

55. If A, B, and C are matrices and $A(B + C)$ is defined, then B must have the same size as C and the number of columns of A must be equal to the number of rows of B.

56. If A is a 2×4 matrix and B is a matrix such that ABA is defined, then the size of B must be 4×2.

2.5 Solutions to Self-Check Exercises

1. We compute

$$\begin{bmatrix} 1 & 3 & 0 \\ 2 & 4 & -1 \end{bmatrix} \begin{bmatrix} 3 & 1 & 4 \\ 2 & 0 & 3 \\ 1 & 2 & -1 \end{bmatrix}$$

$$= \begin{bmatrix} 1(3) + 3(2) + 0(1) & 1(1) + 3(0) + 0(2) & 1(4) + 3(3) + 0(-1) \\ 2(3) + 4(2) - 1(1) & 2(1) + 4(0) - 1(2) & 2(4) + 4(3) - 1(-1) \end{bmatrix}$$

$$= \begin{bmatrix} 9 & 1 & 13 \\ 13 & 0 & 21 \end{bmatrix}$$

2. Let

$$A = \begin{bmatrix} 0 & 1 & -2 \\ 2 & -1 & 3 \\ 1 & 0 & 4 \end{bmatrix} \quad X = \begin{bmatrix} x \\ y \\ z \end{bmatrix} \quad B = \begin{bmatrix} 1 \\ 0 \\ 7 \end{bmatrix}$$

Then the given system may be written as the matrix equation

$$AX = B$$

3. Write

$$B = \begin{bmatrix} 54 \\ 113 \\ 112 \\ 70 \end{bmatrix} \begin{array}{l} \text{AT\&T} \\ \text{TWX} \\ \text{IBM} \\ \text{GM} \end{array}$$

and compute

$$AB = \begin{array}{c} \text{Ash} \\ \text{Joan} \end{array} \begin{bmatrix} 2000 & 1000 & 500 & 5000 \\ 1000 & 2500 & 2000 & 0 \end{bmatrix} \begin{bmatrix} 54 \\ 113 \\ 112 \\ 70 \end{bmatrix}$$

$$= \begin{bmatrix} 627,000 \\ 560,500 \end{bmatrix} \begin{array}{l} \text{Ash} \\ \text{Joan} \end{array}$$

We conclude that Ash's stock holdings were worth $627,000 and Joan's stock holdings were worth $560,500 on June 1.

USING TECHNOLOGY

■ Matrix Multiplication

Graphing Utility

A graphing utility can be used to perform matrix multiplication.

EXAMPLE 1 Let

$$A = \begin{bmatrix} 1.2 & 3.1 & -1.4 \\ 2.7 & 4.2 & 3.4 \end{bmatrix}$$

$$B = \begin{bmatrix} 0.8 & 1.2 & 3.7 \\ 6.2 & -0.4 & 3.3 \end{bmatrix}$$

$$C = \begin{bmatrix} 1.2 & 2.1 & 1.3 \\ 4.2 & -1.2 & 0.6 \\ 1.4 & 3.2 & 0.7 \end{bmatrix}$$

Find (a) AC and (b) $(1.1A + 2.3B)C$.

Solution First, we enter the matrices A, B, and C into the calculator.

a. Using matrix operations, we enter the expression $A*C$. We obtain the matrix

$$\begin{bmatrix} 12.5 & -5.68 & 2.44 \\ 25.64 & 11.51 & 8.41 \end{bmatrix}$$

(You need to scroll the display on the screen to obtain the complete matrix.)

b. Using matrix operations, we enter the expression $(1.1A + 2.3B)C$. We obtain the matrix

$$\begin{bmatrix} 39.464 & 21.536 & 12.689 \\ 52.078 & 67.999 & 32.55 \end{bmatrix}$$

Excel

We use the **MMULT** function in Excel to perform matrix multiplication.

EXAMPLE 2 Let

$$A = \begin{bmatrix} 1.2 & 3.1 & -1.4 \\ 2.7 & 4.2 & 3.4 \end{bmatrix} \quad B = \begin{bmatrix} 0.8 & 1.2 & 3.7 \\ 6.2 & -0.4 & 3.3 \end{bmatrix} \quad C = \begin{bmatrix} 1.2 & 2.1 & 1.3 \\ 4.2 & -1.2 & 0.6 \\ 1.4 & 3.2 & 0.7 \end{bmatrix}$$

Find (a) AC and (b) $(1.1A + 2.3B)C$.

Note: Boldfaced words/characters in a box (for example, **Enter**) indicate that an action (click, select, or press) is required. Words/characters printed blue (for example, Chart sub-type:) indicate words/characters that appear on the screen. Words/characters printed in a typewriter font (for example, =(−2/3)*A2+2) indicate words/characters that need to be typed and entered.

Solution

a. *First, enter the matrices A, B, and C onto a spreadsheet* (Figure T1).

	A	B	C	D	E	F	G
1		A				B	
2	1.2	3.1	−1.4		0.8	1.2	3.7
3	2.7	4.2	3.4		6.2	−0.4	3.3
4							
5		C					
6	1.2	2.1	1.3				
7	4.2	−1.2	0.6				
8	1.4	3.2	0.7				

FIGURE T1
Spreadsheet showing the matrices *A, B,* and *C*

Second, compute AC. Highlight the cells that will contain the matrix product *AC*, which has order 2 × 3. Type =MMULT(, highlight the cells in matrix *A*, type **,**, highlight the cells in matrix *C*, type **)**, and press **Ctrl-Shift-Enter**. The matrix product *AC* shown in Figure T2 will appear on your spreadsheet.

	A	B	C
10		AC	
11	12.5	−5.68	2.44
12	25.64	11.51	8.41

FIGURE T2
The matrix product *AC*

b. *Compute* $(1.1A + 2.3B)C$. Highlight the cells that will contain the matrix product $(1.1A + 2.3B)C$. Next, type =MMULT(1.1*, highlight the cells in matrix *A*, type +2.3*, highlight the cells in matrix *B*, type **,**, highlight the cells in matrix *C*, type **)**, and then press **Ctrl-Shift-Enter**. The matrix product shown in Figure T3 will appear on your spreadsheet.

	A	B	C
13		(1.1A+2.3B)C	
14	39.464	21.536	12.689
15	52.078	67.999	32.55

FIGURE T3
The matrix product $(1.1A + 2.3B)C$

TECHNOLOGY EXERCISES

In Exercises 1–8, refer to the following matrices and perform the indicated operations. Round your answers to two decimal places.

$$A = \begin{bmatrix} 1.2 & 3.1 & -1.2 & 4.3 \\ 7.2 & 6.3 & 1.8 & -2.1 \\ 0.8 & 3.2 & -1.3 & 2.8 \end{bmatrix}$$

$$B = \begin{bmatrix} 0.7 & 0.3 & 1.2 & -0.8 \\ 1.2 & 1.7 & 3.5 & 4.2 \\ -3.3 & -1.2 & 4.2 & 3.2 \end{bmatrix}$$

$$C = \begin{bmatrix} 0.8 & 7.1 & 6.2 \\ 3.3 & -1.2 & 4.8 \\ 1.3 & 2.8 & -1.5 \\ 2.1 & 3.2 & -8.4 \end{bmatrix}$$

1. AC

2. CB

3. $(A + B)C$

4. $(2A + 3B)C$

5. $(2A - 3.1B)C$

6. $C(2.1A + 3.2B)$

7. $(4.1A + 2.7B)1.6C$

8. $2.5C(1.8A - 4.3B)$

In Exercises 9–12, refer to the following matrices and perform the indicated operations. Round your answers to two decimal places.

$$A = \begin{bmatrix} 2 & 5 & -4 & 2 & 8 \\ 6 & 7 & 2 & 9 & 6 \\ 4 & 5 & 4 & 4 & 4 \\ 9 & 6 & 8 & 3 & 2 \end{bmatrix} \quad B = \begin{bmatrix} 2 & 6 & 7 & 5 \\ 3 & 4 & 6 & 2 \\ -5 & 8 & 4 & 3 \\ 8 & 6 & 9 & 5 \\ 4 & 7 & 8 & 8 \end{bmatrix}$$

$$C = \begin{bmatrix} 6.2 & 7.3 & -4.0 & 7.1 & 9.3 \\ 4.8 & 6.5 & 8.4 & -6.3 & 8.4 \\ 5.4 & 3.2 & 6.3 & 9.1 & -2.8 \\ 8.2 & 7.3 & 6.5 & 4.1 & 9.8 \\ 10.3 & 6.8 & 4.8 & -9.1 & 20.4 \end{bmatrix}$$

$$D = \begin{bmatrix} 4.6 & 3.9 & 8.4 & 6.1 & 9.8 \\ 2.4 & -6.8 & 7.9 & 11.4 & 2.9 \\ 7.1 & 9.4 & 6.3 & 5.7 & 4.2 \\ 3.4 & 6.1 & 5.3 & 8.4 & 6.3 \\ 7.1 & -4.2 & 3.9 & -6.4 & 7.1 \end{bmatrix}$$

9. Find AB and BA.

10. Find CD and DC. Is $CD = DC$?

11. Find $AC + AD$.

12. Find
  a. AC
  b. AD
  c. $A(C + D)$
  d. Is $A(C + D) = AC + AD$?

2.6 The Inverse of a Square Matrix

The Inverse of a Square Matrix

In this section, we discuss a procedure for finding the inverse of a matrix and show how the inverse can be used to help us solve a system of linear equations. The inverse of a matrix also plays a central role in the Leontief input–output model, which we will discuss in Section 2.7.

Recall that if a is a nonzero real number, then there exists a unique real number a^{-1} $\left(\text{that is, } \frac{1}{a}\right)$ such that

$$a^{-1}a = \left(\frac{1}{a}\right)(a) = 1$$

The use of the (multiplicative) inverse of a real number enables us to solve algebraic equations of the form

$$ax = b \qquad\qquad \textbf{(12)}$$

For if $a \neq 0$, then $a^{-1} = \frac{1}{a}$. Multiplying both sides of (12) by a^{-1}, we have

$$a^{-1}(ax) = a^{-1}b$$
$$\left(\frac{1}{a}\right)(ax) = \frac{1}{a}(b)$$
$$x = \frac{b}{a}$$

For example, since the inverse of 2 is $2^{-1} = \frac{1}{2}$, we can solve the equation

$$2x = 5$$

by multiplying both sides of the equation by $2^{-1} = \frac{1}{2}$, giving

$$2^{-1}(2x) = 2^{-1} \cdot 5$$
$$x = \frac{5}{2}$$

We can use a similar procedure to solve the matrix equation

$$AX = B$$

where A, X, and B are matrices of the proper sizes. To do this we need the matrix equivalent of the inverse of a real number. Such a matrix, whenever it exists, is called the **inverse of a matrix**.

> **Inverse of a Matrix**
> Let A be a square matrix of size n. A square matrix A^{-1} of size n such that
> $$A^{-1}A = AA^{-1} = I_n$$
> is called the inverse of A.

Let's show that the matrix

$$A = \begin{bmatrix} 1 & 2 \\ 3 & 4 \end{bmatrix}$$

has the matrix

$$A^{-1} = \begin{bmatrix} -2 & 1 \\ \frac{3}{2} & -\frac{1}{2} \end{bmatrix}$$

as its inverse.

EXPLORE & DISCUSS

In defining the inverse of a matrix A, why is it necessary to require that A be a square matrix?

Since

$$AA^{-1} = \begin{bmatrix} 1 & 2 \\ 3 & 4 \end{bmatrix}\begin{bmatrix} -2 & 1 \\ \frac{3}{2} & -\frac{1}{2} \end{bmatrix} = \begin{bmatrix} 1 & 0 \\ 0 & 1 \end{bmatrix} = I$$

$$A^{-1}A = \begin{bmatrix} -2 & 1 \\ \frac{3}{2} & -\frac{1}{2} \end{bmatrix}\begin{bmatrix} 1 & 2 \\ 3 & 4 \end{bmatrix} = \begin{bmatrix} 1 & 0 \\ 0 & 1 \end{bmatrix} = I$$

we see that A^{-1} is the inverse of A, as asserted.

Not every square matrix has an inverse. A square matrix that has an inverse is said to be **nonsingular**. A matrix that does not have an inverse is said to be **singular**. An example of a singular matrix is given by

$$B = \begin{bmatrix} 0 & 1 \\ 0 & 0 \end{bmatrix}$$

If B had an inverse given by

$$B^{-1} = \begin{bmatrix} a & b \\ c & d \end{bmatrix}$$

where a, b, c, and d are some appropriate numbers, then by the definition of an inverse we would have $BB^{-1} = I$; that is,

$$\begin{bmatrix} 0 & 1 \\ 0 & 0 \end{bmatrix}\begin{bmatrix} a & b \\ c & d \end{bmatrix} = \begin{bmatrix} 1 & 0 \\ 0 & 1 \end{bmatrix}$$

$$\begin{bmatrix} c & d \\ 0 & 0 \end{bmatrix} = \begin{bmatrix} 1 & 0 \\ 0 & 1 \end{bmatrix}$$

which implies that $0 = 1$—an impossibility! This contradiction shows that B does not have an inverse.

A Method for Finding the Inverse of a Square Matrix

The methods of Section 2.5 can be used to find the inverse of a nonsingular matrix. To discover such an algorithm, let's find the inverse of the matrix

$$A = \begin{bmatrix} 1 & 2 \\ -1 & 3 \end{bmatrix}$$

Suppose A^{-1} exists and is given by

$$A^{-1} = \begin{bmatrix} a & b \\ c & d \end{bmatrix}$$

where a, b, c, and d are to be determined. By the definition of an inverse, we have $AA^{-1} = I$; that is,

$$\begin{bmatrix} 1 & 2 \\ -1 & 3 \end{bmatrix}\begin{bmatrix} a & b \\ c & d \end{bmatrix} = \begin{bmatrix} 1 & 0 \\ 0 & 1 \end{bmatrix}$$

which simplifies to

$$\begin{bmatrix} a + 2c & b + 2d \\ -a + 3c & -b + 3d \end{bmatrix} = \begin{bmatrix} 1 & 0 \\ 0 & 1 \end{bmatrix}$$

But this matrix equation is equivalent to the two systems of linear equations

$$\left.\begin{array}{r} a + 2c = 1 \\ -a + 3c = 0 \end{array}\right\} \quad \text{and} \quad \left.\begin{array}{r} b + 2d = 0 \\ -b + 3d = 1 \end{array}\right\}$$

with augmented matrices given by

$$\begin{bmatrix} 1 & 2 & | & 1 \\ -1 & 3 & | & 0 \end{bmatrix} \quad \text{and} \quad \begin{bmatrix} 1 & 2 & | & 0 \\ -1 & 3 & | & 1 \end{bmatrix}$$

Note that the matrices of coefficients of the two systems are identical. This suggests that we solve the two systems of simultaneous linear equations by writing the following augmented matrix, which we obtain by joining the coefficient matrix and the two columns of constants:

$$\begin{bmatrix} 1 & 2 & | & 1 & 0 \\ -1 & 3 & | & 0 & 1 \end{bmatrix}$$

Using the Gauss–Jordan elimination method, we obtain the following sequence of equivalent matrices:

$$\begin{bmatrix} 1 & 2 & | & 1 & 0 \\ -1 & 3 & | & 0 & 1 \end{bmatrix} \xrightarrow{R_2 + R_1} \begin{bmatrix} 1 & 2 & | & 1 & 0 \\ 0 & 5 & | & 1 & 1 \end{bmatrix} \xrightarrow{\frac{1}{5}R_2}$$

$$\begin{bmatrix} 1 & 2 & | & 1 & 0 \\ 0 & 1 & | & \frac{1}{5} & \frac{1}{5} \end{bmatrix} \xrightarrow{R_1 - 2R_2} \begin{bmatrix} 1 & 0 & | & \frac{3}{5} & -\frac{2}{5} \\ 0 & 1 & | & \frac{1}{5} & \frac{1}{5} \end{bmatrix}$$

Thus, $a = \frac{3}{5}$, $c = \frac{1}{5}$, $b = -\frac{2}{5}$, and $d = \frac{1}{5}$, giving

$$A^{-1} = \begin{bmatrix} \frac{3}{5} & -\frac{2}{5} \\ \frac{1}{5} & \frac{1}{5} \end{bmatrix}$$

The following computations verify that A^{-1} is indeed the inverse of A:

$$\begin{bmatrix} 1 & 2 \\ -1 & 3 \end{bmatrix}\begin{bmatrix} \frac{3}{5} & -\frac{2}{5} \\ \frac{1}{5} & \frac{1}{5} \end{bmatrix} = \begin{bmatrix} 1 & 0 \\ 0 & 1 \end{bmatrix} = \begin{bmatrix} \frac{3}{5} & -\frac{2}{5} \\ \frac{1}{5} & \frac{1}{5} \end{bmatrix}\begin{bmatrix} 1 & 2 \\ -1 & 3 \end{bmatrix}$$

The preceding example suggests a general algorithm for computing the inverse of a square matrix of size n when it exists.

Finding the Inverse of a Matrix
Given the $n \times n$ matrix A:

1. Adjoin the $n \times n$ identity matrix I to obtain the augmented matrix

$$[A \mid I]$$

2. Use a sequence of row operations to reduce $[A \mid I]$ to the form

$$[I \mid B]$$

if possible.

Then the matrix B is the inverse of A.

Note   Although matrix multiplication is not generally commutative, it is possible to prove that if A has an inverse and $AB = I$, then $BA = I$ also. Hence, to verify that B is the inverse of A, it suffices to show that $AB = I$.   ■

EXAMPLE 1   Find the inverse of the matrix

$$A = \begin{bmatrix} 2 & 1 & 1 \\ 3 & 2 & 1 \\ 2 & 1 & 2 \end{bmatrix}$$

Solution   We form the augmented matrix

$$\begin{bmatrix} 2 & 1 & 1 & \big| & 1 & 0 & 0 \\ 3 & 2 & 1 & \big| & 0 & 1 & 0 \\ 2 & 1 & 2 & \big| & 0 & 0 & 1 \end{bmatrix}$$

and use the Gauss–Jordan elimination method to reduce it to the form $[I \mid B]$:

$$\begin{bmatrix} 2 & 1 & 1 & \big| & 1 & 0 & 0 \\ 3 & 2 & 1 & \big| & 0 & 1 & 0 \\ 2 & 1 & 2 & \big| & 0 & 0 & 1 \end{bmatrix} \xrightarrow{R_1 - R_2} \begin{bmatrix} -1 & -1 & 0 & \big| & 1 & -1 & 0 \\ 3 & 2 & 1 & \big| & 0 & 1 & 0 \\ 2 & 1 & 2 & \big| & 0 & 0 & 1 \end{bmatrix}$$

$$\xrightarrow[\substack{R_2 + 3R_1 \\ R_3 + 2R_1}]{-R_1} \begin{bmatrix} 1 & 1 & 0 & \big| & -1 & 1 & 0 \\ 0 & -1 & 1 & \big| & 3 & -2 & 0 \\ 0 & -1 & 2 & \big| & 2 & -2 & 1 \end{bmatrix}$$

$$\xrightarrow[\substack{-R_2 \\ R_3 - R_2}]{R_1 + R_2} \begin{bmatrix} 1 & 0 & 1 & \big| & 2 & -1 & 0 \\ 0 & 1 & -1 & \big| & -3 & 2 & 0 \\ 0 & 0 & 1 & \big| & -1 & 0 & 1 \end{bmatrix}$$

$$\xrightarrow[R_2 + R_3]{R_1 - R_3} \begin{bmatrix} 1 & 0 & 0 & \big| & 3 & -1 & -1 \\ 0 & 1 & 0 & \big| & -4 & 2 & 1 \\ 0 & 0 & 1 & \big| & -1 & 0 & 1 \end{bmatrix}$$

The inverse of A is the matrix

$$A^{-1} = \begin{bmatrix} 3 & -1 & -1 \\ -4 & 2 & 1 \\ -1 & 0 & 1 \end{bmatrix}$$

We leave it to you to verify these results.   ■

Example 2 illustrates what happens to the reduction process when a matrix A does *not* have an inverse.

EXAMPLE 2   Find the inverse of the matrix

$$A = \begin{bmatrix} 1 & 2 & 3 \\ 2 & 1 & 2 \\ 3 & 3 & 5 \end{bmatrix}$$

Solution We form the augmented matrix

$$\begin{bmatrix} 1 & 2 & 3 & | & 1 & 0 & 0 \\ 2 & 1 & 2 & | & 0 & 1 & 0 \\ 3 & 3 & 5 & | & 0 & 0 & 1 \end{bmatrix}$$

and use the Gauss–Jordan elimination method:

$$\begin{bmatrix} 1 & 2 & 3 & | & 1 & 0 & 0 \\ 2 & 1 & 2 & | & 0 & 1 & 0 \\ 3 & 3 & 5 & | & 0 & 0 & 1 \end{bmatrix} \xrightarrow[R_3 - 3R_1]{R_2 - 2R_1} \begin{bmatrix} 1 & 2 & 3 & | & 1 & 0 & 0 \\ 0 & -3 & -4 & | & -2 & 1 & 0 \\ 0 & -3 & -4 & | & -3 & 0 & 1 \end{bmatrix}$$

$$\xrightarrow[R_3 - R_2]{-R_2} \begin{bmatrix} 1 & 2 & 3 & | & 1 & 0 & 0 \\ 0 & 3 & 4 & | & 2 & -1 & 0 \\ 0 & 0 & 0 & | & -1 & -1 & 1 \end{bmatrix}$$

Since the entries in the last row of the 3×3 submatrix that comprises the left-hand side of the augmented matrix just obtained are all equal to zero, the latter cannot be reduced to the form $[I \mid B]$. Accordingly, we draw the conclusion that A is singular—that is, does not have an inverse. ■

EXPLORE & DISCUSS

Explain in terms of solutions to systems of linear equations why the final augmented matrix in Example 2 implies that A has no inverse. *Hint*: See the discussion on pages 138–139.

More generally, we have the following criterion for determining when the inverse of a matrix does not exist.

Matrices That Have No Inverses
If there is a row to the left of the vertical line in the augmented matrix containing all zeros, then the matrix does not have an inverse.

A Formula for the Inverse of a 2 × 2 Matrix

Before turning to some applications, we show an alternative method that employs a formula for finding the inverse of a 2×2 matrix. This method will prove useful in many situations; we will see an application in Example 5. The derivation of this formula is left as an exercise (Exercise 48).

Formula for the Inverse of a 2 × 2 Matrix
Let

$$A = \begin{bmatrix} a & b \\ c & d \end{bmatrix}$$

Suppose $D = ad - bc$ is not equal to zero. Then A^{-1} exists and is given by

$$A^{-1} = \frac{1}{D} \begin{bmatrix} d & -b \\ -c & a \end{bmatrix} \tag{13}$$

Note   As an aid to memorizing the formula, note that D is the product of the elements along the main diagonal minus the product of the elements along the other diagonal:

$$\begin{bmatrix} a & b \\ c & d \end{bmatrix} \qquad D = ad - bc$$

Main diagonal

EXPLORE & DISCUSS

Suppose A is a square matrix with the property that one of its rows is a nonzero constant multiple of another row. What can you say about the existence or nonexistence of A^{-1}? Explain your answer.

Next, the matrix

$$\begin{bmatrix} d & -b \\ -c & a \end{bmatrix}$$

is obtained by interchanging a and d and reversing the signs of b and c. Finally, A^{-1} is obtained by dividing this matrix by D. ∎

EXAMPLE 3   Find the inverse of

$$A = \begin{bmatrix} 1 & 2 \\ 3 & 4 \end{bmatrix}$$

Solution   We first compute $D = (1)(4) - (2)(3) = 4 - 6 = -2$. Next, we write the matrix

$$\begin{bmatrix} 4 & -2 \\ -3 & 1 \end{bmatrix}$$

Finally, dividing this matrix by D, we obtain

$$A^{-1} = \frac{1}{-2} \begin{bmatrix} 4 & -2 \\ -3 & 1 \end{bmatrix} = \begin{bmatrix} -2 & 1 \\ \frac{3}{2} & -\frac{1}{2} \end{bmatrix}$$

∎

Solving Systems of Equations with Inverses

We now show how the inverse of a matrix may be used to solve certain systems of linear equations in which the number of equations in the system is equal to the number of variables. For simplicity, let's illustrate the process for a system of three linear equations in three variables:

$$\begin{aligned} a_{11}x_1 + a_{12}x_2 + a_{13}x_3 &= c_1 \\ a_{21}x_1 + a_{22}x_2 + a_{23}x_3 &= c_2 \\ a_{31}x_1 + a_{32}x_2 + a_{33}x_3 &= c_3 \end{aligned} \qquad (14)$$

Let's write

$$A = \begin{bmatrix} a_{11} & a_{12} & a_{13} \\ a_{21} & a_{22} & a_{23} \\ a_{31} & a_{32} & a_{33} \end{bmatrix} \qquad X = \begin{bmatrix} x_1 \\ x_2 \\ x_3 \end{bmatrix} \qquad B = \begin{bmatrix} b_1 \\ b_2 \\ b_3 \end{bmatrix}$$

You should verify that System (14) of linear equations may be written in the form of the matrix equation

$$AX = B \qquad (15)$$

If A is nonsingular, then the method of this section may be used to compute A^{-1}. Next, multiplying both sides of Equation (15) by A^{-1} (on the left), we obtain

$$A^{-1}AX = A^{-1}B \quad \text{or} \quad IX = A^{-1}B \quad \text{or} \quad X = A^{-1}B$$

the desired solution to the problem.

In the case of a system of n equations with n unknowns, we have the following more general result.

Using Inverses to Solve Systems of Equations
If $AX = B$ is a linear system of n equations in n unknowns and if A^{-1} exists, then

$$X = A^{-1}B$$

is the unique solution of the system.

The use of inverses to solve systems of equations is particularly advantageous when we are required to solve more than one system of equations, $AX = B$, involving the same coefficient matrix, A, and different matrices of constants, B. As you will see in Examples 4 and 5, we need to compute A^{-1} just once in each case.

EXAMPLE 4 Solve the following systems of linear equations:

a. $\begin{aligned} 2x + y + z &= 1 \\ 3x + 2y + z &= 2 \\ 2x + y + 2z &= -1 \end{aligned}$
   b. $\begin{aligned} 2x + y + z &= 2 \\ 3x + 2y + z &= -3 \\ 2x + y + 2z &= 1 \end{aligned}$

Solution We may write the given systems of equations in the form
$$AX = B \quad \text{and} \quad AX = C$$
respectively, where

$$A = \begin{bmatrix} 2 & 1 & 1 \\ 3 & 2 & 1 \\ 2 & 1 & 2 \end{bmatrix} \quad X = \begin{bmatrix} x \\ y \\ z \end{bmatrix} \quad B = \begin{bmatrix} 1 \\ 2 \\ -1 \end{bmatrix} \quad C = \begin{bmatrix} 2 \\ -3 \\ 1 \end{bmatrix}$$

The inverse of the matrix A,

$$A^{-1} = \begin{bmatrix} 3 & -1 & -1 \\ -4 & 2 & 1 \\ -1 & 0 & 1 \end{bmatrix}$$

was found in Example 1. Using this result, we find that the solution of the first system (a) is

$$X = A^{-1}B = \begin{bmatrix} 3 & -1 & -1 \\ -4 & 2 & 1 \\ -1 & 0 & 1 \end{bmatrix}\begin{bmatrix} 1 \\ 2 \\ -1 \end{bmatrix}$$

$$= \begin{bmatrix} (3)(1) + (-1)(2) + (-1)(-1) \\ (-4)(1) + (2)(2) + (1)(-1) \\ (-1)(1) + (0)(2) + (1)(-1) \end{bmatrix} = \begin{bmatrix} 2 \\ -1 \\ -2 \end{bmatrix}$$

or $x = 2$, $y = -1$, and $z = -2$.

The solution of the second system (b) is

$$X = A^{-1}C = \begin{bmatrix} 3 & -1 & -1 \\ -4 & 2 & 1 \\ -1 & 0 & 1 \end{bmatrix} \begin{bmatrix} 2 \\ -3 \\ 1 \end{bmatrix} = \begin{bmatrix} 8 \\ -13 \\ -1 \end{bmatrix}$$

or $x = 8$, $y = -13$, and $z = -1$.   ∎

APPLIED EXAMPLE 5   Capital Expenditure Planning   The management of Checkers Rent-A-Car plans to expand its fleet of rental cars for the next quarter by purchasing compact and full-size cars. The average cost of a compact car is $10,000, and the average cost of a full-size car is $24,000.

a. If a total of 800 cars is to be purchased with a budget of $12 million, how many cars of each size will be acquired?
b. If the predicted demand calls for a total purchase of 1000 cars with a budget of $14 million, how many cars of each type will be acquired?

Solution   Let x and y denote the number of compact and full-size cars to be purchased. Furthermore, let n denote the total number of cars to be acquired and b the amount of money budgeted for the purchase of these cars. Then,

$$\begin{aligned} x + \quad\quad y &= n \\ 10{,}000x + 24{,}000y &= b \end{aligned}$$

This system of two equations in two variables may be written in the matrix form

$$AX = B$$

where

$$A = \begin{bmatrix} 1 & 1 \\ 10{,}000 & 24{,}000 \end{bmatrix} \quad X = \begin{bmatrix} x \\ y \end{bmatrix} \quad B = \begin{bmatrix} n \\ b \end{bmatrix}$$

Therefore,

$$X = A^{-1}B$$

Since A is a 2×2 matrix, its inverse may be found by using Formula (13). We find $D = (1)(24{,}000) - (1)(10{,}000) = 14{,}000$, so

$$A^{-1} = \frac{1}{14{,}000} \begin{bmatrix} 24{,}000 & -1 \\ -10{,}000 & 1 \end{bmatrix} = \begin{bmatrix} \frac{24{,}000}{14{,}000} & -\frac{1}{14{,}000} \\ -\frac{10{,}000}{14{,}000} & \frac{1}{14{,}000} \end{bmatrix}$$

Thus,

$$X = \begin{bmatrix} \frac{12}{7} & -\frac{1}{14{,}000} \\ -\frac{5}{7} & \frac{1}{14{,}000} \end{bmatrix} \begin{bmatrix} n \\ b \end{bmatrix}$$

a. Here, $n = 800$ and $b = 12{,}000{,}000$, so

$$X = A^{-1}B = \begin{bmatrix} \frac{12}{7} & -\frac{1}{14{,}000} \\ -\frac{5}{7} & \frac{1}{14{,}000} \end{bmatrix} \begin{bmatrix} 800 \\ 12{,}000{,}000 \end{bmatrix} = \begin{bmatrix} 514.3 \\ 285.7 \end{bmatrix}$$

Therefore, 514 compact cars and 286 full-size cars will be acquired in this case.

b. Here, $n = 1000$ and $b = 14{,}000{,}000$, so

$$X = A^{-1}B = \begin{bmatrix} \frac{12}{7} & -\frac{1}{14{,}000} \\ -\frac{5}{7} & \frac{1}{14{,}000} \end{bmatrix} \begin{bmatrix} 1000 \\ 14{,}000{,}000 \end{bmatrix} = \begin{bmatrix} 714.3 \\ 285.7 \end{bmatrix}$$

Therefore, 714 compact cars and 286 full-size cars will be purchased in this case. ∎

2.6 Self-Check Exercises

1. Find the inverse of the matrix

$$A = \begin{bmatrix} 2 & 1 & -1 \\ 1 & 1 & -1 \\ -1 & -2 & 3 \end{bmatrix}$$

if it exists.

2. Solve the system of linear equations

$$\begin{aligned} 2x + y - z &= b_1 \\ x + y - z &= b_2 \\ -x - 2y + 3z &= b_3 \end{aligned}$$

where (a) $b_1 = 5$, $b_2 = 4$, $b_3 = -8$ and (b) $b_1 = 2$, $b_2 = 0$, $b_3 = 5$, by finding the inverse of the coefficient matrix.

3. Grand Canyon Tours offers air and ground scenic tours of the Grand Canyon. Tickets for the $7\frac{1}{2}$-hr tour cost $169 for an adult and $129 for a child, and each tour group is limited to 19 people. On three recent fully booked tours, total receipts were $2931 for the first tour, $3011 for the second tour, and $2771 for the third tour. Determine how many adults and how many children were in each tour.

Solutions to Self-Check Exercises 2.6 can be found on page 148.

2.6 Concept Questions

1. What is the inverse of a matrix A?

2. Explain how you would find the inverse of a nonsingular matrix.

3. Give the formula for the inverse of the 2×2 matrix

$$A = \begin{bmatrix} a & b \\ c & d \end{bmatrix}$$

4. Explain how the inverse of a matrix can be used to solve a system of n linear equations in n unknowns. Can the method work for a system of m linear equations in n unknowns with $m \neq n$? Explain.

2.6 Exercises

In Exercises 1–4, show that the matrices are inverses of each other by showing that their product is the identity matrix I.

1. $\begin{bmatrix} 1 & -3 \\ 1 & -2 \end{bmatrix}$ and $\begin{bmatrix} -2 & 3 \\ -1 & 1 \end{bmatrix}$

2. $\begin{bmatrix} 4 & 5 \\ 2 & 3 \end{bmatrix}$ and $\begin{bmatrix} \frac{3}{2} & -\frac{5}{2} \\ -1 & 2 \end{bmatrix}$

3. $\begin{bmatrix} 3 & 2 & 3 \\ 2 & 2 & 1 \\ 2 & 1 & 1 \end{bmatrix}$ and $\begin{bmatrix} -\frac{1}{3} & -\frac{1}{3} & \frac{4}{3} \\ 0 & 1 & -1 \\ \frac{2}{3} & -\frac{1}{3} & -\frac{2}{3} \end{bmatrix}$

4. $\begin{bmatrix} 2 & 4 & -2 \\ -4 & -6 & 1 \\ 3 & 5 & -1 \end{bmatrix}$ and $\begin{bmatrix} \frac{1}{2} & -3 & -4 \\ -\frac{1}{2} & 2 & 3 \\ -1 & 1 & 2 \end{bmatrix}$

In Exercises 5–16, find the inverse of the matrix, if it exists. Verify your answer.

5. $\begin{bmatrix} 2 & 5 \\ 1 & 3 \end{bmatrix}$ **6.** $\begin{bmatrix} 2 & 3 \\ 3 & 5 \end{bmatrix}$

7. $\begin{bmatrix} 3 & -3 \\ -2 & 2 \end{bmatrix}$ **8.** $\begin{bmatrix} 4 & 2 \\ 6 & 3 \end{bmatrix}$

9. $\begin{bmatrix} 2 & -3 & -4 \\ 0 & 0 & -1 \\ 1 & -2 & 1 \end{bmatrix}$ **10.** $\begin{bmatrix} 1 & -1 & 3 \\ 2 & 1 & 2 \\ -2 & -2 & 1 \end{bmatrix}$

11. $\begin{bmatrix} 4 & 2 & 2 \\ -1 & -3 & 4 \\ 3 & -1 & 6 \end{bmatrix}$ **12.** $\begin{bmatrix} 1 & 2 & 0 \\ -3 & 4 & -2 \\ -5 & 0 & -2 \end{bmatrix}$

13. $\begin{bmatrix} 1 & 4 & -1 \\ 2 & 3 & -2 \\ -1 & 2 & 3 \end{bmatrix}$ **14.** $\begin{bmatrix} 3 & -2 & 7 \\ -2 & 1 & 4 \\ 6 & -5 & 8 \end{bmatrix}$

15. $\begin{bmatrix} 1 & 1 & -1 & 1 \\ 2 & 1 & 1 & 0 \\ 2 & 1 & 0 & 1 \\ 2 & -1 & -1 & 3 \end{bmatrix}$ **16.** $\begin{bmatrix} 1 & 1 & 2 & 3 \\ 2 & 3 & 0 & -1 \\ 0 & 2 & -1 & 1 \\ 1 & 2 & 1 & 1 \end{bmatrix}$

In Exercises 17–24, (a) write a matrix equation that is equivalent to the system of linear equations and (b) solve the system using the inverses found in Exercises 5–16.

17. $2x + 5y = 3$
$x + 3y = 2$
(See Exercise 5.)

18. $2x + 3y = 5$
$3x + 5y = 8$
(See Exercise 6.)

19. $2x - 3y - 4z = 4$
$-z = 3$
$x - 2y + z = -8$
(See Exercise 9.)

20. $x_1 - x_2 + 3x_3 = 2$
$2x_1 + x_2 + 2x_3 = 2$
$-2x_1 - 2x_2 + x_3 = 3$
(See Exercise 10.)

21. $x + 4y - z = 3$
$2x + 3y - 2z = 1$
$-x + 2y + 3z = 7$
(See Exercise 13.)

22. $3x_1 - 2x_2 + 7x_3 = 6$
$-2x_1 + x_2 + 4x_3 = 4$
$6x_1 - 5x_2 + 8x_3 = 4$
(See Exercise 14.)

23. $x_1 + x_2 - x_3 + x_4 = 6$
$2x_1 + x_2 + x_3 = 4$
$2x_1 + x_2 + x_4 = 7$
$2x_1 - x_2 - x_3 + 3x_4 = 9$
(See Exercise 15.)

24. $x_1 + x_2 + 2x_3 + 3x_4 = 4$
$2x_1 + 3x_2 - x_4 = 11$
$2x_2 - x_3 + x_4 = 7$
$x_1 + 2x_2 + x_3 + x_4 = 6$
(See Exercise 16.)

In Exercises 25–32, (a) write each system of equations as a matrix equation and (b) solve the system of equations by using the inverse of the coefficient matrix.

25.
$$x + 2y = b_1$$
$$2x - y = b_2$$
where (i) $b_1 = 14$, $b_2 = 5$
and (ii) $b_1 = 4$, $b_2 = -1$

26.
$$3x - 2y = b_1$$
$$4x + 3y = b_2$$
where (i) $b_1 = -6$, $b_2 = 10$
and (ii) $b_1 = 3$, $b_2 = -2$

27.
$$x + 2y + z = b_1$$
$$x + y + z = b_2$$
$$3x + y + z = b_3$$
where (i) $b_1 = 7$, $b_2 = 4$, $b_3 = 2$
and (ii) $b_1 = 5$, $b_2 = -3$, $b_3 = -1$

28.
$$x_1 + x_2 + x_3 = b_1$$
$$x_1 - x_2 + x_3 = b_2$$
$$x_1 - 2x_2 - x_3 = b_3$$
where (i) $b_1 = 5$, $b_2 = -3$, $b_3 = -1$
and (ii) $b_1 = 1$, $b_2 = 4$, $b_3 = -2$

29.
$$3x + 2y - z = b_1$$
$$2x - 3y + z = b_2$$
$$x - y - z = b_3$$
where (i) $b_1 = 2$, $b_2 = -2$, $b_3 = 4$
and (ii) $b_1 = 8$, $b_2 = -3$, $b_3 = 6$

30.
$$2x_1 + x_2 + x_3 = b_1$$
$$x_1 - 3x_2 + 4x_3 = b_2$$
$$-x_1 + x_3 = b_3$$
where (i) $b_1 = 1$, $b_2 = 4$, $b_3 = -3$
and (ii) $b_1 = 2$, $b_2 = -5$, $b_3 = 0$

31.
$$x_1 + x_2 + x_3 + x_4 = b_1$$
$$x_1 - x_2 - x_3 + x_4 = b_2$$
$$x_2 + 2x_3 + 2x_4 = b_3$$
$$x_1 + 2x_2 + x_3 - 2x_4 = b_4$$
where (i) $b_1 = 1$, $b_2 = -1$, $b_3 = 4$, $b_4 = 0$
and (ii) $b_1 = 2$, $b_2 = 8$, $b_3 = 4$, $b_4 = -1$

32.
$$x_1 + x_2 + 2x_3 + x_4 = b_1$$
$$4x_1 + 5x_2 + 9x_3 + x_4 = b_2$$
$$3x_1 + 4x_2 + 7x_3 + x_4 = b_3$$
$$2x_1 + 3x_2 + 4x_3 + 2x_4 = b_4$$
where (i) $b_1 = 3$, $b_2 = 6$, $b_3 = 5$, $b_4 = 7$
and (ii) $b_1 = 1$, $b_2 = -1$, $b_3 = 0$, $b_4 = -4$

33. Let

$$A = \begin{bmatrix} 2 & 3 \\ -4 & -5 \end{bmatrix}$$

a. Find A^{-1}.   **b.** Show that $(A^{-1})^{-1} = A$.

34. Let

$$A = \begin{bmatrix} 6 & -4 \\ -4 & 3 \end{bmatrix} \quad \text{and} \quad B = \begin{bmatrix} 3 & -5 \\ 4 & -7 \end{bmatrix}$$

a. Find AB, A^{-1}, and B^{-1}.
b. Show that $(AB)^{-1} = B^{-1}A^{-1}$.

35. Let

$$A = \begin{bmatrix} 2 & -5 \\ 1 & -3 \end{bmatrix} \quad B = \begin{bmatrix} 4 & 3 \\ 1 & 1 \end{bmatrix} \quad C = \begin{bmatrix} 2 & 3 \\ -2 & 1 \end{bmatrix}$$

a. Find ABC, A^{-1}, B^{-1}, and C^{-1}.
b. Show that $(ABC)^{-1} = C^{-1}B^{-1}A^{-1}$.

36. TICKET REVENUES   Rainbow Harbor Cruises charges $16/adult and $8/child for a round-trip ticket. The records show that, on a certain weekend, 1000 people took the cruise on Saturday and 800 people took the cruise on Sunday. The total receipts for Saturday were $12,800 and the total receipts for Sunday were $9,600. Determine how many adults and children took the cruise on Saturday and on Sunday.

37. PRICING   BelAir Publishing publishes a deluxe leather edition and a standard edition of its Daily Organizer. The company's marketing department estimates that x copies of the deluxe edition and y copies of the standard edition will be demanded per month when the unit prices are p dollars and q dollars, respectively, where x, y, p, and q are related by the following system of linear equations:

$$5x + y = 1000(70 - p)$$
$$x + 3y = 1000(40 - q)$$

Find the monthly demand for the deluxe edition and the standard edition when the unit prices are set according to the following schedules:
a. $p = 50$ and $q = 25$ **b.** $p = 45$ and $q = 25$
c. $p = 45$ and $q = 20$

38. NUTRITION/DIET PLANNING   Bob, a nutritionist who works for the University Medical Center, has been asked to prepare special diets for two patients, Susan and Tom. Bob has decided that Susan's meals should contain at least 400 mg of calcium, 20 mg of iron, and 50 mg of vitamin C, whereas Tom's meals should contain at least 350 mg of calcium, 15 mg of iron, and 40 mg of vitamin C. Bob has also decided that the meals are to be prepared from three basic foods: food A, food B, and food C. The special nutritional contents of these foods are summarized in the accompanying table. Find how many ounces of each type of food should be used in a meal so that the minimum requirements of calcium, iron, and vitamin C are met for each patient's meals.

| | Contents (mg/oz) | | |
	Calcium	Iron	Vitamin C
Food A	30	1	2
Food B	25	1	5
Food C	20	2	4

39. AGRICULTURE   Jackson Farms has allotted a certain amount of land for cultivating soybeans, corn, and wheat. Cultivating 1 acre of soybeans requires 2 labor-hours, and cultivating 1 acre of corn or wheat requires 6 labor-hours. The cost of seeds for 1 acre of soybeans is $12, for 1 acre of corn is $20, and for 1 acre of wheat is $8. If all resources are to be used, how many acres of each crop should be cultivated if the following hold?
a. 1000 acres of land are allotted, 4400 labor-hours are available, and $13,200 is available for seeds.
b. 1200 acres of land are allotted, 5200 labor-hours are available, and $16,400 is available for seeds.

40. MIXTURE PROBLEM—FERTILIZER   Lawnco produces three grades of commercial fertilizers. A 100-lb bag of grade A fertilizer contains 18 lb of nitrogen, 4 lb of phosphate, and 5 lb of potassium. A 100-lb bag of grade B fertilizer contains 20 lb of nitrogen and 4 lb each of phosphate and potassium. A 100-lb bag of grade C fertilizer contains 24 lb of nitrogen, 3 lb of phosphate, and 6 lb of potassium. How many 100-lb bags of each of the three grades of fertilizers should Lawnco produce if
a. 26,400 lb of nitrogen, 4900 lb of phosphate, and 6200 lb of potassium are available and all the nutrients are used?
b. 21,800 lb of nitrogen, 4200 lb of phosphate, and 5300 lb of potassium are available and all the nutrients are used?

41. INVESTMENT CLUBS   A private investment club has a certain amount of money earmarked for investment in stocks. To arrive at an acceptable overall level of risk, the stocks that management is considering have been classified into three categories: high-risk, medium-risk, and low-risk. Management estimates that high-risk stocks will have a rate of return of 15%/year; medium-risk stocks, 10%/year; and low-risk stocks, 6%/year. The members have decided that the investment in low-risk stocks should be equal to the sum of the investments in the stocks of the other two categories. Determine how much the club should invest in each type of stock in each of the following scenarios. (In all cases, assume that the entire sum available for investment is invested.)
a. The club has $200,000 to invest, and the investment goal is to have a return of $20,000/year on the total investment.
b. The club has $220,000 to invest, and the investment goal is to have a return of $22,000/year on the total investment.
c. The club has $240,000 to invest, and the investment goal is to have a return of $22,000/year on the total investment.

42. RESEARCH FUNDING   The Carver Foundation funds three nonprofit organizations engaged in alternate-energy research activities. From past data, the proportion of funds spent by each organization in research on solar energy, energy from harnessing the wind, and energy from the motion of ocean tides is given in the accompanying table.

	Proportion of Money Spent		
	Solar	**Wind**	**Tides**
Organization I	0.6	0.3	0.1
Organization II	0.4	0.3	0.3
Organization III	0.2	0.6	0.2

Find the amount awarded to each organization if the total amount spent by all three organizations on solar, wind, and tidal research is

a. $9.2 million, $9.6 million, and $5.2 million, respectively.
b. $8.2 million, $7.2 million, and $3.6 million, respectively.

43. Find the value(s) of k such that

$$A = \begin{bmatrix} 1 & 2 \\ k & 3 \end{bmatrix}$$

has an inverse. What is the inverse of A?
Hint: Use Formula 13.

44. Find the value(s) of k such that

$$A = \begin{bmatrix} 1 & 0 & 1 \\ -2 & 1 & k \\ -1 & 2 & k^2 \end{bmatrix}$$

has an inverse.
Hint: Find the value(s) of k such that the augmented matrix $[A \mid I]$ can be reduced to the form $[I \mid B]$.

In Exercises 45–47, determine whether the statement is true or false. If it is true, explain why it is true. If it is false, give an example to show why it is false.

45. If A is a square matrix with inverse A^{-1} and c is a nonzero real number, then

$$(cA)^{-1} = \left(\frac{1}{c}\right)A^{-1}$$

46. The matrix

$$A = \begin{bmatrix} a & b \\ c & d \end{bmatrix}$$

has an inverse if and only if $ad - bc = 0$.

47. If A^{-1} does not exist, then the system $AX = B$ of n linear equations in n unknowns does not have a unique solution.

48. Let

$$A = \begin{bmatrix} a & b \\ c & d \end{bmatrix}$$

a. Find A^{-1}.
b. Find the necessary condition for A to be nonsingular.
c. Verify that $AA^{-1} = A^{-1}A = I$.

2.6 Solutions to Self-Check Exercises

1. We form the augmented matrix

$$\begin{bmatrix} 2 & 1 & -1 & | & 1 & 0 & 0 \\ 1 & 1 & -1 & | & 0 & 1 & 0 \\ -1 & -2 & 3 & | & 0 & 0 & 1 \end{bmatrix}$$

and row-reduce as follows:

$$\begin{bmatrix} 2 & 1 & -1 & | & 1 & 0 & 0 \\ 1 & 1 & -1 & | & 0 & 1 & 0 \\ -1 & -2 & 3 & | & 0 & 0 & 1 \end{bmatrix} \xrightarrow{R_1 \leftrightarrow R_2}$$

$$\begin{bmatrix} 1 & 1 & -1 & | & 0 & 1 & 0 \\ 2 & 1 & -1 & | & 1 & 0 & 0 \\ -1 & -2 & 3 & | & 0 & 0 & 1 \end{bmatrix} \begin{smallmatrix} R_2 - 2R_1 \\ \xrightarrow{\hspace{1cm}} \\ R_3 + R_1 \end{smallmatrix}$$

$$\begin{bmatrix} 1 & 1 & -1 & | & 0 & 1 & 0 \\ 0 & -1 & 1 & | & 1 & -2 & 0 \\ 0 & -1 & 2 & | & 0 & 1 & 1 \end{bmatrix} \begin{smallmatrix} R_1 + R_2 \\ \xrightarrow{\hspace{1cm}} \\ -R_2 \\ R_3 - R_2 \end{smallmatrix}$$

$$\begin{bmatrix} 1 & 0 & 0 & | & 1 & -1 & 0 \\ 0 & 1 & -1 & | & -1 & 2 & 0 \\ 0 & 0 & 1 & | & -1 & 3 & 1 \end{bmatrix} \xrightarrow{R_2 + R_3}$$

$$\begin{bmatrix} 1 & 0 & 0 & | & 1 & -1 & 0 \\ 0 & 1 & 0 & | & -2 & 5 & 1 \\ 0 & 0 & 1 & | & -1 & 3 & 1 \end{bmatrix}$$

From the preceding results, we see that

$$A^{-1} = \begin{bmatrix} 1 & -1 & 0 \\ -2 & 5 & 1 \\ -1 & 3 & 1 \end{bmatrix}$$

2. a. We write the systems of linear equations in the matrix form

$$AX = B_1$$

where

$$A = \begin{bmatrix} 2 & 1 & -1 \\ 1 & 1 & -1 \\ -1 & -2 & 3 \end{bmatrix} \quad X = \begin{bmatrix} x \\ y \\ z \end{bmatrix} \quad B_1 = \begin{bmatrix} 5 \\ 4 \\ -8 \end{bmatrix}$$

Now, using the results of Exercise 1, we have

$$X = \begin{bmatrix} x \\ y \\ z \end{bmatrix} = A^{-1}B_1 = \begin{bmatrix} 1 & -1 & 0 \\ -2 & 5 & 1 \\ -1 & 3 & 1 \end{bmatrix} \begin{bmatrix} 5 \\ 4 \\ -8 \end{bmatrix} = \begin{bmatrix} 1 \\ 2 \\ -1 \end{bmatrix}$$

Therefore, $x = 1$, $y = 2$, and $z = -1$.

b. Here A and X are as in part (a), but

$$B_2 = \begin{bmatrix} 2 \\ 0 \\ 5 \end{bmatrix}$$

Therefore,

$$X = \begin{bmatrix} x \\ y \\ z \end{bmatrix} = A^{-1}B_2 = \begin{bmatrix} 1 & -1 & 0 \\ -2 & 5 & 1 \\ -1 & 3 & 1 \end{bmatrix} \begin{bmatrix} 2 \\ 0 \\ 5 \end{bmatrix} = \begin{bmatrix} 2 \\ 1 \\ 3 \end{bmatrix}$$

or $x = 2$, $y = 1$, and $z = 3$.

3. Let x denote the number of adults and y the number of children on a tour. Since the tours are filled to capacity, we have

$$x + y = 19$$

Next, since the total receipts for the first tour were \$2931 we have

$$169x + 129y = 2931$$

Therefore, the number of adults and the number of children in the first tour is found by solving the system of linear equations

$$\begin{aligned} x + \quad y &= \quad 19 \\ 169x + 129y &= 2931 \end{aligned} \quad \textbf{(a)}$$

Similarly, we see that the number of adults and the number of children in the second and third tours are found by solving the systems

$$\begin{aligned} x + \quad y &= \quad 19 \\ 169x + 129y &= 3011 \end{aligned} \quad \textbf{(b)}$$

$$\begin{aligned} x + \quad y &= \quad 19 \\ 169x + 129y &= 2771 \end{aligned} \quad \textbf{(c)}$$

These systems may be written in the form

$$AX = B_1 \qquad AX = B_2 \qquad AX = B_3$$

where

$$A = \begin{bmatrix} 1 & 1 \\ 169 & 129 \end{bmatrix} \quad X = \begin{bmatrix} x \\ y \end{bmatrix}$$

$$B_1 = \begin{bmatrix} 19 \\ 2931 \end{bmatrix} \quad B_2 = \begin{bmatrix} 19 \\ 3011 \end{bmatrix} \quad B_3 = \begin{bmatrix} 19 \\ 2771 \end{bmatrix}$$

To solve these systems, we first find A^{-1}. Using Formula (13), we obtain

$$A^{-1} = \begin{bmatrix} -\frac{129}{40} & \frac{1}{40} \\ \frac{169}{40} & -\frac{1}{40} \end{bmatrix}$$

Then, solving each system, we find

$$X = \begin{bmatrix} x \\ y \end{bmatrix} = A^{-1}B_1$$

$$= \begin{bmatrix} -\frac{129}{40} & \frac{1}{40} \\ \frac{169}{40} & -\frac{1}{40} \end{bmatrix} \begin{bmatrix} 19 \\ 2931 \end{bmatrix} = \begin{bmatrix} 12 \\ 7 \end{bmatrix} \quad \textbf{(a)}$$

$$X = \begin{bmatrix} x \\ y \end{bmatrix} = A^{-1}B_2$$

$$= \begin{bmatrix} -\frac{129}{40} & \frac{1}{40} \\ \frac{169}{40} & -\frac{1}{40} \end{bmatrix} \begin{bmatrix} 19 \\ 3011 \end{bmatrix}$$

$$= \begin{bmatrix} 14 \\ 5 \end{bmatrix} \quad \textbf{(b)}$$

$$X = \begin{bmatrix} x \\ y \end{bmatrix} = A^{-1}B_3$$

$$= \begin{bmatrix} -\frac{129}{40} & \frac{1}{40} \\ \frac{169}{40} & -\frac{1}{40} \end{bmatrix} \begin{bmatrix} 19 \\ 2771 \end{bmatrix} = \begin{bmatrix} 8 \\ 11 \end{bmatrix} \quad \textbf{(c)}$$

We conclude that there were
a. 12 adults and 7 children on the first tour.
b. 14 adults and 5 children on the second tour.
c. 8 adults and 11 children on the third tour.

USING TECHNOLOGY

▬ Finding the Inverse of a Square Matrix

Graphing Utility

A graphing utility can be used to find the inverse of a square matrix.

EXAMPLE 1   Use a graphing utility to find the inverse of

$$\begin{bmatrix} 1 & 3 & 5 \\ -2 & 2 & 4 \\ 5 & 1 & 3 \end{bmatrix}$$

Solution   We first enter the given matrix as

$$A = \begin{bmatrix} 1 & 3 & 5 \\ -2 & 2 & 4 \\ 5 & 1 & 3 \end{bmatrix}$$

Then, recalling the matrix A and using the $\boxed{x^{-1}}$ key, we find

$$A^{-1} = \begin{bmatrix} 0.1 & -0.2 & 0.1 \\ 1.3 & -1.1 & -0.7 \\ -0.6 & 0.7 & 0.4 \end{bmatrix}$$

EXAMPLE 2   Use a graphing utility to solve the system

$$\begin{aligned} x + 3y + 5z &= 4 \\ -2x + 2y + 4z &= 3 \\ 5x + \ y + 3z &= 2 \end{aligned}$$

by using the inverse of the coefficient matrix.

Solution   The given system can be written in the matrix form $AX = B$, where

$$A = \begin{bmatrix} 1 & 3 & 5 \\ -2 & 2 & 4 \\ 5 & 1 & 3 \end{bmatrix} \qquad X = \begin{bmatrix} x \\ y \\ z \end{bmatrix} \qquad B = \begin{bmatrix} 4 \\ 3 \\ 2 \end{bmatrix}$$

The solution is $X = A^{-1}B$. Entering the matrices A and B in the graphing utility and using the matrix multiplication capability of the utility gives the output shown in Figure T1—that is, $x = 0$, $y = 0.5$, and $z = 0.5$.

```
[A]⁻¹ [B]

                    [ [0 ]
                      [.5]
                      [.5] ]
Ans →
```

FIGURE T1
The TI-83 screen showing $A^{-1}B$

Excel

We use the function **MINVERSE** to find the inverse of a square matrix using Excel.

EXAMPLE 3   Find the inverse of

$$A = \begin{bmatrix} 1 & 3 & 5 \\ -2 & 2 & 4 \\ 5 & 1 & 3 \end{bmatrix}$$

Solution

1. Enter the elements of matrix A onto a spreadsheet (Figure T2).
2. Compute the inverse of the matrix A: Highlight the cells that will contain the inverse matrix A^{-1}, type = MINVERSE (, highlight the cells containing matrix A, type), and press   Ctrl-Shift-Enter . The desired matrix will appear in your spreadsheet (Figure T2).

	A	B	C
1		Matrix A	
2	1	3	5
3	-2	2	4
5	5	1	3
6			
7		Matrix A⁻¹	
8	0.1	-0.2	0.1
9	1.3	-1.1	-0.7
10	-0.6	0.7	0.4

FIGURE T2
Matrix A and its inverse, matrix A^{-1}

EXAMPLE 4   Solve the system

$$x + 3y + 5z = 4$$
$$-2x + 2y + 4z = 3$$
$$5x + y + 3z = 2$$

by using the inverse of the coefficient matrix.

Solution   The given system can be written in the matrix form $AX = B$, where

$$A = \begin{bmatrix} 1 & 3 & 5 \\ -2 & 2 & 4 \\ 5 & 1 & 3 \end{bmatrix} \quad X = \begin{bmatrix} x \\ y \\ z \end{bmatrix} \quad B = \begin{bmatrix} 4 \\ 3 \\ 2 \end{bmatrix}$$

The solution is $X = A^{-1}B$.

Note: Boldfaced words/characters enclosed in a box (for example, Enter) indicate that an action (click, select, or press) is required. Words/characters printed in blue (for example, Chart sub-type:) indicate words/characters on the screen. Words/characters printed in a typewriter font (for example, =(-2/3)*A2+2) indicate words/characters that need to be typed and entered.

(continued)

	A
12	Matrix X
13	5.55112E−17
14	0.5
15	0.5

FIGURE T3
Matrix X gives the solution to the problem

1. *Enter the matrix B on a spreadsheet.*
2. *Compute $A^{-1}B$.* Highlight the cells that will contain the matrix X, and then type `=MMULT(`, highlight the cells in the matrix A^{-1}, type `,`, highlight the cells in the matrix B, type `)`, and press ┃**Ctrl-Shift-Enter**┃. (*Note*: The matrix A^{-1} was found in Example 3.) The matrix X shown in Figure T3 will appear on your spreadsheet. Thus, $x = 0$, $y = 0.5$, and $z = 0.5$. ■

TECHNOLOGY EXERCISES

In Exercises 1–6, find the inverse of the matrix. Round your answers to two decimal places.

1. $\begin{bmatrix} 1.2 & 3.1 & -2.1 \\ 3.4 & 2.6 & 7.3 \\ -1.2 & 3.4 & -1.3 \end{bmatrix}$ 2. $\begin{bmatrix} 4.2 & 3.7 & 4.6 \\ 2.1 & -1.3 & -2.3 \\ 1.8 & 7.6 & -2.3 \end{bmatrix}$

3. $\begin{bmatrix} 1.1 & 2.3 & 3.1 & 4.2 \\ 1.6 & 3.2 & 1.8 & 2.9 \\ 4.2 & 1.6 & 1.4 & 3.2 \\ 1.6 & 2.1 & 2.8 & 7.2 \end{bmatrix}$

4. $\begin{bmatrix} 2.1 & 3.2 & -1.4 & -3.2 \\ 6.2 & 7.3 & 8.4 & 1.6 \\ 2.3 & 7.1 & 2.4 & -1.3 \\ -2.1 & 3.1 & 4.6 & 3.7 \end{bmatrix}$

5. $\begin{bmatrix} 2 & -1 & 3 & 2 & 4 \\ 3 & 2 & -1 & 4 & 1 \\ 3 & 2 & 6 & 4 & -1 \\ 2 & 1 & -1 & 4 & 2 \\ 3 & 4 & 2 & 5 & 6 \end{bmatrix}$

6. $\begin{bmatrix} 1 & 4 & 2 & 3 & 1.4 \\ 6 & 2.4 & 5 & 1.2 & 3 \\ 4 & 1 & 2 & 3 & 1.2 \\ -1 & 2 & -3 & 4 & 2 \\ 1.1 & 2.2 & 3 & 5.1 & 4 \end{bmatrix}$

In Exercises 7–10, solve the system of linear equations by first writing the system in the form $AX = B$ and then solving the resulting system by using A^{-1}. Round your answers to two decimal places.

7. $2x - 3y + 4z = 2.4$
$3x + 2y - 7z = -8.1$
$x + 4y - 2z = 10.2$

8. $3.2x - 4.7y + 3.2z = 7.1$
$2.1x + 2.6y + 6.2z = 8.2$
$5.1x - 3.1y - 2.6z = -6.5$

9. $3x_1 - 2x_2 + 4x_3 - 8x_4 = 8$
$2x_1 + 3x_2 - 2x_3 + 6x_4 = 4$
$3x_1 + 2x_2 - 6x_3 - 7x_4 = -2$
$4x_1 - 7x_2 + 4x_3 + 6x_4 = 22$

10. $1.2x_1 + 2.1x_2 - 3.2x_3 + 4.6x_4 = 6.2$
$3.1x_1 - 1.2x_2 + 4.1x_3 - 3.6x_4 = -2.2$
$1.8x_1 + 3.1x_2 - 2.4x_3 + 8.1x_4 = 6.2$
$2.6x_1 - 2.4x_2 + 3.6x_3 - 4.6x_4 = 3.6$

2.7 Leontief Input–Output Model (Optional)

▪ Input–Output Analysis

One of the many important applications of matrix theory to the field of economics is the study of the relationship between industrial production and consumer demand. At the heart of this analysis is the Leontief input–output model pioneered by Wassily Leontief, who was awarded a Nobel Prize in economics in 1973 for his contributions to the field.

To illustrate this concept, let's consider an oversimplified economy consisting of three sectors: agriculture (*A*), manufacturing (*M*), and service (*S*). In general, part of the output of one sector is absorbed by another sector through interindustry purchases, with the excess available to fulfill consumer demands. The relationship governing both intraindustrial and interindustrial sales and purchases is conveniently represented by means of an **input–output matrix**:

$$
\begin{array}{c}
\text{Input} \\
\text{(amount used in production)}
\end{array}
\quad
\begin{array}{cc}
& \begin{array}{c}\text{Output (amount produced)}\\ \begin{matrix} A & M & S \end{matrix}\end{array} \\
\begin{matrix} A \\ M \\ S \end{matrix} &
\begin{bmatrix}
0.2 & 0.2 & 0.1 \\
0.2 & 0.4 & 0.1 \\
0.1 & 0.2 & 0.3
\end{bmatrix}
\end{array}
\qquad (16)
$$

The first column (read from top to bottom) tells us that the production of 1 unit of agricultural products requires the consumption of 0.2 unit of agricultural products, 0.2 unit of manufactured goods, and 0.1 unit of services. The second column tells us that the production of 1 unit of manufactured products requires the consumption of 0.2 unit of agricultural products, 0.4 unit of manufactured products, and 0.2 unit of services. Finally, the third column tells us that the production of 1 unit of services requires the consumption of 0.1 unit each of agricultural goods and manufactured products and 0.3 unit of services.

APPLIED EXAMPLE 1   Input–Output Analysis   Refer to the input–output matrix (16).

a. If the units are measured in millions of dollars, determine the amount of agricultural products consumed in the production of $100 million worth of manufactured goods.

b. Determine the dollar amount of manufactured products required to produce $200 million worth of all goods and services in the economy.

Solution

a. The production of 1 unit—that is, $1 million worth of manufactured goods— requires the consumption of 0.2 unit of agricultural products. Thus, the amount of agricultural products consumed in the production of $100 million worth of manufactured goods is given by (100)(0.2), or $20 million.

b. The amount of manufactured goods required to produce 1 unit of all goods and services in the economy is given by adding the numbers of the second row of the input–output matrix—that is, 0.2 + 0.4 + 0.1, or 0.7 unit. Therefore, the production of $200 million worth of all goods and services in the economy requires 200(0.7), or $140 million worth, of manufactured products. ▪

Next, suppose the total output of goods of the agriculture and manufacturing sectors and the total output from the service sector of the economy are given by x, y, and z units, respectively. What is the value of agricultural products consumed in the internal process of producing this total output of various goods and services?

To answer this question, we first note, by examining the input–output matrix

$$
\begin{array}{cc}
 & \begin{array}{ccc} & \text{Output} & \\ A & M & S \end{array} \\
\text{Input} \quad \begin{array}{c} A \\ M \\ S \end{array} & \begin{bmatrix} 0.2 & 0.2 & 0.1 \\ 0.2 & 0.4 & 0.1 \\ 0.1 & 0.2 & 0.3 \end{bmatrix}
\end{array}
$$

that 0.2 unit of agricultural products is required to produce 1 unit of agricultural products, so the amount of agricultural goods required to produce x units of agricultural products is given by $0.2x$ unit. Next, again referring to the input–output matrix, we see that 0.2 unit of agricultural products is required to produce 1 unit of manufactured products, so the requirement for producing y units of the latter is $0.2y$ unit of agricultural products. Finally, we see that 0.1 unit of agricultural goods is required to produce 1 unit of services, so the value of agricultural products required to produce z units of services is $0.1z$ unit. Thus, the total amount of agricultural products required to produce the total output of goods and services in the economy is

$$0.2x + 0.2y + 0.1z$$

units. In a similar manner, we see that the total amount of manufactured products and the total value of services required to produce the total output of goods and services in the economy are given by

$$0.2x + 0.4y + 0.1z$$
$$0.1x + 0.2y + 0.3z$$

respectively.

These results could also be obtained using matrix multiplication. To see this, write the total output of goods and services x, y, and z as a 3×1 matrix

$$X = \begin{bmatrix} x \\ y \\ z \end{bmatrix} \qquad \text{Gross production matrix}$$

The matrix X is called the **gross production matrix**. Letting A denote the input–output matrix, we have

$$A = \begin{bmatrix} 0.2 & 0.2 & 0.1 \\ 0.2 & 0.4 & 0.1 \\ 0.1 & 0.2 & 0.3 \end{bmatrix} \qquad \text{Input–output matrix}$$

Then, the product

$$
\begin{aligned}
AX &= \begin{bmatrix} 0.2 & 0.2 & 0.1 \\ 0.2 & 0.4 & 0.1 \\ 0.1 & 0.2 & 0.3 \end{bmatrix} \begin{bmatrix} x \\ y \\ z \end{bmatrix} \\
&= \begin{bmatrix} 0.2x + 0.2y + 0.1z \\ 0.2x + 0.4y + 0.1z \\ 0.1x + 0.2y + 0.3z \end{bmatrix} \qquad \text{Internal consumption matrix}
\end{aligned}
$$

is a 3×1 matrix whose entries represent the respective values of the agricultural products, manufactured products, and services consumed in the internal process of production. The matrix AX is referred to as the **internal consumption matrix**.

Now, since X gives the total production of goods and services in the economy and AX, as we have just seen, gives the amount of products and services consumed in the production of these goods and services, it follows that the 3×1 matrix $X - AX$ gives the net output of goods and services that is exactly enough to satisfy consumer demands. Letting matrix D represent these consumer demands, we are led to the following matrix equation:

$$X - AX = D$$
$$(I - A)X = D$$

where I is the 3×3 identity matrix.

Assuming that the inverse of $(I - A)$ exists, multiplying both sides of the last equation by $(I - A)^{-1}$ yields

$$X = (I - A)^{-1}D$$

Leontief Input–Output Model

In a **Leontief input–output model**, the matrix equation giving the net output of goods and services needed to satisfy consumer demand is

$$
\underset{\substack{\text{Total} \\ \text{output}}}{X} \quad - \quad \underset{\substack{\text{Internal} \\ \text{consumption}}}{AX} \quad = \quad \underset{\substack{\text{Consumer} \\ \text{demand}}}{D}
$$

where X is the total output matrix, A is the input–output matrix, and D is the matrix representing consumer demand.

The solution to this equation is

$$X = (I - A)^{-1}D \qquad \text{Assuming that } (I - A)^{-1} \text{ exists} \qquad \textbf{(17)}$$

which gives the amount of goods and services that must be produced to satisfy consumer demand.

Equation (17) gives us a means of finding the amount of goods and services to be produced in order to satisfy a given level of consumer demand, as illustrated by the following example.

APPLIED EXAMPLE 2 Input–Output Model for a Three-Sector Economy For the three-sector economy with input–output matrix given by (16), which is reproduced here:

$$\begin{bmatrix} 0.2 & 0.2 & 0.1 \\ 0.2 & 0.4 & 0.1 \\ 0.1 & 0.2 & 0.3 \end{bmatrix} \qquad \text{Each unit equals \$1 million.}$$

a. Find the gross output of goods and services needed to satisfy a consumer demand of \$100 million worth of agricultural products, \$80 million worth of manufactured products, and \$50 million worth of services.

b. Find the value of the goods and services consumed in the internal process of production in order to meet this gross output.

Solution

a. We are required to determine the gross production matrix

$$X = \begin{bmatrix} x \\ y \\ z \end{bmatrix}$$

where x, y, and z denote the value of the agricultural products, the manufactured products, and services. The matrix representing the consumer demand is given by

$$D = \begin{bmatrix} 100 \\ 80 \\ 50 \end{bmatrix}$$

Next, we compute

$$I - A = \begin{bmatrix} 1 & 0 & 0 \\ 0 & 1 & 0 \\ 0 & 0 & 1 \end{bmatrix} - \begin{bmatrix} 0.2 & 0.2 & 0.1 \\ 0.2 & 0.4 & 0.1 \\ 0.1 & 0.2 & 0.3 \end{bmatrix} = \begin{bmatrix} 0.8 & -0.2 & -0.1 \\ -0.2 & 0.6 & -0.1 \\ -0.1 & -0.2 & 0.7 \end{bmatrix}$$

Using the method of Section 2.6, we find (to two decimal places)

$$(I - A)^{-1} = \begin{bmatrix} 1.43 & 0.57 & 0.29 \\ 0.54 & 1.96 & 0.36 \\ 0.36 & 0.64 & 1.57 \end{bmatrix}$$

Finally, using Equation (17), we find

$$X = (I - A)^{-1}D = \begin{bmatrix} 1.43 & 0.57 & 0.29 \\ 0.54 & 1.96 & 0.36 \\ 0.36 & 0.64 & 1.57 \end{bmatrix} \begin{bmatrix} 100 \\ 80 \\ 50 \end{bmatrix} = \begin{bmatrix} 203.1 \\ 228.8 \\ 165.7 \end{bmatrix}$$

To fulfill consumer demand, \$203 million worth of agricultural products, \$229 million worth of manufactured products, and \$166 million worth of services should be produced.

b. The amount of goods and services consumed in the internal process of production is given by AX or, equivalently, by $X - D$. In this case it is more convenient to use the latter, which gives the required result of

$$\begin{bmatrix} 203.1 \\ 228.8 \\ 165.7 \end{bmatrix} - \begin{bmatrix} 100 \\ 80 \\ 50 \end{bmatrix} = \begin{bmatrix} 103.1 \\ 148.8 \\ 115.7 \end{bmatrix}$$

or \$103 million worth of agricultural products, \$149 million worth of manufactured products, and \$116 million worth of services.   ∎

APPLIED EXAMPLE 3   An Input–Output Model for a Three-Product Company   TKK Corporation, a large conglomerate, has three subsidiaries engaged in producing raw rubber, manufacturing tires, and manufacturing other rubber-based goods. The production of 1 unit of raw rubber requires the consumption of 0.08 unit of rubber, 0.04 unit of tires, and 0.02 unit of other rubber-based goods. To produce 1 unit of tires requires 0.6 unit of raw rubber, 0.02 unit of tires, and 0 units of other rubber-based goods. To produce 1 unit of other rubber-based goods requires 0.3 unit of raw rubber, 0.01 unit of tires, and 0.06 unit of other rubber-based goods. Market research indicates that the demand for the following year will be $200 million for raw rubber, $800 million for tires, and $120 million for other rubber-based products. Find the level of production for each subsidiary in order to satisfy this demand.

Solution   View the corporation as an economy having three sectors and with an input–output matrix given by

$$A = \begin{array}{c} \\ \text{Raw rubber} \\ \text{Tires} \\ \text{Goods} \end{array} \begin{array}{ccc} \text{Raw rubber} & \text{Tires} & \text{Goods} \end{array} \\ \begin{bmatrix} 0.08 & 0.60 & 0.30 \\ 0.04 & 0.02 & 0.01 \\ 0.02 & 0 & 0.06 \end{bmatrix}$$

Using Equation (17), we find that the required level of production is given by

$$X = \begin{bmatrix} x \\ y \\ z \end{bmatrix} = (I - A)^{-1}D$$

where x, y, and z denote the outputs of raw rubber, tires, and other rubber-based goods and where

$$D = \begin{bmatrix} 200 \\ 800 \\ 120 \end{bmatrix}$$

Now,

$$I - A = \begin{bmatrix} 0.92 & -0.60 & -0.30 \\ -0.04 & 0.98 & -0.01 \\ -0.02 & 0 & 0.94 \end{bmatrix}$$

You are asked to verify that

$$(I - A)^{-1} = \begin{bmatrix} 1.13 & 0.69 & 0.37 \\ 0.05 & 1.05 & 0.03 \\ 0.02 & 0.02 & 1.07 \end{bmatrix} \quad \text{See Exercise 7.}$$

Therefore,

$$X = (I - A)^{-1}D = \begin{bmatrix} 1.13 & 0.69 & 0.37 \\ 0.05 & 1.05 & 0.03 \\ 0.02 & 0.02 & 1.07 \end{bmatrix} \begin{bmatrix} 200 \\ 800 \\ 120 \end{bmatrix} = \begin{bmatrix} 822.4 \\ 853.6 \\ 148.4 \end{bmatrix}$$

To fulfill the predicted demand, $822 million worth of raw rubber, $854 million worth of tires, and $148 million worth of other rubber-based goods should be produced.

2.7　Self-Check Exercises

1. Solve the matrix equation $(I - A)X = D$ for x and y, given that

$$A = \begin{bmatrix} 0.4 & 0.1 \\ 0.2 & 0.2 \end{bmatrix} \quad X = \begin{bmatrix} x \\ y \end{bmatrix} \quad D = \begin{bmatrix} 50 \\ 10 \end{bmatrix}$$

2. A simple economy consists of two sectors: agriculture (A) and transportation (T). The input–output matrix for this economy is given by

$$A = \begin{matrix} A \\ T \end{matrix} \begin{bmatrix} 0.4 & 0.1 \\ 0.2 & 0.2 \end{bmatrix}$$
$$\begin{matrix} & A & T \end{matrix}$$

a. Find the gross output of agricultural products needed to satisfy a consumer demand for $50 million worth of agricultural products and $10 million worth of transportation.

b. Find the value of agricultural products and transportation consumed in the internal process of production in order to meet the gross output.

Solutions to Self-Check Exercises 2.7 can be found on page 160.

2.7　Concept Questions

1. What do the quantities X, AX, and D represent in the matrix equation $X - AX = D$ for a Leontief input–output model?

2. What is the solution to the matrix equation $X - AX = D$? Does the solution to this equation always exist? Why or why not?

2.7　Exercises

1. **AN INPUT–OUTPUT MATRIX FOR A THREE-SECTOR ECONOMY**　A simple economy consists of three sectors: agriculture (A), manufacturing (M), and transportation (T). The input–output matrix for this economy is given by

$$\begin{matrix} & A & M & T \end{matrix}$$
$$\begin{matrix} A \\ M \\ T \end{matrix} \begin{bmatrix} 0.4 & 0.1 & 0.1 \\ 0.1 & 0.4 & 0.3 \\ 0.2 & 0.2 & 0.2 \end{bmatrix}$$

a. If the units are measured in millions of dollars, determine the amount of agricultural products consumed in the production of $100 million worth of manufactured goods.
b. Determine the dollar amount of manufactured products required to produce $200 million worth of all goods in the economy.
c. Which sector consumes the greatest amount of agricultural products in the production of a unit of goods in that sector? The least?

2. **AN INPUT–OUTPUT MATRIX FOR A FOUR-SECTOR ECONOMY**　The relationship governing the intraindustrial and interindustrial sales and purchases of four basic industries—agriculture (A), manufacturing (M), transportation (T), and energy (E)—of a certain economy is given by the following input–output matrix.

$$\begin{matrix} & A & M & T & E \end{matrix}$$
$$\begin{matrix} A \\ M \\ T \\ E \end{matrix} \begin{bmatrix} 0.3 & 0.2 & 0 & 0.1 \\ 0.2 & 0.3 & 0.2 & 0.1 \\ 0.2 & 0.2 & 0.1 & 0.3 \\ 0.1 & 0.2 & 0.3 & 0.2 \end{bmatrix}$$

a. How many units of energy are required to produce 1 unit of manufactured goods?
b. How many units of energy are required to produce 3 units of all goods in the economy?
c. Which sector of the economy is least dependent on the cost of energy?
d. Which sector of the economy has the smallest intraindustry purchases (sales)?

In Exercises 3–6, solve the matrix equation $(I - A)X = D$ for the matrices A and D.

3. $A = \begin{bmatrix} 0.4 & 0.2 \\ 0.3 & 0.1 \end{bmatrix}$ and $D = \begin{bmatrix} 10 \\ 12 \end{bmatrix}$

4. $A = \begin{bmatrix} 0.2 & 0.3 \\ 0.5 & 0.2 \end{bmatrix}$ and $D = \begin{bmatrix} 4 \\ 8 \end{bmatrix}$

5. $A = \begin{bmatrix} 0.5 & 0.2 \\ 0.2 & 0.5 \end{bmatrix}$ and $D = \begin{bmatrix} 10 \\ 20 \end{bmatrix}$

6. $A = \begin{bmatrix} 0.6 & 0.2 \\ 0.1 & 0.4 \end{bmatrix}$ and $D = \begin{bmatrix} 8 \\ 12 \end{bmatrix}$

7. Let

$$A = \begin{bmatrix} 0.08 & 0.60 & 0.30 \\ 0.04 & 0.02 & 0.01 \\ 0.02 & 0 & 0.06 \end{bmatrix}$$

Show that

$$(I - A)^{-1} = \begin{bmatrix} 1.13 & 0.69 & 0.37 \\ 0.05 & 1.05 & 0.03 \\ 0.02 & 0.02 & 1.07 \end{bmatrix}$$

8. AN INPUT–OUTPUT MODEL FOR A TWO-SECTOR ECONOMY A simple economy consists of two industries: agriculture and manufacturing. The production of 1 unit of agricultural products requires the consumption of 0.2 unit of agricultural products and 0.3 unit of manufactured goods. The production of 1 unit of manufactured products requires the consumption of 0.4 unit of agricultural products and 0.3 unit of manufactured goods.

a. Find the gross output of goods needed to satisfy a consumer demand for $100 million worth of agricultural products and $150 million worth of manufactured products.

b. Find the value of the goods consumed in the internal process of production in order to meet the gross output.

9. Rework Exercise 8 if the consumer demand for the output of agricultural goods and the consumer demand for manufactured products are $120 million and $140 million, respectively.

10. Refer to Example 3. Suppose the demand for raw rubber increases by 10%, the demand for tires increases by 20%, and the demand for other rubber-based products decreases by 10%. Find the level of production for each subsidiary in order to meet this demand.

11. AN INPUT–OUTPUT MODEL FOR A THREE-SECTOR ECONOMY Consider the economy of Exercise 1, consisting of three sectors: agriculture (A), manufacturing (M), and transportation (T), with an input–output matrix given by

$$\begin{array}{c} \\ A \\ M \\ T \end{array} \begin{array}{ccc} A & M & T \\ \begin{bmatrix} 0.4 & 0.1 & 0.1 \\ 0.1 & 0.4 & 0.3 \\ 0.2 & 0.2 & 0.2 \end{bmatrix} \end{array}$$

a. Find the gross output of goods needed to satisfy a consumer demand for $200 million worth of agricultural products, $100 million worth of manufactured products, and $60 million worth of transportation.

b. Find the value of goods and transportation consumed in the internal process of production in order to meet this gross output.

12. AN INPUT–OUTPUT MODEL FOR A THREE-SECTOR ECONOMY Consider a simple economy consisting of three sectors: food, clothing, and shelter. The production of 1 unit of food requires the consumption of 0.4 unit of food, 0.2 unit of clothing, and 0.2 unit of shelter. The production of 1 unit of clothing requires the consumption of 0.1 unit of food, 0.2 unit of clothing, and 0.3 unit of shelter. The production of 1 unit of shelter requires the consumption of 0.3 unit of food, 0.1 unit of clothing, and 0.1 unit of shelter. Find the level of production for each sector in order to satisfy the demand for $100 million worth of food, $30 million worth of clothing, and $250 million worth of shelter.

In Exercises 13–16, matrix *A* is an input–output matrix associated with an economy, and matrix *D* (units in millions of dollars) is a demand vector. In each problem, find the final outputs of each industry such that the demands of both industry and the open sector are met.

13. $A = \begin{bmatrix} 0.4 & 0.2 \\ 0.3 & 0.5 \end{bmatrix}$ and $D = \begin{bmatrix} 12 \\ 24 \end{bmatrix}$

14. $A = \begin{bmatrix} 0.1 & 0.4 \\ 0.3 & 0.2 \end{bmatrix}$ and $D = \begin{bmatrix} 5 \\ 10 \end{bmatrix}$

15. $A = \begin{bmatrix} \frac{1}{5} & \frac{2}{5} & \frac{1}{5} \\ \frac{1}{2} & 0 & \frac{1}{2} \\ 0 & \frac{1}{5} & 0 \end{bmatrix}$ and $D = \begin{bmatrix} 10 \\ 5 \\ 15 \end{bmatrix}$

16. $A = \begin{bmatrix} 0.2 & 0.4 & 0.1 \\ 0.3 & 0.2 & 0.1 \\ 0.1 & 0.2 & 0.2 \end{bmatrix}$ and $D = \begin{bmatrix} 6 \\ 8 \\ 10 \end{bmatrix}$

2.7   Solutions to Self-Check Exercises

1. Multiplying both sides of the given equation on the left by $(I - A)^{-1}$, we see that

$$X = (I - A)^{-1}D$$

Now,

$$I - A = \begin{bmatrix} 1 & 0 \\ 0 & 1 \end{bmatrix} - \begin{bmatrix} 0.4 & 0.1 \\ 0.2 & 0.2 \end{bmatrix} = \begin{bmatrix} 0.6 & -0.1 \\ -0.2 & 0.8 \end{bmatrix}$$

Next, we use the Gauss–Jordan procedure to compute $(I - A)^{-1}$ (to two decimal places):

$$\begin{bmatrix} 0.6 & -0.1 & | & 1 & 0 \\ -0.2 & 0.8 & | & 0 & 1 \end{bmatrix} \xrightarrow{\frac{1}{0.6}R_1}$$

$$\begin{bmatrix} 1 & -0.17 & | & 1.67 & 0 \\ -0.2 & 0.8 & | & 0 & 1 \end{bmatrix} \xrightarrow{R_2 + 0.2R_1}$$

$$\begin{bmatrix} 1 & -0.17 & | & 1.67 & 0 \\ 0 & 0.77 & | & 0.33 & 1 \end{bmatrix} \xrightarrow{\frac{1}{0.77}R_2}$$

$$\begin{bmatrix} 1 & -0.17 & | & 1.67 & 0 \\ 0 & 1 & | & 0.43 & 1.30 \end{bmatrix} \xrightarrow{R_1 + 0.17R_2}$$

$$\begin{bmatrix} 1 & 0 & | & 1.74 & 0.22 \\ 0 & 1 & | & 0.43 & 1.30 \end{bmatrix}$$

giving

$$(I - A)^{-1} = \begin{bmatrix} 1.74 & 0.22 \\ 0.43 & 1.30 \end{bmatrix}$$

Therefore,

$$X = \begin{bmatrix} x \\ y \end{bmatrix} = (I - A)^{-1}D = \begin{bmatrix} 1.74 & 0.22 \\ 0.43 & 1.30 \end{bmatrix}\begin{bmatrix} 50 \\ 10 \end{bmatrix} = \begin{bmatrix} 89.2 \\ 34.5 \end{bmatrix}$$

or $x = 89.2$ and $y = 34.5$.

2. a. Let

$$X = \begin{bmatrix} x \\ y \end{bmatrix}$$

denote the gross production matrix, where x denotes the value of the agricultural products and y the value of transportation. Also, let

$$D = \begin{bmatrix} 50 \\ 10 \end{bmatrix}$$

denote the consumer demand. Then

$$(I - A)X = D$$

or, equivalently,

$$X = (I - A)^{-1}D$$

Using the results of Exercise 1, we find that $x = 89.2$ and $y = 34.5$. That is, in order to fulfill consumer demands, $89.2 million worth of agricultural products must be produced and $34.5 million worth of transportation services must be used.

b. The amount of agricultural products consumed and transportation services used is given by

$$X - D = \begin{bmatrix} 89.2 \\ 34.5 \end{bmatrix} - \begin{bmatrix} 50 \\ 10 \end{bmatrix} = \begin{bmatrix} 39.2 \\ 24.5 \end{bmatrix}$$

or $39.2 million worth of agricultural products and $24.5 million worth of transportation services.

USING TECHNOLOGY

■ The Leontief Input–Output Model

Graphing Utility

Since the solution to a problem involving a Leontief input–output model often involves several matrix operations, a graphing utility can be used to facilitate the necessary computations.

APPLIED EXAMPLE 1 Suppose that the input–output matrix associated with an economy is given by A and that the matrix D is a demand vector, where

$$A = \begin{bmatrix} 0.2 & 0.4 & 0.15 \\ 0.3 & 0.1 & 0.4 \\ 0.25 & 0.4 & 0.2 \end{bmatrix} \quad \text{and} \quad D = \begin{bmatrix} 20 \\ 15 \\ 40 \end{bmatrix}$$

Find the final outputs of each industry such that the demands of both industry and the open sector are met.

Solution   First, we enter the matrices I (the identity matrix), A, and D. We are required to compute the output matrix $X = (I - A)^{-1}D$. Using the matrix operations of the graphing utility, we find

$$X = (I - A)^{-1}*D = \begin{bmatrix} 110.28 \\ 116.95 \\ 142.94 \end{bmatrix}$$

Hence the final outputs of the first, second, and third industries are 110.28, 116.95, and 142.94 units, respectively. ∎

Excel

Here we show how to solve a problem involving a Leontief input–output model using matrix operations on a spreadsheet.

APPLIED EXAMPLE 2 Suppose that the input–output matrix associated with an economy is given by matrix A and that the matrix D is a demand vector, where

$$A = \begin{bmatrix} 0.2 & 0.4 & 0.15 \\ 0.3 & 0.1 & 0.4 \\ 0.25 & 0.4 & 0.2 \end{bmatrix} \quad \text{and} \quad D = \begin{bmatrix} 20 \\ 15 \\ 40 \end{bmatrix}$$

Find the final outputs of each industry such that the demands of both industry and the open sector are met.

Solution

1. *Enter the elements of the matrix A and D onto a spreadsheet* (Figure T1).

	A	B	C	D	E
1		Matrix A			Matrix D
2	0.2	0.4	0.15		20
3	0.3	0.1	0.4		15
4	0.25	0.4	0.2		40

FIGURE T1
Spreadsheet showing matrix A and matrix D

2. *Find* $(I - A)^{-1}$. Enter the elements of the 3×3 identity matrix I onto a spreadsheet. Highlight the cells that will contain the matrix $(I - A)^{-1}$. Type =MINVERSE(, highlight the cells containing the matrix I; type −, highlight the

(*continued*)

cells containing the matrix A; type $)$, and press $\boxed{\textbf{Ctrl-Shift-Enter}}$. These results are shown in Figure T2.

	A	B	C
6		Matrix I	
7	1	0	0
8	0	1	0
9	0	0	1
10			
11		Matrix $(I-A)^{-1}$	
12	2.151777	1.460134	1.133525
13	1.306436	2.315082	1.402498
14	1.325648	1.613833	2.305476

FIGURE T2
Matrix I and matrix $(I-A)^{-1}$

3. *Compute* $(I - A)^{-1} * D$. Highlight the cells that will contain the matrix $(I - A)^{-1} * D$. Type $=$MMULT$($, highlight the cells containing the matrix $(I - A)^{-1}$, type $,$, highlight the cells containing the matrix D, type $)$, and press $\boxed{\textbf{Ctrl-Shift-Enter}}$. The resulting matrix is shown in Figure T3. So, the final outputs of the first, second, and third industries are 110.28, 116.95, and 142.94, respectively.

	A
16	Matrix $(I-A)^{-1}$ *D
17	110.2786
18	116.9549
19	142.9395

FIGURE T3
Matrix $(I-A)^{-1}*D$

TECHNOLOGY EXERCISES

In Exercises 1–4, A is an input–output matrix associated with an economy and D (in units of dollars) is a demand vector. Find the final outputs of each industry such that the demands of both industry and the open sector are met.

1.
$$A = \begin{bmatrix} 0.3 & 0.2 & 0.4 & 0.1 \\ 0.2 & 0.1 & 0.2 & 0.3 \\ 0.3 & 0.1 & 0.2 & 0.3 \\ 0.4 & 0.2 & 0.1 & 0.2 \end{bmatrix} \text{ and } D = \begin{bmatrix} 40 \\ 60 \\ 70 \\ 20 \end{bmatrix}$$

2.
$$A = \begin{bmatrix} 0.12 & 0.31 & 0.40 & 0.05 \\ 0.31 & 0.22 & 0.12 & 0.20 \\ 0.18 & 0.32 & 0.05 & 0.15 \\ 0.32 & 0.14 & 0.22 & 0.05 \end{bmatrix} \text{ and } D = \begin{bmatrix} 50 \\ 20 \\ 40 \\ 60 \end{bmatrix}$$

3.
$$A = \begin{bmatrix} 0.2 & 0.2 & 0.3 & 0.05 \\ 0.1 & 0.1 & 0.2 & 0.3 \\ 0.3 & 0.2 & 0.1 & 0.4 \\ 0.2 & 0.05 & 0.2 & 0.1 \end{bmatrix} \text{ and } D = \begin{bmatrix} 25 \\ 30 \\ 50 \\ 40 \end{bmatrix}$$

4.
$$A = \begin{bmatrix} 0.2 & 0.4 & 0.3 & 0.1 \\ 0.1 & 0.2 & 0.1 & 0.3 \\ 0.2 & 0.1 & 0.4 & 0.05 \\ 0.3 & 0.1 & 0.2 & 0.05 \end{bmatrix} \text{ and } D = \begin{bmatrix} 40 \\ 20 \\ 30 \\ 60 \end{bmatrix}$$

Note: Boldfaced words/characters in a box (for example, $\boxed{\textbf{Enter}}$) indicate that an action (click, select, or press) is required. Words/characters printed blue (for example, Chart sub-type:) indicate words/characters on the screen. Words/characters printed in a typewriter font (for example, $=$(−2/3)*A2+2) indicate words/characters that need to be typed and entered.

CHAPTER 2 **Summary of Principal Formulas and Terms**

 FORMULAS

1. Laws for matrix addition	
a. Commutative law	$A + B = B + A$
b. Associative law	$(A + B) + C = A + (B + C)$
2. Laws for matrix multiplication	
a. Associative law	$(AB)C = A(BC)$
b. Distributive law	$A(B + C) = AB + AC$
3. Inverse of a 2×2 matrix	If $\quad A = \begin{bmatrix} a & b \\ c & d \end{bmatrix}$ and $\quad D = ad - bc \neq 0$ then $\quad A^{-1} = \dfrac{1}{D}\begin{bmatrix} d & -b \\ -c & a \end{bmatrix}$
4. Solution of system $AX = B$ (A nonsingular)	$X = A^{-1}B$

TERMS

system of linear equations (72)

solution of a system of linear equations (72)

parameter (74)

dependent system (74)

inconsistent system (74)

Gauss–Jordan elimination method (80)

equivalent system (80)

coefficient matrix (83)

augmented matrix (83)

row-reduced form of a matrix (84)

row operations (85)

unit column (85)

pivoting (85)

size of a matrix (109)

matrix (109)

row matrix (109)

column matrix (109)

square matrix (109)

transpose of a matrix (113)

scalar (113)

scalar product (113)

matrix product (122)

identity matrix (126)

inverse of a matrix (137)

nonsingular matrix (138)

singular matrix (138)

input–output matrix (153)

gross production matrix (154)

internal consumption matrix (155)

Leontief input–output model (155)

CHAPTER 2 **Concept Review Questions**

Fill in the blanks.

1. a. Two lines in the plane can intersect at (a) exactly _____ point, (b) infinitely _____ points, or (c) at _____ point.

   b. A system of two linear equations in two variables can have (a) exactly _____ solution, (b) infinitely _____ solutions, or (c) _____ solution.

2. To find the point(s) of intersection of two lines, we solve the system of _____ describing the two lines.

3. The row operations used in the Gauss–Jordan elimination method are denoted by _____, _____, and _____. The use of each of these operations does not alter the _____ of the system of linear equations.

4. a. A system of linear equations with fewer equations than variables cannot have a/an _____ solution.

b. A system of linear equations with at least as many equations as variables may have _____ solution, _____ _____ solutions, or a _____ solution.

5. Two matrices are equal provided they have the same _____ and their corresponding _____ are equal.

6. Two matrices may be added (subtracted) together if they both have the same _____. To add or subtract two matrices, we add or subtract their _____ entries.

7. The transpose of a/an _____ matrix with elements a_{ij} is the matrix of size _____ with entries _____.

8. The scalar product of a matrix A by the scalar c is the matrix _____ obtained by multiplying each entry of A by _____.

9. a. For the product AB of two matrices A and B to be defined, the number of _____ of A must be equal to the number of _____ of B.

b. If A is an $m \times n$ matrix and B is an $n \times p$ matrix, then the size of AB is _____.

10. a. If the products and sums are defined for the matrices A, B, and C, then the associative law states that $(AB)C =$ ____; the distributive law states that $A(B + C) =$ ____.

b. If I is an identity matrix of order n, then $IA = A$ if A is any matrix of order _____.

11. A matrix A is nonsingular if there exists a matrix A^{-1} such that _____ = _____ = I. If A^{-1} does not exist, then A is said to be _____.

12. A system of n linear equations in n variables written in the form $AX = B$ has a unique solution given by $X =$ _____ if A has an inverse.

CHAPTER 2 **Review Exercises**

In Exercises 1–4, perform the operations if possible.

1. $\begin{bmatrix} 1 & 2 \\ -1 & 3 \\ 2 & 1 \end{bmatrix} + \begin{bmatrix} 1 & 0 \\ 0 & 1 \\ 1 & 2 \end{bmatrix}$

2. $\begin{bmatrix} -1 & 2 \\ 3 & 4 \end{bmatrix} - \begin{bmatrix} 1 & 2 \\ 5 & -2 \end{bmatrix}$

3. $\begin{bmatrix} -3 & 2 & 1 \end{bmatrix} \begin{bmatrix} 2 & 1 \\ -1 & 0 \\ 2 & 1 \end{bmatrix}$

4. $\begin{bmatrix} 1 & 3 & 2 \\ -1 & 2 & 3 \end{bmatrix} \begin{bmatrix} 1 \\ 4 \\ 2 \end{bmatrix}$

In Exercises 5–8, find the values of the variables.

5. $\begin{bmatrix} 1 & x \\ y & 3 \end{bmatrix} = \begin{bmatrix} z & 2 \\ 3 & w \end{bmatrix}$

6. $\begin{bmatrix} 3 & x \\ y & 3 \end{bmatrix} \begin{bmatrix} 1 \\ 2 \end{bmatrix} = \begin{bmatrix} 7 \\ 4 \end{bmatrix}$

7. $\begin{bmatrix} 3 & a+3 \\ -1 & b \\ c+1 & d \end{bmatrix} = \begin{bmatrix} 3 & 6 \\ e+2 & 4 \\ -1 & 2 \end{bmatrix}$

8. $\begin{bmatrix} x & 3 & 1 \\ 0 & y & 2 \end{bmatrix} \begin{bmatrix} 1 & 1 \\ 3 & z \\ 4 & 2 \end{bmatrix} = \begin{bmatrix} 12 & 4 \\ 2 & 2 \end{bmatrix}$

In Exercises 9–16, compute the expressions if possible, given that

$$A = \begin{bmatrix} 1 & 3 & 1 \\ -2 & 1 & 3 \\ 4 & 0 & 2 \end{bmatrix}$$

$$B = \begin{bmatrix} 2 & 1 & 3 \\ -2 & -1 & -1 \\ 1 & 4 & 2 \end{bmatrix}$$

$$C = \begin{bmatrix} 3 & -1 & 2 \\ 1 & 6 & 4 \\ 2 & 1 & 3 \end{bmatrix}$$

9. $2A + 3B$ **10.** $3A - 2B$

11. $2(3A)$ **12.** $2(3A - 4B)$

13. $A(B - C)$ **14.** $AB + AC$

15. $A(BC)$ **16.** $\frac{1}{2}(CA - CB)$

In Exercises 17–24, solve the system of linear equations using the Gauss–Jordan elimination method.

17. $2x - 3y = 5$
    $3x + 4y = -1$

18. $3x + 2y = 3$
    $2x - 4y = -14$

19. $x - y + 2z = 5$
    $3x + 2y + z = 10$
    $2x - 3y - 2z = -10$

20. $3x - 2y + 4z = 16$
    $2x + y - 2z = -1$
    $x + 4y - 8z = -18$

21. $3x - 2y + 4z = 11$
    $2x - 4y + 5z = 4$
    $x + 2y - z = 10$

22. $x - 2y + 3z + 4w = 17$
    $2x + y - 2z - 3w = -9$
    $3x - y + 2z - 4w = 0$
    $4x + 2y - 3z + w = -2$

23. $3x - 2y + z = 4$
    $x + 3y - 4z = -3$
    $2x - 3y + 5z = 7$
    $x - 8y + 9z = 10$

24. $2x - 3y + z = 10$
    $3x + 2y - 2z = -2$
    $x - 3y - 4z = -7$
    $4x + y - z = 4$

In Exercises 25–32, find the inverse of the matrix (if it exists).

25. $A = \begin{bmatrix} 3 & 1 \\ 1 & 2 \end{bmatrix}$

26. $A = \begin{bmatrix} 2 & 4 \\ 1 & 6 \end{bmatrix}$

27. $A = \begin{bmatrix} 3 & 4 \\ 2 & 2 \end{bmatrix}$

28. $A = \begin{bmatrix} 2 & 4 \\ 1 & -2 \end{bmatrix}$

29. $A = \begin{bmatrix} 2 & 3 & 1 \\ 1 & -1 & 2 \\ 1 & 2 & 1 \end{bmatrix}$

30. $A = \begin{bmatrix} 1 & 2 & 4 \\ 2 & 1 & 3 \\ -1 & 0 & 2 \end{bmatrix}$

31. $A = \begin{bmatrix} 1 & 2 & 4 \\ 3 & 1 & 2 \\ 1 & 0 & -6 \end{bmatrix}$

32. $A = \begin{bmatrix} 2 & 1 & -3 \\ 1 & 2 & -4 \\ 3 & 1 & -2 \end{bmatrix}$

In Exercises 33–36, compute the value of the expressions if possible, given that

$$A = \begin{bmatrix} 1 & 2 \\ -1 & 2 \end{bmatrix} \quad B = \begin{bmatrix} 3 & 1 \\ 4 & 2 \end{bmatrix} \quad C = \begin{bmatrix} 1 & 1 \\ -1 & 2 \end{bmatrix}$$

33. $(A^{-1}B)^{-1}$

34. $(ABC)^{-1}$

35. $(2A - C)^{-1}$

36. $(A + B)^{-1}$

In Exercises 37–40, write each system of linear equations in the form $AX = C$. Find A^{-1} and use the result to solve the system.

37. $2x + 3y = -8$
    $x - 2y = 3$

38. $x - 3y = -1$
    $2x + 4y = 8$

39. $x - 2y + 4z = 13$
    $2x + 3y - 2z = 0$
    $x + 4y - 6z = -15$

40. $2x - 3y + 4z = 17$
    $x + 2y - 4z = -7$
    $3x - y + 2z = 14$

41. **GASOLINE SALES** Gloria Newburg operates three self-service gasoline stations in different parts of town. On a certain day, station A sold 600 gal of premium, 800 gal of super, 1000 gal of regular gasoline, and 700 gal of diesel fuel; station B sold 700 gal of premium, 600 gal of super, 1200 gal of regular gasoline, and 400 gal of diesel fuel; station C sold 900 gal of premium, 700 gal of super, 1400 gal of regular gasoline, and 800 gal of diesel fuel. Assume that the price of gasoline was $2.60/gal for premium, $2.40/gal for super, and $2.20/gal for regular and that diesel fuel sold for $2.50/gal. Use matrix algebra to find the total revenue at each station.

42. **COMMON STOCK TRANSACTIONS** Jack Spaulding bought 10,000 shares of stock X, 20,000 shares of stock Y, and 30,000 shares of stock Z at a unit price of $20, $30, and $50 per share, respectively. Six months later, the closing prices of stocks X, Y, and Z were $22, $35, and $51 per share, respectively. Jack made no other stock transactions during the period in question. Compare the value of Jack's stock holdings at the time of purchase and 6 mo later.

43. **MACHINE SCHEDULING** Desmond Jewelry wishes to produce three types of pendants: type A, type B, and type C. To manufacture a type-A pendant requires 2 min on machines I and II and 3 min on machine III. A type-B pendant requires 2 min on machine I, 3 min on machine II, and 4 min on machine III. A type-C pendant requires 3 min on machine I, 4 min on machine II, and 3 min on machine III. There are $3\frac{1}{2}$ hr available on machine I, $4\frac{1}{2}$ hr available on machine II, and 5 hr available on machine III. How many pendants of each type should Desmond make in order to use all the available time?

44. **PETROLEUM PRODUCTION** Wildcat Oil Company has two refineries, one located in Houston and the other in Tulsa. The Houston refinery ships 60% of its petroleum to a Chicago distributor and 40% of its petroleum to a Los Angeles distributor. The Tulsa refinery ships 30% of its petroleum to the Chicago distributor and 70% of its petroleum to the Los Angeles distributor. Assume that, over the year, the Chicago distributor received 240,000 gal of petroleum and the Los Angeles distributor received 460,000 gal of petroleum. Find the amount of petroleum produced at each of Wildcat's refineries.

CHAPTER 2   **Before Moving On . . .**

1. Solve the following system of linear equations, using the Gauss–Jordan elimination method:

$$2x + y - z = -1$$
$$x + 3y + 2z = 2$$
$$3x + 3y - 3z = -5$$

2. Find the solution(s), if it exists, of the system of linear equations whose augmented matrix in reduced form follows.

a. $\begin{bmatrix} 1 & 0 & 0 & | & 2 \\ 0 & 1 & 0 & | & -3 \\ 0 & 0 & 1 & | & 1 \end{bmatrix}$ b. $\begin{bmatrix} 1 & 0 & 0 & | & 3 \\ 0 & 1 & 0 & | & 0 \\ 0 & 0 & 0 & | & 1 \end{bmatrix}$

c. $\begin{bmatrix} 1 & 0 & 0 & | & 2 \\ 0 & 1 & 3 & | & 1 \\ 0 & 0 & 0 & | & 0 \end{bmatrix}$ d. $\begin{bmatrix} 1 & 0 & 0 & 0 & | & 0 \\ 0 & 1 & 0 & 0 & | & 0 \\ 0 & 0 & 1 & 0 & | & 0 \\ 0 & 0 & 0 & 1 & | & 0 \end{bmatrix}$

e. $\begin{bmatrix} 1 & 0 & -1 & | & 2 \\ 0 & 1 & 2 & | & 3 \end{bmatrix}$

3. Solve each system of linear equations using the Gauss–Jordan elimination method.

a. $x + 2y = 3$
   $3x - y = -5$
   $4x + y = -2$

b. $x - 2y + 4z = 2$
   $3x + y - 2z = 1$

4. Let

$$A = \begin{bmatrix} 1 & -2 & 4 \\ 3 & 0 & 1 \end{bmatrix} \qquad B = \begin{bmatrix} 1 & -1 & 2 \\ 3 & 1 & -1 \\ 2 & 1 & 0 \end{bmatrix}$$

$$C = \begin{bmatrix} 2 & -2 \\ 1 & 1 \\ 3 & 4 \end{bmatrix}$$

Find (a) AB, (b) $(A + C^T)B$, and (c) $C^T B - AB^T$.

5. Find A^{-1} if

$$A = \begin{bmatrix} 2 & 1 & 2 \\ 0 & -1 & 3 \\ 1 & 1 & 0 \end{bmatrix}$$

6. Solve the system

$$2x \qquad + z = 4$$
$$2x + y - z = -1$$
$$3x + y - z = 0$$

by first writing it in the matrix form $AX = B$ and then finding A^{-1}.

3

Linear Programming: A Geometric Approach

How should the aircraft engines be shipped? Curtis-Roe Aviation Industries manufactures jet engines in two different locations. These engines are to be shipped to the company's two main assembly plants. In Example 3, page 178, we will show how many engines should be produced and shipped from each manufacturing plant to each assembly plant in order to minimize shipping costs.

© Michael Melford/The Image Bank/Getty Images

MANY PRACTICAL PROBLEMS involve maximizing or minimizing a function subject to certain constraints. For example, we may wish to maximize a profit function subject to certain limitations on the amount of material and labor available. Maximization or minimization problems that can be formulated in terms of a *linear* objective function and constraints in the form of linear inequalities are called *linear programming problems*. In this chapter we look at linear programming problems involving two variables. These problems are amenable to geometric analysis, and the method of solution introduced here will shed much light on the basic nature of a linear programming problem.

3.1   Graphing Systems of Linear Inequalities in Two Variables

Graphing Linear Inequalities

In Chapter 1, we saw that a linear equation in two variables x and y

$$ax + by + c = 0 \qquad \text{\small a, b not both equal to zero}$$

has a *solution set* that may be exhibited graphically as points on a straight line in the xy-plane. We now show that there is also a simple graphical representation for **linear inequalities** in two variables:

$$ax + by + c < 0 \qquad ax + by + c \le 0$$
$$ax + by + c > 0 \qquad ax + by + c \ge 0$$

Before turning to a general procedure for graphing such inequalities, let's consider a specific example. Suppose we wish to graph

$$2x + 3y < 6 \tag{1}$$

We first graph the equation $2x + 3y = 6$, which is obtained by replacing the given inequality "<" with an equality "=" (Figure 1).

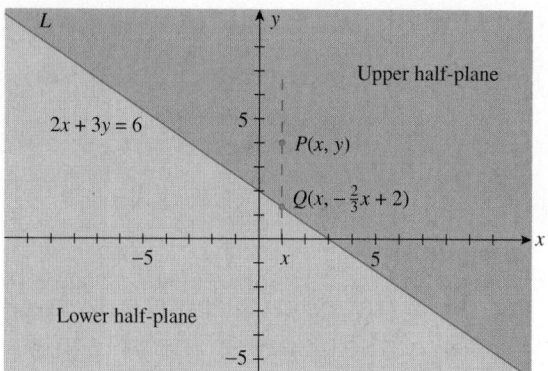

FIGURE 1
A straight line divides the xy-plane into two half-planes.

Observe that this line divides the xy-plane into two half-planes: an upper half-plane and a lower half-plane. Let's show that the upper half-plane is the graph of the linear inequality

$$2x + 3y > 6 \tag{2}$$

whereas the lower half-plane is the graph of the linear inequality

$$2x + 3y < 6 \tag{3}$$

To see this, let's write Equations (2) and (3) in the equivalent forms

$$y > -\frac{2}{3}x + 2 \tag{4}$$

and

$$y < -\frac{2}{3}x + 2 \tag{5}$$

The equation of the line itself is

$$y = -\frac{2}{3}x + 2 \tag{6}$$

Now pick any point $P(x, y)$ lying above the line L. Let Q be the point lying on L and directly below P (see Figure 1). Since Q lies on L, its coordinates must satisfy Equation (6). In other words, Q has representation $Q(x, -\frac{2}{3}x + 2)$. Comparing the y-coordinates of P and Q and recalling that P lies above Q, so that its y-coordinate must be larger than that of Q, we have

$$y > -\frac{2}{3}x + 2$$

But this inequality is just Inequality (4) or, equivalently, Inequality (2). Similarly, we can show that any point lying below L must satisfy Inequality (5) and therefore (3).

This analysis shows that the lower half-plane provides a solution to our problem (Figure 2). (The dashed line shows that the points on L do not belong to the solution set.) Observe that the two half-planes in question are mutually exclusive; that is, they do not have any points in common. Because of this, there is an alternative and easier method of determining the solution to the problem.

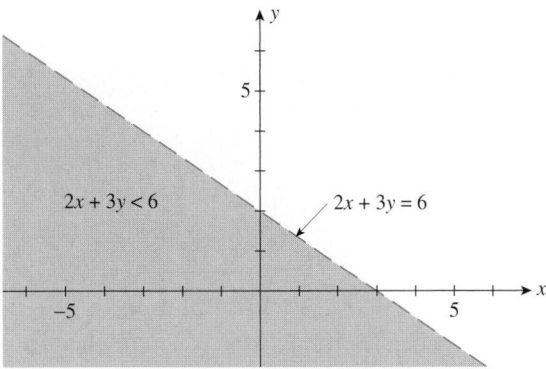

FIGURE 2
The set of points lying below the dashed line satisfies the given inequality.

To determine the required half-plane, let's pick *any* point lying in one of the half-planes. For simplicity, pick the origin $(0, 0)$, which lies in the lower half-plane. Substituting $x = 0$ and $y = 0$ (the coordinates of this point) into the given Inequality (1), we find

$$2(0) + 3(0) < 6$$

or $0 < 6$, which is certainly true. This tells us that the required half-plane is the one containing the test point—namely, the lower half-plane.

Next, let's see what happens if we choose the point $(2, 3)$, which lies in the upper half-plane. Substituting $x = 2$ and $y = 3$ into the given inequality, we find

$$2(2) + 3(3) < 6$$

or $13 < 6$, which is false. This tells us that the upper half-plane is *not* the required half-plane, as expected. Note, too, that no point $P(x, y)$ lying on the line constitutes a solution to our problem, given the *strict* inequality $<$.

This discussion suggests the following procedure for graphing a linear inequality in two variables.

> **Procedure for Graphing Linear Inequalities**
> 1. Draw the graph of the equation obtained for the given inequality by replacing the inequality sign with an equal sign. Use a dashed or dotted line if the problem involves a strict inequality, $<$ or $>$. Otherwise, use a solid line to indicate that the line itself constitutes part of the solution.
> 2. Pick a test point lying in one of the half-planes determined by the line sketched in step 1 and substitute the values of x and y into the given inequality. Use the origin whenever possible.
> 3. If the inequality is satisfied, the graph of the inequality includes the half-plane containing the test point. Otherwise, the solution includes the half-plane not containing the test point.

EXAMPLE 1   Determine the solution set for the inequality $2x + 3y \geq 6$.

Solution   Replacing the inequality $\geq$ with an equality $=$, we obtain the equation $2x + 3y = 6$, whose graph is the straight line shown in Figure 3. Instead of a

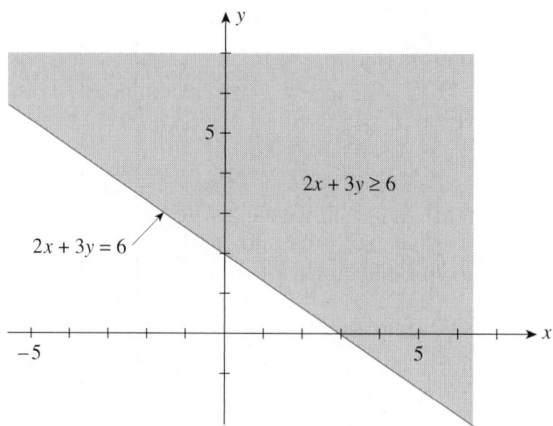

FIGURE 3
The set of points lying on the line and in the upper half-plane satisfies the given inequality.

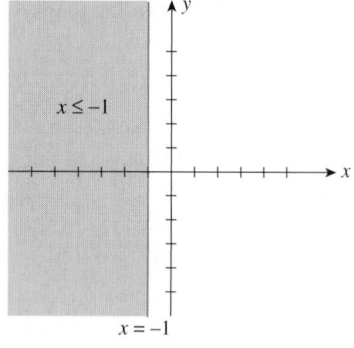

FIGURE 4
The set of points lying on the line $x = -1$ and in the left half-plane satisfies the given inequality.

dashed line as before, we use a solid line to show that all points on the line are also solutions to the problem. Picking the origin as our test point, we find $2(0) + 3(0) \geq 6$, or $0 \geq 6$, which is impossible. So we conclude that the solution set is made up of the half-plane not containing the origin, including (in this case) the line given by $2x + 3y = 6$.   ■

EXAMPLE 2   Graph $x \leq -1$.

Solution   The graph of $x = -1$ is the vertical line shown in Figure 4. Picking the origin $(0, 0)$ as a test point, we find $0 \leq -1$, which is false. Therefore, the required solution is the *left* half-plane, which does not contain the origin.   ■

EXAMPLE 3   Graph $x - 2y > 0$.

Solution   We first graph the equation $x - 2y = 0$, or $y = \frac{1}{2}x$ (Figure 5). Since the origin lies on the line, we may not use it as a test point. (Why?) Let's pick $(1, 2)$ as a test point. Substituting $x = 1$ and $y = 2$ into the given inequality, we find $1 - 2(2) > 0$, or $-3 > 0$, which is false. Therefore, the required solution is the half-plane that does not contain the test point—namely, the lower half-plane.

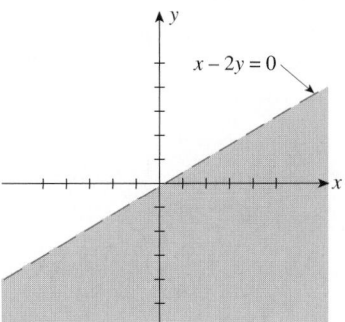

FIGURE 5
The set of points in the lower half-plane satisfies $x - 2y > 0$.

Graphing Systems of Linear Inequalities

By the **solution set of a system of linear inequalities** in the two variables x and y we mean the set of all points (x, y) satisfying each inequality of the system. The graphical solution of such a system may be obtained by graphing the solution set for each inequality independently and then determining the region in common with each solution set.

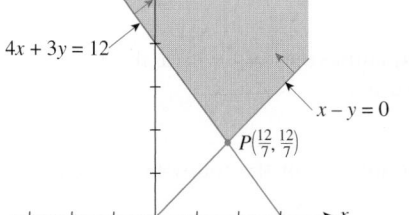

FIGURE 6
The set of points in the shaded area satisfies the system
$$4x + 3y \geq 12$$
$$x - y \leq 0$$

EXAMPLE 4 Determine the solution set for the system

$$4x + 3y \geq 12$$
$$x - y \leq 0$$

Solution Proceeding as in the previous examples, you should have no difficulty locating the half-planes determined by each of the linear inequalities that make up the system. These half-planes are shown in Figure 6. The intersection of the two half-planes is the shaded region. A point in this region is an element of the solution set for the given system. The point P, the intersection of the two straight lines determined by the equations, is found by solving the simultaneous equations

$$4x + 3y = 12$$
$$x - y = 0$$

EXAMPLE 5 Sketch the solution set for the system

$$x \geq 0$$
$$y \geq 0$$
$$x + y - 6 \leq 0$$
$$2x + y - 8 \leq 0$$

Solution The first inequality in the system defines the right half-plane—all points to the right of the y-axis plus all points lying on the y-axis itself. The second inequality in the system defines the upper half-plane, including the x-axis. The half-planes defined by the third and fourth inequalities are indicated by arrows in Figure 7. Thus, the required region—the intersection of the four half-planes defined by the four inequalities in the given system of linear inequalities—is the shaded region. The point P is found by solving the simultaneous equations $x + y - 6 = 0$ and $2x + y - 8 = 0$.

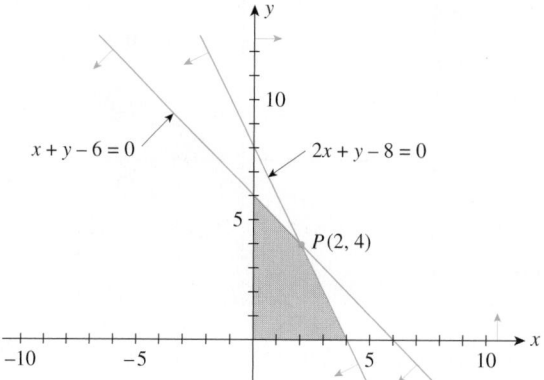

FIGURE 7
The set of points in the shaded region, including the x- and y-axes, satisfies the given inequalities.

The solution set found in Example 5 is an example of a bounded set. Observe that the set can be enclosed by a circle. For example, if you draw a circle of radius 10 with center at the origin, you will see that the set lies entirely inside the circle. On the other hand, the solution set of Example 4 cannot be enclosed by a circle and is said to be unbounded.

Bounded and Unbounded Solution Sets
The solution set of a system of linear inequalities is bounded if it can be enclosed by a circle. Otherwise, it is unbounded.

EXAMPLE 6   Determine the graphical solution set for the following system of linear inequalities:

$$\begin{aligned} 2x + \ y &\geq 50 \\ x + 2y &\geq 40 \\ x &\geq 0 \\ y &\geq 0 \end{aligned}$$

Solution   The required solution set is the unbounded region shown in Figure 8.

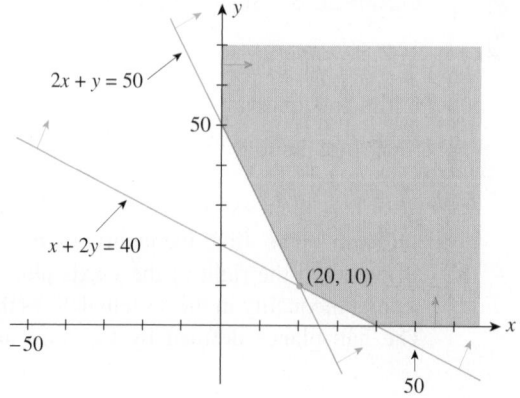

FIGURE 8
The solution set is an unbounded region.

3.1 Self-Check Exercises

1. Determine graphically the solution set for the following system of inequalities:

$$x + 2y \leq 10$$
$$5x + 3y \leq 30$$
$$x \geq 0, y \geq 0$$

2. Determine graphically the solution set for the following system of inequalities:

$$5x + 3y \geq 30$$
$$x - 3y \leq 0$$
$$x \geq 2$$

Solutions to Self-Check Exercises 3.1 can be found on page 175.

3.1 Concept Questions

1. **a.** What is the difference, geometrically, between the solution set of $ax + by < c$ and the solution set of $ax + by \leq c$?
   b. Describe the set that is obtained by intersecting the solution set of $ax + by \leq c$ with the solution set of $ax + by \geq c$.

2. **a.** What is the solution set of a system of linear inequalities?
   b. How do you find the graphical solution of a system of linear inequalities?

3.1 Exercises

In Exercises 1–10, find the graphical solution of each inequality.

1. $4x - 8 < 0$

2. $3y + 2 > 0$

3. $x - y \leq 0$

4. $3x + 4y \leq -2$

5. $x \leq -3$

6. $y \geq -1$

7. $2x + y \leq 4$

8. $-3x + 6y \geq 12$

9. $4x - 3y \leq -24$

10. $5x - 3y \geq 15$

In Exercises 11–18, write a system of linear inequalities that describes the shaded region.

11.

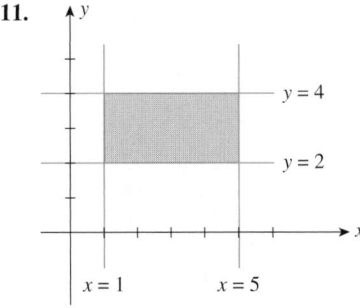

12.

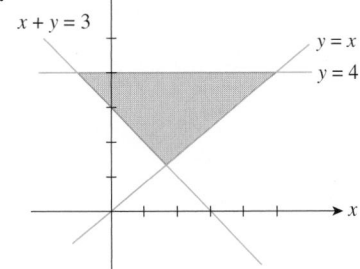

13.

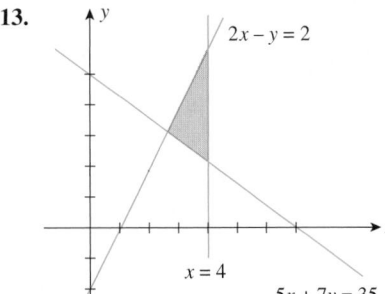

14.

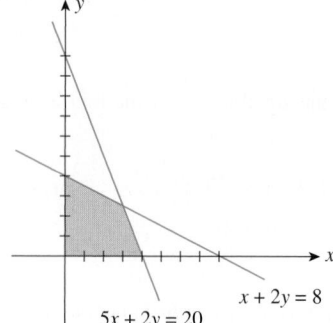

$x + 2y = 8$

$5x + 2y = 20$

15.

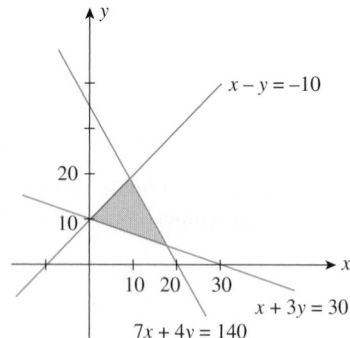

$x - y = -10$

$x + 3y = 30$

$7x + 4y = 140$

16.

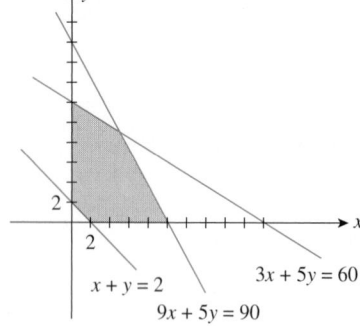

$x + y = 2$

$3x + 5y = 60$

$9x + 5y = 90$

17.

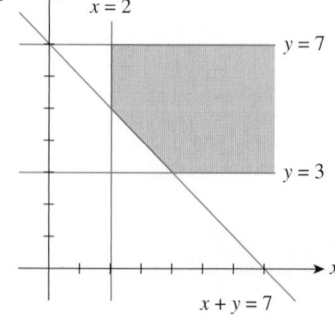

$x = 2$

$y = 7$

$y = 3$

$x + y = 7$

18.

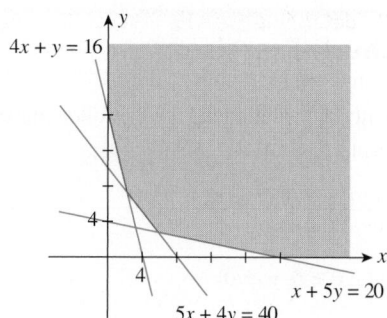

$4x + y = 16$

$x + 5y = 20$

$5x + 4y = 40$

In Exercises 19–36, determine graphically the solution set for each system of inequalities and indicate whether the solution set is bounded or unbounded.

19. $\begin{aligned} 2x + 4y &> 16 \\ -x + 3y &\geq 7 \end{aligned}$

20. $\begin{aligned} 3x - 2y &> -13 \\ -x + 2y &> 5 \end{aligned}$

21. $\begin{aligned} x - y &\leq 0 \\ 2x + 3y &\geq 10 \end{aligned}$

22. $\begin{aligned} x + y &\geq -2 \\ 3x - y &\leq 6 \end{aligned}$

23. $\begin{aligned} x + 2y &\geq 3 \\ 2x + 4y &\leq -2 \end{aligned}$

24. $\begin{aligned} 2x - y &\geq 4 \\ 4x - 2y &< -2 \end{aligned}$

25. $\begin{aligned} x + y &\leq 6 \\ 0 \leq x &\leq 3 \\ y &\geq 0 \end{aligned}$

26. $\begin{aligned} 4x - 3y &\leq 12 \\ 5x + 2y &\leq 10 \\ x \geq 0, y &\geq 0 \end{aligned}$

27. $\begin{aligned} 3x - 6y &\leq 12 \\ -x + 2y &\leq 4 \\ x \geq 0, y &\geq 0 \end{aligned}$

28. $\begin{aligned} x + y &\geq 20 \\ x + 2y &\geq 40 \\ x \geq 0, y &\geq 0 \end{aligned}$

29. $\begin{aligned} 3x - 7y &\geq -24 \\ x + 3y &\geq 8 \\ x \geq 0, y &\geq 0 \end{aligned}$

30. $\begin{aligned} 3x + 4y &\geq 12 \\ 2x - y &\geq -2 \\ 0 \leq y &\leq 3 \\ x &\geq 0 \end{aligned}$

31. $\begin{aligned} x + 2y &\geq 3 \\ 5x - 4y &\leq 16 \\ 0 \leq y &\leq 2 \\ x &\geq 0 \end{aligned}$

32. $\begin{aligned} x + y &\leq 4 \\ 2x + y &\leq 6 \\ 2x - y &\geq -1 \\ x \geq 0, y &\geq 0 \end{aligned}$

33. $\begin{aligned} 6x + 5y &\leq 30 \\ 3x + y &\geq 6 \\ x + y &\geq 4 \\ x \geq 0, y &\geq 0 \end{aligned}$

34. $\begin{aligned} 6x + 7y &\leq 84 \\ 12x - 11y &\leq 18 \\ 6x - 7y &\leq 28 \\ x \geq 0, y &\geq 0 \end{aligned}$

35. $\begin{aligned} x - y &\geq -6 \\ x - 2y &\leq -2 \\ x + 2y &\geq 6 \\ x - 2y &\geq -14 \\ x \geq 0, y &\geq 0 \end{aligned}$

36. $\begin{aligned} x - 3y &\geq -18 \\ 3x - 2y &\geq 2 \\ x - 3y &\leq -4 \\ 3x - 2y &\leq 16 \\ x \geq 0, y &\geq 0 \end{aligned}$

In Exercises 37–40, determine whether the statement is true or false. If it is true, explain why it is true. If it is false, give an example to show why it is false.

37. The solution set of a linear inequality involving two variables is either a half plane or a straight line.

38. The solution set of the inequality $ax + by + c \leq 0$ is either a left half-plane or a lower half-plane.

39. The solution set of a system of linear inequalities in two variables is bounded if it can be enclosed by a rectangle.

40. The solution set of the system

$$ax + by \leq e$$
$$cx + dy \leq f$$
$$x \geq 0, y \geq 0$$

where a, b, c, d, e, and f are positive real numbers, is a bounded set.

3.1 Solutions to Self-Check Exercises

1. The required solution set is shown in the following figure:

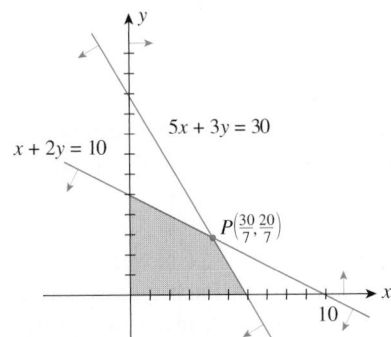

The point P is found by solving the system of equations

$$x + 2y = 10$$
$$5x + 3y = 30$$

Solving the first equation for x in terms of y gives

$$x = 10 - 2y$$

Substituting this value of x into the second equation of the system gives

$$
\begin{aligned}
5(10 - 2y) + 3y &= 30 \\
50 - 10y + 3y &= 30 \\
-7y &= -20
\end{aligned}
$$

so $y = \frac{20}{7}$. Substituting this value of y into the expression for x found earlier, we obtain

$$x = 10 - 2\left(\frac{20}{7}\right) = \frac{30}{7}$$

giving the point of intersection as $\left(\frac{30}{7}, \frac{20}{7}\right)$.

2. The required solution set is shown in the following figure:

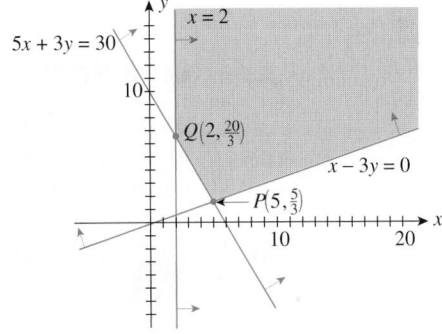

To find the coordinates of P, we solve the system

$$5x + 3y = 30$$
$$x - 3y = 0$$

Solving the second equation for x in terms of y and substituting this value of x in the first equation gives

$$5(3y) + 3y = 30$$

or $y = \frac{5}{3}$. Substituting this value of y into the second equation gives $x = 5$. Next, the coordinates of Q are found by solving the system

$$5x + 3y = 30$$
$$x = 2$$

yielding $x = 2$ and $y = \frac{20}{3}$.

3.2   Linear Programming Problems

In many business and economic problems we are asked to optimize (maximize or minimize) a function subject to a system of equalities or inequalities. The function to be optimized is called the objective function. Profit functions and cost functions are examples of objective functions. The system of equalities or inequalities to which the objective function is subjected reflects the constraints (for example, limitations on resources such as materials and labor) imposed on the solution(s) to the problem. Problems of this nature are called **mathematical programming problems**. In particular, problems in which both the objective function and the constraints are expressed as linear equations or inequalities are called linear programming problems.

> **Linear Programming Problem**
>
> A linear programming problem consists of a linear objective function to be maximized or minimized subject to certain constraints in the form of linear equations or inequalities.

■ A Maximization Problem

As an example of a linear programming problem in which the objective function is to be maximized, let's consider the following simplified version of a production problem involving two variables.

APPLIED EXAMPLE 1   A Production Problem   Ace Novelty wishes to produce two types of souvenirs: type A and type B. Each type-A souvenir will result in a profit of $1, and each type-B souvenir will result in a profit of $1.20. To manufacture a type-A souvenir requires 2 minutes on machine I and 1 minute on machine II. A type-B souvenir requires 1 minute on machine I and 3 minutes on machine II. There are 3 hours available on machine I and 5 hours available on machine II. How many souvenirs of each type should Ace make in order to maximize its profit?

Solution   As a first step toward the mathematical formulation of this problem, we tabulate the given information (see Table 1).

TABLE 1

	Type A	Type B	Time Available
Machine I	2 min	1 min	180 min
Machine II	1 min	3 min	300 min
Profit/Unit	$1	$1.20	

Let x be the number of type-A souvenirs and y the number of type-B souvenirs to be made. Then, the total profit P (in dollars) is given by

$$P = x + 1.2y$$

which is the objective function to be maximized.

The total amount of time that machine I is used is given by $2x + y$ minutes and must not exceed 180 minutes. Thus, we have the inequality

$$2x + y \leq 180$$

Similarly, the total amount of time that machine II is used is $x + 3y$ minutes, which cannot exceed 300 minutes, so we are led to the inequality

$$x + 3y \leq 300$$

Finally, neither x nor y can be negative, so

$$x \geq 0$$
$$y \geq 0$$

To summarize, the problem at hand is one of maximizing the objective function $P = x + 1.2y$ subject to the system of inequalities

$$2x + \ y \leq 180$$
$$x + 3y \leq 300$$
$$x \geq 0$$
$$y \geq 0$$

The solution to this problem will be completed in Example 1, Section 3.3. ■

■ Minimization Problems

In the following linear programming problem, the objective function is to be minimized.

APPLIED EXAMPLE 2 A Nutrition Problem A nutritionist advises an individual who is suffering from iron and vitamin-B deficiency to take at least 2400 milligrams (mg) of iron, 2100 mg of vitamin B_1 (thiamine), and 1500 mg of vitamin B_2 (riboflavin) over a period of time. Two vitamin pills are suitable, brand A and brand B. Each brand-A pill costs 6 cents and contains 40 mg of iron, 10 mg of vitamin B_1, and 5 mg of vitamin B_2. Each brand-B pill costs 8 cents and contains 10 mg of iron and 15 mg each of vitamins B_1 and B_2 (Table 2). What combination of pills should the individual purchase in order to meet the minimum iron and vitamin requirements at the lowest cost?

TABLE 2

	Brand A	Brand B	Minimum Requirement
Iron	40 mg	10 mg	2400 mg
Vitamin B_1	10 mg	15 mg	2100 mg
Vitamin B_2	5 mg	15 mg	1500 mg
Cost/Pill	6¢	8¢	

Solution Let x be the number of brand-A pills and y the number of brand-B pills to be purchased. The cost C (in cents) is given by

$$C = 6x + 8y$$

and is the objective function to be minimized.

The amount of iron contained in x brand-A pills and y brand-B pills is given by $40x + 10y$ mg, and this must be greater than or equal to 2400 mg. This translates into the inequality

$$40x + 10y \geq 2400$$

Similar considerations involving the minimum requirements of vitamins B_1 and B_2 lead to the inequalities

$$10x + 15y \geq 2100$$
$$5x + 15y \geq 1500$$

respectively. Thus, the problem here is to minimize $C = 6x + 8y$ subject to

$$40x + 10y \geq 2400$$
$$10x + 15y \geq 2100$$
$$5x + 15y \geq 1500$$
$$x \geq 0, y \geq 0$$

The solution to this problem will be completed in Example 2, Section 3.3.   ■

APPLIED EXAMPLE 3 A Transportation Problem Curtis-Roe Aviation Industries has two plants, I and II, that produce the Zephyr jet engines used in their light commercial airplanes. The maximum production capacities of these two plants are 100 units and 110 units per month, respectively. The engines are shipped to two of Curtis-Roe's main assembly plants, A and B. The shipping costs (in dollars) per engine from plants I and II to the main assembly plants A and B are as follows:

From	To Assembly Plant A	B
Plant I	100	60
Plant II	120	70

In a certain month, assembly plant A needs 80 engines whereas assembly plant B needs 70 engines. Find how many engines should be shipped from each plant to each main assembly plant if shipping costs are to be kept to a minimum.

Solution Let x denote the number of engines shipped from plant I to assembly plant A, and let y denote the number of engines shipped from plant I to assembly plant B. Since the requirements of assembly plants A and B are 80 and 70 engines, respectively, the number of engines shipped from plant II to assembly plants A and B are $(80 - x)$ and $(70 - y)$, respectively. These numbers may be displayed in a schematic. With the aid of the accompanying schematic and the shipping cost schedule, we find that the total shipping cost incurred by Curtis-Roe is given by

$$C = 100x + 60y + 120(80 - x) + 70(70 - y)$$
$$= 14{,}500 - 20x - 10y$$

Next, the production constraints on plants I and II lead to the inequalities

$$x + y \leq 100$$
$$(80 - x) + (70 - y) \leq 110$$

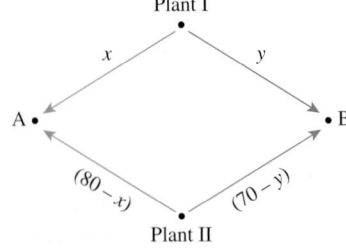

The last inequality simplifies to

$$x + y \geq 40$$

Also, the requirements of the two main assembly plants lead to the inequalities

$$x \geq 0 \qquad y \geq 0 \qquad 80 - x \geq 0 \qquad 70 - y \geq 0$$

The last two may be written as $x \leq 80$ and $y \leq 70$.

Summarizing, we have the following linear programming problem: Minimize the objective (cost) function $C = 14{,}500 - 20x - 10y$ subject to the constraints

$$x + y \geq 40$$
$$x + y \leq 100$$
$$x \leq 80$$
$$y \leq 70$$

where $x \geq 0$ and $y \geq 0$.

You will be asked to complete the solution to this problem in Exercise 43, Section 3.3.

APPLIED EXAMPLE 4   A Warehouse Problem   Acrosonic manufactures its model F loudspeaker systems in two separate locations, plant I and plant II. The output at plant I is at most 400 per month, whereas the output at plant II is at most 600 per month. These loudspeaker systems are shipped to three warehouses that serve as distribution centers for the company. For the warehouses to meet their orders, the minimum monthly requirements of warehouses A, B, and C are 200, 300, and 400 systems, respectively. Shipping costs from plant I to warehouses A, B, and C are $20, $8, and $10 per loudspeaker system, respectively, and shipping costs from plant II to each of these warehouses are $12, $22, and $18, respectively. What should the shipping schedule be if Acrosonic wishes to meet the requirements of the distribution centers and at the same time keep its shipping costs to a minimum?

Solution   The respective shipping costs (in dollars) per loudspeaker system may be tabulated as in Table 3. Letting x_1 denote the number of loudspeaker systems shipped from plant I to warehouse A, x_2 the number shipped from plant I to warehouse B, and so on leads to Table 4.

TABLE 3

Plant	Warehouse A	B	C
I	20	8	10
II	12	22	18

TABLE 4

Plant	Warehouse A	B	C	Max. Prod.
I	x_1	x_2	x_3	400
II	x_4	x_5	x_6	600
Min. Req.	200	300	400	

From Tables 3 and 4 we see that the cost of shipping x_1 loudspeaker systems from plant I to warehouse A is $20x_1$, the cost of shipping x_2 loudspeaker systems from plant I to warehouse B is $\$8x_2$, and so on. Thus, the total monthly shipping cost incurred by Acrosonic is given by

$$C = 20x_1 + 8x_2 + 10x_3 + 12x_4 + 22x_5 + 18x_6$$

Next, the production constraints on plants I and II lead to the inequalities

$$x_1 + x_2 + x_3 \leq 400$$
$$x_4 + x_5 + x_6 \leq 600$$

(see Table 4). Also, the minimum requirements of each of the three warehouses lead to the three inequalities

$$x_1 + x_4 \geq 200$$
$$x_2 + x_5 \geq 300$$
$$x_3 + x_6 \geq 400$$

Summarizing, we have the following linear programming problem:

Minimize $C = 20x_1 + 8x_2 + 10x_3 + 12x_4 + 22x_5 + 18x_6$

subject to

$$x_1 + x_2 + x_3 \leq 400$$
$$x_4 + x_5 + x_6 \leq 600$$
$$x_1 + x_4 \geq 200$$
$$x_2 + x_5 \geq 300$$
$$x_3 + x_6 \geq 400$$
$$x_1 \geq 0, x_2 \geq 0, \ldots, x_6 \geq 0$$

The solution to this problem will be completed in Section 4.2, Example 5.   ■

3.2   Self-Check Exercise

Gino Balduzzi, proprietor of Luigi's Pizza Palace, allocates $9000 a month for advertising in two newspapers, the *City Tribune* and the *Daily News*. The *City Tribune* charges $300 for a certain advertisement, whereas the *Daily News* charges $100 for the same ad. Gino has stipulated that the ad is to appear in at least 15 but no more than 30 editions of the *Daily News* per month. The *City Tribune* has a daily circulation of 50,000, and the *Daily News* has a circulation of 20,000. Under these condi-

tions, determine how many ads Gino should place in each newspaper in order to reach the largest number of readers. Formulate but do not solve the problem. (The solution to this problem can be found in Exercise 3 of Solutions to Self-Check Exercises 3.3.)

The solution to Self-Check Exercise 3.2 can be found on page 184.

3.2   Concept Questions

1. What is a linear programming problem?

2. Suppose you are asked to formulate a linear programming problem in two variables x and y. How would you express the fact that x and y are nonnegative? Why are these conditions often required in practical problems?

3. What is the difference between a maximization linear programming problem and a minimization linear programming problem?

3.2 Exercises

Formulate but do not solve each of the following exercises as a linear programming problem. You will be asked to solve these problems later.

1. MANUFACTURING—PRODUCTION SCHEDULING A company manufactures two products, A and B, on two machines, I and II. It has been determined that the company will realize a profit of $3 on each unit of product A and a profit of $4 on each unit of product B. To manufacture a unit of product A requires 6 min on machine I and 5 min on machine II. To manufacture a unit of product B requires 9 min on machine I and 4 min on machine II. There are 5 hr of machine time available on machine I and 3 hr of machine time available on machine II in each work shift. How many units of each product should be produced in each shift to maximize the company's profit?

2. MANUFACTURING—PRODUCTION SCHEDULING National Business Machines manufactures two models of fax machines: A and B. Each model A costs $100 to make, and each model B costs $150. The profits are $30 for each model A and $40 for each model B fax machine. If the total number of fax machines demanded per month does not exceed 2500 and the company has earmarked no more than $600,000/month for manufacturing costs, how many units of each model should National make each month in order to maximize its monthly profits?

3. MANUFACTURING—PRODUCTION SCHEDULING Kane Manufacturing has a division that produces two models of fireplace grates, model A and model B. To produce each model A grate requires 3 lb of cast iron and 6 min of labor. To produce each model B grate requires 4 lb of cast iron and 3 min of labor. The profit for each model A grate is $2.00, and the profit for each model B grate is $1.50. If 1000 lb of cast iron and 20 hr of labor are available for the production of grates per day, how many grates of each model should the division produce per day in order to maximize Kane's profits?

4. MANUFACTURING—PRODUCTION SCHEDULING Refer to Exercise 3. Because of a backlog of orders on model A grates, the manager of Kane Manufacturing has decided to produce at least 150 of these models a day. Operating under this additional constraint, how many grates of each model should Kane produce to maximize profit?

5. FINANCE—ALLOCATION OF FUNDS Madison Finance has a total of $20 million earmarked for homeowner and auto loans. On the average, homeowner loans have a 10% annual rate of return whereas auto loans yield a 12% annual rate of return. Management has also stipulated that the total amount of homeowner loans should be greater than or equal to 4 times the total amount of automobile loans. Determine the total amount of loans of each type Madison should extend to each category in order to maximize its returns.

6. INVESTMENTS—ASSET ALLOCATION A financier plans to invest up to $500,000 in two projects. Project A yields a return of 10% on the investment whereas project B yields a return of 15% on the investment. Because the investment in project B is riskier than the investment in project A, the financier has decided that the investment in project B should not exceed 40% of the total investment. How much should she invest in each project in order to maximize the return on her investment?

7. MANUFACTURING—PRODUCTION SCHEDULING Acoustical Company manufactures a CD storage cabinet that can be bought fully assembled or as a kit. Each cabinet is processed in the fabrications department and the assembly department. If the fabrication department only manufactures fully assembled cabinets, then it can produce 200 units/day; and if it only manufactures kits, it can produce 200 units/day. If the assembly department only produces fully assembled cabinets, then it can produce 100 units/day; but if it only produces kits, then it can produce 300 units/day. Each fully assembled cabinet contributes $50 to the profits of the company whereas each kit contributes $40 to its profits. How many fully assembled units and how many kits should the company produce per day in order to maximize its profits?

8. AGRICULTURE—CROP PLANNING A farmer plans to plant two crops, A and B. The cost of cultivating crop A is $40/acre whereas that of crop B is $60/acre. The farmer has a maximum of $7400 available for land cultivation. Each acre of crop A requires 20 labor-hours, and each acre of crop B requires 25 labor-hours. The farmer has a maximum of 3300 labor-hours available. If she expects to make a profit of $150/acre on crop A and $200/acre on crop B, how many acres of each crop should she plant in order to maximize her profit?

9. MINING—PRODUCTION Perth Mining Company operates two mines for the purpose of extracting gold and silver. The Saddle Mine costs $14,000/day to operate, and it yields 50 oz of gold and 3000 oz of silver each day. The Horseshoe Mine costs $16,000/day to operate, and it yields 75 oz of gold and 1000 oz of silver each day. Company management has set a target of at least 650 oz of gold and 18,000 oz of silver. How many days should each mine be operated so that the target can be met at a minimum cost?

10. TRANSPORTATION Deluxe River Cruises operates a fleet of river vessels. The fleet has two types of vessels: A type-A vessel has 60 deluxe cabins and 160 standard cabins, whereas a type-B vessel has 80 deluxe cabins and 120 standard cabins. Under a charter agreement with Odyssey Travel Agency, Deluxe River Cruises is to provide Odyssey with a minimum of 360 deluxe and 680 standard cabins for their 15-day cruise in May. It costs $44,000 to operate a type-A

vessel and $54,000 to operate a type-B vessel for that period. How many of each type vessel should be used in order to keep the operating costs to a minimum?

11. NUTRITION—DIET PLANNING A nutritionist at the Medical Center has been asked to prepare a special diet for certain patients. She has decided that the meals should contain a minimum of 400 mg of calcium, 10 mg of iron, and 40 mg of vitamin C. She has further decided that the meals are to be prepared from foods A and B. Each ounce of food A contains 30 mg of calcium, 1 mg of iron, 2 mg of vitamin C, and 2 mg of cholesterol. Each ounce of food B contains 25 mg of calcium, 0.5 mg of iron, 5 mg of vitamin C, and 5 mg of cholesterol. Find how many ounces of each type of food should be used in a meal so that the cholesterol content is minimized and the minimum requirements of calcium, iron, and vitamin C are met.

12. SOCIAL PROGRAMS PLANNING AntiFam, a hunger-relief organization, has earmarked between $2 and $2.5 million (inclusive) for aid to two African countries, country A and country B. Country A is to receive between $1 and $1.5 million (inclusive), and country B is to receive at least $0.75 million. It has been estimated that each dollar spent in country A will yield an effective return of $.60, whereas a dollar spent in country B will yield an effective return of $.80. How should the aid be allocated if the money is to be utilized most effectively according to these criteria?

Hint: If x and y denote the amount of money to be given to country A and country B, respectively, then the objective function to be maximized is $P = 0.6x + 0.8y$.

13. ADVERTISING Everest Deluxe World Travel has decided to advertise in the Sunday editions of two major newspapers in town. These advertisements are directed at three groups of potential customers. Each advertisement in newspaper I is seen by 70,000 group-A customers, 40,000 group-B customers, and 20,000 group-C customers. Each advertisement in newspaper II is seen by 10,000 group-A, 20,000 group-B, and 40,000 group-C customers. Each advertisement in newspaper I costs $1000, and each advertisement in newspaper II costs $800. Everest would like their advertisements to be read by at least 2 million people from group A, 1.4 million people from group B, and 1 million people from group C. How many advertisements should Everest place in each newspaper to achieve its advertisement goals at a minimum cost?

14. MANUFACTURING—SHIPPING COSTS TMA manufactures 37-in. high definition LCD televisions in two separate locations, location I and location II. The output at location I is at most 6000 televisions/month, whereas the output at location II is at most 5000 televisions/month. TMA is the main supplier of televisions to Pulsar Corporation, its holding company, which has priority in having all its requirements met. In a certain month, Pulsar placed orders for 3000 and 4000 televisions to be shipped to two of its factories located in city A and city B, respectively. The shipping costs (in dollars) per

television from the two TMA plants to the two Pulsar factories are as follows:

From TMA	To Pulsar Factories City A	City B
Location I	$6	$4
Location II	$8	$10

Find a shipping schedule that meets the requirements of both companies while keeping costs to a minimum.

15. INVESTMENTS—ASSET ALLOCATION A financier plans to invest up to $2 million in three projects. She estimates that project A will yield a return of 10% on her investment, project B will yield a return of 15% on her investment, and project C will yield a return of 20% on her investment. Because of the risks associated with the investments, she decided to put not more than 20% of her total investment in project C. She also decided that her investments in projects B and C should not exceed 60% of her total investment. Finally, she decided that her investment in project A should be at least 60% of her investments in projects B and C. How much should the financier invest in each project if she wishes to maximize the total returns on her investments?

16. INVESTMENTS—ASSET ALLOCATION Ashley has earmarked at most $250,000 for investment in three mutual funds: a money market fund, an international equity fund, and a growth-and-income fund. The money market fund has a rate of return of 6%/year, the international equity fund has a rate of return of 10%/year, and the growth-and-income fund has a rate of return of 15%/year. Ashley has stipulated that no more than 25% of her total portfolio should be in the growth-and-income fund and that no more than 50% of her total portfolio should be in the international equity fund. To maximize the return on her investment, how much should Ashley invest in each type of fund?

17. MANUFACTURING—PRODUCTION SCHEDULING A company manufactures products A, B, and C. Each product is processed in three departments: I, II, and III. The total available labor-hours per week for departments I, II, and III are 900, 1080, and 840, respectively. The time requirements (in hours per unit) and profit per unit for each product are as follows:

	Product A	Product B	Product C
Dept. I	2	1	2
Dept. II	3	1	2
Dept. III	2	2	1
Profit	$18	$12	$15

How many units of each product should the company produce in order to maximize its profit?

18. ADVERTISING As part of a campaign to promote its annual clearance sale, the Excelsior Company decided to buy television advertising time on Station KAOS. Excelsior's advertising budget is $102,000. Morning time costs $3000/minute, afternoon time costs $1000/minute, and evening (prime) time costs $12,000/minute. Because of previous commitments, KAOS cannot offer Excelsior more than 6 min of prime time or more than a total of 25 min of advertising time over the 2 weeks in which the commercials are to be run. KAOS estimates that morning commercials are seen by 200,000 people, afternoon commercials are seen by 100,000 people, and evening commercials are seen by 600,000 people. How much morning, afternoon, and evening advertising time should Excelsior buy in order to maximize exposure of its commercials?

19. MANUFACTURING—PRODUCTION SCHEDULING Custom Office Furniture Company is introducing a new line of executive desks made from a specially selected grade of walnut. Initially, three different models—A, B, and C—are to be marketed. Each model A desk requires $1\frac{1}{4}$ hr for fabrication, 1 hr for assembly, and 1 hr for finishing; each model B desk requires $1\frac{1}{2}$ hr for fabrication, 1 hr for assembly, and 1 hr for finishing; each model C desk requires $1\frac{1}{2}$ hr, $\frac{3}{4}$ hr, and $\frac{1}{2}$ hr for fabrication, assembly, and finishing, respectively. The profit on each model A desk is $26, the profit on each model B desk is $28, and the profit on each model C desk is $24. The total time available in the fabrication department, the assembly department, and the finishing department in the first month of production is 310 hr, 205 hr, and 190 hr, respectively. To maximize Custom's profit, how many desks of each model should be made in the month?

20. MANUFACTURING—SHIPPING COSTS Acrosonic of Example 4 also manufactures a model G loudspeaker system in plants I and II. The output at plant I is at most 800 systems/month whereas the output at plant II is at most 600/month. These loudspeaker systems are also shipped to the three warehouses —A, B, and C—whose minimum monthly requirements are 500, 400, and 400, respectively. Shipping costs from plant I to warehouse A, warehouse B, and warehouse C are $16, $20, and $22 per system, respectively, and shipping costs from plant II to each of these warehouses are $18, $16, and $14 per system, respectively. What shipping schedule will enable Acrosonic to meet the warehouses' requirements and at the same time keep its shipping costs to a minimum?

21. MANUFACTURING—SHIPPING COSTS Steinwelt Piano manufactures uprights and consoles in two plants, plant I and plant II. The output of plant I is at most 300/month whereas the output of plant II is at most 250/month. These pianos are shipped to three warehouses that serve as distribution centers for the company. To fill current and projected future orders, warehouse A requires a minimum of 200 pianos/month, warehouse B requires at least 150 pianos/month, and warehouse C requires at least 200 pianos/month. The shipping cost of each piano from plant I to warehouse A, warehouse

B, and warehouse C is $60, $60, and $80, respectively, and the shipping cost of each piano from plant II to warehouse A, warehouse B, and warehouse C is $80, $70, and $50, respectively. What shipping schedule will enable Steinwelt to meet the warehouses' requirements while keeping shipping costs to a minimum?

22. MANUFACTURING—PREFABRICATED HOUSING PRODUCTION Boise Lumber has decided to enter the lucrative prefabricated housing business. Initially, it plans to offer three models: standard, deluxe, and luxury. Each house is prefabricated and partially assembled in the factory, and the final assembly is completed on site. The dollar amount of building material required, the amount of labor required in the factory for prefabrication and partial assembly, the amount of on-site labor required, and the profit per unit are as follows:

	Standard Model	Deluxe Model	Luxury Model
Material	$6,000	$8,000	$10,000
Factory Labor (hr)	240	220	200
On-site Labor (hr)	180	210	300
Profit	$3,400	$4,000	$5,000

For the first year's production, a sum of $8.2 million is budgeted for the building material; the number of labor-hours available for work in the factory (for prefabrication and partial assembly) is not to exceed 218,000 hr; and the amount of labor for on-site work is to be less than or equal to 237,000 labor-hours. Determine how many houses of each type Boise should produce (market research has confirmed that there should be no problems with sales) in order to maximize its profit from this new venture.

23. PRODUCTION—JUICE PRODUCTS CalJuice Company has decided to introduce three fruit juices made from blending two or more concentrates. These juices will be packaged in 2-qt (64-oz) cartons. One carton of pineapple–orange juice requires 8 oz each of pineapple and orange juice concentrates. One carton of orange–banana juice requires 12 oz of orange juice concentrate and 4 oz of banana pulp concentrate. Finally, one carton of pineapple–orange–banana juice requires 4 oz of pineapple juice concentrate, 8 oz of orange juice concentrate, and 4 oz of banana pulp. The company has decided to allot 16,000 oz of pineapple juice concentrate, 24,000 oz of orange juice concentrate, and 5000 oz of banana pulp concentrate for the initial production run. The company has also stipulated that the production of pineapple–orange–banana juice should not exceed 800 cartons. Its profit on one carton of pineapple–orange juice is $1.00, its profit on one carton of orange–banana juice is $.80, and its profit on one carton of pineapple–orange–banana juice is $.90. To realize a maximum profit, how many cartons of each blend should the company produce?

24. MANUFACTURING—COLD FORMULA PRODUCTION Beyer Pharmaceutical produces three kinds of cold formulas: formula I, formula II, and formula III. It takes 2.5 hr to produce 1000 bottles of formula I, 3 hr to produce 1000 bottles of formula II, and 4 hr to produce 1000 bottles of formula III. The profits for each 1000 bottles of formula I, formula II, and formula III are $180, $200, and $300, respectively. For a certain production run, there are enough ingredients on hand to make at most 9000 bottles of formula I, 12,000 bottles of formula II, and 6000 bottles of formula III. Furthermore, the time for the production run is limited to a maximum of 70 hr. How many bottles of each formula should be produced in this production run so that the profit is maximized?

In Exercises 25 and 26, determine whether the statement is true or false. If it is true, explain why it is true. If it is false, give an example to show why it is false.

25. The problem

$$\text{Maximize} \quad P = xy$$
$$\text{subject to} \quad 2x + 3y \le 12$$
$$2x + y \le 8$$
$$x \ge 0, y \ge 0$$

is a linear programming problem.

26. The problem

$$\text{Minimize} \quad C = 2x + 3y$$
$$\text{subject to} \quad 2x + 3y \le 6$$
$$x - y = 0$$
$$x \ge 0, y \ge 0$$

is a linear programming problem.

3.2 Solution to Self-Check Exercise

Let x denote the number of ads to be placed in the *City Tribune* and y the number to be placed in the *Daily News*. The total cost for placing x ads in the *City Tribune* and y ads in the *Daily News* is $300x + 100y$ dollars, and since the monthly budget is $9000, we must have

$$300x + 100y \le 9000$$

Next, the condition that the ad must appear in at least 15 but no more than 30 editions of the *Daily News* translates into the inequalities

$$y \ge 15$$
$$y \le 30$$

Finally, the objective function to be maximized is

$$P = 50,000x + 20,000y$$

To summarize, we have the following linear programming problem:

$$\text{Maximize} \quad P = 50,000x + 20,000y$$
$$\text{subject to} \quad 300x + 100y \le 9000$$
$$y \ge 15$$
$$y \le 30$$
$$x \ge 0, y \ge 0$$

3.3 Graphical Solution of Linear Programming Problems

The Graphical Method

Linear programming problems in two variables have relatively simple geometric interpretations. For example, the system of linear constraints associated with a two-dimensional linear programming problem, unless it is inconsistent, defines a planar region whose boundary is composed of straight-line segments and/or half-lines. Such problems are therefore amenable to graphical analysis.

Consider the following two-dimensional linear programming problem:

$$\begin{aligned} \text{Maximize} \quad & P = 3x + 2y \\ \text{subject to} \quad & 2x + 3y \le 12 \\ & 2x + y \le 8 \\ & x \ge 0, y \ge 0 \end{aligned} \tag{7}$$

The system of linear inequalities (7) defines the planar region S shown in Figure 9. Each point in S is a candidate for the solution of the problem at hand and is referred to as a **feasible solution**. The set S itself is referred to as a **feasible set**. Our goal is to find, from among all the points in the set S, the point(s) that optimizes the objective function P. Such a feasible solution is called an **optimal solution** and constitutes the solution to the linear programming problem under consideration.

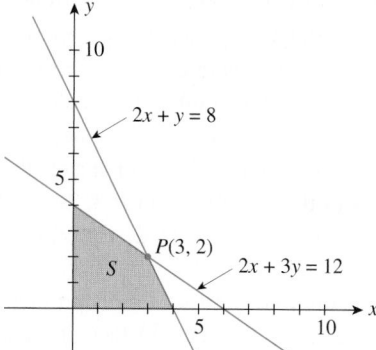

FIGURE 9
Each point in the feasible set S is a candidate for the optimal solution.

As noted earlier, each point $P(x, y)$ in S is a candidate for the optimal solution to the problem at hand. For example, the point $(1, 3)$ is easily seen to lie in S and is therefore in the running. The value of the objective function P at the point $(1, 3)$ is given by $P = 3(1) + 2(3) = 9$. Now, if we could compute the value of P corresponding to each point in S, then the point(s) in S that gave the largest value to P would constitute the solution set sought. Unfortunately, in most problems the number of candidates either is too large or, as in this problem, is infinite. Thus, this method is at best unwieldy and at worst impractical.

Let's turn the question around. Instead of asking for the value of the objective function P at a feasible point, let's assign a value to the objective function P and ask whether there are feasible points that would correspond to the given value of P. Toward this end, suppose we assign a value of 6 to P. Then the objective function P becomes $3x + 2y = 6$, a linear equation in x and y, and thus it has a graph that is a straight line L_1 in the plane. In Figure 10 we have drawn the graph of this straight line superimposed on the feasible set S.

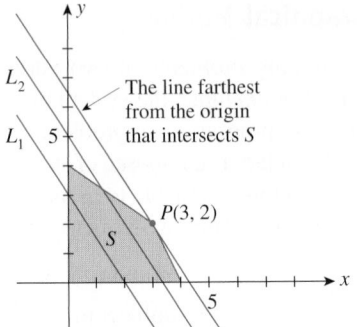

FIGURE 10
A family of parallel lines that intersect the feasible set S

It is clear that each point on the straight-line segment given by the intersection of the straight line L_1 and the feasible set S corresponds to the given value, 6, of P. For this reason the line L_1 is called an **isoprofit line**. Let's repeat the process, this time assigning a value of 10 to P. We obtain the equation $3x + 2y = 10$ and the line L_2 (see Figure 10), which suggests that there are feasible points that correspond to a larger value of P. Observe that the line L_2 is parallel to the line L_1 because both lines have slope equal to $-\frac{3}{2}$, which is easily seen by casting the corresponding equations in the slope-intercept form.

In general, by assigning different values to the objective function, we obtain a family of parallel lines, each with slope equal to $-\frac{3}{2}$. Furthermore, a line corresponding to a larger value of P lies farther away from the origin than one with a smaller value of P. The implication is clear. To obtain the optimal solution(s) to the problem at hand, find the straight line, from this family of straight lines, that is farthest from the origin and still intersects the feasible set S. The required line is the one that passes through the point $P(3, 2)$ (see Figure 10), so the solution to the problem is given by $x = 3$, $y = 2$, resulting in a maximum value of $P = 3(3) + 2(2) = 13$.

That the optimal solution to this problem was found to occur at a vertex of the feasible set S is no accident. In fact, the result is a consequence of the following basic theorem on linear programming, which we state without proof.

THEOREM 1

Linear Programming

If a linear programming problem has a solution then it must occur at a vertex, or corner point, of the feasible set S associated with the problem.
Furthermore, if the objective function P is optimized at two adjacent vertices of S, then it is optimized at every point on the line segment joining these vertices, in which case there are infinitely many solutions to the problem.

Theorem 1 tells us that our search for the solution(s) to a linear programming problem may be restricted to the examination of the set of vertices of the feasible set S associated with the problem. Since a feasible set S has finitely many vertices, the theorem suggests that the solution(s) to the linear programming problem may be found by inspecting the values of the objective function P at these vertices.

Although Theorem 1 sheds some light on the nature of the solution of a linear programming problem, it does not tell us when a linear programming problem has a solution. The following theorem states some conditions that guarantee when a linear programming problem has a solution.

THEOREM 2

Existence of a Solution

Suppose we are given a linear programming problem with a feasible set S and an objective function $P = ax + by$.

a. If S is bounded, then P has both a maximum and a minimum value on S.
b. If S is unbounded and both a and b are nonnegative, then P has a minimum value on S provided that the constraints defining S include the inequalities $x \geq 0$ and $y \geq 0$.
c. If S is the empty set, then the linear programming problem has no solution; that is, P has neither a maximum nor a minimum value.

The **method of corners**, a simple procedure for solving linear programming problems based on Theorem 1, follows.

The Method of Corners

1. Graph the feasible set.
2. Find the coordinates of all corner points (vertices) of the feasible set.
3. Evaluate the objective function at each corner point.
4. Find the vertex that renders the objective function a maximum (minimum). If there is only one such vertex, then this vertex constitutes a unique solution to the problem. If the objective function is maximized (minimized) at two adjacent corner points of S, there are infinitely many optimal solutions given by the points on the line segment determined by these two vertices.

APPLIED EXAMPLE 1   Maximizing Profits We are now in a position to complete the solution to the production problem posed in Example 1, Section 3.2. Recall that the mathematical formulation led to the following linear programming problem:

$$\begin{aligned}
\text{Maximize} \quad & P = x + 1.2y \\
\text{subject to} \quad & 2x + y \leq 180 \\
& x + 3y \leq 300 \\
& x \geq 0, y \geq 0
\end{aligned}$$

Solution   The feasible set S for the problem is shown in Figure 11.

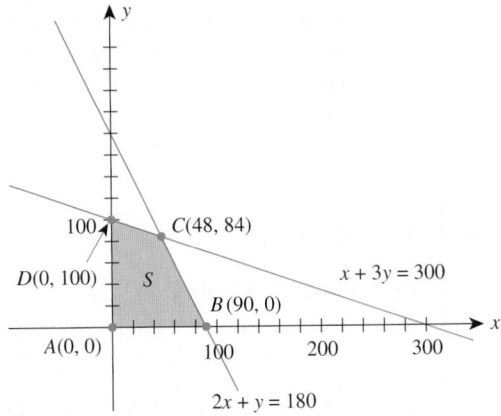

FIGURE 11
The corner point that yields the maximum profit is $C(48, 84)$.

The vertices of the feasible set are $A(0, 0)$, $B(90, 0)$, $C(48, 84)$, and $D(0, 100)$. The values of P at these vertices may be tabulated as follows:

Vertex	$P = x + 1.2y$
$A(0, 0)$	0
$B(90, 0)$	90
$C(48, 84)$	148.8
$D(0, 100)$	120

From the table, we see that the maximum of $P = x + 1.2y$ occurs at the vertex $(48, 84)$ and has a value of 148.8. Recalling what the symbols x, y, and P represent, we conclude that Ace Novelty would maximize its profit (a figure of $148.80) by producing 48 type-A souvenirs and 84 type-B souvenirs.

EXPLORE & DISCUSS

Consider the linear programming problem

$$\text{Maximize} \quad P = 4x + 3y$$
$$\text{subject to} \quad 2x + \ y \le 10$$
$$2x + 3y \le 18$$
$$x \ge 0, y \ge 0$$

1. Sketch the feasible set S for the linear programming problem.
2. Draw the isoprofit lines superimposed on S corresponding to $P = 12, 16, 20,$ and 24, and show that these lines are parallel to each other.
3. Show that the solution to the linear programming problem is $x = 3$ and $y = 4$. Is this result the same as that found using the method of corners?

   APPLIED EXAMPLE 2   A Nutrition Problem   Complete the solution of the nutrition problem posed in Example 2, Section 3.2.

Solution   Recall that the mathematical formulation of the problem led to the following linear programming problem in two variables:

$$\text{Minimize} \quad C = 6x + 8y$$
$$\text{subject to} \quad 40x + 10y \geq 2400$$
$$10x + 15y \geq 2100$$
$$5x + 15y \geq 1500$$
$$x \geq 0, y \geq 0$$

The feasible set S defined by the system of constraints is shown in Figure 12.

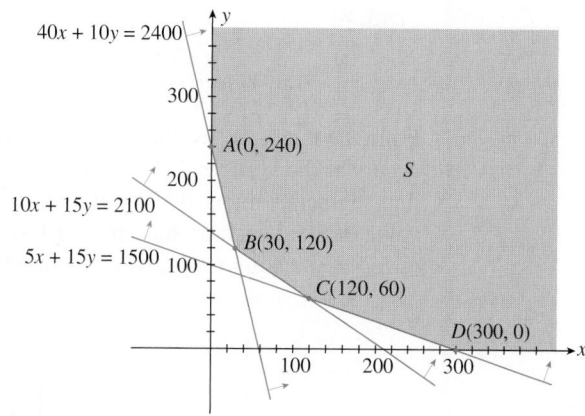

FIGURE 12
The corner point that yields the minimum cost is $B(30, 120)$.

The vertices of the feasible set S are $A(0, 240)$, $B(30, 120)$, $C(120, 60)$, and $D(300, 0)$. The values of the objective function C at these vertices are given in the following table:

Vertex	$C = 6x + 8y$
$A(0, 240)$	1920
$B(30, 120)$	1140
$C(120, 60)$	1200
$D(300, 0)$	1800

From the table, we can see that the minimum for the objective function $C = 6x + 8y$ occurs at the vertex $B(30, 120)$ and has a value of 1140. Thus, the individual should purchase 30 brand-A pills and 120 brand-B pills at a minimum cost of $11.40. ▪

EXAMPLE 3   A Linear Programming Problem with Multiple Solutions
Find the maximum and minimum of $P = 2x + 3y$ subject to the following system of linear inequalities:

$$2x + 3y \leq 30$$
$$y - x \leq 5$$
$$x + y \geq 5$$
$$x \leq 10$$
$$x \geq 0, y \geq 0$$

Solution   The feasible set S is shown in Figure 13. The vertices of the feasible set S are $A(5, 0)$, $B(10, 0)$, $C(10, \frac{10}{3})$, $D(3, 8)$, and $E(0, 5)$. The values of the objective function P at these vertices are given in the following table:

Vertex	$P = 2x + 3y$
$A(5, 0)$	10
$B(10, 0)$	20
$C(10, \frac{10}{3})$	30
$D(3, 8)$	30
$E(0, 5)$	15

From the table, we see that the maximum for the objective function $P = 2x + 3y$ occurs at the vertices $C(10, \frac{10}{3})$ and $D(3, 8)$. This tells us that every point on the line segment joining the points $C(10, \frac{10}{3})$ and $D(3, 8)$ maximizes P, giving it a value of 30 at each of these points. From the table, it is also clear that P is minimized at the point $(5, 0)$, where it attains a value of 10.

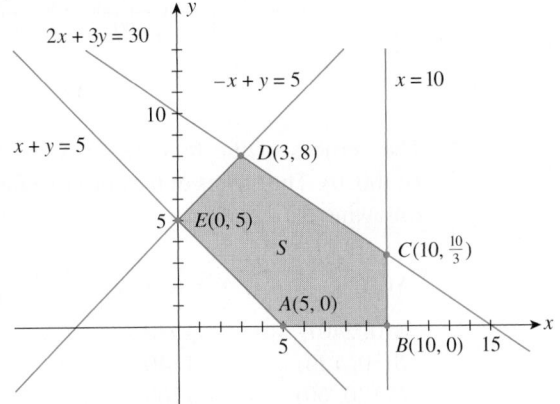

FIGURE 13
Every point lying on the line segment joining C and D maximizes P.

EXPLORE & DISCUSS

Consider the linear programming problem

$$\text{Maximize} \quad P = 2x + 3y$$
$$\text{subject to} \quad 2x + y \le 10$$
$$2x + 3y \le 18$$
$$x \ge 0, y \ge 0$$

1. Sketch the feasible set S for the linear programming problem.
2. Draw the isoprofit lines superimposed on S corresponding to $P = 6, 8, 12$, and 18, and show that these lines are parallel to each other.
3. Show that there are infinitely many solutions to the problem. Is this result as predicted by the method of corners?

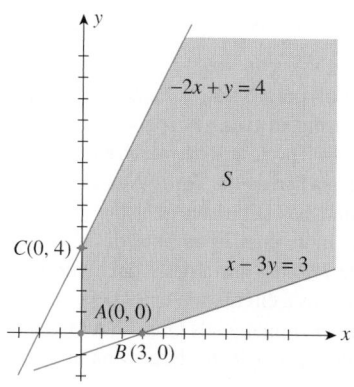

FIGURE 14
This maximization problem has no solution because the feasible set is unbounded.

We close this section by examining two situations in which a linear programming problem may have no solution.

EXAMPLE 4   An Unbounded Linear Programming Problem with No Solution   Solve the following linear programming problem:

$$\text{Maximize} \quad P = x + 2y$$
$$\text{subject to} \quad -2x + y \le 4$$
$$x - 3y \le 3$$
$$x \ge 0, y \ge 0$$

Solution   The feasible set S for this problem is shown in Figure 14. Since the set S is unbounded (both x and y can take on arbitrarily large positive values), we see that we can make P as large as we please by making x and y large enough. This problem has no solution. The problem is said to be unbounded. ▪

EXAMPLE 5   An Infeasible Linear Programming Problem   Solve the following linear programming problem:

$$\text{Maximize} \quad P = x + 2y$$
$$\text{subject to} \quad x + 2y \le 4$$
$$2x + 3y \ge 12$$
$$x \ge 0, y \ge 0$$

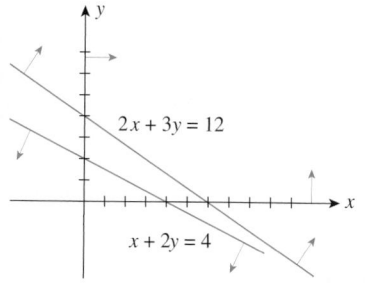

FIGURE 15
This problem is inconsistent because there is no point that satisfies all given inequalities.

Solution   The half-planes described by the constraints (inequalities) have no points in common (Figure 15). Hence there are no feasible points and the problem has no solution. In this situation, we say that the problem is **infeasible**, or **inconsistent**. (These situations are unlikely to occur in well-posed problems arising from practical applications of linear programming.) ▪

The method of corners is particularly effective in solving two-variable linear programming problems with a small number of constraints, as the preceding examples have amply demonstrated. Its effectiveness, however, decreases rapidly as the number of variables and/or constraints increases. For example, it may be shown that a linear programming problem in three variables and five constraints may have up to ten feasible corner points. The determination of the feasible corner points calls for the solution of ten 3×3 systems of linear equations and then the verification—by the substitution of each of these solutions into the system of constraints—to see if it is, in fact, a feasible point. When the number of variables and constraints goes up to five and ten, respectively (still a very small system from the standpoint of applications in economics), the number of vertices to be found and checked for feasible corner points increases dramatically to 252, and each of these vertices is found by solving a 5×5 linear system! For this reason, the method of corners is seldom used to solve linear programming problems; its redeeming value lies in the fact that much insight is gained into the nature of the solutions of linear programming problems through its use in solving two-variable problems.

3.3 Self-Check Exercises

1. Use the method of corners to solve the following linear programming problem:

$$\text{Maximize} \quad P = 4x + 5y$$
$$\text{subject to} \quad x + 2y \le 10$$
$$5x + 3y \le 30$$
$$x \ge 0, y \ge 0$$

2. Use the method of corners to solve the following linear programming problem:

$$\text{Minimize} \quad C = 5x + 3y$$
$$\text{subject to} \quad 5x + 3y \ge 30$$
$$x - 3y \le 0$$
$$x \ge 2$$

3. Gino Balduzzi, proprietor of Luigi's Pizza Palace, allocates $9000 a month for advertising in two newspapers, the *City Tribune* and the *Daily News*. The *City Tribune* charges $300 for a certain advertisement, whereas the *Daily News* charges $100 for the same ad. Gino has stipulated that the ad is to appear in at least 15 but no more than 30 editions of the *Daily News* per month. The *City Tribune* has a daily circulation of 50,000, and the *Daily News* has a circulation of 20,000. Under these conditions, determine how many ads Gino should place in each newspaper in order to reach the largest number of readers.

Solutions to Self-Check Exercises 3.3 can be found on page 197.

3.3 Concept Questions

1. **a.** What is the feasible set associated with a linear programming problem?
   b. What is a feasible solution of a linear programming problem?
   c. What is an optimal solution of a linear programming problem?

2. Describe the method of corners.

3.3 Exercises

In Exercises 1–6, find the optimal (maximum and/or minimum) value(s) of the objective function on the feasible set S.

1. $Z = 2x + 3y$

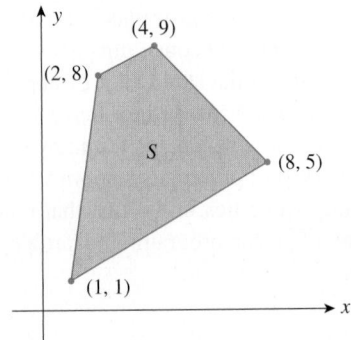

2. $Z = 3x - y$

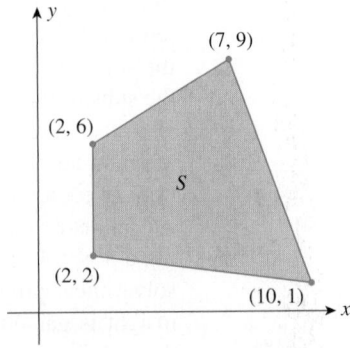

3. $Z = 3x + 4y$

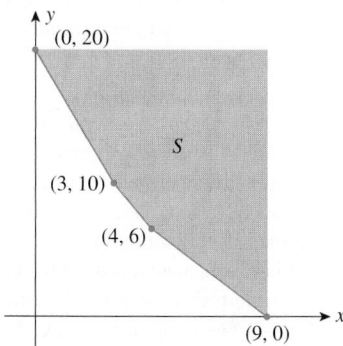

4. $Z = 7x + 9y$

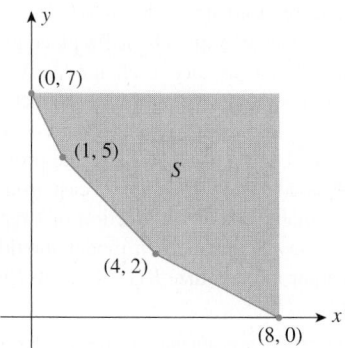

5. $Z = x + 4y$

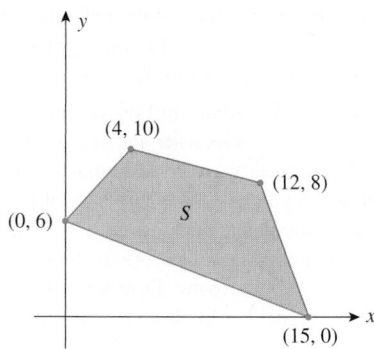

6. $Z = 3x + 2y$

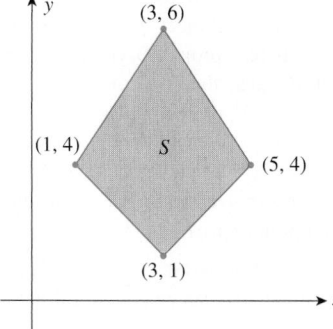

In Exercises 7–28, solve each linear programming problem by the method of corners.

7. Maximize $P = 2x + 3y$
subject to $\quad x + y \le 6$
$\qquad\qquad\qquad x \le 3$
$\qquad\qquad x \ge 0, y \ge 0$

8. Maximize $P = x + 2y$
subject to $\quad x + y \le 4$
$\qquad\qquad 2x + y \le 5$
$\qquad\qquad x \ge 0, y \ge 0$

9. Maximize $P = 2x + y$ subject to the constraints of Exercise 8.

10. Maximize $P = 4x + 2y$
subject to $\quad x + y \le 8$
$\qquad\qquad 2x + y \le 10$
$\qquad\qquad x \ge 0, y \ge 0$

11. Maximize $P = x + 8y$ subject to the constraints of Exercise 10.

12. Maximize $P = 3x - 4y$
subject to $\quad x + 3y \le 15$
$\qquad\qquad 4x + \ y \le 16$
$\qquad\qquad x \ge 0, y \ge 0$

13. Maximize $P = x + 3y$
subject to $\quad 2x + y \le 6$
$\qquad\qquad\ x + y \le 4$
$\qquad\qquad\qquad x \le 1$
$\qquad\qquad x \ge 0, y \ge 0$

14. Maximize $P = 2x + 5y$
subject to $\quad 2x + \ y \le 16$
$\qquad\qquad 2x + 3y \le 24$
$\qquad\qquad\qquad y \le 6$
$\qquad\qquad x \ge 0, y \ge 0$

15. Minimize $C = 3x + 4y$
subject to $\quad x + \ y \ge 3$
$\qquad\qquad x + 2y \ge 4$
$\qquad\qquad x \ge 0, y \ge 0$

16. Minimize $C = 2x + 4y$ subject to the constraints of Exercise 15.

17. Minimize $C = 3x + 6y$
subject to $\quad x + 2y \ge 40$
$\qquad\qquad x + \ y \ge 30$
$\qquad\qquad x \ge 0, y \ge 0$

18. Minimize $C = 3x + y$ subject to the constraints of Exercise 17.

19. Minimize $C = 2x + 10y$
subject to $\quad 5x + 2y \ge 40$
$\qquad\qquad\ x + 2y \ge 20$
$\qquad\qquad y \ge 3, x \ge 0$

20. Minimize $C = 2x + 5y$
subject to $\quad 4x + y \geq 40$
$\qquad 2x + y \geq 30$
$\qquad x + 3y \geq 30$
$\qquad x \geq 0, y \geq 0$

21. Minimize $C = 10x + 15y$
subject to $\quad x + y \leq 10$
$\qquad 3x + y \geq 12$
$\qquad -2x + 3y \geq 3$
$\qquad x \geq 0, y \geq 0$

22. Maximize $P = 2x + 5y$ subject to the constraints of Exercise 21.

23. Maximize $P = 3x + 4y$
subject to $\quad x + 2y \leq 50$
$\qquad 5x + 4y \leq 145$
$\qquad 2x + y \geq 25$
$\qquad y \geq 5, x \geq 0$

24. Maximize $P = 4x - 3y$ subject to the constraints of Exercise 23.

25. Maximize $P = 2x + 3y$
subject to $\quad x + y \leq 48$
$\qquad x + 3y \geq 60$
$\qquad 9x + 5y \leq 320$
$\qquad x \geq 10, y \geq 0$

26. Minimize $C = 5x + 3y$ subject to the constraints of Exercise 25.

27. Find the maximum and minimum of $P = 10x + 12y$ subject to

$$5x + 2y \geq 63$$
$$x + y \geq 18$$
$$3x + 2y \leq 51$$
$$x \geq 0, y \geq 0$$

28. Find the maximum and minimum of $P = 4x + 3y$ subject to

$$3x + 5y \geq 20$$
$$3x + y \leq 16$$
$$-2x + y \leq 1$$
$$x \geq 0, y \geq 0$$

The problems in Exercises 29–42 correspond to those in Exercises 1–14, Section 3.2. Use the results of your previous work to help you solve these problems.

29. MANUFACTURING—PRODUCTION SCHEDULING   A company manufactures two products, A and B, on two machines, I and II. It has been determined that the company will realize a profit of $3/unit of product A and a profit of $4/unit of product B. To manufacture a unit of product A requires 6 min on machine I and 5 min on machine II. To manufacture a unit of product B requires 9 min on machine I and 4 min on machine II.

There are 5 hr of machine time available on machine I and 3 hr of machine time available on machine II in each work shift. How many units of each product should be produced in each shift to maximize the company's profits? What is the optimal profit?

30. MANUFACTURING—PRODUCTION SCHEDULING   National Business Machines manufactures two models of fax machines: A and B. Each model A costs $100 to make, and each model B costs $150. The profits are $30 for each model A and $40 for each model B fax machine. If the total number of fax machines demanded per month does not exceed 2500 and the company has earmarked no more than $600,000/month for manufacturing costs, how many units of each model should National make each month in order to maximize its monthly profits? What is the optimal profit?

31. MANUFACTURING—PRODUCTION SCHEDULING   Kane Manufacturing has a division that produces two models of fireplace grates, model A and model B. To produce each model A grate requires 3 lb of cast iron and 6 min of labor. To produce each model B grate requires 4 lb of cast iron and 3 min of labor. The profit for each model A grate is $2.00, and the profit for each model B grate is $1.50. If 1000 lb of cast iron and 20 labor-hours are available for the production of fireplace grates per day, how many grates of each model should the division produce in order to maximize Kane's profits? What is the optimal profit?

32. MANUFACTURING—PRODUCTION SCHEDULING   Refer to Exercise 31. Because of a backlog of orders for model A grates, Kane's manager had decided to produce at least 150 of these models a day. Operating under this additional constraint, how many grates of each model should Kane produce to maximize profit? What is the optimal profit?

33. FINANCE—ALLOCATION OF FUNDS   Madison Finance has a total of $20 million earmarked for homeowner and auto loans. On the average, homeowner loans have a 10% annual rate of return whereas auto loans yield a 12% annual rate of return. Management has also stipulated that the total amount of homeowner loans should be greater than or equal to 4 times the total amount of automobile loans. Determine the total amount of loans of each type that Madison should extend to each category in order to maximize its returns. What are the optimal returns?

34. INVESTMENTS—ASSET ALLOCATION   A financier plans to invest up to $500,000 in two projects. Project A yields a return of 10% on the investment whereas project B yields a return of 15% on the investment. Because the investment in project B is riskier than the investment in project A, the financier has decided that the investment in project B should not exceed 40% of the total investment. How much should she invest in each project in order to maximize the return on her investment? What is the maximum return?

35. MANUFACTURING—PRODUCTION SCHEDULING   Acoustical manufactures a CD storage cabinet that can be bought fully assembled or as a kit. Each cabinet is processed in the fabrications department and the assembly department. If the fabrication department only manufactures fully assembled cabinets, then it can produce 200 units/day; and if it only manufactures kits, it can produce 200 units/day. If the assembly department produces only fully assembled cabinets, then it can produce 100 units/day; but if it produces only kits, then it can produce 300 units/day. Each fully assembled cabinet contributes $50 to the profits of the company whereas each kit contributes $40 to its profits. How many fully assembled units and how many kits should the company produce per day in order to maximize its profits? What is the optimal profit?

36. AGRICULTURE—CROP PLANNING A farmer plans to plant two crops, A and B. The cost of cultivating crop A is $40/acre whereas that of crop B is $60/acre. The farmer has a maximum of $7400 available for land cultivation. Each acre of crop A requires 20 labor-hours, and each acre of crop B requires 25 labor-hours. The farmer has a maximum of 3300 labor-hours available. If she expects to make a profit of $150/acre on crop A and $200/acre on crop B, how many acres of each crop should she plant in order to maximize her profit? What is the optimal profit?

37. MINING—PRODUCTION Perth Mining Company operates two mines for the purpose of extracting gold and silver. The Saddle Mine costs $14,000/day to operate, and it yields 50 oz of gold and 3000 oz of silver each day. The Horseshoe Mine costs $16,000/day to operate, and it yields 75 oz of gold and 1000 oz of silver each day. Company management has set a target of at least 650 oz of gold and 18,000 oz of silver. How many days should each mine be operated so that the target can be met at a minimum cost? What is the minimum cost?

38. TRANSPORTATION Deluxe River Cruises operates a fleet of river vessels. The fleet has two types of vessels: A type-A vessel has 60 deluxe cabins and 160 standard cabins, whereas a type-B vessel has 80 deluxe cabins and 120 standard cabins. Under a charter agreement with Odyssey Travel Agency, Deluxe River Cruises is to provide Odyssey with a minimum of 360 deluxe and 680 standard cabins for their 15-day cruise in May. It costs $44,000 to operate a type-A vessel and $54,000 to operate a type-B vessel for that period. How many of each type vessel should be used in order to keep the operating costs to a minimum? What is the minimum cost?

39. NUTRITION—DIET PLANNING A nutritionist at the Medical Center has been asked to prepare a special diet for certain patients. She has decided that the meals should contain a minimum of 400 mg of calcium, 10 mg of iron, and 40 mg of vitamin C. She has further decided that the meals are to be prepared from foods A and B. Each ounce of food A contains 30 mg of calcium, 1 mg of iron, 2 mg of vitamin C, and 2 mg of cholesterol. Each ounce of food B contains 25 mg of calcium, 0.5 mg of iron, 5 mg of vitamin C, and 5 mg of cholesterol. Find how many ounces of each type of food should be used in a meal so that the cholesterol content is minimized

and the minimum requirements of calcium, iron, and vitamin C are met.

40. SOCIAL PROGRAMS PLANNING AntiFam, a hunger-relief organization, has earmarked between $2 and $2.5 million (inclusive) for aid to two African countries, country A and country B. Country A is to receive between $1 and $1.5 million (inclusive), and country B is to receive at least $0.75 million. It has been estimated that each dollar spent in country A will yield an effective return of $.60, whereas a dollar spent in country B will yield an effective return of $.80. How should the aid be allocated if the money is to be utilized most effectively according to these criteria?
Hint: If x and y denote the amount of money to be given to country A and country B, respectively, then the objective function to be maximized is $P = 0.6x + 0.8y$.

41. ADVERTISING Everest Deluxe World Travel has decided to advertise in the Sunday editions of two major newspapers in town. These advertisements are directed at three groups of potential customers. Each advertisement in newspaper I is seen by 70,000 group-A customers, 40,000 group-B customers, and 20,000 group-C customers. Each advertisement in newspaper II is seen by 10,000 group-A, 20,000 group-B, and 40,000 group-C customers. Each advertisement in newspaper I costs $1000, and each advertisement in newspaper II costs $800. Everest would like their advertisements to be read by at least 2 million people from group A, 1.4 million people from group B, and 1 million people from group C. How many advertisements should Everest place in each newspaper to achieve its advertising goals at a minimum cost? What is the minimum cost?
Hint: Use different scales for drawing the feasible set.

42. MANUFACTURING—SHIPPING COSTS TMA manufactures 37-in. high definition LCD televisions in two separate locations, locations I and II. The output at location I is at most 6000 televisions/month, whereas the output at location II is at most 5000 televisions/month. TMA is the main supplier of televisions to the Pulsar Corporation, its holding company, which has priority in having all its requirements met. In a certain month, Pulsar placed orders for 3000 and 4000 televisions to be shipped to two of its factories located in city A and city B, respectively. The shipping costs (in dollars) per television from the two TMA plants to the two Pulsar factories are as follows:

From TMA	To Pulsar Factories City A	City B
Location I	$6	$4
Location II	$8	$10

Find a shipping schedule that meets the requirements of both companies while keeping costs to a minimum.

43. Complete the solution to Example 3, Section 3.2.

44. MANUFACTURING—PRODUCTION SCHEDULING Bata Aerobics manufactures two models of steppers used for aerobic exercises.

Manufacturing each luxury model requires 10 lb of plastic and 10 min of labor. Manufacturing each standard model requires 16 lb of plastic and 8 min of labor. The profit for each luxury model is $40, and the profit for each standard model is $30. If 6000 lb of plastic and 60 labor-hours are available for the production of the steppers per day, how many steppers of each model should Bata produce each day in order to maximize its profits? What is the optimal profit?

45. INVESTMENT PLANNING Patricia has at most $30,000 to invest in securities in the form of corporate stocks. She has narrowed her choices to two groups of stocks: growth stocks that she assumes will yield a 15% return (dividends and capital appreciation) within a year and speculative stocks that she assumes will yield a 25% return (mainly in capital appreciation) within a year. Determine how much she should invest in each group of stocks in order to maximize the return on her investments within a year if she has decided to invest at least 3 times as much in growth stocks as in speculative stocks.

46. VETERINARY SCIENCE A veterinarian has been asked to prepare a diet for a group of dogs to be used in a nutrition study at the School of Animal Science. It has been stipulated that each serving should be no larger than 8 oz and must contain at least 29 units of nutrient I and 20 units of nutrient II. The vet has decided that the diet may be prepared from two brands of dog food: brand A and brand B. Each ounce of brand A contains 3 units of nutrient I and 4 units of nutrient II. Each ounce of brand B contains 5 units of nutrient I and 2 units of nutrient II. Brand A costs 3 cents/ounce and brand B costs 4 cents/ounce. Determine how many ounces of each brand of dog food should be used per serving to meet the given requirements at a minimum cost.

47. MARKET RESEARCH Trendex, a telephone survey company, has been hired to conduct a television-viewing poll among urban and suburban families in the Los Angeles area. The client has stipulated that a maximum of 1500 families is to be interviewed. At least 500 urban families must be interviewed, and at least half of the total number of families interviewed must be from the suburban area. For this service, Trendex will be paid $6000 plus $8 for each completed interview. From previous experience, Trendex has determined that it will incur an expense of $4.40 for each successful interview with an urban family and $5 for each successful interview with a suburban family. How many urban and suburban families should Trendex interview in order to maximize its profit?

In Exercises 48–51, determine whether the statement is true or false. If it is true, explain why it is true. If it is false, give an example to show why it is false.

48. An optimal solution of a linear programming problem is a feasible solution, but a feasible solution of a linear programming problem need not be an optimal solution.

49. A linear programming problem can have exactly three (optimal) solutions.

50. If a maximization problem has no solution, then the feasible set associated with the linear programming problem must be unbounded.

51. Suppose you are given the following linear programming problem: Maximize $P = ax + by$ on the unbounded feasible set S shown in the accompanying figure.

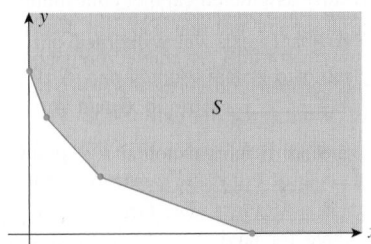

a. If $a > 0$ or $b > 0$, then the linear programming problem has no optimal solution.

b. If $a \le 0$ and $b \le 0$, then the linear programming problem has at least one optimal solution.

52. Suppose you are given the following linear programming problem: Maximize $P = ax + by$, where $a > 0$ and $b > 0$, on the feasible set S shown in the accompanying figure.

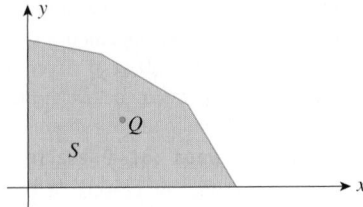

Explain, without using Theorem 1, why the optimal solution of the linear programming problem cannot occur at the point Q.

53. Suppose you are given the following linear programming problem: Maximize $P = ax + by$, where $a > 0$ and $b > 0$, on the feasible set S shown in the accompanying figure.

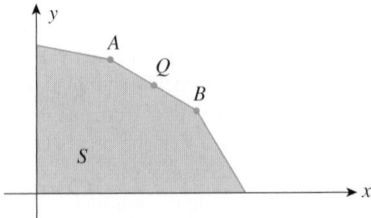

Explain, without using Theorem 1, why the optimal solution of the linear programming problem cannot occur at the point

Q unless the problem has infinitely many solutions lying along the line segment joining the vertices A and B.
Hint: Let $A(x_1, y_1)$ and $B(x_2, y_2)$. Then $Q(\bar{x}, \bar{y})$, where $\bar{x} = x_1 + (x_2 - x_1)t$ and $\bar{y} = y_2 + (y_2 - y_1)t$ with $0 < t < 1$. Study the value of P at and near Q.

54. Consider the linear programming problem

$$\text{Maximize} \quad P = 2x + 7y$$
$$\text{subject to} \quad 2x + y \geq 8$$
$$x + y \geq 6$$
$$x \geq 0, y \geq 0$$

a. Sketch the feasible set S.
b. Find the corner points of S.
c. Find the values of P at the corner points of S found in part (b).
d. Show that the linear programming problem has no (optimal) solution. Does this contradict Theorem 1?

55. Consider the linear programming problem

$$\text{Minimize} \quad C = -2x + 5y$$
$$\text{subject to} \quad x + y \leq 3$$
$$2x + y \leq 4$$
$$5x + 8y \geq 40$$
$$x \geq 0, y \geq 0$$

a. Sketch the feasible set.
b. Find the solution(s) of the linear programming problem, if it exists.

3.3 Solutions to Self-Check Exercises

1. The feasible set S for the problem was graphed in the solution to Exercise 1, Self-Check Exercises 3.1. It is reproduced in the accompanying figure.

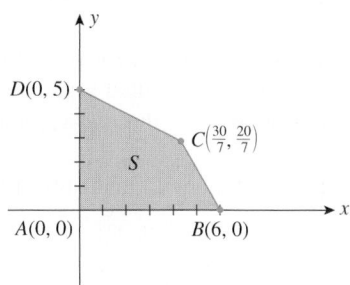

The values of the objective function P at the vertices of S are summarized in the accompanying table.

Vertex	$P = 4x + 5y$
$A(0, 0)$	0
$B(6, 0)$	24
$C(\frac{30}{7}, \frac{20}{7})$	$\frac{220}{7} = 31\frac{3}{7}$
$D(0, 5)$	25

From the table, we see that the maximum for the objective function P is attained at the vertex $C(\frac{30}{7}, \frac{20}{7})$. Therefore, the solution to the problem is $x = \frac{30}{7}$, $y = \frac{20}{7}$, and $P = 31\frac{3}{7}$.

2. The feasible set S for the problem was graphed in the solution to Exercise 2, Self-Check Exercises 3.1. It is reproduced in the accompanying figure.

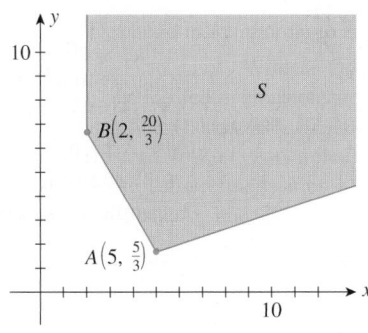

Evaluating the objective function $C = 5x + 3y$ at each corner point, we obtain the table

Vertex	$C = 5x + 3y$
$(5, \frac{5}{3})$	30
$B(2, \frac{20}{3})$	30

We conclude that (i) the objective function is minimized at every point on the line segment joining the points $(5, \frac{5}{3})$ and $(2, \frac{20}{3})$, and (ii) the minimum value of C is 30.

3. Refer to Self-Check Exercise 3.2. The problem is to maximize $P = 50{,}000x + 20{,}000y$ subject to

$$300x + 100y \leq 9000$$
$$y \geq 15$$
$$y \leq 30$$
$$x \geq 0, y \geq 0$$

The feasible set S for the problem is shown in the accompanying figure.

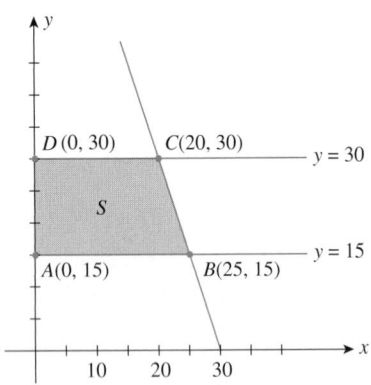

Evaluating the objective function $P = 50{,}000x + 20{,}000y$ at each vertex of S, we obtain

Vertex	$P = 50{,}000x + 20{,}000y$
$A(0, 15)$	300,000
$B(25, 15)$	1,550,000
$C(20, 30)$	1,600,000
$D(0, 30)$	600,000

From the table, we see that P is maximized when $x = 20$ and $y = 30$. Therefore, Gino should place 20 ads in the *City Tribune* and 30 in the *Daily News*.

3.4   Sensitivity Analysis (Optional)

In this section we investigate how changes in the parameters of a linear programming problem affect its optimal solution. This type of analysis is called **sensitivity analysis.** As in the previous sections, we restrict our analysis to the two-variable case, which is amenable to graphical analysis.

Recall the production problem posed in Example 1, Section 3.2, and solved in Example 1, Section 3.3:

Maximize	$P = x + 1.2y$	Objective function
subject to	$2x + y \leq 180$	Constraint 1
	$x + 3y \leq 300$	Constraint 2
	$x \geq 0, y \geq 0$	

where x denotes the number of type-A souvenirs and y denotes the number of type-B souvenirs to be made. The optimal solution of this problem is $x = 48$, $y = 84$ (corresponding to the point C), and $P = 148.8$ (Figure 16).

The following questions arise in connection with this production problem.

1. How do changes made to the coefficients of the objective function affect the optimal solution?

2. How do changes made to the constants on the right-hand side of the constraints affect the optimal solution?

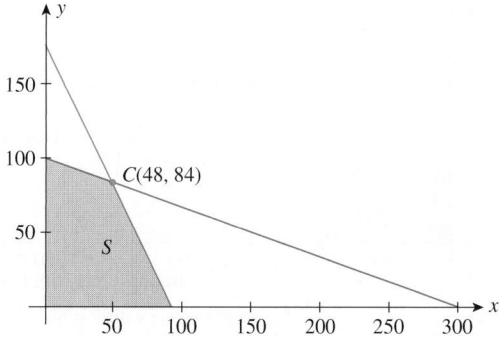

FIGURE 16
The optimal solution occurs at the point $C(48, 84)$.

■ Changes in the Coefficients of the Objective Function

In the production problem under consideration, the objective function is $P = x + 1.2y$. The coefficient of x, which is 1, tells us that the contribution to the profit for each type-A souvenir is $1.00. The coefficient of y, 1.2, tells us that the contribution to the profit for each type-B souvenir is $1.20. Now suppose the contribution to the profit for each type-B souvenir remains fixed at $1.20 per souvenir. By how much can the contribution to the profit for each type-A souvenir vary without affecting the current optimal solution?

To answer this question, suppose the contribution to the profit of each type-A souvenir is c so that

$$P = cx + 1.2y \tag{8}$$

We need to determine the range of values of c such that the solution remains optimal.

We begin by rewriting Equation (8) for the isoprofit line in the slope-intercept form. Thus,

$$y = -\frac{c}{1.2}x + \frac{P}{1.2} \tag{9}$$

The slope of the isoprofit line is $-c/1.2$. If the slope of the isoprofit line exceeds that of the line associated with constraint 2, then the optimal solution shifts from point C to point D (see Figure 17 on page 200).

On the other hand, if the slope of the isoprofit line is *less than or equal to* the slope of the line associated with constraint 2, then the optimal solution remains unaffected. (You may verify that $-\frac{1}{3}$ is the slope of the line associated with constraint 2 by writing the equation $x + 3y = 300$ in the slope-intercept form.) In other words, we must have

$$-\frac{c}{1.2} \leq -\frac{1}{3}$$

$$\frac{c}{1.2} \geq \frac{1}{3} \qquad \text{Multiplying each side by } -1 \text{ reverses the inequality sign.}$$

$$c \geq \frac{1.2}{3} = 0.4$$

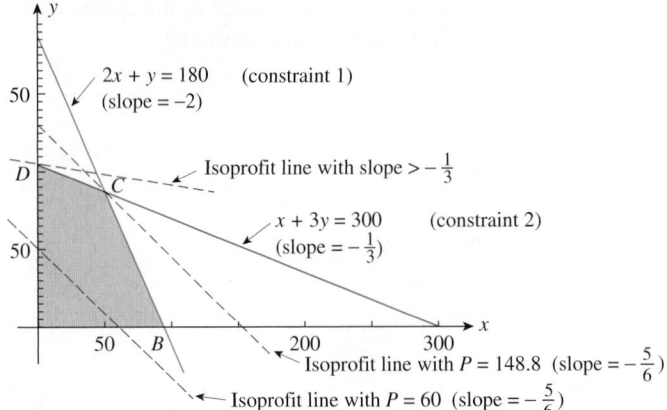

FIGURE 17
Increasing the slope of the isoprofit line $P = cx + 1.2y$ beyond $-\frac{1}{3}$ shifts the optimal solution from point C to point D.

A similar analysis shows that if the slope of the isoprofit line is less than that of the line associated with constraint 1, then the optimal solution shifts from point C to point B. Since the slope of the line associated with constraint 1 is -2, we see that point C will remain optimal provided that the slope of the isoprofit line is *greater than or equal to* -2; that is, if

$$-\frac{c}{1.2} \geq -2$$

$$\frac{c}{1.2} \leq 2$$

$$c \leq 2.4$$

Thus, we have shown that if $0.4 \leq c \leq 2.4$, then the optimal solution obtained previously remains unaffected.

This result tells us that if the contribution to the profit of each type-A souvenir lies between \$.40 and \$2.40, then Ace Novelty should still make 48 type-A souvenirs and 84 type-B souvenirs. Of course, the profit of the company will change with a change in the value of c—it's the product mix that stays the same. For example, if the contribution to the profit of a type-A souvenir is \$1.50, then the profit of the company will be \$172.80. (See Exercise 1.) Incidentally, our analysis shows that the parameter c is not a sensitive parameter.

We leave it as an exercise for you to show that, with the contribution to the profit of type-A souvenirs held constant at \$1.00 per souvenir, the contribution to each type-B souvenir can vary between \$.50 and \$3.00 without affecting the product mix for the optimal solution (see Exercise 1).

APPLIED EXAMPLE 1   Profit Function Analysis   Kane Manufacturing has a division that produces two models of grates, model A and model B. To produce each model A grate requires 3 pounds of cast iron and 6 minutes of labor. To produce each model B grate requires 4 pounds of cast iron and 3 minutes of labor. The profit for each model A grate is \$2.00, and the profit for each model B grate is \$1.50. Available for grate production each day are 1000 pounds of cast iron and 20 labor-hours. Because of an excess inventory of model A grates, management has decided to limit the production of model A grates to no more than 180 grates per day.

a. Use the method of corners to determine the number of grates of each model Kane should produce in order to maximize its profits.

b. Find the range of values that the contribution to the profit of a model A grate can assume without changing the optimal solution.

c. Find the range of values that the contribution to the profit of a model B grate can assume without changing the optimal solution.

Solution

a. Let x denote the number of model A grates and y the number of model B grates produced. Then verify that we are led to the following linear programming problem:

$$\text{Maximize} \quad P = 2x + 1.5y$$

$$
\begin{aligned}
\text{subject to} \quad & 3x + 4y \leq 1000 && \text{Constraint 1} \\
& 6x + 3y \leq 1200 && \text{Constraint 2} \\
& x \leq 180 && \text{Constraint 3} \\
& x \geq 0, y \geq 0
\end{aligned}
$$

The graph of the feasible set S is shown in Figure 18.

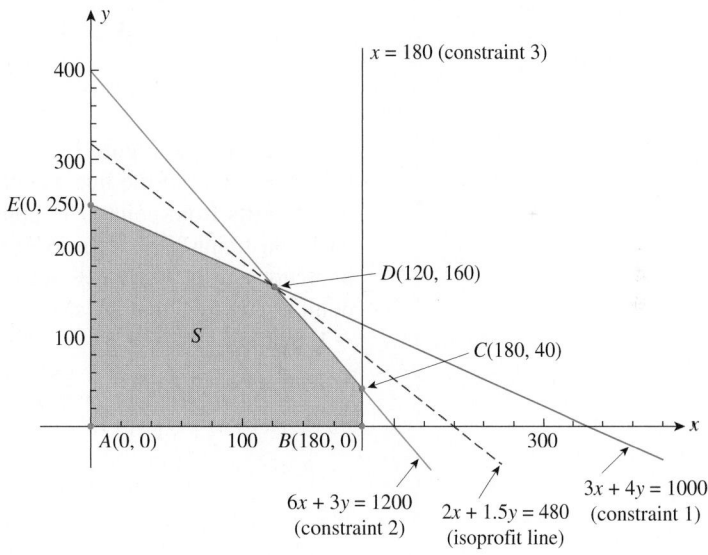

FIGURE 18
The shaded region is the feasible set S. Also shown are the lines of the equations associated with the constraints.

From the following table of values,

Vertex	$P = 2x + 1.5y$
$A(0, 0)$	0
$B(180, 0)$	360
$C(180, 40)$	420
$D(120, 160)$	480
$E(0, 250)$	375

we see that the maximum of $P = 2x + 1.5y$ occurs at the vertex $D(120, 160)$ with a value of 480. Thus, Kane realizes a maximum profit of $480 per day by producing 120 model A grates and 160 model B grates each day.

b. Let c (in dollars) denote the contribution to the profit of a model A grate. Then $P = cx + 1.5y$ or, upon solving for y,

$$y = -\frac{c}{1.5}x + \frac{P}{1.5}$$
$$= \left(-\frac{2}{3}c\right)x + \frac{2}{3}P$$

Referring to Figure 18 on page 201, you can see that if the slope of the iso-profit line is greater than the slope of the line associated with constraint 1, then the optimal solution will shift from point D to point E. Thus, for the optimal solution to remain unaffected, the slope of the isoprofit line must be less than or equal to the slope of the line associated with constraint 1. But the slope of the line associated with constraint 1 is $-\frac{3}{4}$, which you can see by rewriting the equation $3x + 4y = 1000$ in the slope-intercept form $y = -\frac{3}{4}x + 250$. Since the slope of the isoprofit line is $-2c/3$, we must have

$$-\frac{2c}{3} \le -\frac{3}{4}$$
$$\frac{2c}{3} \ge \frac{3}{4}$$
$$c \ge \left(\frac{3}{4}\right)\left(\frac{3}{2}\right) = \frac{9}{8} = 1.125$$

Again referring to Figure 18, you can see that if the slope of the isoprofit line is less than that of the line associated with constraint 2, then the optimal solution shifts from point D to point C. Since the slope of the line associated with constraint 2 is -2 (rewrite the equation $6x + 3y = 1200$ in the slope-intercept form $y = -2x + 400$), we see that the optimal solution remains at point D provided that the slope of the isoprofit line is greater than or equal to -2; that is,

$$-\frac{2c}{3} \ge -2$$
$$\frac{2c}{3} \le 2$$
$$c \le (2)\left(\frac{3}{2}\right) = 3$$

We conclude that the contribution to the profit of a model A grate can assume values between $1.125 and $3.00 without changing the optimal solution.

c. Let c (in dollars) denote the contribution to the profit of a model B grate. Then

$$P = 2x + cy$$

or, upon solving for y,

$$y = -\frac{2}{c}x + \frac{P}{c}$$

An analysis similar to that performed in part (b) with respect to constraint 1 shows that the optimal solution will remain in effect provided that

$$-\frac{2}{c} \leq -\frac{3}{4}$$

$$\frac{2}{c} \geq \frac{3}{4}$$

$$c \leq 2\left(\frac{4}{3}\right) = \frac{8}{3} = 2\frac{2}{3}$$

Performing an analysis with respect to constraint 2 shows that the optimal solution will remain in effect provided that

$$-\frac{2}{c} \geq -2$$

$$\frac{2}{c} \leq 2$$

$$c \geq 1$$

Thus, the contribution to the profit of a model B grate can assume values between $1.00 and $2.67 without changing the optimal solution. ■

▬ Changes to the Constants on the Right-Hand Side of the Constraint Inequalities

Let's return to the production problem posed at the beginning of this section:

$$\begin{aligned}
\text{Maximize} \quad & P = x + 1.2y \\
\text{subject to} \quad & 2x + y \leq 180 \quad \text{Constraint 1}\\
& x + 3y \leq 300 \quad \text{Constraint 2}\\
& x \geq 0, y \geq 0
\end{aligned}$$

Now suppose the time available on machine I is changed from 180 minutes to $(180 + h)$ minutes, where h is a real number. Then the constraint on machine I is changed to

$$2x + y \leq 180 + h$$

Observe that the line with equation $2x + y = 180 + h$ is parallel to the line $2x + y = 180$ associated with the original constraint 1.

As you can see from Figure 19 on page 204, the result of adding the constant h to the right-hand side of constraint 1 is to shift the current optimal solution from the point C to the new optimal solution occurring at the point C'. To find the coordinates of C', we observe that C' is the point of intersection of the lines with equations

$$2x + y = 180 + h \quad \text{and} \quad x + 3y = 300$$

Thus, the coordinates of the point are found by solving the system of linear equations

$$\begin{aligned}
2x + y &= 180 + h \\
x + 3y &= 300
\end{aligned}$$

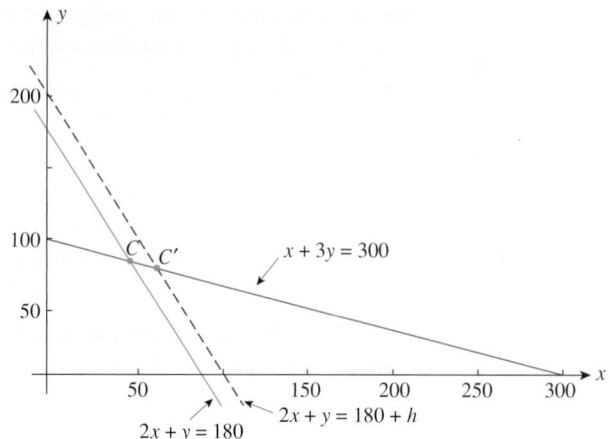

FIGURE 19
The lines with equations $2x + y = 180$ and $2x + y = 180 + h$ are parallel to each other.

The solutions are

$$x = \frac{3}{5}(80 + h) \qquad \text{and} \qquad y = \frac{1}{5}(420 - h) \qquad \textbf{(10)}$$

The nonnegativity of x implies that

$$\frac{3}{5}(80 + h) \geq 0$$

$$80 + h \geq 0$$

$$h \geq -80$$

Next, the nonnegativity of y implies that

$$\frac{1}{5}(420 - h) \geq 0$$

$$420 - h \geq 0$$

$$h \leq 420$$

Thus, h must satisfy the inequalities $-80 \leq h \leq 420$. Our computations reveal that a meaningful solution will require that the time available for machine I must range between $(180 - 80)$ and $(180 + 420)$ minutes—that is, between 100 and 600 minutes. Under these conditions, Ace Novelty should produce $\frac{3}{5}(80 + h)$ type-A souvenirs and $\frac{1}{5}(420 - h)$ type-B souvenirs.

For example, if Ace Novelty can manage to increase the time available on machine I by 10 minutes, then it should produce $\frac{3}{5}(80 + 10)$, or 54, type-A souvenirs and $\frac{1}{5}(420 - 10)$, or 82, type-B souvenirs; the resulting profit is

$$P = x + 1.2y = 54 + (1.2)(82) = 152.4$$

or \$152.40.

We leave it as an exercise for you to show that if the time available on machine II is changed from 300 minutes to $(300 + k)$ minutes with no change in the maximum capacity for machine I, then k must satisfy the inequalities $-210 \leq k \leq 240$. Thus, for a meaningful solution to the problem, the time available on machine II must lie between 90 and 540 min (see Exercise 2). Furthermore, in this case, Ace Novelty should produce $\frac{1}{5}(240 - k)$ type-A souvenirs and $\frac{1}{5}(420 + 2k)$ type-B souvenirs (see Exercise 2).

◾ Shadow Prices

We have just seen that if Ace Novelty could increase the maximum available time on machine I by 10 minutes, then the profit would increase from the original optimal value of $148.80 to $152.40. In this case, finding the extra time on machine I proved beneficial to the company. More generally, to study the economic benefits that can be derived from increasing its resources, a company looks at the shadow prices associated with the respective resources. More specifically, we define the shadow price for the ith resource (associated with the ith constraint of the linear programming problem) to be the amount by which the value of the objective function is improved—increased in a maximization problem and decreased in a minimization problem—if the right-hand side of the ith constraint is increased by 1 unit.

In the Ace Novelty example discussed earlier, we showed that if the right-hand side of constraint 1 is increased by h units then the optimal solution is given by (10):

$$x = \frac{3}{5}(80 + h) \quad \text{and} \quad y = \frac{1}{5}(420 - h)$$

The resulting profit is calculated as follows:

$$P = x + 1.2y$$

$$= x + \frac{6}{5}y$$

$$= \frac{3}{5}(80 + h) + \left(\frac{6}{5}\right)\left(\frac{1}{5}\right)(420 - h)$$

$$= \frac{3}{25}(1240 + 3h)$$

Upon setting $h = 1$, we find

$$P = \frac{3}{25}(1240 + 3)$$

$$= 149.16$$

Since the optimal profit for the original problem is $148.80, we see that the shadow price for the first resource is $149.16 - 148.80$, or $.36. To summarize, Ace Novelty's profit increases at the rate of $.36 per 1-minute increase in the time available on machine I.

We leave it as an exercise for you to show that the shadow price for resource 2 (associated with constraint 2) is $.28 (see Exercise 2).

APPLIED EXAMPLE 2   Shadow Prices   Consider the problem posed in Example 1:

$$\begin{aligned} \text{Maximize} \quad & P = 2x + 1.5y \\ \text{subject to} \quad & 3x + 4y \leq 1000 \quad &\text{Constraint 1} \\ & 6x + 3y \leq 1200 \quad &\text{Constraint 2} \\ & x \leq 180 \quad &\text{Constraint 3} \\ & x \geq 0, y \geq 0 \end{aligned}$$

a. Find the range of values that resource 1 (the constant on the right-hand side of constraint 1) can assume.

b. Find the shadow price for resource 1.

TITLE Land Use Planner
INSTITUTION City of Burien

As a Land Use Planner for the city of Burien, Washington, I assist property owners every day in the development of their land. By definition, land use planners develop plans and recommend policies for managing land use. To do this, I must take into account many existing and potential factors, such as public transportation, zoning laws, and other municipal laws. By using the basic ideas of linear programming, I work with the property owners to figure out maximum and minimum use requirements for each individual situation. Then, I am able to review and evaluate proposals for land use plans and prepare recommendations. All this is necessary to process an application for a land development permit.

Here's how it works. A property owner will come to me who wants to start a business on a vacant commercially zoned piece of property. First, we would have a discussion to find out what type of commercial zoning the property is in and whether or not the use is permitted or would require additional land use review. If the use is permitted and no further land use review is required, I would let the applicant know what criteria would have to be met and shown on building plans. At this point the applicant will begin working with their building contractor, architect, or engineer and landscape architect to meet the zoning code criteria. Once the applicant has worked with one or more of these professionals, building plans can be submitted for review. Then, they are routed to several different departments (building, engineer, public works, and the fire department). Because I am the land use planner for the project, one set of plans is routed to my desk for review.

During this review, I determine whether or not the zoning requirements have been met in order to make a final determination of the application. These zoning requirements are assessed by asking the applicant to give us a site plan showing lot area measurements, building and impervious surface coverage calculations, and building setbacks, just to name a few. Additionally, I would have to determine the parking requirements. How many off-street parking spaces are required? What are the isle widths? Is there enough room for backing space? Then, I would look at the landscaping requirements. Plans would need to be drawn up by a landscape architect and list specifics about the location, size, and types of plants that will be used.

By weighing all of these factors and measurements, I am able to determine the viability of a land development project. The basic ideas of linear programming are, fundamentally, at the heart of this determination and are key to the day-to-day choices I must make in my profession.

Solution

a. Suppose the right-hand side of constraint 1 is replaced by $1000 + h$, where h is a real number. Then the new optimal solution occurs at the point D' (Figure 20).

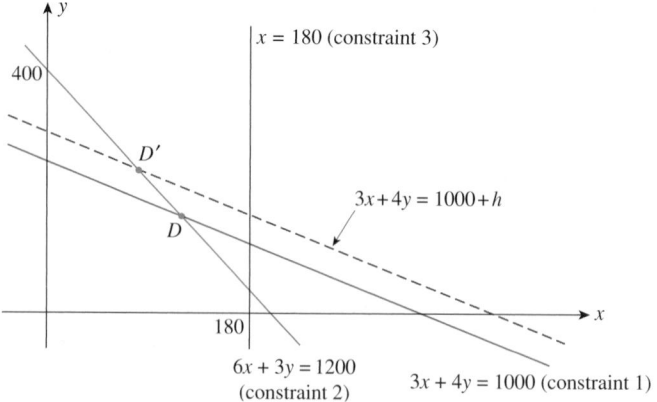

FIGURE 20
As the amount of resource 1 changes, the point at which the optimal solution occurs shifts from D to D'.

To find the coordinates of D', we solve the system

$$3x + 4y = 1000 + h$$
$$6x + 3y = 1200$$

Multiplying the first equation by -2 and then adding the resulting equation to the second equation gives

$$-5y = -800 - 2h$$

$$y = \frac{2}{5}(400 + h)$$

Substituting this value of y into the second equation in the system gives

$$6x + \frac{6}{5}(400 + h) = 1200$$

$$x + \frac{1}{5}(400 + h) = 200$$

$$x = \frac{1}{5}(600 - h)$$

The nonnegativity of y implies that $h \geq -400$, and the nonnegativity of x implies that $h \leq 600$. But constraint 3 dictates that x must also satisfy

$$x = \frac{1}{5}(600 - h) \leq 180$$

$$600 - h \leq 900$$

$$-h \leq 300$$

$$h \geq -300$$

Therefore, h must satisfy $-300 \leq h \leq 600$. This tells us that the amount of resource 1 must lie between $1000 - 300$, or 700, and $1000 + 600$, or 1600—that is, between 700 and 1600 pounds.

b. If we set $h = 1$ in part (a), we obtain

$$x = \frac{1}{5}(600 - 1) = \frac{599}{5}$$

$$y = \frac{2}{5}(400 + 1) = \frac{802}{5}$$

Therefore, the profit realized at this level of production is

$$P = 2x + \frac{3}{2}y = 2\left(\frac{599}{5}\right) + \frac{3}{2}\left(\frac{802}{5}\right)$$

$$= \frac{2401}{5} = 480.2$$

Since the original optimal profit is \$480 (see Example 1), we see that the shadow price for resource 1 is \$.20. ◼

If you examine Figure 20, you can see that increasing resource 3 (the constant on the right-hand side of constraint 3) has no effect on the optimal solution $D(120, 160)$ of the problem at hand. In other words, an increase in the resource associated with constraint 3 has no economic benefit for Kane Manufacturing. The shadow price for this resource is *zero*. There is a *surplus* of this resource. Hence we say that the constraint $x \leq 180$ is not binding on the optimal solution $D(120, 160)$.

On the other hand, constraints 1 and 2, which *hold with equality* at the optimal solution $D(120, 160)$, are said to be **binding constraints**. The objective function cannot be increased without increasing these resources. They have *positive* shadow prices.

■ Importance of Sensitivity Analysis

We conclude this section by pointing out the importance of sensitivity analysis in solving real-world problems. The values of the parameters in these problems may change. For example, the management of Ace Novelty might wish to increase the price of a type-A souvenir because of increased demand for the product, or they might want to see how a change in the time available on machine I affects the (optimal) profit of the company.

When a parameter of a linear programming problem is changed, it is true that one need only re-solve the problem to obtain a new solution to the problem. But since a real-world linear programming problem often involves thousands of parameters, the amount of work involved in finding a new solution is prohibitive. Another disadvantage in using this approach is that it often takes many trials with different values of a parameter in order to see their effect on the optimal solution of the problem. Thus, a more analytical approach such as that discussed earlier is desirable.

Returning to the discussion of Ace Novelty, our analysis of the changes in the coefficients of the objective (profit) function suggests that if management decides to raise the price of a type-A souvenir, it can do so with the assurance that the optimal solution holds as long as the new price leaves the contribution to the profit of a type-A souvenir between $.40 and $2.40. There is no need to re-solve the linear programming problem for each new price being considered. Also, our analysis of the changes in the parameters on the right-hand side of the constraints suggests, for example, that a meaningful solution to the problem requires that the time available for machine I lie in the range between 100 and 600 minutes. Furthermore, the analysis tells us how to compute the increase (decrease) in the optimal profit when the resource is adjusted by using the shadow price associated with that constraint. Again, there is no need to re-solve the linear programming problem each time a change in the resource available is anticipated.

Using Technology examples and exercises that are solved using Excel's Solver can be found on pages 239–242 and 257–260.

3.4   Self-Check Exercises

Consider the linear programming problem:

Maximize $P = 2x + 4y$
subject to $2x + 5y \le 19$ Constraint 1
                     $3x + 2y \le 12$ Constraint 2
                     $x \ge 0, y \ge 0$

1. Use the method of corners to solve this problem.

2. Find the range of values that the coefficient of x can assume without changing the optimal solution.

3. Find the range of values that resource 1 (the constant on the right-hand side of constraint 1) can assume without changing the optimal solution.

4. Find the shadow price for resource 1.

5. Identify the binding and nonbinding constraints.

Solutions to Self-Check Exercises can be found on page 211.

3.4   Concept Questions

1. Suppose $P = 3x + 4y$ is the objective function in a linear programming (maximization) problem, where x denotes the number of units of product A and y denotes the number of units of product B to be made. What does the coefficient of x represent? The coefficient of y?

2. Given the linear programming problem

$$\text{Maximize} \quad P = 3x + 4y$$
$$\text{subject to} \quad x + y \le 4 \quad \text{Resource 1}$$
$$2x + y \le 5 \quad \text{Resource 2}$$

a. Write the inequality that represents an increase of h units in resource 1.

b. Write the inequality that represents an increase of k units in resource 2.

3. Explain the meaning of (a) a shadow price and (b) a binding constraint.

3.4   Exercises

1. Refer to the production problem discussed on pages 198–200.

a. Show that the optimal solution holds if the contribution to the profit of a type-B souvenir lies between $.50 and $3.00.

b. Show that if the contribution to the profit of a type-A souvenir is $1.50 (with the contribution to the profit of a type-B souvenir held at $1.20), then the optimal profit of the company will be $172.80.

c. What will be the optimal profit of the company if the contribution to the profit of a type-B souvenir is $2.00 (with the contribution to the profit of a type-A souvenir held at $1.00)?

2. Refer to the production problem discussed on pages 203–205.

a. Show that, for a meaningful solution, the time available on machine II must lie between 90 and 540 min.

b. Show that, if the time available on machine II is changed from 300 min to $(300 + k)$ min, with no change in the maximum capacity for machine I, then Ace Novelty's profit is maximized by producing $\frac{1}{5}(240 - k)$ type-A souvenirs and $\frac{1}{5}(420 - 2k)$ type-B souvenirs, where $-210 \le k \le 240$.

c. Show that the shadow price for resource 2 (associated with constraint 2) is $.28.

3. Refer to Example 2.

a. Find the range of values that resource 2 can assume.

b. By how much can the right-hand side of constraint 3 be changed such that the current optimal solution still holds?

4. Refer to Example 2.
   a. Find the shadow price for resource 2.
   b. Identify the binding and nonbinding constraints.

In Exercises 5–10, you are given a linear programming problem.
a. Use the method of corners to solve the problem.
b. Find the range of values that the coefficient of x can assume without changing the optimal solution.
c. Find the range of values that resource 1 (requirement 1) can assume.
d. Find the shadow price for resource 1 (requirement 1).
e. Identify the binding and nonbinding constraints.

5. Maximize $P = 3x + 4y$
   subject to $2x + 3y \le 12$ Resource 1
   $2x + y \le 8$ Resource 2
   $x \ge 0, y \ge 0$

6. Maximize $P = 2x + 5y$
   subject to $x + 3y \le 15$ Resource 1
   $4x + y \le 16$ Resource 2
   $x \ge 0, y \ge 0$

7. Minimize $C = 2x + 5y$
   subject to $x + 2y \ge 4$ Requirement 1
   $x + y \ge 3$ Requirement 2
   $x \ge 0, y \ge 0$

8. Minimize $C = 3x + 4y$
   subject to $x + 3y \ge 8$ Requirement 1
   $x + y \ge 4$ Requirement 2
   $x \ge 0, y \ge 0$

9. Maximize $P = 4x + 3y$
subject to $5x + 3y \leq 30$ Resource 1
             $2x + 3y \leq 21$ Resource 2
                   $x \leq 4$ Resource 3
             $x \geq 0, y \geq 0$

10. Maximize $P = 4x + 5y$
subject to $x + y \leq 30$ Resource 1
            $x + 2y \leq 40$ Resource 2
                  $x \leq 25$ Resource 3
            $x \geq 0, y \geq 0$

11. MANUFACTURING—PRODUCTION SCHEDULING   A company manufactures two products, A and B, on machines I and II. The company will realize a profit of $3/unit of product A and a profit of $4/unit of product B. Manufacturing 1 unit of product A requires 6 min on machine I and 5 min on machine II. Manufacturing 1 unit of product B requires 9 min on machine I and 4 min on machine II. There are 5 hr of time available on machine I and 3 hr of time available on machine II in each work shift.

a. How many units of each product should be produced in each shift to maximize the company's profit?

b. Find the range of values that the contribution to the profit of 1 unit of product A can assume without changing the optimal solution.

c. Find the range of values that the resource associated with the time constraint on machine I can assume.

d. Find the shadow price for the resource associated with the time constraint on machine I.

12. AGRICULTURE—CROP PLANNING   A farmer plans to plant two crops, A and B. The cost of cultivating crop A is $40/acre whereas that of crop B is $60/acre. The farmer has a maximum of $7400 available for land cultivation. Each acre of crop A requires 20 labor-hours, and each acre of crop B requires 25 labor-hours. The farmer has a maximum of 3300 labor-hours available. If he expects to make a profit of $150/acre on crop A and $200/acre on crop B, how many acres of each crop should he plant in order to maximize his profit?

a. Find the range of values that the contribution to the profit of an acre of crop A can assume without changing the optimal solution.

b. Find the range of values that the resource associated with the constraint on the available land can assume.

c. Find the shadow price for the resource associated with the constraint on the available land.

13. MINING—PRODUCTION   Perth Mining Company operates two mines for the purpose of extracting gold and silver. The Saddle Mine costs $14,000/day to operate, and it yields 50 oz of gold and 3000 oz of silver per day. The Horseshoe Mine costs $16,000/day to operate, and it yields 75 oz of gold and 1000 ounces of silver per day. Company management has set a target of at least 650 oz of gold and 18,000 oz of silver.

a. How many days should each mine be operated so that the target can be met at a minimum cost?

b. Find the range of values that the Saddle Mine's daily operating cost can assume without changing the optimal solution.

c. Find the range of values that the requirement for gold can assume.

d. Find the shadow price for the requirement for gold.

14. TRANSPORTATION   Deluxe River Cruises operates a fleet of river vessels. The fleet has two types of vessels: a type-A vessel has 60 deluxe cabins and 160 standard cabins, whereas a type-B vessel has 80 deluxe cabins and 120 standard cabins. Under a charter agreement with the Odyssey Travel Agency, Deluxe River Cruises is to provide Odyssey with a minimum of 360 deluxe and 680 standard cabins for their 15-day cruise in May. It costs $44,000 to operate a type-A vessel and $54,000 to operate a type-B vessel for that period.

a. How many of each type of vessel should be used in order to keep the operating costs to a minimum?

b. Find the range of values that the cost of operating a type-A vessel can assume without changing the optimal solution.

c. Find the range of values that the requirement for deluxe cabins can assume.

d. Find the shadow price for the requirement for deluxe cabins.

15. MANUFACTURING—PRODUCTION SCHEDULING   Soundex produces two models of satellite radios. Model A requires 15 min of work on assembly line I and 10 min of work on assembly line II. Model B requires 10 min of work on assembly line I and 12 min of work on assembly line II. At most 25 hr of assembly time on line I and 22 hr of assembly time on line II are available each day. Soundex anticipates a profit of $12 on model A and $10 on model B. Because of previous overproduction, management decides to limit the production of model A satellite radios to no more than 80/day.

a. To maximize Soundex's profit, how many satellite radios of each model should be produced each day?

b. Find the range of values that the contribution to the profit of a model A satellite radio can assume without changing the optimal solution.

c. Find the range of values that the resource associated with the time constraint on machine I can assume.

d. Find the shadow price for the resource associated with the time constraint on machine I.

e. Identify the binding and nonbinding constraints.

16. MANUFACTURING   Refer to Exercise 15.

a. If the contribution to the profit of a model A satellite radio is changed to $8.50/radio, will the original optimal solution still hold? What will be the optimal profit?

b. If the contribution to the profit of a model A satellite radio is changed to $14.00/radio, will the original optimal solution still hold? What will be the optimal profit?

17. MANUFACTURING—PRODUCTION SCHEDULING Kane Manufacturing has a division that produces two models of fireplace grates, model A and model B. To produce each model A grate requires 3 lb of cast iron and 6 min of labor. To produce each model B grate requires 4 lb of cast iron and 3 min of labor. The profit for each model A grate is $2, and the profit for each model B grate is $1.50. 1000 lb of cast iron and 20 labor-hours are available for the production of grates each day. Because of an excess inventory of model B grates, management has decided to limit the production of model B grates to no more than 200 grates per day. How many grates of each model should the division produce daily to maximize Kane's profits?

a. Use the method of corners to solve the problem.

b. Find the range of values that the coefficient of x can assume without changing the optimal solution.

c. Find the range of values that the resource for cast iron can assume without changing the optimal solution.

d. Find the shadow price for the resource for cast iron.

e. Identify the binding and nonbinding constraints.

18. MANUFACTURING Refer to Exercise 17.

a. If the contribution to the profit of a model A grate is changed to $1.75/grate, will the original optimal solution still hold? What will be the new optimal solution?

b. If the contribution to the profit of a model A grate is changed to $2.50/grate, will the original optimal solution still hold? What will be the new optimal solution?

3.4 Exercises

1. The feasible set for the problem is shown in the accompanying figure.

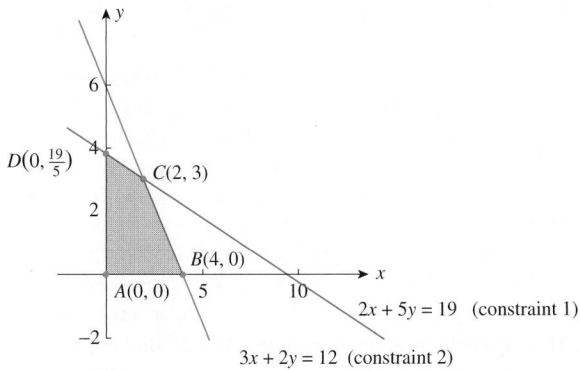

Evaluating the objective function $P = 2x + 4y$ at each feasible corner point, we obtain the following table:

Vertex	$P = 2x + 4y$
$A(0, 0)$	0
$B(4, 0)$	8
$C(2, 3)$	16
$D(0, \frac{19}{5})$	$15\frac{1}{5}$

We conclude that the maximum value of P is 16 attained at the point $(2, 3)$.

2. Assume that $P = cx + 4y$. Then

$$y = -\frac{c}{4}x + \frac{P}{4}$$

The slope of the isoprofit line is $-\dfrac{c}{4}$ and must be less than or equal to the slope of the line associated with constraint 1; that is,

$$-\frac{c}{4} \leq -\frac{2}{5}$$

Solving, we find $c \geq \frac{8}{5}$. A similar analysis shows that the slope of the isoprofit line must be greater than or equal to the slope of the line associated with constraint 2; that is,

$$-\frac{c}{4} \geq -\frac{3}{2}$$

Solving, we find $c \leq 6$. Thus, we have shown that if $1.6 \leq c \leq 6$ then the optimal solution obtained previously remains unaffected.

3. Suppose the right-hand side of constraint 1 is replaced by $19 + h$, where h is a real number. Then the new optimal solution occurs at the point whose coordinates are found by solving the system

$$2x + 5y = 19 + h$$
$$3x + 2y = 12$$

Multiplying the second equation by -5 and adding the resulting equation to 2 times the first equation, we obtain

$$-11x = -60 + 2(19 + h) = -22 + 2h$$

$$x = 2 - \frac{2}{11}h$$

Substituting this value of x into the second equation in the system gives

$$3\left(2 - \frac{2}{11}h\right) + 2y = 12$$

$$2y = 12 - 6 + \frac{6}{11}h$$

$$y = 3 + \frac{3}{11}h$$

The nonnegativity of x implies $2 - \frac{2}{11}h \geq 0$, or $h \leq 11$. The nonnegativity of y implies $3 + \frac{3}{11}h \geq 0$, or $h \geq -11$. Therefore, h must satisfy $-11 \leq h \leq 11$. This tells us that the amount used of resource 1 must lie between $19 - 11$ and $19 + 11$—that is, between 8 and 30.

4. If we set $h = 1$ in Exercise 3, we find that $x = \frac{20}{11}$ and $y = \frac{36}{11}$. Therefore, for these values of x and y,

$$P = 2\left(\frac{20}{11}\right) + 4\left(\frac{36}{11}\right) = \frac{184}{11} = 16\frac{8}{11}$$

Since the original optimal value of P is 16, we see that the shadow price for resource 1 is $\frac{8}{11}$.

5. Since both constraints hold with equality of the optimal solution $C(2, 3)$, they are binding constraints.

CHAPTER 3   Summary of Principal Terms

 TERMS

solution set of a system of linear inequalities (171)	feasible solution (185)	sensitivity analysis (198)
bounded solution set (172)	feasible set (185)	shadow price (205)
unbounded solution set (172)	optimal solution (185)	binding constraint (207)
objective function (176)	isoprofit line (186)	
linear programming problem (176)	method of corners (187)	

CHAPTER 3   Concept Review Questions

Fill in the blanks.

1. **a.** The solution set of the inequality $ax + by < c$ is a/an _____ _____ that does not include the _____ with equation $ax + by = c$.
   b. If $ax + by < c$ describes the lower half-plane, then the inequality _____ describes the lower half-plane together with the line having equation _____.

2. **a.** The solution set of a system of linear inequalities in the two variables x and y is the set of all _____ satisfying _____ inequality of the system.
   b. The solution set of a system of linear inequalities is _____ if it can be _____ by a circle.

3. A linear programming problem consists of a linear function, called a/an _____ _____ to be _____ or _____ subject to constraints in the form of _____ equations or _____.

4. **a.** If a linear programming problem has a solution, then it must occur at a/an _____ _____ of the feasible set.
   b. If the objective function of a linear programming problem is optimized at two adjacent vertices of the feasible set, then it is optimized at every point on the _____ segment joining these vertices.

5. In sensitivity analysis, we investigate how changes in the _____ of a linear programming problem affect the _____ solution.

6. The shadow price for the ith _____ is the _____ by which the _____ of the objective function is _____ if the right-hand side of the ith constraint is _____ by 1 unit.

In Exercises 1 and 2, find the optimal value(s) of the objective function on the feasible set S.

1. $Z = 2x + 3y$

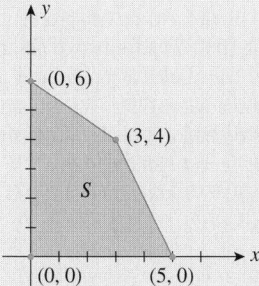

2. $Z = 4x + 3y$

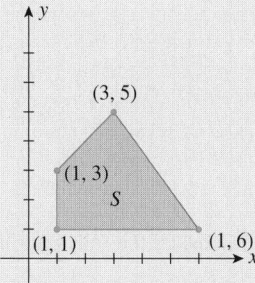

In Exercises 3–12, use the method of corners to solve the linear programming problem.

3. Maximize $P = 3x + 5y$
subject to $2x + 3y \leq 12$
$x + y \leq 5$
$x \geq 0, y \geq 0$

4. Maximize $P = 2x + 3y$
subject to $2x + y \leq 12$
$x - 2y \leq 1$
$x \geq 0, y \geq 0$

5. Minimize $C = 2x + 5y$
subject to $x + 3y \geq 15$
$4x + y \geq 16$
$x \geq 0, y \geq 0$

6. Minimize $C = 3x + 4y$
subject to $2x + y \geq 4$
$2x + 5y \geq 10$
$x \geq 0, y \geq 0$

7. Maximize $P = 3x + 2y$
subject to $2x + y \leq 16$
$2x + 3y \leq 36$
$4x + 5y \geq 28$
$x \geq 0, y \geq 0$

8. Maximize $P = 6x + 2y$
subject to $x + 2y \leq 12$
$x + y \leq 8$
$2x - 3y \geq 6$
$x \geq 0, y \geq 0$

9. Minimize $C = 2x + 7y$
subject to $3x + 5y \geq 45$
$3x + 10y \geq 60$
$x \geq 0, y \geq 0$

10. Minimize $C = 4x + y$
subject to $6x + y \geq 18$
$2x + y \geq 10$
$x + 4y \geq 12$
$x \geq 0, y \geq 0$

11. Find the maximum and minimum of $Q = x + y$ subject to

$$5x + 2y \geq 20$$
$$x + 2y \geq 8$$
$$x + 4y \leq 22$$
$$x \geq 0, y \geq 0$$

12. Find the maximum and minimum of $Q = 2x + 5y$ subject to

$$x + y \geq 4$$
$$-x + y \leq 6$$
$$x + 3y \leq 30$$
$$x \leq 12$$
$$x \geq 0, y \geq 0$$

13. FINANCIAL ANALYSIS An investor has decided to commit no more than $80,000 to the purchase of the common stocks of two companies, company A and company B. He has also estimated that there is a chance of at most a 1% capital loss on his investment in company A and a chance of at most a 4% loss on his investment in company B, and he has decided that these losses should not exceed $2000. On the other hand, he expects to make a 14% profit from his investment in company A and a 20% profit from his investment in company B. Determine how much he should invest in the stock of each company in order to maximize his investment returns.

14. MANUFACTURING—PRODUCTION SCHEDULING Soundex produces two models of satellite radios. Model A requires 15 min of work on assembly line I and 10 min of work on assembly line II. Model B requires 10 min of work on assembly line I and 12 min of work on assembly line II. At most, 25 labor-hours of assembly time on line I and 22 labor-hours of assembly time on line II are available each day. It is anticipated that Soundex will realize a profit of $12 on model A and $10 on model B. How many satellite radios of each model should be produced each day in order to maximize Soundex's profit?

15. MANUFACTURING—PRODUCTION SCHEDULING Kane Manufacturing has a division that produces two models of grates, model A and model B. To produce each model A grate requires 3 lb of cast iron and 6 min of labor. To produce each model B grate requires 4 lb of cast iron and 3 min of labor. The profit for each model A grate is $2.00, and the profit for

each model B grate is $1.50. Available for grate production each day are 1000 lb of cast iron and 20 labor-hours. Because of a backlog of orders for model B grates, Kane's manager has decided to produce at least 180 model B grates/day. How many grates of each model should Kane produce to maximize its profits?

16. MINIMIZING SHIPPING COSTS A manufacturer of projection TVs must ship a total of at least 1000 TVs to its two central warehouses. Each warehouse can hold a maximum of 750 TVs. The first warehouse already has 150 TVs on hand, whereas the second has 50 TVs on hand. It costs $40 to ship a TV to the first warehouse, and it costs $80 to ship a TV to the second warehouse. How many TVs should be shipped to each warehouse to minimize the cost?

CHAPTER 3   Before Moving On . . .

1. Determine graphically the solution set for the following systems of inequalities.

 a. $2x + y \leq 10$
    $x + 3y \leq 15$
    $x \leq 4$
    $x \geq 0, y \geq 0$

 b. $2x + y \geq 8$
    $2x + 3y \geq 15$
    $x \geq 0$
    $y \geq 2$

2. Find the maximum and minimum values of $Z = 3x - y$ on the following feasible set.

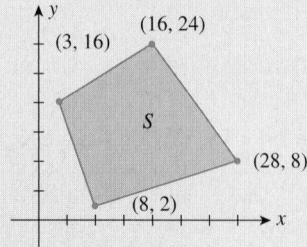

3. Maximize $P = x + 3y$
   subject to $2x + 3y \leq 11$
               $3x + 7y \leq 24$
               $x \geq 0, y \geq 0$

4. Minimize $C = 4x + y$
   subject to $2x + y \geq 10$
               $2x + 3y \geq 24$
               $x + 3y \geq 15$
               $x \geq 0, y \geq 0$

5. **Sensitivity Analysis (Optional)** Consider the following linear programming problem:

$$\text{Maximize} \quad P = 2x + 3y$$
$$\text{subject to} \quad x + 2y \leq 16$$
$$3x + 2y \leq 24$$
$$x \geq 0, y \geq 0$$

 a. Solve the problem.
 b. Find the range of values that the coefficient of x can assume without changing the optimal solution.
 c. Find the range of values that resource 1 (requirement 1) can assume.
 d. Find the shadow price for resource 1.
 e. Identify the binding and nonbinding constraints.

4

Linear Programming: An Algebraic Approach

How much profit? The Ace Novelty Company produces three types of souvenirs. Each type requires a certain amount of time on each of three different machines. Each machine may be operated for a certain amount of time per day. In Example 5, page 229, we will determine how many souvenirs of each type Ace Novelty should make per day in order to maximize its daily profits.

© Roy Gumpel/Stone/Getty Images

THE GEOMETRIC APPROACH introduced in the previous chapter may be used to solve linear programming problems involving two or even three variables. But for linear programming problems involving more than two variables, an algebraic approach is preferred. One such technique, the *simplex method*, was developed by George Dantzig in the late 1940s and remains in wide use to this day.

We begin Chapter 4 by developing the simplex method for solving *standard maximization problems*. We then see how, thanks to the principle of duality discovered by the great mathematician John von Neumann, this method can be used to solve a restricted class of *standard minimization problems*. Finally, we see how the simplex method can be adapted to solve *nonstandard problems*—problems that do not belong to the other aforementioned categories.

4.1 The Simplex Method: Standard Maximization Problems

The Simplex Method

As mentioned in Chapter 3, the method of corners is not suitable for solving linear programming problems when the number of variables or constraints is large. Its major shortcoming is that a knowledge of all the corner points of the feasible set S associated with the problem is required. What we need is a method of solution that is based on a judicious selection of the corner points of the feasible set S, thereby reducing the number of points to be inspected. One such technique, called the *simplex method*, was developed in the late 1940s by George Dantzig and is based on the Gauss–Jordan elimination method. The simplex method is readily adaptable to the computer, which makes it ideally suitable for solving linear programming problems involving large numbers of variables and constraints.

Basically, the simplex method is an iterative procedure; that is, it is repeated over and over again. Beginning at some initial feasible solution (a corner point of the feasible set S, usually the origin), each iteration brings us to another corner point of S with an improved (but certainly no worse) value of the objective function. The iteration is terminated when the optimal solution is reached (if it exists).

In this section we describe the simplex method for solving a large class of problems that are referred to as standard maximization problems.

Before stating a formal procedure for solving standard linear programming problems based on the simplex method, let's consider the following analysis of a two-variable problem. The ensuing discussion will clarify the general procedure and at the same time enhance our understanding of the simplex method by examining the motivation that led to the steps of the procedure.

A Standard Linear Programming Problem

A **standard maximization problem** is one in which

1. The objective function is to be maximized.
2. All the variables involved in the problem are nonnegative.
3. Each linear constraint may be written so that the expression involving the variables is less than or equal to a nonnegative constant.

Consider the linear programming problem presented at the beginning of Section 3.3:

$$\text{Maximize} \quad P = 3x + 2y \qquad \qquad \textbf{(1)}$$
$$\text{subject to} \quad 2x + 3y \le 12$$
$$2x + y \le 8 \qquad \qquad \textbf{(2)}$$
$$x \ge 0, y \ge 0$$

You can easily verify that this is a standard maximization problem. The feasible set S associated with this problem is reproduced in Figure 1, where we have labeled the four feasible corner points $A(0, 0)$, $B(4, 0)$, $C(3, 2)$, and $D(0, 4)$. Recall that the optimal solution to the problem occurs at the corner point $C(3, 2)$.

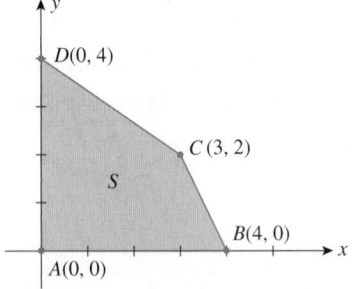

FIGURE 1
The optimal solution occurs at $C(3, 2)$.

As a first step in the solution using the simplex method, we replace the system of inequality constraints (2) with a system of equality constraints. This may be accomplished by using nonnegative variables called **slack variables**. Let's begin by considering the inequality

$$2x + 3y \leq 12$$

Observe that the left-hand side of this equation is always less than or equal to the right-hand side. Therefore, by adding a nonnegative variable u to the left-hand side to compensate for this difference, we obtain the equality

$$2x + 3y + u = 12$$

For example, if $x = 1$ and $y = 1$ [you can see by referring to Figure 1 that the point $(1, 1)$ is a feasible point of S], then $u = 7$. Thus,

$$2(1) + 3(1) + 7 = 12$$

If $x = 2$ and $y = 1$ [the point $(2, 1)$ is also a feasible point of S], then $u = 5$. Thus,

$$2(2) + 3(1) + 5 = 12$$

The variable u is a slack variable.

Similarly, the inequality $2x + y \leq 8$ is converted into the equation $2x + y + v = 8$ through the introduction of the slack variable v. System (2) of linear inequalities may now be viewed as the system of linear equations

$$
\begin{aligned}
2x + 3y + u \quad\;\;\; &= 12 \\
2x + \;y \quad\;\; + v &= \;8
\end{aligned}
$$

where x, y, u, and v are all nonnegative.

Finally, rewriting the objective function (1) in the form $-3x - 2y + P = 0$, where the coefficient of P is $+1$, we are led to the following system of linear equations:

$$
\begin{aligned}
2x + 3y + u \qquad\qquad\;\;\; &= 12 \\
2x + \;y \qquad + v \qquad\;\; &= \;8 \qquad\qquad \textbf{(3)} \\
-3x - 2y \qquad\qquad + P &= \;0
\end{aligned}
$$

Since System (3) consists of three linear equations in the five variables x, y, u, v, and P, we may solve for three of the variables in terms of the other two. Thus, there are infinitely many solutions to this system expressible in terms of two parameters. Our linear programming problem is now seen to be equivalent to the following: From among all the solutions of System (3) for which x, y, u, and v are nonnegative (such solutions are called **feasible solutions**), determine the solution(s) that renders P a maximum.

The augmented matrix associated with System (3) is

Nonbasic variables ⸺ Basic variables
Column of constants

$$
\begin{array}{ccccc}
x & y & u & v & P \\
\end{array}
$$

$$
\left[
\begin{array}{ccccc|c}
2 & 3 & 1 & 0 & 0 & 12 \\
2 & 1 & 0 & 1 & 0 & 8 \\
-3 & -2 & 0 & 0 & 1 & 0
\end{array}
\right] \qquad \textbf{(4)}
$$

Observe that each of the u-, v-, and P-columns of the augmented matrix (4) is a unit column (see page 85). The variables associated with unit columns are called basic variables; all other variables are called nonbasic variables.

Now, the configuration of the augmented matrix (4) suggests that we solve for the basic variables u, v, and P in terms of the nonbasic variables x and y, obtaining

$$\begin{aligned} u &= 12 - 2x - 3y \\ v &= 8 - 2x - y \\ P &= 3x + 2y \end{aligned} \tag{5}$$

Of the infinitely many feasible solutions obtainable by assigning arbitrary nonnegative values to the parameters x and y, a particular solution is obtained by letting $x = 0$ and $y = 0$. In fact, this solution is given by

$$x = 0 \qquad y = 0 \qquad u = 12 \qquad v = 8 \qquad P = 0$$

Such a solution, obtained by setting the nonbasic variables equal to zero, is called a basic solution of the system. This particular solution corresponds to the corner point $A(0, 0)$ of the feasible set associated with the linear programming problem (see Figure 1). Observe that $P = 0$ at this point.

Now, if the value of P cannot be increased, we have found the optimal solution to the problem at hand. To determine whether the value of P can in fact be improved, let's turn our attention to the objective function in (1). Since both the coefficients of x and y are positive, the value of P can be improved by increasing x and/or y—that is, by moving away from the origin. Note that we arrive at the same conclusion by observing that the last row of the augmented matrix (4) contains entries that are *negative*. (Compare the original objective function, $P = 3x + 2y$, with the rewritten objective function, $-3x - 2y + P = 0$.)

Continuing our quest for an optimal solution, our next task is to determine whether it is more profitable to increase the value of x or that of y (increasing x and y simultaneously is more difficult). Since the coefficient of x is greater than that of y, a unit increase in the x-direction will result in a greater increase in the value of the objective function P than a unit increase in the y-direction. Thus, we should increase the value of x while holding y constant. How much can x be increased while holding $y = 0$? Upon setting $y = 0$ in the first two equations of (5), we see that

$$\begin{aligned} u &= 12 - 2x \\ v &= 8 - 2x \end{aligned} \tag{6}$$

Since u must be nonnegative, the first equation of (6) implies that x cannot exceed $\frac{12}{2}$, or 6. The second equation of (6) and the nonnegativity of v implies that x cannot exceed $\frac{8}{2}$, or 4. Thus, we conclude that x can be increased by at most 4.

Now, if we set $y = 0$ and $x = 4$ in System (5), we obtain the solution

$$x = 4 \qquad y = 0 \qquad u = 4 \qquad v = 0 \qquad P = 12$$

which is a basic solution to System (3), this time with y and v as nonbasic variables. (Recall that the nonbasic variables are precisely the variables that are set equal to zero.)

Let's see how this basic solution may be found by working with the augmented matrix of the system. Since x is to replace v as a basic variable, our aim is to find an augmented matrix that is equivalent to the matrix (4) and has a configuration in which the x-column is in the unit form

$$\begin{bmatrix} 0 \\ 1 \\ 0 \end{bmatrix}$$

replacing what is presently the form of the v-column in (4). This may be accomplished by pivoting about the circled number 2.

$$\begin{array}{ccccc} x & y & u & v & P & \text{Const.} \\ \begin{bmatrix} 2 & 3 & 1 & 0 & 0 & 12 \\ ② & 1 & 0 & 1 & 0 & 8 \\ -3 & -2 & 0 & 0 & 1 & 0 \end{bmatrix} \xrightarrow{\frac{1}{2}R_2} \begin{bmatrix} 2 & 3 & 1 & 0 & 0 & 12 \\ ① & \frac{1}{2} & 0 & \frac{1}{2} & 0 & 4 \\ -3 & -2 & 0 & 0 & 1 & 0 \end{bmatrix} \end{array} \quad (7)$$

$$\begin{array}{ccccc} & x & y & u & v & P & \text{Const.} \\ \begin{array}{c} \xrightarrow{R_1 - 2R_2} \\ \xrightarrow{R_3 + 3R_2} \end{array} & \begin{bmatrix} 0 & 2 & 1 & -1 & 0 & 4 \\ 1 & \frac{1}{2} & 0 & \frac{1}{2} & 0 & 4 \\ 0 & -\frac{1}{2} & 0 & \frac{3}{2} & 1 & 12 \end{bmatrix} \end{array} \quad (8)$$

Using (8), we now solve for the basic variables x, u, and P in terms of the non-basic variables y and v, obtaining

$$x = 4 - \frac{1}{2}y - \frac{1}{2}v$$

$$u = 4 - 2y + v$$

$$P = 12 + \frac{1}{2}y - \frac{3}{2}v$$

Setting the nonbasic variables y and v equal to zero gives

$$x = 4 \qquad y = 0 \qquad u = 4 \qquad v = 0 \qquad P = 12$$

as before.

We have now completed one iteration of the simplex procedure, and our search has brought us from the feasible corner point $A(0, 0)$, where $P = 0$, to the feasible corner point $B(4, 0)$, where P attained a value of 12, which is certainly an improvement! (See Figure 2.)

Before going on, let's introduce the following terminology. The circled element 2 in the first augmented matrix of (7), which was to be converted into a 1, is called a *pivot element*. The column containing the pivot element is called the *pivot column*. The pivot column is associated with a nonbasic variable that is to be converted to a basic variable. Note that *the last entry in the pivot column is the negative number with the largest absolute value to the left of the vertical line in the last row—* precisely the criterion for choosing the direction of maximum increase in P.

The row containing the pivot element is called the *pivot row*. The pivot row can also be found by dividing each positive number in the pivot column into the corresponding number in the last column (the column of constants). *The pivot row is the one with the smallest ratio.* In the augmented matrix (7), the pivot row is the second row because the ratio $\frac{8}{2}$, or 4, is less than the ratio $\frac{12}{2}$, or 6. (Compare this with the

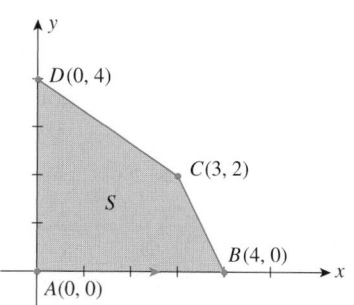

FIGURE 2
One iteration has taken us from $A(0, 0)$, where $P = 0$, to $B(4, 0)$, where $P = 12$.

earlier analysis pertaining to the determination of the largest permissible increase in the value of x.)

The following is a summary of the procedure for selecting the pivot element.

Selecting the Pivot Element

1. *Select the pivot column:* Locate the most negative entry to the left of the vertical line in the last row. The column containing this entry is the **pivot column**. (If there is more than one such column, choose any one.)
2. *Select the pivot row:* Divide each positive entry in the pivot column into its corresponding entry in the column of constants. The **pivot row** is the row corresponding to the smallest ratio thus obtained. (If there is more than one such entry, choose any one.)
3. The **pivot element** is the element common to both the pivot column and the pivot row.

Continuing with the solution to our problem, we observe that the last row of the augmented matrix (8) contains a negative number—namely, $-\frac{1}{2}$. This indicates that P is not maximized at the feasible corner point $B(4, 0)$, so another iteration is required. Without once again going into a detailed analysis, we proceed immediately to the selection of a pivot element. In accordance with the rules, we perform the necessary row operations as follows:

$$
\text{Pivot row} \rightarrow
\begin{array}{ccccc|c}
x & y & u & v & P & \\
0 & ② & 1 & -1 & 0 & 4 \\
1 & \frac{1}{2} & 0 & \frac{1}{2} & 0 & 4 \\
0 & -\frac{1}{2} & 0 & \frac{3}{2} & 1 & 12
\end{array}
\qquad
\begin{array}{c}
\text{Ratio} \\
\frac{4}{2} = 2 \\
\frac{4}{1/2} = 8
\end{array}
$$

$$\uparrow \text{ Pivot column}$$

$$
\xrightarrow{\frac{1}{2}R_1}
\begin{array}{ccccc|c}
x & y & u & v & P & \\
0 & ① & \frac{1}{2} & -\frac{1}{2} & 0 & 2 \\
1 & \frac{1}{2} & 0 & \frac{1}{2} & 0 & 4 \\
0 & -\frac{1}{2} & 0 & \frac{3}{2} & 1 & 12
\end{array}
$$

$$
\begin{array}{c}
R_2 - \frac{1}{2}R_1 \\
\xrightarrow{\hspace{1.5cm}} \\
R_3 + \frac{1}{2}R_1
\end{array}
\begin{array}{ccccc|c}
x & y & u & v & P & \\
0 & 1 & \frac{1}{2} & -\frac{1}{2} & 0 & 2 \\
1 & 0 & -\frac{1}{4} & \frac{3}{4} & 0 & 3 \\
0 & 0 & \frac{1}{4} & \frac{5}{4} & 1 & 13
\end{array}
$$

Interpreting the last augmented matrix in the usual fashion, we find the basic solution $x = 3$, $y = 2$, and $P = 13$. Since there are no negative entries in the last row, the solution is optimal and P cannot be increased further. The optimal solution is the feasible corner point $C(3, 2)$ (Figure 3). Observe that this agrees with the solution we found using the method of corners in Section 3.3.

Having seen how the simplex method works, let's list the steps involved in the procedure. The first step is to set up the initial **simplex tableau**.

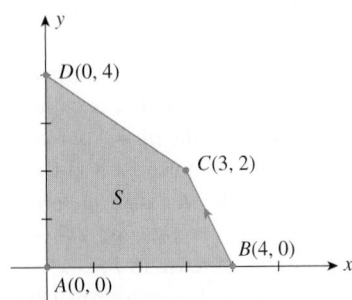

FIGURE 3
The next iteration has taken us from $B(4, 0)$, where $P = 12$, to $C(3, 2)$, where $P = 13$.

Setting Up the Initial Simplex Tableau

1. Transform the system of linear inequalities into a system of linear equations by introducing slack variables.

2. Rewrite the objective function

$$P = c_1 x_1 + c_2 x_2 + \cdots + c_n x_n$$

in the form

$$-c_1 x_1 - c_2 x_2 - \cdots - c_n x_n + P = 0$$

where all the variables are on the left and the coefficient of P is $+1$. Write this equation below the equations of step 1.

3. Write the augmented matrix associated with this system of linear equations.

EXAMPLE 1 Set up the initial simplex tableau for the linear programming problem posed in Example 1, Section 3.2.

Solution The problem at hand is to maximize

$$P = x + 1.2y$$

or, equivalently,

$$P = x + \frac{6}{5}y$$

subject to

$$
\begin{aligned}
2x + y &\le 180 \\
x + 3y &\le 300 \\
x \ge 0, y &\ge 0
\end{aligned}
\tag{9}
$$

This is a standard maximization problem and may be solved by the simplex method. Since System (9) has two linear inequalities (other than $x \ge 0$, $y \ge 0$), we introduce the two slack variables u and v to convert it to a system of linear equations:

$$
\begin{aligned}
2x + y + u \phantom{{}+v} &= 180 \\
x + 3y \phantom{{}+u} + v &= 300
\end{aligned}
$$

Next, by rewriting the objective function in the form

$$-x - \frac{6}{5}y + P = 0$$

where the coefficient of P is $+1$, and placing it below the system of equations, we obtain the system of linear equations

$$
\begin{aligned}
2x + y + u \phantom{{}+v} \phantom{{}+P} &= 180 \\
x + 3y \phantom{{}+u} + v \phantom{{}+P} &= 300 \\
-x - \frac{6}{5}y \phantom{{}+u} \phantom{{}+v} + P &= 0
\end{aligned}
$$

The initial simplex tableau associated with this system is

x	y	u	v	P	Constant
2	1	1	0	0	180
1	3	0	1	0	300
-1	$-\frac{6}{5}$	0	0	1	0

Before completing the solution to the problem posed in Example 1, let's summarize the main steps of the **simplex method**.

The Simplex Method
1. *Set up the initial simplex tableau.*
2. *Determine whether the optimal solution has been reached by examining all entries in the last row to the left of the vertical line.*
   a. If all the entries are nonnegative, the optimal solution has been reached. Proceed to step 4.
   b. If there are one or more negative entries, the optimal solution has not been reached. Proceed to step 3.
3. *Perform the pivot operation.* Locate the pivot element and convert it to a 1 by dividing all the elements in the pivot row by the pivot element. Using row operations, convert the pivot column into a unit column by adding suitable multiples of the pivot row to each of the other rows as required. Return to step 2.
4. *Determine the optimal solution(s).* The value of the variable heading each unit column is given by the entry lying in the column of constants in the row containing the 1. The variables heading columns not in unit form are assigned the value zero.

EXAMPLE 2   Complete the solution to the problem discussed in Example 1.

Solution   The first step in our procedure, setting up the initial simplex tableau, was completed in Example 1. We continue with Step 2.

Step 2   *Determine whether the optimal solution has been reached.* First, refer to the initial simplex tableau:

x	y	u	v	P	Constant
2	1	1	0	0	180
1	3	0	1	0	300
-1	$-\frac{6}{5}$	0	0	1	0

(10)

Since there are negative entries in the last row of the initial simplex tableau, the initial solution is not optimal. We proceed to Step 3.

Step 3   *Perform the following iterations.* First, locate the pivot element:
   a. Since the entry $-\frac{6}{5}$ is the most negative entry to the left of the vertical line in the last row of the initial simplex tableau, the second column in the tableau is the pivot column.

b. Divide each positive number of the pivot column into the corresponding entry in the column of constants and compare the ratios thus obtained. We see that the ratio $\frac{300}{3}$ is less than the ratio $\frac{180}{1}$, so row 2 in the tableau is the pivot row.

c. The entry 3 lying in the pivot column and the pivot row is the pivot element.

Next, we convert this pivot element into a 1 by multiplying all the entries in the pivot row by $\frac{1}{3}$. Then, using elementary row operations, we complete the conversion of the pivot column into a unit column. The details of the iteration are recorded as follows:

	x	y	u	v	P	Constant	Ratio
	2	1	1	0	0	180	$\frac{180}{1} = 180$
Pivot row →	1	③	0	1	0	300	$\frac{300}{3} = 100$
	-1	$-\frac{6}{5}$	0	0	1	0	

↑ Pivot column

	x	y	u	v	P	Constant
$\frac{1}{3}R_2$ →	2	1	1	0	0	180
	$\frac{1}{3}$	①	0	$\frac{1}{3}$	0	100
	-1	$-\frac{6}{5}$	0	0	1	0

	x	y	u	v	P	Constant
$R_1 - R_2$ →	$\frac{5}{3}$	0	1	$-\frac{1}{3}$	0	80
$R_3 + \frac{6}{5}R_2$	$\frac{1}{3}$	1	0	$\frac{1}{3}$	0	100
	$-\frac{3}{5}$	0	0	$\frac{2}{5}$	1	120

(11)

This completes one iteration. The last row of the simplex tableau contains a negative number, so an optimal solution has not been reached. Therefore, we repeat the iterative step once again, as follows:

	x	y	u	v	P	Constant	Ratio
Pivot row →	⑤⁄₃	0	1	$-\frac{1}{3}$	0	80	$\frac{80}{5/3} = 48$
	$\frac{1}{3}$	1	0	$\frac{1}{3}$	0	100	$\frac{100}{1/3} = 300$
	$-\frac{3}{5}$	0	0	$\frac{2}{5}$	1	120	

↑ Pivot column

	x	y	u	v	P	Constant
$\frac{3}{5}R_1$ →	①	0	$\frac{3}{5}$	$-\frac{1}{5}$	0	48
	$\frac{1}{3}$	1	0	$\frac{1}{3}$	0	100
	$-\frac{3}{5}$	0	0	$\frac{2}{5}$	1	120

	x	y	u	v	P	Constant
$R_2 - \frac{1}{3}R_1$	1	0	$\frac{3}{5}$	$-\frac{1}{5}$	0	48
$\xrightarrow{\quad}$ $R_3 + \frac{3}{5}R_1$	0	1	$-\frac{1}{5}$	$\frac{2}{5}$	0	84
	0	0	$\frac{9}{25}$	$\frac{7}{25}$	1	$148\frac{4}{5}$

(12)

The last row of the simplex tableau (12) contains no negative numbers, so we conclude that the optimal solution has been reached.

Step 4　*Determine the optimal solution.* Locate the basic variables in the final tableau. In this case the basic variables (those heading unit columns) are x, y, and P. The value assigned to the basic variable x is the number 48, which is the entry lying in the column of constants and in row 1 (the row that contains the 1).

x	y	u	v	P	Constant
①	0	$\frac{3}{5}$	$-\frac{1}{5}$	0	48 ←
0	①	$-\frac{1}{5}$	$\frac{2}{5}$	0	84 ←
0	0	$\frac{9}{25}$	$\frac{7}{25}$	①	$148\frac{4}{5}$ ←

Similarly, we conclude that $y = 84$ and $P = 148.8$. Next, we note that the variables u and v are nonbasic and are accordingly assigned the values $u = 0$ and $v = 0$. These results agree with those obtained in Example 1, Section 3.3.

EXAMPLE 3

$$\text{Maximize}\quad P = 2x + 2y + z$$
$$\text{subject to}\quad 2x + y + 2z \le 14$$
$$2x + 4y + z \le 26$$
$$x + 2y + 3z \le 28$$
$$x \ge 0, y \ge 0, z \ge 0$$

Solution　Introducing the slack variables u, v, and w and rewriting the objective function in the standard form gives the system of linear equations

$$2x + y + 2z + u \qquad\qquad = 14$$
$$2x + 4y + z \qquad + v \qquad = 26$$
$$x + 2y + 3z \qquad\qquad + w \qquad = 28$$
$$-2x - 2y - z \qquad\qquad\qquad + P = 0$$

The initial simplex tableau is given by

x	y	z	u	v	w	P	Constant
2	1	2	1	0	0	0	14
2	4	1	0	1	0	0	26
1	2	3	0	0	1	0	28
-2	-2	-1	0	0	0	1	0

Since the most negative entry in the last row (-2) occurs twice, we may choose either the x- or the y-column as the pivot column. Choosing the x-column as the pivot column and proceeding with the first iteration, we obtain the following sequence of tableaus:

	x	y	z	u	v	w	P	Constant	Ratio
Pivot row $\rightarrow$	②	1	2	1	0	0	0	14	$\frac{14}{2} = 7$
	2	4	1	0	1	0	0	26	$\frac{26}{2} = 13$
	1	2	3	0	0	1	0	28	$\frac{28}{1} = 28$
	-2	-2	-1	0	0	0	1	0	

Pivot column (under x)

	x	y	z	u	v	w	P	Constant
$\frac{1}{2}R_1$	①	$\frac{1}{2}$	1	$\frac{1}{2}$	0	0	0	7
$\longrightarrow$	2	4	1	0	1	0	0	26
	1	2	3	0	0	1	0	28
	-2	-2	-1	0	0	0	1	0

	x	y	z	u	v	w	P	Constant
	1	$\frac{1}{2}$	1	$\frac{1}{2}$	0	0	0	7
$R_2 - 2R_1$	0	3	-1	-1	1	0	0	12
$R_3 - R_1$	0	$\frac{3}{2}$	2	$-\frac{1}{2}$	0	1	0	21
$R_4 + 2R_1$	0	-1	1	1	0	0	1	14

Since there is a negative number in the last row of the simplex tableau, we perform another iteration, as follows:

	x	y	z	u	v	w	P	Constant	Ratio
	1	$\frac{1}{2}$	1	$\frac{1}{2}$	0	0	0	7	$\frac{7}{1/2} = 14$
Pivot row $\rightarrow$	0	③	-1	-1	1	0	0	12	$\frac{12}{3} = 4$
	0	$\frac{3}{2}$	2	$-\frac{1}{2}$	0	1	0	21	$\frac{21}{3/2} = 14$
	0	-1	1	1	0	0	1	14	

Pivot column (under y)

	x	y	z	u	v	w	P	Constant
	1	$\frac{1}{2}$	1	$\frac{1}{2}$	0	0	0	7
$\frac{1}{3}R_2$	0	①	$-\frac{1}{3}$	$-\frac{1}{3}$	$\frac{1}{3}$	0	0	4
$\longrightarrow$	0	$\frac{3}{2}$	2	$-\frac{1}{2}$	0	1	0	21
	0	-1	1	1	0	0	1	14

	x	y	z	u	v	w	P	Constant
	1	0	$\frac{7}{6}$	$\frac{2}{3}$	$-\frac{1}{6}$	0	0	5
$R_1 - \frac{1}{2}R_2$	0	1	$-\frac{1}{3}$	$-\frac{1}{3}$	$\frac{1}{3}$	0	0	4
$R_3 - \frac{3}{2}R_2$	0	0	$\frac{5}{2}$	0	$-\frac{1}{2}$	1	0	15
$R_4 + R_2$	0	0	$\frac{2}{3}$	$\frac{2}{3}$	$\frac{1}{3}$	0	1	18

All entries in the last row are nonnegative, so we have reached the optimal solution. We conclude that $x = 5$, $y = 4$, $z = 0$, $u = 0$, $v = 0$, $w = 15$, and $P = 18$. ∎

EXPLORE & DISCUSS

Consider the linear programming problem

$$\text{Maximize} \quad P = x + 2y$$
$$\text{subject to} \quad -2x + y \le 4$$
$$x - 3y \le 3$$
$$x \ge 0, \, y \ge 0$$

1. Sketch the feasible set S for the linear programming problem and explain why the problem has an unbounded solution.

2. Use the simplex method to solve the problem as follows:
   a. Perform one iteration on the initial simplex tableau. Interpret your result. Indicate the point on S corresponding to this (nonoptimal) solution.
   b. Show that the simplex procedure breaks down when you attempt to perform another iteration by demonstrating that there is no pivot element.
   c. Describe what happens if you violate the rule for finding the pivot element by allowing the ratios to be negative and proceeding with the iteration.

The following example is constructed to illustrate the geometry associated with the simplex method when used to solve a problem in three-dimensional space. We sketch the feasible set for the problem and show the path dictated by the simplex method in arriving at the optimal solution for the problem. The use of a calculator will help in the arithmetic operations if you wish to verify the steps.

EXAMPLE 4

$$\text{Maximize} \quad P = 20x + 12y + 18z$$
$$\text{subject to} \quad 3x + y + 2z \le 9$$
$$2x + 3y + z \le 8$$
$$x + 2y + 3z \le 7$$
$$x \ge 0, \, y \ge 0, \, z \ge 0$$

Solution   Introducing the slack variables u, v, and w and rewriting the objective function in standard form gives the following system of linear equations:

$$3x + y + 2z + u \qquad\qquad = 9$$
$$2x + 3y + z \qquad + v \qquad\quad = 8$$
$$x + 2y + 3z \qquad\qquad + w \quad = 7$$
$$-20x - 12y - 18z \qquad\qquad\quad + P = 0$$

The initial simplex tableau is given by

x	y	z	u	v	w	P	Constant
3	1	2	1	0	0	0	9
2	3	1	0	1	0	0	8
1	2	3	0	0	1	0	7
−20	−12	−18	0	0	0	1	0

Since the most negative entry in the last row (-20) occurs in the x-column, we choose the x-column as the pivot column. Proceeding with the first iteration, we obtain the following sequence of tableaus:

	x	y	z	u	v	w	P	Constant	Ratio
Pivot row →	③	1	2	1	0	0	0	9	$\frac{9}{3} = 3$
	2	3	1	0	1	0	0	8	$\frac{8}{2} = 4$
	1	2	3	0	0	1	0	7	$\frac{7}{1} = 7$
	-20	-12	-18	0	0	0	1	0	

↑ Pivot column

	x	y	z	u	v	w	P	Constant
$\frac{1}{3}R_1$ ⟶	①	$\frac{1}{3}$	$\frac{2}{3}$	$\frac{1}{3}$	0	0	0	3
	2	3	1	0	1	0	0	8
	1	2	3	0	0	1	0	7
	-20	-12	-18	0	0	0	1	0

	x	y	z	u	v	w	P	Constant	Ratio
$R_2 - 2R_1$	1	$\frac{1}{3}$	$\frac{2}{3}$	$\frac{1}{3}$	0	0	0	3	9
$R_3 - R_1$ ⟶	0	⑦⁄₃	$-\frac{1}{3}$	$-\frac{2}{3}$	1	0	0	2	$\frac{6}{7}$
$R_4 + 20R_1$ Pivot row	0	$\frac{5}{3}$	$\frac{7}{3}$	$-\frac{1}{3}$	0	1	0	4	$\frac{12}{5}$
	0	$-\frac{16}{3}$	$-\frac{14}{3}$	$\frac{20}{3}$	0	0	1	60	

↑ Pivot column

After one iteration we are at the point $(3, 0, 0)$ with $P = 60$. (See Figure 4 on page 228.) Since the most negative entry in the last row is $-\frac{16}{3}$, we choose the y-column as the pivot column. Proceeding with this iteration, we obtain

	x	y	z	u	v	w	P	Constant
$\frac{3}{7}R_2$ ⟶	1	$\frac{1}{3}$	$\frac{2}{3}$	$\frac{1}{3}$	0	0	0	3
	0	①	$-\frac{1}{7}$	$-\frac{2}{7}$	$\frac{3}{7}$	0	0	$\frac{6}{7}$
	0	$\frac{5}{3}$	$\frac{7}{3}$	$-\frac{1}{3}$	0	1	0	4
	0	$-\frac{16}{3}$	$-\frac{14}{3}$	$\frac{20}{3}$	0	0	1	60

	x	y	z	u	v	w	P	Constant	Ratio
$R_1 - \frac{1}{3}R_2$	1	0	$\frac{5}{7}$	$\frac{3}{7}$	$-\frac{1}{7}$	0	0	$\frac{19}{7}$	$\frac{19}{5}$
$R_3 - \frac{5}{3}R_2$ ⟶	0	1	$-\frac{1}{7}$	$-\frac{2}{7}$	$\frac{3}{7}$	0	0	$\frac{6}{7}$	—
$R_4 + \frac{16}{3}R_2$	0	0	⑱⁄₇	$\frac{1}{7}$	$-\frac{5}{7}$	1	0	$\frac{18}{7}$	1
	0	0	$-\frac{38}{7}$	$\frac{36}{7}$	$\frac{16}{7}$	0	1	$64\frac{4}{7}$	

↑ Pivot column

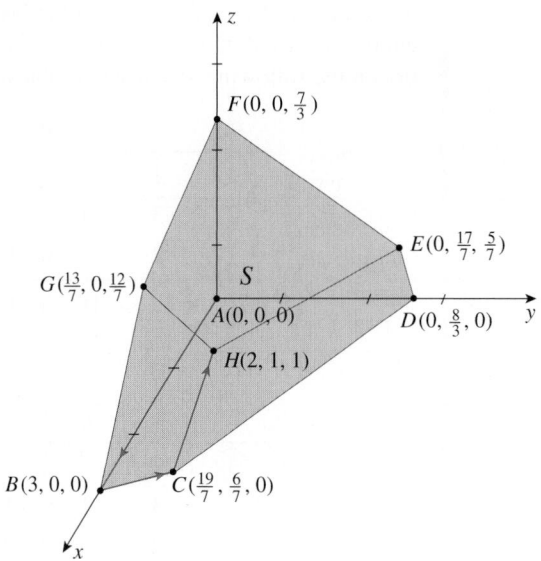

FIGURE 4
The simplex method brings us from the point A to the point H, at which the objective function is maximized.

The second iteration brings us to the point $\left(\frac{19}{7}, \frac{6}{7}, 0\right)$ with $P = 64\frac{4}{7}$. (See Figure 4.) Since there is a negative number in the last row of the simplex tableau, we perform another iteration, as follows:

	x	y	z	u	v	w	P	Constant
$\xrightarrow{\frac{7}{18}R_3}$	1	0	$\frac{5}{7}$	$\frac{3}{7}$	$-\frac{1}{7}$	0	0	$\frac{19}{7}$
	0	1	$-\frac{1}{7}$	$-\frac{2}{7}$	$\frac{3}{7}$	0	0	$\frac{6}{7}$
	0	0	①	$\frac{1}{18}$	$-\frac{5}{18}$	$\frac{7}{18}$	0	1
	0	0	$-\frac{38}{7}$	$\frac{36}{7}$	$\frac{16}{7}$	0	1	$64\frac{4}{7}$

	x	y	z	u	v	w	P	Constant
$R_1 - \frac{5}{7}R_3$	1	0	0	$\frac{7}{18}$	$\frac{1}{18}$	$-\frac{5}{18}$	0	2
$R_2 + \frac{1}{7}R_3$	0	1	0	$-\frac{5}{18}$	$\frac{7}{18}$	$\frac{1}{18}$	0	1
$\xrightarrow{\quad}$ $R_4 + \frac{38}{7}R_3$	0	0	1	$\frac{1}{18}$	$-\frac{5}{18}$	$\frac{7}{18}$	0	1
	0	0	0	$\frac{49}{9}$	$\frac{7}{9}$	$\frac{19}{9}$	1	70

All entries in the last row are nonnegative, so we have reached the optimal solution. We conclude that $x = 2$, $y = 1$, $z = 1$, $u = 0$, $v = 0$, $w = 0$, and $P = 70$.

The feasible set S for the problem is the hexahedron shown in Figure 5. It is the intersection of the half-spaces determined by the planes P_1, P_2, and P_3 with equations $3x + y + 2z = 9$, $2x + 3y + z = 8$, $x + 2y + 3z = 7$, respectively, and the coordinate planes $x = 0$, $y = 0$, and $z = 0$. That portion of the figure showing the feasible set S is shown in Figure 4. Observe that the first iteration of the simplex method brings us from $A(0, 0, 0)$ with $P = 0$ to $B(3, 0, 0)$ with $P = 60$. The second iteration brings us from $B(3, 0, 0)$ to $C\left(\frac{19}{7}, \frac{6}{7}, 0\right)$ with $P = 64\frac{4}{7}$, and the third iteration brings us from $C\left(\frac{19}{7}, \frac{6}{7}, 0\right)$ to the point $H(2, 1, 1)$ with an optimal value of 70 for P.

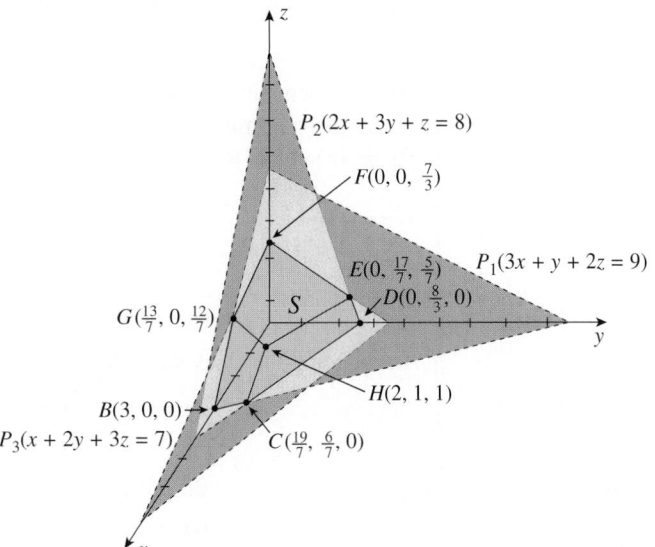

FIGURE 5
The feasible set S is obtained from the intersection of the half-spaces determined by P_1, P_2, and P_3 with the coordinate planes $x = 0$, $y = 0$, and $z = 0$.

APPLIED EXAMPLE 5   Production Planning   Ace Novelty Company has determined that the profit for each type-A, type-B, and type-C souvenir that it plans to produce is \$6, \$5, and \$4, respectively. To manufacture a type-A souvenir requires 2 minutes on machine I, 1 minute on machine II, and 2 minutes on machine III. A type-B souvenir requires 1 minute on machine I, 3 minutes on machine II, and 1 minute on machine III. A type-C souvenir requires 1 minute on machine I and 2 minutes on each of machines II and III. Each day there are 3 hours available on machine I, 5 hours available on machine II, and 4 hours available on machine III for manufacturing these souvenirs. How many souvenirs of each type should Ace Novelty make per day in order to maximize its profit? (Compare with Example 1, Section 2.1.)

Solution   The given information is tabulated as follows:

	Type A	Type B	Type C	Time Available (min)
Machine I	2	1	1	180
Machine II	1	3	2	300
Machine III	2	1	2	240
Profit per Unit	\$6	\$5	\$4	

Let x, y, and z denote the respective numbers of type-A, type-B, and type-C souvenirs to be made. The total amount of time that machine I is used is given by $2x + y + z$ minutes and must not exceed 180 minutes. Thus, we have the inequality

$$2x + y + z \leq 180$$

Similar considerations on the use of machines II and III lead to the inequalities

$$x + 3y + 2z \leq 300$$
$$2x + y + 2z \leq 240$$

The profit resulting from the sale of the souvenirs produced is given by

$$P = 6x + 5y + 4z$$

The mathematical formulation of this problem leads to the following standard linear programming problem: Maximize the objective (profit) function $P = 6x + 5y + 4z$ subject to

$$2x + y + z \leq 180$$
$$x + 3y + 2z \leq 300$$
$$2x + y + 2z \leq 240$$
$$x \geq 0, y \geq 0, z \geq 0$$

Introducing the slack variables u, v, and w gives the system of linear equations

$$2x + y + z + u \qquad\qquad = 180$$
$$x + 3y + 2z \qquad + v \qquad\quad = 300$$
$$2x + y + 2z \qquad\qquad + w \quad = 240$$
$$-6x - 5y - 4z \qquad\qquad\quad + P = 0$$

The tableaus resulting from the use of the simplex algorithm are

	x	y	z	u	v	w	P	Constant
Pivot row →	②	1	1	1	0	0	0	180
	1	3	2	0	1	0	0	300
	2	1	2	0	0	1	0	240
	−6	−5	−4	0	0	0	1	0

Ratio
$\frac{180}{2} = 90$
$\frac{300}{1} = 300$
$\frac{240}{2} = 120$

↑ Pivot column

$\frac{1}{2}R_1 \longrightarrow$

	x	y	z	u	v	w	P	Constant
	①	$\frac{1}{2}$	$\frac{1}{2}$	$\frac{1}{2}$	0	0	0	90
	1	3	2	0	1	0	0	300
	2	1	2	0	0	1	0	240
	−6	−5	−4	0	0	0	1	0

$\begin{array}{l} R_2 - R_1 \\ R_3 - 2R_1 \\ R_4 + 6R_1 \end{array} \longrightarrow$ Pivot row

	x	y	z	u	v	w	P	Constant
	1	$\frac{1}{2}$	$\frac{1}{2}$	$\frac{1}{2}$	0	0	0	90
	0	⑤⁄₂	$\frac{3}{2}$	$-\frac{1}{2}$	1	0	0	210
	0	0	1	−1	0	1	0	60
	0	−2	−1	3	0	0	1	540

Ratio
$\frac{90}{1/2} = 180$
$\frac{210}{5/2} = 84$

↑ Pivot column

$\frac{2}{5}R_2 \longrightarrow$

	x	y	z	u	v	w	P	Constant
	1	$\frac{1}{2}$	$\frac{1}{2}$	$\frac{1}{2}$	0	0	0	90
	0	①	$\frac{3}{5}$	$-\frac{1}{5}$	$\frac{2}{5}$	0	0	84
	0	0	1	−1	0	1	0	60
	0	−2	−1	3	0	0	1	540

$$
\begin{array}{c}
R_1 - \frac{1}{2}R_2 \\
\xrightarrow{\hspace{1.5cm}} \\
R_4 + 2R_2
\end{array}
\quad
\begin{array}{ccccccc|c}
x & y & z & u & v & w & P & \text{Constant} \\
\hline
1 & 0 & \frac{1}{5} & \frac{3}{5} & -\frac{1}{5} & 0 & 0 & 48 \\
0 & 1 & \frac{3}{5} & -\frac{1}{5} & \frac{2}{5} & 0 & 0 & 84 \\
0 & 0 & 1 & -1 & 0 & 1 & 0 & 60 \\
\hline
0 & 0 & \frac{1}{5} & \frac{13}{5} & \frac{4}{5} & 0 & 1 & 708
\end{array}
$$

From the final simplex tableau, we read off the solution

$$x = 48 \qquad y = 84 \qquad z = 0 \qquad u = 0 \qquad v = 0 \qquad w = 60 \qquad P = 708$$

Thus, in order to maximize its profit, Ace Novelty should produce 48 type-A souvenirs, 84 type-B souvenirs, and no type-C souvenirs. The resulting profit is $708 per day. The value of the slack variable $w = 60$ tells us that 1 hour of the available time on machine III is left unused.   ■

Interpreting Our Results   Let's compare the results obtained here with those obtained in Example 7, Section 2.2. Recall that, to use all available machine time on each of the three machines, Ace Novelty had to produce 36 type-A, 48 type-B, and 60 type-C souvenirs. This would have resulted in a profit of $696. Example 5 shows how, through the optimal use of equipment, a company can boost its profit while reducing machine wear!

▬ Problems with Multiple Solutions and Problems with No Solutions

As we saw in Section 3.3, a linear programming problem may have infinitely many solutions. We also saw that a linear programming problem may have no solution. How do we spot each of these phenomena when using the simplex method to solve a problem?

A linear programming problem may have infinitely many solutions if and only if the last row to the left of the vertical line of the final simplex tableau has a zero in a column that is not a unit column. Also, a linear programming problem will have *no* solution if the simplex method breaks down at some stage. For example, if at some stage there are no nonnegative ratios in our computation, then the linear programming problem has no solution (see Exercise 42).

EXPLORE & DISCUSS

Consider the linear programming problem

$$\text{Maximize} \quad P = 4x + 6y$$
$$\text{subject to} \quad 2x + y \le 10$$
$$2x + 3y \le 18$$
$$x \ge 0, y \ge 0$$

1. Sketch the feasible set for the linear programming problem.

2. Use the method of corners to show that there are infinitely many optimal solutions. What are they?

3. Use the simplex method to solve the problem as follows.
   a. Perform one iteration on the initial simplex tableau and conclude that you have arrived at an optimal solution. What is the value of P, and where is it attained? Compare this result with that obtained in step 2.
   b. Observe that the tableau obtained in part (a) indicates that there are infinitely many solutions (see the comment on multiple solutions on page 231). Now perform another iteration on the simplex tableau using the x-column as the pivot column. Interpret the final tableau.

4.1 Self-Check Exercises

1. Solve the following linear programming problem by the simplex method:

$$\text{Maximize} \quad P = 2x + 3y + 6z$$
$$\text{subject to} \quad 2x + 3y + z \le 10$$
$$x + y + 2z \le 8$$
$$2y + 3z \le 6$$
$$x \ge 0, y \ge 0, z \ge 0$$

2. The LaCrosse Iron Works makes two models of cast-iron fireplace grates, model A and model B. Producing one model A

grate requires 20 lb of cast iron and 20 min of labor, whereas producing one model B grate requires 30 lb of cast iron and 15 min of labor. The profit for a model A grate is $6, and the profit for a model B grate is $8. There are 7200 lb of cast iron and 100 labor-hours available each week. Because of a surplus from the previous week, the proprietor has decided that he should make no more than 150 units of model A grates this week. Determine how many of each model he should make in order to maximize his profits.

Solutions to Self-Check Exercises 4.1 can be found on page 237.

4.1 Concept Questions

1. Give the three characteristics of a standard maximization linear programming problem.

2. a. When the initial simplex tableau is set up, how is the system of linear inequalities transformed into a system of linear equations? How is the objective function $P = c_1x_1 + c_2x_2 + \cdots + c_nx_n$ rewritten?

   b. If you are given a simplex tableau, how do you determine whether the optimal solution has been reached?

3. In the simplex method, how is a pivot column selected? A pivot row? A pivot element?

4.1 Exercises

In Exercises 1–10, determine whether the given simplex tableau is in final form. If so, find the solution to the associated regular linear programming problem. If not, find the pivot element to be used in the next iteration of the simplex method.

1.

x	y	u	v	P	Constant
0	1	$\frac{5}{7}$	$-\frac{1}{7}$	0	$\frac{20}{7}$
1	0	$-\frac{3}{7}$	$\frac{2}{7}$	0	$\frac{30}{7}$
0	0	$\frac{13}{7}$	$\frac{3}{7}$	1	$\frac{220}{7}$

2.

x	y	u	v	P	Constant
1	1	1	0	0	6
1	0	-1	1	0	2
3	0	5	0	1	30

3.

x	y	u	v	P	Constant
0	$\frac{1}{2}$	1	$-\frac{1}{2}$	0	2
1	$\frac{1}{2}$	0	$\frac{1}{2}$	0	4
0	$-\frac{1}{2}$	0	$\frac{3}{2}$	1	12

4.

x	y	z	u	v	w	P	Constant
3	0	5	1	1	0	0	28
2	1	3	0	1	0	0	16
2	0	8	0	3	0	1	48

5.

x	y	z	u	v	w	P	Constant
1	$-\frac{1}{3}$	0	$\frac{1}{3}$	0	$-\frac{2}{3}$	0	$\frac{1}{3}$
0	2	0	0	1	1	0	6
0	$\frac{2}{3}$	1	$\frac{1}{3}$	0	$\frac{1}{3}$	0	$\frac{13}{3}$
0	4	0	1	0	2	1	17

6.

x	y	z	u	v	w	P	Constant
$\frac{1}{2}$	0	$\frac{1}{4}$	1	$-\frac{1}{4}$	0	0	$\frac{19}{2}$
$\frac{1}{2}$	1	$\frac{3}{4}$	0	$\frac{1}{4}$	0	0	$\frac{21}{2}$
2	0	3	0	0	1	0	30
-1	0	$-\frac{1}{2}$	6	$\frac{3}{2}$	0	1	63

7.

x	y	z	s	t	u	v	P	Constant
$\frac{5}{2}$	3	0	1	0	0	-4	0	46
1	0	0	0	1	0	0	0	9
0	1	0	0	0	1	0	0	12
0	0	1	0	0	0	1	0	6
-180	-200	0	0	0	0	300	1	1800

8.

x	y	z	s	t	u	v	P	Constant
1	0	0	$\frac{2}{5}$	0	$-\frac{6}{5}$	$-\frac{8}{5}$	0	4
0	0	0	$-\frac{2}{5}$	1	$\frac{6}{5}$	$\frac{8}{5}$	0	5
0	1	0	0	0	1	0	0	12
0	0	1	0	0	0	1	0	6
0	0	0	72	0	-16	12	1	4920

9.

x	y	z	u	v	P	Constant
1	0	$\frac{3}{5}$	0	$\frac{1}{5}$	0	30
0	1	$-\frac{19}{5}$	1	$-\frac{3}{5}$	0	10
0	0	$\frac{26}{5}$	0	0	1	60

10.

x	y	z	u	v	w	P	Constant
0	$\frac{1}{2}$	0	1	$-\frac{1}{2}$	0	0	2
1	$\frac{1}{2}$	1	0	$\frac{1}{2}$	0	0	13
2	$\frac{1}{2}$	0	0	$-\frac{3}{2}$	1	0	4
-1	3	0	0	1	0	1	26

In Exercises 11–25, solve each linear programming problem by the simplex method.

11. Maximize $P = 3x + 4y$
subject to $x + y \le 4$
$2x + y \le 5$
$x \ge 0, y \ge 0$

12. Maximize $P = 5x + 3y$
subject to $x + y \le 80$
$3x \le 90$
$x \ge 0, y \ge 0$

13. Maximize $P = 10x + 12y$
subject to $x + 2y \le 12$
$3x + 2y \le 24$
$x \ge 0, y \ge 0$

14. Maximize $P = 5x + 4y$
subject to $3x + 5y \le 78$
$4x + y \le 36$
$x \ge 0, y \ge 0$

15. Maximize $P = 4x + 6y$
subject to $3x + y \le 24$
$2x + y \le 18$
$x + 3y \le 24$
$x \ge 0, y \ge 0$

16. Maximize $\quad P = 15x + 12y$
subject to $\quad x + y \le 12$
$\quad 3x + y \le 30$
$\quad 10x + 7y \le 70$
$\quad x \ge 0, y \ge 0$

17. Maximize $\quad P = 3x + 4y + 5z$
subject to $\quad x + y + z \le 8$
$\quad 3x + 2y + 4z \le 24$
$\quad x \ge 0, y \ge 0, z \ge 0$

18. Maximize $\quad P = 3x + 3y + 4z$
subject to $\quad x + y + 3z \le 15$
$\quad 4x + 4y + 3z \le 65$
$\quad x \ge 0, y \ge 0, z \ge 0$

19. Maximize $\quad P = 3x + 4y + z$
subject to $\quad 3x + 10y + 5z \le 120$
$\quad 5x + 2y + 8z \le 6$
$\quad 8x + 10y + 3z \le 105$
$\quad x \ge 0, y \ge 0, z \ge 0$

20. Maximize $\quad P = x + 2y - z$
subject to $\quad 2x + y + z \le 14$
$\quad 4x + 2y + 3z \le 28$
$\quad 2x + 5y + 5z \le 30$
$\quad x \ge 0, y \ge 0, z \ge 0$

21. Maximize $\quad P = 4x + 6y + 5z$
subject to $\quad x + y + z \le 20$
$\quad 2x + 4y + 3z \le 42$
$\quad 2x + 3z \le 30$
$\quad x \ge 0, y \ge 0, z \ge 0$

22. Maximize $\quad P = x + 4y - 2z$
subject to $\quad 3x + y - z \le 80$
$\quad 2x + y - z \le 40$
$\quad -x + y + z \le 80$
$\quad x \ge 0, y \ge 0, z \ge 0$

23. Maximize $\quad P = 12x + 10y + 5z$
subject to $\quad 2x + y + z \le 10$
$\quad 3x + 5y + z \le 45$
$\quad 2x + 5y + z \le 40$
$\quad x \ge 0, y \ge 0, z \ge 0$

24. Maximize $\quad P = 2x + 6y + 6z$
subject to $\quad 2x + y + 3z \le 10$
$\quad 4x + y + 2z \le 56$
$\quad 6x + 4y + 3z \le 126$
$\quad 2x + y + z \le 32$
$\quad x \ge 0, y \ge 0, z \ge 0$

25. Maximize $\quad P = 24x + 16y + 23z$
subject to $\quad 2x + y + 2z \le 7$
$\quad 2x + 3y + z \le 8$
$\quad x + 2y + 3z \le 7$
$\quad x \ge 0, y \ge 0, z \ge 0$

26. Rework Example 3 using the y-column as the pivot column in the first iteration of the simplex method.

27. Show that the following linear programming problem

Maximize $\quad P = 2x + 2y - 4z$
subject to $\quad 3x + 3y - 2z \le 100$
$\quad 5x + 5y + 3z \le 150$
$\quad x \ge 0, y \ge 0, z \ge 0$

has optimal solutions $x = 30$, $y = 0$, $z = 0$, $P = 60$ and $x = 0$, $y = 30$, $z = 0$, $P = 60$.

28. MANUFACTURING—PRODUCTION SCHEDULING   A company manufactures two products, A and B, on two machines, I and II. It has been determined that the company will realize a profit of $3/unit of product A and a profit of $4/unit of product B. To manufacture 1 unit of product A requires 6 min on machine I and 5 min on machine II. To manufacture 1 unit of product B requires 9 min on machine I and 4 min on machine II. There are 5 hr of machine time available on machine I and 3 hr of machine time available on machine II in each work shift. How many units of each product should be produced in each shift to maximize the company's profit? What is the largest profit the company can realize? Is there any time left unused on the machines?

29. MANUFACTURING—PRODUCTION SCHEDULING   National Business Machines Corporation manufactures two models of fax machines: A and B. Each model A costs $100 to make, and each model B costs $150. The profits are $30 for each model A and $40 for each model B fax machine. If the total number of fax machines demanded each month does not exceed 2500 and the company has earmarked no more than $600,000/month for manufacturing costs, find how many units of each model National should make each month in order to maximize its monthly profits. What is the largest monthly profit the company can make?

30. MANUFACTURING—PRODUCTION SCHEDULING   Kane Manufacturing has a division that produces two models of hibachis, model A and model B. To produce each model A hibachi requires 3 lb of cast iron and 6 min of labor. To produce each model B hibachi requires 4 lb of cast iron and 3 min of labor. The profit for each model A hibachi is $2, and the profit for each model B hibachi is $1.50. If 1000 lb of cast iron and 20 labor-hours are available for the production of hibachis each day, how many hibachis of each model should the division produce in order to maximize Kane's profits? What is the largest profit the company can realize? Is there any raw material left over?

31. AGRICULTURE—CROP PLANNING   A farmer has 150 acres of land suitable for cultivating crops A and B. The cost of cultivating crop A is $40/acre whereas that of crop B is $60/acre. The farmer has a maximum of $7400 available for land cultivation. Each acre of crop A requires 20 labor-hours, and each acre of crop B requires 25 labor-hours. The farmer has a maximum of 3300 labor-hours available. If he expects to make a profit of $150/acre on crop A and $200/acre on crop B, how many acres of each crop should he plant in order to maximize his profit? What is the largest profit the farmer can realize? Are there any resources left over?

32. INVESTMENTS—ASSET ALLOCATION   A financier plans to invest up to $500,000 in two projects. Project A yields a return of 10% on the investment whereas project B yields a return of 15% on the investment. Because the investment in project B is riskier than the investment in project A, the financier has decided that the investment in project B should not exceed 40% of the total investment. How much should she invest in each project in order to maximize the return on her investment? What is the maximum return?

33. INVESTMENTS—ASSET ALLOCATION   Ashley has earmarked at most $250,000 for investment in three mutual funds: a money market fund, an international equity fund, and a growth-and-income fund. The money market fund has a rate of return of 6%/year, the international equity fund has a rate of return of 10%/year, and the growth-and-income fund has a rate of return of 15%/year. Ashley has stipulated that no more than 25% of her total portfolio should be in the growth-and-income fund and that no more than 50% of her total portfolio should be in the international equity fund. To maximize the return on her investment, how much should Ashley invest in each type of fund? What is the maximum return?

34. MANUFACTURING—PRODUCTION SCHEDULING   A company manufactures products A, B, and C. Each product is processed in three departments: I, II, and III. The total available labor-hours per week for departments I, II, and III are 900, 1080, and 840, respectively. The time requirements (in hours per unit) and profit per unit for each product are as follows:

	Product A	Product B	Product C
Dept. I	2	1	2
Dept. II	3	1	2
Dept. III	2	2	1
Profit	$18	$12	$15

How many units of each product should the company produce in order to maximize its profit? What is the largest profit the company can realize? Are there any resources left over?

35. ADVERTISING—TELEVISION COMMERCIALS   As part of a campaign to promote its annual clearance sale, Excelsior Company decided to buy television advertising time on Station KAOS.

Excelsior's television advertising budget is $102,000. Morning time costs $3000/minute, afternoon time costs $1000/minute, and evening (prime) time costs $12,000/minute. Because of previous commitments, KAOS cannot offer Excelsior more than 6 min of prime time or more than a total of 25 min of advertising time over the 2 weeks in which the commercials are to be run. KAOS estimates that morning commercials are seen by 200,000 people, afternoon commercials are seen by 100,000 people, and evening commercials are seen by 600,000 people. How much morning, afternoon, and evening advertising time should Excelsior buy to maximize exposure of its commercials?

36. INVESTMENTS—ASSET ALLOCATION   Sharon has a total of $200,000 to invest in three types of mutual funds: growth, balanced, and income funds. Growth funds have a rate of return of 12%/year, balanced funds have a rate of return of 10%/year, and income funds have a return of 6%/year. The growth, balanced, and income mutual funds are assigned risk factors of 0.1, 0.06, and 0.02, respectively. Sharon has decided that at least 50% of her total portfolio is to be in income funds and at least 25% in balanced funds. She has also decided that the average risk factor for her investment should not exceed 0.05. How much should Sharon invest in each type of fund in order to realize a maximum return on her investment? What is the maximum return?
Hint: The average risk factor for the investment is given by $0.1x + 0.06y + 0.02z \leq 0.05(x + y + z)$.

37. MANUFACTURING—PRODUCTION CONTROL   Custom Office Furniture is introducing a new line of executive desks made from a specially selected grade of walnut. Initially, three models—A, B, and C—are to be marketed. Each model A desk requires $1\frac{1}{4}$ hr for fabrication, 1 hr for assembly, and 1 hr for finishing; each model B desk requires $1\frac{1}{2}$ hr for fabrication, 1 hr for assembly, and 1 hr for finishing; each model C desk requires $1\frac{1}{2}$ hr, $\frac{3}{4}$ hr, and $\frac{1}{2}$ hr for fabrication, assembly, and finishing, respectively. The profit on each model A desk is $26, the profit on each model B desk is $28, and the profit on each model C desk is $24. The total time available in the fabrication department, the assembly department, and the finishing department in the first month of production is 310 hr, 205 hr, and 190 hr, respectively. To maximize Custom's profit, how many desks of each model should be made in the month? What is the largest profit the company can realize? Are there any resources left over?

38. MANUFACTURING—PREFABRICATED HOUSING PRODUCTION   Boise Lumber has decided to enter the lucrative prefabricated housing business. Initially, it plans to offer three models: standard, deluxe, and luxury. Each house is prefabricated and partially assembled in the factory, and the final assembly is completed on site. The dollar amount of building material required, the amount of labor required in the factory for prefabrication and partial assembly, the amount of on-site labor required, and the profit per unit are as follows:

	Standard Model	Deluxe Model	Luxury Model
Material	$6,000	$8,000	$10,000
Factory Labor (hr)	240	220	200
On-site Labor (hr)	180	210	300
Profit	$3,400	$4,000	$5,000

For the first year's production, a sum of $8,200,000 is budgeted for the building material; the number of labor-hours available for work in the factory (for prefabrication and partial assembly) is not to exceed 218,000 hr; and the amount of labor for on-site work is to be less than or equal to 237,000 labor-hours. Determine how many houses of each type Boise should produce (market research has confirmed that there should be no problems with sales) to maximize its profit from this new venture.

39. MANUFACTURING—COLD FORMULA PRODUCTION   Beyer Pharmaceutical produces three kinds of cold formulas: I, II, and III. It takes 2.5 hr to produce 1000 bottles of formula I, 3 hr to produce 1000 bottles of formula II, and 4 hr to produce 1000 bottles of formula III. The profits for each 1000 bottles of formula I, formula II, and formula III are $180, $200, and $300, respectively. Suppose, for a certain production run, there are enough ingredients on hand to make at most 9000 bottles of formula I, 12,000 bottles of formula II, and 6000 bottles of formula III. Furthermore, suppose the time for the production run is limited to a maximum of 70 hr. How many bottles of each formula should be produced in this production run so that the profit is maximized? What is the maximum profit realizable by the company? Are there any resources left over?

40. PRODUCTION—JUICE PRODUCTS   CalJuice Company has decided to introduce three fruit juices made from blending two or more concentrates. These juices will be packaged in 2-qt (64-oz) cartons. One carton of pineapple–orange juice requires 8 oz each of pineapple and orange juice concentrates. One carton of orange–banana juice requires 12 oz of orange juice concentrate and 4 oz of banana pulp concentrate. Finally, one carton of pineapple–orange–banana juice requires 4 oz of pineapple juice concentrate, 8 oz of orange juice concentrate, and 4 oz of banana pulp. The company has decided to allot 16,000 oz of pineapple juice concentrate, 24,000 oz of orange juice concentrate, and 5000 oz of banana pulp concentrate for the initial production run. The company has also stipulated that the production of pineapple–orange–banana juice should not exceed 800 cartons. Its profit on one carton of pineapple–orange juice is $1.00, its profit on one carton of orange–banana juice is $.80, and its profit on one carton of pineapple–orange–banana juice is $.90. To realize a maximum profit, how many cartons of each blend should the company produce? What is the largest profit it can realize? Are there any concentrates left over?

41. INVESTMENTS—ASSET ALLOCATION   A financier plans to invest up to $2 million in three projects. She estimates that project

A will yield a return of 10% on her investment, project B will yield a return of 15% on her investment, and project C will yield a return of 20% on her investment. Because of the risks associated with the investments, she decided to put not more than 20% of her total investment in project C. She also decided that her investments in projects B and C should not exceed 60% of her total investment. Finally, she decided that her investment in project A should be at least 60% of her investments in projects B and C. How much should the financier invest in each project if she wishes to maximize the total returns on her investments? What is the maximum amount she can expect to make from her investments?

42. Consider the linear programming problem

$$\text{Maximize} \quad P = 3x + 2y$$
$$\text{subject to} \quad x - y \leq 3$$
$$x \leq 2$$
$$x \geq 0, y \geq 0$$

a. Sketch the feasible set for the linear programming problem.
b. Show that the linear programming problem is unbounded.
c. Solve the linear programming problem using the simplex method. How does the method break down?
d. Explain why the result in part (c) implies that no solution exists for the linear programming problem.

In Exercises 43–46, determine whether the statement is true or false. If it is true, explain why it is true. If it is false, give an example to show why it is false.

43. If at least one of the coefficients $a_1, a_2, \ldots, a_n$ of the objective function $P = a_1x_1 + a_2x_2 + \cdots + a_nx_n$ is positive, then $(0, 0, \ldots, 0)$ cannot be the optimal solution of the standard (maximization) linear programming problem.

44. Choosing the pivot row by requiring that the ratio associated with that row be the smallest ensures that the iteration will not take us from a feasible point to a nonfeasible point.

45. Choosing the pivot column by requiring that it be the column associated with the most negative entry to the left of the vertical line in the last row of the simplex tableau ensures that the iteration will result in the greatest increase or, at worse, no decrease in the objective function.

46. If, at any stage of an iteration of the simplex method, it is not possible to compute the ratios (division by zero) or the ratios are negative, then we can conclude that the standard linear programming problem may have no solution.

4.1 Solutions to Self-Check Exercises

1. Introducing the slack variables u, v, and w, we obtain the system of linear equations

$$
\begin{aligned}
2x + 3y + z + u &= 10 \\
x + y + 2z + v &= 8 \\
2y + 3z + w &= 6 \\
-2x - 3y - 6z + P &= 0
\end{aligned}
$$

The initial simplex tableau and the successive tableaus resulting from the use of the simplex procedure follow:

	x	y	z	u	v	w	P	Constant		Ratio
	2	3	1	1	0	0	0	10		$\frac{10}{1} = 10$
	1	1	2	0	1	0	0	8		$\frac{8}{2} = 4$
Pivot row →	0	2	③	0	0	1	0	6		$\frac{6}{3} = 2$
	−2	−3	−6	0	0	0	1	0		

$\xrightarrow{\;\frac{1}{3}R_3\;}$

↑ Pivot column

	x	y	z	u	v	w	P	Constant
	2	3	1	1	0	0	0	10
	1	1	2	0	1	0	0	8
	0	$\frac{2}{3}$	①	0	0	$\frac{1}{3}$	0	2
	−2	−3	−6	0	0	0	1	0

$\xrightarrow[\substack{R_2 - 2R_3 \\ R_4 + 6R_3}]{R_1 - R_3}$

	x	y	z	u	v	w	P	Constant		Ratio
	2	$\frac{7}{3}$	0	1	0	$-\frac{1}{3}$	0	8		$\frac{8}{2} = 4$
Pivot row →	①	$-\frac{1}{3}$	0	0	1	$-\frac{2}{3}$	0	4		$\frac{4}{1} = 4$
	0	$\frac{2}{3}$	1	0	0	$\frac{1}{3}$	0	2		—
	−2	1	0	0	0	2	1	12		

$\xrightarrow[\substack{R_4 + 2R_2}]{R_1 - 2R_2}$

↑ Pivot column

x	y	z	u	v	w	P	Constant
0	3	0	1	−2	1	0	0
1	$-\frac{1}{3}$	0	0	1	$-\frac{2}{3}$	0	4
0	$\frac{2}{3}$	1	0	0	$\frac{1}{3}$	0	2
0	$\frac{1}{3}$	0	0	2	$\frac{2}{3}$	1	20

All entries in the last row are nonnegative, and the tableau is final. We conclude that $x = 4$, $y = 0$, $z = 2$, and $P = 20$.

2. Let x denote the number of model A grates and y the number of model B grates to be made this week. Then the profit function to be maximized is given by

$$P = 6x + 8y$$

The limitations on the availability of material and labor may be expressed by the linear inequalities

$$20x + 30y \le 7200 \quad \text{or} \quad 2x + 3y \le 720$$
$$20x + 15y \le 6000 \quad \text{or} \quad 4x + 3y \le 1200$$

Finally, the condition that no more than 150 units of model A grates be made each week may be expressed by the linear inequality

$$x \le 150$$

Thus, we are led to the following linear programming problem:

$$
\begin{aligned}
\text{Maximize} \quad & P = 6x + 8y \\
\text{subject to} \quad & 2x + 3y \le 720 \\
& 4x + 3y \le 1200 \\
& x \le 150 \\
& x \ge 0, \, y \ge 0
\end{aligned}
$$

To solve this problem, we introduce slack variables u, v, and w and use the simplex method, obtaining the following sequence of simplex tableaus:

	x	y	u	v	w	P	Constant		Ratio
Pivot row →	2	③	1	0	0	0	720		$\frac{720}{3} = 240$
	4	3	0	1	0	0	1200		$\frac{1200}{3} = 400$
	1	0	0	0	1	0	150		—
	−6	−8	0	0	0	1	0		

↑ Pivot column

	x	y	u	v	w	P	Constant
	$\frac{2}{3}$	①	$\frac{1}{3}$	0	0	0	240
	4	3	0	1	0	0	1200
	1	0	0	0	1	0	150
	−6	−8	0	0	0	1	0

$\xrightarrow{\;\frac{1}{3}R_1\;}$

	x	y	u	v	w	P	Constant		Ratio
	$\frac{2}{3}$	1	$\frac{1}{3}$	0	0	0	240		$\frac{240}{2/3} = 360$
	2	0	−1	1	0	0	480		$\frac{480}{2} = 240$
Pivot row →	①	0	0	0	1	0	150		$\frac{150}{1} = 150$
	$-\frac{2}{3}$	0	$\frac{8}{3}$	0	0	1	1920		

$\xrightarrow[\substack{R_2 - 3R_1 \\ R_4 + 8R_1}]{}$

↑ Pivot column

	x	y	u	v	w	P	Constant
	0	1	$\frac{1}{3}$	0	$-\frac{2}{3}$	0	140
	0	0	−1	1	−2	0	180
	1	0	0	0	1	0	150
	0	0	$\frac{8}{3}$	0	$\frac{2}{3}$	1	2020

$\xrightarrow[\substack{R_1 - \frac{2}{3}R_3 \\ R_2 - 2R_3 \\ R_4 + \frac{2}{3}R_3}]{}$

The last tableau is final, and we see that $x = 150$, $y = 140$, and $P = 2020$. Therefore, LaCrosse should make 150 model A grates and 140 model B grates this week. The profit will be $2020.

USING TECHNOLOGY

■ The Simplex Method: Solving Maximization Problems

Graphing Utility

A graphing utility can be used to solve a linear programming problem by the simplex method, as illustrated in Example 1.

EXAMPLE 1   (Refer to Example 5, Section 4.1.) The problem reduces to the following linear programming problem:

$$\text{Maximize} \quad P = 6x + 5y + 4z$$
$$\text{subject to} \quad 2x + y + z \le 180$$
$$x + 3y + 2z \le 300$$
$$2x + y + 2z \le 240$$
$$x \ge 0, y \ge 0, z \ge 0$$

With u, v, and w as slack variables, we are led to the following sequence of simplex tableaus, where the first tableau is entered as the matrix A:

	x	y	z	u	v	w	P	Constant
Pivot row →	②	1	1	1	0	0	0	180
	1	3	2	0	1	0	0	300
	2	1	2	0	0	1	0	240
	−6	−5	−4	0	0	0	1	0

Pivot column (under x)

Ratio: $\frac{180}{2} = 90$; $\frac{300}{1} = 300$; $\frac{240}{2} = 120$

$\xrightarrow{\ast\mathbf{row}\left(\frac{1}{2}, A, 1\right) \blacktriangleright B}$

	x	y	z	u	v	w	P	Constant
	①	0.5	0.5	0.5	0	0	0	90
	1	3	2	0	1	0	0	300
	2	1	2	0	0	1	0	240
	−6	−5	−4	0	0	0	1	0

$\xrightarrow{\ast\mathbf{row}+(-1, B, 1, 2) \blacktriangleright C}$
$\xrightarrow{\ast\mathbf{row}+(-2, C, 1, 3) \blacktriangleright B}$
$\xrightarrow{\ast\mathbf{row}+(6, B, 1, 4) \ \ \blacktriangleright C}$

	x	y	z	u	v	w	P	Constant
	1	0.5	0.5	0.5	0	0	0	90
Pivot row →	0	②.5	1.5	−0.5	1	0	0	210
	0	0	1	−1	0	1	0	60
	0	−2	−1	3	0	0	1	540

Pivot column (under y)

Ratio: $\frac{90}{0.5} = 180$; $\frac{210}{2.5} = 84$

$\xrightarrow{\ast\mathbf{row}\left(\frac{1}{2.5}, C, 2\right) \blacktriangleright B}$

x	y	z	u	v	w	P	Constant
1	0.5	0.5	0.5	0	0	0	90
0	①	0.6	−0.2	0.4	0	0	84
0	0	1	−1	0	1	0	60
0	−2	−1	3	0	0	1	540

$$\xrightarrow{\substack{*(\text{row}+(-0.5, B, 2, 1) \blacktriangleright C \\ *\text{row}+(2, C, 2, 4) \quad \blacktriangleright B}}$$

x	y	z	u	v	w	P	Constant
1	0	0.2	0.6	−0.2	0	0	48
0	1	0.6	−0.2	0.4	0	0	84
0	0	1	−1	0	1	0	60
0	0	0.2	2.6	0.8	0	1	708

The final simplex tableau is the same as the one obtained earlier. We see that $x = 48$, $y = 84$, $z = 0$, and $P = 708$. Hence Ace Novelty should produce 48 type-A souvenirs, 84 type-B souvenirs, and no type-C souvenirs—resulting in a profit of $708 per day. ∎

Excel

Solver is an Excel add-in that is used to solve linear programming problems. When you start the Excel program, check the *Tools* menu for the *Solver* command. If it is not there, you will need to install it. (Check your manual for installation instructions.)

EXAMPLE 2   Solve the following linear programming problem:

$$\begin{aligned} \text{Maximize} \quad & P = 6x + 5y + 4z \\ \text{subject to} \quad & 2x + y + z \leq 180 \\ & x + 3y + 2z \leq 300 \\ & 2x + y + 2z \leq 240 \\ & x \geq 0, y \geq 0, z \geq 0 \end{aligned}$$

Solution

1. *Enter the data for the linear programming problem onto a spreadsheet.* Enter the labels shown in column A and the variables with which we are working under Decision Variables in cells B4:B6, as shown in Figure T1. This optional step will help us organize our work.

Note: Boldfaced words/characters enclosed in a box (for example, **Enter**) indicate that an action (click, select, or press) is required. Words/characters printed blue (for example, Chart sub-type:) indicate words/characters that appear on the screen. Words/characters printed in a typewriter font (for example, =(−2/3)*A2+2) indicate words/characters that need to be typed and entered.

(continued)

	A	B	C	D	E	F	G	H	I
1	Maximization Problem								
2									
3	Decision Variables								
4			x						
5			y						
6			z						
7									
8	Objective Function		0						
9									
10	Constraints								
11			0	<=	180				
12			0	<=	300				
13			0	<=	240				

Formulas for indicated cells
C8: $= 6*C4 + 5*C5 + 4*C6$
C11: $= 2*C4 + C5 + C6$
C12: $= C4 + 3*C5 + 2*C6$
C13: $= 2*C4 + C5 + 2*C6$

FIGURE T1
Setting up the spreadsheet for Solver

For the moment, the cells that will contain the values of the variables (C4:C6) are left blank. In C8 type the formula for the objective function: =6*C4+5*C5+4*C6. In C11 type the formula for the left-hand side of the first constraint: =2*C4+C5+C6. In C12 type the formula for the left-hand side of the second constraint: =C4+3*C5+2*C6. In C13 type the formula for the left-hand side of the third constraint: =2*C4+C5+2*C6. Zeros will then appear in cell B8 and cells C11:C13. In cells D11:D13, type **<=** to indicate that each constraint is of the form $\leq$. Finally, in cells E11:E13, type the right-hand value of each constraint—in this case, **180**, **300**, and **240**, respectively. Note that we need not enter the nonnegativity constraints $x \geq 0$, $y \geq 0$, and $z \geq 0$. The resulting spreadsheet is shown in Figure T1, where the formulas that were entered for the objective function and the constraints are shown in the comment box.

2. *Use Solver to solve the problem.* Click **Tools** on the menu bar and then click   **Solver** . The Solver Parameters dialog box will appear.

   a. The pointer will be in the Set Target Cell: box (refer to Figure T2). Highlight the cell on your spreadsheet containing the formula for the objective function—in this case, C8.

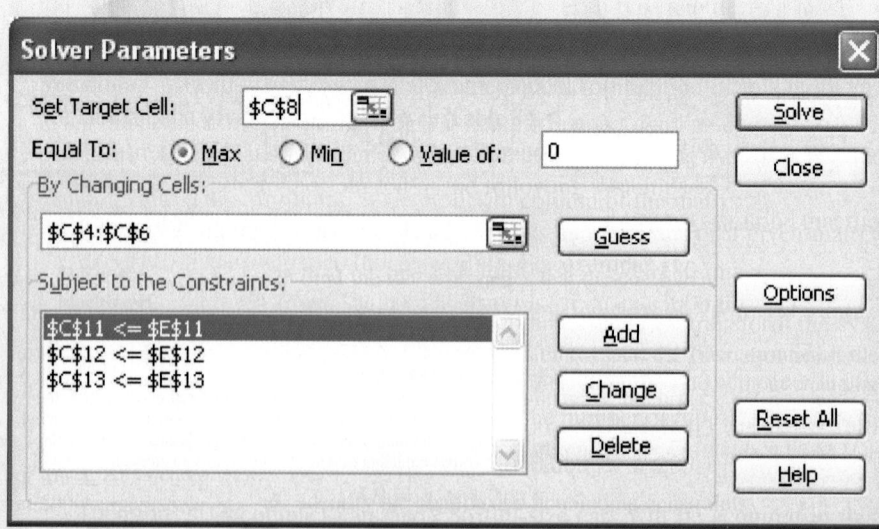

FIGURE T2
The completed Solver Parameters dialog box

Then, next to Equal To: select $\boxed{\text{Max}}$. Select the $\boxed{\textbf{By Changing Cells:}}$ box and highlight the cells in your spreadsheet that will contain the values of the variables—in this case, C4:C6. Select the $\boxed{\textbf{Subject to the Constraints:}}$ box and then click $\boxed{\textbf{Add}}$. The Add Constraint dialog box will appear (Figure T3).

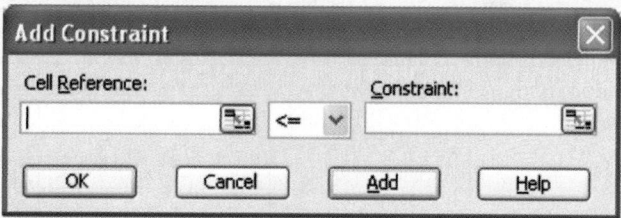

FIGURE T3
The Add Constraint dialog box

b. The pointer will appear in the Cell Reference: box. Highlight the cells on your spreadsheet that contain the formula for the left-hand side of the first constraint—in this case, C11. Next, select the symbol for the appropriate constraint—in this case, $\boxed{\text{<=}}$. Select the $\boxed{\textbf{Constraint:}}$ box and highlight the value of the right-hand side of the first constraint on your spreadsheet—in this case, 180. Click $\boxed{\textbf{Add}}$ and then follow the same procedure to enter the second and third constraint. Click $\boxed{\textbf{OK}}$. The resulting Solver Parameters dialog box shown in Figure T2 will appear.

c. In the Solver Parameters dialog box, click $\boxed{\textbf{Options}}$ (see Figure T2). In the Solver Options dialog box that appears, select $\boxed{\textbf{Assume Linear Model}}$ and $\boxed{\textbf{Assume Non-Negative}}$ constraints (Figure T4). Click $\boxed{\textbf{OK}}$.

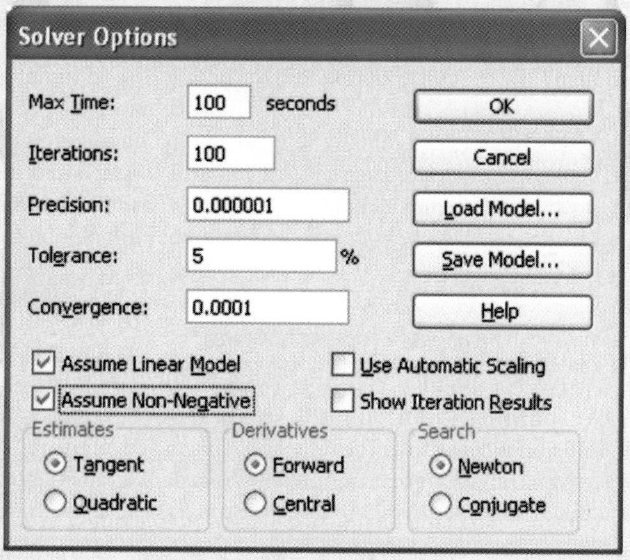

FIGURE T4
The Solver Options dialog box

(continued)

d. In the Solver Parameters dialog box that appears (see Figure T2), click [**Solve**] . A Solver Results dialog box will then appear and at the same time the answers will appear on your spreadsheet (Figure T5).

	A	B	C	D	E
1	Maximization Problem				
2					
3	Decision Variables				
4		x	48		
5		y	84		
6		z	0		
7					
8	Objective Function		708		
9					
10	Constraints				
11			180	<=	180
12			300	<=	300
13			240	<=	240

FIGURE T5
Completed spreadsheet after using Solver

3. *Read off your answers.* From the spreadsheet, we see that the objective function attains a maximum value of 708 (cell C8) when $x = 48$, $y = 84$, and $z = 0$ (cells C4:C6). ∎

TECHNOLOGY EXERCISES

Solve the linear programming problems.

1. Maximize $P = 2x + 3y + 4z + 2w$
subject to $x + 2y + 3z + 2w \le 6$
$2x + 4y + z - w \le 4$
$3x + 2y - 2z + 3w \le 12$
$x \ge 0, y \ge 0, z \ge 0, w \ge 0$

2. Maximize $P = 3x + 2y + 2z + w$
subject to $2x + y - z + 2w \le 8$
$2x - y + 2z + 3w \le 20$
$x + y + z + 2w \le 8$
$4x - 2y + z + 3w \le 24$
$x \ge 0, y \ge 0, z \ge 0, w \ge 0$

3. Maximize $P = x + y + 2z + 3w$
subject to $3x + 6y + 4z + 2w \le 12$
$x + 4y + 8z + 4w \le 16$
$2x + y + 4z + w \le 10$
$x \ge 0, y \ge 0, z \ge 0, w \ge 0$

4. Maximize $P = 2x + 4y + 3z + 5w$
subject to $x - 2y + 3z + 4w \le 8$
$2x + 2y + 4z + 6w \le 12$
$3x + 2y + z + 5w \le 10$
$2x + 8y - 2z + 6w \le 24$
$x \ge 0, y \ge 0, z \ge 0, w \ge 0$

4.2 The Simplex Method: Standard Minimization Problems

■ Minimization with ≤ Constraints

In the last section we developed a procedure, called the simplex method, for solving standard linear programming problems. Recall that a standard maximization problem satisfies three conditions:

1. The objective function is to be maximized.
2. All the variables involved are nonnegative.
3. Each linear constraint may be written so that the expression involving the variables is less than or equal to a nonnegative constant.

In this section we see how the simplex method may be used to solve certain classes of problems that are not necessarily standard maximization problems. In particular, we see how a modified procedure may be used to solve problems involving the minimization of objective functions.

We begin by considering the class of linear programming problems that calls for the minimization of objective functions but otherwise satisfies Conditions 2 and 3 for standard maximization problems. The method used to solve these problems is illustrated in the following example.

EXAMPLE 1

$$\begin{aligned}
\text{Minimize} \quad & C = -2x - 3y \\
\text{subject to} \quad & 5x + 4y \le 32 \\
& x + 2y \le 10 \\
& x \ge 0, y \ge 0
\end{aligned}$$

Solution   This problem involves the minimization of the objective function and is accordingly *not* a standard maximization problem. Note, however, that all other conditions for a standard maximization problem hold true. To solve a problem of this type, we observe that minimizing the objective function C is equivalent to maximizing the objective function $P = -C$. Thus, the solution to this problem may be found by solving the following associated standard maximization problem: Maximize $P = 2x + 3y$ subject to the given constraints. Using the simplex method with u and v as slack variables, we obtain the following sequence of simplex tableaus:

	x	y	u	v	P	Constant
	5	4	1	0	0	32
Pivot row →	1	②	0	1	0	10
	-2	-3	0	0	1	0

Ratio
$\frac{32}{4} = 8$
$\frac{10}{2} = 5$

↑
Pivot column

	x	y	u	v	P	Constant
	5	4	1	0	0	32
$\frac{1}{2}R_2$ →	$\frac{1}{2}$	①	0	$\frac{1}{2}$	0	5
	-2	-3	0	0	1	0

EXPLORE & DISCUSS

Refer to Example 1.

1. Sketch the feasible set S for the linear programming problem.

2. Solve the problem using the method of corners.

3. Indicate on S the corner points corresponding to each iteration of the simplex procedure and trace the path leading to the optimal solution.

Pivot row →

	x	y	u	v	P	Constant		Ratio
	③	0	1	-2	0	12		$\frac{12}{3} = 4$
$\xrightarrow{\substack{R_1 - 4R_2 \\ R_3 + 3R_2}}$	$\frac{1}{2}$	1	0	$\frac{1}{2}$	0	5		$\frac{5}{1/2} = 10$
	$-\frac{1}{2}$	0	0	$\frac{3}{2}$	1	15		

↑ Pivot column

	x	y	u	v	P	Constant
$\xrightarrow{\frac{1}{3}R_1}$	①	0	$\frac{1}{3}$	$-\frac{2}{3}$	0	4
	$\frac{1}{2}$	1	0	$\frac{1}{2}$	0	5
	$-\frac{1}{2}$	0	0	$\frac{3}{2}$	1	15

	x	y	u	v	P	Constant
$\xrightarrow{\substack{R_2 - \frac{1}{2}R_1 \\ R_3 + \frac{1}{2}R_1}}$	1	0	$\frac{1}{3}$	$-\frac{2}{3}$	0	4
	0	1	$-\frac{1}{6}$	$\frac{5}{6}$	0	3
	0	0	$\frac{1}{6}$	$\frac{7}{6}$	1	17

The last tableau is in final form. The solution to the standard maximization problem associated with the given linear programming problem is $x = 4$, $y = 3$, and $P = 17$, so the required solution is given by $x = 4$, $y = 3$, and $C = -17$. You may verify that the solution is correct by using the method of corners. ■

The Dual Problem

Another special class of linear programming problems we encounter in practical applications is characterized by the following conditions:

1. The objective function is to be *minimized*.
2. All the variables involved are nonnegative.
3. Each linear constraint may be written so that the expression involving the variables is *greater than or equal to* a constant.

Such problems are called **standard minimization problems**.

A convenient method for solving this type of problem is based on the following observation. Each maximization linear programming problem is associated with a minimization problem, and vice versa. For the purpose of identification, the given problem is called the **primal problem**; the problem related to it is called the **dual problem**. The following example illustrates the technique for constructing the dual of a given linear programming problem.

EXAMPLE 2 Write the dual problem associated with the following problem:

$$\left. \begin{aligned} \text{Minimize the objective function } C &= 6x + 8y \\ \text{subject to} \quad 40x + 10y &\geq 2400 \\ 10x + 15y &\geq 2100 \\ 5x + 15y &\geq 1500 \\ x \geq 0, y &\geq 0 \end{aligned} \right\} \begin{array}{c} \text{Primal} \\ \text{problem} \end{array}$$

Solution   We first write down the following tableau for the given primal problem:

x	y	Constant
40	10	2400
10	15	2100
5	15	1500
6	8	

Next, we interchange the columns and rows of the foregoing tableau and head the three columns of the resulting array with the three variables u, v, and w, obtaining the tableau

u	v	w	Constant
40	10	5	6
10	15	15	8
2400	2100	1500	

Interpreting the last tableau as if it were part of the initial simplex tableau for a standard maximization problem—with the exception that the signs of the coefficients pertaining to the objective function are not reversed—we construct the required dual problem as follows:

Maximize the objective function $P = 2400u + 2100v + 1500w$

subject to $40u + 10v + 5w \leq 6$

$10u + 15v + 15w \leq 8$

where $u \geq 0$, $v \geq 0$, and $w \geq 0$.

Dual problem

The connection between the solution of the primal problem and that of the dual problem is given by the following theorem. The theorem, attributed to John von Neumann (1903–1957), is stated without proof.

THEOREM 1

The Fundamental Theorem of Duality

A primal problem has a solution if and only if the corresponding dual problem has a solution. Furthermore, if a solution exists, then:

a. The objective functions of both the primal and the dual problem attain the same optimal value.

b. The optimal solution to the primal problem appears under the slack variables in the last row of the final simplex tableau associated with the dual problem.

Armed with this theorem, we will solve the problem posed in Example 2.

EXAMPLE 3   Complete the solution to the problem posed in Example 2.

Solution   Observe that the dual problem associated with the given (primal) problem is a standard maximization problem. The solution may thus be found using the simplex algorithm. Introducing the slack variables x and y, we obtain the system of linear equations

$$
\begin{aligned}
40u + 10v + 5w + x &= 6 \\
10u + 15v + 15w + y &= 8 \\
-2400u - 2100v - 1500w + P &= 0
\end{aligned}
$$

Continuing with the simplex algorithm, we obtain the following sequence of simplex tableaus:

	u	v	w	x	y	P	Constant
Pivot row $\rightarrow$	(40)	10	5	1	0	0	6
	10	15	15	0	1	0	8
	-2400	-2100	-1500	0	0	1	0

Ratio: $\frac{6}{40} = \frac{3}{20}$, $\frac{8}{10} = \frac{4}{5}$

Pivot column $\uparrow$

	u	v	w	x	y	P	Constant
$\frac{1}{40}R_1 \longrightarrow$	(1)	$\frac{1}{4}$	$\frac{1}{8}$	$\frac{1}{40}$	0	0	$\frac{3}{20}$
	10	15	15	0	1	0	8
	-2400	-2100	-1500	0	0	1	0

	u	v	w	x	y	P	Constant
$\frac{R_2 - 10R_1}{R_3 + 2400R_1} \longrightarrow$	1	$\frac{1}{4}$	$\frac{1}{8}$	$\frac{1}{40}$	0	0	$\frac{3}{20}$
	0	$\left(\frac{25}{2}\right)$	$\frac{55}{4}$	$-\frac{1}{4}$	1	0	$\frac{13}{2}$
	0	-1500	-1200	60	0	1	360

Ratio: $\frac{3/20}{1/4} = \frac{3}{5}$, $\frac{13/2}{25/2} = \frac{13}{25}$

	u	v	w	x	y	P	Constant
$\frac{2}{25}R_2 \longrightarrow$	1	$\frac{1}{4}$	$\frac{1}{8}$	$\frac{1}{40}$	0	0	$\frac{3}{20}$
	0	(1)	$\frac{11}{10}$	$-\frac{1}{50}$	$\frac{2}{25}$	0	$\frac{13}{25}$
	0	-1500	-1200	60	0	1	360

	u	v	w	x	y	P	Constant
$\frac{R_1 - \frac{1}{4}R_2}{R_3 + 1500R_2} \longrightarrow$	1	0	$-\frac{3}{20}$	$\frac{3}{100}$	$-\frac{1}{50}$	0	$\frac{1}{50}$
	0	1	$\frac{11}{10}$	$-\frac{1}{50}$	$\frac{2}{25}$	0	$\frac{13}{25}$
	0	0	450	30	120	1	1140

Solution for the primal problem

The last tableau is final. The fundamental theorem of duality tells us that the solution to the primal problem is $x = 30$ and $y = 120$ with a minimum value for C of 1140. Observe that the solution to the dual (maximization) problem may be read from the simplex tableau in the usual manner: $u = \frac{1}{50}$, $v = \frac{13}{25}$, $w = 0$, and $P = 1140$. Note that the maximum value of P is equal to the minimum value of

C, as guaranteed by the fundamental theorem of duality. The solution to the primal problem agrees with the solution of the same problem solved in Section 3.3, Example 2, using the method of corners.   ■

Notes

1. We leave it to you to demonstrate that the dual of a standard minimization problem is always a standard maximization problem provided that the coefficients of the objective function in the primal problem are all nonnegative. Such problems can always be solved by applying the simplex method to solve the dual problem.
2. Standard minimization problems in which the coefficients of the objective function are not all nonnegative do not necessarily have a dual problem that is a standard maximization problem.   ■

EXAMPLE 4

$$\begin{aligned} \text{Minimize} \quad & C = 3x + 2y \\ \text{subject to} \quad & 8x + y \ge 80 \\ & 8x + 5y \ge 240 \\ & x + 5y \ge 100 \\ & x \ge 0, y \ge 0 \end{aligned}$$

Solution   We begin by writing the dual problem associated with the given primal problem. First, we write down the following tableau for the primal problem:

x	y	Constant
8	1	80
8	5	240
1	5	100
3	2	

Next, interchanging the columns and rows of this tableau and heading the three columns of the resulting array with the three variables, u, v, and w, we obtain the tableau

u	v	w	Constant
8	8	1	3
1	5	5	2
80	240	100	

Interpreting the last tableau as if it were part of the initial simplex tableau for a standard maximization problem—with the exception that the signs of the coefficients pertaining to the objective function are not reversed—we construct the dual problem as follows: Maximize the objective function $P = 80u + 240v + 100w$ subject to the constraints

$$\begin{aligned} 8u + 8v + w &\le 3 \\ u + 5v + 5w &\le 2 \end{aligned}$$

where $u \geq 0$, $v \geq 0$, and $w \geq 0$. Having constructed the dual problem, which is a standard maximization problem, we now solve it using the simplex method. Introducing the slack variables x and y, we obtain the system of linear equations

$$
\begin{aligned}
8u + 8v + w + x \qquad\qquad &= 3 \\
u + 5v + 5w \qquad + y \qquad &= 2 \\
-80u - 240v - 100w \qquad\qquad + P &= 0
\end{aligned}
$$

Continuing with the simplex algorithm, we obtain the following sequence of simplex tableaus:

	u	v	w	x	y	P	Constant	Ratio
Pivot row →	8	⑧	1	1	0	0	3	$\frac{3}{8}$
	1	5	5	0	1	0	2	$\frac{2}{5}$
	−80	−240	−100	0	0	1	0	

↑
Pivot column

	u	v	w	x	y	P	Constant
$\frac{1}{8}R_1 \longrightarrow$	1	①	$\frac{1}{8}$	$\frac{1}{8}$	0	0	$\frac{3}{8}$
	1	5	5	0	1	0	2
	−80	−240	−100	0	0	1	0

	u	v	w	x	y	P	Constant	Ratio
	1	1	$\frac{1}{8}$	$\frac{1}{8}$	0	0	$\frac{3}{8}$	3
$\begin{array}{c}R_2 - 5R_1 \\ R_3 + 240R_1\end{array} \longrightarrow$ Pivot row	−4	0	$\frac{35}{8}$	$-\frac{5}{8}$	1	0	$\frac{1}{8}$	$\frac{1}{35}$
	160	0	−70	30	0	1	90	

↑
Pivot column

	u	v	w	x	y	P	Constant
$\frac{8}{35}R_2 \longrightarrow$	1	1	$\frac{1}{8}$	$\frac{1}{8}$	0	0	$\frac{3}{8}$
	$-\frac{32}{35}$	0	①	$-\frac{1}{7}$	$\frac{8}{35}$	0	$\frac{1}{35}$
	160	0	−70	30	0	1	90

	u	v	w	x	y	P	Constant
$\begin{array}{c}R_1 - \frac{1}{8}R_2 \\ R_3 + 70R_2\end{array} \longrightarrow$	$\frac{39}{35}$	1	0	$\frac{1}{7}$	$-\frac{1}{35}$	0	$\frac{13}{35}$
	$-\frac{32}{35}$	0	1	$-\frac{1}{7}$	$\frac{8}{35}$	0	$\frac{1}{35}$
	96	0	0	20	16	1	92

Solution for the
primal problem

The last tableau is final. The fundamental theorem of duality tells us that the solution to the primal problem is $x = 20$ and $y = 16$ with a minimum value for C of 92. ∎

Our last example illustrates how the warehouse problem posed in Section 3.2 may be solved by duality.

 APPLIED EXAMPLE 5   A Warehouse Problem   Complete the solution to the warehouse problem given in Section 3.2, Example 4 (page 179).

Minimize

$$C = 20x_1 + 8x_2 + 10x_3 + 12x_4 + 22x_5 + 18x_6 \qquad \textbf{(13)}$$

subject to

$$
\begin{aligned}
x_1 + x_2 + x_3 & \le 400 \\
x_4 + x_5 + x_6 &\le 600 \\
x_1 + x_4 & \ge 200 \\
x_2 + x_5 & \ge 300 \\
x_3 + x_6 &\ge 400 \\
x_1 \ge 0,\, x_2 \ge 0, \dots,\, x_6 &\ge 0
\end{aligned}
\qquad \textbf{(14)}
$$

Solution   Upon multiplying each of the first two inequalities of (14) by -1, we obtain the following equivalent system of constraints in which each of the expressions involving the variables is greater than or equal to a constant:

$$
\begin{aligned}
-x_1 - x_2 - x_3 &\ge -400 \\
-x_4 - x_5 - x_6 &\ge -600 \\
x_1 + x_4 &\ge 200 \\
x_2 + x_5 &\ge 300 \\
x_3 + x_6 &\ge 400 \\
x_1 \ge 0,\, x_2 \ge 0, \dots,\, x_6 &\ge 0
\end{aligned}
$$

The problem may now be solved by duality. First, we write the array of numbers:

x_1	x_2	x_3	x_4	x_5	x_6	Constant
-1	-1	-1	0	0	0	-400
0	0	0	-1	-1	-1	-600
1	0	0	1	0	0	200
0	1	0	0	1	0	300
0	0	1	0	0	1	400
20	8	10	12	22	18	

Interchanging the rows and columns of this array of numbers and heading the five columns of the resulting array of numbers by the variables u_1, u_2, u_3, u_4, and u_5, we obtain

u_1	u_2	u_3	u_4	u_5	Constant
-1	0	1	0	0	20
-1	0	0	1	0	8
-1	0	0	0	1	10
0	-1	1	0	0	12
0	-1	0	1	0	22
0	-1	0	0	1	18
-400	-600	200	300	400	

from which we construct the associated dual problem: Maximize $P = -400u_1 - 600u_2 + 200u_3 + 300u_4 + 400u_5$ subject to

$$
\begin{aligned}
-u_1 \quad + u_3 \quad &\le 20 \\
-u_1 \quad + u_4 \quad &\le 8 \\
-u_1 \quad + u_5 &\le 10 \\
-u_2 + u_3 \quad &\le 12 \\
-u_2 \quad + u_4 \quad &\le 22 \\
-u_2 \quad + u_5 &\le 18 \\
u_1 \ge 0,\, u_2 \ge 0,\, \ldots,\, u_5 &\ge 0
\end{aligned}
$$

Solving the standard maximization problem by the simplex algorithm, we obtain the following sequence of tableaus ($x_1, x_2, \ldots, x_6$ are slack variables):

u_1	u_2	u_3	u_4	u_5	x_1	x_2	x_3	x_4	x_5	x_6	P	Constant	Ratio
-1	0	1	0	0	1	0	0	0	0	0	0	20	—
-1	0	0	1	0	0	1	0	0	0	0	0	8	—
-1	0	0	0	①	0	0	1	0	0	0	0	10	10
0	-1	1	0	0	0	0	0	1	0	0	0	12	—
0	-1	0	1	0	0	0	0	0	1	0	0	22	—
0	-1	0	0	1	0	0	0	0	0	1	0	18	18
400	600	-200	-300	-400	0	0	0	0	0	0	1	0	

Pivot row → (third row)

Pivot column ↑ (u_5)

u_1	u_2	u_3	u_4	u_5	x_1	x_2	x_3	x_4	x_5	x_6	P	Constant	Ratio
-1	0	1	0	0	1	0	0	0	0	0	0	20	—
-1	0	0	①	0	0	1	0	0	0	0	0	8	8
-1	0	0	0	1	0	0	1	0	0	0	0	10	—
0	-1	1	0	0	0	0	0	1	0	0	0	12	—
0	-1	0	1	0	0	0	0	0	1	0	0	22	22
1	-1	0	0	0	0	0	-1	0	0	1	0	8	—
0	600	-200	-300	0	0	0	400	0	0	0	1	4000	

Pivot row → (second row)

$\dfrac{R_6 - R_3}{R_7 + 400R_3}$

Pivot column ↑ (u_4)

u_1	u_2	u_3	u_4	u_5	x_1	x_2	x_3	x_4	x_5	x_6	P	Constant	Ratio
-1	0	1	0	0	1	0	0	0	0	0	0	20	—
-1	0	0	1	0	0	1	0	0	0	0	0	8	—
-1	0	0	0	1	0	0	1	0	0	0	0	10	—
0	-1	1	0	0	0	0	0	1	0	0	0	12	—
1	-1	0	0	0	0	-1	0	0	1	0	0	14	14
①	-1	0	0	0	0	0	-1	0	0	1	0	8	8
-300	600	-200	0	0	0	300	400	0	0	0	1	6400	

$\dfrac{R_5 - R_2}{R_7 + 300R_2}$

Pivot row → (sixth row)

Pivot column ↑ (u_1)

	u_1	u_2	u_3	u_4	u_5	x_1	x_2	x_3	x_4	x_5	x_6	P	Constant		Ratio
	0	−1	1	0	0	1	0	−1	0	0	1	0	28		28
	0	−1	0	1	0	0	1	−1	0	0	1	0	16		—
	0	−1	0	0	1	0	0	0	0	0	1	0	18		—
	0	−1	①	0	0	0	0	0	1	0	0	0	12		12
	0	0	0	0	0	0	−1	1	0	1	−1	0	6		—
	1	−1	0	0	0	0	0	−1	0	0	1	0	8		—
	0	300	−200	0	0	0	300	100	0	0	300	1	8800		

$R_1 + R_6$
$R_2 + R_6$
$R_3 + R_6$
$R_5 - R_6$
$R_7 + 300R_6$ Pivot row

↑ Pivot column

	u_1	u_2	u_3	u_4	u_5	x_1	x_2	x_3	x_4	x_5	x_6	P	Constant
	0	0	0	0	0	1	0	−1	−1	0	1	0	16
	0	−1	0	1	0	0	1	−1	0	0	1	0	16
	0	−1	0	0	1	0	0	0	0	0	1	0	18
	0	−1	1	0	0	0	0	0	1	0	0	0	12
	0	0	0	0	0	0	−1	1	0	1	−1	0	6
	1	−1	0	0	0	0	0	−1	0	0	1	0	8
	0	100	0	0	0	0	300	100	200	0	300	1	11,200

$R_1 - R_4$
$R_7 + 200R_4$

The last tableau is final, and we find that

$$x_1 = 0 \qquad x_2 = 300 \qquad x_3 = 100 \qquad x_4 = 200$$
$$x_5 = 0 \qquad x_6 = 300 \qquad P = 11,200$$

Thus, to minimize shipping costs, Acrosonic should ship 300 loudspeaker systems from plant I to warehouse B, 100 systems from plant I to warehouse C, 200 systems from plant II to warehouse A, and 300 systems from plant II to warehouse C. The company's total shipping cost is $11,200. ∎

4.2 Self-Check Exercises

1. Write the dual problem associated with the following problem:

Minimize $C = 2x + 5y$
subject to $4x + y \geq 40$
$2x + y \geq 30$
$x + 3y \geq 30$
$x \geq 0, y \geq 0$

2. Solve the primal problem posed in Exercise 1.

Solutions to Self-Check Exercises 4.2 can be found on page 254.

4.2 Concept Questions

1. Suppose you are given the linear programming problem

$$\text{Minimize} \quad C = -3x - 5y$$
$$\text{subject to} \quad 5x + 2y \le 30$$
$$x + 3y \le 21$$
$$x \ge 0, y \ge 0$$

Give the associated standard maximization problem that you would use to solve this linear programming problem via the simplex method.

2. Give three characteristics of a standard minimization linear programming problem.

3. What is the primal problem associated with a standard minimization linear programming problem? The dual problem?

4. a. What does the fundamental theorem of duality tell us about the existence of a solution to a primal problem?
   b. How are the optimal values of the primal and dual problems related?
   c. Given the final simplex tableau associated with a dual problem, how would you determine the optimal solution to the associated primal problem?

4.2 Exercises

In Exercises 1–6, use the technique developed in this section to solve the minimization problem.

1. Minimize $C = -2x + y$
   subject to $x + 2y \le 6$
   $3x + 2y \le 12$
   $x \ge 0, y \ge 0$

2. Minimize $C = -2x - 3y$
   subject to $3x + 4y \le 24$
   $7x - 4y \le 16$
   $x \ge 0, y \ge 0$

3. Minimize $C = -3x - 2y$ subject to the constraints of Exercise 2.

4. Minimize $C = x - 2y + z$
   subject to $x - 2y + 3z \le 10$
   $2x + y - 2z \le 15$
   $2x + y + 3z \le 20$
   $x \ge 0, y \ge 0, z \ge 0$

5. Minimize $C = 2x - 3y - 4z$
   subject to $-x + 2y - z \le 8$
   $x - 2y + 2z \le 10$
   $2x + 4y - 3z \le 12$
   $x \ge 0, y \ge 0, z \ge 0$

6. Minimize $C = -3x - 2y - z$ subject to the constraints of Exercise 5.

In Exercises 7–10, you are given the final simplex tableau for the dual problem. Give the solution to the primal problem and to the associated dual problem.

7. Problem: Minimize $C = 8x + 12y$
   subject to $x + 3y \ge 2$
   $2x + 2y \ge 3$
   $x \ge 0, y \ge 0$

Final tableau:

u	v	x	y	P	Constant
0	1	$\frac{3}{4}$	$-\frac{1}{4}$	0	3
1	0	$-\frac{1}{2}$	$\frac{1}{2}$	0	2
0	0	$\frac{5}{4}$	$\frac{1}{4}$	1	13

8. Problem: Minimize $C = 3x + 2y$
   subject to $5x + y \ge 10$
   $2x + 2y \ge 12$
   $x + 4y \ge 12$
   $x \ge 0, y \ge 0$

Final tableau:

u	v	w	x	y	P	Constant
1	0	$-\frac{3}{4}$	$\frac{1}{4}$	$-\frac{1}{4}$	0	$\frac{1}{4}$
0	1	$\frac{19}{8}$	$-\frac{1}{8}$	$\frac{5}{8}$	0	$\frac{7}{8}$
0	0	9	1	5	1	13

9. Problem: Minimize $C = 10x + 3y + 10z$

subject to $2x + y + 5z \geq 20$

$4x + y + z \geq 30$

$x \geq 0, y \geq 0, z \geq 0$

Final tableau:

u	v	x	y	z	P	Constant
0	1	$\frac{1}{2}$	-1	0	0	2
1	0	$-\frac{1}{2}$	2	0	0	1
0	0	2	-9	1	0	3
0	0	5	10	0	1	80

10. Problem: Minimize $C = 2x + 3y$

subject to $x + 4y \geq 8$

$x + y \geq 5$

$2x + y \geq 7$

$x \geq 0, y \geq 0$

Final tableau:

u	v	w	x	y	P	Constant
0	1	$\frac{7}{3}$	$\frac{4}{3}$	$-\frac{1}{3}$	0	$\frac{5}{3}$
1	0	$-\frac{1}{3}$	$-\frac{1}{3}$	$\frac{1}{3}$	0	$\frac{1}{3}$
0	0	2	4	1	1	11

In Exercises 11–20, construct the dual problem associated with the primal problem. Solve the primal problem.

11. Minimize $C = 2x + 5y$

subject to $x + 2y \geq 4$

$3x + 2y \geq 6$

$x \geq 0, y \geq 0$

12. Minimize $C = 3x + 2y$

subject to $2x + 3y \geq 90$

$3x + 2y \geq 120$

$x \geq 0, y \geq 0$

13. Minimize $C = 6x + 4y$

subject to $6x + y \geq 60$

$2x + y \geq 40$

$x + y \geq 30$

$x \geq 0, y \geq 0$

14. Minimize $C = 10x + y$

subject to $4x + y \geq 16$

$x + 2y \geq 12$

$x \geq 2$

$x \geq 0, y \geq 0$

15. Minimize $C = 200x + 150y + 120z$

subject to $20x + 10y + z \geq 10$

$x + y + 2z \geq 20$

$x \geq 0, y \geq 0, z \geq 0$

16. Minimize $C = 40x + 30y + 11z$

subject to $2x + y + z \geq 8$

$x + y - z \geq 6$

$x \geq 0, y \geq 0, z \geq 0$

17. Minimize $C = 6x + 8y + 4z$

subject to $x + 2y + 2z \geq 10$

$2x + y + z \geq 24$

$x + y + z \geq 16$

$x \geq 0, y \geq 0, z \geq 0$

18. Minimize $C = 12x + 4y + 8z$

subject to $2x + 4y + z \geq 6$

$3x + 2y + 2z \geq 2$

$4x + y + z \geq 2$

$x \geq 0, y \geq 0, z \geq 0$

19. Minimize $C = 30x + 12y + 20z$

subject to $2x + 4y + 3z \geq 6$

$6x + z \geq 2$

$6y + 2z \geq 4$

$x \geq 0, y \geq 0, z \geq 0$

20. Minimize $C = 8x + 6y + 4z$

subject to $2x + 3y + z \geq 6$

$x + 2y - 2z \geq 4$

$x + y + 2z \geq 2$

$x \geq 0, y \geq 0, z \geq 0$

21. TRANSPORTATION Deluxe River Cruises operates a fleet of river vessels. The fleet has two types of vessels: A type-A vessel has 60 deluxe cabins and 160 standard cabins, whereas a type-B vessel has 80 deluxe cabins and 120 standard cabins. Under a charter agreement with Odyssey Travel Agency, Deluxe River Cruises is to provide Odyssey with a minimum of 360 deluxe and 680 standard cabins for their 15-day cruise in May. It costs $44,000 to operate a type-A vessel and $54,000 to operate a type-B vessel for that period. How many of each type vessel should be used in order to keep the operating costs to a minimum? What is the minimum cost?

22. SHIPPING COSTS Acrosonic manufactures a model G loudspeaker system in plants I and II. The output at plant I is at most 800/month, and the output at plant II is at most 600/month. Model G loudspeaker systems are also shipped to the three warehouses—A, B, and C—whose minimum monthly requirements are 500, 400, and 400 systems, respectively. Shipping costs from plant I to warehouse A, warehouse B, and warehouse C are $16, $20, and $22 per loudspeaker system, respectively, and shipping costs from plant II to each of these warehouses are $18, $16, and $14, respectively. What shipping schedule will enable Acrosonic to meet the requirements of the warehouses while keeping its shipping costs to a minimum? What is the minimum cost?

23. ADVERTISING   Everest Deluxe World Travel has decided to advertise in the Sunday editions of two major newspapers in town. These advertisements are directed at three groups of potential customers. Each advertisement in newspaper I is seen by 70,000 group-A customers, 40,000 group-B customers, and 20,000 group-C customers. Each advertisement in newspaper II is seen by 10,000 group-A, 20,000 group-B, and 40,000 group-C customers. Each advertisement in newspaper I costs $1000, and each advertisement in newspaper II costs $800. Everest would like their advertisements to be read by at least 2 million people from group A, 1.4 million people from group B, and 1 million people from group C. How many advertisements should Everest place in each newspaper to achieve its advertising goals at a minimum cost? What is the minimum cost?

24. SHIPPING COSTS   Steinwelt Piano manufactures uprights and consoles in two plants, plant I and plant II. The output of plant I is at most 300/month, and the output of plant II is at most 250/month. These pianos are shipped to three warehouses that serve as distribution centers for Steinwelt. To fill current and projected future orders, warehouse A requires a minimum of 200 pianos/month, warehouse B requires at least 150 pianos/month, and warehouse C requires at least 200 pianos/month. The shipping cost of each piano from plant I to warehouse A, warehouse B, and warehouse C is $60, $60, and $80, respectively, and the shipping cost of each piano from plant II to warehouse A, warehouse B, and warehouse C is $80, $70, and $50, respectively. What shipping schedule will enable Steinwelt to meet the requirements of the warehouses while keeping the shipping costs to a minimum? What is the minimum cost?

25. NUTRITION—DIET PLANNING   The owner of the Health JuiceBar wishes to prepare a low-calorie fruit juice with a high vitamin-A and -C content by blending orange juice and pink grapefruit juice. Each glass of the blended juice is to contain at least 1200 International Units (IU) of vitamin A and 200 IU of vitamin C. One ounce of orange juice contains 60 IU of vitamin A, 16 IU of vitamin C, and 14 calories; each ounce of pink grapefruit juice contains 120 IU of vitamin A, 12 IU of vitamin C, and 11 calories. How many ounces of each juice should a glass of the blend contain if it is to meet the minimum vitamin requirements while containing a minimum number of calories?

26. PRODUCTION CONTROL   An oil company operates two refineries in a certain city. Refinery I has an output of 200, 100, and 100 barrels of low-, medium-, and high-grade oil per day, respectively. Refinery II has an output of 100, 200, and 600 barrels of low-, medium-, and high-grade oil per day, respectively. The company wishes to produce at least 1000, 1400, and 3000 barrels of low-, medium-, and high-grade oil to fill an order. If it costs $200/day to operate refinery I and $300/day to operate refinery II, determine how many days each refinery should be operated in order to meet the production requirements at minimum cost to the company. What is the minimum cost?

In Exercises 27 and 28, determine whether the statement is true or false. If it is true, explain why it is true. If it is false, give an example to show why it is false.

27. If a standard minimization linear programming problem has a unique solution, then so does the corresponding maximization problem with objective function $P = -C$, where $C = a_1x_1 + a_2x_2 + \cdots + a_nx_n$ is the objective function for the minimization problem.

28. The optimal value attained by the objective function of the primal problem may be different from that attained by the objective function of the dual problem.

4.2 Solutions to Self-Check Exercises

1. We first write down the following tableau for the given (primal) problem:

x	y	Constant
4	1	40
2	1	30
1	3	30
2	5	0

Next, we interchange the columns and rows of the tableau and head the three columns of the resulting array with the three variables, u, v, and w, obtaining the tableau

u	v	w	Constant
4	2	1	2
1	1	3	5
40	30	30	0

Interpreting the last tableau as if it were the initial tableau for a standard linear programming problem—with the exception that the signs of the coefficients pertaining to the objective function are not reversed—we construct the required dual problem as follows:

$$\text{Maximize} \quad P = 40u + 30v + 30w$$
$$\text{subject to} \quad 4u + 2v + w \le 2$$
$$u + v + 3w \le 5$$
$$u \ge 0, v \ge 0, w \ge 0$$

2. We introduce slack variables x and y to obtain the system of linear equations

$$4u + 2v + w + x \qquad = 2$$
$$u + v + 3w \qquad + y \qquad = 5$$
$$-40u - 30v - 30w \qquad + P = 0$$

Using the simplex algorithm, we obtain the sequence of simplex tableaus

	u	v	w	x	y	P	Constant	
Pivot row →	④	2	1	1	0	0	2	Ratio: $\frac{2}{4} = \frac{1}{2}$ $\frac{1}{4}R_1$
	1	1	3	0	1	0	5	$\frac{5}{1} = 5$
	-40	-30	-30	0	0	1	0	

↑ Pivot column

u	v	w	x	y	P	Constant	
①	$\frac{1}{2}$	$\frac{1}{4}$	$\frac{1}{4}$	0	0	$\frac{1}{2}$	$R_2 - R_1$
1	1	3	0	1	0	5	$R_3 + 40R_1$
-40	-30	-30	0	0	1	0	

	u	v	w	x	y	P	Constant	
	1	$\frac{1}{2}$	$\frac{1}{4}$	$\frac{1}{4}$	0	0	$\frac{1}{2}$	Ratio: $\frac{1/2}{1/4} = 2$ $\frac{4}{11}R_2$
Pivot row →	0	$\frac{1}{2}$	⑪⁄₄	$-\frac{1}{4}$	1	0	$\frac{9}{2}$	$\frac{9/2}{11/4} = \frac{18}{11}$
	0	-10	-20	10	0	1	20	

↑ Pivot column

u	v	w	x	y	P	Constant	
1	$\frac{1}{2}$	$\frac{1}{4}$	$\frac{1}{4}$	0	0	$\frac{1}{2}$	$R_1 - \frac{1}{4}R_2$
0	$\frac{2}{11}$	①	$-\frac{1}{11}$	$\frac{4}{11}$	0	$\frac{18}{11}$	$R_3 + 20R_2$
0	-10	-20	10	0	1	20	

	u	v	w	x	y	P	Constant	
Pivot row →	1	⑤⁄₁₁	0	$\frac{3}{11}$	$-\frac{1}{11}$	0	$\frac{1}{11}$	Ratio: $\frac{1/11}{5/11} = \frac{1}{5}$ $\frac{11}{5}R_1$
	0	$\frac{2}{11}$	1	$-\frac{1}{11}$	$\frac{4}{11}$	0	$\frac{18}{11}$	$\frac{18/11}{2/11} = 9$
	0	$-\frac{70}{11}$	0	$\frac{90}{11}$	$\frac{80}{11}$	1	$\frac{580}{11}$	

↑ Pivot column

u	v	w	x	y	P	Constant	
$\frac{11}{5}$	①	0	$\frac{3}{5}$	$-\frac{1}{5}$	0	$\frac{1}{5}$	$R_2 - \frac{2}{11}R_1$
0	$\frac{2}{11}$	1	$-\frac{1}{11}$	$\frac{4}{11}$	0	$\frac{18}{11}$	$R_3 + \frac{70}{11}R_1$
0	$-\frac{70}{11}$	0	$\frac{90}{11}$	$\frac{80}{11}$	1	$\frac{580}{11}$	

u	v	w	x	y	P	Constant
$\frac{11}{5}$	1	0	$\frac{3}{5}$	$-\frac{1}{5}$	0	$\frac{1}{5}$
$-\frac{2}{5}$	0	1	$-\frac{1}{5}$	$\frac{2}{5}$	0	$\frac{8}{5}$
14	0	0	12	6	1	54

Solution for the primal problem

The last tableau is final, and the solution to the primal problem is $x = 12$ and $y = 6$ with a minimum value for C of 54.

USING TECHNOLOGY

■ The Simplex Method: Solving Minimization Problems

Graphing Utility

A graphing utility can be used to solve minimization problems using the simplex method.

EXAMPLE 1

$$\text{Minimize} \quad C = 2x + 3y$$
$$\text{subject to} \quad 8x + y \ge 80$$
$$3x + 2y \ge 100$$
$$x + 4y \ge 80$$
$$x \ge 0, y \ge 0$$

(continued)

Solution   We begin by writing the dual problem associated with the given primal problem. From the tableau for the primal problem

x	y	Constant
8	1	80
3	2	100
1	4	80
2	3	

we find—upon changing the columns and rows of this tableau and heading the three columns of the resulting array with the variables u, v, and w—the tableau

u	v	w	Constant
8	3	1	2
1	2	4	3
80	100	80	

This tells us that the dual problem is

$$\text{Maximize} \quad P = 80u + 100v + 80w$$
$$\text{subject to} \quad 8u + 3v + \ w \le 2$$
$$u + 2v + 4w \le 3$$
$$u \ge 0, v \ge 0, w \ge 0$$

To solve this standard maximization problem, we proceed as follows:

	u	v	w	x	y	P	Constant	Ratio
Pivot row →	8	③	1	1	0	0	2	$\frac{2}{3}$
	1	2	4	0	1	0	3	$\frac{3}{2}$
	-80	-100	-80	0	0	1	0	

$*\mathbf{row}(\frac{1}{3}, A, 1) \blacktriangleright B \longrightarrow$

↑ Pivot column

u	v	w	x	y	P	Constant
2.67	①	0.33	0.33	0	0	0.67
1	2	4	0	1	0	3
-80	-100	-80	0	0	1	0

$*\mathbf{row}+(-2, B, 1, 2) \blacktriangleright C \longrightarrow$
$*\mathbf{row}+(100, C, 1, 3) \blacktriangleright B$

	u	v	w	x	y	P	Constant	Ratio
	2.67	1	0.33	0.33	0	0	0.67	2
Pivot row →	-4.33	0	③.33	-0.67	1	0	1.67	0.5
	186.67	0	-46.67	33.33	0	1	66.67	

$*\mathbf{row}(\frac{1}{3.33}, B, 2) \blacktriangleright C \longrightarrow$

↑ Pivot column

u	v	w	x	y	P	Constant
2.67	1	0.33	0.33	0	0	0.67
-1.30	0	1	-0.2	0.3	0	0.5
186.67	0	-46.67	33.33	0	1	66.67

$*\mathbf{row}+(-0.33, C, 2, 1) \blacktriangleright B \longrightarrow$
$*\mathbf{row}+(46.67, B, 2, 3) \ \blacktriangleright C$

u	v	w	x	y	P	Constant
3.1	1	0	0.4	−0.1	0	0.50
−1.3	0	1	−0.2	0.3	0	0.50
125.93	0	0.05	23.99	14.02	1	90.03

Solution for the
primal problem

From the last tableau, we see that $x = 23.99$, $y = 14.02$, and the minimum value of C is 90.03.

Excel

EXAMPLE 2

$$\text{Minimize} \quad C = 2x + 3y$$
$$\text{subject to} \quad 8x + y \geq 80$$
$$3x + 2y \geq 100$$
$$x + 4y \geq 80$$
$$x \geq 0, y \geq 0$$

Solution   We use Solver as outlined in Example 2, pages 239–242, to obtain the spreadsheet shown in Figure T1. (In this case, select $\boxed{\text{Min}}$ next to Equal to: instead of Max because this is a minimization problem. Also select $\boxed{\text{>=}}$ in the Add Constraint dialog box because the inequalities in the problem are of the form $\geq$.) From the spreadsheet, we read off the solution: $x = 24$, $y = 14$, and $C = 90$.

	A	B	C	D	E	F	G	H	I
1	Minimization Problem								
2					Formulas for indicated cells				
3	Decision Variables				C8: = 2*C4 + 3*C5				
4		x	24		C11: = 8*C4 + C5				
5		y	14		C12: = 3*C4 + 2*C5				
6					C13: = C4 + 4*C5				
7									
8	Objective Function		90						
9									
10	Constraints								
11			206	>=	80				
12			100	>=	100				
13			80	>=	80				

FIGURE T1
Completed spreadsheet after using Solver

EXAMPLE 3   Sensitivity Analysis*   Solve the following linear problem:

$$\text{Maximize} \quad P = 2x + 4y$$
$$\text{subject to} \quad 2x + 5y \leq 19 \qquad \text{Constraint 1}$$
$$3x + 2y \leq 12 \qquad \text{Constraint 2}$$
$$x \geq 0, y \geq 0$$

Note: Boldfaced words/characters enclosed in a box (for example, $\boxed{\text{Enter}}$) indicate that an action (click, select, or press) is required. Words/characters printed blue (for example, Chart sub-type) indicate words/characters that appear on the screen. Words/characters printed in a typewriter font (for example, =(−2/3)*A2+2) indicate words/characters that need to be typed and entered.

*For those who have completed Section 3.4.

(*continued*)

a. Use Solver to solve the problem.

b. Find the range of values that the coefficient of *x* can assume without changing the optimal solution.

c. Find the range of values that resource 1 (the constant on the right-hand side of constraint 1) can assume without changing the optimal solution.

d. Find the shadow price for resource 1.

e. Identify the binding and nonbinding constraints.

Solution

a. Use Solver as outlined in Example 2, pages 239–242, to obtain the spreadsheet shown in Figure T2.

	A	B	C	D	E	F	G	H	I
1	Maximization Problem								
2									
3	Decision Variables								
4		x	2						
5		y	3						
6									
7									
8	Objective Function		16						
9									
10	Constraints								
11			19	<=	19				
12			12	<=	12				

Formulas for indicated cells
C8: = 2∗C4 + 4∗C5
C11: = 2∗C4 + 5∗C5
C12: = 3∗C4 + 2∗C5

FIGURE T2
Completed spreadsheet for a maximization problem

b. In the Solver Results dialog box, hold down Ctrl while selecting **Answer** and **Sensitivity** under Reports, and then click **OK** . By clicking the **Answer Report 1** and **Sensitivity Report 1** tabs that appear at the bottom of your worksheet, you can obtain the reports shown in Figures T3a and T3b.

Target Cell (Max)

Cell	Name	Original Value	Final Value
$C$8	Objective Function	0	16

Adjustable Cells

Cell	Name	Original Value	Final Value
$C$4	x	0	2
$C$5	y	0	3

Constraints

Cell	Name	Cell Value	Formula	Status	Slack
$C$11		19	$C$11<=$E$11	Binding	0
$C$12		12	$C$12<=$E$12	Binding	0

FIGURE T3
The Solver reports

(a) The Answer Report

Adjustable Cells

Cell	Name	Final Value	Reduced Cost	Objective Coefficient	Allowable Increase	Allowable Decrease
$C$4	x	2	0	2	4	0.4
$C$5	y	3	0	4	1	2.666666667

Constraints

Cell	Name	Final Value	Shadow Price	Constraint R.H. Side	Allowable Increase	Allowable Decrease
$C$11		19	0.727272727	19	11	11
$C$12		12	0.181818182	12	16.5	4.4

FIGURE T3
(*continued*)

(b) The Sensitivity Report

From the sensitivity report, we see that the value of the Objective Coefficient for x is 2, the Allowable Increase for the coefficient is 4, and the Allowable Decrease for the coefficient is 0.4. Thus, the coefficient can vary from 1.6 to 6 without affecting the optimal solution.

c. From the sensitivity report, we see that the Final Value of the constraint for resource 1 is 19 and the Allowable Increase and Allowable Decrease for this value is 11. Thus, the value of the constraint must lie between $19 - 11$ and $19 + 11$—that is, between 8 and 30.

d. From the sensitivity report, we see that the Shadow Price for resource 1 is 0.727272727.

e. From the answer report, we conclude that both Constraints are binding.   ■

TECHNOLOGY EXERCISES

In Exercises 1–4, solve the linear programming problem by the simplex method.

1. Minimize $C = x + y + 3z$
subject to $2x + y + 3z \geq 6$
$x + 2y + 4z \geq 8$
$3x + y - 2z \geq 4$
$x \geq 0, y \geq 0, z \geq 0$

2. Minimize $C = 2x + 4y + z$
subject to $x + 2y + 4z \geq 7$
$3x + y - z \geq 6$
$x + 4y + 2z \geq 24$
$x \geq 0, y \geq 0, z \geq 0$

3. Minimize $C = x + 1.2y + 3.5z$
subject to $2x + 3y + 5z \geq 12$
$3x + 1.2y - 2.2z \geq 8$
$1.2x + 3y + 1.8z \geq 14$
$x \geq 0, y \geq 0, z \geq 0$

4. Minimize $C = 2.1x + 1.2y + z$
subject to $x + y - z \geq 5.2$
$x - 2.1y + 4.2z \geq 8.4$
$x \geq 0, y \geq 0, z \geq 0$

Exercises 5–8 are for Excel users only. You are given a linear programming problem:

a. Use Solver to solve the problem.
b. Find the range of values that the coefficient of *x* can assume without changing the optimal solution and the range of values that the coefficient of *y* can assume without changing the optimal solution.
c. Find the range of values that each resource (requirement) can assume.
d. Find the shadow price for each resource (requirement).
e. Identify the binding and nonbinding constraints.

5. Maximize $P = 3x + 4y$
subject to $2x + 3y \leq 12$ Resource 1
$2x + y \leq 8$ Resource 2
$x \geq 0, y \geq 0$

(*continued*)

6. Maximize $P = 2x + 5y$
    subject to $x + 3y \le 15$ Resource 1
              $4x + y \le 16$ Resource 2
              $x \ge 0, y \ge 0$

7. Minimize $C = 2x + 5y$
    subject to $x + 2y \ge 4$ Requirement 1
              $x + y \ge 3$ Requirement 2
              $x \ge 0, \; y \ge 0$

8. Minimize $C = 3x + 4y$
    subject to $x + 3y \ge 8$ Requirement 1
              $x + y \ge 4$ Requirement 2
              $x \ge 0, \; y \ge 0$

4.3   The Simplex Method: Nonstandard Problems (Optional)

Section 4.1 showed how we can use the simplex method to solve standard maximization problems; Section 4.2 showed how, thanks to duality, we can use it to solve standard minimization problems provided that the coefficients in the objective function are all nonnegative.

In this section, we see how the simplex method can be incorporated into a method for solving **nonstandard problems**—problems that do not fall into either of the two previous categories. We begin by recalling the characteristics of standard problems.

Standard Maximization Problem

1. The objective function is to be maximized.
2. All variables involved in the problem are nonnegative.
3. Each linear constraint may be written so that the expression involving the variables is less than or equal to a nonnegative constant.

Standard Minimization Problem (Restricted Version—See Condition 4)

1. The objective function is to be minimized.
2. All variables involved in the problem are nonnegative.
3. Each linear constraint may be written so that the expression involving the variables is greater than or equal to a constant.
4. *All coefficients in the objective function are nonnegative.*

Note   Recall that if all coefficients in the objective function are nonnegative, then a standard minimization problem can be solved by using the simplex method to solve the associated dual problem.   ■

We now give some examples of linear programming problems that do not fit into these two categories of problems.

EXAMPLE 1   Explain why the following linear programming problem is *not* a standard maximization problem.

$$\text{Maximize} \quad P = x + 2y$$
$$\text{subject to} \quad 4x + 3y \le 18$$
$$-x + 3y \ge 3$$
$$x \ge 0, y \ge 0$$

Solution   This is not a standard maximization problem because the second constraint,

$$-x + 3y \ge 3$$

violates condition 3. Observe that multiplying both sides of this inequality by -1 yields

$$x - 3y \le -3$$ Recall that multiplying both sides of an inequality by a negative number reverses the inequality sign.

Now the last equation still violates condition 3 because the constant on the right is *negative*.   ■

Observe that the constraints in Example 1 involve both *less than or equal to constraints* ($\le$) and *greater than or equal to constraints* ($\ge$). Such constraints are called **mixed constraints**. We will solve the problem posed in Example 1 later.

EXAMPLE 2   Explain why the following linear programming problem is *not* a restricted standard minimization problem.

$$\text{Minimize} \quad C = 2x - 3y$$
$$\text{subject to} \quad x + y \le 5$$
$$x + 3y \ge 9$$
$$-2x + y \le 2$$
$$x \ge 0, y \ge 0$$

Solution   Observe that the coefficients in the objective function C are not all nonnegative. Therefore, the problem is not a restricted standard minimization problem. By constructing the dual problem, you can convince yourself that the latter is not a standard maximization problem and thus cannot be solved using the methods described in Sections 4.1 and 4.2. Again, we will solve this problem later.   ■

EXAMPLE 3   Explain why the following linear programming problem is *not* a standard maximization problem. Show that it cannot be rewritten as a restricted standard minimization problem.

$$\text{Maximize} \quad P = x + 2y$$
$$\text{subject to} \quad 2x + 3y \le 12$$
$$-x + 3y = 3$$
$$x \ge 0, y \ge 0$$

Solution   The constraint equation $-x + 3y = 3$ is equivalent to the two inequalities

$$-x + 3y \leq 3 \quad \text{and} \quad -x + 3y \geq 3$$

If we multiply both sides of the second inequality by -1, then it can be written in the form

$$x - 3y \leq -3$$

Therefore, the two given constraints are equivalent to the three constraints

$$
\begin{aligned}
2x + 3y &\leq 12 \\
-x + 3y &\leq 3 \\
x - 3y &\leq -3
\end{aligned}
$$

The third inequality violates Condition 3 for a standard maximization problem. Next, we see that the given problem is equivalent to the following:

$$
\begin{aligned}
\text{Minimize} \quad & C = -x - 2y \\
\text{subject to} \quad & -2x - 3y \geq -12 \\
& x - 3y \geq -3 \\
& -x + 3y \geq 3 \\
& x \geq 0, y \geq 0
\end{aligned}
$$

Since the coefficients of the objective function are not all nonnegative, we conclude that the given problem cannot be rewritten as a restricted standard minimization problem. You will be asked to solve this problem in Exercise 11.   ∎

▬ The Simplex Method for Solving Nonstandard Problems

To describe a technique for solving nonstandard problems, let's consider the problem of Example 1:

$$
\begin{aligned}
\text{Maximize} \quad & P = x + 2y \\
\text{subject to} \quad & 4x + 3y \leq 18 \\
& -x + 3y \geq 3 \\
& x \geq 0, y \geq 0
\end{aligned}
$$

As a first step, we rewrite the constraints so that the second constraint involves a $\leq$ constraint. As in Example 1, we obtain

$$
\begin{aligned}
4x + 3y &\leq 18 \\
x - 3y &\leq -3 \\
x \geq 0, y &\geq 0
\end{aligned}
$$

Disregarding the fact that the constant on the right of the second inequality is negative, let's attempt to solve the problem using the simplex method for problems in standard form. Introducing the slack variables u and v gives the system of linear equations

$$
\begin{aligned}
4x + 3y + u \quad\quad\quad &= 18 \\
x - 3y \quad + v \quad\quad &= -3 \\
-x - 2y \quad\quad\quad + P &= 0
\end{aligned}
$$

The initial simplex tableau is

x	y	u	v	P	Constant
4	3	1	0	0	18
1	-3	0	1	0	-3
-1	-2	0	0	1	0

Interpreting the tableau in the usual fashion, we see that

$$x = 0 \qquad y = 0 \qquad u = 18 \qquad v = -3$$

Since the value of the slack variable v is negative, we see that this cannot be a feasible solution (remember, all variables must be nonnegative). In fact, you can see from Figure 6 that the point $(0, 0)$ does not lie in the feasible set associated with the given problem. Since we must start from a feasible point when using the simplex method for problems in standard form, we see that this method is not applicable at this juncture.

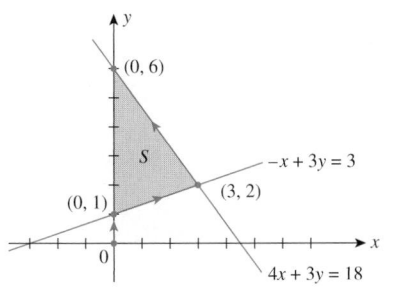

FIGURE 6
S is the feasible set for the problem.

Let's find a way to bring us from the nonfeasible point $(0, 0)$ to *any* feasible point, after which we can switch to the simplex method for problems in standard form. This can be accomplished by pivoting as follows. Referring to the tableau

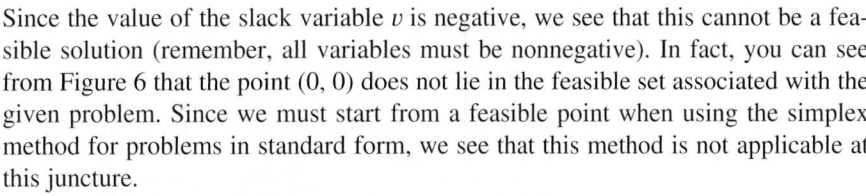

notice the negative number -3 lying in the column of constants and above the lower horizontal line. Locate any negative number to the left of this number (there must always be at least one such number if the problem has a solution). For the problem under consideration there is only one such number, the number -3 in the y-column. This column is designated as the pivot column. To find the pivot element, we form the *positive* ratios of the numbers in the column of constants to the corresponding numbers in the pivot column (above the last row). The pivot row is the row corresponding to the smallest ratio, and the pivot element is the element common to both the pivot row and the pivot column (the number circled in the foregoing tableau). Pivoting about this element, we have

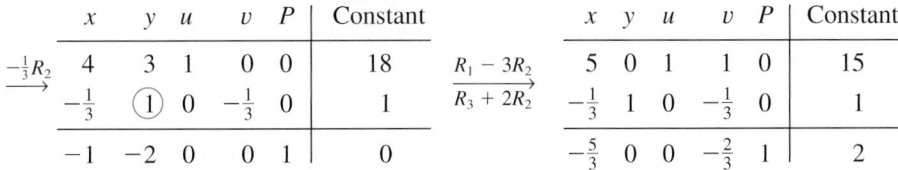

Interpreting the last tableau in the usual fashion, we see that

$$x = 0 \qquad y = 1 \qquad u = 15 \qquad v = 0 \qquad P = 2$$

Observe that the point $(0, 1)$ is a feasible point (see Figure 6). Our iteration has brought us from a nonfeasible point to a feasible point in one iteration. Observe, too, that all the constants in the column of constants are now nonnegative, reflecting the fact that $(0, 1)$ is a feasible point, as we have just noted.

We can now use the simplex method for problems in standard form to complete the solution to our problem.

	x	y	u	v	P	Constant	Ratio
Pivot row →	⑤	0	1	1	0	15	3
	$-\frac{1}{3}$	1	0	$-\frac{1}{3}$	0	1	—
	$-\frac{5}{3}$	0	0	$-\frac{2}{3}$	1	2	

↑ Pivot column

$\xrightarrow{\frac{1}{5}R_1}$

	x	y	u	v	P	Constant
	①	0	$\frac{1}{5}$	$\frac{1}{5}$	0	3
	$-\frac{1}{3}$	1	0	$-\frac{1}{3}$	0	1
	$-\frac{5}{3}$	0	0	$-\frac{2}{3}$	1	2

Pivot row

$\xrightarrow[R_3 + \frac{5}{3}R_1]{R_2 + \frac{1}{3}R_1}$

	x	y	u	v	P	Constant	Ratio
→	1	0	$\frac{1}{5}$	⑤$\frac{1}{5}$	0	3	15
	0	1	$\frac{1}{15}$	$-\frac{4}{15}$	0	2	—
	0	0	$\frac{1}{3}$	$-\frac{1}{3}$	1	7	

↑ Pivot column

$\xrightarrow{5R_1}$

x	y	u	v	P	Constant
5	0	1	①	0	15
0	1	$\frac{1}{15}$	$-\frac{4}{15}$	0	2
0	0	$\frac{1}{3}$	$-\frac{1}{3}$	1	7

$\xrightarrow[R_3 + \frac{1}{3}R_1]{R_2 + \frac{4}{15}R_1}$

x	y	u	v	P	Constant
5	0	1	1	0	15
$\frac{4}{3}$	1	$\frac{1}{3}$	0	0	6
$\frac{5}{3}$	0	$\frac{2}{3}$	0	1	12

All entries in the last row are nonnegative and so the tableau is final. We see that the optimal solution is

$$x = 0 \qquad y = 6 \qquad u = 0 \qquad v = 15 \qquad P = 12$$

Observe that the maximum of P occurs at $(0, 6)$ (see Figure 6). The arrows indicate the path that our search for the maximum of P has taken us on.

Before looking at further examples, let's summarize the method for solving nonstandard problems.

The Simplex Method for Solving Nonstandard Problems

1. If necessary, rewrite the problem as a maximization problem (recall that minimizing C is equivalent to maximizing $-C$).
2. If necessary, rewrite all constraints (except $x \geq 0, y \geq 0, z \geq 0, \ldots$) using less than or equal to ($\leq$) inequalities.
3. Introduce slack variables and set up the initial simplex tableau.
4. Scan the upper part of the column of constants of the tableau for negative entries.
   a. If there are no negative entries, complete the solution using the simplex method for problems in standard form.
   b. If there are negative entries, proceed to step 5.
5. a. Pick any negative entry in the row in which a negative entry in the column of constants occurs. The column containing this entry is the pivot column.
   b. Compute the positive ratios of the numbers in the column of constants to the corresponding numbers in the pivot column (above the last row). The pivot row corresponds to the smallest ratio. The intersection of the pivot column and the pivot row determines the pivot element.
   c. Pivot the tableau about the pivot element. Then return to step 4.

We now apply the method to solve the nonstandard problem posed in Example 2.

EXAMPLE 4   Solve the problem of Example 2:

$$\text{Minimize} \quad C = 2x - 3y$$
$$\text{subject to} \quad x + y \leq 5$$
$$x + 3y \geq 9$$
$$-2x + y \leq 2$$
$$x \geq 0, y \geq 0$$

Solution   We first rewrite the problem as a maximization problem with constraints using $\leq$, which gives the following equivalent problem:

$$\text{Maximize} \quad P = -C = -2x + 3y$$
$$\text{subject to} \quad x + y \leq 5$$
$$-x - 3y \leq -9$$
$$-2x + y \leq 2$$
$$x \geq 0, y \geq 0$$

Introducing slack variables u, v, and w and following the procedure for solving nonstandard problems outlined previously, we obtain the following sequence of tableaus:

	x	y	u	v	w	P	Constant	Ratio	
	1	1	1	0	0	0	5	$\frac{5}{1} = 5$	Column 1 could
	−1	−3	0	1	0	0	−9	$\frac{-9}{-3} = 3$	have been chosen
Pivot row →	−2	①	0	0	1	0	2	$\frac{2}{1} = 2$	as the pivot column as well.
	2	−3	0	0	0	1	0		

Pivot column

	x	y	u	v	w	P	Constant
	3	0	1	0	-1	0	3
	$\boxed{-7}$	0	0	1	3	0	-3
	-2	1	0	0	1	0	2
	-4	0	0	0	3	1	6

$\xrightarrow[\begin{array}{c}R_1 - R_3\\ R_2 + 3R_3\\ R_4 + 3R_3\end{array}]{}$ Pivot row

↑ Pivot column

Ratio
$\frac{3}{3} = 1$
$\frac{-3}{-7} = \frac{3}{7}$

	x	y	u	v	w	P	Constant
	3	0	1	0	-1	0	3
	$\boxed{1}$	0	0	$-\frac{1}{7}$	$-\frac{3}{7}$	0	$\frac{3}{7}$
	-2	1	0	0	1	0	2
	-4	0	0	0	3	1	6

$\xrightarrow{-\frac{1}{7}R_2}$

	x	y	u	v	w	P	Constant
	0	0	1	$\boxed{\frac{3}{7}}$	$\frac{2}{7}$	0	$\frac{12}{7}$
	1	0	0	$-\frac{1}{7}$	$-\frac{3}{7}$	0	$\frac{3}{7}$
	0	1	0	$-\frac{2}{7}$	$\frac{1}{7}$	0	$\frac{20}{7}$
	0	0	0	$-\frac{4}{7}$	$\frac{9}{7}$	1	$\frac{54}{7}$

$\xrightarrow[\begin{array}{c}R_1 - 3R_2\\ R_3 + 2R_2\\ R_4 + 4R_2\end{array}]{}$ Pivot row

↑ Pivot column

Ratio
4
—
—

We now use the simplex method for problems in standard form to complete the problem.

	x	y	u	v	w	P	Constant
	0	0	$\frac{7}{3}$	$\boxed{1}$	$\frac{2}{3}$	0	4
	1	0	0	$-\frac{1}{7}$	$-\frac{3}{7}$	0	$\frac{3}{7}$
	0	1	0	$-\frac{2}{7}$	$\frac{1}{7}$	0	$\frac{20}{7}$
	0	0	0	$-\frac{4}{7}$	$\frac{9}{7}$	1	$\frac{54}{7}$

$\xrightarrow{\frac{7}{3}R_1}$

	x	y	u	v	w	P	Constant
	0	0	$\frac{7}{3}$	1	$\frac{2}{3}$	0	4
	1	0	$\frac{1}{3}$	0	$-\frac{1}{3}$	0	1
	0	1	$\frac{2}{3}$	0	$\frac{1}{3}$	0	4
	0	0	$\frac{4}{3}$	0	$\frac{5}{3}$	1	10

$\xrightarrow[\begin{array}{c}R_2 + \frac{1}{7}R_1\\ R_3 + \frac{2}{7}R_1\\ R_4 + \frac{4}{7}R_1\end{array}]{}$

All the entries in the last row are nonnegative and hence the tableau is final. We see that the optimal solution is

$$x = 1 \qquad y = 4 \qquad u = 0 \qquad v = 4 \qquad w = 0 \qquad C = -P = -10$$

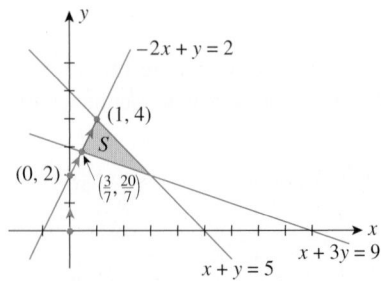

FIGURE 7
The feasible set S and the path leading from the initial nonfeasible point $(0, 0)$ to the optimal point $(1, 4)$

The feasible set S for this problem is shown in Figure 7. The path leading from the nonfeasible initial point $(0, 0)$ to the optimal point $(1, 4)$ goes through the nonfeasible point $(0, 2)$ and the feasible point $(\frac{3}{7}, \frac{20}{7})$, in that order.   ■

APPLIED EXAMPLE 5 **Production Planning** Rockford manufactures two models of exercise bicycles—a standard model and a deluxe model—in two separate plants, plant I and plant II. The maximum output at plant I is 1200 per month, and the maximum output at plant II is 1000 per month. The profit per bike for standard and deluxe models manufactured at plant I is $40 and $60, respectively; the profit per bike for standard and deluxe models manufactured at plant II is $45 and $50, respectively.

For the month of May, Rockford received an order for 1000 standard models and 800 deluxe models. If prior commitments dictate that the number of deluxe models manufactured at plant I not exceed the number of standard models manufactured there by more than 200, find how many of each model should be produced at each plant so as to satisfy the order and at the same time maximize Rockford's profit.

Solution Let x and y denote the number of standard and deluxe models to be manufactured at plant I. Since the number of standard and deluxe models required are 1000 and 800, respectively, we see that the number of standard and deluxe models to be manufactured at plant II are $(1000 - x)$ and $(800 - y)$, respectively. Rockford's profit will then be

$$P = 40x + 60y + 45(1000 - x) + 50(800 - y)$$
$$= 85,000 - 5x + 10y$$

Since the maximum output of plant I is 1200, we have the constraint

$$x + y \leq 1200$$

Similarly, since the maximum output of plant II is 1000, we have

$$(1000 - x) + (800 - y) \leq 1000$$

or, equivalently,

$$-x - y \leq -800$$

Finally, the additional constraints placed on the production schedule at plants I and II translate into the inequalities

$$y - x \leq 200$$
$$x \leq 1000$$
$$y \leq 800$$

To summarize, the problem at hand is the following nonstandard problem:

$$\begin{aligned} \text{Maximize} \quad & P = 85,000 - 5x + 10y \\ \text{subject to} \quad & x + y \leq 1200 \\ & -x - y \leq -800 \\ & -x + y \leq 200 \\ & x \leq 1000 \\ & y \leq 800 \\ & x \geq 0, y \geq 0 \end{aligned}$$

Let's introduce the slack variables, u v, w, r, and s. Using the simplex method for nonstandard problems, we obtain the following sequence of tableaus:

	x	y	u	v	w	r	s	P	Constant		Ratio
	1	1	1	0	0	0	0	0	1,200		$\frac{1200}{1} = 1200$
Pivot row →	(−1)	−1	0	1	0	0	0	0	−800		$\frac{-800}{-1} = 800$
	−1	1	0	0	1	0	0	0	200		—
	1	0	0	0	0	1	0	0	1,000		$\frac{1000}{1} = 1000$
	0	1	0	0	0	0	1	0	800		—
	5	−10	0	0	0	0	0	1	85,000		

↑ Pivot column

	x	y	u	v	w	r	s	P	Constant
	1	1	1	0	0	0	0	0	1,200
$\xrightarrow{-R_2}$	(1)	1	0	−1	0	0	0	0	800
	−1	1	0	0	1	0	0	0	200
	1	0	0	0	0	1	0	0	1,000
	0	1	0	0	0	0	1	0	800
	5	−10	0	0	0	0	0	1	85,000

	x	y	u	v	w	r	s	P	Constant		Ratio
	0	0	1	1	0	0	0	0	400		—
$\begin{matrix}R_1 - R_2\\R_3 + R_2\end{matrix}$	1	1	0	−1	0	0	0	0	800		$\frac{800}{1} = 800$
$\xrightarrow{}$ →	0	(2)	0	−1	1	0	0	0	1,000		$\frac{1000}{2} = 500$
$\begin{matrix}R_4 - R_2\\R_6 - 5R_2\end{matrix}$	0	−1	0	1	0	1	0	0	200		—
Pivot row	0	1	0	0	0	0	1	0	800		$\frac{800}{1} = 800$
	0	−15	0	5	0	0	0	1	81,000		

We now use the simplex method for standard problems.

↑ Pivot column

	x	y	u	v	w	r	s	P	Constant
	0	0	1	1	0	0	0	0	400
	1	1	0	−1	0	0	0	0	800
$\xrightarrow{\frac{1}{2}R_3}$	0	(1)	0	$-\frac{1}{2}$	$\frac{1}{2}$	0	0	0	500
	0	−1	0	1	0	1	0	0	200
	0	1	0	0	0	0	1	0	800
	0	−15	0	5	0	0	0	1	81,000

	x	y	u	v	w	r	s	P	Constant		Ratio
Pivot row →	0	0	1	(1)	0	0	0	0	400		$\frac{400}{1} = 400$
$\begin{matrix}R_2 - R_3\\R_4 + R_3\end{matrix}$	1	0	0	$-\frac{1}{2}$	$-\frac{1}{2}$	0	0	0	300		—
$\xrightarrow{}$	0	1	0	$-\frac{1}{2}$	$\frac{1}{2}$	0	0	0	500		—
$\begin{matrix}R_5 - R_3\\R_6 + 15R_3\end{matrix}$	0	0	0	$\frac{1}{2}$	$\frac{1}{2}$	1	0	0	700		$\frac{700}{1/2} = 1400$
	0	0	0	$\frac{1}{2}$	$-\frac{1}{2}$	0	1	0	300		$\frac{300}{1/2} = 600$
	0	0	0	$-\frac{5}{2}$	$\frac{15}{2}$	0	0	1	88,500		

↑ Pivot column

	x	y	u	v	w	r	s	P	Constant
	0	0	1	1	0	0	0	0	400
$R_2 + \frac{1}{2}R_1$	1	0	$\frac{1}{2}$	0	$-\frac{1}{2}$	0	0	0	500
$R_3 + \frac{1}{2}R_1$	0	1	$\frac{1}{2}$	0	$\frac{1}{2}$	0	0	0	700
$\xrightarrow{} R_4 - \frac{1}{2}R_1$	0	0	$-\frac{1}{2}$	0	$\frac{1}{2}$	1	0	0	500
$R_5 - \frac{1}{2}R_1$	0	0	$-\frac{1}{2}$	0	$-\frac{1}{2}$	0	1	0	100
$R_6 + \frac{5}{2}R_1$									
	0	0	$\frac{5}{2}$	0	$\frac{15}{2}$	0	0	1	89,500

All entries in the last row are nonnegative, and the tableau is final. We see that $x = 500$, $y = 700$, and $P = 89,500$. This tells us that plant I should manufacture 500 standard and 700 deluxe exercise bicycles and that plant II should manufacture $(1000 - 500)$, or 500, standard and $(800 - 700)$, or 100, deluxe models. Rockford's profit will then be $89,500. ∎

EXPLORE & DISCUSS

Refer to Example 5.

1. Sketch the feasible set S for the linear programming problem.

2. Solve the problem using the method of corners.

3. Indicate on S the points (both nonfeasible and feasible) corresponding to each iteration of the simplex method, and trace the path leading to the optimal solution.

4.3 Self-Check Exercises

1. Solve the following nonstandard problem using the method of this section:

$$\text{Maximize} \quad P = 2x + 3y$$
$$\text{subject to} \quad x + y \leq 40$$
$$-x + 2y \leq -10$$
$$x \geq 0, y \geq 0$$

2. A farmer has 150 acres of land suitable for cultivating crops A and B. The cost of cultivating crop A is $40/acre and that of crop B is $60/acre. The farmer has a maximum of $7400 available for land cultivation. Each acre of crop A requires 20 labor-hours, and each acre of crop B requires 25 labor-hours. The farmer has a maximum of 3300 labor-hours available. She has also decided that she will cultivate at least 70 acres of crop A. If she expects to make a profit of $150/acre on crop A and $200/acre on crop B, how many acres of each crop should she plant in order to maximize her profit?

Solutions to Self-Check Exercises 4.3 can be found on page 272.

4.3 Concept Questions

1. Explain why the following linear programming problem is *not* a standard maximization problem.

$$\text{Maximize} \quad P = 2x + 3y$$
$$\text{subject to} \quad 4x + 2y \leq 20$$
$$-2x + 3y \geq 5$$
$$x \geq 0, y \geq 0$$

2. Explain why the following linear programming problem is *not* a restricted standard minimization problem.

$$\text{Minimize} \quad C = 3x - y$$
$$\text{subject to} \quad x + y \geq 5$$
$$3x + 5y \geq 16$$
$$x \geq 0, y \geq 0$$

3. Explain why the following linear programming problem is *not* a standard maximization problem.

$$\text{Maximize} \quad P = x + 3y$$
$$\text{subject to} \quad 3x + 4y \leq 16$$
$$2x - 3y = 5$$
$$x \geq 0, y \geq 0$$

Can it be rewritten as a restricted standard minimization problem? Why or why not?

4.3   Exercises

In Exercises 1–4, rewrite each linear programming problem as a maximization problem with constraints involving inequalities of the form ≤ (with the exception of the inequalities $x \geq 0$, $y \geq 0$, and $z \geq 0$).

1. Minimize $C = 2x - 3y$

subject to $3x + 5y \geq 20$
$3x + y \leq 16$
$-2x + y \leq 1$
$x \geq 0, y \geq 0$

2. Minimize $C = 2x + 3y$

subject to $x + y \leq 10$
$x + 2y \geq 12$
$2x + y \geq 12$
$x \geq 0, y \geq 0$

3. Minimize $C = 5x + 10y + z$

subject to $2x + y + z \geq 4$
$x + 2y + 2z \geq 2$
$2x + 4y + 3z \leq 12$
$x \geq 0, y \geq 0, z \geq 0$

4. Maximize $P = 2x + y - 2z$

subject to $x + 2y + z \geq 10$
$3x + 4y + 2z \geq 5$
$2x + 5y + 12z \leq 20$
$x \geq 0, y \geq 0, z \geq 0$

In Exercises 5–20, use the method of this section to solve each linear programming problem.

5. Maximize $P = x + 2y$

subject to $2x + 5y \leq 20$
$x - 5y \leq -5$
$x \geq 0, y \geq 0$

6. Maximize $P = 2x + 3y$

subject to $x + 2y \leq 8$
$x - y \leq -2$
$x \geq 0, y \geq 0$

7. Minimize $C = -2x + y$

subject to $x + 2y \leq 6$
$3x + 2y \leq 12$
$x \geq 0, y \geq 0$

8. Minimize $C = -2x + 3y$

subject to $x + 3y \leq 60$
$2x + y \geq 45$
$x \leq 40$
$x \geq 0, y \geq 0$

9. Maximize $P = x + 4y$

subject to $x + 3y \leq 6$
$-2x + 3y \leq -6$
$x \geq 0, y \geq 0$

10. Maximize $P = 5x + y$

subject to $2x + y \leq 8$
$-x + y \geq 2$
$x \geq 0, y \geq 0$

11. Maximize $P = x + 2y$

subject to $2x + 3y \leq 12$
$-x + 3y = 3$
$x \geq 0, y \geq 0$

12. Minimize $C = x + 2y$

subject to $4x + 7y \leq 70$
$2x + y = 20$
$x \geq 0, y \geq 0$

13. Maximize $P = 5x + 4y + 2z$
   subject to $x + 2y + 3z \leq 24$
   $x - y + z \geq 6$
   $x \geq 0, y \geq 0, z \geq 0$

14. Maximize $P = x - 2y + z$
   subject to $2x + 3y + 2z \leq 12$
   $x + 2y - 3z \geq 6$
   $x \geq 0, y \geq 0, z \geq 0$

15. Minimize $C = x - 2y + z$
   subject to $x - 2y + 3z \leq 10$
   $2x + y - 2z \leq 15$
   $2x + y + 3z \leq 20$
   $x \geq 0, y \geq 0, z \geq 0$

16. Minimize $C = 2x - 3y + 4z$
   subject to $-x + 2y - z \leq 8$
   $x - 2y + 2z \leq 10$
   $2x + 4y - 3z \leq 12$
   $x \geq 0, y \geq 0, z \geq 0$

17. Maximize $P = 2x + y + z$
   subject to $x + 2y + 3z \leq 28$
   $2x + 3y - z \leq 6$
   $x - 2y + z \geq 4$
   $x \geq 0, y \geq 0, z \geq 0$

18. Minimize $C = 2x - y + 3z$
   subject to $2x + y + z \geq 2$
   $x + 3y + z \geq 6$
   $2x + y + 2z \leq 12$
   $x \geq 0, y \geq 0, z \geq 0$

19. Maximize $P = x + 2y + 3z$
   subject to $x + 2y + z \leq 20$
   $3x + y \leq 30$
   $2x + y + z = 10$
   $x \geq 0, y \geq 0, z \geq 0$

20. Minimize $C = 3x + 2y + z$
   subject to $x + 2y + z \leq 20$
   $3x + y \leq 30$
   $2x + y + z = 10$
   $x \geq 0, y \geq 0, z \geq 0$

21. AGRICULTURE—CROP PLANNING   A farmer has 150 acres of land suitable for cultivating crops A and B. The cost of cultivating crop A is $40/acre and that of crop B is $60/acre. The farmer has a maximum of $7400 available for land cultivation. Each acre of crop A requires 20 labor-hours, and each acre of crop B requires 25 labor-hours. The farmer has a maximum of 3300 labor-hours available. He has also decided that he will cultivate at least 80 acres of crop A. If he expects to make a profit of $150/acre on crop A and $200/acre on crop B, how many acres of each crop should he plant in order to maximize his profit?

22. VETERINARY SCIENCE   A veterinarian has been asked to prepare a diet for a group of dogs to be used in a nutrition study at the School of Animal Science. It has been stipulated that each serving should be no larger than 8 oz and must contain at least 29 units of nutrient I and 20 units of nutrient II. The vet has decided that the diet may be prepared from two brands of dog food: brand A and brand B. Each ounce of brand A contains 3 units of nutrient I and 4 units of nutrient II. Each ounce of brand B contains 5 units of nutrient I and 2 units of nutrient II. Brand A costs 3 cents/ounce and brand B costs 4 cents/ounce. Determine how many ounces of each brand of dog food should be used per serving to meet the given requirements at a minimum cost.

23. FINANCE—ALLOCATION OF FUNDS   The First Street branch of Capitol Bank has a sum of $60 million earmarked for home and commercial-development loans. The bank expects to realize an 8% annual rate of return on the home loans and a 6% annual rate of return on the commercial-development loans. Management has decided that the total amount of home loans is to be greater than or equal to 3 times the total amount of commercial-development loans. Owing to prior commitments, at least $10 million of the funds has been designated for commercial-development loans. Determine the amount of each type of loan the bank should extend in order to maximize its returns.

24. FINANCE—INVESTMENTS   Natsano has at most $50,000 to invest in the common stocks of two companies. He estimates that an investment in company A will yield a return of 10% whereas an investment in company B, which he feels is a riskier investment, will yield a return of 20%. If he decides that his investment in the stocks of company A is to exceed his investment in the stocks of company B by at least $20,000, determine how much he should invest in the stocks of each company in order to maximize the returns on his investment.

25. MANUFACTURING—PRODUCTION SCHEDULING   A company manufactures products A, B, and C. Each product is processed in three departments: I, II, and III. The total available labor-hours per week for departments I, II, and III are 900, 1080, and 840, respectively. The time requirements (in hours/unit) and the profit per unit for each product are as follows:

	Product A	Product B	Product C
Dept. I	2	1	2
Dept. II	3	1	2
Dept. III	2	2	1
Profit	$18	$12	$15

If management decides that the number of units of product B manufactured must equal or exceed the number of units of products A and C manufactured, how many units of each product should the company produce in order to maximize its profit?

26. MANUFACTURING—PRODUCTION SCHEDULING   Wayland Company manufactures two models of its twin-size futons, standard and deluxe, in two locations, I and II. The maximum output at location I is 600/week whereas the maximum output at location II is 400/week. The profit per futon for standard and deluxe models manufactured at location I is $30 and $20, respectively; the profit per futon for standard and deluxe models manufactured at location II is $34 and $18, respectively. For a certain week, the company has received an order for 600 standard models and 300 deluxe models. If prior commitments dictate that the number of deluxe models manufactured at location II not exceed the number of standard models manufactured there by more than 50, find how many of each model should be manufactured at each location so as to satisfy the order and at the same time maximize Wayland's profit.

27. NUTRITION—DIET PLANNING   A nutritionist at the Medical Center has been asked to prepare a special diet for certain patients. He has decided that the meals should contain a minimum of 400 mg of calcium, 10 mg of iron, and 40 mg of vitamin C. He has further decided that the meals are to be prepared from foods A and B. Each ounce of food A contains 30 mg of calcium, 1 mg of iron, 2 mg of vitamin C, and 2 mg of cholesterol. Each ounce of food B contains 25 mg of calcium, 0.5 mg of iron, 5 mg of vitamin C, and 5 mg of cholesterol. Use the method of this section to determine how many ounces of each type of food the nutritionist should use in a meal so that the cholesterol content is minimized and the minimum requirements of calcium, iron, and vitamin C are met.

28. MANUFACTURING—SHIPPING COSTS   Steinwelt Piano manufactures uprights and consoles in two plants, plant I and plant II. The output of plant I is at most 300/month, whereas the output of plant II is at most 250/month. These pianos are shipped to three warehouses that serve as distribution centers for the company. To fill current and projected orders, warehouse A requires a minimum of 200 pianos/month, warehouse B requires at least 150 pianos/month, and warehouse C requires at least 200 pianos/month. The shipping cost of each piano from plant I to warehouse A, warehouse B, and warehouse C is $60, $60, and $80, respectively, and the shipping cost of each piano from plant II to warehouse A, warehouse B, and warehouse C is $80, $70, and $50, respectively. Use the method of this section to determine the shipping schedule that will enable Steinwelt to meet the warehouses' requirements while keeping the shipping costs to a minimum.

4.3 Solutions to Self-Check Exercises

1. We are given the problem

$$\text{Maximize} \quad P = 2x + 3y$$
$$\text{subject to} \quad x + y \le 40$$
$$-x + 2y \le -10$$
$$x \ge 0, y \ge 0$$

Using the method of this section and introducing the slack variables u and v, we obtain the following tableaus:

	x	y	u	v	P	Constant	Ratio
	1	1	1	0	0	40	$\frac{40}{1} = 40$
Pivot row →	(−1)	2	0	1	0	−10	$\frac{-10}{-1} = 10$
	−2	−3	0	0	1	0	

Pivot column ↑

	x	y	u	v	P	Constant
	1	1	1	0	0	40
$-R_2 \longrightarrow$	①	−2	0	−1	0	10
	−2	−3	0	0	1	0

Pivot row →

	x	y	u	v	P	Constant	Ratio
$R_1 - R_2 \longrightarrow$	0	③	1	1	0	30	$\frac{30}{3} = 10$
$R_3 + 2R_2$	1	−2	0	−1	0	10	—
	0	−7	0	−2	1	20	

Pivot column ↑

	x	y	u	v	P	Constant
$\frac{1}{3}R_1 \longrightarrow$	0	①	$\frac{1}{3}$	$\frac{1}{3}$	0	10
	1	−2	0	−1	0	10
	0	−7	0	−2	1	20

	x	y	u	v	P	Constant
$R_2 + 2R_1 \longrightarrow$	0	1	$\frac{1}{3}$	$\frac{1}{3}$	0	10
$R_3 + 7R_1$	1	0	$\frac{2}{3}$	$-\frac{1}{3}$	0	30
	0	0	$\frac{7}{3}$	$\frac{1}{3}$	1	90

The last tableau is in final form, so we have obtained the following solution

$$x = 30 \qquad y = 10 \qquad u = 0 \qquad v = 0 \qquad P = 90$$

2. Let x denote the number of acres of crop A to be cultivated and y the number of acres of crop B to be cultivated. Since there is a total of 150 acres of land available for cultivation, we have $x + y \leq 150$. Next, the restriction on the amount of money available for land cultivation implies that $40x + 60y \leq 7400$. Similarly, the restriction on the amount of time available for labor implies that $20x + 25y \leq 3300$. Also, since the farmer will cultivate at least 70 acres of crop A, $x \geq 70$. Since the profit on each acre of crop A is \$150 and the profit on each acre of crop B is \$200, we see that the profit realizable by the farmer is $P = 150x + 200y$. Summarizing, we have the following linear programming problem:

$$\text{Maximize} \quad P = 150x + 200y$$
$$\text{subject to} \quad x + y \leq 150$$
$$40x + 60y \leq 7400$$
$$20x + 25y \leq 3300$$
$$x \geq 70$$
$$x \geq 0, y \geq 0$$

To solve this nonstandard problem, we rewrite the fourth inequality in the form

$$-x \leq -70$$

Using the method of this section with u, v, w, and z as slack variables, we have the following tableaus:

	x	y	u	v	w	z	P	Constant	Ratio
	1	1	1	0	0	0	0	150	$\frac{150}{1} = 150$
	40	60	0	1	0	0	0	7,400	$\frac{7400}{40} = 185$
	20	25	0	0	1	0	0	3,300	$\frac{3300}{20} = 165$
Pivot row →	(−1)	0	0	0	0	1	0	−70	$\frac{-70}{-1} = 70$
	−150	−200	0	0	0	0	1	0	

↑ Pivot column

$-R_4$ →

	x	y	u	v	w	z	P	Constant
	1	1	1	0	0	0	0	150
	40	60	0	1	0	0	0	7,400
	20	25	0	0	1	0	0	3,300
	(1)	0	0	0	0	−1	0	70
	−150	−200	0	0	0	0	1	0

$R_1 - R_4$, $R_2 - 40R_4$, $R_3 - 20R_4$, $R_5 + 150R_4$ →

	x	y	u	v	w	z	P	Constant	Ratio
	0	1	1	0	0	1	0	80	$\frac{80}{1} = 80$
	0	60	0	1	0	40	0	4,600	$\frac{4600}{60} = 76\frac{2}{3}$
	0	(25)	0	0	1	20	0	1,900	$\frac{1900}{25} = 76$
	1	0	0	0	0	−1	0	70	—
Pivot row	0	−200	0	0	0	−150	1	10,500	

↑ Pivot column

$\frac{1}{25}R_3$ →

	x	y	u	v	w	z	P	Constant
	0	1	1	0	0	1	0	80
	0	60	0	1	0	40	0	4,600
	0	(1)	0	0	$\frac{1}{25}$	$\frac{4}{5}$	0	76
	1	0	0	0	0	−1	0	70
	0	−200	0	0	0	−150	1	10,500

$R_1 - R_3$, $R_2 - 60R_3$, $R_5 + 200R_3$ →

	x	y	u	v	w	z	P	Constant
	0	0	1	0	$-\frac{1}{25}$	$\frac{1}{5}$	0	4
	0	0	0	1	$-\frac{12}{5}$	−8	0	40
	0	1	0	0	$\frac{1}{25}$	$\frac{4}{5}$	0	76
	1	0	0	0	0	−1	0	70
	0	0	0	0	8	10	1	25,700

This last tableau is in final form, and the solution is

$$x = 70 \quad y = 76 \quad u = 4 \quad v = 40$$
$$w = 0 \quad z = 0 \quad P = 25,700$$

Thus, by cultivating 70 acres of crop A and 76 acres of crop B, the farmer will attain a maximum profit of \$25,700.

CHAPTER 4   Summary of Principal Terms

TERMS

standard maximization problem (216)

slack variable (217)

basic variable (218)

nonbasic variable (218)

basic solution (218)

pivot column (220)

pivot row (220)

pivot element (220)

simplex tableau (220)

simplex method (222)

standard minimization problem (244)

primal problem (244)

dual problem (244)

nonstandard problem (260)

mixed constraints (261)

CHAPTER 4   Concept Review Questions

Fill in the blanks.

1. In a standard maximization problem: the objective function is to be _____; all the variables involved in the problem are _____; and each linear constraint may be written so that the expression involving the variables is _____ _____ or _____ _____ a nonnegative constant.

2. In setting up the initial simplex tableau, we first transform the system of linear inequalities into a system of linear _____, using _____ _____; the objective function is rewritten so that it has the form _____ and then is placed _____ the system of linear equations obtained earlier. Finally, the initial simplex tableau is the

_____ matrix associated with this system of linear equations.

3. In a standard minimization problem: the objective function is to be _____; all the variables involved in the problem are _____; and each linear constraint may be written so that the expression involving the variables is _____ _____ or _____ _____ a constant.

4. The fundamental theorem of duality states that a primal problem has a solution if and only if the corresponding _____ problem has a solution. If a solution exists, then the _____ function of both the primal and dual problem attain the same _____ _____.

CHAPTER 4   Review Exercises

In Exercises 1–12, use the simplex method to solve each linear programming problem.

1. Maximize $P = 3x + 4y$
   subject to $x + 3y \le 15$
   $4x + \;\; y \le 16$
   $x \ge 0, y \ge 0$

2. Maximize $P = 2x + 5y$
   subject to $2x + \;\; y \le 16$
   $2x + 3y \le 24$
   $y \le 6$
   $x \ge 0, y \ge 0$

3. Maximize $P = 2x + 3y + 5z$
   subject to $x + 2y + 3z \le 12$
   $x - 3y + 2z \le 10$
   $x \ge 0, y \ge 0, z \ge 0$

4. Maximize $P = x + 2y + 3z$
   subject to $2x + \;\; y + \;\; z \le 14$
   $3x + 2y + 4z \le 24$
   $2x + 5y - 2z \le 10$
   $x \ge 0, y \ge 0, z \ge 0$

5. Minimize $C = 3x + 2y$
   subject to $2x + 3y \ge 6$
   $2x + \;\; y \ge 4$
   $x \ge 0, y \ge 0$

6. Minimize $C = x + 2y$
   subject to $3x + \;\; y \ge 12$
   $x + 4y \ge 16$
   $x \ge 0, y \ge 0$

7. Minimize $C = 24x + 18y + 24z$
   subject to $3x + 2y + \;\; z \ge 4$
   $x + \;\; y + 3z \ge 6$
   $x \ge 0, y \ge 0, z \ge 0$

8. Minimize $C = 4x + 2y + 6z$
   subject to $x + 2y + \;\; z \ge 4$
   $2x + \;\; y + 2z \ge 2$
   $3x + 2y + \;\; z \ge 3$
   $x \ge 0, y \ge 0, z \ge 0$

9. Maximize $P = 3x - 4y$
   subject to $x + y \le 45$
   $x - 2y \ge 10$
   $x \ge 0, y \ge 0$

10. Minimize $C = 2x + 3y$
   subject to $x + y \le 10$
   $x + 2y \ge 12$
   $2x + y \ge 12$
   $x \ge 0, y \ge 0$

11. Maximize $P = 2x + 3y$
   subject to $2x + 5y \le 20$
   $-x + 5y \ge 5$
   $x \ge 0, y \ge 0$

12. Minimize $C = -3x - 4y$
   subject to $x + y \le 45$
   $x + y \ge 15$
   $x \le 30$
   $y \le 25$
   $x \ge 0, y \ge 0$

13. MINING—PRODUCTION   Perth Mining Company operates two mines for the purpose of extracting gold and silver. The Saddle Mine costs $14,000/day to operate, and it yields 50 oz of gold and 3000 oz of silver each day. The Horseshoe Mine costs $16,000/day to operate, and it yields 75 oz of gold and 1000 oz of silver each day. Company management has set a target of at least 650 oz of gold and 18,000 oz of silver. How many days should each mine be operated at so that the target can be met at a minimum cost to the company? What is the minimum cost?

14. INVESTMENT ANALYSIS   Jorge has decided to invest at most $100,000 in securities in the form of corporate stocks. He has classified his options into three groups of stocks: blue-chip stocks that he assumes will yield a 10% return (dividends and capital appreciation) within a year, growth stocks that he assumes will yield a 15% return within a year, and speculative stocks that he assumes will yield a 20% return (mainly due to capital appreciation) within a year. Because of the relative risks involved in his investment, Jorge has further decided that no more than 30% of his investment should be in growth and speculative stocks and at least 50% of his investment should be in blue-chip and speculative stocks. Determine how much Jorge should invest in each group of stocks in the hope of maximizing the return on his investments.

15. MAXIMIZING PROFITS   A company manufactures three products, A, B, and C, on two machines, I and II. It has been determined that the company will realize a profit of $4/unit of product A, $6/unit of product B, and $8/unit of product C. Manufacturing a unit of product A requires 9 min on machine I and 6 min on machine II; manufacturing a unit of product B requires 12 min on machine I and 6 min on machine II; manufacturing a unit of product C requires 18 min on machine I and 10 min on machine II. There are 6 hr of machine time available on machine I and 4 hr of machine time available on machine II in each work shift. How many units of each product should be produced in each shift in order to maximize the company's profit?

16. INVESTMENT ANALYSIS   Sandra has at most $200,000 to invest in stocks, bonds, and money market funds. She expects annual yields of 15%, 10%, and 8%, respectively, on these investments. If Sandra wants at least $50,000 to be invested in money market funds and requires that the amount invested in bonds be greater than or equal to the sum of her investments in stocks and money market funds, determine how much she should invest in each vehicle in order to maximize the return on her investments.

CHAPTER 4   **Before Moving On . . .**

1. Consider the following linear programming problem:

   Maximize $P = x + 2y - 3z$
   subject to $2x + y - z \le 3$
   $x - 2y + 3z \le 1$
   $3x + 2y + 4z \le 17$
   $x \ge 0, y \ge 0, z \ge 0$

Write the initial simplex tableau for the problem and identify the pivot element to be used in the first iteration of the simplex method.

2. The following simplex tableau is in final form. Find the solution to the linear programming problem associated with this tableau.

x	y	z	u	v	w	P	Constant
0	$\frac{1}{2}$	0	1	$-\frac{1}{2}$	0	0	2
0	$\frac{1}{4}$	1	0	$\frac{5}{4}$	$-\frac{1}{2}$	0	11
1	$\frac{1}{4}$	0	0	$-\frac{3}{4}$	$\frac{1}{2}$	0	2
0	$\frac{13}{4}$	0	0	$\frac{1}{4}$	$\frac{1}{2}$	1	28

3. Using the simplex method, solve the following linear programming problem:

   Maximize $P = 5x + 2y$
   subject to $4x + 3y \le 30$
   $2x - 3y \le 6$
   $x \ge 0, y \ge 0$

4. Using the simplex method, solve the following linear programming problem:

$$\begin{aligned} \text{Minimize} \quad & C = x + 2y \\ \text{subject to} \quad & x + y \geq 3 \\ & 2x + 3y \geq 6 \\ & x \geq 0, y \geq 0 \end{aligned}$$

5. **Mixed Constraints (Optional)** Using the simplex method, solve the following linear programming problem:

$$\begin{aligned} \text{Maximize} \quad & P = 2x + y \\ \text{subject to} \quad & 2x + 5y \leq 20 \\ & 4x + 3y \geq 16 \\ & x \geq 0, y \geq 0 \end{aligned}$$

5 Mathematics of Finance

© Kim Kulish/Corbis

How much will the home mortgage payment be? The Blakelys received a bank loan to help finance the purchase of a house. They have agreed to repay the loan in equal monthly installments over a certain period of time. In Example 2, page 310, we will show how to determine the size of the monthly installment so that the loan is fully amortized at the end of the term.

INTEREST THAT IS periodically added to the principal and thereafter itself earns interest is called *compound interest*. We begin this chapter by deriving the *compound interest formula*, which gives the amount of money accumulated when an initial amount of money is invested in an account for a fixed term and earns compound interest.

An *annuity* is a sequence of payments made at regular intervals. We derive formulas giving the *future value of an annuity* (what you end up with) and the *present value of an annuity* (the lump sum that, when invested now, will yield the same future value as that of the annuity). Then, using these formulas, we answer questions involving the amortization of certain types of installment loans and questions involving *sinking funds* (funds that are set up to be used for a specific purpose at a future date).

5.1   Compound Interest

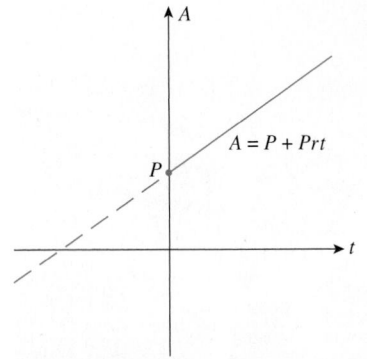

FIGURE 1
The accumulated amount is a linear function of *t*.

■ Simple Interest

A natural application of linear functions to the business world is found in the computation of **simple interest**—interest that is computed on the original principal only. Thus, if *I* denotes the interest on a principal *P* (in dollars) at an interest rate of *r* per year for *t* years, then we have

$$I = Prt$$

The **accumulated amount** *A*, the sum of the principal and interest after *t* years, is given by

$$A = P + I = P + Prt$$
$$= P(1 + rt)$$

and is a linear function of *t* (see Exercise 40 at the end of this section). In business applications, we are normally interested only in the case where *t* is positive, so only that part of the line that lies in Quadrant I is of interest to us (Figure 1).

Simple Interest Formulas		
Interest:	$I = Prt$	**(1a)**
Accumulated amount:	$A = P(1 + rt)$	**(1b)**

EXAMPLE 1 A bank pays simple interest at the rate of 8% per year for certain deposits. If a customer deposits $1000 and makes no withdrawals for 3 years, what is the total amount on deposit at the end of 3 years? What is the interest earned in that period of time?

Solution   Using Equation (1b) with $P = 1000$, $r = 0.08$, and $t = 3$, we see that the total amount on deposit at the end of 3 years is given by

$$A = P(1 + rt)$$
$$= 1000[1 + (0.08)(3)] = 1240$$

or $1240.
     The interest earned over the 3-year period is given by

$$I = Prt \qquad \text{Use (1a).}$$
$$= 1000(0.08)(3) = 240$$

or $240.   ■

EXPLORING WITH TECHNOLOGY

Refer to Example 1. Use a graphing utility to plot the graph of the function $A = 1000(1 + 0.08t)$, using the viewing window $[0, 10] \times [0, 2000]$.

1. What is the *A*-intercept of the straight line, and what does it represent?

2. What is the slope of the straight line, and what does it represent? (See Exercise 40.)

EXAMPLE 2 An amount of $2000 is invested in a 10-year trust fund that pays 6% annual simple interest. What is the total amount of the trust fund at the end of 10 years?

Solution The total amount of the trust fund at the end of 10 years is given by

$$A = P(1 + rt)$$
$$= 2000[1 + (0.06)(10)] = 3200$$

or $3200. ∎

■ Compound Interest

In contrast to simple interest, **compound interest** is earned interest that is periodically added to the principal and thereafter itself earns interest at the same rate. To find a formula for the accumulated amount, let's consider a numerical example. Suppose $1000 (the principal) is deposited in a bank for a term of 3 years, earning interest at the rate of 8% per year (called the **nominal**, or **stated**, **rate**) compounded annually. Then, using Equation (1b) with $P = 1000$, $r = 0.08$, and $t = 1$, we see that the accumulated amount at the end of the first year is

$$A_1 = P(1 + rt)$$
$$= 1000[1 + (0.08)(1)] = 1000(1.08) = 1080$$

or $1080.

To find the accumulated amount A_2 at the end of the second year, we use (1b) once again, this time with $P = A_1$. (Remember, the principal *and* interest now earn interest over the second year.) We obtain

$$A_2 = P(1 + rt) = A_1(1 + rt)$$
$$= 1000[1 + 0.08(1)][1 + 0.08(1)]$$
$$= 1000[1 + 0.08]^2 = 1000(1.08)^2 = 1166.40$$

or $1166.40.

Finally, the accumulated amount A_3 at the end of the third year is found using (1b) with $P = A_2$, giving

$$A_3 = P(1 + rt) = A_2(1 + rt)$$
$$= 1000[1 + 0.08(1)]^2[1 + 0.08(1)]$$
$$= 1000[1 + 0.08]^3 = 1000(1.08)^3 \approx 1259.71$$

or approximately $1259.71.

If you reexamine our calculations, you will see that the accumulated amounts at the end of each year have the following form.

First year: $A_1 = 1000(1 + 0.08)$, or $A_1 = P(1 + r)$

Second year: $A_2 = 1000(1 + 0.08)^2$, or $A_2 = P(1 + r)^2$

Third year: $A_3 = 1000(1 + 0.08)^3$, or $A_3 = P(1 + r)^3$

These observations suggest the following general result: If P dollars is invested over a term of t years, earning interest at the rate of r per year compounded annually, then the accumulated amount is

$$A = P(1 + r)^t \qquad (2)$$

Formula (2) was derived under the assumption that interest was compounded *annually*. In practice, however, interest is usually compounded more than once a year. The interval of time between successive interest calculations is called the **conversion period.**

If interest at a nominal rate of r per year is compounded m times a year on a principal of P dollars, then the simple interest rate per conversion period is

$$i = \frac{r}{m} \qquad \frac{\text{Annual interest rate}}{\text{Periods per year}}$$

For example, if the nominal interest rate is 8% per year ($r = 0.08$) and interest is compounded quarterly ($m = 4$), then

$$i = \frac{r}{m} = \frac{0.08}{4} = 0.02$$

or 2% per period.

To find a general formula for the accumulated amount when a principal of P dollars is deposited in a bank for a term of t years and earns interest at the (nominal) rate of r per year compounded m times per year, we proceed as before, using (1b) repeatedly with the interest rate $i = r/m$. We see that the accumulated amount at the end of each period is

First period: $A_1 = P(1 + i)$

Second period: $A_2 = A_1(1 + i)$ $= [P(1 + i)](1 + i) = P(1 + i)^2$

Third period: $A_3 = A_2(1 + i)$ $= [P(1 + i)^2](1 + i) = P(1 + i)^3$

$\vdots$ $\vdots$

nth period: $A_n = A_{n-1}(1 + i) = [P(1 + i)^{n-1}](1 + i) = P(1 + i)^n$

There are $n = mt$ periods in t years (number of conversion periods times the term). Hence the accumulated amount at the end of t years is given by

$$A = P(1 + i)^n$$

Compound Interest Formula (Accumulated Amount)

$$A = P(1 + i)^n \tag{3}$$

where $i = \dfrac{r}{m}$, $n = mt$, and

A = Accumulated amount at the end of n conversion periods

P = Principal

r = Nominal interest rate per year

m = Number of conversion periods per year

t = Term (number of years)

EXPLORING WITH TECHNOLOGY

Let $A_1(t)$ denote the accumulated amount of $100 earning simple interest at the rate of 10% per year over t years, and let $A_2(t)$ denote the accumulated amount of $100 earning interest at the rate of 10% per year compounded monthly over t years.

1. Find expressions for $A_1(t)$ and $A_2(t)$.

2. Use a graphing utility to plot the graphs of A_1 and A_2 on the same set of axes, using the viewing window $[0, 20] \times [0, 800]$.

3. Comment on the growth of $A_1(t)$ and $A_2(t)$ by referring to the graphs of A_1 and A_2.

EXAMPLE 3 Find the accumulated amount after 3 years if $1000 is invested at 8% per year compounded (a) annually, (b) semiannually, (c) quarterly, (d) monthly, and (e) daily.

Solution

a. Here, $P = 1000$, $r = 0.08$, and $m = 1$. Thus, $i = r = 0.08$ and $n = 3$, so Equation (3) gives

$$A = 1000(1 + 0.08)^3$$
$$\approx 1259.71$$

or $1259.71.

b. Here, $P = 1000$, $r = 0.08$, and $m = 2$. Thus, $i = \frac{0.08}{2}$ and $n = (3)(2) = 6$, so Equation (3) gives

$$A = 1000\left(1 + \frac{0.08}{2}\right)^6$$
$$\approx 1265.32$$

or $1265.32.

c. In this case, $P = 1000$, $r = 0.08$, and $m = 4$. Thus, $i = \frac{0.08}{4}$ and $n = (3)(4) = 12$, so Equation (3) gives

$$A = 1000\left(1 + \frac{0.08}{4}\right)^{12}$$
$$\approx 1268.24$$

or $1268.24

d. Here, $P = 1000$, $r = 0.08$, and $m = 12$. Thus, $i = \frac{0.08}{12}$ and $n = (3)(12) = 36$, so Equation (3) gives

$$A = 1000\left(1 + \frac{0.08}{12}\right)^{36}$$
$$\approx 1270.24$$

or $1270.24.

e. Here, $P = 1000$, $r = 0.08$, $m = 365$, and $t = 3$. Thus, $i = \frac{0.08}{365}$ and $n = (3)(365) = 1095$, so Equation (3) gives

$$A = 1000\left(1 + \frac{0.08}{365}\right)^{1095}$$
$$\approx 1271.22$$

or $1271.22. These results are summarized in Table 1.

TABLE 1

Nominal Rate, r	Conversion Period	Interest Rate/ Conversion Period	Initial Investment	Accumulated Amount
8%	Annual ($m = 1$)	8%	$1000	$1259.71
8	Semiannual ($m = 2$)	4	1000	1265.32
8	Quarterly ($m = 4$)	2	1000	1268.24
8	Monthly ($m = 12$)	2/3	1000	1270.24
8	Daily ($m = 365$)	8/365	1000	1271.22

EXPLORING WITH TECHNOLOGY

Investments that are allowed to grow over time can increase in value surprisingly fast. Consider the potential growth of $10,000 if earnings are reinvested. More specifically, suppose $A_1(t)$, $A_2(t)$, $A_3(t)$, $A_4(t)$, and $A_5(t)$ denote the accumulated values of an investment of $10,000 over a term of t years and earning interest at the rate of 4%, 6%, 8%, 10%, and 12% per year compounded annually.

1. Find expressions for $A_1(t)$, $A_2(t)$, . . . , $A_5(t)$.

2. Use a graphing utility to plot the graphs of $A_1, A_2, \ldots, A_5$ on the same set of axes, using the viewing window $[0, 20] \times [0, 100{,}000]$.

3. Use TRACE to find $A_1(20)$, $A_2(20)$, . . . , $A_5(20)$ and then interpret your results.

■ Continuous Compounding of Interest

One question that arises naturally in the study of compound interest is: What happens to the accumulated amount over a fixed period of time if the interest is computed more and more frequently?

Intuition suggests that the more often interest is compounded, the larger the accumulated amount will be. This is confirmed by the results of Example 3, where we found that the accumulated amounts did in fact increase when we increased the number of conversion periods per year.

This leads us to another question: Does the accumulated amount keep growing, or does it approach a fixed number when the interest is computed more and more frequently over a fixed period of time?

To answer this question, let's look again at the compound interest formula:

$$A = P(1 + i)^n = P\left(1 + \frac{r}{m}\right)^{mt} \tag{4}$$

Recall that m is the number of conversion periods per year. So to find an answer to our question, we should let m get larger and larger in (4). If we let $u = \frac{m}{r}$ so that $m = ru$, then (4) becomes

$$A = P\left(1 + \frac{1}{u}\right)^{urt} \qquad \frac{r}{m} = \frac{1}{u}$$

$$= P\left[\left(1 + \frac{1}{u}\right)^u\right]^{rt} \qquad \text{Since } a^{xy} = (a^x)^y$$

TABLE 2

u	$\left(1 + \dfrac{1}{u}\right)^u$
10	2.59374
100	2.70481
1000	2.71692
10,000	2.71815
100,000	2.71827
1,000,000	2.71828

Now let's see what happens to the expression

$$\left(1 + \frac{1}{u}\right)^u$$

as u gets larger and larger. From Table 2 you can see that, as u increases,

$$\left(1 + \frac{1}{u}\right)^u$$

seems to approach the number 2.71828 (we have rounded all our calculations to five decimal places).

It can be shown—although we will not do so here—that as u gets larger and larger, the value of the expression $(1 + \frac{1}{u})^u$ approaches the irrational number 2.71828. . . , which we denote by e. (See the Exploring with Technology exercise that follows.)

EXPLORING WITH TECHNOLOGY

To obtain a visual confirmation of the fact that the expression $(1 + \frac{1}{u})^u$ approaches the number $e = 2.71828 \ldots$ as u gets larger and larger, plot the graph of $f(x) = (1 + \frac{1}{x})^x$ in a suitable viewing window and observe that $f(x)$ approaches 2.71828. . . as x gets larger and larger. Use ZOOM and TRACE to find the value of $f(x)$ for large values of x.

Using this result, we can see that, as m gets larger and larger, A approaches $P(e)^{rt} = Pe^{rt}$. In this situation, we say that interest is *compounded continuously*. Let's summarize this important result.

Continuous Compound Interest Formula

$$A = Pe^{rt} \qquad\qquad (5)$$

where

P = Principal

r = Annual interest rate compounded continuously

t = Time in years

A = Accumulated amount at the end of t years

EXAMPLE 4 Find the accumulated amount after 3 years if $1000 is invested at 8% per year compounded (a) daily (assume a 365-day year) and (b) continuously.

Solution

a. Use Formula (3) with $P = 1000$, $r = 0.08$, $m = 365$, and $t = 3$. Thus, $i = 0.08/365$ and $n = (365)(3) = 1095$, so that

$$A = 1000\left(1 + \frac{0.08}{365}\right)^{(365)(3)} \approx 1271.22$$

or $1271.22.

b. Here we use Formula (5) with $P = 1000$, $r = 0.08$, and $t = 3$, obtaining

$$A = 1000e^{(0.08)(3)}$$
$$\approx 1271.25$$

or $1271.25.

Observe that the accumulated amounts corresponding to interest compounded daily and interest compounded continuously differ by very little. The continuous compound interest formula is a very important tool in theoretical work in financial analysis.

Effective Rate of Interest

Example 3 showed that the interest actually earned on an investment depends on the frequency with which the interest is compounded. Thus the stated, or nominal, rate of 8% per year does not reflect the actual rate at which interest is earned. This suggests that we need to find a common basis for comparing interest rates. One such way of comparing interest rates is provided by the use of the *effective rate*. The effective rate is the *simple* interest rate that would produce the same accumulated amount in 1 year as the nominal rate compounded m times a year. The effective rate is also called the **effective annual yield**.

To derive a relationship between the nominal interest rate, r per year compounded m times, and its corresponding effective rate, R per year, let's assume an initial investment of P dollars. Then the accumulated amount after 1 year at a simple interest rate of R per year is

$$A = P(1 + R)$$

Also, the accumulated amount after 1 year at an interest rate of r per year compounded m times a year is

$$A = P(1 + i)^n = P\left(1 + \frac{r}{m}\right)^m \qquad \text{Since } i = \frac{r}{m}$$

Equating the two expressions gives

$$P(1 + R) = P\left(1 + \frac{r}{m}\right)^m$$

$$1 + R = \left(1 + \frac{r}{m}\right)^m \qquad \text{Divide both sides by } P.$$

If we solve the preceding equation for R, we obtain the following formula for computing the effective rate of interest.

Effective Rate of Interest Formula

$$r_{\text{eff}} = \left(1 + \frac{r}{m}\right)^m - 1 \qquad\qquad (6)$$

where

$r_{\text{eff}} = $ Effective rate of interest

$r = $ Nominal interest rate per year

$m = $ Number of conversion periods per year

EXAMPLE 5 Find the effective rate of interest corresponding to a nominal rate of 8% per year compounded (a) annually, (b) semiannually, (c) quarterly, (d) monthly, and (e) daily.

Solution

a. The effective rate of interest corresponding to a nominal rate of 8% per year compounded annually is, of course, given by 8% per year. This result is also confirmed by using Equation (6) with $r = 0.08$ and $m = 1$. Thus,

$$r_{eff} = (1 + 0.08) - 1 = 0.08$$

b. Let $r = 0.08$ and $m = 2$. Then Equation (6) yields

$$r_{eff} = \left(1 + \frac{0.08}{2}\right)^2 - 1$$
$$= (1.04)^2 - 1$$
$$= 0.0816$$

so the effective rate is 8.16% per year.

c. Let $r = 0.08$ and $m = 4$. Then Equation (6) yields

$$r_{eff} = \left(1 + \frac{0.08}{4}\right)^4 - 1$$
$$= (1.02)^4 - 1$$
$$\approx 0.08243$$

so the corresponding effective rate in this case is 8.243% per year.

d. Let $r = 0.08$ and $m = 12$. Then Equation (6) yields

$$r_{eff} = \left(1 + \frac{0.08}{12}\right)^{12} - 1$$
$$\approx 0.08300$$

so the corresponding effective rate in this case is 8.3% per year.

e. Let $r = 0.08$ and $m = 365$. Then Equation (6) yields

$$r_{eff} = \left(1 + \frac{0.08}{365}\right)^{365} - 1$$
$$\approx 0.08328$$

so the corresponding effective rate in this case is 8.328% per year. ▪

If the effective rate of interest r_{eff} is known, then the accumulated amount after t years on an investment of P dollars may be more readily computed by using the formula

$$A = P(1 + r_{eff})^t$$

The 1968 Truth in Lending Act passed by Congress requires that the effective rate of interest be disclosed in all contracts involving interest charges. The passage of this act has benefited consumers because they now have a common basis for comparing the various nominal rates quoted by different financial institutions. Furthermore, knowing the effective rate enables consumers to compute the actual charges involved in a transaction. Thus, if the effective rates of interest found in

---

EXPLORE & DISCUSS

Recall the effective rate of interest formula:

$$r_{eff} = \left(1 + \frac{r}{m}\right)^m - 1$$

1. Show that

$$r = m[(1 + r_{eff})^{1/m} - 1]$$

2. A certificate of deposit (CD) is known to have an effective rate of 5.3%. If interest is compounded monthly, find the nominal rate of interest by using the result of part 1.

Example 4 were known, then the accumulated values of Example 3 could have been readily found (see Table 3).

TABLE 3

Nominal Rate, r	Frequency of Interest Payment	Effective Rate	Initial Investment	Accumulated Amount after 3 Years	
8%	Annually	8%	$1000	$1000(1 + 0.08)^3$	≈ $1259.71
8	Semiannually	8.16	1000	$1000(1 + 0.0816)^3$	≈ 1265.32
8	Quarterly	8.243	1000	$1000(1 + 0.08243)^3$	≈ 1268.23
8	Monthly	8.300	1000	$1000(1 + 0.08300)^3$	≈ 1270.24
8	Daily	8.328	1000	$1000(1 + 0.08328)^3$	≈ 1271.22

Present Value

Let's return to the compound interest Formula (3), which expresses the accumulated amount at the end of n periods when interest at the rate of r is compounded m times a year. The principal P in (3) is often referred to as the **present value**, and the accumulated value A is called the **future value**, since it is realized at a future date. In certain instances an investor may wish to determine how much money he should invest now, at a fixed rate of interest, so that he will realize a certain sum at some future date. This problem may be solved by expressing P in terms of A. Thus, from (3) we find

$$P = A(1 + i)^{-n}$$

Here, as before, $i = r/m$, where m is the number of conversion periods per year.

Present Value Formula for Compound Interest

$$P = A(1 + i)^{-n} \tag{7}$$

EXAMPLE 6   How much money should be deposited in a bank paying interest at the rate of 6% per year compounded monthly so that, at the end of 3 years, the accumulated amount will be $20,000?

Solution   Here, $r = 0.06$ and $m = 12$, so $i = \frac{0.06}{12}$ and $n = (3)(12) = 36$. Thus, the problem is to determine P given that $A = 20,000$. Using Equation (7), we obtain

$$P = 20,000\left(1 + \frac{0.06}{12}\right)^{-36}$$
$$\approx 16,713$$

or $16,713.

EXAMPLE 7   Find the present value of $49,158.60 due in 5 years at an interest rate of 10% per year compounded quarterly.

Solution Using (7) with $r = 0.1$ and $m = 4$, so that $i = \frac{0.1}{4}$, $n = (4)(5) = 20$, and $A = 49{,}158.6$, we obtain

$$P = (49{,}158.6)\left(1 + \frac{0.1}{4}\right)^{-20} \approx 30{,}000.07$$

or approximately \$30,000.

If we solve Formula (5) for P, we have

$$A = Pe^{rt}$$

and

$$P = Ae^{-rt} \qquad\qquad \textbf{(8)}$$

which gives the present value in terms of the future (accumulated) value for the case of continuous compounding.

APPLIED EXAMPLE 8   Real Estate Investment Blakely Investment Company owns an office building located in the commercial district of a city. As a result of the continued success of an urban renewal program, local business is enjoying a miniboom. The market value of Blakely's property is

$$V(t) = 300{,}000e^{\sqrt{t}/2}$$

where $V(t)$ is measured in dollars and t is the time in years from the present. If the expected rate of inflation is 9% compounded continuously for the next 10 years, find an expression for the present value $P(t)$ of the market price of the property that will be valid for the next 10 years. Compute $P(7)$, $P(8)$, and $P(9)$ and then interpret your results.

Solution Using Formula (8) with $A = V(t)$ and $r = 0.09$, we find that the present value of the market price of the property t years from now is

$$\begin{aligned}
P(t) &= V(t)e^{-0.09t} \\
&= 300{,}000e^{-0.09t + \sqrt{t}/2} \qquad (0 \le t \le 10)
\end{aligned}$$

Letting $t = 7$, 8, and 9, respectively, we find that

$$P(7) = 300{,}000e^{-0.09(7) + \sqrt{7}/2} \approx 599{,}837, \text{ or } \$599{,}837$$
$$P(8) = 300{,}000e^{-0.09(8) + \sqrt{8}/2} \approx 600{,}640, \text{ or } \$600{,}640$$
$$P(9) = 300{,}000e^{-0.09(9) + \sqrt{9}/2} \approx 598{,}115, \text{ or } \$598{,}115$$

From the results of these computations, we see that the present value of the property's market price seems to decrease after a certain period of growth. This suggests that there is an optimal time for the owners to sell. You can show that the highest present value of the property's market value is \$600,779, and that it occurs at time $t \approx 7.72$ years, by sketching the graph of the function P.

The returns on certain investments such as zero coupon certificates of deposit (CDs) and zero coupon bonds are compared by quoting the time it takes for each investment to triple, or even quadruple. These calculations make use of the compound interest Formula (3).

APPLIED EXAMPLE 9   Investment Options   Jane has narrowed her investment options down to two:

1. Purchase a CD that matures in 12 years and pays interest upon maturity at the rate of 10% per year compounded daily (assume 365 days in a year).
2. Purchase a zero coupon CD that will triple her investment in the same period.

Which option will optimize her investment?

Solution   Let's compute the accumulated amount under option 1. Here,

$$r = 0.10 \qquad m = 365 \qquad t = 12$$

so $n = 12(365) = 4380$ and $i = 0.10/365$. The accumulated amount at the end of 12 years (after 4380 conversion periods) is

$$A = P\left(1 + \frac{0.10}{365}\right)^{4380} \approx 3.32P$$

or $3.32P. If Jane chooses option 2, the accumulated amount of her investment after 12 years will be $3P. Therefore, she should choose option 1. ∎

APPLIED EXAMPLE 10   IRAs   Moesha has an Individual Retirement Account (IRA) with a brokerage firm. Her money is invested in a money market mutual fund that pays interest on a daily basis. Over a 2-year period in which no deposits or withdrawals were made, her account grew from $4500 to $5268.24. Find the effective rate at which Moesha's account was earning interest over that period (assume 365 days in a year).

Solution   Let r_{eff} denote the required effective rate of interest. We have

$$5268.24 = 4500(1 + r_{eff})^2$$
$$(1 + r_{eff})^2 = 1.17072$$
$$1 + r_{eff} \approx 1.081998 \qquad \text{Take the square root on both sides.}$$

or $r_{eff} = 0.081998$. Therefore, the effective rate was 8.20% per year. ∎

5.1   Self-Check Exercises

1. Find the present value of $20,000 due in 3 yr at an interest rate of 12%/yr compounded monthly.

2. Paul is a retiree living on Social Security and the income from his investment. Currently, his $100,000 investment in a 1-yr CD is yielding 10.6% interest compounded daily. If he rein-

vests the principal ($100,000) on the due date of the CD in another 1-yr CD paying 9.2% interest compounded daily, find the net decrease in his yearly income from his investment.

Solutions to Self-Check Exercises 5.1 can be found on page 292.

5.1   Concept Questions

1. Explain the difference between simple interest and compound interest.

2. What is the difference between the accumulated amount (future value) and the present value of an investment?

3. What is the effective rate of interest?

5.1 Exercises

1. Find the simple interest on a $500 investment made for 2 yr at an interest rate of 8%/yr. What is the accumulated amount?

2. Find the simple interest on a $1000 investment made for 3 yr at an interest rate of 5%/yr. What is the accumulated amount?

3. Find the accumulated amount at the end of 9 mo on an $800 deposit in a bank paying simple interest at a rate of 6%/yr.

4. Find the accumulated amount at the end of 8 mo on a $1200 bank deposit paying simple interest at a rate of 7%/yr.

5. If the accumulated amount is $1160 at the end of 2 yr and the simple rate of interest is 8%/yr, then what is the principal?

6. A bank deposit paying simple interest at the rate of 5%/yr grew to a sum of $3100 in 10 mo. Find the principal.

7. How many days will it take for a sum of $1000 to earn $20 interest if it is deposited in a bank paying ordinary simple interest at the rate of 5%/yr? (Use a 365-day year.)

8. How many days will it take for a sum of $1500 to earn $25 interest if it is deposited in a bank paying 5%/yr? (Use a 365-day year.)

9. A bank deposit paying simple interest grew from an initial sum of $1000 to a sum of $1075 in 9 mo. Find the interest rate.

10. Determine the simple interest rate at which $1200 will grow to $1250 in 8 mo.

In Exercises 11–20, find the accumulated amount A if the principal P is invested at the interest rate of r per year for t years.

11. $P = \$1000$, $r = 7\%$, $t = 8$, compounded annually

12. $P = \$1000$, $r = 8\frac{1}{2}\%$, $t = 6$, compounded annually

13. $P = \$2500$, $r = 7\%$, $t = 10$, compounded semiannually

14. $P = \$2500$, $r = 9\%$, $t = 10\frac{1}{2}$, compounded semiannually

15. $P = \$12,000$, $r = 8\%$, $t = 10\frac{1}{2}$, compounded quarterly

16. $P = \$42,000$, $r = 7\frac{3}{4}\%$, $t = 8$, compounded quarterly

17. $P = \$150,000$, $r = 14\%$, $t = 4$, compounded monthly

18. $P = \$180,000$, $r = 9\%$, $t = 6\frac{1}{4}$, compounded monthly

19. $P = \$150,000$, $r = 12\%$, $t = 3$, compounded daily

20. $P = \$200,000$, $r = 8\%$, $t = 4$, compounded daily

In Exercises 21–24, find the effective rate corresponding to the given nominal rate.

21. 10%/yr compounded semiannually

22. 9%/yr compounded quarterly

23. 8%/yr compounded monthly

24. 8%/yr compounded daily

In Exercises 25–28, find the present value of $40,000 due in 4 years at the given rate of interest.

25. 8%/yr compounded semiannually

26. 8%/yr compounded quarterly

27. 7%/yr compounded monthly

28. 9%/yr compounded daily

29. Find the accumulated amount after 4 yr if $5000 is invested at 8%/yr compounded continuously.

30. Find the accumulated amount after 6 yr if $6500 is invested at 7%/yr compounded continuously.

31. Find the interest rate needed for an investment of $5000 to grow to an amount of $6000 in 3 yr if interest is compounded continuously.

32. Find the interest rate needed for an investment of $4000 to double in 5 yr if interest is compounded continuously.

33. How long will it take an investment of $6000 to grow to $7000 if the investment earns interest at the rate of $7\frac{1}{2}\%$ compounded continuously?

34. How long will it take an investment of $8000 to double if the investment earns interest at the rate of 8% compounded continuously?

35. CONSUMER DECISIONS Mitchell has been given the option of either paying his $300 bill now or settling it for $306 after 1 mo. If he chooses to pay after 1 mo, find the simple interest rate at which he would be charged.

36. COURT JUDGMENT Jennifer was awarded damages of $150,000 in a successful lawsuit she brought against her employer 5 years ago. Interest (simple) on the judgment accrues at the rate of 12%/yr from the date of filing. If the case were settled today, how much would Jennifer receive in the final judgment?

37. BRIDGE LOANS To help finance the purchase of a new house, the Abdullahs have decided to apply for a short-term loan (a bridge loan) in the amount of $120,000 for a term of 3 mo. If the bank charges simple interest at the rate of 12%/yr, how much will the Abdullahs owe the bank at the end of the term?

38. **CORPORATE BONDS** David owns $20,000 worth of 10-yr bonds of Ace Corporation. These bonds pay interest every 6 mo at the rate of 7%/yr (simple interest). How much income will David receive from this investment every 6 mo? How much interest will David receive over the life of the bonds?

39. **MUNICIPAL BONDS** Maya paid $10,000 for a 7-yr bond issued by a city. She received interest amounting to $3500 over the life of the bonds. What rate of (simple) interest did the bond pay?

40. Write Equation (1b) in the slope-intercept form and interpret the meaning of the slope and the A-intercept in terms of r and P.
    Hint: Refer to Figure 1.

41. **HOSPITAL COSTS** If the cost of a semiprivate room in a hospital was $480/day 5 years ago and hospital costs have risen at the rate of 8%/yr since that time, what rate would you expect to pay for a semiprivate room today?

42. **FAMILY FOOD EXPENDITURE** Today a typical family of four spends $600/mo for food. If inflation occurs at the rate of 3%/yr over the next 6 yr, how much should the typical family of four expect to spend for food 6 yr from now?

43. **HOUSING APPRECIATION** The Kwans are planning to buy a house 4 years from now. Housing experts in their area have estimated that the cost of a home will increase at a rate of 5%/yr during that period. If this economic prediction holds true, how much can the Kwans expect to pay for a house that currently costs $210,000?

44. **ELECTRICITY CONSUMPTION** A utility company in a western city of the United States expects the consumption of electricity to increase by 8%/yr during the next decade, due mainly to the expected increase in population. If consumption does increase at this rate, find the amount by which the utility company will have to increase its generating capacity in order to meet the needs of the area at the end of the decade.

45. **PENSION FUNDS** The managers of a pension fund have invested $1.5 million in U.S. government certificates of deposit that pay interest at the rate of 5.5%/yr compounded semiannually over a period of 10 yr. At the end of this period, how much will the investment be worth?

46. **RETIREMENT FUNDS** Five and a half years ago, Chris invested $10,000 in a retirement fund that grew at the rate of 10.82%/yr compounded quarterly. What is his account worth today?

47. **MUTUAL FUNDS** Jodie invested $15,000 in a mutual fund 4 years ago. If the fund grew at the rate of 9.8%/yr compounded monthly, what would Jodie's account be worth today?

48. **TRUST FUNDS** A young man is the beneficiary of a trust fund established for him 21 years ago at his birth. If the original amount placed in trust was $10,000, how much will he receive if the money has earned interest at the rate of

8%/yr compounded annually? Compounded quarterly? Compounded monthly?

49. **INVESTMENT PLANNING** Find how much money should be deposited in a bank paying interest at the rate of 8.5%/yr compounded quarterly so that, at the end of 5 years, the accumulated amount will be $40,000.

50. **PROMISSORY NOTES** An individual purchased a 4-yr, $10,000 promissory note with an interest rate of 8.5%/yr compounded semiannually. How much did the note cost?

51. **FINANCING A COLLEGE EDUCATION** The parents of a child have just come into a large inheritance and wish to establish a trust fund for her college education. If they estimate that they will need $100,000 in 13 years, how much should they set aside in the trust now if they can invest the money at $8\frac{1}{2}$%/yr compounded (a) annually, (b) semiannually, and (c) quarterly?

52. **INVESTMENTS** Anthony invested a sum of money 5 years ago in a savings account that has since paid interest at the rate of 8%/yr compounded quarterly. His investment is now worth $22,289.22. How much did he originally invest?

53. **RATE COMPARISONS** In the last 5 years, Bendix Mutual Fund grew at the rate of 10.4%/yr compounded quarterly. Over the same period, Acme Mutual Fund grew at the rate of 10.6%/yr compounded semiannually. Which mutual fund has a better rate of return?

54. **RATE COMPARISONS** Fleet Street Savings Bank pays interest at the rate of 4.25%/yr compounded weekly in a savings account, whereas Washington Bank pays interest at the rate of 4.125%/yr compounded daily (assume a 365-day year). Which bank offers a better rate of interest?

55. **LOAN CONSOLIDATION** The proprietors of The Coachmen Inn secured two loans from Union Bank: one for $8000 due in 3 yr and one for $15,000 due in 6 yr, both at an interest rate of 10%/yr compounded semiannually. The bank has agreed to allow the two loans to be consolidated into one loan payable in 5 yr at the same interest rate. What amount will the proprietors of the inn be required to pay the bank at the end of 5 yr?

56. **EFFECTIVE RATE OF INTEREST** Find the effective rate of interest corresponding to a nominal rate of 9%/yr compounded annually, semiannually, quarterly, and monthly.

57. **HOUSING APPRECIATION** Georgia purchased a house in 2000 for $200,000. In 2006, she sold the house and made a net profit of $56,000. Find the effective annual rate of return on her investment over the 6-yr period.

58. **COMMON STOCK TRANSACTION** Steven purchased 1000 shares of a certain stock for $25,250 (including commissions). He sold the shares 2 yr later and received $32,100 after deducting commissions. Find the effective annual rate of return on his investment over the 2-yr period.

59. **ZERO COUPON BONDS** Nina purchased a zero coupon bond for $6724.53. The bond matures in 7 yr and has a face value of

$10,000. Find the effective rate of interest for the bond if interest is compounded semiannually.

Hint: Assume that the purchase price of the bond is the initial investment and that the face value of the bond is the accumulated amount.

60. **ZERO COUPON BONDS** Juan is contemplating buying a zero coupon bond that matures in 10 years and has a face value of $10,000. If the bond yields a return of 5.25%/yr, how much should Juan pay for the bond?

61. **MONEY MARKET MUTUAL FUNDS** Carlos invested $5000 in a money market mutual fund that pays interest on a daily basis. The balance in his account at the end of 8 months (245 days) was $5170.42. Find the effective rate at which Carlos's account earned interest over this period (assume a 365-day year).

62. **REVENUE GROWTH OF A HOME THEATER BUSINESS** Maxwell started a home theater business in 2004. The revenue of his company for that year was $240,000. The revenue grew by 20% in 2005 and by 30% in 2006. Maxwell projected that the revenue growth for his company in the next 3 years will be at least 25%/yr. How much does Maxwell expect his minimum revenue to be for 2009?

63. **ONLINE RETAIL SALES** Online retail sales stood at $23.5 billion for the year 2000. For the next 2 years, they grew by 33.2% and 27.8% per year, respectively. For the next 6 years, online retail sales are projected to grow at 30.5%, 19.9%, 24.3%, 14.0%, 17.6%, and 10.5% per year, respectively. What are the projected online sales for 2008?

Source: Jupiter Research

64. **PURCHASING POWER** The inflation rates in the U.S. economy for 2000 through 2003 are 3.4%, 2.8%, 1.6%, and 2.3%, respectively. What is the purchasing power of a dollar at the beginning of 2004 compared to that at the beginning of 2000?

Source: U.S. Census Bureau

65. **INVESTMENT OPTIONS** Investment A offers a 10% return compounded semiannually, and investment B offers a 9.75% return compounded continuously. Which investment has a higher rate of return over a 4-yr period?

66. **EFFECT OF INFLATION ON SALARIES** Leonard's current annual salary is $45,000. Ten years from now, how much will he need to earn in order to retain his present purchasing power if the rate of inflation over that period is 3%/yr? Assume that inflation is continuously compounded.

67. **SAVING FOR COLLEGE** Having received a large inheritance, Jing-mei's parents wish to establish a trust for her college education. If 7 yr from now they need an estimated $70,000, how much should they set aside in trust now, if they invest the money at 10.5% compounded quarterly? Continuously?

68. **PENSIONS** Maria, who is now 50 years old, is employed by a firm that guarantees her a pension of $40,000/year at age 65. What is the present value of her first year's pension if infla-

tion over the next 15 years is 6%? 8%? 12%? Assume that inflation is continuously compounded.

69. **REAL ESTATE INVESTMENTS** An investor purchased a piece of waterfront property. Because of the development of a marina in the vicinity, the market value of the property is expected to increase according to the rule

$$V(t) = 80,000e^{\sqrt{t}/2}$$

where $V(t)$ is measured in dollars and t is the time (in yr) from the present. If the inflation rate is expected to be 9% compounded continuously for the next 8 yr, find an expression for the present value $P(t)$ of the property's market price valid for the next 8 yr. What is $P(t)$ expected to be in 4 yr?

70. The simple interest formula $A = P(1 + rt)$ [Formula (1b)] can be written in the form $A = Prt + P$, which is the slope-intercept form of a straight line with slope Pr and A-intercept P.
   a. Describe the family of straight lines obtained by keeping the value of r fixed and allowing the value of P to vary. Interpret your results.
   b. Describe the family of straight lines obtained by keeping the value of P fixed and allowing the value of r to vary. Interpret your results.

71. **EFFECTIVE RATE OF INTEREST** Suppose an initial investment of $P grows to an accumulated amount of $A in t yr. Show that the effective rate (annual effective yield) is

$$r_{eff} = (A/P)^{1/t} - 1$$

72. **EFFECTIVE RATE OF INTEREST** Martha invested $40,000 in a boutique 5 years ago. Her investment is worth $70,000 today. What is the effective rate (annual effective yield) of her investment?

Hint: See Exercise 71.

In Exercises 73–76, determine whether the statement is true or false. If it is true, explain why it is true. If it is false, give an example to show why it is false.

73. When simple interest is used, the accumulated amount is a linear function of t.

74. If compound interest is converted annually, then the accumulated amount after t yr is the same as the accumulated amount under simple interest over t yr.

75. If interest is compounded annually, then the effective rate is the same as the nominal rate.

76. Susan's salary increased from $50,000/year to $60,000/year over a 5-year period. Therefore, Susan received annual increases of 5% over that period.

5.1   Solutions to Self-Check Exercises

1. Using Equation (7) with $A = 20,000$, $r = 0.12$, and $m = 12$ so that $i = 0.12/12$ and $n = (3)(12) = 36$, we find the required present value to be

$$P = 20,000\left(1 + \frac{0.12}{12}\right)^{-36} \approx 13,978.50$$

or $13,978.50

2. The accumulated amount of Paul's current investment is found by using Equation (3) with $P = 100,000$, $r = 0.106$, and $m = 365$. Thus, $i = 0.106/365$ and $n = 365$, so the required accumulated amount is given by

$$A = 100,000\left(1 + \frac{0.106}{365}\right)^{365} \approx 111,180.48$$

or $111,180.48. Next, we compute the accumulated amount of Paul's reinvestment. Now using (3) with $P = 100,000$, $r = 0.092$, and $m = 365$ so that $i = 0.092/365$ and $n = 365$, we find the required accumulated amount in this case to be

$$\overline{A} = 100,000\left(1 + \frac{0.092}{365}\right)^{365}$$

or $109,635.21. Therefore, Paul can expect to experience a net decrease in yearly income of

$$111,180.48 - 109,635.21$$

or $1545.27.

USING TECHNOLOGY

■ **Finding the Accumulated Amount of an Investment, the Effective Rate of Interest, and the Present Value of an Investment**

Graphing Utility

Some graphing utilities have built-in routines for solving problems involving the mathematics of finance. For example, the TI-83 incorporates several functions that can be used to solve the problems that are encountered in Sections 5.1–5.3. The step-by-step procedures for using these functions can be found on our Companion Website.

```
N   = 120
I%  = 10
PV  = −5000
PMT = 0
■ FV = 13535.20745
P/Y = 12
C/Y = 12
PMT : END    BEGIN
```

FIGURE T1
The TI-83 screen showing the future value (FV) of an investment

EXAMPLE 1   Finding the Accumulated Amount of an Investment Find the accumulated amount after 10 years if $5000 is invested at a rate of 10% per year compounded monthly.

Solution Using the TI-83 TVM SOLVER with the following inputs,

$$N = 120 \qquad \text{(10)(12)}$$

$$I\% = 10$$

$$PV = -5000 \qquad \text{Recall that an investment is an outflow.}$$

$$PMT = 0$$

$$FV = 0$$

$$P/Y = 12 \qquad \text{The number of payments each year}$$

$$C/Y = 12 \qquad \text{The number of conversion periods each year}$$

$$PMT:END \; BEGIN$$

we obtain the display shown in Figure T1. We conclude that the required accumulated amount is $13,535.21.

EXAMPLE 2 **Finding the Effective Rate of Interest** Find the effective rate of interest corresponding to a nominal rate of 10% per year compounded quarterly.

Solution Here we use the **Eff** function of the TI-83 calculator to obtain the result shown in Figure T2. The required effective rate is approximately 10.38% per year.

▶ Eff (10, 4)
                10.38128906

FIGURE T2
The TI-83 screen showing the effective rate of interest (Eff)

EXAMPLE 3 **Finding the Present Value of an Investment** Find the present value of $20,000 due in 5 years if the interest rate is 7.5% per year compounded daily.

Solution Using the TI-83 TVM SOLVER with the following inputs,

$$\begin{aligned}
N &= 1825 \qquad \text{(5)(365)}\\
I\% &= 7.5\\
PV &= 0\\
PMT &= 0\\
FV &= 20000\\
P/Y &= 365 \qquad \text{The number of payments each year}\\
C/Y &= 365 \qquad \text{The number of conversions each year}\\
&\text{PMT:END BEGIN}
\end{aligned}$$

we obtain the display shown in Figure T3. We see that the required present value is approximately $13,746.32. Note that PV is negative because an investment is an outflow (money is paid out).

```
N    = 1825
I%   = 7.5
■ PV   = −13746.3151
PMT  = 0
FV   = 20000
P/Y  = 365
C/Y  = 365
PMT : END    BEGIN
```

FIGURE T3
The TI-83 screen showing the present value (PV) of an investment

Excel

Excel has many built-in functions for solving problems involving the mathematics of finance. Here we illustrate the use of the FV (future value), EFFECT (effective rate), and the PV (present value) functions to solve problems of the type that we have encountered in Section 5.1.

EXAMPLE 4 **Finding the Accumulated Amount of an Investment** Find the accumulated amount after 10 years if $5000 is invested at a rate of 10% per year compounded monthly.

(continued)

Solution   Here we are computing the future value of a lump-sum investment, so we use the FV (future value) function. Select $\boxed{f_x}$ from the toolbar to obtain the Insert Function dialog box. Then select $\boxed{\text{Financial}}$ from the Or select a category: list box. Next, select $\boxed{\text{FV}}$ under Select a function: and click $\boxed{\text{OK}}$. The Function Arguments dialog box will appear (see Figure T4). In our example, the mouse cursor is in the edit box headed by Type, so a definition of that term appears near the bottom of the box. Figure T4 shows the entries for each edit box in our example.

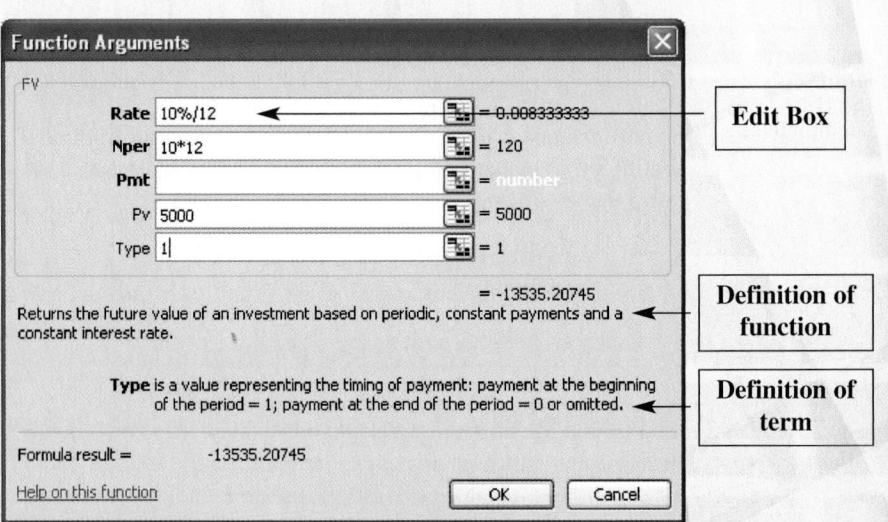

FIGURE T4
Excel's dialog box for computing the future value (FV) of an investment

Note that the entry for Nper is given by the total number of periods for which the investment earns interest. The Pmt box is left blank since no money is added to the original investment. The Pv entry is 5000. The entry for Type is a 1 because the lump-sum payment is made at the beginning of the investment period. The answer, −$13,535.21, is shown at the bottom of the dialog box. It is negative because an investment is considered to be an outflow of money (money is paid out). (Click $\boxed{\text{OK}}$ and the answer will also appear on your spreadsheet.)   ∎

EXAMPLE 5   Finding the Effective Rate of Interest   Find the effective rate of interest corresponding to a nominal rate of 10% per year compounded quarterly.

Solution   Here we use the EFFECT function to compute the effective rate of interest. Accessing this function from the Insert Function dialog box and making the required entries, we obtain the Function Arguments dialog box shown in Figure T5. The required effective rate is approximately 10.38% per year.

Note: Boldfaced words/characters enclosed in a box (for example, $\boxed{\text{Enter}}$) indicate that an action (click, select, or press) is required. Words/characters printed blue (for example, Chart sub-type:) indicate words/characters appearing on the screen. Words/characters in a typewriter font (for example, =(−2/3)*A2+2) indicate words/characters that need to be typed and entered.

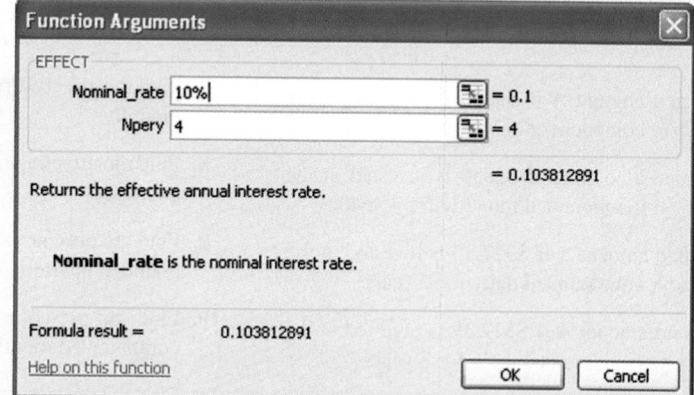

FIGURE T5
Excel's dialog box for the effective rate
of interest function (EFFECT)

EXAMPLE 6   Finding the Present Value of an Investment   Find the present value of $20,000 due in 5 years if the interest rate is 7.5% per year compounded daily.

Solution   We use the PV function to compute the present value of a lump-sum investment. Accessing this function from the Insert Function dialog box and making the required entries, we obtain the PV dialog box shown in Figure T6. Once again, the Pmt edit box is left blank since no additional money is added to the original investment. The Fv entry is 20000. The answer is negative because an investment is considered to be an outflow of money (money is paid out). We deduce that the required amount is $13,746.32.

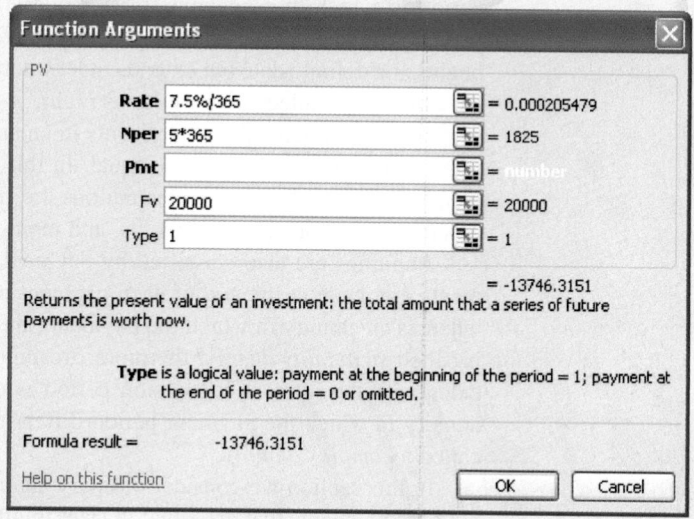

FIGURE T6
Excel dialog box for the present value
function (PV)

(continued)

TECHNOLOGY EXERCISES

1. Find the accumulated amount A if \$5000 is invested at the interest rate of $5\frac{3}{8}\%$/yr compounded monthly for 3 years.

2. Find the accumulated amount A if \$2850 is invested at the interest rate of $6\frac{5}{8}\%$/yr compounded monthly for 4 years.

3. Find the accumulated amount A if \$327.35 is invested at the interest rate of $5\frac{1}{3}\%$/yr compounded daily for 7 years.

4. Find the accumulated amount A if \$327.35 is invested at the interest rate of $6\frac{7}{8}\%$/yr compounded daily for 8 years.

5. Find the effective rate corresponding to $8\frac{2}{3}\%$/yr compounded quarterly.

6. Find the effective rate corresponding to $10\frac{5}{8}\%$/yr compounded monthly.

7. Find the effective rate corresponding to $9\frac{3}{4}\%$/yr compounded monthly.

8. Find the effective rate corresponding to $4\frac{3}{8}\%$/yr compounded quarterly.

9. Find the present value of \$38,000 due in 3 yr at $8\frac{1}{4}\%$/yr compounded quarterly.

10. Find the present value of \$150,000 due in 5 yr at $9\frac{3}{8}\%$/yr compounded monthly.

11. Find the present value of \$67,456 due in 3 yr at $7\frac{7}{8}\%$/yr compounded monthly.

12. Find the present value of \$111,000 due in 5 yr at $11\frac{5}{8}\%$/yr compounded monthly.

5.2 Annuities

■ **Future Value of an Annuity**

An **annuity** is a sequence of payments made at regular time intervals. The time period in which these payments are made is called the **term** of the annuity. Depending on whether the term is given by a *fixed time interval*—a time interval that begins at a definite date but extends indefinitely—or one that is not fixed in advance, an annuity is called an **annuity certain**, a *perpetuity*, or a *contingent annuity*, respectively. In general, the payments in an annuity need not be equal, but in many important applications they are equal. In this section we assume that annuity payments are equal. Examples of annuities are regular deposits to a savings account, monthly home mortgage payments, and monthly insurance payments.

Annuities are also classified by payment dates. An annuity in which the payments are made at the *end* of each payment period is called an **ordinary annuity**, whereas an annuity in which the payments are made at the beginning of each period is called an *annuity due*. Furthermore, an annuity in which the payment period coincides with the interest conversion period is called a **simple annuity**, whereas an annuity in which the payment period differs from the interest conversion period is called a *complex annuity*.

In this section we consider ordinary annuities that are certain and simple, with periodic payments that are equal in size. In other words, we study annuities that are subject to the following conditions:

1. The terms are given by fixed time intervals.
2. The periodic payments are equal in size.
3. The payments are made at the *end* of the payment periods.
4. The payment periods coincide with the interest conversion periods.

To find a formula for the accumulated amount S of an annuity, suppose a sum of $100 is paid into an account at the end of each quarter over a period of 3 years. Furthermore, suppose the account earns interest on the deposit at the rate of 8% per year, compounded quarterly. Then, the first payment of $100 made at the end of the first quarter earns interest at the rate of 8% per year compounded four times a year (or 8/4 = 2% per quarter) over the remaining 11 quarters and therefore, by the compound interest formula, has an accumulated amount of

$$100\left(1 + \frac{0.08}{4}\right)^{11} \quad \text{or} \quad 100(1 + 0.02)^{11}$$

dollars at the end of the term of the annuity (Figure 2).

The second payment of $100 made at the end of the second quarter earns interest at the same rate over the remaining 10 quarters and therefore has an accumulated amount of

$$100(1 + 0.02)^{10}$$

dollars at the end of the term of the annuity, and so on. The last payment earns no interest because it is due at the end of the term. The amount of the annuity is obtained by adding all the terms in Figure 2. Thus,

$$S = 100 + 100(1 + 0.02) + 100(1 + 0.02)^2 + \cdots + 100(1 + 0.02)^{11}$$

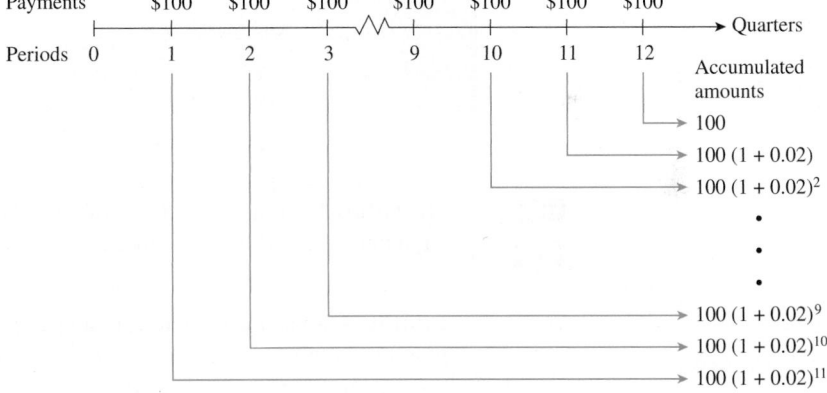

FIGURE 2
The sum of the accumulated amounts is the amount of the annuity.

The sum on the right is the sum of the first n terms of a *geometric progression* with first term R and common ratio $(1 + i)$. We will show in Section 5.4 that the sum S can be written in the more compact form

$$S = 100\left[\frac{(1 + 0.02)^{12} - 1}{0.02}\right]$$

$$\approx 1341.21$$

or approximately $1341.21.

To find a general formula for the accumulated amount S of an annuity, suppose a sum of R is paid into an account at the end of each period for n periods and that the account earns interest at the rate of i per period. Then, proceeding as we did with the numerical example, we obtain

$$S = R + R(1 + i) + R(1 + i)^2 + \cdots + R(1 + i)^{n-1}$$

$$= R\left[\frac{(1 + i)^n - 1}{i}\right] \tag{9}$$

The expression inside the brackets is commonly denoted by $s_{\overline{n}|i}$ (read "s angle n at i") and is called the **compound-amount factor.** Extensive tables have been constructed that give values of $s_{\overline{n}|i}$ for different values of i and n (see e.g. Table 1 on our Companion Website). In terms of the compound-amount factor,

$$S = Rs_{\overline{n}|i} \tag{10}$$

The quantity S in Equations (9) and (10) is realizable at some future date and is accordingly called the future value of an annuity.

Future Value of an Annuity

The **future value S of an annuity** of n payments of R dollars each, paid at the end of each investment period into an account that earns interest at the rate of i per period, is

$$S = R\left[\frac{(1 + i)^n - 1}{i}\right]$$

EXAMPLE 1   Find the amount of an ordinary annuity consisting of 12 monthly payments of $100 that earn interest at 12% per year compounded monthly.

Solution   Since i is the interest rate per *period* and since interest is compounded monthly in this case, we have $i = \frac{0.12}{12} = 0.01$. Using Equation (9) with $R = 100$, $n = 12$, and $i = 0.01$, we have

$$S = \frac{100[(1.01)^{12} - 1]}{0.01}$$

$$\approx 1268.25 \qquad \text{Use a calculator.}$$

or $1268.25. The same result is obtained by observing that

$$S = 100s_{\overline{12}|0.01}$$

$$= 100(12.6825)$$

$$= 1268.25 \qquad \text{Use Table 1 from the Website.} \qquad \blacksquare$$

EXPLORE & DISCUSS

Future Value *S* of an Annuity Due

1. Consider an annuity satisfying conditions 1, 2, and 4 on page 296 but with condition 3 replaced by the condition that payments are made at the *beginning* of the payment periods. By using an argument similar to that used to establish Formula (9), show that the future value *S* of an annuity due of *n* payments of *R* dollars each, paid at the beginning of each investment into an account that earns interest at the rate of *i* per period, is

$$S = R(1 + i)\left[\frac{(1 + i)^n - 1}{i}\right]$$

2. Use the result of part 1 to see how large your nest egg will be at age 65 if you start saving $4000 annually at age 30, assuming a 10% average annual return; if you start saving at 35; if you start saving at 40. [Moral of the story: It is never too early to start saving!]

 EXPLORING WITH TECHNOLOGY

Refer to the preceding Explore & Discuss problem.

1. Show that if $R = 4000$ and $i = 0.1$, then $S = 44,000[(1.1)^n - 1]$. Using a graphing utility, plot the graph of $f(x) = 44,000[(1.1)^x - 1]$, using the viewing window $[0, 40] \times [0, 1,200,000]$.

2. Verify the results of part 1 by evaluating $f(35)$, $f(30)$, and $f(25)$ using the EVAL function.

Present Value of an Annuity

In certain instances, you may want to determine the current value *P* of a sequence of equal periodic payments that will be made over a certain period of time. After each payment is made, the new balance continues to earn interest at some nominal rate. The amount *P* is referred to as the present value of an annuity.

To derive a formula for determining the present value *P* of an annuity, we may argue as follows. The amount *P* invested now and earning interest at the rate of *i* per period will have an accumulated value of $P(1 + i)^n$ at the end of *n* periods. But this must be equal to the future value of the annuity *S* given by Formula (9). Therefore, equating the two expressions, we have

$$P(1 + i)^n = R\left[\frac{(1 + i)^n - 1}{i}\right]$$

Multiplying both sides of this equation by $(1 + i)^{-n}$ gives

$$P = R(1 + i)^{-n}\left[\frac{(1 + i)^n - 1}{i}\right]$$

$$= R\left[\frac{(1 + i)^n(1 + i)^{-n} - (1 + i)^{-n}}{i}\right] \qquad (1 + i)^n(1 + i)^{-n} = 1$$

$$= R\left[\frac{1 - (1 + i)^{-n}}{i}\right]$$

$$= Ra_{\overline{n}|i}$$

where the factor $a_{\overline{n}|i}$ (read "a angle n at i") represents the expression inside the brackets. Extensive tables have also been constructed giving values of $a_{\overline{n}|i}$ for different values of i and n (see Table 1 on the Companion Website).

Present Value of an Annuity

The **present value P of an annuity** consisting of n payments of R dollars each, paid at the end of each investment period into an account that earns interest at the rate of i per period, is

$$P = R\left[\frac{1 - (1 + i)^{-n}}{i}\right] \tag{11}$$

EXAMPLE 2   Find the present value of an ordinary annuity consisting of 24 monthly payments of \$100 each and earning interest at 9% per year compounded monthly.

Solution   Here, $R = 100$, $i = \frac{r}{m} = \frac{0.09}{12} = 0.0075$, and $n = 24$, so by Formula (11) we have

$$P = \frac{100\left[1 - (1.0075)^{-24}\right]}{0.0075}$$

$$\approx 2188.91$$

or \$2188.91. The same result may be obtained by using Table 1 from the Companion Website. Thus,

$$P = 100a_{\overline{24}|0.0075}$$

$$= 100(21.8891)$$

$$= 2188.91$$ ◾

APPLIED EXAMPLE 3   Saving for a College Education   As a savings program toward Alberto's college education, his parents decide to deposit \$100 at the end of every month into a bank account paying interest at the rate of 6% per year compounded monthly. If the savings program began when Alberto was 6 years old, how much money would have accumulated by the time he turns 18?

Solution   By the time the child turns 18, the parents would have made 144 deposits into the account. Thus, $n = 144$. Furthermore, we have $R = 100$, $r = 0.06$, and $m = 12$, so $i = \frac{0.06}{12} = 0.005$. Using Equation (9), we find that the amount of money that would have accumulated is given by

$$S = \frac{100\left[(1.005)^{144} - 1\right]}{0.005}$$

$$\approx 21{,}015$$

or $21,015.

APPLIED EXAMPLE 4   Financing a Car   After making a down payment of $4000 for an automobile, Murphy paid $400 per month for 36 months with interest charged at 12% per year compounded monthly on the unpaid balance. What was the original cost of the car? What portion of Murphy's total car payments went toward interest charges?

Solution   The loan taken up by Murphy is given by the present value of the annuity

$$P = \frac{400\left[1 - (1.01)^{-36}\right]}{0.01} = 400a_{\overline{36}|0.01}$$

$$\approx 12{,}043$$

or $12,043. Therefore, the original cost of the automobile is $16,043 ($12,043 plus the $4000 down payment). The interest charges paid by Murphy are given by $(36)(400) - 12{,}043$, or $2,357.

One important application of annuities is in the area of tax planning. During the 1980s, Congress created many tax-sheltered retirement savings plans, such as Individual Retirement Accounts (IRAs), Keogh plans, and Simplified Employee Pension (SEP) plans. These plans are examples of annuities in which the individual is allowed to make contributions (which are often tax deductible) to an investment account. The amount of the contribution is limited by congressional legislation. The taxes on the contributions and/or the interest accumulated in these accounts are deferred until the money is withdrawn—ideally during retirement, when tax brackets should be lower. In the interim period, the individual has the benefit of tax-free growth on his or her investment.

Suppose, for example, you are eligible to make a fully deductible contribution to an IRA and you are in a marginal tax bracket of 28%. Additionally, suppose you receive a year-end bonus of $2000 from your employer and have the option of depositing the $2000 into either an IRA or a regular savings account, where both accounts earn interest at an effective annual rate of 8% per year. If you choose to invest your bonus in a regular savings account, you will first have to pay taxes on the $2000, leaving $1440 to invest. At the end of 1 year, you will also have to pay taxes on the interest earned, leaving you with

Accumulated amount	−	Tax on interest	=	Net amount
1555.20	−	32.26	=	1522.94

or $1522.94.

On the other hand, if you put the money into the IRA account, the entire sum will earn interest, and at the end of 1 year you will have (1.08)($2000), or $2160, in your account. Of course, you will still have to pay taxes on this money when you withdraw it, but you will have gained the advantage of tax-free growth of the larger principal over the years. The disadvantage of this option is that if you withdraw the money before you reach the age of $59\frac{1}{2}$, you will be liable for taxes on both your contributions and the interest earned *and* you will also have to pay a 10% penalty.

Note   In practice, the size of the contributions an individual might make to the various retirement plans might vary from year to year. Also, he or she might make the contributions at different payment periods. To simplify our discussion, we will consider examples in which fixed payments are made at regular intervals.   ■

APPLIED EXAMPLE 5   IRA Accounts   Caroline is planning to make a contribution of $2000 on January 31 of each year into an IRA earning interest at an effective rate of 9% per year. After she makes her 25th payment on January 31 of the year following her retirement, how much will she have in her IRA?

Solution   The amount of money Caroline will have after her 25th payment into her account is found by using Equation (9) with $R = 2000$, $r = 0.09$, and $m = 1$, so $i = r/m = 0.09$ and $n = 25$. The required amount is given by

$$S = \frac{2000\left[(1.09)^{25} - 1\right]}{0.09}$$
$$\approx 169,401.79$$

or $169,401.79.   ■

After-tax-deferred annuities are another type of investment vehicle that allows an individual to build assets for retirement, college funds, or other future needs. The advantage gained in this type of investment is that the tax on the accumulated interest is deferred to a later date. Note that in this type of investment the contributions themselves are not tax deductible. At first glance, the advantage thus gained may seem to be relatively inconsequential, but its true effect is illustrated by the next example.

APPLIED EXAMPLE 6   Investment Analysis   Both Clark and Colby are salaried individuals, 45 years of age, who are saving for their retirement 20 years from now. Both Clark and Colby are also in the 28% marginal tax bracket. Clark makes a $1000 contribution annually on December 31 into a savings account earning an effective rate of 8% per year. At the same time, Colby makes a $1000 annual payment to an insurance company for an after-tax-deferred annuity. The annuity also earns interest at an effective rate of 8% per year. (Assume that both men remain in the same tax bracket throughout this period, and disregard state income taxes.)
a. Calculate how much each man will have in his investment account at the end of 20 years.
b. Compute the interest earned on each account.

c. Show that even if the interest on Colby's investment were subjected to a tax of 28% upon withdrawal of his investment at the end of 20 years, the net accumulated amount of his investment would still be greater than that of Clark's.

Solution

a. Because Clark is in the 28% marginal tax bracket, the net yield for his investment is $(0.72)(8)$, or 5.76%, per year.

Using Formula (9) with $R = 1000$, $r = 0.0576$, and $m = 1$, so that $i = 0.0576$ and $n = 20$, we see that Clark's investment will be worth

$$S = \frac{1000[(1 + 0.0576)^{20} - 1]}{0.0576}$$

$$\approx 35{,}850.49$$

or $35,850.49 at his retirement.

Colby has a tax-sheltered investment with an effective yield of 8% per year. Using Formula (9) with $R = 1000$, $r = 0.08$, and $m = 1$, so that $i = 0.08$ and $n = 20$, we see that Colby's investment will be worth

$$S = \frac{1000[(1 + 0.08)^{20} - 1]}{0.08}$$

$$\approx 45{,}761.96$$

or $45,761.96 at his retirement.

b. Each man will have paid $20(1000)$, or $20,000, into his account. Therefore, the total interest earned in Clark's account will be $(35{,}850.49 - 20{,}000)$, or $15,850.49, whereas the total interest earned in Colby's account will be $(45{,}761.96 - 20{,}000)$, or $25,761.96.

c. From part (b) we see that the total interest earned in Colby's account will be $25,761.96. If it were taxed at 28%, he would still end up with $(0.72)(25{,}761.96)$, or $18,548.61. This is larger than the total interest of $15,850.49 earned by Clark.

5.2 Self-Check Exercises

1. Phyliss opened an IRA on January 31, 1990, with a contribution of $2000. She plans to make a contribution of $2000 thereafter on January 31 of each year until her retirement in the year 2009 (20 payments). If the account earns interest at the rate of 8%/yr compounded yearly, how much will Phyliss have in her account when she retires?

2. Denver Wildcatting Company has an immediate need for a loan. In an agreement worked out with its banker, Denver assigns its royalty income of $4800/month for the next 3 years from certain oil properties to the bank, with the first payment due at the end of the first month. If the bank charges interest at the rate of 9%/yr compounded monthly, what is the amount of the loan negotiated between the parties?

Solutions to Self-Check Exercises 5.2 can be found on page 305.

5.2 Concept Questions

1. In an ordinary annuity, is the term fixed or variable? Are the periodic payments all of the same size, or do they vary in size? Are the payments made at the beginning or the end of the payment period? Do the payment periods coincide with the interest conversion periods?

2. What is the difference between an ordinary annuity and an annuity due?

3. What is the future value of an annuity? Give an example.

4. What is the present value of an annuity? Give an example.

5.2   Exercises

In Exercises 1–8, find the amount (future value) of each ordinary annuity.

1. $1000 a year for 10 years at 10%/yr compounded annually

2. $1500 per semiannual period for 8 years at 9%/yr compounded semiannually

3. $1800 per quarter for 6 years at 8%/yr compounded quarterly

4. $500 per semiannual period for 12 years at 11%/yr compounded semiannually

5. $600 per quarter for 9 years at 12%/yr compounded quarterly

6. $150 per month for 15 years at 10%/yr compounded monthly

7. $200/month for $20\frac{1}{4}$ years at 9%/yr compounded monthly

8. $100/week for $7\frac{1}{2}$ years at 7.5%/yr compounded weekly

In Exercises 9–14, find the present value of each ordinary annuity.

9. $5000 a year for 8 years at 8%/yr compounded annually

10. $1200 per semiannual period for 6 years at 10%/yr compounded semiannually

11. $4000 a year for 5 years at 9%/yr compounded yearly

12. $3000 per semiannual period for 6 years at 11%/yr compounded semiannually

13. $800 per quarter for 7 years at 12%/yr compounded quarterly

14. $150 per month for 10 years at 8%/yr compounded monthly

15. **IRAs**   If a merchant deposits $1500 annually at the end of each tax year in an IRA account paying interest at the rate of 8%/yr compounded annually, how much will she have in her account at the end of 25 years?

16. **SAVINGS ACCOUNTS**   If Jackson deposits $100 at the end of each month in a savings account earning interest at the rate of 8%/yr compounded monthly, how much will he have on deposit in his savings account at the end of 6 years, assuming that he makes no withdrawals during that period?

17. **SAVINGS ACCOUNTS**   Linda has joined a "Christmas Fund Club" at her bank. At the end of every month, December through October inclusive, she will make a deposit of $40 in her fund. If the money earns interest at the rate of 7%/yr compounded monthly, how much will she have in her account on December 1 of the following year?

18. **KEOGH ACCOUNTS**   Robin, who is self-employed, contributes $5000/yr into a Keogh account. How much will he have in the account after 25 years if the account earns interest at the rate of 8.5%/yr compounded yearly?

19. **RETIREMENT PLANNING**   As a fringe benefit for the past 12 years, Colin's employer has contributed $100 at the end of each month into an employee retirement account for Colin that pays interest at the rate of 7%/yr compounded monthly. Colin has also contributed $2000 at the end of each of the last 8 years into an IRA that pays interest at the rate of 9%/yr compounded yearly. How much does Colin have in his retirement fund at this time?

20. **SAVINGS ACCOUNTS**   The Pirerras are planning to go to Europe 3 years from now and have agreed to set aside $150 each month for their trip. If they deposit this money at the end of each month into a savings account paying interest at the rate of 8%/yr compounded monthly, how much money will be in their travel fund at the end of the third year?

21. **INVESTMENT ANALYSIS**   Karen has been depositing $150 at the end of each month in a tax-free retirement account since she was 25. Matt, who is the same age as Karen, started depositing $250 at the end of each month in a tax-free retirement account when he was 35. Assuming that both accounts have been and will be earning interest at the rate of 5%/yr compounded monthly, who will end up with the larger retirement account at the age of 65?

22. **INVESTMENT ANALYSIS**   Luis has $150,000 in his retirement account at his present company. Because he is assuming a position with another company, Luis is planning to "roll over" his assets to a new account. Luis also plans to put $3000/quarter into the new account until his retirement 20 years from now. If the account earns interest at the rate of 8%/yr compounded quarterly, how much will Luis have in his account at the time of his retirement?
Hint: Use the compound interest formula and the annuity formula.

23. **AUTO LEASING**   The Betzes have leased an auto for 2 years at $450/month. If money is worth 9%/yr compounded monthly, what is the equivalent cash payment (present value) of this annuity?

24. **AUTO FINANCING**   Lupé made a down payment of $4000 toward the purchase of a new car. To pay the balance of the purchase price, she has secured a loan from her bank at the rate of 12%/yr compounded monthly. Under the terms of her finance agreement, she is required to make payments of $420/month for 36 months. What is the cash price of the car?

25. **INSTALLMENT PLANS**   Pierce Publishing sells encyclopedias under two payment plans: cash or installment. Under the installment plan, the customer pays $22/month over 3 years

with interest charged on the balance at a rate of 18%/yr compounded monthly. Find the cash price for a set of encyclopedias if it is equivalent to the price paid by a customer using the installment plan.

26. LOTTERY PAYOUTS   A state lottery commission pays the winner of the "Million Dollar" lottery 20 installments of $50,000/year. The commission makes the first payment of $50,000 immediately and the other $n = 19$ payments at the end of each of the next 19 years. Determine how much money the commission should have in the bank initially to guarantee the payments, assuming that the balance on deposit with the bank earns interest at the rate of 8%/yr compounded yearly.
Hint: Find the present value of an annuity.

27. PURCHASING A HOME   The Johnsons have accumulated a nest egg of $40,000 that they intend to use as a down payment toward the purchase of a new house. Because their present gross income has placed them in a relatively high tax bracket, they have decided to invest a minimum of $2400/month in monthly payments (to take advantage of their tax deductions) toward the purchase of their house. However, because of other financial obligations, their monthly payments should not exceed $3000. If local mortgage rates are 7.5%/yr compounded monthly for a conventional 30-year mortgage, what is the price range of houses that they should consider?

28. PURCHASING A HOME   Refer to Exercise 27. If local mortgage rates were increased to 8%, how would this affect the price range of houses that the Johnsons should consider?

29. PURCHASING A HOME   Refer to Exercise 27. If the Johnsons decide to secure a 15-yr mortgage instead of a 30-yr mortgage, what is the price range of houses they should consider when the local mortgage rate for this type of loan is 7%?

30. SAVINGS PLAN   Lauren plans to deposit $5000 into a bank account at the beginning of next month and $200/month into

the same account at the end of that month and at the end of each subsequent month for the next 5 years. If her bank pays interest at the rate of 6%/yr compounded monthly, how much will Lauren have in her account at the end of 5 years? (Assume she makes no withdrawals during the 5-yr period.)

31. FINANCIAL PLANNING   Joe plans to deposit $200 at the end of each month into a bank account for a period of 2 years, after which he plans to deposit $300 at the end of each month into the same account for another 3 years. If the bank pays interest at the rate of 6%/yr compounded monthly, how much will Joe have in his account by the end of 5 years? (Assume no withdrawals are made during the 5-yr period.)

32. INVESTMENT ANALYSIS   From age 25 to age 40, Jessica deposited $200 at the end of each month into a tax-free retirement account. She made no withdrawals or further contributions until age 65. Alex made deposits of $300 into his tax-free retirement account from age 40 to age 65. If both accounts earned interest at the rate of 5%/yr compounded monthly, who ends up with a bigger nest egg upon reaching the age of 65?
Hint: Use both the annuity formula and the compound interest formula.

In Exercises 33 and 34, determine whether the statement is true or false. If it is true, explain why it is true. If it is false, give an example to show why it is false.

33. The future value of an annuity can be found by adding together all the payments that are paid into the account.

34. If the future value of an annuity consisting of n payments of R dollars each—paid at the end of each investment period into an account that earns interest at the rate of i per period —is S dollars, then

$$R = \frac{iS}{(1+i)^n - 1}$$

5.2   Solutions to Self-Check Exercises

1. The amount Phyliss will have in her account when she retires may be found by using Formula (9) with $R = 2000$, $r = 0.08$, and $m = 1$, so that $i = r = 0.08$ and $n = 20$. Thus,

$$S = \frac{2000[(1.08)^{20} - 1]}{0.08}$$

$$\approx 91,523.93$$

or $91,523.93.

2. We want to find the present value of an ordinary annuity consisting of 36 monthly payments of $4800 each and earning interest at 9%/yr compounded monthly. Using Formula (11) with $R = 4800$ and $m = 12$, so that $i = r/m = 0.09/12 = 0.0075$ and $n = (12)(3) = 36$, we find

$$P = \frac{4800[1 - (1.0075)^{-36}]}{0.0075} \approx 150,944.67$$

or $150,944.67, the amount of the loan negotiated.

USING TECHNOLOGY

▬ Finding the Amount of an Annuity

Graphing Utility

As mentioned in Using Technology, Section 5.1, the TI-83 can facilitate the solution of problems in finance. We continue to exploit its versatility in this section.

EXAMPLE 1   Finding the Future Value of an Annuity   Find the amount of an ordinary annuity of 36 quarterly payments of $220 each that earn interest at the rate of 10% per year compounded quarterly.

Solution   We use the TI-83 TVM SOLVER with the following inputs:

$$N = 36$$
$$I\% = 10$$
$$PV = 0$$
$$PMT = -220 \qquad \text{Recall that a payment is an outflow.}$$
$$FV = 0$$
$$P/Y = 4 \qquad \text{The number of payments each year}$$
$$C/Y = 4 \qquad \text{The number of conversion periods each year}$$
$$\text{PMT:END BEGIN}$$

The result is displayed in Figure T1. We deduce that the desired amount is $12,606.31.

```
N    = 36
I%   = 10
PV   = 0
PMT  = -220
FV   = 12606.31078
P/Y  = 4
C/Y  = 4
PMT : END    BEGIN
```

FIGURE T1
The TI-83 screen showing the future value (FV) of an annuity

EXAMPLE 2   Finding the Present Value of an Annuity   Find the present value of an ordinary annuity consisting of 48 monthly payments of $300 each and earning interest at the rate of 9% per year compounded monthly.

Solution   We use the TI-83 TVM SOLVER with the following inputs:

$$N = 48$$
$$I\% = 9$$
$$PV = 0$$
$$PMT = -300 \qquad \text{A payment is an outflow.}$$
$$FV = 0$$
$$P/Y = 12 \qquad \text{The number of payments each year}$$
$$C/Y = 12 \qquad \text{The number of conversion periods each year}$$
$$\text{PMT:END BEGIN}$$

The output is displayed in Figure T2. We see that the required present value of the annuity is \$12,055.43.

```
N    = 48
I%   = 9
PV   = 12055.43457
PMT  = −300
FV   = 0
P/Y  = 12
C/Y  = 12
PMT : END    BEGIN
```

FIGURE T2
The TI-83 screen showing the present value (PV) of an ordinary annuity

Excel

Now we show how Excel can be used to solve financial problems involving annuities.

EXAMPLE 3   Finding the Future Value of an Annuity   Find the amount of an ordinary annuity of 36 quarterly payments of \$220 each that earn interest at the rate of 10% per year compounded quarterly.

Solution   Here we are computing the future value of a series of equal payments, so we use the FV (future value) function. As before, we access the Insert Function dialog box to obtain the Function Arguments dialog box. After making each of the required entries, we obtain the dialog box shown in Figure T3.

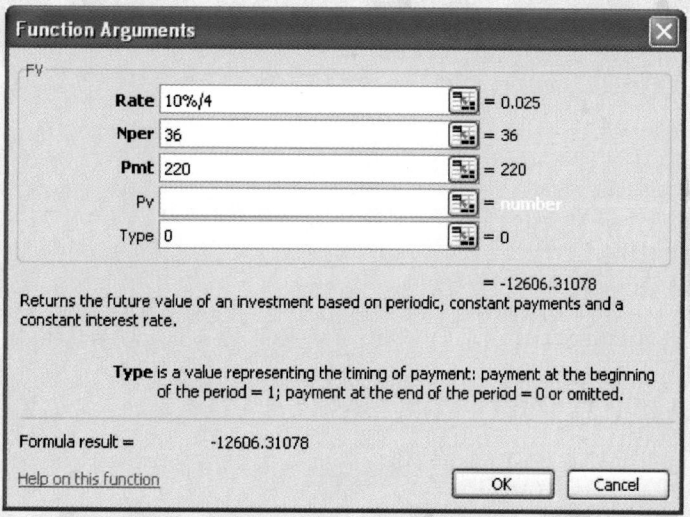

FIGURE T3
Excel's dialog box for the future value (FV) of an annuity

Note that a 0 is entered in the Type edit box because payments are made at the end of each payment period. Once again, the answer is negative because cash is paid out. We deduce that the desired amount is \$12,606.31.

Note: Boldfaced words/characters enclosed in a box (for example, **Enter**) indicate that an action (click, select, or press) is required. Words/characters printed blue (for example, Chart sub-type:) indicate words/characters that appear on the screen. Words/characters printed in a typewriter font (for example, =(−2/3)*A2+2) indicate words/characters that need to be typed and entered.

(*continued*)

EXAMPLE 4   Finding the Present Value of an Annuity   Find the present value of an ordinary annuity consisting of 48 monthly payments of $300 each and earning interest at the rate of 9% per year compounded monthly.

Solution   Here we use the PV function to compute the present value of an annuity. Accessing the PV (present value) function from the Insert Function dialog box and making the required entries, we obtain the PV dialog box shown in Figure T4. We see that the required present value of the annuity is $12,055.43.

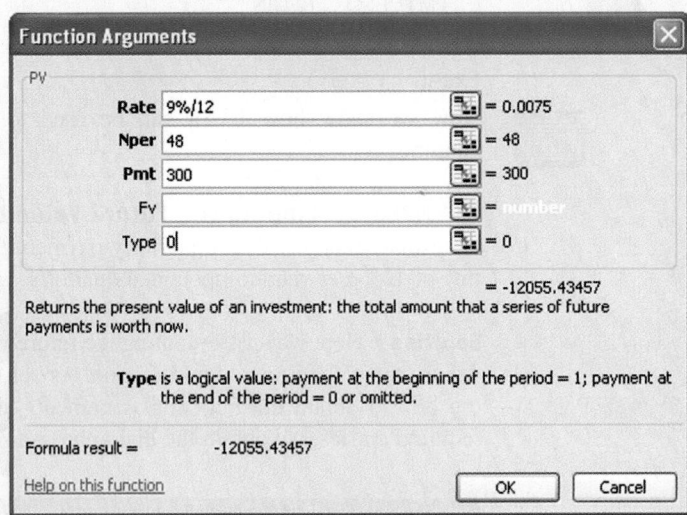

FIGURE T4
Excel's dialog box for computing the present value (PV) of an annuity

---

TECHNOLOGY EXERCISES

1. Find the amount of an ordinary annuity of 20 payments of $2500/quarter at $7\frac{1}{4}$%/yr compounded quarterly.

2. Find the amount of an ordinary annuity of 24 payments of $1790/quarter at $8\frac{3}{4}$%/yr compounded quarterly.

3. Find the amount of an ordinary annuity of $120/month for 5 years at $6\frac{3}{8}$%/yr compounded monthly.

4. Find the amount of an ordinary annuity of $225/month for 6 years at $7\frac{5}{8}$%/yr compounded monthly.

5. Find the present value of an ordinary annuity of $4500/semi-annual period for 5 years earning interest at 9%/yr compounded semiannually.

6. Find the present value of an ordinary annuity of $2100/quarter for 7 years earning interest at $7\frac{1}{8}$%/yr compounded quarterly.

7. Find the present value of an ordinary annuity of $245/month for 6 years earning interest at $8\frac{3}{8}$%/yr compounded monthly.

8. Find the present value of an ordinary annuity of $185/month for 12 years earning interest at $6\frac{5}{8}$%/yr compounded monthly.

5.3 Amortization and Sinking Funds

◼ Amortization of Loans

The annuity formulas derived in Section 5.2 may be used to answer questions involving the amortization of certain types of installment loans. For example, in a typical housing loan the mortgagor makes periodic payments toward reducing his indebtedness to the lender, who charges interest at a fixed rate on the unpaid portion of the debt. In practice, the borrower is required to repay the lender in periodic installments, usually of the same size and over a fixed term, so that the loan (principal plus interest charges) is amortized at the end of the term.

By thinking of the monthly loan repayments R as the payments in an annuity, we see that the original amount of the loan is given by P, the present value of the annuity. From Equation (11), Section 5.2, we have

$$P = R\left[\frac{1 - (1 + i)^{-n}}{i}\right] = Ra_{\overline{n}|i} \tag{12}$$

A question a financier might ask is: How much should the monthly installment be so that a loan will be amortized at the end of the term of the loan? To answer this question, we simply solve (12) for R in terms of P, obtaining

$$R = \frac{Pi}{1 - (1 + i)^{-n}} = \frac{P}{a_{\overline{n}|i}}$$

Amortization Formula

The periodic payment R on a loan of P dollars to be amortized over n periods with interest charged at the rate of i per period is

$$R = \frac{Pi}{1 - (1 + i)^{-n}} \tag{13}$$

EXAMPLE 1 A sum of $50,000 is to be repaid over a 5-year period through equal installments made at the end of each year. If an interest rate of 8% per year is charged on the unpaid balance and interest calculations are made at the end of each year, determine the size of each installment so that the loan (principal plus interest charges) is amortized at the end of 5 years. Verify the result by displaying the amortization schedule.

Solution Substituting $P = 50,000$, $i = r = 0.08$ (here, $m = 1$), and $n = 5$ into Formula (13), we obtain

$$R = \frac{(50,000)(0.08)}{1 - (1.08)^{-5}} \approx 12,522.82$$

giving the required yearly installment as $12,522.82.

TABLE 4

An Amortization Schedule

End of Period	Interest Charged	Repayment Made	Payment Toward Principal	Outstanding Principal
0	—	—	—	$50,000.00
1	$4,000.00	$12,522.82	$ 8,522.82	41,477.18
2	3,318.17	12,522.82	9,204.65	32,272.53
3	2,581.80	12,522.82	9,941.02	22,331.51
4	1,786.52	12,522.82	10,736.30	11,595.21
5	927.62	12,522.82	11,595.20	0.01

The amortization schedule is presented in Table 4. The outstanding principal at the end of 5 years is, of course, zero. (The figure of $.01 in Table 4 is the result of round-off errors.) Observe that initially the larger portion of the repayment goes toward payment of interest charges, but as time goes by more and more of the payment goes toward repayment of the principal.

Financing a Home

APPLIED EXAMPLE 2   Home Mortgage Payments   The Blakelys borrowed $120,000 from a bank to help finance the purchase of a house. The bank charges interest at a rate of 9% per year on the unpaid balance, with interest computations made at the end of each month. The Blakelys have agreed to repay the loan in equal monthly installments over 30 years. How much should each payment be if the loan is to be amortized at the end of the term?

Solution   Here, $P = 120,000$, $i = \frac{r}{m} = \frac{0.09}{12} = 0.0075$, and $n = (30)(12) = 360$. Using Formula (13) we find that the size of each monthly installment required is given by

$$R = \frac{(120,000)(0.0075)}{1 - (1.0075)^{-360}}$$

$$\approx 965.55$$

or $965.55.

APPLIED EXAMPLE 3   Home Equity   Teresa and Raul purchased a house 10 years ago for $200,000. They made a down payment of 20% of the purchase price and secured a 30-year conventional home mortgage at 9% per year on the unpaid balance. The house is now worth $380,000. How much equity do Teresa and Raul have in their house now (after making 120 monthly payments)?

Solution   Since the down payment was 20%, we know that they secured a loan of 80% of $200,000, or $160,000. Furthermore, using Formula (13) with $P = 160,000$, $i = \frac{r}{m} = \frac{0.09}{12} = 0.0075$ and $n = (30)(12) = 360$, we determine that their monthly installment is

$$R = \frac{(160,000)(0.0075)}{1 - (1.0075)^{-360}}$$

$$\approx 1287.40$$

or $1287.40.

After 120 monthly payments have been made, the outstanding principal is given by the sum of the present values of the remaining installments (that is, $360 - 120 = 240$ installments). But this sum is just the present value of an annuity with $n = 240$, $R = 1287.40$, and $i = 0.0075$. Using Formula (11), we find

$$P = 1287.40 \left[\frac{1 - (1 + 0.0075)^{-240}}{0.0075} \right]$$

$$\approx 143,088.01$$

or approximately $143,088. Therefore, Teresa and Raul have an equity of $380,000 - 143,088$, that is, $236,912. ■

EXPLORE & DISCUSS and EXPLORING WITH TECHNOLOGY

1. Consider the amortization Formula (13):

$$R = \frac{Pi}{1 - (1 + i)^{-n}}$$

Suppose you know the values of R, P, and n and you wish to determine i. Explain why you can accomplish this task by finding the point of intersection of the graphs of the functions

$$y_1 = R \quad \text{and} \quad y_2 = \frac{Pi}{1 - (1 + i)^{-n}}$$

2. Thalia knows that her monthly repayment on her 30-year conventional home loan of $150,000 is $1100.65 per month. Help Thalia determine the interest rate for her loan by verifying or executing the following steps:

a. Plot the graphs of

$$y_1 = 1100.65 \quad \text{and} \quad y_2 = \frac{150,000x}{1 - (1 + x)^{-360}}$$

using the viewing window $[0, 0.01] \times [0, 1200]$.

b. Use the ISECT (intersection) function of the graphing utility to find the point of intersection of the graphs of part (a). Explain why this gives the value of i.

c. Compute r from the relationship $r = 12i$.

EXPLORE & DISCUSS and EXPLORING WITH TECHNOLOGY

1. Suppose you secure a home mortgage loan of $P with an interest rate of r per year to be amortized over t years through monthly installments of $R. Show that, after N installments, your outstanding principal is given by

$$B(N) = P\left[\frac{(1 + i)^n - (1 + i)^N}{(1 + i)^n - 1}\right] \qquad (0 \le N \le n)$$

Hint: $B(N) = R\left[\dfrac{1 - (1 + i)^{-n+N}}{i}\right]$. To see this, study Example 3, page 310. Replace R using Formula (13).

2. Refer to Example 3, page 310. Using the result of part 1, show that Teresa and Raul's outstanding balance after making N payments is

$$E(N) = \frac{160,000(1.0075^{360} - 1.0075^N)}{1.0075^{360} - 1} \qquad (0 \le N \le 360)$$

3. Using a graphing utility, plot the graph of

$$E(x) = \frac{160,000(1.0075^{360} - 1.0075^x)}{1.0075^{360} - 1}$$

using the viewing window $[0, 360] \times [0, 160,000]$.

4. Referring to the graph in part 3, observe that the outstanding principal drops off slowly in the early years and accelerates quickly to zero toward the end of the loan. Can you explain why?

5. How long does it take Teresa and Raul to repay half of the loan of $160,000? Hint: See the previous Explore & Discuss and Exploring with Technology box.

APPLIED EXAMPLE 4   Home Affordability   The Jacksons have determined that, after making a down payment, they could afford at most $1000 for a monthly house payment. The bank charges interest at the rate of 9.6% per year on the unpaid balance, with interest computations made at the end of each month. If the loan is to be amortized in equal monthly installments over 30 years, what is the maximum amount that the Jacksons can borrow from the bank?

Solution   Here, $i = \frac{r}{m} = \frac{0.096}{12} = 0.008$, $n = (30)(12) = 360$, and $R = 1000$; we are required to find P. From Equation (11), we have

$$P = \frac{R\left[1 - (1 + i)^{-n}\right]}{i}$$

Substituting the numerical values for R, n, and i into this expression for P, we obtain

$$P = \frac{1000\left[1 - (1.008)^{-360}\right]}{0.008} \approx 117,902$$

Therefore, the Jacksons can borrow at most $117,902.

TITLE   Director of Finance

INSTITUTION   Thomson Higher Education

At Thomson Higher Education, we publish hundreds of college textbooks a year, and it is my job to oversee all the financial aspects of our division. One aspect of my position is that I must keep abreast of the rules and regulations that are set forth by the U.S. Securities and Exchange Commission (SEC). In an effort to apply newly designed rules of revenue recognition from the Sarbanes–Oxley Act of 2002, we review the investment costs for each program that we publish and amortize that investment to reflect the trend in the way each title sells. So, if we spend one million dollars on a program, we will take the *Sum of the Years' Digits* formula and determine the amount we will amortize when the book publishes. Since company policy requires that we amortize these costs over three years, we would amortize $500,000 in the first year, $333,000 in the second year and $167,000 in the final year.

The reason we use SOYD rather than straight lining the amortization is due to the pattern of sales. When a title is first published, the supply is limited exclusively to what the publisher sells. Therefore, no other supply exists and the publisher sells the highest quantity of titles within the life of the edition. In the second year, other avenues, such as used books, are available to the customer. Because the supply is higher but demand is the same, the number sold by the publisher is less. The further out these sales go (four or five years), the number of units the publisher sells decreases. So if the publisher straight-lined the development costs over three years, the amortization would be the same every year and would not reflect the selling patterns.

Many titles also have an increasing number of media components. One of the most popular components is the online homework system. Since these are considered internally developed software systems, they must amortize under similar conditions but are considered a direct expense as opposed to a cost of goods sold (COGS). Corporate accounting rules stipulate that these types of costs begin amortizing immediately after they go live rather than at the end of their copyright year, which is different from the rules of bookplate amortization.

Sinking Funds

Sinking funds are another important application of the annuity formulas. Simply stated, a **sinking fund** is an account that is set up for a specific purpose at some future date. For example, an individual might establish a sinking fund for the purpose of discharging a debt at a future date. A corporation might establish a sinking fund in order to accumulate sufficient capital to replace equipment that is expected to be obsolete at some future date.

By thinking of the amount to be accumulated by a specific date in the future as the future value of an annuity [Equation (9), Section 5.2], we can answer questions about a large class of sinking fund problems.

APPLIED EXAMPLE 5   Sinking Fund   The proprietor of Carson Hardware has decided to set up a sinking fund for the purpose of purchasing a truck in 2 years' time. It is expected that the truck will cost $30,000. If the fund earns 10% interest per year compounded quarterly, determine the size of each (equal) quarterly installment the proprietor should pay into the fund. Verify the result by displaying the schedule.

313

Solution   The problem at hand is to find the size of each quarterly payment R of an annuity given that its future value is $S = 30,000$, the interest earned per conversion period is $i = \frac{r}{m} = \frac{0.1}{4} = 0.025$, and the number of payments is $n = (2)(4) = 8$. The formula for the annuity,

$$S = R\left[\frac{(1 + i)^n - 1}{i}\right]$$

when solved for R yields

$$R = \frac{iS}{(1 + i)^n - 1} \tag{14}$$

or, equivalently,

$$R = \frac{S}{s_{\overline{n}|i}}$$

Substituting the appropriate numerical values for i, S, and n into Equation (14), we obtain the desired quarterly payment

$$R = \frac{(0.025)(30,000)}{(1.025)^8 - 1} \approx 3434.02$$

or $3434.02. Table 5 shows the required schedule.

TABLE 5

A Sinking Fund Schedule

End of Period	Deposit Made	Interest Earned	Addition to Fund	Accumulated Amount in Fund
1	$3,434.02	0	$3,434.02	$ 3,434.02
2	3,434.02	$ 85.85	3,519.87	6,953.89
3	3,434.02	173.85	3,607.87	10,561.76
4	3,434.02	264.04	3,698.06	14,259.82
5	3,434.02	356.50	3,790.52	18,050.34
6	3,434.02	451.26	3,885.28	21,935.62
7	3,434.02	548.39	3,982.41	25,918.03
8	3,434.02	647.95	4,081.97	30,000.00

The formula derived in this last example is restated as follows.

Sinking Fund Payment
The periodic payment R required to accumulate a sum of S dollars over n periods with interest charged at the rate of i per period is

$$R = \frac{iS}{(1 + i)^n - 1}$$

5.3 Self-Check Exercises

1. The Mendozas wish to borrow $100,000 from a bank to help finance the purchase of a house. Their banker has offered the following plans for their consideration. In plan I, the Mendozas have 30 years to repay the loan in monthly installments with interest on the unpaid balance charged at 10.5%/yr compounded monthly. In plan II, the loan is to be repaid in monthly installments over 15 years with interest on the unpaid balance charged at 9.75%/yr compounded monthly.
   a. Find the monthly repayment for each plan.
   b. What is the difference in total payments made under each plan?

2. Harris, a self-employed individual who is 46 years old, is setting up a defined-benefit retirement plan. If he wishes to have $250,000 in this retirement account by age 65, what is the size of each yearly installment he will be required to make into a savings account earning interest at $8\frac{1}{4}$%/yr? (Assume that Harris is eligible to make each of the 20 required contributions.)

Solutions to Self-Check Exercises 5.3 can be found on page 318.

5.3 Concept Questions

1. Write the amortization formula.
   a. If P and i are fixed and n is allowed to increase, what will happen to R?
   b. Interpret the result of part (a).

2. Using the formula for computing a sinking fund payment, show that if the number of payments into a sinking fund increases, then the size of the periodic payment into the sinking fund decreases.

5.3 Exercises

In Exercises 1–8, find the periodic payment R required to amortize a loan of P dollars over t years with interest earned at the rate of r%/year compounded m times a year.

1. $P = 100,000, r = 8, t = 10, m = 1$

2. $P = 40,000, r = 3, t = 15, m = 2$

3. $P = 5000, r = 4, t = 3, m = 4$

4. $P = 16,000, r = 9, t = 4, m = 12$

5. $P = 25,000, r = 3, t = 12, m = 4$

6. $P = 80,000, r = 10.5, t = 15, m = 12$

7. $P = 80,000, r = 10.5, t = 30, m = 12$

8. $P = 100,000, r = 10.5, t = 25, m = 12$

In Exercises 9–14, find the periodic payment R required to accumulate a sum of S dollars over t years with interest earned at the rate of r%/year compounded m times a year.

9. $S = 20,000, r = 4, t = 6, m = 2$

10. $S = 40,000, r = 4, t = 9, m = 4$

11. $S = 100,000, r = 4.5, t = 20, m = 6$

12. $S = 120,000, r = 4.5, t = 30, m = 6$

13. $S = 250,000, r = 10.5, t = 25, m = 12$

14. $S = 350,000, r = 7.5, t = 10, m = 12$

15. Suppose payments were made at the end of each quarter into an ordinary annuity earning interest at the rate of 10%/yr compounded quarterly. If the future value of the annuity after 5 years is $50,000, what was the size of each payment?

16. Suppose payments were made at the end of each month into an ordinary annuity earning interest at the rate of 9%/yr compounded monthly. If the future value of the annuity after 10 years is $60,000, what was the size of each payment?

17. Suppose payments will be made for $6\frac{1}{2}$ years at the end of each semiannual period into an ordinary annuity earning interest at the rate of 7.5%/yr compounded semiannually. If the present value of the annuity is $35,000, what should be the size of each payment?

18. Suppose payments will be made for $9\frac{1}{4}$ years at the end of each month into an ordinary annuity earning interest at the rate of 6.25%/yr compounded monthly. If the present value of the annuity is $42,000, what should be the size of each payment?

19. **LOAN AMORTIZATION** A sum of $100,000 is to be repaid over a 10-year period through equal installments made at the end of each year. If an interest rate of 10%/yr is charged on the unpaid balance and interest calculations are made at the end of each year, determine the size of each installment so that the loan (principal plus interest charges) is amortized at the end of 10 years.

20. **LOAN AMORTIZATION** What monthly payment is required to amortize a loan of $30,000 over 10 years if interest at the rate of 12%/yr is charged on the unpaid balance and interest calculations are made at the end of each month?

21. **HOME MORTGAGES** Complete the following table, which shows the monthly payments on a $100,000, 30-year mortgage at the interest rates shown. Use this information to answer the following questions.

Amount of Mortgage, $	Interest Rate, %	Monthly Payment, $
100,000	7	665.30
100,000	8	. . .
100,000	9	. . .
100,000	10	. . .
100,000	11	. . .
100,000	12	1028.61

a. What is the difference in monthly payments between a $100,000, 30-yr mortgage secured at 7%/yr and one secured at 10%/yr?

b. Use the table to calculate the monthly mortgage payments on a $150,000 mortgage at 10%/yr over 30 yr and a $50,000 mortgage at 10%/yr over 30 yr.

22. **FINANCING A HOME** The Flemings secured a bank loan of $96,000 to help finance the purchase of a house. The bank charges interest at a rate of 9%/yr on the unpaid balance, and interest computations are made at the end of each month. The Flemings have agreed to repay the loan in equal monthly installments over 25 years. What should be the size of each repayment if the loan is to be amortized at the end of the term?

23. **FINANCING A CAR** The price of a new car is $16,000. Assume that an individual makes a down payment of 25% toward the purchase of the car and secures financing for the balance at the rate of 10%/yr compounded monthly.

a. What monthly payment will she be required to make if the car is financed over a period of 36 mo? Over a period of 48 mo?

b. What will the interest charges be if she elects the 36-mo plan? The 48-mo plan?

24. **FINANCIAL ANALYSIS** A group of private investors purchased a condominium complex for $2 million. They made an initial down payment of 10% and obtained financing for the balance. If the loan is to be amortized over 15 years at an inter-

est rate of 12%/yr compounded quarterly, find the required quarterly payment.

25. **FINANCING A HOME** The Taylors have purchased a $270,000 house. They made an initial down payment of $30,000 and secured a mortgage with interest charged at the rate of 8%/yr on the unpaid balance. Interest computations are made at the end of each month. If the loan is to be amortized over 30 yr, what monthly payment will the Taylors be required to make? What is their equity (disregarding appreciation) after 5 yr? After 10 yr? After 20 yr?

26. **FINANCIAL PLANNING** Jessica wants to accumulate $10,000 by the end of 5 yr in a special bank account, which she had opened for this purpose. To achieve this goal, Jessica plans to deposit a fixed sum of money into the account at the end of the month over the 5-yr period. If the bank pays interest at the rate of 5%/yr compounded monthly, how much does she have to deposit each month into her account?

27. **SINKING FUNDS** A city has $2.5 million worth of school bonds that are due in 20 years and has established a sinking fund to retire this debt. If the fund earns interest at the rate of 7%/yr compounded annually, what amount must be deposited annually in this fund?

28. **TRUST FUNDS** Carl is the beneficiary of a $20,000 trust fund set up for him by his grandparents. Under the terms of the trust, he is to receive the money over a 5-yr period in equal installments at the end of each year. If the fund earns interest at the rate of 9%/yr compounded annually, what amount will he receive each year?

29. **SINKING FUNDS** Lowell Corporation wishes to establish a sinking fund to retire a $200,000 debt that is due in 10 years. If the investment will earn interest at the rate of 9%/yr compounded quarterly, find the amount of the quarterly deposit that must be made in order to accumulate the required sum.

30. **SINKING FUNDS** The management of Gibraltar Brokerage Services anticipates a capital expenditure of $20,000 in 3 years' time for the purpose of purchasing new fax machines and has decided to set up a sinking fund to finance this purchase. If the fund earns interest at the rate of 10%/yr compounded quarterly, determine the size of each (equal) quarterly installment that should be deposited in the fund.

31. **RETIREMENT ACCOUNTS** Andrea, a self-employed individual, wishes to accumulate a retirement fund of $250,000. How much should she deposit each month into her retirement account, which pays interest at the rate of 8.5%/yr compounded monthly, to reach her goal upon retirement 25 years from now?

32. **STUDENT LOANS** Joe secured a loan of $12,000 3 years ago from a bank for use toward his college expenses. The bank charges interest at the rate of 4%/yr compounded monthly on his loan. Now that he has graduated from college, Joe wishes to repay the loan by amortizing it through monthly payments over 10 years at the same interest rate. Find the size of the monthly payments he will be required to make.

33. RETIREMENT ACCOUNTS Robin wishes to accumulate a sum of $450,000 in a retirement account by the time of her retirement 30 years from now. If she wishes to do this through monthly payments into the account that earn interest at the rate of 10%/yr compounded monthly, what should be the size of each payment?

34. FINANCING COLLEGE EXPENSES Yumi's grandparents presented her with a gift of $20,000 when she was 10 yr old to be used for her college education. Over the next 7 yr, until she turned 17, Yumi's parents had invested her money in a tax-free account that had yielded interest at the rate of 5.5%/yr compounded monthly. Upon turning 17, Yumi now plans to withdraw her funds in equal annual installments over the next 4 yr, starting at age 18. If the college fund is expected to earn interest at the rate of 6%/yr, compounded annually, what will be the size of each installment?

35. IRAs Martin has deposited $375 in his IRA at the end of each quarter for the past 20 years. His investment has earned interest at the rate of 8%/yr compounded quarterly over this period. Now, at age 60, he is considering retirement. What quarterly payment will he receive over the next 15 years? (Assume that the money is earning interest at the same rate and that payments are made at the end of each quarter.) If he continues working and makes quarterly payments of the same amount in his IRA account until age 65, what quarterly payment will he receive from his fund upon retirement over the following 10 years?

36. FINANCING A CAR Darla purchased a new car during a special sales promotion by the manufacturer. She secured a loan from the manufacturer in the amount of $16,000 at a rate of 7.9%/yr compounded monthly. Her bank is now charging 11.5%/yr compounded monthly for new car loans. Assuming that each loan would be amortized by 36 equal monthly installments, determine the amount of interest she would have paid at the end of 3 years for each loan. How much less will she have paid in interest payments over the life of the loan by borrowing from the manufacturer instead of her bank?

37. FINANCING A HOME The Sandersons are planning to refinance their home. The outstanding principal on their original loan is $100,000 and was to be amortized in 240 equal monthly installments at an interest rate of 10%/yr compounded monthly. The new loan they expect to secure is to be amortized over the same period at an interest rate of 7.8%/yr compounded monthly. How much less can they expect to pay over the life of the loan in interest payments by refinancing the loan at this time?

38. INVESTMENT ANALYSIS Since he was 22 years old, Ben has been depositing $200 at the end of each month into a tax-free retirement account earning interest at the rate of 6.5%/yr compounded monthly. Larry, who is the same age as Ben, decided to open a tax-free retirement account 5 years after Ben opened his. If Larry's account earns interest at the same

rate as Ben's, determine how much Larry should deposit each month into his account so that both men will have the same amount of money in their accounts at age 65.

39. FINANCING A HOME Eight years ago, Kim secured a bank loan of $180,000 to help finance the purchase of a house. The mortgage was for a term of 30 years, with an interest rate of 9.5%/yr compounded monthly on the unpaid balance to be amortized through monthly payments. What is the outstanding principal on Kim's house now?

40. BALLOON PAYMENT MORTGAGE Olivia plans to secure a 5-yr balloon mortgage of $200,000 toward the purchase of a condominium. Her monthly payment for the 5 yr is calculated on the basis of a 30-yr conventional mortgage at the rate of 6%/yr compounded monthly. At the end of the 5 yr, Olivia is required to pay the balance owed (the "balloon" payment). What will be her monthly payment, and what will be her balloon payment?

41. FINANCING A HOME Sarah secured a bank loan of $200,000 for the purchase of a house. The mortgage is to be amortized through monthly payments for a term of 15 yr, with an interest rate of 6%/yr compounded monthly on the unpaid balance. She plans to sell her house in 5 yr. How much will Sarah still owe on her house?

42. HOME REFINANCING Five years ago, Diane secured a bank loan of $300,000 to help finance the purchase of a loft in the San Francisco Bay area. The term of the mortgage was 30 yr, and the interest rate was 9%/yr compounded monthly on the unpaid balance. Because the interest rate for a conventional 30-yr home mortgage has now dropped to 7%/yr compounded monthly, Diane is thinking of refinancing her property.
 a. What is Diane's current monthly mortgage payment?
 b. What is Diane's current outstanding principal?
 c. If Diane decides to refinance her property by securing a 30-yr home mortgage loan in the amount of the current outstanding principal at the prevailing interest rate of 7%/yr compounded monthly, what will be her monthly mortgage payment?
 d. How much less would Diane's monthly mortgage payment be if she refinances?

43. HOME REFINANCING Four years ago, Emily secured a bank loan of $200,000 to help finance the purchase of an apartment in Boston. The term of the mortgage is 30 yr, and the interest rate is 9.5%/yr compounded monthly. Because the interest rate for a conventional 30-yr home mortgage has now dropped to 6.75%/yr compounded monthly, Emily is thinking of refinancing her property.
 a. What is Emily's current monthly mortgage payment?
 b. What is Emily's current outstanding principal?
 c. If Emily decides to refinance her property by securing a 30-yr home mortgage loan in the amount of the current outstanding principal at the prevailing interest rate of

6.75%/yr compounded monthly, what will be her monthly mortgage payment?

d. How much less would Emily's monthly mortgage payment be if she refinances?

44. **ADJUSTABLE-RATE MORTGAGE**   Three years ago, Samantha secured an adjustable-rate mortgage (ARM) loan to help finance the purchase of a house. The amount of the original loan was $150,000 for a term of 30 years, with interest at the rate of 7.5%/yr compounded monthly. Currently the interest rate is 7%/yr compounded monthly, and Samantha's monthly payments are due to be recalculated. What will be her new monthly payment?
Hint: Calculate her current outstanding principal. Then, to amortize the loan in the next 27 years, determine the monthly payment based on the current interest rate.

45. **FINANCING A HOME**   After making a down payment of $25,000, the Meyers need to secure a loan of $280,000 to purchase a certain house. Their bank's current rate for 25-yr home loans is 11%/yr compounded monthly. The owner has offered to finance the loan at 9.8%/yr compounded monthly. Assuming that both loans would be amortized over a 25-yr

period by 300 equal monthly installments, determine the difference in the amount of interest the Meyers would pay by choosing the seller's financing rather than their bank's.

46. **REFINANCING A HOME**   The Martinezes are planning to refinance their home. The outstanding balance on their original loan is $150,000. Their finance company has offered them two options:

Option A: A fixed-rate mortgage at an interest rate of 7.5%/yr compounded monthly, payable over a 30-yr period in 360 equal monthly installments.

Option B: A fixed-rate mortgage at an interest rate of 7.25%/yr compounded monthly, payable over a 15-yr period in 180 equal monthly installments. (Assume that there are no additional finance charges.)

a. Find the monthly payment required to amortize each of these loans over the life of the loan.

b. How much interest would the Martinezes save if they chose the 15-yr mortgage instead of the 30-yr mortgage?

5.3   Solutions to Self-Check Exercises

1. a. We use Equation (13) in each instance. Under plan I,

$$P = 100{,}000 \qquad i = \frac{r}{m} = \frac{0.105}{12} = 0.00875$$

$$n = (30)(12) = 360$$

Therefore, the size of each monthly repayment under plan I is

$$R = \frac{100{,}000(0.00875)}{1 - (1.00875)^{-360}}$$

$$\approx 914.74$$

or $914.74.

Under plan II,

$$P = 100{,}000 \qquad i = \frac{r}{m} = \frac{0.0975}{12} = 0.008125$$

$$n = (15)(12) = 180$$

Therefore, the size of each monthly repayment under plan II is

$$R = \frac{100{,}000(0.008125)}{1 - (1.008125)^{-180}}$$

$$\approx 1059.36$$

or $1059.36.

b. Under plan I, the total amount of repayments will be

$$(360)(914.74) = 329{,}306.40 \qquad \text{Number of payments} \atop \text{× the size of each installment}$$

or $329,306.40. Under plan II, the total amount of repayments will be

$$(180)(1059.36) = 190{,}684.80$$

or $190,684.80. Therefore, the difference in payments is

$$329{,}306.40 - 190{,}684.80 = 138{,}621.60$$

or $138,621.60.

2. We use Equation (14) with

$$S = 250{,}000$$

$$i = r = 0.0825 \qquad \text{Since } m = 1$$

$$n = 20$$

giving the required size of each installment as

$$R = \frac{(0.0825)(250{,}000)}{(1.0825)^{20} - 1}$$

$$\approx 5313.59$$

or $5313.59.

USING TECHNOLOGY

■ Amortizing a Loan

Graphing Utility

Here we use the TI-83 TVM SOLVER function to help us solve problems involving amortization and sinking funds.

 APPLIED EXAMPLE 1 Finding the Payment to Amortize a Loan The Wongs are considering obtaining a preapproved 30-year loan of $120,000 to help finance the purchase of a house. The mortgage company charges interest at the rate of 8% per year on the unpaid balance, with interest computations made at the end of each month. What will be the monthly installments if the loan is amortized at the end of the term?

Solution We use the TI-83 TVM SOLVER with the following inputs:

$$N = 360 \qquad \text{(30)(12)}$$
$$I\% = 8$$
$$PV = 120000$$
$$PMT = 0$$
$$FV = 0$$
$$P/Y = 12 \qquad \text{The number of payments each year}$$
$$C/Y = 12 \qquad \text{The number of conversion periods each year}$$

PMT:END BEGIN

```
N    = 360
I%   = 8
PV   = 120000
■ PMT = −880.51748...
FV   = 0
P/Y  = 12
C/Y  = 12
PMT : END    BEGIN
```

FIGURE T1
The TI-83 screen showing the monthly installment, PMT

From the output shown in Figure T1, we see that the required payment is $880.52.

 APPLIED EXAMPLE 2 Finding the Payment in a Sinking Fund Heidi wishes to establish a retirement account that will be worth $500,000 in 20 years' time. She expects that the account will earn interest at the rate of 11% per year compounded monthly. What should be the monthly contribution into her account each month?

Solution We use the TI-83 TVM SOLVER with the following inputs:

$$N = 240 \qquad \text{(20)(12)}$$
$$I\% = 11$$
$$PV = 0$$
$$PMT = 0$$
$$FV = 500000$$
$$P/Y = 12 \qquad \text{The number of payments each year}$$
$$C/Y = 12 \qquad \text{The number of conversion periods each year}$$

PMT:END BEGIN

```
N    = 240
I%   = 11
PV   = 0
■ PMT = −577.60862...
FV   = 500000
P/Y  = 12
C/Y  = 12
PMT : END    BEGIN
```

FIGURE T2
The TI-83 screen showing the monthly payment, PMT

The result is displayed in Figure T2. We see that Heidi's monthly contribution should be $577.61. (*Note:* The display for PMT is negative because it is an outflow.)

(continued)

Excel

Here we use Excel to help us solve problems involving amortization and sinking funds.

APPLIED EXAMPLE 3   Finding the Payment to Amortize a Loan   The Wongs are considering a preapproved 30-year loan of $120,000 to help finance the purchase of a house. The mortgage company charges interest at the rate of 8% per year on the unpaid balance, with interest computations made at the end of each month. What will be the monthly installments if the loan is amortized at the end of the term?

Solution   We use the PMT function to solve this problem. Accessing this function from the Insert Function dialog box and making the required entries, we obtain the Function Arguments dialog box shown in Figure T3. We see that the desired result is $880.52. (Recall that cash you pay out is represented by a negative number.)

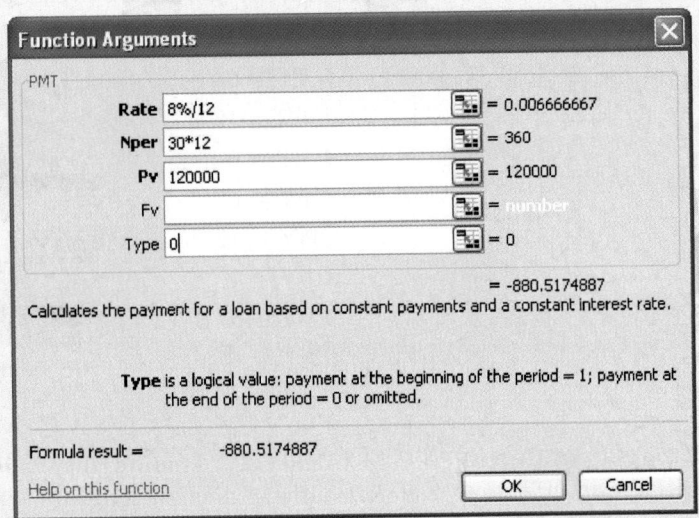

FIGURE T3
Excel's dialog box giving the payment function, PMT

APPLIED EXAMPLE 4   Finding the Payment in a Sinking Fund   Heidi wishes to establish a retirement account that will be worth $500,000 in 20 years' time. She expects that the account will earn interest at the rate of 11% per year compounded monthly. What should be the monthly contribution into her account each month?

Solution   As in Example 3, we use the PMT function, but this time we are given the future value of the investment. Accessing the PMT function from the Insert Function dialog box and making the required entries, we obtain the Function

Note: Words/characters printed blue (for example, Chart sub-type:) indicate words/characters on the screen.

Arguments dialog box shown in Figure T4. We see that Heidi's monthly contribution should be $577.61. (Note that the value for PMT is negative because it is an outflow.)

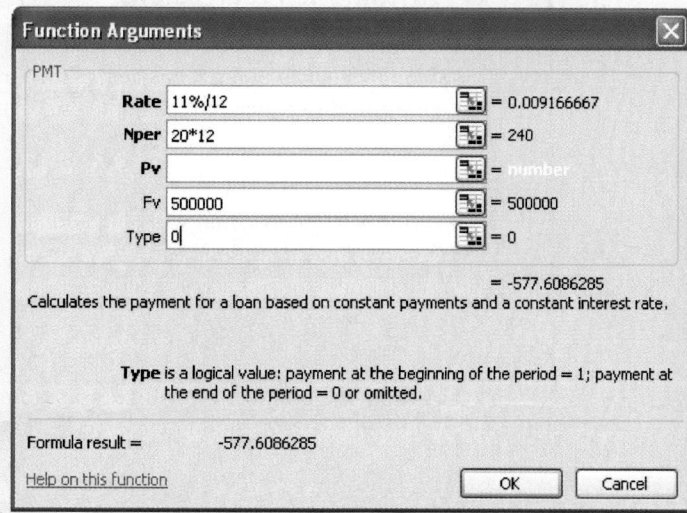

FIGURE T4
Excel's dialog box giving the payment function, PMT

TECHNOLOGY EXERCISES

1. Find the periodic payment required to amortize a loan of $55,000 over 120 months with interest earned at the rate of $6\frac{5}{8}\%$/yr compounded monthly.

2. Find the periodic payment required to amortize a loan of $178,000 over 180 months with interest earned at the rate of $7\frac{1}{8}\%$/yr compounded monthly.

3. Find the periodic payment required to amortize a loan of $227,000 over 360 months with interest earned at the rate of $8\frac{1}{8}\%$/yr compounded monthly.

4. Find the periodic payment required to amortize a loan of $150,000 over 360 months with interest earned at the rate of $7\frac{3}{8}\%$/yr compounded monthly.

5. Find the periodic payment required to accumulate $25,000 over 12 quarters with interest earned at the rate of $4\frac{3}{8}\%$/yr compounded quarterly.

6. Find the periodic payment required to accumulate $50,000 over 36 quarters with interest earned at the rate of $3\frac{7}{8}\%$/yr compounded quarterly.

7. Find the periodic payment required to accumulate $137,000 over 120 months with interest earned at the rate of $4\frac{3}{4}\%$/yr compounded monthly.

8. Find the periodic payment required to accumulate $144,000 over 120 months with interest earned at the rate of $4\frac{5}{8}\%$/yr compounded monthly.

9. A loan of $120,000 is to be repaid over a 10-year period through equal installments made at the end of each year. If an interest rate of 8.5%/yr is charged on the unpaid balance and interest calculations are made at the end of each year, determine the size of each installment so that the loan is amortized at the end of 10 years. Verify the result by displaying the amortization schedule.

10. A loan of $265,000 is to be repaid over an 8-year period through equal installments made at the end of each year. If an interest rate of 7.4%/yr is charged on the unpaid balance and interest calculations are made at the end of each year, determine the size of each installment so that the loan is amortized at the end of 8 years. Verify the result by displaying the amortization schedule.

5.4 Arithmetic and Geometric Progressions (Optional)

Arithmetic Progressions

An **arithmetic progression** is a sequence of numbers in which each term after the first is obtained by adding a constant d to the preceding term. The constant d is called the **common difference.** For example, the sequence

$$3, 6, 9, 12, \ldots$$

is an arithmetic progression with the common difference equal to 3.

Observe that an arithmetic progression is completely determined if the first term and the common difference are known. In fact, if

$$a_1, a_2, a_3, \ldots, a_n, \ldots$$

is an arithmetic progression with the first term given by a and common difference given by d, then by definition we have

$$
\begin{aligned}
a_1 &= a \\
a_2 &= a_1 + d = a + d \\
a_3 &= a_2 + d = (a + d) + d = a + 2d \\
a_4 &= a_3 + d = (a + 2d) + d = a + 3d \\
&\vdots \\
a_n &= a_{n-1} + d = a + (n - 2)d + d = a + (n - 1)d
\end{aligned}
$$

Thus, we see that the nth term of an arithmetic progression with first term a and common difference d is given by

$$a_n = a + (n - 1)d \qquad \textbf{(15)}$$

---

nth Term of an Arithmetic Progression
The nth term of an arithmetic progression with first term a and common difference d is given by

$$a_n = a + (n - 1)d$$

EXAMPLE 1 Find the twelfth term of the arithmetic progression

$$2, 7, 12, 17, 22, \ldots$$

Solution The first term of the arithmetic progression is $a_1 = a = 2$, and the common difference is $d = 5$; so, upon setting $n = 12$ in Equation (15), we find

$$a_{12} = 2 + (12 - 1)5 = 57 \qquad \blacksquare$$

EXAMPLE 2 Write the first five terms of an arithmetic progression whose third and eleventh terms are 21 and 85, respectively.

Solution Using Equation (15), we obtain

$$a_3 = a + 2d = 21$$
$$a_{11} = a + 10d = 85$$

Subtracting the first equation from the second gives $8d = 64$, or $d = 8$. Substituting this value of d into the first equation yields $a + 16 = 21$, or $a = 5$. Thus, the required arithmetic progression is given by the sequence

$$5, 13, 21, 29, 37, \ldots$$ ∎

Let S_n denote the sum of the first n terms of an arithmetic progression with first term $a_1 = a$ and common difference d. Then

$$S_n = a + (a + d) + (a + 2d) + \cdots + [a + (n - 1)d] \qquad \textbf{(16)}$$

Rewriting the expression for S_n with the terms in reverse order gives

$$S_n = [a + (n - 1)d] + [a + (n - 2)d] + \cdots + (a + d) + a \qquad \textbf{(17)}$$

Adding Equations (16) and (17), we obtain

$$2S_n = [2a + (n - 1)d] + [2a + (n - 1)d]$$
$$+ \cdots + [2a + (n - 1)d]$$
$$= n[2a + (n - 1)d]$$
$$S_n = \frac{n}{2}[2a + (n - 1)d]$$

Sum of Terms in an Arithmetic Progression
The sum of the first n terms of an arithmetic progression with first term a and common difference d is given by

$$S_n = \frac{n}{2}[2a + (n - 1)d] \qquad \textbf{(18)}$$

EXAMPLE 3 Find the sum of the first 20 terms of the arithmetic progression of Example 1.

Solution Letting $a = 2$, $d = 5$, and $n = 20$ in Equation (18), we obtain

$$S_{20} = \frac{20}{2}[2 \cdot 2 + 19 \cdot 5] = 990$$ ∎

APPLIED EXAMPLE 4 Company Sales Madison Electric Company had sales of \$200,000 in its first year of operation. If the sales increased by \$30,000 per year thereafter, find Madison's sales in the fifth year and its total sales over the first 5 years of operation.

Solution Madison's yearly sales follow an arithmetic progression, with the first term given by $a = 200{,}000$ and the common difference given by $d = 30{,}000$. The sales in the fifth year are found by using Equation (15) with $n = 5$. Thus,

$$a_5 = 200{,}000 + (5 - 1)30{,}000 = 320{,}000$$

or \$320,000.

Madison's total sales over the first 5 years of operation are found by using (18) with $n = 5$. Thus,

$$S_5 = \frac{5}{2}[2(200{,}000) + (5 - 1)30{,}000]$$
$$= 1{,}300{,}000$$

or $1,300,000.

Geometric Progressions

A **geometric progression** is a sequence of numbers in which each term after the first is obtained by multiplying the preceding term by a constant r. The constant r is called the **common ratio**.

A geometric progression is completely determined if the first term and the common ratio are known. Thus, if

$$a_1, a_2, a_3, \ldots, a_n, \ldots$$

is a geometric progression with the first term given by a and common ratio given by r, then by definition we have

$$a_1 = a$$
$$a_2 = a_1 r = ar$$
$$a_3 = a_2 r = ar^2$$
$$a_4 = a_3 r = ar^3$$
$$\vdots$$
$$a_n = a_{n-1} r = ar^{n-1}$$

Thus, we see that the nth term of a geometric progression with first term a and common ratio r is given by

$$a_n = ar^{n-1} \tag{19}$$

nth Term of a Geometric Progression
The nth term of a geometric progression with first term a and common ratio r is given by

$$a_n = ar^{n-1}$$

EXAMPLE 5   Find the eighth term of a geometric progression whose first five terms are 162, 54, 18, 6, and 2.

Solution   The common ratio is found by taking the ratio of any term other than the first to the preceding term. Taking the ratio of the fourth term to the third term, for example, gives $r = \frac{6}{18} = \frac{1}{3}$. To find the eighth term of the geometric progression, use Formula (19) with $a = 162$, $r = \frac{1}{3}$, and $n = 8$, obtaining

$$a_8 = 162\left(\frac{1}{3}\right)^7$$

$$= \frac{2}{27}$$

EXAMPLE 6   Find the tenth term of a geometric progression whose third term is 16 and whose seventh term is 1.

Solution   Using Equation (19) with $n = 3$ and $n = 7$, respectively, yields

$$a_3 = ar^2 = 16$$
$$a_7 = ar^6 = 1$$

Dividing a_7 by a_3 gives

$$\frac{ar^6}{ar^2} = \frac{1}{16}$$

from which we obtain $r^4 = \frac{1}{16}$, or $r = \frac{1}{2}$. Substituting this value of r into the expression for a_3, we obtain

$$a\left(\frac{1}{2}\right)^2 = 16 \quad \text{or} \quad a = 64$$

Finally, using (19) once again with $a = 64$, $r = \frac{1}{2}$, and $n = 10$ gives

$$a_{10} = 64\left(\frac{1}{2}\right)^9 = \frac{1}{8}$$

To find the sum of the first n terms of a geometric progression with the first term $a_1 = a$ and common ratio r, denote the required sum by S_n. Then,

$$S_n = a + ar + ar^2 + \cdots + ar^{n-2} + ar^{n-1} \tag{20}$$

Upon multiplying (20) by r, we obtain

$$rS_n = ar + ar^2 + ar^3 + \cdots + ar^{n-1} + ar^n \tag{21}$$

Subtracting (21) from (20) gives

$$S_n - rS_n = a - ar^n$$
$$(1 - r)S_n = a(1 - r^n)$$

If $r \neq 1$, we may divide both sides of the last equation by $(1 - r)$, obtaining

$$S_n = \frac{a(1 - r^n)}{(1 - r)}$$

If $r = 1$, then (20) gives

$$S_n = a + a + a + \cdots + a \quad \text{\scriptsize n terms}$$
$$= na$$

Thus,

$$S_n = \begin{cases} \dfrac{a(1 - r^n)}{1 - r} & \text{if } r \neq 1 \\ na & \text{if } r = 1 \end{cases}$$

Sum of Terms in a Geometric Progression
The sum of the first n terms of a geometric progression with first term a and common ratio r is given by

$$S_n = \begin{cases} \dfrac{a(1 - r^n)}{1 - r} & \text{if } r \neq 1 \\ na & \text{if } r = 1 \end{cases} \tag{22}$$

EXAMPLE 7   Find the sum of the first six terms of the following geometric progression:

$$3, 6, 12, 24, \ldots$$

Solution   Here, $a = 3$, $r = \frac{6}{3} = 2$, and $n = 6$, so Formula (22) gives

$$S_6 = \frac{3(1 - 2^6)}{1 - 2} = 189$$

APPLIED EXAMPLE 8   Company Sales   Michaelson Land Development Company had sales of $1 million in its first year of operation. If sales increased by 10% per year thereafter, find Michaelson's sales in the fifth year and its total sales over the first 5 years of operation.

Solution   Michaelson's yearly sales follow a geometric progression, with the first term given by $a = 1{,}000{,}000$ and the common ratio given by $r = 1.1$. The sales in the fifth year are found by using Formula (19) with $n = 5$. Thus,

$$a_5 = 1{,}000{,}000(1.1)^4 = 1{,}464{,}100$$

or $1,464,100.

Michaelson's total sales over the first 5 years of operation are found by using Equation (22) with $n = 5$. Thus,

$$S_5 = \frac{1{,}000{,}000[1 - (1.1)^5]}{1 - 1.1}$$
$$= 6{,}105{,}100$$

or $6,105,100.

Double Declining-Balance Method of Depreciation

In Section 1.3, we discussed the straight-line, or linear, method of depreciating an asset. Linear depreciation assumes that the asset depreciates at a constant rate. For certain assets (such as machines) whose market values drop rapidly in the early years of usage and thereafter less rapidly, another method of depreciation called the **double declining-balance method** is often used. In practice, a business firm normally employs the double declining-balance method for depreciating such assets for a certain number of years and then switches over to the linear method.

To derive an expression for the book value of an asset being depreciated by the double declining-balance method, let C (in dollars) denote the original cost of the asset and let the asset be depreciated over N years. Using this method, the amount depreciated each year is $\frac{2}{N}$ times the value of the asset at the beginning of that year. Thus, the amount by which the asset is depreciated in its first year of use is given by $\frac{2C}{N}$, so if $V(1)$ denotes the book value of the asset at the end of the first year then

$$V(1) = C - \frac{2C}{N} = C\left(1 - \frac{2}{N}\right)$$

Next, if $V(2)$ denotes the book value of the asset at the end of the second year, then a similar argument leads to

$$V(2) = C\left(1 - \frac{2}{N}\right) - C\left(1 - \frac{2}{N}\right)\frac{2}{N}$$
$$= C\left(1 - \frac{2}{N}\right)\left(1 - \frac{2}{N}\right)$$
$$= C\left(1 - \frac{2}{N}\right)^2$$

Continuing, we find that if $V(n)$ denotes the book value of the asset at the end of n years, then the terms $C, V(1), V(2), \ldots, V(n)$ form a geometric progression with first term C and common ratio $\left(1 - \frac{2}{N}\right)$. Consequently, the nth term, $V(n)$, is given by

$$V(n) = C\left(1 - \frac{2}{N}\right)^n \qquad (1 \leq n \leq N) \tag{23}$$

Also, if $D(n)$ denotes the amount by which the asset has been depreciated by the end of the nth year, then

$$D(n) = C - C\left(1 - \frac{2}{N}\right)^n$$
$$= C\left[1 - \left(1 - \frac{2}{N}\right)^n\right] \tag{24}$$

APPLIED EXAMPLE 9   Depreciation of Equipment   A tractor purchased at a cost of $60,000 is to be depreciated by the double declining-balance method over 10 years. What is the book value of the tractor at the end of 5 years? By what amount has the tractor been depreciated by the end of the fifth year?

Solution   We have $C = 60,000$ and $N = 10$. Thus, using Formula (23) with $n = 5$ gives the book value of the tractor at the end of 5 years as

$$V(5) = 60,000\left(1 - \frac{2}{10}\right)^5$$
$$= 60,000\left(\frac{4}{5}\right)^5 = 19,660.80$$

or $19,660.80.

The amount by which the tractor has been depreciated by the end of the fifth year is given by

$$60{,}000 - 19{,}660.80 = 40{,}339.20$$

or $40,339.20. You may verify the last result by using Equation (24) directly. ■

EXPLORING WITH TECHNOLOGY

A tractor purchased at a cost of $60,000 is to be depreciated over 10 years with a residual value of $0. Using the double declining-balance method, its value at the end of n years is $V_1(n) = 60{,}000(0.8)^n$ dollars. Using straight-line depreciation, its value at the end of n years is $V_2(n) = 60{,}000 - 6000n$. Use a graphing utility to sketch the graphs of V_1 and V_2 in the viewing window $[0, 10] \times [0, 70{,}000]$. Comment on the relative merits of each method of depreciation.

5.4   Self-Check Exercises

1. Find the sum of the first five terms of the geometric progression with first term -24 and common ratio $-\frac{1}{2}$.

2. Office equipment purchased for $75,000 is to be depreciated by the double declining-balance method over 5 years. Find the book value at the end of 3 years.

3. Derive the formula for the future value of an annuity [Equation (9), Section 5.2].

Solutions to Self-Check Exercises 5.4 can be found on page 330.

5.4   Concept Questions

1. Suppose an arithmetic progression has first term a and common difference d.
   a. What is the formula for the nth term of this progression?
   b. What is the formula for the sum of the first n terms of this progression?

2. Suppose a geometric progression has first term a and common ratio r.
   a. What is the formula for the nth term of this progression?
   b. What is the formula for the sum of the first n terms of this progression?

5.4   Exercises

In Exercises 1–4, find the nth term of the arithmetic progression that has the given values of a, d, and n.

1. $a = 6, d = 3, n = 9$

2. $a = -5, d = 3, n = 7$

3. $a = -15, d = \dfrac{3}{2}, n = 8$

4. $a = 1.2, d = 0.4, n = 98$

5. Find the first five terms of the arithmetic progression whose fourth and eleventh terms are 30 and 107, respectively.

6. Find the first five terms of the arithmetic progression whose seventh and twenty-third terms are -5 and -29, respectively.

7. Find the seventh term of the arithmetic progression $x, x + y, x + 2y, \ldots$.

8. Find the eleventh term of the arithmetic progression $a + b, 2a, 3a - b, \ldots$.

9. Find the sum of the first 15 terms of the arithmetic progression $4, 11, 18, \ldots$.

10. Find the sum of the first 20 terms of the arithmetic progression $5, -1, -7, \ldots$.

11. Find the sum of the odd integers between 14 and 58.

12. Find the sum of the even integers between 21 and 99.

13. Find $f(1) + f(2) + f(3) + \cdots + f(20)$, given that $f(x) = 3x - 4$.

14. Find $g(1) + g(2) + g(3) + \cdots + g(50)$, given that $g(x) = 12 - 4x$.

15. Show that Equation (18) can be written as

$$S_n = \frac{n}{2}(a + a_n)$$

where a_n represents the last term of an arithmetic progression. Use this formula to find:
a. The sum of the first 11 terms of the arithmetic progression whose first and eleventh terms are 3 and 47, respectively.
b. The sum of the first 20 terms of the arithmetic progression whose first and twentieth terms are 5 and -33, respectively.

16. SALES GROWTH Moderne Furniture Company had sales of $1,500,000 during its first year of operation. If the sales increased by $160,000/year thereafter, find Moderne's sales in the fifth year and its total sales over the first 5 years of operation.

17. EXERCISE PROGRAM As part of her fitness program, Karen has taken up jogging. If she jogs 1 mi the first day and increases her daily run by 1/4 mi every week, how long will it take her to reach her goal of 10 mi/day?

18. COST OF DRILLING A 100-foot oil well is to be drilled. The cost of drilling the first foot is $10.00, and the cost of drilling each additional foot is $4.50 more than that of the preceding foot. Find the cost of drilling the entire 100 feet.

19. CONSUMER DECISIONS Kunwoo wishes to go from the airport to his hotel, which is 25 miles away. The taxi rate is $1.00 for the first mile and $.60 for each additional mile. The airport limousine also goes to his hotel and charges a flat rate of $7.50. How much money will the tourist save by taking the airport limousine?

20. SALARY COMPARISONS Markeeta, a recent college graduate, received two job offers. Company A offered her an initial salary of $48,800 with guaranteed annual increases of $2000/year for the first 5 years. Company B offered an initial salary of $50,400 with guaranteed annual increases of $1500 per year for the first 5 years.
a. Which company is offering a higher salary for the fifth year of employment?
b. Which company is offering more money for the first 5 years of employment?

21. SUM-OF-THE-YEARS'-DIGITS METHOD OF DEPRECIATION One of the methods that the Internal Revenue Service allows for computing depreciation of certain business property is the sum-of-the-years'-digits method. If a property valued at C dollars has an estimated useful life of N years and a salvage value of S dollars, then the amount of depreciation D_n allowed during the nth year is given by

$$D_n = (C - S)\frac{N - (n - 1)}{S_N} \qquad (0 \leq n \leq N)$$

where S_N is the sum of the first N positive integers representing the estimated useful life of the property. Thus,

$$S_N = 1 + 2 + \cdots + N = \frac{N(N + 1)}{2}$$

a. Verify that the sum of the arithmetic progression $S_N = 1 + 2 + \cdots + N$ is given by

$$\frac{N(N + 1)}{2}$$

b. If office furniture worth $6000 is to be depreciated by this method over $N = 10$ years and the salvage value of the furniture is $500, find the depreciation for the third year by computing D_3.

22. SUM-OF-THE-YEARS'-DIGITS METHOD OF DEPRECIATION Refer to Example 2, Section 1.3. The amount of depreciation allowed for a printing machine, which has an estimated useful life of 5 years and an initial value of $100,000 (with no salvage value), was $20,000/year using the straight-line method of depreciation. Determine the amount of depreciation that would be allowed for the first year if the printing machine were depreciated using the sum-of-the-years'-digits method described in Exercise 21. Which method would result in a larger depreciation of the asset in its first year of use?

In Exercises 23–28, determine which of the sequences are geometric progressions. For each geometric progression, find the seventh term and the sum of the first seven terms.

23. 4, 8, 16, 32, . . .

24. $1, -\dfrac{1}{2}, \dfrac{1}{4}, -\dfrac{1}{8}, \ldots$

25. $\dfrac{1}{2}, -\dfrac{3}{8}, \dfrac{1}{4}, -\dfrac{9}{64}, \ldots$

26. 0.004, 0.04, 0.4, 4, . . .

27. 243, 81, 27, 9, . . .

28. $-1, 1, 3, 5, \ldots$

29. Find the twentieth term and sum of the first 20 terms of the geometric progression $-3, 3, -3, 3, \ldots$.

30. Find the twenty-third term in a geometric progression having the first term $a = 0.1$ and ratio $r = 2$.

31. POPULATION GROWTH It has been projected that the population of a certain city in the southwest will increase by 8% during each of the next 5 years. If the current population is 200,000, what is the expected population in 5 years?

32. SALES GROWTH Metro Cable TV had sales of $2,500,000 in its first year of operation. If thereafter the sales increased by 12% of the previous year, find the sales of the company in the fifth year and the total sales over the first 5 years of operation.

33. COLAs   Suppose the cost-of-living index had increased by 3% during each of the past 6 years and that a member of the EUW Union had been guaranteed an annual increase equal to 2% above the cost-of-living index over that period. What would be the present salary of a union member whose salary 6 years ago was $42,000?

34. Savings Plans   The parents of a 9-yr-old boy have agreed to deposit $10 in their son's bank account on his 10th birthday and to double the size of their deposit every year thereafter until his 18th birthday.
   a. How much will they have to deposit on his 18th birthday?
   b. How much will they have deposited by his 18th birthday?

35. Salary Comparisons   A Stenton Printing Co. employee whose current annual salary is $48,000 has the option of taking an annual raise of 8%/yr for the next 4 years or a fixed annual raise of $4000/year. Which option would be more profitable to him considering his total earnings over the 4-yr period?

36. Bacteria Growth   A culture of a certain bacteria is known to double in number every 3 hr. If the culture has an initial count of 20, what will be the population of the culture at the end of 24 hr?

37. Trust Funds   Sarah is the recipient of a trust fund that she will receive over a period of 6 years. Under the terms of the trust, she is to receive $10,000 the first year and each succeeding annual payment is to be increased by 15%.
   a. How much will she receive during the sixth year?
   b. What is the total amount of the six payments she will receive?

In Exercises 38–40, find the book value of office equipment purchased at a cost C at the end of the nth year if it is to be depreciated by the double declining-balance method over 10 years. Assume a salvage value of $0.

38. $C = \$20,000$, $n = 4$ **39.** $C = \$150,000$, $n = 8$

40. $C = \$80,000$, $n = 7$

41. Double Declining-Balance Method of Depreciation   Restaurant equipment purchased at a cost of $150,000 is to be depreciated by the double declining-balance method over 10 years. What is the book value of the equipment at the end of 6 years? By what amount has the equipment been depreciated at the end of the sixth year?

42. Double Declining-Balance Method of Depreciation   Refer to Exercise 22. Recall that a printing machine with an estimated useful life of 5 years and an initial value of $100,000 (and no salvage value) was to be depreciated. At the end of the first year, the amount of depreciation allowed was $20,000 using the straight-line method and $33,333 using the sum-of-the-years'-digits method. Determine the amount of depreciation that would be allowed for the first year if the printing machine were depreciated by the double declining-balance method. Which of these three methods would result in the largest depreciation of the printing machine at the end of its first year of use?

In Exercises 43 and 44, determine whether the statement is true or false. If it is true, explain why it is true. If it is false, give an example to show why it is false.

43. If $a_1, a_2, a_3, \ldots, a_n$ and $b_1, b_2, b_3, \ldots, b_n$ are arithmetic progressions, then $a_1 + b_1, a_2 + b_2, a_3 + b_3, \ldots, a_n + b_n$ is also an arithmetic progression.

44. If $a_1, a_2, a_3, \ldots, a_n$ and $b_1, b_2, b_3, \ldots, b_n$ are geometric progressions, then $a_1b_1, a_2b_2, a_3b_3, \ldots, a_nb_n$ is also a geometric progression.

5.4   Solutions to Self-Check Exercises

1. Use Equation (22) with $a = -24$ and $r = -\frac{1}{2}$, obtaining

$$S_5 = \frac{-24\left[1 - \left(-\frac{1}{2}\right)^5\right]}{1 - \left(-\frac{1}{2}\right)}$$

$$= \frac{-24\left(1 + \frac{1}{32}\right)}{\frac{3}{2}} = -\frac{33}{2}$$

2. Use Equation (23) with $C = 75,000$, $N = 5$, and $n = 3$, giving the book value of the office equipment at the end of 3 years as

$$V(3) = 75,000\left(1 - \frac{2}{5}\right)^3 = 16,200$$

or $16,200.

3. We have

$$S = R + R(1 + i) + R(1 + i)^2 + \cdots + R(1 + i)^{n-1}$$

The sum on the right is easily seen to be the sum of the first n terms of a geometric progression with first term R and common ratio $(1 + i)$, so by virtue of Formula (22) we obtain

$$S = \frac{R[1 - (1 + i)^n]}{1 - (1 + i)} = R\left[\frac{(1 + i)^n - 1}{i}\right]$$

Summary of Principal Formulas and Terms

FORMULAS

1. Simple interest (accumulated amount)	$A = P(1 + rt)$
2. Compound interest	
a. Accumulated amount	$A = P(1 + i)^n$
b. Present value	$P = A(1 + i)^{-n}$
c. Interest rate per conversion period	$i = \dfrac{r}{m}$
d. Number of conversion periods	$n = mt$
3. Continuous compound interest	
a. Accumulated amount	$A = Pe^{rt}$
b. Present value	$P = Ae^{-rt}$
4. Effective rate of interest	$r_{\text{eff}} = \left(1 + \dfrac{r}{m}\right)^m - 1$
5. Annuities	
a. Future value	$S = R\left[\dfrac{(1 + i)^n - 1}{i}\right]$
b. Present value	$P = R\left[\dfrac{1 - (1 + i)^{-n}}{i}\right]$
6. Amortization payment	$R = \dfrac{Pi}{1 - (1 + i)^{-n}}$
7. Sinking fund payment	$R = \dfrac{iS}{(1 + i)^n - 1}$

TERMS

simple interest (278)

accumulated amount (278)

compound interest (279)

nominal rate (stated rate) (279)

conversion period (280)

effective rate (284)

present value (286)

future value (286)

annuity (296)

annuity certain (296)

ordinary annuity (296)

simple annuity (296)

future value of an annuity (298)

present value of an annuity (300)

sinking fund (313)

Concept Review Questions

Fill in the blanks.

1. a. Simple interest is computed on the _____ principal only. The formula for simple interest is $A =$ _____.

b. In calculations using compound interest, earned interest is periodically added to the principal and thereafter itself earns _____. The formula for compound interest is $A =$ _____. Solving this equation for P gives the present value formula for compound interest as $P =$ _____.

2. The effective rate of interest is the _____ interest rate that would produce the same accumulated amount in _____ year as the _____ rate compounded _____ times a year. The formula for calculating the effective rate is $r_{\text{eff}} =$ _____.

3. A sequence of payments made at regular time intervals is called a/an _____; if the payments are made at the end of each payment period, then it is called a/an _____ _____; if the payment period coincides with the interest conversion period, then it is called a/an _____.

4. a. The future value of an annuity is $S =$ _____.
   b. The present value of an annuity is $P =$ _____.

5. The periodic payment R on a loan of P dollars to be amortized over n periods with interest charged at the rate of i per period is $R =$ _____.

6. A sinking fund is an account that is set up for a specific purpose at some _____ date. The periodic payment R required to accumulate a sum of S dollars over n periods with interest charged at the rate of i per period is $R =$ _____.

7. An arithmetic progression is a sequence of numbers in which each term after the first is obtained by adding a/an _____ _____ to the preceding term. The nth term of an arithmetic progression is $a_n =$ _____. The sum of the first n terms of an arithmetic progression is $S_n =$ _____.

8. A geometric progression is a sequence of numbers in which each term after the first is obtained by multiplying the preceding term by a/an _____ _____. The nth term of a geometric progression is $a_n =$ _____. If $r \neq 1$, the sum of the first n terms of a geometric progression is $S_n =$ _____.

CHAPTER 5 Review Exercises

1. Find the accumulated amount after 4 years if $5000 is invested at 10%/yr compounded (a) annually, (b) semiannually, (c) quarterly, and (d) monthly.

2. Find the accumulated amount after 8 years if $12,000 is invested at 6.5%/yr compounded (a) annually, (b) semiannually, (c) quarterly, and (d) monthly.

3. Find the effective rate of interest corresponding to a nominal rate of 12%/yr compounded (a) annually, (b) semiannually, (c) quarterly, and (d) monthly.

4. Find the effective rate of interest corresponding to a nominal rate of 11.5%/yr compounded (a) annually, (b) semiannually, (c) quarterly, and (d) monthly.

5. Find the present value of $41,413 due in 5 years at an interest rate of 6.5%/yr compounded quarterly.

6. Find the present value of $64,540 due in 6 years at an interest rate of 8%/yr compounded monthly.

7. Find the amount (future value) of an ordinary annuity of $150/quarter for 7 years at 8%/yr compounded quarterly.

8. Find the future value of an ordinary annuity of $120/month for 10 years at 9%/yr compounded monthly.

9. Find the present value of an ordinary annuity of 36 payments of $250 each made monthly and earning interest at 9%/yr compounded monthly.

10. Find the present value of an ordinary annuity of 60 payments of $5000 each made quarterly and earning interest at 8%/yr compounded quarterly.

11. Find the payment R needed to amortize a loan of $22,000 at 8.5%/yr compounded monthly with 36 monthly installments over a period of 3 years.

12. Find the payment R needed to amortize a loan of $10,000 at 9.2%/yr compounded monthly with 36 monthly installments over a period of 3 years.

13. Find the payment R needed to accumulate $18,000 with 48 monthly installments over a period of 4 years at an interest rate of 6%/yr compounded monthly.

14. Find the payment R needed to accumulate $15,000 with 60 monthly installments over a period of 5 years at an interest rate of 7.2%/yr compounded monthly.

15. Find the rate of interest per year compounded on a daily basis that is equivalent to 7.2%/yr compounded monthly.

16. Find the rate of interest per year compounded on a daily basis that is equivalent to 9.6%/yr compounded monthly.

17. INVESTMENT RETURN A hotel was purchased by a conglomerate for $4.5 million and sold 5 years later for $8.2 million. Find the annual rate of return (compounded continuously).

18. Find the present value of $119,346 due in 4 years at an interest rate of 10%/yr compounded continuously.

19. COMPANY SALES JCN Media had sales of $1,750,000 in the first year of operation. If the sales increased by 14%/yr thereafter, find the company's sales in the fourth year and the total sales over the first 4 years of operation.

20. **CDs** The manager of a money market fund has invested $4.2 million in certificates of deposit that pay interest at the rate of 5.4%/yr compounded quarterly over a period of 5 yr. How much will the investment be worth at the end of 5 yr?

21. **SAVINGS ACCOUNTS** Emily deposited $2000 into a bank account 5 years ago. The bank paid interest at the rate of 8%/yr compounded weekly. What is Emily's account worth today?

22. **SAVINGS ACCOUNTS** Kim invested a sum of money 4 years ago in a savings account that has since paid interest at the rate of 6.5%/yr compounded monthly. Her investment is now worth $19,440.31. How much did she originally invest?

23. **SAVINGS ACCOUNTS** Andrew withdrew $5986.09 from a savings account, which he closed this morning. The account had earned interest at the rate of 6%/yr compounded continuously during the 3-yr period that the money was on deposit. How much did Andrew originally deposit into the account?

24. **MUTUAL FUNDS** Juan invested $24,000 in a mutual fund 5 years ago. Today his investment is worth $34,616. Find the effective annual rate of return on his investment over the 5-year period.

25. **COLLEGE SAVINGS PROGRAM** The Blakes have decided to start a monthly savings program in order to provide for their son's college education. How much should they deposit at the end of each month in a savings account earning interest at the rate of 8%/yr compounded monthly so that, at the end of the tenth year, the accumulated amount will be $40,000?

26. **RETIREMENT ACCOUNTS** Mai Lee has contributed $200 at the end of each month into her company's employee retirement account for the past 10 yr. Her employer has matched her contribution each month. If the account has earned interest at the rate of 8%/yr compounded monthly over the 10-yr period, determine how much Mai Lee now has in her retirement account.

27. **AUTOMOBILE LEASING** Maria has leased an auto for 4 years at $300/month. If money is worth 5%/yr compounded monthly, what is the equivalent cash payment (present value) of this annuity? (Assume that the payments are made at the end of each month.)

28. **INSTALLMENT FINANCING** Peggy made a down payment of $400 toward the purchase of new furniture. To pay the balance of the purchase price, she has secured a loan from her bank at 12%/yr compounded monthly. Under the terms of her finance agreement, she is required to make payments of $75.32 at the end of each month for 24 mo. What was the purchase price of the furniture?

29. **HOME FINANCING** The Turners have purchased a house for $150,000. They made an initial down payment of $30,000 and secured a mortgage with interest charged at the rate of 9%/yr on the unpaid balance. (Interest computations are made at the end of each month.) Assume the loan is amortized over 30 yr.
   a. What monthly payment will the Turners be required to make?
   b. What will be their total interest payment?
   c. What will be their equity (disregard depreciation) after 10 yr?

30. **HOME FINANCING** Refer to Exercise 29. If the loan is amortized over 15 yr:
   a. What monthly payment will the Turners be required to make?
   b. What will be their total interest payment?
   c. What will be their equity (disregard depreciation) after 10 yr?

31. **SINKING FUNDS** The management of a corporation anticipates a capital expenditure of $500,000 in 5 years for the purpose of purchasing replacement machinery. To finance this purchase, a sinking fund that earns interest at the rate of 10%/yr compounded quarterly will be set up. Determine the amount of each (equal) quarterly installment that should be deposited in the fund. (Assume that the payments are made at the end of each quarter.)

32. **SINKING FUNDS** The management of a condominium association anticipates a capital expenditure of $120,000 in 2 yr for the purpose of painting the exterior of the condominium. To pay for this maintenance, a sinking fund will be set up that will earn interest at the rate of 5.8%/yr compounded monthly. Determine the amount of each (equal) monthly installment the association will be required to deposit into the fund at the end of each month for the next 2 yr.

33. **CREDIT CARD PAYMENTS** The outstanding balance on Bill's credit-card account is $3200. The bank issuing the credit card is charging 18.6%/yr compounded monthly. If Bill decides to pay off this balance in equal monthly installments at the end of each month for the next 18 months, how much will be his monthly payment? What is the effective rate of interest the bank is charging Bill?

34. **FINANCIAL PLANNING** Matt's parents have agreed to contribute $250/month toward the rent for his apartment in his junior year in college. The plan is for Matt's parents to deposit a lump sum in Matt's bank account on August 1 and then have Matt withdraw $250 on the first of each month starting on September 1 and ending on May 1 the following year. If the bank pays interest on the balance at the rate of 5%/yr compounded monthly, how much should Matt's parents deposit into his account?

CHAPTER 5   Before Moving On . . .

1. Find the accumulated amount at the end of 3 years if $2000 is deposited in an account paying interest at the rate of 8%/yr compounded monthly.

2. Find the effective rate of interest corresponding to a nominal rate of 6%/yr compounded daily.

3. Find the future value of an ordinary annuity of $800/week for 10 years at 6%/yr compounded weekly.

4. Find the monthly payment required to amortize a loan of $100,000 over 10 years with interest earned at the rate of 8%/yr compounded monthly.

5. Find the weekly payment required to accumulate a sum of $15,000 over 6 years with interest earned at the rate of 10%/yr compounded weekly.

6. **Arithmetic and Geometric Progression (Optional)**
   a. Find the sum of the first ten terms of the arithmetic progression 3, 7, 11, 15, 19, ...
   b. Find the sum of the first eight terms of the geometric progression $\frac{1}{2}$, 2, 4, 8, ...

6 Sets and Counting

© Jason Homa/The Image Bank/Getty Images

What are the investment options? An investor has decided to purchase shares of stock from a recommended list of aerospace, energy development, and electronics companies. In Example 5, page 355, we will determine how many ways the investor may select a group of three companies from the list.

WE OFTEN DEAL with well-defined collections of objects called *sets*. In this chapter, we see how sets can be combined algebraically to yield other sets. We also look at some techniques for determining the number of elements in a set and for determining the number of ways the elements of a set can be arranged or combined. These techniques enable us to solve many practical problems, as you will see throughout the chapter.

6.1   Sets and Set Operations

■   Set Terminology and Notation

We often deal with collections of different kinds of objects. For example, in conducting a study of the distribution of the weights of newborn infants, we might consider the collection of all infants born in the Massachusetts General Hospital during 2006. In a study of the fuel consumption of compact cars, we might be interested in the collection of compact cars manufactured by General Motors in the 2006 model year. Such collections are examples of sets. More specifically, a **set** is a well-defined collection of objects. Thus, a set is not just any collection of objects; a set must be well defined in the sense that if we are given an object, then we should be able to determine whether or not it belongs to the collection.

The objects of a set are called the **elements**, or *members*, **of a set** and are usually denoted by lowercase letters $a, b, c, \ldots$; the sets themselves are usually denoted by uppercase letters $A, B, C, \ldots$. The elements of a set may be displayed by listing each element between braces. For example, using **roster notation**, the set A consisting of the first three letters of the English alphabet is written

$$A = \{a, b, c\}$$

The set B of all letters of the alphabet may be written

$$B = \{a, b, c, \ldots, z\}$$

Another notation commonly used is **set-builder notation**. Here, a rule is given that describes the definite property or properties an object x must satisfy to qualify for membership in the set. Using this notation, the set B is written as

$$B = \{x \mid x \text{ is a letter of the English alphabet}\}$$

and is read "B is the set of all elements x such that x is a letter of the English alphabet."

If a is an element of a set A, we write $a \in A$ and read "a belongs to A" or "a is an element of A." If, however, the element a does not belong to the set A, then we write $a \notin A$ and read "a does not belong to A." For example, if $A = \{1, 2, 3, 4, 5\}$, then $3 \in A$ but $6 \notin A$.

EXPLORE & DISCUSS

1. Let A denote the collection of all the days in August 2006 in which the average daily temperature in San Francisco was approximately 75°F. Is A a set? Explain your answer.

2. Let B denote the collection of all the days in August 2006 in which the average daily temperature in San Francisco was between 73.5°F and 81.2°F, inclusive. Is B a set? Explain your answer.

Set Equality

Two sets A and B are **equal**, written $A = B$, if and only if they have exactly the same elements.

EXAMPLE 1   Let A, B, and C be the sets

$$A = \{a, e, i, o, u\}$$
$$B = \{a, i, o, e, u\}$$
$$C = \{a, e, i, o\}$$

Then, $A = B$ since they both contain exactly the same elements. Note that the order in which the elements are displayed is immaterial. Also, $A \neq C$ since $u \in A$ but $u \notin C$. Similarly, we conclude that $B \neq C$.   ■

Subset

If every element of a set A is also an element of a set B, then we say that A is a subset of B and write $A \subseteq B$.

By this definition, two sets A and B are equal if and only if (1) $A \subseteq B$ and (2) $B \subseteq A$. You may verify this (see Exercise 66).

EXAMPLE 2   Referring to Example 1, we find that $C \subseteq B$ since every element of C is also an element of B. Also, if D is the set

$$D = \{a, e, i, o, x\}$$

then D is not a subset of A, written $D \nsubseteq A$, since $x \in D$ but $x \notin A$. Observe that $A \nsubseteq D$ as well, since $u \in A$ but $u \notin D$.   ■

If A and B are sets such that $A \subseteq B$ but $A \neq B$, then we say that A is a **proper subset** of B. In other words, a set A is a proper subset of a set B, written $A \subset B$, if (1) $A \subseteq B$ and (2) there exists at least one element in B that is not in A. The latter condition states that the set A is properly "smaller" than the set B.

EXAMPLE 3   Let $A = \{1, 2, 3, 4, 5, 6\}$ and $B = \{2, 4, 6\}$. Then B is a proper subset of A because (1) $B \subseteq A$, which is easily verified, and (2) there exists at least one element in A that is not in B—for example, the element 1.   ■

 Notice that when we are referring to sets and subsets we use the symbols $\subset$, $\subseteq$, $\supset$, and $\supseteq$ to express the idea of "containment." However, when we wish to show that an element is contained in a set, we use the symbol $\in$ to express the idea of "membership." Thus, in Example 3, we would write $1 \in A$ and *not* $\{1\} \in A$.

Empty Set

The set that contains no elements is called the empty set and is denoted by $\varnothing$.

The empty set, $\varnothing$, is a subset of every set. To see this, observe that $\varnothing$ has no elements and thus contains no element that is not also in A.

VIDEO

EXAMPLE 4   List all subsets of the set $A = \{a, b, c\}$.

Solution   There is one subset consisting of no elements—namely, the empty set $\varnothing$. Next, observe that there are three subsets consisting of one element,

$$\{a\}, \{b\}, \{c\}$$

three subsets consisting of two elements,

$$\{a, b\}, \{a, c\}, \{b, c\}$$

and one subset consisting of three elements, the set A itself. Therefore, the subsets of A are

$$\varnothing, \{a\}, \{b\}, \{c\}, \{a, b\}, \{a, c\}, \{b, c\}, \{a, b, c\}$$ ∎

In contrast with the empty set, we have, at the other extreme, the notion of a largest, or universal, set. A **universal set** is the set of all elements of interest in a particular discussion. It is the largest in the sense that all sets considered in the discussion of the problem are subsets of the universal set. Of course, different universal sets are associated with different problems, as shown in Example 5.

EXAMPLE 5
a. If the problem at hand is to determine the ratio of female to male students in a college, then a logical choice of a universal set is the set consisting of the whole student body of the college.
b. If the problem is to determine the ratio of female to male students in the business department of the college in part (a), then the set of all students in the business department may be chosen as the universal set.   ∎

A visual representation of sets is realized through the use of **Venn diagrams**, which are of considerable help in understanding the concepts introduced earlier as well as in solving problems involving sets. The universal set U is represented by a rectangle, and subsets of U are represented by regions lying inside the rectangle.

EXAMPLE 6   Use Venn diagrams to illustrate the following statements:
a. The sets A and B are equal.
b. The set A is a proper subset of the set B.
c. The sets A and B are not subsets of each other.

Solution   The respective Venn diagrams are shown in Figure 1a–c.

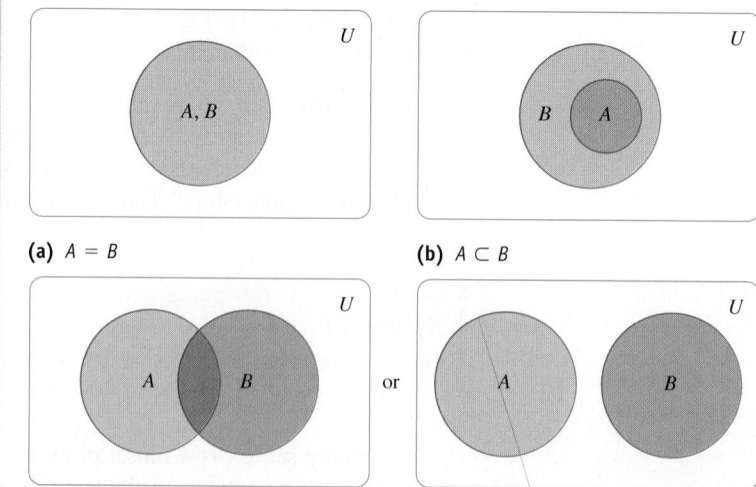

(a) $A = B$

(b) $A \subset B$

(c) $A \nsubseteq B$ and $B \nsubseteq A$

FIGURE 1

Set Operations

Having introduced the concept of a set, our next task is to consider operations on sets—that is, to consider ways in which sets may be combined to yield other sets. These operations enable us to combine sets in much the same way the operations of addition and multiplication enable us to combine numbers to obtain other numbers. In what follows, all sets are assumed to be subsets of a given universal set U.

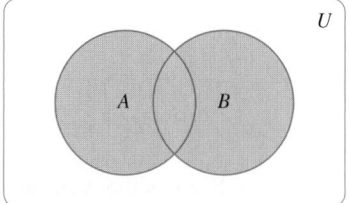

FIGURE 2
Set union $A \cup B$

Set Union

Let A and B be sets. The **union** of A and B, written $A \cup B$, is the set of all elements that belong to either A or B or both.

$$A \cup B = \{x \mid x \in A \text{ or } x \in B \text{ or both}\}$$

The shaded portion of the Venn diagram (Figure 2) depicts the set $A \cup B$.

EXAMPLE 7 If $A = \{a, b, c\}$ and $B = \{a, c, d\}$, then $A \cup B = \{a, b, c, d\}$.

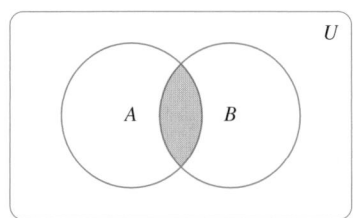

FIGURE 3
Set intersection $A \cap B$

Set Intersection

Let A and B be sets. The set of elements in common with the sets A and B, written $A \cap B$, is called the **intersection** of A and B.

$$A \cap B = \{x \mid x \in A \text{ and } x \in B\}$$

The shaded portion of the Venn diagram (Figure 3) depicts the set $A \cap B$.

EXAMPLE 8 Let $A = \{a, b, c\}$ and $B = \{a, c, d\}$. Then $A \cap B = \{a, c\}$. (Compare this result with Example 7.)

EXAMPLE 9 Let $A = \{1, 3, 5, 7, 9\}$ and $B = \{2, 4, 6, 8, 10\}$. Then $A \cap B = \varnothing$.

The two sets of Example 9 have null intersection. In general, the sets A and B are said to be **disjoint** if they have no elements in common—that is, if $A \cap B = \varnothing$.

EXAMPLE 10 Let U be the set of all students in the classroom. If $M = \{x \in U \mid x \text{ is male}\}$ and $F = \{x \in U \mid x \text{ is female}\}$, then $F \cap M = \varnothing$ and so F and M are disjoint.

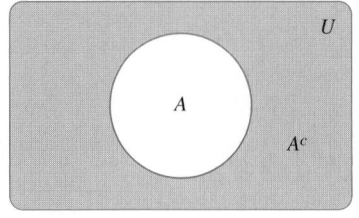

FIGURE 4
Set complementation

Complement of a Set

If U is a universal set and A is a subset of U, then the set of all elements in U that are not in A is called the **complement** of A and is denoted A^c.

$$A^c = \{x \mid x \in U, x \notin A\}$$

The shaded portion of the Venn diagram (Figure 4) shows the set A^c.

EXPLORE & DISCUSS

Let A, B, and C be nonempty subsets of a set U.

1. Suppose $A \cap B \neq \varnothing$, $A \cap C \neq \varnothing$, and $B \cap C \neq \varnothing$. Can you conclude that $A \cap B \cap C \neq \varnothing$? Explain your answer with an example.

2. Suppose $A \cap B \cap C \neq \varnothing$. Can you conclude that $A \cap B \neq \varnothing$, $A \cap C \neq \varnothing$, and $B \cap C \neq \varnothing$ simultaneously? Explain your answer.

EXAMPLE 11   Let $U = \{1, 2, 3, 4, 5, 6, 7, 8, 9, 10\}$ and $A = \{2, 4, 6, 8, 10\}$. Then $A^c = \{1, 3, 5, 7, 9\}$.   ◾

The following rules hold for the operation of complementation. See whether you can verify them.

Set Complementation

If U is a universal set and A is a subset of U, then

a. $U^c = \varnothing$ **b.** $\varnothing^c = U$ **c.** $(A^c)^c = A$

d. $A \cup A^c = U$ **e.** $A \cap A^c = \varnothing$

The following rules govern the operations on sets.

Set Operations

Let U be a universal set. If A, B, and C are arbitrary subsets of U, then

$$A \cup B = B \cup A \qquad \text{Commutative law for union}$$

$$A \cap B = B \cap A \qquad \text{Commutative law for intersection}$$

$$A \cup (B \cup C) = (A \cup B) \cup C \qquad \text{Associative law for union}$$

$$A \cap (B \cap C) = (A \cap B) \cap C \qquad \text{Associative law for intersection}$$

$$A \cup (B \cap C)$$
$$= (A \cup B) \cap (A \cup C) \qquad \text{Distributive law for union}$$

$$A \cap (B \cup C)$$
$$= (A \cap B) \cup (A \cap C) \qquad \text{Distributive law for intersection}$$

Two additional rules, referred to as De Morgan's laws, govern the operations on sets.

De Morgan's Laws

Let A and B be sets. Then

$$(A \cup B)^c = A^c \cap B^c \tag{1}$$

$$(A \cap B)^c = A^c \cup B^c \tag{2}$$

Equation (1) states that the complement of the union of two sets is equal to the intersection of their complements. Equation (2) states that the complement of the intersection of two sets is equal to the union of their complements.

We will not prove De Morgan's laws here, but the plausibility of (2) is illustrated in the following example.

EXAMPLE 12 Using Venn diagrams, show that

$$(A \cap B)^c = A^c \cup B^c$$

Solution $(A \cap B)^c$ is the set of elements in U but not in $A \cap B$ and is thus the shaded region shown in Figure 5. Next, A^c and B^c are shown in Figure 6a–b. Their union, $A^c \cup B^c$, is easily seen to be equivalent to $(A \cap B)^c$ by referring once again to Figure 5.

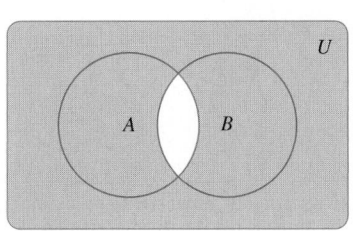

FIGURE 5
$(A \cap B)^c$

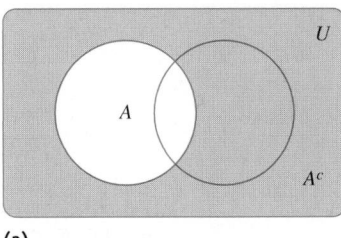

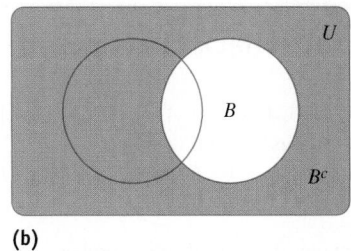

FIGURE 6
$A^c \cup B^c$ is the set obtained by joining (a) and (b).

(a)   (b)

EXAMPLE 13 Let $U = \{1, 2, 3, 4, 5, 6, 7, 8, 9, 10\}$, $A = \{1, 2, 4, 8, 9\}$, and $B = \{3, 4, 5, 6, 8\}$. Verify by direct computation that $(A \cup B)^c = A^c \cap B^c$.

Solution $A \cup B = \{1, 2, 3, 4, 5, 6, 8, 9\}$, so $(A \cup B)^c = \{7, 10\}$. Moreover, $A^c = \{3, 5, 6, 7, 10\}$ and $B^c = \{1, 2, 7, 9, 10\}$, so $A^c \cap B^c = \{7, 10\}$. The required result follows.

APPLIED EXAMPLE 14 Automobile Options Let U denote the set of all cars in a dealer's lot, and let

$A = \{x \in U \mid x \text{ is equipped with automatic transmission}\}$
$B = \{x \in U \mid x \text{ is equipped with air conditioning}\}$
$C = \{x \in U \mid x \text{ is equipped with side air bags}\}$

Find an expression in terms of A, B, and C for each of the following sets:
a. The set of cars with at least one of the given options
b. The set of cars with exactly one of the given options
c. The set of cars with automatic transmission and side air bags but no air conditioning

Solution
a. The set of cars with at least one of the given options is $A \cup B \cup C$ (Figure 7a).
b. The set of cars with automatic transmission only is given by $A \cap B^c \cap C^c$. Similarly, we find that the set of cars with air conditioning only is given by $B \cap C^c \cap A^c$, while the set of cars with side air bags only is given by $C \cap A^c \cap B^c$. Thus, the set of cars with exactly one of the given options is $(A \cap B^c \cap C^c) \cup (B \cap C^c \cap A^c) \cup (C \cap A^c \cap B^c)$ (Figure 7b).

c. The set of cars with automatic transmission and side air bags but no air conditioning is given by $A \cap C \cap B^c$ (Figure 7c).

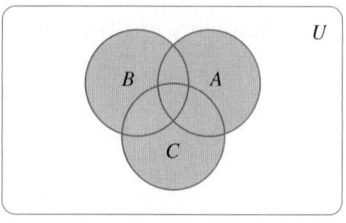

(a) The set of cars with at least one option

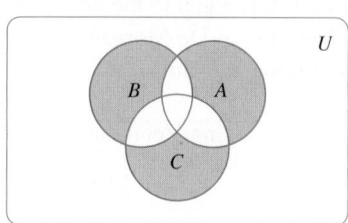

(b) The set of cars with exactly one option

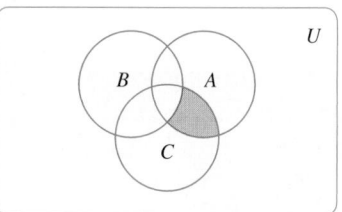

(c) The set of cars with automatic transmission and side air bags but no air conditioning

FIGURE 7

6.1 Self-Check Exercises

1. Let $U = \{1, 2, 3, 4, 5, 6, 7\}$, $A = \{1, 2, 3\}$, $B = \{3, 4, 5, 6\}$, and $C = \{2, 3, 4\}$. Find the following sets:
   a. A^c **b.** $A \cup B$ **c.** $B \cap C$
   d. $(A \cup B) \cap C$ **e.** $(A \cap B) \cup C$ **f.** $A^c \cap (B \cup C)^c$

2. Let U denote the set of all members of the House of Representatives. Let

$$D = \{x \in U \mid x \text{ is a Democrat}\}$$
$$R = \{x \in U \mid x \text{ is a Republican}\}$$

$$F = \{x \in U \mid x \text{ is a female}\}$$
$$L = \{x \in U \mid x \text{ is a lawyer by training}\}$$

Describe each of the following sets in words.
   a. $D \cap F$ **b.** $F^c \cap R$ **c.** $D \cap F \cap L^c$

Solutions to Self-Check Exercises 6.1 can be found on page 345.

6.1 Concept Questions

1. a. What is a set? Give an example.
   b. When are two sets equal? Give an example of two equal sets.
   c. What is the empty set?

2. What can you say about two sets A and B such that
   a. $A \cup B \subseteq A$ **b.** $A \cup B = \varnothing$
   c. $A \cap B = B$ **d.** $A \cap B = \varnothing$

3. a. If $A \subset B$, what can you say about the relationship between A^c and B^c?
   b. If $A^c = \varnothing$, what can you say about A?

6.1 Exercises

In Exercises 1–4, write the set in set-builder notation.

1. The set of gold medalists in the 2006 Winter Olympic Games

2. The set of football teams in the NFL

3. $\{3, 4, 5, 6, 7\}$

4. $\{1, 3, 5, 7, 9, 11, \ldots, 39\}$

In Exercises 5–8, list the elements of the set in roster notation.

5. $\{x \mid x \text{ is a digit in the number } 352,646\}$

6. $\{x \mid x \text{ is a letter in the word } HIPPOPOTAMUS\}$

7. $\{x \mid 2 - x = 4; x \text{ an integer}\}$

8. $\{x \mid 2 - x = 4; x \text{ a fraction}\}$

In Exercises 9–14, state whether the statements are true or false.

9. a. $\{a, b, c\} = \{c, a, b\}$ **b.** $A \in A$

10. a. $\varnothing \in A$ **b.** $A \subset A$

11. a. $0 \in \varnothing$ **b.** $0 = \varnothing$

12. a. $\{\varnothing\} = \varnothing$ **b.** $\{a, b\} \in \{a, b, c\}$

13. $\{$Chevrolet, Pontiac, Buick$\} \subset \{x \mid x$ is a division of General Motors$\}$

14. $\{x \mid x$ is a silver medalist in the 2006 Winter Olympic Games$\} = \varnothing$

In Exercises 15 and 16, let $A = \{1, 2, 3, 4, 5\}$. Determine whether the statements are true or false.

15. a. $2 \in A$ **b.** $A \subseteq \{2, 4, 6\}$

16. a. $0 \in A$ **b.** $\{1, 3, 5\} \in A$

17. Let $A = \{1, 2, 3\}$. Which of the following sets are equal to A?
a. $\{2, 1, 3\}$ **b.** $\{3, 2, 1\}$
c. $\{0, 1, 2, 3\}$

18. Let $A = \{a, e, l, t, r\}$. Which of the following sets are equal to A?
a. $\{x \mid x$ is a letter of the word *later*$\}$
b. $\{x \mid x$ is a letter of the word *latter*$\}$
c. $\{x \mid x$ is a letter of the word *relate*$\}$

19. List all subsets of the following sets:
a. $\{1, 2\}$ **b.** $\{1, 2, 3\}$ **c.** $\{1, 2, 3, 4\}$

20. List all subsets of the set $A = \{$IBM, U.S. Steel, Union Carbide, Boeing$\}$. Which of these are proper subsets of A?

In Exercises 21–24, find the smallest possible set (i.e., the set with the least number of elements) that contains the given sets as subsets.

21. $\{1, 2\}, \{1, 3, 4\}, \{4, 6, 8, 10\}$

22. $\{1, 2, 4\}, \{a, b\}$

23. $\{$Jill, John, Jack$\}, \{$Susan, Sharon$\}$

24. $\{$GM, Ford, Chrysler$\}, \{$Daimler-Benz, Volkswagen$\}, \{$Toyota, Nissan$\}$

25. Use Venn diagrams to represent the following relationships:
a. $A \subset B$ and $B \subset C$
b. $A \subset U$ and $B \subset U$, where A and B have no elements in common
c. The sets A, B, and C are equal.

26. Let U denote the set of all students who applied for admission to the freshman class at Faber College for the upcoming academic year, and let

$A = \{x \in U \mid x$ is a successful applicant$\}$

$B = \{x \in U \mid x$ is a female student who enrolled in the freshman class$\}$

$C = \{x \in U \mid x$ is a male student who enrolled in the freshman class$\}$

a. Use Venn diagrams to represent the sets U, A, B, and C.
b. Determine whether the following statements are true or false.
i. $A \subseteq B$ **ii.** $B \subset A$ **iii.** $C \subset B$

In Exercises 27 and 28, shade the portion of the accompanying figure that represents each set.

27. a. $A \cap B^c$
b. $A^c \cap B$

28. a. $A^c \cap B^c$
b. $(A \cup B)^c$

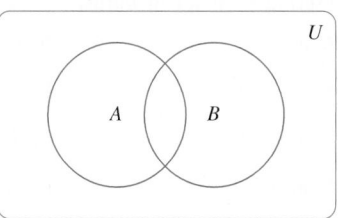

In Exercises 29–32, shade the portion of the accompanying figure that represents each set.

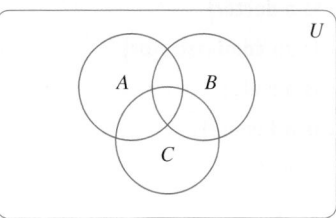

29. a. $A \cup B \cup C$ **b.** $A \cap B \cap C$

30. a. $A \cap B \cap C^c$ **b.** $A^c \cap B \cap C$

31. a. $A^c \cap B^c \cap C^c$ **b.** $(A \cup B)^c \cap C$

32. a. $A \cup (B \cap C)^c$ **b.** $(A \cup B \cup C)^c$

In Exercises 33–36, let $U = \{1, 2, 3, 4, 5, 6, 7, 8, 9, 10\}$, $A = \{1, 3, 5, 7, 9\}$, $B = \{2, 4, 6, 8, 10\}$, and $C = \{1, 2, 4, 5, 8, 9\}$. List the elements of each set.

33. a. A^c **b.** $B \cup C$ **c.** $C \cup C^c$

34. a. $C \cap C^c$ **b.** $(A \cap C)^c$ **c.** $A \cup (B \cap C)$

35. a. $(A \cap B) \cup C$ **b.** $(A \cup B \cup C)^c$
c. $(A \cap B \cap C)^c$

36. a. $A^c \cap (B \cap C^c)$ **b.** $(A \cup B^c) \cup (B \cap C^c)$
c. $(A \cup B)^c \cap C^c$

In Exercises 37 and 38, determine whether the pairs of sets are disjoint.

37. a. $\{1, 2, 3, 4\}, \{4, 5, 6, 7\}$
 b. $\{a, c, e, g\}, \{b, d, f\}$

38. a. $\varnothing, \{1, 3, 5\}$
 b. $\{0, 1, 3, 4\}, \{0, 2, 5, 7\}$

In Exercises 39–42, let U denote the set of all employees at Universal Life Insurance Company and let

$$T = \{x \in U \mid x \text{ drinks tea}\}$$
$$C = \{x \in U \mid x \text{ drinks coffee}\}$$

Describe each set in words.

39. a. T^c **b.** C^c

40. a. $T \cup C$ **b.** $T \cap C$

41. a. $T \cap C^c$ **b.** $T^c \cap C$

42. a. $T^c \cap C^c$ **b.** $(T \cup C)^c$

In Exercises 43–46, let U denote the set of all employees in a hospital. Let

$$N = \{x \in U \mid x \text{ is a nurse}\}$$
$$D = \{x \in U \mid x \text{ is a doctor}\}$$
$$A = \{x \in U \mid x \text{ is an administrator}\}$$
$$M = \{x \in U \mid x \text{ is a male}\}$$
$$F = \{x \in U \mid x \text{ is a female}\}$$

Describe each set in words.

43. a. D^c **b.** N^c

44. a. $N \cup D$ **b.** $N \cap M$

45. a. $D \cap M^c$ **b.** $D \cap A$

46. a. $N \cap F$ **b.** $(D \cup N)^c$

In Exercises 47 and 48, let U denote the set of all senators in Congress and let

$$D = \{x \in U \mid x \text{ is a Democrat}\}$$
$$R = \{x \in U \mid x \text{ is a Republican}\}$$
$$F = \{x \in U \mid x \text{ is a female}\}$$
$$L = \{x \in U \mid x \text{ is a lawyer}\}$$

Write the set that represents each statement.

47. a. The set of all Democrats who are female
 b. The set of all Republicans who are male and are not lawyers

48. a. The set of all Democrats who are female or are lawyers
 b. The set of all senators who are not Democrats or are lawyers

In Exercises 49 and 50, let U denote the set of all students in the business college of a certain university. Let

$$A = \{x \in U \mid x \text{ had taken a course in accounting}\}$$
$$B = \{x \in U \mid x \text{ had taken a course in economics}\}$$
$$C = \{x \in U \mid x \text{ had taken a course in marketing}\}$$

Write the set that represents each statement.

49. a. The set of students who have not had a course in economics
 b. The set of students who have had courses in accounting and economics
 c. The set of students who have had courses in accounting and economics but not marketing

50. a. The set of students who have had courses in economics but not courses in accounting or marketing
 b. The set of students who have had at least one of the three courses
 c. The set of students who have had all three courses

In Exercises 51 and 52, refer to the following diagram, where U is the set of all tourists surveyed over a 1-week period in London and where

$$A = \{x \in U \mid x \text{ has taken the underground [subway]}\}$$
$$B = \{x \in U \mid x \text{ has taken a cab}\}$$
$$C = \{x \in U \mid x \text{ has taken a bus}\}$$

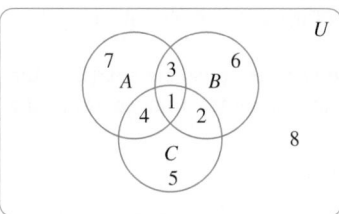

Express the indicated regions in set notation and in words.

51. a. Region 1
 b. Regions 1 and 4 together
 c. Regions 4, 5, 7, and 8 together

52. a. Region 3
 b. Regions 4 and 6 together
 c. Regions 5, 6, and 7 together

In Exercises 53–58, use Venn diagrams to illustrate each statement.

53. $A \subset A \cup B; B \subset A \cup B$ **54.** $A \cap B \subset A; A \cap B \subset B$

55. $A \cup (B \cup C) = (A \cup B) \cup C$

56. $A \cap (B \cap C) = (A \cap B) \cap C$

57. $A \cap (B \cup C) = (A \cap B) \cup (A \cap C)$

58. $(A \cup B)^c = A^c \cap B^c$

In Exercises 59 and 60, let

$$U = \{1, 2, 3, 4, 5, 6, 7, 8, 9, 10\}$$
$$A = \{1, 3, 5, 7, 9\}, B = \{1, 2, 4, 7, 8\}$$
$$C = \{2, 4, 6, 8\}$$

Verify each equation by direct computation.

59. a. $A \cup (B \cup C) = (A \cup B) \cup C$
  b. $A \cap (B \cap C) = (A \cap B) \cap C$

60. a. $A \cap (B \cup C) = (A \cap B) \cup (A \cap C)$
  b. $(A \cup B)^c = A^c \cap B^c$

In Exercises 61–64, refer to the accompanying figure and list the points that belong to each set.

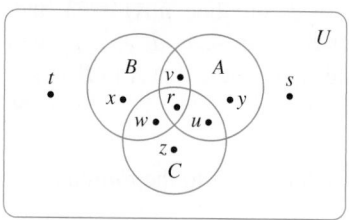

61. a. $A \cup B$ **b.** $A \cap B$

62. a. $A \cap (B \cup C)$ **b.** $(B \cap C)^c$

63. a. $(B \cup C)^c$ **b.** A^c

64. a. $(A \cap B) \cap C^c$ **b.** $(A \cup B \cup C)^c$

65. Suppose $A \subset B$ and $B \subset C$, where A and B are any two sets. What conclusion can be drawn regarding the sets A and C?

66. Verify the assertion that two sets A and B are equal if and only if (1) $A \subseteq B$ and (2) $B \subseteq A$.

In Exercises 67–72, determine whether the statement is true or false. If it is true, explain why it is true. If it is false, give an example to show why it is false.

67. A set is never a subset of itself.

68. A proper subset of a set is itself a subset of the set but not vice versa.

69. If $A \cup B = \varnothing$, then $A = \varnothing$ and $B = \varnothing$.

70. If $A \cap B = \varnothing$, then either $A = \varnothing$ and $B = \varnothing$ or both A and B are empty.

71. $(A \cup A^c)^c = \varnothing$

72. If $A \subseteq B$, then $A \cap B = A$.

6.1   Solutions to Self-Check Exercises

1. a. A^c is the set of all elements in U but not in A. Therefore,

$$A^c = \{4, 5, 6, 7\}$$

b. $A \cup B$ consists of all elements in A and/or B. Hence

$$A \cup B = \{1, 2, 3, 4, 5, 6\}$$

c. $B \cap C$ is the set of all elements in both B and C. Therefore,

$$B \cap C = \{3, 4\}$$

d. Using the result from part (b), we find

$$(A \cup B) \cap C = \{1, 2, 3, 4, 5, 6\} \cap \{2, 3, 4\}$$
$$= \{2, 3, 4\}$$

e. First, we compute

$$A \cap B = \{3\}$$

Next, since $(A \cap B) \cup C$ is the set of all elements in $(A \cap B)$ and/or C, we conclude that

$$(A \cap B) \cup C = \{3\} \cup \{2, 3, 4\}$$
$$= \{2, 3, 4\}$$

f. From part (a), we have $A^c = \{4, 5, 6, 7\}$. Next, we compute

$$B \cup C = \{3, 4, 5, 6\} \cup \{2, 3, 4\}$$
$$= \{2, 3, 4, 5, 6\}$$

from which we deduce that

$$(B \cup C)^c = \{1, 7\} \qquad \text{The set of elements in } U \text{ but not in } B \cup C$$

Finally, using these results, we obtain

$$A^c \cap (B \cup C)^c = \{4, 5, 6, 7\} \cap \{1, 7\} = \{7\}$$

2. a. $D \cap F$ denotes the set of all elements in both D and F. Since an element in D is a Democrat and an element in F is a female representative, we see that $D \cap F$ is the set of all female Democrats in the House of Representatives.
  b. Since F^c is the set of male representatives and R is the set of Republicans, it follows that $F^c \cap R$ is the set of male Republicans in the House of Representatives.
  c. L^c is the set of representatives who are not lawyers by training. Therefore, $D \cap F \cap L^c$ is the set of female Democratic representatives who are not lawyers by training.

6.2   The Number of Elements in a Finite Set

▬ Counting the Elements in a Set

The solution to some problems in mathematics calls for finding the number of elements in a set. Such problems are called **counting problems** and constitute a field of study known as **combinatorics**. Our study of combinatorics is restricted to the results that will be required for our work in probability later on.

The number of elements in a finite set is determined by simply counting the elements in the set. If A is a set, then $n(A)$ denotes the number of elements in A. For example, if

$$A = \{1, 2, 3, \ldots, 20\} \qquad B = \{a, b\} \qquad C = \{8\}$$

then $n(A) = 20$, $n(B) = 2$, and $n(C) = 1$.

The empty set has no elements in it, so $n(\emptyset) = 0$. Another result that is easily seen to be true is the following: If A and B are disjoint sets, then

$$n(A \cup B) = n(A) + n(B) \qquad\qquad (3)$$

> **EXAMPLE 1**   If $A = \{a, c, d\}$ and $B = \{b, e, f, g\}$, then $n(A) = 3$ and $n(B) = 4$, so $n(A) + n(B) = 7$. However, $A \cup B = \{a, b, c, d, e, f, g\}$ and $n(A \cup B) = 7$. Thus, Equation (3) holds true in this case. Note that $A \cap B = \emptyset$. ▪

In the general case, A and B need not be disjoint, which leads us to the formula

$$\boxed{n(A \cup B) = n(A) + n(B) - n(A \cap B)} \qquad\qquad (4)$$

To see this, we observe that the set $A \cup B$ may be viewed as the union of three mutually disjoint sets with x, y, and z elements, respectively (Figure 8). This figure shows that

$$n(A \cup B) = x + y + z$$

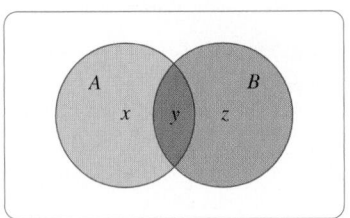

FIGURE 8
$n(A \cup B) = x + y + z$

Also,

$$n(A) = x + y \quad \text{and} \quad n(B) = y + z$$

so

$$\begin{aligned} n(A) + n(B) &= (x + y) + (y + z) \\ &= (x + y + z) + y \\ &= n(A \cup B) + n(A \cap B) \qquad {\scriptstyle n(A \cap B) = y} \end{aligned}$$

Solving for $n(A \cup B)$, we obtain

$$n(A \cup B) = n(A) + n(B) - n(A \cap B)$$

which is the desired result.

> **EXAMPLE 2**   Let $A = \{a, b, c, d, e\}$ and $B = \{b, d, f, h\}$. Verify Equation (4) directly.

Solution

$$A \cup B = \{a, b, c, d, e, f, h\} \quad \text{so} \quad n(A \cup B) = 7$$
$$A \cap B = \{b, d\} \quad \text{so} \quad n(A \cap B) = 2$$

Furthermore,

$$n(A) = 5 \quad \text{and} \quad n(B) = 4$$

so

$$n(A) + n(B) - n(A \cap B) = 5 + 4 - 2 = 7 = n(A \cup B) \qquad \blacksquare$$

APPLIED EXAMPLE 3   Consumer Surveys   In a survey of 100 coffee drinkers, it was found that 70 take sugar, 60 take cream, and 50 take both sugar and cream with their coffee. How many coffee drinkers take sugar or cream with their coffee?

Solution   Let U denote the set of 100 coffee drinkers surveyed, and let

$$A = \{x \in U \,|\, x \text{ takes sugar}\}$$
$$B = \{x \in U \,|\, x \text{ takes cream}\}$$

Then, $n(A) = 70$, $n(B) = 60$, and $n(A \cap B) = 50$. The set of coffee drinkers who take sugar or cream with their coffee is given by $A \cup B$. Using (4), we find

$$n(A \cup B) = n(A) + n(B) - n(A \cap B)$$
$$= 70 + 60 - 50 = 80$$

Thus, 80 out of the 100 coffee drinkers surveyed take cream or sugar with their coffee. $\qquad \blacksquare$

EXPLORE & DISCUSS

Prove Formula (5), using an argument similar to that used to prove Formula (4). Another proof is outlined in Exercise 40 on page 352.

An equation similar to (4) may be derived for the case that involves any finite number of finite sets. For example, a relationship involving the number of elements in the sets A, B, and C is given by

$$\boxed{\begin{aligned} n(A \cup B \cup C) = \; & n(A) + n(B) + n(C) - n(A \cap B) \\ & - n(A \cap C) - n(B \cap C) + n(A \cap B \cap C) \end{aligned}} \qquad (5)$$

As useful as equations such as (5) are, in practice it is often easier to attack a problem directly with the aid of Venn diagrams, as shown by the following example.

APPLIED EXAMPLE 4   Marketing Surveys   A leading cosmetics manufacturer advertises its products in three magazines: *Cosmopolitan*, *McCall's*, and the *Ladies Home Journal*. A survey of 500 customers by the manufacturer reveals the following information:

180 learned of its products from *Cosmopolitan*.

200 learned of its products from *McCall's*.

192 learned of its products from the *Ladies Home Journal*.

84 learned of its products from *Cosmopolitan* and *McCall's*.

52 learned of its products from *Cosmopolitan* and the *Ladies Home Journal*.

64 learned of its products from *McCall's* and the *Ladies Home Journal*.

38 learned of its products from all three magazines.

How many of the customers saw the manufacturer's advertisement in
a. At least one magazine?
b. Exactly one magazine?

Solution Let U denote the set of all customers surveyed, and let

$C = \{x \in U \mid x \text{ learned of the products from } Cosmopolitan\}$

$M = \{x \in U \mid x \text{ learned of the products from } McCall's\}$

$L = \{x \in U \mid x \text{ learned of the products from the } Ladies\ Home\ Journal\}$

The result that 38 customers learned of the products from all three magazines translates into $n(C \cap M \cap L) = 38$ (Figure 9a). Next, the result that 64 learned of the products from *McCall's* and the *Ladies Home Journal* translates into $n(M \cap L) = 64$. This leaves

$$64 - 38 = 26$$

who learned of the products from only *McCall's* and the *Ladies Home Journal* (Figure 9b). Similarly, $n(C \cap L) = 52$, so

$$52 - 38 = 14$$

learned of the products from only *Cosmopolitan* and the *Ladies Home Journal*, and $n(C \cap M) = 84$, so

$$84 - 38 = 46$$

learned of the products from only *Cosmopolitan* and *McCall's*. These numbers appear in the appropriate regions in Figure 9b.

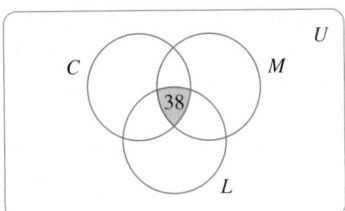

(a) All three magazines   (b) Two or more magazines

FIGURE 9

Continuing, we have $n(L) = 192$, so the number who learned of the products from the *Ladies Home Journal* only is given by

$$192 - 14 - 38 - 26 = 114$$

(Figure 10). Similarly, $n(M) = 200$, so

$$200 - 46 - 38 - 26 = 90$$

learned of the products from only *McCall's*, and $n(C) = 180$, so

$$180 - 14 - 38 - 46 = 82$$

learned of the products from only *Cosmopolitan*. Finally,

$$500 - (90 + 26 + 114 + 14 + 82 + 46 + 38) = 90$$

learned of the products from other sources.

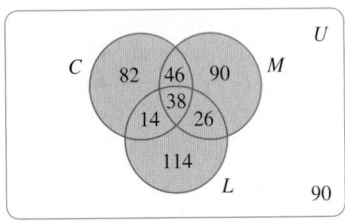

FIGURE 10
At least one magazine

We are now in a position to answer questions (a) and (b).

a. Referring to Figure 10, we see that the number of customers who learned of the products from at least one magazine is given by

$$n(C \cup M \cup L) = 90 + 26 + 114 + 14 + 82 + 46 + 38 = 410$$

b. The number of customers who learned of the products from exactly one magazine (Figure 11) is given by

$$n(L \cap C^c \cap M^c) + n(M \cap C^c \cap L^c) + n(C \cap L^c \cap M^c)$$
$$= 114 + 90 + 82 = 286$$

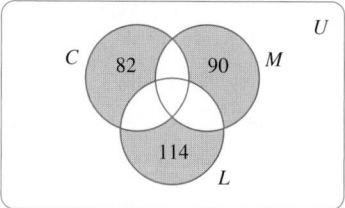

FIGURE 11
Exactly one magazine

6.2 Self-Check Exercises

1. Let A and B be subsets of a universal set U and suppose that $n(U) = 100$, $n(A) = 60$, $n(B) = 40$, and $n(A \cap B) = 20$. Compute:
   a. $n(A \cup B)$ **b.** $n(A \cap B^c)$ **c.** $n(A^c \cap B)$

2. In a survey of 1000 readers of *Video Magazine*, it was found that 900 own at least one VCR in the VHS format, 240 own at least one VCR in the S-VHS format, and 160 own VCRs in both formats. How many of the readers surveyed own VCRs in the VHS format only? How many of the readers surveyed do not own a VCR in either format?

Solutions to Self-Check Exercises 6.2 can be found on page 352.

6.2 Concept Questions

1. **a.** If A and B are sets with $A \cap B = \emptyset$, what can you say about $n(A) + n(B)$? Explain.
   b. If A and B are sets satisfying $n(A \cup B) \neq n(A) + n(B)$, what can you say about $A \cap B$? Explain.

2. Let A and B be subsets of U, the universal set, and suppose $A \cap B = \emptyset$. Is it true that $n(A) - n(B) = n(B^c) - n(A^c)$? Explain.

6.2 Exercises

In Exercises 1 and 2, verify the equation

$$n(A \cup B) = n(A) + n(B)$$

for the given disjoint sets.

1. $A = \{a, e, i, o, u\}$ and $B = \{g, h, k, l, m\}$

2. $A = \{x \mid x$ is a whole number between 0 and 4$\}$
   $B = \{x \mid x$ is a negative integer greater than $-4\}$

3. Let $A = \{2, 4, 6, 8\}$ and $B = \{6, 7, 8, 9, 10\}$. Compute:
   a. $n(A)$ **b.** $n(B)$
   c. $n(A \cup B)$ **d.** $n(A \cap B)$

4. Let $U = \{1, 2, 3, 4, 5, 6, 7, a, b, c, d, e\}$. If $A = \{1, 2, a, e\}$ and $B = \{1, 2, 3, 4, a, b, c\}$, find:
   a. $n(A^c)$
   b. $n(A \cap B^c)$
   c. $n(A \cup B^c)$
   d. $n(A^c \cap B^c)$

5. Verify directly that $n(A \cup B) = n(A) + n(B) - n(A \cap B)$ for the sets in Exercise 3.

6. Let $A = \{a, e, i, o, u\}$ and $B = \{b, d, e, o, u\}$. Verify by direct computation that $n(A \cup B) = n(A) + n(B) - n(A \cap B)$.

7. If $n(A) = 15$, $n(A \cap B) = 5$, and $n(A \cup B) = 30$, then what is $n(B)$?

8. If $n(A) = 10$, $n(A \cup B) = 15$, and $n(B) = 8$, then what is $n(A \cap B)$?

In Exercises 9 and 10, let A and B be subsets of a universal set U and suppose $n(U) = 200$, $n(A) = 100$, $n(B) = 80$, and $n(A \cap B) = 40$. Compute:

9. a. $n(A \cup B)$　　b. $n(A^c)$　　c. $n(A \cap B^c)$

10. a. $n(A^c \cap B)$　　b. $n(B^c)$　　c. $n(A^c \cap B^c)$

11. Find $n(A \cup B)$ given that $n(A) = 6$, $n(B) = 10$, and $n(A \cap B) = 3$.

12. If $n(B) = 6$, $n(A \cup B) = 14$, and $n(A \cap B) = 3$, find $n(A)$.

13. If $n(A) = 4$, $n(B) = 5$, and $n(A \cup B) = 9$, find $n(A \cap B)$.

14. If $n(A) = 16$, $n(B) = 16$, $n(C) = 14$, $n(A \cap B) = 6$, $n(A \cap C) = 5$, $n(B \cap C) = 6$, and $n(A \cup B \cup C) = 31$, find $n(A \cap B \cap C)$.

15. If $n(A) = 12$, $n(B) = 12$, $n(A \cap B) = 5$, $n(A \cap C) = 5$, $n(B \cap C) = 4$, $n(A \cap B \cap C) = 2$, and $n(A \cup B \cup C) = 25$, find $n(C)$.

16. A survey of 1000 subscribers to the *Los Angeles Times* revealed that 900 people subscribe to the daily morning edition and 500 subscribe to both the daily and the Sunday editions. How many subscribe to the Sunday edition? How many subscribe to the Sunday edition only?

17. On a certain day, the Wilton County Jail had 190 prisoners. Of these, 130 were accused of felonies and 121 were accused of misdemeanors. How many prisoners were accused of both a felony and a misdemeanor?

18. Of 100 clock radios sold recently in a department store, 70 had FM circuitry and 90 had AM circuitry. How many radios had both FM and AM circuitry? How many could receive FM transmission only? How many could receive AM transmission only?

19. CONSUMER SURVEYS In a survey of 120 consumers conducted in a shopping mall, 80 consumers indicated that they buy brand A of a certain product, 68 buy brand B, and 42 buy both brands. How many consumers participating in the survey buy
   a. At least one of these brands?
   b. Exactly one of these brands?
   c. Only brand A?
   d. None of these brands?

20. CONSUMER SURVEYS In a survey of 200 members of a local sports club, 100 members indicated that they plan to attend the next Summer Olympic Games, 60 indicated that they plan to attend the next Winter Olympic Games, and 40 indicated that they plan to attend both games. How many members of the club plan to attend
   a. At least one of the two games?
   b. Exactly one of the games?
   c. The Summer Olympic Games only?
   d. None of the games?

21. INVESTING In a poll conducted among 200 active investors, it was found that 120 use discount brokers, 126 use full-service brokers, and 64 use both discount and full-service brokers. How many investors
   a. Use at least one kind of broker?
   b. Use exactly one kind of broker?
   c. Use only discount brokers?
   d. Don't use a broker?

22. COMMUTER TRENDS Of 50 employees of a store located in downtown Boston, 18 people take the subway to work, 12 take the bus, and 7 take both the subway and the bus. How many employees
   a. Take the subway or the bus to work?
   b. Take only the bus to work?
   c. Take either the bus or the subway to work?
   d. Get to work by some other means?

23. CONSUMER SURVEYS In a survey of 200 households regarding the ownership of desktop and laptop computers, the following information was obtained:

   120 households own only desktop computers.

   10 households own only laptop computers.

   40 households own neither desktop nor laptop computers.

   How many households own both desktop and laptop computers?

24. CONSUMER SURVEYS In a survey of 400 households regarding the ownership of VCRs and DVD players, the following data was obtained:

   360 households own one or more VCRs.

   170 households own one or more VCRs and one or more DVD players.

   19 households do not own a VCR or a DVD player.

   How many households own only one or more DVD players?

In Exercises 25–28, let *A*, *B*, and *C* be subsets of a universal set *U* and suppose $n(U) = 100$, $n(A) = 28$, $n(B) = 30$, $n(C) = 34$, $n(A \cap B) = 8$, $n(A \cap C) = 10$, $n(B \cap C) = 15$, and $n(A \cap B \cap C) = 5$. Compute:

25. a. $n(A \cup B \cup C)$ b. $n(A^c \cap B \cap C)$

26. a. $n[A \cap (B \cup C)]$ b. $n[A \cap (B \cup C)^c]$

27. a. $n(A^c \cap B^c \cap C^c)$ b. $n[A^c \cap (B \cup C)]$

28. a. $n[A \cup (B \cap C)]$ b. $n(A^c \cap B^c \cap C^c)^c$

29. ECONOMIC SURVEYS   A survey of the opinions of 10 leading economists in a certain country showed that, because oil prices were expected to drop in that country over the next 12 months,

7 had lowered their estimate of the consumer inflation rate.

8 had raised their estimate of the gross national product (GNP) growth rate.

2 had lowered their estimate of the consumer inflation rate but had not raised their estimate of the GNP growth rate.

How many economists had both lowered their estimate of the consumer inflation rate and raised their estimate of the GNP growth rate for that period?

30. STUDENT DROPOUT RATE   Data released by the Department of Education regarding the rate (percentage) of ninth-grade students who don't graduate showed that, out of 50 states,

12 states had an increase in the dropout rate during the past 2 yr.

15 states had a dropout rate of at least 30% during the past 2 yr.

21 states had an increase in the dropout rate and/or a dropout rate of at least 30% during the past 2 yr.

a. How many states had both a dropout rate of at least 30% and an increase in the dropout rate over the 2-yr period?
b. How many states had a dropout rate that was less than 30% but that had increased over the 2-yr period?

31. STUDENT READING HABITS   A survey of 100 college students who frequent the reading lounge of a university revealed the following results:

40 read *Time*.

30 read *Newsweek*.

25 read *U.S. News & World Report*.

15 read *Time* and *Newsweek*.

12 read *Time* and *U.S. News & World Report*.

10 read *Newsweek* and *U.S. News & World Report*.

4 read all three magazines.

How many of the students surveyed read
a. At least one magazine?
b. Exactly one magazine?
c. Exactly two magazines?
d. None of these magazines?

32. SAT SCORES   Results of a Department of Education survey of SAT test scores in 22 states showed that

10 states had an average composite test score of at least 1000 during the past 3 yr.

15 states had an increase of at least 10 points in the average composite score during the past 3 yr.

8 states had both an average composite SAT score of at least 1000 and an increase in the average composite score of at least 10 points during the past 3 yr.

a. How many of the 22 states had composite scores of less than 1000 and showed an increase of at least 10 points over the 3-yr period?
b. How many of the 22 states had composite scores of at least 1000 and did not show an increase of at least 10 points over the 3-yr period?

33. CONSUMER SURVEYS   The 120 consumers of Exercise 19 were also asked about their buying preferences concerning another product that is sold in the market under three labels. The results were:

12 buy only those sold under label A.

25 buy only those sold under label B.

26 buy only those sold under label C.

15 buy only those sold under labels A and B.

10 buy only those sold under labels A and C.

12 buy only those sold under labels B and C.

8 buy the product sold under all three labels.

How many of the consumers surveyed buy the product sold under
a. At least one of the three labels?
b. Labels A and B but not C?
c. Label A?
d. None of these labels?

34. STUDENT SURVEYS   To help plan the number of meals to be prepared in a college cafeteria, a survey was conducted and the following data were obtained:

130 students ate breakfast.

180 students ate lunch.

275 students ate dinner.

68 students ate breakfast and lunch.

112 students ate breakfast and dinner.

90 students ate lunch and dinner.

58 students ate all three meals.

How many of the students ate
a. At least one meal in the cafeteria?
b. Exactly one meal in the cafeteria?
c. Only dinner in the cafeteria?
d. Exactly two meals in the cafeteria?

35. INVESTMENTS In a survey of 200 employees of a company regarding their 401(k) investments, the following data were obtained:

141 had investments in stock funds.

91 had investments in bond funds.

60 had investments in money market funds.

47 had investments in stock funds and bond funds.

36 had investments in stock funds and money market funds.

36 had investments in bond funds and money market funds.

5 had investments in some other vehicle.

a. How many of the employees surveyed had investments in all three types of funds?
b. How many of the employees had investments in stock funds only?

36. NEWSPAPER SUBSCRIPTIONS In a survey of 300 individual investors regarding subscriptions to the *New York Times* (*NYT*), *Wall Street Journal* (*WSJ*), and *USA Today* (*UST*), the following data were obtained:

122 subscribe to the *NYT*.

150 subscribe to the *WSJ*.

62 subscribe to the *UST*.

38 subscribe to the *NYT* and *WSJ*.

28 subscribe to the *WSJ* and *UST*.

20 subscribe to the *NYT* and *UST*.

36 do not subscribe to any of these newspapers.

a. How many of the individual investors surveyed subscribe to all three newspapers?
b. How many subscribe to only one of these newspapers?

In Exercises 37–39, determine whether the statement is true or false. If it is true, explain why it is true. If it is false, give an example to show why it is false.

37. If $A \cap B \neq \varnothing$, then $n(A \cup B) \neq n(A) + n(B)$.

38. If $A \subseteq B$, then $n(B) = n(A) + n(A^c \cap B)$.

39. If $n(A \cup B) = n(A) + n(B)$, then $A \cap B = \varnothing$.

40. Derive Equation (5).
Hint: Equation (4) may be written as $n(D \cup E) = n(D) + n(E) - n(D \cap E)$. Now, put $D = A \cup B$ and $E = C$. Use (4) again if necessary.

6.2 Solutions to Self-Check Exercises

1. Refer to the following Venn diagram:

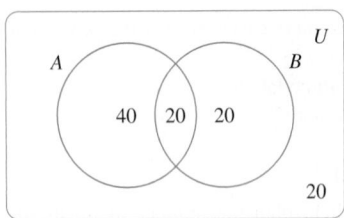

Using this result we see that
a. $n(A \cup B) = 40 + 20 + 20 = 80$
b. $n(A \cap B^c) = 40$
c. $n(A^c \cap B) = 20$

2. Let U denote the set of all readers surveyed, and let

$A = \{x \in U \,|\, x \text{ owns at least one VCR in the VHS format}\}$
$B = \{x \in U \,|\, x \text{ owns at least one VCR in the S-VHS format}\}$

Then, the result that 160 of the readers own VCRs in both formats means that $n(A \cap B) = 160$. Also, $n(A) = 900$ and

$n(B) = 240$. Using this information, we obtain the following Venn diagram:

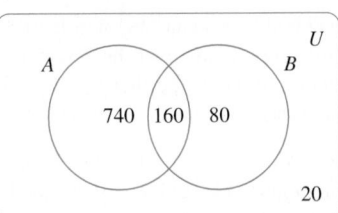

From the Venn diagram we see that the number of readers who own VCRs in the VHS format only is given by

$$n(A \cap B^c) = 740$$

The number of readers who do not own a VCR in either format is given by

$$n(A^c \cap B^c) = 20$$

6.3 The Multiplication Principle

The Fundamental Principle of Counting

The solution of certain problems requires more sophisticated counting techniques than those developed in the previous section. We look at some such techniques in this and the following section. We begin by stating a fundamental principle of counting called the **multiplication principle**.

> ### The Multiplication Principle
> Suppose there are m ways of performing a task T_1 and n ways of performing a task T_2. Then, there are mn ways of performing the task T_1 followed by the task T_2.

EXAMPLE 1 Three trunk roads connect town A and town B, and two trunk roads connect town B and town C.
a. Use the multiplication principle to find the number of ways a journey from town A to town C via town B may be completed.
b. Verify part (a) directly by exhibiting all possible routes.

Solution
a. Since there are three ways of performing the first task (going from town A to town B) followed by two ways of performing the second task (going from town B to town C), the multiplication principle says that there are $3 \cdot 2$, or 6, ways to complete a journey from town A to town C via town B.
b. Label the trunk roads connecting town A and town B with the Roman numerals I, II, and III, and label the trunk roads connecting town B and town C with the lowercase letters a and b. A schematic of this is shown in Figure 12. Then the routes from town A to town C via town B may be exhibited

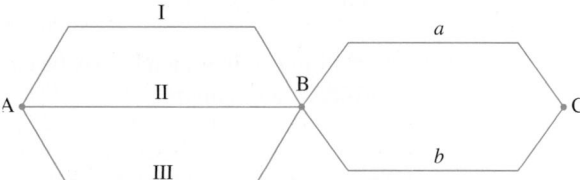

FIGURE 12
Roads from town A to town C

with the aid of a **tree diagram** (Figure 13). If we follow all of the branches from the initial point A to the right-hand edge of the tree, we obtain the six routes represented by six ordered pairs:

$$(\text{I}, a), (\text{I}, b), (\text{II}, a), (\text{II}, b), (\text{III}, a), (\text{III}, b)$$

where (I, a) means that the journey from town A to town B is made on trunk road I with the rest of the journey from town B to town C to be completed on trunk road a, and so forth.

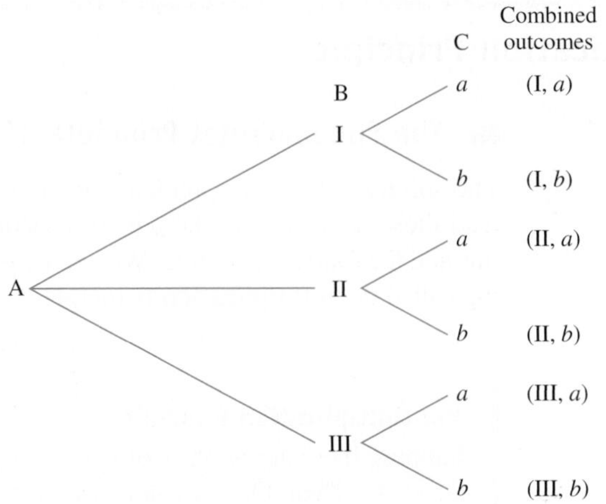

Combined
C   outcomes

I	a	(I, a)
	b	(I, b)
II	a	(II, a)
	b	(II, b)
III	a	(III, a)
	b	(III, b)

FIGURE 13
Tree diagram displaying the possible routes from town A to town C

EXPLORE & DISCUSS

One way of gauging the performance of an airline is to track the arrival times of its flights. Suppose we denote by E, O, and L, a flight that arrives early, on time, or late, respectively.

1. Use a tree diagram to exhibit the possible outcomes when you track two successive flights of the airline. How many outcomes are there?

2. How many outcomes are there if you track three successive flights? Justify your answer.

EXAMPLE 2   Diners at Angelo's Spaghetti Bar may select their entree from 6 varieties of pasta and 28 choices of sauce. How many such combinations are there that consist of 1 variety of pasta and 1 kind of sauce?

Solution   There are 6 ways of choosing a pasta followed by 28 ways of choosing a sauce, so by the multiplication principle, there are $6 \cdot 28$, or 168, combinations of this pasta dish. ∎

The multiplication principle may be easily extended, which leads to the **generalized multiplication principle**.

Generalized Multiplication Principle

Suppose a task T_1 can be performed in N_1 ways, a task T_2 can be performed in N_2 ways, . . . , and, finally, a task T_n can be performed in N_n ways. Then, the number of ways of performing the tasks $T_1, T_2, \ldots, T_n$ in succession is given by the product

$$N_1 N_2 \cdots N_n$$

We now illustrate the application of the generalized multiplication principle to several diverse situations.

EXAMPLE 3   A coin is tossed 3 times, and the sequence of heads and tails is recorded.
  a. Use the generalized multiplication principle to determine the number of possible outcomes of this activity.
  b. Exhibit all the sequences by means of a tree diagram.

Solution
  a. The coin may land in two ways. Therefore, in three tosses the number of outcomes (sequences) is given by $2 \cdot 2 \cdot 2$, or 8.
  b. Let H and T denote the outcomes "a head" and "a tail," respectively. Then the required sequences may be obtained as shown in Figure 14, giving the sequence as HHH, HHT, HTH, HTT, THH, THT, TTH, and TTT.

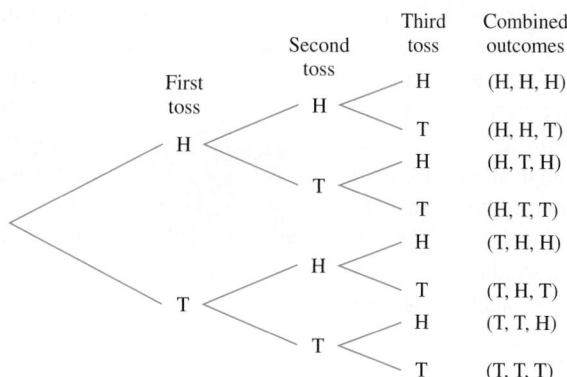

FIGURE 14
Tree diagram displaying possible outcomes of three consecutive coin tosses

APPLIED EXAMPLE 4   Combination Locks   A combination lock is unlocked by dialing a sequence of numbers: first to the left, then to the right, and to the left again. If there are ten digits on the dial, determine the number of possible combinations.

Solution   There are ten choices for the first number, followed by ten for the second and ten for the third, so by the generalized multiplication principle there are $10 \cdot 10 \cdot 10$, or 1000, possible combinations.

APPLIED EXAMPLE 5   Investment Options   An investor has decided to purchase shares in the stock of three companies: one engaged in aerospace activities, one involved in energy development, and one involved in electronics. After some research, the account executive of a brokerage firm has recommended that the investor consider stock from five aerospace companies, three energy development companies, and four electronics companies. In how many ways may the investor select the group of three companies from the executive's list?

PORTFOLIO Stephanie Molina

TITLE Computer Crimes Detective
INSTITUTION Maricopa County Sheriff's Office

Working as a detective in the computer crimes division of the Maricopa County Sheriff's Office, I find applied mathematics techniques play a significant role in my job when I search for evidence contained on computer hard drives and other forms of media. To obtain evidence, I am required to have a working knowledge of certain applied mathematics skills so that I can effectively communicate with the computer forensic analyst who will be decoding the evidence. In order to conduct an effective investigation, I am also required to understand this data in a wide variety of formats. With this information, I can work with the analyst to reconstruct data that may play a significant roll in determining events that occurred pertaining to a crime.

During the course of an investigation, I have to look at the data not only in text, but also in code. Using this view, the analyst can decipher different file types and possible evidence in unallocated space throughout the hard drive. This unallocated space can contain deleted files that may contain potential evidence. The analyst also has to decode files by hand and, at this point, recognizing patterns among the files becomes very important. From here, we can derive an algorithm to define those patterns. By producing an algorithm, it makes it possible to write a program that will decode the files for you.

For example, there was a case that involved a suspect that was receiving files through a mail server. This suspect was then opening the files and deleting the email. Members of my computer forensic laboratory and I viewed these files in their original code to try and discover any patterns or inconsistencies within the code to find a solution to the problem. We did find a clue buried within the code. We then derived an algorithm defining its pattern. By inputting the algorithm, we could then extract the files from the coded data.

While I do not have a solid background in computer science or, even, mathematics, my knowledge of applied mathematics helps me to understand the procedures involved in obtaining evidence. Best of all, I am able to clearly convey my needs to the forensic analysts in my department.

Solution The investor has five choices for selecting an aerospace company, three choices for selecting an energy development company, and four choices for selecting an electronics company. Therefore, by the generalized multiplication principle, there are $5 \cdot 3 \cdot 4$, or 60, ways in which she can select a group of three companies, one from each industry group. ∎

APPLIED EXAMPLE 6 Travel Options Tom is planning to leave for New York City from Washington, D.C., on Monday morning and has decided that he will either fly or take the train. There are five flights and two trains departing for New York City from Washington that morning. When he returns on Sunday afternoon, Tom plans to either fly or hitch a ride with a friend. There are two flights departing from New York City to Washington that afternoon. In how many ways can Tom complete this round trip?

Solution There are seven ways Tom can go from Washington, D.C., to New York City (five by plane and two by train). On the return trip, Tom can travel in three ways (two by plane and one by car). Therefore, by the multiplication principle, Tom can complete the round trip in $7 \cdot 3$, or 21, ways. ∎

6.3 Self-Check Exercises

1. Encore Travel offers a "Theater Week in London" package originating from New York City. There is a choice of eight flights departing from New York City each week, a choice of five hotel accommodations, and a choice of one complimentary ticket to one of eight shows. How many such travel packages can one choose from?

2. The Café Napolean offers a dinner special on Wednesdays consisting of a choice of two entrées (beef bourguignon and chicken basquaise); one dinner salad; one french roll; a choice of three vegetables; a choice of a carafe of burgundy, rosé, or chablis wine; a choice of coffee or tea; and a choice of six french pastries for dessert. How many combinations of dinner specials are there?

Solutions to Self-Check Exercises 6.3 can be found on page 359.

6.3 Concept Questions

1. Explain the multiplication principle and illustrate it with a diagram.

2. Given the following tree diagram for an activity, what are the possible outcomes?

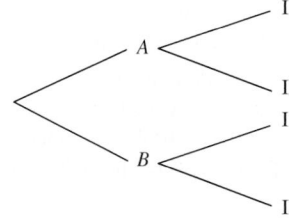

6.3 Exercises

1. RENTAL RATES Lynbrook West, an apartment complex financed by the State Housing Finance Agency, consists of one-, two-, three-, and four-bedroom units. The rental rate for each type of unit—low, moderate, or market—is determined by the income of the tenant. How many different rates are there?

2. COMMUTER PASSES Five different types of monthly commuter passes are offered by a city's local transit authority for three different groups of passengers: youths, adults, and senior citizens. How many different kinds of passes must be printed each month?

3. BLACKJACK In the game of blackjack, a 2-card hand consisting of an ace and either a face card or a 10 is called a "blackjack." If a standard 52-card deck is used, determine how many blackjack hands can be dealt. (A "face card" is a jack, queen, or king.)

4. COIN TOSSES A coin is tossed 4 times and the sequence of heads and tails is recorded.

a. Use the generalized multiplication principle to determine the number of outcomes of this activity.
b. Exhibit all the sequences by means of a tree diagram.

5. WARDROBE SELECTION A female executive selecting her wardrobe purchased two blazers, four blouses, and three skirts in coordinating colors. How many ensembles consisting of a blazer, a blouse, and a skirt can she create from this collection?

6. COMMUTER OPTIONS Four commuter trains and three express buses depart from city A to city B in the morning, and three commuter trains and three express buses operate on the return trip in the evening. In how many ways can a commuter from city A to city B complete a daily round trip via bus and/or train?

7. PSYCHOLOGY EXPERIMENTS A psychologist has constructed the following maze for use in an experiment. The maze is constructed so that a rat must pass through a series of one-way doors. How many different paths are there from start to finish?

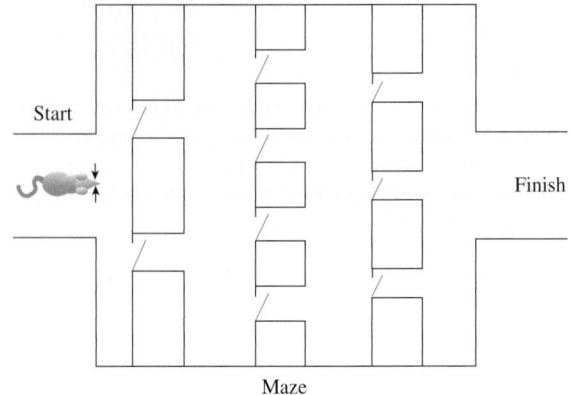

Start

Finish

Maze

8. **UNION BARGAINING ISSUES** In a survey conducted by a union, members were asked to rate the importance of the following issues: (1) job security, (2) increased fringe benefits, and (3) improved working conditions. Five different responses were allowed for each issue. Among completed surveys, how many different responses to this survey were possible?

9. **HEALTH-CARE PLAN OPTIONS** A new state employee is offered a choice of ten basic health plans, three dental plans, and two vision care plans. How many different health-care plans are there to choose from if one plan is selected from each category?

10. **CODE WORDS** How many three-letter code words can be constructed from the first ten letters of the Greek alphabet if no repetitions are allowed?

11. **SOCIAL SECURITY NUMBERS** A Social Security number has nine digits. How many Social Security numbers are possible?

12. **SERIAL NUMBERS** Computers manufactured by a certain company have a serial number consisting of a letter of the alphabet followed by a four-digit number. If all the serial numbers of this type have been used, how many sets have already been manufactured?

13. **COMPUTER DATING** A computer dating service uses the results of its compatibility survey for arranging dates. The survey consists of 50 questions, each having five possible answers. How many different responses are possible if every question is answered?

14. **AUTOMOBILE SELECTION** An automobile manufacturer has three different subcompact cars in the line. Customers selecting one of these cars have a choice of three engine sizes, four body styles, and three color schemes. How many different selections can a customer make?

15. **MENU SELECTIONS** Two soups, five entrées, and three desserts are listed on the "Special" menu at the Neptune Restaurant. How many different selections consisting of one soup, one entrée, and one dessert can a customer choose from this menu?

16. **TELEVISION-VIEWING POLLS** An opinion poll is to be conducted among cable TV viewers. Six multiple-choice questions, each with four possible answers, will be asked. In how many different ways can a viewer complete the poll if exactly one response is given to each question?

17. **ATM CARDS** To gain access to his account, a customer using an automatic teller machine (ATM) must enter a four-digit code. If repetition of the same four digits is not allowed (for example, 5555), how many possible combinations are there?

18. **POLITICAL POLLS** An opinion poll was conducted by the Morris Polling Group. Respondents were classified according to their sex (M or F), political affiliation (D, I, R), and the region of the country in which they reside (NW, W, C, S, E, NE).
    a. Use the generalized multiplication principle to determine the number of possible classifications.
    b. Construct a tree diagram to exhibit all possible classifications of females.

19. **LICENSE PLATE NUMBERS** Over the years, the state of California has used different combinations of letters of the alphabet and digits on its automobile license plates.
    a. At one time, license plates were issued that consisted of three letters followed by three digits. How many different license plates can be issued under this arrangement?
    b. Later on, license plates were issued that consisted of three digits followed by three letters. How many different license plates can be issued under this arrangement?

20. **LICENSE PLATE NUMBERS** In recent years, the state of California issued license plates using a combination of one letter of the alphabet followed by three digits, followed by another three letters of the alphabet. How many different license plates can be issued using this configuration?

21. **EXAMS** An exam consists of ten true-or-false questions. Assuming that every question is answered, in how many different ways can a student complete the exam? In how many ways may the exam be completed if a student may leave some questions unanswered because, say, a penalty is assessed for each incorrect answer?

22. **WARRANTY NUMBERS** A warranty identification number for a certain product consists of a letter of the alphabet followed by a five-digit number. How many possible identification numbers are there if the first digit of the five-digit number must be nonzero?

23. **LOTTERIES** In a state lottery, there are 15 finalists eligible for the Big Money Draw. In how many ways can the first, second, and third prizes be awarded if no ticket holder may win more than one prize?

24. **TELEPHONE NUMBERS**
    a. How many seven-digit telephone numbers are possible if the first digit must be nonzero?

b. How many international direct-dialing numbers are possible if each number consists of a three-digit area code (the first digit of which must be nonzero) and a number of the type described in part (a)?

25. SLOT MACHINES A "lucky dollar" is one of the nine symbols printed on each reel of a slot machine with three reels. A player receives one of various payouts whenever one or more "lucky dollars" appear in the window of the machine. Find the number of winning combinations for which the machine gives a payoff.
Hint: (a) Compute the number of ways in which the nine symbols on the first, second, and third wheels can appear in the window slot and (b) compute the number of ways in which the eight symbols other than the "lucky dollar" can appear in the window slot. The difference $(a - b)$ is the number of ways in which the "lucky dollar" can appear in the window slot. Why?

26. STAFFING Student Painters, which specializes in painting the exterior of residential buildings, has five people available to be organized into two-person and three-person teams.

a. In how many ways can the two-person team be formed?
b. In how many ways can the three-person team be formed?
c. In how many ways can the company organize the available people into two- or three-person teams?

In Exercises 27 and 28, determine whether the statement is true or false. If it is true, explain why it is true. If it is false, give an example to show why it is false.

27. There are 32 three-digit odd numbers that can be formed from the digits 1, 2, 3, and 4.

28. If there are six toppings available, then the number of different pizzas that can be made is 2^5, or 32, pizzas.

6.3 Solutions to Self-Check Exercises

1. A tourist has a choice of eight flights, five hotel accommodations, and eight tickets. By the generalized multiplication principle, there are $8 \cdot 5 \cdot 8$, or 320, travel packages.

2. There is a choice of two entrées, one dinner salad, one french roll, a choice of three vegetables, a choice of three wines, a

choice of two nonalcoholic beverages, and a choice of six pastries. Therefore, by the generalized multiplication principle, there are $2 \cdot 1 \cdot 1 \cdot 3 \cdot 3 \cdot 2 \cdot 6$, or 216, combinations of dinner specials.

6.4 Permutations and Combinations

Permutations

In this section, we apply the generalized multiplication principle to the solution of two types of counting problems. Both types involve determining the number of ways the elements of a set may be arranged, and both play an important role in the solution of problems in probability.

We begin by considering the permutations of a set. Specifically, given a set of distinct objects, a **permutation** of the set is an arrangement of these objects in a *definite order*. To see why the order in which objects are arranged is important in certain practical situations, suppose the winning number for the first prize in a raffle is 9237. Then the number 2973, although it contains the same digits as the winning number, cannot be a first-prize winner (Figure 15). Here, the four objects—the numbers 9, 2, 3, and 7—are arranged in a different order; one arrangement is associated with the winning number for the first prize, and the other is not.

FIGURE 15
The same digits appear on each ticket, but the order of the digits is different.

EXAMPLE 1   Let $A = \{a, b, c\}$.
a. Find the number of permutations of A.
b. List all the permutations of A with the aid of a tree diagram.

Solution
a. Each permutation of A consists of a sequence of the three letters a, b, c. Therefore, we may think of such a sequence as being constructed by filling in each of the three blanks

$$\underline{\quad} \quad \underline{\quad} \quad \underline{\quad}$$

with one of the three letters. Now, there are three ways in which we may fill the first blank—we may choose a, b, or c. Having selected a letter for the first blank, there are two letters left for the second blank. Finally, there is but one way left to fill the third blank. Schematically, we have

$$\underline{3} \quad \underline{2} \quad \underline{1}$$

Invoking the generalized multiplication principle, we conclude that there are $3 \cdot 2 \cdot 1$, or 6, permutations of the set A.
b. The tree diagram associated with this problem appears in Figure 16, and the six permutations of A are abc, acb, bac, bca, cab, and cba.

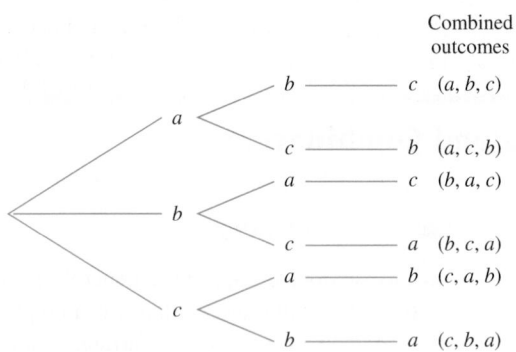

FIGURE 16
Permutations of three objects

Note   Notice that, when the possible outcomes are listed in the tree diagram in Example 1, order is taken into account. Thus, (a, b, c) and (a, c, b) are two different arrangements.

EXAMPLE 2   Find the number of ways a baseball team consisting of nine people can arrange themselves in a line for a group picture.

Solution   We want to determine the number of permutations of the nine members of the baseball team. Each permutation in this situation consists of an arrangement of the nine team members in a line. The nine positions can be represented by nine blanks. Thus,

Position   1   2   3   4   5   6   7   8   9

There are nine ways to choose from among the nine players to fill the first position. When that position is filled, eight players are left, which gives us eight ways to fill the second position. Proceeding in a similar manner, we find that there are seven ways to fill the third position, and so on. Schematically, we have

Number of ways to
fill each position

Invoking the generalized multiplication principle, we conclude that there are $9 \cdot 8 \cdot 7 \cdot 6 \cdot 5 \cdot 4 \cdot 3 \cdot 2 \cdot 1$, or 362,880, ways the baseball team can be arranged for the picture.

 Whenever we are asked to determine the number of ways the objects of a set can be arranged in a line, order is important. For example, if we take a picture of two baseball players, A and B, then the two players can line up for the picture in two ways, AB or BA, and the two pictures will be different.

Pursuing the same line of argument used in solving the problems in the last two examples, we can derive an expression for the number of ways of permuting a set A of n distinct objects taken n at a time. In fact, each permutation may be viewed as being obtained by filling each of n blanks with one and only one element from the set. There are n ways of filling the first blank, followed by $(n-1)$ ways of filling the second blank, and so on. Thus, by the generalized multiplication principle, there are

$$n(n-1)(n-2)\cdots\cdot 3 \cdot 2 \cdot 1$$

ways of permuting the elements of the set A.

Before stating this result formally, let's introduce a notation that will enable us to write in a compact form many of the expressions that follow. We use the symbol $n!$ (read "*n*-factorial") to denote the product of the first n natural numbers.

---

n-Factorial

For any natural number n,

$$n! = n(n-1)(n-2)\cdots\cdot 3 \cdot 2 \cdot 1$$
$$0! = 1$$

For example,

$$1! = 1$$
$$2! = 2 \cdot 1 = 2$$
$$3! = 3 \cdot 2 \cdot 1 = 6$$
$$4! = 4 \cdot 3 \cdot 2 \cdot 1 = 24$$
$$5! = 5 \cdot 4 \cdot 3 \cdot 2 \cdot 1 = 120$$
$$\vdots$$
$$10! = 10 \cdot 9 \cdot 8 \cdot 7 \cdot 6 \cdot 5 \cdot 4 \cdot 3 \cdot 2 \cdot 1 = 3,628,800$$

Using this notation, we may express *the number of permutations of n distinct objects taken n at a time, P(n, n)*, as

$$P(n, n) = n!$$

In many situations, we are interested in determining the number of ways of permuting n distinct objects taken r at a time, where $r \le n$. To derive a formula for computing the number of ways of permuting a set consisting of n distinct objects taken r at a time, we observe that each such permutation may be viewed as being obtained by filling each of r blanks with precisely one element from the set. Now there are n ways of filling the first blank, followed by $(n - 1)$ ways of filling the second blank, and so on. Finally, there are $(n - r + 1)$ ways of filling the rth blank. We may represent this argument schematically:

Number of ways	n	$n - 1$	$n - 2$	$\cdots$	$n - r + 1$
Position	1st	2nd	3rd		rth

Using the generalized multiplication principle, we conclude that *the number of ways of permuting n distinct objects taken r at a time, P(n, r), is given by*

$$P(n, r) = \underbrace{n(n - 1)(n - 2) \cdots (n - r + 1)}_{r \text{ factors}}$$

Since

$$n(n - 1)(n - 2) \cdots (n - r + 1)$$

$$= [n(n - 1)(n - 2) \cdots (n - r + 1)] \cdot \underbrace{\frac{[(n - r)(n - r - 1) \cdot \cdots \cdot 3 \cdot 2 \cdot 1]}{[(n - r)(n - r - 1) \cdot \cdots \cdot 3 \cdot 2 \cdot 1]}}_{\text{Here we are multiplying by 1.}}$$

$$= \frac{[n(n - 1)(n - 2) \cdots (n - r + 1)][(n - r)(n - r - 1) \cdot \cdots \cdot 3 \cdot 2 \cdot 1]}{[(n - r)(n - r - 1) \cdot \cdots \cdot 3 \cdot 2 \cdot 1]}$$

$$= \frac{n!}{(n - r)!}$$

we have the following formula.

Permutations of *n* Distinct Objects

The number of *permutations* of n distinct objects taken r at a time is

$$P(n, r) = \frac{n!}{(n - r)!} \qquad \qquad \textbf{(6)}$$

Note   When $r = n$, Equation (6) reduces to

$$P(n, n) = \frac{n!}{0!} = \frac{n!}{1} = n! \qquad \text{Note that } 0! = 1.$$

In other words, the number of permutations of a set of n distinct objects, taken all together, is $n!$.

EXAMPLE 3 Compute (a) $P(4, 4)$ and (b) $P(4, 2)$ and interpret your results.

Solution

a. $P(4, 4) = \dfrac{4!}{(4-4)!} = \dfrac{4!}{0!} = \dfrac{4!}{1} = \dfrac{4 \cdot 3 \cdot 2 \cdot 1}{1} = 24$ Recall that $0! = 1$.

This gives the number of permutations of four objects taken four at a time.

b. $P(4, 2) = \dfrac{4!}{(4-2)!} = \dfrac{4!}{2!} = \dfrac{4 \cdot 3 \cdot 2 \cdot 1}{2 \cdot 1} = 12$

This is the number of permutations of four objects taken two at a time. ∎

EXAMPLE 4 Let $A = \{a, b, c, d\}$.

a. Use Equation (6) to compute the number of permutations of the set A taken two at a time.

b. Display the permutations of part (a) with the aid of a tree diagram.

Solution

a. Here, $n = 4$ and $r = 2$, so the required number of permutations is given by

$$P(4, 2) = \frac{4!}{(4-2)!} = \frac{4!}{2!} = \frac{4 \cdot 3 \cdot 2 \cdot 1}{2 \cdot 1} = 4 \cdot 3$$
$$= 12$$

b. The tree diagram associated with the problem is shown in Figure 17, and the permutations of A taken two at a time are

$$ab,\ ac,\ ad,\ ba,\ bc,\ bd,\ ca,\ cb,\ cd,\ da,\ db,\ dc$$ ∎

Combined outcomes

a	b	(a, b)
	c	(a, c)
	d	(a, d)
b	a	(b, a)
	c	(b, c)
	d	(b, d)
c	a	(c, a)
	b	(c, b)
	d	(c, d)
d	a	(d, a)
	b	(d, b)
	c	(d, c)

FIGURE 17
Permutations of four objects taken two at a time

EXAMPLE 5 Find the number of ways a chairman, a vice-chairman, a secretary, and a treasurer can be chosen from a committee of eight members.

Solution The problem is equivalent to finding the number of permutations of eight distinct objects taken four at a time. Therefore, there are

$$P(8, 4) = \frac{8!}{(8-4)!} = \frac{8!}{4!} = 8 \cdot 7 \cdot 6 \cdot 5 = 1680$$

ways of choosing the four officials from the committee of eight members. ∎

The permutations considered thus far have been those involving sets of *distinct* objects. In many situations we are interested in finding the number of permutations of a set of objects in which not all of the objects are distinct.

Permutations of n Objects, Not All Distinct
Given a set of n objects in which n_1 objects are alike and of one kind, n_2 objects are alike and of another kind, . . . , and n_r objects are alike and of yet another kind, so that

$$n_1 + n_2 + \cdots + n_r = n$$

then the number of permutations of these n objects taken n at a time is given by

$$\frac{n!}{n_1!\, n_2! \cdots n_r!} \tag{7}$$

To establish Equation (7), let's denote the number of such permutations by x. Now, if we *think* of the n_1 objects as being distinct, then they may be permuted in $n_1!$ ways. Similarly, if we *think* of the n_2 objects as being distinct, then they may be permuted in $n_2!$ ways, and so on. Therefore, if we *think* of the n objects as being distinct, then, by the generalized multiplication principle, there are $x \cdot n_1! \cdot n_2! \cdot \cdots \cdot n_r!$ permutations of these objects. But, the number of permutations of a set of n distinct objects taken n at a time is just equal to $n!$. Therefore, we have

$$x(n_1! \cdot n_2! \cdot \cdots \cdot n_r!) = n!$$

from which we deduce that

$$x = \frac{n!}{n_1! \, n_2! \cdots n_r!}$$

EXAMPLE 6 Find the number of permutations that can be formed from all the letters in the word *ATLANTA*.

Solution There are seven objects (letters) involved, so $n = 7$. However, three of them are alike and of one kind (the three *A*s), while two of them are alike and of another kind (the two *T*s); hence, in this case we have $n_1 = 3$, $n_2 = 2$, $n_3 = 1$, and $n_4 = 1$. Therefore, using Formula (7), there are

$$\frac{7!}{3! \, 2! \, 1! \, 1!} = \frac{7 \cdot 6 \cdot 5 \cdot 4 \cdot 3 \cdot 2 \cdot 1}{3 \cdot 2 \cdot 1 \cdot 2 \cdot 1} = 420$$

required permutations.

APPLIED EXAMPLE 7 Management Decisions Weaver and Kline, a stock brokerage firm, has received nine inquiries regarding new accounts. In how many ways can these inquiries be directed to three of the firm's account executives if each account executive is to handle three inquiries?

Solution If we think of the nine inquiries as being slots arranged in a row with inquiry 1 on the left and inquiry 9 on the right, then the problem can be thought of as one of filling each slot with a business card from an account executive. Then nine business cards would be used, of which three are alike and of one kind, three are alike and of another kind, and three are alike and of yet another kind. Thus, using (7) with $n = 9$ and $n_1 = n_2 = n_3 = 3$, there are

$$\frac{9!}{3! \, 3! \, 3!} = \frac{9 \cdot 8 \cdot 7 \cdot 6 \cdot 5 \cdot 4 \cdot 3 \cdot 2 \cdot 1}{3 \cdot 2 \cdot 1 \cdot 3 \cdot 2 \cdot 1 \cdot 3 \cdot 2 \cdot 1} = 1680$$

ways of assigning the inquiries.

Combinations

Until now we have dealt with permutations of a set—that is, with arrangements of the objects of the set in which the *order* of the elements is taken into consideration. In many situations one is interested in determining the number of ways of selecting r objects from a set of n objects without any regard to the order in which the objects are selected. Such a subset is called a **combination**.

For example, if one is interested in knowing the number of 5-card poker hands that can be dealt from a standard deck of 52 cards, then the order in which the poker hand is dealt is unimportant (Figure 18). In this situation, we are interested in determining the number of combinations of 5 cards (objects) selected from a deck (set) of 52 cards (objects). (We will solve this problem in Example 10.)

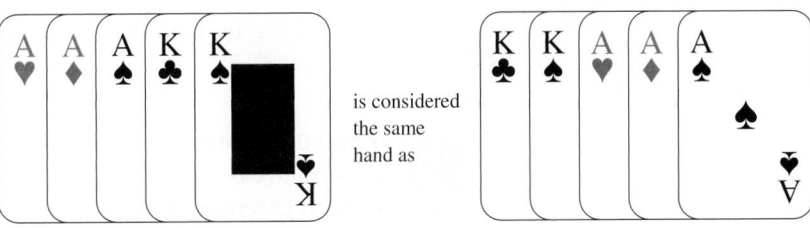

is considered the same hand as

FIGURE 18

To derive a formula for determining the number of combinations of n objects taken r at a time, written

$$C(n, r) \quad \text{or} \quad \binom{n}{r}$$

we observe that each of the $C(n, r)$ combinations of r objects can be permuted in $r!$ ways (Figure 19).

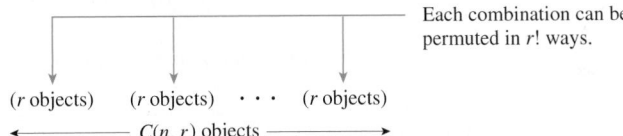

Each combination can be permuted in $r!$ ways.

$(r$ objects) $(r$ objects) $\cdots$ $(r$ objects)

$\longleftarrow$ $C(n, r)$ objects $\longrightarrow$

FIGURE 19

Thus, by the multiplication principle, the product $r!\, C(n, r)$ gives the number of permutations of n objects taken r at a time; that is,

$$r!\, C(n, r) = P(n, r)$$

from which we find

$$C(n, r) = \frac{P(n, r)}{r!}$$

or, using Equation (6),

$$C(n, r) = \frac{n!}{r!\,(n - r)!}$$

Combinations of n Objects

The number of combinations of n distinct objects taken r at a time is given by

$$C(n, r) = \frac{n!}{r!\,(n - r)!} \qquad \text{(where } r \leq n\text{)} \qquad \textbf{(8)}$$

EXAMPLE 8   Compute and interpret the results of (a) $C(4, 4)$ and (b) $C(4, 2)$.

Solution

a. $C(4, 4) = \dfrac{4!}{4!\,(4 - 4)!} = \dfrac{4!}{4!\, 0!} = 1$ Recall that $0! = 1$.

This gives 1 as the number of combinations of four distinct objects taken four at a time.

b. $C(4, 2) = \dfrac{4!}{2!\,(4-2)!} = \dfrac{4!}{2!\,2!} = \dfrac{4 \cdot 3}{2} = 6$

This gives 6 as the number of combinations of four distinct objects taken two at a time. ■

 APPLIED EXAMPLE 9   Committee Selection   A Senate investigation subcommittee of four members is to be selected from a Senate committee of ten members. Determine the number of ways this can be done.

Solution   The order in which the members of the subcommittee are selected is unimportant and so the number of ways of choosing the subcommittee is given by $C(10, 4)$, the number of combinations of ten objects taken four at a time. Hence, there are

$$C(10, 4) = \frac{10!}{4!\,(10-4)!} = \frac{10!}{4!\,6!} = \frac{10 \cdot 9 \cdot 8 \cdot 7}{4 \cdot 3 \cdot 2 \cdot 1} = 210$$

ways of choosing such a subcommittee. ■

Note   Remember, a combination is a selection of objects *without* regard to order. Thus, in Example 9, we used a combination formula rather than a permutation formula to solve the problem because the order of selection was not important; that is, it did not matter whether a member of the subcommittee was selected first, second, third, or fourth. ■

APPLIED EXAMPLE 10   Poker   How many poker hands of 5 cards can be dealt from a standard deck of 52 cards?

Solution   The order in which the 5 cards are dealt is not important. The number of ways of dealing a poker hand of 5 cards from a standard deck of 52 cards is given by $C(52, 5)$, the number of combinations of 52 objects taken five at a time. Thus, there are

$$\begin{aligned}
C(52, 5) &= \frac{52!}{5!\,(52-5)!} = \frac{52!}{5!\,47!} \\
&= \frac{52 \cdot 51 \cdot 50 \cdot 49 \cdot 48}{5 \cdot 4 \cdot 3 \cdot 2 \cdot 1} \\
&= 2{,}598{,}960
\end{aligned}$$

ways of dealing such a poker hand. ■

The next several examples show that solving a counting problem often involves the repeated application of Equation (6) and/or (8), possibly in conjunction with the multiplication principle.

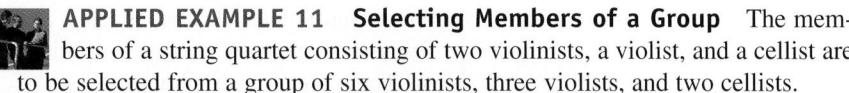

APPLIED EXAMPLE 11   Selecting Members of a Group The members of a string quartet consisting of two violinists, a violist, and a cellist are to be selected from a group of six violinists, three violists, and two cellists.

a. In how many ways can the string quartet be formed?

b. In how many ways can the string quartet be formed if one of the violinists is to be designated as the first violinist and the other is to be designated as the second violinist?

Solution

a. Since the order in which each musician is selected is not important, we use combinations. The violinists may be selected in $C(6, 2)$, or 15, ways; the violist may be selected in $C(3, 1)$, or 3, ways; and the cellist may be selected in $C(2, 1)$, or 2, ways. By the multiplication principle, there are $15 \cdot 3 \cdot 2$, or 90, ways of forming the string quartet.

b. The order in which the violinists are selected is important here. Consequently, the number of ways of selecting the violinists is given by $P(6, 2)$, or 30, ways. The number of ways of selecting the violist and the cellist remain, of course, 3 and 2, respectively. Therefore, the number of ways in which the string quartet may be formed is given by $30 \cdot 3 \cdot 2$, or 180, ways. ∎

Note The solution of Example 11 involves both a permutation and a combination. When we select two violinists from six violinists, order is not important, and we use a combination formula to solve the problem. However, when one of the violinists is designated as a first violinist, order is important, and we use a permutation formula to solve the problem. ∎

APPLIED EXAMPLE 12   Investment Options Refer to Example 5, page 355. Suppose the investor has decided to purchase shares in the stocks of two aerospace companies, two energy development companies, and two electronics companies. In how many ways may the investor select the group of six companies for the investment from the recommended list of five aerospace companies, three energy development companies, and four electronics companies?

Solution There are $C(5, 2)$ ways in which the investor may select the aerospace companies, $C(3, 2)$ ways in which she may select the companies involved in energy development, and $C(4, 2)$ ways in which she may select the electronics companies as investments. By the generalized multiplication principle, there are

$$C(5, 2)C(3, 2)C(4, 2) = \frac{5!}{2!\,3!} \cdot \frac{3!}{2!\,1!} \cdot \frac{4!}{2!\,2!}$$

$$= \frac{5 \cdot 4}{2} \cdot 3 \cdot \frac{4 \cdot 3}{2} = 180$$

ways of selecting the group of six companies for her investment. ∎

APPLIED EXAMPLE 13   Scheduling Performances The Futurists, a rock group, are planning a concert tour with performances to be given in five

cities: San Francisco, Los Angeles, San Diego, Denver, and Las Vegas. In how many ways can they arrange their itinerary if
a. There are no restrictions?
b. The three performances in California must be given consecutively?

Solution
a. The order is important here, and we see that there are

$$P(5, 5) = 5! = 120$$

ways of arranging their itinerary.
b. First, note that there are $P(3, 3)$ ways of choosing between performing in California and in the two cities outside that state. Next, there are $P(3, 3)$ ways of arranging their itinerary in the three cities in California. Therefore, by the multiplication principle, there are

$$P(3, 3)P(3, 3) = \frac{3!}{(3 - 3)!} \cdot \frac{3!}{(3 - 3)!} = (6)(6) = 36$$

ways of arranging their itinerary.　◼

APPLIED EXAMPLE 14　U.N. Security Council Voting　The U.N. Security Council consists of 5 permanent members and 10 nonpermanent members. Decisions made by the council require 9 votes for passage. However, any permanent member may veto a measure and thus block its passage. Assuming there are no abstentions, in how many ways can a measure be passed if all 15 members of the Council vote?

Solution　If a measure is to be passed, then all 5 permanent members must vote for passage of that measure. This can be done in $C(5, 5)$, or 1, way.

Next, observe that since 9 votes are required for passage of a measure, *at least* 4 of the 10 nonpermanent members must also vote for its passage. To determine the number of ways this can be done, notice that there are $C(10, 4)$ ways in which exactly 4 of the nonpermanent members can vote for passage of a measure, $C(10, 5)$ ways in which exactly 5 of them can vote for passage of a measure, and so on. Finally, there are $C(10, 10)$ ways in which all 10 nonpermanent members can vote for passage of a measure. Hence, there are

$$C(10, 4) + C(10, 5) + \cdots + C(10, 10)$$

ways in which at least 4 of the 10 nonpermanent members can vote for a measure. So, by the multiplication principle, there are

$$C(5, 5)[C(10, 4) + C(10, 5) + \cdots + C(10, 10)]$$
$$= (1)\left[\frac{10!}{4! \, 6!} + \frac{10!}{5! \, 5!} + \cdots + \frac{10!}{10! \, 0!}\right]$$
$$= (1)(210 + 252 + 210 + 120 + 45 + 10 + 1) = 848$$

ways a measure can be passed.　◼

6.4 Self-Check Exercises

1. Evaluate:
   a. 5!   **b.** $C(7, 4)$ **c.** $P(6, 2)$

2. A space shuttle crew consists of a shuttle commander, a pilot, three engineers, a scientist, and a civilian. The shuttle commander and pilot are to be chosen from 8 candidates, the three engineers from 12 candidates, the scientist from 5 candidates,

and the civilian from 2 candidates. How many such space shuttle crews can be formed?

Solutions to Self-Check Exercises 6.4 can be found on page 372.

6.4 Concept Questions

1. **a.** What is a permutation of a set of distinct objects?
   b. How many permutations of a set of five distinct objects taken three at a time are there?

2. Given a set of ten objects in which three are alike and of one kind, three are alike and of yet another kind, and four are alike and of yet another kind, what is the formula for computing the permutation of these ten objects taken ten at a time?

3. **a.** What is a combination of a set of n distinct objects taken r at a time?
   b. How many combinations are there of six distinct objects taken three at a time?

6.4 Exercises

In Exercises 1–22, evaluate the given expression.

1. $3(5!)$

2. $2(7!)$

3. $\dfrac{5!}{2!\,3!}$

4. $\dfrac{6!}{4!\,2!}$

5. $P(5, 5)$

6. $P(6, 6)$

7. $P(5, 2)$

8. $P(5, 3)$

9. $P(n, 1)$

10. $P(k, 2)$

11. $C(6, 6)$

12. $C(8, 8)$

13. $C(7, 4)$

14. $C(9, 3)$

15. $C(5, 0)$

16. $C(6, 5)$

17. $C(9, 6)$

18. $C(10, 3)$

19. $C(n, 2)$

20. $C(7, r)$

21. $P(n, n - 2)$

22. $C(n, n - 2)$

In Exercises 23–30, classify each problem according to whether it involves a permutation or a combination.

23. In how many ways can the letters of the word *GLACIER* be arranged?

24. A four-member executive committee is to be formed from a twelve-member board of directors. In how many ways can it be formed?

25. As part of a quality-control program, 3 cell phones are selected at random for testing from each of 100 cell phones produced by the manufacturer. In how many ways can this test batch be chosen?

26. How many three-digit numbers can be formed using the numerals in the set {3, 2, 7, 9} if repetition is not allowed?

27. In how many ways can nine different books be arranged on a shelf?

28. A member of a book club wishes to purchase two books from a selection of eight books recommended for a certain month. In how many ways can she choose them?

29. How many five-card poker hands can be dealt consisting of three queens and a pair?

30. In how many ways can a six-letter security password be formed from letters of the alphabet if no letter is repeated?

31. How many four-letter permutations can be formed from the first four letters of the alphabet?

32. How many three-letter permutations can be formed from the first five letters of the alphabet?

33. In how many ways can four students be seated in a row of four seats?

34. In how many ways can five people line up at a checkout counter in a supermarket?

35. How many different batting orders can be formed for a nine-member baseball team?

36. In how many ways can the names of six candidates for political office be listed on a ballot?

37. In how many ways can a member of a hiring committee select 3 of 12 job applicants for further consideration?

38. In how many ways can an investor select four mutual funds for his investment portfolio from a recommended list of eight mutual funds?

39. Find the number of distinguishable permutations that can be formed from the letters of the word *ANTARCTICA*.

40. Find the number of distinguishable permutations that can be formed from the letters of the word *PHILIPPINES*.

41. MANAGEMENT DECISIONS In how many ways can a supermarket chain select 3 out of 12 possible sites for the construction of new supermarkets?

42. BOOK SELECTIONS A student is given a reading list of ten books from which he must select two for an outside reading requirement. In how many ways can he make his selections?

43. QUALITY CONTROL In how many ways can a quality-control engineer select a sample of 3 microprocessors for testing from a batch of 100 microprocessors?

44. STUDY GROUPS A group of five students studying for a bar exam had formed a study group. Each member of the group will be responsible for preparing a study outline for one of five courses. In how many different ways can the five courses be assigned to the members of the group?

45. TELEVISION PROGRAMMING In how many ways can a television-programming director schedule six different commercials in the six time slots allocated to commercials during a 1-hr program?

46. WAITING LINES Seven people arrive at the ticket counter of a cinema at the same time. In how many ways can they line up to purchase their tickets?

47. MANAGEMENT DECISIONS Weaver and Kline, a stock brokerage firm, has received six inquiries regarding new accounts. In how many ways can these inquiries be directed to its twelve account executives if each executive handles no more than one inquiry?

48. CAR POOLS A company car that has a seating capacity of six is to be used by six employees who have formed a car pool. If only four of these employees can drive, how many possible seating arrangements are there for the group?

49. BOOK DISPLAYS At a college library exhibition of faculty publications, three mathematics books, four social science books, and three biology books will be displayed on a shelf. (Assume that none of the books is alike.)

a. In how many ways can the ten books be arranged on the shelf?

b. In how many ways can the ten books be arranged on the shelf if books on the same subject matter are placed together?

50. SEATING In how many ways can four married couples attending a concert be seated in a row of eight seats if

a. There are no restrictions?

b. Each married couple is seated together?

c. The members of each sex are seated together?

51. NEWSPAPER ADVERTISEMENTS Four items from five different departments of Metro Department Store will be featured in a one-page newspaper advertisement, as shown in the following diagram:

Advertisement

1	2	3	4
5	6	7	8
9	10	11	12
13	14	15	16
17	18	19	20

a. In how many different ways can the 20 featured items be arranged on the page?

b. If items from the same department must be in the same row, how many arrangements are possible?

52. MANAGEMENT DECISIONS C & J Realty has received twelve inquiries from prospective home buyers. In how many ways can the inquiries be directed to four of the firm's real estate agents if each agent handles three inquiries?

53. SPORTS In the women's tennis tournament at Wimbledon, two finalists, A and B, are competing for the title, which will be awarded to the first player to win two sets. In how many different ways can the match be completed?

54. SPORTS In the men's tennis tournament at Wimbledon, two finalists, A and B, are competing for the title, which will be awarded to the first player to win three sets. In how many different ways can the match be completed?

55. U.N. VOTING Refer to Example 14. In how many ways can a measure be passed if two particular permanent and two particular nonpermanent members of the Council abstain from voting?

56. JURY SELECTION In how many different ways can a panel of 12 jurors and 2 alternate jurors be chosen from a group of 30 prospective jurors?

57. TEACHING ASSISTANTSHIPS Twelve graduate students have applied for three available teaching assistantships. In how many ways can the assistantships be awarded among these applicants if

a. No preference is given to any student?

b. One particular student must be awarded an assistantship?

c. The group of applicants includes seven men and five women and it is stipulated that at least one woman must be awarded an assistantship?

58. EXAMS A student taking an examination is required to answer 10 out of 15 questions.

a. In how many ways can the 10 questions be selected?

b. In how many ways can the 10 questions be selected if exactly 2 of the first 3 questions must be answered?

59. CONTRACT BIDDING UBS Television Company is considering bids submitted by seven different firms for three different contracts. In how many ways can the contracts be awarded among these firms if no firm is to receive more than two contracts?

60. SENATE COMMITTEES In how many ways can a subcommittee of four be chosen from a Senate committee of five Democrats and four Republicans if

a. All members are eligible?

b. The subcommittee must consist of two Republicans and two Democrats?

61. COURSE SELECTION A student planning her curriculum for the upcoming year must select one of five business courses, one of three mathematics courses, two of six elective courses, and either one of four history courses or one of three social science courses. How many different curricula are available for her consideration?

62. PERSONNEL SELECTION JCL Computers has five vacancies in its executive trainee program. In how many ways can the company select five trainees from a group of ten female and ten male applicants if the vacancies

a. May be filled by any combination of men and women?

b. Must be filled by two men and three women?

63. DRIVERS' TESTS A state Motor Vehicle Department requires learners to pass a written test on the motor vehicle laws of the state. The exam consists of ten true-or-false questions, of which eight must be answered correctly to qualify for a permit. In how many different ways can a learner who answers all the questions on the exam qualify for a permit?

64. QUALITY CONTROL Goodman Tire has 32 tires of a particular size and grade in stock, 2 of which are defective. If a set of 4 tires is to be selected,

a. How many different selections can be made?

b. How many different selections can be made that do not include any defective tires?

A list of poker hands ranked in order from the highest to the lowest is shown in the following table, along with a description and example of each hand. Use the table to answer Exercises 65–70.

Hand	Description	Example
Straight flush	5 cards in sequence in the same suit	A ♥ 2 ♥ 3 ♥ 4 ♥ 5 ♥
Four of a kind	4 cards of the same rank and any other card	K ♥ K ♦ K ♠ K ♣ 2 ♥
Full house	3 of a kind and a pair	3 ♥ 3 ♦ 3 ♣ 7 ♥ 7 ♦
Flush	5 cards of the same suit that are not all in sequence	5 ♥ 6 ♥ 9 ♥ J ♥ K ♥
Straight	5 cards in sequence but not all of the same suit	10 ♥ J ♦ Q ♣ K ♠ A ♥
Three of a kind	3 cards of the same rank and 2 unmatched cards	K ♥ K ♦ K ♠ 2 ♥ 4 ♦
Two pair	2 cards of the same rank and 2 cards of any other rank with an unmatched card	K ♥ K ♦ 2 ♥ 2 ♠ 4 ♣
One pair	2 cards of the same rank and 3 unmatched cards	K ♥ K ♦ 5 ♥ 2 ♠ 4 ♥

If a 5-card poker hand is dealt from a well-shuffled deck of 52 cards, how many different hands consist of the following:

65. A straight flush? (Note that an ace may be played as either a high or a low card in a straight sequence—that is, A, 2, 3, 4, 5 or 10, J, Q, K, A. Hence, there are ten possible sequences for a straight in one suit.)

66. A straight (but not a straight flush)?

67. A flush (but not a straight flush)?

68. Four of a kind?

69. A full house?

70. Two pairs?

71. BUS ROUTING The following is a schematic diagram of a city's street system between the points *A* and *B*. The City Transit Authority is in the process of selecting a route from *A* to *B* along which to provide bus service. If the company's intention is to keep the route as short as possible, how many routes must be considered?

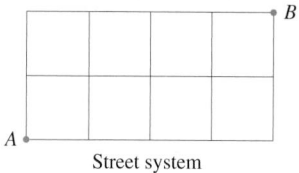

Street system

72. Sports In the World Series, one National League team and one American League team compete for the title, which is awarded to the first team to win four games. In how many different ways can the series be completed?

73. Voting Quorums A quorum (minimum) of 6 voting members is required at all meetings of the Curtis Townhomes Owners Association. If there is a total of 12 voting members in the group, find the number of ways this quorum can be formed.

74. Circular Permutations Suppose n distinct objects are arranged in a circle. Show that the number of (different) circular arrangements of the n objects is $(n - 1)!$.
Hint: Consider the arrangement of the five letters A, B, C, D, and E in the accompanying figure. The permutations $ABCDE$, $BCDEA$, $CDEAB$, $DEABC$, and $EABCD$ are not distinguishable. Generalize this observation to the case of n objects.

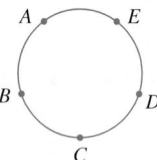

75. Refer to Exercise 74. In how many ways can five TV commentators be seated at a round table for a discussion?

76. Refer to Exercise 74. In how many ways can four men and four women be seated at a round table at a dinner party if each guest is seated between members of the opposite sex?

77. At the end of Section 3.3, we mentioned that solving a linear programming problem in three variables and five constraints requires that we solve 56 3 × 3 systems of linear equations. Verify this assertion.

78. Refer to Exercise 77. Show that, in order to solve a linear programming problem in five variables and ten constraints, we must solve 3003 5 × 5 systems of linear equations. This assertion was also made at the end of Section 3.3.

In Exercises 79–82, determine whether the statement is true or false. If it is true, explain why it is true. If it is false, give an example to show why it is false.

79. The number of permutations of n distinct objects taken all together is $n!$

80. $P(n, r) = r! \, C(n, r)$

81. The number of combinations of n objects taken $n - r$ at a time is the same as the number taken r at a time.

82. If a set of n objects consists of r elements of one kind and $n - r$ elements of another kind, then the number of permutations of the n objects taken all together is $P(n, r)$.

6.4 Solutions to Self-Check Exercises

1. a. $5! = 5 \cdot 4 \cdot 3 \cdot 2 \cdot 1 = 120$

  b. $C(7, 4) = \dfrac{7!}{4! \, 3!} = \dfrac{7 \cdot 6 \cdot 5}{3 \cdot 2 \cdot 1} = 35$

  c. $P(6, 2) = \dfrac{6!}{4!} = 6 \cdot 5 = 30$

2. There are $P(8, 2)$ ways of picking the shuttle commander and pilot (the order *is* important here), $C(12, 3)$ ways of picking the engineers (the order is not important here), $C(5, 1)$ ways of picking the scientist, and $C(2, 1)$ ways of picking the civilian. By the multiplication principle, there are

$$P(8, 2) \cdot C(12, 3) \cdot C(5, 1) \cdot C(2,1)$$
$$= \frac{8!}{6!} \cdot \frac{12!}{9! \, 3!} \cdot \frac{5!}{4! \, 1!} \cdot \frac{2!}{1! \, 1!}$$
$$= \frac{(8)(7)(12)(11)(10)(5)(2)}{(3)(2)}$$
$$= 123{,}200$$

ways a crew can be selected.

USING TECHNOLOGY

■ Evaluating $n!$, $P(n, r)$, and $C(n, r)$

Graphing Utility

A graphing utility can be used to calculate factorials, permutations, and combinations with relative ease. A graphing utility is therefore an indispensable tool in solving counting problems involving large numbers of objects. Here we use the **nPr** (permutation) and **nCr** (combination) functions of a graphing utility.

EXAMPLE 1 Use a graphing utility to find (a) 12!, (b) $P(52, 5)$, and (c) $C(38, 10)$.

Solution

a. Using the factorial function, we find that $12! = 479,001,600$.
b. Using the **nPr** function, we have

$$P(52, 5) = 52 \textbf{ nPr } 5 = 311,875,200$$

c. Using the **nCr** function, we obtain

$$C(38, 10) = 38 \textbf{ nCr } 10 = 472,733,756$$

Excel

Excel has built-in functions for calculating factorials, permutations, and combinations.

EXAMPLE 2 Use Excel to calculate
a. 12!
b. $P(52, 5)$
c. $C(38, 10)$

Solution

a. In cell A1, enter `=FACT(12)` and press Shift-Enter . The number 479001600 will appear.
b. In cell A2, enter `=PERMUT(52,5)` and press Shift-Enter . The number 311875200 will appear.
c. In cell A3, enter `=COMBIN(38,10)` and press Shift-Enter . The number 472733756 will appear.

Note: Boldfaced words/characters enclosed in a box (for example, Enter) indicate that an action (click, select, or press) is required. Words/characters printed blue (for example, Chart sub-type:) indicate words/characters that appear on the screen. Words/characters printed in a typewriter font (for example, `=(-2/3)*A2+2)`) indicate words/characters that need to be typed and entered.

(continued)

TECHNOLOGY EXERCISES

In Exercises 1–10, evaluate the expression.

1. $15!$

2. $20!$

3. $4(18!)$

4. $\dfrac{30!}{18!}$

5. $P(52, 7)$

6. $P(24, 8)$

7. $C(52, 7)$

8. $C(26, 8)$

9. $P(10, 4)C(12, 6)$

10. $P(20, 5)C(9, 3)C(8, 4)$

11. A mathematics professor uses a computerized test bank to prepare her final exam. If 25 different problems are available for the first three exam questions, 40 different problems available for the next five questions, and 30 different problems available for the last two questions, how many different ten-question exams can she prepare? (Assume that the order of the questions within each group is not important.)

12. S & S Brokerage has received 100 inquiries from prospective clients. In how many ways can the inquiries be directed to five of the firm's brokers if each broker handles 20 inquiries?

CHAPTER 6 **Summary of Principal Formulas and Terms**

FORMULAS

1. Commutative laws	$A \cup B = B \cup A$ $A \cap B = B \cap A$
2. Associative laws	$A \cup (B \cup C) = (A \cup B) \cup C$ $A \cap (B \cap C) = (A \cap B) \cap C$
3. Distributive laws	$A \cup (B \cap C)$ $\quad = (A \cup B) \cap (A \cup C)$ $A \cap (B \cup C)$ $\quad = (A \cap B) \cup (A \cap C)$
4. De Morgan's laws	$(A \cup B)^c = A^c \cap B^c$ $(A \cap B)^c = A^c \cup B^c$
5. Number of elements in the union of two finite sets	$n(A \cup B) = n(A) + n(B)$ $\quad\quad - n(A \cap B)$
6. Permutation of n distinct objects, taken r at a time	$P(n, r) = \dfrac{n!}{(n - r)!}$
7. Permutation of n objects, not all distinct, taken n at a time	$\dfrac{n!}{n_1! n_2! \cdots n_r!}$
8. Combination of n distinct objects, taken r at a time	$C(n, r) = \dfrac{n!}{r!(n - r)!}$

TERMS

set (336)

element of a set (336)

roster notation (336)

set-builder notation (336)

set equality (336)

subset (337)

empty set (337)

universal set (338)

Venn diagram (338)

set union (339)

set intersection (339)

set complementation (339)

multiplication principle (353)

generalized multiplication principle (354)

permutation (359)

n-factorial (361)

combination (364)

CHAPTER 6 Concept Review Questions

Fill in the blanks.

1. A well-defined collection of objects is called a/an _____. These objects are also called _____ of the _____.

2. Two sets having exactly the same elements are said to be _____.

3. If every element of a set A is also an element of a set B, then A is a/an _____ of B.

4. **a.** The empty set $\varnothing$ is the set containing _____ elements.
   b. The universal set is the set containing _____ elements.

5. **a.** The set of all elements in A and/or B is called the _____ of A and B.
   b. The set of all elements in A and B is called the _____ of A and B.

6. The set of all elements in U that are not in A is called the _____ of A.

7. Applying De Morgan's law, we can write $(A \cup B \cup C)^c =$ _____.

8. An arrangement of a set of distinct objects in a definite order is called a/an _____; an arrangement in which the order is not important is a/an _____.

CHAPTER 6 Review Exercises

In Exercises 1–4, list the elements of each set in roster notation.

1. $\{x \mid 3x - 2 = 7;\ x \text{ an integer}\}$

2. $\{x \mid x \text{ is a letter of the word } TALLAHASSEE\}$

3. The set whose elements are the even numbers between 3 and 11

4. $\{x \mid (x - 3)(x + 4) = 0;\ x \text{ a negative integer}\}$

Let $A = \{a, c, e, r\}$. In Exercises 5–8, determine whether the set is equal to A.

5. $\{r, e, c, a\}$

6. $\{x \mid x \text{ is a letter of the word } career\}$

7. $\{x \mid x \text{ is a letter of the word } racer\}$

8. $\{x \mid x \text{ is a letter of the word } cares\}$

In Exercises 9–12, shade the portion of the accompanying figure that represents the given set.

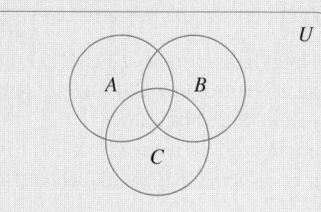

9. $A \cup (B \cap C)$

10. $(A \cap B \cap C)^c$

11. $A^c \cap B^c \cap C^c$

12. $A^c \cap (B^c \cup C^c)$

Let $U = \{a, b, c, d, e\}$, $A = \{a, b\}$, $B = \{b, c, d\}$, and $C = \{a, d, e\}$. In Exercises 13–16, verify the equation by direct computation.

13. $A \cup (B \cup C) = (A \cup B) \cup C$

14. $A \cap (B \cap C) = (A \cap B) \cap C$

15. $A \cap (B \cup C) = (A \cap B) \cup (A \cap C)$

16. $A \cup (B \cap C) = (A \cup B) \cap (A \cup C)$

Let $U = \{$all participants in a consumer-behavior survey conducted by a national polling group$\}$

$A = \{$consumers who avoided buying a product because it is not recyclable$\}$

$B = \{$consumers who used cloth rather than disposable diapers$\}$

$C = \{$consumers who boycotted a company's products because of their record on the environment$\}$

$D = \{$consumers who voluntarily recycled their garbage$\}$

In Exercises 17–20, describe each set in words.

17. $A \cap C$

18. $A \cup D$

19. $B^c \cap D$

20. $C^c \cup D^c$

Let A and B be subsets of a universal set U and suppose $n(U) = 350$, $n(A) = 120$, $n(B) = 80$, and $n(A \cap B) = 50$. In Exercises 21–26, find the number of elements in each set.

21. $n(A \cup B)$

22. $n(A^c)$

23. $n(B^c)$

24. $n(A^c \cap B)$

25. $n(A \cap B^c)$

26. $n(A^c \cap B^c)$

In Exercises 27–30, evaluate each quantity.

27. $C(20, 18)$

28. $P(9, 7)$

29. $C(5, 3) \cdot P(4, 2)$

30. $4 \cdot P(5, 3) \cdot C(7, 4)$

31. CREDIT-CARD COMPARISONS   A comparison of five major credit cards showed that

3 offered cash advances.

3 offered extended payments for all goods and services purchased.

2 required an annual fee of less than $35.

2 offered both cash advances and extended payments.

1 offered extended payments and had an annual fee less than $35.

No card had an annual fee less than $35 and offered both cash advances and extended payments.

How many cards had an annual fee less than $35 and offered cash advances? (Assume that every card had at least one of the three mentioned features.)

32. STUDENT SURVEYS   The Department of Foreign Languages of a liberal arts college conducted a survey of its recent graduates to determine the foreign language courses they had taken while undergraduates at the college. Of the 480 graduates,

200 had at least 1 yr of Spanish.

178 had at least 1 yr of French.

140 had at least 1 yr of German.

33 had at least 1 yr of Spanish and French.

24 had at least 1 yr of Spanish and German.

18 had at least 1 yr of French and German.

3 had at least 1 yr of all three languages.

How many of the graduates had
a. At least 1 yr of at least one of the three languages?
b. At least 1 yr of exactly one of the three languages?
c. Less than 1 yr of any of the three languages?

33. In how many ways can six different compact discs be arranged on a shelf?

34. In how many ways can three pictures be selected from a group of six different pictures?

35. Find the number of distinguishable permutations that can be formed from the letters of each word.
a. *CINCINNATI*   **b.** *HONOLULU*

36. How many three-digit numbers can be formed from the numerals in the set $\{1, 2, 3, 4, 5\}$ if
a. Repetition of digits is not allowed?
b. Repetition of digits is allowed?

37. INVESTMENTS   In a survey conducted by Helena, a financial consultant, it was revealed that of her 400 clients

300 own stocks.

180 own bonds.

160 own mutual funds.

110 own both stocks and bonds.

120 own both stocks and mutual funds.

90 own both bonds and mutual funds.

How many of Helena's clients own stocks, bonds, and mutual funds?

38. POKER   From a standard 52-card deck, how many 5-card poker hands can be dealt consisting of
a. Five clubs?   **b.** Three kings and one pair?

39. ELECTIONS   In an election being held by the Associated Students Organization, there are six candidates for president, four for vice president, five for secretary, and six for treasurer. How many different possible outcomes are there for this election?

40. TEAM SELECTION   There are eight seniors and six juniors in the Math Club at Jefferson High School. In how many ways can a math team consisting of four seniors and two juniors be selected from the members of the Math Club?

41. SEATING ARRANGEMENTS   In how many ways can seven students be assigned seats in a row containing seven desks if
a. There are no restrictions?
b. Two of the students must not be seated next to each other?

42. QUALITY CONTROL   From a shipment of 60 CPUs, 5 of which are defective, a sample of 4 CPUs is selected at random.
a. In how many different ways can the sample be selected?
b. How many samples contain 3 defective CPUs?
c. How many samples do not contain any defective CPUs?

43. RANDOM SAMPLES   A sample of 4 balls is to be selected at random from an urn containing 15 balls numbered 1 to 15. If 6 balls are green, 5 are white, and 4 are black, then:
a. How many different samples can be selected?
b. How many samples can be selected that contain at least 1 white ball?

CHAPTER 6 Before Moving On . . .

1. Let $U = \{a, b, c, d, e, f, g\}$, $A = \{a, d, f, g\}$, $B = \{d, f, g\}$, and $C = \{b, c, e, f\}$. Find
   a. $A \cap (B \cup C)$
   b. $(A \cap C) \cup (B \cup C)$
   c. A^c

2. Let A, B, and C be subsets of a universal set U and suppose that $n(U) = 120$, $n(A) = 20$, $n(A \cap B) = 10$, $n(A \cap C) = 11$, $n(B \cap C) = 9$, and $n(A \cap B \cap C) = 4$. Find $n[A \cap (B \cup C)^c]$.

3. In how many ways can four compact discs be selected from six different compact discs?

4. From a standard 52-card deck, how many 5-card poker hands can be dealt consisting of 3 deuces and 2 face cards?

5. There are six seniors and five juniors in the Chess Club at Madison High School. In how many ways can a team consisting of three seniors and two juniors be selected from the members of the Chess Club?

7 Probability

© Michael Rosenfeld/Stone/Getty Images

Where did the defective picture tube come from? Picture tubes for the Pulsar 19-inch color television sets are manufactured in three locations and then shipped to the main plant of Vista Vision for final assembly. Each location produces a certain number of the picture tubes with different degrees of reliability. In Example 1, page 430, we will determine the likelihood that a defective picture tube is manufactured in a particular location.

THE SYSTEMATIC STUDY of probability began in the 17th century, when certain aristocrats wanted to discover superior strategies to use in the gaming rooms of Europe. Some of the best mathematicians of the period were engaged in this pursuit. Since then, probability has evolved in virtually every sphere of human endeavor in which an element of uncertainty is present.

We begin by introducing some of the basic terminology used in the study of the subject. Then, in Section 7.2, we give the technical meaning of the term *probability*. The rest of this chapter is devoted to the development of techniques for computing the probabilities of the occurrence of certain events.

7.1   Experiments, Sample Spaces, and Events

▬ Terminology

A number of specialized terms are used in the study of probability. We begin by defining the term *experiment*.

Experiment
An **experiment** is an activity with observable results.

The results of the experiment are called the **outcomes** of the experiment. Three examples of experiments are the following:

- Tossing a coin and observing whether it falls "heads" or "tails"
- Casting a die and observing which of the numbers 1, 2, 3, 4, 5, or 6 shows up
- Testing a spark plug from a batch of 100 spark plugs and observing whether or not it is defective

In our discussion of experiments, we use the following terms:

Sample Point, Sample Space, and Event
Sample point: An outcome of an experiment
Sample space: The set consisting of all possible sample points of an experiment
Event: A subset of a sample space of an experiment

The sample space of an experiment is a universal set whose elements are precisely the outcomes, or the sample points, of the experiment; the events of the experiment are the subsets of the universal set. A sample space associated with an experiment that has a finite number of possible outcomes (sample points) is called a **finite sample space**.

Since the events of an experiment are subsets of a universal set (the sample space of the experiment), we may use the results for set theory given in Chapter 6 to help us study probability. The event B is said to **occur** in a trial of an experiment whenever B contains the observed outcome. We begin by explaining the roles played by the empty set and a universal set when viewed as events associated with an experiment. The empty set, $\varnothing$, is called the *impossible event*; it cannot occur because $\varnothing$ has no elements (outcomes). Next, the universal set S is referred to as the *certain event*; it must occur because S contains all the outcomes of the experiment.

This terminology is illustrated in the next several examples.

EXAMPLE 1   Describe the sample space associated with the experiment of tossing a coin and observing whether it falls "heads" or "tails." What are the events of this experiment?

Solution The two outcomes are "heads" and "tails," and the required sample space is given by $S = \{H, T\}$, where H denotes the outcome "heads" and T denotes the outcome "tails." The events of the experiment, the subsets of S, are

$$\varnothing, \{H\}, \{T\}, S$$

Note that we have included the impossible event, $\varnothing$, and the certain event, S. ■

Since the events of an experiment are subsets of the sample space of the experiment, we may talk about the union and intersection of any two events; we can also consider the complement of an event with respect to the sample space.

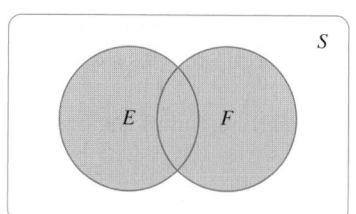

$E \cup F$

(a) The union of two events

Union of Two Events
The **union** of the **two events** E and F is the event $E \cup F$.

Thus, the event $E \cup F$ contains the set of outcomes of E and/or F.

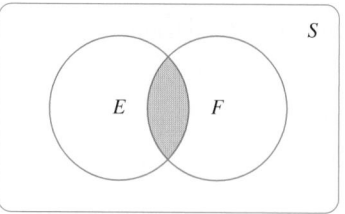

$E \cap F$

(b) The intersection of two events

Intersection of Two Events
The **intersection** of the **two events** E and F is the event $E \cap F$.

Thus, the event $E \cap F$ contains the set of outcomes of E and F.

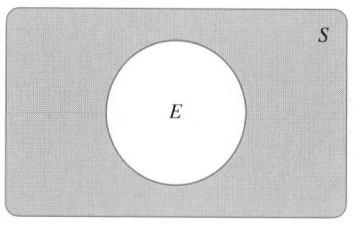

E^c

(c) The complement of the event E

FIGURE 1

Complement of an Event
The **complement** of an **event** E is the event E^c.

Thus, the event E^c is the set containing all the outcomes in the sample space S that are not in E.

Venn diagrams depicting the union, intersection, and complementation of events are shown in Figure 1. These concepts are illustrated in the following example.

EXAMPLE 2 Consider the experiment of casting a die and observing the number that falls uppermost. Let $S = \{1, 2, 3, 4, 5, 6\}$ denote the sample space of the experiment and $E = \{2, 4, 6\}$ and $F = \{1, 3\}$ be events of this experiment. Compute (a) $E \cup F$, (b) $E \cap F$, and (c) F^c. Interpret your results.

Solution
a. $E \cup F = \{1, 2, 3, 4, 6\}$ and is the event that the outcome of the experiment is a 1, a 2, a 3, a 4, or a 6.
b. $E \cap F = \varnothing$ is the impossible event; the number appearing uppermost when a die is cast cannot be both even and odd at the same time.
c. $F^c = \{2, 4, 5, 6\}$ is precisely the event that the event F does not occur. ■

If two events cannot occur at the same time, they are said to be mutually exclusive. Using set notation, we have the following definition.

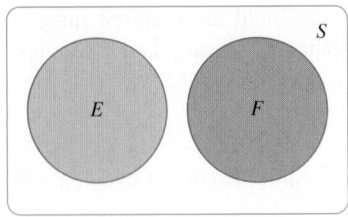

FIGURE 2
Mutually exclusive events

EXPLORE & DISCUSS

1. Suppose E and F are two complementary events. Must E and F be mutually exclusive? Explain your answer.

2. Suppose E and F are mutually exclusive events. Must E and F be complementary? Explain your answer.

FIGURE 3

Mutually Exclusive Events
E and F are **mutually exclusive** if $E \cap F = \emptyset$.

As before, we may use Venn diagrams to illustrate these events. In this case, the two mutually exclusive events are depicted as two nonintersecting circles (Figure 2).

EXAMPLE 3 An experiment consists of tossing a coin three times and observing the resulting sequence of "heads" and "tails."
a. Describe the sample space S of the experiment.
b. Determine the event E that exactly two heads appear.
c. Determine the event F that at least one head appears.

Solution
a. The sample points may be obtained with the aid of a tree diagram (Figure 3).

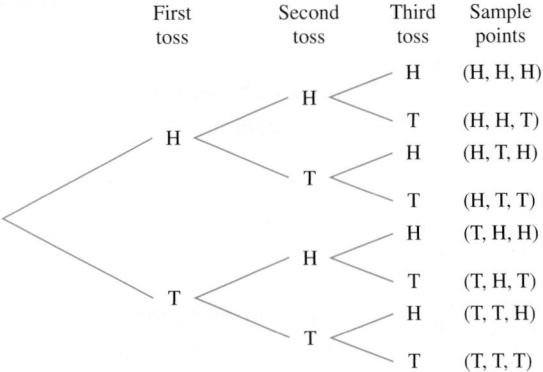

The required sample space S is given by

$$S = \{HHH, HHT, HTH, HTT, THH, THT, TTH, TTT\}$$

b. By scanning the sample space S obtained in part (a), we see that the outcomes in which exactly two heads appear are given by the event

$$E = \{HHT, HTH, THH\}$$

c. Proceeding as in part (b), we find

$$F = \{HHH, HHT, HTH, HTT, THH, THT, TTH\}$$

EXAMPLE 4 An experiment consists of casting a pair of dice and observing the number that falls uppermost on each die.
a. Describe an appropriate sample space S for this experiment.
b. Determine the events $E_2, E_3, E_4, \ldots, E_{12}$ that the sum of the numbers falling uppermost is 2, 3, 4, $\ldots$, 12, respectively.

Solution
a. We may represent each outcome of the experiment by an ordered pair of numbers, the first representing the number that appears uppermost on the first die

and the second representing the number that appears uppermost on the second die. To distinguish between the two dice, think of the first die as being red and the second as being green. Since there are six possible outcomes for each die, the multiplication principle implies that there are $6 \cdot 6$, or 36, elements in the sample space:

$$
\begin{aligned}
S = \{&(1, 1), (1, 2), (1, 3), (1, 4), (1, 5), (1, 6), \\
&(2, 1), (2, 2), (2, 3), (2, 4), (2, 5), (2, 6), \\
&(3, 1), (3, 2), (3, 3), (3, 4), (3, 5), (3, 6), \\
&(4, 1), (4, 2), (4, 3), (4, 4), (4, 5), (4, 6), \\
&(5, 1), (5, 2), (5, 3), (5, 4), (5, 5), (5, 6), \\
&(6, 1), (6, 2), (6, 3), (6, 4), (6, 5), (6, 6)\}
\end{aligned}
$$

b. With the aid of the results of part (a), we obtain the required list of events, shown in Table 1.

TABLE 1

Sum of Uppermost Numbers	Event
2	$E_2 = \{(1, 1)\}$
3	$E_3 = \{(1, 2), (2, 1)\}$
4	$E_4 = \{(1, 3), (2, 2), (3, 1)\}$
5	$E_5 = \{(1, 4), (2, 3), (3, 2), (4, 1)\}$
6	$E_6 = \{(1, 5), (2, 4), (3, 3), (4, 2), (5, 1)\}$
7	$E_7 = \{(1, 6), (2, 5), (3, 4), (4, 3), (5, 2), (6, 1)\}$
8	$E_8 = \{(2, 6), (3, 5), (4, 4), (5, 3), (6, 2)\}$
9	$E_9 = \{(3, 6), (4, 5), (5, 4), (6, 3)\}$
10	$E_{10} = \{(4, 6), (5, 5), (6, 4)\}$
11	$E_{11} = \{(5, 6), (6, 5)\}$
12	$E_{12} = \{(6, 6)\}$

APPLIED EXAMPLE 5 Movie Attendance The manager of a local cinema records the number of patrons attending a first-run movie at the 1 p.m. screening. The theater has a seating capacity of 500.
a. What is an appropriate sample space for this experiment?
b. Describe the event E that fewer than 50 people attend the screening.
c. Describe the event F that the theater is more than half full at the screening.

Solution
a. The number of patrons at the screening (the outcome) could run from 0 to 500. Therefore, a sample space for this experiment is

$$
S = \{0, 1, 2, 3, \ldots, 500\}
$$

b. $E = \{0, 1, 2, 3, \ldots, 49\}$
c. $F = \{251, 252, 253, \ldots, 500\}$

APPLIED EXAMPLE 6   Family Composition   An experiment consists of studying the composition of a three-child family in which the children were born at different times.

a. Describe an appropriate sample space S for this experiment.
b. Describe the event E that there are two girls and a boy in the family.
c. Describe the event F that the oldest child is a girl.
d. Describe the event G that the oldest child is a girl and the youngest child is a boy.

Solution

a. The sample points of the experiment may be obtained with the aid of the tree diagram shown in Figure 4, where b denotes a boy and g denotes a girl.

First child	Second child	Third child	Sample points
		b	$b\,b\,b$
b	b	g	$b\,b\,g$
	g	b	$b\,g\,b$
		g	$b\,g\,g$
	b	b	$g\,b\,b$
g		g	$g\,b\,g$
	g	b	$g\,g\,b$
		g	$g\,g\,g$

FIGURE 4
Tree diagram for three-child families

We see from the tree diagram that the required sample space is given by

$$S = \{bbb, bbg, bgb, bgg, gbb, gbg, ggb, ggg\}$$

Using the tree diagram, we find that:

b. $E = \{bgg, gbg, ggb\}$
c. $F = \{gbb, gbg, ggb, ggg\}$
d. $G = \{gbb, ggb\}$

The next example shows that sample spaces may be infinite.

APPLIED EXAMPLE 7   Testing New Products   EverBrite is developing a high-amperage, high-capacity battery as a source for powering electric cars. The battery is tested by installing it in a prototype electric car and running the car with a fully charged battery on a test track at a constant speed of 55 mph until the car runs out of power. The distance covered by the car is then observed.

a. What is the sample space for this experiment?
b. Describe the event E that the driving range under test conditions is less than 150 miles.
c. Describe the event F that the driving range is between 200 and 250 miles, inclusive.

Solution

a. Since the distance d covered by the car in any run may be given by any nonnegative number, the sample space S is given by

$$S = \{d \mid d \geq 0\}$$

b. The event E is given by

$$E = \{d \,|\, d < 150\}$$

c. The event F is given by

$$F = \{d \,|\, 200 \leq d \leq 250\}$$

7.1 Self-Check Exercises

1. A sample of three apples taken from Cavallero's Fruit Stand is examined to determine whether the apples are good or rotten.
   a. What is an appropriate sample space for this experiment?
   b. Describe the event E that exactly one of the apples picked is rotten.
   c. Describe the event F that the first apple picked is rotten.

2. Refer to Self-Check Exercise 1.
   a. Find $E \cup F$.
   b. Find $E \cap F$.
   c. Find F^c.
   d. Are the events E and F mutually exclusive?

Solutions to Self-Check Exercises 7.1 can be found on page 388.

7.1 Concept Questions

1. Explain what is meant by an experiment. Give an example. For the example you have chosen, describe (a) a sample point, (b) the sample space, and (c) an event of the experiment.

2. What does it mean for two events to be mutually exclusive? Give an example of two mutually exclusive events E and F. How can you prove that they are mutually exclusive?

7.1 Exercises

In Exercises 1–6, let $S = \{a, b, c, d, e, f\}$ be a sample space of an experiment and let $E = \{a, b\}$, $F = \{a, d, f\}$, and $G = \{b, c, e\}$ be events of this experiment.

1. Find the events $E \cup F$ and $E \cap F$.

2. Find the events $F \cup G$ and $F \cap G$.

3. Find the events F^c and $E \cap G^c$.

4. Find the events E^c and $F^c \cap G$.

5. Are the events E and F mutually exclusive?

6. Are the events $E \cup F$ and $E \cap F^c$ mutually exclusive?

In Exercises 7–14, let $S = \{1, 2, 3, 4, 5, 6\}$, $E = \{2, 4, 6\}$, $F = \{1, 3, 5\}$, and $G = \{5, 6\}$.

7. Find the event $E \cup F \cup G$.

8. Find the event $E \cap F \cap G$.

9. Find the event $(E \cup F \cup G)^c$.

10. Find the event $(E \cap F \cap G)^c$.

11. Are the events E and F mutually exclusive?

12. Are the events F and G mutually exclusive?

13. Are the events E and F complementary?

14. Are the events F and G complementary?

In Exercises 15–20, let S be any sample space and let E, F, and G be any three events associated with the experiment. Describe the events using the symbols $\cup$, $\cap$, and ^{c}.

15. The event that E and/or F occurs

16. The event that both E and F occur

17. The event that G does not occur

18. The event that E but not F occurs

19. The event that none of the events E, F, and G occurs

20. The event that E occurs but neither of the events F or G occurs

21. Consider the sample space S of Example 4, page 382.
  a. Determine the event that the number that falls uppermost on the first die is greater than the number that falls uppermost on the second die.
  b. Determine the event that the number that falls uppermost on the second die is double the number that falls on the first die.

22. Consider the sample space S of Example 4, page 382.
  a. Determine the event that the sum of the numbers falling uppermost is less than or equal to 7.
  b. Determine the event that the number falling uppermost on one die is a 4 and the number falling uppermost on the other die is greater than 4.

23. Let $S = \{a, b, c\}$ be a sample space of an experiment with outcomes a, b, and c. List all the events of this experiment.

24. Let $S = \{1, 2, 3\}$ be a sample space associated with an experiment.
  a. List all events of this experiment.
  b. How many subsets of S contain the number 3?
  c. How many subsets of S contain either the number 2 or the number 3?

25. An experiment consists of selecting a card from a standard deck of playing cards and noting whether it is black (B) or red (R).
  a. Describe an appropriate sample space for this experiment.
  b. What are the events of this experiment?

26. An experiment consists of selecting a letter at random from the letters in the word *MASSACHUSETTS* and observing the outcomes.
  a. What is an appropriate sample space for this experiment?
  b. Describe the event "the letter selected is a vowel."

27. An experiment consists of tossing a coin, casting a die, and observing the outcomes.
  a. Describe an appropriate sample space for this experiment.
  b. Describe the event "a head is tossed and an even number is cast."

28. An experiment consists of spinning the hand of the numbered disc shown in the following figure and then observing the region in which the pointer stops. (If the needle stops on a line, the result is discounted and the needle is spun again.)

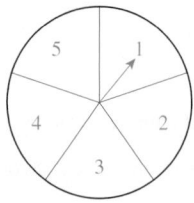

  a. What is an appropriate sample space for this experiment?
  b. Describe the event "the spinner points to the number 2."
  c. Describe the event "the spinner points to an odd number."

29. A die is cast and the number that falls uppermost is observed. Let E denote the event that the number shown is a 2 and let F denote the event that the number shown is an even number.
  a. Are the events E and F mutually exclusive?
  b. Are the events E and F complementary?

30. A die is cast and the number that falls uppermost is observed. Let E denote the event that the number shown is even, and let F denote the event that the number is an odd number.
  a. Are the events E and F mutually exclusive?
  b. Are the events E and F complementary?

31. QUALITY CONTROL   A sample of three transistors taken from a local electronics store was examined to determine whether the transistors were defective (d) or nondefective (n). What is an appropriate sample space for this experiment?

32. BLOOD TYPING   Human blood is classified by the presence or absence of three main antigens (A, B, and Rh). When a blood specimen is typed, the presence of the A and/or B antigen is indicated by listing the letter A and/or the letter B. If neither the A nor B antigen is present, the letter O is used. The presence or absence of the Rh antigen is indicated by the symbols $+$ or $-$, respectively. Thus, if a blood specimen is classified as AB^+, it contains the A and the B antigens as well as the Rh antigen. Similarly, O^- blood contains none of the three antigens. Using this information, determine the sample space corresponding to the different blood groups.

33. GAME SHOWS   In a television game show, the winner is asked to select three prizes from five different prizes, A, B, C, D, and E.
  a. Describe a sample space of possible outcomes (order is not important).
  b. How many points are there in the sample space corresponding to a selection that includes A?
  c. How many points are there in the sample space corresponding to a selection that includes A and B?
  d. How many points are there in the sample space corresponding to a selection that includes either A or B?

34. AUTOMATIC TELLERS   The manager of a local bank observes how long it takes a customer to complete his transactions at the automatic bank teller.
  a. Describe an appropriate sample space for this experiment.
  b. Describe the event that it takes a customer between 2 and 3 minutes to complete his transactions at the automatic bank teller.

35. COMMON STOCKS   Robin purchased shares of a machine tool company and shares of an airline company. Let E be the event that the shares of the machine tool company increase in value over the next 6 mo, and let F be the event that the shares of the airline company increase in value over the next 6 mo. Using the symbols $\cup$, $\cap$, and ^{c}, describe the following events.

a. The shares in the machine tool company do not increase in value.

b. The shares in both the machine tool company and the airline company do not increase in value.

c. The shares of at least one of the two companies increase in value.

d. The shares of only one of the two companies increase in value.

36. **CUSTOMER SERVICE SURVEYS**   The customer service department of Universal Instruments, manufacturer of the Galaxy home computer, conducted a survey among customers who had returned their purchase registration cards. Purchasers of its deluxe model home computer were asked to report the length of time (t) in days before service was required.

a. Describe a sample space corresponding to this survey.

b. Describe the event E that a home computer required service before a period of 90 days had elapsed.

c. Describe the event F that a home computer did not require service before a period of 1 year had elapsed.

37. **ASSEMBLY-TIME STUDIES**   A time study was conducted by the production manager of Vista Vision to determine the length of time in minutes required by an assembly worker to complete a certain task during the assembly of its Pulsar color television sets.

a. Describe a sample space corresponding to this time study.

b. Describe the event E that an assembly worker took 2 minutes or less to complete the task.

c. Describe the event F that an assembly worker took more than 2 minutes to complete the task.

38. **POLITICAL POLLS**   An opinion poll is conducted among a state's electorate to determine the relationship between their income levels and their stands on a proposition aimed at reducing state income taxes. Voters are classified as belonging to either the low-, middle-, or upper-income group. They are asked whether they favor, oppose, or are undecided about the proposition. Let the letters L, M, and U represent the low-, middle-, and upper-income groups, respectively, and let the letters f, o, and u represent the responses—favor, oppose, and undecided, respectively.

a. Describe a sample space corresponding to this poll.

b. Describe the event E_1 that a respondent favors the proposition.

c. Describe the event E_2 that a respondent opposes the proposition and does not belong to the low-income group.

d. Describe the event E_3 that a respondent does not favor the proposition and does not belong to the upper-income group.

39. **QUALITY CONTROL**   As part of a quality-control procedure, an inspector at Bristol Farms randomly selects ten eggs from each consignment of eggs he receives and records the number of broken eggs.

a. What is an appropriate sample space for this experiment?

b. Describe the event E that at most three eggs are broken.

c. Describe the event F that at least five eggs are broken.

40. **POLITICAL POLLS**   In the opinion poll of Exercise 38, the voters were also asked to indicate their political affiliations—Democrat, Republican, or Independent. As before, let the letters L, M, and U represent the low-, middle-, and upper-income groups, respectively, and let the letters D, R, and I represent Democrat, Republican, and Independent, respectively.

a. Describe a sample space corresponding to this poll.

b. Describe the event E_1 that a respondent is a Democrat.

c. Describe the event E_2 that a respondent belongs to the upper-income group and is a Republican.

d. Describe the event E_3 that a respondent belongs to the middle-income group and is not a Democrat.

41. **SHUTTLE BUS USAGE**   A certain airport hotel operates a shuttle bus service between the hotel and the airport. The maximum capacity of a bus is 20 passengers. On alternate trips of the shuttle bus over a period of 1 week, the hotel manager kept a record of the number of passengers arriving at the hotel in each bus.

a. What is an appropriate sample space for this experiment?

b. Describe the event E that a shuttle bus carried fewer than ten passengers.

c. Describe the event F that a shuttle bus arrived with a full capacity.

42. **SPORTS**   Eight players, A, B, C, D, E, F, G, and H, are competing in a series of elimination matches of a tennis tournament in which the winner of each preliminary match will advance to the semifinals and the winner of the semifinals will advance to the finals. An outline of the scheduled matches follows. Describe a sample space listing the possible participants in the finals.

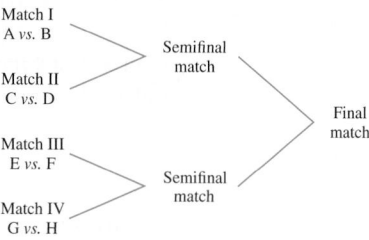

43. An experiment consists of selecting a card at random from a well-shuffled 52-card deck. Let E denote the event that an ace is drawn and let F denote the event that a spade is drawn. Show that $n(E \cup F) = n(E) + n(F) - n(E \cap F)$.

44. Let S be a sample space for an experiment. Show that if E is any event of an experiment, then E and E^c are mutually exclusive.

45. Let S be a sample space for an experiment and let E and F be events of this experiment. Show that the events $E \cup F$ and $E^c \cap F^c$ are mutually exclusive.
Hint: Use De Morgan's law.

46. Let S be a sample space of an experiment with n outcomes. Determine the number of events of this experiment.

In Exercises 47 and 48, determine whether the statement is true or false. If it is true, explain why it is true. If it is false, give an example to show why it is false.

47. If E and F are mutually exclusive and E and G are mutually exclusive, then F and G are mutually exclusive.

48. The numbers 1, 2, and 3 are written separately on three pieces of paper. These slips of paper are then placed in a bowl. If you draw two slips from the bowl, one at a time and without replacement, then the sample space for this experiment consists of six elements.

7.1 Solutions to Self-Check Exercises

1. a. Let g denote a good apple and r a rotten apple. Thus, the required sample points may be obtained with the aid of a tree diagram (compare with Example 3). The required sample space is given by

$$S = \{ggg, ggr, grg, grr, rgg, rgr, rrg, rrr\}$$

b. By scanning the sample space S obtained in part (a), we identify the outcomes in which exactly one apple is rotten. We find

$$E = \{ggr, grg, rgg\}$$

c. Proceeding as in part (b), we find

$$F = \{rgg, rgr, rrg, rrr\}$$

2. Using the results of Self-Check Exercise 1, we find:
a. $E \cup F = \{ggr, grg, rgg, rgr, rrg, rrr\}$
b. $E \cap F = \{rgg\}$
c. F^c is the set of outcomes in S but not in F. Thus,

$$F^c = \{ggg, ggr, grg, grr\}$$

d. Since $E \cap F \neq \emptyset$, we conclude that E and F are not mutually exclusive.

7.2 Definition of Probability

▬ Finding the Probability of an Event

Let's return to the coin-tossing experiment. The sample space of this experiment is given by $S = \{H, T\}$, where the sample points H and T correspond to the two possible outcomes, "heads" or "tails." If the coin is *unbiased*, then there is *one chance out of two* of obtaining a head (or a tail) and we say that the *probability* of tossing a head (tail) is $\frac{1}{2}$, abbreviated

$$P(H) = \frac{1}{2} \quad \text{and} \quad P(T) = \frac{1}{2}$$

An alternative method of obtaining the values of $P(H)$ and $P(T)$ is based on continued experimentation and does not depend on the assumption that the two outcomes are equally likely. Table 2 summarizes the results of such an exercise.

Observe that the relative frequencies (column 3) differ considerably when the number of trials is small, but as the number of trials becomes very large, the relative frequency approaches the number .5. This result suggests that we assign to $P(H)$ the value $\frac{1}{2}$, as before.

More generally, consider an experiment that may be repeated over and over again under independent and similar conditions. Suppose that in n trials an event E

TABLE 2

As the Number of Trials Increases, the Relative Frequency Approaches .5

Number of Tosses, n	Number of Heads, m	Relative Frequency of Heads, m/n
10	4	.4000
100	58	.5800
1,000	492	.4920
10,000	5,034	.5034
20,000	10,024	.5012
40,000	20,032	.5008

occurs m times. We call the ratio m/n the **relative frequency** of the event E after n repetitions. If this relative frequency approaches some value $P(E)$ as n becomes larger and larger, then $P(E)$ is called the **empirical probability** of E. Thus, the probability $P(E)$ of an event occurring is a measure of the proportion of the time that the event E will occur in the long run. Observe that this method of computing the probability of a head occurring is effective even in the case when a biased coin is used in the experiment. The relative frequency distribution is often referred to as an observed or **empirical probability distribution**.

The **probability of an event** is a number that lies between 0 and 1. In general, the larger the probability of an event, the more likely the event will occur. Thus, an event with a probability of .8 is more likely to occur than an event with a probability of .6. An event with a probability of $\frac{1}{2}$, or .5, has a "fifty-fifty," or equal, chance of occurring.

Now suppose we are given an experiment and wish to determine the probabilities associated with certain events of the experiment. This problem could be solved by computing $P(E)$ directly for each event E of interest. In practice, however, the number of events that we may be interested in is usually quite large, so this approach is not satisfactory.

The following approach is particularly suitable when the sample space of an experiment is finite.* Let S be a finite sample space with n outcomes; that is,

$$S = \{s_1, s_2, s_3, \ldots, s_n\}$$

Then the events

$$\{s_1\}, \{s_2\}, \{s_3\}, \ldots, \{s_n\}$$

that consist of exactly one point are called **elementary**, or **simple**, **events** of the experiment. They are elementary in the sense that any (nonempty) event of the experiment may be obtained by taking a finite union of suitable elementary events. The simple events of an experiment are also **mutually exclusive**; that is, given any two simple events of the experiment, only one can occur.

By assigning probabilities to each of the simple events, we obtain the results shown in Table 3. This table is called a **probability distribution** for the experiment. The function P, which assigns a probability to each of the simple events, is called a **probability function**.

TABLE 3

A Probability Distribution

Simple Event	Probability*
$\{s_1\}$	$P(s_1)$
$\{s_2\}$	$P(s_2)$
$\{s_3\}$	$P(s_3)$
⋮	⋮
$\{s_n\}$	$P(s_n)$

*For simplicity, we use the notation $P(s_i)$ instead of the technically more correct $P(\{s_i\})$.

*For the remainder of the chapter, we assume that all sample spaces are finite.

The numbers $P(s_1), P(s_2), \ldots, P(s_n)$ have the following properties:

1. $0 \leq P(s_i) \leq 1$ $i = 1, 2, \ldots, n$
2. $P(s_1) + P(s_2) + \cdots + P(s_n) = 1$
3. $P(\{s_i\} \cup \{s_j\}) = P(s_i) + P(s_j)$ $(i \neq j)$ $i = 1, 2, \ldots, n; j = 1, 2, \ldots, n$

The first property simply states that the probability of a simple event must be between 0 and 1 inclusive. The second property states that the sum of the probabilities of all simple events of the sample space is 1. This follows from the fact that the event S is certain to occur. The third property states that the probability of the union of two mutually exclusive events is given by the sum of their probabilities.

As we saw earlier, there is no unique method for assigning probabilities to the simple events of an experiment. In practice, the methods used to determine these probabilities may range from theoretical considerations of the problem on the one extreme to the reliance on "educated guesses" on the other.

Sample spaces in which the outcomes are equally likely are called **uniform sample spaces**. Assigning probabilities to the simple events in these spaces is relatively easy.

Probability of an Event in a Uniform Sample Space
If

$$S = \{s_1, s_2, \ldots, s_n\}$$

is the sample space for an experiment in which the outcomes are equally likely, then we assign the probabilities

$$P(s_1) = P(s_2) = \cdots = P(s_n) = \frac{1}{n}$$

to each of the simple events $s_1, s_2, \ldots, s_n$.

EXAMPLE 1   A fair die is cast, and the number that falls uppermost is observed. Determine the probability distribution for the experiment.

Solution   The sample space for the experiment is $S = \{1, 2, 3, 4, 5, 6\}$, and the simple events are accordingly given by the sets $\{1\}, \{2\}, \{3\}, \{4\}, \{5\},$ and $\{6\}$. Since the die is assumed to be fair, the six outcomes are equally likely. We therefore assign a probability of $\frac{1}{6}$ to each of the simple events and obtain the probability distribution shown in Table 4.   ■

TABLE 4

A Probability Distribution

Simple Event	Probability
$\{1\}$	$\frac{1}{6}$
$\{2\}$	$\frac{1}{6}$
$\{3\}$	$\frac{1}{6}$
$\{4\}$	$\frac{1}{6}$
$\{5\}$	$\frac{1}{6}$
$\{6\}$	$\frac{1}{6}$

EXPLORE & DISCUSS

You suspect that a die is biased.

1. Describe a method you might use to prove your assertion.

2. How would you assign the probability to each outcome 1 through 6 of an experiment that consists of casting the die and observing the number that lands uppermost?

The next example shows how the *relative frequency* interpretation of probability lends itself to the computation of probabilities.

TABLE 5

Data Obtained during 200 Test Runs of an Electric Car

Distance Covered in Miles, x	Frequency of Occurrence
$0 < x \leq 50$	4
$50 < x \leq 100$	10
$100 < x \leq 150$	30
$150 < x \leq 200$	100
$200 < x \leq 250$	40
$250 < x$	16

TABLE 6

A Probability Distribution

Simple Event	Probability
$\{s_1\}$	.02
$\{s_2\}$	.05
$\{s_3\}$	.15
$\{s_4\}$	.50
$\{s_5\}$	.20
$\{s_6\}$	.08

APPLIED EXAMPLE 2   Testing New Products   Refer to Example 7, Section 7.1. The data shown in Table 5 were obtained in tests involving 200 test runs. Each run was made with a fully charged battery.

a. Describe an appropriate sample space for this experiment.
b. Find the empirical probability distribution for this experiment.

Solution

a. Let s_1 denote the outcome that the distance covered by the car does not exceed 50 miles; let s_2 denote the outcome that the distance covered by the car is greater than 50 miles but does not exceed 100 miles, and so on. Finally, let s_6 denote the outcome that the distance covered by the car is greater than 250 miles. Then, the required sample space is given by

$$S = \{s_1, s_2, s_3, s_4, s_5, s_6\}$$

b. To compute the empirical probability distribution for the experiment, we turn to the relative frequency interpretation of probability. Accepting the inaccuracies inherent in a relatively small number of trials (200 runs), we take the probability of s_1 occurring as

$$P(s_1) = \frac{\text{Number of trials in which } s_1 \text{ occurs}}{\text{Total number of trials}}$$

$$= \frac{4}{200} = .02$$

In a similar manner, we assign probabilities to the other simple events, obtaining the probability distribution shown in Table 6.   ■

We are now in a position to give a procedure for computing the probability $P(E)$ of an arbitrary event E of an experiment.

Finding the Probability of an Event E

1. Determine a sample space S associated with the experiment.
2. Assign probabilities to the simple events of S.
3. If $E = \{s_1, s_2, s_3, \ldots, s_n\}$, where $\{s_1\}, \{s_2\}, \{s_3\}, \ldots, \{s_n\}$ are simple events, then

$$P(E) = P(s_1) + P(s_2) + P(s_3) + \cdots + P(s_n)$$

If E is the empty set, $\varnothing$, then $P(E) = 0$.

The principle stated in step 3 is called the **addition principle** and is a consequence of Property 3 of the probability function (page 390). This principle allows us to find the probabilities of all other events once the probabilities of the simple events are known.

The addition rule applies *only* to the addition of probabilities of simple events.

APPLIED EXAMPLE 3   Casting Dice   A pair of fair dice is cast.
a. Calculate the probability that the two dice show the same number.
b. Calculate the probability that the sum of the numbers of the two dice is 6.

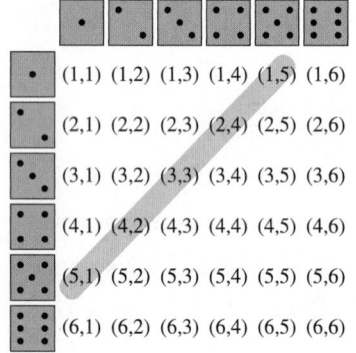

FIGURE 5
The event that the two dice show the same number

FIGURE 6
The event that the sum of the numbers on the two dice is 6

Solution　From the results of Example 4, page 382, we see that the sample space S of the experiment consists of 36 outcomes:

$$S = \{(1, 1), (1, 2), \ldots, (6, 5), (6, 6)\}$$

Since both dice are fair, each of the 36 outcomes is equally likely. Accordingly, we assign the probability of $\frac{1}{36}$ to each simple event. We are now in a position to answer the questions posed.

a. The event that the two dice show the same number is given by

$$E = \{(1, 1), (2, 2), (3, 3), (4, 4), (5, 5), (6, 6)\}$$

(Figure 5). Therefore, by the addition principle, the probability that the two dice show the same number is given by

$$
\begin{aligned}
P(E) &= P[(1, 1)] + P[(2, 2)] + \cdots + P[(6, 6)] \\
&= \frac{1}{36} + \frac{1}{36} + \cdots + \frac{1}{36} \quad \text{Six terms} \\
&= \frac{1}{6}
\end{aligned}
$$

b. The event that the sum of the numbers of the two dice is 6 is given by

$$E_6 = \{(1, 5), (2, 4), (3, 3), (4, 2), (5, 1)\}$$

(Figure 6). Therefore, the probability that the sum of the numbers on the two dice is 6 is given by

$$
\begin{aligned}
P(E_6) &= P[(1, 5)] + P[(2, 4)] + P[(3, 3)] + P[(4, 2)] + P[(5, 1)] \\
&= \frac{1}{36} + \frac{1}{36} + \cdots + \frac{1}{36} \quad \text{Five terms} \\
&= \frac{5}{36}
\end{aligned}
$$

APPLIED EXAMPLE 4　Testing New Products　Consider the experiment by EverBrite in Example 2. What is the probability that the prototype car will travel more than 150 miles on a fully charged battery?

Solution　Using the results of Example 2, we see that the event that the car will travel more than 150 miles on a fully charged battery is given by $E = \{s_4, s_5, s_6\}$. Therefore, the probability that the car will travel more than 150 miles on one charge is given by

$$P(E) = P(s_4) + P(s_5) + P(s_6)$$

or, using the probability distribution for the experiment obtained in Example 2,

$$P(E) = .50 + .20 + .08 = .78$$

7.2　Self-Check Exercises

1. A biased die was cast repeatedly, and the results of the experiment are summarized in the following table:

Outcome	1	2	3	4	5	6
Frequency of Occurrence	142	173	158	175	162	190

Using the relative frequency interpretation of probability, find the empirical probability distribution for this experiment.

2. In an experiment conducted to study the effectiveness of an eye-level third brake light in the prevention of rear-end collisions, 250 of the 500 highway patrol cars of a certain state were equipped with such lights. At the end of the 1-yr trial period, the records revealed that for those equipped with a third brake light there were 14 incidents of rear-end collision. There were 22 such incidents involving the cars not equipped with the accessory. Based on these data, what is the probability that

a highway patrol car equipped with a third brake light will be rear-ended within a 1-yr period? What is the probability that a car not so equipped will be rear-ended within a 1-yr period?

Solutions to Self-Check Exercises 7.2 can be found on page 397.

7.2 Concept Questions

1. Define (a) a probability distribution and (b) a probability function. Give examples of each.

2. If $S = \{s_1, s_2, \ldots, s_n\}$ is the sample space for an experiment in which the outcomes are equally likely, what is the probability of each of the simple events $s_1, s_2, \ldots, s_n$? What do we call this type of sample space?

3. Suppose $E = \{s_1, s_2, s_3, \ldots, s_n\}$, where E is an event of an experiment and $\{s_1\}, \{s_2\}, \{s_3\}, \ldots, \{s_n\}$ are simple events. If E is nonempty, what is $P(E)$? If E is empty, what is $P(E)$?

7.2 Exercises

In Exercises 1–8, list the simple events associated with each experiment.

1. A nickel and a dime are tossed, and the result of heads or tails is recorded for each coin.

2. A card is selected at random from a standard 52-card deck, and its suit—hearts (h), diamonds (d), spades (s), or clubs (c)—is recorded.

3. OPINION POLLS An opinion poll is conducted among a group of registered voters. Their political affiliation—Democrat (D), Republican (R), or Independent (I)—and their sex—male (m) or female (f)—are recorded.

4. QUALITY CONTROL As part of a quality-control procedure, eight circuit boards are checked, and the number of defectives is recorded.

5. MOVIE ATTENDANCE In a survey conducted to determine whether movie attendance is increasing (i), decreasing (d), or holding steady (s) among various sectors of the population, participants are classified as follows:

   Group 1: Those aged 10–19

   Group 2: Those aged 20–29

   Group 3: Those aged 30–39

   Group 4: Those aged 40–49

   Group 5: Those aged 50 and over

   The response of each participant and his or her age group are recorded.

6. DURABLE GOODS ORDERS Data concerning durable goods orders are obtained each month by an economist. A record is kept for a 1-yr period of any increase (i), decrease (d), or unchanged movement (u) in the number of durable goods orders for each month as compared with the number of such orders in the same month of the previous year.

7. BLOOD TYPES Blood tests are given as a part of the admission procedure at the Monterey Garden Community Hospital. The blood type of each patient (A, B, AB, or O) and the presence or absence of the Rh factor in each patient's blood (Rh^+ or Rh^-) are recorded.

8. METEOROLOGY A meteorologist preparing a weather map classifies the expected average temperature in each of five neighboring states for the upcoming week as follows:
   a. More than 10° below average
   b. Normal to 10° below average
   c. Higher than normal to 10° above average
   d. More than 10° above average
   Using each state's abbreviation and the categories—(a), (b), (c), and (d)—the meteorologist records these data.

9. GRADE DISTRIBUTIONS The grade distribution for a certain class is shown in the following table. Find the probability distribution associated with these data.

Grade	A	B	C	D	F
Frequency of Occurrence	4	10	18	6	2

10. **Blood Types**   The percentage of the general population that has each blood type is shown in the following table. Determine the probability distribution associated with these data.

Blood Type	A	B	AB	O
Population, %	41	12	3	44

11. **Traffic Surveys**   The number of cars entering a tunnel leading to an airport in a major city over a period of 200 peak hours was observed, and the following data were obtained:

Cars, x	Frequency of Occurrence
$0 < x \le 200$	15
$200 < x \le 400$	20
$400 < x \le 600$	35
$600 < x \le 800$	70
$800 < x \le 1000$	45
$x > 1000$	15

a. Describe an appropriate sample space for this experiment.
b. Find the empirical probability distribution for this experiment.

12. **Arrival Times**   The arrival times of the 8 a.m. Boston-based commuter train as observed in the suburban town of Sharon over 120 weekdays is summarized below:

Arrival Time, x	Frequency of Occurrence
7:56 a.m. $< x \le$ 7:58 a.m.	4
7:58 a.m. $< x \le$ 8:00 a.m.	18
8:00 a.m. $< x \le$ 8:02 a.m.	50
8:02 a.m. $< x \le$ 8:04 a.m.	32
8:04 a.m. $< x \le$ 8:06 a.m.	9
8:06 a.m. $< x \le$ 8:08 a.m.	4
8:08 a.m. $< x \le$ 8:10 a.m.	3

a. Describe an appropriate sample space for this experiment.
b. Find the empirical probability distribution for this experiment.

13. **Same-Sex Marriage**   In a *Los Angeles Times* poll of 1936 California residents conducted in February 2004, the following question was asked: Do you favor or oppose an amendment to the U.S. Constitution barring same-sex marriage? The following results were obtained.

Opinion	Favor	Oppose	Don't know
Respondents	910	891	135

Determine the empirical probability distribution associated with these data.
Source: The Field Poll, *Los Angeles Times*

14. **E-mail Services**   The number of subscribers to five leading e-mail services is shown in the accompanying table.

Company	A	B	C
Subscribers	300,000	200,000	120,000

Company	D	E
Subscribers	80,000	60,000

Find the empirical probability distribution associated with these data.

15. **Political Views**   In a poll conducted among 2000 college freshmen to ascertain the political views of college students, the accompanying data were obtained. Determine the empirical probability distribution associated with these data.

Political Views	A	B	C	D	E
Respondents	52	398	1140	386	24

A: Far left
B: Liberal
C: Middle of the road
D: Conservative
E: Far right

16. **Product Surveys**   The accompanying data were obtained from a survey of 1500 Americans who were asked: How safe are American-made consumer products? Determine the empirical probability distribution associated with these data.

Rating	A	B	C	D	E
Respondents	285	915	225	30	45

A: Very safe
B: Somewhat safe
C: Not too safe
D: Not safe at all
E: Don't know

17. **Assembly-Time Studies**   The results of a time study conducted by the production manager of Ace Novelty are shown in the accompanying table, where the number of action figures produced each quarter hour during an 8-hour workday has been tabulated. Find the empirical probability distribution associated with this experiment.

Figures Produced (in dozens)	Frequency of Occurrence
30	4
31	0
32	6
33	8
34	6
35	4
36	4

18. SERVICE-UTILIZATION STUDIES Metro Telephone Company compiled the accompanying information during a service utilization study pertaining to the number of customers using their Dial-the-Time service from 7 a.m. to 9 a.m. on a certain weekday morning. Using these data, find the empirical probability distribution associated with the experiment.

Calls Received/Minute	Frequency of Occurrence
10	6
11	15
12	12
13	3
14	12
15	36
16	24
17	0
18	6
19	6

19. CORRECTIVE LENS USE According to Mediamark Research, 84 million out of 179 million adults in the United States correct their vision by using prescription eyeglasses, bifocals, or contact lenses. (Some respondents use more than one type.) What is the probability that an adult selected at random from the adult population uses corrective lenses?

Source: Mediamark Research

20. TRAFFIC DEATHS A study of deaths in car crashes from 1986 to 2002 revealed the following data on deaths in crashes by day of the week.

Day of the Week	Sunday	Monday	Tuesday	Wednesday
Average Number of Deaths	132	98	95	98

Day of the Week	Thursday	Friday	Saturday
Average Number of Deaths	105	133	158

Find the empirical probability distribution associated with these data.

Source: Insurance Institute for Highway Safety

21. FAMILY INCOME According to the 2000 U.S. Census Bureau, the income distribution of households and families was as follows:

Income ($)	0–24,999	25,000–49,999	50,000–74,999
Households and Families:	30,261,220	30,965,514	20,540,604

Income ($)	75,000–99,999	100,000–124,999	125,000–149,999
Households and Families:	10,779,245	5,491,526	2,656,300

Income ($)	150,000–199,999	200,000 or more
Households and Families:	2,322,038	2,502,675

Find the empirical probability distribution associated with these data.

Source: U.S. Census Bureau

22. CORRECTIONAL SUPERVISION A study conducted by the Corrections Department of a certain state revealed that 163,605 people out of a total adult population of 1,778,314 were under correctional supervision (on probation, parole, or in jail). What is the probability that a person selected at random from the adult population in that state is under correctional supervision?

23. LIGHTNING DEATHS According to data obtained from the National Weather Service, 376 of the 439 people killed by lightning in the United States between 1985 and 1992 were men. (Job and recreational habits of men make them more vulnerable to lightning.) Assuming that this trend holds in the future, what is the probability that a person killed by lightning
a. Is a male?   b. Is a female?

24. QUALITY CONTROL One light bulb is selected at random from a lot of 120 light bulbs, of which 5% are defective. What is the probability that the light bulb selected is defective?

25. EFFORTS TO STOP SHOPLIFTING According to a survey of 176 retailers, 46% of them use electronic tags as protection against shoplifting and employee theft. If one of these retailers is selected at random, what is the probability that he or she uses electronic tags as antitheft devices?

26. If a ball is selected at random from an urn containing three red balls, two white balls, and five blue balls, what is the probability that it will be a white ball?

27. If a card is drawn at random from a standard 52-card deck, what is the probability that the card drawn is
a. A diamond?   b. A black card?
c. An ace?

28. A pair of fair dice is cast. What is the probability that
a. The sum of the numbers shown uppermost is less than 5?
b. At least one 6 is cast?

29. **TRAFFIC LIGHTS**   What is the probability of arriving at a traffic light when it is red if the red signal is lit for 30 sec, the yellow signal for 5 sec, and the green signal for 45 sec?

30. **ROULETTE**   What is the probability that a roulette ball will come to rest on an even number other than 0 or 00? (Assume that there are 38 equally likely outcomes consisting of the numbers 1–36, 0, and 00.)

31. Refer to Exercise 9. What is the probability that a student selected at random from this class received a passing grade (D or better)?

32. Refer to Exercise 11. What is the probability that more than 600 cars will enter the airport tunnel during a peak hour?

33. **DISPOSITION OF CRIMINAL CASES**   Of the 98 first-degree murder cases from 2002 through the first half of 2004 in the Suffolk superior court, 9 cases were thrown out of the system, 62 cases were plea-bargained, and 27 cases went to trial. What is the probability that a case selected at random
    a. Was settled through plea bargaining?
    b. Went to trial?
    Source: Boston Globe

34. **SWEEPSTAKES**   In a sweepstakes sponsored by Gemini Paper Products, 100,000 entries have been received. If 1 grand prize, 5 first prizes, 25 second prizes, and 500 third prizes are to be awarded, what is the probability that a person who has submitted one entry will win
    a. The grand prize?
    b. A prize?

35. A pair of fair dice is cast, and the sum of the two numbers falling uppermost is observed. The probability of obtaining a sum of 2 is the same as that of obtaining a 7 since there is only one way of getting a 2—namely, by each die showing a 1; and there is only one way of obtaining a 7—namely, by one die showing a 3 and the other die showing a 4. What is wrong with this argument?

In Exercises 36–39, determine whether the given experiment has a sample space with equally likely outcomes.

36. A loaded die is cast, and the number appearing uppermost on the die is recorded.

37. Two fair dice are cast, and the sum of the numbers appearing uppermost is recorded.

38. A ball is selected at random from an urn containing six black balls and six red balls, and the color of the ball is recorded.

39. A weighted coin is thrown, and the outcome of heads or tails is recorded.

40. Let $S = \{s_1, s_2, s_3, s_4, s_5, s_6\}$ be the sample space associated with an experiment having the following probability distribution:

Outcome	s_1	s_2	s_3	s_4	s_5	s_6
Probability	$\frac{1}{12}$	$\frac{1}{4}$	$\frac{1}{12}$	$\frac{1}{6}$	$\frac{1}{3}$	$\frac{1}{12}$

Find the probability of the event:
a. $A = \{s_1, s_3\}$
b. $B = \{s_2, s_4, s_5, s_6\}$
c. $C = S$

41. Consider the composition of a three-child family in which the children were born at different times. Assume that a girl is as likely as a boy at each birth. What is the probability that
    a. There are two girls and a boy in the family?
    b. The oldest child is a girl?
    c. The oldest child is a girl and the youngest child is a boy?

42. Let $S = \{s_1, s_2, s_3, s_4, s_5\}$ be the sample space associated with an experiment having the following probability distribution:

Outcome	s_1	s_2	s_3	s_4	s_5
Probability	$\frac{1}{14}$	$\frac{3}{14}$	$\frac{6}{14}$	$\frac{2}{14}$	$\frac{2}{14}$

Find the probability of the event:
a. $A = \{s_1, s_2, s_4\}$ **b.** $B = \{s_1, s_5\}$
c. $C = S$

43. **AIRLINE SAFETY**   In an attempt to study the leading causes of airline crashes, the following data were compiled from records of airline crashes from 1959 to 1994 (excluding sabotage and military action).

Primary Factor	Accidents
Flight crew	327
Airplane	49
Maintenance	14
Weather	22
Airport/air traffic control	19
Miscellaneous/other	15

Assume that you have just learned of an airline crash and that the data give a generally good indication of the causes of airline crashes. Give an estimate of the probability that the primary cause of the crash was due to pilot error or bad weather.
Source: National Transportation Safety Board

44. **POLITICAL POLLS**   An opinion poll was conducted among a group of registered voters in a certain state concerning a proposition aimed at limiting state and local taxes. Results of the poll indicated that 35% of the voters favored the proposition, 32% were against it, and the remaining group were undecided. If the results of the poll are assumed to be representative of the opinions of the state's electorate, what is the probability that a registered voter selected at random from the electorate
    a. Favors the proposition?
    b. Is undecided about the proposition?

In Exercises 45 and 46, determine whether the statement is true or false. If it is true, explain why it is true. If it is false, give an example to show why it is false.

45. If $S = \{s_1, s_2, \ldots, s_n\}$ is a uniform sample space with n outcomes, then $0 \leq P(s_1) + P(s_2) + \cdots + P(s_n) \leq 1$.

46. Let $S = \{s_1, s_2, \ldots, s_n\}$ be a uniform sample space for an experiment. If $n \geq 5$ and $E = \{s_1, s_2, s_5\}$, then $P(E) = 3/n$.

7.2 Solutions to Self-Check Exercises

1.

$$P(1) = \frac{\text{Number of trials in which a 1 appears uppermost}}{\text{Total number of trials}}$$

$$= \frac{142}{1000}$$

$$= .142$$

Similarly, we compute $P(2), \ldots, P(6)$, obtaining the following probability distribution:

Outcome	1	2	3	4	5	6
Probability	.142	.173	.158	.175	.162	.190

2. The probability that a highway patrol car equipped with a third brake light will be rear-ended within a 1-yr period is given by

$$\frac{\substack{\text{Number of rear-end collisions involving} \\ \text{cars equipped with a third brake light}}}{\text{Total number of such cars}} = \frac{14}{250} = .056$$

The probability that a highway patrol car not equipped with a third brake light will be rear-ended within a 1-yr period is given by

$$\frac{\substack{\text{Number of rear-end collisions involving cars} \\ \text{not equipped with a third brake light}}}{\text{Total number of such cars}} = \frac{22}{250} = .088$$

7.3 Rules of Probability

■ Properties of the Probability Function and Their Applications

In this section, we examine some of the properties of the probability function and look at the role they play in solving certain problems. We begin by looking at the generalization of the three properties of the probability function, which were stated for simple events in the last section. Let S be a sample space of an experiment and suppose E and F are events of the experiment. We have:

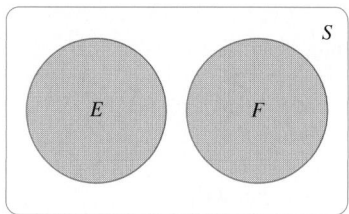

FIGURE 7
If E and F are mutually exclusive events, then $P(E \cup F) = P(E) + P(F)$.

Property 1. $P(E) \geq 0$ for any E.
Property 2. $P(S) = 1$.
Property 3. If E and F are mutually exclusive (that is, only one of them can occur or, equivalently, $E \cap F = \varnothing$), then

$$P(E \cup F) = P(E) + P(F)$$

(Figure 7).

Property 3 may be easily extended to the case involving any finite number of mutually exclusive events. Thus, if $E_1, E_2, \ldots, E_n$ are mutually exclusive events, then

$$P(E_1 \cup E_2 \cup \cdots \cup E_n) = P(E_1) + P(E_2) + \cdots + P(E_n)$$

TABLE 7

Probability Distribution

Score, x	Probability
$x > 700$	.01
$600 < x \le 700$	.07
$500 < x \le 600$	.19
$400 < x \le 500$	.23
$300 < x \le 400$	.31
$x \le 300$	.19

APPLIED EXAMPLE 1   SAT Verbal Scores The superintendent of a metropolitan school district has estimated the probabilities associated with the SAT verbal scores of students from that district. The results are shown in Table 7. If a student is selected at random, what is the probability that his or her SAT verbal score will be

a. More than 400?
b. Less than or equal to 500?
c. Greater than 400 but less than or equal to 600?

Solution   Let A, B, C, D, E, and F denote, respectively, the event that the score is greater than 700, greater than 600 but less than or equal to 700, greater than 500 but less than or equal to 600, and so forth. Then these events are mutually exclusive. Therefore,

a. The probability that the student's score will be more than 400 is given by

$$P(D \cup C \cup B \cup A) = P(D) + P(C) + P(B) + P(A)$$
$$= .23 + .19 + .07 + .01$$
$$= .5$$

b. The probability that the student's score will be less than or equal to 500 is given by

$$P(D \cup E \cup F) = P(D) + P(E) + P(F)$$
$$= .23 + .31 + .19 = .73$$

c. The probability that the student's score will be greater than 400 but less than or equal to 600 is given by

$$P(C \cup D) = P(C) + P(D)$$
$$= .19 + .23 = .42$$

Property 3 holds if and only if E and F are mutually exclusive. In the general case, we have the following rule.

> **Property 4. Addition Rule**
> If E and F are any two events of an experiment, then
> $$P(E \cup F) = P(E) + P(F) - P(E \cap F)$$

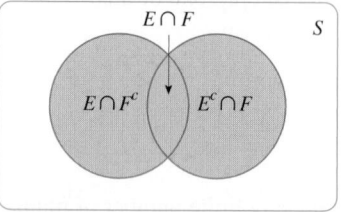

FIGURE 8
$E \cup F = (E \cap F^c) \cup (E \cap F) \cup$
$(E^c \cap F)$

To derive this property, refer to Figure 8. Observe that we can write

$$E = (E \cap F^c) \cup (E \cap F) \quad \text{and} \quad F = (E^c \cap F) \cup (E \cap F)$$

as a union of disjoint sets. Therefore,

$$P(E) = P(E \cap F^c) + P(E \cap F) \quad \text{or} \quad P(E \cap F^c) = P(E) - P(E \cap F)$$

and

$$P(F) = P(E^c \cap F) + P(E \cap F) \quad \text{or} \quad P(E^c \cap F) = P(F) - P(E \cap F)$$

Finally, since $E \cup F = (E \cap F^c) \cup (E \cap F) \cup (E^c \cap F)$ is a union of disjoint sets, we have

$$
\begin{aligned}
P(E \cup F) &= P(E \cap F^c) + P(E \cap F) + P(E^c \cap F) \\
&= P(E) - P(E \cap F) + P(E \cap F) + P(F) - P(E \cap F) \quad \text{\small Use the earlier} \\
&= P(E) + P(F) - P(E \cap F) \qquad\qquad\qquad\qquad\qquad\qquad\text{\small results.}
\end{aligned}
$$

Note   Observe that if E and F are mutually exclusive—that is, if $E \cap F = \varnothing$—then the equation of Property 4 reduces to that of Property 3. In other words, if E and F are mutually exclusive events, then $P(E \cup F) = P(E) + P(F)$. If E and F are not mutually exclusive events, then $P(E \cup F) = P(E) + P(F) - P(E \cap F)$. ■

EXAMPLE 2   A card is drawn from a well-shuffled deck of 52 playing cards. What is the probability that it is an ace or a spade?

Solution   Let E denote the event that the card drawn is an ace and let F denote the event that the card drawn is a spade. Then,

$$
P(E) = \frac{4}{52} \quad \text{and} \quad P(F) = \frac{13}{52}
$$

Furthermore, E and F are not mutually exclusive events. In fact, $E \cap F$ is the event that the card drawn is an ace of spades. Consequently,

$$
P(E \cap F) = \frac{1}{52}
$$

The event that a card drawn is an ace or a spade is $E \cup F$, with probability given by

$$
\begin{aligned}
P(E \cup F) &= P(E) + P(F) - P(E \cap F) \\
&= \frac{4}{52} + \frac{13}{52} - \frac{1}{52} = \frac{16}{52} = \frac{4}{13}
\end{aligned}
$$

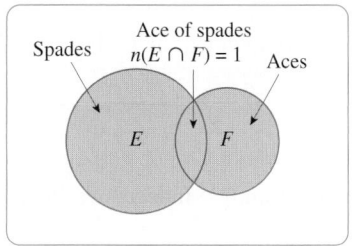

FIGURE 9
$P(E \cup F) = P(E) + P(F) - P(E \cap F)$

(Figure 9). This result, of course, can be obtained by arguing that 16 of the 52 cards are either spades or aces of other suits. ■

EXPLORE & DISCUSS

Let E, F, and G be any three events of an experiment. Use Formula (5) of Section 6.2 to show that

$$
\begin{aligned}
P(E \cup F \cup G) = P(E) + P(F) + P(G) &- P(E \cap F) - P(E \cap G) \\
&- P(F \cap G) + P(E \cap F \cap G)
\end{aligned}
$$

If E, F, and G are mutually exclusive, what is $P(E \cup F \cup G)$?

APPLIED EXAMPLE 3   Quality Control   The quality-control department of Vista Vision, manufacturer of the Pulsar 42-inch plasma TV, has determined from records obtained from the company's service centers that 3% of the sets sold experience video problems, 1% experience audio problems, and 0.1%

experience both video as well as audio problems before the expiration of the 90-day warranty. Find the probability that a plasma TV purchased by a consumer will experience video or audio problems before the warranty expires.

Solution Let E denote the event that a plasma TV purchased will experience video problems within 90 days and let F denote the event that a plasma TV purchased will experience audio problems within 90 days. Then,

$$P(E) = .03 \qquad P(F) = .01 \qquad P(E \cap F) = .001$$

The event that a plasma TV purchased will experience video problems or audio problems before the warranty expires is $E \cup F$, and the probability of this event is given by

$$P(E \cup F) = P(E) + P(F) - P(E \cap F)$$
$$= .03 + .01 - .001$$
$$= .039$$

(Figure 10). ∎

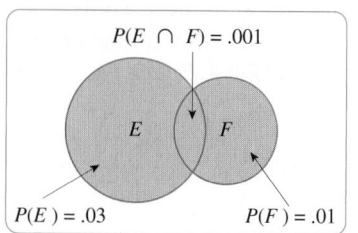

FIGURE 10
$P(E \cup F) = P(E) + P(F) - P(E \cap F)$

Here is another property of a probability function that is of considerable aid in computing the probability of an event.

> **Property 5. Rule of Complements**
> If E is an event of an experiment and E^c denotes the complement of E, then
> $$P(E^c) = 1 - P(E)$$

Property 5 is an immediate consequence of Properties 2 and 3. Indeed, we have $E \cup E^c = S$ and $E \cap E^c = \varnothing$, so

$$1 = P(S) = P(E \cup E^c) = P(E) + P(E^c)$$

and, therefore,

$$P(E^c) = 1 - P(E)$$

APPLIED EXAMPLE 4   Refer to Example 3. What is the probability that a Pulsar 42-inch plasma TV bought by a consumer will *not* experience video or audio difficulties before the warranty expires?

Solution Let E denote the event that a plasma TV bought by a consumer will experience video or audio difficulties before the warranty expires. Then, the event that the plasma TV will not experience either problem before the warranty expires is given by E^c, with probability

$$P(E^c) = 1 - P(E)$$
$$= 1 - .039$$
$$= .961$$

TITLE   Owner and Broker

INSTITUTION   Good and Associates

The insurance business is all about probabilities. Maybe your car will be stolen, maybe it won't; maybe you'll be in an accident, maybe you won't; maybe someone will slip on your steps and sue you; maybe they won't. Naturally, we hope that nothing bad happens to you or your family, but common sense and historical statistics tell us that it might.

All of these have probabilities associated with them. For example, if you own a Toyota Camry, there is a much higher probability your car will be stolen than if you drive a Dodge Caravan. As a matter of fact, in recent statistics listing the top ten stolen cars in America, the Toyota Camry was not only in first place; various model years were also in 2nd, 3rd, and 8th place! And the Honda Accord was also pretty popular with thieves—during the same year, different model years of the Accord were in 5th, 7th, 9th, and 10th place on the top-ten stolen list.

In terms of safety statistics, we've found that accidents and injuries tend to vary quite a bit from one make and model to another. With other factors being equal, a person who drives a Pontiac Firebird is much more likely to be involved in an accident than someone who drives a Buick Park Avenue. For example, a 23-year-old, unmarried male with a good driving record who uses his Firebird for commuting and pleasure driving in Philadelphia, Pennsylvania, would be charged an annual insurance premium over 30% higher than the same person driving a Buick Park Avenue would pay for the same personal injury and liability coverage.

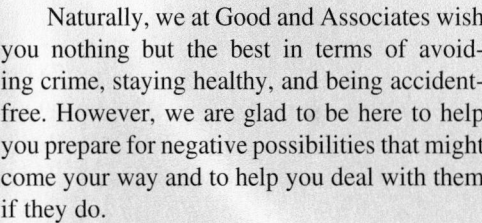

Naturally, we at Good and Associates wish you nothing but the best in terms of avoiding crime, staying healthy, and being accident-free. However, we are glad to be here to help you prepare for negative possibilities that might come your way and to help you deal with them if they do.

▬ Computations Involving the Rules of Probability

We close this section by looking at two additional examples that illustrate the rules of probability.

EXAMPLE 5   Let E and F be two mutually exclusive events and suppose that $P(E) = .1$ and $P(F) = .6$. Compute:
a. $P(E \cap F)$ **b.** $P(E \cup F)$ **c.** $P(E^c)$
d. $P(E^c \cap F^c)$ **e.** $P(E^c \cup F^c)$

Solution

a. Since the events E and F are mutually exclusive—that is, $E \cap F = \varnothing$—we have $P(E \cap F) = 0$.

b. $P(E \cup F) = P(E) + P(F)$ Since E and F are mutually exclusive

$$= .1 + .6$$

$$= .7$$

c. $P(E^c) = 1 - P(E)$ Property 5

$$= 1 - .1$$

$$= .9$$

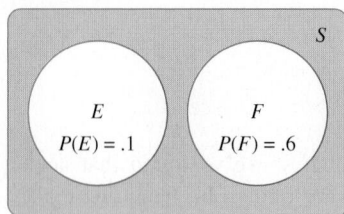

FIGURE 11
$P(E^c \cap F^c) = P[(E \cup F)^c]$

d. Observe that, by De Morgan's law, $E^c \cap F^c = (E \cup F)^c$. Hence

$$P(E^c \cap F^c) = P[(E \cup F)^c] \qquad \text{See Figure 11.}$$

$$= 1 - P(E \cup F) \qquad \text{Property 5}$$

$$= 1 - .7 \qquad \text{Use the result of part (b).}$$

$$= .3$$

e. Again using De Morgan's law, we find

$$P(E^c \cup F^c) = P[(E \cap F)^c]$$

$$= 1 - P(E \cap F)$$

$$= 1 - 0 \qquad \text{Use the result of part (a).}$$

$$= 1$$

EXAMPLE 6 Let E and F be two events of an experiment with sample space S. Suppose $P(E) = .2$, $P(F) = .1$, and $P(E \cap F) = .05$. Compute:
a. $P(E \cup F)$
b. $P(E^c \cap F^c)$
c. $P(E^c \cap F)$ *Hint:* Draw a Venn diagram.

Solution
a. $P(E \cup F) = P(E) + P(F) - P(E \cap F)$ Property 4
$$= .2 + .1 - .05$$
$$= .25$$

b. Using De Morgan's law, we have

$$P(E^c \cap F^c) = P[(E \cup F)^c]$$

$$= 1 - P(E \cup F) \qquad \text{Property 5}$$

$$= 1 - .25 \qquad \text{Use the result of part (a).}$$

$$= .75$$

c. From the Venn diagram describing the relationship between E, F, and S (Figure 12), we have

$$P(E^c \cap F) = .05 \qquad \text{The shaded subset is the event } E^c \cap F.$$

This result may also be obtained by using the relationship

$$P(E^c \cap F) = P(F) - P(E \cap F)$$

$$= .1 - .05$$

$$= .05$$

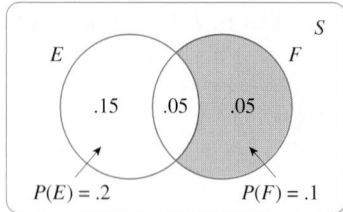

FIGURE 12
$P(E^c \cap F)$: the probability that the event F, but not the event E, will occur

as before.

7.3 Self-Check Exercises

1. Let E and F be events of an experiment with sample space S. Suppose $P(E) = .4$, $P(F) = .5$, and $P(E \cap F) = .1$. Compute:
   a. $P(E \cup F)$ **b.** $P(E \cap F^c)$

2. Susan Garcia wishes to sell or lease a condominium through a realty company. The realtor estimates that the probability of finding a buyer within a month of the date the property is listed for sale or lease is .3, the probability of finding a lessee is .8, and the probability of finding both a buyer and a lessee is .1. Determine the probability that the property will be sold or leased within 1 month from the date the property is listed for sale or lease.

Solutions to Self-Check Exercises 7.3 can be found on page 407.

7.3 Concept Questions

1. Suppose S is a sample space of an experiment, E and F are events of the experiment, and P is a probability function. Give the meaning of each of the following statements:
   a. $P(E) = 0$ **b.** $P(F) = 0.5$ **c.** $P(S) = 1$
   d. $P(E \cup F) = P(E) + P(F) - P(E \cap F)$

2. Give an example, based on a real-life situation, illustrating the property $P(E^c) = 1 - P(E)$, where E is an event and E^c is the complement of E.

7.3 Exercises

A pair of dice is cast, and the number that appears uppermost on each die is observed. In Exercises 1–6, refer to this experiment and find the probability of the given event.

1. The sum of the numbers is an even number.

2. The sum of the numbers is either 7 or 11.

3. A pair of 1s is thrown.

4. A double is thrown.

5. One die shows a 6, and the other is a number less than 3.

6. The sum of the numbers is at least 4.

An experiment consists of selecting a card at random from a 52-card deck. In Exercises 7–12, refer to this experiment and find the probability of the event.

7. A king of diamonds is drawn.

8. A diamond or a king is drawn.

9. A face card (i.e., a jack, queen, or king) is drawn.

10. A red face card is drawn.

11. An ace is not drawn.

12. A black face card is not drawn.

13. Five hundred people have purchased raffle tickets. What is the probability that a person holding one ticket will win the first prize? What is the probability that he or she will not win the first prize?

14. **TV HOUSEHOLDS** The results of a recent television survey of American TV households revealed that 87 out of every 100 TV households have at least one remote control. What is the probability that a randomly selected TV household does not have at least one remote control?

In Exercises 15–24, explain why the statement is incorrect.

15. The sample space associated with an experiment is given by $S = \{a, b, c\}$, where $P(a) = .3$, $P(b) = .4$, and $P(c) = .4$.

16. The probability that a bus will arrive late at the Civic Center is .35, and the probability that it will be on time or early is .60.

17. A person participates in a weekly office pool in which he has one chance in ten of winning the purse. If he participates for 5 weeks in succession, the probability of winning at least one purse is $\frac{5}{10}$.

18. The probability that a certain stock will increase in value over a period of 1 week is .6. Therefore, the probability that the stock will decrease in value is .4.

19. A red die and a green die are tossed. The probability that a 6 will appear uppermost on the red die is $\frac{1}{6}$, and the probability that a 1 will appear uppermost on the green die is $\frac{1}{6}$. Hence, the probability that the red die will show a 6 or the green die will show a 1 is $\frac{1}{6} + \frac{1}{6}$.

20. Joanne, a high school senior, has applied for admission to four colleges, A, B, C, and D. She has estimated that the probability that she will be accepted for admission by A, B, C, and D is .5, .3, .1, and .08, respectively. Thus, the probability that she will be accepted for admission by at least one college is $P(A) + P(B) + P(C) + P(D) = .5 + .3 + .1 + .08 = .98$.

21. The sample space associated with an experiment is given by $S = \{a, b, c, d, e\}$. The events $E = \{a, b\}$ and $F = \{c, d\}$ are mutually exclusive. Hence, the events E^c and F^c are mutually exclusive.

22. A 5-card poker hand is dealt from a 52-card deck. Let A denote the event that a flush is dealt and let B be the event that a straight is dealt. Then the events A and B are mutually exclusive.

23. Mark Owens, an optician, estimates that the probability that a customer coming into his store will purchase one or more pairs of glasses but not contact lenses is .40, and the probability that he will purchase one or more pairs of contact lenses but not glasses is .25. Hence, Owens concludes that the probability that a customer coming into his store will purchase neither a pair of glasses nor a pair of contact lenses is .35.

24. There are eight grades in Garfield Elementary School. If a student is selected at random from the school, then the probability that the student is in the first grade is $\frac{1}{8}$.

25. Let E and F be two events that are mutually exclusive, and suppose $P(E) = .2$ and $P(F) = .5$. Compute
  a. $P(E \cap F)$ **b.** $P(E \cup F)$
  c. $P(E^c)$ **d.** $P(E^c \cap F^c)$

26. Let E and F be two events of an experiment with sample space S. Suppose $P(E) = .6$, $P(F) = .4$, and $P(E \cap F) = .2$. Compute
  a. $P(E \cup F)$ **b.** $P(E^c)$
  c. $P(F^c)$ **d.** $P(E^c \cap F)$

27. Let $S = \{s_1, s_2, s_3, s_4\}$ be the sample space associated with an experiment having the probability distribution shown in the accompanying table. If $A = \{s_1, s_2\}$ and $B = \{s_1, s_3\}$, find
  a. $P(A)$, $P(B)$ **b.** $P(A^c)$, $P(B^c)$
  c. $P(A \cap B)$ **d.** $P(A \cup B)$

Outcome	Probability
s_1	$\frac{1}{8}$
s_2	$\frac{3}{8}$
s_3	$\frac{1}{4}$
s_4	$\frac{1}{4}$

28. Let $S = \{s_1, s_2, s_3, s_4, s_5, s_6\}$ be the sample space associated with an experiment having the probability distribution shown in the accompanying table. If $A = \{s_1, s_2\}$ and $B = \{s_1, s_5, s_6\}$, find
  a. $P(A)$, $P(B)$ **b.** $P(A^c)$, $P(B^c)$
  c. $P(A \cap B)$ **d.** $P(A \cup B)$
  e. $P(A^c \cap B^c)$ **f.** $P(A^c \cup B^c)$

Outcome	Probability
s_1	$\frac{1}{3}$
s_2	$\frac{1}{8}$
s_3	$\frac{1}{6}$
s_4	$\frac{1}{6}$
s_5	$\frac{1}{12}$
s_6	$\frac{1}{8}$

29. TEACHER ATTITUDES A nonprofit organization conducted a survey of 2140 metropolitan-area teachers regarding their beliefs about educational problems. The following data were obtained:

900 said that lack of parental support is a problem.

890 said that abused or neglected children are problems.

680 said that malnutrition or students in poor health is a problem.

120 said that lack of parental support and abused or neglected children are problems.

110 said that lack of parental support and malnutrition or poor health are problems.

140 said that abused or neglected children and malnutrition or poor health are problems.

40 said that lack of parental support, abuse or neglect, and malnutrition or poor health are problems.

What is the probability that a teacher selected at random from this group said that lack of parental support is the only problem hampering a student's schooling?
Hint: Draw a Venn diagram.

30. COURSE ENROLLMENTS Among 500 freshmen pursuing a business degree at a university, 320 are enrolled in an economics course, 225 are enrolled in a mathematics course, and 140 are enrolled in both an economics and a mathematics course. What is the probability that a freshman selected at random from this group is enrolled in
  a. An economics and/or a mathematics course?
  b. Exactly one of these two courses?
  c. Neither an economics course nor a mathematics course?

31. CONSUMER SURVEYS A leading manufacturer of kitchen appliances advertised its products in two magazines: *Good*

Housekeeping and the *Ladies Home Journal*. A survey of 500 customers revealed that 140 learned of its products from *Good Housekeeping*, 130 learned of its products from the *Ladies Home Journal*, and 80 learned of its products from both magazines. What is the probability that a person selected at random from this group saw the manufacturer's advertisement in
a. Both magazines?
b. At least one of the two magazines?
c. Exactly one magazine?

32. ROLLOVER DEATHS   The following table gives the number of people killed in rollover crashes in various types of vehicles in 2002:

Types of Vehicles	Cars	Pickups	SUVs	Vans
Deaths	4768	2742	2448	698

Find the empirical probability distribution associated with these data. If a fatality due to a rollover crash in 2002 is picked at random, what is the probability that the victim was in
a. A car?   b. An SUV?   c. A pickup or an SUV?
Source: National Highway Traffic Safety Administration

33. INVESTMENT IN TECHNOLOGY   One hundred sixty top regional executives were asked: "Do you plan to invest more or less in computers and information technology in the coming year?" The results of the poll follow.

Answer	Respondents
Less	27
The same	66
More	61
No answer	6

If one of the participants in the poll is selected at random, what is the probability that he or she said they would invest
a. More in computers and information technology in the coming year?
b. The same or less in computers and information technology in the coming year?
Source: Greater Boston Chamber of Commerce

34. IN-FLIGHT SERVICE   In a survey conducted in November 2002 of 1400 international business travelers concerning in-flight service over the past few years, the following information was obtained:

Comments on Quality of Service	Respondents
Has remained the same from two years ago	630
Has diminished over that time frame	406
Has improved over that time frame	336
Weren't sure	28

If a person in the survey is chosen at random, what is the probability that he or she has rated the in-flight service as
a. Remaining the same or improved over the time frame in question?
b. Remaining the same or diminished over the time frame in question?
Source: American Express

35. SWITCHING JOBS   Two hundred workers were asked: "Would a better economy lead you to switch jobs?" The results of the survey follow.

Answer	Very likely	Somewhat likely	Somewhat unlikely	Very unlikely	Don't know
Respondents	40	28	26	104	2

If a worker is chosen at random, what is the probability that he or she
a. Is very unlikely to switch jobs?
b. Is somewhat likely or very likely to switch jobs?
Source: Accountemps

36. 401(K) INVESTORS   According to a study conducted in 2003 concerning the participation, by age, of 401(k) investors, the following data were obtained:

Age	20s	30s	40s	50s	60s
Percent	11	28	32	22	7

a. What percent of 401(k) investors are in their 20s or 60s?
b. What percent of 401(k) investors are under the age of 50?
Source: Investment Company Institute

37. ASSET ALLOCATION   When asked by a 30-yr-old bachelor for professional advice concerning the asset allocation of his $200,000 nest egg, the financial-planning firm of Sagemark Consulting suggested the following:

Asset	Percent
U.S. mid/small cap stocks	22
U.S. large cap stocks	46
International stocks	17
Bonds/income	8
REITS	5
Cash	2

a. According to Sagemark Consulting, what percent of the bachelor's portfolio should be invested in U.S. large cap or U.S. mid/small cap stocks?
b. According to Sagemark Consulting, what percent of the bachelor's portfolio should be invested in vehicles other than bonds/incomes or REITS?
Source: Sagemark Consulting

38. DOWNLOADING MUSIC   The following table, compiled in 2004, gives the percent of music downloaded from the United States and other countries by U.S. users.

Country	U.S.	Germany	Canada	Italy	U.K.	France	Japan	Other
Percent	45.1	16.5	6.9	6.1	4.2	3.8	2.5	14.9

a. Verify that the table does give a probability distribution for the experiment.

b. What is the probability that a user who downloads music, selected at random, obtained it from either the United States or Canada?

c. What is the probability that a U.S. user who downloads music, selected at random, does not obtain it from Italy, the United Kingdom (U.K.), or France?

Source: Felix Oberholtzer-Gee and Koleman Strumpf

39. Assembly-Time Studies A time study was conducted by the production manager of Universal Instruments to determine how much time it took an assembly worker to complete a certain task during the assembly of its Galaxy home computers. Results of the study indicated that 20% of the workers were able to complete the task in less than 3 min, 60% of the workers were able to complete the task in 4 min or less, and 10% of the workers required more than 5 min to complete the task. If an assembly-line worker is selected at random from this group, what is the probability that:

a. He or she will be able to complete the task in 5 min or less?

b. He or she will not be able to complete the task within 4 min?

c. The time taken for the worker to complete the task will be between 3 and 4 min (inclusive)?

40. Plans to Keep Cars In a survey conducted to see how long Americans keep their cars, 2000 automobile owners were asked how long they plan to keep their present cars. The results of the survey follow:

Years Car Is Kept, x	Respondents
$0 \le x < 1$	60
$1 \le x < 3$	440
$3 \le x < 5$	360
$5 \le x < 7$	340
$7 \le x < 10$	240
$10 \le x$	560

Find the probability distribution associated with these data. What is the probability that an automobile owner selected at random from those surveyed plans to keep his or her present car

a. Less than five years?

b. Three or more years?

41. Gun-Control Laws A poll was conducted among 250 residents of a certain city regarding tougher gun-control laws. The results of the poll are shown in the table:

	Own Only a Handgun	Own Only a Rifle	Own a Handgun and a Rifle	Own Neither	Total
Favor Tougher Laws	0	12	0	138	150
Oppose Tougher Laws	58	5	25	0	88
No Opinion	0	0	0	12	12
Total	58	17	25	150	250

If one of the participants in this poll is selected at random, what is the probability that he or she

a. Favors tougher gun-control laws?

b. Owns a handgun?

c. Owns a handgun but not a rifle?

d. Favors tougher gun-control laws and does not own a handgun?

42. Risk of an Airplane Crash According to a study of Western-built commercial jets involved in crashes from 1988 to 1998, the percent of airplane crashes that occur at each stage of flight are as follows:

Phase	Percent
On ground, taxiing	4
During takeoff	10
Climbing to cruise altitude	19
En route	5
Descent and approach	31
Landing	31

If one of the doomed flights in the period 1988–1998 is picked at random, what is the probability that it crashed

a. While taxiing on the ground or while en route?

b. During takeoff or landing?

If the study is indicative of airplane crashes in general, when is the risk of a plane crash the highest?

Source: National Transportation and Safety Board

43. Suppose the probability that Bill can solve a problem is p_1 and the probability that Mike can solve it is p_2. Show that the probability that Bill and Mike working independently can solve the problem is $p_1 + p_2 - p_1 p_2$.

44. Fifty raffle tickets are numbered 1 through 50, and one of them is drawn at random. What is the probability that the number is a multiple of 5 or 7? Consider the following "solution": Since 10 tickets bear numbers that are multiples of 5 and since 7 tickets bear numbers that are multiples of 7, we conclude that the required probability is

$$\frac{10}{50} + \frac{7}{50} = \frac{17}{50}$$

What is wrong with this argument? What is the correct answer?

In Exercises 45–48, determine whether the statement is true or false. If it is true, explain why it is true. If it is false, give an example to show why it is false.

45. If A is a subset of B and $P(B) = 0$, then $P(A) = 0$.

46. If A is a subset of B, then $P(A) \le P(B)$.

47. If $E_1, E_2, \ldots, E_n$ are events of an experiment, then $P(E_1 \cup E_2 \cup \cdots \cup E_n) = P(E_1) + P(E_2) + \cdots + P(E_n)$.

48. If E is an event of an experiment, then $P(E) + P(E^c) = 1$.

7.3 Solutions to Self-Check Exercises

1. a. Using Property 4, we find

$$P(E \cup F) = P(E) + P(F) - P(E \cap F)$$
$$= .4 + .5 - .1$$
$$= .8$$

b. From the accompanying Venn diagram, in which the subset $E \cap F^c$ is shaded, we see that

$$P(E \cap F^c) = .3$$

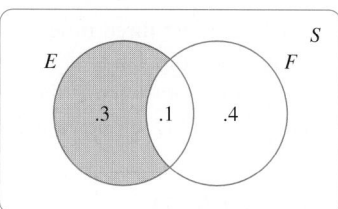

The result may also be obtained by using the relationship

$$P(E \cap F^c) = P(E) - P(E \cap F)$$
$$= .4 - .1 = .3$$

2. Let E denote the event that the property will be sold within 1 month of the date it is listed for sale or lease and let F denote the event that the property will be leased within the same time period. Then,

$$P(E) = .3 \qquad P(F) = .8 \qquad P(E \cap F) = .1$$

The probability of the event that the property will be sold or leased within 1 month of the date it is listed for sale or lease is given by

$$P(E \cup F) = P(E) + P(F) - P(E \cap F)$$
$$= .3 + .8 - .1 = 1$$

—that is, a certainty.

7.4 Use of Counting Techniques in Probability

■ Further Applications of Counting Techniques

As we have seen many times before, a problem in which the underlying sample space has a small number of elements may be solved by first determining all such sample points. However, for problems involving sample spaces with a large number of sample points, this approach is neither practical nor desirable.

In this section, we see how the counting techniques studied in Chapter 6 may be employed to help us solve problems in which the associated sample spaces contain large numbers of sample points. In particular, we restrict our attention to the study of uniform sample spaces—that is, sample spaces in which the outcomes are equally likely. For such spaces we have the following result.

Computing the Probability of an Event in a Uniform Sample Space
Let S be a uniform sample space and let E be any event. Then

$$P(E) = \frac{\text{Number of favorable outcomes in } E}{\text{Number of possible outcomes in } S} = \frac{n(E)}{n(S)} \qquad \textbf{(1)}$$

EXAMPLE 1 An unbiased coin is tossed six times. What is the probability that the coin will land heads
a. Exactly three times?
b. At most three times?
c. On the first and the last toss?

Solution

a. Each outcome of the experiment may be represented as a sequence of heads and tails. Using the generalized multiplication principle, we see that the number of outcomes of this experiment is given by 2^6, or 64. Let E denote the event that the coin lands heads exactly three times. Since there are $C(6, 3)$ ways this can occur, we see that the required probability is

$$P(E) = \frac{n(E)}{n(S)} = \frac{C(6, 3)}{64} = \frac{\dfrac{6!}{3!\,3!}}{64} \qquad \text{\small S, sample space of the experiment}$$

$$= \frac{\dfrac{6 \cdot 5 \cdot 4}{3 \cdot 2}}{64} = \frac{20}{64} = \frac{5}{16} = .3125$$

b. Let F denote the event that the coin lands heads at most three times. Then $n(F)$ is given by the sum of the number of ways the coin lands heads zero times (no heads!), the number of ways it lands heads exactly once, the number of ways it lands heads exactly twice, and the number of ways it lands heads exactly three times. That is,

$$n(F) = C(6, 0) + C(6, 1) + C(6, 2) + C(6, 3)$$

$$= \frac{6!}{0!\,6!} + \frac{6!}{1!\,5!} + \frac{6!}{2!\,4!} + \frac{6!}{3!\,3!}$$

$$= 1 + 6 + \frac{(6)(5)}{2} + \frac{(6)(5)(4)}{(3)(2)} = 42$$

Therefore, the required probability is

$$P(F) = \frac{n(F)}{n(S)} = \frac{42}{64} = \frac{21}{32} \approx .66$$

c. Let F denote the event that the coin lands heads on the first and the last toss. Then $n(F) = 1 \cdot 2 \cdot 2 \cdot 2 \cdot 2 \cdot 1 = 2^4$, so the probability that this event occurs is

$$P(F) = \frac{2^4}{2^6}$$

$$= \frac{1}{2^2}$$

$$= \frac{1}{4}$$

■

EXAMPLE 2   Two cards are selected at random from a well-shuffled pack of 52 playing cards. What is the probability that
a. They are both aces?   **b.** Neither of them is an ace?

Solution

a. The experiment consists of selecting 2 cards from a pack of 52 playing cards. Since the order in which the cards are selected is immaterial, the sample points are combinations of 52 cards taken 2 at a time. Now there are $C(52, 2)$ ways of selecting 52 cards taken 2 at a time, so the number of elements in the sample space S is given by $C(52, 2)$. Next, we observe that there are $C(4, 2)$ ways of selecting 2 aces from the 4 in the deck. Therefore, if E denotes the event that the cards selected are both aces, then

$$P(E) = \frac{n(E)}{n(S)}$$

$$= \frac{C(4, 2)}{C(52, 2)} = \frac{\dfrac{4!}{2!\,2!}}{\dfrac{52!}{2!\,50!}}$$

$$= \frac{1}{221}$$

b. Let F denote the event that neither of the two cards selected is an ace. Since there are $C(48, 2)$ ways of selecting two cards, neither of which is an ace, we find that

$$P(F) = \frac{n(F)}{n(S)} = \frac{C(48, 2)}{C(52, 2)} = \frac{\dfrac{48!}{2!\,46!}}{\dfrac{52!}{2!\,50!}} = \frac{48 \cdot 47}{2} \cdot \frac{2}{52 \cdot 51}$$

$$= \frac{188}{221}$$

APPLIED EXAMPLE 3 Quality Control A bin in the hi-fi department of Building 20, a bargain outlet, contains 100 blank cassette tapes, of which 10 are known to be defective. If a customer selects 6 of these cassette tapes, determine the probability

a. That 2 of them are defective.

b. That at least 1 of them is defective.

Solution

a. There are $C(100, 6)$ ways of selecting a set of 6 cassette tapes from the 100, and this gives $n(S)$, the number of outcomes in the sample space associated with the experiment. Next, we observe that there are $C(10, 2)$ ways of selecting a set of 2 defective cassette tapes from the 10 defective cassette tapes and $C(90, 4)$ ways of selecting a set of 4 nondefective cassette tapes from the 90 nondefective cassette tapes (Figure 13). Thus, by the multiplication principle, there are $C(10, 2) \cdot C(90, 4)$ ways of selecting 2 defective and 4 nondefective cassette tapes. Therefore, the probability of selecting 6 cassette tapes, of which 2 are defective, is given by

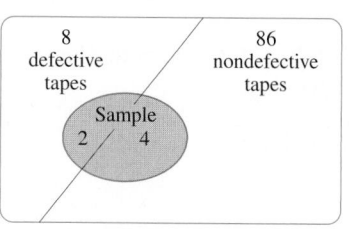

FIGURE 13
A sample of 6 tapes selected from 90 nondefective tapes and 10 defective tapes

$$\frac{C(10, 2) \cdot C(90, 4)}{C(100, 6)} = \frac{\dfrac{10!}{2!\,8!}\,\dfrac{90!}{4!\,86!}}{\dfrac{100!}{6!\,94!}}$$

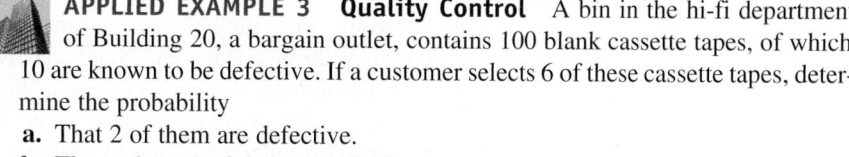

$$= \frac{10 \cdot 9}{2} \cdot \frac{90 \cdot 89 \cdot 88 \cdot 87}{4 \cdot 3 \cdot 2 \cdot 1} \cdot \frac{6 \cdot 5 \cdot 4 \cdot 3 \cdot 2 \cdot 1}{100 \cdot 99 \cdot 98 \cdot 97 \cdot 96 \cdot 95}$$

$$\approx .096$$

b. Let E denote the event that none of the cassette tapes selected is defective. Then E^c gives the event that at least 1 of the cassette tapes is defective. By the rule of complements,

$$P(E^c) = 1 - P(E)$$

To compute $P(E)$, we observe that there are $C(90, 6)$ ways of selecting a set of 6 cassette tapes that are nondefective. Therefore,

$$P(E) = \frac{C(90, 6)}{C(100, 6)}$$

$$P(E^c) = 1 - \frac{C(90, 6)}{C(100, 6)}$$

$$= 1 - \frac{\dfrac{90!}{6! \, 84!}}{\dfrac{100!}{6! \, 94!}}$$

$$= 1 - \frac{90 \cdot 89 \cdot 88 \cdot 87 \cdot 86 \cdot 85}{6 \cdot 5 \cdot 4 \cdot 3 \cdot 2 \cdot 1} \cdot \frac{6 \cdot 5 \cdot 4 \cdot 3 \cdot 2 \cdot 1}{100 \cdot 99 \cdot 98 \cdot 97 \cdot 96 \cdot 95}$$

$$\approx .48$$

The Birthday Problem

EXAMPLE 4   A group of five people is selected at random. What is the probability that at least two of them have the same birthday?

Solution   For simplicity, we assume that none of the five people was born on February 29 of a leap year. Since the five people were selected at random, we may also assume that each of them is equally likely to have any of the 365 days of a year as his or her birthday. If we let A, B, C, D, and F represent the five people, then an outcome of the experiment may be represented by (a, b, c, d, f), where the numbers $a, b, c, d,$ and f give the birthdays of A, B, C, D, and F, respectively.

We first observe that since there are 365 possibilities for each of the dates a, $b, c, d,$ and f, the multiplication principle implies that there are

$$\boxed{365} \cdot \boxed{365} \cdot \boxed{365} \cdot \boxed{365} \cdot \boxed{365}$$
$$\;\;\;a \qquad\quad b \qquad\quad c \qquad\quad d \qquad\quad f$$

or 365^5 outcomes of the experiment. Therefore,

$$n(S) = 365^5$$

where S denotes the sample space of the experiment.

Next, let E denote the event that two or more of the five people have the same birthday. It is now necessary to compute $P(E)$. However, a direct computation of $P(E)$ is relatively difficult. It is much easier to compute $P(E^c)$, where E^c is the event that no two of the five people have the same birthday, and then use the relation

$$P(E) = 1 - P(E^c)$$

To compute $P(E^c)$, observe that there are 365 ways (corresponding to the 365 dates) on which A's birthday can occur, followed by 364 ways on which B's birthday could occur if B were not to have the same birthday as A, and so on. Therefore, by the generalized multiplication principle,

$$n(E^c) = \underset{\substack{\text{A's} \\ \text{birthday}}}{365} \cdot \underset{\substack{\text{B's} \\ \text{birthday}}}{364} \cdot \underset{\substack{\text{C's} \\ \text{birthday}}}{363} \cdot \underset{\substack{\text{D's} \\ \text{birthday}}}{362} \cdot \underset{\substack{\text{F's} \\ \text{birthday}}}{361}$$

Thus,

$$P(E^c) = \frac{n(E^c)}{n(S)}$$

$$= \frac{365 \cdot 364 \cdot 363 \cdot 362 \cdot 361}{365^5}$$

$$P(E) = 1 - P(E^c)$$

$$= 1 - \frac{365 \cdot 364 \cdot 363 \cdot 362 \cdot 361}{365^5}$$

$$\approx .027$$

We can extend the result obtained in Example 4 to the general case involving r people. In fact, if E denotes the event that at least two of the r people have the same birthday, an argument similar to that used in Example 4 leads to the result

$$P(E) = 1 - \frac{365 \cdot 364 \cdot 363 \cdots (365 - r + 1)}{365^r}$$

By letting r take on the values 5, 10, 15, 20, . . . , 50, in turn, we obtain the probabilities that at least 2 of 5, 10, 15, 20, . . . , 50 people, respectively, have the same birthday. These results are summarized in Table 8.

The results show that, in a group of 23 randomly selected people, the chances are greater than 50% that at least 2 of them will have the same birthday. In a group of 50 people, it is an excellent bet that at least 2 people in the group will have the same birthday.

TABLE 8

Probability That at Least Two People in a Randomly Selected Group of *r* People Have the Same Birthday

r	P(E)
5	.027
10	.117
15	.253
20	.411
22	.476
23	.507
25	.569
30	.706
40	.891
50	.970

EXPLORE & DISCUSS

During an episode of the *Tonight Show*, a talk show host related "The Birthday Problem" to the audience—noting that, in a group of 50 or more people, probabilists have calculated that the probability of at least 2 people having the same birthday is very high. To illustrate this point, he proceeded to conduct his own experiment. A person selected at random from the audience was asked to state his birthday. The host then asked if anyone in the audience had the same birthday. The response was negative. He repeated the experiment. Once again, the response was negative. These results, observed the host, were contrary to expectations. In a later episode of the show, the host explained why this experiment had been improperly conducted. Explain why the host failed to illustrate the point he was trying to make in the earlier episode.

7.4 Self-Check Exercises

1. Four balls are selected at random without replacement from an urn containing 10 white balls and 8 red balls. What is the probability that all the chosen balls are white?

2. A box contains 20 microchips, of which 4 are substandard. If 2 of the chips are taken from the box, what is the probability that they are both substandard?

Solutions to Self-Check Exercises 7.4 can be found on page 414.

7.4 Concept Questions

1. What is the probability of an event E in a uniform sample space S?

2. Suppose we want to find the probability that at least two people in a group of six randomly selected people have the same birthday.

   a. If S denotes the sample space of this experiment, what is $n(S)$?
   b. If E is the event that two or more of the six people in the group have the same birthday, explain how you would use $P(E^c)$ to determine $P(E)$.

7.4 Exercises

An unbiased coin is tossed five times. In Exercises 1–4, find the probability of the given event.

1. The coin lands heads all five times.

2. The coin lands heads exactly once.

3. The coin lands heads at least once.

4. The coin lands heads more than once.

Two cards are selected at random without replacement from a well-shuffled deck of 52 playing cards. In Exercises 5–8, find the probability of the given event.

5. A pair is drawn.   6. A pair is not drawn.

7. Two black cards are drawn.

8. Two cards of the same suit are drawn.

Four balls are selected at random without replacement from an urn containing three white balls and five blue balls. In Exercises 9–12, find the probability of the given event.

9. Two of the balls are white and two are blue.

10. All of the balls are blue.

11. Exactly three of the balls are blue.

12. Two or three of the balls are white.

Assume that the probability of a boy being born is the same as the probability of a girl being born. In Exercises 13–16, find the probability that a family with three children will have the given composition.

13. Two boys and one girl

14. At least one girl

15. No girls

16. The two oldest children are girls.

17. An exam consists of ten true-or-false questions. If a student guesses at every answer, what is the probability that he or she will answer exactly six questions correctly?

18. PERSONNEL SELECTION Jacobs & Johnson, an accounting firm, employs 14 accountants, of whom 8 are CPAs. If a delegation of 3 accountants is randomly selected from the firm to attend a conference, what is the probability that 3 CPAs will be selected?

19. QUALITY CONTROL Two light bulbs are selected at random from a lot of 24, of which 4 are defective. What is the probability that
   a. Both of the light bulbs are defective?
   b. At least 1 of the light bulbs is defective?

20. A customer at Cavallaro's Fruit Stand picks a sample of 3 oranges at random from a crate containing 60 oranges, of which 4 are rotten. What is the probability that the sample contains 1 or more rotten oranges?

21. QUALITY CONTROL A shelf in the Metro Department Store contains 80 colored ink cartridges for a popular ink-jet printer. Six of the cartridges are defective. If a customer selects 2 cartridges at random from the shelf, what is the probability that
   a. Both are defective?
   b. At least 1 is defective?

22. QUALITY CONTROL Electronic baseball games manufactured by Tempco Electronics are shipped in lots of 24. Before shipping, a quality-control inspector randomly selects a sample of 8 from each lot for testing. If the sample contains any defective games, the entire lot is rejected. What is the probability that a lot containing exactly 2 defective games will still be shipped?

23. PERSONNEL SELECTION The City Transit Authority plans to hire 12 new bus drivers. From a group of 100 qualified applicants, of which 60 are men and 40 are women, 12 names are to be selected by lot. Suppose that Mary and John Lewis are among the 100 qualified applicants.

a. What is the probability that Mary's name will be selected? That both Mary's and John's names will be selected?

b. If it is stipulated that an equal number of men and women are to be selected (6 men from the group of 60 men and 6 women from the group of 40 women), what is the probability that Mary's name will be selected? That Mary's and John's names will be selected?

24. **PUBLIC HOUSING** The City Housing Authority has received 50 applications from qualified applicants for eight low-income apartments. Three of the apartments are on the north side of town, and five are on the south side. If the apartments are to be assigned by means of a lottery, what is the probability that
   a. A specific qualified applicant will be selected for one of these apartments?
   b. Two specific qualified applicants will be selected for apartments on the same side of town?

25. A student studying for a vocabulary test knows the meanings of 12 words from a list of 20 words. If the test contains 10 words from the study list, what is the probability that at least 8 of the words on the test are words that the student knows?

26. **DRIVING TESTS** Four different written driving tests are administered by the Motor Vehicle Department. One of these four tests is selected at random for each applicant for a driver's license. If a group consisting of two women and three men apply for a license, what is the probability that
   a. Exactly two of the five will take the same test?
   b. The two women will take the same test?

27. **BRAND SELECTION** A druggist wishes to select three brands of aspirin to sell in his store. He has five major brands to choose from: A, B, C, D, and E. If he selects the three brands at random, what is the probability that he will select
   a. Brand B?
   b. Brands B and C?
   c. At least one of the two brands B and C?

28. **BLACKJACK** In the game of blackjack, a 2-card hand consisting of an ace and a face card or a 10 is called a blackjack.
   a. If a player is dealt 2 cards from a standard deck of 52 well-shuffled cards, what is the probability that the player will receive a blackjack?
   b. If a player is dealt 2 cards from 2 well-shuffled standard decks, what is the probability that the player will receive a blackjack?

29. **SLOT MACHINES** Refer to Exercise 25, Section 6.3, where the "lucky dollar" slot machine was described. What is the probability that the three "lucky dollar" symbols will appear in the window of the slot machine?

30. **ROULETTE** In 1959 a world record was set for the longest run on an ungaffed (fair) roulette wheel at the El San Juan Hotel in Puerto Rico. The number 10 appeared six times in a row.

What is the probability of the occurrence of this event? (Assume that there are 38 equally likely outcomes consisting of the numbers 1–36, 0, and 00.)

In "The Numbers Game," a state lottery, four numbers are drawn with replacement from an urn containing the digits 0–9, inclusive. In Exercises 31–34, find the probability of a ticket holder having the indicated winning ticket.

31. All four digits in exact order (the grand prize)

32. Two specified, consecutive digits in exact order (the first two digits, the middle two digits, or the last two digits)

33. One specified digit in exact order (the first, second, third, or fourth digit)

34. All four digits in any order (including the other winning tickets)

A list of poker hands, ranked in order from the highest to the lowest, is shown in the accompanying table along with a description and example of each hand. Use the table to answer Exercises 35–40.

Hand	Description	Example
Straight flush	5 cards in sequence in the same suit	A ♥ 2 ♥ 3 ♥ 4 ♥ 5 ♥
Four of a kind	4 cards of the same rank and any other card	K ♥ K ♦ K ♠ K ♣ 2 ♥
Full house	3 of a kind and a pair	3 ♥ 3 ♦ 3 ♣ 7 ♥ 7 ♦
Flush	5 cards of the same suit that are not all in sequence	5 ♥ 6 ♥ 9 ♥ J ♥ K ♥
Straight	5 cards in sequence but not all of the same suit	10 ♥ J ♦ Q ♣ K ♠ A ♥
Three of a kind	3 cards of the same rank and 2 unmatched cards	K ♥ K ♦ K ♠ 2 ♥ 4 ♦
Two pair	2 cards of the same rank and 2 cards of any other rank with an unmatched card	K ♥ K ♦ 2 ♥ 2 ♠ 4 ♣
One pair	2 cards of the same rank and 3 unmatched cards	K ♥ K ♦ 5 ♥ 2 ♠ 4 ♥

If a 5-card poker hand is dealt from a well-shuffled deck of 52 cards, what is the probability of being dealt the given hand?

35. A straight flush (Note that an ace may be played as either a high or low card in a straight sequence—that is, A, 2, 3, 4, 5 or 10, J, Q, K, A. Hence, there are ten possible sequences for a straight in one suit.)

36. A straight (but not a straight flush)

37. A flush (but not a straight flush)

38. Four of a kind

39. A full house

40. Two pairs

41. ZODIAC SIGNS There are 12 signs of the Zodiac: Aries, Taurus, Gemini, Cancer, Leo, Virgo, Libra, Scorpio, Sagittarius, Capricorn, Aquarius, and Pisces. Each sign corresponds to a different calendar period of approximately 1 month. Assuming that a person is just as likely to be born under one sign as another, what is the probability that in a group of five people at least two of them
  a. Have the same sign?
  b. Were born under the sign of Aries?

42. BIRTHDAY PROBLEM What is the probability that at least two of the nine justices of the U.S. Supreme Court have the same birthday?

43. BIRTHDAY PROBLEM Fifty people are selected at random. What is the probability that none of the people in this group have the same birthday?

44. BIRTHDAY PROBLEM There were 42 different presidents of the United States from 1789 through 2000. What is the probability that at least two of them had the same birthday? Compare your calculation with the facts by checking an almanac or some other source.

7.4 Solutions to Self-Check Exercises

1. The probability that all 4 balls selected are white is given by

The number of ways of selecting 4 white balls from the 10 in the urn
───
The number of ways of selecting any 4 balls from the 18 balls in the urn

$$= \frac{C(10, 4)}{C(18, 4)}$$

$$= \frac{\dfrac{10!}{4!\, 6!}}{\dfrac{18!}{4!\, 14!}}$$

$$= \frac{10 \cdot 9 \cdot 8 \cdot 7}{4 \cdot 3 \cdot 2} \cdot \frac{4 \cdot 3 \cdot 2}{18 \cdot 17 \cdot 16 \cdot 15}$$

$$\approx .069$$

2. The probability that both chips are substandard is given by

The number of ways of choosing any 2 of the 4 substandard chips
───
The number of ways of choosing any 2 of the 20 chips

$$= \frac{C(4, 2)}{C(20, 2)}$$

$$= \frac{\dfrac{4!}{2!\, 2!}}{\dfrac{20!}{2!\, 18!}}$$

$$= \frac{4 \cdot 3}{2} \cdot \frac{2}{20 \cdot 19}$$

$$\approx .032$$

7.5 Conditional Probability and Independent Events

▰ Conditional Probability

Three cities, A, B, and C, are vying to play host to the Summer Olympic Games in 2008. If each city has the same chance of winning the right to host the Games, then the probability of city A hosting the Games is $\frac{1}{3}$. Now suppose city B decides to pull out of contention because of fiscal problems. Then it would seem that city A's chances of playing host will increase. In fact, if each of the two remaining cities have equal chances of winning, then the probability of city A playing host to the Games is $\frac{1}{2}$.

In general, the probability of an event is affected by the occurrence of other events and/or by the knowledge of information relevant to the event. Basically, the injection of conditions into a problem modifies the underlying sample space of the original problem. This in turn leads to a change in the probability of the event.

EXAMPLE 1   Two cards are drawn without replacement from a well-shuffled deck of 52 playing cards.
 a. What is the probability that the first card drawn is an ace?
 b. What is the probability that the second card drawn is an ace given that the first card drawn was not an ace?
 c. What is the probability that the second card drawn is an ace given that the first card drawn was an ace?

Solution
 a. The sample space here consists of 52 equally likely outcomes, 4 of which are aces. Therefore, the probability that the first card drawn is an ace is $\frac{4}{52}$.
 b. Having drawn the first card, there are 51 cards left in the deck. In other words, for the second phase of the experiment, we are working in a *reduced* sample space. If the first card drawn was not an ace, then this modified sample space of 51 points contains 4 "favorable" outcomes (the 4 aces), so the probability that the second card drawn is an ace is given by $\frac{4}{51}$.
 c. If the first card drawn was an ace, then there are 3 aces left in the deck of 51 playing cards, so the probability that the second card drawn is an ace is given by $\frac{3}{51}$.   ∎

Observe that in Example 1 the occurrence of the first event reduces the size of the original sample space. The information concerning the first card drawn also leads us to the consideration of modified sample spaces: In part (b) the deck contained 4 aces, and in part (c) the deck contained 3 aces.

The probability found in part (b) or (c) of Example 1 is known as a **conditional probability**, since it is the probability of an event occurring given that another event has already occurred. For example, in part (b) we computed the probability of the event that the second card drawn is an ace *given that* the first card drawn was not an ace. In general, given two events A and B of an experiment, under certain circumstances one may compute the probability of the event B given that the event A has already occurred. This probability, denoted by $P(B|A)$, is called the **conditional probability of B given A**.

A formula for computing the conditional probability of B given A may be discovered with the aid of a Venn diagram. Consider an experiment with a uniform sample space S, and suppose A and B are two events of the experiment (Figure 14).

The condition that the event A has occurred tells us that the possible outcomes of the experiment in the second phase are restricted to those outcomes (elements) in the set A. In other words, we may work with the reduced sample space A instead of the original sample space S in the experiment. Next we observe that, with respect to the reduced sample space A, the outcomes favorable to the event B are precisely those elements in the set $A \cap B$. Consequently, the conditional probability of B given A is

$$P(B|A) = \frac{\text{Number of elements in } A \cap B}{\text{Number of elements in } A}$$

$$= \frac{n(A \cap B)}{n(A)} \qquad n(A) \neq 0$$

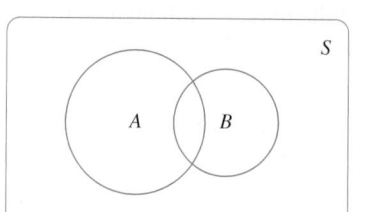

(a) Original sample space

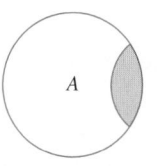

(b) Reduced sample space A. The shaded area is $A \cap B$.

FIGURE 14

Dividing the numerator and the denominator by $n(S)$, the number of elements in S, we have

$$P(B|A) = \frac{\dfrac{n(A \cap B)}{n(S)}}{\dfrac{n(A)}{n(S)}}$$

which is equivalent to the following formula.

Conditional Probability of an Event

If A and B are events in an experiment and $P(A) \neq 0$, then the conditional probability that the event B will occur given that the event A has already occurred is

$$P(B|A) = \frac{P(A \cap B)}{P(A)} \tag{2}$$

EXAMPLE 2   A pair of fair dice is cast. What is the probability that the sum of the numbers falling uppermost is 7 if it is known that one of the numbers is a 5?

Solution   Let A denote the event that the sum of the numbers falling uppermost is 7 and let B denote the event that one of the numbers is a 5. From the results of Example 4, Section 7.1, we find that

$$A = \{(6, 1), (5, 2), (4, 3), (3, 4), (2, 5), (1, 6)\}$$
$$B = \{(5, 1), (5, 2), (5, 3), (5, 4), (5, 5), (5, 6),$$
$$(1, 5), (2, 5), (3, 5), (4, 5), (6, 5)\}$$

so that

$$A \cap B = \{(5, 2), (2, 5)\}$$

(Figure 15). Since the dice are fair, each outcome of the experiment is equally likely; therefore,

$$P(A \cap B) = \frac{2}{36} \quad \text{and} \quad P(B) = \frac{11}{36} \qquad \text{Recall that } n(S) = 36.$$

Thus, the probability that the sum of the numbers falling uppermost is 7 given that one of the numbers is a 5 is, by virtue of Equation (2),

$$P(A|B) = \frac{\dfrac{2}{36}}{\dfrac{11}{36}} = \frac{2}{11}$$

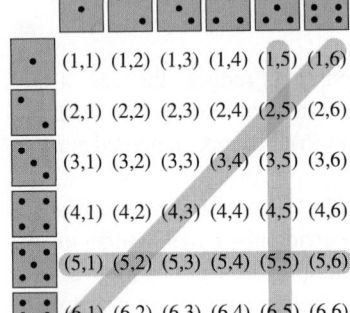

(1,1) (1,2) (1,3) (1,4) (1,5) (1,6)

(2,1) (2,2) (2,3) (2,4) (2,5) (2,6)

(3,1) (3,2) (3,3) (3,4) (3,5) (3,6)

(4,1) (4,2) (4,3) (4,4) (4,5) (4,6)

(5,1) (5,2) (5,3) (5,4) (5,5) (5,6)

(6,1) (6,2) (6,3) (6,4) (6,5) (6,6)

FIGURE 15
$A \cap B = \{(5, 2), (2, 5)\}$

APPLIED EXAMPLE 3   Color Blindness   In a test conducted by the U.S. Army, it was found that of 1000 new recruits (600 men and 400 women), 50 of the men and 4 of the women were red-green color-blind. Given that a recruit selected at random from this group is red-green color-blind, what is the probability that the recruit is a male?

Solution   Let C denote the event that a randomly selected subject is red-green color-blind and let M denote the event that the subject is a male recruit. Since 54 out of the 1000 subjects are color-blind, we may take

$$P(C) = \frac{54}{1000} = .054$$

Therefore, by Equation (2), the probability that a subject is male given that the subject is red-green color-blind is

$$P(M|C) = \frac{P(M \cap C)}{P(C)}$$

$$= \frac{.05}{.054} = \frac{25}{27}$$

EXPLORE & DISCUSS

Let A and B be events in an experiment and let $P(A) \neq 0$. In n trials, the event A occurs m times, the event B occurs k times, and the events A and B occur together l times.

1. Explain why it makes good sense to call the ratio l/m the conditional relative frequency of the event B given the event A.

2. Show that the relative frequencies l/m, m/n, and l/n satisfy the equation

$$\frac{l}{m} = \frac{\dfrac{l}{n}}{\dfrac{m}{n}}$$

3. Explain why the result of part 2 suggests that Formula (2)

$$P(B|A) = \frac{P(A \cap B)}{P(A)} \qquad [P(A) \neq 0]$$

is plausible.

In certain problems, the probability of an event B occurring given that A has occurred, written $P(B|A)$, is known, and we wish to find the probability of A and B occurring. The solution to such a problem is facilitated by the use of the following formula:

Product Rule

$$P(A \cap B) = P(A) \cdot P(B|A) \tag{3}$$

This formula is obtained from (2) by multiplying both sides of the equation by $P(A)$. We illustrate the use of the product rule in the next several examples.

 APPLIED EXAMPLE 4   Seniors with Driver's Licenses   There are 300 seniors in Jefferson High School, of which 140 are males. It is known that 80% of the males and 60% of the females have their driver's license. If a student is selected at random from this senior class, what is the probability that the student is
a. A male and has a driver's license?
b. A female who does not have a driver's license?

Solution

a. Let M denote the event that the student is a male and let D denote the event that the student has a driver's license. Then,

$$P(M) = \frac{140}{300} \quad \text{and} \quad P(D|M) = .8$$

The event that the student selected at random is a male and has a driver's license is $M \cap D$ and, by the product rule, the probability of this event occurring is given by

$$P(M \cap D) = P(M) \cdot P(D|M)$$

$$= \left(\frac{140}{300}\right)(.8) = \frac{28}{75}$$

b. Let F denote the event that the student is a female. Then D^c is the event that the student does not have a driver's license. We have

$$P(F) = \frac{160}{300} \quad \text{and} \quad P(D^c|F) = 1 - .6 = .4$$

Note that we have used the rule of complements in the computation of $P(D^c|F)$. The event that the student selected at random is a female and does not have a driver's license is $F \cap D^c$ and so, by the product rule, the probability of this event occurring is given by

$$P(F \cap D^c) = P(F) \cdot P(D^c|F)$$

$$= \left(\frac{160}{300}\right)(.4) = \frac{16}{75}$$

EXAMPLE 5   Two cards are drawn without replacement from a well-shuffled deck of 52 playing cards. What is the probability that the first card drawn is an ace and the second card drawn is a face card?

Solution   Let A denote the event that the first card drawn is an ace and let F denote the event that the second card drawn is a face card. Then $P(A) = \frac{4}{52}$. After drawing the first card, there are 51 cards left in the deck, of which 12 are face cards. Therefore, the probability of drawing a face card given that the first card drawn was an ace is given by

$$P(F|A) = \frac{12}{51}$$

By the product rule, the probability that the first card drawn is an ace and the second card drawn is a face card is given by

$$P(A \cap F) = P(A) \cdot P(F|A)$$

$$= \left(\frac{4}{52}\right)\left(\frac{12}{51}\right) = \frac{4}{221}$$

The product rule may be generalized to the case involving any finite number of events. For example, in the case involving the three events E, F, and G, it may be shown that

$$P(E \cap F \cap G) = P(E) \cdot P(F|E) \cdot P(G|E \cap F) \tag{4}$$

■ More on Tree Diagrams

Formula (4) and its generalizations may be used to help us solve problems that involve finite stochastic processes. More specifically, a **finite stochastic process** is an experiment consisting of a finite number of stages in which the outcomes and associated probabilities of each stage depend on the outcomes and associated probabilities of the preceding stages.

We can use tree diagrams to help us solve problems involving finite stochastic processes. Consider, for example, the experiment consisting of drawing 2 cards without replacement from a well-shuffled deck of 52 playing cards. What is the probability that the second card drawn is a face card?

We may think of this experiment as a stochastic process with two stages. The events associated with the first stage are F, that the card drawn is a face card, and F^c, that the card drawn is not a face card. Since there are 12 face cards, we have

$$P(F) = \frac{12}{52} \quad \text{and} \quad P(F^c) = 1 - \frac{12}{52} = \frac{40}{52}$$

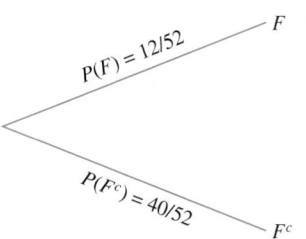

FIGURE 16
F is the probability that a face card is drawn.

The outcomes of this trial, together with the associated probabilities, may be represented along two branches of a tree diagram as shown in Figure 16.

In the second trial, we again have two events: G, that the card drawn is a face card, and G^c, that the card drawn is not a face card. But the outcome of the second trial depends on the outcome of the first trial. For example, if the first card drawn was a face card, then the event G that the second card drawn is a face card has probability given by the *conditional probability* $P(G|F)$. Since the occurrence of a face card in the first draw leaves 11 face cards in a deck of 51 cards for the second draw, we see that

$$P(G|F) = \frac{11}{51} \qquad \text{The probability of drawing a face card given that a face card has already been drawn}$$

Similarly, the occurrence of a face card in the first draw leaves 40 that are other than face cards in a deck of 51 cards for the second draw. Therefore, the probability of

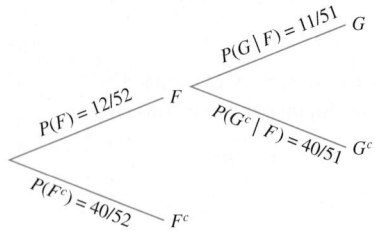

FIGURE 17
G is the probability that the second
card drawn is a face card.

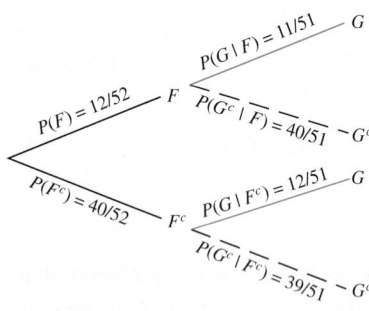

FIGURE 18
Tree diagram showing the two trials of
the experiment

drawing other than a face card in the second draw given that the first card drawn is
a face card is

$$P(G^c|F) = \frac{40}{51}$$

Using these results, we extend the tree diagram of Figure 16 by displaying another
two branches of the tree growing from its upper branch (Figure 17).

To complete the tree diagram, we compute $P(G|F^c)$ and $P(G^c|F^c)$, the condi-
tional probabilities that the second card drawn is a face card and other than a face
card, respectively, given that the first card drawn is not a face card. We find that

$$P(G|F^c) = \frac{12}{51} \quad \text{and} \quad P(G^c|F^c) = \frac{39}{51}$$

This leads to the completion of the tree diagram, shown in Figure 18, where the
branches of the tree that lead to the two outcomes of interest have been highlighted.

Having constructed the tree diagram associated with the problem, we are now
in a position to answer the question posed earlier: "What is the probability of the second card being a face card?" Observe that Figure 18 shows the two ways in which a
face card may result in the second draw—namely, the two Gs on the extreme right
of the diagram.

Now, by the product rule, the probability that the second card drawn is a
face card and the first card drawn is a face card (this is represented by the upper
branch) is

$$P(G \cap F) = P(F) \cdot P(G|F)$$

Similarly, the probability that the second card drawn is a face card and the first card
drawn is other than a face card (this corresponds to the other branch) is

$$P(G \cap F^c) = P(F^c) \cdot P(G|F^c)$$

Observe that each of these probabilities is obtained by taking the *product of the
probabilities appearing on the respective branch.* Since $G \cap F$ and $G \cap F^c$ are
mutually exclusive events (why?), the probability that the second card drawn is a
face card is given by

$$P(G \cap F) + P(G \cap F^c) = P(F) \cdot P(G|F) + P(F^c) \cdot P(G|F^c)$$

or, upon replacing the probabilities on the right of the expression by their numerical
values,

$$P(G \cap F) + P(G \cap F^c) = \left(\frac{12}{52}\right)\left(\frac{11}{51}\right) + \left(\frac{40}{52}\right)\left(\frac{12}{51}\right)$$

$$= \frac{3}{13}$$

APPLIED EXAMPLE 6   Quality Control The picture tubes for the
Pulsar 19-inch color television sets are manufactured in three locations and
then shipped to the main plant of Vista Vision for final assembly. Plants A, B,
and C supply 50%, 30%, and 20%, respectively, of the picture tubes used by the
company. The quality-control department of the company has determined that
1% of the picture tubes produced by plant A are defective, whereas 2% of the picture tubes produced by plants B and C are defective. What is the probability that
a randomly selected Pulsar 19-inch color television set will have a defective picture tube?

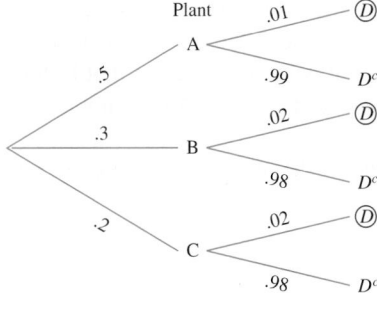

FIGURE 19
Tree diagram showing the probabilities of producing defective picture tubes at each plant

Solution Let A, B, and C denote the events that the set chosen has a picture tube manufactured in plant A, plant B, and plant C, respectively. Also, let D denote the event that a set has a defective picture tube. Using the given information, we draw the tree diagram shown in Figure 19. (The events that result in a set with a defective picture tube being selected are circled.) Taking the product of the probabilities along each branch leading to such an event and then adding them yields the probability that a set chosen at random has a defective picture tube. Thus, the required probability is given by

$$(.5)(.01) + (.3)(.02) + (.2)(.02) = .005 + .006 + .004$$
$$= .015 \qquad \blacksquare$$

APPLIED EXAMPLE 7   Quality Control A box contains eight 9-volt batteries, of which two are known to be defective. The batteries are selected one at a time without replacement and tested until a nondefective one is found. What is the probability that the number of batteries tested is (a) one, (b) two, and (c) three?

Solution We may view this experiment as a multistage process with up to three stages. In the first stage, a battery is selected with a probability of $\frac{6}{8}$ of its being nondefective and a probability of $\frac{2}{8}$ of its being defective. If the battery selected is good, the experiment is terminated. Otherwise, a second battery is selected with probabilities of $\frac{6}{7}$ and $\frac{1}{7}$, respectively, of its being nondefective and defective. If the second battery selected is good, the experiment is terminated. Otherwise, a third battery is selected with probabilities of 1 and 0, respectively, of its being nondefective and defective. The tree diagram associated with this experiment is shown in Figure 20, where N denotes the event that the battery selected is non-defective and D denotes the event that the battery selected is defective.

FIGURE 20
In this experiment, batteries are selected until a nondefective one is found.

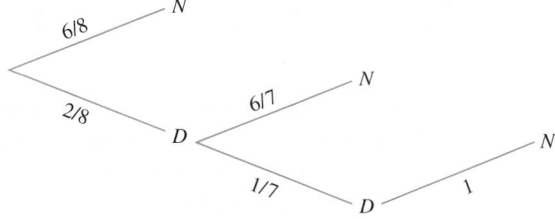

With the aid of the tree diagram we see that (a) the probability that only one battery is selected is $\frac{6}{8}$, (b) the probability that two batteries are selected is $\left(\frac{2}{8}\right)\left(\frac{6}{7}\right)$, or $\frac{3}{14}$, and (c) the probability that three batteries are selected is $\left(\frac{2}{8}\right)\left(\frac{1}{7}\right)(1) = \frac{1}{28}$. $\qquad \blacksquare$

Independent Events

Let's return to the experiment of drawing 2 cards in succession without replacement from a well-shuffled deck of 52 playing cards as considered in Example 5. Let E denote the event that the first card drawn is not a face card and let F denote the event that the second card drawn is a face card. It is intuitively clear that the events E and F are *not* independent of each other, because whether or not the first card drawn is a face card affects the likelihood that the second card drawn is a face card.

Next, let's consider the experiment of tossing a coin twice and observing the outcomes: If H denotes the event that the first toss produces "heads" and T denotes the event that the second toss produces "tails," then it is intuitively clear that H and T *are* independent of each other because the outcome of the first toss does not affect the outcome of the second.

In general, two events A and B are independent if the outcome of one does not affect the outcome of the other. Thus, we have

---

Independent Events
If A and B are **independent events**, then

$$P(A|B) = P(A) \quad \text{and} \quad P(B|A) = P(B)$$

Using the product rule, we can find a simple test to determine the independence of two events. Suppose that A and B are independent and that $P(A) \neq 0$ and $P(B) \neq 0$. Then

$$P(B|A) = P(B)$$

Thus, by the product rule, we have

$$P(A \cap B) = P(A) \cdot P(B|A) = P(A) \cdot P(B)$$

Conversely, if this equation holds then it can be seen that $P(B|A) = P(B)$; that is, A and B are independent. Accordingly, we have the following test for the independence of two events.

Test for the Independence of Two Events
Two events A and B are independent if and only if

$$P(A \cap B) = P(A) \cdot P(B) \tag{5}$$

 Do not confuse *independent* events with *mutually exclusive* events. The former pertains to how the occurrence of one event affects the occurrence of another event, whereas the latter pertains to the question of whether the events can occur at the same time.

EXAMPLE 8   Consider the experiment consisting of tossing a fair coin twice and observing the outcomes. Show that the event of "heads" in the first toss and "tails" in the second toss are independent events.

Solution   Let A denote the event that the outcome of the first toss is a *head*, and let B denote the event that the outcome of the second toss is a *tail*. The sample space of the experiment is

$$S = \{(\text{H, H}), (\text{H, T}), (\text{T, H}), (\text{T, T})\}$$
$$A = \{(\text{H, H}), (\text{H, T})\}$$
$$B = \{(\text{H, T}), (\text{T, T})\}$$

so that

$$A \cap B = \{(\text{H, T})\}$$

Next, we compute

$$P(A \cap B) = \frac{1}{4} \qquad P(A) = \frac{1}{2} \qquad P(B) = \frac{1}{2}$$

and observe that Equation (5) is satisfied in this case. Hence A and B are independent events, as we set out to show. ∎

APPLIED EXAMPLE 9   Medical Surveys   A survey conducted by an independent agency for the National Lung Society found that, of 2000 women, 680 were heavy smokers and 50 had emphysema. Of those who had emphysema, 42 were also heavy smokers. Using the data in this survey, determine whether the events "being a heavy smoker" and "having emphysema" are independent events.

Solution   Let A denote the event that a woman is a heavy smoker and let B denote the event that a woman has emphysema. Then, the probability that a woman is a heavy smoker and has emphysema is given by

$$P(A \cap B) = \frac{42}{2000} = .021$$

Next,

$$P(A) = \frac{680}{2000} = .34 \quad \text{and} \quad P(B) = \frac{50}{2000} = .025$$

so that

$$P(A) \cdot P(B) = (.34)(.025) = .0085$$

Since $P(A \cap B) \neq P(A) \cdot P(B)$, we conclude that A and B are not independent events. ∎

The solution of many practical problems involves more than two independent events. In such cases we use the following result.

EXPLORE & DISCUSS

Let E and F be independent events in a sample space S. Are E^c and F^c independent?

Independence of More Than Two Events
If $E_1, E_2, \ldots, E_n$ are independent events, then

$$P(E_1 \cap E_2 \cap \cdots \cap E_n) = P(E_1) \cdot P(E_2) \cdot \cdots \cdot P(E_n) \qquad \textbf{(6)}$$

Formula (6) states that the probability of the simultaneous occurrence of n independent events is equal to the product of the probabilities of the n events.

It is important to note that the mere requirement that the n events $E_1, E_2, \ldots, E_n$ satisfy (6) is not sufficient to guarantee that the n events are indeed independent. However, a criterion does exist for determining the independence of n events and may be found in more advanced texts on probability.

EXAMPLE 10   It is known that the three events A, B, and C are independent and that $P(A) = .2$, $P(B) = .4$, and $P(C) = .5$. Compute:
a. $P(A \cap B)$ **b.** $P(A \cap B \cap C)$

Solution   Using Formulas (5) and (6), we find
a. $P(A \cap B) = P(A) \cdot P(B)$
$$= (.2)(.4) = .08$$
b. $P(A \cap B \cap C) = P(A) \cdot P(B) \cdot P(C)$
$$= (.2)(.4)(.5) = .04$$

APPLIED EXAMPLE 11   Quality Control   The Acrosonic model F loudspeaker system has four loudspeaker components: a woofer, a midrange, a tweeter, and an electrical crossover. The quality-control manager of Acrosonic has determined that on the average 1% of the woofers, 0.8% of the midranges, and 0.5% of the tweeters are defective, while 1.5% of the electrical crossovers are defective. Determine the probability that a loudspeaker system selected at random as it comes off the assembly line (and before final inspection) is not defective. Assume that the defects in the manufacturing of the components are unrelated.

Solution   Let A, B, C, and D denote, respectively, the events that the woofer, the midrange, the tweeter, and the electrical crossover are defective. Then,

$$P(A) = .01 \qquad P(B) = .008 \qquad P(C) = .005 \qquad P(D) = .015$$

and the probabilities of the corresponding complementary events are

$$P(A^c) = .99 \qquad P(B^c) = .992 \qquad P(C^c) = .995 \qquad P(D^c) = .985$$

The event that a loudspeaker system selected at random is not defective is given by $A^c \cap B^c \cap C^c \cap D^c$. Because the events A, B, C, and D (and therefore also A^c, B^c, C^c, and D^c) are assumed to be independent, we find that the required probability is given by

$$P(A^c \cap B^c \cap C^c \cap D^c) = P(A^c) \cdot P(B^c) \cdot P(C^c) \cdot P(D^c)$$
$$= (.99)(.992)(.995)(.985)$$
$$\approx .96$$

7.5   Self-Check Exercises

1. Let A and B be events in a sample space S such that $P(A) = .4$, $P(B) = .8$, and $P(A \cap B) = .3$. Find:
   a. $P(A|B)$ **b.** $P(B|A)$

2. According to a survey cited in *Newsweek*, 29.7% of married survey respondents who were married between the ages of 20 and 22 (inclusive), 26.9% of those married between the ages of 23 and 27, and 45.1% of those married at age 28 or older said that "their marriage was less than 'very happy'." Suppose that a survey respondent from each of the three age groups was selected at random. What is the probability that all three respondents said that they were "less than very happy"?

Source: Marc Bain, *Newsweek.*

Solutions to Self-Check Exercises 7.5 can be found on page 429.

7.5 Concept Questions

1. What is conditional probability? Illustrate the concept with an example.

2. If A and B are events in an experiment and $P(A) \neq 0$, then what is the formula for computing $P(B|A)$?

3. If A and B are events in an experiment and the conditional probability $P(B|A)$ is known, give the formula that can be used to compute the event that both A and B will occur.

4. **a.** What is the test for determining the independence of two events?
   b. What is the difference between mutually exclusive events and independent events?

7.5 Exercises

1. Let A and B be two events in a sample space S such that $P(A) = .6$, $P(B) = .5$, and $P(A \cap B) = .2$. Find:
   a. $P(A|B)$　　　　**b.** $P(B|A)$

2. Let A and B be two events in a sample space S such that $P(A) = .4$, $P(B) = .6$, and $P(A \cap B) = .3$. Find:
   a. $P(A|B)$　　　　**b.** $P(B|A)$

3. Let A and B be two events in a sample space S such that $P(A) = .6$ and $P(B|A) = .5$. Find $P(A \cap B)$.

4. Let A and B be the events described in Exercise 1. Find:
   a. $P(A|B^c)$　　　　**b.** $P(B|A^c)$
   Hint: $(A \cap B^c) \cup (A \cap B) = A$.

In Exercises 5–8, determine whether the events A and B are independent.

5. $P(A) = .3$, $P(B) = .6$, $P(A \cap B) = .18$

6. $P(A) = .6$, $P(B) = .8$, $P(A \cap B) = .2$

7. $P(A) = .5$, $P(B) = .7$, $P(A \cup B) = .85$

8. $P(A^c) = .3$, $P(B^c) = .4$, $P(A \cap B) = .42$

9. If A and B are independent events, $P(A) = .4$, and $P(B) = .6$, find
   a. $P(A \cap B)$　　　　**b.** $P(A \cup B)$

10. If A and B are independent events, $P(A) = .35$, and $P(B) = .45$, find
    a. $P(A \cap B)$　　　　**b.** $P(A \cup B)$

11. The accompanying tree diagram represents an experiment consisting of two trials:

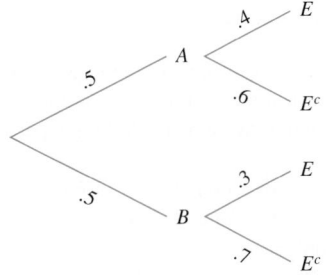

Use the diagram to find
   a. $P(A)$　　　　　　**b.** $P(E|A)$
   c. $P(A \cap E)$　　　**d.** $P(E)$
   e. Does $P(A \cap E) = P(A) \cdot P(E)$?
   f. Are A and E independent events?

12. The accompanying tree diagram represents an experiment consisting of two trials. Use the diagram to find
    a. $P(A)$　　　　　　**b.** $P(E|A)$
    c. $P(A \cap E)$　　　**d.** $P(E)$
    e. Does $P(A \cap E) = P(A) \cdot P(E)$?
    f. Are A and E independent events?

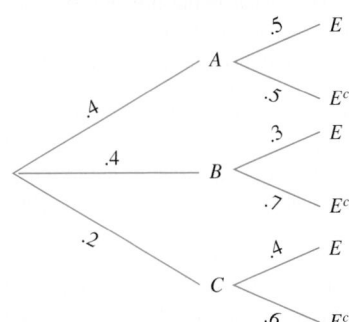

13. An experiment consists of two trials. The outcomes of the first trial are A and B with probabilities of occurring equal to .4 and .6. There are also two outcomes, C and D, in the second trial with probabilities of .3 and .7. Draw a tree diagram representing this experiment. Use this diagram to find
    a. $P(A)$ **b.** $P(C|A)$
    c. $P(A \cap C)$ **d.** $P(C)$
    e. Does $P(A \cap C) = P(A) \cdot P(C)$?
    f. Are A and C independent events?

14. An experiment consists of two trials. The outcomes of the first trial are A, B, and C, with probabilities of occurring equal to .2, .5, and .3, respectively. The outcomes of the second trial are E and F, with probabilities of occurring equal to .6 and .4. Draw a tree diagram representing this experiment. Use this diagram to find
    a. $P(B)$ **b.** $P(F|B)$
    c. $P(B \cap F)$ **d.** $P(F)$
    e. Does $P(B \cap F) = P(B) \cdot P(F)$?
    f. Are B and F independent events?

15. A pair of fair dice is cast. Let E denote the event that the number falling uppermost in the first die is 5, and let F denote the event that the sum of the numbers falling uppermost is 10.
    a. Compute $P(F)$.   **b.** Compute $P(E \cap F)$.
    c. Compute $P(F|E)$.   **d.** Compute $P(E)$.
    e. Are E and F independent events?

16. A pair of fair dice is cast. Let E denote the event that the number falling uppermost in the first die is 4, and let F denote the event that the sum of the numbers falling uppermost is 6.
    a. Compute $P(F)$.   **b.** Compute $P(E \cap F)$.
    c. Compute $P(F|E)$.   **d.** Compute $P(E)$.
    e. Are E and F independent events?

17. A pair of fair dice is cast. What is the probability that the sum of the numbers falling uppermost is less than 9, given that one of the numbers is a 6?

18. A pair of fair dice is cast. What is the probability that the number landing uppermost on the first die is a 4 if it is known that the sum of the numbers landing uppermost is 7?

19. A pair of fair dice is cast. Let E denote the event that the number landing uppermost on the first die is a 3, and let F denote the event that the sum of the numbers landing uppermost is 7. Determine whether E and F are independent events.

20. A pair of fair dice is cast. Let E denote the event that the number landing uppermost on the first die is a 3, and let F denote the event that the sum of the numbers landing uppermost is 6. Determine whether E and F are independent events.

21. A card is drawn from a well-shuffled deck of 52 playing cards. Let E denote the event that the card drawn is black and let F denote the event that the card drawn is a spade. Determine whether E and F are independent events. Give an intuitive explanation for your answer.

22. A card is drawn from a well-shuffled deck of 52 playing cards. Let E denote the event that the card drawn is an ace and let F denote the event that the card drawn is a diamond. Determine whether E and F are independent events. Give an intuitive explanation for your answer.

23. **PRODUCT RELIABILITY**   The probability that a battery will last 10 hr or more is .80, and the probability that it will last 15 hr or more is .15. Given that a battery has lasted 10 hr, find the probability that it will last 15 hr or more.

24. Two cards are drawn without replacement from a well-shuffled deck of 52 playing cards.
    a. What is the probability that the first card drawn is a heart?
    b. What is the probability that the second card drawn is a heart if the first card drawn was not a heart?
    c. What is the probability that the second card drawn is a heart if the first card drawn was a heart?

25. Five black balls and four white balls are placed in an urn. Two balls are then drawn in succession. What is the probability that the second ball drawn is a white ball if
    a. The second ball is drawn without replacing the first?
    b. The first ball is replaced before the second is drawn?

26. **AUDITING TAX RETURNS**   A tax specialist has estimated the probability that a tax return selected at random will be audited is .02. Furthermore, he estimates the probability that an audited return will result in additional assessments being levied on the taxpayer is .60. What is the probability that a tax return selected at random will result in additional assessments being levied on the taxpayer?

27. **STUDENT ENROLLMENT**   At a certain medical school, $\frac{1}{7}$ of the students are from a minority group. Of those students who belong to a minority group, $\frac{1}{3}$ are black.
    a. What is the probability that a student selected at random from this medical school is black?
    b. What is the probability that a student selected at random from this medical school is black if it is known that the student is a member of a minority group?

28. **EDUCATIONAL LEVEL OF VOTERS**   In a survey of 1000 eligible voters selected at random, it was found that 80 had a college degree. Additionally, it was found that 80% of those who had a college degree voted in the last presidential election, whereas 55% of the people who did not have a college degree voted in the last presidential election. Assuming that the poll is representative of all eligible voters, find the probability that an eligible voter selected at random
    a. Had a college degree and voted in the last presidential election.

b. Did not have a college degree and did not vote in the last presidential election.

c. Voted in the last presidential election.

d. Did not vote in the last presidential election.

29. Three cards are drawn without replacement from a well-shuffled deck of 52 playing cards. What is the probability that the third card drawn is a diamond?

30. A coin is tossed three times. What is the probability that the coin will land heads

  a. At least twice?

  b. On the second toss, given that heads were thrown on the first toss?

  c. On the third toss, given that tails were thrown on the first toss?

31. In a three-child family, what is the probability that all three children are girls given that one of the children is a girl? (Assume that the probability of a boy being born is the same as the probability of a girl being born.)

32. QUALITY CONTROL An automobile manufacturer obtains the microprocessors used to regulate fuel consumption in its automobiles from three microelectronic firms: A, B, and C. The quality-control department of the company has determined that 1% of the microprocessors produced by firm A are defective, 2% of those produced by firm B are defective, and 1.5% of those produced by firm C are defective. Firms A, B, and C supply 45%, 25%, and 30%, respectively, of the microprocessors used by the company. What is the probability that a randomly selected automobile manufactured by the company will have a defective microprocessor?

33. CAR THEFT Figures obtained from a city's police department seem to indicate that, of all motor vehicles reported as stolen, 64% were stolen by professionals whereas 36% were stolen by amateurs (primarily for joy rides). Of those vehicles presumed stolen by professionals, 24% were recovered within 48 hr, 16% were recovered after 48 hr, and 60% were never recovered. Of those vehicles presumed stolen by amateurs, 38% were recovered within 48 hr, 58% were recovered after 48 hr, and 4% were never recovered.

  a. Draw a tree diagram representing these data.

  b. What is the probability that a vehicle stolen by a professional in this city will be recovered within 48 hr?

  c. What is the probability that a vehicle stolen in this city will never be recovered?

34. HOUSING LOANS The chief loan officer of La Crosse Home Mortgage Company summarized the housing loans extended by the company in 2003 according to type and term of the loan. Her list shows that 70% of the loans were fixed-rate mortgages (F), 25% were adjustable-rate mortgages (A), and 5% belong to some other category (O) (mostly second trust-deed loans and loans extended under the graduated payment plan). Of the fixed-rate mortgages, 80% were 30-yr loans and 20% were 15-yr loans; of the adjustable-rate mortgages, 40% were 30-yr loans and 60% were 15-yr loans; finally, of

the other loans extended, 30% were 20-yr loans, 60% were 10-yr loans, and 10% were for a term of 5 yr or less.

  a. Draw a tree diagram representing these data.

  b. What is the probability that a home loan extended by La Crosse has an adjustable rate and is for a term of 15 yr?

  c. What is the probability that a home loan extended by La Crosse is for a term of 15 yr?

35. COLLEGE ADMISSIONS The admissions office of a private university released the following admission data for the preceding academic year: From a pool of 3900 male applicants, 40% were accepted by the university and 40% of these subsequently enrolled. Additionally, from a pool of 3600 female applicants, 45% were accepted by the university and 40% of these subsequently enrolled. What is the probability that

  a. A male applicant will be accepted by and subsequently will enroll in the university?

  b. A student who applies for admissions will be accepted by the university?

  c. A student who applies for admission will be accepted by the university and subsequently will enroll?

36. QUALITY CONTROL A box contains two defective Christmas tree lights that have been inadvertently mixed with eight nondefective lights. If the lights are selected one at a time without replacement and tested until both defective lights are found, what is the probability that both defective lights will be found after three trials?

37. QUALITY CONTROL It is estimated that 0.80% of a large consignment of eggs in a certain supermarket is broken.

  a. What is the probability that a customer who randomly selects a dozen of these eggs receives at least one broken egg?

  b. What is the probability that a customer who selects these eggs at random will have to check three cartons before finding a carton without any broken eggs? (Each carton contains a dozen eggs.)

38. STUDENT FINANCIAL AID The accompanying data were obtained from the financial aid office of a certain university:

	Receiving Financial Aid	Not Receiving Financial Aid	Total
Undergraduates	4,222	3,898	8,120
Graduates	1,879	731	2,610
Total	6,101	4,629	10,730

Let A be the event that a student selected at random from this university is an undergraduate student, and let B be the event that a student selected at random is receiving financial aid.

  a. Find each of the following probabilities: $P(A)$, $P(B)$, $P(A \cap B)$, $P(B|A)$, and $P(B|A^c)$.

  b. Are the events A and B independent events?

39. EMPLOYEE EDUCATION AND INCOME The personnel department of Franklin National Life Insurance Company compiled the accompanying data regarding the income and education of its employees:

	Income $50,000 or Below	Income Above $50,000
Noncollege Graduate	2040	840
College Graduate	400	720

Let A be the event that a randomly chosen employee has a college degree and B the event that the chosen employee's income is more than $50,000.
a. Find each of the following probabilities: $P(A)$, $P(B)$, $P(A \cap B)$, $P(B|A)$, and $P(B|A^c)$.
b. Are the events A and B independent events?

40. Two cards are drawn without replacement from a well-shuffled deck of 52 cards. Let A be the event that the first card drawn is a heart, and let B be the event that the second card drawn is a red card. Show that the events A and B are dependent events.

41. MEDICAL RESEARCH A nationwide survey conducted by the National Cancer Society revealed the following information. Of 10,000 people surveyed, 3200 were "heavy coffee drinkers" and 160 had cancer of the pancreas. Of those who had cancer of the pancreas, 132 were heavy coffee drinkers. Using the data in this survey, determine whether the events "being a heavy coffee drinker" and "having cancer of the pancreas" are independent events.

42. SWITCHING INTERNET SERVICE PROVIDERS (ISPs) According to a survey conducted in 2004 of 1000 American adults with Internet access, one in four households plan to switch ISPs in the next 6 months. Of those who plan to switch, 1% of the households are likely to switch to a satellite connection, 27% to digital subscriber line (DSL), 28% to cable modem, 35% to dial-up modem, and 9% don't know which service provider they will switch to.
a. If a person participating in the survey is chosen at random, what is the probability that he or she will switch to a dial-up modem connection?
b. If a person in the survey has planned to switch ISPs, what is the probability that he or she will upgrade to high-speed service (satellite, DSL, or cable)?

Source: Ipsos-Insight

43. RELIABILITY OF SECURITY SYSTEMS Before being allowed to enter a maximum-security area at a military installation, a person must pass three identification tests: a voice-pattern test, a fingerprint test, and a handwriting test. If the reliability of the first test is 97%, the reliability of the second test is 98.5%, and that of the third is 98.5%, what is the probability that this security system will allow an improperly identified person to enter the maximum-security area?

44. RELIABILITY OF A HOME THEATER SYSTEM In a home theater system, the probability that the video components need repair within 1 yr is .01, the probability that the electronic components need repair within 1 yr is .005, and the probability that the audio components need repair within 1 yr is .001. As-

suming the probabilities are independent, find the probability that
a. At least one of these components will need repair within 1 yr.
b. Exactly one of these components will need repair within 1 yr.

45. PROBABILITY OF TRANSPLANT REJECTION The independent probabilities that the three patients who are scheduled to receive kidney transplants at General Hospital will suffer rejection are $\frac{1}{2}$, $\frac{1}{3}$, and $\frac{1}{10}$. Find the probabilities that
a. At least one patient will suffer rejection.
b. Exactly two patients will suffer rejection.

46. QUALITY CONTROL Copykwik has four photocopy machines: A, B, C, and D. The probability that a given machine will break down on a particular day is

$$P(A) = \frac{1}{50} \quad P(B) = \frac{1}{60} \quad P(C) = \frac{1}{75} \quad P(D) = \frac{1}{40}$$

Assuming independence, what is the probability on a particular day that
a. All four machines will break down?
b. None of the machines will break down?
c. Exactly one machine will break down?

47. PRODUCT RELIABILITY The proprietor of Cunningham's Hardware Store has decided to install floodlights on the premises as a measure against vandalism and theft. If the probability is .01 that a certain brand of floodlight will burn out within a year, find the minimum number of floodlights that must be installed to ensure that the probability that at least one of them will remain functional within the year is at least .99999. (Assume that the floodlights operate independently.)

48. Let E be any event in a sample space S.
a. Are E and S independent? Explain your answer.
b. Are E and $\emptyset$ independent? Explain your answer.

49. Suppose the probability that an event will occur in one trial is p. Show that the probability that the event will occur at least once in n independent trials is $1 - (1 - p)^n$.

50. Let E and F be mutually exclusive events and suppose $P(F) \neq 0$. Find $P(E|F)$ and interpret your result.

51. Let E and F be events such that $F \subset E$. Find $P(E|F)$ and interpret your result.

52. Suppose that A and B are mutually exclusive events and that $P(A \cup B) \neq 0$. What is $P(A|A \cup B)$?

In Exercises 53–56, determine whether the statement is true or false. If it is true, explain why it is true. If it is false, give an example to show why it is false.

53. If A and B are mutually exclusive and $P(B) \neq 0$, then $P(A|B) = 0$.

54. If A is an event of an experiment, then $P(A|A^c) \neq 0$.

55. If A and B are events of an experiment, then
$$P(A \cap B) = P(A|B) \cdot P(B) = P(B|A) \cdot P(A)$$

56. If A and B are independent events with $P(A) \neq 0$ and $P(B) \neq 0$, then $A \cap B \neq \emptyset$.

7.5   Solutions to Self-Check Exercises

1. a. $P(A|B) = \dfrac{P(A \cap B)}{P(B)}$ **b.** $P(B|A) = \dfrac{P(A \cap B)}{P(A)}$

$= \dfrac{.3}{.8} = \dfrac{3}{8}$ $= \dfrac{.3}{.4} = \dfrac{3}{4}$

2. Let A, B, and C denote the events that a respondent who was married between the ages of 20 and 22, between the ages of 23 and 27, and at age 28 or older (respectively) said that they were "less than very happy." Then the probability of each of these events occurring is $P(A) = .297$, $P(B) = .269$, and $P(C) = .451$. So the probability that all three of the respondents said that they were "less than very happy" is

$$P(A) \cdot P(B) \cdot P(C) = (.297)(.269)(.451) = .036$$

7.6   Bayes' Theorem

◼ A Posteriori Probabilities

Suppose three machines, A, B, and C, produce similar engine components. Machine A produces 45% of the total components, machine B produces 30%, and machine C, 25%. For the usual production schedule, 6% of the components produced by machine A do not meet established specifications; for machine B and machine C, the corresponding figures are 4% and 3%. One component is selected at random from the total output and is found to be defective. What is the probability that the component selected was produced by machine A?

The answer to this question is found by calculating the probability *after* the outcomes of the experiment have been observed. Such probabilities are called **a posteriori probabilities** as opposed to **a priori probabilities**—probabilities that give the likelihood that an event *will* occur, the subject of the last several sections.

Returning to the example under consideration, we need to determine the a posteriori probability for the event that the component selected was produced by machine A. Toward this end, let A, B, and C denote the event that a component is produced by machine A, machine B, and machine C, respectively. We may represent this experiment with a Venn diagram (Figure 21).

The three mutually exclusive events A, B, and C form a **partition** of the sample space S; that is, aside from being mutually exclusive, their union is precisely S. The event D that a component is defective is the shaded area. Again referring to Figure 21, we see that

1. The event D may be expressed as

$$D = (A \cap D) \cup (B \cap D) \cup (C \cap D)$$

2. The event that a component is defective and is produced by machine A is given by $A \cap D$.

Thus, the a posteriori probability that a defective component selected was produced by machine A is given by

$$P(A|D) = \frac{n(A \cap D)}{n(D)}$$

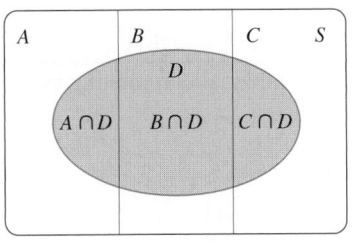

FIGURE 21
D is the event that a defective component is produced by machine A, machine B, or machine C.

Upon dividing both the numerator and the denominator by $n(S)$ and observing that the events $A \cap D$, $B \cap D$, and $C \cap D$ are mutually exclusive, we obtain

$$P(A|D) = \frac{P(A \cap D)}{P(D)}$$

$$= \frac{P(A \cap D)}{P(A \cap D) + P(B \cap D) + P(C \cap D)} \qquad (7)$$

Next, using the product rule, we may express

$$P(A \cap D) = P(A) \cdot P(D|A)$$
$$P(B \cap D) = P(B) \cdot P(D|B)$$
$$P(C \cap D) = P(C) \cdot P(D|C)$$

so that Equation (7) may be expressed in the form

$$P(A|D) = \frac{P(A) \cdot P(D|A)}{P(A) \cdot P(D|A) + P(B) \cdot P(D|B) + P(C) \cdot P(D|C)} \qquad (8)$$

which is a special case of a result known as **Bayes' theorem**.

Observe that the expression on the right of (8) involves the probabilities $P(A)$, $P(B)$, and $P(C)$ as well as the conditional probabilities $P(D|A)$, $P(D|B)$, and $P(D|C)$, all of which may be calculated in the usual fashion. In fact, by displaying these quantities on a tree diagram, we obtain Figure 22. We may compute the required probability by substituting the relevant quantities into (8), or we may make use of the following device:

$$P(A|D) = \frac{\text{Product of probabilities along the branch through } A}{\text{Sum of products of the probabilities along each branch terminating at } D}$$

In either case, we obtain

$$P(A|D) = \frac{(.45)(.06)}{(.45)(.06) + (.3)(.04) + (.25)(.03)}$$

$$\approx .58$$

Before looking at any further examples, let's state the general form of Bayes' theorem.

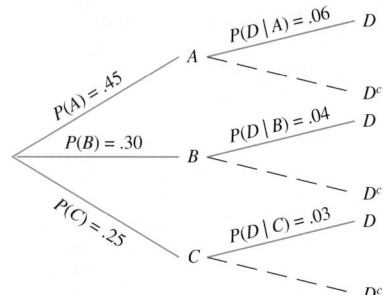

FIGURE 22
A tree diagram displaying the probabilities that a defective component is produced by machine A, machine B, or machine C

Bayes' Theorem

Let $A_1, A_2, \ldots, A_n$ be a partition of a sample space S, and let E be an event of the experiment such that $P(E) \neq 0$. Then the a posteriori probability $P(A_i|E)$ $(1 \leq i \leq n)$ is given by

$$P(A_i|E) = \frac{P(A_i) \cdot P(E|A_i)}{P(A_1) \cdot P(E|A_1) + P(A_2) \cdot P(E|A_2) + \cdots + P(A_n) \cdot P(E|A_n)} \qquad (9)$$

APPLIED EXAMPLE 1 Quality Control The picture tubes for the Pulsar 19-inch color television sets are manufactured in three locations and then shipped to the main plant of Vista Vision for final assembly. Plants A, B, and C supply 50%, 30%, and 20%, respectively, of the picture tubes used by

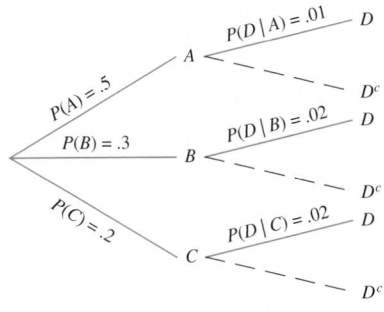

FIGURE 23

$P(C|D)$

$$= \frac{\text{Product of probabilities of branches to } D \text{ through } C}{\text{Sum of product of probabilities of branches leading to } D}$$

Vista Vision. The quality-control department of the company has determined that 1% of the picture tubes produced by plant A are defective, whereas 2% of the picture tubes produced by plants B and C are defective. If a Pulsar 19-inch color television set is selected at random and the picture tube is found to be defective, what is the probability that the picture tube was manufactured in plant C? (Compare with Example 6, page 420.)

Solution Let A, B, and C denote the event that the set chosen has a picture tube manufactured in plant A, plant B, and plant C, respectively. Also, let D denote the event that a set has a defective picture tube. Using the given information, we may draw the tree diagram shown in Figure 23. Next, using Formula (9), we find that the required a posteriori probability is given by

$$P(C|D) = \frac{P(C) \cdot P(D|C)}{P(A) \cdot P(D|A) + P(B) \cdot P(D|B) + P(C) \cdot P(D|C)}$$

$$= \frac{(.2)(.02)}{(.5)(.01) + (.3)(.02) + (.2)(.02)}$$

$$\approx .27$$

APPLIED EXAMPLE 2   Income Distributions A study was conducted in a large metropolitan area to determine the annual incomes of married couples in which the husbands were the sole providers and of those in which the husbands and wives were both employed. Table 9 gives the results of this study.

TABLE 9

Annual Family Income, $	Married Couples, %	Income Group with Both Spouses Working, %
125,000 and over	4	65
100,000–124,999	10	73
75,000–99,999	21	68
50,000–74,999	24	63
30,000–49,999	30	43
Under 30,000	11	28

a. What is the probability that a couple selected at random from this area has two incomes?

b. If a randomly chosen couple has two incomes, what is the probability that the annual income of this couple is over $125,000?

c. If a randomly chosen couple has two incomes, what is the probability that the annual income of this couple is greater than $49,999?

Solution Let A denote the event that the annual income of the couple is $125,000 and over; let B denote the event that the annual income is between $100,000 and $124,999; let C denote the event that the annual income is between $75,000 and $99,999; and so on. Finally, let F denote the event that the annual income is less than $30,000 and let T denote the event that both spouses work. The probabilities of the occurrence of these events are displayed in Figure 24.

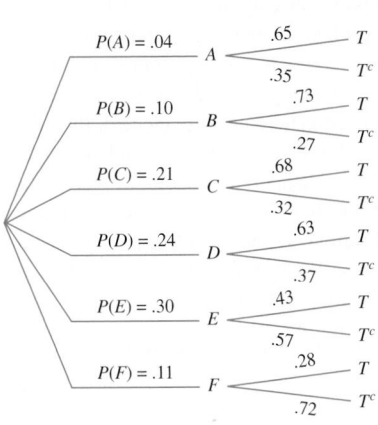

FIGURE 24

a. The probability that a couple selected at random from this group has two incomes is given by

$$P(T) = P(A) \cdot P(T|A) + P(B) \cdot P(T|B) + P(C) \cdot P(T|C)$$
$$+ P(D) \cdot P(T|D) + P(E) \cdot P(T|E) + P(F) \cdot P(T|F)$$
$$= (.04)(.65) + (.10)(.73) + (.21)(.68) + (.24)(.63)$$
$$+ (.30)(.43) + (.11)(.28)$$
$$= .5528$$

b. Using the results of part (a) and Bayes' theorem, we find that the probability that a randomly chosen couple has an annual income over $125,000, given that both spouses are working, is

$$P(A|T) = \frac{P(A) \cdot P(T|A)}{P(T)} = \frac{(.04)(.65)}{.5528}$$
$$\approx .047$$

c. The probability that a randomly chosen couple has an annual income greater than $49,999, given that both spouses are working, is

$$P(A|T) + P(B|T) + P(C|T) + P(D|T)$$
$$= \frac{P(A) \cdot P(T|A) + P(B) \cdot P(T|B) + P(C) \cdot P(T|C) + P(D) \cdot P(T|D)}{P(T)}$$
$$= \frac{(.04)(.65) + (.1)(.73) + (.21)(.68) + (.24)(.63)}{.5528}$$
$$\approx .711$$

7.6   Self-Check Exercises

1. The accompanying tree diagram represents a two-stage experiment. Use the diagram to find $P(B|D)$.

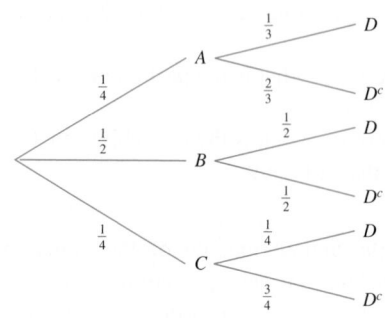

2. In a recent presidential election, it was estimated that the probability that the Republican candidate would be elected was $\frac{3}{5}$ and therefore the probability that the Democratic candidate would be elected was $\frac{2}{5}$ (the two Independent candidates were given little chance of being elected). It was also estimated that if the Republican candidate were elected, then the probability that research for a new manned bomber would continue was $\frac{4}{5}$. But if the Democratic candidate were successful, then the probability that the research would continue was $\frac{3}{10}$. Research was terminated shortly after the successful presidential candidate took office. What is the probability that the Republican candidate won that election?

Solutions to Self-Check Exercises 7.6 can be found on page 438.

7.6 Concept Questions

1. What are a priori probabilities and a posteriori probabilities? Give an example of each.

2. Suppose the events A, B, and C form the partition of a sample space S, and suppose E is an event of an experiment such that $P(E) \neq 0$. Use Bayes' theorem to write the formula for the a posteriori probability $P(A|E)$.

3. Refer to Concept Question 2. If E is the event that a product was produced in factory A, factory B, or factory C and $P(E) \neq 0$, what does $P(A|E)$ represent?

7.6 Exercises

In Exercises 1–3, refer to the accompanying Venn diagram. An experiment in which the three mutually exclusive events A, B, and C form a partition of the uniform sample space S is depicted in the diagram.

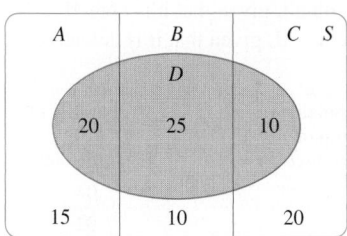

1. Using the information given in the Venn diagram, draw a tree diagram illustrating the probabilities of the events A, B, C, and D.

2. Find: **a.** $P(D)$ **b.** $P(A|D)$

3. Find: **a.** $P(D^c)$ **b.** $P(B|D^c)$

In Exercises 4–6, refer to the accompanying Venn diagram. An experiment in which the three mutually exclusive events A, B, and C form a partition of the uniform sample space S is depicted in the diagram.

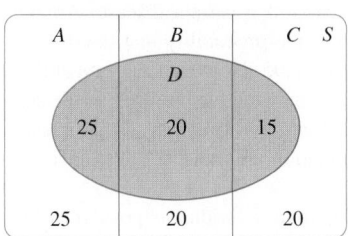

4. Using the information given in the Venn diagram, draw a tree diagram illustrating the probabilities of the events A, B, C, and D.

5. Find: **a.** $P(D)$ **b.** $P(B|D)$

6. Find: **a.** $P(D^c)$ **b.** $P(B|D^c)$

7. The accompanying tree diagram represents a two-stage experiment. Use the diagram to find

    a. $P(A) \cdot P(D|A)$ **b.** $P(B) \cdot P(D|B)$

    c. $P(A|D)$

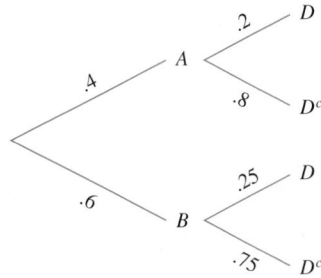

8. The accompanying tree diagram represents a two-stage experiment. Use the diagram to find

    a. $P(A) \cdot P(D|A)$ **b.** $P(B) \cdot P(D|B)$

    c. $P(A|D)$

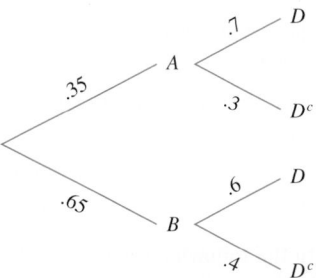

9. The accompanying tree diagram represents a two-stage experiment. Use the diagram to find

    a. $P(A) \cdot P(D|A)$ **b.** $P(B) \cdot P(D|B)$

    c. $P(C) \cdot P(D|C)$ **d.** $P(A|D)$

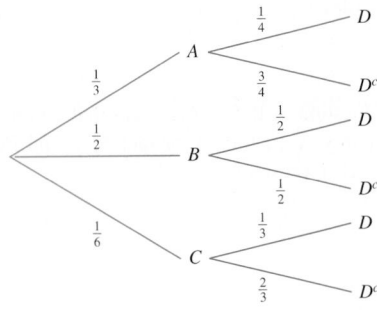

10. The accompanying tree diagram represents a two-stage experiment. Use this diagram to find
   a. $P(A \cap D)$ **b.** $P(B \cap D)$ **c.** $P(C \cap D)$ **d.** $P(D)$
   e. Verify:

$$P(A|D) = \frac{P(A \cap D)}{P(D)}$$

$$= \frac{P(A) \cdot P(D|A)}{P(A) \cdot P(D|A) + P(B) \cdot P(D|B) + P(C) \cdot P(D|C)}$$

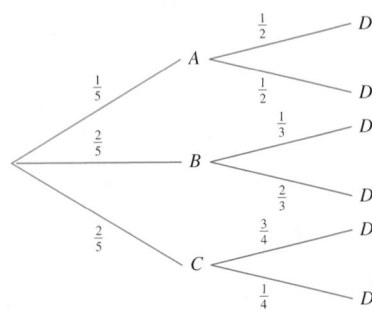

In Exercises 11–14, refer to the following experiment: Two cards are drawn in succession without replacement from a standard deck of 52 cards.

11. What is the probability that the first card is a heart given that the second card is a heart?

12. What is the probability that the first card is a heart given that the second card is a diamond?

13. What is the probability that the first card is a jack given that the second card is an ace?

14. What is the probability that the first card is a face card given that the second card is an ace?

In Exercises 15–18, refer to the following experiment: Urn A contains four white and six black balls. Urn B contains three white and five black balls. A ball is drawn from urn A and then transferred to urn B. A ball is then drawn from urn B.

15. Represent the probabilities associated with this two-stage experiment in the form of a tree diagram.

16. What is the probability that the transferred ball was white given that the second ball drawn was white?

17. What is the probability that the transferred ball was black given that the second ball drawn was white?

18. What is the probability that the transferred ball was black given that the second ball drawn was black?

19. Politics   The 1992 U.S. Senate was composed of 57 Democrats and 43 Republicans. Of the Democrats, 38 served in the military, whereas 28 of the Republicans had seen military service. If a senator selected at random had served in the military, what is the probability that he was Republican? *Note:* No congresswoman had served in the military.

20. Quality Control   Jansen Electronics has four machines that produce an identical component for use in its videocassette players. The proportion of the components produced by each machine and the probability of that component being defective are shown in the accompanying table. What is the probability that a component selected at random
   a. Is defective?
   b. Was produced by machine I, given that it is defective?
   c. Was produced by machine II, given that it is defective?

Machine	Proportion of Components Produced	Probability of Defective Component
I	.15	.04
II	.30	.02
III	.35	.02
IV	.20	.03

21. An experiment consists of randomly selecting one of three coins, tossing it, and observing the outcome—heads or tails. The first coin is a two-headed coin, the second is a biased coin such that $P(H) = .75$, and the third is a fair coin.
   a. What is the probability that the coin that is tossed will show heads?
   b. If the coin selected shows heads, what is the probability that this coin is the fair coin?

22. Reliability of Medical Tests   A medical test has been designed to detect the presence of a certain disease. Among those who have the disease, the probability that the disease will be detected by the test is .95. However, the probability that the test will erroneously indicate the presence of the disease in those who do not actually have it is .04. It is estimated that 4% of the population who take this test have the disease.
   a. If the test administered to an individual is positive, what is the probability that the person actually has the disease?
   b. If an individual takes the test twice and both times the test is positive, what is the probability that the person actually has the disease?

23. RELIABILITY OF MEDICAL TESTS   Refer to Exercise 22. Suppose 20% of the people who were referred to a clinic for the test did in fact have the disease. If the test administered to an individual from this group is positive, what is the probability that the person actually has the disease?

24. QUALITY CONTROL   A desk lamp produced by Luminar was found to be defective. The company has three factories where the lamps are manufactured. The percentage of the total number of desk lamps produced by each factory and the probability that a lamp manufactured by that factory is defective are shown in the accompanying table. What is the probability that the defective lamp was manufactured in factory III?

Factory	Percentage of Total Production	Probability of Defective Component
I	.35	.015
II	.35	.01
III	.30	.02

25. AUTO-ACCIDENT RATES   An insurance company has compiled the accompanying data relating the age of drivers and the accident rate (the probability of being involved in an accident during a 1-yr period) for drivers within that group:

Age Group	Percentage of Insured Drivers	Accident Rate
Under 25	.16	.055
25–44	.40	.025
45–64	.30	.02
65 and over	.14	.04

a. What is the probability that an insured driver will be involved in an accident during a particular 1-yr period?
b. What is the probability that an insured driver who is involved in an accident is under 25?

26. SEAT-BELT COMPLIANCE   Data compiled by the Highway Patrol Department regarding the use of seat belts by drivers in a certain area after the passage of a compulsory seat-belt law are shown in the accompanying table.

Drivers	Percentage of Drivers in Group	Percentage of Group Stopped for Moving Violation
Group I (using seat belts)	.64	.002
Group II (not using seat belts)	.36	.005

If a driver in that area is stopped for a moving violation, what is the probability that he or she
a. Will have a seat belt on?
b. Will not have a seat belt on?

27. MEDICAL RESEARCH   Based on data obtained from the National Institute of Dental Research, it has been determined that 42% of 12-yr-olds have never had a cavity, 34% of 13-yr-olds have never had a cavity, and 28% of 14-yr-olds have never had a cavity. Suppose a child is selected at random from a group of 24 junior high school students that includes six 12-yr-olds, eight 13-yr-olds, and ten 14-yr-olds. If this child does not have a cavity, what is the probability that this child is 14 yrs old?

28. VOTING PATTERNS   In a recent senatorial election, 50% of the voters in a certain district were registered as Democrats, 35% were registered as Republicans, and 15% were registered as Independents. The incumbent Democratic senator was reelected over her Republican and Independent opponents. Exit polls indicated that she gained 75% of the Democratic vote, 25% of the Republican vote, and 30% of the Independent vote. Assuming that the exit poll is accurate, what is the probability that a vote for the incumbent was cast by a registered Republican?

29. CRIME RATES   Data compiled by the Department of Justice on the number of people arrested in a certain year for serious crimes (murder, forcible rape, robbery, and so on) revealed that 89% were male and 11% were female. Of the males, 30% were under 18, whereas 27% of the females arrested were under 18.
a. What is the probability that a person arrested for a serious crime in that year was under 18?
b. If a person arrested for a serious crime in that year is known to be under 18, what is the probability that the person is female?

30. OPINION POLLS   A survey involving 600 Democrats, 400 Republicans, and 200 Independents asked the question: "Do you favor or oppose eliminating taxes on dividends paid to shareholders?" The following results were obtained:

Answer	Democrats	Republicans	Independents
Favor	29%	66%	48%
Opposed	71%	34%	52%

If a randomly chosen respondent in the survey answered "favor," what is the probability that he or she is an Independent?

Source: TechnoMetrica Market Intelligence

31. OPINION POLLS   A poll was conducted among 500 registered voters in a certain area regarding their position on a national lottery to raise revenue for the government. The results of the poll are shown in the accompanying table.

Sex	Percentage of Voters Polled	Percentage Favoring Lottery	Percentage Not Favoring Lottery	Percentage Expressing No Opinion
Male	.51	.62	.32	.06
Female	.49	.68	.28	.04

What is the probability that a registered voter who
a. Favored a national lottery was a woman?
b. Expressed no opinion regarding the lottery was a woman?

32. **SELECTION OF SUPREME COURT JUDGES**   In a past presidential election, it was estimated that the probability that the Republican candidate would be elected was $\frac{3}{5}$, and therefore the probability that the Democratic candidate would be elected was $\frac{2}{5}$ (the two Independent candidates were given little chance of being elected). It was also estimated that if the Republican candidate were elected, the probabilities that a conservative, moderate, or liberal judge would be appointed to the Supreme Court (one retirement was expected during the presidential term) were $\frac{1}{2}$, $\frac{1}{3}$, and $\frac{1}{6}$, respectively. If the Democratic candidate were elected, the probabilities that a conservative, moderate, or liberal judge would be appointed to the Supreme Court would be $\frac{1}{8}$, $\frac{3}{8}$, and $\frac{1}{2}$, respectively. A conservative judge was appointed to the Supreme Court during the presidential term. What is the probability that the Democratic candidate was elected?

33. **PERSONNEL SELECTION**   Applicants for temporary office work at Carter Temporary Help Agency who have successfully completed a typing test are then placed in suitable positions by Nancy Dwyer and Darla Newberg. Employers who hire temporary help through the agency return a card indicating satisfaction or dissatisfaction with the work performance of those hired. From past experience it is known that 80% of the employees placed by Nancy are rated as satisfactory and 70% of those placed by Darla are rated as satisfactory. Darla places 55% of the temporary office help at the agency and Nancy the remaining 45%. If a Carter office worker is rated unsatisfactory, what is the probability that he or she was placed by Darla?

34. **COLLEGE MAJORS**   The Office of Admissions and Records of a large western university released the accompanying information concerning the contemplated majors of its freshman class:

Major	Percentage of Freshmen Choosing This Major	Percentage of Females	Percentage of Males
Business	.24	.38	.62
Humanities	.08	.60	.40
Education	.08	.66	.34
Social science	.07	.58	.42
Natural sciences	.09	.52	.48
Other	.44	.48	.52

a. What is the probability that a student selected at random from the freshman class is a female?
b. What is the probability that a business student selected at random from the freshman class is a male?
c. What is the probability that a female student selected at random from the freshman class is majoring in business?

35. **MEDICAL DIAGNOSES**   A study was conducted among a certain group of union members whose health insurance policies required second opinions prior to surgery. Of those members whose doctors advised them to have surgery, 20% were informed by a second doctor that no surgery was needed. Of these, 70% took the second doctor's opinion and did not go through with the surgery. Of the members who were advised to have surgery by both doctors, 95% went through with the surgery. What is the probability that a union member who had surgery was advised to do so by a second doctor?

36. **AGE DISTRIBUTION OF RENTERS**   A study conducted by the Metro Housing Agency in a midwestern city revealed the following information concerning the age distribution of renters within the city.

Age	Percentage of Adult Population	Percentage of Group Who Are Renters
21–44	.51	.58
45–64	.31	.45
65 and over	.18	.60

a. What is the probability that an adult selected at random from this population is a renter?
b. If a renter is selected at random, what is the probability that he or she is in the 21–44 age bracket?
c. If a renter is selected at random, what is the probability that he or she is 45 years of age or older?

37. **PERSONAL HABITS**   There were 80 male guests at a party. The number of men in each of four age categories is given in the following table. The table also gives the probability that a man in the respective age category will keep his paper money in order of denomination.

Age	Number of Men	Percentage of Men Who Keep Paper Money in Order
21–34	25	.9
35–44	30	.61
45–54	15	.80
55 and over	10	.80

A man's wallet was retrieved and the paper money in it was kept in order of denomination. What is the probability that the wallet belonged to a male guest between the ages of 35 and 44?

38. **THE SOCIAL LADDER**   The following table summarizes the results of a poll conducted with 1154 adults.

Annual Household Income, $	Respondents within That Income Range, %	Percent of Respondents Who Call Themselves		
		Rich	Middle Class	Poor
Less than 15,000	11.2	0	24	76
15,000–29,999	18.6	3	60	37
30,000–49,999	24.5	0	86	14
50,000–74,999	21.9	2	90	8
75,000 and higher	23.8	5	91	4

a. What is the probability that a respondent chosen at random calls himself or herself middle class?

b. If a randomly chosen respondent calls himself or herself middle class, what is the probability that the annual household income of that individual is between $30,000 and $49,999, inclusive?

c. If a randomly chosen respondent calls himself or herself middle class, what is the probability that the individual's income is either less than or equal to $29,999 or greater than or equal to $50,000?

Source: New York Times/CBS News; Wall Street Journal Almanac

39. OPINION POLLS A survey involving 400 likely Democratic voters and 300 likely Republican voters asked the question: "Do you support or oppose legislation that would require registration of all handguns?" The following results were obtained:

Answer	Democrats, %	Republicans, %
Support	77	59
Oppose	14	31
Don't know/refused	9	10

If a randomly chosen respondent in the survey answered "Oppose," what is the probability that he or she is a likely Democratic voter?

40. OPINION POLLS A survey involving 400 likely Democratic voters and 300 likely Republican voters asked the question: "Do you support or oppose legislation that would require trigger locks on guns, to prevent misuse by children?" The following results were obtained:

Answer	Democrats, %	Republicans, %
Support	88	71
Oppose	7	20
Don't know/refused	5	9

If a randomly chosen respondent in the survey answered "Support," what is the probability that he or she is a likely Republican voter?

41. VOTER TURNOUT BY INCOME Voter turnout drops steadily as income level declines. The following table gives the percent of eligible voters in a certain city, categorized by income, who responded with "Did not vote" in the 2000 presidential election. The table also gives the number of eligible voters in the city, categorized by income.

Income (percentile)	Percent Who "Did Not Vote"	Eligible Voters
0–16	52	4,000
17–33	31	11,000
34–67	30	17,500
68–95	14	12,500
96–100	12	5,000

If an eligible voter from this city who had voted in the election is selected at random, what is the probability that this person had an income in the 17–33 percentile?

Source: The National Election Studies

42. VOTER TURNOUT BY PROFESSION The following table gives the percent of eligible voters grouped according to profession who responded with "Voted" in the 2000 presidential election. The table also gives the percent of people in a survey categorized by their profession.

Profession	Percent Who Voted	Percent in Each Profession
Professionals	84	12
White collar	73	24
Blue collar	66	32
Unskilled	57	10
Farmers	68	8
Housewives	66	14

If an eligible voter who participated in the survey and voted in the election is selected at random, what is the probability that this person is a housewife?

Source: The National Election Studies

7.6 Solutions to Self-Check Exercises

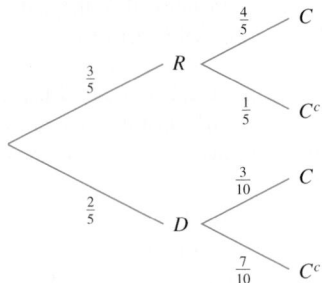

1. Using the probabilities given in the tree diagram and Bayes' theorem, we have

$$P(B|D) = \frac{P(B) \cdot P(D|B)}{P(A) \cdot P(D|A) + P(B) \cdot P(D|B) + P(C) \cdot P(D|C)}$$

$$= \frac{(\frac{1}{2})(\frac{1}{2})}{(\frac{1}{4})(\frac{1}{3}) + (\frac{1}{2})(\frac{1}{2}) + (\frac{1}{4})(\frac{1}{4})} = \frac{12}{19}$$

2. Let R and D, respectively, denote the event that the Republican and the Democratic candidate won the presidential election. Then, $P(R) = \frac{3}{5}$ and $P(D) = \frac{2}{5}$. Also, let C denote the event that research for the new manned bomber would continue. These data may be exhibited as in the accompanying tree diagram:

Using Bayes' theorem, we find that the probability that the Republican candidate had won the election is given by

$$P(R|C^c) = \frac{P(R) \cdot P(C^c|R)}{P(R) \cdot P(C^c|R) + P(D) \cdot P(C^c|D)}$$

$$= \frac{(\frac{3}{5})(\frac{1}{5})}{(\frac{3}{5})(\frac{1}{5}) + (\frac{2}{5})(\frac{7}{10})} = \frac{3}{10}$$

CHAPTER 7   Summary of Principal Formulas and Terms

 FORMULAS

1. Probability of an event in a uniform sample space	$P(E) = \dfrac{n(E)}{n(S)}$
2. Probability of the union of two mutually exclusive events	$P(E \cup F) = P(E) + P(F)$
3. Addition rule	$P(E \cup F) = P(E) + P(F) - P(E \cap F)$
4. Rule of complements	$P(E^c) = 1 - P(E)$
5. Conditional probability	$P(B|A) = \dfrac{P(A \cap B)}{P(A)}$
6. Product rule	$P(A \cap B) = P(A) \cdot P(B|A)$
7. Test for independence	$P(A \cap B) = P(A) \cdot P(B)$

TERMS

experiment (380)
outcome (380)
sample point (380)
sample space (380)
event (380)
finite sample space (380)
union of two events (381)
intersection of two events (381)

complement of an event (381)
mutually exclusive events (382)
relative frequency (389)
empirical probability (389)
probability of an event (389)
elementary (simple) event (389)
probability distribution (389)
probability function (389)

uniform sample space (390)
addition principle (391)
conditional probability (415)
finite stochastic process (419)
independent events (422)
Bayes' theorem (430)

CHAPTER 7 Concept Review Questions

Fill in the blanks.

1. An activity with observable results is called a/an _____; an outcome of an experiment is called a _____ point, and the set consisting of all possible sample points of an experiment is called a sample _____; a subset of a sample space of an experiment is called a/an _____.

2. The events E and F are mutually exclusive if $E \cap F =$ _____.

3. A sample space in which the outcomes are equally likely is called a/an _____ sample space; if such a space contains n simple events, then the probability of each simple event is _____.

4. The probability of the event B given that the event A has already occurred is called the _____ probability of B given A.

5. If the outcome of one event does not depend on the other, then the two events are said to be _____.

6. The probability of an event after the outcomes of an experiment have been observed is called a/an _____ _____.

CHAPTER 7 Review Exercises

1. Let E and F be two mutually exclusive events, and suppose $P(E) = .4$ and $P(F) = .2$. Compute
   a. $P(E \cap F)$ **b.** $P(E \cup F)$
   c. $P(E^c)$ **d.** $P(E^c \cap F^c)$
   e. $P(E^c \cup F^c)$

2. Let E and F be two events of an experiment with sample space S. Suppose $P(E) = .3$, $P(F) = .2$, and $P(E \cap F) = .15$. Compute
   a. $P(E \cup F)$
   b. $P(E^c \cap F^c)$
   c. $P(E^c \cap F)$

3. A die is loaded, and it has been determined that the probability distribution associated with the experiment of casting the die and observing which number falls uppermost is given by

Simple Event	Probability
{1}	.20
{2}	.12
{3}	.16
{4}	.18
{5}	.15
{6}	.19

   a. What is the probability of the number being even?
   b. What is the probability of the number being either a 1 or a 6?
   c. What is the probability of the number being less than 4?

4. An urn contains six red, five black, and four green balls. If two balls are selected at random without replacement from the urn, what is the probability that a red ball and a black ball will be selected?

5. QUALITY CONTROL The quality-control department of Starr Communications, a manufacturer of video-game DVDs, has determined from records that 1.5% of the DVDs sold have video defects, 0.8% have audio defects, and 0.4% have both audio and video defects. What is the probability that a DVD purchased by a customer
   a. Will have a video or audio defect?
   b. Will not have a video or audio defect?

6. Let E and F be two events, and suppose that $P(E) = .35$, $P(F) = .55$, and $P(E \cup F) = .70$. Find $P(E|F)$.

The accompanying tree diagram represents an experiment consisting of two trials. In Exercises 7–11, use the diagram to find the given probability.

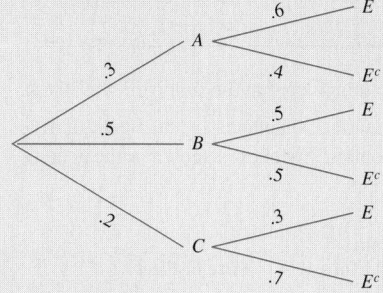

7. $P(A \cap E)$ 8. $P(B \cap E)$

9. $P(C \cap E)$ 10. $P(A|E)$

11. $P(E)$

12. An experiment consists of tossing a fair coin three times and observing the outcomes. Let A be the event that at least one

head is thrown and B the event that at most two tails are thrown.
   a. Find $P(A)$.   **b.** Find $P(B)$.
   c. Are A and B independent events?

13. QUALITY CONTROL   In a group of 20 ballpoint pens on a shelf in the stationery department of Metro Department Store, 2 are known to be defective. If a customer selects 3 of these pens, what is the probability that
   a. At least 1 is defective?
   b. No more than 1 is defective?

14. Five people are selected at random. What is the probability that none of the people in this group were born on the same day of the week?

15. A pair of fair dice is cast. What is the probability that the sum of the numbers falling uppermost is 8 if it is known that the two numbers are different?

16. A fair die is cast three times. What is the probability that it shows an even number in the first toss, an odd number in the second toss, and a 1 on the third toss? Assume that the outcomes of the tosses are independent.

17. A fair die is cast, a fair coin is tossed, and a card is drawn from a standard deck of 52 playing cards. Assuming these events are independent, what is the probability that the number falling uppermost on the die is a 6, the coin shows a tail, and the card drawn is a face card?

Three cards are drawn at random without replacement from a standard deck of 52 playing cards. In Exercises 18–22, find the probability of each of the given events.

18. All three cards are aces.

19. All three cards are face cards.

20. The second and third cards are red.

21. The second card is black, given that the first card was red.

22. The second card is a club, given that the first card was black.

23. AIRFONE USAGE   The number of planes in the fleets of five leading airlines that contain Airfones is shown in the accompanying table:

Airline	Planes with Airfones	Size of Fleet
A	50	295
B	40	325
C	31	167
D	29	50
E	25	248

   a. If a plane is selected at random from airline A, what is the probability that it contains an Airfone?
   b. If a plane is selected at random from the entire fleet of the five airlines, what is the probability that it contains an Airfone?

24. FLEX-TIME   Of 320 male and 280 female employees at the home office of Gibraltar Insurance Company, 160 of the men and 190 of the women are on flex-time (flexible working hours). Given that an employee selected at random from this group is on flex-time, what is the probability that the employee is a man?

25. QUALITY CONTROL   In a manufacturing plant, three machines, A, B, and C, produce 40%, 35%, and 25%, respectively, of the total production. The company's quality-control department has determined that 1% of the items produced by machine A, 1.5% of the items produced by machine B, and 2% of the items produced by machine C are defective. If an item is selected at random and found to be defective, what is the probability that it was produced by machine B?

26. COLLEGE ADMISSIONS   Applicants who wish to be admitted to a certain professional school in a large university are required to take a screening test devised by an educational testing service. From past results, the testing service has estimated that 70% of all applicants are eligible for admission and that 92% of those who are eligible for admission pass the exam, whereas 12% of those who are ineligible for admission pass the exam. Using these results, what is the probability that an applicant for admission
   a. Passed the exam?
   b. Passed the exam but was actually ineligible?

27. COMMUTING TIMES   Bill commutes to work in the business district of Boston. He takes the train $\frac{3}{5}$ of the time and drives $\frac{2}{5}$ of the time (when he visits clients). If he takes the train, then he gets home by 6:30 p.m. 85% of the time; if he drives, then he gets home by 6:30 p.m. 60% of the time. If Bill gets home by 6:30 p.m., what is the probability that he drove to work?

28. CUSTOMER SURVEYS   The sales department of Thompson Drug Company released the accompanying data concerning the sales of a certain pain reliever manufactured by the company.

Pain Reliever	Percentage of Drug Sold	Percentage of Group Sold in Extra-Strength Dosage
Group I (capsule form)	.57	.38
Group II (tablet form)	.43	.31

If a customer purchased the extra-strength dosage of this drug, what is the probability that it was in capsule form?

CHAPTER 7 **Before Moving On . . .**

1. Let $S = \{s_1, s_2, s_3, s_4, s_5, s_6\}$ be the sample space associated with an experiment having the following probability distribution:

Outcome	s_1	s_2	s_3	s_4	s_5	s_6
Probability	$\frac{1}{12}$	$\frac{2}{12}$	$\frac{3}{12}$	$\frac{2}{12}$	$\frac{3}{12}$	$\frac{1}{12}$

Find the probability of the event $A = \{s_1, s_3, s_6\}$.

2. A card is drawn from a well-shuffled 52-card deck. What is the probability that the card drawn is a deuce or a face card?

3. Let E and F be events of an experiment with sample space S. Suppose $P(E) = .5$, $P(F) = .6$, and $P(E \cap F) = .2$. Compute:
   a. $P(E \cup F)$ b. $P(E \cap F^c)$

4. Suppose A and B are independent events with $P(A) = .3$ and $P(B) = .6$. Find $P(A \cup B)$.

5. The accompanying tree diagram represents a two-stage experiment. Use the diagram to find $P(A|D)$.

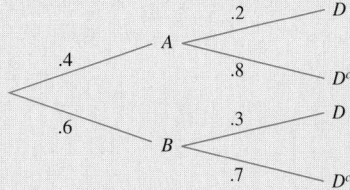

8

Probability Distributions and Statistics

Where do we go from here? Some students in the top 10% of this senior class will further their education at one of the campuses of the state university system. In Example 3, page 500, we will determine the minimum grade point average a senior needs to be eligible for admission to a state university.

© Andy Sacks/Stone/Getty Images

STATISTICS IS THAT branch of mathematics concerned with the collection, analysis, and interpretation of data. In Sections 8.1–8.3, we take a look at descriptive statistics; here, our interest lies in the description and presentation of data in the form of tables and graphs. In the rest of the chapter, we briefly examine inductive statistics, and we see how mathematical tools such as those developed in Chapter 7 may be used in conjunction with these data to help us draw certain conclusions and make forecasts.

8.1   Distributions of Random Variables

▬ Random Variables

In many situations, it is desirable to assign numerical values to the outcomes of an experiment. For example, if an experiment consists of casting a die and observing the face that lands uppermost, then it is natural to assign the numbers 1, 2, 3, 4, 5, and 6, respectively, to the outcomes *one, two, three, four, five,* and *six* of the experiment. If we let X denote the outcome of the experiment, then X assumes one of these numbers. Because the values assumed by X depend on the outcomes of a chance experiment, the outcome X is referred to as a random variable.

> **Random Variable**
>
> A random variable is a rule that assigns a number to each outcome of a chance experiment.

More precisely, a random variable is a function with domain given by the set of outcomes of a chance experiment and range contained in the set of real numbers.

EXAMPLE 1   A coin is tossed three times. Let the random variable X denote the number of heads that occur in the three tosses.
a. List the outcomes of the experiment; that is, find the domain of the function X.
b. Find the value assigned to each outcome of the experiment by the random variable X.
c. Find the event comprising the outcomes to which a value of 2 has been assigned by X. This event is written $(X = 2)$ and is the event consisting of the outcomes in which two heads occur.

Solution
a. From the results of Example 3, Section 7.1 (page 382), we see that the set of outcomes of the experiment is given by the sample space

$$S = \{HHH, HHT, HTH, THH, HTT, THT, TTH, TTT\}$$

b. The outcomes of the experiment are displayed in the first column of Table 1. The corresponding value assigned to each such outcome by the random variable X (the number of heads) appears in the second column.
c. With the aid of Table 1, we see that the event $(X = 2)$ is given by the set

$$\{HHT, HTH, THH\}$$

EXAMPLE 2   A coin is tossed repeatedly until a head occurs. Let the random variable Y denote the number of coin tosses in the experiment. What are the values of Y?

Solution   The outcomes of the experiment make up the infinite set

$$S = \{H, TH, TTH, TTTH, TTTTH, \ldots\}$$

These outcomes of the experiment are displayed in the first column of Table 2. The corresponding values assumed by the random variable Y (the number of tosses) appear in the second column.

TABLE 1

Number of Heads in Three Coin Tosses

Outcome	Value of X
HHH	3
HHT	2
HTH	2
THH	2
HTT	1
THT	1
TTH	1
TTT	0

TABLE 2

Number of Coin Tosses before Heads Appear

Outcome	Value of Y
H	1
TH	2
TTH	3
TTTH	4
TTTTH	5
⋮	⋮

APPLIED EXAMPLE 3   Product Reliability   A disposable flashlight is turned on until its battery runs out. Let the random variable Z denote the length (in hours) of the life of the battery. What values may Z assume?

Solution   The values assumed by Z may be any nonnegative real numbers; that is, the possible values of Z comprise the interval $0 \leq Z < \infty$.   ∎

One advantage of working with random variables—rather than working directly with the outcomes of an experiment—is that random variables are functions that may be added, subtracted, and multiplied. Because of this, results developed in the field of algebra and other areas of mathematics may be used freely to help us solve problems in probability and statistics.

A random variable is classified into three categories depending on the set of values it assumes. A random variable is called **finite discrete** if it assumes only finitely many values. For example, the random variable X of Example 1 is finite discrete because it may assume values only from the finite set $\{0, 1, 2, 3\}$ of numbers. Next, a random variable is said to be **infinite discrete** if it takes on infinitely many values, which may be arranged in a sequence. For example, the random variable Y of Example 2 is infinite discrete because it assumes values from the set $\{1, 2, 3, 4, 5, \ldots\}$, which has been arranged in the form of an infinite sequence. Finally, a random variable is called **continuous** if the values it may assume comprise an interval of real numbers. For example, the random variable Z of Example 3 is continuous because the values it may assume comprise the interval of nonnegative real numbers. For the remainder of this section, *all random variables will be assumed to be finite discrete*.

■ Probability Distributions of Random Variables

In Section 7.2, we learned how to construct the probability distribution for an experiment. There, the probability distribution took the form of a table that gave the probabilities associated with the outcomes of an experiment. Since the random variable associated with an experiment is related to the outcomes of the experiment, it is clear that we should be able to construct a probability distribution associated with the *random variable* rather than one associated with the outcomes of the experiment. Such a distribution is called the **probability distribution of a random variable** and may be given in the form of a formula or displayed in a table that gives the distinct (numerical) values of the random variable X and the probabilities associated with these values. Thus, if $x_1, x_2, \ldots, x_n$ are the values assumed by the random variable X with associated probabilities $P(X = x_1), P(X = x_2), \ldots, P(X = x_n)$, respectively, then the required probability distribution of the random variable X, where $p_i = P(X = x_i)$, $i = 1, 2, \ldots, n$, may be expressed in the form of the table shown in Table 3.

In the next several examples we illustrate the construction of probability distributions.

TABLE 3

Probability Distribution for the Random Variable X

x	$P(X = x)$
x_1	p_1
x_2	p_2
x_3	p_3
$\vdots$	$\vdots$
x_n	p_n

EXAMPLE 4   Find the probability distribution of the random variable associated with the experiment of Example 1.

Solution   From the results of Example 1, we see that the values assumed by the random variable X are 0, 1, 2, and 3, corresponding to the events of 0, 1, 2, and

TABLE 4

Probability Distribution

x	$P(X = x)$
0	$\frac{1}{8}$
1	$\frac{3}{8}$
2	$\frac{3}{8}$
3	$\frac{1}{8}$

TABLE 5

Probability Distribution for the Random Variable That Gives the Sum of the Faces of Two Dice

x	$P(X = x)$
2	$\frac{1}{36}$
3	$\frac{2}{36}$
4	$\frac{3}{36}$
5	$\frac{4}{36}$
6	$\frac{5}{36}$
7	$\frac{6}{36}$
8	$\frac{5}{36}$
9	$\frac{4}{36}$
10	$\frac{3}{36}$
11	$\frac{2}{36}$
12	$\frac{1}{36}$

TABLE 6

Probability Distribution

x	$P(X = x)$
0	.03
1	.15
2	.27
3	.20
4	.13
5	.10
6	.07
7	.03
8	.02

3 heads occurring, respectively. Referring to Table 1 once again, we see that the outcome associated with the event $(X = 0)$ is given by the set {TTT}. Consequently, the probability associated with the random variable X when it assumes the value 0 is given by

$$P(X = 0) = \frac{1}{8} \qquad \text{Note that } n(S) = 8.$$

Next, observe that the event $(X = 1)$ is given by the set {HTT, THT, TTH}, so

$$P(X = 1) = \frac{3}{8}$$

In a similar manner, we may compute $P(X = 2)$ and $P(X = 3)$, which gives the probability distribution shown in Table 4. ∎

EXAMPLE 5   Let X denote the random variable that gives the sum of the faces that fall uppermost when two fair dice are cast. Find the probability distribution of X.

Solution   The values assumed by the random variable X are 2, 3, 4, ... , 12, corresponding to the events $E_2, E_3, E_4, \ldots , E_{12}$ (see Example 4, Section 7.1). Next, the probabilities associated with the random variable X when X assumes the values 2, 3, 4, ... , 12 are precisely the probabilities $P(E_2), P(E_3), \ldots , P(E_{12})$, respectively, and may be computed in much the same way as the solution to Example 3, Section 7.2. Thus,

$$P(X = 2) = P(E_2) = \frac{1}{36}$$

$$P(X = 3) = P(E_3) = \frac{2}{36}$$

and so on. The required probability distribution of X is given in Table 5. ∎

APPLIED EXAMPLE 6   Waiting Lines   The following data give the number of cars observed waiting in line at the beginning of 2-minute intervals between 3 p.m. and 5 p.m. on a certain Friday at the drive-in teller of Westwood Savings Bank and the corresponding frequency of occurrence. Find the probability distribution of the random variable X, where X denotes the number of cars observed waiting in line.

Cars	0	1	2	3	4	5	6	7	8
Frequency of Occurrence	2	9	16	12	8	6	4	2	1

Solution   Dividing each number in the last row of the given table by 60 (the sum of these numbers) gives the respective probabilities (here, we use the relative frequency interpretation of probability) associated with the random variable X when X assumes the values 0, 1, 2, ... , 8. For example,

$$P(X = 0) = \frac{2}{60} \approx .03$$

$$P(X = 1) = \frac{9}{60} \approx .15$$

and so on. The resulting probability distribution is shown in Table 6. ∎

Histograms

A probability distribution of a random variable may be exhibited graphically by means of a **histogram**. To construct a histogram of a particular probability distribution, first locate the values of the random variable on a number line. Then, above each such number, erect a rectangle with width 1 and height equal to the probability associated with that value of the random variable. For example, the histogram of the probability distribution appearing in Table 4 is shown in Figure 1. The histograms of the probability distributions of Examples 5 and 6 are constructed in a similar manner and are displayed in Figures 2 and 3, respectively.

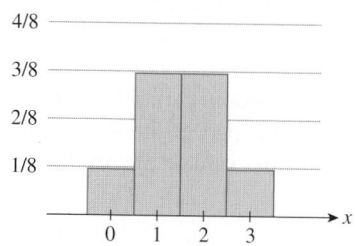

FIGURE 1
Histogram showing the probability distribution for the number of heads occurring in three coin tosses

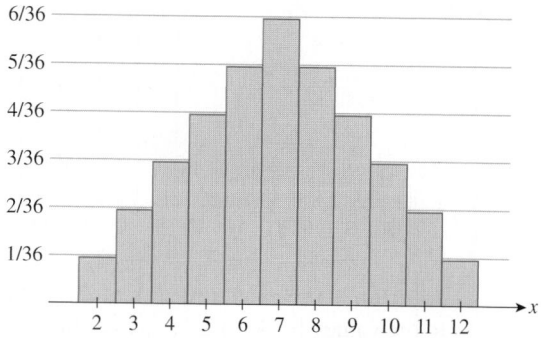

FIGURE 2
Histogram showing the probability distribution for the sum of the faces of two dice

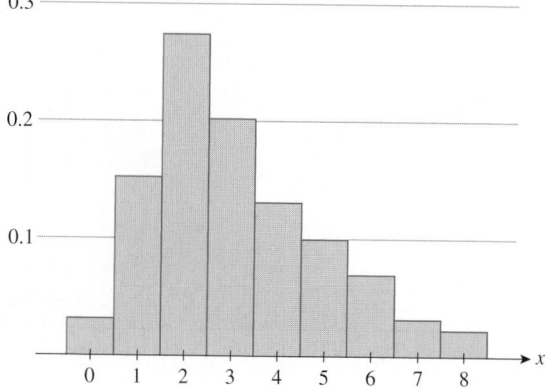

FIGURE 3
Histogram showing the probability distribution for the number of cars waiting in line

Observe that in each histogram, the area of a rectangle associated with a value of a random variable X gives precisely the probability associated with the value of X. This follows because each such rectangle, by construction, has width 1 and height corresponding to the probability associated with the value of the random variable. Another consequence arising from the method of construction of a histogram is that *the probability associated with more than one value of the random variable X is given by the sum of the areas of the rectangles associated with those values of X.* For example, in the coin-tossing experiment of Example 1, the event of obtaining at least two heads, which corresponds to the event $(X = 2)$ or $(X = 3)$, is given by

$$P(X = 2) + P(X = 3)$$

and may be obtained from the histogram depicted in Figure 1 by adding the areas associated with the values 2 and 3 of the random variable X. We obtain

$$P(X = 2) + P(X = 3) = (1)\left(\frac{3}{8}\right) + (1)\left(\frac{1}{8}\right) = \frac{1}{2}$$

This result provides us with a method of computing the probabilities of events directly from the knowledge of a histogram of the probability distribution of the random variable associated with the experiment.

EXAMPLE 7   Suppose the probability distribution of a random variable X is represented by the histogram shown in Figure 4. Identify that part of the histogram whose area gives the probability $P(10 \leq X \leq 20)$. Do not evaluate the result.

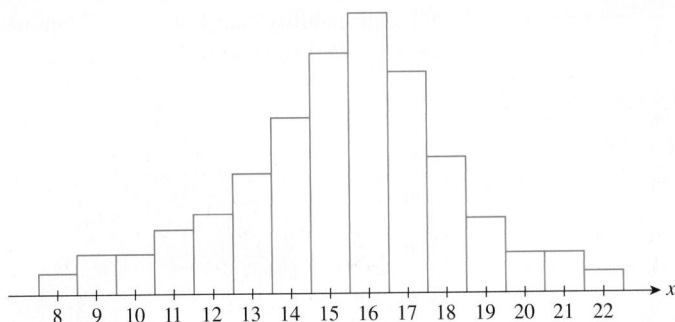

FIGURE 4

Solution   The event $(10 \leq X \leq 20)$ is the event consisting of outcomes related to the values 10, 11, 12, . . . , 20 of the random variable X. The probability of this event $P(10 \leq X \leq 20)$ is therefore given by the shaded area of the histogram in Figure 5.

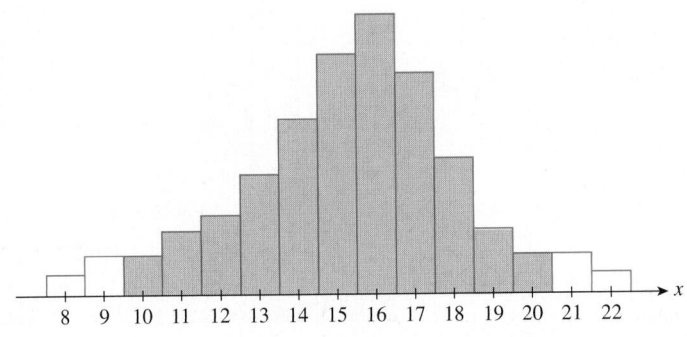

FIGURE 5
$P(10 \leq X \leq 20)$

8.1   Self-Check Exercises

1. Three balls are selected at random without replacement from an urn containing four black balls and five white balls. Let the random variable X denote the number of black balls drawn.
   a. List the outcomes of the experiment.
   b. Find the value assigned to each outcome of the experiment by the random variable X.
   c. Find the event consisting of the outcomes to which a value of 2 has been assigned by X.

2. The following data, extracted from the records of Dover Public Library, give the number of books borrowed by the library's members over a 1-mo period:

Books	0	1	2	3	4	5	6	7	8
Frequency of Occurrence	780	300	412	205	98	54	57	30	6

   a. Find the probability distribution of the random variable X, where X denotes the number of books checked out over a 1-mo period by a randomly chosen member.
   b. Draw the histogram representing this probability distribution.

Solutions to Self-Check Exercises 8.1 can be found on page 451.

8.1 Concept Questions

1. What is a random variable? Give an example.

2. Give an example of (a) a finite discrete random variable, (b) an infinite discrete random variable, and (c) a continuous random variable.

3. Suppose you are given the probability distribution for a random variable X. Explain how you would construct a histogram for this probability distribution. What does the area of each rectangle in the histogram represent?

8.1 Exercises

1. Three balls are selected at random without replacement from an urn containing four green balls and six red balls. Let the random variable X denote the number of green balls drawn.
  a. List the outcomes of the experiment.
  b. Find the value assigned to each outcome of the experiment by the random variable X.
  c. Find the event consisting of the outcomes to which a value of 3 has been assigned by X.

2. A coin is tossed four times. Let the random variable X denote the number of tails that occur.
  a. List the outcomes of the experiment.
  b. Find the value assigned to each outcome of the experiment by the random variable X.
  c. Find the event consisting of the outcomes to which a value of 2 has been assigned by X.

3. A die is cast repeatedly until a 6 falls uppermost. Let the random variable X denote the number of times the die is cast. What are the values that X may assume?

4. Cards are selected one at a time without replacement from a well-shuffled deck of 52 cards until an ace is drawn. Let X denote the random variable that gives the number of cards drawn. What values may X assume?

5. Let X denote the random variable that gives the sum of the faces that fall uppermost when two fair dice are cast. Find $P(X = 7)$.

6. Two cards are drawn from a well-shuffled deck of 52 playing cards. Let X denote the number of aces drawn. Find $P(X = 2)$.

In Exercises 7–12, give the range of values that the random variable X may assume and classify the random variable as finite discrete, infinite discrete, or continuous.

7. X = The number of times a die is thrown until a 2 appears

8. X = The number of defective watches in a sample of eight watches

9. X = The distance a commuter travels to work

10. X = The number of hours a child watches television on a given day

11. X = The number of times an accountant takes the CPA examination before passing

12. X = The number of boys in a four-child family

13. The probability distribution of the random variable X is shown in the accompanying table:

x	-10	-5	0	5	10	15	20
$P(X = x)$	.20	.15	.05	.1	.25	.1	.15

Find
  a. $P(X = -10)$ **b.** $P(X \geq 5)$
  c. $P(-5 \leq X \leq 5)$ **d.** $P(X \leq 20)$

14. The probability distribution of the random variable X is shown in the accompanying table:

x	-5	-3	-2	0	2	3
$P(X = x)$	.17	.13	.33	.16	.11	.10

Find
  a. $P(X \leq 0)$ **b.** $P(X \leq -3)$
  c. $P(-2 \leq X \leq 2)$

15. Suppose a probability distribution of a random variable X is represented by the accompanying histogram. Shade that part of the histogram whose area gives the probability $P(17 \leq X \leq 20)$.

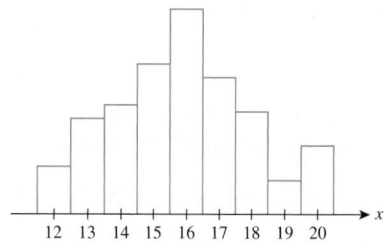

16. EXAMS   An examination consisting of ten true-or-false questions was taken by a class of 100 students. The probability distribution of the random variable X, where X denotes the number of questions answered correctly by a randomly chosen student, is represented by the accompanying histogram. The rectangle with base centered on the number 8 is missing. What should be the height of this rectangle?

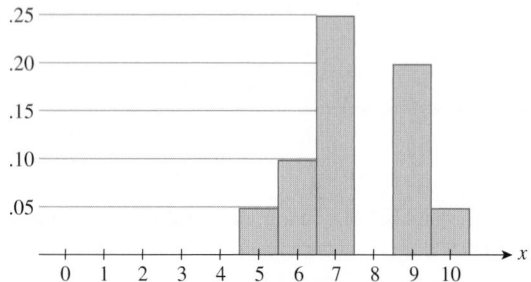

17. Two dice are cast. Let the random variable X denote the number that falls uppermost on the first die, and let Y denote the number that falls uppermost on the second die.
    a. Find the probability distributions of X and Y.
    b. Find the probability distribution of $X + Y$.

18. DISTRIBUTION OF FAMILIES BY SIZE   A survey was conducted by the Public Housing Authority in a certain community among 1000 families to determine the distribution of families by size. The results follow:

Family Size	2	3	4	5	6	7	8
Frequency of Occurrence	350	200	245	125	66	10	4

    a. Find the probability distribution of the random variable X, where X denotes the number of persons in a randomly chosen family.
    b. Draw the histogram corresponding to the probability distribution found in part (a).

19. WAITING LINES   The accompanying data were obtained in a study conducted by the manager of SavMore Supermarket. In this study, the number of customers waiting in line at the express checkout at the beginning of each 3-min interval between 9 a.m. and 12 noon on Saturday was observed.

Customers	0	1	2	3	4
Frequency of Occurrence	1	4	2	7	14

Customers	5	6	7	8	9	10
Frequency of Occurrence	8	10	6	3	4	1

    a. Find the probability distribution of the random variable X, where X denotes the number of customers observed waiting in line.
    b. Draw the histogram representing the probability distribution.

20. MONEY MARKET RATES   The rates paid by 30 financial institutions on a certain day for money market deposit accounts are shown in the accompanying table:

Rate, %	6	6.25	6.55	6.56
Institutions	1	7	7	1

Rate, %	6.58	6.60	6.65	6.85
Institutions	1	8	3	2

    Let the random variable X denote the interest paid by a randomly chosen financial institution on its money market deposit accounts and find the probability distribution associated with these data.

21. TELEVISION PILOTS   After the private screening of a new television pilot, audience members were asked to rate the new show on a scale of 1 to 10 (10 being the highest rating). From a group of 140 people, the following responses were obtained:

Rating	1	2	3	4	5	6	7	8	9	10
Frequency of Occurrence	1	4	3	11	23	21	28	29	16	4

    Let the random variable X denote the rating given to the show by a randomly chosen audience member. Find the probability distribution associated with these data.

22. U.S. POPULATION BY AGE   The following table gives the 2002 age distribution of the U.S. population:

Age (in years)	Under 5	5–19	20–24	25–44	45–64	65 and over
Number (in thousands)	19,527	59,716	18,611	83,009	66,088	33,590

    Let the random variable X denote a randomly chosen age group within the population. Find the probability distribution associated with these data.
    Source: U.S. Census Bureau

In Exercises 23 and 24, determine whether the statement is true or false. If it is true, explain why it is true. If it is false, give an example to show why it is false.

23. Suppose X is a finite discrete random variable assuming the values $x_1, x_2, \ldots, x_n$ and associated probabilities $p_1, p_2, \ldots, p_n$. Then $p_1 + p_2 + \cdots + p_n = 1$.

24. The area of a histogram associated with a probability distribution is a number between 0 and 1.

8.1 Solutions to Self-Check Exercises

1. a. Using the accompanying tree diagram, we see that the outcomes of the experiment are

$$S = \{BBB, BBW, BWB, BWW, WBB, WBW, WWB, WWW\}$$

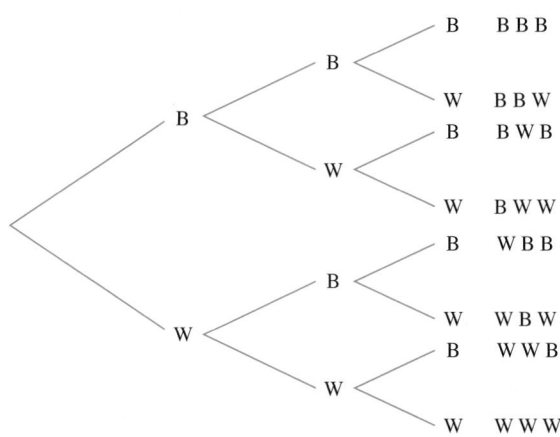

table by 1942 (the sum of these numbers) to obtain the probabilities associated with the random variable X when X takes on the values 0, 1, 2, 3, 4, 5, 6, 7, and 8. For example,

$$P(X = 0) = \frac{780}{1942} \approx .402$$

$$P(X = 1) = \frac{300}{1942} \approx .154$$

The required probability distribution and histogram follow.

x	0	1	2	3	4
$P(X = x)$	.402	.154	.212	.106	.050

x	5	6	7	8
$P(X = x)$	.028	.029	.015	.003

b. Using the results of part (a), we obtain the values assigned to the outcomes of the experiment as follows:

Outcome	BBB	BBW	BWB	BWW
Value	3	2	2	1

Outcome	WBB	WBW	WWB	WWW
Value	2	1	1	0

b.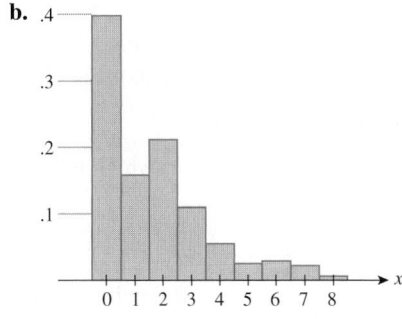

c. The required event is {BBW, BWB, WBB}.

2. a. We divide each number in the bottom row of the given

USING TECHNOLOGY

Graphing a Histogram

Graphing Utility

A graphing utility can be used to plot the histogram for a given set of data, as illustrated by the following example.

 APPLIED EXAMPLE 1 A survey of 90,000 households conducted in 1995 revealed the following percent of women who wear the given shoe size.

Shoe Size	<5	5–5½	6–6½	7–7½	8–8½	9–9½	10–10½	>10½
Women, %	1	5	15	27	29	14	7	2

Source: Footwear Market Insights survey

(continued)

a. Plot a histogram for the given data.

b. What percent of women in the survey wear size $7–7\frac{1}{2}$ or $8–8\frac{1}{2}$ shoes?

Solution

a. Let X denote the random variable taking on the values 1 through 8, where 1 corresponds to a shoe size less than 5, 2 corresponds to a shoe size of $5–5\frac{1}{2}$, and so on. Enter the values of X as $x_1 = 1, x_2 = 2, \ldots, x_8 = 8$ and the corresponding values of Y as $y_1 = 1, y_2 = 5, \ldots, y_8 = 2$. Then using the **DRAW** function from the Statistics menu, we draw the histogram shown in Figure T1.

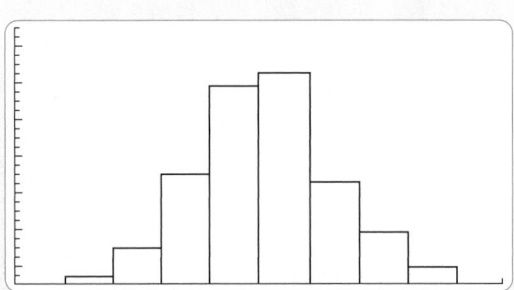

FIGURE T1
The histogram for the given data, using the viewing window $[0, 9] \times [0, 35]$

b. The probability that a woman participating in the survey wears size $7–7\frac{1}{2}$ or $8–8\frac{1}{2}$ shoes is given by

$$P(X = 4) + P(X = 5) = .27 + .29 = .56$$

which tells us that 56% of the women wear either size $7–7\frac{1}{2}$ or $8–8\frac{1}{2}$ shoes. ∎

Excel

Excel can be used to plot the histogram for a given set of data, as illustrated by the following example.

APPLIED EXAMPLE 2   A survey of 90,000 households conducted in 1995 revealed the following percent of women who wear the given shoe size.

Shoe Size	<5	$5–5\frac{1}{2}$	$6–6\frac{1}{2}$	$7–7\frac{1}{2}$	$8–8\frac{1}{2}$	$9–9\frac{1}{2}$	$10–10\frac{1}{2}$	$>10\frac{1}{2}$
Women, %	1	5	15	27	29	14	7	2

Source: Footwear Market Insights survey

a. Plot a histogram for the given data.

b. What percent of women in the survey wear size $7–7\frac{1}{2}$ or $8–8\frac{1}{2}$ shoes?

Solution

a. Let X denote the random variable taking on the values 1 through 8, where 1 corresponds to a shoe size less than 5, 2 corresponds to a shoe size of $5–5\frac{1}{2}$, and so on. Next, enter the given data in columns A and B onto a spreadsheet, as shown in Figure T2. Highlight the data in column B and select $\boxed{\Sigma}$ from the toolbar. The

	A	B	C
	X	Frequency	Probability
1			
2	1	1	0.01
3	2	5	0.05
4	3	15	0.15
5	4	27	0.27
6	5	29	0.29
7	6	14	0.14
8	7	7	0.07
9	8	2	0.02
10		100	

FIGURE T2
Completed spreadsheet for Example 2

Note: Boldfaced words/characters enclosed in a box (for example, $\boxed{\textbf{Enter}}$) indicate an action (click, select, or press) is required. Words/characters printed blue (for example, Chart sub-type:) indicate words/characters that appear on the screen. Words/characters printed in a typewriter font (for example, =(−2/3)*A2+2) indicate words/characters that need to be typed and entered.

sum of the numbers in this column (100) will appear in cell B10. In cell C2, type =B2/100 and then press Enter . To extend the formula to cell C9, move the pointer to the small black box at the lower right corner of cell C2. Drag the black + that appears (at the lower right corner of cell C2) through cell C9 and then release it. The probability distribution shown in cells C2 to C9 will then appear on your spreadsheet. Then highlight the data in the Probability column and select Chart Wizard from the toolbar. Select Column under Chart type: and click Next twice. Under the Titles tab, enter Histogram, X, and Probability in the appropriate boxes. Under the Legend tab, click the Show legend box to delete the check mark. Then click Finish . The histogram shown in Figure T3 will appear.

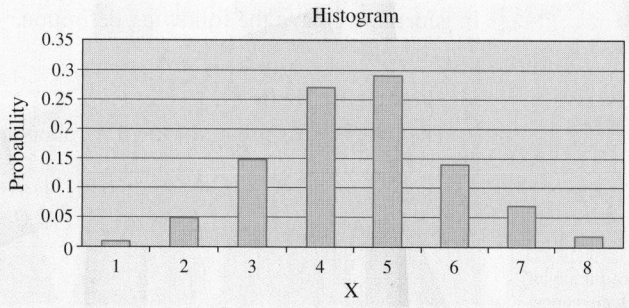

FIGURE T3
The histogram for the random variable X

b. The probability that a woman participating in the survey wears size $7-7\frac{1}{2}$ or $8-8\frac{1}{2}$ shoes is given by

$$P(X = 4) + P(X = 5) = .27 + .29 = .56$$

which tells us that 56% of the women wear either size $7-7\frac{1}{2}$ or $8-8\frac{1}{2}$ shoes. ◾

TECHNOLOGY EXERCISES

1. Graph the histogram associated with the data given in Table 1, page 444. Compare your graph with that given in Figure 1, page 447.
2. Graph the histogram associated with the data given in Exercise 18, page 450.
3. Graph the histogram associated with the data given in Exercise 19, page 450.
4. Graph the histogram associated with the data given in Exercise 21, page 450.

8.2   Expected Value

■ Mean

The average value of a set of numbers is a familiar notion to most people. For example, to compute the average of the four numbers

$$12, 16, 23, 37$$

we simply add these numbers and divide the resulting sum by 4, giving the required average as

$$\frac{12 + 16 + 23 + 37}{4} = \frac{88}{4} = 22$$

In general, we have the following definition.

> **Average, or Mean**
> The average, or mean, of the n numbers
>
> $$x_1, x_2, \ldots, x_n$$
>
> is $\bar{x}$ (read "x bar"), where
>
> $$\bar{x} = \frac{x_1 + x_2 + \cdots + x_n}{n}$$

TABLE 7

Cars	Frequency of Occurrence
0	2
1	9
2	16
3	12
4	8
5	6
6	4
7	2
8	1

APPLIED EXAMPLE 1   Waiting Times   Refer to Example 6, Section 8.1. Find the average number of cars waiting in line at the bank's drive-in teller at the beginning of each 2-minute interval during the period in question.

Solution   The number of cars, together with its corresponding frequency of occurrence, are reproduced in Table 7. Observe that the number 0 (of cars) occurs twice, the number 1 occurs 9 times, and so on. There are altogether

$$2 + 9 + 16 + 12 + 8 + 6 + 4 + 2 + 1 = 60$$

numbers to be averaged. Therefore, the required average is given by

$$\frac{(0 \cdot 2) + (1 \cdot 9) + (2 \cdot 16) + (3 \cdot 12) + (4 \cdot 8) + (5 \cdot 6) + (6 \cdot 4) + (7 \cdot 2) + (8 \cdot 1)}{60} \approx 3.1 \qquad \textbf{(1)}$$

or approximately 3.1 cars.   ■

■ Expected Value

Let's reconsider the expression in Equation (1) that gives the average of the frequency distribution shown in Table 7. Dividing each term by the denominator, the expression may be rewritten in the form

$$0 \cdot \left(\frac{2}{60}\right) + 1 \cdot \left(\frac{9}{60}\right) + 2 \cdot \left(\frac{16}{60}\right) + 3 \cdot \left(\frac{12}{60}\right) + 4 \cdot \left(\frac{8}{60}\right) + 5 \cdot \left(\frac{6}{60}\right)$$

$$+ 6 \cdot \left(\frac{4}{60}\right) + 7 \cdot \left(\frac{2}{60}\right) + 8 \cdot \left(\frac{1}{60}\right)$$

Observe that each term in the sum is a product of two factors; the first factor is the value assumed by the random variable X, where X denotes the number of cars waiting in line, and the second factor is just the probability associated with that value of the random variable. This observation suggests the following general method for calculating the expected value (that is, the average or mean) of a random variable X that assumes a finite number of values from the knowledge of its probability distribution.

> ### Expected Value of a Random Variable X
> Let X denote a random variable that assumes the values $x_1, x_2, \ldots, x_n$ with associated probabilities $p_1, p_2, \ldots, p_n$, respectively. Then the **expected value** of X, $E(X)$, is given by
>
> $$E(X) = x_1 p_1 + x_2 p_2 + \cdots + x_n p_n \qquad (2)$$

Note The numbers $x_1, x_2, \ldots, x_n$ may be positive, zero, or negative. For example, such a number will be positive if it represents a profit and negative if it represents a loss. ■

TABLE 8

Probability Distribution

x	$P(X = x)$
0	.03
1	.15
2	.27
3	.20
4	.13
5	.10
6	.07
7	.03
8	.02

APPLIED EXAMPLE 2 Waiting Times Re-solve Example 1 by using the probability distribution associated with the experiment, which is reproduced in Table 8.

Solution Let X denote the number of cars waiting in line. Then the average number of cars waiting in line is given by the expected value of X—that is, by

$$E(X) = (0)(0.3) + (1)(.15) + (2)(.27) + (3)(.20) + (4)(.13)$$
$$+ (5)(.10) + (6)(.07) + (7)(.03) + (8)(.02)$$
$$= 3.1 \text{ cars}$$

which agrees with the earlier result. ■

The expected value of a random variable X is a measure of the central tendency of the probability distribution associated with X. In repeated trials of an experiment with random variable X, the average of the observed values of X gets closer and closer to the expected value of X as the number of trials gets larger and larger. Geometrically, the expected value of a random variable X has the following simple interpretation: If a laminate is made of the histogram of a probability distribution associated with a random variable X, then the expected value of X corresponds to the point on the base of the laminate at which the latter will balance perfectly when the point is directly over a fulcrum (Figure 6).

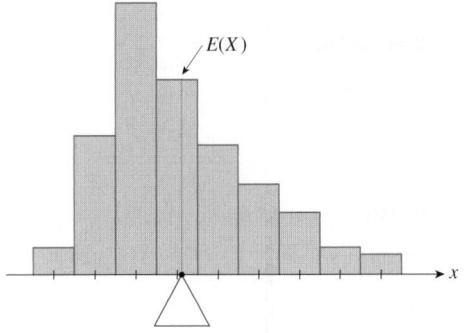

FIGURE 6
Expected value of a random variable X

TABLE 9

Probability Distribution

x	$P(X = x)$
2	$\frac{1}{36}$
3	$\frac{2}{36}$
4	$\frac{3}{36}$
5	$\frac{4}{36}$
6	$\frac{5}{36}$
7	$\frac{6}{36}$
8	$\frac{5}{36}$
9	$\frac{4}{36}$
10	$\frac{3}{36}$
11	$\frac{2}{36}$
12	$\frac{1}{36}$

FIGURE 7
Histogram showing the probability distribution for the sum of the faces of two dice

EXAMPLE 3 Let X denote the random variable that gives the sum of the faces that fall uppermost when two fair dice are cast. Find the expected value, $E(X)$, of X.

Solution The probability distribution of X, reproduced in Table 9, was found in Example 5, Section 8.1. Using this result, we find

$$E(X) = 2\left(\frac{1}{36}\right) + 3\left(\frac{2}{36}\right) + 4\left(\frac{3}{36}\right) + 5\left(\frac{4}{36}\right) + 6\left(\frac{5}{36}\right) + 7\left(\frac{6}{36}\right)$$
$$+ 8\left(\frac{5}{36}\right) + 9\left(\frac{4}{36}\right) + 10\left(\frac{3}{36}\right) + 11\left(\frac{2}{36}\right) + 12\left(\frac{1}{36}\right)$$
$$= 7$$

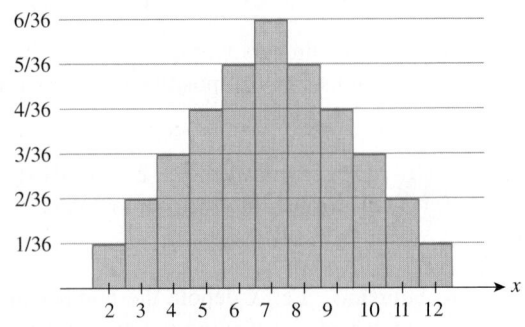

Note that, because of the symmetry of the histogram of the probability distribution with respect to the vertical line $x = 7$, the result could have been obtained by merely inspecting Figure 7. ■

The next example shows how we can use the concept of expected value to help us make the best investment decision.

APPLIED EXAMPLE 4 Expected Profit A group of private investors intends to purchase one of two motels currently being offered for sale in a certain city. The terms of sale of the two motels are similar, although the Regina Inn has 52 rooms and is in a slightly better location than the Merlin Motor Lodge, which has 60 rooms. Records obtained for each motel reveal that the occupancy rates, with corresponding probabilities, during the May–September tourist season are as shown in the following tables.

Regina Inn					
Occupancy Rate	.80	.85	.90	.95	1.00
Probability	.19	.22	.31	.23	.05

Merlin Motor Lodge						
Occupancy Rate	.75	.80	.85	.90	.95	1.00
Probability	.35	.21	.18	.15	.09	.02

The average profit per day for each occupied room at the Regina Inn is $10, whereas the average profit per day for each occupied room at the Merlin Motor Lodge is $9.

a. Find the average number of rooms occupied per day at each motel.

b. If the investors' objective is to purchase the motel that generates the higher daily profit, which motel should they purchase? (Compare the expected daily profit of the two motels.)

Solution

a. Let X denote the occupancy rate at the Regina Inn. Then the average daily occupancy rate at the Regina Inn is given by the expected value of X—that is, by

$$E(X) = (.80)(.19) + (.85)(.22) + (.90)(.31)$$
$$+ (.95)(.23) + (1.00)(.05)$$
$$= .8865$$

The average number of rooms occupied per day at the Regina is

$$(.8865)(52) \approx 46.1$$

or approximately 46.1 rooms. Similarly, letting Y denote the occupancy rate at the Merlin Motor Lodge, we have

$$E(Y) = (.75)(.35) + (.80)(.21) + (.85)(.18) + (.90)(.15)$$
$$+ (.95)(.09) + (1.00)(.02)$$
$$= .8240$$

The average number of rooms occupied per day at the Merlin is

$$(.8240)(60) \approx 49.4$$

or approximately 49.4 rooms.

b. The expected daily profit at the Regina is given by

$$(46.1)(10) = 461$$

or $461. The expected daily profit at the Merlin is given by

$$(49.4)(9) \approx 445$$

or $445. From these results we conclude that the investors should purchase the Regina Inn, which is expected to yield a higher daily profit. ▪

APPLIED EXAMPLE 5   Raffles The Island Club is holding a fund-raising raffle. Ten thousand tickets have been sold for $2 each. There will be a first prize of $3000, 3 second prizes of $1000 each, 5 third prizes of $500 each, and 20 consolation prizes of $100 each. Letting X denote the net winnings (that is, winnings less the cost of the ticket) associated with the tickets, find $E(X)$. Interpret your results.

Solution   The values assumed by X are $(0 - 2), (100 - 2), (500 - 2), (1000 - 2)$, and $(3000 - 2)$—that is, $-2, 98, 498, 998$, and 2998—which correspond, respectively, to the value of a losing ticket, a consolation prize, a third prize, and so on.

TABLE 10

Probability Distribution for a Raffle

x	$P(X = x)$
-2	.9971
98	.0020
498	.0005
998	.0003
2998	.0001

The probability distribution of X may be calculated in the usual manner and appears in Table 10. Using the table, we find

$$E(X) = (-2)(.9971) + 98(.0020) + 498(.0005)$$
$$+ 998(.0003) + 2998(.0001)$$
$$= -0.95$$

This expected value gives the long-run average loss (negative gain) of a holder of one ticket; that is, if one participated in such a raffle by purchasing one ticket each time, in the long run, one may expect to lose, on the average, 95 cents per raffle. ◼

APPLIED EXAMPLE 6   Roulette   In the game of roulette as played in Las Vegas casinos, the wheel is divided into 38 compartments numbered 1 through 36, 0, and 00. One-half of the numbers 1 through 36 are red, the other half black, and 0 and 00 are green (Figure 8). Of the many types of bets that may be placed, one type involves betting on the outcome of the color of the winning number. For example, one may place a certain sum of money on *red*. If the winning number is red, one wins an amount equal to the bet placed and loses the bet otherwise. Find the expected value of the winnings on a $1 bet placed on *red*.

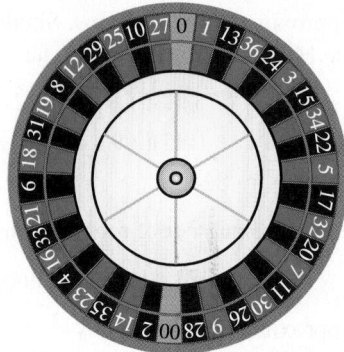

FIGURE 8
Roulette wheel

Solution   Let X be a random variable whose values are 1 and -1, which correspond to a win and a loss. The probabilities associated with the values 1 and -1 are $\frac{18}{38}$ and $\frac{20}{38}$, respectively. Therefore, the expected value is given by

$$E(X) = 1\left(\frac{18}{38}\right) + (-1)\left(\frac{20}{38}\right) = -\frac{2}{38}$$
$$\approx -0.053$$

Thus, if one places a $1 bet on *red* over and over again, one may expect to lose, on the average, approximately 5 cents per bet in the long run. ◼

Examples 5 and 6 illustrate games that are not "fair." Of course, most participants in such games are aware of this fact and participate in them for other reasons. In a fair game, neither party has an advantage, a condition that translates into the condition that $E(X) = 0$, where X takes on the values of a player's winnings.

APPLIED EXAMPLE 7   Fair Games   Mike and Bill play a card game with a standard deck of 52 cards. Mike selects a card from a well-shuffled deck and receives A dollars from Bill if the card selected is a diamond; otherwise, Mike pays Bill a dollar. Determine the value of A if the game is to be fair.

Solution Let X denote a random variable whose values are associated with Mike's winnings. Then X takes on the value A with probability $P(X = A) = \frac{1}{4}$ (since there are 13 diamonds in the deck) if Mike wins and takes on the value -1 with probability $P(X = -1) = \frac{3}{4}$ if Mike loses. Since the game is to be a fair one, the expected value $E(X)$ of Mike's winnings must be equal to zero; that is,

$$E(X) = A\left(\frac{1}{4}\right) + (-1)\left(\frac{3}{4}\right) = 0$$

Solving this equation for A gives $A = 3$. Thus, the card game will be fair if Bill makes a $3 payoff for a winning bet of $1 placed by Mike. ∎

▪ Odds

In everyday parlance, the probability of the occurrence of an event is often stated in terms of the *odds in favor of* (or *odds against*) the occurrence of the event. For example, one often hears statements such as "the odds that the Dodgers will win the World Series this season are 7 to 5" and "the odds that it will not rain tomorrow are 3 to 2." We will return to these examples later. But first, let us look at a definition that ties together these two concepts.

Odds In Favor Of and Odds Against

If $P(E)$ is the probability of an event E occurring, then

1. The odds in favor of E occurring are

$$\frac{P(E)}{1 - P(E)} = \frac{P(E)}{P(E^c)} \qquad [P(E) \neq 1] \qquad \textbf{(3a)}$$

2. The odds against E occurring are

$$\frac{1 - P(E)}{P(E)} = \frac{P(E^c)}{P(E)} \qquad [P(E) \neq 0] \qquad \textbf{(3b)}$$

Notes

1. The odds in favor of the occurrence of an event are given by the ratio of the probability of the event occurring to the probability of the event not occurring. The odds against the occurrence of an event are given by the reciprocal of the odds in favor of the occurrence of the event.

2. Whenever possible, odds are expressed as ratios of whole numbers. If the odds in favor of E are a/b, we say the odds in favor of E are a to b. If the odds against E occurring are b/a, we say the odds against E are b to a. ∎

EXAMPLE 8 Find the odds in favor of winning a bet on *red* in American roulette. What are the odds against winning a bet on *red*?

Solution The probability of winning a bet here—the probability that the ball lands in a red compartment—is given by $P = \frac{18}{38}$. Therefore, using Formula (3a), we see that the odds in favor of winning a bet on *red* are

$$\frac{P(E)}{1 - P(E)} = \frac{\frac{18}{38}}{1 - \frac{18}{38}} \qquad \text{\textit{E}, event of winning a bet on \textit{red}}$$

$$= \frac{\frac{18}{38}}{\frac{38 - 18}{38}}$$

$$= \frac{18}{38} \cdot \frac{38}{20}$$

$$= \frac{18}{20} = \frac{9}{10}$$

or 9 to 10. Next, using (3b), we see that the odds against winning a bet on *red* are $\frac{10}{9}$, or 10 to 9.   ■

Now, suppose the odds in favor of the occurrence of an event are a to b. Then, (3a) gives

$$\frac{a}{b} = \frac{P(E)}{1 - P(E)}$$

$$a[1 - P(E)] = bP(E) \qquad \text{Cross-multiply}$$

$$a - aP(E) = bP(E)$$

$$a = (a + b)P(E)$$

$$P(E) = \frac{a}{a + b}$$

which leads us to the following result.

Probability of an Event (Given the Odds)
If the odds in favor of an event E occurring are a to b, then the probability of E occurring is

$$P(E) = \frac{a}{a + b} \qquad\qquad\qquad \textbf{(4)}$$

Formula (4) is often used to determine subjective probabilities, as the next example shows.

EXAMPLE 9   Consider each of the following statements.
a. "The odds that the Dodgers will win the World Series this season are 7 to 5."
b. "The odds that it will not rain tomorrow are 3 to 2."

Express each of these odds as a probability of the event occurring.

Solution
a. Using Formula (4) with $a = 7$ and $b = 5$ gives the required probability as

$$\frac{7}{7 + 5} = \frac{7}{12} \approx .5833$$

b. Here, the event is that it will not rain tomorrow. Using (4) with $a = 3$ and $b = 2$, we conclude that the probability that it will not rain tomorrow is

$$\frac{3}{3 + 2} = \frac{3}{5} = .6$$ ■

TITLE   Senior Project Director
INSTITUTION   GfK ARBOR

GfK ARBOR, LLC is a full-service, custom marketing research and consulting firm. We develop and apply advanced research methodologies and analyses to a wide array of marketing and marketing research problems. We have provided services to many of the largest corporations in the United States as well as many other countries around the world across a wide array of industries. Statistics play a big part in helping our clients find solutions to their marketing problems.

A manufacturer of a brand of juice wanted to know how their brand was performing overall and how various factors in the marketplace were affecting consumers' perceptions and usage of their brand as well as the juice category overall. This juice brand was beginning a new advertising campaign, introducing new packaging and flavors, and had reduced the juice content of their product.

A year-long tracking study was conducted over the Internet with a sample of juice users. Consumers were asked their opinions, perceptions, and consumption of various brands of juice.

Using statistics, we were able to determine with certainty whether the juice brand's overall performance had improved or declined over time. Opinion ratings and brand awareness as well as consumption levels from before the new advertising campaign, packaging, flavors, and product formulation occurred were compared to levels after these events. Oftentimes we see movement or change over time, but these changes or differences must be statistically significant for us to say there has in fact been a change.

The information obtained from this research gave the juice manufacturer direction for future advertising and the ability to tailor their advertising to include elements that appeal to consumers and motivate them to purchase this particular brand of juice. The manufacturer was able to make informed decisions about whether or not to continue with the new packaging and flavors and if modifications were needed. We were able to determine that the reduced juice content did not have detrimental effects on consumer opinions or consumption of the brand.

EXPLORE & DISCUSS

In the movie *Casino*, the executive of the Tangiers Casino, Sam Rothstein (Robert DeNiro), fired the manager of the slot machines in the casino after three gamblers hit three "million dollar" jackpots in a span of 20 minutes. Rothstein claimed that it was a scam and that somebody had gotten into those machines to set the wheels. He was especially annoyed at the slot machine manager's assertion that there was no way to determine this. According to Rothstein the odds of hitting a jackpot in a four-wheel machine is 1 in $1\frac{1}{2}$ million, and the probability of hitting three jackpots in a row is "in the billions." "It cannot happen! It will not happen!" To see why Mr. Rothstein was so indignant, find the odds of hitting the jackpots in three of the machines in quick succession and comment on the likelihood of this happening.

■ Median and Mode

In addition to the mean, there are two other measures of central tendency of a set of numerical data: the median and the mode of a set of numbers.

Median
The median of a set of numbers arranged in increasing or decreasing order is (a) the middle number if there is an odd number of entries or (b) the mean of the two middle numbers if there is an even number of entries.

APPLIED EXAMPLE 10   Commuting Times

a. The times, in minutes, Susan took to go to work on nine consecutive working days were

$$46 \quad 42 \quad 49 \quad 40 \quad 52 \quad 48 \quad 45 \quad 43 \quad 50$$

What is the median of her morning commute times?

b. The times, in minutes, Susan took to return home from work on eight consecutive working days were

$$37 \quad 36 \quad 39 \quad 37 \quad 34 \quad 38 \quad 41 \quad 40$$

What is the median of her evening commute times?

Solution

a. Arranging the numbers in increasing order, we have

$$40 \quad 42 \quad 43 \quad 45 \quad 46 \quad 48 \quad 49 \quad 50 \quad 52$$

Here we have an odd number of entries with the middle number equal to 46, and this gives the required median.

b. Arranging the numbers in increasing order, we have

$$34 \quad 36 \quad 37 \quad 37 \quad 38 \quad 39 \quad 40 \quad 41$$

Here the number of entries is even, and the required median is

$$\frac{37 + 38}{2} = 37.5$$

Mode

The **mode** of a set of numbers is the number in the set that occurs most frequently.

Note

A set may have no mode, a unique mode, or more than one mode.

EXAMPLE 11   Find the mode, if there is one, of the given set of numbers.
a. 1, 2, 3, 4, 6   **b.** 2, 3, 3, 4, 6, 8   **c.** 2, 3, 3, 3, 4, 4, 4, 8

Solution

a. The set has no mode because there isn't a number that occurs more frequently than the others.

b. The mode is 3 because it occurs more frequently than the others.

c. The modes are 3 and 4 because each number occurs three times.

Finally, observe that of the three measures of central tendency of a set of numerical data, only the mean is suitable in work that requires mathematical computations.

8.2 Self-Check Exercises

1. Find the expected value of a random variable X having the following probability distribution:

x	-4	-3	-1	0	1	2
$P(X = x)$	.10	.20	.25	.10	.25	.10

2. The developer of Shoreline Condominiums has provided the following estimate of the probability that 20, 25, 30, 35, 40, 45, or 50 of the townhouses will be sold within the first month they are offered for sale.

Units	20	25	30	35	40	45	50
Probability	.05	.10	.30	.25	.15	.10	.05

How many townhouses can the developer expect to sell within the first month they are put on the market?

Solutions to Self-Check Exercises 8.2 can be found on page 467.

8.2 Concept Questions

1. What is the expected value of a random variable? Give an example.

2. What is a fair game? Is the game of roulette as played in American casinos a fair game? Why?

3. a. If the probability of an event E occurring is $P(E)$, what are the odds in favor of E occurring?
   b. If the odds in favor of an event occurring are a to b, what is the probability of E occurring?

8.2 Exercises

1. During the first year at a university that uses a 4-point grading system, a freshman took ten 3-credit courses and received two As, three Bs, four Cs, and one D.
   a. Compute this student's grade-point average.
   b. Let the random variable X denote the number of points corresponding to a given letter grade. Find the probability distribution of the random variable X and compute $E(X)$, the expected value of X.

2. Records kept by the chief dietitian at the university cafeteria over a 30-week period show the following weekly consumption of milk (in gallons).

Milk	200	205	210	215	220
Weeks	3	4	6	5	4

Milk	225	230	235	240
Weeks	3	2	2	1

   a. Find the average number of gallons of milk consumed per week in the cafeteria.
   b. Let the random variable X denote the number of gallons of milk consumed in a week at the cafeteria. Find the probability distribution of the random variable X and compute $E(X)$, the expected value of X.

3. Find the expected value of a random variable X having the following probability distribution:

x	-5	-1	0	1	5	8
$P(X = x)$	.12	.16	.28	.22	.12	.1

4. Find the expected value of a random variable X having the following probability distribution:

x	0	1	2	3	4	5
$P(X = x)$	$\frac{1}{8}$	$\frac{1}{4}$	$\frac{3}{16}$	$\frac{1}{4}$	$\frac{1}{16}$	$\frac{1}{8}$

5. The daily earnings X of an employee who works on a commission basis are given by the following probability distribution. Find the employee's expected earnings.

x (in $)	0	25	50	75
$P(X = x)$	.07	.12	.17	.14

x (in $)	100	125	150
$P(X = x)$	.28	.18	.04

6. In a four-child family, what is the expected number of boys? (Assume that the probability of a boy being born is the same as the probability of a girl being born.)

7. Based on past experience, the manager of the VideoRama Store has compiled the following table, which gives the probabilities that a customer who enters the VideoRama Store will buy 0, 1, 2, 3, or 4 videocassettes. How many videocassettes can a customer entering this store be expected to buy?

Video-cassettes	0	1	2	3	4
Probability	.42	.36	.14	.05	.03

8. If a sample of three batteries is selected from a lot of ten, of which two are defective, what is the expected number of defective batteries?

9. AUTO ACCIDENTS The number of accidents that occur at a certain intersection known as "Five Corners" on a Friday afternoon between the hours of 3 p.m. and 6 p.m., along with the corresponding probabilities, are shown in the following table. Find the expected number of accidents during the period in question.

Accidents	0	1	2	3	4
Probability	.935	.03	.02	.01	.005

10. EXPECTED DEMAND The owner of a newsstand in a college community estimates the weekly demand for a certain magazine as follows:

Quantity Demanded	10	11	12	13	14	15
Probability	.05	.15	.25	.30	.20	.05

Find the number of issues of the magazine that the newsstand owner can expect to sell per week.

11. EXPECTED PRODUCT RELIABILITY A bank has two automatic tellers at its main office and two at each of its three branches. The number of machines that break down on a given day, along with the corresponding probabilities, are shown in the following table.

Machines That Break Down	0	1	2	3	4
Probability	.43	.19	.12	.09	.04

Machines That Break Down	5	6	7	8
Probability	.03	.03	.02	.05

Find the expected number of machines that will break down on a given day.

12. EXPECTED SALES The management of the Cambridge Company has projected the sales of its products (in millions of dollars) for the upcoming year, with the associated probabilities shown in the following table:

Sales	20	22	24	26	28	30
Probability	.05	.1	.35	.3	.15	.05

What does the management expect the sales to be next year?

13. INTEREST-RATE PREDICTION A panel of 50 economists was asked to predict the average prime interest rate for the upcoming year. The results of the survey follow:

Interest Rate, %	4.9	5.0	5.1	5.2	5.3	5.4
Economists	3	8	12	14	8	5

Based on this survey, what does the panel expect the average prime interest rate to be next year?

14. UNEMPLOYMENT RATES A panel of 64 economists was asked to predict the average unemployment rate for the upcoming year. The results of the survey follow:

Unemployment Rate, %	4.5	4.6	4.7	4.8	4.9	5.0	5.1
Economists	2	4	8	20	14	12	4

Based on this survey, what does the panel expect the average unemployment rate to be next year?

15. LOTTERIES In a lottery, 5000 tickets are sold for $1 each. One first prize of $2000, 1 second prize of $500, 3 third prizes of $100, and 10 consolation prizes of $25 are to be awarded. What are the expected net earnings of a person who buys one ticket?

16. LIFE INSURANCE PREMIUMS A man wishes to purchase a 5-yr term-life insurance policy that will pay the beneficiary $20,000 in the event that the man's death occurs during the next 5 yr. Using life insurance tables, he determines that the probability that he will live another 5 yr is .96. What is the minimum amount that he can expect to pay for his premium?
Hint: The minimum premium occurs when the insurance company's expected profit is zero.

17. LIFE INSURANCE PREMIUMS A woman purchased a $10,000, 1-yr term-life insurance policy for $130. Assuming that the probability that she will live another year is .992, find the company's expected gain.

18. LIFE INSURANCE POLICIES As a fringe benefit, Dennis Taylor receives a $25,000 life insurance policy from his employer. The probability that Dennis will live another year is .9935. If he purchases the same coverage for himself, what is the minimum amount that he can expect to pay for the policy?

19. EXPECTED PROFIT A buyer for Discount Fashions, an outlet for women's apparel, is considering the purchase of a batch of clothing for $64,000. She estimates that the company will be able to sell it for $80,000, $75,000, or $70,000 with probabilities of .30, .60, and .10, respectively. Based on these estimates, what will be the company's expected gross profit?

20. INVESTMENT ANALYSIS The proprietor of Midland Construction Company has to decide between two projects. He estimates that the first project will yield a profit of $180,000 with a probability of .7 or a profit of $150,000 with a probability of .3; the second project will yield a profit of $220,000 with a probability of .6 or a profit of $80,000 with a probability of .4. Which project should the proprietor choose if he wants to maximize his expected profit?

21. CABLE TELEVISION   The management of MultiVision, a cable TV company, intends to submit a bid for the cable television rights in one of two cities, A or B. If the company obtains the rights to city A, the probability of which is .2, the estimated profit over the next 10 yr is $10 million; if the company obtains the rights to city B, the probability of which is .3, the estimated profit over the next 10 yr is $7 million. The cost of submitting a bid for rights in city A is $250,000 and that in city B is $200,000. By comparing the expected profits for each venture, determine whether the company should bid for the rights in city A or city B.

22. EXPECTED AUTO SALES   Roger Hunt intends to purchase one of two car dealerships currently for sale in a certain city. Records obtained from each of the two dealers reveal that their weekly volume of sales, with corresponding probabilities, are as follows:

Dahl Motors

Cars Sold/Week	5	6	7	8
Probability	.05	.09	.14	.24

Cars Sold/Week	9	10	11	12
Probability	.18	.14	.11	.05

Farthington Auto Sales

Cars Sold/Week	5	6	7	8	9	10
Probability	.08	.21	.31	.24	.10	.06

The average profit/car at Dahl Motors is $362, and the average profit/car at Farthington Auto Sales is $436.
a. Find the average number of cars sold each week at each dealership.
b. If Roger's objective is to purchase the dealership that generates the higher weekly profit, which dealership should he purchase? (Compare the expected weekly profit for each dealership.)

23. EXPECTED HOME SALES   Sally Leonard, a real estate broker, is relocating in a large metropolitan area where she has received job offers from realty company A and realty company B. The number of houses she expects to sell in a year at each firm and the associated probabilities are shown in the following tables.

Company A

Houses Sold	12	13	14	15	16
Probability	.02	.03	.05	.07	.07

Houses Sold	17	18	19	20
Probability	.16	.17	.13	.11

Houses Sold	21	22	23	24
Probability	.09	.06	.03	.01

Company B

Houses Sold	6	7	8	9	10
Probability	.01	.04	.07	.06	.11

Houses Sold	11	12	13	14
Probability	.12	.19	.17	.13

Houses Sold	15	16	17	18
Probability	.04	.03	.02	.01

The average price of a house in the locale of company A is $308,000, whereas the average price of a house in the locale of company B is $474,000. If Sally will receive a 3% commission on sales at both companies, which job offer should she accept to maximize her expected yearly commission?

24. INVESTMENT ANALYSIS   Bob, the proprietor of Midway Lumber, bases his projections for the annual revenues of the company on the performance of the housing market. He rates the performance of the market as very strong, strong, normal, weak, and very weak. For the next year, Bob estimates that the probabilities for these outcomes are .18, .27, .42, .10, and .03, respectively. He also thinks that the revenues corresponding to these outcomes are $20, $18.8, $16.2, $14, and $12 million, respectively. What is Bob's expected revenue for next year?

25. REVENUE PROJECTION   Maria sees the growth of her business for the upcoming year as being tied to the gross domestic product (GDP). She believes that her business will grow (or contract) at the rate of 5%, 4.5%, 3%, 0%, or −0.5% per year if the GDP grows (or contracts) at the rate of between 2 and 2.5%, between 1.5 and 2%, between 1 and 1.5%, between 0 and 1%, and between −1 and 0%, respectively. Maria has decided to assign a probability of .12, .24, .40, .20, and .04, respectively, to each outcome. At what rate does Maria expect her business to grow next year?

26. WEATHER PREDICTIONS   Suppose the probability that it will rain tomorrow is .3.
a. What are the odds that it will rain tomorrow?
b. What are the odds that it will not rain tomorrow?

27. ROULETTE   In American roulette, as described in Example 6, a player may bet on a split (two adjacent numbers). In this case, if the player bets $1 and either number comes up, the player wins $17 and gets his $1 back. If neither comes up, he loses his $1 bet. Find the expected value of the winnings on a $1 bet placed on a split.

28. ROULETTE   If a player placed a $1 bet on *red* and a $1 bet on *black* in a single play in American roulette, what would be the expected value of his winnings?

29. ROULETTE   In European roulette, the wheel is divided into 37 compartments numbered 1 through 36 and 0. (In American roulette there are 38 compartments numbered 1 through 36, 0, and 00.) Find the expected value of the winnings on a $1 bet placed on *red* in European roulette.

30. The probability of an event E occurring is .8. What are the odds in favor of E occurring? What are the odds against E occurring?

31. The probability of an event E not occurring is .6. What are the odds in favor of E occurring? What are the odds against E occurring?

32. The odds in favor of an event E occurring are 9 to 7. What is the probability of E occurring?

33. The odds against an event E occurring are 2 to 3. What is the probability of E not occurring?

34. ODDS Carmen, a computer sales representative, feels that the odds are 8 to 5 that she will clinch the sale of a mini-computer to a certain company. What is the (subjective) probability that Carmen will make the sale?

35. SPORTS Steffi feels that the odds in favor of her winning her tennis match tomorrow are 7 to 5. What is the (subjective) probability that she will win her match tomorrow?

36. SPORTS If a sports forecaster states that the odds of a certain boxer winning a match are 4 to 3, what is the (subjective) probability that the boxer will win the match?

37. ODDS Bob, the proprietor of Midland Lumber, feels that the odds in favor of a business deal going through are 9 to 5. What is the (subjective) probability that this deal will *not* materialize?

38. ROULETTE
a. Show that, for any number c,

$$E(cX) = cE(X)$$

b. Use this result to find the expected loss if a gambler bets $300 on *red* in a single play in American roulette.
Hint: Use the results of Example 6.

39. EXAM SCORES In an examination given to a class of 20 students, the following test scores were obtained:

40 45 50 50 55 60 60 75 75 80
80 85 85 85 85 90 90 95 95 100

a. Find the mean (or average) score, the mode, and the median score.
b. Which of these three measures of central tendency do you think is the least representative of the set of scores?

40. WAGE RATES The frequency distribution of the hourly wage rates (in dollars) among blue-collar workers in a certain factory is given in the following table. Find the mean (or average) wage rate, the mode, and the median wage rate of these workers.

Wage Rate	10.70	10.80	10.90	11.00	11.10	11.20
Frequency	60	90	75	120	60	45

41. WAITING TIMES Refer to Example 6, Section 8.1. Find the median of the number of cars waiting in line at the bank's drive-in teller at the beginning of each 2-minute interval during the period in question. Compare you answer to the mean obtained in Example 1, Section 8.2.

42. SAN FRANCISCO WEATHER The normal daily minimum temperature in degrees Fahrenheit for the months of January through December in San Francisco follows:

46.2 48.4 48.6 49.2 50.7 52.5
53.1 54.2 55.8 54.8 51.5 47.2

Find the average and the median daily temperature in San Francisco for these months.
Source: San Francisco Convention and Visitors Bureau

43. WEIGHT OF POTATO CHIPS The weights, in ounces, of ten packages of potato chips are

16.1 16 15.8 16 15.9 16.1 15.9 16 16 16.2

Find the average, the median, and the mode of these weights.

44. BOSTON WEATHER The relative humidity, in percent, in the morning for the months of January through December in Boston follows:

68 67 69 69 71 73
74 76 79 77 74 70

Find the average, the median, and the mode of these humidity readings.
Source: National Weather Service Forecast Office

In Exercises 45 and 46, determine whether the statement is true or false. If it is true, explain why it is true. If it is false, give an example to show why it is false.

45. A game between two persons is fair if the expected value to both persons is zero.

46. If the odds in favor of an event E occurring are a to b, then the probability of E^c occurring is $b/(a + b)$.

8.2 Solutions to Self-Check Exercises

1. $E(X) = (-4)(.10) + (-3)(.20) + (-1)(.25)$
$\quad\quad\quad + (0)(.10) + (1)(.25) + (2)(.10)$
$\quad\quad = -0.8$

2. Let X denote the number of townhouses that will be sold within 1 mo of being put on the market. Then, the number of townhouses the developer expects to sell within 1 mo is given by the expected value of X—that is, by

$$E(X) = 20(.05) + 25(.10) + 30(.30) + 35(.25)$$
$$\quad\quad + 40(.15) + 45(.10) + 50(.05)$$
$$\quad\quad = 34.25$$

or 34 townhouses.

8.3 Variance and Standard Deviation

Variance

The mean, or expected value, of a random variable enables us to express an important property of the probability distribution associated with the random variable in terms of a single number. But the knowledge of the location, or central tendency, of a probability distribution alone is usually not enough to give a reasonably accurate picture of the probability distribution. Consider, for example, the two probability distributions whose histograms appear in Figure 9. Both distributions have the same expected value, or mean, of $\mu = 4$ (the Greek letter μ is read "mu"). Note that the probability distribution with the histogram shown in Figure 9a is closely concentrated about its mean μ, whereas the one with the histogram shown in Figure 9b is widely dispersed or spread about its mean.

As another example, suppose David Horowitz, host of the popular television show *The Consumer Advocate*, decides to demonstrate the accuracy of the weights of two popular brands of potato chips. Ten packages of potato chips of each brand are selected at random and weighed carefully. The results are as follows:

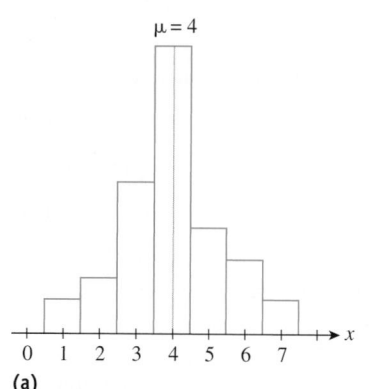

$\mu = 4$

(a)

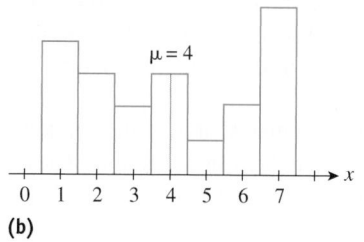

$\mu = 4$

(b)

FIGURE 9
The histograms of two probability distributions

Weight in Ounces									
Brand A	16.1	16	15.8	16	15.9	16.1	15.9	16	16	16.2
Brand B	16.3	15.7	15.8	16.2	15.9	16.1	15.7	16.2	16	16.1

In Example 3, we will verify that the mean weights for each of the two brands is 16 ounces. However, a cursory examination of the data now shows that the weights of the brand B packages exhibit much greater dispersion about the mean than those of brand A.

One measure of the degree of dispersion, or spread, of a probability distribution about its mean is given by the variance of the random variable associated with the probability distribution. A probability distribution with a small spread about its mean will have a small variance, whereas one with a larger spread will have a larger variance. Thus, the variance of the random variable associated with the probability distribution whose histogram appears in Figure 9a is smaller than the variance of the random variable associated with the probability distribution whose histogram is shown in Figure 9b (see Example 1). Also, as we will see in Example 3, the variance of the random variable associated with the weights of the brand A potato chips is smaller than that of the random variable associated with the weights of the brand B potato chips. (This observation was made earlier.)

We now define the variance of a random variable.

Variance of a Random Variable X

Suppose a random variable has the probability distribution

x	x_1	x_2	x_3	$\cdots$	x_n
$P(X = x)$	p_1	p_2	p_3	$\cdots$	p_n

and expected value

$$E(X) = \mu$$

Then the **variance** of the random variable X is

$$\text{Var}(X) = p_1(x_1 - \mu)^2 + p_2(x_2 - \mu)^2 + \cdots + p_n(x_n - \mu)^2 \quad \textbf{(5)}$$

Let's look a little closer at Equation (5). First, note that the numbers

$$x_1 - \mu, x_2 - \mu, \ldots, x_n - \mu \quad \textbf{(6)}$$

measure the **deviations** of $x_1, x_2, \ldots, x_n$ from μ, respectively. Thus, the numbers

$$(x_1 - \mu)^2, (x_2 - \mu)^2, \ldots, (x_n - \mu)^2 \quad \textbf{(7)}$$

measure the squares of the deviations of $x_1, x_2, \ldots, x_n$ from μ, respectively. Next, by multiplying each of the numbers in (7) by the probability associated with each value of the random variable X, the numbers are weighted accordingly so that their sum is a measure of the variance of X about its mean. An attempt to define the variance of a random variable about its mean in a similar manner using the deviations in (6), rather than their squares, would not be fruitful since some of the deviations may be positive whereas others may be negative and hence (because of cancellations) the sum will not give a satisfactory measure of the variance of the random variable.

EXAMPLE 1   Find the variance of the random variable X and of the random variable Y whose probability distributions are shown in the following table. These are the probability distributions associated with the histograms shown in Figure 9a–b.

x	$P(X = x)$	y	$P(Y = y)$
1	.05	1	.2
2	.075	2	.15
3	.2	3	.1
4	.375	4	.15
5	.15	5	.05
6	.1	6	.1
7	.05	7	.25

Solution   The mean of the random variable X is given by

$$\mu_X = (1)(.05) + (2)(.075) + (3)(.2) + (4)(.375) + (5)(.15)$$
$$+ (6)(.1) + (7)(.05)$$
$$= 4$$

Therefore, using Equation (5) and the data from the probability distribution of X, we find that the variance of X is given by

$$\text{Var}(X) = (.05)(1 - 4)^2 + (.075)(2 - 4)^2 + (.2)(3 - 4)^2$$
$$+ (.375)(4 - 4)^2 + (.15)(5 - 4)^2$$
$$+ (.1)(6 - 4)^2 + (.05)(7 - 4)^2$$
$$= 1.95$$

Next, we find that the mean of the random variable Y is given by

$$\mu_Y = (1)(.2) + (2)(.15) + (3)(.1) + (4)(.15) + (5)(.05)$$
$$+ (6)(.1) + (7)(.25)$$
$$= 4$$

and so the variance of Y is given by

$$\text{Var}(Y) = (.2)(1 - 4)^2 + (.15)(2 - 4)^2 + (.1)(3 - 4)^2$$
$$+ (.15)(4 - 4)^2 + (.05)(5 - 4)^2$$
$$+ (.1)(6 - 4)^2 + (.25)(7 - 4)^2$$
$$= 5.2$$

Note that $\text{Var}(X)$ is smaller than $\text{Var}(Y)$, which confirms our earlier observations about the spread (or dispersion) of the probability distribution of X and Y, respectively. ◼

■ Standard Deviation

Because Equation (5), which gives the variance of the random variable X, involves the squares of the deviations, the unit of measurement of $\text{Var}(X)$ is the square of the unit of measurement of the values of X. For example, if the values assumed by the random variable X are measured in units of a gram, then $\text{Var}(X)$ will be measured in units involving the *square* of a gram. To remedy this situation, one normally works with the square root of $\text{Var}(X)$ rather than $\text{Var}(X)$ itself. The former is called the standard deviation of X.

Standard Deviation of a Random Variable X
The **standard deviation** of a random variable X, σ (pronounced "sigma"), is defined by

$$\sigma = \sqrt{\text{Var}(X)}$$
$$= \sqrt{p_1(x_1 - \mu)^2 + p_2(x_2 - \mu)^2 + \cdots + p_n(x_n - \mu)^2} \qquad \textbf{(8)}$$

where $x_1, x_2, \ldots, x_n$ denote the values assumed by the random variable X and $p_1 = P(X = x_1), p_2 = P(X = x_2), \ldots, p_n = P(X = x_n)$.

EXAMPLE 2   Find the standard deviations of the random variables X and Y of Example 1.

Solution   From the results of Example 1, we have $\text{Var}(X) = 1.95$ and $\text{Var}(Y) = 5.2$. Taking their respective square roots, we have

$$\sigma_X = \sqrt{1.95}$$
$$\approx 1.40$$
$$\sigma_Y = \sqrt{5.2}$$
$$\approx 2.28 \qquad \blacksquare$$

APPLIED EXAMPLE 3   Packaging   Let X and Y denote the random variables whose values are the weights of the brand A and brand B potato chips, respectively (see page 467). Compute the means and standard deviations of X and Y and interpret your results.

Solution   The probability distributions of X and Y may be computed from the given data as follows:

Brand A			Brand B		
x	Relative Frequency of Occurrence	$P(X = x)$	y	Relative Frequency of Occurrence	$P(Y = y)$
15.8	1	.1	15.7	2	.2
15.9	2	.2	15.8	1	.1
16.0	4	.4	15.9	1	.1
16.1	2	.2	16.0	1	.1
16.2	1	.1	16.1	2	.2
			16.2	2	.2
			16.3	1	.1

The means of X and Y are given by

$$\mu_X = (.1)(15.8) + (.2)(15.9) + (.4)(16.0) + (.2)(16.1)$$
$$+ (.1)(16.2)$$
$$= 16$$
$$\mu_Y = (.2)(15.7) + (.1)(15.8) + (.1)(15.9) + (.1)(16.0)$$
$$+ (.2)(16.1) + (.2)(16.2) + (.1)(16.3)$$
$$= 16$$

Therefore,

$$\text{Var}(X) = (.1)(15.8 - 16)^2 + (.2)(15.9 - 16)^2 + (.4)(16 - 16)^2$$
$$+ (.2)(16.1 - 16)^2 + (.1)(16.2 - 16)^2$$
$$= 0.012$$
$$\text{Var}(Y) = (.2)(15.7 - 16)^2 + (.1)(15.8 - 16)^2 + (.1)(15.9 - 16)^2$$
$$+ (.1)(16 - 16)^2 + (.2)(16.1 - 16)^2 + (.2)(16.2 - 16)^2$$
$$+ (.1)(16.3 - 16)^2$$
$$= 0.042$$

so that the standard deviations are

$$\sigma_X = \sqrt{\text{Var}(X)}$$
$$= \sqrt{0.012}$$
$$\approx 0.11$$
$$\sigma_Y = \sqrt{\text{Var}(Y)}$$
$$= \sqrt{0.042}$$
$$\approx 0.20$$

The mean of X and that of Y are both equal to 16. Therefore, the average weight of a package of potato chips of either brand is 16 ounces. However, the standard deviation of Y is greater than that of X. This tells us that the weights of the packages of brand B potato chips are more widely dispersed about the common mean of 16 than are those of brand A.

EXPLORE & DISCUSS

Suppose the mean weight of m packages of brand A potato chips is μ_1 and the standard deviation from the mean of their weight distribution is σ_1. Also suppose the mean weight of n packages of brand B potato chips is μ_2 and the standard deviation from the mean of their weight distribution is σ_2.

1. Show that the mean of the weights of packages of brand A and brand B combined is

$$\mu = \frac{m\mu_1 + n\mu_2}{m + n}$$

2. If $\mu_1 = \mu_2$, show that the standard deviation from the mean of the combined-weight distribution is

$$\sigma = \left[\frac{m\sigma_1^2 + n\sigma_2^2}{m + n} \right]^{1/2}$$

3. Refer to Example 3, page 470. Using the results of parts 1 and 2, find the mean and the standard deviation of the combined-weight distribution.

▣ Chebychev's Inequality

The standard deviation of a random variable X may be used in statistical estimations. For example, the following result, derived by the Russian mathematician P. L. Chebychev (1821–1894), gives the proportion of the values of X lying within k standard deviations of the expected value of X.

Chebychev's Inequality

Let X be a random variable with expected value μ and standard deviation σ. Then the probability that a randomly chosen outcome of the experiment lies between $\mu - k\sigma$ and $\mu + k\sigma$ is at least $1 - (1/k^2)$; that is,

$$P(\mu - k\sigma \leq X \leq \mu + k\sigma) \geq 1 - \frac{1}{k^2} \tag{9}$$

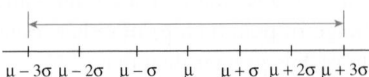

FIGURE 10
At least 75% of the outcomes fall within this interval

FIGURE 11
At least 89% of the outcomes fall within this interval

To shed some light on this result, let's take $k = 2$ in Inequality (9) and compute

$$P(\mu - 2\sigma \leq X \leq \mu + 2\sigma) \geq 1 - \frac{1}{2^2} = 1 - \frac{1}{4} = .75$$

This tells us that at least 75% of the outcomes of the experiment lie within 2 standard deviations of the mean (Figure 10). Taking $k = 3$ in Formula (9), we have

$$P(\mu - 3\sigma \leq X \leq \mu + 3\sigma) \geq 1 - \frac{1}{3^2} = 1 - \frac{1}{9} = \frac{8}{9} \approx .89$$

This tells us that at least 89% of the outcomes of the experiment lie within 3 standard deviations of the mean (Figure 11).

EXAMPLE 4   A probability distribution has a mean of 10 and a standard deviation of 1.5. Use Chebychev's inequality to estimate the probability that an outcome of the experiment lies between 7 and 13.

Solution   Here, $\mu = 10$ and $\sigma = 1.5$. Next, to determine the value of k, note that $\mu - k\sigma = 7$ and $\mu + k\sigma = 13$. Substituting the appropriate values for μ and σ, we find $k = 2$. Using Chebychev's Inequality (9), we see that the probability that an outcome of the experiment lies between 7 and 13 is given by

$$P(7 \leq X \leq 13) \geq 1 - \left(\frac{1}{2^2}\right)$$

$$= \frac{3}{4}$$

—that is, at least 75%.   ∎

Note   The results of Example 4 tell us that at least 75% of the outcomes of the experiment lie between $10 - 2\sigma$ and $10 + 2\sigma$—that is, between 7 and 13.   ∎

APPLIED EXAMPLE 5   Industrial Accidents   Great Northwest Lumber Company employs 400 workers in its mills. It has been estimated that X, the random variable measuring the number of mill workers who have industrial accidents during a 1-year period, is distributed with a mean of 40 and a standard deviation of 6. Using Chebychev's Inequality (9), estimate the probability that the number of workers who will have an industrial accident over a 1-year period is between 30 and 50, inclusive.

Solution   Here, $\mu = 40$ and $\sigma = 6$. We wish to estimate $P(30 \leq X \leq 50)$. To use Chebychev's Inequality (9), we first determine the value of k from the equation

$$\mu - k\sigma = 30 \quad \text{or} \quad \mu + k\sigma = 50$$

Since $\mu = 40$ and $\sigma = 6$ in this case, we see that k satisfies

$$40 - 6k = 30 \quad \text{and} \quad 40 + 6k = 50$$

from which we deduce that $k = \frac{5}{3}$. Thus, the probability that the number of mill workers who will have an industrial accident during a 1-year period is between 30 and 50 is given by

$$P(30 \le X \le 50) \ge 1 - \frac{1}{(\frac{5}{3})^2}$$

$$= \frac{16}{25}$$

—that is, at least 64%.

8.3 Self-Check Exercises

1. Compute the mean, variance, and standard deviation of the random variable X with probability distribution as follows:

x	-4	-3	-1	0	2	5
$P(X = x)$	.1	.1	.2	.3	.1	.2

2. James recorded the following travel times (the length of time in minutes it took him to drive to work) on 10 consecutive days:

   55   50   52   48   50   52   46   48   50   51

Calculate the mean and standard deviation of the random variable X associated with these data.

Solutions to Self-Check Exercises 8.3 can be found on page 477.

8.3 Concept Questions

1. a. What is the variance of a random variable X?
    b. What is the standard deviation of a random variable X?

2. What does Chebychev's inequality measure?

8.3 Exercises

In Exercises 1–6, the probability distribution of a random variable X is given. Compute the mean, variance, and standard deviation of X.

1.

x	1	2	3	4
$P(X = x)$	.4	.3	.2	.1

2.

x	-4	-2	0	2	4
$P(X = x)$	.1	.2	.3	.1	.3

3.

x	-2	-1	0	1	2
$P(X = x)$	1/16	4/16	6/16	4/16	1/16

4.

x	10	11	12	13	14	15
$P(X = x)$	1/8	2/8	1/8	2/8	1/8	1/8

5.

x	430	480	520	565	580
$P(X = x)$	.1	.2	.4	.2	.1

6.

x	-198	-195	-193	-188	-185
$P(X = x)$	.15	.30	.10	.25	.20

7. The following histograms represent the probability distributions of the random variables X and Y. Determine by inspection which probability distribution has the larger variance.

(a)
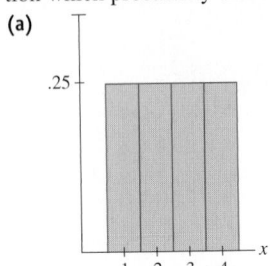

(b)
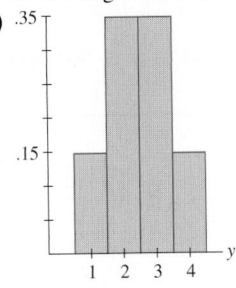

8. The following histograms represent the probability distributions of the random variables X and Y.

(a)
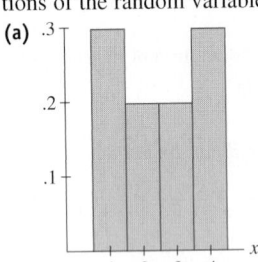

(b)
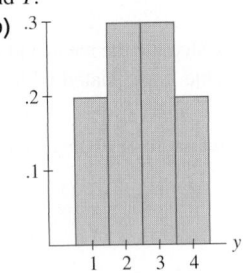

Determine by inspection which probability distribution has the larger variance.

In Exercises 9 and 10, find the variance of the probability distribution for the histogram shown.

9.
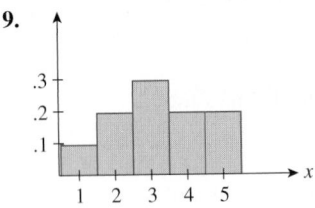

10.
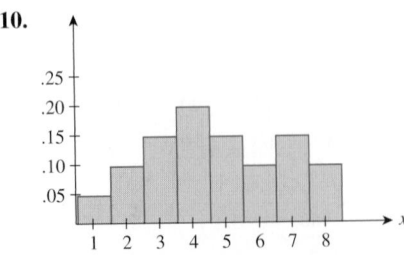

11. An experiment consists of casting an eight-sided die (numbered 1 through 8) and observing the number that appears uppermost. Find the mean and variance of this experiment.

12. DRIVING AGE REQUIREMENTS   The minimum age requirement for a regular driver's license differs from state to state. The frequency distribution for this age requirement in the 50 states is given in the following table:

Minimum Age	15	16	17	18	19	21
Frequency of Occurrence	1	15	4	28	1	1

a. Describe a random variable X that is associated with these data.
b. Find the probability distribution for the random variable X.
c. Compute the mean, variance, and standard deviation of X.

13. BIRTHRATES   The birthrates in the United States for the years 1991–2000 are given in the following table. (The birthrate is the number of live births/1000 population.)

Year	1991	1992	1993	1994
Birthrate	16.3	15.9	15.5	15.2

Year	1995	1996	1997
Birthrate	14.8	14.7	14.5

Year	1998	1999	2000
Birthrate	14.6	14.5	14.7

a. Describe a random variable X that is associated with these data.
b. Find the probability distribution for the random variable X.
c. Compute the mean, variance, and standard deviation of X.
Source: National Center for Health Statistics

14. INVESTMENT ANALYSIS   Paul Hunt is considering two business ventures. The anticipated returns (in thousands of dollars) of each venture are described by the following probability distributions:

Venture A

Earnings	Probability
−20	.3
40	.4
50	.3

Venture B

Earnings	Probability
−15	.2
30	.5
40	.3

a. Compute the mean and variance for each venture.
b. Which investment would provide Paul with the higher expected return (the greater mean)?
c. In which investment would the element of risk be less (that is, which probability distribution has the smaller variance)?

15. **INVESTMENT ANALYSIS** Rosa Walters is considering investing $10,000 in two mutual funds. The anticipated returns from price appreciation and dividends (in hundreds of dollars) are described by the following probability distributions:

Mutual Fund A

Returns	Probability
−4	.2
8	.5
10	.3

Mutual Fund B

Returns	Probability
−2	.2
6	.4
8	.4

a. Compute the mean and variance associated with the returns for each mutual fund.
b. Which investment would provide Rosa with the higher expected return (the greater mean)?
c. In which investment would the element of risk be less (that is, which probability distribution has the smaller variance)?

16. The distribution of the number of chocolate chips (x) in a cookie is shown in the following table. Find the mean and the variance of the number of chocolate chips in a cookie.

x	0	1	2
$P(X = x)$	.01	.03	.05

x	3	4	5
$P(X = x)$	.11	.13	.24

x	6	7	8
$P(X = x)$	.22	.16	.05

17. Formula (5) can also be expressed in the form

$$\text{Var}(X) = (p_1 x_1^2 + p_2 x_2^2 + \cdots + p_n x_n^2) - \mu^2$$

Find the variance of the distribution of Exercise 1 using this formula.

18. Find the variance of the distribution of Exercise 16 using the formula

$$\text{Var}(X) = (p_1 x_1^2 + p_2 x_2^2 + \cdots + p_n x_n^2) - \mu^2$$

19. **HOUSING PRICES** A survey was conducted by the market research department of the National Real Estate Company among 500 prospective buyers in a large metropolitan area to determine the maximum price a prospective buyer would be willing to pay for a house. From the data collected, the distribution that follows was obtained. Compute the mean, variance, and standard deviation of the maximum price x (in thousands of dollars) that these buyers were willing to pay for a house.

Maximum Price Considered, x	$P(X = x)$
280	$\frac{10}{500}$
290	$\frac{20}{500}$
300	$\frac{75}{500}$
310	$\frac{85}{500}$
320	$\frac{70}{500}$
350	$\frac{90}{500}$
380	$\frac{90}{500}$
400	$\frac{55}{500}$
450	$\frac{5}{500}$

20. **VOLKSWAGEN'S REVENUE** The revenue of Volkswagen (in billions of euros) for the five quarters beginning with the first quarter of 2003 are summarized in the following table:

	2003				2004
Quarter	Q1	Q2	Q3	Q4	Q1
Revenue	20.6	22.1	21.3	21.8	21.9

Find the average quarterly revenue of Volkswagen for the five quarters in question. What is the standard deviation?
Source: Company Reports

21. **EXAM SCORES** The following table gives the scores of 30 students in a mathematics examination:

Scores	90–99	80–89	70–79	60–69	50–59
Students	4	8	12	4	2

Find the mean and the standard deviation of the distribution of the given data.
Hint: Assume that all scores lying within a group interval take the middle value of that group.

22. **MARITAL STATUS OF MEN** The number of married men (in thousands) between the ages of 20 and 44 in the United States in 1998 is given in the following table:

Age	20–24	25–29	30–34	35–39	40–44
Men	1332	4219	6345	7598	7633

Find the mean and the standard deviation of the given data.
Hint: See the hint for Exercise 21.
Source: U.S. Census Bureau

23. **HOURS WORKED IN SOME COUNTRIES** The number of average hours worked per year per worker in the United States and five European countries in 2002 is given in the following table:

Country	U.S.	Spain	Great Britain	France	West Germany	Norway
Average Hours Worked	1815	1807	1707	1545	1428	1342

Find the average of the average hours worked, per worker, in 2002 for workers in the six countries. What is the standard deviation?

Source: Office of Economic Cooperation and Development

24. **AMERICANS WITHOUT HEALTH INSURANCE** The number of Americans without health insurance, in millions, from 1995 through 2002 is summarized in the following table:

Year	1995	1996	1997	1998	1999	2000	2001	2002
Americans	40.7	41.8	43.5	44.5	40.2	39.9	41.2	43.6

Find the average number of Americans without health insurance in the period from 1995 through 2002. What is the standard deviation?

Source: U.S. Census Bureau

25. **ACCESS TO CAPITAL** One of the key determinants to economic growth is access to capital. Using 54 variables to create an index of 1–7, with 7 being best possible access to capital, Milken Institute ranked the following as the top ten nations (although technically Hong Kong is not a nation) by the ability of their entrepreneurs to gain access to capital:

Country	Hong Kong	Netherlands	U.K.	Singapore	Switzerland
Index	5.7	5.59	5.57	5.56	5.55

Country	U.S.	Australia	Finland	Germany	Denmark
Index	5.55	5.31	5.24	5.23	5.22

Find the mean of the indices of the top ten nations. What is the standard deviation?

Source: Milken Institute

26. **ACCESS TO CAPITAL** Refer to Exercise 25. Milken Institute also ranked the following as the ten worst-performing nations by the ability of their entrepreneurs to gain access to capital:

Country	Peru	Mexico	Bulgaria	Brazil	Indonesia
Index	3.76	3.7	3.66	3.5	3.46

Country	Colombia	Turkey	Argentina	Venezuela	Russia
Index	3.46	3.43	3.2	2.88	2.19

Find the mean of the indices of the ten worst-performing nations. What is the standard deviation?

Source: Milken Institute

27. **SALES OF VEHICLES** The seasonally adjusted annualized sales rate for U.S. cars and light trucks, in millions of units, for May 2003 through April 2004 are given in the following tables:

2003

M	J	J	A	S	O	N	D
16.5	16.5	17.0	18.5	17.0	16.0	17.0	18.0

2004

J	F	M	A
16.3	16.5	16.8	16.5

What is the average monthly annualized sales rate for U.S. motor vehicles for the period in question? What is the standard deviation, and what does it say about the monthly sales?

Source: Autodata

28. **ELECTION TURNOUT** The percent of the voting age population who cast ballots in presidential election years from 1932 through 2000 are given in the following table:

Election Year	1932	1936	1940	1944	1948	1952	1956	1960	1964
Turnout, %	53	57	59	56	51	62	59	59	62

Election Year	1968	1972	1976	1980	1984	1988	1992	1996	2000
Turnout %	61	55	54	53	53	50	55	49	51

Find the mean and the standard deviation of the given data.

Source: Federal Election Commission

29. A probability distribution has a mean of 42 and a standard deviation of 2. Use Chebychev's inequality to estimate the probability that an outcome of the experiment lies between
    a. 38 and 46.   **b.** 32 and 52.

30. A probability distribution has a mean of 20 and a standard deviation of 3. Use Chebychev's inequality to estimate the probability that an outcome of the experiment lies between
    a. 15 and 25.   **b.** 10 and 30.

31. A probability distribution has a mean of 50 and a standard deviation of 1.4. Use Chebychev's inequality to find the value of c that guarantees the probability is at least 96% that an outcome of the experiment lies between $50 - c$ and $50 + c$.

32. Suppose X is a random variable with mean μ and standard deviation σ. If a large number of trials is observed, at least what percentage of these values is expected to lie between $\mu - 2\sigma$ and $\mu + 2\sigma$?

33. **PRODUCT RELIABILITY** The deluxe model hair dryer produced by Roland Electric has a mean expected lifetime of 24 mo with a standard deviation of 3 mo. Find the probability that one of these hair dryers will last between 20 and 28 mo.

34. **PRODUCT RELIABILITY** A Christmas tree light has an expected life of 200 hr and a standard deviation of 2 hr.
    a. Estimate the probability that one of these Christmas tree lights will last between 190 hr and 210 hr.
    b. Suppose 150,000 of these Christmas tree lights are used by a large city as part of its Christmas decorations. Estimate the number of lights that will require replacement between 180 hr and 220 hr of use.

35. **STARTING SALARIES** The mean annual starting salary of a new graduate in a certain profession is $52,000 with a standard deviation of $500. What is the probability that the starting

salary of a new graduate in this profession will be between $50,000 and $54,000?

36. QUALITY CONTROL   Sugar packaged by a certain machine has a mean weight of 5 lb and a standard deviation of 0.02 lb. For what values of c can the manufacturer of the machinery claim that the sugar packaged by this machine has a weight between $5 - c$ and $5 + c$ lb with probability at least 96%?

In Exercises 37 and 38, determine whether the statement is true or false. If it is true, explain why it is true. If it is false, give an example to show why it is false.

37. Both the variance and the standard deviation of a random variable measure the spread of a probability distribution.

38. Chebychev's inequality is useless when $k \leq 1$.

8.3   Solutions to Self-Check Exercises

1. The mean of the random variable X is

$$\mu = (-4)(.1) + (-3)(.1) + (-1)(.2)$$
$$+ (0)(.3) + (2)(.1) + (5)(.2)$$
$$= 0.3$$

The variance of X is

$$\text{Var}(X) = (.1)(-4 - 0.3)^2 + (.1)(-3 - 0.3)^2$$
$$+ (.2)(-1 - 0.3)^2 + (.3)(0 - 0.3)^2$$
$$+ (.1)(2 - 0.3)^2 + (.2)(5 - 0.3)^2$$
$$= 8.01$$

The standard deviation of X is

$$\sigma = \sqrt{\text{Var}(X)} = \sqrt{8.01} \approx 2.83$$

2. We first compute the probability distribution of X from the given data as follows:

x	Relative Frequency of Occurrence	$P(X = x)$
46	1	.1
48	2	.2
50	3	.3
51	1	.1
52	2	.2
55	1	.1

The mean of X is

$$\mu = (.1)(46) + (.2)(48) + (.3)(50)$$
$$+ (.1)(51) + (.2)(52) + (.1)(55)$$
$$= 50.2$$

The variance of X is

$$\text{Var}(X) = (.1)(46 - 50.2)^2 + (.2)(48 - 50.2)^2$$
$$+ (.3)(50 - 50.2)^2 + (.1)(51 - 50.2)^2$$
$$+ (.2)(52 - 50.2)^2 + (.1)(55 - 50.2)^2$$
$$= 5.76$$

from which we deduce the standard deviation

$$\sigma = \sqrt{5.76}$$
$$= 2.4$$

USING TECHNOLOGY

Finding the Mean and Standard Deviation

The calculation of the mean and standard deviation of a random variable is facilitated by the use of a graphing utility.

APPLIED EXAMPLE 1   Age Distribution of Company Directors   A survey conducted in 1995 of the Fortune 1000 companies revealed the following age distribution of the company directors:

Age	20–25	25–30	30–35	35–40	40–45	45–50	50–55
Directors	1	6	28	104	277	607	1142

Age	55–60	60–65	65–70	70–75	75–80	80–85	85–90
Directors	1413	1424	494	159	62	31	5

Source: Directorship

a. Plot a histogram for the given data.
b. Find the mean age and the standard deviation of the company directors.

Solution

a. Let X denote the random variable taking on the values 1 through 14, where 1 corresponds to the age bracket 20–25, 2 corresponds to the age bracket 25–30, and so on. Enter the values of X as $x_1 = 1, x_2 = 2, \ldots, x_{14} = 14$ and the corresponding values of Y as $y_1 = 1, y_2 = 6, \ldots, y_{14} = 5$. Then using the DRAW function from the Statistics menu of a graphing utility, we obtain the histogram shown in Figure T1.

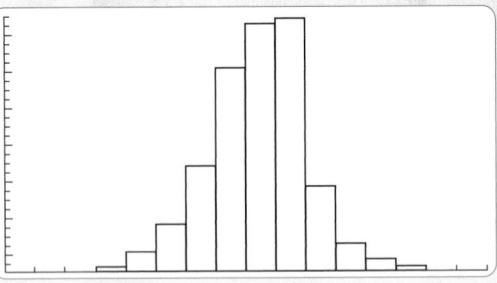

FIGURE T1
The histogram for the given data, using the viewing window $[0, 16] \times [0, 1500]$

b. Using the appropriate function from the Statistics menu, we find that $\bar{x} = 7.9193$ and $\sigma x = 1.6378$; that is, the mean of X is $\mu \approx 7.9$ and the standard deviation is $\sigma \approx 1.6$. Thus, the average age of the directors is in the 55–60-year-old bracket.

TECHNOLOGY EXERCISES

1. a. Graph the histogram associated with the random variable X in Example 1, page 468.
  b. Find the mean and the standard deviation for these data.

2. a. Graph the histogram associated with the random variable Y in Example 1, page 468.
  b. Find the mean and the standard deviation for these data.

3. a. Graph the histogram associated with the data given in Exercise 12, page 474.
  b. Find the mean and the standard deviation for these data.

4. a. Graph the histogram associated with the data given in Exercise 16, page 475.
  b. Find the mean and the standard deviation for these data.

5. A sugar refiner uses a machine to pack sugar in 5-lb cartons. To check the machine's accuracy, cartons are selected at random and weighed. The results follow:

4.98	5.02	4.96	4.97	5.03
4.96	4.98	5.01	5.02	5.06
4.97	5.04	5.04	5.01	4.99
4.98	5.04	5.01	5.03	5.05
4.96	4.97	5.02	5.04	4.97
5.03	5.01	5.00	5.01	4.98

a. Describe a random variable X that is associated with these data.
b. Find the probability distribution for the random variable X.
c. Compute the mean and standard deviation of X.

6. The scores of 25 students in a mathematics examination are as follows:

90	85	74	92	68	94	66
87	85	70	72	68	73	72
69	66	58	70	74	88	90
98	71	75	68			

a. Describe a random variable X that is associated with these data.
b. Find the probability distribution for the random variable X.
c. Compute the mean and standard deviation of X.

7. HEIGHTS OF WOMEN The following data, obtained from the records of the Westwood Health Club, give the heights (to the nearest inch) of 200 female members of the club:

Height	62	$62\frac{1}{2}$	63	$63\frac{1}{2}$	64	$64\frac{1}{2}$	65	$65\frac{1}{2}$	66
Frequency	2	3	4	8	11	20	32	30	18

Height	$66\frac{1}{2}$	67	$67\frac{1}{2}$	68	$68\frac{1}{2}$	69	$69\frac{1}{2}$	70	$70\frac{1}{2}$	71
Frequency	18	16	8	10	5	5	4	3	2	1

a. Plot a histogram for the given data.
b. Find the mean and the standard deviation (from the mean).

8. AGE DISTRIBUTION IN A TOWN The following table gives the distribution of the ages (in years) of the residents (in hundreds) of the town of Monroe who are under the age of 40:

Age	0–3	4–7	8–11	12–15	16–19
Residents	30	42	50	60	50

Age	20–23	24–27	28–31	32–35	36–39
Residents	41	50	45	42	34

Let X denote the random variable taking on the values 1 through 10, where 1 corresponds to the range 0–3, ..., and 10 corresponds to the range 36–39.
a. Plot a histogram for the given data.
b. Find the mean and the standard deviation.

8.4   The Binomial Distribution

▬ Bernoulli Trials

An important class of experiments have (or may be viewed as having) two out-comes. For example, in a coin-tossing experiment, the two outcomes are *heads* and *tails*. In the card game played by Mike and Bill (Example 7, Section 8.2), one may view the selection of a diamond as a *win* (for Mike) and the selection of a card of another suit as a *loss* for Mike. For a third example, consider an experiment in which a person is inoculated with a flu vaccine. Here, the vaccine may be classified as being "effective" or "ineffective" with respect to that particular person.

In general, experiments with two outcomes are called **Bernoulli trials**, or **binomial trials**. It is standard practice to label one of the outcomes of a binomial trial a *success* and the other a *failure*. For example, in a coin-tossing experiment, the outcome a *head* may be called a success, in which case the outcome a *tail* is called a failure. Note that by using the terms *success* and *failure* in this way, we depart from their usual connotations.

A sequence of Bernoulli (binomial) trials is called a binomial experiment. More precisely, we have the following definition:

Binomial Experiment

A **binomial experiment** has the following properties:

1. The number of trials in the experiment is fixed.
2. There are two outcomes of the experiment: "success" and "failure."
3. The probability of success in each trial is the same.
4. The trials are independent of each other.

In a binomial experiment, it is customary to denote the probability of a success by the letter p and the probability of a failure by the letter q. Because the event of a success and the event of a failure are complementary events, we have the relationship

$$p + q = 1$$

or, equivalently,

$$q = 1 - p$$

The properties of a binomial experiment are illustrated in the following example.

EXAMPLE 1   A fair die is cast four times. Compute the probability of obtaining exactly one 6 in the four throws.

Solution   There are four trials in this experiment. Each trial consists of casting the die once and observing the face that lands uppermost. We may view each trial as an experiment with two outcomes: a success (S) if the face that lands upper-most is a 6 and a failure (F) if it is any of the other five numbers. Letting p and q

denote the probability of success and failure, respectively, of a single trial of the experiment, we find that

$$p = \frac{1}{6} \quad \text{and} \quad q = 1 - \frac{1}{6} = \frac{5}{6}$$

Furthermore, we may assume that the trials of this experiment are independent. Thus, we have a binomial experiment.

With the aid of the multiplication principle, we see that the experiment has 2^4, or 16, outcomes. We can obtain these outcomes by constructing the tree diagram associated with the experiment (see Table 11, where the outcomes are listed according to the number of successes). From the table, we see that the event of obtaining exactly one success in four trials is given by

$$E = \{\text{SFFF, FSFF, FFSF, FFFS}\}$$

with probability given by

$$P(E) = P(\text{SFFF}) + P(\text{FSFF}) + P(\text{FFSF}) + P(\text{FFFS}) \tag{10}$$

TABLE 11

0 Success	1 Success	2 Successes	3 Successes	4 Successes
FFFF	SFFF	SSFF	SSSF	SSSS
	FSFF	SFSF	SSFS	
	FFSF	SFFS	SFSS	
	FFFS	FSSF	FSSS	
		FSFS		
		FFSS		

Since the trials (throws) are independent, the terms on the right-hand side of Equation (10) may be computed as follows:

$$P(\text{SFFF}) = P(\text{S})P(\text{F})P(\text{F})P(\text{F}) = p \cdot q \cdot q \cdot q = pq^3$$
$$P(\text{FSFF}) = P(\text{F})P(\text{S})P(\text{F})P(\text{F}) = q \cdot p \cdot q \cdot q = pq^3$$
$$P(\text{FFSF}) = P(\text{F})P(\text{F})P(\text{S})P(\text{F}) = q \cdot q \cdot p \cdot q = pq^3$$
$$P(\text{FFFS}) = P(\text{F})P(\text{F})P(\text{F})P(\text{S}) = q \cdot q \cdot q \cdot p = pq^3$$

Therefore, upon substituting these values in (10), we obtain

$$P(E) = pq^3 + pq^3 + pq^3 + pq^3 = 4pq^3$$
$$= 4\left(\frac{1}{6}\right)\left(\frac{5}{6}\right)^3 \approx .386$$

Probabilities in Bernoulli Trials

Let's reexamine the computations performed in the last example. There it was found that the probability of obtaining exactly one success in a binomial experiment with four independent trials with probability of success in a single trial p is given by

$$P(E) = 4pq^3 \qquad (\text{where } q = 1 - p) \tag{11}$$

Observe that the coefficient 4 of pq^3 appearing in Equation (11) is precisely the number of outcomes of the experiment with exactly one success and three failures, the outcomes being

$$\text{SFFF}\quad \text{FSFF}\quad \text{FFSF}\quad \text{FFFS}$$

Another way of obtaining this coefficient is to think of the outcomes as arrangements of the letters S and F. Then, the number of ways of selecting one position for S from four possibilities is given by

$$C(4, 1) = \frac{4!}{1!\,(4-1)!}$$
$$= 4$$

Next, observe that, because the trials are independent, each of the four outcomes of the experiment has the same probability, given by

$$pq^3$$

where the exponents 1 and 3 of p and q, respectively, correspond to exactly one success and three failures in the trials that make up each outcome.

As a result of the foregoing discussion, we may write (11) as

$$P(E) = C(4, 1)pq^3 \tag{12}$$

We are also in a position to generalize this result. Suppose that in a binomial experiment the probability of success in any trial is p. What is the probability of obtaining exactly x successes in n independent trials? We start by counting the number of outcomes of the experiment, each of which has exactly x successes. Now, one such outcome involves x successive successes followed by $(n-x)$ failures—that is,

$$\underbrace{\text{SS} \cdots \text{S}}_{x}\, \underbrace{\text{FF} \cdots \text{F}}_{n-x} \tag{13}$$

The other outcomes, each of which has exactly x successes, are obtained by rearranging the Ss (x of them) and Fs ($n-x$ of them). There are $C(n, x)$ ways of arranging these letters. Next, arguing as in Example 1, we see that each such outcome has probability given by

$$p^x q^{n-x}$$

For example, for the outcome (13) we find

$$P(\underbrace{\text{SS} \cdots \text{S}}_{x}\, \underbrace{\text{FF} \cdots \text{F}}_{(n-x)}) = \underbrace{P(\text{S})P(\text{S}) \cdots P(\text{S})}_{x}\, \underbrace{P(\text{F})P(\text{F}) \cdots P(\text{F})}_{(n-x)}$$
$$= \underbrace{pp \cdots p}_{x}\, \underbrace{qq \cdots q}_{n-x}$$
$$= p^x q^{n-x}$$

Let's now state this important result formally.

Computation of Probabilities in Bernoulli Trials

In a binomial experiment in which the probability of success in any trial is p, the probability of exactly x successes in n independent trials is given by

$$C(n, x)p^x q^{n-x}$$

If we let X be the random variable that gives the number of successes in a binomial experiment, then the probability of exactly x successes in n independent trials may be written

$$P(X = x) = C(n, x)p^x q^{n-x} \qquad (x = 0, 1, 2, \ldots, n) \tag{14}$$

The random variable X is called a **binomial random variable**, and the probability distribution of X is called a **binomial distribution**.

EXAMPLE 2   A fair die is cast five times. If a 1 or a 6 lands uppermost in a trial, then the throw is considered a success. Otherwise, the throw is considered a failure.
a. Find the probability of obtaining exactly 0, 1, 2, 3, 4, and 5 successes, respectively, in this experiment.
b. Using the results obtained in the solution to part (a), construct the binomial distribution for this experiment and draw the histogram associated with it.

Solution
a. This is a binomial experiment with X, the binomial random variable, taking on each of the values 0, 1, 2, 3, 4, and 5 corresponding to exactly 0, 1, 2, 3, 4, and 5 successes, respectively, in five trials. Since the die is fair, the probability of a 1 or a 6 landing uppermost in any trial is given by $p = 2/6 = 1/3$, from which it also follows that $q = 1 - p = 2/3$. Finally, $n = 5$ since there are five trials (throws of the die) in this experiment. Using Equation (14), we find that the required probabilities are

$$P(X = 0) = C(5, 0)\left(\frac{1}{3}\right)^0\left(\frac{2}{3}\right)^5 = \frac{5!}{0!\,5!} \cdot 1 \cdot \frac{32}{243} \approx .132$$

$$P(X = 1) = C(5, 1)\left(\frac{1}{3}\right)^1\left(\frac{2}{3}\right)^4 = \frac{5!}{1!\,4!} \cdot \frac{16}{243} \approx .329$$

$$P(X = 2) = C(5, 2)\left(\frac{1}{3}\right)^2\left(\frac{2}{3}\right)^3 = \frac{5!}{2!\,3!} \cdot \frac{8}{243} \approx .329$$

$$P(X = 3) = C(5, 3)\left(\frac{1}{3}\right)^3\left(\frac{2}{3}\right)^2 = \frac{5!}{3!\,2!} \cdot \frac{4}{243} \approx .165$$

$$P(X = 4) = C(5, 4)\left(\frac{1}{3}\right)^4\left(\frac{2}{3}\right)^1 = \frac{5!}{4!\,1!} \cdot \frac{2}{243} \approx .041$$

$$P(X = 5) = C(5, 5)\left(\frac{1}{3}\right)^5\left(\frac{2}{3}\right)^0 = \frac{5!}{5!\,0!} \cdot \frac{1}{243} \approx .004$$

b. Using these results, we find the required binomial distribution associated with this experiment given in Table 12. Next, we use this table to construct the histogram associated with the probability distribution (Figure 12). ∎

TABLE 12

Probability Distribution

x	$P(X = x)$
0	.132
1	.329
2	.329
3	.165
4	.041
5	.004

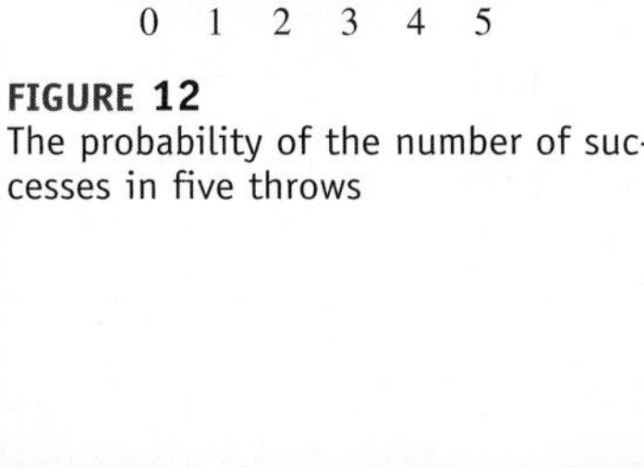

FIGURE 12
The probability of the number of successes in five throws

EXAMPLE 3   A fair die is cast five times. If a 1 or a 6 lands uppermost in a trial, then the throw is considered a success. Use the results from Example 2 to answer the following questions:
a. What is the probability of obtaining 0 or 1 success in the experiment?
b. What is the probability of obtaining at least 1 success in the experiment?

Solution   Interpreting the probability associated with the random variable X, when X assumes the value $X = a$, as the area of the rectangle centered about $X = a$ (Figure 12), we find that

a. The probability of obtaining 0 or 1 success in the experiment is given by

$$P(X = 0) + P(X = 1) = .132 + .329 = .461$$

b. The probability of obtaining at least 1 success in the experiment is given by

$$P(X = 1) + P(X = 2) + P(X = 3) + P(X = 4) + P(X = 5)$$
$$= .329 + .329 + .165 + .041 + .004$$
$$= .868$$

EXPLORE & DISCUSS

Consider the equation

$$P(X = x) = C(n, x)p^x q^{n-x}$$

for the binomial distribution.

1. Construct the histogram with $n = 5$ and $p = .2$; the histogram with $n = 5$ and $p = .5$; and the histogram with $n = 5$ and $p = .8$.
2. Comment on the shape of the histograms and give an interpretation.

The following formulas (which we state without proof) will be useful in solving problems that involve binomial experiments.

Mean, Variance, and Standard Deviation of a Random Variable X

If X is a binomial random variable associated with a binomial experiment consisting of n trials with probability of success p and probability of failure q, then the **mean** (expected value), **variance**, and **standard deviation** of X are

$$\mu = E(X) = np \tag{15a}$$

$$\mathrm{Var}(X) = npq \tag{15b}$$

$$\sigma_X = \sqrt{npq} \tag{15c}$$

EXAMPLE 4   For the experiment in Examples 2 and 3, compute the mean, the variance, and the standard deviation of X by (a) using Formulas (15a), (15b), and (15c) and (b) using the definition of each term (Sections 8.2 and 8.3).

Solution

a. We use (15a), (15b), and (15c), with $p = \frac{1}{3}$, $q = \frac{2}{3}$, and $n = 5$, obtaining

$$\mu = E(X) = (5)\left(\frac{1}{3}\right) = \frac{5}{3} \approx 1.67$$

$$\mathrm{Var}(X) = (5)\left(\frac{1}{3}\right)\left(\frac{2}{3}\right) = \frac{10}{9} \approx 1.11$$

$$\sigma_X = \sqrt{\mathrm{Var}(X)} = \sqrt{1.11} \approx 1.05$$

We leave it to you to interpret the results.

b. Using the definition of expected value (Section 8.2) and the values of the probability distribution shown in Table 12, we find that

$$\mu = E(X) = (0)(.132) + (1)(.329) + (2)(.329)$$
$$+ (3)(.165) + (4)(.041) + (5)(.004)$$
$$\approx 1.67$$

which agrees with the result obtained in part (a). Next, using the definition of variance and $\mu = 1.67$, we find that

$$Var(X) = (.132)(-1.67)^2 + (.329)(-0.67)^2 + (.329)(0.33)^2$$
$$+ (.165)(1.33)^2 + (.041)(2.33)^2 + (.004)(3.33)^2$$
$$\approx 1.11$$
$$\sigma_X \approx \sqrt{Var(X)}$$
$$\approx \sqrt{1.11} \approx 1.05$$

which again agrees with the preceding results.

We close this section by looking at several examples involving binomial experiments. In working through these examples, you may use a calculator or you may consult Table 1, Appendix B.

APPLIED EXAMPLE 5   Quality Control   A division of Solaron manufactures photovoltaic cells to use in the company's solar energy converters. It is estimated that 5% of the cells manufactured are defective. If a random sample of 20 is selected from a large lot of cells manufactured by the company, what is the probability that it will contain at most 2 defective cells?

Solution   We may view this as a binomial experiment. To see this, first note that a fixed number of trials ($n = 20$) corresponds to the selection of exactly 20 photovoltaic cells. Second, observe that there are exactly two outcomes in the experiment, defective ("success") and nondefective ("failure"). Third, the probability of success in each trial is .05 ($p = .05$) and the probability of failure in each trial is .95 ($q = .95$). This assumption is justified by virtue of the fact that the lot from which the cells are selected is "large," so the removal of a few cells will not appreciably affect the percentage of defective cells in the lot in each successive trial. Finally, the trials are independent of each other—once again because of the lot size.

Letting X denote the number of defective cells, we find that the probability of finding at most 2 defective cells in the sample of 20 is given by

$$P(X = 0) + P(X = 1) + P(X = 2)$$
$$= C(20, 0)(.05)^0(.95)^{20} + C(20, 1)(.05)^1(.95)^{19}$$
$$+ C(20, 2)(.05)^2(.95)^{18}$$
$$\approx .3585 + .3774 + .1887$$
$$= .9246$$

Thus, for lots of photovoltaic cells manufactured by Solaron, approximately 92% of the samples will have at most 2 defective cells; equivalently, approximately 8% of the samples will contain more than 2 defective cells.

APPLIED EXAMPLE 6   Success of Heart Transplants   The probability that a heart transplant performed at the Medical Center is successful (that is, the patient survives 1 year or more after undergoing such an operation) is .7. Of six patients who have recently undergone such an operation, what is the probability that, 1 year from now,

a. None of the heart recipients will be alive?
b. Exactly three will be alive?
c. At least three will be alive?
d. All will be alive?

Solution   Here, $n = 6, p = .7$, and $q = .3$. Let X denote the number of successful operations. Then:

a. The probability that no heart recipients will be alive after 1 year is given by

$$P(X = 0) = C(6, 0)(.7)^0(.3)^6$$
$$= \frac{6!}{0!\,6!} \cdot 1 \cdot (.3)^6$$
$$\approx .0007$$

b. The probability that exactly three will be alive after 1 year is given by

$$P(X = 3) = C(6, 3)(.7)^3(.3)^3$$
$$= \frac{6!}{3!\,3!}(.7)^3(.3)^3$$
$$\approx .19$$

c. The probability that at least three will be alive after 1 year is given by

$$P(X = 3) + P(X = 4) + P(X = 5) + P(X = 6)$$
$$= C(6, 3)(.7)^3(.3)^3 + C(6, 4)(.7)^4(.3)^2$$
$$+ C(6, 5)(.7)^5(.3)^1 + C(6, 6)(.7)^6(.3)^0$$
$$= \frac{6!}{3!\,3!}(.7)^3(.3)^3 + \frac{6!}{4!\,2!}(.7)^4(.3)^2 + \frac{6!}{5!\,1!}(.7)^5(.3)^1$$
$$+ \frac{6!}{6!\,0!}(.7)^6 \cdot 1$$
$$\approx .93$$

d. The probability that all will be alive after 1 year is given by

$$P(X = 6) = C(6, 6)(.7)^6(.3)^0 = \frac{6!}{6!\,0!}(.7)^6$$
$$\approx .12$$

APPLIED EXAMPLE 7   Quality Control   PAR Bearings manufactures ball bearings packaged in lots of 100 each. The company's quality-control department has determined that 2% of the ball bearings manufactured do not meet the specifications imposed by a buyer. Find the average number of ball bearings per package that fail to meet the buyer's specification.

Solution   The experiment under consideration is binomial. The average number of ball bearings per package that fail to meet with the specifications is therefore given by the expected value of the associated binomial random variable. Using (15a), we find that

$$\mu = E(X) = np = (100)(.02) = 2$$

substandard ball bearings in a package of 100.

8.4 Self-Check Exercises

1. A binomial experiment consists of four independent trials. The probability of success in each trial is .2.
   a. Find the probability of obtaining exactly 0, 1, 2, 3, and 4 successes, respectively, in this experiment.
   b. Construct the binomial distribution and draw the histogram associated with this experiment.
   c. Compute the mean and the standard deviation of the random variable associated with this experiment.

2. A survey shows that 60% of the households in a large metropolitan area have microwave ovens. If ten households are selected at random, what is the probability that five or fewer of these households have microwave ovens?

Solutions to Self-Check Exercises 8.4 can be found on page 490.

8.4 Concept Questions

1. Suppose that you are given a Bernoulli experiment.
   a. How many outcomes are there in the experiment?
   b. Can the number of trials in the experiment vary, or is it fixed?
   c. Are the trials in the experiment dependent?
   d. If the probability of success in any trial is p, what is the probability of exactly x successes in n independent trials?

2. Give the formula for the mean, variance, and standard deviation of X, where X is a binomial random variable associated with a binomial experiment consisting of n trials with probability of success p and probability of failure q.

8.4 Exercises

In Exercises 1–6, determine whether the experiment is a binomial experiment. Justify your answer.

1. Casting a fair die three times and observing the number of times a 6 is thrown

2. Casting a fair die and observing the number of times the die is thrown until a 6 appears uppermost

3. Casting a fair die three times and observing the number that appears uppermost

4. A card is selected from a deck of 52 cards, and its color is observed. A second card is then drawn (without replacement), and its color is observed.

5. Recording the number of accidents that occur at a given intersection on four clear days and one rainy day

6. Recording the number of hits a baseball player, whose batting average is .325, gets after being up to bat five times

In Exercises 7–10, find $C(n, x)p^x q^{n-x}$ for the given values of n, x, and p.

7. $n = 4, x = 2, p = \dfrac{1}{3}$

8. $n = 6, x = 4, p = \dfrac{1}{4}$

9. $n = 5, x = 3, p = .2$

10. $n = 6, x = 5, p = .4$

In Exercises 11–16, use the formula $C(n, x)p^x q^{n-x}$ to determine the probability of the given event.

11. The probability of exactly no successes in five trials of a binomial experiment in which $p = \frac{1}{3}$

12. The probability of exactly three successes in six trials of a binomial experiment in which $p = \frac{1}{2}$

13. The probability of at least three successes in six trials of a binomial experiment in which $p = \frac{1}{2}$

14. The probability of no successful outcomes in six trials of a binomial experiment in which $p = \frac{1}{3}$

15. The probability of no failures in five trials of a binomial experiment in which $p = \frac{1}{3}$

16. The probability of at least one failure in five trials of a binomial experiment in which $p = \frac{1}{3}$

17. A fair die is cast four times. Calculate the probability of obtaining exactly two 6s.

18. Let X be the number of successes in five independent trials of a binomial experiment in which the probability of success is $p = \frac{2}{5}$. Find:
    a. $P(X = 4)$ **b.** $P(2 \leq X \leq 4)$

19. A binomial experiment consists of five independent trials. The probability of success in each trial is .4.
    a. Find the probability of obtaining exactly 0, 1, 2, 3, 4, and 5 successes, respectively, in this experiment.
    b. Construct the binomial distribution and draw the histogram associated with this experiment.
    c. Compute the mean and the standard deviation of the random variable associated with this experiment.

20. Let the random variable X denote the number of girls in a five-child family. If the probability of a female birth is .5,
    a. Find the probability of 0, 1, 2, 3, 4, and 5 girls in a five-child family.
    b. Construct the binomial distribution and draw the histogram associated with this experiment.
    c. Compute the mean and the standard deviation of the random variable X.

21. The probability that a fuse produced by a certain manufacturing process will be defective is $\frac{1}{50}$. Is it correct to infer from this statement that there is at most 1 defective fuse in each lot of 50 produced by this process? Justify your answer.

22. SPORTS If the probability that a certain tennis player will serve an ace is $\frac{1}{4}$, what is the probability that he will serve exactly two aces out of five serves?

23. SPORTS If the probability that a certain tennis player will serve an ace is .15, what is the probability that she will serve at least two aces out of five serves?

24. SALES PREDICTIONS From experience, the manager of Kramer's Book Mart knows that 40% of the people who are browsing in the store will make a purchase. What is the probability that, among ten people who are browsing in the store, at least three will make a purchase?

25. CUSTOMER SERVICES Mayco, a mail-order department store, has six telephone lines available for customers who wish to place their orders. If the probability is $\frac{1}{4}$ that any one of the six telephone lines is engaged during business hours, find the probability that all six lines will be in use when a customer calls to place an order.

26. RESTAURANT VIOLATIONS OF THE HEALTH CODE Suppose 30% of the restaurants in a certain part of a town are in violation of the health code. If a health inspector randomly selects five of the restaurants for inspection, what is the probability that
    a. None of the restaurants are in violation of the health code?
    b. Just one of the restaurants is in violation of the health code?
    c. At least two of the restaurants are in violation of the health code?

27. ADVERTISEMENTS An advertisement for brand A chicken noodle soup claims that 60% of all consumers prefer brand A over brand B, the chief competitor's product. To test this claim, David Horowitz, host of *The Consumer Advocate*, selected ten people at random from the audience. After tasting both soups, each person was asked to state his or her preference. Assuming the company's claim is correct, find the probability that
    a. The company's claim was supported by the experiment; that is, six or more people stated a preference for brand A.
    b. The company's claim was not supported by the experiment; that is, fewer than six people stated a preference for brand A.

28. VOTERS In a certain congressional district, it is known that 40% of the registered voters classify themselves as conservatives. If ten registered voters are selected at random from this district, what is the probability that four of them will be conservatives?

29. VIOLATIONS OF THE BUILDING CODE Suppose that one third of the new buildings in a town are in violation of the building code. If a building inspector inspects five of the buildings, find the probability that
    a. The first three buildings will pass the inspection and the remaining two will fail the inspection.
    b. Just three of the buildings will pass inspection.

30. EXAMS A biology quiz consists of eight multiple-choice questions. Five must be answered correctly to receive a passing grade. If each question has five possible answers, of

which only one is correct, what is the probability that a student who guesses at random on each question will pass the examination?

31. BLOOD TYPES   It is estimated that one third of the general population has blood type A^+. If a sample of nine people is selected at random, what is the probability that
    a. Exactly three of them have blood type A^+?
    b. At most three of them have blood type A^+?

32. EXAMS   A psychology quiz consists of ten true-or-false questions. If a student knows the correct answer to six of the questions but determines the answers to the remaining questions by flipping a coin, what is the probability that she will obtain a score of at least 90%?

33. QUALITY CONTROL   The probability that a DVD player produced by VCA Television is defective is estimated to be .02. If a sample of ten players is selected at random, what is the probability that the sample contains
    a. No defectives?   b. At most two defectives?

34. QUALITY CONTROL   As part of its quality-control program, the video-game DVDs produced by Starr Communications are subjected to a final inspection before shipment. A sample of six DVDs is selected at random from each lot of DVDs produced, and the lot is rejected if the sample contains one or more defective DVDs. If 1.5% of the DVDs produced by Starr is defective, find the probability that a shipment will be accepted.

35. ROBOT RELIABILITY   An automobile-manufacturing company uses ten industrial robots as welders on its assembly line. On a given working day, the probability that a robot will be inoperative is .05. What is the probability that, on a given working day,
    a. Exactly two robots are inoperative?
    b. More than two robots are inoperative?

36. ENGINE FAILURES   The probability that an airplane engine will fail in a transcontinental flight is .001. Assuming that engine failures are independent of each other, what is the probability that, on a certain transcontinental flight, a four-engine plane will experience
    a. Exactly one engine failure?
    b. Exactly two engine failures?
    c. More than two engine failures? (*Note:* In this event, the airplane will crash.)

37. QUALITY CONTROL   The manager of Toy World has decided to accept a shipment of electronic games if none of a random sample of 20 is found to be defective.

a. What is the probability that he will accept the shipment if 10% of the electronic games is defective?
b. What is the probability that he will accept the shipment if 5% of the electronic games is defective?

38. QUALITY CONTROL   Refer to Exercise 37. If the manager's criterion for accepting shipment is that there be no more than 1 defective electronic game in a random sample of 20, what is the probability that he will accept the shipment if 10% of the electronic games is defective?

39. QUALITY CONTROL   Refer to Exercise 37. If the manager of the store changes his sample size to 10 and decides to accept shipment only if none of the sampled games is defective, what is the probability that he will accept the shipment if 10% of the games is defective?

40. How many times must a person toss a coin for the chances of obtaining at least one head to be 99% or better?

41. DRUG TESTING   A new drug has been found to be effective in treating 75% of the people afflicted by a certain disease. If the drug is administered to 500 people who have this disease, what are the mean and the standard deviation of the number of people for whom the drug can be expected to be effective?

42. COLLEGE GRADUATES   At a certain university, the probability that an entering freshman will graduate within 4 yr is .6. From an incoming class of 2000 freshmen, find
    a. The expected number of students who will graduate within 4 yr.
    b. The standard deviation of the number of students who will graduate within 4 yr.

In Exercises 43–46, determine whether the statement is true or false. If it is true, explain why it is true. If it is false, give an example to show why it is false.

43. In a binomial experiment, the number of outcomes of the experiment may be any finite number.

44. In a binomial experiment with $n = 3$, $P(X = 1 \text{ or } 2) = 3pq$.

45. If the probability that a batter gets a hit is $\frac{1}{4}$, then the batter is sure to get a hit if she bats four times.

46. The histogram associated with a binomial distribution is symmetric with respect to $x = \frac{n}{2}$ if $p = \frac{1}{2}$.

8.4 Solutions to Self-Check Exercises

1. a. We use Formula (14) with $n = 4$, $p = .2$, and $q = 1 - .2 = .8$, obtaining

$$P(X = 0) = C(4, 0)(.2)^0(.8)^4$$
$$= \frac{4!}{0!\,4!} \cdot 1 \cdot (.8)^4 \approx .410$$

$$P(X = 1) = C(4, 1)(.2)^1(.8)^3$$
$$= \frac{4!}{1!\,3!}(.2)(.8)^3 \approx .410$$

$$P(X = 2) = C(4, 2)(.2)^2(.8)^2$$
$$= \frac{4!}{2!\,2!}(.2)^2(.8)^2 \approx .154$$

$$P(X = 3) = C(4, 3)(.2)^3(.8)^1$$
$$= \frac{4!}{3!\,1!}(.2)^3(.8) \approx .026$$

$$P(X = 4) = C(4, 4)(.2)^4(.8)^0$$
$$= \frac{4!}{4!\,0!}(.2)^4 \cdot 1 \approx .002$$

b. The required binomial distribution and histogram are as follows:

x	$P(X = x)$
0	.410
1	.410
2	.154
3	.026
4	.002

c. The mean is

$$\mu = E(X) = np = (4)(.2)$$
$$= 0.8$$

and the standard deviation is

$$\sigma = \sqrt{npq} = \sqrt{(4)(.2)(.8)}$$
$$= 0.8$$

2. This is a binomial experiment with $n = 10$, $p = .6$, and $q = .4$. Let X denote the number of households that have microwave ovens. Then, the probability that five or fewer households have microwave ovens is given by

$$P(X = 0) + P(X = 1) + P(X = 2) + P(X = 3)$$
$$+ P(X = 4) + P(X = 5)$$
$$= C(10, 0)(.6)^0(.4)^{10} + C(10, 1)(.6)^1(.4)^9$$
$$+ C(10, 2)(.6)^2(.4)^8 + C(10, 3)(.6)^3(.4)^7$$
$$+ C(10, 4)(.6)^4(.4)^6 + C(10, 5)(.6)^5(.4)^5$$
$$\approx 0 + .002 + .011 + .042 + .111 + .201$$
$$\approx .37$$

8.5 The Normal Distribution

■ Probability Density Functions

The probability distributions discussed in the preceding sections were all associated with finite random variables—that is, random variables that take on finitely many values. Such probability distributions are referred to as *finite probability distributions*. In this section, we consider probability distributions associated with a continuous random variable—that is, a random variable that may take on any value lying in an interval of real numbers. Such probability distributions are called **continuous probability distributions**.

Unlike a finite probability distribution, which may be exhibited in the form of a table, a continuous probability distribution is defined by a function f whose domain coincides with the interval of values taken on by the random variable associated with the experiment. Such a function *f* is called the **probability density function** associated with the probability distribution, and it has the following properties:

1. *f*(*x*) is nonnegative for all values of *x*.
2. The area of the region between the graph of *f* and the *x*-axis is equal to 1 (Figure 13).

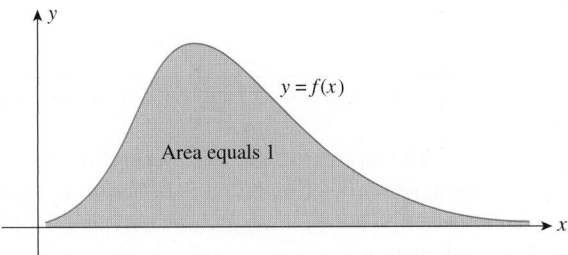

FIGURE 13
A probability density function

Now suppose we are given a continuous probability distribution defined by a probability density function *f*. Then, the probability that the random variable *X* assumes a value in an interval *a* < *x* < *b* is given by the area of the region between the graph of *f* and the *x*-axis from *x* = *a* to *x* = *b* (Figure 14). We denote the value of this probability by $P(a < X < b)$.* Observe that property 2 of the probability density function states that the probability that a continuous random variable takes on a value lying in its range is 1, a certainty, which is expected. Note the analogy between the areas under the probability density curves and the areas of the histograms associated with finite probability distributions (see Section 8.1).

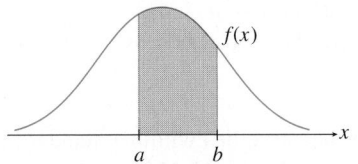

FIGURE 14
$P(a < X < b)$ is given by the area of the shaded region.

Normal Distributions

The mean μ and the standard deviation σ of a continuous probability distribution have roughly the same meanings as the mean and standard deviation of a finite probability distribution. Thus, the mean of a continuous probability distribution is a measure of the central tendency of the probability distribution, and the standard deviation of the probability distribution measures its spread about its mean. Both of these numbers will play an important role in the following discussion.

For the remainder of this section, we will discuss a special class of continuous probability distributions known as **normal distributions**. The normal distribution is without doubt the most important of all the probability distributions. Many phenomena—such as the heights of people in a given population, the weights of newborn infants, the IQs of college students, the actual weights of 16-ounce packages of cereals, and so on—have probability distributions that are normal. The normal distribution also provides us with an accurate approximation to the distributions of many random variables associated with random-sampling problems. In fact, in the next section we will see how a normal distribution may be used to approximate a binomial distribution under certain conditions.

The graph of a normal distribution, which is bell shaped, is called a **normal curve** (Figure 15).

---

*Because the area under one point of the graph of *f* is equal to zero, we see that $P(a < X < b) = P(a < X \leq b) = P(a \leq X < b) = P(a \leq X \leq b)$.

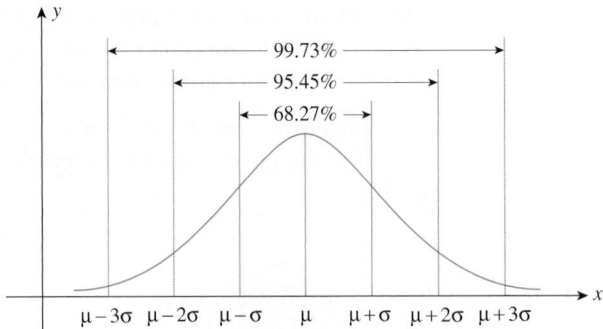

FIGURE 15
A normal curve

The normal curve (and therefore the corresponding normal distribution) is completely determined by its mean μ and standard deviation σ. In fact, the normal curve has the following characteristics, which are described in terms of these two parameters.*

1. The curve has a peak at $x = \mu$.
2. The curve is symmetric with respect to the vertical line $x = \mu$.
3. The curve always lies above the x-axis but approaches the x-axis as x extends indefinitely in either direction.
4. The area under the curve is 1.
5. For any normal curve, 68.27% of the area under the curve lies within 1 standard deviation of the mean (that is, between $\mu - \sigma$ and $\mu + \sigma$), 95.45% of the area lies within 2 standard deviations of the mean, and 99.73% of the area lies within 3 standard deviations of the mean.

Figure 16 shows two normal curves with different means μ_1 and μ_2 but the same deviation. Next, Figure 17 shows two normal curves with the same mean but different standard deviations σ_1 and σ_2. (Which number is smaller?)

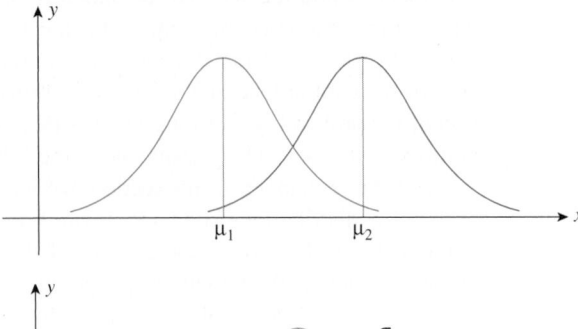

FIGURE 16
Two normal curves that have the same standard deviation but different means

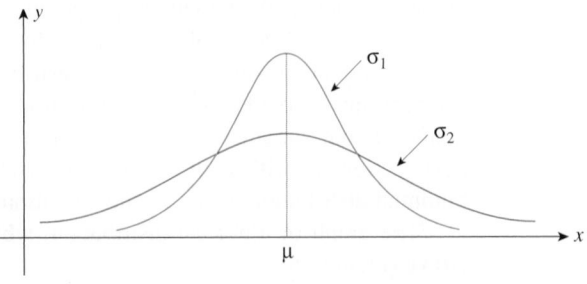

FIGURE 17
Two normal curves that have the same mean but different standard deviations

---

*The probability density function associated with this normal curve is given by

$$y = \frac{1}{\sigma\sqrt{2\pi}} e^{-(1/2)[(x-\mu)/\sigma]^2}$$

but the direct use of this formula will not be required in our discussion of the normal distribution.

In general, the mean μ of a normal distribution determines where the center of the curve is located, whereas the standard deviation σ of a normal distribution determines the sharpness (or flatness) of the curve.

As this discussion reveals, there are infinitely many normal curves corresponding to different choices of the parameters μ and σ that characterize such curves. Fortunately, any normal curve may be transformed into any other normal curve (as we will see later), so in the study of normal curves it suffices to single out one such particular curve for special attention. The normal curve with mean $\mu = 0$ and standard deviation $\sigma = 1$ is called the **standard normal curve**. The corresponding distribution is called the **standard normal distribution**. The random variable itself is called the **standard normal variable** and is commonly denoted by Z.

EXPLORING WITH TECHNOLOGY

Consider the probability density function

$$f(x) = \frac{1}{\sqrt{2\pi}} e^{-x^2/2}$$

which is the formula given in the footnote on page 492 with $\mu = 0$ and $\sigma = 1$.

1. Use a graphing utility to plot the graph of f, using the viewing window $[-4, 4] \times [0, 0.5]$.

2. Use the numerical integration function of a graphing utility to find the area of the region under the graph of f on the intervals $[-1, 1]$, $[-2, 2]$, and $[-3, 3]$, thereby verifying property 5 of normal distributions for the special case where $\mu = 0$ and $\sigma = 1$.

■ Computations of Probabilities Associated with Normal Distributions

Areas under the standard normal curve have been extensively computed and tabulated. Table 2, Appendix B, gives the areas of the regions under the standard normal curve to the left of the number z; these areas correspond, of course, to probabilities of the form $P(Z < z)$ or $P(Z \le z)$. The next several examples illustrate the use of this table in computations involving the probabilities associated with the standard normal variable.

EXAMPLE 1   Let Z be the standard normal variable. Make a sketch of the appropriate region under the standard normal curve, and then find the values of
a. $P(Z < 1.24)$ **b.** $P(Z > 0.5)$ **c.** $P(0.24 < Z < 1.48)$
d. $P(-1.65 < Z < 2.02)$

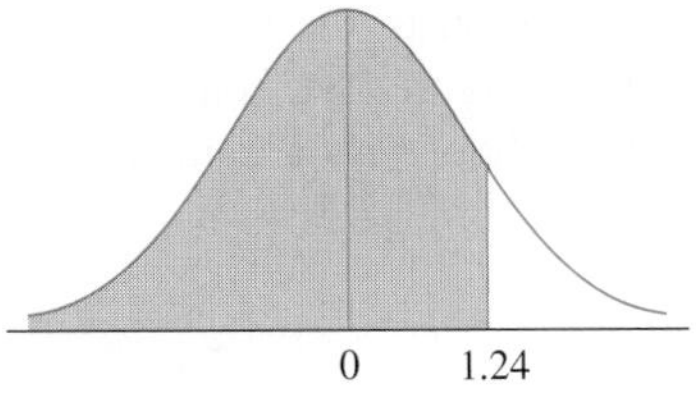

Solution
a. The region under the standard normal curve associated with the probability $P(Z < 1.24)$ is shown in Figure 18. To find the area of the required region using Table 2, Appendix B, we first locate the number 1.2 in the column and the number 0.04 in the row, both headed by z, and read off the number 0.8925 appearing in the body of the table. Thus,

$$P(Z < 1.24) = .8925$$

FIGURE 18
$P(Z < 1.24)$

0   1.24

b. The region under the standard normal curve associated with the probability $P(Z > 0.5)$ is shown in Figure 19a. Observe, however, that the required area is, by virtue of the symmetry of the standard normal curve, equal to the shaded area shown in Figure 19b. Thus,

$$P(Z > 0.5) = P(Z < -0.5)$$
$$= .3085$$

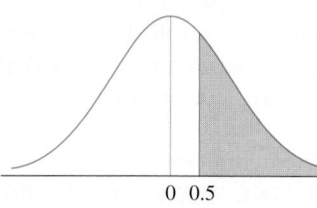 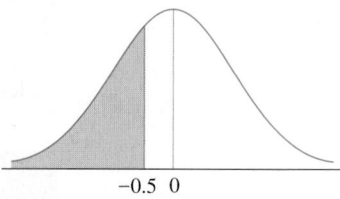

FIGURE 19

(a) $P(Z > 0.5)$ **(b)** $P(Z < -0.5)$

c. The probability $P(0.24 < Z < 1.48)$ is equal to the shaded area shown in Figure 20. This area is obtained by subtracting the area under the curve to the left of $z = 0.24$ from the area under the curve to the left of $z = 1.48$; that is

$$P(0.24 < Z < 1.48) = P(Z < 1.48) - P(Z < 0.24)$$
$$= .9306 - .5948$$
$$= .3358$$

d. The probability $P(-1.65 < Z < 2.02)$ is given by the shaded area shown in Figure 21. We have

$$P(-1.65 < Z < 2.02) = P(Z < 2.02) - P(Z < -1.65)$$
$$= .9783 - .0495$$
$$= .9288$$
∎

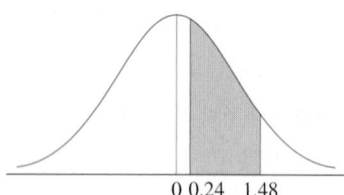

FIGURE 20
$P(0.24 < Z < 1.48)$

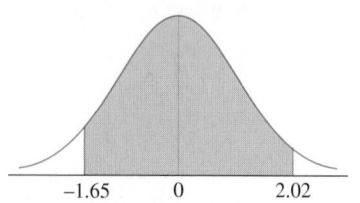

FIGURE 21
$P(-1.65 < Z < 2.02)$

EXAMPLE 2   Let Z be the standard normal variable. Find the value of z if z satisfies
a. $P(Z < z) = .9474$ **b.** $P(Z > z) = .9115$
c. $P(-z < Z < z) = .7888$

Solution
a. Refer to Figure 22. We want the value of Z such that the area of the region under the standard normal curve and to the left of $Z = z$ is .9474. Locating the number .9474 in Table 2, Appendix B, and reading back, we find that $z = 1.62$.
b. Since $P(Z > z)$, or equivalently, the area of the region to the right of z is greater than 0.5, it follows that z must be negative (Figure 23); hence $-z$ is positive. Furthermore, the area of the region to the right of z is the same as the area of the region to the left of $-z$. Therefore,

$$P(Z > z) = P(Z < -z)$$
$$= .9115$$

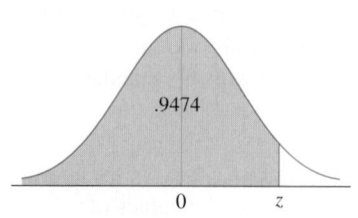

FIGURE 22
$P(Z < z) = .9474$

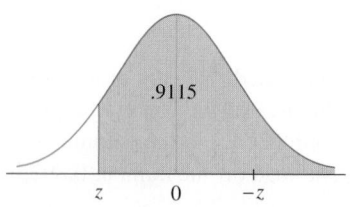

FIGURE 23
$P(Z > z) = .9115$

Looking up the table, we find $-z = 1.35$ and so $z = -1.35$.

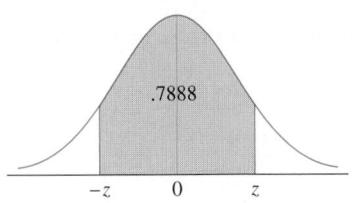

FIGURE 24
$P(-z < Z < z) = .7888$

c. The region associated with $P(-z < Z < z)$ is shown in Figure 24. Observe that, by symmetry, the area of this region is just double that of the area of the region between $Z = 0$ and $Z = z$; that is,

$$P(-z < Z < z) = 2P(0 < Z < z)$$

$$P(0 < Z < z) = P(Z < z) - \frac{1}{2}$$

(Figure 25). Therefore,

$$\frac{1}{2}P(-z < Z < z) = P(Z < z) - \frac{1}{2}$$

or, solving for $P(Z < z)$,

$$P(Z < z) = \frac{1}{2} + \frac{1}{2}P(-z < Z < z)$$

$$= \frac{1}{2}(1 + .7888)$$

$$= .8944$$

Consulting the table, we find $z = 1.25$.

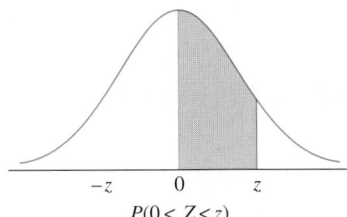 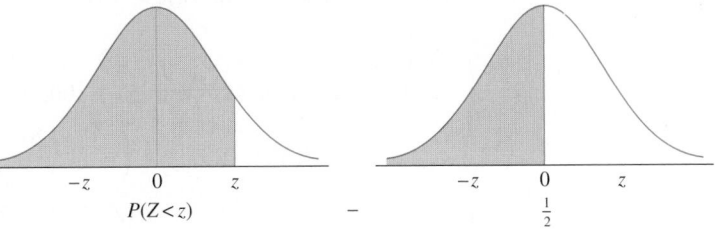

FIGURE 25

We now turn our attention to the computation of probabilities associated with normal distributions whose means and standard deviations are not necessarily equal to 0 and 1, respectively. As mentioned earlier, any normal curve may be transformed into the standard normal curve. In particular, it may be shown that if X is a normal random variable with mean μ and standard deviation σ, then it can be transformed into the standard normal random variable Z by means of the substitution

$$Z = \frac{X - \mu}{\sigma}$$

The area of the region under the normal curve (with random variable X) between $x = a$ and $x = b$ is *equal* to the area of the region under the standard normal curve between $z = (a - \mu)/\sigma$ and $z = (b - \mu)/\sigma$. In terms of probabilities associated with these distributions, we have

$$P(a < X < b) = P\left(\frac{a - \mu}{\sigma} < Z < \frac{b - \mu}{\sigma}\right) \tag{16}$$

(Figure 26). Similarly, we have

$$P(X < b) = P\left(Z < \frac{b - \mu}{\sigma}\right) \tag{17}$$

$$P(X > a) = P\left(Z > \frac{a - \mu}{\sigma}\right) \tag{18}$$

Thus, with the help of Equations (16)–(18), computations of probabilities associated with any normal distribution may be reduced to the computations of areas of regions under the standard normal curve.

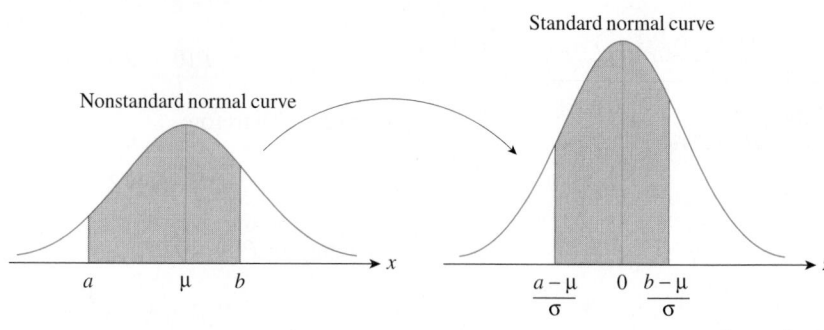

FIGURE 26

Area under the curve between a and b =   Area under the curve between $\dfrac{a-\mu}{\sigma}$ and $\dfrac{b-\mu}{\sigma}$

EXAMPLE 3   Suppose X is a normal random variable with $\mu = 100$ and $\sigma = 20$. Find the values of

a. $P(X < 120)$ **b.** $P(X > 70)$ **c.** $P(75 < X < 110)$

Solution

a. Using Formula (17) with $\mu = 100$, $\sigma = 20$, and $b = 120$, we have

$$P(X < 120) = P\left(Z < \frac{120 - 100}{20}\right)$$
$$= P(Z < 1) = .8413 \qquad \text{Use the table of values of } Z.$$

b. Using Formula (18) with $\mu = 100$, $\sigma = 20$, and $a = 70$, we have

$$P(X > 70)$$
$$= P\left(Z > \frac{70 - 100}{20}\right)$$
$$= P(Z > -1.5) = P(Z < 1.5) = .9332 \qquad \text{Use the table of values of } Z.$$

c. Using Formula (16) with $\mu = 100$, $\sigma = 20$, $a = 75$, and $b = 110$, we have

$$P(75 < X < 110)$$
$$= P\left(\frac{75 - 100}{20} < Z < \frac{110 - 100}{20}\right)$$
$$= P(-1.25 < Z < 0.5)$$
$$= P(Z < 0.5) - P(Z < -1.25) \qquad \text{See Figure 27.}$$
$$= .6915 - .1056 = .5859 \qquad \text{Use the table of values of } Z.$$

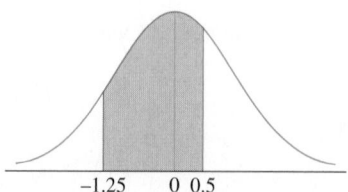

FIGURE 27

8.5 Self-Check Exercises

1. Let Z be a standard normal variable.
  a. Find the value of $P(-1.2 < Z < 2.1)$ by first making a sketch of the appropriate region under the standard normal curve.
  b. Find the value of z if z satisfies $P(-z < Z < z) = .8764$.

2. Let X be a normal random variable with $\mu = 80$ and $\sigma = 10$. Find the values of
  a. $P(X < 100)$ **b.** $P(X > 60)$ **c.** $P(70 < X < 90)$

Solutions to Self-Check Exercises 8.5 can be found on page 498.

8.5 Concept Questions

1. Consider the following normal curve with mean μ and standard deviation σ:

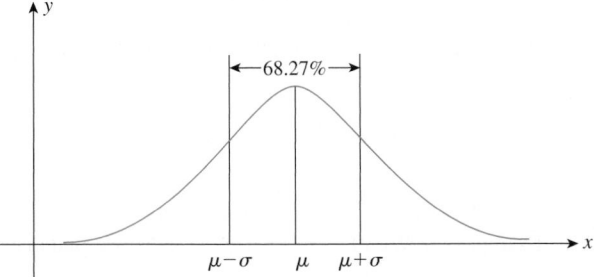

  a. What is the x-coordinate of the peak of the curve?
  b. What can you say about the symmetry of the curve?
  c. Does the curve always lie above the x-axis? What happens to the curve as x extends indefinitely to the left or right?
  d. What is the value of the area under the curve?
  e. Between what values does 68.27% of the area under the curve lie?

2. a. What is the difference between a normal curve and a standard normal curve?
  b. If X is a normal random variable with mean μ and standard deviation σ, write $P(a < X < b)$ in terms of the probabilities associated with the standard normal random variable Z.

8.5 Exercises

In Exercises 1–6, find the value of the probability of the standard normal variable Z corresponding to the shaded area under the standard normal curve.

1. $P(Z < 1.45)$

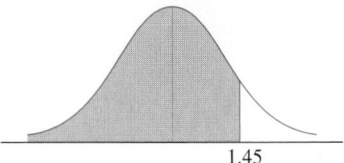

2. $P(Z > 1.11)$

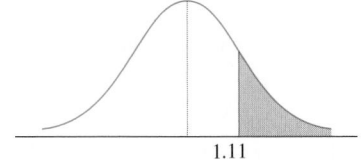

3. $P(Z < -1.75)$

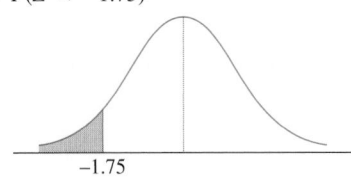

4. $P(0.3 < Z < 1.83)$

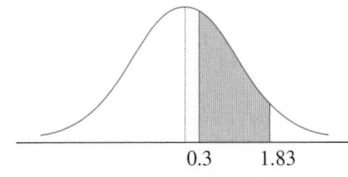

5. $P(-1.32 < Z < 1.74)$

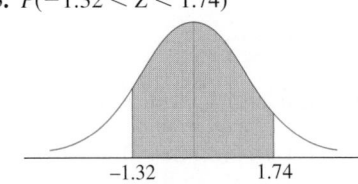

6. $P(-2.35 < Z < -0.51)$

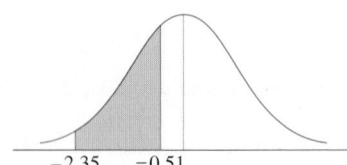

$$-2.35 \quad -0.51$$

In Exercises 7–14, (a) make a sketch of the area under the standard normal curve corresponding to the probability and (b) find the value of the probability of the standard normal variable Z corresponding to this area.

7. $P(Z < 1.37)$ **8.** $P(Z > 2.24)$

9. $P(Z < -0.65)$ **10.** $P(0.45 < Z < 1.75)$

11. $P(Z > -1.25)$ **12.** $P(-1.48 < Z < 1.54)$

13. $P(0.68 < Z < 2.02)$ **14.** $P(-1.41 < Z < -0.24)$

15. Let Z be the standard normal variable. Find the values of z if z satisfies
  a. $P(Z < z) = .8907$ **b.** $P(Z < z) = .2090$

16. Let Z be the standard normal variable. Find the values of z if z satisfies
  a. $P(Z > z) = .9678$ **b.** $P(-z < Z < z) = .8354$

17. Let Z be the standard normal variable. Find the values of z if z satisfies
  a. $P(Z > -z) = .9713$ **b.** $P(Z < -z) = .9713$

18. Suppose X is a normal random variable with $\mu = 380$ and $\sigma = 20$. Find the value of
  a. $P(X < 405)$ **b.** $P(400 < X < 430)$ **c.** $P(X > 400)$

19. Suppose X is a normal random variable with $\mu = 50$ and $\sigma = 5$. Find the value of
  a. $P(X < 60)$ **b.** $P(X > 43)$ **c.** $P(46 < X < 58)$

20. Suppose X is a normal random variable with $\mu = 500$ and $\sigma = 75$. Find the value of
  a. $P(X < 750)$ **b.** $P(X > 350)$ **c.** $P(400 < X < 600)$

8.5   Solutions to Self-Check Exercises

1. a. The probability $P(-1.2 < Z < 2.1)$ is given by the shaded area in the accompanying figure:

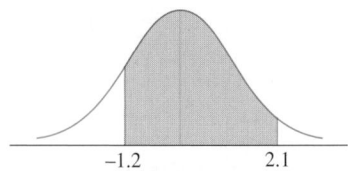

$$-1.2 \qquad 2.1$$

We have

$$P(-1.2 < Z < 2.1) = P(Z < 2.1) - P(Z < -1.2)$$
$$= .9821 - .1151$$
$$= .867$$

b. The region associated with $P(-z < Z < z)$ is shown in the accompanying figure:

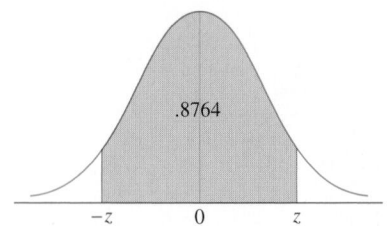

$$.8764$$
$$-z \qquad 0 \qquad z$$

Observe that we have the following relationship:

$$P(Z < z) = \frac{1}{2}[1 + P(-z < Z < z)]$$

(see Example 2c). With $P(-z < Z < z) = .8764$, we find that

$$P(Z < z) = \frac{1}{2}(1 + .8764)$$
$$= .9382$$

Consulting the table, we find $z = 1.54$.

2. Using the transformation (16) and the table of values of Z, we have

a. $P(X < 100) = P\left(Z < \dfrac{100 - 80}{10}\right)$
$$= P(Z < 2)$$
$$= .9772$$

b. $P(X > 60) = P\left(Z > \dfrac{60 - 80}{10}\right)$
$$= P(Z > -2)$$
$$= P(Z < 2)$$
$$= .9772$$

c. $P(70 < X < 90) = P\left(\dfrac{70 - 80}{10} < Z < \dfrac{90 - 80}{10}\right)$
$$= P(-1 < Z < 1)$$
$$= P(Z < 1) - P(Z < -1)$$
$$= .8413 - .1587$$
$$= .6826$$

8.6 Applications of the Normal Distribution

Applications Involving Normal Random Variables

In this section we look at some applications involving the normal distribution.

APPLIED EXAMPLE 1   Birth Weights of Infants   The medical records of infants delivered at the Kaiser Memorial Hospital show that the infants' birth weights in pounds are normally distributed with a mean of 7.4 and a standard deviation of 1.2. Find the probability that an infant selected at random from among those delivered at the hospital weighed more than 9.2 pounds at birth.

Solution   Let X be the normal random variable denoting the birth weights of infants delivered at the hospital. Then, the probability that an infant selected at random has a birth weight of more than 9.2 pounds is given by $P(X > 9.2)$. To compute $P(X > 9.2)$, we use Formula (18), Section 8.5, with $\mu = 7.4$, $\sigma = 1.2$, and $a = 9.2$. We find that

$$P(X > 9.2) = P\left(Z > \frac{9.2 - 7.4}{1.2}\right) \qquad P(X > a) = P\left(Z > \frac{a - \mu}{\sigma}\right)$$
$$= P(Z > 1.5)$$
$$= P(Z < -1.5)$$
$$= .0668$$

Thus, the probability that an infant delivered at the hospital weighs more than 9.2 pounds is .0668.

APPLIED EXAMPLE 2   Packaging   Idaho Natural Produce Corporation ships potatoes to its distributors in bags whose weights are normally distributed with a mean weight of 50 pounds and standard deviation of 0.5 pound. If a bag of potatoes is selected at random from a shipment, what is the probability that it weighs

a. More than 51 pounds?
b. Less than 49 pounds?
c. Between 49 and 51 pounds?

Solution   Let X denote the weight of potatoes packed by the company. Then the mean and standard deviation of X are $\mu = 50$ and $\sigma = 0.5$, respectively.

a. The probability that a bag selected at random weighs more than 51 pounds is given by

$$P(X > 51) = P\left(Z > \frac{51 - 50}{0.5}\right) \qquad P(X > a) = P\left(Z > \frac{a - \mu}{\sigma}\right)$$
$$= P(Z > 2)$$
$$= P(Z < -2)$$
$$= .0228$$

b. The probability that a bag selected at random weighs less than 49 pounds is given by

$$P(X < 49) = P\left(Z < \frac{49 - 50}{0.5}\right) \qquad P(X < b) = P\left(Z < \frac{b - \mu}{\sigma}\right)$$
$$= P(Z < -2)$$
$$= .0228$$

c. The probability that a bag selected at random weighs between 49 and 51 pounds is given by

$$P(49 < X < 51) \qquad\qquad P(a < X < b)$$
$$= P\left(\frac{49 - 50}{0.5} < Z < \frac{51 - 50}{0.5}\right) \qquad = P\left(\frac{a - \mu}{\sigma} < Z < \frac{b - \mu}{\sigma}\right)$$
$$= P(-2 < Z < 2)$$
$$= P(Z < 2) - P(Z < -2)$$
$$= .9772 - .0228$$
$$= .9544 \qquad\qquad\qquad\qquad\qquad\qquad \blacksquare$$

APPLIED EXAMPLE 3 College Admissions The grade-point average (GPA) of the senior class of Jefferson High School is normally distributed with a mean of 2.7 and a standard deviation of 0.4 point. If a senior in the top 10% of his or her class is eligible for admission to any of the nine campuses of the state university system, what is the minimum GPA that a senior should have to ensure eligibility for university admission?

Solution Let X denote the GPA of a randomly selected senior at Jefferson High School, and let x denote the minimum GPA that will ensure his or her eligibility for admission to the university. Since only the top 10% are eligible for admission, x must satisfy the equation

$$P(X \geq x) = .1$$

Using Formula (18), Section 8.5, with $\mu = 2.7$ and $\sigma = 0.4$, we find that

$$P(X \geq x) = P\left(Z \geq \frac{x - 2.7}{0.4}\right) = .1 \qquad P(X > a) = P\left(Z > \frac{a - \mu}{\sigma}\right)$$

This is equivalent to the equation

$$P\left(Z \leq \frac{x - 2.7}{0.4}\right) = .9 \qquad \text{Why?}$$

Consulting Table 2, Appendix B, we find that

$$\frac{x - 2.7}{0.4} = 1.28$$

Upon solving for x, we obtain

$$x = (1.28)(0.4) + 2.7$$
$$\approx 3.2$$

Thus, to ensure eligibility for admission to one of the nine campuses of the state university system, a senior at Jefferson High School should have a minimum of 3.2 GPA.   ▪

■ Approximating Binomial Distributions

As mentioned in the last section, one important application of the normal distribution is that it provides us with an accurate approximation of other continuous probability distributions. Here we show how a binomial distribution may be approximated by a suitable normal distribution. This technique leads to a convenient and simple solution to certain problems involving binomial probabilities.

Recall that a binomial distribution is a probability distribution of the form

$$P(X = x) = C(n, x)p^x q^{n-x} \qquad x = 0, 1, 2, \ldots, n \qquad \textbf{(19)}$$

(See Section 8.4.) For small values of n, the arithmetic computations of the binomial probabilities may be done with relative ease. However, if n is large, then the work involved becomes prodigious, even when tables of $P(X = x)$ are available. For example, if $n = 50$, $p = .3$, and $q = .7$, then the probability of ten or more successes is given by

$$P(X \geq 10) = P(X = 10) + P(X = 11) + \cdots + P(X = 50)$$

$$= \frac{50!}{10!\,40!}(.3)^{10}(.7)^{40} + \frac{50!}{11!\,39!}(.3)^{11}(.7)^{39} + \cdots + \frac{50!}{50!\,0!}(.3)^{50}(.7)^0$$

To see how the normal distribution helps us in such situations, let's consider a coin-tossing experiment. Suppose a fair coin is tossed 20 times and we wish to compute the probability of obtaining 10 or more heads. The solution to this problem may be obtained, of course, by computing

$$P(X \geq 10) = P(X = 10) + P(X = 11) + \cdots + P(X = 20)$$

The inconvenience of this approach for solving the problem at hand has already been pointed out. As an alternative solution, let's begin by interpreting the solution in terms of finding the area of suitable rectangles of the histogram for the distribution associated with the problem. We may use Formula (19) to compute the probability of obtaining exactly x heads in 20 coin tosses. The results lead to the binomial distribution displayed in Table 13.

Using the data from the table, we next construct the histogram for the distribution (Figure 28). The probability of obtaining 10 or more heads in 20 coin tosses

TABLE 13

Probability Distribution

x	$P(X = x)$
0	.0000
1	.0000
2	.0002
3	.0011
4	.0046
5	.0148
6	.0370
7	.0739
8	.1201
9	.1602
10	.1762
11	.1602
12	.1201
.	.
.	.
.	.
20	.0000

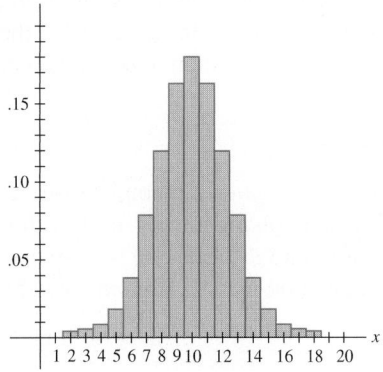

FIGURE 28
Histogram showing the probability of obtaining x heads in 20 coin tosses

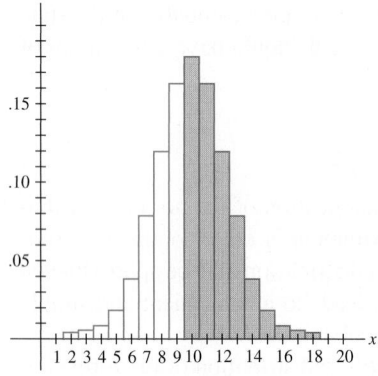

FIGURE 29
The shaded area gives the probability of obtaining ten or more heads in 20 coin tosses.

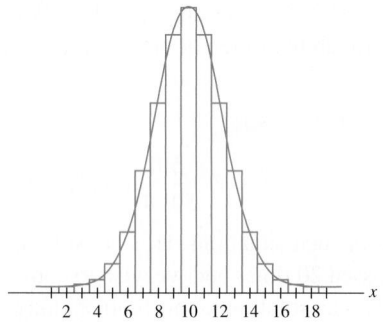

FIGURE 30
Normal curve superimposed on the histogram for a binomial distribution

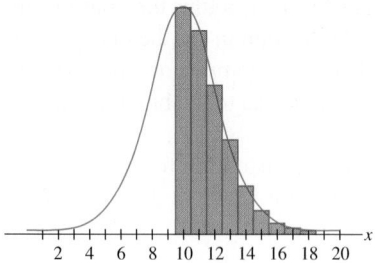

FIGURE 31
$P(X \geq 10)$ is approximated by the area under the normal curve.

is equal to the sum of the areas of the shaded rectangles of the histogram of the binomial distribution shown in Figure 29.

Next, observe that the shape of the histogram suggests that the binomial distribution under consideration may be approximated by a suitable normal distribution. Since the mean and standard deviation of the binomial distribution are given by

$$\mu = np$$
$$= (20)(.5) = 10$$
$$\sigma = \sqrt{npq}$$
$$= \sqrt{(20)(.5)(.5)}$$
$$= 2.24$$

respectively (see Section 8.4), the natural choice of a normal curve for this purpose is one with a mean of 10 and standard deviation of 2.24. Figure 30 shows such a normal curve superimposed on the histogram of the binomial distribution.

The good fit suggests that the sum of the areas of the rectangles representing $P(X \geq 10)$, the probability of obtaining 10 or more heads in 20 coin tosses, may be approximated by the area of an appropriate region under the normal curve. To determine this region, let's note that the base of the portion of the histogram representing the required probability extends from $x = 9.5$ on, since the base of the leftmost rectangle is centered at $x = 10$ and the base of each rectangle has length 1 (Figure 31). Therefore, the required region under the normal curve should also have $x \geq 9.5$. Letting Y denote the continuous normal variable, we obtain

$$P(X \geq 10) \approx P(Y \geq 9.5)$$
$$= P(Y > 9.5)$$
$$= P\left(Z > \frac{9.5 - 10}{2.24}\right) \qquad P(X > a) = P\left(Z > \frac{a - \mu}{\sigma}\right)$$
$$= P(Z > -0.22)$$
$$= P(Z < 0.22)$$
$$= .5871 \qquad \text{Use the table of values of } Z.$$

The exact value of $P(X \geq 10)$ may be found by computing

$$P(X = 10) + P(X = 11) + \cdots + P(X = 20)$$

in the usual fashion and is equal to .5881. Thus, the normal distribution with suitably chosen mean and standard deviation does provide us with a good approximation of the binomial distribution.

In the general case, the following result, which is a special case of the *central limit theorem*, guarantees the accuracy of the approximation of a binomial distribution by a normal distribution under certain conditions.

THEOREM 1

Suppose we are given a binomial distribution associated with a binomial experiment involving n trials, each with a probability of success p and probability of failure q. Then, if n is large and p is not close to 0 or 1, the binomial distribution may be approximated by a normal distribution with

$$\mu = np \quad \text{and} \quad \sigma = \sqrt{npq}$$

Note   It can be shown that if both np and nq are greater than 5, then the error resulting from this approximation is negligible.   ▪

◼ Applications Involving Binomial Random Variables

Next, we look at some applications involving binomial random variables.

APPLIED EXAMPLE 4   Quality Control   An automobile manufacturer receives the microprocessors used to regulate fuel consumption in its automobiles in shipments of 1000 each from a certain supplier. It has been estimated that, on the average, 1% of the microprocessors manufactured by the supplier are defective. Determine the probability that more than 20 of the microprocessors in a single shipment are defective.

Solution   Let X denote the number of defective microprocessors in a single shipment. Then X has a binomial distribution with $n = 1000$, $p = .01$, and $q = .99$, so

$$\mu = (1000)(.01) = 10$$
$$\sigma = \sqrt{(1000)(.01)(.99)}$$
$$\approx 3.15$$

Approximating the binomial distribution by a normal distribution with a mean of 10 and a standard deviation of 3.15, we find that the probability that more than 20 microprocessors in a shipment are defective is given by

$$P(X > 20) \approx P(Y > 20.5) \qquad \text{Where } Y \text{ denotes the normal random variable}$$
$$= P\left(Z > \frac{20.5 - 10}{3.15}\right) \qquad P(X > a) = P\left(Z > \frac{a - \mu}{\sigma}\right)$$
$$= P(Z > 3.33)$$
$$= P(Z < -3.33)$$
$$= .0004$$

In other words, approximately 0.04% of the shipments containing 1000 microprocessors each will contain more than 20 defective units.   ▪

APPLIED EXAMPLE 5   Heart Transplant Survival Rate   The probability that a heart transplant performed at the Medical Center is successful (that is, the patient survives 1 year or more after undergoing the surgery) is .7. Of 100 patients who have undergone such an operation, what is the probability that
a. Fewer than 75 will survive 1 year or more after the operation?
b. Between 80 and 90, inclusive, will survive 1 year or more after the operation?

Solution   Let X denote the number of patients who survive 1 year or more after undergoing a heart transplant at the Medical Center; then X is a binomial random variable. Also, $n = 100$, $p = .7$, and $q = .3$, so

$$\mu = (100)(.7) = 70$$
$$\sigma = \sqrt{(100)(.7)(.3)}$$
$$\approx 4.58$$

Approximating the binomial distribution by a normal distribution with a mean of 70 and a standard deviation of 4.58, we find, upon letting Y denote the associated normal random variable:

a. The probability that fewer than 75 patients will survive 1 year or more is given by

$$P(X < 75) \approx P(Y < 74.5) \qquad \text{Why?}$$

$$= P\left(Z < \frac{74.5 - 70}{4.58}\right) \qquad P(X < b) = P\left(Z < \frac{b - \mu}{\sigma}\right)$$

$$= P(Z < 0.98)$$

$$= .8365$$

b. The probability that between 80 and 90, inclusive, of the patients will survive 1 year or more is given by

$$P(80 \le X \le 90)$$

$$\approx P(79.5 < Y < 90.5)$$

$$= P\left(\frac{79.5 - 70}{4.58} < Z < \frac{90.5 - 70}{4.58}\right) \qquad \begin{array}{l} P(a < X < b) \\ = P\left(\frac{a - \mu}{\sigma} < Z < \frac{b - \mu}{\sigma}\right) \end{array}$$

$$= P(2.07 < Z < 4.48)$$

$$= P(Z < 4.48) - P(Z < 2.07)$$

$$= 1 - .9808 \qquad \text{Note: } P(Z < 4.48) \approx 1$$

$$= .0192$$

8.6 Self-Check Exercises

1. The serum cholesterol levels in milligrams/decaliter (mg/dL) in a current Mediterranean population are found to be normally distributed with a mean of 160 and a standard deviation of 50. Scientists at the National Heart, Lung, and Blood Institute consider this pattern ideal for a minimal risk of heart attacks. Find the percentage of the population having blood cholesterol levels between 160 and 180 mg/dL.

2. It has been estimated that 4% of the luggage manufactured by The Luggage Company fails to meet the standards established by the company and is sold as "seconds" to discount and outlet stores. If 500 bags are produced, what is the probability that more than 30 will be classified as "seconds"?

Solutions to Self-Check Exercises 8.6 can be found on page 506.

8.6 Concept Questions

1. What does the central limit theorem allow us to do?

2. Suppose a binomial distribution is associated with a binomial experiment involving n trials, each with a probability of success p and probability of failure q, and suppose n and p satisfy the other conditions given in the central limit theorem. What are the formulas for μ and σ that can be used to approximate this binomial distribution by a normal distribution?

8.6 Exercises

1. MEDICAL RECORDS   The medical records of infants delivered at Kaiser Memorial Hospital show that the infants' lengths at birth (in inches) are normally distributed with a mean of 20 and a standard deviation of 2.6. Find the probability that an infant selected at random from among those delivered at the hospital measures
   a. More than 22 in.   **b.** Less than 18 in.
   c. Between 19 and 21 in.

2. FACTORY WORKERS' WAGES   According to the data released by the Chamber of Commerce of a certain city, the weekly wages of factory workers are normally distributed with a mean of $600 and a standard deviation of $50. What is the probability that a factory worker selected at random from the city makes a weekly wage
   a. Of less than $600?   **b.** Of more than $760?
   c. Between $550 and $650?

3. PRODUCT RELIABILITY   TKK Products manufactures electric light bulbs in the 50-, 60-, 75-, and 100-watt range. Laboratory tests show that the lives of these light bulbs are normally distributed with a mean of 750 hr and a standard deviation of 75 hr. What is the probability that a TKK light bulb selected at random will burn
   a. For more than 900 hr?
   b. For less than 600 hr?
   c. Between 750 and 900 hr?
   d. Between 600 and 800 hr?

4. EDUCATION   On average, a student takes 100 words/minute midway through an advanced court reporting course at the American Institute of Court Reporting. Assuming that the dictation speeds of the students are normally distributed and that the standard deviation is 20 words/minute, what is the probability that a student randomly selected from the course can take dictation at a speed
   a. Of more than 120 words/minute?
   b. Between 80 and 120 words/minute?
   c. Of less than 80 words/minute?

5. IQs   The IQs of students at Wilson Elementary School were measured recently and found to be normally distributed with a mean of 100 and a standard deviation of 15. What is the probability that a student selected at random will have an IQ
   a. Of 140 or higher?   **b.** Of 120 or higher?
   c. Between 100 and 120?   **d.** Of 90 or less?

6. PRODUCT RELIABILITY   The tread lives of the Super Titan radial tires under normal driving conditions are normally distributed with a mean of 40,000 mi and a standard deviation of 2000 mi. What is the probability that a tire selected at random will have a tread life of more than 35,000 mi? If four new tires are installed in a car and they experience even wear, determine the probability that all four tires still have useful tread lives after 35,000 mi of driving.

7. FEMALE FACTORY WORKERS' WAGES   According to data released by the Chamber of Commerce of a certain city, the weekly wages (in dollars) of female factory workers are normally distributed with a mean of 575 and a standard deviation of 50. Find the probability that a female factory worker selected at random from the city makes a weekly wage of $550 to $650.

8. CIVIL SERVICE EXAMS   To be eligible for further consideration, applicants for certain civil service positions must first pass a written qualifying examination on which a score of 70 or more must be obtained. In a recent examination it was found that the scores were normally distributed with a mean of 60 points and a standard deviation of 10 points. Determine the percentage of applicants who passed the written qualifying examination.

9. WARRANTIES   The general manager of the service department of MCA Television has estimated that the time that elapses between the dates of purchase and the dates on which the 19-in. sets manufactured by the company first require service is normally distributed with a mean of 22 mo and a standard deviation of 4 mo. If the company gives a 1-yr warranty on parts and labor for these sets, determine the percentage of sets manufactured and sold by the company that may require service before the warranty period runs out.

10. GRADE DISTRIBUTIONS   The scores on an economics examination are normally distributed with a mean of 72 and a standard deviation of 16. If the instructor assigns a grade of A to 10% of the class, what is the lowest score a student may have and still obtain an A?

11. GRADE DISTRIBUTIONS   The scores on a sociology examination are normally distributed with a mean of 70 and a standard deviation of 10. If the instructor assigns As to 15%, Bs to 25%, Cs to 40%, Ds to 15%, and Fs to 5% of the class, find the cutoff points for these grades.

12. HIGHWAY SPEEDS   The speeds (in mph) of motor vehicles on a certain stretch of Route 3A as clocked at a certain place along the highway are normally distributed with a mean of 64.2 mph and a standard deviation of 8.44 mph. What is the probability that a motor vehicle selected at random is traveling at
   a. More than 65 mph?   **b.** Less than 60 mph?
   c. Between 65 and 70 mph?

In Exercises 13–24, use the appropriate normal distributions to approximate the resulting binomial distributions.

13. A coin is weighted so that the probability of obtaining a head in a single toss is .4. If the coin is tossed 25 times, what is the probability of obtaining
    a. Fewer than 10 heads?
    b. Between 10 and 12 heads, inclusive?
    c. More than 15 heads?

14. A fair coin is tossed 20 times. What is the probability of obtaining
  a. Fewer than 8 heads?   **b.** More than 6 heads?
  c. Between 6 and 10 heads inclusive?

15. SPORTS A marksman's chance of hitting a target with each of his shots is 60%. If he fires 30 shots, what is the probability of his hitting the target
  a. At least 20 times?   **b.** Fewer than 10 times?
  c. Between 15 and 20 times, inclusive?

16. SPORTS A basketball player has a 75% chance of making a free throw. What is the probability of her making 100 or more free throws in 120 trials?

17. QUALITY CONTROL The manager of C & R Clothiers, a major manufacturer of men's shirts, has determined that 3% of C & R's shirts do not meet with company standards and are sold as "seconds" to discount and outlet stores. What is the probability that, in a day's production of 200 dozen shirts, fewer than 10 dozen will be classified as "seconds"?

18. TELEMARKETING Jorge sells magazine subscriptions over the phone. He estimates that the probability of his making a sale with each attempt is .12. What is the probability of Jorge making more than 10 sales if he makes 80 calls?

19. INDUSTRIAL ACCIDENTS Colorado Mining and Mineral has 800 employees engaged in its mining operations. It has been estimated that the probability of a worker meeting with an accident during a 1-yr period is .1. What is the probability that more than 70 workers will meet with an accident during the 1-yr period?

20. QUALITY CONTROL PAR Bearings is the principal supplier of ball bearings for the Sperry Gyroscope Company. It has been determined that 6% of the ball bearings shipped are rejected because they fail to meet tolerance requirements. What is the probability that a shipment of 200 ball bearings contains more than 10 rejects?

21. DRUG TESTING An experiment was conducted to test the effectiveness of a new drug in treating a certain disease. The drug was administered to 50 mice that had been previously exposed to the disease. It was found that 35 mice subsequently recovered from the disease. It has been determined that the natural recovery rate from the disease is 0.5.
  a. Determine the probability that 35 or more of the mice not treated with the drug would recover from the disease.
  b. Using the results obtained in part (a), comment on the effectiveness of the drug in the treatment of the disease.

22. LOAN DELINQUENCIES The manager of Madison Finance Company has estimated that, because of a recession year, 5% of its 400 loan accounts will be delinquent. If the manager's estimate is correct, what is the probability that 25 or more of the accounts will be delinquent?

23. CRUISE SHIP BOOKINGS Because of late cancellations, Neptune Lines, an operator of cruise ships, has a policy of accepting more reservations than there are accommodations available. From experience, 8% of the bookings for the 90-day around-the-world cruise on the S.S. *Drion*, which has accommodations for 2000 passengers, are subsequently canceled. If the management of Neptune Lines has decided, for public relations reasons, that a person who has made a reservation should have a probability of .99 of obtaining accommodation on the ship, determine the largest number of reservations that should be taken for this cruise on the S.S. *Drion*.

24. THEATER BOOKINGS Preview Showcase, a research firm, screens pilots of new TV shows before a randomly selected audience and then solicits their opinions of the shows. Based on past experience, 20% of those who receive complimentary tickets are "no-shows." The theater has a seating capacity of 500. Management has decided, for public relations reasons, that a person who has been solicited for a screening should have a probability of .99 of being seated. How many tickets should the company send out to prospective viewers for each screening?

8.6 Solutions to Self-Check Exercises

1. Let X be the normal random variable denoting the serum cholesterol levels in mg/dL in the current Mediterranean population under consideration. Then, the percentage of the population having blood cholesterol levels between 160 and 180 mg/dL is given by $P(160 < X < 180)$. To compute $P(160 < X < 180)$, we use Formula (16), Section 8.5, with $\mu = 160$, $\sigma = 50$, $a = 160$, and $b = 180$. We find

$$P(160 < X < 180) = P\left(\frac{160 - 160}{50} < Z < \frac{180 - 160}{50}\right)$$
$$= P(0 < Z < 0.4)$$
$$= P(Z < 0.4) - P(Z < 0)$$
$$= .6554 - .5000$$
$$= .1554$$

Thus, approximately 15.5% of the population has blood cholesterol levels between 160 and 180 mg/dL.

2. Let X denote the number of substandard bags in the production. Then X has a binomial distribution with $n = 500$, $p = .04$, and $q = .96$, so

$$\mu = (500)(.04) = 20$$
$$\sigma = \sqrt{(500)(.04)(.96)} = 4.38$$

Approximating the binomial distribution by a normal distribution with a mean of 20 and standard deviation of 4.38, we find that the probability that more than 30 bags in the production of 500 will be substandard is given by

$$P(X > 30) \approx P(Y > 30.5)$$

Where Y denotes the normal random variable

$$= P\left(Z > \frac{30.5 - 20}{4.38}\right)$$
$$= P(Z > 2.40)$$
$$= P(Z < -2.40)$$
$$= .0082$$

or approximately 0.8%.

 CHAPTER 8 **Summary of Principal Formulas and Terms**

FORMULAS

1. Mean of n numbers	$\bar{x} = \dfrac{x_1 + x_2 + \cdots + x_n}{n}$
2. Expected value	$E(X) = x_1 p_1 + x_2 p_2 + \cdots + x_n p_n$
3. Odds in favor of E occurring	$\dfrac{P(E)}{P(E^c)}$
4. Odds against E occurring	$\dfrac{P(E^c)}{P(E)}$
5. Probability of an event occurring given the odds	$\dfrac{a}{a + b}$
6. Variance of a random variable	$\mathrm{Var}(X) = p_1(x_1 - \mu)^2 + p_2(x_2 - \mu)^2 + \cdots + p_n(x_n - \mu)^2$
7. Standard deviation of a random variable	$\sigma = \sqrt{\mathrm{Var}(X)}$
8. Chebychev's inequality	$P(\mu - k\sigma \leq X \leq \mu + k\sigma)$ $\geq 1 - \dfrac{1}{k^2}$
9. Probability of x successes in n Bernoulli trials	$C(n, x)p^x q^{n-x}$
10. Binomial random variable: Mean Variance Standard deviation	$\mu = E(X) = np$ $\mathrm{Var}(X) = npq$ $\sigma_x = \sqrt{npq}$

TERMS

CHAPTER 8 Concept Review Questions

Fill in the blanks.

1. A rule that assigns a number to each outcome of a chance experiment is called a/an _____ variable.

2. If a random variable assumes only finitely many values, then it is called _____ discrete; if it takes on infinitely many values that can be arranged in a sequence, then it is called _____ discrete; if it takes on all real numbers in an interval, then it is said to be _____.

3. The expected value of a random variable X is given by the _____ of the products of the values assumed by the random variable and its associated probabilities. For example, if X assumes the values -2, 3, and 4 with associated probabilities $\frac{1}{2}$, $\frac{1}{4}$, and $\frac{1}{4}$, then its expected value is _____.

4. **a.** If the probability of an event E occurring is $P(E)$, then the odds in favor of E occurring are _____.
   b. If the odds in favor of an event E occurring are a to b, then the probability of E occurring is _____.

5. Suppose a random variable X takes on the values $x_1, x_2, \ldots,$ x_n with probabilities $p_1, p_2, \ldots, p_n$ and has a mean of μ. Then the variance of X is _____ and the standard deviation of X is _____.

6. In a binomial experiment, the number of trials is _____, there are exactly _____ outcomes of the experiment, the probability of "success" in each trial is the _____, and the trials are _____ of each other.

7. A probability distribution that is associated with a continuous random variable is called a/an _____ probability distribution. Such a probability distribution is defined by a/an _____ _____ _____ whose domain is the _____ of values taken on by the random variable associated with the experiment.

8. A binomial distribution may be approximated by a/an _____ distribution with $\mu = np$ and $\sigma = \sqrt{npq}$ if n is _____ and p is not close to _____ or _____.

CHAPTER 8 Review Exercises

1. Three balls are selected at random without replacement from an urn containing three white balls and four blue balls. Let the random variable X denote the number of blue balls drawn.
   a. List the outcomes of this experiment.
   b. Find the value assigned to each outcome of this experiment by the random variable X.
   c. Find the probability distribution of the random variable associated with this experiment.
   d. Draw the histogram representing this distribution.

2. A man purchased a $25,000, 1-yr term-life insurance policy for $375. Assuming that the probability that he will live for another year is .989, find the company's expected gain.

3. The probability distribution of a random variable X is shown in the following table:

x	$P(X = x)$
0	.1
1	.1
2	.2
3	.3
4	.2
5	.1

   a. Compute $P(1 \leq X \leq 4)$.
   b. Compute the mean and standard deviation of X.

4. A binomial experiment consists of four trials in which the probability of success in any one trial is $\frac{2}{5}$.
   a. Construct the probability distribution for the experiment.
   b. Compute the mean and standard deviation of the probability distribution.

In Exercises 5–8, let Z be the standard normal variable. Make a rough sketch of the appropriate region under the standard normal curve and find the probability.

5. $P(Z < 0.5)$

6. $P(Z < -0.75)$

7. $P(-0.75 < Z < 0.5)$

8. $P(-0.42 < Z < 0.66)$

In Exercises 9–12, let Z be the standard normal variable. Find z if z satisfies the given value.

9. $P(Z < z) = .9922$

10. $P(Z < z) = .1469$

11. $P(Z > z) = .9788$

12. $P(-z < Z < z) = .8444$

In Exercises 13–16, let X be a normal random variable with $\mu = 10$ and $\sigma = 2$. Find the value of the given probability.

13. $P(X < 11)$

14. $P(X > 8)$

15. $P(7 < X < 9)$

16. $P(6.5 < X < 11.5)$

17. **Sports** If the probability that a bowler will bowl a strike is .7, what is the probability that he will get exactly two strikes in four attempts? At least two strikes in four attempts?

18. **Heights of Women** The heights of 4000 women who participated in a recent survey were found to be normally distributed with a mean of 64.5 in. and a standard deviation of 2.5 in. What percentage of these women have heights of 67 in. or greater?

19. **Heights of Women** Refer to Exercise 18. Use Chebychev's inequality to estimate the probability that the height of a woman who participated in the survey will fall within 2 standard deviations of the mean—that is, that her height will be between 59.5 and 69.5 in.

20. **Marital Status of Women** The number of single women (in thousands) between the ages of 20 and 44 in the United States in 1998 is given in the following table:

Age	20–24	25–29	30–34	35–39	40–44
Women	6178	3689	2219	1626	1095

Find the mean and the standard deviation of the given data. Hint: Assume that all values lying within a group interval take the middle value of that group.

Source: U.S. Census Bureau

21. **Quality Control** The proprietor of a hardware store will accept a shipment of ceramic wall tiles if no more than 2 of a random sample of 20 are found to be defective. What is the probability that he will accept shipment if exactly 10% of the tiles in a certain shipment is defective?

22. **Drug Effectiveness** An experimental drug has been found to be effective in treating 15% of the people afflicted by a certain disease. If the drug is administered to 800 people who have this disease, what are the mean and standard deviation of the number of people for whom the drug can be expected to be effective?

23. **Quality Control** Dayton Iron Works manufactures steel rods to a specification of 1-in. diameter. These rods are accepted by the buyer if they fall within the tolerance limits of 0.995 and 1.005. Assuming that the diameter of the rods is normally distributed about a mean of 1 in. and a standard deviation of 0.002 in., estimate the percentage of rods that will be rejected by the buyer.

24. **Coin Tosses** A coin is biased so that the probability of it landing heads is .6. If the coin is tossed 100 times, what is the probability that heads will appear more than 50 times in the 100 tosses?

25. **Quality Control** A division of Solaron Corporation manufactures photovoltaic cells for use in the company's solar energy converters. It is estimated that 5% of the cells manufactured are defective. In a batch of 200 cells manufactured by the company, what is the probability that it will contain at most 20 defective units?

CHAPTER 8 Before Moving On . . .

1. The values taken on by a random variable X and the frequency of their occurrence are shown in the following table. Find the probability distribution of X.

x	−3	−2	0	1	2	3
Frequency of Occurrence	4	8	20	24	16	8

2. The probability distribution of the random variable X is shown in the following table. Find (a) $P(X \le 0)$ and (b) $P(-4 \le X \le 1)$.

x	−4	−3	−1	0	1	3
P(X = x)	.06	.14	.32	.28	.12	.08

3. Find the mean, variance, and standard deviation of a random variable X having the following probability distribution:

x	−3	−1	0	1	3	5
P(X = x)	.08	.24	.32	.16	.12	.08

4. A binomial experiment consists of four independent trials, and the probability of success in each trial is 0.3.
   a. Find the probability of obtaining 0, 1, 2, 3, and 4 successes, respectively.
   b. Compute the mean and standard deviation of the random variable associated with this experiment.

5. Let X be a normal random variable with $\mu = 60$ and $\sigma = 5$. Find the values of (a) $P(X < 70)$, (b) $P(X > 50)$, and (c) $P(50 < X < 70)$.

6. A fair coin is tossed 30 times. Using the appropriate normal distribution to approximate a binomial distribution, find the probability of obtaining (a) fewer than 10 heads, (b) between 12 and 16 heads, inclusive, and (c) more than 20 heads.

9 Markov Chains

© Joseph Sohm; Chromosohm/Corbis

Will the flowers be red? A certain species of plant produces red, pink, or white flowers, depending on its genetic makeup. In Example 4, page 538, we will show that if the offspring of two plants are crossed successively with plants of a certain genetic makeup only, then in the long run all the flowers produced by the plants will be red.

IN THIS CHAPTER, we look at an important application of mathematics that is based primarily on matrix theory and the theory of probability. *Markov chains*, a relatively recent development in the field of mathematics, have widespread applications in many practical areas.

9.1 Markov Chains

■ Transitional Probabilities

A finite stochastic process, you may recall, is an experiment consisting of a finite number of stages in which the outcomes and associated probabilities at each stage depend on the outcomes and associated probabilities of the *preceding stages*. In this chapter, we are concerned with a special class of stochastic processes—namely, those in which the probabilities associated with the outcomes at any stage of the experiment depend only on the outcomes of the *preceding stage*. Such a process is called a **Markov process**, or a **Markov chain**, named after the Russian mathematician A. A. Markov (1856–1922).

The outcome at any stage of the experiment in a Markov process is called the **state** of the experiment. In particular, the outcome at the current stage of the experiment is called the **current state** of the process. Here is a typical problem involving a Markov chain:

Starting from one state of a process (the current state), determine the probability that the process will be at a particular state at some future time.

APPLIED EXAMPLE 1   Common Stocks   An analyst at Weaver and Kline, a stock brokerage firm, observes that the closing price of the preferred stock of an airline company over a short span of time depends only on its previous closing price. At the end of each trading day, he makes a note of the stock's performance for that day, recording the closing price as "higher," "unchanged," or "lower" according to whether the stock closes higher, unchanged, or lower than the previous day's closing price. This sequence of observations may be viewed as a Markov chain. ■

The transition from one state to another in a Markov chain may be studied with the aid of tree diagrams, as in the next example.

APPLIED EXAMPLE 2   Common Stocks   Refer to Example 1. If on a certain day the stock's closing price is higher than that of the previous day, then the probability that it closes higher, unchanged, or lower on the next trading day is .2, .3, and .5, respectively. Next, if the stock's closing price is unchanged from the previous day, then the probability that it closes higher, unchanged, or lower on the next trading day is .5, .2, and .3, respectively. Finally, if the stock's closing price is lower than that of the previous day, then the probability that it closes higher, unchanged, or lower on the next trading day is .4, .4, and .2, respectively. With the aid of tree diagrams, describe the transition between states and the probabilities associated with these transitions.

Solution   The Markov chain being described has three states: higher, unchanged, and lower. If the current state is higher, then the transition to the other states from this state may be displayed by constructing a tree diagram in which the associated probabilities are shown on the appropriate branches (Figure 1). Tree diagrams describing the transition from each of the other two possible current states, unchanged and lower, to the other states are constructed in a similar manner.

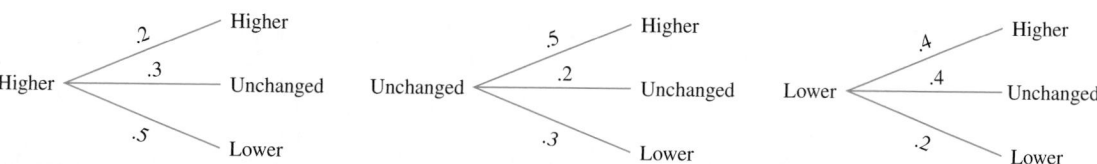

FIGURE 1
Tree diagrams showing transition probabilities between states

The probabilities encountered in this example are called **transition probabilities** because they are associated with the transition from one state to the next in the Markov process. These transition probabilities may be conveniently represented in the form of a matrix. Suppose for simplicity that we have a Markov chain with three possible outcomes at each stage of the experiment. Let's refer to these outcomes as state 1, state 2, and state 3. Then the transition probabilities associated with the transition from state 1 to each of the states 1, 2, and 3 in the next phase of the experiment are precisely the respective conditional probabilities that the outcome is state 1, state 2, and state 3 *given* that the outcome state 1 has already occurred. In short, the desired transition probabilities are $P(\text{state } 1 | \text{state } 1)$, $P(\text{state } 2 | \text{state } 1)$, and $P(\text{state } 3 | \text{state } 1)$, respectively. Thus, we write

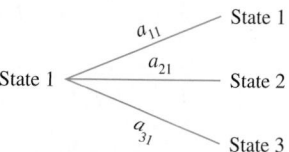

$$a_{11} = P(\text{state } 1 | \text{state } 1)$$
$$a_{21} = P(\text{state } 2 | \text{state } 1)$$
$$a_{31} = P(\text{state } 3 | \text{state } 1)$$

Note that the first subscript in this notation refers to the state in the next stage of the experiment and that the second subscript refers to the current state. Using a tree diagram, we have the following representation:

$$
\text{State 1} <
\begin{array}{l}
\xrightarrow{a_{11}} \text{State 1} \\
\xrightarrow{a_{21}} \text{State 2} \\
\xrightarrow{a_{31}} \text{State 3}
\end{array}
$$

Similarly, the transition probabilities associated with the transition from state 2 and state 3 to each of the states 1, 2, and 3 are

$$a_{12} = P(\text{state } 1 | \text{state } 2) \quad \text{and} \quad a_{13} = P(\text{state } 1 | \text{state } 3)$$
$$a_{22} = P(\text{state } 2 | \text{state } 2) \qquad\qquad a_{23} = P(\text{state } 2 | \text{state } 3)$$
$$a_{32} = P(\text{state } 3 | \text{state } 2) \qquad\qquad a_{33} = P(\text{state } 3 | \text{state } 3)$$

These observations lead to the following matrix representation of the transition probabilities:

$$\text{Next state}\quad\begin{array}{c}\\\text{State 1}\\\text{State 2}\\\text{State 3}\end{array}\overset{\begin{array}{ccc}\text{State 1}&\text{State 2}&\text{State 3}\end{array}}{\begin{bmatrix}a_{11}&a_{12}&a_{13}\\a_{21}&a_{22}&a_{23}\\a_{31}&a_{32}&a_{33}\end{bmatrix}}$$

Current state

EXAMPLE 3   Use a matrix to represent the transition probabilities obtained in Example 2.

Solution   There are three states at each stage of the Markov chain under consideration. Letting state 1, state 2, and state 3 denote the states "higher," "unchanged," and "lower," respectively, we find that

$$a_{11} = .2 \quad a_{21} = .3 \quad a_{31} = .5$$

and so on. Hence the required matrix representation is given by

$$T = \begin{bmatrix} .2 & .5 & .4 \\ .3 & .2 & .4 \\ .5 & .3 & .2 \end{bmatrix}$$

The matrix obtained in Example 3 is a transition matrix. In the general case, we have the following definition.

Transition Matrix

A **transition matrix** associated with a Markov chain with n states is an $n \times n$ matrix T with entries a_{ij} $(1 \leq i \leq n; 1 \leq j \leq n)$

Current state

$$T = \begin{array}{c}\text{Next}\\\text{state}\end{array}\begin{array}{c}\\\text{State 1}\\\text{State 2}\\\vdots\\\text{State }i\\\vdots\\\text{State }n\end{array}\overset{\begin{array}{cccccc}\text{State 1}&\text{State 2}&\cdots&\text{State }j&\cdots&\text{State }n\end{array}}{\begin{bmatrix}a_{11}&a_{12}&\cdots&a_{1j}&\cdots&a_{1n}\\a_{21}&a_{22}&\cdots&a_{2j}&\cdots&a_{2n}\\\vdots&\vdots&&\vdots&&\vdots\\a_{i1}&a_{i2}&\cdots&a_{ij}&\cdots&a_{in}\\\vdots&\vdots&&\vdots&&\vdots\\a_{n1}&a_{n2}&\cdots&a_{nj}&\cdots&a_{nn}\end{bmatrix}}$$

having the following properties:

1. $a_{ij} \geq 0$ for all i and j.
2. The sum of the entries in each column of T is 1.

Since $a_{ij} = P(\text{state } i \,|\, \text{state } j)$ is the probability of the occurrence of an event, it must be nonnegative, and this is precisely what property 1 implies. Property 2 follows from the fact that the transition from any one of the current states must terminate in one of the n states in the next stage of the experiment. Any square matrix satisfying properties 1 and 2 is referred to as a **stochastic matrix**.

EXPLORE & DISCUSS

Let

$$A = \begin{bmatrix} p & q \\ 1-p & 1-q \end{bmatrix} \quad \text{and} \quad B = \begin{bmatrix} r & s \\ 1-r & 1-s \end{bmatrix}$$

be two 2×2 stochastic matrices, where $0 \le p \le 1$, $0 \le q \le 1$, $0 \le r \le 1$, and $0 \le s \le 1$.

1. Show that AB is a 2×2 stochastic matrix.

2. Use the result of part (a) to explain why $A^2, A^3, \ldots, A^n$, where n is a positive integer, are also 2×2 stochastic matrices.

One advantage of representing the transition probabilities in the form of a matrix is that we may use the results from matrix theory to help us solve problems involving Markov processes, as we will see in the next several sections.

For simplicity, let's now consider the following Markov process where each stage of the experiment has precisely two possible states.

APPLIED EXAMPLE 4   Urban–Suburban Population Flow   Because of the continued successful implementation of an urban renewal program, it is expected that each year 3% of the population currently residing in the city will move to the suburbs and 6% of the population currently residing in the suburbs will move into the city. At present, 65% of the total population of the metropolitan area live in the city itself, while the remaining 35% live in the suburbs. Assuming that the total population of the metropolitan area remains constant, what will be the distribution of the population 1 year from now?

Solution   This problem may be solved with the aid of a tree diagram and the techniques of Chapter 7. The required tree diagram describing this process is shown in Figure 2. Using the method of Section 7.5, we find the probability that a person selected at random will be a city dweller 1 year from now is given by

$$(.65)(.97) + (.35)(.06) = .6515$$

In a similar manner, we find that the probability that a person selected at random will reside in the suburbs 1 year from now is given by

$$(.65)(.03) + (.35)(.94) = .3485$$

Current population · Population 1 year later

.65 — City
.97 — City
.03 — Suburbs
.35 — Suburbs
.06 — City
.94 — Suburbs

FIGURE 2
Tree diagram showing a Markov process with two states: living in the city and living in the suburbs

Thus, the population of the area 1 year from now may be expected to be distributed as follows: 65.15% living in the city and 34.85% residing in the suburbs.

Let's reexamine the solution to this problem. As noted previously, the process under consideration may be viewed as a Markov chain with two possible states at each stage of the experiment: "living in the city" (state 1) and "living in the suburbs" (state 2). The transition matrix associated with this Markov chain is

$$T = \begin{array}{c} \\ \text{State 1} \\ \text{State 2} \end{array} \overset{\text{State 1 \quad State 2}}{\begin{bmatrix} .97 & .06 \\ .03 & .94 \end{bmatrix}} \qquad \text{Transition matrix}$$

Next, observe that the initial (current) probability distribution of the population may be summarized in the form of a column vector of dimension 2 (that is, a 2×1 matrix). Thus,

$$X_0 = \begin{array}{c} \text{State 1} \\ \text{State 2} \end{array} \begin{bmatrix} .65 \\ .35 \end{bmatrix} \qquad \text{Initial-state matrix}$$

Using the results of Example 4, we may write the population distribution 1 year later as

$$X_1 = \begin{array}{c} \text{State 1} \\ \text{State 2} \end{array} \begin{bmatrix} .6515 \\ .3485 \end{bmatrix} \qquad \text{Distribution after 1 year}$$

You may now verify that

$$TX_0 = \begin{bmatrix} .97 & .06 \\ .03 & .94 \end{bmatrix} \begin{bmatrix} .65 \\ .35 \end{bmatrix} = \begin{bmatrix} .6515 \\ .3485 \end{bmatrix} = X_1$$

so this problem may be solved using matrix multiplication.

APPLIED EXAMPLE 5   Urban–Suburban Population Flow   Refer to Example 4. What is the population distribution of the city after 2 years? After 3 years?

Solution   Let X_2 be the column vector representing the probability population distribution of the metropolitan area after 2 years. We may view X_1, the vector representing the probability population distribution of the metropolitan area after 1 year, as representing the "initial" probability distribution in this part of our calculation. Thus,

$$X_2 = TX_1 = \begin{bmatrix} .97 & .06 \\ .03 & .94 \end{bmatrix} \begin{bmatrix} .6515 \\ .3485 \end{bmatrix} = \begin{bmatrix} .6529 \\ .3471 \end{bmatrix}$$

The vector representing the probability distribution of the metropolitan area after 3 years is given by

$$X_3 = TX_2 = \begin{bmatrix} .97 & .06 \\ .03 & .94 \end{bmatrix} \begin{bmatrix} .6529 \\ .3471 \end{bmatrix} = \begin{bmatrix} .6541 \\ .3459 \end{bmatrix}$$

That is, after 3 years, the population will be distributed as follows: 65.41% will live in the city and 34.59% will live in the suburbs.

Distribution Vectors

Observe that, in the foregoing computations, we have $X_1 = TX_0$, $X_2 = TX_1 = T^2X_0$, and $X_3 = TX_2 = T^3X_0$. These results are easily generalized. To see this, suppose we have a Markov process in which there are n possible states at each stage of the experiment. Suppose further that the probability of the system initially being in state 1, state 2, . . . , state n is given by $p_1, p_2, \ldots, p_n$, respectively. This distribution may be represented as an n-dimensional vector,

$$X_0 = \begin{bmatrix} p_1 \\ p_2 \\ \vdots \\ p_n \end{bmatrix}$$

called a **distribution vector**. If T represents the $n \times n$ transition matrix associated with the Markov process, then the probability distribution of the system after m observations is given by

$$X_m = T^m X_0 \tag{1}$$

APPLIED EXAMPLE 6   Taxicab Movement   To keep track of the location of its cabs, Zephyr Cab has divided a town into three zones: zone I, zone II, and zone III. Zephyr's management has determined from company records that, of the passengers picked up in zone I, 60% are discharged in the same zone, 30% are discharged in zone II, and 10% are discharged in zone III. Of those picked up in zone II, 40% are discharged in zone I, 30% are discharged in zone II, and 30% are discharged in zone III. Of those picked up in zone III, 30% are discharged in zone I, 30% are discharged in zone II, and 40% are discharged in zone III. At the beginning of the day, 80% of the cabs are in zone I, 15% are in zone II, and 5% are in zone III. A taxi without a passenger will cruise within the zone it is currently in until a pickup is made.

a. Find the transition matrix for the Markov chain that describes the successive locations of a cab.
b. What is the distribution of the cabs after all of them have made one pickup and discharge?
c. What is the distribution of the cabs after all of them have made two pickups and discharges?

Solution   Let zone I, zone II, and zone III correspond to state 1, state 2, and state 3 of the Markov chain.

a. The required transition matrix is given by

$$T = \begin{bmatrix} .6 & .4 & .3 \\ .3 & .3 & .3 \\ .1 & .3 & .4 \end{bmatrix}$$

b. The initial distribution vector associated with the problem is

$$X_0 = \begin{bmatrix} .8 \\ .15 \\ .05 \end{bmatrix}$$

If X_1 denotes the distribution vector after one observation—that is, after all the cabs have made one pickup and discharge—then

$$X_1 = TX_0$$
$$= \begin{bmatrix} .6 & .4 & .3 \\ .3 & .3 & .3 \\ .1 & .3 & .4 \end{bmatrix} \begin{bmatrix} .8 \\ .15 \\ .05 \end{bmatrix} = \begin{bmatrix} .555 \\ .3 \\ .145 \end{bmatrix}$$

That is, 55.5% of the cabs are in zone I, 30% are in zone II, and 14.5% are in zone III.

c. Let X_2 denote the distribution vector after all the cabs have made two pickups and discharges. Then

$$X_2 = TX_1$$
$$= \begin{bmatrix} .6 & .4 & .3 \\ .3 & .3 & .3 \\ .1 & .3 & .4 \end{bmatrix} \begin{bmatrix} .555 \\ .3 \\ .145 \end{bmatrix} = \begin{bmatrix} .4965 \\ .3 \\ .2035 \end{bmatrix}$$

That is, 49.65% of the cabs are in zone I, 30% are in zone II, and 20.35% are in zone III. You should verify that the same result may be obtained by computing T^2X_0. ∎

Note In this simplified model, we do not take into consideration variable demand and variable delivery time. ∎

9.1 Self-Check Exercises

1. Three supermarkets serve a certain section of a city. During the upcoming year, supermarket A is expected to retain 80% of its customers, lose 5% of its customers to supermarket B, and lose 15% to supermarket C. Supermarket B is expected to retain 90% of its customers and lose 5% of its customers to each of supermarkets A and C. Supermarket C is expected to retain 75% of its customers, lose 10% to supermarket A, and lose 15% to supermarket B. Construct the transition matrix for the Markov chain that describes the change in the market share of the three supermarkets.

2. Refer to Self-Check Exercise 1. Currently the market shares of supermarket A, supermarket B, and supermarket C are 0.4, 0.3, and 0.3, respectively.
   a. Find the initial distribution vector for this Markov chain.
   b. What share of the market will be held by each supermarket after 1 year? Assuming that the trend continues, what will be the market share after 2 years?

Solutions to Self-Check Exercises 9.1 can be found on page 522.

9.1 Concept Questions

1. What is a finite stochastic process? What can you say about the finite stochastic processes in a Markov chain?

2. Define the following terms for a Markov chain:
   a. State
   b. Current state
   c. Transition probabilities

3. Consider a transition matrix T for a Markov chain with entries a_{ij}, where $1 \leq i \leq n$ and $1 \leq j \leq n$.
   a. If there are n states associated with the Markov chain, then what is the size of the matrix T?
   b. Describe the probability that each entry represents. Can an entry be negative?
   c. What is the sum of the entries in each column of T?

9.1 Exercises

In Exercises 1–10, determine which of the matrices are stochastic.

1. $\begin{bmatrix} .4 & .7 \\ .6 & .3 \end{bmatrix}$

2. $\begin{bmatrix} .8 & .2 \\ .3 & .7 \end{bmatrix}$

3. $\begin{bmatrix} \frac{1}{4} & \frac{1}{8} \\ \frac{3}{4} & \frac{7}{8} \end{bmatrix}$

4. $\begin{bmatrix} \frac{1}{3} & 0 & \frac{1}{2} \\ \frac{1}{2} & 1 & 0 \\ \frac{1}{4} & 0 & \frac{1}{2} \end{bmatrix}$

5. $\begin{bmatrix} .3 & .2 & .4 \\ .4 & .7 & .3 \\ .3 & .1 & .2 \end{bmatrix}$

6. $\begin{bmatrix} \frac{1}{3} & \frac{1}{4} & \frac{1}{2} \\ \frac{1}{3} & 0 & -\frac{1}{2} \\ \frac{1}{4} & \frac{3}{4} & \frac{1}{2} \end{bmatrix}$

7. $\begin{bmatrix} .1 & .4 & .3 \\ .7 & .2 & .1 \\ .2 & .4 & .6 \end{bmatrix}$

8. $\begin{bmatrix} 1 & 0 & 0 \\ 0 & 0 & 1 \\ 0 & 1 & 0 \end{bmatrix}$

9. $\begin{bmatrix} .2 & .3 \\ .3 & .1 \\ .5 & .6 \end{bmatrix}$

10. $\begin{bmatrix} .5 & .2 & .3 \\ .2 & .3 & .2 \\ .3 & .4 & .1 \\ 0 & .1 & .4 \end{bmatrix}$

11. The transition matrix for a Markov process is given by

$$T = \begin{matrix} & \text{State} \\ & \begin{matrix} 1 & \ \ 2 \end{matrix} \\ \begin{matrix} \text{State 1} \\ \text{State 2} \end{matrix} & \begin{bmatrix} .3 & .6 \\ .7 & .4 \end{bmatrix} \end{matrix}$$

   a. What does the entry $a_{11} = .3$ represent?
   b. Given that the outcome state 1 has occurred, what is the probability that the next outcome of the experiment will be state 2?

   c. If the initial-state distribution vector is given by

$$X_0 = \begin{matrix} \text{State 1} \\ \text{State 2} \end{matrix} \begin{bmatrix} .4 \\ .6 \end{bmatrix}$$

   find TX_0, the probability distribution of the system after one observation.

12. The transition matrix for a Markov process is given by

$$T = \begin{matrix} & \text{State} \\ & \begin{matrix} 1 & \ \ 2 \end{matrix} \\ \begin{matrix} \text{State 1} \\ \text{State 2} \end{matrix} & \begin{bmatrix} \frac{1}{6} & \frac{2}{3} \\ \frac{5}{6} & \frac{1}{3} \end{bmatrix} \end{matrix}$$

   a. What does the entry $a_{22} = \frac{1}{3}$ represent?
   b. Given that the outcome state 1 has occurred, what is the probability that the next outcome of the experiment will be state 2?
   c. If the initial-state distribution vector is given by

$$X_0 = \begin{matrix} \text{State 1} \\ \text{State 2} \end{matrix} \begin{bmatrix} \frac{1}{4} \\ \frac{3}{4} \end{bmatrix}$$

   find TX_0, the probability distribution of the system after one observation.

13. The transition matrix for a Markov process is given by

$$T = \begin{matrix} & \text{State} \\ & \begin{matrix} 1 & \ \ 2 \end{matrix} \\ \begin{matrix} \text{State 1} \\ \text{State 2} \end{matrix} & \begin{bmatrix} .6 & .2 \\ .4 & .8 \end{bmatrix} \end{matrix}$$

   and the initial-state distribution vector is given by

$$X_0 = \begin{matrix} \text{State 1} \\ \text{State 2} \end{matrix} \begin{bmatrix} .5 \\ .5 \end{bmatrix}$$

Find TX_0 and interpret your result with the aid of a tree diagram.

14. The transition matrix for a Markov process is given by

$$T = \begin{matrix} & \text{State} \\ & 1 \quad 2 \\ \text{State 1} \\ \text{State 2} \end{matrix} \begin{bmatrix} \frac{1}{2} & \frac{3}{4} \\ \frac{1}{2} & \frac{1}{4} \end{bmatrix}$$

and the initial-state distribution vector is given by

$$X_0 = \begin{matrix} \text{State 1} \\ \text{State 2} \end{matrix} \begin{bmatrix} \frac{1}{3} \\ \frac{2}{3} \end{bmatrix}$$

Find TX_0 and interpret your result with the aid of a tree diagram.

In Exercises 15–18, find X_2 (the probability distribution of the system after two observations) for the distribution vector X_0 and the transition matrix T.

15. $X_0 = \begin{bmatrix} .6 \\ .4 \end{bmatrix}$, $T = \begin{bmatrix} .4 & .8 \\ .6 & .2 \end{bmatrix}$

16. $X_0 = \begin{bmatrix} \frac{1}{2} \\ \frac{1}{2} \\ 0 \end{bmatrix}$, $T = \begin{bmatrix} \frac{1}{2} & \frac{1}{3} & \frac{1}{2} \\ 0 & \frac{1}{3} & \frac{1}{4} \\ \frac{1}{2} & \frac{1}{3} & \frac{1}{4} \end{bmatrix}$

17. $X_0 = \begin{bmatrix} \frac{1}{4} \\ \frac{1}{2} \\ \frac{1}{4} \end{bmatrix}$, $T = \begin{bmatrix} \frac{1}{4} & \frac{1}{4} & \frac{1}{2} \\ \frac{1}{4} & \frac{1}{2} & \frac{1}{2} \\ \frac{1}{2} & \frac{1}{4} & 0 \end{bmatrix}$

18. $X_0 = \begin{bmatrix} .25 \\ .40 \\ .35 \end{bmatrix}$, $T = \begin{bmatrix} .1 & .1 & .3 \\ .8 & .7 & .2 \\ .1 & .2 & .5 \end{bmatrix}$

19. PSYCHOLOGY EXPERIMENTS A psychologist conducts an experiment in which a mouse is placed in a T-maze, where it has a choice at the T-junction of turning left and receiving a reward (cheese) or turning right and receiving a mild electric shock (see accompanying figure). At the end of each trial, a record is kept of the mouse's response. It is observed that the mouse is as likely to turn left (state 1) as right (state 2) during the first trial. In subsequent trials, however, the observation is made that if the mouse had turned left in the previous trial, then on the next trial the probability that it will turn left is .8, whereas the probability that it will turn right is .2. If the mouse had turned right in the previous trial, then the probability that it will turn right on the next trial is .1, whereas the probability that it will turn left is .9.

a. Using a tree diagram, describe the transitions between states and the probabilities associated with these transitions.

b. Represent the transition probabilities obtained in part (a) in terms of a matrix.

c. What is the initial-state probability vector?

d. Use the results of parts (b) and (c) to find the probability that a mouse will turn left on the second trial.

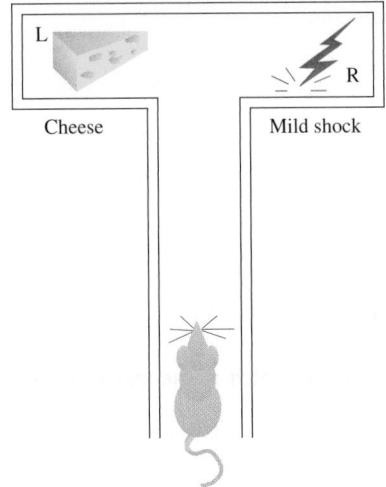

Cheese   Mild shock

20. SMALL-TOWN REVIVAL At the beginning of 1990, the population of a certain state was 55.4% rural and 44.6% urban. Based on past trends, it is expected that 10% of the population currently residing in the rural areas will move into the urban areas—while 17% of the population currently residing in the urban areas will move into the rural areas—in the next decade. What was the population distribution in that state at the beginning of 2000?

21. POLITICAL POLLS Morris Polling conducted a poll 6 months before an election in a state in which a Democrat and a Republican were running for governor and found that 60% of the voters intended to vote for the Republican and 40% intended to vote for the Democrat. In a poll conducted 3 months later, it was found that 70% of those who had earlier stated a preference for the Republican candidate still maintained that preference whereas 30% of these voters now preferred the Democratic candidate. Of those who had earlier stated a preference for the Democrat, 80% still maintained that preference whereas 20% now preferred the Republican candidate.

a. If the election were held at this time, who would win?

b. Assuming that this trend continues, which candidate is expected to win the election?

22. COMMUTER TRENDS Within a large metropolitan area, 20% of the commuters currently use the public transportation system, whereas the remaining 80% commute via automobile. The city has recently revitalized and expanded its public transportation system. It is expected that, 6 months from now, 30% of those who are now commuting to work via automobile will switch to public transportation and 70% will continue to commute via automobile. At the same time, it is expected that 20% of those now using public transportation will commute via automobile and 80% will continue to use public transportation.

a. Construct the transition matrix for the Markov chain that describes the change in the mode of transportation used by these commuters.

b. Find the initial distribution vector for this Markov chain.

c. What percentage of the commuters are expected to use public transportation 6 months from now?

23. Refer to Example 6. If the initial distribution vector for the location of the taxis is

$$X_0 = \begin{matrix} \text{Zone I} \\ \text{Zone II} \\ \text{Zone III} \end{matrix} \begin{bmatrix} .6 \\ .2 \\ .2 \end{bmatrix}$$

what will be the distribution after all of them have made one pickup and discharge?

24. URBAN–SUBURBAN POPULATION FLOW Refer to Example 4. If the initial probability distribution is

$$X_0 = \begin{matrix} \text{City} \\ \text{Suburb} \end{matrix} \begin{bmatrix} .80 \\ .20 \end{bmatrix}$$

what will be the population distribution of the city after 1 year? After 2 years?

25. MARKET SHARE At a certain university, three bookstores—the University Bookstore, the Campus Bookstore, and the Book Mart—currently serve the university community. From a survey conducted at the beginning of the fall quarter, it was found that the University Bookstore and the Campus Bookstore each had 40% of the market, whereas the Book Mart had 20% of the market. Each quarter, the University Bookstore retains 80% of its customers but loses 10% to the Campus Bookstore and 10% to the Book Mart. The Campus Bookstore retains 75% of its customers but loses 10% to the University Bookstore and 15% to the Book Mart. The Book Mart retains 90% of its customers but loses 5% to the University Bookstore and 5% to the Campus Bookstore. If these trends continue, what percent of the market will each store have at the beginning of the second quarter? The third quarter?

26. MARKET SHARE OF AUTO MANUFACTURERS In a study of the domestic market share of the three major automobile manufacturers A, B, and C in a certain country, it was found that their current market shares were 60%, 30%, and 10%, respectively. It was also found that, of the customers who bought a car manufactured by A, 75% would again buy a car manufactured by A, 15% would buy a car manufactured by B, and 10% would buy a car manufactured by C. Of the customers who bought a car manufactured by B, 90% would again buy a car manufactured by B, whereas 5% each would buy cars

manufactured by A and C, respectively. Finally, of the customers who bought a car manufactured by C, 85% would again buy a car manufactured by C, 5% would buy a car manufactured by A, and 10% would buy a car manufactured by B. Assuming that these sentiments reflect the buying habits of customers in the future, determine the market share that will be held by each manufacturer after the next two model years.

27. COLLEGE MAJORS Records compiled by the Admissions Office at a state university indicating the percentage of students who change their major each year are shown in the following transition matrix. Of the freshmen now at the university, 30% have chosen their major field in business, 30% in the humanities, 20% in education, and 20% in the natural sciences and other fields. Assuming that this trend continues, find the percent of these students that will be majoring in each of the given areas in their senior year.
Hint: Find $T^3 X_0$.

	Bus.	Hum.	Educ.	Nat. sci. and others
Business	.80	.10	.20	.10
Humanities	.10	.70	.10	.05
Education	.05	.10	.60	.05
Nat. sci. and others	.05	.10	.10	.80

28. HOMEOWNERS' CHOICE OF ENERGY A study conducted by the Urban Energy Commission in a large metropolitan area indicates the probabilities that homeowners within the area will use certain heating fuels or solar energy during the next 10 years as the major source of heat for their homes. The transition matrix representing the transition probabilities from one state to another is

	Elec.	Gas	Oil	Solar
Electricity	.70	0	0	0
Natural gas	.15	.90	.20	.05
Fuel oil	.05	.02	.75	0
Solar energy	.10	.08	.05	.95

Among homeowners within the area, 20% currently use electricity, 35% use natural gas, 40% use oil, and 5% use solar energy as the major source of heat for their homes. What is the expected distribution of the homeowners that will be using each type of heating fuel or solar energy within the next decade?

In Exercises 29 and 30, determine whether the statement is true or false. If it is true, explain why it is true. If it is false, give an example to show why it is false.

29. A Markov chain is a process in which the outcomes at any stage of the experiment depend on the outcomes of the preceding stages.

30. The sum of the entries in each column of a transition matrix must not exceed 1.

9.1   Solutions to Self-Check Exercises

1. The required transition matrix is

$$T = \begin{bmatrix} .80 & .05 & .10 \\ .05 & .90 & .15 \\ .15 & .05 & .75 \end{bmatrix}$$

2. **a.** The initial distribution vector is

$$X_0 = \begin{bmatrix} .4 \\ .3 \\ .3 \end{bmatrix}$$

b. The vector representing the market share of each supermarket after 1 year is

$$X_1 = TX_0$$

$$= \begin{bmatrix} .80 & .05 & .10 \\ .05 & .90 & .15 \\ .15 & .05 & .75 \end{bmatrix} \begin{bmatrix} .4 \\ .3 \\ .3 \end{bmatrix} = \begin{bmatrix} .365 \\ .335 \\ .3 \end{bmatrix}$$

That is, after 1 year supermarket A will command a 36.5% market share, supermarket B will have a 33.5% share, and supermarket C will have a 30% market share.

The vector representing the market share of the supermarkets after 2 years is

$$X_2 = TX_1$$

$$= \begin{bmatrix} .80 & .05 & .10 \\ .05 & .90 & .15 \\ .15 & .05 & .75 \end{bmatrix} \begin{bmatrix} .365 \\ .335 \\ .3 \end{bmatrix} = \begin{bmatrix} .3388 \\ .3648 \\ .2965 \end{bmatrix}$$

That is, 2 years later the market shares of supermarkets A, B, and C will be 33.88%, 36.48%, and 29.65%, respectively.

USING TECHNOLOGY

■ Finding Distribution Vectors

Since the computation of the probability distribution of a system after a certain number of observations involves matrix multiplication, a graphing utility may be used to facilitate the work.

EXAMPLE 1   Consider the problem posed in Example 6, page 517, where

$$T = \begin{bmatrix} .6 & .4 & .3 \\ .3 & .3 & .3 \\ .1 & .3 & .4 \end{bmatrix} \quad \text{and} \quad X_0 = \begin{bmatrix} .8 \\ .15 \\ .05 \end{bmatrix}$$

Verify that

$$X_2 = \begin{bmatrix} .4965 \\ .3 \\ .2035 \end{bmatrix}$$

as obtained in that example.

Solution   First, we enter the matrix X_0 as the matrix A and the matrix T as the matrix B. Then, performing the indicated multiplication, we find that

$$B^2 * A = \begin{bmatrix} .4965 \\ .3 \\ .2035 \end{bmatrix}$$

That is,

$$X_2 = T^2 X_0 = \begin{bmatrix} .4965 \\ .3 \\ .2035 \end{bmatrix}$$

as was to be shown. ∎

TECHNOLOGY EXERCISES

In Exercises 1 and 2, find X_5 (the probability distribution of the system after five observations) for the distribution vector X_0 and the transition matrix T.

1. $X_0 = \begin{bmatrix} .2 \\ .3 \\ .2 \\ .1 \\ .2 \end{bmatrix}$, $T = \begin{bmatrix} .2 & .2 & .3 & .2 & .1 \\ .1 & .2 & .1 & .2 & .1 \\ .3 & .4 & .1 & .3 & .3 \\ .2 & .1 & .2 & .2 & .2 \\ .2 & .1 & .3 & .1 & .3 \end{bmatrix}$

2. $X_0 = \begin{bmatrix} .1 \\ .2 \\ .2 \\ .3 \\ .2 \end{bmatrix}$, $T = \begin{bmatrix} .3 & .2 & .1 & .3 & .1 \\ .2 & .1 & .2 & .1 & .2 \\ .1 & .2 & .3 & .2 & .2 \\ .1 & .3 & .2 & .3 & .2 \\ .3 & .2 & .2 & .1 & .3 \end{bmatrix}$

3. Refer to Exercise 26 on page 521. Using the same data, determine the market share that will be held by each manufacturer five model years after the study began.

4. Refer to Exercise 25 on page 521. Using the same data, determine the expected market share that each store will hold at the beginning of the fourth quarter.

9.2 Regular Markov Chains

▬ Steady-State Distribution Vectors

In the last section, we derived a formula for computing the likelihood that a physical system will be in any one of the possible states associated with each stage of a Markov process describing the system. In this section we use this formula to help us investigate the long-term trends of certain Markov processes.

APPLIED EXAMPLE 1   Educational Status of Women A survey conducted by the National Commission on the Educational Status of Women reveals that 70% of the daughters of women who have completed 2 or more years of college have themselves completed 2 or more years of college, whereas only 20% of the daughters of women with less than 2 years of college education have themselves completed 2 or more years of college. If this trend continues, determine, in the long run, the percentage of women in the population who will have completed at least 2 years of college given that currently only 20% have completed at least 2 years of college.

Solution   This problem may be viewed as a Markov process with two possible states: "completed 2 or more years of college" (state 1) and "completed fewer than 2 years of college" (state 2). The transition matrix associated with this Markov chain is given by

$$T = \begin{bmatrix} .7 & .2 \\ .3 & .8 \end{bmatrix}$$

The initial distribution vector is given by

$$X_0 = \begin{bmatrix} .2 \\ .8 \end{bmatrix}$$

To study the long-term trend pertaining to this particular aspect of the educational status of women, let's compute $X_1, X_2, \ldots$, the distribution vectors associated with the Markov process under consideration. These vectors give the percentage of women with 2 or more years of college and that of women with fewer than 2 years of college after one generation, after two generations, and so on. With the aid of Formula (1), Section 9.1, we find (to four decimal places) that

After one generation $$X_1 = TX_0 = \begin{bmatrix} .7 & .2 \\ .3 & .8 \end{bmatrix} \begin{bmatrix} .2 \\ .8 \end{bmatrix} = \begin{bmatrix} .3 \\ .7 \end{bmatrix}$$

After two generations $$X_2 = TX_1 = \begin{bmatrix} .7 & .2 \\ .3 & .8 \end{bmatrix} \begin{bmatrix} .3 \\ .7 \end{bmatrix} = \begin{bmatrix} .35 \\ .65 \end{bmatrix}$$

After three generations $$X_3 = TX_2 = \begin{bmatrix} .7 & .2 \\ .3 & .8 \end{bmatrix} \begin{bmatrix} .35 \\ .65 \end{bmatrix} = \begin{bmatrix} .375 \\ .625 \end{bmatrix}$$

Proceeding further, we obtain the following sequence of vectors:

$$X_4 = \begin{bmatrix} .3875 \\ .6125 \end{bmatrix}$$

$$X_5 = \begin{bmatrix} .3938 \\ .6062 \end{bmatrix}$$

$$X_6 = \begin{bmatrix} .3969 \\ .6031 \end{bmatrix}$$

$$X_7 = \begin{bmatrix} .3984 \\ .6016 \end{bmatrix}$$

$$X_8 = \begin{bmatrix} .3992 \\ .6008 \end{bmatrix}$$

$$X_9 = \begin{bmatrix} .3996 \\ .6004 \end{bmatrix}$$

After ten generations $$X_{10} = \begin{bmatrix} .3998 \\ .6002 \end{bmatrix}$$

The results of these computations show that, as m increases, the probability distribution vector X_m approaches the probability distribution vector

$$\begin{bmatrix} .4 \\ .6 \end{bmatrix} \quad \text{or} \quad \begin{bmatrix} \frac{2}{5} \\ \frac{3}{5} \end{bmatrix}$$

Such a vector is called the **limiting**, or **steady-state, distribution vector** for the system. We interpret these results as follows. Initially, 20% of the women in the population have completed 2 or more years of college whereas 80% have completed fewer than 2 years of college. After one generation, the former share has increased to 30% of the population and the latter has dropped to 70% of the population. The trend continues, and eventually 40% of all women in future generations will have completed 2 or more years of college whereas 60% will have completed fewer than 2 years of college. ▪

In order to explain the foregoing result, let's analyze Formula (1), Section 9.1, more closely. The initial distribution vector X_0 is a constant; that is, it remains fixed throughout our computation of $X_1, X_2, \ldots$. It appears reasonable, therefore, to conjecture that this phenomenon is a result of the behavior of the powers, T^m, of the transition matrix T. Pursuing this line of investigation, we compute

$$T^2 = \begin{bmatrix} .7 & .2 \\ .3 & .8 \end{bmatrix}\begin{bmatrix} .7 & .2 \\ .3 & .8 \end{bmatrix} = \begin{bmatrix} .55 & .3 \\ .45 & .7 \end{bmatrix}$$

$$T^3 = \begin{bmatrix} .7 & .2 \\ .3 & .8 \end{bmatrix}\begin{bmatrix} .55 & .3 \\ .45 & .7 \end{bmatrix} = \begin{bmatrix} .475 & .35 \\ .525 & .65 \end{bmatrix}$$

Proceeding further, we obtain the following sequence of matrices:

$$T^4 = \begin{bmatrix} .4375 & .375 \\ .5625 & .625 \end{bmatrix} \quad T^5 = \begin{bmatrix} .4188 & .3875 \\ .5813 & .6125 \end{bmatrix}$$

$$T^6 = \begin{bmatrix} .4094 & .3938 \\ .5906 & .6062 \end{bmatrix} \quad T^7 = \begin{bmatrix} .4047 & .3969 \\ .5953 & .6031 \end{bmatrix}$$

$$T^8 = \begin{bmatrix} .4023 & .3984 \\ .5977 & .6016 \end{bmatrix} \quad T^9 = \begin{bmatrix} .4012 & .3992 \\ .5988 & .6008 \end{bmatrix}$$

$$T^{10} = \begin{bmatrix} .4006 & .3996 \\ .5994 & .6004 \end{bmatrix} \quad T^{11} = \begin{bmatrix} .4003 & .3998 \\ .5997 & .6002 \end{bmatrix}$$

These results show that the powers T^m of the transition matrix T tend toward a fixed matrix as m gets larger and larger. In this case, the "limiting matrix" is the matrix

$$L = \begin{bmatrix} .40 & .40 \\ .60 & .60 \end{bmatrix} \quad \text{or} \quad \begin{bmatrix} \frac{2}{5} & \frac{2}{5} \\ \frac{3}{5} & \frac{3}{5} \end{bmatrix}$$

Such a matrix is called the **steady-state matrix** for the system. Thus, as suspected, the long-term behavior of a Markov process such as the one in this example depends on the behavior of the limiting matrix of the powers of the transition matrix—the steady-state matrix for the system. In view of this, the long-term (steady-state) distribution vector for this problem may be found by computing the product

$$LX_0 = \begin{bmatrix} .40 & .40 \\ .60 & .60 \end{bmatrix} \begin{bmatrix} .2 \\ .8 \end{bmatrix} = \begin{bmatrix} .40 \\ .60 \end{bmatrix}$$

which agrees with the result obtained earlier.

Next, since the transition matrix T in this situation seems to have a stabilizing effect over the long term, we are led to wonder whether the steady state would be reached regardless of the initial state of the system. To answer this question, suppose the initial distribution vector is

$$X_0 = \begin{bmatrix} p \\ 1 - p \end{bmatrix}$$

Then, as before, the steady-state distribution vector is given by

$$LX_0 = \begin{bmatrix} .40 & .40 \\ .60 & .60 \end{bmatrix} \begin{bmatrix} p \\ 1 - p \end{bmatrix} = \begin{bmatrix} .40 \\ .60 \end{bmatrix}$$

Thus, the steady state is reached regardless of the initial state of the system!

■ Regular Markov Chains

The transition matrix T of Example 1 has several important properties, which we emphasized in the foregoing discussion. First, the sequence $T, T^2, T^3, \ldots$ approaches a steady-state matrix in which the rows of the limiting matrix are all equal and all entries are positive. A matrix T having this property is called a *regular* Markov chain.

> **Regular Markov Chain**
> A stochastic matrix T is a **regular Markov chain** if the sequence
>
> $$T, T^2, T^3, \ldots$$
>
> approaches a steady-state matrix in which the rows of the limiting matrix are all equal and all the entries are positive.

It can be shown that *a stochastic matrix T is regular if and only if some power of T has entries that are all positive.* Second, as in the case of Example 1, a Markov chain with a regular transition matrix has a steady-state distribution vector whose elements coincide with those of a row (since they are all the same) of the steady-state matrix; thus, this steady-state distribution vector is always reached regardless of the initial distribution vector.

We will return to computations involving regular Markov chains, but for the moment let's see how one may determine whether a given matrix is indeed regular.

EXAMPLE 2 Determine which of the following matrices are regular.

a. $\begin{bmatrix} .7 & .2 \\ .3 & .8 \end{bmatrix}$ **b.** $\begin{bmatrix} .4 & 1 \\ .6 & 0 \end{bmatrix}$ **c.** $\begin{bmatrix} 0 & 1 \\ 1 & 0 \end{bmatrix}$

Solution

a. Since all the entries of the matrix are positive, the given matrix is regular. Note that this is the transition matrix of Example 1.

b. In this case, one of the entries of the given matrix is equal to zero. Let's compute

$$\begin{bmatrix} .4 & 1 \\ .6 & 0 \end{bmatrix}^2 = \begin{bmatrix} .4 & 1 \\ .6 & 0 \end{bmatrix}\begin{bmatrix} .4 & 1 \\ .6 & 0 \end{bmatrix} = \begin{bmatrix} .76 & .4 \\ .24 & .6 \end{bmatrix}$$

$$\uparrow$$
All entries are positive.

Since the second power of the matrix has entries that are all positive, we conclude that the given matrix is in fact regular.

c. Denote the given matrix by A. Then,

$$A = \begin{bmatrix} 0 & 1 \\ 1 & 0 \end{bmatrix}$$

$$A^2 = \begin{bmatrix} 0 & 1 \\ 1 & 0 \end{bmatrix}\begin{bmatrix} 0 & 1 \\ 1 & 0 \end{bmatrix} = \begin{bmatrix} 1 & 0 \\ 0 & 1 \end{bmatrix}$$

$$A^3 = \begin{bmatrix} 0 & 1 \\ 1 & 0 \end{bmatrix}\begin{bmatrix} 1 & 0 \\ 0 & 1 \end{bmatrix} = \begin{bmatrix} 0 & 1 \\ 1 & 0 \end{bmatrix}$$

Not all entries are positive

> **EXPLORE & DISCUSS**
>
> Find the set of all 2×2 stochastic matrices with elements that are either 0 or 1.

Observe that $A^3 = A$. It therefore follows that $A^4 = A^2$, $A^5 = A$, and so on. In other words, any power of A must coincide with either A or A^2. Since not all entries of A and A^2 are positive, the same is true of any power of A. We thus conclude that the given matrix is not regular. ∎

We now return to the study of regular Markov chains. In Example 1 we found the steady-state distribution vector associated with a regular Markov chain by studying the limiting behavior of a sequence of distribution vectors. Alternatively, as pointed out in the subsequent discussion, the steady-state distribution vector may also be obtained by first determining the steady-state matrix associated with the regular Markov chain.

Fortunately, there is a relatively simple procedure for finding the steady-state distribution vector associated with a regular Markov process. It does not involve the rather tedious computations required to obtain the sequences in Example 1. The procedure follows.

> **Finding the Steady-State Distribution Vector**
> Let T be a regular stochastic matrix. Then the steady-state distribution vector X may be found by solving the vector equation
>
> $$TX = X$$
>
> together with the condition that the sum of the elements of the vector X be equal to 1.

A justification of the foregoing procedure is given in Exercise 29.

VIDEO

EXAMPLE 3   Find the steady-state distribution vector for the regular Markov chain whose transition matrix is

$$T = \begin{bmatrix} .7 & .2 \\ .3 & .8 \end{bmatrix} \qquad \text{See Example 1.}$$

Solution   Let

$$X = \begin{bmatrix} x \\ y \end{bmatrix}$$

be the steady-state distribution vector associated with the Markov process, where the numbers x and y are to be determined. The condition $TX = X$ translates into the matrix equation

$$\begin{bmatrix} .7 & .2 \\ .3 & .8 \end{bmatrix} \begin{bmatrix} x \\ y \end{bmatrix} = \begin{bmatrix} x \\ y \end{bmatrix}$$

or, equivalently, the system of linear equations

$$0.7x + 0.2y = x$$
$$0.3x + 0.8y = y$$

But each of the equations that make up this system of equations is equivalent to the single equation

$$0.3x - 0.2y = 0 \qquad \begin{matrix} 0.7x - x + 0.2y = 0 \\ 0.3x + 0.8y - y = 0 \end{matrix}$$

Next, the condition that the elements of X add up to 1 gives

$$x + y = 1$$

Thus, the simultaneous fulfillment of these two conditions implies that x and y are the solutions of the system

$$0.3x - 0.2y = 0$$
$$x + \quad y = 1$$

Solving the first equation for x, we obtain

$$x = \frac{2}{3}y$$

which, upon substitution into the second, yields

$$\frac{2}{3}y + y = 1$$

$$y = \frac{3}{5}$$

Hence $x = \dfrac{2}{5}$, and the required steady-state distribution vector is given by

$$X = \begin{bmatrix} \frac{2}{5} \\ \frac{3}{5} \end{bmatrix}$$

which agrees with the result obtained earlier.   ∎

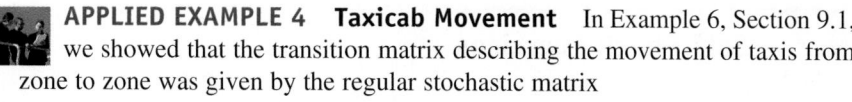

APPLIED EXAMPLE 4   Taxicab Movement   In Example 6, Section 9.1, we showed that the transition matrix describing the movement of taxis from zone to zone was given by the regular stochastic matrix

$$T = \begin{bmatrix} .6 & .4 & .3 \\ .3 & .3 & .3 \\ .1 & .3 & .4 \end{bmatrix}$$

Use this information to determine the long-term distribution of the taxis in the three zones.

Solution   Let

$$X = \begin{bmatrix} x \\ y \\ z \end{bmatrix}$$

be the steady-state distribution vector associated with the Markov process under consideration, where x, y, and z are to be determined. The condition $TX = X$ translates into the matrix equation

$$\begin{bmatrix} .6 & .4 & .3 \\ .3 & .3 & .3 \\ .1 & .3 & .4 \end{bmatrix} \begin{bmatrix} x \\ y \\ z \end{bmatrix} = \begin{bmatrix} x \\ y \\ z \end{bmatrix}$$

or, equivalently, the system of linear equations

$$0.6x + 0.4y + 0.3z = x$$
$$0.3x + 0.3y + 0.3z = y$$
$$0.1x + 0.3y + 0.4z = z$$

This system simplifies to

$$4x - 4y - 3z = 0$$
$$3x - 7y + 3z = 0$$
$$x + 3y - 6z = 0$$

Since $x + y + z = 1$ as well, we are required to solve the system

$$x + y + z = 1$$
$$4x - 4y - 3z = 0$$
$$3x - 7y + 3z = 0$$
$$x + 3y - 6z = 0$$

Using the Gauss–Jordan elimination procedure of Chapter 2, we find that

$$x = \frac{33}{70} \qquad y = \frac{3}{10} \qquad z = \frac{8}{35}$$

or $x \approx 0.47$, $y = 0.30$, and $z \approx 0.23$. Thus, in the long run, 47% of the taxis will be in zone I, 30% in zone II, and 23% in zone III.

9.2   Self-Check Exercises

1. Find the steady-state distribution vector for the regular Markov chain whose transition matrix is

$$T = \begin{bmatrix} .5 & .8 \\ .5 & .2 \end{bmatrix}$$

2. Three supermarkets serve a certain section of a city. During the year, supermarket A is expected to retain 80% of its customers, lose 5% of its customers to supermarket B, and lose 15% to supermarket C. Supermarket B is expected to retain 90% of its customers and lose 5% to each of supermarket A and supermarket C. Supermarket C is expected to retain 75% of its customers, lose 10% to supermarket A, and lose 15% to supermarket B. If these trends continue, what will be the market share of each supermarket in the long run?

Solutions to Self-Check Exercises 9.2 can be found on page 532.

9.2   Concept Questions

1. Explain (a) a steady-state distribution vector, (b) a steady-state matrix, and (c) a regular Markov chain.

2. How do you find the steady-state distribution vector once you are given a regular stochastic matrix T associated with a Markov process?

9.2   Exercises

In Exercises 1–8, determine which of the matrices are regular.

1. $\begin{bmatrix} \frac{2}{5} & \frac{3}{4} \\ \frac{3}{5} & \frac{1}{4} \end{bmatrix}$

2. $\begin{bmatrix} 0 & .3 \\ 1 & .7 \end{bmatrix}$

3. $\begin{bmatrix} 1 & .8 \\ 0 & .2 \end{bmatrix}$

4. $\begin{bmatrix} \frac{1}{3} & 0 \\ \frac{2}{3} & 1 \end{bmatrix}$

5. $\begin{bmatrix} \frac{1}{2} & \frac{3}{4} & 0 \\ \frac{1}{2} & 0 & \frac{1}{2} \\ 0 & \frac{1}{4} & \frac{1}{2} \end{bmatrix}$

6. $\begin{bmatrix} 1 & .3 & .1 \\ 0 & .4 & .8 \\ 0 & .3 & .1 \end{bmatrix}$

7. $\begin{bmatrix} .7 & .2 & .3 \\ .3 & .8 & .3 \\ 0 & 0 & .4 \end{bmatrix}$

8. $\begin{bmatrix} 0 & 0 & \frac{1}{4} \\ 1 & 0 & 0 \\ 0 & 1 & \frac{3}{4} \end{bmatrix}$

In Exercises 9–16, find the steady-state vector for the given transition matrix.

9. $\begin{bmatrix} \frac{1}{3} & \frac{1}{4} \\ \frac{2}{3} & \frac{3}{4} \end{bmatrix}$

10. $\begin{bmatrix} \frac{4}{5} & \frac{3}{5} \\ \frac{1}{5} & \frac{2}{5} \end{bmatrix}$

11. $\begin{bmatrix} .5 & .2 \\ .5 & .8 \end{bmatrix}$

12. $\begin{bmatrix} .9 & 1 \\ .1 & 0 \end{bmatrix}$

13. $\begin{bmatrix} 0 & \frac{1}{8} & 1 \\ 1 & \frac{5}{8} & 0 \\ 0 & \frac{1}{4} & 0 \end{bmatrix}$

14. $\begin{bmatrix} .6 & .3 & 0 \\ .4 & .4 & .6 \\ 0 & .3 & .4 \end{bmatrix}$

15. $\begin{bmatrix} .2 & 0 & .3 \\ 0 & .6 & .4 \\ .8 & .4 & .3 \end{bmatrix}$

16. $\begin{bmatrix} .1 & .2 & .3 \\ .1 & .2 & .3 \\ .8 & .6 & .4 \end{bmatrix}$

17. **PSYCHOLOGY EXPERIMENTS**   A psychologist conducts an experiment in which a mouse is placed in a T-maze, where it has a choice at the T-junction of turning left and receiving a reward (cheese) or turning right and receiving a mild shock. At the end of each trial a record is kept of the mouse's response. It is observed that the mouse is as likely to turn left (state 1) as right (state 2) during the first trial. In subsequent trials, however, the observation is made that if the mouse had turned left in the previous trial, then the probability that it will turn left in the next trial is .8 whereas the probability that it will turn right is .2. If the mouse had turned right in the previous trial, then the probability that it will turn right in the next trial is .1 whereas the probability that it will turn left is .9. In the long run, what percentage of the time will the mouse turn left at the T-junction?

18. **COMMUTER TRENDS**   Within a large metropolitan area, 20% of the commuters currently use the public transportation system and the remaining 80% commute via automobile. The city has recently revitalized and expanded its public transportation system. It is expected that, 6 months from now, 30% of those who are now commuting to work via automobile will switch to public transportation and 70% will continue to commute via automobile. At the same time, it is expected

that 20% of those now using public transportation will commute via automobile and 80% will continue to use public transportation. In the long run, what percent of the commuters will be using public transportation?

19. **ONE- AND TWO-INCOME FAMILIES** The following transition matrix was constructed from data compiled over a 10-year period by Manpower, Inc., in a statewide study of married couples in which at least one spouse was working. The matrix gives the transitional probabilities for one and two wage earners among married couples.

		Current State	
		1 Wage Earner	2 Wage Earners
Next	1 Wage Earner	.72	.12
State	2 Wage Earners	.28	.88

At the present time, 48% of the married couples (in which at least one spouse is working) have one wage earner and 52% have two wage earners. Assuming that this trend continues, what will be the distribution of one- and two-wage earner families among married couples in this area 10 years from now? Over the long run?

20. **PROFESSIONAL WOMEN** From data compiled over a 5-year period by *Women's Daily* in a study of the number of women in the professions, the following transition matrix was constructed. It gives the transitional probabilities for the number of men and women in the professions.

		Current State	
		Men	Women
Next State	Men	.95	.04
	Women	.05	.96

At the beginning of 1986, 52.9% of professional jobs were held by men. If this trend continues, what percent of professional jobs will be held by women in the long run?

21. **BUYING TRENDS OF HOME BUYERS** From data collected by the Association of Realtors of a certain city, the following transition matrix was obtained. The matrix describes the buying pattern of home buyers who buy single-family homes (S) or condominiums (C).

		Current State	
		S	C
Next State	S	.85	.35
	C	.15	.65

Currently, 80% of the homeowners live in single-family homes whereas 20% live in condominiums. If this trend continues, what will be the percent of homeowners in this city who will own single-family homes and condominiums 2 years from now? In the long run?

22. **HOMEOWNERS' CHOICE OF ENERGY** A study conducted by the Urban Energy Commission in a large metropolitan area indicates the probabilities that homeowners within the area will use certain heating fuels or solar energy during the next 10 years as the major source of heat for their homes. The following transition matrix represents the transition probabilities from one state to another:

	Elec.	Gas	Oil	Solar
Electricity	.70	0	.10	0
Natural gas	.15	.90	.10	.05
Fuel oil	.05	.02	.75	.05
Solar energy	.10	.08	.05	.90

Among the homeowners within the area, 20% currently use electricity, 35% use natural gas, 40% use oil, and 5% use solar energy as their major source of heat for their homes. In the long run, what percent of homeowners within the area will be using solar energy as their major source of heating fuel?

23. **NETWORK NEWS VIEWERSHIP** A television poll was conducted among regular viewers of the national news in a certain region where the three national networks share the same time slot for the evening news. Results of the poll indicate that 30% of the viewers watch the ABC evening news, 40% watch the CBS evening news, and 30% watch the NBC evening news. It was also found that, of those viewers who watched the ABC evening news during one week, 80% would again watch the ABC evening news during the next week, 10% would watch the CBS news, and 10% would watch the NBC news. Of those viewers who watched the CBS evening news during one week, 85% would again watch the CBS evening news during the next week, 10% would watch the ABC news, and 5% would watch the NBC news. Of those viewers who watched the NBC evening news during one week, 85% would again watch the NBC news during the next week, 10% would watch ABC, and 5% would watch CBS.
   a. What share of the audience consisting of regular viewers of the national news will each network command after two weeks?
   b. In the long run, what share of the audience will each network command?

24. **NETWORK NEWS VIEWERSHIP** Refer to Exercise 23. If the initial distribution vector is

$$X_0 = \begin{matrix} \text{ABC} \\ \text{CBS} \\ \text{NBC} \end{matrix} \begin{bmatrix} .40 \\ .40 \\ .20 \end{bmatrix}$$

what share of the audience will each network command in the long run?

25. **GENETICS** In a certain species of roses, a plant with genotype (genetic makeup) *AA* has red flowers, a plant with genotype *Aa* has pink flowers, and a plant with genotype *aa* has white flowers. Here *A* is the dominant gene and *a* is the recessive gene for color. If a plant with one genotype is crossed with another plant, then the color of the offspring's flowers is determined by the genotype of the parent plants. If a plant of

each genotype is crossed with a pink-flowered plant, then the transition matrix used to determine the color of the offspring's flowers is given by

$$
\begin{array}{c}
\hspace{3.5cm} \text{Parent} \\
\hspace{2.7cm} \text{Red} \quad \text{Pink} \quad \text{White}
\end{array}
$$

		Red	Pink	White
Offspring	Red (AA)	$\frac{1}{2}$	$\frac{1}{4}$	0
	Pink (Aa) or (aA)	$\frac{1}{2}$	$\frac{1}{2}$	$\frac{1}{2}$
	White (aa)	0	$\frac{1}{4}$	$\frac{1}{2}$

If the offspring of each generation are crossed only with pink-flowered plants, in the long run what percentage of the plants will have red flowers? Pink flowers? White flowers?

26. **MARKET SHARE OF AUTO MANUFACTURERS** In a study of the domestic market share of the three major automobile manufacturers A, B, and C in a certain country, it was found that, of the customers who bought a car manufactured by A, 75% would again buy a car manufactured by A, 15% would buy a car manufactured by B, and 10% would buy a car manufactured by C. Of the customers who bought a car manufactured by B, 90% would again buy a car manufactured by B, whereas 5% each would buy cars manufactured by A and C, respectively. Finally, of the customers who bought a car manufactured by C, 85% would again buy a car manufactured by C, 5% would buy a car manufactured by A, and 10% would buy a car manufactured by B. Assuming that these sentiments reflect the buying habits of customers in future model years, determine the market share that will be held by each manufacturer in the long run.

In Exercises 27 and 28, determine whether the statement is true or false. If it is true, explain why it is true. If it is false, give an example to show why it is false.

27. A stochastic matrix T is a regular Markov chain if the powers of T approach a fixed matrix whose rows are all equal.

28. To find the steady-state distribution vector X, we solve the system

$$
\begin{cases}
TX = X \\
x_1 + x_2 + \cdots + x_n = 1
\end{cases}
$$

where T is the regular stochastic matrix associated with the Markov process and

$$
X = \begin{bmatrix} x_1 \\ x_2 \\ \vdots \\ x_n \end{bmatrix}
$$

29. Let T be a regular stochastic matrix. Show that the steady-state distribution vector X may be found by solving the vector equation $TX = X$ together with the condition that the sum of the elements of X be equal to 1.
    Hint: Take the initial distribution to be X, the steady-state distribution vector. Then, when n is large, $X \approx T^n X$. (Why?) Multiply both sides of the last equation by T (on the left) and consider the resulting equation when n is large.

9.2 Solutions to Self-Check Exercises

1. Let

$$
X = \begin{bmatrix} x \\ y \end{bmatrix}
$$

be the steady-state distribution vector associated with the Markov process, where the numbers x and y are to be determined. The condition $TX = X$ translates into the matrix equation

$$
\begin{bmatrix} .5 & .8 \\ .5 & .2 \end{bmatrix} \begin{bmatrix} x \\ y \end{bmatrix} = \begin{bmatrix} x \\ y \end{bmatrix}
$$

which is equivalent to the system of linear equations

$$
0.5x + 0.8y = x
$$
$$
0.5x + 0.2y = y
$$

Each equation in the system is equivalent to the equation

$$
0.5x - 0.8y = 0
$$

Next, the condition that the elements of X add up to 1 gives

$$
x + y = 1
$$

Thus, the simultaneous fulfillment of these two conditions implies that x and y are the solutions of the system

$$
0.5x - 0.8y = 0
$$
$$
x + \quad y = 1
$$

Solving the first equation for x, we obtain

$$
x = \frac{8}{5}y
$$

which, upon substitution into the second, yields

$$
\frac{8}{5}y + y = 1
$$
$$
y = \frac{5}{13}
$$

Therefore, $x = \dfrac{8}{13}$, and the required steady-state distribution vector is

$$\begin{bmatrix} \frac{8}{13} \\ \frac{5}{13} \end{bmatrix}$$

2. The transition matrix for the Markov process under consideration is

$$T = \begin{bmatrix} .80 & .05 & .10 \\ .05 & .90 & .15 \\ .15 & .05 & .75 \end{bmatrix}$$

Now, let

$$X = \begin{bmatrix} x \\ y \\ z \end{bmatrix}$$

be the steady-state distribution vector associated with the Markov process under consideration, where x, y, and z are to be determined. The condition $TX = X$ is

$$\begin{bmatrix} .80 & .05 & .10 \\ .05 & .90 & .15 \\ .15 & .05 & .75 \end{bmatrix} \begin{bmatrix} x \\ y \\ z \end{bmatrix} = \begin{bmatrix} x \\ y \\ z \end{bmatrix}$$

or, equivalently, the system of linear equations

$$\begin{aligned} 0.80x + 0.05y + 0.10z &= x \\ 0.05x + 0.90y + 0.15z &= y \\ 0.15x + 0.05y + 0.75z &= z \end{aligned}$$

This system simplifies to

$$\begin{aligned} 4x - y - 2z &= 0 \\ x - 2y + 3z &= 0 \\ 3x + y - 5z &= 0 \end{aligned}$$

Since $x + y + z = 1$ as well, we are required to solve the system

$$\begin{aligned} 4x - y - 2z &= 0 \\ x - 2y + 3z &= 0 \\ 3x + y - 5z &= 0 \\ x + y + z &= 1 \end{aligned}$$

Using the Gauss–Jordan elimination procedure, we find

$$x = \frac{1}{4} \qquad y = \frac{1}{2} \qquad z = \frac{1}{4}$$

Therefore, in the long run, supermarkets A and C will each have one-quarter of the customers and supermarket B will have half the customers.

USING TECHNOLOGY

■ Finding the Long-Term Distribution Vector

The problem of finding the long-term distribution vector for a regular Markov chain ultimately rests on the problem of solving a system of linear equations. As such, the **rref** or equivalent function of a graphing utility proves indispensable, as the following example shows.

EXAMPLE 1 Find the steady-state distribution vector for the regular Markov chain whose transition matrix is

$$T = \begin{bmatrix} .4 & .2 & .1 \\ .3 & .4 & .5 \\ .3 & .4 & .4 \end{bmatrix}$$

Solution Let

$$\begin{bmatrix} x \\ y \\ z \end{bmatrix}$$

(continued)

be the steady-state distribution vector, where x, y, and z are to be determined. The condition $TX = X$ translates into the matrix equation

$$\begin{bmatrix} .4 & .2 & .1 \\ .3 & .4 & .5 \\ .3 & .4 & .4 \end{bmatrix} \begin{bmatrix} x \\ y \\ z \end{bmatrix} = \begin{bmatrix} x \\ y \\ z \end{bmatrix}$$

or, equivalently, the system of linear equations

$$0.4x + 0.2y + 0.1z = x$$
$$0.3x + 0.4y + 0.5z = y$$
$$0.3x + 0.4y + 0.4z = z$$

Since $x + y + z = 1$, we are required to solve the system

$$-0.6x + 0.2y + 0.1z = 0$$
$$0.3x - 0.6y + 0.5z = 0$$
$$0.3x + 0.4y - 0.6z = 0$$
$$x + y + z = 1$$

Entering this system into the graphing calculator as the augmented matrix

$$A = \begin{bmatrix} -.6 & .2 & .1 & | & 0 \\ .3 & -.6 & .5 & | & 0 \\ .3 & .4 & -.6 & | & 0 \\ 1 & 1 & 1 & | & 1 \end{bmatrix}$$

and then using the **rref** function, we obtain the equivalent system (to two decimal places)

$$\begin{bmatrix} 1 & 0 & 0 & | & .20 \\ 0 & 1 & 0 & | & .42 \\ 0 & 0 & 1 & | & .38 \\ 0 & 0 & 0 & | & 0 \end{bmatrix}$$

Hence $x \approx 0.20$, $y \approx 0.42$, and $z \approx 0.38$, so the required steady-state distribution vector is

$$\begin{bmatrix} .20 \\ .42 \\ .38 \end{bmatrix}$$

TECHNOLOGY EXERCISES

In Exercises 1 and 2, find the steady-state vector for the matrix T.

1.
$$T = \begin{bmatrix} .2 & .2 & .3 & .2 & .1 \\ .1 & .2 & .1 & .2 & .1 \\ .3 & .4 & .1 & .3 & .3 \\ .2 & .1 & .2 & .2 & .2 \\ .2 & .1 & .3 & .1 & .3 \end{bmatrix}$$

2.
$$T = \begin{bmatrix} .3 & .2 & .1 & .3 & .1 \\ .2 & .1 & .2 & .1 & .2 \\ .1 & .2 & .3 & .2 & .2 \\ .1 & .3 & .2 & .3 & .2 \\ .3 & .2 & .2 & .1 & .3 \end{bmatrix}$$

3. Verify that the steady-state vector for Example 4, page 529, is

$$X = \begin{bmatrix} .47 \\ .30 \\ .23 \end{bmatrix}$$

9.3 Absorbing Markov Chains

Absorbing Markov Chains

In this section, we investigate the long-term trends of a certain class of Markov chains that involve transition matrices that are not regular. In particular, we study Markov chains in which the transition matrices, known as *absorbing* stochastic matrices, have the special properties to be described presently.

Consider the stochastic matrix

$$\begin{bmatrix} 1 & 0 & .2 & 0 \\ 0 & 1 & .3 & 1 \\ 0 & 0 & .5 & 0 \\ 0 & 0 & 0 & 0 \end{bmatrix}$$

associated with a Markov process. Interpreting it in the usual fashion, we see that after one observation the probability is 1 (a certainty) that an object previously in state 1 will remain in state 1. Similarly, we see that an object previously in state 2 must remain in state 2. Next, we find that an object previously in state 3 has a probability of .2 of going to state 1, a probability of .3 of going to state 2, a probability of .5 of remaining in state 3, and no chance of going to state 4. Finally, an object previously in state 4 must, after one observation, end up in state 2.

This stochastic matrix exhibits certain special characteristics. First, as observed earlier, an object in state 1 or state 2 must stay in state 1 or state 2, respectively. Such states are called absorbing states. In general, an **absorbing state** is one from which it is impossible for an object to leave. To identify the absorbing states of a stochastic matrix, we simply examine each column of the matrix. If column i has a 1 in the a_{ii} position (that is, on the main diagonal of the matrix) and zeros elsewhere in that column, then and only then is state i an absorbing state.

Second, observe that states 3 and 4, although not absorbing states, have the property that an object in each of these states has a possibility of going to an absorbing state. For example, an object currently in state 3 has a probability of .2 of ending up in state 1, an absorbing state, and an object in state 4 must end up in state 2, also an absorbing state, after one transition.

Absorbing Stochastic Matrix

An **absorbing stochastic matrix** has the following properties:

1. There is at least one absorbing state.
2. It is possible to go from any nonabsorbing state to an absorbing state in one or more stages.

A Markov chain is said to be an **absorbing Markov chain** if the transition matrix associated with the process is an absorbing stochastic matrix.

EXAMPLE 1   Determine whether the following matrices are absorbing stochastic matrices.

a.
$$\begin{bmatrix} .7 & 0 & .1 & 0 \\ 0 & 1 & .5 & 0 \\ .3 & 0 & .2 & 0 \\ 0 & 0 & .2 & 1 \end{bmatrix}$$
b.
$$\begin{bmatrix} 1 & 0 & 0 & 0 \\ 0 & 1 & 0 & 0 \\ 0 & 0 & .5 & .4 \\ 0 & 0 & .5 & .6 \end{bmatrix}$$

Solution

a. States 2 and 4 are both absorbing states. Furthermore, even though state 1 is not an absorbing state, there is a possibility (with probability .3) that an object may go from this state to state 3. State 3 itself is nonabsorbing, but an object in that state has a probability of .5 of going to the absorbing state 2 and a probability of .2 of going to the absorbing state 4. Thus, the given matrix is an absorbing stochastic matrix.

b. States 1 and 2 are absorbing states. However, it is impossible for an object to go from the nonabsorbing states 3 and 4 to either or both of the absorbing states. Thus, the given matrix is not an absorbing stochastic matrix. ■

Given an absorbing stochastic matrix, it is always possible, by suitably reordering the states if necessary, to rewrite it so that the absorbing states appear first. Then, the resulting matrix can be partitioned into four submatrices,

Absorbing   Nonabsorbing

$$\left[\begin{array}{c|c} I & S \\ \hline O & R \end{array}\right]$$

where I is an identity matrix whose order is determined by the number of absorbing states and O is a zero matrix. The submatrices R and S correspond to the nonabsorbing states. As an example, the absorbing stochastic matrix of Example 1(a)

$$\begin{array}{c} \\ 1 \\ 2 \\ 3 \\ 4 \end{array} \begin{array}{cccc} 1 & 2 & 3 & 4 \\ \left[\begin{array}{cccc} .7 & 0 & .1 & 0 \\ 0 & 1 & .5 & 0 \\ .3 & 0 & .2 & 0 \\ 0 & 0 & .2 & 1 \end{array}\right] \end{array} \quad \text{may be written as} \quad \begin{array}{c} \\ 4 \\ 2 \\ 1 \\ 3 \end{array} \begin{array}{cccc} 4 & 2 & 1 & 3 \\ \left[\begin{array}{cc|cc} 1 & 0 & 0 & .2 \\ 0 & 1 & 0 & .5 \\ \hline 0 & 0 & .7 & .1 \\ 0 & 0 & .3 & .2 \end{array}\right] \end{array}$$

upon reordering the states as indicated.

APPLIED EXAMPLE 2   Gambler's Ruin   John has decided to risk $2 in the following game of chance. He places a $1 bet on each repeated play of the game in which the probability of his winning $1 is .4, and he continues to play until he has accumulated a total of $3 or has lost all of his money. Write the transition matrix for the related absorbing Markov chain.

Solution   There are four states in this Markov chain, which correspond to John accumulating a total of $0, $1, $2, and $3. The first and last states listed are absorbing states and so we will list these states first, resulting in the transition matrix

$$
\begin{array}{c}
\overbrace{}^{\text{Absorbing}}\ \overbrace{}^{\text{Nonabsorbing}} \\
\begin{array}{cccc} \$0 & \$3 & \$1 & \$2 \end{array} \\
\begin{array}{c} \$0 \\ \$3 \\ \$1 \\ \$2 \end{array}
\begin{bmatrix}
1 & 0 & .6 & 0 \\
0 & 1 & 0 & .4 \\
0 & 0 & 0 & .6 \\
0 & 0 & .4 & 0
\end{bmatrix}
\end{array}
$$

which is constructed as follows. Since the state "$0" is an absorbing state, we see that $a_{11} = 1$ and $a_{21} = a_{31} = a_{41} = 0$. Similarly, the state "$3" is an absorbing state, so $a_{22} = 1$ and $a_{12} = a_{32} = a_{42} = 0$. To construct the column corresponding to the nonabsorbing state "$1," we note that there is a probability of .6 (John loses) in going from an accumulated amount of $1 to $0, so $a_{13} = .6$; $a_{23} = a_{33} = 0$ because it is not feasible to go from an accumulated amount of $1 to either an accumulated amount of $3 or $1 in one transition (play). Finally, there is a probability of .4 (John wins) in going from an accumulated amount of $1 to an accumulated amount of $2, so $a_{43} = .4$. The last column of the transition matrix is constructed by reasoning in a similar manner. ∎

The following question arises in connection with the last example: If John continues to play the game as originally planned, what is the probability that he will depart from the game victorious—that is, leave with an accumulated amount of $3?

To answer this question, we must look at the long-term trend of the relevant Markov chain. Taking a cue from our work in the last section, we may compute the powers of the transition matrix associated with the Markov chain. Just as in the case of regular stochastic matrices, it turns out that the powers of an absorbing stochastic matrix approach a steady-state matrix. However, instead of demonstrating this, we use the following result, which we state without proof, for computing the steady-state matrix.

Finding the Steady-State Matrix for an Absorbing Stochastic Matrix
Suppose an absorbing stochastic matrix A has been partitioned into submatrices

$$
A = \left[\begin{array}{c|c} I & S \\ \hline O & R \end{array}\right]
$$

Then the *steady-state matrix* of A is given by

$$
\left[\begin{array}{c|c} I & S(I - R)^{-1} \\ \hline O & O \end{array}\right]
$$

where the order of the identity matrix appearing in the expression $(I - R)^{-1}$ is chosen to have the same order as R.

APPLIED EXAMPLE 3 Gambler's Ruin (continued) Refer to Example 2. If John continues to play the game until he has accumulated a sum of $3 or until he has lost all of his money, what is the probability that he will accumulate $3?

Solution   The transition matrix associated with the Markov process is

$$A = \begin{bmatrix} 1 & 0 & .6 & 0 \\ 0 & 1 & 0 & .4 \\ \hline 0 & 0 & 0 & .6 \\ 0 & 0 & .4 & 0 \end{bmatrix}$$

(see Example 2). We need to find the steady-state matrix of A. In this case,

$$R = \begin{bmatrix} 0 & .6 \\ .4 & 0 \end{bmatrix} \quad \text{and} \quad S = \begin{bmatrix} .6 & 0 \\ 0 & .4 \end{bmatrix}$$

so

$$I - R = \begin{bmatrix} 1 & 0 \\ 0 & 1 \end{bmatrix} - \begin{bmatrix} 0 & .6 \\ .4 & 0 \end{bmatrix} = \begin{bmatrix} 1 & -.6 \\ -.4 & 1 \end{bmatrix}$$

Using the formula in Section 2.6 for finding the inverse of a 2×2 matrix yields

$$(I - R)^{-1} = \begin{bmatrix} 1.32 & .79 \\ .53 & 1.32 \end{bmatrix}$$

and so

$$S(I - R)^{-1} = \begin{bmatrix} .6 & 0 \\ 0 & .4 \end{bmatrix} \begin{bmatrix} 1.32 & .79 \\ .53 & 1.32 \end{bmatrix} = \begin{bmatrix} .79 & .47 \\ .21 & .53 \end{bmatrix}$$

Therefore, the required steady-state matrix of A is given by

$$\left[\begin{array}{c|c} I & S(I-R)^{-1} \\ \hline O & O \end{array} \right] = \begin{array}{c} \$0 \\ \$3 \\ \$1 \\ \$2 \end{array} \begin{bmatrix} 1 & 0 & .79 & .47 \\ 0 & 1 & .21 & .53 \\ \hline 0 & 0 & 0 & 0 \\ 0 & 0 & 0 & 0 \end{bmatrix}$$

with columns labeled $\$0 \quad \$3 \quad \$1 \quad \$2$.

From this result we see that, if John starts with $2, the probability is .53 that he will leave the game with an accumulated amount of $3—that is, that he wins $1. ■

Our last example shows an application of Markov chains in the field of genetics.

EXPLORE & DISCUSS

Consider the stochastic matrix

$$A = \begin{bmatrix} 1 & 0 & a \\ 0 & 1 & b \\ 0 & 0 & 1-a-b \end{bmatrix}$$

where a and b satisfy $0 < a < 1$, $0 < b < 1$, and $0 < a + b < 1$.

1. Find the steady-state matrix.

2. What is the probability that state 3 will be absorbed in state 2?

APPLIED EXAMPLE 4   Genetics   In a certain species of flowers, a plant of genotype (genetic makeup) AA has red flowers, a plant of genotype Aa has pink flowers, and a plant of genotype aa has white flowers, where A is the dominant gene and a is the recessive gene for color. If a plant of one genotype is crossed with another plant, then the color of the offspring's flowers is determined by the genotype of the parent plants. If the offspring are crossed successively with plants of genotype AA only, show that in the long run all the flowers produced by the plants will be red.

Solution   First, let's construct the transition matrix associated with the resulting Markov chain. In crossing a plant of genotype AA with another of the same genotype AA, the offspring will inherit one dominant gene from each parent and thus will have genotype *AA*. Therefore, the probabilities of the offspring being genotype *AA*, *Aa*, and *aa* are 1, 0, and 0, respectively.

Next, in crossing a plant of genotype *AA* with one of genotype *Aa*, the probability of the offspring having genotype *AA* (inheriting an *A* gene from the first parent and an *A* from the second) is $\frac{1}{2}$; the probability of the offspring having genotype *Aa* (inheriting an *A* gene from the first parent and an *a* gene from the second parent) is $\frac{1}{2}$; finally, the probability of the offspring being of genotype *aa* is 0 since this is clearly impossible.

A similar argument shows that, when a plant of genotype *AA* is crossed with one of genotype *aa*, the probabilities of the offspring having genotype *AA*, *Aa*, and *aa* are 0, 1, and 0, respectively.

The required transition matrix is thus given by

$$
\begin{array}{c}
\text{Absorbing state} \\
\downarrow \\
\begin{array}{ccc} AA & Aa & aa \end{array} \\
T = \begin{array}{c} AA \\ Aa \\ aa \end{array}
\begin{bmatrix}
1 & \frac{1}{2} & 0 \\
0 & \frac{1}{2} & 1 \\
0 & 0 & 0
\end{bmatrix}
\end{array}
$$

Observe that the state *AA* is an absorbing state. Furthermore, it is possible to go from each of the other two nonabsorbing states to the absorbing state *AA*. Thus, the Markov chain is an absorbing Markov chain. To determine the long-term effects of this experiment, let's compute the steady-state matrix of *T*. Partitioning *T* in the usual manner, we find

$$
T = \begin{bmatrix}
1 & \vdots & \frac{1}{2} & 0 \\
\hline
0 & \vdots & \frac{1}{2} & 1 \\
0 & \vdots & 0 & 0
\end{bmatrix}
$$

so that

$$
R = \begin{bmatrix} \frac{1}{2} & 1 \\ 0 & 0 \end{bmatrix} \quad \text{and} \quad S = \begin{bmatrix} \frac{1}{2} & 0 \end{bmatrix}
$$

Next, we compute

$$
I - R = \begin{bmatrix} 1 & 0 \\ 0 & 1 \end{bmatrix} - \begin{bmatrix} \frac{1}{2} & 1 \\ 0 & 0 \end{bmatrix} = \begin{bmatrix} \frac{1}{2} & -1 \\ 0 & 1 \end{bmatrix}
$$

and, using the formula for finding the inverse of a 2×2 matrix from Section 2.6,

$$
(I - R)^{-1} = \begin{bmatrix} 2 & 2 \\ 0 & 1 \end{bmatrix}
$$

Thus,

$$
S(I - R)^{-1} = \begin{bmatrix} \frac{1}{2} & 0 \end{bmatrix} \begin{bmatrix} 2 & 2 \\ 0 & 1 \end{bmatrix} = \begin{bmatrix} 1 & 1 \end{bmatrix}
$$

Therefore, the steady-state matrix of *T* is given by

$$
\begin{bmatrix}
I & \vdots & S(I - R)^{-1} \\
\hline
O & \vdots & O
\end{bmatrix} =
\begin{array}{c}
\begin{array}{ccc} AA & Aa & aa \end{array} \\
\begin{array}{c} AA \\ Aa \\ aa \end{array}
\begin{bmatrix}
1 & \vdots & 1 & 1 \\
\hline
0 & \vdots & 0 & 0 \\
0 & \vdots & 0 & 0
\end{bmatrix}
\end{array}
$$

Interpreting the steady-state matrix of T, we see that the long-term result of crossing the offspring with plants of genotype AA leads only to the absorbing state AA. In other words, such a procedure will result in the production of plants that will bear only red flowers, as we set out to demonstrate. ∎

9.3   Self-Check Exercises

1. Let

$$T = \begin{bmatrix} .2 & 0 & 0 \\ .3 & 1 & .6 \\ .5 & 0 & .4 \end{bmatrix}$$

a. Show that T is an absorbing stochastic matrix.
b. Rewrite T so that the absorbing states appear first, partition the resulting matrix, and identify the submatrices R and S.
c. Compute the steady-state matrix of T.

2. There is a trend toward increased use of computer-aided transcription (CAT) and electronic recording (ER) as alternatives to manual transcription (MT) of court proceedings by court stenographers in a certain state. Suppose the following sto-

chastic matrix is the transition matrix associated with the Markov process:

$$T = \begin{array}{c} \\ \text{CAT} \\ \text{ER} \\ \text{MT} \end{array} \begin{array}{c} \begin{array}{ccc} \text{CAT} & \text{ER} & \text{MT} \end{array} \\ \begin{bmatrix} 1 & .3 & .2 \\ 0 & .6 & .3 \\ 0 & .1 & .5 \end{bmatrix} \end{array}$$

Determine the probability that a court now using electronic recording or manual transcribing of its proceedings will eventually change to CAT.

Solutions to Self-Check Exercises 9.3 can be found on page 542.

9.3   Concept Questions

1. What is an absorbing stochastic matrix?

2. Suppose the absorbing stochastic matrix A has been partitioned into submatrices

$$\left[\begin{array}{c|c} I & S \\ \hline O & R \end{array} \right]$$

Write the expression representing the steady-state matrix of A. (Assume that I and R have the same order.)

9.3   Exercises

In Exercises 1–8, determine whether the given matrix is an absorbing stochastic matrix.

1. $\begin{bmatrix} \frac{2}{5} & 0 \\ \frac{3}{5} & 1 \end{bmatrix}$

2. $\begin{bmatrix} 1 & 0 \\ 0 & 1 \end{bmatrix}$

3. $\begin{bmatrix} 1 & .5 & 0 \\ 0 & 0 & 1 \\ 0 & .5 & 0 \end{bmatrix}$

4. $\begin{bmatrix} 1 & 0 & 0 \\ 0 & .7 & .2 \\ 0 & .3 & .8 \end{bmatrix}$

5. $\begin{bmatrix} \frac{1}{8} & 0 & 0 \\ \frac{1}{4} & 1 & 0 \\ \frac{5}{8} & 0 & 1 \end{bmatrix}$

6. $\begin{bmatrix} 1 & 0 & 0 & 0 \\ 0 & \frac{5}{8} & 0 & \frac{1}{6} \\ 0 & \frac{1}{8} & 1 & 0 \\ 0 & \frac{1}{4} & 0 & \frac{5}{6} \end{bmatrix}$

7. $\begin{bmatrix} 1 & 0 & .3 & 0 \\ 0 & 1 & .2 & 0 \\ 0 & 0 & .1 & .5 \\ 0 & 0 & .4 & .5 \end{bmatrix}$

8. $\begin{bmatrix} 1 & 0 & 0 & 0 \\ 0 & 1 & 0 & 0 \\ 0 & 0 & .2 & .6 \\ 0 & 0 & .8 & .4 \end{bmatrix}$

In Exercises 9–14, rewrite each absorbing stochastic matrix so that the absorbing states appear first, partition the resulting matrix, and identify the submatrices R and S.

9. $\begin{bmatrix} .6 & 0 \\ .4 & 1 \end{bmatrix}$

10. $\begin{bmatrix} \frac{1}{4} & 0 & 0 \\ \frac{1}{4} & 1 & 0 \\ \frac{1}{2} & 0 & 1 \end{bmatrix}$

11. $\begin{bmatrix} 0 & .2 & 0 \\ .5 & .4 & 0 \\ .5 & .4 & 1 \end{bmatrix}$ **12.** $\begin{bmatrix} .5 & 0 & .3 \\ 0 & 1 & .1 \\ .5 & 0 & .6 \end{bmatrix}$

13. $\begin{bmatrix} .4 & .2 & 0 & 0 \\ .2 & .3 & 0 & 0 \\ 0 & .3 & 1 & 0 \\ .4 & .2 & 0 & 1 \end{bmatrix}$ **14.** $\begin{bmatrix} .1 & 0 & 0 & 0 \\ .2 & 1 & 0 & .2 \\ .3 & 0 & 1 & 0 \\ .4 & 0 & 0 & .8 \end{bmatrix}$

In Exercises 15–24, compute the steady-state matrix of each stochastic matrix.

15. $\begin{bmatrix} .55 & 0 \\ .45 & 1 \end{bmatrix}$ **16.** $\begin{bmatrix} \frac{3}{5} & 0 \\ \frac{2}{5} & 1 \end{bmatrix}$

17. $\begin{bmatrix} 1 & .2 & .3 \\ 0 & .4 & .2 \\ 0 & .4 & .5 \end{bmatrix}$ **18.** $\begin{bmatrix} \frac{1}{5} & 0 & 0 \\ 0 & 1 & \frac{3}{8} \\ \frac{4}{5} & 0 & \frac{5}{8} \end{bmatrix}$

19. $\begin{bmatrix} \frac{1}{2} & 0 & \frac{1}{3} & 0 \\ \frac{1}{2} & 1 & 0 & 0 \\ 0 & 0 & \frac{2}{3} & 0 \\ 0 & 0 & 0 & 1 \end{bmatrix}$ **20.** $\begin{bmatrix} 1 & \frac{1}{8} & \frac{1}{3} & 0 \\ 0 & \frac{1}{8} & 0 & 0 \\ 0 & \frac{1}{4} & \frac{2}{3} & 0 \\ 0 & \frac{1}{2} & 0 & 1 \end{bmatrix}$

21. $\begin{bmatrix} 1 & 0 & \frac{1}{4} & \frac{1}{3} \\ 0 & 1 & \frac{1}{4} & \frac{1}{3} \\ 0 & 0 & \frac{1}{2} & 0 \\ 0 & 0 & 0 & \frac{1}{3} \end{bmatrix}$ **22.** $\begin{bmatrix} 1 & 0 & .2 & .1 \\ 0 & 1 & .4 & .2 \\ 0 & 0 & 0 & .4 \\ 0 & 0 & .4 & .3 \end{bmatrix}$

23. $\begin{bmatrix} 1 & 0 & 0 & .2 & .1 \\ 0 & 1 & 0 & .1 & .2 \\ 0 & 0 & 1 & .3 & .1 \\ 0 & 0 & 0 & .2 & .2 \\ 0 & 0 & 0 & .2 & .4 \end{bmatrix}$ **24.** $\begin{bmatrix} 1 & 0 & \frac{1}{4} & \frac{1}{3} & 0 \\ 0 & 1 & 0 & \frac{1}{3} & \frac{1}{2} \\ 0 & 0 & \frac{1}{4} & \frac{1}{3} & 0 \\ 0 & 0 & \frac{1}{2} & 0 & \frac{1}{2} \\ 0 & 0 & 0 & 0 & 0 \end{bmatrix}$

25. GASOLINE CONSUMPTION   As more and more old cars are taken off the road and replaced by late models that use unleaded fuel, the consumption of leaded gasoline will continue to drop. Suppose the transition matrix

$$A = \begin{matrix} & \begin{matrix} \text{L} & \text{UL} \end{matrix} \\ \begin{matrix} \text{L} \\ \text{UL} \end{matrix} & \begin{bmatrix} .80 & 0 \\ .20 & 1 \end{bmatrix} \end{matrix}$$

describes this Markov process, where L denotes leaded gasoline and UL denotes unleaded gasoline.

  a. Show that A is an absorbing stochastic matrix and rewrite it so that the absorbing state appears first. Partition the resulting matrix and identify the submatrices R and S.
  b. Compute the steady-state matrix of A and interpret your results.

26. Diane has decided to play the following game of chance. She places a $1 bet on each repeated play of a game in which the

probability of her winning $1 is .5. She has further decided to continue playing the game until she has either accumulated a total of $3 or has lost all her money. What is the probability that Diane will eventually leave the game a winner if she started with a capital of $1? Of $2?

27. Refer to Exercise 26. Suppose Diane has decided to stop playing only after she has accumulated a sum of $4 or has lost all her money. All other conditions being the same, what is the probability that Diane will leave the game a winner if she started with a capital of $1? Of $2? Of $3?

28. USE OF AUTOMATED OFFICE EQUIPMENT   Because of the proliferation of more affordable automated office equipment, more and more companies are turning to them as replacements for obsolete equipment. The following transition matrix describes the Markov process. Here, E stands for electric typewriters, W stands for electric typewriters with some form of word processing capabilities, and C stands for computers with word processing software.

$$A = \begin{matrix} & \begin{matrix} \text{E} & \text{W} & \text{C} \end{matrix} \\ \begin{matrix} \text{E} \\ \text{W} \\ \text{C} \end{matrix} & \begin{bmatrix} .10 & 0 & 0 \\ .70 & .60 & 0 \\ .20 & .40 & 1 \end{bmatrix} \end{matrix}$$

  a. Show that A is an absorbing stochastic matrix and rewrite it so that the absorbing state appears first. Partition the resulting matrix and identify the submatrices R and S.
  b. Compute the steady-state matrix of A and interpret your results.

29. EDUCATION RECORDS   The registrar of Computronics Institute has compiled the following statistics on the progress of the school's students in their 2-yr computer programming course leading to an associate degree. Of beginning students in a particular year, 75% successfully complete their first year of study and move on to the second year whereas 25% drop out of the program; of second-year students in a particular year, 90% go on to graduate at the end of the year whereas 10% drop out of the program.

  a. Construct the transition matrix associated with this Markov process.
  b. Compute the steady-state matrix.
  c. Determine the probability that a beginning student enrolled in the program will complete the course successfully.

30. EDUCATION RECORDS   The registrar of a law school has compiled the following statistics on the progress of the school's students working toward the LLB degree. Of the first-year students in a particular year, 85% successfully complete their course of studies and move on to the second year whereas 15% drop out of the program; of the second-year students in a particular year, 92% go on to the third year whereas 8% drop out of the program; of the third-year students in a particular year, 98% go on to graduate at the end of the year whereas 2% drop out of the program.

a. Construct the transition matrix associated with the Markov process.
b. Find the steady-state matrix.
c. Determine the probability that a beginning law student enrolled in the program will go on to graduate.

In Exercises 31 and 32, determine whether the statement is true or false. If it is true, explain why it is true. If it is false, give an example to show why it is false.

31. An absorbing stochastic matrix need not contain an absorbing state.

32. In partitioning an absorbing matrix into subdivisions,

$$A = \left[\begin{array}{c|c} I & S \\ \hline O & R \end{array}\right]$$

the identity matrix I is chosen to have the same order as R.

33. GENETICS Refer to Example 4. If the offspring are crossed successively with plants of genotype aa only, show that in the long run all the flowers produced by the plants will be white.

9.3 Solutions to Self-Check Exercises

1. a. State 2 is an absorbing state. States 1 and 3 are not absorbing, but each has a possibility (with probability .3 and .6) that an object may go from these states to state 2. Therefore, the matrix T is an absorbing stochastic matrix.
b. Denoting the states as indicated, we rewrite

$$\begin{array}{c} \\ 1 \\ 2 \\ 3 \end{array}\begin{array}{ccc} 1 & 2 & 3 \\ \left[\begin{array}{ccc} .2 & 0 & 0 \\ .3 & 1 & .6 \\ .5 & 0 & .4 \end{array}\right] \end{array}$$

in the form

$$\begin{array}{c} \\ 2 \\ 3 \\ 1 \end{array}\begin{array}{ccc} 2 & 3 & 1 \\ \left[\begin{array}{c|cc} 1 & .6 & .3 \\ \hline 0 & .4 & .5 \\ 0 & 0 & .2 \end{array}\right] \end{array}$$

We see that

$$S = [.6 \quad .3] \quad \text{and} \quad R = \left[\begin{array}{cc} .4 & .5 \\ 0 & .2 \end{array}\right]$$

c. We compute

$$I - R = \left[\begin{array}{cc} 1 & 0 \\ 0 & 1 \end{array}\right] - \left[\begin{array}{cc} .4 & .5 \\ 0 & .2 \end{array}\right] = \left[\begin{array}{cc} .6 & -.5 \\ 0 & .8 \end{array}\right]$$

and, using the formula for finding the inverse of a 2×2 matrix from Section 2.6,

$$(I - R)^{-1} = \left[\begin{array}{cc} 1.67 & 1.04 \\ 0 & 1.25 \end{array}\right]$$

and so

$$S(I - R)^{-1} = [.6 \quad .3]\left[\begin{array}{cc} 1.67 & 1.04 \\ 0 & 1.25 \end{array}\right] = [1 \quad 1]$$

Therefore, the steady-state matrix of T is

$$\left[\begin{array}{c|cc} 1 & 1 & 1 \\ \hline 0 & 0 & 0 \\ 0 & 0 & 0 \end{array}\right]$$

2. We want to compute the steady-state matrix of T. Note that T is in the form

$$\left[\begin{array}{c|c} I & S \\ \hline O & R \end{array}\right]$$

where

$$S = [.3 \quad .2] \quad \text{and} \quad R = \left[\begin{array}{cc} .6 & .3 \\ .1 & .5 \end{array}\right]$$

We compute

$$I - R = \left[\begin{array}{cc} 1 & 0 \\ 0 & 1 \end{array}\right] - \left[\begin{array}{cc} .6 & .3 \\ .1 & .5 \end{array}\right] = \left[\begin{array}{cc} .4 & -.3 \\ -.1 & .5 \end{array}\right]$$

and, using the inverse formula from Section 2.6,

$$(I - R)^{-1} = \left[\begin{array}{cc} 2.94 & 1.76 \\ 0.59 & 2.36 \end{array}\right]$$

so

$$S(I - R)^{-1} = [.3 \quad .2]\left[\begin{array}{cc} 2.94 & 1.76 \\ 0.59 & 2.36 \end{array}\right] = [1 \quad 1]$$

Therefore, the steady-state matrix of T is

$$\begin{array}{c} \\ \text{CAT} \\ \text{ER} \\ \text{MT} \end{array}\begin{array}{ccc} \text{CAT} & \text{ER} & \text{MT} \\ \left[\begin{array}{c|cc} 1 & 1 & 1 \\ \hline 0 & 0 & 0 \\ 0 & 0 & 0 \end{array}\right] \end{array}$$

Interpreting the steady-state matrix of T, we see that in the long run all courts in this state will use computer-aided transcription.

CHAPTER 9 **Summary of Principal Formulas and Terms**

FORMULA

Steady-state matrix for an absorbing stochastic matrix	If $A = \begin{bmatrix} I & \vdots & S \\ \cdots & + & \cdots \\ O & \vdots & R \end{bmatrix}$ then the steady-state matrix of A is $\begin{bmatrix} I & \vdots & S(I - R)^{-1} \\ \cdots & + & \cdots \\ O & \vdots & O \end{bmatrix}$

TERMS

Markov chain (process) (512)

transition matrix (514)

stochastic matrix (515)

steady-state (limiting) distribution vector (525)

steady-state matrix (525)

regular Markov chain (526)

absorbing state (535)

absorbing stochastic matrix (535)

absorbing Markov chain (535)

CHAPTER 9 **Concept Review Questions**

Fill in the blanks.

1. A Markov chain is a stochastic process in which the _____ associated with the outcomes at any stage of the experiment depend only on the outcomes of the _____ stage.

2. The outcome at any stage of the experiment in a Markov process is called the _____ of the experiment; the outcome at the current stage of the experiment is called the current _____.

3. The probabilities in a Markov chain are called _____ probabilities because they are associated with the transition from one state to the next in the Markov process.

4. A transition matrix associated with a Markov chain with n states is a/an _____ matrix T with entries satisfying the following conditions: (a) all entries are _____ and (b) the sum of the entries in each column of T is _____.

5. If the probability distribution vector X_N associated with a Markov process approaches a fixed vector as N gets larger and larger, then the latter is called the steady-state _____ vector for the system. To find this vector, we are led to finding the limit of T^m, which (if it exists) is called the _____ matrix.

6. A stochastic matrix T is a/an _____ Markov chain if T^n approaches a steady-state matrix in which the _____ of the limiting matrix are all _____ and all the entries are _____. To find the steady-state distribution vector X, we solve the vector equation _____ together with the condition that the sum of the _____ of the vector X is equal to _____.

7. In an absorbing stochastic matrix (a) there is at least one _____ state, a state in which it is impossible for an object to _____ and (b) it is possible to go from any nonabsorbing state to an absorbing state in one or more _____.

CHAPTER 9 Review Exercises

In Exercises 1–4, determine which of the following are regular stochastic matrices.

1. $\begin{bmatrix} 1 & -2 \\ 0 & -8 \end{bmatrix}$

2. $\begin{bmatrix} .3 & 1 \\ .7 & 0 \end{bmatrix}$

3. $\begin{bmatrix} \frac{1}{2} & 0 & \frac{1}{3} \\ 0 & 0 & \frac{1}{3} \\ \frac{1}{2} & 1 & \frac{1}{3} \end{bmatrix}$

4. $\begin{bmatrix} .3 & 0 & .5 \\ .2 & 1 & 0 \\ .1 & 0 & .5 \end{bmatrix}$

In Exercises 5 and 6, find X_2, (the probability distribution of the system after two observations) for the distribution vector X_0 and the transition matrix T.

5. $X_0 = \begin{bmatrix} \frac{1}{2} \\ \frac{1}{2} \\ 0 \end{bmatrix}$, $T = \begin{bmatrix} 0 & \frac{1}{4} & \frac{3}{5} \\ \frac{2}{5} & \frac{1}{2} & \frac{1}{5} \\ \frac{3}{5} & \frac{1}{4} & \frac{1}{5} \end{bmatrix}$

6. $X_0 = \begin{bmatrix} .35 \\ .25 \\ .40 \end{bmatrix}$, $T = \begin{bmatrix} .2 & .1 & .3 \\ .5 & .4 & .4 \\ .3 & .5 & .3 \end{bmatrix}$

In Exercises 7–10, determine whether the given matrix is an absorbing stochastic matrix.

7. $\begin{bmatrix} 1 & .6 & .1 \\ 0 & .2 & .6 \\ 0 & .2 & .3 \end{bmatrix}$

8. $\begin{bmatrix} .3 & .2 & .1 \\ .7 & .5 & .3 \\ 0 & .3 & .6 \end{bmatrix}$

9. $\begin{bmatrix} .32 & .22 & .44 \\ .68 & .78 & .56 \\ 0 & 0 & 0 \end{bmatrix}$

10. $\begin{bmatrix} .31 & .35 & 0 \\ .32 & .40 & 0 \\ .37 & .25 & 1 \end{bmatrix}$

In Exercises 11–14, find the steady-state matrix for the given transition matrix.

11. $\begin{bmatrix} .6 & .3 \\ .4 & .7 \end{bmatrix}$

12. $\begin{bmatrix} .5 & .4 \\ .5 & .6 \end{bmatrix}$

13. $\begin{bmatrix} .6 & .4 & .3 \\ .2 & .2 & .2 \\ .2 & .4 & .5 \end{bmatrix}$

14. $\begin{bmatrix} .1 & .2 & .6 \\ .3 & .4 & .2 \\ .6 & .4 & .2 \end{bmatrix}$

15. URBANIZATION OF FARMLAND   A study conducted by the State Department of Agriculture in a Sunbelt state reveals an increasing trend toward urbanization of the farmland within the state. Ten years ago, 50% of the land within the state was used for agricultural purposes (A), 15% had been urbanized (U), and the remaining 35% was neither agricultural nor urban (N). Since that time, 10% of the agricultural land has been converted to urban land, 5% has been used for other purposes, and the remaining 85% is still agricultural. Of the urban land, 95% has remained urban and 5% of it has been used for nonagricultural purposes. Of the land that was neither agricultural nor urban, 10% has been converted to agricultural land, 5% has been urbanized, and the remaining 85% remains unchanged.

    a. Construct the transition matrix for the Markov chain that describes the shift in land use within the state.

    b. Find the probability vector describing the distribution of land within the state 10 years ago.

    c. Assuming that this trend continues, find the probability vector describing the distribution of land within the state 10 years from now.

16. AUTOMOBILE SURVEY   *Auto Trend* magazine conducted a survey among automobile owners in a certain area of the country to determine what type of car they now own and what type of car they expect to own 4 years from now. For purposes of classification, automobiles mentioned in the survey were placed into three categories: large, intermediate, and small. Results of the survey follow:

		Present car		
		Large	Intermediate	Small
Future car	Large	.3	.1	.1
	Intermediate	.3	.5	.2
	Small	.4	.4	.7

Assuming that these results indicate the long-term buying trend of car owners within the area, what will be the distribution of cars (relative to size) in this area over the long run?

Before Moving On . . .

1. The transition matrix for a Markov process is

$$T = \begin{matrix} & \text{State} \\ & \begin{matrix} 1 & 2 \end{matrix} \\ \begin{matrix} \text{State 1} \\ \text{State 2} \end{matrix} & \begin{bmatrix} .3 & .4 \\ .7 & .6 \end{bmatrix} \end{matrix}$$

and the initial-state distribution vector is

$$X_0 = \begin{matrix} \text{State 1} \\ \text{State 2} \end{matrix} \begin{bmatrix} .6 \\ .4 \end{bmatrix}$$

Find X_2.

2. Find the steady-state vector for the transition matrix

$$T = \begin{bmatrix} \frac{1}{3} & \frac{1}{4} \\ \frac{2}{3} & \frac{3}{4} \end{bmatrix}$$

3. Compute the steady-state matrix of the absorbing stochastic matrix

$$\begin{bmatrix} \frac{1}{3} & 0 & 0 \\ 0 & 1 & \frac{1}{4} \\ \frac{2}{3} & 0 & \frac{3}{4} \end{bmatrix}$$

Precalculus Review

How much money is needed to purchase at least 100,000 shares of the Starr Communications Company? Corbyco, a giant conglomerate, wishes to purchase a minimum of 100,000 shares of the company. In Example 4, page 570, you will see how Corbyco's management determines how much money they will need for the acquisition.

PhotoAlto/Getty Images

THIS CHAPTER contains a brief review of the algebra you will need in the study of calculus. To solve many calculus problems, you will often need to simplify certain algebraic expressions before moving on to the next step. So, it is important that you are familiar with this process. This chapter also contains a short review of inequalities and absolute value; their uses range from describing the domains of functions to formulating practical problems.

10.1 Exponents and Radicals

■ Exponents and Radicals

This section reviews the properties of exponents and radicals. Recall that if b is any real number and n is a positive integer, then the expression b^n (read "b to the power n") is defined as the number

$$b^n = \underbrace{b \cdot b \cdot b \cdot \cdots \cdot b}_{n \text{ factors}}$$

The number b is called the **base**, and the superscript n is called the **power** of the exponential expression b^n. For example,

$$2^5 = 2 \cdot 2 \cdot 2 \cdot 2 \cdot 2 = 32 \qquad \text{and} \qquad \left(\frac{2}{3}\right)^3 = \left(\frac{2}{3}\right)\left(\frac{2}{3}\right)\left(\frac{2}{3}\right) = \frac{8}{27}$$

If $b \neq 0$, we define

$$b^0 = 1$$

For example, $2^0 = 1$ and $(-\pi)^0 = 1$, but the expression 0^0 is undefined.

Next, recall that if n is a positive integer then the expression $b^{1/n}$ is defined to be the number that, when raised to the nth power, is equal to b. Thus,

$$(b^{1/n})^n = b$$

Such a number, if it exists, is called the ***nth root of b***, also written $\sqrt[n]{b}$.

 Observe that the nth root of a negative number is not defined when n is even. For example, the square root of -2 is not defined because there is no real number b such that $b^2 = -2$. Also, given a number b, more than one number might satisfy our definition of the nth root. For example, both 3 and -3 squared equal 9, and each is a square root of 9. So, to avoid ambiguity, we define $b^{1/n}$ to be the positive nth root of b whenever it exists. Thus, $\sqrt{9} = 9^{1/2} = 3$.

Next, recall that if p/q (p, q positive integers with $q \neq 0$) is a rational number in lowest terms, then the expression $b^{p/q}$ is defined as the number $(b^{1/q})^p$ or, equivalently, $\sqrt[q]{b^p}$, whenever it exists. Expressions involving negative rational exponents are taken care of by the definition

$$b^{-p/q} = \frac{1}{b^{p/q}}$$

Examples are

$$2^{3/2} = (2^{1/2})^3 \approx (1.4142)^3 \approx 2.8283$$

and

$$4^{-5/2} = \frac{1}{4^{5/2}} = \frac{1}{(4^{1/2})^5} = \frac{1}{2^5} = \frac{1}{32}$$

The rules defining the exponential expression a^n, where $a > 0$ for all rational values of n, are given in Table 1.

TABLE 1

Definitions of Exponents

Definition of a^n ($a > 0$)	Example	Definition of a^n ($a > 0$)	Example
Integer exponent: If n is a positive integer, then $$a^n = a \cdot a \cdot a \cdots a$$ (n factors of a)	$2^5 = 2 \cdot 2 \cdot 2 \cdot 2 \cdot 2$ (five factors) $= 32$	**Fractional exponent** **a.** If n is a positive integer, then $$a^{1/n} \quad \text{or} \quad \sqrt[n]{a}$$ denotes the nth root of a.	$16^{1/2} = \sqrt{16}$ $= 4$
Zero exponent: If n is equal to zero, then $$a^0 = 1$$ (0^0 is not defined.)	$7^0 = 1$	**b.** If m and n are positive integers, then $$a^{m/n} = \sqrt[n]{a^m} = (\sqrt[n]{a})^m$$	$8^{2/3} = (\sqrt[3]{8})^2$ $= 4$
Negative exponent: If n is a positive integer, then $$a^{-n} = \frac{1}{a^n} \quad (a \neq 0)$$	$6^{-2} = \frac{1}{6^2}$ $= \frac{1}{36}$	**c.** If m and n are positive integers, then $$a^{-m/n} = \frac{1}{a^{m/n}} \quad (a \neq 0)$$	$9^{-3/2} = \frac{1}{9^{3/2}}$ $= \frac{1}{27}$

The first three definitions in Table 1 are also valid for negative values of a, whereas the fourth definition is valid for negative values of a only when n is odd. Thus,

$$(-8)^{1/3} = \sqrt[3]{-8} = -2 \qquad \text{\footnotesize n is odd.}$$

$$(-8)^{1/2} \text{ has no real value} \qquad \text{\footnotesize n is even.}$$

Finally, note that it can be shown that a^n has meaning for *all* real numbers n. For example, using a pocket calculator with a $\boxed{\text{y}^\text{x}}$, we see that $2^{\sqrt{2}} \approx 2.665144$.

The five laws of exponents are listed in Table 2.

TABLE 2

Laws of Exponents

Law	Example
1. $a^m \cdot a^n = a^{m+n}$	$x^2 \cdot x^3 = x^{2+3} = x^5$
2. $\dfrac{a^m}{a^n} = a^{m-n} \quad (a \neq 0)$	$\dfrac{x^7}{x^4} = x^{7-4} = x^3$
3. $(a^m)^n = a^{m \cdot n}$	$(x^4)^3 = x^{4 \cdot 3} = x^{12}$
4. $(ab)^n = a^n \cdot b^n$	$(2x)^4 = 2^4 \cdot x^4 = 16x^4$
5. $\left(\dfrac{a}{b}\right)^n = \dfrac{a^n}{b^n} \quad (b \neq 0)$	$\left(\dfrac{x}{2}\right)^3 = \dfrac{x^3}{2^3} = \dfrac{x^3}{8}$

The laws in Table 2 are valid for any real numbers a, b, m, and n whenever the quantities are defined.

 Remember, $(x^2)^3 \neq x^5$. The correct equation is $(x^2)^3 = x^{2 \cdot 3} = x^6$.

The next several examples illustrate the use of the laws of exponents.

EXAMPLE 1   Simplify the following expressions.

a. $(3x^2)(4x^3)$ **b.** $\dfrac{16^{5/4}}{16^{1/2}}$ **c.** $(6^{2/3})^3$ **d.** $(x^3 y^{-2})^{-2}$ **e.** $\left(\dfrac{y^{3/2}}{x^{1/4}}\right)^{-2}$

Solution

a. $(3x^2)(4x^3) = 12x^{2+3} = 12x^5$ 　　　　　　　　　　　Law 1

b. $\dfrac{16^{5/4}}{16^{1/2}} = 16^{5/4 - 1/2} = 16^{3/4} = (\sqrt[4]{16})^3 = 2^3 = 8$ 　　Law 2

c. $(6^{2/3})^3 = 6^{2/3 \cdot 3} = 6^{6/3} = 6^2 = 36$ 　　　　　　　　Law 3

d. $(x^3 y^{-2})^{-2} = (x^3)^{-2}(y^{-2})^{-2} = x^{(3)(-2)} y^{(-2)(-2)} = x^{-6} y^4 = \dfrac{y^4}{x^6}$ 　Law 4

e. $\left(\dfrac{y^{3/2}}{x^{1/4}}\right)^{-2} = \dfrac{y^{(3/2)(-2)}}{x^{(1/4)(-2)}} = \dfrac{y^{-3}}{x^{-1/2}} = \dfrac{x^{1/2}}{y^3}$ 　　　　　Law 5 ∎

▬ Simplifying Radicals

We can also use the laws of exponents to simplify expressions involving radicals, as illustrated in the next example.

EXAMPLE 2   Simplify the following expressions. (Assume that x, y, and n are positive.)

a. $\sqrt[4]{16x^4 y^8}$ **b.** $\sqrt{12m^3 n} \cdot \sqrt{3m^5 n}$ **c.** $\dfrac{\sqrt[3]{-27x^6}}{\sqrt[3]{8y^3}}$

Solution

a. $\sqrt[4]{16x^4 y^8} = (16x^4 y^8)^{1/4} = 16^{1/4} \cdot x^{4/4} y^{8/4} = 2xy^2$

b. $\sqrt{12m^3 n} \cdot \sqrt{3m^5 n} = \sqrt{36m^8 n^2} = (36m^8 n^2)^{1/2} = 36^{1/2} \cdot m^{8/2} n^{2/2} = 6m^4 n$

c. $\dfrac{\sqrt[3]{-27x^6}}{\sqrt[3]{8y^3}} = \dfrac{(-27x^6)^{1/3}}{(8y^3)^{1/3}} = \dfrac{-27^{1/3} x^{6/3}}{8^{1/3} y^{3/3}} = -\dfrac{3x^2}{2y}$ ∎

When a radical appears in the numerator or denominator of an algebraic expression, we often try to simplify the expression by eliminating the radical from the numerator or denominator. This process, called **rationalization**, is illustrated in the next two examples.

EXAMPLE 3   Rationalize the denominators:

a. $\dfrac{3x}{2\sqrt{x}}$ **b.** $\sqrt[3]{\dfrac{x}{y}}$

Solution

a. $\dfrac{3x}{2\sqrt{x}} = \dfrac{3x}{2\sqrt{x}} \cdot \dfrac{\sqrt{x}}{\sqrt{x}} = \dfrac{3x\sqrt{x}}{2\sqrt{x^2}} = \dfrac{3x\sqrt{x}}{2x} = \dfrac{3}{2}\sqrt{x}$

b. $\sqrt[3]{\dfrac{x}{y}} = \sqrt[3]{\dfrac{x}{y}} \cdot \sqrt[3]{\dfrac{y^2}{y^2}} = \dfrac{\sqrt[3]{xy^2}}{\sqrt[3]{y^3}} = \dfrac{\sqrt[3]{xy^2}}{y}$

EXAMPLE 4   Rationalize the numerators:

a. $\dfrac{3\sqrt{x}}{2x}$ b. $\sqrt[3]{\dfrac{x^2}{yz^3}}$

Solution

a. $\dfrac{3\sqrt{x}}{2x} = \dfrac{3\sqrt{x}}{2x} \cdot \dfrac{\sqrt{x}}{\sqrt{x}} = \dfrac{3\sqrt{x^2}}{2x\sqrt{x}} = \dfrac{3x}{2x\sqrt{x}} = \dfrac{3}{2\sqrt{x}}$

b. $\sqrt[3]{\dfrac{x^2}{yz^3}} \cdot \sqrt[3]{\dfrac{x}{x}} = \dfrac{\sqrt[3]{x^3}}{\sqrt[3]{xyz^3}} = \dfrac{x}{z\sqrt[3]{xy}}$

10.1   Exercises

In Exercises 1–16, evaluate the given expression.

1. $27^{2/3}$
2. $8^{-4/3}$
3. $\left(\dfrac{1}{\sqrt{3}}\right)^0$
4. $(7^{1/2})^4$
5. $\left[\left(\dfrac{1}{8}\right)^{1/3}\right]^{-2}$
6. $\left[\left(-\dfrac{1}{3}\right)^2\right]^{-3}$
7. $\left(\dfrac{7^{-5}\cdot 7^2}{7^{-2}}\right)^{-1}$
8. $\left(\dfrac{9}{16}\right)^{-1/2}$
9. $(125^{2/3})^{-1/2}$
10. $\sqrt[3]{2^6}$
11. $\dfrac{\sqrt{32}}{\sqrt{8}}$
12. $\sqrt[3]{\dfrac{-8}{27}}$
13. $\dfrac{16^{5/8}16^{1/2}}{16^{7/8}}$
14. $\left(\dfrac{9^{-3}\cdot 9^5}{9^{-2}}\right)^{-1/2}$
15. $16^{1/4}\cdot(8)^{-1/3}$
16. $\dfrac{6^{2.5}\cdot 6^{-1.9}}{6^{-1.4}}$

In Exercises 17–26, determine whether the statement is true or false. Give a reason for your choice.

17. $x^4 + 2x^4 = 3x^4$
18. $3^2\cdot 2^2 = 6^2$
19. $x^3\cdot 2x^2 = 2x^6$
20. $3^3 + 3 = 3^4$
21. $\dfrac{2^{4x}}{1^{3x}} = 2^{4x-3x}$
22. $(2^2\cdot 3^2)^2 = 6^4$
23. $\dfrac{1}{4^{-3}} = \dfrac{1}{64}$
24. $\dfrac{4^{3/2}}{2^4} = \dfrac{1}{2}$
25. $(1.2^{1/2})^{-1/2} = 1$
26. $5^{2/3}\cdot(25)^{2/3} = 25$

In Exercises 27–32, rewrite the expression using positive exponents only.

27. $(xy)^{-2}$
28. $3s^{1/3}\cdot s^{-7/3}$
29. $\dfrac{x^{-1/3}}{x^{1/2}}$
30. $\sqrt{x^{-1}}\cdot\sqrt{9x^{-3}}$
31. $12^0(s+t)^{-3}$
32. $(x-y)(x^{-1}+y^{-1})$

In Exercises 33–48, simplify the expression.

33. $\dfrac{x^{7/3}}{x^{-2}}$
34. $(49x^{-2})^{-1/2}$
35. $(x^2y^{-3})(x^{-5}y^3)$
36. $\dfrac{5x^6y^3}{2x^2y^7}$

37. $\dfrac{x^{3/4}}{x^{-1/4}}$

38. $\left(\dfrac{x^3 y^2}{z^2}\right)^2$

39. $\left(\dfrac{x^3}{-27y^{-6}}\right)^{-2/3}$

40. $\left(\dfrac{e^x}{e^{x-2}}\right)^{-1/2}$

41. $\left(\dfrac{x^{-3}}{y^{-2}}\right)^2\left(\dfrac{y}{x}\right)^4$

42. $\dfrac{(r^n)^4}{r^{5-2n}}$

43. $\sqrt[3]{x^{-2}} \cdot \sqrt{4x^5}$

44. $\sqrt{81x^6 y^{-4}}$

45. $-\sqrt[4]{16x^4 y^8}$

46. $\sqrt[3]{x^{3a+b}}$

47. $\sqrt[6]{64x^8 y^3}$

48. $\sqrt[3]{27r^6} \cdot \sqrt{s^2 t^4}$

In Exercises 49–52, use the fact that $2^{1/2} \approx 1.414$ and $3^{1/2} \approx 1.732$ to evaluate the expression without using a calculator.

49. $2^{3/2}$

50. $8^{1/2}$

51. $9^{3/4}$

52. $6^{1/2}$

In Exercises 53–56, use the fact that $10^{1/2} \approx 3.162$ and $10^{1/3} \approx 2.154$ to evaluate the expression without using a calculator.

53. $10^{3/2}$

54. $1000^{3/2}$

55. $10^{2.5}$

56. $(0.0001)^{-1/3}$

In Exercises 57–62, rationalize the denominator of the expression.

57. $\dfrac{3}{2\sqrt{x}}$

58. $\dfrac{3}{\sqrt{xy}}$

59. $\dfrac{2y}{\sqrt{3y}}$

60. $\dfrac{5x^2}{\sqrt{3x}}$

61. $\dfrac{1}{\sqrt[3]{x}}$

62. $\sqrt{\dfrac{2x}{y}}$

In Exercises 63–68, rationalize the numerator of the expression.

63. $\dfrac{2\sqrt{x}}{3}$

64. $\dfrac{\sqrt[3]{x}}{24}$

65. $\sqrt{\dfrac{2y}{x}}$

66. $\sqrt[3]{\dfrac{2x}{3y}}$

67. $\dfrac{\sqrt[3]{x^2 z}}{y}$

68. $\dfrac{\sqrt[3]{x^2 y}}{2x}$

10.2   Algebraic Expressions

▪ Operations with Algebraic Expressions

In calculus we often work with algebraic expressions such as

$$2x^{4/3} - x^{1/3} + 1 \qquad 2x^2 - x - \frac{2}{\sqrt{x}} \qquad \frac{3xy + 2}{x + 1} \qquad 2x^3 + 2x + 1$$

An algebraic expression of the form ax^n, where the coefficient a is a real number and n is a nonnegative integer, is called a **monomial**, meaning that it consists of one term. For example, $7x^2$ is a monomial. A **polynomial** is a monomial or the sum of two or more monomials. For example,

$$x^2 + 4x + 4 \qquad x^3 + 5 \qquad x^4 + 3x^2 + 3 \qquad x^2 y + xy + y$$

are all polynomials.

Adding and Subtracting Algebraic Expressions   Constant terms and terms containing the same variable factor are called **like**, or **similar**, **terms**. Like terms may be combined by adding or subtracting their numerical coefficients. For example,

$$3x + 7x = 10x \quad \text{and} \quad \frac{1}{2}xy + 3xy = \frac{7}{2}xy$$

The distributive property of the real number system,

$$ab + ac = a \cdot (b + c)$$

is used to justify this procedure.

To add or subtract two or more algebraic expressions, first remove the parentheses and then combine like terms. The resulting expression is written in order of decreasing degree from left to right.

 *

EXAMPLE 1

a. $(3x^4 + 10x^3 + 6x^2 + 10x + 3) + (2x^4 + 10x^3 + 6x^2 + 4x)$

$\quad = 3x^4 + 2x^4 + 10x^3 + 10x^3 + 6x^2 + 6x^2 + 10x$

$\qquad + 4x + 3$ 　　　　　　　　　　　Remove parentheses.

$\quad = 5x^4 + 20x^3 + 12x^2 + 14x + 3$ 　　Combine like terms.

b. $(2x^4 + 3x^3 + 4x + 6) - (3x^4 + 9x^3 + 3x^2)$

$\quad = 2x^4 + 3x^3 + 4x + 6 - 3x^4 - 9x^3 - 3x^2$ 　　Remove parentheses.

$\quad = 2x^4 - 3x^4 + 3x^3 - 9x^3 - 3x^2 + 4x + 6$

$\quad = -x^4 - 6x^3 - 3x^2 + 4x + 6$ 　　　　　Combine like terms.

c. $2\sqrt{x} + 7y - \dfrac{1}{2}\sqrt{x} + 3y = 2\sqrt{x} - \dfrac{1}{2}\sqrt{x} + 7y + 3y = \dfrac{3}{2}\sqrt{x} + 10y$

d. $\left(2x^{3/2} - x^{1/2} + \dfrac{2}{3}x^{-1/3} \right) - \left(3x^{3/2} + \dfrac{1}{2}x^{1/2} - \dfrac{1}{3}x^{-1/3} \right)$

$\quad = 2x^{3/2} - x^{1/2} + \dfrac{2}{3}x^{-1/3} - 3x^{3/2} - \dfrac{1}{2}x^{1/2} + \dfrac{1}{3}x^{-1/3}$ 　　Remove parentheses.

$\quad = 2x^{3/2} - 3x^{3/2} - x^{1/2} - \dfrac{1}{2}x^{1/2} + \dfrac{2}{3}x^{-1/3} + \dfrac{1}{3}x^{-1/3}$

$\quad = -x^{3/2} - \dfrac{3}{2}x^{1/2} + x^{-1/3}$ 　　　　　　　Combine like terms.

e. $2t^3 - \{t^2 - [t - (2t - 1)] + 4\}$

$\quad = 2t^3 - \{t^2 - [t - 2t + 1] + 4\}$

$\quad = 2t^3 - \{t^2 - [-t + 1] + 4\}$ 　　Remove parentheses and combine like terms within brackets.

$\quad = 2t^3 - \{t^2 + t - 1 + 4\}$ 　　Remove brackets.

$\quad = 2t^3 - \{t^2 + t + 3\}$ 　　Combine like terms within braces.

$\quad = 2t^3 - t^2 - t - 3$ 　　Remove braces. ◼

An algebraic expression is said to be **simplified** if none of its terms are similar. Observe that when the algebraic expression in Example 1e was simplified, the innermost grouping symbols were removed first; that is, the parentheses ( ) were removed first, the brackets [ ] second, and the braces { } third.

Multiplying Algebraic Expressions　When we multiply algebraic expressions, each term of one algebraic expression is multiplied by each term of the other. The resulting algebraic expression is then simplified.

---

*The symbol ▨ indicates that these examples were selected from the calculus portion of the text in order to help you review the algebraic computations you will actually be using in calculus.

EXAMPLE 2　Expand the expression $(x + 3)(x - 2)$.

Solution

$$(x + 3)(x - 2) = x(x - 2) + 3(x - 2)$$
$$= x^2 - 2x + 3x - 6$$
$$= x^2 + x - 6$$ ∎

EXAMPLE 3　Perform the indicated operations.

a. $(3x - 4)(3x^2 - 2x + 3)$　　　**b.** $(x^2 + 1)(3x^2 + 10x + 3)$
c. $(x^2 + x + 1)(x^2 + x + 1)$　　**d.** $(e^t + e^{-t})e^t - e^t(e^t - e^{-t})$

Solution

a. $(3x - 4)(3x^2 - 2x + 3)$
$$= 3x\,(3x^2 - 2x + 3) - 4(3x^2 - 2x + 3)$$
$$= 9x^3 - 6x^2 + 9x - 12x^2 + 8x - 12$$
$$= 9x^3 - 18x^2 + 17x - 12$$
b. $(x^2 + 1)(3x^2 + 10x + 3)$
$$= x^2(3x^2 + 10x + 3) + 1(3x^2 + 10x + 3)$$
$$= 3x^4 + 10x^3 + 3x^2 + 3x^2 + 10x + 3$$
$$= 3x^4 + 10x^3 + 6x^2 + 10x + 3$$
c. $(x^2 + x + 1)(x^2 + x + 1)$
$$= x^2(x^2 + x + 1) + x(x^2 + x + 1) + 1(x^2 + x + 1)$$
$$= x^4 + x^3 + x^2 + x^3 + x^2 + x + x^2 + x + 1$$
$$= x^4 + 2x^3 + 3x^2 + 2x + 1$$
d. $(e^t + e^{-t})e^t - e^t(e^t - e^{-t})$
$$= e^{2t} + e^0 - e^{2t} + e^0$$
$$= e^{2t} - e^{2t} + e^0 + e^0$$
$$= 1 + 1 \qquad \text{Recall that } e^0 = 1.$$
$$= 2$$ ∎

Certain product formulas that are frequently used in algebraic computations are given in Table 3.

TABLE 3

Product Formulas

Formula	Example
$(a + b)^2 = a^2 + 2ab + b^2$	$(2x + 3y)^2 = (2x)^2 + 2(2x)(3y) + (3y)^2$ $= 4x^2 + 12xy + 9y^2$
$(a - b)^2 = a^2 - 2ab + b^2$	$(4x - 2y)^2 = (4x)^2 - 2(4x)(2y) + (2y)^2$ $= 16x^2 - 16xy + 4y^2$
$(a + b)(a - b) = a^2 - b^2$	$(2x + y)(2x - y) = (2x)^2 - (y)^2$ $= 4x^2 - y^2$

Factoring

Factoring is the process of expressing an algebraic expression as a product of other algebraic expressions. For example, by applying the distributive property, we may write

$$3x^2 - x = x(3x - 1)$$

The first step in factoring an algebraic expression is to check whether it contains any common terms. If it does, the greatest common term is then factored out. For example, the common factor of the algebraic expression $2a^2x + 4ax + 6a$ is $2a$ because

$$2a^2x + 4ax + 6a = 2a \cdot ax + 2a \cdot 2x + 2a \cdot 3 = 2a(ax + 2x + 3)$$

EXAMPLE 4   Factor out the greatest common factor in each of the following expressions.

a. $-0.3t^2 + 3t$ **b.** $6a^4b^4c - 3a^3b^2c - 9a^2b^2$ **c.** $2x^{3/2} - 3x^{1/2}$
d. $2ye^{xy^2} + 2xy^3e^{xy^2}$ **e.** $4x(x + 1)^{1/2} - 2x^2(\frac{1}{2})(x + 1)^{-1/2}$

Solution

a. $-0.3t^2 + 3t = -0.3t(t - 10)$
b. $6a^4b^4c - 3a^3b^2c - 9a^2b^2 = 3a^2b^2(2a^2b^2c - ac - 3)$
c. $2x^{3/2} - 3x^{1/2} = x^{1/2}(2x - 3)$
d. $2ye^{xy^2} + 2xy^3e^{xy^2} = 2ye^{xy^2}(1 + xy^2)$
e. $4x(x + 1)^{1/2} - 2x^2(\frac{1}{2})(x + 1)^{-1/2} = 4x(x + 1)^{1/2} - x^2(x + 1)^{-1/2}$
$$= x(x + 1)^{-1/2}[4(x + 1)^{1/2}(x + 1)^{1/2} - x]$$
$$= x(x + 1)^{-1/2}[4(x + 1) - x]$$
$$= x(x + 1)^{-1/2}(4x + 4 - x) = x(x + 1)^{-1/2}(3x + 4)$$

Here we select $(x + 1)^{-1/2}$ as the common factor because it is "contained" in each algebraic term. In particular, observe that

$$(x + 1)^{-1/2}(x + 1)^{1/2}(x + 1)^{1/2} = (x + 1)^{1/2}$$ ∎

Sometimes an algebraic expression may be factored by regrouping and rearranging its terms and then factoring out a common term. This technique is illustrated in Example 5.

EXAMPLE 5   Factor:

a. $2ax + 2ay + bx + by$ **b.** $3x\sqrt{y} - 4 - 2\sqrt{y} + 6x$

Solution

a. First, factor the common term $2a$ from the first two terms and the common term b from the last two terms. Thus,

$$2ax + 2ay + bx + by = 2a(x + y) + b(x + y)$$

Since $(x + y)$ is common to both terms of the polynomial, we may factor it out. Hence,

$$2a(x + y) + b(x + y) = (x + y)(2a + b)$$

b. $3x\sqrt{y} - 4 - 2\sqrt{y} + 6x = 3x\sqrt{y} - 2\sqrt{y} + 6x - 4$

$$= \sqrt{y}(3x - 2) + 2(3x - 2)$$
$$= (3x - 2)(\sqrt{y} + 2) \qquad \blacksquare$$

■ Factoring Polynomials

A polynomial that is **prime** over a given set of numbers cannot be expressed as a product of two polynomials of positive degree with coefficients belonging to that set. For example, $x^2 + 2x + 2$ is prime over the set of polynomials with integral coefficients because it cannot be expressed as the product of two polynomials of positive degree that have integral coefficients.

The first step in factoring a polynomial is to find the common factors. The next step is to express the polynomial as the product of a constant and/or one or more prime polynomials.

Certain product formulas that are useful in factoring binomials and trinomials are listed in Table 4.

TABLE 4

Formula	Example
Difference of two squares $x^2 - y^2 = (x + y)(x - y)$	$x^2 - 36 = (x + 6)(x - 6)$ $8x^2 - 2y^2 = 2(4x^2 - y^2)$ $\qquad\qquad = 2(2x + y)(2x - y)$ $9 - a^6 = (3 + a^3)(3 - a^3)$
Perfect-square trinomial $x^2 + 2xy + y^2 = (x + y)^2$ $x^2 - 2xy + y^2 = (x - y)^2$	$x^2 + 8x + 16 = (x + 4)^2$ $4x^2 - 4xy + y^2 = (2x - y)^2$
Sum of two cubes $x^3 + y^3 = (x + y)(x^2 - xy + y^2)$	$z^3 + 27 = z^3 + (3)^3$ $\qquad\qquad = (z + 3)(z^2 - 3z + 9)$
Difference of two cubes $x^3 - y^3 = (x - y)(x^2 + xy + y^2)$	$8x^3 - y^6 = (2x)^3 - (y^2)^3$ $\qquad\qquad = (2x - y^2)(4x^2 + 2xy^2 + y^4)$

The factors of the second-degree polynomial with integral coefficients

$$px^2 + qx + r$$

are $(ax + b)(cx + d)$, where $ac = p$, $ad + bc = q$, and $bd = r$. Since only a limited number of choices are possible, we use a trial-and-error method to factor polynomials having this form.

For example, to factor $x^2 - 2x - 3$, we first observe that the only possible first-degree terms are

$$(x \quad)(x \quad) \qquad \text{\small Since the coefficient of } x^2 \text{ is 1}$$

Next, we observe that the product of the constant term is (-3). This gives us the following possible factors:

$$(x - 1)(x + 3)$$
$$(x + 1)(x - 3)$$

Looking once again at the polynomial $x^2 - 2x - 3$, we see that the coefficient of x is -2. Checking to see which set of factors yields -2 for the coefficient of x, we find that

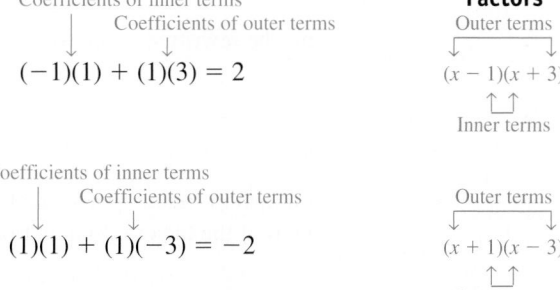

Coefficients of inner terms
Coefficients of outer terms

$$(-1)(1) + (1)(3) = 2$$

Factors
Outer terms
$(x - 1)(x + 3)$
Inner terms

Coefficients of inner terms
Coefficients of outer terms

$$(1)(1) + (1)(-3) = -2$$

Outer terms
$(x + 1)(x - 3)$
Inner terms

We thus conclude that the correct factorization is

$$x^2 - 2x - 3 = (x + 1)(x - 3)$$

With practice, you will soon find that you can perform many of these steps mentally and will no longer need to write out each step.

EXAMPLE 6   Factor:

a. $3x^2 + 4x - 4$ **b.** $3x^2 - 6x - 24$

Solution

a. Using trial and error, we find that the correct factorization is

$$3x^2 + 4x - 4 = (3x - 2)(x + 2)$$

b. Since each term has the common factor 3, we have

$$3x^2 - 6x - 24 = 3(x^2 - 2x - 8)$$

Using the trial-and-error method of factorization, we find that

$$x^2 - 2x - 8 = (x - 4)(x + 2)$$

Therefore,

$$3x^2 - 6x - 24 = 3(x - 4)(x + 2)$$

Roots of Polynomial Equations

A polynomial equation of degree n in the variable x is an equation of the form

$$a_n x^n + a_{n-1}x^{n-1} + \cdots + a_0 = 0$$

where n is a nonnegative integer and $a_0, a_1, \ldots, a_n$ are real numbers with $a_n \neq 0$. For example, the equation

$$-2x^5 + 8x^3 - 6x^2 + 3x + 1 = 0$$

is a polynomial equation of degree 5 in x.

The **roots of a polynomial equation** are precisely the values of x that satisfy the given equation.* To find the roots of a polynomial equation, first factor the polynomial and then solve the resulting equation. For example, the polynomial equation

$$x^3 - 3x^2 + 2x = 0$$

may be rewritten in the form

$$x(x^2 - 3x + 2) = 0$$

or

$$x(x - 1)(x - 2) = 0$$

Since the product of two real numbers can be equal to zero if and only if one (or both) of the factors is equal to zero, we have

$$x = 0, \quad x - 1 = 0, \quad \text{or} \quad x - 2 = 0$$

from which we see that the desired roots are $x = 0$, 1, and 2.

■ The Quadratic Formula

In general, the problem of finding the roots of a polynomial equation is a difficult one. But the roots of a quadratic equation (a polynomial equation of degree 2) are easily found either by factoring or by using the following quadratic formula.

Quadratic Formula

The solutions of the equation $ax^2 + bx + c = 0$ $(a \neq 0)$ are given by

$$x = \frac{-b \pm \sqrt{b^2 - 4ac}}{2a}$$

EXAMPLE 7 Solve each of the following quadratic equations.

a. $2x^2 + 5x - 12 = 0$ **b.** $x^2 = -3x + 8$

Solution

a. The equation is in standard form, with $a = 2$, $b = 5$, and $c = -12$. Using the quadratic formula, we find

$$x = \frac{-b \pm \sqrt{b^2 - 4ac}}{2a} = \frac{-5 \pm \sqrt{5^2 - 4(2)(-12)}}{2(2)}$$

$$= \frac{-5 \pm \sqrt{121}}{4} = \frac{-5 \pm 11}{4}$$

$$= -4 \quad \text{or} \quad \frac{3}{2}$$

---

*In this book, we are interested only in the *real* roots of an equation.

This equation can also be solved by factoring. Thus,

$$2x^2 + 5x - 12 = (2x - 3)(x + 4) = 0$$

from which we see that the desired roots are $x = 3/2$ or $x = -4$, as obtained previously.

b. We first rewrite the given equation in the standard form $x^2 + 3x - 8 = 0$, from which we see that $a = 1$, $b = 3$, and $c = -8$. Using the quadratic formula, we find

$$x = \frac{-b \pm \sqrt{b^2 - 4ac}}{2a} = \frac{-3 \pm \sqrt{3^2 - 4(1)(-8)}}{2(1)}$$

$$= \frac{-3 \pm \sqrt{41}}{2}$$

That is, the solutions are

$$\frac{-3 + \sqrt{41}}{2} \approx 1.7 \quad \text{and} \quad \frac{-3 - \sqrt{41}}{2} \approx -4.7$$

In this case, the quadratic formula proves quite handy!

10.2 Exercises

In Exercises 1–22, perform the indicated operations and then simplify each expression.

1. $(7x^2 - 2x + 5) + (2x^2 + 5x - 4)$

2. $(3x^2 + 5xy + 2y) + (4 - 3xy - 2x^2)$

3. $(5y^2 - 2y + 1) - (y^2 - 3y - 7)$

4. $3(2a - b) - 4(b - 2a)$

5. $x - \{2x - [-x - (1 - x)]\}$

6. $3x^2 - \{x^2 + 1 - x[x - (2x - 1)]\} + 2$

7. $\left(\frac{1}{3} - 1 + e\right) - \left(-\frac{1}{3} - 1 + e^{-1}\right)$

8. $-\frac{3}{4}y - \frac{1}{4}x + 100 + \frac{1}{2}x + \frac{1}{4}y - 120$

9. $3\sqrt{8} + 8 - 2\sqrt{y} + \frac{1}{2}\sqrt{x} - \frac{3}{4}\sqrt{y}$

10. $\frac{8}{9}x^2 + \frac{2}{3}x + \frac{16}{3}x^2 - \frac{16}{3}x - 2x + 2$

11. $(x + 8)(x - 2)$

12. $(5x + 2)(3x - 4)$

13. $(a + 5)^2$

14. $(3a - 4b)^2$

15. $(x + 2y)^2$

16. $(6 - 3x)^2$

17. $(2x + y)(2x - y)$

18. $(3x + 2)(2 - 3x)$

19. $(x^2 - 1)(2x) - x^2(2x)$

20. $(x^{1/2} + 1)\left(\frac{1}{2}x^{-1/2}\right) - (x^{1/2} - 1)\left(\frac{1}{2}x^{-1/2}\right)$

21. $2(t + \sqrt{t})^2 - 2t^2$

22. $2x^2 + (-x + 1)^2$

In Exercises 23–30, factor out the greatest common factor from each expression.

23. $4x^5 - 12x^4 - 6x^3$

24. $4x^2y^2z - 2x^5y^2 + 6x^3y^2z^2$

25. $7a^4 - 42a^2b^2 + 49a^3b$

26. $3x^{2/3} - 2x^{1/3}$

27. $e^{-x} - xe^{-x}$

28. $2ye^{xy^2} + 2xy^3e^{xy^2}$

29. $2x^{-5/2} - \frac{3}{2}x^{-3/2}$

30. $\frac{1}{2}\left(\frac{2}{3}u^{3/2} - 2u^{1/2}\right)$

In Exercises 31–44, factor each expression.

31. $6ac + 3bc - 4ad - 2bd$

32. $3x^3 - x^2 + 3x - 1$

33. $4a^2 - b^2$

34. $12x^2 - 3y^2$

35. $10 - 14x - 12x^2$ **36.** $x^2 - 2x - 15$

37. $3x^2 - 6x - 24$ **38.** $3x^2 - 4x - 4$

39. $12x^2 - 2x - 30$ **40.** $(x + y)^2 - 1$

41. $9x^2 - 16y^2$ **42.** $8a^2 - 2ab - 6b^2$

43. $x^6 + 125$ **44.** $x^3 - 27$

In Exercises 45–52, perform the indicated operations and then simplify the algebraic expression.

45. $(x^2 + y^2)x - xy(2y)$ **46.** $2kr(R - r) - kr^2$

47. $2(x - 1)(2x + 2)^3[4(x - 1) + (2x + 2)]$

48. $5x^2(3x^2 + 1)^4(6x) + (3x^2 + 1)^5(2x)$

49. $4(x - 1)^2(2x + 2)^3(2) + (2x + 2)^4(2)(x - 1)$

50. $(x^2 + 1)(4x^3 - 3x^2 + 2x) - (x^4 - x^3 + x^2)(2x)$

51. $(x^2 + 2)^2[5(x^2 + 2)^2 - 3](2x)$

52. $(x^2 - 4)(x^2 + 4)(2x + 8) - (x^2 + 8x - 4)(4x^3)$

In Exercises 53–58, find the real roots of each equation by factoring.

53. $x^2 + x - 12 = 0$ **54.** $3x^2 - x - 4 = 0$

55. $4t^2 + 2t - 2 = 0$ **56.** $-6x^2 + x + 12 = 0$

57. $\frac{1}{4}x^2 - x + 1 = 0$ **58.** $\frac{1}{2}a^2 + a - 12 = 0$

In Exercises 59–64, use the quadratic formula to solve the given quadratic equation.

59. $4x^2 + 5x - 6 = 0$ **60.** $3x^2 - 4x + 1 = 0$

61. $8x^2 - 8x - 3 = 0$ **62.** $x^2 - 6x + 6 = 0$

63. $2x^2 + 4x - 3 = 0$ **64.** $2x^2 + 7x - 15 = 0$

10.3   Algebraic Fractions

▬ Rational Expressions

Quotients of polynomials are called **rational expressions**. Examples of rational expressions are

$$\frac{6x - 1}{2x + 3} \qquad \frac{3x^2y^3 - 2xy}{4x} \qquad \frac{2}{5ab}$$

Since rational expressions are quotients in which the variables represent real numbers, the properties of the real numbers apply to rational expressions as well, and operations with rational fractions are performed in the same manner as operations with arithmetic fractions. For example, using the properties of the real number system, we may write

$$\frac{ac}{bc} = \frac{a}{b} \cdot \frac{c}{c} = \frac{a}{b} \cdot 1 = \frac{a}{b}$$

where a, b, and c are any real numbers and b and c are not zero.

Similarly, using the same properties of real numbers, we may write

$$\frac{(x + 2)(x - 3)}{(x - 2)(x - 3)} = \frac{x + 2}{x - 2} \qquad (x \neq 2, 3)$$

after "canceling" the common factors.

 $\frac{3 + 4x}{3} \neq 1 + 4x$ is an example of incorrect cancellation. Instead we need to

write $\frac{3 + 4x}{3} = \frac{3}{3} + \frac{4x}{3} = 1 + \frac{4x}{3}$.

A rational expression is simplified, or in lowest terms, when the numerator and denominator have no common factors other than 1 and -1 and the expression contains no negative exponents.

EXAMPLE 1 Simplify the following expressions.

a. $\dfrac{x^2 + 2x - 3}{x^2 + 4x + 3}$ **b.** $\dfrac{3 - 4x - 4x^2}{2x - 1}$

c. $\dfrac{[(t^2 + 4)(2t - 4) - (t^2 - 4t + 4)(2t)]}{(t^2 + 4)^2}$

d. $\dfrac{(1 + x^2)^2(-2x) - (1 - x^2)2(1 + x^2)(2x)}{(1 + x^2)^4}$

e. $\dfrac{\frac{1}{4}(x + h)^2 - \frac{1}{4}x^2}{h}$

Solution

a. $\dfrac{x^2 + 2x - 3}{x^2 + 4x + 3} = \dfrac{(x + 3)(x - 1)}{(x + 3)(x + 1)} = \dfrac{x - 1}{x + 1}$

b. $\dfrac{3 - 4x - 4x^2}{2x - 1} = \dfrac{(1 - 2x)(3 + 2x)}{2x - 1}$

$\qquad\qquad = \dfrac{-(2x - 1)(2x + 3)}{2x - 1} \qquad 1 - 2x = -(2x - 1)$

$\qquad\qquad = -(2x + 3)$

c. $\dfrac{[(t^2 + 4)(2t - 4) - (t^2 - 4t + 4)(2t)]}{(t^2 + 4)^2}$

$\qquad = \dfrac{2t^3 - 4t^2 + 8t - 16 - 2t^3 + 8t^2 - 8t}{(t^2 + 4)^2}$ Carry out the indicated multiplication.

$\qquad = \dfrac{4t^2 - 16}{(t^2 + 4)^2}$ Combine like terms.

$\qquad = \dfrac{4(t^2 - 4)}{(t^2 + 4)^2}$ Factor.

d. $\dfrac{(1 + x^2)^2(-2x) - (1 - x^2)2(1 + x^2)(2x)}{(1 + x^2)^4}$

$\qquad = \dfrac{(1 + x^2)(2x)[(1 + x^2)(-1) - 2(1 - x^2)]}{(1 + x^2)^4}$ Factor out $(1 + x^2)(2x)$.

$\qquad = \dfrac{2x[-1 - x^2 - 2 + 2x^2]}{(1 + x^2)^3}$ Divide numerator and denominator by $(1 + x^2)$.

$\qquad = \dfrac{2x(x^2 - 3)}{(1 + x^2)^3}$ Combine like terms.

e. $\dfrac{\frac{1}{4}(x + h)^2 - \frac{1}{4}x^2}{h} = \dfrac{\frac{1}{4}x^2 + \frac{1}{2}xh + \frac{1}{4}h^2 - \frac{1}{4}x^2}{h}$

$\qquad\qquad = \dfrac{h(\frac{1}{2}x + \frac{1}{4}h)}{h}$ Square the binomial and combine like terms.

$\qquad\qquad = \dfrac{1}{2}x + \dfrac{1}{4}h$ Divide numerator and denominator by h.

TABLE 5

Operation	Example

If P, Q, R, and S are polynomials, then

Multiplication

$$\frac{P}{Q} \cdot \frac{R}{S} = \frac{PR}{QS} \qquad (Q, S \neq 0)$$

$$\frac{2x}{y} \cdot \frac{(x + 1)}{(y - 1)} = \frac{2x(x + 1)}{y(y - 1)} = \frac{2x^2 + 2x}{y^2 - y}$$

Division

$$\frac{P}{Q} \div \frac{R}{S} = \frac{P}{Q} \cdot \frac{S}{R} = \frac{PS}{QR} \qquad (Q, R, S \neq 0)$$

$$\frac{x^2 + 3}{y} \div \frac{y^2 + 1}{x} = \frac{x^2 + 3}{y} \cdot \frac{x}{y^2 + 1} = \frac{x^3 + 3x}{y^3 + y}$$

The operations of multiplication and division are performed with algebraic fractions in the same manner as with arithmetic fractions (Table 5).

When rational expressions are multiplied and divided, the resulting expressions should be simplified.

EXAMPLE 2   Perform the indicated operations and then simplify.

a. $\dfrac{2x - 8}{x + 2} \cdot \dfrac{x^2 + 4x + 4}{x^2 - 16}$ **b.** $\dfrac{x^2 - 6x + 9}{3x + 12} \div \dfrac{x^2 - 9}{6x^2 + 18x}$

Solution

a.
$$\frac{2x - 8}{x + 2} \cdot \frac{x^2 + 4x + 4}{x^2 - 16}$$
$$= \frac{2(x - 4)}{x + 2} \cdot \frac{(x + 2)^2}{(x + 4)(x - 4)}$$
$$= \frac{2(x - 4)(x + 2)(x + 2)}{(x + 2)(x + 4)(x - 4)} \qquad \text{Cancel the common factors } (x + 2)(x - 4).$$
$$= \frac{2(x + 2)}{x + 4}$$

b.
$$\frac{x^2 - 6x + 9}{3x + 12} \div \frac{x^2 - 9}{6x^2 + 18x}$$
$$= \frac{x^2 - 6x + 9}{3x + 12} \cdot \frac{6x^2 + 18x}{x^2 - 9}$$
$$= \frac{(x - 3)^2}{3(x + 4)} \cdot \frac{6x(x + 3)}{(x + 3)(x - 3)}$$
$$= \frac{(x - 3)(x - 3)(6x)(x + 3)}{3(x + 4)(x + 3)(x - 3)} \qquad \text{Cancel the common factors } 3(x + 3)(x - 3).$$
$$= \frac{2x(x - 3)}{x + 4}$$

For rational expressions, the operations of addition and subtraction are performed by finding a common denominator of the fractions and then adding or subtracting the fractions. Table 6 shows the rules for fractions with equal denominators.

To add or subtract fractions that have different denominators, first find a com-

TABLE 6

Operation	Example
If P, Q, and R are polynomials, then	
Addition	
$\dfrac{P}{R} + \dfrac{Q}{R} = \dfrac{P + Q}{R}$ $\quad (R \neq 0)$	$\dfrac{2x}{x + 2} + \dfrac{6x}{x + 2} = \dfrac{2x + 6x}{x + 2} = \dfrac{8x}{x + 2}$
Subtraction	
$\dfrac{P}{R} - \dfrac{Q}{R} = \dfrac{P - Q}{R}$ $\quad (R \neq 0)$	$\dfrac{3y}{y - x} - \dfrac{y}{y - x} = \dfrac{3y - y}{y - x} = \dfrac{2y}{y - x}$

mon denominator, preferably the least common denominator. Then carry out the indicated operations following the procedure illustrated in Table 6.

To find the least common denominator (LCD) of two or more rational expressions:

1. *Find the prime factors of each denominator.*
2. *Form the product of the different prime factors that occur in the denominators. Each prime factor in this product should be raised to the highest power of that factor appearing in the denominators.*

$$\frac{x}{2 + y} \neq \frac{x}{2} + \frac{x}{y}$$

EXAMPLE 3

a. $\dfrac{3x + 4}{4x} + \dfrac{4y - 2}{3y}$ $\qquad$ **b.** $\dfrac{2x}{x^2 + 1} + \dfrac{6(3x^2)}{x^3 + 2}$ $\qquad$ **c.** $\dfrac{1}{x + h} - \dfrac{1}{x}$

Solution

a. $\dfrac{3x + 4}{4x} + \dfrac{4y - 2}{3y}$

$\qquad = \dfrac{3x + 4}{4x} \cdot \dfrac{3y}{3y} + \dfrac{4y - 2}{3y} \cdot \dfrac{4x}{4x}$ $\qquad$ LCD $= (4x)(3y) = 12xy$

$\qquad = \dfrac{9xy + 12y}{12xy} + \dfrac{16xy - 8x}{12xy}$

$\qquad = \dfrac{25xy - 8x + 12y}{12xy}$

b. $\dfrac{2x}{x^2 + 1} + \dfrac{6(3x^2)}{x^3 + 2}$

$\qquad = \dfrac{2x(x^3 + 2) + 6(3x^2)(x^2 + 1)}{(x^2 + 1)(x^3 + 2)}$ $\qquad$ LCD $= (x^2 + 1)(x^3 + 2)$

$\qquad = \dfrac{2x^4 + 4x + 18x^4 + 18x^2}{(x^2 + 1)(x^3 + 2)}$

$\qquad = \dfrac{20x^4 + 18x^2 + 4x}{(x^2 + 1)(x^3 + 2)}$

$\qquad = \dfrac{2x(10x^3 + 9x + 2)}{(x^2 + 1)(x^3 + 2)}$

c. $\dfrac{1}{x+h} - \dfrac{1}{x} = \dfrac{1}{x+h} \cdot \dfrac{x}{x} - \dfrac{1}{x} \cdot \dfrac{x+h}{x+h}$ LCD $= (x)(x+h)$

$\qquad\qquad\qquad = \dfrac{x}{x(x+h)} - \dfrac{x+h}{x(x+h)}$

$\qquad\qquad\qquad = \dfrac{x-x-h}{x(x+h)}$

$\qquad\qquad\qquad = \dfrac{-h}{x(x+h)}$

■ Other Algebraic Fractions

The techniques used to simplify rational expressions may also be used to simplify algebraic fractions in which the numerator and denominator are not polynomials, as illustrated in Example 4.

EXAMPLE 4   Simplify:

a. $\dfrac{1 + \dfrac{1}{x+1}}{x - \dfrac{4}{x}}$ **b.** $\dfrac{x^{-1} + y^{-1}}{x^{-2} - y^{-2}}$

Solution

a. $\dfrac{1 + \dfrac{1}{x+1}}{x - \dfrac{4}{x}} = \dfrac{1 \cdot \dfrac{x+1}{x+1} + \dfrac{1}{x+1}}{x \cdot \dfrac{x}{x} - \dfrac{4}{x}} = \dfrac{\dfrac{x+1+1}{x+1}}{\dfrac{x^2-4}{x}}$

$\qquad\qquad = \dfrac{x+2}{x+1} \cdot \dfrac{x}{x^2-4} = \dfrac{x+2}{x+1} \cdot \dfrac{x}{(x+2)(x-2)}$

$\qquad\qquad = \dfrac{x}{(x+1)(x-2)}$

b. $\dfrac{x^{-1} + y^{-1}}{x^{-2} - y^{-2}} = \dfrac{\dfrac{1}{x} + \dfrac{1}{y}}{\dfrac{1}{x^2} - \dfrac{1}{y^2}} = \dfrac{\dfrac{y+x}{xy}}{\dfrac{y^2-x^2}{x^2y^2}}$ $x^{-n} = \dfrac{1}{x^n}$

$\qquad\qquad = \dfrac{y+x}{xy} \cdot \dfrac{x^2y^2}{y^2-x^2} = \dfrac{y+x}{xy} \cdot \dfrac{(xy)^2}{(y+x)(y-x)}$

$\qquad\qquad = \dfrac{xy}{y-x}$

EXAMPLE 5   Show that the algebraic expression

$$\dfrac{3 + \dfrac{8}{x} - \dfrac{4}{x^2}}{2 + \dfrac{4}{x} - \dfrac{5}{x^2}}$$

is equal to

$$\frac{3x^2 + 8x - 4}{2x^2 + 4x - 5}$$

(We will work with the first form of this algebraic expression in Chapter 11 when we evaluate limits.)

Solution   The LCD of both the numerator and denominator is x^2. Therefore,

$$\frac{3 + \dfrac{8}{x} - \dfrac{4}{x^2}}{2 + \dfrac{4}{x} - \dfrac{5}{x^2}} = \frac{\dfrac{3(x^2) + 8(x) - 4}{x^2}}{\dfrac{2(x^2) + 4(x) - 5}{x^2}}$$

$$= \frac{3x^2 + 8x - 4}{x^2} \cdot \frac{x^2}{2x^2 + 4x - 5}$$

$$= \frac{3x^2 + 8x - 4}{2x^2 + 4x - 5}$$

EXAMPLE 6   Perform the given operations and then simplify.

a. $\dfrac{x^2(2x^2 + 1)^{1/2}}{x - 1} \cdot \dfrac{4x^3 - 6x^2 + x - 2}{x(x - 1)(2x^2 + 1)}$ **b.** $\dfrac{12x^2}{\sqrt{2x^2 + 3}} + 6\sqrt{2x^2 + 3}$

Solution

a. $\dfrac{x^2(2x^2 + 1)^{1/2}}{x - 1} \cdot \dfrac{4x^3 - 6x^2 + x - 2}{x(x - 1)(2x^2 + 1)}$

$$= \frac{x(4x^3 - 6x^2 + x - 2)}{(x - 1)^2(2x^2 + 1)^{1 - 1/2}}$$

$$= \frac{x(4x^3 - 6x^2 + x - 2)}{(x - 1)^2(2x^2 + 1)^{1/2}}$$

b. We first rewrite the given expression without radicals and then proceed to simplify the resulting expression.

$$\frac{12x^2}{\sqrt{2x^2 + 3}} + 6\sqrt{2x^2 + 3} = \frac{12x^2}{(2x^2 + 3)^{1/2}} + 6(2x^2 + 3)^{1/2}$$

$$= \frac{12x^2 + 6(2x^2 + 3)^{1/2}(2x^2 + 3)^{1/2}}{(2x^2 + 3)^{1/2}}$$

$$= \frac{12x^2 + 6(2x^2 + 3)}{(2x^2 + 3)^{1/2}}$$

$$= \frac{24x^2 + 18}{(2x^2 + 3)^{1/2}} = \frac{6(4x^2 + 3)}{\sqrt{2x^2 + 3}}$$

Rationalizing Algebraic Fractions

When the denominator of an algebraic fraction contains sums or differences involving radicals, we may **rationalize the denominator**—that is, transform the fraction into an equivalent one with a denominator that does not contain radicals. In doing so, we make use of the fact that

$$(\sqrt{a} + \sqrt{b})(\sqrt{a} - \sqrt{b}) = (\sqrt{a})^2 - (\sqrt{b})^2$$
$$= a - b$$

This procedure is illustrated in Example 7.

EXAMPLE 7 Rationalize the denominator of $\dfrac{1}{1 + \sqrt{x}}$.

Solution Upon multiplying the numerator and the denominator by $(1 - \sqrt{x})$, we obtain

$$\frac{1}{1 + \sqrt{x}} = \frac{1}{1 + \sqrt{x}} \cdot \frac{1 - \sqrt{x}}{1 - \sqrt{x}}$$
$$= \frac{1 - \sqrt{x}}{1 - (\sqrt{x})^2}$$
$$= \frac{1 - \sqrt{x}}{1 - x}$$

In other situations, it may be necessary to **rationalize the numerator** of an algebraic expression. In calculus, for example, one encounters the following type of problem.

EXAMPLE 8 Rationalize the numerator of $\dfrac{\sqrt{1 + h} - 1}{h}$.

Solution

$$\frac{\sqrt{1 + h} - 1}{h} = \frac{\sqrt{1 + h} - 1}{h} \cdot \frac{\sqrt{1 + h} + 1}{\sqrt{1 + h} + 1}$$
$$= \frac{(\sqrt{1 + h})^2 - (1)^2}{h(\sqrt{1 + h} + 1)}$$
$$= \frac{1 + h - 1}{h(\sqrt{1 + h} + 1)} \qquad (\sqrt{1 + h})^2 = \sqrt{1 + h} \cdot \sqrt{1 + h} = 1 + h$$
$$= \frac{h}{h(\sqrt{1 + h} + 1)}$$
$$= \frac{1}{\sqrt{1 + h} + 1}$$

10.3 Exercises

In Exercises 1-10, simplify the given expression.

1. $\dfrac{x^2 + x - 2}{x^2 - 4}$

2. $\dfrac{2a^2 - 3ab - 9b^2}{2ab^2 + 3b^3}$

3. $\dfrac{12t^2 + 12t + 3}{4t^2 - 1}$

4. $\dfrac{x^3 + 2x^2 - 3x}{-2x^2 - x + 3}$

5. $\dfrac{(4x - 1)(3) - (3x + 1)(4)}{(4x - 1)^2}$

6. $\dfrac{(1 + x^2)^2(2) - 2x(2)(1 + x^2)(2x)}{(1 + x^2)^4}$

7. $\dfrac{(2x + 3)(1) - (x + 1)(2)}{(2x + 3)^2}$

8. $\dfrac{(x^2 - 1)^4(2x) - x^2(4)(x^2 - 1)^3(2x)}{(x^2 - 1)^8}$

9. $\dfrac{(e^x + 1)e^x - e^x(2)(e^x + 1)(e^x)}{(e^x + 1)^4}$

10. $\dfrac{(u^3 - 1)(3)(u^2 + u - 1)^2(2u + 1) - (u^2 + u - 1)^3(3u^2)}{(u^3 - 1)^2}$

29. $\dfrac{4x^2}{2\sqrt{2x^2 + 7}} + \sqrt{2x^2 + 7}$

30. $\dfrac{x^3}{\sqrt{8 - 3x^2}}(-6x) + \sqrt{8 - 3x^2}(6x^2)$

31. $\dfrac{2x(x + 1)^{-1/2} - (x + 1)^{1/2}}{x^2}$

32. $\dfrac{(x^2 + 1)^{1/2} - 2x^2(x^2 + 1)^{-1/2}}{1 - x^2}$

33. $\dfrac{(2x + 1)^{1/2} - (x + 2)(2x + 1)^{-1/2}}{2x + 1}$

34. $\dfrac{2(2x - 3)^{1/3} - (x - 1)(2x - 3)^{-2/3}}{(2x - 3)^{2/3}}$

In Exercises 11–34, perform the indicated operations and then simplify each expression.

11. $\dfrac{2a^2 - 2b^2}{b - a} \cdot \dfrac{4a + 4b}{a^2 + 2ab + b^2}$

12. $\dfrac{x^2 - 6x + 9}{x^2 - x - 6} \cdot \dfrac{3x + 6}{2x^2 - 7x + 3}$

13. $\dfrac{3x^2 + 2x - 1}{2x + 6} \div \dfrac{x^2 - 1}{x^2 + 2x - 3}$

14. $\dfrac{3x^2 - 4xy - 4y^2}{x^2y} \div \dfrac{(2y - x)^2}{x^3y}$

15. $\dfrac{58}{3(3t + 2)} + \dfrac{1}{3}$

16. $\dfrac{a + 1}{3a} + \dfrac{b - 2}{5b}$

17. $\dfrac{2x}{2x - 1} - \dfrac{3x}{2x + 5}$

18. $\dfrac{-xe^x}{x + 1} + e^x$

19. $\left(x + \dfrac{1}{x}\right)(x^2 - 1)$

20. $\left(2x - \dfrac{3}{x}\right)(x^3 + 1)$

21. $\dfrac{4}{x^2 - 9} - \dfrac{5}{x^2 - 6x + 9}$

22. $\dfrac{x}{1 - x} + \dfrac{2x + 3}{x^2 - 1}$

23. $2 + \dfrac{1}{a + 2} - \dfrac{2a}{a - 2}$

24. $x - \dfrac{x^2}{x + 2} + \dfrac{2}{x - 2}$

25. $\dfrac{1 + \dfrac{1}{x}}{1 - \dfrac{1}{x}}$

26. $\dfrac{\dfrac{1}{x} + \dfrac{1}{y}}{1 - \dfrac{1}{xy}}$

27. $\dfrac{x^{-2} - y^{-2}}{x + y}$

28. $\dfrac{x^{-3} - y^{-3}}{x^{-1} - y^{-1}}$

In Exercises 35–38, show that expressions (a) and (b) are equivalent.

35. a. $\dfrac{ax}{x + b}$ b. $\dfrac{a}{1 + \dfrac{b}{x}}$

36. a. $\dfrac{x + 3}{x^2 - x}$ b. $\dfrac{\dfrac{1}{x} + \dfrac{3}{x^2}}{1 - \dfrac{1}{x}}$

37. a. $\dfrac{x^4 + 1}{x^3 - 1}$ b. $\dfrac{x + \dfrac{1}{x^3}}{1 - \dfrac{1}{x^3}}$

38. a. $\sqrt{\dfrac{4x^2 + 2x - 1}{x^2 + 2}}$ b. $\sqrt{\dfrac{4 + \dfrac{2}{x} - \dfrac{1}{x^2}}{1 + \dfrac{2}{x^2}}}$

In Exercises 39–44, rationalize the denominator of each expression.

39. $\dfrac{1}{\sqrt{3} - 1}$

40. $\dfrac{1}{\sqrt{x} + 5}$

41. $\dfrac{1}{\sqrt{x} - \sqrt{y}}$

42. $\dfrac{a}{1 - \sqrt{a}}$

43. $\dfrac{\sqrt{a} + \sqrt{b}}{\sqrt{a} - \sqrt{b}}$

44. $\dfrac{2\sqrt{a} + \sqrt{b}}{2\sqrt{a} - \sqrt{b}}$

In Exercises 45–50, rationalize the numerator of each expression.

45. $\dfrac{\sqrt{x}}{3}$

46. $\dfrac{\sqrt[3]{y}}{x}$

47. $\dfrac{1 - \sqrt{3}}{3}$

48. $\dfrac{\sqrt{x} - 1}{x}$

49. $\dfrac{1 + \sqrt{x + 2}}{\sqrt{x + 2}}$

50. $\dfrac{\sqrt{x + 3} - \sqrt{x}}{3}$

10.4 Inequalities and Absolute Value

▬ Intervals

The system of real numbers and some of its properties were mentioned briefly in Section 1.1. In much of our later work, we restrict our attention to certain subsets of the set of real numbers. For example, if x denotes the number of cars rolling off an assembly line each day in an automobile assembly plant, then x must be nonnegative —that is, $x \geq 0$. Taking this example one step further, suppose management decides that the daily production must not exceed 200 cars. Then x must satisfy the inequality $0 \leq x \leq 200$.

More generally, we will be interested in the following subsets of real numbers: open intervals, closed intervals, and half-open intervals. The set of all real numbers that lie *strictly* between two fixed numbers a and b is called an **open interval** (a, b). It consists of all real numbers x that satisfy the inequalities $a < x < b$, and it is called "open" because neither of its endpoints is included in the interval. A **closed interval** contains *both* of its endpoints. Thus, the set of all real numbers x that satisfy the inequalities $a \leq x \leq b$ is the closed interval $[a, b]$. Notice that brackets are used to indicate that the endpoints are included in this interval. **Half-open intervals** contain only *one* of their endpoints. Thus, the interval $[a, b)$ is the set of all real numbers x that satisfy $a \leq x < b$, whereas the interval $(a, b]$ is described by the inequalities $a < x \leq b$. Examples of these **finite intervals** are illustrated in Table 7.

In addition to finite intervals, we will encounter **infinite intervals**. Examples of infinite intervals are the half-lines (a, ∞), $[a, \infty)$, $(-\infty, a)$, and $(-\infty, a]$ defined by the set of all real numbers that satisfy $x > a$, $x \geq a$, $x < a$, and $x \leq a$, respectively. The symbol ∞, called *infinity*, is not a real number. It is used here only for notational purposes in conjunction with the definition of infinite intervals. The notation

TABLE 7

Finite Intervals

Interval		Graph	Example	
Open	(a, b)		$(-2, 1)$	
Closed	$[a, b]$		$[-1, 2]$	
Half-open	$(a, b]$		$\left(\frac{1}{2}, 3\right]$	
Half-open	$[a, b)$		$\left[-\frac{1}{2}, 3\right)$	

TABLE 8

Infinite Intervals

Interval	Graph	Example	
(a, ∞)		$(2, \infty)$	
$[a, \infty)$		$[-1, \infty)$	
$(-\infty, a)$		$(-\infty, 1)$	
$(-\infty, a]$		$(-\infty, -\frac{1}{2}]$	

$(-\infty, \infty)$ is used for the set of all real numbers x since, by definition, the inequalities $-\infty < x < \infty$ hold for any real number x. Infinite intervals are illustrated in Table 8.

■ Properties of Inequalities

In practical applications, intervals are often found by solving one or more inequalities involving a variable. In such situations, the following properties may be used to advantage.

Properties of Inequalities

If a, b, and c are any real numbers, then

		Example
Property 1	If $a < b$ and $b < c$, then $a < c$.	$2 < 3$ and $3 < 8$, so $2 < 8$.
Property 2	If $a < b$, then $a + c < b + c$.	$-5 < -3$, so $-5 + 2 < -3 + 2$; that is, $-3 < -1$.
Property 3	If $a < b$ and $c > 0$, then $ac < bc$.	$-5 < -3$, and since $2 > 0$ we have $(-5)(2) < (-3)(2)$; that is, $-10 < -6$.
Property 4	If $a < b$ and $c < 0$, then $ac > bc$.	$-2 < 4$, and since $-3 < 0$ we have $(-2)(-3) > (4)(-3)$; that is, $6 > -12$.

Similar properties hold if each inequality sign, $<$, between a and b is replaced by $\geq$, $>$, or $\leq$.

A real number is a *solution of an inequality* involving a variable if a true statement is obtained when the variable is replaced by that number. The set of all real numbers satisfying the inequality is called the *solution set*.

EXAMPLE 1   Find the set of real numbers that satisfy $-1 \leq 2x - 5 < 7$.

Solution   Add 5 to each member of the given double inequality, obtaining

$$4 \leq 2x < 12$$

Next, multiply each member of the resulting double inequality by $\frac{1}{2}$, yielding

$$2 \leq x < 6$$

Thus, the solution is the set of all values of x lying in the interval $[2, 6)$. ■

EXAMPLE 2   Solve the inequality $x^2 + 2x - 8 < 0$.

Solution   Observe that $x^2 + 2x - 8 = (x + 4)(x - 2)$, so the given inequality is equivalent to the inequality $(x + 4)(x - 2) < 0$. Since the product of two real numbers is negative if and only if the two numbers have opposite signs, we solve the inequality $(x + 4)(x - 2) < 0$ by studying the signs of the two factors $x + 4$ and $x - 2$. Now, $x + 4 > 0$ when $x > -4$, and $x + 4 < 0$ when $x < -4$. Similarly, $x - 2 > 0$ when $x > 2$, and $x - 2 < 0$ when $x < 2$. These results are summarized graphically in Figure 1.

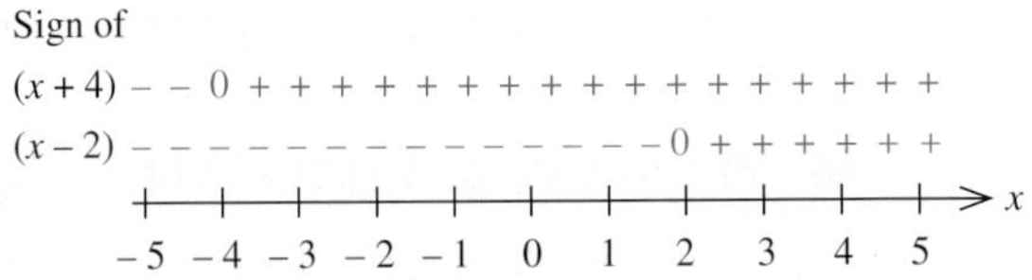

FIGURE 1
Sign diagram for $(x + 4)(x - 2)$

Sign of
$(x+4)$ $-$ $-$ 0 $+$ $+$ $+$ $+$ $+$ $+$ $+$ $+$ $+$ $+$ $+$ $+$ $+$ $+$ $+$
$(x-2)$ $-$ $-$ $-$ $-$ $-$ $-$ $-$ $-$ $-$ $-$ $-$ $-$ $-$ 0 $+$ $+$ $+$ $+$ $+$ $+$

-5 -4 -3 -2 -1 0 1 2 3 4 5

From Figure 1 we see that the two factors $x + 4$ and $x - 2$ have opposite signs when and only when x lies strictly between -4 and 2. Therefore, the required solution is the interval $(-4, 2)$. ■

EXAMPLE 3   Solve the inequality $\dfrac{x + 1}{x - 1} \geq 0$.

Solution   The quotient $(x + 1)/(x - 1)$ is strictly positive if and only if both the numerator and the denominator have the same sign. The signs of $x + 1$ and $x - 1$ are shown in Figure 2.

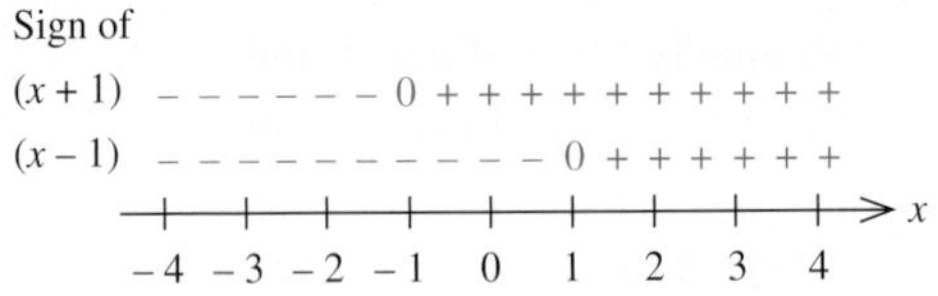

FIGURE 2
Sign diagram for $\dfrac{x + 1}{x - 1}$

Sign of
$(x+1)$ $-$ $-$ $-$ $-$ $-$ $-$ 0 $+$ $+$ $+$ $+$ $+$ $+$ $+$ $+$
$(x-1)$ $-$ $-$ $-$ $-$ $-$ $-$ $-$ $-$ $-$ 0 $+$ $+$ $+$ $+$ $+$ $+$

-4 -3 -2 -1 0 1 2 3 4

From Figure 2 we see that $x + 1$ and $x - 1$ have the same sign when and only when $x < -1$ or $x > 1$. The quotient $(x + 1)/(x - 1)$ is equal to zero when and only when $x = -1$. Therefore, the required solution is the set of all x in the intervals $(-\infty, -1]$ and $(1, \infty)$. ■

APPLIED EXAMPLE 4 Stock Purchase   The management of Corbyco, a giant conglomerate, has estimated that x thousand dollars is needed to purchase

$$100{,}000(-1 + \sqrt{1 + 0.001x})$$

shares of common stock of Starr Communications. Determine how much money Corbyco needs to purchase at least 100,000 shares of Starr's stock.

Solution   The amount of cash Corbyco needs to purchase at least 100,000 shares is found by solving the inequality

$$100{,}000(-1 + \sqrt{1 + 0.001x}) \geq 100{,}000$$

Proceeding, we find

$$-1 + \sqrt{1 + 0.001x} \geq 1$$
$$\sqrt{1 + 0.001x} \geq 2$$
$$1 + 0.001x \geq 4 \qquad \text{Square both sides.}$$
$$0.001x \geq 3$$
$$x \geq 3000$$

so Corbyco needs at least $3,000,000. ∎

Absolute Value

> **Absolute Value**
> The **absolute value** of a number a is denoted by $|a|$ and is defined by
> $$|a| = \begin{cases} a & \text{if } a \geq 0 \\ -a & \text{if } a < 0 \end{cases}$$

Since $-a$ is a positive number when a is negative, it follows that the absolute value of a number is always nonnegative. For example, $|5| = 5$ and $|-5| = -(-5) = 5$. Geometrically, $|a|$ is the distance between the origin and the point on the number line that represents the number a (Figure 3).

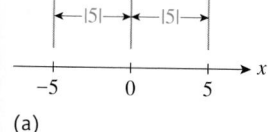

(a)

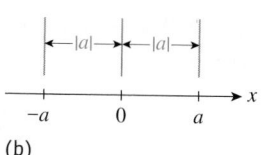

(b)

FIGURE 3
The absolute value of a number

> **Absolute Value Properties**
> If a and b are any real numbers, then
>
> | | | **Example** |
> |---|---|---|
> | **Property 5** | $\|-a\| = \|a\|$ | $\|-3\| = -(-3) = 3 = \|3\|$ |
> | **Property 6** | $\|ab\| = \|a\|\|b\|$ | $\|(2)(-3)\| = \|-6\| = 6 = (2)(3)$ |
> | | | $= \|2\|\|-3\|$ |
> | **Property 7** | $\left\|\dfrac{a}{b}\right\| = \dfrac{\|a\|}{\|b\|} \quad (b \neq 0)$ | $\left\|\dfrac{(-3)}{(-4)}\right\| = \dfrac{\|3\|}{\|4\|} = \dfrac{3}{4} = \dfrac{\|-3\|}{\|-4\|}$ |
> | **Property 8** | $\|a + b\| \leq \|a\| + \|b\|$ | $\|8 + (-5)\| = \|3\| = 3$ |
> | | | $\leq \|8\| + \|-5\|$ |
> | | | $= 13$ |

Property 8 is called the **triangle inequality**.

EXAMPLE 5   Evaluate each expression.

a. $|\pi - 5| + 3$ **b.** $|\sqrt{3} - 2| + |2 - \sqrt{3}|$

Solution

a. Since $\pi - 5 < 0$, we see that $|\pi - 5| = -(\pi - 5)$. Therefore,

$$|\pi - 5| + 3 = -(\pi - 5) + 3 = 8 - \pi$$

b. Since $\sqrt{3} - 2 < 0$, we see that $|\sqrt{3} - 2| = -(\sqrt{3} - 2)$. Next, observe that $2 - \sqrt{3} > 0$, so $|2 - \sqrt{3}| = 2 - \sqrt{3}$. Therefore,

$$|\sqrt{3} - 2| + |2 - \sqrt{3}| = -(\sqrt{3} - 2) + (2 - \sqrt{3})$$
$$= 4 - 2\sqrt{3} = 2(2 - \sqrt{3}) \qquad ■$$

EXAMPLE 6   Solve the inequalities $|x| \leq 5$ and $|x| \geq 5$.

Solution   First, we consider the inequality $|x| \leq 5$. If $x \geq 0$ then $|x| = x$, so $|x| \leq 5$ implies $x \leq 5$ in this case. On the other hand, if $x < 0$ then $|x| = -x$, so $|x| \leq 5$ implies $-x \leq 5$ or $x \geq -5$. Thus, $|x| \leq 5$ means $-5 \leq x \leq 5$ (Figure 4a). To obtain an alternative solution, observe that $|x|$ is the distance from the point x to zero, so the inequality $|x| \leq 5$ implies immediately that $-5 \leq x \leq 5$.

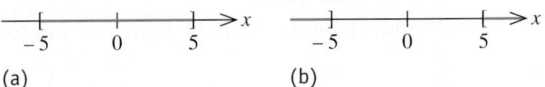

FIGURE 4   (a)   (b)

Next, the inequality $|x| \geq 5$ states that the distance from x to zero is greater than or equal to 5. This observation yields the result $x \geq 5$ or $x \leq -5$ (Figure 4b).   ■

EXAMPLE 7   Solve the inequality $|2x - 3| \leq 1$.

Solution   The inequality $|2x - 3| \leq 1$ is equivalent to the inequalities $-1 \leq 2x - 3 \leq 1$ (see Example 6). Thus, $2 \leq 2x \leq 4$ and $1 \leq x \leq 2$. The solution is therefore given by the set of all x in the interval $[1, 2]$ (Figure 5).

FIGURE 5
$|2x - 3| \leq 1$

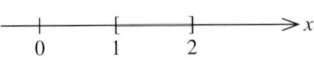

10.4   Exercises

In Exercises 1–4, determine whether the statement is true or false.

1. $-3 < -20$

2. $-5 \leq -5$

3. $\dfrac{2}{3} > \dfrac{5}{6}$

4. $-\dfrac{5}{6} < -\dfrac{11}{12}$

In Exercises 5–10, show the given interval on a number line.

5. $(3, 6)$

6. $(-2, 5]$

7. $[-1, 4)$

8. $\left[-\dfrac{6}{5}, -\dfrac{1}{2}\right]$

9. $(0, \infty)$

10. $(-\infty, 5]$

In Exercises 11–28, find the values of x that satisfy the inequality (inequalities).

11. $2x + 4 < 8$

12. $-6 > 4 + 5x$

13. $-4x \geq 20$

14. $-12 \leq -3x$

15. $-6 < x - 2 < 4$

16. $0 \leq x + 1 \leq 4$

17. $x + 1 > 4$ or $x + 2 < -1$

18. $x + 1 > 2$ or $x - 1 < -2$

19. $x + 3 > 1$ and $x - 2 < 1$

20. $x - 4 \le 1$ and $x + 3 > 2$

21. $(x + 3)(x - 5) \le 0$ **22.** $(2x - 4)(x + 2) \ge 0$

23. $(2x - 3)(x - 1) \ge 0$ **24.** $(3x - 4)(2x + 2) \le 0$

25. $\dfrac{x + 3}{x - 2} \ge 0$ **26.** $\dfrac{2x - 3}{x + 1} \ge 4$

27. $\dfrac{x - 2}{x - 1} \le 2$ **28.** $\dfrac{2x - 1}{x + 2} \le 4$

In Exercises 29–38, evaluate the given expression.

29. $|-6 + 2|$ **30.** $4 + |-4|$

31. $\dfrac{|-12 + 4|}{|16 - 12|}$ **32.** $\left|\dfrac{0.2 - 1.4}{1.6 - 2.4}\right|$

33. $\sqrt{3}|-2| + 3|-\sqrt{3}|$ **34.** $|-1| + \sqrt{2}|-2|$

35. $|\pi - 1| + 2$ **36.** $|\pi - 6| - 3$

37. $|\sqrt{2} - 1| + |3 - \sqrt{2}|$

38. $|2\sqrt{3} - 3| - |\sqrt{3} - 4|$

In Exercises 39–44, suppose that a and b are real numbers other than zero and that $a > b$. State whether the inequality is true or false.

39. $b - a > 0$ **40.** $\dfrac{a}{b} > 1$

41. $a^2 > b^2$ **42.** $\dfrac{1}{a} > \dfrac{1}{b}$

43. $a^3 > b^3$ **44.** $-a < -b$

In Exercises 45–50, determine whether the statement is true for all real numbers a and b.

45. $|-a| = a$ **46.** $|b^2| = b^2$

47. $|a - 4| = |4 - a|$ **48.** $|a + 1| = |a| + 1$

49. $|a + b| = |a| + |b|$ **50.** $|a - b| = |a| - |b|$

51. DRIVING RANGE OF A CAR An advertisement for a certain car states that the EPA fuel economy is 20 mpg city and 27 mpg highway and that the car's fuel-tank capacity is 18.1 gal. Assuming ideal driving conditions, determine the driving range for the car from these data.

52. Find the minimum cost C (in dollars), given that

$$5(C - 25) \ge 1.75 + 2.5C$$

53. Find the maximum profit P (in dollars) given that

$$6(P - 2500) \le 4(P + 2400)$$

54. CELSIUS AND FAHRENHEIT TEMPERATURES The relationship between Celsius (°C) and Fahrenheit (°F) temperatures is given by the formula

$$C = \frac{5}{9}(F - 32)$$

  a. If the temperature range for Montreal during the month of January is $-15° < °C < -5°$, find the range in degrees Fahrenheit in Montreal for the same period.
  b. If the temperature range for New York City during the month of June is $63° < °F < 80°$, find the range in degrees Celsius in New York City for the same period.

55. MEETING SALES TARGETS A salesman's monthly commission is 15% on all sales over $24,000. If his goal is to make a commission of at least $6000 per month, what minimum monthly sales figures must he attain?

56. MARKUP ON A CAR The markup on a used car was at least 30% of its current wholesale price. If the car was sold for $5600, what was the maximum wholesale price?

57. QUALITY CONTROL PAR Manufacturing Company manufactures steel rods. Suppose the rods ordered by a customer are manufactured to a specification of 0.5 in. and are acceptable only if they are within the tolerance limits of 0.49 in. and 0.51 in. Letting x denote the diameter of a rod, write an inequality using absolute value to express a criterion involving x that must be satisfied in order for a rod to be acceptable.

58. QUALITY CONTROL The diameter x (in inches) of a batch of ball bearings manufactured by PAR Manufacturing Company satisfies the inequality

$$|x - 0.1| \le 0.01$$

What is the smallest diameter a ball bearing in the batch can have? The largest diameter?

59. MEETING PROFIT GOALS A manufacturer of a certain commodity has estimated that her profit in thousands of dollars is given by the expression

$$-6x^2 + 30x - 10$$

where x (in thousands) is the number of units produced. What production range will enable the manufacturer to realize a profit of at least $14,000 on the commodity?

60. DISTRIBUTION OF INCOMES The distribution of income in a certain city can be described by the mathematical model $y = (2.8 \cdot 10^{11})(x)^{-1.5}$, where y is the number of families with an income of x or more dollars.
  a. How many families in this city have an income of $20,000 or more?
  b. How many families have an income of $40,000 or more?
  c. How many families have an income of $100,000 or more?

CHAPTER 10 Summary of Principal Formulas and Terms

FORMULAS

1. Product formulas	$(a + b)^2 = a^2 + 2ab + b^2$ $(a - b)^2 = a^2 - 2ab + b^2$ $(a + b)(a - b) = a^2 - b^2$
2. Quadratic formula	$x = \dfrac{-b \pm \sqrt{b^2 - 4ac}}{2a}$

TERMS

roots of a polynomial equation (558)   half-open interval (568)   absolute value (571)

open interval (568)   finite interval (568)   triangle inequality (571)

closed interval (568)   infinite interval (568)

CHAPTER 10 Review Exercises

In Exercises 1–6, evaluate the given expression.

1. $\left(\dfrac{9}{4}\right)^{3/2}$

2. $\dfrac{5^6}{5^4}$

3. $(3 \cdot 4)^{-2}$

4. $(-8)^{5/3}$

5. $\dfrac{(3 \cdot 2^{-3})(4 \cdot 3^5)}{2 \cdot 9^3}$

6. $\dfrac{3\sqrt[3]{54}}{\sqrt[3]{18}}$

In Exercises 7–14, simplify the expression.

7. $\dfrac{4(x^2 + y)^3}{x^2 + y}$

8. $\dfrac{a^6 b^{-5}}{(a^3 b^{-2})^{-3}}$

9. $\dfrac{\sqrt[4]{16x^5 yz}}{\sqrt[4]{81xyz^5}}$

10. $(2x^3)(-3x^{-2})\left(\dfrac{1}{6}x^{-1/2}\right)$

11. $\left(\dfrac{3xy^2}{4x^3 y}\right)^{-2}\left(\dfrac{3xy^3}{2x^2}\right)^3$

12. $(-3a^2 b^3)^2 (2a^{-1}b^{-2})^{-1}$

13. $\sqrt[3]{81x^5 y^{10}} \ \sqrt[3]{9xy^2}$

14. $\left(\dfrac{-x^{1/2}y^{2/3}}{x^{1/3}y^{3/4}}\right)^6$

In Exercises 15–20, factor the expression.

15. $-2\pi^2 r^3 + 100\pi r^2$

16. $2v^3 w + 2vw^3 + 2u^2 vw$

17. $16 - x^2$

18. $12t^3 - 6t^2 - 18t$

19. $-2x^2 - 4x + 6$

20. $12x^2 - 92x + 120$

In Exercises 21–24, perform the indicated operations and then simplify the expression.

21. $\dfrac{(t + 6)(60) - (60t + 180)}{(t + 6)^2}$

22. $\dfrac{6x}{2(3x^2 + 2)} + \dfrac{1}{4(x + 2)}$

23. $\dfrac{2}{3}\left(\dfrac{4x}{2x^2 - 1}\right) + 3\left(\dfrac{3}{3x - 1}\right)$

24. $\dfrac{-2x}{\sqrt{x + 1}} + 4\sqrt{x + 1}$

In Exercises 25–28, solve the equation by factoring.

25. $8x^2 + 2x - 3 = 0$

26. $-6x^2 - 10x + 4 = 0$

27. $-x^3 - 2x^2 + 3x = 0$

28. $2x^4 + x^2 = 1$

In Exercises 29–32, find the values of x that satisfy the given inequalities.

29. $-x + 3 \le 2x + 9$

30. $-2 \le 3x + 1 \le 7$

31. $x - 3 > 2$ or $x + 3 < -1$

32. $2x^2 > 50$

In Exercises 33–36, evaluate each expression.

33. $|-5 + 7| + |-2|$ **34.** $\left| \dfrac{5 - 12}{-4 - 3} \right|$

35. $|2\pi - 6| - \pi$ **36.** $|\sqrt{3} - 4| + |4 - 2\sqrt{3}|$

In Exercises 37–40, find the value(s) of x that satisfy the expression.

37. $2x^2 + 3x - 2 \le 0$ **38.** $\dfrac{1}{x + 2} > 2$

39. $|2x - 3| < 5$ **40.** $\left| \dfrac{x + 1}{x - 1} \right| = 5$

41. Rationalize the numerator:

$$\frac{\sqrt{x} - 1}{x - 1}$$

42. Rationalize the denominator:

$$\frac{\sqrt{x} - 1}{2\sqrt{x}}$$

In Exercises 43 and 44, use the quadratic formula to solve the given quadratic equation.

43. $x^2 - 2x - 5 = 0$ **44.** $2x^2 + 8x + 7 = 0$

45. Find the minimum cost C (in dollars), given that

$$2(1.5C + 80) \le 2(2.5C - 20)$$

46. Find the maximum revenue R (in dollars), given that

$$12(2R - 320) \le 4(3R + 240)$$

CHAPTER 10   **Before Moving On . . .**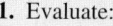

1. Evaluate:
   a. $|\pi - 2\sqrt{3}| - |\sqrt{3} - \sqrt{2}|$

   b. $\left[\left(-\dfrac{1}{3} \right)^{-3} \right]^{1/3}$

2. Simplify:
   a. $\sqrt[3]{64x^6} \cdot \sqrt{9y^2x^6}$ **b.** $\left(\dfrac{a^{-3}}{b^{-4}} \right)^2 \left(\dfrac{b}{a} \right)^{-3}$

3. Rationalize the denominator:
   a. $\dfrac{2x}{3\sqrt{y}}$ **b.** $\dfrac{x}{\sqrt{x} - 4}$

4. Perform each operation and simplify:
   a. $\dfrac{(x^2 + 1)(\frac{1}{2}x^{-1/2}) - x^{1/2}(2x)}{(x^2 + 1)^2}$

   b. $-\dfrac{3x}{\sqrt{x + 2}} + 3\sqrt{x + 2}$

5. Rationalize the numerator: $\dfrac{\sqrt{x} + \sqrt{y}}{\sqrt{x} - \sqrt{y}}$.

6. Factor:
   a. $12x^3 - 10x^2 - 12x$ **b.** $2bx - 2by + 3cx - 3cy$

7. Solve each equation:
   a. $12x^2 - 9x - 3 = 0$ **b.** $3x^2 - 5x + 1 = 0$

8. Find the values of x that satisfy $(3x + 2)(2x - 3) \le 0$.

11 Functions, Limits, and the Derivative

How does the change in the demand for a certain make of tires affect the unit price of the tires? The management of the Titan Tire Company has determined the demand function that relates the unit price of its Super Titan tires to the quantity demanded. In Example 7, page 667–668, you will see how this function can be used to compute the rate of change of the unit price of the Super Titan tires with respect to the quantity demanded.

© Chris Collins/Corbis

IN THIS CHAPTER, we review the concept of a *function* and begin the study of differential calculus. Historically, differential calculus was developed in response to the problem of finding the tangent line to an arbitrary curve. But it quickly became apparent that solving this problem provided mathematicians with a method for solving many practical problems involving the rate of change of one quantity with respect to another. The basic tool used in differential calculus is the *derivative* of a function. The concept of the derivative is based, in turn, on a more fundamental notion—that of the *limit* of a function.

11.1 Functions and Their Graphs

■ Functions

The notion of a *function* was introduced in Section 1.3, where we were concerned primarily with the special class of functions known as linear functions. In the study of calculus, we will be dealing with a more general class of functions called *nonlinear functions*. First, however, let's restate the definition of a function.

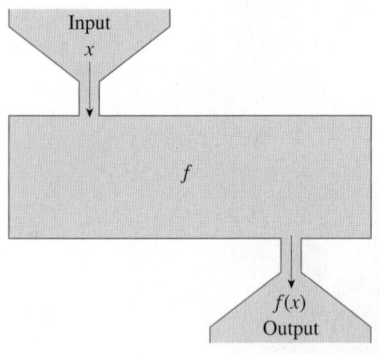

FIGURE 1
A function machine

> **Function**
> A **function** is a rule that assigns to each element in a set A one and only one element in a set B.

The set A is called the **domain** of the function. It is customary to denote a function by a letter of the alphabet, such as the letter f. If x is an element in the domain of a function f, then the element in B that f associates with x is written $f(x)$ (read "f of x") and is called the value of f at x. The set comprising all the values assumed by $y = f(x)$ as x takes on all possible values in its domain is called the **range** of the function f.

We can think of a function f as a machine. The domain is the set of inputs (raw material) for the machine, the rule describes how the input is to be processed, and the value(s) of the function are the outputs of the machine (Figure 1).

We can also think of a function f as a mapping in which an element x in the domain of f is mapped onto a unique element $f(x)$ in B (Figure 2).

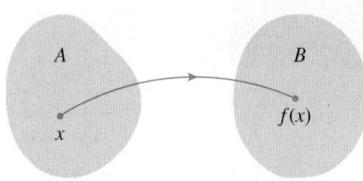

FIGURE 2
The function f viewed as a mapping

Notes

1. The output $f(x)$ associated with an input x is unique. To appreciate the importance of this uniqueness property, consider a rule that associates with each item x in a department store its selling price y. Then, each x must correspond to *one and only one y*. Notice, however, that different x's may be associated with the same y. In the context of the present example, this says that different items may have the same price.
2. Although the sets A and B that appear in the definition of a function may be quite arbitrary, in this book they will denote sets of real numbers.   ■

In general, to evaluate a function at a specific value of x, we replace x with that value, as illustrated in Examples 1 and 2.

EXAMPLE 1   Let the function f be defined by the rule $f(x) = 2x^2 - x + 1$. Find:

a. $f(1)$ **b.** $f(-2)$ **c.** $f(a)$ **d.** $f(a + h)$

Solution

a. $f(1) = 2(1)^2 - (1) + 1 = 2 - 1 + 1 = 2$
b. $f(-2) = 2(-2)^2 - (-2) + 1 = 8 + 2 + 1 = 11$
c. $f(a) = 2(a)^2 - (a) + 1 = 2a^2 - a + 1$
d. $f(a + h) = 2(a + h)^2 - (a + h) + 1 = 2a^2 + 4ah + 2h^2 - a - h + 1$ ▪

APPLIED EXAMPLE 2 Profit Functions ThermoMaster manufactures an indoor–outdoor thermometer at its Mexican subsidiary. Management esti-mates that the profit (in dollars) realizable by ThermoMaster in the manufacture and sale of x thermometers per week is

$$P(x) = -0.001x^2 + 8x - 5000$$

Find ThermoMaster's weekly profit if its level of production is (a) 1000 ther-mometers per week and (b) 2000 thermometers per week.

Solution

a. The weekly profit when the level of production is 1000 units per week is found by evaluating the profit function P at $x = 1000$. Thus,

$$P(1000) = -0.001(1000)^2 + 8(1000) - 5000 = 2000$$

or $2000.
b. When the level of production is 2000 units per week, the weekly profit is given by

$$P(2000) = -0.001(2000)^2 + 8(2000) - 5000 = 7000$$

or $7000. ▪

■ Determining the Domain of a Function

Suppose we are given the function $y = f(x)$.* Then, the variable x is called the **inde-pendent variable**. The variable y, whose value depends on x, is called the **dependent variable**.

To determine the domain of a function, we need to find what restrictions, if any, are to be placed on the independent variable x. In many practical applications, the domain of a function is dictated by the nature of the problem, as illustrated in Example 3.

APPLIED EXAMPLE 3 Packaging An open box is to be made from a rectangular piece of cardboard 16 inches long and 10 inches wide by cutting away identical squares (x inches by x inches) from each corner and folding up the

---

*It is customary to refer to a function f as $f(x)$ or by the equation $y = f(x)$ defining it.

resulting flaps (Figure 3). Find an expression that gives the volume V of the box as a function of x. What is the domain of the function?

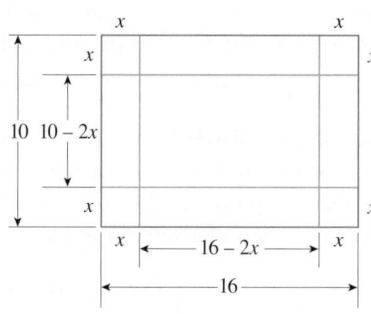 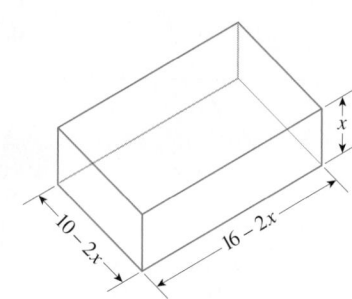

FIGURE 3

(a) The box is constructed by cutting x-by-x-inch squares from each corner.

(b) The dimensions of the resulting box are $(10 - 2x)''$ by $(16 - 2x)''$ by x''.

Solution   The length of the box is $(16 - 2x)$ inches, the width is $(10 - 2x)$ inches, and the height is x inches, so the volume of the box (in cubic inches) is given by

$$V = f(x) = (16 - 2x)(10 - 2x)x \qquad \text{Length} \cdot \text{width} \cdot \text{height}$$
$$= (160 - 52x + 4x^2)x$$
$$= 4x^3 - 52x^2 + 160x$$

Since the length of each side of the box must be greater than or equal to zero, we see that

$$16 - 2x \geq 0 \qquad 10 - 2x \geq 0 \qquad x \geq 0$$

simultaneously; that is,

$$x \leq 8 \qquad x \leq 5 \qquad x \geq 0$$

All three inequalities are satisfied simultaneously provided that $0 \leq x \leq 5$. Thus, the domain of the function f is the interval $[0, 5]$.   ∎

In general, if a function is defined by a rule relating x to $f(x)$ without specific mention of its domain, it is understood that the domain will consist of all values of x for which $f(x)$ is a real number. In this connection, you should keep in mind that (1) division by zero is not permitted and (2) the square root of a negative number is not defined.

EXAMPLE 4   Find the domain of each function.

a. $f(x) = \sqrt{x - 1}$ **b.** $f(x) = \dfrac{1}{x^2 - 4}$ **c.** $f(x) = x^2 + 3$

Solution

a. Since the square root of a negative number is undefined, it is necessary that $x - 1 \geq 0$. The inequality is satisfied by the set of real numbers $x \geq 1$. Thus, the domain of f is the interval $[1, \infty)$.

b. The only restriction on x is that $x^2 - 4$ be different from zero since division by zero is not allowed. But $(x^2 - 4) = (x + 2)(x - 2) = 0$ if $x = -2$ or $x = 2$. Thus, the domain of f in this case consists of the intervals $(-\infty, -2)$, $(-2, 2)$, and $(2, \infty)$.

c. Here, any real number satisfies the equation, so the domain of f is the set of all real numbers. ∎

Graphs of Functions

If f is a function with domain A, then corresponding to each real number x in A there is precisely one real number $f(x)$. We can also express this fact by using **ordered pairs** of real numbers. Write each number x in A as the first member of an ordered pair and each number $f(x)$ corresponding to x as the second member of the ordered pair. This gives exactly one ordered pair $(x, f(x))$ for each x in A. This observation leads to an **alternative definition of a function f**:

Function (Alternative Definition)
A function f with domain A is the set of all ordered pairs $(x, f(x))$ where x belongs to A.

Observe that the condition that there be one and only one number $f(x)$ corresponding to each number x in A translates into the requirement that *no two ordered pairs have the same first number.*

Since ordered pairs of real numbers correspond to points in the plane, we have found a way to exhibit a function graphically.

Graph of a Function of One Variable
The **graph of a function f** is the set of all points (x, y) in the xy-plane such that x is in the domain of f and $y = f(x)$.

Figure 4 shows the graph of a function f. Observe that the y-coordinate of the point (x, y) on the graph of f gives the height of that point (the distance above the x-axis), if $f(x)$ is positive. If $f(x)$ is negative, then $-f(x)$ gives the depth of the point (x, y) (the distance below the x-axis). Also, observe that the domain of f is a set of real numbers lying on the x-axis, whereas the range of f lies on the y-axis.

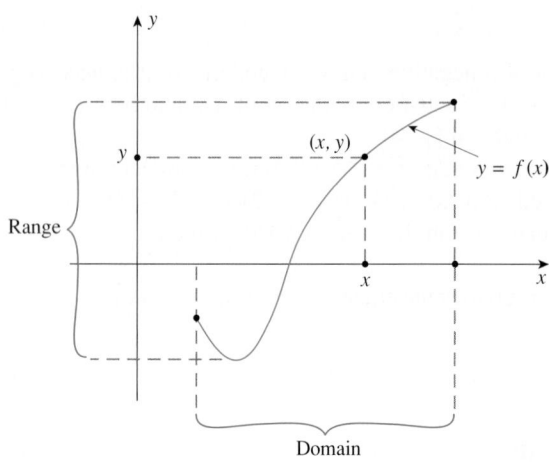

FIGURE 4
The graph of f

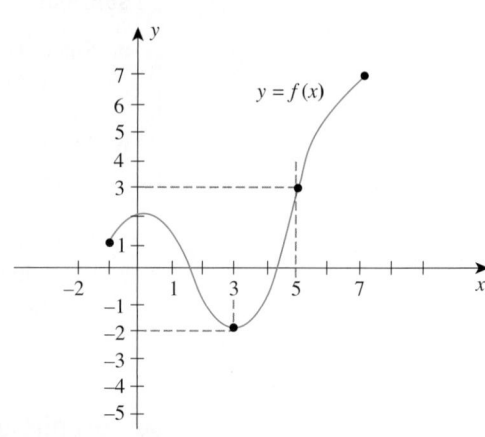

FIGURE 5

EXAMPLE 5   The graph of a function f is shown in Figure 5.

a. What is the value of $f(3)$? The value of $f(5)$?
b. What is the height or depth of the point $(3, f(3))$ from the x-axis? The point $(5, f(5))$ from the x-axis?
c. What is the domain of f? The range of f?

Solution

a. From the graph of f, we see that $y = -2$ when $x = 3$ and conclude that $f(3) = -2$. Similarly, we see that $f(5) = 3$.
b. Since the point $(3, -2)$ lies below the x-axis, we see that the depth of the point $(3, f(3))$ is $-f(3) = -(-2) = 2$ units below the x-axis. The point $(5, f(5))$ lies above the x-axis and is located at a height of $f(5)$, or 3 units above the x-axis.
c. Observe that x may take on all values between $x = -1$ and $x = 7$, inclusive, and so the domain of f is $[-1, 7]$. Next, observe that as x takes on all values in the domain of f, $f(x)$ takes on all values between -2 and 7, inclusive. (You can easily see this by running your index finger along the x-axis from $x = -1$ to $x = 7$ and observing the corresponding values assumed by the y-coordinate of each point of the graph of f.) Therefore, the range of f is $[-2, 7]$.   ◼

Much information about the graph of a function can be gained by plotting a few points on its graph. Later on we will develop more systematic and sophisticated techniques for graphing functions.

EXAMPLE 6   Sketch the graph of the function defined by the equation $y = x^2 + 1$. What is the range of f?

Solution   The domain of the function is the set of all real numbers. By assigning several values to the variable x and computing the corresponding values for y, we obtain the following solutions to the equation $y = x^2 + 1$:

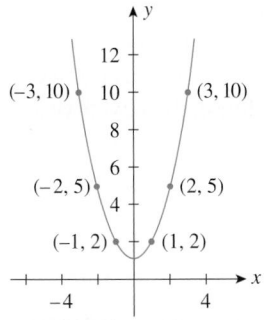

FIGURE 6
The graph of $y = x^2 + 1$ is a parabola.

x	-3	-2	-1	0	1	2	3
y	10	5	2	1	2	5	10

By plotting these points and then connecting them with a smooth curve, we obtain the graph of $y = f(x)$, which is a parabola (Figure 6). To determine the range of f, we observe that $x^2 \geq 0$, if x is any real number and so $x^2 + 1 \geq 1$ for all real numbers x. We conclude that the range of f is $[1, \infty)$. The graph of f confirms this result visually.   ■

EXPLORING WITH TECHNOLOGY

Let $f(x) = x^2$.

1. Plot the graphs of $F(x) = x^2 + c$ on the same set of axes for $c = -2, -1, -\frac{1}{2}, 0, \frac{1}{2}, 1, 2$.

2. Plot the graphs of $G(x) = (x + c)^2$ on the same set of axes for $c = -2, -1, -\frac{1}{2}, 0, \frac{1}{2}, 1, 2$.

3. Plot the graphs of $H(x) = cx^2$ on the same set of axes for $c = -2, -1, -\frac{1}{2}, -\frac{1}{4}, 0, \frac{1}{4}, \frac{1}{2}, 1, 2$.

4. Study the family of graphs in parts 1–3 and describe the relationship between the graph of a function f and the graphs of the functions defined by (a) $y = f(x) + c$, (b) $y = f(x + c)$, and (c) $y = cf(x)$, where c is a constant.

A function that is defined by more than one rule is called a **piecewise-defined function**.

EXAMPLE 7   Sketch the graph of the function f defined by

$$f(x) = \begin{cases} -x & \text{if } x < 0 \\ \sqrt{x} & \text{if } x \geq 0 \end{cases}$$

Solution   The function f is defined in a piecewise fashion on the set of all real numbers. In the subdomain $(-\infty, 0)$, the rule for f is given by $f(x) = -x$. The equation $y = -x$ is a linear equation in the slope-intercept form (with slope -1 and intercept 0). Therefore, the graph of f corresponding to the subdomain $(-\infty, 0)$ is the half line shown in Figure 7. Next, in the subdomain $[0, \infty)$, the rule for f is given by $f(x) = \sqrt{x}$. The values of $f(x)$ corresponding to $x = 0, 1, 2, 3, 4, 9$, and 16 are shown in the following table:

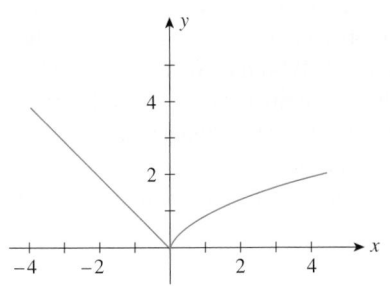

FIGURE 7
The graph of $y = f(x)$ is obtained by graphing $y = -x$ over $(-\infty, 0)$ and $y = \sqrt{x}$ over $[0, \infty)$.

x	0	1	2	3	4	9	16
$f(x)$	0	1	$\sqrt{2}$	$\sqrt{3}$	2	3	4

Using these values, we sketch the graph of the function f as shown in Figure 7.   ■

APPLIED EXAMPLE 8 Bank Deposits Madison Finance Company plans to open two branch offices 2 years from now in two separate locations: an industrial complex and a newly developed commercial center in the city. As a result of these expansion plans, Madison's total deposits during the next 5 years are expected to grow in accordance with the rule

$$f(x) = \begin{cases} \sqrt{2x} + 20 & \text{if } 0 \leq x \leq 2 \\ \dfrac{1}{2}x^2 + 20 & \text{if } 2 < x \leq 5 \end{cases}$$

where $y = f(x)$ gives the total amount of money (in millions of dollars) on deposit with Madison in year x ($x = 0$ corresponds to the present). Sketch the graph of the function f.

Solution The function f is defined in a piecewise fashion on the interval $[0, 5]$. In the subdomain $[0, 2]$, the rule for f is given by $f(x) = \sqrt{2x} + 20$. The values of $f(x)$ corresponding to $x = 0$, 1, and 2 may be tabulated as follows:

x	0	1	2
$f(x)$	20	21.4	22

Next, in the subdomain $(2, 5]$, the rule for f is given by $f(x) = \frac{1}{2}x^2 + 20$. The values of $f(x)$ corresponding to $x = 3$, 4, and 5 are shown in the following table:

x	3	4	5
$f(x)$	24.5	28	32.5

Using the values of $f(x)$ in this table, we sketch the graph of the function f as shown in Figure 8. ∎

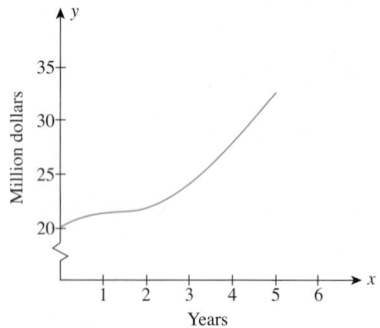

FIGURE 8
We obtain the graph of the function $y = f(x)$ by graphing $y = \sqrt{2x} + 20$ over $[0, 2]$ and $y = \frac{1}{2}x^2 + 20$ over $(2, 5]$.

The Vertical-Line Test

Although it is true that every function f of a variable x has a graph in the xy-plane, it is not true that every curve in the xy-plane is the graph of a function. For example, consider the curve depicted in Figure 9. This is the graph of the equation $y^2 = x$. In general, the **graph of an equation** is the set of all ordered pairs (x, y) that satisfy the given equation. Observe that the points $(9, -3)$ and $(9, 3)$ both lie on the curve. This implies that the number $x = 9$ is associated with *two* numbers: $y = -3$ and $y = 3$. But this clearly violates the uniqueness property of a function. Thus, we conclude that the curve under consideration cannot be the graph of a function.

This example suggests the following **vertical-line test** for determining when a curve is the graph of a function.

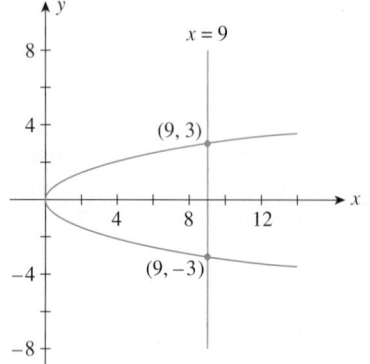

FIGURE 9
Since a vertical line passes through the curve at more than one point, we deduce that it is *not* the graph of a function.

Vertical-Line Test
A curve in the xy-plane is the graph of a function $y = f(x)$ if and only if each vertical line intersects it in at most one point.

EXAMPLE 9   Determine which of the curves shown in Figure 10 are the graphs of functions of x.

Solution   The curves depicted in Figure 10a, c, and d are graphs of functions because each curve satisfies the requirement that each vertical line intersects the curve in at most one point. Note that the vertical line shown in Figure 10c does *not* intersect the graph because the point on the x-axis through which this line passes does not lie in the domain of the function. The curve depicted in Figure 10b is *not* the graph of a function because the vertical line shown there intersects the graph at three points.

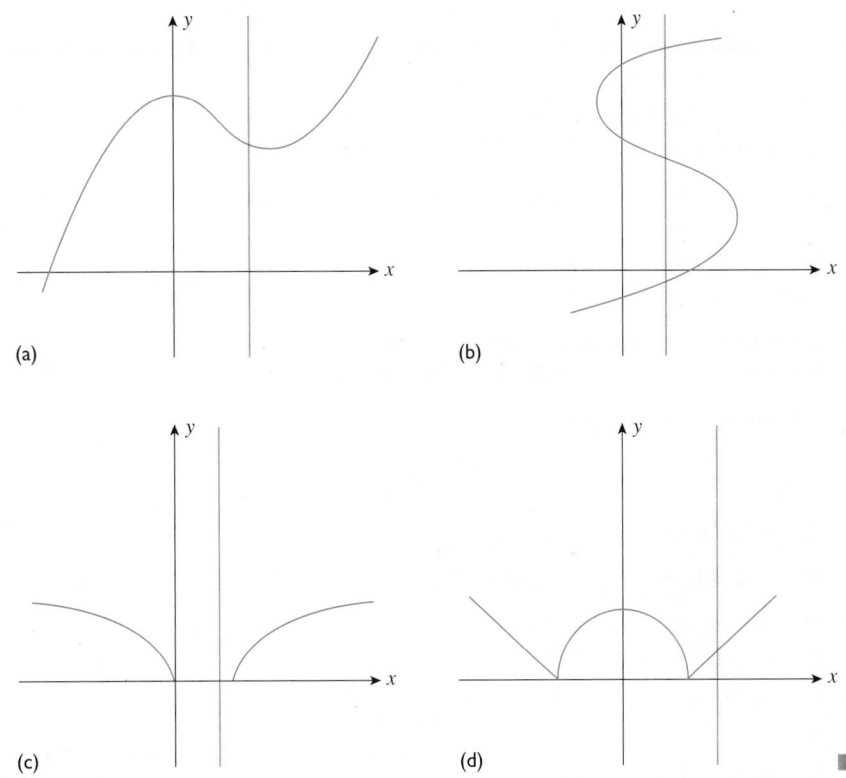

(a)

(b)

(c)

(d)

FIGURE 10
The vertical-line test can be used to determine which of these curves are graphs of functions.

11.1   Self-Check Exercises

1. Let f be the function defined by

$$f(x) = \frac{\sqrt{x+1}}{x}$$

a. Find the domain of f.   **c.** Compute $f(a + h)$.
b. Compute $f(3)$.

2. Statistics show that more and more motorists are pumping their own gas. The following function gives self-serve sales as a percent of all U.S. gas sales:

$$f(t) = \begin{cases} 6t + 17 & \text{if } 0 \le t \le 6 \\ 15.98(t-6)^{1/4} + 53 & \text{if } 6 < t \le 20 \end{cases}$$

Here t is measured in years, with $t = 0$ corresponding to the beginning of 1974.

a. Sketch the graph of the function f.
b. What percent of all gas sales at the beginning of 1978 were self-serve? At the beginning of 1994?

Source: Amoco Corporation

3. Let $f(x) = \sqrt{2x+1} + 2$. Determine whether the point $(4, 6)$ lies on the graph of f.

Solutions to Self-Check Exercises 11.1 can be found on page 591.

11.1 Concept Questions

1. **a.** What is a function?
   b. What is the domain of a function? The range of a function?
   c. What is an independent variable? A dependent variable?

2. **a.** What is the graph of a function? Use a drawing to illustrate the graph, the domain, and the range of a function.
   b. If you are given a curve in the xy-plane, how can you tell if the graph is that of a function f defined by $y = f(x)$?

11.1 Exercises

1. Let f be the function defined by $f(x) = 5x + 6$. Find $f(3)$, $f(-3), f(a), f(-a)$, and $f(a + 3)$.

2. Let f be the function defined by $f(x) = 4x - 3$. Find $f(4)$, $f(\frac{1}{4}), f(0), f(a)$, and $f(a + 1)$.

3. Let g be the function defined by $g(x) = 3x^2 - 6x - 3$. Find $g(0), g(-1), g(a), g(-a)$, and $g(x + 1)$.

4. Let h be the function defined by $h(x) = x^3 - x^2 + x + 1$. Find $h(-5), h(0), h(a)$, and $h(-a)$.

5. Let f be the function defined by $f(x) = 2x + 5$. Find $f(a + h), f(-a), f(a^2), f(a - 2h)$, and $f(2a - h)$.

6. Let g be the function defined by $g(x) = -x^2 + 2x$. Find $g(a + h), g(-a), g(\sqrt{a}), a + g(a)$, and $\dfrac{1}{g(a)}$.

7. Let s be the function defined by $s(t) = \dfrac{2t}{t^2 - 1}$. Find $s(4), s(0), s(a), s(2 + a)$, and $s(t + 1)$.

8. Let g be the function defined by $g(u) = (3u - 2)^{3/2}$. Find $g(1), g(6), g(\frac{11}{3})$, and $g(u + 1)$.

9. Let f be the function defined by $f(t) = \dfrac{2t^2}{\sqrt{t - 1}}$. Find $f(2)$, $f(a), f(x + 1)$, and $f(x - 1)$.

10. Let f be the function defined by $f(x) = 2 + 2\sqrt{5 - x}$. Find $f(-4), f(1), f(\frac{11}{4})$, and $f(x + 5)$.

11. Let f be the function defined by
$$f(x) = \begin{cases} x^2 + 1 & \text{if } x \le 0 \\ \sqrt{x} & \text{if } x > 0 \end{cases}$$
Find $f(-2), f(0)$, and $f(1)$.

12. Let g be the function defined by
$$g(x) = \begin{cases} -\dfrac{1}{2}x + 1 & \text{if } x < 2 \\ \sqrt{x - 2} & \text{if } x \ge 2 \end{cases}$$
Find $g(-2), g(0), g(2)$, and $g(4)$.

13. Let f be the function defined by
$$f(x) = \begin{cases} -\dfrac{1}{2}x^2 + 3 & \text{if } x < 1 \\ 2x^2 + 1 & \text{if } x \ge 1 \end{cases}$$
Find $f(-1), f(0), f(1)$, and $f(2)$.

14. Let f be the function defined by
$$f(x) = \begin{cases} 2 + \sqrt{1 - x} & \text{if } x \le 1 \\ \dfrac{1}{1 - x} & \text{if } x > 1 \end{cases}$$
Find $f(0), f(1)$, and $f(2)$.

15. Refer to the graph of the function f in the following figure.

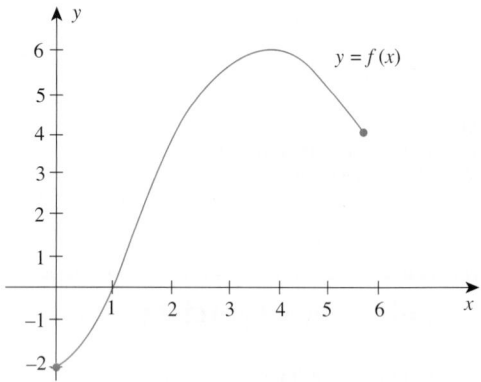

a. Find the value of $f(0)$.
b. Find the value of x for which (i) $f(x) = 3$ and (ii) $f(x) = 0$.
c. Find the domain of f.
d. Find the range of f.

16. Refer to the graph of the function f in the following figure.

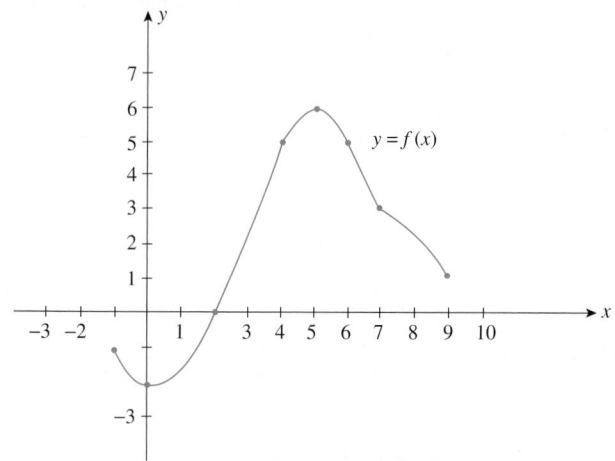

a. Find the value of $f(7)$.
b. Find the values of x corresponding to the point(s) on the graph of f located at a height of 5 units from the x-axis.
c. Find the point on the x-axis at which the graph of f crosses it. What is the value of $f(x)$ at this point?
d. Find the domain and range of f.

In Exercises 17–20, determine whether the point lies on the graph of the function.

17. $(2, \sqrt{3}); g(x) = \sqrt{x^2 - 1}$

18. $(3, 3); f(x) = \dfrac{x + 1}{\sqrt{x^2 + 7}} + 2$

19. $(-2, -3); f(t) = \dfrac{|t - 1|}{t + 1}$

20. $\left(-3, -\dfrac{1}{13}\right); h(t) = \dfrac{|t + 1|}{t^3 + 1}$

In Exercises 21–34, find the domain of the function.

21. $f(x) = x^2 + 3$

22. $f(x) = 7 - x^2$

23. $f(x) = \dfrac{3x + 1}{x^2}$

24. $g(x) = \dfrac{2x + 1}{x - 1}$

25. $f(x) = \sqrt{x^2 + 1}$

26. $f(x) = \sqrt{x - 5}$

27. $f(x) = \sqrt{5 - x}$

28. $g(x) = \sqrt{2x^2 + 3}$

29. $f(x) = \dfrac{x}{x^2 - 1}$

30. $f(x) = \dfrac{1}{x^2 + x - 2}$

31. $f(x) = (x + 3)^{3/2}$

32. $g(x) = 2(x - 1)^{5/2}$

33. $f(x) = \dfrac{\sqrt{1 - x}}{x^2 - 4}$

34. $f(x) = \dfrac{\sqrt{x - 1}}{(x + 2)(x - 3)}$

35. Let f be a function defined by the rule $f(x) = x^2 - x - 6$.
a. Find the domain of f.
b. Compute $f(x)$ for $x = -3, -2, -1, 0, \frac{1}{2}, 1, 2, 3$.
c. Use the results obtained in parts (a) and (b) to sketch the graph of f.

36. Let f be a function defined by the rule $f(x) = 2x^2 + x - 3$.
a. Find the domain of f.
b. Compute $f(x)$ for $x = -3, -2, -1, -\frac{1}{2}, 0, 1, 2, 3$.
c. Use the results obtained in parts (a) and (b) to sketch the graph of f.

In Exercises 37–48, sketch the graph of the function with the given rule. Find the domain and range of the function.

37. $f(x) = 2x^2 + 1$

38. $f(x) = 9 - x^2$

39. $f(x) = 2 + \sqrt{x}$

40. $g(x) = 4 - \sqrt{x}$

41. $f(x) = \sqrt{1 - x}$

42. $f(x) = \sqrt{x - 1}$

43. $f(x) = |x| - 1$

44. $f(x) = |x| + 1$

45. $f(x) = \begin{cases} x & \text{if } x < 0 \\ 2x + 1 & \text{if } x \geq 0 \end{cases}$

46. $f(x) = \begin{cases} 4 - x & \text{if } x < 2 \\ 2x - 2 & \text{if } x \geq 2 \end{cases}$

47. $f(x) = \begin{cases} -x + 1 & \text{if } x \leq 1 \\ x^2 - 1 & \text{if } x > 1 \end{cases}$

48. $f(x) = \begin{cases} -x - 1 & \text{if } x < -1 \\ 0 & \text{if } -1 \leq x \leq 1 \\ x + 1 & \text{if } x > 1 \end{cases}$

In Exercises 49–56, use the vertical-line test to determine whether the graph represents y as a function of x.

49.   **50.**

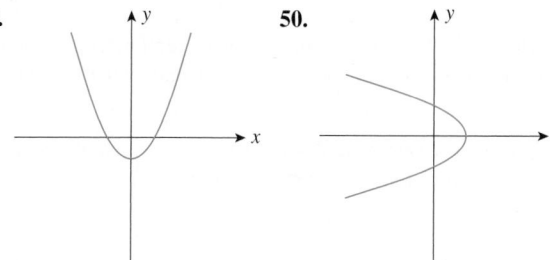

51.
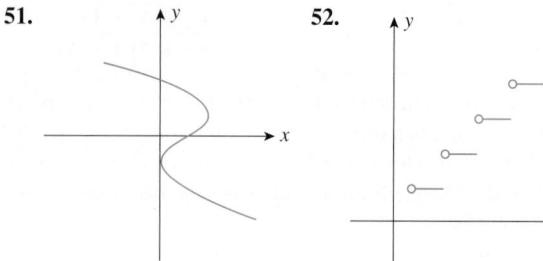

52.

53.
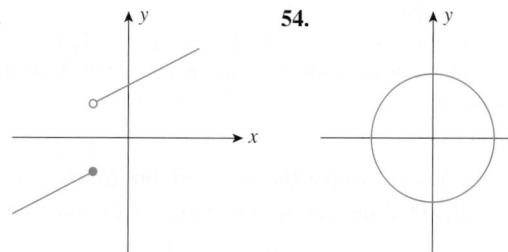

54.

55.
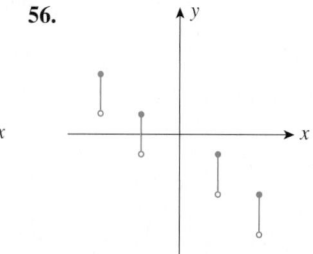

56.

57. The circumference of a circle is given by $C(r) = 2\pi r$, where r is the radius of the circle. What is the circumference of a circle with a 5-in. radius?

58. The volume of a sphere of radius r is given by $V(r) = \frac{4}{3}\pi r^3$. Compute $V(2.1)$ and $V(2)$. What does the quantity $V(2.1) - V(2)$ measure?

59. GROWTH OF A CANCEROUS TUMOR   The volume of a spherical cancerous tumor is given by the function

$$V(r) = \frac{4}{3}\pi r^3$$

where r is the radius of the tumor in centimeters. By what factor is the volume of the tumor increased if its radius is doubled?

60. GROWTH OF A CANCEROUS TUMOR   The surface area of a spherical cancerous tumor is given by the function

$$S(r) = 4\pi r^2$$

where r is the radius of the tumor in centimeters. After extensive chemotherapy treatment, the surface area of the tumor is reduced by 75%. What is the radius of the tumor after treatment?

61. SALES OF PRERECORDED MUSIC   The following graphs show the sales y of prerecorded music (in billions of dollars) by format as a function of time t (in years), with $t = 0$ corresponding to 1985.

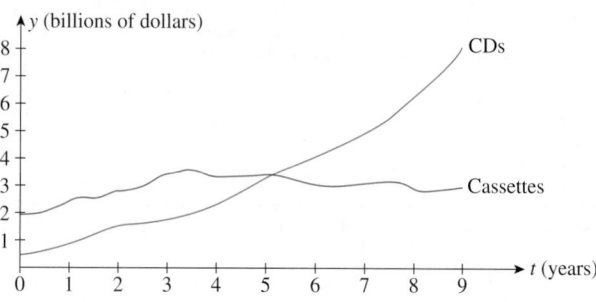

a. In what years were the sales of prerecorded cassettes greater than those of prerecorded CDs?
b. In what years were the sales of prerecorded CDs greater than those of prerecorded cassettes?
c. In what year were the sales of prerecorded cassettes the same as those of prerecorded CDs? Estimate the level of sales in each format at that time.

Source: Recording Industry Association of America

62. THE GENDER GAP   The following graph shows the ratio of women's earnings to men's from 1960 through 2000.

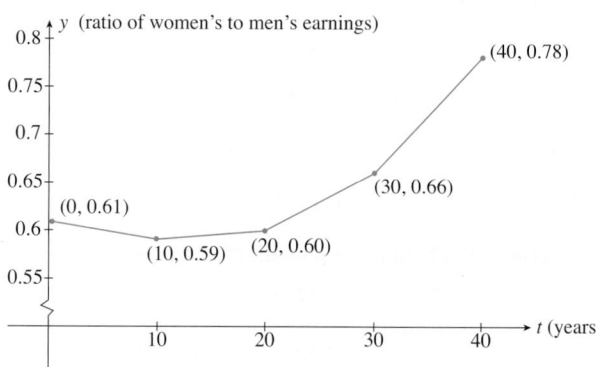

a. Write the rule for the function f giving the ratio of women's earnings to men's in year t, with $t = 0$ corresponding to 1960.
Hint: The function f is defined piecewise and is linear over each of four subintervals.
b. In what decade(s) was the gender gap expanding? Shrinking?
c. Refer to part (b). How fast was the gender gap (the ratio/year) expanding or shrinking in each of these decades?

Source: U.S. Bureau of Labor Statistics

63. CLOSING THE GENDER GAP IN EDUCATION   The following graph shows the ratio of bachelor's degrees earned by women to men from 1960 through 1990.

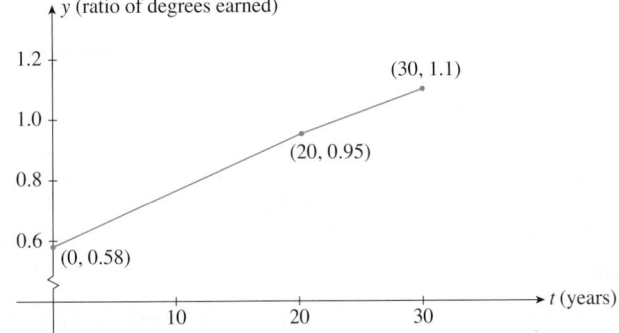

a. Write the rule for the function f giving the ratio of bachelor's degrees earned by women to men in year t, with $t = 0$ corresponding to 1960.
   Hint: The function f is defined piecewise and is linear over each of two subintervals.
b. How fast was the ratio changing in the period from 1960 to 1980? From 1980 to 1990?
c. In what year (approximately) was the number of bachelor's degrees earned by women equal for the first time to that earned by men?

Source: Department of Education

64. CONSUMPTION FUNCTION   The consumption function in a certain economy is given by the equation

$$C(y) = 0.75y + 6$$

where $C(y)$ is the personal consumption expenditure, y is the disposable personal income, and both $C(y)$ and y are measured in billions of dollars. Find $C(0)$, $C(50)$, and $C(100)$.

65. SALES TAXES   In a certain state, the sales tax T on the amount of taxable goods is 6% of the value of the goods purchased (x), where both T and x are measured in dollars.
a. Express T as a function of x.
b. Find $T(200)$ and $T(5.65)$.

66. SURFACE AREA OF A SINGLE-CELLED ORGANISM   The surface area S of a single-celled organism may be found by multiplying 4π times the square of the radius r of the cell. Express S as a function of r.

67. FRIEND'S RULE   Friend's rule, a method for calculating pediatric drug dosages, is based on a child's age. If a denotes the adult dosage (in milligrams) and if t is the age of the child (in years), then the child's dosage is given by

$$D(t) = \frac{2}{25}ta$$

If the adult dose of a substance is 500 mg, how much should a 4-yr-old child receive?

68. COLAs   Social Security recipients receive an automatic cost-of-living adjustment (COLA) once each year. Their monthly benefit is increased by the amount that consumer prices increased during the preceding year. Suppose that consumer prices increased by 5.3% during the preceding year.
a. Express the adjusted monthly benefit of a Social Security recipient as a function of his or her current monthly benefit.
b. If Harrington's monthly Social Security benefit is now $620, what will be his adjusted monthly benefit?

69. BROADBAND INTERNET HOUSEHOLDS   The number of U.S. broadband Internet households stood at 20 million at the beginning of 2002 and is projected to grow at the rate of 7.5 million households per year for the next 6 yr.
a. Find a function $f(t)$ giving the projected U.S. broadband Internet households (in millions) in year t, where $t = 0$ corresponds to the beginning of 2002.
   Hint: The graph of f is a straight line.
b. What is the projected size of U.S. broadband Internet households at the beginning of 2008?

Source: Strategy Analytics Inc.

70. COST OF RENTING A TRUCK   Ace Truck leases its 10-ft box truck at $30/day and $.45/mi, whereas Acme Truck leases a similar truck at $25/day and $.50/mi.
a. Find the daily cost of leasing from each company as a function of the number of miles driven.
b. Sketch the graphs of the two functions on the same set of axes.
c. Which company should a customer rent a truck from for 1 day if she plans to drive at most 70 mi and wishes to minimize her cost?

71. LINEAR DEPRECIATION   A new machine was purchased by National Textile for $120,000. For income tax purposes, the machine is depreciated linearly over 10 yr; that is, the book value of the machine decreases at a constant rate, so that at the end of 10 yr the book value is zero.
a. Express the book value of the machine (V) as a function of the age, in years, of the machine (n).
b. Sketch the graph of the function in part (a).
c. Find the book value of the machine at the end of the sixth year.
d. Find the rate at which the machine is being depreciated each year.

72. LINEAR DEPRECIATION   Refer to Exercise 71. An office building worth $1 million when completed in 1986 was depreciated linearly over 50 yr. What was the book value of the building in 2001? What will be the book value in 2005? In 2009? (Assume that the book value of the building will be zero at the end of the 50th year.)

73. BOYLE'S LAW As a consequence of Boyle's law, the pressure P of a fixed sample of gas held at a constant temperature is related to the volume V of the gas by the rule

$$P = f(V) = \frac{k}{V}$$

where k is a constant. What is the domain of the function f? Sketch the graph of the function f.

74. POISEUILLE'S LAW According to a law discovered by the 19th-century physician Poiseuille, the velocity (in centimeters/second) of blood r cm from the central axis of an artery is given by

$$v(r) = k(R^2 - r^2)$$

where k is a constant and R is the radius of the artery. Suppose that for a certain artery, $k = 1000$ and $R = 0.2$ so that $v(r) = 1000(0.04 - r^2)$.
a. What is the domain of the function $v(r)$?
b. Compute $v(0)$, $v(0.1)$, and $v(0.2)$ and interpret your results.
c. Sketch the graph of the function v on the interval $[0, 0.2]$.
d. What can you say about the velocity of blood as we move away from the central axis toward the artery wall?

75. CANCER SURVIVORS The number of living Americans who have had a cancer diagnosis has increased drastically since 1971. In part, this is due to more testing for cancer and better treatment for some cancers. In part, it is because the population is older, and cancer is largely a disease of the elderly. The number of cancer survivors (in millions) between 1975 ($t = 0$) and 2000 ($t = 25$) is approximately

$$N(t) = 0.0031t^2 + 0.16t + 3.6 \qquad (0 \le t \le 25)$$

a. How many living Americans had a cancer diagnosis in 1975? In 2000?
b. Assuming the trend continued, how many cancer survivors were there in 2005?
Source: National Cancer Institute

76. WORKER EFFICIENCY An efficiency study conducted for Elektra Electronics showed that the number of "Space Commander" walkie-talkies assembled by the average worker t hr after starting work at 8:00 a.m. is given by

$$N(t) = -t^3 + 6t^2 + 15t \qquad (0 \le t \le 4)$$

How many walkie-talkies can an average worker be expected to assemble between 8:00 and 9:00 a.m.? Between 9:00 and 10:00 a.m.?

77. POLITICS Political scientists have discovered the following empirical rule, known as the "cube rule," which gives the relationship between the proportion of seats in the House of Representatives won by Democratic candidates $s(x)$ and the proportion of popular votes x received by the Democratic presidential candidate:

$$s(x) = \frac{x^3}{x^3 + (1 - x)^3} \qquad (0 \le x \le 1)$$

Compute $s(0.6)$ and interpret your result.

78. PREVALENCE OF ALZHEIMER'S PATIENTS Based on a study conducted in 1997, the percent of the U.S. population by age afflicted with Alzheimer's disease is given by the function

$$P(x) = 0.0726x^2 + 0.7902x + 4.9623 \qquad (0 \le x \le 25)$$

where x is measured in years, with $x = 0$ corresponding to age 65. What percent of the U.S. population at age 65 is expected to have Alzheimer's disease? At age 90?
Source: Alzheimer's Association

79. REGISTERED VEHICLES IN MASSACHUSETTS The number of registered vehicles (in millions) in Massachusetts between 1991 ($t = 0$) and 2003 ($t = 12$) is approximately

$$N(t) = -0.0014t^3 + 0.027t^2 - 0.008t + 4.1 \qquad (0 \le t \le 12)$$

a. How many registered vehicles were there in Massachusetts in 1991?
b. How many registered vehicles where there in Massachusetts in 2003?
Source: Mass. Registry of Motor Vehicles

80. POSTAL REGULATIONS In 2006 the postage for first-class mail was raised to 39¢ for the first ounce or fraction thereof and 24¢ for each additional ounce or fraction thereof. Any parcel not exceeding 12 oz may be sent by first-class mail. Letting x denote the weight of a parcel in ounces and $f(x)$ the postage in cents, complete the following description of the "postage function" f:

$$f(x) = \begin{cases} 39 & \text{if } 0 < x \le 1 \\ 63 & \text{if } 1 < x \le 2 \\ \vdots \\ ? & \text{if } 11 < x \le 12 \end{cases}$$

a. What is the domain of f?
b. Sketch the graph of f.

81. HARBOR CLEANUP The amount of solids discharged from the MWRA (Massachusetts Water Resources Authority) sewage treatment plant on Deer Island (near Boston Harbor) is given by the function

$$f(t) = \begin{cases} 130 & \text{if } 0 \le t \le 1 \\ -30t + 160 & \text{if } 1 < t \le 2 \\ 100 & \text{if } 2 < t \le 4 \\ -5t^2 + 25t + 80 & \text{if } 4 < t \le 6 \\ 1.25t^2 - 26.25t + 162.5 & \text{if } 6 < t \le 10 \end{cases}$$

where $f(t)$ is measured in tons/day and t is measured in years, with $t = 0$ corresponding to 1989.
a. What amount of solids were discharged per day in 1989? In 1992? In 1996?
b. Sketch the graph of f.
Source: Metropolitan District Commission

82. RISING MEDIAN AGE Increased longevity and the aging of the baby boom generation—those born between 1946 and 1965—are the primary reasons for a rising median age. The median age (in years) of the U.S. population from 1900 through 2000 is approximated by the function

$$f(t) = \begin{cases} 1.3t + 22.9 & \text{if } 0 \le t \le 3 \\ -0.7t^2 + 7.2t + 11.5 & \text{if } 3 < t \le 7 \\ 2.6t + 9.4 & \text{if } 7 < t \le 10 \end{cases}$$

where t is measured in decades, with $t = 0$ corresponding to the beginning of 1900.

a. What was the median age of the U.S. population at the beginning of 1900? At the beginning of 1950? At the beginning of 1990?

b. Sketch the graph of f.

Source: U.S. Census Bureau

In Exercises 83–86, determine whether the statement is true or false. If it is true, explain why it is true. If it is false, give an example to show why it is false.

83. If $a = b$, then $f(a) = f(b)$.

84. If $f(a) = f(b)$, then $a = b$.

85. If f is a function, then $f(a + b) = f(a) + f(b)$.

86. A vertical line must intersect the graph of $y = f(x)$ at exactly one point.

11.1 Solutions to Self-Check Exercises

1. a. The expression under the radical sign must be nonnegative, so $x + 1 \ge 0$ or $x \ge -1$. Also, $x \ne 0$ because division by zero is not permitted. Therefore, the domain of f is $[-1, 0) \cup (0, \infty)$.

b. $f(3) = \dfrac{\sqrt{3 + 1}}{3} = \dfrac{\sqrt{4}}{3} = \dfrac{2}{3}$

c. $f(a + h) = \dfrac{\sqrt{(a + h) + 1}}{a + h} = \dfrac{\sqrt{a + h + 1}}{a + h}$

2. a. For t in the subdomain $[0, 6]$, the rule for f is given by $f(t) = 6t + 17$. The equation $y = 6t + 17$ is a linear equation, so that portion of the graph of f is the line segment joining the points $(0, 17)$ and $(6, 53)$. Next, in the subdomain $(6, 20]$, the rule for f is given by $f(t) = 15.98(t - 6)^{1/4} + 53$. Using a calculator, we construct the following table of values of $f(t)$ for selected values of t.

t	6	8	10	12	14	16	18	20
$f(t)$	53	72	75.6	78	79.9	81.4	82.7	83.9

We have included $t = 6$ in the table, although it does not lie in the subdomain of the function under consideration, in order to help us obtain a better sketch of that portion of the graph of f in the subdomain $(6, 20]$. The graph of f follows:

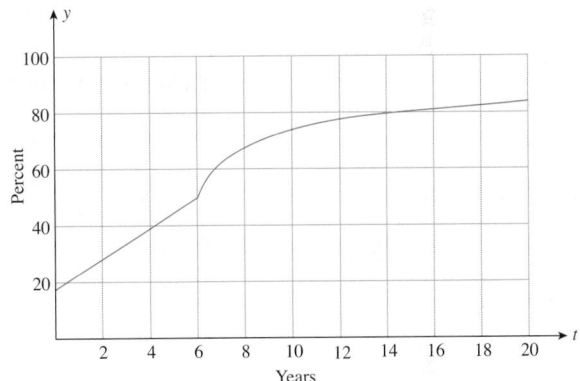

b. The percent of all self-serve gas sales at the beginning of 1978 is found by evaluating f at $t = 4$. Since this point lies in the interval $[0, 6]$, we use the rule $f(t) = 6t + 17$ and find

$$f(4) = 6(4) + 17 = 41$$

giving 41% as the required figure. The percent of all self-serve gas sales at the beginning of 1994 is given by

$$f(20) = 15.98(20 - 6)^{1/4} + 53 \approx 83.9$$

or approximately 83.9%.

3. A point (x, y) lies on the graph of the function f if and only if the coordinates satisfy the equation $y = f(x)$. Now,

$$f(4) = \sqrt{2(4) + 1} + 2 = \sqrt{9} + 2 = 5 \ne 6$$

and we conclude that the given point does *not* lie on the graph of f.

USING TECHNOLOGY

Graphing a Function

Most of the graphs of functions in this book can be plotted with the help of a graphing utility. Furthermore, a graphing utility can be used to analyze the nature of a function. However, the amount and accuracy of the information obtained using a graphing utility depend on the experience and sophistication of the user. As you progress through this book, you will see that the more knowledge of calculus you gain, the more effective the graphing utility will prove to be as a tool in problem solving.

Finding a Suitable Viewing Window

The first step in plotting the graph of a function with a graphing utility is to select a suitable viewing window. We usually do this by experimenting. For example, you might first plot the graph using the *standard viewing window* $[-10, 10]$ by $[-10, 10]$. If necessary, you then might adjust the viewing window by enlarging it or reducing it to obtain a sufficiently complete view of the graph or at least the portion of the graph that is of interest.

EXAMPLE 1 Plot the graph of $f(x) = 2x^2 - 4x - 5$ in the standard viewing window.

Solution The graph of f, shown in Figure T1a, is a parabola. From our previous work (Example 6, Section 11.1), we know that the figure does give a good view of the graph.

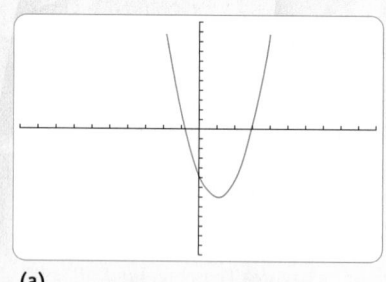

(a)

WINDOW	
X_{min}	$= -10$
X_{max}	$= 10$
X_{scl}	$= 1$
Y_{min}	$= -10$
Y_{max}	$= 10$
Y_{scl}	$= 1$
X_{res}	$= 1$

(b)

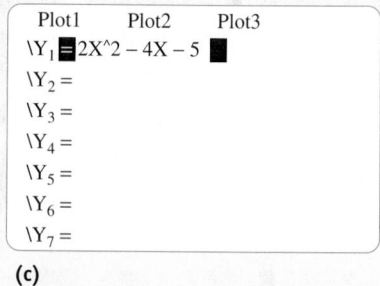

(c)

FIGURE T1
(a) The graph of $f(x) = 2x^2 - 4x - 5$ on $[-10, 10] \times [-10, 10]$; (b) the TI-83 window screen for (a); (c) the TI-83 equation screen

EXAMPLE 2 Let $f(x) = x^3(x - 3)^4$.

a. Plot the graph of f in the standard viewing window.
b. Plot the graph of f in the window $[-1, 5] \times [-40, 40]$.

Solution

a. The graph of f in the standard viewing window is shown in Figure T2a. Since the graph does not appear to be complete, we need to adjust the viewing window.

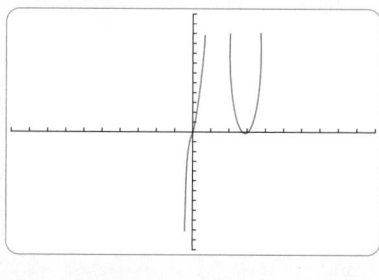

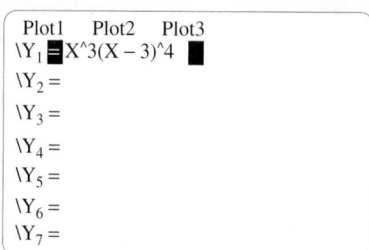

FIGURE T2
(a) An incomplete sketch of
$f(x) = x^3(x - 3)^4$ on $[-10, 10] \times$
$[-10, 10]$; (b) the TI-83 equation screen

(a)

(b)

b. The graph of f in the window $[-1, 5] \times [-40, 40]$, shown in Figure T3a, is an improvement over the previous graph. (Later we will be able to show that the figure does in fact give a rather complete view of the graph of f.)

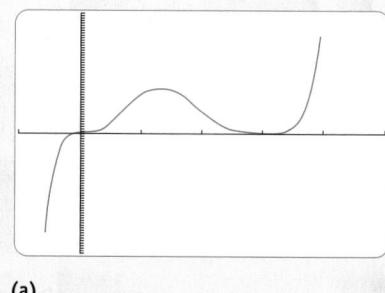

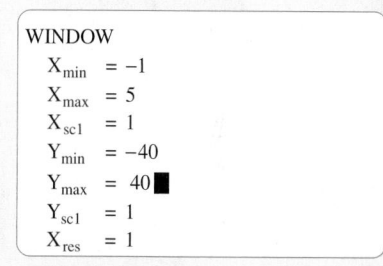

FIGURE T3
(a) A complete sketch of $f(x) =$
$x^3(x - 3)^4$ is shown using the window
$[-1, 5] \times [-40, 40]$; (b) the TI-83
window screen

(a)

(b)

Evaluating a Function

A graphing utility can be used to find the value of a function with minimal effort, as the next example shows.

EXAMPLE 3   Let $f(x) = x^3 - 4x^2 + 4x + 2$.

a. Plot the graph of f in the standard viewing window.
b. Find $f(3)$ and verify your result by direct computation.
c. Find $f(4.215)$.

Solution

a. The graph of f is shown in Figure T4a.

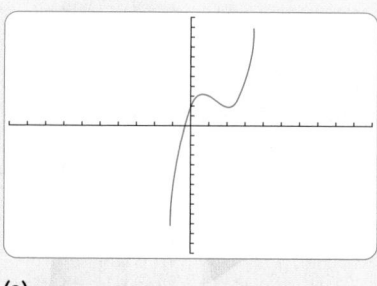

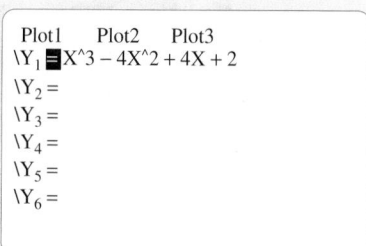

FIGURE T4
(a) The graph of $f(x) = x^3 - 4x^2 +$
$4x + 2$ in the standard viewing window;
(b) the TI-83 equation screen

(a)

(b)

(continued)

b. Using the evaluation function of the graphing utility and the value 3 for x, we find $y = 5$. This result is verified by computing

$$f(3) = 3^3 - 4(3^2) + 4(3) + 2 = 27 - 36 + 12 + 2 = 5$$

c. Using the evaluation function of the graphing utility and the value 4.215 for x, we find $y = 22.679738375$. Thus, $f(4.215) = 22.679738375$. The efficacy of the graphing utility is clearly demonstrated here!

 APPLIED EXAMPLE 4 Number of Alzheimer's Patients   The number of Alzheimer's patients in the United States is given by

$$f(t) = -0.0277t^4 + 0.3346t^3 - 1.1261t^2 + 1.7575t + 3.7745 \qquad (0 \le t \le 6)$$

where $f(t)$ is measured in millions and t is measured in decades, with $t = 0$ corresponding to the beginning of 1990.

a. Use a graphing utility to plot the graph of f in the viewing window $[0, 7] \times [0, 12]$.
b. What is the anticipated number of Alzheimer's patients in the United States at the beginning of 2010 ($t = 2$)? At the beginning of 2030 ($t = 4$)?
Source: Alzheimer's Association

Solution

a. The graph of f in the viewing rectangle $[0, 7] \times [0, 12]$ is shown in Figure T5a.

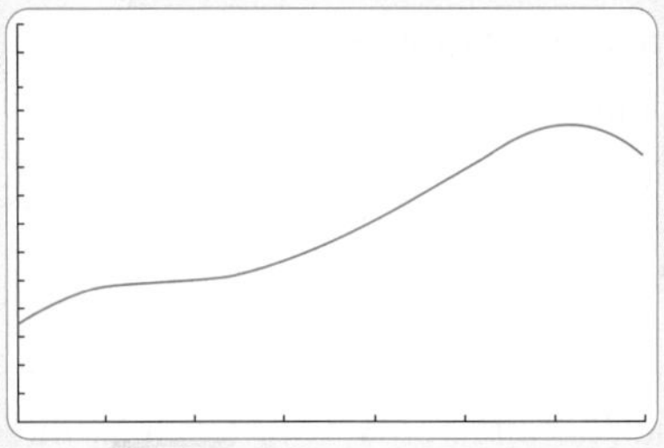

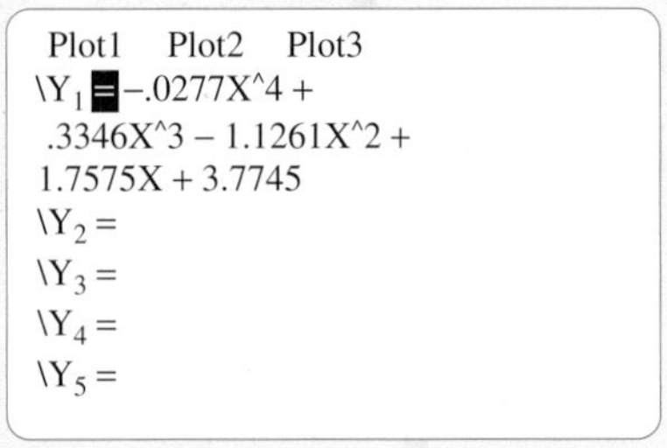

```
Plot1    Plot2    Plot3
\Y1 ■-.0277X^4 +
   .3346X^3 - 1.1261X^2 +
   1.7575X + 3.7745
\Y2 =
\Y3 =
\Y4 =
\Y5 =
```

(a)   (b)

FIGURE T5
(a) The graph of f in the viewing window $[0, 7] \times [0, 12]$; (b) the TI-83 equation screen

b. Using the evaluation function of the graphing utility and the value 2 for x, we see that the anticipated number of Alzheimer's patients at the beginning of 2010 is given by

$$f(2) = 5.0187$$

or approximately 5 million. The anticipated number of Alzheimer's patients at the beginning of 2030 is given by

$$f(4) = 7.1101$$

or approximately 7.1 million.

TECHNOLOGY EXERCISES

In Exercises 1–4, plot the graph of the function f in the standard viewing window.

1. $f(x) = 2x^2 - 16x + 29$

2. $f(x) = -2.01x^3 + 1.21x^2 - 0.78x + 1$

3. $f(x) = 0.2x^4 - 2.1x^2 + 1$

4. $f(x) = \dfrac{\sqrt{x} + 1}{\sqrt{x} - 1}$

In Exercises 5–10, plot the graph of the function f in (a) the standard viewing window and (b) the indicated window.

5. $f(x) = 2x^2 - 32x + 125; [5, 15] \times [-5, 10]$

6. $f(x) = x^3 - 20x^2 + 8x - 10; [-20, 20] \times [-1200, 100]$

7. $f(x) = x^4 - 2x^2 + 8; [-2, 2] \times [6, 10]$

8. $f(x) = \dfrac{4}{x^2 - 8}; [-5, 5] \times [-5, 5]$

9. $f(x) = x\sqrt{4 - x^2}; [-3, 3] \times [-2, 2]$

10. $f(x) = x - 2\sqrt{x}; [0, 20] \times [-2, 10]$

In Exercises 11–16, plot the graph of the function f in an appropriate viewing window. (Note: The answer is not unique.)

11. $f(x) = 2x^3 - 10x^2 + 5x - 10$

12. $f(x) = -x^3 + 5x^2 - 14x + 20$

13. $f(x) = 2x^4 - 3x^3 + 5x^2 - 20x + 40$

14. $f(x) = -2x^4 + 5x^2 - 4$

15. $f(x) = \dfrac{x^3}{x^3 + 1}$ 16. $f(x) = \dfrac{2x^4 - 3x}{x^2 - 1}$

In Exercises 17–20, use the evaluation function of your graphing utility to find the value of f at the given value of x. Verify your result by direct computation.

17. $f(x) = -3x^3 + 5x^2 - 2x + 8; x = -1$

18. $f(x) = 2x^4 - 3x^3 + 2x^2 + x - 5; x = 2$

19. $f(x) = \dfrac{x^4 - 3x^2}{x - 2}; x = 1$ 20. $f(x) = \dfrac{\sqrt{x^2 - 1}}{3x + 4}; x = 2$

In Exercises 21–24, use the evaluation function of your graphing utility to find the value of f at the indicated value of x. Express your answer accurate to four decimal places.

21. $f(x) = 3x^3 - 2x^2 + x - 4; x = 2.145$

22. $f(x) = 5x^4 - 2x^2 + 8x - 3; x = 1.28$

23. $f(x) = \dfrac{2x^3 - 3x + 1}{3x - 2}; x = 2.41$

24. $f(x) = \sqrt{2x^2 + 1} + \sqrt{3x^2 - 1}; x = 0.62$

25. **MANUFACTURING CAPACITY**   Data show that the annual increase in manufacturing capacity between 1994 and 2000 is given by

$$f(t) = 0.0094t^3 - 0.4266t^2 + 2.7489t + 5.54 \qquad (0 \le t \le 6)$$

where $f(t)$ is expressed as a percent and t is measured in years, with $t = 0$ corresponding to the beginning of 1994.
   a. Plot the graph of f in the viewing window $[0, 8] \times [0, 12]$.
   b. What was the annual increase in manufacturing capacity at the beginning of 1996 ($t = 2$)? At the beginning of 2000 ($t = 6$)?

Source: Federal Reserve

26. **DECLINE OF UNION MEMBERSHIP**   The total union membership as a percent of the private workforce is given by

$$f(t) = 0.00017t^4 - 0.00921t^3 + 0.15437t^2$$
$$- 1.360723t + 16.8028 \qquad (0 \le t \le 10)$$

where t is measured in years, with $t = 0$ corresponding to the beginning of 1983.
   a. Plot the graph of f in the viewing window $[0, 11] \times [8, 20]$.
   b. What was the total union membership as a percent of the private force at the beginning of 1986? At the beginning of 1993?

Source: American Federation of Labor and Congress of Industrial Organizations

27. **KEEPING WITH THE TRAFFIC FLOW**   By driving at a speed to match the prevailing traffic speed, you decrease the chances of an accident. According to data obtained in a university study, the number of accidents/100 million vehicle

(continued)

miles, y, is related to the deviation from the mean speed, x, in mph by

$$y = 1.05x^3 - 21.95x^2 + 155.9x - 327.3 \qquad (6 \le x \le 11)$$

a. Plot the graph of y in the viewing window $[6, 11] \times [20, 150]$.
b. What is the number of accidents/100 million vehicle miles if the deviation from the mean speed is 6 mph, 8 mph, and 11 mph?

Source: University of Virginia School of Engineering and Applied Science

28. **SAFE DRIVERS**   The fatality rate in the United States (per 100 million miles traveled) by age of driver (in years) is given by the function

$$f(x) = 0.00000304x^4 - 0.0005764x^3 + 0.04105x^2$$
$$- 1.30366x + 16.579 \qquad (18 \le x \le 82)$$

a. Plot the graph of f in the viewing window $[18, 82] \times [0, 8]$.
b. What is the fatality rate for 18-yr-old drivers? For 50-yr-old drivers? For 80-yr-old drivers?

Source: National Highway Traffic Safety Administration

11.2 The Algebra of Functions

▬ The Sum, Difference, Product, and Quotient of Functions

Let $S(t)$ and $R(t)$ denote, respectively, the federal government's spending and revenue at any time t, measured in billions of dollars. The graphs of these functions for the period between 1990 and 2000 are shown in Figure 11.

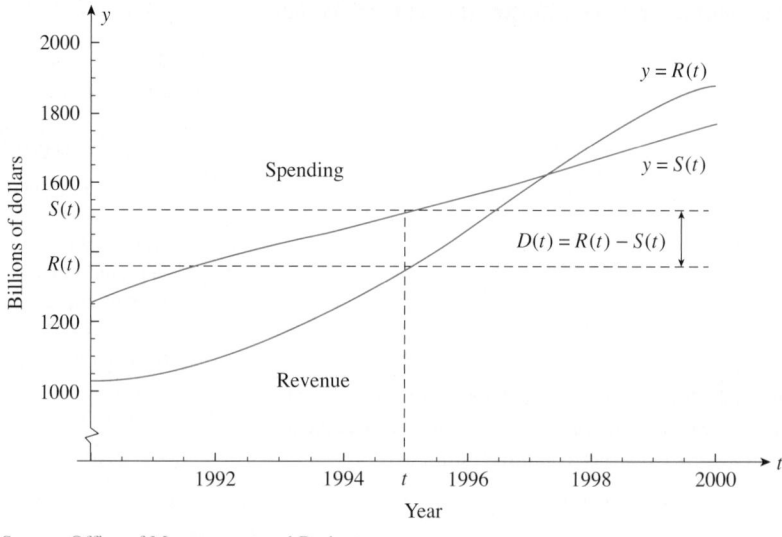

FIGURE 11
$R(t) - S(t)$ gives the federal budget deficit (surplus) at any time t.

Source: Office of Management and Budget

The difference $R(t) - S(t)$ gives the deficit (surplus) in billions of dollars at any time t if $R(t) - S(t)$ is negative (positive). This observation suggests that we can define a function D whose value at any time t is given by $R(t) - S(t)$. The function D, the *difference* of the two functions R and S, is written $D = R - S$ and may be called the "deficit (surplus) function" since it gives the budget deficit or surplus at any time t. It has the same domain as the functions S and R. The graph of the function D is shown in Figure 12.

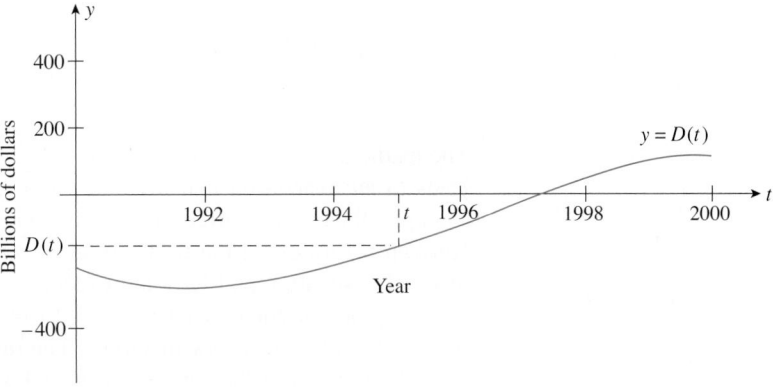

FIGURE 12
The graph of $D(t)$

Source: Office of Management and Budget

Most functions are built up from other, generally simpler functions. For example, we may view the function $f(x) = 2x + 4$ as the sum of the two functions $g(x) = 2x$ and $h(x) = 4$. The function $g(x) = 2x$ may in turn be viewed as the product of the functions $p(x) = 2$ and $q(x) = x$.

In general, given the functions f and g, we define the sum $f + g$, the difference $f - g$, the product fg, and the quotient f/g of f and g as follows.

The Sum, Difference, Product, and Quotient of Functions

Let f and g be functions with domains A and B, respectively. Then the **sum** $f + g$, **difference** $f - g$, and **product** fg of f and g are functions with domain $A \cap B^*$ and rule given by

$$(f + g)(x) = f(x) + g(x) \qquad \text{Sum}$$
$$(f - g)(x) = f(x) - g(x) \qquad \text{Difference}$$
$$(fg)(x) = f(x)g(x) \qquad \text{Product}$$

The **quotient** f/g of f and g has domain $A \cap B$ excluding all numbers x such that $g(x) = 0$ and rule given by

$$\left(\frac{f}{g}\right)(x) = \frac{f(x)}{g(x)} \qquad \text{Quotient}$$

---

*$A \cap B$ is read "A intersected with B" and denotes the set of all points common to both A and B.

EXAMPLE 1 Let $f(x) = \sqrt{x + 1}$ and $g(x) = 2x + 1$. Find the sum s, the difference d, the product p, and the quotient q of the functions f and g.

Solution Since the domain of f is $A = [-1, \infty)$ and the domain of g is $B = (-\infty, \infty)$, we see that the domain of s, d, and p is $A \cap B = [-1, \infty)$. The rules follow.

$$s(x) = (f + g)(x) = f(x) + g(x) = \sqrt{x + 1} + 2x + 1$$
$$d(x) = (f - g)(x) = f(x) - g(x) = \sqrt{x + 1} - (2x + 1) = \sqrt{x + 1} - 2x - 1$$
$$p(x) = (fg)(x) = f(x)g(x) = \sqrt{x + 1}(2x + 1) = (2x + 1)\sqrt{x + 1}$$

The quotient function q has rule

$$q(x) = \left(\frac{f}{g}\right)(x) = \frac{f(x)}{g(x)} = \frac{\sqrt{x+1}}{2x+1}$$

Its domain is $[-1, \infty)$ together with the restriction $x \neq -\frac{1}{2}$. We denote this by $[-1, -\frac{1}{2}) \cup (-\frac{1}{2}, \infty)$.   ■

The mathematical formulation of a problem arising from a practical situation often leads to an expression that involves the combination of functions. Consider, for example, the costs incurred in operating a business. Costs that remain more or less constant regardless of the firm's level of activity are called **fixed costs**. Examples of fixed costs are rental fees and executive salaries. On the other hand, costs that vary with production or sales are called **variable costs**. Examples of variable costs are wages and costs of raw materials. The **total cost** of operating a business is thus given by the *sum* of the variable costs and the fixed costs, as illustrated in the next example.

APPLIED EXAMPLE 2   Cost Functions   Suppose Puritron, a manufacturer of water filters, has a monthly fixed cost of $10,000 and a variable cost of

$$-0.0001x^2 + 10x \qquad (0 \leq x \leq 40{,}000)$$

dollars, where x denotes the number of filters manufactured per month. Find a function C that gives the total monthly cost incurred by Puritron in the manufacture of x filters.

Solution   Puritron's monthly fixed cost is always $10,000, regardless of the level of production, and it is described by the constant function $F(x) = 10{,}000$. Next, the variable cost is described by the function $V(x) = -0.0001x^2 + 10x$. Since the total cost incurred by Puritron at any level of production is the sum of the variable cost and the fixed cost, we see that the required total cost function is given by

$$\begin{aligned} C(x) &= V(x) + F(x) \\ &= -0.0001x^2 + 10x + 10{,}000 \qquad (0 \leq x \leq 40{,}000) \end{aligned}$$ ■

Next, the **total profit** realized by a firm in operating a business is the *difference* between the total revenue realized and the total cost incurred; that is,

$$P(x) = R(x) - C(x)$$

APPLIED EXAMPLE 3 Profit Functions   Refer to Example 2. Suppose the total revenue realized by Puritron from the sale of x water filters is given by the total revenue function

$$R(x) = -0.0005x^2 + 20x \qquad (0 \leq x \leq 40{,}000)$$

a. Find the total profit function—that is, the function that describes the total profit Puritron realizes in manufacturing and selling x water filters per month.
b. What is the profit when the level of production is 10,000 filters per month?

Solution

a. The total profit realized by Puritron in manufacturing and selling x water filters per month is the difference between the total revenue realized and the total cost incurred. Thus, the required total profit function is given by

$$P(x) = R(x) - C(x)$$
$$= (-0.0005x^2 + 20x) - (-0.0001x^2 + 10x + 10,000)$$
$$= -0.0004x^2 + 10x - 10,000$$

b. The profit realized by Puritron when the level of production is 10,000 filters per month is

$$P(10,000) = -0.0004(10,000)^2 + 10(10,000) - 10,000 = 50,000$$

or \$50,000 per month.   ∎

Composition of Functions

Another way to build up a function from other functions is through a process known as the *composition of functions*. Consider, for example, the function h, whose rule is given by $h(x) = \sqrt{x^2 - 1}$. Let f and g be functions defined by the rules $f(x) = x^2 - 1$ and $g(x) = \sqrt{x}$. Evaluating the function g at the point $f(x)$ [remember that for each real number x in the domain of f, $f(x)$ is simply a real number], we find that

$$g(f(x)) = \sqrt{f(x)} = \sqrt{x^2 - 1}$$

which is just the rule defining the function h!

In general, the composition of a function g with a function f is defined as follows.

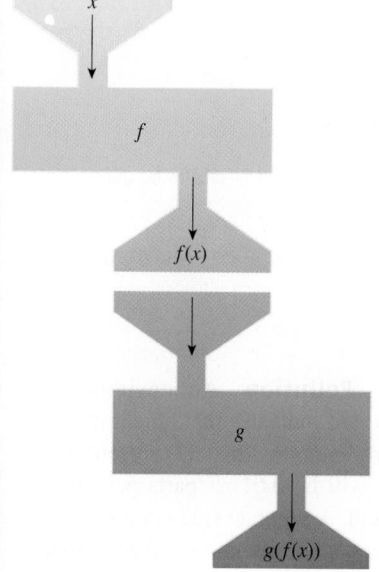

FIGURE 13
The composite function $h = g \circ f$ viewed as a machine

The Composition of Two Functions

Let f and g be functions. Then the composition of g and f is the function $g \circ f$ defined by

$$(g \circ f)(x) = g(f(x))$$

The domain of $g \circ f$ is the set of all x in the domain of f such that $f(x)$ lies in the domain of g.

The function $g \circ f$ (read "g circle f") is also called a **composite function**. The interpretation of the function $h = g \circ f$ as a machine is illustrated in Figure 13 and its interpretation as a mapping is shown in Figure 14.

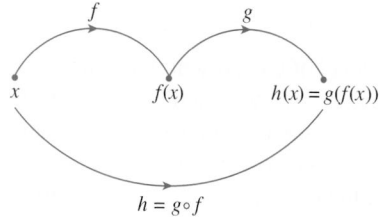

FIGURE 14
The function $h = g \circ f$ viewed as a mapping

EXAMPLE 4   Let $f(x) = x^2 - 1$ and $g(x) = \sqrt{x} + 1$. Find:

a. The rule for the composite function $g \circ f$.
b. The rule for the composite function $f \circ g$.

Solution

a. To find the rule for the composite function $g \circ f$, evaluate the function g at $f(x)$. We obtain

$$(g \circ f)(x) = g(f(x)) = \sqrt{f(x)} + 1 = \sqrt{x^2 - 1} + 1$$

b. To find the rule for the composite function $f \circ g$, evaluate the function f at $g(x)$. Thus,

$$(f \circ g)(x) = f(g(x)) = (g(x))^2 - 1 = (\sqrt{x} + 1)^2 - 1$$
$$= x + 2\sqrt{x} + 1 - 1 = x + 2\sqrt{x}$$

 Example 4 reminds us that in general $g \circ f$ is different from $f \circ g$, so care must be taken when finding the rule for a composite function.

EXPLORE & DISCUSS

Let $f(x) = \sqrt{x} + 1$ for $x \geq 0$ and let $g(x) = (x - 1)^2$ for $x \geq 1$.

1. Show that $(g \circ f)(x)$ and $(f \circ g)(x) = x$. (*Note:* The function g is said to be the *inverse* of f and vice versa.)

2. Plot the graphs of f and g together with the straight line $y = x$. Describe the relationship between the graphs of f and g.

 APPLIED EXAMPLE 5 Automobile Pollution An environmental impact study conducted for the city of Oxnard indicates that, under existing environmental protection laws, the level of carbon monoxide (CO) present in the air due to pollution from automobile exhaust will be $0.01x^{2/3}$ parts per million when the number of motor vehicles is x thousand. A separate study conducted by a state government agency estimates that t years from now the number of motor vehicles in Oxnard will be $0.2t^2 + 4t + 64$ thousand.

a. Find an expression for the concentration of CO in the air due to automobile exhaust t years from now.
b. What will be the level of concentration 5 years from now?

Solution

a. The level of CO present in the air due to pollution from automobile exhaust is described by the function $g(x) = 0.01x^{2/3}$, where x is the number (in thousands) of motor vehicles. But the number of motor vehicles x (in thousands) t years from now may be estimated by the rule $f(t) = 0.2t^2 + 4t + 64$. Therefore, the concentration of CO due to automobile exhaust t years from now is given by

$$C(t) = (g \circ f)(t) = g(f(t)) = 0.01(0.2t^2 + 4t + 64)^{2/3}$$

parts per million.

b. The level of concentration 5 years from now will be

$$C(5) = 0.01[0.2(5)^2 + 4(5) + 64]^{2/3}$$
$$= (0.01)89^{2/3} \approx 0.20$$

or approximately 0.20 parts per million. ∎

11.2 Self-Check Exercises

1. Let f and g be functions defined by the rules

$$f(x) = \sqrt{x} + 1 \quad \text{and} \quad g(x) = \frac{x}{1 + x}$$

respectively. Find the rules for
a. The sum s, the difference d, the product p, and the quotient q of f and g.
b. The composite functions $f \circ g$ and $g \circ f$.

2. Health-care spending per person by the private sector includes payments by individuals, corporations, and their insurance companies and is approximated by the function

$$f(t) = 2.48t^2 + 18.47t + 509 \qquad (0 \le t \le 6)$$

where $f(t)$ is measured in dollars and t is measured in years, with $t = 0$ corresponding to the beginning of 1994. The cor-responding government spending—including expenditures for Medicaid, Medicare, and other federal, state, and local government public health care—is

$$g(t) = -1.12t^2 + 29.09t + 429 \qquad (0 \le t \le 6)$$

where t has the same meaning as before.
a. Find a function that gives the difference between private and government health-care spending per person at any time t.
b. What was the difference between private and government expenditures per person at the beginning of 1995? At the beginning of 2000?
Source: Health Care Financing Administration

Solutions to Self-Check Exercises 11.2 can be found on page 604.

11.2 Concept Questions

1. a. Explain what is meant by the sum, difference, product, and quotient of the functions f and g with domains A and B, respectively.
b. If $f(2) = 3$ and $g(2) = -2$, what is $(f + g)(2)$? $(f - g)(2)$? $(fg)(2)$? $(f/g)(2)$?

2. a. What is the composition of the functions f and g? The functions g and f?
b. If $f(2) = 3$ and $g(3) = 8$, what is $(g \circ f)(2)$? Can you con-clude from the given information what $(f \circ g)(3)$ is? Explain.

11.2 Exercises

In Exercises 1–8, let $f(x) = x^3 + 5$, $g(x) = x^2 - 2$, and $h(x) = 2x + 4$. Find the rule for each function.

1. $f + g$ **2.** $f - g$ **3.** fg **4.** gf

5. $\dfrac{f}{g}$ **6.** $\dfrac{f - g}{h}$ **7.** $\dfrac{fg}{h}$ **8.** fgh

In Exercises 9–18, let $f(x) = x - 1$, $g(x) = \sqrt{x + 1}$, and $h(x) = 2x^3 - 1$. Find the rule for each function.

9. $f + g$ **10.** $g - f$ **11.** fg **12.** gf

13. $\dfrac{g}{h}$ **14.** $\dfrac{h}{g}$ **15.** $\dfrac{fg}{h}$ **16.** $\dfrac{fh}{g}$

17. $\dfrac{f - h}{g}$ **18.** $\dfrac{gh}{g - f}$

In Exercises 19–24, find the functions $f + g$, $f - g$, fg, and f/g.

19. $f(x) = x^2 + 5; g(x) = \sqrt{x} - 2$

20. $f(x) = \sqrt{x - 1}; g(x) = x^3 + 1$

21. $f(x) = \sqrt{x + 3}; g(x) = \dfrac{1}{x - 1}$

22. $f(x) = \dfrac{1}{x^2 + 1}; g(x) = \dfrac{1}{x^2 - 1}$

23. $f(x) = \dfrac{x + 1}{x - 1}; g(x) = \dfrac{x + 2}{x - 2}$

24. $f(x) = x^2 + 1; g(x) = \sqrt{x + 1}$

In Exercises 25–30, find the rules for the composite functions $f \circ g$ and $g \circ f$.

25. $f(x) = x^2 + x + 1; g(x) = x^2$

26. $f(x) = 3x^2 + 2x + 1; g(x) = x + 3$

27. $f(x) = \sqrt{x} + 1; g(x) = x^2 - 1$

28. $f(x) = 2\sqrt{x} + 3; g(x) = x^2 + 1$

29. $f(x) = \dfrac{x}{x^2 + 1}; g(x) = \dfrac{1}{x}$

30. $f(x) = \sqrt{x + 1}; g(x) = \dfrac{1}{x - 1}$

In Exercises 31–34, evaluate $h(2)$, where $h = g \circ f$.

31. $f(x) = x^2 + x + 1; g(x) = x^2$

32. $f(x) = \sqrt[3]{x^2 - 1}; g(x) = 3x^3 + 1$

33. $f(x) = \dfrac{1}{2x + 1}; g(x) = \sqrt{x}$

34. $f(x) = \dfrac{1}{x - 1}; g(x) = x^2 + 1$

In Exercises 35–42, find functions f and g such that $h = g \circ f$. (*Note:* The answer is *not* unique.)

35. $h(x) = (2x^3 + x^2 + 1)^5$ **36.** $h(x) = (3x^2 - 4)^{-3}$

37. $h(x) = \sqrt{x^2 - 1}$ **38.** $h(x) = (2x - 3)^{3/2}$

39. $h(x) = \dfrac{1}{x^2 - 1}$ **40.** $h(x) = \dfrac{1}{\sqrt{x^2 - 4}}$

41. $h(x) = \dfrac{1}{(3x^2 + 2)^{3/2}}$

42. $h(x) = \dfrac{1}{\sqrt{2x + 1}} + \sqrt{2x + 1}$

In Exercises 43–46, find $f(a + h) - f(a)$ for each function. Simplify your answer.

43. $f(x) = 3x + 4$ **44.** $f(x) = -\dfrac{1}{2}x + 3$

45. $f(x) = 4 - x^2$ **46.** $f(x) = x^2 - 2x + 1$

In Exercises 47–52, find and simplify

$$\frac{f(a + h) - f(a)}{h} \quad (h \neq 0)$$

for each function.

47. $f(x) = x^2 + 1$ **48.** $f(x) = 2x^2 - x + 1$

49. $f(x) = x^3 - x$ **50.** $f(x) = 2x^3 - x^2 + 1$

51. $f(x) = \dfrac{1}{x}$ **52.** $f(x) = \sqrt{x}$

53. Restaurant Revenue Nicole owns and operates two restaurants. The revenue of the first restaurant at time t is $f(t)$ dollars, and the revenue of the second restaurant at time t is $g(t)$ dollars. What does the function $F(t) = f(t) + g(t)$ represent?

54. Birthrate of Endangered Species The birthrate of an endangered species of whales in year t is $f(t)$ whales/year. This species of whales is dying at the rate of $g(t)$ whales/year in year t. What does the function $F(t) = f(t) - g(t)$ represent?

55. Value of an Investment The number of IBM shares that Nancy owns is given by $f(t)$. The price per share of the stock of IBM at time t is $g(t)$ dollars. What does the function $f(t)g(t)$ represent?

56. Production Costs The total cost incurred by time t in the production of a certain commodity is $f(t)$ dollars. The number of products produced by time t is $g(t)$ units. What does the function $f(t)/g(t)$ represent?

57. Carbon Monoxide Pollution The number of cars running in the business district of a town at time t is given by $f(t)$. Carbon monoxide pollution coming from these cars is given by $g(x)$ parts per million, where x is the number of cars being operated in the district. What does the function $g \circ f$ represent?

58. Effect of Advertising on Revenue The revenue of Leisure Travel is given by $f(x)$ dollars, where x is the dollar amount spent by the company on advertising. The amount spent by Leisure at time t on advertising is given by $g(t)$ dollars. What does the function $f \circ g$ represent?

59. Manufacturing Costs TMI, a manufacturer of blank audiocassette tapes, has a monthly fixed cost of $12,100 and a variable cost of $.60/tape. Find a function C that gives the total cost incurred by TMI in the manufacture of x tapes/month.

60. Spam Messages The total number of email messages per day (in billions) between 2003 and 2007 was estimated to be

$$f(t) = 1.54t^2 + 7.1t + 31.4 \quad (0 \leq t \leq 4)$$

where t is measured in years, with $t = 0$ corresponding to 2003. Over the same period, the total number of spam messages per day (in billions) was estimated to be

$$g(t) = 1.21t^2 + 6t + 14.5 \quad (0 \leq t \leq 4)$$

a. Find the rule for the function $D = f - g$. Compute $D(4)$ and explain what it measures.

b. Find the rule for the function $P = f/g$. Compute $P(4)$ and explain what it means.

Source: Technology Review

61. **GLOBAL SUPPLY OF PLUTONIUM** The global stockpile of plutonium for military applications between 1990 ($t = 0$) and 2003 ($t = 13$) stood at a constant 267 tons. On the other hand, the global stockpile of plutonium for civilian use was

$$2t^2 + 46t + 733$$

tons in year t over the same period.
a. Find the function f giving the global stockpile of plutonium for military use from 1990 through 2003 and the function g giving the global stockpile of plutonium for civilian use over the same period.
b. Find the function h giving the total global stockpile of plutonium between 1990 and 2003.
c. What was the total global stockpile of plutonium in 2003?

Source: Institute for Science and International Security

62. **COST OF PRODUCING PDAS** Apollo manufactures PDAs at a variable cost of

$$V(x) = 0.000003x^3 - 0.03x^2 + 200x$$

dollars, where x denotes the number of units manufactured per month. The monthly fixed cost attributable to the division that produces these PDAs is $100,000. Find a function C that gives the total cost incurred by the manufacture of x PDAs. What is the total cost incurred in producing 2000 units/month?

63. **PROFIT FROM SALE OF PDAS** Refer to Exercise 62. Suppose the total revenue realized by Apollo from the sale of x PDAs is given by the total revenue function

$$R(x) = -0.1x^2 + 500x \qquad (0 \le x \le 5000)$$

where $R(x)$ is measured in dollars.
a. Find the total profit function.
b. What is the profit when 1500 units are produced and sold each month?

64. **PROFIT FROM SALE OF PAGERS** A division of Chapman Corporation manufactures a pager. The weekly fixed cost for the division is $20,000, and the variable cost for producing x pagers/week is

$$V(x) = 0.000001x^3 - 0.01x^2 + 50x$$

dollars. The company realizes a revenue of

$$R(x) = -0.02x^2 + 150x \qquad (0 \le x \le 7500)$$

dollars from the sale of x pagers/week.
a. Find the total cost function.
b. Find the total profit function.
c. What is the profit for the company if 2000 units are produced and sold each week?

65. **OVERCROWDING OF PRISONS** The 1980s saw a trend toward old-fashioned punitive deterrence as opposed to the more liberal penal policies and community-based corrections popular in the 1960s and early 1970s. As a result, prisons became more crowded, and the gap between the number of people in prison and the prison capacity widened. The number of prisoners (in thousands) in federal and state prisons is approximated by the function

$$N(t) = 3.5t^2 + 26.7t + 436.2 \qquad (0 \le t \le 10)$$

where t is measured in years, with $t = 0$ corresponding to 1983. The number of inmates for which prisons were designed is given by

$$C(t) = 24.3t + 365 \qquad (0 \le t \le 10)$$

where $C(t)$ is measured in thousands and t has the same meaning as before.
a. Find an expression that shows the gap between the number of prisoners and the number of inmates for which the prisons were designed at any time t.
b. Find the gap at the beginning of 1983 and at the beginning of 1986.

Source: U.S. Department of Justice

66. **EFFECT OF MORTGAGE RATES ON HOUSING STARTS** A study prepared for the National Association of Realtors estimated that the number of housing starts per year over the next 5 yr will be

$$N(r) = \frac{7}{1 + 0.02r^2}$$

million units, where r (percent) is the mortgage rate. Suppose the mortgage rate over the next t mo is

$$r(t) = \frac{10t + 150}{t + 10} \qquad (0 \le t \le 24)$$

percent/year.
a. Find an expression for the number of housing starts per year as a function of t, t mo from now.
b. Using the result from part (a), determine the number of housing starts at present, 12 mo from now, and 18 mo from now.

67. **HOTEL OCCUPANCY RATE** The occupancy rate of the all-suite Wonderland Hotel, located near an amusement park, is given by the function

$$r(t) = \frac{10}{81}t^3 - \frac{10}{3}t^2 + \frac{200}{9}t + 55 \qquad (0 \le t \le 11)$$

where t is measured in months and $t = 0$ corresponds to the beginning of January. Management has estimated that the monthly revenue (in thousands of dollars) is approximated by the function

$$R(r) = -\frac{3}{5000}r^3 + \frac{9}{50}r^2 \qquad (0 \le r \le 100)$$

where r (percent) is the occupancy rate.

a. What is the hotel's occupancy rate at the beginning of January? At the beginning of June?

b. What is the hotel's monthly revenue at the beginning of January? At the beginning of June?
Hint: Compute $R(r(0))$ and $R(r(5))$.

68. HOUSING STARTS AND CONSTRUCTION JOBS The president of a major housing construction firm reports that the number of construction jobs (in millions) created is given by

$$N(x) = 1.42x$$

where x denotes the number of housing starts. Suppose the number of housing starts in the next t mo is expected to be

$$x(t) = \frac{7(t + 10)^2}{(t + 10)^2 + 2(t + 15)^2}$$

million units/year. Find an expression for the number of jobs created per month in the next t mo. How many jobs will have been created 6 mo and 12 mo from now?

In Exercises 69–72, determine whether the statement is true or false. If it is true, explain why it is true. If it is false, give an example to show why it is false.

69. If f and g are functions with domain D, then $f + g = g + f$.

70. If $g \circ f$ is defined at $x = a$, then $f \circ g$ must also be defined at $x = a$.

71. If f and g are functions, then $f \circ g = g \circ f$.

72. If f is a function, then $f \circ f = f^2$.

11.2 Solutions to Self-Check Exercises

1. a. $s(x) = f(x) + g(x) = \sqrt{x} + 1 + \dfrac{x}{1 + x}$

$d(x) = f(x) - g(x) = \sqrt{x} + 1 - \dfrac{x}{1 + x}$

$p(x) = f(x)g(x) = (\sqrt{x} + 1) \cdot \dfrac{x}{1 + x} = \dfrac{x(\sqrt{x} + 1)}{1 + x}$

$q(x) = \dfrac{f(x)}{g(x)} = \dfrac{\sqrt{x} + 1}{\dfrac{x}{1 + x}} = \dfrac{(\sqrt{x} + 1)(1 + x)}{x}$

b. $(f \circ g)(x) = f(g(x)) = \sqrt{\dfrac{x}{1 + x} + 1}$

$(g \circ f)(x) = g(f(x)) = \dfrac{\sqrt{x} + 1}{1 + (\sqrt{x} + 1)} = \dfrac{\sqrt{x} + 1}{\sqrt{x} + 2}$

2. a. The difference between private and government health-care spending per person at any time t is given by the function d with the rule

$d(t) = f(t) - g(t) = (2.48t^2 + 18.47t + 509)$
$\qquad - (-1.12t^2 + 29.09t + 429)$
$\qquad = 3.6t^2 - 10.62t + 80$

b. The difference between private and government expenditures per person at the beginning of 1995 is given by

$$d(1) = 3.6(1)^2 - 10.62(1) + 80$$

or $72.98/person.

The difference between private and government expenditures per person at the beginning of 2000 is given by

$$d(6) = 3.6(6)^2 - 10.62(6) + 80$$

or $145.88/person.

11.3 Functions and Mathematical Models

▬ Mathematical Models

One of the fundamental goals in this book is to show how mathematics and, in particular, calculus can be used to solve real-world problems such as those arising from the world of business and the social, life, and physical sciences. You have already seen some of these problems earlier. Here are a few more examples of real-world phenomena that we will analyze in this and ensuing chapters.

▪ The average U.S. credit card debt (p. 613).

▪ The population growth in the fastest-growing metropolitan area in the United States (p. 620).

▪ The increase in the revenue collected by the IRS from the alternative minimum tax (p. 805).

■ The prevalence of Alzheimer's patients in the United States (p. 590).
■ The solvency of the Social Security retirement and disability trust funds (p. 823).
■ The concentration of glucose in the bloodstream (p. 903).

Before going on, let's look at a mathematical model. This model, which is derived from data using the least-squares technique, is used to project the growth of the number of people enrolled in health maintenance organizations (HMOs). In Using Technology on pages 618–621, you can see more examples of mathematical models that are constructed from raw data.

APPLIED EXAMPLE 1 HMO Membership The number of people (in millions) enrolled in HMOs from 1994 through 2002 is given in the following table:

Year	1994	1995	1996	1997	1998	1999	2000	2001	2002
People	45.4	50.6	58.7	67.0	76.4	81.3	80.9	80.0	74.2

The **scatter plot** (graph of the data) associated with this data is shown in Figure 15a, where $t = 0$ corresponds to 1994. A mathematical model giving the approximate number of people, $N(t)$ (in millions), enrolled in HMOs during this period is

$$N(t) = 0.030915t^4 - 0.67974t^3 + 3.704t^2 + 1.63t + 45.5 \qquad (0 \le t \le 8)$$

The graph of N is shown in Figure 15b.

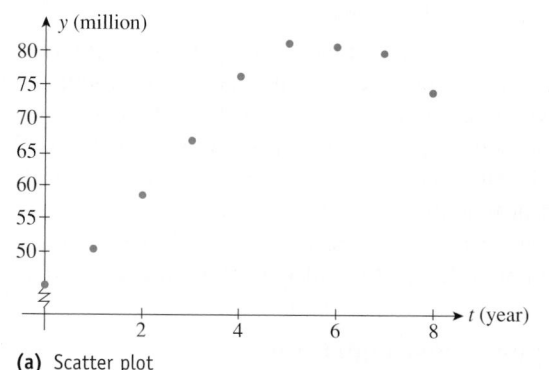

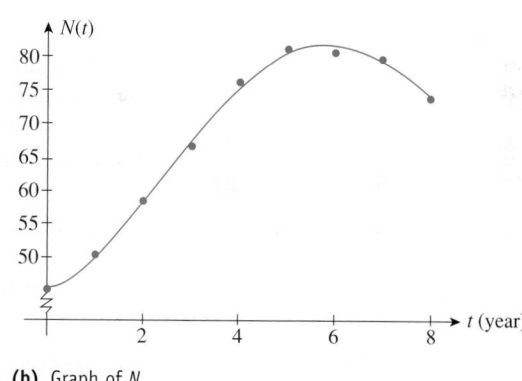

FIGURE 15   (a) Scatter plot   (b) Graph of N

a. Use the model to estimate the number of people enrolled in HMOs in 2000. How does this number compare with the actual number?
b. Assume that the trend continued and use the model to predict the number of people enrolled in HMOs in 2003.

Source: Group Health Association of America

Solution

a. The number of people that were enrolled in HMOs at the beginning of 2000 ($t = 6$) is given by

$$N(6) = 0.030915(6)^4 - 0.67974(6)^3 + 3.704(6)^2 + 1.63(6) + 45.5 \approx 81.87$$

or approximately 81.9 million. This number is very close to the actual number of 80.9 million people.

b. Assuming that the trend continued, the number of people enrolled in HMOs in 2003 ($t = 9$) was given by

$$N(9) = 0.030915(9)^4 - 0.67974(9)^3 + 3.704(9)^2 + 1.63(9) + 45.5 \approx 67.50$$

or approximately 67.5 million.   ∎

Polynomial Functions

We begin by recalling a special class of functions, polynomial functions.

> ### Polynomial Function
> A **polynomial function** of degree n is a function of the form
>
> $$f(x) = a_0 x^n + a_1 x^{n-1} + \cdots + a_{n-1}x + a_n \qquad (a_0 \neq 0)$$
>
> where $a_0, a_1, \ldots, a_n$ are constants and n is a nonnegative integer.

For example, the functions

$$f(x) = 4x^5 - 3x^4 + x^2 - x + 8$$
$$g(x) = 0.001x^3 - 2x^2 + 20x + 400$$

are polynomial functions of degrees 5 and 3, respectively. Observe that a polynomial function is defined everywhere so that it has domain $(-\infty, \infty)$.

A polynomial function of degree 1 ($n = 1$)

$$f(x) = a_0 x + a_1 \qquad (a_0 \neq 0)$$

is the equation of a straight line in the slope-intercept form with slope $m = a_0$ and y-intercept $b = a_1$ (see Section 1.2). For this reason, a polynomial function of degree 1 is called a **linear function**. For example, the linear function $f(x) = 2x + 3$ may be written as a linear equation in x and y—namely, $y = 2x + 3$ or $2x - y + 3 = 0$. Conversely, the linear equation $2x - 3y + 4 = 0$ can be solved for y in terms of x to yield the linear function $y = f(x) = \frac{2}{3}x + \frac{4}{3}$.

A polynomial function of degree 2 is referred to as a **quadratic function**. A polynomial function of degree 3 is called a **cubic function**, and so on.

Rational and Power Functions

Another important class of functions is rational functions. A **rational function** is simply the quotient of two polynomials. Examples of rational functions are

$$F(x) = \frac{3x^3 + x^2 - x + 1}{x - 2}$$
$$G(x) = \frac{x^2 + 1}{x^2 - 1}$$

In general, a rational function has the form

$$R(x) = \frac{f(x)}{g(x)}$$

where $f(x)$ and $g(x)$ are polynomial functions. Since division by zero is not allowed, we conclude that the domain of a rational function is the set of all real numbers except the zeros of g—that is, the roots of the equation $g(x) = 0$. Thus, the domain of the function F is the set of all numbers except $x = 2$, whereas the domain of the function G is the set of all numbers except those that satisfy $x^2 - 1 = 0$, or $x = \pm 1$.

Functions of the form

$$f(x) = x^r$$

where r is any real number, are called **power functions**. We encountered examples of power functions earlier in our work. For example, the functions

$$f(x) = \sqrt{x} = x^{1/2} \quad \text{and} \quad g(x) = \frac{1}{x^2} = x^{-2}$$

are power functions.

Many of the functions we will encounter later will involve combinations of the functions introduced here. For example, the following functions may be viewed as combinations of such functions:

$$f(x) = \sqrt{\frac{1 - x^2}{1 + x^2}}$$

$$g(x) = \sqrt{x^2 - 3x + 4}$$

$$h(x) = (1 + 2x)^{1/2} + \frac{1}{(x^2 + 2)^{3/2}}$$

As with polynomials of degree 3 or greater, analyzing the properties of these functions is facilitated by using the tools of calculus, to be developed later.

In the next example, we use a power function to construct a model that describes the driving costs of a car.

APPLIED EXAMPLE 2 Driving Costs   A study of driving costs based on the 2002 Ford Taurus SEL found the following average costs (car payments, gas, insurance, upkeep, and depreciation), measured in cents per mile.

Miles/year, x	5000	10,000	15,000	20,000
Cost/mile, y (¢)	80.0	60.0	49.8	44.9

A mathematical model (using least-squares techniques) giving the average cost in cents per mile is

$$C(x) = \frac{157.6}{x^{0.421}}$$

where x (in thousands) denotes the number of miles the car is driven in 1 year. The scatter plot associated with this data and the graph of C are shown in Figure 16. Using this model, estimate the average cost of driving the 2002 Ford Taurus SEL 8000 miles per year and 18,000 miles per year.

Solution   The average cost for driving the car 8000 miles per year is

$$C(8) = \frac{157.6}{8^{0.421}} \approx 65.67$$

or approximately 65.7¢/mile. The average cost for driving it 18,000 miles per year is

$$C(18) = \frac{157.6}{18^{0.421}} \approx 46.68$$

or approximately 46.7¢/mile.

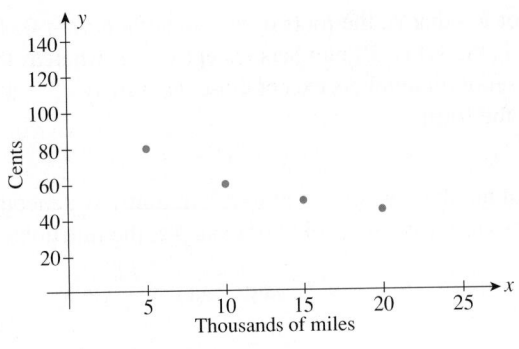

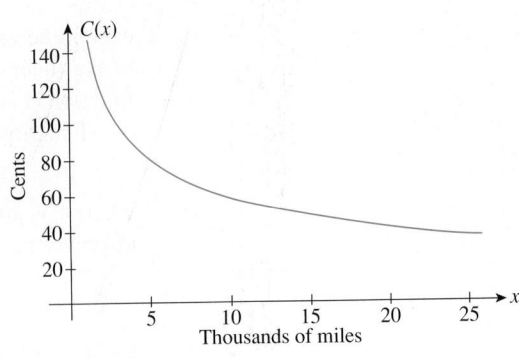

FIGURE 16
(a) The scatter plot and (b) the graph of the model for driving costs

(a) Scatter plot

Source: American Automobile Association

(b) Graph of *C*

Some Economic Models

In the remainder of this section, we look at some examples of *nonlinear* supply and **demand functions**. Before proceeding, however, you might want to review the concepts of supply and demand equations and of market equilibrium (see Sections 1.3 and 1.4).

EXAMPLE 3 The demand function for a certain brand of videocassette is given by

$$p = d(x) = -0.01x^2 - 0.2x + 8$$

where p is the wholesale unit price in dollars and x is the quantity demanded each week, measured in units of a thousand. Sketch the corresponding demand curve. Above what price will there be no demand? What is the maximum quantity demanded per week?

Solution The given function is quadratic, and its graph, which appears in Figure 17, may be sketched using the methods just developed. The p-intercept, 8, gives the wholesale unit price above which there will be no demand. To obtain the maximum quantity demanded, we set $p = 0$, which gives

$$-0.01x^2 - 0.2x + 8 = 0$$
$$x^2 + 20x - 800 = 0 \qquad \text{(Multiplying both sides of the equation by } -100)$$

or

$$(x + 40)(x - 20) = 0$$

That is, $x = -40$ or $x = 20$. Since x must be nonnegative, we reject the root $x = -40$. Thus, the maximum number of videocassettes demanded per week is 20,000.

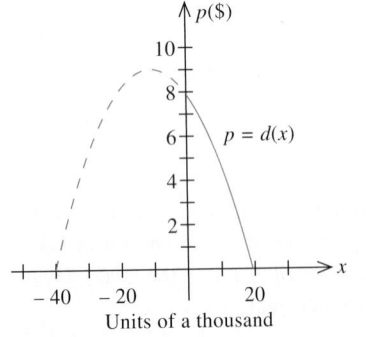

FIGURE 17
A demand curve

EXAMPLE 4 The supply function for a certain brand of videocassette is given by

$$p = s(x) = 0.01x^2 + 0.1x + 3$$

where p is the unit wholesale price in dollars and x stands for the quantity that will be made available in the market by the supplier, measured in units of a thou-

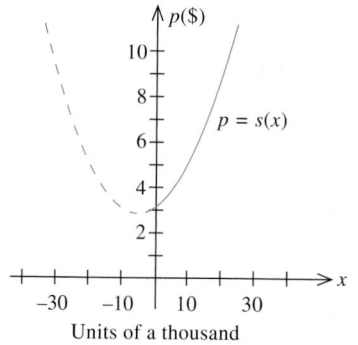

FIGURE 18
A supply curve

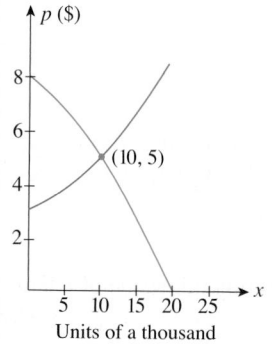

FIGURE 19
The supply curve and the demand curve
intersect at the point (10, 5).

sand. Sketch the corresponding supply curve. What is the lowest price at which the supplier will make the videocassettes available in the market?

Solution   Figure 18 is a sketch of the supply curve. The *p*-intercept, 3, gives the lowest price at which the supplier will make the videocassettes available in the market.

APPLIED EXAMPLE 5   Supply-Demand   The demand function for a certain brand of videocassette is given by

$$p = d(x) = -0.01x^2 - 0.2x + 8$$

and the corresponding supply function is given by

$$p = s(x) = 0.01x^2 + 0.1x + 3$$

where *p* is expressed in dollars and *x* is measured in units of a thousand. Find the equilibrium quantity and price.

Solution   We solve the following system of equations:

$$p = -0.01x^2 - 0.2x + 8$$
$$p = 0.01x^2 + 0.1x + 3$$

Substituting the first equation into the second yields

$$-0.01x^2 - 0.2x + 8 = 0.01x^2 + 0.1x + 3$$

which is equivalent to

$$0.02x^2 + 0.3x - 5 = 0$$
$$2x^2 + 30x - 500 = 0$$
$$x^2 + 15x - 250 = 0$$
$$(x + 25)(x - 10) = 0$$

Thus, $x = -25$ or $x = 10$. Since *x* must be nonnegative, the root $x = -25$ is rejected. Therefore, the equilibrium quantity is 10,000 videocassettes. The equilibrium price is given by

$$p = 0.01(10)^2 + 0.1(10) + 3 = 5$$

or $5 per videocassette (Figure 19).

EXPLORING WITH TECHNOLOGY

1. a. Use a graphing utility to plot the straight lines L_1 and L_2 with equations $y = 2x - 1$ and $y = 2.1x + 3$, respectively, on the same set of axes, using the standard viewing window. Do the lines appear to intersect?
   b. Plot the straight lines L_1 and L_2, using the viewing window $[-100, 100] \times [-100, 100]$. Do the lines appear to intersect? Can you find the point of intersection using TRACE and ZOOM? Using the "intersection" function of your graphing utility?
   c. Find the point of intersection of L_1 and L_2 algebraically.
   d. Comment on the effectiveness of the methods of solutions in parts (b) and (c).

2. a. Use a graphing utility to plot the straight lines L_1 and L_2 with equations $y = 3x - 2$ and $y = -2x + 3$, respectively, on the same set of axes, using the standard viewing window. Then use TRACE and ZOOM to find the point of intersection of L_1 and L_2. Repeat using the "intersection" function of your graphing utility.
   b. Find the point of intersection of L_1 and L_2 algebraically.
   c. Comment on the effectiveness of the methods.

Constructing Mathematical Models

We close this section by showing how some mathematical models can be constructed using elementary geometric and algebraic arguments.

The following guidelines can be used to construct mathematical models.

Guidelines for Constructing Mathematical Models

1. Assign a letter to each variable mentioned in the problem. If appropriate, draw and label a figure.
2. Find an expression for the quantity sought.
3. Use the conditions given in the problem to write the quantity sought as a function f of one variable. Note any restrictions to be placed on the domain of f from physical considerations of the problem.

APPLIED EXAMPLE 6 Enclosing an Area The owner of the Rancho Los Feliz has 3000 yards of fencing with which to enclose a rectangular piece of grazing land along the straight portion of a river. Fencing is not required along the river. Letting x denote the width of the rectangle, find a function f in the variable x giving the area of the grazing land if she uses all of the fencing (Figure 20).

Solution

1. This information was given.

2. The area of the rectangular grazing land is $A = xy$. Next, observe that the amount of fencing is $2x + y$ and this must be equal to 3000 since all the fencing is used; that is,

$$2x + y = 3000$$

3. From the equation we see that $y = 3000 - 2x$. Substituting this value of y into the expression for A gives

$$A = xy = x(3000 - 2x) = 3000x - 2x^2$$

Finally, observe that both x and y must be nonnegative since they represent the width and length of a rectangle, respectively. Thus, $x \geq 0$ and $y \geq 0$. But the latter is equivalent to $3000 - 2x \geq 0$, or $x \leq 1500$. So the required function is $f(x) = 3000x - 2x^2$ with domain $0 \leq x \leq 1500$.   ∎

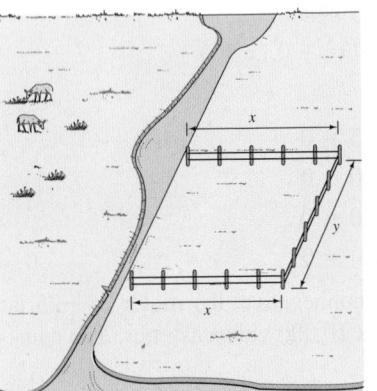

FIGURE 20
The rectangular grazing land has width x and length y.

Note Observe that if we view the function $f(x) = 3000x - 2x^2$ strictly as a mathematical entity, then its domain is the set of all real numbers. But physical considerations dictate that its domain should be restricted to the interval [0, 1500].   ∎

APPLIED EXAMPLE 7 Charter-Flight Revenue If exactly 200 people sign up for a charter flight, Leisure World Travel Agency charges $300 per person. However, if more than 200 people sign up for the flight (assume this is the case), then each fare is reduced by $1 for each additional person. Letting x denote the number of passengers above 200, find a function giving the revenue realized by the company.

Solution

1. This information was given.

2. If there are x passengers above 200, then the number of passengers signing up for the flight is $200 + x$. Furthermore, the fare will be $(300 - x)$ dollars per passenger.

3. The revenue will be

$$R = (200 + x)(300 - x) \qquad \text{Number of passengers} \times$$
$$= -x^2 + 100x + 60{,}000 \qquad \text{the fare per passenger}$$

Clearly, x must be nonnegative, and $300 - x \geq 0$, or $x \leq 300$. So the required function is $f(x) = -x^2 + 100x + 60{,}000$ with domain $[0, 300]$. ■

11.3 Self-Check Exercises

1. Thomas Young has suggested the following rule for calculating the dosage of medicine for children from ages 1 to 12 yr. If a denotes the adult dosage (in milligrams) and t is the age of the child (in years), then the child's dosage is given by

$$D(t) = \frac{at}{t + 12}$$

If the adult dose of a substance is 500 mg, how much should a 4-yr-old child receive?

2. The demand function for Mrs. Baker's cookies is given by

$$d(x) = -\frac{2}{15}x + 4$$

where $d(x)$ is the wholesale price in dollars/pound and x is the quantity demanded each week, measured in thousands of pounds. The supply function for the cookies is given by

$$s(x) = \frac{1}{75}x^2 + \frac{1}{10}x + \frac{3}{2}$$

where $s(x)$ is the wholesale price in dollars/pound and x is the quantity, in thousands of pounds, that will be made available in the market each week by the supplier.

a. Sketch the graphs of the functions d and s.
b. Find the equilibrium quantity and price.

Solutions to Self-Check Exercises 11.3 can be found on page 617.

11.3 Concept Questions

1. Describe mathematical modeling in your own words.

2. Define (a) a polynomial function and (b) a rational function. Give an example of each.

3. a. What is a demand function? A supply function?

b. What is market equilibrium? Describe how you would go about finding the equilibrium quantity and equilibrium price given the demand and supply equations associated with a commodity.

11.3 Exercises

In Exercises 1–6, determine whether the given function is a polynomial function, a rational function, or some other function. State the degree of each polynomial function.

1. $f(x) = 3x^6 - 2x^2 + 1$

2. $f(x) = \dfrac{x^2 - 9}{x - 3}$

3. $G(x) = 2(x^2 - 3)^3$

4. $H(x) = 2x^{-3} + 5x^{-2} + 6$

5. $f(t) = 2t^2 + 3\sqrt{t}$

6. $f(r) = \dfrac{6r}{r^3 - 8}$

7. Disposable Income   Economists define the *disposable annual income* for an individual by the equation $D = (1 - r)T$, where T is the individual's total income and r is the net rate at which he or she is taxed. What is the disposable income for an individual whose income is $60,000 and whose net tax rate is 28%?

8. Worker Efficiency   An efficiency study showed that the average worker at Delphi Electronics assembled cordless telephones at the rate of

$$f(t) = -\frac{3}{2}t^2 + 6t + 10 \qquad (0 \le t \le 4)$$

phones/hour, t hr after starting work during the morning shift. At what rate does the average worker assemble telephones 2 hr after starting work?

9. **EFFECT OF ADVERTISING ON SALES** The quarterly profit of Cunningham Realty depends on the amount of money x spent on advertising/quarter according to the rule

$$P(x) = -\frac{1}{8}x^2 + 7x + 30 \qquad (0 \le x \le 50)$$

where $P(x)$ and x are measured in thousands of dollars. What is Cunningham's profit when its quarterly advertising budget is $28,000?

10. **SPENDING ON MEDICAL DEVICES** The U.S. market size for medical devices (in billions of dollars) in year t, from 1999 ($t = 0$) through 2005 ($t = 6$) is approximated by the function

$$S(t) = 0.288t^2 + 3.03t + 45.9 \qquad (0 \le t \le 6)$$

What was the U.S. market size for medical devices in 1999? In 2005?

Source: Frost and Sullivan

11. **SOLAR POWER** More and more businesses and homeowners are installing solar panels on their roofs to draw energy from the Sun's rays. According to the U.S. Department of Energy, the solar cell kilowatt-hour used in the United States (in millions) is projected to be

$$S(t) = 0.73t^2 + 15.8t + 2.7 \qquad (0 \le t \le 8)$$

in year t, with $t = 0$ corresponding to 2000. What is the projected solar cell kilowatt-hour use in the United States for 2006? For 2008?

Source: U.S. Department of Energy

12. **SALES OF DIGITAL TVS** The number of homes with digital TVs is expected to grow according to the function

$$f(t) = 0.1714t^2 + 0.6657t + 0.7143 \qquad (0 \le t \le 6)$$

where t is measured in years, with $t = 0$ corresponding to the beginning of 2000, and $f(t)$ is measured in millions of homes.
a. How many homes had digital TVs at the beginning of 2000?
b. How many homes had digital TVs at the beginning of 2005?

Source: Consumer Electronics Manufacturers Association

13. **AGING DRIVERS** The number of fatalities due to car crashes, based on the number of miles driven, begins to climb after the driver is past age 65. Aside from declining ability as one ages, the older driver is more fragile. The number of fatalities per 100 million vehicle miles driven is approximately

$$N(x) = 0.0336x^3 - 0.118x^2 + 0.215x + 0.7 \qquad (0 \le x \le 7)$$

where x denotes the age group of drivers, with $x = 0$ corresponding to those aged 50–54, $x = 1$ corresponding to those

aged 55–59, $x = 2$ corresponding to those aged 60–64, . . . , and $x = 7$ corresponding to those aged 85–89. What is the fatality rate per 100 million vehicle miles driven for an average driver in the 50–54 age group? In the 85–89 age group?

Source: U.S. Department of Transportation

14. **TESTOSTERONE USE** Fueled by the promotion of testosterone as an antiaging elixir, use of the hormone by middle-age and older men grew dramatically. The total number of prescriptions for testosterone from 1999 through 2002 is given by

$$N(t) = -35.8t^3 + 202t^2 + 87.8t + 648 \qquad (0 \le t \le 3)$$

where $N(t)$ is measured in thousands and t is measured in years, with $t = 0$ corresponding to 1999. Find the total number of prescriptions for testosterone in 1999, 2000, 2001, and 2002.

Source: IMS Health

15. **BLACKBERRY SUBSCRIBERS** According to a study conducted in 2004, the number of subscribers of BlackBerry, the handheld email devices manufactured by Research in Motion Ltd., is expected to be

$$N(t) = -0.0675t^4 + 0.5083t^3 - 0.893t^2 + 0.66t + 0.32$$
$$(0 \le t \le 4)$$

where $N(t)$ is measured in millions and t in years, with $t = 0$ corresponding to the beginning of 2002.
a. How many BlackBerry subscribers were there at the beginning of 2002?
b. What is the projected number of BlackBerry subscribers at the beginning of 2006?

Source: ThinkEquity Partners

16. **CHIP SALES** The worldwide flash memory chip sales (in billions of dollars) is projected to be

$$S(t) = 4.3(t + 2)^{0.94} \qquad (0 \le t \le 6)$$

where t is measured in years, with $t = 0$ corresponding to 2002. Flash chips are used in cell phones, digital cameras, and other products.
a. What were the worldwide flash memory chip sales in 2002?
b. What are the projected sales for 2008?

Source: Web-Feet Research Inc.

17. **OUTSOURCING OF JOBS** According to a study conducted in 2003, the total number of U.S. jobs (in millions) that are projected to leave the country by year t, where $t = 0$ corresponds to 2000, is

$$N(t) = 0.0018425(t + 5)^{2.5} \qquad (0 \le t \le 15)$$

How many jobs will be outsourced by 2005? By 2010?

Source: Forrester Research

18. **SELLING PRICE OF DVD RECORDERS** The rise of digital music and the improvement to the DVD format are part of the rea-

sons why the average selling price of standalone DVD recorders will drop in the coming years. The function

$$A(t) = \frac{699}{(t+1)^{0.94}} \quad (0 \le t \le 5)$$

gives the projected average selling price (in dollars) of standalone DVD recorders in year t, where $t = 0$ corresponds to the beginning of 2002. What was the average selling price of standalone DVD recorders at the beginning of 2002? What will it be at the beginning of 2007?

Source: Consumer Electronics Association

19. **E-MAIL USAGE**   The number of international emailings per day (in millions) is approximated by the function

$$f(t) = 38.57t^2 - 24.29t + 79.14 \quad (0 \le t \le 4)$$

where t is measured in years, with $t = 0$ corresponding to the beginning of 1998.
a. Sketch the graph of f.
b. How many international emailings per day were there at the beginning of 2002?

Source: Pioneer Consulting

20. **DOCUMENT MANAGEMENT**   The size (measured in millions of dollars) of the document-management business is described by the function

$$f(t) = 0.22t^2 + 1.4t + 3.77 \quad (0 \le t \le 6)$$

where t is measured in years, with $t = 0$ corresponding to the beginning of 1996.
a. Sketch the graph of f.
b. What was the size of the document-management business at the beginning of 2002?

Source: Sun Trust Equitable Securities

21. **REACTION OF A FROG TO A DRUG**   Experiments conducted by A. J. Clark suggest that the response $R(x)$ of a frog's heart muscle to the injection of x units of acetylcholine (as a percent of the maximum possible effect of the drug) may be approximated by the rational function

$$R(x) = \frac{100x}{b+x} \quad (x \ge 0)$$

where b is a positive constant that depends on the particular frog.
a. If a concentration of 40 units of acetylcholine produces a response of 50% for a certain frog, find the "response function" for this frog.
b. Using the model found in part (a), find the response of the frog's heart muscle when 60 units of acetylcholine are administered.

22. **WALKING VERSUS RUNNING**   The oxygen consumption (in milliliter/pound/minute) for a person walking at x mph is approximated by the function

$$f(x) = \frac{5}{3}x^2 + \frac{5}{3}x + 10 \quad (0 \le x \le 9)$$

whereas the oxygen consumption for a runner at x mph is approximated by the function

$$g(x) = 11x + 10 \quad (4 \le x \le 9)$$

a. Sketch the graphs of f and g.
b. At what speed is the oxygen consumption the same for a walker as it is for a runner? What is the level of oxygen consumption at that speed?
c. What happens to the oxygen consumption of the walker and the runner at speeds beyond that found in part (b)?

Source: William McArdley, Frank Katch, and Victor Katch, *Exercise Physiology*

23. **CREDIT CARD DEBT**   Following the introduction in 1950 of the nation's first credit card, the Diners Club Card, credit cards have proliferated over the years. More than 720 different cards are now used at more than 4 million locations in the United States. The average U.S. credit card debt (per household) in thousands of dollars is approximately given by

$$D(t) = \begin{cases} 4.77(1+t)^{0.2676} & \text{if } 0 \le t \le 2 \\ 5.6423t^{0.1818} & \text{if } 2 < t \le 6 \end{cases}$$

where t is measured in years, with $t = 0$ corresponding to the beginning of 1994. What was the average U.S. credit card debt (per household) at the beginning of 1994? At the beginning of 1996? At the beginning of 1999?

Source: David Evans and Richard Schmalensee, *Paying with Plastic: The Digital Revolution in Buying and Borrowing*

24. **OBESE CHILDREN IN THE UNITED STATES**   The percent of obese children aged 12–19 in the United States is approximately

$$P(t) = \begin{cases} 0.04t + 4.6 & \text{if } 0 \le t < 10 \\ -0.01005t^2 + 0.945t - 3.4 & \text{if } 10 \le t \le 30 \end{cases}$$

where t is measured in years, with $t = 0$ corresponding to the beginning of 1970. What was the percent of obese children aged 12–19 at the beginning of 1970? At the beginning of 1985? At the beginning of 2000?

Source: Centers for Disease Control

25. **U.S. NUTRITIONAL SUPPLEMENTS MARKET**   The size of the U.S. nutritional supplements market from 1999 through 2003 is approximated by the function

$$A(t) = 16.4(t+1)^{0.1} \quad (0 \le t \le 4)$$

where $A(t)$ is measured in billions of dollars and t is measured in years, with $t = 0$ corresponding to 1999.
a. Compute $A(0)$, $A(1)$, $A(2)$, $A(3)$, and $A(4)$. Interpret your results.
b. Use the results of part (a) to sketch the graph of A.

Source: Nutrition Business Journal

26. **PRICE OF IVORY**   According to the World Wildlife Fund, a group in the forefront of the fight against illegal ivory trade, the price of ivory (in dollars/kilo) compiled from a variety of legal and black market sources is approximated by the function

$$f(t) = \begin{cases} 8.37t + 7.44 & \text{if } 0 \le t \le 8 \\ 2.84t + 51.68 & \text{if } 8 < t \le 30 \end{cases}$$

where t is measured in years, with $t = 0$ corresponding to the beginning of 1970.

a. Sketch the graph of the function f.

b. What was the price of ivory at the beginning of 1970? At the beginning of 1990?

Source: World Wildlife Fund

27. **SENIOR CITIZENS' HEALTH CARE** According to a study, the out-of-pocket cost to senior citizens for health care, $f(t)$ (as a percent of income), in year t where $t = 0$ corresponds to 1977, is given by

$$y = \begin{cases} \dfrac{2}{7}t + 12 & \text{if } 0 \le t \le 7 \\ t + 7 & \text{if } 7 < t \le 10 \\ \dfrac{3}{5}t + 11 & \text{if } 10 < t \le 20 \end{cases}$$

a. Sketch the graph of f.

b. What was the out-of-pocket cost, as a percent of income, to senior citizens for health care in 1982? In 1992?

Source: Senate Select Committee on Aging, AARP

28. **WORKING-AGE POPULATION** The ratio of working-age population to the elderly in the United States (including projections after 2000) is given by

$$f(t) = \begin{cases} 4.1 & \text{if } 0 \le t < 5 \\ -0.03t + 4.25 & \text{if } 5 \le t < 15 \\ -0.075t + 4.925 & \text{if } 15 \le t \le 35 \end{cases}$$

with $t = 0$ corresponding to the beginning of 1995.

a. Sketch the graph of f.

b. What was the ratio at the beginning of 2005? What will be the ratio at the beginning of 2020?

c. Over what years is the ratio constant?

d. Over what years is the decline of the ratio greatest?

Source: U.S. Census Bureau

29. **PRICE OF AUTOMOBILE PARTS** For years, automobile manufacturers had a monopoly on the replacement-parts market, particularly for sheet metal parts such as fenders, doors, and hoods, the parts most often damaged in a crash. Beginning in the late 1970s, however, competition appeared on the scene. In a report conducted by an insurance company to study the effects of the competition, the price of an OEM (original equipment manufacturer) fender for a particular 1983 model car was found to be

$$f(t) = \frac{110}{\frac{1}{2}t + 1} \qquad (0 \le t \le 2)$$

where $f(t)$ is measured in dollars and t is in years. Over the same period of time, the price of a non-OEM fender for the car was found to be

$$g(t) = 26\left(\frac{1}{4}t^2 - 1\right)^2 + 52 \qquad (0 \le t \le 2)$$

where $g(t)$ is also measured in dollars. Find a function $h(t)$ that gives the difference in price between an OEM fender

and a non-OEM fender. Compute $h(0)$, $h(1)$, and $h(2)$. What does the result of your computation seem to say about the price gap between OEM and non-OEM fenders over the 2 yr?

30. **SALES OF DVD PLAYERS VS. VCRS** The sales of DVD players in year t (in millions of units) is given by the function

$$f(t) = 5.6(1 + t) \qquad (0 \le t \le 3)$$

where $t = 0$ corresponds to 2001. Over the same period, the sales of VCRs (in millions of units) is given by

$$g(t) = \begin{cases} -9.6t + 22.5 & \text{if } 0 \le t \le 1 \\ -0.5t + 13.4 & \text{if } 1 < t \le 2 \\ -7.8t + 28 & \text{if } 2 < t \le 3 \end{cases}$$

a. Show that more VCRs than DVD players were sold in 2001.

b. When did the sales of DVD players first exceed those of VCRs?

Source: Popular Science

31. **HOTEL OCCUPANCY RATE** A forecast released by PricewaterhouseCoopers in June of 2004 predicted the occupancy rate of U.S. hotels between 2001 ($t = 0$) and 2005 ($t = 4$) to be

$$P(t) = \begin{cases} -0.9t + 59.8 & \text{if } 0 \le t < 1 \\ 0.3t + 58.6 & \text{if } 1 \le t < 2 \\ 56.79t^{0.06} & \text{if } 2 \le t \le 4 \end{cases}$$

percent.

a. Compute $P(0)$, $P(1)$, $P(2)$, $P(3)$, and $P(4)$.

b. Sketch the graph of P.

c. What was the predicted occupancy rate of hotels for 2004?

Source: PricewaterhouseCoopers LLP Hospitality & Leisure Research

32. **IMMIGRATION TO THE UNITED STATES** The immigration to the United States from Europe, as a percent of the total immigration, is approximately

$$P(t) = 0.767t^3 - 0.636t^2 - 19.17t + 52.7 \qquad (0 \le t \le 4)$$

where t is measured in decades, with $t = 0$ corresponding to the decade of the 1950s.

a. Complete the table:

t	0	1	2	3	4
$P(t)$					

b. Use the result of part (a) to sketch the graph of P.

c. Use the result of part (b) to estimate the decade when the immigration, as a percent of the total immigration, was the greatest and the smallest.

Source: Jeffrey Williamson, Harvard University

For the demand equations in Exercises 33–36, where x represents the quantity demanded in units of a thousand and p is the unit price in dollars, (a) sketch the demand curve and (b) determine the quantity demanded when the unit price is set at $p.

33. $p = -x^2 + 16; p = 7$ **34.** $p = -x^2 + 36; p = 11$

35. $p = \sqrt{18 - x^2}; p = 3$ **36.** $p = \sqrt{9 - x^2}; p = 2$

For the supply equations in Exercises 37–40, where x is the quantity supplied in units of a thousand and p is the unit price in dollars, (a) sketch the supply curve and (b) determine the price at which the supplier will make 2000 units of the commodity available in the market.

37. $p = x^2 + 16x + 40$ **38.** $p = 2x^2 + 18$

39. $p = x^3 + 2x + 3$ **40.** $p = x^3 + x + 10$

41. DEMAND FOR SMOKE ALARMS   The demand function for the Sentinel smoke alarm is given by

$$p = \frac{30}{0.02x^2 + 1} \qquad (0 \le x \le 10)$$

where x (measured in units of a thousand) is the quantity demanded per week and p is the unit price in dollars.
a. Sketch the graph of the demand function.
b. What is the unit price that corresponds to a quantity demanded of 10,000 units?

42. DEMAND FOR COMMODITIES   Assume that the demand function for a certain commodity has the form

$$p = \sqrt{-ax^2 + b} \qquad (a \ge 0, b \ge 0)$$

where x is the quantity demanded, measured in units of a thousand and p is the unit price in dollars. Suppose the quantity demanded is 6000 ($x = 6$) when the unit price is $8.00 and 8000 ($x = 8$) when the unit price is $6.00. Determine the demand equation. What is the quantity demanded when the unit price is set at $7.50?

43. SUPPLY FUNCTIONS   The supply function for the Luminar desk lamp is given by

$$p = 0.1x^2 + 0.5x + 15$$

where x is the quantity supplied (in thousands) and p is the unit price in dollars.
a. Sketch the graph of the supply function.
b. What unit price will induce the supplier to make 5000 lamps available in the marketplace?

44. SUPPLY FUNCTIONS   Suppliers of satellite radios will market 10,000 units when the unit price is $20 and 62,500 units when the unit price is $35. Determine the supply function if it is known to have the form

$$p = a\sqrt{x} + b \qquad (a \ge 0, b \ge 0)$$

where x is the quantity supplied and p is the unit price in dollars. Sketch the graph of the supply function. What unit price will induce the supplier to make 40,000 satellite radios available in the marketplace?

For each pair of supply and demand equations in Exercises 45–48, where x represents the quantity demanded in units of a thousand and p the unit price in dollars, find the equilibrium quantity and the equilibrium price.

45. $p = -x^2 - 2x + 100$ and $p = 8x + 25$

46. $p = -2x^2 + 80$ and $p = 15x + 30$

47. $p = 60 - 2x^2$ and $p = x^2 + 9x + 30$

48. $11p + 3x - 66 = 0$ and $2p^2 + p - x = 10$

49. MARKET EQUILIBRIUM   The weekly demand and supply functions for Sportsman 5×7 tents are given by

$$p = -0.1x^2 - x + 40$$
$$p = 0.1x^2 + 2x + 20$$

respectively, where p is measured in dollars and x is measured in units of a hundred. Find the equilibrium quantity and price.

50. MARKET EQUILIBRIUM   The management of Titan Tire Company has determined that the weekly demand and supply functions for their Super Titan tires are given by

$$p = 144 - x^2$$
$$p = 48 + \frac{1}{2}x^2$$

respectively, where p is measured in dollars and x is measured in units of a thousand. Find the equilibrium quantity and price.

51. ENCLOSING AN AREA   Patricia wishes to have a rectangular-shaped garden in her backyard. She has 80 ft of fencing with which to enclose her garden. Letting x denote the width of the garden, find a function f in the variable x giving the area of the garden. What is its domain?

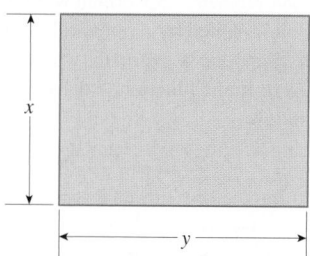

52. ENCLOSING AN AREA   Patricia's neighbor, Juanita, also wishes to have a rectangular-shaped garden in her backyard. But Juanita wants her garden to have an area of 250 ft². Letting x denote the width of the garden, find a function f in the variable x giving the length of the fencing required to construct the garden. What is the domain of the function?

Hint: Refer to the figure for Exercise 51. The amount of fencing required is equal to the perimeter of the rectangle, which is twice the width plus twice the length of the rectangle.

53. **PACKAGING**   By cutting away identical squares from each corner of a rectangular piece of cardboard and folding up the resulting flaps, an open box may be made. If the cardboard is 15 in. long and 8 in. wide and the square cutaways have dimensions of x in. by x in., find a function giving the volume of the resulting box.

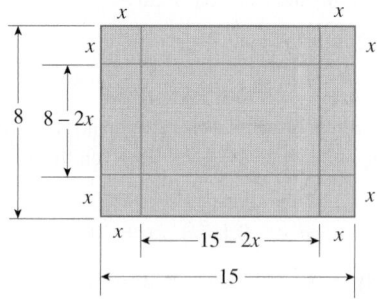

54. **CONSTRUCTION COSTS**   A rectangular box is to have a square base and a volume of 20 ft³. The material for the base costs 30¢/ft², the material for the sides costs 10¢/ft², and the material for the top costs 20¢/ft². Letting x denote the length of one side of the base, find a function in the variable x giving the cost of constructing the box.

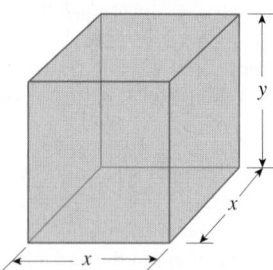

55. **AREA OF A NORMAN WINDOW**   A Norman window has the shape of a rectangle surmounted by a semicircle (see the accompanying figure). Suppose a Norman window is to have a perimeter of 28 ft; find a function in the variable x giving the area of the window.

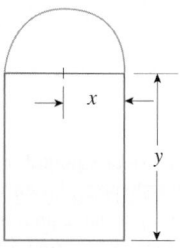

56. **YIELD OF AN APPLE ORCHARD**   An apple orchard has an average yield of 36 bushels of apples/tree if tree density is 22 trees/acre. For each unit increase in tree density, the yield decreases by 2 bushels/tree. Letting x denote the number of trees beyond 22/acre, find a function in x that gives the yield of apples.

57. **BOOK DESIGN**   A book designer has decided that the pages of a book should have 1-in. margins at the top and bottom and $\frac{1}{2}$-in. margins on the sides. She further stipulated that each page should have an area of 50 in.². Find a function in the variable x, giving the area of the printed page. What is the domain of the function?

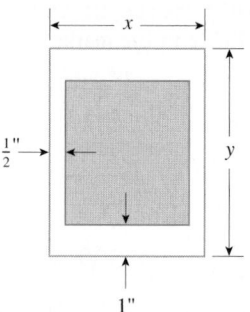

58. **PROFIT OF A VINEYARD**   Phillip, the proprietor of a vineyard, estimates that if 10,000 bottles of wine were produced this season, then the profit would be $5/bottle. But if more than 10,000 bottles were produced, then the profit/bottle would drop by $0.0002 for each additional bottle sold. Assume at least 10,000 bottles of wine are produced and sold and let x denote the number of bottles produced and sold above 10,000.
   a. Find a function P giving the profit in terms of x.
   b. What is the profit Phillip can expect from the sale of 16,000 bottles of wine from his vineyard?

59. **CHARTER REVENUE**   The owner of a luxury motor yacht that sails among the 4000 Greek islands charges $600/person/day if exactly 20 people sign up for the cruise. However, if more than 20 people sign up (up to the maximum capacity of 90) for the cruise, then each fare is reduced by $4 for each additional passenger. Assume at least 20 people sign up for the cruise and let x denote the number of passengers above 20.
   a. Find a function R giving the revenue/day realized from the charter.
   b. What is the revenue/day if 60 people sign up for the cruise?
   c. What is the revenue/day if 80 people sign up for the cruise?

In Exercises 60–63, determine whether the statement is true or false. If it is true, explain why it is true. If it is false, give an example to show why it is false.

60. A polynomial function is a sum of constant multiples of power functions.

61. A polynomial function is a rational function, but the converse is false.

62. If $r > 0$, then the power function $f(x) = x^r$ is defined for all values of x.

63. The function $f(x) = 2^x$ is a power function.

11.3   Solutions to Self-Check Exercises

1. Since the adult dose of the substance is 500 mg, $a = 500$; thus, the rule in this case is

$$D(t) = \frac{500t}{t + 12}$$

A 4-yr-old should receive

$$D(4) = \frac{500(4)}{4 + 12}$$

or 125 mg of the substance.

2. a. The graphs of the functions d and s are shown in the following figure:

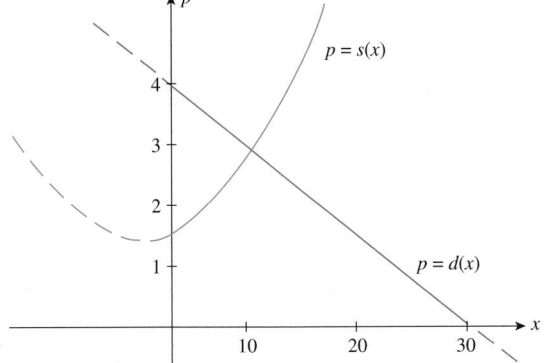

b. Solve the following system of equations:

$$p = -\frac{2}{15}x + 4$$
$$p = \frac{1}{75}x^2 + \frac{1}{10}x + \frac{3}{2}$$

Substituting the first equation into the second yields

$$\frac{1}{75}x^2 + \frac{1}{10}x + \frac{3}{2} = -\frac{2}{15}x + 4$$
$$\frac{1}{75}x^2 + \left(\frac{1}{10} + \frac{2}{15}\right)x - \frac{5}{2} = 0$$
$$\frac{1}{75}x^2 + \frac{7}{30}x - \frac{5}{2} = 0$$

Multiplying both sides of the last equation by 150, we have

$$2x^2 + 35x - 375 = 0$$
$$(2x - 15)(x + 25) = 0$$

Thus, $x = -25$ or $x = 15/2 = 7.5$. Since x must be nonnegative, we take $x = 7.5$, and the equilibrium quantity is 7500 lb. The equilibrium price is given by

$$p = -\frac{2}{15}\left(\frac{15}{2}\right) + 4$$

or \$3/lb.

USING TECHNOLOGY

■ Finding the Points of Intersection of Two Graphs and Modeling

A graphing utility can be used to find the point(s) of intersection of the graphs of two functions.

EXAMPLE 1   Find the points of intersection of the graphs of

$$f(x) = 0.3x^2 - 1.4x - 3 \quad \text{and} \quad g(x) = -0.4x^2 + 0.8x + 6.4$$

Solution   The graphs of both f and g in the standard viewing window are shown in Figure T1a. Using TRACE and ZOOM or the function for finding the points of intersection of two graphs on your graphing utility, we find the point(s) of inter-

(*continued*)

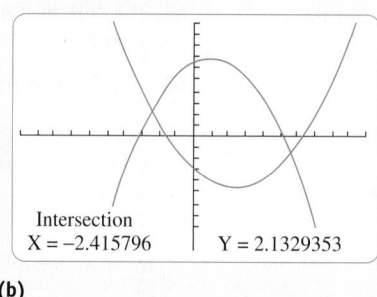
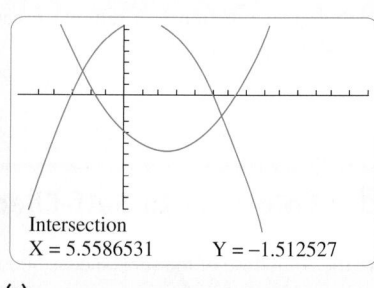

(a)   **(b)**   **(c)**

FIGURE T1
(a) The graphs of f and g in the standard viewing window; (b) and (c) the TI-83 intersection screens

section, accurate to four decimal places, to be $(-2.4158, 2.1329)$ (Figure T1b) and $(5.5587, -1.5125)$ (Figure T1c).

EXAMPLE 2   Consider the demand and supply functions

$$p = d(x) = -0.01x^2 - 0.2x + 8 \quad \text{and} \quad p = s(x) = 0.01x^2 + 0.1x + 3$$

of Example 5 in Section 11.3.

a. Plot the graphs of d and s in the viewing window $[0, 15] \times [0, 10]$.
b. Verify that the equilibrium point is $(10, 5)$, as obtained in Example 5.

Solution

a. The graphs of d and s are shown in Figure T2a.

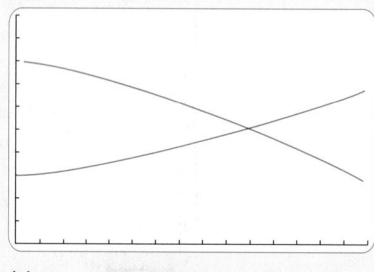
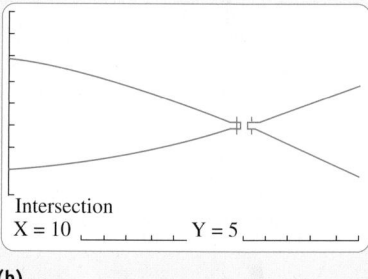

(a)   **(b)**

FIGURE T2
(a) The graphs of d and s in the window $[0, 15] \times [0, 10]$; (b) the TI-83 intersection screen

b. Using **TRACE** and **ZOOM** or the function for finding the point of intersection of two graphs, we see that $x = 10$ and $y = 5$ (Figure T2b), so the equilibrium point is $(10, 5)$, as obtained before.

Constructing Mathematical Models from Raw Data

A graphing utility can sometimes be used to construct mathematical models from sets of data. For example, if the points corresponding to the given data are scattered about a straight line, then use **LINR** (linear regression) from the **STAT CALC** (statistical calculation) menu of the graphing utility to obtain a function (model) that approximates the data at hand. If the points seem to be scattered along a parabola (the graph of a quadratic function), then use **P2REG** (second-order polynomial regression), and so on. (These are functions on the TI-83 calculator.)

APPLIED EXAMPLE 3 Indian Gaming Industry   The following data gives the estimated gross revenues (in billions of dollars) from the Indian gaming industries from 1990 ($t = 0$) to 1997 ($t = 7$).

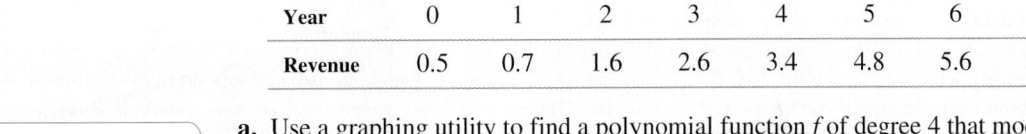

Year	0	1	2	3	4	5	6	7
Revenue	0.5	0.7	1.6	2.6	3.4	4.8	5.6	6.8

a. Use a graphing utility to find a polynomial function f of degree 4 that models the data.
b. Plot the graph of the function f, using the viewing window $[0, 8] \times [0, 10]$.
c. Use the function evaluation capability of the graphing utility to compute $f(0)$, $f(1), \ldots, f(7)$ and compare these values with the original data.

Source: Christiansen/Cummings Associates

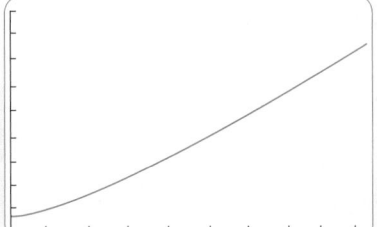

FIGURE T3
The graph of f in the viewing window
$[0, 8] \times [0, 10]$

Solution

a. Choosing **P4REG** (fourth-order polynomial regression) from the **STAT CALC** (statistical calculations) menu of a graphing utility, we find

$$f(t) = 0.00379t^4 - 0.06616t^3 + 0.41667t^2 - 0.07291t + 0.48333$$

b. The graph of f is shown in Figure T3.
c. The required values, which compare favorably with the given data, follow:

t	0	1	2	3	4	5	6	7
$f(t)$	0.5	0.8	1.5	2.5	3.6	4.6	5.7	6.8

TECHNOLOGY EXERCISES

In Exercises 1–6, find the points of intersection of the graphs of the functions. Express your answer accurate to four decimal places.

1. $f(x) = 1.2x + 3.8$; $g(x) = -0.4x^2 + 1.2x + 7.5$

2. $f(x) = 0.2x^2 - 1.3x - 3$; $g(x) = -1.3x + 2.8$

3. $f(x) = 0.3x^2 - 1.7x - 3.2$; $g(x) = -0.4x^2 + 0.9x + 6.7$

4. $f(x) = -0.3x^2 + 0.6x + 3.2$; $g(x) = 0.2x^2 - 1.2x - 4.8$

5. $f(x) = 0.3x^3 - 1.8x^2 + 2.1x - 2$; $g(x) = 2.1x - 4.2$

6. $f(x) = -0.2x^3 + 1.2x^2 - 1.2x + 2$; $g(x) = -0.2x^2 + 0.8x + 2.1$

7. MARKET EQUILIBRIUM The monthly demand and supply functions for a certain brand of wall clock are given by

$$p = -0.2x^2 - 1.2x + 50$$
$$p = 0.1x^2 + 3.2x + 25$$

respectively, where p is measured in dollars and x is measured in units of a hundred.
a. Plot the graphs of both functions in an appropriate viewing window.
b. Find the equilibrium quantity and price.

8. MARKET EQUILIBRIUM The quantity demanded x (in units of a hundred) of Mikado miniature cameras/week is related to the unit price p (in dollars) by

$$p = -0.2x^2 + 80$$

The quantity x (in units of a hundred) that the supplier is willing to make available in the market is related to the unit price p (in dollars) by

$$p = 0.1x^2 + x + 40$$

a. Plot the graphs of both functions in an appropriate viewing window.
b. Find the equilibrium quantity and price.

In Exercises 9–18, use the STAT CALC menu to construct a mathematical model associated with the given data.

9. SALES OF DSPS The projected sales (in billions of dollars) of digital signal processors (DSPs) and the scatter plot for these data follow:

Year	1997	1998	1999	2000	2001	2002
Sales	3.1	4	5	6.2	8	10

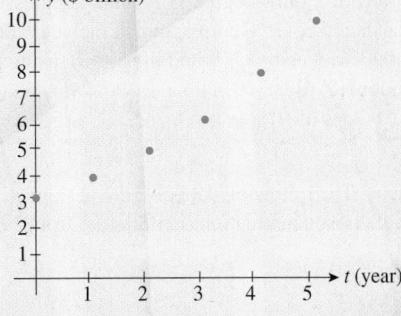

(continued)

a. Use **P2REG** to find a second-degree polynomial regression model for the data. Let $t = 0$ correspond to 1997.

b. Plot the graph of the function f found in part (a), using the viewing window $[0, 5] \times [0, 12]$.

c. Compute the values of $f(t)$ for $t = 0, 1, 2, 3, 4$, and 5. How does your model compare with the given data?

Source: A. G. Edwards & Sons, Inc.

10. **ANNUAL RETAIL SALES** Annual retail sales in the United States from 1990 through the year 2000 (in billions of dollars) are given in the following table:

Year	1990	1991	1992	1993	1994	1995
Sales	471.6	485.4	519.2	553.4	595	625.5

Year	1996	1997	1998	1999	2000
Sales	656.6	685.6	727.2	781.7	877.7

a. Let $t = 0$ correspond to 1990 and use **P2REG** to find a second-degree polynomial regression model based on the given data.

b. Plot the graph of the function f found in part (a) using the viewing window $[0, 10] \times [0, 1000]$.

c. Compute $f(0), f(5)$, and $f(10)$. Compare these values with the given data.

Source: National Retail Federation

11. **DIGITAL TV SHIPMENTS** The estimated number of digital TV shipments between the year 2000 and 2006 (in millions of units) and the scatter plot for these data follow:

Year	2000	2001	2002	2003	2004	2005	2006
Units Shipped	0.63	1.43	2.57	4.1	6	8.1	10

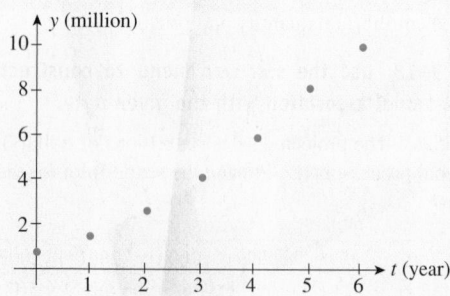

a. Use **P3REG** to find a third-degree polynomial regression model for the data. Let $t = 0$ correspond to the year 2000.

b. Plot the graph of the function f found in part (a), using the viewing window $[0, 6] \times [0, 11]$.

c. Compute the values of $f(t)$ for $t = 0, 1, 2, 3, 4, 5$, and 6.

Source: Consumer Electronics Manufacturers Association

12. **PRISON POPULATION** The following data give the past, present, and projected U.S. prison population (in millions) from 1980 through 2005:

Year	1980	1985	1990	1995	2000	2005
Population	0.52	0.77	1.18	1.64	2.23	3.20

a. Let $t = 0$ correspond to the beginning of 1980, where t is measured in 5-yr intervals, and use **P2REG** to find a second-degree polynomial regression model based on the given data.

b. Plot the graph of the function f found in part (a), using the viewing window $[0, 5] \times [0, 3.5]$.

c. Compute $f(0), f(1), f(2), f(3), f(4)$, and $f(5)$. Compare these values with the given data.

13. **FEDERAL DEBT** According to data obtained from the Congressional Budget Office, the federal debt (in trillions of dollars) for selected years from 1990 through 2010, with figures after 2000 being estimates, is given in the following table:

Year	1990	1997	2000	2010
Debt	2.55	3.84	3.41	1.70

a. Use **P3REG** to find a third-degree polynomial regression model for the data, letting $t = 0$ correspond to the beginning of 1990.

b. Plot the graph of the function f found in part (a).

c. Compare the values of f at $t = 0, 7, 10$, and 20, with the given data.

Source: Congressional Budget Office

14. **POPULATION GROWTH IN CLARK COUNTY** Clark County in Nevada —dominated by greater Las Vegas—is the fastest-growing metropolitan area in the United States. The population of the county from 1970 through 2000 is given in the following table:

Year	1970	1980	1990	2000
Population	273,288	463,087	741,459	1,375,765

a. Use **P3REG** to find a third-degree polynomial regression model for the data. Let t be measured in decades, with $t = 0$ corresponding to the beginning of 1970.

b. Plot the graph of the function f found in part (a), using the viewing window $[0, 3] \times [0, 1,500,000]$.

c. Compare the values of f at $t = 0, 1, 2$, and 3, with the given data.

Source: U.S. Census Bureau

15. **CABLE AD REVENUE** The revenues (in billions of dollars) from cable advertisement for the years 1995 through 2000 follow:

Year	1995	1996	1997	1998	1999	2000
Revenue	5.1	6.6	8.1	9.4	11.1	13.7

a. Use **P3REG** to find a third-degree polynomial regression model for the data. Let $t = 0$ correspond to 1995.

b. Plot the graph of the function f found in part (a), using the viewing window $[0, 6] \times [0, 14]$.

c. Compare the values of f at $t = 1, 2, 3, 4$, and 5, with the given data.

Source: National Cable Television Association

16. ONLINE SHOPPING The following data give the revenue per year (in billions of dollars) from Internet shopping:

Year	1997	1998	1999	2000	2001
Revenue	2.4	5	8	12	17.4

a. Use **P3REG** to find a third-degree polynomial regression model for the data. Let $t = 0$ correspond to 1997.
b. Plot the graph of the function f found in part (a), using the viewing window $[0, 4] \times [0, 20]$.
c. Compare the values of f at $t = 0, 1, 2, 3,$ and 4, with the given data.

Source: Forrester Research, Inc.

17. MARIJUANA ARRESTS The number of arrests (in thousands) for marijuana sales and possession in New York City from 1992 through 1997 and the scatter plot for these data follow:

Year	1992	1993	1994	1995	1996	1997
Arrests	5.0	5.8	8.8	11.7	18.5	27.5

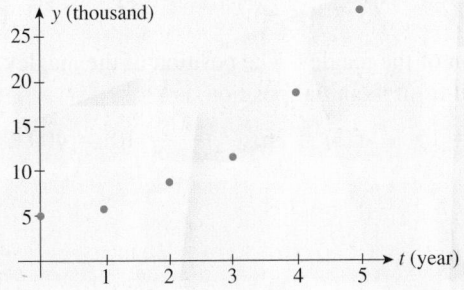

a. Use **P4REG** to find a fourth-degree polynomial regression model for the data. Let $t = 0$ correspond to 1992.
b. Plot the graph of the function f found in part (a), using the viewing window $[0, 5] \times [0, 30]$.
c. Compare the values of f at $t = 0, 1, 2, 3, 4,$ and 5, with the given data.

Source: New York State Division of Criminal Justice Services

18. ONLINE SPENDING The following data give the worldwide spending (in billions of dollars) on the Web from 1997 through 2002.

Year	1997	1998	1999	2000	2001	2002
Spending	5.0	10.5	20.5	37.5	60	95

a. Use **P4REG** to find a fourth-degree polynomial regression model for the data. Let $t = 0$ correspond to 1997.
b. Plot the graph of the function f found in part (a), using the viewing window $[0, 5] \times [0, 100]$.

Source: International Data Corporation

11.4 Limits

■ Introduction to Calculus

Historically, the development of calculus by Isaac Newton (1642–1727) and Gottfried Wilhelm Leibniz (1646–1716) resulted from the investigation of the following problems:

1. Finding the tangent line to a curve at a given point on the curve (Figure 21a)
2. Finding the area of a planar region bounded by an arbitrary curve (Figure 21b)

The tangent-line problem might appear to be unrelated to any practical applications of mathematics, but as you will see later, the problem of finding the *rate of change* of one quantity with respect to another is mathematically equivalent to the geometric problem of finding the slope of the *tangent line* to a curve at a given point on the curve. It is precisely the discovery of the relationship between these two problems that spurred the development of calculus in the 17th century and made it such an indispensable tool for solving practical problems. The following are a few examples of such problems:

- Finding the velocity of an object
- Finding the rate of change of a bacteria population with respect to time

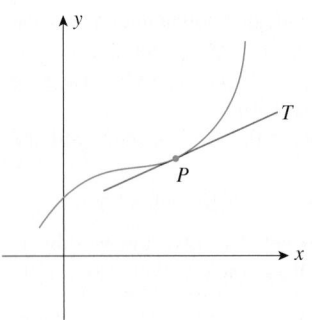

(a) What is the slope of the tangent line *T* at point *P*?

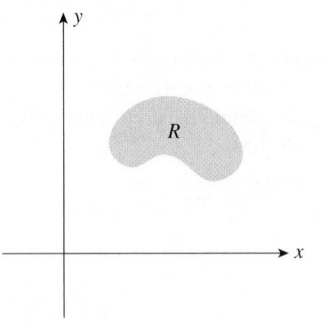

(b) What is the area of the region *R*?

FIGURE 21

- Finding the rate of change of a company's profit with respect to time
- Finding the rate of change of a travel agency's revenue with respect to the agency's expenditure for advertising

The study of the tangent-line problem led to the creation of *differential calculus*, which relies on the concept of the *derivative* of a function. The study of the area problem led to the creation of *integral calculus*, which relies on the concept of the *antiderivative*, or *integral*, of a function. (The derivative of a function and the integral of a function are intimately related, as you will see in Section 15.4.) Both the derivative of a function and the integral of a function are defined in terms of a more fundamental concept—the limit—our next topic.

■ A Real-Life Example

From data obtained in a test run conducted on a prototype of a maglev (magnetic levitation train), which moves along a straight monorail track, engineers have determined that the position of the maglev (in feet) from the origin at time *t* is given by

$$s = f(t) = 4t^2 \qquad (0 \le t \le 30) \tag{1}$$

where *f* is called the **position function** of the maglev. The position of the maglev at time $t = 0, 1, 2, 3, \ldots, 10$, measured from its initial position, is

$$f(0) = 0 \qquad f(1) = 4 \qquad f(2) = 16 \qquad f(3) = 36, \ldots \qquad f(10) = 400$$

feet (Figure 22).

FIGURE 22
A maglev moving along an elevated monorail track

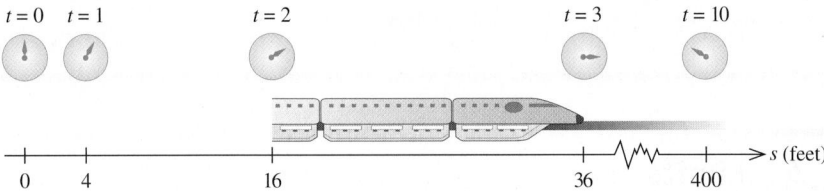

Suppose we want to find the velocity of the maglev at *t* = 2. This is just the velocity of the maglev as shown on its speedometer at that precise instant of time. Offhand, calculating this quantity using only Equation (1) appears to be an impossible task; but consider what quantities we *can* compute using this relationship. Obviously, we can compute the position of the maglev at any time *t* as we did earlier for some selected values of *t*. Using these values, we can then compute the *average velocity* of the maglev over an interval of time. For example, the average velocity of the train over the time interval [2, 4] is given by

$$\frac{\text{Distance covered}}{\text{Time elapsed}} = \frac{f(4) - f(2)}{4 - 2}$$
$$= \frac{4(4^2) - 4(2^2)}{2}$$
$$= \frac{64 - 16}{2} = 24$$

or 24 feet/second.

Although this is not quite the velocity of the maglev at *t* = 2, it does provide us with an approximation of its velocity at that time.

Can we do better? Intuitively, the smaller the time interval we pick (with $t = 2$ as the left endpoint), the better the average velocity over that time interval will approximate the actual velocity of the maglev at $t = 2$.*

Now, let's describe this process in general terms. Let $t > 2$. Then, the average velocity of the maglev over the time interval $[2, t]$ is given by

$$\frac{f(t) - f(2)}{t - 2} = \frac{4t^2 - 4(2^2)}{t - 2} = \frac{4(t^2 - 4)}{t - 2} \tag{2}$$

By choosing the values of t closer and closer to 2, we obtain a sequence of numbers that give the average velocities of the maglev over smaller and smaller time intervals. As we observed earlier, this sequence of numbers should approach the *instantaneous velocity* of the train at $t = 2$.

Let's try some sample calculations. Using Equation (2) and taking the sequence $t = 2.5, 2.1, 2.01, 2.001,$ and 2.0001, which approaches 2, we find

The average velocity over $[2, 2.5]$ is $\dfrac{4(2.5^2 - 4)}{2.5 - 2} = 18$, or 18 feet/second

The average velocity over $[2, 2.1]$ is $\dfrac{4(2.1^2 - 4)}{2.1 - 2} = 16.4$, or 16.4 feet/second

and so forth. These results are summarized in Table 1.

TABLE 1

		t approaches 2 from the right.			
t	2.5	2.1	2.01	2.001	2.0001
Average Velocity over $[2, t]$	18	16.4	16.04	16.004	16.0004

Average velocity approaches 16 from the right.

From Table 1, we see that the average velocity of the maglev seems to approach the number 16 as it is computed over smaller and smaller time intervals. These computations suggest that the instantaneous velocity of the train at $t = 2$ is 16 feet/second.

Note Notice that we cannot obtain the instantaneous velocity for the maglev at $t = 2$ by substituting $t = 2$ into Equation (2) because this value of t is not in the domain of the average velocity function. ■

■ Intuitive Definition of a Limit

Consider the function g defined by

$$g(t) = \frac{4(t^2 - 4)}{t - 2}$$

which gives the average velocity of the maglev [see Equation (2)]. Suppose we are required to determine the value that $g(t)$ approaches as t approaches the (fixed) number 2. If we take the sequence of values of t approaching 2 from the right-hand side, as we did earlier, we see that $g(t)$ approaches the number 16. Similarly, if we take a

*Actually, any interval containing $t = 2$ will do.

TABLE 2

	t approaches 2 from the left.				
t	1.5	1.9	1.99	1.999	1.9999
$g(t)$	14	15.6	15.96	15.996	15.9996

$g(t)$ approaches 16 from the left.

sequence of values of t approaching 2 from the left, such as $t = 1.5, 1.9, 1.99, 1.999$, and 1.9999, we obtain the results shown in Table 2.

Observe that $g(t)$ approaches the number 16 as t approaches 2—this time from the left-hand side. In other words, as t approaches 2 from *either* side of 2, $g(t)$ approaches 16. In this situation, we say that the limit of $g(t)$ as t approaches 2 is 16, written

$$\lim_{t \to 2} g(t) = \lim_{t \to 2} \frac{4(t^2 - 4)}{t - 2} = 16$$

The graph of the function g, shown in Figure 23, confirms this observation.

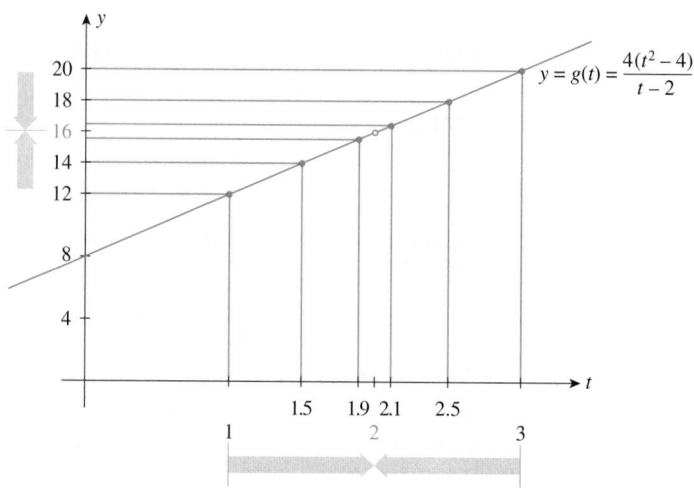

FIGURE 23
As t approaches $t = 2$ from either direction, $g(t)$ approaches $y = 16$.

Observe that the point $t = 2$ is not in the domain of the function g [for this reason, the point $(2, 16)$ is missing from the graph of g]. This, however, is inconsequential because the value, if any, of $g(t)$ at $t = 2$ plays no role in computing the limit.

This example leads to the following informal definition.

Limit of a Function
The function f has the **limit** L as x approaches a, written

$$\lim_{x \to a} f(x) = L$$

if the value of $f(x)$ can be made as close to the number L as we please by taking x sufficiently close to (but not equal to) a.

EXPLORING WITH TECHNOLOGY

1. Use a graphing utility to plot the graph of

$$g(x) = \frac{4(x^2 - 4)}{x - 2}$$

in the viewing window $[0, 3] \times [0, 20]$.

2. Use ZOOM and TRACE to describe what happens to the values of $g(x)$ as x approaches 2, first from the right and then from the left.

3. What happens to the y-value when you try to evaluate $g(x)$ at $x = 2$? Explain.

4. Reconcile your results with those of the preceding example.

■ Evaluating the Limit of a Function

Let's now consider some examples involving the computation of limits.

EXAMPLE 1 Let $f(x) = x^3$ and evaluate $\lim_{x \to 2} f(x)$.

Solution The graph of f is shown in Figure 24. You can see that $f(x)$ can be made as close to the number 8 as we please by taking x sufficiently close to 2. Therefore,

$$\lim_{x \to 2} x^3 = 8$$

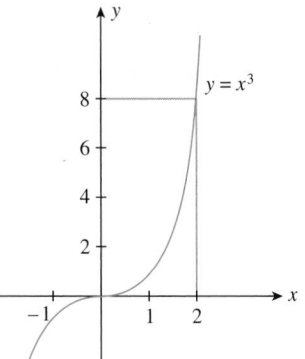

FIGURE 24
$f(x)$ is close to 8 whenever x is close to 2.

EXAMPLE 2 Let

$$g(x) = \begin{cases} x + 2 & \text{if } x \neq 1 \\ 1 & \text{if } x = 1 \end{cases}$$

Evaluate $\lim_{x \to 1} g(x)$.

Solution The domain of g is the set of all real numbers. From the graph of g shown in Figure 25, we see that $g(x)$ can be made as close to 3 as we please by taking x sufficiently close to 1. Therefore,

$$\lim_{x \to 1} g(x) = 3$$

Observe that $g(1) = 1$, which is not equal to the limit of the function g as x approaches 1. [Once again, the value of $g(x)$ at $x = 1$ has no bearing on the existence or value of the limit of g as x approaches 1.]

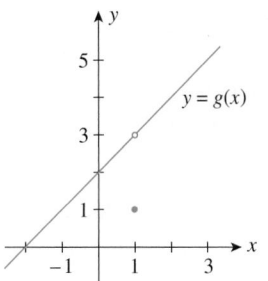

FIGURE 25
$\lim_{x \to 1} g(x) = 3$

EXAMPLE 3 Evaluate the limit of the following functions as x approaches the indicated point.

a. $f(x) = \begin{cases} -1 & \text{if } x < 0 \\ 1 & \text{if } x \geq 0 \end{cases}; \; x = 0$ **b.** $g(x) = \frac{1}{x^2}; \; x = 0$

Solution The graphs of the functions f and g are shown in Figure 26.

a. Referring to Figure 26a, we see that no matter how close x is to zero, $f(x)$ takes on the values 1 or -1, depending on whether x is positive or negative. Thus, there is no *single* real number L that $f(x)$ approaches as x approaches zero. We conclude that the limit of $f(x)$ does *not* exist as x approaches zero.

b. Referring to Figure 26b, we see that as x approaches zero (from either side), $g(x)$ increases without bound and thus does not approach any specific real

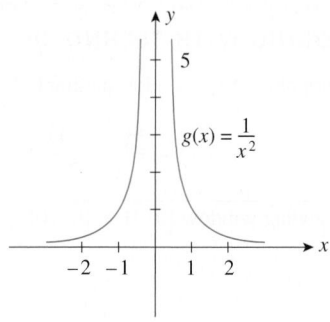

FIGURE 26

(a) $\lim\limits_{x \to 0} f(x)$ does not exist.   **(b)** $\lim\limits_{x \to 0} g(x)$ does not exist.

number. We conclude, accordingly, that the limit of $g(x)$ does *not* exist as x approaches zero.   ∎

EXPLORE & DISCUSS

Consider the graph of the function h whose graph is depicted in the following figure.

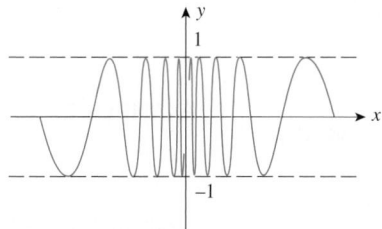

It has the property that as x approaches zero from either the right or the left, the curve oscillates more and more frequently between the lines $y = -1$ and $y = 1$.

1. Explain why $\lim\limits_{x \to 0} h(x)$ does not exist.

2. Compare this function with those in Example 3. More specifically, discuss the different ways the functions fail to have a limit at $x = 0$.

Until now, we have relied on knowing the actual values of a function or the graph of a function near $x = a$ to help us evaluate the limit of the function $f(x)$ as x approaches a. The following properties of limits, which we list without proof, enable us to evaluate limits of functions algebraically.

THEOREM 1

Properties of Limits
Suppose

$$\lim_{x \to a} f(x) = L \qquad \text{and} \qquad \lim_{x \to a} g(x) = M$$

Then,

1. $\lim\limits_{x \to a}[f(x)]^r = \left[\lim\limits_{x \to a} f(x)\right]^r = L^r$ r, a real number

2. $\lim\limits_{x \to a} cf(x) = c \lim\limits_{x \to a} f(x) = cL$ c, a real number

3. $\lim\limits_{x\to a}[f(x) \pm g(x)] = \lim\limits_{x\to a} f(x) \pm \lim\limits_{x\to a} g(x) = L \pm M$

4. $\lim\limits_{x\to a}[f(x)g(x)] = [\lim\limits_{x\to a} f(x)][\lim\limits_{x\to a} g(x)] = LM$

5. $\lim\limits_{x\to a}\dfrac{f(x)}{g(x)} = \dfrac{\lim\limits_{x\to a} f(x)}{\lim\limits_{x\to a} g(x)} = \dfrac{L}{M}$ Provided that $M \neq 0$

EXAMPLE 4   Use Theorem 1 to evaluate the following limits.

a. $\lim\limits_{x\to 2} x^3$ **b.** $\lim\limits_{x\to 4} 5x^{3/2}$ **c.** $\lim\limits_{x\to 1}(5x^4 - 2)$

d. $\lim\limits_{x\to 3} 2x^3\sqrt{x^2 + 7}$ **e.** $\lim\limits_{x\to 2}\dfrac{2x^2 + 1}{x + 1}$

Solution

a. $\lim\limits_{x\to 2} x^3 = \left[\lim\limits_{x\to 2} x\right]^3$ Property 1

$= 2^3 = 8$ $\lim\limits_{x\to 2} x = 2$

b. $\lim\limits_{x\to 4} 5x^{3/2} = 5[\lim\limits_{x\to 4} x^{3/2}]$ Property 2

$= 5(4)^{3/2} = 40$

c. $\lim\limits_{x\to 1}(5x^4 - 2) = \lim\limits_{x\to 1} 5x^4 - \lim\limits_{x\to 1} 2$ Property 3

To evaluate $\lim\limits_{x\to 1} 2$, observe that the constant function $g(x) = 2$ has value 2 for all values of x. Therefore, $g(x)$ must approach the limit 2 as x approaches $x = 1$ (or any other point for that matter!). Therefore,

$$\lim\limits_{x\to 1}(5x^4 - 2) = 5(1)^4 - 2 = 3$$

d. $\lim\limits_{x\to 3} 2x^3\sqrt{x^2 + 7} = 2\lim\limits_{x\to 3} x^3\sqrt{x^2 + 7}$ Property 2

$= 2\lim\limits_{x\to 3} x^3 \lim\limits_{x\to 3}\sqrt{x^2 + 7}$ Property 4

$= 2(3)^3\sqrt{3^2 + 7}$ Property 1

$= 2(27)\sqrt{16} = 216$

e. $\lim\limits_{x\to 2}\dfrac{2x^2 + 1}{x + 1} = \dfrac{\lim\limits_{x\to 2}(2x^2 + 1)}{\lim\limits_{x\to 2}(x + 1)}$ Property 5

$= \dfrac{2(2)^2 + 1}{2 + 1} = \dfrac{9}{3} = 3$

▬ Indeterminate Forms

Let's emphasize once again that Property 5 of limits is valid only when the limit of the function that appears in the denominator is not equal to zero at the number in question.

If the numerator has a limit different from zero and the denominator has a limit equal to zero, then the limit of the quotient does not exist at the number in question. This is the case with the function $g(x) = 1/x^2$ in Example 3b. Here, as x approaches zero, the numerator approaches 1 but the denominator approaches zero, so the quotient becomes arbitrarily large. Thus, as observed earlier, the limit does not exist.

Next, consider

$$\lim_{x \to 2} \frac{4(x^2 - 4)}{x - 2}$$

which we evaluated earlier by looking at the values of the function for x near $x = 2$. If we attempt to evaluate this expression by applying Property 5 of limits, we see that both the numerator and denominator of the function

$$\frac{4(x^2 - 4)}{x - 2}$$

approach zero as x approaches 2; that is, we obtain an expression of the form 0/0. In this event, we say that the limit of the quotient $f(x)/g(x)$ as x approaches 2 has the **indeterminate form 0/0**.

We need to evaluate limits of this type when we discuss the derivative of a function, a fundamental concept in the study of calculus. As the name suggests, the meaningless expression 0/0 does not provide us with a solution to our problem. One strategy that can be used to solve this type of problem follows.

Strategy for Evaluating Indeterminate Forms

1. Replace the given function with an appropriate one that takes on the same values as the original function everywhere except at $x = a$.

2. Evaluate the limit of this function as x approaches a.

Examples 5 and 6 illustrate this strategy.

EXAMPLE 5   Evaluate:

$$\lim_{x \to 2} \frac{4(x^2 - 4)}{x - 2}$$

Solution   Since both the numerator and the denominator of this expression approach zero as x approaches 2, we have the indeterminate form 0/0. We rewrite

$$\frac{4(x^2 - 4)}{x - 2} = \frac{4(x - 2)(x + 2)}{(x - 2)}$$

which, upon canceling the common factors, is equivalent to $4(x + 2)$, provided $x \neq 2$. Next, we replace $4(x^2 - 4)/(x - 2)$ with $4(x + 2)$ and find that

$$\lim_{x \to 2} \frac{4(x^2 - 4)}{x - 2} = \lim_{x \to 2} 4(x + 2) = 16$$

The graphs of the functions

$$f(x) = \frac{4(x^2 - 4)}{x - 2} \quad \text{and} \quad g(x) = 4(x + 2)$$

are shown in Figure 27. Observe that the graphs are identical except when $x = 2$. The function g is defined for all values of x and, in particular, its value at $x = 2$ is $g(2) = 4(2 + 2) = 16$. Thus, the point $(2, 16)$ is on the graph of g. However, the function f is not defined at $x = 2$. Since $f(x) = g(x)$ for all values of x except $x = 2$, it follows that the graph of f must look exactly like the graph of g, with the exception that the point $(2, 16)$ is missing from the graph of f. This illustrates graphically why we can evaluate the limit of f by evaluating the limit of the "equivalent" function g.

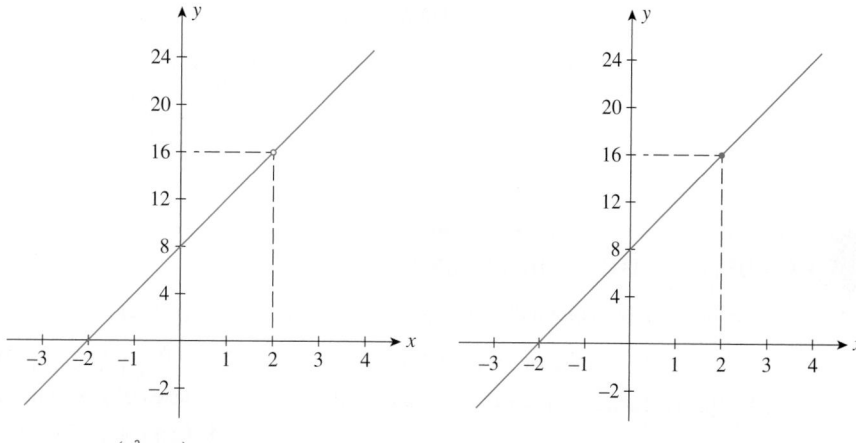

FIGURE 27
The graphs of $f(x)$ and $g(x)$ are identical except at the point (2, 16).

(a) $f(x) = \dfrac{4(x^2 - 4)}{x - 2}$

(b) $g(x) = 4(x + 2)$

Note Notice that the limit in Example 5 is the same limit that we evaluated earlier when we discussed the instantaneous velocity of a maglev at a specified time.

EXPLORING WITH TECHNOLOGY

1. Use a graphing utility to plot the graph of

$$f(x) = \frac{4(x^2 - 4)}{x - 2}$$

in the viewing window $[0, 3] \times [0, 20]$. Then use ZOOM and TRACE to find

$$\lim_{x \to 2} \frac{4(x^2 - 4)}{x - 2}$$

2. Use a graphing utility to plot the graph of $g(x) = 4(x + 2)$ in the viewing window $[0, 3] \times [0, 20]$. Then use ZOOM and TRACE to find $\lim_{x \to 2} 4(x + 2)$. What happens to the y-value when you try to evaluate $f(x)$ at $x = 2$? Explain.

3. Can you distinguish between the graphs of f and g?

4. Reconcile your results with those of Example 5.

EXAMPLE 6 Evaluate:

$$\lim_{h \to 0} \frac{\sqrt{1 + h} - 1}{h}$$

Solution Letting h approach zero, we obtain the indeterminate form 0/0. Next, we rationalize the numerator of the quotient (see page 566) by multiplying both the numerator and the denominator by the expression $(\sqrt{1 + h} + 1)$, obtaining

$$\frac{\sqrt{1 + h} - 1}{h} = \frac{(\sqrt{1 + h} - 1)(\sqrt{1 + h} + 1)}{h(\sqrt{1 + h} + 1)}$$

$$= \frac{1 + h - 1}{h(\sqrt{1 + h} + 1)} \qquad (\sqrt{a} - \sqrt{b})(\sqrt{a} + \sqrt{b}) = a - b$$

$$= \frac{h}{h(\sqrt{1 + h} + 1)}$$

$$= \frac{1}{\sqrt{1 + h} + 1}$$

Therefore,

$$\lim_{h \to 0} \frac{\sqrt{1+h}-1}{h} = \lim_{h \to 0} \frac{1}{\sqrt{1+h}+1}$$

$$= \frac{1}{\sqrt{1}+1} = \frac{1}{2} \qquad \blacksquare$$

EXPLORING WITH TECHNOLOGY

1. Use a graphing utility to plot the graph of
$g(x) = \dfrac{\sqrt{1+x}-1}{x}$ in the viewing window
$[-1, 2] \times [0, 1]$. Then use **ZOOM** and **TRACE** to find
$\displaystyle\lim_{x \to 0} \frac{\sqrt{1+x}-1}{x}$ by observing the values of $g(x)$ as x
approaches zero from the left and from the right.

2. Use a graphing utility to plot the graph of
$f(x) = \dfrac{1}{\sqrt{1+x}+1}$ in the viewing window

$[-1, 2] \times [0, 1]$. Then use **ZOOM** and **TRACE** to find
$\displaystyle\lim_{x \to 0} \frac{1}{\sqrt{1+x}+1}$. What happens to the y-value when x
takes on the value zero? Explain.

3. Can you distinguish between the graphs of f and g?

4. Reconcile your results with those of Example 6.

■ Limits at Infinity

Up to now we have studied the limit of a function as x approaches a (finite) number a. There are occasions, however, when we want to know whether $f(x)$ approaches a unique number as x increases without bound. Consider, for example, the function P, giving the number of fruit flies (*Drosophila*) in a container under controlled laboratory conditions, as a function of a time t. The graph of P is shown in Figure 28. You can see from the graph of P that, as t increases without bound (gets larger and larger), $P(t)$ approaches the number 400. This number, called the *carrying capacity* of the environment, is determined by the amount of living space and food available, as well as other environmental factors.

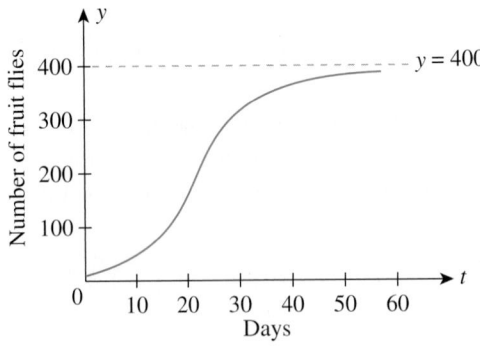

FIGURE 28
The graph of $P(t)$ gives the population of fruit flies in a laboratory experiment.

As another example, suppose we are given the function

$$f(x) = \frac{2x^2}{1+x^2}$$

and we want to determine what happens to $f(x)$ as x gets larger and larger. Picking the sequence of numbers 1, 2, 5, 10, 100, and 1000 and computing the corresponding values of $f(x)$, we obtain the following table of values:

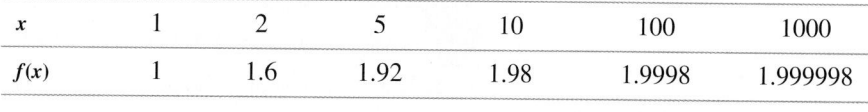

x	1	2	5	10	100	1000
$f(x)$	1	1.6	1.92	1.98	1.9998	1.999998

From the table, we see that as x gets larger and larger, $f(x)$ gets closer and closer to 2. The graph of the function f shown in Figure 29 confirms this observation. We call the line $y = 2$ a **horizontal asymptote.*** In this situation, we say that the limit of the function $f(x)$ as x increases without bound is 2, written

$$\lim_{x \to \infty} \frac{2x^2}{1 + x^2} = 2$$

In the general case, the following definition for a **limit of a function at infinity** is applicable.

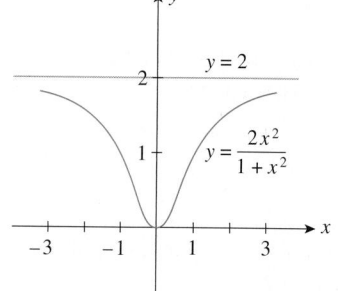

FIGURE 29

The graph of $y = \dfrac{2x^2}{1 + x^2}$ has a horizontal asymptote at $y = 2$.

Limit of a Function at Infinity

The function f has the limit L as x increases without bound (or, as x approaches infinity), written

$$\lim_{x \to \infty} f(x) = L$$

if $f(x)$ can be made arbitrarily close to L by taking x large enough.

Similarly, the function f has the limit M as x decreases without bound (or as x approaches negative infinity), written

$$\lim_{x \to -\infty} f(x) = M$$

if $f(x)$ can be made arbitrarily close to M by taking x to be negative and sufficiently large in absolute value.

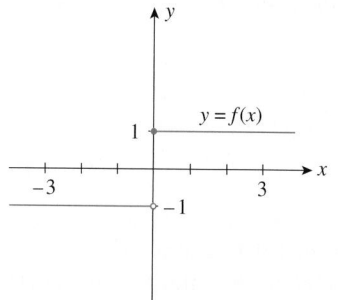

(a) $\lim_{x \to \infty} f(x) = 1$ and $\lim_{x \to -\infty} f(x) = -1$

EXAMPLE 7   Let f and g be the functions

$$f(x) = \begin{cases} -1 & \text{if } x < 0 \\ 1 & \text{if } x \geq 0 \end{cases} \quad \text{and} \quad g(x) = \frac{1}{x^2}$$

Evaluate:

a. $\lim_{x \to \infty} f(x)$ and $\lim_{x \to -\infty} f(x)$ **b.** $\lim_{x \to \infty} g(x)$ and $\lim_{x \to -\infty} g(x)$

Solution   The graphs of $f(x)$ and $g(x)$ are shown in Figure 30. Referring to the graphs of the respective functions, we see that

a. $\lim_{x \to \infty} f(x) = 1$ and $\lim_{x \to -\infty} f(x) = -1$ **b.** $\lim_{x \to \infty} \frac{1}{x^2} = 0$ and $\lim_{x \to -\infty} \frac{1}{x^2} = 0$

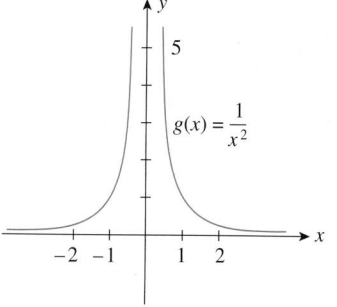

(b) $\lim_{x \to \infty} g(x) = 0$ and $\lim_{x \to -\infty} g(x) = 0$

FIGURE 30

All the properties of limits listed in Theorem 1 are valid when a is replaced by ∞ or $-\infty$. In addition, we have the following property for the limit at infinity.

THEOREM 2

For all $n > 0$, $\lim_{x \to \infty} \dfrac{1}{x^n} = 0$ and $\lim_{x \to -\infty} \dfrac{1}{x^n} = 0$

provided that $\dfrac{1}{x^n}$ is defined.

---

*We will discuss asymptotes in greater detail in Section 13.3.

EXPLORING WITH TECHNOLOGY

1. Use a graphing utility to plot the graphs of

$$y_1 = \frac{1}{x^{0.5}} \qquad y_2 = \frac{1}{x} \qquad y_3 = \frac{1}{x^{1.5}}$$

in the viewing window $[0, 200] \times [0, 0.5]$. What can you say about $\lim_{x \to \infty} \frac{1}{x^n}$ if $n = 0.5$, $n = 1$, and $n = 1.5$? Are these results predicted by Theorem 2?

2. Use a graphing utility to plot the graphs of

$$y_1 = \frac{1}{x} \qquad \text{and} \qquad y_2 = \frac{1}{x^{5/3}}$$

in the viewing window $[-50, 0] \times [-0.5, 0]$. What can you say about $\lim_{x \to -\infty} \frac{1}{x^n}$ if $n = 1$ and $n = \frac{5}{3}$? Are these results predicted by Theorem 2?
Hint: To graph y_2, write it in the form y2 = 1/(x^(1/3))^5.

We often use the following technique to evaluate the limit at infinity of a rational function: *Divide the numerator and denominator of the expression by x^n, where n is the highest power present in the denominator of the expression.*

EXAMPLE 8 Evaluate

$$\lim_{x \to \infty} \frac{x^2 - x + 3}{2x^3 + 1}$$

Solution Since the limits of both the numerator and the denominator do not exist as x approaches infinity, the property pertaining to the limit of a quotient (Property 5) is not applicable. Let's divide the numerator and denominator of the rational expression by x^3, obtaining

$$\lim_{x \to \infty} \frac{x^2 - x + 3}{2x^3 + 1} = \lim_{x \to \infty} \frac{\dfrac{1}{x} - \dfrac{1}{x^2} + \dfrac{3}{x^3}}{2 + \dfrac{1}{x^3}}$$

$$= \frac{0 - 0 + 0}{2 + 0} = \frac{0}{2} \qquad \text{Use Theorem 2.}$$

$$= 0 \qquad \blacksquare$$

EXAMPLE 9 Let

$$f(x) = \frac{3x^2 + 8x - 4}{2x^2 + 4x - 5}$$

Compute $\lim_{x \to \infty} f(x)$ if it exists.

Solution Again, we see that Property 5 is not applicable. Dividing the numerator and the denominator by x^2, we obtain

$$\lim_{x\to\infty}\frac{3x^2+8x-4}{2x^2+4x-5}=\lim_{x\to\infty}\frac{3+\dfrac{8}{x}-\dfrac{4}{x^2}}{2+\dfrac{4}{x}-\dfrac{5}{x^2}}$$

$$=\frac{\lim\limits_{x\to\infty}3+8\lim\limits_{x\to\infty}\dfrac{1}{x}-4\lim\limits_{x\to\infty}\dfrac{1}{x^2}}{\lim\limits_{x\to\infty}2+4\lim\limits_{x\to\infty}\dfrac{1}{x}-5\lim\limits_{x\to\infty}\dfrac{1}{x^2}}$$

$$=\frac{3+0-0}{2+0-0}\qquad\text{Use Theorem 2.}$$

$$=\frac{3}{2}$$

EXAMPLE 10 Let $f(x)=\dfrac{2x^3-3x^2+1}{x^2+2x+4}$ and evaluate:

a. $\lim\limits_{x\to\infty}f(x)$ **b.** $\lim\limits_{x\to-\infty}f(x)$

Solution

a. Dividing the numerator and the denominator of the rational expression by x^2, we obtain

$$\lim_{x\to\infty}\frac{2x^3-3x^2+1}{x^2+2x+4}=\lim_{x\to\infty}\frac{2x-3+\dfrac{1}{x^2}}{1+\dfrac{2}{x}+\dfrac{4}{x^2}}$$

Since the numerator becomes arbitrarily large whereas the denominator approaches 1 as x approaches infinity, we see that the quotient $f(x)$ gets larger and larger as x approaches infinity. In other words, the limit does not exist. We indicate this by writing

$$\lim_{x\to\infty}\frac{2x^3-3x^2+1}{x^2+2x+4}=\infty$$

b. Once again, dividing both the numerator and the denominator by x^2, we obtain

$$\lim_{x\to-\infty}\frac{2x^3-3x^2+1}{x^2+2x+4}=\lim_{x\to-\infty}\frac{2x-3+\dfrac{1}{x^2}}{1+\dfrac{2}{x}+\dfrac{4}{x^2}}$$

In this case, the numerator becomes arbitrarily large in magnitude but negative in sign, whereas the denominator approaches 1 as x approaches negative infinity. Therefore, the quotient $f(x)$ decreases without bound, and the limit does not exist. We indicate this by writing

$$\lim_{x\to-\infty}\frac{2x^3-3x^2+1}{x^2+2x+4}=-\infty$$

Example 11 gives an application of the concept of the limit of a function at infinity.

APPLIED EXAMPLE 11 Average Cost Functions Custom Office makes a line of executive desks. It is estimated that the total cost of making x Senior Executive Model desks is $C(x) = 100x + 200{,}000$ dollars per year, so the average cost of making x desks is given by

$$\overline{C}(x) = \frac{C(x)}{x} = \frac{100x + 200{,}000}{x} = 100 + \frac{200{,}000}{x}$$

dollars per desk. Evaluate $\lim_{x\to\infty} \overline{C}(x)$ and interpret your results.

Solution

$$\lim_{x\to\infty} \overline{C}(x) = \lim_{x\to\infty}\left(100 + \frac{200{,}000}{x}\right)$$

$$= \lim_{x\to\infty} 100 + \lim_{x\to\infty} \frac{200{,}000}{x} = 100$$

A sketch of the graph of the function $\overline{C}(x)$ appears in Figure 31. The result we obtained is fully expected if we consider its economic implications. Note that as the level of production increases, the fixed cost per desk produced, represented by the term $(200{,}000/x)$, drops steadily. The average cost should approach a constant unit cost of production—$100 in this case. ∎

FIGURE 31
As the level of production increases, the average cost approaches $100 per desk.

EXPLORE & DISCUSS

Consider the graph of the function f depicted in the following figure:

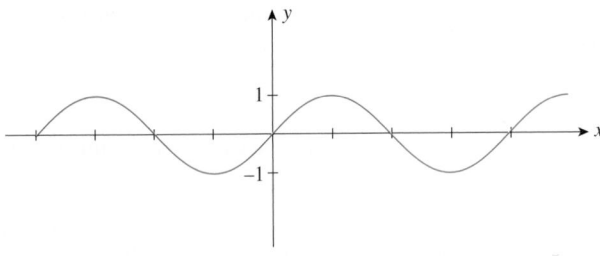

It has the property that the curve oscillates between $y = -1$ and $y = 1$ indefinitely in either direction.

1. Explain why $\lim_{x\to -\infty} f(x)$ and $\lim_{x\to\infty} f(x)$ do not exist.
2. Compare this function with those of Example 10. More specifically, discuss the different ways each function fails to have a limit at infinity or minus infinity.

11.4   Self-Check Exercises

1. Find the indicated limit if it exists.

   a. $\lim_{x \to 3} \dfrac{\sqrt{x^2 + 7} + \sqrt{3x - 5}}{x + 2}$

   b. $\lim_{x \to -1} \dfrac{x^2 - x - 2}{2x^2 - x - 3}$

2. The average cost per disc (in dollars) incurred by Herald

Records in pressing x compact discs (CDs) is given by the average cost function

$$\overline{C}(x) = 1.8 + \frac{3000}{x}$$

Evaluate $\lim_{x \to \infty} \overline{C}(x)$ and interpret your result.

Solutions to Self-Check Exercises 11.4 can be found on page 639.

11.4   Concept Questions

1. Explain what is meant by the statement $\lim_{x \to 2} f(x) = 3$.

2. **a.** If $\lim_{x \to 3} f(x) = 5$, what can you say about $f(3)$? Explain.

   b. If $f(2) = 6$, what can you say about $\lim_{x \to 2} f(x)$? Explain.

3. Evaluate and state the property of limits that you use at each step.

   a. $\lim_{x \to 4} \sqrt{x}(2x^2 + 1)$ **b.** $\lim_{x \to 1} \left(\dfrac{2x^2 + x + 5}{x^4 + 1} \right)^{3/2}$

4. What is an indeterminate form? Illustrate with an example.

5. Explain in your own words the meaning of $\lim_{x \to \infty} f(x) = L$ and $\lim_{x \to -\infty} f(x) = M$.

11.4   Exercises

In Exercises 1–8, use the graph of the given function f to determine $\lim_{x \to a} f(x)$ at the indicated value of a, if it exists.

1.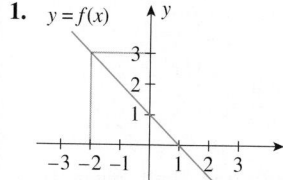
$a = -2$

2.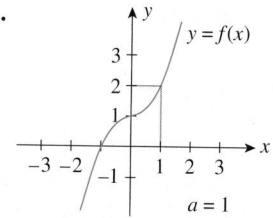
$a = 1$

3. **4.**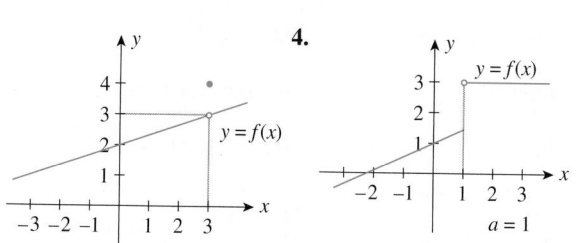
$a = 3$ $a = 1$

5.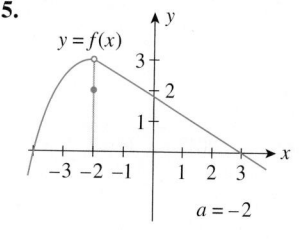
$a = -2$

6.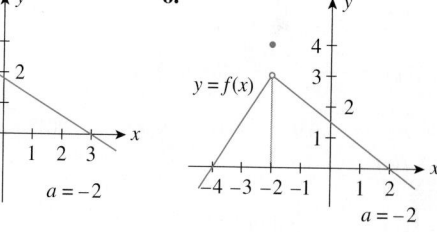
$a = -2$

7.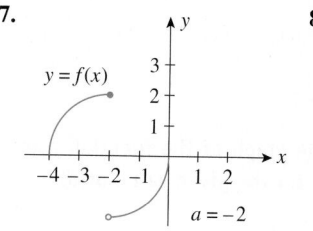
$a = -2$

8.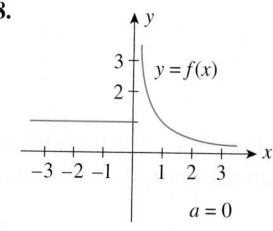
$a = 0$

In Exercises 9–16, complete the table by computing $f(x)$ at the given values of x. Use these results to estimate the indicated limit (if it exists).

9. $f(x) = x^2 + 1;\ \lim\limits_{x\to 2} f(x)$

x	1.9	1.99	1.999	2.001	2.01	2.1
$f(x)$						

10. $f(x) = 2x^2 - 1;\ \lim\limits_{x\to 1} f(x)$

x	0.9	0.99	0.999	1.001	1.01	1.1
$f(x)$						

11. $f(x) = \dfrac{|x|}{x};\ \lim\limits_{x\to 0} f(x)$

x	−0.1	−0.01	−0.001	0.001	0.01	0.1
$f(x)$						

12. $f(x) = \dfrac{|x-1|}{x-1};\ \lim\limits_{x\to 1} f(x)$

x	0.9	0.99	0.999	1.001	1.01	1.1
$f(x)$						

13. $f(x) = \dfrac{1}{(x-1)^2};\ \lim\limits_{x\to 1} f(x)$

x	0.9	0.99	0.999	1.001	1.01	1.1
$f(x)$						

14. $f(x) = \dfrac{1}{x-2};\ \lim\limits_{x\to 2} f(x)$

x	1.9	1.99	1.999	2.001	2.01	2.1
$f(x)$						

15. $f(x) = \dfrac{x^2 + x - 2}{x - 1};\ \lim\limits_{x\to 1} f(x)$

x	0.9	0.99	0.999	1.001	1.01	1.1
$f(x)$						

16. $f(x) = \dfrac{x - 1}{x - 1};\ \lim\limits_{x\to 1} f(x)$

x	0.9	0.99	0.999	1.001	1.01	1.1
$f(x)$						

In Exercises 17–22, sketch the graph of the function f and evaluate $\lim\limits_{x\to a} f(x)$, if it exists, for the given value of a.

17. $f(x) = \begin{cases} x - 1 & \text{if } x \le 0 \\ -1 & \text{if } x > 0 \end{cases}$ $(a = 0)$

18. $f(x) = \begin{cases} x - 1 & \text{if } x \le 3 \\ -2x + 8 & \text{if } x > 3 \end{cases}$ $(a = 3)$

19. $f(x) = \begin{cases} x & \text{if } x < 1 \\ 0 & \text{if } x = 1 \\ -x + 2 & \text{if } x > 1 \end{cases}$ $(a = 1)$

20. $f(x) = \begin{cases} -2x + 4 & \text{if } x < 1 \\ 4 & \text{if } x = 1 \\ x^2 + 1 & \text{if } x > 1 \end{cases}$ $(a = 1)$

21. $f(x) = \begin{cases} |x| & \text{if } x \ne 0 \\ 1 & \text{if } x = 0 \end{cases}$ $(a = 0)$

22. $f(x) = \begin{cases} |x - 1| & \text{if } x \ne 1 \\ 0 & \text{if } x = 1 \end{cases}$ $(a = 1)$

In Exercises 23–40, find the indicated limit.

23. $\lim\limits_{x\to 2} 3$

24. $\lim\limits_{x\to -2} -3$

25. $\lim\limits_{x\to 3} x$

26. $\lim\limits_{x\to -2} -3x$

27. $\lim\limits_{x\to 1} (1 - 2x^2)$

28. $\lim\limits_{t\to 3} (4t^2 - 2t + 1)$

29. $\lim\limits_{x\to 1} (2x^3 - 3x^2 + x + 2)$

30. $\lim\limits_{x\to 0} (4x^5 - 20x^2 + 2x + 1)$

31. $\lim\limits_{s\to 0} (2s^2 - 1)(2s + 4)$

32. $\lim\limits_{x\to 2} (x^2 + 1)(x^2 - 4)$

33. $\lim\limits_{x\to 2} \dfrac{2x + 1}{x + 2}$

34. $\lim\limits_{x\to 1} \dfrac{x^3 + 1}{2x^3 + 2}$

35. $\lim\limits_{x\to 2} \sqrt{x + 2}$

36. $\lim\limits_{x\to -2} \sqrt[3]{5x + 2}$

37. $\lim\limits_{x\to -3} \sqrt{2x^4 + x^2}$

38. $\lim\limits_{x\to 2} \sqrt{\dfrac{2x^3 + 4}{x^2 + 1}}$

39. $\lim\limits_{x\to -1} \dfrac{\sqrt{x^2 + 8}}{2x + 4}$

40. $\lim\limits_{x\to 3} \dfrac{x\sqrt{x^2 + 7}}{2x - \sqrt{2x + 3}}$

In Exercises 41–48, find the indicated limit given that $\lim\limits_{x\to a} f(x) = 3$ and $\lim\limits_{x\to a} g(x) = 4$.

41. $\lim\limits_{x\to a} [f(x) - g(x)]$

42. $\lim\limits_{x\to a} 2f(x)$

43. $\lim\limits_{x\to a} [2f(x) - 3g(x)]$

44. $\lim\limits_{x\to a} [f(x)g(x)]$

45. $\lim\limits_{x\to a} \sqrt{g(x)}$

46. $\lim\limits_{x\to a} \sqrt[3]{5f(x) + 3g(x)}$

47. $\lim\limits_{x\to a} \dfrac{2f(x) - g(x)}{f(x)g(x)}$

48. $\lim\limits_{x\to a} \dfrac{g(x) - f(x)}{f(x) + \sqrt{g(x)}}$

In Exercises 49–62, find the indicated limit, if it exists.

49. $\lim\limits_{x \to 1} \dfrac{x^2 - 1}{x - 1}$

50. $\lim\limits_{x \to -2} \dfrac{x^2 - 4}{x + 2}$

51. $\lim\limits_{x \to 0} \dfrac{x^2 - x}{x}$

52. $\lim\limits_{x \to 0} \dfrac{2x^2 - 3x}{x}$

53. $\lim\limits_{x \to -5} \dfrac{x^2 - 25}{x + 5}$

54. $\lim\limits_{b \to -3} \dfrac{b + 1}{b + 3}$

55. $\lim\limits_{x \to 1} \dfrac{x}{x - 1}$

56. $\lim\limits_{x \to 2} \dfrac{x + 2}{x - 2}$

57. $\lim\limits_{x \to -2} \dfrac{x^2 - x - 6}{x^2 + x - 2}$

58. $\lim\limits_{z \to 2} \dfrac{z^3 - 8}{z - 2}$

59. $\lim\limits_{x \to 1} \dfrac{\sqrt{x} - 1}{x - 1}$

Hint: Multiply by $\dfrac{\sqrt{x} + 1}{\sqrt{x} + 1}$.

60. $\lim\limits_{x \to 4} \dfrac{x - 4}{\sqrt{x} - 2}$

Hint: See Exercise 59.

61. $\lim\limits_{x \to 1} \dfrac{x - 1}{x^3 + x^2 - 2x}$

62. $\lim\limits_{x \to -2} \dfrac{4 - x^2}{2x^2 + x^3}$

In Exercises 63–68, use the graph of the function f to determine $\lim\limits_{x \to \infty} f(x)$ and $\lim\limits_{x \to -\infty} f(x)$, if they exist.

63.
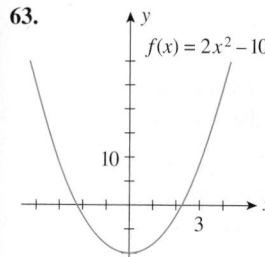
$f(x) = 2x^2 - 10$

64.
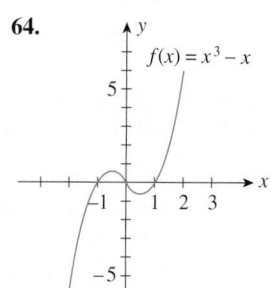
$f(x) = x^3 - x$

65.
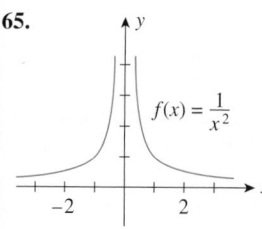
$f(x) = \dfrac{1}{x^2}$

66.
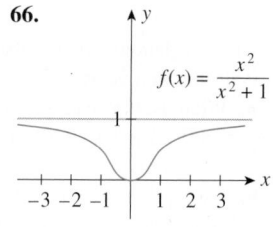
$f(x) = \dfrac{x^2}{x^2 + 1}$

67.
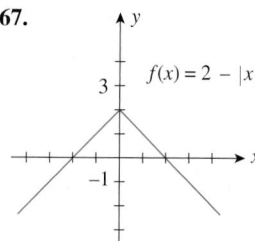
$f(x) = 2 - |x|$

68.
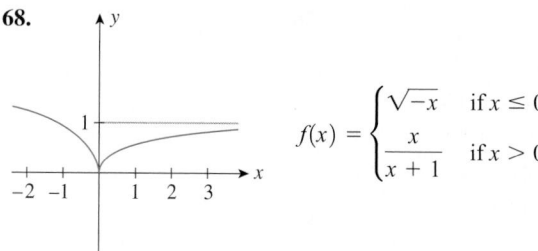
$f(x) = \begin{cases} \sqrt{-x} & \text{if } x \le 0 \\ \dfrac{x}{x + 1} & \text{if } x > 0 \end{cases}$

In Exercises 69–72, complete the table by computing $f(x)$ at the given values of x. Use the results to guess at the indicated limits, if they exist.

69. $f(x) = \dfrac{1}{x^2 + 1}$; $\lim\limits_{x \to \infty} f(x)$ and $\lim\limits_{x \to -\infty} f(x)$

x	1	10	100	1000
$f(x)$				

x	-1	-10	-100	-1000
$f(x)$				

70. $f(x) = \dfrac{2x}{x + 1}$; $\lim\limits_{x \to \infty} f(x)$ and $\lim\limits_{x \to -\infty} f(x)$

x	1	10	100	1000
$f(x)$				

x	-5	-10	-100	-1000
$f(x)$				

71. $f(x) = 3x^3 - x^2 + 10$; $\lim\limits_{x \to \infty} f(x)$ and $\lim\limits_{x \to -\infty} f(x)$

x	1	5	10	100	1000
$f(x)$					

x	-1	-5	-10	-100	-1000
$f(x)$					

72. $f(x) = \dfrac{|x|}{x}$; $\lim\limits_{x\to\infty} f(x)$ and $\lim\limits_{x\to-\infty} f(x)$

x	1	10	100	-1	-10	-100
$f(x)$						

In Exercises 73–80, find the indicated limits, if they exist.

73. $\lim\limits_{x\to\infty} \dfrac{3x+2}{x-5}$

74. $\lim\limits_{x\to-\infty} \dfrac{4x^2-1}{x+2}$

75. $\lim\limits_{x\to-\infty} \dfrac{3x^3+x^2+1}{x^3+1}$

76. $\lim\limits_{x\to\infty} \dfrac{2x^2+3x+1}{x^4-x^2}$

77. $\lim\limits_{x\to-\infty} \dfrac{x^4+1}{x^3-1}$

78. $\lim\limits_{x\to\infty} \dfrac{4x^4-3x^2+1}{2x^4+x^3+x^2+x+1}$

79. $\lim\limits_{x\to\infty} \dfrac{x^5-x^3+x-1}{x^6+2x^2+1}$

80. $\lim\limits_{x\to\infty} \dfrac{2x^2-1}{x^3+x^2+1}$

81. Toxic Waste A city's main well was recently found to be contaminated with trichloroethylene, a cancer-causing chemical, as a result of an abandoned chemical dump leaching chemicals into the water. A proposal submitted to city council members indicates that the cost, measured in millions of dollars, of removing $x\%$ of the toxic pollutant is given by

$$C(x) = \frac{0.5x}{100-x} \quad (0 < x < 100)$$

a. Find the cost of removing 50%, 60%, 70%, 80%, 90%, and 95% of the pollutant.
b. Evaluate

$$\lim_{x\to 100} \frac{0.5x}{100-x}$$

and interpret your result.

82. A Doomsday Situation The population of a certain breed of rabbits introduced into an isolated island is given by

$$P(t) = \frac{72}{9-t} \quad (0 \le t < 9)$$

where t is measured in months.
a. Find the number of rabbits present in the island initially (at $t = 0$).
b. Show that the population of rabbits is increasing without bound.
c. Sketch the graph of the function P.
(*Comment:* This phenomenon is referred to as a *doomsday situation*.)

83. Average Cost The average cost/disc in dollars incurred by Herald Records in pressing x DVDs is given by the average cost function

$$\overline{C}(x) = 2.2 + \frac{2500}{x}$$

Evaluate $\lim\limits_{x\to\infty} \overline{C}(x)$ and interpret your result.

84. Concentration of a Drug in the Bloodstream The concentration of a certain drug in a patient's bloodstream t hr after injection is given by

$$C(t) = \frac{0.2t}{t^2+1} \text{ mg/cm}^3$$

Evaluate $\lim\limits_{t\to\infty} C(t)$ and interpret your result.

85. Box-Office Receipts The total worldwide box-office receipts for a long-running blockbuster movie are approximated by the function

$$T(x) = \frac{120x^2}{x^2+4}$$

where $T(x)$ is measured in millions of dollars and x is the number of months since the movie's release.
a. What are the total box-office receipts after the first month? The second month? The third month?
b. What will the movie gross in the long run (when x is very large)?

86. Population Growth A major corporation is building a 4325-acre complex of homes, offices, stores, schools, and churches in the rural community of Glen Cove. As a result of this development, the planners have estimated that Glen Cove's population (in thousands) t yr from now will be given by

$$P(t) = \frac{25t^2+125t+200}{t^2+5t+40}$$

a. What is the current population of Glen Cove?
b. What will be the population in the long run?

87. Driving Costs A study of driving costs of 1992 model subcompact (four-cylinder) cars found that the average cost (car payments, gas, insurance, upkeep, and depreciation), measured in cents/mile, is approximated by the function

$$C(x) = \frac{2010}{x^{2.2}} + 17.80$$

where x denotes the number of miles (in thousands) the car is driven in a year.
a. What is the average cost of driving a subcompact car 5000 mi/yr? 10,000 mi/yr? 15,000 mi/yr? 20,000 mi/yr? 25,000 mi/yr?
b. Use part (a) to sketch the graph of the function C.
c. What happens to the average cost as the number of miles driven increases without bound?

Source: American Automobile Association

88. PHOTOSYNTHESIS The rate of production R in photosynthesis is related to the light intensity I by the function

$$R(I) = \frac{aI}{b + I^2}$$

where a and b are positive constants.

a. Taking $a = b = 1$, compute $R(I)$ for $I = 0, 1, 2, 3, 4,$ and 5.

b. Evaluate $\lim_{I \to \infty} R(I)$.

c. Use the results of parts (a) and (b) to sketch the graph of R. Interpret your results.

In Exercises 89–94, determine whether the statement is true or false. If it is true, explain why it is true. If it is false, give an example to show why it is false.

89. If $\lim_{x \to a} f(x)$ exists, then f is defined at $x = a$.

90. If $\lim_{x \to 0} f(x) = 4$ and $\lim_{x \to 0} g(x) = 0$, then $\lim_{x \to 0} f(x)g(x) = 0$.

91. If $\lim_{x \to 2} f(x) = 3$ and $\lim_{x \to 2} g(x) = 0$, then $\lim_{x \to 2} [f(x)]/[g(x)]$ does not exist.

92. If $\lim_{x \to 3} f(x) = 0$ and $\lim_{x \to 3} g(x) = 0$, then $\lim_{x \to 3} [f(x)]/[g(x)]$ does not exist.

93. $\lim_{x \to 2} \left(\dfrac{x}{x + 1} + \dfrac{3}{x - 1} \right) = \lim_{x \to 2} \dfrac{x}{x + 1} + \lim_{x \to 2} \dfrac{3}{x - 1}$

94. $\lim_{x \to 1} \left(\dfrac{2x}{x - 1} - \dfrac{2}{x - 1} \right) = \lim_{x \to 1} \dfrac{2x}{x - 1} - \lim_{x \to 1} \dfrac{2}{x - 1}$

95. SPEED OF A CHEMICAL REACTION Certain proteins, known as enzymes, serve as catalysts for chemical reactions in living things. In 1913 Leonor Michaelis and L. M. Menten discovered the following formula giving the initial speed V (in moles/liter/second) at which the reaction begins in terms of the amount of substrate x (the substance being acted upon, measured in moles/liters) present:

$$V = \frac{ax}{x + b}$$

where a and b are positive constants. Evaluate

$$\lim_{x \to \infty} \frac{ax}{x + b}$$

and interpret your result.

96. Show by means of an example that $\lim_{x \to a} [f(x) + g(x)]$ may exist even though neither $\lim_{x \to a} f(x)$ nor $\lim_{x \to a} g(x)$ exists. Does this example contradict Theorem 1?

97. Show by means of an example that $\lim_{x \to a} [f(x)g(x)]$ may exist even though neither $\lim_{x \to a} f(x)$ nor $\lim_{x \to a} g(x)$ exists. Does this example contradict Theorem 1?

98. Show by means of an example that $\lim_{x \to a} f(x)/g(x)$ may exist even though neither $\lim_{x \to a} f(x)$ nor $\lim_{x \to a} g(x)$ exists. Does this example contradict Theorem 1?

11.4 Solutions to Self-Check Exercises

1. a. $\lim_{x \to 3} \dfrac{\sqrt{x^2 + 7} + \sqrt{3x - 5}}{x + 2} = \dfrac{\sqrt{9 + 7} + \sqrt{3(3) - 5}}{3 + 2}$

$= \dfrac{\sqrt{16} + \sqrt{4}}{5}$

$= \dfrac{6}{5}$

b. Letting x approach -1 leads to the indeterminate form $0/0$. Thus, we proceed as follows:

$\lim_{x \to -1} \dfrac{x^2 - x - 2}{2x^2 - x - 3} = \lim_{x \to -1} \dfrac{(x + 1)(x - 2)}{(x + 1)(2x - 3)}$

$= \lim_{x \to -1} \dfrac{x - 2}{2x - 3}$ Cancel the common factors.

$= \dfrac{-1 - 2}{2(-1) - 3}$

$= \dfrac{3}{5}$

2. $\lim_{x \to \infty} \overline{C}(x) = \lim_{x \to \infty} \left(1.8 + \dfrac{3000}{x} \right)$

$= \lim_{x \to \infty} 1.8 + \lim_{x \to \infty} \dfrac{3000}{x}$

$= 1.8$

Our computation reveals that, as the production of CDs increases "without bound," the average cost drops and approaches a unit cost of $1.80/disc.

USING TECHNOLOGY

■ **Finding the Limit of a Function**

A graphing utility can be used to help us find the limit of a function, if it exists, as illustrated in the following examples.

EXAMPLE 1 Let $f(x) = \dfrac{x^3 - 1}{x - 1}$.

a. Plot the graph of f in the viewing window $[-2, 2] \times [0, 4]$.

b. Use ZOOM to find $\displaystyle\lim_{x \to 1} \frac{x^3 - 1}{x - 1}$.

c. Verify your result by evaluating the limit algebraically.

Solution

a. The graph of f in the viewing window $[-2, 2] \times [0, 4]$ is shown in Figure T1a.

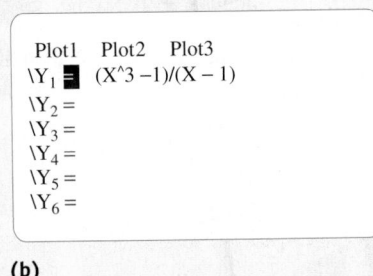

FIGURE T1
(a) The graph of $f(x) = \dfrac{x^3 - 1}{x - 1}$ in the viewing window $[-2, 2] \times [0, 4]$; (b) the TI-83 equation screen

(a)   **(b)**

b. Using ZOOM-IN repeatedly, we see that the y-value approaches 3 as the x-value approaches 1. We conclude, accordingly, that

$$\lim_{x \to 1} \frac{x^3 - 1}{x - 1} = 3$$

c. We compute

$$\lim_{x \to 1} \frac{x^3 - 1}{x - 1} = \lim_{x \to 1} \frac{(x - 1)(x^2 + x + 1)}{x - 1}$$

$$= \lim_{x \to 1} (x^2 + x + 1) = 3$$

Note If you attempt to find the limit in Example 1 by using the evaluation function of your graphing utility to find the value of $f(x)$ when $x = 1$, you will see that the graphing utility does not display the y-value. This happens because $x = 1$ is not in the domain of f.

EXAMPLE 2 Use ZOOM to find $\displaystyle\lim_{x \to 0} (1 + x)^{1/x}$.

Solution We first plot the graph of $f(x) = (1 + x)^{1/x}$ in a suitable viewing window. Figure T2a shows a plot of f in the window $[-1, 1] \times [0, 4]$. Using ZOOM-IN repeatedly, we see that $\displaystyle\lim_{x \to 0} (1 + x)^{1/x} \approx 2.71828$.

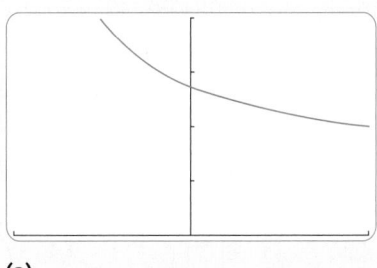

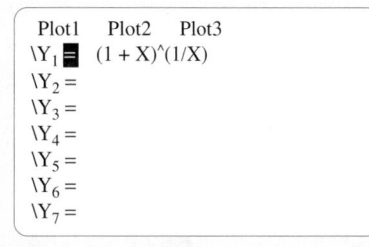

FIGURE T2
(a) The graph of $f(x) = (1 + x)^{1/x}$ in the viewing window $[-1, 1] \times [0, 4]$; (b) the T1-83 equation screen

(a)   **(b)**

The limit of $f(x) = (1 + x)^{1/x}$ as x approaches zero, denoted by the letter e, plays a very important role in the study of mathematics and its applications (see Section 14.5). Thus,

$$\lim_{x \to 0} (1 + x)^{1/x} = e$$

where, as we have just seen, $e \approx 2.71828$.   ■

APPLIED EXAMPLE 3 Oxygen Content of a Pond   When organic waste is dumped into a pond, the oxidation process that takes place reduces the pond's oxygen content. However, given time, nature will restore the oxygen content to its natural level. Suppose the oxygen content t days after the organic waste has been dumped into the pond is given by

$$f(t) = 100 \left(\frac{t^2 + 10t + 100}{t^2 + 20t + 100} \right)$$

percent of its normal level.

a. Plot the graph of f in the viewing window $[0, 200] \times [70, 100]$.
b. What can you say about $f(t)$ when t is very large?
c. Verify your observation in part (b) by evaluating $\lim_{t \to \infty} f(t)$.

Solution

a. The graph of f is shown in Figure T3a.
b. From the graph of f, it appears that $f(t)$ approaches 100 steadily as t gets larger and larger. This observation tells us that eventually the oxygen content of the pond will be restored to its natural level.

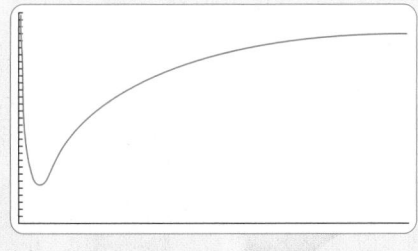

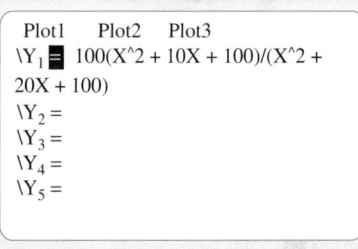

FIGURE T3
(a) The graph of f in the viewing window $[0, 200] \times [70, 100]$; (b) the TI-83 equation screen

(a)   **(b)**

(continued)

c. To verify the observation made in part (b), we compute

$$\lim_{t \to \infty} f(t) = \lim_{t \to \infty} 100\left(\frac{t^2 + 10t + 100}{t^2 + 20t + 100}\right)$$

$$= 100 \lim_{t \to \infty} \left(\frac{1 + \dfrac{10}{t} + \dfrac{100}{t^2}}{1 + \dfrac{20}{t} + \dfrac{100}{t^2}}\right) = 100$$

■

TECHNOLOGY EXERCISES

In Exercises 1–10, find the indicated limit by first plotting the graph of the function in a suitable viewing window and then using the ZOOM-IN feature of the calculator.

1. $\lim\limits_{x \to 1} \dfrac{2x^3 - 2x^2 + 3x - 3}{x - 1}$

2. $\lim\limits_{x \to -2} \dfrac{2x^3 + 3x^2 - x + 2}{x + 2}$

3. $\lim\limits_{x \to -1} \dfrac{x^3 + 1}{x + 1}$

4. $\lim\limits_{x \to -1} \dfrac{x^4 - 1}{x - 1}$

5. $\lim\limits_{x \to 1} \dfrac{x^3 - x^2 - x + 1}{x^3 - 3x + 2}$

6. $\lim\limits_{x \to 2} \dfrac{x^3 + 2x^2 - 16}{2x^3 - x^2 + 2x - 16}$

7. $\lim\limits_{x \to 0} \dfrac{\sqrt{x + 1} - 1}{x}$

8. $\lim\limits_{x \to 0} \dfrac{(x + 4)^{3/2} - 8}{x}$

9. $\lim\limits_{x \to 0} (1 + 2x)^{1/x}$

10. $\lim\limits_{x \to 0} \dfrac{2^x - 1}{x}$

11. Show that $\lim\limits_{x \to 3} \dfrac{2}{x - 3}$ does not exist.

12. Show that $\lim\limits_{x \to 2} \dfrac{x^3 - 2x + 1}{x - 2}$ does not exist.

13. CITY PLANNING A major developer is building a 5000-acre complex of homes, offices, stores, schools, and churches in the rural community of Marlboro. As a result of this development, the planners have estimated that Marlboro's population (in thousands) t yr from now will be given by

$$P(t) = \frac{25t^2 + 125t + 200}{t^2 + 5t + 40}$$

a. Plot the graph of P in the viewing window [0, 50] × [0, 30].

b. What will be the population of Marlboro in the long run?
Hint: Find $\lim\limits_{t \to \infty} P(t)$.

14. AMOUNT OF RAINFALL The total amount of rain (in inches) after t hr during a rainfall is given by

$$T(t) = \frac{0.8t}{t + 4.1}$$

a. Plot the graph of T in the viewing window [0, 30] × [0, 0.8].

b. What is the total amount of rain during this rainfall?
Hint: Find $\lim\limits_{t \to \infty} T(t)$.

11.5 One-Sided Limits and Continuity

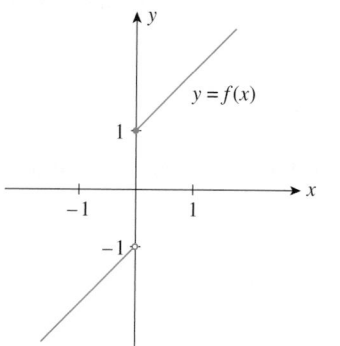

FIGURE 32
The function f does not have a limit as x approaches zero.

■ One-Sided Limits

Consider the function f defined by

$$f(x) = \begin{cases} x - 1 & \text{if } x < 0 \\ x + 1 & \text{if } x \geq 0 \end{cases}$$

From the graph of f shown in Figure 32, we see that the function f does not have a limit as x approaches zero because, no matter how close x is to zero, $f(x)$ takes on values that are close to 1 if x is positive and values that are close to -1 if x is negative. Therefore, $f(x)$ cannot be close to a single number L—no matter how close x is to zero. Now, if we restrict x to be greater than zero (to the right of zero), then we see that $f(x)$ can be made as close to 1 as we please by taking x sufficiently close to zero. In this situation we say that the right-hand limit of f as x approaches zero (from the right) is 1, written

$$\lim_{x \to 0^+} f(x) = 1$$

Similarly, we see that $f(x)$ can be made as close to -1 as we please by taking x sufficiently close to, but to the left of, zero. In this situation we say that the left-hand limit of f as x approaches zero (from the left) is -1, written

$$\lim_{x \to 0^-} f(x) = -1$$

These limits are called **one-sided limits.** More generally, we have the following definitions.

> **One-Sided Limits**
> The function f has the **right-hand limit** L as x approaches a from the right, written
>
> $$\lim_{x \to a^+} f(x) = L$$
>
> if the values of $f(x)$ can be made as close to L as we please by taking x sufficiently close to (but not equal to) a and to the right of a.
> Similarly, the function f has the **left-hand limit** M as x approaches a from the left, written
>
> $$\lim_{x \to a^-} f(x) = M$$
>
> if the values of $f(x)$ can be made as close to M as we please by taking x sufficiently close to (but not equal to) a and to the left of a.

The connection between one-sided limits and the two-sided limit defined earlier is given by the following theorem.

> **THEOREM 3**
> Let f be a function that is defined for all values of x close to $x = a$ with the possible exception of a itself. Then
>
> $$\lim_{x \to a} f(x) = L \quad \text{if and only if} \quad \lim_{x \to a^+} f(x) = \lim_{x \to a^-} f(x) = L$$

Thus, the two-sided limit exists if and only if the one-sided limits exist and are equal.

EXAMPLE 1   Let

$$f(x) = \begin{cases} \sqrt{x} & \text{if } x > 0 \\ -x & \text{if } x \le 0 \end{cases} \quad \text{and} \quad g(x) = \begin{cases} -1 & \text{if } x < 0 \\ 1 & \text{if } x \ge 0 \end{cases}$$

a. Show that $\lim_{x \to 0} f(x)$ exists by studying the one-sided limits of f as x approaches $x = 0$.
b. Show that $\lim_{x \to 0} g(x)$ does not exist.

Solution

a. For $x > 0$, we find

$$\lim_{x \to 0^+} f(x) = \lim_{x \to 0^+} \sqrt{x} = 0$$

and for $x \le 0$

$$\lim_{x \to 0^-} f(x) = \lim_{x \to 0^-} (-x) = 0$$

Thus,

$$\lim_{x \to 0} f(x) = 0$$

(Figure 33a).
b. We have

$$\lim_{x \to 0^-} g(x) = -1 \quad \text{and} \quad \lim_{x \to 0^+} g(x) = 1$$

and since these one-sided limits are not equal, we conclude that $\lim_{x \to 0} g(x)$ does not exist (Figure 33b).

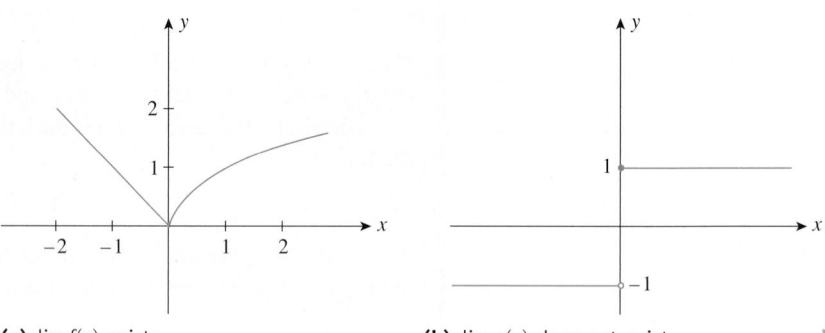

FIGURE 33

(a) $\lim_{x \to 0} f(x)$ exists. **(b)** $\lim_{x \to 0} g(x)$ does not exist.

Continuous Functions

Continuous functions will play an important role throughout most of our study of calculus. Loosely speaking, a function is continuous at a point if the graph of the function at that point is devoid of holes, gaps, jumps, or breaks. Consider, for example, the graph of the function f depicted in Figure 34.

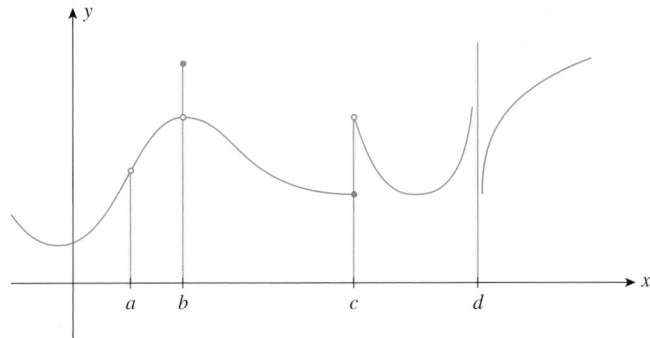

FIGURE 34
The graph of this function is not continuous at $x = a$, $x = b$, $x = c$, and $x = d$.

Let's take a closer look at the behavior of f at or near $x = a$, $x = b$, $x = c$, and $x = d$. First, note that f is not defined at $x = a$; that is, $x = a$ is not in the domain of f, thereby resulting in a "hole" in the graph of f. Next, observe that the value of f at b, $f(b)$, is not equal to the limit of $f(x)$ as x approaches b, resulting in a "jump" in the graph of f at $x = b$. The function f does not have a limit at $x = c$ since the left-hand and right-hand limits of $f(x)$ are not equal, also resulting in a jump in the graph of f at $x = c$. Finally, the limit of f does not exist at $x = d$, resulting in a break in the graph of f. The function f is *discontinuous* at each of these numbers. It is *continuous* everywhere else.

> **Continuity of a Function at a Number**
> A function f is **continuous at a number** $x = a$ if the following conditions are satisfied.
>
> **1.** $f(a)$ is defined.   **2.** $\lim\limits_{x \to a} f(x)$ exists.   **3.** $\lim\limits_{x \to a} f(x) = f(a)$

Thus, a function f is continuous at $x = a$ if the limit of f at $x = a$ exists and has the value $f(a)$. Geometrically, f is continuous at $x = a$ if the proximity of x to a implies the proximity of $f(x)$ to $f(a)$.

If f is not continuous at $x = a$, then f is said to be **discontinuous** at $x = a$. Also, f is **continuous on an interval** if f is continuous at every number in the interval.

Figure 35 depicts the graph of a continuous function on the interval (a, b). Notice that the graph of the function over the stated interval can be sketched without lifting one's pencil from the paper.

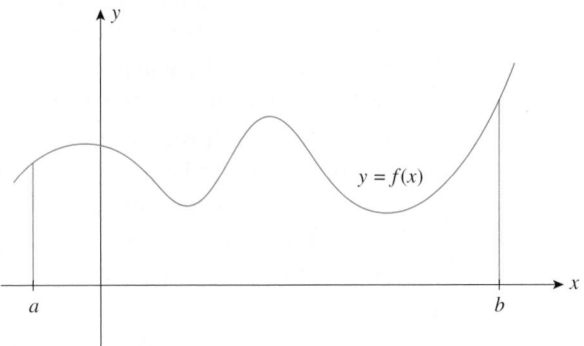

FIGURE 35
The graph of f is continuous on the interval (a, b).

EXAMPLE 2   Find the values of x for which each function is continuous.

a. $f(x) = x + 2$ **b.** $g(x) = \dfrac{x^2 - 4}{x - 2}$ **c.** $h(x) = \begin{cases} x + 2 & \text{if } x \neq 2 \\ 1 & \text{if } x = 2 \end{cases}$

d. $F(x) = \begin{cases} -1 & \text{if } x < 0 \\ 1 & \text{if } x \geq 0 \end{cases}$ **e.** $G(x) = \begin{cases} \dfrac{1}{x} & \text{if } x > 0 \\ -1 & \text{if } x \leq 0 \end{cases}$

The graph of each function is shown in Figure 36.

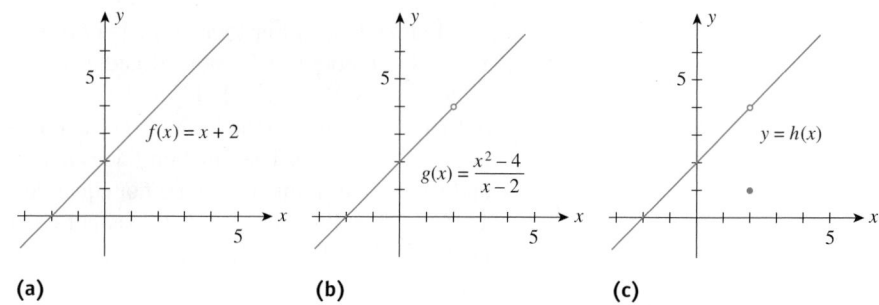

(a)   (b)   (c)

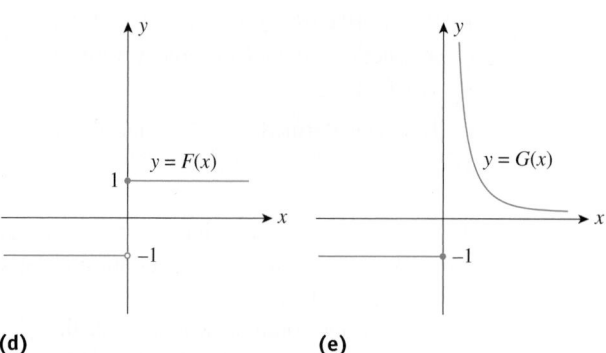

(d)   (e)

FIGURE 36

Solution

a. The function f is continuous everywhere because the three conditions for continuity are satisfied for all values of x.

b. The function g is discontinuous at $x = 2$ because g is not defined at that number. It is continuous everywhere else.

c. The function h is discontinuous at $x = 2$ because the third condition for continuity is violated; the limit of $h(x)$ as x approaches 2 exists and has the value 4, but this limit is not equal to $h(2) = 1$. It is continuous for all other values of x.

d. The function F is continuous everywhere except at $x = 0$, where the limit of $F(x)$ fails to exist as x approaches zero (see Example 3a, Section 11.4).

e. Since the limit of $G(x)$ does not exist as x approaches zero, we conclude that G fails to be continuous at $x = 0$. The function G is continuous everywhere else.

■ Properties of Continuous Functions

The following properties of continuous functions follow directly from the definition of continuity and the corresponding properties of limits. They are stated without proof.

> **Properties of Continuous Functions**
>
> **1.** The constant function $f(x) = c$ is continuous everywhere.
>
> **2.** The identity function $f(x) = x$ is continuous everywhere.
>
> *If f and g are continuous at x = a, then*
>
> **3.** $[f(x)]^n$, where n is a real number, is continuous at $x = a$ whenever it is defined at that number.
>
> **4.** $f \pm g$ is continuous at $x = a$.
>
> **5.** fg is continuous at $x = a$.
>
> **6.** f/g is continuous at $x = a$ provided $g(a) \neq 0$.

Using these properties of continuous functions, we can prove the following results. (A proof is sketched in Exercise 97, pages 655–656.)

> **Continuity of Polynomial and Rational Functions**
>
> **1.** A polynomial function $y = P(x)$ is continuous at every value of x.
>
> **2.** A rational function $R(x) = p(x)/q(x)$ is continuous at every value of x where $q(x) \neq 0$.

EXAMPLE 3   Find the values of x for which each function is continuous.

a. $f(x) = 3x^3 + 2x^2 - x + 10$ **b.** $g(x) = \dfrac{8x^{10} - 4x + 1}{x^2 + 1}$

c. $h(x) = \dfrac{4x^3 - 3x^2 + 1}{x^2 - 3x + 2}$

Solution

a. The function f is a polynomial function of degree 3, so $f(x)$ is continuous for all values of x.

b. The function g is a rational function. Observe that the denominator of g—namely, $x^2 + 1$—is never equal to zero. Therefore, we conclude that g is continuous for all values of x.

c. The function h is a rational function. In this case, however, the denominator of h is equal to zero at $x = 1$ and $x = 2$, which can be seen by factoring it. Thus,

$$x^2 - 3x + 2 = (x - 2)(x - 1)$$

We therefore conclude that h is continuous everywhere except at $x = 1$ and $x = 2$, where it is discontinuous.   ■

Up to this point, most of the applications we have discussed involved functions that are continuous everywhere. In Example 4, we consider an application from the field of educational psychology that involves a discontinuous function.

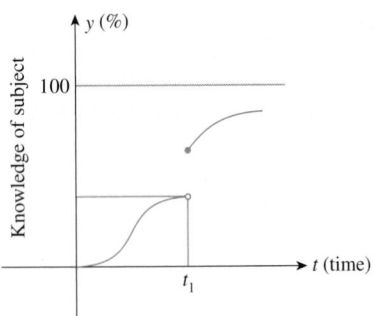

FIGURE 37
A learning curve that is discontinuous at $t = t_1$

APPLIED EXAMPLE 4 **Learning Curves** Figure 37 depicts the learning curve associated with a certain individual. Beginning with no knowledge of the subject being taught, the individual makes steady progress toward understanding it over the time interval $0 \le t < t_1$. In this instance, the individual's progress slows as we approach time t_1 because he fails to grasp a particularly difficult concept. All of a sudden, a breakthrough occurs at time t_1, propelling his knowledge of the subject to a higher level. The curve is discontinuous at t_1. ∎

Intermediate Value Theorem

Let's look again at our model of the motion of the maglev on a straight stretch of track. We know that the train cannot vanish at any instant of time and it cannot skip portions of the track and reappear someplace else. To put it another way, the train cannot occupy the positions s_1 and s_2 without at least, at some time, occupying an intermediate position (Figure 38).

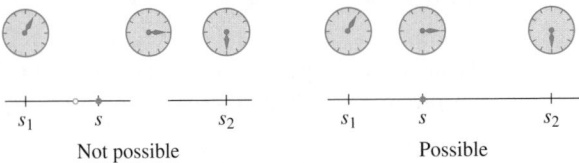

FIGURE 38
The position of the maglev

To state this fact mathematically, recall that the position of the maglev as a function of time is described by

$$f(t) = 4t^2 \qquad (0 \le t \le 10)$$

Suppose the position of the maglev is s_1 at some time t_1 and its position is s_2 at some time t_2 (Figure 39). Then, if s_3 is any number between s_1 and s_2 giving an intermediate position of the maglev, there must be at least one t_3 between t_1 and t_2 giving the time at which the train is at s_3—that is, $f(t_3) = s_3$.

This discussion carries the gist of the intermediate value theorem. The proof of this theorem can be found in most advanced calculus texts.

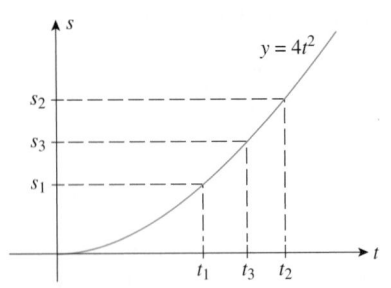

FIGURE 39
If $s_1 \le s_3 \le s_2$, then there must be at least one t_3 ($t_1 \le t_3 \le t_2$) such that $f(t_3) = s_3$.

THEOREM 4

The Intermediate Value Theorem

If f is a continuous function on a closed interval $[a, b]$ and M is any number between $f(a)$ and $f(b)$, then there is at least one number c in $[a, b]$ such that $f(c) = M$ (Figure 40).

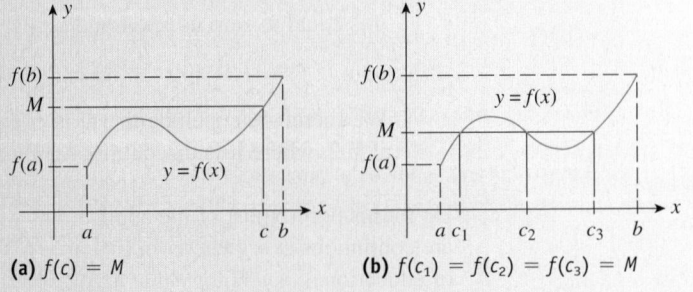

(a) $f(c) = M$

(b) $f(c_1) = f(c_2) = f(c_3) = M$

FIGURE 40

To illustrate the intermediate value theorem, let's look at the example involving the motion of the maglev again (see Figure 22, page 622). Notice that the initial position of the train is $f(0) = 0$ and the position at the end of its test run is $f(10) = 400$. Furthermore, the function f is continuous on $[0, 10]$. So, the intermediate value theorem guarantees that if we arbitrarily pick a number between 0 and 400—say, 100—giving the position of the maglev, there must be a $\bar{t}$ (read "t bar") between 0 and 10 at which time the train is at the position $s = 100$.

To find the value of $\bar{t}$, we solve the equation $f(\bar{t}) = s$, or

$$4\bar{t}^2 = 100$$

giving $\bar{t} = 5$ (t must lie between 0 and 10).

 It is important to remember when we use Theorem 4 that the function f must be continuous. The conclusion of the intermediate value theorem may not hold if f is not continuous (see Exercise 98, page 656).

The next theorem is an immediate consequence of the intermediate value theorem. It not only tells us when a zero of a function f [root of the equation $f(x) = 0$] exists but also provides the basis for a method of approximating it.

THEOREM 5

Existence of Zeros of a Continuous Function

If f is a continuous function on a closed interval $[a, b]$, and if $f(a)$ and $f(b)$ have opposite signs, then there is at least one solution of the equation $f(x) = 0$ in the interval (a, b) (Figure 41).

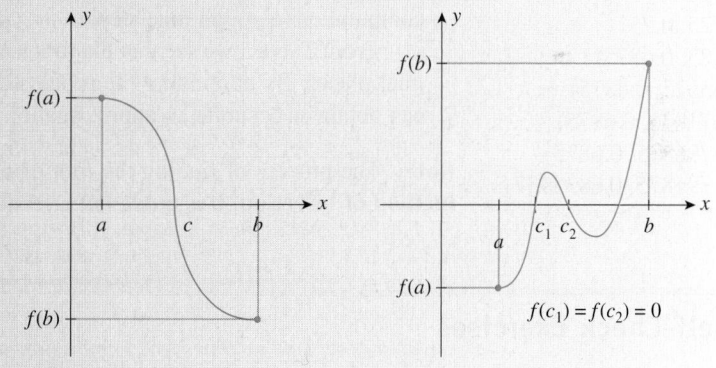

FIGURE 41
If $f(a)$ and $f(b)$ have opposite signs, there must be at least one number c ($a < c < b$) such that $f(c) = 0$.

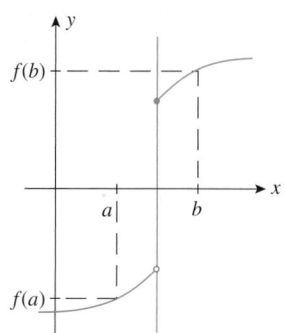

FIGURE 42
$f(a) < 0$ and $f(b) > 0$, but the graph of f does not cross the x-axis between a and b because f is discontinuous.

Geometrically, this property states that if the graph of a continuous function goes from above the x-axis to below the x-axis, or vice versa, it must *cross* the x-axis. This is not necessarily true if the function is discontinuous (Figure 42).

EXAMPLE 5   Let $f(x) = x^3 + x + 1$.

a. Show that f is continuous for all values of x.
b. Compute $f(-1)$ and $f(1)$ and use the results to deduce that there must be at least one number $x = c$, where c lies in the interval $(-1, 1)$ and $f(c) = 0$.

Solution

a. The function f is a polynomial function of degree 3 and is therefore continuous everywhere.

b. $f(-1) = (-1)^3 + (-1) + 1 = -1$ and $f(1) = 1^3 + 1 + 1 = 3$
Since $f(-1)$ and $f(1)$ have opposite signs, Theorem 5 tells us that there must be at least one number $x = c$ with $-1 < c < 1$ such that $f(c) = 0$.　■

The next example shows how the intermediate value theorem can be used to help us find the zero of a function.

EXAMPLE 6　Let $f(x) = x^3 + x - 1$. Since f is a polynomial function, it is continuous everywhere. Observe that $f(0) = -1$ and $f(1) = 1$ so that Theorem 5 guarantees the existence of at least one root of the equation $f(x) = 0$ in $(0, 1)$.*

We can locate the root more precisely by using Theorem 5 once again as follows: Evaluate $f(x)$ at the midpoint of $[0, 1]$, obtaining

$$f(0.5) = -0.375$$

Because $f(0.5) < 0$ and $f(1) > 0$, Theorem 5 now tells us that the root must lie in $(0.5, 1)$.

Repeat the process: Evaluate $f(x)$ at the midpoint of $[0.5, 1]$, which is

$$\frac{0.5 + 1}{2} = 0.75$$

Thus,

$$f(0.75) = 0.1719$$

Because $f(0.5) < 0$ and $f(0.75) > 0$, Theorem 5 tells us that the root is in $(0.5, 0.75)$. This process can be continued. Table 3 summarizes the results of our computations through nine steps.

From Table 3 we see that the root is approximately 0.68, accurate to two decimal places. By continuing the process through a sufficient number of steps, we can obtain as accurate an approximation to the root as we please.　■

TABLE 3

Step	Root of $f(x) = 0$ Lies in
1	$(0, 1)$
2	$(0.5, 1)$
3	$(0.5, 0.75)$
4	$(0.625, 0.75)$
5	$(0.625, 0.6875)$
6	$(0.65625, 0.6875)$
7	$(0.671875, 0.6875)$
8	$(0.6796875, 0.6875)$
9	$(0.6796875, 0.6835937)$

Note　The process of finding the root of $f(x) = 0$ used in Example 6 is called the **method of bisection**. It is crude but effective.　■

11.5　Self-Check Exercises

1. Evaluate $\lim_{x \to -1^-} f(x)$ and $\lim_{x \to -1^+} f(x)$, where

$$f(x) = \begin{cases} 1 & \text{if } x < -1 \\ 1 + \sqrt{x + 1} & \text{if } x \ge -1 \end{cases}$$

Does $\lim_{x \to -1} f(x)$ exist?

2. Determine the values of x for which the given function is discontinuous. At each number where f is discontinuous, indicate which condition(s) for continuity are violated. Sketch the graph of the function.

a. $f(x) = \begin{cases} -x^2 + 1 & \text{if } x \le 1 \\ x - 1 & \text{if } x > 1 \end{cases}$

b. $g(x) = \begin{cases} -x + 1 & \text{if } x < -1 \\ 2 & \text{if } -1 < x \le 1 \\ -x + 3 & \text{if } x > 1 \end{cases}$

Solutions to Self-Check Exercises 11.5 can be found on page 656.

*It can be shown that f has precisely one zero in $(0, 1)$ (see Exercise 103, Section 13.1).

11.5   Concept Questions

1. Explain what is meant by the statement $\lim_{x \to 3^-} f(x) = 2$ and $\lim_{x \to 3^+} f(x) = 4$.

2. Suppose $\lim_{x \to 1^-} f(x) = 3$ and $\lim_{x \to 1^+} f(x) = 4$.
   a. What can you say about $\lim_{x \to 1} f(x)$? Explain.
   b. What can you say about $f(1)$? Explain.

3. Explain what it means for a function f to be continuous (a) at a number a and (b) on an interval I.

4. Determine whether each function f is continuous or discontinuous. Explain your answer.

   a. $f(t)$ gives the altitude of an airplane at time t.
   b. $f(t)$ measures the total amount of rainfall at time t at the Municipal Airport.
   c. $f(s)$ measures the fare as a function of the distance s for taking a cab from Kennedy Airport to downtown Manhattan.
   d. $f(t)$ gives the interest rate charged by a financial institution at time t.

5. Explain the intermediate value theorem in your own words.

11.5   Exercises

In Exercises 1–8, use the graph of the function f to find $\lim_{x \to a^-} f(x)$, $\lim_{x \to a^+} f(x)$, and $\lim_{x \to a} f(x)$ at the indicated value of a, if the limit exists.

1.
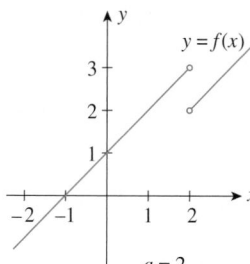

2.
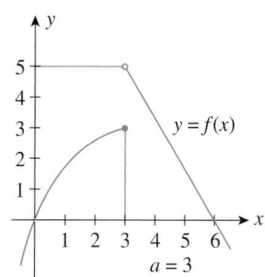

3.
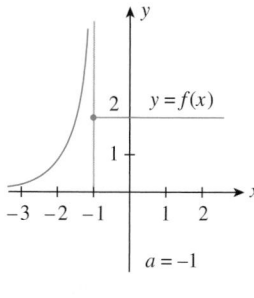

4.
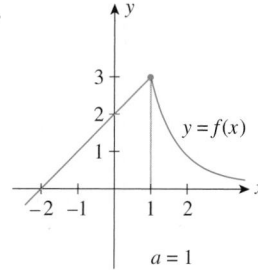

5.
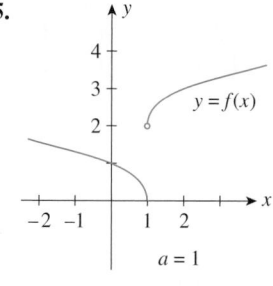

6.
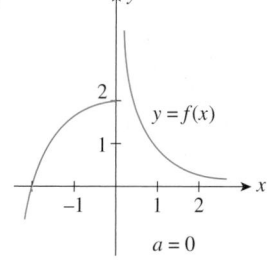

7.
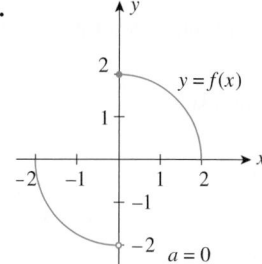

8.
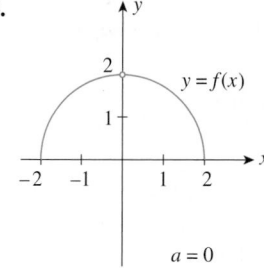

In Exercises 9–14, refer to the graph of the function f and determine whether each statement is true or false.

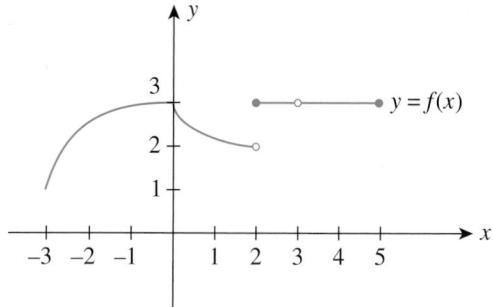

9. $\lim_{x \to -3^+} f(x) = 1$

10. $\lim_{x \to 0^-} f(x) = f(0)$

11. $\lim_{x \to 2^-} f(x) = 2$

12. $\lim_{x \to 2^+} f(x) = 3$

13. $\lim_{x \to 3^-} f(x)$ does not exist.

14. $\lim_{x \to 5^-} f(x) = 3$

In Exercises 15–20, refer to the graph of the function f and determine whether each statement is true or false.

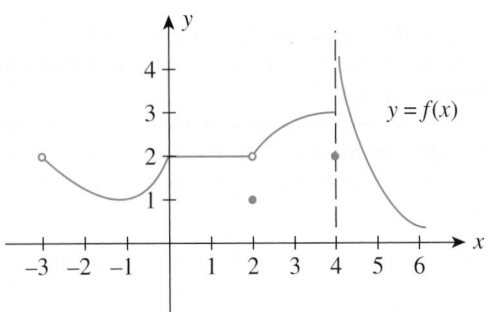

15. $\lim\limits_{x \to -3^+} f(x) = 2$

16. $\lim\limits_{x \to 0} f(x) = 2$

17. $\lim\limits_{x \to 2} f(x) = 1$

18. $\lim\limits_{x \to 4^-} f(x) = 3$

19. $\lim\limits_{x \to 4^+} f(x)$ does not exist.

20. $\lim\limits_{x \to 4} f(x) = 2$

In Exercises 21–38, find the indicated one-sided limit, if it exists.

21. $\lim\limits_{x \to 1^+} (2x + 4)$

22. $\lim\limits_{x \to 1^-} (3x - 4)$

23. $\lim\limits_{x \to 2^-} \dfrac{x - 3}{x + 2}$

24. $\lim\limits_{x \to 1^+} \dfrac{x + 2}{x + 1}$

25. $\lim\limits_{x \to 0^+} \dfrac{1}{x}$

26. $\lim\limits_{x \to 0^-} \dfrac{1}{x}$

27. $\lim\limits_{x \to 0^+} \dfrac{x - 1}{x^2 + 1}$

28. $\lim\limits_{x \to 2^+} \dfrac{x + 1}{x^2 - 2x + 3}$

29. $\lim\limits_{x \to 0^+} \sqrt{x}$

30. $\lim\limits_{x \to 2^+} 2\sqrt{x - 2}$

31. $\lim\limits_{x \to -2^+} (2x + \sqrt{2 + x})$

32. $\lim\limits_{x \to -5^+} x(1 + \sqrt{5 + x})$

33. $\lim\limits_{x \to 1^-} \dfrac{1 + x}{1 - x}$

34. $\lim\limits_{x \to 1^+} \dfrac{1 + x}{1 - x}$

35. $\lim\limits_{x \to 2^-} \dfrac{x^2 - 4}{x - 2}$

36. $\lim\limits_{x \to -3^+} \dfrac{\sqrt{x + 3}}{x^2 + 1}$

37. $\lim\limits_{x \to 0^+} f(x)$ and $\lim\limits_{x \to 0^-} f(x)$, where

$$f(x) = \begin{cases} 2x & \text{if } x < 0 \\ x^2 & \text{if } x \geq 0 \end{cases}$$

38. $\lim\limits_{x \to 0^+} f(x)$ and $\lim\limits_{x \to 0^-} f(x)$, where

$$f(x) = \begin{cases} -x + 1 & \text{if } x \leq 0 \\ 2x + 3 & \text{if } x > 0 \end{cases}$$

In Exercises 39–44, determine the values of x, if any, at which each function is discontinuous. At each number where f is discontinuous, state the condition(s) for continuity that are violated.

39.

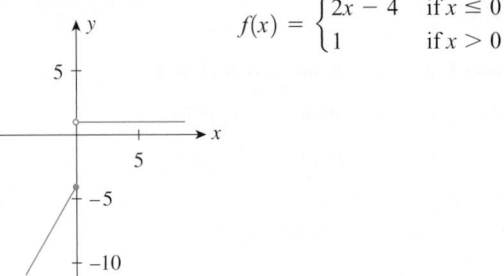

$$f(x) = \begin{cases} 2x - 4 & \text{if } x \leq 0 \\ 1 & \text{if } x > 0 \end{cases}$$

40.

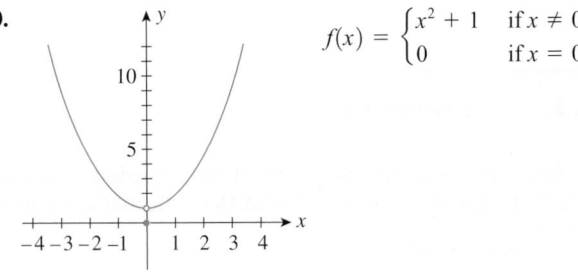

$$f(x) = \begin{cases} x^2 + 1 & \text{if } x \neq 0 \\ 0 & \text{if } x = 0 \end{cases}$$

41.

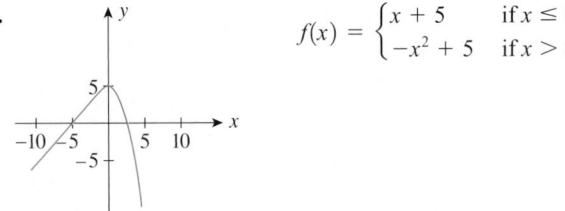

$$f(x) = \begin{cases} x + 5 & \text{if } x \leq 0 \\ -x^2 + 5 & \text{if } x > 0 \end{cases}$$

42.

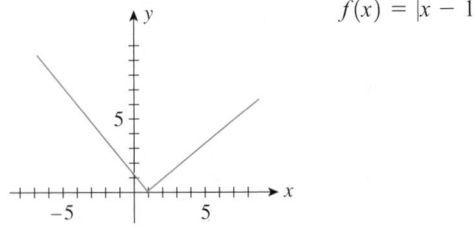

$$f(x) = |x - 1|$$

43.

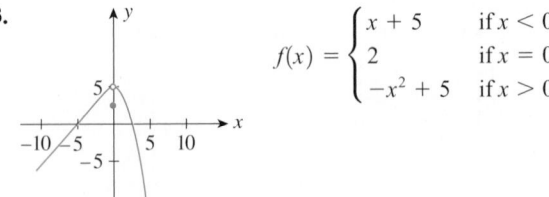

$$f(x) = \begin{cases} x + 5 & \text{if } x < 0 \\ 2 & \text{if } x = 0 \\ -x^2 + 5 & \text{if } x > 0 \end{cases}$$

44.

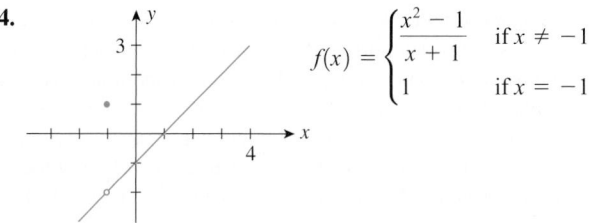

$$f(x) = \begin{cases} \dfrac{x^2 - 1}{x + 1} & \text{if } x \neq -1 \\ 1 & \text{if } x = -1 \end{cases}$$

In Exercises 45–56, find the values of x for which each function is continuous.

45. $f(x) = 2x^2 + x - 1$

46. $f(x) = x^3 - 2x^2 + x - 1$

47. $f(x) = \dfrac{2}{x^2 + 1}$

48. $f(x) = \dfrac{x}{2x^2 + 1}$

49. $f(x) = \dfrac{2}{2x - 1}$

50. $f(x) = \dfrac{x + 1}{x - 1}$

51. $f(x) = \dfrac{2x + 1}{x^2 + x - 2}$

52. $f(x) = \dfrac{x - 1}{x^2 + 2x - 3}$

53. $f(x) = \begin{cases} x & \text{if } x \leq 1 \\ 2x - 1 & \text{if } x > 1 \end{cases}$

54. $f(x) = \begin{cases} -2x + 1 & \text{if } x < 0 \\ x^2 + 1 & \text{if } x \geq 0 \end{cases}$

55. $f(x) = |x + 1|$

56. $f(x) = \dfrac{|x - 1|}{x - 1}$

In Exercises 57–60, determine all values of x at which the function is discontinuous.

57. $f(x) = \dfrac{2x}{x^2 - 1}$

58. $f(x) = \dfrac{1}{(x - 1)(x - 2)}$

59. $f(x) = \dfrac{x^2 - 2x}{x^2 - 3x + 2}$

60. $f(x) = \dfrac{x^2 - 3x + 2}{x^2 - 2x}$

61. THE POSTAGE FUNCTION The graph of the "postage function" for 2006,

$$f(x) = \begin{cases} 39 & \text{if } 0 < x \leq 1 \\ 63 & \text{if } 1 < x \leq 2 \\ \vdots & \\ 303 & \text{if } 11 < x \leq 12 \end{cases}$$

where x denotes the weight of a parcel in ounces and $f(x)$ the postage in cents, is shown in the accompanying figure. Determine the values of x for which f is discontinuous.

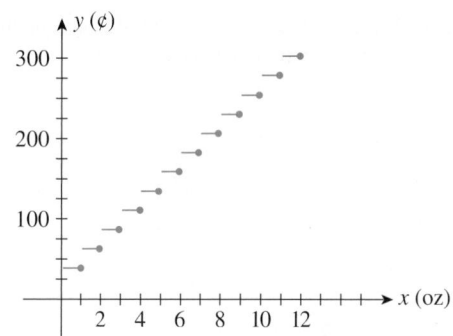

62. INVENTORY CONTROL As part of an optimal inventory policy, the manager of an office supply company orders 500 reams of photocopy paper every 20 days. The accompanying graph shows the *actual* inventory level of paper in an office supply store during the first 60 business days of 2007. Determine the values of t for which the "inventory function" is discontinuous and give an interpretation of the graph.

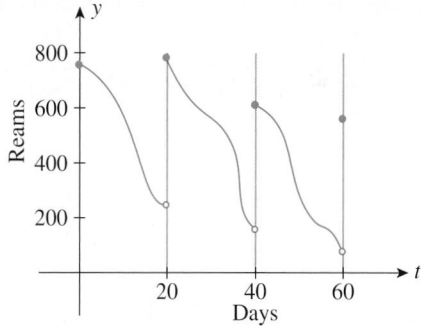

63. LEARNING CURVES The following graph describes the progress Michael made in solving a problem correctly during a mathematics quiz. Here, y denotes the percent of work completed, and x is measured in minutes. Give an interpretation of the graph.

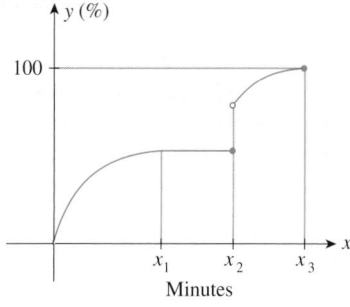

64. AILING FINANCIAL INSTITUTIONS Franklin Savings and Loan acquired two ailing financial institutions in 2006. One of them was acquired at time $t = T_1$, and the other was acquired at time $t = T_2$ ($t = 0$ corresponds to the beginning of 2006). The following graph shows the total amount of money on deposit with Franklin. Explain the significance of the discontinuities of the function at T_1 and T_2.

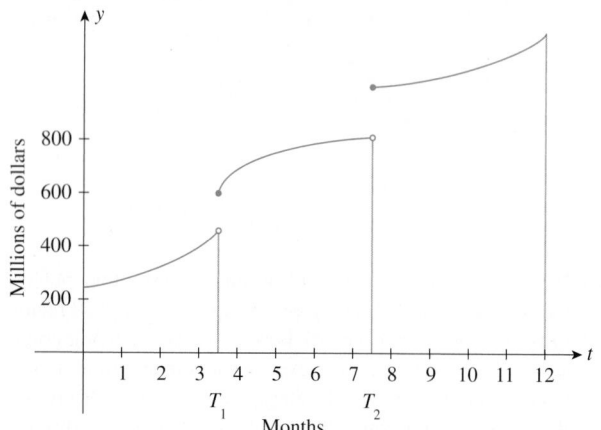

65. **ENERGY CONSUMPTION**   The following graph shows the amount of home heating oil remaining in a 200-gal tank over a 120-day period ($t = 0$ corresponds to October 1). Explain why the function is discontinuous at $t = 40$, 70, 95, and 110.

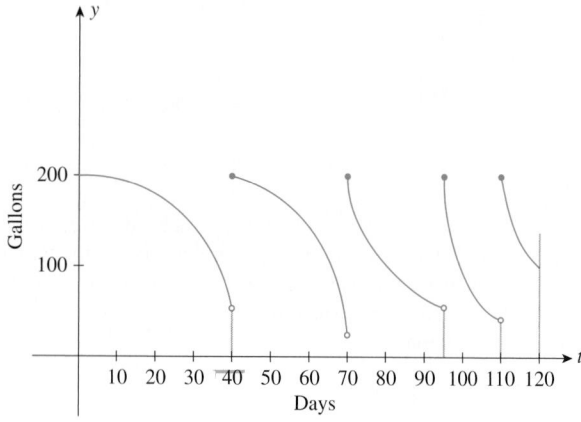

66. **PRIME INTEREST RATE**   The function P, whose graph follows, gives the prime rate (the interest rate banks charge their best corporate customers) as a function of time for the first 32 wk in 1989. Determine the values of t for which P is discontinuous and interpret your results.

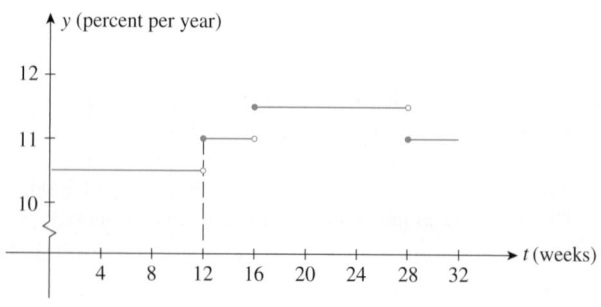

67. **ADMINISTRATION OF AN INTRAVENOUS SOLUTION**   A dextrose solution is being administered to a patient intravenously. The 1-liter (L) bottle holding the solution is removed and replaced by another as soon as the contents drop to approximately 5% of the initial (1-L) amount. The rate of discharge is constant, and it takes 6 hr to discharge 95% of the contents of a full bottle. Draw a graph showing the amount of dextrose solution in a bottle in the IV system over a 24-hr period, assuming that we started with a full bottle.

68. **COMMISSIONS**   The base monthly salary of a salesman working on commission is $12,000. For each $50,000 of sales beyond $100,000, he is paid a $1000 commission. Sketch a graph showing his earnings as a function of the level of his sales x. Determine the values of x for which the function f is discontinuous.

69. **PARKING FEES**   The fee charged per car in a downtown parking lot is $2.00 for the first half hour and $1.00 for each additional half hour or part thereof, subject to a maximum of $10.00. Derive a function f relating the parking fee to the length of time a car is left in the lot. Sketch the graph of f and determine the values of x for which the function f is discontinuous.

70. **COMMODITY PRICES**   The function that gives the cost of a certain commodity is defined by

$$C(x) = \begin{cases} 5x & \text{if } 0 < x < 10 \\ 4x & \text{if } 10 \le x < 30 \\ 3.5x & \text{if } 30 \le x < 60 \\ 3.25x & \text{if } x \ge 60 \end{cases}$$

where x is the number of pounds of a certain commodity sold and $C(x)$ is measured in dollars. Sketch the graph of the function C and determine the values of x for which the function C is discontinuous.

71. **WEISS'S LAW**   According to Weiss's law of excitation of tissue, the strength S of an electric current is related to the time t the current takes to excite tissue by the formula

$$S(t) = \frac{a}{t} + b \qquad (t > 0)$$

where a and b are positive constants.
a. Evaluate $\lim_{t \to 0^+} S(t)$ and interpret your result.
b. Evaluate $\lim_{t \to \infty} S(t)$ and interpret your result.

(*Note:* The limit in part (b) is called the threshold strength of the current. Why?)

72. **ENERGY EXPENDED BY A FISH**   Suppose a fish swimming a distance of L ft at a speed of v ft/sec relative to the water and against a current flowing at the rate of u ft/sec ($u < v$) expends a total energy given by

$$E(v) = \frac{aLv^3}{v - u}$$

where E is measured in foot-pounds (ft-lb) and a is a constant.
a. Evaluate $\lim_{v \to u^+} E(v)$ and interpret your result.
b. Evaluate $\lim_{v \to \infty} E(v)$ and interpret your result.

73. Let

$$f(x) = \begin{cases} x + 2 & \text{if } x \leq 1 \\ kx^2 & \text{if } x > 1 \end{cases}$$

Find the value of k that will make f continuous on $(-\infty, \infty)$.

74. Let

$$f(x) = \begin{cases} \dfrac{x^2 - 4}{x + 2} & \text{if } x \neq -2 \\ k & \text{if } x = -2 \end{cases}$$

For what value of k will f be continuous on $(-\infty, \infty)$?

75. a. Suppose f is continuous at a and g is discontinuous at a. Is the sum $f + g$ discontinuous at a? Explain.
  b. Suppose f and g are both discontinuous at a. Is the sum $f + g$ necessarily discontinuous at a? Explain.

76. a. Suppose f is continuous at a and g is discontinuous at a. Is the product fg necessarily discontinuous at a? Explain.
  b. Suppose f and g are both discontinuous at a. Is the product fg necessarily discontinuous at a? Explain.

In Exercises 77–80, (a) show that the function f is continuous for all values of x in the interval $[a, b]$ and (b) prove that f must have at least one zero in the interval (a, b) by showing that $f(a)$ and $f(b)$ have opposite signs.

77. $f(x) = x^2 - 6x + 8$; $a = 1, b = 3$

78. $f(x) = 2x^3 - 3x^2 - 36x + 14$; $a = 0, b = 1$

79. $f(x) = x^3 - 2x^2 + 3x + 2$; $a = -1, b = 1$

80. $f(x) = 2x^{5/3} - 5x^{4/3}$; $a = 14, b = 16$

In Exercises 81–82, use the intermediate value theorem to find the value of c such that $f(c) = M$.

81. $f(x) = x^2 - 4x + 6$ on $[0, 3]$; $M = 4$

82. $f(x) = x^2 - x + 1$ on $[-1, 4]$; $M = 7$

83. Use the method of bisection (see Example 6) to find the root of the equation $x^5 + 2x - 7 = 0$ accurate to two decimal places.

84. Use the method of bisection (see Example 6) to find the root of the equation $x^3 - x + 1 = 0$ accurate to two decimal places.

85. Falling Object   Joan is looking straight out a window of an apartment building at a height of 32 ft from the ground. A boy throws a tennis ball straight up by the side of the building where the window is located. Suppose the height of the ball (measured in feet) from the ground at time t is $h(t) = 4 + 64t - 16t^2$.
  a. Show that $h(0) = 4$ and $h(2) = 68$.
  b. Use the intermediate value theorem to conclude that the ball must cross Joan's line of sight at least once.
  c. At what time(s) does the ball cross Joan's line of sight? Interpret your results.

86. Oxygen Content of a Pond   The oxygen content t days after organic waste has been dumped into a pond is given by

$$f(t) = 100 \left(\frac{t^2 + 10t + 100}{t^2 + 20t + 100} \right)$$

percent of its normal level.
  a. Show that $f(0) = 100$ and $f(10) = 75$.
  b. Use the intermediate value theorem to conclude that the oxygen content of the pond must have been at a level of 80% at some time.
  c. At what time(s) is the oxygen content at the 80% level? Hint: Use the quadratic formula.

In Exercises 87–93, determine whether the statement is true or false. If it is true, explain why it is true. If it is false, give an example to show why it is false.

87. If $f(2) = 4$, then $\lim\limits_{x \to 2} f(x) = 4$.

88. If $\lim\limits_{x \to 0} f(x) = 3$, then $f(0) = 3$.

89. Suppose the function f is defined on the interval $[a, b]$. If $f(a)$ and $f(b)$ have the same sign, then f has no zero in $[a, b]$.

90. If $\lim\limits_{x \to a} f(x) = L$, then $\lim\limits_{x \to a^+} f(x) - \lim\limits_{x \to a^-} f(x) \neq 0$.

91. If $\lim\limits_{x \to a^-} f(x) = L$ and $\lim\limits_{x \to a^+} f(x) = L$, then $f(a) = L$.

92. If $\lim\limits_{x \to a} f(x) = L$ and $g(a) = M$, then $\lim\limits_{x \to a} f(x)g(x) = LM$.

93. If f is continuous for all $x \neq 0$ and $f(0) = 0$, then $\lim\limits_{x \to 0} f(x) = 0$.

94. Suppose f is continuous on $[a, b]$ and $f(a) < f(b)$. If M is a number that lies outside the interval $[f(a), f(b)]$, then there does not exist a number $a < c < b$ such that $f(c) = M$. Does this contradict the intermediate value theorem?

95. Let $f(x) = x - \sqrt{1 - x^2}$.
  a. Show that f is continuous for all values of x in the interval $[-1, 1]$.
  b. Show that f has at least one zero in $[-1, 1]$.
  c. Find the zeros of f in $[-1, 1]$ by solving the equation $f(x) = 0$.

96. Let $f(x) = \dfrac{x^2}{x^2 + 1}$.
  a. Show that f is continuous for all values of x.
  b. Show that $f(x)$ is nonnegative for all values of x.
  c. Show that f has a zero at $x = 0$. Does this contradict Theorem 5?

97. a. Prove that a polynomial function $y = P(x)$ is continuous at every number x. Follow these steps:
  (1) Use Properties 2 and 3 of continuous functions to establish that the function $g(x) = x^n$, where n is a positive integer, is continuous everywhere.
  (2) Use Properties 1 and 5 to show that $f(x) = cx^n$, where c is a constant and n is a positive integer, is continuous everywhere.
  (3) Use Property 4 to complete the proof of the result.

b. Prove that a rational function $R(x) = p(x)/q(x)$ is continuous at every point x, where $q(x) \neq 0$.

Hint: Use the result of part (a) and Property 6.

98. Show that the conclusion of the intermediate value theorem does not hold if f is discontinuous on $[a, b]$.

11.5   Solutions to Self-Check Exercises

1. For $x < -1$, $f(x) = 1$, and so

$$\lim_{x \to -1^-} f(x) = \lim_{x \to -1^-} 1 = 1$$

For $x \geq -1$, $f(x) = 1 + \sqrt{x + 1}$, and so

$$\lim_{x \to -1^+} f(x) = \lim_{x \to -1^+} (1 + \sqrt{x + 1}) = 1$$

Since the left-hand and right-hand limits of f exist as x approaches $x = -1$ and both are equal to 1, we conclude that

$$\lim_{x \to -1} f(x) = 1$$

2. a. The graph of f follows:

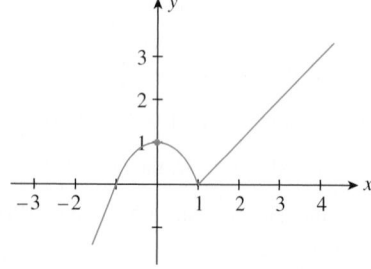

We see that f is continuous everywhere.

b. The graph of g follows:

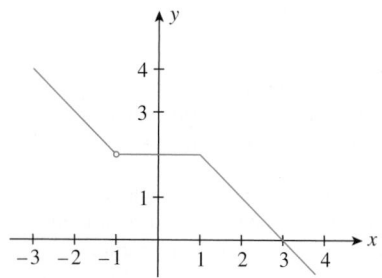

Since g is not defined at $x = -1$, it is discontinuous there. It is continuous everywhere else.

USING TECHNOLOGY

■ Finding the Points of Discontinuity of a Function

You can very often recognize the points of discontinuity of a function f by examining its graph. For example, Figure T1a shows the graph of $f(x) = x/(x^2 - 1)$ obtained using a graphing utility. It is evident that f is discontinuous at $x = -1$ and $x = 1$. This observation is also borne out by the fact that both these points are not in the domain of f.

FIGURE T1
(a) The graph of $f(x) = \dfrac{x}{x^2 - 1}$ in the viewing window $[-4, 4] \times [-10, 10]$; (b) the TI-83 equation screen

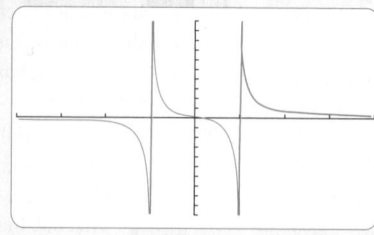

(a)

Plot1   Plot2   Plot3
\Y$_1$ ■ X/(X^2 – 1)
\Y$_2$ =
\Y$_3$ =
\Y$_4$ =
\Y$_5$ =
\Y$_6$ =
\Y$_7$ =

(b)

Consider the function

$$g(x) = \frac{2x^3 + x^2 - 7x - 6}{x^2 - x - 2}$$

Using a graphing utility, we obtain the graph of g shown in Figure T2a.

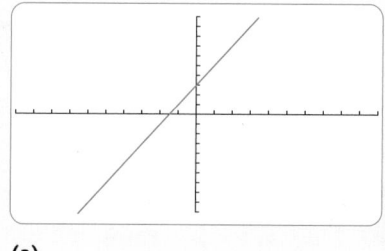

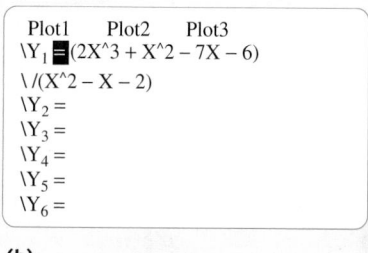

FIGURE T2
(a) The graph of
$$g(x) = \frac{2x^3 + x^2 - 7x - 6}{x^2 - x - 2}$$ in the
standard viewing window; (b) the TI-83
equation screen.

(a)

(b)

An examination of this graph does not reveal any points of discontinuity. However, if we factor both the numerator and the denominator of the rational expression, we see that

$$g(x) = \frac{(x + 1)(x - 2)(2x + 3)}{(x + 1)(x - 2)}$$

$$= 2x + 3$$

provided $x \neq -1$ and $x \neq 2$, so that its graph in fact looks like that shown in Figure T3.

This example shows the limitation of the graphing utility and reminds us of the importance of studying functions analytically!

Graphing Functions Defined Piecewise

The following example illustrates how to plot the graphs of functions defined in a piecewise manner on a graphing utility.

EXAMPLE 1 Plot the graph of

$$f(x) = \begin{cases} x + 1 & \text{if } x \leq 1 \\ \dfrac{2}{x} & \text{if } x > 1 \end{cases}$$

FIGURE T3
The graph of g has holes at $(-1, 1)$ and $(2, 7)$.

Solution We enter the function

$$y1 = (x + 1)(x \leq 1) + (2/x)(x > 1)$$

Figure T4 shows the graph of the function in the viewing window $[-5, 5] \times [-2, 4]$.

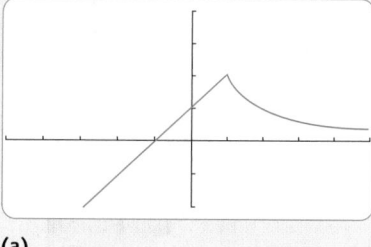

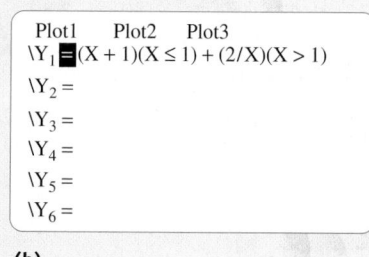

FIGURE T4
(a) The graph of f in the viewing window
$[-5, 5] \times [-2, 4]$; (b) the TI-83 equation
screen.

(a)

(b)

(continued)

APPLIED EXAMPLE 2 TV Viewing Patterns The percent of U.S. households, $P(t)$, watching television during weekdays between the hours of 4 p.m. and 4 a.m. is given by

$$P(t) = \begin{cases} 0.01354t^4 - 0.49375t^3 + 2.58333t^2 + 3.8t + 31.60704 & \text{if } 0 \le t \le 8 \\ 1.35t^2 - 33.05t + 208 & \text{if } 8 < t \le 12 \end{cases}$$

where t is measured in hours, with $t = 0$ corresponding to 4 p.m. Plot the graph of P in the viewing window $[0, 12] \times [0, 80]$.

Source: A. C. Nielsen Co.

Solution   We enter the function

$$y1 = (.01354x^4 - .49375x^3 + 2.58333x^2 + 3.8x + 31.60704)(x \ge 0)(x \le 8)$$
$$+ (1.35x^2 - 33.05x + 208)(x > 8)(x \le 12)$$

Figure T5a shows the graph of P.

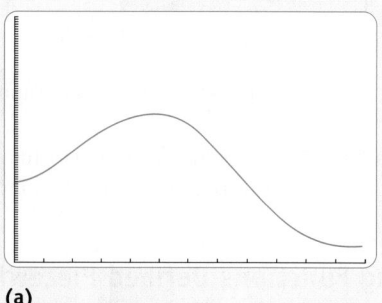

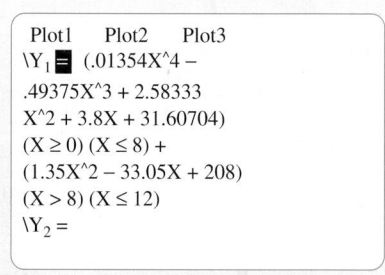

FIGURE T5
(a) The graph of P in the viewing window $[0, 12] \times [0, 80]$; (b) the TI-83 equation screen

(a)

(b)

TECHNOLOGY EXERCISES

In Exercises 1–10, plot the graph of f and find the points of discontinuity of f. Then use analytical means to verify your observation and find all numbers where f is discontinuous.

1. $f(x) = \dfrac{2}{x^2 - x}$

2. $f(x) = \dfrac{2x + 1}{x^2 + x - 2}$

3. $f(x) = \dfrac{\sqrt{x}}{x^2 - x - 2}$

4. $f(x) = \dfrac{3}{\sqrt{x}(x + 1)}$

5. $f(x) = \dfrac{6x^3 + x^2 - 2x}{2x^2 - x}$

6. $f(x) = \dfrac{2x^3 - x^2 - 13x - 6}{2x^2 - 5x - 3}$

7. $f(x) = \dfrac{2x^4 - 3x^3 - 2x^2}{2x^2 - 3x - 2}$

8. $f(x) = \dfrac{6x^4 - x^3 + 5x^2 - 1}{6x^2 - x - 1}$

9. $f(x) = \dfrac{x^3 + x^2 - 2x}{x^4 + 2x^3 - x - 2}$
   Hint: $x^4 + 2x^3 - x - 2 = (x^3 - 1)(x + 2)$

10. $f(x) = \dfrac{x^3 - x}{x^{4/3} - x + x^{1/3} - 1}$
   Hint: $x^{4/3} - x + x^{1/3} - 1 = (x^{1/3} - 1)(x + 1)$

   Can you explain why part of the graph is missing?

In Exercises 11–14, plot the graph of f in the indicated viewing window.

11. $f(x) = \begin{cases} -1 & \text{if } x \le 1 \\ x + 1 & \text{if } x > 1 \end{cases}; [-5, 5] \times [-2, 8]$

12. $f(x) = \begin{cases} \dfrac{1}{3}x^2 - 2x & \text{if } x \le 3 \\ -x + 6 & \text{if } x > 3 \end{cases}; [0, 7] \times [-5, 5]$

13. $f(x) = \begin{cases} 2 & \text{if } x \le 0 \\ \sqrt{4 - x^2} & \text{if } x > 0 \end{cases}; [-2, 2] \times [-4, 4]$

14. $f(x) = \begin{cases} -x^2 + x + 2 & \text{if } x \le 1 \\ 2x^3 - x^2 - 4 & \text{if } x > 1 \end{cases}; [-4, 4] \times [-5, 5]$

15. FLIGHT PATH OF A PLANE   The function

$$f(x) = \begin{cases} 0 & \text{if } 0 \le x < 1 \\ \begin{aligned} -0.00411523x^3 + 0.0679012x^2 \\ -0.123457x + 0.0596708 \end{aligned} & \text{if } 1 \le x < 10 \\ 1.5 & \text{if } 10 \le x \le 100 \end{cases}$$

where both x and $f(x)$ are measured in units of 1000 ft, describes the flight path of a plane taking off from the origin and climbing to an altitude of 15,000 ft. Plot the graph of f to visualize the trajectory of the plane.

16. HOME SHOPPING INDUSTRY   According to industry sources, revenue from the home shopping industry for the years since its inception may be approximated by the function

$$R(t) = \begin{cases} -0.03t^3 + 0.25t^2 - 0.12t & \text{if } 0 \le t \le 3 \\ 0.57t - 0.63 & \text{if } 3 < t \le 11 \end{cases}$$

where $R(t)$ measures the revenue in billions of dollars and t is measured in years, with $t = 0$ corresponding to the beginning of 1984. Plot the graph of R.

Source: Paul Kagan Associates

11.6 The Derivative

■ An Intuitive Example

We mentioned in Section 11.4 that the problem of finding the *rate of change* of one quantity with respect to another is mathematically equivalent to the problem of finding the *slope of the tangent line* to a curve at a given point on the curve. Before going on to establish this relationship, let's show its plausibility by looking at it from an intuitive point of view.

Consider the motion of the maglev discussed in Section 11.4. Recall that the position of the maglev at any time t is given by

$$s = f(t) = 4t^2 \qquad (0 \le t \le 30)$$

where s is measured in feet and t in seconds. The graph of the function f is sketched in Figure 43.

Observe that the graph of f rises slowly at first but more rapidly as t increases, reflecting the fact that the speed of the maglev is increasing with time. This observation suggests a relationship between the speed of the maglev at any time t and the *steepness* of the curve at the point corresponding to this value of t. Thus, it would appear that we can solve the problem of finding the speed of the maglev at any time if we can find a way to measure the steepness of the curve at any point on the curve.

To discover a yardstick that will measure the steepness of a curve, consider the graph of a function f such as the one shown in Figure 44a. Think of the curve as representing a stretch of roller coaster track (Figure 44b). When the car is at the point P on the curve, a passenger sitting erect in the car and looking straight ahead will have a line of sight that is parallel to the line T, the tangent to the curve at P.

As Figure 44a suggests, the steepness of the curve—that is, the rate at which y is increasing or decreasing with respect to x—is given by the slope of the tangent line to the graph of f at the point $P(x, f(x))$. But for now we will show how this relationship can be used to estimate the rate of change of a function from its graph.

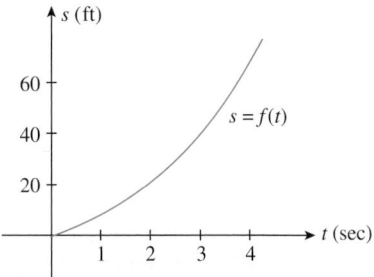

FIGURE 43
Graph showing the position s of a maglev at time t

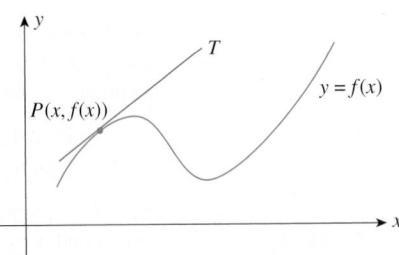
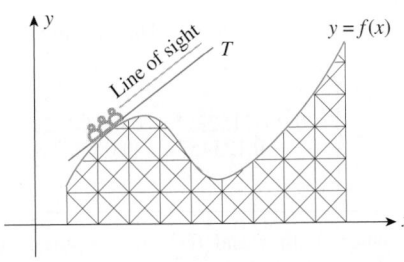

FIGURE 44

(a) *T* is the tangent line to the curve at *P*.   **(b)** *T* is parallel to the line of sight.

APPLIED EXAMPLE 1 Social Security Beneficiaries The graph of the function $y = N(t)$, shown in Figure 45, gives the number of Social Security beneficiaries from the beginning of 1990 ($t = 0$) through the year 2045 ($t = 55$).

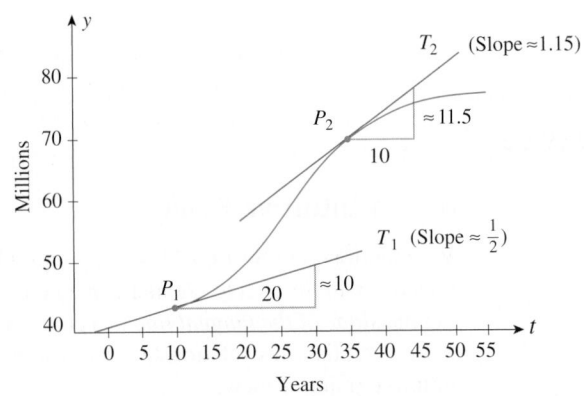

FIGURE 45
The number of Social Security beneficiaries from 1990 through 2045. We can use the slope of the tangent line at the indicated points to estimate the rate at which the number of Social Security beneficiaries will be changing.

Use the graph of $y = N(t)$ to estimate the rate at which the number of Social Security beneficiaries was growing at the beginning of the year 2000 ($t = 10$). How fast will the number be growing at the beginning of 2025 ($t = 35$)? [Assume that the rate of change of the function N at any value of t is given by the slope of the tangent line at the point $P(t, N(t))$.]
Source: Social Security Administration

Solution From the figure, we see that the slope of the tangent line T_1 to the graph of $y = N(t)$ at $P_1(10, 44.7)$ is approximately 0.5. This tells us that the quantity y is increasing at the rate of $\frac{1}{2}$ unit per unit increase in t, when $t = 10$. In other words, at the beginning of the year 2000, the number of Social Security beneficiaries was increasing at the rate of approximately 0.5 million, or 500,000, per year.

The slope of the tangent line T_2 at $P_2(35, 71.9)$ is approximately 1.15. This tells us that at the beginning of 2025 the number of Social Security beneficiaries will be growing at the rate of approximately 1.15 million, or 1,150,000, per year.   ■

Slope of a Tangent Line

In Example 1 we answered the questions raised by drawing the graph of the function *N* and estimating the position of the tangent lines. Ideally, however, we would like to solve a problem analytically whenever possible. To do this we need a precise definition of the slope of a tangent line to a curve.

To define the tangent line to a curve C at a point P on the curve, fix P and let Q be any point on C distinct from P (Figure 46). The straight line passing through P and Q is called a **secant line**.

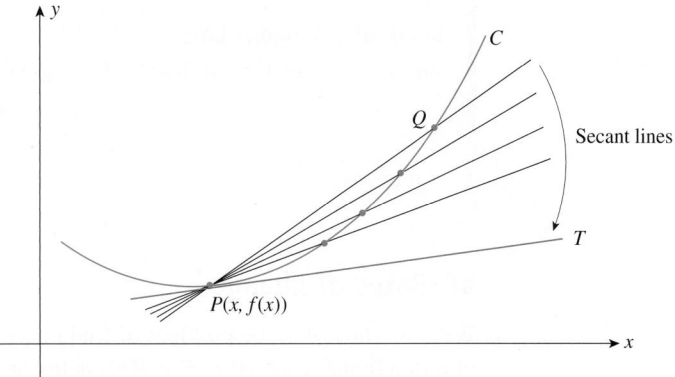

FIGURE 46
As Q approaches P along the curve C, the secant lines approach the tangent line T.

Now, as the point Q is allowed to move toward P along the curve, the secant line through P and Q rotates about the fixed point P and approaches a fixed line through P. This fixed line, which is the limiting position of the secant lines through P and Q as Q approaches P, is the **tangent line to the graph of f at the point P**.

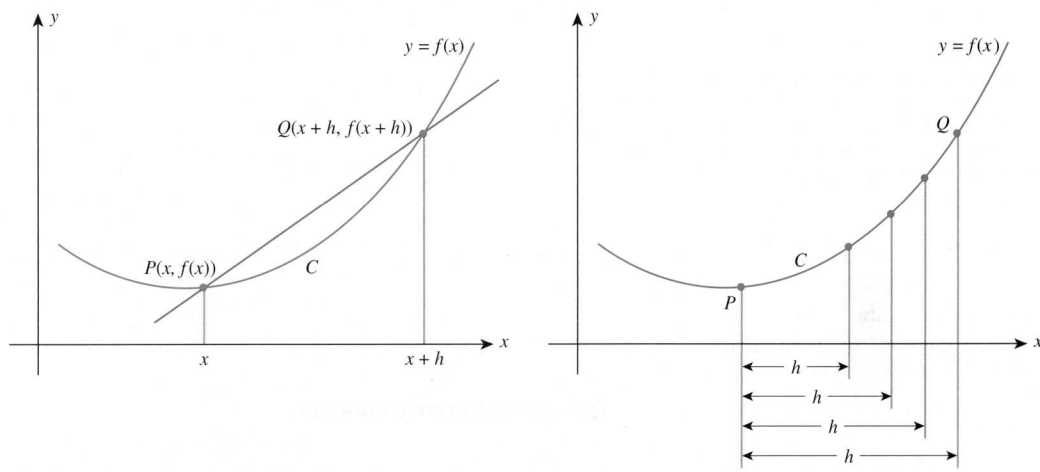

FIGURE 47   **(a)** The points $P(x, f(x))$ and $Q(x + h, f(x + h))$ **(b)** As h approaches zero, Q approaches P.

We can describe the process more precisely as follows. Suppose the curve C is the graph of a function f defined by $y = f(x)$. Then the point P is described by $P(x, f(x))$ and the point Q by $Q(x + h, f(x + h))$, where h is some appropriate nonzero number (Figure 47a). Observe that we can make Q approach P along the curve C by letting h approach zero (Figure 47b).

Next, using the formula for the slope of a line, we can write the slope of the secant line passing through $P(x, f(x))$ and $Q(x + h, f(x + h))$ as

$$\frac{f(x + h) - f(x)}{(x + h) - x} = \frac{f(x + h) - f(x)}{h} \tag{3}$$

As observed earlier, Q approaches P, and therefore the secant line through P and Q approaches the tangent line T as h approaches zero. Consequently, we might expect that the slope of the secant line would approach the slope of the tangent line T as h approaches zero. This leads to the following definition.

> **Slope of a Tangent Line**
> The slope of the tangent line to the graph of f at the point $P(x, f(x))$ is given by
> $$\lim_{h \to 0} \frac{f(x + h) - f(x)}{h} \tag{4}$$
> if it exists.

■ Rates of Change

We now show that the problem of finding the slope of the tangent line to the graph of a function f at the point $P(x, f(x))$ is mathematically equivalent to the problem of finding the rate of change of f at x. To see this, suppose we are given a function f that describes the relationship between the two quantities x and y—that is, $y = f(x)$. The number $f(x + h) - f(x)$ measures the change in y that corresponds to a change h in x (Figure 48).

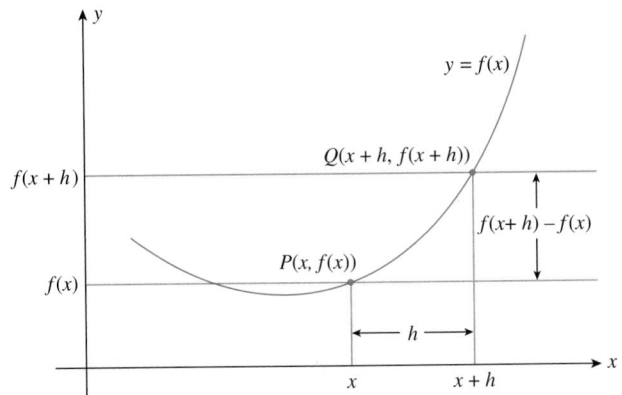

FIGURE 48
$f(x + h) - f(x)$ is the change in y that corresponds to a change h in x.

Then, the **difference quotient**

$$\frac{f(x + h) - f(x)}{h} \tag{5}$$

measures the **average rate of change of y with respect to x** over the interval $[x, x + h]$. For example, if y measures the position of a car at time x, then quotient (5) gives the average velocity of the car over the time interval $[x, x + h]$.

Observe that the difference quotient (5) is the same as (3). We conclude that the difference quotient (5) also measures the slope of the secant line that passes through the two points $P(x, f(x))$ and $Q(x + h, f(x + h))$ lying on the graph of $y = f(x)$. Next, by taking the limit of the difference quotient (5) as h goes to zero—that is, by evaluating

$$\lim_{h \to 0} \frac{f(x + h) - f(x)}{h} \tag{6}$$

we obtain the **rate of change of f at x**. For example, if y measures the position of a car at time x, then the limit (6) gives the velocity of the car at time x. For emphasis, the rate of change of a function f at x is often called the **instantaneous rate of**

change of f at x. This distinguishes it from the average rate of change of f, which is computed over an *interval* $[x, x + h]$ rather than at a *number* x.

Observe that the limit (6) is the same as (4). Therefore, the limit of the difference quotient also measures the slope of the tangent line to the graph of $y = f(x)$ at the point $(x, f(x))$. The following summarizes this discussion.

Average and Instantaneous Rates of Change

The **average rate of change** of f over the interval $[x, x + h]$ or **slope of the secant line** to the graph of f through the points $(x, f(x))$ and $(x + h, f(x + h))$ is

$$\frac{f(x + h) - f(x)}{h} \tag{7}$$

The **instantaneous rate of change** of f at x or **slope of the tangent line** to the graph of f at $(x, f(x))$ is

$$\lim_{h \to 0} \frac{f(x + h) - f(x)}{h} \tag{8}$$

EXPLORE & DISCUSS

Explain the difference between the average rate of change of a function and the instantaneous rate of change of a function.

The Derivative

The limit (4) or (8), which measures both the slope of the tangent line to the graph of $y = f(x)$ at the point $P(x, f(x))$ and the (instantaneous) rate of change of f at x, is given a special name: the **derivative of f at x.**

Derivative of a Function

The derivative of a function f with respect to x is the function f' (read "f prime"),

$$f'(x) = \lim_{h \to 0} \frac{f(x + h) - f(x)}{h} \tag{9}$$

The domain of f' is the set of all x where the limit exists.

Thus, the derivative of a function f is a function f' that gives the slope of the tangent line to the graph of f at *any* point $(x, f(x))$ and also the rate of change of f at x (Figure 49).

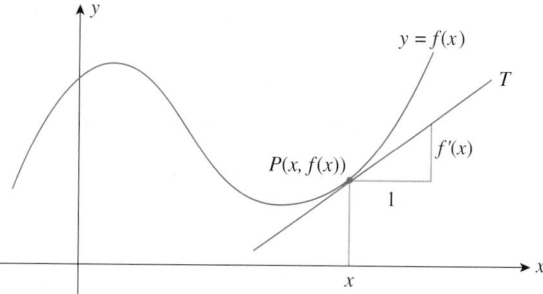

FIGURE 49
The slope of the tangent line at $P(x, f(x))$ is $f'(x)$; f changes at the rate of $f'(x)$ units per unit change in x at x.

Other notations for the derivative of f include:

$$D_x f(x) \qquad \text{Read "}d \text{ sub } x \text{ of } f \text{ of } x\text{"}$$

$$\frac{dy}{dx} \qquad \text{Read "}d \, y \, d \, x\text{"}$$

$$y' \qquad \text{Read "}y \text{ prime"}$$

The last two are used when the rule for f is written in the form $y = f(x)$.

The calculation of the derivative of f is facilitated using the following four-step process.

Four-Step Process for Finding $f'(x)$

1. Compute $f(x + h)$.
2. Form the difference $f(x + h) - f(x)$.
3. Form the quotient $\dfrac{f(x + h) - f(x)}{h}$.
4. Compute $f'(x) = \displaystyle\lim_{h \to 0} \dfrac{f(x + h) - f(x)}{h}$.

EXAMPLE 2   Find the slope of the tangent line to the graph of $f(x) = 3x + 5$ at any point $(x, f(x))$.

Solution   The slope of the tangent line at any point on the graph of f is given by the derivative of f at x. To find the derivative, we use the four-step process:

Step 1 $f(x + h) = 3(x + h) + 5 = 3x + 3h + 5$

Step 2 $f(x + h) - f(x) = (3x + 3h + 5) - (3x + 5) = 3h$

Step 3 $\dfrac{f(x + h) - f(x)}{h} = \dfrac{3h}{h} = 3$

Step 4 $f'(x) = \displaystyle\lim_{h \to 0} \dfrac{f(x + h) - f(x)}{h} = \lim_{h \to 0} 3 = 3$

We expect this result since the tangent line to any point on a straight line must coincide with the line itself and therefore must have the same slope as the line. In this case, the graph of f is a straight line with slope 3.   ∎

EXAMPLE 3   Let $f(x) = x^2$.

a. Find $f'(x)$.
b. Compute $f'(2)$ and interpret your result.

Solution

a. To find $f'(x)$, we use the four-step process:

Step 1 $f(x + h) = (x + h)^2 = x^2 + 2xh + h^2$

Step 2 $f(x + h) - f(x) = x^2 + 2xh + h^2 - x^2 = 2xh + h^2 = h(2x + h)$

Step 3 $\dfrac{f(x + h) - f(x)}{h} = \dfrac{h(2x + h)}{h} = 2x + h$

Step 4 $f'(x) = \displaystyle\lim_{h \to 0} \dfrac{f(x + h) - f(x)}{h} = \lim_{h \to 0} (2x + h) = 2x$

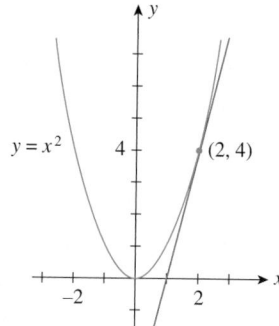

$y = x^2$ 4   (2, 4)

−2   2

FIGURE 50
The tangent line to the graph of
$f(x) = x^2$ at (2, 4)

b. $f'(2) = 2(2) = 4$. This result tells us that the slope of the tangent line to the graph of f at the point (2, 4) is 4. It also tells us that the function f is changing at the rate of 4 units per unit change in x at $x = 2$. The graph of f and the tangent line at (2, 4) are shown in Figure 50. ∎

EXPLORING WITH TECHNOLOGY

1. Consider the function $f(x) = x^2$ of Example 3. Suppose we want to compute $f'(2)$, using Equation (9). Thus,

$$f'(2) = \lim_{h \to 0} \frac{f(2 + h) - f(2)}{h} = \lim_{h \to 0} \frac{(2 + h)^2 - 2^2}{h}$$

Use a graphing utility to plot the graph of

$$g(x) = \frac{(2 + x)^2 - 4}{x}$$

in the viewing window $[-3, 3] \times [-2, 6]$.

2. Use ZOOM and TRACE to find $\lim_{h \to 0} g(x)$.

3. Explain why the limit found in part 2 is $f'(2)$.

EXAMPLE 4   Let $f(x) = x^2 - 4x$.

a. Find $f'(x)$.
b. Find the point on the graph of f where the tangent line to the curve is horizontal.
c. Sketch the graph of f and the tangent line to the curve at the point found in part (b).
d. What is the rate of change of f at this point?

Solution

a. To find $f'(x)$, we use the four-step process:

Step 1 $f(x + h) = (x + h)^2 - 4(x + h) = x^2 + 2xh + h^2 - 4x - 4h$

Step 2 $f(x + h) - f(x) = x^2 + 2xh + h^2 - 4x - 4h - (x^2 - 4x)$
$= 2xh + h^2 - 4h = h(2x + h - 4)$

Step 3 $\dfrac{f(x + h) - f(x)}{h} = \dfrac{h(2x + h - 4)}{h} = 2x + h - 4$

Step 4 $f'(x) = \lim_{h \to 0} \dfrac{f(x + h) - f(x)}{h} = \lim_{h \to 0} (2x + h - 4) = 2x - 4$

b. At a point on the graph of f where the tangent line to the curve is horizontal and hence has slope zero, the derivative f' of f is zero. Accordingly, to find such point(s) we set $f'(x) = 0$, which gives $2x - 4 = 0$, or $x = 2$. The corresponding value of y is given by $y = f(2) = -4$, and the required point is $(2, -4)$.
c. The graph of f and the tangent line are shown in Figure 51.
d. The rate of change of f at $x = 2$ is zero. ∎

$y = x^2 - 4x$

−1   1   2   3   5

−2

−3

T $y = -4$

(2, −4)

FIGURE 51
The tangent line to the graph of
$y = x^2 - 4x$ at (2, −4) is $y = -4$.

EXPLORE & DISCUSS

Can the tangent line to the graph of a function intersect the graph at more than one point? Explain your answer using illustrations.

EXAMPLE 5   Let $f(x) = \dfrac{1}{x}$.

a. Find $f'(x)$.
b. Find the slope of the tangent line T to the graph of f at the point where $x = 1$.
c. Find an equation of the tangent line T in part (b).

Solution

a. To find $f'(x)$, we use the four-step process:

Step 1 $f(x + h) = \dfrac{1}{x + h}$

Step 2 $f(x + h) - f(x) = \dfrac{1}{x + h} - \dfrac{1}{x} = \dfrac{x - (x + h)}{x(x + h)} = -\dfrac{h}{x(x + h)}$

Step 3 $\dfrac{f(x + h) - f(x)}{h} = -\dfrac{h}{x(x + h)} \cdot \dfrac{1}{h} = -\dfrac{1}{x(x + h)}$

Step 4 $f'(x) = \lim\limits_{h \to 0} \dfrac{f(x + h) - f(x)}{h} = \lim\limits_{h \to 0} -\dfrac{1}{x(x + h)} = -\dfrac{1}{x^2}$

b. The slope of the tangent line T to the graph of f where $x = 1$ is given by $f'(1) = -1$.

c. When $x = 1$, $y = f(1) = 1$ and T is tangent to the graph of f at the point $(1, 1)$. From part (b), we know that the slope of T is -1. Thus, an equation of T is

$$y - 1 = -1(x - 1)$$
$$y = -x + 2$$

(Figure 52).

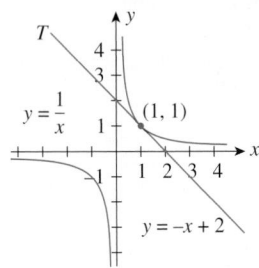

FIGURE 52
The tangent line to the graph of $f(x) = 1/x$ at $(1, 1)$

EXPLORING WITH TECHNOLOGY

1. Use the results of Example 5 to draw the graph of $f(x) = 1/x$ and its tangent line at the point $(1, 1)$ by plotting the graphs of $y_1 = 1/x$ and $y_2 = -x + 2$ in the viewing window $[-4, 4] \times [-4, 4]$.

2. Some graphing utilities draw the tangent line to the graph of a function at a given point automatically—you need only specify the function and give the x-coordinate of the point of tangency. If your graphing utility has this feature, verify the result of part 1 without finding an equation of the tangent line.

EXPLORE & DISCUSS

Consider the following alternative approach to the definition of the derivative of a function: Let h be a positive number and suppose $P(x - h, f(x - h))$ and $Q(x + h, f(x + h))$ are two points on the graph of f.

1. Give a geometric and a physical interpretation of the quotient

$$\frac{f(x + h) - f(x - h)}{2h}$$

Make a sketch to illustrate your answer.

2. Give a geometric and a physical interpretation of the limit

$$\lim\limits_{h \to 0} \frac{f(x + h) - f(x - h)}{2h}$$

Make a sketch to illustrate your answer.

3. Explain why it makes sense to define

$$f'(x) = \lim\limits_{h \to 0} \frac{f(x + h) - f(x - h)}{2h}$$

4. Using the definition given in part 3, formulate a four-step process for finding $f'(x)$ similar to that given on page 664 and use it to find the derivative of $f(x) = x^2$. Compare your answer with that obtained in Example 3 on page 664.

APPLIED EXAMPLE 6 Average Velocity of a Car   Suppose the distance (in feet) covered by a car moving along a straight road t seconds after starting from rest is given by the function $f(t) = 2t^2$ $(0 \leq t \leq 30)$.

a. Calculate the average velocity of the car over the time intervals [22, 23], [22, 22.1], and [22, 22.01].
b. Calculate the (instantaneous) velocity of the car when $t = 22$.
c. Compare the results obtained in part (a) with that obtained in part (b).

Solution

a. We first compute the average velocity (average rate of change of f) over the interval $[t, t + h]$ using Formula (7). We find

$$\frac{f(t + h) - f(t)}{h} = \frac{2(t + h)^2 - 2t^2}{h}$$

$$= \frac{2t^2 + 4th + 2h^2 - 2t^2}{h}$$

$$= 4t + 2h$$

Next, using $t = 22$ and $h = 1$, we find that the average velocity of the car over the time interval [22, 23] is

$$4(22) + 2(1) = 90$$

or 90 feet per second. Similarly, using $t = 22$, $h = 0.1$, and $h = 0.01$, we find that its average velocities over the time intervals [22, 22.1] and [22, 22.01] are 88.2 and 88.02 feet per second, respectively.

b. Using the limit (8), we see that the instantaneous velocity of the car at any time t is given by

$$\lim_{h \to 0} \frac{f(t + h) - f(t)}{h} = \lim_{h \to 0} (4t + 2h) \qquad \text{Use the results from part (a).}$$

$$= 4t$$

In particular, the velocity of the car 22 seconds from rest ($t = 22$) is given by

$$v = 4(22)$$

or 88 feet per second.

c. The computations in part (a) show that, as the time intervals over which the average velocity of the car are computed become smaller and smaller, the average velocities over these intervals do approach 88 feet per second, the instantaneous velocity of the car at $t = 22$. ■

APPLIED EXAMPLE 7 Demand for Tires   The management of Titan Tire Company has determined that the weekly demand function of their Super Titan tires is given by

$$p = f(x) = 144 - x^2$$

where p is measured in dollars and x is measured in units of a thousand (Figure 53).

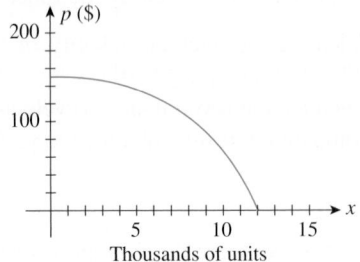

FIGURE 53
The graph of the demand function
$p = 144 - x^2$

a. Find the average rate of change in the unit price of a tire if the quantity demanded is between 5000 and 6000 tires, between 5000 and 5100 tires, and between 5000 and 5010 tires.

b. What is the instantaneous rate of change of the unit price when the quantity demanded is 5000 units?

Solution

a. The average rate of change of the unit price of a tire if the quantity demanded is between x and $x + h$ is

$$\frac{f(x + h) - f(x)}{h} = \frac{[144 - (x + h)^2] - (144 - x^2)}{h}$$

$$= \frac{144 - x^2 - 2xh - h^2 - 144 + x^2}{h}$$

$$= -2x - h$$

To find the average rate of change of the unit price of a tire when the quantity demanded is between 5000 and 6000 tires (that is, over the interval $[5, 6]$), we take $x = 5$ and $h = 1$, obtaining

$$-2(5) - 1 = -11$$

or $-\$11$ per 1000 tires. (Remember, x is measured in units of a thousand.) Similarly, taking $h = 0.1$ and $h = 0.01$ with $x = 5$, we find that the average rates of change of the unit price when the quantities demanded are between 5000 and 5100 and between 5000 and 5010 are $-\$10.10$ and $-\$10.01$ per 1000 tires, respectively.

b. The instantaneous rate of change of the unit price of a tire when the quantity demanded is x units is given by

$$\lim_{h \to 0} \frac{f(x + h) - f(x)}{h} = \lim_{h \to 0} (-2x - h) \qquad \text{Use the results from part (a).}$$

$$= -2x$$

In particular, the instantaneous rate of change of the unit price per tire when the quantity demanded is 5000 is given by $-2(5)$, or $-\$10$ per 1000 tires. ■

The derivative of a function provides us with a tool for measuring the rate of change of one quantity with respect to another. Table 4 lists several other applications involving this limit.

TABLE 4

Applications Involving Rate of Change

x Stands for	*y* Stands for	$\dfrac{f(a+h)-f(a)}{h}$ Measures	$\displaystyle\lim_{h\to 0}\dfrac{f(a+h)-f(a)}{h}$ Measures
Time	**Concentration of a drug** in the bloodstream at time *x*	Average rate of change in the concentration of the drug over the time interval $[a, a+h]$	Instantaneous rate of change in the concentration of the drug in the bloodstream at time $x = a$
Number of items sold	**Revenue** at a sales level of *x* units	Average rate of change in the revenue when the sales level is between $x = a$ and $x = a + h$	Instantaneous rate of change in the revenue when the sales level is *a* units
Time	**Volume of sales** at time *x*	Average rate of change in the volume of sales over the time interval $[a, a+h]$	Instantaneous rate of change in the volume of sales at time $x = a$
Time	**Population** of *Drosophila* (fruit flies) at time *x*	Average rate of growth of the fruit fly population over the time interval $[a, a+h]$	Instantaneous rate of change of the fruit fly population at time $x = a$
Temperature in a chemical reaction	**Amount of product formed in the chemical reaction** when the temperature is *x* degrees	Average rate of formation of chemical product over the temperature range $[a, a+h]$	Instantaneous rate of formation of chemical product when the temperature is *a* degrees

▬ Differentiability and Continuity

In practical applications, one encounters continuous functions that fail to be **differentiable**—that is, do not have a derivative—at certain values in the domain of the function *f*. It can be shown that a continuous function *f* fails to be differentiable at $x = a$ when the graph of *f* makes an abrupt change of direction at $(a, f(a))$. We call such a point a "corner." A function also fails to be differentiable at a point where the tangent line is vertical since the slope of a vertical line is undefined. These cases are illustrated in Figure 54.

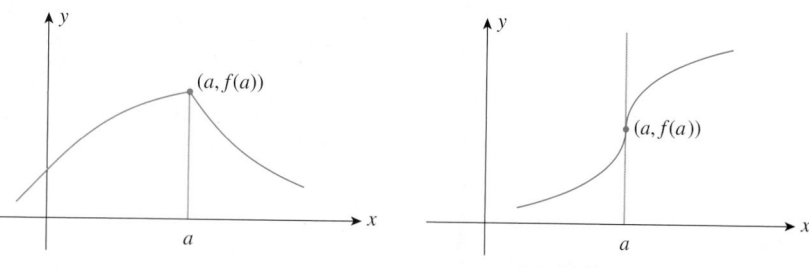

(a) The graph makes an abrupt change of direction at $x = a$.

(b) The slope at $x = a$ is undefined.

FIGURE 54

The next example illustrates a function that is not differentiable at a point.

APPLIED EXAMPLE 8 Wages   Mary works at the B&O department store, where, on a weekday, she is paid $8 an hour for the first 8 hours and $12 an hour for overtime. The function

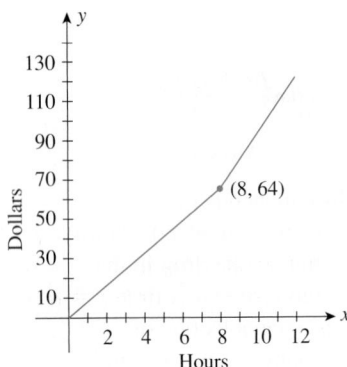

FIGURE 55
The function f is not differentiable at $(8, 64)$.

$$f(x) = \begin{cases} 8x & \text{if } 0 \le x \le 8 \\ 12x - 32 & \text{if } 8 < x \end{cases}$$

gives Mary's earnings on a weekday in which she worked x hours. Sketch the graph of the function f and explain why it is not differentiable at $x = 8$.

Solution   The graph of f is shown in Figure 55. Observe that the graph of f has a corner at $x = 8$ and consequently is not differentiable at $x = 8$.   ∎

We close this section by mentioning the connection between the continuity and the differentiability of a function at a given value $x = a$ in the domain of f. By reexamining the function of Example 8, it becomes clear that f is continuous everywhere and, in particular, when $x = 8$. This shows that in general the continuity of a function at $x = a$ does not necessarily imply the differentiability of the function at that number. The converse, however, is true: If a function f is differentiable at $x = a$, then it is continuous there.

Differentiability and Continuity
If a function is differentiable at $x = a$, then it is continuous at $x = a$.

For a proof of this result, see Exercise 60, page 675.

EXPLORE & DISCUSS

Suppose a function f is differentiable at $x = a$. Can there be two tangent lines to the graphs of f at the point $(a, f(a))$? Explain your answer.

EXPLORING WITH TECHNOLOGY

1. Use a graphing utility to plot the graph of $f(x) = x^{1/3}$ in the viewing window $[-2, 2] \times [-2, 2]$.

2. Use a graphing utility to draw the tangent line to the graph of f at the point $(0, 0)$. Can you explain why the process breaks down?

EXAMPLE 9   Figure 56 depicts a portion of the graph of a function. Explain why the function fails to be differentiable at each of the numbers $x = a, b, c, d, e, f,$ and g.

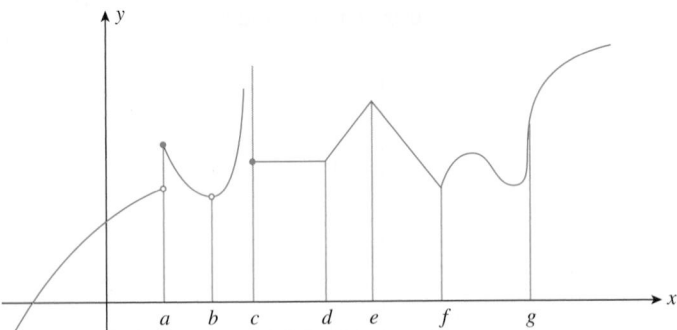

FIGURE 56
The graph of this function is not differentiable at the numbers a–g.

Solution   The function fails to be differentiable at $x = a, b,$ and c because it is discontinuous at each of these numbers. The derivative of the function does not exist at $x = d, e,$ and f because it has a kink at each point on the graph corresponding to these numbers. Finally, the function is not differentiable at $x = g$ because the tangent line is vertical at $(g, f(g))$.   ∎

11.6   Self-Check Exercises

1. Let $f(x) = -x^2 - 2x + 3$.
   a. Find the derivative f' of f, using the definition of the derivative.
   b. Find the slope of the tangent line to the graph of f at the point $(0, 3)$.
   c. Find the rate of change of f when $x = 0$.
   d. Find an equation of the tangent line to the graph of f at the point $(0, 3)$.
   e. Sketch the graph of f and the tangent line to the curve at the point $(0, 3)$.

2. The losses (in millions of dollars) due to bad loans extended chiefly in agriculture, real estate, shipping, and energy by the Franklin Bank are estimated to be

$$A = f(t) = -t^2 + 10t + 30 \qquad (0 \leq t \leq 10)$$

where t is the time in years ($t = 0$ corresponds to the beginning of 1994). How fast were the losses mounting at the beginning of 1997? At the beginning of 1999? At the beginning of 2001?

Solutions to Self-Check Exercises 11.6 can be found on page 676.

11.6   Concept Questions

1. Let $P(2, f(2))$ and $Q(2 + h, f(2 + h))$ be points on the graph of a function f.
   a. Find an expression for the slope of the secant line passing through P and Q.
   b. Find an expression for the slope of the tangent line passing through P.

2. Refer to Question 1.
   a. Find an expression for the average rate of change of f over the interval $[2, 2 + h]$.
   b. Find an expression for the instantaneous rate of change of f at 2.

   c. Compare your answers for part (a) and (b) with those of Exercise 1.

3. a. Give a geometric and a physical interpretation of the expression
$$\frac{f(x + h) - f(x)}{h}$$
   b. Give a geometric and a physical interpretation of the expression
$$\lim_{h \to 0} \frac{f(x + h) - f(x)}{h}$$

4. Under what conditions does a function fail to have a derivative at a number? Illustrate your answer with sketches.

11.6   Exercises

1. AVERAGE WEIGHT OF AN INFANT   The following graph shows the weight measurements of the average infant from the time of birth ($t = 0$) through age 2 ($t = 24$). By computing the slopes of the respective tangent lines, estimate the rate of change of the average infant's weight when $t = 3$ and when $t = 18$. What is the average rate of change in the average infant's weight over the first year of life?

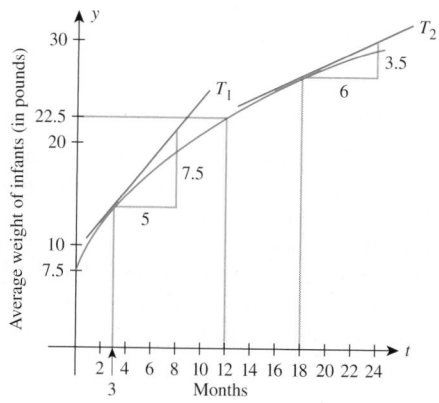

2. FORESTRY   The following graph shows the volume of wood produced in a single-species forest. Here $f(t)$ is measured in cubic meters/hectare and t is measured in years. By computing the slopes of the respective tangent lines, estimate the rate at which the wood grown is changing at the beginning of year 10 and at the beginning of year 30.

Source: The Random House Encyclopedia

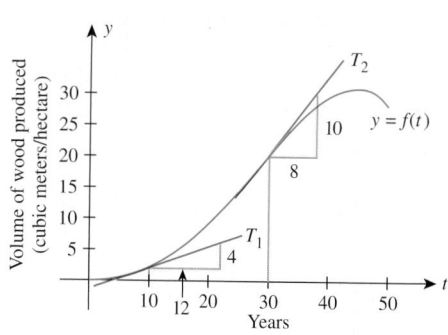

3. **TV-VIEWING PATTERNS**   The following graph shows the percent of U.S. households watching television during a 24-hr period on a weekday ($t = 0$ corresponds to 6 a.m.). By computing the slopes of the respective tangent lines, estimate the rate of change of the percent of households watching television at 4 p.m. and 11 p.m.

Source: A. C. Nielsen Company

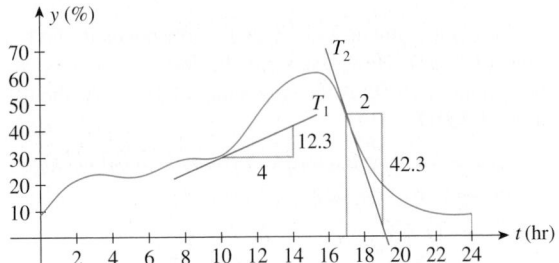

4. **CROP YIELD**   Productivity and yield of cultivated crops are often reduced by insect pests. The following graph shows the relationship between the yield of a certain crop, $f(x)$, as a function of the density of aphids x. (Aphids are small insects that suck plant juices.) Here, $f(x)$ is measured in kilograms/4000 square meters, and x is measured in hundreds of aphids/bean stem. By computing the slopes of the respective tangent lines, estimate the rate of change of the crop yield with respect to the density of aphids when that density is 200 aphids/bean stem and when it is 800 aphids/bean stem.

Source: The Random House Encyclopedia

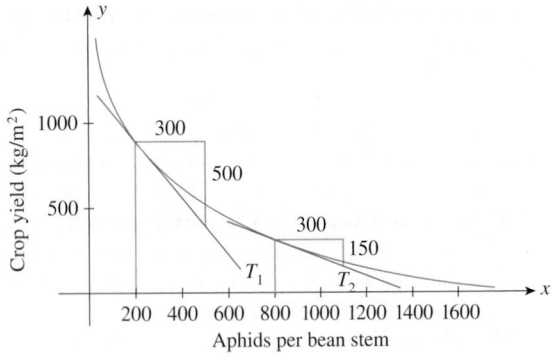

5. The position of car A and car B, starting out side by side and traveling along a straight road, is given by $s = f(t)$ and $s = g(t)$, respectively, where s is measured in feet and t is measured in seconds (see the accompanying figure).

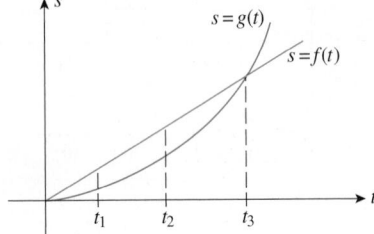

a. Which car is traveling faster at t_1?
b. What can you say about the speed of the cars at t_2?
   Hint: Compare tangent lines.
c. Which car is traveling faster at t_3?
d. What can you say about the positions of the cars at t_3?

6. The velocity of car A and car B, starting out side by side and traveling along a straight road, is given by $v = f(t)$ and $v = g(t)$, respectively, where v is measured in feet/second and t is measured in seconds (see the accompanying figure).

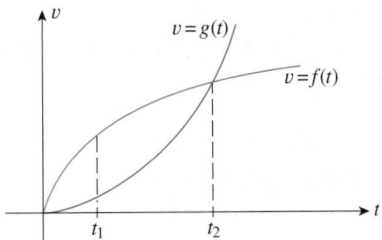

a. What can you say about the velocity and acceleration of the two cars at t_1? (Acceleration is the rate of change of velocity.)
b. What can you say about the velocity and acceleration of the two cars at t_2?

7. **EFFECT OF A BACTERICIDE ON BACTERIA**   In the following figure, $f(t)$ gives the population P_1 of a certain bacteria culture at time t after a portion of bactericide A was introduced into the population at $t = 0$. The graph of g gives the population P_2 of a similar bacteria culture at time t after a portion of bactericide B was introduced into the population at $t = 0$.
a. Which population is decreasing faster at t_1?
b. Which population is decreasing faster at t_2?

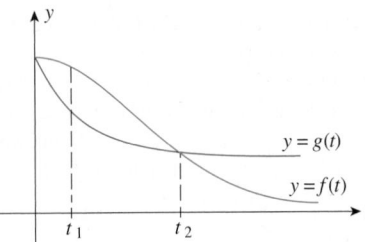

c. Which bactericide is more effective in reducing the population of bacteria in the short run? In the long run?

8. **MARKET SHARE**   The following figure shows the devastating effect the opening of a new discount department store had on an established department store in a small town. The revenue of the discount store at time t (in months) is given by $f(t)$ million dollars, whereas the revenue of the established department store at time t is given by $g(t)$ million dollars. Answer the following questions by giving the value of t at which the specified event took place.

a. The revenue of the established department store is decreasing at the slowest rate.

b. The revenue of the established department store is decreasing at the fastest rate.

c. The revenue of the discount store first overtakes that of the established store.

d. The revenue of the discount store is increasing at the fastest rate.

In Exercises 9–16, use the four-step process to find the slope of the tangent line to the graph of the given function at any point.

9. $f(x) = 13$

10. $f(x) = -6$

11. $f(x) = 2x + 7$

12. $f(x) = 8 - 4x$

13. $f(x) = 3x^2$

14. $f(x) = -\frac{1}{2}x^2$

15. $f(x) = -x^2 + 3x$

16. $f(x) = 2x^2 + 5x$

In Exercises 17–22, find the slope of the tangent line to the graph of each function at the given point and determine an equation of the tangent line.

17. $f(x) = 2x + 7$ at $(2, 11)$

18. $f(x) = -3x + 4$ at $(-1, 7)$

19. $f(x) = 3x^2$ at $(1, 3)$

20. $f(x) = 3x - x^2$ at $(-2, -10)$

21. $f(x) = -\frac{1}{x}$ at $\left(3, -\frac{1}{3}\right)$

22. $f(x) = \frac{3}{2x}$ at $\left(1, \frac{3}{2}\right)$

23. Let $f(x) = 2x^2 + 1$.
  a. Find the derivative f' of f.
  b. Find an equation of the tangent line to the curve at the point $(1, 3)$.
  c. Sketch the graph of f.

24. Let $f(x) = x^2 + 6x$.
  a. Find the derivative f' of f.
  b. Find the point on the graph of f where the tangent line to the curve is horizontal.
  Hint: Find the value of x for which $f'(x) = 0$.

c. Sketch the graph of f and the tangent line to the curve at the point found in part (b).

25. Let $f(x) = x^2 - 2x + 1$.
  a. Find the derivative f' of f.
  b. Find the point on the graph of f where the tangent line to the curve is horizontal.
  c. Sketch the graph of f and the tangent line to the curve at the point found in part (b).
  d. What is the rate of change of f at this point?

26. Let $f(x) = \dfrac{1}{x - 1}$.
  a. Find the derivative f' of f.
  b. Find an equation of the tangent line to the curve at the point $\left(-1, -\frac{1}{2}\right)$.
  c. Sketch the graph of f and the tangent line to the curve at $\left(-1, -\frac{1}{2}\right)$.

27. Let $y = f(x) = x^2 + x$.
  a. Find the average rate of change of y with respect to x in the interval from $x = 2$ to $x = 3$, from $x = 2$ to $x = 2.5$, and from $x = 2$ to $x = 2.1$.
  b. Find the (instantaneous) rate of change of y at $x = 2$.
  c. Compare the results obtained in part (a) with that of part (b).

28. Let $y = f(x) = x^2 - 4x$.
  a. Find the average rate of change of y with respect to x in the interval from $x = 3$ to $x = 4$, from $x = 3$ to $x = 3.5$, and from $x = 3$ to $x = 3.1$.
  b. Find the (instantaneous) rate of change of y at $x = 3$.
  c. Compare the results obtained in part (a) with that of part (b).

29. VELOCITY OF A CAR Suppose the distance s (in feet) covered by a car moving along a straight road after t sec is given by the function $f(t) = 2t^2 + 48t$.
  a. Calculate the average velocity of the car over the time intervals $[20, 21]$, $[20, 20.1]$, and $[20, 20.01]$.
  b. Calculate the (instantaneous) velocity of the car when $t = 20$.
  c. Compare the results of part (a) with that of part (b).

30. VELOCITY OF A BALL THROWN INTO THE AIR A ball is thrown straight up with an initial velocity of 128 ft/sec, so that its height (in feet) after t sec is given by $s(t) = 128t - 16t^2$.
  a. What is the average velocity of the ball over the time intervals $[2, 3]$, $[2, 2.5]$, and $[2, 2.1]$?
  b. What is the instantaneous velocity at time $t = 2$?
  c. What is the instantaneous velocity at time $t = 5$? Is the ball rising or falling at this time?
  d. When will the ball hit the ground?

31. During the construction of a high-rise building, a worker accidentally dropped his portable electric screwdriver from a height of 400 ft. After t sec, the screwdriver had fallen a distance of $s = 16t^2$ ft.
  a. How long did it take the screwdriver to reach the ground?
  b. What was the average velocity of the screwdriver between the time it was dropped and the time it hit the ground?

c. What was the velocity of the screwdriver at the time it hit the ground?

32. A hot-air balloon rises vertically from the ground so that its height after t sec is $h = \frac{1}{2}t^2 + \frac{1}{2}t$ ft $(0 \leq t \leq 60)$.
  a. What is the height of the balloon at the end of 40 sec?
  b. What is the average velocity of the balloon between $t = 0$ and $t = 40$?
  c. What is the velocity of the balloon at the end of 40 sec?

33. At a temperature of 20°C, the volume V (in liters) of 1.33 g of O_2 is related to its pressure p (in atmospheres) by the formula $V = 1/p$.
  a. What is the average rate of change of V with respect to p as p increases from $p = 2$ to $p = 3$?
  b. What is the rate of change of V with respect to p when $p = 2$?

34. COST OF PRODUCING SURFBOARDS   The total cost $C(x)$ (in dollars) incurred by Aloha Company in manufacturing x surfboards a day is given by
$$C(x) = -10x^2 + 300x + 130 \qquad (0 \leq x \leq 15)$$
  a. Find $C'(x)$.
  b. What is the rate of change of the total cost when the level of production is ten surfboards a day?

35. EFFECT OF ADVERTISING ON PROFIT   The quarterly profit (in thousands of dollars) of Cunningham Realty is given by
$$P(x) = -\frac{1}{3}x^2 + 7x + 30 \qquad (0 \leq x \leq 50)$$
where x (in thousands of dollars) is the amount of money Cunningham spends on advertising per quarter.
  a. Find $P'(x)$.
  b. What is the rate of change of Cunningham's quarterly profit if the amount it spends on advertising is $10,000/quarter $(x = 10)$ and $30,000/quarter $(x = 30)$?

36. DEMAND FOR TENTS   The demand function for Sportsman 5×7 tents is given by
$$p = f(x) = -0.1x^2 - x + 40$$
where p is measured in dollars and x is measured in units of a thousand.
  a. Find the average rate of change in the unit price of a tent if the quantity demanded is between 5000 and 5050 tents; between 5000 and 5010 tents.
  b. What is the rate of change of the unit price if the quantity demanded is 5000?

37. A COUNTRY'S GDP   The gross domestic product (GDP) of a certain country is projected to be
$$N(t) = t^2 + 2t + 50 \qquad (0 \leq t \leq 5)$$
billion dollars t yr from now. What will be the rate of change of the country's GDP 2 yr and 4 yr from now?

38. GROWTH OF BACTERIA   Under a set of controlled laboratory conditions, the size of the population of a certain bacteria culture at time t (in minutes) is described by the function
$$P = f(t) = 3t^2 + 2t + 1$$
Find the rate of population growth at $t = 10$ min.

In Exercises 39–44, let x and $f(x)$ represent the given quantities. Fix $x = a$ and let h be a small positive number. Give an interpretation of the quantities
$$\frac{f(a+h) - f(a)}{h} \quad \text{and} \quad \lim_{h \to 0} \frac{f(a+h) - f(a)}{h}$$

39. x denotes time and $f(x)$ denotes the population of seals at time x.

40. x denotes time and $f(x)$ denotes the prime interest rate at time x.

41. x denotes time and $f(x)$ denotes a country's industrial production.

42. x denotes the level of production of a certain commodity, and $f(x)$ denotes the total cost incurred in producing x units of the commodity.

43. x denotes altitude and $f(x)$ denotes atmospheric pressure.

44. x denotes the speed of a car (in mph), and $f(x)$ denotes the fuel economy of the car measured in miles per gallon (mpg).

In each of Exercises 45–50, the graph of a function is shown. For each function, state whether or not (a) $f(x)$ has a limit at $x = a$, (b) $f(x)$ is continuous at $x = a$, and (c) $f(x)$ is differentiable at $x = a$. Justify your answers.

45.

46.

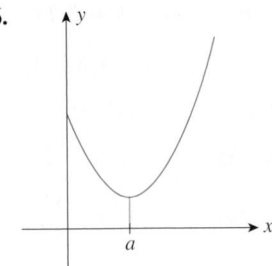

47.

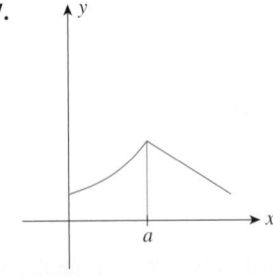

48.

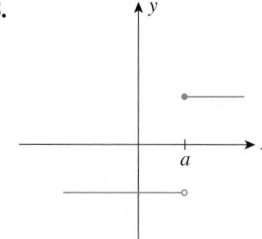

49.

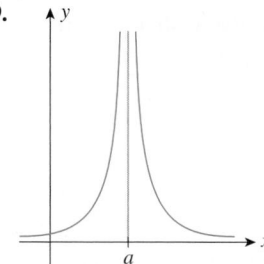

50.

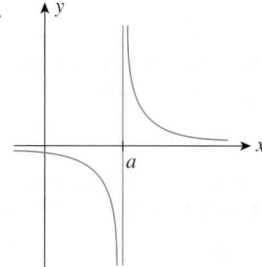

51. The distance s (in feet) covered by a motorcycle traveling in a straight line and starting from rest in t sec is given by the function

$$s(t) = -0.1t^3 + 2t^2 + 24t$$

Calculate the motorcycle's average velocity over the time interval $[2, 2 + h]$ for $h = 1, 0.1, 0.01, 0.001, 0.0001$, and 0.00001 and use your results to guess at the motorcycle's instantaneous velocity at $t = 2$.

52. The daily total cost $C(x)$ incurred by Trappee and Sons for producing x cases of TexaPep hot sauce is given by

$$C(x) = 0.000002x^3 + 5x + 400$$

Calculate

$$\frac{C(100 + h) - C(100)}{h}$$

for $h = 1, 0.1, 0.01, 0.001$, and 0.0001 and use your results to estimate the rate of change of the total cost function when the level of production is 100 cases/day.

In Exercises 53 and 54, determine whether the statement is true or false. If it is true, explain why it is true. If it is false, give an example to show why it is false.

53. If f is continuous at $x = a$, then f is differentiable at $x = a$.

54. If f is continuous at $x = a$ and g is differentiable at $x = a$, then $\lim_{x \to a} f(x)g(x) = f(a)g(a)$.

55. Sketch the graph of the function $f(x) = |x + 1|$ and show that the function does not have a derivative at $x = -1$.

56. Sketch the graph of the function $f(x) = 1/(x - 1)$ and show that the function does not have a derivative at $x = 1$.

57. Let

$$f(x) = \begin{cases} x^2 & \text{if } x \le 1 \\ ax + b & \text{if } x > 1 \end{cases}$$

Find the values of a and b so that f is continuous and has a derivative at $x = 1$. Sketch the graph of f.

58. Sketch the graph of the function $f(x) = x^{2/3}$. Is the function continuous at $x = 0$? Does $f'(0)$ exist? Why or why not?

59. Prove that the derivative of the function $f(x) = |x|$ for $x \ne 0$ is given by

$$f'(x) = \begin{cases} 1 & \text{if } x > 0 \\ -1 & \text{if } x < 0 \end{cases}$$

Hint: Recall the definition of the absolute value of a number.

60. Show that if a function f is differentiable at $x = a$, then f must be continuous at that number.
Hint: Write

$$f(x) - f(a) = \left[\frac{f(x) - f(a)}{x - a} \right](x - a)$$

Use the product rule for limits and the definition of the derivative to show that

$$\lim_{x \to a} [f(x) - f(a)] = 0$$

11.6 Solutions to Self-Check Exercises

1. a. $f'(x)$

$$= \lim_{h \to 0} \frac{f(x+h) - f(x)}{h}$$

$$= \lim_{h \to 0} \frac{[-(x+h)^2 - 2(x+h) + 3] - (-x^2 - 2x + 3)}{h}$$

$$= \lim_{h \to 0} \frac{-x^2 - 2xh - h^2 - 2x - 2h + 3 + x^2 + 2x - 3}{h}$$

$$= \lim_{h \to 0} \frac{h(-2x - h - 2)}{h}$$

$$= \lim_{h \to 0} (-2x - h - 2) = -2x - 2$$

b. From the result of part (a), we see that the slope of the tangent line to the graph of f at any point $(x, f(x))$ is given by

$$f'(x) = -2x - 2$$

In particular, the slope of the tangent line to the graph of f at $(0, 3)$ is

$$f'(0) = -2$$

c. The rate of change of f when $x = 0$ is given by $f'(0) = -2$, or -2 units/unit change in x.

d. Using the result from part (b), we see that an equation of the required tangent line is

$$y - 3 = -2(x - 0)$$
$$y = -2x + 3$$

e.

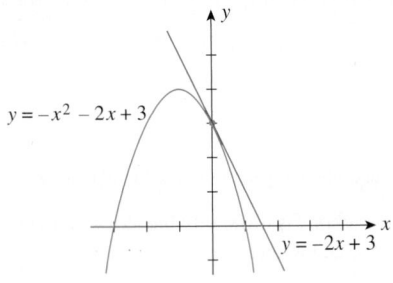

$y = -x^2 - 2x + 3$

$y = -2x + 3$

2. The rate of change of the losses at any time t is given by

$f'(t)$

$$= \lim_{h \to 0} \frac{f(t+h) - f(t)}{h}$$

$$= \lim_{h \to 0} \frac{[-(t+h)^2 + 10(t+h) + 30] - (-t^2 + 10t + 30)}{h}$$

$$= \lim_{h \to 0} \frac{-t^2 - 2th - h^2 + 10t + 10h + 30 + t^2 - 10t - 30}{h}$$

$$= \lim_{h \to 0} \frac{h(-2t - h + 10)}{h}$$

$$= \lim_{h \to 0} (-2t - h + 10)$$

$$= -2t + 10$$

Therefore, the rate of change of the losses suffered by the bank at the beginning of 1997 ($t = 3$) was

$$f'(3) = -2(3) + 10 = 4$$

In other words, the losses were increasing at the rate of $4 million/year. At the beginning of 1999 ($t = 5$),

$$f'(5) = -2(5) + 10 = 0$$

and we see that the growth in losses due to bad loans was zero at this point. At the beginning of 2001 ($t = 7$),

$$f'(7) = -2(7) + 10 = -4$$

and we conclude that the losses were decreasing at the rate of $4 million/year.

USING TECHNOLOGY

■ Graphing a Function and Its Tangent Line

We can use a graphing utility to plot the graph of a function f and the tangent line at any point on the graph.

EXAMPLE 1 Let $f(x) = x^2 - 4x$.

a. Find an equation of the tangent line to the graph of f at the point $(3, -3)$.

b. Plot both the graph of f and the tangent line found in part (a) on the same set of axes.

Solution

a. The slope of the tangent line at any point on the graph of f is given by $f'(x)$. But from Example 4 (page 665) we find $f'(x) = 2x - 4$. Using this result, we see that the slope of the required tangent line is

$$f'(3) = 2(3) - 4 = 2$$

Finally, using the point-slope form of the equation of a line, we find that an equation of the tangent line is

$$y - (-3) = 2(x - 3)$$
$$y + 3 = 2x - 6$$
$$y = 2x - 9$$

b. Figure T1a shows the graph of f in the standard viewing window and the tangent line of interest.

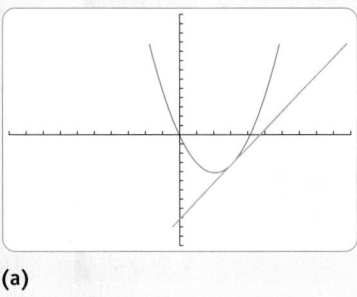

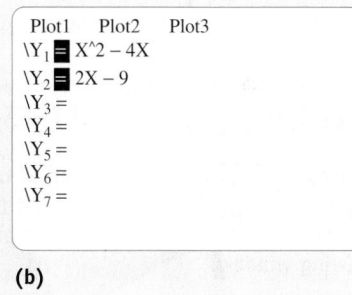

FIGURE T1
(a) The graph of $f(x) = x^2 - 4x$ and the tangent line $y = 2x - 9$ in the standard viewing window; (b) the TI-83 equation screen.

(a)   (b)

Note Some graphing utilities will draw both the graph of a function f and the tangent line to the graph of f at a specified point when the function and the specified value of x are entered.

▬ Finding the Derivative of a Function at a Given Point

The numerical derivative operation of a graphing utility can be used to give an approximate value of the derivative of a function for a given value of x.

EXAMPLE 2 Let $f(x) = \sqrt{x}$.

a. Use the numerical derivative operation of a graphing utility to find the derivative of f at $(4, 2)$.
b. Find an equation of the tangent line to the graph of f at $(4, 2)$.
c. Plot the graph of f and the tangent line on the same set of axes.

Solution

a. Using the numerical derivative operation of a graphing utility, we find that

$$f'(4) = \frac{1}{4}$$

(Figure T2).

```
nDeriv(X^.5, X, 4)
         .250000002
```

FIGURE T2
The TI-83 numerical derivative screen.

(continued)

b. An equation of the required tangent line is

$$y - 2 = \frac{1}{4}(x - 4)$$

$$y = \frac{1}{4}x + 1$$

c. Figure T3a shows the graph of f and the tangent line in the viewing window $[0, 15] \times [0, 4]$.

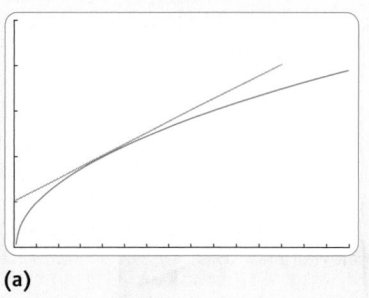

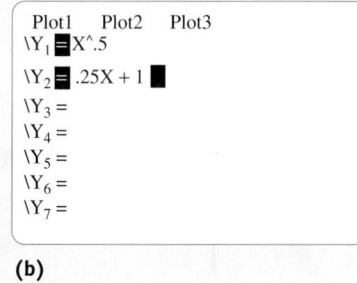

FIGURE T3
(a) The graph of $f(x) = \sqrt{x}$ and the tangent line $y = \frac{1}{4}x + 1$ in the viewing window $[0, 15] \times [0, 4]$; (b) the TI-83 equation screen

(a)   **(b)**

TECHNOLOGY EXERCISES

In Exercises 1–6, (a) find an equation of the tangent line to the graph of f at the indicated point and (b) plot the graph of f and the tangent line on the same set of axes. Use a suitable viewing window.

1. $f(x) = 4x - 3$; $(2, 5)$

2. $f(x) = 2x^2 + x$; $(-2, 6)$

3. $f(x) = 2x^2 + x - 3$; $(2, 7)$

4. $f(x) = x + \dfrac{1}{x}$; $(1, 2)$

5. $f(x) = \sqrt{x}$; $(4, 2)$

6. $f(x) = \dfrac{1}{\sqrt{x}}$; $\left(4, \dfrac{1}{2}\right)$

In Exercises 7–12, (a) use the numerical derivative operation to find the derivative of f for the given value of x (to two desired places of accuracy), (b) find an equation of the tangent line to the graph of f at the indicated point, and (c) plot the graph of f and the tangent line on the same set of axes. Use a suitable viewing window.

7. $f(x) = x^3 + x + 1$; $x = 1$; $(1, 3)$

8. $f(x) = x^4 - 3x^2 + 1$; $x = 2$; $(2, 5)$

9. $f(x) = x - \sqrt{x}$; $x = 4$; $(4, 2)$

10. $f(x) = \dfrac{1}{x + 1}$; $x = 1$; $\left(1, \dfrac{1}{2}\right)$

11. $f(x) = x\sqrt{x^2 + 1}$; $x = 2$; $(2, 2\sqrt{5})$

12. $f(x) = \dfrac{x}{\sqrt{x^2 + 1}}$; $x = 1$; $\left(1, \dfrac{\sqrt{2}}{2}\right)$

13. DRIVING COSTS   The average cost of owning and operating a car in the United States from 1991 through 2001 is approximated by the function

$$C(t) = 0.06t^2 + 0.74t + 37.3 \qquad (0 \le t \le 11)$$

where $C(t)$ is measured in cents/mile and t is measured in years, with $t = 0$ corresponding to the beginning of 1991.
a. Plot the graph of C in the viewing window $[0, 10] \times [35, 52]$.
b. What was the average cost of driving a car at the beginning of 1995?
c. How fast was the average cost of driving a car changing at the beginning of 1995?
Source: Automobile Association of America

14. ANNUAL RETAIL SALES   The annual retail sales in the United States from 1990 through the year 2000 (in millions of dollars) are approximated by the function

$$S(t) = 1.9152t^2 + 18.7176t + 473.8 \qquad (0 \le t \le 11)$$

where t is measured in years, with $t = 0$ corresponding to 1990.
a. Plot the graph of S in the viewing window $[0, 10] \times [0, 1000]$.
b. What were the annual retail sales in the United States in 1999 ($t = 9$)?
c. Approximately, how fast were the retail sales changing in 1999 ($t = 9$)?
Source: U.S. Census Bureau

CHAPTER 11   Summary of Principal Formulas and Terms

FORMULAS

1. Average rate of change of f over $[x, x+h]$ or Slope of the secant line to the graph of f through $(x, f(x))$ and $(x + h, f(x + h))$ or Difference quotient	$\dfrac{f(x + h) - f(x)}{h}$
2. Instantaneous rate of change of f at $(x, f(x))$ or Slope of tangent line to the graph of f at $(x, f(x))$ at x or Derivative of f	$\displaystyle\lim_{h \to 0} \dfrac{f(x + h) - f(x)}{h}$

TERMS

function (578)

domain (578)

range (578)

independent variable (579)

dependent variable (579)

ordered pairs (581)

function (alternative definition) (581)

graph of a function (581)

graph of an equation (584)

vertical-line test (584)

composite function (599)

polynomial function (606)

linear function (606)

quadratic function (606)

cubic function (606)

rational function (606)

power function (607)

limit of a function (624)

indeterminate form (628)

limit of a function at infinity (631)

right-hand limit of a function (643)

left-hand limit of a function (643)

continuity of a function at a number (645)

secant line (661)

tangent line to the graph of f (661)

differentiable function (669)

CHAPTER 11   Concept Review Questions

Fill in the blanks.

1. If f is a function from the set A to the set B, then A is called the _____ of f, and the set of all values of $f(x)$ as x takes on all possible values in A is called the _____ of f. The range of f is contained in the set _____.

2. The graph of a function is the set of all points (x, y) in the xy-plane such that x is in the _____ of f and $y =$ _____. The vertical-line test states that a curve in the xy-plane is the graph of a function $y = f(x)$ if and only if each _____ line intersects it in at most one _____.

3. If f and g are functions with domains A and B, respectively, then (a) $(f \pm g)(x) =$ _____, (b) $(fg)(x) =$ _____, and (c) $\left(\dfrac{f}{g}\right)(x) =$ _____. The domain of $f + g$ is _____. The domain of $\dfrac{f}{g}$ is _____ with the additional condition that $g(x)$ is never _____.

4. The composition of g and f is the function with rule $(g \circ f)(x) =$ _____. Its domain is the set of all x in the domain of _____ such that _____ lies in the domain of _____.

5. a. A polynomial function of degree n is a function of the form _____.

b. A polynomial function of degree 1 is called a _____ function; one of degree 2 is called a _____ function; one of degree 3 is called a _____ function.

c. A rational function is a/an _____ of two _____.

d. A power function has the form $f(x) =$ _____.

6. The statement $\displaystyle\lim_{x \to a} f(x) = L$ means that there is a number _____ such that the values of _____ can be made as close to _____ as we please by taking x sufficiently close to _____.

7. If $\lim\limits_{x\to a} f(x) = L$ and $\lim\limits_{x\to a} g(x) = M$, then

  a. $\lim\limits_{x\to a} [f(x)]^r =$ _____, where r is a real number.

  b. $\lim\limits_{x\to a} [f(x) \pm g(x)] =$ _____ .

  c. $\lim\limits_{x\to a} [f(x)g(x)] =$ _____ .

  d. $\lim\limits_{x\to a} \dfrac{f(x)}{g(x)} =$ _____ provided that _____.

8. a. The statement $\lim\limits_{x\to\infty} f(x) = L$ means that $f(x)$ can be made arbitrarily close to _____ by taking _____ large enough.

  b. The statement $\lim\limits_{x\to-\infty} f(x) = M$ means that $f(x)$ can be made arbitrarily close to _____ by taking x to be _____ and sufficiently large in _____ value.

9. a. The statement $\lim\limits_{x\to a^+} f(x) = L$ is similar to the statement $\lim\limits_{x\to a} f(x) = L$, but here x is required to lie to the _____ of a.

  b. The statement $\lim\limits_{x\to a^-} f(x) = L$ is similar to the statement $\lim\limits_{x\to a} f(x) = L$, but here x is required to lie to the _____ of a.

  c. $\lim\limits_{x\to a} f(x) = L$ if and only if both $\lim\limits_{x\to a^-} f(x) =$ _____ and $\lim\limits_{x\to a^+} f(x) =$ _____ .

10. a. If $f(a)$ is defined, $\lim\limits_{x\to a} f(x)$ exists and $\lim\limits_{x\to a} f(x) = f(a)$, then f is _____ at a.

  b. If f is not continuous at a, then it is _____ at a.

  c. f is continuous on an interval I if f is continuous at _____ number in the interval.

11. a. If f and g are continuous at a, then $f \pm g$ and fg are continuous at _____. Also, $\frac{f}{g}$ is continuous at _____, provided _____ $\neq 0$.

  b. A polynomial function is continuous _____.

  c. A rational function $R = \frac{P}{Q}$ is continuous everywhere except at values of x where _____ $= 0$.

12. a. Suppose f is continuous on $[a, b]$ and $f(a) < M < f(b)$. Then the intermediate value theorem guarantees the existence of at least one number c in _____ such that _____.

  b. If f is continuous on $[a, b]$ and $f(a)f(b) < 0$, then there must be at least one solution of the equation _____ in the interval _____.

13. a. The tangent line at $P(a, f(a))$ to the graph of f is the line passing through P and having slope _____.

  b. If the slope of the tangent line at $P(a, f(a))$ is m, then an equation of the tangent line at P is _____.

14. a. The slope of the secant line passing through $P(a, f(a))$ and $Q(a + h, f(a + h))$ and the average rate of change of f over the interval $[a, a + h]$ are both given by _____.

  b. The slope of the tangent line at $P(a, f(a))$ and the instantaneous rate of change of f at a are both given by _____.

CHAPTER 11   Review Exercises

1. Find the domain of each function:

  a. $f(x) = \sqrt{9 - x}$ **b.** $f(x) = \dfrac{x + 3}{2x^2 - x - 3}$

2. Let $f(x) = 3x^2 + 5x - 2$. Find:

  a. $f(-2)$ **b.** $f(a + 2)$

  c. $f(2a)$ **d.** $f(a + h)$

3. Let $y^2 = 2x + 1$.

  a. Sketch the graph of this equation.

  b. Is y a function of x? Why?

  c. Is x a function of y? Why?

4. Sketch the graph of the function defined by

$$f(x) = \begin{cases} x + 1 & \text{if } x < 1 \\ -x^2 + 4x - 1 & \text{if } x \geq 1 \end{cases}$$

5. Let $f(x) = 1/x$ and $g(x) = 2x + 3$. Find:

  a. $f(x)g(x)$ **b.** $f(x)/g(x)$

  c. $f(g(x))$ **d.** $g(f(x))$

In Exercises 6–19, find the indicated limits, if they exist.

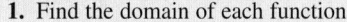

6. $\lim\limits_{x\to 0} (5x - 3)$ **7.** $\lim\limits_{x\to 1} (x^2 + 1)$

8. $\lim\limits_{x\to -1} (3x^2 + 4)(2x - 1)$

9. $\lim\limits_{x\to 3} \dfrac{x - 3}{x + 4}$ **10.** $\lim\limits_{x\to 2} \dfrac{x + 3}{x^2 - 9}$

11. $\lim\limits_{x\to -2} \dfrac{x^2 - 2x - 3}{x^2 + 5x + 6}$ **12.** $\lim\limits_{x\to 3} \sqrt{2x^3 - 5}$

13. $\lim\limits_{x\to 3} \dfrac{4x - 3}{\sqrt{x + 1}}$ **14.** $\lim\limits_{x\to 1^+} \dfrac{x - 1}{x(x - 1)}$

15. $\lim\limits_{x\to 1^-} \dfrac{\sqrt{x} - 1}{x - 1}$ **16.** $\lim\limits_{x\to\infty} \dfrac{x^2}{x^2 - 1}$

17. $\lim\limits_{x\to -\infty} \dfrac{x + 1}{x}$ **18.** $\lim\limits_{x\to\infty} \dfrac{3x^2 + 2x + 4}{2x^2 - 3x + 1}$

19. $\lim\limits_{x\to -\infty} \dfrac{x^2}{x + 1}$

20. Sketch the graph of the function

$$f(x) = \begin{cases} 2x - 3 & \text{if } x \leq 2 \\ -x + 3 & \text{if } x > 2 \end{cases}$$

and evaluate $\lim\limits_{x\to a^+} f(x)$, $\lim\limits_{x\to a^-} f(x)$, and $\lim\limits_{x\to a} f(x)$ at the point $a = 2$, if the limits exist.

21. Sketch the graph of the function

$$f(x) = \begin{cases} 4 - x & \text{if } x \le 2 \\ x + 2 & \text{if } x > 2 \end{cases}$$

and evaluate $\lim\limits_{x \to a^+} f(x)$, $\lim\limits_{x \to a^-} f(x)$, and $\lim\limits_{x \to a} f(x)$ at the point $a = 2$, if the limits exist.

In Exercises 22–25, determine all values of x for which each function is discontinuous.

22. $g(x) = \begin{cases} x + 3 & \text{if } x \ne 2 \\ 0 & \text{if } x = 2 \end{cases}$

23. $f(x) = \dfrac{3x + 4}{4x^2 - 2x - 2}$

24. $f(x) = \begin{cases} \dfrac{1}{(x + 1)^2} & \text{if } x \ne -1 \\ 2 & \text{if } x = -1 \end{cases}$

25. $f(x) = \dfrac{|2x|}{x}$

26. Let $y = x^2 + 2$.
   a. Find the average rate of change of y with respect to x in the intervals $[1, 2]$, $[1, 1.5]$, and $[1, 1.1]$.
   b. Find the (instantaneous) rate of change of y at $x = 1$.

27. Use the definition of the derivative to find the slope of the tangent line to the graph of the function $f(x) = 3x + 5$ at any point $P(x, f(x))$ on the graph.

28. Use the definition of the derivative to find the slope of the tangent line to the graph of the function $f(x) = -1/x$ at any point $P(x, f(x))$ on the graph.

29. Use the definition of the derivative to find the slope of the tangent line to the graph of the function $f(x) = \frac{3}{2}x + 5$ at the point $(-2, 2)$ and determine an equation of the tangent line.

30. Use the definition of the derivative to find the slope of the tangent line to the graph of the function $f(x) = -x^2$ at the point $(2, -4)$ and determine an equation of the tangent line.

31. The graph of the function f is shown in the accompanying figure.
   a. Is f continuous at $x = a$? Why?
   b. Is f differentiable at $x = a$? Justify your answers.

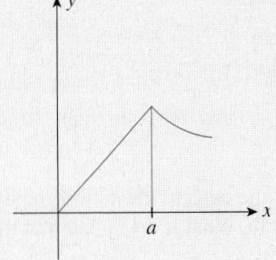

32. SALES OF CLOCK RADIOS Sales of a certain stereo clock radio are approximated by the relationship $S(x) = 6000x + 30,000$ $(0 \le x \le 5)$, where $S(x)$ denotes the number of clock radios sold in year x ($x = 0$ corresponds to the year 2002). Find the number of clock radios expected to be sold in 2006.

33. SALES OF A COMPANY A company's total sales (in millions of dollars) are approximately linear as a function of time (in years). Sales in 2001 were $2.4 million, whereas sales in 2006 amounted to $7.4 million.
   a. Find an equation that gives the company's sales as a function of time.
   b. What were the sales in 2004?

34. PROFIT FUNCTIONS A company has a fixed cost of $30,000 and a production cost of $6 for each unit it manufactures. A unit sells for $10.
   a. What is the cost function?
   b. What is the revenue function?
   c. What is the profit function?
   d. Compute the profit (loss) corresponding to production levels of 6000, 8000, and 12,000 units, respectively.

35. Find the point of intersection of the two straight lines having the equations $y = \frac{3}{4}x + 6$ and $3x - 2y + 3 = 0$.

36. The cost and revenue functions for a certain firm are given by $C(x) = 12x + 20,000$ and $R(x) = 20x$, respectively. Find the company's profit function.

37. MARKET EQUILIBRIUM Given the demand equation $3x + p - 40 = 0$ and the supply equation $2x - p + 10 = 0$, where p is the unit price in dollars and x represents the quantity in units of a thousand, determine the equilibrium quantity and the equilibrium price.

38. CLARK'S RULE Clark's rule is a method for calculating pediatric drug dosages based on a child's weight. If a denotes the adult dosage (in milligrams) and if w is the weight of the child (in pounds), then the child's dosage is given by

$$D(w) = \frac{aw}{150}$$

If the adult dose of a substance is 500 mg, how much should a child who weighs 35 lb receive?

39. REVENUE FUNCTIONS The revenue (in dollars) realized by Apollo from the sale of its ink-jet printers is given by

$$R(x) = -0.1x^2 + 500x$$

where x denotes the number of units manufactured each month. What is Apollo's revenue when 1000 units are produced?

40. REVENUE FUNCTIONS The monthly revenue R (in hundreds of dollars) realized in the sale of Royal electric shavers is related to the unit price p (in dollars) by the equation

$$R(p) = -\frac{1}{2}p^2 + 30p$$

Find the revenue when an electric shaver is priced at $30.

41. **HEALTH CLUB MEMBERSHIP** The membership of the newly opened Venus Health Club is approximated by the function

$$N(x) = 200(4 + x)^{1/2} \qquad (1 \le x \le 24)$$

where $N(x)$ denotes the number of members x mo after the club's grand opening. Find $N(0)$ and $N(12)$ and interpret your results.

42. **POPULATION GROWTH** A study prepared for a Sunbelt town's Chamber of Commerce projected that the population of the town in the next 3 yr will grow according to the rule

$$P(x) = 50,000 + 30x^{3/2} + 20x$$

where $P(x)$ denotes the population x mo from now. By how much will the population increase during the next 9 mo? During the next 16 mo?

43. **THURSTONE LEARNING CURVE** Psychologist L. L. Thurstone discovered the following model for the relationship between the learning time T and the length of a list n:

$$T = f(n) = An\sqrt{n - b}$$

where A and b are constants that depend on the person and the task. Suppose that, for a certain person and a certain task, $A = 4$ and $b = 4$. Compute $f(4), f(5), \ldots, f(12)$ and use this information to sketch the graph of the function f. Interpret your results.

44. **FORECASTING SALES** The annual sales of Crimson Drug Store are expected to be given by

$$S_1(t) = 2.3 + 0.4t$$

million dollars t yr from now, whereas the annual sales of Cambridge Drug Store are expected to be given by

$$S_2(t) = 1.2 + 0.6t$$

million dollars t yr from now. When will the annual sales of Cambridge first surpass the annual sales of Crimson?

45. **MARKET EQUILIBRIUM** The monthly demand and supply functions for the Luminar desk lamp are given by

$$p = d(x) = -1.1x^2 + 1.5x + 40$$
$$p = s(x) = 0.1x^2 + 0.5x + 15$$

respectively, where p is measured in dollars and x in units of a thousand. Find the equilibrium quantity and price.

46. **OIL SPILLS** The oil spilling from the ruptured hull of a grounded tanker spreads in all directions in calm waters. Suppose the area polluted is a circle of radius r and the radius is increasing at the rate of 2 ft/sec.
   a. Find a function f giving the area polluted in terms of r.
   b. Find a function g giving the radius of the polluted area in terms of t.
   c. Find a function h giving the area polluted in terms of t.
   d. What is the size of the polluted area 30 sec after the hull was ruptured?

47. **FILM CONVERSION PRICES** PhotoMart transfers movie films to CDs. The fees charged for this service are shown in the following table. Find a function C relating the cost $C(x)$ to the number of feet x of film transferred. Sketch the graph of the function C and discuss its continuity.

Length of Film in Feet, x	Price ($) for Conversion
$1 \le x \le 100$	5.00
$100 < x \le 200$	9.00
$200 < x \le 300$	12.50
$300 < x \le 400$	15.00
$x > 400$	$7 + 0.02x$

48. **AVERAGE PRICE OF A COMMODITY** The average cost (in dollars) of producing x units of a certain commodity is given by

$$\overline{C}(x) = 20 + \frac{400}{x}$$

Evaluate $\lim_{x \to \infty} \overline{C}(x)$ and interpret your results.

CHAPTER 11   **Before Moving On . . .**

1. Let

$$f(x) = \begin{cases} -2x + 1 & -1 \le x < 0 \\ x^2 + 2 & 0 \le x \le 2 \end{cases}$$

Find (a) $f(-1)$, (b) $f(0)$, and (c) $f(\frac{3}{2})$.

2. Let $f(x) = \frac{1}{x + 1}$ and $g(x) = x^2 + 1$. Find the rules for (a) $f + g$, (b) fg, (c) $f \circ g$, and (d) $g \circ f$.

3. Postal regulations specify that a parcel sent by parcel post may have a combined length and girth of no more than 108 in. Suppose a rectangular package that has a square cross section of x in. $\times$ x in. is to have a combined length and girth of exactly 108 in. Find a function in terms of x giving the volume of the package.
   Hint: The length plus the girth is $4x + h$ (see the accompanying figure).

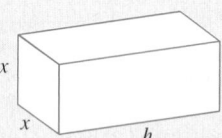

4. Find $\lim_{x \to -1} \dfrac{x^2 + 4x + 3}{x^2 + 3x + 2}$.

5. Let

$$f(x) = \begin{cases} x^2 - 1 & -2 \le x < 1 \\ x^3 & 1 \le x \le 2 \end{cases}$$

Find (a) $\lim_{x \to 1^-} f(x)$ and (b) $\lim_{x \to 1^+} f(x)$. Is f continuous at $x = 1$? Explain.

6. Find the slope of the tangent line to the graph of $x^2 - 3x + 1$ at the point $(1, -1)$. What is an equation of the tangent line?

12 Differentiation

© Theo Allofs/Corbis

How is a pond's oxygen content affected by organic waste? In Example 7, page 702, you will see how to find the rate at which oxygen is being restored to the pond after organic waste has been dumped into it.

THIS CHAPTER GIVES several rules that will greatly simplify the task of finding the derivative of a function, thus enabling us to study how fast one quantity is changing with respect to another in many real-world situations. For example, we will be able to find how fast the population of an endangered species of whales grows after certain conservation measures have been implemented, how fast an economy's consumer price index (CPI) is changing at any time, and how fast the time taken to learn the items on a list changes with respect to the length of a list. We also see how these rules of differentiation facilitate the study of marginal analysis, the study of the rate of change of economic quantities. Finally, we introduce the notion of the differential of a function. Using differentials is a relatively easy way of approximating the change in one quantity due to a small change in a related quantity.

12.1 Basic Rules of Differentiation

▬ Four Basic Rules

The method used in Chapter 11 for computing the derivative of a function is based on a faithful interpretation of the definition of the derivative as the limit of a quotient. Thus, to find the rule for the derivative f' of a function f, we first computed the difference quotient

$$\frac{f(x + h) - f(x)}{h}$$

and then evaluated its limit as h approached zero. As you have probably observed, this method is tedious even for relatively simple functions.

The main purpose of this chapter is to derive certain rules that will simplify the process of finding the derivative of a function. Throughout this book, we will use the notation

$$\frac{d}{dx}[f(x)] \qquad \text{Read "d, d x of f of x"}$$

to mean "the derivative of f with respect to x at x." In stating the rules of differentiation, we assume that the functions f and g are differentiable.

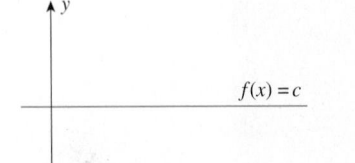

FIGURE 1
The slope of the tangent line to the graph of $f(x) = c$, where c is a constant, is zero.

Rule 1: Derivative of a Constant

$$\frac{d}{dx}(c) = 0 \qquad (c, \text{ a constant})$$

The derivative of a constant function is equal to zero.

We can see this from a geometric viewpoint by recalling that the graph of a constant function is a straight line parallel to the x-axis (Figure 1). Since the tangent line to a straight line at any point on the line coincides with the straight line itself, its slope [as given by the derivative of $f(x) = c$] must be zero. We can also use the definition of the derivative to prove this result by computing

$$f'(x) = \lim_{h \to 0} \frac{f(x + h) - f(x)}{h}$$
$$= \lim_{h \to 0} \frac{c - c}{h}$$
$$= \lim_{h \to 0} 0 = 0$$

EXAMPLE 1

a. If $f(x) = 28$, then

$$f'(x) = \frac{d}{dx}(28) = 0$$

b. If $f(x) = -2$, then

$$f'(x) = \frac{d}{dx}(-2) = 0$$

Rule 2: The Power Rule

If n is any real number, then $\dfrac{d}{dx}(x^n) = nx^{n-1}$.

Let's verify the power rule for the special case $n = 2$. If $f(x) = x^2$, then

$$f'(x) = \frac{d}{dx}(x^2) = \lim_{h \to 0} \frac{f(x + h) - f(x)}{h}$$

$$= \lim_{h \to 0} \frac{(x + h)^2 - x^2}{h}$$

$$= \lim_{h \to 0} \frac{x^2 + 2xh + h^2 - x^2}{h}$$

$$= \lim_{h \to 0} \frac{2xh + h^2}{h} = \lim_{h \to 0}(2x + h) = 2x$$

as we set out to show.

The proof of the power rule for the general case is not easy to prove and will be omitted. However, you will be asked to prove the rule for the special case $n = 3$ in Exercise 77, page 694.

EXAMPLE 2

a. If $f(x) = x$, then

$$f'(x) = \frac{d}{dx}(x) = 1 \cdot x^{1-1} = x^0 = 1$$

b. If $f(x) = x^8$, then

$$f'(x) = \frac{d}{dx}(x^8) = 8x^7$$

c. If $f(x) = x^{5/2}$, then

$$f'(x) = \frac{d}{dx}(x^{5/2}) = \frac{5}{2}x^{3/2}$$

To differentiate a function whose rule involves a radical, we first rewrite the rule using fractional powers. The resulting expression can then be differentiated using the power rule.

EXAMPLE 3 Find the derivative of the following functions:

a. $f(x) = \sqrt{x}$ **b.** $g(x) = \dfrac{1}{\sqrt[3]{x}}$

Solution

a. Rewriting $\sqrt{x}$ in the form $x^{1/2}$, we obtain

$$f'(x) = \frac{d}{dx}(x^{1/2})$$

$$= \frac{1}{2}x^{-1/2} = \frac{1}{2x^{1/2}} = \frac{1}{2\sqrt{x}}$$

b. Rewriting $\dfrac{1}{\sqrt[3]{x}}$ in the form $x^{-1/3}$, we obtain

$$g'(x) = \frac{d}{dx}(x^{-1/3})$$

$$= -\frac{1}{3}x^{-4/3} = -\frac{1}{3x^{4/3}}$$

◼

Rule 3: Derivative of a Constant Multiple of a Function

$$\frac{d}{dx}[cf(x)] = c\frac{d}{dx}[f(x)] \qquad \text{(c, a constant)}$$

The derivative of a constant times a differentiable function is equal to the constant times the derivative of the function.

This result follows from the following computations.

If $g(x) = cf(x)$, then

$$g'(x) = \lim_{h \to 0} \frac{g(x + h) - g(x)}{h} = \lim_{h \to 0} \frac{cf(x + h) - cf(x)}{h}$$

$$= c\lim_{h \to 0} \frac{f(x + h) - f(x)}{h}$$

$$= cf'(x)$$

EXAMPLE 4

a. If $f(x) = 5x^3$, then

$$f'(x) = \frac{d}{dx}(5x^3) = 5\frac{d}{dx}(x^3)$$

$$= 5(3x^2) = 15x^2$$

b. If $f(x) = \dfrac{3}{\sqrt{x}}$, then

$$f'(x) = \frac{d}{dx}(3x^{-1/2})$$

$$= 3\left(-\frac{1}{2}x^{-3/2}\right) = -\frac{3}{2x^{3/2}}$$

◼

Rule 4: The Sum Rule

$$\frac{d}{dx}[f(x) \pm g(x)] = \frac{d}{dx}[f(x)] \pm \frac{d}{dx}[g(x)]$$

The derivative of the sum (difference) of two differentiable functions is equal to the sum (difference) of their derivatives.

This result may be extended to the sum and difference of any finite number of differentiable functions. Let's verify the rule for a sum of two functions.

If $s(x) = f(x) + g(x)$, then

$$s'(x) = \lim_{h \to 0} \frac{s(x + h) - s(x)}{h}$$

$$= \lim_{h \to 0} \frac{[f(x + h) + g(x + h)] - [f(x) + g(x)]}{h}$$

$$= \lim_{h \to 0} \frac{[f(x + h) - f(x)] + [g(x + h) - g(x)]}{h}$$

$$= \lim_{h \to 0} \frac{f(x + h) - f(x)}{h} + \lim_{h \to 0} \frac{g(x + h) - g(x)}{h}$$

$$= f'(x) + g'(x)$$

EXAMPLE 5   Find the derivatives of the following functions:

a. $f(x) = 4x^5 + 3x^4 - 8x^2 + x + 3$ **b.** $g(t) = \dfrac{t^2}{5} + \dfrac{5}{t^3}$

Solution

a. $f'(x) = \dfrac{d}{dx}(4x^5 + 3x^4 - 8x^2 + x + 3)$

$$= \frac{d}{dx}(4x^5) + \frac{d}{dx}(3x^4) - \frac{d}{dx}(8x^2) + \frac{d}{dx}(x) + \frac{d}{dx}(3)$$

$$= 20x^4 + 12x^3 - 16x + 1$$

b. Here, the independent variable is t instead of x, so we differentiate with respect to t. Thus,

$$g'(t) = \frac{d}{dt}\left(\frac{1}{5}t^2 + 5t^{-3}\right) \qquad \text{Rewrite } \frac{1}{t^3} \text{ as } t^{-3}.$$

$$= \frac{2}{5}t - 15t^{-4}$$

$$= \frac{2t^5 - 75}{5t^4} \qquad \text{Rewrite } t^{-4} \text{ as } \frac{1}{t^4} \text{ and simplify.}$$

EXAMPLE 6   Find the slope and an equation of the tangent line to the graph of $f(x) = 2x + 1/\sqrt{x}$ at the point $(1, 3)$.

Solution   The slope of the tangent line at any point on the graph of f is given by

$$f'(x) = \frac{d}{dx}\left(2x + \frac{1}{\sqrt{x}}\right)$$

$$= \frac{d}{dx}(2x + x^{-1/2}) \qquad \text{Rewrite } \frac{1}{\sqrt{x}} \text{ as } \frac{1}{x^{1/2}} = x^{-1/2}.$$

$$= 2 - \frac{1}{2}x^{-3/2} \qquad \text{Use the sum rule.}$$

$$= 2 - \frac{1}{2x^{3/2}}$$

In particular, the slope of the tangent line to the graph of f at $(1, 3)$ (where $x = 1$) is

$$f'(1) = 2 - \frac{1}{2(1^{3/2})} = 2 - \frac{1}{2} = \frac{3}{2}$$

Using the point-slope form of the equation of a line with slope $\frac{3}{2}$ and the point $(1, 3)$, we see that an equation of the tangent line is

$$y - 3 = \frac{3}{2}(x - 1) \qquad y - y_1 = m(x - x_1)$$

or, upon simplification,

$$y = \frac{3}{2}x + \frac{3}{2}$$

(see Figure 2).

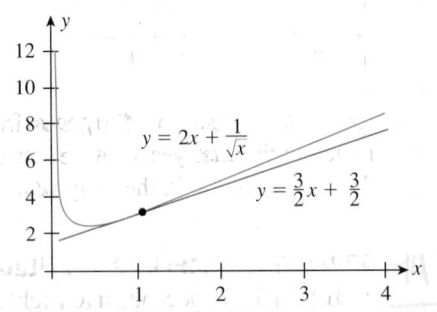

FIGURE 2
The tangent line to the graph of $f(x) = 2x + 1/\sqrt{x}$ at $(1, 3)$.

APPLIED EXAMPLE 7   Conservation of a Species A group of marine biologists at the Neptune Institute of Oceanography recommended that a series of conservation measures be carried out over the next decade to save a certain species of whale from extinction. After implementing the conservation measures, the population of this species is expected to be

$$N(t) = 3t^3 + 2t^2 - 10t + 600 \qquad (0 \le t \le 10)$$

where $N(t)$ denotes the population at the end of year t. Find the rate of growth of the whale population when $t = 2$ and $t = 6$. How large will the whale population be 8 years after implementing the conservation measures?

Solution   The rate of growth of the whale population at any time t is given by

$$N'(t) = 9t^2 + 4t - 10$$

In particular, when $t = 2$ and $t = 6$, we have

$$N'(2) = 9(2)^2 + 4(2) - 10$$
$$= 34$$
$$N'(6) = 9(6)^2 + 4(6) - 10$$
$$= 338$$

Thus, the whale population's rate of growth will be 34 whales per year after 2 years and 338 per year after 6 years.

The whale population at the end of the eighth year will be

$$N(8) = 3(8)^3 + 2(8)^2 - 10(8) + 600$$
$$= 2184 \text{ whales}$$

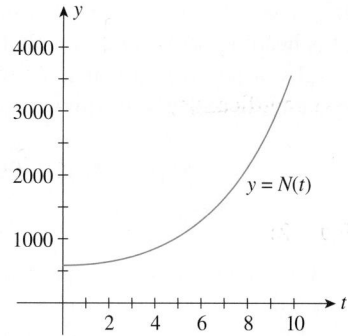

FIGURE 3
The whale population after year t is given by $N(t)$.

The graph of the function N appears in Figure 3. Note the rapid growth of the population in the later years, as the conservation measures begin to pay off, compared with the growth in the early years.

APPLIED EXAMPLE 8 Altitude of a Rocket The altitude of a rocket (in feet) t seconds into flight is given by

$$s = f(t) = -t^3 + 96t^2 + 195t + 5 \qquad (t \geq 0)$$

a. Find an expression v for the rocket's velocity at any time t.
b. Compute the rocket's velocity when $t = 0, 30, 50, 65,$ and 70. Interpret your results.
c. Using the results from the solution to part (b) and the observation that at the highest point in its trajectory the rocket's velocity is zero, find the maximum altitude attained by the rocket.

Solution

a. The rocket's velocity at any time t is given by

$$v = f'(t) = -3t^2 + 192t + 195$$

b. The rocket's velocity when $t = 0, 30, 50, 65,$ and 70 is given by

$$f'(0) = -3(0)^2 + 192(0) + 195 = 195$$
$$f'(30) = -3(30)^2 + 192(30) + 195 = 3255$$
$$f'(50) = -3(50)^2 + 192(50) + 195 = 2295$$
$$f'(65) = -3(65)^2 + 192(65) + 195 = 0$$
$$f'(70) = -3(70)^2 + 192(70) + 195 = -1065$$

or 195, 3255, 2295, 0, and -1065 feet per second (ft/sec).

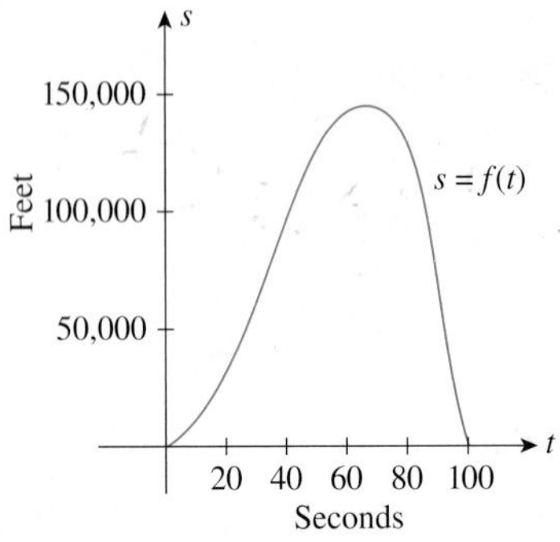

FIGURE 4
The rocket's altitude t seconds into flight is given by $f(t)$.

Thus, the rocket has an initial velocity of 195 ft/sec at $t = 0$ and accelerates to a velocity of 3255 ft/sec at $t = 30$. Fifty seconds into the flight, the rocket's velocity is 2295 ft/sec, which is less than the velocity at $t = 30$. This means that the rocket begins to decelerate after an initial period of acceleration. (Later on we will learn how to determine the rocket's maximum velocity.)

The deceleration continues: The velocity is 0 ft/sec at $t = 65$ and -1065 ft/sec when $t = 70$. This result tells us that 70 seconds into flight the rocket is heading back to Earth with a speed of 1065 ft/sec.

c. The results of part (b) show that the rocket's velocity is zero when $t = 65$. At this instant, the rocket's maximum altitude is

$$s = f(65) = -(65)^3 + 96(65)^2 + 195(65) + 5$$
$$= 143,655 \text{ feet}$$

A sketch of the graph of f appears in Figure 4.

EXPLORING WITH TECHNOLOGY

Refer to Example 8.

1. Use a graphing utility to plot the graph of the velocity function

$$v = f'(t) = -3t^2 + 192t + 195$$

using the viewing window $[0, 120] \times [-5000, 5000]$. Then, using **ZOOM** and **TRACE** or the root-finding capability of your graphing utility, verify that $f'(65) = 0$.

2. Plot the graph of the position function of the rocket

$$s = f(t) = -t^3 + 96t^2 + 195t + 5$$

using the viewing window $[0, 120] \times [0, 150,000]$. Then, using **ZOOM** and **TRACE** repeatedly, verify that the maximum altitude of the rocket is 143,655 feet.

3. Use **ZOOM** and **TRACE** or the root-finding capability of your graphing utility to find when the rocket returns to Earth.

12.1   Self-Check Exercises

1. Find the derivative of each function using the rules of differentiation.
   a. $f(x) = 1.5x^2 + 2x^{1.5}$
   b. $g(x) = 2\sqrt{x} + \dfrac{3}{\sqrt{x}}$

2. Let $f(x) = 2x^3 - 3x^2 + 2x - 1$.
   a. Compute $f'(x)$.
   b. What is the slope of the tangent line to the graph of f when $x = 2$?
   c. What is the rate of change of the function f at $x = 2$?

3. A certain country's gross domestic product (GDP) (in millions of dollars) is described by the function

$$G(t) = -2t^3 + 45t^2 + 20t + 6000 \qquad (0 \le t \le 11)$$

where $t = 0$ corresponds to the beginning of 1995.
   a. At what rate was the GDP changing at the beginning of 2000? At the beginning of 2002? At the beginning of 2005?
   b. What was the average rate of growth of the GDP over the period 2000–2005?

Solutions to Self-Check Exercises 12.1 can be found on page 695.

12.1 Concept Questions

1. State the following rules of differentiation in your own words.
   a. The rule for differentiating a constant function
   b. The power rule
   c. The constant multiple rule
   d. The sum rule

2. If $f'(2) = 3$ and $g'(2) = -2$, find
   a. $h'(2)$ if $h(x) = 2f(x)$
   b. $F'(2)$ if $F(x) = 3f(x) - 4g(x)$

12.1 Exercises

In Exercises 1–34, find the derivative of the function f by using the rules of differentiation.

1. $f(x) = -3$

2. $f(x) = 365$

3. $f(x) = x^5$

4. $f(x) = x^7$

5. $f(x) = x^{2.1}$

6. $f(x) = x^{0.8}$

7. $f(x) = 3x^2$

8. $f(x) = -2x^3$

9. $f(r) = \pi r^2$

10. $f(r) = \dfrac{4}{3}\pi r^3$

11. $f(x) = 9x^{1/3}$

12. $f(x) = \dfrac{5}{4}x^{4/5}$

13. $f(x) = 3\sqrt{x}$

14. $f(u) = \dfrac{2}{\sqrt{u}}$

15. $f(x) = 7x^{-12}$

16. $f(x) = 0.3x^{-1.2}$

17. $f(x) = 5x^2 - 3x + 7$

18. $f(x) = x^3 - 3x^2 + 1$

19. $f(x) = -x^3 + 2x^2 - 6$

20. $f(x) = x^4 - 2x^2 + 5$

21. $f(x) = 0.03x^2 - 0.4x + 10$

22. $f(x) = 0.002x^3 - 0.05x^2 + 0.1x - 20$

23. $f(x) = \dfrac{x^3 - 4x^2 + 3}{x}$

24. $f(x) = \dfrac{x^3 + 2x^2 + x - 1}{x}$

25. $f(x) = 4x^4 - 3x^{5/2} + 2$

26. $f(x) = 5x^{4/3} - \dfrac{2}{3}x^{3/2} + x^2 - 3x + 1$

27. $f(x) = 3x^{-1} + 4x^{-2}$

28. $f(x) = -\dfrac{1}{3}(x^{-3} - x^6)$

29. $f(t) = \dfrac{4}{t^4} - \dfrac{3}{t^3} + \dfrac{2}{t}$

30. $f(x) = \dfrac{5}{x^3} - \dfrac{2}{x^2} - \dfrac{1}{x} + 200$

31. $f(x) = 2x - 5\sqrt{x}$

32. $f(t) = 2t^2 + \sqrt{t^3}$

33. $f(x) = \dfrac{2}{x^2} - \dfrac{3}{x^{1/3}}$

34. $f(x) = \dfrac{3}{x^3} + \dfrac{4}{\sqrt{x}} + 1$

35. Let $f(x) = 2x^3 - 4x$. Find:
   a. $f'(-2)$ **b.** $f'(0)$ **c.** $f'(2)$

36. Let $f(x) = 4x^{5/4} + 2x^{3/2} + x$. Find:
   a. $f'(0)$ **b.** $f'(16)$

In Exercises 37–40, find each limit by evaluating the derivative of a suitable function at an appropriate point.
Hint: Look at the definition of the derivative.

37. $\displaystyle\lim_{h \to 0} \dfrac{(1 + h)^3 - 1}{h}$

38. $\displaystyle\lim_{x \to 1} \dfrac{x^5 - 1}{x - 1}$
   Hint: Let $h = x - 1$.

39. $\displaystyle\lim_{h \to 0} \dfrac{3(2 + h)^2 - (2 + h) - 10}{h}$

40. $\displaystyle\lim_{t \to 0} \dfrac{1 - (1 + t)^2}{t(1 + t)^2}$

In Exercises 41–44, find the slope and an equation of the tangent line to the graph of the function f at the specified point.

41. $f(x) = 2x^2 - 3x + 4;\ (2, 6)$

42. $f(x) = -\dfrac{5}{3}x^2 + 2x + 2;\ \left(-1, -\dfrac{5}{3}\right)$

43. $f(x) = x^4 - 3x^3 + 2x^2 - x + 1;\ (1, 0)$

44. $f(x) = \sqrt{x} + \dfrac{1}{\sqrt{x}};\ \left(4, \dfrac{5}{2}\right)$

45. Let $f(x) = x^3$.
   a. Find the point on the graph of f where the tangent line is horizontal.
   b. Sketch the graph of f and draw the horizontal tangent line.

46. Let $f(x) = x^3 - 4x^2$. Find the point(s) on the graph of f where the tangent line is horizontal.

47. Let $f(x) = x^3 + 1$.
  a. Find the point(s) on the graph of f where the slope of the tangent line is equal to 12.
  b. Find the equation(s) of the tangent line(s) of part (a).
  c. Sketch the graph of f showing the tangent line(s).

48. Let $f(x) = \frac{2}{3}x^3 + x^2 - 12x + 6$. Find the values of x for which:
  a. $f'(x) = -12$ **b.** $f'(x) = 0$
  c. $f'(x) = 12$

49. Let $f(x) = \frac{1}{4}x^4 - \frac{1}{3}x^3 - x^2$. Find the point(s) on the graph of f where the slope of the tangent line is equal to:
  a. $-2x$ **b.** 0 **c.** $10x$

50. A straight line perpendicular to and passing through the point of tangency of the tangent line is called the *normal* to the curve. Find an equation of the tangent line and the normal to the curve $y = x^3 - 3x + 1$ at the point (2, 3).

51. **GROWTH OF A CANCEROUS TUMOR** The volume of a spherical cancerous tumor is given by the function

$$V(r) = \frac{4}{3}\pi r^3$$

where r is the radius of the tumor in centimeters. Find the rate of change in the volume of the tumor when

  a. $r = \frac{2}{3}$ cm   **b.** $r = \frac{5}{4}$ cm

52. **VELOCITY OF BLOOD IN AN ARTERY** The velocity (in centimeters/second) of blood r cm from the central axis of an artery is given by

$$v(r) = k(R^2 - r^2)$$

where k is a constant and R is the radius of the artery (see the accompanying figure). Suppose $k = 1000$ and $R = 0.2$ cm. Find $v(0.1)$ and $v'(0.1)$ and interpret your results.

Blood vessel

53. **SALES OF DIGITAL CAMERAS** According to projections made in 2004, the worldwide shipments of digital point-and-shoot cameras are expected to grow in accordance with the rule

$$N(t) = 16.3t^{0.8766} \qquad (1 \le t \le 6)$$

where $N(t)$ is measured in millions and t is measured in years, with $t = 1$ corresponding to 2001.
  a. How many digital cameras were sold in 2001 ($t = 1$)?
  b. How fast were sales increasing in 2001?
  c. What were the projected sales in 2005?
  d. How fast were the sales projected to grow in 2005?

Source: International Data Corp.

54. **ONLINE BUYERS** As use of the Internet grows, so does the number of consumers who shop online. The number of online buyers, as a percent of net users, is expected to be

$$P(t) = 53t^{0.12} \qquad (1 \le t \le 7)$$

where t is measured in years, with $t = 1$ corresponding to the beginning of 2002.
  a. How many online buyers, as a percent of net users, are there expected to be at the beginning of 2007?
  b. How fast is the number of online buyers, as a percent of net users, expected to be changing at the beginning of 2007?

Source: Strategy Analytics

55. **MARRIED HOUSEHOLDS WITH CHILDREN** The percent of families that were married households with children between 1970 and 2000 is approximately

$$P(t) = \frac{49.6}{t^{0.27}} \qquad (1 \le t \le 4)$$

where t is measured in decades, with $t = 1$ corresponding to 1970.
  a. What percent of families were married households with children in 1970? In 1980? In 1990? In 2000?
  b. How fast was the percent of families that were married households with children changing in 1980? In 1990?

Source: U.S. Census Bureau

56. **EFFECT OF STOPPING ON AVERAGE SPEED** According to data from a study, the average speed of your trip A (in mph) is related to the number of stops/mile you make on the trip x by the equation

$$A = \frac{26.5}{x^{0.45}}$$

Compute dA/dx for $x = 0.25$ and $x = 2$. How is the rate of change of the average speed of your trip affected by the number of stops/mile?

Source: General Motors

57. **PORTABLE PHONES** The percent of the U.S. population with portable phones is projected to be

$$P(t) = 24.4t^{0.34} \qquad (1 \le t \le 10)$$

where t is measured in years, with $t = 1$ corresponding to the beginning of 1998.
  a. What percent of the U.S. population is expected to have portable phones by the beginning of 2006?
  b. How fast is the percent of the U.S. population with portable phones expected to be changing at the beginning of 2006?

Source: BancAmerica Robertson Stephens

58. **DEMAND FUNCTIONS** The demand function for the Luminar desk lamp is given by

$$p = f(x) = -0.1x^2 - 0.4x + 35$$

where x is the quantity demanded (measured in thousands) and p is the unit price in dollars.

a. Find $f'(x)$.

b. What is the rate of change of the unit price when the quantity demanded is 10,000 units ($x = 10$)? What is the unit price at that level of demand?

59. STOPPING DISTANCE OF A RACING CAR During a test by the editors of an auto magazine, the stopping distance s (in feet) of the MacPherson X-2 racing car conformed to the rule

$$s = f(t) = 120t - 15t^2 \qquad (t \geq 0)$$

where t was the time (in seconds) after the brakes were applied.

a. Find an expression for the car's velocity v at any time t.

b. What was the car's velocity when the brakes were first applied?

c. What was the car's stopping distance for that particular test?

Hint: The stopping time is found by setting $v = 0$.

60. SALES OF DSPs The sales of digital signal processors (DSPs) in billions of dollars is projected to be

$$S(t) = 0.14t^2 + 0.68t + 3.1 \qquad (0 \leq t \leq 6)$$

where t is measured in years, with $t = 0$ corresponding to the beginning of 1997.

a. What were the sales of DSPs at the beginning of 1997? What were the sales at the beginning of 2002?

b. How fast was the level of sales increasing at the beginning of 1997? How fast were sales increasing at the beginning of 2002?

Source: World Semiconductor Trade Statistics

61. CHILD OBESITY The percent of obese children, ages 12–19, in the United States has grown dramatically in recent years. The percent of obese children from 1980 through the year 2000 is approximated by the function

$$P(t) = -0.0105t^2 + 0.735t + 5 \qquad (0 \leq t \leq 20)$$

where t is measured in years, with $t = 0$ corresponding to the beginning of 1980.

a. What percent of children were obese at the beginning of 1980? At the beginning of 1990? At the beginning of the year 2000?

b. How fast was the percent of obese children changing at the beginning of 1985? At the beginning of 1990?

Source: Centers for Disease Control and Prevention

62. SPENDING ON MEDICARE Based on the current eligibility requirement, a study conducted in 2004 showed that federal spending on entitlement programs, particularly Medicare, would grow enormously in the future. The study predicted that spending on Medicare, as a percent of the gross domestic product (GDP), will be

$$P(t) = 0.27t^2 + 1.4t + 2.2 \qquad (0 \leq t \leq 5)$$

percent in year t, where t is measured in decades, with $t = 0$ corresponding to 2000.

a. How fast will the spending on Medicare, as a percent of the GDP, be growing in 2010? In 2020?

b. What will the predicted spending on Medicare be in 2010? In 2020?

Source: Congressional Budget Office

63. FISHERIES The total groundfish population on Georges Bank in New England between 1989 and 1999 is approximated by the function

$$f(t) = 5.303t^2 - 53.977t + 253.8 \qquad (0 \leq t \leq 10)$$

where $f(t)$ is measured in thousands of metric tons and t is measured in years, with $t = 0$ corresponding to the beginning of 1989.

a. What was the rate of change of the groundfish population at the beginning of 1994? At the beginning of 1996?

b. Fishing restrictions were imposed on Dec. 7, 1994. Were the conservation measures effective?

Source: New England Fishery Management Council

64. WORKER EFFICIENCY An efficiency study conducted for Elektra Electronics showed that the number of Space Commander walkie-talkies assembled by the average worker t hr after starting work at 8 a.m. is given by

$$N(t) = -t^3 + 6t^2 + 15t$$

a. Find the rate at which the average worker will be assembling walkie-talkies t hr after starting work.

b. At what rate will the average worker be assembling walkie-talkies at 10 a.m.? At 11 a.m.?

c. How many walkie-talkies will the average worker assemble between 10 a.m. and 11 a.m.?

65. CONSUMER PRICE INDEX An economy's consumer price index (CPI) is described by the function

$$I(t) = -0.2t^3 + 3t^2 + 100 \qquad (0 \leq t \leq 10)$$

where $t = 0$ corresponds to 1997.

a. At what rate was the CPI changing in 2002? In 2004? In 2007?

b. What was the average rate of increase in the CPI over the period from 2002 to 2007?

66. EFFECT OF ADVERTISING ON SALES The relationship between the amount of money x that Cannon Precision Instruments spends on advertising and the company's total sales $S(x)$ is given by the function

$$S(x) = -0.002x^3 + 0.6x^2 + x + 500 \qquad (0 \leq x \leq 200)$$

where x is measured in thousands of dollars. Find the rate of change of the sales with respect to the amount of money spent on advertising. Are Cannon's total sales increasing at a faster rate when the amount of money spent on advertising is (a) $100,000 or (b) $150,000?

67. SUPPLY FUNCTIONS The supply function for a certain make of satellite radio is given by

$$p = f(x) = 0.0001x^{5/4} + 10$$

where x is the quantity supplied and p is the unit price in dollars.

a. Find $f'(x)$.

b. What is the rate of change of the unit price if the quantity supplied is 10,000 satellite radios?

68. **POPULATION GROWTH** A study prepared for a Sunbelt town's chamber of commerce projected that the town's population in the next 3 yr will grow according to the rule

$$P(t) = 50,000 + 30t^{3/2} + 20t$$

where $P(t)$ denotes the population t mo from now. How fast will the population be increasing 9 mo and 16 mo from now?

69. **AVERAGE SPEED OF A VEHICLE ON A HIGHWAY** The average speed of a vehicle on a stretch of Route 134 between 6 a.m. and 10 a.m. on a typical weekday is approximated by the function

$$f(t) = 20t - 40\sqrt{t} + 50 \qquad (0 \le t \le 4)$$

where $f(t)$ is measured in mph and t is measured in hours, with $t = 0$ corresponding to 6 a.m.

a. Compute $f'(t)$.

b. What is the average speed of a vehicle on that stretch of Route 134 at 6 a.m.? At 7 a.m.? At 8 a.m.?

c. How fast is the average speed of a vehicle on that stretch of Route 134 changing at 6:30 a.m.? At 7 a.m.? At 8 a.m.?

70. **CURBING POPULATION GROWTH** Five years ago, the government of a Pacific Island state launched an extensive propaganda campaign toward curbing the country's population growth. According to the Census Department, the population (measured in thousands of people) for the following 4 yr was

$$P(t) = -\frac{1}{3}t^3 + 64t + 3000$$

where t is measured in years and $t = 0$ corresponds to the start of the campaign. Find the rate of change of the population at the end of years 1, 2, 3, and 4. Was the plan working?

71. **CONSERVATION OF SPECIES** A certain species of turtle faces extinction because dealers collect truckloads of turtle eggs to be sold as aphrodisiacs. After severe conservation measures are implemented, it is hoped that the turtle population will grow according to the rule

$$N(t) = 2t^3 + 3t^2 - 4t + 1000 \qquad (0 \le t \le 10)$$

where $N(t)$ denotes the population at the end of year t. Find the rate of growth of the turtle population when $t = 2$ and $t = 8$. What will be the population 10 yr after the conservation measures are implemented?

72. **FLIGHT OF A ROCKET** The altitude (in feet) of a rocket t sec into flight is given by

$$s = f(t) = -2t^3 + 114t^2 + 480t + 1 \qquad (t \ge 0)$$

a. Find an expression v for the rocket's velocity at any time t.

b. Compute the rocket's velocity when $t = 0, 20, 40$, and 60. Interpret your results.

c. Using the results from the solution to part (b), find the maximum altitude attained by the rocket.
Hint: At its highest point, the velocity of the rocket is zero.

73. **OBESITY IN AMERICA** The body mass index (BMI) measures body weight in relation to height. A BMI of 25 to 29.9 is considered overweight, a BMI of 30 or more is considered obese, and a BMI of 40 or more is morbidly obese. The percent of the U.S. population that is obese is approximated by the function

$$P(t) = 0.0004t^3 + 0.0036t^2 + 0.8t + 12 \qquad (0 \le t \le 13)$$

where t is measured in years, with $t = 0$ corresponding to the beginning of 1991.

a. What percent of the U.S. population was deemed obese at the beginning of 1991? At the beginning of 2004?

b. How fast was the percent of the U.S. population that is deemed obese changing at the beginning of 1991? At the beginning of 2004?

(*Note:* A formula for calculating the BMI of a person is given in Exercise 27, page 1058.)

Source: Centers for Disease Control and Prevention

74. **HEALTH-CARE SPENDING** Despite efforts at cost containment, the cost of the Medicare program is increasing. Two major reasons for this increase are an aging population and extensive use by physicians of new technologies. Based on data from the Health Care Financing Administration and the U.S. Census Bureau, health-care spending through the year 2000 may be approximated by the function

$$S(t) = 0.02836t^3 - 0.05167t^2 + 9.60881t \\ + 41.9 \qquad (0 \le t \le 35)$$

where $S(t)$ is the spending in billions of dollars and t is measured in years, with $t = 0$ corresponding to the beginning of 1965.

a. Find an expression for the rate of change of health-care spending at any time t.

b. How fast was health-care spending changing at the beginning of 1980? At the beginning of 2000?

c. What was the amount of health-care spending at the beginning of 1980? At the beginning of 2000?

Source: Health Care Financing Administration and U.S. Census Bureau

In Exercises 75 and 76, determine whether the statement is true or false. If it is true, explain why it is true. If it is false, give an example to show why it is false.

75. If f and g are differentiable, then

$$\frac{d}{dx}[2f(x) - 5g(x)] = 2f'(x) - 5g'(x)$$

76. If $f(x) = \pi^x$, then $f'(x) = x\pi^{x-1}$.

77. Prove the power rule (Rule 2) for the special case $n = 3$.

Hint: Compute $\displaystyle \lim_{h \to 0}\left[\frac{(x + h)^3 - x^3}{h}\right]$.

12.1 Solutions to Self-Check Exercises

1. a. $f'(x) = \dfrac{d}{dx}(1.5x^2) + \dfrac{d}{dx}(2x^{1.5})$

$= (1.5)(2x) + (2)(1.5x^{0.5})$

$= 3x + 3\sqrt{x} = 3(x + \sqrt{x})$

b. $g'(x) = \dfrac{d}{dx}(2x^{1/2}) + \dfrac{d}{dx}(3x^{-1/2})$

$= (2)\left(\dfrac{1}{2}x^{-1/2}\right) + (3)\left(-\dfrac{1}{2}x^{-3/2}\right)$

$= x^{-1/2} - \dfrac{3}{2}x^{-3/2}$

$= \dfrac{1}{2}x^{-3/2}(2x - 3) = \dfrac{2x - 3}{2x^{3/2}}$

2. a. $f'(x) = \dfrac{d}{dx}(2x^3) - \dfrac{d}{dx}(3x^2) + \dfrac{d}{dx}(2x) - \dfrac{d}{dx}(1)$

$= (2)(3x^2) - (3)(2x) + 2$

$= 6x^2 - 6x + 2$

b. The slope of the tangent line to the graph of f when $x = 2$ is given by

$$f'(2) = 6(2)^2 - 6(2) + 2 = 14$$

c. The rate of change of f at $x = 2$ is given by $f'(2)$. Using the results of part (b), we see that the required rate of change is 14 units/unit change in x.

3. a. The rate at which the GDP was changing at any time t $(0 < t < 11)$ is given by

$$G'(t) = -6t^2 + 90t + 20$$

In particular, the rates of change of the GDP at the beginning of the years 2000 $(t = 5)$, 2002 $(t = 7)$, and 2005 $(t = 10)$ are given by

$$G'(5) = 320 \qquad G'(7) = 356 \qquad G'(10) = 320$$

respectively—that is, by \$320 million/year, \$356 million/year, and \$320 million/year, respectively.

b. The average rate of growth of the GDP over the period from the beginning of 2000 $(t = 5)$ to the beginning of 2005 $(t = 10)$ is given by

$$\frac{G(10) - G(5)}{10 - 5} = \frac{[-2(10)^3 + 45(10)^2 + 20(10) + 6000]}{5}$$

$$- \frac{[-2(5)^3 + 45(5)^2 + 20(5) + 6000]}{5}$$

$$= \frac{8700 - 6975}{5}$$

or \$345 million/year.

USING TECHNOLOGY

■ Finding the Rate of Change of a Function

We can use the numerical derivative operation of a graphing utility to obtain the value of the derivative at a given value of x. Since the derivative of a function $f(x)$ measures the rate of change of the function with respect to x, the numerical derivative operation can be used to answer questions pertaining to the rate of change of one quantity y with respect to another quantity x, where $y = f(x)$, for a specific value of x.

EXAMPLE 1   Let $y = 3t^3 + 2\sqrt{t}$.

a. Use the numerical derivative operation of a graphing utility to find how fast y is changing with respect to t when $t = 1$.

b. Verify the result of part (a), using the rules of differentiation of this section.

Solution

a. Write $f(t) = 3t^3 + 2\sqrt{t}$. Using the numerical derivative operation of a graphing utility, we find that the rate of change of y with respect to t when $t = 1$ is given by $f'(1) = 10$ (Figure T1).

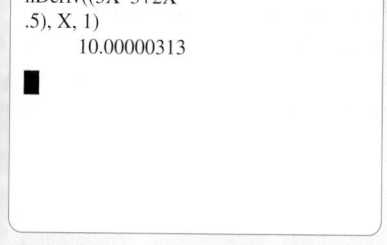

FIGURE T1
The TI-83 numerical derivative screen for computing $f'(1)$

(continued)

b. Here, $f(t) = 3t^3 + 2t^{1/2}$ and

$$f'(t) = 9t^2 + 2\left(\frac{1}{2}t^{-1/2}\right) = 9t^2 + \frac{1}{\sqrt{t}}$$

Using this result, we see that when $t = 1$, y is changing at the rate of

$$f'(1) = 9(1^2) + \frac{1}{\sqrt{1}} = 10$$

units per unit change in t, as obtained earlier.

APPLIED EXAMPLE 2   Fuel Economy of Cars   According to data obtained from the U.S. Department of Energy and the Shell Development Company, a typical car's fuel economy depends on the speed it is driven and is approximated by the function

$$f(x) = 0.00000310315x^4 - 0.000455174x^3$$
$$+ 0.00287869x^2 + 1.25986x \quad (0 \le x \le 75)$$

where x is measured in mph and $f(x)$ is measured in miles per gallon (mpg).

a. Use a graphing utility to graph the function f on the interval $[0, 75]$.
b. Find the rate of change of f when $x = 20$ and when $x = 50$.
c. Interpret your results.
Source: U.S. Department of Energy and the Shell Development Company

Solution

a. The graph is shown in Figure T2.
b. Using the numerical derivative operation of a graphing utility, we see that $f'(20) = .9280996$. The rate of change of f when $x = 50$ is given by $f'(50) = -.3145009995$. (See Figure T3a and T3b.)

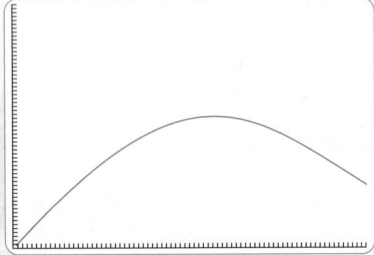

FIGURE T2
The graph of the function f on the interval $[0, 75]$

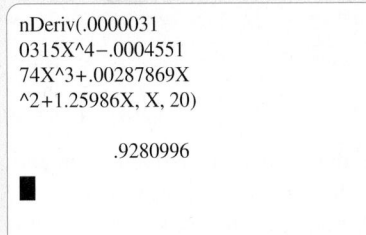

nDeriv(.0000031
0315X^4−.0004551
74X^3+.00287869X
^2+1.25986X, X, 20)

            .9280996

nDeriv(.0000031
0315X^4−.0004551
74X^3+.00287869X
^2+1.25986X, X, 5
0)
            −.3145009995

(a)   **(b)**

FIGURE T3
The TI-83 numerical derivative screen for computing (a) $f'(20)$ and (b) $f'(50)$

c. The results of part (b) tell us that when a typical car is being driven at 20 mph, its fuel economy increases at the rate of approximately 0.9 mpg per 1 mph increase in its speed. At a speed of 50 mph, its fuel economy decreases at the rate of approximately 0.3 mpg per 1 mph increase in its speed.

TECHNOLOGY EXERCISES

In Exercises 1–6, use the numerical derivative operation to find the rate of change of $f(x)$ at the given value of x. Give your answer accurate to four decimal places.

1. $f(x) = 4x^5 - 3x^3 + 2x^2 + 1; x = 0.5$

2. $f(x) = -x^5 + 4x^2 + 3; x = 0.4$

3. $f(x) = x - 2\sqrt{x}; x = 3$

4. $f(x) = \dfrac{\sqrt{x} - 1}{x}; x = 2$

5. $f(x) = x^{1/2} - x^{1/3}$; $x = 1.2$

6. $f(x) = 2x^{5/4} + x$; $x = 2$

7. CARBON MONOXIDE IN THE ATMOSPHERE The projected average global atmospheric concentration of carbon monoxide is approximated by the function

$$f(t) = 0.881443t^4 - 1.45533t^3 + 0.695876t^2 + 2.87801t + 293 \qquad (0 \le t \le 4)$$

where t is measured in 40-yr intervals with $t = 0$ corresponding to the beginning of 1860, and $f(t)$ is measured in parts per million by volume.

a. Plot the graph of f in the viewing window $[0, 4] \times [280, 400]$.

b. Use a graphing utility to estimate how fast the projected average global atmospheric concentration of carbon monoxide was changing at the beginning of the year 1900 ($t = 1$) and at the beginning of 2000 ($t = 3.5$).

Source: Meadows et al. "Beyond the Limits"

8. GROWTH OF HMOs Based on data compiled by the Group Health Association of America, the number of people receiving their care in an HMO (health maintenance organization) from the beginning of 1984 through 1994 is approximated by the function

$$f(t) = 0.0514t^3 - 0.853t^2 + 6.8147t + 15.6524 \qquad (0 \le t \le 11)$$

where $f(t)$ gives the number of people in millions and t is measured in years, with $t = 0$ corresponding to the beginning of 1984.

a. Plot the graph of f in the viewing window $[0, 11] \times [0, 80]$.

b. How fast was the number of people receiving their care in an HMO changing at the beginning of 1992?

Source: Group Health Association of America

9. HOME SALES The average number of days a single-family home remains for sale from listing to accepted offer (in the greater Boston area) is approximated by the function

$$f(t) = 0.0171911t^4 - 0.662121t^3 + 6.18083t^2 - 8.97086t + 53.3357 \qquad (0 \le t \le 12)$$

where t is measured in years, with $t = 0$ corresponding to the beginning of 1984.

a. Plot the graph of f in the viewing window $[0, 12] \times [0, 120]$.

b. How fast was the average number of days a single-family home remained for sale from listing to accepted offer changing at the beginning of 1984 ($t = 0$)? At the beginning of 1988 ($t = 4$)?

Source: Greater Boston Real Estate Board—Multiple Listing Service

10. SPREAD OF HIV The estimated number of children newly infected with HIV through mother-to-child contact worldwide is given by

$$f(t) = -0.2083t^3 + 3.0357t^2 + 44.0476t + 200.2857 \qquad (0 \le t \le 12)$$

where $f(t)$ is measured in thousands and t is measured in years, with $t = 0$ corresponding to the beginning of 1990.

a. Plot the graph of f in the viewing window $[0, 12] \times [0, 800]$.

b. How fast was the estimated number of children newly infected with HIV through mother-to-child contact worldwide increasing at the beginning of the year 2000?

Source: United Nations

11. MANUFACTURING CAPACITY Data show that the annual change in manufacturing capacity between 1994 and 2000 is given by

$$f(t) = 0.009417t^3 - 0.426571t^2 + 2.74894t + 5.54 \qquad (0 \le t \le 6)$$

percent, where t is measured in years, with $t = 0$ corresponding to the beginning of 1994.

a. Plot the graph of f in the viewing window $[0, 8] \times [0, 10]$.

b. How fast was $f(t)$ changing at the beginning of 1996 ($t = 2$)? At the beginning of 1998 ($t = 4$)?

Source: Federal Reserve

12. FISHERIES The total groundfish population on Georges Bank in New England between 1989 and 1999 is approximated by the function

$$f(t) = 5.303t^2 - 53.977t + 253.8 \qquad (0 \le t \le 10)$$

where $f(t)$ is measured in thousands of metric tons and t is measured in years, with $t = 0$ corresponding to the beginning of 1989.

a. Plot the graph of f in the viewing window $[0, 10] \times [100, 250]$.

b. How fast was the total groundfish population changing at the beginning of 1990? At the beginning of 1996?

(*Note:* Fishing restrictions were imposed on December 7, 1994.)

Source: New England Fishery Management Council

12.2 The Product and Quotient Rules

In this section we study two more rules of differentiation: the **product rule** and the **quotient rule**.

■ The Product Rule

The derivative of the product of two differentiable functions is given by the following rule:

> **Rule 5: The Product Rule**
> $$\frac{d}{dx}[f(x)g(x)] = f(x)g'(x) + g(x)f'(x)$$

The derivative of the product of two functions is the first function times the derivative of the second plus the second function times the derivative of the first.

The product rule may be extended to the case involving the product of any finite number of functions (see Exercise 63, p. 706). We prove the product rule at the end of this section.

 The derivative of the product of two functions is *not* given by the product of the derivatives of the functions; that is, in general

$$\frac{d}{dx}[f(x)g(x)] \neq f'(x)g'(x)$$

EXAMPLE 1   Find the derivative of the function

$$f(x) = (2x^2 - 1)(x^3 + 3)$$

Solution   By the product rule,

$$f'(x) = (2x^2 - 1)\frac{d}{dx}(x^3 + 3) + (x^3 + 3)\frac{d}{dx}(2x^2 - 1)$$
$$= (2x^2 - 1)(3x^2) + (x^3 + 3)(4x)$$
$$= 6x^4 - 3x^2 + 4x^4 + 12x$$
$$= 10x^4 - 3x^2 + 12x$$
$$= x(10x^3 - 3x + 12)$$

EXAMPLE 2   Differentiate (that is, find the derivative of) the function

$$f(x) = x^3(\sqrt{x} + 1)$$

Solution   First, we express the function in exponential form, obtaining

$$f(x) = x^3(x^{1/2} + 1)$$

By the product rule,

$$f'(x) = x^3 \frac{d}{dx}(x^{1/2} + 1) + (x^{1/2} + 1)\frac{d}{dx}x^3$$

$$= x^3 \left(\frac{1}{2}x^{-1/2}\right) + (x^{1/2} + 1)(3x^2)$$

$$= \frac{1}{2}x^{5/2} + 3x^{5/2} + 3x^2$$

$$= \frac{7}{2}x^{5/2} + 3x^2$$

Note   We can also solve the problem by first expanding the product before differentiating f. Examples for which this is not possible will be considered in Section 12.3, where the true value of the product rule will be appreciated.

The Quotient Rule

The derivative of the quotient of two differentiable functions is given by the following rule:

Rule 6: The Quotient Rule

$$\frac{d}{dx}\left[\frac{f(x)}{g(x)}\right] = \frac{g(x)f'(x) - f(x)g'(x)}{[g(x)]^2} \qquad (g(x) \neq 0)$$

As an aid to remembering this expression, observe that it has the following form:

$$\frac{d}{dx}\left[\frac{f(x)}{g(x)}\right] = \frac{(\text{Denominator})\begin{pmatrix}\text{Derivative of}\\\text{numerator}\end{pmatrix} - (\text{Numerator})\begin{pmatrix}\text{Derivative of}\\\text{denominator}\end{pmatrix}}{(\text{Square of denominator})}$$

For a proof of the quotient rule, see Exercise 64, page 706.

The derivative of a quotient is *not* equal to the quotient of the derivatives; that is,

$$\frac{d}{dx}\left[\frac{f(x)}{g(x)}\right] \neq \frac{f'(x)}{g'(x)}$$

For example, if $f(x) = x^3$ and $g(x) = x^2$, then

$$\frac{d}{dx}\left[\frac{f(x)}{g(x)}\right] = \frac{d}{dx}\left(\frac{x^3}{x^2}\right) = \frac{d}{dx}(x) = 1$$

which is *not* equal to

$$\frac{f'(x)}{g'(x)} = \frac{\dfrac{d}{dx}(x^3)}{\dfrac{d}{dx}(x^2)} = \frac{3x^2}{2x} = \frac{3}{2}x$$

EXAMPLE 3   Find $f'(x)$ if $f(x) = \dfrac{x}{2x - 4}$.

Solution   Using the quotient rule, we obtain

$$f'(x) = \frac{(2x - 4)\dfrac{d}{dx}(x) - x\dfrac{d}{dx}(2x - 4)}{(2x - 4)^2}$$

$$= \frac{(2x - 4)(1) - x(2)}{(2x - 4)^2}$$

$$= \frac{2x - 4 - 2x}{(2x - 4)^2} = -\frac{4}{(2x - 4)^2}$$

EXAMPLE 4   Find $f'(x)$ if $f(x) = \dfrac{x^2 + 1}{x^2 - 1}$.

Solution   By the quotient rule,

$$f'(x) = \frac{(x^2 - 1)\dfrac{d}{dx}(x^2 + 1) - (x^2 + 1)\dfrac{d}{dx}(x^2 - 1)}{(x^2 - 1)^2}$$

$$= \frac{(x^2 - 1)(2x) - (x^2 + 1)(2x)}{(x^2 - 1)^2}$$

$$= \frac{2x^3 - 2x - 2x^3 - 2x}{(x^2 - 1)^2}$$

$$= -\frac{4x}{(x^2 - 1)^2}$$

EXAMPLE 5   Find $h'(x)$ if $h(x) = \dfrac{\sqrt{x}}{x^2 + 1}$.

Solution   Rewrite $h(x)$ in the form $h(x) = \dfrac{x^{1/2}}{x^2 + 1}$. By the quotient rule, we find

$$h'(x) = \frac{(x^2 + 1)\dfrac{d}{dx}(x^{1/2}) - x^{1/2}\dfrac{d}{dx}(x^2 + 1)}{(x^2 + 1)^2}$$

$$= \frac{(x^2 + 1)(\frac{1}{2}x^{-1/2}) - x^{1/2}(2x)}{(x^2 + 1)^2}$$

$$= \frac{\frac{1}{2}x^{-1/2}(x^2 + 1 - 4x^2)}{(x^2 + 1)^2} \quad \text{\small Factor out } \tfrac{1}{2}x^{-1/2} \\ \text{\small from the numerator.}$$

$$= \frac{1 - 3x^2}{2\sqrt{x}(x^2 + 1)^2}$$

APPLIED EXAMPLE 6   Rate of Change of DVD Sales   The sales (in millions of dollars) of a DVD recording of a hit movie t years from the date of release is given by

$$S(t) = \frac{5t}{t^2 + 1}$$

a. Find the rate at which the sales are changing at time t.
b. How fast are the sales changing at the time the DVDs are released ($t = 0$)? Two years from the date of release?

Solution

a. The rate at which the sales are changing at time t is given by $S'(t)$. Using the quotient rule, we obtain

$$S'(t) = \frac{d}{dt}\left[\frac{5t}{t^2 + 1}\right] = 5\frac{d}{dt}\left[\frac{t}{t^2 + 1}\right]$$

$$= 5\left[\frac{(t^2 + 1)(1) - t(2t)}{(t^2 + 1)^2}\right]$$

$$= 5\left[\frac{t^2 + 1 - 2t^2}{(t^2 + 1)^2}\right] = \frac{5(1 - t^2)}{(t^2 + 1)^2}$$

b. The rate at which the sales are changing at the time the DVDs are released is given by

$$S'(0) = \frac{5(1 - 0)}{(0 + 1)^2} = 5$$

That is, they are increasing at the rate of $5 million per year.

Two years from the date of release, the sales are changing at the rate of

$$S'(2) = \frac{5(1 - 4)}{(4 + 1)^2} = -\frac{3}{5} = -0.6$$

That is, they are decreasing at the rate of $600,000 per year.

The graph of the function S is shown in Figure 5.

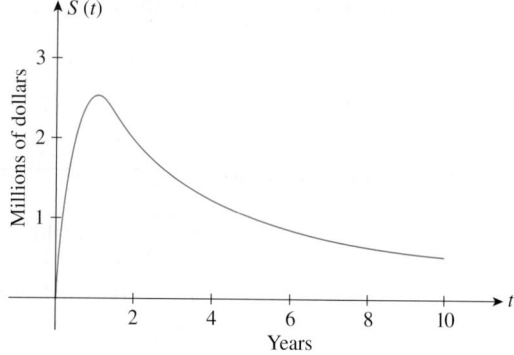

FIGURE 5
After a spectacular rise, the sales begin to taper off.

EXPLORING WITH TECHNOLOGY

Refer to Example 6.

1. Use a graphing utility to plot the graph of the function S, using the viewing window $[0, 10] \times [0, 3]$.

2. Use TRACE and ZOOM to determine the coordinates of the highest point on the graph of S in the interval $[0, 10]$. Interpret your results.

EXPLORE & DISCUSS

Suppose the revenue of a company is given by $R(x) = xp(x)$, where x is the number of units of the product sold at a unit price of $p(x)$ dollars.

1. Compute $R'(x)$ and explain, in words, the relationship between $R'(x)$ and $p(x)$ and/or its derivative.

2. What can you say about $R'(x)$ if $p(x)$ is constant? Is this expected?

APPLIED EXAMPLE 7   Oxygen-Restoration Rate in a Pond   When organic waste is dumped into a pond, the oxidation process that takes place reduces the pond's oxygen content. However, given time, nature will restore the oxygen content to its natural level. Suppose the oxygen content t days after organic waste has been dumped into the pond is given by

$$f(t) = 100\left[\frac{t^2 + 10t + 100}{t^2 + 20t + 100}\right] \qquad (0 < t < \infty)$$

percent of its normal level.

a. Derive a general expression that gives the rate of change of the pond's oxygen level at any time t.

b. How fast is the pond's oxygen content changing 1 day, 10 days, and 20 days after the organic waste has been dumped?

Solution

a. The rate of change of the pond's oxygen level at any time t is given by the derivative of the function f. Thus, the required expression is

$$f'(t) = 100\frac{d}{dt}\left[\frac{t^2 + 10t + 100}{t^2 + 20t + 100}\right]$$

$$= 100\left[\frac{(t^2 + 20t + 100)\dfrac{d}{dt}(t^2 + 10t + 100) - (t^2 + 10t + 100)\dfrac{d}{dt}(t^2 + 20t + 100)}{(t^2 + 20t + 100)^2}\right]$$

$$= 100\left[\frac{(t^2 + 20t + 100)(2t + 10) - (t^2 + 10t + 100)(2t + 20)}{(t^2 + 20t + 100)^2}\right]$$

$$= 100\left[\frac{2t^3 + 10t^2 + 40t^2 + 200t + 200t + 1000 - 2t^3 - 20t^2 - 20t^2 - 200t - 200t - 2000}{(t^2 + 20t + 100)^2}\right]$$

$$= 100\left[\frac{10t^2 - 1000}{(t^2 + 20t + 100)^2}\right]$$

b. The rate at which the pond's oxygen content is changing 1 day after the organic waste has been dumped is given by

$$f'(1) = 100\left[\frac{10 - 1000}{(1 + 20 + 100)^2}\right] \approx -6.76$$

That is, it is dropping at the rate of 6.8% per day. After 10 days, the rate is

$$f'(10) = 100\left[\frac{10(10)^2 - 1000}{(10^2 + 20(10) + 100)^2}\right] = 0$$

That is, it is neither increasing nor decreasing. After 20 days, the rate is

$$f'(20) = 100\left[\frac{10(20)^2 - 1000}{(20^2 + 20(20) + 100)^2}\right] \approx 0.37$$

That is, the oxygen content is increasing at the rate of 0.37% per day, and the restoration process has indeed begun. ∎

■ Verification of the Product Rule

We will now verify the product rule. If $p(x) = f(x)g(x)$, then

$$p'(x) = \lim_{h\to 0}\frac{p(x + h) - p(x)}{h}$$
$$= \lim_{h\to 0}\frac{f(x + h)g(x + h) - f(x)g(x)}{h}$$

By adding $-f(x + h)g(x) + f(x + h)g(x)$ (which is zero!) to the numerator and factoring, we have

$$p'(x) = \lim_{h\to 0}\frac{f(x + h)[g(x + h) - g(x)] + g(x)[f(x + h) - f(x)]}{h}$$
$$= \lim_{h\to 0}\left\{f(x + h)\left[\frac{g(x + h) - g(x)}{h}\right] + g(x)\left[\frac{f(x + h) - f(x)}{h}\right]\right\}$$
$$= \lim_{h\to 0}f(x + h)\left[\frac{g(x + h) - g(x)}{h}\right]$$
$$+ \lim_{h\to 0}g(x)\left[\frac{f(x + h) - f(x)}{h}\right] \qquad \text{By Property 3 of limits}$$
$$= \lim_{h\to 0}f(x + h)\cdot\lim_{h\to 0}\frac{g(x + h) - g(x)}{h}$$
$$+ \lim_{h\to 0}g(x)\cdot\lim_{h\to 0}\frac{f(x + h) - f(x)}{h} \qquad \text{By Property 4 of limits}$$
$$= f(x)g'(x) + g(x)f'(x)$$

Observe that in the second from the last link in the chain of equalities, we have used the fact that $\lim_{h\to 0} f(x + h) = f(x)$ because f is continuous at x.

12.2 Self-Check Exercises

1. Find the derivative of $f(x) = \dfrac{2x + 1}{x^2 - 1}$.

2. What is the slope of the tangent line to the graph of

$$f(x) = (x^2 + 1)(2x^3 - 3x^2 + 1)$$

at the point (2, 25)? How fast is the function f changing when $x = 2$?

3. The total sales of Security Products in its first 2 yr of operation are given by

$$S = f(t) = \frac{0.3t^3}{1 + 0.4t^2} \qquad (0 \le t \le 2)$$

where S is measured in millions of dollars and $t = 0$ corresponds to the date Security Products began operations. How

fast were the sales increasing at the beginning of the company's second year of operation?

Solutions to Self-Check Exercises 12.2 can be found on page 706.

12.2   Concept Questions

1. State the rule of differentiation in your own words.
   a. Product rule   **b.** Quotient rule

2. If $f(1) = 3$, $g(1) = 2$, $f'(1) = -1$, and $g'(1) = 4$, find:
   a. $h'(1)$ if $h(x) = f(x)g(x)$ **b.** $F'(1)$ if $F(x) = \dfrac{f(x)}{g(x)}$

12.2   Exercises

In Exercises 1–30, find the derivative of each function.

1. $f(x) = 2x(x^2 + 1)$

2. $f(x) = 3x^2(x - 1)$

3. $f(t) = (t - 1)(2t + 1)$

4. $f(x) = (2x + 3)(3x - 4)$

5. $f(x) = (3x + 1)(x^2 - 2)$

6. $f(x) = (x + 1)(2x^2 - 3x + 1)$

7. $f(x) = (x^3 - 1)(x + 1)$

8. $f(x) = (x^3 - 12x)(3x^2 + 2x)$

9. $f(w) = (w^3 - w^2 + w - 1)(w^2 + 2)$

10. $f(x) = \dfrac{1}{5}x^5 + (x^2 + 1)(x^2 - x - 1) + 28$

11. $f(x) = (5x^2 + 1)(2\sqrt{x} - 1)$

12. $f(t) = (1 + \sqrt{t})(2t^2 - 3)$

13. $f(x) = (x^2 - 5x + 2)\left(x - \dfrac{2}{x}\right)$

14. $f(x) = (x^3 + 2x + 1)\left(2 + \dfrac{1}{x^2}\right)$

15. $f(x) = \dfrac{1}{x - 2}$

16. $g(x) = \dfrac{3}{2x + 4}$

17. $f(x) = \dfrac{x - 1}{2x + 1}$

18. $f(t) = \dfrac{1 - 2t}{1 + 3t}$

19. $f(x) = \dfrac{1}{x^2 + 1}$

20. $f(u) = \dfrac{u}{u^2 + 1}$

21. $f(s) = \dfrac{s^2 - 4}{s + 1}$

22. $f(x) = \dfrac{x^3 - 2}{x^2 + 1}$

23. $f(x) = \dfrac{\sqrt{x}}{x^2 + 1}$

24. $f(x) = \dfrac{x^2 + 1}{\sqrt{x}}$

25. $f(x) = \dfrac{x^2 + 2}{x^2 + x + 1}$

26. $f(x) = \dfrac{x + 1}{2x^2 + 2x + 3}$

27. $f(x) = \dfrac{(x + 1)(x^2 + 1)}{x - 2}$

28. $f(x) = (3x^2 - 1)\left(x^2 - \dfrac{1}{x}\right)$

29. $f(x) = \dfrac{x}{x^2 - 4} - \dfrac{x - 1}{x^2 + 4}$

30. $f(x) = \dfrac{x + \sqrt{3x}}{3x - 1}$

In Exercises 31–34, suppose f and g are functions that are differentiable at $x = 1$ and that $f(1) = 2$, $f'(1) = -1$, $g(1) = -2$, and $g'(1) = 3$. Find the value of $h'(1)$.

31. $h(x) = f(x)g(x)$

32. $h(x) = (x^2 + 1)g(x)$

33. $h(x) = \dfrac{xf(x)}{x + g(x)}$

34. $h(x) = \dfrac{f(x)g(x)}{f(x) - g(x)}$

In Exercises 35–38, find the derivative of each function and evaluate $f'(x)$ at the given value of x.

35. $f(x) = (2x - 1)(x^2 + 3)$; $x = 1$

36. $f(x) = \dfrac{2x + 1}{2x - 1}$; $x = 2$

37. $f(x) = \dfrac{x}{x^4 - 2x^2 - 1}$; $x = -1$

38. $f(x) = (\sqrt{x} + 2x)(x^{3/2} - x)$; $x = 4$

In Exercises 39–42, find the slope and an equation of the tangent line to the graph of the function f at the specified point.

39. $f(x) = (x^3 + 1)(x^2 - 2)$; $(2, 18)$

40. $f(x) = \dfrac{x^2}{x + 1}$; $\left(2, \dfrac{4}{3}\right)$

41. $f(x) = \dfrac{x + 1}{x^2 + 1}$; $(1, 1)$

42. $f(x) = \dfrac{1 + 2x^{1/2}}{1 + x^{3/2}}; \left(4, \dfrac{5}{9}\right)$

43. Find an equation of the tangent line to the graph of the function $f(x) = (x^3 + 1)(3x^2 - 4x + 2)$ at the point $(1, 2)$.

44. Find an equation of the tangent line to the graph of the function $f(x) = \dfrac{3x}{x^2 - 2}$ at the point $(2, 3)$.

45. Let $f(x) = (x^2 + 1)(2 - x)$. Find the point(s) on the graph of f where the tangent line is horizontal.

46. Let $f(x) = \dfrac{x}{x^2 + 1}$. Find the point(s) on the graph of f where the tangent line is horizontal.

47. Find the point(s) on the graph of the function $f(x) = (x^2 + 6)(x - 5)$ where the slope of the tangent line is equal to -2.

48. Find the point(s) on the graph of the function $f(x) = \dfrac{x + 1}{x - 1}$ where the slope of the tangent line is equal to $-\dfrac{1}{2}$.

49. A straight line perpendicular to and passing through the point of tangency of the tangent line is called the *normal* to the curve. Find the equation of the tangent line and the normal to the curve $y = \dfrac{1}{1 + x^2}$ at the point $(1, \frac{1}{2})$.

50. Concentration of a Drug in the Bloodstream   The concentration of a certain drug in a patient's bloodstream t hr after injection is given by
$$C(t) = \frac{0.2t}{t^2 + 1}$$
   a. Find the rate at which the concentration of the drug is changing with respect to time.
   b. How fast is the concentration changing $\frac{1}{2}$ hr, 1 hr, and 2 hr after the injection?

51. Cost of Removing Toxic Waste   A city's main well was recently found to be contaminated with trichloroethylene, a cancer-causing chemical, as a result of an abandoned chemical dump leaching chemicals into the water. A proposal submitted to the city's council members indicates that the cost, measured in millions of dollars, of removing $x\%$ of the toxic pollutant is given by
$$C(x) = \frac{0.5x}{100 - x}$$
Find $C'(80)$, $C'(90)$, $C'(95)$, and $C'(99)$. What does your result tell you about the cost of removing *all* of the pollutant?

52. Drug Dosages   Thomas Young has suggested the following rule for calculating the dosage of medicine for children 1 to 12 yr old. If a denotes the adult dosage (in milligrams) and if t is the child's age (in years), then the child's dosage is given by
$$D(t) = \frac{at}{t + 12}$$

Suppose the adult dosage of a substance is 500 mg. Find an expression that gives the rate of change of a child's dosage with respect to the child's age. What is the rate of change of a child's dosage with respect to his or her age for a 6-yr-old child? A 10-yr-old child?

53. Effect of Bactericide   The number of bacteria $N(t)$ in a certain culture t min after an experimental bactericide is introduced obeys the rule
$$N(t) = \frac{10{,}000}{1 + t^2} + 2000$$
Find the rate of change of the number of bacteria in the culture 1 min and 2 min after the bactericide is introduced. What is the population of the bacteria in the culture 1 min and 2 min after the bactericide is introduced?

54. Demand Functions   The demand function for the Sicard wristwatch is given by
$$d(x) = \frac{50}{0.01x^2 + 1} \qquad (0 \le x \le 20)$$
where x (measured in units of a thousand) is the quantity demanded per week and $d(x)$ is the unit price in dollars.
   a. Find $d'(x)$.
   b. Find $d'(5)$, $d'(10)$, and $d'(15)$ and interpret your results.

55. Learning Curves   From experience, Emory Secretarial School knows that the average student taking Advanced Typing will progress according to the rule
$$N(t) = \frac{60t + 180}{t + 6} \qquad (t \ge 0)$$
where $N(t)$ measures the number of words/minute the student can type after t wk in the course.
   a. Find an expression for $N'(t)$.
   b. Compute $N'(t)$ for $t = 1$, 3, 4, and 7 and interpret your results.
   c. Sketch the graph of the function N. Does it confirm the results obtained in part (b)?
   d. What will be the average student's typing speed at the end of the 12-wk course?

56. Box-Office Receipts   The total worldwide box-office receipts for a long-running movie are approximated by the function
$$T(x) = \frac{120x^2}{x^2 + 4}$$
where $T(x)$ is measured in millions of dollars and x is the number of years since the movie's release. How fast are the total receipts changing 1 yr, 3 yr, and 5 yr after its release?

57. Formaldehyde Levels   A study on formaldehyde levels in 900 homes indicates that emissions of various chemicals can decrease over time. The formaldehyde level (parts per million) in an average home in the study is given by
$$f(t) = \frac{0.055t + 0.26}{t + 2} \qquad (0 \le t \le 12)$$

where t is the age of the house in years. How fast is the formaldehyde level of the average house dropping when it is new? At the beginning of its fourth year?

Source: Bonneville Power Administration

58. **POPULATION GROWTH** A major corporation is building a 4325-acre complex of homes, offices, stores, schools, and churches in the rural community of Glen Cove. As a result of this development, the planners have estimated that Glen Cove's population (in thousands) t yr from now will be given by

$$P(t) = \frac{25t^2 + 125t + 200}{t^2 + 5t + 40}$$

a. Find the rate at which Glen Cove's population is changing with respect to time.
b. What will be the population after 10 yr? At what rate will the population be increasing when $t = 10$?

In Exercises 59–62, determine whether the statement is true or false. If it is true, explain why it is true. If it is false, give an example to show why it is false.

59. If f and g are differentiable, then

$$\frac{d}{dx}[f(x)g(x)] = f'(x)g'(x)$$

60. If f is differentiable, then

$$\frac{d}{dx}[xf(x)] = f(x) + xf'(x)$$

61. If f is differentiable, then

$$\frac{d}{dx}\left[\frac{f(x)}{x^2}\right] = \frac{f'(x)}{2x}$$

62. If f, g, and h are differentiable, then

$$\frac{d}{dx}\left[\frac{f(x)g(x)}{h(x)}\right] = \frac{f'(x)g(x)h(x) + f(x)g'(x)h(x) - f(x)g(x)h'(x)}{[h(x)]^2}$$

63. Extend the product rule for differentiation to the following case involving the product of three differentiable functions: Let $h(x) = u(x)v(x)w(x)$ and show that $h'(x) = u(x)v(x)w'(x) + u(x)v'(x)w(x) + u'(x)v(x)w(x)$.
Hint: Let $f(x) = u(x)v(x)$, $g(x) = w(x)$, and $h(x) = f(x)g(x)$ and apply the product rule to the function h.

64. Prove the quotient rule for differentiation (Rule 6).
Hint: Let $k(x) = f(x)/g(x)$ and verify the following steps:

a. $$\frac{k(x + h) - k(x)}{h} = \frac{f(x + h)g(x) - f(x)g(x + h)}{hg(x + h)g(x)}$$

b. By adding $[-f(x)g(x) + f(x)g(x)]$ to the numerator and simplifying, show that

$$\frac{k(x + h) - k(x)}{h} = \frac{1}{g(x + h)g(x)}$$

$$\times \left\{ \left[\frac{f(x + h) - f(x)}{h}\right] \cdot g(x) \right.$$

$$\left. - \left[\frac{g(x + h) - g(x)}{h}\right] \cdot f(x) \right\}$$

c. $$k'(x) = \lim_{h \to 0} \frac{k(x + h) - k(x)}{h}$$

$$= \frac{g(x)f'(x) - f(x)g'(x)}{[g(x)]^2}$$

12.2 Solutions to Self-Check Exercises

1. We use the quotient rule to obtain

$$f'(x) = \frac{(x^2 - 1)\dfrac{d}{dx}(2x + 1) - (2x + 1)\dfrac{d}{dx}(x^2 - 1)}{(x^2 - 1)^2}$$

$$= \frac{(x^2 - 1)(2) - (2x + 1)(2x)}{(x^2 - 1)^2}$$

$$= \frac{2x^2 - 2 - 4x^2 - 2x}{(x^2 - 1)^2}$$

$$= \frac{-2x^2 - 2x - 2}{(x^2 - 1)^2}$$

$$= \frac{-2(x^2 + x + 1)}{(x^2 - 1)^2}$$

2. The slope of the tangent line to the graph of f at any point is given by

$$f'(x) = (x^2 + 1)\frac{d}{dx}(2x^3 - 3x^2 + 1)$$

$$+ (2x^3 - 3x^2 + 1)\frac{d}{dx}(x^2 + 1)$$

$$= (x^2 + 1)(6x^2 - 6x) + (2x^3 - 3x^2 + 1)(2x)$$

In particular, the slope of the tangent line to the graph of f when $x = 2$ is

$$f'(2) = (2^2 + 1)[6(2^2) - 6(2)]$$

$$+ [2(2^3) - 3(2^2) + 1][2(2)]$$

$$= 60 + 20 = 80$$

Note that it is not necessary to simplify the expression for $f'(x)$ since we are required only to evaluate the expression at $x = 2$. We also conclude, from this result, that the function f is changing at the rate of 80 units/unit change in x when $x = 2$.

3. The rate at which the company's total sales are changing at any time t is given by

$$S'(t) = \frac{(1 + 0.4t^2)\frac{d}{dt}(0.3t^3) - (0.3t^3)\frac{d}{dt}(1 + 0.4t^2)}{(1 + 0.4t^2)^2}$$

$$= \frac{(1 + 0.4t^2)(0.9t^2) - (0.3t^3)(0.8t)}{(1 + 0.4t^2)^2}$$

Therefore, at the beginning of the second year of operation, Security Products' sales were increasing at the rate of

$$S'(1) = \frac{(1 + 0.4)(0.9) - (0.3)(0.8)}{(1 + 0.4)^2} = 0.520408$$

or \$520,408/year.

USING TECHNOLOGY

■ The Product and Quotient Rules

EXAMPLE 1 Let $f(x) = (2\sqrt{x} + 0.5x)\left(0.3x^3 + 2x - \frac{0.3}{x}\right)$. Find $f'(0.2)$.

Solution Using the numerical derivative operation of a graphing utility, we find

$$f'(0.2) = 6.4797499802$$

See Figure T1.

```
nDeriv((2X^.5+.5
X)(.3X^3+2X−.3/X),
X, .2)
          6.4797499802
■
```

FIGURE T1
The TI-83 numerical derivative screen for computing $f'(0.2)$

APPLIED EXAMPLE 2   Importance of Time in Treating Heart Attacks
According to the American Heart Association, the treatment benefit for heart attacks depends on the time until treatment and is described by the function

$$f(t) = \frac{-16.94t + 203.28}{t + 2.0328} \quad (0 \le t \le 12)$$

where t is measured in hours and $f(t)$ is expressed as a percent.

a. Use a graphing utility to graph the function f using the viewing window $[0, 13] \times [0, 120]$.
b. Use a graphing utility to find the derivative of f when $t = 0$ and $t = 2$.
c. Interpret the results obtained in part (b).

Source: American Heart Association

Solution

a. The graph of f is shown in Figure T2.

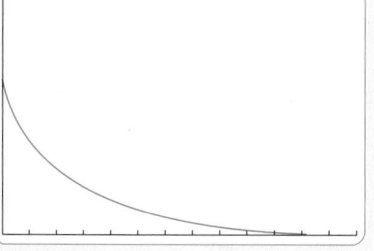

FIGURE T2

(continued)

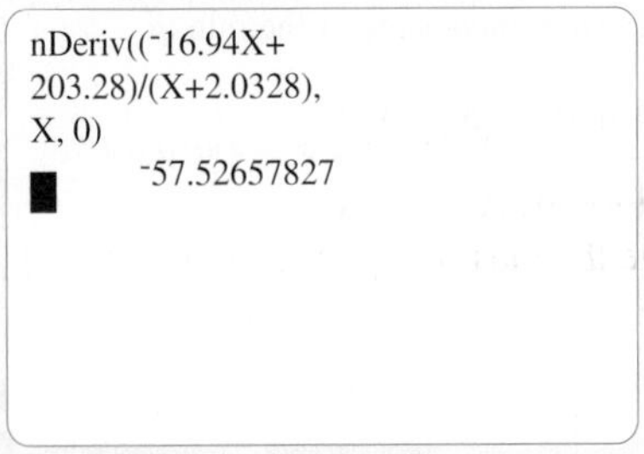

```
nDeriv((⁻16.94X+
203.28)/(X+2.0328),
X, 0)
■          ⁻57.52657827
```

FIGURE T3
The TI-83 numerical derivative screen
for computing $f'(0)$

b. Using the numerical derivative operation of a graphing utility, we find

$$f'(0) \approx -57.5266$$
$$f'(2) \approx -14.6165$$

(see Figure T3).

c. The results of part (b) show that the treatment benefit drops off at the rate of 58% per hour at the time when the heart attack first occurs and falls off at the rate of 15% per hour when the time to treatment is 2 hours. Thus, it is extremely urgent that a patient suffering a heart attack receive medical attention as soon as possible. ■

TECHNOLOGY EXERCISES

In Exercises 1–6, use the numerical derivative operation to find the rate of change of $f(x)$ at the given value of x. Give your answer accurate to four decimal places.

1. $f(x) = (2x^2 + 1)(x^3 + 3x + 4); x = -0.5$

2. $f(x) = (\sqrt{x} + 1)(2x^2 + x - 3); x = 1.5$

3. $f(x) = \dfrac{\sqrt{x} - 1}{\sqrt{x} + 1}; x = 3$

4. $f(x) = \dfrac{\sqrt{x}(x^2 + 4)}{x^3 + 1}; x = 4$

5. $f(x) = \dfrac{\sqrt{x}(1 + x^{-1})}{x + 1}; x = 1$

6. $f(x) = \dfrac{x^2(2 + \sqrt{x})}{1 + \sqrt{x}}; x = 1$

7. New Construction Jobs   The president of a major housing construction company claims that the number of construction jobs

created in the next t mo is given by

$$f(t) = 1.42\left(\frac{7t^2 + 140t + 700}{3t^2 + 80t + 550}\right)$$

where $f(t)$ is measured in millions of jobs/year. At what rate will construction jobs be created 1 yr from now, assuming her projection is correct?

8. Population Growth   A major corporation is building a 4325-acre complex of homes, offices, stores, schools, and churches in the rural community of Glen Cove. As a result of this development, the planners have estimated that Glen Cove's population (in thousands) t yr from now will be given by

$$P(t) = \frac{25t^2 + 125t + 200}{t^2 + 5t + 40}$$

a. What will be the population 10 yr from now?

b. At what rate will the population be increasing 10 yr from now?

12.3 The Chain Rule

This section introduces another rule of differentiation called the **chain rule**. When used in conjunction with the rules of differentiation developed in the last two sections, the chain rule enables us to greatly enlarge the class of functions that we are able to differentiate.

■ The Chain Rule

Consider the function $h(x) = (x^2 + x + 1)^2$. If we were to compute $h'(x)$ using only the rules of differentiation from the previous sections, then our approach might be to expand $h(x)$. Thus,

$$h(x) = (x^2 + x + 1)^2 = (x^2 + x + 1)(x^2 + x + 1)$$
$$= x^4 + 2x^3 + 3x^2 + 2x + 1$$

from which we find

$$h'(x) = 4x^3 + 6x^2 + 6x + 2$$

But what about the function $H(x) = (x^2 + x + 1)^{100}$? The same technique may be used to find the derivative of the function H, but the amount of work involved in this case would be prodigious! Consider, also, the function $G(x) = \sqrt{x^2 + 1}$. For each of the two functions H and G, the rules of differentiation of the previous sections cannot be applied directly to compute the derivatives H' and G'.

Observe that both H and G are **composite functions**; that is, each is composed of, or built up from, simpler functions. For example, the function H is composed of the two simpler functions $f(x) = x^2 + x + 1$ and $g(x) = x^{100}$ as follows:

$$H(x) = g[f(x)] = [f(x)]^{100}$$
$$= (x^2 + x + 1)^{100}$$

In a similar manner, we see that the function G is composed of the two simpler functions $f(x) = x^2 + 1$ and $g(x) = \sqrt{x}$. Thus,

$$G(x) = g[f(x)] = \sqrt{f(x)}$$
$$= \sqrt{x^2 + 1}$$

As a first step toward finding the derivative h' of a composite function $h = g \circ f$ defined by $h(x) = g[f(x)]$, we write

$$u = f(x) \quad \text{and} \quad y = g[f(x)] = g(u)$$

The dependency of h on g and f is illustrated in Figure 6. Since u is a function of x, we may compute the derivative of u with respect to x, if f is a differentiable function, obtaining $du/dx = f'(x)$. Next, if g is a differentiable function of u, we may compute the derivative of g with respect to u, obtaining $dy/du = g'(u)$. Now, since the function h is composed of the function g and the function f, we might suspect that the rule $h'(x)$ for the derivative h' of h will be given by an expression that involves the rules for the derivatives of f and g. But how do we combine these derivatives to yield h'?

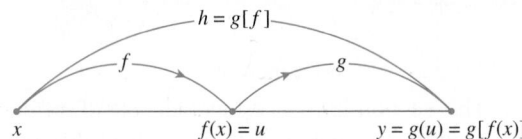

FIGURE 6
The composite function $h(x) = g[f(x)]$

This question can be answered by interpreting the derivative of each function as giving the rate of change of that function. For example, suppose $u = f(x)$ changes three times as fast as x—that is,

$$f'(x) = \frac{du}{dx} = 3$$

And suppose $y = g(u)$ changes twice as fast as u—that is,

$$g'(u) = \frac{dy}{du} = 2$$

Then, we would expect $y = h(x)$ to change six times as fast as x—that is,

$$h'(x) = g'(u)f'(x) = (2)(3) = 6$$

or equivalently,

$$\frac{dy}{dx} = \frac{dy}{du} \cdot \frac{du}{dx} = (2)(3) = 6$$

This observation suggests the following result, which we state without proof.

Rule 7: The Chain Rule
If $h(x) = g[f(x)]$, then

$$h'(x) = \frac{d}{dx} g(f(x)) = g'(f(x))f'(x) \qquad \textbf{(1)}$$

Equivalently, if we write $y = h(x) = g(u)$, where $u = f(x)$, then

$$\frac{dy}{dx} = \frac{dy}{du} \cdot \frac{du}{dx} \qquad \textbf{(2)}$$

Notes

1. If we label the composite function h in the following manner

Inside function
↓
$$h(x) = g[f(x)]$$
↑
Outside function

then $h'(x)$ is just the *derivative* of the "outside function" *evaluated at* the "inside function" times the *derivative* of the "inside function."

2. Equation (2) can be remembered by observing that if we "cancel" the du's, then

$$\frac{dy}{dx} = \frac{dy}{du} \cdot \frac{du}{dx} = \frac{dy}{dx}$$

∎

■ The Chain Rule for Powers of Functions

Many composite functions have the special form $h(x) = g(f(x))$, where g is defined by the rule $g(x) = x^n$ (n, a real number)—that is,

$$h(x) = [f(x)]^n$$

In other words, the function h is given by the power of a function f. The functions

$$h(x) = (x^2 + x + 1)^2 \qquad H = (x^2 + x + 1)^{100} \qquad G = \sqrt{x^2 + 1}$$

discussed earlier are examples of this type of composite function. By using the following corollary of the chain rule, the general power rule, we can find the derivative of this type of function much more easily than by using the chain rule directly.

The General Power Rule

If the function f is differentiable and $h(x) = [f(x)]^n$ (n, a real number), then

$$h'(x) = \frac{d}{dx}[f(x)]^n = n[f(x)]^{n-1}f'(x) \qquad \qquad \text{(3)}$$

To see this, we observe that $h(x) = g(f(x))$, where $g(x) = x^n$, so that, by virtue of the chain rule, we have

$$h'(x) = g'(f(x))f'(x)$$
$$= n[f(x)]^{n-1}f'(x)$$

since $g'(x) = nx^{n-1}$.

EXAMPLE 1   Let $F(x) = (3x + 1)^2$.

a. Find $F'(x)$, using the general power rule.
b. Verify your result without the benefit of the general power rule.

Solution

a. Using the general power rule, we obtain

$$F'(x) = 2(3x + 1)^1 \frac{d}{dx}(3x + 1)$$
$$= 2(3x + 1)(3)$$
$$= 6(3x + 1)$$

b. We first expand $F(x)$. Thus,

$$F(x) = (3x + 1)^2 = 9x^2 + 6x + 1$$

Next, differentiating, we have

$$F'(x) = \frac{d}{dx}(9x^2 + 6x + 1)$$
$$= 18x + 6$$
$$= 6(3x + 1)$$

as before.   ■

EXAMPLE 2　Differentiate the function $G(x) = \sqrt{x^2 + 1}$.

Solution　We rewrite the function $G(x)$ as

$$G(x) = (x^2 + 1)^{1/2}$$

and apply the general power rule, obtaining

$$G'(x) = \frac{1}{2}(x^2 + 1)^{-1/2}\frac{d}{dx}(x^2 + 1)$$

$$= \frac{1}{2}(x^2 + 1)^{-1/2} \cdot 2x = \frac{x}{\sqrt{x^2 + 1}}$$ ■

EXAMPLE 3　Differentiate the function $f(x) = x^2(2x + 3)^5$.

Solution　Applying the product rule followed by the general power rule, we obtain

$$f'(x) = x^2\frac{d}{dx}(2x + 3)^5 + (2x + 3)^5\frac{d}{dx}(x^2)$$

$$= (x^2)5(2x + 3)^4 \cdot \frac{d}{dx}(2x + 3) + (2x + 3)^5(2x)$$

$$= 5x^2(2x + 3)^4(2) + 2x(2x + 3)^5$$

$$= 2x(2x + 3)^4(5x + 2x + 3) = 2x(7x + 3)(2x + 3)^4$$ ■

EXAMPLE 4　Find $f'(x)$ if $f(x) = (2x^2 + 3)^4(3x - 1)^5$.

Solution　Applying the product rule, we have

$$f'(x) = (2x^2 + 3)^4\frac{d}{dx}(3x - 1)^5 + (3x - 1)^5\frac{d}{dx}(2x^2 + 3)^4$$

Next, we apply the general power rule to each term, obtaining

$$f'(x) = (2x^2 + 3)^4 \cdot 5(3x - 1)^4\frac{d}{dx}(3x - 1) + (3x - 1)^5 \cdot 4(2x^2 + 3)^3\frac{d}{dx}(2x^2 + 3)$$
$$= 5(2x^2 + 3)^4(3x - 1)^4 \cdot 3 + 4(3x - 1)^5(2x^2 + 3)^3(4x)$$

Finally, observing that $(2x^2 + 3)^3(3x - 1)^4$ is common to both terms, we can factor and simplify as follows:

$$f'(x) = (2x^2 + 3)^3(3x - 1)^4[15(2x^2 + 3) + 16x(3x - 1)]$$
$$= (2x^2 + 3)^3(3x - 1)^4(30x^2 + 45 + 48x^2 - 16x)$$
$$= (2x^2 + 3)^3(3x - 1)^4(78x^2 - 16x + 45)$$ ■

EXAMPLE 5　Find $f'(x)$ if $f(x) = \dfrac{1}{(4x^2 - 7)^2}$.

Solution　Rewriting $f(x)$ and then applying the general power rule, we obtain

$$f'(x) = \frac{d}{dx}\left[\frac{1}{(4x^2 - 7)^2}\right] = \frac{d}{dx}(4x^2 - 7)^{-2}$$

$$= -2(4x^2 - 7)^{-3}\frac{d}{dx}(4x^2 - 7)$$

$$= -2(4x^2 - 7)^{-3}(8x) = -\frac{16x}{(4x^2 - 7)^3}$$ ■

EXAMPLE 6 Find the slope of the tangent line to the graph of the function

$$f(x) = \left(\frac{2x+1}{3x+2}\right)^3$$

at the point $\left(0, \frac{1}{8}\right)$.

Solution The slope of the tangent line to the graph of f at any point is given by $f'(x)$. To compute $f'(x)$, we use the general power rule followed by the quotient rule, obtaining

$$f'(x) = 3\left(\frac{2x+1}{3x+2}\right)^2 \frac{d}{dx}\left(\frac{2x+1}{3x+2}\right)$$

$$= 3\left(\frac{2x+1}{3x+2}\right)^2 \left[\frac{(3x+2)(2) - (2x+1)(3)}{(3x+2)^2}\right]$$

$$= 3\left(\frac{2x+1}{3x+2}\right)^2 \left[\frac{6x+4-6x-3}{(3x+2)^2}\right]$$

$$= \frac{3(2x+1)^2}{(3x+2)^4}$$

In particular, the slope of the tangent line to the graph of f at $\left(0, \frac{1}{8}\right)$ is given by

$$f'(0) = \frac{3(0+1)^2}{(0+2)^4} = \frac{3}{16}$$

EXPLORING WITH TECHNOLOGY

Refer to Example 6.

1. Use a graphing utility to plot the graph of the function f, using the viewing window $[-2, 1] \times [-1, 2]$. Then draw the tangent line to the graph of f at the point $(0, \frac{1}{8})$.

2. For a better picture, repeat part 1 using the viewing window $[-1, 1] \times [-0.1, 0.3]$.

3. Use the numerical differentiation capability of the graphing utility to verify that the slope of the tangent line at $(0, \frac{1}{8})$ is $\frac{3}{16}$.

APPLIED EXAMPLE 7 Growth in a Health Club Membership The membership of The Fitness Center, which opened a few years ago, is approximated by the function

$$N(t) = 100(64 + 4t)^{2/3} \qquad (0 \le t \le 52)$$

where $N(t)$ gives the number of members at the beginning of week t.

a. Find $N'(t)$.
b. How fast was the center's membership increasing initially ($t = 0$)?
c. How fast was the membership increasing at the beginning of the 40th week?
d. What was the membership when the center first opened? At the beginning of the 40th week?

Solution

a. Using the general power rule, we obtain

$$N'(t) = \frac{d}{dt}\left[100(64 + 4t)^{2/3}\right]$$

$$= 100\frac{d}{dt}(64 + 4t)^{2/3}$$

$$= 100\left(\frac{2}{3}\right)(64 + 4t)^{-1/3}\frac{d}{dt}(64 + 4t)$$

$$= \frac{200}{3}(64 + 4t)^{-1/3}(4)$$

$$= \frac{800}{3(64 + 4t)^{1/3}}$$

b. The rate at which the membership was increasing when the center first opened is given by

$$N'(0) = \frac{800}{3(64)^{1/3}} \approx 66.7$$

or approximately 67 people per week.

c. The rate at which the membership was increasing at the beginning of the 40th week is given by

$$N'(40) = \frac{800}{3(64 + 160)^{1/3}} \approx 43.9$$

or approximately 44 people per week.

d. The membership when the center first opened is given by

$$N(0) = 100(64)^{2/3} = 100(16)$$

or approximately 1600 people. The membership at the beginning of the 40th week is given by

$$N(40) = 100(64 + 160)^{2/3} \approx 3688.3$$

or approximately 3688 people.

EXPLORE & DISCUSS

The profit P of a one-product software manufacturer depends on the number of units of its products sold. The manufacturer estimates that it will sell x units of its product per week. Suppose $P = g(x)$ and $x = f(t)$, where g and f are differentiable functions.

1. Write an expression giving the rate of change of the profit with respect to the number of units sold.

2. Write an expression giving the rate of change of the number of units sold per week.

3. Write an expression giving the rate of change of the profit per week.

APPLIED EXAMPLE 8 Arteriosclerosis Arteriosclerosis begins during childhood when plaque (soft masses of fatty material) forms in the arterial walls, blocking the flow of blood through the arteries and leading to heart attacks, strokes, and gangrene. Suppose the idealized cross section of the aorta is

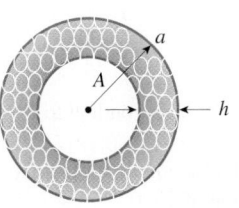

FIGURE 7
Cross section of the aorta

circular with radius a cm and by year t the thickness of the plaque (assume it is uniform) is $h = g(t)$ cm (Figure 7). Then the area of the opening is given by $A = \pi(a - h)^2$ square centimeters (cm²).

Suppose the radius of an individual's artery is 1 cm ($a = 1$) and the thickness of the plaque in year t is given by

$$h = g(t) = 1 - 0.01(10{,}000 - t^2)^{1/2} \text{ cm}$$

Since the area of the arterial opening is given by

$$A = f(h) = \pi(1 - h)^2$$

the rate at which A is changing with respect to time is given by

$$\frac{dA}{dt} = \frac{dA}{dh} \cdot \frac{dh}{dt} = f'(h) \cdot g'(t) \qquad \text{By the chain rule}$$

$$= 2\pi(1 - h)(-1)\left[-0.01\left(\frac{1}{2}\right)(10{,}000 - t^2)^{-1/2}(-2t)\right] \qquad \text{Use the chain rule twice.}$$

$$= -2\pi(1 - h)\left[\frac{0.01t}{(10{,}000 - t^2)^{1/2}}\right]$$

$$= -\frac{0.02\pi(1 - h)t}{\sqrt{10{,}000 - t^2}}$$

For example, when $t = 50$,

$$h = g(50) = 1 - 0.01(10{,}000 - 2500)^{1/2} \approx 0.134$$

so that

$$\frac{dA}{dt} = -\frac{0.02\pi(1 - 0.134)50}{\sqrt{10{,}000 - 2500}} \approx -0.03$$

That is, the area of the arterial opening is decreasing at the rate of 0.03 cm² per year. ∎

EXPLORE & DISCUSS

Suppose the population P of a certain bacteria culture is given by $P = f(T)$, where T is the temperature of the medium. Further, suppose the temperature T is a function of time t in seconds—that is, $T = g(t)$. Give an interpretation of each of the following quantities:

1. $\dfrac{dP}{dT}$ **2.** $\dfrac{dT}{dt}$ **3.** $\dfrac{dP}{dt}$ **4.** $(f \circ g)(t)$ **5.** $f'(g(t))g'(t)$

12.3   Self-Check Exercises

1. Find the derivative of

$$f(x) = -\frac{1}{\sqrt{2x^2 - 1}}$$

2. Suppose the life expectancy at birth (in years) of a female in a certain country is described by the function

$$g(t) = 50.02(1 + 1.09t)^{0.1} \qquad (0 \le t \le 150)$$

where t is measured in years, with $t = 0$ corresponding to the beginning of 1900.

a. What is the life expectancy at birth of a female born at the beginning of 1980? At the beginning of 2000?

b. How fast is the life expectancy at birth of a female born at any time t changing?

Solutions to Self-Check Exercises 12.3 can be found on page 720.

12.3   Concept Questions

1. In your own words, state the chain rule for differentiating the composite function $h(x) = g[f(x)]$.

2. State the general power rule for differentiating the function $h(x) = [f(x)]^n$, where n is a real number.

12.3   Exercises

In Exercises 1–48, find the derivative of each function.

1. $f(x) = (2x - 1)^4$

2. $f(x) = (1 - x)^3$

3. $f(x) = (x^2 + 2)^5$

4. $f(t) = 2(t^3 - 1)^5$

5. $f(x) = (2x - x^2)^3$

6. $f(x) = 3(x^3 - x)^4$

7. $f(x) = (2x + 1)^{-2}$

8. $f(t) = \frac{1}{2}(2t^2 + t)^{-3}$

9. $f(x) = (x^2 - 4)^{3/2}$

10. $f(t) = (3t^2 - 2t + 1)^{3/2}$

11. $f(x) = \sqrt{3x - 2}$

12. $f(t) = \sqrt{3t^2 - t}$

13. $f(x) = \sqrt[3]{1 - x^2}$

14. $f(x) = \sqrt{2x^2 - 2x + 3}$

15. $f(x) = \dfrac{1}{(2x + 3)^3}$

16. $f(x) = \dfrac{2}{(x^2 - 1)^4}$

17. $f(t) = \dfrac{1}{\sqrt{2t - 3}}$

18. $f(x) = \dfrac{1}{\sqrt{2x^2 - 1}}$

19. $y = \dfrac{1}{(4x^4 + x)^{3/2}}$

20. $f(t) = \dfrac{4}{\sqrt[3]{2t^2 + t}}$

21. $f(x) = (3x^2 + 2x + 1)^{-2}$

22. $f(t) = (5t^3 + 2t^2 - t + 4)^{-3}$

23. $f(x) = (x^2 + 1)^3 - (x^3 + 1)^2$

24. $f(t) = (2t - 1)^4 + (2t + 1)^4$

25. $f(t) = (t^{-1} - t^{-2})^3$

26. $f(v) = (v^{-3} + 4v^{-2})^3$

27. $f(x) = \sqrt{x + 1} + \sqrt{x - 1}$

28. $f(u) = (2u + 1)^{3/2} + (u^2 - 1)^{-3/2}$

29. $f(x) = 2x^2(3 - 4x)^4$

30. $h(t) = t^2(3t + 4)^3$

31. $f(x) = (x - 1)^2(2x + 1)^4$

32. $g(u) = (1 + u^2)^5(1 - 2u^2)^8$

33. $f(x) = \left(\dfrac{x + 3}{x - 2}\right)^3$

34. $f(x) = \left(\dfrac{x + 1}{x - 1}\right)^5$

35. $s(t) = \left(\dfrac{t}{2t + 1}\right)^{3/2}$

36. $g(s) = \left(s^2 + \dfrac{1}{s}\right)^{3/2}$

37. $g(u) = \sqrt{\dfrac{u + 1}{3u + 2}}$

38. $g(x) = \sqrt{\dfrac{2x + 1}{2x - 1}}$

39. $f(x) = \dfrac{x^2}{(x^2 - 1)^4}$

40. $g(u) = \dfrac{2u^2}{(u^2 + u)^3}$

41. $h(x) = \dfrac{(3x^2 + 1)^3}{(x^2 - 1)^4}$

42. $g(t) = \dfrac{(2t - 1)^2}{(3t + 2)^4}$

43. $f(x) = \dfrac{\sqrt{2x + 1}}{x^2 - 1}$

44. $f(t) = \dfrac{4t^2}{\sqrt{2t^2 + 2t - 1}}$

45. $g(t) = \dfrac{\sqrt{t + 1}}{\sqrt{t^2 + 1}}$

46. $f(x) = \dfrac{\sqrt{x^2 + 1}}{\sqrt{x^2 - 1}}$

47. $f(x) = (3x + 1)^4(x^2 - x + 1)^3$

48. $g(t) = (2t + 3)^2(3t^2 - 1)^{-3}$

In Exercises 49–54, find $\dfrac{dy}{du}, \dfrac{du}{dx},$ **and** $\dfrac{dy}{dx}.$

49. $y = u^{4/3}$ and $u = 3x^2 - 1$

50. $y = \sqrt{u}$ and $u = 7x - 2x^2$

51. $y = u^{-2/3}$ and $u = 2x^3 - x + 1$

52. $y = 2u^2 + 1$ and $u = x^2 + 1$

53. $y = \sqrt{u} + \dfrac{1}{\sqrt{u}}$ and $u = x^3 - x$

54. $y = \dfrac{1}{u}$ and $u = \sqrt{x} + 1$

55. Suppose $F(x) = g(f(x))$ and $f(2) = 3, f'(2) = -3, g(3) = 5$, and $g'(3) = 4$. Find $F'(2)$.

56. Suppose $h = f \circ g$. Find $h'(0)$ given that $f(0) = 6, f'(5) = -2$, $g(0) = 5$, and $g'(0) = 3$.

57. Suppose $F(x) = f(x^2 + 1)$. Find $F'(1)$ if $f'(2) = 3$.

58. Let $F(x) = f(f(x))$. Does it follow that $F'(x) = [f'(x)]^2$?
Hint: Let $f(x) = x^2$.

59. Suppose $h = g \circ f$. Does it follow that $h' = g' \circ f'$?
Hint: Let $f(x) = x$ and $g(x) = x^2$.

60. Suppose $h = f \circ g$. Show that $h' = (f' \circ g)g'$.

In Exercises 61–64, find an equation of the tangent line to the graph of the function at the given point.

61. $f(x) = (1 - x)(x^2 - 1)^2$; $(2, -9)$

62. $f(x) = \left(\dfrac{x + 1}{x - 1}\right)^2$; $(3, 4)$

63. $f(x) = x\sqrt{2x^2 + 7}$; $(3, 15)$

64. $f(x) = \dfrac{8}{\sqrt{x^2 + 6x}}$; $(2, 2)$

65. TELEVISION VIEWING The number of viewers of a television series introduced several years ago is approximated by the function

$$N(t) = (60 + 2t)^{2/3} \quad (1 \le t \le 26)$$

where $N(t)$ (measured in millions) denotes the number of weekly viewers of the series in the tth week. Find the rate of increase of the weekly audience at the end of week 2 and at the end of week 12. How many viewers were there in week 2? In week 24?

66. OUTSOURCING OF JOBS According to a study conducted in 2003, the total number of U.S. jobs that are projected to leave the country by year t, where $t = 0$ corresponds to 2000, is

$$N(t) = 0.0018425(t + 5)^{2.5} \quad (0 \le t \le 15)$$

where $N(t)$ is measured in millions. How fast will the number of U.S. jobs that are outsourced be changing in 2005? In 2010 $(t = 10)$?

Source: Forrester Research

67. WORKING MOTHERS The percent of mothers who work outside the home and have children younger than age 6 yr is approximated by the function

$$P(t) = 33.55(t + 5)^{0.205} \quad (0 \le t \le 21)$$

where t is measured in years, with $t = 0$ corresponding to the beginning of 1980. Compute $P'(t)$. At what rate was the percent of these mothers changing at the beginning of 2000? What was the percent of these mothers at the beginning of 2000?

Source: U.S. Bureau of Labor Statistics

68. SELLING PRICE OF DVD RECORDERS The rise of digital music and the improvement to the DVD format are some of the reasons why the average selling price of standalone DVD recorders will drop in the coming years. The function

$$A(t) = \frac{699}{(t + 1)^{0.94}} \quad (0 \le t \le 5)$$

gives the projected average selling price (in dollars) of standalone DVD recorders in year t, where $t = 0$ corresponds to the beginning of 2002. How fast was the average selling price of standalone DVD recorders falling at the beginning of 2002? How fast was it falling at the beginning of 2006?

Source: Consumer Electronics Association

69. SOCIALLY RESPONSIBLE FUNDS Since its inception in 1971, socially responsible investments, or SRIs, have yielded returns to investors on par with investments in general. The assets of socially responsible funds (in billions of dollars) from 1991 through 2001 is given by

$$f(t) = 23.7(0.2t + 1)^{1.32} \quad (0 \le t \le 11)$$

where $t = 0$ corresponds to the beginning of 1991.
a. Find the rate at which the assets of SRIs were changing at the beginning of 2000.
b. What were the assets of SRIs at the beginning of 2000?

Source: Thomson Financial Wiesenberger

70. AGING POPULATION The population of Americans age 55 and over as a percent of the total population is approximated by the function

$$f(t) = 10.72(0.9t + 10)^{0.3} \quad (0 \le t \le 20)$$

where t is measured in years, with $t = 0$ corresponding to the year 2000. At what rate was the percent of Americans age 55 and over changing at the beginning of 2000? At what rate will the percent of Americans age 55 and over be changing in 2010? What will be the percent of the population of Americans age 55 and over in 2010?

Source: U.S. Census Bureau

71. CONCENTRATION OF CARBON MONOXIDE (CO) IN THE AIR According to a joint study conducted by Oxnard's Environmental Management Department and a state government agency, the concentration of CO in the air due to automobile exhaust t yr from now is given by

$$C(t) = 0.01(0.2t^2 + 4t + 64)^{2/3}$$

parts per million.

a. Find the rate at which the level of CO is changing with respect to time.

b. Find the rate at which the level of CO will be changing 5 yr from now.

72. **CONTINUING EDUCATION ENROLLMENT** The registrar of Kellogg University estimates that the total student enrollment in the Continuing Education division will be given by

$$N(t) = -\frac{20,000}{\sqrt{1 + 0.2t}} + 21,000$$

where $N(t)$ denotes the number of students enrolled in the division t yr from now. Find an expression for $N'(t)$. How fast is the student enrollment increasing currently? How fast will it be increasing 5 yr from now?

73. **AIR POLLUTION** According to the South Coast Air Quality Management District, the level of nitrogen dioxide, a brown gas that impairs breathing, present in the atmosphere on a certain May day in downtown Los Angeles is approximated by

$$A(t) = 0.03t^3(t - 7)^4 + 60.2 \qquad (0 \le t \le 7)$$

where $A(t)$ is measured in pollutant standard index and t is measured in hours, with $t = 0$ corresponding to 7 a.m.
a. Find $A'(t)$.
b. Find $A'(1)$, $A'(3)$, and $A'(4)$ and interpret your results.

74. **EFFECT OF LUXURY TAX ON CONSUMPTION** Government economists of a developing country determined that the purchase of imported perfume is related to a proposed "luxury tax" by the formula

$$N(x) = \sqrt{10,000 - 40x - 0.02x^2} \qquad (0 \le x \le 200)$$

where $N(x)$ measures the percentage of normal consumption of perfume when a "luxury tax" of $x\%$ is imposed on it. Find the rate of change of $N(x)$ for taxes of 10%, 100%, and 150%.

75. **PULSE RATE OF AN ATHLETE** The pulse rate (the number of heartbeats/minute) of a long-distance runner t sec after leaving the starting line is given by

$$P(t) = \frac{300\sqrt{\frac{1}{2}t^2 + 2t + 25}}{t + 25} \qquad (t \ge 0)$$

Compute $P'(t)$. How fast is the athlete's pulse rate increasing 10 sec, 60 sec, and 2 min into the run? What is her pulse rate 2 min into the run?

76. **THURSTONE LEARNING MODEL** Psychologist L. L. Thurstone suggested the following relationship between learning time T and the length of a list n:

$$T = f(n) = An\sqrt{n - b}$$

where A and b are constants that depend on the person and the task.

a. Compute dT/dn and interpret your result.
b. For a certain person and a certain task, suppose $A = 4$ and $b = 4$. Compute $f'(13)$ and $f'(29)$ and interpret your results.

77. **OIL SPILLS** In calm waters, the oil spilling from the ruptured hull of a grounded tanker spreads in all directions. Assuming that the area polluted is a circle and that its radius is increasing at a rate of 2 ft/sec, determine how fast the area is increasing when the radius of the circle is 40 ft.

78. **ARTERIOSCLEROSIS** Refer to Example 8, page 714. Suppose the radius of an individual's artery is 1 cm and the thickness of the plaque (in centimeters) t yr from now is given by

$$h = g(t) = \frac{0.5t^2}{t^2 + 10} \qquad (0 \le t \le 10)$$

How fast will the arterial opening be decreasing 5 yr from now?

79. **TRAFFIC FLOW** Opened in the late 1950s, the Central Artery in downtown Boston was designed to move 75,000 vehicles a day. The number of vehicles moved per day is approximated by the function

$$x = f(t) = 6.25t^2 + 19.75t + 74.75 \qquad (0 \le t \le 5)$$

where x is measured in thousands and t in decades, with $t = 0$ corresponding to the beginning of 1959. Suppose the average speed of traffic flow in mph is given by

$$S = g(x) = -0.00075x^2 + 67.5 \qquad (75 \le x \le 350)$$

where x has the same meaning as before. What was the rate of change of the average speed of traffic flow at the beginning of 1999? What was the average speed of traffic flow at that time?
Hint: $S = g[f(t)]$.

80. **HOTEL OCCUPANCY RATES** The occupancy rate of the all-suite Wonderland Hotel, located near an amusement park, is given by the function

$$r(t) = \frac{10}{81}t^3 - \frac{10}{3}t^2 + \frac{200}{9}t + 60 \qquad (0 \le t \le 12)$$

where t is measured in months, with $t = 0$ corresponding to the beginning of January. Management has estimated that the monthly revenue (in thousands of dollars/month) is approximated by the function

$$R(r) = -\frac{3}{5000}r^3 + \frac{9}{50}r^2 \qquad (0 \le r \le 100)$$

where r is the occupancy rate.
a. Find an expression that gives the rate of change of Wonderland's occupancy rate with respect to time.
b. Find an expression that gives the rate of change of Wonderland's monthly revenue with respect to the occupancy rate.

c. What is the rate of change of Wonderland's monthly revenue with respect to time at the beginning of January? At the beginning of July?
Hint: Use the chain rule to find $R'(r(0))r'(0)$ and $R'(r(6))r'(6)$.

81. **EFFECT OF HOUSING STARTS ON JOBS** The president of a major housing construction firm claims that the number of construction jobs created is given by

$$N(x) = 1.42x$$

where x denotes the number of housing starts. Suppose the number of housing starts in the next t mo is expected to be

$$x(t) = \frac{7t^2 + 140t + 700}{3t^2 + 80t + 550}$$

million units/year. Find an expression that gives the rate at which the number of construction jobs will be created t mo from now. At what rate will construction jobs be created 1 yr from now?

82. **DEMAND FOR PCs** The quantity demanded per month, x, of a certain make of personal computer (PC) is related to the average unit price, p (in dollars), of PCs by the equation

$$x = f(p) = \frac{100}{9}\sqrt{810,000 - p^2}$$

It is estimated that t mo from now, the average price of a PC will be given by

$$p(t) = \frac{400}{1 + \frac{1}{8}\sqrt{t}} + 200 \qquad (0 \le t \le 60)$$

dollars. Find the rate at which the quantity demanded per month of the PCs will be changing 16 mo from now.

83. **DEMAND FOR WATCHES** The demand equation for the Sicard wristwatch is given by

$$x = f(p) = 10\sqrt{\frac{50 - p}{p}} \qquad (0 < p \le 50)$$

where x (measured in units of a thousand) is the quantity demanded each week and p is the unit price in dollars. Find the rate of change of the quantity demanded of the wristwatches with respect to the unit price when the unit price is $25.

84. **CRUISE SHIP BOOKINGS** The management of Cruise World, operators of Caribbean luxury cruises, expects that the percent of young adults booking passage on their cruises in the years ahead will rise dramatically. They have constructed the following model, which gives the percent of young adult passengers in year t:

$$p = f(t) = 50\left(\frac{t^2 + 2t + 4}{t^2 + 4t + 8}\right) \qquad (0 \le t \le 5)$$

Young adults normally pick shorter cruises and generally spend less on their passage. The following model gives an approximation of the average amount of money R (in dollars) spent per passenger on a cruise when the percent of young adults is p:

$$R(p) = 1000\left(\frac{p + 4}{p + 2}\right)$$

Find the rate at which the price of the average passage will be changing 2 yr from now.

In Exercises 85–88, determine whether the statement is true or false. If it is true, explain why it is true. If it is false, give an example to show why it is false.

85. If f and g are differentiable and $h = f \circ g$, then $h'(x) = f'[g(x)]g'(x)$.

86. If f is differentiable and c is a constant, then

$$\frac{d}{dx}[f(cx)] = cf'(cx).$$

87. If f is differentiable, then

$$\frac{d}{dx}\sqrt{f(x)} = \frac{f'(x)}{2\sqrt{f(x)}}$$

88. If f is differentiable, then

$$\frac{d}{dx}\left[f\left(\frac{1}{x}\right)\right] = f'\left(\frac{1}{x}\right)$$

89. In Section 12.1, we proved that

$$\frac{d}{dx}(x^n) = nx^{n-1}$$

for the special case when $n = 2$. Use the chain rule to show that

$$\frac{d}{dx}(x^{1/n}) = \frac{1}{n}x^{1/n-1}$$

for any nonzero integer n, assuming that $f(x) = x^{1/n}$ is differentiable.
Hint: Let $f(x) = x^{1/n}$ so that $[f(x)]^n = x$. Differentiate both sides with respect to x.

90. With the aid of Exercise 89, prove that

$$\frac{d}{dx}(x^r) = rx^{r-1}$$

for any rational number r.
Hint: Let $r = m/n$, where m and n are integers, with $n \ne 0$, and write $x^r = (x^m)^{1/n}$.

12.3   Solutions to Self-Check Exercises

1. Rewriting, we have

$$f(x) = -(2x^2 - 1)^{-1/2}$$

Using the general power rule, we find

$$f'(x) = -\frac{d}{dx}(2x^2 - 1)^{-1/2}$$

$$= -\left(-\frac{1}{2}\right)(2x^2 - 1)^{-3/2}\frac{d}{dx}(2x^2 - 1)$$

$$= \frac{1}{2}(2x^2 - 1)^{-3/2}(4x)$$

$$= \frac{2x}{(2x^2 - 1)^{3/2}}$$

2. a. The life expectancy at birth of a female born at the beginning of 1980 is given by

$$g(80) = 50.02[1 + 1.09(80)]^{0.1} \approx 78.29$$

or approximately 78 yr. Similarly, the life expectancy at birth of a female born at the beginning of the year 2000 is given by

$$g(100) = 50.02[1 + 1.09(100)]^{0.1} \approx 80.04$$

or approximately 80 yr.

b. The rate of change of the life expectancy at birth of a female born at any time t is given by $g'(t)$. Using the general power rule, we have

$$g'(t) = 50.02\frac{d}{dt}(1 + 1.09t)^{0.1}$$

$$= (50.02)(0.1)(1 + 1.09t)^{-0.9}\frac{d}{dt}(1 + 1.09t)$$

$$= (50.02)(0.1)(1.09)(1 + 1.09t)^{-0.9}$$

$$= 5.45218(1 + 1.09t)^{-0.9}$$

$$= \frac{5.45218}{(1 + 1.09t)^{0.9}}$$

USING TECHNOLOGY

■ Finding the Derivative of a Composite Function

EXAMPLE 1 Find the rate of change of $f(x) = \sqrt{x}(1 + 0.02x^2)^{3/2}$ when $x = 2.1$.

Solution Using the numerical derivative operation of a graphing utility, we find

$$f'(2.1) = 0.5821463392$$

or approximately 0.58 unit per unit change in x. (See Figure T1.)

```
nDeriv(X^.5(1+.0
2X^2)^1.5, X, 2.1)
      .5821463392
■
```

FIGURE T1
The TI-83 numerical derivative screen for computing $f'(2.1)$

APPLIED EXAMPLE 2   Amusement Park Attendance   The management of AstroWorld ("The Amusement Park of the Future") estimates that the total number of visitors (in thousands) to the amusement park t hours after opening time at

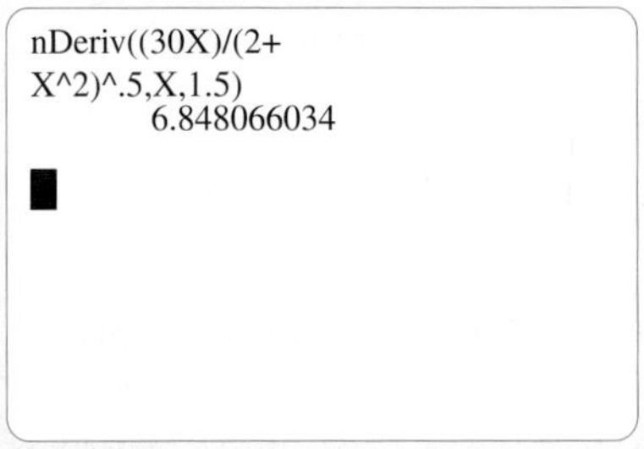

```
nDeriv((30X)/(2+
X^2)^.5,X,1.5)
        6.848066034
■
```

FIGURE T2
The TI-83 numerical derivative screen
for computing $N'(1.5)$

9 a.m. is given by

$$N(t) = \frac{30t}{\sqrt{2 + t^2}}$$

What is the rate at which visitors are admitted to the amusement park at 10:30 a.m.?

Solution   Using the numerical derivative operation of a graphing utility, we find

$$N'(1.5) \approx 6.8481$$

or approximately 6848 visitors per hour. (See Figure T2.)   ■

TECHNOLOGY EXERCISES

In Exercises 1–6, use the numerical derivative operation to find the rate of change of $f(x)$ at the given value of x. Give your answer accurate to four decimal places.

1. $f(x) = \sqrt{x^2 - x^4}$; $x = 0.5$

2. $f(x) = x - \sqrt{1 - x^2}$; $x = 0.4$

3. $f(x) = x\sqrt{1 - x^2}$; $x = 0.2$

4. $f(x) = (x + \sqrt{x^2 + 4})^{3/2}$; $x = 1$

5. $f(x) = \dfrac{\sqrt{1 + x^2}}{x^3 + 2}$; $x = -1$

6. $f(x) = \dfrac{x^3}{1 + (1 + x^2)^{3/2}}$; $x = 3$

7. WORLDWIDE PRODUCTION OF VEHICLES   The worldwide production of vehicles between 1960 and 1990 is given by the function

$$f(t) = 16.5\sqrt{1 + 2.2t} \qquad (0 \le t \le 3)$$

where $f(t)$ is measured in millions and t is measured in decades, with $t = 0$ corresponding to the beginning of 1960. What was the rate of change of the worldwide production of vehicles at the beginning of 1970? At the beginning of 1980?
Source: Automotive News

8. ACCUMULATION YEARS   People from their mid-40s to their mid-50s are in the prime investing years. Demographic studies of this type are of particular importance to financial institutions. The function

$$N(t) = 34.4(1 + 0.32125t)^{0.15} \qquad (0 \le t \le 12)$$

gives the projected number of people in this age group in the United States (in millions) in year t, where $t = 0$ corresponds to the beginning of 1996.

a. How large is this segment of the population projected to be at the beginning of 2005?

b. How fast will this segment of the population be growing at the beginning of 2005?
Source: U.S. Census Bureau

12.4 Marginal Functions in Economics

Marginal analysis is the study of the rate of change of economic quantities. For example, an economist is not merely concerned with the value of an economy's gross domestic product (GDP) at a given time but is equally concerned with the rate at which it is growing or declining. In the same vein, a manufacturer is not only interested in the total cost corresponding to a certain level of production of a commodity but also is interested in the rate of change of the total cost with respect to the level of production, and so on. Let's begin with an example to explain the meaning of the adjective *marginal*, as used by economists.

Cost Functions

 APPLIED EXAMPLE 1   Rate of Change of Cost Functions   Suppose the total cost in dollars incurred each week by Polaraire for manufacturing x refrigerators is given by the total cost function

$$C(x) = 8000 + 200x - 0.2x^2 \qquad (0 \le x \le 400)$$

a. What is the actual cost incurred for manufacturing the 251st refrigerator?
b. Find the rate of change of the total cost function with respect to x when $x = 250$.
c. Compare the results obtained in parts (a) and (b).

Solution

a. The actual cost incurred in producing the 251st refrigerator is the difference between the total cost incurred in producing the first 251 refrigerators and the total cost of producing the first 250 refrigerators:

$$\begin{aligned}
C(251) - C(250) &= [8000 + 200(251) - 0.2(251)^2] \\
&\quad - [8000 + 200(250) - 0.2(250)^2] \\
&= 45{,}599.8 - 45{,}500 \\
&= 99.80
\end{aligned}$$

or $99.80.

b. The rate of change of the total cost function C with respect to x is given by the derivative of C—that is, $C'(x) = 200 - 0.4x$. Thus, when the level of production is 250 refrigerators, the rate of change of the total cost with respect to x is given by

$$\begin{aligned}
C'(250) &= 200 - 0.4(250) \\
&= 100
\end{aligned}$$

or $100.

c. From the solution to part (a), we know that the actual cost for producing the 251st refrigerator is $99.80. This answer is very closely approximated by the answer to part (b), $100. To see why this is so, observe that the difference $C(251) - C(250)$ may be written in the form

$$\frac{C(251) - C(250)}{1} = \frac{C(250 + 1) - C(250)}{1} = \frac{C(250 + h) - C(250)}{h}$$

where $h = 1$. In other words, the difference $C(251) - C(250)$ is precisely the average rate of change of the total cost function C over the interval $[250, 251]$, or, equivalently, the slope of the secant line through the points $(250, 45{,}500)$ and $(251, 45{,}599.8)$. However, the number $C'(250) = 100$ is the instantaneous rate of change of the total cost function C at $x = 250$, or, equivalently, the slope of the tangent line to the graph of C at $x = 250$.

Now when h is small, the average rate of change of the function C is a good approximation to the instantaneous rate of change of the function C, or, equivalently, the slope of the secant line through the points in question is a good approximation to the slope of the tangent line through the point in question. Thus, we may expect

$$C(251) - C(250) = \frac{C(251) - C(250)}{1} = \frac{C(250 + h) - C(250)}{h}$$

$$\approx \lim_{h \to 0} \frac{C(250 + h) - C(250)}{h} = C'(250)$$

which is precisely the case in this example.   ∎

The actual cost incurred in producing an additional unit of a certain commodity given that a plant is already at a certain level of operation is called the **marginal cost.** Knowing this cost is very important to management. As we saw in Example 1, the marginal cost is approximated by the rate of change of the total cost function evaluated at the appropriate point. For this reason, economists have defined the **marginal cost function** to be the derivative of the corresponding total cost function. In other words, if C is a total cost function, then the marginal cost function is defined to be its derivative C'. Thus, the adjective *marginal* is synonymous with *derivative of.*

APPLIED EXAMPLE 2 Marginal Cost Functions A subsidiary of Elektra Electronics manufactures a programmable pocket calculator. Management determined that the daily total cost of producing these calculators (in dollars) is given by

$$C(x) = 0.0001x^3 - 0.08x^2 + 40x + 5000$$

where x stands for the number of calculators produced.

a. Find the marginal cost function.
b. What is the marginal cost when $x = 200, 300, 400,$ and 600?
c. Interpret your results.

Solution

a. The marginal cost function C' is given by the derivative of the total cost function C. Thus,

$$C'(x) = 0.0003x^2 - 0.16x + 40$$

b. The marginal cost when $x = 200, 300, 400,$ and 600 is given by

$$C'(200) = 0.0003(200)^2 - 0.16(200) + 40 = 20$$
$$C'(300) = 0.0003(300)^2 - 0.16(300) + 40 = 19$$
$$C'(400) = 0.0003(400)^2 - 0.16(400) + 40 = 24$$
$$C'(600) = 0.0003(600)^2 - 0.16(600) + 40 = 52$$

or $20, $19, $24,$ and $52, respectively.

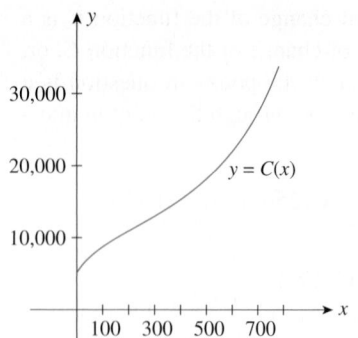

FIGURE 8
The cost of producing x calculators is given by $C(x)$.

c. From the results of part (b), we see that Elektra's actual cost for producing the 201st calculator is approximately $20. The actual cost incurred for producing one additional calculator when the level of production is already 300 calculators is approximately $19, and so on. Observe that when the level of production is already 600 units, the actual cost of producing one additional unit is approximately $52. The higher cost for producing this additional unit when the level of production is 600 units may be the result of several factors, among them excessive costs incurred because of overtime or higher maintenance, production breakdown caused by greater stress and strain on the equipment, and so on. The graph of the total cost function appears in Figure 8. ∎

Average Cost Functions

Let's now introduce another marginal concept closely related to the marginal cost. Let $C(x)$ denote the total cost incurred in producing x units of a certain commodity. Then the **average cost** of producing x units of the commodity is obtained by dividing the total production cost by the number of units produced. This leads to the following definition:

Average Cost Function

Suppose $C(x)$ is a total cost function. Then the **average cost function**, denoted by $\overline{C}(x)$ (read "C bar of x"), is

$$\frac{C(x)}{x} \tag{4}$$

The derivative $\overline{C}'(x)$ of the average cost function, called the **marginal average cost function**, measures the rate of change of the average cost function with respect to the number of units produced.

  APPLIED EXAMPLE 3 Marginal Average Cost Functions The total cost of producing x units of a certain commodity is given by

$$C(x) = 400 + 20x$$

dollars.

a. Find the average cost function $\overline{C}$.
b. Find the marginal average cost function $\overline{C}'$.
c. What are the economic implications of your results?

Solution

a. The average cost function is given by

$$\overline{C}(x) = \frac{C(x)}{x} = \frac{400 + 20x}{x}$$

$$= 20 + \frac{400}{x}$$

b. The marginal average cost function is

$$\overline{C}'(x) = -\frac{400}{x^2}$$

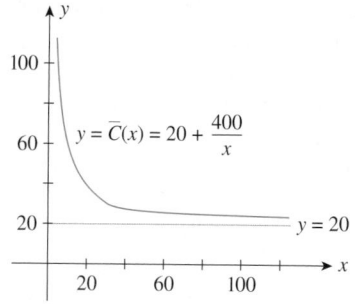

$$y = \overline{C}(x) = 20 + \frac{400}{x}$$

$$y = 20$$

FIGURE 9
As the level of production increases,
the average cost approaches $20.

c. Since the marginal average cost function is negative for all admissible values of x, the rate of change of the average cost function is negative for all $x > 0$; that is, $\overline{C}(x)$ decreases as x increases. However, the graph of $\overline{C}$ always lies above the horizontal line $y = 20$, but it approaches the line since

$$\lim_{x \to \infty} \overline{C}(x) = \lim_{x \to \infty} \left(20 + \frac{400}{x} \right) = 20$$

A sketch of the graph of the function $\overline{C}(x)$ appears in Figure 9. This result is fully expected if we consider the economic implications. Note that as the level of production increases, the fixed cost per unit of production, represented by the term $(400/x)$, drops steadily. The average cost approaches the constant unit of production, which is $20 in this case. ∎

APPLIED EXAMPLE 4 Marginal Average Cost Functions Once again consider the subsidiary of Elektra Electronics. The daily total cost for producing its programmable calculators is given by

$$C(x) = 0.0001x^3 - 0.08x^2 + 40x + 5000$$

dollars, where x stands for the number of calculators produced (see Example 2).

a. Find the average cost function $\overline{C}$.
b. Find the marginal average cost function $\overline{C}'$. Compute $\overline{C}'(500)$.
c. Sketch the graph of the function $\overline{C}$ and interpret the results obtained in parts (a) and (b).

Solution

a. The average cost function is given by

$$\overline{C}(x) = \frac{C(x)}{x} = 0.0001x^2 - 0.08x + 40 + \frac{5000}{x}$$

b. The marginal average cost function is given by

$$\overline{C}'(x) = 0.0002x - 0.08 - \frac{5000}{x^2}$$

Also,

$$\overline{C}'(500) = 0.0002(500) - 0.08 - \frac{5000}{(500)^2} = 0$$

c. To sketch the graph of the function $\overline{C}$, observe that if x is a small positive number, then $\overline{C}(x) > 0$. Furthermore, $\overline{C}(x)$ becomes arbitrarily large as x approaches zero from the right, since the term $(5000/x)$ becomes arbitrarily large as x approaches zero. Next, the result $\overline{C}'(500) = 0$ obtained in part (b) tells us that the tangent line to the graph of the function $\overline{C}$ is horizontal at the point $(500, 35)$ on the graph. Finally, plotting the points on the graph corresponding to, say, $x = 100, 200, 300, \ldots, 900$, we obtain the sketch in Figure 10. As expected, the average cost drops as the level of production increases. But in this case, as opposed to the case in Example 3, the average cost reaches a minimum value of $35, corresponding to a production level of 500, and *increases* thereafter.

This phenomenon is typical in situations where the marginal cost increases from some point on as production increases, as in Example 2. This

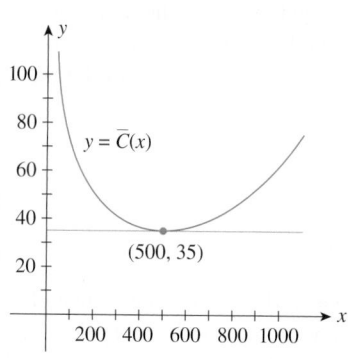

$$y = \overline{C}(x)$$

$$(500, 35)$$

FIGURE 10
The average cost reaches a minimum
of $35 when 500 calculators are
produced.

situation is in contrast to that of Example 3, in which the marginal cost remains constant at any level of production.   ■

EXPLORING WITH TECHNOLOGY

Refer to Example 4.

1. Use a graphing utility to plot the graph of the average cost function

$$\overline{C}(x) = 0.0001x^2 - 0.08x + 40 + \frac{5000}{x}$$

using the viewing window $[0, 1000] \times [0, 100]$. Then, using ZOOM and TRACE, show that the lowest point on the graph of $\overline{C}$ is $(500, 35)$.

2. Draw the tangent line to the graph of $\overline{C}$ at $(500, 35)$. What is its slope? Is this expected?

3. Plot the graph of the marginal average cost function

$$\overline{C}'(x) = 0.0002x - 0.08 - \frac{5000}{x^2}$$

using the viewing window $[0, 2000] \times [-1, 1]$. Then use ZOOM and TRACE to show that the zero of the function $\overline{C}'$ occurs at $x = 500$. Verify this result using the root-finding capability of your graphing utility. Is this result compatible with that obtained in part 2? Explain your answer.

■ Revenue Functions

Recall that a revenue function $R(x)$ gives the revenue realized by a company from the sale of x units of a certain commodity. If the company charges p dollars per unit, then

$$R(x) = px \tag{5}$$

However, the price that a company can command for the product depends on the market in which it operates. If the company is one of many—none of which is able to dictate the price of the commodity—then in this competitive market environment the price is determined by market equilibrium (see Section 11.3). On the other hand, if the company is the sole supplier of the product, then under this monopolistic situation it can manipulate the price of the commodity by controlling the supply. The unit selling price p of the commodity is related to the quantity x of the commodity demanded. This relationship between p and x is called a *demand equation* (see Section 11.3). Solving the demand equation for p in terms of x, we obtain the unit price function f. Thus,

$$p = f(x)$$

and the revenue function R is given by

$$R(x) = px = xf(x)$$

The **marginal revenue** gives the actual revenue realized from the sale of an additional unit of the commodity given that sales are already at a certain level. Following an argument parallel to that applied to the cost function in Example 1, you can convince yourself that the marginal revenue is approximated by $R'(x)$. Thus, we define the **marginal revenue function** to be $R'(x)$, where R is the revenue function. The derivative R' of the function R measures the rate of change of the revenue function.

APPLIED EXAMPLE 5 Marginal Revenue Functions Suppose the relationship between the unit price p in dollars and the quantity demanded x of the Acrosonic model F loudspeaker system is given by the equation

$$p = -0.02x + 400 \qquad (0 \le x \le 20{,}000)$$

a. Find the revenue function R.
b. Find the marginal revenue function R'.
c. Compute $R'(2000)$ and interpret your result.

Solution

a. The revenue function R is given by

$$\begin{aligned} R(x) &= px \\ &= x(-0.02x + 400) \\ &= -0.02x^2 + 400x \quad (0 \le x \le 20{,}000) \end{aligned}$$

b. The marginal revenue function R' is given by

$$R'(x) = -0.04x + 400$$

c.
$$R'(2000) = -0.04(2000) + 400 = 320$$

Thus, the actual revenue to be realized from the sale of the 2001st loudspeaker system is approximately $320.

Profit Functions

Our final example of a marginal function involves the profit function. The profit function P is given by

$$P(x) = R(x) - C(x) \tag{6}$$

where R and C are the revenue and cost functions and x is the number of units of a commodity produced and sold. The **marginal profit function** $P'(x)$ measures the rate of change of the profit function P and provides us with a good approximation of the actual profit or loss realized from the sale of the $(x + 1)$st unit of the commodity (assuming the xth unit has been sold).

APPLIED EXAMPLE 6 Marginal Profit Functions Refer to Example 5. Suppose the cost of producing x units of the Acrosonic model F loudspeaker is

$$C(x) = 100x + 200{,}000$$

dollars.
a. Find the profit function P.
b. Find the marginal profit function P'.
c. Compute $P'(2000)$ and interpret your result.
d. Sketch the graph of the profit function P.

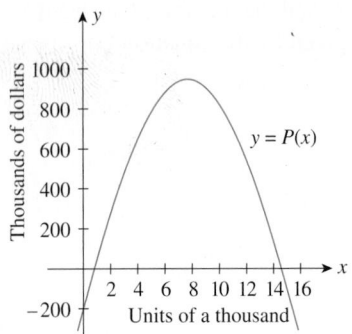

FIGURE 11
The total profit made when x loudspeakers are produced is given by P(x).

Solution

a. From the solution to Example 5a, we have

$$R(x) = -0.02x^2 + 400x$$

Thus, the required profit function P is given by

$$P(x) = R(x) - C(x)$$
$$= (-0.02x^2 + 400x) - (100x + 200,000)$$
$$= -0.02x^2 + 300x - 200,000$$

b. The marginal profit function P' is given by

$$P'(x) = -0.04x + 300$$

c.
$$P'(2000) = -0.04(2000) + 300 = 220$$

Thus, the actual profit realized from the sale of the 2001st loudspeaker system
is approximately \$220.

d. The graph of the profit function P appears in Figure 11.   ∎

Elasticity of Demand

Finally, let's use the marginal concepts introduced in this section to derive an important criterion used by economists to analyze the demand function: elasticity of
demand.

In what follows, it will be convenient to write the demand function f in the form
$x = f(p)$; that is, we will think of the quantity demanded of a certain commodity as
a function of its unit price. Since the quantity demanded of a commodity usually
decreases as its unit price increases, the function f is typically a decreasing function
of p (Figure 12a).

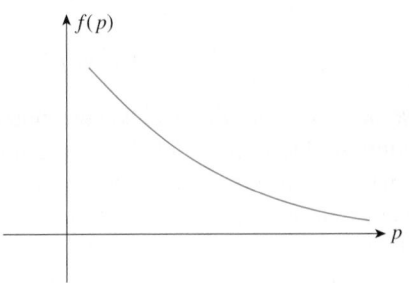

(a) A demand function

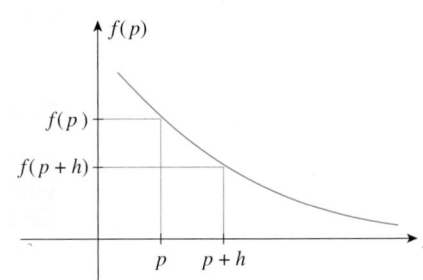

(b) $f(p + h)$ is the quantity demanded
when the unit price increases from p
to $p + h$ dollars.

FIGURE 12

Suppose the unit price of a commodity is increased by h dollars from p dollars
to $(p + h)$ dollars (Figure 12b). Then the quantity demanded drops from $f(p)$ units
to $f(p + h)$ units, a change of $[f(p + h) - f(p)]$ units. The percentage change in
the unit price is

$$\frac{h}{p}(100) \qquad \frac{\text{Change in unit price}}{\text{Price } p}(100)$$

and the corresponding percentage change in the quantity demanded is

$$100 \left[\frac{f(p+h) - f(p)}{f(p)} \right] \qquad \frac{\text{Change in quantity demanded}}{\text{Quantity demanded at price } p}(100)$$

Now, one good way to measure the effect that a percentage change in price has on the percentage change in the quantity demanded is to look at the ratio of the latter to the former. We find

$$\frac{\text{Percentage change in the quantity demanded}}{\text{Percentage change in the unit price}} = \frac{100 \left[\dfrac{f(p+h) - f(p)}{f(p)} \right]}{100 \left(\dfrac{h}{p} \right)}$$

$$= \frac{\dfrac{f(p+h) - f(p)}{h}}{\dfrac{f(p)}{p}}$$

If f is differentiable at p, then

$$\frac{f(p+h) - f(p)}{h} \approx f'(p)$$

when h is small. Therefore, if h is small, then the ratio is approximately equal to

$$\frac{f'(p)}{\dfrac{f(p)}{p}} = \frac{pf'(p)}{f(p)}$$

Economists call the negative of this quantity the elasticity of demand.

Elasticity of Demand

If f is a differentiable demand function defined by $x = f(p)$, then the **elasticity of demand** at price p is given by

$$E(p) = -\frac{pf'(p)}{f(p)} \tag{7}$$

Note   It will be shown later (Section 13.1) that if f is decreasing on an interval, then $f'(p) < 0$ for p in that interval. In light of this, we see that since both p and $f(p)$ are positive, the quantity $\dfrac{pf'(p)}{f(p)}$ is negative. Because economists would rather work with a positive value, the elasticity of demand $E(p)$ is defined to be the negative of this quantity.   ■

APPLIED EXAMPLE 7   Elasticity of Demand   Consider the demand equation

$$p = -0.02x + 400 \qquad (0 \le x \le 20{,}000)$$

which describes the relationship between the unit price in dollars and the quantity demanded x of the Acrosonic model F loudspeaker systems.

a. Find the elasticity of demand $E(p)$.
b. Compute $E(100)$ and interpret your result.
c. Compute $E(300)$ and interpret your result.

Solution

a. Solving the given demand equation for x in terms of p, we find

$$x = f(p) = -50p + 20,000$$

from which we see that

$$f'(p) = -50$$

Therefore,

$$E(p) = -\frac{pf'(p)}{f(p)} = -\frac{p(-50)}{-50p + 20,000}$$

$$= \frac{p}{400 - p}$$

b.
$$E(100) = \frac{100}{400 - 100} = \frac{1}{3}$$

which is the elasticity of demand when $p = 100$. To interpret this result, recall that $E(100)$ is the negative of the ratio of the percentage change in the quantity demanded to the percentage change in the unit price when $p = 100$. Therefore, our result tells us that when the unit price p is set at $100 per speaker, an increase of 1% in the unit price will cause a decrease of approximately 0.33% in the quantity demanded.

c.
$$E(300) = \frac{300}{400 - 300} = 3$$

which is the elasticity of demand when $p = 300$. It tells us that when the unit price is set at $300 per speaker, an increase of 1% in the unit price will cause a decrease of approximately 3% in the quantity demanded. ∎

Economists often use the following terminology to describe demand in terms of elasticity.

Elasticity of Demand

The demand is said to be **elastic** if $E(p) > 1$.

The demand is said to be **unitary** if $E(p) = 1$.

The demand is said to be **inelastic** if $E(p) < 1$.

As an illustration, our computations in Example 7 revealed that demand for Acrosonic loudspeakers is elastic when $p = 300$ but inelastic when $p = 100$. These computations confirm that when demand is elastic, a small percentage change in the unit price will result in a greater percentage change in the quantity demanded; and when demand is inelastic, a small percentage change in the unit price will cause a smaller percentage change in the quantity demanded. Finally, when demand is unitary, a small percentage change in the unit price will result in the same percentage change in the quantity demanded.

We can describe the way revenue responds to changes in the unit price using the notion of elasticity. If the quantity demanded of a certain commodity is related to its unit price by the equation $x = f(p)$, then the revenue realized through the sale of x units of the commodity at a price of p dollars each is

$$R(p) = px = pf(p)$$

The rate of change of the revenue with respect to the unit price p is given by

$$\begin{aligned}
R'(p) &= f(p) + pf'(p) \\
&= f(p)\left[1 + \frac{pf'(p)}{f(p)}\right] \\
&= f(p)[1 - E(p)]
\end{aligned}$$

Now, suppose demand is elastic when the unit price is set at a dollars. Then $E(a) > 1$, and so $1 - E(a) < 0$. Since $f(p)$ is positive for all values of p, we see that

$$R'(a) = f(a)[1 - E(a)] < 0$$

and so $R(p)$ is decreasing at $p = a$. This implies that a small increase in the unit price when $p = a$ results in a decrease in the revenue, whereas a small decrease in the unit price will result in an increase in the revenue. Similarly, you can show that if the demand is inelastic when the unit price is set at a dollars, then a small increase in the unit price will cause the revenue to increase, and a small decrease in the unit price will cause the revenue to decrease. Finally, if the demand is unitary when the unit price is set at a dollars, then $E(a) = 1$ and $R'(a) = 0$. This implies that a small increase or decrease in the unit price will not result in a change in the revenue. The following statements summarize this discussion.

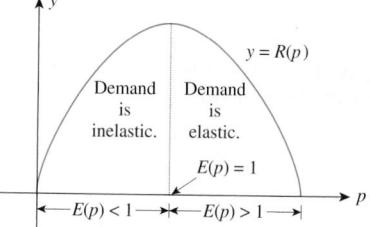

FIGURE 13
The revenue is increasing on an interval where the demand is inelastic, decreasing on an interval where the demand is elastic, and stationary at the point where the demand is unitary.

---

1. If the demand is elastic at p [$E(p) > 1$], then an increase in the unit price will cause the revenue to decrease, whereas a decrease in the unit price will cause the revenue to increase.

2. If the demand is inelastic at p [$E(p) < 1$], then an increase in the unit price will cause the revenue to increase, and a decrease in the unit price will cause the revenue to decrease.

3. If the demand is unitary at p [$E(p) = 1$], then an increase in the unit price will cause the revenue to stay about the same.

These results are illustrated in Figure 13.

Note   As an aid to remembering this, note the following:

1. If demand is elastic, then the change in revenue and the change in the unit price move in opposite directions.
2. If demand is inelastic, then they move in the same direction.   ■

APPLIED EXAMPLE 8   Elasticity of Demand   Refer to Example 7.

a. Is demand elastic, unitary, or inelastic when $p = 100$? When $p = 300$?
b. If the price is \$100, will raising the unit price slightly cause the revenue to increase or decrease?

Solution

a. From the results of Example 7, we see that $E(100) = \frac{1}{3} < 1$ and $E(300) = 3 > 1$. We conclude accordingly that demand is inelastic when $p = 100$ and elastic when $p = 300$.
b. Since demand is inelastic when $p = 100$, raising the unit price slightly will cause the revenue to increase. ◼

12.4   Self-Check Exercises

1. The weekly demand for Pulsar DVD recorders is given by the demand equation

$$p = -0.02x + 300 \qquad (0 \le x \le 15,000)$$

where p denotes the wholesale unit price in dollars and x denotes the quantity demanded. The weekly total cost function associated with manufacturing these recorders is

$$C(x) = 0.000003x^3 - 0.04x^2 + 200x + 70,000$$

dollars.
a. Find the revenue function R and the profit function P.

b. Find the marginal cost function C', the marginal revenue function R', and the marginal profit function P'.
c. Find the marginal average cost function $\overline{C}'$.
d. Compute $C'(3000)$, $R'(3000)$, and $P'(3000)$ and interpret your results.

2. Refer to the preceding exercise. Determine whether the demand is elastic, unitary, or inelastic when $p = 100$ and when $p = 200$.

Solutions to Self-Check Exercises 12.4 can be found on page 735.

12.4   Concept Questions

1. Explain each term in your own words:
   a. Marginal cost function
   b. Average cost function
   c. Marginal average cost function
   d. Marginal revenue function
   e. Marginal profit function

2. **a.** Define the elasticity of demand.
   b. When is the elasticity of demand elastic? Unitary? Inelastic? Explain the meaning of each term.

12.4   Exercises

1. PRODUCTION COSTS   The graph of a typical total cost function $C(x)$ associated with the manufacture of x units of a certain commodity is shown in the following figure.
   a. Explain why the function C is always increasing.
   b. As the level of production x increases, the cost/unit drops so that $C(x)$ increases but at a slower pace. However, a level of production is soon reached at which the cost/unit begins to increase dramatically (due to a shortage of raw material, overtime, breakdown of machinery due to excessive stress and strain) so that $C(x)$ continues to increase at a faster pace. Use the graph of C to find the approximate

level of production x_0 where this occurs.

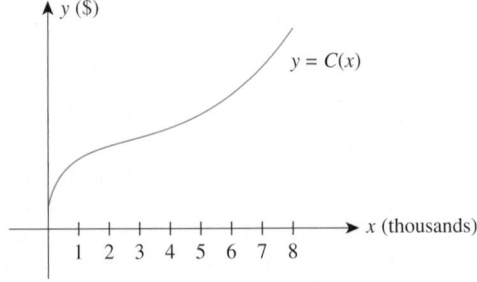

2. PRODUCTION COSTS   The graph of a typical average cost function $A(x) = C(x)/x$, where $C(x)$ is a total cost function associated with the manufacture of x units of a certain commodity is shown in the following figure.
   a. Explain in economic terms why $A(x)$ is large if x is small and why $A(x)$ is large if x is large.
   b. What is the significance of the numbers x_0 and y_0, the x- and y-coordinates of the lowest point on the graph of the function A?

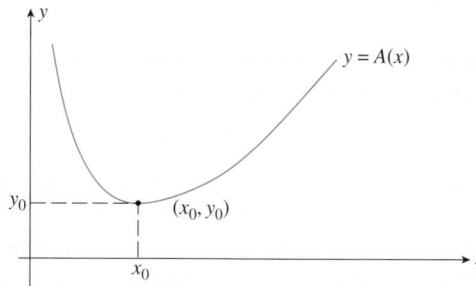

3. MARGINAL COST   The total weekly cost (in dollars) incurred by Lincoln Records in pressing x compact discs is

$$C(x) = 2000 + 2x - 0.0001x^2 \qquad (0 \le x \le 6000)$$

   a. What is the actual cost incurred in producing the 1001st and the 2001st disc?
   b. What is the marginal cost when $x = 1000$ and 2000?

4. MARGINAL COST   A division of Ditton Industries manufactures the Futura model microwave oven. The daily cost (in dollars) of producing these microwave ovens is

$$C(x) = 0.0002x^3 - 0.06x^2 + 120x + 5000$$

   where x stands for the number of units produced.
   a. What is the actual cost incurred in manufacturing the 101st oven? The 201st oven? The 301st oven?
   b. What is the marginal cost when $x = 100$, 200, and 300?

5. MARGINAL AVERAGE COST   Custom Office makes a line of executive desks. It is estimated that the total cost for making x units of their Senior Executive model is

$$C(x) = 100x + 200,000$$

   dollars/year.
   a. Find the average cost function $\overline{C}$.
   b. Find the marginal average cost function $\overline{C}'$.
   c. What happens to $\overline{C}(x)$ when x is very large? Interpret your results.

6. MARGINAL AVERAGE COST   The management of ThermoMaster Company, whose Mexican subsidiary manufactures an indoor–outdoor thermometer, has estimated that the total

weekly cost (in dollars) for producing x thermometers is

$$C(x) = 5000 + 2x$$

   a. Find the average cost function $\overline{C}$.
   b. Find the marginal average cost function $\overline{C}'$.
   c. Interpret your results.

7. Find the average cost function $\overline{C}$ and the marginal average cost function $\overline{C}'$ associated with the total cost function C of Exercise 3.

8. Find the average cost function $\overline{C}$ and the marginal average cost function $\overline{C}'$ associated with the total cost function C of Exercise 4.

9. MARGINAL REVENUE   Williams Commuter Air Service realizes a monthly revenue of

$$R(x) = 8000x - 100x^2$$

   dollars when the price charged per passenger is x dollars.
   a. Find the marginal revenue R'.
   b. Compute $R'(39)$, $R'(40)$, and $R'(41)$.
   c. Based on the results of part (b), what price should the airline charge in order to maximize their revenue?

10. MARGINAL REVENUE   The management of Acrosonic plans to market the ElectroStat, an electrostatic speaker system. The marketing department has determined that the demand for these speakers is

$$p = -0.04x + 800 \qquad (0 \le x \le 20,000)$$

   where p denotes the speaker's unit price (in dollars) and x denotes the quantity demanded.
   a. Find the revenue function R.
   b. Find the marginal revenue function R'.
   c. Compute $R'(5000)$ and interpret your results.

11. MARGINAL PROFIT   Refer to Exercise 10. Acrosonic's production department estimates that the total cost (in dollars) incurred in manufacturing x ElectroStat speaker systems in the first year of production will be

$$C(x) = 200x + 300,000$$

   a. Find the profit function P.
   b. Find the marginal profit function P'.
   c. Compute $P'(5000)$ and $P'(8000)$.
   d. Sketch the graph of the profit function and interpret your results.

12. MARGINAL PROFIT   Lynbrook West, an apartment complex, has 100 two-bedroom units. The monthly profit (in dollars) realized from renting x apartments is

$$P(x) = -10x^2 + 1760x - 50,000$$

   a. What is the actual profit realized from renting the 51st unit, assuming that 50 units have already been rented?

b. Compute the marginal profit when $x = 50$ and compare your results with that obtained in part (a).

13. MARGINAL COST, REVENUE, AND PROFIT   The weekly demand for the Pulsar 25 color LED television is

$$p = 600 - 0.05x \qquad (0 \le x \le 12,000)$$

where p denotes the wholesale unit price in dollars and x denotes the quantity demanded. The weekly total cost function associated with manufacturing the Pulsar 25 is given by

$$C(x) = 0.000002x^3 - 0.03x^2 + 400x + 80,000$$

where $C(x)$ denotes the total cost incurred in producing x sets.
a. Find the revenue function R and the profit function P.
b. Find the marginal cost function C', the marginal revenue function R', and the marginal profit function P'.
c. Compute $C'(2000)$, $R'(2000)$, and $P'(2000)$ and interpret your results.
d. Sketch the graphs of the functions C, R, and P and interpret parts (b) and (c), using the graphs obtained.

14. MARGINAL COST, REVENUE, AND PROFIT   Pulsar also manufactures a series of 19-in. color television sets. The quantity x of these sets demanded each week is related to the wholesale unit price p by the equation

$$p = -0.006x + 180$$

The weekly total cost incurred by Pulsar for producing x sets is

$$C(x) = 0.000002x^3 - 0.02x^2 + 120x + 60,000$$

dollars. Answer the questions in Exercise 13 for these data.

15. MARGINAL AVERAGE COST   Find the average cost function $\overline{C}$ associated with the total cost function C of Exercise 13.
a. What is the marginal average cost function $\overline{C}'$?
b. Compute $\overline{C}'(5000)$ and $\overline{C}'(10,000)$ and interpret your results.
c. Sketch the graph of $\overline{C}$.

16. MARGINAL AVERAGE COST   Find the average cost function $\overline{C}$ associated with the total cost function C of Exercise 14.
a. What is the marginal average cost function $\overline{C}'$?
b. Compute $\overline{C}'(5000)$ and $\overline{C}'(10,000)$ and interpret your results.

17. MARGINAL REVENUE   The quantity of Sicard wristwatches demanded each month is related to the unit price by the equation

$$p = \frac{50}{0.01x^2 + 1} \qquad (0 \le x \le 20)$$

where p is measured in dollars and x in units of a thousand.
a. Find the revenue function R.

b. Find the marginal revenue function R'.
c. Compute $R'(2)$ and interpret your result.

18. MARGINAL PROPENSITY TO CONSUME   The consumption function of the U.S. economy from 1929 to 1941 is

$$C(x) = 0.712x + 95.05$$

where $C(x)$ is the personal consumption expenditure and x is the personal income, both measured in billions of dollars. Find the rate of change of consumption with respect to income, dC/dx. This quantity is called the *marginal propensity to consume*.

19. MARGINAL PROPENSITY TO CONSUME   Refer to Exercise 18. Suppose a certain economy's consumption function is

$$C(x) = 0.873x^{1.1} + 20.34$$

where $C(x)$ and x are measured in billions of dollars. Find the marginal propensity to consume when $x = 10$.

20. MARGINAL PROPENSITY TO SAVE   Suppose $C(x)$ measures an economy's personal consumption expenditure and x the personal income, both in billions of dollars. Then,

$$S(x) = x - C(x) \qquad \text{Income minus consumption}$$

measures the economy's savings corresponding to an income of x billion dollars. Show that

$$\frac{dS}{dx} = 1 - \frac{dC}{dx}$$

The quantity dS/dx is called the *marginal propensity to save*.

21. Refer to Exercise 20. For the consumption function of Exercise 18, find the marginal propensity to save.

22. Refer to Exercise 20. For the consumption function of Exercise 19, find the marginal propensity to save when $x = 10$.

For each demand equation in Exercises 23–28, compute the elasticity of demand and determine whether the demand is elastic, unitary, or inelastic at the indicated price.

23. $x = -\dfrac{5}{4}p + 20; p = 10$

24. $x = -\dfrac{3}{2}p + 9; p = 2$

25. $x + \dfrac{1}{3}p - 20 = 0; p = 30$

26. $0.4x + p - 20 = 0; p = 10$

27. $p = 169 - x^2; p = 29$ 28. $p = 144 - x^2; p = 96$

29. ELASTICITY OF DEMAND   The demand equation for the Roland portable hair dryer is given by

$$x = \frac{1}{5}(225 - p^2) \qquad (0 \le p \le 15)$$

where x (measured in units of a hundred) is the quantity demanded per week and p is the unit price in dollars.

a. Is the demand elastic or inelastic when $p = 8$ and when $p = 10$?

b. When is the demand unitary?

  Hint: Solve $E(p) = 1$ for p.

c. If the unit price is lowered slightly from $10, will the revenue increase or decrease?

d. If the unit price is increased slightly from $8, will the revenue increase or decrease?

30. ELASTICITY OF DEMAND   The management of Titan Tire Company has determined that the quantity demanded x of their Super Titan tires per week is related to the unit price p by the equation

$$x = \sqrt{144 - p} \qquad (0 \le p \le 144)$$

where p is measured in dollars and x in units of a thousand.

a. Compute the elasticity of demand when $p = 63$, 96, and 108.

b. Interpret the results obtained in part (a).

c. Is the demand elastic, unitary, or inelastic when $p = 63$, 96, and 108?

31. ELASTICITY OF DEMAND   The proprietor of the Showplace, a video store, has estimated that the rental price p (in dollars) of prerecorded videodiscs is related to the quantity x rented/day by the demand equation

$$x = \frac{2}{3}\sqrt{36 - p^2} \qquad (0 \le p \le 6)$$

Currently, the rental price is $2/disc.

a. Is the demand elastic or inelastic at this rental price?

b. If the rental price is increased, will the revenue increase or decrease?

32. ELASTICITY OF DEMAND   The quantity demanded each week x (in units of a hundred) of the Mikado digital camera is related to the unit price p (in dollars) by the demand equation

$$x = \sqrt{400 - 5p} \qquad (0 \le p \le 80)$$

a. Is the demand elastic or inelastic when $p = 40$? When $p = 60$?

b. When is the demand unitary?

c. If the unit price is lowered slightly from $60, will the revenue increase or decrease?

d. If the unit price is increased slightly from $40, will the revenue increase or decrease?

33. ELASTICITY OF DEMAND   The demand function for a certain make of exercise bicycle sold exclusively through cable television is

$$p = \sqrt{9 - 0.02x} \qquad (0 \le x \le 450)$$

where p is the unit price in hundreds of dollars and x is the quantity demanded/week. Compute the elasticity of demand and determine the range of prices corresponding to inelastic, unitary, and elastic demand.

  Hint: Solve the equation $E(p) = 1$.

34. ELASTICITY OF DEMAND   The demand equation for the Sicard wristwatch is given by

$$x = 10\sqrt{\frac{50 - p}{p}} \qquad (0 < p \le 50)$$

where x (measured in units of a thousand) is the quantity demanded/week and p is the unit price in dollars. Compute the elasticity of demand and determine the range of prices corresponding to inelastic, unitary, and elastic demand.

In Exercises 35 and 36, determine whether the statement is true or false. If it is true, explain why it is true. If it is false, give an example to show why it is false.

35. If C is a differentiable total cost function, then the marginal average cost function is

$$\overline{C}'(x) = \frac{xC'(x) - C(x)}{x^2}$$

36. If the marginal profit function is positive at $x = a$, then it makes sense to decrease the level of production.

12.4 Solutions to Self-Check Exercises

1. a. $R(x) = px$
$\quad = x(-0.02x + 300)$
$\quad = -0.02x^2 + 300x \qquad (0 \le x \le 15,000)$

$P(x) = R(x) - C(x)$
$\quad = -0.02x^2 + 300x$
$\qquad - (0.000003x^3 - 0.04x^2 + 200x + 70,000)$
$\quad = -0.000003x^3 + 0.02x^2 + 100x - 70,000$

b. $C'(x) = 0.000009x^2 - 0.08x + 200$
$R'(x) = -0.04x + 300$
$P'(x) = -0.000009x^2 + 0.04x + 100$

c. The average cost function is

$$\overline{C}(x) = \frac{C(x)}{x}$$

$$= \frac{0.000003x^3 - 0.04x^2 + 200x + 70,000}{x}$$

$$= 0.000003x^2 - 0.04x + 200 + \frac{70,000}{x}$$

Therefore, the marginal average cost function is

$$\overline{C}'(x) = 0.000006x - 0.04 - \frac{70,000}{x^2}$$

d. Using the results from part (b), we find

$$C'(3000) = 0.000009(3000)^2 - 0.08(3000) + 200$$
$$= 41$$

That is, when the level of production is already 3000 recorders, the actual cost of producing one additional recorder is approximately $41. Next,

$$R'(3000) = -0.04(3000) + 300 = 180$$

That is, the actual revenue to be realized from selling the 3001st recorder is approximately $180. Finally,

$$P'(3000) = -0.000009(3000)^2 + 0.04(3000) + 100$$
$$= 139$$

That is, the actual profit realized from selling the 3001st DVD recorder is approximately $139.

2. We first solve the given demand equation for x in terms of p,

obtaining

$$x = f(p) = -50p + 15,000$$
$$f'(p) = -50$$

Therefore,

$$E(p) = -\frac{pf'(p)}{f(p)} = -\frac{p}{-50p + 15,000}(-50)$$
$$= \frac{p}{300 - p} \quad (0 \le p < 300)$$

Next, we compute

$$E(100) = \frac{100}{300 - 100} = \frac{1}{2} < 1$$

and we conclude that demand is inelastic when $p = 100$. Also,

$$E(200) = \frac{200}{300 - 200} = 2 > 1$$

and we see that demand is elastic when $p = 200$.

12.5 Higher-Order Derivatives

▬ Higher-Order Derivatives

The derivative f' of a function f is also a function. As such, the differentiability of f' may be considered. Thus, the function f' has a derivative f'' at a point x in the domain of f' if the limit of the quotient

$$\frac{f'(x + h) - f'(x)}{h}$$

exists as h approaches zero. In other words, it is the derivative of the first derivative.

The function f'' obtained in this manner is called the **second derivative of** the function f, just as the derivative f' of f is often called the first derivative of f. Continuing in this fashion, we are led to considering the third, fourth, and higher-order derivatives of f whenever they exist. Notations for the first, second, third, and, in general, nth derivatives of a function f at a point x are

$$f'(x), f''(x), f'''(x), \ldots, f^{(n)}(x)$$

or

$$D^1 f(x), D^2 f(x), D^3 f(x), \ldots, D^n f(x)$$

If f is written in the form $y = f(x)$, then the notations for its derivatives are

$$y', y'', y''', \ldots, y^{(n)}$$
$$\frac{dy}{dx}, \frac{d^2y}{dx^2}, \frac{d^3y}{dx^3}, \ldots, \frac{d^ny}{dx^n}$$

or

$$D^1 y, D^2 y, D^3 y, \ldots, D^n y$$

respectively.

EXAMPLE 1   Find the derivatives of all orders of the polynomial function

$$f(x) = x^5 - 3x^4 + 4x^3 - 2x^2 + x - 8$$

Solution   We have

$$f'(x) = 5x^4 - 12x^3 + 12x^2 - 4x + 1$$

$$f''(x) = \frac{d}{dx}f'(x) = 20x^3 - 36x^2 + 24x - 4$$

$$f'''(x) = \frac{d}{dx}f''(x) = 60x^2 - 72x + 24$$

$$f^{(4)}(x) = \frac{d}{dx}f'''(x) = 120x - 72$$

$$f^{(5)}(x) = \frac{d}{dx}f^{(4)}(x) = 120$$

and, in general,

$$f^{(n)}(x) = 0 \qquad (\text{for } n > 5)$$

EXAMPLE 2   Find the third derivative of the function f defined by $y = x^{2/3}$. What is its domain?

Solution   We have

$$y' = \frac{2}{3}x^{-1/3}$$

$$y'' = \left(\frac{2}{3}\right)\left(-\frac{1}{3}\right)x^{-4/3} = -\frac{2}{9}x^{-4/3}$$

737

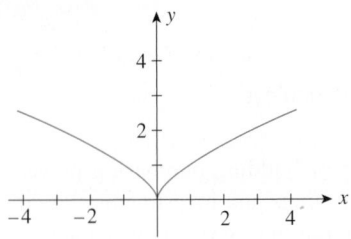

FIGURE 14
The graph of the function $y = x^{2/3}$

so the required derivative is

$$y''' = \left(-\frac{2}{9}\right)\left(-\frac{4}{3}\right)x^{-7/3} = \frac{8}{27}x^{-7/3} = \frac{8}{27x^{7/3}}$$

The common domain of the functions f', f'', and f''' is the set of all real numbers except $x = 0$. The domain of $y = x^{2/3}$ is the set of all real numbers. The graph of the function $y = x^{2/3}$ appears in Figure 14. ∎

Note Always simplify an expression before differentiating it to obtain the next order derivative. ∎

EXAMPLE 3 Find the second derivative of the function $y = (2x^2 + 3)^{3/2}$.

Solution We have, using the general power rule,

$$y' = \frac{3}{2}(2x^2 + 3)^{1/2}(4x) = 6x(2x^2 + 3)^{1/2}$$

Next, using the product rule and then the chain rule, we find

$$y'' = (6x) \cdot \frac{d}{dx}(2x^2 + 3)^{1/2} + \left[\frac{d}{dx}(6x)\right](2x^2 + 3)^{1/2}$$

$$= (6x)\left(\frac{1}{2}\right)(2x^2 + 3)^{-1/2}(4x) + 6(2x^2 + 3)^{1/2}$$

$$= 12x^2(2x^2 + 3)^{-1/2} + 6(2x^2 + 3)^{1/2}$$

$$= 6(2x^2 + 3)^{-1/2}[2x^2 + (2x^2 + 3)]$$

$$= \frac{6(4x^2 + 3)}{\sqrt{2x^2 + 3}}$$

∎

Just as the derivative of a function f at a point x measures the rate of change of the function f at that point, the second derivative of f (the derivative of f') measures the rate of change of the derivative f' of the function f. The third derivative of the function f, f''', measures the rate of change of f'', and so on.

In Chapter 13, we will discuss applications involving the geometric interpretation of the second derivative of a function. The following example gives an interpretation of the second derivative in a familiar role.

APPLIED EXAMPLE 4 Acceleration of a Maglev Refer to the example on page 622. The distance s (in feet) covered by a maglev moving along a straight track t seconds after starting from rest is given by the function $s = 4t^2$ ($0 \le t \le 10$). What is the maglev's acceleration at the end of 30 seconds?

Solution The velocity of the maglev t seconds from rest is given by

$$v = \frac{ds}{dt} = \frac{d}{dt}(4t^2) = 8t$$

The acceleration of the maglev t seconds from rest is given by the rate of change of the velocity of t—that is,

$$a = \frac{d}{dt}v = \frac{d}{dt}\left(\frac{ds}{dt}\right) = \frac{d^2s}{dt^2} = \frac{d}{dt}(8t) = 8$$

or 8 feet per second per second, normally abbreviated 8 ft/sec². ∎

APPLIED EXAMPLE 5 Acceleration and Velocity of a Falling Object
A ball is thrown straight up into the air from the roof of a building. The height of the ball as measured from the ground is given by

$$s = -16t^2 + 24t + 120$$

where s is measured in feet and t in seconds. Find the velocity and acceleration of the ball 3 seconds after it is thrown into the air.

Solution The velocity v and acceleration a of the ball at any time t are given by

$$v = \frac{ds}{dt} = \frac{d}{dt}(-16t^2 + 24t + 120) = -32t + 24$$

and

$$a = \frac{d^2s}{dt^2} = \frac{d}{dt}\left(\frac{ds}{dt}\right) = \frac{d}{dt}(-32t + 24) = -32$$

Therefore, the velocity of the ball 3 seconds after it is thrown into the air is

$$v = -32(3) + 24 = -72$$

That is, the ball is falling downward at a speed of 72 ft/sec. The acceleration of the ball is 32 ft/sec² downward at any time during the motion.

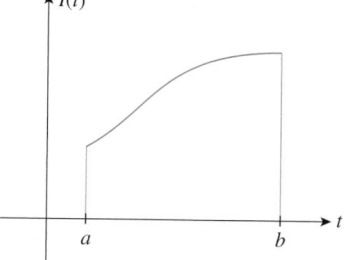

FIGURE 15
The CPI of a certain economy from year a to year b is given by $I(t)$.

Another interpretation of the second derivative of a function—this time from the field of economics—follows. Suppose the consumer price index (CPI) of an economy between the years a and b is described by the function $I(t)$ ($a \le t \le b$) (Figure 15). Then the first derivative of I at $t = c$, $I'(c)$, where $a < c < b$, gives the rate of change of I at c. The quantity

$$\frac{I'(c)}{I(c)}$$

called the *relative rate of change of $I(t)$* with respect to t at $t = c$, measures the *inflation rate* of the economy at $t = c$. The second derivative of I at $t = c$, $I''(c)$, gives the rate of change of I' at $t = c$. Now, it is possible for $I'(t)$ to be positive and $I''(t)$ to be negative at $t = c$ (see Example 6). This tells us that at $t = c$ the economy is experiencing inflation (the CPI is increasing) but the rate at which inflation is growing is in fact decreasing. This is precisely the situation described by an economist or a politician when she claims that "inflation is slowing." One may not jump to the conclusion from the aforementioned quote that prices of goods and services are about to drop!

APPLIED EXAMPLE 6 Inflation Rate of an Economy The function
$$I(t) = -0.2t^3 + 3t^2 + 100 \qquad (0 \le t \le 9)$$
gives the CPI of an economy, where $t = 0$ corresponds to the beginning of 1995.

a. Find the inflation rate at the beginning of 2001 ($t = 6$).
b. Show that inflation was moderating at that time.

Solution

a. We find $I'(t) = -0.6t^2 + 6t$. Next, we compute

$$I'(6) = -0.6(6)^2 + 6(6) = 14.4 \quad \text{and} \quad I(6) = -0.2(6)^3 + 3(6)^2 + 100 = 164.8$$

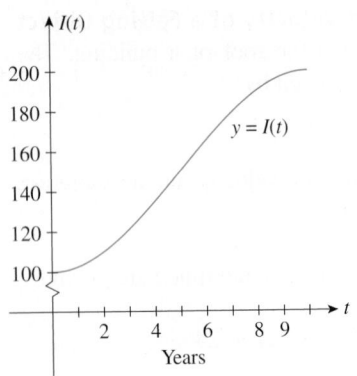

FIGURE 16
The CPI of an economy is given by $I(t)$.

from which we see that the inflation rate is

$$\frac{I'(6)}{I(6)} = \frac{14.4}{164.8} \approx 0.0874$$

or approximately 8.7%.

b. We find

$$I''(t) = \frac{d}{dt}(-0.6t^2 + 6t) = -1.2t + 6$$

Since

$$I''(6) = -1.2(6) + 6 = -1.2$$

we see that I' is indeed decreasing at $t = 6$ and conclude that inflation was moderating at that time (Figure 16). ■

12.5   Self-Check Exercises

1. Find the third derivative of

$$f(x) = 2x^5 - 3x^3 + x^2 - 6x + 10$$

2. Let

$$f(x) = \frac{1}{1 + x}$$

Find $f'(x), f''(x),$ and $f'''(x)$.

3. A certain species of turtles faces extinction because dealers collect truckloads of turtle eggs to be sold as aphrodisiacs.

After severe conservation measures are implemented, it is hoped that the turtle population will grow according to the rule

$$N(t) = 2t^3 + 3t^2 - 4t + 1000 \qquad (0 \le t \le 10)$$

where $N(t)$ denotes the population at the end of year t. Compute $N''(2)$ and $N''(8)$. What do your results tell you about the effectiveness of the program?

Solutions to Self-Check Exercises 12.5 can be found on page 742.

12.5   Concept Questions

1. a. What is the second derivative of a function f?
b. How do you find the second derivative of a function f, assuming that it exists?

2. If $s = f(t)$ gives the position of an object moving on the coordinate line, what do $f'(t)$ and $f''(t)$ measure?

3. If $I(t)$ gives the CPI of an economy, how do you compute the relative rate of change of I with respect to t at time $t = c$?

12.5   Exercises

In Exercises 1–20, find the first and second derivatives of the function.

1. $f(x) = 4x^2 - 2x + 1$

2. $f(x) = -0.2x^2 + 0.3x + 4$

3. $f(x) = 2x^3 - 3x^2 + 1$

4. $g(x) = -3x^3 + 24x^2 + 6x - 64$

5. $h(t) = t^4 - 2t^3 + 6t^2 - 3t + 10$

6. $f(x) = x^5 - x^4 + x^3 - x^2 + x - 1$

7. $f(x) = (x^2 + 2)^5$ **8.** $g(t) = t^2(3t + 1)^4$

9. $g(t) = (2t^2 - 1)^2(3t^2)$

10. $h(x) = (x^2 + 1)^2(x - 1)$

11. $f(x) = (2x^2 + 2)^{7/2}$

12. $h(w) = (w^2 + 2w + 4)^{5/2}$

13. $f(x) = x(x^2 + 1)^2$

14. $g(u) = u(2u - 1)^3$

15. $f(x) = \dfrac{x}{2x + 1}$

16. $g(t) = \dfrac{t^2}{t - 1}$

17. $f(s) = \dfrac{s - 1}{s + 1}$

18. $f(u) = \dfrac{u}{u^2 + 1}$

19. $f(u) = \sqrt{4 - 3u}$

20. $f(x) = \sqrt{2x - 1}$

In Exercises 21–28, find the third derivative of the given function.

21. $f(x) = 3x^4 - 4x^3$

22. $f(x) = 3x^5 - 6x^4 + 2x^2 - 8x + 12$

23. $f(x) = \dfrac{1}{x}$

24. $f(x) = \dfrac{2}{x^2}$

25. $g(s) = \sqrt{3s - 2}$

26. $g(t) = \sqrt{2t + 3}$

27. $f(x) = (2x - 3)^4$

28. $g(t) = \left(\dfrac{1}{2}t^2 - 1\right)^5$

29. ACCELERATION OF A FALLING OBJECT   During the construction of an office building, a hammer is accidentally dropped from a height of 256 ft. The distance (in feet) the hammer falls in t sec is $s = 16t^2$. What is the hammer's velocity when it strikes the ground? What is its acceleration?

30. ACCELERATION OF A CAR   The distance s (in feet) covered by a car after t sec is given by

$$s = -t^3 + 8t^2 + 20t \qquad (0 \le t \le 6)$$

Find a general expression for the car's acceleration at any time t $(0 \le t \le 6)$. Show that the car is decelerating after $2\frac{2}{3}$ sec.

31. CRIME RATES   The number of major crimes committed in Bronxville between 1988 and 1995 is approximated by the function

$$N(t) = -0.1t^3 + 1.5t^2 + 100 \qquad (0 \le t \le 7)$$

where $N(t)$ denotes the number of crimes committed in year t, with $t = 0$ corresponding to 1988. Enraged by the dramatic increase in the crime rate, Bronxville's citizens, with the help of the local police, organized "Neighborhood Crime Watch" groups in early 1992 to combat this menace.
a. Verify that the crime rate was increasing from 1988 through 1995.
Hint: Compute $N'(0), N'(1), \ldots, N'(7)$.
b. Show that the Neighborhood Crime Watch program was working by computing $N''(4), N''(5), N''(6),$ and $N''(7)$.

32. GDP OF A DEVELOPING COUNTRY   A developing country's gross domestic product (GDP) from 1996 to 2004 is approximated by the function

$$G(t) = -0.2t^3 + 2.4t^2 + 60 \qquad (0 \le t \le 8)$$

where $G(t)$ is measured in billions of dollars, with $t = 0$ corresponding to 1996.
a. Compute $G'(0), G'(1), \ldots, G'(8)$.
b. Compute $G''(0), G''(1), \ldots, G''(8)$.
c. Using the results obtained in parts (a) and (b), show that after a spectacular growth rate in the early years, the growth of the GDP cooled off.

33. DISABILITY BENEFITS   The number of persons aged 18–64 receiving disability benefits through Social Security, the Supplemental Security income, or both, from 1990 through 2000 is approximated by the function

$$N(t) = 0.00037t^3 - 0.0242t^2 + 0.52t + 5.3 \qquad (0 \le t \le 10)$$

where $f(t)$ is measured in units of a million and t is measured in years, with $t = 0$ corresponding to the beginning of 1990. Compute $N(8), N'(8),$ and $N''(8)$ and interpret your results.
Source: Social Security Administration

34. OBESITY IN AMERICA   The body mass index (BMI) measures body weight in relation to height. A BMI of 25 to 29.9 is considered overweight, a BMI of 30 or more is considered obese, and a BMI of 40 or more is morbidly obese. The percent of the U.S. population that is obese is approximated by the function

$$P(t) = 0.0004t^3 + 0.0036t^2 + 0.8t + 12 \qquad (0 \le t \le 13)$$

where t is measured in years, with $t = 0$ corresponding to the beginning of 1991. Show that the rate of the rate of change of the percent of the U.S. population that is deemed obese was positive from 1991 to 2004. What does this mean?
Source: Centers for Disease Control and Prevention

35. TEST FLIGHT OF A VTOL   In a test flight of the McCord Terrier, McCord Aviation's experimental VTOL (vertical takeoff and landing) aircraft, it was determined that t sec after liftoff, when the craft was operated in the vertical takeoff mode, its altitude (in feet) was

$$h(t) = \frac{1}{16}t^4 - t^3 + 4t^2 \qquad (0 \le t \le 8)$$

a. Find an expression for the craft's velocity at time t.
b. Find the craft's velocity when $t = 0$ (the initial velocity), $t = 4$, and $t = 8$.
c. Find an expression for the craft's acceleration at time t.
d. Find the craft's acceleration when $t = 0, 4,$ and 8.
e. Find the craft's height when $t = 0, 4,$ and 8.

36. AIR PURIFICATION   During testing of a certain brand of air purifier, the amount of smoke remaining t min after the start of the test was

$$A(t) = -0.00006t^5 + 0.00468t^4 - 0.1316t^3$$
$$+ 1.915t^2 - 17.63t + 100$$

percent of the original amount. Compute $A'(10)$ and $A''(10)$ and interpret your results.

Source: Consumer Reports

37. **AGING POPULATION** The population of Americans age 55 yr and over as a percent of the total population is approximated by the function

$$f(t) = 10.72(0.9t + 10)^{0.3} \qquad (0 \le t \le 20)$$

where t is measured in years, with $t = 0$ corresponding to 2000. Compute $f''(10)$ and interpret your result.

Source: U.S. Census Bureau

38. **WORKING MOTHERS** The percent of mothers who work outside the home and have children younger than age 6 yr is approximated by the function

$$P(t) = 33.55(t + 5)^{0.205} \qquad (0 \le t \le 21)$$

where t is measured in years, with $t = 0$ corresponding to the beginning of 1980. Compute $P''(20)$ and interpret your result.

Source: U.S. Bureau of Labor Statistics

In Exercises 39–43, determine whether the statement is true or false. If it is true, explain why it is true. If it is false, give an example to show why it is false.

39. If the second derivative of f exists at $x = a$, then $f''(a) = [f'(a)]^2$.

40. If $h = fg$ where f and g have second-order derivatives, then

$$h''(x) = f''(x)g(x) + 2f'(x)g'(x) + f(x)g''(x)$$

41. If $f(x)$ is a polynomial function of degree n, then $f^{(n+1)}(x) = 0$.

42. Suppose $P(t)$ represents the population of bacteria at time t and suppose $P'(t) > 0$ and $P''(t) < 0$; then the population is increasing at time t but at a decreasing rate.

43. If $h(x) = f(2x)$, then $h''(x) = 4f''(2x)$.

44. Let f be the function defined by the rule $f(x) = x^{7/3}$. Show that f has first- and second-order derivatives at all points x, and in particular at $x = 0$. Also show that the third derivative of f does *not* exist at $x = 0$.

45. Construct a function f that has derivatives of order up through and including n at a point a but fails to have the $(n + 1)$st derivative there.
Hint: See Exercise 44.

46. Show that a polynomial function has derivatives of all orders.
Hint: Let $P(x) = a_0x^n + a_1x^{n-1} + a_2x^{n-2} + \cdots + a_n$ be a polynomial of degree n, where n is a positive integer and $a_0, a_1, \ldots, a_n$ are constants with $a_0 \neq 0$. Compute $P'(x), P''(x), \ldots$.

12.5   Solutions to Self-Check Exercises

1. $f'(x) = 10x^4 - 9x^2 + 2x - 6$
$f''(x) = 40x^3 - 18x + 2$
$f'''(x) = 120x^2 - 18$

2. We write $f(x) = (1 + x)^{-1}$ and use the general power rule, obtaining

$$f'(x) = (-1)(1 + x)^{-2}\frac{d}{dx}(1 + x) = -(1 + x)^{-2}(1)$$

$$= -(1 + x)^{-2} = -\frac{1}{(1 + x)^2}$$

Continuing, we find

$$f''(x) = -(-2)(1 + x)^{-3}$$

$$= 2(1 + x)^{-3} = \frac{2}{(1 + x)^3}$$

$$f'''(x) = 2(-3)(1 + x)^{-4}$$
$$= -6(1 + x)^{-4}$$
$$= -\frac{6}{(1 + x)^4}$$

3. $N'(t) = 6t^2 + 6t - 4$ and $N''(t) = 12t + 6 = 6(2t + 1)$

Therefore, $N''(2) = 30$ and $N''(8) = 102$. The results of our computations reveal that at the end of year 2, the *rate* of growth of the turtle population is increasing at the rate of 30 turtles/year/year. At the end of year 8, the rate is increasing at the rate of 102 turtles/year/year. Clearly, the conservation measures are paying off handsomely.

USING TECHNOLOGY

▬ Finding the Second Derivative of a Function at a Given Point

Some graphing utilities have the capability of numerically computing the second derivative of a function at a point. If your graphing utility has this capability, use it to work through the examples and exercises of this section.

EXAMPLE 1   Use the (second) numerical derivative operation of a graphing utility to find the second derivative of $f(x) = \sqrt{x}$ when $x = 4$.

Solution   Using the (second) numerical derivative operation, we find

$$f''(4) = \text{der2}(x^.5, x, 4) = -.03125$$

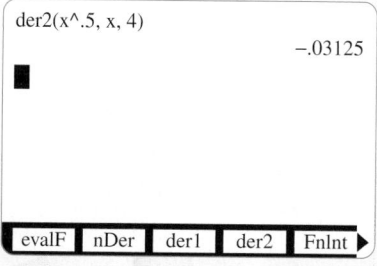

FIGURE T1
The TI-86 second derivative screen for computing $f''(4)$

(Figure T1).

APPLIED EXAMPLE 2   Prevalence of Alzheimer's Patients   The number of Alzheimer's patients in the United States is given by

$$f(t) = -0.02765t^4 + 0.3346t^3 - 1.1261t^2$$
$$+ 1.7575t + 3.7745 \quad (0 \le t \le 6)$$

where $f(t)$ is measured in millions and t is measured in decades, with $t = 0$ corresponding to the beginning of 1990.

a. How fast is the number of Alzheimer's patients in the United States anticipated to be changing at the beginning of 2030?
b. How fast is the rate of change of the number of Alzheimer's patients in the United States anticipated to be changing at the beginning of 2030?
c. Plot the graph of f in the viewing window $[0, 7] \times [0, 12]$.
Source: Alzheimer's Association

Solution

a. Using the numerical derivative operation of a graphing utility, we find that the number of Alzheimer's patients at the beginning of 2030 can be anticipated to be changing at the rate of

$$f'(4) = 1.7311$$

That is, the number is increasing at the rate of approximately 1.7 million patients per decade.

(continued)

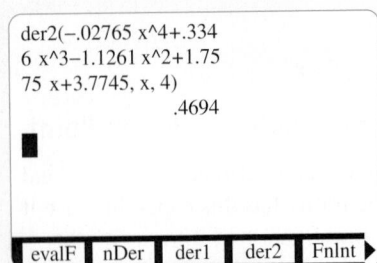

```
der2(−.02765 x^4+.334
6 x^3−1.1261 x^2+1.75
75 x+3.7745, x, 4)
              .4694
■
```

| evalF | nDer | der1 | der2 | FnInt ▶ |

FIGURE T2
The TI-86 second derivative screen for computing $f''(4)$

b. Using the (second) numerical derivative operation of a graphing utility, we find that

$$f''(4) = .4694$$

(Figure T2); that is, the rate of change of the number of Alzheimer's patients is increasing at the rate of approximately 0.5 million patients per decade per decade.

c. Figure T3 shows the graph.

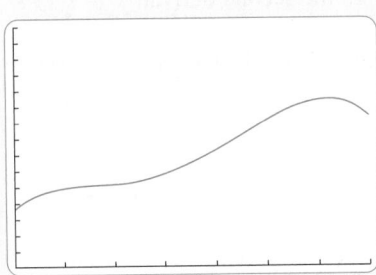

FIGURE T3
The graph of f in the viewing window $[0, 7] \times [0, 12]$

■ TECHNOLOGY EXERCISES

In Exercises 1–8, find the value of the second derivative of f at the given value of x. Express your answer correct to four decimal places.

1. $f(x) = 2x^3 - 3x^2 + 1; x = -1$

2. $f(x) = 2.5x^5 - 3x^3 + 1.5x + 4; x = 2.1$

3. $f(x) = 2.1x^{3.1} - 4.2x^{1.7} + 4.2; x = 1.4$

4. $f(x) = 1.7x^{4.2} - 3.2x^{1.3} + 4.2x - 3.2; x = 2.2$

5. $f(x) = \dfrac{x^2 + 2x - 5}{x^3 + 1}; x = 2.1$

6. $f(x) = \dfrac{x^3 + x + 2}{2x^2 - 5x + 4}; x = 1.2$

7. $f(x) = \dfrac{x^{1/2} + 2x^{3/2} + 1}{2x^{1/2} + 3}; x = 0.5$

8. $f(x) = \dfrac{\sqrt{x} - 1}{2x + \sqrt{x} + 4}; x = 2.3$

9. RATE OF BANK FAILURES The rate at which banks were failing between 1982 and 1994 is given by

$$f(t) = -0.063447t^4 - 1.953283t^3 + 14.632576t^2 \\ - 6.684704t + 47.458874 \quad (0 \le t \le 12)$$

where $f(t)$ is measured in the number of banks/year and t is measured in years, with $t = 0$ corresponding to the beginning of 1982. Compute $f''(6)$ and interpret your results.

Source: Federal Deposit Insurance Corporation

10. MULTIMEDIA SALES Sales in the multimedia market (hardware and software) are given by

$$S(t) = -0.0094t^4 + 0.1204t^3 - 0.0868t^2 \\ + 0.0195t + 3.3325 \quad (0 \le t \le 10)$$

where $S(t)$ is measured in billions of dollars and t is measured in years, with $t = 0$ corresponding to 1990. Compute $S''(7)$ and interpret your results.

Source: Electronics Industries Association

12.6 Implicit Differentiation and Related Rates (Optional)

■ Differentiating Implicitly

Up to now we have dealt with functions expressed in the form $y = f(x)$; that is, the dependent variable y is expressed *explicitly* in terms of the independent variable x. However, not all functions are expressed in this form. Consider, for example, the equation

$$x^2y + y - x^2 + 1 = 0 \tag{8}$$

This equation does express y *implicitly* as a function of x. In fact, solving (8) for y in terms of x, we obtain

$$(x^2 + 1)y = x^2 - 1 \qquad \text{Implicit equation}$$

$$y = f(x) = \frac{x^2 - 1}{x^2 + 1} \qquad \text{Explicit equation}$$

which gives an explicit representation of f.

Next, consider the equation

$$y^4 - y^3 - y + 2x^3 - x = 8$$

When certain restrictions are placed on x and y, this equation defines y as a function of x. But in this instance, we would be hard pressed to find y explicitly in terms of x. The following question arises naturally: How does one go about computing dy/dx in this case?

As it turns out, thanks to the chain rule, a method *does* exist for computing the derivative of a function directly from the implicit equation defining the function. This method is called **implicit differentiation** and is demonstrated in the next several examples.

EXAMPLE 1   Given the equation $y^2 = x$, find $\dfrac{dy}{dx}$.

Solution   Differentiating both sides of the equation with respect to x, we obtain

$$\frac{d}{dx}(y^2) = \frac{d}{dx}(x)$$

To carry out the differentiation of the term $\dfrac{d}{dx}(y^2)$, we note that y is a function of x. Writing $y = f(x)$ to remind us of this fact, we find that

$$\frac{d}{dx}(y^2) = \frac{d}{dx}[f(x)]^2 \qquad \text{Write } y = f(x).$$

$$= 2f(x)f'(x) \qquad \text{Use the chain rule.}$$

$$= 2y\frac{dy}{dx} \qquad \text{Return to using } y \text{ instead of } f(x).$$

Therefore, the equation

$$\frac{d}{dx}(y^2) = \frac{d}{dx}(x)$$

is equivalent to

$$2y\frac{dy}{dx} = 1$$

Solving for $\frac{dy}{dx}$ yields

$$\frac{dy}{dx} = \frac{1}{2y}$$
∎

Before considering other examples, let's summarize the important steps involved in implicit differentiation. (Here we assume that dy/dx exists.)

Finding $\dfrac{dy}{dx}$ by Implicit Differentiation

1. Differentiate both sides of the equation *with respect to x*. (Make sure that the derivative of any term involving y includes the factor dy/dx).
2. Solve the resulting equation for dy/dx in terms of x and y.

EXAMPLE 2   Find $\dfrac{dy}{dx}$ given the equation

$$y^3 - y + 2x^3 - x = 8$$

Solution   Differentiating both sides of the given equation with respect to x, we obtain

$$\frac{d}{dx}(y^3 - y + 2x^3 - x) = \frac{d}{dx}(8)$$

$$\frac{d}{dx}(y^3) - \frac{d}{dx}(y) + \frac{d}{dx}(2x^3) - \frac{d}{dx}(x) = 0$$

Now, recalling that y is a function of x, we apply the chain rule to the first two terms on the left. Thus,

$$3y^2\frac{dy}{dx} - \frac{dy}{dx} + 6x^2 - 1 = 0$$

$$(3y^2 - 1)\frac{dy}{dx} = 1 - 6x^2$$

$$\frac{dy}{dx} = \frac{1 - 6x^2}{3y^2 - 1}$$
∎

EXPLORE AND DISCUSS

Refer to Example 2. Suppose we think of the equation $y^3 - y + 2x^3 - x = 8$ as defining x implicitly as a function of y. Find dx/dy and justify your method of solution.

EXAMPLE 3   Consider the equation $x^2 + y^2 = 4$.

a. Find dy/dx by implicit differentiation.
b. Find the slope of the tangent line to the graph of the function $y = f(x)$ at the point $(1, \sqrt{3})$.
c. Find an equation of the tangent line of part (b).

Solution

a. Differentiating both sides of the equation with respect to x, we obtain

$$\frac{d}{dx}(x^2 + y^2) = \frac{d}{dx}(4)$$

$$\frac{d}{dx}(x^2) + \frac{d}{dx}(y^2) = 0$$

$$2x + 2y\frac{dy}{dx} = 0$$

$$\frac{dy}{dx} = -\frac{x}{y} \qquad (y \neq 0)$$

b. The slope of the tangent line to the graph of the function at the point $(1, \sqrt{3})$ is given by

$$\left.\frac{dy}{dx}\right|_{(1,\sqrt{3})} = \left.-\frac{x}{y}\right|_{(1,\sqrt{3})} = -\frac{1}{\sqrt{3}}$$

(*Note:* This notation is read "dy/dx evaluated at the point $(1, \sqrt{3})$.")

c. An equation of the tangent line in question is found by using the point-slope form of the equation of a line with the slope $m = -1/\sqrt{3}$ and the point $(1, \sqrt{3})$. Thus,

$$y - \sqrt{3} = -\frac{1}{\sqrt{3}}(x - 1)$$

$$\sqrt{3}y - 3 = -x + 1$$

$$x + \sqrt{3}y - 4 = 0$$

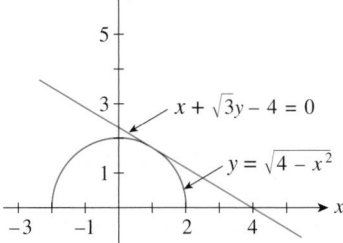

FIGURE 17
The line $x + \sqrt{3}y - 4 = 0$ is tangent to the graph of the function $y = f(x)$.

A sketch of this tangent line is shown in Figure 17.

We can also solve the equation $x^2 + y^2 = 4$ explicitly for y in terms of x. If we do this, we obtain

$$y = \pm\sqrt{4 - x^2}$$

From this, we see that the equation $x^2 + y^2 = 4$ defines the two functions

$$y = f(x) = \sqrt{4 - x^2}$$

$$y = g(x) = -\sqrt{4 - x^2}$$

Since the point $(1, \sqrt{3})$ does not lie on the graph of $y = g(x)$, we conclude that

$$y = f(x) = \sqrt{4 - x^2}$$

is the required function. The graph of f is the upper semicircle shown in Figure 17. ▪

Note The notation

$$\left.\frac{dy}{dx}\right|_{(a, b)}$$

is used to denote the value of dy/dx at the point (a, b). ▪

To find dy/dx at a *specific point* (a, b), differentiate the given equation implicitly with respect to x and then replace x and y by a and b, respectively, *before* solving the equation for dy/dx. This often simplifies the amount of algebra involved.

EXAMPLE 4 Find $\dfrac{dy}{dx}$ given that x and y are related by the equation

$$x^2y^3 + 6x^2 = y + 12$$

and that $y = 2$ when $x = 1$.

Solution Differentiating both sides of the given equation with respect to x, we obtain

$$\frac{d}{dx}(x^2y^3) + \frac{d}{dx}(6x^2) = \frac{d}{dx}(y) + \frac{d}{dx}(12)$$

$$x^2 \cdot \frac{d}{dx}(y^3) + y^3 \cdot \frac{d}{dx}(x^2) + 12x = \frac{dy}{dx} \qquad \text{Use the product rule on } \frac{d}{dx}(x^2y^3).$$

$$3x^2y^2\frac{dy}{dx} + 2xy^3 + 12x = \frac{dy}{dx}$$

Substituting $x = 1$ and $y = 2$ into this equation gives

$$3(1)^2(2)^2\frac{dy}{dx} + 2(1)(2)^3 + 12(1) = \frac{dy}{dx}$$

$$12\frac{dy}{dx} + 16 + 12 = \frac{dy}{dx}$$

and, solving for $\dfrac{dy}{dx}$,

$$\frac{dy}{dx} = -\frac{28}{11}$$

Note that it is not necessary to find an explicit expression for dy/dx.   ∎

Note In Examples 3 and 4, you can verify that the points at which we evaluated dy/dx actually lie on the curve in question by showing that the coordinates of the points satisfy the given equations.   ∎

EXAMPLE 5   Find $\dfrac{dy}{dx}$ given that x and y are related by the equation

$$\sqrt{x^2 + y^2} - x^2 = 5$$

Solution   Differentiating both sides of the given equation with respect to x, we obtain

$$\frac{d}{dx}(x^2 + y^2)^{1/2} - \frac{d}{dx}(x^2) = \frac{d}{dx}(5)$$

Write $\sqrt{x^2 + y^2} = (x^2 + y^2)^{1/2}$.

$$\frac{1}{2}(x^2 + y^2)^{-1/2}\frac{d}{dx}(x^2 + y^2) - 2x = 0$$

Use the general power rule on the first term.

$$\frac{1}{2}(x^2 + y^2)^{-1/2}\left(2x + 2y\frac{dy}{dx}\right) - 2x = 0$$

$$2x + 2y\frac{dy}{dx} = 4x(x^2 + y^2)^{1/2}$$

Transpose $2x$ and multiply both sides by $2(x^2 + y^2)^{1/2}$.

$$2y\frac{dy}{dx} = 4x(x^2 + y^2)^{1/2} - 2x$$

$$\frac{dy}{dx} = \frac{2x\sqrt{x^2 + y^2} - x}{y}$$ ∎

■ Related Rates

Implicit differentiation is a useful technique for solving a class of problems known as **related-rates** problems. The following is a typical related-rates problem: Suppose x and y are two quantities that depend on a third quantity t and we know the relationship between x and y in the form of an equation. Can we find a relationship between dx/dt and dy/dt? In particular, if we know one of the rates of change at a specific value of t—say, dx/dt—can we find the other rate, dy/dt, at that value of t?

VIDEO

   APPLIED EXAMPLE 6   Rate of Change of Housing Starts   A study prepared for the National Association of Realtors estimates that the number of housing starts in the southwest, $N(t)$ (in units of a million), over the next 5 years is related to the mortgage rate $r(t)$ (percent per year) by the equation

$$9N^2 + r = 36$$

What is the rate of change of the number of housing starts with respect to time when the mortgage rate is 11% per year and is increasing at the rate of 1.5% per year?

Solution   We are given that

$$r = 11 \quad \text{and} \quad \frac{dr}{dt} = 1.5$$

at a certain instant of time, and we are required to find dN/dt. First, by substituting $r = 11$ into the given equation, we find

$$9N^2 + 11 = 36$$

$$N^2 = \frac{25}{9}$$

or $N = 5/3$ (we reject the negative root). Next, differentiating the given equation implicitly on both sides with respect to t, we obtain

$$\frac{d}{dt}(9N^2) + \frac{d}{dt}(r) = \frac{d}{dt}(36)$$

$$18N\frac{dN}{dt} + \frac{dr}{dt} = 0 \qquad \text{Use the chain rule on the first term.}$$

Then, substituting $N = 5/3$ and $dr/dt = 1.5$ into this equation gives

$$18\left(\frac{5}{3}\right)\frac{dN}{dt} + 1.5 = 0$$

Solving this equation for dN/dt then gives

$$\frac{dN}{dt} = -\frac{1.5}{30} \approx -0.05$$

Thus, at the instant of time under consideration, the number of housing starts is decreasing at the rate of 50,000 units per year. ∎

APPLIED EXAMPLE 7   Supply-Demand   A major audiotape manufacturer is willing to make x thousand ten-packs of metal alloy audiocassette tapes available in the marketplace each week when the wholesale price is \$$p$ per ten-pack. It is known that the relationship between x and p is governed by the supply equation

$$x^2 - 3xp + p^2 = 5$$

How fast is the supply of tapes changing when the price per ten-pack is \$11, the quantity supplied is 4000 ten-packs, and the wholesale price per ten-pack is increasing at the rate of \$.10 per ten-pack each week?

Solution   We are given that

$$p = 11, \qquad x = 4, \qquad \frac{dp}{dt} = 0.1$$

at a certain instant of time, and we are required to find dx/dt. Differentiating the given equation on both sides with respect to t, we obtain

$$\frac{d}{dt}(x^2) - \frac{d}{dt}(3xp) + \frac{d}{dt}(p^2) = \frac{d}{dt}(5)$$

$$2x\frac{dx}{dt} - 3\left(p\frac{dx}{dt} + x\frac{dp}{dt}\right) + 2p\frac{dp}{dt} = 0 \qquad \text{Use the product rule on the second term.}$$

Substituting the given values of p, x, and dp/dt into the last equation, we have

$$2(4)\frac{dx}{dt} - 3\left[(11)\frac{dx}{dt} + 4(0.1)\right] + 2(11)(0.1) = 0$$

$$8\frac{dx}{dt} - 33\frac{dx}{dt} - 1.2 + 2.2 = 0$$

$$25\frac{dx}{dt} = 1$$

$$\frac{dx}{dt} = 0.04$$

Thus, at the instant of time under consideration the supply of ten-pack audiocassettes is increasing at the rate of (0.04)(1000), or 40, ten-packs per week. ■

In certain related-rates problems, we need to formulate the problem mathematically before analyzing it. The following guidelines can be used to help solve problems of this type.

Solving Related-Rates Problems

1. Assign a variable to each quantity. Draw a diagram if needed.
2. Write the *given* values of the variables and their rates of change with respect to t.
3. Find an equation giving the relationship between the variables.
4. Differentiate both sides of this equation implicitly with respect to t.
5. Replace the variables and their derivatives by the numerical data found in step 2 and solve the equation for the required rate of change.

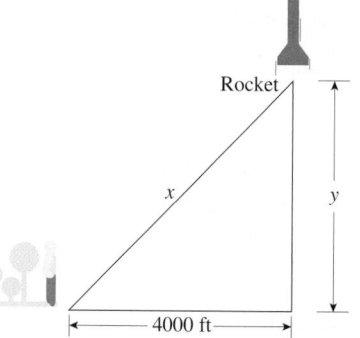

FIGURE 18
The rate at which x is changing with respect to time is related to the rate of change of y with respect to time.

APPLIED EXAMPLE 8 Watching a Rocket Launch   At a distance of 4000 feet from the launch site, a spectator is observing a rocket being launched. If the rocket lifts off vertically and is rising at a speed of 600 feet/second when it is at an altitude of 3000 feet, how fast is the distance between the rocket and the spectator changing at that instant?

Solution

Step 1   Let

$$y = \text{altitude of the rocket}$$
$$x = \text{distance between the rocket and the spectator}$$

at any time t (Figure 18).

Step 2   We are given that at a certain instant of time

$$y = 3000 \quad \text{and} \quad \frac{dy}{dt} = 600$$

and are asked to find dx/dt at that instant.

Step 3   Applying the Pythagorean theorem to the right triangle in Figure 18, we find that

$$x^2 = y^2 + 4000^2$$

Therefore, when $y = 3000$,

$$x = \sqrt{3000^2 + 4000^2} = 5000$$

Step 4   Next, we differentiate the equation $x^2 = y^2 + 4000^2$ with respect to t, obtaining

$$2x\frac{dx}{dt} = 2y\frac{dy}{dt}$$

(Remember, both x and y are functions of t.)

Step 5   Substituting $x = 5000$, $y = 3000$, and $dy/dt = 600$, we find

$$2(5000)\frac{dx}{dt} = 2(3000)(600)$$

$$\frac{dx}{dt} = 360$$

Therefore, the distance between the rocket and the spectator is changing at a rate of 360 feet/second. ■

 Be sure that you do *not* replace the variables in the equation found in Step 3 by their numerical values before differentiating the equation.

EXAMPLE 9   A passenger ship and an oil tanker left port sometime in the morning; the former headed north, and the latter headed east. At noon, the passenger ship was 40 miles from port and sailing at 30 mph, while the oil tanker was 30 miles from port and sailing at 20 mph. How fast was the distance between the two ships changing at that time?

Solution

Step 1   Let

$$x = \text{distance of the oil tanker from port}$$
$$y = \text{distance of the passenger ship from port}$$
$$z = \text{distance between the two ships}$$

See Figure 19.

Step 2   We are given that at noon

$$x = 30 \qquad y = 40 \qquad \frac{dx}{dt} = 20 \qquad \frac{dy}{dt} = 30$$

and we are required to find $\frac{dz}{dt}$ at that time.

Step 3   Applying the Pythagorean theorem to the right triangle in Figure 19, we find that

$$z^2 = x^2 + y^2 \tag{9}$$

In particular, when $x = 30$ and $y = 40$, we have

$$z^2 = 30^2 + 40^2 = 2500 \quad \text{or} \quad z = 50.$$

Step 4   Differentiating (9) implicitly with respect to t, we obtain

$$2z\frac{dz}{dt} = 2x\frac{dx}{dt} + 2y\frac{dy}{dt}$$

$$z\frac{dz}{dt} = x\frac{dx}{dt} + y\frac{dy}{dt}$$

Step 5   Finally, substituting $x = 30$, $y = 40$, $z = 50$, $dx/dt = 20$, and $dy/dt = 30$ into the last equation, we find

$$50\frac{dz}{dt} = (30)(20) + (40)(30) \quad \text{and} \quad \frac{dz}{dt} = 36$$

Therefore, at noon on the day in question, the ships are moving apart at the rate of 36 mph. ■

Passenger ship

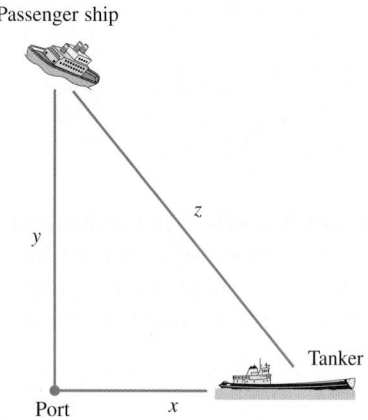

FIGURE 19
We want to find *dz/dt*, the rate at which the distance between the two ships is changing at a certain instant of time.

12.6 Self-Check Exercises

1. Given the equation $x^3 + 3xy + y^3 = 4$, find dy/dx by implicit differentiation.

2. Find an equation of the tangent line to the graph of

$$16x^2 + 9y^2 = 144$$

at the point

$$\left(2, -\frac{4\sqrt{5}}{3}\right)$$

Solutions to Self-Check Exercises 12.6 can be found on page 756.

12.6 Concept Questions

1. **a.** Suppose the equation $F(x, y) = 0$ defines y as a function of x. Explain how implicit differentiation can be used to find dy/dx.
   b. What is the role of the chain rule in implicit differentiation?

2. Suppose the equation $xg(y) + yf(x) = 0$, where f and g are differentiable functions, defines y as a function of x. Find an expression for dy/dx.

3. In your own words, describe what a related-rates problem is.

4. Give the steps that you would use to solve a related-rates problem.

12.6 Exercises

In Exercises 1–8, find the derivative dy/dx (a) by solving each of the implicit equations for y explicitly in terms of x and (b) by differentiating each of the equations implicitly. Show that, in each case, the results are equivalent.

1. $x + 2y = 5$

2. $3x + 4y = 6$

3. $xy = 1$

4. $xy - y - 1 = 0$

5. $x^3 - x^2 - xy = 4$

6. $x^2y - x^2 + y - 1 = 0$

7. $\dfrac{x}{y} - x^2 = 1$

8. $\dfrac{y}{x} - 2x^3 = 4$

In Exercises 9–30, find dy/dx by implicit differentiation.

9. $x^2 + y^2 = 16$

10. $2x^2 + y^2 = 16$

11. $x^2 - 2y^2 = 16$

12. $x^3 + y^3 + y - 4 = 0$

13. $x^2 - 2xy = 6$

14. $x^2 + 5xy + y^2 = 10$

15. $x^2y^2 - xy = 8$

16. $x^2y^3 - 2xy^2 = 5$

17. $x^{1/2} + y^{1/2} = 1$

18. $x^{1/3} + y^{1/3} = 1$

19. $\sqrt{x + y} = x$

20. $(2x + 3y)^{1/3} = x^2$

21. $\dfrac{1}{x^2} + \dfrac{1}{y^2} = 1$

22. $\dfrac{1}{x^3} + \dfrac{1}{y^3} = 5$

23. $\sqrt{xy} = x + y$

24. $\sqrt{xy} = 2x + y^2$

25. $\dfrac{x + y}{x - y} = 3x$

26. $\dfrac{x - y}{2x + 3y} = 2x$

27. $xy^{3/2} = x^2 + y^2$

28. $x^2y^{1/2} = x + 2y^3$

29. $(x + y)^3 + x^3 + y^3 = 0$

30. $(x + y^2)^{10} = x^2 + 25$

In Exercises 31–34, find an equation of the tangent line to the graph of the function f defined by the equation at the indicated point.

31. $4x^2 + 9y^2 = 36$; $(0, 2)$

32. $y^2 - x^2 = 16$; $(2, 2\sqrt{5})$

33. $x^2y^3 - y^2 + xy - 1 = 0$; $(1, 1)$

34. $(x - y - 1)^3 = x$; $(1, -1)$

In Exercises 35–38, find the second derivative d^2y/dx^2 of each of the functions defined implicitly by the equation.

35. $xy = 1$

36. $x^3 + y^3 = 28$

37. $y^2 - xy = 8$

38. $x^{1/3} + y^{1/3} = 1$

39. The volume of a right-circular cylinder of radius r and height h is $V = \pi r^2 h$. Suppose the radius and height of the cylinder are changing with respect to time t.
   a. Find a relationship between dV/dt, dr/dt, and dh/dt.
   b. At a certain instant of time, the radius and height of the cylinder are 2 and 6 in. and are increasing at the rate of

0.1 and 0.3 in./sec, respectively. How fast is the volume of the cylinder increasing?

40. A car leaves an intersection traveling west. Its position 4 sec later is 20 ft from the intersection. At the same time, another car leaves the same intersection heading north so that its position 4 sec later is 28 ft from the intersection. If the speed of the cars at that instant of time is 9 ft/sec and 11 ft/sec, respectively, find the rate at which the distance between the two cars is changing.

41. PRICE-DEMAND   Suppose the quantity demanded weekly of the Super Titan radial tires is related to its unit price by the equation

$$p + x^2 = 144$$

where p is measured in dollars and x is measured in units of a thousand. How fast is the quantity demanded changing when $x = 9$, $p = 63$, and the price/tire is increasing at the rate of $2/week?

42. PRICE-SUPPLY   Suppose the quantity x of Super Titan radial tires made available each week in the marketplace is related to the unit-selling price by the equation

$$p - \frac{1}{2}x^2 = 48$$

where x is measured in units of a thousand and p is in dollars. How fast is the weekly supply of Super Titan radial tires being introduced into the marketplace when $x = 6$, $p = 66$, and the price/tire is decreasing at the rate of $3/week?

43. PRICE-DEMAND   The demand equation for a certain brand of metal alloy audiotape is

$$100x^2 + 9p^2 = 3600$$

where x represents the number (in thousands) of ten-packs demanded each week when the unit price is p. How fast is the quantity demanded increasing when the unit price/ten-pack is $14 and the selling price is dropping at the rate of $.15/ten-pack/week?
Hint: To find the value of x when $p = 14$, solve the equation $100x^2 + 9p^2 = 3600$ for x when $p = 14$.

44. EFFECT OF PRICE ON SUPPLY   Suppose the wholesale price of a certain brand of medium-sized eggs p (in dollars/carton) is related to the weekly supply x (in thousands of cartons) by the equation

$$625p^2 - x^2 = 100$$

If 25,000 cartons of eggs are available at the beginning of a certain week and the price is falling at the rate of 2¢/carton/week, at what rate is the supply falling?
Hint: To find the value of p when $x = 25$, solve the supply equation for p when $x = 25$.

45. SUPPLY-DEMAND   Refer to Exercise 44. If 25,000 cartons of eggs are available at the beginning of a certain week and the supply is falling at the rate of 1000 cartons/week, at what rate is the wholesale price changing?

46. ELASTICITY OF DEMAND   The demand function for a certain make of ink-jet cartridge is

$$p = -0.01x^2 - 0.1x + 6$$

where p is the unit price in dollars and x is the quantity demanded each week, measured in units of a thousand. Compute the elasticity of demand and determine whether the demand is inelastic, unitary, or elastic when $x = 10$.

47. ELASTICITY OF DEMAND   The demand function for a certain brand of compact disc is

$$p = -0.01x^2 - 0.2x + 8$$

where p is the wholesale unit price in dollars and x is the quantity demanded each week, measured in units of a thousand. Compute the elasticity of demand and determine whether the demand is inelastic, unitary, or elastic when $x = 15$.

48. The volume V of a cube with sides of length x in. is changing with respect to time. At a certain instant of time, the sides of the cube are 5 in. long and increasing at the rate of 0.1 in./sec. How fast is the volume of the cube changing at that instant of time?

49. OIL SPILLS   In calm waters, oil spilling from the ruptured hull of a grounded tanker spreads in all directions. If the area polluted is a circle and its radius is increasing at a rate of 2 ft/sec, determine how fast the area is increasing when the radius of the circle is 40 ft.

50. Two ships leave the same port at noon. Ship A sails north at 15 mph, and ship B sails east at 12 mph. How fast is the distance between them changing at 1 p.m.?

51. A car leaves an intersection traveling east. Its position t sec later is given by $x = t^2 + t$ ft. At the same time, another car leaves the same intersection heading north, traveling $y = t^2 + 3t$ ft in t sec. Find the rate at which the distance between the two cars will be changing 5 sec later.

52. At a distance of 50 ft from the pad, a man observes a helicopter taking off from a heliport. If the helicopter lifts off vertically and is rising at a speed of 44 ft/sec when it is at an altitude of 120 ft, how fast is the distance between the helicopter and the man changing at that instant?

53. A spectator watches a rowing race from the edge of a river bank. The lead boat is moving in a straight line that is 120 ft from the river bank. If the boat is moving at a constant speed of 20 ft/sec, how fast is the boat moving away from the spectator when it is 50 ft past her?

54. A boat is pulled toward a dock by means of a rope wound on a drum that is located 4 ft above the bow of the boat. If the rope is being pulled in at the rate of 3 ft/sec, how fast is the boat approaching the dock when it is 25 ft from the dock? (See figure on the next page.)

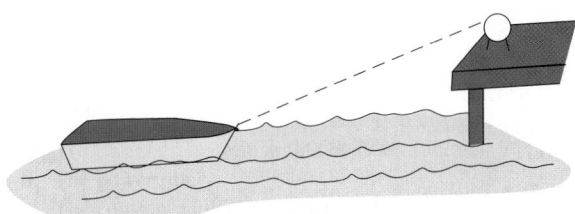

55. Assume that a snowball is in the shape of a sphere. If the snowball melts at a rate that is proportional to its surface area, show that its radius decreases at a constant rate.
Hint: Its volume is $V = (4/3)\pi r^3$, and its surface area is $S = 4\pi r^2$.

56. BLOWING SOAP BUBBLES Carlos is blowing air into a soap bubble at the rate of 8 cm³/sec. Assuming that the bubble is spherical, how fast is its radius changing at the instant of time when the radius is 10 cm? How fast is the surface area of the bubble changing at that instant of time?

57. COAST GUARD PATROL SEARCH MISSION The pilot of a Coast Guard patrol aircraft on a search mission had just spotted a disabled fishing trawler and decided to go in for a closer look. Flying in a straight line at a constant altitude of 1000 ft and at a steady speed of 264 ft/sec, the aircraft passed directly over the trawler. How fast was the aircraft receding from the trawler when it was 1500 ft from it?

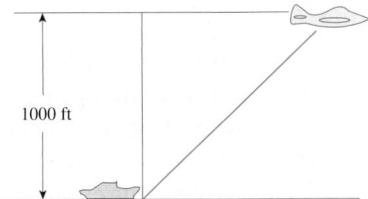

58. A coffee pot in the form of a circular cylinder of radius 4 in. is being filled with water flowing at a constant rate. If the water level is rising at the rate of 0.4 in./sec, what is the rate at which water is flowing into the coffee pot?

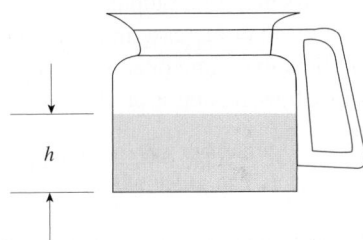

59. A 6-ft tall man is walking away from a street light 18 ft high at a speed of 6 ft/sec. How fast is the tip of his shadow moving along the ground?

60. A 20-ft ladder leaning against a wall begins to slide. How fast is the top of the ladder sliding down the wall at the instant of time when the bottom of the ladder is 12 ft from the wall and sliding away from the wall at the rate of 5 ft/sec?

Hint: Refer to the accompanying figure. By the Pythagorean theorem, $x^2 + y^2 = 400$. Find dy/dt when $x = 12$ and $dx/dt = 5$.

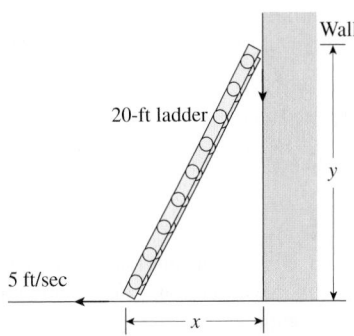

61. The base of a 13-ft ladder leaning against a wall begins to slide away from the wall. At the instant of time when the base is 12 ft from the wall, the base is moving at the rate of 8 ft/sec. How fast is the top of the ladder sliding down the wall at that instant of time?
Hint: Refer to the hint in Problem 60.

62. Water flows from a tank of constant cross-sectional area 50 ft² through an orifice of constant cross-sectional area 1.4 ft² located at the bottom of the tank (see the figure).

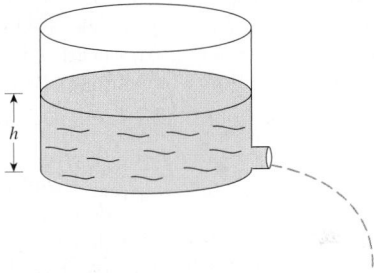

Initially, the height of the water in the tank was 20 ft and its height t sec later is given by the equation

$$2\sqrt{h} + \frac{1}{25}t - 2\sqrt{20} = 0 \qquad (0 \le t \le 50\sqrt{20})$$

How fast was the height of the water decreasing when its height was 8 ft?

In Exercises 63 and 64, determine whether the statement is true or false. If it is true, explain why it is true. If it is false, give an example to show why it is false.

63. If f and g are differentiable and $f(x)g(y) = 0$, then

$$\frac{dy}{dx} = -\frac{f'(x)g(y)}{f(x)g'(y)} \qquad (f(x) \ne 0 \text{ and } g'(y) \ne 0)$$

64. If f and g are differentiable and $f(x) + g(y) = 0$, then

$$\frac{dy}{dx} = -\frac{f'(x)}{g'(y)}$$

12.6 Solutions to Self-Check Exercises

1. Differentiating both sides of the equation with respect to x, we have

$$3x^2 + 3y + 3xy' + 3y^2y' = 0$$
$$(x^2 + y) + (x + y^2)y' = 0$$
$$y' = -\frac{x^2 + y}{x + y^2}$$

2. To find the slope of the tangent line to the graph of the function at any point, we differentiate the equation implicitly with respect to x, obtaining

$$32x + 18yy' = 0$$
$$y' = -\frac{16x}{9y}$$

In particular, the slope of the tangent line at $\left(2, -\dfrac{4\sqrt{5}}{3}\right)$ is

$$m = -\frac{16(2)}{9\left(-\dfrac{4\sqrt{5}}{3}\right)} = \frac{8}{3\sqrt{5}}$$

Using the point-slope form of the equation of a line, we find

$$y - \left(-\frac{4\sqrt{5}}{3}\right) = \frac{8}{3\sqrt{5}}(x - 2)$$
$$y = \frac{8\sqrt{5}}{15}x - \frac{36\sqrt{5}}{15} = \frac{8\sqrt{5}}{15}x - \frac{12\sqrt{5}}{5}$$

12.7 Differentials

The Millers are planning to buy a house in the near future and estimate that they will need a 30-year fixed-rate mortgage of $240,000. If the interest rate increases from the present rate of 7% per year to 7.4% per year between now and the time the Millers decide to secure the loan, approximately how much more per month will their mortgage be? (You will be asked to answer this question in Exercise 44, page 763.)

Questions such as this, in which one wishes to *estimate* the change in the dependent variable (monthly mortgage payment) corresponding to a small change in the independent variable (interest rate per year), occur in many real-life applications. For example:

- An economist would like to know how a small increase in a country's capital expenditure will affect the country's gross domestic output.
- A sociologist would like to know how a small increase in the amount of capital investment in a housing project will affect the crime rate.
- A businesswoman would like to know how raising a product's unit price by a small amount will affect her profit.
- A bacteriologist would like to know how a small increase in the amount of a bactericide will affect a population of bacteria.

To calculate these changes and estimate their effects, we use the *differential* of a function, a concept that will be introduced shortly.

Increments

Let x denote a variable quantity and suppose x changes from x_1 to x_2. This change in x is called the **increment in x** and is denoted by the symbol Δx (read "delta x"). Thus,

$$\Delta x = x_2 - x_1 \quad \text{Final value} - \text{initial value} \tag{10}$$

EXAMPLE 1 Find the increment in x as x changes (a) from 3 to 3.2 and (b) from 3 to 2.7.

Solution

a. Here, $x_1 = 3$ and $x_2 = 3.2$, so

$$\Delta x = x_2 - x_1 = 3.2 - 3 = 0.2$$

b. Here, $x_1 = 3$ and $x_2 = 2.7$. Therefore,

$$\Delta x = x_2 - x_1 = 2.7 - 3 = -0.3$$

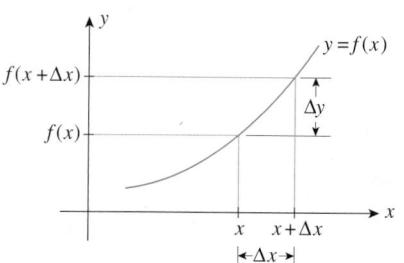

FIGURE 20
An increment of Δx in x induces an increment of $\Delta y = f(x + \Delta x) - f(x)$ in y.

Observe that Δx plays the same role that h played in Section 11.4.

Now, suppose two quantities, x and y, are related by an equation $y = f(x)$, where f is a function. If x changes from x to $x + \Delta x$, then the corresponding change in y is called the **increment in y**. It is denoted by Δy and is defined by

$$\Delta y = f(x + \Delta x) - f(x) \tag{11}$$

(see Figure 20).

EXAMPLE 2 Let $y = x^3$. Find Δx and Δy when x changes (a) from 2 to 2.01 and (b) from 2 to 1.98.

Solution Let $f(x) = x^3$.

a. Here, $\Delta x = 2.01 - 2 = 0.01$. Next,

$$\begin{aligned}
\Delta y &= f(x + \Delta x) - f(x) = f(2.01) - f(2) \\
&= (2.01)^3 - 2^3 = 8.120601 - 8 = 0.120601
\end{aligned}$$

b. Here, $\Delta x = 1.98 - 2 = -0.02$. Next,

$$\begin{aligned}
\Delta y &= f(x + \Delta x) - f(x) = f(1.98) - f(2) \\
&= (1.98)^3 - 2^3 = 7.762392 - 8 = -0.237608
\end{aligned}$$

■ Differentials

We can obtain a relatively quick and simple way of approximating Δy, the change in y due to a small change Δx, by examining the graph of the function f shown in Figure 21.

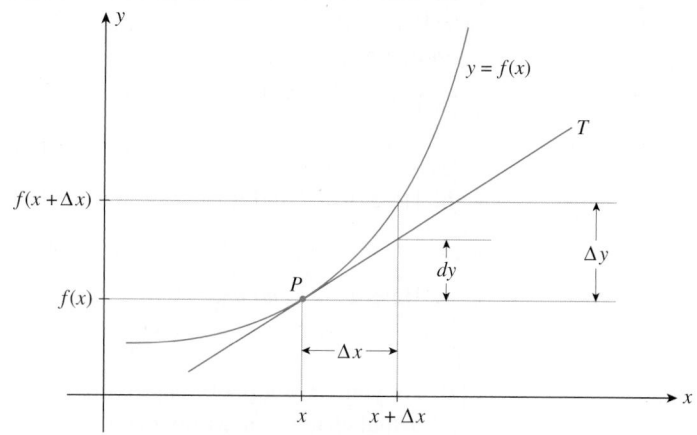

FIGURE 21
If Δx is small, dy is a good approximation of Δy.

Observe that near the point of tangency P, the tangent line T is close to the graph of f. Therefore, if Δx is small, then dy is a good approximation of Δy. We can find an expression for dy as follows: Notice that the slope of T is given by

$$\frac{dy}{\Delta x} \qquad \text{Rise} \div \text{run}$$

However, the slope of T is given by $f'(x)$. Therefore, we have

$$\frac{dy}{\Delta x} = f'(x)$$

or $dy = f'(x)\Delta x$. Thus, we have the approximation

$$\Delta y \approx dy = f'(x)\Delta x$$

in terms of the derivative of f at x. The quantity dy is called the *differential of y*.

The Differential

Let $y = f(x)$ define a differentiable function of x. Then,

1. The **differential** dx of the independent variable x is $dx = \Delta x$.

2. The **differential** dy of the dependent variable y is

$$dy = f'(x)\Delta x = f'(x)\,dx \tag{12}$$

Notes

1. For the independent variable x: There is no difference between Δx and dx—both measure the change in x from x to $x + \Delta x$.
2. For the dependent variable y: Δy measures the *actual* change in y as x changes from x to $x + \Delta x$, whereas dy measures the *approximate* change in y corresponding to the same change in x.
3. The differential dy depends on both x and dx, but for fixed x, dy is a linear function of dx.

EXAMPLE 3   Let $y = x^3$.

a. Find the differential dy of y.
b. Use dy to approximate Δy when x changes from 2 to 2.01.
c. Use dy to approximate Δy when x changes from 2 to 1.98.
d. Compare the results of part (b) with those of Example 2.

Solution

a. Let $f(x) = x^3$. Then,

$$dy = f'(x)\,dx = 3x^2\,dx$$

b. Here, $x = 2$ and $dx = 2.01 - 2 = 0.01$. Therefore,

$$dy = 3x^2\,dx = 3(2)^2(0.01) = 0.12$$

c. Here, $x = 2$ and $dx = 1.98 - 2 = -0.02$. Therefore,

$$dy = 3x^2\,dx = 3(2)^2(-0.02) = -0.24$$

d. As you can see, both approximations 0.12 and -0.24 are quite close to the actual changes of Δy obtained in Example 2: 0.120601 and -0.237608.

Observe how much easier it is to find an approximation to the exact change in a function with the help of the differential, rather than calculating the exact change in the function itself. In the following examples, we take advantage of this fact.

EXAMPLE 4   Approximate the value of $\sqrt{26.5}$ using differentials. Verify your result using the $\boxed{\sqrt{}}$ key on your calculator.

Solution   Since we want to compute the square root of a number, let's consider the function $y = f(x) = \sqrt{x}$. Since 25 is the number nearest 26.5 whose square root is readily recognized, let's take $x = 25$. We want to know the change in y, Δy, as x changes from $x = 25$ to $x = 26.5$, an increase of $\Delta x = 1.5$ units. Using Equation (12), we find

$$\Delta y \approx dy = f'(x)\,\Delta x$$

$$= \left(\frac{1}{2\sqrt{x}} \bigg|_{x=25} \right) \cdot (1.5) = \left(\frac{1}{10} \right)(1.5) = 0.15$$

Therefore,

$$\sqrt{26.5} - \sqrt{25} = \Delta y \approx 0.15$$
$$\sqrt{26.5} \approx \sqrt{25} + 0.15 = 5.15$$

The exact value of $\sqrt{26.5}$, rounded off to five decimal places, is 5.14782. Thus, the error incurred in the approximation is 0.00218.   ∎

APPLIED EXAMPLE 5   The Effect of Speed on Vehicular Operating Cost   The total cost incurred in operating a certain type of truck on a 500-mile trip, traveling at an average speed of v mph, is estimated to be

$$C(v) = 125 + v + \frac{4500}{v}$$

dollars. Find the approximate change in the total operating cost when the average speed is increased from 55 mph to 58 mph.

Solution   With $v = 55$ and $\Delta v = dv = 3$, we find

$$\Delta C \approx dC = C'(v)\,dv = \left(1 - \frac{4500}{v^2} \right) \bigg|_{v=55} (3)$$

$$= \left(1 - \frac{4500}{3025} \right)(3) \approx -1.46$$

so the total operating cost is found to decrease by $1.46. This might explain why so many independent truckers often exceed the 55 mph speed limit.   ∎

APPLIED EXAMPLE 6   The Effect of Advertising on Sales   The relationship between the amount of money x spent by Cannon Precision Instruments on advertising and Cannon's total sales $S(x)$ is given by the function

$$S(x) = -0.002x^3 + 0.6x^2 + x + 500 \qquad (0 \le x \le 200)$$

where x is measured in thousands of dollars. Use differentials to estimate the change in Cannon's total sales if advertising expenditures are increased from $100,000 ($x = 100$) to $105,000 ($x = 105$).

Solution The required change in sales is given by

$$\Delta S \approx dS = S'(100)dx$$
$$= -0.006x^2 + 1.2x + 1|_{x=100} \cdot (5) \qquad dx = 105 - 100 = 5$$
$$= (-60 + 120 + 1)(5) = 305$$

—that is, an increase of $305,000. ∎

APPLIED EXAMPLE 7 **The Rings of Neptune**

$dr = R - r$

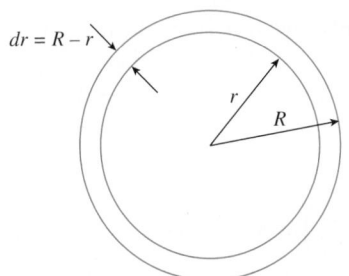

(a) The area of the ring is the circumference of the inner circle times the thickness.

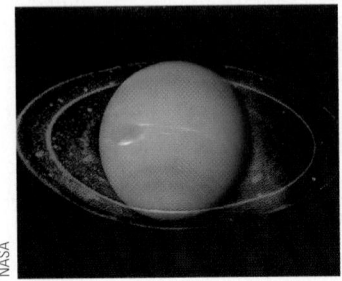

NASA

(b) Neptune and its rings

FIGURE 22

a. A ring has an inner radius of r units and an outer radius of R units, where $(R - r)$ is small in comparison to r (Figure 22a). Use differentials to estimate the area of the ring.

b. Recent observations, including those of *Voyager I* and *II*, showed that Neptune's ring system is considerably more complex than had been believed. For one thing, it is made up of a large number of distinguishable rings rather than one continuous great ring as previously thought (Figure 22b). The outermost ring, 1989N1R, has an inner radius of approximately 62,900 kilometers (measured from the center of the planet), and a radial width of approximately 50 kilometers. Using these data, estimate the area of the ring.

Solution

a. Since the area of a circle of radius x is $A = f(x) = \pi x^2$, we find

$$\pi R^2 - \pi r^2 = f(R) - f(r)$$
$$= \Delta A \qquad \text{Remember, } \Delta A = \text{change in } f \text{ when } x$$
$$\approx dA \qquad \text{changes from } x = r \text{ to } x = R.$$
$$= f'(r)dr$$

where $dr = R - r$. So, we see that the area of the ring is approximately $2\pi r(R - r)$ square units. In words, the area of the ring is approximately equal to

Circumference of the inner circle × Thickness of the ring

b. Applying the results of part (a) with $r = 62,900$ and $dr = 50$, we find that the area of the ring is approximately $2\pi(62,900)(50)$, or 19,760,618 square kilometers, which is roughly 4% of Earth's surface. ∎

Before looking at the next example, we need to familiarize ourselves with some terminology. If a quantity with exact value q is measured or calculated with an error of Δq, then the quantity $\Delta q/q$ is called the *relative error* in the measurement or calculation of q. If the quantity $\Delta q/q$ is expressed as a percentage, it is then called the *percentage error*. Because Δq is approximated by dq, we normally approximate the relative error $\Delta q/q$ by dq/q.

 APPLIED EXAMPLE 8 **Estimating Errors in Measurement** Suppose the radius of a ball-bearing is measured to be 0.5 inch, with a maximum error of ± 0.0002 inch. Then, the relative error in r is

$$\frac{dr}{r} = \frac{\pm 0.0002}{0.5} = \pm 0.0004$$

and the percentage error is $\pm 0.04\%$. ∎

APPLIED EXAMPLE 9 Estimating Errors in Measurement Suppose the side of a cube is measured with a maximum percentage error of 2%. Use differentials to estimate the maximum percentage error in the calculated volume of the cube.

Solution Suppose the side of the cube is x, so its volume is

$$V = x^3$$

We are given that $\left| \dfrac{dx}{x} \right| \leq 0.02$. Now,

$$dV = 3x^2 dx$$

and so

$$\frac{dV}{V} = \frac{3x^2 dx}{x^3} = 3\frac{dx}{x}$$

Therefore,

$$\left| \frac{dV}{V} \right| = 3\left| \frac{dx}{x} \right| \leq 3(0.02) = 0.06$$

and we see that the maximum percentage error in the measurement of the volume of the cube is 6%. ■

Finally, if at some point in reading this section you have a sense of déjà vu, do not be surprised, because the notion of the differential was first used in Section 12.4 (see Example 1). There we took $\Delta x = 1$ since we were interested in finding the marginal cost when the level of production was increased from $x = 250$ to $x = 251$. If we had used differentials, we would have found

$$C(251) - C(250) \approx C'(250)\, dx$$

so that taking $dx = \Delta x = 1$, we have $C(251) - C(250) \approx C'(250)$, which agrees with the result obtained in Example 1. Thus, in Section 12.4, we touched upon the notion of the differential, albeit in the special case in which $dx = 1$.

12.7 Self-Check Exercises

1. Find the differential of $f(x) = \sqrt{x} + 1$.
2. A certain country's government economists have determined that the demand equation for corn in that country is given by

$$p = f(x) = \frac{125}{x^2 + 1}$$

where p is expressed in dollars/bushel and x, the quantity demanded each year, is measured in billions of bushels. The

economists are forecasting a harvest of 6 billion bushels for the year. If the actual production of corn were 6.2 billion bushels for the year instead, what would be the approximate drop in the predicted price of corn/bushel?

Solutions to Self-Check Exercises 12.7 can be found on page 764.

12.7 Concept Questions

1. If $y = f(x)$, what is the differential of x? Write an expression for the differential dy.

2. Let $y = f(x)$. What is the relationship between the actual change in y, Δy, when x changes from x to $x + \Delta x$, and the differential dy of f at x? Illustrate this relationship graphically.

12.7 Exercises

In Exercises 1–14, find the differential of the function.

1. $f(x) = 2x^2$

2. $f(x) = 3x^2 + 1$

3. $f(x) = x^3 - x$

4. $f(x) = 2x^3 + x$

5. $f(x) = \sqrt{x + 1}$

6. $f(x) = \dfrac{3}{\sqrt{x}}$

7. $f(x) = 2x^{3/2} + x^{1/2}$

8. $f(x) = 3x^{5/6} + 7x^{2/3}$

9. $f(x) = x + \dfrac{2}{x}$

10. $f(x) = \dfrac{3}{x - 1}$

11. $f(x) = \dfrac{x - 1}{x^2 + 1}$

12. $f(x) = \dfrac{2x^2 + 1}{x + 1}$

13. $f(x) = \sqrt{3x^2 - x}$

14. $f(x) = (2x^2 + 3)^{1/3}$

15. Let f be the function defined by

$$y = f(x) = x^2 - 1$$

   a. Find the differential of f.
   b. Use your result from part (a) to find the approximate change in y if x changes from 1 to 1.02.
   c. Find the actual change in y if x changes from 1 to 1.02 and compare your result with that obtained in part (b).

16. Let f be the function defined by

$$y = f(x) = 3x^2 - 2x + 6$$

   a. Find the differential of f.
   b. Use your result from part (a) to find the approximate change in y if x changes from 2 to 1.97.
   c. Find the actual change in y if x changes from 2 to 1.97 and compare your result with that obtained in part (b).

17. Let f be the function defined by

$$y = f(x) = \dfrac{1}{x}$$

   a. Find the differential of f.
   b. Use your result from part (a) to find the approximate change in y if x changes from -1 to -0.95.
   c. Find the actual change in y if x changes from -1 to -0.95 and compare your result with that obtained in part (b).

18. Let f be the function defined by

$$y = f(x) = \sqrt{2x + 1}$$

   a. Find the differential of f.
   b. Use your result from part (a) to find the approximate change in y if x changes from 4 to 4.1.

   c. Find the actual change in y if x changes from 4 to 4.1 and compare your result with that obtained in part (b).

In Exercises 19–26, use differentials to approximate the quantity.

19. $\sqrt{10}$

20. $\sqrt{17}$

21. $\sqrt{49.5}$

22. $\sqrt{99.7}$

23. $\sqrt[3]{7.8}$

24. $\sqrt[4]{81.6}$

25. $\sqrt{0.089}$

26. $\sqrt[3]{0.00096}$

27. Use a differential to approximate $\sqrt{4.02} + \dfrac{1}{\sqrt{4.02}}$.

   Hint: Let $f(x) = \sqrt{x} + \dfrac{1}{\sqrt{x}}$ and compute dy with $x = 4$ and $dx = 0.02$.

28. Use a differential to approximate $\dfrac{2(4.98)}{(4.98)^2 + 1}$.

   Hint: Study the hint for Exercise 27.

29. ERROR ESTIMATION The length of each edge of a cube is 12 cm, with a possible error in measurement of 0.02 cm. Use differentials to estimate the error that might occur when the volume of the cube is calculated.

30. ESTIMATING THE AMOUNT OF PAINT REQUIRED A coat of paint of thickness 0.05 cm is to be applied uniformly to the faces of a cube of edge 30 cm. Use differentials to find the approximate amount of paint required for the job.

31. ERROR ESTIMATION A hemisphere-shaped dome of radius 60 ft is to be coated with a layer of rust-proofer before painting. Use differentials to estimate the amount of rust-proofer needed if the coat is to be 0.01 in. thick.
   Hint: The volume of a hemisphere of radius r is $V = \frac{2}{3}\pi r^3$.

32. GROWTH OF A CANCEROUS TUMOR The volume of a spherical cancerous tumor is given by

$$V(r) = \dfrac{4}{3}\pi r^3$$

If the radius of a tumor is estimated at 1.1 cm, with a maximum error in measurement of 0.005 cm, determine the error that might occur when the volume of the tumor is calculated.

33. UNCLOGGING ARTERIES Research done in the 1930s by the French physiologist Jean Poiseuille showed that the resistance R of a blood vessel of length l and radius r is $R = kl/r^4$, where k is a constant. Suppose a dose of the drug TPA increases r by 10%. How will this affect the resistance R? Assume that l is constant.

34. **GROSS DOMESTIC PRODUCT** An economist has determined that a certain country's gross domestic product (GDP) is approximated by the function $f(x) = 640x^{1/5}$, where $f(x)$ is measured in billions of dollars and x is the capital outlay in billions of dollars. Use differentials to estimate the change in the country's GDP if the country's capital expenditure changes from $243 billion to $248 billion.

35. **LEARNING CURVES** The length of time (in seconds) a certain individual takes to learn a list of n items is approximated by

$$f(n) = 4n\sqrt{n-4}$$

Use differentials to approximate the additional time it takes the individual to learn the items on a list when n is increased from 85 to 90 items.

36. **EFFECT OF ADVERTISING ON PROFITS** The relationship between Cunningham Realty's quarterly profits, $P(x)$, and the amount of money x spent on advertising per quarter is described by the function

$$P(x) = -\frac{1}{8}x^2 + 7x + 30 \qquad (0 \le x \le 50)$$

where both $P(x)$ and x are measured in thousands of dollars. Use differentials to estimate the increase in profits when advertising expenditure each quarter is increased from $24,000 to $26,000.

37. **EFFECT OF MORTGAGE RATES ON HOUSING STARTS** A study prepared for the National Association of Realtors estimates that the number of housing starts per year over the next 5 yr will be

$$N(r) = \frac{7}{1 + 0.02r^2}$$

million units, where r (percent) is the mortgage rate. Use differentials to estimate the decrease in the number of housing starts when the mortgage rate is increased from 12% to 12.5%.

38. **SUPPLY-PRICE** The supply equation for a certain brand of radio is given by

$$p = s(x) = 0.3\sqrt{x} + 10$$

where x is the quantity supplied and p is the unit price in dollars. Use differentials to approximate the change in price when the quantity supplied is increased from 10,000 units to 10,500 units.

39. **DEMAND-PRICE** The demand function for the Sentinel smoke alarm is given by

$$p = d(x) = \frac{30}{0.02x^2 + 1}$$

where x is the quantity demanded (in units of a thousand) and p is the unit price in dollars. Use differentials to estimate the change in the price p when the quantity demanded changes from 5000 to 5500 units/week.

40. **SURFACE AREA OF AN ANIMAL** Animal physiologists use the formula

$$S = kW^{2/3}$$

to calculate an animal's surface area (in square meters) from its weight W (in kilograms), where k is a constant that depends on the animal under consideration. Suppose a physiologist calculates the surface area of a horse ($k = 0.1$). If the horse's weight is estimated at 300 kg, with a maximum error in measurement of 0.6 kg, determine the percentage error in the calculation of the horse's surface area.

41. **FORECASTING PROFITS** The management of Trappee and Sons forecast that they will sell 200,000 cases of their TexaPep hot sauce next year. Their annual profit is described by

$$P(x) = -0.000032x^3 + 6x - 100$$

thousand dollars, where x is measured in thousands of cases. If the maximum error in the forecast is 15%, determine the corresponding error in Trappee's profits.

42. **FORECASTING COMMODITY PRICES** A certain country's government economists have determined that the demand equation for soybeans in that country is given by

$$p = f(x) = \frac{55}{2x^2 + 1}$$

where p is expressed in dollars/bushel and x, the quantity demanded each year, is measured in billions of bushels. The economists are forecasting a harvest of 1.8 billion bushels for the year, with a maximum error of 15% in their forecast. Determine the corresponding maximum error in the predicted price per bushel of soybeans.

43. **CRIME STUDIES** A sociologist has found that the number of serious crimes in a certain city each year is described by the function

$$N(x) = \frac{500(400 + 20x)^{1/2}}{(5 + 0.2x)^2}$$

where x (in cents/dollar deposited) is the level of reinvestment in the area in conventional mortgages by the city's ten largest banks. Use differentials to estimate the change in the number of crimes if the level of reinvestment changes from 20¢/dollar deposited to 22¢/dollar deposited.

44. **FINANCING A HOME** The Millers are planning to buy a home in the near future and estimate that they will need a 30-yr fixed-rate mortgage for $240,000. Their monthly payment P (in dollars) can be computed using the formula

$$P = \frac{20,000r}{1 - \left(1 + \dfrac{r}{12}\right)^{-360}}$$

where r is the interest rate per year.

a. Find the differential of P.

b. If the interest rate increases from the present rate of 7%/year to 7.2%/year between now and the time the Millers decide to secure the loan, approximately how much more will their monthly mortgage payment be? How much more will it be if the interest rate increases to 7.3%/year? To 7.4%/year? To 7.5%/year?

45. **INVESTMENTS** Lupé deposits a sum of $10,000 into an account that pays interest at the rate of r/year compounded monthly. Her investment at the end of 10 yr is given by

$$A = 10,000 \left(1 + \frac{r}{12} \right)^{120}$$

a. Find the differential of A.

b. Approximately how much more would Lupé's account be worth at the end of the term if her account paid 8.1%/year instead of 8%/year? 8.2%/year instead of 8%/year? 8.3%/year instead of 8%/year?

46. **KEOGH ACCOUNTS** Ian, who is self-employed, contributes $2000 a month into a Keogh account earning interest at the rate of r/year compounded monthly. At the end of 25 yr, his account will be worth

$$S = \frac{24,000 \left[\left(1 + \dfrac{r}{12} \right)^{300} - 1 \right]}{r}$$

dollars.

a. Find the differential of S.

b. Approximately how much more would Ian's account be worth at the end of 25 yr if his account earned 9.1%/year instead of 9%/year? 9.2%/year instead of 9%/year? 9.3%/year instead of 9%/year?

In Exercises 47 and 48, determine whether the statement is true or false. If it is true, explain why it is true. If it is false, give an example to show why it is false.

47. If $y = ax + b$ where a and b are constants, then $\Delta y = dy$.

48. If $A = f(x)$, then the percentage change in A is

$$\frac{100f'(x)}{f(x)} dx$$

12.7   Solutions to Self-Check Exercises

1. We find

$$f'(x) = \frac{1}{2}x^{-1/2} = \frac{1}{2\sqrt{x}}$$

Therefore, the required differential of f is

$$dy = \frac{1}{2\sqrt{x}} dx$$

2. We first compute the differential

$$dp = -\frac{250x}{(x^2 + 1)^2} dx$$

Next, using Equation (12) with $x = 6$ and $dx = 0.2$, we find

$$\Delta p \approx dp = -\frac{250(6)}{(36 + 1)^2}(0.2) = -0.22$$

or a drop in price of 22¢/bushel.

USING TECHNOLOGY

■ Finding the Differential of a Function

The calculation of the differential of f at a given value of x involves the evaluation of the derivative of f at that point and can be facilitated through the use of the numerical derivative function.

EXAMPLE 1   Use dy to approximate Δy if $y = x^2(2x^2 + x + 1)^{2/3}$ and x changes from 2 to 1.98.

Solution   Let $f(x) = x^2(2x^2 + x + 1)^{2/3}$. Since $dx = 1.98 - 2 = -0.02$, we find the required approximation to be

$$dy = f'(2)(-0.02)$$

But using the numerical derivative operation, we find

$$f'(2) = 30.57581679$$

(see Figure T1). Thus,

$$dy = (-0.02)(30.57581679) = -0.6115163358$$

```
nDeriv(X^2(2X^2+
X+1)^(2/3), X, 2)
        30.57581679
```

FIGURE T1
The TI-83 numerical derivative screen
for computing $f'(2)$

APPLIED EXAMPLE 2   Financing a Home   The Meyers are considering the purchase of a house in the near future and estimate that they will need a loan of $120,000. Based on a 30-year conventional mortgage with an annual interest rate of r, their monthly repayment will be

$$P = \frac{10,000r}{1 - \left(1 + \dfrac{r}{12}\right)^{-360}}$$

dollars. If the interest rate increases from 10% per year to 10.2% per year between now and the time the Meyers decide to secure the loan, approximately how much more will their monthly mortgage payment be?

Solution   Let's write

$$P = f(r) = \frac{10,000r}{1 - \left(1 + \dfrac{r}{12}\right)^{-360}}$$

Then the increase in the mortgage payment will be approximately

$$dP = f'(0.1)\, dr = f'(0.1)(0.002) \qquad \text{Since } dr = 0.102 - 0.1$$
$$= (8867.543909)(0.002) \approx 17.7351 \qquad \text{Use the numerical derivative operation.}$$

or approximately $17.74 per month. (See Figure T2.)

```
nDeriv((10000X)/
(1−(1+X/12)^ −360),
X, .1)
        8867.543909
```

FIGURE T2
The TI-83 numerical derivative screen
for computing $f'(0.1)$

(continued)

TECHNOLOGY EXERCISES

In Exercises 1–6, use *dy* to approximate Δ*y* for the function
y = *f*(*x*) when *x* changes from *x* = *a* to *x* = *b*.

1. $f(x) = 0.21x^7 - 3.22x^4 + 5.43x^2 + 1.42x + 12.42$; $a = 3$, $b = 3.01$

2. $f(x) = \dfrac{0.2x^2 + 3.1}{1.2x + 1.3}$; $a = 2$, $b = 1.96$

3. $f(x) = \sqrt{2.2x^2 + 1.3x + 4}$; $a = 1$, $b = 1.03$

4. $f(x) = x\sqrt{2x^3 - x + 4}$; $a = 2$, $b = 1.98$

5. $f(x) = \dfrac{\sqrt{x^2 + 4}}{x - 1}$; $a = 4$, $b = 4.1$

6. $f(x) = 2.1x^2 + \dfrac{3}{\sqrt{x}} + 5$; $a = 3$, $b = 2.95$

7. CALCULATING MORTGAGE PAYMENTS Refer to Example 2. How much more will the Meyers' mortgage payment be each month if the interest rate increases from 10% to 10.3%/year? To 10.4%/year? To 10.5%/year?

8. ESTIMATING THE AREA OF A RING OF NEPTUNE The ring 1989N2R of the planet Neptune has an inner radius of approximately 53,200 km (measured from the center of the planet) and a radial width of 15 km. Use differentials to estimate the area of the ring.

9. EFFECT OF PRICE INCREASE ON QUANTITY DEMANDED The quantity demanded each week of the Alpha Sports Watch, *x* (in thousands), is related to its unit price of *p* dollars by the equation

$$x = f(p) = 10\sqrt{\dfrac{50 - p}{p}} \qquad (0 \le p \le 50)$$

Use differentials to find the decrease in the quantity of the watches demanded each week if the unit price is increased from $40 to $42.

10. PERIOD OF A COMMUNICATIONS SATELLITE According to Kepler's third law, the period *T* (in days) of a satellite moving in a circular orbit *d* mi above the surface of Earth is given by

$$T = 0.0588\left(1 + \dfrac{d}{3959}\right)^{3/2}$$

Suppose a communications satellite that was moving in a circular orbit 22,000 mi above Earth's surface at one time has, because of friction, dropped down to a new orbit that is 21,500 mi above Earth's surface. Estimate the decrease in the period of the satellite to the nearest $\frac{1}{100}$th hr.

CHAPTER 12 **Summary of Principal Formulas and Terms**

FORMULAS

1. Derivative of a constant	$\dfrac{d}{dx}(c) = 0$ $\quad$ (*c*, a constant)
2. Power rule	$\dfrac{d}{dx}(x^n) = nx^{n-1}$
3. Constant multiple rule	$\dfrac{d}{dx}[cf(x)] = cf'(x)$
4. Sum rule	$\dfrac{d}{dx}[f(x) \pm g(x)] = f'(x) \pm g'(x)$
5. Product rule	$\dfrac{d}{dx}[f(x)g(x)] = f(x)g'(x) + g(x)f'(x)$
6. Quotient rule	$\dfrac{d}{dx}\left[\dfrac{f(x)}{g(x)}\right] = \dfrac{g(x)f'(x) - f(x)g'(x)}{[g(x)]^2}$
7. Chain rule	$\dfrac{d}{dx}g(f(x)) = g'(f(x))f'(x)$
8. General power rule	$\dfrac{d}{dx}[f(x)]^n = n[f(x)]^{n-1}f'(x)$
9. Average cost function	$\overline{C}(x) = \dfrac{C(x)}{x}$
10. Revenue function	$R(x) = px$
11. Profit function	$P(x) = R(x) - C(x)$
12. Elasticity of demand	$E(p) = -\dfrac{pf'(p)}{f(p)}$
13. Differential of *y*	$dy = f'(x)dx$

TERMS

marginal cost (723)

marginal cost function (723)

average cost (724)

marginal average cost function (724)

marginal revenue (726)

marginal revenue function (727)

marginal profit function (727)

elasticity of demand (729)

elastic demand (730)

unitary demand (730)

inelastic demand (730)

second derivative of f (736)

implicit differentiation (745)

related rates (749)

CHAPTER 12 Concept Review Questions

Fill in the blanks.

1. a. If c is a constant, then $\frac{d}{dx}(c) =$ _____.

b. The power rule states that if n is any real number, then $\frac{d}{dx}(x^n) =$ _____.

c. The constant multiple rule states that if c is a constant, then $\frac{d}{dx}[cf(x)] =$ _____.

d. The sum rule states that $\frac{d}{dx}[f(x) \pm g(x)] =$ _____.

2. a. The product rule states that $\frac{d}{dx}[f(x)g(x)] =$ _____.

b. The quotient rule states that $\frac{d}{dx}[f(x)/g(x)] =$ _____.

3. a. The chain rule states that if $h(x) = g[f(x)]$, then $h'(x) =$ _____.

b. The general power rule states that if $h(x) = [f(x)]^n$, then $h'(x) =$ _____.

4. If $C, R, P,$ and $\overline{C}$ denote the total cost function, the total revenue function, the profit function, and the average cost function, respectively, then C' denotes the _____ _____ function, R' denotes the _____ _____ function, P' denotes the _____ _____ function, and $\overline{C}'$ denotes the _____ _____ _____ function.

5. a. If f is a differentiable demand function defined by $x = f(p)$, then the elasticity of demand at price p is given by $E(p) =$ _____.

b. The demand is _____ if $E(p) > 1$; it is _____ if $E(p) = 1$; it is _____ if $E(p) < 1$.

6. Suppose a function $y = f(x)$ is defined implicitly by an equation in x and y. To find $\frac{dy}{dx}$, we differentiate _____ _____ of the equation with respect to x and then solve the resulting equation for $\frac{dy}{dx}$. The derivative of a term involving y includes _____ as a factor.

7. In a related-rates problem, we are given a relationship between x and _____ that depends on a third variable t. Knowing the values of x, y, and $\frac{dx}{dt}$ at a, we want to find _____ at _____.

8. Let $y = f(t)$ and $x = g(t)$. If $x^2 + y^2 = 4$, then $\frac{dx}{dt} =$ _____. If $xy = 1$, then $\frac{dy}{dt} =$ _____.

9. a. If a variable quantity x changes from x_1 to x_2, then the increment in x is $\Delta x =$ _____.

b. If $y = f(x)$ and x changes from x to $x + \Delta x$, then the increment in y is $\Delta y =$ _____.

10. If $y = f(x)$, where f is a differentiable function, then the differential dx of x is $dx =$ _____, where _____ is an increment in _____, and the differential dy of y is $dy =$ _____.

CHAPTER 12 Review Exercises

In Exercises 1–30, find the derivative of the function.

1. $f(x) = 3x^5 - 2x^4 + 3x^2 - 2x + 1$

2. $f(x) = 4x^6 + 2x^4 + 3x^2 - 2$

3. $g(x) = -2x^{-3} + 3x^{-1} + 2$

4. $f(t) = 2t^2 - 3t^3 - t^{-1/2}$

5. $g(t) = 2t^{-1/2} + 4t^{-3/2} + 2$

6. $h(x) = x^2 + \dfrac{2}{x}$

7. $f(t) = t + \dfrac{2}{t} + \dfrac{3}{t^2}$

8. $g(s) = 2s^2 - \dfrac{4}{s} + \dfrac{2}{\sqrt{s}}$

9. $h(x) = x^2 - \dfrac{2}{x^{3/2}}$

10. $f(x) = \dfrac{x + 1}{2x - 1}$

11. $g(t) = \dfrac{t^2}{2t^2 + 1}$

12. $h(t) = \dfrac{\sqrt{t}}{\sqrt{t} + 1}$

13. $f(x) = \dfrac{\sqrt{x} - 1}{\sqrt{x} + 1}$

14. $f(t) = \dfrac{t}{2t^2 + 1}$

15. $f(x) = \dfrac{x^2(x^2 + 1)}{x^2 - 1}$

16. $f(x) = (2x^2 + x)^3$

17. $f(x) = (3x^3 - 2)^8$

18. $h(x) = (\sqrt{x} + 2)^5$ 19. $f(t) = \sqrt{2t^2 + 1}$

20. $g(t) = \sqrt[3]{1 - 2t^3}$ 21. $s(t) = (3t^2 - 2t + 5)^{-2}$

22. $f(x) = (2x^3 - 3x^2 + 1)^{-3/2}$

23. $h(x) = \left(x + \dfrac{1}{x}\right)^2$ 24. $h(x) = \dfrac{1 + x}{(2x^2 + 1)^2}$

25. $h(t) = (t^2 + t)^4(2t^2)$ 26. $f(x) = (2x + 1)^3(x^2 + x)^2$

27. $g(x) = \sqrt{x}(x^2 - 1)^3$ 28. $f(x) = \dfrac{x}{\sqrt{x^3 + 2}}$

29. $h(x) = \dfrac{\sqrt{3x + 2}}{4x - 3}$ 30. $f(t) = \dfrac{\sqrt{2t + 1}}{(t + 1)^3}$

In Exercises 31–36, find the second derivative of the function.

31. $f(x) = 2x^4 - 3x^3 + 2x^2 + x + 4$

32. $g(x) = \sqrt{x} + \dfrac{1}{\sqrt{x}}$ 33. $h(t) = \dfrac{t}{t^2 + 4}$

34. $f(x) = (x^3 + x + 1)^2$ 35. $f(x) = \sqrt{2x^2 + 1}$

36. $f(t) = t(t^2 + 1)^3$

In Exercises 37–42, find *dy/dx* by implicit differentiation.

37. $6x^2 - 3y^2 = 9$ 38. $2x^3 - 3xy = 4$

39. $y^3 + 3x^2 = 3y$ 40. $x^2 + 2x^2y^2 + y^2 = 10$

41. $x^2 - 4xy - y^2 = 12$

42. $3x^2y - 4xy + x - 2y = 6$

43. Find the differential of $f(x) = x^2 + \dfrac{1}{x^2}$.

44. Find the differential of $f(x) = \dfrac{1}{\sqrt{x^3 + 1}}$.

45. Let f be the function defined by $f(x) = \sqrt{2x^2 + 4}$.
   a. Find the differential of f.
   b. Use your result from part (a) to find the approximate change in $y = f(x)$ if x changes from 4 to 4.1.
   c. Find the actual change in y if x changes from 4 to 4.1 and compare your result with that obtained in part (b).

46. Use a differential to approximate $\sqrt[3]{26.8}$.

47. Let $f(x) = 2x^3 - 3x^2 - 16x + 3$.
   a. Find the points on the graph of f at which the slope of the tangent line is equal to -4.
   b. Find the equation(s) of the tangent line(s) of part (a).

48. Let $f(x) = \frac{1}{3}x^3 + \frac{1}{2}x^2 - 4x + 1$.
   a. Find the points on the graph of f at which the slope of the tangent line is equal to -2.
   b. Find the equation(s) of the tangent line(s) of part (a).

49. Find an equation of the tangent line to the graph of $y = \sqrt{4 - x^2}$ at the point $(1, \sqrt{3})$.

50. Find an equation of the tangent line to the graph of $y = x(x + 1)^5$ at the point $(1, 32)$.

51. Find the third derivative of the function

$$f(x) = \dfrac{1}{2x - 1}$$

What is its domain?

52. The demand equation for a certain product is $2x + 5p - 60 = 0$, where p is the unit price and x is the quantity demanded of the product. Find the elasticity of demand and determine whether the demand is elastic or inelastic, at the indicated prices.
   a. $p = 3$ **b.** $p = 6$ **c.** $p = 9$

53. The demand equation for a certain product is

$$x = \dfrac{25}{\sqrt{p}} - 1$$

where p is the unit price and x is the quantity demanded for the product. Compute the elasticity of demand and determine the range of prices corresponding to inelastic, unitary, and elastic demand.

54. The demand equation for a certain product is $x = 100 - 0.01p^2$.
   a. Is the demand elastic, unitary, or inelastic when $p = 40$?
   b. If the price is \$40, will raising the price slightly cause the revenue to increase or decrease?

55. The demand equation for a certain product is

$$p = 9\sqrt[3]{1000 - x}$$

   a. Is the demand elastic, unitary, or inelastic when $p = 60$?
   b. If the price is \$60, will raising the price slightly cause the revenue to increase or decrease?

56. **ADULT OBESITY** In the United States, the percent of adults (age 20–74) classified as obese held steady through the 1960s and 1970s at around 14% but began to rise rapidly during the 1980s and 1990s. This rise in adult obesity coincided with the period when an increasing number of Americans began eating more sugar and fats. The function

$$P(t) = 0.01484t^2 + 0.446t + 15 \qquad (0 \le t \le 22)$$

gives the percent of obese adults from 1978 ($t = 0$) through the year 2000 ($t = 22$).
   a. What percent of adults were obese in 1978? In 2000?
   b. How fast was the percent of obese adults increasing in 1980 ($t = 2$)? In 1998 ($t = 20$)?

   Source: Journal of the American Medical Association

57. **CABLE TV SUBSCRIBERS** The number of subscribers to CNC Cable Television in the town of Randolph is approximated by the function

$$N(x) = 1000(1 + 2x)^{1/2} \qquad (1 \le x \le 30)$$

where $N(x)$ denotes the number of subscribers to the service in the xth week. Find the rate of increase in the number of subscribers at the end of the 12th week.

58. **Cost of Wireless Phone Calls**   As cellular phone usage continues to soar, the airtime costs have dropped. The average price per minute of use (in cents) is projected to be

$$f(t) = 31.88(1 + t)^{-0.45} \qquad (0 \le t \le 6)$$

where t is measured in years and $t = 0$ corresponds to the beginning of 1998. Compute $f'(t)$. How fast was the average price/minute of use changing at the beginning of 2000? What was the average price/minute of use at the beginning of 2000?

Source: Cellular Telecommunications Industry Association

59. **Male Life Expectancy**   Suppose the life expectancy of a male at birth in a certain country is described by the function

$$f(t) = 46.9(1 + 1.09t)^{0.1} \qquad (0 \le t \le 150)$$

where t is measured in years, with $t = 0$ corresponding to the beginning of 1900. How long can a male born at the beginning of 2000 in that country expect to live? What is the rate of change of the life expectancy of a male born in that country at the beginning of 2000?

60. **Cost of Producing DVDs**   The total weekly cost in dollars incurred by Herald Media Corp. in producing x DVDs is given by the total cost function

$$C(x) = 2500 + 2.2x \qquad (0 \le x \le 8000)$$

a. What is the marginal cost when $x = 1000$ and 2000?
b. Find the average cost function $\bar{C}$ and the marginal average cost function $\bar{C}'$.

c. Using the results from part (b), show that the average cost incurred by Herald in pressing a DVD approaches $2.20/disc when the level of production is high enough.

61. **Demand for Cordless Phones**   The marketing department of Telecon has determined that the demand for their cordless phones obeys the relationship

$$p = -0.02x + 600 \qquad (0 \le x \le 30,000)$$

where p denotes the phone's unit price (in dollars) and x denotes the quantity demanded.
a. Find the revenue function R.
b. Find the marginal revenue function R'.
c. Compute $R'(10,000)$ and interpret your result.

62. **Demand for Photocopying Machines**   The weekly demand for the LectroCopy photocopying machine is given by the demand equation

$$p = 2000 - 0.04x \qquad (0 \le x \le 50,000)$$

where p denotes the wholesale unit price in dollars and x denotes the quantity demanded. The weekly total cost function for manufacturing these copiers is given by

$$C(x) = 0.000002x^3 - 0.02x^2 + 1000x + 120,000$$

where $C(x)$ denotes the total cost incurred in producing x units.
a. Find the revenue function R, the profit function P, and the average cost function $\bar{C}$.
b. Find the marginal cost function C', the marginal revenue function R', the marginal profit function P', and the marginal average cost function $\bar{C}'$.
c. Compute $C'(3000)$, $R'(3000)$, and $P'(3000)$.
d. Compute $C'(5000)$ and $\bar{C}'(8000)$ and interpret your results.

CHAPTER 12   **Before Moving On . . .**

1. Find the derivative of $f(x) = 2x^3 - 3x^{1/3} + 5x^{-2/3}$.

2. Differentiate $g(x) = x\sqrt{2x^2 - 1}$.

3. Find $\dfrac{dy}{dx}$ if $y = \dfrac{2x + 1}{x^2 + x + 1}$.

4. Find the first three derivatives of $f(x) = \dfrac{1}{\sqrt{x + 1}}$.

5. Find $\dfrac{dy}{dx}$ given that $xy^2 - x^2y + x^3 = 4$.

6. Let $y = x\sqrt{x^2 + 5}$.
   a. Find the differential of y.
   b. If x changes from $x = 2$ to $x = 2.01$, what is the approximate change in y?

13 Applications of the Derivative

© NASA

What is the maximum altitude and the maximum velocity attained by the rocket? In Example 7, page 830, you will see how the techniques of calculus can be used to help answer these questions.

THIS CHAPTER FURTHER explores the power of the derivative as a tool to help analyze the properties of functions. The information obtained can then be used to accurately sketch graphs of functions. We also see how the derivative is used in solving a large class of optimization problems, including finding what level of production will yield a maximum profit for a company, finding what level of production will result in minimal cost to a company, finding the maximum height attained by a rocket, finding the maximum velocity at which air is expelled when a person coughs, and a host of other problems.

13.1   Applications of the First Derivative

■ Determining the Intervals Where a Function Is Increasing or Decreasing

According to a study by the U.S. Department of Energy and the Shell Development Company, a typical car's fuel economy as a function of its speed is described by the graph shown in Figure 1. Observe that the fuel economy $f(x)$ in miles per gallon (mpg) improves as x, the vehicle's speed in miles per hour (mph), increases from 0 to 42, and then drops as the speed increases beyond 42 mph. We use the terms *increasing* and *decreasing* to describe the behavior of a function as we move from left to right along its graph.

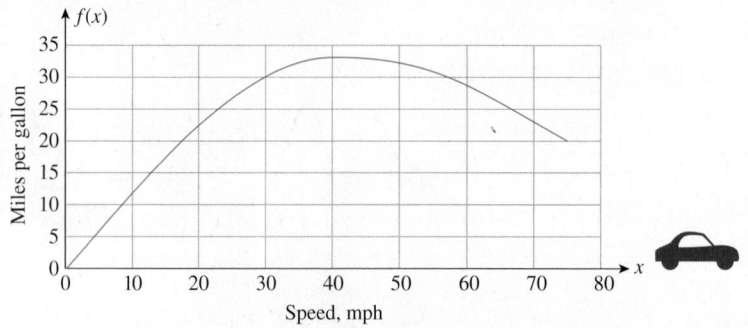

FIGURE 1
A typical car's fuel economy improves as the speed at which it is driven increases from 0 mph to 42 mph and drops at speeds greater than 42 mph.

Source: U.S. Department of Energy and Shell Development Co.

More precisely, we have the following definitions.

Increasing and Decreasing Functions
A function f is **increasing** on an interval (a, b) if for any two numbers x_1 and x_2 in (a, b), $f(x_1) < f(x_2)$ whenever $x_1 < x_2$ (Figure 2a).
A function f is **decreasing** on an interval (a, b) if for any two numbers x_1 and x_2 in (a, b), $f(x_1) > f(x_2)$ whenever $x_1 < x_2$ (Figure 2b).

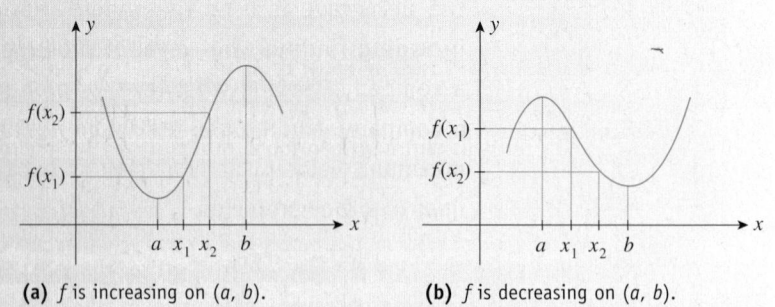

(a) f is increasing on (a, b).

(b) f is decreasing on (a, b).

FIGURE 2

We say that f is *increasing at a number c* if there exists an interval (a, b) containing c such that f is increasing on (a, b). Similarly, we say that f is *decreasing at a number c* if there exists an interval (a, b) containing c such that f is decreasing on (a, b).

Since the rate of change of a function at $x = c$ is given by the derivative of the function at that number, the derivative lends itself naturally to being a tool for determining the intervals where a differentiable function is increasing or decreasing.

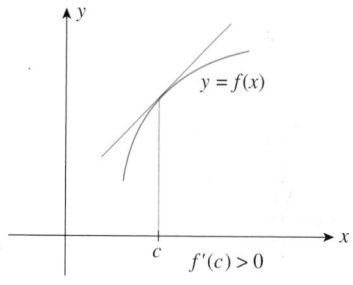

(a) f is increasing at $x = c$.

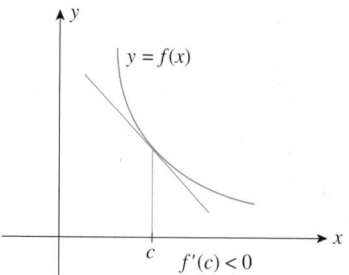

(b) f is decreasing at $x = c$.
FIGURE 3

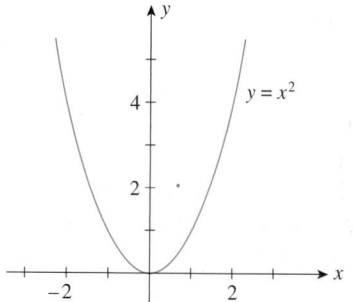

FIGURE 4
The graph of f falls on $(-\infty, 0)$ where $f'(x) < 0$ and rises on $(0, \infty)$ where $f'(x) > 0$.

Indeed, as we saw in Chapter 11, the derivative of a function at a number measures both the slope of the tangent line to the graph of the function at the point on the graph of f corresponding to that number and the rate of change of the function at that number. In fact, at a number where the derivative is positive, the slope of the tangent line to the graph is positive, and the function is increasing. At a number where the derivative is negative, the slope of the tangent line to the graph is negative, and the function is decreasing (Figure 3).

These observations lead to the following important theorem, which we state without proof.

THEOREM 1

a. If $f'(x) > 0$ for each value of x in an interval (a, b), then f is increasing on (a, b).

b. If $f'(x) < 0$ for each value of x in an interval (a, b), then f is decreasing on (a, b).

c. If $f'(x) = 0$ for each value of x in an interval (a, b), then f is constant on (a, b).

EXAMPLE 1   Find the interval where the function $f(x) = x^2$ is increasing and the interval where it is decreasing.

Solution   The derivative of $f(x) = x^2$ is $f'(x) = 2x$. Since

$$f'(x) = 2x > 0 \quad \text{if } x > 0 \qquad \text{and} \qquad f'(x) = 2x < 0 \quad \text{if } x < 0$$

f is increasing on the interval $(0, \infty)$ and decreasing on the interval $(-\infty, 0)$ (Figure 4). ∎

Recall that the graph of a continuous function cannot have any breaks. As a consequence, a continuous function cannot change sign unless it equals zero for some value of x. (See Theorem 5, page 649.) This observation suggests the following procedure for determining the sign of the derivative f' of a function f, and hence the intervals where the function f is increasing and where it is decreasing.

Determining the Intervals Where a Function Is Increasing or Decreasing

1. Find all values of x for which $f'(x) = 0$ or f' is discontinuous and identify the open intervals determined by these numbers.

2. Select a test number c in each interval found in step 1 and determine the sign of $f'(c)$ in that interval.

    a. If $f'(c) > 0$, f is increasing on that interval.

    b. If $f'(c) < 0$, f is decreasing on that interval.

EXAMPLE 2   Determine the intervals where the function $f(x) = x^3 - 3x^2 - 24x + 32$ is increasing and where it is decreasing.

Solution

1. The derivative of f is

$$f'(x) = 3x^2 - 6x - 24 = 3(x + 2)(x - 4)$$

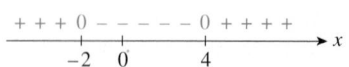

FIGURE 5
Sign diagram for f'

and it is continuous everywhere. The zeros of $f'(x)$ are $x = -2$ and $x = 4$, and these numbers divide the real line into the intervals $(-\infty, -2)$, $(-2, 4)$, and $(4, \infty)$.

2. To determine the sign of $f'(x)$ in the intervals $(-\infty, -2)$, $(-2, 4)$, and $(4, \infty)$, compute $f'(x)$ at a convenient test point in each interval. The results are shown in the following table.

Interval	Test Point c	$f'(c)$	Sign of $f'(x)$
$(-\infty, -2)$	-3	21	$+$
$(-2, 4)$	0	-24	$-$
$(4, \infty)$	5	21	$+$

Using these results, we obtain the sign diagram shown in Figure 5. We conclude that f is increasing on the intervals $(-\infty, -2)$ and $(4, \infty)$ and is decreasing on the interval $(-2, 4)$. Figure 6 shows the graph of f.

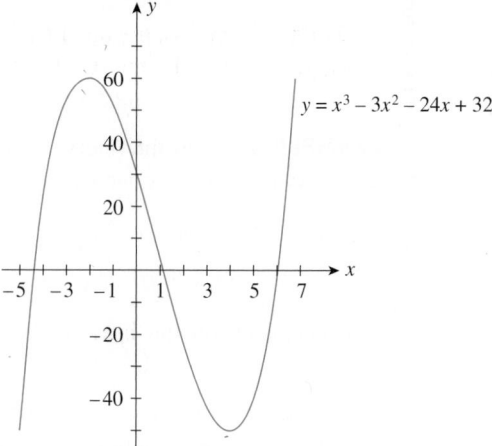

$y = x^3 - 3x^2 - 24x + 32$

FIGURE 6
The graph of f rises on $(-\infty, -2)$, falls on $(-2, 4)$, and rises again on $(4, \infty)$.

Note Do not be concerned with how the graphs in this section are obtained. We will learn how to sketch these graphs later. However, if you are familiar with the use of a graphing utility, you may go ahead and verify each graph.

EXPLORING WITH TECHNOLOGY

Refer to Example 2.

1. Use a graphing utility to plot the graphs of

$$f(x) = x^3 - 3x^2 - 24x + 32$$

and its derivative function

$$f'(x) = 3x^2 - 6x - 24$$

using the viewing window $[-10, 10] \times [-50, 70]$.

2. By looking at the graph of f', determine the intervals where $f'(x) > 0$ and the intervals where $f'(x) < 0$. Next, look at the graph of f and determine the intervals where it is increasing and the intervals where it is decreasing. Describe the relationship. Is it what you expected?

EXAMPLE 3   Find the interval where the function $f(x) = x^{2/3}$ is increasing and the interval where it is decreasing.

Solution

1. The derivative of f is

$$f'(x) = \frac{2}{3}x^{-1/3} = \frac{2}{3x^{1/3}}$$

The function f' is not defined at $x = 0$, so f' is discontinuous there. It is continuous everywhere else. Furthermore, f' is not equal to zero anywhere. The number 0 divides the real line (the domain of f) into the intervals $(-\infty, 0)$ and $(0, \infty)$.

2. Pick a test point (say, $x = -1$) in the interval $(-\infty, 0)$ and compute

$$f'(-1) = -\frac{2}{3}$$

Since $f'(-1) < 0$, we see that $f'(x) < 0$ on $(-\infty, 0)$. Next, we pick a test point (say, $x = 1$) in the interval $(0, \infty)$ and compute

$$f'(1) = \frac{2}{3}$$

Since $f'(1) > 0$, we see that $f'(x) > 0$ on $(0, \infty)$. Figure 7 shows these results in the form of a sign diagram.

We conclude that f is decreasing on the interval $(-\infty, 0)$ and increasing on the interval $(0, \infty)$. The graph of f, shown in Figure 8, confirms these results.   ∎

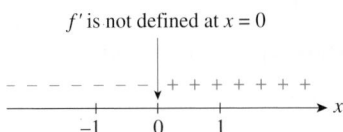

FIGURE 7
Sign diagram for f'

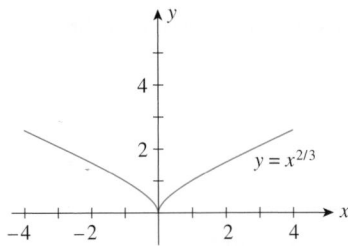

FIGURE 8
f decreases on $(-\infty, 0)$ and increases on $(0, \infty)$.

EXAMPLE 4   Find the intervals where the function $f(x) = x + \dfrac{1}{x}$ is increasing and where it is decreasing.

Solution

1. The derivative of f is

$$f'(x) = 1 - \frac{1}{x^2} = \frac{x^2 - 1}{x^2}$$

Since f' is not defined at $x = 0$, it is discontinuous there. Furthermore, $f'(x)$ is equal to zero when $x^2 - 1 = 0$ or $x = \pm 1$. These values of x partition the domain of f' into the open intervals $(-\infty, -1)$, $(-1, 0)$, $(0, 1)$, and $(1, \infty)$, where the sign of f' is different from zero.

2. To determine the sign of f' in each of these intervals, we compute $f'(x)$ at the test points $x = -2$, $-\frac{1}{2}$, $\frac{1}{2}$, and 2, respectively, obtaining $f'(-2) = \frac{3}{4}$, $f'(-\frac{1}{2}) = -3$, $f'(\frac{1}{2}) = -3$, and $f'(2) = \frac{3}{4}$. From the sign diagram for f' (Figure 9), we conclude that f is increasing on $(-\infty, -1)$ and $(1, \infty)$ and decreasing on $(-1, 0)$ and $(0, 1)$.

The graph of f appears in Figure 10. Note that f' does not change sign as we move across $x = 0$. (Compare this with Example 3.)   ∎

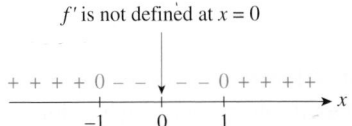

FIGURE 9
f' does not change sign as we move across $x = 0$.

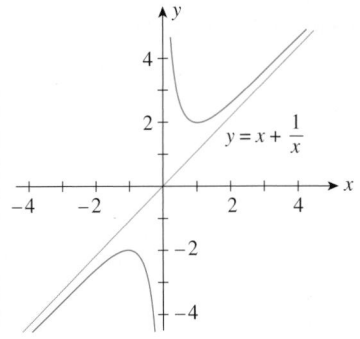

FIGURE 10
The graph of f rises on $(-\infty, -1)$, falls on $(-1, 0)$ and $(0, 1)$, and rises again on $(1, \infty)$.

Example 4 reminds us that we must *not* automatically conclude that the derivative f' must change sign when we move across a number where f' is discontinuous or a zero of f'.

EXPLORE & DISCUSS

Consider the profit function P associated with a certain commodity defined by

$$P(x) = R(x) - C(x) \qquad (x \geq 0)$$

where R is the revenue function, C is the total cost function, and x is the number of units of the product produced and sold.

1. Find an expression for $P'(x)$.

2. Find relationships in terms of the derivatives of R and C so that

   a. P is increasing at $x = a$.

   b. P is decreasing at $x = a$.

   c. P is neither increasing nor decreasing at $x = a$.
   Hint: Recall that the derivative of a function at $x = a$ measures the rate of change of the function at that number.

3. Explain the results of part 2 in economic terms.

EXPLORING WITH TECHNOLOGY

1. Use a graphing utility to sketch the graphs of $f(x) = x^3 - ax$ for $a = -2, -1, 0, 1,$ and 2, using the viewing window $[-2, 2] \times [-2, 2]$.

2. Use the results of part 1 to guess at the values of a so that f is increasing on $(-\infty, \infty)$.

3. Prove your conjecture analytically.

Relative Extrema

Besides helping us determine where the graph of a function is increasing and decreasing, the first derivative may be used to help us locate certain "high points" and "low points" on the graph of f. Knowing these points is invaluable in sketching the graphs of functions and solving optimization problems. These "high points" and "low points" correspond to the *relative (local) maxima* and *relative minima* of a function. They are so called because they are the highest or the lowest points when compared with points nearby.

The graph shown in Figure 11 gives the U.S. budget deficit from 1980 ($t = 0$) to 1991. The relative maxima and the relative minima of the function f are indicated on the graph.

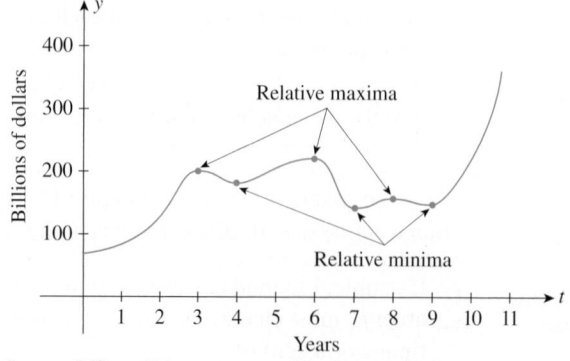

FIGURE 11
U.S. budget deficit from 1980 to 1991

Source: Office of Management and Budget

More generally, we have the following definition:

> **Relative Maximum**
>
> A function f has a **relative maximum** at $x = c$ if there exists an open interval (a, b) containing c such that $f(x) \le f(c)$ for all x in (a, b).

Geometrically, this means that there is *some* interval containing $x = c$ such that no point on the graph of f with its x-coordinate in that interval can lie above the point $(c, f(c))$; that is, $f(c)$ is the largest value of $f(x)$ in some interval around $x = c$. Figure 12 depicts the graph of a function f that has a relative maximum at $x = x_1$ and another at $x = x_3$.

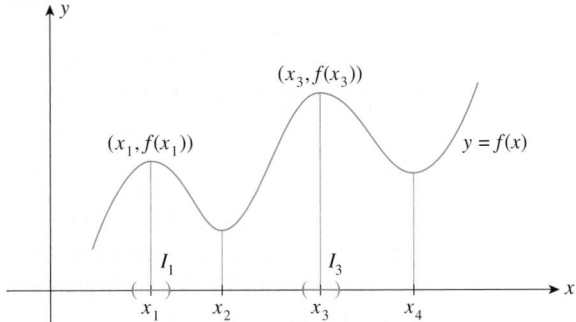

FIGURE 12
f has a relative maximum at $x = x_1$ and at $x = x_3$.

Observe that all the points on the graph of f with x-coordinates in the interval I_1 containing x (shown in blue) lie on or below the point $(x_1, f(x_1))$. This is also true for the point $(x_3, f(x_3))$ and the interval I_3. Thus, even though there are points on the graph of f that are "higher" than the points $(x_1, f(x_1))$ and $(x_3, f(x_3))$, the latter points are "highest" relative to points in their respective neighborhoods (intervals). Points on the graph of a function f that are "highest" and "lowest" with respect to *all* points in the domain of f will be studied in Section 13.4.

The definition of the relative minimum of a function parallels that of the relative maximum of a function.

> **Relative Minimum**
>
> A function f has a **relative minimum** at $x = c$ if there exists an open interval (a, b) containing c such that $f(x) \ge f(c)$ for all x in (a, b).

The graph of the function f, depicted in Figure 12, has a relative minimum at $x = x_2$ and another at $x = x_4$.

▪ Finding the Relative Extrema

We refer to the relative maximum and relative minimum of a function as the **relative extrema** of that function. As a first step in our quest to find the relative extrema of a function, we consider functions that have derivatives at such points. Suppose that f is a function that is differentiable in some interval (a, b) that contains a number c and that f has a relative maximum at $x = c$ (Figure 13a).

Observe that the slope of the tangent line to the graph of f must change from positive to negative as we move across $x = c$ from left to right. Therefore, the tangent line to the graph of f at the point $(c, f(c))$ must be horizontal; that is, $f'(c) = 0$ (Figure 13a).

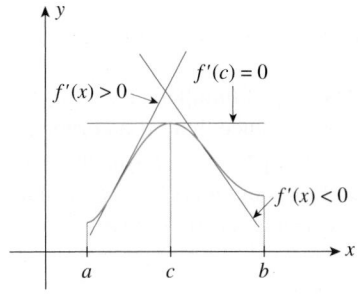

(a) f has a relative maximum at $x = c$.

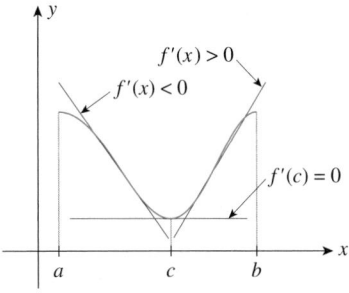

(b) f has a relative minimum at $x = c$.
FIGURE 13

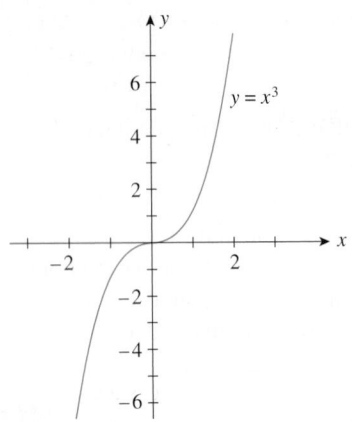

FIGURE 14
$f'(0) = 0$, but f does not have a relative extremum at $(0, 0)$.

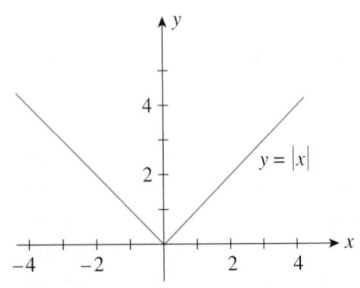

(a)

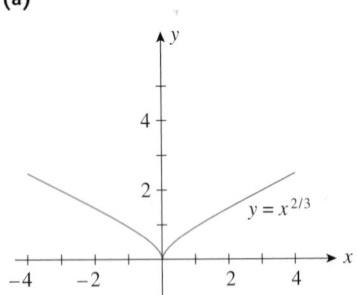

(b)
FIGURE 15
Each of these functions has a relative extremum at $(0, 0)$, but the derivative does not exist there.

Using a similar argument, it may be shown that the derivative f' of a differentiable function f must also be equal to zero at $x = c$, where f has a relative minimum (Figure 13b).

This analysis reveals an important characteristic of the relative extrema of a differentiable function f: *At any number c where f has a relative extremum,* $f'(c) = 0$.

 Before we develop a procedure for finding such numbers, a few words of caution are in order. First, this result tells us that if a differentiable function f has a relative extremum at a number $x = c$, then $f'(c) = 0$. The converse of this statement—if $f'(c) = 0$ at $x = c$, then f must have a relative extremum at that number—is *not* true. Consider, for example, the function $f(x) = x^3$. Here, $f'(x) = 3x^2$, so $f'(0) = 0$. Yet, f has neither a relative maximum nor a relative minimum at $x = 0$ (Figure 14).

Second, our result assumes that the function is differentiable and thus has a derivative at a number that gives rise to a relative extremum. The functions $f(x) = |x|$ and $g(x) = x^{2/3}$ demonstrate that a relative extremum of a function may exist at a number at which the derivative does not exist. Both these functions fail to be differentiable at $x = 0$, but each has a relative minimum there. Figure 15 shows the graphs of these functions. Note that the slopes of the tangent lines change from negative to positive as we move across $x = 0$, just as in the case of a function that is differentiable at a value of x that gives rise to a relative minimum.

We refer to a number in the domain of f that *may* give rise to a relative extremum as a critical number.

Critical Number of f
A **critical number** of a function f is any number x in the domain of f such that $f'(x) = 0$ or $f'(x)$ does not exist.

Figure 16 depicts the graph of a function that has critical numbers at $x = a, b, c, d,$ and e. Observe that $f'(x) = 0$ at $x = a, b,$ and c. Next, since there is a corner at $x = d, f'(x)$ does not exist there. Finally, $f'(x)$ does not exist at $x = e$ because the tangent line there is vertical. Also, observe that the critical numbers $x = a, b,$ and d give rise to relative extrema of f, whereas the critical numbers $x = c$ and $x = e$ do not.

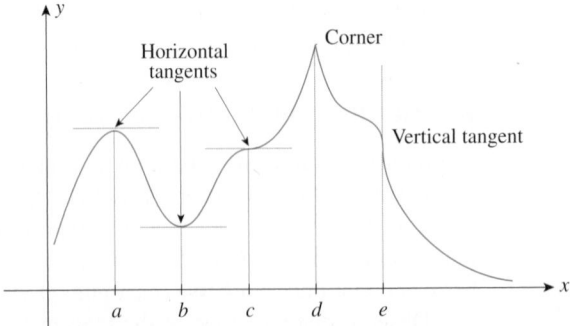

FIGURE 16
Critical numbers of f

Having defined what a critical number is, we can now state a formal procedure for finding the relative extrema of a continuous function that is differentiable everywhere except at isolated values of x. Incorporated into the procedure is the so-called **first derivative test**, which helps us determine whether a number gives rise to a relative maximum or a relative minimum of the function f.

The First Derivative Test

> **Procedure for Finding Relative Extrema of a Continuous Function f**
>
> 1. Determine the critical numbers of f.
> 2. Determine the sign of $f'(x)$ to the left and right of each critical number.
>   a. If $f'(x)$ changes sign from *positive* to *negative* as we move across a critical number c, then $f(c)$ is a relative maximum.
>   b. If $f'(x)$ changes sign from *negative* to *positive* as we move across a critical number c, then $f(c)$ is a relative minimum.
>   c. If $f'(x)$ does not change sign as we move across a critical number c, then $f(c)$ is not a relative extremum.

EXAMPLE 5 Find the relative maxima and relative minima of the function $f(x) = x^2$.

Solution The derivative of $f(x) = x^2$ is given by $f'(x) = 2x$. Setting $f'(x) = 0$ yields $x = 0$ as the only critical number of f. Since

$$f'(x) < 0 \quad \text{if } x < 0 \qquad \text{and} \qquad f'(x) > 0 \quad \text{if } x > 0$$

we see that $f'(x)$ changes sign from negative to positive as we move across the critical number 0. Thus, we conclude that $f(0) = 0$ is a relative minimum of f (Figure 17).

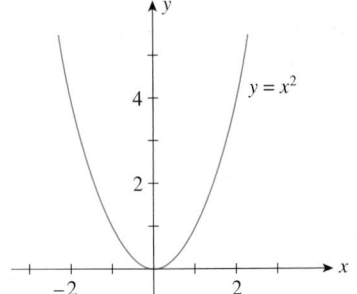

FIGURE 17
f has a relative minimum at $x = 0$.

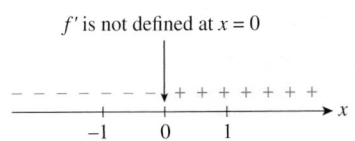

f' is not defined at $x = 0$

FIGURE 18
Sign diagram for f'

EXAMPLE 6 Find the relative maxima and relative minima of the function $f(x) = x^{2/3}$ (see Example 3).

Solution The derivative of f is $f'(x) = \frac{2}{3}x^{-1/3}$. As noted in Example 3, f' is not defined at $x = 0$, is continuous everywhere else, and is not equal to zero in its domain. Thus, $x = 0$ is the only critical number of the function f.

The sign diagram obtained in Example 3 is reproduced in Figure 18. We can see that the sign of $f'(x)$ changes from negative to positive as we move across $x = 0$ from left to right. Thus, an application of the first derivative test tells us that $f(0) = 0$ is a relative minimum of f (Figure 19).

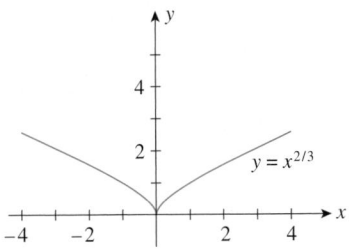

FIGURE 19
f has a relative minimum at $x = 0$.

EXPLORE & DISCUSS

Recall that the average cost function $\overline{C}$ is defined by

$$\overline{C} = \frac{C(x)}{x}$$

where $C(x)$ is the total cost function and x is the number of units of a commodity manufactured (see Section 12.4).

1. Show that

$$\overline{C}'(x) = \frac{C'(x) - \overline{C}(x)}{x} \qquad (x > 0)$$

2. Use the result of part 1 to conclude that $\overline{C}$ is decreasing for values of x at which $C'(x) < \overline{C}(x)$. Find similar conditions for which $\overline{C}$ is increasing and for which $\overline{C}$ is constant.

3. Explain the results of part 2 in economic terms.

EXAMPLE 7   Find the relative maxima and relative minima of the function

$$f(x) = x^3 - 3x^2 - 24x + 32$$

Solution   The derivative of f is

$$f'(x) = 3x^2 - 6x - 24 = 3(x + 2)(x - 4)$$

and it is continuous everywhere. The zeros of $f'(x)$, $x = -2$ and $x = 4$, are the only critical numbers of the function f. The sign diagram for f' is shown in Figure 20. Examine the two critical numbers $x = -2$ and $x = 4$ for a relative extremum using the first derivative test and the sign diagram for f':

1. *The critical number* -2: Since the function $f'(x)$ changes sign from positive to negative as we move across $x = -2$ from left to right, we conclude that a relative maximum of f occurs at $x = -2$. The value of $f(x)$ when $x = -2$ is

$$f(-2) = (-2)^3 - 3(-2)^2 - 24(-2) + 32 = 60$$

2. *The critical number* 4: $f'(x)$ changes sign from negative to positive as we move across $x = 4$ from left to right, so $f(4) = -48$ is a relative minimum of f. The graph of f appears in Figure 21.

FIGURE 20
Sign diagram for f'

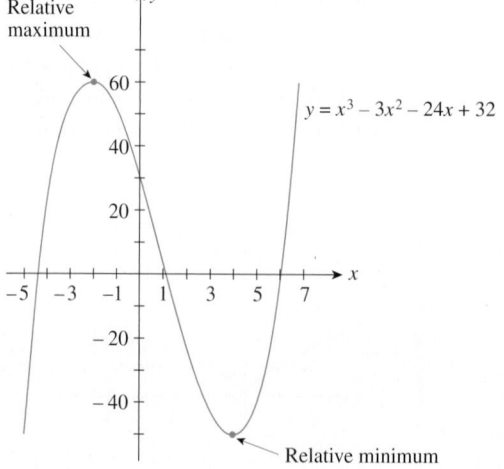

FIGURE 21
f has a relative maximum at $x = -2$ and a relative minimum at $x = 4$.

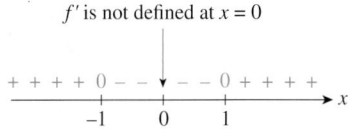

FIGURE 22
$x = 0$ is not a critical number because f is not defined at $x = 0$.

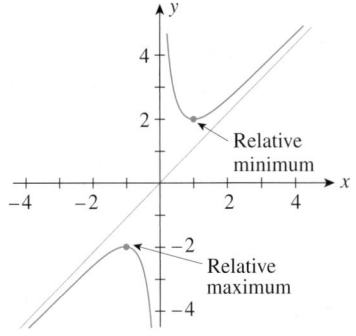

FIGURE 23
$f(x) = x + \dfrac{1}{x}$

EXAMPLE 8   Find the relative maxima and the relative minima of the function

$$f(x) = x + \frac{1}{x}$$

Solution   The derivative of f is

$$f'(x) = 1 - \frac{1}{x^2} = \frac{x^2 - 1}{x^2} = \frac{(x + 1)(x - 1)}{x^2}$$

Since f' is equal to zero at $x = -1$ and $x = 1$, these are critical numbers for the function f. Next, observe that f' is discontinuous at $x = 0$. However, because f is *not defined at that number*, $x = 0$ does not qualify as a critical number of f. Figure 22 shows the sign diagram for f'.

Since $f'(x)$ changes sign from positive to negative as we move across $x = -1$ from left to right, the first derivative test implies that $f(-1) = -2$ is a relative maximum of the function f. Next, $f'(x)$ changes sign from negative to positive as we move across $x = 1$ from left to right, so $f(1) = 2$ is a relative minimum of the function f. The graph of f appears in Figure 23. Note that this function has a relative maximum that lies below its relative minimum.   ■

EXPLORING WITH TECHNOLOGY

Refer to Example 8.

1. Use a graphing utility to plot the graphs of $f(x) = x + 1/x$ and its derivative function $f'(x) = 1 - 1/x^2$, using the viewing window $[-4, 4] \times [-8, 8]$.

2. By studying the graph of f', determine the critical numbers of f. Next, note the sign of $f'(x)$ immediately to the left and to the right of each critical number. What can you conclude about each critical number? Are your conclusions borne out by the graph of f?

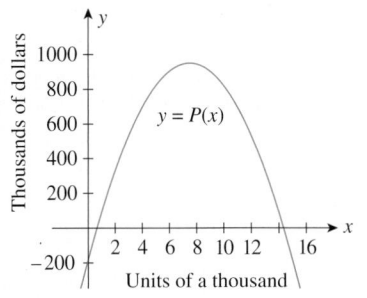

FIGURE 24
The profit function is increasing on $(0, 7500)$ and decreasing on $(7500, \infty)$.

APPLIED EXAMPLE 9   Profit Functions   The profit function of Acrosonic Company is given by

$$P(x) = -0.02x^2 + 300x - 200{,}000$$

dollars, where x is the number of Acrosonic model F loudspeaker systems produced. Find where the function P is increasing and where it is decreasing.

Solution   The derivative P' of the function P is

$$P'(x) = -0.04x + 300 = -0.04(x - 7500)$$

Thus, $P'(x) = 0$ when $x = 7500$. Furthermore, $P'(x) > 0$ for x in the interval $(0, 7500)$, and $P'(x) < 0$ for x in the interval $(7500, \infty)$. This means that the profit function P is increasing on $(0, 7500)$ and decreasing on $(7500, \infty)$ (Figure 24).   ■

APPLIED EXAMPLE 10   Crime Rates   The number of major crimes committed in the city of Bronxville from 1998 to 2005 is approximated by the function

$$N(t) = -0.1t^3 + 1.5t^2 + 100 \qquad (0 \le t \le 7)$$

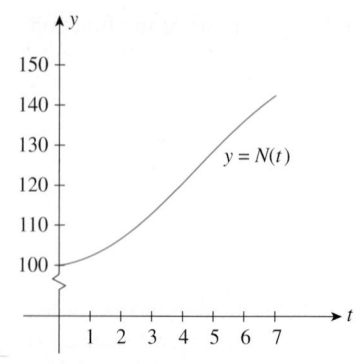

FIGURE 25
The number of crimes, $N(t)$, is increasing over the 7-year interval.

where $N(t)$ denotes the number of crimes committed in year t, with $t = 0$ corresponding to the beginning of 1998. Find where the function N is increasing and where it is decreasing.

Solution   The derivative N' of the function N is

$$N'(t) = -0.3t^2 + 3t = -0.3t(t - 10)$$

Since $N'(t) > 0$ for t in the interval $(0, 7)$, the function N is increasing throughout that interval (Figure 25). ∎

13.1   Self-Check Exercises

1. Find the intervals where the function $f(x) = \frac{2}{3}x^3 - x^2 - 12x + 3$ is increasing and the intervals where it is decreasing.

2. Find the relative extrema of $f(x) = \dfrac{x^2}{1 - x^2}$.

Solutions to Self-Check Exercises 13.1 can be found on page 788.

13.1   Concept Questions

1. Explain each of the following:
  a. f is increasing on an interval I.
  b. f is decreasing on an interval I.

2. Describe a procedure for determining where a function is increasing and where it is decreasing.

3. Explain each term: **(a)** relative maximum and **(b)** relative minimum.

4. a. What is a critical number of a function f?
  b. Explain the role of a critical number in determining the relative extrema of a function.

5. Describe the first derivative test and describe a procedure for finding the relative extrema of a function.

13.1   Exercises

In Exercises 1–8, you are given the graph of a function f. Determine the intervals where f is increasing, constant, or decreasing.

1.

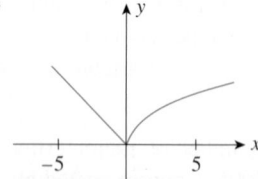

2.

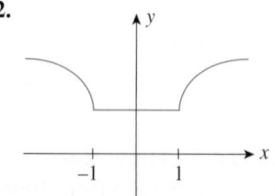

3.

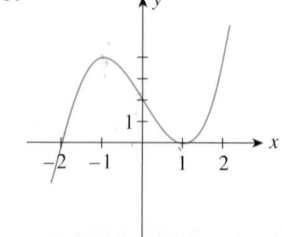

4.

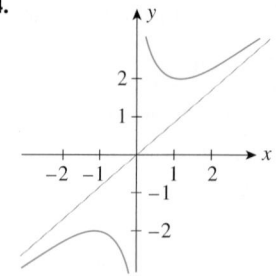

5.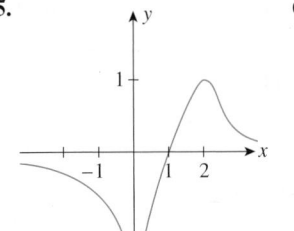

6.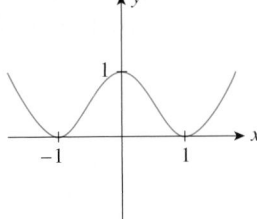

7.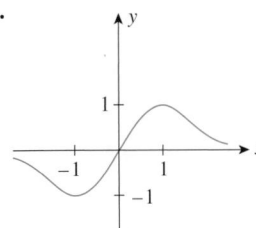

8.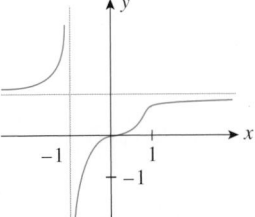

9. THE BOSTON MARATHON The graph of the function f shown in the accompanying figure gives the elevation of that part of the Boston Marathon course that includes the notorious Heartbreak Hill. Determine the intervals (stretches of the course) where the function f is increasing (the runner is laboring), where it is constant (the runner is taking a breather), and where it is decreasing (the runner is coasting).

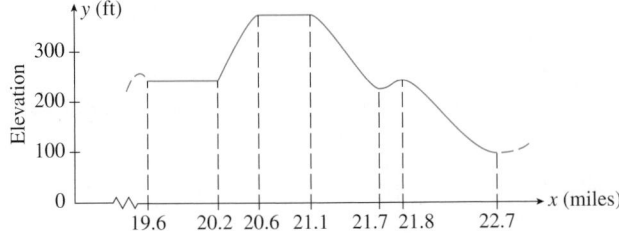

10. AIRCRAFT STRUCTURAL INTEGRITY Among the important factors in determining the structural integrity of an aircraft is its age. Advancing age makes planes more likely to crack. The graph of the function f, shown in the accompanying figure, is referred to as a "bathtub curve" in the airline industry. It gives the fleet damage rate (damage due to corrosion, accident, and metal fatigue) of a typical fleet of commercial aircraft as a function of the number of years of service.

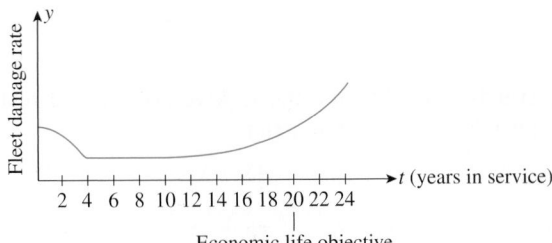

a. Determine the interval where f is decreasing. This corresponds to the time period when the fleet damage rate is dropping as problems are found and corrected during the initial "shakedown" period.

b. Determine the interval where f is constant. After the initial shakedown period, planes have few structural problems, and this is reflected by the fact that the function is constant on this interval.

c. Determine the interval where f is increasing. Beyond the time period mentioned in part (b), the function is increasing—reflecting an increase in structural defects due mainly to metal fatigue.

In Exercises 11–34, find the interval(s) where each function is increasing and the interval(s) where it is decreasing.

11. $f(x) = 3x + 5$

12. $f(x) = 4 - 5x$

13. $f(x) = x^2 - 3x$

14. $f(x) = 2x^2 + x + 1$

15. $g(x) = x - x^3$

16. $f(x) = x^3 - 3x^2$

17. $g(x) = x^3 + 3x^2 + 1$

18. $f(x) = x^3 - 3x + 4$

19. $f(x) = \dfrac{1}{3}x^3 - 3x^2 + 9x + 20$

20. $f(x) = \dfrac{2}{3}x^3 - 2x^2 - 6x - 2$

21. $h(x) = x^4 - 4x^3 + 10$

22. $g(x) = x^4 - 2x^2 + 4$

23. $f(x) = \dfrac{1}{x - 2}$

24. $h(x) = \dfrac{1}{2x + 3}$

25. $h(t) = \dfrac{t}{t - 1}$

26. $g(t) = \dfrac{2t}{t^2 + 1}$

27. $f(x) = x^{3/5}$

28. $f(x) = x^{2/3} + 5$

29. $f(x) = \sqrt{x + 1}$

30. $f(x) = (x - 5)^{2/3}$

31. $f(x) = \sqrt{16 - x^2}$

32. $g(x) = x\sqrt{x + 1}$

33. $f(x) = \dfrac{x^2 - 1}{x}$

34. $h(x) = \dfrac{x^2}{x - 1}$

In Exercises 35–42, you are given the graph of a function f. Determine the relative maxima and relative minima, if any.

35.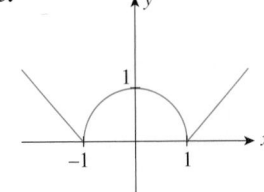

36.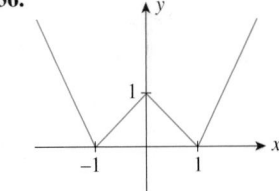

37.

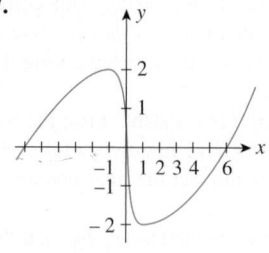

38.

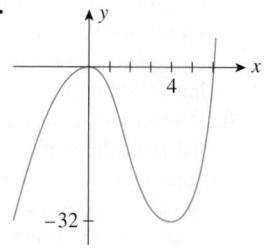

39.

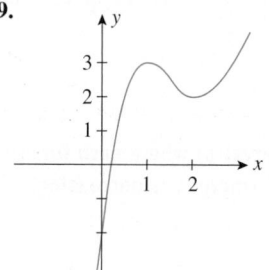

40.

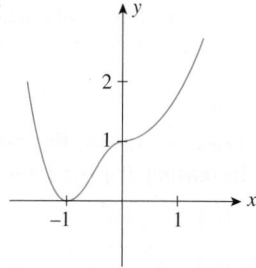

41.

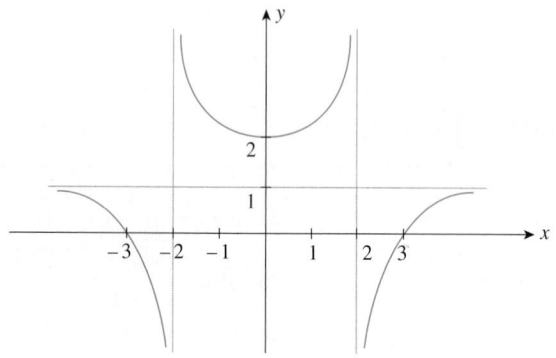

42.

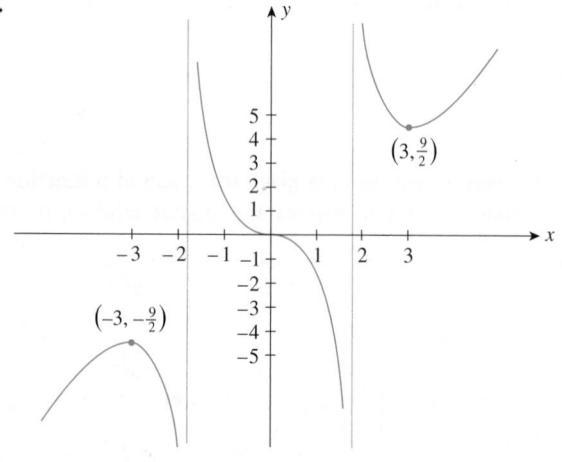

In Exercises 43–46, match the graph of the function with the graph of its derivative in (a)–(d).

43.

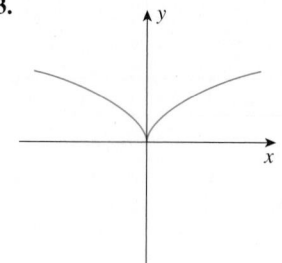

44.

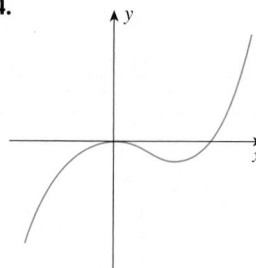

45.

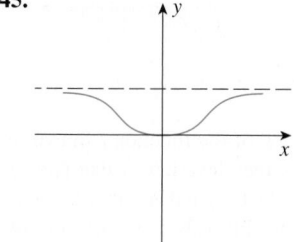

46.

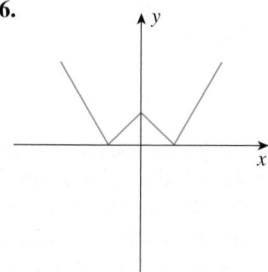

(a)

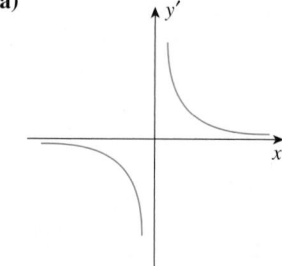

(b)

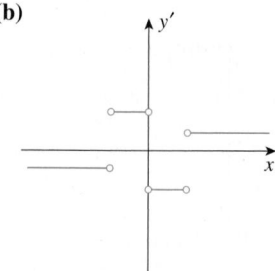

(c)

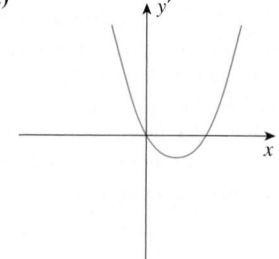

(d)

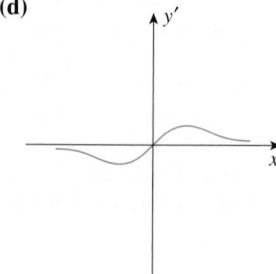

In Exercises 47–68, find the relative maxima and relative minima, if any, of each function.

47. $f(x) = x^2 - 4x$ **48.** $g(x) = x^2 + 3x + 8$

49. $h(t) = -t^2 + 6t + 6$ **50.** $f(x) = \dfrac{1}{2}x^2 - 2x + 4$

51. $f(x) = x^{5/3}$ **52.** $f(x) = x^{2/3} + 2$

53. $g(x) = x^3 - 3x^2 + 4$ **54.** $f(x) = x^3 - 3x + 6$

55. $f(x) = \dfrac{1}{2}x^4 - x^2$

56. $h(x) = \dfrac{1}{2}x^4 - 3x^2 + 4x - 8$

57. $F(x) = \dfrac{1}{3}x^3 - x^2 - 3x + 4$

58. $F(t) = 3t^5 - 20t^3 + 20$

59. $g(x) = x^4 - 4x^3 + 8$

60. $f(x) = 3x^4 - 2x^3 + 4$

61. $g(x) = \dfrac{x+1}{x}$ **62.** $h(x) = \dfrac{x}{x+1}$

63. $f(x) = x + \dfrac{9}{x} + 2$

64. $g(x) = 2x^2 + \dfrac{4000}{x} + 10$

65. $f(x) = \dfrac{x}{1+x^2}$ **66.** $g(x) = \dfrac{x}{x^2-1}$

67. $f(x) = (x-1)^{2/3}$ **68.** $g(x) = x\sqrt{x-4}$

69. A stone is thrown straight up from the roof of an 80-ft building. The distance (in feet) of the stone from the ground at any time t (in seconds) is given by

$$h(t) = -16t^2 + 64t + 80$$

When is the stone rising, and when is it falling? If the stone were to miss the building, when would it hit the ground? Sketch the graph of h.
Hint: The stone is on the ground when $h(t) = 0$.

70. Profit Functions The Mexican subsidiary of ThermoMaster manufactures an indoor–outdoor thermometer. Management estimates that the profit (in dollars) realizable by the company for the manufacture and sale of x units of thermometers each week is

$$P(x) = -0.001x^2 + 8x - 5000$$

Find the intervals where the profit function P is increasing and the intervals where P is decreasing.

71. Prevalence of Alzheimer's Patients Based on a study conducted in 1997, the percent of the U.S. population by age afflicted with Alzheimer's disease is given by the function

$$P(x) = 0.0726x^2 + 0.7902x + 4.9623 \quad (0 \le x \le 25)$$

where x is measured in years, with $x = 0$ corresponding to age 65. Show that P is an increasing function of x on the interval $(0, 25)$. What does your result tell you about the

relationship between Alzheimer's disease and age for the population that is age 65 and over?
Source: Alzheimer's Association

72. Growth of Managed Services Almost half of companies let other firms manage some of their Web operations—a practice called Web hosting. Managed services—monitoring a customer's technology services—is the fastest growing part of Web hosting. Managed services sales are expected to grow in accordance with the function

$$f(t) = 0.469t^2 + 0.758t + 0.44 \quad (0 \le t \le 6)$$

where $f(t)$ is measured in billions of dollars and t is measured in years, with $t = 0$ corresponding to 1999.
a. Find the interval where f is increasing and the interval where f is decreasing.
b. What does your result tell you about sales in managed services from 1999 through 2005?
Source: International Data Corp.

73. Flight of a Rocket The height (in feet) attained by a rocket t sec into flight is given by the function

$$h(t) = -\dfrac{1}{3}t^3 + 16t^2 + 33t + 10 \quad (t \ge 0)$$

When is the rocket rising, and when is it descending?

74. Environment of Forests Following the lead of the National Wildlife Federation, the Department of the Interior of a South American country began to record an index of environmental quality that measured progress and decline in the environmental quality of its forests. The index for the years 1984 through 1994 is approximated by the function

$$I(t) = \dfrac{1}{3}t^3 - \dfrac{5}{2}t^2 + 80 \quad (0 \le t \le 10)$$

where $t = 0$ corresponds to 1984. Find the intervals where the function I is increasing and the intervals where it is decreasing. Interpret your results.

75. Average Speed of a Highway Vehicle The average speed of a vehicle on a stretch of Route 134 between 6 a.m. and 10 a.m. on a typical weekday is approximated by the function

$$f(t) = 20t - 40\sqrt{t} + 50 \quad (0 \le t \le 4)$$

where $f(t)$ is measured in miles per hour and t is measured in hours, with $t = 0$ corresponding to 6 a.m. Find the interval where f is increasing and the interval where f is decreasing and interpret your results.

76. Average Cost The average cost (in dollars) incurred by Lincoln Records each week in pressing x compact discs is given by

$$\overline{C}(x) = -0.0001x + 2 + \dfrac{2000}{x} \quad (0 < x \le 6000)$$

Show that $\overline{C}(x)$ is always decreasing over the interval $(0, 6000)$.

77. **Web Hosting**   Refer to Exercise 72. Sales in the Web-hosting industry are projected to grow in accordance with the function

$$f(t) = -0.05t^3 + 0.56t^2 + 5.47t + 7.5 \qquad (0 \le t \le 6)$$

where $f(t)$ is measured in billions of dollars and t is measured in years, with $t = 0$ corresponding to 1999.
a. Find the interval where f is increasing and the interval where f is decreasing.
Hint: Use the quadratic formula.
b. What does your result tell you about sales in the Web-hosting industry from 1999 through 2005?

Source: International Data Corp.

78. **Cellular Phone Revenue**   According to a study conducted in 1997, the revenue (in millions of dollars) in the U.S. cellular phone market in the next 6 yr is approximated by the function

$$R(t) = 0.03056t^3 - 0.45357t^2 + 4.81111t + 31.7 \quad (0 \le t \le 6)$$

where t is measured in years, with $t = 0$ corresponding to 1997.
a. Find the interval where R is increasing and the interval where R is decreasing.
b. What does your result tell you about the revenue in the cellular phone market in the years under consideration?
Hint: Use the quadratic formula.

Source: Paul Kagan Associates, Inc.

79. **Elderly Workforce**   The percent of men 65 years and older in the workforce from 1970 through the year 2000 is approximated by the function

$$P(t) = 0.00093t^3 - 0.018t^2 - 0.51t + 25 \qquad (0 \le t \le 30)$$

where t is measured in years, with $t = 0$ corresponding to the beginning of 1970.
a. Find the interval where P is decreasing and the interval where P is increasing.
b. Interpret the results of part (a).

Source: U.S. Census Bureau

80. **Medical School Applicants**   According to a study from the American Medical Association, the number of medical school applicants from academic year 1997–1998 ($t = 0$) through the academic year 2002–2003 is approximated by the function

$$N(t) = -0.0333t^3 + 0.47t^2 - 3.8t + 47 \qquad (0 \le t \le 5)$$

where $N(t)$ measured in thousands.
a. Show that the number of medical school applicants had been declining over the period in question.
Hint: Use the quadratic formula.
b. What was the largest number of medical school applicants in any one academic year for the period in question? In what academic year did that occur?

Source: Journal of the American Medical Association

81. **Sales of Functional Food Products**   The sales of functional food products—those that promise benefits beyond basic nutrition—have risen sharply in recent years. The sales (in billions of dollars) of foods and beverages with herbal and other additives is approximated by the function

$$S(t) = 0.46t^3 - 2.22t^2 + 6.21t + 17.25 \qquad (0 \le t \le 4)$$

where t is measured in years, with $t = 0$ corresponding to the beginning of 1997. Show that S is increasing on the interval $[0, 4]$.
Hint: Use the quadratic formula.

Source: Frost & Sullivan

82. **Projected Retirement Funds**   Based on data from the Central Provident Fund of a certain country (a government agency similar to the Social Security Administration), the estimated cash in the fund in 1995 is given by

$$A(t) = -96.6t^4 + 403.6t^3$$
$$+ 660.9t^2 + 250 \qquad (0 \le t \le 5)$$

where $A(t)$ is measured in billions of dollars and t is measured in decades, with $t = 0$ corresponding to 1995. Find the interval where A is increasing and the interval where A is decreasing and interpret your results.
Hint: Use the quadratic formula.

83. **Spending on Fiber-Optic Links**   U.S. telephone company spending on fiber-optic links to homes and businesses from 2001 to 2006 is projected to be

$$S(t) = -2.315t^3 + 34.325t^2 + 1.32t + 23 \qquad (0 \le t \le 5)$$

billion dollars in year t, where t is measured in years with $t = 0$ corresponding to 2001. Show that $S'(t) > 0$ for all t in the interval $[0, 5]$. What conclusion can you draw from this result?
Hint: Use the quadratic formula.

Source: RHK Inc.

84. **Air Pollution**   According to the South Coast Air Quality Management District, the level of nitrogen dioxide, a brown gas that impairs breathing, present in the atmosphere on a certain May day in downtown Los Angeles is approximated by

$$A(t) = 0.03t^3(t - 7)^4 + 60.2 \qquad (0 \le t \le 7)$$

where $A(t)$ is measured in pollutant standard index (PSI) and t is measured in hours, with $t = 0$ corresponding to 7 a.m. At what time of day is the air pollution increasing, and at what time is it decreasing?

85. **Drug Concentration in the Blood**   The concentration (in milligrams/cubic centimeter) of a certain drug in a patient's body t hr after injection is given by

$$C(t) = \frac{t^2}{2t^3 + 1} \qquad (0 \le t \le 4)$$

When is the concentration of the drug increasing, and when is it decreasing?

86. **AGE OF DRIVERS IN CRASH FATALITIES** The number of crash fatalities per 100,000 vehicle miles of travel (based on 1994 data) is approximated by the model

$$f(x) = \frac{15}{0.08333x^2 + 1.91667x + 1} \quad (0 \le x \le 11)$$

where x is the age of the driver in years, with $x = 0$ corresponding to age 16. Show that f is decreasing on $(0, 11)$ and interpret your result.

Source: National Highway Traffic Safety Administration

87. **AIR POLLUTION** The amount of nitrogen dioxide, a brown gas that impairs breathing, present in the atmosphere on a certain May day in the city of Long Beach is approximated by

$$A(t) = \frac{136}{1 + 0.25(t - 4.5)^2} + 28 \quad (0 \le t \le 11)$$

where $A(t)$ is measured in pollutant standard index (PSI) and t is measured in hours, with $t = 0$ corresponding to 7 a.m. Find the intervals where A is increasing and where A is decreasing and interpret your results.

88. **PRISON OVERCROWDING** The 1980s saw a trend toward old-fashioned punitive deterrence as opposed to the more liberal penal policies and community-based corrections popular in the 1960s and early 1970s. As a result, prisons became more crowded, and the gap between the number of people in prison and the prison capacity widened. The number of prisoners (in thousands) in federal and state prisons is approximated by the function

$$N(t) = 3.5t^2 + 26.7t + 436.2 \quad (0 \le t \le 10)$$

where t is measured in years, with $t = 0$ corresponding to 1984. The number of inmates for which prisons were designed is given by

$$C(t) = 24.3t + 365 \quad (0 \le t \le 10)$$

where $C(t)$ is measured in thousands and t has the same meaning as before. Show that the gap between the number of prisoners and the number for which the prisons were designed has been widening at any time t.
Hint: First, write a function G that gives the gap between the number of prisoners and the number for which the prisons were designed at any time t. Then show that $G'(t) > 0$ for all values of t in the interval $(0, 10)$.

Source: U.S. Department of Justice

89. **U.S. NURSING SHORTAGE** The demand for nurses between 2000 and 2015 is estimated to be

$$D(t) = 0.0007t^2 + 0.0265t + 2 \quad (0 \le t \le 15)$$

where $D(t)$ is measured in millions and $t = 0$ corresponds to the year 2000. The supply of nurses over the same time period is estimated to be

$$S(t) = -0.0014t^2 + 0.0326t + 1.9 \quad (0 \le t \le 15)$$

where $S(t)$ is also measured in millions.

a. Find an expression $G(t)$ giving the gap between the demand and supply of nurses over the period in question.
b. Find the interval where G is decreasing and where it is increasing. Interpret your result.
c. Find the relative extrema of G. Interpret your result.

Source: U.S. Department of Health and Human Services

In Exercises 90–95, determine whether the statement is true or false. If it is true, explain why it is true. If it is false, give an example to show why it is false.

90. If f is decreasing on (a, b), then $f'(x) < 0$ for each x in (a, b).

91. If f and g are both increasing on (a, b), then $f + g$ is increasing on (a, b).

92. If f and g are both decreasing on (a, b), then $f - g$ is decreasing on (a, b).

93. If $f(x)$ and $g(x)$ are positive on (a, b) and both f and g are increasing on (a, b), then fg is increasing on (a, b).

94. If $f'(c) = 0$, then f has a relative maximum or a relative minimum at $x = c$.

95. If f has a relative minimum at $x = c$, then $f'(c) = 0$.

96. Using Theorem 1, verify that the linear function $f(x) = mx + b$ is (a) increasing everywhere if $m > 0$, (b) decreasing everywhere if $m < 0$, and (c) constant if $m = 0$.

97. Show that the function $f(x) = x^3 + x + 1$ has no relative extrema on $(-\infty, \infty)$.

98. Let

$$f(x) = \begin{cases} -3x & \text{if } x < 0 \\ 2x + 4 & \text{if } x \ge 0 \end{cases}$$

a. Compute $f'(x)$ and show that it changes sign from negative to positive as we move across $x = 0$.
b. Show that f does not have a relative minimum at $x = 0$. Does this contradict the first derivative test? Explain your answer.

99. Let

$$f(x) = \begin{cases} -x^2 + 3 & \text{if } x \ne 0 \\ 2 & \text{if } x = 0 \end{cases}$$

a. Compute $f'(x)$ and show that it changes sign from positive to negative as we move across $x = 0$.
b. Show that f does not have a relative maximum at $x = 0$. Does this contradict the first derivative test? Explain your answer.

100. Let

$$f(x) = \begin{cases} \dfrac{1}{x^2} & \text{if } x > 0 \\ x^2 & \text{if } x \le 0 \end{cases}$$

a. Compute $f'(x)$ and show that it does not change sign as we move across $x = 0$.

b. Show that f has a relative minimum at $x = 0$. Does this contradict the first derivative test? Explain your answer.

101. Show that the quadratic function

$$f(x) = ax^2 + bx + c \qquad (a \neq 0)$$

has a relative extremum when $x = -b/2a$. Also, show that the relative extremum is a relative maximum if $a < 0$ and a relative minimum if $a > 0$.

102. Show that the cubic function

$$f(x) = ax^3 + bx^2 + cx + d \qquad (a \neq 0)$$

has no relative extremum if and only if $b^2 - 3ac \leq 0$.

103. Refer to Example 6, page 650.

a. Show that f is increasing on the interval $(0, 1)$.

b. Show that $f(0) = -1$ and $f(1) = 1$ and use the result of part (a) together with the intermediate value theorem to conclude that there is exactly one root of $f(x) = 0$ in $(0, 1)$.

104. Show that the function

$$f(x) = \frac{ax + b}{cx + d}$$

does not have a relative extremum if $ad - bc \neq 0$. What can you say about f if $ad - bc = 0$?

13.1 Solutions to Self-Check Exercises

1. The derivative of f is

$$f'(x) = 2x^2 - 2x - 12 = 2(x + 2)(x - 3)$$

and it is continuous everywhere. The zeros of $f'(x)$ are $x = -2$ and $x = 3$. The sign diagram of f' is shown in the accompanying figure. We conclude that f is increasing on the intervals $(-\infty, -2)$ and $(3, \infty)$ and decreasing on the interval $(-2, 3)$.

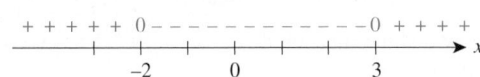

and it is continuous everywhere except at $x = \pm 1$. Since $f'(x)$ is equal to zero at $x = 0$, $x = 0$ is a critical number of f. Next, observe that $f'(x)$ is discontinuous at $x = \pm 1$, but since these numbers are not in the domain of f, they do not qualify as critical numbers of f. Finally, from the sign diagram of f' shown in the accompanying figure, we conclude that $f(0) = 0$ is a relative minimum of f.

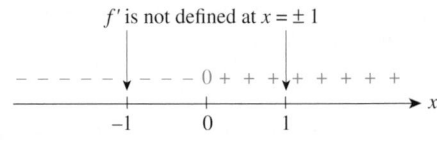

2. The derivative of f is

$$f'(x) = \frac{(1 - x^2)\dfrac{d}{dx}(x^2) - x^2\dfrac{d}{dx}(1 - x^2)}{(1 - x^2)^2}$$

$$= \frac{(1 - x^2)(2x) - x^2(-2x)}{(1 - x^2)^2} = \frac{2x}{(1 - x^2)^2}$$

USING TECHNOLOGY

■ Using the First Derivative to Analyze a Function

A graphing utility is an effective tool for analyzing the properties of functions. This is especially true when we also bring into play the power of calculus, as the following examples show.

EXAMPLE 1   Let $f(x) = 2.4x^4 - 8.2x^3 + 2.7x^2 + 4x + 1$.

a. Use a graphing utility to plot the graph of f.

b. Find the intervals where f is increasing and the intervals where f is decreasing.

c. Find the relative extrema of f.

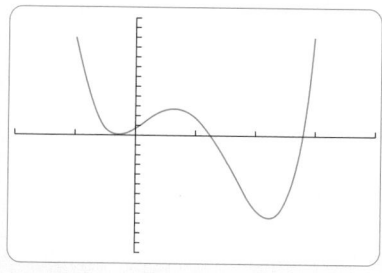

FIGURE T1
The graph of f in the viewing window $[-2, 4] \times [-10, 10]$

Solution

a. The graph of f in the viewing window $[-2, 4] \times [-10, 10]$ is shown in Figure T1.
b. We compute

$$f'(x) = 9.6x^3 - 24.6x^2 + 5.4x + 4$$

and observe that f' is continuous everywhere, so the critical numbers of f occur at values of x where $f'(x) = 0$. To solve this last equation, observe that $f'(x)$ is a *polynomial function* of degree 3. The easiest way to solve the polynomial equation

$$9.6x^3 - 24.6x^2 + 5.4x + 4 = 0$$

is to use the function on a graphing utility for solving polynomial equations. (Not all graphing utilities have this function.) You can also use TRACE and ZOOM, but this will not give the same accuracy without a much greater effort.
    We find

$$x_1 \approx 2.22564943249 \qquad x_2 \approx 0.63272944121 \qquad x_3 \approx -0.295878873696$$

Referring to Figure T1, we conclude that f is decreasing on $(-\infty, -0.2959)$ and $(0.6327, 2.2256)$ (correct to four decimal places) and f is increasing on $(-0.2959, 0.6327)$ and $(2.2256, \infty)$.
c. Using the evaluation function of a graphing utility, we find the value of f at each of the critical numbers found in part (b). Upon referring to Figure T1 once again, we see that $f(x_3) \approx 0.2836$ and $f(x_1) \approx -8.2366$ are relative minimum values of f and $f(x_2) \approx 2.9194$ is a relative maximum value of f. ∎

Note The equation $f'(x) = 0$ in Example 1 is a polynomial equation, and so it is easily solved using the function for solving polynomial equations. We could also solve the equation using the function for finding the roots of equations, but that would require much more work. For equations that are *not* polynomial equations, however, our only choice is to use the function for finding the roots of equations. ∎

If the derivative of a function is difficult to compute or simplify and we do not require great precision in the solution, we can find the relative extrema of the function using a combination of ZOOM and TRACE. This technique, which does not require the use of the derivative of f, is illustrated in the following example.

EXAMPLE 2 Let $f(x) = x^{1/3}(x^2 + 1)^{-3/2}3^{-x}$.

a. Use a graphing utility to plot the graph of f.*
b. Find the relative extrema of f.

Solution

a. The graph of f in the viewing window $[-4, 2] \times [-2, 1]$ is shown in Figure T2.
b. From the graph of f in Figure T2, we see that f has relative maxima when $x \approx -2$ and $x \approx 0.25$ and a relative minimum when $x \approx -0.75$. To obtain a better approximation of the first relative maximum, we zoom-in with the cursor at approximately the point on the graph corresponding to $x \approx -2$. Then, using TRACE, we see that a relative maximum occurs when $x \approx -1.76$ with value

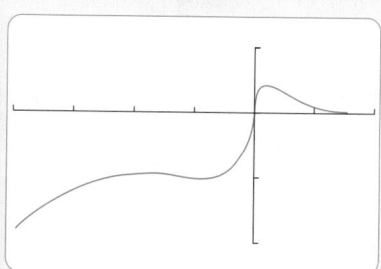

FIGURE T2
The graph of f in the viewing window $[-4, 2] \times [-2, 1]$

*Functions of the form $f(x) = 3^{-x}$ are called *exponential functions*, and we will study them in greater detail in Chapter 14.

(continued)

$y \approx -1.01$. Similarly, we find the other relative maximum where $x \approx 0.20$ with value $y \approx 0.44$. Repeating the procedure, we find the relative minimum at $x \approx -0.86$ and $y \approx -1.07$. ∎

You can also use the "minimum" and "maximum" functions of a graphing utility to find the relative extrema of the function. See the Web site for the procedure.

Finally, we comment that if you have access to a computer and software such as Derive, Maple, or Mathematica, then symbolic differentiation will yield the derivative $f'(x)$ of any differentiable function. This software will also solve the equation $f'(x) = 0$ with ease. Thus, the use of a computer will simplify even more greatly the analysis of functions.

TECHNOLOGY EXERCISES

In Exercises 1–4, find (a) the intervals where f is increasing and the intervals where f is decreasing and (b) the relative extrema of f. Express your answers accurate to four decimal places.

1. $f(x) = 3.4x^4 - 6.2x^3 + 1.8x^2 + 3x - 2$

2. $f(x) = 1.8x^4 - 9.1x^3 + 5x - 4$

3. $f(x) = 2x^5 - 5x^3 + 8x^2 - 3x + 2$

4. $f(x) = 3x^5 - 4x^2 + 3x - 1$

In Exercises 5–8, use the ZOOM and TRACE features to find (a) the intervals where f is increasing and the intervals where f is decreasing and (b) the relative extrema of f. Express your answers accurate to two decimal places.

5. $f(x) = (2x + 1)^{1/3}(x^2 + 1)^{-2/3}$

6. $f(x) = [x^2(x^3 - 1)]^{1/3} + \dfrac{1}{x}$

7. $f(x) = x - \sqrt{1 - x^2}$

8. $f(x) = \dfrac{\sqrt{x}(x^2 - 1)^2}{x - 2}$

9. **Rate of Bank Failures** The rate at which banks were failing between 1982 and 1994 was estimated to be

$$f(t) = 0.063447t^4 - 1.953283t^3 + 14.632576t^2$$
$$- 6.684704t + 47.458874 \qquad (0 \le t \le 12)$$

where $f(t)$ is the number of banks failing each year and t is measured in years, with $t = 0$ corresponding to the beginning of 1982.
   a. Plot the graph of f in the viewing window $[0, 13] \times [0, 220]$.
   b. Determine the intervals where f is increasing and where f is decreasing and interpret your result.
   c. Find the relative maximum of f and interpret your result.
   Source: Federal Deposit Insurance Corporation

10. **Cargo Volume** The volume of cargo moved in the port of New York/New Jersey from 1991 through 2002 is modeled by the function

$$V(t) = 0.0094665t^3 - 0.052775t^2 + 0.39895t + 13.7$$
$$(0 \le t \le 11)$$

where $V(t)$ is measured in millions of tons and t is measured in years, with $t = 0$ corresponding to 1991.
   a. Plot the graph of V in the viewing window $[0, 11] \times [0, 25]$.
   b. Where is V increasing? What does this tell us?
   c. Verify the result of part (b) analytically.
   Source: Port Authority of New York/New Jersey

11. **Manufacturing Capacity** Data show that the annual increase in manufacturing capacity between 1994 and 2000 is given by

$$f(t) = 0.009417t^3 - 0.426571t^2 + 2.74894t + 5.54$$
$$(0 \le t \le 6)$$

percent where t is measured in years, with $t = 0$ corresponding to the beginning of 1994.
   a. Plot the graph of f in the viewing window $[0, 6] \times [0, 11]$.
   b. Determine the interval where f is increasing and the interval where f is decreasing and interpret your result.
   Source: Federal Reserve

12. **Surgeries in Physicians' Offices** Driven by technological advances and financial pressures, the number of surgeries performed in physicians' offices nationwide has been increasing over the years. The function

$$f(t) = -0.00447t^3 + 0.09864t^2 + 0.05192t + 0.8$$
$$(0 \le t \le 15)$$

gives the number of surgeries (in millions) performed in physicians' offices in year t, with $t = 0$ corresponding to the beginning of 1986.
   a. Plot the graph of f in the viewing window $[0, 15] \times [0, 10]$.

b. Prove that *f* is increasing on the interval [0, 15].
  Hint: Show that *f′* is positive on the interval.
Source: SMG Marketing Group

13. **MORNING TRAFFIC RUSH** The speed of traffic flow on a certain stretch of Route 123 between 6 a.m. and 10 a.m. on a typical weekday is approximated by the function

$$f(t) = 20t - 40\sqrt{t} + 52 \qquad (0 \le t \le 4)$$

where *f(t)* is measured in miles per hour and *t* is measured in hours, with *t* = 0 corresponding to 6 a.m. Find the interval where *f* is increasing, the interval where *f* is decreasing, and the relative extrema of *f*. Interpret your results.

14. **AIR POLLUTION** The amount of nitrogen dioxide, a brown gas that impairs breathing, present in the atmosphere on a certain May day in the city of Long Beach, is approximated by

$$A(t) = \frac{136}{1 + 0.25(t - 4.5)^2} + 28 \qquad (0 \le t \le 11)$$

where *A(t)* is measured in pollutant standard index (PSI) and *t* is measured in hours, with *t* = 0 corresponding to 7 a.m. When is the PSI increasing and when is it decreasing? At what time is the PSI highest, and what is its value at that time?

13.2 Applications of the Second Derivative

■ Determining the Intervals of Concavity

Consider the graphs shown in Figure 26, which give the estimated population of the world and of the United States through the year 2000. Both graphs are rising, indicating that both the U.S. population and the world population continued to increase through the year 2000. But observe that the graph in Figure 26a opens upward, whereas the graph in Figure 26b opens downward. What is the significance of this? To answer this question, let's look at the slopes of the tangent lines to various points on each graph (Figure 27).

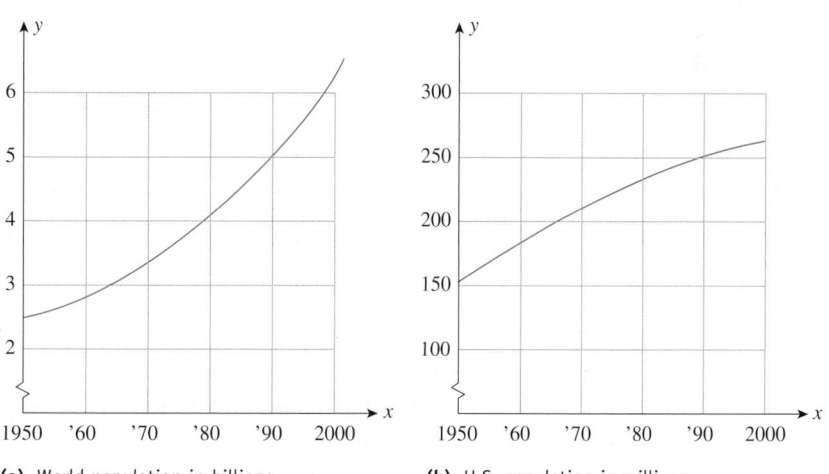

(a) World population in billions   **(b)** U.S. population in millions

FIGURE 26   *Source:* U.S. Department of Commerce and Worldwatch Institute

In Figure 27a, we see that the slopes of the tangent lines to the graph are increasing as we move from left to right. Since the slope of the tangent line to the graph at a point on the graph measures the rate of change of the function at that point, we conclude that the world population was not only increasing through the year 2000 but was also increasing at an *increasing* pace. A similar analysis of Figure 27b reveals that the U.S. population was increasing, but at a *decreasing* pace.

(a) Slopes of tangent lines are increasing.   **(b)** Slopes of tangent lines are decreasing.

The shape of a curve can be described using the notion of concavity.

> **Concavity of a Function f**
> Let the function f be differentiable on an interval (a, b). Then,
> **1.** f is **concave upward** on (a, b) if f' is increasing on (a, b).
> **2.** f is **concave downward** on (a, b) if f' is decreasing on (a, b).

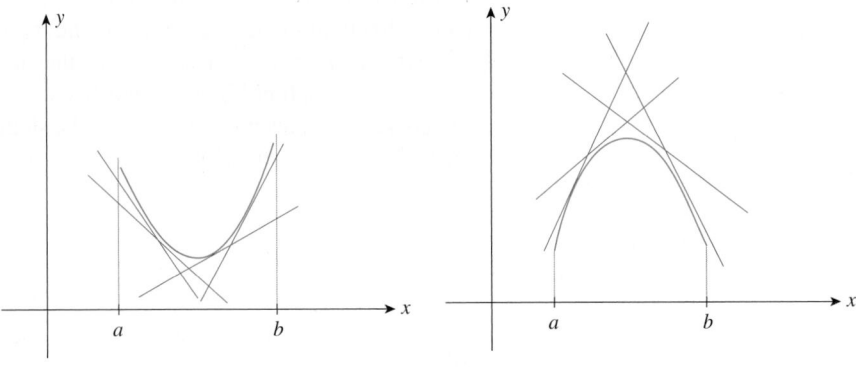

(a) f is concave upward on (a, b).   **(b)** f is concave downward on (a, b).

Geometrically, a curve is concave upward if it lies above its tangent lines (Figure 28a). Similarly, a curve is concave downward if it lies below its tangent lines (Figure 28b).

We also say that f is *concave upward at a number c* if there exists an interval (a, b) containing c in which f is concave upward. Similarly, we say that f is *concave downward at a number c* if there exists an interval (a, b) containing c in which f is concave downward.

If a function f has a second derivative f'', we can use f'' to determine the intervals of concavity of the function. Recall that $f''(x)$ measures the rate of change of the slope $f'(x)$ of the tangent line to the graph of f at the point $(x, f(x))$. Thus, if $f''(x) > 0$ on an interval (a, b), then the slopes of the tangent lines to the graph of f are increasing on (a, b), and so f is concave upward on (a, b). Similarly, if $f''(x) < 0$ on (a, b), then f is concave downward on (a, b). These observations suggest the following theorem.

THEOREM 2

a. If $f''(x) > 0$ for each value of x in (a, b), then f is concave upward on (a, b).

b. If $f''(x) < 0$ for each value of x in (a, b), then f is concave downward on (a, b).

The following procedure, based on the conclusions of Theorem 2, may be used to determine the intervals of concavity of a function.

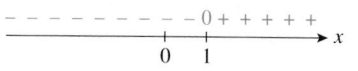

FIGURE 29
Sign diagram for f''

Determining the Intervals of Concavity of f

1. Determine the values of x for which f'' is zero or where f'' is not defined, and identify the open intervals determined by these numbers.

2. Determine the sign of f'' in each interval found in step 1. To do this, compute $f''(c)$, where c is any conveniently chosen test number in the interval.

   a. If $f''(c) > 0$, f is concave upward on that interval.

   b. If $f''(c) < 0$, f is concave downward on that interval.

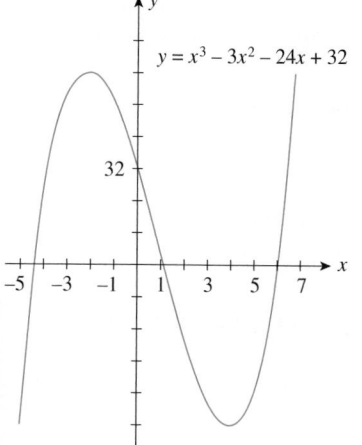

FIGURE 30
f is concave downward on $(-\infty, 1)$ and concave upward on $(1, \infty)$.

EXAMPLE 1   Determine where the function $f(x) = x^3 - 3x^2 - 24x + 32$ is concave upward and where it is concave downward.

Solution   Here,

$$f'(x) = 3x^2 - 6x - 24$$
$$f''(x) = 6x - 6 = 6(x - 1)$$

and f'' is defined everywhere. Setting $f''(x) = 0$ gives $x = 1$. The sign diagram of f'' appears in Figure 29. We conclude that f is concave downward on the interval $(-\infty, 1)$ and is concave upward on the interval $(1, \infty)$. Figure 30 shows the graph of f.   ■

EXPLORING WITH TECHNOLOGY

Refer to Example 1.

1. Use a graphing utility to plot the graph of $f(x) = x^3 - 3x^2 - 24x + 32$ and its second derivative $f''(x) = 6x - 6$ using the viewing window $[-10, 10] \times [-80, 90]$.

2. By studying the graph of f'', determine the intervals where $f''(x) > 0$ and the intervals where $f''(x) < 0$. Next, look at the graph of f and determine the intervals where the graph of f is concave upward and the intervals where the graph of f is concave downward. Are these observations what you might have expected?

EXAMPLE 2   Determine the intervals where the function $f(x) = x + \dfrac{1}{x}$ is concave upward and where it is concave downward.

f'' is not defined at $x = 0$

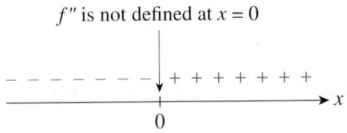

FIGURE 31
The sign diagram for f''

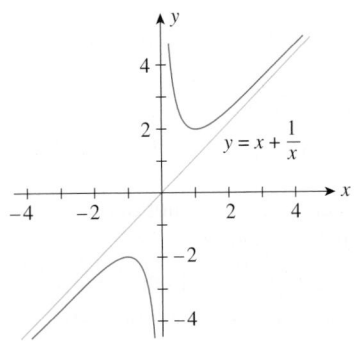

FIGURE 32
f is concave downward on $(-\infty, 0)$ and concave upward on $(0, \infty)$.

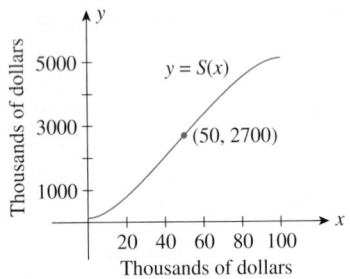

FIGURE 33
The graph of S has a point of inflection at (50, 2700).

FIGURE 34
At each point of inflection, the graph of a function crosses its tangent line.

Solution   We have

$$f'(x) = 1 - \frac{1}{x^2}$$

$$f''(x) = \frac{2}{x^3}$$

We deduce from the sign diagram for f'' (Figure 31) that the function f is concave downward on the interval $(-\infty, 0)$ and concave upward on the interval $(0, \infty)$. The graph of f is sketched in Figure 32.   ∎

Inflection Points

Figure 33 shows the total sales S of a manufacturer of automobile air conditioners versus the amount of money x that the company spends on advertising its product. Notice that the graph of the continuous function $y = S(x)$ changes concavity—from upward to downward—at the point (50, 2700). This point is called an inflection point of S. To understand the significance of this inflection point, observe that the total sales increase rather slowly at first, but as more money is spent on advertising, the total sales increase rapidly. This rapid increase reflects the effectiveness of the company's ads. However, a point is soon reached after which any additional advertising expenditure results in increased sales but at a slower rate of increase. This point, commonly known as the *point of diminishing returns,* is the point of inflection of the function S. We will return to this example later.

Let's now state formally the definition of an inflection point.

Inflection Point
A point on the graph of a continuous function f where the tangent line exists and where the concavity changes is called an **inflection point.**

Observe that the graph of a function crosses its tangent line at a point of inflection (Figure 34).

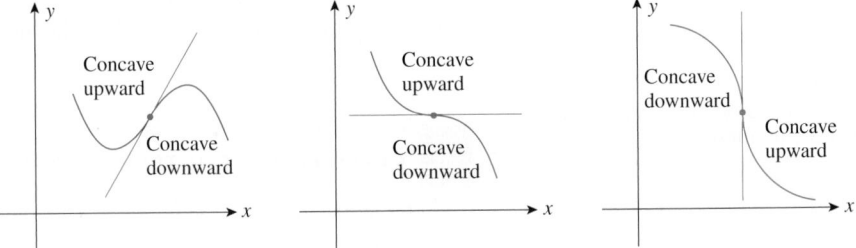

The following procedure may be used to find inflection points.

Finding Inflection Points

1. Compute $f''(x)$.

2. Determine the numbers in the domain of f for which $f''(x) = 0$ or $f''(x)$ does not exist.

3. Determine the sign of $f''(x)$ to the left and right of each number c found in step 2. If there is a change in the sign of $f''(x)$ as we move across $x = c$, then $(c, f(c))$ is an inflection point of f.

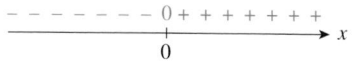

FIGURE 35
Sign diagram for f''

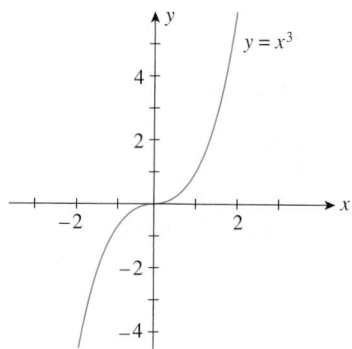

FIGURE 36
f has an inflection point at $(0, 0)$.

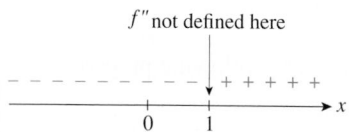

FIGURE 37
Sign diagram for f''

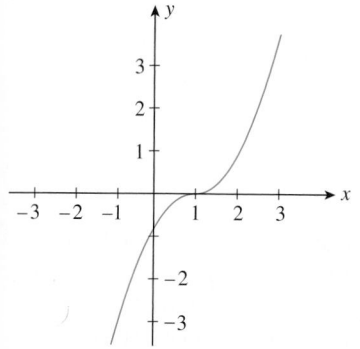

FIGURE 38
f has an inflection point at $(1, 0)$.

The numbers determined in step 2 are only *candidates* for the inflection points of f. For example, you can easily verify that $f''(0) = 0$ if $f(x) = x^4$, but a sketch of the graph of f will show that $(0, 0)$ is *not* an inflection point of f.

EXAMPLE 3   Find the points of inflection of the function $f(x) = x^3$.

Solution

$$f'(x) = 3x^2$$
$$f''(x) = 6x$$

Observe that f'' is continuous everywhere and is zero if $x = 0$. The sign diagram of f'' is shown in Figure 35. From this diagram, we see that $f''(x)$ changes sign as we move across $x = 0$. Thus, the point $(0, 0)$ is an inflection point of the function f (Figure 36). ∎

EXAMPLE 4   Determine the intervals where the function $f(x) = (x - 1)^{5/3}$ is concave upward and where it is concave downward and find the inflection points of f.

Solution   The first derivative of f is

$$f'(x) = \frac{5}{3}(x - 1)^{2/3}$$

and the second derivative of f is

$$f''(x) = \frac{10}{9}(x - 1)^{-1/3} = \frac{10}{9(x - 1)^{1/3}}$$

We see that f'' is not defined at $x = 1$. Furthermore, $f''(x)$ is not equal to zero anywhere. The sign diagram of f'' is shown in Figure 37. From the sign diagram, we see that f is concave downward on $(-\infty, 1)$ and concave upward on $(1, \infty)$. Next, since $x = 1$ does lie in the domain of f, our computations also reveal that the point $(1, 0)$ is an inflection point of f (Figure 38). ∎

EXAMPLE 5   Determine the intervals where the function

$$f(x) = \frac{1}{x^2 + 1}$$

is concave upward and where it is concave downward and find the inflection points of f.

Solution   The first derivative of f is

$$f'(x) = \frac{d}{dx}(x^2 + 1)^{-1} = -2x(x^2 + 1)^{-2} \qquad \text{Rewrite the original function and use the general power rule.}$$

$$= -\frac{2x}{(x^2 + 1)^2}$$

Next, using the quotient rule, we find

$$f''(x) = \frac{(x^2 + 1)^2(-2) + (2x)2(x^2 + 1)(2x)}{(x^2 + 1)^4}$$

$$= \frac{(x^2 + 1)[-2(x^2 + 1) + 8x^2]}{(x^2 + 1)^4} = \frac{(x^2 + 1)(6x^2 - 2)}{(x^2 + 1)^4}$$

$$= \frac{2(3x^2 - 1)}{(x^2 + 1)^3} \qquad \text{Cancel the common factors.}$$

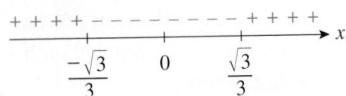

FIGURE 39
Sign diagram for f''

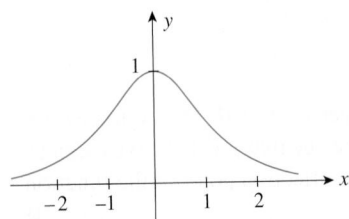

FIGURE 40

The graph of $f(x) = \dfrac{1}{x^2 + 1}$ is concave upward on $(-\infty, -\sqrt{3}/3) \cup (\sqrt{3}/3, \infty)$ and concave downward on $(-\sqrt{3}/3, \sqrt{3}/3)$.

Observe that f'' is continuous everywhere and is zero if

$$3x^2 - 1 = 0$$
$$x^2 = \frac{1}{3}$$

or $x = \pm\sqrt{3}/3$. The sign diagram for f'' is shown in Figure 39. From the sign diagram for f'', we see that f is concave upward on $(-\infty, -\sqrt{3}/3) \cup (\sqrt{3}/3, \infty)$ and concave downward on $(-\sqrt{3}/3, \sqrt{3}/3)$. Also, observe that $f''(x)$ changes sign as we move across the numbers $x = -\sqrt{3}/3$ and $x = \sqrt{3}/3$. Since

$$f\left(-\frac{\sqrt{3}}{3}\right) = \frac{1}{\frac{1}{3} + 1} = \frac{3}{4} \quad \text{and} \quad f\left(\frac{\sqrt{3}}{3}\right) = \frac{3}{4}$$

we see that the points $(-\sqrt{3}/3, 3/4)$ and $(\sqrt{3}/3, 3/4)$ are inflection points of f. The graph of f is shown in Figure 40.   ■

EXPLORE & DISCUSS

1. Suppose $(c, f(c))$ is an inflection point of f. Can you conclude that f has no relative extremum at $x = c$? Explain your answer.

2. True or false: A polynomial function of degree 3 has exactly one inflection point.
   Hint: Study the function $f(x) = ax^3 + bx^2 + cx + d \ (a \neq 0)$.

The next example uses an interpretation of the first and second derivatives to help us sketch a graph of a function.

EXAMPLE 6   Sketch the graph of a function having the following properties:

$$f(-1) = 4$$
$$f(0) = 2$$
$$f(1) = 0$$
$$f'(-1) = 0$$
$$f'(1) = 0$$
$$f'(x) > 0 \quad \text{on } (-\infty, -1) \cup (1, \infty)$$
$$f'(x) < 0 \quad \text{on } (-1, 1)$$
$$f''(x) < 0 \quad \text{on } (-\infty, 0)$$
$$f''(x) > 0 \quad \text{on } (0, \infty)$$

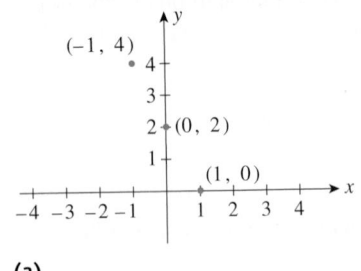

(a)

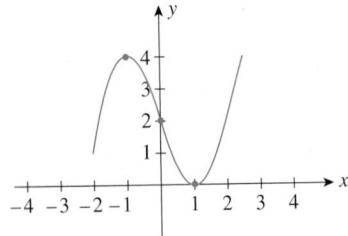

(b)
FIGURE 41

Solution   First, we plot the points $(-1, 4)$, $(0, 2)$, and $(1, 0)$ that lie on the graph of f. Since $f'(-1) = 0$ and $f'(1) = 0$, the tangent lines at the points $(-1, 4)$ and $(1, 0)$ are horizontal. Since $f'(x) > 0$ on $(-\infty, -1)$ and $f'(x) < 0$ on $(-1, 1)$, we see that f has a relative maximum at the point $(-1, 4)$. Also, $f'(x) < 0$ on $(-1, 1)$ and $f'(x) > 0$ on $(1, \infty)$ implies that f has a relative minimum at the point $(1, 0)$ (Figure 41a).

Since $f''(x) < 0$ on $(-\infty, 0)$ and $f''(x) > 0$ on $(0, \infty)$, we see that the point $(0, 2)$ is an inflection point. Finally, we complete the graph making use of the fact that f is increasing on $(-\infty, -1) \cup (1, \infty)$, where it is given that $f'(x) > 0$, and f is decreasing on $(-1, 1)$, where $f'(x) < 0$. Also, make sure that f is concave downward on $(-\infty, 0)$ and concave upward on $(0, \infty)$ (Figure 41b).   ■

Examples 7 and 8 illustrate familiar interpretations of the significance of the inflection point of a function.

APPLIED EXAMPLE 7   Effect of Advertising on Sales   The total sales S (in thousands of dollars) of Arctic Air Corporation, a manufacturer of automobile air conditioners, is related to the amount of money x (in thousands of dollars) the company spends on advertising its products by the formula

$$S = -0.01x^3 + 1.5x^2 + 200 \qquad (0 \le x \le 100)$$

Find the inflection point of the function S.

Solution   The first two derivatives of S are given by

$$S' = -0.03x^2 + 3x$$
$$S'' = -0.06x + 3$$

Setting $S'' = 0$ gives $x = 50$. So $(50, S(50))$ is the only candidate for an inflection point of S. Moreover, since

$$S'' > 0 \quad \text{for} \quad x < 50$$

and

$$S'' < 0 \quad \text{for} \quad x > 50$$

the point $(50, 2700)$ is an inflection point of the function S. The graph of S appears in Figure 42. Notice that this is the graph of the function we discussed earlier.

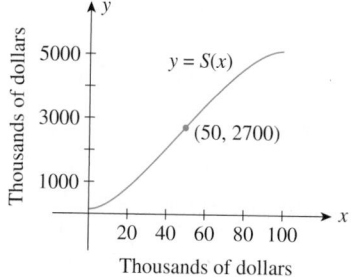

FIGURE 42
The graph of $S(x)$ has a point of inflection at (50, 2700).

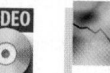

APPLIED EXAMPLE 8   Consumer Price Index   An economy's consumer price index (CPI) is described by the function

$$I(t) = -0.2t^3 + 3t^2 + 100 \qquad (0 \le t \le 9)$$

where $t = 0$ corresponds to the year 1995. Find the point of inflection of the function I and discuss its significance.

Solution   The first two derivatives of I are given by

$$I'(t) = -0.6t^2 + 6t$$
$$I''(t) = -1.2t + 6 = -1.2(t - 5)$$

Setting $I''(t) = 0$ gives $t = 5$. So $(5, I(5))$ is the only candidate for an inflection point of I. Next, we observe that

$$I'' > 0 \quad \text{for} \quad t < 5$$
$$I'' < 0 \quad \text{for} \quad t > 5$$

so the point $(5, 150)$ is an inflection point of I. The graph of I is sketched in Figure 43.

Since the second derivative of I measures the rate of change of the inflation rate, our computations reveal that the rate of inflation had in fact peaked at $t = 5$. Thus, relief actually began at the beginning of 2000.

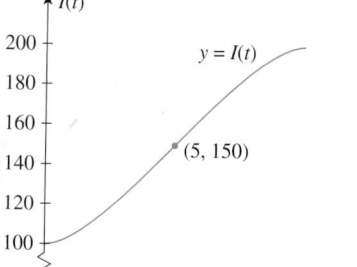

FIGURE 43
The graph of $I(t)$ has a point of inflection at (5, 150).

The Second Derivative Test

We now show how the second derivative f'' of a function f can be used to help us determine whether a critical number of f gives rise to a relative extremum of f. Figure 44a shows the graph of a function that has a relative maximum at $x = c$.

Observe that f is concave downward at that number. Similarly, Figure 44b shows that at a relative minimum of f the graph is concave upward. But from our previous work, we know that f is concave downward at $x = c$ if $f''(c) < 0$ and f is concave upward at $x = c$ if $f''(c) > 0$. These observations suggest the following alternative procedure for determining whether a critical number of f gives rise to a relative extremum of f. This result is called the **second derivative test** and is applicable when f'' exists.

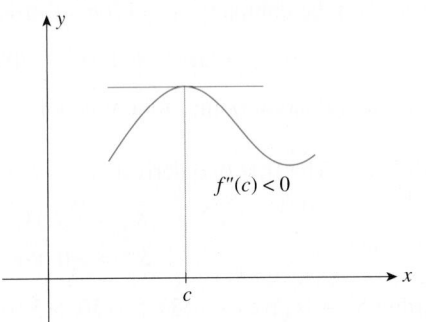
$f''(c) < 0$

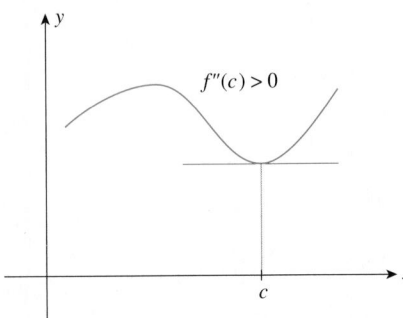
$f''(c) > 0$

(a) f has a relative maximum at $x = c$.   **(b)** f has a relative minimum at $x = c$.

FIGURE 44

The Second Derivative Test

1. Compute $f'(x)$ and $f''(x)$.
2. Find all the critical numbers of f at which $f'(x) = 0$.
3. Compute $f''(c)$ for each such critical number c.
   a. If $f''(c) < 0$, then f has a relative maximum at c.
   b. If $f''(c) > 0$, then f has a relative minimum at c.
   c. If $f''(c) = 0$, the test fails; that is, it is inconclusive.

Note   The second derivative test does not yield a conclusion if $f''(c) = 0$ or if $f''(c)$ does not exist. In other words, $x = c$ may give rise to a relative extremum or an inflection point (see Exercise 102, page 807). In such cases, you should revert to the first derivative test. ∎

EXAMPLE 9   Determine the relative extrema of the function

$$f(x) = x^3 - 3x^2 - 24x + 32$$

using the second derivative test. (See Example 7, Section 13.1.)

Solution   We have

$$f'(x) = 3x^2 - 6x - 24 = 3(x + 2)(x - 4)$$

so $f'(x) = 0$ gives $x = -2$ and $x = 4$, the critical numbers of f, as in Example 7. Next, we compute

$$f''(x) = 6x - 6 = 6(x - 1)$$

Since

$$f''(-2) = 6(-2 - 1) = -18 < 0$$

the second derivative test implies that $f(-2) = 60$ is a relative maximum of f.

Also,

$$f''(4) = 6(4 - 1) = 18 > 0$$

and the second derivative test implies that $f(4) = -48$ is a relative minimum of f, which confirms the results obtained earlier.   ∎

EXPLORE & DISCUSS

Suppose a function f has the following properties:

1. $f''(x) > 0$ for all x in an interval (a, b).

2. There is a number c between a and b such that $f'(c) = 0$.

What special property can you ascribe to the point $(c, f(c))$? Answer the question if Property 1 is replaced by the property that $f''(x) < 0$ for all x in (a, b).

Comparing the First and Second Derivative Tests

Notice that both the first derivative test and the second derivative test are used to classify the critical numbers of f. What are the pros and cons of the two tests? Since the second derivative test is applicable only when f'' exists, it is less versatile than the first derivative test. For example, it cannot be used to locate the relative minimum $f(0) = 0$ of the function $f(x) = x^{2/3}$.

Furthermore, the second derivative test is inconclusive when f'' is equal to zero at a critical number of f, whereas the first derivative test always yields positive conclusions. The second derivative test is also inconvenient to use when f'' is difficult to compute. On the plus side, if f'' is computed easily, then we use the second derivative test since it involves just the evaluation of f'' at the critical number(s) of f. Also, the conclusions of the second derivative test are important in theoretical work.

We close this section by summarizing the different roles played by the first derivative f' and the second derivative f'' of a function f in determining the properties of the graph of f. The first derivative f' tells us where f is increasing and where f is decreasing, whereas the second derivative f'' tells us where f is concave upward and where f is concave downward. These different properties of f are reflected by the signs of f' and f'' in the interval of interest. The following table shows the general characteristics of the function f for various possible combinations of the signs of f' and f'' in the interval (a, b).

Signs of f' and f''	Properties of the Graph of f	General Shape of the Graph of f
$f'(x) > 0$ $f''(x) > 0$	f increasing f concave upward	
$f'(x) > 0$ $f''(x) < 0$	f increasing f concave downward	
$f'(x) < 0$ $f''(x) > 0$	f decreasing f concave upward	
$f'(x) < 0$ $f''(x) < 0$	f decreasing f concave downward	

13.2   Self-Check Exercises

1. Determine where the function $f(x) = 4x^3 - 3x^2 + 6$ is concave upward and where it is concave downward.

2. Using the second derivative test, if applicable, find the relative extrema of the function $f(x) = 2x^3 - \frac{1}{2}x^2 - 12x - 10$.

3. A certain country's gross domestic product (GDP) (in millions of dollars) in year t is described by the function

$$G(t) = -2t^3 + 45t^2 + 20t + 6000 \qquad (0 \le t \le 11)$$

where $t = 0$ corresponds to the beginning of 1995. Find the inflection point of the function G and discuss its significance.

Solutions to Self-Check Exercises 13.2 can be found on page 807.

13.2   Concept Questions

1. Explain what it means for a function f to be (a) concave upward and (b) concave downward on an open interval I. Given that f has a second derivative on I (except at isolated numbers), how do you determine where the graph of f is concave upward and where it is concave downward?

2. What is an inflection point of the graph of a function f? How do you find the inflection point(s) of the graph of a function f whose rule is given?

3. State the second derivative test. What are the pros and cons of using the first derivative test and the second derivative test?

13.2   Exercises

In Exercises 1–8, you are given the graph of a function f. Determine the intervals where f is concave upward and where it is concave downward. Also, find all inflection points of f, if any.

1.

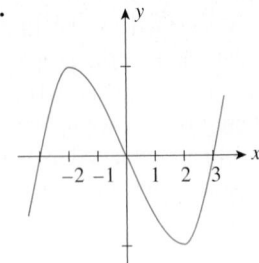

2.

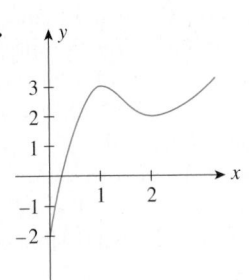

3.

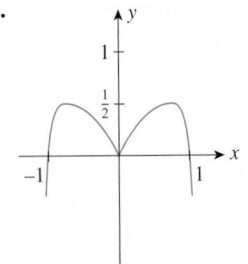

4.

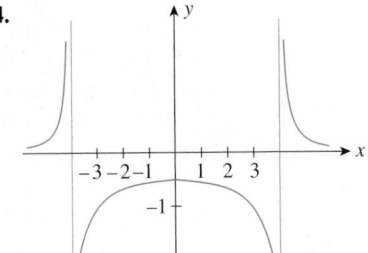

5.

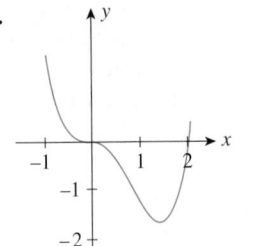

6.

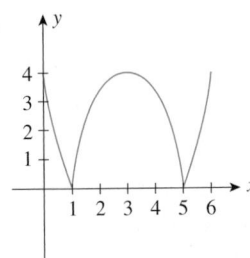

7.

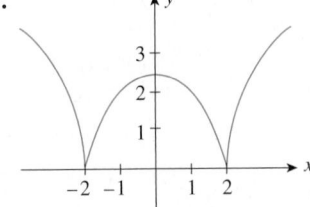

8.

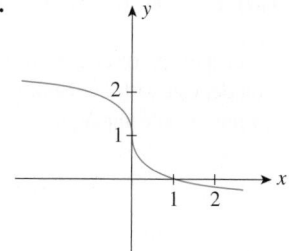

In Exercises 9–12, determine which graph—(a), (b), or (c)—is the graph of the function f with the specified properties.

9. $f(2) = 1, f'(2) > 0$, and $f''(2) < 0$

(a)

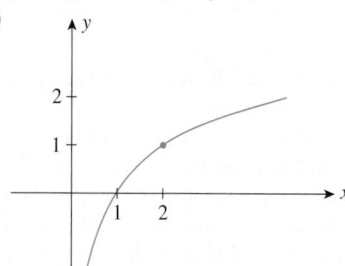

(b)

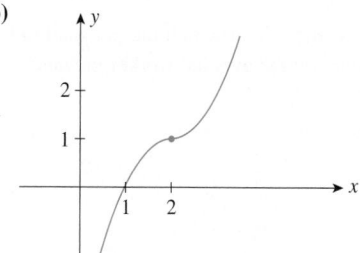

(c)

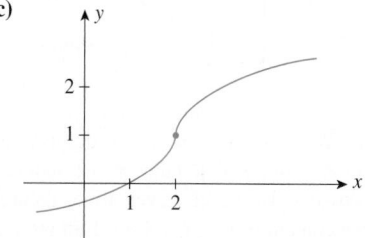

10. $f(1) = 2, f'(x) > 0$ on $(-\infty, 1) \cup (1, \infty)$, and $f''(1) = 0$

(a)

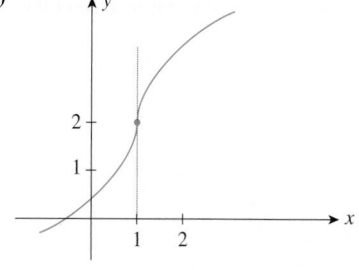

(b)

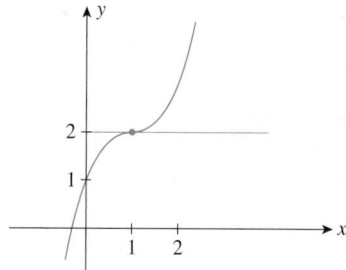

(c)

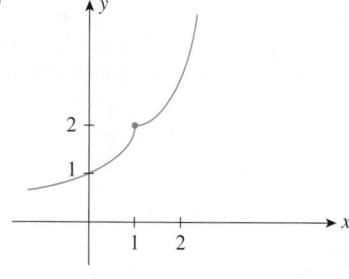

11. $f'(0)$ is undefined, f is decreasing on $(-\infty, 0)$, f is concave downward on $(0, 3)$, and f has an inflection point at $x = 3$.

(a)

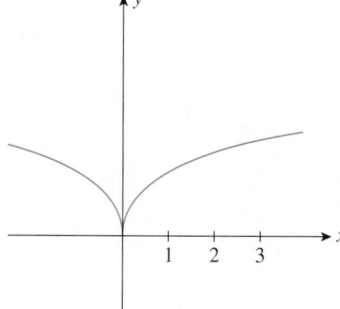

(b)

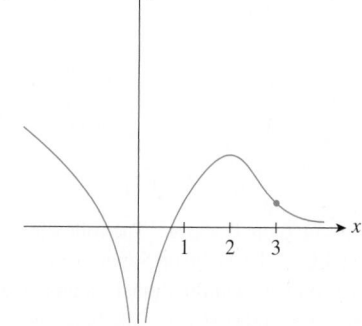

(c)

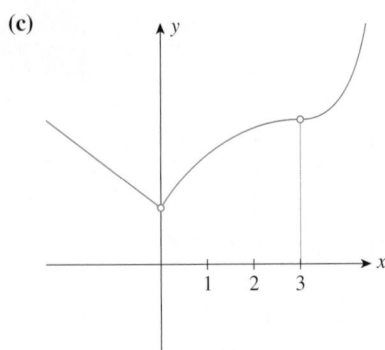

12. f is decreasing on $(-\infty, 2)$ and increasing on $(2, \infty)$, f is concave upward on $(1, \infty)$, and f has inflection points at $x = 0$ and $x = 1$.

(a)

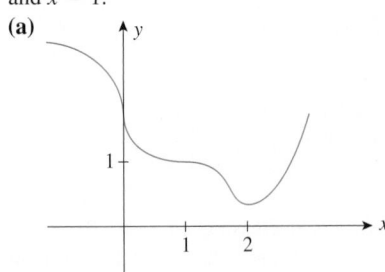

(b)

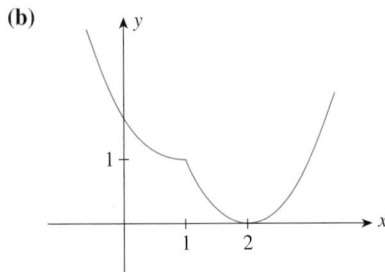

(c)

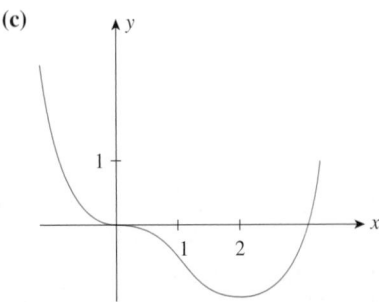

13. EFFECT OF ADVERTISING ON BANK DEPOSITS The following graphs were used by the CEO of the Madison Savings Bank to illustrate what effect a projected promotional campaign would have on its deposits over the next year. The functions D_1 and D_2 give the projected amount of money on deposit with the bank over the next 12 mo with and without the proposed promotional campaign, respectively.

a. Determine the signs of $D_1'(t)$, $D_2'(t)$, $D_1''(t)$, and $D_2''(t)$ on the interval $(0, 12)$.

b. What can you conclude about the rate of change of the growth rate of the money on deposit with the bank with and without the proposed promotional campaign?

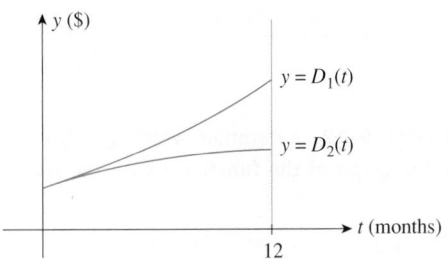

14. ASSEMBLY TIME OF A WORKER In the following graph, $N(t)$ gives the number of personal radios assembled by the average worker by the tth hr, where $t = 0$ corresponds to 8 a.m. and $0 \le t \le 4$. The point P is an inflection point of N.

a. What can you say about the rate of change of the rate of the number of personal radios assembled by the average worker between 8 a.m. and 10 a.m.? Between 10 a.m. and 12 a.m.?

b. At what time is the rate at which the personal radios are being assembled by the average worker greatest?

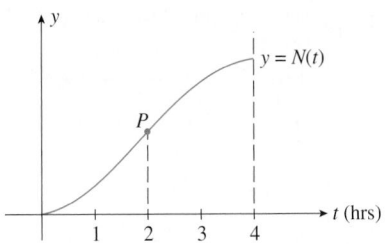

15. WATER POLLUTION When organic waste is dumped into a pond, the oxidation process that takes place reduces the pond's oxygen content. However, given time, nature will restore the oxygen content to its natural level. In the following graph, $P(t)$ gives the oxygen content (as a percent of its normal level) t days after organic waste has been dumped into the pond. Explain the significance of the inflection point Q.

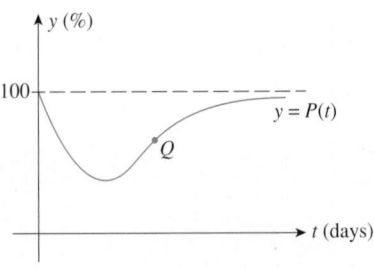

16. SPREAD OF A RUMOR   Initially, a handful of students heard a rumor on campus. The rumor spread and, after t hr, the number had grown to $N(t)$. The graph of the function N is shown in the following figure:

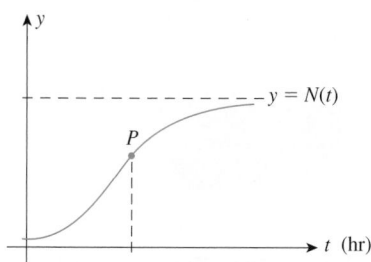

Describe the spread of the rumor in terms of the speed it was spread. In particular, explain the significance of the inflection point P of the graph of N.

In Exercises 17–20, show that the function is concave upward wherever it is defined.

17. $f(x) = 4x^2 - 12x + 7$

18. $g(x) = x^4 + \frac{1}{2}x^2 + 6x + 10$

19. $f(x) = \frac{1}{x^4}$ 20. $g(x) = -\sqrt{4 - x^2}$

In Exercises 21–40, determine where the function is concave upward and where it is concave downward.

21. $f(x) = 2x^2 - 3x + 4$ 22. $g(x) = -x^2 + 3x + 4$

23. $f(x) = x^3 - 1$ 24. $g(x) = x^3 - x$

25. $f(x) = x^4 - 6x^3 + 2x + 8$

26. $f(x) = 3x^4 - 6x^3 + x - 8$

27. $f(x) = x^{4/7}$ 28. $f(x) = \sqrt[3]{x}$

29. $f(x) = \sqrt{4 - x}$ 30. $g(x) = \sqrt{x - 2}$

31. $f(x) = \frac{1}{x - 2}$ 32. $g(x) = \frac{x}{x + 1}$

33. $f(x) = \frac{1}{2 + x^2}$ 34. $g(x) = \frac{x}{1 + x^2}$

35. $h(t) = \frac{t^2}{t - 1}$ 36. $f(x) = \frac{x + 1}{x - 1}$

37. $g(x) = x + \frac{1}{x^2}$ 38. $h(r) = -\frac{1}{(r - 2)^2}$

39. $g(t) = (2t - 4)^{1/3}$ 40. $f(x) = (x - 2)^{2/3}$

In Exercises 41–52, find the inflection point(s), if any, of each function.

41. $f(x) = x^3 - 2$ 42. $g(x) = x^3 - 6x$

43. $f(x) = 6x^3 - 18x^2 + 12x - 15$

44. $g(x) = 2x^3 - 3x^2 + 18x - 8$

45. $f(x) = 3x^4 - 4x^3 + 1$

46. $f(x) = x^4 - 2x^3 + 6$

47. $g(t) = \sqrt[3]{t}$ 48. $f(x) = \sqrt[5]{x}$

49. $f(x) = (x - 1)^3 + 2$ 50. $f(x) = (x - 2)^{4/3}$

51. $f(x) = \frac{2}{1 + x^2}$ 52. $f(x) = 2 + \frac{3}{x}$

In Exercises 53–68, find the relative extrema, if any, of each function. Use the second derivative test, if applicable.

53. $f(x) = -x^2 + 2x + 4$ 54. $g(x) = 2x^2 + 3x + 7$

55. $f(x) = 2x^3 + 1$ 56. $g(x) = x^3 - 6x$

57. $f(x) = \frac{1}{3}x^3 - 2x^2 - 5x - 10$

58. $f(x) = 2x^3 + 3x^2 - 12x - 4$

59. $g(t) = t + \frac{9}{t}$ 60. $f(t) = 2t + \frac{3}{t}$

61. $f(x) = \frac{x}{1 - x}$ 62. $f(x) = \frac{2x}{x^2 + 1}$

63. $f(t) = t^2 - \frac{16}{t}$ 64. $g(x) = x^2 + \frac{2}{x}$

65. $g(s) = \frac{s}{1 + s^2}$ 66. $g(x) = \frac{1}{1 + x^2}$

67. $f(x) = \frac{x^4}{x - 1}$ 68. $f(x) = \frac{x^2}{x^2 + 1}$

In Exercises 69–74, sketch the graph of a function having the given properties.

69. $f(2) = 4, f'(2) = 0, f''(x) < 0$ on $(-\infty, \infty)$

70. $f(2) = 2, f'(2) = 0, f'(x) > 0$ on $(-\infty, 2), f'(x) > 0$ on $(2, \infty), f''(x) < 0$ on $(-\infty, 2), f''(x) > 0$ on $(2, \infty)$

71. $f(-2) = 4, f(3) = -2, f'(-2) = 0, f'(3) = 0, f'(x) > 0$ on $(-\infty, -2) \cup (3, \infty), f'(x) < 0$ on $(-2, 3)$, inflection point at $(1, 1)$

72. $f(0) = 0, f'(0)$ does not exist, $f''(x) < 0$ if $x \neq 0$

73. $f(0) = 1, f'(0) = 0, f(x) > 0$ on $(-\infty, \infty), f''(x) < 0$ on $(-\sqrt{2}/2, \sqrt{2}/2), f''(x) > 0$ on $(-\infty, -\sqrt{2}/2) \cup (\sqrt{2}/2, \infty)$

74. f has domain $[-1, 1], f(-1) = -1, f(-\frac{1}{2}) = -2, f'(-\frac{1}{2}) = 0, f''(x) > 0$ on $(-1, 1)$

75. DEMAND FOR RNS   The following graph gives the total number of help-wanted ads for RNs (registered nurses) in 22 cities over the last 12 mo as a function of time t (t measured in months).
   a. Explain why $N'(t)$ is positive on the interval $(0, 12)$.
   b. Determine the signs of $N''(t)$ on the interval $(0, 6)$ and the interval $(6, 12)$.

c. Interpret the results of part (b).

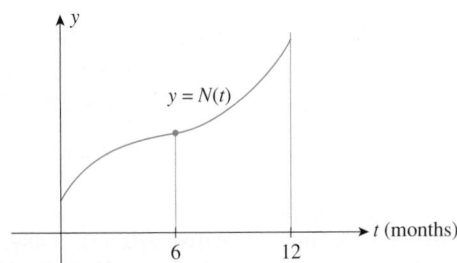

76. EFFECT OF BUDGET CUTS ON DRUG-RELATED CRIMES The graphs below were used by a police commissioner to illustrate what effect a budget cut would have on crime in the city. The number $N_1(t)$ gives the projected number of drug-related crimes in the next 12 mo. The number $N_2(t)$ gives the projected number of drug-related crimes in the same time frame if next year's budget is cut.
a. Explain why $N_1'(t)$ and $N_2'(t)$ are both positive on the interval $(0, 12)$.
b. What are the signs of $N_1''(t)$ and $N_2''(t)$ on the interval $(0, 12)$?
c. Interpret the results of part (b).

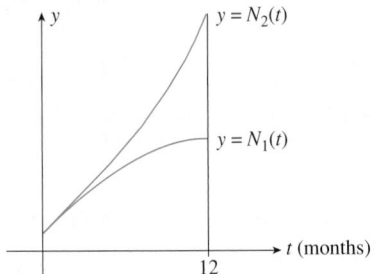

77. In the following figure, water is poured into the vase at a constant rate (in appropriate units), and the water level rises to a height of $f(t)$ units at time t as measured from the base of the vase. The graph of f follows. Explain the shape of the curve in terms of its concavity. What is the significance of the inflection point?

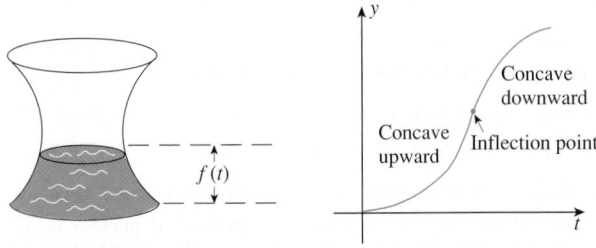

78. In the following figure, water is poured into an urn at a constant rate (in appropriate units), and the water level rises to a height of $f(t)$ units at time t as measured from the base of the

urn. Sketch the graph of f and explain its shape, indicating where it is concave upward and concave downward. Indicate the inflection point on the graph and explain its significance.
Hint: Study Exercise 77.

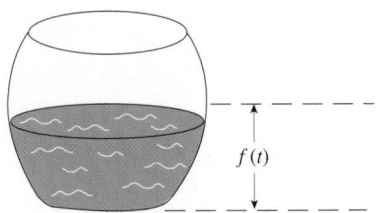

79. EFFECT OF SMOKING BANS The sales (in billions of dollars) in restaurants and bars in California from 1993 ($t = 0$) through 2000 ($t = 7$) are approximated by the function

$$S(t) = 0.195t^2 + 0.32t + 23.7 \qquad (0 \le t \le 7)$$

a. Show that the sales in restaurants and bars continued to rise after smoking bans were implemented in restaurants in 1995 and in bars in 1998.
Hint: Show that S is increasing in the interval $(2, 7)$.
b. What can you say about the rate at which the sales were rising after smoking bans were implemented?
Source: California Board of Equalization

80. DIGITAL TELEVISION SALES Since their introduction into the market in the late 1990s, the sales of digital televisions, including high-definition television sets, have slowly gathered momentum. The model

$$S(t) = 0.164t^2 + 0.85t + 0.3 \qquad (0 \le t \le 4)$$

describes the sales of digital television sets (in billions of dollars) between 1999 ($t = 0$) and 2003 ($t = 4$).
a. Find $S'(t)$ and $S''(t)$.
b. Use the results of part (a) to conclude that the sales of digital TVs were increasing between 1999 and 2003 and that the sales were increasing at an increasing rate over that time interval.
Source: Consumer Electronics Association

81. WORKER EFFICIENCY An efficiency study conducted for Elektra Electronics showed that the number of Space Commander walkie-talkies assembled by the average worker t hr after starting work at 8 a.m. is given by

$$N(t) = -t^3 + 6t^2 + 15t \qquad (0 \le t \le 4)$$

At what time during the morning shift is the average worker performing at peak efficiency?

82. FLIGHT OF A ROCKET The altitude (in feet) of a rocket t sec into flight is given by

$$s = f(t) = -t^3 + 54t^2 + 480t + 6 \qquad (t \ge 0)$$

Find the point of inflection of the function f and interpret your result. What is the maximum velocity attained by the rocket?

83. BUSINESS SPENDING ON TECHNOLOGY   In a study conducted in 2003, business spending on technology (in billions of dollars) from 2000 through 2005 was projected to be

$$S(t) = -1.88t^3 + 30.33t^2 - 76.14t + 474 \qquad (0 \le t \le 5)$$

where t is measured in years, with $t = 0$ corresponding to 2000. Show that the graph of S is concave upward on the interval $(0, 5)$. What does this result tell you about the rate of business spending on technology over the period in question?

Source: Quantit Economic Group

84. ALTERNATIVE MINIMUM TAX   Congress created the alternative minimum tax (AMT) in the late 1970s to ensure that wealthy people paid their fair share of taxes. But because of quirks in the law, even middle-income taxpayers have started to get hit with the tax. The AMT (in billions of dollars) projected to be collected by the IRS from 2001 through 2010 is

$$f(t) = 0.0117t^3 + 0.0037t^2 + 0.7563t + 4.1 \qquad (0 \le t \le 9)$$

where t is measured in years, with $t = 0$ corresponding to 2001.
a. Show that f is increasing on the interval $(0, 9)$. What does this result tell you about the projected amount of AMT paid over the years in question?
b. Show that f' is increasing on the interval $(0, 9)$. What conclusion can you draw from this result concerning the rate of growth at which the AMT is paid over the years in question?

Source: U.S. Congress Joint Economic Committee

85. EFFECT OF ADVERTISING ON HOTEL REVENUE   The total annual revenue R of the Miramar Resorts Hotel is related to the amount of money x the hotel spends on advertising its services by the function

$$R(x) = -0.003x^3 + 1.35x^2 + 2x + 8000 \qquad (0 \le x \le 400)$$

where both R and x are measured in thousands of dollars.
a. Find the interval where the graph of R is concave upward and the interval where the graph of R is concave downward. What is the inflection point of R?
b. Would it be more beneficial for the hotel to increase its advertising budget slightly when the budget is $140,000 or when it is $160,000?

86. FORECASTING PROFITS   As a result of increasing energy costs, the growth rate of the profit of the 4-yr old Venice Glassblowing Company has begun to decline. Venice's management, after consulting with energy experts, decides to implement certain energy-conservation measures aimed at cutting energy bills. The general manager reports that, according to his calculations, the growth rate of Venice's profit should be on the increase again within 4 yr. If Venice's profit (in hundreds of dollars) t yr from now is given by the function

$$P(t) = t^3 - 9t^2 + 40t + 50 \qquad (0 \le t \le 8)$$

determine whether the general manager's forecast will be accurate.
Hint: Find the inflection point of the function P and study the concavity of P.

87. OUTSOURCING   The amount (in billions of dollars) spent by the top 15 U.S. financial institutions on IT (information technology) offshore outsourcing is projected to be

$$A(t) = 0.92(t + 1)^{0.61} \qquad (0 \le t \le 4)$$

where t is measured in years, with $t = 0$ corresponding to 2004.
a. Show that A is increasing on $(0, 4)$ and interpret your result.
b. Show that A is concave downward on $(0, 4)$. Interpret your result.

Source: Tower Group

88. SALES OF MOBILE PROCESSORS   The rising popularity of notebook computers is fueling the sales of mobile PC processors. In a study conducted in 2003, the sales of these chips (in billions of dollars) was projected to be

$$S(t) = 6.8(t + 1.03)^{0.49} \qquad (0 \le t \le 4)$$

where t is measured in years, with $t = 0$ corresponding to 2003.
a. Show that S is increasing on the interval $(0, 4)$ and interpret your result.
b. Show that the graph of S is concave downward on the interval $(0, 4)$. Interpret your result.

Source: International Data Corp.

89. DRUG SPENDING   Medicaid spending on drugs in Massachusetts started slowing down in part after the state demanded that patients use more generic drugs and limited the range of drugs available to the program. The annual pharmacy spending (in millions of dollars) from 1999 through 2004 is given by

$$S(t) = -1.806t^3 + 10.238t^2 + 93.35t + 583 \qquad (0 \le t \le 5)$$

where t is measured in years with $t = 0$ corresponding to 1999. Find the inflection point of S and interpret your result.

Source: MassHealth

90. PC SHIPMENTS   In a study conducted in 2003, it was projected that worldwide PC shipments (in millions) through 2005 will be given by

$$N(t) = 0.75t^3 - 1.5t^2 + 8.25t + 133 \qquad (0 \le t \le 4)$$

where t is measured in years, with $t = 0$ corresponding to 2001.
a. Determine the intervals where N is concave upward and where it is concave downward.
b. Find the inflection point of N and interpret your result.

Source: International Data Corporation

91. Google's Revenue   The revenue for Google from 1999 ($t = 0$) through 2003 ($t = 4$) is approximated by the function

$$R(t) = 24.975t^3 - 49.81t^2 + 41.25t + 0.2 \qquad (0 \le t \le 4)$$

where $R(t)$ is measured in millions of dollars.
  a. Find $R'(t)$ and $R''(t)$.
  b. Show that $R'(t) > 0$ for all t in the interval $(0, 4)$ and interpret your result.
  Hint: Use the quadratic formula.
  c. Find the inflection point of R and interpret your result.
  Source: Company Report

92. Population Growth in Clark County   Clark County in Nevada—dominated by greater Las Vegas—is the fastest-growing metropolitan area in the United States. The population of the county from 1970 through 2000 is approximated by the function

$$P(t) = 44560t^3 - 89394t^2 + 234633t + 273288 \quad (0 \le t \le 4)$$

where t is measured in decades, with $t = 0$ corresponding to the beginning of 1970.
  a. Show that the population of Clark County was always increasing over the time period in question.
  Hint: Show that $P'(t) > 0$ for all t in the interval $(0, 4)$.
  b. Show that the population of Clark County was increasing at the slowest pace some time toward the middle of August 1976.
  Hint: Find the inflection point of P in the interval $(0, 4)$.
  Source: U.S. Census Bureau

93. Air Pollution   The level of ozone, an invisible gas that irritates and impairs breathing, present in the atmosphere on a certain May day in the city of Riverside was approximated by

$$A(t) = 1.0974t^3 - 0.0915t^4 \qquad (0 \le t \le 11)$$

where $A(t)$ is measured in pollutant standard index (PSI) and t is measured in hours, with $t = 0$ corresponding to 7 a.m. Use the second derivative test to show that the function A has a relative maximum at approximately $t = 9$. Interpret your results.

94. Cash Reserves at Blue Cross and Blue Shield   Based on company financial reports, the cash reserves of Blue Cross and Blue Shield as of the beginning of year t is approximated by the function

$$R(t) = -1.5t^4 + 14t^3 - 25.4t^2 + 64t + 290 \qquad (0 \le t \le 6)$$

where $R(t)$ is measured in millions of dollars and t is measured in years, with $t = 0$ corresponding to the beginning of 1998.
  a. Find the inflection points of R.
  Hint: Use the quadratic formula.
  b. Use the result of part (a) to show that the cash reserves of the company was growing at the greatest rate at the beginning of 2002.
  Source: Blue Cross and Blue Shield

95. Women's Soccer   Starting with the youth movement that took hold in the 1970s and buoyed by the success of the U.S. national women's team in international competition in recent years, girls and women have taken to soccer in ever-growing numbers. The function

$$N(t) = -0.9307t^3 + 74.04t^2 + 46.8667t + 3967 \quad (0 \le t \le 16)$$

gives the number of participants in women's soccer in year t, with $t = 0$ corresponding to the beginning of 1985.
  a. Verify that the number of participants in women's soccer had been increasing from 1985 through 2000.
  Hint: Use the quadratic formula.
  b. Show that the number of participants in women's soccer had been increasing at an increasing rate from 1985 through 2000.
  Hint: Show that the sign of N'' is positive on the interval in question.
  Source: NCCA News

96. Dependency Ratio   The share of the world population that is over 60 years of age compared to the rest of the working population in the world is of concern to economists. An increasing dependency ratio means that there will be fewer workers to support an aging population. The dependency ratio over the next century is forecast to be

$$R(t) = 0.00731t^4 - 0.174t^3 + 1.528t^2 + 0.48t + 19.3$$
$$(0 \le t \le 10)$$

in year t, where t is measured in decades with $t = 0$ corresponding to 2000.
  a. Show that the dependency ratio will be increasing at the fastest pace around 2052.
  Hint: Use the quadratic formula.
  b. What will the dependency ratio be at that time?
  Source: International Institute for Applied Systems Analysis

In Exercises 97–100, determine whether the statement is true or false. If it is true, explain why it is true. If it is false, give an example to show why it is false.

97. If the graph of f is concave upward on (a, b), then the graph of $-f$ is concave downward on (a, b).

98. If the graph of f is concave upward on (a, c) and concave downward on (c, b), where $a < c < b$, then f has an inflection point at $(c, f(c))$.

99. If c is a critical number of f where $a < c < b$ and $f''(x) < 0$ on (a, b), then f has a relative maximum at $x = c$.

100. A polynomial function of degree n ($n \ge 3$) can have at most $(n - 2)$ inflection points.

101. Show that the quadratic function

$$f(x) = ax^2 + bx + c \qquad (a \ne 0)$$

is concave upward if $a > 0$ and concave downward if $a < 0$. Thus, by examining the sign of the coefficient of x^2, one can tell immediately whether the parabola opens upward or downward.

102. Consider the functions $f(x) = x^3$, $g(x) = x^4$, and $h(x) = -x^4$.

a. Show that $x = 0$ is a critical number of each of the functions f, g, and h.
b. Show that the second derivative of each of the functions f, g, and h equals zero at $x = 0$.
c. Show that f has neither a relative maximum nor a relative minimum at $x = 0$, that g has a relative minimum at $x = 0$, and that h has a relative maximum at $x = 0$.

13.2 Solutions to Self-Check Exercises

1. We first compute

$$f'(x) = 12x^2 - 6x$$
$$f''(x) = 24x - 6 = 6(4x - 1)$$

Observe that f'' is continuous everywhere and has a zero at $x = \frac{1}{4}$. The sign diagram of f'' is shown in the accompanying figure.

$$- - - - - - 0 + + + + + +$$
$$\xrightarrow{0 \frac{1}{4} } x$$

From the sign diagram for f'', we see that f is concave upward on $(\frac{1}{4}, \infty)$ and concave downward on $(-\infty, \frac{1}{4})$.

2. First, we find the critical numbers of f by solving the equation

$$f'(x) = 6x^2 - x - 12 = 0$$

That is,

$$(3x + 4)(2x - 3) = 0$$

giving $x = -\frac{4}{3}$ and $x = \frac{3}{2}$. Next, we compute

$$f''(x) = 12x - 1$$

Since

$$f''\left(-\frac{4}{3}\right) = 12\left(-\frac{4}{3}\right) - 1 = -17 < 0$$

the second derivative test implies that $f(-\frac{4}{3}) = \frac{10}{27}$ is a relative maximum of f. Also,

$$f''\left(\frac{3}{2}\right) = 12\left(\frac{3}{2}\right) - 1 = 17 > 0$$

and we see that $f(\frac{3}{2}) = -\frac{179}{8}$ is a relative minimum.

3. We compute the second derivative of G. Thus,

$$G'(t) = -6t^2 + 90t + 20$$
$$G''(t) = -12t + 90$$

Now, G'' is continuous everywhere, and $G''(t) = 0$, where $t = \frac{15}{2}$, giving $t = \frac{15}{2}$ as the only candidate for an inflection point of G. Since $G''(t) > 0$ for $t < \frac{15}{2}$ and $G''(t) < 0$ for $t > \frac{15}{2}$, we see that $(\frac{15}{2}, \frac{15,675}{2})$ is an inflection point of G. The results of our computations tell us that the country's GDP was increasing most rapidly at the beginning of July 2002.

USING TECHNOLOGY

■ Finding the Inflection Points of a Function

A graphing utility can be used to find the inflection points of a function and hence the intervals where the graph of the function is concave upward and the intervals where it is concave downward. Some graphing utilities have an operation for finding inflection points directly. For example, both the TI-85 and TI-86 graphing calculators have this capability. If your graphing utility has this capability, use it to work through the example and exercises in this section.

EXAMPLE 1 Let $f(x) = 2.5x^5 - 12.4x^3 + 4.2x^2 - 5.2x + 4$.

a. Use a graphing utility to plot the graph of f.
b. Find the inflection points of f.
c. Find the intervals where f is concave upward and where it is concave downward.

(continued)

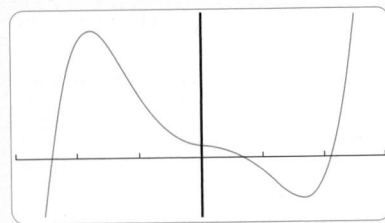

FIGURE T1
The graph of f in the viewing window
$[-3, 3] \times [-25, 60]$

Solution

a. The graph of f, using the viewing window $[-3, 3] \times [-25, 60]$, is shown in Figure T1.

b. We describe here the procedure using the TI-85. See the Web site for instructions for using the TI-86. From Figure T1 we see that f has three inflection points—one occurring at the point where the x-coordinate is approximately -1, another at the point where $x \approx 0$, and the third at the point where $x \approx 1$. To find the first inflection point, we use the inflection operation, moving the cursor to the point on the graph of f where $x \approx -1$. We obtain the point $(-1.2728, 34.6395)$ (accurate to four decimal places). Next, setting the cursor near $x = 0$ yields the inflection point $(0.1139, 3.4440)$. Finally, with the cursor set at $x = 1$, we obtain the third inflection point $(1.1589, -10.4594)$. (See Figure T2a–c.)

c. From the results of part (b), we see that f is concave upward on the intervals $(-1.2728, 0.1139)$ and $(1.1589, \infty)$ and concave downward on $(-\infty, -1.2728)$ and $(0.1139, 1.1589)$.

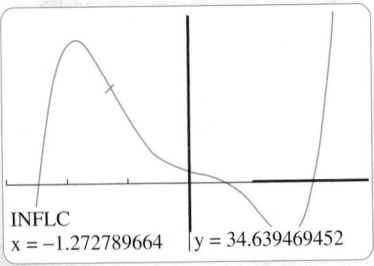

INFLC
x = -1.272789664   y = 34.639469452

(a)

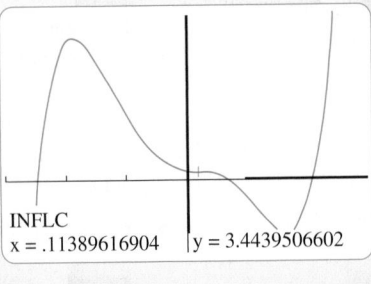

INFLC
x = .11389616904   y = 3.4439506602

(b)

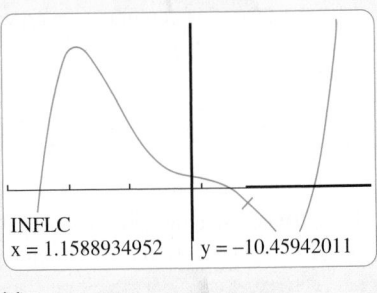

INFLC
x = 1.1588934952   y = -10.45942011

(c)

FIGURE T2
The TI-85 inflection point screens showing the points (a) $(-1.2728, 34.6395)$, (b) $(0.1139, 3.4440)$, and (c) $(1.1589, -10.4594)$

TECHNOLOGY EXERCISES

In Exercises 1–8, find (a) the intervals where f is concave upward and the intervals where f is concave downward and (b) the inflection points of f. Express your answers accurate to four decimal places.

1. $f(x) = 1.8x^4 - 4.2x^3 + 2.1x + 2$

2. $f(x) = -2.1x^4 + 3.1x^3 + 2x^2 - x + 1.2$

3. $f(x) = 1.2x^5 - 2x^4 + 3.2x^3 - 4x + 2$

4. $f(x) = -2.1x^5 + 3.2x^3 - 2.2x^2 + 4.2x - 4$

5. $f(x) = x^3(x^2 + 1)^{-1/3}$ **6.** $f(x) = x^2(x^3 - 1)^3$

7. $f(x) = \dfrac{x^2 - 1}{x^3}$ **8.** $f(x) = \dfrac{x + 1}{\sqrt{x}}$

9. GROWTH OF HMOS   Data show that the number of people receiving their care in an HMO (health maintenance organization) from the beginning of 1984 through 1994 is approximated by the function

$$f(t) = 0.0514t^3 - 0.853t^2 + 6.8147t + 15.6524 \quad (0 < t \le 11)$$

where $f(t)$ gives the number of people in millions and t is measured in years, with $t = 0$ corresponding to the beginning of 1984.

a. Plot the graph of f in the viewing window $[0, 12] \times [0, 120]$.

b. Find the points of inflection of f.

c. At what time in the given time interval was the number of people receiving their care at an HMO increasing at the slowest rate?

Source: Group Health Association of America

10. MANUFACTURING CAPACITY   Data show that the annual increase in manufacturing capacity between 1988 and 1994 is given by

$$f(t) = 0.0388889t^3 - 0.283333t^2 + 0.477778t + 2.04286 \quad (0 \le t \le 6)$$

percent where t is measured in years, with $t = 0$ corresponding to the beginning of 1988.

a. Plot the graph of f in the viewing window $[0, 8] \times [0, 4]$.
b. Find the point of inflection and interpret your result.

Source: Federal Reserve

11. **TIME ON THE MARKET** The average number of days a single-family home remains for sale from listing to accepted offer (in the greater Boston area) is approximated by the function

$$f(t) = 0.0171911t^4 - 0.662121t^3 + 6.18083t^2$$
$$- 8.97086t + 53.3357 \qquad (0 \le t \le 10)$$

where t is measured in years, with $t = 0$ corresponding to the beginning of 1984.
a. Plot the graph of f in the viewing window $[0, 12] \times [0, 120]$.
b. Find the points of inflection and interpret your result.

Source: Greater Boston Real Estate Board—Multiple Listing Service

12. **MULTIMEDIA SALES** Sales in the multimedia market (hardware and software) are approximated by the function

$$S(t) = -0.0094t^4 + 0.1204t^3 - 0.0868t^2$$
$$+ 0.0195t + 3.3325 \qquad (0 \le t \le 10)$$

where $S(t)$ is measured in billions of dollars and t is measured in years, with $t = 0$ corresponding to 1990.
a. Plot the graph of S in the viewing window $[0, 12] \times [0, 25]$.
b. Find the inflection point of S and interpret your result.

Source: Electronic Industries Association

13. **SURGERIES IN PHYSICIANS' OFFICES** Driven by technological advances and financial pressures, the number of surgeries performed in physicians' offices nationwide has been increasing over the years. The function

$$f(t) = -0.00447t^3 + 0.09864t^2 + 0.05192t + 0.8$$
$$(0 \le t \le 15)$$

gives the number of surgeries (in millions) performed in physicians' offices in year t, with $t = 0$ corresponding to the beginning of 1986.
a. Plot the graph of f in the viewing window $[0, 15] \times [0, 10]$.
b. At what time in the period under consideration is the number of surgeries performed in physicians' offices increasing at the fastest rate?

Source: SMG Marketing Group

14. **COMPUTER SECURITY** The number of computer-security incidents, including computer viruses and intrusions, in which the same tool or exploit is used by an intruder, from 1999 through 2003 is approximated by the function

$$N(t) = 2.0417t^4 - 16.083t^3 + 44.46t^2 - 19.42t + 10$$
$$(0 \le t \le 4)$$

where $N(t)$ is measured in thousands and t is measured in years, with $t = 0$ corresponding to 1999.
a. Show that the number of computer-security incidents was always increasing between 2000 and 2003.
Hint: Plot the graph of N' and show that it always lies above the t-axis for $1 \le t \le 4$.
b. Show that between 2000 and 2001, the number of computer-security incidents was increasing at the fastest rate in the middle of 2000 and that between 2001 and 2003 the number of incidents was increasing at the slowest rate in the middle of 2001.
Hint: Study the nature of the inflection points of N.

Source: CERT Coordination Center

13.3 Curve Sketching

■ A Real-Life Example

As we have seen on numerous occasions, the graph of a function is a useful aid for visualizing the function's properties. From a practical point of view, the graph of a function also gives, at one glance, a complete summary of all the information captured by the function.

Consider, for example, the graph of the function giving the Dow-Jones Industrial Average (DJIA) on Black Monday, October 19, 1987 (Figure 45). Here, $t = 0$ corresponds to 8:30 a.m., when the market was open for business, and $t = 7.5$ corresponds to 4 p.m., the closing time. The following information may be gleaned from studying the graph.

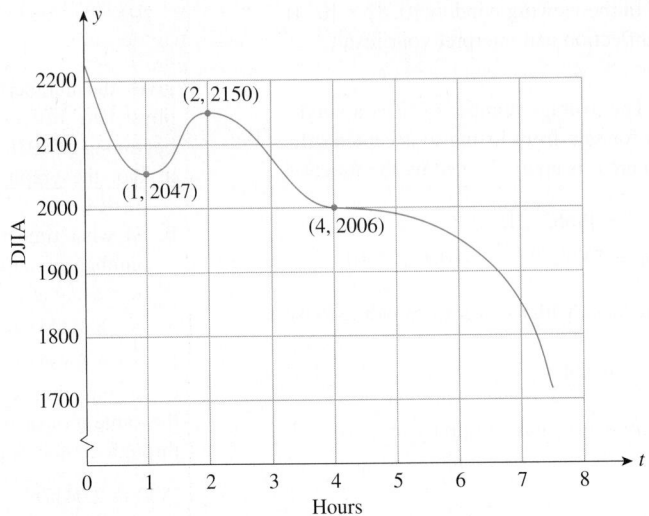

FIGURE 45
The Dow-Jones Industrial Average on
Black Monday.

Source: Wall Street Journal

The graph is *decreasing* rapidly from $t = 0$ to $t = 1$, reflecting the sharp drop in the index in the first hour of trading. The point $(1, 2047)$ is a *relative minimum* point of the function, and this turning point coincides with the start of an aborted recovery. The short-lived rally, represented by the portion of the graph that is *increasing* on the interval $(1, 2)$, quickly fizzled out at $t = 2$ (10:30 a.m.). The *relative maximum* point $(2, 2150)$ marks the highest point of the recovery. The function is decreasing in the rest of the interval. The point $(4, 2006)$ is an *inflection point* of the function; it shows that there was a temporary respite at $t = 4$ (12:30 p.m.). However, selling pressure continued unabated, and the DJIA continued to fall until the closing bell. Finally, the graph also shows that the index opened at the high of the day [$f(0) = 2247$ is the *absolute maximum* of the function] and closed at the low of the day [$f(\frac{15}{2}) = 1739$ is the *absolute minimum* of the function], a drop of 508 points!*

Before we turn our attention to the actual task of sketching the graph of a function, let's look at some properties of graphs that will be helpful in this connection.

▄ Vertical Asymptotes

Before going on, you might want to review the material on one-sided limits and the limit at infinity of a function (Sections 11.4 and 11.5).

Consider the graph of the function

$$f(x) = \frac{x + 1}{x - 1}$$

shown in Figure 46. Observe that $f(x)$ increases without bound (tends to infinity) as x approaches $x = 1$ from the right; that is,

$$\lim_{x \to 1^+} \frac{x + 1}{x - 1} = \infty$$

You can verify this by taking a sequence of values of x approaching $x = 1$ from the right and looking at the corresponding values of $f(x)$.

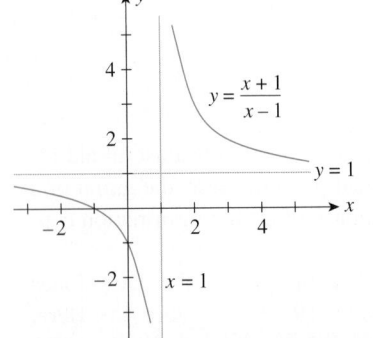

FIGURE 46
The graph of f has a vertical asymptote
at $x = 1$.

*Absolute maxima and absolute minima of functions are covered in Section 13.4.

Here is another way of looking at the situation: Observe that if x is a number that is a little larger than 1, then both $(x + 1)$ and $(x - 1)$ are positive, so $(x + 1)/(x - 1)$ is also positive. As x approaches $x = 1$, the numerator $(x + 1)$ approaches the number 2, but the denominator $(x - 1)$ approaches zero, so the quotient $(x + 1)/(x - 1)$ approaches infinity, as observed earlier. The line $x = 1$ is called a vertical asymptote of the graph of f.

For the function $f(x) = (x + 1)/(x - 1)$, you can show that

$$\lim_{x \to 1^-} \frac{x + 1}{x - 1} = -\infty$$

and this tells us how $f(x)$ approaches the asymptote $x = 1$ from the left.

More generally, we have the following definition:

Vertical Asymptote

The line $x = a$ is a **vertical asymptote** of the graph of a function f if either

$$\lim_{x \to a^+} f(x) = \infty \quad \text{or} \quad -\infty$$

or

$$\lim_{x \to a^-} f(x) = \infty \quad \text{or} \quad -\infty$$

Note   Although a vertical asymptote of a graph is not part of the graph, it serves as a useful aid for sketching the graph.   ∎

For rational functions

$$f(x) = \frac{P(x)}{Q(x)}$$

there is a simple criterion for determining whether the graph of f has any vertical asymptotes.

Finding Vertical Asymptotes of Rational Functions

Suppose f is a rational function

$$f(x) = \frac{P(x)}{Q(x)}$$

where P and Q are polynomial functions. Then, the line $x = a$ is a vertical asymptote of the graph of f if $Q(a) = 0$ but $P(a) \neq 0$.

For the function

$$f(x) = \frac{x + 1}{x - 1}$$

considered earlier, $P(x) = x + 1$ and $Q(x) = x - 1$. Observe that $Q(1) = 0$ but $P(1) = 2 \neq 0$, so $x = 1$ is a vertical asymptote of the graph of f.

EXAMPLE 1   Find the vertical asymptotes of the graph of the function

$$f(x) = \frac{x^2}{4 - x^2}$$

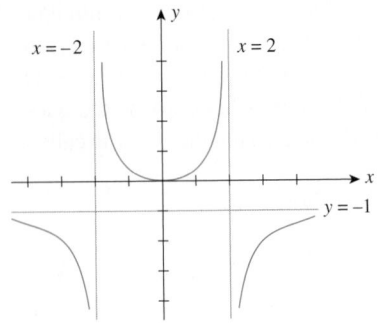

FIGURE 47
$x = -2$ and $x = 2$ are vertical asymptotes of the graph of f.

Solution   The function f is a rational function with $P(x) = x^2$ and $Q(x) = 4 - x^2$. The zeros of Q are found by solving

$$4 - x^2 = 0$$

—that is,

$$(2 - x)(2 + x) = 0$$

giving $x = -2$ and $x = 2$. These are candidates for the vertical asymptotes of the graph of f. Examining $x = -2$, we compute $P(-2) = (-2)^2 = 4 \neq 0$, and we see that $x = -2$ is indeed a vertical asymptote of the graph of f. Similarly, we find $P(2) = 2^2 = 4 \neq 0$, and so $x = 2$ is also a vertical asymptote of the graph of f. The graph of f sketched in Figure 47 confirms these results.   ∎

 Recall that in order for the line $x = a$ to be a vertical asymptote of the graph of a rational function f, *only* the denominator of $f(x)$ must be equal to zero at $x = a$. If *both* $P(a)$ and $Q(a)$ are equal to zero, then $x = a$ need *not* be a vertical asymptote. For example, look at the function

$$f(x) = \frac{4(x^2 - 4)}{x - 2}$$

whose graph appears in Figure 27a, page 629.

▪ Horizontal Asymptotes

Let's return to the function f defined by

$$f(x) = \frac{x + 1}{x - 1}$$

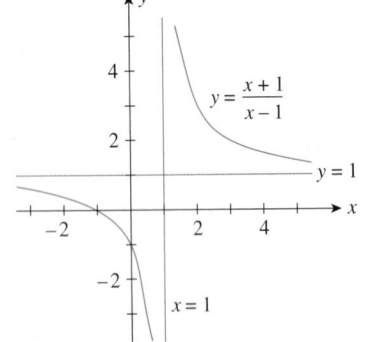

FIGURE 48
The graph of f has a horizontal asymptote at $y = 1$.

(Figure 48).
Observe that $f(x)$ approaches the horizontal line $y = 1$ as x approaches infinity, and, in this case, $f(x)$ approaches $y = 1$ as x approaches minus infinity as well. The line $y = 1$ is called a horizontal asymptote of the graph of f. More generally, we have the following definition:

Horizontal Asymptote
The line $y = b$ is a **horizontal asymptote** of the graph of a function f if either

$$\lim_{x \to \infty} f(x) = b \quad \text{or} \quad \lim_{x \to -\infty} f(x) = b$$

For the function

$$f(x) = \frac{x + 1}{x - 1}$$

we see that

$$\lim_{x \to \infty} \frac{x + 1}{x - 1} = \lim_{x \to \infty} \frac{1 + \frac{1}{x}}{1 - \frac{1}{x}} \qquad \text{Divide numerator and denominator by } x.$$

$$= 1$$

Also,

$$\lim_{x \to -\infty} \frac{x+1}{x-1} = \lim_{x \to -\infty} \frac{1+\frac{1}{x}}{1-\frac{1}{x}}$$

$$= 1$$

In either case, we conclude that $y = 1$ is a horizontal asymptote of the graph of f, as observed earlier.

EXAMPLE 2 Find the horizontal asymptotes of the graph of the function

$$f(x) = \frac{x^2}{4 - x^2}$$

Solution We compute

$$\lim_{x \to \infty} \frac{x^2}{4 - x^2} = \lim_{x \to \infty} \frac{1}{\frac{4}{x^2} - 1} \qquad \text{Divide numerator and denominator by } x^2.$$

$$= -1$$

and so $y = -1$ is a horizontal asymptote, as before. (Similarly, $\lim_{x \to -\infty} f(x) = -1$, as well.) The graph of f sketched in Figure 49 confirms this result.

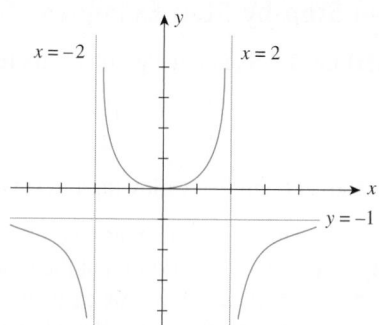

FIGURE 49
The graph of f has a horizontal asymptote at $y = -1$.

We next state an important property of polynomial functions.

> A polynomial function has no vertical or horizontal asymptotes.

To see this, note that a polynomial function $P(x)$ can be written as a rational function with denominator equal to 1. Thus,

$$P(x) = \frac{P(x)}{1}$$

Since the denominator is never equal to zero, P has no vertical asymptotes. Next, if P is a polynomial of degree greater than or equal to 1, then

$$\lim_{x \to \infty} P(x) \quad \text{and} \quad \lim_{x \to -\infty} P(x)$$

are either infinity or minus infinity; that is, they do not exist. Therefore, P has no horizontal asymptotes.

In the last two sections, we saw how the first and second derivatives of a function are used to reveal various properties of the graph of a function f. We now show how this information can be used to help us sketch the graph of f. We begin by giving a general procedure for curve sketching.

> **A Guide to Curve Sketching**
>
> **1.** Determine the domain of f.
>
> **2.** Find the x- and y-intercepts of f.*
>
> **3.** Determine the behavior of f for large absolute values of x.
>
> **4.** Find all horizontal and vertical asymptotes of f.
>
> **5.** Determine the intervals where f is increasing and where f is decreasing.
>
> **6.** Find the relative extrema of f.
>
> **7.** Determine the concavity of f.
>
> **8.** Find the inflection points of f.
>
> **9.** Plot a few additional points to help further identify the shape of the graph of f and sketch the graph.
>
> ---
>
> *The equation $f(x) = 0$ may be difficult to solve, in which case one may decide against finding the x-intercepts or to use technology, if available, for assistance.

We now illustrate the techniques of curve sketching with several examples.

Two Step-by-Step Examples

EXAMPLE 3   Sketch the graph of the function

$$y = f(x) = x^3 - 6x^2 + 9x + 2$$

Solution   Obtain the following information on the graph of f.

1. The domain of f is the interval $(-\infty, \infty)$.

2. By setting $x = 0$, we find that the y-intercept is 2. The x-intercept is found by setting $y = 0$, which in this case leads to a cubic equation. Since the solution is not readily found, we will not use this information.

3. Since

$$\lim_{x \to -\infty} f(x) = \lim_{x \to -\infty} (x^3 - 6x^2 + 9x + 2) = -\infty$$
$$\lim_{x \to \infty} f(x) = \lim_{x \to \infty} (x^3 - 6x^2 + 9x + 2) = \infty$$

we see that f decreases without bound as x decreases without bound and that f increases without bound as x increases without bound.

4. Since f is a polynomial function, there are no asymptotes.

5.
$$f'(x) = 3x^2 - 12x + 9 = 3(x^2 - 4x + 3)$$
$$= 3(x - 3)(x - 1)$$

Setting $f'(x) = 0$ gives $x = 1$ or $x = 3$. The sign diagram for f' shows that f is increasing on the intervals $(-\infty, 1)$ and $(3, \infty)$ and decreasing on the interval $(1, 3)$ (Figure 50).

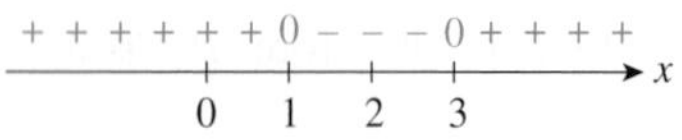

6. From the results of step 5, we see that $x = 1$ and $x = 3$ are critical numbers of f. Furthermore, f' changes sign from positive to negative as we move across $x = 1$, so a relative maximum of f occurs at $x = 1$. Similarly, we see that a

$+ + + + + 0 - - - 0 + + + +$

$\longrightarrow x$

$\quad 0 \quad 1 \quad 2 \quad 3$

FIGURE 50
Sign diagram for f'

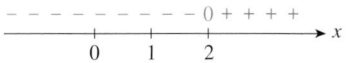

FIGURE 51
Sign diagram for f''

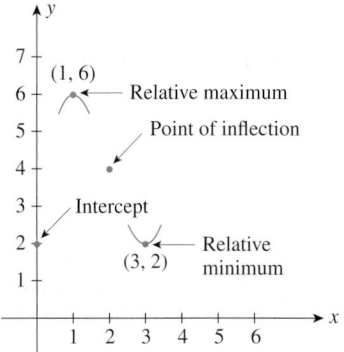

FIGURE 52
We first plot the intercept, the relative extrema, and the inflection point.

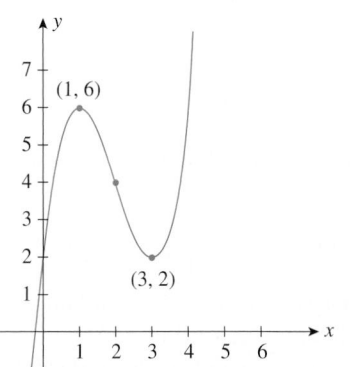

FIGURE 53
The graph of $y = x^3 - 6x^2 + 9x + 2$

relative minimum of f occurs at $x = 3$. Now,

$$f(1) = 1 - 6 + 9 + 2 = 6$$
$$f(3) = 3^3 - 6(3)^2 + 9(3) + 2 = 2$$

so $f(1) = 6$ is a relative maximum of f and $f(3) = 2$ is a relative minimum of f.

7. $$f''(x) = 6x - 12 = 6(x - 2)$$

which is equal to zero when $x = 2$. The sign diagram of f'' shows that f is concave downward on the interval $(-\infty, 2)$ and concave upward on the interval $(2, \infty)$ (Figure 51).

8. From the results of step 7, we see that f'' changes sign as we move across $x = 2$. Next,

$$f(2) = 2^3 - 6(2)^2 + 9(2) + 2 = 4$$

and so the required inflection point of f is $(2, 4)$.

Summarizing, we have the following:

> Domain: $(-\infty, \infty)$
>
> Intercept: $(0, 2)$
>
> $\lim_{x \to -\infty} f(x)$; $\lim_{x \to \infty} f(x)$: $-\infty$; ∞
>
> Asymptotes: None
>
> Intervals where f is ↗ or ↘: ↗ on $(-\infty, 1) \cup (3, \infty)$; ↘ on $(1, 3)$
>
> Relative extrema: Relative maximum at $(1, 6)$; relative minimum at $(3, 2)$
>
> Concavity: Downward on $(-\infty, 2)$; upward on $(2, \infty)$
>
> Point of inflection: $(2, 4)$

In general, it is a good idea to start graphing by plotting the intercept(s), relative extrema, and inflection point(s) (Figure 52). Then, using the rest of the information, we complete the graph of f, as sketched in Figure 53.   ■

EXPLORE & DISCUSS

The average price of gasoline at the pump over a 3-month period, during which there was a temporary shortage of oil, is described by the function f defined on the interval $[0, 3]$. During the first month, the price was increasing at an increasing rate. Starting with the second month, the good news was that the rate of increase was slowing down, although the price of gas was still increasing. This pattern continued until the end of the second month. The price of gas peaked at the end of $t = 2$ and began to fall at an increasing rate until $t = 3$.

1. Describe the signs of $f'(t)$ and $f''(t)$ over each of the intervals $(0, 1)$, $(1, 2)$, and $(2, 3)$.

2. Make a sketch showing a plausible graph of f over $[0, 3]$.

EXAMPLE 4 Sketch the graph of the function

$$y = f(x) = \frac{x + 1}{x - 1}$$

Solution Obtain the following information:

1. f is undefined when $x = 1$, so the domain of f is the set of all real numbers other than $x = 1$.
2. Setting $y = 0$ gives -1, the x-intercept of f. Next, setting $x = 0$ gives -1 as the y-intercept of f.
3. Earlier we found that

$$\lim_{x \to \infty} \frac{x + 1}{x - 1} = 1 \quad \text{and} \quad \lim_{x \to -\infty} \frac{x + 1}{x - 1} = 1$$

(see pp. 812–813). Consequently, we see that $f(x)$ approaches the line $y = 1$ as $|x|$ becomes arbitrarily large. For $x > 1$, $f(x) > 1$ and $f(x)$ approaches the line $y = 1$ from above. For $x < 1$, $f(x) < 1$, so $f(x)$ approaches the line $y = 1$ from below.

4. The straight line $x = 1$ is a vertical asymptote of the graph of f. Also, from the results of step 3, we conclude that $y = 1$ is a horizontal asymptote of the graph of f.

5. $$f'(x) = \frac{(x - 1)(1) - (x + 1)(1)}{(x - 1)^2} = -\frac{2}{(x - 1)^2}$$

and is discontinuous at $x = 1$. The sign diagram of f' shows that $f'(x) < 0$ whenever it is defined. Thus, f is decreasing on the intervals $(-\infty, 1)$ and $(1, \infty)$ (Figure 54).

6. From the results of step 5, we see that there are no critical numbers of f since $f'(x)$ is never equal to zero for any value of x in the domain of f.

7. $$f''(x) = \frac{d}{dx}[-2(x - 1)^{-2}] = 4(x - 1)^{-3} = \frac{4}{(x - 1)^3}$$

The sign diagram of f'' shows immediately that f is concave downward on the interval $(-\infty, 1)$ and concave upward on the interval $(1, \infty)$ (Figure 55).

8. From the results of step 7, we see that there are no candidates for inflection points of f since $f''(x)$ is never equal to zero for any value of x in the domain of f. Hence, f has no inflection points.

Summarizing, we have the following:

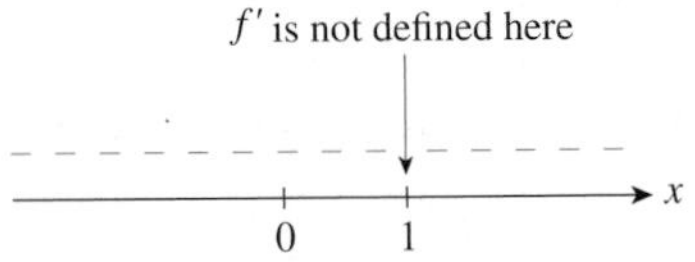

Domain: $(-\infty, 1) \cup (1, \infty)$

Intercepts: $(0, -1)$; $(-1, 0)$

$\lim_{x \to -\infty} f(x)$; $\lim_{x \to \infty} f(x)$: 1; 1

Asymptotes: $x = 1$ is a vertical asymptote
$y = 1$ is a horizontal asymptote

Intervals where f is ↗ or ↘: ↘ on $(-\infty, 1) \cup (1, \infty)$

Relative extrema: None

Concavity: Downward on $(-\infty, 1)$; upward on $(1, \infty)$

Points of inflection: None

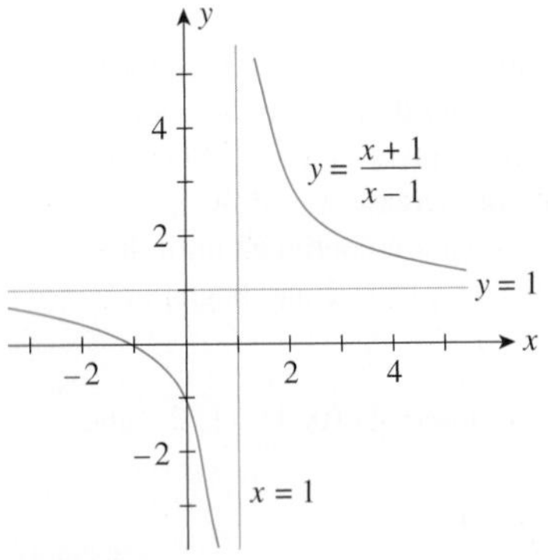

The graph of f is sketched in Figure 56.

---

f' is not defined here

FIGURE 54
The sign diagram for f'

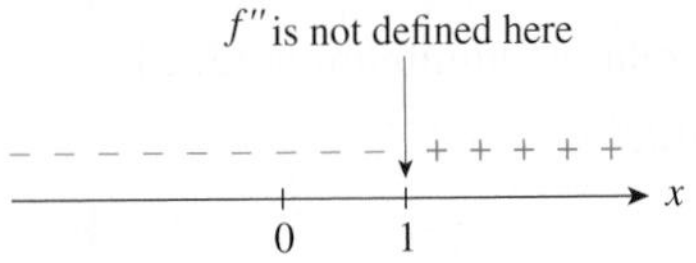

f'' is not defined here

FIGURE 55
The sign diagram for f''

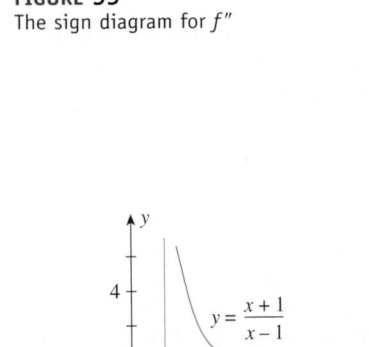

FIGURE 56
The graph of f has a horizontal asymptote at $y = 1$ and a vertical asymptote at $x = 1$.

13.3 Self-Check Exercises

1. Find the horizontal and vertical asymptotes of the graph of the function

$$f(x) = \frac{2x^2}{x^2 - 1}$$

2. Sketch the graph of the function

$$f(x) = \frac{2}{3}x^3 - 2x^2 - 6x + 4$$

Solutions to Self-Check Exercises 13.3 can be found on page 821.

13.3 Concept Questions

1. Explain the following terms in your own words:
  a. Vertical asymptote   **b.** Horizontal asymptote

2. a. How many vertical asymptotes can the graph of a function *f* have? Explain using graphs.
  b. How many horizontal asymptotes can the graph of a function *f* have? Explain using graphs.

3. How do you find the vertical asymptotes of a rational function?

4. Give a procedure for sketching the graph of a function.

13.3 Exercises

In Exercises 1–10, find the horizontal and vertical asymptotes of the graph.

1.

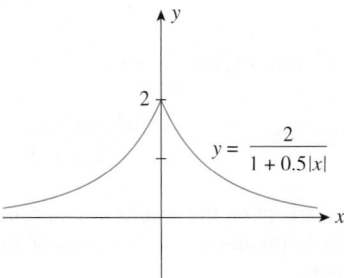

$$y = \frac{2}{1 + 0.5|x|}$$

2.

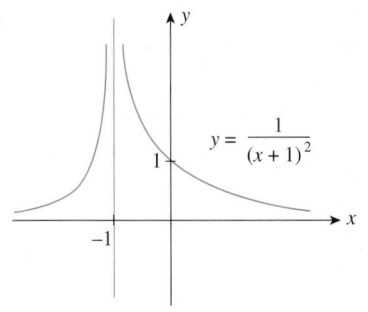

$$y = \frac{1}{(x+1)^2}$$

3.

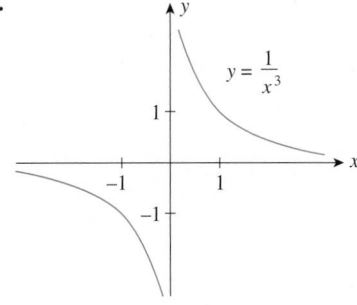

$$y = \frac{1}{x^3}$$

4.

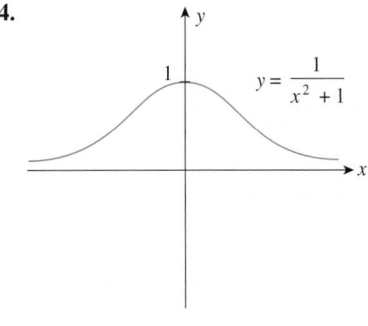

$$y = \frac{1}{x^2 + 1}$$

5.

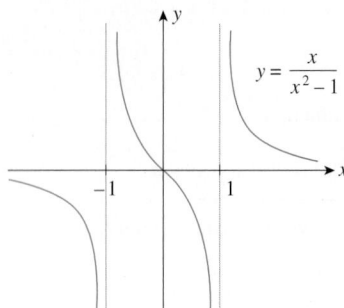

$$y = \frac{x}{x^2 - 1}$$

6.

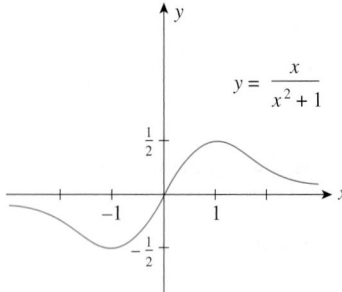

$$y = \frac{x}{x^2 + 1}$$

7.

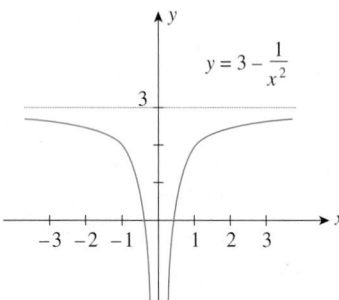

$$y = 3 - \frac{1}{x^2}$$

8.

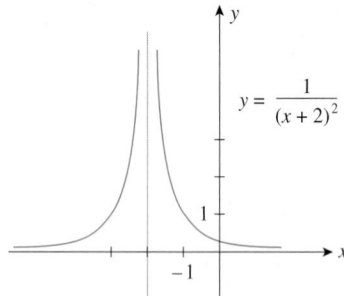

$$y = \frac{1}{(x + 2)^2}$$

9.

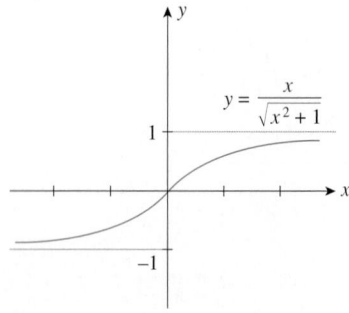

$$y = \frac{x}{\sqrt{x^2 + 1}}$$

10.

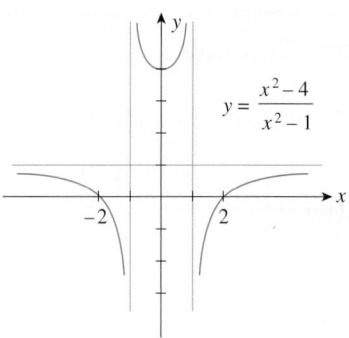

$$y = \frac{x^2 - 4}{x^2 - 1}$$

In Exercises 11–28, find the horizontal and vertical asymptotes of the graph of the function. (You need not sketch the graph.)

11. $f(x) = \dfrac{1}{x}$

12. $f(x) = \dfrac{1}{x + 2}$

13. $f(x) = -\dfrac{2}{x^2}$

14. $g(x) = \dfrac{1}{1 + 2x^2}$

15. $f(x) = \dfrac{x - 1}{x + 1}$

16. $g(t) = \dfrac{t + 1}{2t - 1}$

17. $h(x) = x^3 - 3x^2 + x + 1$

18. $g(x) = 2x^3 + x^2 + 1$

19. $f(t) = \dfrac{t^2}{t^2 - 9}$

20. $g(x) = \dfrac{x^3}{x^2 - 4}$

21. $f(x) = \dfrac{3x}{x^2 - x - 6}$

22. $g(x) = \dfrac{2x}{x^2 + x - 2}$

23. $g(t) = 2 + \dfrac{5}{(t - 2)^2}$

24. $f(x) = 1 + \dfrac{2}{x - 3}$

25. $f(x) = \dfrac{x^2 - 2}{x^2 - 4}$

26. $h(x) = \dfrac{2 - x^2}{x^2 + x}$

27. $g(x) = \dfrac{x^3 - x}{x(x + 1)}$

28. $f(x) = \dfrac{x^4 - x^2}{x(x - 1)(x + 2)}$

In Exercises 29 and 30, you are given the graphs of two functions f and g. One function is the derivative function of the other. Identify each of them.

29.

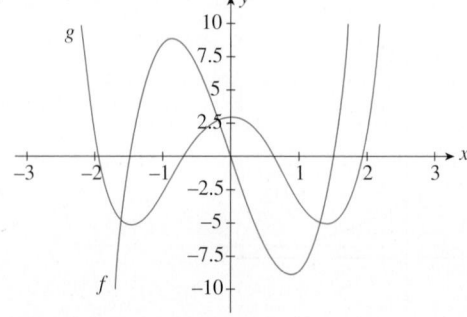

30.

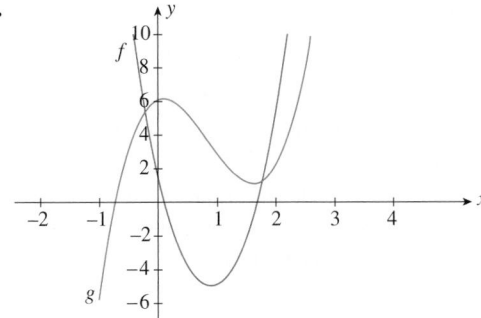

31. TERMINAL VELOCITY   A skydiver leaps from the gondola of a hot-air balloon. As she free-falls, air resistance, which is proportional to her velocity, builds up to a point where it balances the force due to gravity. The resulting motion may be described in terms of her velocity as follows: Starting at rest (zero velocity), her velocity increases and approaches a constant velocity, called the *terminal velocity*. Sketch a graph of her velocity v versus time t.

32. SPREAD OF A FLU EPIDEMIC   Initially, 10 students at a junior high school contracted influenza. The flu spread over time, and the total number of students who eventually contracted the flu approached but never exceeded 200. Let $P(t)$ denote the number of students who had contracted the flu after t days, where P is an appropriate function.
   a. Make a sketch of the graph of P. (Your answer will *not* be unique.)
   b. Where is the function increasing?
   c. Does P have a horizontal asymptote? If so, what is it?
   d. Discuss the concavity of P. Explain its significance.
   e. Is there an inflection point on the graph of P? If so, explain its significance.

In Exercises 33–36, use the information summarized in the table to sketch the graph of f.

33. $f(x) = x^3 - 3x^2 + 1$

Domain: $(-\infty, \infty)$
Intercept: y-intercept: 1
Asymptotes: None
Intervals where f is $\nearrow$ and $\searrow$: $\nearrow$ on $(-\infty, 0) \cup (2, \infty)$;
              $\searrow$ on $(0, 2)$
Relative extrema: Rel. max. at $(0, 1)$; rel. min. at $(2, -3)$
Concavity: Downward on $(-\infty, 1)$; upward on $(1, \infty)$
Point of inflection: $(1, -1)$

---

34. $f(x) = \dfrac{1}{9}(x^4 - 4x^3)$

Domain: $(-\infty, \infty)$
Intercepts: x-intercepts: 0, 4; y-intercept: 0
Asymptotes: None
Intervals where f is $\nearrow$ and $\searrow$: $\nearrow$ on $(3, \infty)$;
              $\searrow$ on $(-\infty, 3)$

Relative extrema: Rel. min. at $(3, -3)$
Concavity: Downward on $(0, 2)$;
              upward on $(-\infty, 0) \cup (2, \infty)$
Points of inflection: $(0, 0)$ and $\left(2, -\frac{16}{9}\right)$

---

35. $f(x) = \dfrac{4x - 4}{x^2}$

Domain: $(-\infty, 0) \cup (0, \infty)$
Intercept: x-intercept: 1
Asymptotes: x-axis and y-axis
Intervals where f is $\nearrow$ and $\searrow$: $\nearrow$ on $(0, 2)$;
              $\searrow$ on $(-\infty, 0) \cup (2, \infty)$
Relative extrema: Rel. max. at $(2, 1)$
Concavity: Downward on $(-\infty, 0) \cup (0, 3)$;
              upward on $(3, \infty)$
Point of inflection: $\left(3, \frac{8}{9}\right)$

---

36. $f(x) = x - 3x^{1/3}$

Domain: $(-\infty, \infty)$
Intercepts: x-intercepts: $\pm 3\sqrt{3}, 0$
Asymptotes: None
Intervals where f is $\nearrow$ and $\searrow$: $\nearrow$ on $(-\infty, -1) \cup (1, \infty)$;
              $\searrow$ on $(-1, 1)$
Relative extrema: Rel. max. at $(-1, 2)$; rel. min. at $(1, -2)$
Concavity: Downward on $(-\infty, 0)$; upward on $(0, \infty)$
Point of inflection: $(0, 0)$

In Exercises 37–60, sketch the graph of the function, using the curve-sketching guide of this section.

37. $g(x) = 4 - 3x - 2x^3$ **38.** $f(x) = x^2 - 2x + 3$

39. $h(x) = x^3 - 3x + 1$ **40.** $f(x) = 2x^3 + 1$

41. $f(x) = -2x^3 + 3x^2 + 12x + 2$

42. $f(t) = 2t^3 - 15t^2 + 36t - 20$

43. $h(x) = \dfrac{3}{2}x^4 - 2x^3 - 6x^2 + 8$

44. $f(t) = 3t^4 + 4t^3$

45. $f(t) = \sqrt{t^2 - 4}$ **46.** $f(x) = \sqrt{x^2 + 5}$

47. $g(x) = \dfrac{1}{2}x - \sqrt{x}$ **48.** $f(x) = \sqrt[3]{x^2}$

49. $g(x) = \dfrac{2}{x - 1}$ **50.** $f(x) = \dfrac{1}{x + 1}$

51. $h(x) = \dfrac{x + 2}{x - 2}$ **52.** $g(x) = \dfrac{x}{x - 1}$

53. $f(t) = \dfrac{t^2}{1 + t^2}$ **54.** $g(x) = \dfrac{x}{x^2 - 4}$

55. $g(t) = -\dfrac{t^2 - 2}{t - 1}$ **56.** $f(x) = \dfrac{x^2 - 9}{x^2 - 4}$

57. $g(t) = \dfrac{t^2}{t^2 - 1}$ **58.** $h(x) = \dfrac{1}{x^2 - x - 2}$

59. $h(x) = (x - 1)^{2/3} + 1$ **60.** $g(x) = (x + 2)^{3/2} + 1$

61. Cost of Removing Toxic Pollutants   A city's main well was recently found to be contaminated with trichloroethylene (a cancer-causing chemical) as a result of an abandoned chemical dump that leached chemicals into the water. A proposal submitted to the city council indicated that the cost, measured in millions of dollars, of removing $x\%$ of the toxic pollutants is given by

$$C(x) = \frac{0.5x}{100 - x}$$

a. Find the vertical asymptote of $C(x)$.
b. Is it possible to remove 100% of the toxic pollutant from the water?

62. Average Cost of Producing DVDs   The average cost per disc (in dollars) incurred by Herald Media Corporation in pressing x DVDs is given by the average cost function

$$\overline{C}(x) = 2.2 + \frac{2500}{x}$$

a. Find the horizontal asymptote of $\overline{C}(x)$.
b. What is the limiting value of the average cost?

63. Concentration of a Drug in the Bloodstream   The concentration (in milligrams/cubic centimeter) of a certain drug in a patient's bloodstream t hr after injection is given by

$$C(t) = \frac{0.2t}{t^2 + 1}$$

a. Find the horizontal asymptote of $C(t)$.
b. Interpret your result.

64. Effect of Enzymes on Chemical Reactions   Certain proteins, known as enzymes, serve as catalysts for chemical reactions in living things. In 1913 Leonor Michaelis and L. M. Menten discovered the following formula giving the initial speed V (in moles/liter/second) at which the reaction begins in terms of the amount of substrate x (the substance that is being acted upon, measured in moles/liter):

$$V = \frac{ax}{x + b}$$

where a and b are positive constants.
a. Find the horizontal asymptote of V.
b. What does the result of part (a) tell you about the initial speed at which the reaction begins, if the amount of substrate is very large?

65. GDP of a Developing Country   A developing country's gross domestic product (GDP) from 1997 to 2005 is approximated by the function

$$G(t) = -0.2t^3 + 2.4t^2 + 60 \qquad (0 \le t \le 8)$$

where $G(t)$ is measured in billions of dollars, with $t = 0$ corresponding to 1997. Sketch the graph of the function G and interpret your results.

66. Worker Efficiency   An efficiency study showed that the total number of cordless telephones assembled by an average worker at Delphi Electronics t hr after starting work at 8 a.m. is given by

$$N(t) = -\frac{1}{2}t^3 + 3t^2 + 10t \qquad (0 \le t \le 4)$$

Sketch the graph of the function N and interpret your results.

67. Concentration of a Drug in the Bloodstream   The concentration (in millimeters/cubic centimeter) of a certain drug in a patient's bloodstream t hr after injection is given by

$$C(t) = \frac{0.2t}{t^2 + 1}$$

Sketch the graph of the function C and interpret your results.

68. Oxygen Content of a Pond   When organic waste is dumped into a pond, the oxidation process that takes place reduces the pond's oxygen content. However, given time, nature will restore the oxygen content to its natural level. Suppose the oxygen content t days after organic waste has been dumped into the pond is given by

$$f(t) = 100\left(\frac{t^2 - 4t + 4}{t^2 + 4}\right) \qquad (0 \le t < \infty)$$

percent of its normal level. Sketch the graph of the function f and interpret your results.

69. Box-Office Receipts   The total worldwide box-office receipts for a long-running movie are approximated by the function

$$T(x) = \frac{120x^2}{x^2 + 4}$$

where $T(x)$ is measured in millions of dollars and x is the number of years since the movie's release. Sketch the graph of the function T and interpret your results.

70. Cost of Removing Toxic Pollutants   Refer to Exercise 61. The cost, measured in millions of dollars, of removing $x\%$ of a toxic pollutant is given by

$$C(x) = \frac{0.5x}{100 - x}$$

Sketch the graph of the function C and interpret your results.

13.3 Solutions to Self-Check Exercises

1. Since

$$\lim_{x \to \infty} \frac{2x^2}{x^2 - 1} = \lim_{x \to \infty} \frac{2}{1 - \dfrac{1}{x^2}}$$

Divide the numerator
and denominator by x^2.

$$= 2$$

we see that $y = 2$ is a horizontal asymptote. Next, since

$$x^2 - 1 = (x + 1)(x - 1) = 0$$

implies $x = -1$ or $x = 1$, these are candidates for the vertical asymptotes of f. Since the numerator of f is not equal to zero for $x = -1$ or $x = 1$, we conclude that $x = -1$ and $x = 1$ are vertical asymptotes of the graph of f.

2. We obtain the following information on the graph of f.
(1) The domain of f is the interval $(-\infty, \infty)$.
(2) By setting $x = 0$, we find the y-intercept is 4.
(3) Since

$$\lim_{x \to -\infty} f(x) = \lim_{x \to -\infty} \left(\frac{2}{3}x^3 - 2x^2 - 6x + 4 \right) = -\infty$$

$$\lim_{x \to \infty} f(x) = \lim_{x \to \infty} \left(\frac{2}{3}x^3 - 2x^2 - 6x + 4 \right) = \infty$$

we see that $f(x)$ decreases without bound as x decreases without bound and that $f(x)$ increases without bound as x increases without bound.
(4) Since f is a polynomial function, there are no asymptotes.
(5)
$$f'(x) = 2x^2 - 4x - 6 = 2(x^2 - 2x - 3)$$
$$= 2(x + 1)(x - 3)$$

Setting $f'(x) = 0$ gives $x = -1$ or $x = 3$. The accompanying sign diagram for f' shows that f is increasing on the intervals $(-\infty, -1)$ and $(3, \infty)$ and decreasing on $(-1, 3)$.

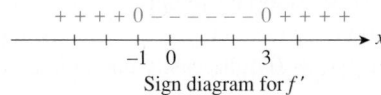
Sign diagram for f'

(6) From the results of step 5, we see that $x = -1$ and $x = 3$ are critical numbers of f. Furthermore, the sign diagram of f' tells us that $x = -1$ gives rise to a relative maximum of f and $x = 3$ gives rise to a relative minimum of f. Now,

$$f(-1) = \frac{2}{3}(-1)^3 - 2(-1)^2 - 6(-1) + 4 = \frac{22}{3}$$

$$f(3) = \frac{2}{3}(3)^3 - 2(3)^2 - 6(3) + 4 = -14$$

so $f(-1) = \frac{22}{3}$ is a relative maximum of f and $f(3) = -14$ is a relative minimum of f.
(7)
$$f''(x) = 4x - 4 = 4(x - 1)$$
which is equal to zero when $x = 1$. The accompanying sign diagram of f'' shows that f is concave downward on the interval $(-\infty, 1)$ and concave upward on the interval $(1, \infty)$.

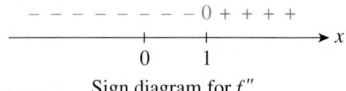
Sign diagram for f''

(8) From the results of step 7, we see that $x = 1$ is the only candidate for an inflection point of f. Since $f''(x)$ changes sign as we move across the point $x = 1$ and

$$f(1) = \frac{2}{3}(1)^3 - 2(1)^2 - 6(1) + 4 = -\frac{10}{3}$$

we see that the required inflection point is $(1, -\frac{10}{3})$.
(9) Summarizing this information, we have the following:

Domain: $(-\infty, \infty)$

Intercept: $(0, 4)$
$\lim_{x \to -\infty} f(x)$; $\lim_{x \to \infty} f(x)$: $-\infty$; ∞
Asymptotes: None

Intervals where f is $\nearrow$ or $\searrow$: $\nearrow$ on $(-\infty, -1) \cup (3, \infty)$; $\searrow$ on $(-1, 3)$

Relative extrema: Rel. max. at $(-1, \frac{22}{3})$; rel. min. at $(3, -14)$

Concavity: Downward on $(-\infty, 1)$; upward on $(1, \infty)$

Point of inflection: $(1, -\frac{10}{3})$

The graph of f is sketched in the accompanying figure.

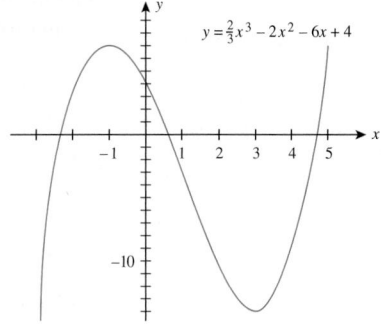

USING TECHNOLOGY

■ Analyzing the Properties of a Function

One of the main purposes of studying Section 13.3 is to see how the many concepts of calculus come together to paint a picture of a function. The techniques of graphing also play a very practical role. For example, using the techniques of graphing developed in Section 13.3, you can tell if the graph of a function generated by a graphing utility is reasonably complete. Furthermore, these techniques can often reveal details that are missing from a graph.

EXAMPLE 1   Consider the function $f(x) = 2x^3 - 3.5x^2 + x - 10$. A plot of the graph of f in the standard viewing window is shown in Figure T1. Since the domain of f is the interval $(-\infty, \infty)$, we see that Figure T1 does not reveal the part of the graph to the left of the y-axis. This suggests that we enlarge the viewing window accordingly. Figure T2 shows the graph of f in the viewing window $[-10, 10] \times [-20, 10]$.

The behavior of f for large values of f

$$\lim_{x \to -\infty} f(x) = -\infty \quad \text{and} \quad \lim_{x \to \infty} f(x) = \infty$$

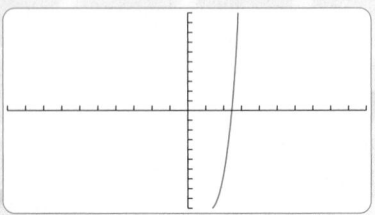

FIGURE T1
The graph of f in the standard viewing window

suggests that this viewing window has captured a sufficiently complete picture of f. Next, an analysis of the first derivative of f,

$$f'(x) = 6x^2 - 7x + 1 = (6x - 1)(x - 1)$$

reveals that f has critical values at $x = \frac{1}{6}$ and $x = 1$. In fact, a sign diagram of f' shows that f has a relative maximum at $x = \frac{1}{6}$ and a relative minimum at $x = 1$, details that are not revealed in the graph of f shown in Figure T2. To examine this portion of the graph of f, we use, say, the viewing window $[-1, 2] \times [-11, -9]$. The resulting graph of f is shown in Figure T3, which certainly reveals the hitherto missing details! Thus, through an interaction of calculus and a graphing utility, we are able to obtain a good picture of the properties of f.   ■

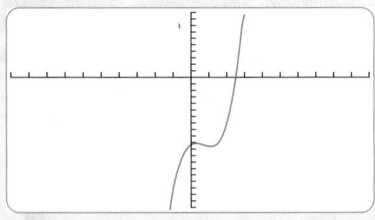

FIGURE T2
The graph of f in the viewing window $[-10, 10] \times [-20, 10]$

Finding x-Intercepts

As noted in Section 13.3, it is not always easy to find the x-intercepts of the graph of a function. But this information is very important in applications. By using the function for solving polynomial equations or the function for finding the roots of an equation, we can solve the equation $f(x) = 0$ quite easily and hence yield the x-intercepts of the graph of a function.

EXAMPLE 2   Let $f(x) = x^3 - 3x^2 + x + 1.5$.

a. Use the function for solving polynomial equations on a graphing utility to find the x-intercepts of the graph of f.

b. Use the function for finding the roots of an equation on a graphing utility to find the x-intercepts of the graph of f.

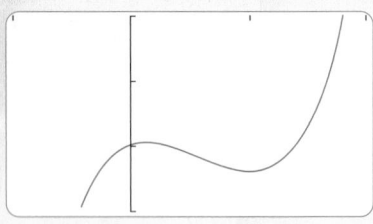

FIGURE T3
The graph of f in the viewing window $[-1, 2] \times [-11, -9]$

Solution

a. Observe that f is a polynomial function of degree 3, and so we may use the function for solving polynomial equations to solve the equation $x^3 - 3x^2 + x + 1.5 = 0$ [$f(x) = 0$]. We find that the solutions (x-intercepts) are

$$x_1 \approx -0.525687120865 \qquad x_2 \approx 1.2586520225 \qquad x_3 \approx 2.26703509836$$

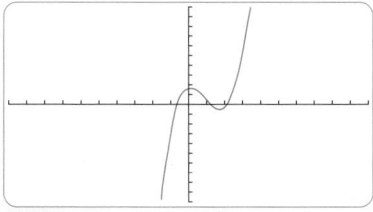

FIGURE T4
The graph of $f(x) = x^3 - 3x^2 + x + 1.5$

b. Using the graph of f (Figure T4), we see that $x_1 \approx -0.5$, $x_2 \approx 1$, and $x_3 \approx 2$. Using the function for finding the roots of an equation on a graphing utility, and these values of x as initial guesses, we find

$$x_1 \approx -0.5256871209 \qquad x_2 \approx 1.2586520225 \qquad x_3 \approx 2.2670350984 \qquad ■$$

Note The function for solving polynomial equations on a graphing utility will solve a polynomial equation $f(x) = 0$, where f is a polynomial function. The function for finding the roots of a polynomial, however, will solve equations $f(x) = 0$ even if f is not a polynomial. ■

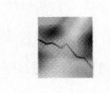

APPLIED EXAMPLE 3 Solvency of Social Security Fund Unless payroll taxes are increased significantly and/or benefits are scaled back drastically, it is a matter of time before the current Social Security system goes broke. Data show that the assets of the system—the Social Security "trust fund"—may be approximated by

$$f(t) = -0.0129t^4 + 0.3087t^3 + 2.1760t^2 + 62.8466t + 506.2955 \qquad (0 \le t \le 35)$$

where $f(t)$ is measured in millions of dollars and t is measured in years, with $t = 0$ corresponding to 1995.

a. Use a graphing calculator to sketch the graph of f.
b. Based on this model, when can the Social Security system be expected to go broke?

Source: Social Security Administration

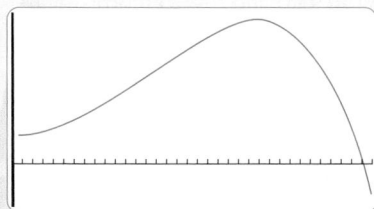

FIGURE T5
The graph of $f(t)$

Solution

a. The graph of f in the window $[0, 35] \times [-1000, 3500]$ is shown in Figure T5.
b. Using the function for finding the roots on a graphing utility, we find that $y = 0$ when $t \approx 34.1$, and this tells us that the system is expected to go broke around 2029. ■

TECHNOLOGY EXERCISES

In Exercises 1–4, use the method of Example 1 to analyze the function. (*Note:* Your answers will *not* be unique.)

1. $f(x) = 4x^3 - 4x^2 + x + 10$

2. $f(x) = x^3 + 2x^2 + x - 12$

3. $f(x) = \frac{1}{2}x^4 + x^3 + \frac{1}{2}x^2 - 10$

4. $f(x) = 2.25x^4 - 4x^3 + 2x^2 + 2$

In Exercises 5–10, find the x-intercepts of the graph of f. Give your answer accurate to four decimal places.

5. $f(x) = 0.2x^3 - 1.2x^2 + 0.8x + 2.1$

6. $f(x) = -0.5x^3 + 1.7x^2 - 1.2$

7. $f(x) = 0.3x^4 - 1.2x^3 + 0.8x^2 + 1.1x - 2$

8. $f(x) = -0.2x^4 + 0.8x^3 - 2.1x + 1.2$

9. $f(x) = 2x^2 - \sqrt{x + 1} - 3$

10. $f(x) = x - \sqrt{1 - x^2}$

13.4 Optimization I

■ Absolute Extrema

The graph of the function f in Figure 57 shows the average age of cars in use in the United States from the beginning of 1946 ($t = 0$) to the beginning of 2002 ($t = 56$). Observe that the highest average age of cars in use during this period is 9 years, whereas the lowest average age of cars in use during the same period is $5\frac{1}{2}$ years. The number 9, the largest value of $f(t)$ for all values of t in the interval [0, 56] (the domain of f), is called the *absolute maximum value of f* on that interval. The number $5\frac{1}{2}$, the smallest value of $f(t)$ for all values of t in [0, 56], is called the *absolute minimum value of f* on that interval. Notice, too, that the absolute maximum value of f is attained at the endpoint $t = 0$ of the interval, whereas the absolute minimum value of f is attained at the points $t = 12$ (corresponding to 1958) and $t = 23$ (corresponding to 1969) that lie within the interval (0, 56).

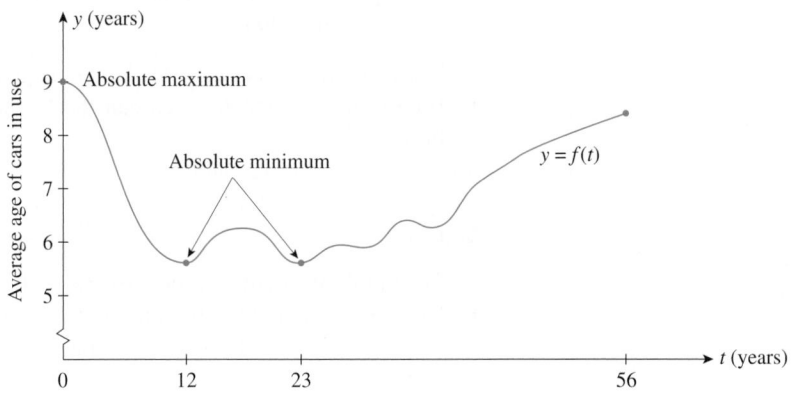

FIGURE 57
$f(t)$ gives the average age of cars in use in year t, t in [0, 56].

Source: American Automobile Association

(Incidentally, it is interesting to note that 1946 marked the first year of peace following World War II, and the two years, 1958 and 1969, marked the end of two periods of prosperity in recent U.S. history!)

A precise definition of the **absolute extrema** (absolute maximum or absolute minimum) of a function follows.

> **The Absolute Extrema of a Function f**
> If $f(x) \leq f(c)$ for all x in the domain of f, then $f(c)$ is called the **absolute maximum value** of f.
> If $f(x) \geq f(c)$ for all x in the domain of f, then $f(c)$ is called the **absolute minimum value** of f.

Figure 58 shows the graphs of several functions and gives the absolute maximum and absolute minimum of each function, if they exist.

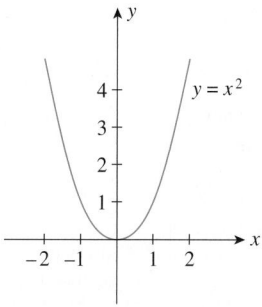

(a) $f(0) = 0$ is the absolute minimum of f; f has no absolute maximum.

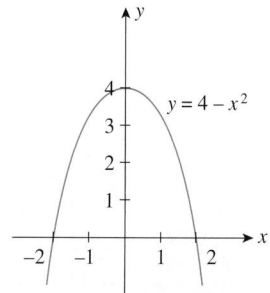

(b) $f(0) = 4$ is the absolute maximum of f; f has no absolute minimum.

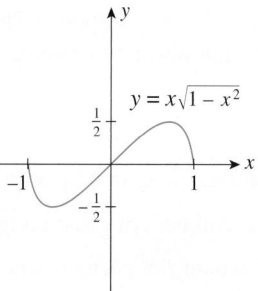

(c) $f(\sqrt{2}/2) = 1/2$ is the absolute maximum of f; $f(-\sqrt{2}/2) = -1/2$ is the absolute minimum of f.

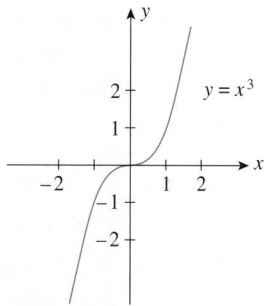

(d) f has no absolute extrema.

FIGURE 58

Absolute Extrema on a Closed Interval

As the preceding examples show, a continuous function defined on an arbitrary interval does not always have an absolute maximum or an absolute minimum. But an important case arises often in practical applications in which both the absolute maximum and the absolute minimum of a function are guaranteed to exist. This occurs where a continuous function is defined on a *closed* interval. Let's state this important result in the form of a theorem, whose proof we will omit.

THEOREM 3

If a function f is continuous on a closed interval $[a, b]$, then f has both an absolute maximum value and an absolute minimum value on $[a, b]$.

Observe that if an absolute extremum of a continuous function f occurs at a point in an open interval (a, b), then it must be a relative extremum of f and hence its x-coordinate must be a critical number of f. Otherwise, the absolute extremum of f must occur at one or both of the endpoints of the interval $[a, b]$. A typical situation is illustrated in Figure 59.

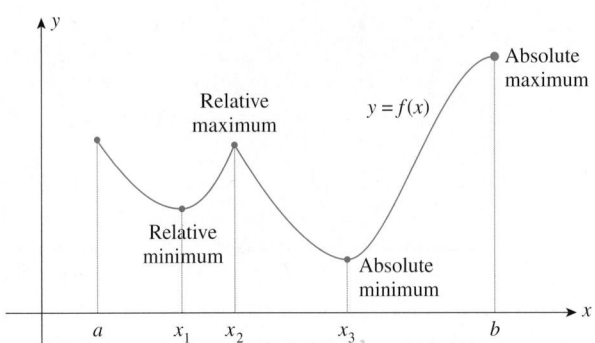

FIGURE 59
The relative minimum of f at x_3 is the absolute minimum of f. The right endpoint b of the interval $[a, b]$ gives rise to the absolute maximum value $f(b)$ of f.

Here x_1, x_2, and x_3 are critical numbers of f. The absolute minimum of f occurs at x_3, which lies in the open interval (a, b) and is a critical number of f. The absolute maximum of f occurs at b, an endpoint. This observation suggests the following procedure for finding the absolute extrema of a continuous function on a closed interval.

Finding the Absolute Extrema of f on a Closed Interval

1. Find the critical numbers of f that lie in (a, b).

2. Compute the value of f at each critical number found in step 1 and compute $f(a)$ and $f(b)$.

3. The absolute maximum value and absolute minimum value of f will correspond to the largest and smallest numbers, respectively, found in step 2.

EXAMPLE 1 Find the absolute extrema of the function $F(x) = x^2$ defined on the interval $[-1, 2]$.

Solution The function F is continuous on the closed interval $[-1, 2]$ and differentiable on the open interval $(-1, 2)$. The derivative of F is

$$F'(x) = 2x$$

so 0 is the only critical number of F. Next, evaluate $F(x)$ at $x = -1$, $x = 0$, and $x = 2$. Thus,

$$F(-1) = 1 \qquad F(0) = 0 \qquad F(2) = 4$$

It follows that 0 is the absolute minimum value of F and 4 is the absolute maximum value of F. The graph of F, in Figure 60, confirms our results. ∎

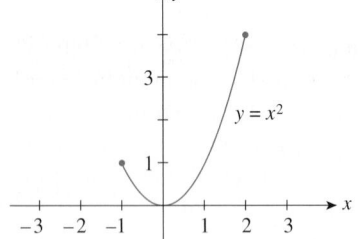

FIGURE 60
F has an absolute minimum value of 0 and an absolute maximum value of 4.

EXAMPLE 2 Find the absolute extrema of the function

$$f(x) = x^3 - 2x^2 - 4x + 4$$

defined on the interval $[0, 3]$.

Solution The function f is continuous on the closed interval $[0, 3]$ and differentiable on the open interval $(0, 3)$. The derivative of f is

$$f'(x) = 3x^2 - 4x - 4 = (3x + 2)(x - 2)$$

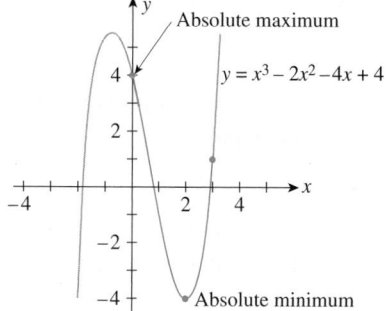

FIGURE 61
f has an absolute maximum value of 4 and an absolute minimum value of -4.

and it is equal to zero when $x = -\frac{2}{3}$ and $x = 2$. Since $x = -\frac{2}{3}$ lies outside the interval $[0, 3]$, it is dropped from further consideration, and $x = 2$ is seen to be the sole critical number of f. Next, we evaluate $f(x)$ at the critical number of f as well as the endpoints of f, obtaining

$$f(0) = 4 \qquad f(2) = -4 \qquad f(3) = 1$$

From these results, we conclude that -4 is the absolute minimum value of f and 4 is the absolute maximum value of f. The graph of f, which appears in Figure 61, confirms our results. Observe that the absolute maximum of f occurs at the endpoint $x = 0$ of the interval $[0, 3]$, while the absolute minimum of f occurs at $x = 2$, which lies in the interval $(0, 3)$.

EXPLORING WITH TECHNOLOGY

Let $f(x) = x^3 - 2x^2 - 4x + 4$. (This is the function of Example 2.)

1. Use a graphing utility to plot the graph of f, using the viewing window $[0, 3] \times [-5, 5]$. Use TRACE to find the absolute extrema of f on the interval $[0, 3]$ and thus verify the results obtained analytically in Example 2.

2. Plot the graph of f, using the viewing window $[-2, 1] \times [-5, 6]$. Use ZOOM and TRACE to find the absolute extrema of f on the interval $[-2, 1]$. Verify your results analytically.

EXAMPLE 3 Find the absolute maximum and absolute minimum values of the function $f(x) = x^{2/3}$ on the interval $[-1, 8]$.

Solution The derivative of f is

$$f'(x) = \frac{2}{3}x^{-1/3} = \frac{2}{3x^{1/3}}$$

Note that f' is not defined at $x = 0$, is continuous everywhere else, and does not equal zero for all x. Therefore, 0 is the only critical number of f. Evaluating $f(x)$ at $x = -1, 0$, and 8, we obtain

$$f(-1) = 1 \qquad f(0) = 0 \qquad f(8) = 4$$

We conclude that the absolute minimum value of f is 0, attained at $x = 0$, and the absolute maximum value of f is 4, attained at $x = 8$ (Figure 62).

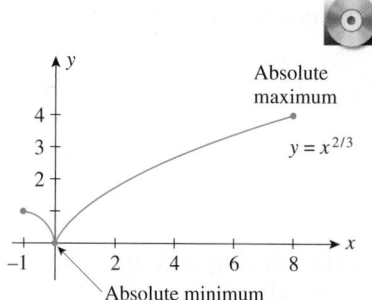

FIGURE 62
f has an absolute minimum value of $f(0) = 0$ and an absolute maximum value of $f(8) = 4$.

Many real-world applications call for finding the absolute maximum value or the absolute minimum value of a given function. For example, management is interested in finding what level of production will yield the maximum profit for a company; a farmer is interested in finding the right amount of fertilizer to maximize crop yield; a doctor is interested in finding the maximum concentration of a drug in a patient's body and the time at which it occurs; and an engineer is interested in finding the dimension of a container with a specified shape and volume that can be constructed at a minimum cost.

APPLIED EXAMPLE 4 Maximizing Profits Acrosonic's total profit (in dollars) from manufacturing and selling x units of their model F loudspeaker systems is given by

$$P(x) = -0.02x^2 + 300x - 200,000 \qquad (0 \leq x \leq 20,000)$$

How many units of the loudspeaker system must Acrosonic produce to maximize its profits?

Solution   To find the absolute maximum of P on $[0, 20,000]$, first find the critical points of P on the interval $(0, 20,000)$. To do this, compute

$$P'(x) = -0.04x + 300$$

Solving the equation $P'(x) = 0$ gives $x = 7500$. Next, evaluate $P(x)$ at $x = 7500$ as well as the endpoints $x = 0$ and $x = 20,000$ of the interval $[0, 20,000]$, obtaining

$$P(0) = -200,000$$
$$P(7500) = 925,000$$
$$P(20,000) = -2,200,000$$

From these computations we see that the absolute maximum value of the function P is 925,000. Thus, by producing 7500 units, Acrosonic will realize a maximum profit of $925,000. The graph of P is sketched in Figure 63. ◼

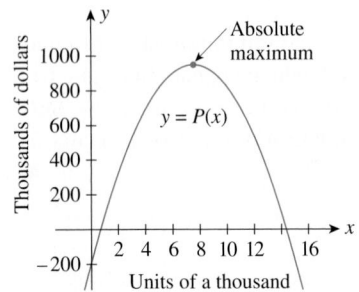

FIGURE 63
P has an absolute maximum at
(7500, 925,000).

EXPLORE & DISCUSS

Recall that the total profit function P is defined as $P(x) = R(x) - C(x)$, where R is the total revenue function, C is the total cost function, and x is the number of units of a product produced and sold. (Assume all derivatives exist.)

1. Show that at the level of production x_0 that yields the maximum profit for the company, the following two conditions are satisfied:

$$R'(x_0) = C'(x_0) \quad \text{and} \quad R''(x_0) < C''(x_0)$$

2. Interpret the two conditions in part 1 in economic terms and explain why they make sense.

APPLIED EXAMPLE 5   Trachea Contraction during a Cough   When a person coughs, the trachea (windpipe) contracts, allowing air to be expelled at a maximum velocity. It can be shown that during a cough the velocity v of airflow is given by the function

$$v = f(r) = kr^2(R - r)$$

where r is the trachea's radius (in centimeters) during a cough, R is the trachea's normal radius (in centimeters), and k is a positive constant that depends on the length of the trachea. Find the radius r for which the velocity of airflow is greatest.

Solution   To find the absolute maximum of f on $[0, R]$, first find the critical numbers of f on the interval $(0, R)$. We compute

$$f'(r) = 2kr(R - r) - kr^2 \quad \text{Use the product rule.}$$
$$= -3kr^2 + 2kRr = kr(-3r + 2R)$$

Setting $f'(r) = 0$ gives $r = 0$ or $r = \frac{2}{3}R$, and so $\frac{2}{3}R$ is the sole critical number of f ($r = 0$ is an endpoint). Evaluating $f(r)$ at $r = \frac{2}{3}R$, as well as at the endpoints

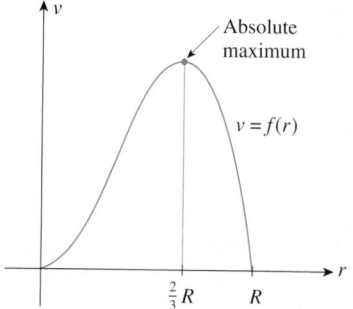

FIGURE 64
The velocity of airflow is greatest when the radius of the contracted trachea is $\frac{2}{3}R$.

$r = 0$ and $r = R$, we obtain

$$f(0) = 0$$

$$f\left(\frac{2}{3}R\right) = \frac{4k}{27}R^3$$

$$f(R) = 0$$

from which we deduce that the velocity of airflow is greatest when the radius of the contracted trachea is $\frac{2}{3}R$—that is, when the radius is contracted by approximately 33%. The graph of the function f is shown in Figure 64. ∎

EXPLORE & DISCUSS

Prove that if a cost function $C(x)$ is concave upward $[C''(x) > 0]$, then the level of production that will result in the smallest average production cost occurs when

$$\overline{C}(x) = C'(x)$$

—that is, when the average cost $\overline{C}(x)$ is equal to the marginal cost $C'(x)$.

Hints:

1. Show that

$$\overline{C}'(x) = \frac{xC'(x) - C(x)}{x^2}$$

so that the critical number of the function $\overline{C}$ occurs when

$$xC'(x) - C(x) = 0$$

2. Show that at a critical number of $\overline{C}$

$$\overline{C}''(x) = \frac{C''(x)}{x}$$

Use the second derivative test to reach the desired conclusion.

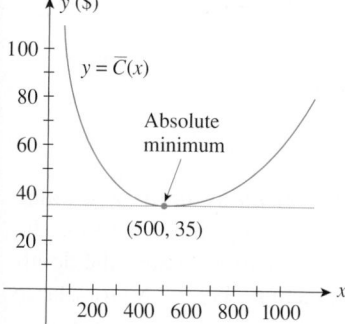

FIGURE 65
The minimum average cost is $35 per unit.

APPLIED EXAMPLE 6 Minimizing Average Cost The daily average cost function (in dollars per unit) of Elektra Electronics is given by

$$\overline{C}(x) = 0.0001x^2 - 0.08x + 40 + \frac{5000}{x} \qquad (x > 0)$$

where x stands for the number of programmable calculators that Elektra produces. Show that a production level of 500 units per day results in a minimum average cost for the company.

Solution The domain of the function $\overline{C}$ is the interval $(0, \infty)$, which is not closed. To solve the problem, we resort to the graphical method. Using the techniques of graphing from the last section, we sketch the graph of $\overline{C}$ (Figure 65). Now,

$$\overline{C}'(x) = 0.0002x - 0.08 - \frac{5000}{x^2}$$

Substituting the given value of x, 500, into $\overline{C}'(x)$ gives $\overline{C}'(500) = 0$, so 500 is a critical number of $\overline{C}$. Next,

$$\overline{C}''(x) = 0.0002 + \frac{10{,}000}{x^3}$$

Thus,

$$\overline{C}''(500) = 0.0002 + \frac{10{,}000}{(500)^3} > 0$$

and by the second derivative test, a relative minimum of the function $\overline{C}$ occurs at 500. Furthermore, $\overline{C}''(x) > 0$ for $x > 0$, which implies that the graph of $\overline{C}$ is concave upward everywhere, so the relative minimum of $\overline{C}$ must be the absolute minimum of $\overline{C}$. The minimum average cost is given by

$$\overline{C}(500) = 0.0001(500)^2 - 0.08(500) + 40 + \frac{5000}{500}$$

$$= 35$$

or $35 per unit. ∎

EXPLORING WITH TECHNOLOGY

Refer to the preceding Explore & Discuss and Example 6.

1. Using a graphing utility, plot the graphs of

$$\overline{C}(x) = 0.0001x^2 - 0.08x + 40 + \frac{5000}{x}$$

$$C'(x) = 0.0003x^2 - 0.16x + 40$$

using the viewing window $[0, 1000] \times [0, 150]$.

Note: $C(x) = 0.0001x^3 - 0.08x^2 + 40x + 5000$ (Why?)

2. Find the point of intersection of the graphs of $\overline{C}$ and C' and thus verify the assertion in the Explore & Discuss for the special case studied in Example 6.

APPLIED EXAMPLE 7   Flight of a Rocket   The altitude (in feet) of a rocket t seconds into flight is given by

$$s = f(t) = -t^3 + 96t^2 + 195t + 5 \qquad (t \geq 0)$$

a. Find the maximum altitude attained by the rocket.
b. Find the maximum velocity attained by the rocket.

Solution

a. The maximum altitude attained by the rocket is given by the largest value of the function f in the closed interval $[0, T]$, where T denotes the time the rocket impacts Earth. We know that such a number exists because the dominant term in the expression for the continuous function f is $-t^3$. So for t large enough, the value of $f(t)$ must change from positive to negative and, in particular, it must attain the value 0 for some T.

  To find the absolute maximum of f, compute

$$f'(t) = -3t^2 + 192t + 195$$

$$= -3(t - 65)(t + 1)$$

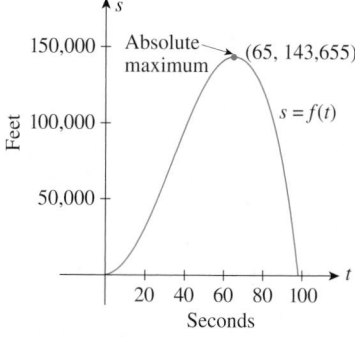

FIGURE 66
The maximum altitude of the rocket is
143,655 feet.

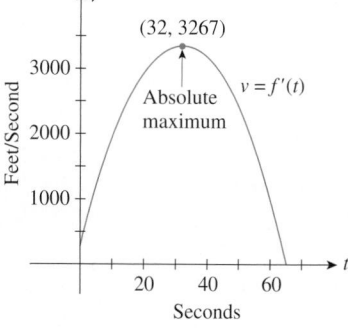

FIGURE 67
The maximum velocity of the rocket is
3267 feet per second.

and solve the equation $f'(t) = 0$, obtaining $t = -1$ and $t = 65$. Ignore $t = -1$ since it lies outside the interval $[0, T]$. This leaves the critical number 65 of f. Continuing, we compute

$$f(0) = 5 \qquad f(65) = 143{,}655 \qquad f(T) = 0$$

and conclude, accordingly, that the absolute maximum value of f is 143,655. Thus, the maximum altitude of the rocket is 143,655 feet, attained 65 seconds into flight. The graph of f is sketched in Figure 66.

b. To find the maximum velocity attained by the rocket, find the largest value of the function that describes the rocket's velocity at any time t—namely,

$$v = f'(t) = -3t^2 + 192t + 195 \qquad (t \geq 0)$$

We find the critical number of v by setting $v' = 0$. But

$$v' = -6t + 192$$

and the critical number of v is 32. Since

$$v'' = -6 < 0$$

the second derivative test implies that a relative maximum of v occurs at $t = 32$. Our computation has in fact clarified the property of the "velocity curve." Since $v'' < 0$ everywhere, the velocity curve is concave downward everywhere. With this observation, we assert that the relative maximum must in fact be the absolute maximum of v. The maximum velocity of the rocket is given by evaluating v at $t = 32$,

$$f'(32) = -3(32)^2 + 192(32) + 195$$

or 3267 feet per second. The graph of the velocity function v is sketched in Figure 67.

13.4   Self-Check Exercises

1. Let $f(x) = x - 2\sqrt{x}$
  a. Find the absolute extrema of f on the interval $[0, 9]$.
  b. Find the absolute extrema of f.

2. Find the absolute extrema of $f(x) = 3x^4 + 4x^3 + 1$ on $[-2, 1]$.

3. The operating rate (expressed as a percent) of factories, mines, and utilities in a certain region of the country on the tth day of 2006 is given by the function

$$f(t) = 80 + \frac{1200t}{t^2 + 40{,}000} \qquad (0 \leq t \leq 250)$$

On which day of the first 250 days of 2006 was the manufacturing capacity operating rate highest?

Solutions to Self-Check Exercises 13.4 can be found on page 836.

13.4   Concept Questions

1. Explain the following terms: (a) absolute maximum and (b) absolute minimum.

2. Describe the procedure for finding the absolute extrema of a continuous function on a closed interval.

13.4   Exercises

In Exercises 1–8, you are given the graph of a function f defined on the indicated interval. Find the absolute maximum and the absolute minimum of f, if they exist.

1.

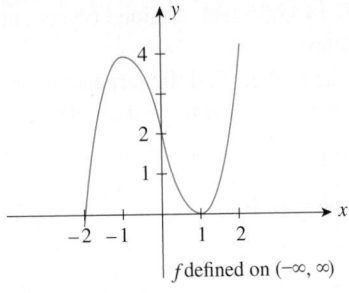

f defined on $(-\infty, \infty)$

2.

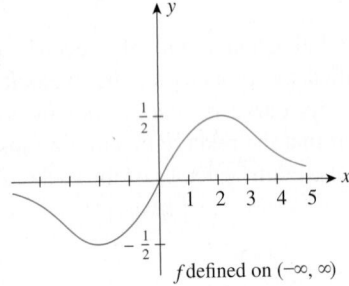

f defined on $(-\infty, \infty)$

3.

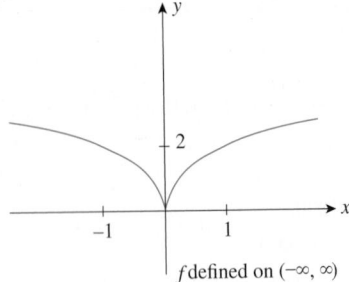

f defined on $(-\infty, \infty)$

4.

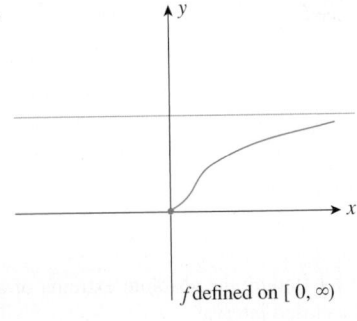

f defined on $[\,0, \infty)$

5.

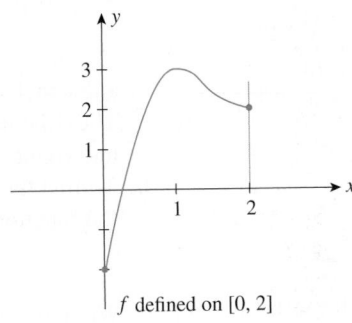

f defined on $[0, 2]$

6.

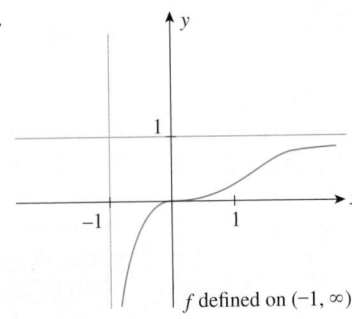

f defined on $(-1, \infty)$

7.

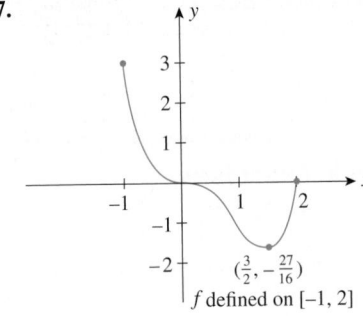

$\left(\frac{3}{2}, -\frac{27}{16}\right)$

f defined on $[-1, 2]$

8.

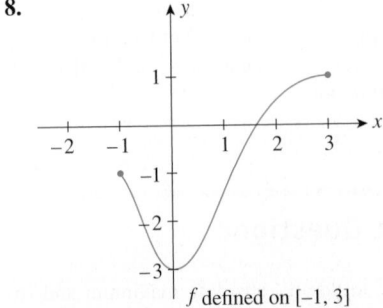

f defined on $[-1, 3]$

In Exercises 9–38, find the absolute maximum value and the absolute minimum value, if any, of each function.

9. $f(x) = 2x^2 + 3x - 4$ **10.** $g(x) = -x^2 + 4x + 3$

11. $h(x) = x^{1/3}$ **12.** $f(x) = x^{2/3}$

13. $f(x) = \dfrac{1}{1 + x^2}$ **14.** $f(x) = \dfrac{x}{1 + x^2}$

15. $f(x) = x^2 - 2x - 3$ on $[-2, 3]$

16. $g(x) = x^2 - 2x - 3$ on $[0, 4]$

17. $f(x) = -x^2 + 4x + 6$ on $[0, 5]$

18. $f(x) = -x^2 + 4x + 6$ on $[3, 6]$

19. $f(x) = x^3 + 3x^2 - 1$ on $[-3, 2]$

20. $g(x) = x^3 + 3x^2 - 1$ on $[-3, 1]$

21. $g(x) = 3x^4 + 4x^3$ on $[-2, 1]$

22. $f(x) = \dfrac{1}{2}x^4 - \dfrac{2}{3}x^3 - 2x^2 + 3$ on $[-2, 3]$

23. $f(x) = \dfrac{x + 1}{x - 1}$ on $[2, 4]$ **24.** $g(t) = \dfrac{t}{t - 1}$ on $[2, 4]$

25. $f(x) = 4x + \dfrac{1}{x}$ on $[1, 3]$ **26.** $f(x) = 9x - \dfrac{1}{x}$ on $[1, 3]$

27. $f(x) = \dfrac{1}{2}x^2 - 2\sqrt{x}$ on $[0, 3]$

28. $g(x) = \dfrac{1}{8}x^2 - 4\sqrt{x}$ on $[0, 9]$

29. $f(x) = \dfrac{1}{x}$ on $(0, \infty)$ **30.** $g(x) = \dfrac{1}{x + 1}$ on $(0, \infty)$

31. $f(x) = 3x^{2/3} - 2x$ on $[0, 3]$

32. $g(x) = x^2 + 2x^{2/3}$ on $[-2, 2]$

33. $f(x) = x^{2/3}(x^2 - 4)$ on $[-1, 2]$

34. $f(x) = x^{2/3}(x^2 - 4)$ on $[-1, 3]$

35. $f(x) = \dfrac{x}{x^2 + 2}$ on $[-1, 2]$

36. $f(x) = \dfrac{1}{x^2 + 2x + 5}$ on $[-2, 1]$

37. $f(x) = \dfrac{x}{\sqrt{x^2 + 1}}$ on $[-1, 1]$

38. $g(x) = x\sqrt{4 - x^2}$ on $[0, 2]$

39. A stone is thrown straight up from the roof of an 80-ft building. The height (in feet) of the stone at any time t (in seconds), measured from the ground, is given by

$$h(t) = -16t^2 + 64t + 80$$

What is the maximum height the stone reaches?

40. MAXIMIZING PROFITS Lynbrook West, an apartment complex, has 100 two-bedroom units. The monthly profit (in dollars) realized from renting out x apartments is given by

$$P(x) = -10x^2 + 1760x - 50,000$$

To maximize the monthly rental profit, how many units should be rented out? What is the maximum monthly profit realizable?

41. SENIORS IN THE WORKFORCE The percent of men, age 65 and above, in the workforce from 1950 ($t = 0$) through 2000 ($t = 50$) is approximately

$$P(t) = 0.0135t^2 - 1.126t + 41.2 \qquad (0 \le t \le 50)$$

Show that the percent of men, age 65 and above, in the workforce in the period of time under consideration was smallest around mid-September 1991. What is that percent?
Source: U.S. Census Bureau

42. FLIGHT OF A ROCKET The altitude (in feet) attained by a model rocket t sec into flight is given by the function

$$h(t) = -\frac{1}{3}t^3 + 4t^2 + 20t + 2 \qquad (t \ge 0)$$

Find the maximum altitude attained by the rocket.

43. FEMALE SELF-EMPLOYED WORKFORCE Data show that the number of nonfarm, full-time, self-employed women can be approximated by

$$N(t) = 0.81t - 1.14\sqrt{t} + 1.53 \qquad (0 \le t \le 6)$$

where $N(t)$ is measured in millions and t is measured in 5-yr intervals, with $t = 0$ corresponding to the beginning of 1963. Determine the absolute extrema of the function N on the interval $[0, 6]$. Interpret your results.
Source: U.S. Department of Labor

44. AVERAGE SPEED OF A VEHICLE ON A HIGHWAY The average speed of a vehicle on a stretch of Route 134 between 6 a.m. and 10 a.m. on a typical weekday is approximated by the function

$$f(t) = 20t - 40\sqrt{t} + 50 \qquad (0 \le t \le 4)$$

where $f(t)$ is measured in miles per hour and t is measured in hours, with $t = 0$ corresponding to 6 a.m. At what time of the morning commute is the traffic moving at the slowest rate? What is the average speed of a vehicle at that time?

45. MAXIMIZING PROFITS The management of Trappee and Sons, producers of the famous TexaPep hot sauce, estimate that their profit (in dollars) from the daily production and sale of x cases (each case consisting of 24 bottles) of the hot sauce is given by

$$P(x) = -0.000002x^3 + 6x - 400$$

What is the largest possible profit Trappee can make in 1 day?

46. MAXIMIZING PROFITS The quantity demanded each month of the Walter Serkin recording of Beethoven's *Moonlight Sonata*, manufactured by Phonola Record Industries, is related to the price/compact disc. The equation

$$p = -0.00042x + 6 \qquad (0 \le x \le 12{,}000)$$

where p denotes the unit price in dollars and x is the number of discs demanded, relates the demand to the price. The total monthly cost (in dollars) for pressing and packaging x copies of this classical recording is given by

$$C(x) = 600 + 2x - 0.00002x^2 \qquad (0 \le x \le 20{,}000)$$

To maximize its profits, how many copies should Phonola produce each month?
Hint: The revenue is $R(x) = px$, and the profit is $P(x) = R(x) - C(x)$.

47. MAXIMIZING PROFIT A manufacturer of tennis rackets finds that the total cost $C(x)$ (in dollars) of manufacturing x rackets/day is given by $C(x) = 400 + 4x + 0.0001x^2$. Each racket can be sold at a price of p dollars, where p is related to x by the demand equation $p = 10 - 0.0004x$. If all rackets that are manufactured can be sold, find the daily level of production that will yield a maximum profit for the manufacturer.

48. MAXIMIZING PROFIT The weekly demand for the Pulsar 25-in. color console television is given by the demand equation

$$p = -0.05x + 600 \qquad (0 \le x \le 12{,}000)$$

where p denotes the wholesale unit price in dollars and x denotes the quantity demanded. The weekly total cost function associated with manufacturing these sets is given by

$$C(x) = 0.000002x^3 - 0.03x^2 + 400x + 80{,}000$$

where $C(x)$ denotes the total cost incurred in producing x sets. Find the level of production that will yield a maximum profit for the manufacturer.
Hint: Use the quadratic formula.

49. MAXIMIZING PROFIT A division of Chapman Corporation manufactures a pager. The weekly fixed cost for the division is $20,000, and the variable cost for producing x pagers/week is

$$V(x) = 0.000001x^3 - 0.01x^2 + 50x$$

dollars. The company realizes a revenue of

$$R(x) = -0.02x^2 + 150x \qquad (0 \le x \le 7500)$$

dollars from the sale of x pagers/week. Find the level of production that will yield a maximum profit for the manufacturer.
Hint: Use the quadratic formula.

50. MINIMIZING AVERAGE COST Suppose the total cost function for manufacturing a certain product is $C(x) = 0.2(0.01x^2 + 120)$ dollars, where x represents the number of units produced. Find the level of production that will minimize the average cost.

51. MINIMIZING PRODUCTION COSTS The total monthly cost (in dollars) incurred by Cannon Precision Instruments for manufacturing x units of the model M1 camera is given by the function

$$C(x) = 0.0025x^2 + 80x + 10{,}000$$

a. Find the average cost function $\overline{C}$.
b. Find the level of production that results in the smallest average production cost.
c. Find the level of production for which the average cost is equal to the marginal cost.
d. Compare the result of part (c) with that of part (b).

52. MINIMIZING PRODUCTION COSTS The daily total cost (in dollars) incurred by Trappee and Sons for producing x cases of TexaPep hot sauce is given by the function

$$C(x) = 0.000002x^3 + 5x + 400$$

Using this function, answer the questions posed in Exercise 51.

53. MAXIMIZING REVENUE Suppose the quantity demanded per week of a certain dress is related to the unit price p by the demand equation $p = \sqrt{800 - x}$, where p is in dollars and x is the number of dresses made. To maximize the revenue, how many dresses should be made and sold each week?
Hint: $R(x) = px$.

54. MAXIMIZING REVENUE The quantity demanded each month of the Sicard wristwatch is related to the unit price by the equation

$$p = \frac{50}{0.01x^2 + 1} \qquad (0 \le x \le 20)$$

where p is measured in dollars and x is measured in units of a thousand. To yield a maximum revenue, how many watches must be sold?

55. OXYGEN CONTENT OF A POND When organic waste is dumped into a pond, the oxidation process that takes place reduces the pond's oxygen content. However, given time, nature will restore the oxygen content to its natural level. Suppose the oxygen content t days after organic waste has been dumped into the pond is given by

$$f(t) = 100 \left[\frac{t^2 - 4t + 4}{t^2 + 4} \right] \qquad (0 \le t \le \infty)$$

percent of its normal level.
a. When is the level of oxygen content lowest?
b. When is the rate of oxygen regeneration greatest?

56. AIR POLLUTION The amount of nitrogen dioxide, a brown gas that impairs breathing, present in the atmosphere on a certain May day in the city of Long Beach is approximated by

$$A(t) = \frac{136}{1 + 0.25(t - 4.5)^2} + 28 \qquad (0 \le t \le 11)$$

where $A(t)$ is measured in pollutant standard index (PSI) and t is measured in hours, with $t = 0$ corresponding to 7 a.m. Determine the time of day when the pollution is at its highest level.

57. **MAXIMIZING REVENUE**   The average revenue is defined as the function

$$\overline{R}(x) = \frac{R(x)}{x} \qquad (x > 0)$$

Prove that if a revenue function $R(x)$ is concave downward $[R''(x) < 0]$, then the level of sales that will result in the largest average revenue occurs when $\overline{R}(x) = R'(x)$.

58. **VELOCITY OF BLOOD**   According to a law discovered by the 19th-century physician Jean Louis Marie Poiseuille, the velocity (in centimeters/second) of blood r cm from the central axis of an artery is given by

$$v(r) = k(R^2 - r^2)$$

where k is a constant and R is the radius of the artery. Show that the velocity of blood is greatest along the central axis.

59. **GDP OF A DEVELOPING COUNTRY**   A developing country's gross domestic product (GDP) from 1997 to 2005 is approximated by the function

$$G(t) = -0.2t^3 + 2.4t^2 + 60 \qquad (0 \le t \le 8)$$

where $G(t)$ is measured in billions of dollars and $t = 0$ corresponds to 1997. Show that the growth rate of the country's GDP was maximal in 2001.

60. **CRIME RATES**   The number of major crimes committed in the city of Bronxville between 1997 and 2004 is approximated by the function

$$N(t) = -0.1t^3 + 1.5t^2 + 100 \qquad (0 \le t \le 7)$$

where $N(t)$ denotes the number of crimes committed in year t ($t = 0$ corresponds to 1997). Enraged by the dramatic increase in the crime rate, the citizens of Bronxville, with the help of the local police, organized "Neighborhood Crime Watch" groups in early 2001 to combat this menace. Show that the growth in the crime rate was maximal in 2002, giving credence to the claim that the Neighborhood Crime Watch program was working.

61. **FOREIGN-BORN RESIDENTS**   The percent of foreign-born residents in the United States from 1910 through 2000 is approximated by the function

$$P(t) = 0.04363t^3 - 0.267t^2 - 1.59t + 14.7 \qquad (0 \le t \le 9)$$

where t is measured in decades, with $t = 0$ corresponding to 1910. Show that the percentage of foreign-born residents was lowest in early 1970.
Hint: Use the quadratic formula.

Source: Journal of American Medical Association

62. **AVERAGE PRICES OF HOMES**   The average annual price of single-family homes in Massachusetts between 1990 and 2002 is approximated by the function

$$P(t) = -0.183t^3 + 4.65t^2 - 17.3t + 200 \qquad (0 \le t \le 12)$$

where $P(t)$ is measured in thousands of dollars and t is measured in years, with $t = 0$ corresponding to 1990. In what year was the average annual price of single-family homes in Massachusetts lowest? What was the approximate lowest average annual price?
Hint: Use the quadratic formula.

Source: Massachusetts Association of Realtors

63. **OFFICE RENTS**   After the economy softened, the sky-high office space rents of the late 1990s started to come down to earth. The function R gives the approximate price per square foot in dollars, $R(t)$, of prime space in Boston's Back Bay and Financial District from 1997 ($t = 0$) through 2002, where

$$R(t) = -0.711t^3 + 3.76t^2 + 0.2t + 36.5 \qquad (0 \le t \le 5)$$

Show that the office space rents peaked at about the middle of 2000. What was the highest office space rent during the period in question?
Hint: Use the quadratic formula.

Source: Meredith & Grew Inc./Oncor

64. **WORLD POPULATION**   The total world population is forecast to be

$$P(t) = 0.00074t^3 - 0.0704t^2 + 0.89t + 6.04 \qquad (0 \le t \le 10)$$

in year t, where t is measured in decades with $t = 0$ corresponding to 2000 and $P(t)$ is measured in billions.
a. Show that the world population is forecast to peak around 2071.
   Hint: Use the quadratic formula.
b. What will the population peak at?

Source: International Institute for Applied Systems Analysis

65. **VENTURE-CAPITAL INVESTMENT**   Venture-capital investment increased dramatically in the late 1990s but came to a screeching halt after the dot-com bust. The venture-capital investment (in billions of dollars) from 1995 ($t = 0$) through 2003 ($t = 8$) is approximated by the function

$$C(t) = \begin{cases} 0.6t^2 + 2.4t + 7.6 & 0 \le t < 3 \\ 3t^2 + 18.8t - 63.2 & 3 \le t < 5 \\ -3.3167t^3 + 80.1t^2 - 642.583t + 1730.8025 & 5 \le t \le 8 \end{cases}$$

a. In what year did venture-capital investment peak over the period under consideration? What was the amount of that investment?
b. In what year was the venture-capital investment lowest over this period? What was the amount of that investment?
   Hint: Find the absolute extrema of C on each of the closed intervals [0, 3], [3, 5], and [5, 8].

Sources: Venture One; Ernst & Young

66. ENERGY EXPENDED BY A FISH   It has been conjectured that a fish swimming a distance of L ft at a speed of v ft/sec relative to the water and against a current flowing at the rate of u ft/sec ($u < v$) expends a total energy given by

$$E(v) = \frac{aLv^3}{v - u}$$

where E is measured in foot-pounds (ft-lb) and a is a constant. Find the speed v at which the fish must swim in order to minimize the total energy expended. (*Note:* This result has been verified by biologists.)

67. REACTION TO A DRUG   The strength of a human body's reaction R to a dosage D of a certain drug is given by

$$R = D^2\left(\frac{k}{2} - \frac{D}{3}\right)$$

where k is a positive constant. Show that the maximum reaction is achieved if the dosage is k units.

68. Refer to Exercise 67. Show that the rate of change in the reaction R with respect to the dosage D is maximal if $D = k/2$.

In Exercises 69–72, determine whether the statement is true or false. If it is true, explain why it is true. If it is false, give an example to show why it is false.

69. If f is defined on a closed interval $[a, b]$, then f has an absolute maximum value.

70. If f is continuous on an open interval (a, b), then f does not have an absolute minimum value.

71. If f is not continuous on the closed interval $[a, b]$, then f cannot have an absolute maximum value.

72. If $f''(x) < 0$ on (a, b) and $f'(c) = 0$ where $a < c < b$, then $f(c)$ is the absolute maximum value of f on $[a, b]$.

73. Let f be a constant function—that is, let $f(x) = c$, where c is some real number. Show that every number a gives rise to an absolute maximum and, at the same time, an absolute minimum of f.

74. Show that a polynomial function defined on the interval $(-\infty, \infty)$ cannot have both an absolute maximum and an absolute minimum unless it is a constant function.

75. One condition that must be satisfied before Theorem 3 (p. 825) is applicable is that the function f must be continuous on the closed interval $[a, b]$. Define a function f on the closed interval $[-1, 1]$ by

$$f(x) = \begin{cases} \dfrac{1}{x} & \text{if } x \in [-1, 1] \quad (x \neq 0) \\ 0 & \text{if } x = 0 \end{cases}$$

a. Show that f is not continuous at $x = 0$.
b. Show that $f(x)$ does not attain an absolute maximum or an absolute minimum on the interval $[-1, 1]$.
c. Confirm your results by sketching the function f.

76. One condition that must be satisfied before Theorem 3 (page 825) is applicable is that the interval on which f is defined must be a closed interval $[a, b]$. Define a function f on the *open* interval $(-1, 1)$ by $f(x) = x$. Show that f does not attain an absolute maximum or an absolute minimum on the interval $(-1, 1)$.
Hint: What happens to $f(x)$ if x is close to but not equal to $x = -1$? If x is close to but not equal to $x = 1$?

13.4   Solutions to Self-Check Exercises

1. a. The function f is continuous in its domain and differentiable in the interval $(0, 9)$. The derivative of f is

$$f'(x) = 1 - x^{-1/2} = \frac{x^{1/2} - 1}{x^{1/2}}$$

and it is equal to zero when $x = 1$. Evaluating $f(x)$ at the endpoints $x = 0$ and $x = 9$ and at the critical number 1 of f, we have

$$f(0) = 0 \qquad f(1) = -1 \qquad f(9) = 3$$

From these results, we see that -1 is the absolute minimum value of f and 3 is the absolute maximum value of f.

b. In this case, the domain of f is the interval $[0, \infty)$, which is not closed. Therefore, we resort to the graphic method. Using the techniques of graphing, we sketch in the accompanying figure the graph of f.

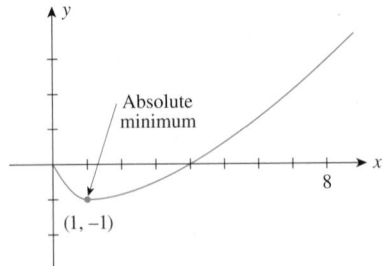

The graph of f shows that -1 is the absolute minimum value of f, but f has no absolute maximum since $f(x)$ increases without bound as x increases without bound.

2. The function f is continuous on the interval $[-2, 1]$. It is also differentiable on the open interval $(-2, 1)$. The derivative of f is

$$f'(x) = 12x^3 + 12x^2 = 12x^2(x + 1)$$

and it is continuous on $(-2, 1)$. Setting $f'(x) = 0$ gives -1 and 0 as critical numbers of f. Evaluating $f(x)$ at these critical numbers of f as well as at the endpoints of the interval $[-2, 1]$, we obtain

$$f(-2) = 17 \quad f(-1) = 0 \quad f(0) = 1 \quad f(1) = 8$$

From these results, we see that 0 is the absolute minimum value of f and 17 is the absolute maximum value of f.

3. The problem is solved by finding the absolute maximum of the function f on $[0, 250]$. Differentiating $f(t)$, we obtain

$$f'(t) = \frac{(t^2 + 40{,}000)(1200) - 1200t(2t)}{(t^2 + 40{,}000)^2}$$

$$= \frac{-1200(t^2 - 40{,}000)}{(t^2 + 40{,}000)^2}$$

Upon setting $f'(t) = 0$ and solving the resulting equation, we obtain $t = -200$ or 200. Since -200 lies outside the interval $[0, 250]$, we are interested only in the critical number 200 of f. Evaluating $f(t)$ at $t = 0$, $t = 200$, and $t = 250$, we find

$$f(0) = 80 \quad f(200) = 83 \quad f(250) = 82.93$$

We conclude that the manufacturing capacity operating rate was the highest on the 200th day of 2006—that is, a little past the middle of July 2006.

USING TECHNOLOGY

■ Finding the Absolute Extrema of a Function

Some graphing utilities have a function for finding the absolute maximum and the absolute minimum values of a continuous function on a closed interval. If your graphing utility has this capability, use it to work through the example and exercises of this section.

EXAMPLE 1 Let $f(x) = \dfrac{2x + 4}{(x^2 + 1)^{3/2}}$.

a. Use a graphing utility to plot the graph of f in the viewing window $[-3, 3] \times [-1, 5]$.
b. Find the absolute maximum and absolute minimum values of f on the interval $[-3, 3]$. Express your answers accurate to four decimal places.

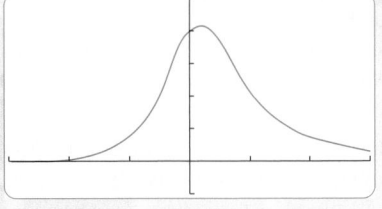

FIGURE T1
The graph of f in the viewing window
$[-3, 3] \times [-1, 5]$

Solution

a. The graph of f is shown in Figure T1.
b. Using the function on a graphing utility for finding the absolute minimum value of a continuous function on a closed interval, we find the absolute minimum value of f to be -0.0632. Similarly, using the function for finding the absolute maximum value, we find the absolute maximum value to be 4.1593. ■

Note Some graphing utilities will enable you to find the absolute minimum and absolute maximum values of a continuous function on a closed interval without having to graph the function. ■

(continued)

TECHNOLOGY EXERCISES

In Exercises 1–6, find the absolute maximum and the absolute minimum values of f in the given interval using the method of Example 1. Express your answers accurate to four decimal places.

1. $f(x) = 3x^4 - 4.2x^3 + 6.1x - 2; [-2, 3]$

2. $f(x) = 2.1x^4 - 3.2x^3 + 4.1x^2 + 3x - 4; [-1, 2]$

3. $f(x) = \dfrac{2x^3 - 3x^2 + 1}{x^2 + 2x - 8}; [-3, 1]$

4. $f(x) = \sqrt{x}(x^3 - 4)^2; [0.5, 1]$

5. $f(x) = \dfrac{x^3 - 1}{x^2}; [1, 3]$

6. $f(x) = \dfrac{x^3 - x^2 + 1}{x - 2}; [1, 3]$

7. RATE OF BANK FAILURES The estimated rate at which banks were failing between 1982 and 1994 is given by

$$f(t) = 0.063447t^4 - 1.953283t^3 + 14.632576t^2$$
$$- 6.684704t + 47.458874 \quad (0 \le t \le 12)$$

where $f(t)$ is the number of banks per year and t is measured in years, with $t = 0$ corresponding to the beginning of 1982.
a. Plot the graph of f in the viewing window $[0, 12] \times [0, 220]$.
b. What is the highest rate of bank failures during the period in question?
Source: Federal Deposit Insurance Corporation

8. TIME ON THE MARKET The average number of days a single-family home remains for sale from listing to accepted offer (in the greater Boston area) is approximated by the function

$$f(t) = 0.0171911t^4 - 0.662121t^3 + 6.18083t^2$$
$$- 8.97086t + 53.3357 \quad (0 \le t \le 10)$$

where t is measured in years, with $t = 0$ corresponding to the beginning of 1984.
a. Plot the graph of f in the viewing window $[0, 12] \times [0, 120]$.
b. Find the absolute maximum value and the absolute minimum value of f in the interval $[0, 12]$. Interpret your results.
Source: Greater Boston Real Estate Board—Multiple Listing Service

9. USE OF DIESEL ENGINES Diesel engines are popular in cars in Europe, where fuel prices are high. The percent of new vehicles in Western Europe equipped with diesel engines is approximated by the function

$$f(t) = 0.3t^4 - 2.58t^3 + 8.11t^2 - 7.71t + 23.75 \quad (0 \le t \le 4)$$

where t is measured in years, with $t = 0$ corresponding to the beginning of 1996.
a. Use a graphing utility to sketch the graph of f on $[0, 4] \times [0, 40]$.
b. What was the lowest percent of new vehicles equipped with diesel engines for the period in question?
Source: German Automobile Industry Association

10. FEDERAL DEBT According to data obtained from the Congressional Budget Office, the national debt (in trillions of dollars) is given by the function

$$f(t) = 0.001532t^3 - 0.0588t^2$$
$$+ 0.5208t + 2.55 \quad (0 \le t \le 20)$$

where t is measured in years, with $t = 0$ corresponding to the beginning of 1990.
a. Sketch the graph of f, using the viewing window $[0, 20] \times [0, 4]$.
b. Estimate (to the nearest year) when the federal debt was at the highest level over the period under consideration. What was that level?
Source: Congressional Budget Office

11. SICKOUTS In a sickout by pilots of American Airlines in February 1999, the number of canceled flights from February 6 ($t = 0$) through February 14 ($t = 8$) is approximated by the function

$$N(t) = 1.2576t^4 - 26.357t^3 + 127.98t^2 + 82.3t + 43 \quad (0 \le t \le 8)$$

where t is measured in days. The sickout ended after the union was threatened with millions of dollars of fines.
a. Show that the number of canceled flights was increasing at the fastest rate on February 8.
b. Estimate the maximum number of canceled flights in a day during the sickout.
Source: Associated Press

12. 401(K) INVESTORS The average account balance of a 401(k) investor from 1996 through 2002 is approximately

$$A(t) = 0.28636t^4 - 3.4864t^3 + 11.689t^2 - 6.08t + 37.6 \quad (0 \le t \le 6)$$

where $A(t)$ is measured in thousands of dollars and t is measured in years with $t = 0$ corresponding to the beginning of 1996.
a. Plot the graph of A, using the viewing window $[0, 6] \times [0, 60]$.
b. When was the average account balance lowest in the period under consideration? When was it highest?
c. What were the lowest average account balance and the highest average account balance during the period under consideration?
Source: Investment Company Institute

13. **Why SS Benefits May Exceed Payroll Taxes**   Unless payroll taxes are increased significantly and/or benefits are scaled back drastically, it is only a matter of time before the current Social Security system goes broke. Data show that the assets of the system—the Social Security "trust fund"—may be approximated by

$$f(t) = -0.0129t^4 + 0.3087t^3 + 2.1760t^2$$
$$+ 62.8466t + 506.2955 \qquad (0 \le t \le 35)$$

where $f(t)$ is measured in millions of dollars and t is measured in years, with $t = 0$ corresponding to the beginning of 1995.
a. Use a graphing utility to sketch the graph of f.
b. Based on this model, when will the Social Security system start to pay out more benefits than it gets in payroll taxes?

Source: Board of Trustees, Social Security Administration

14. **Food Stamp Recipients**   The number of people in the United States receiving food stamps from 1988 through 1998 is approximated by the function

$$f(t) = 0.00944t^4 - 0.2311t^3 + 1.516t^2$$
$$- 1.35t + 18.6 \qquad (0 \le t \le 10)$$

where $f(t)$ is measured in millions of people and t is measured in years, with $t = 0$ corresponding to the fiscal year ending in September 1988.
a. Plot the graph of f, using the viewing window $[0, 10] \times [0, 28]$.
b. Based on this model, in what year did the number of food stamps recipients peak?

Source: U.S. Department of Agriculture

13.5 Optimization II

Section 13.4 outlined how to find the solution to certain optimization problems in which the objective function is given. In this section, we consider problems in which we are required to first find the appropriate function to be optimized. The following guidelines will be useful for solving these problems.

Guidelines for Solving Optimization Problems

1. Assign a letter to each variable mentioned in the problem. If appropriate, draw and label a figure.
2. Find an expression for the quantity to be optimized.
3. Use the conditions given in the problem to write the quantity to be optimized as a function f of *one* variable. Note any restrictions to be placed on the domain of f from physical considerations of the problem.
4. Optimize the function f over its domain using the methods of Section 13.4.

Note   In carrying out step 4, remember that if the function f to be optimized is continuous on a closed interval, then the absolute maximum and absolute minimum of f are, respectively, the largest and smallest values of $f(x)$ on the set composed of the critical numbers of f and the endpoints of the interval. If the domain of f is not a closed interval, then we resort to the graphical method.

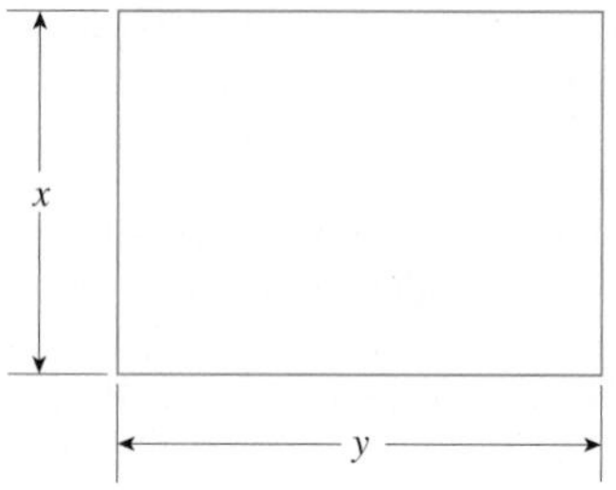

■ Maximization Problems

APPLIED EXAMPLE 1 Fencing a Garden A man wishes to have a rectangular-shaped garden in his backyard. He has 50 feet of fencing with which to enclose his garden. Find the dimensions for the largest garden he can have if he uses all of the fencing.

FIGURE 68
What is the maximum rectangular area that can be enclosed with 50 feet of fencing?

Solution

Step 1   Let x and y denote the dimensions (in feet) of two adjacent sides of the garden (Figure 68) and let A denote its area.

Step 2   The area of the garden

$$A = xy \tag{1}$$

is the quantity to be maximized.

Step 3   The perimeter of the rectangle, $(2x + 2y)$ feet, must equal 50 feet. Therefore, we have the equation

$$2x + 2y = 50$$

Next, solving this equation for y in terms of x yields

$$y = 25 - x \tag{2}$$

which, when substituted into Equation (1), gives

$$A = x(25 - x)$$
$$= -x^2 + 25x$$

(Remember, the function to be optimized must involve just one variable.) Since the sides of the rectangle must be nonnegative, we must have $x \geq 0$ and $y = 25 - x \geq 0$; that is, we must have $0 \leq x \leq 25$. Thus, the problem is reduced to that of finding the absolute maximum of $A = f(x) = -x^2 + 25x$ on the closed interval $[0, 25]$.

Step 4   Observe that f is continuous on $[0, 25]$, so the absolute maximum value of f must occur at the endpoint(s) of the interval or at the critical number(s) of f. The derivative of the function A is given by

$$A' = f'(x) = -2x + 25$$

Setting $A' = 0$ gives

$$-2x + 25 = 0$$

or 12.5, as the critical number of A. Next, we evaluate the function $A = f(x)$ at $x = 12.5$ and at the endpoints $x = 0$ and $x = 25$ of the interval $[0, 25]$, obtaining

$$f(0) = 0 \qquad f(12.5) = 156.25 \qquad f(25) = 0$$

We see that the absolute maximum value of the function f is 156.25. From Equation (2) we see that $y = 12.5$ when $x = 12.5$. Thus, the garden of maximum area (156.25 square feet) is a square with sides of length 12.5 feet.   ■

APPLIED EXAMPLE 2   Packaging By cutting away identical squares from each corner of a rectangular piece of cardboard and folding up the resulting flaps, the cardboard may be turned into an open box. If the cardboard is 16 inches long and 10 inches wide, find the dimensions of the box that will yield the maximum volume.

Solution

Step 1   Let x denote the length (in inches) of one side of each of the identical squares to be cut out of the cardboard (Figure 69) and let V denote the volume of the resulting box.

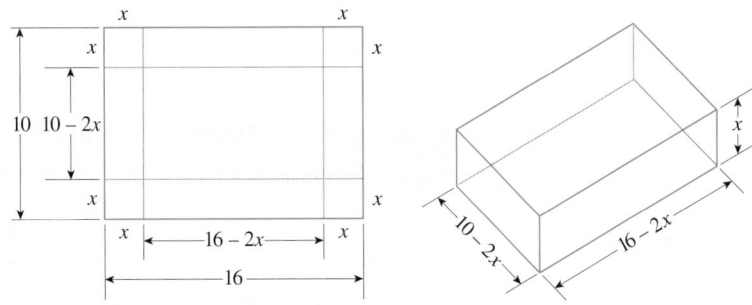

FIGURE 69
The dimensions of the open box are $(16 - 2x)''$ by $(10 - 2x)''$ by x''.

Step 2   The dimensions of the box are $(16 - 2x)$ inches by $(10 - 2x)$ inches by x inches. Therefore, its volume (in cubic inches),

$$V = (16 - 2x)(10 - 2x)x$$
$$= 4(x^3 - 13x^2 + 40x) \qquad \text{Expand the expression.}$$

is the quantity to be maximized.

Step 3   Since each side of the box must be nonnegative, x must satisfy the inequalities $x \geq 0$, $16 - 2x \geq 0$, and $10 - 2x \geq 0$. This set of inequalities is satisfied if $0 \leq x \leq 5$. Thus, the problem at hand is equivalent to that of finding the absolute maximum of

$$V = f(x) = 4(x^3 - 13x^2 + 40x)$$

on the closed interval $[0, 5]$.

Step 4   Observe that f is continuous on $[0, 5]$, so the absolute maximum value of f must be attained at the endpoint(s) or at the critical number(s) of f.
    Differentiating $f(x)$, we obtain

$$f'(x) = 4(3x^2 - 26x + 40)$$
$$= 4(3x - 20)(x - 2)$$

Upon setting $f'(x) = 0$ and solving the resulting equation for x, we obtain $x = \frac{20}{3}$ or $x = 2$. Since $\frac{20}{3}$ lies outside the interval $[0, 5]$, it is no longer considered, and we are interested only in the critical number 2 of f. Next, evaluating $f(x)$ at $x = 0$, $x = 5$ (the endpoints of the interval $[0, 5]$), and $x = 2$, we obtain

$$f(0) = 0 \qquad f(2) = 144 \qquad f(5) = 0$$

Thus, the volume of the box is maximized by taking $x = 2$. The dimensions of the box are $12'' \times 6'' \times 2''$, and the volume is 144 cubic inches. ∎

EXPLORING WITH TECHNOLOGY

Refer to Example 2.

1. Use a graphing utility to plot the graph of

$$f(x) = 4(x^3 - 13x^2 + 40x)$$

using the viewing window $[0, 5] \times [0, 150]$. Explain what happens to $f(x)$ as x increases from $x = 0$ to $x = 5$ and give a physical interpretation.

2. Using ZOOM and TRACE, find the absolute maximum of f on the interval $[0, 5]$ and thus verify the solution for Example 2 obtained analytically.

APPLIED EXAMPLE 3 Optimal Subway Fare   A city's Metropolitan Transit Authority (MTA) operates a subway line for commuters from a certain suburb to the downtown metropolitan area. Currently, an average of 6000 passengers a day take the trains, paying a fare of $3.00 per ride. The board of the MTA, contemplating raising the fare to $3.50 per ride in order to generate a larger revenue, engages the services of a consulting firm. The firm's study reveals that for each $.50 increase in fare, the ridership will be reduced by an average of 1000 passengers a day. Thus, the consulting firm recommends that MTA stick to the current fare of $3.00 per ride, which already yields a maximum revenue. Show that the consultants are correct.

Solution

Step 1   Let x denote the number of passengers per day, p denote the fare per ride, and R be MTA's revenue.

Step 2   To find a relationship between x and p, observe that the given data imply that when $x = 6000$, $p = 3$, and when $x = 5000$, $p = 3.50$. Therefore, the points $(6000, 3)$ and $(5000, 3.50)$ lie on a straight line. (Why?) To find the linear relationship between p and x, use the point-slope form of the equation of a straight line. Now, the slope of the line is

$$m = \frac{3.50 - 3}{5000 - 6000} = -0.0005$$

Therefore, the required equation is

$$p - 3 = -0.0005(x - 6000)$$
$$= -0.0005x + 3$$
$$p = -0.0005x + 6$$

Therefore, the revenue

$$R = f(x) = xp = -0.0005x^2 + 6x \qquad \text{Number of riders} \times \text{unit fare}$$

is the quantity to be maximized.

Step 3   Since both p and x must be nonnegative, we see that $0 \leq x \leq 12{,}000$, and the problem is that of finding the absolute maximum of the function f on the closed interval $[0, 12{,}000]$.

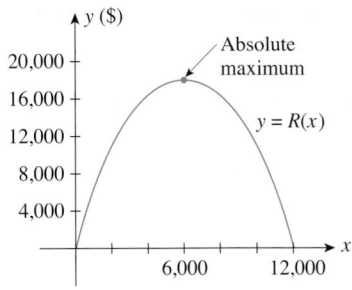

FIGURE 70
f has an absolute maximum of 18,000 when $x = 6000$.

Step 4 Observe that f is continuous on [0, 12,000]. To find the critical number of R, we compute

$$f'(x) = -0.001x + 6$$

and set it equal to zero, giving $x = 6000$. Evaluating the function f at $x = 6000$, as well as at the endpoints $x = 0$ and $x = 12,000$, yields

$$f(0) = 0$$
$$f(6000) = 18,000$$
$$f(12,000) = 0$$

We conclude that a maximum revenue of $18,000 per day is realized when the ridership is 6000 per day. The optimum price of the fare per ride is therefore $3.00, as recommended by the consultants. The graph of the revenue function R is shown in Figure 70. ∎

Minimization Problems

APPLIED EXAMPLE 4   Packaging Betty Moore Company requires that its corned beef hash containers have a capacity of 54 cubic inches, have the shape of right circular cylinders, and be made of aluminum. Determine the radius and height of the container that requires the least amount of metal.

Solution

Step 1 Let the radius and height of the container be r and h inches, respectively, and let S denote the surface area of the container (Figure 71).

Step 2 The amount of aluminum used to construct the container is given by the total surface area of the cylinder. Now, the area of the base and the top of the cylinder are each πr^2 square inches and the area of the side is $2\pi rh$ square inches. Therefore,

$$S = 2\pi r^2 + 2\pi rh \tag{3}$$

is the quantity to be minimized.

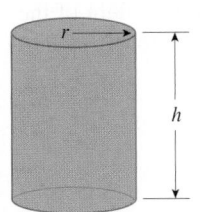

FIGURE 71
We want to minimize the amount of material used to construct the container.

Step 3 The requirement that the volume of a container be 54 cubic inches implies that

$$\pi r^2 h = 54 \tag{4}$$

Solving Equation (4) for h, we obtain

$$h = \frac{54}{\pi r^2} \tag{5}$$

which, when substituted into (3), yields

$$S = 2\pi r^2 + 2\pi r \left(\frac{54}{\pi r^2} \right)$$
$$= 2\pi r^2 + \frac{108}{r}$$

Clearly, the radius r of the container must satisfy the inequality $r > 0$. The problem now is reduced to finding the absolute minimum of the function $S = f(r)$ on the interval $(0, \infty)$.

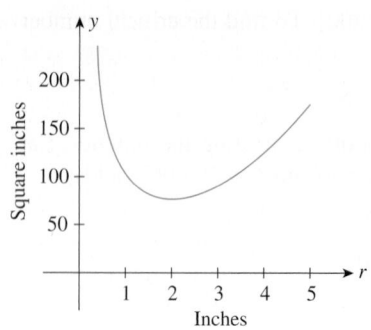

FIGURE 72
The total surface area of the right cylindrical container is graphed as a function of r.

Step 4   Using the curve-sketching techniques of Section 13.3, we obtain the graph of f in Figure 72.

To find the critical number of f, we compute

$$S' = 4\pi r - \frac{108}{r^2}$$

and solve the equation $S' = 0$ for r:

$$4\pi r - \frac{108}{r^2} = 0$$

$$4\pi r^3 - 108 = 0$$

$$r^3 = \frac{27}{\pi}$$

$$r = \frac{3}{\sqrt[3]{\pi}} \approx 2 \tag{6}$$

Next, let's show that this value of r gives rise to the absolute minimum of f. To show this, we first compute

$$S'' = 4\pi + \frac{216}{r^3}$$

Since $S'' > 0$ for $r = 3/\sqrt[3]{\pi}$, the second derivative test implies that the value of r in Equation (6) gives rise to a relative minimum of f. Finally, this relative minimum of f is also the absolute minimum of f since f is always concave upward ($S'' > 0$ for all $r > 0$). To find the height of the given container, we substitute the value of r given in (6) into (5). Thus,

$$h = \frac{54}{\pi r^2} = \frac{54}{\pi \left(\dfrac{3}{\pi^{1/3}} \right)^2}$$

$$= \frac{54\pi^{2/3}}{(\pi)9}$$

$$= \frac{6}{\pi^{1/3}} = \frac{6}{\sqrt[3]{\pi}}$$

$$= 2r$$

We conclude that the required container has a radius of approximately 2 inches and a height of approximately 4 inches, or twice the size of the radius.   ■

■ An Inventory Problem

One problem faced by many companies is that of controlling the inventory of goods carried. Ideally, the manager must ensure that the company has sufficient stock to meet customer demand at all times. At the same time, she must make sure that this is accomplished without overstocking (incurring unnecessary storage costs) and also without having to place orders too frequently (incurring reordering costs).

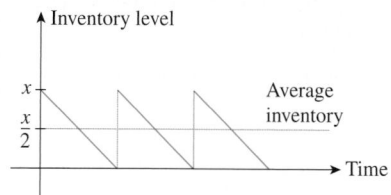

FIGURE 73
As each lot is depleted, the new lot arrives. The average inventory level is $x/2$ if x is the lot size.

APPLIED EXAMPLE 5   Inventory Control and Planning Dixie Import-Export is the sole agent for the Excalibur 250-cc motorcycle. Management estimates that the demand for these motorcycles is 10,000 per year and that they will sell at a uniform rate throughout the year. The cost incurred in ordering each shipment of motorcycles is $10,000, and the cost per year of storing each motorcycle is $200.

Dixie's management faces the following problem: Ordering too many motorcycles at one time ties up valuable storage space and increases the storage cost. On the other hand, placing orders too frequently increases the ordering costs. How large should each order be, and how often should orders be placed, to minimize ordering and storage costs?

Solution Let x denote the number of motorcycles in each order (the lot size). Then, assuming that each shipment arrives just as the previous shipment has been sold, the average number of motorcycles in storage during the year is $x/2$. You can see that this is the case by examining Figure 73. Thus, Dixie's storage cost for the year is given by $200(x/2)$, or $100x$ dollars.

Next, since the company requires 10,000 motorcycles for the year and since each order is for x motorcycles, the number of orders required is

$$\frac{10,000}{x}$$

This gives an ordering cost of

$$10,000\left(\frac{10,000}{x}\right) = \frac{100,000,000}{x}$$

dollars for the year. Thus, the total yearly cost incurred by Dixie, which includes the ordering and storage costs attributed to the sale of these motorcycles, is given by

$$C(x) = 100x + \frac{100,000,000}{x}$$

The problem is reduced to finding the absolute minimum of the function C in the interval $(0, 10,000]$. To accomplish this, we compute

$$C'(x) = 100 - \frac{100,000,000}{x^2}$$

Setting $C'(x) = 0$ and solving the resulting equation, we obtain $x = \pm 1000$. Since the number -1000 is outside the domain of the function C, it is rejected, leaving 1000 as the only critical number of C. Next, we find

$$C''(x) = \frac{200,000,000}{x^3}$$

Since $C''(1000) > 0$, the second derivative test implies that the critical number 1000 is a relative minimum of the function C (Figure 74). Also, since $C''(x) > 0$ for all x in $(0, 10,000]$, the function C is concave upward everywhere so that $x = 1000$ also gives the absolute minimum of C. Thus, to minimize the ordering and storage costs, Dixie should place 10,000/1000, or 10, orders a year, each for a shipment of 1000 motorcycles. ∎

FIGURE 74
C has an absolute minimum at (1000, 200,000).

13.5   Self-Check Exercises

1. A man wishes to have an enclosed vegetable garden in his backyard. If the garden is to be a rectangular area of 300 ft², find the dimensions of the garden that will minimize the amount of fencing material needed.

2. The demand for the Super Titan tires is 1,000,000/year. The setup cost for each production run is $4000, and the manufac-

turing cost is $20/tire. The cost of storing each tire over the year is $2. Assuming uniformity of demand throughout the year and instantaneous production, determine how many tires should be manufactured per production run in order to keep the production cost to a minimum.

Solutions to Self-Check Exercises 13.5 can be found on page 850.

13.5   Concept Questions

1. If the domain of a function f is not a closed interval, how would you find the absolute extrema of f, if they exist?

2. Refer to Example 4 (page 843). In the solution given in the example, we solved for h in terms of r, resulting in a function of r, which we then optimized with respect to r. Write S in terms of h and re-solve the problem. Which choice is better?

13.5   Exercises

1. **ENCLOSING THE LARGEST AREA** The owner of the Rancho Los Feliz has 3000 yd of fencing material with which to enclose a rectangular piece of grazing land along the straight portion of a river. If fencing is not required along the river, what are the dimensions of the largest area that he can enclose? What is this area?

2. **ENCLOSING THE LARGEST AREA** Refer to Exercise 1. As an alternative plan, the owner of the Rancho Los Feliz might use the 3000 yd of fencing material to enclose the rectangular piece of grazing land along the straight portion of the river and then subdivide it by means of a fence running parallel to the sides. Again, no fencing is required along the river. What are the dimensions of the largest area that can be enclosed? What is this area? (See the accompanying figure.)

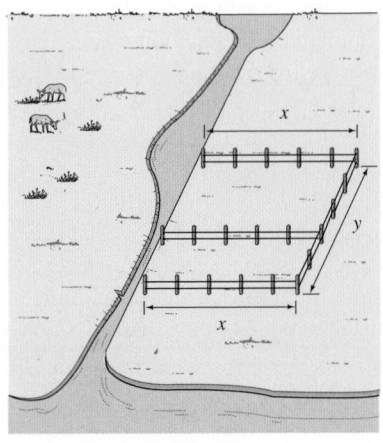

3. **MINIMIZING CONSTRUCTION COSTS** The management of the UNICO department store has decided to enclose an 800-ft² area outside the building for displaying potted plants and flowers. One side will be formed by the external wall of the store, two sides will be constructed of pine boards, and the fourth side will be made of galvanized steel fencing material. If the pine board fencing costs $6/running foot and the steel fencing costs $3/running foot, determine the dimensions of the enclosure that can be erected at minimum cost.

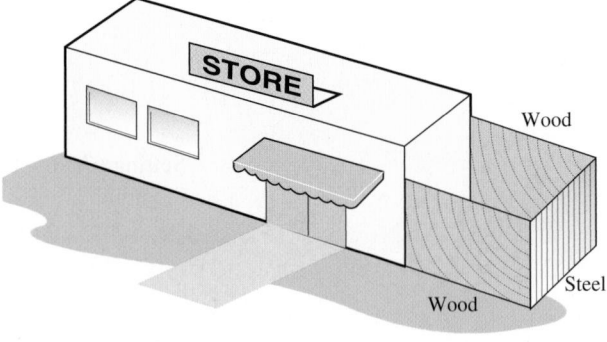

4. **PACKAGING** By cutting away identical squares from each corner of a rectangular piece of cardboard and folding up the resulting flaps, an open box may be made. If the cardboard is 15 in. long and 8 in. wide, find the dimensions of the box that will yield the maximum volume.

5. **METAL FABRICATION** If an open box is made from a tin sheet 8 in. square by cutting out identical squares from each corner

and bending up the resulting flaps, determine the dimensions of the largest box that can be made.

6. **MINIMIZING PACKAGING COSTS**   If an open box has a square base and a volume of 108 in.^{3} and is constructed from a tin sheet, find the dimensions of the box, assuming a minimum amount of material is used in its construction.

7. **MINIMIZING PACKAGING COSTS**   What are the dimensions of a closed rectangular box that has a square cross section, a capacity of 128 in.^{3}, and is constructed using the least amount of material?

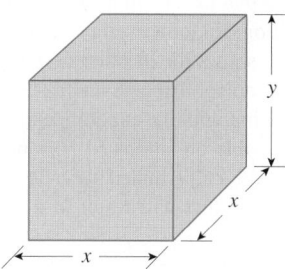

8. **MINIMIZING PACKAGING COSTS**   A rectangular box is to have a square base and a volume of 20 ft^3. If the material for the base costs 30¢/square foot, the material for the sides costs 10¢/square foot, and the material for the top costs 20¢/square foot, determine the dimensions of the box that can be constructed at minimum cost. (Refer to the figure for Exercise 7.)

9. **PARCEL POST REGULATIONS**   Postal regulations specify that a parcel sent by parcel post may have a combined length and girth of no more than 108 in. Find the dimensions of a rectangular package that has a square cross section and the largest volume that may be sent through the mail. What is the volume of such a package?
Hint: The length plus the girth is $4x + h$ (see the accompanying figure).

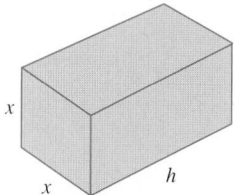

10. **BOOK DESIGN**   A book designer has decided that the pages of a book should have 1-in. margins at the top and bottom and $\frac{1}{2}$-in. margins on the sides. She further stipulated that each page should have an area of 50 in.^{2} (see the accompanying figure). Determine the page dimensions that will result in the maximum printed area on the page.

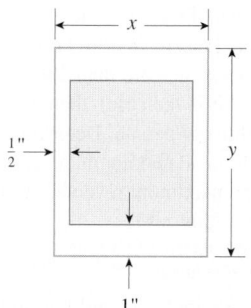

11. **PARCEL POST REGULATIONS**   Postal regulations specify that a parcel sent by parcel post may have a combined length and girth of no more than 108 in. Find the dimensions of the cylindrical package of greatest volume that may be sent through the mail. What is the volume of such a package? Compare with Exercise 9.
Hint: The length plus the girth is $2\pi r + l$.

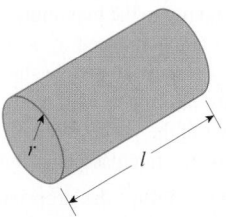

12. **MINIMIZING COSTS**   For its beef stew, Betty Moore Company uses aluminum containers that have the form of right circular cylinders. Find the radius and height of a container if it has a capacity of 36 in.^{3} and is constructed using the least amount of metal.

13. **PRODUCT DESIGN**   The cabinet that will enclose the Acrosonic model D loudspeaker system will be rectangular and will have an internal volume of 2.4 ft^3. For aesthetic reasons, it has been decided that the height of the cabinet is to be 1.5 times its width. If the top, bottom, and sides of the cabinet are constructed of veneer costing 40¢/square foot and the front (ignore the cutouts in the baffle) and rear are constructed of particle board costing 20¢/square foot, what are the dimensions of the enclosure that can be constructed at a minimum cost?

14. **DESIGNING A NORMAN WINDOW**   A Norman window has the shape of a rectangle surmounted by a semicircle (see the accompanying figure). If a Norman window is to have a perimeter of 28 ft, what should its dimensions be in order to allow the maximum amount of light through the window?

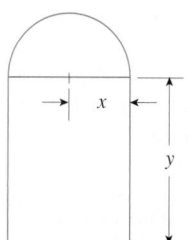

15. OPTIMAL CHARTER-FLIGHT FARE   If exactly 200 people sign up for a charter flight, Leisure World Travel Agency charges $300/person. However, if more than 200 people sign up for the flight (assume this is the case), then each fare is reduced by $1 for each additional person. Determine how many passengers will result in a maximum revenue for the travel agency. What is the maximum revenue? What would be the fare per passenger in this case?
Hint: Let x denote the number of passengers above 200. Show that the revenue function R is given by $R(x) = (200 + x)(300 - x)$.

16. MAXIMIZING YIELD   An apple orchard has an average yield of 36 bushels of apples/tree if tree density is 22 trees/acre. For each unit increase in tree density, the yield decreases by 2 bushels/tree. How many trees should be planted in order to maximize the yield?

17. CHARTER REVENUE   The owner of a luxury motor yacht that sails among the 4000 Greek islands charges $600/person/day if exactly 20 people sign up for the cruise. However, if more than 20 people sign up (up to the maximum capacity of 90) for the cruise, then each fare is reduced by $4 for each additional passenger. Assuming at least 20 people sign up for the cruise, determine how many passengers will result in the maximum revenue for the owner of the yacht. What is the maximum revenue? What would be the fare/passenger in this case?

18. PROFIT OF A VINEYARD   Phillip, the proprietor of a vineyard, estimates that the first 10,000 bottles of wine produced this season will fetch a profit of $5/bottle. However, the profit from each bottle beyond 10,000 drops by $0.0002 for each additional bottle sold. Assuming at least 10,000 bottles of wine are produced and sold, what is the maximum profit? What would be the price/bottle in this case?

19. STRENGTH OF A BEAM   A wooden beam has a rectangular cross section of height h in. and width w in. (see the accompanying figure). The strength S of the beam is directly proportional to its width and the square of its height. What are the dimensions of the cross section of the strongest beam that can be cut from a round log of diameter 24 in.?
Hint: $S = kh^2w$, where k is a constant of proportionality.

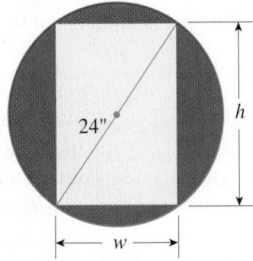

20. DESIGNING A GRAIN SILO   A grain silo has the shape of a right circular cylinder surmounted by a hemisphere (see the accompanying figure). If the silo is to have a capacity of 504π ft^3, find the radius and height of the silo that requires the least amount of material to construct.
Hint: The volume of the silo is $\pi r^2 h + \frac{2}{3}\pi r^3$, and the surface area (including the floor) is $\pi(3r^2 + 2rh)$.

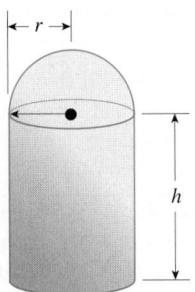

21. MINIMIZING COST OF LAYING CABLE   In the following diagram, S represents the position of a power relay station located on a straight coast, and E shows the location of a marine biology experimental station on an island. A cable is to be laid connecting the relay station with the experimental station. If the cost of running the cable on land is $1.50/running foot and the cost of running the cable under water is $2.50/running foot, locate the point P that will result in a minimum cost (solve for x).

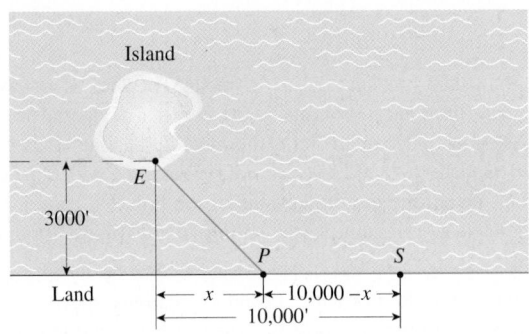

22. STORING RADIOACTIVE WASTE   A cylindrical container for storing radioactive waste is to be constructed from lead and have a thickness of 6 in. (see the accompanying figure). If the volume of the outside cylinder is to be 16π ft^3, find the radius and the height of the inside cylinder that will result in a container of maximum storage capacity.

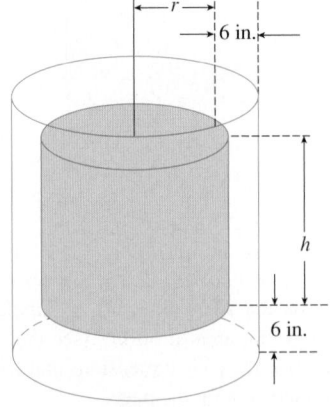

Hint: Show that the storage capacity (inside volume) is given by

$$V(r) = \pi r^2 \left[\frac{16}{(r + \frac{1}{2})^2} - 1 \right] \qquad \left(0 \le r \le \frac{7}{2} \right)$$

23. **FLIGHTS OF BIRDS** During daylight hours, some birds fly more slowly over water than over land because some of their energy is expended in overcoming the downdrafts of air over open bodies of water. Suppose a bird that flies at a constant speed of 4 mph over water and 6 mph over land starts its journey at the point E on an island and ends at its nest N on the shore of the mainland, as shown in the accompanying figure. Find the location of the point P that allows the bird to complete its journey in the minimum time (solve for x).

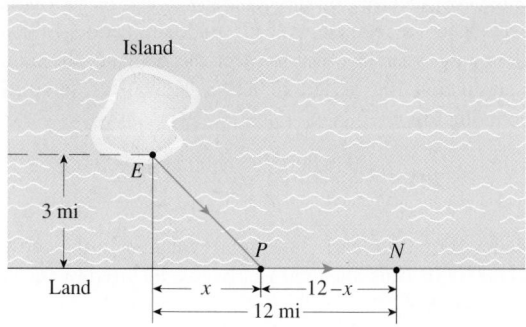

24. **OPTIMAL SPEED OF A TRUCK** A truck gets $600/x$ mpg when driven at a constant speed of x mph (between 50 and 70 mph). If the price of fuel is \$3/gallon and the driver is paid \$18/hour, at what speed between 50 and 70 mph is it most economical to drive?

25. **RACETRACK DESIGN** The accompanying figure depicts a racetrack with ends that are semicircular in shape. The length of the track is 1760 ft ($\frac{1}{3}$ mi). Find l and r so that the area enclosed by the rectangular region of the racetrack is as large as possible. What is the area enclosed by the track in this case?

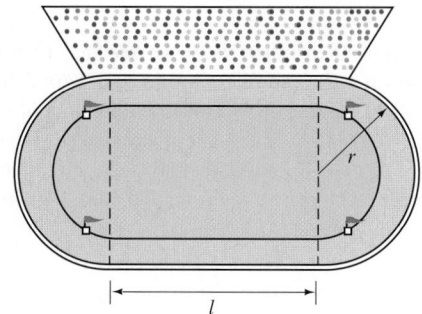

26. **INVENTORY CONTROL AND PLANNING** The demand for motorcycle tires imported by Dixie Import-Export is 40,000/year and may be assumed to be uniform throughout the year. The cost of ordering a shipment of tires is \$400, and the cost of storing each tire for a year is \$2. Determine how many tires should be in each shipment if the ordering and storage costs are to be minimized. (Assume that each shipment arrives just as the previous one has been sold.)

27. **INVENTORY CONTROL AND PLANNING** McDuff Preserves expects to bottle and sell 2,000,000 32-oz jars of jam. The company orders its containers from Consolidated Bottle Company. The cost of ordering a shipment of bottles is \$200, and the cost of storing each empty bottle for a year is \$.40. How many orders should McDuff place per year and how many bottles should be in each shipment if the ordering and storage costs are to be minimized? (Assume that each shipment of bottles is used up before the next shipment arrives.)

28. **INVENTORY CONTROL AND PLANNING** Neilsen Cookie Company sells its assorted butter cookies in containers that have a net content of 1 lb. The estimated demand for the cookies is 1,000,000 1-lb containers. The setup cost for each production run is \$500, and the manufacturing cost is \$.50 for each container of cookies. The cost of storing each container of cookies over the year is \$.40. Assuming uniformity of demand throughout the year and instantaneous production, how many containers of cookies should Neilsen produce per production run in order to minimize the production cost?
Hint: Following the method of Example 5, show that the total production cost is given by the function

$$C(x) = \frac{500,000,000}{x} + 0.2x + 500,000$$

Then minimize the function C on the interval $(0, 1,000,000)$.

29. **INVENTORY CONTROL AND PLANNING** A company expects to sell D units of a certain product per year. Sales are assumed to be at a steady rate with no shortages allowed. Each time an order for the product is placed, an ordering cost of K dollars is incurred. Each item costs p dollars, and the holding cost is h dollars per item per year.
   a. Show that the inventory cost (the combined ordering cost, purchasing cost, and holding cost) is

   $$C(x) = \frac{KD}{x} + pD + \frac{hx}{2} \qquad (x > 0)$$

   where x is the order quantity (the number of items in each order).
   b. Use the result of part (a) to show that the inventory cost is minimized if

   $$x = \sqrt{\frac{2KD}{h}}$$

   This quantity is called the *economic order quantity* (EOQ).

30. **INVENTORY CONTROL AND PLANNING** Refer to Exercise 29. The Camera Store sells 960 Yamaha A35 digital cameras per year. Each time an order for cameras is placed with the manufacturer, an ordering cost of \$10 is incurred. The store pays \$80 for each camera, and the cost for holding a camera (mainly due to the opportunity cost incurred in tying up caital in inventory) is \$12/yr. Assume that the cameras sel¹ uniform rate and no shortages are allowed.
   a. What is the EOQ?
   b. How many orders will be placed each year?
   c. What is the interval between orders?

13.5   Solutions to Self-Check Exercises

1. Let x and y (measured in feet) denote the length and width of the rectangular garden.

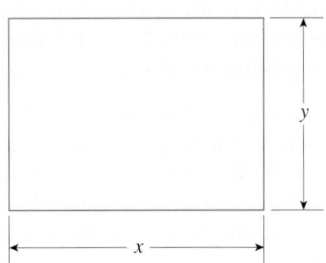

Since the area is to be 300 ft^2, we have

$$xy = 300$$

Next, the amount of fencing to be used is given by the perimeter, and this quantity is to be minimized. Thus, we want to minimize

$$2x + 2y$$

or, since $y = 300/x$ (obtained by solving for y in the first equation), we see that the expression to be minimized is

$$f(x) = 2x + 2\left(\frac{300}{x}\right)$$
$$= 2x + \frac{600}{x}$$

for positive values of x. Now

$$f'(x) = 2 - \frac{600}{x^2}$$

Setting $f'(x) = 0$ yields $x = -\sqrt{300}$ or $x = \sqrt{300}$. We consider only the critical number $\sqrt{300}$ since $-\sqrt{300}$ lies outside the interval $(0, \infty)$. We then compute

$$f''(x) = \frac{1200}{x^3}$$

Since

$$f''(300) > 0$$

the second derivative test implies that a relative minimum of f occurs at $x = \sqrt{300}$. In fact, since $f''(x) > 0$ for all x in $(0, \infty)$, we conclude that $x = \sqrt{300}$ gives rise to the absolute minimum of f. The corresponding value of y, obtained by substituting this value of x into the equation $xy = 300$, is $y = \sqrt{300}$. Therefore, the required dimensions of the vegetable garden are approximately 17.3 ft $\times$ 17.3 ft.

2. Let x denote the number of tires in each production run. Then, the average number of tires in storage is $x/2$, so the storage cost incurred by the company is $2(x/2)$, or x dollars. Next, since the company needs to manufacture 1,000,000 tires for the year in order to meet the demand, the number of production runs is $1,000,000/x$. This gives setup costs amounting to

$$4000\left(\frac{1,000,000}{x}\right) = \frac{4,000,000,000}{x}$$

dollars for the year. The total manufacturing cost is $20,000,000. Thus, the total yearly cost incurred by the company is given by

$$C(x) = x + \frac{4,000,000,000}{x} + 20,000,000$$

Differentiating $C(x)$, we find

$$C'(x) = 1 - \frac{4,000,000,000}{x^2}$$

Setting $C'(x) = 0$ gives 63,246 as the critical number in the interval (0, 1,000,000). Next, we find

$$C''(x) = \frac{8,000,000,000}{x^3}$$

Since $C''(x) > 0$ for all $x > 0$, we see that C is concave upward for all $x > 0$. Furthermore, $C''(63,246) > 0$ implies that $x = 63,246$ gives rise to a relative minimum of C (by the second derivative test). Since C is always concave upward for $x > 0$, $x = 63,246$ gives the absolute minimum of C. Therefore, the company should manufacture 63,246 tires in each production run.

Summary of Principal Terms

TERMS

increasing function (772)	first derivative test (779)	horizontal asymptote (812)
decreasing function (772)	concave upward (792)	absolute extrema (824)
relative maximum (777)	concave downward (792)	absolute maximum value (824)
relative minimum (777)	inflection point (794)	absolute minimum value (824)
relative extrema (777)	second derivative test (798)	
critical number (778)	vertical asymptote (811)	

Concept Review Questions

Fill in the blanks.

1. a. A function f is increasing on an interval I, if for any two numbers x_1 and x_2 in I, $x_1 < x_2$ implies that _____.
b. A function f is decreasing on an interval I, if for any two numbers x_1 and x_2 in I, $x_1 < x_2$ implies that _____.

2. a. If f is differentiable on an open interval (a, b) and $f'(x) > 0$ on (a, b), then f is _____ on (a, b).
b. If f is differentiable on an open interval (a, b) and _____ on (a, b), then f is decreasing on (a, b).
c. If $f'(x) = 0$ for each value of x in the interval (a, b), then f is _____ on (a, b).

3. a. A function f has a relative maximum at c if there exists an open interval (a, b) containing c such that _____ for all x in (a, b).
b. A function f has a relative minimum at c if there exists an open interval (a, b) containing c such that _____ for all x in (a, b).

4. a. A critical number of a function f is any number in the _____ of f at which $f'(c)$ _____ or $f'(c)$ does *not* _____.
b. If f has a relative extremum at c, then c must be a _____ _____ of f.
c. If c is a critical number of f, then f may or may not have a _____ _____ at c.

5. a. A differentiable function f is concave upward on an interval I if _____ is increasing on I.
b. If f has a second derivative on an open interval I and $f''(x)$ _____ on I, then the graph of f is concave upward on I.

c. If the graph of a continuous function f has a tangent line at $P(c, f(c))$ and the graph of f changes _____ at P, then P is called an inflection point of the graph of f.
d. Suppose f has a continuous second derivative on an interval (a, b), containing a critical number c of f. If $f''(c) < 0$, then f has a _____ _____ at c. If $f''(c) = 0$, then f may or may not have a _____ _____ at c.

6. The line $x = a$ is a vertical asymptote of the graph f if at least one of the following is true: $\lim_{x \to a^+} f(x) = $ _____ or $\lim_{x \to a^-} f(x) = $ _____.

7. For a rational function $f(x) = \dfrac{P(x)}{Q(x)}$, the line $x = a$ is a vertical asymptote of the graph of f if $Q(a) = $ _____ but $P(a) \neq $ _____.

8. The line $y = b$ is a horizontal asymptote of the graph of a function f if either $\lim_{x \to \infty} f(x) = $ _____ or $\lim_{x \to -\infty} f(x) = $ _____.

9. a. A function f has an absolute maximum at c if _____ for all x in the domain D of f. The number $f(c)$ is called the _____ _____ of f on D.
b. A function f has a relative minimum at c if _____ for all values of x in some _____ _____ containing c.

10. The extreme value theorem states that if f is _____ on the closed interval $[a, b]$, then f has both an _____ maximum value and an _____ minimum value on $[a, b]$.

CHAPTER 13 | Review Exercises

In Exercises 1–10, (a) find the intervals where the function f is increasing and where it is decreasing, (b) find the relative extrema of f, (c) find the intervals where f is concave upward and where it is concave downward, and (d) find the inflection points, if any, of f.

1. $f(x) = \dfrac{1}{3}x^3 - x^2 + x - 6$

2. $f(x) = (x - 2)^3$

3. $f(x) = x^4 - 2x^2$

4. $f(x) = x + \dfrac{4}{x}$

5. $f(x) = \dfrac{x^2}{x - 1}$

6. $f(x) = \sqrt{x - 1}$

7. $f(x) = (1 - x)^{1/3}$

8. $f(x) = x\sqrt{x - 1}$

9. $f(x) = \dfrac{2x}{x + 1}$

10. $f(x) = \dfrac{-1}{1 + x^2}$

In Exercises 11–18, obtain as much information as possible on each function. Then use this information to sketch the graph of the function.

11. $f(x) = x^2 - 5x + 5$

12. $f(x) = -2x^2 - x + 1$

13. $g(x) = 2x^3 - 6x^2 + 6x + 1$

14. $g(x) = \dfrac{1}{3}x^3 - x^2 + x - 3$

15. $h(x) = x\sqrt{x - 2}$

16. $h(x) = \dfrac{2x}{1 + x^2}$

17. $f(x) = \dfrac{x - 2}{x + 2}$

18. $f(x) = x - \dfrac{1}{x}$

In Exercises 19–22, find the horizontal and vertical asymptotes of the graph of each function. Do not sketch the graph.

19. $f(x) = \dfrac{1}{2x + 3}$

20. $f(x) = \dfrac{2x}{x + 1}$

21. $f(x) = \dfrac{5x}{x^2 - 2x - 8}$

22. $f(x) = \dfrac{x^2 + x}{x(x - 1)}$

In Exercises 23–32, find the absolute maximum value and the absolute minimum value, if any, of the function.

23. $f(x) = 2x^2 + 3x - 2$

24. $g(x) = x^{2/3}$

25. $g(t) = \sqrt{25 - t^2}$

26. $f(x) = \dfrac{1}{3}x^3 - x^2 + x + 1$ on $[0, 2]$

27. $h(t) = t^3 - 6t^2$ on $[2, 5]$

28. $g(x) = \dfrac{x}{x^2 + 1}$ on $[0, 5]$

29. $f(x) = x - \dfrac{1}{x}$ on $[1, 3]$

30. $h(t) = 8t - \dfrac{1}{t^2}$ on $[1, 3]$

31. $f(s) = s\sqrt{1 - s^2}$ on $[-1, 1]$

32. $f(x) = \dfrac{x^2}{x - 1}$ on $[-1, 3]$

33. **MAXIMIZING PROFITS** Odyssey Travel Agency's monthly profit (in thousands of dollars) depends on the amount of money x (in thousands of dollars) spent on advertising each month according to the rule

$$P(x) = -x^2 + 8x + 20$$

To maximize its monthly profits, what should be Odyssey's monthly advertising budget?

34. **ONLINE HOTEL RESERVATIONS** The online lodging industry is expected to grow dramatically. In a study conducted in 1999, analysts projected the U.S. online travel spending for lodging to be approximately

$$f(t) = 0.157t^2 + 1.175t + 2.03 \qquad (0 \le t \le 6)$$

billion dollars, where t is measured in years, with $t = 0$ corresponding to 1999.
a. Show that f is increasing on the interval $(0, 6)$.
b. Show that the graph of f is concave upward on $(0, 6)$.
c. What do your results from parts (a) and (b) tell you about the growth of online travel spending over the years in question?
Source: International Data Corp.

35. **SALES OF CAMERA PHONES** Camera phones, virtually nonexistent a few years ago, are quickly gaining in popularity. The function

$$N(t) = 8.125t^2 + 24.625t + 18.375 \qquad (0 \le t \le 3)$$

gives the projected worldwide shipments of camera phones (in millions of units) in year t, with $t = 0$ corresponding to 2002.
a. Find $N'(t)$. What does this say about the sales of camera phones between 2002 and 2005?
b. Find $N''(t)$. What does this say about the rate of the rate of sales of camera phones between 2002 and 2005?
Source: In-Stat/MDR

36. **EFFECT OF ADVERTISING ON SALES** The total sales S of Cannon Precision Instruments is related to the amount of money x that Cannon spends on advertising its products by the function

$$S(x) = -0.002x^3 + 0.6x^2 + x + 500 \qquad (0 \le x \le 200)$$

where S and x are measured in thousands of dollars. Find the inflection point of the function S and discuss its significance.

37. **COST OF PRODUCING CALCULATORS** A subsidiary of Elektra Electronics manufactures graphing calculators. Management determines that the daily cost $C(x)$ (in dollars) of producing these calculators is

$$C(x) = 0.0001x^3 - 0.08x^2 + 40x + 5000$$

where x is the number of calculators produced. Find the inflection point of the function C and interpret your result.

38. **INDEX OF ENVIRONMENTAL QUALITY** The Department of the Interior of an African country began to record an index of environmental quality to measure progress or decline in the environmental quality of its wildlife. The index for the years 1984 through 1994 is approximated by the function

$$I(t) = \frac{50t^2 + 600}{t^2 + 10} \qquad (0 \le t \le 10)$$

a. Compute $I'(t)$ and show that $I(t)$ is decreasing on the interval $(0, 10)$.
b. Compute $I''(t)$. Study the concavity of the graph of I.
c. Sketch the graph of I.
d. Interpret your results.

39. **MAXIMIZING PROFITS** The weekly demand for DVDs manufactured by Herald Media Corporation is given by

$$p = -0.0005x^2 + 60$$

where p denotes the unit price in dollars and x denotes the quantity demanded. The weekly total cost function associated with producing these discs is given by

$$C(x) = -0.001x^2 + 18x + 4000$$

where $C(x)$ denotes the total cost (in dollars) incurred in pressing x discs. Find the production level that will yield a maximum profit for the manufacturer.
Hint: Use the quadratic formula.

40. **MAXIMIZING PROFITS** The estimated monthly profit (in dollars) realizable by Cannon Precision Instruments for manufacturing and selling x units of its model M1 digital camera is

$$P(x) = -0.04x^2 + 240x - 10{,}000$$

To maximize its profits, how many cameras should Cannon produce each month?

41. **MINIMIZING AVERAGE COST** The total monthly cost (in dollars) incurred by Carlota Music in manufacturing x units of its Professional Series guitars is given by the function

$$C(x) = 0.001x^2 + 100x + 4000$$

a. Find the average cost function $\overline{C}$.
b. Determine the production level that will result in the smallest average production cost.

42. **WORKER EFFICIENCY** The average worker at Wakefield Avionics can assemble

$$N(t) = -2t^3 + 12t^2 + 2t \qquad (0 \le t \le 4)$$

ready-to-fly radio-controlled model airplanes t hr into the 8 a.m. to 12 noon morning shift. At what time during this shift is the average worker performing at peak efficiency?

43. **SENIOR WORKFORCE** The percent of women 65 years and older in the workforce from 1970 through the year 2000 is approximated by the function

$$P(t) = -0.0002t^3 + 0.018t^2 - 0.36t + 10 \qquad (0 \le t \le 30)$$

where t is measured in years, with $t = 0$ corresponding to the beginning of 1970.
a. Find the interval where P is decreasing and the interval where P is increasing.
b. Find the absolute minimum of P.
c. Interpret the results of parts (a) and (b).
Source: U.S. Census Bureau

44. **SPREAD OF A CONTAGIOUS DISEASE** The incidence (number of new cases/day) of a contagious disease spreading in a population of M people is given by

$$R(x) = kx(M - x)$$

where k is a positive constant and x denotes the number of people already infected. Show that the incidence R is greatest when half the population is infected.

45. **MAXIMIZING THE VOLUME OF A BOX** A box with an open top is to be constructed from a square piece of cardboard, 10 in. wide, by cutting out a square from each of the four corners and bending up the sides. What is the maximum volume of such a box?

46. **MINIMIZING CONSTRUCTION COSTS** A man wishes to construct a cylindrical barrel with a capacity of 32π ft^3. The cost/square foot of the material for the side of the barrel is half that of the cost/square foot for the top and bottom. Help him find the dimensions of the barrel that can be constructed at a minimum cost in terms of material used.

47. **PACKAGING** You wish to construct a closed rectangular box that has a volume of 4 ft^3. The length of the base of the box will be twice as long as its width. The material for the top

and bottom of the box costs 30¢/square foot. The material for the sides of the box costs 20¢/square foot. Find the dimensions of the least expensive box that can be constructed.

48. **INVENTORY CONTROL AND PLANNING** Lehen Vinters imports a certain brand of beer. The demand, which may be assumed to be uniform, is 800,000 cases/year. The cost of ordering a shipment of beer is $500, and the cost of storing each case of beer for a year is $2. Determine how many cases of beer should be in each shipment if the ordering and storage costs are to be kept at a minimum. (Assume that each shipment of beer arrives just as the previous one has been sold.)

49. In what interval is the quadratic function

$$f(x) = ax^2 + bx + c \qquad (a \neq 0)$$

increasing? In what interval is f decreasing?

50. Let

$$f(x) = \begin{cases} x^3 + 1 & \text{if } x \neq 0 \\ 2 & \text{if } x = 0 \end{cases}$$

a. Compute $f'(x)$ and show that it does not change sign as we move across $x = 0$.

b. Show that f has a relative maximum at $x = 0$. Does this contradict the first derivative test? Explain your answer.

CHAPTER 13   **Before Moving On . . .**

1. Find the interval(s) where $f(x) = \dfrac{x^2}{1 - x}$ is increasing and where it is decreasing.

2. Find the relative maxima and relative minima, if any, of $f(x) = 2x^2 - 12x^{1/3}$.

3. Find the intervals where $f(x) = \frac{1}{3}x^3 - \frac{1}{4}x^2 - \frac{1}{2}x + 1$ is concave upward, the intervals where f is concave downward, and the inflection point(s) of f.

4. Sketch the graph of $f(x) = 2x^3 - 9x^2 + 12x - 1$.

5. Find the absolute maximum and absolute minimum values of $f(x) = 2x^3 + 3x^2 - 1$ on the interval $[-2, 3]$.

6. An open bucket in the form of a right circular cylinder is to be constructed with a capacity of 1 ft³. Find the radius and height of the cylinder if the amount of material used is minimal.

14 Exponential and Logarithmic Functions

© Andrew Brookes/Corbis

How many bacteria will there be in a culture at the end of a certain period of time? How fast will the bacteria population be growing at the end of that time? Example 1, page 894, answers these questions.

THE EXPONENTIAL FUNCTION is, without doubt, the most important function in mathematics and its applications. After a brief introduction to the exponential function and its *inverse*, the logarithmic function, we learn how to differentiate such functions. This lays the foundation for exploring the many applications involving exponential functions. For example, we look at the role played by exponential functions in studying the growth of a bacteria population in the laboratory, the way radioactive matter decays, the rate at which a factory worker learns a certain process, and the rate at which a communicable disease is spread over time.

14.1 Exponential Functions

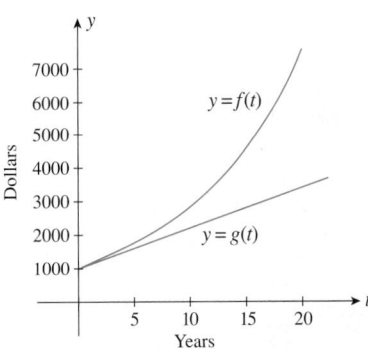

FIGURE 1
Under continuous compounding, a sum of money grows exponentially.

■ Exponential Functions and Their Graphs

Suppose you deposit a sum of $1000 in an account earning interest at the rate of 10% per year *compounded continuously* (the way most financial institutions compute interest). Then, the accumulated amount at the end of t years ($0 \leq t \leq 20$) is described by the function f, whose graph appears in Figure 1. This function is called an *exponential function*. Observe that the graph of f rises rather slowly at first but very rapidly as time goes by. For purposes of comparison, we have also shown the graph of the function $y = g(t) = 1000(1 + 0.10t)$, giving the accumulated amount for the same principal ($1000) but earning *simple* interest at the rate of 10% per year. The moral of the story: It is never too early to save.

Exponential functions play an important role in many real-world applications, as you will see throughout this chapter.

Observe that whenever b is a positive number and n is any real number, the expression b^n is a real number. This enables us to define an exponential function as follows:

Exponential Function
The function defined by

$$f(x) = b^x \qquad (b > 0, b \neq 1)$$

is called an **exponential function with base b and exponent x**. The domain of f is the set of all real numbers.

For example, the exponential function with base 2 is the function

$$f(x) = 2^x$$

with domain $(-\infty, \infty)$. The values of $f(x)$ for selected values of x follow:

$$f(3) = 2^3 = 8 \qquad f\left(\frac{3}{2}\right) = 2^{3/2} = 2 \cdot 2^{1/2} = 2\sqrt{2} \qquad f(0) = 2^0 = 1$$

$$f(-1) = 2^{-1} = \frac{1}{2} \qquad f\left(-\frac{2}{3}\right) = 2^{-2/3} = \frac{1}{2^{2/3}} = \frac{1}{\sqrt[3]{4}}$$

Computations involving exponentials are facilitated by the laws of exponents. These laws were stated in Section 10.1, and you might want to review the material there. For convenience, however, we will restate these laws.

Laws of Exponents
Let a and b be positive numbers and let x and y be real numbers. Then,

1. $b^x \cdot b^y = b^{x+y}$ 4. $(ab)^x = a^x b^x$

2. $\dfrac{b^x}{b^y} = b^{x-y}$ 5. $\left(\dfrac{a}{b}\right)^x = \dfrac{a^x}{b^x}$

3. $(b^x)^y = b^{xy}$

The use of the laws of exponents is illustrated in the next example.

EXAMPLE 1

a. $16^{7/4} \cdot 16^{-1/2} = 16^{7/4-1/2} = 16^{5/4} = 2^5 = 32$ Law 1

b. $\dfrac{8^{5/3}}{8^{-1/3}} = 8^{5/3-(-1/3)} = 8^2 = 64$ Law 2

c. $(64^{4/3})^{-1/2} = 64^{(4/3)(-1/2)} = 64^{-2/3}$

$$= \frac{1}{64^{2/3}} = \frac{1}{(64^{1/3})^2} = \frac{1}{4^2} = \frac{1}{16}$$ Law 3

d. $(16 \cdot 81)^{-1/4} = 16^{-1/4} \cdot 81^{-1/4} = \dfrac{1}{16^{1/4}} \cdot \dfrac{1}{81^{1/4}} = \dfrac{1}{2} \cdot \dfrac{1}{3} = \dfrac{1}{6}$ Law 4

e. $\left(\dfrac{3^{1/2}}{2^{1/3}}\right)^4 = \dfrac{3^{4/2}}{2^{4/3}} = \dfrac{9}{2^{4/3}}$ Law 5

EXAMPLE 2   Let $f(x) = 2^{2x-1}$. Find the value of x for which $f(x) = 16$.

Solution   We want to solve the equation

$$2^{2x-1} = 16 = 2^4$$

But this equation holds if and only if

$$2x - 1 = 4 \qquad b^m = b^n \Rightarrow m = n$$

giving $x = \dfrac{5}{2}$.

Exponential functions play an important role in mathematical analysis. Because of their special characteristics, they are some of the most useful functions and are found in virtually every field where mathematics is applied. To mention a few examples: Under ideal conditions the number of bacteria present at any time t in a culture may be described by an exponential function of t; radioactive substances decay over time in accordance with an "exponential" law of decay; money left on fixed deposit and earning compound interest grows exponentially; and some of the most important distribution functions encountered in statistics are exponential.

Let's begin our investigation into the properties of exponential functions by studying their graphs.

EXAMPLE 3   Sketch the graph of the exponential function $y = 2^x$.

Solution   First, as discussed earlier, the domain of the exponential function $y = f(x) = 2^x$ is the set of real numbers. Next, putting $x = 0$ gives $y = 2^0 = 1$, the y-intercept of f. There is no x-intercept since there is no value of x for which $y = 0$. To find the range of f, consider the following table of values:

x	-5	-4	-3	-2	-1	0	1	2	3	4	5
y	$\frac{1}{32}$	$\frac{1}{16}$	$\frac{1}{8}$	$\frac{1}{4}$	$\frac{1}{2}$	1	2	4	8	16	32

We see from these computations that 2^x decreases and approaches zero as x decreases without bound and that 2^x increases without bound as x increases without bound. Thus, the range of f is the interval $(0, \infty)$—that is, the set of positive real numbers. Finally, we sketch the graph of $y = f(x) = 2^x$ in Figure 2.

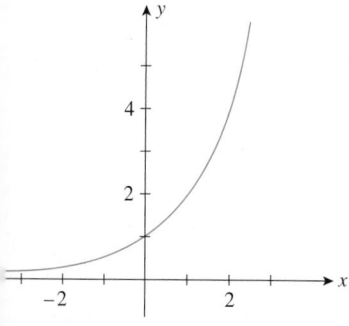

FIGURE 2
The graph of $y = 2^x$

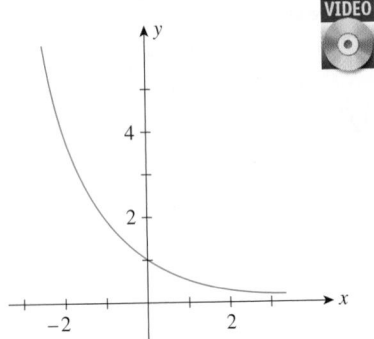

FIGURE 3
The graph of $y = \left(\dfrac{1}{2}\right)^x$

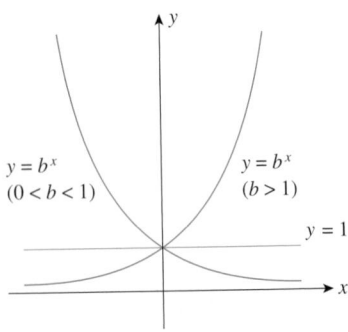

FIGURE 4
$y = b^x$ is an increasing function of x if $b > 1$, a constant function if $b = 1$, and a decreasing function if $0 < b < 1$.

m	$\left(1 + \dfrac{1}{m}\right)^m$
10	2.59374
100	2.70481
1000	2.71692
10,000	2.71815
100,000	2.71827
1,000,000	2.71828

TABLE 1

EXAMPLE 4 Sketch the graph of the exponential function $y = (1/2)^x$.

Solution The domain of the exponential function $y = (1/2)^x$ is the set of all real numbers. The y-intercept is $(1/2)^0 = 1$; there is no x-intercept since there is no value of x for which $y = 0$. From the following table of values

x	-5	-4	-3	-2	-1	0	1	2	3	4	5
y	32	16	8	4	2	1	$\dfrac{1}{2}$	$\dfrac{1}{4}$	$\dfrac{1}{8}$	$\dfrac{1}{16}$	$\dfrac{1}{32}$

we deduce that $(1/2)^x = 1/2^x$ increases without bound as x decreases without bound and that $(1/2)^x$ decreases and approaches zero as x increases without bound. Thus, the range of f is the interval $(0, \infty)$. The graph of $y = f(x) = (1/2)^x$ is sketched in Figure 3. ∎

The functions $y = 2^x$ and $y = (1/2)^x$, whose graphs you studied in Examples 3 and 4, are special cases of the exponential function $y = f(x) = b^x$, obtained by setting $b = 2$ and $b = 1/2$, respectively. In general, the exponential function $y = b^x$ with $b > 1$ has a graph similar to $y = 2^x$, whereas the graph of $y = b^x$ for $0 < b < 1$ is similar to that of $y = (1/2)^x$ (Exercises 27 and 28 on page 860). When $b = 1$, the function $y = b^x$ reduces to the constant function $y = 1$. For comparison, the graphs of all three functions are sketched in Figure 4.

Properties of the Exponential Function
The exponential function $y = b^x$ ($b > 0, b \neq 1$) has the following properties:

1. Its domain is $(-\infty, \infty)$.
2. Its range is $(0, \infty)$.
3. Its graph passes through the point $(0, 1)$.
4. It is continuous on $(-\infty, \infty)$.
5. It is increasing on $(-\infty, \infty)$ if $b > 1$ and decreasing on $(-\infty, \infty)$ if $b < 1$.

The Base e

Exponential functions to the base e, where e is an irrational number whose value is 2.7182818. . . . , play an important role in both theoretical and applied problems. It can be shown, although we will not do so here, that

$$e = \lim_{m \to \infty} \left(1 + \frac{1}{m}\right)^m \tag{1}$$

However, you may convince yourself of the plausibility of this definition of the number e by examining Table 1, which may be constructed with the help of a calculator.

EXPLORING WITH TECHNOLOGY

To obtain a visual confirmation of the fact that the expression $(1 + 1/m)^m$ approaches the number $e = 2.71828 \ldots$ as m increases without bound, plot the graph of $f(x) = (1 + 1/x)^x$ in a suitable viewing window and observe that $f(x)$ approaches $2.71828 \ldots$ as x increases without bound. Use ZOOM and TRACE to find the value of $f(x)$ for large values of x.

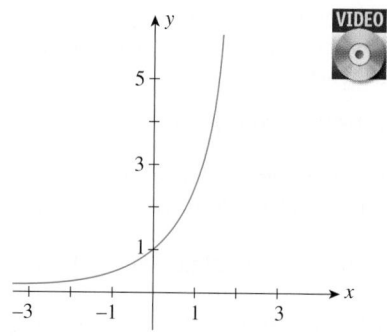

FIGURE 5
The graph of $y = e^x$

EXAMPLE 5   Sketch the graph of the function $y = e^x$.

Solution   Since $e > 1$, it follows from our previous discussion that the graph of $y = e^x$ is similar to the graph of $y = 2^x$ (see Figure 2). With the aid of a calculator, we obtain the following table:

x	-3	-2	-1	0	1	2	3
y	0.05	0.14	0.37	1	2.72	7.39	20.09

The graph of $y = e^x$ is sketched in Figure 5.   ∎

Next, we consider another exponential function to the base e that is closely related to the previous function and is particularly useful in constructing models that describe "exponential decay."

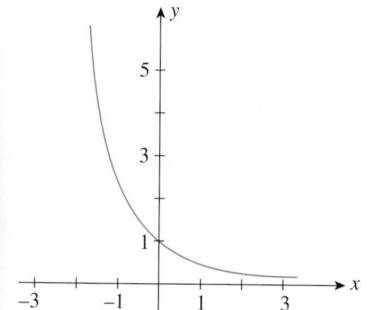

FIGURE 6
The graph of $y = e^{-x}$

EXAMPLE 6   Sketch the graph of the function $y = e^{-x}$.

Solution   Since $e > 1$, it follows that $0 < 1/e < 1$, so $f(x) = e^{-x} = 1/e^x = (1/e)^x$ is an exponential function with base less than 1. Therefore, it has a graph similar to that of the exponential function $y = (1/2)^x$. As before, we construct the following table of values of $y = e^{-x}$ for selected values of x:

x	-3	-2	-1	0	1	2	3
y	20.09	7.39	2.72	1	0.37	0.14	0.05

Using this table, we sketch the graph of $y = e^{-x}$ in Figure 6.   ∎

14.1   Self-Check Exercises

1. Solve the equation $2^{2x+1} \cdot 2^{-3} = 2^{x-1}$.

2. Sketch the graph of $y = e^{0.4x}$.

Solutions to Self-Check Exercises 14.1 can be found on page 861.

14.1   Concept Questions

1. Define the exponential function f with base b and exponent x. What restrictions, if any, are placed on b?

2. For the exponential function $y = b^x$ ($b > 0$, $b \neq 1$), state (a) its domain and range, (b) its y-intercept, (c) where it is continuous, and (d) where it is increasing and where it is decreasing for the case $b > 1$ and the case $b < 1$.

14.1   Exercises

In Exercises 1–8, evaluate the expression.

1. a. $4^{-3} \cdot 4^5$ **b.** $3^{-3} \cdot 3^6$

2. a. $(2^{-1})^3$ **b.** $(3^{-2})^3$

3. a. $9(9)^{-1/2}$ **b.** $5(5)^{-1/2}$

4. a. $\left[\left(-\dfrac{1}{2} \right)^3 \right]^{-2}$ **b.** $\left[\left(-\dfrac{1}{3} \right)^2 \right]^{-3}$

5. a. $\dfrac{(-3)^4(-3)^5}{(-3)^8}$　　**b.** $\dfrac{(2^{-4})(2^6)}{2^{-1}}$

6. a. $3^{1/4} \cdot 9^{-5/8}$　　**b.** $2^{3/4} \cdot 4^{-3/2}$

7. a. $\dfrac{5^{3.3} \cdot 5^{-1.6}}{5^{-0.3}}$　　**b.** $\dfrac{4^{2.7} \cdot 4^{-1.3}}{4^{-0.4}}$

8. a. $\left(\dfrac{1}{16}\right)^{-1/4}\left(\dfrac{27}{64}\right)^{-1/3}$　　**b.** $\left(\dfrac{8}{27}\right)^{-1/3}\left(\dfrac{81}{256}\right)^{-1/4}$

In Exercises 9–16, simplify the expression.

9. a. $(64x^9)^{1/3}$　　**b.** $(25x^3y^4)^{1/2}$

10. a. $(2x^3)(-4x^{-2})$　　**b.** $(4x^{-2})(-3x^5)$

11. a. $\dfrac{6a^{-5}}{3a^{-3}}$　　**b.** $\dfrac{4b^{-4}}{12b^{-6}}$

12. a. $y^{-3/2}y^{5/3}$　　**b.** $x^{-3/5}x^{8/3}$

13. a. $(2x^3y^2)^3$　　**b.** $(4x^2y^2z^3)^2$

14. a. $(x^{r/s})^{s/r}$　　**b.** $(x^{-b/a})^{-a/b}$

15. a. $\dfrac{5^0}{(2^{-3}x^{-3}y^2)^2}$　　**b.** $\dfrac{(x+y)(x-y)}{(x-y)^0}$

16. a. $\dfrac{(a^m \cdot a^{-n})^{-2}}{(a^{m+n})^2}$　　**b.** $\left(\dfrac{x^{2n-2}y^{2n}}{x^{5n+1}y^{-n}}\right)^{1/3}$

In Exercises 17–26, solve the equation for x.

17. $6^{2x} = 6^4$　　**18.** $5^{-x} = 5^3$

19. $3^{3x-4} = 3^5$　　**20.** $10^{2x-1} = 10^{x+3}$

21. $(2.1)^{x+2} = (2.1)^5$　　**22.** $(-1.3)^{x-2} = (-1.3)^{2x+1}$

23. $8^x = \left(\dfrac{1}{32}\right)^{x-2}$　　**24.** $3^{x-x^2} = \dfrac{1}{9^x}$

25. $3^{2x} - 12 \cdot 3^x + 27 = 0$

26. $2^{2x} - 4 \cdot 2^x + 4 = 0$

In Exercises 27–36, sketch the graphs of the given functions on the same axes.

27. $y = 2^x$, $y = 3^x$, and $y = 4^x$

28. $y = \left(\dfrac{1}{2}\right)^x$, $y = \left(\dfrac{1}{3}\right)^x$, and $y = \left(\dfrac{1}{4}\right)^x$

29. $y = 2^{-x}$, $y = 3^{-x}$, and $y = 4^{-x}$

30. $y = 4^{0.5x}$ and $y = 4^{-0.5x}$

31. $y = 4^{0.5x}$, $y = 4^x$, and $y = 4^{2x}$

32. $y = e^x$, $y = 2e^x$, and $y = 3e^x$

33. $y = e^{0.5x}$, $y = e^x$, and $y = e^{1.5x}$

34. $y = e^{-0.5x}$, $y = e^{-x}$, and $y = e^{-1.5x}$

35. $y = 0.5e^{-x}$, $y = e^{-x}$, and $y = 2e^{-x}$

36. $y = 1 - e^{-x}$ and $y = 1 - e^{-0.5x}$

37. DISABILITY RATES Because of medical technology advances, the disability rates for people over 65 have been dropping rather dramatically. The function

$$R(t) = 26.3e^{-0.016t} \qquad (0 \le t \le 18)$$

gives the disability rate $R(t)$, in percent, for people over age 65 from 1982 ($t = 0$) through 2000, where t is measured in years.
  a. What was the disability rate in 1982? In 1986? In 1994? In 2000?
  b. Sketch the graph of R.
  Source: Frost and Sullivan

38. TRACKING WITH GPS Employers are increasingly turning to GPS (global positioning system) technology to keep track of their fleet vehicles. In a study conducted in 2004, the estimated number of automatic vehicle trackers installed on fleet vehicles in the United States is

$$N(t) = 0.6e^{0.17t} \qquad (0 \le t \le 5)$$

where $N(t)$ is measured in millions and t is measured in years, with $t = 0$ corresponding to 2000.
  a. What was the number of automatic vehicles trackers installed in the year 2000? How many were projected to be installed in 2005?
  b. Sketch the graph of N.
  Source: C. J. Driscoll Associates

39. GROWTH OF WEB SITES According to a study conducted in 2000, the projected number of Web addresses (in billions) is approximated by the function

$$N(t) = 0.45e^{0.5696t} \qquad (0 \le t \le 5)$$

where t is measured in years, with $t = 0$ corresponding to 1997.
  a. Complete the following table by finding the number of Web addresses in each year:

Year	0	1	2	3	4	5
Number of Web Addresses (billions)						

  b. Sketch the graph of N.

40. MARRIED HOUSEHOLDS The percent of families that were married households between 1970 and 2000 is approximately

$$P(t) = 86.9e^{-0.05t} \qquad (0 \le t \le 3)$$

where t is measured in decades, with $t = 0$ corresponding to 1970.
  a. What percent of families were married households in 1970? In 1980? In 1990? In 2000?
  b. Sketch the graph of P.
  Source: U.S. Census Bureau

where $x(t)$ is measured in grams/cubic centimeter (g/cm³).
a. What is the initial concentration of the drug in the organ?
b. What is the concentration of the drug in the organ after 20 sec?
c. What will be the concentration of the drug in the organ in the long run?
d. Sketch the graph of x.

44. ABSORPTION OF DRUGS Jane took 100 mg of a drug in the morning and another 100 mg of the same drug at the same time the following morning. The amount of the drug in her body t days after the first dosage was taken is given by

$$A(t) = \begin{cases} 100e^{-1.4t} & \text{if } 0 \le t < 1 \\ 100(1 + e^{1.4})e^{-1.4t} & \text{if } t \ge 1 \end{cases}$$

a. What was the amount of drug in Jane's body immediately after taking the second dose? After 2 days? In the long run?
b. Sketch the graph of A.

In Exercises 45–48, determine whether the statement is true or false. If it is true, explain why it is true. If it is false, give an example to show why it is false.

45. $(x^2 + 1)^3 = x^6 + 1$

46. $e^{xy} = e^x e^y$

47. If $x < y$, then $e^x < e^y$.

48. If $0 < b < 1$ and $x < y$, then $b^x > b^y$.

41. ALTERNATIVE MINIMUM TAX The alternative minimum tax was created in 1969 to prevent the very wealthy from using creative deductions and shelters to avoid having to pay anything to the Internal Revenue Service. But it has increasingly hit the middle class. The number of taxpayers subject to an alternative minimum tax is projected to be

$$N(t) = \frac{35.5}{1 + 6.89e^{-0.8674t}} \qquad (0 \le t \le 6)$$

where $N(t)$ is measured in millions and t is measured in years, with $t = 0$ corresponding to 2004. What is the projected number of taxpayers subjected to an alternative minimum tax in 2010?
Source: Brookings Institution

42. ABSORPTION OF DRUGS The concentration of a drug in an organ at any time t (in seconds) is given by

$$C(t) = \begin{cases} 0.3t - 18(1 - e^{-t/60}) & \text{if } 0 \le t \le 20 \\ 18e^{-t/60} - 12e^{-(t-20)/60} & \text{if } t > 20 \end{cases}$$

where $C(t)$ is measured in grams/cubic centimeter (g/cm³).
a. What is the initial concentration of the drug in the organ?
b. What is the concentration of the drug in the organ after 10 sec?
c. What is the concentration of the drug in the organ after 30 sec?
d. What will be the concentration of the drug in the long run?

43. ABSORPTION OF DRUGS The concentration of a drug in an organ at any time t (in seconds) is given by

$$x(t) = 0.08 + 0.12(1 - e^{-0.02t})$$

14.1 Solutions to Self-Check Exercises

1. $2^{2x+1} \cdot 2^{-3} = 2^{x-1}$

$$\frac{2^{2x+1}}{2^{x-1}} \cdot 2^{-3} = 1 \qquad \text{Divide both sides by } 2^{x-1}.$$

$$2^{(2x+1)-(x-1)-3} = 1$$

$$2^{x-1} = 1$$

This is true if and only if $x - 1 = 0$ or $x = 1$.

2. We first construct the following table of values:

x	-3	-2	-1	0	1	2	3	4
$y = e^{0.4x}$	0.3	0.4	0.7	1	1.5	2.2	3.3	5

Next, we plot these points and join them by a smooth curve to obtain the graph of f shown in the accompanying figure.

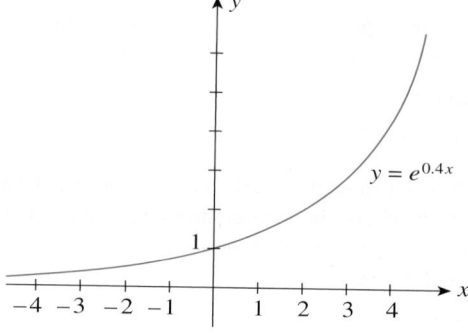

USING TECHNOLOGY

Although the proof is outside the scope of this book, it can be proved that an exponential function of the form $f(x) = b^x$, where $b > 1$, will ultimately grow faster than the power function $g(x) = x^n$ for *any* positive real number n. To give a visual demonstration of this result for the special case of the exponential function $f(x) = e^x$, we can use a graphing utility to plot the graphs of both f and g (for selected values of n) on the same set of axes in an appropriate viewing window and observe that the graph of f ultimately lies above that of g.

EXAMPLE 1   Use a graphing utility to plot the graphs of (a) $f(x) = e^x$ and $g(x) = x^3$ on the same set of axes in the viewing window $[0, 6] \times [0, 250]$ and (b) $f(x) = e^x$ and $g(x) = x^5$ in the viewing window $[0, 20] \times [0, 1{,}000{,}000]$.

Solution

a. The graphs of $f(x) = e^x$ and $g(x) = x^3$ in the viewing window $[0, 6] \times [0, 250]$ are shown in Figure T1a.

b. The graphs of $f(x) = e^x$ and $g(x) = x^5$ in the viewing window $[0, 20] \times [0, 1{,}000{,}000]$ are shown in Figure T1b.

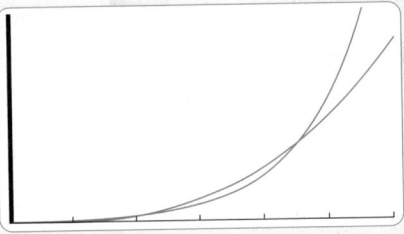
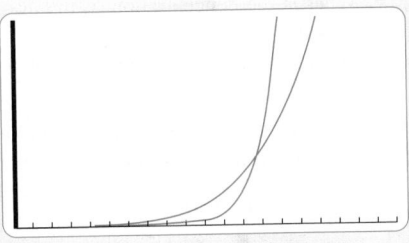

(a) The graphs of $f(x) = e^x$ and $g(x) = x^3$ in the viewing window $[0, 6] \times [0, 250]$

(b) The graphs of $f(x) = e^x$ and $g(x) = x^5$ in the viewing window $[0, 20] \times [0, 1{,}000{,}000]$

FIGURE T1

In the exercises that follow, you are asked to use a graphing utility to reveal the properties of exponential functions.

TECHNOLOGY EXERCISES

In Exercises 1 and 2, plot the graphs of the functions f and g on the same set of axes in the specified viewing window.

1. $f(x) = e^x$ and $g(x) = x^2$; $[0, 4] \times [0, 30]$

2. $f(x) = e^x$ and $g(x) = x^4$; $[0, 15] \times [0, 20{,}000]$

In Exercises 3 and 4, plot the graphs of the functions f and g on the same set of axes in an appropriate viewing window to demonstrate that f ultimately grows faster than g. (*Note:* Your answer will *not* be unique.)

3. $f(x) = 2^x$ and $g(x) = x^{2.5}$

4. $f(x) = 3^x$ and $g(x) = x^3$

5. Plot the graphs of $f(x) = 2^x$, $g(x) = 3^x$, and $h(x) = 4^x$ on the same set of axes in the viewing window $[0, 5] \times [0, 100]$. Comment on the relationship between the base b and the growth of the function $f(x) = b^x$.

6. Plot the graphs of $f(x) = (1/2)^x$, $g(x) = (1/3)^x$, and $h(x) = (1/4)^x$ on the same set of axes in the viewing window $[0, 4] \times [0, 1]$. Comment on the relationship between the base b and the growth of the function $f(x) = b^x$.

7. Plot the graphs of $f(x) = e^x$, $g(x) = 2e^x$, and $h(x) = 3e^x$ on the same set of axes in the viewing window $[-3, 3] \times [0, 10]$. Comment on the role played by the constant k in the graph of $f(x) = ke^x$.

8. Plot the graphs of $f(x) = -e^x$, $g(x) = -2e^x$, and $h(x) = -3e^x$ on the same set of axes in the viewing window $[-3, 3] \times [-10, 0]$. Comment on the role played by the constant k in the graph of $f(x) = ke^x$.

9. Plot the graphs of $f(x) = e^{0.5x}$, $g(x) = e^x$, and $h(x) = e^{1.5x}$ on the same set of axes in the viewing window $[-2, 2] \times [0, 4]$. Comment on the role played by the constant k in the graph of $f(x) = e^{kx}$.

10. Plot the graphs of $f(x) = e^{-0.5x}$, $g(x) = e^{-x}$, and $h(x) = e^{-1.5x}$ on the same set of axes in the viewing window $[-2, 2] \times [0, 4]$. Comment on the role played by the constant k in the graph of $f(x) = e^{kx}$.

11. **ABSORPTION OF DRUGS** The concentration of a drug in an organ at any time t (in seconds) is given by

$$x(t) = 0.08 + 0.12(1 - e^{-0.02t})$$

where $x(t)$ is measured in grams/cubic centimeter (g/cm³).
   a. Plot the graph of the function x in the viewing window $[0, 200] \times [0, 0.2]$.
   b. What is the initial concentration of the drug in the organ?
   c. What is the concentration of the drug in the organ after 20 sec?
   d. What will be the concentration of the drug in the organ in the long run?

12. **ABSORPTION OF DRUGS** Jane took 100 mg of a drug in the morning and another 100 mg of the same drug at the same time the following morning. The amount of the drug in her body t days after the first dosage was taken is given by

$$A(t) = \begin{cases} 100e^{-1.4t} & \text{if } 0 \le t < 1 \\ 100(1 + e^{1.4})e^{-1.4t} & \text{if } t \ge 1 \end{cases}$$

   a. Plot the graph of the function A in the viewing window $[0, 5] \times [0, 140]$.
   b. Verify the results of Exercise 44, page 861.

13. **ABSORPTION OF DRUGS** The concentration of a drug in an organ at any time t (in seconds) is given by

$$C(t) = \begin{cases} 0.3t - 18(1 - e^{-t/60}) & \text{if } 0 \le t \le 20 \\ 18e^{-t/60} - 12e^{-(t-20)/60} & \text{if } t > 20 \end{cases}$$

where $C(t)$ is measured in grams/cubic centimeter (g/cm³).
   a. Plot the graph of the function C in the viewing window $[0, 120] \times [0, 1]$.
   b. How long after the drug is first introduced will it take for the concentration of the drug to reach a peak?
   c. How long after the concentration of the drug has peaked will it take for the concentration of the drug to fall back to 0.5 g/cm³?
   Hint: Plot the graphs of $y_1 = C(x)$ and $y_2 = 0.5$ and use the ISECT function of your graphing utility.

14.2 Logarithmic Functions

■ Logarithms

You are already familiar with exponential equations of the form

$$b^y = x \qquad (b > 0, b \ne 1)$$

where the variable x is expressed in terms of a real number b and a variable y. But what about solving this same equation for y? You may recall from your study of algebra that the number y is called the **logarithm of x to the base b** and is denoted by **$\log_b x$**. It is the power to which the base b must be raised in order to obtain the number x.

Logarithm of x to the Base b

$$y = \log_b x \quad \text{if and only if} \quad x = b^y \qquad (x > 0)$$

 Observe that the logarithm $\log_b x$ is defined only for positive values of x.

EXAMPLE 1

a. $\log_{10} 100 = 2$ since $100 = 10^2$
b. $\log_5 125 = 3$ since $125 = 5^3$

c. $\log_3 \dfrac{1}{27} = -3$ since $\dfrac{1}{27} = \dfrac{1}{3^3} = 3^{-3}$

d. $\log_{20} 20 = 1$ since $20 = 20^1$

EXAMPLE 2 Solve each of the following equations for x.

a. $\log_3 x = 4$ **b.** $\log_{16} 4 = x$ **c.** $\log_x 8 = 3$

Solution

a. By definition, $\log_3 x = 4$ implies $x = 3^4 = 81$.
b. $\log_{16} 4 = x$ is equivalent to $4 = 16^x = (4^2)^x = 4^{2x}$, or $4^1 = 4^{2x}$, from which we deduce that

$$2x = 1 \qquad b^m = b^n \Rightarrow m = n$$
$$x = \frac{1}{2}$$

c. Referring once again to the definition, we see that the equation $\log_x 8 = 3$ is equivalent to

$$8 = 2^3 = x^3$$
$$x = 2 \qquad a^m = b^m \Rightarrow a = b$$

The two widely used systems of logarithms are the system of **common logarithms**, which uses the number 10 as the base, and the system of **natural logarithms**, which uses the irrational number $e = 2.71828\ldots$ as the base. Also, it is standard practice to write **log** for $\log_{10}$ and **ln** for $\log_e$.

Logarithmic Notation

$$\log x = \log_{10} x \qquad \text{Common logarithm}$$
$$\ln x = \log_e x \qquad \text{Natural logarithm}$$

The system of natural logarithms is widely used in theoretical work. Using natural logarithms rather than logarithms to other bases often leads to simpler expressions.

Laws of Logarithms

Computations involving logarithms are facilitated by the following **laws of logarithms**.

Laws of Logarithms
If m and n are positive numbers, then

1. $\log_b mn = \log_b m + \log_b n$
2. $\log_b \dfrac{m}{n} = \log_b m - \log_b n$
3. $\log_b m^n = n \log_b m$
4. $\log_b 1 = 0$
5. $\log_b b = 1$

Do not confuse the expression log m/n (Law 2) with the expression log m/log n. For example,

$$\log \frac{100}{10} = \log 100 - \log 10 = 2 - 1 = 1 \neq \frac{\log 100}{\log 10} = \frac{2}{1} = 2$$

You will be asked to prove these laws in Exercises 70–72 on page 872. Their derivations are based on the definition of a logarithm and the corresponding laws of exponents. The following examples illustrate the properties of logarithms.

EXAMPLE 3

a. $\log(2 \cdot 3) = \log 2 + \log 3$ **b.** $\ln \dfrac{5}{3} = \ln 5 - \ln 3$

c. $\log \sqrt{7} = \log 7^{1/2} = \dfrac{1}{2} \log 7$ **d.** $\log_5 1 = 0$

e. $\log_{45} 45 = 1$

EXAMPLE 4   Given that $\log 2 \approx 0.3010$, $\log 3 \approx 0.4771$, and $\log 5 \approx 0.6990$, use the laws of logarithms to find

a. log 15   **b.** log 7.5   **c.** log 81   **d.** log 50

Solution

a. Note that $15 = 3 \cdot 5$, so by Law 1 for logarithms,

$$\begin{aligned}
\log 15 &= \log 3 \cdot 5 \\
&= \log 3 + \log 5 \\
&\approx 0.4771 + 0.6990 \\
&= 1.1761
\end{aligned}$$

b. Observing that $7.5 = 15/2 = (3 \cdot 5)/2$, we apply Laws 1 and 2, obtaining

$$\begin{aligned}
\log 7.5 &= \log \frac{(3)(5)}{2} \\
&= \log 3 + \log 5 - \log 2 \\
&\approx 0.4771 + 0.6990 - 0.3010 \\
&= 0.8751
\end{aligned}$$

c. Since $81 = 3^4$, we apply Law 3 to obtain

$$\begin{aligned}
\log 81 &= \log 3^4 \\
&= 4 \log 3 \\
&\approx 4(0.4771) \\
&= 1.9084
\end{aligned}$$

d. We write $50 = 5 \cdot 10$ and find

$$\begin{aligned}
\log 50 &= \log(5)(10) \\
&= \log 5 + \log 10 \\
&\approx 0.6990 + 1 \qquad \text{Use Law 5} \\
&= 1.6990
\end{aligned}$$

EXAMPLE 5   Expand and simplify the following expressions:

a. $\log_3 x^2 y^3$ **b.** $\log_2 \dfrac{x^2 + 1}{2^x}$ **c.** $\ln \dfrac{x^2 \sqrt{x^2 - 1}}{e^x}$

Solution

a.
$$
\begin{aligned}
\log_3 x^2 y^3 &= \log_3 x^2 + \log_3 y^3 && \text{Law 1} \\
&= 2 \log_3 x + 3 \log_3 y && \text{Law 3}
\end{aligned}
$$

b.
$$
\begin{aligned}
\log_2 \frac{x^2 + 1}{2^x} &= \log_2(x^2 + 1) - \log_2 2^x && \text{Law 2} \\
&= \log_2(x^2 + 1) - x \log_2 2 && \text{Law 3} \\
&= \log_2(x^2 + 1) - x && \text{Law 5}
\end{aligned}
$$

c.
$$
\begin{aligned}
\ln \frac{x^2 \sqrt{x^2 - 1}}{e^x} &= \ln \frac{x^2 (x^2 - 1)^{1/2}}{e^x} && \text{Rewrite} \\
&= \ln x^2 + \ln(x^2 - 1)^{1/2} - \ln e^x && \text{Laws 1 and 2} \\
&= 2 \ln x + \frac{1}{2} \ln(x^2 - 1) - x \ln e && \text{Law 3} \\
&= 2 \ln x + \frac{1}{2} \ln(x^2 - 1) - x && \text{Law 5}
\end{aligned}
$$

Logarithmic Functions and Their Graphs

The definition of a logarithm implies that if b and n are positive numbers and b is different from 1, then the expression $\log_b n$ is a real number. This enables us to define a logarithmic function as follows:

> **Logarithmic Function**
> The function defined by
> $$ f(x) = \log_b x \qquad (b > 0, b \neq 1) $$
> is called the **logarithmic function with base b**. The domain of f is the set of all positive numbers.

One easy way to obtain the graph of the logarithmic function $y = \log_b x$ is to construct a table of values of the logarithm (base b). However, another method—and a more instructive one—is based on exploiting the intimate relationship between logarithmic and exponential functions.

If a point (u, v) lies on the graph of $y = \log_b x$, then

$$ v = \log_b u $$

But we can also write this equation in exponential form as

$$ u = b^v $$

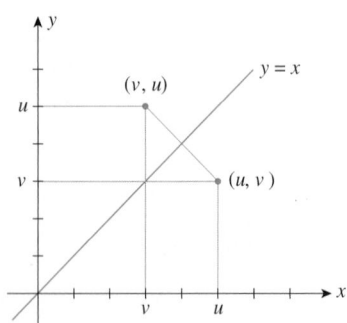

FIGURE 7
The points (u, v) and (v, u) are mirror reflections of each other.

So the point (v, u) also lies on the graph of the function $y = b^x$. Let's look at the relationship between the points (u, v) and (v, u) and the line $y = x$ (Figure 7). If we think of the line $y = x$ as a mirror, then the point (v, u) is the mirror reflection of the point (u, v). Similarly, the point (u, v) is a mirror reflection of the point (v, u). We can take

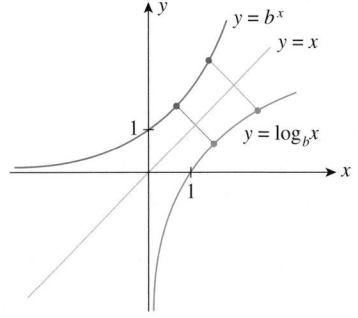

FIGURE 8
The graphs of $y = b^x$ and $y = \log_b x$ are mirror reflections of each other.

advantage of this relationship to help us draw the graph of logarithmic functions. For example, if we wish to draw the graph of $y = \log_b x$, where $b > 1$, then we need only draw the mirror reflection of the graph of $y = b^x$ with respect to the line $y = x$ (Figure 8).

You may discover the following properties of the logarithmic function by taking the reflection of the graph of an appropriate exponential function (Exercises 33 and 34 on page 871).

Properties of the Logarithmic Function

The logarithmic function $y = \log_b x$ $(b > 0, b \neq 1)$ has the following properties:

1. Its domain is $(0, \infty)$.

2. Its range is $(-\infty, \infty)$.

3. Its graph passes through the point $(1, 0)$.

4. It is continuous on $(0, \infty)$.

5. It is increasing on $(0, \infty)$ if $b > 1$ and decreasing on $(0, \infty)$ if $b < 1$.

EXAMPLE 6 Sketch the graph of the function $y = \ln x$.

Solution We first sketch the graph of $y = e^x$. Then, the required graph is obtained by tracing the mirror reflection of the graph of $y = e^x$ with respect to the line $y = x$ (Figure 9). ∎

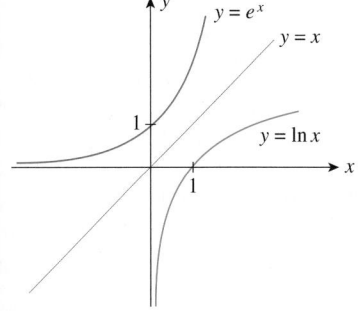

FIGURE 9
The graph of $y = \ln x$ is the mirror reflection of the graph of $y = e^x$.

Properties Relating the Exponential and Logarithmic Functions

We made use of the relationship that exists between the exponential function $f(x) = e^x$ and the logarithmic function $g(x) = \ln x$ when we sketched the graph of g in Example 6. This relationship is further described by the following properties, which are an immediate consequence of the definition of the logarithm of a number.

Properties Relating e^x and $\ln x$

$$e^{\ln x} = x \qquad (x > 0) \qquad \textbf{(2)}$$

$$\ln e^x = x \qquad \text{(for any real number } x) \qquad \textbf{(3)}$$

(Try to verify these properties.)

From Properties 2 and 3, we conclude that the composite function

$$(f \circ g)(x) = f[g(x)]$$
$$= e^{\ln x} = x$$
$$(g \circ f)(x) = g[f(x)]$$
$$= \ln e^x = x$$

Thus,

$$f[g(x)] = g[f(x)]$$
$$= x$$

Any two functions f and g that satisfy this relationship are said to be **inverses** of each other. Note that the function f undoes what the function g does, and vice versa, so the composition of the two functions in any order results in the identity function $F(x) = x$.

The relationships expressed in Equations (2) and (3) are useful in solving equations that involve exponentials and logarithms.

EXPLORING WITH TECHNOLOGY

You can demonstrate the validity of Properties 2 and 3, which state that the exponential function $f(x) = e^x$ and the logarithmic function $g(x) = \ln x$ are inverses of each other as follows:

1. Sketch the graph of $(f \circ g)(x) = e^{\ln x}$, using the viewing window $[0, 10] \times [0, 10]$. Interpret the result.

2. Sketch the graph of $(g \circ f)(x) = \ln e^x$, using the standard viewing window. Interpret the result.

EXAMPLE 7   Solve the equation $2e^{x+2} = 5$.

Solution   We first divide both sides of the equation by 2 to obtain

$$e^{x+2} = \frac{5}{2} = 2.5$$

Next, taking the natural logarithm of each side of the equation and using Equation (3), we have

$$\ln e^{x+2} = \ln 2.5$$
$$x + 2 = \ln 2.5$$
$$x = -2 + \ln 2.5$$
$$\approx -1.08$$

EXPLORE & DISCUSS

Consider the equation $y = y_0 b^{kx}$, where y_0 and k are positive constants and $b > 0$, $b \neq 1$. Suppose we want to express y in the form $y = y_0 e^{px}$. Use the laws of logarithms to show that $p = k \ln b$ and hence that $y = y_0 e^{(k \ln b)x}$ is an alternative form of $y = y_0 b^{kx}$ using the base e.

EXAMPLE 8   Solve the equation $5 \ln x + 3 = 0$.

Solution   Adding -3 to both sides of the equation leads to

$$5 \ln x = -3$$
$$\ln x = -\frac{3}{5} = -0.6$$

and so

$$e^{\ln x} = e^{-0.6}$$

Using Equation (2), we conclude that

$$x = e^{-0.6}$$
$$\approx 0.55$$

The next two examples show how logarithms can be used to solve problems involving compound interest.

EXAMPLE 9 How long will it take $10,000 to grow to $15,000 if the investment earns an interest rate of 12% per year compounded quarterly?

Solution Using the compound interest formula with $A = 15,000$, $P = 10,000$, $r = 0.12$, and $m = 4$, we obtain

$$15,000 = 10,000\left(1 + \frac{0.12}{4}\right)^{4t}$$

$$(1.03)^{4t} = \frac{15,000}{10,000} = 1.5$$

Taking the logarithm on each side of the equation gives

$$\ln(1.03)^{4t} = \ln 1.5$$
$$4t \ln 1.03 = \ln 1.5 \qquad \log_b m^n = n \log_b m$$
$$4t = \frac{\ln 1.5}{\ln 1.03}$$
$$t = \frac{\ln 1.5}{4 \ln 1.03} \approx 3.43$$

So it will take approximately 3.4 years for the investment to grow from $10,000 to $15,000.

EXAMPLE 10 Find the interest rate needed for an investment of $10,000 to grow to an amount of $18,000 in 5 years if the interest is compounded monthly.

Solution Using the compound interest formula with $A = 18,000$, $P = 10,000$, $m = 12$, and $t = 5$, we obtain

$$18,000 = 10,000\left(1 + \frac{r}{12}\right)^{12(5)}$$

Dividing both sides of the equation by 10,000 gives

$$\frac{18,000}{10,000} = \left(1 + \frac{r}{12}\right)^{60}$$

or, upon simplification,

$$\left(1 + \frac{r}{12}\right)^{60} = 1.8$$

Now, we take the logarithm on each side of the equation, obtaining

$$\ln\left(1 + \frac{r}{12}\right)^{60} = \ln 1.8$$

$$60\ln\left(1 + \frac{r}{12}\right) = \ln 1.8$$

$$\ln\left(1 + \frac{r}{12}\right) = \frac{\ln 1.8}{60} = 0.009796$$

$$\left(1 + \frac{r}{12}\right) = e^{0.009796} \qquad \text{By Property 2}$$

$$= 1.009844$$

and

$$\frac{r}{12} = 1.009844 - 1$$

$$r = 0.1181$$

or 11.81% per year. ∎

14.2 Self-Check Exercises

1. Sketch the graph of $y = 3^x$ and $y = \log_3 x$ on the same set of axes.

2. Solve the equation $3e^{x+1} - 2 = 4$.

Solutions to Self-Check Exercises 14.2 can be found on page 873.

14.2 Concept Questions

1. **a.** Define $y = \log_b x$.
   b. Define the logarithmic function f with base b. What restrictions, if any, are placed on b?

2. For the logarithmic function $y = \log_b x$ ($b > 0$, $b \neq 1$), state (a) its domain and range, (b) its x-intercept, (c) where it is continuous, and (d) where it is increasing and where it is decreasing for the case $b > 1$ and the case $b < 1$.

3. **a.** If $x > 0$, what is $e^{\ln x}$?
   b. If x is any real number, what is $\ln e^x$?

14.2 Exercises

In Exercises 1–10, express each equation in logarithmic form.

1. $2^6 = 64$

2. $3^5 = 243$

3. $3^{-2} = \dfrac{1}{9}$

4. $5^{-3} = \dfrac{1}{125}$

5. $\left(\dfrac{1}{3}\right)^1 = \dfrac{1}{3}$

6. $\left(\dfrac{1}{2}\right)^{-4} = 16$

7. $32^{3/5} = 8$

8. $81^{3/4} = 27$

9. $10^{-3} = 0.001$

10. $16^{-1/4} = 0.5$

In Exercises 11–16, use the facts that log 3 = 0.4771 and log 4 = 0.6021 to find the value of each logarithm.

11. log 12

12. $\log \dfrac{3}{4}$

13. log 16

14. $\log \sqrt{3}$

15. log 48

16. $\log \dfrac{1}{300}$

In Exercises 17–20, write the expression as the logarithm of a single quantity.

17. $2 \ln a + 3 \ln b$

18. $\dfrac{1}{2} \ln x + 2 \ln y - 3 \ln z$

19. $\ln 3 + \dfrac{1}{2} \ln x + \ln y - \dfrac{1}{3} \ln z$

20. $\ln 2 + \dfrac{1}{2} \ln(x + 1) - 2 \ln(1 + \sqrt{x})$

In Exercises 21–28, use the laws of logarithms to expand and simplify the expression.

21. $\log x(x + 1)^4$

22. $\log x(x^2 + 1)^{-1/2}$

23. $\log \dfrac{\sqrt{x + 1}}{x^2 + 1}$

24. $\ln \dfrac{e^x}{1 + e^x}$

25. $\ln xe^{-x^2}$

26. $\ln x(x + 1)(x + 2)$

27. $\ln \dfrac{x^{1/2}}{x^2\sqrt{1 + x^2}}$

28. $\ln \dfrac{x^2}{\sqrt{x}(1 + x)^2}$

In Exercises 29–32, sketch the graph of the equation.

29. $y = \log_3 x$

30. $y = \log_{1/3} x$

31. $y = \ln 2x$

32. $y = \ln \dfrac{1}{2}x$

In Exercises 33 and 34, sketch the graphs of the equations on the same coordinate axes.

33. $y = 2^x$ and $y = \log_2 x$

34. $y = e^{3x}$ and $y = \ln 3x$

In Exercises 35–44, use logarithms to solve the equation for t.

35. $e^{0.4t} = 8$

36. $\dfrac{1}{3}e^{-3t} = 0.9$

37. $5e^{-2t} = 6$

38. $4e^{t-1} = 4$

39. $2e^{-0.2t} - 4 = 6$

40. $12 - e^{0.4t} = 3$

41. $\dfrac{50}{1 + 4e^{0.2t}} = 20$

42. $\dfrac{200}{1 + 3e^{-0.3t}} = 100$

43. $A = Be^{-t/2}$

44. $\dfrac{A}{1 + Be^{t/2}} = C$

45. Find the interest rate needed for an investment of $5000 to grow to an amount of $7500 in 3 yr if interest is compounded monthly.

46. Find the interest rate needed for an investment of $5000 to grow to an amount of $7500 in 3 yr if interest is compounded quarterly.

47. Find the interest rate needed for an investment of $5000 to grow to an amount of $8000 in 4 yr if interest is compounded semiannually.

48. Find the interest rate needed for an investment of $5000 to grow to an amount of $5500 in 6 mo if interest is compounded monthly.

49. Find the interest rate needed for an investment of $2000 to double in 5 yr if interest is compounded annually.

50. Find the interest rate needed for an investment of $2000 to triple in 5 yr if interest is compounded monthly.

51. How long will it take $5000 to grow to $6500 if the investment earns interest at the rate of 12%/year compounded monthly?

52. How long will it take $12,000 to grow to $15,000 if the investment earns interest at the rate of 8%/year compounded monthly?

53. How long will it take an investment of $2000 to double if the investment earns interest at the rate of 9%/year compounded monthly?

54. How long will it take an investment of $5000 to triple if the investment earns interest at the rate of 8%/year compounded daily?

55. BLOOD PRESSURE   A normal child's systolic blood pressure may be approximated by the function

$$p(x) = m(\ln x) + b$$

where $p(x)$ is measured in millimeters of mercury, x is measured in pounds, and m and b are constants. Given that $m = 19.4$ and $b = 18$, determine the systolic blood pressure of a child who weighs 92 lb.

56. MAGNITUDE OF EARTHQUAKES   On the Richter scale, the magnitude R of an earthquake is given by the formula

$$R = \log \dfrac{I}{I_0}$$

where I is the intensity of the earthquake being measured and I_0 is the standard reference intensity.

a. Express the intensity I of an earthquake of magnitude $R = 5$ in terms of the standard intensity I_0.

b. Express the intensity I of an earthquake of magnitude $R = 8$ in terms of the standard intensity I_0. How many times greater is the intensity of an earthquake of magnitude 8 than one of magnitude 5?

c. In modern times, the greatest loss of life attributable to an earthquake occurred in eastern China in 1976. Known as the Tangshan earthquake, it registered 8.2 on the Richter scale. How does the intensity of this earthquake compare with the intensity of an earthquake of magnitude $R = 5$?

57. **SOUND INTENSITY**   The relative loudness of a sound D of intensity I is measured in decibels (db), where

$$D = 10 \log \frac{I}{I_0}$$

and I_0 is the standard threshold of audibility.
  a. Express the intensity I of a 30-db sound (the sound level of normal conversation) in terms of I_0.
  b. Determine how many times greater the intensity of an 80-db sound (rock music) is than that of a 30-db sound.
  c. Prolonged noise above 150 db causes immediate and permanent deafness. How does the intensity of a 150-db sound compare with the intensity of an 80-db sound?

58. **BAROMETRIC PRESSURE**   Halley's law states that the barometric pressure (in inches of mercury) at an altitude of x mi above sea level is approximated by the equation

$$p(x) = 29.92e^{-0.2x} \qquad (x \geq 0)$$

If the barometric pressure as measured by a hot-air balloonist is 20 in. of mercury, what is the balloonist's altitude?

59. **HEIGHT OF TREES**   The height (in feet) of a certain kind of tree is approximated by

$$h(t) = \frac{160}{1 + 240e^{-0.2t}}$$

where t is the age of the tree in years. Estimate the age of an 80-ft tree.

60. **NEWTON'S LAW OF COOLING**   The temperature of a cup of coffee t min after it is poured is given by

$$T = 70 + 100e^{-0.0446t}$$

where T is measured in degrees Fahrenheit.
  a. What is the temperature of the coffee when it was poured?
  b. When will the coffee be cool enough to drink (say, 120°F)?

61. **LENGTHS OF FISH**   The length (in centimeters) of a typical Pacific halibut t yr old is approximately

$$f(t) = 200(1 - 0.956e^{-0.18t})$$

Suppose a Pacific halibut caught by Mike measures 140 cm. What is its approximate age?

62. **ABSORPTION OF DRUGS**   The concentration of a drug in an organ at any time t (in seconds) is given by

$$x(t) = 0.08(1 - e^{-0.02t})$$

where $x(t)$ is measured in grams/cubic centimeter (g/cm³).
  a. How long would it take for the concentration of the drug in the organ to reach 0.02 g/cm³?

  b. How long would it take for the concentration of the drug in the organ to reach 0.04 g/cm³?

63. **ABSORPTION OF DRUGS**   The concentration of a drug in an organ at any time t (in seconds) is given by

$$x(t) = 0.08 + 0.12e^{-0.02t}$$

where $x(t)$ is measured in grams/cubic centimeter (g/cm³).
  a. How long would it take for the concentration of the drug in the organ to reach 0.18 g/cm³?
  b. How long would it take for the concentration of the drug in the organ to reach 0.16 g/cm³?

64. **FORENSIC SCIENCE**   Forensic scientists use the following law to determine the time of death of accident or murder victims. If T denotes the temperature of a body t hr after death, then

$$T = T_0 + (T_1 - T_0)(0.97)^t$$

where T_0 is the air temperature and T_1 is the body temperature at the time of death. John Doe was found murdered at midnight in his house, when the room temperature was 70°F and his body temperature was 80°F. When was he killed? Assume that the normal body temperature is 98.6°F.

In Exercises 65–68, determine whether the statement is true or false. If it is true, explain why it is true. If it is false, give an example to show why it is false.

65. $(\ln x)^3 = 3 \ln x$ for all x in $(0, \infty)$.

66. $\ln a - \ln b = \ln(a - b)$ for all positive real numbers a and b.

67. The function $f(x) = \dfrac{1}{\ln x}$ is continuous on $(1, \infty)$.

68. The function $f(x) = \ln|x|$ is continuous for all $x \neq 0$.

69. a. Given that $2^x = e^{kx}$, find k.
    b. Show that, in general, if b is a nonnegative real number, then any equation of the form $y = b^x$ may be written in the form $y = e^{kx}$, for some real number k.

70. Use the definition of a logarithm to prove
    a. $\log_b mn = \log_b m + \log_b n$
    b. $\log_b \dfrac{m}{n} = \log_b m - \log_b n$
    Hint: Let $\log_b m = p$ and $\log_b n = q$. Then, $b^p = m$ and $b^q = n$.

71. Use the definition of a logarithm to prove

$$\log_b m^n = n \log_b m$$

72. Use the definition of a logarithm to prove
    a. $\log_b 1 = 0$
    b. $\log_b b = 1$

14.2 Solutions to Self-Check Exercises

1. First, sketch the graph of $y = 3^x$ with the help of the following table of values:

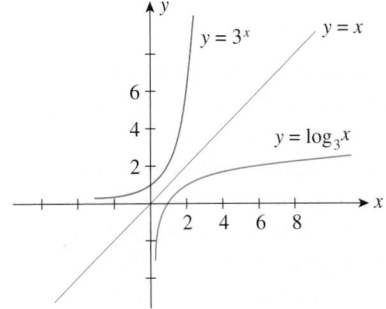

x	-3	-2	-1	0	1	2	3
$y = 3^x$	1/27	1/9	1/3	0	3	9	27

Next, take the mirror reflection of this graph with respect to the line $y = x$ to obtain the graph of $y = \log_3 x$.

2.
$$3e^{x+1} - 2 = 4$$
$$3e^{x+1} = 6$$
$$e^{x+1} = 2$$

$\ln e^{x+1} = \ln 2$ Take the logarithm of both sides.

$(x + 1)\ln e = \ln 2$ Law 3

$x + 1 = \ln 2$ Law 5

$x = \ln 2 - 1$

≈ -0.3069

14.3 Differentiation of Exponential Functions

■ The Derivative of the Exponential Function

To study the effects of budget deficit-reduction plans at different income levels, it is important to know the income distribution of American families. Based on data from the House Budget Committee, the House Ways and Means Committee, and the U.S. Census Bureau, the graph of f shown in Figure 10 gives the number of American families y (in millions) as a function of their annual income x (in thousands of dollars) in 1990.

Observe that the graph of f rises very quickly and then tapers off. From the graph of f, you can see that the bulk of American families earned less than $100,000 per year. In fact, 95% of U.S. families earned less than $102,358 per year in 1990. (We will refer to this model again in Using Technology at the end of this section.)

To analyze mathematical models involving exponential and logarithmic functions in greater detail, we need to develop rules for computing the derivative of these functions. We begin by looking at the rule for computing the derivative of the exponential function.

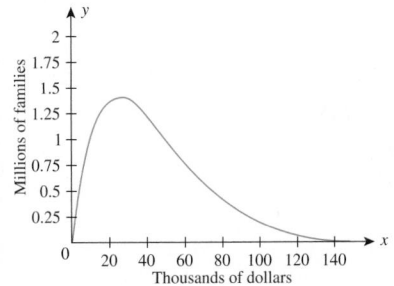

FIGURE 10
The graph of f shows the number of families versus their annual income.

Source: House Budget Committee, House Ways and Means Committee, and U.S. Census Bureau

---

Rule 1: Derivative of the Exponential Function
$$\frac{d}{dx}(e^x) = e^x$$

Thus, the derivative of the exponential function with base e is equal to the function itself. To demonstrate the validity of this rule, we compute

TABLE 2

h	$\dfrac{e^h - 1}{h}$
0.1	1.0517
0.01	1.0050
0.001	1.0005
−0.1	0.9516
−0.01	0.9950
−0.001	0.9995

$$
\begin{aligned}
f'(x) &= \lim_{h \to 0} \frac{f(x + h) - f(x)}{h} \\
&= \lim_{h \to 0} \frac{e^{x+h} - e^x}{h} \\
&= \lim_{h \to 0} \frac{e^x(e^h - 1)}{h} \qquad \text{Write } e^{x+h} = e^x e^h \text{ and factor.} \\
&= e^x \lim_{h \to 0} \frac{e^h - 1}{h} \qquad \text{Why?}
\end{aligned}
$$

To evaluate

$$
\lim_{h \to 0} \frac{e^h - 1}{h}
$$

let's refer to Table 2, which is constructed with the aid of a calculator. From the table, we see that

$$
\lim_{h \to 0} \frac{e^h - 1}{h} = 1
$$

(Although a rigorous proof of this fact is possible, it is beyond the scope of this book. Also see Example 1, Using Technology, page 884.) Using this result, we conclude that

$$f'(x) = e^x \cdot 1 = e^x$$

as we set out to show.

EXAMPLE 1 Find the derivative of each of the following functions:

a. $f(x) = x^2 e^x$ **b.** $g(t) = (e^t + 2)^{3/2}$

Solution

a. The product rule gives

$$f'(x) = \frac{d}{dx}(x^2 e^x) = x^2 \frac{d}{dx}(e^x) + e^x \frac{d}{dx}(x^2)$$
$$= x^2 e^x + e^x(2x) = xe^x(x + 2)$$

b. Using the general power rule, we find

$$g'(t) = \frac{3}{2}(e^t + 2)^{1/2}\frac{d}{dt}(e^t + 2) = \frac{3}{2}(e^t + 2)^{1/2}e^t = \frac{3}{2}e^t(e^t + 2)^{1/2}$$

EXPLORING WITH TECHNOLOGY

Consider the exponential function $f(x) = b^x$ ($b > 0$, $b \neq 1$).

1. Use the definition of the derivative of a function to show that

$$f'(x) = b^x \cdot \lim_{h \to 0} \frac{b^h - 1}{h}$$

2. Use the result of part 1 to show that

$$\frac{d}{dx}(2^x) = 2^x \cdot \lim_{h \to 0} \frac{2^h - 1}{h}$$
$$\frac{d}{dx}(3^x) = 3^x \cdot \lim_{h \to 0} \frac{3^h - 1}{h}$$

3. Use the technique in Using Technology, page 884, to show that (to two decimal places)

$$\lim_{h \to 0} \frac{2^h - 1}{h} = 0.69 \quad \text{and} \quad \lim_{h \to 0} \frac{3^h - 1}{h} = 1.10$$

4. Conclude from the results of parts 2 and 3 that

$$\frac{d}{dx}(2^x) \approx (0.69)2^x \quad \text{and} \quad \frac{d}{dx}(3^x) \approx (1.10)3^x$$

Thus,

$$\frac{d}{dx}(b^x) = k \cdot b^x$$

where k is an appropriate constant.

(continued)

5. The results of part 4 suggest that, for convenience, we pick the base b, where $2 < b < 3$, so that $k = 1$. This value of b is $e \approx 2.718281828. \ldots$ Thus,

$$\frac{d}{dx}(e^x) = e^x$$

This is why we prefer to work with the exponential function $f(x) = e^x$.

Applying the Chain Rule to Exponential Functions

To enlarge the class of exponential functions to be differentiated, we appeal to the chain rule to obtain the following rule for differentiating composite functions of the form $h(x) = e^{f(x)}$. An example of such a function is $h(x) = e^{x^2 - 2x}$. Here, $f(x) = x^2 - 2x$.

> **Rule 2: Chain Rule for Exponential Functions**
> If $f(x)$ is a differentiable function, then
>
> $$\frac{d}{dx}(e^{f(x)}) = e^{f(x)}f'(x)$$

To see this, observe that if $h(x) = g[f(x)]$, where $g(x) = e^x$, then by virtue of the chain rule,

$$h'(x) = g'(f(x))f'(x) = e^{f(x)}f'(x)$$

since $g'(x) = e^x$.

As an aid to remembering the chain rule for exponential functions, observe that it has the following form:

$$\frac{d}{dx}(e^{f(x)}) = e^{f(x)} \cdot \text{derivative of exponent}$$
$$\underset{\text{Same}}{\upharpoonleft \qquad \upharpoonright}$$

EXAMPLE 2   Find the derivative of each of the following functions:

a. $f(x) = e^{2x}$ **b.** $y = e^{-3x}$ **c.** $g(t) = e^{2t^2 + t}$

Solution

a. $f'(x) = e^{2x}\dfrac{d}{dx}(2x) = e^{2x} \cdot 2 = 2e^{2x}$

b. $\dfrac{dy}{dx} = e^{-3x}\dfrac{d}{dx}(-3x) = -3e^{-3x}$

c. $g'(t) = e^{2t^2 + t} \cdot \dfrac{d}{dt}(2t^2 + t) = (4t + 1)e^{2t^2 + t}$

EXAMPLE 3   Differentiate the function $y = xe^{-2x}$.

Solution   Using the product rule, followed by the chain rule, we find

$$\frac{dy}{dx} = x\frac{d}{dx}e^{-2x} + e^{-2x}\frac{d}{dx}(x)$$

$$= xe^{-2x}\frac{d}{dx}(-2x) + e^{-2x} \qquad \text{Use the chain rule on the first term.}$$

$$= -2xe^{-2x} + e^{-2x}$$

$$= e^{-2x}(1 - 2x)$$

EXAMPLE 4   Differentiate the function $g(t) = \dfrac{e^t}{e^t + e^{-t}}$.

Solution   Using the quotient rule, followed by the chain rule, we find

$$g'(t) = \frac{(e^t + e^{-t})\dfrac{d}{dt}(e^t) - e^t\dfrac{d}{dt}(e^t + e^{-t})}{(e^t + e^{-t})^2}$$

$$= \frac{(e^t + e^{-t})e^t - e^t(e^t - e^{-t})}{(e^t + e^{-t})^2}$$

$$= \frac{e^{2t} + 1 - e^{2t} + 1}{(e^t + e^{-t})^2} \qquad e^0 = 1$$

$$= \frac{2}{(e^t + e^{-t})^2}$$

EXAMPLE 5   In Section 14.5, we will discuss some practical applications of the exponential function

$$Q(t) = Q_0e^{kt}$$

where Q_0 and k are positive constants and $t \in [0, \infty)$. A quantity $Q(t)$ growing according to this law experiences exponential growth. Show that for a quantity $Q(t)$ experiencing exponential growth, the rate of growth of the quantity $Q'(t)$ at any time t is directly proportional to the amount of the quantity present.

Solution   Using the chain rule for exponential functions, we compute the derivative Q' of the function Q. Thus,

$$Q'(t) = Q_0e^{kt}\frac{d}{dt}(kt)$$

$$= Q_0e^{kt}(k)$$

$$= kQ_0e^{kt}$$

$$= kQ(t) \qquad Q(t) = Q_0e^{kt}$$

which is the desired conclusion.

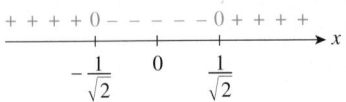

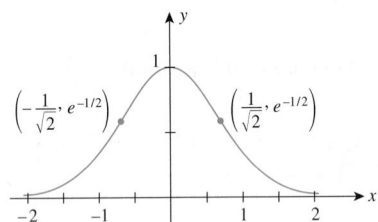

FIGURE 11
Sign diagram for f''

FIGURE 12
The graph of $y = e^{-x^2}$ has two inflection points.

EXAMPLE 6   Find the inflection points of the function $f(x) = e^{-x^2}$.

Solution   The first derivative of f is

$$f'(x) = -2xe^{-x^2}$$

Differentiating $f'(x)$ with respect to x yields

$$f''(x) = (-2x)(-2xe^{-x^2}) - 2e^{-x^2}$$
$$= 2e^{-x^2}(2x^2 - 1)$$

Setting $f''(x) = 0$ gives

$$2e^{-x^2}(2x^2 - 1) = 0$$

Since e^{-x^2} never equals zero for any real value of x, we see that $x = \pm 1/\sqrt{2}$ are the only candidates for inflection points of f. The sign diagram of f'', shown in Figure 11, tells us that both $x = -1/\sqrt{2}$ and $x = 1/\sqrt{2}$ give rise to inflection points of f.
    Next,

$$f\left(-\frac{1}{\sqrt{2}}\right) = f\left(\frac{1}{\sqrt{2}}\right) = e^{-1/2}$$

and the inflection points of f are $(-1/\sqrt{2}, e^{-1/2})$ and $(1/\sqrt{2}, e^{-1/2})$. The graph of f appears in Figure 12.

Our final example involves finding the absolute maximum of an exponential function.

APPLIED EXAMPLE 7   Optimal Market Price   Refer to Example 8, Section 5.1. The present value of the market price of the Blakely Office Building is given by

$$P(t) = 300{,}000e^{-0.09t + \sqrt{t}/2} \qquad (0 \leq t \leq 10)$$

Find the optimal present value of the building's market price.

Solution   To find the maximum value of P over $[0, 10]$, we compute

$$P'(t) = 300{,}000e^{-0.09t + \sqrt{t}/2}\frac{d}{dt}\left(-0.09t + \frac{1}{2}t^{1/2}\right)$$
$$= 300{,}000e^{-0.09t + \sqrt{t}/2}\left(-0.09 + \frac{1}{4}t^{-1/2}\right)$$

Setting $P'(t) = 0$ gives

$$-0.09 + \frac{1}{4t^{1/2}} = 0$$

since $e^{-0.09t + \sqrt{t}/2}$ is never zero for any value of t. Solving this equation, we find

$$\frac{1}{4t^{1/2}} = 0.09$$

$$t^{1/2} = \frac{1}{4(0.09)}$$

$$= \frac{1}{0.36}$$

$$t = \left(\frac{1}{0.36}\right)^2 \approx 7.72$$

the sole critical number of the function P. Finally, evaluating $P(t)$ at the critical number as well as at the endpoints of $[0, 10]$, we have

t	0	7.72	10
$P(t)$	300,000	600,779	592,838

We conclude, accordingly, that the optimal present value of the property's market price is $600,779 and that this will occur 7.72 years from now.   ∎

14.3   Self-Check Exercises

1. Let $f(x) = xe^{-x}$.
  a. Find the first and second derivatives of f.
  b. Find the relative extrema of f.
  c. Find the inflection points of f.

2. An industrial asset is being depreciated at a rate so that its book value t yr from now will be

$$V(t) = 50{,}000e^{-0.4t}$$

dollars. How fast will the book value of the asset be changing 3 yr from now?

Solutions to Self-Check Exercises 14.3 can be found on page 883.

14.3   Concept Questions

1. State the rule for differentiating (a) $f(x) = e^x$ and (b) $g(x) = e^{f(x)}$, where f is a differentiable function.

2. Let $f(x) = e^{kx}$.
  a. Compute $f'(x)$.
  b. Use the result to deduce the behavior of f for the case $k > 0$ and the case $k < 0$.

14.3   Exercises

In Exercises 1–28, find the derivative of the function.

1. $f(x) = e^{3x}$

2. $f(x) = 3e^x$

3. $g(t) = e^{-t}$

4. $f(x) = e^{-2x}$

5. $f(x) = e^x + x$

6. $f(x) = 2e^x - x^2$

7. $f(x) = x^3e^x$

8. $f(u) = u^2e^{-u}$

9. $f(x) = \dfrac{2e^x}{x}$

10. $f(x) = \dfrac{x}{e^x}$

11. $f(x) = 3(e^x + e^{-x})$

12. $f(x) = \dfrac{e^x + e^{-x}}{2}$

13. $f(w) = \dfrac{e^w + 1}{e^w}$

14. $f(x) = \dfrac{e^x}{e^x + 1}$

15. $f(x) = 2e^{3x-1}$

16. $f(t) = 4e^{3t+2}$

17. $h(x) = e^{-x^2}$

18. $f(x) = e^{x^2-1}$

19. $f(x) = 3e^{-1/x}$

20. $f(x) = e^{1/(2x)}$

21. $f(x) = (e^x + 1)^{25}$

22. $f(x) = (4 - e^{-3x})^3$

23. $f(x) = e^{\sqrt{x}}$

24. $f(x) = -e^{-\sqrt{2t}}$

25. $f(x) = (x - 1)e^{3x+2}$

26. $f(s) = (s^2 + 1)e^{-s^2}$

27. $f(x) = \dfrac{e^x - 1}{e^x + 1}$

28. $g(t) = \dfrac{e^{-t}}{1 + t^2}$

In Exercises 29–32, find the second derivative of the function.

29. $f(x) = e^{-4x} + 2e^{3x}$

30. $f(t) = 3e^{-2t} - 5e^{-t}$

31. $f(x) = 2xe^{3x}$

32. $f(t) = t^2 e^{-2t}$

33. Find an equation of the tangent line to the graph of $y = e^{2x-3}$ at the point $\left(\frac{3}{2}, 1\right)$.

34. Find an equation of the tangent line to the graph of $y = e^{-x^2}$ at the point $(1, 1/e)$.

35. Determine the intervals where the function $f(x) = e^{-x^2/2}$ is increasing and where it is decreasing.

36. Determine the intervals where the function $f(x) = x^2 e^{-x}$ is increasing and where it is decreasing.

37. Determine the intervals of concavity for the function $f(x) = \dfrac{e^x - e^{-x}}{2}$.

38. Determine the intervals of concavity for the function $f(x) = xe^x$.

39. Find the inflection point of the function $f(x) = xe^{-2x}$.

40. Find the inflection point(s) of the function $f(x) = 2e^{-x^2}$.

41. Find the equations of the tangent lines to the graph of $f(x) = e^{-x^2}$ at its inflection points.

42. Find an equation of the tangent line to the graph of $f(x) = xe^{-x}$ at its inflection point.

In Exercises 43–46, find the absolute extrema of the function.

43. $f(x) = e^{-x^2}$ on $[-1, 1]$

44. $h(x) = e^{x^2-4}$ on $[-2, 2]$

45. $g(x) = (2x - 1)e^{-x}$ on $[0, \infty)$

46. $f(x) = xe^{-x^2}$ on $[0, 2]$

In Exercises 47–50, use the curve-sketching guidelines of Chapter 13, page 814, to sketch the graph of the function.

47. $f(t) = e^t - t$

48. $h(x) = \dfrac{e^x + e^{-x}}{2}$

49. $f(x) = 2 - e^{-x}$

50. $f(x) = \dfrac{3}{1 + e^{-x}}$

51. PERCENTAGE OF POPULATION RELOCATING   Based on data obtained from the Census Bureau, the manager of Plymouth Van Lines estimates that the percent of the total population relocating in year t ($t = 0$ corresponds to the year 1960) may be approximated by the formula

$$P(t) = 20.6e^{-0.009t} \qquad (0 \leq t \leq 35)$$

Compute $P'(10)$, $P'(20)$, and $P'(30)$ and interpret your results.

52. ONLINE BANKING   In a study prepared in 2000, the percent of households using online banking was projected to be

$$f(t) = 1.5e^{0.78t} \qquad (0 \leq t \leq 4)$$

where t is measured in years, with $t = 0$ corresponding to the beginning of 2000.
a. What was the projected percent of households using online banking at the beginning of 2003?
b. How fast was the projected percent of households using online banking changing at the beginning of 2003?
c. How fast was the rate of the projected percent of households using online banking changing at the beginning of 2003?
Hint: We want $f''(3)$. Why?

Source: Online Banking Report

53. OVER-100 POPULATION   Based on data obtained from the Census Bureau, the number of Americans over age 100 is expected to be

$$P(t) = 0.07e^{0.54t} \qquad (0 \leq t \leq 4)$$

where $P(t)$ is measured in millions and t is measured in decades, with $t = 0$ corresponding to the beginning of 2000.
a. What was the population of Americans over age 100 at the beginning of 2000? What will it be at the beginning of 2030?
b. How fast was the population of Americans over age 100 changing at the beginning of 2000? How fast will it be changing at the beginning of 2030?

Source: U.S. Census Bureau

54. WORLD POPULATION GROWTH   After its fastest rate of growth ever during the 1980s and 1990s, the rate of growth of world population is expected to slow dramatically in the 21st century. The function

$$G(t) = 1.58e^{-0.213t}$$

gives the projected annual average percent population growth/decade in the tth decade, with $t = 1$ corresponding to 2000.

a. What will the projected annual average population growth rate be in 2020 ($t = 3$)?

b. How fast will the projected annual average population growth rate be changing in 2020?

Source: U.S. Census Bureau

55. **LOANS AT JAPANESE BANKS** The total loans outstanding at all Japanese banks have been declining in recent years. The function

$$L(t) = 4.6e^{-0.04t} \qquad (0 \le t \le 6)$$

gives the approximate total loans outstanding from 1998 ($t = 0$) through 2004 in trillions of dollars.

a. What were the total loans outstanding in 1998? In 2004?

b. How fast were the total loans outstanding declining in 1998? In 2004?

c. Show that the total loans outstanding were declining but at a slower rate over the years between 1998 and 2004.
Hint: Look at L'' on the interval $(0, 6)$.

Source: Bank of Japan

56. **ENERGY CONSUMPTION OF APPLIANCES** The average energy consumption of the typical refrigerator/freezer manufactured by York Industries is approximately

$$C(t) = 1486e^{-0.073t} + 500 \qquad (0 \le t \le 20)$$

kilowatt-hours (kWh) per year, where t is measured in years, with $t = 0$ corresponding to 1972.

a. What was the average energy consumption of the York refrigerator/freezer at the beginning of 1972?

b. Prove that the average energy consumption of the York refrigerator/freezer is decreasing over the years in question.

c. All refrigerator/freezers manufactured as of January 1, 1990, must meet the 950-kWh/year maximum energy-consumption standard set by the National Appliance Conservation Act. Show that the York refrigerator/freezer satisfies this requirement.

57. **SALES PROMOTION** The Lady Bug, a women's clothing chain store, found that t days after the end of a sales promotion the volume of sales was given by

$$S(t) = 20{,}000(1 + e^{-0.5t}) \qquad (0 \le t \le 5)$$

dollars.

a. Find the rate of change of The Lady Bug's sales volume when $t = 1$, $t = 2$, $t = 3$, and $t = 4$.

b. After how many days will the sales volume drop below $27,400?

58. **BLOOD ALCOHOL LEVEL** The percentage of alcohol in a person's bloodstream t hr after drinking 8 fluid oz of whiskey is given by

$$A(t) = 0.23te^{-0.4t} \qquad (0 \le t \le 12)$$

a. What is the percentage of alcohol in a person's bloodstream after $\frac{1}{2}$ hr? After 8 hr?

b. How fast is the percentage of alcohol in a person's bloodstream changing after $\frac{1}{2}$ hr? After 8 hr?

Source: Encyclopedia Britannica

59. **POLIO IMMUNIZATION** Polio, a once-feared killer, declined markedly in the United States in the 1950s after Jonas Salk developed the inactivated polio vaccine and mass immunization of children took place. The number of polio cases in the United States from the beginning of 1959 to the beginning of 1963 is approximated by the function

$$N(t) = 5.3e^{0.095t^2 - 0.85t} \qquad (0 \le t \le 4)$$

where $N(t)$ gives the number of polio cases (in thousands) and t is measured in years with $t = 0$ corresponding to the beginning of 1959.

a. Show that the function N is decreasing over the time interval under consideration.

b. How fast was the number of polio cases decreasing at the beginning of 1959? At the beginning of 1962? (*Comment:* Following the introduction of the oral vaccine developed by Dr. Albert B. Sabin in 1963, polio in the United States has, for all practical purposes, been eliminated.)

60. **MARGINAL REVENUE** The unit selling price p (in dollars) and the quantity demanded (in pairs) of a certain brand of women's gloves is given by the demand equation

$$p = 100e^{-0.0001x} \qquad (0 \le x \le 20{,}000)$$

a. Find the revenue function R.
Hint: $R(x) = px$.

b. Find the marginal revenue function R'.

c. What is the marginal revenue when $x = 10$?

61. **MAXIMIZING REVENUE** Refer to Exercise 60. How many pairs of the gloves must be sold to yield a maximum revenue? What will be the maximum revenue?

62. **PRICE OF PERFUME** The monthly demand for a certain brand of perfume is given by the demand equation

$$p = 100e^{-0.0002x} + 150$$

where p denotes the retail unit price (in dollars) and x denotes the quantity (in 1-oz bottles) demanded.

a. Find the rate of change of the price per bottle when $x = 1000$ and when $x = 2000$.

b. What is the price per bottle when $x = 1000$? When $x = 2000$?

63. PRICE OF WINE The monthly demand for a certain brand of table wine is given by the demand equation

$$p = 240\left(1 - \frac{3}{3 + e^{-0.0005x}}\right)$$

where p denotes the wholesale price per case (in dollars) and x denotes the number of cases demanded.
a. Find the rate of change of the price per case when $x = 1000$.
b. What is the price per case when $x = 1000$?

64. SPREAD OF AN EPIDEMIC During a flu epidemic, the total number of students on a state university campus who had contracted influenza by the xth day was given by

$$N(x) = \frac{3000}{1 + 99e^{-x}} \qquad (x \geq 0)$$

a. How many students had influenza initially?
b. Derive an expression for the rate at which the disease was being spread and prove that the function N is increasing on the interval $(0, \infty)$.
c. Sketch the graph of N. What was the total number of students who contracted influenza during that particular epidemic?

65. WEIGHTS OF CHILDREN The Ehrenberg equation

$$W = 2.4e^{1.84h}$$

gives the relationship between the height (in meters) and the average weight W (in kilograms) for children between 5 and 13 yr of age.
a. What is the average weight of a 10-yr-old child who stands at 1.6 m tall?
b. Use differentials to estimate the change in the average weight of a 10-yr-old child whose height increases from 1.6 m to 1.65 m.

66. MAXIMUM OIL PRODUCTION It has been estimated that the total production of oil from a certain oil well is given by

$$T(t) = -1000(t + 10)e^{-0.1t} + 10,000$$

thousand barrels t yr after production has begun. Determine the year when the oil well will be producing at maximum capacity.

67. OPTIMAL SELLING TIME The present value of a piece of waterfront property purchased by an investor is given by the function

$$P(t) = 80,000e^{\sqrt{t}/2 - 0.09t} \qquad (0 \leq t \leq 8)$$

Determine the optimal time (based on present value) for the investor to sell the property. What is the property's optimal present value?

68. BLOOD ALCOHOL LEVEL Refer to Exercise 58, p. 881. At what time after drinking the alcohol is the percentage of alcohol in

the person's bloodstream at its highest level? What is that level?

69. OIL USED TO FUEL PRODUCTIVITY A study on worldwide oil use was prepared for a major oil company. The study predicted that the amount of oil used to fuel productivity in a certain country is given by

$$f(t) = 1.5 + 1.8te^{-1.2t} \qquad (0 \leq t \leq 4)$$

where $f(t)$ denotes the number of barrels per \$1000 of economic output and t is measured in decades ($t = 0$ corresponds to 1965). Compute $f'(0), f'(1), f'(2)$, and $f'(3)$ and interpret your results.

70. PRICE OF A COMMODITY The price of a certain commodity in dollars per unit at time t (measured in weeks) is given by $p = 18 - 3e^{-2t} - 6e^{-t/3}$.
a. What is the price of the commodity at $t = 0$?
b. How fast is the price of the commodity changing at $t = 0$?
c. Find the equilibrium price of the commodity.
Hint: It is given by $\lim_{t \to \infty} p$.

71. PRICE OF A COMMODITY The price of a certain commodity in dollars per unit at time t (measured in weeks) is given by $p = 8 + 4e^{-2t} + te^{-2t}$.
a. What is the price of the commodity at $t = 0$?
b. How fast is the price of the commodity changing at $t = 0$?
c. Find the equilibrium price of the commodity.
Hint: It's given by $\lim_{t \to \infty} p$. Also, use the fact that $\lim_{t \to \infty} te^{-2t} = 0$.

72. ABSORPTION OF DRUGS A liquid carries a drug into an organ of volume V cm^3 at the rate of a cm^3/sec and leaves at the same rate. The concentration of the drug in the entering liquid is c g/cm^3. Letting $x(t)$ denote the concentration of the drug in the organ at any time t, we have $x(t) = c(1 - e^{-at/V})$.
a. Show that x is an increasing function on $(0, \infty)$.
b. Sketch the graph of x.

73. ABSORPTION OF DRUGS Refer to Exercise 72. Suppose the maximum concentration of the drug in the organ must *not* exceed m g/cm^3, where $m < c$. Show that the liquid must not be allowed to enter the organ for a time longer than

$$T = \left(\frac{V}{a}\right) \ln\left(\frac{c}{c - m}\right)$$

minutes.

74. CONCENTRATION OF A DRUG IN THE BLOODSTREAM The concentration of a drug in the bloodstream t sec after injection into a muscle is given by

$$y = c(e^{-bt} - e^{-at}) \qquad (\text{for } t \geq 0)$$

where a, b, and c are positive constants, with $a > b$.
a. Find the time at which the concentration is maximal.
b. Find the time at which the concentration of the drug in the bloodstream is decreasing most rapidly.

75. **ABSORPTION OF DRUGS** The concentration of a drug in an organ at any time t (in seconds) is given by

$$C(t) = \begin{cases} 0.3t - 18(1 - e^{-t/60}) & \text{if } 0 \le t \le 20 \\ 18e^{-t/60} - 12e^{-(t-20)/60} & \text{if } t > 20 \end{cases}$$

where $C(t)$ is measured in grams/cubic centimeter (g/cm^3).
a. How fast is the concentration of the drug in the organ changing after 10 sec?
b. How fast is the concentration of the drug in the organ changing after 30 sec?
c. When will the concentration of the drug in the organ reach a maximum?
d. What is the maximum drug concentration in the organ?

76. **ABSORPTION OF DRUGS** Jane took 100 mg of a drug in the morning and another 100 mg of the same drug at the same time the following morning. The amount of the drug in her body t days after the first dosage was taken is given by

$$A(t) = \begin{cases} 100e^{-1.4t} & \text{if } 0 \le t < 1 \\ 100(1 + e^{1.4})e^{-1.4t} & \text{if } t \ge 1 \end{cases}$$

a. How fast was the amount of drug in Jane's body changing after 12 hr $(t = \frac{1}{2})$? After 2 days?
b. When was the amount of drug in Jane's body a maximum?
c. What was the maximum amount of drug in Jane's body?

In Exercises 77–80, determine whether the statement is true or false. If it is true, explain why it is true. If it is false, give an example to show why it is false.

77. If $f(x) = 3^x$, then $f'(x) = x \cdot 3^{x-1}$.

78. If $f(x) = e^{\pi}$, then $f'(x) = e^{\pi}$.

79. If $f(x) = \pi^x$, then $f'(x) = \pi^x$.

80. If $x^2 + e^y = 10$, then $y' = \dfrac{-2x}{e^y}$.

14.3 Solutions to Self-Check Exercises

1. **a.** Using the product rule, we obtain

$$f'(x) = x\frac{d}{dx}e^{-x} + e^{-x}\frac{d}{dx}x$$

$$= -xe^{-x} + e^{-x} = (1 - x)e^{-x}$$

Using the product rule once again, we obtain

$$f''(x) = (1 - x)\frac{d}{dx}e^{-x} + e^{-x}\frac{d}{dx}(1 - x)$$

$$= (1 - x)(-e^{-x}) + e^{-x}(-1)$$

$$= -e^{-x} + xe^{-x} - e^{-x} = (x - 2)e^{-x}$$

b. Setting $f'(x) = 0$ gives

$$(1 - x)e^{-x} = 0$$

Since $e^{-x} \ne 0$, we see that $1 - x = 0$, and this gives 1 as the only critical number of f. The sign diagram of f' shown in the accompanying figure tells us that the point $(1, e^{-1})$ is a relative maximum of f.

c. Setting $f''(x) = 0$ gives $x - 2 = 0$, so $x = 2$ is a candidate for an inflection point of f. The sign diagram of f'' (see the accompanying figure) shows that $(2, 2e^{-2})$ is an inflection point of f.

2. The rate change of the book value of the asset t yr from now is

$$V'(t) = 50,000\frac{d}{dt}e^{-0.4t}$$

$$= 50,000(-0.4)e^{-0.4t} = -20,000e^{-0.4t}$$

Therefore, 3 yr from now the book value of the asset will be changing at the rate of

$$V'(3) = -20,000e^{-0.4(3)} = -20,000e^{-1.2} \approx -6023.88$$

—that is, decreasing at the rate of approximately \$6024/year.

USING TECHNOLOGY

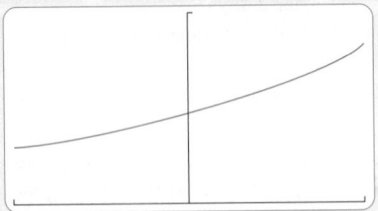

FIGURE T1
The graph of f in the viewing window
$[-1, 1] \times [0, 2]$

EXAMPLE 1   At the beginning of Section 14.3, we demonstrated via a table of values of $(e^h - 1)/h$ for selected values of h the plausibility of the result

$$\lim_{h \to 0} \frac{e^h - 1}{h} = 1$$

To obtain a visual confirmation of this result, we plot the graph of

$$f(x) = \frac{e^x - 1}{x}$$

in the viewing window $[-1, 1] \times [0, 2]$ (Figure T1). From the graph of f, we see that $f(x)$ appears to approach 1 as x approaches 0.

The numerical derivative function of a graphing utility will yield the derivative of an exponential or logarithmic function for any value of x, just as it did for algebraic functions.*

---

*The rules for differentiating logarithmic functions will be covered in Section 14.4. However, the exercises given here can be done without using these rules.

TECHNOLOGY EXERCISES

In Exercises 1–6, use the numerical derivative operation to find the rate of change of $f(x)$ at the given value of x. Give your answer accurate to four decimal places.

1. $f(x) = x^3 e^{-1/x};\ x = -1$

2. $f(x) = (\sqrt{x} + 1)^{3/2} e^{-x};\ x = 0.5$

3. $f(x) = x^3 \sqrt{\ln x};\ x = 2$

4. $f(x) = \dfrac{\sqrt{x} \ln x}{x + 1};\ x = 3.2$

5. $f(x) = e^{-x} \ln(2x + 1);\ x = 0.5$

6. $f(x) = \dfrac{e^{-\sqrt{x}}}{\ln(x^2 + 1)};\ x = 1$

7. AN EXTINCTION SITUATION   The number of saltwater crocodiles in a certain area of northern Australia is given by

$$P(t) = \frac{300e^{-0.024t}}{5e^{-0.024t} + 1}$$

a. How many crocodiles were in the population initially?
b. Show that $\lim_{t \to \infty} P(t) = 0$.
c. Plot the graph of P in the viewing window $[0, 200] \times [0, 70]$.
(*Comment:* This phenomenon is referred to as an *extinction situation.*)

8. INCOME OF AMERICAN FAMILIES   Based on data, it is estimated that the number of American families y (in millions) who earned x thousand dollars in 1990 is related by the equation

$$y = 0.1584xe^{-0.0000016x^3 + 0.00011x^2 - 0.04491x} \qquad (x > 0)$$

a. Plot the graph of the equation in the viewing window $[0, 150] \times [0, 2]$.
b. How fast is y changing with respect to x when $x = 10$? When $x = 50$? Interpret your results.

Source: House Budget Committee, House Ways and Means Committee, and U.S. Census Bureau

9. WORLD POPULATION GROWTH   Based on data obtained in a study, the world population (in billions) is approximated by the function

$$f(t) = \frac{12}{1 + 3.74914e^{-1.42804t}} \qquad (0 \le t \le 4)$$

where t is measured in half centuries, with $t = 0$ corresponding to the beginning of 1950.
a. Plot the graph of f in the viewing window $[0, 5] \times [0, 14]$.
b. How fast was the world population expected to increase at the beginning of 2000?

Source: United Nations Population Division

10. LOAN AMORTIZATION   The Sotos plan to secure a loan of $160,000 to purchase a house. They are considering a conventional 30-yr home mortgage at 9%/year on the unpaid balance. It can be shown that the Sotos will have an outstanding principal of

$$B(x) = \frac{160,000(1.0075^{360} - 1.0075^{x})}{1.0075^{360} - 1}$$

dollars after making x monthly payments of $1287.40.

a. Plot the graph of $B(x)$, using the viewing window $[0, 360]$ $\times$ $[0, 160{,}000]$.

b. Compute $B(0)$ and $B'(0)$ and interpret your results; compute $B(180)$ and $B'(180)$ and interpret your results.

11. **INCREASE IN JUVENILE OFFENDERS**   The number of youths aged 15 to 19 increased by 21% between 1994 and 2005, pushing up the crime rate. According to the National Council on Crime and Delinquency, the number of violent crime arrests of juveniles under age 18 in year t is given by

$$f(t) = -0.438t^2 + 9.002t + 107 \qquad (0 \le t \le 13)$$

where $f(t)$ is measured in thousands and t in years, with $t = 0$ corresponding to 1989. According to the same source, if trends like inner-city drug use and wider availability of guns continues, then the number of violent crime arrests of juveniles under age 18 in year t is given by

$$g(t) = \begin{cases} -0.438t^2 + 9.002t + 107 & \text{if } 0 \le t < 4 \\ 99.456e^{0.07824t} & \text{if } 4 \le t \le 13 \end{cases}$$

where $g(t)$ is measured in thousands and $t = 0$ corresponds to 1989.

a. Compute $f(11)$ and $g(11)$ and interpret your results.

b. Compute $f'(11)$ and $g'(11)$ and interpret your results.

Source: National Council on Crime and Delinquency

12. **INCREASING CROP YIELDS**   If left untreated on bean stems, aphids (small insects that suck plant juices) will multiply at an increasing rate during the summer months and reduce productivity and crop yield of cultivated crops. But if the aphids are treated in mid-June, the numbers decrease sharply to less than 100/bean stem, allowing for steep rises in crop yield. The function

$$F(t) = \begin{cases} 62e^{1.152t} & \text{if } 0 \le t < 1.5 \\ 349e^{-1.324(t-1.5)} & \text{if } 1.5 \le t \le 3 \end{cases}$$

gives the number of aphids on a typical bean stem at time t, where t is measured in months, with $t = 0$ corresponding to the beginning of May.

a. How many aphids are there on a typical bean stem at the beginning of June $(t = 1)$? At the beginning of July $(t = 2)$?

b. How fast is the population of aphids changing at the beginning of June? At the beginning of July?

Source: The Random House Encyclopedia

13. **PERCENT OF FEMALES IN THE LABOR FORCE**   Based on data from the U.S. Census Bureau, the chief economist of Manpower, Inc., constructed the following formula giving the percent of the total female population in the civilian labor force, $P(t)$, at the beginning of the tth decade ($t = 0$ corresponds to the year 1900):

$$P(t) = \frac{74}{1 + 2.6e^{-0.166t + 0.04536t^2 - 0.0066t^3}} \qquad (0 \le t \le 11)$$

Assume this trend continued for the rest of the twentieth century.

a. What was the percent of the total female population in the civilian labor force at the beginning of 2000?

b. What was the growth rate of the percent of the total female population in the civilian labor force at the beginning of 2000?

Source: U.S. Census Bureau

14.4 Differentiation of Logarithmic Functions

■ The Derivative of ln x

Let's now turn our attention to the differentiation of logarithmic functions.

Rule 3: Derivative of ln x

$$\frac{d}{dx} \ln |x| = \frac{1}{x} \qquad (x \ne 0)$$

To derive Rule 3, suppose $x > 0$ and write $f(x) = \ln x$ in the equivalent form

$$x = e^{f(x)}$$

Differentiating both sides of the equation with respect to x, we find, using the chain rule,

$$1 = e^{f(x)} \cdot f'(x)$$

from which we see that

$$f'(x) = \frac{1}{e^{f(x)}}$$

or, since $e^{f(x)} = x$,

$$f'(x) = \frac{1}{x}$$

as we set out to show. You are asked to prove the rule for the case $x < 0$ in Exercise 71, page 892.

EXAMPLE 1   Find the derivative of each function:

a. $f(x) = x \ln x$ **b.** $g(x) = \dfrac{\ln x}{x}$

Solution

a. Using the product rule, we obtain

$$f'(x) = \frac{d}{dx}(x \ln x) = x \frac{d}{dx}(\ln x) + (\ln x) \frac{d}{dx}(x)$$

$$= x\left(\frac{1}{x}\right) + \ln x = 1 + \ln x$$

b. Using the quotient rule, we obtain

$$g'(x) = \frac{x \dfrac{d}{dx}(\ln x) - (\ln x)\dfrac{d}{dx}(x)}{x^2} = \frac{x\left(\dfrac{1}{x}\right) - \ln x}{x^2} = \frac{1 - \ln x}{x^2}$$

EXPLORE & DISCUSS

You can derive the formula for the derivative of $f(x) = \ln x$ directly from the definition of the derivative, as follows.

1. Show that

$$f'(x) = \lim_{h \to 0} \frac{f(x+h) - f(x)}{h} = \lim_{h \to 0} \ln\left(1 + \frac{h}{x}\right)^{1/h}$$

2. Put $m = x/h$ and note that $m \to \infty$ as $h \to 0$. Then, $f'(x)$ can be written in the form

$$f'(x) = \lim_{m \to \infty} \ln\left(1 + \frac{1}{m}\right)^{m/x}$$

3. Finally, use both the fact that the natural logarithmic function is continuous and the definition of the number e to show that

$$f'(x) = \frac{1}{x} \ln\left[\lim_{m \to \infty}\left(1 + \frac{1}{m}\right)^m\right] = \frac{1}{x}$$

■ The Chain Rule for Logarithmic Functions

To enlarge the class of logarithmic functions to be differentiated, we appeal once more to the chain rule to obtain the following rule for differentiating composite functions of the form $h(x) = \ln f(x)$, where $f(x)$ is assumed to be a positive differentiable function.

> **Rule 4: Chain Rule for Logarithmic Functions**
> If $f(x)$ is a differentiable function, then
> $$\frac{d}{dx}[\ln f(x)] = \frac{f'(x)}{f(x)} \qquad [f(x) > 0]$$

To see this, observe that $h(x) = g[f(x)]$, where $g(x) = \ln x$ $(x > 0)$. Since $g'(x) = 1/x$, we have, using the chain rule,

$$h'(x) = g'(f(x))f'(x)$$
$$= \frac{1}{f(x)}f'(x) = \frac{f'(x)}{f(x)}$$

Observe that in the special case $f(x) = x$, $h(x) = \ln x$, so the derivative of h is, by Rule 3, given by $h'(x) = 1/x$.

EXAMPLE 2   Find the derivative of the function $f(x) = \ln(x^2 + 1)$.

Solution   Using Rule 4, we see immediately that

$$f'(x) = \frac{\dfrac{d}{dx}(x^2 + 1)}{x^2 + 1} = \frac{2x}{x^2 + 1}$$

When differentiating functions involving logarithms, the rules of logarithms may be used to advantage, as shown in Examples 3 and 4.

EXAMPLE 3   Differentiate the function $y = \ln[(x^2 + 1)(x^3 + 2)^6]$.

Solution   We first rewrite the given function using the properties of logarithms:

$$y = \ln[(x^2 + 1)(x^3 + 2)^6]$$
$$= \ln(x^2 + 1) + \ln(x^3 + 2)^6 \qquad \ln mn = \ln m + \ln n$$
$$= \ln(x^2 + 1) + 6\ln(x^3 + 2) \qquad \ln m^n = n\ln m$$

Differentiating and using Rule 4, we obtain

$$y' = \frac{\dfrac{d}{dx}(x^2 + 1)}{x^2 + 1} + \frac{6\dfrac{d}{dx}(x^3 + 2)}{x^3 + 2}$$
$$= \frac{2x}{x^2 + 1} + \frac{6(3x^2)}{x^3 + 2} = \frac{2x}{x^2 + 1} + \frac{18x^2}{x^3 + 2}$$

EXPLORING WITH TECHNOLOGY

Use a graphing utility to plot the graphs of $f(x) = \ln x$; its first derivative function, $f'(x) = 1/x$; and its second derivative function $f''(x) = -1/x^2$, using the same viewing window $[0, 4] \times [-3, 3]$.

1. Describe the properties of the graph of f revealed by studying the graph of $f'(x)$. What can you say about the rate of increase of f for large values of x?

2. Describe the properties of the graph of f revealed by studying the graph of $f''(x)$. What can you say about the concavity of f for large values of x?

EXAMPLE 4   Find the derivative of the function $g(t) = \ln(t^2 e^{-t^2})$.

Solution   Here again, to save a lot of work, we first simplify the given expression using the properties of logarithms. We have

$$
\begin{aligned}
g(t) &= \ln(t^2 e^{-t^2}) \\
&= \ln t^2 + \ln e^{-t^2} \qquad &&\ln mn = \ln m + \ln n \\
&= 2 \ln t - t^2 \qquad &&\ln m^n = n \ln m \quad \text{and} \quad \ln e = 1
\end{aligned}
$$

Therefore,

$$
g'(t) = \frac{2}{t} - 2t = \frac{2(1 - t^2)}{t}
$$

Logarithmic Differentiation

As we saw in the last two examples, the task of finding the derivative of a given function can be made easier by first applying the laws of logarithms to simplify the function. We now illustrate a process called **logarithmic differentiation**, which not only simplifies the calculation of the derivatives of certain functions but also enables us to compute the derivatives of functions we could not otherwise differentiate using the techniques developed thus far.

EXAMPLE 5   Differentiate $y = x(x + 1)(x^2 + 1)$, using logarithmic differentiation.

Solution   First, we take the natural logarithm on both sides of the given equation, obtaining

$$
\ln y = \ln x(x + 1)(x^2 + 1)
$$

Next, we use the properties of logarithms to rewrite the right-hand side of this equation, obtaining

$$
\ln y = \ln x + \ln(x + 1) + \ln(x^2 + 1)
$$

If we differentiate both sides of this equation, we have

$$
\begin{aligned}
\frac{d}{dx} \ln y &= \frac{d}{dx} \left[\ln x + \ln(x + 1) + \ln(x^2 + 1) \right] \\
&= \frac{1}{x} + \frac{1}{x + 1} + \frac{2x}{x^2 + 1} \qquad \text{Use Rule 4.}
\end{aligned}
$$

To evaluate the expression on the left-hand side, note that y is a function of x. Therefore, writing $y = f(x)$ to remind us of this fact, we have

$$\frac{d}{dx} \ln y = \frac{d}{dx} \ln[f(x)] \qquad \text{Write } y = f(x).$$

$$= \frac{f'(x)}{f(x)} \qquad \text{Use Rule 4.}$$

$$= \frac{y'}{y} \qquad \text{Return to using } y \text{ instead of } f(x).$$

Therefore, we have

$$\frac{y'}{y} = \frac{1}{x} + \frac{1}{x+1} + \frac{2x}{x^2+1}$$

Finally, solving for y', we have

$$y' = y\left(\frac{1}{x} + \frac{1}{x+1} + \frac{2x}{x^2+1}\right)$$

$$= x(x+1)(x^2+1)\left(\frac{1}{x} + \frac{1}{x+1} + \frac{2x}{x^2+1}\right)$$

Before considering other examples, let's summarize the important steps involved in logarithmic differentiation.

Finding $\dfrac{dy}{dx}$ by Logarithmic Differentiation

1. Take the natural logarithm on both sides of the equation and use the properties of logarithms to write any "complicated expression" as a sum of simpler terms.

2. Differentiate both sides of the equation with respect to x.

3. Solve the resulting equation for $\dfrac{dy}{dx}$.

EXAMPLE 6   Differentiate $y = x^2(x-1)(x^2+4)^3$.

Solution   Taking the natural logarithm on both sides of the given equation and using the laws of logarithms, we obtain

$$\ln y = \ln x^2(x-1)(x^2+4)^3$$

$$= \ln x^2 + \ln(x-1) + \ln(x^2+4)^3$$

$$= 2\ln x + \ln(x-1) + 3\ln(x^2+4)$$

Differentiating both sides of the equation with respect to x, we have

$$\frac{d}{dx} \ln y = \frac{y'}{y} = \frac{2}{x} + \frac{1}{x-1} + 3 \cdot \frac{2x}{x^2+4}$$

Finally, solving for y', we have

$$y' = y\left(\frac{2}{x} + \frac{1}{x-1} + \frac{6x}{x^2+4}\right)$$

$$= x^2(x-1)(x^2+4)^3\left(\frac{2}{x} + \frac{1}{x-1} + \frac{6x}{x^2+4}\right)$$

EXAMPLE 7   Find the derivative of $f(x) = x^x$ $(x > 0)$.

Solution   A word of caution! This function is neither a power function nor an exponential function. Taking the natural logarithm on both sides of the equation gives

$$\ln f(x) = \ln x^x = x \ln x$$

Differentiating both sides of the equation with respect to x, we obtain

$$\frac{f'(x)}{f(x)} = x \frac{d}{dx} \ln x + (\ln x) \frac{d}{dx} x$$

$$= x \left(\frac{1}{x} \right) + \ln x$$

$$= 1 + \ln x$$

Therefore,

$$f'(x) = f(x)(1 + \ln x) = x^x(1 + \ln x)$$

EXPLORING WITH TECHNOLOGY

Refer to Example 7.

1. Use a graphing utility to plot the graph of $f(x) = x^x$, using the viewing window $[0, 2] \times [0, 2]$. Then use ZOOM and TRACE to show that

$$\lim_{x \to 0^+} f(x) = 1$$

2. Use the results of part 1 and Example 7 to show that $\lim_{x \to 0^+} f'(x) = -\infty$. Justify your answer.

14.4   Self-Check Exercises

1. Find an equation of the tangent line to the graph of $f(x) = x \ln(2x + 3)$ at the point $(-1, 0)$.

2. Use logarithmic differentiation to compute y', given $y = (2x + 1)^3(3x + 4)^5$.

Solutions to Self-Check Exercises 14.4 can be found on page 892.

14.4   Concept Questions

1. State the rule for differentiating (a) $f(x) = \ln|x|$ $(x \neq 0)$, and $g(x) = \ln f(x)$ $[f(x) > 0]$, where f is a differentiable function.

2. Explain the technique of logarithmic differentiation.

14.4   Exercises

In Exercises 1–32, find the derivative of the function.

1. $f(x) = 5 \ln x$

2. $f(x) = \ln 5x$

3. $f(x) = \ln(x + 1)$

4. $g(x) = \ln(2x + 1)$

5. $f(x) = \ln x^8$

6. $h(t) = 2 \ln t^5$

7. $f(x) = \ln \sqrt{x}$

8. $f(x) = \ln(\sqrt{x} + 1)$

9. $f(x) = \ln \dfrac{1}{x^2}$

10. $f(x) = \ln \dfrac{1}{2x^3}$

11. $f(x) = \ln(4x^2 - 6x + 3)$

12. $f(x) = \ln(3x^2 - 2x + 1)$

13. $f(x) = \ln \dfrac{2x}{x+1}$

14. $f(x) = \ln \dfrac{x+1}{x-1}$

15. $f(x) = x^2 \ln x$

16. $f(x) = 3x^2 \ln 2x$

17. $f(x) = \dfrac{2 \ln x}{x}$

18. $f(x) = \dfrac{3 \ln x}{x^2}$

19. $f(u) = \ln(u - 2)^3$

20. $f(x) = \ln(x^3 - 3)^4$

21. $f(x) = \sqrt{\ln x}$

22. $f(x) = \sqrt{\ln x + x}$

23. $f(x) = (\ln x)^3$

24. $f(x) = 2(\ln x)^{3/2}$

25. $f(x) = \ln(x^3 + 1)$

26. $f(x) = \ln \sqrt{x^2 - 4}$

27. $f(x) = e^x \ln x$

28. $f(x) = e^x \ln \sqrt{x + 3}$

29. $f(t) = e^{2t} \ln(t + 1)$

30. $g(t) = t^2 \ln(e^{2t} + 1)$

31. $f(x) = \dfrac{\ln x}{x}$

32. $g(t) = \dfrac{t}{\ln t}$

In Exercises 33–36, find the second derivative of the function.

33. $f(x) = \ln 2x$

34. $f(x) = \ln(x + 5)$

35. $f(x) = \ln(x^2 + 2)$

36. $f(x) = (\ln x)^2$

In Exercises 37–46, use logarithmic differentiation to find the derivative of the function.

37. $y = (x + 1)^2(x + 2)^3$

38. $y = (3x + 2)^4(5x - 1)^2$

39. $y = (x - 1)^2(x + 1)^3(x + 3)^4$

40. $y = \sqrt{3x + 5}(2x - 3)^4$

41. $y = \dfrac{(2x^2 - 1)^5}{\sqrt{x + 1}}$

42. $y = \dfrac{\sqrt{4 + 3x^2}}{\sqrt[3]{x^2 + 1}}$

43. $y = 3^x$

44. $y = x^{x+2}$

45. $y = (x^2 + 1)^x$

46. $y = x^{\ln x}$

47. Find an equation of the tangent line to the graph of $y = x \ln x$ at the point $(1, 0)$.

48. Find an equation of the tangent line to the graph of $y = \ln x^2$ at the point $(2, \ln 4)$.

49. Determine the intervals where the function $f(x) = \ln x^2$ is increasing and where it is decreasing.

50. Determine the intervals where the function $f(x) = \dfrac{\ln x}{x}$ is increasing and where it is decreasing.

51. Determine the intervals of concavity for the function $f(x) = x^2 + \ln x^2$.

52. Determine the intervals of concavity for the function $f(x) = \dfrac{\ln x}{x}$.

53. Find the inflection points of the function $f(x) = \ln(x^2 + 1)$.

54. Find the inflection points of the function $f(x) = x^2 \ln x$.

55. Find an equation of the tangent line to the graph of $f(x) = x^2 + 2 \ln x$ at its inflection point.

56. Find an equation of the tangent line to the graph of $f(x) = e^{x/2} \ln x$ at its inflection point.
Hint: Show that $(1, 0)$ is the only inflection point of f.

57. Find the absolute extrema of the function $f(x) = x - \ln x$ on $\left[\frac{1}{2}, 3\right]$.

58. Find the absolute extrema of the function $g(x) = \dfrac{x}{\ln x}$ on $[2, 5]$.

59. STRAIN ON VERTEBRAE The strain (percent of compression) on the lumbar vertebral disks in an adult human as a function of the load x (in kilograms) is given by

$$f(x) = 7.2956 \ln(0.0645012x^{0.95} + 1)$$

What is the rate of change of the strain with respect to the load when the load is 100 kg? When the load is 500 kg?
Source: Benedek and Villars, *Physics with Illustrative Examples from Medicine and Biology*

60. HEIGHTS OF CHILDREN For children between the ages of 5 and 13 years old, the Ehrenberg equation

$$\ln W = \ln 2.4 + 1.84h$$

gives the relationship between the weight W (in kilograms) and the height h (in meters) of a child. Use differentials to estimate the change in the weight of a child who grows from 1 m to 1.1 m.

61. YAHOO! IN EUROPE Yahoo! is putting more emphasis on Western Europe, where the number of online households is expected to grow steadily. In a study conducted in 2004, the number of online households (in millions) in Western Europe was projected to be

$$N(t) = 34.68 + 23.88 \ln(1.05t + 5.3) \qquad (0 \le t \le 2)$$

where $t = 0$ corresponds to the beginning of 2004.
a. What was the projected number of online households in Western Europe at the beginning of 2005?
b. How fast was the projected number of online households in Western Europe increasing at the beginning of 2005?
Source: Jupiter Research

62. ONLINE BUYERS The number of online buyers in Western Europe is expected to grow steadily in the coming years. The function

$$P(t) = 28.5 + 14.42 \ln t \qquad (1 \le t \le 7)$$

gives the estimated online buyers as a percent of the total population, where t is measured in years with $t = 1$ corresponding to 2001.
a. What was the percent of online buyers in 2001 ($t = 1$)? How fast was it changing in 2001?
b. What was the percent of online buyers expected to be in 2006 ($t = 6$)? How fast was it expected to be changing in 2006?

Source: Jupiter Research

63. **ABSORPTION OF LIGHT**   When light passes through a window glass, some of it is absorbed. It can be shown that if $r\%$ of the light is absorbed by a glass of thickness w, then the percent of light absorbed by a piece of glass of thickness nw is

$$A(n) = 100\left[1 - \left(1 - \frac{r}{100}\right)^n\right] \qquad (0 \le r \le 100)$$

a. Show that A is an increasing function of n on $(0, \infty)$ if $0 < r < 100$.
Hint: Use logarithmic differentiation.
b. Sketch the graph of A for the special case where $r = 10$.
c. Evaluate $\lim_{n \to \infty} A(n)$ and interpret your result.

64. **LAMBERT'S LAW OF ABSORPTION**   Lambert's law of absorption states that the light intensity $I(x)$ (in calories/square centimeter/second) at a depth of x m as measured from the surface of a material is given by $I = I_0 a^x$, where I_0 and a are positive constants.
a. Find the rate of change of the light intensity with respect to x at a depth of x m from the surface of the material.
b. Using the result of part (a), conclude that the rate of change $I'(x)$ at a depth of x m is proportional to $I(x)$. What is the constant of proportion?

65. **MAGNITUDE OF EARTHQUAKES**   On the Richter scale, the magnitude R of an earthquake is given by the formula

$$R = \log \frac{I}{I_0}$$

where I is the intensity of the earthquake being measured and I_0 is the standard reference intensity.
a. What is the magnitude of an earthquake that has intensity 1 million times that of I_0?
b. Suppose an earthquake is measured with a magnitude of 6 on the Richter scale with an error of at most 2%. Use differentials to find the error in the intensity of the earthquake.
Hint: Observe that $I = I_0 10^R$ and use logarithmic differentiation.

66. **WEBER–FECHNER LAW**   The Weber–Fechner law

$$R = k \ln \frac{S}{S_0}$$

where k is a positive constant, describes the relationship between a stimulus S and the resulting response R. Here, S_0, a positive constant, is the threshold level.
a. Show that $R = 0$ if the stimulus is at the threshold level S_0.
b. The derivative dR/dS is the *sensitivity* corresponding to the stimulus level S and measures the capability to detect small changes in the stimulus level. Show that dR/dS is inversely proportional to S and interpret your result.

In Exercises 67 and 68, use the guidelines on page 814 to sketch the graph of the given function.

67. $f(x) = \ln(x - 1)$
68. $f(x) = 2x - \ln x$

In Exercises 69 and 70, determine whether the statement is true or false. If it is true, explain why it is true. If it is false, give an example to show why it is false.

69. If $f(x) = \ln 5$, then $f'(x) = \frac{1}{5}$.

70. If $f(x) = \ln a^x$, then $f'(x) = \ln a$.

71. Prove that $\frac{d}{dx} \ln|x| = \frac{1}{x}$ $(x \ne 0)$ for the case $x < 0$.

72. Use the definition of the derivative to show that

$$\lim_{x \to 0} \frac{\ln(x + 1)}{x} = 1$$

14.4 Solutions to Self-Check Exercises

1. The slope of the tangent line to the graph of f at any point $(x, f(x))$ lying on the graph of f is given by $f'(x)$. Using the product rule, we find

$$f'(x) = \frac{d}{dx}[x \ln(2x + 3)]$$
$$= x\frac{d}{dx}\ln(2x + 3) + \ln(2x + 3) \cdot \frac{d}{dx}(x)$$
$$= x\left(\frac{2}{2x + 3}\right) + \ln(2x + 3) \cdot 1$$
$$= \frac{2x}{2x + 3} + \ln(2x + 3)$$

In particular, the slope of the tangent line to the graph of f at the point $(-1, 0)$ is

$$f'(-1) = \frac{-2}{-2 + 3} + \ln 1 = -2$$

Therefore, using the point-slope form of the equation of a line, we see that a required equation is

$$y - 0 = -2(x + 1)$$
$$y = -2x - 2$$

2. Taking the logarithm on both sides of the equation gives

$$\ln y = \ln(2x + 1)^3(3x + 4)^5$$
$$= \ln(2x + 1)^3 + \ln(3x + 4)^5$$
$$= 3\ln(2x + 1) + 5\ln(3x + 4)$$

Differentiating both sides of the equation with respect to x, keeping in mind that y is a function of x, we obtain

$$\frac{d}{dx}(\ln y) = \frac{y'}{y} = 3 \cdot \frac{2}{2x + 1} + 5 \cdot \frac{3}{3x + 4}$$

$$= 3\left(\frac{2}{2x + 1} + \frac{5}{3x + 4}\right)$$

and

$$y' = 3(2x + 1)^3(3x + 4)^5\left(\frac{2}{2x + 1} + \frac{5}{3x + 4}\right)$$

14.5 Exponential Functions as Mathematical Models

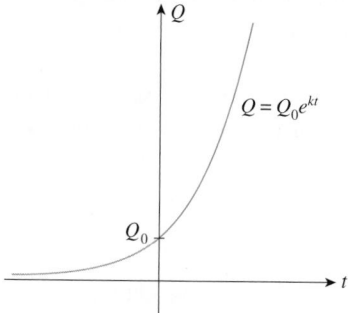

FIGURE 13
Exponential growth

Exponential Growth

Many problems arising from practical situations can be described mathematically in terms of exponential functions or functions closely related to the exponential function. In this section we look at some applications involving exponential functions from the fields of the life and social sciences.

In Section 14.1, we saw that the exponential function $f(x) = b^x$ is an increasing function when $b > 1$. In particular, the function $f(x) = e^x$ shares this property. From this result, one may deduce that the function $Q(t) = Q_0 e^{kt}$, where Q_0 and k are positive constants, has the following properties:

1. $Q(0) = Q_0$
2. $Q(t)$ increases "rapidly" without bound as t increases without bound (Figure 13).

Property 1 follows from the computation

$$Q(0) = Q_0 e^0 = Q_0$$

Next, to study the rate of change of the function $Q(t)$, we differentiate it with respect to t, obtaining

$$Q'(t) = \frac{d}{dt}(Q_0 e^{kt})$$
$$= Q_0 \frac{d}{dt}(e^{kt})$$
$$= kQ_0 e^{kt}$$
$$= kQ(t) \tag{4}$$

Since $Q(t) > 0$ (because Q_0 is assumed to be positive) and $k > 0$, we see that $Q'(t) > 0$ and so $Q(t)$ is an increasing function of t. Our computation has in fact shed more light on an important property of the function $Q(t)$. Equation (4) says that the rate of increase of the function $Q(t)$ is proportional to the amount $Q(t)$ of the quantity present at time t. The implication is that as $Q(t)$ increases, so does the *rate of increase* of $Q(t)$, resulting in a very rapid increase in $Q(t)$ as t increases without bound.

Thus, the exponential function

$$Q(t) = Q_0 e^{kt} \qquad (0 \le t < \infty) \tag{5}$$

provides us with a mathematical model of a quantity $Q(t)$ that is initially present in the amount of $Q(0) = Q_0$ and whose rate of growth at any time t is directly proportional to the amount of the quantity present at time t. Such a quantity is said to exhibit **exponential growth**, and the constant k of proportionality is called the **growth constant**. Interest earned on a fixed deposit when compounded continuously exhibits exponential growth. Other examples of unrestricted exponential growth follow.

APPLIED EXAMPLE 1 Growth of Bacteria Under ideal laboratory conditions, the number of bacteria in a culture grows in accordance with the law $Q(t) = Q_0 e^{kt}$, where Q_0 denotes the number of bacteria initially present in the culture, k is a constant determined by the strain of bacteria under consideration, and t is the elapsed time measured in hours. Suppose 10,000 bacteria are present initially in the culture and 60,000 present 2 hours later.

a. How many bacteria will there be in the culture at the end of 4 hours?
b. What is the rate of growth of the population after 4 hours?

Solution

a. We are given that $Q(0) = Q_0 = 10{,}000$, so $Q(t) = 10{,}000 e^{kt}$. Next, the fact that 60,000 bacteria are present 2 hours later translates into $Q(2) = 60{,}000$. Thus,

$$60{,}000 = 10{,}000 e^{2k}$$
$$e^{2k} = 6$$

Taking the natural logarithm on both sides of the equation, we obtain

$$\ln e^{2k} = \ln 6$$
$$2k = \ln 6 \qquad \text{Since } \ln e = 1$$
$$k \approx 0.8959$$

Thus, the number of bacteria present at any time t is given by

$$Q(t) = 10{,}000 e^{0.8959t}$$

In particular, the number of bacteria present in the culture at the end of 4 hours is given by

$$Q(4) = 10{,}000 e^{0.8959(4)}$$
$$\approx 360{,}029$$

b. The rate of growth of the bacteria population at any time t is given by

$$Q'(t) = kQ(t)$$

Thus, using the result from part (a), we find that the rate at which the population is growing at the end of 4 hours is

$$Q'(4) = kQ(4)$$
$$\approx (0.8959)(360{,}029)$$
$$\approx 322{,}550$$

or approximately 322,550 bacteria per hour.

Exponential Decay

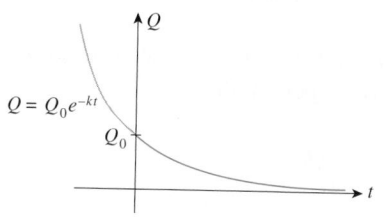

$Q = Q_0 e^{-kt}$

FIGURE 14
Exponential decay

In contrast to exponential growth, a quantity exhibits **exponential decay** if it decreases at a rate that is directly proportional to its size. Such a quantity may be described by the exponential function

$$Q(t) = Q_0 e^{-kt} \qquad t \in [0, \infty) \tag{6}$$

where the positive constant Q_0 measures the amount present initially ($t = 0$) and k is some suitable positive number, called the **decay constant**. The choice of this number is determined by the nature of the substance under consideration. The graph of this function is sketched in Figure 14.

To verify the properties ascribed to the function $Q(t)$, we simply compute

$$Q(0) = Q_0 e^0 = Q_0$$

$$Q'(t) = \frac{d}{dt}(Q_0 e^{-kt})$$

$$= Q_0 \frac{d}{dt}(e^{-kt})$$

$$= -kQ_0 e^{-kt} = -kQ(t)$$

APPLIED EXAMPLE 2   Radioactive Decay   Radioactive substances decay exponentially. For example, the amount of radium present at any time t obeys the law $Q(t) = Q_0 e^{-kt}$, where Q_0 is the initial amount present and k is a suitable positive constant. The **half-life of a radioactive substance** is the time required for a given amount to be reduced by one-half. Now, it is known that the half-life of radium is approximately 1600 years. Suppose initially there are 200 milligrams of pure radium. Find the amount left after t years. What is the amount left after 800 years?

Solution   The initial amount of radium present is 200 milligrams, so $Q(0) = Q_0 = 200$. Thus, $Q(t) = 200e^{-kt}$. Next, the datum concerning the half-life of radium implies that $Q(1600) = 100$, and this gives

$$100 = 200e^{-1600k}$$

$$e^{-1600k} = \frac{1}{2}$$

Taking the natural logarithm on both sides of this equation yields

$$-1600k \ln e = \ln \frac{1}{2}$$

$$-1600k = \ln \frac{1}{2} \qquad \ln e = 1$$

$$k = -\frac{1}{1600} \ln \left(\frac{1}{2}\right) = 0.0004332$$

Therefore, the amount of radium left after t years is

$$Q(t) = 200e^{-0.0004332t}$$

In particular, the amount of radium left after 800 years is

$$Q(800) = 200e^{-0.0004332(800)} \approx 141.42$$

or approximately 141 milligrams.

APPLIED EXAMPLE 3   Carbon-14 Decay   Carbon 14, a radioactive iso-tope of carbon, has a half-life of 5770 years. What is its decay constant?

Solution   We have $Q(t) = Q_0 e^{-kt}$. Since the half-life of the element is 5770 years, half of the substance is left at the end of that period; that is,

$$Q(5770) = Q_0 e^{-5770k} = \frac{1}{2} Q_0$$

$$e^{-5770k} = \frac{1}{2}$$

Taking the natural logarithm on both sides of this equation, we have

$$\ln e^{-5770k} = \ln \frac{1}{2}$$

$$-5770k = -0.693147$$

$$k \approx 0.00012$$

Carbon-14 dating is a well-known method used by anthropologists to establish the age of animal and plant fossils. This method assumes that the proportion of carbon 14 (C-14) present in the atmosphere has remained constant over the past 50,000 years. Professor Willard Libby, recipient of the Nobel Prize in chemistry in 1960, proposed this theory.

The amount of C-14 in the tissues of a living plant or animal is constant. However, when an organism dies, it stops absorbing new quantities of C-14, and the amount of C-14 in the remains diminishes because of the natural decay of the radioactive substance. Thus, the approximate age of a plant or animal fossil can be determined by measuring the amount of C-14 present in the remains.

APPLIED EXAMPLE 4   Carbon-14 Dating   A skull from an archeolog-ical site has one-tenth the amount of C-14 that it originally contained. Determine the approximate age of the skull.

Solution   Here,

$$Q(t) = Q_0 e^{-kt}$$
$$= Q_0 e^{-0.00012t}$$

where Q_0 is the amount of C-14 present originally and k, the decay constant, is equal to 0.00012 (see Example 3). Since $Q(t) = (1/10)Q_0$, we have

$$\frac{1}{10} Q_0 = Q_0 e^{-0.00012t}$$

$$\ln \frac{1}{10} = -0.00012t \qquad \text{Take the natural logarithm on both sides.}$$

$$t = \frac{\ln \frac{1}{10}}{-0.00012}$$

$$\approx 19,200$$

or approximately 19,200 years.

Learning Curves

The next example shows how the exponential function may be applied to describe certain types of learning processes. Consider the function

$$Q(t) = C - Ae^{-kt}$$

where C, A, and k are positive constants. To sketch the graph of the function Q, observe that its y-intercept is given by $Q(0) = C - A$. Next, we compute

$$Q'(t) = kAe^{-kt}$$

Since both k and A are positive, we see that $Q'(t) > 0$ for all values of t. Thus, $Q(t)$ is an increasing function of t. Also,

$$\lim_{t \to \infty} Q(t) = \lim_{t \to \infty} (C - Ae^{-kt})$$
$$= \lim_{t \to \infty} C - \lim_{t \to \infty} Ae^{-kt}$$
$$= C$$

so $y = C$ is a horizontal asymptote of Q. Thus, $Q(t)$ increases and approaches the number C as t increases without bound. The graph of the function Q is shown in Figure 15, where that part of the graph corresponding to the negative values of t is drawn with a gray line since, in practice, one normally restricts the domain of the function to the interval $[0, \infty)$.

Observe that $Q(t)$ $(t > 0)$ increases rather rapidly initially but that the rate of increase slows down considerably after a while. To see this, we compute

$$\lim_{t \to \infty} Q'(t) = \lim_{t \to \infty} kAe^{-kt} = 0$$

This behavior of the graph of the function Q closely resembles the learning pattern experienced by workers engaged in highly repetitive work. For example, the productivity of an assembly-line worker increases very rapidly in the early stages of the training period. This productivity increase is a direct result of the worker's training and accumulated experience. But the rate of increase of productivity slows as time goes by, and the worker's productivity level approaches some fixed level due to the limitations of the worker and the machine. Because of this characteristic, the graph of the function $Q(t) = C - Ae^{-kt}$ is often called a **learning curve**.

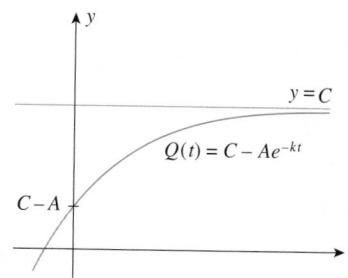

FIGURE 15
A learning curve

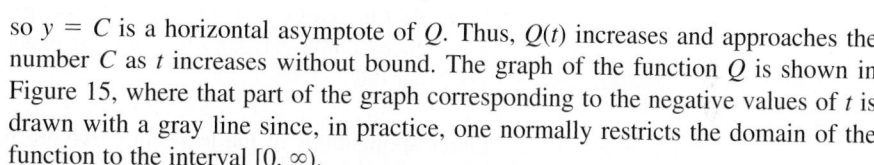

APPLIED EXAMPLE 5   Assembly Time The Camera Division of Eastman Optical produces a 35-mm single-lens reflex camera. Eastman's training department determines that after completing the basic training program, a new, previously inexperienced employee will be able to assemble

$$Q(t) = 50 - 30e^{-0.5t}$$

model F cameras per day, t months after the employee starts work on the assembly line.

a. How many model F cameras can a new employee assemble per day after basic training?

b. How many model F cameras can an employee with 1 month of experience assemble per day? An employee with 2 months of experience? An employee with 6 months of experience?

c. How many model F cameras can the average experienced employee assemble per day?

Solution

a. The number of model F cameras a new employee can assemble is given by

$$Q(0) = 50 - 30 = 20$$

b. The number of model F cameras that an employee with 1 month of experience, 2 months of experience, and 6 months of experience can assemble per day is given by

$$Q(1) = 50 - 30e^{-0.5} \approx 31.80$$
$$Q(2) = 50 - 30e^{-1} \approx 38.96$$
$$Q(6) = 50 - 30e^{-3} \approx 48.51$$

or approximately 32, 39, and 49, respectively.

c. As t increases without bound, $Q(t)$ approaches 50. Hence, the average experienced employee can ultimately be expected to assemble 50 model F cameras per day.   ∎

Other applications of the learning curve are found in models that describe the dissemination of information about a product or the velocity of an object dropped into a viscous medium.

Logistic Growth Functions

Our last example of an application of exponential functions to the description of natural phenomena involves the **logistic** (also called the **S-shaped,** or **sigmoidal**) **curve**, which is the graph of the function

$$Q(t) = \frac{A}{1 + Be^{-kt}}$$

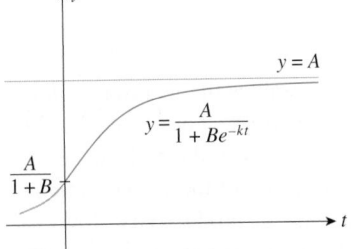

FIGURE 16
A logistic curve

where A, B, and k are positive constants. The function Q is called a logistic growth function, and the graph of the function Q is sketched in Figure 16.

Observe that $Q(t)$ increases rather rapidly for small values of t. In fact, for small values of t, the logistic curve resembles an exponential growth curve. However, the *rate of growth* of $Q(t)$ decreases quite rapidly as t increases and $Q(t)$ approaches the number A as t increases without bound.

Thus, the logistic curve exhibits both the property of rapid growth of the exponential growth curve as well as the "saturation" property of the learning curve. Because of these characteristics, the logistic curve serves as a suitable mathematical model for describing many natural phenomena. For example, if a small number of rabbits were introduced to a tiny island in the South Pacific, the rabbit population might be expected to grow very rapidly at first, but the growth rate would decrease quickly as overcrowding, scarcity of food, and other environmental factors affected it. The population would eventually stabilize at a level compatible with the life-support capacity of the environment. This level, given by A, is called the *carrying capacity* of the environment. Models describing the spread of rumors and epidemics are other examples of the application of the logistic curve.

APPLIED EXAMPLE 6   Spread of Flu   The number of soldiers at Fort MacArthur who contracted influenza after t days during a flu epidemic is approximated by the exponential model

$$Q(t) = \frac{5000}{1 + 1249e^{-kt}}$$

If 40 soldiers contracted the flu by day 7, find how many soldiers contracted the flu by day 15.

Solution   The given information implies that

$$Q(7) = \frac{5000}{1 + 1249e^{-7k}} = 40$$

Thus,

$$40(1 + 1249e^{-7k}) = 5000$$

$$1 + 1249e^{-7k} = \frac{5000}{40} = 125$$

$$e^{-7k} = \frac{124}{1249}$$

$$-7k = \ln \frac{124}{1249}$$

$$k = -\frac{\ln \frac{124}{1249}}{7} \approx 0.33$$

Therefore, the number of soldiers who contracted the flu after t days is given by

$$Q(t) = \frac{5000}{1 + 1249e^{-0.33t}}$$

In particular, the number of soldiers who contracted the flu by day 15 is given by

$$Q(15) = \frac{5000}{1 + 1249e^{-15(0.33)}}$$

$$\approx 508$$

or approximately 508 soldiers.   ◼

EXPLORING WITH TECHNOLOGY

Refer to Example 6.

1. Use a graphing utility to plot the graph of the function Q, using the viewing window $[0, 40] \times [0, 5000]$.

2. Find how long it takes for the first 1000 soldiers to contract the flu.
   Hint: Plot the graphs of $y_1 = Q(t)$ and $y_2 = 1000$ and find the point of intersection of the two graphs.

14.5   Self-Check Exercise

Suppose the population (in millions) of a country at any time t grows in accordance with the rule

$$P = \left(P_0 + \frac{I}{k}\right)e^{kt} - \frac{I}{k}$$

where P denotes the population at any time t, k is a constant reflecting the natural growth rate of the population, I is a constant giving the (constant) rate of immigration into the country, and P_0

is the total population of the country at time $t = 0$. The population of the United States in 1980 ($t = 0$) was 226.5 million. If the natural growth rate is 0.8% annually ($k = 0.008$) and net immigration is allowed at the rate of half a million people per year ($I = 0.5$) until 2010, what is the expected population of the United States in 2005?

The solution to Self-Check Exercise 14.5 can be found on page 903.

14.5   Concept Questions

1. Give the model for unrestricted exponential growth and the model for exponential decay. What effect does the magnitude of the growth (decay) constant have on the growth (decay) of a quantity?

2. What is the half-life of a radioactive substance?

3. What is the logistic growth function? What are its characteristics?

14.5   Exercises

1. **EXPONENTIAL GROWTH**   Given that a quantity $Q(t)$ is described by the exponential growth function

$$Q(t) = 400e^{0.05t}$$

where t is measured in minutes, answer the following questions:
   a. What is the growth constant?
   b. What quantity is present initially?
   c. Complete the following table of values:

t	0	10	20	100	1000
Q					

2. **EXPONENTIAL DECAY**   Given that a quantity $Q(t)$ exhibiting exponential decay is described by the function

$$Q(t) = 2000e^{-0.06t}$$

where t is measured in years, answer the following questions:
   a. What is the decay constant?
   b. What quantity is present initially?
   c. Complete the following table of values:

t	0	5	10	20	100
Q					

3. **GROWTH OF BACTERIA**   The growth rate of the bacterium *Escherichia coli*, a common bacterium found in the human intestine, is proportional to its size. Under ideal laboratory conditions, when this bacterium is grown in a nutrient broth medium, the number of cells in a culture doubles approximately every 20 min.

   a. If the initial cell population is 100, determine the function $Q(t)$ that expresses the exponential growth of the number of cells of this bacterium as a function of time t (in minutes).
   b. How long will it take for a colony of 100 cells to increase to a population of 1 million?
   c. If the initial cell population were 1000, how would this alter our model?

4. **WORLD POPULATION**   The world population at the beginning of 1990 was 5.3 billion. Assume that the population continues to grow at the rate of approximately 2%/year and find the function $Q(t)$ that expresses the world population (in billions) as a function of time t (in years), with $t = 0$ corresponding to the beginning of 1990.

   a. Using this function, complete the following table of values and sketch the graph of the function Q.

Year	1990	1995	2000	2005
World Population				

Year	2010	2015	2020	2025
World Population				

   b. Find the estimated rate of growth in 2010.

5. **WORLD POPULATION**   Refer to Exercise 4.
   a. If the world population continues to grow at the rate of approximately 2%/year, find the length of time t_0 required for the world population to triple in size.

b. Using the time t_0 found in part (a), what would be the world population if the growth rate were reduced to 1.8%/yr?

6. RESALE VALUE A certain piece of machinery was purchased 3 yr ago by Garland Mills for $500,000. Its present resale value is $320,000. Assuming that the machine's resale value decreases exponentially, what will it be 4 yr from now?

7. ATMOSPHERIC PRESSURE If the temperature is constant, then the atmospheric pressure P (in pounds/square inch) varies with the altitude above sea level h in accordance with the law

$$P = p_0 e^{-kh}$$

where p_0 is the atmospheric pressure at sea level and k is a constant. If the atmospheric pressure is 15 lb/in.^{2} at sea level and 12.5 lb/in.^{2} at 4000 ft, find the atmospheric pressure at an altitude of 12,000 ft. How fast is the atmospheric pressure changing with respect to altitude at an altitude of 12,000 ft?

8. RADIOACTIVE DECAY The radioactive element polonium decays according to the law

$$Q(t) = Q_0 \cdot 2^{-(t/140)}$$

where Q_0 is the initial amount and the time t is measured in days. If the amount of polonium left after 280 days is 20 mg, what was the initial amount present?

9. RADIOACTIVE DECAY Phosphorus 32 (P-32) has a half-life of 14.2 days. If 100 g of this substance are present initially, find the amount present after t days. What amount will be left after 7.1 days? How fast is P-32 decaying when $t = 7.1$?

10. NUCLEAR FALLOUT Strontium 90 (Sr-90), a radioactive isotope of strontium, is present in the fallout resulting from nuclear explosions. It is especially hazardous to animal life, including humans, because, upon ingestion of contaminated food, it is absorbed into the bone structure. Its half-life is 27 yr. If the amount of Sr-90 in a certain area is found to be four times the "safe" level, find how much time must elapse before an "acceptable level" is reached.

11. CARBON-14 DATING Wood deposits recovered from an archeological site contain 20% of the C-14 they originally contained. How long ago did the tree from which the wood was obtained die?

12. CARBON-14 DATING Skeletal remains of the so-called Pittsburgh Man, unearthed in Pennsylvania, had lost 82% of the C-14 they originally contained. Determine the approximate age of the bones.

13. LEARNING CURVES The American Court Reporting Institute finds that the average student taking Advanced Machine Shorthand, an intensive 20-wk course, progresses according to the function

$$Q(t) = 120(1 - e^{-0.05t}) + 60 \qquad (0 \le t \le 20)$$

where $Q(t)$ measures the number of words (per minute) of

dictation that the student can take in machine shorthand after t wk in the course. Sketch the graph of the function Q and answer the following questions:
a. What is the beginning shorthand speed for the average student in this course?
b. What shorthand speed does the average student attain halfway through the course?
c. How many words per minute can the average student take after completing this course?

14. EFFECT OF ADVERTISING ON SALES Metro Department Store found that t wk after the end of a sales promotion the volume of sales was given by a function of the form

$$S(t) = B + Ae^{-kt} \qquad (0 \le t \le 4)$$

where $B = 50,000$ and is equal to the average weekly volume of sales before the promotion. The sales volumes at the end of the first and third weeks were $83,515 and $65,055, respectively. Assume that the sales volume is decreasing exponentially.
a. Find the decay constant k.
b. Find the sales volume at the end of the fourth week.
c. How fast is the sales volume dropping at the end of the fourth week?

15. DEMAND FOR COMPUTERS Universal Instruments found that the monthly demand for its new line of Galaxy Home Computers t mo after placing the line on the market was given by

$$D(t) = 2000 - 1500e^{-0.05t} \qquad (t > 0)$$

Graph this function and answer the following questions:
a. What is the demand after 1 mo? After 1 yr? After 2 yr? After 5 yr?
b. At what level is the demand expected to stabilize?
c. Find the rate of growth of the demand after the tenth month.

16. RELIABILITY OF COMPUTER CHIPS The percent of a certain brand of computer chips that will fail after t yr of use is estimated to be

$$P(t) = 100(1 - e^{-0.1t})$$

a. What percent of this brand of computer chips are expected to be usable after 3 yr?
b. Evaluate $\lim_{t \to \infty} P(t)$. Did you expect this result?

17. LENGTHS OF FISH The length (in centimeters) of a typical Pacific halibut t yr old is approximately

$$f(t) = 200(1 - 0.956e^{-0.18t})$$

a. What is the length of a typical 5-yr-old Pacific halibut?
b. How fast is the length of a typical 5-yr-old Pacific halibut increasing?
c. What is the maximum length a typical Pacific halibut can attain?

18. SPREAD OF AN EPIDEMIC   During a flu epidemic, the number of children in the Woodbridge Community School System who contracted influenza after t days was given by

$$Q(t) = \frac{1000}{1 + 199e^{-0.8t}}$$

a. How many children were stricken by the flu after the first day?
b. How many children had the flu after 10 days?
c. How many children eventually contracted the disease?

19. LAY TEACHERS AT ROMAN CATHOLIC SCHOOLS   The change from religious to lay teachers at Roman Catholic schools has been partly attributed to the decline in the number of women and men entering religious orders. The percent of teachers who are lay teachers is given by

$$f(t) = \frac{98}{1 + 2.77e^{-t}} \qquad (0 \le t \le 4)$$

where t is measured in decades, with $t = 0$ corresponding to the beginning of 1960.
a. What percent of teachers were lay teachers at the beginning of 1990?
b. How fast was the percent of lay teachers changing at the beginning of 1990?
c. Find the year when the percent of lay teachers was increasing most rapidly.
Sources: National Catholic Education Association and the Department of Education

20. GROWTH OF A FRUIT FLY POPULATION   On the basis of data collected during an experiment, a biologist found that the growth of a fruit fly (*Drosophila*) with a limited food supply could be approximated by the exponential model

$$N(t) = \frac{400}{1 + 39e^{-0.16t}}$$

where t denotes the number of days since the beginning of the experiment.
a. What was the initial fruit fly population in the experiment?
b. What was the maximum fruit fly population that could be expected under this laboratory condition?
c. What was the population of the fruit fly colony on the 20th day?
d. How fast was the population changing on the 20th day?

21. DEMOGRAPHICS   The number of citizens aged 45–64 is projected to be

$$P(t) = \frac{197.9}{1 + 3.274e^{-0.0361t}} \qquad (0 \le t \le 20)$$

where $P(t)$ is measured in millions and t is measured in years, with $t = 0$ corresponding to the beginning of 1990. People belonging to this age group are the targets of

insurance companies that want to sell them annuities. What is the projected population of citizens aged 45–64 in 2010?
Source: K. G. Securities

22. POPULATION GROWTH IN THE 21ST CENTURY   The U.S. population is approximated by the function

$$P(t) = \frac{616.5}{1 + 4.02e^{-0.5t}}$$

where $P(t)$ is measured in millions of people and t is measured in 30-yr intervals, with $t = 0$ corresponding to 1930. What is the expected population of the United States in 2020 ($t = 3$)?

23. DISSEMINATION OF INFORMATION   Three hundred students attended the dedication ceremony of a new building on a college campus. The president of the traditionally female college announced a new expansion program, which included plans to make the college coeducational. The number of students who learned of the new program t hr later is given by the function

$$f(t) = \frac{3000}{1 + Be^{-kt}}$$

If 600 students on campus had heard about the new program 2 hr after the ceremony, how many students had heard about the policy after 4 hr? How fast was the news spreading 4 hr after the ceremony?

24. PRICE OF A COMMODITY   The unit price of a certain commodity is given by

$$p = f(t) = 6 + 4e^{-2t}$$

where p is measured in dollars and t is measured in months.
a. Show that f is decreasing on $(0, \infty)$.
b. Show that the graph of f is concave upward on $(0, \infty)$.
c. Evaluate $\lim_{t \to \infty} f(t)$. (*Note:* This value is called the *equilibrium price* of the commodity, and in this case, we have *price stability.*)
d. Sketch the graph of f.

25. CHEMICAL MIXTURES   Two chemicals react to form another chemical. Suppose the amount of the chemical formed in time t (in hours) is given by

$$x(t) = \frac{15\left[1 - \left(\frac{2}{3}\right)^{3t}\right]}{1 - \frac{1}{4}\left(\frac{2}{3}\right)^{3t}}$$

where $x(t)$ is measured in pounds. How many pounds of the chemical are formed eventually?
Hint: You need to evaluate $\lim_{t \to \infty} x(t)$.

26. VON BERTALANFFY GROWTH FUNCTION   The length (in centimeters) of a common commercial fish is approximated by the von Bertalanffy growth function

$$f(t) = a(1 - be^{-kt})$$

where a, b, and k are positive constants.

a. Show that f is increasing on the interval $(0, \infty)$.
b. Show that the graph of f is concave downward on $(0, \infty)$.
c. Show that $\lim_{t \to \infty} f(t) = a$.
d. Use the results of parts (a)–(c) to sketch the graph of f.

27. **ABSORPTION OF DRUGS** The concentration of a drug in grams/cubic centimeter (g/cm³) t min after it has been injected into the bloodstream is given by

$$C(t) = \frac{k}{b - a}(e^{-at} - e^{-bt})$$

where a, b, and k are positive constants, with $b > a$.
a. At what time is the concentration of the drug the greatest?
b. What will be the concentration of the drug in the long run?

28. **CONCENTRATION OF GLUCOSE IN THE BLOODSTREAM** A glucose solution is administered intravenously into the bloodstream at a constant rate of r mg/hr. As the glucose is being administered, it is converted into other substances and removed from the bloodstream. Suppose the concentration of the glucose solution at time t is given by

$$C(t) = \frac{r}{k} - \left[\left(\frac{r}{k}\right) - C_0 \right]e^{-kt}$$

where C_0 is the concentration at time $t = 0$ and k is a positive constant. Assuming that $C_0 < r/k$, evaluate $\lim_{t \to \infty} C(t)$.
a. What does your result say about the concentration of the glucose solution in the long run?
b. Show that the function C is increasing on $(0, \infty)$.

c. Show that the graph of C is concave downward on $(0, \infty)$.
d. Sketch the graph of the function C.

29. **RADIOACTIVE DECAY** A radioactive substance decays according to the formula

$$Q(t) = Q_0 e^{-kt}$$

where $Q(t)$ denotes the amount of the substance present at time t (measured in years), Q_0 denotes the amount of the substance present initially, and k (a positive constant) is the decay constant.
a. Show that half-life of the substance is $\bar{t} = \ln 2/k$.
b. Suppose a radioactive substance decays according to the formula

$$Q(t) = 20e^{-0.0001238t}$$

How long will it take for the substance to decay to half the original amount?

30. **GOMPERTZ GROWTH CURVE** Consider the function

$$Q(t) = Ce^{-Ae^{-kt}}$$

where $Q(t)$ is the size of a quantity at time t and A, C, and k are positive constants. The graph of this function, called the *Gompertz growth curve*, is used by biologists to describe restricted population growth.
a. Show that the function Q is always increasing.
b. Find the time t at which the growth rate $Q'(t)$ is increasing most rapidly.
    Hint: Find the inflection point of Q.
c. Show that $\lim_{t \to \infty} Q(t) = C$ and interpret your result.

14.5 Solution to Self-Check Exercise

We are given that $P_0 = 226.5$, $k = 0.008$, and $I = 0.5$. So

$$P = \left(226.5 + \frac{0.5}{0.008}\right)e^{0.008t} - \frac{0.5}{0.008}$$

$$= 289e^{0.008t} - 62.5$$

Therefore, the expected population in 2005 is given by

$$P(25) = 289e^{0.2} - 62.5$$

$$\approx 290.5$$

or approximately 290.5 million.

USING TECHNOLOGY

■ **Analyzing Mathematical Models**

We can use a graphing utility to analyze the mathematical models encountered in this section.

APPLIED EXAMPLE 1 **Households with Microwaves** The number of households with microwave ovens increased greatly in the 1980s and 1990s. The percent

(continued)

of households with microwave ovens from 1981 through 1999 is given by

$$f(t) = \frac{87}{1 + 4.209e^{-0.3727t}} \qquad (0 \le t \le 18)$$

where t is measured in years, with $t = 0$ corresponding to the beginning of 1981.

a. Use a graphing utility to plot the graph of f on the interval $[0, 18]$.
b. What percent of households owned microwave ovens at the beginning of 1984? At the beginning of 1994?
c. At what rate was the ownership of microwave ovens increasing at the beginning of 1984? At the beginning of 1994?
d. At what time was the increase in ownership of microwave ovens greatest?

Source: Energy Information Agency

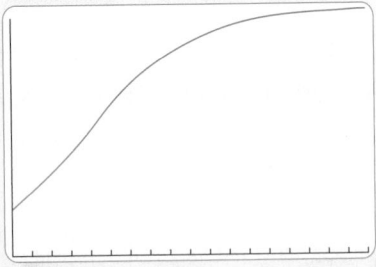

FIGURE T1
The graph of f in the viewing window
$[0, 18] \times [0, 100]$

Solution

a. The graph of f using the viewing window $[0, 18] \times [0, 100]$ is shown in Figure T1.
b. Using the evaluation function, we find $f(3) \approx 36.6$ and $f(13) \approx 84.2$ and conclude that the percent of households owning microwave ovens in 1984 was 36.6 and that in 1994 was 84.2.
c. Using the numerical derivative operation, we find $f'(3) \approx 7.9$ and $f'(13) \approx 1.0$. We conclude that the ownership of microwave ovens was increasing at the rate of 7.9% per year in 1984 and at the rate of 1% per year in 1994.
d. We see that the inflection point of f occurs at $(3.9, 43.5)$. Thus, the largest increase in ownership of microwave ovens occurred at the time when $t \approx 3.9$, or near the end of 1984.

TECHNOLOGY EXERCISES

1. **ONLINE BANKING** In a study prepared in 2000, the percent of households using online banking was projected to be

$$f(t) = 1.5e^{0.78t} \qquad (0 \le t \le 4)$$

where t is measured in years, with $t = 0$ corresponding to the beginning of 2000.
a. Plot the graph of f, using the viewing window $[0, 4] \times [0, 40]$.
b. How fast was the projected percent of households using online banking changing at the beginning of 2003?
c. How fast was the rate of the projected percent of households using online banking changing at the beginning of 2003?
Hint: We want $f''(3)$. Why?

Source: Online Banking Report

2. **NEWTON'S LAW OF COOLING** The temperature of a cup of coffee t min after it is poured is given by

$$T = 70 + 100e^{-0.0446t}$$

where T is measured in degrees Fahrenheit.

a. Plot the graph of T, using the viewing window $[0, 30] \times [0, 200]$.
b. When will the coffee be cool enough to drink (say, $120°$)?
Hint: Use the ISECT function.

3. **AIR TRAVEL** Air travel has been rising dramatically in the past 30 yr. In a study conducted in 2000, the FAA projected further exponential growth for air travel through 2010. The function

$$f(t) = 666e^{0.0413t} \qquad (0 \le t \le 10)$$

gives the number of passengers (in millions) in year t, with $t = 0$ corresponding to 2000.
a. Plot the graph of f, using the viewing window $[0, 10] \times [0, 1000]$.
b. How many air passengers were there in 2000? What was the projected number of air passengers for 2005?
c. What was the rate of change of the number of air passengers in 2005?

Source: Federal Aviation Administration

4. LENGTHS OF FISH The length (in centimeters) of a typical Pacific halibut t yr old is approximately

$$f(t) = 200(1 - 0.956e^{-0.18t})$$

a. Plot the graph of f, using the viewing window $[0, 10] \times [0, 200]$.

b. Suppose a Pacific halibut caught by Mike measures 140 cm. What is its approximate age?

5. LAY TEACHERS AT ROMAN CATHOLIC SCHOOLS The change from religious to lay teachers at Roman Catholic schools has been partly attributed to the decline in the number of women and men entering religious orders. The percent of teachers who are lay teachers is given by

$$f(t) = \frac{98}{1 + 2.77e^{-t}} \qquad (0 \le t \le 4)$$

where t is measured in decades, with $t = 0$ corresponding to the beginning of 1960.

a. Plot the graph of f, using the viewing window $[0, 4] \times [0, 100]$.

b. What percent of teachers were lay teachers at the beginning of 1990?

c. How fast was the percent of lay teachers changing at the beginning of 1990?

d. Find the year when the percent of lay teachers was increasing most rapidly.

Sources: National Catholic Education Association and the Department of Education

6. GROWTH OF A FRUIT FLY POPULATION On the basis of data collected during an experiment, a biologist found that the growth of a fruit fly (*Drosophila*) with a limited food supply could be approximated by the exponential model

$$N(t) = \frac{400}{1 + 39e^{-0.16t}}$$

where t denotes the number of days since the beginning of the experiment.

a. Plot the graph of f using the viewing window $[0, 50] \times [0, 400]$.

b. What was the initial fruit fly population in the experiment?

c. What was the maximum fruit fly population that could be expected under these laboratory conditions?

d. What was the population of the fruit fly colony on the 20th day?

e. How fast was the population changing on the 20th day?

7. POPULATION GROWTH IN THE 21ST CENTURY The U.S. population is approximated by the function

$$P(t) = \frac{616.5}{1 + 4.02e^{-0.5t}}$$

where $P(t)$ is measured in millions of people and t is measured in 30-yr intervals, with $t = 0$ corresponding to 1930.

a. Plot the graph of f, using the viewing window $[0, 4] \times [0, 650]$.

b. What is the expected population of the United States in 2020 ($t = 3$)?

c. What is the expected rate of growth of the U.S. population in 2020?

8. TIME RATE OF GROWTH OF A TUMOR The rate at which a tumor grows, with respect to time, is given by

$$R = Ax \ln\frac{B}{x} \qquad (\text{for } 0 < x < B)$$

where A and B are positive constants and x is the radius of the tumor.

a. Plot the graph of R for the case $A = B = 10$.

b. Find the radius of the tumor when the tumor is growing most rapidly with respect to time.

9. ABSORPTION OF DRUGS The concentration of a drug in an organ at any time t (in seconds) is given by

$$C(t) = \begin{cases} 0.3t - 18(1 - e^{-t/60}) & \text{if } 0 \le t \le 20 \\ 18e^{-t/60} - 12e^{-(t-20)/60} & \text{if } t > 20 \end{cases}$$

where $C(t)$ is measured in grams/cubic centimeter (g/cm³).

a. Plot the graph of f, using the viewing window $[0, 120] \times [0, 1]$.

b. What is the initial concentration of the drug in the organ?

c. What is the concentration of the drug in the organ after 10 sec?

d. What is the concentration of the drug in the organ after 30 sec?

e. What will be the concentration of the drug in the long run?

10. ANNUITIES At the time of retirement, Christine expects to have a sum of $500,000 in her retirement account. Assuming that the account pays interest at the rate of 5%/year compounded continuously, her accountant pointed out to her that if she made withdrawals amounting to x dollars per year ($x > 25,000$), then the time required to deplete her savings would be T years, where

$$T = f(x) = 20 \ln\left(\frac{x}{x - 25,000}\right) \qquad (x > 25,000)$$

a. Plot the graph of f, using the viewing window $[25,000, 50,000] \times [0, 100]$.

b. How much should Christine plan to withdraw from her retirement account each year if she wants it to last for 25 yr?

c. Evaluate $\lim\limits_{x \to 25,000^+} f(x)$. Is the result expected? Explain.

d. Evaluate $\lim\limits_{x \to \infty} f(x)$. Is the result expected? Explain.

CHAPTER 14   **Summary of Principal Formulas and Terms**

⚡ FORMULAS

1. Exponential function with base b	$y = b^x$
2. The number e	$e = \lim\limits_{m \to \infty} \left(1 + \dfrac{1}{m}\right)^m = 2.71828 \ldots$
3. Exponential function with base e	$y = e^x$
4. Logarithmic function with base b	$y = \log_b x$
5. Logarithmic function with base e	$y = \ln x$
6. Inverse properties of $\ln x$ and e	$\ln e^x = x$ and $e^{\ln x} = x$
7. Derivative of the exponential function	$\dfrac{d}{dx}(e^x) = e^x$
8. Chain rule for exponential functions	$\dfrac{d}{dx}(e^u) = e^u \dfrac{du}{dx}$
9. Derivative of the logarithmic function	$\dfrac{d}{dx}\ln|x| = \dfrac{1}{x}$
10. Chain rule for logarithmic functions	$\dfrac{d}{dx}(\ln u) = \dfrac{1}{u}\dfrac{du}{dx}$

⚡ TERMS

common logarithm (864)

natural logarithm (864)

logarithmic differentiation (888)

exponential growth (894)

growth constant (894)

exponential decay (895)

decay constant (895)

half-life of a radioactive substance (895)

logistic growth function (898)

CHAPTER 14   **Concept Review Questions**

Fill in the blanks.

1. The function $f(x) = x^b$ (b, a real number) is called a/an _____ function, whereas the function $g(x) = b^x$, where $b >$ _____ and $b \neq$ _____, is called a/an _____ function.

2. a. The domain of the function $y = 3^x$ is _____, and its range is _____.

   b. The graph of the function $y = 0.3^x$ passes through the point _____ and is decreasing on _____.

3. a. If $b > 0$ and $b \neq 1$, then the logarithmic function $y = \log_b x$ has domain _____ and range _____; its graph passes through the point _____.

   b. The graph of $y = \log_b x$ is decreasing if b _____ and increasing if b _____.

4. a. If $x > 0$, then $e^{\ln x} =$ _____.

   b. If x is any real number, then $\ln e^x =$ _____.

5. a. If $g(x) = e^{f(x)}$, where f is a differentiable function, then $g'(x) =$ _____.

   b. If $g(x) = \ln f(x)$, where $f(x) > 0$ is differentiable, then $g'(x) =$ _____.

6. a. In the unrestricted exponential growth model $Q = Q_0 e^{kt}$, Q_0 represents the quantity present _____, and k is called the _____ constant.

   b. In the exponential decay model $Q = Q_0 e^{-kt}$, k is called the _____ constant.

   c. The half-life of a radioactive substance is the _____ required for a substance to decay to _____ _____ of its original amount.

7. a. For the model $Q(t) = C - Ae^{-kt}$ describing a learning curve, $y = C$ is a/an _____ _____ of the graph of Q. The value of $Q(t)$ never exceeds _____.

b. For the logistic growth model $Q(t) = \dfrac{A}{1 + Be^{-kt}}$, $y = A$ is

a/an _____ _____ of the graph of Q. If the quantity $Q(t)$ is initially smaller than A, then $Q(t)$ will eventually approach _____ as t increases; the number A represents the life-support capacity of the environment and is called the _____ _____ of the environment.

CHAPTER 14 Review Exercises

1. Sketch on the same set of coordinate axes the graphs of the exponential functions defined by the equations.

a. $y = 2^{-x}$

b. $y = \left(\dfrac{1}{2}\right)^x$

In Exercises 2 and 3, express each equation in logarithmic form.

2. $\left(\dfrac{2}{3}\right)^{-3} = \dfrac{27}{8}$

3. $16^{-3/4} = 0.125$

In Exercises 4 and 5, solve each equation for x.

4. $\log_4(2x + 1) = 2$

5. $\ln(x - 1) + \ln 4 = \ln(2x + 4) - \ln 2$

In Exercises 6–8, given that $\ln 2 = x$, $\ln 3 = y$, and $\ln 5 = z$, express each of the given logarithmic values in terms of x, y, and z.

6. $\ln 30$

7. $\ln 3.6$

8. $\ln 75$

9. Sketch the graph of the function $y = \log_2(x + 3)$.

10. Sketch the graph of the function $y = \log_3(x + 1)$.

In Exercises 11–28, find the derivative of the function.

11. $f(x) = xe^{2x}$

12. $f(t) = \sqrt{t}e^t + t$

13. $g(t) = \sqrt{t}e^{-2t}$

14. $g(x) = e^x\sqrt{1 + x^2}$

15. $y = \dfrac{e^{2x}}{1 + e^{-2x}}$

16. $f(x) = e^{2x^2-1}$

17. $f(x) = xe^{-x^2}$

18. $g(x) = (1 + e^{2x})^{3/2}$

19. $f(x) = x^2e^x + e^x$

20. $g(t) = t \ln t$

21. $f(x) = \ln(e^{x^2} + 1)$

22. $f(x) = \dfrac{x}{\ln x}$

23. $f(x) = \dfrac{\ln x}{x + 1}$

24. $y = (x + 1)e^x$

25. $y = \ln(e^{4x} + 3)$

26. $f(r) = \dfrac{re^r}{1 + r^2}$

27. $f(x) = \dfrac{\ln x}{1 + e^x}$

28. $g(x) = \dfrac{e^{x^2}}{1 + \ln x}$

29. Find the second derivative of the function $y = \ln(3x + 1)$.

30. Find the second derivative of the function $y = x \ln x$.

31. Find $h'(0)$ if $h(x) = g(f(x))$, $g(x) = x + \dfrac{1}{x}$, and $f(x) = e^x$.

32. Find $h'(1)$ if $h(x) = g(f(x))$, $g(x) = \dfrac{x + 1}{x - 1}$, and $f(x) = \ln x$.

33. Use logarithmic differentiation to find the derivative of $f(x) = (2x^3 + 1)(x^2 + 2)^3$.

34. Use logarithmic differentiation to find the derivative of $f(x) = \dfrac{x(x^2 - 2)^2}{(x - 1)}$.

35. Find an equation of the tangent line to the graph of $y = e^{-2x}$ at the point $(1, e^{-2})$.

36. Find an equation of the tangent line to the graph of $y = xe^{-x}$ at the point $(1, e^{-1})$.

37. Sketch the graph of the function $f(x) = xe^{-2x}$.

38. Sketch the graph of the function $f(x) = x^2 - \ln x$.

39. Find the absolute extrema of the function $f(t) = te^{-t}$.

40. Find the absolute extrema of the function

$$g(t) = \dfrac{\ln t}{t}$$

on $[1, 2]$.

41. CONSUMER PRICE INDEX At an annual inflation rate of 7.5%, how long will it take the Consumer Price Index (CPI) to double?

42. GROWTH OF BACTERIA A culture of bacteria that initially contained 2000 bacteria has a count of 18,000 bacteria after 2 hr.

a. Determine the function $Q(t)$ that expresses the exponential growth of the number of cells of this bacterium as a function of time t (in minutes).

b. Find the number of bacteria present after 4 hr.

43. RADIOACTIVE DECAY The radioactive element radium has a half-life of 1600 yr. What is its decay constant?

44. DEMAND FOR DVD PLAYERS VCA Television found that the monthly demand for its new line of DVD players t mo after placing the players on the market is given by

$$D(t) = 4000 - 3000e^{-0.06t} \quad (t \geq 0)$$

Graph this function and answer the following questions:
a. What was the demand after 1 mo? After 1 yr? After 2 yr?
b. At what level is the demand expected to stabilize?

45. FLU EPIDEMIC During a flu epidemic, the number of students at a certain university who contracted influenza after t days could be approximated by the exponential model

$$Q(t) = \frac{3000}{1 + 499e^{-kt}}$$

If 90 students contracted the flu by day 10, how many students contracted the flu by day 20?

46. U.S. INFANT MORTALITY RATE The U.S. infant mortality rate (per 1000 live births) is approximated by the function

$$N(t) = 12.5e^{-0.0294t} \quad (0 \leq t \leq 21)$$

where t is measured in years, with $t = 0$ corresponding to 1980.
a. What was the mortality rate in 1980? In 1990? In 2000?
b. Sketch the graph of N.

Source: U.S. Department of Health and Human Services

47. ABSORPTION OF DRUGS The concentration of a drug in an organ at any time t (in seconds) is given by

$$x(t) = 0.08(1 - e^{-0.02t})$$

where $x(t)$ is measured in grams/cubic centimeter (g/cm^3).
a. What is the initial concentration of the drug in the organ?
b. What is the concentration of the drug in the organ after 30 sec?
c. What will be the concentration of the drug in the organ in the long run?
d. Sketch the graph of x.

CHAPTER 14 **Before Moving On . . .**

1. Solve the equation $\dfrac{100}{1 + 2e^{0.3t}} = 40$ for t.

2. Find the slope of the tangent line to the graph of $f(x) = e^{\sqrt{x}}$.

3. Find the rate at which $y = x \ln(x^2 + 1)$ is changing at $x = 1$.

4. Find the second derivative of $y = e^{2x} \ln 3x$.

5. The temperature of a cup of coffee at time t (in minutes) is

$$T(t) = 70 + ce^{-kt}$$

Initially, the temperature of the coffee was 200°F. Three minutes later, it was 180°. When will the temperature of the coffee be 150°F?

15 Integration

How much will the solar cell panels cost? The head of Soloron Corporation's research and development department has projected that the cost of producing solar cell panels will drop at a certain rate in the next several years. In Example 7, page 928, you will see how this information can be used to predict the cost of solar cell panels in the coming years.

© Nikolay Zurek/Taxi/Getty Images

DIFFERENTIAL CALCULUS IS concerned with the problem of finding the rate of change of one quantity with respect to another. In this chapter, we begin the study of the other branch of calculus, known as integral calculus. Here we are interested in precisely the opposite problem: If we know the rate of change of one quantity with respect to another, can we find the relationship between the two quantities? The principal tool used in the study of integral calculus is the *antiderivative* of a function, and we develop rules for antidifferentiation, or *integration*, as the process of finding the antiderivative is called. We also show that a link is established between differential and integral calculus—via the fundamental theorem of calculus.

15.1 Antiderivatives and the Rules of Integration

■ Antiderivatives

Let's return, once again, to the example involving the motion of the maglev (Figure 1).

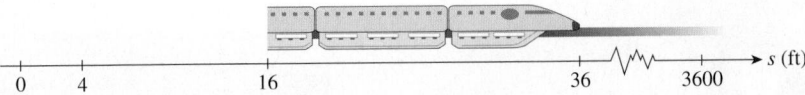

In Chapter 11, we discussed the following problem:

If we know the position of the maglev at any time t, can we find its velocity at time t?

As it turns out, if the position of the maglev is described by the position function f, then its velocity at any time t is given by $f'(t)$. Here f'—the velocity function of the maglev—is just the derivative of f.

Now, in Chapters 15 and 16, we will consider precisely the opposite problem:

If we know the velocity of the maglev at any time t, can we find its position at time t?

Stated another way, if we know the velocity function f' of the maglev, can we find its position function f?

To solve this problem, we need the concept of an antiderivative of a function.

> **Antiderivative**
> A function F is an **antiderivative** of f on an interval I if $F'(x) = f(x)$ for all x in I.

Thus, an antiderivative of a function f is a function F whose derivative is f. For example, $F(x) = x^2$ is an antiderivative of $f(x) = 2x$ because

$$F'(x) = \frac{d}{dx}(x^2) = 2x = f(x)$$

and $F(x) = x^3 + 2x + 1$ is an antiderivative of $f(x) = 3x^2 + 2$ because

$$F'(x) = \frac{d}{dx}(x^3 + 2x + 1) = 3x^2 + 2 = f(x)$$

EXAMPLE 1 Let $F(x) = \frac{1}{3}x^3 - 2x^2 + x - 1$. Show that F is an antiderivative of $f(x) = x^2 - 4x + 1$.

Solution Differentiating the function F, we obtain

$$F'(x) = x^2 - 4x + 1 = f(x)$$

and the desired result follows. ■

EXAMPLE 2 Let $F(x) = x$, $G(x) = x + 2$, and $H(x) = x + C$, where C is a constant. Show that F, G, and H are all antiderivatives of the function f defined by $f(x) = 1$.

Solution Since

$$F'(x) = \frac{d}{dx}(x) = 1 = f(x)$$

$$G'(x) = \frac{d}{dx}(x + 2) = 1 = f(x)$$

$$H'(x) = \frac{d}{dx}(x + C) = 1 = f(x)$$

we see that F, G, and H are indeed antiderivatives of f.

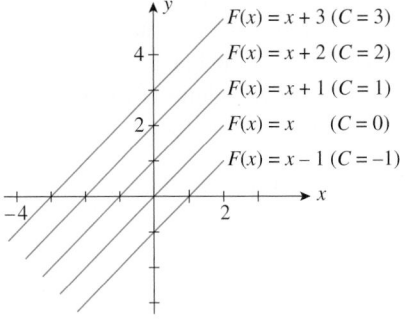

FIGURE 2
The graphs of some antiderivatives of $f(x) = 1$

Example 2 shows that once an antiderivative G of a function f is known, then another antiderivative of f may be found by adding an arbitrary constant to the function G. The following theorem states that no function other than one obtained in this manner can be an antiderivative of f. (We omit the proof.)

THEOREM 1

Let G be an antiderivative of a function f. Then, every antiderivative F of f must be of the form $F(x) = G(x) + C$, where C is a constant.

Returning to Example 2, we see that there are infinitely many antiderivatives of the function $f(x) = 1$. We obtain each one by specifying the constant C in the function $F(x) = x + C$. Figure 2 shows the graphs of some of these antiderivatives for selected values of C. These graphs constitute part of a family of infinitely many parallel straight lines, each having a slope equal to 1. This result is expected since there are infinitely many curves (straight lines) with a given slope equal to 1. The antiderivatives $F(x) = x + C$ (C, a constant) are precisely the functions representing this family of straight lines.

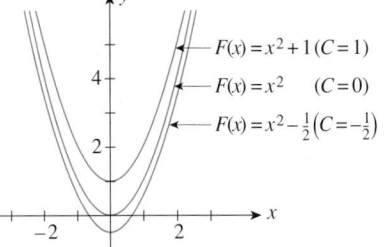

FIGURE 3
The graphs of some antiderivatives of $f(x) = 2x$

EXAMPLE 3 Prove that the function $G(x) = x^2$ is an antiderivative of the function $f(x) = 2x$. Write a general expression for the antiderivatives of f.

Solution Since $G'(x) = 2x = f(x)$, we have shown that $G(x) = x^2$ is an antiderivative of $f(x) = 2x$. By Theorem 1, every antiderivative of the function $f(x) = 2x$ has the form $F(x) = x^2 + C$, where C is some constant. The graphs of a few of the antiderivatives of f are shown in Figure 3.

EXPLORING WITH TECHNOLOGY

Let $f(x) = x^2 - 1$.

1. Show that $F(x) = \frac{1}{3}x^3 - x + C$, where C is an arbitrary constant, is an antiderivative of f.

2. Use a graphing utility to plot the graphs of the antiderivatives of f corresponding to $C = -2$, $C = -1$, $C = 0$, $C = 1$, and $C = 2$ on the same set of axes, using the viewing window $[-4, 4] \times [-4, 4]$.

3. If your graphing utility has the capability, draw the tangent line to each of the graphs in part 2 at the point whose x-coordinate is 2. What can you say about this family of tangent lines?

4. What is the slope of a tangent line in this family? Explain how you obtained your answer.

▬ The Indefinite Integral

The process of finding all antiderivatives of a function is called **antidifferentiation**, or **integration**. We use the symbol $\int$, called an **integral sign**, to indicate that the operation of integration is to be performed on some function f. Thus,

$$\int f(x)\, dx = F(x) + C$$

[read "the indefinite integral of $f(x)$ with respect to x equals $F(x)$ plus C"] tells us that the **indefinite integral** of f is the family of functions given by $F(x) + C$, where $F'(x) = f(x)$. The function f to be integrated is called the **integrand**, and the constant C is called a **constant of integration**. The expression dx following the integrand $f(x)$ reminds us that the operation is performed with respect to x. If the independent variable is t, we write $\int f(t)\, dt$ instead. In this sense both t and x are "dummy variables."

Using this notation, we can write the results of Examples 2 and 3 as

$$\int 1\, dx = x + C \quad \text{and} \quad \int 2x\, dx = x^2 + K$$

where C and K are arbitrary constants.

▬ Basic Integration Rules

Our next task is to develop some rules for finding the indefinite integral of a given function f. Because integration and differentiation are reverse operations, we discover many of the rules of integration by first making an "educated guess" at the antiderivative F of the function f to be integrated. Then this result is verified by demonstrating that $F' = f$.

> **Rule 1: The Indefinite Integral of a Constant**
>
> $$\int k\, dx = kx + C \qquad (k, \text{ a constant})$$

To prove this result, observe that

$$F'(x) = \frac{d}{dx}(kx + C) = k$$

EXAMPLE 4   Find each of the following indefinite integrals:

a. $\int 2\, dx$ **b.** $\int \pi^2\, dx$

Solution   Each of the integrands has the form $f(x) = k$, where k is a constant. Applying Rule 1 in each case yields

a. $\int 2\, dx = 2x + C$ **b.** $\int \pi^2\, dx = \pi^2 x + C$ ■

Next, from the rule of differentiation,

$$\frac{d}{dx} x^n = nx^{n-1}$$

we obtain the following rule of integration.

Rule 2: The Power Rule

$$\int x^n \, dx = \frac{1}{n + 1} x^{n+1} + C \qquad (n \neq -1)$$

An antiderivative of a power function is another power function obtained from the integrand by increasing its power by 1 and dividing the resulting expression by the new power.

To prove this result, observe that

$$F'(x) = \frac{d}{dx} \left(\frac{1}{n + 1} x^{n+1} + C \right)$$
$$= \frac{n + 1}{n + 1} x^n$$
$$= x^n$$
$$= f(x)$$

EXAMPLE 5   Find each of the following indefinite integrals:

a. $\displaystyle\int x^3 \, dx$ **b.** $\displaystyle\int x^{3/2} \, dx$ **c.** $\displaystyle\int \frac{1}{x^{3/2}} \, dx$

Solution   Each integrand is a power function with exponent $n \neq -1$. Applying Rule 2 in each case yields the following results:

a. $\displaystyle\int x^3 \, dx = \frac{1}{4} x^4 + C$

b. $\displaystyle\int x^{3/2} \, dx = \frac{1}{\frac{5}{2}} x^{5/2} + C = \frac{2}{5} x^{5/2} + C$

c. $\displaystyle\int \frac{1}{x^{3/2}} \, dx = \int x^{-3/2} \, dx = \frac{1}{-\frac{1}{2}} x^{-1/2} + C = -2x^{-1/2} + C = -\frac{2}{x^{1/2}} + C$

These results may be verified by differentiating each of the antiderivatives and showing that the result is equal to the corresponding integrand.   ∎

The next rule tells us that a constant factor may be moved through an integral sign.

Rule 3: The Indefinite Integral of a Constant Multiple of a Function

$$\int cf(x) \, dx = c \int f(x) \, dx \qquad (c, \text{ a constant})$$

The indefinite integral of a constant multiple of a function is equal to the constant multiple of the indefinite integral of the function.

This result follows from the corresponding rule of differentiation (see Rule 3, Section 12.1).

⚠ Only a constant can be "moved out" of an integral sign. For example, it would be incorrect to write

$$\int x^2 \, dx = x^2 \int 1 \, dx$$

In fact, $\int x^2 \, dx = \frac{1}{3} x^3 + C$, whereas $x^2 \int 1 \, dx = x^2(x + C) = x^3 + Cx^2$.

EXAMPLE 6   Find each of the following indefinite integrals:

a. $\displaystyle\int 2t^3 \, dt$ **b.** $\displaystyle\int -3x^{-2} \, dx$

Solution   Each integrand has the form $cf(x)$, where c is a constant. Applying Rule 3, we obtain:

a. $\displaystyle\int 2t^3 \, dt = 2\int t^3 \, dt = 2\left(\frac{1}{4}t^4 + K\right) = \frac{1}{2}t^4 + 2K = \frac{1}{2}t^4 + C$

where $C = 2K$. From now on, we will write the constant of integration as C, since any nonzero multiple of an arbitrary constant is an arbitrary constant.

b. $\displaystyle\int -3x^{-2} \, dx = -3\int x^{-2} \, dx = (-3)(-1)x^{-1} + C = \frac{3}{x} + C$ ■

Rule 4: The Sum Rule

$$\int [f(x) + g(x)] \, dx = \int f(x) \, dx + \int g(x) \, dx$$

$$\int [f(x) - g(x)] \, dx = \int f(x) \, dx - \int g(x) \, dx$$

The indefinite integral of a sum (difference) of two integrable functions is equal to the sum (difference) of their indefinite integrals.

This result is easily extended to the case involving the sum and difference of any finite number of functions. As in Rule 3, the proof of Rule 4 follows from the corresponding rule of differentiation (see Rule 4, Section 12.1).

EXAMPLE 7   Find the indefinite integral

$$\int (3x^5 + 4x^{3/2} - 2x^{-1/2}) \, dx$$

Solution   Applying the extended version of Rule 4, we find that

$$\int (3x^5 + 4x^{3/2} - 2x^{-1/2}) \, dx$$

$$= \int 3x^5 \, dx + \int 4x^{3/2} \, dx - \int 2x^{-1/2} \, dx$$

$$= 3\int x^5 \, dx + 4\int x^{3/2} \, dx - 2\int x^{-1/2} \, dx \qquad \text{Rule 3}$$

$$= (3)\left(\frac{1}{6}\right)x^6 + (4)\left(\frac{2}{5}\right)x^{5/2} - (2)(2)x^{1/2} + C \qquad \text{Rule 2}$$

$$= \frac{1}{2}x^6 + \frac{8}{5}x^{5/2} - 4x^{1/2} + C \qquad\qquad ■$$

Observe that we have combined the three constants of integration, which arise from evaluating the three indefinite integrals, to obtain one constant C. After all, the sum of three arbitrary constants is also an arbitrary constant.

Rule 5: The Indefinite Integral of the Exponential Function

$$\int e^x \, dx = e^x + C$$

The indefinite integral of the exponential function with base e is equal to the function itself (except, of course, for the constant of integration).

EXAMPLE 8 Find the indefinite integral

$$\int (2e^x - x^3) \, dx$$

Solution We have

$$\int (2e^x - x^3) \, dx = \int 2e^x \, dx - \int x^3 \, dx$$

$$= 2 \int e^x \, dx - \int x^3 \, dx$$

$$= 2e^x - \frac{1}{4}x^4 + C \qquad \blacksquare$$

The last rule of integration in this section covers the integration of the function $f(x) = x^{-1}$. Remember that this function constituted the only exceptional case in the integration of the power function $f(x) = x^n$ (see Rule 2).

Rule 6: The Indefinite Integral of the Function $f(x) = x^{-1}$

$$\int x^{-1} \, dx = \int \frac{1}{x} \, dx = \ln|x| + C \qquad (x \neq 0)$$

To prove Rule 6, observe that

$$\frac{d}{dx} \ln|x| = \frac{1}{x} \qquad \text{See Rule 3, Section 14.4.}$$

EXAMPLE 9 Find the indefinite integral

$$\int \left(2x + \frac{3}{x} + \frac{4}{x^2} \right) dx$$

Solution

$$\int \left(2x + \frac{3}{x} + \frac{4}{x^2} \right) dx = \int 2x \, dx + \int \frac{3}{x} \, dx + \int \frac{4}{x^2} \, dx$$

$$= 2 \int x \, dx + 3 \int \frac{1}{x} \, dx + 4 \int x^{-2} \, dx$$

$$= 2 \left(\frac{1}{2} \right) x^2 + 3 \ln|x| + 4(-1)x^{-1} + C$$

$$= x^2 + 3 \ln|x| - \frac{4}{x} + C \qquad \blacksquare$$

■ Differential Equations

Let's return to the problem posed at the beginning of the section: *Given the derivative of a function, f', can we find the function f?* As an example, suppose we are given the function

$$f'(x) = 2x - 1 \qquad (1)$$

and we wish to find $f(x)$. From what we now know, we can find f by integrating Equation (1). Thus,

$$f(x) = \int f'(x)\, dx = \int (2x - 1)\, dx = x^2 - x + C \qquad (2)$$

where C is an arbitrary constant. Thus, infinitely many functions have the derivative f', each differing from the other by a constant.

Equation (1) is called a differential equation. In general, a **differential equation** is an equation that involves the derivative or differential of an unknown function. [In the case of Equation (1), the unknown function is f.] A **solution** of a differential equation is any function that satisfies the differential equation. Thus, Equation (2) gives *all* the solutions of the differential Equation (1), and it is, accordingly, called the **general solution** of the differential equation $f'(x) = 2x - 1$.

The graphs of $f(x) = x^2 - x + C$ for selected values of C are shown in Figure 4. These graphs have one property in common: For any fixed value of x, the tangent lines to these graphs have the same slope. This follows because any member of the family $f(x) = x^2 - x + C$ must have the same slope at x—namely, $2x - 1$!

Although there are infinitely many solutions to the differential equation $f'(x) = 2x - 1$, we can obtain a **particular solution** by specifying the value the function must assume at a certain value of x. For example, suppose we stipulate that the function f under consideration must satisfy the condition $f(1) = 3$ or, equivalently, the graph of f must pass through the point $(1, 3)$. Then, using the condition on the general solution $f(x) = x^2 - x + C$, we find that

$$f(1) = 1 - 1 + C = 3$$

and $C = 3$. Thus, the particular solution is $f(x) = x^2 - x + 3$ (see Figure 4).

The condition $f(1) = 3$ is an example of an initial condition. More generally, an **initial condition** is a condition imposed on the value of f at $x = a$.

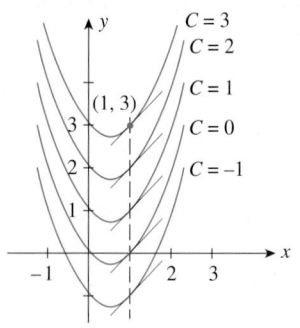

FIGURE 4
The graphs of some of the functions having the derivative $f'(x) = 2x - 1$. Observe that the slopes of the tangent lines to the graphs are the same for a fixed value of x.

■ Initial Value Problems

An **initial value problem** is one in which we are required to find a function satisfying (1) a differential equation and (2) one or more initial conditions. The following are examples of initial value problems.

EXAMPLE 10   Find the function f if it is known that

$$f'(x) = 3x^2 - 4x + 8 \quad \text{and} \quad f(1) = 9$$

Solution   We are required to solve the initial value problem

$$\left. \begin{array}{l} f'(x) = 3x^2 - 4x + 8 \\ f(1) = 9 \end{array} \right\}$$

Integrating the function f', we find

$$f(x) = \int f'(x)\, dx$$

$$= \int (3x^2 - 4x + 8)\, dx$$

$$= x^3 - 2x^2 + 8x + C$$

Using the condition $f(1) = 9$, we have

$$9 = f(1) = 1^3 - 2(1)^2 + 8(1) + C = 7 + C \quad \text{or} \quad C = 2$$

Therefore, the required function f is given by $f(x) = x^3 - 2x^2 + 8x + 2$.   ∎

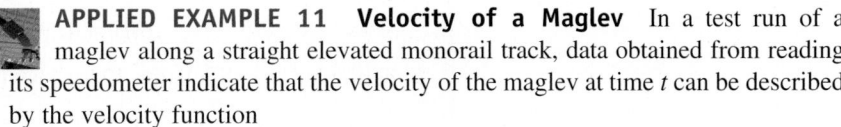

APPLIED EXAMPLE 11   Velocity of a Maglev   In a test run of a maglev along a straight elevated monorail track, data obtained from reading its speedometer indicate that the velocity of the maglev at time t can be described by the velocity function

$$v(t) = 8t \qquad (0 \le t \le 30)$$

Find the position function of the maglev. Assume that initially the maglev is located at the origin of a coordinate line.

Solution   Let $s(t)$ denote the position of the maglev at any time t ($0 \le t \le 30$). Then, $s'(t) = v(t)$. So, we have the initial value problem

$$\left. \begin{array}{l} s'(t) = 8t \\ s(0) = 0 \end{array} \right\}$$

Integrating both sides of the differential equation $s'(t) = 8t$, we obtain

$$s(t) = \int s'(t)\, dt = \int 8t\, dt = 4t^2 + C$$

where C is an arbitrary constant. To evaluate C, we use the initial condition $s(0) = 0$ to write

$$s(0) = 4(0) + C = 0 \quad \text{or} \quad C = 0$$

Therefore, the required position function is $s(t) = 4t^2$ ($0 \le t \le 30$).   ∎

VIDEO

APPLIED EXAMPLE 12   Magazine Circulation   The current circulation of the *Investor's Digest* is 3000 copies per week. The managing editor of the weekly projects a growth rate of

$$4 + 5t^{2/3}$$

copies per week, t weeks from now, for the next 3 years. Based on her projection, what will the circulation of the digest be 125 weeks from now?

Solution   Let $S(t)$ denote the circulation of the digest t weeks from now. Then $S'(t)$ is the rate of change in the circulation in the tth week and is given by

$$S'(t) = 4 + 5t^{2/3}$$

Furthermore, the current circulation of 3000 copies per week translates into the initial condition $S(0) = 3000$. Integrating the differential equation with respect to t gives

$$S(t) = \int S'(t)\,dt = \int (4 + 5t^{2/3})\,dt$$

$$= 4t + 5\left(\frac{t^{5/3}}{\frac{5}{3}}\right) + C = 4t + 3t^{5/3} + C$$

To determine the value of C, we use the condition $S(0) = 3000$ to write

$$S(0) = 4(0) + 3(0) + C = 3000$$

which gives $C = 3000$. Therefore, the circulation of the digest t weeks from now will be

$$S(t) = 4t + 3t^{5/3} + 3000$$

In particular, the circulation 125 weeks from now will be

$$S(125) = 4(125) + 3(125)^{5/3} + 3000 = 12{,}875$$

copies per week.   ∎

15.1   Self-Check Exercises

1. Evaluate $\int \left(\dfrac{1}{\sqrt{x}} - \dfrac{2}{x} + 3e^x \right) dx.$

2. Find the rule for the function f given that (1) the slope of the tangent line to the graph of f at any point $P(x, f(x))$ is given by the expression $3x^2 - 6x + 3$ and (2) the graph of f passes through the point $(2, 9)$.

3. Suppose United Motors' share of the new cars sold in a cer-

tain country is changing at the rate of

$$f(t) = -0.01875t^2 + 0.15t - 1.2 \qquad (0 \le t \le 12)$$

percent at year t ($t = 0$ corresponds to the beginning of 1994). The company's market share at the beginning of 1994 was 48.4%. What was United Motors' market share at the beginning of 2006?

Solutions to Self-Check Exercises 15.1 can be found on page 923.

15.1   Concept Questions

1. What is an antiderivative? Give an example.

2. If $f'(x) = g'(x)$ for all x in an interval I, what is the relationship between f and g?

3. What is the difference between an antiderivative of f and the indefinite integral of f?

15.1   Exercises

In Exercises 1–4, verify directly that F is an antiderivative of f.

1. $F(x) = \dfrac{1}{3}x^3 + 2x^2 - x + 2; f(x) = x^2 + 4x - 1$

2. $F(x) = xe^x + \pi; f(x) = e^x(1 + x)$

3. $F(x) = \sqrt{2x^2 - 1}; f(x) = \dfrac{2x}{\sqrt{2x^2 - 1}}$

4. $F(x) = x \ln x - x; f(x) = \ln x$

In Exercises 5–8, (a) verify that G is an antiderivative of f, (b) find all antiderivatives of f, and (c) sketch the graphs of a few of the family of antiderivatives found in part (b).

5. $G(x) = 2x; f(x) = 2$
6. $G(x) = 2x^2; f(x) = 4x$

7. $G(x) = \dfrac{1}{3}x^3; f(x) = x^2$
8. $G(x) = e^x; f(x) = e^x$

In Exercises 9–50, find the indefinite integral.

9. $\int 6\,dx$

10. $\int \sqrt{2}\,dx$

11. $\int x^3\,dx$

12. $\int 2x^5\,dx$

13. $\int x^{-4}\,dx$

14. $\int 3t^{-7}\,dt$

15. $\int x^{2/3}\,dx$

16. $\int 2u^{3/4}\,du$

17. $\int x^{-5/4}\,dx$

18. $\int 3x^{-2/3}\,dx$

19. $\int \dfrac{2}{x^2}\,dx$

20. $\int \dfrac{1}{3x^5}\,dx$

21. $\int \pi\sqrt{t}\,dt$

22. $\int \dfrac{3}{\sqrt{t}}\,dt$

23. $\int (3 - 2x)\,dx$

24. $\int (1 + u + u^2)\,du$

25. $\int (x^2 + x + x^{-3})\,dx$

26. $\int (0.3t^2 + 0.02t + 2)\,dt$

27. $\int 4e^x\,dx$

28. $\int (1 + e^x)\,dx$

29. $\int (1 + x + e^x)\,dx$

30. $\int (2 + x + 2x^2 + e^x)\,dx$

31. $\int \left(4x^3 - \dfrac{2}{x^2} - 1\right)dx$

32. $\int \left(6x^3 + \dfrac{3}{x^2} - x\right)dx$

33. $\int (x^{5/2} + 2x^{3/2} - x)\,dx$

34. $\int (t^{3/2} + 2t^{1/2} - 4t^{-1/2})\,dt$

35. $\int \left(\sqrt{x} + \dfrac{3}{\sqrt{x}}\right)dx$

36. $\int \left(\sqrt[3]{x^2} - \dfrac{1}{x^2}\right)dx$

37. $\int \left(\dfrac{u^3 + 2u^2 - u}{3u}\right)du$

Hint: $\dfrac{u^3 + 2u^2 - u}{3u} = \dfrac{1}{3}u^2 + \dfrac{2}{3}u - \dfrac{1}{3}$

38. $\int \dfrac{x^4 - 1}{x^2}\,dx$

Hint: $\dfrac{x^4 - 1}{x^2} = x^2 - x^{-2}$

39. $\int (2t + 1)(t - 2)\,dt$

40. $\int u^{-2}(1 - u^2 + u^4)\,du$

41. $\int \dfrac{1}{x^2}(x^4 - 2x^2 + 1)\,dx$

42. $\int \sqrt{t}\,(t^2 + t - 1)\,dt$

43. $\int \dfrac{ds}{(s + 1)^{-2}}$

44. $\int \left(\sqrt{x} + \dfrac{3}{x} - 2e^x\right)dx$

45. $\int (e^t + t^e)\,dt$

46. $\int \left(\dfrac{1}{x^2} - \dfrac{1}{\sqrt[3]{x^2}} + \dfrac{1}{\sqrt{x}}\right)dx$

47. $\int \left(\dfrac{x^3 + x^2 - x + 1}{x^2}\right)dx$

Hint: Simplify the integrand first.

48. $\int \dfrac{t^3 + \sqrt[3]{t}}{t^2}\,dt$

Hint: Simplify the integrand first.

49. $\int \dfrac{(\sqrt{x} - 1)^2}{x^2}\,dx$

Hint: Simplify the integrand first.

50. $\int (x + 1)^2\left(1 - \dfrac{1}{x}\right)dx$

Hint: Simplify the integrand first.

In Exercises 51–58, find $f(x)$ by solving the initial value problem.

51. $f'(x) = 2x + 1; f(1) = 3$

52. $f'(x) = 3x^2 - 6x; f(2) = 4$

53. $f'(x) = 3x^2 + 4x - 1; f(2) = 9$

54. $f'(x) = \dfrac{1}{\sqrt{x}}; f(4) = 2$

55. $f'(x) = 1 + \dfrac{1}{x^2}; f(1) = 2$

56. $f'(x) = e^x - 2x; f(0) = 2$

57. $f'(x) = \dfrac{x + 1}{x}; f(1) = 1$

58. $f'(x) = 1 + e^x + \dfrac{1}{x}; f(1) = 3 + e$

In Exercises 59–62, find the function f given that the slope of the tangent line to the graph of f at any point $(x, f(x))$ is $f'(x)$ and that the graph of f passes through the given point.

59. $f'(x) = \dfrac{1}{2}x^{-1/2}; (2, \sqrt{2})$

60. $f'(t) = t^2 - 2t + 3; (1, 2)$

61. $f'(x) = e^x + x; (0, 3)$ **62.** $f'(x) = \dfrac{2}{x} + 1; (1, 2)$

63. BANK DEPOSITS Madison Finance opened two branches on September 1 ($t = 0$). Branch A is located in an established industrial park, and branch B is located in a fast-growing new development. The net rate at which money was deposited into branch A and branch B in the first 180 business days is given by the graphs of f and g, respectively (see the figure). Which branch has a larger amount on deposit at the end of 180 business days? Justify your answer.

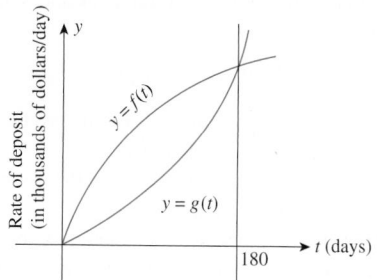

64. VELOCITY OF A CAR Two cars, side by side, start from rest and travel along a straight road. The velocity of car A is given by $v = f(t)$, and the velocity of car B is given by $v = g(t)$. The graphs of f and g are shown in the figure on the next page. Are the cars still side by side after T sec? If not, which car is ahead of the other? Justify your answer.

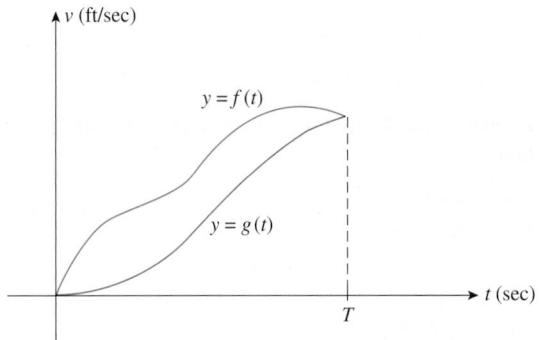

65. **VELOCITY OF A CAR** The velocity of a car (in feet/second) t sec after starting from rest is given by the function

$$f(t) = 2\sqrt{t} \qquad (0 \le t \le 30)$$

Find the car's position at any time t.

66. **VELOCITY OF A MAGLEV** The velocity (in feet/second) of a maglev is

$$v(t) = 0.2t + 3 \qquad (0 \le t \le 120)$$

At $t = 0$, it is at the station. Find the function giving the position of the maglev at time t, assuming that the motion takes place along a straight stretch of track.

67. **COST OF PRODUCING CLOCKS** Lorimar Watch Company manufactures travel clocks. The daily marginal cost function associated with producing these clocks is

$$C'(x) = 0.000009x^2 - 0.009x + 8$$

where $C'(x)$ is measured in dollars/unit and x denotes the number of units produced. Management has determined that the daily fixed cost incurred in producing these clocks is $120. Find the total cost incurred by Lorimar in producing the first 500 travel clocks/day.

68. **REVENUE FUNCTIONS** The management of Lorimar Watch Company has determined that the daily marginal revenue function associated with producing and selling their travel clocks is given by

$$R'(x) = -0.009x + 12$$

where x denotes the number of units produced and sold and $R'(x)$ is measured in dollars/unit.
a. Determine the revenue function $R(x)$ associated with producing and selling these clocks.
b. What is the demand equation that relates the wholesale unit price with the quantity of travel clocks demanded?

69. **PROFIT FUNCTIONS** Cannon Precision Instruments makes an automatic electronic flash with Thyrister circuitry. The estimated marginal profit associated with producing and selling these electronic flashes is

$$-0.004x + 20$$

dollars/unit/month when the production level is x units per month. Cannon's fixed cost for producing and selling these electronic flashes is $16,000/month. At what level of production does Cannon realize a maximum profit? What is the maximum monthly profit?

70. **COST OF PRODUCING GUITARS** Carlota Music Company estimates that the marginal cost of manufacturing its Professional Series guitars is

$$C'(x) = 0.002x + 100$$

dollars/month when the level of production is x guitars/month. The fixed costs incurred by Carlota are $4000/month. Find the total monthly cost incurred by Carlota in manufacturing x guitars/month.

71. **QUALITY CONTROL** As part of a quality-control program, the chess sets manufactured by Jones Brothers are subjected to a final inspection before packing. The rate of increase in the number of sets checked per hour by an inspector t hr into the 8 a.m. to 12 noon morning shift is approximately

$$N'(t) = -3t^2 + 12t + 45 \qquad (0 \le t \le 4)$$

a. Find an expression $N(t)$ that approximates the number of sets inspected at the end of t hours.
Hint: $N(0) = 0$.
b. How many sets does the average inspector check during a morning shift?

72. **MEASURING TEMPERATURE** The temperature on a certain day as measured at the airport of a city is changing at the rate of

$$T'(t) = 0.15t^2 - 3.6t + 14.4 \qquad (0 \le t \le 4)$$

°F/hr, where t is measured in hours, with $t = 0$ corresponding to 6 a.m. The temperature at 6 a.m. was 24°F.
a. Find an expression giving the temperature T at the airport at any time between 6 a.m. and 10 a.m.
b. What was the temperature at 10 a.m.?

73. **SATELLITE RADIO SUBSCRIPTIONS** Based on data obtained by polling automobile buyers, the number of subscribers of satellite radios is expected to grow at the rate of

$$r(t) = -0.375t^2 + 2.1t + 2.45 \qquad (0 \le t \le 5)$$

million subscribers/year between 2003 ($t = 0$) and 2008 ($t = 5$). The number of satellite radio subscribers at the beginning of 2003 was 1.5 million.
a. Find an expression giving the number of satellite radio subscribers in year t ($0 \le t \le 5$).
b. Based on this model, what will be the number of satellite radio subscribers in 2008?
Source: Carmel Group

74. **GENETICALLY MODIFIED CROPS** The total number of acres of genetically modified crops grown worldwide from 1997

through 2003 was changing at the rate of

$$R(t) = 2.718t^2 - 19.86t + 50.18 \qquad (0 \le t \le 6)$$

million acres/year. The total number of acres of such crops grown in 1997 ($t = 0$) was 27.2 million acres. How many acres of genetically modified crops were grown worldwide in 2003?

Source: International Services for the Acquisition of Agri-biotech Applications

75. CREDIT CARD DEBT   The average credit card debt per U.S. household between 1990 ($t = 0$) and 2003 ($t = 13$) was growing at the rate of approximately

$$D(t) = -4.479t^2 + 69.8t + 279.5 \qquad (0 \le t \le 13)$$

dollars per year. The average credit card debt per U.S. household stood at \$2917 in 1990.
   a. Find an expression giving the approximate average credit card debt per U.S. household in year t ($0 \le t \le 13$).
   b. Use the result of part (a) to estimate the average credit card debt per U.S. household in 2003.

Source: Encore Capital Group

76. DVD SALES   The total number of DVDs sold to U.S. dealers for rental and sale from 1999 through 2003 grew at the rate of approximately

$$R(t) = -0.03t^2 + 0.218t - 0.032 \qquad (0 \le t \le 4)$$

billion units/year, where t is measured in years with $t = 0$ corresponding to 1999. The total number of DVDs sold as of 1999 was 0.1 billion units.
   a. Find an expression giving the total number of DVDs sold by year t ($0 \le t \le 4$).
   b. What was the total number of DVDs sold by 2003?

Source: Adams Media

77. GASTRIC BYPASS SURGERIES   One method of weight loss gaining in popularity is stomach-reducing surgery. It is generally reserved for people at least 100 lb overweight because the procedure carries a serious risk of death or complications. According to the American Society of Bariatric Surgery, the number of morbidly obese patients undergoing the procedure was increasing at the rate of

$$R(t) = 9.399t^2 - 13.4t + 14.07 \qquad (0 \le t \le 3)$$

thousands/year, where $t = 0$ corresponds to 2000. The number of gastric bypass surgeries performed in 2000 was 36.7 thousand.
   a. Find an expression giving the number of gastric bypass surgeries performed in year t ($0 \le t \le 3$).
   b. Use the result of part (a) to find the number of gastric bypass surgeries performed in 2003.

Source: American Society for Bariatric Surgery

78. ONLINE AD SALES   According to a study conducted in 2004, the share of online advertisement, worldwide, as a percent of

the total ad market, is expected to grow at the rate of

$$R(t) = -0.033t^2 + 0.3428t + 0.07 \qquad (0 \le t \le 6)$$

percent/year at time t (in years), with $t = 0$ corresponding to the beginning of 2000. The online ad market at the beginning of 2000 was 2.9% of the total ad market.
   a. What is the projected online ad market share at any time t?
   b. What is the projected online ad market share at the beginning of 2005?

Source: Jupiter Media Metrix, Inc.

79. HEALTH-CARE COSTS   The average out-of-pocket costs for beneficiaries in traditional Medicare (including premiums, cost sharing, and prescription drugs not covered by Medicare) is projected to grow at the rate of

$$C'(t) = 12.288t^2 - 150.5594t + 695.23$$

dollars/year, where t is measured in 5-yr intervals, with $t = 0$ corresponding to 2000. The out-of-pocket costs for beneficiaries in 2000 were \$3142.
   a. Find an expression giving the average out-of-pocket costs for beneficiaries in year t.
   b. What is the projected average out-of-pocket costs for beneficiaries in 2010?

Source: The Urban Institute

80. BALLAST DROPPED FROM A BALLOON   A ballast is dropped from a stationary hot-air balloon that is hovering at an altitude of 400 ft. Its velocity after t sec is $-32t$ ft/sec.
   a. Find the height $h(t)$ of the ballast from the ground at time t.
      Hint: $h'(t) = -32t$ and $h(0) = 400$.
   b. When will the ballast strike the ground?
   c. Find the velocity of the ballast when it hits the ground.

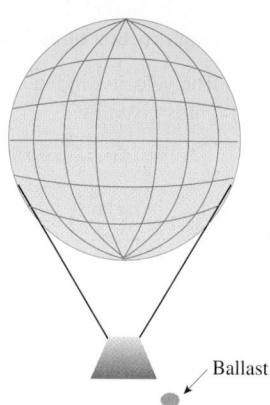
Ballast

81. CABLE TV SUBSCRIBERS   A study conducted by TeleCable estimates that the number of cable TV subscribers will grow at the rate of

$$100 + 210t^{3/4}$$

new subscribers/month t mo from the start date of the service. If 5000 subscribers signed up for the service before the starting date, how many subscribers will there be 16 mo from that date?

82. **AIR POLLUTION** On an average summer day, the level of carbon monoxide (CO) in a city's air is 2 parts per million (ppm). An environmental protection agency's study predicts that, unless more stringent measures are taken to protect the city's atmosphere, the CO concentration present in the air will increase at the rate of

$$0.003t^2 + 0.06t + 0.1$$

ppm/year t yr from now. If no further pollution-control efforts are made, what will be the CO concentration on an average summer day 5 yr from now?

83. **OZONE POLLUTION** The rate of change of the level of ozone, an invisible gas that is an irritant and impairs breathing, present in the atmosphere on a certain May day in the city of Riverside is given by

$$R(t) = 3.2922t^2 - 0.366t^3 \qquad (0 < t < 11)$$

(measured in pollutant standard index/hour). Here, t is measured in hours, with $t = 0$ corresponding to 7 a.m. Find the ozone level $A(t)$ at any time t, assuming that at 7 a.m. it is zero.
Hint: $A'(t) = R(t)$ and $A(0) = 0$.
Source: Los Angeles Times

84. **POPULATION GROWTH** The development of AstroWorld ("The Amusement Park of the Future") on the outskirts of a city will increase the city's population at the rate of

$$4500\sqrt{t} + 1000$$

people/year t yr from the start of construction. The population before construction is 30,000. Determine the projected population 9 yr after construction of the park has begun.

85. **FLIGHT OF A ROCKET** The velocity, in feet/second, of a rocket t sec into vertical flight is given by

$$v(t) = -3t^2 + 192t + 120$$

Find an expression $h(t)$ that gives the rocket's altitude, in feet, t sec after liftoff. What is the altitude of the rocket 30 sec after liftoff?
Hint: $h'(t) = v(t)$; $h(0) = 0$.

86. **SURFACE AREA OF A HUMAN** Empirical data suggest that the surface area of a 180-cm-tall human body changes at the rate of

$$S'(W) = 0.131773W^{-0.575}$$

square meters/kilogram, where W is the weight of the body in kilograms. If the surface area of a 180-cm-tall human

body weighing 70 kg is 1.886277 m^2, what is the surface area of a human body of the same height weighing 75 kg?

87. **OUTPATIENT SERVICE COMPANIES** The number of Medicare-certified home-health-care agencies (70% are freestanding, and 30% are owned by a hospital or other large facility) has been declining at the rate of

$$0.186e^{-0.02t} \qquad (0 \le t \le 14)$$

thousand agencies per year, between 1988 ($t = 0$) and 2002 ($t = 14$). The number of such agencies stood at 9.3 thousand units in 1988.
a. Find an expression giving the number of health-care agencies in year t.
b. What was the number of health-care agencies in 2002?
c. If this model held true through 2005, how many care agencies were there in 2005?
Source: Centers for Medicare and Medicaid Services

88. **HEIGHTS OF CHILDREN** According to the Jenss model for predicting the height of preschool children, the rate of growth of a typical preschool child is

$$R(t) = 25.8931e^{-0.993t} + 6.39 \qquad \left(\frac{1}{4} \le t \le 6\right)$$

cm/yr, where t is measured in years. The height of a typical 3-mo-old preschool child is 60.2952 cm.
a. Find a model for predicting the height of a typical preschool child at age t.
b. Use the result of part (a) to estimate the height of a typical 1-yr-old child.

89. **BLOOD FLOW IN AN ARTERY** Nineteenth-century physician Jean Louis Marie Poiseuille discovered that the rate of change of the velocity of blood r cm from the central axis of an artery (in centimeters/second/centimeter) is given by

$$a(r) = -kr$$

where k is a constant. If the radius of an artery is R cm, find an expression for the velocity of blood as a function of r (see the accompanying figure).
Hint: $v'(r) = a(r)$ and $v(R) = 0$. (Why?)

Blood vessel

90. **ACCELERATION OF A CAR** A car traveling along a straight road at 66 ft/sec accelerated to a speed of 88 ft/sec over a distance of 440 ft. What was the acceleration of the car, assuming it was constant?

91. DECELERATION OF A CAR What constant deceleration would a car moving along a straight road have to be subjected to if it were brought to rest from a speed of 88 ft/sec in 9 sec? What would be the stopping distance?

92. CARRIER LANDING A pilot lands a fighter aircraft on an aircraft carrier. At the moment of touchdown, the speed of the aircraft is 160 mph. If the aircraft is brought to a complete stop in 1 sec and the deceleration is assumed to be constant, find the number of g's the pilot is subjected to during landing (1 g = 32 ft/sec²).

93. CROSSING THE FINISH LINE After rounding the final turn in the bell lap, two runners emerged ahead of the pack. When runner A is 200 ft from the finish line, his speed is 22 ft/sec, a speed that he maintains until he crosses the line. At that instant of time, runner B, who is 20 ft behind runner A and running at a speed of 20 ft/sec, begins to spurt. Assuming that runner B sprints with a constant acceleration, what minimum acceleration will enable him to cross the finish line ahead of runner A?

94. DRAINING A TANK A tank has a constant cross-sectional area of 50 ft² and an orifice of constant cross-sectional area of $\frac{1}{2}$ ft² located at the bottom of the tank (see the accompanying figure).

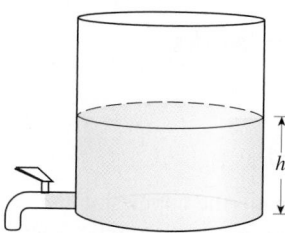

If the tank is filled with water to a height of h ft and allowed to drain, then the height of the water decreases at a rate that

is described by the equation

$$\frac{dh}{dt} = -\frac{1}{25}\left(\sqrt{20} - \frac{t}{50}\right) \qquad (0 \le t \le 50\sqrt{20})$$

Find an expression for the height of the water at any time t if its height initially is 20 ft.

95. AMOUNT OF RAINFALL During a thunderstorm, rain was falling at the rate of

$$\frac{8}{(t+4)^2} \qquad (0 \le t \le 2)$$

in./hr.

a. Find an expression giving the total amount of rainfall after t hr.
Hint: The total amount of rainfall at $t = 0$ is zero.
b. How much rain had fallen after 1 hr? After 2 hr?

96. LAUNCHING A FIGHTER AIRCRAFT A fighter aircraft is launched from the deck of a Nimitz-class aircraft carrier with the help of a steam catapult. If the aircraft is to attain a takeoff speed of at least 240 ft/sec after traveling 800 ft along the flight deck, find the minimum acceleration it must be subjected to, assuming it is constant.

In Exercises 97–100, determine whether the statement is true or false. If it is true, explain why it is true. If it is false, give an example to show why it is false.

97. If F and G are antiderivatives of f on an interval I, then $F(x) = G(x) + C$ on I.

98. If F is an antiderivative of f on an interval I, then $\int f(x)\,dx = F(x)$.

99. If f and g are integrable, then $\int [2f(x) - 3g(x)]\,dx = 2\int f(x)\,dx - 3\int g(x)\,dx$.

100. If f and g are integrable, then $\int f(x)g(x)\,dx = [\int f(x)\,dx][\int g(x)\,dx]$.

15.1 Solutions to Self-Check Exercises

1. $\displaystyle \int\left(\frac{1}{\sqrt{x}} - \frac{2}{x} + 3e^x\right)dx = \int\left(x^{-1/2} - \frac{2}{x} + 3e^x\right)dx$

$$= \int x^{-1/2}\,dx - 2\int\frac{1}{x}\,dx + 3\int e^x\,dx$$

$$= 2x^{1/2} - 2\ln|x| + 3e^x + C$$

$$= 2\sqrt{x} - 2\ln|x| + 3e^x + C$$

2. The slope of the tangent line to the graph of the function f at any point $P(x, f(x))$ is given by the derivative f' of f. Thus, the first condition implies that

$$f'(x) = 3x^2 - 6x + 3$$

which, upon integration, yields

$$f(x) = \int (3x^2 - 6x + 3)\,dx$$

$$= x^3 - 3x^2 + 3x + k$$

where k is the constant of integration.
    To evaluate k, we use the initial condition (2), which implies that $f(2) = 9$, or

$$9 = f(2) = 2^3 - 3(2)^2 + 3(2) + k$$

or $k = 7$. Hence, the required rule of definition of the

function f is

$$f(x) = x^3 - 3x^2 + 3x + 7$$

3. Let $M(t)$ denote United Motors' market share at year t. Then,

$$M(t) = \int f(t)\, dt$$

$$= \int (-0.01875t^2 + 0.15t - 1.2)\, dt$$

$$= -0.00625t^3 + 0.075t^2 - 1.2t + C$$

To determine the value of C, we use the initial condition

$M(0) = 48.4$, obtaining $C = 48.4$. Therefore,

$$M(t) = -0.00625t^3 + 0.075t^2 - 1.2t + 48.4$$

In particular, United Motors' market share of new cars at the beginning of 2006 is given by

$$M(12) = -0.00625(12)^3 + 0.075(12)^2$$
$$- 1.2(12) + 48.4 = 34$$

or 34%.

15.2 Integration by Substitution

In Section 15.1, we developed certain rules of integration that are closely related to the corresponding rules of differentiation in Chapters 12 and 14. In this section, we introduce a method of integration called the **method of substitution**, which is related to the chain rule for differentiating functions. When used in conjunction with the rules of integration developed earlier, the method of substitution is a powerful tool for integrating a large class of functions.

▬ How the Method of Substitution Works

Consider the indefinite integral

$$\int 2(2x + 4)^5\, dx \tag{3}$$

One way of evaluating this integral is to expand the expression $(2x + 4)^5$ and then integrate the resulting integrand term by term. As an alternative approach, let's see if we can simplify the integral by making a change of variable. Write

$$u = 2x + 4$$

with differential

$$du = 2\, dx$$

If we formally substitute these quantities into Equation (3), we obtain

$$\int 2(2x + 4)^5\, dx = \int (2x + 4)^5 (2\, dx) = \int u^5\, du$$

$$\uparrow$$
Rewrite $\qquad\qquad\qquad$ $\begin{cases} u = 2x + 4 \\ du = 2\, dx \end{cases}$

Now, the last integral involves a power function and is easily evaluated using Rule 2 of Section 15.1. Thus,

$$\int u^5\, du = \frac{1}{6} u^6 + C$$

Therefore, using this result and replacing u by $u = 2x + 4$, we obtain

$$\int 2(2x + 4)^5\, dx = \frac{1}{6}(2x + 4)^6 + C$$

We can verify that the foregoing result is indeed correct by computing

$$\frac{d}{dx}\left[\frac{1}{6}(2x+4)^6\right] = \frac{1}{6}\cdot 6(2x+4)^5(2) \qquad \text{Use the chain rule.}$$
$$= 2(2x+4)^5$$

and observing that the last expression is just the integrand of (3).

■ The Method of Integration by Substitution

To see why the approach used in evaluating the integral in (3) is successful, write

$$f(x) = x^5 \quad \text{and} \quad g(x) = 2x + 4$$

Then, $g'(x) = 2\,dx$. Furthermore, the integrand of (3) is just the composition of f and g. Thus,

$$(f \circ g)(x) = f(g(x))$$
$$= [g(x)]^5 = (2x+4)^5$$

Therefore, (3) can be written as

$$\int f(g(x))g'(x)\,dx \qquad\qquad \textbf{(4)}$$

Next, let's show that an integral having the form (4) can always be written as

$$\int f(u)\,du \qquad\qquad \textbf{(5)}$$

Suppose F is an antiderivative of f. By the chain rule, we have

$$\frac{d}{dx}[F(g(x))] = F'(g(x))g'(x)$$

Therefore,

$$\int F'(g(x))g'(x)\,dx = F(g(x)) + C$$

Letting $F' = f$ and making the substitution $u = g(x)$, we have

$$\int f(g(x))g'(x)\,dx = F(u) + C = \int F'(u)\,du = \int f(u)\,du$$

as we wished to show. Thus, if the transformed integral is readily evaluated, as is the case with the integral (3), then the method of substitution will prove successful.

Before we look at more examples, let's summarize the steps involved in integration by substitution.

Integration by Substitution

Step 1   Let $u = g(x)$, where $g(x)$ is part of the integrand, usually the "inside function" of the composite function $f(g(x))$.

Step 2   Find $du = g'(x)\,dx$.

Step 3   Use the substitution $u = g(x)$ and $du = g'(x)\,dx$ to convert the *entire* integral into one involving *only u*.

Step 4   Evaluate the resulting integral.

Step 5   Replace u by $g(x)$ to obtain the final solution as a function of x.

Note   Sometimes we need to consider different choices of g for the substitution $u = g(x)$ in order to carry out Step 3 and/or Step 4. ∎

EXAMPLE 1   Find $\int 2x(x^2 + 3)^4\, dx$.

Solution

Step 1   Observe that the integrand involves the composite function $(x^2 + 3)^4$ with "inside function" $g(x) = x^2 + 3$. So, we choose $u = x^2 + 3$.

Step 2   Find $du = 2x\, dx$.

Step 3   Making the substitution $u = x^2 + 3$ and $du = 2x\, dx$, we obtain

$$\int 2x(x^2 + 3)^4\, dx = \int \underset{\substack{\uparrow \\ \text{Rewrite}}}{(x^2 + 3)^4(2x\, dx)} = \int u^4\, du$$

an integral involving only the variable u.

Step 4   Evaluate

$$\int u^4\, du = \frac{1}{5} u^5 + C$$

Step 5   Replacing u by $x^2 + 3$, we obtain

$$\int 2x(x^2 + 3)^4\, dx = \frac{1}{5}(x^2 + 3)^5 + C$$ ∎

EXAMPLE 2   Find $\int 3\sqrt{3x + 1}\, dx$.

Solution

Step 1   The integrand involves the composite function $\sqrt{3x + 1}$ with "inside function" $g(x) = 3x + 1$. So, let $u = 3x + 1$.

Step 2   Find $du = 3\, dx$.

Step 3   Making the substitution $u = 3x + 1$ and $du = 3\, dx$, we obtain

$$\int 3\sqrt{3x + 1}\, dx = \int \sqrt{3x + 1}(3\, dx) = \int \sqrt{u}\, du$$

an integral involving only the variable u.

Step 4   Evaluate

$$\int \sqrt{u}\, du = \int u^{1/2}\, du = \frac{2}{3} u^{3/2} + C$$

Step 5   Replacing u by $3x + 1$, we obtain

$$\int 3\sqrt{3x + 1}\, dx = \frac{2}{3}(3x + 1)^{3/2} + C$$ ∎

EXAMPLE 3   Find $\int x^2(x^3 + 1)^{3/2}\, dx$.

Solution

Step 1   The integrand contains the composite function $(x^3 + 1)^{3/2}$ with "inside function" $g(x) = x^3 + 1$. So, let $u = x^3 + 1$.

Step 2   Find $du = 3x^2\, dx$.

Step 3   Making the substitution $u = x^3 + 1$ and $du = 3x^2\, dx$, or $x^2\, dx = \frac{1}{3}\, du$, we obtain

$$\int x^2(x^3 + 1)^{3/2}\, dx = \int (x^3 + 1)^{3/2}(x^2\, dx)$$

$$= \int u^{3/2}\left(\frac{1}{3}\, du\right) = \frac{1}{3}\int u^{3/2}\, du$$

an integral involving only the variable u.

Step 4   We evaluate

$$\frac{1}{3}\int u^{3/2}\, du = \frac{1}{3} \cdot \frac{2}{5} u^{5/2} + C = \frac{2}{15} u^{5/2} + C$$

Step 5   Replacing u by $x^3 + 1$, we obtain

$$\int x^2(x^3 + 1)^{3/2}\, dx = \frac{2}{15}(x^3 + 1)^{5/2} + C$$ ∎

EXPLORE & DISCUSS

Let $f(x) = x^2(x^3 + 1)^{3/2}$. Using the result of Example 3, we see that an antiderivative of f is $F(x) = \frac{2}{15}(x^3 + 1)^{5/2}$. However, in terms of u (where $u = x^3 + 1$), an antiderivative of f is $G(u) = \frac{2}{15} u^{5/2}$. Compute $F(2)$. Next, suppose we want to compute $F(2)$ using the function G instead. At what value of u should you evaluate $G(u)$ in order to obtain the desired result? Explain your answer.

In the remaining examples, we drop the practice of labeling the steps involved in evaluating each integral.

EXAMPLE 4   Find $\int e^{-3x}\, dx$.

Solution   Let $u = -3x$ so that $du = -3\, dx$, or $dx = -\frac{1}{3}\, du$. Then,

$$\int e^{-3x}\, dx = \int e^u\left(-\frac{1}{3}\, du\right) = -\frac{1}{3}\int e^u\, du$$

$$= -\frac{1}{3} e^u + C = -\frac{1}{3} e^{-3x} + C$$ ∎

EXAMPLE 5   Find $\displaystyle\int \frac{x}{3x^2 + 1}\, dx$.

Solution   Let $u = 3x^2 + 1$. Then, $du = 6x\, dx$, or $x\, dx = \frac{1}{6}\, du$. Making the appropriate substitutions, we have

$$\int \frac{x}{3x^2 + 1}\, dx = \int \frac{\frac{1}{6}}{u}\, du$$

$$= \frac{1}{6}\int \frac{1}{u}\, du$$

$$= \frac{1}{6}\ln|u| + C$$

$$= \frac{1}{6}\ln(3x^2 + 1) + C \quad \text{Since } 3x^2 + 1 > 0$$ ∎

EXAMPLE 6   Find $\int \dfrac{(\ln x)^2}{2x}\,dx$.

Solution   Let $u = \ln x$. Then,

$$du = \frac{d}{dx}(\ln x)\,dx = \frac{1}{x}\,dx$$

$$\int \frac{(\ln x)^2}{2x}\,dx = \frac{1}{2}\int \frac{(\ln x)^2}{x}\,dx$$

$$= \frac{1}{2}\int u^2\,du$$

$$= \frac{1}{6}u^3 + C$$

$$= \frac{1}{6}(\ln x)^3 + C$$ ∎

EXPLORE & DISCUSS

Suppose $\int f(u)\,du = F(u) + C$.

1. Show that $\int f(ax + b)\,dx = \dfrac{1}{a}F(ax + b) + C$.

2. How can you use this result to facilitate the evaluation of integrals such as $\int (2x + 3)^5\,dx$ and $\int e^{3x-2}\,dx$? Explain your answer.

Examples 7 and 8 show how the method of substitution can be used in practical situations.

APPLIED EXAMPLE 7   Cost of Producing Solar Cell Panels   In 1990 the head of the research and development department of Soloron Corporation claimed that the cost of producing solar cell panels would drop at the rate of

$$\frac{58}{(3t + 2)^2} \qquad (0 \le t \le 10)$$

dollars per peak watt for the next t years, with $t = 0$ corresponding to the beginning of 1990. (A peak watt is the power produced at noon on a sunny day.) In 1990 the panels, which are used for photovoltaic power systems, cost \$10 per peak watt. Find an expression giving the cost per peak watt of producing solar cell panels at the beginning of year t. What was the cost at the beginning of 2000?

Solution   Let $C(t)$ denote the cost per peak watt for producing solar cell panels at the beginning of year t. Then,

$$C'(t) = -\frac{58}{(3t + 2)^2}$$

Integrating, we find that

$$C(t) = \int \frac{-58}{(3t + 2)^2}\,dt$$

$$= -58 \int (3t + 2)^{-2}\,dt$$

Let $u = 3t + 2$ so that

$$du = 3\, dt \quad \text{or} \quad dt = \frac{1}{3}\, du$$

Then,

$$C(t) = -58\left(\frac{1}{3}\right)\int u^{-2}\, du$$

$$= -\frac{58}{3}(-1)u^{-1} + k$$

$$= \frac{58}{3(3t + 2)} + k$$

where k is an arbitrary constant. To determine the value of k, note that the cost per peak watt of producing solar cell panels at the beginning of 1990 ($t = 0$) was 10, or $C(0) = 10$. This gives

$$C(0) = \frac{58}{3(2)} + k = 10$$

or $k = \frac{1}{3}$. Therefore, the required expression is given by

$$C(t) = \frac{58}{3(3t + 2)} + \frac{1}{3}$$

$$= \frac{58 + (3t + 2)}{3(3t + 2)} = \frac{3t + 60}{3(3t + 2)}$$

$$= \frac{t + 20}{3t + 2}$$

The cost per peak watt for producing solar cell panels at the beginning of 2000 is given by

$$C(10) = \frac{10 + 20}{3(10) + 2} \approx 0.94$$

or approximately \$0.94 per peak watt.

EXPLORING WITH TECHNOLOGY

Refer to Example 7.

1. Use a graphing utility to plot the graph of

$$C(t) = \frac{t + 20}{3t + 2}$$

   using the viewing window $[0, 10] \times [0, 5]$. Then, use the numerical differentiation capability of the graphing utility to compute $C'(10)$.

2. Plot the graph of

$$C'(t) = -\frac{58}{(3t + 2)^2}$$

   using the viewing window $[0, 10] \times [-10, 0]$. Then, use the evaluation capability of the graphing utility to find $C'(10)$. Is this value of $C'(10)$ the same as that obtained in part 1? Explain your answer.

APPLIED EXAMPLE 8   Computer Sales Projections   A study prepared by the marketing department of Universal Instruments forecasts that, after its new line of Galaxy Home Computers is introduced into the market, sales will grow at the rate of

$$2000 - 1500e^{-0.05t} \qquad (0 \le t \le 60)$$

units per month. Find an expression that gives the total number of computers that will sell t months after they become available on the market. How many computers will Universal sell in the first year they are on the market?

Solution   Let $N(t)$ denote the total number of computers that may be expected to be sold t months after their introduction in the market. Then, the rate of growth of sales is given by $N'(t)$ units per month. Thus,

$$N'(t) = 2000 - 1500e^{-0.05t}$$

so that

$$N(t) = \int (2000 - 1500e^{-0.05t})\, dt$$

$$= \int 2000\, dt - 1500 \int e^{-0.05t}\, dt$$

Upon integrating the second integral by the method of substitution, we obtain

$$N(t) = 2000t + \frac{1500}{0.05} e^{-0.05t} + C \qquad \text{Let } u = -0.05t,$$
$$\text{then } du = -0.05\, dt.$$

$$= 2000t + 30{,}000e^{-0.05t} + C$$

To determine the value of C, note that the number of computers sold at the end of month 0 is nil, so $N(0) = 0$. This gives

$$N(0) = 30{,}000 + C = 0 \qquad \text{Since } e^0 = 1$$

or $C = -30{,}000$. Therefore, the required expression is given by

$$N(t) = 2000t + 30{,}000e^{-0.05t} - 30{,}000$$

$$= 2000t + 30{,}000(e^{-0.05t} - 1)$$

The number of computers that Universal can expect to sell in the first year is given by

$$N(12) = 2000(12) + 30{,}000(e^{-0.05(12)} - 1)$$

$$\approx 10{,}464$$

15.2   Self-Check Exercises

1. Evaluate $\int \sqrt{2x + 5}\, dx$.

2. Evaluate $\int \dfrac{x^2}{(2x^3 + 1)^{3/2}}\, dx$.

3. Evaluate $\int xe^{2x^2 - 1}\, dx$.

4. According to a joint study conducted by Oxnard's Environ-

mental Management Department and a state government agency, the concentration of carbon monoxide (CO) in the air due to automobile exhaust is increasing at the rate given by

$$f(t) = \frac{8(0.1t + 1)}{300(0.2t^2 + 4t + 64)^{1/3}}$$

parts per million (ppm) per year t. Currently, the CO concentration due to automobile exhaust is 0.16 ppm. Find an expression giving the CO concentration t yr from now.

Solutions to Self-Check Exercises 15.2 can be found on page 933.

15.2 Concept Questions

1. Explain how the method of substitution works by showing the steps used to find $\int f(g(x))g'(x)\,dx$.

2. Explain why the method of substitution works for the integral $\int xe^{-x^2}\,dx$, but not for the integral $\int e^{-x^2}\,dx$.

15.2 Exercises

In Exercises 1–50, find the indefinite integral.

1. $\displaystyle \int 4(4x+3)^4\,dx$

2. $\displaystyle \int 4x(2x^2+1)^7\,dx$

3. $\displaystyle \int (x^3-2x)^2(3x^2-2)\,dx$

4. $\displaystyle \int (3x^2-2x+1)(x^3-x^2+x)^4\,dx$

5. $\displaystyle \int \frac{4x}{(2x^2+3)^3}\,dx$

6. $\displaystyle \int \frac{3x^2+2}{(x^3+2x)^2}\,dx$

7. $\displaystyle \int 3t^2\sqrt{t^3+2}\,dt$

8. $\displaystyle \int 3t^2(t^3+2)^{3/2}\,dt$

9. $\displaystyle \int (x^2-1)^9 x\,dx$

10. $\displaystyle \int x^2(2x^3+3)^4\,dx$

11. $\displaystyle \int \frac{x^4}{1-x^5}\,dx$

12. $\displaystyle \int \frac{x^2}{\sqrt{x^3-1}}\,dx$

13. $\displaystyle \int \frac{2}{x-2}\,dx$

14. $\displaystyle \int \frac{x^2}{x^3-3}\,dx$

15. $\displaystyle \int \frac{0.3x-0.2}{0.3x^2-0.4x+2}\,dx$

16. $\displaystyle \int \frac{2x^2+1}{0.2x^3+0.3x}\,dx$

17. $\displaystyle \int \frac{x}{3x^2-1}\,dx$

18. $\displaystyle \int \frac{x^2-1}{x^3-3x+1}\,dx$

19. $\displaystyle \int e^{-2x}\,dx$

20. $\displaystyle \int e^{-0.02x}\,dx$

21. $\displaystyle \int e^{2-x}\,dx$

22. $\displaystyle \int e^{2t+3}\,dt$

23. $\displaystyle \int xe^{-x^2}\,dx$

24. $\displaystyle \int x^2 e^{x^3-1}\,dx$

25. $\displaystyle \int (e^x-e^{-x})\,dx$

26. $\displaystyle \int (e^{2x}+e^{-3x})\,dx$

27. $\displaystyle \int \frac{e^x}{1+e^x}\,dx$

28. $\displaystyle \int \frac{e^{2x}}{1+e^{2x}}\,dx$

29. $\displaystyle \int \frac{e^{\sqrt{x}}}{\sqrt{x}}\,dx$

30. $\displaystyle \int \frac{e^{-1/x}}{x^2}\,dx$

31. $\displaystyle \int \frac{e^{3x}+x^2}{(e^{3x}+x^3)^3}\,dx$

32. $\displaystyle \int \frac{e^x-e^{-x}}{(e^x+e^{-x})^{3/2}}\,dx$

33. $\displaystyle \int e^{2x}(e^{2x}+1)^3\,dx$

34. $\displaystyle \int e^{-x}(1+e^{-x})\,dx$

35. $\displaystyle \int \frac{\ln 5x}{x}\,dx$

36. $\displaystyle \int \frac{(\ln u)^3}{u}\,du$

37. $\displaystyle \int \frac{1}{x\ln x}\,dx$

38. $\displaystyle \int \frac{1}{x(\ln x)^2}\,dx$

39. $\displaystyle \int \frac{\sqrt{\ln x}}{x}\,dx$

40. $\displaystyle \int \frac{(\ln x)^{7/2}}{x}\,dx$

41. $\displaystyle \int \left(xe^{x^2}-\frac{x}{x^2+2}\right)dx$

42. $\displaystyle \int \left(xe^{-x^2}+\frac{e^x}{e^x+3}\right)dx$

43. $\displaystyle \int \frac{x+1}{\sqrt{x}-1}\,dx$

Hint: Let $u=\sqrt{x}-1$.

44. $\displaystyle \int \frac{e^{-u}-1}{e^{-u}+u}\,du$

Hint: Let $v=e^{-u}+u$.

45. $\displaystyle \int x(x-1)^5\,dx$

Hint: $u=x-1$ implies $x=u+1$.

46. $\displaystyle \int \frac{t}{t+1}\,dt$

Hint: $\dfrac{t}{t+1}=1-\dfrac{1}{t+1}$.

47. $\displaystyle \int \frac{1-\sqrt{x}}{1+\sqrt{x}}\,dx$

Hint: Let $u=1+\sqrt{x}$.

48. $\displaystyle \int \frac{1+\sqrt{x}}{1-\sqrt{x}}\,dx$

Hint: Let $u=1-\sqrt{x}$.

49. $\displaystyle \int v^2(1-v)^6\,dv$

Hint: Let $u=1-v$.

50. $\displaystyle \int x^3(x^2+1)^{3/2}\,dx$

Hint: Let $u=x^2+1$.

In Exercises 51–54, find the function f given that the slope of the tangent line to the graph of f at any point $(x, f(x))$ is $f'(x)$ and that the graph of f passes through the given point.

51. $f'(x)=5(2x-1)^4$; $(1,3)$

52. $f'(x)=\dfrac{3x^2}{2\sqrt{x^3-1}}$; $(1,1)$

53. $f'(x) = -2xe^{-x^2+1};\ (1, 0)$

54. $f'(x) = 1 - \dfrac{2x}{x^2 + 1};\ (0, 2)$

55. CABLE TELEPHONE SUBSCRIBERS   The number of cable telephone subscribers stood at 3.2 million at the beginning of 2004 ($t = 0$). For the next 5 yr, the number was projected to grow at the rate of

$$R(t) = 3.36(t + 1)^{0.05} \qquad (0 \le t \le 4)$$

million subscribers/year. If the projection holds true, how many cable telephone subscribers will there be at the beginning of 2008 ($t = 4$)?

Source: Sanford C. Bernstein

56. TV VIEWERS: NEWSMAGAZINE SHOWS   The number of viewers of a weekly TV newsmagazine show, introduced in the 2002 season, has been increasing at the rate of

$$3\left(2 + \frac{1}{2}t\right)^{-1/3} \qquad (1 \le t \le 6)$$

million viewers/year in its tth year on the air. The number of viewers of the program during its first year on the air is given by $9(5/2)^{2/3}$ million. Find how many viewers were expected in the 2007 season.

57. STUDENT ENROLLMENT   The registrar of Kellogg University estimates that the total student enrollment in the Continuing Education division will grow at the rate of

$$N'(t) = 2000(1 + 0.2t)^{-3/2}$$

students/year t yr from now. If the current student enrollment is 1000, find an expression giving the total student enrollment t yr from now. What will be the student enrollment 5 yr from now?

58. SUPPLY: WOMEN'S BOOTS   The rate of change of the unit price p (in dollars) of Apex women's boots is given by

$$p'(x) = \frac{240}{(5 - x)^2}$$

where x is the number of pairs in units of a hundred that the supplier will make available in the market daily when the unit price is $\$p$/pair. Find the supply equation for these boots if the quantity the supplier is willing to make available is 200 pairs daily ($x = 2$) when the unit price is $\$50$/pair.

59. DEMAND: WOMEN'S BOOTS   The rate of change of the unit price p (in dollars) of Apex women's boots is given by

$$p'(x) = \frac{-250\,x}{(16 + x^2)^{3/2}}$$

where x is the quantity demanded daily in units of a hundred. Find the demand function for these boots if the quantity demanded daily is 300 pairs ($x = 3$) when the unit price is $\$50$/pair.

60. POPULATION GROWTH   The population of a certain city is projected to grow at the rate of

$$r(t) = 400\left(1 + \frac{2t}{24 + t^2}\right) \qquad (0 \le t \le 5)$$

people/year, t years from now. The current population is 60,000. What will be the population 5 yr from now?

61. OIL SPILL   In calm waters, the oil spilling from the ruptured hull of a grounded tanker forms an oil slick that is circular in shape. If the radius r of the circle is increasing at the rate of

$$r'(t) = \frac{30}{\sqrt{2t + 4}}$$

feet/minute t min after the rupture occurs, find an expression for the radius at any time t. How large is the polluted area 16 min after the rupture occurred?
Hint: $r(0) = 0$.

62. LIFE EXPECTANCY OF A FEMALE   Suppose in a certain country the life expectancy at birth of a female is changing at the rate of

$$g'(t) = \frac{5.45218}{(1 + 1.09t)^{0.9}}$$

years/year. Here, t is measured in years, with $t = 0$ corresponding to the beginning of 1900. Find an expression $g(t)$ giving the life expectancy at birth (in years) of a female in that country if the life expectancy at the beginning of 1900 is 50.02 yr. What is the life expectancy at birth of a female born in 2000 in that country?

63. AVERAGE BIRTH HEIGHT OF BOYS   Using data collected at Kaiser Hospital, pediatricians estimate that the average height of male children changes at the rate of

$$h'(t) = \frac{52.8706e^{-0.3277t}}{(1 + 2.449e^{-0.3277t})^2}$$

in./yr, where the child's height $h(t)$ is measured in inches and t, the child's age, is measured in years, with $t = 0$ corresponding to the age at birth. Find an expression $h(t)$ for the average height of a boy at age t if the height at birth of an average child is 19.4 in. What is the height of an average 8-yr-old boy?

64. LEARNING CURVES   The average student enrolled in the 20-wk Court Reporting I course at the American Institute of Court Reporting progresses according to the rule

$$N'(t) = 6e^{-0.05t} \qquad (0 \le t \le 20)$$

where $N'(t)$ measures the rate of change in the number of words/minute of dictation the student takes in machine shorthand after t wk in the course. Assuming that the average student enrolled in this course begins with a dictation speed of 60 words/minute, find an expression $N(t)$ that gives the dictation speed of the student after t wk in the course.

65. AMOUNT OF GLUCOSE IN THE BLOODSTREAM   Suppose a patient is given a continuous intravenous infusion of glucose at a constant rate of r mg/min. Then, the rate at which the amount of glucose in the bloodstream is changing at time t due to this infusion is given by

$$A'(t) = re^{-at}$$

mg/min, where a is a positive constant associated with the rate at which excess glucose is eliminated from the bloodstream and is dependent on the patient's metabolism rate. Derive an expression for the amount of glucose in the bloodstream at time t.
Hint: $A(0) = 0$.

66. **CONCENTRATION OF A DRUG IN AN ORGAN**   A drug is carried into an organ of volume V cm^3 by a liquid that enters the organ at

the rate of a cm^3/sec and leaves it at the rate of b cm^3/sec. The concentration of the drug in the liquid entering the organ is c g/cm^3. If the concentration of the drug in the organ at time t is increasing at the rate of

$$x'(t) = \frac{1}{V}(ac - bx_0)e^{-bt/V}$$

g/cm^3/sec, and the concentration of the drug in the organ initially is x_0 g/cm^3, show that the concentration of the drug in the organ at time t is given by

$$x(t) = \frac{ac}{b} + \left(x_0 - \frac{ac}{b}\right)e^{-bt/V}$$

15.2 Solutions to Self-Check Exercises

1. Let $u = 2x + 5$. Then, $du = 2\,dx$, or $dx = \frac{1}{2}\,du$. Making the appropriate substitutions, we have

$$\int \sqrt{2x + 5}\,dx = \int \sqrt{u}\left(\frac{1}{2}\,du\right) = \frac{1}{2}\int u^{1/2}\,du$$

$$= \frac{1}{2}\left(\frac{2}{3}\right)u^{3/2} + C$$

$$= \frac{1}{3}(2x + 5)^{3/2} + C$$

2. Let $u = 2x^3 + 1$, so that $du = 6x^2\,dx$, or $x^2\,dx = \frac{1}{6}\,du$. Making the appropriate substitutions, we have

$$\int \frac{x^2}{(2x^3 + 1)^{3/2}}\,dx = \int \frac{\left(\frac{1}{6}\right)du}{u^{3/2}} = \frac{1}{6}\int u^{-3/2}\,du$$

$$= \left(\frac{1}{6}\right)(-2)u^{-1/2} + C$$

$$= -\frac{1}{3}(2x^3 + 1)^{-1/2} + C$$

$$= -\frac{1}{3\sqrt{2x^3 + 1}} + C$$

3. Let $u = 2x^2 - 1$, so that $du = 4x\,dx$, or $x\,dx = \frac{1}{4}\,du$. Then,

$$\int xe^{2x^2 - 1}\,dx = \frac{1}{4}\int e^u\,du$$

$$= \frac{1}{4}e^u + C$$

$$= \frac{1}{4}e^{2x^2 - 1} + C$$

4. Let $C(t)$ denote the CO concentration in the air due to automobile exhaust t yr from now. Then,

$$C'(t) = f(t) = \frac{8(0.1t + 1)}{300(0.2t^2 + 4t + 64)^{1/3}}$$

$$= \frac{8}{300}(0.1t + 1)(0.2t^2 + 4t + 64)^{-1/3}$$

Integrating, we find

$$C(t) = \int \frac{8}{300}(0.1t + 1)(0.2t^2 + 4t + 64)^{-1/3}\,dt$$

$$= \frac{8}{300}\int (0.1t + 1)(0.2t^2 + 4t + 64)^{-1/3}\,dt$$

Let $u = 0.2t^2 + 4t + 64$, so that $du = (0.4t + 4)\,dt = 4(0.1t + 1)\,dt$, or

$$(0.1t + 1)\,dt = \frac{1}{4}\,du$$

Then,

$$C(t) = \frac{8}{300}\left(\frac{1}{4}\right)\int u^{-1/3}\,du$$

$$= \frac{1}{150}\left(\frac{3}{2}u^{2/3}\right) + k$$

$$= 0.01(0.2t^2 + 4t + 64)^{2/3} + k$$

where k is an arbitrary constant. To determine the value of k, we use the condition $C(0) = 0.16$, obtaining

$$C(0) = 0.16 = 0.01(64)^{2/3} + k$$

$$0.16 = 0.16 + k$$

$$k = 0$$

Therefore,

$$C(t) = 0.01(0.2t^2 + 4t + 64)^{2/3}$$

15.3 Area and the Definite Integral

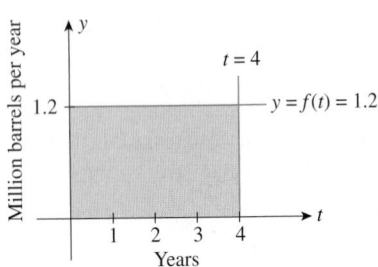

FIGURE 5
The total petroleum consumption is
given by the area of the rectangular
region.

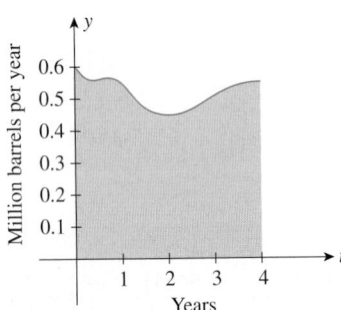

FIGURE 6
The daily petroleum consumption is
given by the "area" of the shaded
region.

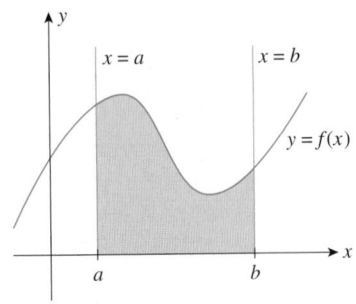

FIGURE 7
The area under the graph of f on $[a, b]$

An Intuitive Look

Suppose a certain state's annual rate of petroleum consumption over a 4-year period
is constant and is given by the function

$$f(t) = 1.2 \qquad (0 \le t \le 4)$$

where t is measured in years and $f(t)$ in millions of barrels per year. Then, the state's
total petroleum consumption over the period of time in question is

$$(1.2)(4 - 0) \qquad \text{Rate of consumption} \times \text{Time elapsed}$$

or 4.8 million barrels. If you examine the graph of f shown in Figure 5, you will see
that this total is just the area of the rectangular region bounded above by the graph
of f, below by the t-axis, and to the left and right by the vertical lines $t = 0$ (the
y-axis) and $t = 4$, respectively.

Figure 6 shows the actual petroleum consumption of a certain New England
state over a 4-year period from 1990 ($t = 0$) to 1994 ($t = 4$). Observe that the rate
of consumption is not constant; that is, the function f is not a constant function. What
is the state's total petroleum consumption over this 4-year period? It seems reasonable to conjecture that it is given by the "area" of the region bounded above by the
graph of f, below by the t-axis, and to the left and right by the vertical lines $t = 0$
and $t = 4$, respectively.

This example raises two questions:

1. What is the "area" of the region shown in Figure 6?
2. How do we compute this area?

The Area Problem

The preceding example touches on the second fundamental problem in calculus:
Calculate the area of the region bounded by the graph of a nonnegative function f,
the x-axis, and the vertical lines $x = a$ and $x = b$ (Figure 7). This area is called the
area under the graph of f on the interval $[a, b]$, or from a to b.

Defining Area—Two Examples

Just as we used the slopes of secant lines (quantities that we could compute) to help
us define the slope of the tangent line to a point on the graph of a function, we now
adopt a parallel approach and use the areas of rectangles (quantities that we can compute) to help us define the area under the graph of a function. We begin by looking
at a specific example.

EXAMPLE 1 Let $f(x) = x^2$ and consider the region R under the graph of f on the
interval $[0, 1]$ (Figure 8a). To obtain an approximation of the area of R, let's construct four nonoverlapping rectangles as follows: Divide the interval $[0, 1]$ into
four subintervals

$$\left[0, \frac{1}{4}\right], \quad \left[\frac{1}{4}, \frac{1}{2}\right], \quad \left[\frac{1}{2}, \frac{3}{4}\right], \quad \left[\frac{3}{4}, 1\right]$$

of equal length $\frac{1}{4}$. Next, construct four rectangles with these subintervals as bases

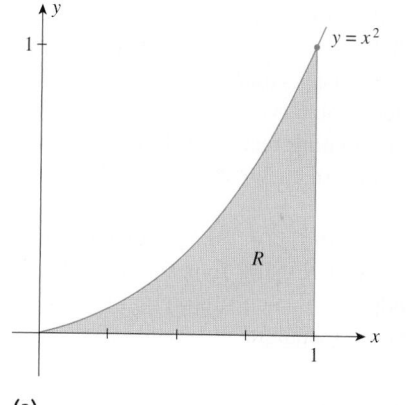

(a)

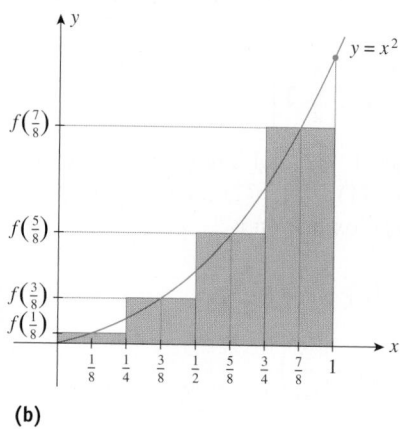

(b)

FIGURE 8
The area of the region under the graph of f on $[0, 1]$ in (a) is approximated by the sum of the areas of the four rectangles in (b).

FIGURE 9
As n increases, the number of rectangles increases, and the approximation improves.

and with heights given by the values of the function at the midpoints

$$\frac{1}{8}, \quad \frac{3}{8}, \quad \frac{5}{8}, \quad \frac{7}{8}$$

of each subinterval. Then, each of these rectangles has width $\frac{1}{4}$ and height

$$f\left(\frac{1}{8}\right), \quad f\left(\frac{3}{8}\right), \quad f\left(\frac{5}{8}\right), \quad f\left(\frac{7}{8}\right)$$

respectively (Figure 8b).

   If we approximate the area A of R by the sum of the areas of the four rectangles, we obtain

$$A \approx \frac{1}{4}f\left(\frac{1}{8}\right) + \frac{1}{4}f\left(\frac{3}{8}\right) + \frac{1}{4}f\left(\frac{5}{8}\right) + \frac{1}{4}f\left(\frac{7}{8}\right)$$

$$= \frac{1}{4}\left[f\left(\frac{1}{8}\right) + f\left(\frac{3}{8}\right) + f\left(\frac{5}{8}\right) + f\left(\frac{7}{8}\right)\right]$$

$$= \frac{1}{4}\left[\left(\frac{1}{8}\right)^2 + \left(\frac{3}{8}\right)^2 + \left(\frac{5}{8}\right)^2 + \left(\frac{7}{8}\right)^2\right] \quad \text{Recall that } f(x) = x^2.$$

$$= \frac{1}{4}\left(\frac{1}{64} + \frac{9}{64} + \frac{25}{64} + \frac{49}{64}\right) = \frac{21}{64}$$

or approximately 0.328125 square unit.   ■

Following the procedure of Example 1, we can obtain approximations of the area of the region R using any number n of rectangles ($n = 4$ in Example 1). Figure 9a shows the approximation of the area A of R using 8 rectangles ($n = 8$), and Figure 9b shows the approximation of the area A of R using 16 rectangles.

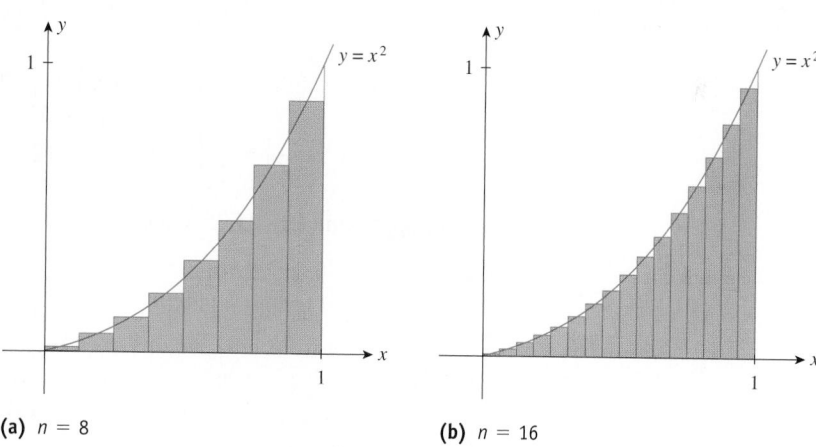

(a) $n = 8$ | **(b)** $n = 16$

   These figures suggest that the approximations seem to get better as n increases. This is borne out by the results given in Table 1, which were obtained using a computer.

TABLE 1

Number of Rectangles, n	4	8	16	32	64	100	200
Approximation of A	0.328125	0.332031	0.333008	0.333252	0.333313	0.333325	0.333331

Our computations seem to suggest that the approximations approach the number $\frac{1}{3}$ as n gets larger and larger. This result suggests that we *define* the area of the region under the graph of $f(x) = x^2$ on the interval [0, 1] to be $\frac{1}{3}$ square unit.

In Example 1, we chose the *midpoint* of each subinterval as the point at which to evaluate $f(x)$ to obtain the height of the approximating rectangle. Let's consider another example, this time choosing the *left endpoint* of each subinterval.

EXAMPLE 2   Let R be the region under the graph of $f(x) = 16 - x^2$ on the interval [1, 3]. Find an approximation of the area A of R using four subintervals of [1, 3] of equal length and picking the left endpoint of each subinterval to evaluate $f(x)$ to obtain the height of the approximating rectangle.

Solution   The graph of f is sketched in Figure 10a. Since the length of [1, 3] is 2, we see that the length of each subinterval is $\frac{2}{4}$, or $\frac{1}{2}$. Therefore, the four subintervals are

$$\left[1, \frac{3}{2}\right], \quad \left[\frac{3}{2}, 2\right], \quad \left[2, \frac{5}{2}\right], \quad \left[\frac{5}{2}, 3\right]$$

The left endpoints of these subintervals are $1, \frac{3}{2}, 2$, and $\frac{5}{2}$, respectively, so the heights of the approximating rectangles are $f(1)$, $f(\frac{3}{2})$, $f(2)$, and $f(\frac{5}{2})$, respectively (Figure 10b). Therefore, the required approximation is

$$A \approx \frac{1}{2}f(1) + \frac{1}{2}f\left(\frac{3}{2}\right) + \frac{1}{2}f(2) + \frac{1}{2}f\left(\frac{5}{2}\right)$$

$$= \frac{1}{2}\left[f(1) + f\left(\frac{3}{2}\right) + f(2) + f\left(\frac{5}{2}\right)\right]$$

$$= \frac{1}{2}\left\{[16 - (1)^2] + \left[16 - \left(\frac{3}{2}\right)^2\right]\right.$$

$$\left. + [16 - (2)^2] + \left[16 - \left(\frac{5}{2}\right)^2\right]\right\} \quad \text{Recall that } f(x) = 16 - x^2.$$

$$= \frac{1}{2}\left(15 + \frac{55}{4} + 12 + \frac{39}{4}\right) = \frac{101}{4}$$

or approximately 25.25 square units.

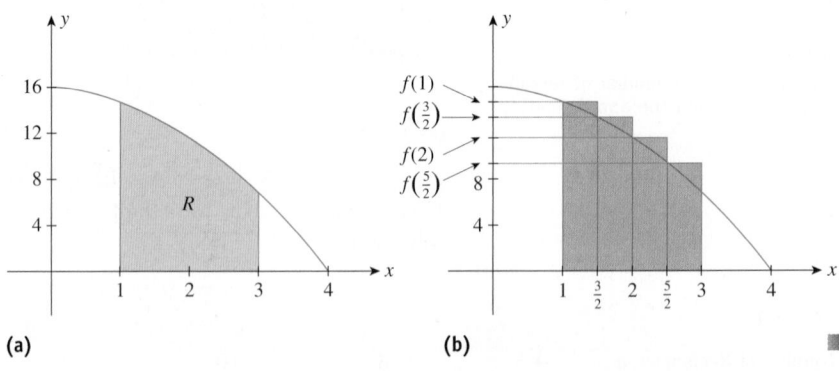

FIGURE 10
The area of R in (a) is approximated by the sum of the areas of the four rectangles in (b).

(a)   (b)

Table 2 shows the approximations of the area A of the region R of Example 2 when n rectangles are used for the approximation and the heights of the approximating rectangles are found by evaluating $f(x)$ at the left endpoints.

TABLE 2

Number of Rectangles, n	4	10	100	1,000	10,000	50,000	100,000
Approximation of A	25.2500	24.1200	23.4132	23.3413	23.3341	23.3335	23.3334

Once again, we see that the approximations seem to approach a unique number as n gets larger and larger—this time the number is $23\frac{1}{3}$. This result suggests that we *define* the area of the region under the graph of $f(x) = 16 - x^2$ on the interval $[1, 3]$ to be $23\frac{1}{3}$ square units.

◼ Defining Area—The General Case

Examples 1 and 2 point the way to defining the area A under the graph of an arbitrary but continuous and nonnegative function f on an interval $[a, b]$ (Figure 11a).

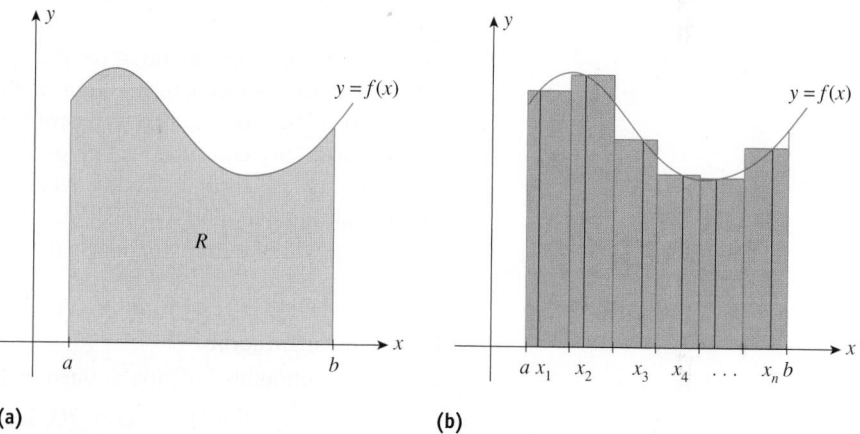

FIGURE 11
The area of the region under the graph of f on $[a, b]$ in (a) is approximated by the sum of the areas of the n rectangles shown in (b).

(a)   **(b)**

Divide the interval $[a, b]$ into n subintervals of equal length $\Delta x = (b - a)/n$. Next, pick n arbitrary points $x_1, x_2, \ldots, x_n$, called *representative points*, from the first, second, $\ldots$, and nth subintervals, respectively (Figure 11b). Then, approximating the area A of the region R by the n rectangles of width Δx and heights $f(x_1)$, $f(x_2), \ldots, f(x_n)$, so that the areas of the rectangles are $f(x_1)\Delta x, f(x_2)\Delta x, \ldots, f(x_n)\Delta x$, we have

$$A \approx f(x_1)\Delta x + f(x_2)\Delta x + \cdots + f(x_n)\Delta x$$

The sum on the right-hand side of this expression is called a **Riemann sum** in honor of the German mathematician Bernhard Riemann (1826–1866). Now, as the earlier examples seem to suggest, the Riemann sum will approach a unique number as n becomes arbitrarily large.* We define this number to be the area A of the region R.

---

*Even though we chose the representative points to be the midpoints of the subintervals in Example 1 and the left endpoints in Example 2, it can be shown that each of the respective sums will always approach a unique number as n approaches infinity.

The Area under the Graph of a Function

Let f be a nonnegative continuous function on $[a, b]$. Then, the area of the region under the graph of f is

$$A = \lim_{n\to\infty} [f(x_1) + f(x_2) + \cdots + f(x_n)]\Delta x \tag{6}$$

where $x_1, x_2, \ldots, x_n$ are arbitrary points in the n subintervals of $[a, b]$ of equal width $\Delta x = (b - a)/n$.

■ The Definite Integral

As we have just seen, the area under the graph of a continuous *nonnegative* function f on an interval $[a, b]$ is defined by the limit of the Riemann sum

$$\lim_{n\to\infty} [f(x_1)\Delta x + f(x_2)\Delta x + \cdots + f(x_n)\Delta x]$$

We now turn our attention to the study of limits of Riemann sums involving functions that are not necessarily nonnegative. Such limits arise in many applications of calculus.

For example the calculation of the distance covered by a body traveling along a straight line involves evaluating a limit of this form. The computation of the total revenue realized by a company over a certain time period, the calculation of the total amount of electricity consumed in a typical home over a 24-hour period, the average concentration of a drug in a body over a certain interval of time, and the volume of a solid—all involve limits of this type.

We begin with the following definition.

The Definite Integral

Let f be a continuous function defined on $[a, b]$. If

$$\lim_{n\to\infty} [f(x_1)\Delta x + f(x_2)\Delta x + \cdots + f(x_n)\Delta x]$$

exists for all choices of representative points $x_1, x_2, \ldots, x_n$ in the n subintervals of $[a, b]$ of equal width $\Delta x = (b - a)/n$, then this limit is called the **definite integral** of f from a to b and is denoted by $\int_a^b f(x)\,dx$. Thus,

$$\int_a^b f(x)\,dx = \lim_{n\to\infty} [f(x_1)\Delta x + f(x_2)\Delta x + \cdots + f(x_n)\Delta x] \tag{7}$$

The number a is the **lower limit of integration**, and the number b is the **upper limit of integration**.

Notes

1. If f is nonnegative, then the limit in (7) is the same as the limit in (6); therefore, the definite integral gives the area under the graph of f on $[a, b]$.
2. The limit in (7) is denoted by the integral sign $\int$ because, as we will see later, the definite integral and the antiderivative of a function f are related.
3. It is important to realize that the definite integral $\int_a^b f(x)\,dx$ is a *number*, whereas

the indefinite integral $\int f(x)\,dx$ represents a *family of functions* (the antiderivatives of f).

4. If the limit in (7) exists, we say that f is **integrable** on the interval $[a, b]$. ■

■ When Is a Function Integrable?

The following theorem, which we state without proof, guarantees that a continuous function is integrable.

> **Integrability of a Function**
>
> Let f be continuous on $[a, b]$. Then, f is integrable on $[a, b]$; that is, the definite integral $\int_a^b f(x)\,dx$ exists.

■ Geometric Interpretation of the Definite Integral

If f is nonnegative and integrable on $[a, b]$, then we have the following geometric interpretation of the definite integral $\int_a^b f(x)\,dx$.

> **Geometric Interpretation of $\int_a^b f(x)\,dx$ for $f(x) \geq 0$ on $[a, b]$**
>
> If f is nonnegative and continuous on $[a, b]$, then
>
> $$\int_a^b f(x)\,dx \qquad\qquad (8)$$
>
> is equal to the area of the region under the graph of f on $[a, b]$ (Figure 12).

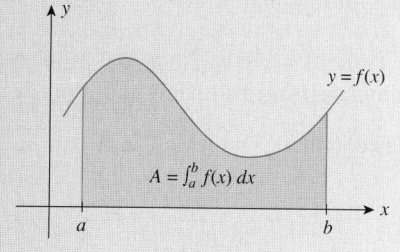

FIGURE 12
If $f(x) \geq 0$ on $[a, b]$, then $\int_a^b f(x)\,dx =$ area under the graph of f on $[a, b]$.

EXPLORE & DISCUSS

Suppose f is nonpositive [that is, $f(x) \leq 0$] and continuous on $[a, b]$. Explain why the area of the region below the x-axis and above the graph of f is given by $-\int_a^b f(x)\,dx$.

Next, let's extend our geometric interpretation of the definite integral to include the case where f assumes both positive as well as negative values on $[a, b]$. Consider a typical Riemann sum of the function f,

$$f(x_1)\Delta x + f(x_2)\Delta x + \cdots + f(x_n)\Delta x$$

corresponding to a partition of $[a, b]$ into n subintervals of equal width $(b - a)/n$, where $x_1, x_2, \ldots, x_n$ are representative points in the subintervals. The sum consists of n terms in which a positive term corresponds to the area of a rectangle of height $f(x_k)$ (for some positive integer k) lying above the x-axis and a negative term corresponds to the area of a rectangle of height $-f(x_k)$ lying below the x-axis. (See Figure 13, which depicts a situation with $n = 6$.)

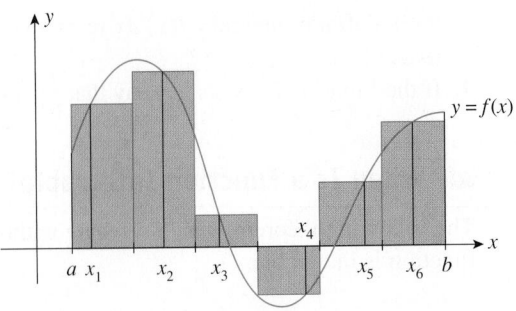

FIGURE 13
The positive terms in the Riemann sum are associated with the areas of the rectangles that lie above the x-axis, and the negative terms are associated with the areas of those that lie below the x-axis.

As n gets larger and larger, the sums of the areas of the rectangles lying above the x-axis seem to give a better and better approximation of the area of the region lying above the x-axis (Figure 14). Similarly, the sums of the areas of those rectangles lying below the x-axis seem to give a better and better approximation of the area of the region lying below the x-axis.

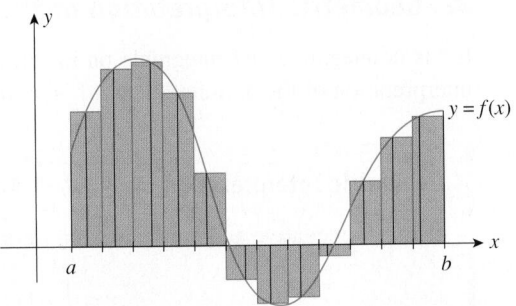

FIGURE 14
As n gets larger, the approximations get better. Here, $n = 12$ and we are approximating with twice as many rectangles as in Figure 13.

These observations suggest the following geometric interpretation of the definite integral for an arbitrary continuous function on an interval $[a, b]$.

Geometric Interpretation of $\int_a^b f(x)\, dx$ on $[a, b]$

If f is continuous on $[a, b]$, then

$$\int_a^b f(x)\, dx$$

is equal to the area of the region above $[a, b]$ minus the area of the region below $[a, b]$ (Figure 15).

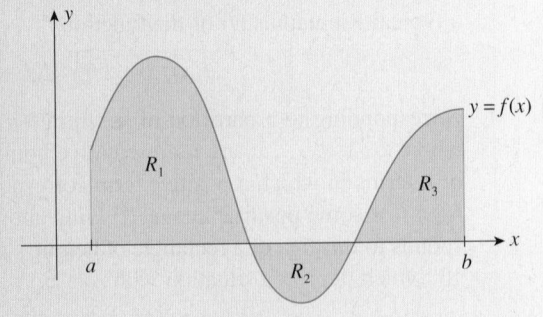

FIGURE 15
$\int_a^b f(x)\, dx$ = Area of R_1 − Area of R_2 + Area of R_3

15.3 Self-Check Exercise

Find an approximation of the area of the region R under the graph of $f(x) = 2x^2 + 1$ on the interval $[0, 3]$, using four subintervals of $[0, 3]$ of equal length and picking the midpoint of each subinterval as a representative point.

The solution to Self-Check Exercise 15.3 can be found on page 943.

15.3 Concept Questions

1. Explain how you would define the area of the region under the graph of a nonnegative continuous function f on the interval $[a, b]$.

2. Define the definite integral of a continuous function on the interval $[a, b]$. Give a geometric interpretation of $\int_a^b f(x)\, dx$

for the case where (a) f is nonnegative on $[a, b]$ and (b) f assumes both positive as well as negative values on $[a, b]$. Illustrate your answers graphically.

15.3 Exercises

In Exercises 1 and 2, find an approximation of the area of the region R under the graph of f by computing the Riemann sum of f corresponding to the partition of the interval into the subintervals shown in the accompanying figures. In each case, use the midpoints of the subintervals as the representative points.

1.

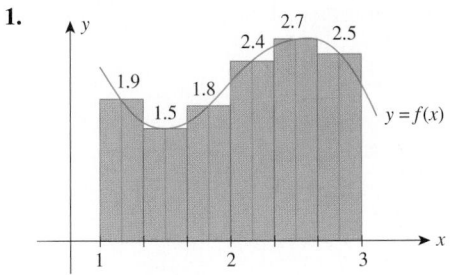

2.

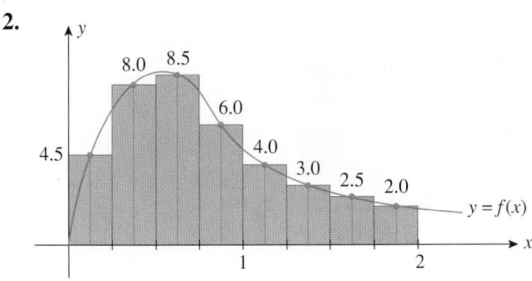

3. Let $f(x) = 3x$.
   a. Sketch the region R under the graph of f on the interval $[0, 2]$ and find its exact area using geometry.

   b. Use a Riemann sum with four subintervals of equal length ($n = 4$) to approximate the area of R. Choose the representative points to be the left endpoints of the subintervals.
   c. Repeat part (b) with eight subintervals of equal length ($n = 8$).
   d. Compare the approximations obtained in parts (b) and (c) with the exact area found in part (a). Do the approximations improve with larger n?

4. Repeat Exercise 3, choosing the representative points to be the right endpoints of the subintervals.

5. Let $f(x) = 4 - 2x$.
   a. Sketch the region R under the graph of f on the interval $[0, 2]$ and find its exact area using geometry.
   b. Use a Riemann sum with five subintervals of equal length ($n = 5$) to approximate the area of R. Choose the representative points to be the left endpoints of the subintervals.
   c. Repeat part (b) with ten subintervals of equal length ($n = 10$).
   d. Compare the approximations obtained in parts (b) and (c) with the exact area found in part (a). Do the approximations improve with larger n?

6. Repeat Exercise 5, choosing the representative points to be the right endpoints of the subintervals.

7. Let $f(x) = x^2$ and compute the Riemann sum of f over the interval $[2, 4]$, using
   a. Two subintervals of equal length ($n = 2$).
   b. Five subintervals of equal length ($n = 5$).
   c. Ten subintervals of equal length ($n = 10$).

In each case, choose the representative points to be the mid-points of the subintervals.

d. Can you guess at the area of the region under the graph of f on the interval [2, 4]?

8. Repeat Exercise 7, choosing the representative points to be the left endpoints of the subintervals.

9. Repeat Exercise 7, choosing the representative points to be the right endpoints of the subintervals.

10. Let $f(x) = x^3$ and compute the Riemann sum of f over the interval [0, 1], using
   a. Two subintervals of equal length ($n = 2$).
   b. Five subintervals of equal length ($n = 5$).
   c. Ten subintervals of equal length ($n = 10$).
   In each case, choose the representative points to be the mid-points of the subintervals.
   d. Can you guess at the area of the region under the graph of f on the interval [0, 1]?

11. Repeat Exercise 10, choosing the representative points to be the left endpoints of the subintervals.

12. Repeat Exercise 10, choosing the representative points to be the right endpoints of the subintervals.

In Exercises 13–16, find an approximation of the area of the region R under the graph of the function f on the interval $[a, b]$. In each case, use n subintervals and choose the representative points as indicated.

13. $f(x) = x^2 + 1$; [0, 2]; $n = 5$; midpoints

14. $f(x) = 4 - x^2$; [−1, 2]; $n = 6$; left endpoints

15. $f(x) = \dfrac{1}{x}$; [1, 3]; $n = 4$; right endpoints

16. $f(x) = e^x$; [0, 3]; $n = 5$; midpoints

17. REAL ESTATE   Figure (a) shows a vacant lot with a 100-ft frontage in a development. To estimate its area, we introduce a coordinate system so that the x-axis coincides with the edge of the straight road forming the lower boundary of the property, as shown in Figure (b). Then, thinking of the upper boundary of the property as the graph of a continuous function f over the interval [0, 100], we see that the problem is mathematically equivalent to that of finding the area under the graph of f on [0, 100]. To estimate the area of the lot using a Riemann sum, we divide the interval [0, 100] into five equal subintervals of length 20 ft. Then, using survey-or's equipment, we measure the distance from the midpoint of each of these subintervals to the upper boundary of the property. These measurements give the values of $f(x)$ at

$x = $ 10, 30, 50, 70, and 90. What is the approximate area of the lot?

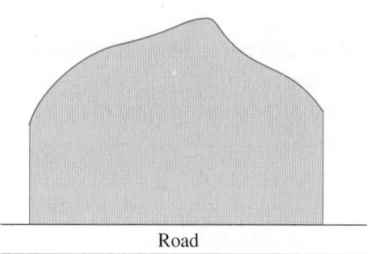

(a)

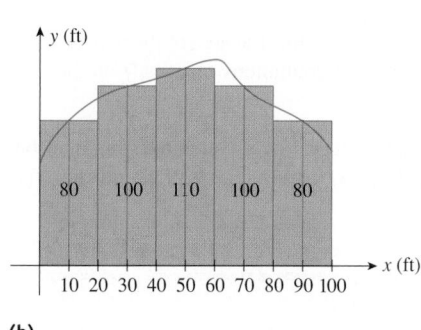

(b)

18. REAL ESTATE   Use the technique of Exercise 17 to obtain an estimate of the area of the vacant lot shown in the accompanying figures.

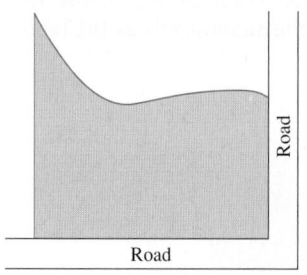

(a)

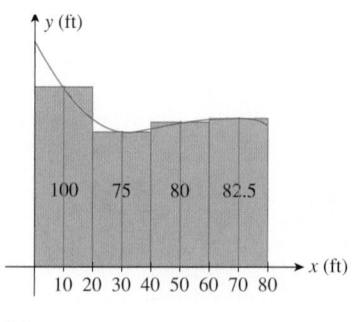

(b)

15.3 Solution to Self-Check Exercise

The length of each subinterval is $\frac{3}{4}$. Therefore, the four subintervals are

$$\left[0, \frac{3}{4}\right], \quad \left[\frac{3}{4}, \frac{3}{2}\right], \quad \left[\frac{3}{2}, \frac{9}{4}\right], \quad \left[\frac{9}{4}, 3\right]$$

The representative points are $\frac{3}{8}$, $\frac{9}{8}$, $\frac{15}{8}$, and $\frac{21}{8}$, respectively. Therefore, the required approximation is

$$
\begin{aligned}
A &= \frac{3}{4} f\left(\frac{3}{8}\right) + \frac{3}{4} f\left(\frac{9}{8}\right) + \frac{3}{4} f\left(\frac{15}{8}\right) + \frac{3}{4} f\left(\frac{21}{8}\right) \\
&= \frac{3}{4}\left[f\left(\frac{3}{8}\right) + f\left(\frac{9}{8}\right) + f\left(\frac{15}{8}\right) + f\left(\frac{21}{8}\right) \right] \\
&= \frac{3}{4}\left\{ \left[2\left(\frac{3}{8}\right)^2 + 1\right] + \left[2\left(\frac{9}{8}\right)^2 + 1\right] + \left[2\left(\frac{15}{8}\right)^2 + 1\right] \right. \\
&\quad \left. + \left[2\left(\frac{21}{8}\right)^2 + 1\right] \right\} \\
&= \frac{3}{4}\left(\frac{41}{32} + \frac{113}{32} + \frac{257}{32} + \frac{473}{32}\right) = \frac{663}{32}
\end{aligned}
$$

or approximately 20.72 square units.

15.4 The Fundamental Theorem of Calculus

■ The Fundamental Theorem of Calculus

In Section 15.3, we defined the definite integral of an arbitrary continuous function on an interval $[a, b]$ as a limit of Riemann sums. Calculating the value of a definite integral by actually taking the limit of such sums is tedious and in most cases impractical. It is important to realize that the numerical results we obtained in Examples 1 and 2 of Section 15.3 were *approximations* of the respective areas of the regions in question, even though these results enabled us to *conjecture* what the actual areas might be. Fortunately, there is a much better way of finding the exact value of a definite integral.

The following theorem shows how to evaluate the definite integral of a continuous function provided we can find an antiderivative of that function. Because of its importance in establishing the relationship between differentiation and integration, this theorem—discovered independently by Sir Isaac Newton (1642–1727) in England and Gottfied Wilhelm Leibniz (1646–1716) in Germany—is called the **fundamental theorem of calculus**.

THEOREM 2

The Fundamental Theorem of Calculus

Let f be continuous on $[a, b]$. Then,

$$\int_a^b f(x)\,dx = F(b) - F(a) \tag{9}$$

where F is any antiderivative of f; that is, $F'(x) = f(x)$.

We will explain why this theorem is true at the end of this section.

When applying the fundamental theorem of calculus, it is convenient to use the notation

$$F(x)\Big|_a^b = F(b) - F(a)$$

For example, using this notation, Equation (9) is written

$$\int_a^b f(x)\, dx = F(x)\Big|_a^b = F(b) - F(a)$$

-EXAMPLE 1   Let R be the region under the graph of $f(x) = x$ on the interval $[1, 3]$. Use the fundamental theorem of calculus to find the area A of R and verify your result by elementary means.

Solution   The region R is shown in Figure 16a. Since f is nonnegative on $[1, 3]$, the area of R is given by the definite integral of f from 1 to 3; that is,

$$A = \int_1^3 x\, dx$$

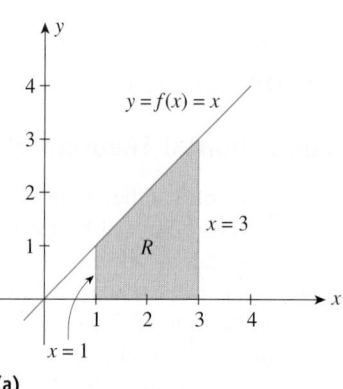
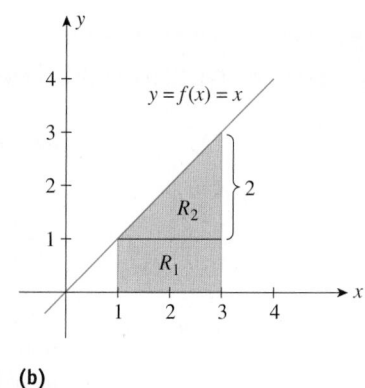

(a)   (b)

FIGURE 16
The area of R can be computed in two different ways.

To evaluate the definite integral, observe that an antiderivative of $f(x) = x$ is $F(x) = \frac{1}{2}x^2 + C$, where C is an arbitrary constant. Therefore, by the fundamental theorem of calculus, we have

$$A = \int_1^3 x\, dx = \frac{1}{2}x^2 + C\Big|_1^3$$

$$= \left(\frac{9}{2} + C\right) - \left(\frac{1}{2} + C\right) = 4 \text{ square units}$$

To verify this result by elementary means, observe that the area A is the area of the rectangle R_1 (width $\times$ height) plus the area of the triangle R_2 ($\frac{1}{2}$ base $\times$ height) (see Figure 16b); that is,

$$2(1) + \frac{1}{2}(2)(2) = 2 + 2 = 4$$

which agrees with the result obtained earlier.   ■

Observe that in evaluating the definite integral in Example 1, the constant of integration "dropped out." This is true in general, for if $F(x) + C$ denotes an antiderivative of some function f, then

$$F(x) + C\Big|_a^b = [F(b) + C] - [F(a) + C]$$
$$= F(b) + C - F(a) - C$$
$$= F(b) - F(a)$$

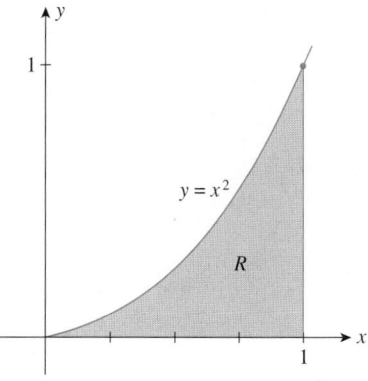

FIGURE 17
The area of R is $\int_0^1 x^2\,dx = \frac{1}{3}$.

With this fact in mind, we may, in all future computations involving the evaluation of a definite integral, drop the constant of integration from our calculations.

■ Finding the Area under a Curve

Having seen how effective the fundamental theorem of calculus is in helping us find the area of simple regions, we now use it to find the area of more complicated regions.

EXAMPLE 2 In Section 15.3, we conjectured that the area of the region R under the graph of $f(x) = x^2$ on the interval $[0, 1]$ was $\frac{1}{3}$ square unit. Use the fundamental theorem of calculus to verify this conjecture.

Solution The region R is reproduced in Figure 17. Observe that f is nonnegative on $[0, 1]$, so the area of R is given by $A = \int_0^1 x^2\,dx$. Since an antiderivative of $f(x) = x^2$ is $F(x) = \frac{1}{3}x^3$, we see, using the fundamental theorem of calculus, that

$$A = \int_0^1 x^2\,dx = \frac{1}{3}x^3 \Big|_0^1 = \frac{1}{3}(1) - \frac{1}{3}(0) = \frac{1}{3}\text{ square unit}$$

as we wished to show. ■

Note It is important to realize that the value, $\frac{1}{3}$, is by definition the exact value of the area of R. ■

EXAMPLE 3 Find the area of the region R under the graph of $y = x^2 + 1$ from $x = -1$ to $x = 2$.

Solution The region R under consideration is shown in Figure 18. Using the fundamental theorem of calculus, we find that the required area is

$$\int_{-1}^2 (x^2 + 1)\,dx = \left(\frac{1}{3}x^3 + x\right)\Big|_{-1}^2$$

$$= \left[\frac{1}{3}(8) + 2\right] - \left[\frac{1}{3}(-1)^3 + (-1)\right] = 6$$

or 6 square units. ■

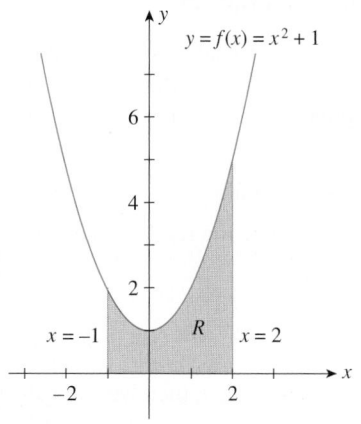

FIGURE 18
The area of R is $\int_{-1}^2 (x^2 + 1)\,dx$.

■ Evaluating Definite Integrals

In Examples 4 and 5, we use the rules of integration of Section 15.1 to help us evaluate the definite integrals.

EXAMPLE 4 Evaluate $\int_1^3 (3x^2 + e^x)\,dx$.

Solution

$$\int_1^3 (3x^2 + e^x)\,dx = x^3 + e^x \Big|_1^3$$

$$= (27 + e^3) - (1 + e) = 26 + e^3 - e$$ ■

EXAMPLE 5 Evaluate $\displaystyle\int_1^2 \left(\frac{1}{x} - \frac{1}{x^2} \right) dx$.

Solution

$$\int_1^2 \left(\frac{1}{x} - \frac{1}{x^2} \right) dx = \int_1^2 \left(\frac{1}{x} - x^{-2} \right) dx$$

$$= \ln|x| + \frac{1}{x} \Big|_1^2$$

$$= \left(\ln 2 + \frac{1}{2} \right) - (\ln 1 + 1)$$

$$= \ln 2 - \frac{1}{2} \qquad \text{Recall, } \ln 1 = 0.$$

EXPLORE & DISCUSS

Consider the definite integral $\displaystyle\int_{-1}^1 \frac{1}{x^2}\, dx$.

1. Show that a formal application of Equation (9) leads to

$$\int_{-1}^1 \frac{1}{x^2}\, dx = -\frac{1}{x} \Big|_{-1}^1 = -1 - 1 = -2$$

2. Observe that $f(x) = 1/x^2$ is positive at each value of x in $[-1, 1]$ where it is defined. Therefore, one might expect that the definite integral with integrand f has a positive value, if it exists.

3. Explain this apparent contradiction in the result (1) and the observation (2).

■ The Definite Integral as a Measure of Net Change

In real-world applications, we are often interested in the net change of a quantity over a period of time. For example, suppose P is a function giving the population, $P(t)$, of a city at time t. Then the *net change* in the population over the period from $t = a$ to $t = b$ is given by

$$P(b) - P(a) \qquad \text{Population at } t = b \text{ minus population at } t = a$$

If P has a continuous derivative P' in $[a, b]$, then we can invoke the fundamental theorem of calculus to write

$$P(b) - P(a) = \int_a^b P'(t)\, dt \qquad P \text{ is an antiderivative of } P'.$$

Thus, if we know the *rate of change* of the population at any time t, then we can calculate the net change in the population from $t = a$ to $t = b$ by evaluating an appropriate definite integral.

APPLIED EXAMPLE 6 Population Growth in Clark County Clark County in Nevada—dominated by Las Vegas—is the fastest-growing metropolitan area in the United States. From 1970 through 2000, the population was growing at the rate of

$$R(t) = 133680t^2 - 178788t + 234633 \qquad (0 \le t \le 3)$$

people per decade, where $t = 0$ corresponds to the beginning of 1970. What was the net change in the population over the decade from 1980 to 1990?

Source: U.S. Census Bureau

Solution   The net change in the population over the decade from 1980 ($t = 1$) to 1990 ($t = 2$) is given by $P(2) - P(1)$, where P denotes the population in the county at time t. But $P' = R$, and so

$$
\begin{aligned}
P(2) - P(1) &= \int_1^2 P'(t)\, dt = \int_1^2 R(t)\, dt \\
&= \int_1^2 (133680t^2 - 178788t + 234633)\, dt \\
&= 44560t^3 - 89394t^2 + 234633t \Big|_1^2 \\
&= [44560(2)^3 - 89394(2)^2 + 234633(2)] \\
&\quad - [44560 - 89394 + 234633] \\
&= 278371
\end{aligned}
$$

and so the net change is 278,371 people.   ∎

More generally, we have the following result. We assume that f has a continuous derivative, even though, the integrability of f' is sufficient.

Net Change Formula

The net change in a function f over an interval $[a, b]$ is given by

$$ f(b) - f(a) = \int_a^b f'(x)\, dx \tag{10} $$

provided f' is continuous on $[a, b]$.

As another example of the net change of a function, let's consider the following example.

APPLIED EXAMPLE 7   Production Costs   The management of Staedtler Office Equipment has determined that the daily marginal cost function associated with producing battery-operated pencil sharpeners is given by

$$ C'(x) = 0.000006x^2 - 0.006x + 4 $$

where $C'(x)$ is measured in dollars per unit and x denotes the number of units produced. Management has also determined that the daily fixed cost incurred in producing these pencil sharpeners is $100. Find Staedtler's daily total cost for producing (a) the first 500 units and (b) the 201st through 400th units.

Solution

a. Since $C'(x)$ is the marginal cost function, its antiderivative $C(x)$ is the total cost function. The daily fixed cost incurred in producing the pencil sharpeners is $C(0)$ dollars. Since the daily fixed cost is given as $100, we have $C(0) = 100$. We are required to find $C(500)$. Let's compute $C(500) - C(0)$, the net change in the total cost function $C(x)$ over the interval $[0, 500]$. Using the fundamental theorem of calculus, we find

$$C(500) - C(0) = \int_0^{500} C'(x)\, dx$$

$$= \int_0^{500} (0.000006x^2 - 0.006x + 4)\, dx$$

$$= 0.000002x^3 - 0.003x^2 + 4x \Big|_0^{500}$$

$$= [0.000002(500)^3 - 0.003(500)^2 + 4(500)]$$
$$- [0.000002(0)^3 - 0.003(0)^2 + 4(0)]$$

$$= 1500$$

Therefore, $C(500) = 1500 + C(0) = 1500 + 100 = 1600$, so the total cost incurred daily by Staedtler in producing 500 pencil sharpeners is $1600.
b. The daily total cost incurred by Staedtler in producing the 201st through 400th units of battery-operated pencil sharpeners is given by

$$C(400) - C(200) = \int_{200}^{400} C'(x)\, dx$$

$$= \int_{200}^{400} (0.000006x^2 - 0.006x + 4)\, dx$$

$$= 0.000002x^3 - 0.003x^2 + 4x \Big|_{200}^{400}$$

$$= [0.000002(400)^3 - 0.003(400)^2 + 4(400)]$$
$$- [0.000002(200)^3 - 0.003(200)^2 + 4(200)]$$

$$= 552$$

or $552.

Since $C'(x)$ is nonnegative for x in the interval $(0, \infty)$, we have the following geometric interpretation of the two definite integrals in Example 7: $\int_0^{500} C'(x)\, dx$ is the area of the region under the graph of the function C' from $x = 0$ to $x = 500$, shown in Figure 19a, and $\int_{200}^{400} C'(x)\, dx$ is the area of the region from $x = 200$ to $x = 400$, shown in Figure 19b.

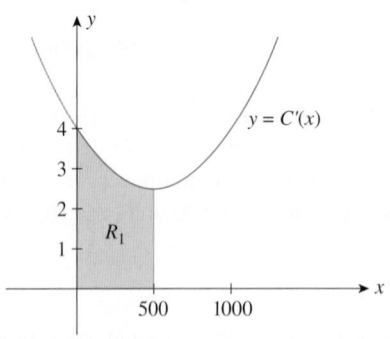

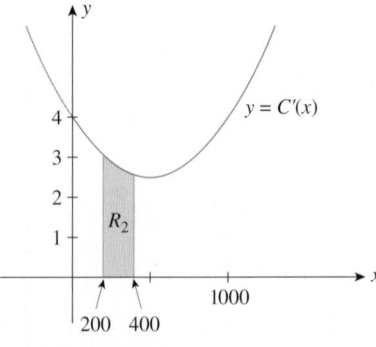

FIGURE 19

(a) Area of $R_1 = \int_0^{500} C'(x)\, dx$

(b) Area of $R_2 = \int_{200}^{400} C'(x)\, dx$

APPLIED EXAMPLE 8   Assembly Time of Workers An efficiency study conducted for Elektra Electronics showed that the rate at which Space Commander walkie-talkies are assembled by the average worker t hours after starting work at 8 a.m. is given by the function

$$f(t) = -3t^2 + 12t + 15 \qquad (0 \le t \le 4)$$

Determine how many walkie-talkies can be assembled by the average worker in the first hour of the morning shift.

Solution   Let $N(t)$ denote the number of walkie-talkies assembled by the average worker t hours after starting work in the morning shift. Then, we have

$$N'(t) = f(t) = -3t^2 + 12t + 15$$

Therefore, the number of units assembled by the average worker in the first hour of the morning shift is

$$
\begin{aligned}
N(1) - N(0) &= \int_0^1 N'(t)\,dt = \int_0^1 (-3t^2 + 12t + 15)\,dt \\
&= -t^3 + 6t^2 + 15t \,\Big|_0^1 = -1 + 6 + 15 \\
&= 20
\end{aligned}
$$

or 20 units.   ■

EXPLORING WITH TECHNOLOGY

You can demonstrate graphically that $\int_0^x t\,dt = \frac{1}{2}x^2$ as follows:

1. Plot the graphs of $y1 = \text{fnInt}\,(t, t, 0, x) = \int_0^x t\,dt$ and $y2 = \frac{1}{2}x^2$ on the same set of axes, using the viewing window $[-5, 5] \times [0, 10]$.

2. Compare the graphs of $y1$ and $y2$ and draw the desired conclusion.

APPLIED EXAMPLE 9   Projected Demand for Electricity A certain city's rate of electricity consumption is expected to grow exponentially with a growth constant of $k = 0.04$. If the present rate of consumption is 40 million kilowatt-hours (kWh) per year, what should be the total production of electricity over the next 3 years in order to meet the projected demand?

Solution   If $R(t)$ denotes the expected rate of consumption of electricity t years from now, then

$$R(t) = 40e^{0.04t}$$

million kWh per year. Next, if $C(t)$ denotes the expected total consumption of electricity over a period of t years, then

$$C'(t) = R(t)$$

Therefore, the total consumption of electricity expected over the next 3 years is given by

---

EXPLORE & DISCUSS

The definite integral $\int_{-3}^{3} \sqrt{9 - x^2}\,dx$ cannot be evaluated using the fundamental theorem of calculus because the method of this section does not enable us to find an antiderivative of the integrand. But the integral can be evaluated by interpreting it as the area of a certain plane region. What is the region? And what is the value of the integral?

$$\int_0^3 C'(t)\, dt = \int_0^3 40e^{0.04t}\, dt$$

$$= \frac{40}{0.04}\, e^{0.04t}\bigg|_0^3$$

$$= 1000(e^{0.12} - 1)$$

$$= 127.5$$

or 127.5 million kWh, the amount that must be produced over the next 3 years in order to meet the demand.   ■

■ Validity of the Fundamental Theorem of Calculus

To demonstrate the plausibility of the fundamental theorem of calculus for the case where f is nonnegative on an interval $[a, b]$, let's define an "area function" A as follows. Let $A(t)$ denote the area of the region R under the graph of $y = f(x)$ from $x = a$ to $x = t$, where $a \leq t \leq b$ (Figure 20).

If h is a small positive number, then $A(t + h)$ is the area of the region under the graph of $y = f(x)$ from $x = a$ to $x = t + h$. Therefore, the difference

$$A(t + h) - A(t)$$

is the area under the graph of $y = f(x)$ from $x = t$ to $x = t + h$ (Figure 21).

Now, the area of this last region can be approximated by the area of the rectangle of width h and height $f(t)$—that is, by the expression $h \cdot f(t)$ (Figure 22). Thus,

$$A(t + h) - A(t) \approx h \cdot f(t)$$

where the approximations improve as h is taken to be smaller and smaller.

Dividing both sides of the foregoing relationship by h, we obtain

$$\frac{A(t + h) - A(t)}{h} \approx f(t)$$

Taking the limit as h approaches zero, we find, by the definition of the derivative, that the left-hand side is

$$\lim_{h \to 0} \frac{A(t + h) - A(t)}{h} = A'(t)$$

The right-hand side, which is independent of h, remains constant throughout the limiting process. Because the approximation becomes exact as h approaches zero, we find that

$$A'(t) = f(t)$$

Since the foregoing equation holds for all values of t in the interval $[a, b]$, we have shown that the *area function* A is an antiderivative of the function $f(x)$. By Theorem 1 of Section 15.1, we conclude that $A(x)$ must have the form

$$A(x) = F(x) + C$$

where F is any antiderivative of f and C is a constant. To determine the value of C, observe that $A(a) = 0$. This condition implies that

$$A(a) = F(a) + C = 0$$

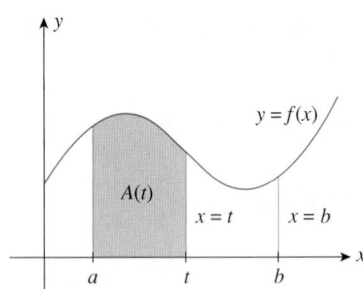

FIGURE 20
$A(t) =$ area under the graph of f from $x = a$ to $x = t$

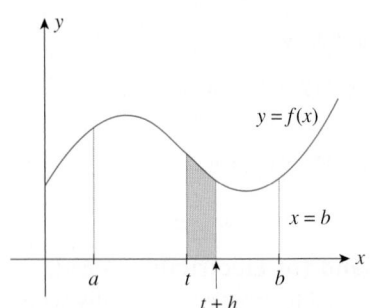

FIGURE 21
$A(t + h) - A(t) =$ area under the graph of f from $x = t$ to $x = t + h$

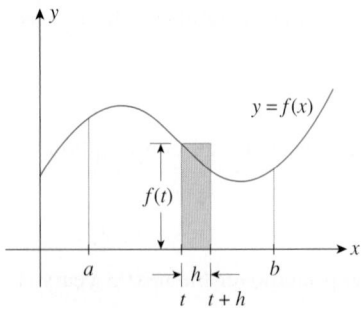

FIGURE 22
The area of the rectangle is $h \cdot f(t)$.

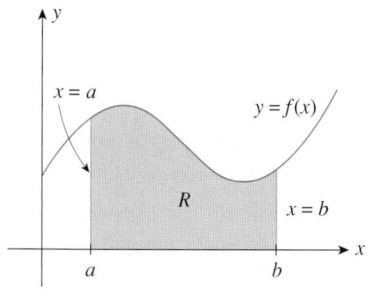

FIGURE 23
The area of R is given by $A(b)$.

or $C = -F(a)$. Next, since the area of the region R is $A(b)$ (Figure 23), we see that the required area is

$$A(b) = F(b) + C$$
$$= F(b) - F(a)$$

Since the area of the region R is

$$\int_a^b f(x)\, dx$$

we have

$$\int_a^b f(x)\, dx = F(b) - F(a)$$

as we set out to show.

15.4 Self-Check Exercises

1. Evaluate $\int_0^2 (x + e^x)\, dx$.
2. The daily marginal profit function associated with producing and selling TexaPep hot sauce is

$$P'(x) = -0.000006x^2 + 6$$

where x denotes the number of cases (each case contains 24 bottles) produced and sold daily and $P'(x)$ is measured in dollars/units. The fixed cost is $400.

a. What is the total profit realizable from producing and selling 1000 cases of TexaPep per day?
b. What is the additional profit realizable if the production and sale of TexaPep is increased from 1000 to 1200 cases/day?

Solutions to Self-Check Exercises 15.4 can be found on page 953.

15.4 Concept Questions

1. State the fundamental theorem of calculus.

2. State the net change formula and use it to answer the following questions:
   a. If a company generates income at the rate of R dollars/day, explain what $\int_a^b R(t)\,dt$ measures, where a and b are measured in days with $a < b$.

   b. If a private jet airplane consumes fuel at the rate of R gal/min, write an integral giving the net fuel consumption by the airplane between times $t = a$ and $t = b$ ($a < b$), where t is measured in minutes.

15.4 Exercises

In Exercises 1–4, find the area of the region under the graph of the function f on the interval $[a, b]$, using the fundamental theorem of calculus. Then verify your result using geometry.

1. $f(x) = 2;\ [1, 4]$ **2.** $f(x) = 4;\ [-1, 2]$

3. $f(x) = 2x;\ [1, 3]$ **4.** $f(x) = -\dfrac{1}{4}x + 1;\ [1, 4]$

In Exercises 5–16, find the area of the region under the graph of the function f on the interval $[a, b]$.

5. $f(x) = 2x + 3;\ [-1, 2]$ **6.** $f(x) = 4x - 1;\ [2, 4]$

7. $f(x) = -x^2 + 4;\ [-1, 2]$ **8.** $f(x) = 4x - x^2;\ [0, 4]$

9. $f(x) = \dfrac{1}{x};\ [1, 2]$ **10.** $f(x) = \dfrac{1}{x^2};\ [2, 4]$

11. $f(x) = \sqrt{x}; [1, 9]$ **12.** $f(x) = x^3; [1, 3]$

13. $f(x) = 1 - \sqrt[3]{x}; [-8, -1]$ **14.** $f(x) = \dfrac{1}{\sqrt{x}}; [1, 9]$

15. $f(x) = e^x; [0, 2]$ **16.** $f(x) = e^x - x; [1, 2]$

In Exercises 17–40, evaluate the definite integral.

17. $\displaystyle\int_2^4 3\,dx$ **18.** $\displaystyle\int_{-1}^2 -2\,dx$

19. $\displaystyle\int_1^3 (2x + 3)\,dx$ **20.** $\displaystyle\int_{-1}^0 (4 - x)\,dx$

21. $\displaystyle\int_{-1}^3 2x^2\,dx$ **22.** $\displaystyle\int_0^2 8x^3\,dx$

23. $\displaystyle\int_{-2}^2 (x^2 - 1)\,dx$ **24.** $\displaystyle\int_1^4 \sqrt{u}\,du$

25. $\displaystyle\int_1^8 4x^{1/3}\,dx$ **26.** $\displaystyle\int_1^4 2x^{-3/2}\,dx$

27. $\displaystyle\int_0^1 (x^3 - 2x^2 + 1)\,dx$ **28.** $\displaystyle\int_1^2 (t^5 - t^3 + 1)\,dt$

29. $\displaystyle\int_2^4 \dfrac{1}{x}\,dx$ **30.** $\displaystyle\int_1^3 \dfrac{2}{x}\,dx$

31. $\displaystyle\int_0^4 x(x^2 - 1)\,dx$ **32.** $\displaystyle\int_0^2 (x - 4)(x - 1)\,dx$

33. $\displaystyle\int_1^3 (t^2 - t)^2\,dt$ **34.** $\displaystyle\int_{-1}^1 (x^2 - 1)^2\,dx$

35. $\displaystyle\int_{-3}^{-1} \dfrac{1}{x^2}\,dx$ **36.** $\displaystyle\int_1^2 \dfrac{2}{x^3}\,dx$

37. $\displaystyle\int_1^4 \left(\sqrt{x} - \dfrac{1}{\sqrt{x}}\right)dx$ **38.** $\displaystyle\int_0^1 \sqrt{2x}(\sqrt{x} + \sqrt{2})\,dx$

39. $\displaystyle\int_1^4 \dfrac{3x^3 - 2x^2 + 4}{x^2}\,dx$ **40.** $\displaystyle\int_1^2 \left(1 + \dfrac{1}{u} + \dfrac{1}{u^2}\right)du$

41. MARGINAL COST A division of Ditton Industries manufactures a deluxe toaster oven. Management has determined that the daily marginal cost function associated with producing these toaster ovens is given by

$$C'(x) = 0.0003x^2 - 0.12x + 20$$

where $C'(x)$ is measured in dollars/unit and x denotes the number of units produced. Management has also determined that the daily fixed cost incurred in the production is $800.

a. Find the total cost incurred by Ditton in producing the first 300 units of these toaster ovens per day.

b. What is the total cost incurred by Ditton in producing the 201st through 300th units/day?

42. MARGINAL REVENUE The management of Ditton Industries has determined that the daily marginal revenue function associated with selling x units of their deluxe toaster ovens is given by

$$R'(x) = -0.1x + 40$$

where $R'(x)$ is measured in dollars/unit.

a. Find the daily total revenue realized from the sale of 200 units of the toaster oven.

b. Find the additional revenue realized when the production (and sales) level is increased from 200 to 300 units.

43. MARGINAL PROFIT Refer to Exercise 41. The daily marginal profit function associated with the production and sales of the deluxe toaster ovens is known to be

$$P'(x) = -0.0003x^2 + 0.02x + 20$$

where x denotes the number of units manufactured and sold daily and $P'(x)$ is measured in dollars/unit.

a. Find the total profit realizable from the manufacture and sale of 200 units of the toaster ovens per day.
   Hint: $P(200) - P(0) = \int_0^{200} P'(x)\,dx$, $P(0) = -800$.

b. What is the additional daily profit realizable if the production and sale of the toaster ovens are increased from 200 to 220 units/day?

44. EFFICIENCY STUDIES Tempco Electronics, a division of Tempco Toys, manufactures an electronic football game. An efficiency study showed that the rate at which the games are assembled by the average worker t hr after starting work at 8 a.m. is

$$-\dfrac{3}{2}t^2 + 6t + 20 \qquad (0 \le t \le 4)$$

units/hour.

a. Find the total number of games the average worker can be expected to assemble in the 4-hr morning shift.

b. How many units can the average worker be expected to assemble in the first hour of the morning shift? In the second hour of the morning shift?

45. SPEEDBOAT RACING In a recent pretrial run for the world water speed record, the velocity of the *Sea Falcon II* t sec after firing the booster rocket was given by

$$v(t) = -t^2 + 20t + 440 \qquad (0 \le t \le 20)$$

ft/sec. Find the distance covered by the boat over the 20-sec period after the booster rocket was activated.
Hint: The distance is given by $\int_0^{20} v(t)\,dt$.

46. POCKET COMPUTERS Annual sales (in millions of units) of pocket computers are expected to grow in accordance with the function

$$f(t) = 0.18t^2 + 0.16t + 2.64 \qquad (0 \le t \le 6)$$

where t is measured in years, with $t = 0$ corresponding to 1997. How many pocket computers were sold over the 6-yr period between the beginning of 1997 and the end of 2002?
Source: Dataquest, Inc.

47. SINGLE FEMALE-HEADED HOUSEHOLDS WITH CHILDREN The percent of families with children that are headed by single females grew at the rate of

$$R(t) = 0.8499t^2 - 3.872t + 5 \qquad (0 \le t \le 3)$$

households/decade between 1970 ($t = 0$) and 2000 ($t = 3$). The number of such households stood at 5.6% of all families in 1970.

a. Find an expression giving the percent of these households in the tth decade.

b. If the trend continued, estimate the percent of these households in 2010.

c. What was the net increase in the percent of these households from 1970 to 2000?

Source: U.S. Census Bureau

48. Air Purification   To test air purifiers, engineers ran a purifier in a smoke-filled 10-ft × 20-ft room. While conducting a test for a certain brand of air purifier, it was determined that the amount of smoke in the room was decreasing at the rate of

$$R(t) = 0.00032t^4 - 0.01872t^3 + 0.3948t^2 - 3.83t + 17.63 \qquad (0 \le t \le 20)$$

percent of the (original) amount of the smoke per minute, t min after the start of the test. How much smoke was left in the room 5 min after the start of the test? Ten minutes after the start of the test?

Source: Consumer Reports

49. Senior Citizens   The population aged 65 years old and older (in millions) from 2000 to 2050 is projected to be

$$f(t) = \frac{85}{1 + 1.859e^{-0.66t}} \qquad (0 \le t \le 5)$$

where t is measured in decades, with $t = 0$ corresponding to 2000. What will be the average population aged 65 years and older over the years from 2000 to 2030?
Hint: The average population is given by $(1/3)\int_0^3 f(t)\,dt$. Multiply the integrand by $e^{0.66t}/e^{0.66t}$ and then use the method of substitution.

50. Blood Flow   Consider an artery of length L cm and radius R cm. Using Poiseuille's law (page 590), it can be shown that

the rate at which blood flows through the artery (measured in cubic centimeters/second) is given by

$$V = \int_0^R \frac{k}{L} x(R^2 - x^2)\,dx$$

where k is a constant. Find an expression for V that does *not* involve an integral.
Hint: Use the substitution $u = R^2 - x^2$.

51. Find the area of the region bounded by the graph of the function $f(x) = x^4 - 2x^2 + 2$, the x-axis, and the lines $x = a$ and $x = b$, where $a < b$ and a and b are the x-coordinates of the relative maximum point and a relative minimum point of f, respectively.

52. Find the area of the region bounded by the graph of the function $f(x) = (x + 1)/\sqrt{x}$, the x-axis, and the lines $x = a$ and $x = b$ where a and b are, respectively, the x-coordinates of the relative minimum point and the inflection point of f.

In Exercises 53–56, determine whether the statement is true or false. If it is true, explain why it is true. If it is false, give an example to show why it is false.

53. $\int_{-1}^{1} \frac{1}{x^3}\,dx = -\frac{1}{2x^2}\Big|_{-1}^{1} = -\frac{1}{2} - \left(-\frac{1}{2}\right) = 0$

54. $\int_{-1}^{1} \frac{1}{x}\,dx = \ln|x|\,\Big|_{-1}^{1} = \ln|1| - \ln|-1| = \ln 1 - \ln 1 = 0$

55. $\int_0^2 (1 - x)\,dx$ gives the area of the region under the graph of $f(x) = 1 - x$ on the interval $[0, 2]$.

56. The total revenue realized in selling the first 5000 units of a product is given by

$$\int_0^{500} R'(x)\,dx = R(500) - R(0)$$

where $R(x)$ is the total revenue.

15.4 Solutions to Self-Check Exercises

1. $\int_0^2 (x + e^x)\,dx = \frac{1}{2}x^2 + e^x\,\Big|_0^2$

$= \left[\frac{1}{2}(2)^2 + e^2\right] - \left[\frac{1}{2}(0) + e^0\right]$

$= 2 + e^2 - 1$

$= e^2 + 1$

2. a. We want $P(1000)$, but

$P(1000) - P(0) = \int_0^{1000} P'(x)\,dx = \int_0^{1000} (-0.000006x^2 + 6)\,dx$

$= -0.000002x^3 + 6x\,\Big|_0^{1000}$

$= -0.000002(1000)^3 + 6(1000)$

$= 4000$

So, $P(1000) = 4000 + P(0) = 4000 - 400$, or \$3600/day $[P(0) = -C(0)]$.

b. The additional profit realizable is given by

$\int_{1000}^{1200} P'(x)\,dx = -0.000002x^3 + 6x\,\Big|_{1000}^{1200}$

$= [-0.000002(1200)^3 + 6(1200)]$
$\quad - [-0.000002(1000)^3 + 6(1000)]$

$= 3744 - 4000$

$= -256$

That is, the company sustains a loss of \$256/day if production is increased to 1200 cases/day.

USING TECHNOLOGY

■ Evaluating Definite Integrals

Some graphing utilities have an operation for finding the definite integral of a function. If your graphing utility has this capability, use it to work through the example and exercises of this section.

EXAMPLE 1 Use the numerical integral operation of a graphing utility to evaluate

$$\int_{-1}^{2} \frac{2x + 4}{(x^2 + 1)^{3/2}} \, dx$$

Solution Using the numerical integral operation of a graphing utility, we find

$$\int_{-1}^{2} \frac{2x + 4}{(x^2 + 1)^{3/2}} \, dx = \text{fnInt}((2x + 4)/(x^2 + 1)^1.5, x, -1, 2) \approx 6.92592226 \quad ■$$

TECHNOLOGY EXERCISES

In Exercises 1–4, find the area of the region under the graph of f on the interval $[a, b]$. Express your answer to four decimal places.

1. $f(x) = 0.002x^5 + 0.032x^4 - 0.2x^2 + 2; [-1.1, 2.2]$

2. $f(x) = x\sqrt{x^3 + 1}; [1, 2]$

3. $f(x) = \sqrt{x}e^{-x}; [0, 3]$

4. $f(x) = \dfrac{\ln x}{\sqrt{1 + x^2}}; [1, 2]$

In Exercises 5–10, evaluate the definite integral.

5. $\displaystyle\int_{-1.2}^{2.3} (0.2x^4 - 0.32x^3 + 1.2x - 1) \, dx$

6. $\displaystyle\int_{1}^{3} x(x^4 - 1)^{3.2} \, dx$

7. $\displaystyle\int_{0}^{2} \frac{3x^3 + 2x^2 + 1}{2x^2 + 3} \, dx$ **8.** $\displaystyle\int_{1}^{2} \frac{\sqrt{x} + 1}{2x^2 + 1} \, dx$

9. $\displaystyle\int_{0}^{2} \frac{e^x}{\sqrt{x^2 + 1}} \, dx$ **10.** $\displaystyle\int_{1}^{3} e^{-x} \ln(x^2 + 1) \, dx$

11. Rework Exercise 48, Exercises 15.4.

12. Rework Exercise 49, Exercises 15.4.

13. THE GLOBAL EPIDEMIC The number of AIDS-related deaths/year in the United States is given by the function

$$f(t) = -53.254t^4 + 673.7t^3 - 2801.07t^2$$
$$+ 8833.379t + 20,000 \quad (0 \le t \le 9)$$

where $t = 0$ corresponds to the beginning of 1988. Find the total number of AIDS-related deaths in the United States between the beginning of 1988 and the end of 1996.
Source: Centers for Disease Control

14. MARIJUANA ARRESTS The number of arrests for marijuana sales and possession in New York City grew at the rate of approximately

$$f(t) = 0.0125t^4 - 0.01389t^3 + 0.55417t^2$$
$$+ 0.53294t + 4.95238 \quad (0 \le t \le 5)$$

thousand/year, where t is measured in years, with $t = 0$ corresponding to the beginning of 1992. Find the approximate number of marijuana arrests in the city from the beginning of 1992 to the end of 1997.
Source: State Division of Criminal Justice Services

15. POPULATION GROWTH The population of a certain city is projected to grow at the rate of $18\sqrt{t + 1} \ln \sqrt{t + 1}$ thousand people/year t yr from now. If the current population is 800,000, what will be the population 45 yr from now?

15.5 Evaluating Definite Integrals

This section continues our discussion of the applications of the fundamental theorem of calculus.

▬ Properties of the Definite Integral

Before going on, we list the following useful properties of the definite integral, some of which parallel the rules of integration of Section 15.1.

Properties of the Definite Integral

Let f and g be integrable functions; then,

1. $\displaystyle\int_a^a f(x)\, dx = 0$

2. $\displaystyle\int_a^b f(x)\, dx = -\int_b^a f(x)\, dx$

3. $\displaystyle\int_a^b cf(x)\, dx = c\int_a^b f(x)\, dx \qquad (c, \text{ a constant})$

4. $\displaystyle\int_a^b \left[f(x) \pm g(x)\right] dx = \int_a^b f(x)\, dx \pm \int_a^b g(x)\, dx$

5. $\displaystyle\int_a^b f(x)\, dx = \int_a^c f(x)\, dx + \int_c^b f(x)\, dx \qquad (a < c < b)$

Property 5 states that if c is a number lying between a and b so that the interval $[a, b]$ is divided into the intervals $[a, c]$ and $[c, b]$, then the integral of f over the interval $[a, b]$ may be expressed as the sum of the integral of f over the interval $[a, c]$ and the integral of f over the interval $[c, b]$.

Property 5 has the following geometric interpretation when f is nonnegative. By definition

$$\int_a^b f(x)\, dx$$

is the area of the region under the graph of $y = f(x)$ from $x = a$ to $x = b$ (Figure 24). Similarly, we interpret the definite integrals

$$\int_a^c f(x)\, dx \quad \text{and} \quad \int_c^b f(x)\, dx$$

as the areas of the regions under the graph of $y = f(x)$ from $x = a$ to $x = c$ and from $x = c$ to $x = b$, respectively. Since the two regions do not overlap, we see that

$$\int_a^b f(x)\, dx = \int_a^c f(x)\, dx + \int_c^b f(x)\, dx$$

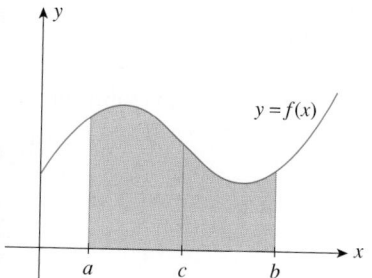

FIGURE 24
$\int_a^b f(x)\, dx = \int_a^c f(x)\, dx + \int_c^b f(x)\, dx$

▬ The Method of Substitution for Definite Integrals

Our first example shows two approaches generally used when evaluating a definite integral using the method of substitution.

EXAMPLE 1   Evaluate $\int_0^4 x\sqrt{9+x^2}\,dx$.

Solution

Method 1   We first find the corresponding indefinite integral:

$$I = \int x\sqrt{9+x^2}\,dx$$

Make the substitution $u = 9 + x^2$ so that

$$du = \frac{d}{dx}(9+x^2)\,dx$$

$$= 2x\,dx$$

$$x\,dx = \frac{1}{2}\,du \qquad \text{Divide both sides by 2.}$$

Then,

$$I = \int \frac{1}{2}\sqrt{u}\,du = \frac{1}{2}\int u^{1/2}\,du$$

$$= \frac{1}{3}u^{3/2} + C = \frac{1}{3}(9+x^2)^{3/2} + C \qquad \begin{array}{l}\text{Substitute}\\ 9+x^2 \text{ for } u.\end{array}$$

Using this result, we now evaluate the given definite integral:

$$\int_0^4 x\sqrt{9+x^2}\,dx = \frac{1}{3}(9+x^2)^{3/2}\Big|_0^4$$

$$= \frac{1}{3}\left[(9+16)^{3/2} - 9^{3/2}\right]$$

$$= \frac{1}{3}(125 - 27) = \frac{98}{3} = 32\tfrac{2}{3}$$

Method 2   *Changing the Limits of Integration.* As before, we make the substitution

$$u = 9 + x^2 \tag{11}$$

so that

$$du = 2x\,dx$$

$$x\,dx = \frac{1}{2}\,du$$

Next, observe that the given definite integral is evaluated *with respect to x* with the range of integration given by the interval [0, 4]. If we perform the integration *with respect to u* via the substitution (11), then we must adjust the range of integration to reflect the fact that the integration is being performed with respect to the new variable u. To determine the proper range of integration, note that when $x = 0$, Equation (11) implies that

$$u = 9 + 0^2 = 9$$

which gives the required lower limit of integration with respect to u. Similarly, when $x = 4$,

$$u = 9 + 16 = 25$$

is the required upper limit of integration with respect to u. Thus, the range of integration when the integration is performed with respect to u is given by the interval $[9, 25]$. Therefore, we have

$$\int_0^4 x\sqrt{9 + x^2}\, dx = \int_9^{25} \frac{1}{2}\sqrt{u}\, du = \frac{1}{2}\int_9^{25} u^{1/2}\, du$$

$$= \frac{1}{3} u^{3/2}\Big|_9^{25} = \frac{1}{3}\left(25^{3/2} - 9^{3/2}\right)$$

$$= \frac{1}{3}(125 - 27) = \frac{98}{3} = 32\tfrac{2}{3}$$

which agrees with the result obtained using Method 1.

When you use the method of substitution, make sure you adjust the limits of integration to reflect integrating with respect to the new variable u.

EXPLORING WITH TECHNOLOGY

Refer to Example 1. You can confirm the results obtained there by using a graphing utility as follows:

1. Use the numerical integration operation of the graphing utility to evaluate

$$\int_0^4 x\sqrt{9 + x^2}\, dx$$

2. Evaluate $\dfrac{1}{2}\displaystyle\int_9^{25} \sqrt{u}\, du$.

3. Conclude that $\displaystyle\int_0^4 x\sqrt{9 + x^2}\, dx = \frac{1}{2}\int_9^{25} \sqrt{u}\, du$.

EXAMPLE 2 Evaluate $\int_0^2 x e^{2x^2}\, dx$.

Solution Let $u = 2x^2$ so that $du = 4x\, dx$, or $x\, dx = \tfrac{1}{4}\, du$. When $x = 0$, $u = 0$, and when $x = 2$, $u = 8$. This gives the lower and upper limits of integration with respect to u. Making the indicated substitutions, we find

$$\int_0^2 x e^{2x^2}\, dx = \int_0^8 \frac{1}{4} e^u\, du = \frac{1}{4} e^u\Big|_0^8 = \frac{1}{4}\left(e^8 - 1\right)$$

EXAMPLE 3 Evaluate $\displaystyle\int_0^1 \frac{x^2}{x^3 + 1}\, dx$.

Solution Let $u = x^3 + 1$ so that $du = 3x^2\, dx$, or $x^2\, dx = \tfrac{1}{3}\, du$. When $x = 0$, $u = 1$, and when $x = 1$, $u = 2$. This gives the lower and upper limits of integration with respect to u. Making the indicated substitutions, we find

$$\int_0^1 \frac{x^2}{x^3 + 1}\, dx = \frac{1}{3}\int_1^2 \frac{du}{u} = \frac{1}{3} \ln|u|\Big|_1^2$$

$$= \frac{1}{3}(\ln 2 - \ln 1) = \frac{1}{3}\ln 2$$

Finding the Area under a Curve

EXAMPLE 4   Find the area of the region R under the graph of $f(x) = e^{(1/2)x}$ from $x = -1$ to $x = 1$.

Solution   The region R is shown in Figure 25. Its area is given by

$$A = \int_{-1}^{1} e^{(1/2)x} \, dx$$

To evaluate this integral, we make the substitution

$$u = \frac{1}{2}x$$

so that

$$du = \frac{1}{2} \, dx$$

$$dx = 2 \, du$$

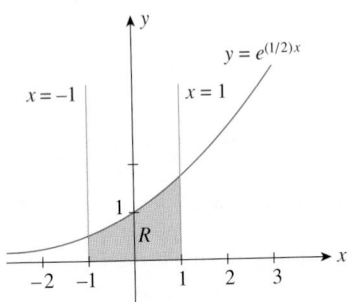

FIGURE 25
Area of $R = \int_{-1}^{1} e^{(1/2)x} \, dx$

When $x = -1$, $u = -\frac{1}{2}$, and when $x = 1$, $u = \frac{1}{2}$. Making the indicated substitutions, we obtain

$$A = \int_{-1}^{1} e^{(1/2)x} \, dx = 2 \int_{-1/2}^{1/2} e^u \, du$$

$$= 2e^u \Big|_{-1/2}^{1/2} = 2(e^{1/2} - e^{-1/2})$$

or approximately 2.08 square units.

EXPLORE & DISCUSS

Let f be a function defined piecewise by the rule

$$f(x) = \begin{cases} \sqrt{x} & \text{if } 0 \le x \le 1 \\ \dfrac{1}{x} & \text{if } 1 < x \le 2 \end{cases}$$

How would you use Property 5 of definite integrals to find the area of the region under the graph of f on $[0, 2]$? What is the area?

Average Value of a Function

The *average value* of a function over an interval provides us with an application of the definite integral. Recall that the average value of a set of n numbers is the number

$$\frac{y_1 + y_2 + \cdots + y_n}{n}$$

Now, suppose f is a continuous function defined on $[a, b]$. Let's divide the interval $[a, b]$ into n subintervals of equal length $(b - a)/n$. Choose points $x_1, x_2, \ldots, x_n$ in the first, second, $\ldots$, and nth subintervals, respectively. Then, the average value of the numbers $f(x_1), f(x_2), \ldots, f(x_n)$, given by

$$\frac{f(x_1) + f(x_2) + \cdots + f(x_n)}{n}$$

is an approximation of the average of all the values of $f(x)$ on the interval $[a, b]$. This expression can be written in the form

$$\frac{(b-a)}{(b-a)}\left[f(x_1)\cdot\frac{1}{n} + f(x_2)\cdot\frac{1}{n} + \cdots + f(x_n)\cdot\frac{1}{n}\right]$$

$$= \frac{1}{b-a}\left[f(x_1)\cdot\frac{b-a}{n} + f(x_2)\cdot\frac{b-a}{n} + \cdots + f(x_n)\cdot\frac{b-a}{n}\right]$$

$$= \frac{1}{b-a}[f(x_1)\Delta x + f(x_2)\Delta x + \cdots + f(x_n)\Delta x] \tag{12}$$

As n gets larger and larger, the expression (12) approximates the average value of $f(x)$ over $[a, b]$ with increasing accuracy. But the sum inside the brackets in (12) is a Riemann sum of the function f over $[a, b]$. In view of this, we have

$$\lim_{n\to\infty}\left[\frac{f(x_1) + f(x_2) + \cdots + f(x_n)}{n}\right]$$

$$= \frac{1}{b-a}\lim_{n\to\infty}[f(x_1)\Delta x + f(x_2)\Delta x + \cdots + f(x_n)\Delta x]$$

$$= \frac{1}{b-a}\int_a^b f(x)\,dx$$

This discussion motivates the following definition.

> **The Average Value of a Function**
> Suppose f is integrable on $[a, b]$. Then the **average value** of f over $[a, b]$ is
> $$\frac{1}{b-a}\int_a^b f(x)\,dx$$

EXAMPLE 5   Find the average value of the function $f(x) = \sqrt{x}$ over the interval $[0, 4]$.

Solution   The required average value is given by

$$\frac{1}{4-0}\int_0^4 \sqrt{x}\,dx = \frac{1}{4}\int_0^4 x^{1/2}\,dx$$

$$= \frac{1}{6}x^{3/2}\Big|_0^4 = \frac{1}{6}(4^{3/2})$$

$$= \frac{4}{3}$$

APPLIED EXAMPLE 6   Automobile Financing   The interest rates charged by Madison Finance on auto loans for used cars over a certain 6-month period in 2000 are approximated by the function

$$r(t) = -\frac{1}{12}t^3 + \frac{7}{8}t^2 - 3t + 12 \qquad (0 \le t \le 6)$$

where t is measured in months and $r(t)$ is the annual percentage rate. What is the average rate on auto loans extended by Madison over the 6-month period?

Solution   The average rate over the 6-month period in question is given by

$$\frac{1}{6-0} \int_0^6 \left(-\frac{1}{12} t^3 + \frac{7}{8} t^2 - 3t + 12 \right) dt$$

$$= \frac{1}{6} \left(-\frac{1}{48} t^4 + \frac{7}{24} t^3 - \frac{3}{2} t^2 + 12t \right) \Big|_0^6$$

$$= \frac{1}{6} \left[-\frac{1}{48} (6^4) + \frac{7}{24} (6^3) - \frac{3}{2} (6^2) + 12(6) \right]$$

$$= 9$$

or 9% per year. ◼

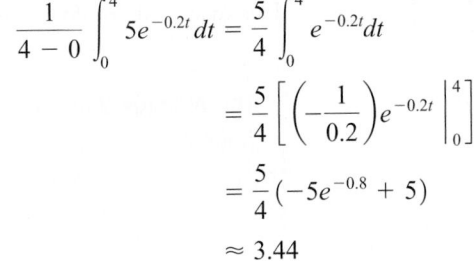

APPLIED EXAMPLE 7   Drug Concentration in a Body   The amount of a certain drug in a patient's body t days after it has been administered is

$$C(t) = 5e^{-0.2t}$$

units. Determine the average amount of the drug present in the patient's body for the first 4 days after the drug has been administered.

Solution   The average amount of the drug present in the patient's body for the first 4 days after it has been administered is given by

$$\frac{1}{4-0} \int_0^4 5e^{-0.2t} dt = \frac{5}{4} \int_0^4 e^{-0.2t} dt$$

$$= \frac{5}{4} \left[\left(-\frac{1}{0.2} \right) e^{-0.2t} \Big|_0^4 \right]$$

$$= \frac{5}{4} (-5e^{-0.8} + 5)$$

$$\approx 3.44$$

or approximately 3.44 units. ◼

We now give a geometric interpretation of the average value of a function f over an interval $[a, b]$. Suppose $f(x)$ is nonnegative so that the definite integral

$$\int_a^b f(x) \, dx$$

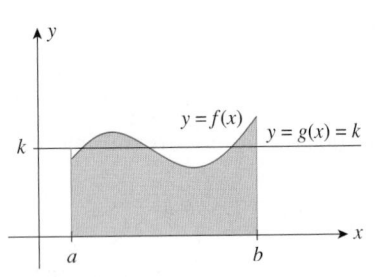

FIGURE 26
The average value of f over $[a, b]$ is k.

gives the area under the graph of f from $x = a$ to $x = b$ (Figure 26). Observe that, in general, the "height" $f(x)$ varies from point to point. Can we replace $f(x)$ by a constant function $g(x) = k$ (which has constant height) such that the areas under each of the two functions f and g are the same? If so, since the area under the graph of g from $x = a$ to $x = b$ is $k(b - a)$, we have

$$k(b - a) = \int_a^b f(x) \, dx$$

$$k = \frac{1}{b - a} \int_a^b f(x) \, dx$$

so that k is the average value of f over $[a, b]$. Thus, the average value of a function f over an interval $[a, b]$ is the height of a rectangle with base of length $(b - a)$ that has the same area as that of the region under the graph of f from $x = a$ to $x = b$.

15.5   Self-Check Exercises

1. Evaluate $\int_0^2 \sqrt{2x + 5}\, dx$.

2. Find the average value of the function $f(x) = 1 - x^2$ over the interval $[-1, 2]$.

3. The median price of a house in a southwestern state between January 1, 2000, and January 1, 2005, is approximated by the function

$$f(t) = t^3 - 7t^2 + 17t + 280 \qquad (0 \le t \le 5)$$

where $f(t)$ is measured in thousands of dollars and t is expressed in years, with $t = 0$ corresponding to the beginning of 2000. Determine the average median price of a house over that time interval.

Solutions to Self-Check Exercises 15.5 can be found on page 964.

15.5   Concept Questions

1. Describe two approaches used to evaluate a definite integral using the method of substitution. Illustrate with the integral $\int_0^1 x^2(x^3 + 1)^2\, dx$.

2. Define the average value of a function f over an interval $[a, b]$. Give a geometric interpretation.

15.5   Exercises

In Exercises 1–28, evaluate the definite integral.

1. $\displaystyle\int_0^2 x(x^2 - 1)^3\, dx$

2. $\displaystyle\int_0^1 x^2(2x^3 - 1)^4\, dx$

3. $\displaystyle\int_0^1 x\sqrt{5x^2 + 4}\, dx$

4. $\displaystyle\int_1^3 x\sqrt{3x^2 - 2}\, dx$

5. $\displaystyle\int_0^2 x^2(x^3 + 1)^{3/2}\, dx$

6. $\displaystyle\int_1^5 (2x - 1)^{5/2}\, dx$

7. $\displaystyle\int_0^1 \frac{1}{\sqrt{2x + 1}}\, dx$

8. $\displaystyle\int_0^2 \frac{x}{\sqrt{x^2 + 5}}\, dx$

9. $\displaystyle\int_1^2 (2x - 1)^4\, dx$

10. $\displaystyle\int_1^2 (2x + 4)(x^2 + 4x - 8)^3\, dx$

11. $\displaystyle\int_{-1}^1 x^2(x^3 + 1)^4\, dx$

12. $\displaystyle\int_1^2 \left(x^3 + \frac{3}{4}\right)(x^4 + 3x)^{-2}\, dx$

13. $\displaystyle\int_1^5 x\sqrt{x - 1}\, dx$

14. $\displaystyle\int_1^4 x\sqrt{x + 1}\, dx$

Hint: Let $u = x + 1$.

15. $\displaystyle\int_0^2 xe^{x^2}\, dx$

16. $\displaystyle\int_0^1 e^{-x}\, dx$

17. $\displaystyle\int_0^1 (e^{2x} + x^2 + 1)\, dx$

18. $\displaystyle\int_0^2 (e^t - e^{-t})\, dt$

19. $\displaystyle\int_{-1}^1 xe^{x^2+1}\, dx$

20. $\displaystyle\int_0^4 \frac{e^{\sqrt{x}}}{\sqrt{x}}\, dx$

21. $\displaystyle\int_3^6 \frac{2}{x - 2}\, dx$

22. $\displaystyle\int_0^1 \frac{x}{1 + 2x^2}\, dx$

23. $\displaystyle\int_1^2 \frac{x^2 + 2x}{x^3 + 3x^2 - 1}\, dx$

24. $\displaystyle\int_0^1 \frac{e^x}{1 + e^x}\, dx$

25. $\displaystyle\int_1^2 \left(4e^{2u} - \frac{1}{u}\right) du$

26. $\displaystyle\int_1^2 \left(1 + \frac{1}{x} + e^x\right) dx$

27. $\displaystyle\int_1^2 \left(2e^{-4x} - \frac{1}{x^2}\right) dx$

28. $\displaystyle\int_1^2 \frac{\ln x}{x}\, dx$

In Exercises 29–38, find the average value of the function f over the indicated interval $[a, b]$.

29. $f(x) = 2x + 3;\ [0, 2]$

30. $f(x) = 8 - x;\ [1, 4]$

31. $f(x) = 2x^2 - 3;\ [1, 3]$

32. $f(x) = 4 - x^2;\ [-2, 3]$

33. $f(x) = x^2 + 2x - 3;\ [-1, 2]$

34. $f(x) = x^3;\ [-1, 1]$

35. $f(x) = \sqrt{2x + 1};\ [0, 4]$

36. $f(x) = e^{-x};\ [0, 4]$

37. $f(x) = xe^{x^2};\ [0, 2]$

38. $f(x) = \dfrac{1}{x + 1};\ [0, 2]$

39. **WORLD PRODUCTION OF COAL**   A study proposed in 1980 by researchers from the major producers and consumers of the world's coal concluded that coal could and must play an important role in fueling global economic growth over the next 20 yr. The world production of coal in 1980 was 3.5 bil-

lion metric tons. If output increased at the rate of $3.5e^{0.05t}$ billion metric tons/year in year t ($t = 0$ corresponding to 1980), determine how much coal was produced worldwide between 1980 and the end of the 20th century.

40. **NEWTON'S LAW OF COOLING**   A bottle of white wine at room temperature (68°F) is placed in a refrigerator at 4 p.m. Its temperature after t hr is changing at the rate of

$$-18e^{-0.6t}$$

°F/hr. By how many degrees will the temperature of the wine have dropped by 7 p.m.? What will the temperature of the wine be at 7 p.m.?

41. **NET INVESTMENT FLOW**   The net investment flow (rate of capital formation) of the giant conglomerate LTF incorporated is projected to be

$$t\sqrt{\frac{1}{2}t^2 + 1}$$

million dollars/year in year t. Find the accruement on the company's capital stock in the second year.
Hint: The amount is given by

$$\int_1^2 t\sqrt{\frac{1}{2}t^2 + 1}\, dt$$

42. **OIL PRODUCTION**   Based on a preliminary report by a geological survey team, it is estimated that a newly discovered oil field can be expected to produce oil at the rate of

$$R(t) = \frac{600t^2}{t^3 + 32} + 5 \qquad (0 \le t \le 20)$$

thousand barrels/year, t yr after production begins. Find the amount of oil that the field can be expected to yield during the first 5 yr of production, assuming that the projection holds true.

43. **DEPRECIATION: DOUBLE DECLINING-BALANCE METHOD**   Suppose a tractor purchased at a price of $60,000 is to be depreciated by the *double declining-balance method* over a 10-yr period. It can be shown that the rate at which the book value will be decreasing is given by

$$R(t) = 13388.61e^{-0.22314t} \qquad (0 \le t \le 10)$$

dollars/year at year t. Find the amount by which the book value of the tractor will depreciate over the first 5 yr of its life.

44. **VELOCITY OF A CAR**   A car moves along a straight road in such a way that its velocity (in feet/second) at any time t (in seconds) is given by

$$v(t) = 3t\sqrt{16 - t^2} \qquad (0 \le t \le 4)$$

Find the distance traveled by the car in the 4 sec from $t = 0$ to $t = 4$.

45. **AVERAGE VELOCITY OF A TRUCK**   A truck traveling along a straight road has a velocity (in feet/second) at time t (in seconds) given by

$$v(t) = \frac{1}{12}t^2 + 2t + 44 \qquad (0 \le t \le 5)$$

What is the average velocity of the truck over the time interval from $t = 0$ to $t = 5$?

46. **MEMBERSHIP IN CREDIT UNIONS**   Credit unions in Massachusetts have grown remarkably in recent years. Their tax-exempt status allows them to offer deposit and loan rates that are often more favorable than those offered by banks. The membership in Massachusetts credit unions grew at the rate of

$$R(t) = -0.0039t^2 + 0.0374t + 0.0046 \qquad (0 \le t \le 9)$$

million members/year between 1994 ($t = 0$) and 2003 ($t = 9$). Find the average rate of growth of membership in Massachusetts credit unions over the period in question.
Source: Massachusetts Credit Union League

47. **WHALE POPULATION**   A group of marine biologists estimates that if certain conservation measures are implemented, the population of an endangered species of whale will be

$$N(t) = 3t^3 + 2t^2 - 10t + 600 \qquad (0 \le t \le 10)$$

where $N(t)$ denotes the population at the end of year t. Find the average population of the whales over the next 10 yr.

48. **U.S. CITIZENS 65 AND OLDER**   The number of U.S. citizens 65 and older from 1900 through 2050 is estimated to be growing at the rate of

$$R(t) = 0.063t^2 - 0.48t + 3.87 \qquad (0 \le t \le 15)$$

million people/decade, where t is measured in decades and $t = 0$ corresponds to 1900. Show that the average rate of growth of U.S. citizens 65 and older between 2000 and 2050 will be growing at twice the rate of that between 1950 and 2000.
Source: American Heart Association

49. **AVERAGE YEARLY SALES**   The sales of Universal Instruments in the first t yr of its operation are approximated by the function

$$S(t) = t\sqrt{0.2t^2 + 4}$$

where $S(t)$ is measured in millions of dollars. What were Universal's average yearly sales over its first 5 yr of operation?

50. **CABLE TV SUBSCRIBERS**   The manager of TeleStar Cable Service estimates that the total number of subscribers to the service in a certain city t yr from now will be

$$N(t) = -\frac{40,000}{\sqrt{1 + 0.2t}} + 50,000$$

Find the average number of cable television subscribers over the next 5 yr if this prediction holds true.

51. Refer to Exercise 44. Find the average velocity of the car over the time interval [0, 4].

52. CONCENTRATION OF A DRUG IN THE BLOODSTREAM The concentration of a certain drug in a patient's bloodstream t hr after injection is

$$C(t) = \frac{0.2t}{t^2 + 1}$$

mg/cm^3. Determine the average concentration of the drug in the patient's bloodstream over the first 4 hr after the drug is injected.

53. AVERAGE PRICE OF A COMMODITY The price of a certain commodity in dollars/unit at time t (measured in weeks) is given by

$$p = 18 - 3e^{-2t} - 6e^{-t/3}$$

What is the average price of the commodity over the 5-wk period from $t = 0$ to $t = 5$?

54. FLOW OF BLOOD IN AN ARTERY According to a law discovered by 19th-century physician Jean Louis Marie Poiseuille, the velocity of blood (in centimeters/second) r cm from the central axis of an artery is given by

$$v(r) = k(R^2 - r^2)$$

where k is a constant and R is the radius of the artery. Find the average velocity of blood along a radius of the artery (see the accompanying figure).

Hint: Evaluate $\frac{1}{R} \int_0^R v(r)\, dr$.

Blood vessel

55. WASTE DISPOSAL When organic waste is dumped into a pond, the oxidization process that takes place reduces the pond's oxygen content. However, in time, nature will restore the oxygen content to its natural level. Suppose that the oxygen content t days after organic waste has been dumped into a pond is given by

$$f(t) = 100\left(\frac{t^2 + 10t + 100}{t^2 + 20t + 100}\right)$$

percent of its normal level. Find the average content of oxygen in the pond over the first 10 days after organic waste has been dumped into it.
Hint: Show that

$$\frac{t^2 + 10t + 100}{t^2 + 20t + 100} = 1 - \frac{10}{t + 10} + \frac{100}{(t + 10)^2}$$

56. VELOCITY OF A FALLING HAMMER During the construction of a high-rise apartment building, a construction worker accidently drops a hammer that falls vertically a distance of h ft. The velocity of the hammer after falling a distance of x ft is $v = \sqrt{2gx}$ ft/sec ($0 \le x \le h$). Show that the average velocity of the hammer over this path is $\bar{v} = \frac{2}{3}\sqrt{2gh}$ ft/sec.

57. Prove Property 1 of the definite integral.
Hint: Let F be an antiderivative of f and use the definition of the definite integral.

58. Prove Property 2 of the definite integral.
Hint: See Exercise 57.

59. Verify by direct computation that

$$\int_1^3 x^2\, dx = -\int_3^1 x^2\, dx$$

60. Prove Property 3 of the definite integral.
Hint: See Exercise 57.

61. Verify by direct computation that

$$\int_1^9 2\sqrt{x}\, dx = 2\int_1^9 \sqrt{x}\, dx$$

62. Verify by direct computation that

$$\int_0^1 (1 + x - e^x)\, dx = \int_0^1 dx + \int_0^1 x\, dx - \int_0^1 e^x\, dx$$

What properties of the definite integral are demonstrated in this exercise?

63. Verify by direct computation that

$$\int_0^3 (1 + x^3)\, dx = \int_0^1 (1 + x^3)\, dx + \int_1^3 (1 + x^3)\, dx$$

What property of the definite integral is demonstrated here?

64. Verify by direct computation that

$$\int_0^3 (1 + x^3)\, dx$$

$$= \int_0^1 (1 + x^3)\, dx + \int_1^2 (1 + x^3)\, dx + \int_2^3 (1 + x^3)\, dx$$

hence showing that Property 5 may be extended.

65. Evaluate $\int_3^3 (1 + \sqrt{x})e^{-x}\, dx$.

66. Evaluate $\int_3^0 f(x)\, dx$, given that $\int_0^3 f(x)\, dx = 4$.

67. Given that $\int_{-1}^2 f(x)\, dx = -2$ and $\int_{-1}^2 g(x)\, dx = 3$, evaluate
a. $\int_{-1}^2 [2f(x) + g(x)]\, dx$
b. $\int_{-1}^2 [g(x) - f(x)]\, dx$
c. $\int_{-1}^2 [2f(x) - 3g(x)]\, dx$

68. Given that $\int_{-1}^2 f(x)\, dx = 2$ and $\int_0^2 f(x)\, dx = 3$, evaluate
a. $\int_{-1}^0 f(x)\, dx$
b. $\int_0^2 f(x)\, dx - \int_{-1}^0 f(x)\, dx$

In Exercises 69–74, determine whether the statement is true or false. If it is true, explain why it is true. If it is false, give an example to show why it is false.

69. $\displaystyle\int_{2}^{2}\frac{e^x}{\sqrt{1+x}}\,dx = 0$

70. $\displaystyle\int_{1}^{3}\frac{dx}{x-2} = -\int_{3}^{1}\frac{dx}{x-2}$

71. $\displaystyle\int_{0}^{1}x\sqrt{x+1}\,dx = \sqrt{x+1}\int_{0}^{1}x\,dx = \frac{1}{2}x^2\sqrt{x+1}\Big|_{0}^{1} = \frac{\sqrt{2}}{2}$

72. If f' is continuous on $[0, 2]$, then $\int_{0}^{2}f'(x)\,dx = f(2) - f(0)$.

73. If f and g are continuous on $[a, b]$ and k is a constant, then

$$\int_{a}^{b}[kf(x) + g(x)]\,dx = k\int_{a}^{b}f(x)\,dx + \int_{a}^{b}g(x)\,dx$$

74. If f is continuous on $[a, b]$ and $a < c < b$, then

$$\int_{b}^{c}f(x)\,dx = \int_{a}^{c}f(x)\,dx - \int_{a}^{b}f(x)\,dx$$

15.5 Solutions to Self-Check Exercises

1. Let $u = 2x + 5$. Then, $du = 2\,dx$, or $dx = \frac{1}{2}\,du$. Also, when $x = 0$, $u = 5$, and when $x = 2$, $u = 9$. Therefore,

$$\begin{aligned}\int_{0}^{2}\sqrt{2x+5}\,dx &= \int_{0}^{2}(2x+5)^{1/2}\,dx \\ &= \frac{1}{2}\int_{5}^{9}u^{1/2}\,du \\ &= \left(\frac{1}{2}\right)\left(\frac{2}{3}u^{3/2}\right)\Big|_{5}^{9} \\ &= \frac{1}{3}[9^{3/2} - 5^{3/2}] \\ &= \frac{1}{3}(27 - 5\sqrt{5})\end{aligned}$$

2. The required average value is given by

$$\begin{aligned}\frac{1}{2-(-1)}\int_{-1}^{2}(1-x^2)\,dx &= \frac{1}{3}\int_{-1}^{2}(1-x^2)\,dx \\ &= \frac{1}{3}\left(x - \frac{1}{3}x^3\right)\Big|_{-1}^{2} \\ &= \frac{1}{3}\left[\left(2 - \frac{8}{3}\right) - \left(-1 + \frac{1}{3}\right)\right] = 0\end{aligned}$$

3. The average median price of a house over the stated time interval is given by

$$\begin{aligned}\frac{1}{5-0}\int_{0}^{5}&(t^3 - 7t^2 + 17t + 280)\,dt \\ &= \frac{1}{5}\left(\frac{1}{4}t^4 - \frac{7}{3}t^3 + \frac{17}{2}t^2 + 280t\right)\Big|_{0}^{5} \\ &= \frac{1}{5}\left[\frac{1}{4}(5)^4 - \frac{7}{3}(5)^3 + \frac{17}{2}(5)^2 + 280(5)\right] \\ &= 295.417\end{aligned}$$

or $295,417.

USING TECHNOLOGY

■ Evaluating Definite Integrals for Piecewise-Defined Functions

We continue using graphing utilities to find the definite integral of a function. But here we will make use of Property 5 of the properties of the definite integral (p. 955).

EXAMPLE 1 Use the numerical integral operation of a graphing utility to evaluate

$$\int_{-1}^{2}f(x)\,dx$$

where

$$f(x) = \begin{cases} -x^2 & \text{if } x < 0 \\ \sqrt{x} & \text{if } x \geq 0 \end{cases}$$

Solution   Using Property 5 of the definite integral, we can write

$$\int_{-1}^{2} f(x)\, dx = \int_{-1}^{0} -x^2\, dx + \int_{0}^{2} x^{1/2}\, dx$$

Using a graphing utility, we find

$$\int_{-1}^{2} f(x)\, dx = \text{fnInt}(-x\string^2, x, -1, 0) + \text{fnInt}(x\string^0.5, x, 0, 2)$$

$$\approx -0.333333 + 1.885618$$

$$= 1.552285$$

TECHNOLOGY EXERCISES

In Exercises 1–4, use Property 5 of the properties of the definite integral (page 955) to evaluate the definite integral accurate to six decimal places.

1. $\int_{-1}^{2} f(x)\, dx$, where

$$f(x) = \begin{cases} 2.3x^3 - 3.1x^2 + 2.7x + 3 & \text{if } x < 1 \\ -1.7x^2 + 2.3x + 4.3 & \text{if } x \geq 1 \end{cases}$$

2. $\int_{0}^{3} f(x)\, dx$, where $f(x) = \begin{cases} \dfrac{\sqrt{x}}{1+x^2} & \text{if } 0 \leq x < 1 \\ 0.5e^{-0.1x^2} & \text{if } x \geq 1 \end{cases}$

3. $\int_{-2}^{2} f(x)\, dx$, where $f(x) = \begin{cases} x^4 - 2x^2 + 4 & \text{if } x < 0 \\ 2\ln(x + e^2) & \text{if } x \geq 0 \end{cases}$

4. $\int_{-2}^{6} f(x)\, dx$, where

$$f(x) = \begin{cases} 2x^3 - 3x^2 + x + 2 & \text{if } x < -1 \\ \sqrt{3x+4} - 5 & \text{if } -1 \leq x \leq 4 \\ x^2 - 3x - 5 & \text{if } x > 4 \end{cases}$$

5. AIDS IN MASSACHUSETTS   The rate of growth (and decline) of the number of AIDS cases diagnosed in Massachusetts from the beginning of 1989 ($t = 0$) through the end of 1997 ($t = 8$) is approximated by the function

$$f(t) = \begin{cases} 69.83333t^2 + 30.16667t + 1000 & \text{if } 0 \leq t < 3 \\ 1719 & \text{if } 3 \leq t < 4 \\ -28.79167t^3 + 491.37500t^2 & \text{if } 4 \leq t \leq 8 \\ \quad -2985.083333t + 7640 \end{cases}$$

where $f(t)$ is measured in the number of cases/year. Estimate the total number of AIDS cases diagnosed in Massachusetts from the beginning of 1989 through the end of 1997.

Source: Massachusetts Department of Health

6. CROP YIELD   If left untreated on bean stems, aphids (small insects that suck plant juices) will multiply at an increasing rate during the summer months and reduce productivity and crop yield of cultivated crops. But if the aphids are treated in mid-June, the numbers decrease sharply to less than 100/bean stem, allowing for steep rises in crop yield. The function

$$F(t) = \begin{cases} 62e^{1.152t} & \text{if } 0 \leq t < 1.5 \\ 349e^{-1.324(t-1.5)} & \text{if } 1.5 \leq t \leq 3 \end{cases}$$

gives the number of aphids on a typical bean stem at time t, where t is measured in months, $t = 0$ corresponding to the beginning of May. Find the average number of aphids on a typical bean stem over the period from the beginning of May to the beginning of August.

7. ABSORPTION OF DRUGS   Jane took 100 mg of a drug in the morning and another 100 mg of the same drug at the same time the following morning. The amount of the drug in her body t days after the first dosage was taken is given by

$$A(t) = \begin{cases} 100e^{-1.4t} & \text{if } 0 \leq t < 1 \\ 100(1 + e^{1.4})e^{-1.4t} & \text{if } t \geq 1 \end{cases}$$

Find the average amount of the drug in Jane's body over the first 2 days.

8. ABSORPTION OF DRUGS   The concentration of a drug in an organ at any time t (in seconds) is given by

$$C(t) = \begin{cases} 0.3t - 18(1 - e^{-t/60}) & \text{if } 0 \leq t \leq 20 \\ 18e^{-t/60} - 12e^{-(t-20)/60} & \text{if } t > 20 \end{cases}$$

where $C(t)$ is measured in grams/cubic centimeter (g/cm^3). Find the average concentration of the drug in the organ over the first 30 sec after it is administered.

15.6 Area between Two Curves

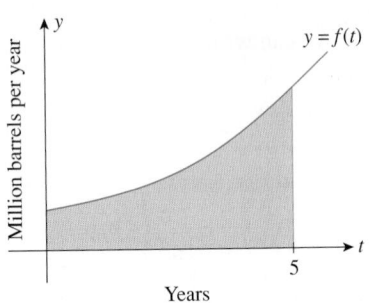

FIGURE 27
At a rate of consumption of $f(t)$ million barrels per year, the total petroleum consumption is given by the area of the region under the graph of f.

Suppose a certain country's petroleum consumption is expected to grow at the rate of $f(t)$ million barrels per year, t years from now, for the next 5 years. Then, the country's total petroleum consumption over the period of time in question is given by the area under the graph of f on the interval $[0, 5]$ (Figure 27).

Next, suppose that because of the implementation of certain energy-conservation measures, the rate of growth of petroleum consumption is expected to be $g(t)$ million barrels per year instead. Then, the country's projected total petroleum consumption over the 5-year period is given by the area under the graph of g on the interval $[0, 5]$ (Figure 28).

Therefore, the area of the shaded region S lying between the graphs of f and g on the interval $[0, 5]$ (Figure 29) gives the amount of petroleum that would be saved over the 5-year period because of the conservation measures.

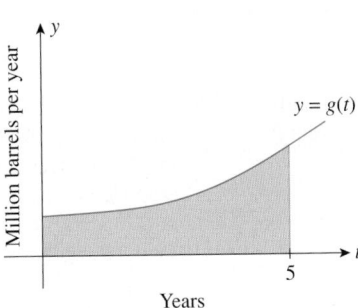

FIGURE 28
At a rate of consumption of $g(t)$ million barrels per year, the total petroleum consumption is given by the area of the region under the graph of g.

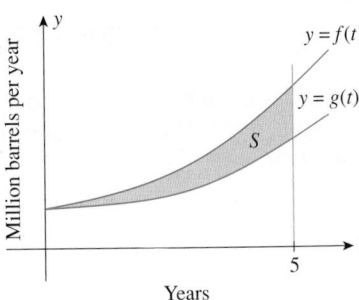

FIGURE 29
The area of S gives the amount of petroleum that would be saved over the 5-year period.

But the area of S is given by

Area under the graph of f on $[a, b]$ − Area under the graph of g on $[a, b]$

$$= \int_0^5 f(t)\, dt - \int_0^5 g(t)\, dt$$

$$= \int_0^5 [f(t) - g(t)]\, dt \qquad \text{By Property 4, Section 15.5}$$

This example shows that some practical problems can be solved by finding the area of a region between two curves, which in turn can be found by evaluating an appropriate definite integral.

Finding the Area between Two Curves

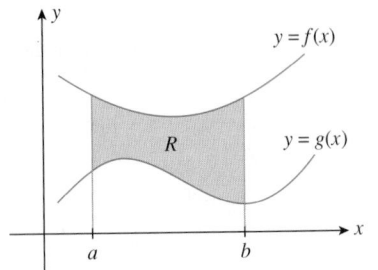

FIGURE 30
Area of $R = \int_a^b [f(x) - g(x)]\, dx$

We now turn our attention to the general problem of finding the area of a plane region bounded both above and below by the graphs of functions. First, consider the situation in which the graph of one function lies above that of another. More specifically, let R be the region in the xy-plane (Figure 30) that is bounded above by the graph of a continuous function f, below by a continuous function g where $f(x) \geq g(x)$ on $[a, b]$, and to the left and right by the vertical lines $x = a$ and $x = b$, respectively. From the figure, we see that

$$\text{Area of } R = \text{Area under } f(x) - \text{Area under } g(x)$$

$$= \int_a^b f(x)\, dx - \int_a^b g(x)\, dx$$

$$= \int_a^b [f(x) - g(x)]\, dx$$

upon using Property 4 of the definite integral.

The Area between Two Curves

Let f and g be continuous functions such that $f(x) \geq g(x)$ on the interval $[a, b]$. Then, the area of the region bounded above by $y = f(x)$ and below by $y = g(x)$ on $[a, b]$ is given by

$$\int_a^b [f(x) - g(x)]\, dx \qquad \textbf{(13)}$$

Even though we assumed that both f and g were nonnegative in the derivation of (13), it may be shown that this equation is valid if f and g are not nonnegative (see Exercise 55). Also, observe that if $g(x)$ is 0 for all x—that is, when the lower boundary of the region R is the x-axis—Equation (13) gives the area of the region under the curve $y = f(x)$ from $x = a$ to $x = b$, as we would expect.

EXAMPLE 1   Find the area of the region bounded by the x-axis, the graph of $y = -x^2 + 4x - 8$, and the lines $x = -1$ and $x = 4$.

Solution   The region R under consideration is shown in Figure 31. We can view R as the region bounded above by the graph of $f(x) = 0$ (the x-axis) and below by the graph of $g(x) = -x^2 + 4x - 8$ on $[-1, 4]$. Therefore, the area of R is given by

$$\int_a^b [f(x) - g(x)]\, dx = \int_{-1}^4 [0 - (-x^2 + 4x - 8)]\, dx$$

$$= \int_{-1}^4 (x^2 - 4x + 8)\, dx$$

$$= \frac{1}{3}x^3 - 2x^2 + 8x \Big|_{-1}^4$$

$$= \left[\frac{1}{3}(64) - 2(16) + 8(4)\right] - \left[\frac{1}{3}(-1) - 2(1) + 8(-1)\right]$$

$$= 31\tfrac{2}{3}$$

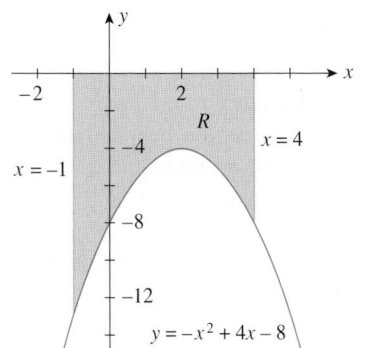

FIGURE 31
Area of $R = -\int_{-1}^4 g(x)\, dx$

or $31\tfrac{2}{3}$ square units.   ∎

EXAMPLE 2   Find the area of the region R bounded by the graphs of

$$f(x) = 2x - 1 \quad \text{and} \quad g(x) = x^2 - 4$$

and the vertical lines $x = 1$ and $x = 2$.

Solution   We first sketch the graphs of the functions $f(x) = 2x - 1$ and

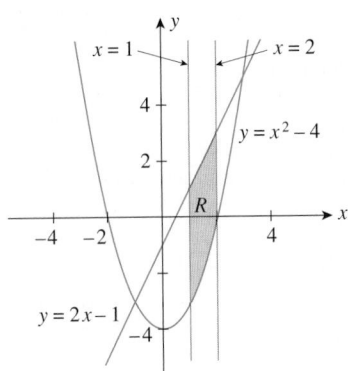

FIGURE 32
Area of $R = \int_1^2 [f(x) - g(x)]\, dx$

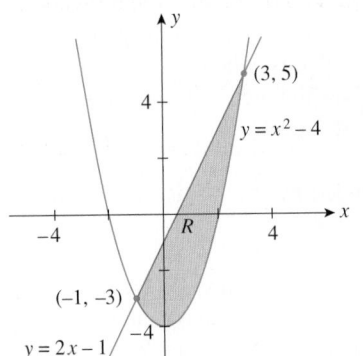

FIGURE 33
Area of $R = \int_{-1}^3 [f(x) - g(x)]\, dx$

$g(x) = x^2 - 4$ and the vertical lines $x = 1$ and $x = 2$, and then we identify the region R whose area is to be calculated (Figure 32).

Since the graph of f always lies above that of g for x in the interval $[1, 2]$, we see by Equation (13) that the required area is given by

$$
\begin{aligned}
\int_1^2 [f(x) - g(x)]\, dx &= \int_1^2 [(2x - 1) - (x^2 - 4)]\, dx \\
&= \int_1^2 (-x^2 + 2x + 3)\, dx \\
&= -\frac{1}{3}x^3 + x^2 + 3x \Big|_1^2 \\
&= \left(-\frac{8}{3} + 4 + 6\right) - \left(-\frac{1}{3} + 1 + 3\right) = \frac{11}{3}
\end{aligned}
$$

or $\frac{11}{3}$ square units.

EXAMPLE 3   Find the area of the region R that is completely enclosed by the graphs of the functions

$$f(x) = 2x - 1 \quad \text{and} \quad g(x) = x^2 - 4$$

Solution   The region R is shown in Figure 33. First, we find the points of intersection of the two curves. To do this, we solve the system that consists of the two equations $y = 2x - 1$ and $y = x^2 - 4$. Equating the two values of y gives

$$
\begin{aligned}
x^2 - 4 &= 2x - 1 \\
x^2 - 2x - 3 &= 0 \\
(x + 1)(x - 3) &= 0
\end{aligned}
$$

so $x = -1$ or $x = 3$. That is, the two curves intersect when $x = -1$ and $x = 3$.

Observe that we could also view the region R as the region bounded above by the graph of the function $f(x) = 2x - 1$, below by the graph of the function $g(x) = x^2 - 4$, and to the left and right by the vertical lines $x = -1$ and $x = 3$, respectively.

Next, since the graph of the function f always lies above that of the function g on $[-1, 3]$, we can use (13) to compute the desired area:

$$
\begin{aligned}
\int_a^b [f(x) - g(x)]\, dx &= \int_{-1}^3 [(2x - 1) - (x^2 - 4)]\, dx \\
&= \int_{-1}^3 (-x^2 + 2x + 3)\, dx \\
&= -\frac{1}{3}x^3 + x^2 + 3x \Big|_{-1}^3 \\
&= (-9 + 9 + 9) - \left(\frac{1}{3} + 1 - 3\right) = \frac{32}{3} \\
&= 10\frac{2}{3}
\end{aligned}
$$

or $10\frac{2}{3}$ square units.

EXAMPLE 4   Find the area of the region R bounded by the graphs of the functions

$$f(x) = x^2 - 2x - 1 \quad \text{and} \quad g(x) = -e^x - 1$$

and the vertical lines $x = -1$ and $x = 1$.

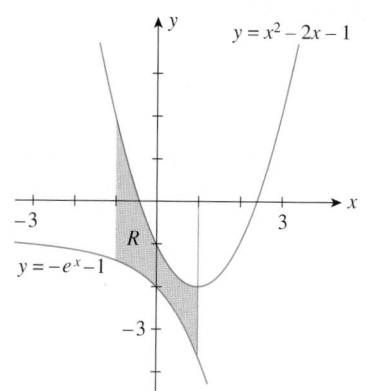

FIGURE 34
Area of $R = \int_{-1}^{1} [f(x) - g(x)]\, dx$

Solution   The region R is shown in Figure 34. Since the graph of the function f always lies above that of the function g, the area of the region R is given by

$$
\int_a^b [f(x) - g(x)]\, dx = \int_{-1}^{1} [(x^2 - 2x - 1) - (-e^x - 1)]\, dx
$$

$$
= \int_{-1}^{1} (x^2 - 2x + e^x)\, dx
$$

$$
= \frac{1}{3}x^3 - x^2 + e^x \Big|_{-1}^{1}
$$

$$
= \left(\frac{1}{3} - 1 + e \right) - \left(-\frac{1}{3} - 1 + e^{-1} \right)
$$

$$
= \frac{2}{3} + e - \frac{1}{e}, \text{ or approximately 3.02 square units} \quad \blacksquare
$$

Equation (13), which gives the area of the region between the curves $y = f(x)$ and $y = g(x)$ for $a \le x \le b$, is valid when the graph of the function f lies above that of the function g over the interval $[a, b]$. Example 5 shows how to use (13) to find the area of a region when the latter condition does not hold.

EXAMPLE 5   Find the area of the region bounded by the graph of the function $f(x) = x^3$, the x-axis, and the lines $x = -1$ and $x = 1$.

Solution   The region R under consideration can be thought of as composed of the two subregions R_1 and R_2, as shown in Figure 35.

Recall that the x-axis is represented by the function $g(x) = 0$. Since $g(x) \ge f(x)$ on $[-1, 0]$, we see that the area of R_1 is given by

$$
\int_a^b [g(x) - f(x)]\, dx = \int_{-1}^{0} (0 - x^3)\, dx = -\int_{-1}^{0} x^3\, dx
$$

$$
= -\frac{1}{4}x^4 \Big|_{-1}^{0} = 0 - \left(-\frac{1}{4} \right) = \frac{1}{4}
$$

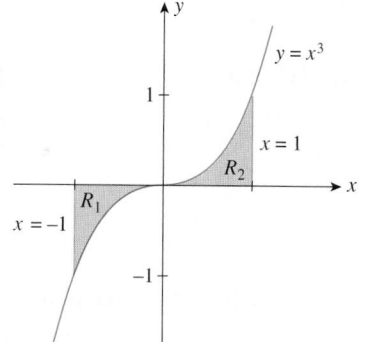

FIGURE 35
Area of R_1 = Area of R_2

To find the area of R_2, we observe that $f(x) \ge g(x)$ on $[0, 1]$, so it is given by

$$
\int_a^b [f(x) - g(x)]\, dx = \int_0^1 (x^3 - 0)\, dx = \int_0^1 x^3\, dx
$$

$$
= \frac{1}{4}x^4 \Big|_0^1 = \left(\frac{1}{4} \right) - 0 = \frac{1}{4}
$$

Therefore, the area of R is $\frac{1}{4} + \frac{1}{4}$, or $\frac{1}{2}$, square units.

By making use of symmetry, we could have obtained the same result by computing

$$
-2\int_{-1}^{0} x^3\, dx \qquad \text{or} \qquad 2\int_0^1 x^3\, dx
$$

as you may verify. $\blacksquare$

EXPLORE & DISCUSS

A function is *even* if it satisfies the condition $f(-x) = f(x)$, and it is *odd* if it satisfies the condition $f(-x) = -f(x)$. Show that the graph of an even function is symmetric with respect to the y-axis while the graph of an odd function is symmetric with respect to the origin. Explain why

$$
\int_{-a}^{a} f(x)\, dx = 2\int_0^a f(x)\, dx \quad \text{if } f \text{ is even}
$$

$$
\int_{-a}^{a} f(x)\, dx = 0 \quad \text{if } f \text{ is odd}
$$

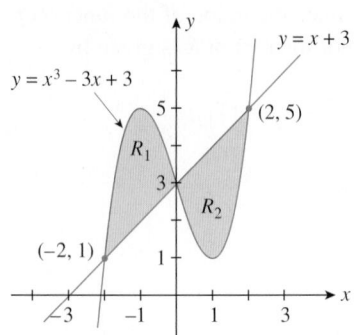

FIGURE 36
Area of R_1 + Area of R_2

$$= \int_{-2}^{0} [f(x) - g(x)] \, dx$$
$$+ \int_{0}^{2} [g(x) - f(x)] \, dx$$

EXAMPLE 6   Find the area of the region completely enclosed by the graphs of the functions

$$f(x) = x^3 - 3x + 3 \quad \text{and} \quad g(x) = x + 3$$

Solution   First, sketch the graphs of $y = x^3 - 3x + 3$ and $y = x + 3$ and then identify the required region R. We can view the region R as being composed of the two subregions R_1 and R_2, as shown in Figure 36. By solving the equations $y = x + 3$ and $y = x^3 - 3x + 3$ simultaneously, we find the points of intersection of the two curves. Equating the two values of y, we have

$$x^3 - 3x + 3 = x + 3$$
$$x^3 - 4x = 0$$
$$x(x^2 - 4) = 0$$
$$x(x + 2)(x - 2) = 0$$
$$x = 0, -2, 2$$

Hence, the points of intersection of the two curves are $(-2, 1)$, $(0, 3)$, and $(2, 5)$.
    For $-2 \le x \le 0$, we see that the graph of the function f lies above that of the function g, so the area of the region R_1 is, by virtue of (13),

$$\int_{-2}^{0} [(x^3 - 3x + 3) - (x + 3)] \, dx = \int_{-2}^{0} (x^3 - 4x) \, dx$$
$$= \frac{1}{4}x^4 - 2x^2 \Big|_{-2}^{0}$$
$$= -(4 - 8)$$
$$= 4$$

or 4 square units. For $0 \le x \le 2$, the graph of the function g lies above that of the function f, and the area of R_2 is given by

$$\int_{0}^{2} [(x + 3) - (x^3 - 3x + 3)] \, dx = \int_{0}^{2} (-x^3 + 4x) \, dx$$
$$= -\frac{1}{4}x^4 + 2x^2 \Big|_{0}^{2}$$
$$= -4 + 8$$
$$= 4$$

or 4 square units. Therefore, the required area is the sum of the area of the two regions $R_1 + R_2$—that is, $4 + 4$, or 8 square units.   ∎

APPLIED EXAMPLE 7   Conservation of Oil   In a 1999 study for a developing country's Economic Development Board, government economists and energy experts concluded that if the Energy Conservation Bill were implemented in 2000, the country's oil consumption for the next 5 years would be expected to grow in accordance with the model

$$R(t) = 20e^{0.05t}$$

where t is measured in years ($t = 0$ corresponding to the year 2000) and $R(t)$ in millions of barrels per year. Without the government-imposed conservation measures, however, the expected rate of growth of oil consumption would be given by

$$R_1(t) = 20e^{0.08t}$$

millions of barrels per year. Using these models, determine how much oil would have been saved from 2000 through 2005 if the bill had been implemented.

Solution   Under the Energy Conservation Bill, the total amount of oil that would have been consumed between 2000 and 2005 is given by

$$\int_0^5 R(t)\, dt = \int_0^5 20e^{0.05t}\, dt \tag{14}$$

Without the bill, the total amount of oil that would have been consumed between 2000 and 2005 is given by

$$\int_0^5 R_1(t)\, dt = \int_0^5 20e^{0.08t}\, dt \tag{15}$$

Equation (14) may be interpreted as the area of the region under the curve $y = R(t)$ from $t = 0$ to $t = 5$. Similarly, we interpret (15) as the area of the region under the curve $y = R_1(t)$ from $t = 0$ to $t = 5$. Furthermore, note that the graph of $y = R_1(t) = 20e^{0.08t}$ always lies on or above the graph of $y = R(t) = 20e^{0.05t}$ ($t \geq 0$). Thus, the area of the shaded region S in Figure 37 shows the amount of oil that would have been saved from 2000 to 2005 if the Energy Conservation Bill had been implemented. But the area of the region S is given by

$$\int_0^5 [R_1(t) - R(t)]\, dt = \int_0^5 [20e^{0.08t} - 20e^{0.05t}]\, dt$$

$$= 20 \int_0^5 (e^{0.08t} - e^{0.05t})\, dt$$

$$= 20 \left(\frac{e^{0.08t}}{0.08} - \frac{e^{0.05t}}{0.05} \right) \Big|_0^5$$

$$= 20 \left[\left(\frac{e^{0.4}}{0.08} - \frac{e^{0.25}}{0.05} \right) - \left(\frac{1}{0.08} - \frac{1}{0.05} \right) \right]$$

$$\approx 9.3$$

or approximately 9.3 square units. Thus, the amount of oil that would have been saved is 9.3 million barrels.   ■

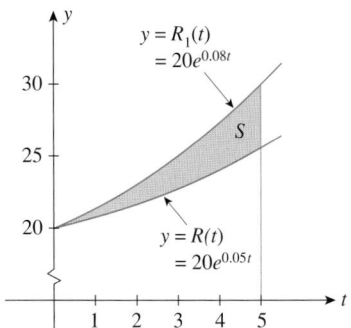

FIGURE 37
Area of $S = \int_0^5 [R_1(t) - R(t)]\, dt$

EXPLORING WITH TECHNOLOGY

Refer to Example 7. Suppose we want to construct a mathematical model giving the amount of oil saved from 2000 through the year 2000 + x, where $x \geq 0$. For example, in Example 7, $x = 5$.

1. Show that this model is given by

$$F(x) = \int_0^x [R_1(t) - R(t)]\, dt$$
$$= 250e^{0.08x} - 400e^{0.05x} + 150$$

   Hint: You may find it helpful to use some of the results of Example 7.
2. Use a graphing utility to plot the graph of F, using the viewing window $[0, 10] \times [0, 50]$.
3. Find $F(5)$ and thus confirm the result of Example 7.
4. What is the main advantage of this model?

15.6   Self-Check Exercises

1. Find the area of the region bounded by the graphs of $f(x) = x^2 + 2$ and $g(x) = 1 - x$ and the vertical lines $x = 0$ and $x = 1$.

2. Find the area of the region completely enclosed by the graphs of $f(x) = -x^2 + 6x + 5$ and $g(x) = x^2 + 5$.

3. The management of Kane Corporation, which operates a chain of hotels, expects its profits to grow at the rate of $1 + t^{2/3}$ million dollars/year t yr from now. However, with renovations and improvements of existing hotels and proposed acquisitions of new hotels, Kane's profits are expected to grow at the rate of $t - 2\sqrt{t} + 4$ million dollars/year in the next decade. What additional profits are expected over the next 10 yr if the group implements the proposed plans?

Solutions to Self-Check Exercises 15.6 can be found on page 976.

15.6   Concept Questions

1. Suppose f and g are continuous functions such that $f(x) \geq g(x)$ on the interval $[a, b]$. Write an integral giving the area of the region bounded above by the graph of f, below by the graph of g, and on the left and right by the lines $x = a$ and $x = b$.

2. Write an expression in terms of definite integrals giving the area of the shaded region in the following figure:

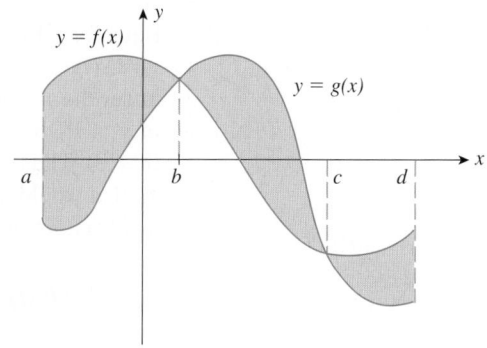

15.6   Exercises

In Exercises 1–8, find the area of the shaded region.

1.

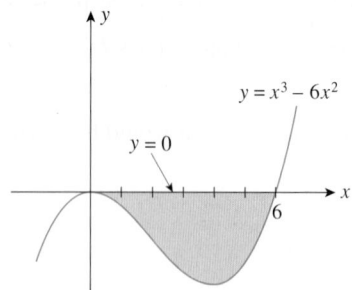

2.

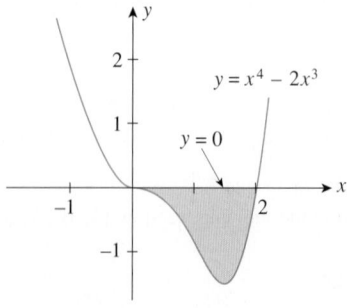

3.

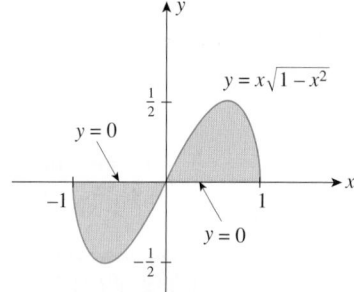

4.

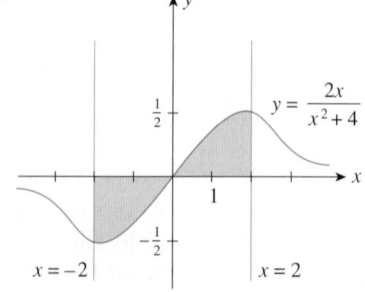

5.

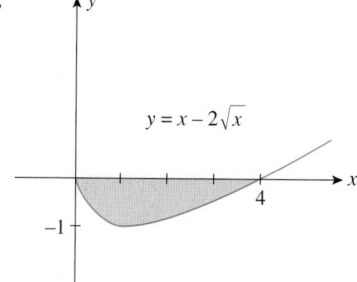

$y = x - 2\sqrt{x}$

6.

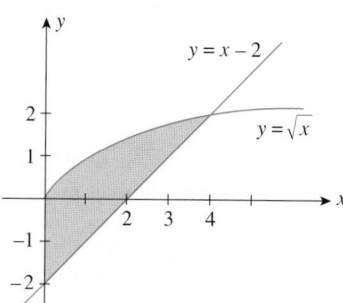

$y = x - 2$

$y = \sqrt{x}$

7.

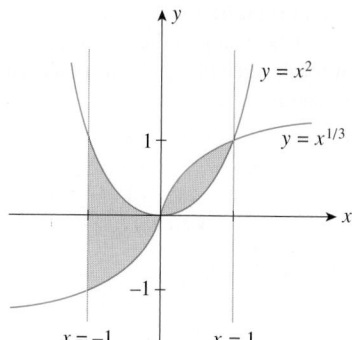

$y = x^2$

$y = x^{1/3}$

$x = -1$ $x = 1$

8.

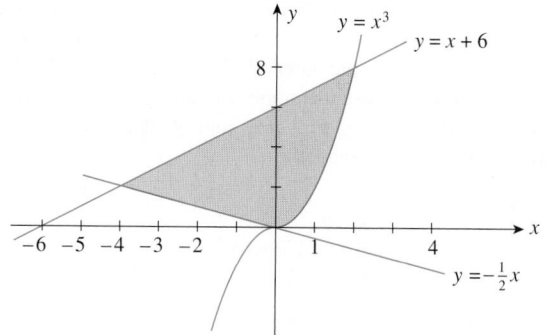

$y = x^3$

$y = x + 6$

$y = -\frac{1}{2}x$

In Exercises 9–16, sketch the graph and find the area of the region bounded below by the graph of each function and above by the *x*-axis from *x* = *a* to *x* = *b*.

9. $f(x) = -x^2$; $a = -1, b = 2$

10. $f(x) = x^2 - 4$; $a = -2, b = 2$

11. $f(x) = x^2 - 5x + 4$; $a = 1, b = 3$

12. $f(x) = x^3$; $a = -1, b = 0$

13. $f(x) = -1 - \sqrt{x}$; $a = 0, b = 9$

14. $f(x) = \frac{1}{2}x - \sqrt{x}$; $a = 0, b = 4$

15. $f(x) = -e^{(1/2)x}$; $a = -2, b = 4$

16. $f(x) = -xe^{-x^2}$; $a = 0, b = 1$

In Exercises 17–26, sketch the graphs of the functions *f* and *g* and find the area of the region enclosed by these graphs and the vertical lines *x* = *a* and *x* = *b*.

17. $f(x) = x^2 + 3, g(x) = 1$; $a = 1, b = 3$

18. $f(x) = x + 2, g(x) = x^2 - 4$; $a = -1, b = 2$

19. $f(x) = -x^2 + 2x + 3, g(x) = -x + 3$; $a = 0, b = 2$

20. $f(x) = 9 - x^2, g(x) = 2x + 3$; $a = -1, b = 1$

21. $f(x) = x^2 + 1, g(x) = \frac{1}{3}x^3$; $a = -1, b = 2$

22. $f(x) = \sqrt{x}, g(x) = -\frac{1}{2}x - 1$; $a = 1, b = 4$

23. $f(x) = \frac{1}{x}, g(x) = 2x - 1$; $a = 1, b = 4$

24. $f(x) = x^2, g(x) = \frac{1}{x^2}$; $a = 1, b = 3$

25. $f(x) = e^x, g(x) = \frac{1}{x}$; $a = 1, b = 2$

26. $f(x) = x, g(x) = e^{2x}$; $a = 1, b = 3$

In Exercises 27–34, sketch the graph and find the area of the region bounded by the graph of the function *f* and the lines *y* = 0, *x* = *a*, and *x* = *b*.

27. $f(x) = x$; $a = -1, b = 2$

28. $f(x) = x^2 - 2x$; $a = -1, b = 1$

29. $f(x) = -x^2 + 4x - 3$; $a = -1, b = 2$

30. $f(x) = x^3 - x^2$; $a = -1, b = 1$

31. $f(x) = x^3 - 4x^2 + 3x$; $a = 0, b = 2$

32. $f(x) = 4x^{1/3} + x^{4/3}$; $a = -1, b = 8$

33. $f(x) = e^x - 1$; $a = -1, b = 3$

34. $f(x) = xe^{x^2}$; $a = 0, b = 2$

In Exercises 35–42, sketch the graph and find the area of the region completely enclosed by the graphs of the given functions *f* and *g*.

35. $f(x) = x + 2$ and $g(x) = x^2 - 4$

36. $f(x) = -x^2 + 4x$ and $g(x) = 2x - 3$

37. $f(x) = x^2$ and $g(x) = x^3$

38. $f(x) = x^3 + 2x^2 - 3x$ and $g(x) = 0$

39. $f(x) = x^3 - 6x^2 + 9x$ and $g(x) = x^2 - 3x$

40. $f(x) = \sqrt{x}$ and $g(x) = x^2$

41. $f(x) = x\sqrt{9 - x^2}$ and $g(x) = 0$

42. $f(x) = 2x$ and $g(x) = x\sqrt{x + 1}$

43. EFFECT OF ADVERTISING ON REVENUE In the accompanying figure, the function f gives the rate of change of Odyssey Travel's revenue with respect to the amount x it spends on advertising with their current advertising agency. By engaging the services of a different advertising agency, it is expected that Odyssey's revenue will grow at the rate given by the function g. Give an interpretation of the area A of the region S and find an expression for A in terms of a definite integral involving f and g.

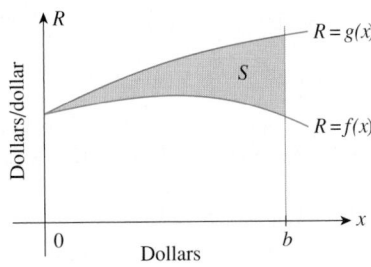

44. PULSE RATE DURING EXERCISE In the accompanying figure, the function f gives the rate of increase of an individual's pulse rate when he walked a prescribed course on a treadmill 6 mo ago. The function g gives the rate of increase of his pulse rate when he recently walked the same prescribed course. Give an interpretation of the area A of the region S and find an expression for A in terms of a definite integral involving f and g.

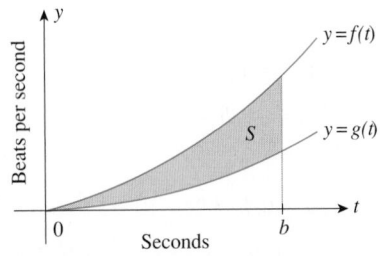

45. OIL PRODUCTION SHORTFALL Energy experts disagree about when global oil production will begin to decline. In the following figure, the function f gives the annual world oil production in billions of barrels from 1980 to 2050, according to the Department of Energy projection. The function g gives the world oil production in billions of barrels per year over the same period, according to longtime petroleum geologist Colin Campbell. Find an expression in terms of the definite integrals involving f and g, giving the shortfall in the total oil production over the period in question heeding Campbell's dire warnings.

Source: U.S. Department of Energy; Colin Campbell

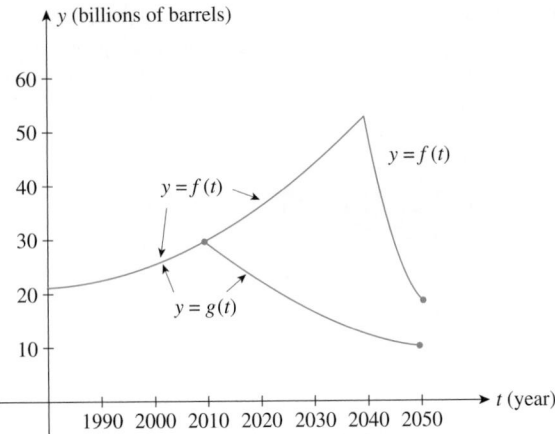

46. AIR PURIFICATION To study the effectiveness of air purifiers in removing smoke, engineers ran each purifier in a smoke-filled 10-ft × 20-ft room. In the accompanying figure, the function f gives the rate of change of the smoke level/minute, t min after the start of the test, when a brand A purifier is used. The function g gives the rate of change of the smoke level/minute when a brand B purifier is used.

a. Give an interpretation of the area of the region S.

b. Find an expression for the area of S in terms of a definite integral involving f and g.

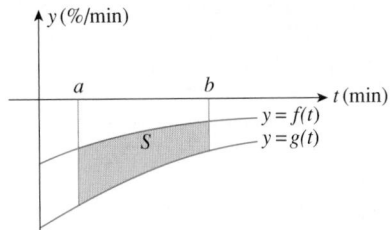

47. Two cars start out side by side and travel along a straight road. The velocity of car 1 is $f(t)$ ft/sec, the velocity of car 2 is $g(t)$ ft/sec over the interval $[0, T]$, and $0 < T_1 < T$. Furthermore, suppose the graphs of f and g are as depicted in the accompanying figure. Let A_1 and A_2 denote the areas of the regions (shown shaded).

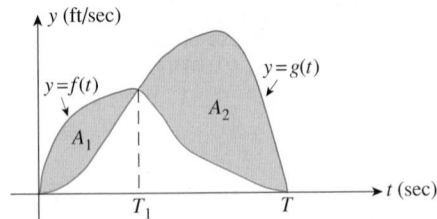

a. Write the number

$$\int_{T_1}^{T} [g(t) - f(t)]\, dt - \int_{0}^{T_1} [f(t) - g(t)]\, dt$$

in terms of A_1 and A_2.

b. What does the number obtained in part (a) represent?

48. The rate of change of the revenue of company A over the (time) interval $[0, T]$ is $f(t)$ dollars/week, whereas the rate of change of the revenue of company B over the same period is $g(t)$ dollars/week. The graphs of f and g are depicted in the accompanying figure. Find an expression in terms of definite integrals involving f and g giving the additional revenue that company B will have over company A in the period $[0, T]$.

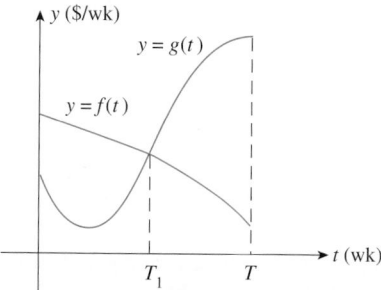

49. TURBO-CHARGED ENGINE VS. STANDARD ENGINE In tests conducted by *Auto Test Magazine* on two identical models of the Phoenix Elite—one equipped with a standard engine and the other with a turbo-charger—it was found that the acceleration of the former is given by

$$a = f(t) = 4 + 0.8t \qquad (0 \le t \le 12)$$

ft/sec/sec, t sec after starting from rest at full throttle, whereas the acceleration of the latter is given by

$$a = g(t) = 4 + 1.2t + 0.03t^2 \qquad (0 \le t \le 12)$$

ft/sec/sec. How much faster is the turbo-charged model moving than the model with the standard engine at the end of a 10-sec test run at full throttle?

50. ALTERNATIVE ENERGY SOURCES Because of the increasingly important role played by coal as a viable alternative energy source, the production of coal has been growing at the rate of

$$3.5e^{0.05t}$$

billion metric tons/year t yr from 1980 (which corresponds to $t = 0$). Had it not been for the energy crisis, the rate of production of coal since 1980 might have been only

$$3.5e^{0.01t}$$

billion metric tons/year t yr from 1980. Determine how much additional coal was produced between 1980 and the end of the century as an alternate energy source.

51. EFFECT OF TV ADVERTISING ON CAR SALES Carl Williams, the proprietor of Carl Williams Auto Sales, estimates that with extensive television advertising, car sales over the next several years could be increasing at the rate of

$$5e^{0.3t}$$

thousand cars/year t yr from now, instead of at the current rate of

$$(5 + 0.5t^{3/2})$$

thousand cars/year t yr from now. Find how many more cars Carl expects to sell over the next 5 yr by implementing his advertising plans.

52. POPULATION GROWTH In an endeavor to curb population growth in a Southeast Asian island state, the government has decided to launch an extensive propaganda campaign. Without curbs, the government expects the rate of population growth to have been

$$60e^{0.02t}$$

thousand people/year t yr from now, over the next 5 yr. However, successful implementation of the proposed campaign is expected to result in a population growth rate of

$$-t^2 + 60$$

thousand people/year t yr from now, over the next 5 yr. Assuming that the campaign is mounted, how many fewer people will there be in that country 5 yr from now then there would have been if no curbs had been imposed?

In Exercises 53 and 54, determine whether the statement is true or false. If it is true, explain why it is true. If it is false, give an example to show why it is false.

53. If f and g are continuous on $[a, b]$ and either $f(x) \ge g(x)$ for all x in $[a, b]$ or $f(x) \le g(x)$ for all x in $[a, b]$, then the area of the region bounded by the graphs of f and g and the vertical lines $x = a$ and $x = b$ is given by $\int_a^b |f(x) - g(x)|\, dx$.

54. The area of the region bounded by the graphs of $f(x) = 2 - x$ and $g(x) = 4 - x^2$ and the vertical lines $x = 0$ and $x = 2$ is given by $\int_0^2 [f(x) - g(x)]\, dx$.

55. Show that the area of a region R bounded above by the graph of a function f and below by the graph of a function g from $x = a$ to $x = b$ is given by

$$\int_a^b [f(x) - g(x)]\, dx$$

Hint: The validity of the formula was verified earlier for the case when both f and g were nonnegative. Now, let f and g be two functions such that $f(x) \ge g(x)$ for $a \le x \le b$. Then, there exists some nonnegative constant c such that the curves $y = f(x) + c$ and $y = g(x) + c$ are translated in the y-direction in such a way that the region R' has the same area as the region R (see the accompanying figures). Show that the area of R' is given by

$$\int_a^b \{[f(x) + c] - [g(x) + c]\}\, dx = \int_a^b [f(x) - g(x)]\, dx$$

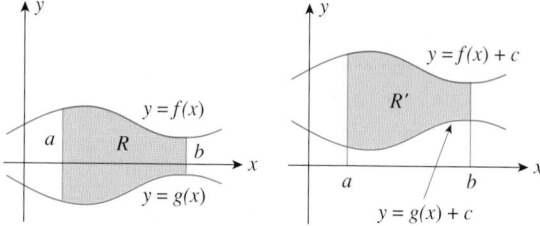

15.6 Solutions to Self-Check Exercises

1. The region in question is shown in the accompanying figure. Since the graph of the function f lies above that of the function g for $0 \le x \le 1$, we see that the required area is given by

$$\int_0^1 [(x^2 + 2) - (1 - x)]\,dx = \int_0^1 (x^2 + x + 1)\,dx$$
$$= \frac{1}{3}x^3 + \frac{1}{2}x^2 + x \Big|_0^1$$
$$= \frac{1}{3} + \frac{1}{2} + 1$$
$$= \frac{11}{6}$$

or $\frac{11}{6}$ square units.

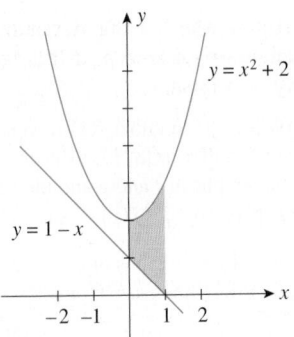

2. The region in question is shown in the accompanying figure. To find the points of intersection of the two curves, we solve the equations

$$-x^2 + 6x + 5 = x^2 + 5$$
$$2x^2 - 6x = 0$$
$$2x(x - 3) = 0$$

giving $x = 0$ or $x = 3$. Therefore, the points of intersection are $(0, 5)$ and $(3, 14)$.

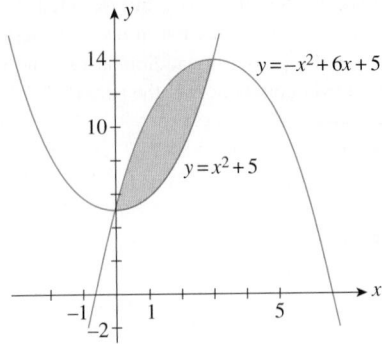

Since the graph of f always lies above that of g for $0 \le x \le 3$, we see that the required area is given by

$$\int_0^3 [(-x^2 + 6x + 5) - (x^2 + 5)]\,dx = \int_0^3 (-2x^2 + 6x)\,dx$$
$$= -\frac{2}{3}x^3 + 3x^2 \Big|_0^3$$
$$= -18 + 27$$
$$= 9$$

or 9 square units.

3. The additional profits realizable over the next 10 yr are given by

$$\int_0^{10} [(t - 2\sqrt{t} + 4) - (1 + t^{2/3})]\,dt$$
$$= \int_0^{10} (t - 2t^{1/2} + 3 - t^{2/3})\,dt$$
$$= \frac{1}{2}t^2 - \frac{4}{3}t^{3/2} + 3t - \frac{3}{5}t^{5/3} \Big|_0^{10}$$
$$= \frac{1}{2}(10)^2 - \frac{4}{3}(10)^{3/2} + 3(10) - \frac{3}{5}(10)^{5/3}$$
$$\approx 9.99$$

or approximately $10 million.

USING TECHNOLOGY

■ Finding the Area between Two Curves

The numerical integral operation can also be used to find the area between two curves. We do this by using the numerical integral operation to evaluate an appropriate definite integral or the sum (difference) of appropriate definite integrals. In the following example, the intersection operation is also used to advantage to help us find the limits of integration.

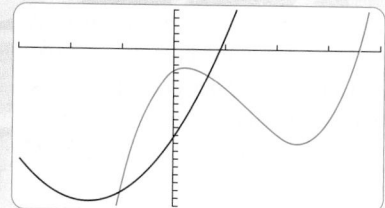

FIGURE T1
The region R is completely enclosed by the graphs of f and g.

EXAMPLE 1   Use a graphing utility to find the area of the smaller region R that is completely enclosed by the graphs of the functions

$$f(x) = 2x^3 - 8x^2 + 4x - 3 \quad \text{and} \quad g(x) = 3x^2 + 10x - 11$$

Solution   The graphs of f and g in the viewing window $[-3, 4] \times [-20, 5]$ are shown in Figure T1.

Using the intersection operation of a graphing utility, we find the x-coordinates of the points of intersection of the two graphs to be approximately -1.04 and 0.65, respectively. Since the graph of f lies above that of g on the interval $[-1.04, 0.65]$, we see that the area of R is given by

$$A = \int_{-1.04}^{0.65} [(2x^3 - 8x^2 + 4x - 3) - (3x^2 + 10x - 11)]\, dx$$

$$= \int_{-1.04}^{0.65} (2x^3 - 11x^2 - 6x + 8)\, dx$$

Using the numerical integral function of a graphing utility, we find $A \approx 9.87$, and so the area of R is approximately 9.87 square units. ■

TECHNOLOGY EXERCISES

In Exercises 1–6, (a) plot the graphs of the functions f and g and (b) find the area of the region enclosed by these graphs and the vertical lines $x = a$ and $x = b$. Express your answers accurate to four decimal places.

1. $f(x) = x^3(x - 5)^4$, $g(x) = 0$; $a = 1$, $b = 3$

2. $f(x) = x - \sqrt{1 - x^2}$, $g(x) = 0$; $a = -\dfrac{1}{2}$, $b = \dfrac{1}{2}$

3. $f(x) = x^{1/3}(x + 1)^{1/2}$, $g(x) = x^{-1}$; $a = 1.2$, $b = 2$

4. $f(x) = 2$, $g(x) = \ln(1 + x^2)$; $a = -1$, $b = 1$

5. $f(x) = \sqrt{x}$, $g(x) = \dfrac{x^2 - 3}{x^2 + 1}$; $a = 0$, $b = 3$

6. $f(x) = \dfrac{4}{x^2 + 1}$, $g(x) = x^4$; $a = -1$, $b = 1$

In Exercises 7–12, (a) plot the graphs of the functions f and g and (b) find the area of the region totally enclosed by the graphs of these functions.

7. $f(x) = 2x^3 - 8x^2 + 4x - 3$ and $g(x) = -3x^2 + 10x - 10$

8. $f(x) = x^4 - 2x^2 + 2$ and $g(x) = 4 - 2x^2$

9. $f(x) = 2x^3 - 3x^2 + x + 5$ and $g(x) = e^{2x} - 3$

10. $f(x) = \dfrac{1}{2}x^2 - 3$ and $g(x) = \ln x$

11. $f(x) = xe^{-x}$ and $g(x) = x - 2\sqrt{x}$

12. $f(x) = e^{-x^2}$ and $g(x) = x^4$

13. Refer to Example 1. Find the area of the larger region that is completely enclosed by the graphs of the functions f and g.

15.7 Applications of the Definite Integral to Business and Economics

In this section, we consider several applications of the definite integral in the fields of business and economics.

■ Consumers' and Producers' Surplus

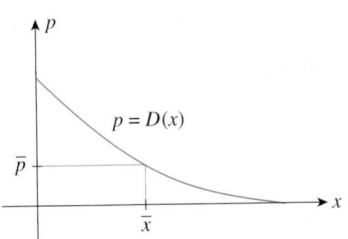

FIGURE 38
$D(x)$ is a demand function.

We begin by deriving a formula for computing the consumers' surplus. Suppose $p = D(x)$ is the demand function that relates the unit price p of a commodity to the quantity x demanded of it. Furthermore, suppose a fixed unit market price $\bar{p}$ has been established for the commodity and corresponding to this unit price the quantity demanded is $\bar{x}$ units (Figure 38). Then, those consumers who would be willing to pay a unit price higher than $\bar{p}$ for the commodity would in effect experience a savings. This difference between what the consumers *would* be willing to pay for $\bar{x}$ units of the commodity and what they *actually* pay for them is called the **consumers' surplus**.

To derive a formula for computing the consumers' surplus, divide the interval $[0, \bar{x}]$ into n subintervals, each of length $\Delta x = \bar{x}/n$, and denote the right endpoints of these subintervals by $x_1, x_2, \ldots, x_n = \bar{x}$ (Figure 39).

We observe in Figure 39 that there are consumers who would pay a unit price of at least $D(x_1)$ dollars for the first Δx units of the commodity instead of the market price of $\bar{p}$ dollars per unit. The savings to these consumers is approximated by

$$D(x_1)\Delta x - \bar{p}\Delta x = [D(x_1) - \bar{p}]\Delta x$$

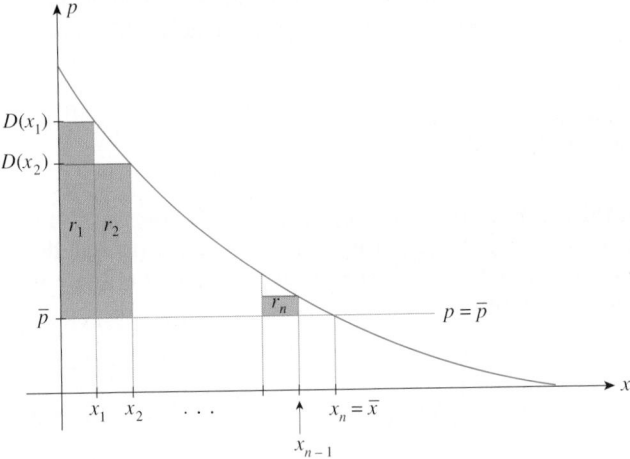

FIGURE 39
Approximating consumers' surplus by the sum of the rectangles $r_1, r_2, \ldots, r_n$

which is the area of the rectangle r_1. Pursuing the same line of reasoning, we find that the savings to the consumers who would be willing to pay a unit price of at least $D(x_2)$ dollars for the next Δx units (from x_1 through x_2) of the commodity, instead of the market price of $\bar{p}$ dollars per unit, is approximated by

$$D(x_2)\Delta x - \bar{p}\Delta x = [D(x_2) - \bar{p}]\Delta x$$

Continuing, we approximate the total savings to the consumers in purchasing $\bar{x}$ units of the commodity by the sum

$$[D(x_1) - \bar{p}]\Delta x + [D(x_2) - \bar{p}]\Delta x + \cdots + [D(x_n) - \bar{p}]\Delta x$$
$$= [D(x_1) + D(x_2) + \cdots + D(x_n)]\Delta x - \underbrace{[\bar{p}\Delta x + \bar{p}\Delta x + \cdots + \bar{p}\Delta x]}_{n \text{ terms}}$$

$$= [D(x_1) + D(x_2) + \cdots + D(x_n)]\Delta x - n\bar{p}\Delta x$$
$$= [D(x_1) + D(x_2) + \cdots + D(x_n)]\Delta x - \bar{p}\,\bar{x}$$

Now, the first term in the last expression is the Riemann sum of the demand function $p = D(x)$ over the interval $[0, \bar{x}]$ with representative points $x_1, x_2, \ldots, x_n$. Letting n approach infinity, we obtain the following formula for the consumers' surplus CS.

> **Consumers' Surplus**
> The consumers' surplus is given by
> $$CS = \int_0^{\bar{x}} D(x)\,dx - \bar{p}\,\bar{x} \qquad (16)$$
> where D is the demand function, $\bar{p}$ is the unit market price, and $\bar{x}$ is the quantity sold.

The consumer's surplus is given by the area of the region bounded above by the demand curve $p = D(x)$ and below by the straight line $p = \bar{p}$ from $x = 0$ to $x = \bar{x}$ (Figure 40). We can also see this if we rewrite Equation (16) in the form

$$\int_0^{\bar{x}} [D(x) - \bar{p}]\,dx$$

and interpret the result geometrically.

Analogously, we can derive a formula for computing the producers' surplus. Suppose $p = S(x)$ is the supply equation that relates the unit price p of a certain commodity to the quantity x that the supplier will make available in the market at that price.

Again, suppose a fixed market price $\bar{p}$ has been established for the commodity and, corresponding to this unit price, a quantity of $\bar{x}$ units will be made available in the market by the supplier (Figure 41). Then, the suppliers who would be willing to make the commodity available at a lower price stand to gain from the fact that the market price is set as such. The difference between what the suppliers actually receive and what they would be willing to receive is called the **producers' surplus**. Proceeding in a manner similar to the derivation of the equation for computing the consumers' surplus, we find that the producers' surplus PS is defined as follows:

> **Producers' Surplus**
> The producers' surplus is given by
> $$PS = \bar{p}\,\bar{x} - \int_0^{\bar{x}} S(x)\,dx \qquad (17)$$
> where S is the supply function, $\bar{p}$ is the unit market price, and $\bar{x}$ is the quantity supplied.

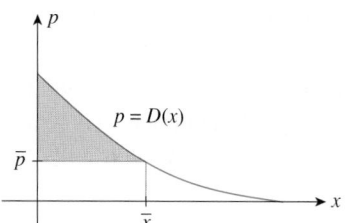

FIGURE 40
Consumers' surplus

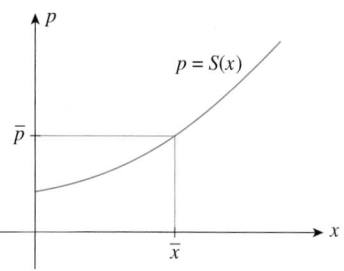

FIGURE 41
$S(x)$ is a supply function.

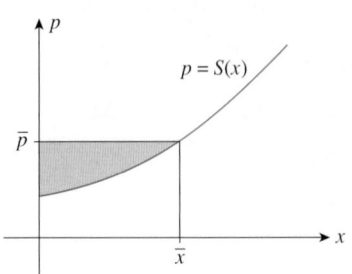

FIGURE 42
Producers' surplus

Geometrically, the producers' surplus is given by the area of the region bounded above by the straight line $p = \bar{p}$ and below by the supply curve $p = S(x)$ from $x = 0$ to $x = \bar{x}$ (Figure 42).

We can also show that the last statement is true by converting Equation (17) to the form

$$\int_0^{\bar{x}} [\bar{p} - S(x)]\, dx$$

and interpreting the definite integral geometrically.

APPLIED EXAMPLE 1   Consumers' and Producers' Surplus   The demand function for a certain make of 10-speed bicycle is given by

$$p = D(x) = -0.001x^2 + 250$$

where p is the unit price in dollars and x is the quantity demanded in units of a thousand. The supply function for these bicycles is given by

$$p = S(x) = 0.0006x^2 + 0.02x + 100$$

where p stands for the unit price in dollars and x stands for the number of bicycles that the supplier will put on the market, in units of a thousand. Determine the consumers' surplus and the producers' surplus if the market price of a bicycle is set at the equilibrium price.

Solution   Recall that the equilibrium price is the unit price of the commodity when market equilibrium occurs. We determine the equilibrium price by solving for the point of intersection of the demand curve and the supply curve (Figure 43). To solve the system of equations

$$p = -0.001x^2 \qquad\quad + 250$$
$$p = 0.0006x^2 + 0.02x + 100$$

we simply substitute the first equation into the second, obtaining

$$0.0006x^2 + 0.02x + 100 = -0.001x^2 + 250$$
$$0.0016x^2 + 0.02x - 150 = 0$$
$$16x^2 + 200x - 1,500,000 = 0$$
$$2x^2 + 25x - 187,500 = 0$$

Factoring this last equation, we obtain

$$(2x + 625)(x - 300) = 0$$

Thus, $x = -625/2$ or $x = 300$. The first number lies outside the interval of interest, so we are left with the solution $x = 300$, with a corresponding value of

$$p = -0.001(300)^2 + 250 = 160$$

Thus, the equilibrium point is (300, 160); that is, the equilibrium quantity is 300,000, and the equilibrium price is $160. Setting the market price at $160 per unit and using Formula (16) with $\bar{p} = 160$ and $\bar{x} = 300$, we find that the

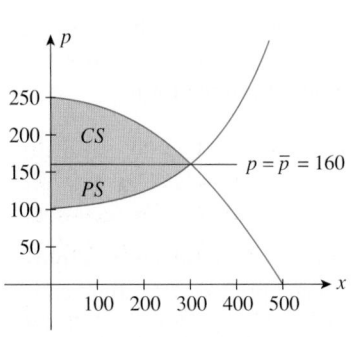

FIGURE 43
Consumers' surplus and producers' surplus when market price = equilibrium price

consumers' surplus is given by

$$CS = \int_0^{300} (-0.001x^2 + 250)\,dx - (160)(300)$$

$$= \left(-\frac{1}{3000}x^3 + 250x\right)\Big|_0^{300} - 48{,}000$$

$$= -\frac{300^3}{3000} + (250)(300) - 48{,}000$$

$$= 18{,}000$$

or $18,000,000. (Recall that x is measured in units of a thousand.) Next, using (17), we find that the producers' surplus is given by

$$PS = (160)(300) - \int_0^{300} (0.0006x^2 + 0.02x + 100)\,dx$$

$$= 48{,}000 - (0.0002x^3 + 0.01x^2 + 100x)\Big|_0^{300}$$

$$= 48{,}000 - [(0.0002)(300)^3 + (0.01)(300)^2 + 100(300)]$$

$$= 11{,}700$$

or $11,700,000. ∎

The Future and Present Value of an Income Stream

Suppose a firm generates a stream of income over a period of time—for example, the revenue generated by a large chain of retail stores over a 5-year period. As the income is realized, it is reinvested and earns interest at a fixed rate. The **accumulated future income stream** over the 5-year period is the amount of money the firm ends up with at the end of that period.

The definite integral can be used to determine this accumulated, or total, future income stream over a period of time. The total future value of an income stream gives us a way to measure the value of such a stream. To find the **total future value of an income stream**, suppose

$R(t) =$ Rate of income generation at any time t Dollars per year

$r =$ Interest rate compounded continuously

$T =$ Term   In years

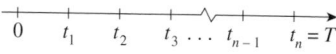

FIGURE 44
The time interval [0, T] is partitioned into n subintervals.

Let's divide the time interval $[0, T]$ into n subintervals of equal length $\Delta t = T/n$ and denote the right endpoints of these intervals by $t_1, t_2, \ldots, t_n = T$, as shown in Figure 44.

If R is a continuous function on $[0, T]$, then $R(t)$ will not differ by much from $R(t_1)$ in the subinterval $[0, t_1]$ provided that the subinterval is small (which is true if n is large). Therefore, the income generated over the time interval $[0, t_1]$ is approximately

$$R(t_1)\Delta t \qquad \text{Constant rate of income} \times \text{Length of time}$$

dollars. The future value of this amount, T years from now, calculated as if it were earned at time t_1, is

$$[R(t_1)\Delta t]e^{r(T-t_1)} \qquad \text{Equation (5), Section 5.1}$$

dollars. Similarly, the income generated over the time interval $[t_1, t_2]$ is approximately $P(t_2)\Delta t$ dollars and has a future value, T years from now, of approximately

$$[R(t_2)\Delta t]e^{r(T-t_2)}$$

dollars. Therefore, the sum of the future values of the income stream generated over the time interval $[0, T]$ is approximately

$$R(t_1)e^{r(T-t_1)}\Delta t + R(t_2)e^{r(T-t_2)}\Delta t + \cdots + R(t_n)e^{r(T-t_n)}\Delta t$$
$$= e^{rT}[R(t_1)e^{-rt_1}\Delta t + R(t_2)e^{-rt_2}\Delta t + \cdots + R(t_n)e^{-rt_n}\Delta t]$$

dollars. But this sum is just the Riemann sum of the function $e^{rT}R(t)e^{-rt}$ over the interval $[0, T]$ with representative points $t_1, t_2, \ldots, t_n$. Letting n approach infinity, we obtain the following result.

Accumulated or Total Future Value of an Income Stream

The accumulated, or total, future value after T years of an income stream of $R(t)$ dollars per year, earning interest at the rate of r per year compounded continuously, is given by

$$A = e^{rT}\int_0^T R(t)e^{-rt}\,dt \qquad \textbf{(18)}$$

APPLIED EXAMPLE 2   Income Stream   Crystal Car Wash recently bought an automatic car-washing machine that is expected to generate $40,000 in revenue per year, t years from now, for the next 5 years. If the income is reinvested in a business earning interest at the rate of 12% per year compounded continuously, find the total accumulated value of this income stream at the end of 5 years.

Solution   We are required to find the total future value of the given income stream after 5 years. Using Equation (18) with $R(t) = 40,000$, $r = 0.12$, and $T = 5$, we see that the required value is given by

$$e^{0.12(5)}\int_0^5 40,000e^{-0.12t}\,dt$$

$$= e^{0.6}\left[-\frac{40,000}{0.12}e^{-0.12t}\right]\Big|_0^5 \qquad \begin{array}{l}\text{Integrate using the}\\ \text{substitution } u = -0.12t.\end{array}$$

$$= -\frac{40,000e^{0.6}}{0.12}\left(e^{-0.6} - 1\right) \approx 274,039.60$$

or approximately $274,040.   ∎

Another way of measuring the value of an income stream is by considering its present value. The **present value of an income stream** of $R(t)$ dollars per year over a term of T years, earning interest at the rate of r per year compounded continuously, is the principal P that will yield the same accumulated value as the income stream itself when P is invested today for a period of T years at the same rate of interest.

In other words,

$$Pe^{rT} = e^{rT} \int_0^T R(t)e^{-rt}\, dt$$

Dividing both sides of the equation by e^{rT} gives the following result.

Present Value of an Income Stream
The present value of an income stream of $R(t)$ dollars per year, earning interest at the rate of r per year compounded continuously, is given by

$$PV = \int_0^T R(t)e^{-rt}\, dt \qquad\qquad \textbf{(19)}$$

APPLIED EXAMPLE 3   Investment Analysis The owner of a local cinema is considering two alternative plans for renovating and improving the theater. Plan A calls for an immediate cash outlay of $250,000, whereas plan B requires an immediate cash outlay of $180,000. It has been estimated that adopting plan A would result in a net income stream generated at the rate of

$$f(t) = 630{,}000$$

dollars per year, whereas adopting plan B would result in a net income stream generated at the rate of

$$g(t) = 580{,}000$$

dollars per year for the next 3 years. If the prevailing interest rate for the next 5 years is 10% per year, which plan will generate a higher net income by the end of 3 years?

Solution Since the initial outlay is $250,000, we find—using Equation (19) with $R(t) = 630{,}000$, $r = 0.1$, and $T = 3$—that the present value of the net income under plan A is given by

$$\int_0^3 630{,}000 e^{-0.1t}\, dt - 250{,}000$$

$$= \frac{630{,}000}{-0.1} e^{-0.1t} \Big|_0^3 - 250{,}000 \qquad \text{\small Integrate using the substitution } u = -0.1t.$$

$$= -6{,}300{,}000 e^{-0.3} + 6{,}300{,}000 - 250{,}000$$

$$\approx 1{,}382{,}845$$

or approximately $1,382,845.

To find the present value of the net income under plan B, we use (19) with $R(t) = 580{,}000$, $r = 0.1$, and $T = 3$, obtaining

$$\int_0^3 580{,}000 e^{-0.1t}\, dt - 180{,}000$$

dollars. Proceeding as in the previous computation, we see that the required value is $1,323,254 (see Exercise 8, page 988).

Comparing the present value of each plan, we conclude that plan A would generate a higher net income by the end of 3 years. ∎

Note The function R in Example 3 is a constant function. If R is not a constant function, then we may need more sophisticated techniques of integration to evaluate the integral in (19). Exercises 16.1 and 16.2 contain problems of this type. ∎

The Amount and Present Value of an Annuity

An annuity is a sequence of payments made at regular time intervals. The time period in which these payments are made is called the *term* of the annuity. Although the payments need not be equal in size, they are equal in many important applications, and we will assume that they are equal in our discussion. Examples of annuities are regular deposits to a savings account, monthly home mortgage payments, and monthly insurance payments.

The **amount of an annuity** is the sum of the payments plus the interest earned. A formula for computing the amount of an annuity A can be derived with the help of (18). Let

$$P = \text{Size of each payment in the annuity}$$
$$r = \text{Interest rate compounded continuously}$$
$$T = \text{Term of the annuity (in years)}$$
$$m = \text{Number of payments per year}$$

The payments into the annuity constitute a constant income stream of $R(t) = mP$ dollars per year. With this value of $R(t)$, (18) yields

$$A = e^{rT} \int_0^T R(t)e^{-rt}\,dt = e^{rT} \int_0^T mPe^{-rt}\,dt$$

$$= mPe^{rT}\left[-\frac{e^{-rt}}{r} \right]\Bigg|_0^T = mPe^{rT}\left[-\frac{e^{-rT}}{r} + \frac{1}{r} \right]$$

$$= \frac{mP}{r}(e^{rT} - 1) \qquad \text{Since } e^{rT} \cdot e^{-rT} = 1$$

This leads us to the following formula.

Amount of an Annuity
The amount of an annuity is

$$A = \frac{mP}{r}(e^{rT} - 1) \tag{20}$$

where P, r, T, and m are as defined earlier.

APPLIED EXAMPLE 4 IRAs On January 1, 1990, Marcus Chapman deposited $2000 into an Individual Retirement Account (IRA) paying interest at the rate of 10% per year compounded continuously. Assuming that he deposits $2000 annually into the account, how much will he have in his IRA at the beginning of the year 2006?

Solution We use (20), with $P = 2000$, $r = 0.1$, $T = 16$, and $m = 1$, obtaining

$$A = \frac{2000}{0.1}(e^{1.6} - 1)$$

$$\approx 79{,}060.65$$

Thus, Marcus will have approximately $79,061 in his account at the beginning of the year 2006.

 EXPLORING WITH TECHNOLOGY

Refer to Example 4. Suppose Marcus wishes to know how much he will have in his IRA at any time in the future, not just at the beginning of 2006, as you were asked to compute in the example.

1. Using Formula (18) and the relevant data from Example 4, show that the required amount at any time x (x measured in years, $x > 0$) is given by

$$A = f(x) = 20{,}000(e^{0.1x} - 1)$$

2. Use a graphing utility to plot the graph of f, using the viewing window $[0, 30] \times [2000, 400{,}000]$.

3. Using ZOOM and TRACE, or using the function evaluation capability of your graphing utility, use the result of part 2 to verify the result obtained in Example 4. Comment on the advantage of the mathematical model found in part 1.

Using (19), we can derive the following formula for the present value of an annuity.

Present Value of an Annuity

The present value of an annuity is given by

$$PV = \frac{mP}{r}\left(1 - e^{-rT}\right) \tag{21}$$

where P, r, T, and m are as defined earlier.

APPLIED EXAMPLE 5   Sinking Funds   Tomas Perez, the proprietor of a hardware store, wants to establish a fund from which he will withdraw $1000 per month for the next 10 years. If the fund earns interest at the rate of 9% per year compounded continuously, how much money does he need to establish the fund?

Solution   We want to find the present value of an annuity with $P = 1000$, $r = 0.09$, $T = 10$, and $m = 12$. Using Equation (21), we find

$$PV = \frac{12{,}000}{0.09}\left(1 - e^{-(0.09)(10)}\right)$$

$$\approx 79{,}124.05$$

Thus, Tomas needs approximately $79,124 to establish the fund.

Lorentz Curves and Income Distributions

One method used by economists to study the distribution of income in a society is based on the **Lorentz curve**, named after American statistician M. D. Lorentz. To describe the Lorentz curve, let $f(x)$ denote the proportion of the total income received by the poorest $100x\%$ of the population for $0 \leq x \leq 1$. Using this terminology, $f(0.3) = 0.1$ simply states that the lowest 30% of the income recipients receive 10% of the total income.

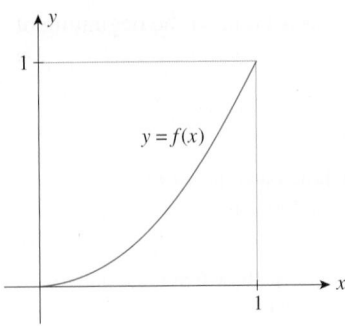

FIGURE 45
A Lorentz curve

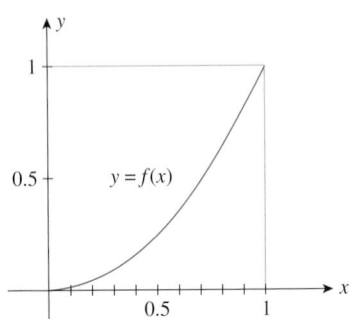

FIGURE 46
The Lorentz curve $f(x) = \dfrac{19}{20}x^2 + \dfrac{1}{20}x$

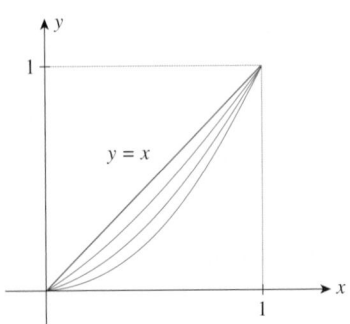

FIGURE 47
The closer the Lorentz curve is to the line, the more equitable the income distribution.

The function f has the following properties:

1. The domain of f is $[0, 1]$.
2. The range of f is $[0, 1]$.
3. $f(0) = 0$ and $f(1) = 1$.
4. $f(x) \le x$ for every x in $[0, 1]$.
5. f is increasing on $[0, 1]$.

The first two properties follow from the fact that both x and $f(x)$ are fractions of a whole. Property 3 is a statement that 0% of the income recipients receive 0% of the total income and 100% of the income recipients receive 100% of the total income. Property 4 follows from the fact that the lowest $100x\%$ of the income recipients cannot receive more than $100x\%$ of the total income. A typical Lorentz curve is shown in Figure 45.

 APPLIED EXAMPLE 6 Lorentz Curves A developing country's income distribution is described by the function

$$f(x) = \frac{19}{20}x^2 + \frac{1}{20}x$$

a. Sketch the Lorentz curve for the given function.
b. Compute $f(0.2)$ and $f(0.8)$ and interpret your results.

Solution

a. The Lorentz curve is shown in Figure 46.

b.
$$f(0.2) = \frac{19}{20}(0.2)^2 + \frac{1}{20}(0.2) = 0.048$$

Thus, the lowest 20% of the people receive 4.8% of the total income.

$$f(0.8) = \frac{19}{20}(0.8)^2 + \frac{1}{20}(0.8) = 0.648$$

Thus, the lowest 80% of the people receive 64.8% of the total income.

Next, let's consider the Lorentz curve described by the function $y = f(x) = x$. Since exactly $100x\%$ of the total income is received by the lowest $100x\%$ of income recipients, the line $y = x$ is called the **line of complete equality**. For example, 10% of the total income is received by the lowest 10% of income recipients, 20% of the total income is received by the lowest 20% of income recipients, and so on. Now, it is evident that the closer a Lorentz curve is to this line, the more equitable the income distribution is among the income recipients. But the proximity of a Lorentz curve to the line of complete equality is reflected by the area between the Lorentz curve and the line $y = x$ (Figure 47). The closer the curve is to the line, the smaller the enclosed area.

This observation suggests that we may define a number, called the coefficient of inequality of a Lorentz curve, as the ratio of the area between the line of complete equality and the Lorentz curve to the area under the line of complete equality. Since the area under the line of complete equality is $\frac{1}{2}$, we see that the coefficient of inequality is given by the following formula.

> **Coefficient of Inequality of a Lorentz Curve**
> The coefficient of inequality, or **Gini index**, of a Lorentz curve is
>
> $$L = 2 \int_0^1 [x - f(x)]\, dx \tag{22}$$

The coefficient of inequality is a number between 0 and 1. For example, a coefficient of zero implies that the income distribution is perfectly uniform.

APPLIED EXAMPLE 7   Income Distributions   In a study conducted by a certain country's Economic Development Board with regard to the income distribution of certain segments of the country's workforce, it was found that the Lorentz curves for the distribution of income of medical doctors and of movie actors are described by the functions

$$f(x) = \frac{14}{15}x^2 + \frac{1}{15}x \quad \text{and} \quad g(x) = \frac{5}{8}x^4 + \frac{3}{8}x$$

respectively. Compute the coefficient of inequality for each Lorentz curve. Which profession has a more equitable income distribution?

Solution   The required coefficients of inequality are, respectively,

$$L_1 = 2\int_0^1 \left[x - \left(\frac{14}{15}x^2 + \frac{1}{15}x\right)\right] dx = 2\int_0^1 \left(\frac{14}{15}x - \frac{14}{15}x^2\right) dx$$

$$= \frac{28}{15}\int_0^1 (x - x^2)\, dx = \frac{28}{15}\left(\frac{1}{2}x^2 - \frac{1}{3}x^3\right)\Big|_0^1$$

$$= \frac{14}{45} \approx 0.311$$

$$L_2 = 2\int_0^1 \left[x - \left(\frac{5}{8}x^4 + \frac{3}{8}x\right)\right] dx = 2\int_0^1 \left(\frac{5}{8}x - \frac{5}{8}x^4\right) dx$$

$$= \frac{5}{4}\int_0^1 (x - x^4)\, dx = \frac{5}{4}\left(\frac{1}{2}x^2 - \frac{1}{5}x^5\right)\Big|_0^1$$

$$= \frac{15}{40} \approx 0.375$$

We conclude that in this country the incomes of medical doctors are more evenly distributed than the incomes of movie actors.   ■

15.7   Self-Check Exercise

The demand function for a certain make of exercise bicycle that is sold exclusively through cable television is

$$p = d(x) = \sqrt{9 - 0.02x}$$

where p is the unit price in hundreds of dollars and x is the quantity demanded/week. The corresponding supply function is given by

$$p = s(x) = \sqrt{1 + 0.02x}$$

where p has the same meaning as before and x is the number of exercise bicycles the supplier will make available at price p. Determine the consumers' surplus and the producers' surplus if the unit price is set at the equilibrium price.

The solution to Self-Check Exercise 15.7 can be found on page 990.

15.7   Concept Questions

1. a. Define consumers' surplus. Give a formula for computing it.

b. Define producers' surplus. Give a formula for computing it.

2. a. Define the accumulated (future) value of an income stream. Give a formula for computing it.

b. Define the present value of an income stream. Give a formula for computing it.

3. Define the amount of an annuity. Give a formula for computing it.

4. Explain the following terms: (a) Lorentz curve (b) Coefficient of inequality of a Lorentz curve.

15.7   Exercises

1. CONSUMERS' SURPLUS   The demand function for a certain make of replacement cartridges for a water purifier is given by

$$p = -0.01x^2 - 0.1x + 6$$

where p is the unit price in dollars and x is the quantity demanded each week, measured in units of a thousand. Determine the consumers' surplus if the market price is set at \$4/cartridge.

2. CONSUMERS' SURPLUS   The demand function for a certain brand of CD is given by

$$p = -0.01x^2 - 0.2x + 8$$

where p is the wholesale unit price in dollars and x is the quantity demanded each week, measured in units of a thousand. Determine the consumers' surplus if the wholesale market price is set at \$5/disc.

3. CONSUMERS' SURPLUS   It is known that the quantity demanded of a certain make of portable hair dryer is x hundred units/week and the corresponding wholesale unit price is

$$p = \sqrt{225 - 5x}$$

dollars. Determine the consumers' surplus if the wholesale market price is set at \$10/unit.

4. PRODUCERS' SURPLUS   The supplier of the portable hair dryers in Exercise 3 will make x hundred units of hair dryers available in the market when the wholesale unit price is

$$p = \sqrt{36 + 1.8x}$$

dollars. Determine the producers' surplus if the wholesale market price is set at \$9/unit.

5. PRODUCERS' SURPLUS   The supply function for the CDs of Exercise 2 is given by

$$p = 0.01x^2 + 0.1x + 3$$

where p is the unit wholesale price in dollars and x stands for the quantity that will be made available in the market by the supplier, measured in units of a thousand. Determine the producers' surplus if the wholesale market price is set at the equilibrium price.

6. CONSUMERS' AND PRODUCERS' SURPLUS   The management of the Titan Tire Company has determined that the quantity demanded x of their Super Titan tires/week is related to the unit price p by the relation

$$p = 144 - x^2$$

where p is measured in dollars and x is measured in units of a thousand. Titan will make x units of the tires available in the market if the unit price is

$$p = 48 + \frac{1}{2}x^2$$

dollars. Determine the consumers' surplus and the producers' surplus when the market unit price is set at the equilibrium price.

7. CONSUMERS' AND PRODUCERS' SURPLUS   The quantity demanded x (in units of a hundred) of the Mikado miniature cameras/week is related to the unit price p (in dollars) by

$$p = -0.2x^2 + 80$$

and the quantity x (in units of a hundred) that the supplier is willing to make available in the market is related to the unit price p (in dollars) by

$$p = 0.1x^2 + x + 40$$

If the market price is set at the equilibrium price, find the consumers' surplus and the producers' surplus.

8. Refer to Example 3, page 983. Verify that

$$\int_0^3 580{,}000e^{-0.1t}\, dt - 180{,}000 \approx 1{,}323{,}254$$

9. PRESENT VALUE OF AN INVESTMENT   Suppose an investment is expected to generate income at the rate of

$$R(t) = 200,000$$

dollars/year for the next 5 yr. Find the present value of this investment if the prevailing interest rate is 8%/year compounded continuously.

10. FRANCHISES   Camille purchased a 15-yr franchise for a computer outlet store that is expected to generate income at the rate of

$$R(t) = 400,000$$

dollars/year. If the prevailing interest rate is 10%/year compounded continuously, find the present value of the franchise.

11. THE AMOUNT OF AN ANNUITY   Find the amount of an annuity if $250/month is paid into it for a period of 20 yr, earning interest at the rate of 8%/year compounded continuously.

12. THE AMOUNT OF AN ANNUITY   Find the amount of an annuity if $400/month is paid into it for a period of 20 yr, earning interest at the rate of 8%/year compounded continuously.

13. THE AMOUNT OF AN ANNUITY   Aiso deposits $150/month in a savings account paying 8%/year compounded continuously. Estimate the amount that will be in his account after 15 yr.

14. CUSTODIAL ACCOUNTS   The Armstrongs wish to establish a custodial account to finance their children's education. If they deposit $200 monthly for 10 yr in a savings account paying 9%/year compounded continuously, how much will their savings account be worth at the end of this period?

15. IRA ACCOUNTS   Refer to Example 4, page 984. Suppose Marcus makes his IRA payment on April 1, 1990, and annually thereafter. If interest is paid at the same initial rate, approximately how much will Marcus have in his account at the beginning of the year 2006?

16. PRESENT VALUE OF AN ANNUITY   Estimate the present value of an annuity if payments are $800 monthly for 12 yr and the account earns interest at the rate of 10%/year compounded continuously.

17. PRESENT VALUE OF AN ANNUITY   Estimate the present value of an annuity if payments are $1200 monthly for 15 yr and the account earns interest at the rate of 10%/year compounded continuously.

18. LOTTERY PAYMENTS   A state lottery commission pays the winner of the "Million Dollar" lottery 20 annual installments of $50,000 each. If the prevailing interest rate is 8%/year compounded continuously, find the present value of the winning ticket.

19. REVERSE ANNUITY MORTGAGES   Sinclair wishes to supplement his retirement income by $300/month for the next 10 yr. He plans to obtain a reverse annuity mortgage (RAM) on his home to meet this need. Estimate the amount of the mortgage he will require if the prevailing interest rate is 12%/year compounded continuously.

20. REVERSE ANNUITY MORTGAGE   Refer to Exercise 19. Leah wishes to supplement her retirement income by $400/month for the next 15 yr by obtaining a RAM. Estimate the amount of the mortgage she will require if the prevailing interest rate is 9%/year compounded continuously.

21. LORENTZ CURVES   A certain country's income distribution is described by the function

$$f(x) = \frac{15}{16}x^2 + \frac{1}{16}x$$

a. Sketch the Lorentz curve for this function.
b. Compute $f(0.4)$ and $f(0.9)$ and interpret your results.

22. LORENTZ CURVES   In a study conducted by a certain country's Economic Development Board, it was found that the Lorentz curve for the distribution of income of college teachers was described by the function

$$f(x) = \frac{13}{14}x^2 + \frac{1}{14}x$$

and that of lawyers by the function

$$g(x) = \frac{9}{11}x^4 + \frac{2}{11}x$$

a. Compute the coefficient of inequality for each Lorentz curve.
b. Which profession has a more equitable income distribution?

23. LORENTZ CURVES   A certain country's income distribution is described by the function

$$f(x) = \frac{14}{15}x^2 + \frac{1}{15}x$$

a. Sketch the Lorentz curve for this function.
b. Compute $f(0.3)$ and $f(0.7)$.

24. LORENTZ CURVES   In a study conducted by a certain country's Economic Development Board, it was found that the Lorentz curve for the distribution of income of stockbrokers was described by the function

$$f(x) = \frac{11}{12}x^2 + \frac{1}{12}x$$

and that of high school teachers by the function

$$g(x) = \frac{5}{6}x^2 + \frac{1}{6}x$$

a. Compute the coefficient of inequality for each Lorentz curve.
b. Which profession has a more equitable income distribution?

15.7   Solution to Self-Check Exercise

We find the equilibrium price and equilibrium quantity by solving the system of equations

$$p = \sqrt{9 - 0.02x}$$
$$p = \sqrt{1 + 0.02x}$$

simultaneously. Substituting the first equation into the second, we have

$$\sqrt{9 - 0.02x} = \sqrt{1 + 0.02x}$$

Squaring both sides of the equation then leads to

$$9 - 0.02x = 1 + 0.02x$$
$$x = 200$$

Therefore,

$$p = \sqrt{9 - 0.02(200)}$$
$$= \sqrt{5} \approx 2.24$$

The equilibrium price is $224, and the equilibrium quantity is 200. The consumers' surplus is given by

$$CS = \int_0^{200} \sqrt{9 - 0.02x}\, dx - (2.24)(200)$$

$$= \int_0^{200} (9 - 0.02x)^{1/2}\, dx - 448$$

$$= -\frac{1}{0.02}\left(\frac{2}{3}\right)(9 - 0.02x)^{3/2}\Big|_0^{200} - 448 \qquad \text{Integrate by substitution.}$$

$$= -\frac{1}{0.03}(5^{3/2} - 9^{3/2}) - 448$$

$$\approx 79.32$$

or approximately $7932.

Next, the producers' surplus is given by

$$PS = (2.24)(200) - \int_0^{200} \sqrt{1 + 0.02x}\, dx$$

$$= 448 - \int_0^{200} (1 + 0.02x)^{1/2}\, dx$$

$$= 448 - \frac{1}{0.02}\left(\frac{2}{3}\right)(1 + 0.02x)^{3/2}\Big|_0^{200}$$

$$= 448 - \frac{1}{0.03}(5^{3/2} - 1)$$

$$\approx 108.66$$

or approximately $10,866.

USING TECHNOLOGY

■ Business and Economic Applications/Technology Exercises

1. Re-solve Example 1, Section 15.7, using a graphing utility.
   Hint: Use the intersection operation to find the equilibrium quantity and the equilibrium price. Use the numerical integral operation to evaluate the definite integral.

2. Re-solve Exercise 7, Section 15.7, using a graphing utility.
   Hint: See Exercise 1.

3. The demand function for a certain brand of travel alarm clocks is given by

$$p = -0.01x^2 - 0.3x + 10$$

where p is the wholesale unit price in dollars and x is the quantity demanded each month, measured in units of a thousand. The supply function for this brand of clocks is given by

$$p = -0.01x^2 + 0.2x + 4$$

where p has the same meaning as before and x is the quantity, in thousands, the supplier will make available in the marketplace per month. Determine the consumers' surplus and the producers' surplus when the market unit price is set at the equilibrium price.

4. The quantity demanded of a certain make of compact disc organizer is x thousand units per week, and the corresponding wholesale unit price is

$$p = \sqrt{400 - 8x}$$

dollars. The supplier of the organizers will make x thousand units available in the market when the unit wholesale price is

$$p = 0.02x^2 + 0.04x + 5$$

dollars. Determine the consumers' surplus and the producers' surplus when the market unit price is set at the equilibrium price.

5. Investment A is expected to generate income at the rate of

$$R_1(t) = 50,000 + 10,000\sqrt{t}$$

dollars/year for the next 5 years and investment B is expected to generate income at the rate of

$$R_2(t) = 50,000 + 6000t$$

dollars/year over the same period of time. If the prevailing interest rate for the next 5 years is 10%/year, which investment will generate a higher net income by the end of 5 years?

6. Investment A is expected to generate income at the rate of

$$R_1(t) = 40,000 + 5000t + 100t^2$$

dollars/year for the next 10 years and investment B is expected to generate income at the rate of

$$R_2(t) = 60,000 + 2000t$$

dollars/year over the same period of time. If the prevailing interest rate for the next 10 years is 8%/year, which investment will generate a higher net income by the end of 10 years?

CHAPTER 15 **Summary of Principal Formulas and Terms**

www **FORMULAS**

1. Indefinite integral of a constant	$\int k\, du = ku + C$
2. Power rule	$\int u^n\, du = \dfrac{u^{n+1}}{n+1} + C \quad (n \neq -1)$
3. Constant multiple rule	$\int k f(u)\, du = k \int f(u)\, du$ (k, a constant)
4. Sum rule	$\int [f(u) \pm g(u)]\, du$ $= \int f(u)\, du \pm \int g(u)\, du$

5. Indefinite integral of the exponential function	$\int e^u \, du = e^u + C$
6. Indefinite integral of $f(u) = \dfrac{1}{u}$	$\int \dfrac{du}{u} = \ln|u| + C$
7. Method of substitution	$\int f'(g(x))g'(x) \, dx = \int f'(u) \, du$
8. Definite integral as the limit of a sum	$\int_a^b f(x) \, dx = \lim_{n \to \infty} S_n,$ where S_n is a Riemann sum
9. Fundamental theorem of calculus	$\int_a^b f(x) \, dx = F(b) - F(a), F'(x) = f(x)$
10. Average value of f over $[a, b]$	$\dfrac{1}{b-a} \int_a^b f(x) \, dx$
11. Area between two curves	$\int_a^b [f(x) - g(x)] \, dx, f(x) \geq g(x)$
12. Consumers' surplus	$CS = \int_0^{\bar{x}} D(x) \, dx - \bar{p}\,\bar{x}$
13. Producers' surplus	$PS = \bar{p}\,\bar{x} - \int_0^{\bar{x}} S(x) \, dx$
14. Accumulated (future) value of an income stream	$A = e^{rT} \int_0^T R(t)e^{-rt} \, dt$
15. Present value of an income stream	$PV = \int_0^T R(t)e^{-rt} \, dt$
16. Amount of an annuity	$A = \dfrac{mP}{r}(e^{rT} - 1)$
17. Present value of an annuity	$PV = \dfrac{mP}{r}(1 - e^{-rT})$
18. Coefficient of inequality of a Lorentz curve	$L = 2 \int_0^1 [x - f(x)] \, dx$

TERMS

antiderivative (910)
antidifferentiation (912)
integration (912)
indefinite integral (912)
integrand (912)

constant of integration (912)
differential equation (916)
initial value problem (916)
Riemann sum (937)
definite integral (938)

lower limit of integration (938)
upper limit of integration (938)
Lorentz curve (985)
line of complete equality (986)

CHAPTER 15 Concept Review Questions

Fill in the blanks.

1. a. A function F is an antiderivative of f on an interval, if _____ for all x in I.

b. If F is an antiderivative of f on an interval I, then every antiderivative of f on I has the form _____.

2. a. $\int cf(x)\,dx =$ _____

b. $\int [f(x) \pm g(x)]\,dx =$ _____

3. a. A differential equation is an equation that involves the derivative or differential of a/an _____ function.

b. A solution of a differential equation on an interval I is any _____ that satisfies the differential equation.

4. If we let $u = g(x)$, then $du =$ _____, and the substitution transforms the integral $\int f(g(x))g'(x)\,dx$ into the integral _____ involving only u.

5. a. If f is continuous and nonnegative on an interval $[a, b]$, then the area of the region under the graph of f on $[a, b]$ is given by _____.

b. If f is continuous on an interval $[a, b]$, then $\int_a^b f(x)\,dx$ is equal to the area(s) of the regions lying above the x-axis and bounded by the graph of f on $[a, b]$ _____ the area(s) of the regions lying below the x-axis and bounded by the graph of f on $[a, b]$.

6. a. The fundamental theorem of calculus states that if f is continuous on $[a, b]$, then $\int_a^b f(x)\,dx =$ _____, where F is a/an _____ of f.

b. The net change in a function f over an interval $[a, b]$ is given by $f(b) - f(a) =$ _____, provided f' is continuous on $[a, b]$.

7. a. If f is continuous on $[a, b]$, then the average value of f over $[a, b]$ is the number _____.

b. If f is a continuous and nonnegative function on $[a, b]$, then the average value of f over $[a, b]$ may be thought of as the _____ of the rectangle with base lying on the interval $[a, b]$ and having the same _____ as the region under the graph of f on $[a, b]$.

8. If f and g are continuous on $[a, b]$ and $f(x) \geq g(x)$ for all x in $[a, b]$, then the area of the region between the graphs of f and g and the vertical lines $x = a$ and $x = b$ is $A =$ _____.

9. a. The consumers' surplus is given by $CS =$ _____.

b. The producers' surplus is given by $PS =$ _____.

10. a. The accumulated value after T years of an income stream of $R(t)$ dollars/year, earning interest of r/year compounded continuously, is given by $A =$ _____.

b. The present value of an income stream is given by $PV =$ _____.

11. The amount of an annuity is $A =$ _____.

12. The coefficient of inequality of a Lorentz curve is $L =$ _____.

CHAPTER 15 Review Exercises

In Exercises 1–20, find each indefinite integral.

1. $\int (x^3 + 2x^2 - x)\,dx$

2. $\int \left(\frac{1}{3}x^3 - 2x^2 + 8\right)dx$

3. $\int \left(x^4 - 2x^3 + \frac{1}{x^2}\right)dx$

4. $\int (x^{1/3} - \sqrt{x} + 4)\,dx$

5. $\int x(2x^2 + x^{1/2})\,dx$

6. $\int (x^2 + 1)(\sqrt{x} - 1)\,dx$

7. $\int \left(x^2 - x + \frac{2}{x} + 5\right)dx$

8. $\int \sqrt{2x + 1}\,dx$

9. $\int (3x - 1)(3x^2 - 2x + 1)^{1/3}\,dx$

10. $\int x^2(x^3 + 2)^{10}\,dx$

11. $\int \frac{x - 1}{x^2 - 2x + 5}\,dx$

12. $\int 2e^{-2x}\,dx$

13. $\int \left(x + \frac{1}{2}\right)e^{x^2+x+1}\,dx$

14. $\int \frac{e^{-x} - 1}{(e^{-x} + x)^2}\,dx$

15. $\int \frac{(\ln x)^5}{x}\,dx$

16. $\int \frac{\ln x^2}{x}\,dx$

17. $\int x^3(x^2 + 1)^{10}\,dx$

18. $\int x\sqrt{x + 1}\,dx$

19. $\int \frac{x}{\sqrt{x - 2}}\,dx$

20. $\int \frac{3x}{\sqrt{x + 1}}\,dx$

In Exercises 21–32, evaluate each definite integral.

21. $\displaystyle\int_0^1 (2x^3 - 3x^2 + 1)\, dx$

22. $\displaystyle\int_0^2 (4x^3 - 9x^2 + 2x - 1)\, dx$

23. $\displaystyle\int_1^4 (\sqrt{x} + x^{-3/2})\, dx$ **24.** $\displaystyle\int_0^1 20x(2x^2 + 1)^4\, dx$

25. $\displaystyle\int_{-1}^0 12(x^2 - 2x)(x^3 - 3x^2 + 1)^3\, dx$

26. $\displaystyle\int_4^7 x\sqrt{x - 3}\, dx$ **27.** $\displaystyle\int_0^2 \frac{x}{x^2 + 1}\, dx$

28. $\displaystyle\int_0^1 \frac{dx}{(5 - 2x)^2}$ **29.** $\displaystyle\int_0^2 \frac{4x}{\sqrt{1 + 2x^2}}\, dx$

30. $\displaystyle\int_0^2 xe^{(-1/2)x^2}\, dx$ **31.** $\displaystyle\int_{-1}^0 \frac{e^{-x}}{(1 + e^{-x})^2}\, dx$

32. $\displaystyle\int_1^e \frac{\ln x}{x}\, dx$

In Exercises 33–36, find the function f given that the slope of the tangent line to the graph at any point $(x, f(x))$ is $f'(x)$ and that the graph of f passes through the given point.

33. $f'(x) = 3x^2 - 4x + 1;\ (1, 1)$

34. $f'(x) = \dfrac{x}{\sqrt{x^2 + 1}};\ (0, 1)$

35. $f'(x) = 1 - e^{-x};\ (0, 2)$

36. $f'(x) = \dfrac{\ln x}{x};\ (1, -2)$

37. Let $f(x) = -2x^2 + 1$ and compute the Riemann sum of f over the interval $[1, 2]$ by partitioning the interval into five subintervals of the same length ($n = 5$), where the points p_i ($1 \le i \le 5$) are taken to be the *right* endpoints of the respective subintervals.

38. MARGINAL COST FUNCTIONS   The management of National Electric has determined that the daily marginal cost function associated with producing their automatic drip coffeemakers is given by

$$C'(x) = 0.00003x^2 - 0.03x + 20$$

where $C'(x)$ is measured in dollars/unit and x denotes the number of units produced. Management has also determined that the daily fixed cost incurred in producing these coffeemakers is $500. What is the total cost incurred by National in producing the first 400 coffeemakers/day?

39. MARGINAL REVENUE FUNCTIONS   Refer to Exercise 38. Management has also determined that the daily marginal revenue function associated with producing and selling their coffeemakers is given by

$$R'(x) = -0.03x + 60$$

where x denotes the number of units produced and sold and $R'(x)$ is measured in dollars/unit.
a. Determine the revenue function $R(x)$ associated with producing and selling these coffeemakers.
b. What is the demand equation relating the wholesale unit price to the quantity of coffeemakers demanded?

40. COMPUTER RESALE VALUE   Franklin National Life Insurance Company purchased new computers for $200,000. If the rate at which the computers' resale value changes is given by the function

$$V'(t) = 3800(t - 10)$$

where t is the length of time since the purchase date and $V'(t)$ is measured in dollars/year, find an expression $V(t)$ that gives the resale value of the computers after t yr. How much would the computers cost after 6 yr?

41. PROJECTION TV SALES   The marketing department of Vista Vision forecasts that sales of their new line of projection television systems will grow at the rate of

$$3000 - 2000e^{-0.04t} \qquad (0 \le t \le 24)$$

units/month once they are introduced into the market. Find an expression giving the total number of the projection television systems that Vista may expect to sell t mo from the time they are put on the market. How many units of the television systems can Vista expect to sell during the first year?

42. COMMUTER TRENDS   Due to the increasing cost of fuel, the manager of the City Transit Authority estimates that the number of commuters using the city subway system will increase at the rate of

$$3000(1 + 0.4t)^{-1/2} \qquad (0 \le t \le 36)$$

per month t mo from now. If 100,000 commuters are currently using the system, find an expression giving the total number of commuters who will be using the subway t mo from now. How many commuters will be using the subway 6 mo from now?

43. SALES: LOUDSPEAKERS   Sales of the Acrosonic model F loudspeaker systems have been growing at the rate of

$$f'(t) = 2000(3 - 2e^{-t})$$

units/yr, where t denotes the number of years these loudspeaker systems have been on the market. Determine the number of loudspeaker systems that were sold in the first 5 yr after they appeared on the market.

44. MARGINAL COST FUNCTIONS The management of a division of Ditton Industries has determined that the daily marginal cost function associated with producing their hot-air corn poppers is given by

$$C'(x) = 0.00003x^2 - 0.03x + 10$$

where $C'(x)$ is measured in dollars/unit and x denotes the number of units manufactured. Management has also determined that the daily fixed cost incurred in producing these corn poppers is $600. Find the total cost incurred by Ditton in producing the first 500 corn poppers.

45. U.S. CENSUS The number of Americans aged 45–54 (which stood at 25 million at the beginning of 1990) grew at the rate of

$$R(t) = 0.00933t^3 + 0.019t^2 - 0.10833t + 1.3467$$

million people/year, t yr from the beginning of 1990. How many Americans aged 45 to 54 were added to the population between 1990 and the year 2000?
Source: U.S. Census Bureau

46. WORLD COAL PRODUCTION In 1980 the world produced 3.5 billion metric tons of coal. If output increased at the rate of

$$3.5e^{0.04t}$$

billion metric tons/year in year t ($t = 0$ corresponds to 1980), determine how much coal was produced worldwide between 1980 and the end of 1985.

47. Find the area of the region under the curve $y = 3x^2 + 2x + 1$ from $x = -1$ to $x = 2$.

48. Find the area of the region under the curve $y = e^{2x}$ from $x = 0$ to $x = 2$.

49. Find the area of the region bounded by the graph of the function $y = 1/x^2$, the x-axis, and the lines $x = 1$ and $x = 3$.

50. Find the area of the region bounded by the curve $y = -x^2 - x + 2$ and the x-axis.

51. Find the area of the region bounded by the graphs of the functions $f(x) = e^x$ and $g(x) = x$ and the vertical lines $x = 0$ and $x = 2$.

52. Find the area of the region that is completely enclosed by the graphs of $f(x) = x^4$ and $g(x) = x$.

53. Find the area of the region between the curve $y = x(x - 1)(x - 2)$ and the x-axis.

54. OIL PRODUCTION Based on current production techniques, the rate of oil production from a certain oil well t yr from now is

estimated to be

$$R_1(t) = 100e^{0.05t}$$

thousand barrels/year. Based on a new production technique, however, it is estimated that the rate of oil production from that oil well t yr from now will be

$$R_2(t) = 100e^{0.08t}$$

thousand barrels/year. Determine how much additional oil will be produced over the next 10 yr if the new technique is adopted.

55. Find the average value of the function

$$f(x) = \frac{x}{\sqrt{x^2 + 16}}$$

over the interval [0, 3].

56. AVERAGE TEMPERATURE The temperature (in °F) in Boston over a 12-hr period on a certain December day was given by

$$T = -0.05t^3 + 0.4t^2 + 3.8t + 5.6 \qquad (0 \le t \le 12)$$

where t is measured in hours, with $t = 0$ corresponding to 6 a.m. Determine the average temperature on that day over the 12-hr period from 6 a.m. to 6 p.m.

57. DEMAND FOR DIGITAL CAMCORDER TAPES The demand function for a brand of blank digital camcorder tapes is given by

$$p = -0.01x^2 - 0.2x + 23$$

where p is the wholesale unit price in dollars and x is the quantity demanded each week, measured in units of a thousand. Determine the consumers' surplus if the wholesale unit price is $8/tape.

58. CONSUMERS' AND PRODUCERS' SURPLUS The quantity demanded x (in units of a hundred) of the Sportsman 5 × 7 tents, per week, is related to the unit price p (in dollars) by the relation

$$p = -0.1x^2 - x + 40$$

The quantity x (in units of a hundred) that the supplier is willing to make available in the market is related to the unit price by the relation

$$p = 0.1x^2 + 2x + 20$$

If the market price is set at the equilibrium price, find the consumers' surplus and the producers' surplus.

59. RETIREMENT ACCOUNT SAVINGS Chi-Tai plans to deposit $4000/year in his Keogh Retirement Account. If interest is compounded continuously at the rate of 8%/year, how much will he have in his retirement account after 20 yr?

60. INSTALLMENT CONTRACTS Glenda sold her house under an installment contract whereby the buyer gave her a down payment of $9000 and agreed to make monthly payments of $925/month for 30 yr. If the prevailing interest rate is

12%/year compounded continuously, find the present value of the purchase price of the house.

61. PRESENT VALUE OF A FRANCHISE Alicia purchased a 10-yr franchise for a health spa that is expected to generate income at the rate of

$$P(t) = 80,000$$

dollars/year. If the prevailing interest rate is 10%/year compounded continuously, find the present value of the franchise.

62. INCOME DISTRIBUTION OF A COUNTRY A certain country's income distribution is described by the function

$$f(x) = \frac{17}{18}x^2 + \frac{1}{18}x$$

a. Sketch the Lorentz curve for this function.
b. Compute $f(0.3)$ and $f(0.6)$ and interpret your results.
c. Compute the coefficient of inequality for this Lorentz curve.

63. POPULATION GROWTH The population of a certain Sunbelt city, currently 80,000, is expected to grow exponentially in the next 5 yr with a growth constant of 0.05. If the prediction comes true, what will be the average population of the city over the next 5 yr?

CHAPTER 15 **Before Moving On . . .**

1. Find $\int \left(2x^3 + \sqrt{x} + \dfrac{2}{x} - \dfrac{2}{\sqrt{x}} \right) dx$.

2. Find f if $f'(x) = e^x + x$ and $f(0) = 2$.

3. Find $\int \dfrac{x}{\sqrt{x^2 + 1}} \, dx$.

4. Evaluate $\int_0^1 x\sqrt{2 - x^2} \, dx$

5. Find the area of the region completely enclosed by the graphs of $y = x^2 - 1$ and $y = 1 - x$.

Additional Topics in Integration

What is the area of the oil spill caused by a grounded tanker? In Example 5, page 1020, you will see how to determine the area of the oil spill.

© Lawson Wood/Corbis

BESIDES THE BASIC rules of integration developed in Chapter 15, there are more sophisticated techniques for finding the antiderivatives of functions. We begin this chapter by looking at the method of integration by parts. We then look at a technique of integration that involves using tables of integrals that have been compiled for this purpose. We also look at numerical methods of integration, which enable us to obtain approximate solutions to definite integrals, especially those whose exact value cannot be found otherwise. More specifically, we study the trapezoidal rule and Simpson's rule. Numerical integration methods are especially useful when the integrand is known only at discrete points. Finally, we learn how to evaluate integrals in which the intervals of integration are unbounded. Such integrals, called *improper integrals*, play an important role in the study of probability, the last topic of this chapter.

16.1 Integration by Parts

▬ The Method of Integration by Parts

Integration by parts is another technique of integration that, like the method of substitution discussed in Chapter 15, is based on a corresponding rule of differentiation. In this case, the rule of differentiation is the product rule, which asserts that if f and g are differentiable functions, then

$$\frac{d}{dx}\left[f(x)g(x)\right] = f(x)g'(x) + g(x)f'(x) \tag{1}$$

If we integrate both sides of Equation (1) with respect to x, we obtain

$$\int \frac{d}{dx} f(x)g(x)\, dx = \int f(x)g'(x)\, dx + \int g(x)f'(x)\, dx$$

$$f(x)g(x) = \int f(x)g'(x)\, dx + \int g(x)f'(x)\, dx$$

This last equation, which may be written in the form

$$\int f(x)g'(x)\, dx = f(x)g(x) - \int g(x)f'(x)\, dx \tag{2}$$

is called the formula for **integration by parts**. This formula is useful since it enables us to express one indefinite integral in terms of another that may be easier to evaluate. Formula (2) may be simplified by letting

$$u = f(x) \qquad dv = g'(x)\, dx$$
$$du = f'(x)\, dx \qquad v = g(x)$$

giving the following version of the formula for integration by parts.

Integration by Parts Formula

$$\int u\, dv = uv - \int v\, du \tag{3}$$

EXAMPLE 1   Evaluate $\int xe^x\, dx$.

Solution   No method of integration developed thus far enables us to evaluate the given indefinite integral in its present form. Therefore, we attempt to write it in terms of an indefinite integral that will be easier to evaluate. Let's use the integration by parts Formula (3) by letting

$$u = x \quad \text{and} \quad dv = e^x\, dx$$

so that

$$du = dx \quad \text{and} \quad v = e^x$$

Therefore,

$$\int xe^x \, dx = \int u \, dv$$
$$= uv - \int v \, du$$
$$= xe^x - \int e^x \, dx$$
$$= xe^x - e^x + C$$
$$= (x - 1)e^x + C$$

The success of the method of integration by parts depends on the proper choice of u and dv. For example, if we had chosen

$$u = e^x \quad \text{and} \quad dv = x \, dx$$

in the last example, then

$$du = e^x \, dx \quad \text{and} \quad v = \frac{1}{2}x^2$$

Thus, (3) would have yielded

$$\int xe^x \, dx = \int u \, dv$$
$$= uv - \int v \, du$$
$$= \frac{1}{2}x^2 e^x - \int \frac{1}{2}x^2 e^x \, dx$$

Since the indefinite integral on the right-hand side of this equation is not readily evaluated (it is in fact more complicated than the original integral!), choosing u and dv as shown has not helped us evaluate the given indefinite integral.

In general, we can use the following guidelines.

Guidelines for Choosing u and dv

Choose u and dv so that

1. du is simpler than u.

2. dv is easy to integrate.

EXAMPLE 2 Evaluate $\int x \ln x \, dx$.

Solution Letting

$$u = \ln x \quad \text{and} \quad dv = x \, dx$$

we have

$$du = \frac{1}{x} \, dx \quad \text{and} \quad v = \frac{1}{2}x^2$$

Therefore,

$$\int x \ln x \, dx = \int u \, dv = uv - \int v \, du$$

$$= \frac{1}{2} x^2 \ln x - \int \frac{1}{2} x^2 \cdot \left(\frac{1}{x} \right) dx$$

$$= \frac{1}{2} x^2 \ln x - \frac{1}{2} \int x \, dx$$

$$= \frac{1}{2} x^2 \ln x - \frac{1}{4} x^2 + C$$

$$= \frac{1}{4} x^2 (2 \ln x - 1) + C$$

EXAMPLE 3   Evaluate $\displaystyle\int \frac{xe^x}{(x+1)^2} \, dx$.

Solution   Let

$$u = xe^x \quad \text{and} \quad dv = \frac{1}{(x+1)^2} \, dx$$

Then,

$$du = (xe^x + e^x) \, dx = e^x(x+1) \, dx \quad \text{and} \quad v = -\frac{1}{x+1}$$

Therefore,

$$\int \frac{xe^x}{(x+1)^2} \, dx = \int u \, dv = uv - \int v \, du$$

$$= xe^x \left(\frac{-1}{x+1} \right) - \int \left(-\frac{1}{x+1} \right) e^x(x+1) \, dx$$

$$= -\frac{xe^x}{x+1} + \int e^x \, dx$$

$$= -\frac{xe^x}{x+1} + e^x + C$$

$$= \frac{e^x}{x+1} + C$$

The next example shows that repeated applications of the technique of integration by parts are sometimes required to evaluate an integral.

EXAMPLE 4   Find $\displaystyle\int x^2 e^x \, dx$.

Solution   Let

$$u = x^2 \quad \text{and} \quad dv = e^x \, dx$$

so that

$$du = 2x \, dx \quad \text{and} \quad v = e^x$$

Therefore,

$$\int x^2 e^x dx = \int u \, dv = uv - \int v \, du$$

$$= x^2 e^x - \int e^x(2x) \, dx = x^2 e^x - 2 \int x e^x \, dx$$

To complete the solution of the problem, we need to evaluate the integral

$$\int x e^x \, dx$$

But this integral may be found using integration by parts. In fact, you will recognize that this integral is precisely that of Example 1. Using the results obtained there, we now find

$$\int x^2 e^x \, dx = x^2 e^x - 2[(x - 1)e^x] + C = e^x(x^2 - 2x + 2) + C \qquad \blacksquare$$

EXPLORE & DISCUSS

1. Use the method of integration by parts to derive the formula

$$\int x^n e^{ax} \, dx = \frac{1}{a} x^n e^{ax} - \frac{n}{a} \int x^{n-1} e^{ax} \, dx$$

   where n is a positive integer and a is a real number.
2. Use the formula of part 1 to evaluate

$$\int x^3 e^x \, dx.$$

   Hint: You may find the results of Example 4 helpful.

 APPLIED EXAMPLE 5 Oil Production The estimated rate at which oil will be produced from a certain oil well t years after production has begun is given by

$$R(t) = 100te^{-0.1t}$$

thousand barrels per year. Find an expression that describes the total production of oil at the end of year t.

Solution Let $T(t)$ denote the total production of oil from the well at the end of year t ($t \geq 0$). Then, the rate of oil production will be given by $T'(t)$ thousand barrels per year. Thus,

$$T'(t) = R(t) = 100te^{-0.1t}$$

so

$$T(t) = \int 100te^{-0.1t} \, dt$$

$$= 100 \int te^{-0.1t} \, dt$$

We use the technique of integration by parts to evaluate this integral. Let

$$u = t \quad \text{and} \quad dv = e^{-0.1t}\, dt$$

so that

$$du = dt \quad \text{and} \quad v = -\frac{1}{0.1}e^{-0.1t} = -10e^{-0.1t}$$

Therefore,

$$
\begin{aligned}
T(t) &= 100\left[-10te^{-0.1t} + 10\int e^{-0.1t}\, dt\right] \\
&= 100[-10te^{-0.1t} - 100e^{-0.1t}] + C \\
&= -1000e^{-0.1t}(t + 10) + C
\end{aligned}
$$

To determine the value of C, note that the total quantity of oil produced at the end of year 0 is nil, so $T(0) = 0$. This gives

$$T(0) = -1000(10) + C = 0$$
$$C = 10{,}000$$

Thus, the required production function is given by

$$T(t) = -1000e^{-0.1t}(t + 10) + 10{,}000$$

EXPLORING WITH TECHNOLOGY

Refer to Example 5.

1. Use a graphing utility to plot the graph of

$$T(t) = -1000e^{-0.1t}(t + 10) + 10{,}000$$

using the viewing window $[0, 100] \times [0, 12{,}000]$. Use TRACE to see what happens when t is very large.
2. Verify the result of part 1 by evaluating $\lim\limits_{t \to \infty} T(t)$. Interpret your results.

16.1 Self-Check Exercises

1. Evaluate $\displaystyle\int x^2 \ln x\, dx$.

2. Since the inauguration of Ryan's Express at the beginning of 2002, the number of passengers (in millions) flying on this commuter airline has been growing at the rate of

$$R(t) = 0.1 + 0.2te^{-0.4t}$$

passengers/year ($t = 0$ corresponds to the beginning of 2002). Assuming that this trend continues through 2006, determine how many passengers will have flown on Ryan's Express by that time.

Solutions to Self-Check Exercises 16.1 can be found on page 1005.

16.1 Concept Questions

1. Write the formula for integration by parts.

2. Explain how you would choose u and dv when using the integration by parts formula. Illustrate your answer with $\int x^2 e^{-x}\, dx$. What happens if you reverse your choices of u and dv?

16.1   Exercises

In Exercises 1–26, find each indefinite integral.

1. $\int xe^{2x}\,dx$

2. $\int xe^{-x}\,dx$

3. $\int xe^{x/4}\,dx$

4. $\int 6xe^{3x}\,dx$

5. $\int (e^x - x)^2\,dx$

6. $\int (e^{-x} + x)^2\,dx$

7. $\int (x + 1)e^x\,dx$

8. $\int (x - 3)e^{3x}\,dx$

9. $\int x(x + 1)^{-3/2}\,dx$

10. $\int x(x + 4)^{-2}\,dx$

11. $\int x\sqrt{x - 5}\,dx$

12. $\int \dfrac{x}{\sqrt{2x + 3}}\,dx$

13. $\int x \ln 2x\,dx$

14. $\int x^2 \ln 2x\,dx$

15. $\int x^3 \ln x\,dx$

16. $\int \sqrt{x} \ln x\,dx$

17. $\int \sqrt{x} \ln \sqrt{x}\,dx$

18. $\int \dfrac{\ln x}{\sqrt{x}}\,dx$

19. $\int \dfrac{\ln x}{x^2}\,dx$

20. $\int \dfrac{\ln x}{x^3}\,dx$

21. $\int \ln x\,dx$

Hint: Let $u = \ln x$ and $dv = dx$.

22. $\int \ln(x + 1)\,dx$

23. $\int x^2 e^{-x}\,dx$

Hint: Integrate by parts twice.

24. $\int e^{-\sqrt{x}}\,dx$

Hint: First, make the substitution $u = \sqrt{x}$; then, integrate by parts.

25. $\int x(\ln x)^2\,dx$

Hint: Integrate by parts twice.

26. $\int x \ln(x + 1)\,dx$

Hint: First, make the substitution $u = x + 1$; then, integrate by parts.

In Exercises 27–32, evaluate each definite integral by using the method of integration by parts.

27. $\displaystyle\int_0^{\ln 2} xe^x\,dx$

28. $\displaystyle\int_0^2 xe^{-x}\,dx$

29. $\displaystyle\int_1^4 \ln x\,dx$

30. $\displaystyle\int_1^2 x \ln x\,dx$

31. $\displaystyle\int_0^2 xe^{2x}\,dx$

32. $\displaystyle\int_0^1 x^2 e^{-x}\,dx$

33. Find the function f given that the slope of the tangent line to the graph of f at any point $(x, f(x))$ is xe^{-2x} and that the graph passes through the point $(0, 3)$.

34. Find the function f given that the slope of the tangent line to the graph of f at any point $(x, f(x))$ is $x\sqrt{x + 1}$ and that the graph passes through the point $(3, 6)$.

35. Find the area of the region under the graph of $f(x) = \ln x$ from $x = 1$ to $x = 5$.

36. Find the area of the region under the graph of $f(x) = xe^{-x}$ from $x = 0$ to $x = 3$.

37. VELOCITY OF A DRAGSTER   The velocity of a dragster t sec after leaving the starting line is

$$100te^{-0.2t}$$

ft/sec. What is the distance covered by the dragster in the first 10 sec of its run?

38. PRODUCTION OF STEAM COAL   In keeping with the projected increase in worldwide demand for steam coal, the boiler-firing fuel used for generating electricity, the management of Consolidated Mining has decided to step up its mining operations. Plans call for increasing the yearly production of steam coal by

$$2te^{-0.05t}$$

million metric tons/year for the next 20 yr. The current yearly production is 20 million metric tons. Find a function that describes Consolidated's total production of steam coal at the end of t yr. How much coal will Consolidated have produced over the next 20 yr if this plan is carried out?

39. CONCENTRATION OF A DRUG IN THE BLOODSTREAM   The concentration (in milligrams/milliliter) of a certain drug in a patient's bloodstream t hr after it has been administered is given by $C(t) = 3te^{-t/3}$ mg/mL. Find the average concentration of the drug in the patient's bloodstream over the first 12 hr after administration.

40. ALCOHOL-RELATED TRAFFIC ACCIDENTS   As a result of increasingly stiff laws aimed at reducing the number of alcohol-related traffic accidents in a certain state, preliminary data indicate that the number of such accidents has been changing at the rate of

$$R(t) = -10 - te^{0.1t}$$

accidents/month t mo after the laws took effect. There were 982 alcohol-related accidents for the year before the enactment of the laws. Determine how many alcohol-related accidents were expected during the first year the laws were in effect.

41. COMPACT DISC SALES   Sales of the latest recording by Brittania, a British rock group, are currently $2te^{-0.1t}$ units/week (each unit representing 10,000 CDs), where t denotes the number of weeks since the recording's release. Find an expression that gives the total number of CDs sold as a function of t.

42. AVERAGE PRICE OF A COMMODITY   The price of a certain commodity in dollars/unit at time t (measured in weeks) is given by

$$p = 8 + 4e^{-2t} + te^{-2t}$$

What is the average price of the commodity over the 4-wk period from $t = 0$ to $t = 4$?

43. RATE OF RETURN ON AN INVESTMENT Suppose an investment is expected to generate income at the rate of

$$P(t) = 30{,}000 + 800t$$

dollars/year for the next 5 yr. Find the present value of this investment if the prevailing interest rate is 8%/year compounded continuously.
Hint: Use Formula (19), Section 15.7 (page 983).

44. PRESENT VALUE OF A FRANCHISE Tracy purchased a 15-yr franchise for a computer outlet store that is expected to generate income at the rate of

$$P(t) = 50{,}000 + 3000t$$

dollars/year. If the prevailing interest rate is 10%/year compounded continuously, find the present value of the franchise.
Hint: Use Formula (19), Section 15.7 (page 983).

45. GROWTH OF HMOS The membership of the Cambridge Community Health Plan (a health maintenance organization) is projected to grow at the rate of $9\sqrt{t + 1}\ln\sqrt{t + 1}$ thousand people/year, t yr from now. If the HMO's current membership is 50,000, what will be the membership 5 yr from now?

46. A MIXTURE PROBLEM Two tanks are connected in tandem as shown in the following figure. Each tank contains 60 gal of water. Starting at time $t = 0$, brine containing 3 lb/gal of salt flows into tank 1 at the rate of 2 gal/min. The mixture then enters and leaves tank 2 at the same rate. The mixtures in both tanks are stirred uniformly. It can be shown that the amount of salt in tank 2 after t min is given by

$$A(t) = 180(1 - e^{-t/30}) - 6te^{-t/30}$$

where $A(t)$ is measured in pounds.

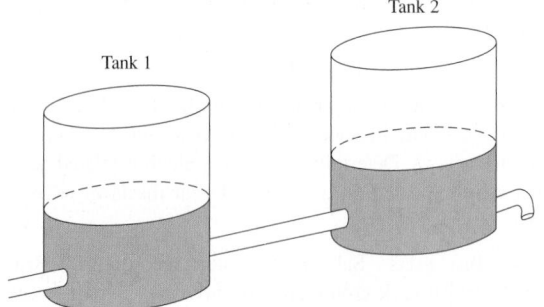

Tank 1

Tank 2

a. What is the initial amount of salt in tank 2?
b. What is the amount of salt in tank 2 after 3 hr (180 min)?
c. What is the average amount of salt in tank 2 over the first 3 hr?

47. DIFFUSION A cylindrical membrane with inner radius r_1 cm and outer radius r_2 cm containing a chemical solution is introduced into a salt bath with constant concentration c_2 moles/liter (see the accompanying figure).

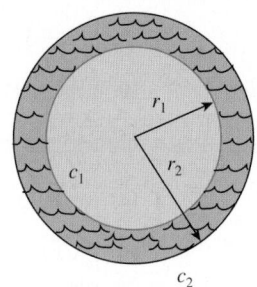

If the concentration of the chemical inside the membrane is kept constant at a different concentration of c_1 moles/liter, then the concentration of the chemical across the membrane will be given by

$$c(r) = \left(\frac{c_1 - c_2}{\ln r_1 - \ln r_2}\right)(\ln r - \ln r_2) + c_2 \qquad (r_1 < r < r_2)$$

moles/liter. Find the average concentration of the chemical across the membrane from $r = r_1$ to $r = r_2$.

48. Suppose f'' is continuous on $[1, 3]$ and $f(1) = 2$, $f(3) = -1$, $f'(1) = 2$, and $f'(3) = 5$. Evaluate $\displaystyle\int_1^3 xf''(x)\,dx$.

In Exercises 49 and 50, determine whether the statement is true or false. If it is true, explain why it is true. If it is false, give an example to show why it is false.

49. $\displaystyle\int u\,dv + \int v\,du = uv$

50. $\displaystyle\int e^x g'(x)\,dx = e^x g(x) - \int e^x g(x)\,dx$

16.1 Solutions to Self-Check Exercises

1. Let $u = \ln x$ and $dv = x^2 \, dx$ so that $du = \dfrac{1}{x} \, dx$ and $v = \dfrac{1}{3} x^3$.
Therefore,

$$\int x^2 \ln x \, dx = \int u \, dv = uv - \int v \, du$$

$$= \frac{1}{3} x^3 \ln x - \int \frac{1}{3} x^2 \, dx$$

$$= \frac{1}{3} x^3 \ln x - \frac{1}{9} x^3 + C$$

$$= \frac{1}{9} x^3 (3 \ln x - 1) + C$$

2. If $N(t)$ denote the total number of passengers who will have flown on Ryan's Express by the end of year t. Then $N'(t) = R(t)$, so that

$$N(t) = \int R(t) \, dt$$

$$= \int (0.1 + 0.2 t e^{-0.4t}) \, dt$$

$$= \int 0.1 \, dt + 0.2 \int t e^{-0.4t} \, dt$$

We now use the technique of integration by parts on the second integral. Letting $u = t$ and $dv = e^{-0.4t} \, dt$, we have

$$du = dt \quad \text{and} \quad v = -\frac{1}{0.4} e^{-0.4t} = -2.5 e^{-0.4t}$$

Therefore,

$$N(t) = 0.1t + 0.2 \left(-2.5 t e^{-0.4t} + 2.5 \int e^{-0.4t} \, dt \right)$$

$$= 0.1t - 0.5 t e^{-0.4t} - \frac{0.5}{0.4} e^{-0.4t} + C$$

$$= 0.1t - 0.5(t + 2.5) e^{-0.4t} + C$$

To determine the value of C, note that $N(0) = 0$, which gives

$$N(0) = -0.5(2.5) + C = 0$$

$$C = 1.25$$

Therefore,

$$N(t) = 0.1t - 0.5(t + 2.5) e^{-0.4t} + 1.25$$

The number of passengers who will have flown on Ryan's Express by the end of 2006 is given by

$$N(5) = 0.1(5) - 0.5(5 + 2.5) e^{-0.4(5)} + 1.25$$

$$= 1.242493$$

—that is, 1,242,493 passengers.

16.2 Integration Using Tables of Integrals

■ A Table of Integrals

We have studied several techniques for finding an antiderivative of a function. However, useful as they are, these techniques are not always applicable. There are of course numerous other methods for finding an antiderivative of a function. Extensive lists of integration formulas have been compiled based on these methods.

A small sample of the integration formulas that can be found in many mathematical handbooks is given in the following table of integrals. The formulas are grouped according to the basic form of the integrand. Note that it may be necessary to modify the integrand of the integral to be evaluated in order to use one of these formulas.

TABLE OF INTEGRALS

Forms Involving $a + bu$

1. $\displaystyle\int \frac{u\,du}{a+bu} = \frac{1}{b^2}\left(a+bu-a\ln|a+bu|\right)+C$

2. $\displaystyle\int \frac{u^2\,du}{a+bu} = \frac{1}{2b^3}\left[(a+bu)^2-4a(a+bu)+2a^2\ln|a+bu|\right]+C$

3. $\displaystyle\int \frac{u\,du}{(a+bu)^2} = \frac{1}{b^2}\left(\frac{a}{a+bu}+\ln|a+bu|\right)+C$

4. $\displaystyle\int u\sqrt{a+bu}\,du = \frac{2}{15b^2}(3bu-2a)(a+bu)^{3/2}+C$

5. $\displaystyle\int \frac{u\,du}{\sqrt{a+bu}} = \frac{2}{3b^2}(bu-2a)\sqrt{a+bu}+C$

6. $\displaystyle\int \frac{du}{u\sqrt{a+bu}} = \frac{1}{\sqrt{a}}\ln\left|\frac{\sqrt{a+bu}-\sqrt{a}}{\sqrt{a+bu}+\sqrt{a}}\right|+C \qquad (\text{if } a>0)$

Forms Involving $\sqrt{a^2+u^2}$

7. $\displaystyle\int \sqrt{a^2+u^2}\,du = \frac{u}{2}\sqrt{a^2+u^2}+\frac{a^2}{2}\ln|u+\sqrt{a^2+u^2}|+C$

8. $\displaystyle\int u^2\sqrt{a^2+u^2}\,du = \frac{u}{8}(a^2+2u^2)\sqrt{a^2+u^2}-\frac{a^4}{8}\ln|u+\sqrt{a^2+u^2}|+C$

9. $\displaystyle\int \frac{du}{\sqrt{a^2+u^2}} = \ln|u+\sqrt{a^2+u^2}|+C$

10. $\displaystyle\int \frac{du}{u\sqrt{a^2+u^2}} = -\frac{1}{a}\ln\left|\frac{\sqrt{a^2+u^2}+a}{u}\right|+C$

11. $\displaystyle\int \frac{du}{u^2\sqrt{a^2+u^2}} = -\frac{\sqrt{a^2+u^2}}{a^2u}+C$

12. $\displaystyle\int \frac{du}{(a^2+u^2)^{3/2}} = \frac{u}{a^2\sqrt{a^2+u^2}}+C$

Forms Involving $\sqrt{u^2-a^2}$

13. $\displaystyle\int \sqrt{u^2-a^2}\,du = \frac{u}{2}\sqrt{u^2-a^2}-\frac{a^2}{2}\ln|u+\sqrt{u^2-a^2}|+C$

14. $\displaystyle\int u^2\sqrt{u^2-a^2}\,du = \frac{u}{8}(2u^2-a^2)\sqrt{u^2-a^2}-\frac{a^4}{8}\ln|u+\sqrt{u^2-a^2}|+C$

15. $\displaystyle\int \frac{\sqrt{u^2-a^2}}{u^2}\,du = -\frac{\sqrt{u^2-a^2}}{u}+\ln|u+\sqrt{u^2-a^2}|+C$

16. $\displaystyle\int \frac{du}{\sqrt{u^2-a^2}} = \ln|u+\sqrt{u^2-a^2}|+C$

17. $\displaystyle\int \frac{du}{u^2\sqrt{u^2-a^2}} = \frac{\sqrt{u^2-a^2}}{a^2u}+C$

18. $\displaystyle\int \frac{du}{(u^2-a^2)^{3/2}} = -\frac{u}{a^2\sqrt{u^2-a^2}}+C$

TABLE OF INTEGRALS (*Continued*)

Forms Involving $\sqrt{a^2 - u^2}$

19. $\displaystyle\int \frac{\sqrt{a^2 - u^2}}{u}\, du = \sqrt{a^2 - u^2} - a \ln\left| \frac{a + \sqrt{a^2 - u^2}}{u} \right| + C$

20. $\displaystyle\int \frac{du}{u\sqrt{a^2 - u^2}} = -\frac{1}{a} \ln\left| \frac{a + \sqrt{a^2 - u^2}}{u} \right| + C$

21. $\displaystyle\int \frac{du}{u^2\sqrt{a^2 - u^2}} = -\frac{\sqrt{a^2 - u^2}}{a^2 u} + C$

22. $\displaystyle\int \frac{du}{(a^2 - u^2)^{3/2}} = \frac{u}{a^2\sqrt{a^2 - u^2}} + C$

Forms Involving e^{au} and $\ln u$

23. $\displaystyle\int u e^{au}\, du = \frac{1}{a^2}(au - 1)e^{au} + C$

24. $\displaystyle\int u^n e^{au}\, du = \frac{1}{a} u^n e^{au} - \frac{n}{a} \int u^{n-1} e^{au}\, du$

25. $\displaystyle\int \frac{du}{1 + be^{au}} = u - \frac{1}{a} \ln(1 + be^{au}) + C$

26. $\displaystyle\int \ln u\, du = u \ln u - u + C$

27. $\displaystyle\int u^n \ln u\, du = \frac{u^{n+1}}{(n+1)^2}\left[(n+1)\ln u - 1\right] + C \qquad (n \neq -1)$

28. $\displaystyle\int \frac{du}{u \ln u} = \ln|\ln u| + C$

29. $\displaystyle\int (\ln u)^n\, du = u(\ln u)^n - n \int (\ln u)^{n-1}\, du$

◼ Using a Table of Integrals

We now consider several examples that illustrate how the table of integrals can be used to evaluate an integral.

EXAMPLE 1 Use the table of integrals to find $\displaystyle\int \frac{2x\, dx}{\sqrt{3 + x}}$.

Solution We first write

$$\int \frac{2x\, dx}{\sqrt{3 + x}} = 2 \int \frac{x\, dx}{\sqrt{3 + x}}$$

Since $\sqrt{3 + x}$ is of the form $\sqrt{a + bu}$, with $a = 3$, $b = 1$, and $u = x$, we use Formula (5),

$$\int \frac{u\, du}{\sqrt{a + bu}} = \frac{2}{3b^2}(bu - 2a)\sqrt{a + bu} + C$$

obtaining

$$2 \int \frac{x}{\sqrt{3 + x}}\, dx = 2\left[\frac{2}{3(1)}(x - 6)\sqrt{3 + x}\right] + C$$

$$= \frac{4}{3}(x - 6)\sqrt{3 + x} + C$$

◼

EXPLORE & DISCUSS

All formulas given in the table of integrals can be verified by direct computation. Describe a method you would use and apply it to verify a formula of your choice.

EXAMPLE 2   Use the table of integrals to find $\int x^2\sqrt{3 + x^2}\,dx$.

Solution   Observe that if we write 3 as $(\sqrt{3})^2$, then $3 + x^2$ has the form $\sqrt{a^2 + u^2}$, with $a = \sqrt{3}$ and $u = x$. Using Formula (8),

$$\int u^2\sqrt{a^2 + u^2}\,du = \frac{u}{8}(a^2 + 2u^2)\sqrt{a^2 + u^2} - \frac{a^4}{8}\ln|u + \sqrt{a^2 + u^2}| + C$$

we obtain

$$\int x^2\sqrt{3 + x^2}\,dx = \frac{x}{8}(3 + 2x^2)\sqrt{3 + x^2} - \frac{9}{8}\ln|x + \sqrt{3 + x^2}| + C \quad\blacksquare$$

EXAMPLE 3   Use the table of integrals to evaluate

$$\int_3^4 \frac{dx}{x^2\sqrt{50 - 2x^2}}$$

Solution   We first find the indefinite integral

$$I = \int \frac{dx}{x^2\sqrt{50 - 2x^2}}$$

Observe that $\sqrt{50 - 2x^2} = \sqrt{2(25 - x^2)} = \sqrt{2}\sqrt{25 - x^2}$, so we can write I as

$$I = \frac{1}{\sqrt{2}}\int \frac{dx}{x^2\sqrt{25 - x^2}} = \frac{\sqrt{2}}{2}\int \frac{dx}{x^2\sqrt{25 - x^2}} \qquad \frac{1}{\sqrt{2}} = \frac{\sqrt{2}}{2}$$

Next, using Formula (21),

$$\int \frac{du}{u^2\sqrt{a^2 - u^2}} = -\frac{\sqrt{a^2 - u^2}}{a^2 u} + C$$

with $a = 5$ and $u = x$, we find

$$I = \frac{\sqrt{2}}{2}\left[-\frac{\sqrt{25 - x^2}}{25x} \right]$$

$$= -\left(\frac{\sqrt{2}}{50}\right)\frac{\sqrt{25 - x^2}}{x}$$

Finally, using this result, we obtain

$$\int_3^4 \frac{dx}{x^2\sqrt{50 - 2x^2}} = -\frac{\sqrt{2}}{50}\frac{\sqrt{25 - x^2}}{x}\Bigg|_3^4$$

$$= -\frac{\sqrt{2}}{50}\frac{\sqrt{25 - 16}}{4} - \left(-\frac{\sqrt{2}}{50}\frac{\sqrt{25 - 9}}{3}\right)$$

$$= -\frac{3\sqrt{2}}{200} + \frac{2\sqrt{2}}{75} = \frac{7\sqrt{2}}{600} \quad\blacksquare$$

EXAMPLE 4   Use the table of integrals to find $\int e^{2x}\sqrt{5 + 2e^x}\,dx$.

Solution   Let $u = e^x$. Then $du = e^x\,dx$. Therefore, the given integral can be written

$$\int e^x\sqrt{5 + 2e^x}\,(e^x\,dx) = \int u\sqrt{5 + 2u}\,du$$

Using Formula (4),

$$\int u\sqrt{a + bu}\, du = \frac{2}{15b^2}(3bu - 2a)(a + bu)^{3/2} + C$$

with $a = 5$ and $b = 2$, we see that

$$\int u\sqrt{5 + 2u}\, du = \frac{2}{15(4)}(6u - 10)(5 + 2u)^{3/2} + C$$

$$= \frac{1}{15}(3u - 5)(5 + 2u)^{3/2} + C$$

Finally, recalling the substitution $u = e^x$, we find

$$\int e^{2x}\sqrt{5 + 2e^x}\, dx = \frac{1}{15}(3e^x - 5)(5 + 2e^x)^{3/2} + C$$

EXPLORE & DISCUSS

The formulas given in the table of integrals were derived using various techniques, including the method of substitution and the method of integration by parts studied earlier. For example, Formula (1),

$$\int \frac{u\, du}{a + bu} = \frac{1}{b^2}\left[a + bu - a \ln|a + bu|\right] + C$$

can be derived using the method of substitution. Show how this is done.

As illustrated in the next example, we may need to apply a formula more than once in order to evaluate an integral.

EXAMPLE 5   Use the table of integrals to find $\int x^2 e^{(-1/2)x}\, dx$.

Solution   Scanning the table of integrals for a formula involving e^{ax} in the integrand, we are led to Formula (24),

$$\int u^n e^{au}\, du = \frac{1}{a}u^n e^{au} - \frac{n}{a}\int u^{n-1} e^{au}\, du$$

With $n = 2$, $a = -\frac{1}{2}$, and $u = x$, we have

$$\int x^2 e^{(-1/2)x}\, dx = \left(\frac{1}{-\frac{1}{2}}\right)x^2 e^{(-1/2)x} - \frac{2}{(-\frac{1}{2})}\int x e^{(-1/2)x}\, dx$$

$$= -2x^2 e^{(-1/2)x} + 4\int x e^{(-1/2)x}\, dx$$

If we use Formula (24) once again, with $n = 1$, $a = -\frac{1}{2}$, and $u = x$, to evaluate the integral on the right, we obtain

$$\int x^2 e^{(-1/2)x}\, dx = -2x^2 e^{(-1/2)x} + 4\left[\left(\frac{1}{-\frac{1}{2}}\right)x e^{(-1/2)x} - \frac{1}{(-\frac{1}{2})}\int e^{(-1/2)x}\, dx\right]$$

$$= -2x^2 e^{(-1/2)x} + 4\left[-2x e^{(-1/2)x} + 2 \cdot \frac{1}{(-\frac{1}{2})} e^{(-1/2)x}\right] + C$$

$$= -2e^{(-1/2)x}(x^2 + 4x + 8) + C$$

VIDEO

APPLIED EXAMPLE 6 Mortgage Rates A study prepared for the National Association of Realtors estimated that the mortgage rate over the next t months will be

$$r(t) = \frac{8t + 100}{t + 10} \qquad (0 \leq t \leq 24)$$

percent per year. If the prediction holds true, what will be the average mortgage rate over the next 12 months?

Solution   The average mortgage rate over the next 12 months will be given by

$$A = \frac{1}{12 - 0} \int_0^{12} \frac{8t + 100}{t + 10} \, dt = \frac{1}{12} \left[\int_0^{12} \frac{8t}{t + 10} \, dt + \int_0^{12} \frac{100}{t + 10} \, dt \right]$$

$$= \frac{8}{12} \int_0^{12} \frac{t}{t + 10} \, dt + \frac{100}{12} \int_0^{12} \frac{1}{t + 10} \, dt$$

Using Formula (1)

$$\int \frac{u \, du}{a + bu} = \frac{1}{b^2} \left[a + bu - a \ln|a + bu| \right] + C \qquad a = 10, b = 1, u = t$$

to evaluate the first integral, we have

$$A = \left(\frac{2}{3} \right) \left[10 + t - 10 \ln(10 + t) \right] \Big|_0^{12} + \left(\frac{25}{3} \right) \ln(10 + t) \Big|_0^{12}$$

$$= \left(\frac{2}{3} \right) \left[(22 - 10 \ln 22) - (10 - 10 \ln 10) \right] + \left(\frac{25}{3} \right) \left[\ln 22 - \ln 10 \right]$$

$$\approx 9.31$$

or approximately 9.31% per year.

16.2   Self-Check Exercises

1. Use the table of integrals to evaluate

$$\int_0^2 \frac{dx}{(5 - x^2)^{3/2}}$$

2. During a flu epidemic, the number of children in Easton Middle School who contracted influenza t days after the outbreak began was given by

$$N(t) = \frac{200}{1 + 9e^{-0.8t}}$$

Determine the average number of children who contracted the flu in the first 10 days of the epidemic.

Solutions to Self-Check Exercises 16.2 can be found on page 1012.

16.2   Concept Questions

1. Consider the integral $\int \dfrac{\sqrt{2 - x^2}}{x} \, dx$.
   a. Which formula from the table of integrals would you choose to help you find the integral?
   b. Find the integral showing the appropriate substitutions you need to use to make the given integral conform to the formula.

2. Consider the integral $\int_2^3 \dfrac{dx}{\sqrt{2x^2 - 5}}$.
   a. Which formula from the table of integrals would you choose to help you evaluate the integral?
   b. Evaluate the integral.

16.2   Exercises

In Exercises 1–32, use the table of integrals in this section to find each integral.

1. $\displaystyle \int \frac{2x}{2 + 3x}\, dx$

2. $\displaystyle \int \frac{x}{(1 + 2x)^2}\, dx$

3. $\displaystyle \int \frac{3x^2}{2 + 4x}\, dx$

4. $\displaystyle \int \frac{x^2}{3 + x}\, dx$

5. $\displaystyle \int x^2 \sqrt{9 + 4x^2}\, dx$

6. $\displaystyle \int x^2 \sqrt{4 + x^2}\, dx$

7. $\displaystyle \int \frac{dx}{x\sqrt{1 + 4x}}$

8. $\displaystyle \int_0^2 \frac{x + 1}{\sqrt{2 + 3x}}\, dx$

9. $\displaystyle \int_0^2 \frac{dx}{\sqrt{9 + 4x^2}}$

10. $\displaystyle \int \frac{dx}{x\sqrt{4 + 8x^2}}$

11. $\displaystyle \int \frac{dx}{(9 - x^2)^{3/2}}$

12. $\displaystyle \int \frac{dx}{(2 - x^2)^{3/2}}$

13. $\displaystyle \int x^2 \sqrt{x^2 - 4}\, dx$

14. $\displaystyle \int_3^5 \frac{dx}{x^2\sqrt{x^2 - 9}}$

15. $\displaystyle \int \frac{\sqrt{4 - x^2}}{x}\, dx$

16. $\displaystyle \int_0^1 \frac{dx}{(4 - x^2)^{3/2}}$

17. $\displaystyle \int x e^{2x}\, dx$

18. $\displaystyle \int \frac{dx}{1 + e^{-x}}$

19. $\displaystyle \int \frac{dx}{(x + 1)\ln(1 + x)}$

Hint: First use the substitution $u = x + 1$.

20. $\displaystyle \int \frac{x}{(x^2 + 1)\ln(x^2 + 1)}\, dx$

Hint: First use the substitution $u = x^2 + 1$.

21. $\displaystyle \int \frac{e^{2x}}{(1 + 3e^x)^2}\, dx$

22. $\displaystyle \int \frac{e^{2x}}{\sqrt{1 + 3e^x}}\, dx$

23. $\displaystyle \int \frac{3e^x}{1 + e^{(1/2)x}}\, dx$

24. $\displaystyle \int \frac{dx}{1 - 2e^{-x}}$

25. $\displaystyle \int \frac{\ln x}{x(2 + 3\ln x)}\, dx$

26. $\displaystyle \int_1^e (\ln x)^2\, dx$

27. $\displaystyle \int_0^1 x^2 e^x\, dx$

28. $\displaystyle \int x^3 e^{2x}\, dx$

29. $\displaystyle \int x^2 \ln x\, dx$

30. $\displaystyle \int x^3 \ln x\, dx$

31. $\displaystyle \int (\ln x)^3\, dx$

32. $\displaystyle \int (\ln x)^4\, dx$

33. CONSUMERS' SURPLUS   Refer to Section 15.7. The demand function for Apex women's boots is

$$p = \frac{250}{\sqrt{16 + x^2}}$$

where p is the wholesale unit price in dollars and x is the quantity demanded daily, in units of a hundred. Find the consumers' surplus if the wholesale price is set at \$50/pair.

34. PRODUCERS' SURPLUS   Refer to Section 15.7. The supplier of Apex women's boots will make x hundred pairs of the boots available in the market daily when the wholesale unit price is

$$p = \frac{30x}{5 - x}$$

dollars. Find the producers' surplus if the wholesale price is set at \$50/pair.

35. AMUSEMENT PARK ATTENDANCE   The management of Astro-World ("The Amusement Park of the Future") estimates that the number of visitors (in thousands) entering the amusement park t hr after opening time at 9 a.m. is given by

$$R(t) = \frac{60}{(2 + t^2)^{3/2}}$$

per hour. Determine the number of visitors admitted by noon.

36. VOTER REGISTRATION   The number of voters in a certain district of a city is expected to grow at the rate of

$$R(t) = \frac{3000}{\sqrt{4 + t^2}}$$

people/year t yr from now. If the number of voters at present is 20,000, how many voters will be in the district 5 yr from now?

37. GROWTH OF FRUIT FLIES   Based on data collected during an experiment, a biologist found that the number of fruit flies (*Drosophila*) with a limited food supply could be approximated by the exponential model

$$N(t) = \frac{1000}{1 + 24e^{-0.02t}}$$

where t denotes the number of days since the beginning of the experiment. Find the average number of fruit flies in the colony in the first 10 days of the experiment and in the first 20 days.

38. VCR OWNERSHIP   The percent of households that own VCRs is given by

$$P(t) = \frac{68}{1 + 21.67e^{-0.62t}} \qquad (0 \le t \le 12)$$

where t is measured in years, with $t = 0$ corresponding to the beginning of 1981. Find the average percent of households owning VCRs from the beginning of 1981 to the beginning of 1993.

Source: Paul Kroger Associates

39. RECYCLING PROGRAMS   The commissioner of the City of Newton Department of Public Works estimates that the number of people in the city who have been recycling their magazines in year t following the introduction of the recycling program at the beginning of 1990 is

$$N(t) = \frac{100,000}{2 + 3e^{-0.2t}}$$

Find the average number of people who will have recycled their magazines during the first 5 yr since the program was introduced.

40. **FRANCHISES** Elaine purchased a 10-yr franchise for a fast-food restaurant that is expected to generate income at the rate of $R(t) = 250,000 + 2000t^2$ dollars/year, t yr from now. If the prevailing interest rate is 10%/year compounded continuously, find the present value of the franchise.
Hint: Use Formula (19), Section 15.7.

41. **ACCUMULATED VALUE OF AN INCOME STREAM** The revenue of Virtual Reality, a video-game arcade, is generated at the rate

of $R(t) = 20,000t$ dollars. If the revenue is invested t yr from now in a business earning interest at the rate of 15%/year compounded continuously, find the accumulated value of this stream of income at the end of 5 yr.
Hint: Use Formula (19), Section 15.7.

42. **LORENTZ CURVES** In a study conducted by a certain country's Economic Development Board regarding the income distribution of certain segments of the country's workforce, it was found that the Lorentz curve for the distribution of income of college professors is described by the function

$$g(x) = \frac{1}{3}x\sqrt{1 + 8x}$$

Compute the coefficient of inequality of the Lorentz curve.
Hint: Use Formula (22), Section 15.7.

16.2 Solutions to Self-Check Exercises

1. Using Formula (22), page 1007, with $a^2 = 5$ and $u = x$, we see that

$$\int_0^2 \frac{dx}{(5 - x^2)^{3/2}} = \frac{x}{5\sqrt{5 - x^2}}\Big|_0^2$$
$$= \frac{2}{5\sqrt{5 - 4}}$$
$$= \frac{2}{5}$$

2. The average number of children who contracted the flu in the first 10 days of the epidemic is given by

$$A = \frac{1}{10}\int_0^{10} \frac{200}{1 + 9e^{-0.8t}}\, dt = 20\int_0^{10} \frac{dt}{1 + 9e^{-0.8t}}$$
$$= 20\left[t + \frac{1}{0.8}\ln(1 + 9e^{-0.8t})\right]\Big|_0^{10} \quad \substack{\text{Formula (25), } a = -0.8, \\ b = 9, u = t}$$
$$= 20\left[10 + \frac{1}{0.8}\ln(1 + 9e^{-8})\right] - 20\left(\frac{1}{0.8}\right)\ln 10$$
$$\approx 200.07537 - 57.56463$$
$$\approx 143$$

or 143 students.

16.3 Numerical Integration

■ Approximating Definite Integrals

One method of measuring cardiac output is to inject 5 to 10 milligrams (mg) of a dye into a vein leading to the heart. After making its way through the lungs, the dye returns to the heart and is pumped into the aorta, where its concentration is measured at equal time intervals. The graph of the function c in Figure 1 shows the concentration of dye in a person's aorta, measured at 2-second intervals after 5 mg of dye

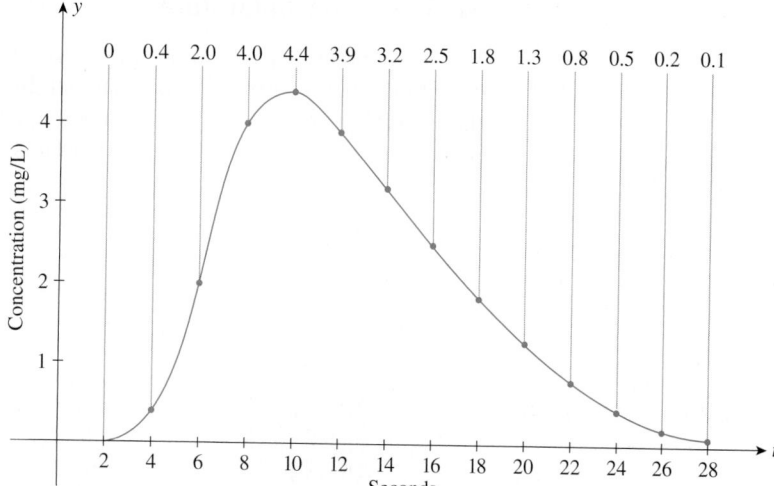

FIGURE 1
The function c gives the concentration of a dye measured at the aorta. The graph is constructed by drawing a smooth curve through a set of discrete points.

have been injected. The person's cardiac output, measured in liters per minute (L/min), is computed using the formula

$$R = \frac{60D}{\displaystyle\int_0^{28} c(t)\, dt} \tag{4}$$

where D is the quantity of dye injected (see Exercise 49, on page 1026).

Now, to use Formula (4), we need to evaluate the definite integral

$$\int_0^{28} c(t)\, dt$$

But we do not have the algebraic rule defining the integrand c for all values of t in $[0, 28]$. In fact, we are given its values only at a set of discrete points in that interval. In situations such as this, the fundamental theorem of calculus proves useless because we cannot find an antiderivative of c. (We will complete the solution to this problem in Example 4.)

Other situations also arise in which an integrable function has an antiderivative that cannot be found in terms of elementary functions (functions that can be expressed as a finite combination of algebraic, exponential, logarithmic, and trigonometric functions). Examples of such functions are

$$f(x) = e^{x^2} \qquad g(x) = x^{-1/2} e^x \qquad h(x) = \frac{1}{\ln x}$$

Riemann sums provide us with a good approximation of a definite integral, provided the number of subintervals in the partitions is large enough. But there are better techniques and formulas, called *quadrature formulas*, that give a more efficient way of computing approximate values of definite integrals. In this section, we look at two rather simple but effective ways of approximating definite integrals.

The Trapezoidal Rule

We assume that $f(x) \geq 0$ on $[a, b]$ in order to simplify the derivation of the trapezoidal rule, but the result is valid without this restriction. We begin by subdividing the interval $[a, b]$ into n subintervals of equal length Δx, by means of the $(n + 1)$ points $x_0 = a, x_1, x_2, \ldots, x_n = b$, where n is a positive integer (Figure 2).

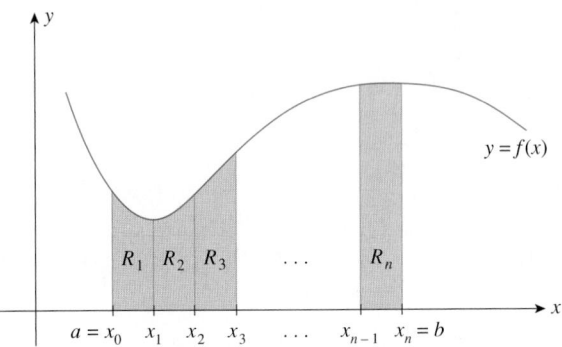

FIGURE 2
The area under the curve is equal to the sum of the n subregions R_1, $R_2, \ldots, R_n$.

Then, the length of each subinterval is given by

$$\Delta x = \frac{b - a}{n}$$

Furthermore, as we saw earlier, we may view the definite integral

$$\int_a^b f(x) \, dx$$

as the area of the region R under the curve $y = f(x)$ between $x = a$ and $x = b$. This area is given by the sum of the areas of the n nonoverlapping subregions $R_1, R_2, \ldots, R_n$, such that R_1 represents the region under the curve $y = f(x)$ from $x = x_0$ to $x = x_1$, and so on.

The basis for the trapezoidal rule lies in the approximation of each of the regions $R_1, R_2, \ldots, R_n$ by a suitable trapezoid. This often leads to a much better approximation than one obtained by means of rectangles (a Riemann sum).

Let's consider the subregion R_1, shown magnified for the sake of clarity in Figure 3. Observe that the area of the region R_1 may be approximated by the trapezoid of width Δx whose parallel sides are of lengths $f(x_0)$ and $f(x_1)$. The area of the trapezoid is given by

$$\left[\frac{f(x_0) + f(x_1)}{2} \right] \Delta x \qquad \text{Average of the lengths of the parallel sides} \times \text{Width}$$

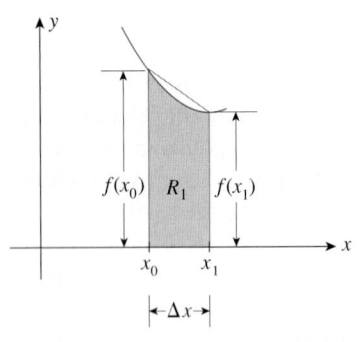

FIGURE 3
The area of R_1 is approximated by the area of the trapezoid.

square units. Similarly, the area of the region R_2 may be approximated by the trapezoid of width Δx and sides of lengths $f(x_1)$ and $f(x_2)$. The area of the trapezoid is given by

$$\left[\frac{f(x_1) + f(x_2)}{2} \right] \Delta x$$

Similarly, we see that the area of the last (*n*th) approximating trapezoid is given by

$$\left[\frac{f(x_{n-1}) + f(x_n)}{2}\right]\Delta x$$

Then, the area of the region *R* is approximated by the sum of the areas of the *n* trapezoids—that is,

$$\left[\frac{f(x_0) + f(x_1)}{2}\right]\Delta x + \left[\frac{f(x_1) + f(x_2)}{2}\right]\Delta x + \cdots + \left[\frac{f(x_{n-1}) + f(x_n)}{2}\right]\Delta x$$

$$= \frac{\Delta x}{2}\left[f(x_0) + f(x_1) + f(x_1) + f(x_2) + \cdots + f(x_{n-1}) + f(x_n)\right]$$

$$= \frac{\Delta x}{2}\left[f(x_0) + 2f(x_1) + 2f(x_2) + \cdots + 2f(x_{n-1}) + f(x_n)\right]$$

Since the area of the region *R* is given by the value of the definite integral we wished to approximate, we are led to the following approximation formula, which is called the **trapezoidal rule**.

Trapezoidal Rule

$$\int_a^b f(x)\,dx \approx \frac{\Delta x}{2}\left[f(x_0) + 2f(x_1) + 2f(x_2) \right.$$

$$\left. + \cdots + 2f(x_{n-1}) + f(x_n)\right] \qquad\qquad \textbf{(5)}$$

where $\Delta x = \dfrac{b - a}{n}$.

The approximation generally improves with larger values of *n*.

VIDEO

EXAMPLE 1   Approximate the value of

$$\int_1^2 \frac{1}{x}\,dx$$

using the trapezoidal rule with $n = 10$. Compare this result with the exact value of the integral.

Solution   Here, $a = 1$, $b = 2$, and $n = 10$, so

$$\Delta x = \frac{b - a}{n} = \frac{1}{10} = 0.1$$

and

$$x_0 = 1 \qquad x_1 = 1.1 \qquad x_2 = 1.2 \qquad x_3 = 1.3, \ldots, x_9 = 1.9 \qquad x_{10} = 2$$

The trapezoidal rule yields

$$\int_1^2 \frac{1}{x}\,dx \approx \frac{0.1}{2}\left[1 + 2\left(\frac{1}{1.1}\right) + 2\left(\frac{1}{1.2}\right) + 2\left(\frac{1}{1.3}\right) + \cdots + 2\left(\frac{1}{1.9}\right) + \frac{1}{2}\right]$$

$$\approx 0.693771$$

In this case, we can easily compute the actual value of the definite integral under

consideration. In fact,

$$\int_1^2 \frac{1}{x}\,dx = \ln x \Big|_1^2 = \ln 2 - \ln 1 = \ln 2$$

$$\approx 0.693147$$

Thus, the trapezoidal rule with $n = 10$ yields a result with an error of 0.000624 to six decimal places. ∎

APPLIED EXAMPLE 2   Consumers' Surplus   The demand function for a certain brand of perfume is given by

$$p = D(x) = \sqrt{10{,}000 - 0.01x^2}$$

where p is the unit price in dollars and x is the quantity demanded each week, measured in ounces. Find the consumers' surplus if the market price is set at $60 per ounce.

Solution   When $p = 60$, we have

$$\sqrt{10{,}000 - 0.01x^2} = 60$$
$$10{,}000 - 0.01x^2 = 3{,}600$$
$$x^2 = 640{,}000$$

or $x = 800$ since x must be nonnegative. Next, using the consumers' surplus formula (page 979) with $\bar{p} = 60$ and $\bar{x} = 800$, we see that the consumers' surplus is given by

$$CS = \int_0^{800} \sqrt{10{,}000 - 0.01x^2}\,dx - (60)(800)$$

It is not easy to evaluate this definite integral by finding an antiderivative of the integrand. Instead, let's use the trapezoidal rule with $n = 10$.

With $a = 0$ and $b = 800$, we find that

$$\Delta x = \frac{b - a}{n} = \frac{800}{10} = 80$$

and

$$x_0 = 0 \qquad x_1 = 80 \qquad x_2 = 160 \qquad x_3 = 240, \ldots, x_9 = 720 \qquad x_{10} = 800$$

so

$$\int_0^{800} \sqrt{10{,}000 - 0.01x^2}\,dx$$

$$\approx \frac{80}{2}\Big[100 + 2\sqrt{10{,}000 - (0.01)(80)^2}$$

$$+ 2\sqrt{10{,}000 - (0.01)(160)^2} + \cdots + 2\sqrt{10{,}000 - (0.01)(720)^2}$$

$$+ \sqrt{10{,}000 - (0.01)(800)^2}\Big]$$

$$= 40(100 + 199.3590 + 197.4234 + 194.1546 + 189.4835$$

$$+ 183.3030 + 175.4537 + 165.6985$$

$$+ 153.6750 + 138.7948 + 60)$$

$$\approx 70{,}293.82$$

Therefore, the consumers' surplus is approximately $70{,}294 - 48{,}000$, or \$22,294. ∎

EXPLORE & DISCUSS

Explain how you would approximate the value of $\int_0^2 f(x)\, dx$ using the trapezoidal rule with $n = 10$, where

$$f(x) = \begin{cases} \sqrt{1 + x^2} & \text{if } 0 \le x \le 1 \\ \dfrac{2}{\sqrt{1 + x^2}} & \text{if } 1 < x \le 2 \end{cases}$$

and find the value.

■ Simpson's Rule

Before stating Simpson's rule, let's review the two rules we have used in approximating a definite integral. Let f be a continuous nonnegative function defined on the interval $[a, b]$. Suppose the interval $[a, b]$ is partitioned by means of the $n + 1$ equally spaced points $x_0 = a, x_1, x_2, \ldots, x_n = b$, where n is a positive integer, so that the length of each subinterval is $\Delta x = (b - a)/n$ (Figure 4).

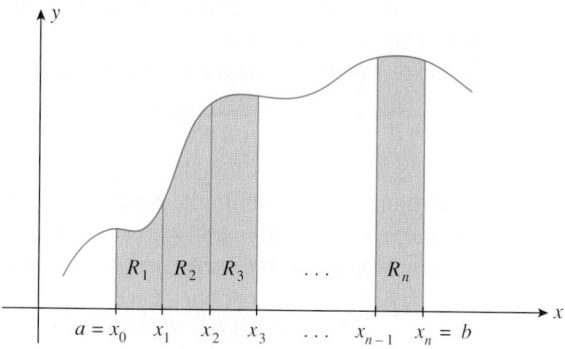

FIGURE 4
The area under the curve is equal to the sum of the n subregions R_1, $R_2, \ldots, R_n$.

Let's concentrate on the portion of the graph of $y = f(x)$ defined on the interval $[x_0, x_2]$. In using a Riemann sum to approximate the definite integral, we are in effect approximating the function $f(x)$ on $[x_0, x_1]$ by the *constant* function $y = f(p_1)$, where p_1 is chosen to be a point in $[x_0, x_1]$; the function $f(x)$ on $[x_1, x_2]$ by the constant function $y = f(p_2)$, where p_2 lies in $[x_1, x_2]$; and so on. Using a Riemann sum, we see that the area of the region under the curve $y = f(x)$ between $x = a$ and $x = b$ is approximated by the area under the approximating "step" function (Figure 5a).

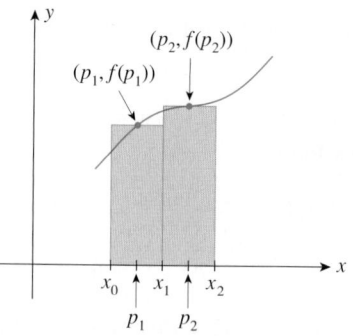

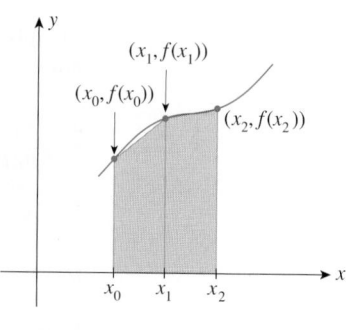

(a) The area under the curve is approximated by the area of the rectangles.

(b) The area under the curve is approximated by the area of the trapezoids.

FIGURE 5

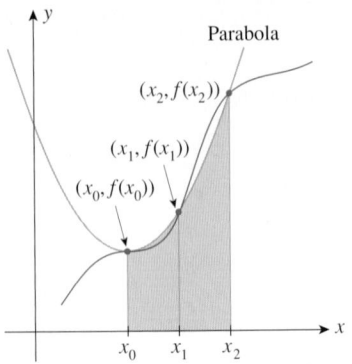

FIGURE 6
Simpson's rule approximates the area under the curve by the area under the parabola.

When we use the trapezoidal rule, we are in effect approximating the function $f(x)$ on the interval $[x_0, x_1]$ by a *linear* function through the two points $(x_0, f(x_0))$ and $(x_1, f(x_1))$; the function $f(x)$ on $[x_1, x_2]$ by a *linear* function through the two points $(x_1, f(x_1))$ and $(x_2, f(x_2))$; and so on. Thus, the trapezoidal rule simply approximates the actual area of the region under the curve $y = f(x)$ from $x = a$ to $x = b$ by the area under the approximating polygonal curve (Figure 5b).

A natural extension of the preceding idea is to approximate portions of the graph of $y = f(x)$ by means of portions of the graphs of second-degree polynomials (parts of parabolas). It can be shown that given any three noncollinear points there is a unique parabola that passes through the given points. Choose the points $(x_0, f(x_0))$, $(x_1, f(x_1))$, and $(x_2, f(x_2))$ corresponding to the first three points of the partition. Then, we can approximate the function $f(x)$ on $[x_0, x_2]$ by means of a quadratic function whose graph contains these three points (Figure 6).

Although we will not do so here, it can be shown that the area under the parabola between $x = x_0$ and $x = x_2$ is given by

$$\frac{\Delta x}{3}\left[f(x_0) + 4f(x_1) + f(x_2)\right]$$

square units. Repeating this argument on the interval $[x_2, x_4]$, we see that the area under the curve between $x = x_2$ and $x = x_4$ is approximated by the area under the parabola between x_2 and x_4—that is, by

$$\frac{\Delta x}{3}\left[f(x_2) + 4f(x_3) + f(x_4)\right]$$

square units. Proceeding, we conclude that if n is even (Why?), then the area under the curve $y = f(x)$ from $x = a$ to $x = b$ may be approximated by the sum of the areas under the $n/2$ approximating parabolas—that is,

$$\frac{\Delta x}{3}\left[f(x_0) + 4f(x_1) + f(x_2)\right] + \frac{\Delta x}{3}\left[f(x_2) + 4f(x_3) + f(x_4)\right] + \cdots$$
$$+ \frac{\Delta x}{3}\left[f(x_{n-2}) + 4f(x_{n-1}) + f(x_n)\right]$$
$$= \frac{\Delta x}{3}\left[f(x_0) + 4f(x_1) + f(x_2) + f(x_2) + 4f(x_3) + f(x_4) + \cdots\right.$$
$$\left. + f(x_{n-2}) + 4f(x_{n-1}) + f(x_n)\right]$$
$$= \frac{\Delta x}{3}\left[f(x_0) + 4f(x_1) + 2f(x_2) + 4f(x_3)\right.$$
$$\left. + 2f(x_4) + \cdots + 4f(x_{n-1}) + f(x_n)\right]$$

The preceding is the derivation of the approximation formula known as **Simpson's rule**.

Simpson's Rule

$$\int_a^b f(x)\, dx \approx \frac{\Delta x}{3}\left[f(x_0) + 4f(x_1) + 2f(x_2) + 4f(x_3) + 2f(x_4)\right.$$
$$\left. + \cdots + 4f(x_{n-1}) + f(x_n)\right] \tag{6}$$

where $\Delta x = \dfrac{b - a}{n}$ and n is even.

In using this rule, remember that n must be even.

EXAMPLE 3 Find an approximation of

$$\int_1^2 \frac{1}{x}\, dx$$

using Simpson's rule with $n = 10$. Compare this result with that of Example 1 and also with the exact value of the integral.

Solution We have $a = 1$, $b = 2$, $f(x) = \dfrac{1}{x}$, and $n = 10$, so

$$\Delta x = \frac{b - a}{n} = \frac{1}{10} = 0.1$$

and

$$x_0 = 1 \qquad x_1 = 1.1 \qquad x_2 = 1.2 \qquad x_3 = 1.3, \ldots, x_9 = 1.9 \qquad x_{10} = 2$$

Simpson's rule yields

$$\int_1^2 \frac{1}{x}\, dx \approx \frac{0.1}{3}\left[f(1) + 4f(1.1) + 2f(1.2) + \cdots + 4f(1.9) + f(2)\right]$$

$$= \frac{0.1}{3}\left[1 + 4\left(\frac{1}{1.1}\right) + 2\left(\frac{1}{1.2}\right) + 4\left(\frac{1}{1.3}\right) + 2\left(\frac{1}{1.4}\right) + 4\left(\frac{1}{1.5}\right)\right.$$

$$\left. + 2\left(\frac{1}{1.6}\right) + 4\left(\frac{1}{1.7}\right) + 2\left(\frac{1}{1.8}\right) + 4\left(\frac{1}{1.9}\right) + \frac{1}{2}\right]$$

$$\approx 0.693150$$

The trapezoidal rule with $n = 10$ yielded an approximation of 0.693771, which is 0.000624 off the value of $\ln 2 \approx 0.693147$ to six decimal places. Simpson's rule yields an approximation with an error of 0.000003, a definite improvement over the trapezoidal rule. ■

APPLIED EXAMPLE 4 Cardiac Output Solve the problem posed at the beginning of this section. Recall that we wished to find a person's cardiac output by using the formula

$$R = \frac{60\, D}{\displaystyle\int_0^{28} c(t)\, dt}$$

where D (the quantity of dye injected) is equal to 5 milligrams and the function c has the graph shown in Figure 7. Use Simpson's rule with $n = 14$ to estimate the value of the integral.

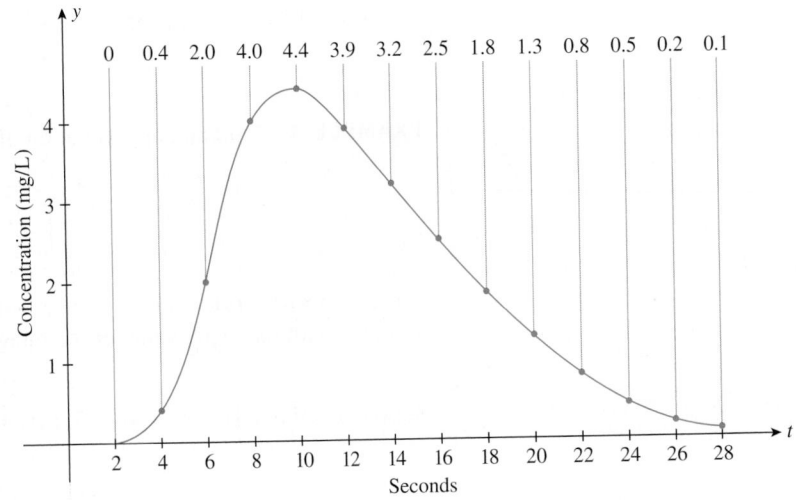

FIGURE 7
The function c gives the concentration of a dye measured at the aorta. The graph is constructed by drawing a smooth curve through a set of discrete points.

Solution Using Simpson's rule with $n = 14$ and $\Delta t = 2$ so that

$$t_0 = 0 \qquad t_1 = 2 \qquad t_2 = 4 \qquad t_3 = 6, \ldots, t_{14} = 28$$

we obtain

$$\int_0^{28} c(t)\, dt \approx \frac{2}{3}[c(0) + 4c(2) + 2c(4) + 4c(6) + \cdots$$

$$+ 4c(26) + c(28)]$$

$$= \frac{2}{3}[0 + 4(0) + 2(0.4) + 4(2.0) + 2(4.0)$$

$$+ 4(4.4) + 2(3.9) + 4(3.2) + 2(2.5) + 4(1.8)$$

$$+ 2(1.3) + 4(0.8) + 2(0.5) + 4(0.2) + 0.1]$$

$$\approx 49.9$$

Therefore, the person's cardiac output is

$$R \approx \frac{60(5)}{49.9} \approx 6.0$$

or 6.0 L/min.

  APPLIED EXAMPLE 5 Oil Spill An oil spill off the coastline was caused by a ruptured tank in a grounded oil tanker. Using aerial photographs, the Coast Guard was able to obtain the dimensions of the oil spill (Figure 8). Using Simpson's rule with $n = 10$, estimate the area of the oil spill.

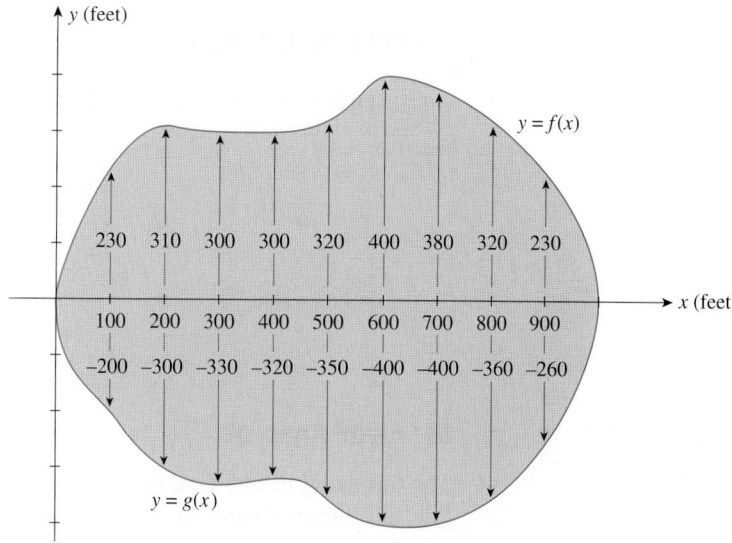

FIGURE 8
Simpson's rule can be used to calculate the area of the oil spill.

Solution We may think of the area affected by the oil spill as the area of the plane region bounded above by the graph of the function $f(x)$ and below by the graph of the function $g(x)$ between $x = 0$ and $x = 1000$ (Figure 8). Then, the required area is given by

$$A = \int_0^{1000} [f(x) - g(x)]\, dx$$

Using Simpson's rule with $n = 10$ and $\Delta x = 100$ so that

$$x_0 = 0 \qquad x_1 = 100 \qquad x_2 = 200, \ldots, x_{10} = 1000$$

we have

$$
\begin{aligned}
A &= \int_0^{1000} [f(x) - g(x)]\, dx \\[4pt]
&\approx \frac{\Delta x}{3} \{ [f(x_0) - g(x_0)] + 4[f(x_1) - g(x_1)] + 2[f(x_2) - g(x_2)] \\
&\quad + \cdots + 4[f(x_9) - g(x_9)] + [f(x_{10}) - g(x_{10})] \} \\[4pt]
&= \frac{100}{3} \{ [0 - 0] + 4[230 - (-200)] + 2[310 - (-300)] \\
&\quad + 4[300 - (-330)] + 2[300 - (-320)] + 4[320 - (-350)] \\
&\quad + 2[400 - (-400)] + 4[380 - (-400)] + 2[320 - (-360)] \\
&\quad + 4[230 - (-260)] + [0 - 0] \} \\[4pt]
&= \frac{100}{3} [0 + 4(430) + 2(610) + 4(630) + 2(620) + 4(670) \\
&\quad + 2(800) + 4(780) + 2(680) + 4(490) + 0] \\[4pt]
&= \frac{100}{3} (17{,}420) \\[4pt]
&\approx 580{,}667
\end{aligned}
$$

or approximately 580,667 square feet.

EXPLORE & DISCUSS

Explain how you would approximate the value of $\int_0^2 f(x)\, dx$ using Simpson's rule with $n = 10$, where

$$f(x) = \begin{cases} \sqrt{1 + x^2} & \text{if } 0 \le x \le 1 \\[2mm] \dfrac{2}{\sqrt{1 + x^2}} & \text{if } 1 < x \le 2 \end{cases}$$

and find the value.

▬ Error Analysis

The following results give the bounds on the errors incurred when the trapezoidal rule and Simpson's rule are used to approximate a definite integral (proof omitted).

Errors in the Trapezoidal and Simpson Approximations
Suppose the definite integral

$$\int_a^b f(x)\, dx$$

is approximated with n subintervals.

1. The *maximum* error incurred in using the trapezoidal rule is

$$\frac{M(b - a)^3}{12n^2} \qquad \text{(7)}$$

where M is a number such that $|f''(x)| \le M$ for all x in $[a, b]$.

2. The *maximum* error incurred in using Simpson's rule is

$$\frac{M(b - a)^5}{180n^4} \qquad \text{(8)}$$

where M is a number such that $|f^{(4)}(x)| \le M$ for all x in $[a, b]$.

Note In many instances, the actual error is less than the upper error bounds given. ▬

EXAMPLE 6 Find bounds on the errors incurred when

$$\int_1^2 \frac{1}{x}\, dx$$

is approximated using (a) the trapezoidal rule and (b) Simpson's rule with $n = 10$. Compare these with the actual errors found in Examples 1 and 3.

Solution

a. Here, $a = 1$, $b = 2$, and $f(x) = 1/x$. Next, to find a value for M, we compute

$$f'(x) = -\frac{1}{x^2} \quad \text{and} \quad f''(x) = \frac{2}{x^3}$$

Since $f''(x)$ is positive and decreasing on $(1, 2)$ (Why?), it attains its maximum value of 2 at $x = 1$, the left endpoint of the interval. Therefore, if we take $M = 2$, then $|f''(x)| \leq 2$. Using (7), we see that the maximum error incurred is

$$\frac{2(2 - 1)^3}{12(10)^2} = \frac{2}{1200} = 0.0016667$$

The actual error found in Example 1, 0.000624, is much less than the upper bound just found.

b. We compute

$$f'''(x) = \frac{-6}{x^4} \quad \text{and} \quad f^{(4)}(x) = \frac{24}{x^5}$$

Since $f^{(4)}(x)$ is positive and decreasing on $(1, 2)$ (just look at $f^{(5)}$ to verify this fact), it attains its maximum at the left endpoint of $[1, 2]$. Now,

$$f^{(4)}(1) = 24$$

and so we may take $M = 24$. Using (8), we obtain the maximum error of

$$\frac{24(2 - 1)^5}{180(10)^4} = 0.0000133$$

The actual error is 0.000003 (see Example 3).

EXPLORE & DISCUSS

Refer to the Explore & Discuss on pages 1017 and 1022. Explain how you would find the maximum error incurred in using (1) the trapezoidal rule and (2) Simpson's rule with $n = 10$ to approximate $\displaystyle\int_0^2 f(x)\, dx$.

16.3 Self-Check Exercises

1. Use the trapezoidal rule and Simpson's rule with $n = 8$ to approximate the value of the definite integral

$$\int_0^2 \frac{2}{\sqrt{1 + x^2}}\, dx$$

2. The graph in the accompanying figure shows the consumption of petroleum in the United States in quadrillion BTU, from 1976 to 1990. Using Simpson's rule with $n = 14$, estimate the average consumption during the 14-yr period.

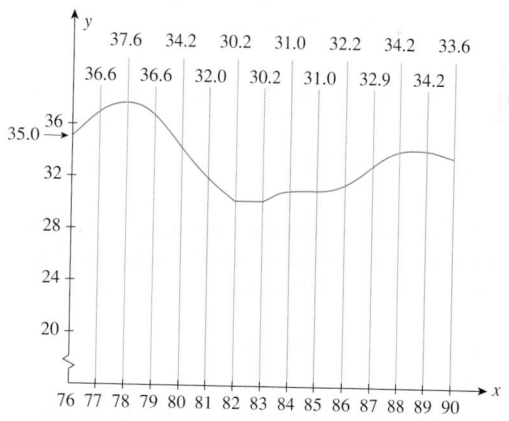

Source: The World Almanac

Solutions to Self-Check Exercises 16.3 can be found on page 1027.

16.3 Concept Questions

1. Explain why n can be odd or even in the trapezoidal rule, but it must be even in Simpson's rule.

2. Explain, without alluding to the error formulas, why the trapezoidal rule gives the exact value of $\int_a^b f(x)\, dx$ if f is a linear function and why Simpson's rule gives the exact value of the integral if f is a quadratic function.

3. Refer to Concept Question 2 and answer the questions using the error formulas for the trapezoidal rule and Simpson's rule.

16.3   Exercises

In Exercises 1–14, use the trapezoidal rule and Simpson's rule to approximate the value of each definite integral. Compare your result with the exact value of the integral.

1. $\displaystyle\int_0^2 x^2\,dx; n = 6$

2. $\displaystyle\int_1^3 (x^2 - 1)\,dx; n = 4$

3. $\displaystyle\int_0^1 x^3\,dx; n = 4$

4. $\displaystyle\int_1^2 x^3\,dx; n = 6$

5. $\displaystyle\int_1^2 \frac{1}{x}\,dx; n = 4$

6. $\displaystyle\int_1^2 \frac{1}{x}\,dx; n = 8$

7. $\displaystyle\int_1^2 \frac{1}{x^2}\,dx; n = 4$

8. $\displaystyle\int_0^1 \frac{1}{1+x}\,dx; n = 4$

9. $\displaystyle\int_0^4 \sqrt{x}\,dx; n = 8$

10. $\displaystyle\int_0^2 x\sqrt{2x^2 + 1}\,dx; n = 6$

11. $\displaystyle\int_0^1 e^{-x}\,dx; n = 6$

12. $\displaystyle\int_0^1 xe^{-x^2}\,dx; n = 6$

13. $\displaystyle\int_1^2 \ln x\,dx; n = 4$

14. $\displaystyle\int_0^1 x\ln(x^2 + 1)\,dx; n = 8$

In Exercises 15–22, use the trapezoidal rule and Simpson's rule to approximate the value of each definite integral.

15. $\displaystyle\int_0^1 \sqrt{1 + x^3}\,dx; n = 4$

16. $\displaystyle\int_0^2 x\sqrt{1 + x^3}\,dx; n = 4$

17. $\displaystyle\int_0^2 \frac{1}{\sqrt{x^3 + 1}}\,dx; n = 4$

18. $\displaystyle\int_0^1 \sqrt{1 - x^2}\,dx; n = 4$

19. $\displaystyle\int_0^2 e^{-x^2}\,dx; n = 4$

20. $\displaystyle\int_0^1 e^{x^2}\,dx; n = 6$

21. $\displaystyle\int_1^2 x^{-1/2}e^x\,dx; n = 4$

22. $\displaystyle\int_2^4 \frac{dx}{\ln x}; n = 6$

In Exercises 23–28, find a bound on the error in approximating each definite integral using (a) the trapezoidal rule and (b) Simpson's rule with *n* intervals.

23. $\displaystyle\int_{-1}^2 x^5\,dx; n = 10$

24. $\displaystyle\int_0^1 e^{-x}\,dx; n = 8$

25. $\displaystyle\int_1^3 \frac{1}{x}\,dx; n = 10$

26. $\displaystyle\int_1^3 \frac{1}{x^2}\,dx; n = 8$

27. $\displaystyle\int_0^2 \frac{1}{\sqrt{1+x}}\,dx; n = 8$

28. $\displaystyle\int_1^3 \ln x\,dx; n = 10$

29. TRIAL RUN OF AN ATTACK SUBMARINE In a submerged trial run of an attack submarine, a reading of the sub's velocity was made every quarter hour, as shown in the accompanying table. Use the trapezoidal rule to estimate the distance traveled by the submarine during the 2-hr period.

Time, t (hr)	0	$\frac{1}{4}$	$\frac{1}{2}$	$\frac{3}{4}$
Velocity, $V(t)$ (mph)	19.5	24.3	34.2	40.5

Time, t (hr)	1	$\frac{5}{4}$	$\frac{3}{2}$	$\frac{7}{4}$	2
Velocity, $V(t)$ (mph)	38.4	26.2	18	16	8

30. REAL ESTATE Cooper Realty is considering development of a time-sharing condominium resort complex along the ocean-front property illustrated in the accompanying graph. To obtain an estimate of the area of this property, measurements of the distances from the edge of a straight road, which defines one boundary of the property, to the corresponding points on the shoreline are made at 100-ft intervals. Using Simpson's rule with $n = 10$, estimate the area of the oceanfront property.

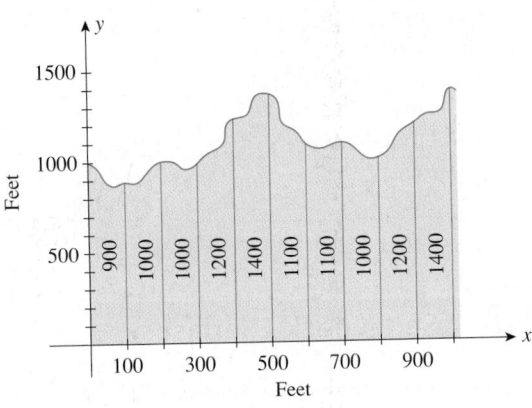

31. FUEL CONSUMPTION OF DOMESTIC CARS Thanks to smaller and more fuel-efficient models, American carmakers have doubled their average fuel economy over a 13-yr period, from 1974 to 1987. The following figure gives the average fuel consumption in miles per gallon (mpg) of domestic-built cars over the period under consideration ($t = 0$ corresponds to the beginning of 1974). Use the trapezoidal rule to estimate the

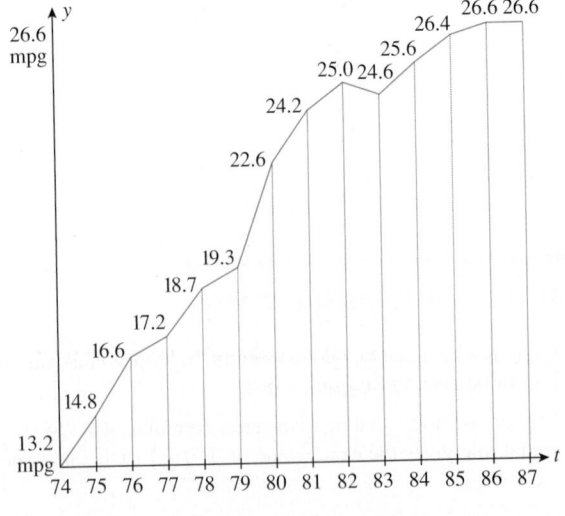

average fuel consumption of the domestic car built during this period.
Hint: Approximate the integral $\frac{1}{13} \int_0^{13} f(t)\, dt$.

32. AVERAGE TEMPERATURE The graph depicted in the following figure shows the daily mean temperatures recorded during one September in Cameron Highlands. Using (a) the trapezoidal rule and (b) Simpson's rule with $n = 10$, estimate the average temperature during that month.

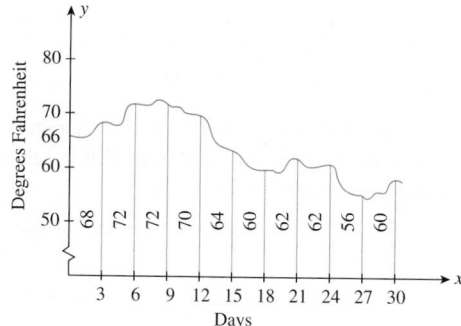

33. U.S. DAILY OIL CONSUMPTION The following table gives the daily consumption of oil in the United States, in millions of barrels, measured in 2-yr intervals, from 1980 through 2000. Use Simpson's rule to estimate the average daily consumption of oil over the period in question.

Year	1980	1982	1984	1986	1988	1990
Consumption	17.1	15.3	15.7	16.3	17.3	17.0

Year	1992	1994	1996	1998	2000
Consumption	17.0	17.7	18.3	18.9	19.7

Source: Office of Transportation Technologies

34. SURFACE AREA OF THE CENTRAL PARK RESERVOIR The reservoir located in Central Park in New York City has the shape depicted in the following figure. The measurements shown were taken at 206-ft intervals. Use Simpson's rule with $n = 10$ to estimate the surface area of the lake.

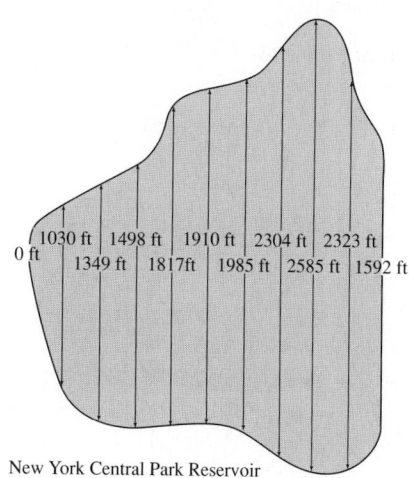

New York Central Park Reservoir

Source: Boston Globe

35. WATER FLOW IN A RIVER At a certain point, a river is 78 ft wide and its depth, measured at 6-ft intervals across the river, is recorded in the following table.

x (ft)	0	6	12	18	24	30	36
y (ft)	0.8	2.6	5.8	6.2	8.2	10.1	10.8

x (ft)	42	48	54	60	66	72	78
y (ft)	9.8	7.6	6.4	5.2	3.9	2.4	1.4

Here, x denotes the distance (in feet) from one bank of the river and y (in feet) is the corresponding depth. If the average rate of flow through this section of the river is 4 ft/sec, use the trapezoidal rule with $n = 13$ to find the rate of the volume of flow of water in the river.
Hint: Volume of flow = rate of flow × area of cross section.

36. CONSUMERS' SURPLUS Refer to Section 15.7. The demand equation for the Sicard wristwatch is given by

$$p = S(x) = \frac{50}{0.01x^2 + 1} \qquad (0 \le x \le 20)$$

where x (measured in units of a thousand) is the quantity demanded per week and p is the unit price in dollars. Use (a) the trapezoidal rule and (b) Simpson's rule (take $n = 8$) to estimate the consumers' surplus if the market price is $25/watch.

37. PRODUCERS' SURPLUS Refer to Section 15.7. The supply function for the CD manufactured by Herald Records is given by

$$p = S(x) = \sqrt{0.01x^2 + 0.11x + 38}$$

where p is the unit wholesale price in dollars and x stands for the quantity that will be made available in the market by the supplier, measured in units of a thousand. Use (a) the trapezoidal rule and (b) Simpson's rule (take $n = 8$) to estimate the producers' surplus if the wholesale price is $8/CD.

38. AIR POLLUTION The amount of nitrogen dioxide, a brown gas that impairs breathing, present in the atmosphere on a certain May day in the city of Long Beach has been approximated by

$$A(t) = \frac{136}{1 + 0.25(t - 4.5)^2} + 28 \qquad (0 \le t \le 11)$$

where $A(t)$ is measured in pollutant standard index (PSI), and t is measured in hours, with $t = 0$ corresponding to 7 a.m. Use the trapezoidal rule with $n = 10$ to estimate the average PSI between 7 a.m. and noon.
Hint: The average value is given by $\frac{1}{5} \int_0^5 A(t)\, dt$.

39. U.S. STRATEGIC PETROLEUM RESERVES According to data from the American Petroleum Institute, the U.S. strategic petroleum reserves from the beginning of 1981 through the beginning of 1990 can be approximated by the function

$$S(t) = \frac{613.7t^2 + 1449.1}{t^2 + 6.3} \qquad (0 \le t \le 9)$$

where $S(t)$ is measured in millions of barrels and t in years, with $t = 0$ corresponding to the beginning of 1981. Using the trapezoidal rule with $n = 9$, estimate the average petroleum reserves from the beginning of 1981 through the beginning of 1990.

Source: American Petroleum Institute

40. **GROWTH OF SERVICE INDUSTRIES** It has been estimated that service industries, which currently make up 30% of the nonfarm workforce in a certain country, will continue to grow at the rate of

$$R(t) = 5e^{1/(t+1)}$$

percent/decade t decades from now. Estimate the percent of the nonfarm workforce in the service industries one decade from now.

Hint: (a) Show that the answer is given by $30 + \int_0^1 5e^{1/(t+1)} \, dt$ and (b) use Simpson's rule with $n = 10$ to approximate the definite integral.

41. **TREAD LIVES OF TIRES** Under normal driving conditions the percent of Super Titan radial tires expected to have a useful tread life of between 30,000 and 40,000 mi is given by

$$P = 100 \int_{30,000}^{40,000} \frac{1}{2000\sqrt{2\pi}} e^{-(1/2)[(x - 40,000)/2000]^2} \, dx$$

Use Simpson's rule with $n = 10$ to estimate P.

42. **LENGTH OF INFANTS AT BIRTH** Medical records of infants delivered at Kaiser Memorial Hospital show that the percent of infants whose length at birth is between 19 and 21 in. is given by

$$P = 100 \int_{19}^{21} \frac{1}{2.6\sqrt{2\pi}} e^{-(1/2)[(x - 20)/2.6]^2} \, dx$$

Use Simpson's rule with $n = 10$ to estimate P.

43. **MEASURING CARDIAC OUTPUT** Eight milligrams of a dye are injected into a vein leading to an individual's heart. The concentration of the dye in the aorta (in mg/L) measured at 2-sec intervals is shown in the accompanying table. Use Simpson's rule and the formula of Example 4 to estimate the person's cardiac output.

t	0	2	4	6	8	10	12
$C(t)$	0	0	2.8	6.1	9.7	7.6	4.8

t	14	16	18	20	22	24
$C(t)$	3.7	1.9	0.8	0.3	0.1	0

44. **ESTIMATING THE FLOW RATE OF A RIVER** A stream is 120 ft wide. The following table gives the depths of the river measured across a section of the river in intervals of 6 ft. Here, x denotes the distance from one bank of the river, and y denotes the corresponding depth (in feet). The average rate of flow of the river across this section of the river is 4.2 ft/sec. Use Simpson's rule with $n = 20$ to estimate the rate of the volume of flow of the river.

x (ft)	0	6	12	18	24	30	36	42	48	54	60
y (ft)	0.8	1.2	3.0	4.1	5.8	6.6	6.8	7.0	7.2	7.4	7.8

x (ft)	66	72	78	84	90	96	102	108	114	120
y (ft)	7.6	7.4	7.0	6.6	6.0	5.1	4.3	3.2	2.2	1.1

In Exercises 45–48, determine whether the statement is true or false. It it is true, explain why it is true. If it is false, give an example to show why it is false.

45. In using the trapezoidal rule, the number of subintervals n must be even.

46. In using Simpson's rule, the number of subintervals n may be chosen to be odd or even.

47. Simpson's rule is more accurate than the trapezoidal rule.

48. If f is a polynomial function of degree less than or equal to 3, then the approximation $\int_a^b f(x) \, dx$ using Simpson's rule is exact.

49. Derive the formula

$$R = \frac{60 \, D}{\int_0^T c(t) \, dt}$$

for calculating the cardiac output of a person in L/min. Here, $c(t)$ is the concentration of dye in the aorta (in mg/L) at time t (in seconds) for t in $[0, T]$, and D is the amount of dye (in mg) injected into a vein leading to the heart.

Hint: Partition the interval $[0, T]$ into n subintervals of equal length Δt. The amount of dye that flows past the measuring point in the aorta during the time interval $[0, \Delta t]$ is approximately $c(t_i)(R\Delta t)/60$ (concentration times volume). Therefore, the total amount of dye measured at the aorta is

$$\frac{[c(t_1)R\Delta t + c(t_2)R\Delta t + \cdots + c(t_n)R\Delta t]}{60} = D$$

Take the limit of the Riemann sum to obtain

$$R = \frac{60 \, D}{\int_0^T c(t) \, dt}$$

16.3 Solutions to Self-Check Exercises

1. We have $x = 0$, $b = 2$, and $n = 8$, so

$$\Delta x = \frac{b - a}{n} = \frac{2}{8} = 0.25$$

and $x_0 = 0$, $x_1 = 0.25$, $x_2 = 0.50$, $x_3 = 0.75, \ldots, x_7 = 1.75$, and $x_8 = 2$. The trapezoidal rule gives

$$\int_0^2 \frac{1}{\sqrt{1 - x^2}}\, dx \approx \frac{0.25}{2}\left[1 + \frac{2}{\sqrt{1 + (0.25)^2}} + \frac{2}{\sqrt{1 + (0.5)^2}} \right.$$

$$+ \cdots + \frac{2}{\sqrt{1 + (1.75)^2}} + \left. \frac{1}{\sqrt{5}} \right]$$

$$\approx 0.125(1 + 1.9403 + 1.7889 + 1.6000 + 1.4142$$

$$+ 1.2494 + 1.1094 + 0.9923 + 0.4472)$$

$$\approx 1.4427$$

Using Simpson's rule with $n = 8$ gives

$$\int_0^2 \frac{1}{\sqrt{1 - x^2}}\, dx \approx \frac{0.25}{3}\left[1 + \frac{4}{\sqrt{1 + (0.25)^2}} + \frac{2}{\sqrt{1 + (0.5)^2}} \right.$$

$$+ \frac{4}{\sqrt{1 + (0.75)^2}} + \cdots + \frac{4}{\sqrt{1 + (1.75)^2}} + \left. \frac{1}{\sqrt{5}} \right]$$

$$\approx \frac{0.25}{3}(1 + 3.8806 + 1.7889 + 3.2000 + 1.4142$$

$$+ 2.4988 + 1.1094 + 1.9846 + 0.4472)$$

$$\approx 1.4436$$

2. The average consumption of petroleum during the 14-yr period is given by

$$\frac{1}{14}\int_0^{14} f(x)\, dx$$

where f is the function describing the given graph. Using Simpson's rule with $a = 0$, $b = 14$, and $n = 14$ so that $\Delta x = 1$ and

$$x_0 = 0 \qquad x_1 = 1 \qquad x_2 = 2, \ldots, x_{14} = 14$$

we have

$$\frac{1}{14}\int_0^{14} f(x)\, dx$$

$$\approx \left(\frac{1}{14}\right)\left(\frac{1}{3}\right)[f(x_0) + 4f(x_1) + 2f(x_2) + 4f(x_3) + \cdots$$

$$+ 4f(x_{13}) + f(x_{14})]$$

$$= \frac{1}{42}[35 + 4(36.6) + 2(37.6) + 4(36.6) + 2(34.2)$$

$$+ 4(32.0) + 2(30.2) + 4(30.2) + 2(31.0) + 4(31.0)$$

$$+ 2(32.2) + 4(32.9) + 2(34.2) + 4(34.2) + 33.6]$$

$$\approx 33.4$$

or approximately 33.4 quadrillion BTU/year.

16.4 Improper Integrals

■ Improper Integrals

All the definite integrals we have encountered have had finite intervals of integration. In many applications, however, we are concerned with integrals that have unbounded intervals of integration. These integrals are called **improper integrals**.

To lead us to the definition of an improper integral of a function f over an infinite interval, consider the problem of finding the area of the region R under the curve $y = f(x) = 1/x^2$ and to the right of the vertical line $x = 1$, as shown in Figure 9. Because the interval over which the integration must be performed is unbounded, the method of integration presented previously cannot be applied directly in solving this problem. However, we can approximate the region R by the definite integral

$$\int_1^b \frac{1}{x^2}\, dx \tag{9}$$

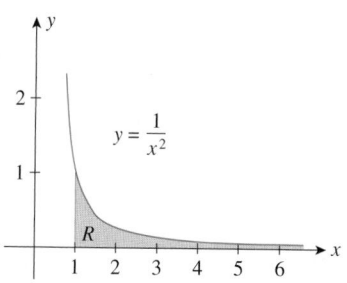

FIGURE 9
The area of the unbounded region R can be approximated by a definite integral.

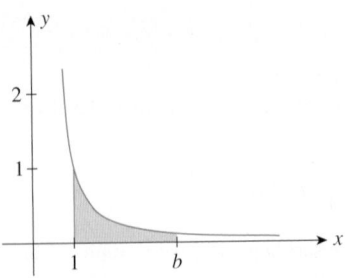

FIGURE 10

Area of shaded region $= \int_1^b \frac{1}{x^2}\, dx$

which gives the area of the region under the curve $y = f(x) = 1/x^2$ from $x = 1$ to $x = b$ (Figure 10). You can see that the approximation of the region R by the definite integral (9) improves as the upper limit of integration, b, becomes larger and larger. Figure 11 illustrates the situation for $b = 2$, 3, and 4, respectively.

This observation suggests that if we define a function $I(b)$ by

$$I(b) = \int_1^b \frac{1}{x^2}\, dx \tag{10}$$

then we can find the area of the required region R by evaluating the limit of $I(b)$ as b tends to infinity; that is, the area of R is given by

$$\lim_{b \to \infty} I(b) = \lim_{b \to \infty} \int_1^b \frac{1}{x^2}\, dx \tag{11}$$

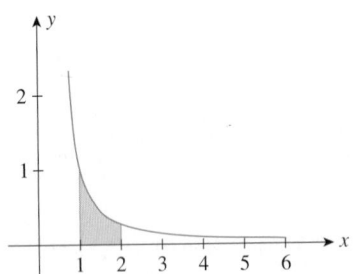

(a) Area of region under the graph of f on [1, 2]

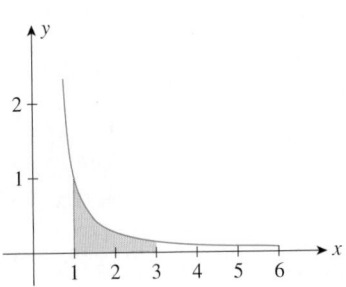

(b) Area of region under the graph of f on [1, 3]

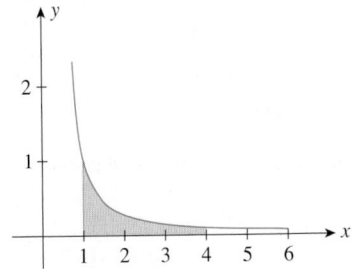

(c) Area of region under the graph of f on [1, 4]

FIGURE 11
As b increases, the approximation of R by the definite integral improves.

EXAMPLE 1

  a. Evaluate the definite integral $I(b)$ in Equation (10).
  b. Compute $I(b)$ for $b = 10$, 100, 1000, 10,000.
  c. Evaluate the limit in Equation (11).
  d. Interpret the results of parts (b) and (c).

Solution

  a. $I(b) = \int_1^b \frac{1}{x^2}\, dx = -\frac{1}{x} \Big|_1^b = -\frac{1}{b} + 1$

  b. From the result of part (a),

$$I(b) = 1 - \frac{1}{b}$$

   Therefore,

$$I(10) = 1 - \frac{1}{10} = 0.9$$

$$I(100) = 1 - \frac{1}{100} = 0.99$$

$$I(1000) = 1 - \frac{1}{1000} = 0.999$$

$$I(10{,}000) = 1 - \frac{1}{10{,}000} = 0.9999$$

c. Once again, using the result of part (a), we find

$$\lim_{b \to \infty} I(b) = \lim_{b \to \infty} \int_1^b \frac{1}{x^2}\, dx$$

$$= \lim_{b \to \infty} \left(1 - \frac{1}{b} \right)$$

$$= 1$$

d. The result of part (c) tells us that the area of the region R is 1 square unit. The results of the computations performed in part (b) reinforce our expectation that $I(b)$ should approach 1, the area of the region R, as b approaches infinity. ■

The preceding discussion and the results of Example 1 suggest that we define the improper integral of a continuous function f over the unbounded interval $[a, \infty)$ as follows.

Improper Integral of f over $[a, \infty)$

Let f be a continuous function on the unbounded interval $[a, \infty)$. Then, the improper integral of f over $[a, \infty)$ is defined by

$$\int_a^\infty f(x)\, dx = \lim_{b \to \infty} \int_a^b f(x)\, dx \qquad (12)$$

if the limit exists.

If the limit exists, the improper integral is said to be **convergent**. An improper integral for which the limit in Equation (12) fails to exist is said to be **divergent**.

EXAMPLE 2   Evaluate $\displaystyle\int_2^\infty \frac{1}{x}\, dx$ if it converges.

Solution

$$\int_2^\infty \frac{1}{x}\, dx = \lim_{b \to \infty} \int_2^b \frac{1}{x}\, dx$$

$$= \lim_{b \to \infty} \ln x \, \Big|_2^b$$

$$= \lim_{b \to \infty} (\ln b - \ln 2)$$

Since $\ln b \to \infty$ as $b \to \infty$, the limit does not exist, and we conclude that the given improper integral is divergent. ■

EXPLORE & DISCUSS

1. Suppose f is continuous and nonnegative on $[0, \infty)$. Furthermore, suppose $\lim_{x \to \infty} f(x) = L$, where L is a positive number. What can you say about the convergence of the improper integral $\int_0^\infty f(x)\, dx$? Explain your answer and illustrate with an example.

2. Suppose f is continuous and nonnegative on $[0, \infty)$ and satisfies the condition $\lim_{x \to \infty} f(x) = 0$. What can you say about $\int_0^\infty f(x)\, dx$? Explain and illustrate your answer with examples.

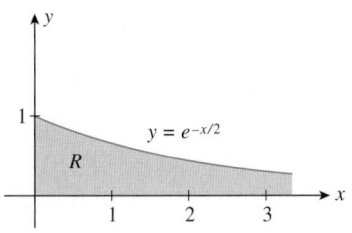

FIGURE 12

Area of $R = \displaystyle\int_0^\infty e^{-x/2}\, dx$

EXAMPLE 3 Find the area of the region R under the curve $y = e^{-x/2}$ for $x \geq 0$.

Solution The region R is shown in Figure 12. Taking $b > 0$, we compute the area of the region under the curve $y = e^{-x/2}$ from $x = 0$ to $x = b$—namely,

$$I(b) = \int_0^b e^{-x/2}\, dx = -2e^{-x/2}\Big|_0^b = -2e^{-b/2} + 2$$

Then, the area of the region R is given by

$$\lim_{b\to\infty} I(b) = \lim_{b\to\infty}(2 - 2e^{-b/2}) = 2 - 2\lim_{b\to\infty}\frac{1}{e^{b/2}}$$
$$= 2$$

or 2 square units. ∎

EXPLORING WITH TECHNOLOGY

You can see how fast the improper integral in Example 3 converges, as follows:

1. Use a graphing utility to plot the graph of $I(b) = 2 - 2e^{-b/2}$, using the viewing window $[0, 50] \times [0, 3]$.
2. Use **TRACE** to follow the values of y for increasing values of x, starting at the origin.

The improper integral defined in Equation (12) has an interval of integration that is unbounded on the right. Improper integrals with intervals of integration that are unbounded on the left also arise in practice and are defined in a similar manner.

Improper Integral of f over $(-\infty, b]$

Let f be a continuous function on the unbounded interval $(-\infty, b]$. Then, the improper integral of f over $(-\infty, b]$ is defined by

$$\int_{-\infty}^{b} f(x)\, dx = \lim_{a\to -\infty} \int_a^b f(x)\, dx \qquad (13)$$

if the limit exists.

In this case, the improper integral is said to be convergent. Otherwise, the improper integral is said to be divergent.

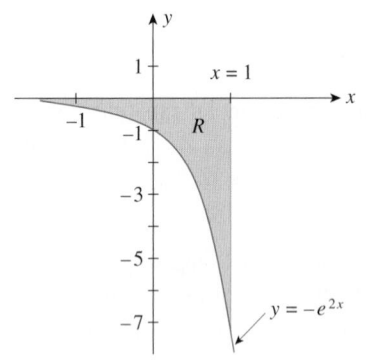

FIGURE 13

Area of $R = -\displaystyle\int_{-\infty}^{1} -e^{2x}\, dx$

EXAMPLE 4 Find the area of the region R bounded above by the x-axis, below by the curve $y = -e^{2x}$, and on the right by the vertical line $x = 1$.

Solution The region R is shown in Figure 13. Taking $a < 1$, compute the area of the region bounded above by the x-axis ($y = 0$), and below by the curve

$y = -e^{2x}$ from $x = a$ to $x = 1$—namely,

$$I(a) = \int_a^1 [0 - (-e^{2x})]\, dx = \int_a^1 e^{2x}\, dx$$

$$= \frac{1}{2} e^{2x} \bigg|_a^1 = \frac{1}{2} e^2 - \frac{1}{2} e^{2a}$$

Then, the area of the required region is given by

$$\lim_{a \to -\infty} I(a) = \lim_{a \to -\infty} \left(\frac{1}{2} e^2 - \frac{1}{2} e^{2a} \right)$$

$$= \frac{1}{2} e^2 - \frac{1}{2} \lim_{a \to -\infty} e^{2a}$$

$$= \frac{1}{2} e^2$$

Another improper integral found in practical applications involves the integration of a function f over the unbounded interval $(-\infty, \infty)$.

> **Improper Integral of f over $(-\infty, \infty)$**
> Let f be a continuous function over the unbounded interval $(-\infty, \infty)$. Let c be any real number and suppose both the improper integrals
>
> $$\int_{-\infty}^c f(x)\, dx \quad \text{and} \quad \int_c^\infty f(x)\, dx$$
>
> are convergent. Then, the improper integral of f over $(-\infty, \infty)$ is defined by
>
> $$\int_{-\infty}^\infty f(x)\, dx = \int_{-\infty}^c f(x)\, dx + \int_c^\infty f(x)\, dx \qquad \textbf{(14)}$$

In this case, we say that the improper integral on the left in Equation (14) is convergent. If either one of the two improper integrals on the right in (14) is divergent, then the improper integral on the left is not defined.

Note Usually, we choose $c = 0$.

EXAMPLE 5 Evaluate the improper integral

$$\int_{-\infty}^\infty x e^{-x^2}\, dx$$

and give a geometric interpretation of the results.

Solution Take the number c in Equation (14) to be 0. Let's first evaluate

$$\int_{-\infty}^0 x e^{-x^2}\, dx = \lim_{a \to -\infty} \int_a^0 x e^{-x^2}\, dx$$

$$= \lim_{a \to -\infty} -\frac{1}{2} e^{-x^2} \bigg|_a^0$$

$$= \lim_{a \to -\infty} \left[-\frac{1}{2} + \frac{1}{2} e^{-a^2} \right] = -\frac{1}{2}$$

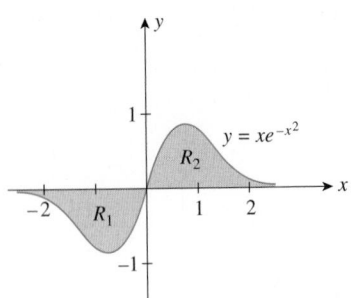

FIGURE 14

$$\int_{-\infty}^{\infty} xe^{-x^2}\, dx$$

$$= \int_{-\infty}^{0} xe^{-x^2}\, dx + \int_{0}^{\infty} xe^{-x^2}\, dx$$

Next, we evaluate

$$\int_{0}^{\infty} xe^{-x^2}\, dx = \lim_{b \to \infty} \int_{0}^{b} xe^{-x^2}\, dx$$

$$= \lim_{b \to \infty} -\frac{1}{2} e^{-x^2} \Big|_{0}^{b}$$

$$= \lim_{b \to \infty} \left(-\frac{1}{2} e^{-b^2} + \frac{1}{2} \right) = \frac{1}{2}$$

Therefore,

$$\int_{-\infty}^{\infty} xe^{-x^2}\, dx = \int_{-\infty}^{0} xe^{-x^2}\, dx + \int_{0}^{\infty} xe^{-x^2}\, dx$$

$$= -\frac{1}{2} + \frac{1}{2}$$

$$= 0$$

The graph of $y = xe^{-x^2}$ is sketched in Figure 14. A glance at the figure tells us that the improper integral

$$\int_{-\infty}^{0} xe^{-x^2}\, dx$$

gives the negative of the area of the region R_1, bounded above by the x-axis, below by the curve $y = xe^{-x^2}$, and on the right by the y-axis ($x = 0$).

However, the improper integral

$$\int_{0}^{\infty} xe^{-x^2}\, dx$$

gives the area of the region R_2 under the curve $y = xe^{-x^2}$ for $x \geq 0$. Since the graph of f is symmetric with respect to the origin, the area of R_1 is equal to the area of R_2. In other words,

$$\int_{-\infty}^{0} xe^{-x^2}\, dx = -\int_{0}^{\infty} xe^{-x^2}\, dx$$

Therefore,

$$\int_{-\infty}^{\infty} xe^{-x^2}\, dx = \int_{-\infty}^{0} xe^{-x^2}\, dx + \int_{0}^{\infty} xe^{-x^2}\, dx$$

$$= -\int_{0}^{\infty} xe^{-x^2}\, dx + \int_{0}^{\infty} xe^{-x^2}\, dx$$

$$= 0$$

as was shown earlier.

Perpetuities

Recall from Section 15.7 that the present value of an annuity is given by

$$PV \approx mP \int_{0}^{T} e^{-rt}\, dt = \frac{mP}{r} (1 - e^{-rT})$$ **(15)**

Now, if the payments of an annuity are allowed to continue indefinitely, we have what is called a **perpetuity**. The present value of a perpetuity may be approximated by the improper integral

$$PV \approx mP \int_{0}^{\infty} e^{-rt}\, dt$$

obtained from Formula (15) by allowing the term of the annuity, T, to approach infinity. Thus,

$$mP \int_0^\infty e^{-rt}\, dt = \lim_{b \to \infty} mP \int_0^b e^{-rt}\, dt$$

$$= mP \lim_{b \to \infty} \int_0^b e^{-rt}\, dt$$

$$= mP \lim_{b \to \infty} \left[-\frac{1}{r} e^{-rt} \Big|_0^b \right]$$

$$= mP \lim_{b \to \infty} \left(-\frac{1}{r} e^{-rb} + \frac{1}{r} \right) = \frac{mP}{r}$$

The Present Value of a Perpetuity
The **present value PV of a perpetuity** is given by

$$PV = \frac{mP}{r} \tag{16}$$

where m is the number of payments per year, P is the size of each payment, and r is the interest rate (compounded continuously).

APPLIED EXAMPLE 6   Endowments   The Robinson family wishes to create a scholarship fund at a college. If a scholarship in the amount of $5000 is awarded annually beginning 1 year from now, find the amount of the endowment they are required to make now. Assume that this fund will earn interest at a rate of 8% per year compounded continuously.

Solution   The amount of the endowment, A, is given by the present value of a perpetuity, with $m = 1$, $P = 5000$, and $r = 0.08$. Using Formula (16), we find

$$A = \frac{(1)(5000)}{0.08}$$

$$= 62{,}500$$

or $62,500.

The improper integral also plays an important role in the study of probability theory, as we will see in Section 16.5.

16.4   Self-Check Exercises

1. Evaluate $\displaystyle\int_{-\infty}^{\infty} \frac{x^3}{(1 + x^4)^{3/2}}\, dx.$

2. Suppose an income stream is expected to continue indefinitely. Then, the present value of such a stream can be calculated from the formula for the present value of an income stream by letting T approach infinity. Thus, the required present value is given by

$$PV = \int_0^\infty P(t) e^{-rt}\, dt$$

Suppose Marcia has an oil well in her backyard that generates a stream of income given by

$$P(t) = 20 e^{-0.02t}$$

where $P(t)$ is expressed in thousands of dollars per year and t

is the time in years from the present. Assuming that the prevailing interest rate in the foreseeable future is 10%/year compounded continuously, what is the present value of the income stream?

Solutions to Self-Check Exercises 16.4 can be found on page 1035.

16.4 Concept Questions

1. a. Define $\displaystyle\int_a^\infty f(x)\, dx$, where f is continuous on $[a, \infty)$.

b. Define $\displaystyle\int_{-\infty}^b f(x)\, dx$, where f is continuous on $(-\infty, b]$.

c. Define $\displaystyle\int_{-\infty}^\infty f(x)\, dx$ where f is continuous on $(-\infty, \infty)$.

2. What is the present value of a perpetuity? Give a formula for computing it.

16.4 Exercises

In Exercises 1–10, find the area of the region under the curve $y = f(x)$ over the indicated interval.

1. $f(x) = \dfrac{2}{x^2}; x \geq 3$

2. $f(x) = \dfrac{2}{x^3}; x \geq 2$

3. $f(x) = \dfrac{1}{(x-2)^2}; x \geq 3$

4. $f(x) = \dfrac{2}{(x+1)^3}; x \geq 0$

5. $f(x) = \dfrac{1}{x^{3/2}}; x \geq 1$

6. $f(x) = \dfrac{3}{x^{5/2}}; x \geq 4$

7. $f(x) = \dfrac{1}{(x+1)^{5/2}}; x \geq 0$

8. $f(x) = \dfrac{1}{(1-x)^{3/2}}; x \leq 0$

9. $f(x) = e^{2x}; x \leq 2$

10. $f(x) = xe^{-x^2}; x \geq 0$

11. Find the area of the region bounded by the x-axis and the graph of the function

$$f(x) = \frac{x}{(1+x^2)^2}$$

12. Find the area of the region bounded by the x-axis and the graph of the function

$$f(x) = \frac{e^x}{(1+e^x)^2}$$

13. Consider the improper integral

$$\int_0^\infty \sqrt{x}\, dx$$

a. Evaluate $I(b) = \displaystyle\int_0^b \sqrt{x}\, dx$.

b. Show that

$$\lim_{b \to \infty} I(b) = \infty$$

thus proving that the given improper integral is divergent.

14. Consider the improper integral

$$\int_1^\infty x^{-2/3}\, dx$$

a. Evaluate $I(b) = \displaystyle\int_1^b x^{-2/3}\, dx$.

b. Show that

$$\lim_{b \to \infty} I(b) = \infty$$

thus proving that the given improper integral is divergent.

In Exercises 15–42, evaluate each improper integral whenever it is convergent.

15. $\displaystyle\int_1^\infty \frac{3}{x^4}\, dx$

16. $\displaystyle\int_1^\infty \frac{1}{x^3}\, dx$

17. $\displaystyle\int_4^\infty \frac{2}{x^{3/2}}\, dx$

18. $\displaystyle\int_1^\infty \frac{1}{\sqrt{x}}\, dx$

19. $\displaystyle\int_1^\infty \frac{4}{x}\, dx$

20. $\displaystyle\int_2^\infty \frac{3}{x}\, dx$

21. $\displaystyle\int_{-\infty}^0 \frac{1}{(x-2)^3}\, dx$

22. $\displaystyle\int_2^\infty \frac{1}{(x+1)^2}\, dx$

23. $\displaystyle\int_1^\infty \frac{1}{(2x-1)^{3/2}}\, dx$

24. $\displaystyle\int_{-\infty}^0 \frac{1}{(4-x)^{3/2}}\, dx$

25. $\displaystyle\int_0^\infty e^{-x}\, dx$

26. $\displaystyle\int_0^\infty e^{-x/2}\, dx$

27. $\displaystyle\int_{-\infty}^0 e^{2x}\, dx$

28. $\displaystyle\int_{-\infty}^0 e^{3x}\, dx$

29. $\displaystyle\int_1^\infty \frac{e^{\sqrt{x}}}{\sqrt{x}}\, dx$

30. $\displaystyle\int_1^\infty \frac{e^{-\sqrt{x}}}{\sqrt{x}}\, dx$

31. $\displaystyle\int_{-\infty}^0 xe^x\, dx$

32. $\displaystyle\int_0^\infty xe^{-2x}\, dx$

33. $\displaystyle\int_{-\infty}^\infty x\, dx$

34. $\displaystyle\int_{-\infty}^\infty x^3\, dx$

35. $\displaystyle\int_{-\infty}^\infty x^3(1+x^4)^{-2}\, dx$

36. $\displaystyle\int_{-\infty}^\infty x(x^2+4)^{-3/2}\, dx$

37. $\displaystyle\int_{-\infty}^\infty xe^{1-x^2}\, dx$

38. $\displaystyle\int_{-\infty}^\infty \left(x-\frac{1}{2}\right)e^{-x^2+x-1}\, dx$

39. $\displaystyle\int_{-\infty}^{\infty} \frac{e^{-x}}{1 + e^{-x}}\, dx$

40. $\displaystyle\int_{-\infty}^{\infty} \frac{xe^{-x^2}}{1 + e^{-x^2}}\, dx$

41. $\displaystyle\int_{e}^{\infty} \frac{1}{x \ln^3 x}\, dx$

42. $\displaystyle\int_{e^2}^{\infty} \frac{1}{x \ln x}\, dx$

43. **THE AMOUNT OF AN ENDOWMENT** A university alumni group wishes to provide an annual scholarship in the amount of $1500 beginning next year. If the scholarship fund will earn an interest rate of 8%/year compounded continuously, find the amount of the endowment the alumni are required to make now.

44. **THE AMOUNT OF AN ENDOWMENT** Mel Thompson wishes to establish a fund to provide a university medical center with an annual research grant of $50,000 beginning next year. If the fund will earn an interest rate of 9%/year compounded continuously, find the amount of the endowment he is required to make now.

45. **PERPETUAL NET INCOME STREAMS** The present value of a perpetual stream of income that flows continually at the rate of $P(t)$ dollars/year is given by the formula

$$PV \approx \int_0^{\infty} P(t)e^{-rt}\, dt$$

where r is the interest rate compounded continuously. Using this formula, find the present value of a perpetual net income stream that is generated at the rate of

$$P(t) = 10{,}000 + 4000t$$

dollars/year.

Hint: $\displaystyle\lim_{b \to \infty} \frac{b}{e^{rb}} = 0$.

46. **ESTABLISHING A TRUST FUND** Becky Wilkinson wants to establish a trust fund that will provide her children and heirs with a perpetual annuity in the amount of

$$P(t) = 20 + t$$

thousand dollars/year beginning next year. If the trust fund will earn an interest rate of 10%/year compounded continuously, find the amount that she must place in the trust fund now.

Hint: Use the formula given in Exercise 45.

In Exercises 47–50, determine whether the statement is true or false. If it is true, explain why it is true. If it is false, give an example to show why it is false.

47. If $\displaystyle\int_a^{\infty} f(x)\, dx$ exists, then $\displaystyle\int_b^{\infty} f(x)\, dx$ exists for every real number $b > a$.

48. If f is continuous on $(-\infty, \infty)$, then
$$\int_{-\infty}^{\infty} f(x)\, dx = \lim_{t \to \infty} \int_{-t}^{t} f(x)\, dx.$$

49. If $\displaystyle\int_{-\infty}^{\infty} f(x)\, dx$ exists, then $\displaystyle\int_0^{\infty} f(x)\, dx$ exists, and
$$\int_{-\infty}^{\infty} f(x)\, dx = 2 \int_0^{\infty} f(x)\, dx.$$

50. If $\displaystyle\int_a^{\infty} f(x)\, dx$ exists, then $\displaystyle\int_{-\infty}^{-a} f(x)\, dx$ exists, and
$$\int_a^{\infty} f(x)\, dx = -\int_{-\infty}^{-a} f(x)\, dx \text{ exists.}$$

51. **CAPITAL VALUE** The capital value (present sale value) CV of property that can be rented on a perpetual basis for R dollars annually is approximated by the formula

$$CV \approx \int_0^{\infty} Re^{-it}\, dt$$

where i is the prevailing continuous interest rate.
a. Show that $CV = R/i$.
b. Find the capital value of property that can be rented at $10,000 annually when the prevailing continuous interest rate is 12%/year.

52. Show that an integral of the form $\displaystyle\int_a^{\infty} e^{-px}\, dx$ is convergent if $p > 0$ and divergent if $p < 0$.

53. Show that an integral of the form $\displaystyle\int_{-\infty}^{b} e^{px}\, dx$ is convergent if $p > 0$ and divergent if $p < 0$.

54. Find the values of p such that $\displaystyle\int_1^{\infty} \frac{1}{x^p}\, dx$ is convergent.

16.4 Solutions to Self-Check Exercises

1. Write

$$\int_{-\infty}^{\infty} \frac{x^3}{(1 + x^4)^{3/2}}\, dx = \int_{-\infty}^{0} \frac{x^3}{(1 + x^4)^{3/2}}\, dx + \int_0^{\infty} \frac{x^3}{(1 + x^4)^{3/2}}\, dx$$

Now,

$$\int_{-\infty}^{0} \frac{x^3}{(1 + x^4)^{3/2}}\, dx = \lim_{a \to -\infty} \int_a^0 x^3(1 + x^4)^{-3/2}\, dx$$

$$= \lim_{a \to -\infty} \frac{1}{4}(-2)(1 + x^4)^{-1/2}\Big|_a^0 \quad \text{Integrate by substitution.}$$

$$= -\frac{1}{2} \lim_{a \to -\infty}\left[1 - \frac{1}{(1 + a^4)^{1/2}}\right]$$

$$= -\frac{1}{2}$$

Similarly, you can show that

$$\int_0^\infty \frac{x^3}{(1+x^4)^{3/2}}\,dx = \frac{1}{2}$$

Therefore,

$$\int_{-\infty}^\infty \frac{x^3}{(1+x^4)^{3/2}}\,dx = -\frac{1}{2} + \frac{1}{2} = 0$$

2. The required present value is given by

$$\begin{aligned}
PV &= \int_0^\infty 20e^{-0.02t}e^{-0.10t}\,dt \\
&= 20\int_0^\infty e^{-0.12t}\,dt \\
&= 20 \lim_{b\to\infty} \int_0^b e^{-0.12t}\,dt \\
&= -\frac{20}{0.12} \lim_{b\to\infty} e^{-0.12t}\Big|_0^b \\
&= -\frac{500}{3} \lim_{b\to\infty} (e^{-0.12b} - 1) \\
&= \frac{500}{3}
\end{aligned}$$

or approximately $166,667.

16.5   Applications of Calculus to Probability

■ Probability Density Functions

In this section we look at an application of the definite integral to the computation of probabilities associated with a continuous random variable. We begin by recalling the properties of a probability density function associated with a continuous random variable x.

Probability Density Function

A **probability density function** of a continuous random variable x in an interval I, where I may be bounded or unbounded, is a nonnegative function f having the following properties.

1. The total area of the region under the graph of f is equal to 1 (Figure 15a).
2. The probability that an observed value of the random variable x lies in the interval $[a, b]$ is given by

$$P(a \le x \le b) = \int_a^b f(x)\,dx$$

(Figure 15b).

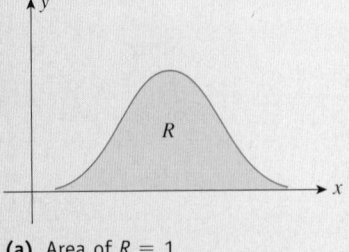

(a) Area of $R = 1$

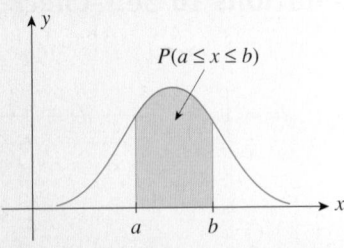

(b) $P(a \le x \le b)$ is the probability that an outcome of an experiment will lie between a and b.

FIGURE 15

A few comments are in order. First, a probability density function of a random variable x may be constructed using methods that range from theoretical considerations of the problem on the one extreme to an interpretation of data associated with the experiment on the other. Second, Property 1 states that the probability that a continuous random variable takes on a value lying in its range is 1, a certainty, which is expected. Third, Property 2 states that the probability that the random variable x assumes a value in an interval $a \leq x \leq b$ is given by the area of the region between the graph of f and the x-axis from $x = a$ to $x = b$. Because the area under one point of the graph of f is equal to zero, we see immediately that $P(a \leq x \leq b) = P(a < x \leq b) = P(a \leq x < b) = P(a < x < b)$.

EXAMPLE 1   Show that each of the following functions satisfies the nonnegativity condition and Property 1 of probability density functions.

a. $f(x) = \dfrac{2}{27} x(x - 1)$ $(1 \leq x \leq 4)$

b. $f(x) = \dfrac{1}{3} e^{(-1/3)x}$ $(0 \leq x < \infty)$

Solution

a. Since the factors x and $(x - 1)$ are both nonnegative, we see that $f(x) \geq 0$ on $[1, 4]$. Next, we compute

$$\int_1^4 \frac{2}{27} x(x - 1)\, dx = \frac{2}{27} \int_1^4 (x^2 - x)\, dx$$

$$= \frac{2}{27} \left(\frac{1}{3} x^3 - \frac{1}{2} x^2 \right) \Big|_1^4$$

$$= \frac{2}{27} \left[\left(\frac{64}{3} - 8 \right) - \left(\frac{1}{3} - \frac{1}{2} \right) \right]$$

$$= \frac{2}{27} \left(\frac{27}{2} \right)$$

$$= 1$$

showing that Property 1 of probability density functions holds as well.

b. First, $f(x) = \frac{1}{3} e^{(-1/3)x} \geq 0$ for all values of x in $[0, \infty)$. Next,

$$\int_0^\infty \frac{1}{3} e^{(-1/3)x}\, dx = \lim_{b \to \infty} \int_0^b \frac{1}{3} e^{(-1/3)x}\, dx$$

$$= \lim_{b \to \infty} - e^{(-1/3)x} \Big|_0^b$$

$$= \lim_{b \to \infty} \left(-e^{(-1/3)b} + 1 \right)$$

$$= 1$$

so the area under the graph of $f(x) = \frac{1}{3} e^{(-1/3)x}$ is equal to 1, as we set out to show. ∎

EXAMPLE 2

a. Determine the value of the constant k so that the function $f(x) = kx^2$ is a probability density function on the interval $[0, 5]$.

b. If x is a continuous random variable with the probability density function given in part (a), compute the probability that x will assume a value between $x = 1$ and $x = 2$.

Solution

a. We compute

$$\int_0^5 kx^2\, dx = k \int_0^5 x^2\, dx$$

$$= \frac{k}{3} x^3 \Big|_0^5$$

$$= \frac{125}{3} k$$

Since this value must be equal to 1, we find that $k = 3/125$.

b. The required probability is given by

$$P(1 \le x \le 2) = \int_1^2 f(x)\, dx = \int_1^2 \frac{3}{125} x^2\, dx$$

$$= \frac{1}{125} x^3 \Big|_1^2 = \frac{1}{125}(8-1)$$

$$= \frac{7}{125}$$

The graph of the probability density function f and the area corresponding to the probability $P(1 \le x \le 2)$ are shown in Figure 16. ∎

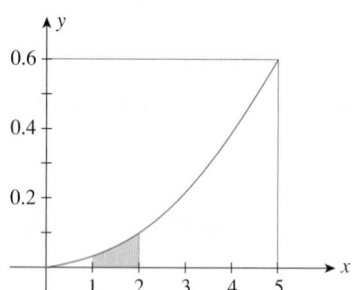

FIGURE 16
$P(1 \le x \le 2)$ for the probability density function $y = \dfrac{3}{125} x^2$

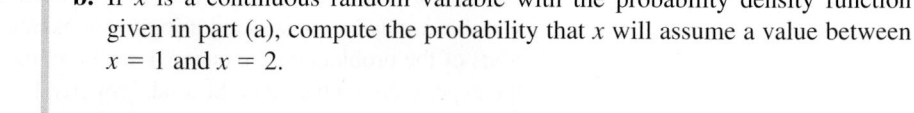

APPLIED EXAMPLE 3 Life Span of Light Bulbs TKK Products manufactures a 200-watt electric light bulb. Laboratory tests show that the life spans of these light bulbs have a distribution described by the probability density function

$$f(x) = .001e^{-.001x} \qquad (0 \le x < \infty)$$

Determine the probability that a light bulb will have a life span of (a) 500 hours or less, (b) more than 500 hours, and (c) more than 1000 hours but less than 1500 hours.

Solution Let x denote the life span of a light bulb.

a. The probability that a light bulb will have a life span of 500 hours or less is given by

$$P(0 \le x \le 500) = \int_0^{500} .001e^{-.001x}\, dx$$

$$= -e^{-.001x} \Big|_0^{500} = -e^{-.5} + 1$$

$$\approx .3935$$

b. The probability that a light bulb will have a life span of more than 500 hours is given by

$$P(x > 500) = \int_{500}^{\infty} .001e^{-.001x}\, dx$$

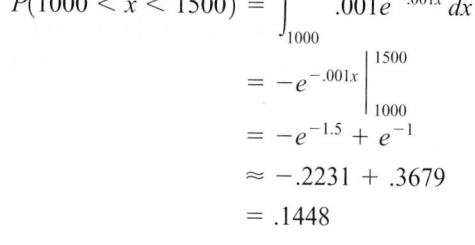

$$= \lim_{b \to \infty} \int_{500}^{b} .001e^{-.001x}\, dx$$

$$= \lim_{b \to \infty} -e^{-.001x}\Big|_{500}^{b}$$

$$= \lim_{b \to \infty} \left(-e^{-.001b} + e^{-.5}\right)$$

$$= e^{-.5} \approx .6065$$

This result may also be obtained by observing that

$$P(x > 500) = 1 - P(x \le 500)$$

$$= 1 - .3935 \qquad \text{Use the result from part (a).}$$

$$\approx .6065$$

c. The probability that a light bulb will have a life span of more than 1000 hours but less than 1500 hours is given by

$$P(1000 < x < 1500) = \int_{1000}^{1500} .001e^{-.001x}\, dx$$

$$= -e^{-.001x}\Big|_{1000}^{1500}$$

$$= -e^{-1.5} + e^{-1}$$

$$\approx -.2231 + .3679$$

$$= .1448$$

The probability density function of Example 3 has the form

$$f(x) = ke^{-kx}$$

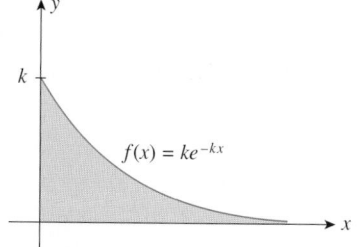

FIGURE 17
The area under the graph of the exponential density function is equal to 1.

where $x \ge 0$ and k is a positive constant. Its graph is shown in Figure 17. This probability function is called an **exponential density function**, and the random variable associated with it is said to be **exponentially distributed**. Exponential random variables are used to represent the life span of electronic components, the duration of telephone calls, the waiting time in a doctor's office, and the time between successive flight arrivals and departures in an airport, to mention but a few applications.

Another probability density function, and the one most widely used, is the **normal density function**, defined by

$$f(x) = \frac{1}{\sigma\sqrt{2\pi}} > e^{-(1/2)[(x-\mu)/\sigma]^2}$$

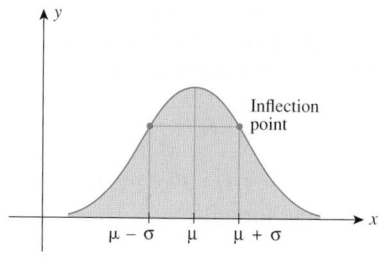

FIGURE 18
The area under the bell-shaped normal distribution curve

where μ and σ are constants. The graph of the normal distribution is bell-shaped (Figure 18). Many phenomena, such as the heights of people in a given population, the weights of newborn infants, the IQs of college students, the actual weights of 16-ounce packages of cereals, and so on, have probability distributions that are normal.

Areas under the standard normal curve (the normal curve with $\mu = 0$ and $\sigma = 1$) have been extensively computed and tabulated. Most problems involving the normal distribution can be solved with the aid of these tables.

■ Expected Value

The average value, or **expected value**, of a discrete variable X that takes on values $x_1, x_2, \ldots, x_n$ with associated probabilities $p_1, p_2, \ldots, p_n$ is defined by

$$E(X) = x_1 p_1 + x_2 p_2 + \cdots + x_n p_n$$

If each of the values $x_1, x_2, \ldots, x_n$ occurs with equal frequency, then $p_1 = p_2 = \cdots = p_n = 1/n$ and

$$\begin{aligned} E(X) &= x_1 \left(\frac{1}{n}\right) + x_2 \left(\frac{1}{n}\right) + \cdots + x_n \left(\frac{1}{n}\right) \\ &= \frac{x_1 + x_2 + \cdots + x_n}{n} \end{aligned}$$

giving the familiar formula for computing the average value of the n numbers $x_1, x_2, \ldots, x_n$.

Now, suppose x is a continuous random variable and f is the probability density function associated with it. For simplicity, let's first assume that $a \leq x \leq b$. Divide the interval $[a, b]$ into n subintervals of equal length $\Delta x = (b - a)/n$ by means of the $(n + 1)$ points $x_0 = a, x_1, x_2, \ldots, x_n = b$ (Figure 19). To find an approximation of the average value, or expected value, of x on the interval $[a, b]$, let's treat x as if it were a discrete random variable that takes on the values $x_1, x_2, \ldots, x_n$ with probabilities $p_1, p_2, \ldots, p_n$. Then,

$$E(x) \approx x_1 p_1 + x_2 p_2 + \cdots + x_n p_n$$

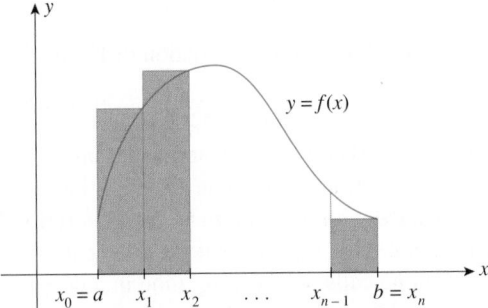

FIGURE 19
Approximating the expected value of a random variable x on $[a, b]$ by a Riemann sum

But p_1 is the probability that x is in the interval $[x_0, x_1]$, and this is just the area under the graph of f from $x = x_0$ to $x = x_1$, which may be approximated by $f(x_1)\Delta x$. The probabilities $p_2, \ldots, p_n$ may be approximated in a similar manner. Thus,

$$E(x) \approx x_1 f(x_1)\Delta x + x_2 f(x_2)\Delta x + \cdots + x_n f(x_n)\Delta x$$

which is seen to be the Riemann sum of the function $g(x) = xf(x)$ over the interval $[a, b]$. Letting n approach infinity, we obtain the following formula:

Expected Value of a Continuous Random Variable

Suppose the function f defined on the interval $[a, b]$ is the probability density function associated with a continuous random variable x. Then, the expected value of x is

$$E(x) = \int_a^b xf(x)\,dx \qquad (17)$$

If either $a = -\infty$ or $b = \infty$, then the integral in (17) becomes an improper integral.

The expected value of a random variable plays an important role in many practical applications. For example, if x represents the life span of a certain brand of electronic components, then the expected value of x gives the average life span of these components. If x measures the waiting time in a doctor's office, then $E(x)$ gives the average waiting time, and so on.

APPLIED EXAMPLE 4 Life Span of Light Bulbs Show that if a continuous random variable x is exponentially distributed with the probability density function

$$f(x) = ke^{-kx} \qquad (0 \le x < \infty)$$

then the expected value $E(x)$ is equal to $1/k$. Using this result, determine the average life span of a 200-watt light bulb manufactured by TKK Products of Example 3.

Solution We compute

$$E(x) = \int_0^\infty xf(x)\,dx$$

$$= \int_0^\infty kxe^{-kx}\,dx$$

$$= k \lim_{b \to \infty} \int_0^b xe^{-kx}\,dx$$

Integrating by parts with

$$u = x \quad \text{and} \quad dv = e^{-kx}\,dx$$

so that

$$du = dx \quad \text{and} \quad v = -\frac{1}{k}e^{-kx}$$

we have

$$E(x) = k \lim_{b \to \infty} \left[-\frac{1}{k} x e^{-kx} \Big|_0^b + \frac{1}{k} \int_0^b e^{-kx} \, dx \right]$$

$$= k \lim_{b \to \infty} \left[-\left(\frac{1}{k}\right) b e^{-kb} - \frac{1}{k^2} e^{-kx} \Big|_0^b \right]$$

$$= k \lim_{b \to \infty} \left[-\left(\frac{1}{k}\right) b e^{-kb} - \frac{1}{k^2} e^{-kb} + \frac{1}{k^2} \right]$$

$$= -\lim_{b \to \infty} \frac{b}{e^{kb}} - \frac{1}{k} \lim_{b \to \infty} \frac{1}{e^{kb}} + \frac{1}{k} \lim_{b \to \infty} 1$$

Now, by taking a sequence of values of b that approaches infinity—for example, $b = 10, 100, 1000, 10,000, \ldots$—we see that, for a fixed k,

$$\lim_{b \to \infty} \frac{b}{e^{kb}} = 0$$

Therefore,

$$E(x) = \frac{1}{k}$$

as we set out to show. Next, since $k = .001$ in Example 3, we see that the average life span of the TKK light bulbs is $1/(.001) = 1000$ hours. ∎

Before considering another example, let's summarize the important result obtained in Example 4.

The Expected Value of an Exponential Density Function
If a continuous random variable x is exponentially distributed with probability density function

$$f(x) = ke^{-kx} \qquad (0 \le x < \infty)$$

then the expected (average) value of x is given by

$$E(x) = \frac{1}{k}$$

APPLIED EXAMPLE 5 Airport Traffic On a typical Monday morning, the time between successive arrivals of planes at Jackson International Airport is an exponentially distributed random variable x with expected value of 10 (minutes).

a. Find the probability density function associated with x.
b. What is the probability that between 6 and 8 minutes will elapse between successive arrivals of planes?
c. What is the probability that the time between successive arrivals of planes will be more than 15 minutes?

TITLE   Associate

INSTITUTION   JPMorgan Chase

As one of the leading financial institutions in the world, JPMorgan Chase & Co. depends on a wide range of mathematical disciplines from statistics to linear programming to calculus. Whether assessing the credit worthiness of a borrower, recommending portfolio investments or pricing an exotic derivative, quantitative understanding is a critical tool in serving the financial needs of clients.

I work in the Fixed-Income Derivatives Strategy group. A derivative in finance is an instrument whose value depends on the price of some other underlying instrument. A simple type of derivative is the forward contract, where two parties agree to a future trade at a specified price. In agriculture, for instance, farmers will often pledge their crops for sale to buyers at an agreed price before even planting the harvest. Depending on the weather, demand and other factors, the actual price may turn out higher or lower. Either the buyer or seller of the forward contract benefits accordingly. The value of the contract changes one-for-one with the actual price. In derivatives lingo, we borrow from calculus and say forward contracts have a delta of one.

Nowadays, the bulk of derivatives deal with interest-rate rather than agricultural risk. The value of any asset with fixed payments over time varies with interest rates. With trillions of dollars in this form, especially government bonds and mortgages, fixed-income derivatives are vital to the economy. As a strategy group, our job is to track and anticipate key drivers and developments in the market using, in significant part, quantitative analysis. Some of the derivatives we look at are of the forward kind, such as interest-rate swaps, where over time you receive fixed-rate payments in exchange for paying a floating-rate or vice-versa. A whole other class of derivatives where statistics and calculus are especially relevant are options.

Whereas forward contracts bind both parties to a future trade, options give the holder the right but not the obligation to trade at a specified time and price. Similar to an insurance policy, the holder of the option pays an upfront premium in exchange for potential gain. Solving this pricing problem requires statistics, stochastic calculus and enough insight to win a Nobel prize. Fortunately for us, this was taken care of by Fischer Black, Myron Scholes, and Robert Merton in the early 1970s (including the 1997 Nobel Prize in Economics for Scholes and Merton). The Black-Scholes differential equation was the first accurate options pricing model, making possible the rapid growth of the derivatives market. Black-Scholes and the many derivatives of it, as it were, continue to be used to this day.

Solution

a. Since x is exponentially distributed, the associated probability density function has the form $f(x) = ke^{-kx}$. Next, since the expected value of x is 10, we see that

$$E(x) = \frac{1}{k} = 10$$

$$k = \frac{1}{10}$$

$$= .1$$

so the required probability density function is

$$f(x) = .1e^{-.1x}$$

b. The probability that between 6 and 8 minutes will elapse between successive arrivals is given by

$$P(6 \le x \le 8) = \int_6^8 .1e^{-.1x}\, dx = -e^{-.1x}\Big|_6^8$$

$$= -e^{-.8} + e^{-.6}$$

$$\approx .10$$

c. The probability that the time between successive arrivals will be more than 15 minutes is given by

$$P(x > 15) = \int_{15}^\infty .1e^{-.1x}\, dx$$

$$= \lim_{b\to\infty} \int_{15}^b .1e^{-.1x}\, dx$$

$$= \lim_{b\to\infty} \left[-e^{-.1x}\Big|_{15}^b \right]$$

$$= \lim_{b\to\infty} \left(-e^{-.1b} + e^{-1.5} \right) = e^{-1.5}$$

$$\approx .22$$

16.5 Self-Check Exercises

1. Determine the value of the constant k such that the function $f(x) = k(4x - x^2)$ is a probability density function on the interval $[0, 4]$.

2. Suppose x is a continuous random variable with the probability density function of Self-Check Exercise 1. Find the probability that x will assume a value between $x = 1$ and $x = 3$.

Solutions to Self-Check Exercises 16.5 can be found on page 1046.

16.5 Concept Questions

1. What is a probability density function of a random variable x on an interval I? Give an example.

2. **a.** What is the expected value of a random variable x associated with a probability density function f defined on $[a, b]$?

b. What is the expected value of x where $f(x) = ke^{-kx}$ $(0 \le x < \infty)$?

16.5 Exercises

In Exercises 1–10, show that the function is a probability density function on the specified interval.

1. $f(x) = \dfrac{2}{32} x;\ (2 \le x \le 6)$

2. $f(x) = \dfrac{2}{9}(3x - x^2);\ (0 \le x \le 3)$

3. $f(x) = \dfrac{3}{8} x^2;\ (0 \le x \le 2)$

4. $f(x) = \dfrac{3}{32}(x - 1)(5 - x);\ (1 \le x \le 5)$

5. $f(x) = 20(x^3 - x^4);\ (0 \le x \le 1)$

6. $f(x) = \dfrac{8}{7x^2};\ (1 \le x \le 8)$

7. $f(x) = \dfrac{3}{14}\sqrt{x};\ (1 \le x \le 4)$

8. $f(x) = \dfrac{12 - x}{72};\ (0 \le x \le 12)$

9. $f(x) = \dfrac{x}{(x^2 + 1)^{3/2}};\ (0 \le x < \infty)$

10. $f(x) = 4xe^{-2x^2};\ (0 \le x < \infty)$

11. a. Determine the value of the constant k so that the function $f(x) = k(4 - x)$ is a probability density function on the interval $[0, 4]$.
  b. If x is a continuous random variable with the probability density function given in part (a), compute the probability that x will assume a value between $x = 1$ and $x = 3$.

12. a. Determine the value of the constant k so that the function $f(x) = k/x^2$ is a probability density function on the interval $[1, 10]$.
  b. If x is a continuous random variable with the probability density function of part (a), compute the probability that x will assume a value between $x = 2$ and $x = 6$.

13. a. Determine the value of the constant k so that the function $f(x) = 2ke^{-kx}$ is a probability density function on the interval $[0, 4]$.
  b. If x is a continuous random variable with the probability density function of part (a), find the probability that x will assume a value between $x = 1$ and $x = 2$.

14. a. Determine the value of the constant k so that the function $f(x) = kxe^{-2x^2}$ is a probability density function on the interval $[0, \infty)$.
  b. If x is a continuous random variable with the probability density function of part (a), find the probability that x will assume a value greater than 1.

15. AVERAGE WAITING TIME FOR PATIENTS The average waiting time for patients arriving at the Newtown Health Clinic between 1 p.m. and 4 p.m. on a weekday is an exponentially distributed random variable x with expected value of 15 min.
  a. Find the probability density function associated with x.
  b. What is the probability that a patient arriving at the clinic between 1 p.m. and 4 p.m. will have to wait between 10 and 12 min?
  c. What is the probability that a patient arriving at the clinic between 1 p.m. and 4 p.m. will have to wait more than 15 min?

16. LIFE SPAN OF A PLANT SPECIES The life span of a certain plant species (in days) is described by the probability density function

$$f(x) = \frac{1}{100} e^{-x/100} \qquad (0 \le x < \infty)$$

a. Find the probability that a plant of this species will live for 100 days or less.
b. Find the probability that a plant of this species will live more than 120 days.
c. Find the probability that a plant of this species will live more than 60 days but less than 140 days.

17. SHOPPING HABITS The amount of time t (in minutes) a shopper spends browsing in the magazine section of a supermarket is a continuous random variable with probability density function

$$f(x) = \frac{2}{25} t \qquad (0 \le t \le 5)$$

How much time is a shopper chosen at random expected to spend in the magazine section?

18. REACTION TIME OF A MOTORIST The amount of time t (in seconds) it takes a motorist to react to a road emergency is a continuous random variable with probability density function

$$f(t) = \frac{9}{4t^3} \qquad (1 \le t \le 3)$$

What is the expected reaction time for a motorist chosen at random?

19. DEMAND FOR BUTTER The quantity demanded x (in thousands of pounds) of a certain brand of butter each week is a continuous random variable with probability density function

$$f(x) = \frac{6}{125} x(5 - x) \qquad (0 \le x \le 5)$$

What is the expected demand for this brand of butter each week?

20. EXPECTED SNOWFALL The amount of snowfall in feet in a remote region of Alaska in the month of January is a continuous random variable with probability density function

$$f(x) = \frac{2}{9} x(3 - x) \qquad (0 \le x \le 3)$$

Find the amount of snowfall one can expect in any given month of January in Alaska.

21. FREQUENCY OF ROAD REPAIRS The number of streets in the downtown section of a certain city that need repairs in a given year is a random variable with a distribution described by the probability density function

$$f(x) = 12x^2(1 - x) \qquad (0 \le x \le 1)$$

Find the probability that at most half of the streets will need repairs in any given year.

22. GAS STATION SALES The amount of gas (in thousands of gallons) Al's Gas Station sells on a typical Monday is a continuous random variable with probability density function

$$f(x) = 4(x - 2)^3 \qquad (2 \le x \le 3)$$

How much gas can the gas station expect to sell each Monday?

23. **LIFE SPAN OF COLOR TELEVISION TUBES** The life span (in years) of a certain brand of color television tube is a continuous random variable with probability density function

$$f(t) = 9(9 + t^2)^{-3/2} \qquad (0 \le t < \infty)$$

How long is one of these color television tubes expected to last?

24. **RELIABILITY OF ROBOTS** National Welding uses industrial robots in some of its assembly-line operations. Management has determined that, on average, a robot breaks down after 1000 hr of use and that the lengths of time between breakdowns are exponentially distributed.
    a. What is the probability that a robot selected at random will break down between 600 and 800 hr of use?
    b. What is the probability that a robot will break down after 1200 hr of use?

25. **EXPRESSWAY TOLLBOOTHS** Suppose the time intervals between arrivals of successive cars at an expressway tollbooth during rush hour are exponentially distributed and that the average time interval between arrivals is 8 sec. Find the probability that the average time interval between arrivals of successive cars is more than 8 sec.

26. **TIME INTERVALS BETWEEN PHONE CALLS** A study conducted by UniMart, a mail-order department store, reveals that the time intervals between incoming telephone calls on its toll-free 800 line between 10 a.m. and 2 p.m. are exponentially distributed and that the average time interval is 30 sec. What is

the probability that the time interval between successive calls is more than 2 min?

27. **RELIABILITY OF MICROPROCESSORS** The microprocessors manufactured by United Motor Works, which are used in automobiles to regulate fuel consumption, are guaranteed against defects for 20,000 mi of use. Tests conducted in the laboratory under simulated driving conditions reveal that the distances driven before the microprocessors break down are exponentially distributed and that the average distance driven before the microprocessors fail is 100,000 mi. What is the probability that a microprocessor selected at random will fail during the warranty period?

In Exercises 28–31, determine whether the statement is true or false. If it is true, explain why it is true. If it is false, give an example to show why it is false.

28. If f is a probability density function defined on $(-\infty, \infty)$ and a and b are real numbers such that $a < b$, then
$$P(x < a) + P(x > b) = 1 - \int_a^b f(x)\, dx$$

29. If $\int_a^b f(x)\, dx = 1$, then f is a probability density function on $[a, b]$.

30. If f is a probability density function of a continuous random variable x in the interval $[a, b]$, then the expected value of x is given by $\int_a^b x^2 f(x)\, dx$.

31. If f is a probability function on an interval $[a, b]$, then f is a probability function on $[c, d]$ for any real numbers c and d satisfying $a < c < d < b$.

16.5 Solutions to Self-Check Exercises

1. We compute

$$\int_0^4 k(4x - x^2)\, dx = k\left(2x^2 - \frac{1}{3}x^3\right)\Big|_0^4$$
$$= k\left(32 - \frac{64}{3}\right)$$
$$= \frac{32}{3}k$$

Since this value must be equal to 1, we see that $k = \dfrac{3}{32}$.

2. The required probability is given by

$$P(1 \le x \le 3) = \int_1^3 f(x)\, dx$$
$$= \int_1^3 \frac{3}{32}(4x - x^2)\, dx$$
$$= \frac{3}{32}\left(2x^2 - \frac{1}{3}x^3\right)\Big|_1^3$$
$$= \frac{3}{32}\left[(18 - 9) - \left(2 - \frac{1}{3}\right)\right] = \frac{11}{16}$$

CHAPTER 16 **Summary of Principal Formulas and Terms**

 **FORMULAS**

1. Integration by parts	$\int u \, dv = uv - \int v \, du$
2. Trapezoidal rule	$\int_a^b f(x) \, dx \approx \dfrac{\Delta x}{2} [f(x_0) + 2f(x_1)$ $+ \, 2f(x_2) + \cdots$ $+ \, 2f(x_{n-1}) + f(x_n)]$ where $\Delta x = \dfrac{b - a}{n}$
3. Simpson's rule	$\int_a^b f(x) \, dx \approx \dfrac{\Delta x}{3} [f(x_0) + 4f(x_1)$ $+ \, 2f(x_2) + 4f(x_3)$ $+ \, 2f(x_4) + \cdots$ $+ \, 4f(x_{n-1}) + f(x_n)]$ where $\Delta x = \dfrac{b - a}{n}$
4. Maximum error for trapezoidal rule	$\dfrac{M(b - a)^3}{12n^2}$, where $\lvert f''(x) \rvert \le M$ $(a \le x \le b)$
5. Maximum error for Simpson's rule	$\dfrac{M(b - a)^5}{180n^4}$, where $\lvert f^{(4)}(x) \rvert \le M$ $(a \le x \le b)$
6. Improper integral of f over $[a, \infty)$	$\int_a^\infty f(x) \, dx = \lim_{b \to \infty} \int_a^b f(x) \, dx$
7. Improper integral of f over $(-\infty, b]$	$\int_{-\infty}^b f(x) \, dx = \lim_{a \to -\infty} \int_a^b f(x) \, dx$
8. Improper integral of f over $(-\infty, \infty)$	$\int_{-\infty}^\infty f(x) \, dx = \int_{-\infty}^c f(x) \, dx$ $+ \int_c^\infty f(x) \, dx$
9. Present value of a perpetuity	$PV = \dfrac{mP}{r}$

 **TERMS**

improper integral (1027)

convergent integral (1029)

divergent integral (1029)

perpetuity (1032)

probability density funtion (1036)

exponential density function (1039)

normal density function (1039)

expected value (1040)

CHAPTER 16 **Concept Review Questions**

Fill in the blanks.

1. The integration by parts formula is obtained by reversing the _____ rule. The formula for integration by parts is $\int u\,dv = $ _____. In choosing u and dv, we want du to be simpler than _____ and dv to be_____ _____ _____.

2. To find $I = \int x\ln(x^2 + 1)\,dx$ using the table of integrals, we need to first use the substitution with $u = $ _____ so that $du = $ _____ to transform I into the integral $I = \frac{1}{2}\int u\ln u\,du$. We then choose Formula _____ to evaluate this integral.

3. The trapezoidal rule states that $\int_a^b f(x)\,dx \approx$ _____ where

$\Delta x = \frac{b-a}{n}$. The error E_n in approximating $\int_a^b f(x)\,dx$ by the trapezoidal rule satisfies $|E_n| \le$ _____.

4. Simpson's rule states that $\int_a^b f(x)\,dx \approx$ _____ where $\Delta x = \frac{b-a}{n}$ and n is _____. The error E_n in approximating $\int_a^b f(x)\,dx$ by Simpson's rule satisfies $|E_n| \le$ _____.

5. The improper integrals $\int_{-\infty}^b f(x)\,dx = $ _____, $\int_a^\infty f(x)\,dx = $ _____, and $\int_{-\infty}^\infty f(x)\,dx = $ _____.

CHAPTER 16 **Review Exercises**

In Exercises 1–6, evaluate the integral.

1. $\int 2xe^{-x}\,dx$

2. $\int xe^{4x}\,dx$

3. $\int \ln 5x\,dx$

4. $\int_1^4 \ln 2x\,dx$

5. $\int_0^1 xe^{-2x}\,dx$

6. $\int_0^2 xe^{2x}\,dx$

7. Find the function f given that the slope of the tangent line to the graph of f at any point $(x, f(x))$ is
$$f'(x) = \frac{\ln x}{\sqrt{x}}$$
and that the graph of f passes through the point $(1, -2)$.

8. Find the function f given that the slope of the tangent line to the graph of f at any point $(x, f(x))$ is
$$f'(x) = xe^{-3x}$$
and that the graph of f passes through the point $(0, 0)$.

In Exercises 9–14, use the table of integrals in Section 16.2 to evaluate the integral.

9. $\int \frac{x^2\,dx}{(3 + 2x)^2}$

10. $\int \frac{2x}{\sqrt{2x + 3}}\,dx$

11. $\int x^2 e^{4x}\,dx$

12. $\int \frac{dx}{(x^2 - 25)^{3/2}}$

13. $\int \frac{dx}{x^2\sqrt{x^2 - 4}}$

14. $\int 8x^3 \ln 2x\,dx$

In Exercises 15–20, evaluate each improper integral whenever it is convergent.

15. $\int_0^\infty e^{-2x}\,dx$

16. $\int_{-\infty}^0 e^{3x}\,dx$

17. $\int_3^\infty \frac{2}{x}\,dx$

18. $\int_2^\infty \frac{1}{(x + 2)^{3/2}}\,dx$

19. $\int_2^\infty \frac{dx}{(1 + 2x)^2}$

20. $\int_1^\infty 3e^{1-x}\,dx$

In Exercises 21–24, use the trapezoidal rule and Simpson's rule to approximate the value of the definite integral.

21. $\int_1^3 \frac{dx}{1 + \sqrt{x}}; n = 4$

22. $\int_0^1 e^{x^2}\,dx; n = 4$

23. $\int_{-1}^1 \sqrt{1 + x^4}\,dx; n = 4$

24. $\int_1^3 \frac{e^x}{x}\,dx; n = 4$

25. Find a bound on the error in approximating the integral $\int_0^1 \frac{dx}{x + 1}$ with $n = 8$ using (a) the trapezoidal rule and (b) Simpson's rule.

26. Show that the function $f(x) = (3/128)(16 - x^2)$ is a probability density function on the interval $[0, 4]$.

27. Show that the function $f(x) = (1/9)x\sqrt{9 - x^2}$ is a probability density function on the interval $[0, 3]$.

28. a. Determine the value of the constant k such that the function $f(x) = kx\sqrt{4 - x^2}$ is a probability function on the interval $[0, 2]$.

  b. If x is a continuous random variable with the probability density function given in part (a), compute the probability that x will assume a value between $x = 1$ and $x = 2$.

29. a. Determine the value of the constant k such that the function $f(x) = k/\sqrt{x}$ is a probability density function on the interval $[1, 4]$.

  b. If x is a continuous random variable with the probability density function given in part (a), compute the probability that x will assume a value between $x = 2$ and $x = 3$.

30. a. Determine the value of the constant k such that the function $f(x) = kx^2(3 - x)$ is a probability density function on the interval $[0, 3]$.

  b. If x is a continuous random variable with the probability density function given in part (a), compute the probability that x will assume a value between $x = 1$ and $x = 2$.

31. LENGTH OF HOSPITAL STAY Records at Centerville Hospital indicate that the length of time in days that a maternity patient stays in the hospital has a probability density function given by

$$P(t) = \frac{1}{4}e^{(-1/4)t}$$

  a. What is the probability that a woman entering the maternity wing will be there more than 6 days?
  b. What is the probability that a woman entering the maternity wing will be there less than 2 days?
  c. What is the average length of time that a woman entering the maternity wing stays in the hospital?

32. PRODUCERS' SURPLUS The supply equation for the GTC Slim-Phone is given by

$$p = 2\sqrt{25 + x^2}$$

where p is the unit price in dollars and x is the quantity demanded per month in units of 10,000. Find the producers' surplus if the market price is $26. Use the table of integrals in Section 16.2 to evaluate the definite integral.

33. COMPUTER GAME SALES The sales of Starr Communication's newest computer game, Laser Beams, are currently

$$te^{-0.05t}$$

units/month (each unit representing 1000 games), where t denotes the number of months since the release of the game. Find an expression that gives the total number of games sold as a function of t. How many games will be sold by the end of the first year?

34. DEMAND FOR COMPUTER SOFTWARE The demand equation for a computer software program is given by

$$p = 2\sqrt{325 - x^2}$$

where p is the unit price in dollars and x is the quantity demanded each month in units of a thousand. Find the consumers' surplus if the market price is $30. Evaluate the definite integral using Simpson's rule with $n = 10$.

35. OIL SPILLS Using aerial photographs, the Coast Guard was able to determine the dimensions of an oil spill along an embankment on a coastline, as shown in the accompanying figure. Using (a) the trapezoidal rule and (b) Simpson's rule with $n = 10$, estimate the area of the oil spill.

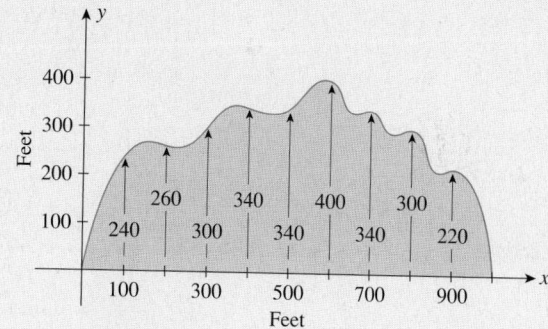

36. SURFACE AREA OF A LAKE A manmade lake located in Lake View Condominiums has the shape depicted in the following figure. The measurements shown were taken at 15-ft intervals. Using Simpson's rule with $n = 10$, estimate the surface area of the lake.

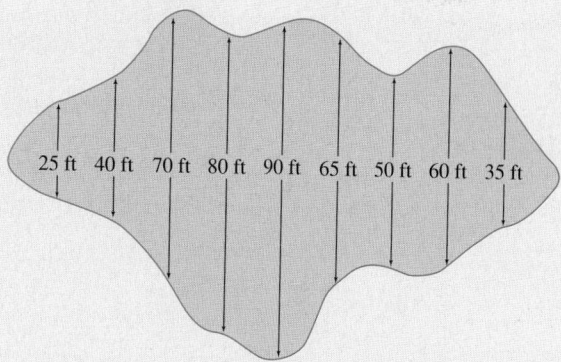

37. PERPETUITIES Lindsey wishes to establish a memorial fund at Newtown Hospital in the amount of $10,000/year beginning next year. If the fund earns interest at a rate of 9%/year compounded continuously, find the amount of endowment that he is required to make now.

CHAPTER 16   Before Moving On . . .

1. Find $\int x^2 \ln x \, dx$.

2. Use the table of integrals to find $\int \dfrac{dx}{x^2\sqrt{8 + 2x^2}}$.

3. Evaluate $\int_2^4 \sqrt{x^2 + 1} \, dx$ using the trapezoidal rule with $n = 5$.

4. Evaluate $\int_1^3 e^{0.2x} \, dx$ using Simpson's rule with $n = 6$.

5. Evaluate $\int_1^\infty e^{-2x} \, dx$

6. Let $f(x) = \dfrac{5}{96} x^{2/3}$ be defined on $[0, 8]$.

    a. Show that f is a probability density function on $[0, 8]$.
    b. Find $P(1 \le x \le 8)$.

17 Calculus of Several Variables

© Richard Cummins/Corbis

What should the dimensions of the new swimming pool be? It will be built in an elliptical area located in the rear of the promenade deck. Subject to this constraint, what are the dimensions of the largest pool that can be built? See Example 5, page 1092, to see how to solve this problem.

UP TO NOW, we have dealt with functions involving one variable. However, many situations involve functions of two or more variables. For example, the Consumer Price Index (CPI) compiled by the Bureau of Labor Statistics depends on the price of more than 95,000 consumer items. To study such relationships, we need the notion of a function of several variables, the first topic in this chapter. Next, generalizing the concept of the derivative of a function of one variable, we study the *partial derivatives* of a function of two or more variables. Using partial derivatives, we study the rate of change of a function with respect to one variable while holding all other variables constant. We then learn how to find the extremum values of a function of several variables. Finally, we generalize the notion of the integral to the case involving a function of two variables.

17.1   Functions of Several Variables

Up to now, our study of calculus has been restricted to functions of one variable. In many practical situations, however, the formulation of a problem results in a mathematical model that involves a function of two or more variables. For example, suppose Ace Novelty determines that the profits are $6, $5, and $4 for three types of souvenirs it produces. Let x, y, and z denote the number of type-A, type-B, and type-C souvenirs to be made; then the company's profit is given by

$$P = 6x + 5y + 4z$$

and P is a function of the three variables, x, y, and z.

■ Functions of Two Variables

Although this chapter deals with real-valued functions of several variables, most of our definitions and results are stated in terms of a function of two variables. One reason for adopting this approach, as you will soon see, is that there is a geometric interpretation for this special case, which serves as an important visual aid. We can then draw upon the experience gained from studying the two-variable case to help us understand the concepts and results connected with the more general case, which, by and large, is just a simple extension of the lower-dimensional case.

A Function of Two Variables

A real-valued **function of two variables** f, consists of

1. A set A of ordered pairs of real numbers (x, y) called the **domain** of the function.
2. A rule that associates with each ordered pair in the domain of f one and only one real number, denoted by $z = f(x, y)$.

The variables x and y are called **independent variables**, and the variable z, which is dependent on the values of x and y, is referred to as a **dependent variable**.

As in the case of a real-valued function of one real variable, the number $z = f(x, y)$ is called the **value of** f at the point (x, y). And, unless specified, the domain of the function f will be taken to be the largest possible set for which the rule defining f is meaningful.

EXAMPLE 1   Let f be the function defined by

$$f(x, y) = x + xy + y^2 + 2$$

Compute $f(0, 0), f(1, 2)$, and $f(2, 1)$.

Solution   We have

$$f(0, 0) = 0 + (0)(0) + 0^2 + 2 = 2$$
$$f(1, 2) = 1 + (1)(2) + 2^2 + 2 = 9$$
$$f(2, 1) = 2 + (2)(1) + 1^2 + 2 = 7$$

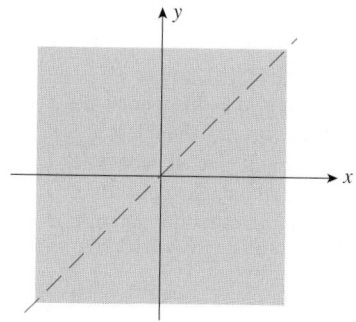

(a) Domain of g

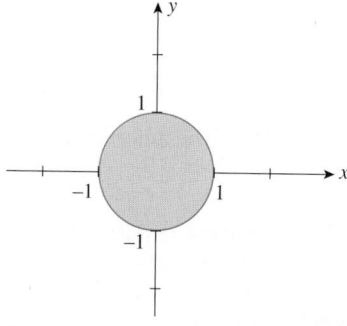

(b) Domain of h

FIGURE 1

The domain of a function of two variables $f(x, y)$ is a set of ordered pairs of real numbers and may therefore be viewed as a subset of the xy-plane.

EXAMPLE 2 Find the domain of each of the following functions.

a. $f(x, y) = x^2 + y^2$ **b.** $g(x, y) = \dfrac{2}{x - y}$ **c.** $h(x, y) = \sqrt{1 - x^2 - y^2}$

Solution

a. $f(x, y)$ is defined for all real values of x and y, so the domain of the function f is the set of all points (x, y) in the xy-plane.
b. $g(x, y)$ is defined for all $x \neq y$, so the domain of the function g is the set of all points in the xy-plane except those lying on the line $y = x$ (Figure 1a).
c. We require that $1 - x^2 - y^2 \geq 0$ or $x^2 + y^2 \leq 1$, which is just the set of all points (x, y) lying on and inside the circle of radius 1 with center at the origin (Figure 1b).

APPLIED EXAMPLE 3 **Revenue Functions** Acrosonic manufactures a bookshelf loudspeaker system that may be bought fully assembled or in a kit. The demand equations that relate the unit prices, p and q, to the quantities demanded weekly, x and y, of the assembled and kit versions of the loudspeaker systems are given by

$$p = 300 - \frac{1}{4}x - \frac{1}{8}y \quad \text{and} \quad q = 240 - \frac{1}{8}x - \frac{3}{8}y$$

a. What is the weekly total revenue function $R(x, y)$?
b. What is the domain of the function R?

Solution

a. The weekly revenue realizable from the sale of x units of the assembled speaker systems at p dollars per unit is given by xp dollars. Similarly, the weekly revenue realizable from the sale of y units of the kits at q dollars per unit is given by yq dollars. Therefore, the weekly total revenue function R is given by

$$R(x, y) = xp + yq$$
$$= x\left(300 - \frac{1}{4}x - \frac{1}{8}y\right) + y\left(240 - \frac{1}{8}x - \frac{3}{8}y\right)$$
$$= -\frac{1}{4}x^2 - \frac{3}{8}y^2 - \frac{1}{4}xy + 300x + 240y$$

b. To find the domain of the function R, let's observe that the quantities x, y, p, and q must be nonnegative. This observation leads to the following system of linear inequalities:

$$300 - \frac{1}{4}x - \frac{1}{8}y \geq 0$$
$$240 - \frac{1}{8}x - \frac{3}{8}y \geq 0$$
$$x \geq 0$$
$$y \geq 0$$

The domain of the function R is sketched in Figure 2.

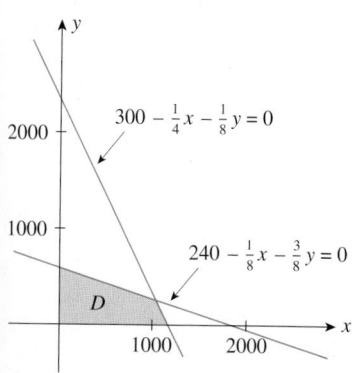

FIGURE 2
The domain of $R(x, y)$

EXPLORE & DISCUSS

Suppose the total profit of a two-product company is given by $P(x, y)$, where x denotes the number of units of the first product produced and sold and y denotes the number of units of the second product produced and sold. Fix $x = a$, where a is a positive number so that (a, y) is in the domain of P. Describe and give an economic interpretation of the function $f(y) = P(a, y)$. Next, fix $y = b$, where b is a positive number so that (x, b) is in the domain of P. Describe and give an economic interpretation of the function $g(x) = P(x, b)$.

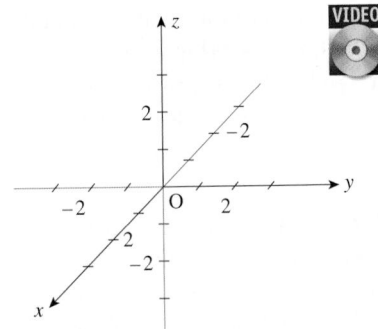

FIGURE 3
The three-dimensional Cartesian coordinate system

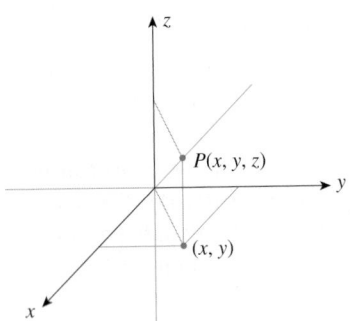

(a) A point in three-dimensional space

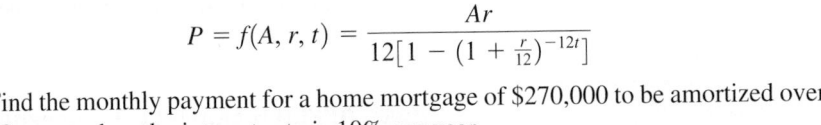

APPLIED EXAMPLE 4   Home Mortgage Payments   The monthly payment that amortizes a loan of A dollars in t years when the interest rate is r per year is given by

$$P = f(A, r, t) = \frac{Ar}{12[1 - (1 + \frac{r}{12})^{-12t}]}$$

Find the monthly payment for a home mortgage of $270,000 to be amortized over 30 years when the interest rate is 10% per year.

Solution   Letting $A = 270{,}000$, $r = 0.1$, and $t = 30$, we find the required monthly payment to be

$$P = f(270{,}000, 0.1, 30) = \frac{270{,}000(0.1)}{12[1 - (1 + \frac{0.1}{12})^{-360}]}$$

$$\approx 2369.44$$

or approximately $2369.44.   ∎

Graphs of Functions of Two Variables

To graph a function of two variables, we need a three-dimensional coordinate system. This is readily constructed by adding a third axis to the plane Cartesian coordinate system in such a way that the three resulting axes are mutually perpendicular and intersect at O. Observe that, by construction, the zeros of the three number scales coincide at the origin of the **three-dimensional Cartesian coordinate system** (Figure 3).

A point in three-dimensional space can now be represented uniquely in this coordinate system by an **ordered triple** of numbers (x, y, z), and, conversely, every ordered triple of real numbers (x, y, z) represents a point in three-dimensional space (Figure 4a). For example, the points $A(2, 3, 4)$, $B(1, -2, -2)$, $C(2, 4, 0)$, and $D(0, 0, 4)$ are shown in Figure 4b.

Now, if $f(x, y)$ is a function of two variables x and y, the domain of f is a subset of the xy-plane. Let $z = f(x, y)$ so that there is one and only one point $(x, y, z) \equiv (x, y, f(x, y))$ associated with each point (x, y) in the domain of f. The totality of all such points makes up the **graph** of the function f and is, except for certain degenerate cases, a surface in three-dimensional space (Figure 5).

In interpreting the graph of a function $f(x, y)$, one often thinks of the value $z = f(x, y)$ of the function at the point (x, y) as the "height" of the point (x, y, z) on the graph of f. If $f(x, y) > 0$, then the point (x, y, z) is $f(x, y)$ units above the xy-plane; if $f(x, y) < 0$, then the point (x, y, z) is $|f(x, y)|$ units below the xy-plane.

(b) Some sample points in three-dimensional space

FIGURE 4

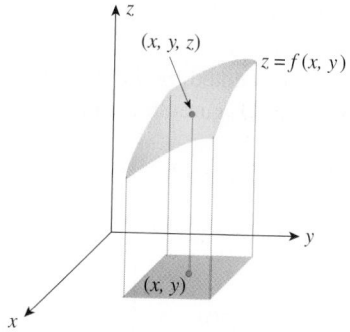

FIGURE 5
The graph of a function in three-dimensional space

FIGURE 6
Two computer-generated graphs of functions of two variables

In general, it is quite difficult to draw the graph of a function of two variables. But techniques have been developed that enable us to generate such graphs with minimum effort, using a computer. Figure 6 shows the computer-generated graphs of two functions.

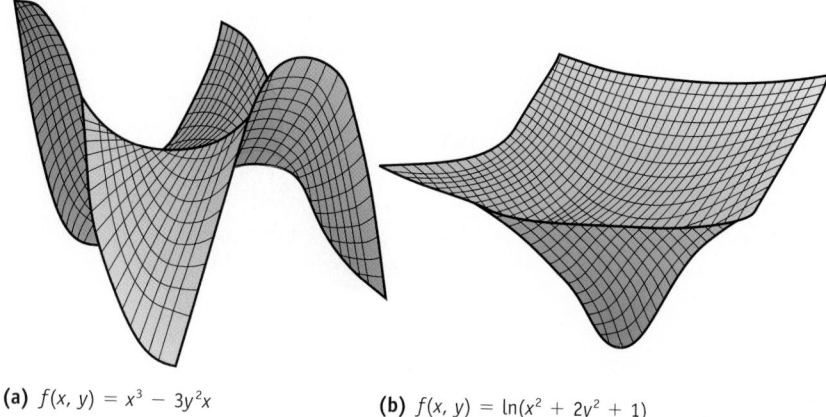

(a) $f(x, y) = x^3 - 3y^2x$ **(b)** $f(x, y) = \ln(x^2 + 2y^2 + 1)$

▬ Level Curves

As mentioned earlier, the graph of a function of two variables is often difficult to sketch, and we will not develop a systematic procedure for sketching it. Instead, we describe a method that is used in constructing topographic maps. This method is relatively easy to apply and conveys sufficient information to enable one to obtain a feel for the graph of the function.

Suppose that $f(x, y)$ is a function of two variables x and y, with a graph as shown in Figure 7. If c is some value of the function f, then the equation $f(x, y) = c$ describes a curve lying on the plane $z = c$ called the **trace** of the graph of f in the plane $z = c$. If this trace is projected onto the xy-plane, the resulting curve in the xy-plane is called a **level curve**. By drawing the level curves corresponding to several admissible values of c, we obtain a **contour map**. Observe that, by construction, every point on a particular level curve corresponds to a point on the surface

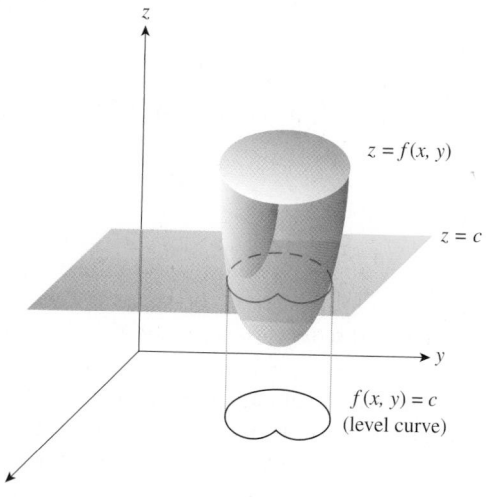

FIGURE 7
The graph of the function $z = f(x, y)$ and its intersection with the plane $z = c$

$z = f(x, y)$ that is a certain fixed distance from the xy-plane. Thus, by elevating or depressing the level curves that make up the contour map in one's mind, it is possible to obtain a feel for the general shape of the surface represented by the function f. Figure 8a shows a part of a mountain range with one peak; Figure 8b is the associated contour map.

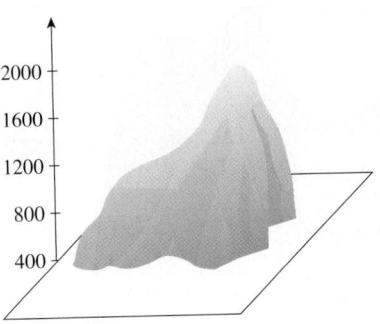

(a) A peak on a mountain range

(b) A contour map for the mountain peak

FIGURE 8

EXAMPLE 5   Sketch a contour map for the function $f(x, y) = x^2 + y^2$.

Solution   The level curves are the graphs of the equation $x^2 + y^2 = c$ for non-negative numbers c. Taking $c = 0, 1, 4, 9$, and 16, for example, we obtain

$$c = \ \ 0: x^2 + y^2 = 0$$
$$c = \ \ 1: x^2 + y^2 = 1$$
$$c = \ \ 4: x^2 + y^2 = 4 = 2^2$$
$$c = \ \ 9: x^2 + y^2 = 9 = 3^2$$
$$c = 16: x^2 + y^2 = 16 = 4^2$$

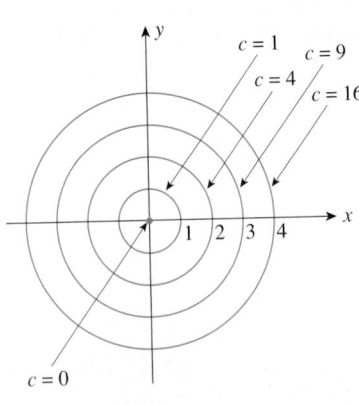

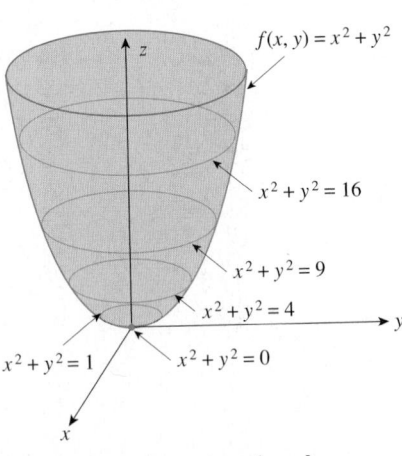

(a) Level curves of $f(x, y) = x^2 + y^2$

(b) The graph of $f(x, y) = x^2 + y^2$

FIGURE 9

The five level curves are concentric circles with center at the origin and radius given by $r = 0, 1, 2, 3$, and 4, respectively (Figure 9a). A sketch of the graph of $f(x, y) = x^2 + y^2$ is included for your reference in Figure 9b.

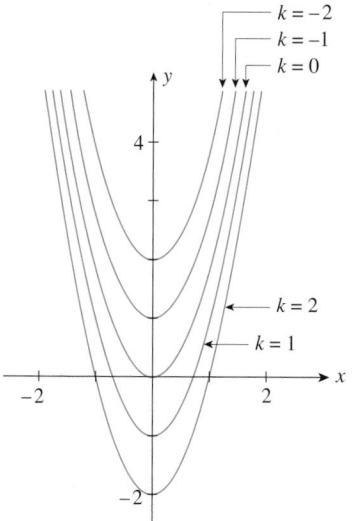

FIGURE 10
Level curves for $f(x, y) = 2x^2 - y$

EXAMPLE 6 Sketch the level curves for the function $f(x, y) = 2x^2 - y$ corresponding to $z = -2, -1, 0, 1$, and 2.

Solution The level curves are the graphs of the equation $2x^2 - y = k$ or $y = 2x^2 - k$ for $k = -2, -1, 0, 1$, and 2. The required level curves are shown in Figure 10.

Level curves of functions of two variables are found in many practical applications. For example, if $f(x, y)$ denotes the temperature at a location within the continental United States with longitude x and latitude y at a certain time of day, then the temperature at the point (x, y) is given by the "height" of the surface, represented by $z = f(x, y)$. In this situation the level curve $f(x, y) = k$ is a curve superimposed on a map of the United States, connecting points having the same temperature at a given time (Figure 11). These level curves are called **isotherms**.

Similarly, if $f(x, y)$ gives the barometric pressure at the location (x, y), then the level curves of the function f are called **isobars**, lines connecting points having the same barometric pressure at a given time.

As a final example, suppose $P(x, y, z)$ is a function of three variables x, y, and z giving the profit realized when x, y, and z units of three products, A, B, and C, respectively, are produced and sold. Then, the equation $P(x, y, z) = k$, where k is a constant, represents a surface in three-dimensional space called a **level surface** of P. In this situation, the level surface represented by $P(x, y, z) = k$ represents the product mix that results in a profit of exactly k dollars. Such a level surface is called an **isoprofit surface**.

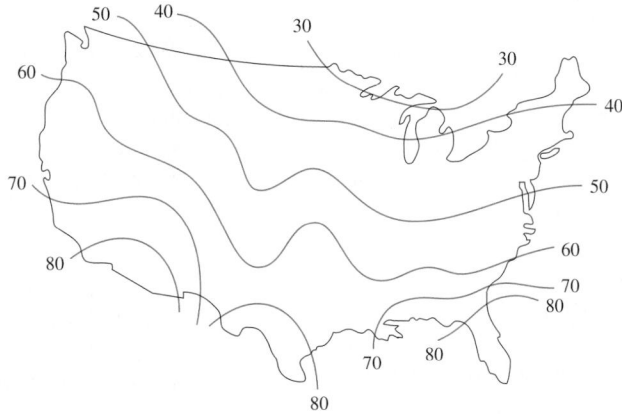

FIGURE 11
Isotherms: curves connecting points that have the same temperature

17.1 Self-Check Exercises

1. Let $f(x, y) = x^2 - 3xy + \sqrt{x + y}$. Compute $f(1, 3)$ and $f(-1, 1)$. Is the point $(-1, 0)$ in the domain of f?

2. Find the domain of $f(x, y) = \dfrac{1}{x} + \dfrac{1}{x - y} - e^{x+y}$.

3. Odyssey Travel Agency has a monthly advertising budget of $20,000. Odyssey's management estimates that if they spend x dollars on newspaper advertising and y dollars on television advertising, then the monthly revenue will be

$$f(x, y) = 30x^{1/4}y^{3/4}$$

dollars. What will be the monthly revenue if Odyssey spends $5000/month on newspaper ads and $15,000/month on television ads? If Odyssey spends $4000/month on newspaper ads and $16,000/month on television ads?

Solutions to Self-Check Exercises 17.1 can be found on page 1060.

17.1 Concept Questions

1. What is a function of two variables? Give an example of a function of two variables and state its rule of definition and domain.

2. If f is a function of two variables, what can you say about the relationship between $f(a, b)$ and $f(c, d)$, if (a, b) is in the domain of f and $c = a$ and $d = b$?

3. Define (a) the graph of $f(x, y)$ and (b) the level curve of f.

17.1 Exercises

1. Let $f(x, y) = 2x + 3y - 4$. Compute $f(0, 0), f(1, 0), f(0, 1)$, $f(1, 2)$, and $f(2, -1)$.

2. Let $g(x, y) = 2x^2 - y^2$. Compute $g(1, 2), g(2, 1), g(1, 1)$, $g(-1, 1)$, and $g(2, -1)$.

3. Let $f(x, y) = x^2 + 2xy - x + 3$. Compute $f(1, 2), f(2, 1)$, $f(-1, 2)$, and $f(2, -1)$.

4. Let $h(x, y) = (x + y)/(x - y)$. Compute $h(0, 1), h(-1, 1)$, $h(2, 1)$, and $h(\pi, -\pi)$.

5. Let $g(s, t) = 3s\sqrt{t} + t\sqrt{s} + 2$. Compute $g(1, 2), g(2, 1)$, $g(0, 4)$, and $g(4, 9)$.

6. Let $f(x, y) = xye^{x^2+y^2}$. Compute $f(0, 0), f(0, 1), f(1, 1)$, and $f(-1, -1)$.

7. Let $h(s, t) = s \ln t - t \ln s$. Compute $h(1, e), h(e, 1)$, and $h(e, e)$.

8. Let $f(u, v) = (u^2 + v^2)e^{uv^2}$. Compute $f(0, 1), f(-1, -1)$, $f(a, b)$, and $f(b, a)$.

9. Let $g(r, s, t) = re^{s/t}$. Compute $g(1, 1, 1), g(1, 0, 1)$, and $g(-1, -1, -1)$.

10. Let $g(u, v, w) = (ue^{vw} + ve^{uw} + we^{uv})/(u^2 + v^2 + w^2)$. Compute $g(1, 2, 3)$ and $g(3, 2, 1)$.

In Exercises 11–18, find the domain of the function.

11. $f(x, y) = 2x + 3y$

12. $g(x, y, z) = x^2 + y^2 + z^2$

13. $h(u, v) = \dfrac{uv}{u - v}$

14. $f(s, t) = \sqrt{s^2 + t^2}$

15. $g(r, s) = \sqrt{rs}$

16. $f(x, y) = e^{-xy}$

17. $h(x, y) = \ln(x + y - 5)$

18. $h(u, v) = \sqrt{4 - u^2 - v^2}$

In Exercises 19–24, sketch the level curves of the function corresponding to each value of z.

19. $f(x, y) = 2x + 3y; z = -2, -1, 0, 1, 2$

20. $f(x, y) = -x^2 + y; z = -2, -1, 0, 1, 2$

21. $f(x, y) = 2x^2 + y; z = -2, -1, 0, 1, 2$

22. $f(x, y) = xy; z = -4, -2, 2, 4$

23. $f(x, y) = \sqrt{16 - x^2 - y^2}; z = 0, 1, 2, 3, 4$

24. $f(x, y) = e^x - y; z = -2, -1, 0, 1, 2$

25. The volume of a cylindrical tank of radius r and height h is given by

$$V = f(r, h) = \pi r^2 h$$

Find the volume of a cylindrical tank of radius 1.5 ft and height 4 ft.

26. **IQs** The IQ (intelligence quotient) of a person whose mental age is m yr and whose chronological age is c yr is defined as

$$f(m, c) = \frac{100m}{c}$$

What is the IQ of a 9-yr-old child who has a mental age of 13.5 yr?

27. **BODY MASS** The body mass index (BMI) is used to identify, evaluate, and treat overweight and obese adults. The BMI value for an adult of weight w (in kilograms) and height h (in meters) is defined to be

$$M = f(w, h) = \frac{w}{h^2}$$

According to federal guidelines, an adult is overweight if he or she has a BMI value between 25 and 29.9 and is "obese" if the value is greater than or equal to 30.

a. What is the BMI of an adult who weighs in at 80 kg and stands 1.8 m tall?

b. What is the maximum weight for an adult of height 1.8 m, who is not classified as overweight or obese?

28. **POISEUILLE'S LAW** Poiseuille's law states that the resistance R, measured in dynes, of blood flowing in a blood vessel of length l and radius r (both in centimeters) is given by

$$R = f(l, r) = \frac{kl}{r^4}$$

where k is the viscosity of blood (in dyne-sec/cm^2). What is the resistance, in terms of k, of blood flowing through an arteriole 4 cm long and of radius 0.1 cm?

29. **REVENUE FUNCTIONS** Country Workshop manufactures both finished and unfinished furniture for the home. The estimated quantities demanded each week of its rolltop desks in

the finished and unfinished versions are x and y units when the corresponding unit prices are

$$p = 200 - \frac{1}{5}x - \frac{1}{10}y$$

$$q = 160 - \frac{1}{10}x - \frac{1}{4}y,$$

dollars, respectively.
a. What is the weekly total revenue function $R(x, y)$?
b. Find the domain of the function R.

30. For the total revenue function $R(x, y)$ of Exercise 29, compute $R(100, 60)$ and $R(60, 100)$. Interpret your results.

31. Revenue Functions   Weston Publishing publishes a deluxe edition and a standard edition of its English language dictionary. Weston's management estimates that the number of deluxe editions demanded is x copies/day and the number of standard editions demanded is y copies/day when the unit prices are

$$p = 20 - 0.005x - 0.001y$$

$$q = 15 - 0.001x - 0.003y$$

dollars, respectively.
a. Find the daily total revenue function $R(x, y)$.
b. Find the domain of the function R.

32. For the total revenue function $R(x, y)$ of Exercise 31, compute $R(300, 200)$ and $R(200, 300)$. Interpret your results.

33. Volume of a Gas   The volume of a certain mass of gas is related to its pressure and temperature by the formula

$$V = \frac{30.9T}{P}$$

where the volume V is measured in liters, the temperature T is measured in degrees Kelvin (obtained by adding $273°$ to the Celsius temperature), and the pressure P is measured in millimeters of mercury pressure.
a. Find the domain of the function V.
b. Calculate the volume of the gas at standard temperature and pressure—that is, when $T = 273$ K and $P = 760$ mm of mercury.

34. Surface Area of a Human Body   An empirical formula by E. F. Dubois relates the surface area S of a human body (in square meters) to its weight W (in kilograms) and its height H (in centimeters). The formula, given by

$$S = 0.007184W^{0.425}H^{0.725}$$

is used by physiologists in metabolism studies.
a. Find the domain of the function S.
b. What is the surface area of a human body that weighs 70 kg and has a height of 178 cm?

35. Production Function   Suppose the output of a certain country is given by

$$f(x, y) = 100x^{3/5}y^{2/5}$$

billion dollars if x billion dollars are spent for labor and y billion dollars are spent on capital. Find the output if the country spent $32 billion on labor and $243 billion on capital.

36. Production Function   Economists have found that the output of a finished product, $f(x, y)$, is sometimes described by the function

$$f(x, y) = ax^b y^{1-b}$$

where x stands for the amount of money expended for labor, y stands for the amount expended on capital, and a and b are positive constants with $0 < b < 1$.
a. If p is a positive number, show that $f(px, py) = pf(x, y)$.
b. Use the result of part (a) to show that if the amount of money expended for labor and capital are both increased by r percent, then the output is also increased by r percent.

37. Arson for Profit   A study of arson for profit was conducted by a team of paid civilian experts and police detectives appointed by the mayor of a large city. It was found that the number of suspicious fires in that city in 2002 was very closely related to the concentration of tenants in the city's public housing and to the level of reinvestment in the area in conventional mortgages by the ten largest banks. In fact, the number of fires was closely approximated by the formula

$$N(x, y) = \frac{100(1000 + 0.03x^2 y)^{1/2}}{(5 + 0.2y)^2} \quad (0 \le x \le 150; 5 \le y \le 35)$$

where x denotes the number of persons/census tract and y denotes the level of reinvestment in the area in cents/dollar deposited. Using this formula, estimate the total number of suspicious fires in the districts of the city where the concentration of public housing tenants was 100/census tract and the level of reinvestment was 20 cents/dollar deposited.

38. Continuously Compounded Interest   If a principal of P dollars is deposited in an account earning interest at the rate of r/year compounded continuously, then the accumulated amount at the end of t yr is given by

$$A = f(P, r, t) = Pe^{rt}$$

dollars. Find the accumulated amount at the end of 3 yr if a sum of $10,000 is deposited in an account earning interest at the rate of 10%/year.

39. Home Mortgages   The monthly payment that amortizes a loan of A dollars in t yr when the interest rate is r per year is given by

$$P = f(A, r, t) = \frac{Ar}{12[1 - (1 + \frac{r}{12})^{-12t}]}$$

a. What is the monthly payment for a home mortgage of $100,000 that will be amortized over 30 yr with an interest rate of 8%/year? An interest rate of 10%/year?

b. Find the monthly payment for a home mortgage of $100,000 that will be amortized over 20 yr with an interest rate of 8%/year.

40. HOME MORTGAGES   Suppose a home buyer secures a bank loan of A dollars to purchase a house. If the interest rate charged is r/year and the loan is to be amortized in t yr, then the principal repayment at the end of i mo is given by

$$B = f(A, r, t, i)$$

$$= A\left[\frac{(1 + \frac{r}{12})^i - 1}{(1 + \frac{r}{12})^{12t} - 1}\right] \qquad (0 \le i \le 12t)$$

Suppose the Blakelys borrow a sum of $280,000 from a bank to help finance the purchase of a house and the bank charges interest at a rate of 6%/year. If the Blakelys agree to repay the loan in equal installments over 30 yr, how much will they owe the bank after the 60th payment (5 yr)? The 240th payment (20 yr)?

41. FORCE GENERATED BY A CENTRIFUGE   A centrifuge is a machine designed for the specific purpose of subjecting materials to a sustained centrifugal force. The actual amount of centrifugal force, F, expressed in dynes (1 gram of force = 980 dynes) is given by

$$F = f(M, S, R) = \frac{\pi^2 S^2 MR}{900}$$

where S is in revolutions per minute (rpm), M is in grams, and R is in centimeters. Show that an object revolving at the rate of 600 rpm in a circle with radius of 10 cm generates a centrifugal force that is approximately 40 times gravity.

42. WILSON LOT-SIZE FORMULA   The Wilson lot-size formula in economics states that the optimal quantity Q of goods for a store to order is given by

$$Q = f(C, N, h) = \sqrt{\frac{2\,CN}{h}}$$

where C is the cost of placing an order, N is the number of items the store sells per week, and h is the weekly holding cost for each item. Find the most economical quantity of 10-speed bicycles to order if it costs the store $20 to place an order, $5 to hold a bicycle for a week, and the store expects to sell 40 bicycles a week.

In Exercises 43–46, determine whether the statement is true or false. If it is true, explain why it is true. If it is false, give an example to show why it is false.

43. If h is a function of x and y, then there are functions f and g of one variable such that

$$h(x, y) = f(x) + g(y)$$

44. If f is a function of x and y and a is a real number, then

$$f(ax, ay) = af(x, y)$$

45. The domain of $f(x, y) = 1/(x^2 - y^2)$ is $\{(x, y) \mid y \ne x\}$.

46. Every point on the level curve $f(x, y) = c$ corresponds to a point on the graph of f that is c units above the xy-plane if $c > 0$ and $|c|$ units below the xy-plane if $c < 0$.

17.1   Solutions to Self-Check Exercises

1. $f(1, 3) = 1^2 - 3(1)(3) + \sqrt{1 + 3} = -6$
   $f(-1, 1) = (-1)^2 - 3(-1)(1) + \sqrt{-1 + 1} = 4$

   The point $(-1, 0)$ is not in the domain of f because the term $\sqrt{x + y}$ is not defined when $x = -1$ and $y = 0$. In fact, the domain of f consists of all real values of x and y that satisfy the inequality $x + y \ge 0$, the shaded half-plane shown in the accompanying figure.

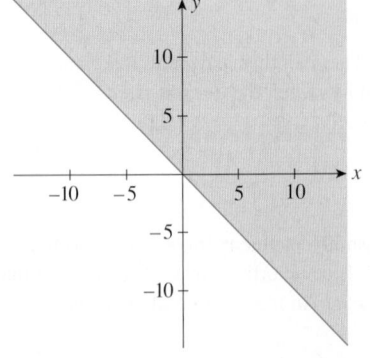

2. Since division by zero is not permitted, we see that $x \ne 0$ and $x - y \ne 0$. Therefore, the domain of f is the set of all points in the xy-plane not containing the y-axis ($x = 0$) and the straight line $x = y$.

3. If Odyssey spends $5000/month on newspaper ads ($x = 5000$) and $15,000/month on television ads ($y = 15,000$), then its monthly revenue will be given by

   $$f(5000, 15,000) = 30(5000)^{1/4}(15,000)^{3/4}$$
   $$\approx 341,926.06$$

   or approximately $341,926. If the agency spends $4000/month on newspaper ads and $16,000/month on television ads, then its monthly revenue will be given by

   $$f(4000, 16,000) = 30(4000)^{1/4}(16,000)^{3/4}$$
   $$\approx 339,411.26$$

   or approximately $339,411.

17.2 Partial Derivatives

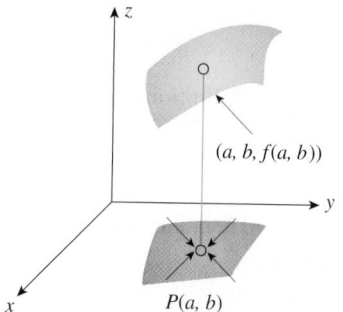

FIGURE 12
We can approach a point in the plane from infinitely many directions.

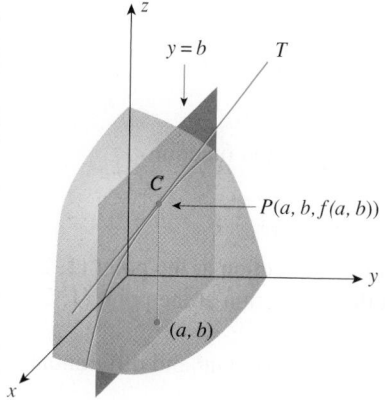

FIGURE 13
The curve C is formed by the intersection of the plane $y = b$ with the surface $z = f(x, y)$.

■ Partial Derivatives

For a function $f(x)$ of one variable x, there is no ambiguity when we speak about the rate of change of $f(x)$ with respect to x since x must be constrained to move along the x-axis. The situation becomes more complicated, however, when we study the rate of change of a function of two or more variables. For example, the domain D of a function of two variables $f(x, y)$ is a subset of the plane (Figure 12), so if $P(a, b)$ is any point in the domain of f, there are infinitely many directions from which one can approach the point P. We may therefore ask for the rate of change of f at P along any of these directions.

However, we will not deal with this general problem. Instead, we will restrict ourselves to studying the rate of change of the function $f(x, y)$ at a point $P(a, b)$ in each of two *preferred directions*—namely, the direction parallel to the x-axis and the direction parallel to the y-axis. Let $y = b$, where b is a constant, so that $f(x, b)$ is a function of the one variable x. Since the equation $z = f(x, y)$ is the equation of a surface, the equation $z = f(x, b)$ is the equation of the curve C on the surface formed by the intersection of the surface and the plane $y = b$ (Figure 13).

Because $f(x, b)$ is a function of one variable x, we may compute the derivative of f with respect to x at $x = a$. This derivative, obtained by keeping the variable y fixed and differentiating the resulting function $f(x, b)$ with respect to x, is called the **first partial derivative of f with respect to x** at (a, b), written

$$\frac{\partial z}{\partial x}(a, b) \quad \text{or} \quad \frac{\partial f}{\partial x}(a, b) \quad \text{or} \quad f_x(a, b)$$

Thus,

$$\frac{\partial z}{\partial x}(a, b) = \frac{\partial f}{\partial x}(a, b) = f_x(a, b) = \lim_{h \to 0} \frac{f(a + h, b) - f(a, b)}{h}$$

provided that the limit exists. The first partial derivative of f with respect to x at (a, b) measures both the slope of the tangent line T to the curve C and the rate of change of the function f in the x-direction when $x = a$ and $y = b$. We also write

$$\left.\frac{\partial f}{\partial x}\right|_{(a, b)} \equiv f_x(a, b)$$

Similarly, we define the **first partial derivative of f with respect to y** at (a, b), written

$$\frac{\partial z}{\partial y}(a, b) \quad \text{or} \quad \frac{\partial f}{\partial y}(a, b) \quad \text{or} \quad f_y(a, b)$$

as the derivative obtained by keeping the variable x fixed and differentiating the resulting function $f(a, y)$ with respect to y. That is,

$$\frac{\partial z}{\partial y}(a, b) = \frac{\partial f}{\partial y}(a, b) = f_y(a, b)$$
$$= \lim_{k \to 0} \frac{f(a, b + k) - f(a, b)}{k}$$

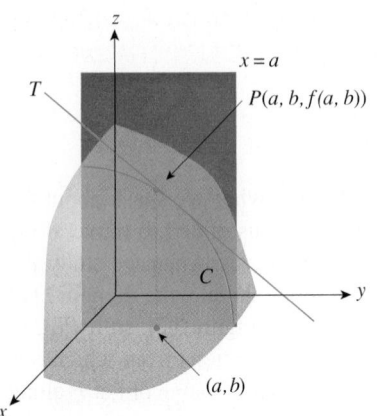

FIGURE 14
The first partial derivative of f with respect to y at (a, b) measures the slope of the tangent line T to the curve C with x held constant.

if the limit exists. The first partial derivative of f with respect to y at (a, b) measures both the slope of the tangent line T to the curve C, obtained by holding x constant (Figure 14), and the rate of change of the function f in the y-direction when $x = a$ and $y = b$. We write

$$\frac{\partial f}{\partial y}\bigg|_{(a, b)} \equiv f_y(a, b)$$

Before looking at some examples, let's summarize these definitions.

First Partial Derivatives of $f(x, y)$

Suppose $f(x, y)$ is a function of the two variables x and y. Then, the **first partial derivative of f** with respect to x at the point (x, y) is

$$\frac{\partial f}{\partial x} = \lim_{h \to 0} \frac{f(x + h, y) - f(x, y)}{h}$$

provided the limit exists. The first partial derivative of f with respect to y at the point (x, y) is

$$\frac{\partial f}{\partial y} = \lim_{k \to 0} \frac{f(x, y + k) - f(x, y)}{k}$$

provided the limit exists.

EXAMPLE 1   Find the partial derivatives $\dfrac{\partial f}{\partial x}$ and $\dfrac{\partial f}{\partial y}$ of the function

$$f(x, y) = x^2 - xy^2 + y^3$$

What is the rate of change of the function f in the x-direction at the point $(1, 2)$? What is the rate of change of the function f in the y-direction at the point $(1, 2)$?

Solution   To compute $\dfrac{\partial f}{\partial x}$, think of the variable y as a constant and differentiate the resulting function of x with respect to x. Let's write

$$f(x, y) = x^2 - xy^2 + y^3$$

where the variable y to be treated as a constant is shown in color. Then,

$$\frac{\partial f}{\partial x} = 2x - y^2$$

To compute $\dfrac{\partial f}{\partial y}$, think of the variable x as being fixed—that is, as a constant—and differentiate the resulting function of y with respect to y. In this case,

$$f(x, y) = x^2 - xy^2 + y^3$$

so that

$$\frac{\partial f}{\partial y} = -2xy + 3y^2$$

The rate of change of the function f in the x-direction at the point $(1, 2)$ is given by

$$f_x(1, 2) = \frac{\partial f}{\partial x}\bigg|_{(1, 2)} = 2(1) - 2^2 = -2$$

That is, f decreases 2 units for each unit increase in the x-direction, y being kept constant ($y = 2$). The rate of change of the function f in the y-direction at the point $(1, 2)$ is given by

$$f_y(1, 2) = \left.\frac{\partial f}{\partial y}\right|_{(1, 2)} = -2(1)(2) + 3(2)^2 = 8$$

That is, f increases 8 units for each unit increase in the y-direction, x being kept constant ($x = 1$). ∎

EXPLORE & DISCUSS

Refer to the Explore & Discuss on page 1054. Suppose management has decided that the projected sales of the first product is a units. Describe how you might help management decide how many units of the second product the company should produce and sell in order to maximize the company's total profit. Justify your method to management. Suppose, however, management feels that b units of the second product can be manufactured and sold. How would you help management decide how many units of the first product to manufacture in order to maximize the company's total profit?

EXAMPLE 2   Compute the first partial derivatives of each function.

a. $f(x, y) = \dfrac{xy}{x^2 + y^2}$ **b.** $g(s, t) = (s^2 - st + t^2)^5$

c. $h(u, v) = e^{u^2 - v^2}$ **d.** $f(x, y) = \ln(x^2 + 2y^2)$

Solution

a. To compute $\dfrac{\partial f}{\partial x}$, think of the variable y as a constant. Thus,

$$f(x, y) = \frac{xy}{x^2 + y^2}$$

so that, upon using the quotient rule, we have

$$\frac{\partial f}{\partial x} = \frac{(x^2 + y^2)y - xy(2x)}{(x^2 + y^2)^2} = \frac{x^2y + y^3 - 2x^2y}{(x^2 + y^2)^2}$$

$$= \frac{y(y^2 - x^2)}{(x^2 + y^2)^2}$$

upon simplification and factorization. To compute $\dfrac{\partial f}{\partial y}$, think of the variable x as a constant. Thus,

$$f(x, y) = \frac{xy}{x^2 + y^2}$$

so that, upon using the quotient rule once again, we obtain

$$\frac{\partial f}{\partial y} = \frac{(x^2 + y^2)x - xy(2y)}{(x^2 + y^2)^2} = \frac{x^3 + xy^2 - 2xy^2}{(x^2 + y^2)^2}$$

$$= \frac{x(x^2 - y^2)}{(x^2 + y^2)^2}$$

b. To compute $\partial g/\partial s$, we treat the variable t as if it were a constant. Thus,

$$g(s, t) = (s^2 - st + t^2)^5$$

Using the general power rule, we find

$$\frac{\partial g}{\partial s} = 5(s^2 - st + t^2)^4 \cdot (2s - t)$$

$$= 5(2s - t)(s^2 - st + t^2)^4$$

To compute $\partial g/\partial t$, we treat the variable s as if it were a constant. Thus,

$$g(s, t) = (s^2 - st + t^2)^5$$

$$\frac{\partial g}{\partial t} = 5(s^2 - st + t^2)^4 (-s + 2t)$$

$$= 5(2t - s)(s^2 - st + t^2)^4$$

c. To compute $\partial h/\partial u$, think of the variable v as a constant. Thus,

$$h(u, v) = e^{u^2 - v^2}$$

Using the chain rule for exponential functions, we have

$$\frac{\partial h}{\partial u} = e^{u^2 - v^2} \cdot 2u$$

$$= 2u e^{u^2 - v^2}$$

Next, we treat the variable u as if it were a constant,

$$h(u, v) = e^{u^2 - v^2}$$

and we obtain

$$\frac{\partial h}{\partial v} = e^{u^2 - v^2} \cdot (-2v)$$

$$= -2v e^{u^2 - v^2}$$

d. To compute $\partial f/\partial x$, think of the variable y as à constant. Thus,

$$f(x, y) = \ln(x^2 + 2y^2)$$

so that the chain rule for logarithmic functions gives

$$\frac{\partial f}{\partial x} = \frac{2x}{x^2 + 2y^2}$$

Next, treating the variable x as if it were a constant, we find

$$f(x, y) = \ln(x^2 + 2y^2)$$

$$\frac{\partial f}{\partial y} = \frac{4y}{x^2 + 2y^2}$$

To compute the partial derivative of a function of several variables with respect to one variable—say, x—we think of the other variables as if they were constants and differentiate the resulting function with respect to x.

EXPLORE & DISCUSS

1. Let (a, b) be a point in the domain of $f(x, y)$. Put $g(x) = f(x, b)$ and suppose g is differentiable at $x = a$. Explain why you can find $f_x(a, b)$ by computing $g'(a)$. How would you go about calculating $f_y(a, b)$ using a similar technique? Give a geometric interpretation of these processes.

2. Let $f(x, y) = x^2y^3 - 3x^2y + 2$. Use the method of Problem 1 to find $f_x(1, 2)$ and $f_y(1, 2)$.

EXAMPLE 3 Compute the first partial derivatives of the function

$$w = f(x, y, z) = xyz - xe^{yz} + x \ln y$$

Solution Here we have a function of three variables, x, y, and z, and we are required to compute

$$\frac{\partial f}{\partial x}, \quad \frac{\partial f}{\partial y}, \quad \frac{\partial f}{\partial z}$$

To compute f_x, we think of the other two variables, y and z, as fixed, and we differentiate the resulting function of x with respect to x, thereby obtaining

$$f_x = yz - e^{yz} + \ln y$$

To compute f_y, we think of the other two variables, x and z, as constants, and we differentiate the resulting function of y with respect to y. We then obtain

$$f_y = xz - xze^{yz} + \frac{x}{y}$$

Finally, to compute f_z, we treat the variables x and y as constants and differentiate the function f with respect to z, obtaining

$$f_z = xy - xye^{yz}$$

 EXPLORING WITH TECHNOLOGY

Refer to the Explore & Discuss on this page. Let

$$f(x, y) = \frac{e^{\sqrt{xy}}}{(1 + xy^2)^{3/2}}$$

1. Compute $g(x) = f(x, 1)$ and use a graphing utility to plot the graph of g in the viewing window $[0, 2] \times [0, 2]$.

2. Use the differentiation operation of your graphing utility to find $g'(1)$ and hence $f_x(1, 1)$.

3. Compute $h(y) = f(1, y)$ and use a graphing utility to plot the graph of g in the viewing window $[0, 2] \times [0, 2]$.

4. Use the differentiation operation of your graphing utility to find $h'(1)$ and hence $f_y(1, 1)$.

▬ The Cobb–Douglas Production Function

For an economic interpretation of the first partial derivatives of a function of two variables, let's turn our attention to the function

$$f(x, y) = ax^b y^{1-b} \qquad (1)$$

where a and b are positive constants with $0 < b < 1$. This function is called the **Cobb–Douglas production function**. Here, x stands for the amount of money expended for labor, y stands for the cost of capital equipment (buildings, machinery, and other tools of production), and the function f measures the output of the finished product (in suitable units) and is called, accordingly, the production function.

The partial derivative f_x is called the **marginal productivity of labor**. It measures the rate of change of production with respect to the amount of money expended for labor, with the level of capital expenditure held constant. Similarly, the partial derivative f_y, called the **marginal productivity of capital**, measures the rate of change of production with respect to the amount expended on capital, with the level of labor expenditure held fixed.

APPLIED EXAMPLE 4 Marginal Productivity A certain country's production in the early years following World War II is described by the function

$$f(x, y) = 30x^{2/3} y^{1/3}$$

units, when x units of labor and y units of capital were used.

a. Compute f_x and f_y.
b. What is the marginal productivity of labor and the marginal productivity of capital when the amounts expended on labor and capital are 125 units and 27 units, respectively?
c. Should the government have encouraged capital investment rather than increasing expenditure on labor to increase the country's productivity?

Solution

a. $f_x = 30 \cdot \dfrac{2}{3} x^{-1/3} y^{1/3} = 20 \left(\dfrac{y}{x} \right)^{1/3}$

   $f_y = 30x^{2/3} \cdot \dfrac{1}{3} y^{-2/3} = 10 \left(\dfrac{x}{y} \right)^{2/3}$

b. The required marginal productivity of labor is given by

$$f_x(125, 27) = 20 \left(\frac{27}{125} \right)^{1/3} = 20 \left(\frac{3}{5} \right)$$

or 12 units per unit increase in labor expenditure (capital expenditure is held constant at 27 units). The required marginal productivity of capital is given by

$$f_y(125, 27) = 10 \left(\frac{125}{27} \right)^{2/3} = 10 \left(\frac{25}{9} \right)$$

or $27\frac{7}{9}$ units per unit increase in capital expenditure (labor outlay is held constant at 125 units).

c. From the results of part (b), we see that a unit increase in capital expenditure resulted in a much faster increase in productivity than a unit increase in labor expenditure would have. Therefore, the government should have encouraged increased spending on capital rather than on labor during the early years of reconstruction.

Substitute and Complementary Commodities

For another application of the first partial derivatives of a function of two variables in the field of economics, let's consider the relative demands of two commodities. We say that the two commodities are substitute (competitive) commodities if a decrease in the demand for one results in an increase in the demand for the other. Examples of substitute commodities are coffee and tea. Conversely, two commodities are referred to as complementary commodities if a decrease in the demand for one results in a decrease in the demand for the other as well. Examples of complementary commodities are automobiles and tires.

We now derive a criterion for determining whether two commodities A and B are substitute or complementary. Suppose the demand equations that relate the quantities demanded, x and y, to the unit prices, p and q, of the two commodities are given by

$$x = f(p, q) \quad \text{and} \quad y = g(p, q)$$

Let's consider the partial derivative $\partial f/\partial p$. Since f is the demand function for commodity A, we see that, for fixed q, f is typically a decreasing function of p—that is, $\partial f/\partial p < 0$. Now, if the two commodities were substitute commodities, then the quantity demanded of commodity B would increase with respect to p—that is, $\partial g/\partial p > 0$. A similar argument with p fixed shows that if A and B are substitute commodities, then $\partial f/\partial q > 0$. Thus, the two commodities A and B are substitute commodities if

$$\frac{\partial f}{\partial q} > 0 \quad \text{and} \quad \frac{\partial g}{\partial p} > 0$$

Similarly, A and B are complementary commodities if

$$\frac{\partial f}{\partial q} < 0 \quad \text{and} \quad \frac{\partial g}{\partial p} < 0$$

Substitute and Complementary Commodities

Two commodities A and B are substitute commodities if

$$\frac{\partial f}{\partial q} > 0 \quad \text{and} \quad \frac{\partial g}{\partial p} > 0 \tag{2}$$

Two commodities A and B are complementary commodities if

$$\frac{\partial f}{\partial q} < 0 \quad \text{and} \quad \frac{\partial g}{\partial p} < 0 \tag{3}$$

 APPLIED EXAMPLE 5   Substitute and Complementary Commodities   Suppose that the daily demand for butter is given by

$$x = f(p, q) = \frac{3q}{1 + p^2}$$

and the daily demand for margarine is given by

$$y = g(p, q) = \frac{2p}{1 + \sqrt{q}} \quad (p > 0, q > 0)$$

where p and q denote the prices per pound (in dollars) of butter and margarine, respectively, and x and y are measured in millions of pounds. Determine whether these two commodities are substitute, complementary, or neither.

Solution We compute

$$\frac{\partial f}{\partial q} = \frac{3}{1 + p^2} \quad \text{and} \quad \frac{\partial g}{\partial p} = \frac{2}{1 + \sqrt{q}}$$

Since

$$\frac{\partial f}{\partial q} > 0 \quad \text{and} \quad \frac{\partial g}{\partial p} > 0$$

for all values of $p > 0$ and $q > 0$, we conclude that butter and margarine are substitute commodities. ■

Second-Order Partial Derivatives

The first partial derivatives $f_x(x, y)$ and $f_y(x, y)$ of a function $f(x, y)$ of the two variables x and y are also functions of x and y. As such, we may differentiate each of the functions f_x and f_y to obtain the **second-order partial derivatives of f** (Figure 15). Thus, differentiating the function f_x with respect to x leads to the second partial derivative

$$f_{xx} \equiv \frac{\partial^2 f}{\partial x^2} = \frac{\partial}{\partial x}(f_x)$$

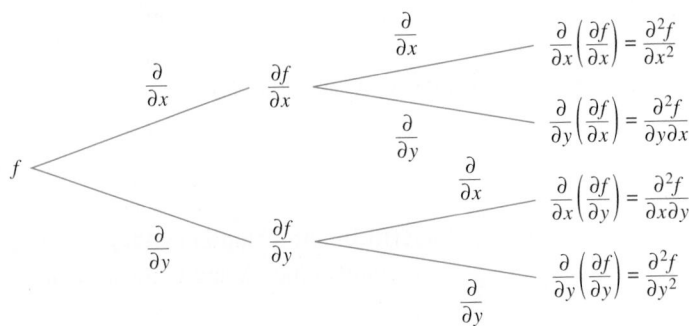

FIGURE 15
A schematic showing the four second-order partial derivatives of f

However, differentiation of f_x with respect to y leads to the second partial derivative

$$f_{xy} \equiv \frac{\partial^2 f}{\partial y \partial x} = \frac{\partial}{\partial y}(f_x)$$

Similarly, differentiation of the function f_y with respect to x and with respect to y leads to

$$f_{yx} \equiv \frac{\partial^2 f}{\partial x \partial y} = \frac{\partial}{\partial x}(f_y)$$

$$f_{yy} \equiv \frac{\partial^2 f}{\partial y^2} = \frac{\partial}{\partial y}(f_y)$$

respectively. Note that, in general, it is not true that $f_{xy} = f_{yx}$, but they are equal if both f_{xy} and f_{yx} are continuous. We might add that this is the case in most practical applications.

EXAMPLE 6 Find the second-order partial derivatives of the function
$$f(x, y) = x^3 - 3x^2y + 3xy^2 + y^2$$

Solution The first partial derivatives of f are
$$f_x = \frac{\partial}{\partial x}(x^3 - 3x^2y + 3xy^2 + y^2)$$
$$= 3x^2 - 6xy + 3y^2$$
$$f_y = \frac{\partial}{\partial y}(x^3 - 3x^2y + 3xy^2 + y^2)$$
$$= -3x^2 + 6xy + 2y$$

Therefore,
$$f_{xx} = \frac{\partial}{\partial x}(f_x) = \frac{\partial}{\partial x}(3x^2 - 6xy + 3y^2)$$
$$= 6x - 6y = 6(x - y)$$
$$f_{xy} = \frac{\partial}{\partial y}(f_x) = \frac{\partial}{\partial y}(3x^2 - 6xy + 3y^2)$$
$$= -6x + 6y = 6(y - x)$$
$$f_{yx} = \frac{\partial}{\partial x}(f_y) = \frac{\partial}{\partial x}(-3x^2 + 6xy + 2y)$$
$$= -6x + 6y = 6(y - x)$$
$$f_{yy} = \frac{\partial}{\partial y}(f_y) = \frac{\partial}{\partial y}(-3x^2 + 6xy + 2y)$$
$$= 6x + 2$$

EXAMPLE 7 Find the second-order partial derivatives of the function
$$f(x, y) = e^{xy^2}$$

Solution We have
$$f_x = \frac{\partial}{\partial x}(e^{xy^2})$$
$$= y^2 e^{xy^2}$$
$$f_y = \frac{\partial}{\partial y}(e^{xy^2})$$
$$= 2xy e^{xy^2}$$

so the required second-order partial derivatives of f are

$$f_{xx} = \frac{\partial}{\partial x}(f_x) = \frac{\partial}{\partial x}(y^2 e^{xy^2})$$
$$= y^4 e^{xy^2}$$
$$f_{xy} = \frac{\partial}{\partial y}(f_x) = \frac{\partial}{\partial y}(y^2 e^{xy^2})$$
$$= 2ye^{xy^2} + 2xy^3 e^{xy^2}$$
$$= 2ye^{xy^2}(1 + xy^2)$$
$$f_{yx} = \frac{\partial}{\partial x}(f_y) = \frac{\partial}{\partial x}(2xye^{xy^2})$$
$$= 2ye^{xy^2} + 2xy^3 e^{xy^2}$$
$$= 2ye^{xy^2}(1 + xy^2)$$
$$f_{yy} = \frac{\partial}{\partial y}(f_y) = \frac{\partial}{\partial y}(2xye^{xy^2})$$
$$= 2xe^{xy^2} + (2xy)(2xy)e^{xy^2}$$
$$= 2xe^{xy^2}(1 + 2xy^2)$$

17.2 Self-Check Exercises

1. Compute the first partial derivatives of $f(x, y) = x^3 - 2xy^2 + y^2 - 8$.

2. Find the first partial derivatives of $f(x, y) = x \ln y + ye^x - x^2$ at $(0, 1)$ and interpret your results.

3. Find the second-order partial derivatives of the function of Self-Check Exercise 1.

4. A certain country's production is described by the function

$$f(x, y) = 60x^{1/3}y^{2/3}$$

when x units of labor and y units of capital are used.

a. What is the marginal productivity of labor and the marginal productivity of capital when the amounts expended on labor and capital are 125 units and 8 units, respectively?

b. Should the government encourage capital investment rather than increased expenditure on labor at this time in order to increase the country's productivity?

Solutions to Self-Check Exercises 17.2 can be found on page 1073.

17.2 Concept Questions

1. a. What is the partial derivative of $f(x, y)$ with respect to x at (a, b)?

   b. Give a geometric interpretation of $f_x(a, b)$ and a physical interpretation of $f_x(a, b)$.

2. a. What are substitute commodities and complementary commodities? Give an example of each.

   b. Suppose $x = f(p, q)$ and $y = g(p, q)$ are demand functions for two commodities A and B, respectively. Give conditions for determining whether A and B are substitute or complementary commodities.

3. List all second-order partial derivatives of f.

17.2 Exercises

In Exercises 1–22, find the first partial derivatives of the function.

1. $f(x, y) = 2x + 3y + 5$

2. $f(x, y) = 2xy$

3. $g(x, y) = 2x^2 + 4y + 1$

4. $f(x, y) = 1 + x^2 + y^2$

5. $f(x, y) = \dfrac{2y}{x^2}$

6. $f(x, y) = \dfrac{x}{1 + y}$

7. $g(u, v) = \dfrac{u - v}{u + v}$

8. $f(x, y) = \dfrac{x^2 - y^2}{x^2 + y^2}$

9. $f(s, t) = (s^2 - st + t^2)^3$

10. $g(s, t) = s^2 t + st^{-3}$

11. $f(x, y) = (x^2 + y^2)^{2/3}$

12. $f(x, y) = x\sqrt{1 + y^2}$

13. $f(x, y) = e^{xy+1}$

14. $f(x, y) = (e^x + e^y)^5$

15. $f(x, y) = x \ln y + y \ln x$

16. $f(x, y) = x^2 e^{y^2}$

17. $g(u, v) = e^u \ln v$

18. $f(x, y) = \dfrac{e^{xy}}{x + y}$

19. $f(x, y, z) = xyz + xy^2 + yz^2 + zx^2$

20. $g(u, v, w) = \dfrac{2uvw}{u^2 + v^2 + w^2}$

21. $h(r, s, t) = e^{rst}$

22. $f(x, y, z) = xe^{y/z}$

In Exercises 23–32, evaluate the first partial derivatives of the function at the given point.

23. $f(x, y) = x^2 y + xy^2;\ (1, 2)$

24. $f(x, y) = x^2 + xy + y^2 + 2x - y;\ (-1, 2)$

25. $f(x, y) = x\sqrt{y} + y^2;\ (2, 1)$

26. $g(x, y) = \sqrt{x^2 + y^2};\ (3, 4)$

27. $f(x, y) = \dfrac{x}{y};\ (1, 2)$

28. $f(x, y) = \dfrac{x + y}{x - y};\ (1, -2)$

29. $f(x, y) = e^{xy};\ (1, 1)$

30. $f(x, y) = e^x \ln y;\ (0, e)$

31. $f(x, y, z) = x^2 yz^3;\ (1, 0, 2)$

32. $f(x, y, z) = x^2 y^2 + z^2;\ (1, 1, 2)$

In Exercises 33–40, find the second-order partial derivatives of the function. In each case, show that the mixed partial derivatives f_{xy} and f_{yx} are equal.

33. $f(x, y) = x^2 y + xy^3$

34. $f(x, y) = x^3 + x^2 y + x + 4$

35. $f(x, y) = x^2 - 2xy + 2y^2 + x - 2y$

36. $f(x, y) = x^3 + x^2 y^2 + y^3 + x + y$

37. $f(x, y) = \sqrt{x^2 + y^2}$

38. $f(x, y) = x\sqrt{y} + y\sqrt{x}$

39. $f(x, y) = e^{-x/y}$

40. $f(x, y) = \ln(1 + x^2 y^2)$

41. PRODUCTIVITY OF A COUNTRY The productivity of a South American country is given by the function

$$f(x, y) = 20x^{3/4} y^{1/4}$$

when x units of labor and y units of capital are used.

a. What is the marginal productivity of labor and the marginal productivity of capital when the amounts expended on labor and capital are 256 units and 16 units, respectively?

b. Should the government encourage capital investment rather than increased expenditure on labor at this time in order to increase the country's productivity?

42. PRODUCTIVITY OF A COUNTRY The productivity of a country in Western Europe is given by the function

$$f(x, y) = 40x^{4/5} y^{1/5}$$

when x units of labor and y units of capital are used.

a. What is the marginal productivity of labor and the marginal productivity of capital when the amounts expended on labor and capital are 32 units and 243 units, respectively?

b. Should the government encourage capital investment rather than increased expenditure on labor at this time in order to increase the country's productivity?

43. LAND PRICES The rectangular region R shown in the figure on the next page represents a city's financial district. The price of land within the district is approximated by the function

$$p(x, y) = 200 - 10\left(x - \frac{1}{2}\right)^2 - 15(y - 1)^2$$

where $p(x, y)$ is the price of land at the point (x, y) in dollars per square foot and x and y are measured in miles. Compute

$$\frac{\partial p}{\partial x}(0, 1) \quad \text{and} \quad \frac{\partial p}{\partial y}(0, 1)$$

and interpret your results.

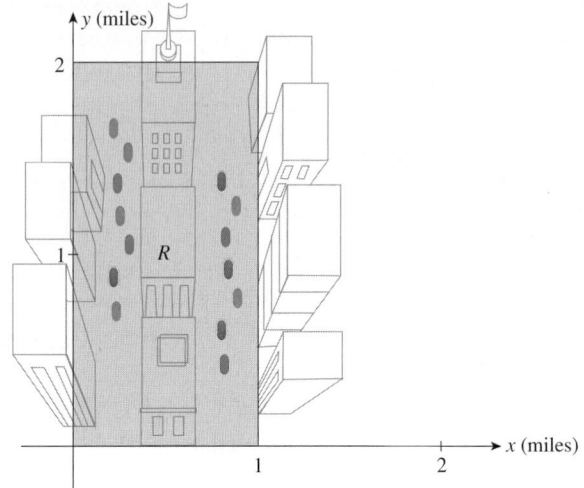

44. COMPLEMENTARY AND SUBSTITUTE COMMODITIES   In a survey conducted by *Home Entertainment* magazine, it was determined that the demand equation for VCRs is given by

$$x = f(p, q) = 10{,}000 - 10p + 0.2q^2$$

and the demand equation for DVD players is given by

$$y = g(p, q) = 5000 + 0.8p^2 - 20q$$

where p and q denote the unit prices (in dollars) for the VCRs and DVD players, respectively, and x and y denote the number of VCRs and DVD players demanded per week. Determine whether these two products are substitute, complementary, or neither.

45. COMPLEMENTARY AND SUBSTITUTE COMMODITIES   In a survey it was determined that the demand equation for VCRs is given by

$$x = f(p, q) = 10{,}000 - 10p - e^{0.5q}$$

The demand equation for blank VCR tapes is given by

$$y = g(p, q) = 50{,}000 - 4000q - 10p$$

where p and q denote the unit prices, respectively, and x and y denote the number of VCRs and the number of blank VCR tapes demanded each week. Determine whether these two products are substitute, complementary, or neither.

46. COMPLEMENTARY AND SUBSTITUTE COMMODITIES   Refer to Exercise 29, Exercises 17.1. Show that the finished and unfinished home furniture manufactured by Country Workshop are substitute commodities.
Hint: Solve the system of equations for x and y in terms of p and q.

47. REVENUE FUNCTIONS   The total weekly revenue (in dollars) of Country Workshop associated with manufacturing and selling their rolltop desks is given by the function

$$R(x, y) = -0.2x^2 - 0.25y^2 - 0.2xy + 200x + 160y$$

where x denotes the number of finished units and y denotes

the number of unfinished units manufactured and sold each week. Compute $\partial R/\partial x$ and $\partial R/\partial y$ when $x = 300$ and $y = 250$. Interpret your results.

48. PROFIT FUNCTIONS   The monthly profit (in dollars) of Bond and Barker Department Store depends on the level of inventory x (in thousands of dollars) and the floor space y (in thousands of square feet) available for display of the merchandise, as given by the equation

$$P(x, y) = -0.02x^2 - 15y^2 + xy$$
$$+ 39x + 25y - 20{,}000$$

Compute $\partial P/\partial x$ and $\partial P/\partial y$ when $x = 4000$ and $y = 150$. Interpret your results. Repeat with $x = 5000$ and $y = 150$.

49. WIND CHILL FACTOR   A formula used by meteorologists to calculate the wind chill temperature (the temperature that you feel in still air that is the same as the actual temperature when the presence of wind is taken into consideration) is

$$T = f(t, s) = 35.74 + 0.6215t - 35.75s^{0.16} + 0.4275ts^{0.16}$$
$$(s \geq 1)$$

where t is the actual air temperature in °F and s is the wind speed in mph.
a. What is the wind chill temperature when the actual air temperature is 32°F and the wind speed is 20 mph?
b. If the temperature is 32°F, by how much approximately will the wind chill temperature change if the wind speed increases from 20 mph to 21 mph?

50. ENGINE EFFICIENCY   The efficiency of an internal combustion engine is given by

$$E = \left(1 - \frac{v}{V}\right)^{0.4}$$

where V and v are the respective maximum and minimum volumes of air in each cylinder.
a. Show that $\partial E/\partial V > 0$ and interpret your result.
b. Show that $\partial E/\partial v < 0$ and interpret your result.

51. VOLUME OF A GAS   The volume V (in liters) of a certain mass of gas is related to its pressure P (in millimeters of mercury) and its temperature T (in degrees Kelvin) by the law

$$V = \frac{30.9T}{P}$$

Compute $\partial V/\partial T$ and $\partial V/\partial P$ when $T = 300$ and $P = 800$. Interpret your results.

52. SURFACE AREA OF A HUMAN BODY   The formula

$$S = 0.007184W^{0.425}H^{0.725}$$

gives the surface area S of a human body (in square meters) in terms of its weight W (in kilograms) and its height H (in centimeters). Compute $\partial S/\partial W$ and $\partial S/\partial H$ when $W = 70$ kg and $H = 180$ cm. Interpret your results.

53. According to the *ideal gas law*, the volume V (in liters) of an ideal gas is related to its pressure P (in pascals) and temperature T (in degrees Kelvin) by the formula

$$V = \frac{kT}{P}$$

where k is a constant. Show that

$$\frac{\partial V}{\partial T} \cdot \frac{\partial T}{\partial P} \cdot \frac{\partial P}{\partial V} = -1$$

54. KINETIC ENERGY OF A BODY   The kinetic energy K of a body of mass m and velocity v is given by

$$K = \frac{1}{2}mv^2$$

Show that $\dfrac{\partial K}{\partial m} \cdot \dfrac{\partial^2 K}{\partial v^2} = K$.

In Exercises 55–58, determine whether the statement is true or false. If it is true, explain why it is true. If it is false, give an example to show why it is false.

55. If $f_x(x, y)$ is defined at (a, b), then $f_y(x, y)$ must also be defined at (a, b).

56. If $f_x(a, b) < 0$, then f is decreasing with respect to x near (a, b).

57. If $f_{xy}(x, y)$ and $f_{yx}(x, y)$ are both continuous for all values of x and y, then $f_{xy} = f_{yx}$ for all values of x and y.

58. If both f_{xy} and f_{yx} are defined at (a, b), then f_{xx} and f_{yy} must be defined at (a, b).

17.2   Solutions to Self-Check Exercises

1. $f_x = \dfrac{\partial f}{\partial x} = 3x^2 - 2y^2$

$f_y = \dfrac{\partial f}{\partial y} = -2x(2y) + 2y$
   $= 2y(1 - 2x)$

2. $f_x = \ln y + ye^x - 2x;\ f_y = \dfrac{x}{y} + e^x$

In particular,

$$f_x(0, 1) = \ln 1 + 1e^0 - 2(0) = 1$$

$$f_y(0, 1) = \frac{0}{1} + e^0 = 1$$

The results tell us that at the point $(0, 1)$, $f(x, y)$ increases 1 unit for each unit increase in the x-direction, y being kept constant; $f(x, y)$ also increases 1 unit for each unit increase in the y-direction, x being kept constant.

3. From the results of Self-Check Exercise 1,

$$f_x = 3x^2 - 2y^2$$

Therefore,

$$f_{xx} = \frac{\partial}{\partial x}(3x^2 - 2y^2) = 6x$$

$$f_{xy} = \frac{\partial}{\partial y}(3x^2 - 2y^2) = -4y$$

Also, from the results of Self-Check Exercise 1,

$$f_y = 2y(1 - 2x)$$

Thus,

$$f_{yx} = \frac{\partial}{\partial x}[2y(1 - 2x)] = -4y$$

$$f_{yy} = \frac{\partial}{\partial y}[2y(1 - 2x)] = 2(1 - 2x)$$

4. a. The marginal productivity of labor when the amounts expended on labor and capital are x and y units, respectively, is given by

$$f_x(x, y) = 60\left(\frac{1}{3}x^{-2/3}\right)y^{2/3} = 20\left(\frac{y}{x}\right)^{2/3}$$

In particular, the required marginal productivity of labor is given by

$$f_x(125, 8) = 20\left(\frac{8}{125}\right)^{2/3} = 20\left(\frac{4}{25}\right)$$

or 3.2 units/unit increase in labor expenditure, capital expenditure being held constant at 8 units. Next, we compute

$$f_y(x, y) = 60x^{1/3}\left(\frac{2}{3}y^{-1/3}\right) = 40\left(\frac{x}{y}\right)^{1/3}$$

and deduce that the required marginal productivity of capital is given by

$$f_y(125, 8) = 40\left(\frac{125}{8}\right)^{1/3} = 40\left(\frac{5}{2}\right)$$

or 100 units/unit increase in capital expenditure, labor expenditure being held constant at 125 units.

b. The results of part (a) tell us that the government should encourage increased spending on capital rather than on labor.

USING TECHNOLOGY

■ Finding Partial Derivatives at a Given Point

Suppose $f(x, y)$ is a function of two variables and we wish to compute

$$f_x(a, b) = \frac{\partial f}{\partial x}\bigg|_{(a, b)}$$

Recall that in computing $\partial f/\partial x$, we think of y as being fixed. But in this situation, we are evaluating $\partial f/\partial x$ at (a, b). Therefore, we set y equal to b. Doing this leads to the function g of one variable, x, defined by

$$g(x) = f(x, b)$$

It follows from the definition of the partial derivative that

$$f_x(a, b) = g'(a)$$

Thus, the value of the partial derivative $\partial f/\partial x$ at a given point (a, b) can be found by evaluating the derivative of a function of one variable. In particular, the latter can be found by using the numerical derivative operation of a graphing utility. We find $f_y(a, b)$ in a similar manner.

EXAMPLE 1 Let $f(x, y) = (1 + xy^2)^{3/2}e^{x^2y}$. Find (a) $f_x(1, 2)$ and (b) $f_y(1, 2)$.

Solution

a. Define $g(x) = f(x, 2) = (1 + 4x)^{3/2}e^{2x^2}$. Using the numerical derivative operation to find $g'(1)$, we obtain

$$f_x(1, 2) = g'(1) \approx 429.585835$$

b. Define $h(y) = f(1, y) = (1 + y^2)^{3/2}e^y$. Using the numerical derivative operation to find $h'(2)$, we obtain

$$f_y(1, 2) = h'(2) \approx 181.7468642$$

TECHNOLOGY EXERCISES

Compute the following at the given point:

$$\frac{\partial f}{\partial x} \quad \text{and} \quad \frac{\partial f}{\partial y}$$

1. $f(x, y) = \sqrt{x}(2 + xy^2)^{1/3}$; $(1, 2)$

2. $f(x, y) = \sqrt{xy}(1 + 2xy)^{2/3}$; $(1, 4)$

3. $f(x, y) = \dfrac{x + y^2}{1 + x^2y}$; $(1, 2)$

4. $f(x, y) = \dfrac{xy^2}{(\sqrt{x} + \sqrt{y})^2}$; $(4, 1)$

5. $f(x, y) = e^{-xy^2}(x + y)^{1/3}$; $(1, 1)$

6. $f(x, y) = \dfrac{\ln(\sqrt{x} + y^2)}{x^2 + y^2}$; $(4, 1)$

17.3 Maxima and Minima of Functions of Several Variables

Maxima and Minima

In Chapter 13, we saw that the solution of a problem often reduces to finding the extreme values of a function of one variable. In practice, however, situations also arise in which a problem is solved by finding the absolute maximum or absolute minimum value of a function of two or more variables.

For example, suppose Scandi Company manufactures computer desks in both assembled and unassembled versions. Its profit P is therefore a function of the number of assembled units, x, and the number of unassembled units, y, manufactured and sold per week; that is, $P = f(x, y)$. A question of paramount importance to the manufacturer is, How many assembled and unassembled desks should the company manufacture per week in order to maximize its weekly profit? Mathematically, the problem is solved by finding the values of x and y that will make $f(x, y)$ a maximum.

In this section we will focus our attention on finding the extrema of a function of two variables. As in the case of a function of one variable, we distinguish between the relative (or local) extrema and the absolute extrema of a function of two variables.

> **Relative Extrema of a Function of Two Variables**
> Let f be a function defined on a region R containing the point (a, b). Then, f has a **relative maximum** at (a, b) if $f(x, y) \leq f(a, b)$ for all points (x, y) that are sufficiently close to (a, b). The number $f(a, b)$ is called a **relative maximum value**. Similarly, f has a **relative minimum** at (a, b), with **relative minimum value** $f(a, b)$ if $f(x, y) \geq f(a, b)$ for all points (x, y) that are sufficiently close to (a, b).

Loosely speaking, f has a relative maximum at (a, b) if the point $(a, b, f(a, b))$ is the highest point on the graph of f when compared with all nearby points. A similar interpretation holds for a relative minimum.

If the inequalities in this last definition hold for *all* points (x, y) in the domain of f, then f has an **absolute maximum** (or **absolute minimum**) at (a, b) with **absolute maximum value** (or **absolute minimum value**) $f(a, b)$. Figure 16 shows

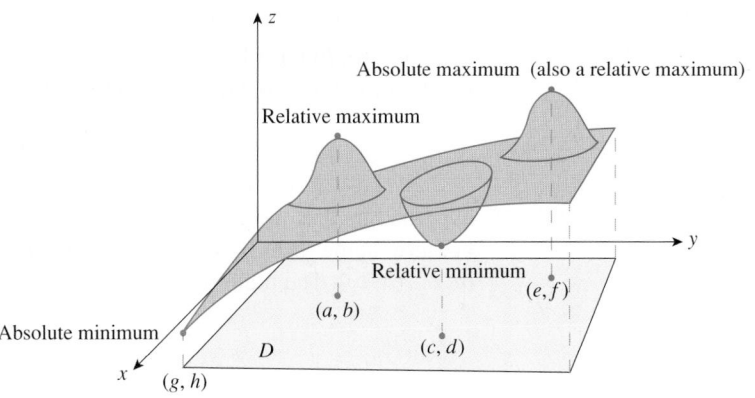

FIGURE 16

the graph of a function with relative maxima at (a, b) and (e, f) and a relative minimum at (c, d). The absolute maximum of f occurs at (e, f) and the absolute minimum of f occurs at (g, h).

Observe that just as in the case of a function of one variable, a relative extremum (relative maximum or relative minimum) may or may not be an absolute extremum.

Now let's turn our attention to the study of relative extrema of a function. Suppose that a differentiable function $f(x, y)$ of two variables has a relative maximum (relative minimum) at a point (a, b) in the domain of f. From Figure 17 it is clear that at the point (a, b) the slope of the "tangent lines" to the surface in any direction must be zero. In particular, this implies that both

$$\frac{\partial f}{\partial x}(a, b) \quad \text{and} \quad \frac{\partial f}{\partial y}(a, b)$$

must be zero.

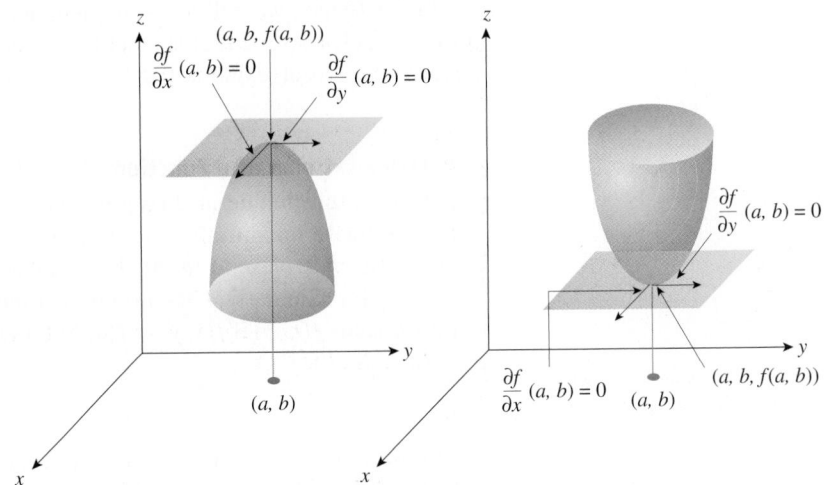

FIGURE 17

(a) f has a relative maximum at (a, b). **(b)** f has a relative minimum at (a, b).

Lest we are tempted to jump to the conclusion that a differentiable function f satisfying both the conditions

$$\frac{\partial f}{\partial x}(a, b) = 0 \quad \text{and} \quad \frac{\partial f}{\partial y}(a, b) = 0$$

at a point (a, b) must have a relative extremum at the point (a, b), let's examine the graph of the function f depicted in Figure 18. Here both

$$\frac{\partial f}{\partial x}(a, b) = 0 \quad \text{and} \quad \frac{\partial f}{\partial y}(a, b) = 0$$

but f has neither a relative maximum nor a relative minimum at the point (a, b) because some nearby points are higher and some are lower than the point $(a, b, f(a, b))$. The point $(a, b, f(a, b))$ is called a **saddle point**.

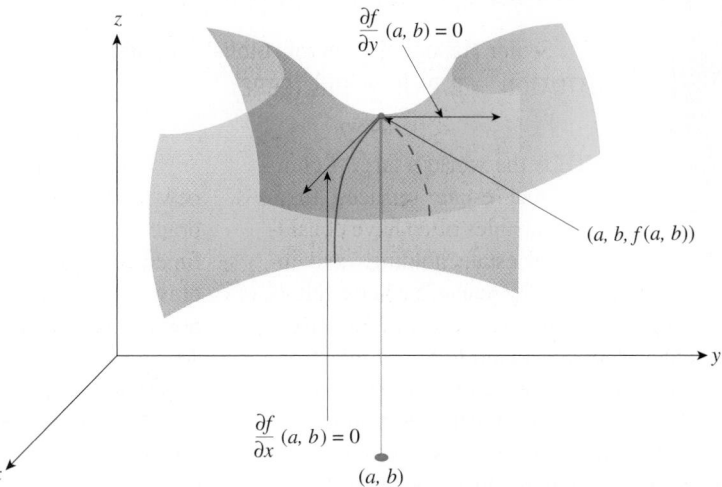

FIGURE 18
The point $(a, b, f(a, b))$ is called a saddle point.

Finally, an examination of the graph of the function f depicted in Figure 19 should convince you that f has a relative maximum at the point (a, b). But both $\partial f/\partial x$ and $\partial f/\partial y$ fail to be defined at (a, b).

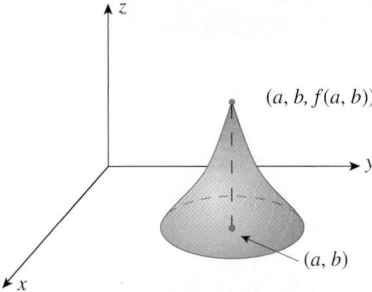

FIGURE 19
f has a relative maximum at (a, b), but neither $\partial f/\partial x$ nor $\partial f/\partial y$ exist at (a, b).

To summarize, a function f of two variables can only have a relative extremum at a point (a, b) in its domain where $\partial f/\partial x$ and $\partial f/\partial y$ both exist and are equal to zero at (a, b) or at least one of the partial derivatives does not exist. As in the case of one variable, we refer to a point in the domain of f that *may* give rise to a relative extremum as a critical point. The precise definition follows.

> **Critical Point of f**
>
> A **critical point** of f is a point (a, b) in the domain of f such that both
>
> $$\frac{\partial f}{\partial x}(a, b) = 0 \quad \text{and} \quad \frac{\partial f}{\partial y}(a, b) = 0$$
>
> or at least one of the partial derivatives does not exist.

To determine the nature of a critical point of a function $f(x, y)$ of two variables, we use the second partial derivatives of f. The resulting test, which helps us classify

these points, is called the **second derivative test** and is incorporated in the following procedure for finding and classifying the relative extrema of f.

Determining Relative Extrema

1. Find the critical points of $f(x, y)$ by solving the system of simultaneous equations

$$f_x = 0$$
$$f_y = 0$$

2. The second derivative test: Let

$$D(x, y) = f_{xx}f_{yy} - f_{xy}^2$$

Then,

a. $D(a, b) > 0$ and $f_{xx}(a, b) < 0$ implies that $f(x, y)$ has a **relative maximum** at the point (a, b).

b. $D(a, b) > 0$ and $f_{xx}(a, b) > 0$ implies that $f(x, y)$ has a **relative minimum** at the point (a, b).

c. $D(a, b) < 0$ implies that $f(x, y)$ has neither a relative maximum nor a relative minimum at the point (a, b).

d. $D(a, b) = 0$ implies that the test is inconclusive, so some other technique must be used to solve the problem.

EXAMPLE 1   Find the relative extrema of the function

$$f(x, y) = x^2 + y^2$$

Solution   We have

$$f_x = 2x$$
$$f_y = 2y$$

To find the critical point(s) of f, we set $f_x = 0$ and $f_y = 0$ and solve the resulting system of simultaneous equations

$$2x = 0$$
$$2y = 0$$

obtaining $x = 0$, $y = 0$, or $(0, 0)$, as the sole critical point of f. Next, we apply the second derivative test to determine the nature of the critical point $(0, 0)$. We compute

$$f_{xx} = 2 \qquad f_{xy} = 0 \qquad f_{yy} = 2$$

and

$$D(x, y) = f_{xx}f_{yy} - f_{xy}^2 = (2)(2) - 0 = 4$$

In particular, $D(0, 0) = 4$. Since $D(0, 0) > 0$ and $f_{xx}(0, 0) = 2 > 0$, we conclude that $f(x, y)$ has a relative minimum at the point $(0, 0)$. The relative minimum value, 0, also happens to be the absolute minimum of f. The graph of the function f, shown in Figure 20, confirms these results.   ■

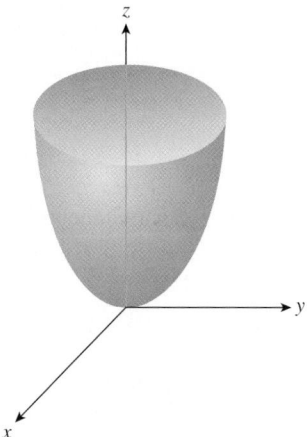

FIGURE 20
The graph of $f(x, y) = x^2 + y^2$

EXAMPLE 2   Find the relative extrema of the function

$$f(x, y) = 3x^2 - 4xy + 4y^2 - 4x + 8y + 4$$

Solution   We have

$$f_x = 6x - 4y - 4$$
$$f_y = -4x + 8y + 8$$

To find the critical points of f, we set $f_x = 0$ and $f_y = 0$ and solve the resulting system of simultaneous equations

$$6x - 4y = \quad 4$$
$$-4x + 8y = -8$$

Multiplying the first equation by 2 and the second equation by 3, we obtain the equivalent system

$$12x - \ 8y = \quad 8$$
$$-12x + 24y = -24$$

Adding the two equations gives $16y = -16$, or $y = -1$. We substitute this value for y into either equation in the system to get $x = 0$. Thus, the only critical point of f is the point $(0, -1)$. Next, we apply the second derivative test to determine whether the point $(0, -1)$ gives rise to a relative extremum of f. We

EXPLORE & DISCUSS

Suppose $f(x, y)$ has a relative extremum (relative maximum or relative minimum) at a point (a, b). Let $g(x) = f(x, b)$ and $h(y) = f(a, y)$. Assuming that f and g are differentiable, explain why $g'(a) = 0$ and $h'(b) = 0$. Explain why these results are equivalent to the conditions $f_x(a, b) = 0$ and $f_y(a, b) = 0$.

VIDEO

EXPLORE & DISCUSS

1. Refer to the second derivative test. Can the condition $f_{xx}(a, b) < 0$ in part 2a be replaced by the condition $f_{yy}(a, b) < 0$? Explain your answer. How about the condition $f_{xx}(a, b) > 0$ in part 2b?
2. Let $f(x, y) = x^4 + y^4$.
   a. Show that $(0, 0)$ is a critical point of f and that $D(0, 0) = 0$.
   b. Explain why f has a relative (in fact, an absolute) minimum at $(0, 0)$. Does this contradict the second derivative test? Explain your answer.

compute

$$f_{xx} = 6 \qquad f_{xy} = -4 \qquad f_{yy} = 8$$

and

$$D(x, y) = f_{xx}f_{yy} - f_{xy}^2 = (6)(8) - (-4)^2 = 32$$

Since $D(0, -1) = 32 > 0$ and $f_{xx}(0, -1) = 6 > 0$, we conclude that $f(x, y)$ has a relative minimum at the point $(0, -1)$. The value of $f(x, y)$ at the point $(0, -1)$ is given by

$$f(0, -1) = 3(0)^2 - 4(0)(-1) + 4(-1)^2 - 4(0) + 8(-1) + 4 = 0 \qquad ∎$$

EXAMPLE 3 Find the relative extrema of the function

$$f(x, y) = 4y^3 + x^2 - 12y^2 - 36y + 2$$

Solution To find the critical points of f, we set $f_x = 0$ and $f_y = 0$ simultaneously, obtaining

$$f_x = 2x = 0$$
$$f_y = 12y^2 - 24y - 36 = 0$$

The first equation implies that $x = 0$. The second equation implies that

$$y^2 - 2y - 3 = 0$$
$$(y + 1)(y - 3) = 0$$

—that is, $y = -1$ or 3. Therefore, there are two critical points of the function f—namely, $(0, -1)$ and $(0, 3)$.

Next, we apply the second derivative test to determine the nature of each of the two critical points. We compute

$$f_{xx} = 2 \qquad f_{xy} = 0 \qquad f_{yy} = 24y - 24 = 24(y - 1)$$

Therefore,

$$D(x, y) = f_{xx}f_{yy} - f_{xy}^2 = 48(y - 1)$$

For the point $(0, -1)$,

$$D(0, -1) = 48(-1 - 1) = -96 < 0$$

Since $D(0, -1) < 0$, we conclude that the point $(0, -1)$ gives a saddle point of f. For the point $(0, 3)$,

$$D(0, 3) = 48(3 - 1) = 96 > 0$$

Since $D(0, 3) > 0$ and $f_{xx}(0, 3) > 0$, we conclude that the function f has a relative minimum at the point $(0, 3)$. Furthermore, since

$$f(0, 3) = 4(3)^3 + (0)^2 - 12(3)^2 - 36(3) + 2$$
$$= -106$$

we see that the relative minimum value of f is -106. ∎

As in the case of a practical optimization problem involving a function of one variable, the solution to an optimization problem involving a function of several variables calls for finding the *absolute* extremum of the function. Determining the absolute extremum of a function of several variables is more difficult than merely finding the relative extrema of the function. However, in many situations, the absolute extremum of a function actually coincides with the largest relative extremum of the function that occurs in the interior of its domain. We assume that the problems considered here belong to this category. Furthermore, the existence of the absolute extremum (solution) of a practical problem is often deduced from the geometric or physical nature of the problem.

APPLIED EXAMPLE 4 Maximizing Profits The total weekly revenue (in dollars) that Acrosonic realizes in producing and selling its bookshelf loudspeaker systems is given by

$$R(x, y) = -\frac{1}{4}x^2 - \frac{3}{8}y^2 - \frac{1}{4}xy + 300x + 240y$$

where x denotes the number of fully assembled units and y denotes the number of kits produced and sold each week. The total weekly cost attributable to the production of these loudspeakers is

$$C(x, y) = 180x + 140y + 5000$$

dollars, where x and y have the same meaning as before. Determine how many assembled units and how many kits Acrosonic should produce per week to maximize its profit.

Solution The contribution to Acrosonic's weekly profit stemming from the production and sale of the bookshelf loudspeaker systems is given by

$$P(x, y) = R(x, y) - C(x, y)$$
$$= \left(-\frac{1}{4}x^2 - \frac{3}{8}y^2 - \frac{1}{4}xy + 300x + 240y\right) - (180x + 140y + 5000)$$
$$= -\frac{1}{4}x^2 - \frac{3}{8}y^2 - \frac{1}{4}xy + 120x + 100y - 5000$$

To find the relative maximum of the profit function $P(x, y)$, we first locate the critical point(s) of P. Setting $P_x(x, y)$ and $P_y(x, y)$ equal to zero, we obtain

$$P_x = -\frac{1}{2}x - \frac{1}{4}y + 120 = 0$$
$$P_y = -\frac{3}{4}y - \frac{1}{4}x + 100 = 0$$

Solving the first of these equations for y yields

$$y = -2x + 480$$

which, upon substitution into the second equation, yields

$$-\frac{3}{4}(-2x + 480) - \frac{1}{4}x + 100 = 0$$
$$6x - 1440 - x + 400 = 0$$
$$x = 208$$

We substitute this value of x into the equation $y = -2x + 480$ to get

$$y = 64$$

Therefore, the function P has the sole critical point $(208, 64)$. To show that the point $(208, 64)$ is a solution to our problem, we use the second derivative test. We compute

$$P_{xx} = -\frac{1}{2} \qquad P_{xy} = -\frac{1}{4} \qquad P_{yy} = -\frac{3}{4}$$

So,

$$D(x, y) = \left(-\frac{1}{2}\right)\left(-\frac{3}{4}\right) - \left(-\frac{1}{4}\right)^2 = \frac{3}{8} - \frac{1}{16} = \frac{5}{16}$$

In particular, $D(208, 64) = 5/16 > 0$.

Since $D(208, 64) > 0$ and $P_{xx}(208, 64) < 0$, the point $(208, 64)$ yields a relative maximum of P. This relative maximum is also the absolute maximum of P. We conclude that Acrosonic can maximize its weekly profit by manufacturing 208 assembled units and 64 kits of their bookshelf loudspeaker systems. The maximum weekly profit realizable from the production and sale of these loudspeaker systems is given by

$$P(208, 64) = -\frac{1}{4}(208)^2 - \frac{3}{8}(64)^2 - \frac{1}{4}(208)(64)$$
$$+ 120(208) + 100(64) - 5000$$
$$= 10{,}680$$

or $10,680.

APPLIED EXAMPLE 5 Locating a Television Relay Station Site A television relay station will serve towns A, B, and C, whose relative locations are shown in Figure 21. Determine a site for the location of the station if the sum of the squares of the distances from each town to the site is minimized.

Solution Suppose the required site is located at the point $P(x, y)$. With the aid of the distance formula, we find that the square of the distance from town A to the site is

$$(x - 30)^2 + (y - 20)^2$$

The respective distances from towns B and C to the site are found in a similar manner, so the sum of the squares of the distances from each town to the site is given by

$$f(x, y) = (x - 30)^2 + (y - 20)^2 + (x + 20)^2$$
$$+ (y - 10)^2 + (x - 10)^2 + (y + 10)^2$$

To find the relative minimum of $f(x, y)$, we first find the critical point(s) of f. Using the chain rule to find $f_x(x, y)$ and $f_y(x, y)$ and setting each equal to zero, we obtain

$$f_x = 2(x - 30) + 2(x + 20) + 2(x - 10) = 6x - 40 = 0$$
$$f_y = 2(y - 20) + 2(y - 10) + 2(y + 10) = 6y - 40 = 0$$

from which we deduce that $\left(\frac{20}{3}, \frac{20}{3}\right)$ is the sole critical point of f. Since

$$f_{xx} = 6 \qquad f_{xy} = 0 \qquad f_{yy} = 6$$

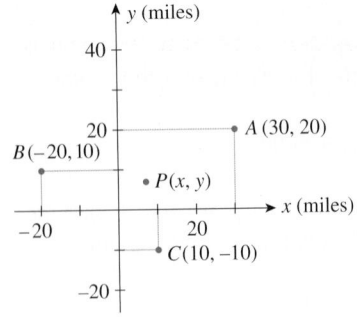

FIGURE 21
Locating a site for a television relay station

we have

$$D(x, y) = f_{xx}f_{yy} - f_{xy}^2 = (6)(6) - 0 = 36$$

Since $D(\frac{20}{3}, \frac{20}{3}) > 0$ and $f_{xx}(\frac{20}{3}, \frac{20}{3}) > 0$, we conclude that the point $(\frac{20}{3}, \frac{20}{3})$ yields a relative minimum of f. Thus, the required site has coordinates $x = \frac{20}{3}$ and $y = \frac{20}{3}$. ∎

17.3 Self-Check Exercises

1. Let $f(x, y) = 2x^2 + 3y^2 - 4xy + 4x - 2y + 3$.
   a. Find the critical point of f.
   b. Use the second derivative test to classify the nature of the critical point.
   c. Find the relative extremum of f, if it exists.

2. Robertson Controls manufactures two basic models of setback thermostats: a standard mechanical thermostat and a deluxe electronic thermostat. Robertson's monthly revenue (in hundreds of dollars) is

$$R(x, y) = -\frac{1}{8}x^2 - \frac{1}{2}y^2 - \frac{1}{4}xy + 20x + 60y$$

where x (in units of a hundred) denotes the number of mechan-

ical thermostats manufactured and y (in units of a hundred) denotes the number of electronic thermostats manufactured each month. The total monthly cost incurred in producing these thermostats is

$$C(x, y) = 7x + 20y + 280$$

hundred dollars. Find how many thermostats of each model Robertson should manufacture each month in order to maximize its profits. What is the maximum profit?

Solutions to Self-Check Exercises 17.3 can be found on page 1085.

17.3 Concept Questions

1. Explain the terms (a) relative maximum of a function $f(x, y)$ and (b) absolute maximum of a function $f(x, y)$.

2. **a.** What is a critical point of a function $f(x, y)$?
   b. Explain the role of a critical point in determining the relative extrema of a function of two variables.

3. Explain how the second derivative test is used to determine the relative extrema of a function of two variables.

17.3 Exercises

In Exercises 1–20, find the critical point(s) of the function. Then use the second derivative test to classify the nature of each point, if possible. Finally, determine the relative extrema of the function.

1. $f(x, y) = 1 - 2x^2 - 3y^2$

2. $f(x, y) = x^2 - xy + y^2 + 1$

3. $f(x, y) = x^2 - y^2 - 2x + 4y + 1$

4. $f(x, y) = 2x^2 + y^2 - 4x + 6y + 3$

5. $f(x, y) = x^2 + 2xy + 2y^2 - 4x + 8y - 1$

6. $f(x, y) = x^2 - 4xy + 2y^2 + 4x + 8y - 1$

7. $f(x, y) = 2x^3 + y^2 - 9x^2 - 4y + 12x - 2$

8. $f(x, y) = 2x^3 + y^2 - 6x^2 - 4y + 12x - 2$

9. $f(x, y) = x^3 + y^2 - 2xy + 7x - 8y + 4$

10. $f(x, y) = 2y^3 - 3y^2 - 12y + 2x^2 - 6x + 2$

11. $f(x, y) = x^3 - 3xy + y^3 - 2$

12. $f(x, y) = x^3 - 2xy + y^2 + 5$

13. $f(x, y) = xy + \dfrac{4}{x} + \dfrac{2}{y}$

14. $f(x, y) = \dfrac{x}{y^2} + xy$

15. $f(x, y) = x^2 - e^{y^2}$

16. $f(x, y) = e^{x^2 - y^2}$

17. $f(x, y) = e^{x^2 + y^2}$

18. $f(x, y) = e^{xy}$

19. $f(x, y) = \ln(1 + x^2 + y^2)$

20. $f(x, y) = xy + \ln x + 2y^2$

21. MAXIMIZING PROFIT   The total weekly revenue (in dollars) of the Country Workshop realized in manufacturing and selling its rolltop desks is given by

$$R(x, y) = -0.2x^2 - 0.25y^2 - 0.2xy + 200x + 160y$$

where x denotes the number of finished units and y denotes the number of unfinished units manufactured and sold each week. The total weekly cost attributable to the manufacture of these desks is given by

$$C(x, y) = 100x + 70y + 4000$$

dollars. Determine how many finished units and how many unfinished units the company should manufacture each week in order to maximize its profit. What is the maximum profit realizable?

22. MAXIMIZING PROFIT   The total daily revenue (in dollars) that Weston Publishing realizes in publishing and selling its English-language dictionaries is given by

$$R(x, y) = -0.005x^2 - 0.003y^2 - 0.002xy$$
$$+ 20x + 15y$$

where x denotes the number of deluxe copies and y denotes the number of standard copies published and sold daily. The total daily cost of publishing these dictionaries is given by

$$C(x, y) = 6x + 3y + 200$$

dollars. Determine how many deluxe copies and how many standard copies Weston should publish each day to maximize its profits. What is the maximum profit realizable?

23. MAXIMUM PRICE   The rectangular region R shown in the accompanying figure represents the financial district of a city. The price of land within the district is approximated by the function

$$p(x, y) = 200 - 10\left(x - \frac{1}{2}\right)^2 - 15(y - 1)^2$$

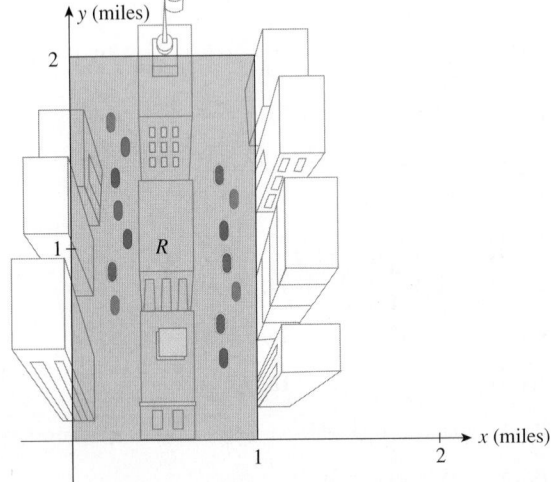

where $p(x, y)$ is the price of land at the point (x, y) in dollars/square foot and x and y are measured in miles. At what point within the financial district is the price of land highest?

24. MAXIMIZING PROFIT   C&G Imports imports two brands of white wine, one from Germany and the other from Italy. The German wine costs \$4/bottle, and the Italian wine costs \$3/bottle. It has been estimated that if the German wine retails at p dollars/bottle and the Italian wine is sold for q dollars/bottle, then

$$2000 - 150p + 100q$$

bottles of the German wine and

$$1000 + 80p - 120q$$

bottles of the Italian wine will be sold each week. Determine the unit price for each brand that will allow C&G to realize the largest possible weekly profit.

25. DETERMINING THE OPTIMAL SITE   An auxiliary electric power station will serve three communities, A, B, and C, whose relative locations are shown in the accompanying figure. Determine where the power station should be located if the sum of the squares of the distances from each community to the site is minimized.

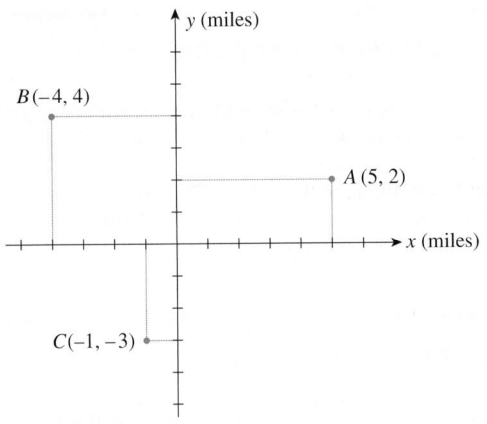

26. PACKAGING   An open rectangular box having a volume of 108 in.^{3} is to be constructed from a tin sheet. Find the dimensions of such a box if the amount of material used in its construction is to be minimal.

Hint: Let the dimensions of the box be x'' by y'' by z''. Then, $xyz = 108$ and the amount of material used is given by $S = xy + 2yz + 2xz$. Show that

$$S = f(x, y) = xy + \frac{216}{x} + \frac{216}{y}$$

Minimize $f(x, y)$.

27. PACKAGING   An open rectangular box having a surface area of 300 in.^{2} is to be constructed from a tin sheet. Find the dimensions of the box if the volume of the box is to be as large as possible. What is the maximum volume?

Hint: Let the dimensions of the box be $x \times y \times z$ (see the figure that follows). Then the surface area is $xy + 2xz + 2yz$, and its volume is xyz.

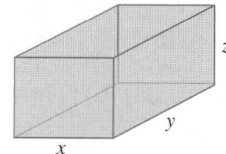

28. PACKAGING Postal regulations specify that the combined length and girth of a parcel sent by parcel post may not exceed 108 in. Find the dimensions of the rectangular package that would have the greatest possible volume under these regulations.

Hint: Let the dimensions of the box be x'' by y'' by z'' (see the figure below). Then, $2x + 2z + y = 108$, and the volume $V = xyz$. Show that

$$V = f(x, y) = 108xz - 2x^2z - 2xz^2$$

Maximize $f(x, z)$.

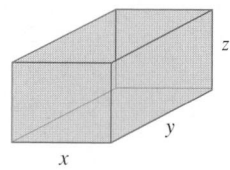

29. MINIMIZING HEATING AND COOLING COSTS A building in the shape of a rectangular box is to have a volume of 12,000 ft³ (see the figure). It is estimated that the annual heating and cooling costs will be \$2/square foot for the top, \$4/square foot for the front and back, and \$3/square foot for the sides. Find the dimensions of the building that will result in a min-

imal annual heating and cooling cost. What is the minimal annual heating and cooling cost?

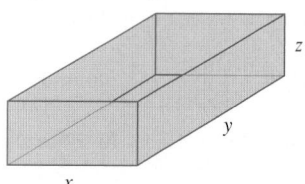

30. PACKAGING An open box having a volume of 48 in.³ is to be constructed. If the box is to include a partition that is parallel to a side of the box, as shown in the figure, and the amount of material used is to be minimal, what should be the dimensions of the box?

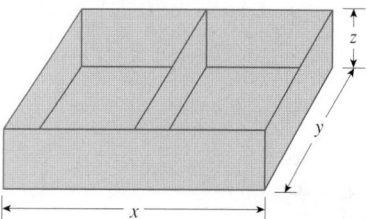

In Exercises 31 and 32, determine whether the statement is true or false. If it is true, explain why it is true. If it is false, give an example to show why it is false.

31. If $f_x(a, b) = 0$ and $f_y(a, b) = 0$, then f must have a relative extremum at (a, b).

32. If (a, b) is a critical point of f and both the conditions $f_{xx}(a, b) < 0$ and $f_{yy}(a, b) < 0$ hold, then f has a relative maximum at (a, b).

17.3 Solutions to Self-Check Exercises

1. a. To find the critical point(s) of f, we solve the system of equations

$$f_x = 4x - 4y + 4 = 0$$
$$f_y = -4x + 6y - 2 = 0$$

obtaining $x = -2$ and $y = -1$. Thus, the only critical point of f is the point $(-2, -1)$.

b. We have $f_{xx} = 4, f_{xy} = -4$, and $f_{yy} = 6$, so

$$D(x, y) = f_{xx}f_{yy} - f_{xy}^2$$
$$= (4)(6) - (-4)^2 = 8$$

Since $D(-2, -1) > 0$ and $f_{xx}(-2, -1) > 0$, we conclude that f has a relative minimum at the point $(-2, -1)$.

c. The relative minimum value of $f(x, y)$ at the point $(-2, -1)$ is

$$f(-2, -1) = 2(-2)^2 + 3(-1)^2 - 4(-2)(-1)$$
$$+ 4(-2) - 2(-1) + 3$$
$$= 0$$

2. Robertson's monthly profit is

$$P(x, y) = R(x, y) - C(x, y)$$
$$= \left(-\frac{1}{8}x^2 - \frac{1}{2}y^2 - \frac{1}{4}xy + 20x + 60y\right) - (7x + 20y + 280)$$
$$= -\frac{1}{8}x^2 - \frac{1}{2}y^2 - \frac{1}{4}xy + 13x + 40y - 280$$

The critical point of P is found by solving the system

$$P_x = -\frac{1}{4}x - \frac{1}{4}y + 13 = 0$$

$$P_y = -\frac{1}{4}x - \quad y + 40 = 0$$

giving $x = 16$ and $y = 36$. Thus, $(16, 36)$ is the critical point of P. Next,

$$P_{xx} = -\frac{1}{4} \qquad P_{xy} = -\frac{1}{4} \qquad P_{yy} = -1$$

and

$$D(x, y) = f_{xx}f_{yy} - f_{xy}^2$$

$$= \left(-\frac{1}{4}\right)(-1) - \left(-\frac{1}{4}\right)^2 = \frac{3}{16}$$

Since $D(16, 36) > 0$ and $P_{xx}(16, 36) < 0$, the point $(16, 36)$ yields a relative maximum of P. We conclude that the monthly profit is maximized by manufacturing 1600 mechanical and 3600 electronic setback thermostats each month. The maximum monthly profit realizable is

$$P(16, 36) = -\frac{1}{8}(16)^2 - \frac{1}{2}(36)^2 - \frac{1}{4}(16)(36)$$

$$+ 13(16) + 40(36) - 280$$

$$= 544$$

or \$54,400.

17.4 Constrained Maxima and Minima and the Method of Lagrange Multipliers

▪ Constrained Relative Extrema

In Section 17.3, we studied the problem of determining the relative extremum of a function $f(x, y)$ without placing any restrictions on the independent variables x and y—except, of course, that the point (x, y) lies in the domain of f. Such a relative extremum of a function f is referred to as an **unconstrained relative extremum** of f. However, in many practical optimization problems, we must maximize or minimize a function in which the independent variables are subjected to certain further constraints.

In this section, we discuss a powerful method for determining the relative extrema of a function $f(x, y)$ whose independent variables x and y are required to satisfy one or more constraints of the form $g(x, y) = 0$. Such a relative extremum of a function f is called a constrained relative extremum of f. We can see the difference between an unconstrained extremum of a function $f(x, y)$ of two variables and a constrained extremum of f, where the independent variables x and y are subjected to a constraint of the form $g(x, y) = 0$, by considering the geometry of the two cases. Figure 22a depicts the graph of a function $f(x, y)$ that has an unconstrained relative minimum at the point $(0, 0)$. However, when the independent variables x and y are subjected to an equality constraint of the form $g(x, y) = 0$, the points (x, y, z) that satisfy both $z = f(x, y)$ and the constraint equation $g(x, y) = 0$ lie on a curve C. Therefore, the constrained relative minimum of f must also lie on C (Figure 22b).

Our first example involves an equality constraint $g(x, y) = 0$ in which we solve for the variable y explicitly in terms of x. In this case we may apply the technique used in Chapter 13 to find the relative extrema of a function of one variable.

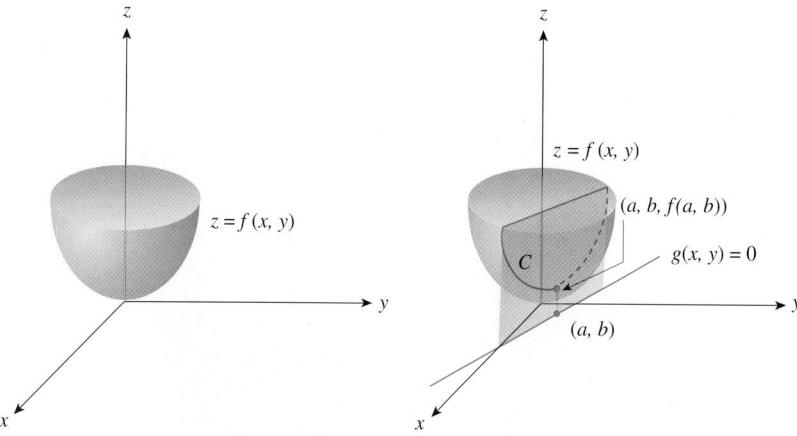

FIGURE 22

(a) $f(x, y)$ has an unconstrained relative extremum at $(0, 0)$.

(b) $f(x, y)$ has a constrained relative extremum at $(a, b, f(a, b))$.

EXAMPLE 1   Find the relative minimum of the function

$$f(x, y) = 2x^2 + y^2$$

subject to the constraint $g(x, y) = x + y - 1 = 0$.

Solution   Solving the constraint equation for y explicitly in terms of x, we obtain $y = -x + 1$. Substituting this value of y into the function $f(x, y) = 2x^2 + y^2$ results in a function of x,

$$h(x) = 2x^2 + (-x + 1)^2 = 3x^2 - 2x + 1$$

The function h describes the curve C lying on the graph of f on which the constrained relative minimum of f occurs. To find this point, use the technique developed in Chapter 13 to determine the relative extrema of a function of one variable:

$$h'(x) = 6x - 2 = 2(3x - 1)$$

Setting $h'(x) = 0$ gives $x = \frac{1}{3}$ as the sole critical point of the function h. Next, we find

$$h''(x) = 6$$

and, in particular,

$$h''\left(\frac{1}{3}\right) = 6 > 0$$

Therefore, by the second derivative test, the point $x = \frac{1}{3}$ gives rise to a relative minimum of h. Substitute this value of x into the constraint equation $x + y - 1 = 0$ to get $y = \frac{2}{3}$. Thus, the point $(\frac{1}{3}, \frac{2}{3})$ gives rise to the required constrained relative minimum of f. Since

$$f\left(\frac{1}{3}, \frac{2}{3}\right) = 2\left(\frac{1}{3}\right)^2 + \left(\frac{2}{3}\right)^2 = \frac{2}{3}$$

the required constrained relative minimum value of f is $\frac{2}{3}$ at the point $(\frac{1}{3}, \frac{2}{3})$. It may be shown that $\frac{2}{3}$ is in fact a constrained absolute minimum value of f (Figure 23).

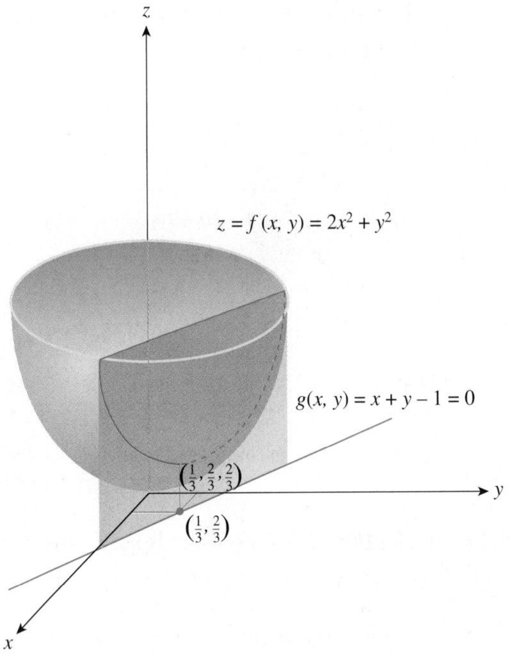

FIGURE 23
f has a constrained absolute minimum of $\frac{2}{3}$ at $\left(\frac{1}{3}, \frac{2}{3}\right)$.

The Method of Lagrange Multipliers

The major drawback of the technique used in Example 1 is that it relies on our ability to solve the constraint equation $g(x, y) = 0$ for y explicitly in terms of x. This is not always an easy task. Moreover, even when we can solve the constraint equation $g(x, y) = 0$ for y explicitly in terms of x, the resulting function of one variable that is to be optimized may turn out to be unnecessarily complicated. Fortunately, an easier method exists. This method, called the **method of Lagrange multipliers** (Joseph Lagrange, 1736–1813), is as follows:

The Method of Lagrange Multipliers
To find the relative extrema of the function $f(x, y)$ subject to the constraint $g(x, y) = 0$ (assuming that these extreme values exist),

1. Form an auxiliary function

$$F(x, y, \lambda) = f(x, y) + \lambda g(x, y)$$

called the Lagrangian function (the variable λ is called the Lagrange multiplier).
2. Solve the system that consists of the equations

$$F_x = 0 \qquad F_y = 0 \qquad F_\lambda = 0$$

for all values of x, y, and λ.
3. The solutions found in step 2 are candidates for the extrema of f.

Let's re-solve Example 1 using the method of Lagrange multipliers.

EXAMPLE 2   Using the method of Lagrange multipliers, find the relative minimum of the function

$$f(x, y) = 2x^2 + y^2$$

subject to the constraint $x + y = 1$.

Solution   Write the constraint equation $x + y = 1$ in the form $g(x, y) = x + y - 1 = 0$. Then, form the Lagrangian function

$$F(x, y, \lambda) = f(x, y) + \lambda g(x, y)$$
$$= 2x^2 + y^2 + \lambda(x + y - 1)$$

To find the critical point(s) of the function F, solve the system composed of the equations

$$F_x = 4x + \lambda = 0$$
$$F_y = 2y + \lambda = 0$$
$$F_\lambda = x + y - 1 = 0$$

Solving the first and second equations in this system for x and y in terms of λ, we obtain

$$x = -\frac{1}{4}\lambda \quad \text{and} \quad y = -\frac{1}{2}\lambda$$

which, upon substitution into the third equation, yields

$$-\frac{1}{4}\lambda - \frac{1}{2}\lambda - 1 = 0 \quad \text{or} \quad \lambda = -\frac{4}{3}$$

Therefore, $x = \frac{1}{3}$ and $y = \frac{2}{3}$, and $\left(\frac{1}{3}, \frac{2}{3}\right)$ affords a constrained minimum of the function f, in agreement with the result obtained earlier. ■

Note   A disadvantage of the method of Lagrange multipliers is that there is no test analogous to the second derivative test mentioned in Section 17.3 for determining whether a critical point of a function of two or more variables leads to a relative maximum or relative minimum (and thus the absolute extrema) of the function. Here we have to rely on the geometric or physical nature of the problem to help us draw the necessary conclusions (see Example 2). ■

The method of Lagrange multipliers may be used to solve a problem involving a function of three or more variables, as illustrated in the next example.

EXAMPLE 3   Use the method of Lagrange multipliers to find the minimum of the function

$$f(x, y, z) = 2xy + 6yz + 8xz$$

subject to the constraint

$$xyz = 12{,}000$$

(*Note:* The existence of the minimum is suggested by the geometry of the problem.)

Solution   Write the constraint equation $xyz = 12{,}000$ in the form $g(x, y, z) = xyz - 12{,}000$. Then, the Lagrangian function is

$$F(x, y, z, \lambda) = f(x, y, z) + \lambda g(x, y, z)$$
$$= 2xy + 6yz + 8xz + \lambda(xyz - 12{,}000)$$

To find the critical point(s) of the function F, we solve the system composed of the equations

$$F_x = 2y + 8z + \lambda yz = 0$$
$$F_y = 2x + 6z + \lambda xz = 0$$
$$F_z = 6y + 8x + \lambda xy = 0$$
$$F_\lambda = xyz - 12{,}000 = 0$$

Solving the first three equations of the system for λ in terms of x, y, and z, we have

$$\lambda = -\frac{2y + 8z}{yz}$$
$$\lambda = -\frac{2x + 6z}{xz}$$
$$\lambda = -\frac{6y + 8x}{xy}$$

Equating the first two expressions for λ leads to

$$\frac{2y + 8z}{yz} = \frac{2x + 6z}{xz}$$
$$2xy + 8xz = 2xy + 6yz$$
$$x = \frac{3}{4}y$$

Next, equating the second and third expressions for λ in the same system yields

$$\frac{2x + 6z}{xz} = \frac{6y + 8x}{xy}$$
$$2xy + 6yz = 6yz + 8xz$$
$$z = \frac{1}{4}y$$

Finally, substituting these values of x and z into the equation $xyz - 12{,}000 = 0$, the fourth equation of the first system of equations, we have

$$\left(\frac{3}{4}y\right)(y)\left(\frac{1}{4}y\right) - 12{,}000 = 0$$
$$y^3 = \frac{(12{,}000)(4)(4)}{3} = 64{,}000$$
$$y = 40$$

The corresponding values of x and z are given by $x = \frac{3}{4}(40) = 30$ and $z = \frac{1}{4}(40) = 10$. Therefore, we see that the point $(30, 40, 10)$ gives the constrained

minimum of f. The minimum value is

$$f(30, 40, 10) = 2(30)(40) + 6(40)(10) + 8(30)(10) = 7200 \qquad \blacksquare$$

 APPLIED EXAMPLE 4 Maximizing Profit Refer to Example 3, Section 17.1. The total weekly profit (in dollars) that Acrosonic realized in producing and selling its bookshelf loudspeaker systems is given by the profit function

$$P(x, y) = -\frac{1}{4}x^2 - \frac{3}{8}y^2 - \frac{1}{4}xy + 120x + 100y - 5000$$

where x denotes the number of fully assembled units and y denotes the number of kits produced and sold per week. Acrosonic's management decides that production of these loudspeaker systems should be restricted to a total of exactly 230 units each week. Under this condition, how many fully assembled units and how many kits should be produced each week to maximize Acrosonic's weekly profit?

Solution The problem is equivalent to the problem of maximizing the function

$$P(x, y) = -\frac{1}{4}x^2 - \frac{3}{8}y^2 - \frac{1}{4}xy + 120x + 100y - 5000$$

subject to the constraint

$$g(x, y) = x + y - 230 = 0$$

The Lagrangian function is

$$F(x, y, \lambda) = P(x, y) + \lambda g(x, y)$$

$$= -\frac{1}{4}x^2 - \frac{3}{8}y^2 - \frac{1}{4}xy + 120x + 100y$$

$$- 5000 + \lambda(x + y - 230)$$

To find the critical point(s) of F, solve the following system of equations:

$$F_x = -\frac{1}{2}x - \frac{1}{4}y + 120 + \lambda = 0$$

$$F_y = -\frac{3}{4}y - \frac{1}{4}x + 100 + \lambda = 0$$

$$F_\lambda = x + y - 230 = 0$$

Solving the first equation of this system for λ, we obtain

$$\lambda = \frac{1}{2}x + \frac{1}{4}y - 120$$

which, upon substitution into the second equation, yields

$$-\frac{3}{4}y - \frac{1}{4}x + 100 + \frac{1}{2}x + \frac{1}{4}y - 120 = 0$$

$$-\frac{1}{2}y + \frac{1}{4}x - 20 = 0$$

Solving the last equation for y gives

$$y = \frac{1}{2}x - 40$$

When we substitute this value of y into the third equation of the system, we have

$$x + \frac{1}{2}x - 40 - 230 = 0$$

$$x = 180$$

The corresponding value of y is $\frac{1}{2}(180) - 40$, or 50. Thus, the required constrained relative maximum of P occurs at the point $(180, 50)$. Again, we can show that the point $(180, 50)$ in fact yields a constrained absolute maximum for P. Thus, Acrosonic's profit is maximized by producing 180 assembled and 50 kit versions of their bookshelf loudspeaker systems. The maximum weekly profit realizable is given by

$$P(180, 50) = -\frac{1}{4}(180)^2 - \frac{3}{8}(50)^2 - \frac{1}{4}(180)(50)$$

$$+ 120(180) + 100(50) - 5000$$

$$= 10{,}312.5$$

or \$10,312.50.   ∎

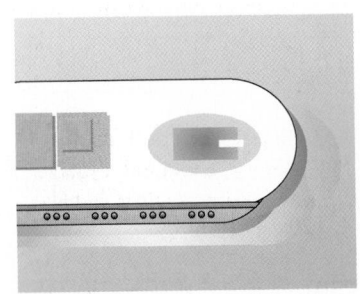

FIGURE 24
A rectangular-shaped pool will be built in the elliptical-shaped poolside area.

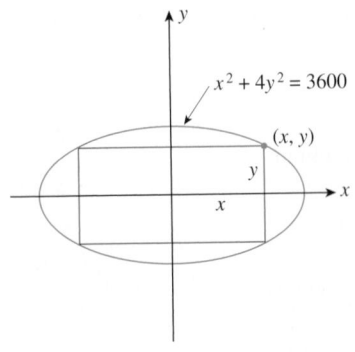

FIGURE 25
We want to find the largest rectangle that can be inscribed in the ellipse described by $x^2 + 4y^2 = 3600$.

APPLIED EXAMPLE 5   Designing a Cruise-Ship Pool   The operators of the *Viking Princess*, a luxury cruise liner, are contemplating the addition of another swimming pool to the ship. The chief engineer has suggested that an area in the form of an ellipse located in the rear of the promenade deck would be suitable for this purpose. This location would provide a poolside area with sufficient space for passenger movement and placement of deck chairs (Figure 24). It has been determined that the shape of the ellipse may be described by the equation $x^2 + 4y^2 = 3600$, where x and y are measured in feet. *Viking*'s operators would like to know the dimensions of the rectangular pool with the largest possible area that would meet these requirements.

Solution   To solve this problem, we need to find the rectangle of largest area that can be inscribed in the ellipse with equation $x^2 + 4y^2 = 3600$. Letting the sides of the rectangle be $2x$ and $2y$ feet, we see that the area of the rectangle is $A = 4xy$ (Figure 25). Furthermore, the point (x, y) must be constrained to lie on the ellipse so that it satisfies the equation $x^2 + 4y^2 = 3600$. Thus, the problem is equivalent to the problem of maximizing the function

$$f(x, y) = 4xy$$

subject to the constraint $g(x, y) = x^2 + 4y^2 - 3600 = 0$. The Lagrangian function is

$$F(x, y, \lambda) = f(x, y) + \lambda g(x, y)$$

$$= 4xy + \lambda(x^2 + 4y^2 - 3600)$$

To find the critical point(s) of F, we solve the following system of equations:

$$F_x = 4y + 2\lambda x = 0$$
$$F_y = 4x + 8\lambda y = 0$$
$$F_\lambda = x^2 + 4y^2 - 3600 = 0$$

Solving the first equation of this system for λ, we obtain

$$\lambda = -\frac{2y}{x}$$

which, upon substitution into the second equation, yields

$$4x + 8\left(-\frac{2y}{x}\right)y = 0 \quad \text{or} \quad x^2 - 4y^2 = 0$$

—that is, $x = \pm 2y$. Substituting these values of x into the third equation of the system, we have

$$4y^2 + 4y^2 - 3600 = 0$$

or, upon solving $y = \pm\sqrt{450} = \pm 15\sqrt{2}$. The corresponding values of x are $\pm 30\sqrt{2}$. Because both x and y must be nonnegative, we have $x = 30\sqrt{2}$ and $y = 15\sqrt{2}$. Thus, the dimensions of the pool with maximum area are $30\sqrt{2}$ feet $\times 60\sqrt{2}$ feet, or approximately 42 feet $\times$ 85 feet. ■

 APPLIED EXAMPLE 6   Cobb–Douglas Production Function   Suppose x units of labor and y units of capital are required to produce

$$f(x, y) = 100x^{3/4}y^{1/4}$$

units of a certain product (recall that this is a Cobb–Douglas production function). If each unit of labor costs $200 and each unit of capital costs $300 and a total of $60,000 is available for production, determine how many units of labor and how many units of capital should be used in order to maximize production.

Solution   The total cost of x units of labor at $200 per unit and y units of capital at $300 per unit is equal to $200x + 300y$ dollars. But $60,000 is budgeted for production, so $200x + 300y = 60,000$, which we rewrite as

$$g(x, y) = 200x + 300y - 60,000 = 0$$

To maximize $f(x, y) = 100x^{3/4}y^{1/4}$ subject to the constraint $g(x, y) = 0$, we form the Lagrangian function

$$F(x, y, \lambda) = f(x, y) + \lambda g(x, y)$$
$$= 100x^{3/4}y^{1/4} + \lambda(200x + 300y - 60,000)$$

To find the critical point(s) of F, we solve the following system of equations:

$$F_x = 75x^{-1/4}y^{1/4} + 200\lambda = 0$$
$$F_y = 25x^{3/4}y^{-3/4} + 300\lambda = 0$$
$$F_\lambda = 200x + 300y - 60,000 = 0$$

Solving the first equation for λ, we have

$$\lambda = -\frac{75x^{-1/4}y^{1/4}}{200} = -\frac{3}{8}\left(\frac{y}{x}\right)^{1/4}$$

which, when substituted into the second equation, yields

$$25\left(\frac{x}{y}\right)^{3/4} + 300\left(-\frac{3}{8}\right)\left(\frac{y}{x}\right)^{1/4} = 0$$

Multiplying the last equation by $\left(\frac{x}{y}\right)^{1/4}$ then gives

$$25\left(\frac{x}{y}\right) - \frac{900}{8} = 0$$

$$x = \left(\frac{900}{8}\right)\left(\frac{1}{25}\right)y = \frac{9}{2}y$$

Substituting this value of x into the third equation of the first system of equations, we have

$$200\left(\frac{9}{2}y\right) + 300y - 60,000 = 0$$

from which we deduce that $y = 50$. Hence, $x = 225$. Thus, maximum production is achieved when 225 units of labor and 50 units of capital are used. ∎

When used in the context of Example 6, the negative of the Lagrange multiplier λ is called the **marginal productivity of money**. That is, if one additional dollar is available for production, then approximately $-\lambda$ units of a product can be produced. Here,

$$\lambda = -\frac{3}{8}\left(\frac{y}{x}\right)^{1/4} = -\frac{3}{8}\left(\frac{50}{225}\right)^{1/4} \approx -0.257$$

so, in this case, the marginal productivity of money is 0.257. For example, if $65,000 is available for production instead of the originally budgeted figure of $60,000, then the maximum production may be boosted from the original

$$f(225, 50) = 100(225)^{3/4}(50)^{1/4}$$

or 15,448 units, to

$$15,448 + 5000(0.257)$$

or 16,733 units.

17.4 Self-Check Exercises

1. Use the method of Lagrange multipliers to find the relative maximum of the function

$$f(x, y) = -2x^2 - y^2$$

subject to the constraint $3x + 4y = 12$.

2. The total monthly profit of Robertson Controls in manufacturing and selling x hundred of its standard mechanical setback thermostats and y hundred of its deluxe electronic setback thermostats each month is given by the total profit function

$$P(x, y) = -\frac{1}{8}x^2 - \frac{1}{2}y^2 - \frac{1}{4}xy + 13x + 40y - 280$$

where P is in hundreds of dollars. If the production of setback thermostats is to be restricted to a total of exactly 4000/month, how many of each model should Robertson manufacture in order to maximize its monthly profits? What is the maximum monthly profit?

Solutions to Self-Check Exercises 17.4 can be found on page 1096.

17.4 Concept Questions

1. What is a constrained relative extremum of a function f?

2. Explain how the method of Lagrange multipliers is used to find the relative extrema of $f(x, y)$ subject to $g(x, y) = 0$.

17.4 Exercises

In Exercises 1–16, use the method of Lagrange multipliers to optimize the function subject to the given constraint.

1. Minimize the function $f(x, y) = x^2 + 3y^2$ subject to the constraint $x + y - 1 = 0$.

2. Minimize the function $f(x, y) = x^2 + y^2 - xy$ subject to the constraint $x + 2y - 14 = 0$.

3. Maximize the function $f(x, y) = 2x + 3y - x^2 - y^2$ subject to the constraint $x + 2y = 9$.

4. Maximize the function $f(x, y) = 16 - x^2 - y^2$ subject to the constraint $x + y - 6 = 0$.

5. Minimize the function $f(x, y) = x^2 + 4y^2$ subject to the constraint $xy = 1$.

6. Minimize the function $f(x, y) = xy$ subject to the constraint $x^2 + 4y^2 = 4$.

7. Maximize the function $f(x, y) = x + 5y - 2xy - x^2 - 2y^2$ subject to the constraint $2x + y = 4$.

8. Maximize the function $f(x, y) = xy$ subject to the constraint $2x + 3y - 6 = 0$.

9. Maximize the function $f(x, y) = xy^2$ subject to the constraint $9x^2 + y^2 = 9$.

10. Minimize the function $f(x, y) = \sqrt{y^2 - x^2}$ subject to the constraint $x + 2y - 5 = 0$.

11. Find the maximum and minimum values of the function $f(x, y) = xy$ subject to the constraint $x^2 + y^2 = 16$.

12. Find the maximum and minimum values of the function $f(x, y) = e^{xy}$ subject to the constraint $x^2 + y^2 = 8$.

13. Find the maximum and minimum values of the function $f(x, y) = xy^2$ subject to the constraint $x^2 + y^2 = 1$.

14. Maximize the function $f(x, y, z) = xyz$ subject to the constraint $2x + 2y + z = 84$.

15. Minimize the function $f(x, y, z) = x^2 + y^2 + z^2$ subject to the constraint $3x + 2y + z = 6$.

16. Find the maximum value of the function $f(x, y, z) = x + 2y - 3z$ subject to the constraint $z = 4x^2 + y^2$.

17. MAXIMIZING PROFIT The total weekly profit (in dollars) realized by Country Workshop in manufacturing and selling its rolltop desks is given by the profit function

$$P(x, y) = -0.2x^2 - 0.25y^2 - 0.2xy$$
$$+ 100x + 90y - 4000$$

where x stands for the number of finished units and y denotes the number of unfinished units manufactured and sold each week. The company's management decides to restrict the manufacture of these desks to a total of exactly 200 units/week. How many finished and how many unfinished units should be manufactured each week to maximize the company's weekly profit?

18. MAXIMIZING PROFIT The total daily profit (in dollars) realized by Weston Publishing in publishing and selling its dictionaries is given by the profit function

$$P(x, y) = -0.005x^2 - 0.003y^2 - 0.002xy$$
$$+ 14x + 12y - 200$$

where x stands for the number of deluxe editions and y denotes the number of standard editions sold daily. Weston's management decides that publication of these dictionaries should be restricted to a total of exactly 400 copies/day. How many deluxe copies and how many standard copies should be published each day to maximize Weston's daily profit?

19. MINIMIZING CONSTRUCTION COSTS The management of UNICO Department Store decides to enclose an 800-ft^2 area outside their building to display potted plants. The enclosed area will be a rectangle, one side of which is provided by the external walls of the store. Two sides of the enclosure will be made of pine board, and the fourth side will be made of galvanized steel fencing material. If the pine board fencing costs $6/running foot and the steel fencing costs $3/running foot, determine the dimensions of the enclosure that will cost the least to erect.

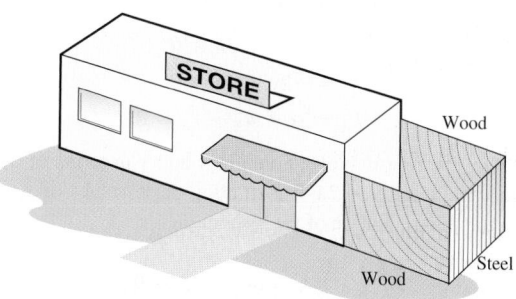

20. **PARCEL POST REGULATIONS** Postal regulations specify that a parcel sent by parcel post may have a combined length and girth of no more than 108 in. Find the dimensions of the cylindrical package of greatest volume that may be sent through the mail. What is the volume of such a package?
Hint: The length plus the girth is $2\pi r + l$, and the volume is $\pi r^2 l$.

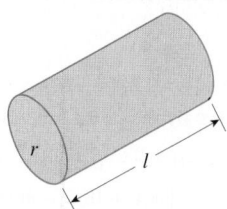

21. **MINIMIZING CONTAINER COSTS** The Betty Moore Company requires that its corned beef hash containers have a capacity of 64 in.^{3}, be right circular cylinders, and be made of a tin alloy. Find the radius and height of the least expensive container that can be made if the metal for the side and bottom costs 4¢/in.^{2} and the metal for the pull-off lid costs $2¢/in.^{2}.
Hint: Let the radius and height of the container be r and h in., respectively. Then, the volume of the container is $\pi r^2 h = 64$, and the cost is given by $C(r, h) = 8\pi rh + 6\pi r^2$.

22. **MINIMIZING CONSTRUCTION COSTS** An open rectangular box is to be constructed from material that costs $3/ft^2 for the bottom and $1/ft^2 for its sides. Find the dimensions of the box of greatest volume that can be constructed for $36.

23. **MINIMIZING CONSTRUCTION COSTS** A closed rectangular box having a volume of 4 ft^3 is to be constructed. If the material for the sides costs $1.00/ft^2 and the material for the top and bottom costs $1.50/ft^2, find the dimensions of the box that can be constructed with minimum cost.

24. **MAXIMIZING SALES** Ross–Simons Company has a monthly advertising budget of $60,000. Their marketing department

estimates that if they spend x dollars on newspaper advertising and y dollars on television advertising, then the monthly sales will be given by

$$z = f(x, y) = 90x^{1/4}y^{3/4}$$

dollars. Determine how much money Ross–Simons should spend on newspaper ads and on television ads each month to maximize its monthly sales.

25. **MAXIMIZING PRODUCTION** John Mills—the proprietor of Mills Engine Company, a manufacturer of model airplane engines—finds that it takes x units of labor and y units of capital to produce

$$f(x, y) = 100x^{3/4}y^{1/4}$$

units of the product. If a unit of labor costs $100, a unit of capital costs $200, and $200,000 is budgeted for production, determine how many units should be expended on labor and how many units should be expended on capital in order to maximize production.

26. Use the method of Lagrange multipliers to solve Exercise 29, Exercises 17.3.

In Exercises 27 and 28, determine whether the statement is true or false. If it is true, explain why it is true. If it is false, give an example to show why it is false.

27. If (a, b) gives rise to a (constrained) relative extremum of f subject to the constraint $g(x, y) = 0$, then (a, b) also gives rise to the unconstrained relative extremum of f.

28. If (a, b) gives rise to a (constrained) relative extremum of f subject to the constraint $g(x, y) = 0$, then $f_x(a, b) = 0$ and $f_y(a, b) = 0$, simultaneously.

17.4   Solutions to Self-Check Exercises

1. Write the constraint equation in the form $g(x, y) = 3x + 4y - 12 = 0$. Then, the Lagrangian function is

$$F(x, y, \lambda) = -2x^2 - y^2 + \lambda(3x + 4y - 12)$$

To find the critical point(s) of F, we solve the system

$$F_x = -4x + 3\lambda = 0$$
$$F_y = -2y + 4\lambda = 0$$
$$F_\lambda = 3x + 4y - 12 = 0$$

Solving the first two equations for x and y in terms of λ, we find $x = \frac{3}{4}\lambda$ and $y = 2\lambda$. Substituting these values of x and y into the third equation of the system yields

$$3\left(\frac{3}{4}\lambda\right) + 4(2\lambda) - 12 = 0$$

or $\lambda = \frac{48}{41}$. Therefore, $x = \left(\frac{3}{4}\right)\left(\frac{48}{41}\right) = \frac{36}{41}$ and $y = 2\left(\frac{48}{41}\right) = \frac{96}{41}$, and we see that the point $\left(\frac{36}{41}, \frac{96}{41}\right)$ gives the constrained maximum of f. The maximum value is

$$f\left(\frac{36}{41}, \frac{96}{41}\right) = -2\left(\frac{36}{41}\right)^2 - \left(\frac{96}{41}\right)^2$$
$$= -\frac{11,808}{1681} = -\frac{288}{41}$$

2. We want to maximize

$$P(x, y) = -\frac{1}{8}x^2 - \frac{1}{2}y^2 - \frac{1}{4}xy + 13x + 40y - 280$$

subject to the constraint

$$g(x, y) = x + y - 40 = 0$$

The Lagrangian function is

$$F(x, y, \lambda) = P(x, y) + \lambda g(x, y)$$

$$= -\frac{1}{8}x^2 - \frac{1}{2}y^2 - \frac{1}{4}xy + 13x$$

$$+ 40y - 280 + \lambda(x + y - 40)$$

To find the critical points of F, solve the following system of equations:

$$F_x = -\frac{1}{4}x - \frac{1}{4}y + 13 + \lambda = 0$$

$$F_y = -\frac{1}{4}x - y + 40 + \lambda = 0$$

$$F_\lambda = x + y - 40 = 0$$

Subtracting the first equation from the second gives

$$-\frac{3}{4}y + 27 = 0 \quad \text{or} \quad y = 36$$

Substituting this value of y into the third equation yields $x = 4$. Therefore, to maximize its monthly profits, Robertson should manufacture 400 standard and 3600 deluxe thermostats. The maximum monthly profit is given by

$$P(4, 36) = -\frac{1}{8}(4)^2 - \frac{1}{2}(36)^2 - \frac{1}{4}(4)(36)$$

$$+ 13(4) + 40(36) - 280$$

$$= 526$$

or $52,600.

17.5 Double Integrals

■ A Geometric Interpretation of the Double Integral

To introduce the notion of the integral of a function of two variables, let's first recall the definition of the definite integral of a continuous function of one variable $y = f(x)$ over the interval $[a, b]$. We first divide the interval $[a, b]$ into n subintervals, each of equal length, by the points $x_0 = a < x_1 < x_2 < \cdots < x_n = b$ and define the **Reimann sum** by

$$S_n = f(p_1)h + f(p_2)h + \cdots + f(p_n)h$$

where $h = (b - a)/n$ and p_i is an arbitrary point in the interval $[x_{i-1}, x_i]$. The definite integral of f over $[a, b]$ is defined as the limit of the Riemann sum S_n as n tends to infinity, whenever it exists. Furthermore, recall that when f is a nonnegative continuous function on $[a, b]$, then the ith term of the Riemann sum, $f(p_i)h$, is an approximation (by the area of a rectangle) of the area under that part of the graph of $y = f(x)$ between $x = x_{i-1}$ and $x = x_i$, so that the Riemann sum S_n provides us with an approximation of the area under the curve $y = f(x)$ from $x = a$ to $x = b$. The integral

$$\int_a^b f(x)\, dx = \lim_{n \to \infty} S_n$$

gives the *actual* area under the curve from $x = a$ to $x = b$.

Now suppose $f(x, y)$ is a continuous function of two variables defined over a region R. For simplicity, we assume for the moment that R is a rectangular region in the plane (Figure 26). Let's construct a Riemann sum for this function over the rectangle R by following a procedure that parallels the case for a function of one variable over an interval I. We begin by observing that the analogue of a *partition* in the two-dimensional case is a rectangular **grid** composed of mn rectangles, each of length h and width k, as a result of partitioning the side of the rectangle R of length $(b - a)$ into m segments and the side of length $(d - c)$ into n segments. By construction

$$h = \frac{b - a}{m} \quad \text{and} \quad k = \frac{d - c}{n}$$

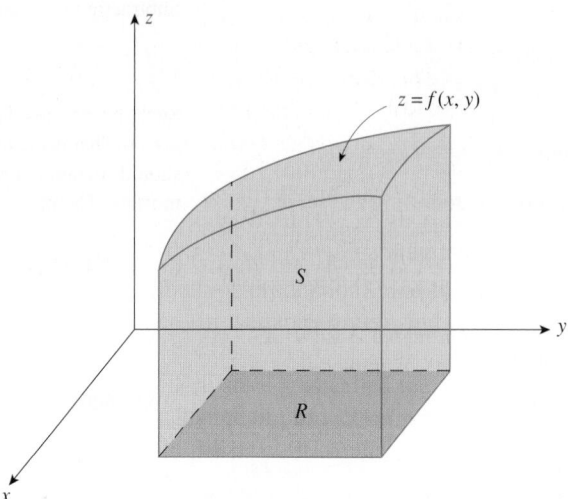

FIGURE 26
$f(x, y)$ is a function defined over a rectangular region R.

A sample grid with $m = 5$ and $n = 4$ is shown in Figure 27.

Let's label the rectangles $R_1, R_2, R_3, \ldots, R_{mn}$. If (x_i, y_i) is *any* point in R_i ($1 \le i \le mn$), then the **Riemann sum of $f(x, y)$ over the region R** is defined as

$$S(m, n) = f(x_1, y_1)hk + f(x_2, y_2)hk + \cdots + f(x_{mn}, y_{mn})hk$$

If the limit of $S(m, n)$ exists as both m and n tend to infinity, we call this limit the value of the **double integral** of $f(x, y)$ **over the region R** and denote it by

$$\iint_R f(x, y)\, dA$$

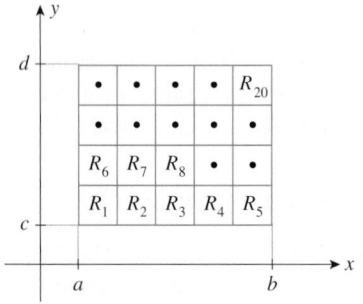

FIGURE 27
Grid with $m = 5$ and $n = 4$

EXPLORE & DISCUSS

Using a geometric interpretation, evaluate

$$\iint_R \sqrt{4 - x^2 - y^2}\, dA$$

where $R = \{(x, y) \mid x^2 + y^2 \le 4\}$.

If $f(x, y)$ is a nonnegative function, then it defines a solid S bounded above by the graph of f and below by the rectangular region R. Furthermore, the solid S is the union of the mn solids bounded above by the graph of f and below by the mn rectangular regions corresponding to the partition of R (Figure 28). The volume of a

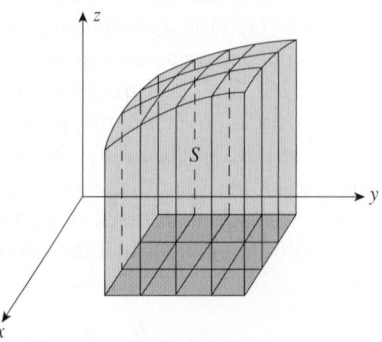

(a) The solid S is the union of mn solids (shown here with $m = 3$ and $n = 4$).

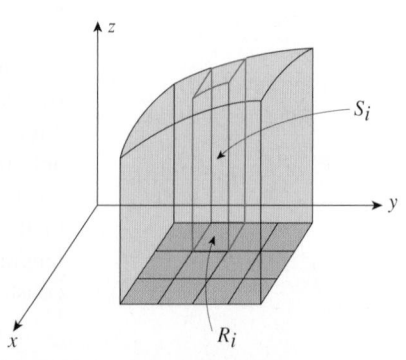

(b) A typical solid S_i is bounded above by the graph of f and lies above R_i.

FIGURE 28

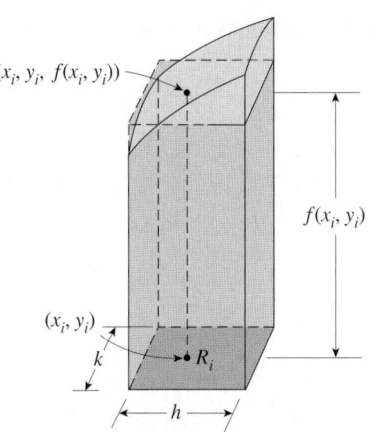

FIGURE 29
The volume of S_i is approximated by the parallelepiped with base R_i and height $f(x_i, y_i)$.

typical solid S_i can be approximated by a parallelepiped with base R_i and height $f(x_i, y_i)$ (Figure 29).

Therefore, the Riemann sum $S(m, n)$ gives us an approximation of the volume of the solid bounded above by the surface $z = f(x, y)$ and below by the plane region R. As both m and n tend to infinity, the Riemann sum $S(m, n)$ approaches the *actual* volume under the solid.

▬ Evaluating a Double Integral over a Rectangular Region

Let's turn our attention to the evaluation of the double integral

$$\iint_R f(x, y) \, dA$$

where R is the rectangular region shown in Figure 26. As in the case of the definite integral of a function of one variable, it turns out that the double integral can be evaluated without our having to first find an appropriate Riemann sum and then take the limit of that sum. Instead, as we will now see, the technique calls for evaluating two single integrals—the so-called **iterated integrals**—in succession, using a process that might be called "antipartial differentiation." The technique is described in the following result, which we state without proof.

Let R be the rectangle defined by the inequalities $a \le x \le b$ and $c \le y \le d$ (see Figure 27). Then,

$$\iint_R f(x, y) \, dA = \int_c^d \left[\int_a^b f(x, y) \, dx \right] dy \tag{4}$$

where the iterated integrals on the right-hand side are evaluated as follows. We first compute the integral

$$\int_a^b f(x, y) \, dx$$

by treating y as if it were a constant and integrating the resulting function of x with respect to x (dx reminds us that we are integrating with respect to x). In this manner we obtain a value for the integral that may contain the variable y. Thus,

$$\int_a^b f(x, y) \, dx = g(y)$$

for some function g. Substituting this value into Equation (4) gives

$$\int_c^d g(y) \, dy$$

which may be integrated in the usual manner.

EXAMPLE 1 Evaluate $\iint_R f(x, y) \, dA$, where $f(x, y) = x + 2y$ and R is the rectangle defined by $1 \le x \le 4$ and $1 \le y \le 2$.

Solution Using Equation (4), we find

$$\iint_R f(x, y) \, dA = \int_1^2 \left[\int_1^4 (x + 2y) \, dx \right] dy$$

To compute

$$\int_1^4 (x + 2y) \, dx$$

we treat y as if it were a constant (remember that dx reminds us that we are integrating with respect to x). We obtain

$$\int_1^4 (x + 2y)\, dx = \frac{1}{2}x^2 + 2xy \Big|_{x=1}^{x=4}$$

$$= \left[\frac{1}{2}(16) + 2(4)y\right] - \left[\frac{1}{2}(1) + 2(1)y\right]$$

$$= \frac{15}{2} + 6y$$

Thus,

$$\iint_R f(x, y)\, dA = \int_1^2 \left(\frac{15}{2} + 6y\right) dy = \left(\frac{15}{2}y + 3y^2\right)\Big|_1^2$$

$$= (15 + 12) - \left(\frac{15}{2} + 3\right) = 16\tfrac{1}{2}$$ ■

Evaluating a Double Integral over a Plane Region

Up to now, we have assumed that the region over which a double integral is to be evaluated is rectangular. In fact, however, it is possible to compute the double integral of functions over rather arbitrary regions. The next theorem, which we state without proof, expands the number of types of regions over which we may integrate.

THEOREM 1

a. Suppose $g_1(x)$ and $g_2(x)$ are continuous functions on $[a, b]$ and the region R is defined by $R = \{(x, y) \mid g_1(x) \le y \le g_2(x);\ a \le x \le b\}$. Then,

$$\iint_R f(x, y)\, dA = \int_a^b \left[\int_{g_1(x)}^{g_2(x)} f(x, y)\, dy\right] dx \tag{5}$$

(Figure 30a).

b. Suppose $h_1(y)$ and $h_2(y)$ are continuous functions on $[c, d]$ and the region R is defined by $R = \{(x, y) \mid h_1(y) \le x \le h_2(y);\ c \le y \le d\}$. Then,

$$\iint_R f(x, y)\, dA = \int_c^d \left[\int_{h_1(y)}^{h_2(y)} f(x, y)\, dx\right] dy \tag{6}$$

(Figure 30b).

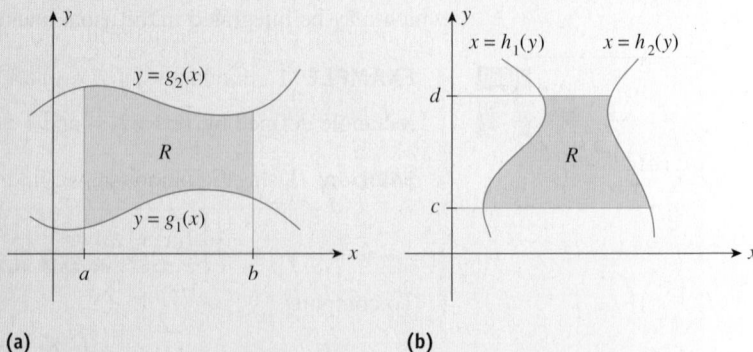

FIGURE 30

(a)   (b)

Notes

1. Observe that in (5) the lower and upper limits of integration with respect to y are given by $y = g_1(x)$ and $y = g_2(x)$. This is to be expected since, for a fixed value of x lying between $x = a$ and $x = b$, y runs between the lower curve defined by $y = g_1(x)$ and the upper curve defined by $y = g_2(x)$ (see Figure 30a). Observe, too, that in the special case when $g_1(x) = c$ and $g_2(x) = d$, the region R is rectangular, and (5) reduces to (4).

2. For a fixed value of y, x runs between $x = h_1(y)$ and $x = h_2(y)$, giving the indicated limits of integration with respect to x in (6) (see Figure 30b).

3. Note that the two curves in Figure 30b are not graphs of functions of x (use the vertical-line test), but they are graphs of functions of y. It is this observation that justifies the approach leading to (6). ∎

We now look at several examples.

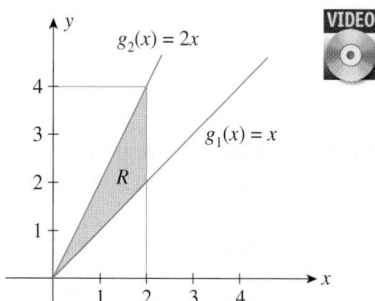

FIGURE 31
R is the region bounded by $g_1(x) = x$ and $g_2(x) = 2x$ for $0 \le x \le 2$.

EXAMPLE 2 Evaluate $\iint_R f(x, y)\, dA$ given that $f(x, y) = x^2 + y^2$ and R is the region bounded by the graphs of $g_1(x) = x$ and $g_2(x) = 2x$ for $0 \le x \le 2$.

Solution The region under consideration is shown in Figure 31. Using Equation (5), we find

$$\iint_R f(x, y)\, dA = \int_0^2 \left[\int_x^{2x} (x^2 + y^2)\, dy \right] dx$$

$$= \int_0^2 \left[\left(x^2 y + \frac{1}{3} y^3 \right) \Big|_x^{2x} \right] dx$$

$$= \int_0^2 \left[\left(2x^3 + \frac{8}{3} x^3 \right) - \left(x^3 + \frac{1}{3} x^3 \right) \right] dx$$

$$= \int_0^2 \frac{10}{3} x^3\, dx = \frac{5}{6} x^4 \Big|_0^2 = 13\tfrac{1}{3}$$ ∎

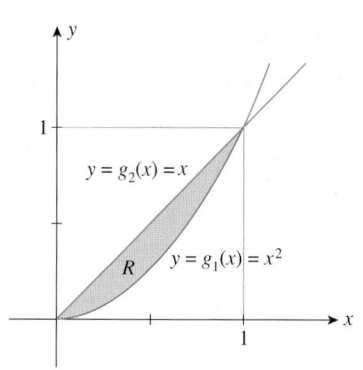

FIGURE 32
R is the region bounded by $y = x^2$ and $y = x$.

EXAMPLE 3 Evaluate $\iint_R f(x, y)\, dA$, where $f(x, y) = xe^y$ and R is the plane region bounded by the graphs of $y = x^2$ and $y = x$.

Solution The region in question is shown in Figure 32. The point of intersection of the two curves is found by solving the equation $x^2 = x$, giving $x = 0$ and $x = 1$. Using Equation (5), we find

$$\iint_R f(x, y)\, dA = \int_0^1 \left[\int_{x^2}^{x} xe^y\, dy \right] dx = \int_0^1 \left[xe^y \Big|_{x^2}^{x} \right] dx$$

$$= \int_0^1 (xe^x - xe^{x^2})\, dx = \int_0^1 xe^x\, dx - \int_0^1 xe^{x^2}\, dx$$

and integrating the first integral on the right-hand side by parts,

$$= \left[(x - 1)e^x - \frac{1}{2} e^{x^2} \right] \Big|_0^1$$

$$= -\frac{1}{2} e - \left(-1 - \frac{1}{2} \right) = \frac{1}{2}(3 - e)$$ ∎

Finding the Volume of a Solid by Double Integrals

As we saw earlier, the double integral

$$\iint\limits_R f(x, y) \, dA$$

gives the volume of the solid bounded by the graph of $f(x, y)$ over the region R.

The Volume of a Solid under a Surface
Let R be a region in the xy-plane and let f be continuous and nonnegative on R. Then, the **volume of the solid under a surface** bounded above by $z = f(x, y)$ and below by R is given by

$$V = \iint\limits_R f(x, y) \, dA$$

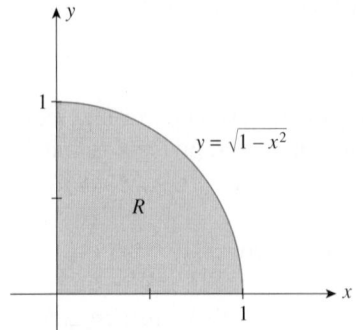

FIGURE 33
The plane region R defined by
$y = \sqrt{1 - x^2} \ (0 \le x \le 1)$

EXAMPLE 4 Find the volume of the solid bounded above by the plane $z = f(x, y) = y$ and below by the plane region R defined by $y = \sqrt{1 - x^2} \ (0 \le x \le 1)$.

Solution The region R is sketched in Figure 33. Observe that $f(x, y) = y \ge 0$ for $(x, y) \in R$. Therefore, the required volume is given by

$$\iint\limits_R y \, dA = \int_0^1 \left[\int_0^{\sqrt{1-x^2}} y \, dy \right] dx = \int_0^1 \left[\frac{1}{2} y^2 \Big|_0^{\sqrt{1-x^2}} \right] dx$$

$$= \int_0^1 \frac{1}{2} (1 - x^2) \, dx = \frac{1}{2} \left(x - \frac{1}{3} x^3 \right) \Big|_0^1 = \frac{1}{3}$$

or $\frac{1}{3}$ cubic unit. The solid is shown in Figure 34. Note that it is not necessary to make a sketch of the solid in order to compute its volume.

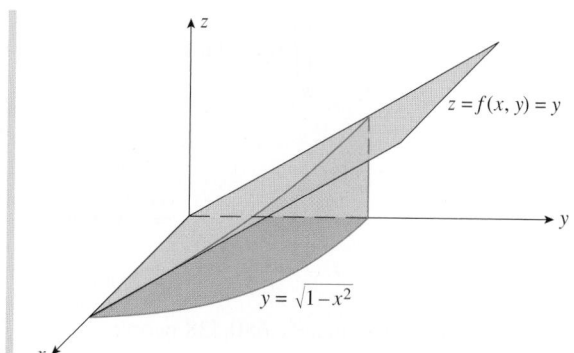

FIGURE 34
The solid bounded above by the plane $z = y$ and below by the plane region defined by $y = \sqrt{1 - x^2}$ $(0 \leq x \leq 1)$

FIGURE 35
The rectangular region R representing a certain district of a city is enclosed by a rectangular grid.

■ Population of a City

Suppose the plane region R represents a certain district of a city and $f(x, y)$ gives the population density (the number of people per square mile) at any point (x, y) in R. Enclose the set R by a rectangle and construct a grid for it in the usual manner. In any rectangular region of the grid that has no point in common with R, set $f(x_i, y_i)hk = 0$ (Figure 35). Then, corresponding to any grid covering the set R, the general term of the Riemann sum $f(x_i, y_i)hk$ (population density times area) gives the number of people living in that part of the city corresponding to the rectangular region R_i. Therefore, the Riemann sum gives an approximation of the number of people living in the district represented by R and, in the limit, the double integral

$$\iint\limits_{R} f(x, y) \, dA$$

gives the actual number of people living in the district under consideration.

APPLIED EXAMPLE 5 Population Density The population density of a certain city is described by the function

$$f(x, y) = 10{,}000e^{-0.2|x| - 0.1|y|}$$

where the origin $(0, 0)$ gives the location of the city hall. What is the population inside the rectangular area described by

$$R = \{(x, y) \mid -10 \leq x \leq 10; \, -5 \leq y \leq 5\}$$

if x and y are in miles? (See Figure 36.)

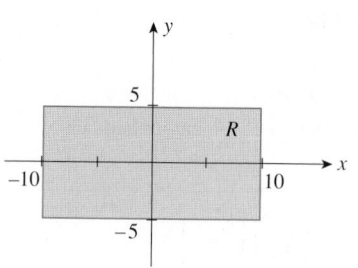

FIGURE 36
The rectangular region R represents a certain district of a city.

Solution By symmetry, it suffices to compute the population in the first quadrant. (Why?) Then, upon observing that in this quadrant

$$f(x, y) = 10{,}000e^{-0.2x - 0.1y} = 10{,}000e^{-0.2x}e^{-0.1y}$$

we see that the population in R is given by

$$\iint\limits_{R} f(x, y)\, dA = 4\int_{0}^{10}\left[\int_{0}^{5} 10{,}000e^{-0.2x}e^{-0.1y}\, dy\right] dx$$

$$= 4\int_{0}^{10}\left[-100{,}000e^{-0.2x}e^{-0.1y}\,\Big|_{0}^{5}\right] dx$$

$$= 400{,}000\left(1 - e^{-0.5}\right)\int_{0}^{10} e^{-0.2x}\, dx$$

$$= 2{,}000{,}000\left(1 - e^{-0.5}\right)\left(1 - e^{-2}\right)$$

or approximately 680,438 people. ∎

EXPLORE & DISCUSS

1. Consider the improper double integral $\iint\limits_{D} f(x, y)\, dA$ of the continuous function f of two variables defined over the plane region

$$D = \{(x, y) \mid 0 \le x < \infty;\ 0 \le y < \infty\}$$

Using the definition of improper integrals of functions of one variable (Section 16.3), explain why it makes sense to define

$$\iint\limits_{D} f(x, y)\, dA = \lim_{N\to\infty}\int_{0}^{N}\left[\lim_{M\to\infty}\int_{0}^{M} f(x, y)\, dx\right] dy$$

$$= \lim_{M\to\infty}\int_{0}^{M}\left[\lim_{N\to\infty}\int_{0}^{N} f(x, y)\, dy\right] dx$$

provided the limits exist.

2. Refer to Example 5. Assuming that the population density of the city is described by

$$f(x, y) = 10{,}000e^{-0.2|x|-0.1|y|}$$

for $-\infty < x < \infty$ and $-\infty < y < \infty$, show that the population outside the rectangular region

$$R = \{(x, y) \mid -10 < x < 10;\ -5 < y \le 5\}$$

of Example 5 is given by

$$4\iint\limits_{D} f(x, y)\, dx\, dy - 680{,}438$$

(recall that 680,438 is the approximate population inside R).

3. Use the results of parts 1 and 2 to determine the population of the city outside the rectangular area R.

■ Average Value of a Function

In Section 15.5, we showed that the average value of a continuous function $f(x)$ over an interval $[a, b]$ is given by

$$\frac{1}{b - a}\int_{a}^{b} f(x)\, dx$$

That is, the average value of a function over $[a, b]$ is the integral of f over $[a, b]$ divided by the length of the interval. An analogous result holds for a function of two variables $f(x, y)$ over a plane region R. To see this, we enclose R by a rectangle and construct a rectangular grid. Let (x_i, y_i) be any point in the rectangle R_i of area hk. Now, the average value of the mn numbers $f(x_1, y_1), f(x_2, y_2), \ldots, f(x_{mn}, y_{mn})$ is given by

$$\frac{f(x_1, y_1) + f(x_2, y_2) + \cdots + f(x_{mn}, y_{mn})}{mn}$$

which can also be written as

$$\frac{hk}{hk}\left[\frac{f(x_1, y_1) + f(x_2, y_2) + \cdots + f(x_{mn}, y_{mn})}{mn}\right]$$

$$= \frac{1}{(mn)\,hk}\left[f(x_1, y_1) + f(x_2, y_2) + \cdots + f(x_{mn}, y_{mn})\right]hk$$

Now the area of R is approximated by the sum of the mn rectangles (*omitting* those having no points in common with R), each of area hk. Note that this is the denominator of the previous expression. Therefore, taking the limit as m and n both tend to infinity, we obtain the following formula for the *average value of $f(x, y)$ over R.*

Average Value of $f(x, y)$ over the Region R

If f is integrable over the plane region R, then its average value over R is given by

$$\frac{\displaystyle\iint_R f(x, y)\, dA}{\text{Area of } R} \quad \text{or} \quad \frac{\displaystyle\iint_R f(x, y)\, dA}{\displaystyle\iint_R dA} \tag{7}$$

Note If we let $f(x, y) = 1$ for all (x, y) in R, then

$$\iint_R f(x, y)\, dA = \iint_R dA = \text{Area of } R \qquad \blacksquare$$

EXAMPLE 6 Find the average value of the function $f(x, y) = xy$ over the plane region defined by $y = e^x \ (0 \le x \le 1)$.

Solution The region R is shown in Figure 37. The area of the region R is given by

$$\int_0^1\left[\int_0^{e^x} dy\right] dx = \int_0^1 \left[y\Big|_0^{e^x}\right] dx$$

$$= \int_0^1 e^x\, dx$$

$$= e^x \Big|_0^1$$

$$= e - 1$$

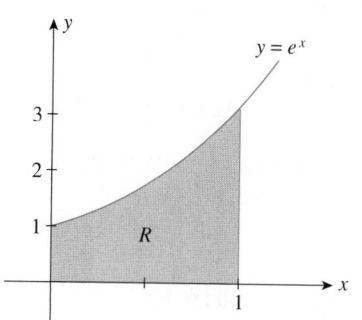

FIGURE 37
The plane region R defined by $y = e^x\ (0 \le x \le 1)$

square units. We would obtain the same result had we viewed the area of this region as the area of the region under the curve $y = e^x$ from $x = 0$ to $x = 1$. Next, we compute

$$\iint\limits_R f(x, y) \, dA = \int_0^1 \left[\int_0^{e^x} xy \, dy \right] dx$$

$$= \int_0^1 \left[\frac{1}{2} xy^2 \Big|_0^{e^x} \right] dx$$

$$= \int_0^1 \frac{1}{2} xe^{2x} \, dx$$

$$= \frac{1}{4} xe^{2x} - \frac{1}{8} e^{2x} \Big|_0^1 \qquad \text{Integrate by parts.}$$

$$= \left(\frac{1}{4} e^2 - \frac{1}{8} e^2 \right) + \frac{1}{8}$$

$$= \frac{1}{8}(e^2 + 1)$$

square units. Therefore, the required average value is given by

$$\frac{\displaystyle\iint\limits_R f(x, y) \, dA}{\displaystyle\iint\limits_R dA} = \frac{\frac{1}{8}(e^2 + 1)}{e - 1} = \frac{e^2 + 1}{8(e - 1)}$$

APPLIED EXAMPLE 7 Population Density (Refer to Example 5.) The population density of a certain city (number of people per square mile) is described by the function

$$f(x, y) = 10,000e^{-0.2|x| - 0.1|y|}$$

where the origin gives the location of the city hall. What is the average population density inside the rectangular area described by

$$R = \{(x, y) \mid -10 \le x \le 10; -5 \le y \le 5\}$$

where x and y are measured in miles?

Solution From the results of Example 5, we know that

$$\iint\limits_R f(x, y) \, dA \approx 680{,}438$$

From Figure 36, we see that the area of the plane rectangular region R is $(20)(10)$, or 200, square miles. Therefore, the average population inside R is

$$\frac{\displaystyle\iint\limits_R f(x, y) \, dA}{\displaystyle\iint\limits_R dA} = \frac{680{,}438}{200} = 3402.19$$

or approximately 3402 people per square mile.

17.5   Self-Check Exercises

1. Evaluate $\iint_R (x + y)\, dA$, where R is the region bounded by the graphs of $g_1(x) = x$ and $g_2(x) = x^{1/3}$.

2. The population density of a coastal town located on an island is described by the function

$$f(x, y) = \frac{5000xe^y}{1 + 2x^2} \qquad (0 \le x \le 4; -2 \le y \le 0)$$

where x and y are measured in miles (see the accompanying figure).

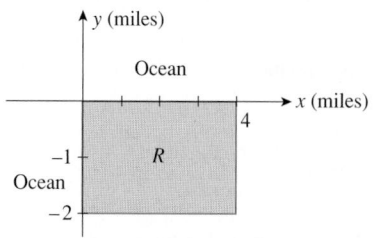

What is the population inside the rectangular area defined by $R = \{(x, y) \mid 0 \le x \le 4; -2 \le y \le 0\}$? What is the average population density in the area?

Solutions to Self-Check Exercises 17.5 can be found on page 1110.

17.5   Concept Questions

1. Give a geometric interpretation of $\iint_R f(x, y)\, dA$, where f is a nonnegative function on the rectangular region R in the xy-plane.

2. What is an iterated integral? How is $\iint_R f(x, y)\, dA$ evaluated in terms of iterated integrals, where R is the rectangular region defined by $a \le x \le b$ and $c \le y \le d$?

3. Suppose g_1 and g_2 are continuous on the interval $[a, b]$ and $R = \{(x, y) \mid g_1(x) \le y \le g_2(x), a \le x \le b\}$, what is $\iint_R f(x, y)\, dA$, where f is a continuous function defined on R?

4. Suppose h_1 and h_2 are continuous on the interval $[c, d]$ and $R = \{(x, y) \mid h_1(y) \le x \le h_2(y), c \le y \le d\}$, what is $\iint_R f(x, y)\, dA$, where f is a continuous function defined on R?

5. What is the average value of $f(x, y)$ over the region R?

17.5   Exercises

In Exercises 1–25, evaluate the double integral $\iint_R f(x, y)\, dA$ for the function $f(x, y)$ and the region R.

1. $f(x, y) = y + 2x$; R is the rectangle defined by $1 \le x \le 2$ and $0 \le y \le 1$.

2. $f(x, y) = x + 2y$; R is the rectangle defined by $-1 \le x \le 2$ and $0 \le y \le 2$.

3. $f(x, y) = xy^2$; R is the rectangle defined by $-1 \le x \le 1$ and $0 \le y \le 1$.

4. $f(x, y) = 12xy^2 + 8y^3$; R is the rectangle defined by $0 \le x \le 1$ and $0 \le y \le 2$.

5. $f(x, y) = \dfrac{x}{y}$; R is the rectangle defined by $-1 \le x \le 2$ and $1 \le y \le e^3$.

6. $f(x, y) = \dfrac{xy}{1 + y^2}$; R is the rectangle defined by $-2 \le x \le 2$ and $0 \le y \le 1$.

7. $f(x, y) = 4xe^{2x^2+y}$; R is the rectangle defined by $0 \le x \le 1$ and $-2 \le y \le 0$.

8. $f(x, y) = \dfrac{y}{x^2} e^{y/x}$; R is the rectangle defined by $1 \le x \le 2$ and $0 \le y \le 1$.

9. $f(x, y) = \ln y$; R is the rectangle defined by $0 \le x \le 1$ and $1 \le y \le e$.

10. $f(x, y) = \dfrac{\ln y}{x}$; R is the rectangle defined by $1 \le x \le e^2$ and $1 \le y \le e$.

11. $f(x, y) = x + 2y$; R is bounded by $x = 0$, $x = 1$, $y = 0$, and $y = x$.

12. $f(x, y) = xy$; R is bounded by $x = 0$, $x = 1$, $y = 0$ and $y = x$.

13. $f(x, y) = 2x + 4y$; R is bounded by $x = 1$, $x = 3$, $y = 0$, and $y = x + 1$.

14. $f(x, y) = 2 - y$; R is bounded by $x = -1$, $x = 1 - y$, $y = 0$, and $y = 2$.

15. $f(x, y) = x + y$; R is bounded by $x = 0$, $x = \sqrt{y}$, $y = 0$, and $y = 4$.

16. $f(x, y) = x^2 y^2$; R is bounded by $x = 0$, $x = 1$, $y = x^2$, and $y = x^3$.

17. $f(x, y) = y$; R is bounded by $x = 0$, $x = \sqrt{4 - y^2}$, $y = 0$, and $y = 2$.

18. $f(x, y) = \dfrac{y}{x^3 + 2}$; R is bounded by $x = 0$, $x = 1$, $y = 0$, and $y = x$.

19. $f(x, y) = 2xe^y$; R is bounded by $x = 0$, $x = 1$, $y = 0$, and $y = x$.

20. $f(x, y) = 2x$; R is bounded by $x = e^{2y}$, $x = y$, $y = 0$, and $y = 1$.

21. $f(x, y) = ye^x$; R is bounded by $y = \sqrt{x}$ and $y = x$.

22. $f(x, y) = xe^{-y^2}$; R is bounded by $x = 0$, $y = x^2$, and $y = 4$.

23. $f(x, y) = e^{y^2}$; R is bounded by $x = 0$, $x = 1$, $y = 2x$, and $y = 2$.

24. $f(x, y) = y$; R is bounded by $x = 1$, $x = e$, $y = 0$, and $y = \ln x$.

25. $f(x, y) = ye^{x^3}$; R is bounded by $x = \dfrac{y}{2}$, $x = 1$, $y = 0$, and $y = 2$.

In Exercises 26–33, use a double integral to find the volume of the solid shown in the figure.

26.

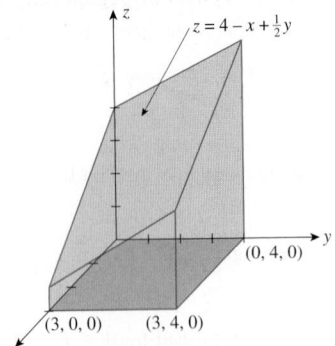

27.

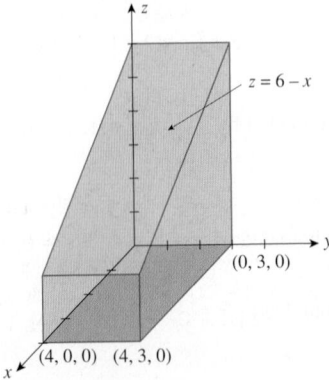

28.

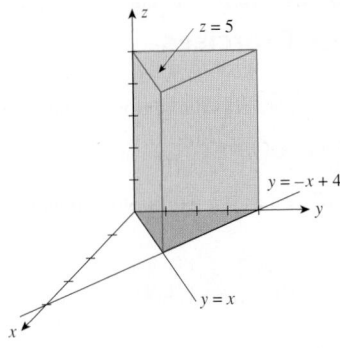

29.

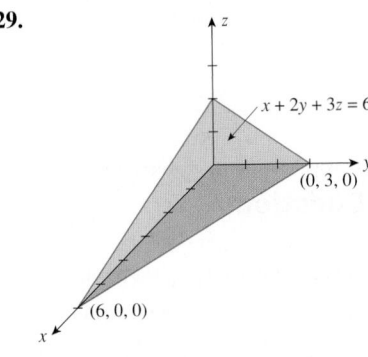

30.

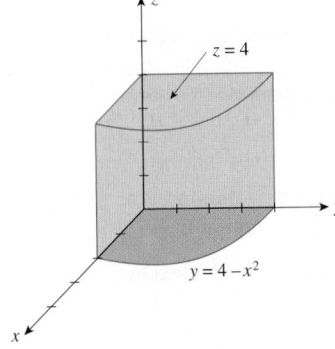

31.

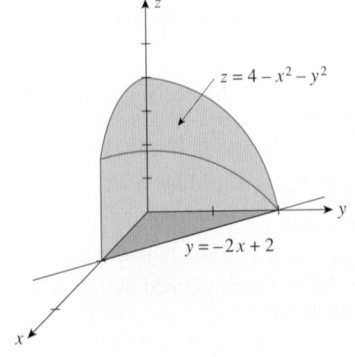

32.

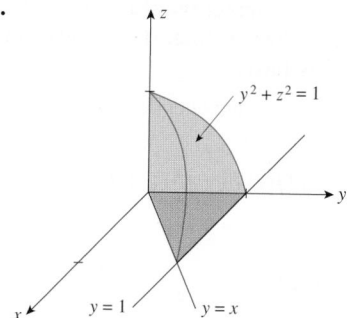

33.

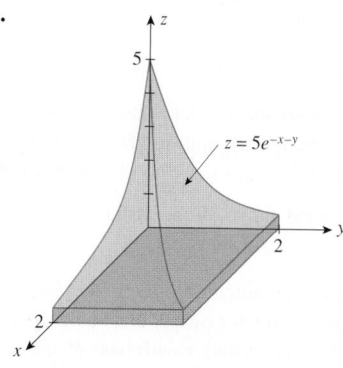

In Exercises 34–41, find the volume of the solid bounded above by the surface $z = f(x, y)$ and below by the plane region R.

34. $f(x, y) = 4 - 2x - y$; $R = \{(x, y) \mid 0 \le x \le 1; 0 \le y \le 2\}$

35. $f(x, y) = 2x + y$; R is the triangle bounded by $y = 2x$, $y = 0$, and $x = 2$.

36. $f(x, y) = x^2 + y^2$; R is the rectangle with vertices $(0, 0)$, $(1, 0)$, $(1, 2)$, and $(0, 2)$.

37. $f(x, y) = e^{x+2y}$; R is the triangle with vertices $(0, 0)$, $(1, 0)$, and $(0, 1)$.

38. $f(x, y) = 2xe^y$; R is the triangle bounded by $y = x$, $y = 2$, and $x = 0$.

39. $f(x, y) = \dfrac{2y}{1 + x^2}$; R is the region bounded by $y = \sqrt{x}$, $y = 0$, and $x = 4$.

40. $f(x, y) = 2x^2y$; R is the region bounded by the graphs of $y = x$ and $y = x^2$.

41. $f(x, y) = x$; R is the region in the first quadrant bounded by the semicircle $y = \sqrt{16 - x^2}$, the x-axis, and the y-axis.

In Exercises 42–47, find the average value of the function $f(x, y)$ over the plane region R.

42. $f(x, y) = 6x^2y^3$; $R = \{(x, y) \mid 0 \le x \le 2; 0 \le y \le 3\}$

43. $f(x, y) = x + 2y$; R is the triangle with vertices $(0, 0)$, $(1, 0)$, and $(1, 1)$.

44. $f(x, y) = xy$; R is the triangle bounded by $y = x$, $y = 2 - x$, and $y = 0$.

45. $f(x, y) = e^{-x^2}$; R is the triangle with vertices $(0, 0)$, $(1, 0)$, and $(1, 1)$.

46. $f(x, y) = xe^y$; R is the triangle with vertices $(0, 0)$, $(1, 0)$, and $(1, 1)$.

47. $f(x, y) = \ln x$; R is the region bounded by the graphs of $y = 2x$ and $y = 0$ from $x = 1$ to $x = 3$.
Hint: Use integration by parts.

48. POPULATION DENSITY The population density of a coastal town is described by the function

$$f(x, y) = \frac{10{,}000e^y}{1 + 0.5|x|} \qquad (-10 \le x \le 10; -4 \le y \le 0)$$

where x and y are measured in miles. Find the population inside the rectangular area described by

$$R = \{(x, y) \mid -5 \le x \le 5; -2 \le y \le 0\}$$

(See the accompanying figure.)

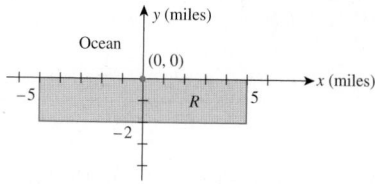

49. AVERAGE POPULATION DENSITY Refer to Exercise 48. Find the average population density inside the rectangular area R.

50. POPULATION DENSITY The population density of a certain city is given by the function

$$f(x, y) = \frac{50{,}000|xy|}{(x^2 + 20)(y^2 + 36)}$$

where the origin $(0, 0)$ gives the location of the government center. Find the population inside the rectangular area described by

$$R = \{(x, y) \mid -15 \le x \le 15; -20 \le y \le 20\}$$

51. AVERAGE PROFIT The Country Workshop's total weekly profit (in dollars) realized in manufacturing and selling its rolltop desks is given by the profit function

$$P(x, y) = -0.2x^2 - 0.25y^2 - 0.2xy + 100x + 90y - 4000$$

where x stands for the number of finished units and y stands for the number of unfinished units manufactured and sold each week. Find the average weekly profit if the number of finished units manufactured and sold varies between 180 and 200 and the number of unfinished units varies between 100 and 120/week.

52. AVERAGE PRICE OF LAND The rectangular region R shown in the accompanying figure represents a city's financial district. The price of land in the district is approximated by the function

$$p(x, y) = 200 - 10\left(x - \frac{1}{2}\right)^2 - 15(y - 1)^2$$

where $p(x, y)$ is the price of land at the point (x, y) in dollars/square foot and x and y are measured in miles. What is the average price of land per square foot in the district?

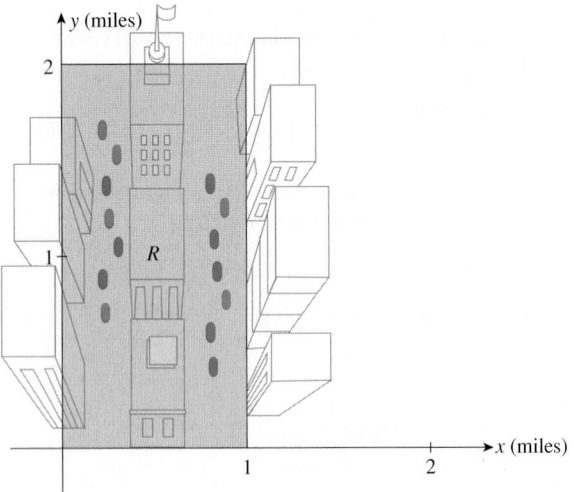

In Exercises 53–56, determine whether the statement is true or false. If it is true, explain why it is true. If it is false, give an example to show why it is false.

53. If $h(x, y) = f(x)g(y)$, where f is continuous on $[a, b]$ and g is continuous on $[c, d]$, then

$$\iint_R h(x, y)\, dA = \left[\int_a^b f(x)\, dx\right]\left[\int_c^d g(y)\, dy\right]$$

where $R = \{(x, y) \mid a \leq x \leq b; c \leq y \leq d\}$.

54. If $\iint_{R_1} f(x, y)\, dA$ exists, where

$$R_1 = \{(x, y) \mid a \leq x \leq b; c \leq y \leq d\}$$

then $\iint_{R_2} f(x, y)\, dA$ exists, where

$$R_2 = \{(x, y) \mid c \leq x \leq d; a \leq y \leq b\}.$$

55. Let R be a region in the xy-plane and let f and g be continuous functions on R that satisfy the condition $f(x, y) \leq g(x, y)$ for all (x, y) in R. Then, $\iint_R [g(x, y) - f(x, y)]\, dA$ gives the volume of the solid bounded above by the surface $z = g(x, y)$ and below by the surface $z = f(x, y)$.

56. Suppose f is nonnegative and integrable over the plane region R. Then, the average value of f over R can be thought of as the (constant) height of the cylinder with base R and volume that is exactly equal to the volume of the solid under the graph of $z = f(x, y)$. (*Note:* The cylinder referred to here has sides perpendicular to R.)

17.5 Solutions to Self-Check Exercises

1. The region R is shown in the accompanying figure. The points of intersection of the two curves are found by solving the equation $x = x^{1/3}$, giving $x = 0$ and $x = 1$. Using Equation (5), we find

$$\iint_R (x + y)\, dA = \int_0^1\left[\int_x^{x^{1/3}} (x + y)\, dy\right] dx$$

$$= \int_0^1 \left[xy + \frac{1}{2}y^2 \Big|_x^{x^{1/3}} \right] dx$$

$$= \int_0^1 \left[\left(x^{4/3} + \frac{1}{2}x^{2/3}\right) - \left(x^2 + \frac{1}{2}x^2\right)\right] dx$$

$$= \int_0^1 \left(x^{4/3} + \frac{1}{2}x^{2/3} - \frac{3}{2}x^2\right) dx$$

$$= \frac{3}{7}x^{7/3} + \frac{3}{10}x^{5/3} - \frac{1}{2}x^3 \Big|_0^1$$

$$= \frac{3}{7} + \frac{3}{10} - \frac{1}{2} = \frac{8}{35}$$

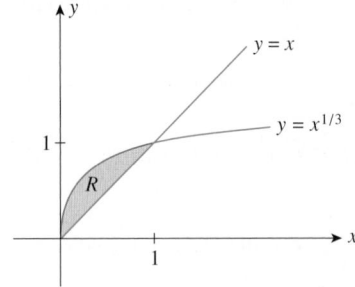

2. The population in R is given by

$$\iint_R f(x, y)\, dA = \int_0^4 \left[\int_{-2}^0 \frac{5000xe^y}{1 + 2x^2}\, dy \right] dx$$

$$= \int_0^4 \left[\frac{5000xe^y}{1 + 2x^2} \Big|_{-2}^0 \right] dx$$

$$= 5000(1 - e^{-2}) \int_0^4 \frac{x}{1 + 2x^2}\, dx$$

$$= 5000(1 - e^{-2}) \left[\frac{1}{4} \ln(1 + 2x^2) \Big|_0^4 \right]$$

$$= 5000(1 - e^{-2}) \left(\frac{1}{4} \right) \ln 33$$

or approximately 3779 people. The average population density inside R is

$$\frac{\displaystyle\iint_R f(x, y)\, dA}{\displaystyle\iint_R dA} = \frac{3779}{(2)(4)}$$

or approximately 472 people/square mile.

Summary of Principal Terms

 **TERMS**

function of two variables (1052)	complementary commodities (1067)	absolute minimum value (1075)
domain (1052)	second-order partial derivative of f (1068)	saddle point (1076)
three-dimensional Cartesian coordinate system (1054)	relative maximum (1075)	critical point (1077)
	relative maximum value (1075)	second derivative test (1078)
level curve (1055)	relative minimum (1075)	constrained relative extremum (1086)
first partial derivatives of f (1062)	relative minimum value (1075)	method of Lagrange multipliers (1088)
Cobb–Douglas production function (1066)	absolute maximum (1075)	Riemann sum (1097)
marginal productivity of labor (1066)	absolute minimum (1075)	double integral (1098)
marginal productivity of capital (1066)	absolute maximum value (1075)	volume of a solid under a surface (1102)
substitute commodities (1067)		

Concept Review Questions

Fill in the blanks.

1. The domain of a function f of two variables is a subset of the _____-plane. The rule of f associates with each _____ _____ in the domain of f one and only one _____ _____, denoted by $z =$ _____.

2. If the function f has rule $z = f(x, y)$, then x and y are called _____ variables, and z is a/an _____ variable. The number z is also called the _____ of f.

3. The graph of a function f of two variables is the set of all points (x, y, z), where _____, and (x, y) is the domain of _____. The graph of a function of two variables is a _____ in three-dimensional space.

4. The trace of the graph of $f(x, y)$ in the plane $z = c$ is the curve with equation _____ lying in the plane $z = c$. The projection of the trace of f in the plane $z = c$ onto the xy-plane is called the _____ _____ of f. The contour map associated with f is obtained by drawing the _____ _____ of f corresponding to several admissible values of _____.

5. The partial derivative $\partial f/\partial x$ of f at (x, y) can be found by thinking of y as a/an _____ _____ in the expression for f, and differentiating this expression with respect to _____ as if it were a function of x alone.

6. The number $f_x(a, b)$ measures the _____ of the tangent line to the curve C obtained by the intersection of the graph of f and the plane $y = b$ at the point ____. It also measures the rate of change of f with respect to ____ at the point (a, b) with y held fixed with value ____.

7. A function $f(x, y)$ has a relative maximum at (a, b) if $f(x, y)$ ____ $f(a, b)$ for all points (x, y) that are sufficiently close to ____. The absolute maximum value of $f(x, y)$ is the number $f(a, b)$ such that $f(x, y)$ ____ $f(a, b)$ for all (x, y) in the ____ of f.

8. A critical point of $f(x, y)$ is a point (a, b) in the ____ of f such that ____ ____ ____ or at least one of the partial derivatives of f does not ____. A critical point of f is a ____ for a relative extremum of f.

9. The method of Lagrange multipliers solves the problem of finding the relative extrema of a function $f(x, y)$ subject to the constraint ____. We first form the Lagrangian function

$F(x, y, \lambda) =$ ____. Then we solve the system consisting of the three equations ____, ____, and ____ for x, y, and λ. These solutions give the critical points that give rise to the relative ____ of f.

10. If $f(x, y)$ is continuous and nonnegative over a region R in the xy-plane and $\iint_R f(x, y)\, dA$ exists, then the double integral gives the ____ of the ____ bounded by the graph of $f(x, y)$ over the region R.

11. The integral $\iint_R f(x, y)\, dA$ is evaluated using ____ integrals. For example, $\iint_R (2x + y^2)\, dA$ where $R = \{(x, y) \mid 0 \le x \le 1;$ $3 \le y \le 5\}$ is equal to $\int_0^1 \int_3^5 (2x + y^2)\, dy\, dx$ or the (iterated) integral ____.

CHAPTER 17   **Review Exercises**

1. Let $f(x, y) = \dfrac{xy}{x^2 + y^2}$. Compute $f(0, 1)$, $f(1, 0)$, and $f(1, 1)$. Does $f(0, 0)$ exist?

2. Let $f(x, y) = \dfrac{xe^y}{1 + \ln xy}$. Compute $f(1, 1)$, $f(1, 2)$, and $f(2, 1)$. Does $f(1, 0)$ exist?

3. Let $h(x, y, z) = xye^z + \dfrac{x}{y}$. Compute $h(1, 1, 0)$, $h(-1, 1, 1)$, and $h(1, -1, 1)$.

4. Find the domain of the function $f(u, v) = \dfrac{\sqrt{u}}{u - v}$.

5. Find the domain of the function $f(x, y) = \dfrac{x - y}{x + y}$.

6. Find the domain of the function $f(x, y) = x\sqrt{y} + y\sqrt{1 - x}$.

7. Find the domain of the function
$$f(x, y, z) = \frac{xy\sqrt{z}}{(1 - x)(1 - y)(1 - z)}$$

In Exercises 8–11, sketch the level curves of the function corresponding to each value of z.

8. $z = f(x, y) = 2x + 3y$; $z = -2, -1, 0, 1, 2$

9. $z = f(x, y) = y - x^2$; $z = -2, -1, 0, 1, 2$

10. $z = f(x, y) = \sqrt{x^2 + y^2}$; $z = 0, 1, 2, 3, 4$

11. $z = f(x, y) = e^{xy}$; $z = 1, 2, 3$

In Exercises 12–21, compute the first partial derivatives of the function.

12. $f(x, y) = x^2y^3 + 3xy^2 + \dfrac{x}{y}$

13. $f(x, y) = x\sqrt{y} + y\sqrt{x}$ 14. $f(u, v) = \sqrt{uv^2 - 2u}$

15. $f(x, y) = \dfrac{x - y}{y + 2x}$ 16. $g(x, y) = \dfrac{xy}{x^2 + y^2}$

17. $h(x, y) = (2xy + 3y^2)^5$ 18. $f(x, y) = (xe^y + 1)^{1/2}$

19. $f(x, y) = (x^2 + y^2)e^{x^2 + y^2}$

20. $f(x, y) = \ln(1 + 2x^2 + 4y^4)$

21. $f(x, y) = \ln\left(1 + \dfrac{x^2}{y^2}\right)$

In Exercises 22–27, compute the second-order partial derivatives of the function.

22. $f(x, y) = x^3 - 2x^2y + y^2 + x - 2y$

23. $f(x, y) = x^4 + 2x^2y^2 - y^4$

24. $f(x, y) = (2x^2 + 3y^2)^3$ 25. $g(x, y) = \dfrac{x}{x + y^2}$

26. $g(x, y) = e^{x^2 + y^2}$ 27. $h(s, t) = \ln\left(\dfrac{s}{t}\right)$

28. Let $f(x, y, z) = x^3y^2z + xy^2z + 3xy - 4z$. Compute $f_x(1, 1, 0)$, $f_y(1, 1, 0)$, and $f_z(1, 1, 0)$ and interpret your results.

In Exercises 29–34, find the critical point(s) of the functions. Then use the second derivative test to classify the nature of each of these points, if possible. Finally, determine the relative extrema of each function.

29. $f(x, y) = 2x^2 + y^2 - 8x - 6y + 4$

30. $f(x, y) = x^2 + 3xy + y^2 - 10x - 20y + 12$

31. $f(x, y) = x^3 - 3xy + y^2$

32. $f(x, y) = x^3 + y^2 - 4xy + 17x - 10y + 8$

33. $f(x, y) = e^{2x^2 + y^2}$

34. $f(x, y) = \ln(x^2 + y^2 - 2x - 2y + 4)$

In Exercises 35–38, use the method of Lagrange multipliers to optimize the function subject to the given constraints.

35. Maximize the function $f(x, y) = -3x^2 - y^2 + 2xy$ subject to the constraint $2x + y = 4$.

36. Minimize the function $f(x, y) = 2x^2 + 3y^2 - 6xy + 4x - 9y + 10$ subject to the constraint $x + y = 1$.

37. Find the maximum and minimum values of the function $f(x, y) = 2x - 3y + 1$ subject to the constraint $2x^2 + 3y^2 - 125 = 0$.

38. Find the maximum and minimum values of the function $f(x, y) = e^{x-y}$ subject to the constraint $x^2 + y^2 = 1$.

In Exercises 39–42, evaluate the double integrals.

39. $f(x, y) = 3x - 2y$; R is the rectangle defined by $2 \le x \le 4$ and $-1 \le y \le 2$.

40. $f(x, y) = e^{-x-2y}$; R is the rectangle defined by $0 \le x \le 2$ and $0 \le y \le 1$.

41. $f(x, y) = 2x^2y$; R is bounded by $x = 0$, $x = 1$, $y = x^2$, and $y = x^3$.

42. $f(x, y) = \dfrac{y}{x}$, R is bounded by $x = 1$, $x = 2$, $y = 1$, and $y = x$.

In Exercises 43 and 44, find the volume of the solid bounded above by the surface $z = f(x, y)$ and below by the plane region R.

43. $f(x, y) = 4x^2 + y^2$; $R = \{0 \le x \le 2; 0 \le y \le 1\}$

44. $f(x, y) = x + y$; R is the region bounded by $y = x^2$, $y = 4x$, and $y = 4$.

45. Find the average value of the function

$$f(x, y) = xy + 1$$

over the plane region R bounded by $y = x^2$ and $y = 2x$.

46. Revenue Functions A division of Ditton Industries makes a 16-speed and a 10-speed electric blender. The company's management estimates that x units of the 16-speed model and y units of the 10-speed model are demanded daily when the unit prices are

$$p = 80 - 0.02x - 0.1y$$
$$q = 60 - 0.1x - 0.05y$$

dollars, respectively.
  a. Find the daily total revenue function $R(x, y)$.
  b. Find the domain of the function R.
  c. Compute $R(100, 300)$ and interpret your result.

47. Demand for CD Players In a survey conducted by *Home Entertainment* magazine, it was determined that the demand equation for CD players is given by

$$x = f(p, q) = 900 - 9p - e^{0.4q}$$

whereas the demand equation for audio CDs is given by

$$y = g(p, q) = 20{,}000 - 3000q - 4p$$

where p and q denote the unit prices (in dollars) for the CD players and audio CDs, respectively, and x and y denote the number of CD players and audio CDs demanded per week. Determine whether these two products are substitute, complementary, or neither.

48. Maximizing Revenue Odyssey Travel Agency's monthly revenue depends on the amount of money x (in thousands of dollars) spent on advertising per month and the number of agents y in its employ in accordance with the rule

$$R(x, y) = -x^2 - 0.5y^2 + xy + 8x + 3y + 20$$

Determine the amount of money the agency should spend per month and the number of agents it should employ in order to maximize its monthly revenue.

49. Minimizing Fencing Costs The owner of the Rancho Grande wants to enclose a rectangular piece of grazing land along the straight portion of a river and then subdivide it using a fence running parallel to the sides. No fencing is required along the river. If the material for the sides costs \$3/running yard and the material for the divider costs \$2/running yard, what will be the dimensions of a 303,750-yd pasture if the cost of the fencing is kept to a minimum?

50. Cobb–Douglas Production Functions The production of Q units of a commodity is related to the amount of labor x and the amount of capital y (in suitable units) expended by the equation

$$Q = f(x, y) = x^{3/4}y^{1/4}$$

If an expenditure of 100 units is available for production, how should it be apportioned between labor and capital so that Q is maximized?
Hint: Use the method of Lagrange multipliers to maximize the function Q subject to the constraint $x + y = 100$.

Before Moving On . . .

1. Find the domain of

$$f(x, y) = \frac{\sqrt{x} + \sqrt{y}}{(1 - x)(2 - y)}$$

2. Find the first- and second-order partial derivatives of $f(x, y) = x^2 y + e^{xy}$.

3. Find the relative extrema, if any, of $f(x, y) = 2x^3 + 2y^3 - 6xy - 5$.

4. Use the method of Lagrange multipliers to find the minimum of $f(x, y) = 3x^2 + 3y^2 + 1$ subject to $x + y = 1$.

5. Evaluate $\iint_R (1 - xy) \, dA$, where R is the region bounded by $x = 0$, $x = 1$, $y = x$, and $y = x^2$.

A The System of Real Numbers

In this appendix, we briefly review the system of real numbers. This system consists of a set of objects called real numbers together with two operations, addition and multiplication, that enable us to combine two or more real numbers to obtain other real numbers. These operations are subjected to certain rules that we will state after first recalling the set of real numbers.

The set of real numbers may be constructed from the set of **natural** (also called counting) numbers

$$N = \{1, 2, 3, \ldots\}$$

by adjoining other objects (numbers) to it. Thus, the set

$$W = \{0, 1, 2, 3, \ldots\}$$

obtained by adjoining the single number 0 to N is called the set of **whole numbers**. By adjoining *negatives* of the numbers 1, 2, 3, . . . to the set W of whole numbers, we obtain the set of **integers**

$$I = \{\ldots, -3, -2, -1, 0, 1, 2, 3, \ldots\}$$

Next, consider the set

$$Q = \left\{ \frac{a}{b} \;\middle|\; a \text{ and } b \text{ are integers with } b \neq 0 \right\}$$

Now, the set I of integers is contained in the set Q of **rational numbers**. To see this, observe that each integer may be written in the form a/b with $b = 1$, thus qualifying as a member of the set Q. The converse, however, is false, for the rational numbers (fractions) such as

$$\frac{1}{2}, \quad \frac{23}{25}, \quad \text{and so on}$$

are clearly not integers.

The sets N, W, I, and Q constructed thus far have

$$N \subset W \subset I \subset Q$$

That is, N is a proper subset of W, W is a proper subset of I, and so on.

Finally, consider the set Ir of all numbers that cannot be expressed in the form a/b, where a, b are integers ($b \neq 0$). The members of this set, called the set of **irrational numbers**, include $\sqrt{2}$, $\sqrt{3}$, π, and so on. The set

$$R = Q \cup Ir$$

which is the set of all rational and irrational numbers, is called the set of **real numbers** (Figure 1).

FIGURE 1
The set of all real numbers consists of the set of rational numbers plus the set of irrational numbers.

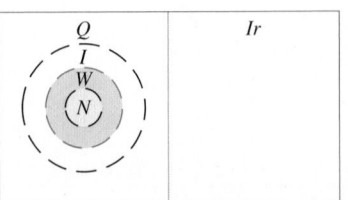

Q = Rationals
I = Integers
W = Whole numbers
N = Natural numbers
Ir = Irrationals

Note the following important representation of real numbers: Every real number has a decimal representation; a rational number has a representation in terms of a repeated decimal. For example,

$$\frac{1}{7} = 0.142857142857142857 \ldots$$

Note that the block of integers 142857 repeats.

On the other hand, the irrational number $\sqrt{2}$ has a representation in terms of a non-repeating decimal. Thus,

$$\sqrt{2} = 1.41421 \ldots$$

As mentioned earlier, any two real numbers may be combined to obtain another real number. The operation of *addition*, written $+$, enables us to combine any two numbers a and b to obtain their sum, denoted by $a + b$. Another operation, called *multiplication*, and written $\cdot$, enables us to combine any two real numbers a and b to form their product, the number $a \cdot b$, or, written more simply, ab. These two operations are subjected to the following rules of operation: Given any three real numbers a, b, and c, we have

I. Under addition

    1. $a + b = b + a$ Commutative law of addition

    2. $a + (b + c) = (a + b) + c$ Associative law of addition

    3. $a + 0 = a$ Identity law of addition

    4. $a + (-a) = 0$ Inverse law of addition

II. Under multiplication

    1. $ab = ba$ Commutative law of multiplication

    2. $a(bc) = (ab)c$ Associative law of multiplication

    3. $a \cdot 1 = a$ Identity law of multiplication

    4. $a(1/a) = 1$ $(a \neq 0)$ Inverse law of multiplication

III. Under addition and multiplication

    1. $a(b + c) = ab + ac$ Distributive law for multiplication with respect to addition

B | Tables

TABLE 1

Binomial Probabilities

n	x	0.05	0.1	0.2	0.3	p 0.4	0.5	0.6	0.7	0.8	0.9	0.95
2	0	0.902	0.810	0.640	0.490	0.360	0.250	0.160	0.090	0.040	0.010	0.002
	1	0.095	0.180	0.320	0.420	0.480	0.500	0.480	0.420	0.320	0.180	0.095
	2	0.002	0.010	0.040	0.090	0.160	0.250	0.360	0.490	0.640	0.810	0.902
3	0	0.857	0.729	0.512	0.343	0.216	0.125	0.064	0.027	0.008	0.001	
	1	0.135	0.243	0.384	0.441	0.432	0.375	0.288	0.189	0.096	0.027	0.007
	2	0.007	0.027	0.096	0.189	0.288	0.375	0.432	0.441	0.384	0.243	0.135
	3		0.001	0.008	0.027	0.064	0.125	0.216	0.343	0.512	0.729	0.857
4	0	0.815	0.656	0.410	0.240	0.130	0.062	0.026	0.008	0.002		
	1	0.171	0.292	0.410	0.412	0.346	0.250	0.154	0.076	0.026	0.004	
	2	0.014	0.049	0.154	0.265	0.346	0.375	0.346	0.265	0.154	0.049	0.014
	3		0.004	0.026	0.076	0.154	0.250	0.346	0.412	0.410	0.292	0.171
	4			0.002	0.008	0.026	0.062	0.130	0.240	0.410	0.656	0.815
5	0	0.774	0.590	0.328	0.168	0.078	0.031	0.010	0.002			
	1	0.204	0.328	0.410	0.360	0.259	0.156	0.077	0.028	0.006		
	2	0.021	0.073	0.205	0.309	0.346	0.312	0.230	0.132	0.051	0.008	0.001
	3	0.001	0.008	0.051	0.132	0.230	0.312	0.346	0.309	0.205	0.073	0.021
	4			0.006	0.028	0.077	0.156	0.259	0.360	0.410	0.328	0.204
	5				0.002	0.010	0.031	0.078	0.168	0.328	0.590	0.774
6	0	0.735	0.531	0.262	0.118	0.047	0.016	0.004	0.001			
	1	0.232	0.354	0.393	0.303	0.187	0.094	0.037	0.010	0.002		
	2	0.031	0.098	0.246	0.324	0.311	0.234	0.138	0.060	0.015	0.001	
	3	0.002	0.015	0.082	0.185	0.276	0.312	0.276	0.185	0.082	0.015	0.002
	4		0.001	0.015	0.060	0.138	0.234	0.311	0.324	0.246	0.098	0.031
	5			0.002	0.010	0.037	0.094	0.187	0.303	0.393	0.354	0.232
	6				0.001	0.004	0.016	0.047	0.118	0.262	0.531	0.735
7	0	0.698	0.478	0.210	0.082	0.028	0.008	0.002				
	1	0.257	0.372	0.367	0.247	0.131	0.055	0.017	0.004			
	2	0.041	0.124	0.275	0.318	0.261	0.164	0.077	0.025	0.004		
	3	0.004	0.023	0.115	0.227	0.290	0.273	0.194	0.097	0.029	0.003	
	4		0.003	0.029	0.097	0.194	0.273	0.290	0.227	0.115	0.023	0.004
	5			0.004	0.025	0.077	0.164	0.261	0.318	0.275	0.124	0.041
	6				0.004	0.017	0.055	0.131	0.247	0.367	0.372	0.257
	7					0.002	0.008	0.028	0.082	0.210	0.478	0.698

TABLE 1 (*continued*)

n	x	p 0.05	0.1	0.2	0.3	0.4	0.5	0.6	0.7	0.8	0.9	0.95
8	0	0.663	0.430	0.168	0.058	0.017	0.004	0.001				
	1	0.279	0.383	0.336	0.198	0.090	0.031	0.008	0.001			
	2	0.051	0.149	0.294	0.296	0.209	0.109	0.041	0.010	0.001		
	3	0.005	0.033	0.147	0.254	0.279	0.219	0.124	0.047	0.009		
	4		0.005	0.046	0.136	0.232	0.273	0.232	0.136	0.046	0.005	
	5			0.009	0.047	0.124	0.219	0.279	0.254	0.147	0.033	0.005
	6			0.001	0.010	0.041	0.109	0.209	0.296	0.294	0.149	0.051
	7				0.001	0.008	0.031	0.090	0.198	0.336	0.383	0.279
	8					0.001	0.004	0.017	0.058	0.168	0.430	0.663
9	0	0.630	0.387	0.134	0.040	0.010	0.002					
	1	0.299	0.387	0.302	0.156	0.060	0.018	0.004				
	2	0.063	0.172	0.302	0.267	0.161	0.070	0.021	0.004			
	3	0.008	0.045	0.176	0.267	0.251	0.164	0.074	0.021	0.003		
	4	0.001	0.007	0.066	0.172	0.251	0.246	0.167	0.074	0.017	0.001	
	5		0.001	0.017	0.074	0.167	0.246	0.251	0.172	0.066	0.007	0.001
	6			0.003	0.021	0.074	0.164	0.251	0.267	0.176	0.045	0.008
	7				0.004	0.021	0.070	0.161	0.267	0.302	0.172	0.063
	8					0.004	0.018	0.060	0.156	0.302	0.387	0.299
	9						0.002	0.010	0.040	0.134	0.387	0.630
10	0	0.599	0.349	0.107	0.028	0.006	0.001					
	1	0.315	0.387	0.268	0.121	0.040	0.010	0.002				
	2	0.075	0.194	0.302	0.233	0.121	0.044	0.011	0.001			
	3	0.010	0.057	0.201	0.267	0.215	0.117	0.042	0.009	0.001		
	4	0.001	0.011	0.088	0.200	0.251	0.205	0.111	0.037	0.006		
	5		0.001	0.026	0.103	0.201	0.246	0.201	0.103	0.026	0.001	
	6			0.006	0.037	0.111	0.205	0.251	0.200	0.088	0.011	0.001
	7			0.001	0.009	0.042	0.117	0.215	0.267	0.201	0.057	0.010
	8				0.001	0.011	0.044	0.121	0.233	0.302	0.194	0.075
	9					0.002	0.010	0.040	0.121	0.268	0.387	0.315
	10						0.001	0.006	0.028	0.107	0.349	0.599
11	0	0.569	0.314	0.086	0.020	0.004						
	1	0.329	0.384	0.236	0.093	0.027	0.005	0.001				
	2	0.087	0.213	0.295	0.200	0.089	0.027	0.005	0.001			
	3	0.014	0.071	0.221	0.257	0.177	0.081	0.023	0.004			
	4	0.001	0.016	0.111	0.220	0.236	0.161	0.070	0.017	0.002		
	5		0.002	0.039	0.132	0.221	0.226	0.147	0.057	0.010		
	6			0.010	0.057	0.147	0.226	0.221	0.132	0.039	0.002	
	7			0.002	0.017	0.070	0.161	0.236	0.220	0.111	0.016	0.001
	8				0.004	0.023	0.081	0.177	0.257	0.221	0.071	0.014
	9				0.001	0.005	0.027	0.089	0.200	0.295	0.213	0.087
	10					0.001	0.005	0.027	0.093	0.236	0.384	0.329
	11							0.004	0.020	0.086	0.314	0.569

TABLE 1 (*continued*)

							p					
n	x	0.05	0.1	0.2	0.3	0.4	0.5	0.6	0.7	0.8	0.9	0.95
12	0	0.540	0.282	0.069	0.014	0.002						
	1	0.341	0.377	0.206	0.071	0.017	0.003					
	2	0.099	0.230	0.283	0.168	0.064	0.016	0.002				
	3	0.017	0.085	0.236	0.240	0.142	0.054	0.012	0.001			
	4	0.002	0.021	0.133	0.231	0.213	0.121	0.042	0.008	0.001		
	5		0.004	0.053	0.158	0.227	0.193	0.101	0.029	0.003		
	6			0.016	0.079	0.177	0.226	0.177	0.079	0.016		
	7			0.003	0.029	0.101	0.193	0.227	0.158	0.053	0.004	
	8			0.001	0.008	0.042	0.121	0.213	0.231	0.133	0.021	0.002
	9				0.001	0.012	0.054	0.142	0.240	0.236	0.085	0.017
	10					0.002	0.016	0.064	0.168	0.283	0.230	0.099
	11						0.003	0.017	0.071	0.206	0.377	0.341
	12							0.002	0.014	0.069	0.282	0.540
13	0	0.513	0.254	0.055	0.010	0.001						
	1	0.351	0.367	0.179	0.054	0.011	0.002					
	2	0.111	0.245	0.268	0.139	0.045	0.010	0.001				
	3	0.021	0.100	0.246	0.218	0.111	0.035	0.006	0.001			
	4	0.003	0.028	0.154	0.234	0.184	0.087	0.024	0.003			
	5		0.006	0.069	0.180	0.221	0.157	0.066	0.014	0.001		
	6		0.001	0.023	0.103	0.197	0.209	0.131	0.044	0.006		
	7			0.006	0.044	0.131	0.209	0.197	0.103	0.023	0.001	
	8			0.001	0.014	0.066	0.157	0.221	0.180	0.069	0.006	
	9				0.003	0.024	0.087	0.184	0.234	0.154	0.028	0.003
	10				0.001	0.006	0.035	0.111	0.218	0.246	0.100	0.021
	11					0.001	0.010	0.045	0.139	0.268	0.245	0.111
	12						0.002	0.011	0.054	0.179	0.367	0.351
	13							0.001	0.010	0.055	0.254	0.513
14	0	0.488	0.229	0.044	0.007	0.001						
	1	0.359	0.356	0.154	0.041	0.007	0.001					
	2	0.123	0.257	0.250	0.113	0.032	0.006	0.001				
	3	0.026	0.114	0.250	0.194	0.085	0.022	0.003				
	4	0.004	0.035	0.172	0.229	0.155	0.061	0.014	0.001			
	5		0.008	0.086	0.196	0.207	0.122	0.041	0.007			
	6		0.001	0.032	0.126	0.207	0.183	0.092	0.023	0.002		
	7			0.009	0.062	0.157	0.209	0.157	0.062	0.009		
	8			0.002	0.023	0.092	0.183	0.207	0.126	0.032	0.001	
	9				0.007	0.041	0.122	0.207	0.196	0.086	0.008	
	10				0.001	0.014	0.061	0.155	0.229	0.172	0.035	0.004
	11					0.003	0.022	0.085	0.194	0.250	0.114	0.026
	12					0.001	0.006	0.032	0.113	0.250	0.257	0.123
	13						0.001	0.007	0.041	0.154	0.356	0.359
	14							0.001	0.007	0.044	0.229	0.488

TABLE 1 (*continued*)

| | | | | | | | *p* | | | | | | |
n	x	0.05	0.1	0.2	0.3	0.4	0.5	0.6	0.7	0.8	0.9	0.95
15	0	0.463	0.206	0.035	0.005							
	1	0.366	0.343	0.132	0.031	0.005						
	2	0.135	0.267	0.231	0.092	0.022	0.003					
	3	0.031	0.129	0.250	0.170	0.063	0.014	0.002				
	4	0.005	0.043	0.188	0.219	0.127	0.042	0.007	0.001			
	5	0.001	0.010	0.103	0.206	0.186	0.092	0.024	0.003			
	6		0.002	0.043	0.147	0.207	0.153	0.061	0.012	0.001		
	7			0.014	0.081	0.177	0.196	0.118	0.035	0.003		
	8			0.003	0.035	0.118	0.196	0.177	0.081	0.014		
	9			0.001	0.012	0.061	0.153	0.207	0.147	0.043	0.002	
	10				0.003	0.024	0.092	0.186	0.206	0.103	0.010	0.001
	11				0.001	0.007	0.042	0.127	0.219	0.188	0.043	0.005
	12					0.002	0.014	0.063	0.170	0.250	0.129	0.031
	13						0.003	0.022	0.092	0.231	0.267	0.135
	14							0.005	0.031	0.132	0.343	0.366
	15								0.005	0.035	0.206	0.463

TABLE 2

The Standard Normal Distribution

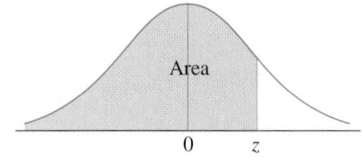

$$F_z(z) = P[Z \leq z]$$

z	0.00	0.01	0.02	0.03	0.04	0.05	0.06	0.07	0.08	0.09
−3.4	0.0003	0.0003	0.0003	0.0003	0.0003	0.0003	0.0003	0.0003	0.0003	0.0002
−3.3	0.0005	0.0005	0.0005	0.0004	0.0004	0.0004	0.0004	0.0004	0.0004	0.0003
−3.2	0.0007	0.0007	0.0006	0.0006	0.0006	0.0006	0.0006	0.0005	0.0005	0.0005
−3.1	0.0010	0.0009	0.0009	0.0009	0.0008	0.0008	0.0008	0.0008	0.0007	0.0007
−3.0	0.0013	0.0013	0.0013	0.0012	0.0012	0.0011	0.0011	0.0011	0.0010	0.0010
−2.9	0.0019	0.0018	0.0017	0.0017	0.0016	0.0016	0.0015	0.0015	0.0014	0.0014
−2.8	0.0026	0.0025	0.0024	0.0023	0.0023	0.0022	0.0021	0.0021	0.0020	0.0019
−2.7	0.0035	0.0034	0.0033	0.0032	0.0031	0.0030	0.0029	0.0028	0.0027	0.0026
−2.6	0.0047	0.0045	0.0044	0.0043	0.0041	0.0040	0.0039	0.0038	0.0037	0.0036
−2.5	0.0062	0.0060	0.0059	0.0057	0.0055	0.0054	0.0052	0.0051	0.0049	0.0048
−2.4	0.0082	0.0080	0.0078	0.0075	0.0073	0.0071	0.0069	0.0068	0.0066	0.0064
−2.3	0.0107	0.0104	0.0102	0.0099	0.0096	0.0094	0.0091	0.0089	0.0087	0.0084
−2.2	0.0139	0.0136	0.0132	0.0129	0.0125	0.0122	0.0119	0.0116	0.0113	0.0110
−2.1	0.0179	0.0174	0.0170	0.0166	0.0162	0.0158	0.0154	0.0150	0.0146	0.0143
−2.0	0.0228	0.0222	0.0217	0.0212	0.0207	0.0202	0.0197	0.0192	0.0188	0.0183
−1.9	0.0287	0.0281	0.0274	0.0268	0.0262	0.0256	0.0250	0.0244	0.0239	0.0233
−1.8	0.0359	0.0352	0.0344	0.0336	0.0329	0.0322	0.0314	0.0307	0.0301	0.0294
−1.7	0.0446	0.0436	0.0427	0.0418	0.0409	0.0401	0.0392	0.0384	0.0375	0.0367
−1.6	0.0548	0.0537	0.0526	0.0516	0.0505	0.0495	0.0485	0.0475	0.0465	0.0455
−1.5	0.0668	0.0655	0.0643	0.0630	0.0618	0.0606	0.0594	0.0582	0.0571	0.0559
−1.4	0.0808	0.0793	0.0778	0.0764	0.0749	0.0735	0.0722	0.0708	0.0694	0.0681
−1.3	0.0968	0.0951	0.0934	0.0918	0.0901	0.0885	0.0869	0.0853	0.0838	0.0823
−1.2	0.1151	0.1131	0.1112	0.1093	0.1075	0.1056	0.1038	0.1020	0.1003	0.0985
−1.1	0.1357	0.1335	0.1314	0.1292	0.1271	0.1251	0.1230	0.1210	0.1190	0.1170
−1.0	0.1587	0.1562	0.1539	0.1515	0.1492	0.1469	0.1446	0.1423	0.1401	0.1379
−0.9	0.1841	0.1814	0.1788	0.1762	0.1736	0.1711	0.1685	0.1660	0.1635	0.1611
−0.8	0.2119	0.2090	0.2061	0.2033	0.2005	0.1977	0.1949	0.1922	0.1894	0.1867
−0.7	0.2420	0.2389	0.2358	0.2327	0.2296	0.2266	0.2236	0.2206	0.2177	0.2148
−0.6	0.2743	0.2709	0.2676	0.2643	0.2611	0.2578	0.2546	0.2514	0.2483	0.2451
−0.5	0.3085	0.3050	0.3015	0.2981	0.2946	0.2912	0.2877	0.2843	0.2810	0.2776
−0.4	0.3446	0.3409	0.3372	0.3336	0.3300	0.3264	0.3228	0.3192	0.3156	0.3121
−0.3	0.3821	0.3783	0.3745	0.3707	0.3669	0.3632	0.3594	0.3557	0.3520	0.3483
−0.2	0.4207	0.4168	0.4129	0.4090	0.4052	0.4013	0.3974	0.3936	0.3897	0.3859
−0.1	0.4602	0.4562	0.4522	0.4483	0.4443	0.4404	0.4364	0.4325	0.4286	0.4247
−0.0	0.5000	0.4960	0.4920	0.4880	0.4840	0.4801	0.4761	0.4721	0.4681	0.4641

TABLE 2 (*continued*)

$$F_z(z) = P[Z \leq z]$$

z	0.00	0.01	0.02	0.03	0.04	0.05	0.06	0.07	0.08	0.09
0.0	0.5000	0.5040	0.5080	0.5120	0.5160	0.5199	0.5239	0.5279	0.5319	0.5359
0.1	0.5398	0.5438	0.5478	0.5517	0.5557	0.5596	0.5636	0.5675	0.5714	0.5753
0.2	0.5793	0.5832	0.5871	0.5910	0.5948	0.5987	0.6026	0.6064	0.6103	0.6141
0.3	0.6179	0.6217	0.6255	0.6293	0.6331	0.6368	0.6406	0.6443	0.6480	0.6517
0.4	0.6554	0.6591	0.6628	0.6664	0.6700	0.6736	0.6772	0.6808	0.6844	0.6879
0.5	0.6915	0.6950	0.6985	0.7019	0.7054	0.7088	0.7123	0.7157	0.7190	0.7224
0.6	0.7257	0.7291	0.7324	0.7357	0.7389	0.7422	0.7454	0.7486	0.7517	0.7549
0.7	0.7580	0.7611	0.7642	0.7673	0.7704	0.7734	0.7764	0.7794	0.7823	0.7852
0.8	0.7881	0.7910	0.7939	0.7967	0.7995	0.8023	0.8051	0.8078	0.8106	0.8133
0.9	0.8159	0.8186	0.8212	0.8238	0.8264	0.8289	0.8315	0.8340	0.8365	0.8389
1.0	0.8413	0.8438	0.8461	0.8485	0.8508	0.8531	0.8554	0.8577	0.8599	0.8621
1.1	0.8643	0.8665	0.8686	0.8708	0.8729	0.8749	0.8770	0.8790	0.8810	0.8830
1.2	0.8849	0.8869	0.8888	0.8907	0.8925	0.8944	0.8962	0.8980	0.8997	0.9015
1.3	0.9032	0.9049	0.9066	0.9082	0.9099	0.9115	0.9131	0.9147	0.9162	0.9177
1.4	0.9192	0.9207	0.9222	0.9236	0.9251	0.9265	0.9278	0.9292	0.9306	0.9319
1.5	0.9332	0.9345	0.9357	0.9370	0.9382	0.9394	0.9406	0.9418	0.9429	0.9441
1.6	0.9452	0.9463	0.9474	0.9484	0.9495	0.9505	0.9515	0.9525	0.9535	0.9545
1.7	0.9554	0.9564	0.9573	0.9582	0.9591	0.9599	0.9608	0.9616	0.9625	0.9633
1.8	0.9641	0.9649	0.9656	0.9664	0.9671	0.9678	0.9686	0.9693	0.9699	0.9706
1.9	0.9713	0.9719	0.9726	0.9732	0.9738	0.9744	0.9750	0.9756	0.9761	0.9767
2.0	0.9772	0.9778	0.9783	0.9788	0.9793	0.9798	0.9803	0.9808	0.9812	0.9817
2.1	0.9821	0.9826	0.9830	0.9834	0.9838	0.9842	0.9846	0.9850	0.9854	0.9857
2.2	0.9861	0.9864	0.9868	0.9871	0.9875	0.9878	0.9881	0.9884	0.9887	0.9890
2.3	0.9893	0.9896	0.9898	0.9901	0.9904	0.9906	0.9909	0.9911	0.9913	0.9916
2.4	0.9918	0.9920	0.9922	0.9925	0.9927	0.9929	0.9931	0.9932	0.9934	0.9936
2.5	0.9938	0.9940	0.9951	0.9943	0.9945	0.9946	0.9948	0.9949	0.9951	0.9952
2.6	0.9953	0.9955	0.9956	0.9957	0.9959	0.9960	0.9961	0.9962	0.9963	0.9964
2.7	0.9965	0.9966	0.9967	0.9968	0.9969	0.9970	0.9971	0.9972	0.9973	0.9974
2.8	0.9974	0.9975	0.9976	0.9977	0.9977	0.9978	0.9979	0.9979	0.9980	0.9981
2.9	0.9981	0.9982	0.9982	0.9983	0.9984	0.9984	0.9985	0.9985	0.9986	0.9986
3.0	0.9987	0.9987	0.9987	0.9988	0.9988	0.9989	0.9989	0.9989	0.9990	0.9990
3.1	0.9990	0.9991	0.9991	0.9991	0.9992	0.9992	0.9992	0.9992	0.9993	0.9993
3.2	0.9993	0.9993	0.9994	0.9994	0.9994	0.9994	0.9994	0.9995	0.9995	0.9995
3.3	0.9995	0.9995	0.9995	0.9996	0.9996	0.9996	0.9996	0.9996	0.9996	0.9997
3.4	0.9997	0.9997	0.9997	0.9997	0.9997	0.9997	0.9997	0.9997	0.9997	0.9998

Answers to Odd-Numbered Exercises

CHAPTER 1

Exercises 1.1, page 7

1. $(3, 3)$; Quadrant I   3. $(2, -2)$; Quadrant IV

5. $(-4, -6)$; Quadrant III   7. A

9. E, F, and G 11. F

13–19. See the following figure.

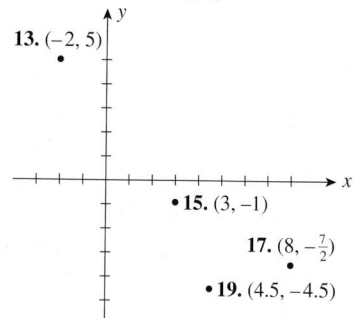

13. $(-2, 5)$

15. $(3, -1)$

17. $(8, -\frac{7}{2})$

19. $(4.5, -4.5)$

21. 5   23. $\sqrt{61}$ 25. $(-8, -6)$ and $(8, -6)$

29. $(x - 2)^2 + (y + 3)^2 = 25$

31. $x^2 + y^2 = 25$

33. $(x - 2)^2 + (y + 3)^2 = 34$ 35. 3400 mi

37. Route 1   39. Model C

41. **a.** $d = 10\sqrt{13}\, t$ **b.** 72.11 mi

43. False

Exercises 1.2, page 19

1. $\frac{1}{2}$ 3. Not defined   5. 5   7. $\frac{5}{6}$

9. $\dfrac{d - b}{c - a}\,(a \neq c)$ 11. **a.** 4   **b.** -8

13. Parallel   15. Perpendicular

17. $a = -5$ 19. $y = -3$

21. e   23. a   25. f   27. $y = 2x - 10$

29. $y = 2$ 31. $y = 3x - 2$ 33. $y = x + 1$

35. $y = 3x + 4$ 37. $y = 5$

39. $y = \frac{1}{2}x; m = \frac{1}{2}; b = 0$

41. $y = \frac{2}{3}x - 3; m = \frac{2}{3}; b = -3$

43. $y = -\frac{1}{2}x + \frac{7}{2}; m = -\frac{1}{2}; b = \frac{7}{2}$

45. $y = \frac{1}{2}x + 3$ 47. $y = -6$ 49. $y = b$

51. $y = \frac{2}{3}x - \frac{2}{3}$ 53. $k = 8$

55.

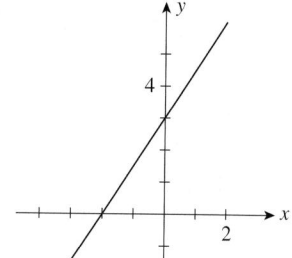

57.

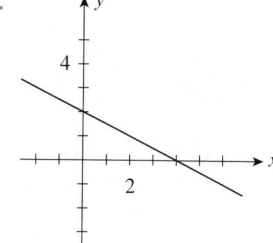

59.

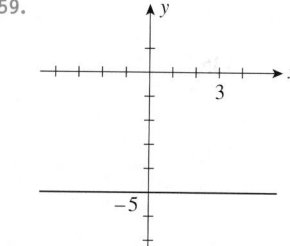

63. $y = -2x - 4$ 65. $y = \frac{1}{8}x - \frac{1}{2}$ 67. Yes

69. **a.**

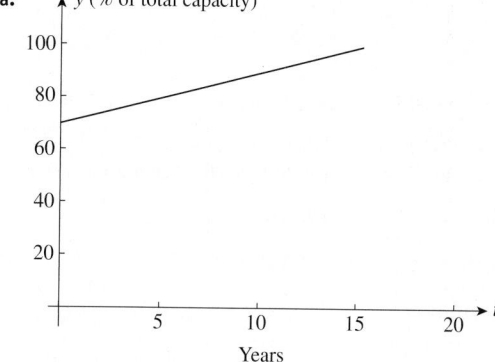

Years

b. 1.9467; 70.082

c. The capacity utilization has been increasing by 1.9467% each year since 1990 when it stood at 70.082%.

d. Shortly after 2005

71. **a.** $y = 0.55x$ **b.** 2000   73. 84.8%

75. a and **b.**

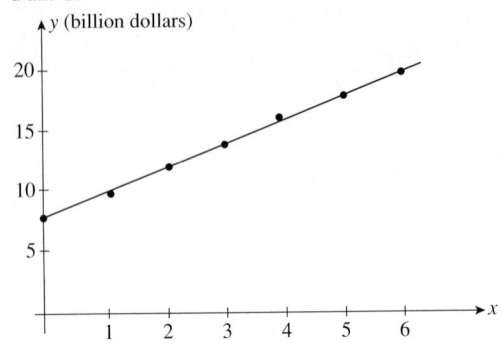

c. $y = 1.82x + 7.9$ **d.** $17 billion; same

77. a and **b.**

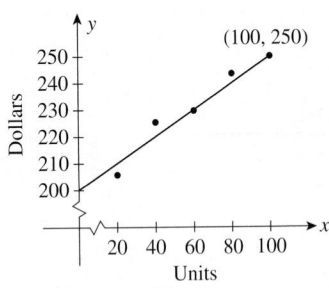

c. $y = \frac{1}{2}x + 200$ **d.** $227

79. a and **b.**

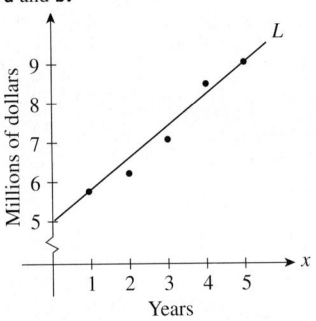

c. $y = 0.8x + 5$ **d.** $12.2 million

81. a. A family of parallel lines having slope m
b. A family of straight lines that pass through the point $(0, b)$

83. False   **85.** True   **87.** True

Using Technology Exercises 1.2, page 27

Graphing Utility

1.

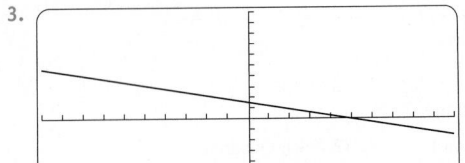

3.

5.

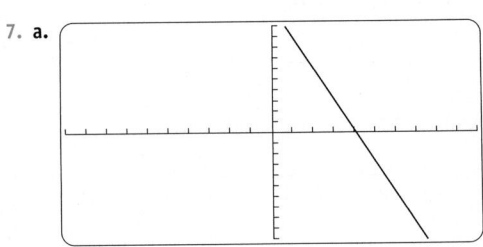

7. a.

b.

9. a.

b.

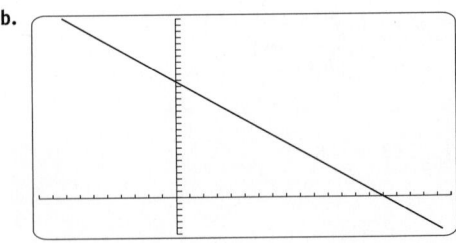

11.

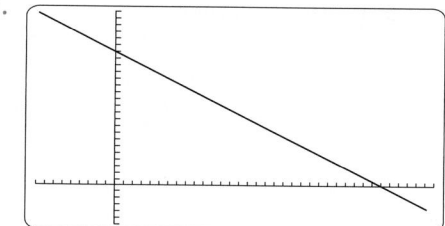

13.

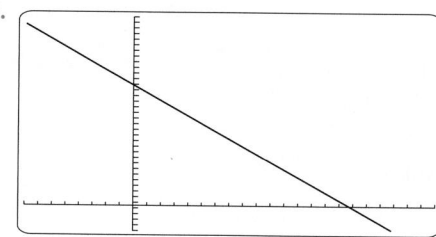

15.

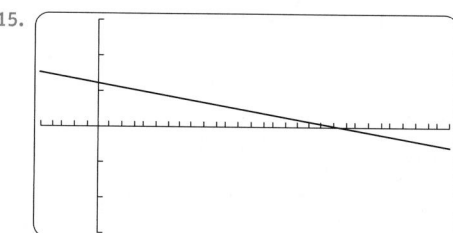

17.

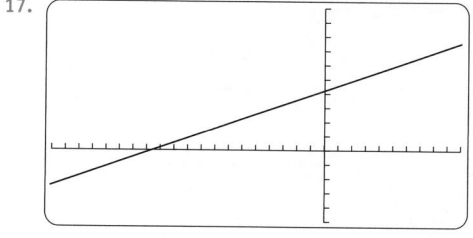

Excel

1.

$3.2x + 2.1y - 6.72 = 0$

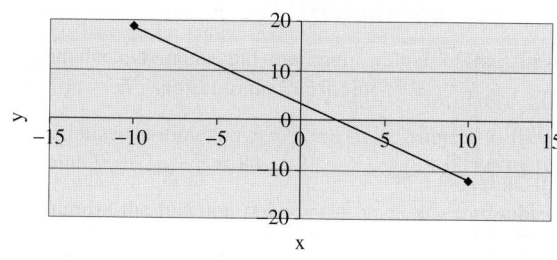

3.

$1.6x + 5.1y = 8.16$

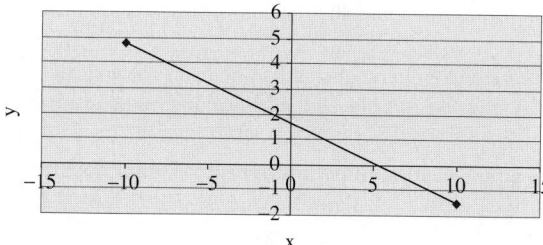

5.

$2.8x = -1.6y + 4.48$

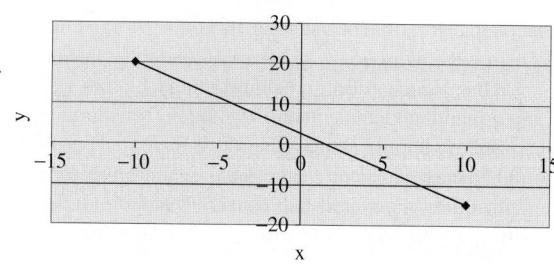

7.

$12.1x + 4.1y - 49.61 = 0$

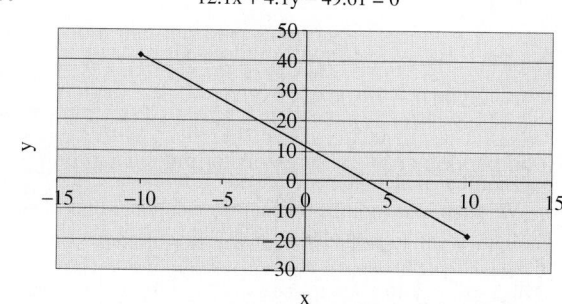

9.

$20x + 16y = 300$

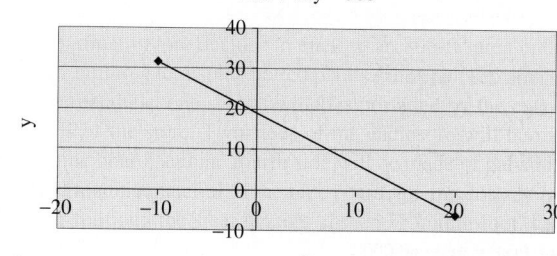

11.

$20x + 30y = 600$

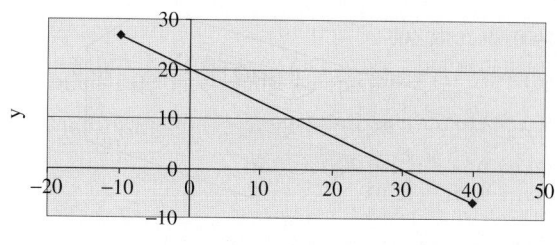

13.

$$22.4x + 16.1y - 352 = 0$$

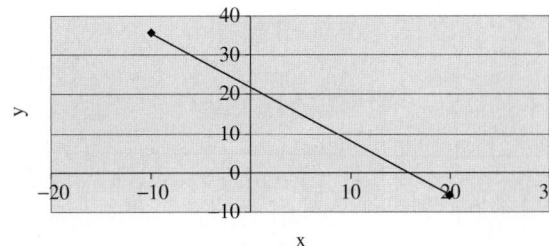

15.

$$1.2x + 20y = 24$$

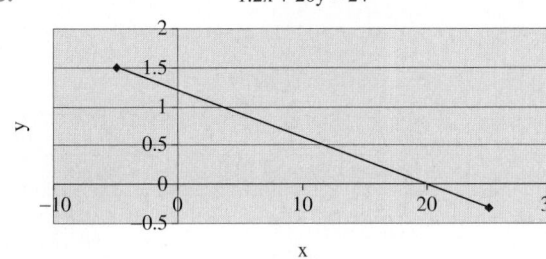

17.

$$-4x + 12y = 50$$

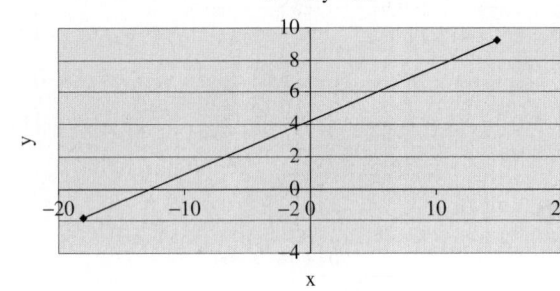

Exercises 1.3, page 35

1. Yes; $y = -\frac{2}{3}x + 2$ 3. Yes; $y = \frac{1}{2}x + 2$

5. Yes; $y = \frac{1}{2}x + \frac{9}{4}$ 7. No   9. No

11. **a.** $C(x) = 8x + 40,000$
    b. $R(x) = 12x$
    c. $P(x) = 4x - 40,000$
    d. Loss of $8000; profit of $8000

13. $m = -1; b = 2$

15. $900,000; $800,000

17. $6 billion; $43.5 billion; $81 billion

19. **a.** $y = 1.053x$ **b.** $1074.06

21. $C(x) = 0.6x + 12,100; R(x) = 1.15x;$
    $P(x) = 0.55x - 12,100$

23. **a.** $12,000/yr   **b.** $V = 60,000 - 12,000t$
    c.

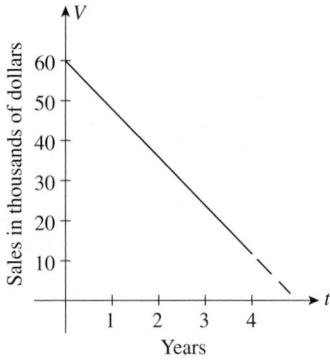

    d. $24,000

25. $900,000; $800,000

27. **a.** $m = a/1.7; b = 0$ **b.** 117.65 mg

29. $f(t) = 7.5t + 20 \ (0 \le t \le 6)$; 65 million

31. **a.** $F = \frac{9}{5}C + 32$ **b.** 68°F   **c.** 21.1°C   33. L_2

35. **a.**

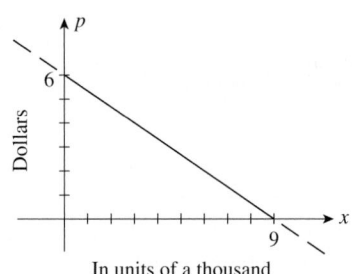

In units of a thousand

    b. 3000

37. **a.**

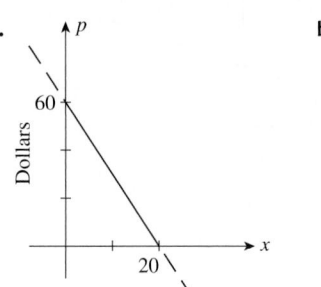

In units of a thousand

    b. 10,000

39. $p = -\frac{3}{40}x + 130$; $130; 1733   41. 2500 units

43. **a.**

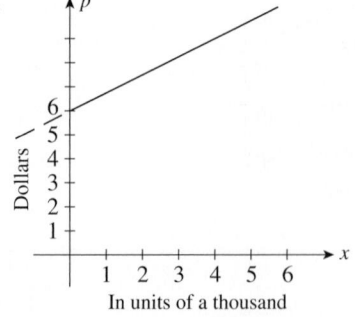

In units of a thousand

    b. 2667 units

45. a.

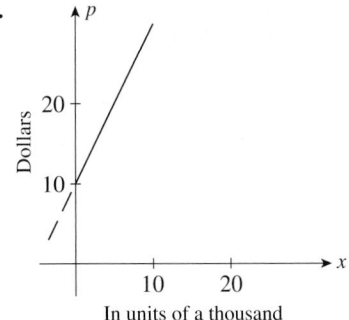

In units of a thousand

b. 2000 units

47. $p = \frac{1}{2}x + 40$ (x is measured in units of a thousand)

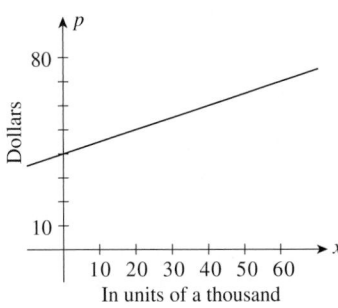

In units of a thousand

60,000 units

49. False

Using Technology Exercises 1.3, page 42

1. 2.2875   **3.** 2.880952381   **5.** 7.2851648352

7. 2.4680851064

Exercises 1.4, page 49

1. (2, 10)   **3.** $\left(4, \frac{2}{3}\right)$ **5.** (−4, −6)

7. 1000 units; $15,000

9. 600 units; $240

11. a.

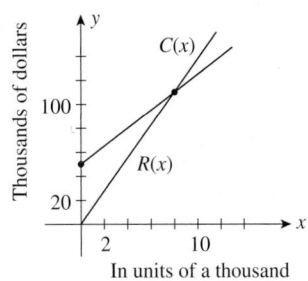

In units of a thousand

b. 8000 units; $112,000

c.

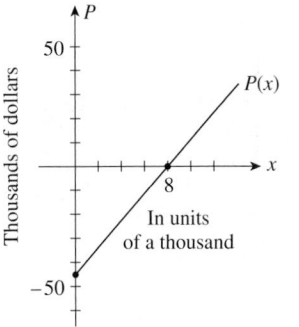

In units
of a thousand

d. (8000, 0)

13. 9259 units; $83,331

15. a. $C_1(x) = 18{,}000 + 15x$
        $C_2(x) = 15{,}000 + 20x$

b.

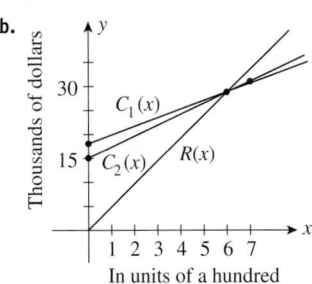

In units of a hundred

c. Machine II; machine II; machine I
d. ($1500); $1500; $4750

17. Middle of 2003

19. a.

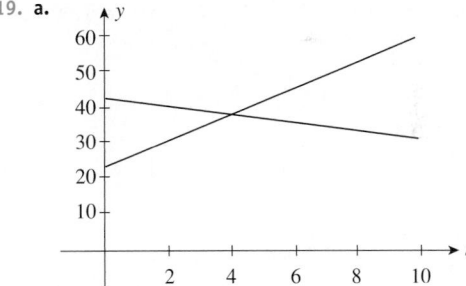

b. Feb. 2005

21. 8000 units; $9   **23.** 2000 units; $18

25. a. $p = -0.08x + 725$ **b.** $p = 0.09x + 300$
c. 2500 DVD players; $525

27. 300 fax machines; $600

29. a. $\dfrac{b - d}{c - a}; \dfrac{bc - ad}{c - a}$
b. If c is increased, x gets smaller and p gets larger.
c. If b is decreased, x decreases and p decreases.

31. True

33. a. $m_1 = m_2$ and $b_2 \neq b_1$ **b.** $m_1 \neq m_2$
    c. $m_1 = m_2$ and $b_1 = b_2$

Using Technology Exercises 1.4, page 53

1. (0.6, 6.2)   **3.** (3.8261, 0.1304)

5. (386.9091, 145.3939)

7. a.

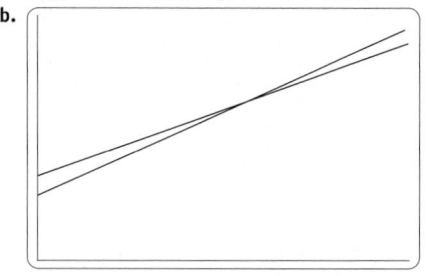

b. (3548.39, 27,996.77)
c.

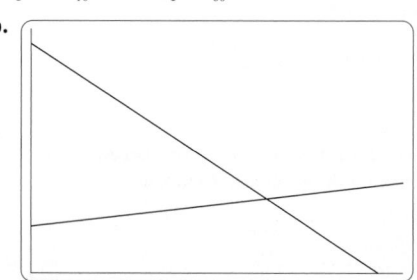

    x-intercept: 3548

9. a. $C_1(x) = 34 + 0.18x; C_2(x) = 28 + 0.22x$
b.

                  c. (150, 61)

d. If the distance driven is less than or equal to 150 mi, rent from Acme Truck Leasing; if the distance driven is more than 150 mi, rent from Ace Truck Leasing.

11. a. $p = -\frac{1}{10}x + 284; p = \frac{1}{60}x + 60$
b.

                  (1920, 92)

c. 1920/wk; $92/radio

Exercises 1.5, page 59

1. a. $y = 2.3x + 1.5$

b.

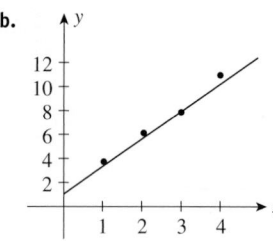

3. a. $y = -0.77x + 5.74$

b.

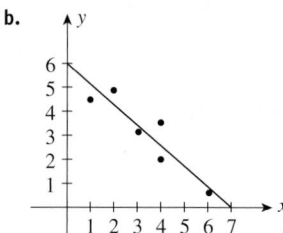

5. a. $y = 1.2x + 2$

b.

7. a. $y = 0.34x - 0.9$

b.

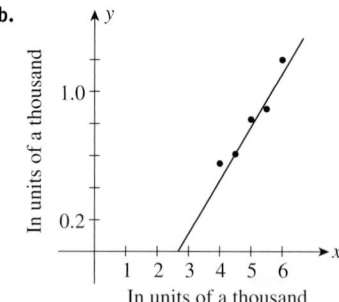

c. 1276 applications

9. a. $y = -2.8x + 440$

b.

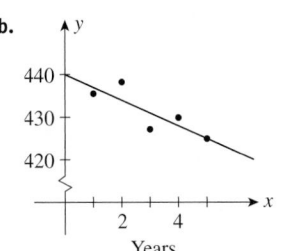

c. 420

Years

11. a. $y = 2.8x + 17.6$ **b.** \$40,000,000

13. a. $y = 22.7t + 124.2$ **b.** \$260.4 billion

15. a. $y = 12.2x + 20.9$ **b.** 106.3 million

17. a. $y = 0.059x + 19.5$ **b.** 21.9 yr   **c.** 21.3 yr

19. a. $y = 1.757x + 7.914$ **b.** \$21.97 billion

21. a. $y = 10x + 90.34$ **b.** 150,340,000

23. a. $y = 46.61x + 495.04$ **b.** \$46.61/buyer

25. a. $y = 2.43x + 78.88$ **b.** \$103,180

27. False   **29.** True

Using Technology Exercises 1.5, page 65

1. $y = 3.8639 + 2.3596x$

3. $y = 3.5525 - 1.1948x$

5. a. $y = 0.55x + 1.17$ **b.** \$5.57 billion

7. a. $y = 13.321x + 72.571$ **b.** 192 million tons

9. a. $y = 14.43x + 212.1$ **b.** 247 trillion cu ft

Chapter 1 Concept Review Questions, page 67

1. Ordered; abscissa (x-coordinate); ordinate (y-coordinate)

2. a. x-; y-;   **b.** third

3. $\sqrt{(c - a)^2 + (d - b)^2}$

4. $(x - a)^2 + (y - b)^2 = r^2$

5. a. $\dfrac{y_2 - y_1}{x_2 - x_1}$ **b.** Undefined   **c.** 0   **d.** Positive

6. $m_1 = m_2$; $m_1 = -\dfrac{1}{m_2}$

7. a. $y - y_1 = m(x - x_1)$; point-slope
b. $y = mx + b$; slope-intercept

8. a. $Ax + By + C = 0$ (A, B, not both zero)   **b.** $-\dfrac{a}{b}$

9. $mx + b$

10. a. Price; demanded; demand
b. Price; supplied; supply

11. Break-even   **12.** Demand; supply

Chapter 1 Review Exercises, page 68

1. 5   **2.** 5   **3.** 5   **4.** 2

5. $x = -2$ **6.** $y = 4$

7. $x + 10y - 38 = 0$ **8.** $y = -\frac{4}{5}x + \frac{12}{5}$

9. $5x - 2y + 18 = 0$ **10.** $y = \frac{3}{4}x + \frac{11}{2}$

11. $y = -\frac{1}{2}x - 3$ **12.** $\frac{3}{5}$; $-\frac{6}{5}$

13. $3x + 4y - 18 = 0$

14. $y = -\frac{3}{5}x + \frac{12}{5}$ **15.** $3x + 2y + 14 = 0$

16.

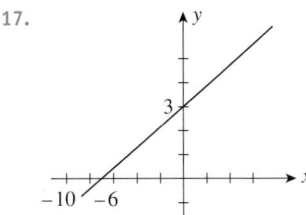

17.

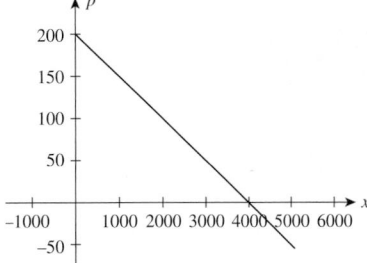

18. 60,000   **19. a.** $f(x) = x + 2.4$ **b.** \$5.4 million

21. b. ≈ 117 mg   **22. a.** \$200,000/yr   **b.** \$4,000,000

23. a. \$22,500/yr   **b.** $V = -22,500t + 300,000$

24. a. $6x + 30,000$ **b.** $10x$ **c.** $4x - 30,000$
d. (\$6,000); \$2000; \$18,000

25. $p = -0.05x + 200$

26. $p = \frac{1}{36}x + \frac{400}{9}$ **27.** $(2, -3)$ **28.** $\left(6, \frac{21}{2}\right)$

29. $(2500, 50,000)$ **30.** 6000; \$22

31. a. $y = 0.25x$ **b.** 1600   **32.** 600; \$80

Chapter 1 Before Moving On, page 69

1.
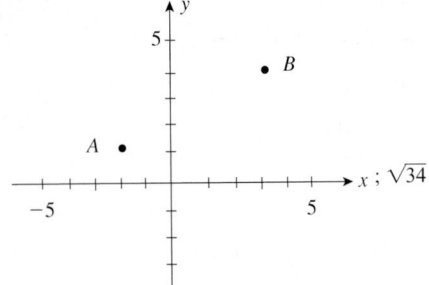

x ; $\sqrt{34}$

2. $y = 3x - 8$ 3. Yes

4. **a.** $18   **b.** $22,000   **c.** $15

5. $(1, \frac{4}{3})$ 6. After 5 yr

CHAPTER 2

Exercises 2.1, page 77

1. Unique solution; $(2, 1)$ 3. No solution

5. Unique solution; $(3, 2)$

7. Infinitely many solutions; $(t, \frac{2}{5}t - 2)$; t, a parameter

9. Unique solution; $(1, -2)$

11. No solution   13. $k = -2$

15. $\begin{aligned} x + y &= 500 \\ 42x + 30y &= 18,600 \end{aligned}$ 17. $\begin{aligned} x + y &= 100 \\ 5x + 6y &= 560 \end{aligned}$

19. $\begin{aligned} x + y &= 1000 \\ 0.5x + 1.5y &= 1300 \end{aligned}$

21. $\begin{aligned} .06x + .08y + .12z &= 21,600 \\ z &= 2x \\ .12z &= .08y \end{aligned}$

23. $\begin{aligned} 18x + 20y + 24z &= 26,400 \\ 4x + 4y + 3z &= 4,900 \\ 5x + 4y + 6z &= 6,200 \end{aligned}$

25. $\begin{aligned} 12,000x + 18,000y + 24,000z &= 1,500,000 \\ x &= 2y \\ x + y + z &= 100 \end{aligned}$

27. $\begin{aligned} 10x + 6y + 8z &= 100 \\ 10x + 12y + 6z &= 100 \\ 5x + 4y + 12z &= 100 \end{aligned}$

29. True

Exercises 2.2, page 91

1. $\begin{bmatrix} 2 & -3 & | & 7 \\ 3 & 1 & | & 4 \end{bmatrix}$

3. $\begin{bmatrix} 0 & -1 & 2 & | & 6 \\ 2 & 2 & -8 & | & 7 \\ 0 & 3 & 4 & | & 0 \end{bmatrix}$

5. $\begin{aligned} 3x + 2y &= -4 \\ x - y &= 5 \end{aligned}$ 7. $\begin{aligned} x + 3y + 2z &= 4 \\ 2x &= 5 \\ 3x - 3y + 2z &= 6 \end{aligned}$

9. Yes   11. No   13. Yes   15. No   17. No

19. $\begin{bmatrix} 1 & 2 & | & 4 \\ 0 & -5 & | & -10 \end{bmatrix}$ 21. $\begin{bmatrix} 1 & -2 & | & -3 \\ 0 & 16 & | & 20 \end{bmatrix}$

23. $\begin{bmatrix} 1 & 2 & 3 & | & 6 \\ 0 & -1 & -5 & | & -7 \\ 0 & -7 & -7 & | & -14 \end{bmatrix}$

25. $\begin{bmatrix} -6 & -11 & 0 & | & -5 \\ 2 & 4 & 1 & | & 3 \\ 1 & -2 & 0 & | & -10 \end{bmatrix}$

27. $\begin{bmatrix} 3 & 9 & | & 6 \\ 2 & 1 & | & 4 \end{bmatrix} \xrightarrow{\frac{1}{3}R_1} \begin{bmatrix} 1 & 3 & | & 2 \\ 2 & 1 & | & 4 \end{bmatrix}$

$\xrightarrow{R_2 - 2R_1} \begin{bmatrix} 1 & 3 & | & 2 \\ 0 & -5 & | & 0 \end{bmatrix} \xrightarrow{-\frac{1}{5}R_2}$

$\begin{bmatrix} 1 & 3 & | & 2 \\ 0 & 1 & | & 0 \end{bmatrix} \xrightarrow{R_1 - 3R_2} \begin{bmatrix} 1 & 0 & | & 2 \\ 0 & 1 & | & 0 \end{bmatrix}$

29. $\begin{bmatrix} 1 & 3 & 1 & | & 3 \\ 3 & 8 & 3 & | & 7 \\ 2 & -3 & 1 & | & -10 \end{bmatrix} \begin{smallmatrix} R_2 - 3R_1 \\ \xrightarrow{\hspace{1cm}} \\ R_3 - 2R_1 \end{smallmatrix}$

$\begin{bmatrix} 1 & 3 & 1 & | & 3 \\ 0 & -1 & 0 & | & -2 \\ 0 & -9 & -1 & | & -16 \end{bmatrix} \xrightarrow{-R_2}$

$\begin{bmatrix} 1 & 3 & 1 & | & 3 \\ 0 & 1 & 0 & | & 2 \\ 0 & -9 & -1 & | & -16 \end{bmatrix} \begin{smallmatrix} R_1 - 3R_2 \\ \xrightarrow{\hspace{1cm}} \\ R_3 + 9R_2 \end{smallmatrix}$

$\begin{bmatrix} 1 & 0 & 1 & | & -3 \\ 0 & 1 & 0 & | & 2 \\ 0 & 0 & -1 & | & 2 \end{bmatrix} \begin{smallmatrix} R_1 + R_3 \\ \xrightarrow{\hspace{1cm}} \\ -R_3 \end{smallmatrix}$

$\begin{bmatrix} 1 & 0 & 0 & | & -1 \\ 0 & 1 & 0 & | & 2 \\ 0 & 0 & 1 & | & -2 \end{bmatrix}$

31. $(2, 0)$ 33. $(-1, 2, -2)$

35. $(4, -2)$ 37. $(-1, 2)$

39. $(\frac{7}{9}, -\frac{1}{9}, -\frac{2}{3})$ 41. $(19, -7, -15)$

43. $(3, 0, 2)$ 45. $(1, -2, 1)$

47. $(-20, -28, 13)$ 49. $(4, -1, 3)$

51. 300 acres of corn, 200 acres of wheat

53. In 100 lb of blended coffee, use 40 lb of the $5/lb coffee and 60 lb of the $6/lb coffee.

55. 200 children and 800 adults

57. $40,000 in a savings account, $120,000 in mutual funds, $80,000 in bonds

59. 400 bags of grade-A fertilizer; 600 bags of grade-B fertilizer; 300 bags of grade-C fertilizer

61. 60 compact, 30 intermediate, and 10 full-size cars

63. 4 oz of food I, 2 oz of food II, 6 oz of food III

65. 240 front orchestra seats, 560 rear orchestra seats, 200 front balcony seats

67. 7 days in London, 4 days in Paris, and 3 days in Rome

69. False

Using Technology Exercises 2.2, page 96

1. $x_1 = 3; x_2 = 1; x_3 = -1; x_4 = 2$

3. $x_1 = 5; x_2 = 4; x_3 = -3; x_4 = -4$

5. $x_1 = 1; x_2 = -1; x_3 = 2; x_4 = 0; x_5 = 3$

Exercises 2.3, page 104

1. a. One solution   **b.** $(3, -1, 2)$

3. a. One solution   **b.** $(2, 4)$

5. a. Infinitely many solutions
b. $(4 - t, -2, t)$; t, a parameter

7. a. No solution

9. a. Infinitely many solutions
b. $(2, -1, 2 - t, t)$; t, a parameter

11. a. Infinitely many solutions
b. $(2 - 3s, 1 + s, s, t)$; s, t, parameters

13. $(2, 1)$ **15.** No solution   **17.** $(1, -1)$

19. $(2 + 2t, t)$; t, a parameter

21. $(\frac{4}{3} - \frac{2}{3}t, t)$; t, a parameter

23. $(-2 + \frac{1}{2}s - \frac{1}{2}t, s, t)$; s, t, parameters

25. $(-1, \frac{17}{7}, \frac{23}{7})$

27. $(1 - \frac{1}{4}s + \frac{1}{4}t, s, t)$; s, t, parameters

29. No solution   **31.** $(2, -1, 4)$

33. $x = 20 + z$, $y = 40 - 2z$; 25 compact cars, 30 mid-sized cars, and 5 full-sized cars; 30 compact cars, 20 mid-sized cars, and 10 full-sized cars

37. a.
$$\begin{aligned} x_1 - x_2 &= 200 \\ x_1 \qquad\quad - x_5 &= 100 \\ - x_2 + x_3 \qquad + x_6 &= 600 \\ - x_3 + x_4 \qquad &= 200 \\ x_4 - x_5 + x_6 &= 700 \end{aligned}$$
b. $x_1 = s + 100; x_2 = s - 100; x_3 = s - t + 500;$
$x_4 = s - t + 700; x_5 = s; x_6 = t$
$(250, 50, 600, 800, 150, 50)$
$(300, 100, 600, 800, 200, 100)$
c. $(300, 100, 700, 900, 200, 0)$

39. $k = 6$; $(-2, 2)$ **41.** False

Using Technology Exercises 2.3, page 107

1. $(1 + t, 2 + t, t)$; t, a parameter

3. $(-\frac{17}{7} + \frac{6}{7}t, 3 - t, -\frac{18}{7} + \frac{1}{7}t, t)$; t, a parameter

5. No solution

Exercises 2.4, page 115

1. 4×4; 4×3; 1×5; 4×1 **3.** 2; 3; 8

5. D; $D^T = \begin{bmatrix} 1 & 3 & -2 & 0 \end{bmatrix}$ **7.** 3×2; 3×2; 3×3; 3×3

9. $\begin{bmatrix} 1 & 6 \\ 6 & -1 \\ 2 & 2 \end{bmatrix}$ **11.** $\begin{bmatrix} 1 & 1 & -4 \\ -1 & -8 & 1 \\ 6 & 3 & 1 \end{bmatrix}$

13. $\begin{bmatrix} 3 & 5 & 9 \\ 4 & 10 & 13 \end{bmatrix}$ **15.** $\begin{bmatrix} 3 & -4 & -16 \\ 17 & -4 & 16 \end{bmatrix}$

17. $\begin{bmatrix} -1.9 & 3.0 & -0.6 \\ 6.0 & 9.6 & 1.2 \end{bmatrix}$

19. $\begin{bmatrix} \frac{7}{2} & 3 & -1 & \frac{10}{3} \\ -\frac{19}{6} & \frac{2}{3} & -\frac{17}{2} & \frac{23}{3} \\ \frac{29}{3} & \frac{17}{6} & -1 & -2 \end{bmatrix}$

21. $x = \frac{5}{2}$, $y = 7$, $z = 2$, and $u = 3$

23. $x = 2$, $y = 2$, $z = -\frac{7}{3}$, and $u = 15$

31. $\begin{bmatrix} 3 \\ 2 \\ -1 \\ 5 \end{bmatrix}$ **33.** $\begin{bmatrix} 1 & 3 & 0 \\ -1 & 4 & 1 \\ 2 & 2 & 0 \end{bmatrix}$

35. $\begin{bmatrix} 220 & 215 & 210 & 205 \\ 220 & 210 & 200 & 195 \\ 215 & 205 & 195 & 190 \end{bmatrix}$

37. a. $D = \begin{bmatrix} 2960 & 1510 & 1150 \\ 1100 & 550 & 490 \\ 1230 & 590 & 470 \end{bmatrix}$

b. $E = \begin{bmatrix} 3256 & 1661 & 1265 \\ 1210 & 605 & 539 \\ 1353 & 649 & 517 \end{bmatrix}$

39.
$$\begin{array}{c|ccc} & 2000 & 2001 & 2002 \\ \hline \text{Mass.} & 6.88 & 7.05 & 7.18 \\ \text{U.S.} & 4.13 & 4.09 & 4.06 \end{array}$$

41.
$$\begin{array}{c|ccc} & \text{White} & \text{Black} & \text{Hispanic} \\ \hline \text{Women} & 81 & 76.1 & 82.2 \\ \text{Men} & 76 & 69.9 & 75.9 \end{array}$$

$$\begin{array}{c|cc} & \text{Women} & \text{Men} \\ \hline \text{White} & 81 & 76 \\ \text{Black} & 76.1 & 69.9 \\ \text{Hispanic} & 82.2 & 75.9 \end{array}$$

43. True   **45.** False

Using Technology Exercises 2.4, page 121

1. $\begin{bmatrix} 15 & 38.75 & -67.5 & 33.75 \\ 51.25 & 40 & 52.5 & -38.75 \\ 21.25 & 35 & -65 & 105 \end{bmatrix}$

3. $\begin{bmatrix} -5 & 6.3 & -6.8 & 3.9 \\ 1 & 0.5 & 5.4 & -4.8 \\ 0.5 & 4.2 & -3.5 & 5.6 \end{bmatrix}$

5. $\begin{bmatrix} 16.44 & -3.65 & -3.66 & 0.63 \\ 12.77 & 10.64 & 2.58 & 0.05 \\ 5.09 & 0.28 & -10.84 & 17.64 \end{bmatrix}$

7. $\begin{bmatrix} 22.2 & -0.3 & -12 & 4.5 \\ 21.6 & 17.7 & 9 & -4.2 \\ 8.7 & 4.2 & -20.7 & 33.6 \end{bmatrix}$

Exercises 2.5, page 129

1. 2×5; not defined   3. 1×1; 7×7

5. $n = s$; $m = t$ 7. $\begin{bmatrix} -1 \\ 3 \end{bmatrix}$ 9. $\begin{bmatrix} 9 \\ -10 \end{bmatrix}$

11. $\begin{bmatrix} 4 & -2 \\ 9 & 13 \end{bmatrix}$ 13. $\begin{bmatrix} 2 & 9 \\ 5 & 16 \end{bmatrix}$ 15. $\begin{bmatrix} 0.57 & 1.93 \\ 0.64 & 1.76 \end{bmatrix}$

17. $\begin{bmatrix} 6 & -3 & 0 \\ -2 & 1 & -8 \\ 4 & -4 & 9 \end{bmatrix}$ 19. $\begin{bmatrix} 5 & 1 & -3 \\ 1 & 7 & -3 \end{bmatrix}$

21. $\begin{bmatrix} -4 & -20 & 4 \\ 4 & 12 & 0 \\ 12 & 32 & 20 \end{bmatrix}$ 23. $\begin{bmatrix} 4 & -3 & 2 \\ 7 & 1 & -5 \end{bmatrix}$

27. $AB = \begin{bmatrix} 10 & 7 \\ 22 & 15 \end{bmatrix}$; $BA = \begin{bmatrix} 5 & 8 \\ 13 & 20 \end{bmatrix}$

31. $A = \begin{bmatrix} -2 & -1 \\ 5 & 2 \end{bmatrix}$ 33. **a.** $A^T = \begin{bmatrix} 2 & 5 \\ 4 & -6 \end{bmatrix}$

35. $AX = B$, where $A = \begin{bmatrix} 2 & -3 \\ 3 & -4 \end{bmatrix}$, $X = \begin{bmatrix} x \\ y \end{bmatrix}$,

and $B = \begin{bmatrix} 7 \\ 8 \end{bmatrix}$

37. $AX = B$, where $A = \begin{bmatrix} 2 & -3 & 4 \\ 0 & 2 & -3 \\ 1 & -1 & 2 \end{bmatrix}$, $X = \begin{bmatrix} x \\ y \\ z \end{bmatrix}$,

and $B = \begin{bmatrix} 6 \\ 7 \\ 4 \end{bmatrix}$

39. $AX = B$, where $A = \begin{bmatrix} -1 & 1 & 1 \\ 2 & -1 & -1 \\ -3 & 2 & 4 \end{bmatrix}$, $X = \begin{bmatrix} x_1 \\ x_2 \\ x_3 \end{bmatrix}$,

and $B = \begin{bmatrix} 0 \\ 2 \\ 4 \end{bmatrix}$

41. **a.** $AB = \begin{bmatrix} 51{,}400 \\ 54{,}200 \end{bmatrix}$

b. The first entry shows that William's total stockholdings are $51,400; the second shows that Michael's stockholdings are $54,200.

43. **a.** $B = \begin{bmatrix} 20{,}000 \\ 22{,}000 \\ 25{,}000 \\ 30{,}000 \end{bmatrix}$

b. $7,160,000 in New York, $2,860,000 in Connecticut, and $1,430,000 in Massachusetts; the total profit is $11,450,000.

45.
$$\begin{array}{ccc} & \text{D} & \text{R} \quad \text{I} \end{array}$$
$$BA = \begin{bmatrix} 41{,}000 & 35{,}000 & 14{,}000 \end{bmatrix}$$

47. $AB = \begin{bmatrix} 1575 & 1590 & 1560 & 975 \\ 410 & 405 & 415 & 270 \\ 215 & 205 & 225 & 155 \end{bmatrix}$

49. $[277.60]$; it represents Cindy's long-distance bill for phone calls to London, Tokyo, and Hong Kong.

51. **a.** $\begin{bmatrix} 8800 \\ 3380 \\ 1020 \end{bmatrix}$ **b.** $\begin{bmatrix} 8800 \\ 3380 \\ 1020 \end{bmatrix}$ **c.** $\begin{bmatrix} 17{,}600 \\ 6{,}760 \\ 2{,}040 \end{bmatrix}$

53. False   55. True

Using Technology Exercises 2.5, page 136

1. $\begin{bmatrix} 18.66 & 15.2 & -12 \\ 24.48 & 41.88 & 89.82 \\ 15.39 & 7.16 & -1.25 \end{bmatrix}$

3. $\begin{bmatrix} 20.09 & 20.61 & -1.3 \\ 44.42 & 71.6 & 64.89 \\ 20.97 & 7.17 & -60.65 \end{bmatrix}$

5. $\begin{bmatrix} 32.89 & 13.63 & -57.17 \\ -12.85 & -8.37 & 256.92 \\ 13.48 & 14.29 & 181.64 \end{bmatrix}$

7. $\begin{bmatrix} 128.59 & 123.08 & -32.50 \\ 246.73 & 403.12 & 481.52 \\ 125.06 & 47.01 & -264.81 \end{bmatrix}$

9. $\begin{bmatrix} 87 & 68 & 110 & 82 \\ 119 & 176 & 221 & 143 \\ 51 & 128 & 142 & 94 \\ 28 & 174 & 174 & 112 \end{bmatrix}$

$\begin{bmatrix} 113 & 117 & 72 & 101 & 90 \\ 72 & 85 & 36 & 72 & 76 \\ 81 & 69 & 76 & 87 & 30 \\ 133 & 157 & 56 & 121 & 146 \\ 154 & 157 & 94 & 127 & 122 \end{bmatrix}$

11. $\begin{bmatrix} 170 & 18.1 & 133.1 & -106.3 & 341.3 \\ 349 & 226.5 & 324.1 & 164 & 506.4 \\ 245.2 & 157.7 & 231.5 & 125.5 & 312.9 \\ 310 & 245.2 & 291 & 274.3 & 354.2 \end{bmatrix}$

Exercises 2.6, page 145

5. $\begin{bmatrix} 3 & -5 \\ -1 & 2 \end{bmatrix}$ 7. Does not exist

9. $\begin{bmatrix} 2 & -11 & -3 \\ 1 & -6 & -2 \\ 0 & -1 & 0 \end{bmatrix}$ 11. Does not exist

13. $\begin{bmatrix} -\frac{13}{10} & \frac{7}{5} & \frac{1}{2} \\ \frac{2}{5} & -\frac{1}{5} & 0 \\ -\frac{7}{10} & \frac{3}{5} & \frac{1}{2} \end{bmatrix}$

15. $\begin{bmatrix} 3 & 4 & -6 & 1 \\ -2 & -3 & 5 & -1 \\ -4 & -4 & 7 & -1 \\ -4 & -5 & 8 & -1 \end{bmatrix}$

17. **a.** $A = \begin{bmatrix} 2 & 5 \\ 1 & 3 \end{bmatrix}; X = \begin{bmatrix} x \\ y \end{bmatrix}; B = \begin{bmatrix} 3 \\ 2 \end{bmatrix}$

   b. $x = -1; y = 1$

19. **a.** $A = \begin{bmatrix} 2 & -3 & -4 \\ 0 & 0 & -1 \\ 1 & -2 & 1 \end{bmatrix}; X = \begin{bmatrix} x \\ y \\ z \end{bmatrix}; B = \begin{bmatrix} 4 \\ 3 \\ -8 \end{bmatrix}$

   b. $x = -1; y = 2; z = -3$

21. **a.** $A = \begin{bmatrix} 1 & 4 & -1 \\ 2 & 3 & -2 \\ -1 & 2 & 3 \end{bmatrix}; X = \begin{bmatrix} x \\ y \\ z \end{bmatrix}; B = \begin{bmatrix} 3 \\ 1 \\ 7 \end{bmatrix}$

   b. $x = 1; y = 1; z = 2$

23. **a.** $A = \begin{bmatrix} 1 & 1 & -1 & 1 \\ 2 & 1 & 1 & 0 \\ 2 & 1 & 0 & 1 \\ 2 & -1 & -1 & 3 \end{bmatrix}; X = \begin{bmatrix} x_1 \\ x_2 \\ x_3 \\ x_4 \end{bmatrix}; B = \begin{bmatrix} 6 \\ 4 \\ 7 \\ 9 \end{bmatrix}$

   b. $x_1 = 1; x_2 = 2; x_3 = 0; x_4 = 3$

25. **b.** **(i)** $x = \frac{24}{5}; y = \frac{23}{5}$ **(ii)** $x = \frac{2}{5}; y = \frac{9}{5}$

27. **b.** **(i)** $x = -1; y = 3; z = 2$
    (ii) $x = 1; y = 8; z = -12$

29. **b.** **(i)** $x = -\frac{2}{17}; y = -\frac{10}{17}; z = -\frac{60}{17}$
    (ii) $x = 1; y = 0; z = -5$

31. **b.** **(i)** $x_1 = 1; x_2 = -4; x_3 = 5; x_4 = -1$
    (ii) $x_1 = 12; x_2 = -24; x_3 = 21; x_4 = -7$

33. **a.** $A^{-1} = \begin{bmatrix} -\frac{5}{2} & -\frac{3}{2} \\ 2 & 1 \end{bmatrix}$

35. **a.** $ABC = \begin{bmatrix} 4 & 10 \\ 2 & 3 \end{bmatrix}; A^{-1} = \begin{bmatrix} 3 & -5 \\ 1 & -2 \end{bmatrix};$

   $B^{-1} = \begin{bmatrix} 1 & -3 \\ -1 & 4 \end{bmatrix}; C^{-1} = \begin{bmatrix} \frac{1}{8} & -\frac{3}{8} \\ \frac{1}{4} & \frac{1}{4} \end{bmatrix}$

37. **a.** 3214; 3929   **b.** 4286; 3571   **c.** 3929; 5357

39. **a.** 400 acres of soybeans; 300 acres of corn; 300 acres of wheat
    b. 500 acres of soybeans; 400 acres of corn; 300 acres of wheat

41. **a.** $80,000 in high-risk stocks; $20,000 in medium-risk stocks; $100,000 in low-risk stocks
    b. $88,000 in high-risk stocks; $22,000 in medium-risk stocks; $110,000 in low-risk stocks
    c. $56,000 in high-risk stocks; $64,000 in medium-risk stocks, $120,000 in low-risk stocks

43. All values of k except $k = \frac{3}{2}$; $\dfrac{1}{3 - 2k}\begin{bmatrix} 3 & -2 \\ -k & 1 \end{bmatrix}$

45. True   47. True

Using Technology Exercises 2.6, page 152

1. $\begin{bmatrix} 0.36 & 0.04 & -0.36 \\ 0.06 & 0.05 & 0.20 \\ -0.19 & 0.10 & 0.09 \end{bmatrix}$

3. $\begin{bmatrix} 0.01 & -0.09 & 0.31 & -0.11 \\ -0.25 & 0.58 & -0.15 & -0.02 \\ 0.86 & -0.42 & 0.07 & -0.37 \\ -0.27 & 0.01 & -0.05 & 0.31 \end{bmatrix}$

5. $\begin{bmatrix} 0.30 & 0.85 & -0.10 & -0.77 & -0.11 \\ -0.21 & 0.10 & 0.01 & -0.26 & 0.21 \\ 0.03 & -0.16 & 0.12 & -0.01 & 0.03 \\ -0.14 & -0.46 & 0.13 & 0.71 & -0.05 \\ 0.10 & -0.05 & -0.10 & -0.03 & 0.11 \end{bmatrix}$

7. $x = 1.2; y = 3.6; z = 2.7$

9. $x_1 = 2.50; x_2 = -0.88; x_3 = 0.70; x_4 = 0.51$

Exercises 2.7, page 158

1. **a.** $10 million   **b.** $160 million
   c. Agricultural; manufacturing and transportation

3. $x = 23.75$ and $y = 21.25$

5. $x = 42.85$ and $y = 57.14$

9. **a.** $318.2 million worth of agricultural goods and $336.4 million worth of manufactured products
   b. $198.2 million worth of agricultural products and $196.4 million worth of manufactured goods

11. **a.** $443.75 million, $381.25 million, and $281.25 million worth of agricultural products, manufactured goods, and transportation, respectively
    b. $243.75 million, $281.25 million, and $221.25 million worth of agricultural products, manufactured goods, and transportation, respectively

13. $45 million and $75 million

15. $34.4 million, $33 million, and $21.6 million

Using Technology Exercises 2.7, page 162

1. The final outputs of the first, second, third, and fourth industries are 602.62, 502.30, 572.57, and 523.46 units, respectively.

3. The final outputs of the first, second, third, and fourth industries are 143.06, 132.98, 188.59, and 125.53 units, respectively.

Chapter 2 Concept Review Questions, page 163

1. **a.** One; many; no   **b.** One; many; no   2. Equations

3. $R_i \longleftrightarrow R_j$; cR_i; $R_i + aR_j$; solution

4. **a.** Unique   **b.** No; infinitely many; unique

5. Size; entries   6. Size; corresponding

7. $m \times n$; $n \times m$; a_{ji} 8. cA; c

9. **a.** Columns; rows   **b.** $m \times p$

10. **a.** $A(BC)$; $AB + AC$ **b.** $n \times r$

11. $A^{-1}A$; AA^{-1}; singular   12. $A^{-1}B$

Chapter 2 Review Exercises, page 164

1. $\begin{bmatrix} 2 & 2 \\ -1 & 4 \\ 3 & 3 \end{bmatrix}$ 2. $\begin{bmatrix} -2 & 0 \\ -2 & 6 \end{bmatrix}$ 3. $[-6 \quad -2]$ 4. $\begin{bmatrix} 17 \\ 13 \end{bmatrix}$

5. $x = 2$; $y = 3$; $z = 1$; $w = 3$ 6. $x = 2$; $y = -2$

7. $a = 3$; $b = 4$; $c = -2$; $d = 2$; $e = -3$

8. $x = -1$; $y = -2$; $z = 1$

9. $\begin{bmatrix} 8 & 9 & 11 \\ -10 & -1 & 3 \\ 11 & 12 & 10 \end{bmatrix}$ 10. $\begin{bmatrix} -1 & 7 & -3 \\ -2 & 5 & 11 \\ 10 & -8 & 2 \end{bmatrix}$

11. $\begin{bmatrix} 6 & 18 & 6 \\ -12 & 6 & 18 \\ 24 & 0 & 12 \end{bmatrix}$ 12. $\begin{bmatrix} -10 & 10 & -18 \\ 4 & 14 & 26 \\ 16 & -32 & -4 \end{bmatrix}$

13. $\begin{bmatrix} -11 & -16 & -15 \\ -4 & -2 & -10 \\ -6 & 14 & 2 \end{bmatrix}$ 14. $\begin{bmatrix} 5 & 20 & 19 \\ -2 & 20 & 8 \\ 26 & 10 & 30 \end{bmatrix}$

15. $\begin{bmatrix} -3 & 17 & 8 \\ -2 & 56 & 27 \\ 74 & 78 & 116 \end{bmatrix}$ 16. $\begin{bmatrix} \frac{3}{2} & -2 & -5 \\ \frac{11}{2} & -1 & 11 \\ \frac{7}{2} & -3 & 0 \end{bmatrix}$

17. $x = 1$; $y = -1$ 18. $x = -1$; $y = 3$

19. $x = 1$; $y = 2$; $z = 3$

20. $(2, 2t - 5, t)$; t, a parameter   21. No solution

22. $x = 1$; $y = -1$; $z = 2$; $w = 2$

23. $x = 1$; $y = 0$; $z = 1$ 24. $x = 2$; $y = -1$; $z = 3$

25. $\begin{bmatrix} \frac{2}{5} & -\frac{1}{5} \\ -\frac{1}{5} & \frac{3}{5} \end{bmatrix}$ 26. $\begin{bmatrix} \frac{3}{4} & -\frac{1}{2} \\ -\frac{1}{8} & \frac{1}{4} \end{bmatrix}$

27. $\begin{bmatrix} -1 & 2 \\ 1 & -\frac{3}{2} \end{bmatrix}$ 28. $\begin{bmatrix} \frac{1}{4} & \frac{1}{2} \\ \frac{1}{8} & -\frac{1}{4} \end{bmatrix}$

29. $\begin{bmatrix} \frac{5}{4} & \frac{1}{4} & -\frac{7}{4} \\ -\frac{1}{4} & -\frac{1}{4} & \frac{3}{4} \\ -\frac{3}{4} & \frac{1}{4} & \frac{5}{4} \end{bmatrix}$ 30. $\begin{bmatrix} -\frac{1}{4} & \frac{1}{2} & -\frac{1}{4} \\ \frac{7}{8} & -\frac{3}{4} & -\frac{5}{8} \\ -\frac{1}{8} & \frac{1}{4} & \frac{3}{8} \end{bmatrix}$

31. $\begin{bmatrix} -\frac{1}{5} & \frac{2}{5} & 0 \\ \frac{2}{3} & -\frac{1}{3} & \frac{1}{3} \\ -\frac{1}{30} & \frac{1}{15} & -\frac{1}{6} \end{bmatrix}$ 32. $\begin{bmatrix} 0 & -\frac{1}{5} & \frac{2}{5} \\ -2 & 1 & 1 \\ -1 & \frac{1}{5} & \frac{3}{5} \end{bmatrix}$

33. $\begin{bmatrix} \frac{3}{2} & 1 \\ -\frac{7}{2} & -1 \end{bmatrix}$ 34. $\begin{bmatrix} \frac{11}{24} & -\frac{7}{8} \\ -\frac{1}{12} & \frac{1}{4} \end{bmatrix}$

35. $\begin{bmatrix} \frac{2}{5} & -\frac{3}{5} \\ \frac{1}{5} & \frac{1}{5} \end{bmatrix}$ 36. $\begin{bmatrix} \frac{4}{7} & -\frac{3}{7} \\ -\frac{3}{7} & \frac{4}{7} \end{bmatrix}$

37. $A^{-1} = \begin{bmatrix} \frac{2}{7} & \frac{3}{7} \\ \frac{1}{7} & -\frac{2}{7} \end{bmatrix}$; $x = -1$; $y = -2$

38. $A^{-1} = \begin{bmatrix} \frac{2}{5} & \frac{3}{10} \\ -\frac{1}{5} & \frac{1}{10} \end{bmatrix}$; $x = 2$; $y = 1$

39. $A^{-1} = \begin{bmatrix} 1 & -\frac{2}{5} & \frac{4}{5} \\ -1 & 1 & -1 \\ -\frac{1}{2} & \frac{3}{5} & -\frac{7}{10} \end{bmatrix}$; $x = 1$; $y = 2$; $z = 4$

40. $A^{-1} = \begin{bmatrix} 0 & \frac{1}{7} & \frac{2}{7} \\ -1 & -\frac{4}{7} & \frac{6}{7} \\ -\frac{1}{2} & -\frac{1}{2} & \frac{1}{2} \end{bmatrix}$; $x = 3$; $y = -1$; $z = 2$

41. $7430, $6900, and $9100

42. $2,300,000; $2,450,000; an increase of $150,000

43. 30 of each type

44. Houston: 100,000 gallons; Tulsa: 600,000 gallons

Chapter 2 Before Moving On, page 166

1. $\left(\frac{2}{3}, -\frac{2}{3}, \frac{5}{3}\right)$

2. **a.** $(2, -3, 1)$ **b.** No solution   **c.** $(2, 1 - 3t, t)$
   d. $(0, 0, 0, 0)$ **e.** $(2 + t, 3 - 2t, t)$

3. **a.** $(-1, 2)$ **b.** $\left(\frac{4}{7}, -\frac{5}{7} + 2t, t\right)$

4. **a.** $\begin{bmatrix} 3 & 1 & 4 \\ 5 & -2 & 6 \end{bmatrix}$ **b.** $\begin{bmatrix} 14 & 3 & 7 \\ 14 & 5 & 1 \end{bmatrix}$ **c.** $\begin{bmatrix} 0 & 5 & 3 \\ 4 & -1 & -11 \end{bmatrix}$

5. $\begin{bmatrix} 3 & -2 & -5 \\ -3 & 2 & 6 \\ -1 & 1 & 2 \end{bmatrix}$ 6. $(1, -1, 2)$

CHAPTER 3

Exercises 3.1, page 173

1.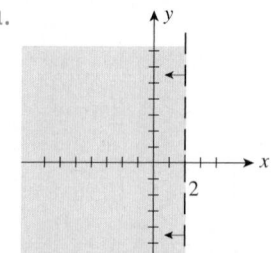

3.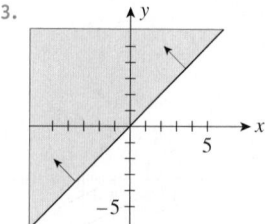

5.

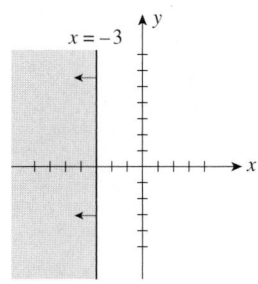

7.

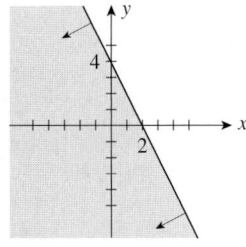

9.

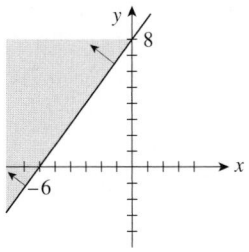

11. $x \geq 1$; $x \leq 5$; $y \geq 2$; $y \leq 4$

13. $2x - y \geq 2$; $5x + 7y \geq 35$; $x \leq 4$

15. $x - y \geq -10$; $7x + 4y \leq 140$; $x + 3y \geq 30$

17. $x + y \geq 7$; $x \geq 2$; $y \geq 3$; $y \leq 7$

19.

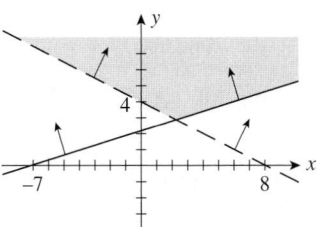

Unbounded

21.

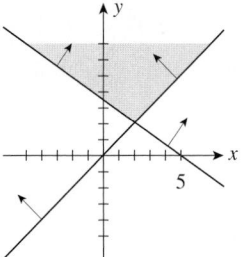

Unbounded

23.

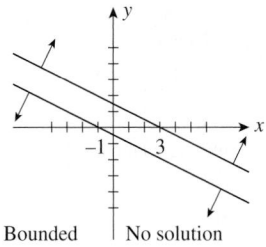

Bounded   No solution

25.

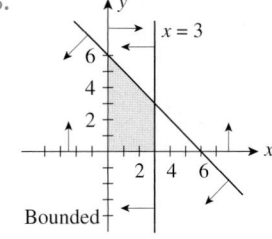

Bounded

27.

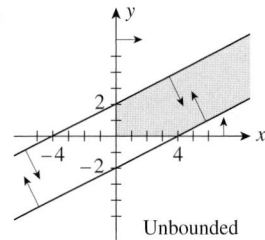

Unbounded

29.

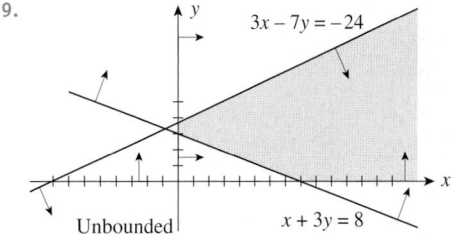

Unbounded

31.

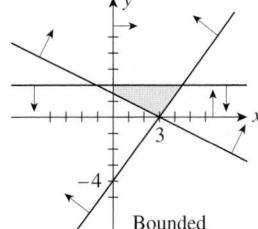

Bounded

33.

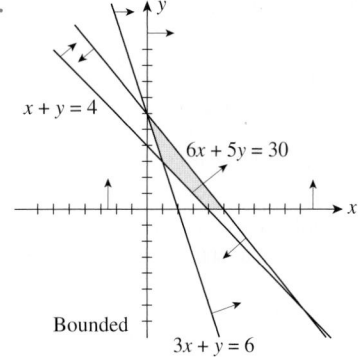

Bounded

35.

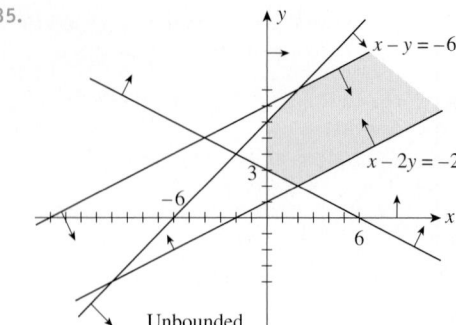

$x - y = -6$

$x - 2y = -2$

Unbounded

37. False   39. True

Exercises 3.2, page 181

1. Maximize $P = 3x + 4y$
   subject to $6x + 9y \leq 300$
   $5x + 4y \leq 180$
   $x \geq 0, y \geq 0$

3. Maximize $P = 2x + 1.5y$
   subject to $3x + 4y \leq 1000$
   $6x + 3y \leq 1200$
   $x \geq 0, y \geq 0$

5. Maximize $P = 0.1x + 0.12y$
   subject to $x + y \leq 20$
   $x - 4y \geq 0$
   $x \geq 0, y \geq 0$

7. Maximize $P = 50x + 40y$
   subject to $\frac{1}{200}x + \frac{1}{200}y \leq 1$
   $\frac{1}{100}x + \frac{1}{300}y \leq 1$
   $x \geq 0, y \geq 0$

9. Minimize $C = 14{,}000x + 16{,}000y$
   subject to $50x + 75y \geq 650$
   $3000x + 1000y \geq 18{,}000$
   $x \geq 0, y \geq 0$

11. Minimize $C = 2x + 5y$
    subject to $30x + 25y \geq 400$
    $x + 0.5y \geq 10$
    $2x + 5y \geq 40$
    $x \geq 0, y \geq 0$

13. Minimize $C = 1000x + 800y$
    subject to $70{,}000x + 10{,}000y \geq 2{,}000{,}000$
    $40{,}000x + 20{,}000y \geq 1{,}400{,}000$
    $20{,}000x + 40{,}000y \geq 1{,}000{,}000$
    $x \geq 0, y \geq 0$

15. Maximize $P = 0.1x + 0.15y + 0.2z$
    subject to $x + y + z \leq 2{,}000{,}000$
    $-2x - 2y + 8z \leq 0$
    $-6x + 4y + 4z \leq 0$
    $-10x + 6y + 6z \leq 0$
    $x \geq 0, y \geq 0, z \geq 0$

17. Maximize $P = 18x + 12y + 15z$
    subject to $2x + y + 2z \leq 900$
    $3x + y + 2z \leq 1080$
    $2x + 2y + z \leq 840$
    $x \geq 0, y \geq 0, z \geq 0$

19. Maximize $P = 26x + 28y + 24z$
    subject to $\frac{5}{4}x + \frac{3}{2}y + \frac{3}{2}z \leq 310$
    $x + y + \frac{3}{4}z \leq 205$
    $x + y + \frac{1}{2}z \leq 190$
    $x \geq 0, y \geq 0, z \geq 0$

21. Minimize $C = 60x_1 + 60x_2 + 80x_3 + 80x_4 + 70x_5 + 50x_6$
    subject to $x_1 + x_2 + x_3 \leq 300$
    $x_4 + x_5 + x_6 \leq 250$
    $x_1 + x_4 \geq 200$
    $x_2 + x_5 \geq 150$
    $x_3 + x_6 \geq 200$
    $x_1 \geq 0, x_2 \geq 0, \ldots, x_6 \geq 0$

23. Maximize $P = x + 0.8y + 0.9z$
    subject to $8x + 4z \leq 16{,}000$
    $8x + 12y + 8z \leq 24{,}000$
    $4y + 4z \leq 5000$
    $z \leq 800$
    $x \geq 0, y \geq 0, z \geq 0$

25. False

Exercises 3.3, page 192

1. Max: 35; min: 5   3. No max. value; min: 27

5. Max: 44; min: 15   7. $x = 0$; $y = 6$; $P = 18$

9. Any point (x, y) lying on the line segment joining $(\frac{5}{2}, 0)$ and $(1, 3)$; $P = 5$

11. $x = 0$; $y = 8$; $P = 64$

13. $x = 0$; $y = 4$; $P = 12$

15. $x = 2$; $y = 1$; $C = 10$

17. Any point (x, y) lying on the line segment joining $(20, 10)$ and $(40, 0)$; $C = 120$

19. $x = 14$; $y = 3$; $C = 58$

21. $x = 3$; $y = 3$; $C = 75$

23. $x = 15$; $y = 17.5$; $P = 115$

25. $x = 10$; $y = 38$; $P = 134$

27. Max: $x = 6$; $y = \frac{33}{2}$; $P = 258$
    Min: $x = 15$; $y = 3$; $P = 186$

29. 20 product A, 20 product B; $140

31. 120 model A, 160 model B; $480

33. $16 million in homeowner loans, $4 million in auto loans; $2.08 million

35. 50 fully assembled units, 150 kits; $8500

37. Saddle Mine: 4 days; Horseshoe Mine: 6 days; $152,000

39. Infinitely many solutions; 10 oz of food A and 4 oz of food B or 20 oz of food A and 0 oz of food B, etc., with a minimum value of 40 mg of cholesterol

41. 30 in newspaper I, 10 in newspaper II; $38,000

43. 80 from I to A, 20 from I to B, 0 from II to A, 50 from II to B

45. a. $22,500 in growth stocks and $7500 in speculative stocks; maximum return; $5250

47. 750 urban, 750 suburban; $10,950

49. False   **51. a.** True   **b.** True

55. a.

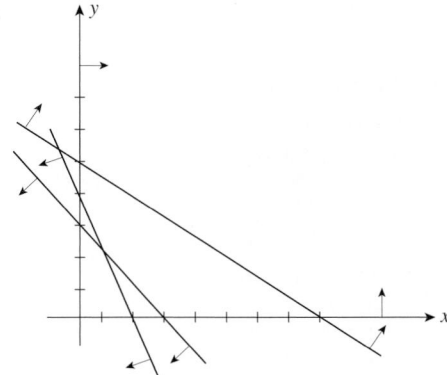

b. No solution

Exercises 3.4, page 209

1. c. $216   **3. a.** Between 750 and 1425
b. It cannot be decreased by more than 60 units.

5. a. $x = 3$; $y = 2$; $P = 17$ **b.** $\frac{8}{3} \le c \le 8$
c. $8 \le b \le 24$ **d.** $\frac{5}{4}$
e. Both constraints are binding.

7. a. $x = 4$; $y = 0$; $C = 8$ **b.** $0 \le c \le \frac{5}{2}$
c. $b \ge 3$ **d.** 2
e. Constraint 1 is binding; constraint 2 is nonbinding.

9. a. $x = 3$; $y = 5$; $P = 27$ **b.** $2 \le c \le 5$
c. $21 \le b \le 33$ **d.** $\frac{2}{3}$
e. The first two constraints are binding; the third is nonbinding.

11. a. 20 units of each product   **b.** $\frac{8}{3} \le c \le 5$
c. $216 \le b \le 405$ **d.** $\frac{8}{21}$

13. a. Operate Saddle Mine for 4 days, Horseshoe Mine for 6 days; minimum cost of $152,000
b. $10,666\frac{2}{3} \le c \le 48,000$ **c.** $300 \le b \le 1350$
d. $194.29

15. a. Produce 60 of each; maximum profit of $1320
b. $8\frac{1}{3} \le c \le 15$ **c.** $1100 \le b \le 1633\frac{1}{3}$
d. $0.55
e. Constraints 1 and 2 are binding; constraint 3 is not.

17. a. 120 model A and 160 model B grates; maximum profit of $480
b. $1.125 \le c \le 3$ **c.** $600 \le b \le 1100$
d. $0.20
e. Constraints 1 and 2 are binding; constraint 3 is nonbinding.

Chapter 3 Concept Review Questions, page 212

1. a. Half plane; line   **b.** $ax + by \le c$; $ax + by = c$

2. a. Points; each   **b.** Bounded; enclosed

3. Objective function; maximized; minimized; linear; inequalities

4. a. Corner point   **b.** Line

5. Parameters; optimal

6. Resource; amount; value; improved; increased

Chapter 3 Review Exercises, page 213

1. Max: 18—any point (x, y) lying on the line segment joining $(0, 6)$ to $(3, 4)$; Min: 0

2. Max: 27; Min: 7   **3.** $x = 0$; $y = 4$; $P = 20$

4. $x = 0$; $y = 12$; $P = 36$

5. $x = 3$; $y = 4$; $C = 26$ **6.** $x = 1.25$; $y = 1.5$; $C = 9.75$

7. $x = 3$; $y = 10$; $P = 29$ **8.** $x = 8$; $y = 0$; $P = 48$

9. $x = 20$; $y = 0$; $C = 40$ **10.** $x = 2$; $y = 6$; and $C = 14$

11. Max: $x = 22$; $y = 0$; $Q = 22$; Min: $x = 3$; $y = \frac{5}{2}$; $Q = \frac{11}{2}$

12. Max: $x = 12$; $y = 6$; $Q = 54$; Min: $x = 4$; $y = 0$; $Q = 8$

13. $40,000 in each company; $P = $13,600$

14. 60 model A clock radios; 60 model B clock radios; $P = 1320

15. 93 model A, 180 model B; $P = 456

16. 600 to Warehouse I and 400 to Warehouse II

Chapter 3 Before Moving On, page 214

1. a.

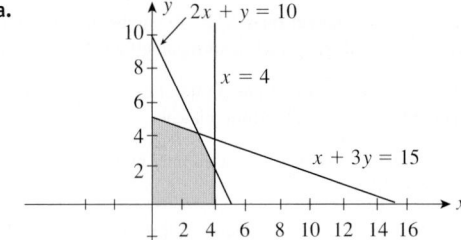

b.

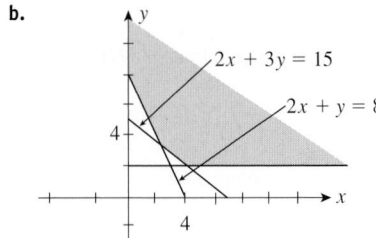

2. Min: $x = 3$, $y = 16$; $C = -7$ **3.** Max; $x = 0$, $y = \frac{24}{7}$; $P = \frac{72}{7}$
  Max: $x = 28$, $y = 8$; $P = 76$

4. Min: $x = 0$, $y = 10$; $C = 10$

5. a. Max: $x = 4$, $y = 6$; $P = 26$ **b.** Between 1.5 and 4.5
  c. Between 8 and 24 **d.** $1.25
  e. Constraints 1 and 2 are binding.

CHAPTER 4

Exercises 4.1, page 233

1. In final form; $x = \frac{30}{7}$, $y = \frac{20}{7}$, $u = 0$, $v = 0$, and $P = \frac{220}{7}$

3. Not in final form; pivot element is $\frac{1}{2}$, lying in the first row, second column.

5. In final form; $x = \frac{1}{3}$, $y = 0$, $z = \frac{13}{3}$; $u = 0$, $v = 6$, $w = 0$, and $P = 17$

7. Not in final form; pivot element is 1, lying in the third row, second column.

9. In final form; $x = 30$, $y = 10$, $z = 0$, $u = 0$, $v = 0$, $P = 60$ and $x = 30$, $y = 0$, $z = 0$, $u = 10$, $v = 0$, $P = 60$

11. $x = 0$, $y = 4$, $u = 0$, $v = 1$, and $P = 16$

13. $x = 6$, $y = 3$, $u = 0$, $v = 0$, and $P = 96$

15. $x = 6$, $y = 6$, $u = 0$, $v = 0$, $w = 0$, and $P = 60$

17. $x = 0$, $y = 4$, $z = 4$, $u = 0$, $v = 0$, and $P = 36$

19. $x = 0$, $y = 3$, $z = 0$, $u = 90$, $v = 0$, $w = 75$, and $P = 12$

21. $x = 15$, $y = 3$, $z = 0$, $u = 2$, $v = 0$, $w = 0$, and $P = 78$

23. $x = \frac{5}{4}$, $y = \frac{15}{2}$, $z = 0$, $u = 0$, $v = \frac{15}{4}$, $w = 0$, and $P = 90$

25. $x = 2$, $y = 1$, $z = 1$, $u = 0$, $v = 0$, $w = 0$, and $P = 87$

29. No model A, 2500 model B; $100,000

31. 65 acres of crop A, 80 acres of crop B; $25,750

33. $62,500 in the money market fund, $125,000 in the international equity fund, $62,500 in the growth-and-income fund; $25,625

35. 22 min of morning advertising time, 3 min of evening advertising time; maximum exposure: 6.2 million viewers

37. 80 units of model A, 80 units of model B, and 60 units of model C; maximum profit: $5760; no

39. 9000 bottles of formula I, 7833 bottles of formula II, 6000 bottles of formula III; maximum profit: $4986.67; Yes, ingredients for 4167 bottles of formula II

41. Project A: $800,000, project B: $800,000, and project C: $400,000; $280,000

43. False **45.** True

Using Technology Exercises 4.1, page 242

1. $x = 1.2$, $y = 0$, $z = 1.6$, $w = 0$, and $P = 8.8$

3. $x = 1.6$, $y = 0$, $z = 0$, $w = 3.6$, and $P = 12.4$

Exercises 4.2, page 252

1. $x = 4$, $y = 0$, and $C = -8$

3. $x = 4$, $y = 3$, and $C = -18$

5. $x = 0$, $y = 13$, $z = 18$, $w = 14$, and $C = -111$

7. $x = \frac{5}{4}$, $y = \frac{1}{4}$, $u = 2$, $v = 3$, and $C = P = 13$

9. $x = 5$, $y = 10$, $z = 0$, $u = 1$, $v = 2$, and $C = P = 80$

11. Maximize $P = 4u + 6v$
  subject to $u + 3v \le 2$
  $2u + 2v \le 5$; $x = 4$, $y = 0$, $C = 8$
  $u \ge 0$, $v \ge 0$

13. Maximize $P = 60u + 40v + 30w$
  subject to $6u + 2v + w \le 6$
  $u + v + w \le 4$; $x = 10$, $y = 20$, $C = 140$
  $u \ge 0$, $v \ge 0$, $w \ge 0$

15. Maximize $P = 10u + 20v$
  subject to $20u + v \le 200$
  $10u + v \le 150$; $x = 0$, $y = 0$, $z = 10$, $C = 1200$
  $u + 2v \le 120$
  $u \ge 0$, $v \ge 0$

17. Maximize $P = 10u + 24v + 16w$
  subject to: $u + 2v + w \le 6$
  $2u + v + w \le 8$; $x = 8$, $y = 0$, $z = 8$, $C = 80$
  $2u + v + w \le 4$
  $u \ge 0$, $v \ge 0$, $w \ge 0$

19. Maximize $P = 6u + 2v + 4w$
  subject to: $2u + 6v \le 30$
  $4u + 6w \le 12$; $x = \frac{1}{3}$, $y = \frac{4}{3}$, $z = 0$, $C = 26$
  $3u + v + 2w \le 20$
  $u \ge 0$, $v \ge 0$, $w \ge 0$

21. 2 type-A vessels; 3 type-B vessels; $250,000

23. 30 in newspaper I; 10 in newspaper II; $38,000

25. 8 oz of orange juice; 6 oz of pink grapefruit juice; 178 calories

27. True

Using Technology Exercises 4.2, page 259

1. $x = \frac{4}{3}$, $y = \frac{10}{3}$, $z = 0$, and $C = \frac{14}{3}$

3. $x = 0.9524$; $y = 4.2857$; $z = 0$; $C = 6.0952$

5. a. $x = 3$, $y = 2$, and $P = 17$ **b.** $\frac{8}{3} \le c_1 \le 8$; $\frac{3}{2} \le c_2 \le \frac{9}{2}$
  c. $8 \le b_1 \le 24$; $4 \le b_2 \le 12$ **d.** $\frac{5}{4}$; $\frac{1}{4}$
  e. Both constraints are binding.

7. a. $x = 4$, $y = 0$, and $C = 8$ **b.** $0 \le c_1 \le \frac{5}{2}$; $4 \le c_2 < \infty$
  c. $3 \le b_1 < \infty$; $-\infty < b_2 \le 4$ **d.** 2; 0
  e. Both constraints are binding.

Exercises 4.3, page 270

1. Maximize $P = -C = -2x + 3y$
  subject to $-3x - 5y \le -20$
  $3x + y \le 16$
  $-2x + y \le 1$
  $x \ge 0$, $y \ge 0$

3. Maximize $P = -C = -5x - 10y - z$
subject to $-2x - y - z \leq -4$
$-x - 2y - 2z \leq -2$
$2x + 4y + 3z \leq 12$
$x \geq 0, y \geq 0, z \geq 0$

5. $x = 5, y = 2$, and $P = 9$

7. $x = 4, y = 0$, and $C = -8$

9. $x = 4, y = \frac{2}{3}$, and $P = \frac{20}{3}$

11. $x = 3, y = 2$, and $P = 7$

13. $x = 24, y = 0, z = 0$, and $P = 120$

15. $x = 0, y = 17, z = 1$, and $C = -33$

17. $x = \frac{46}{7}, y = 0, z = \frac{50}{7}$, and $P = \frac{142}{7}$

19. $x = 0, y = 0, z = 10$, and $P = 30$

21. 80 acres of crop A, 68 acres of crop B; $P = \$25,600$

23. $50 million worth of home loans, $10 million worth of commercial-development loans; maximum return: $4.6 million

25. 0 units of product A, 280 units of product B, 280 units of product C; $P = \$7560$

27. 10 oz of food A, 4 oz of food B, 40 mg of cholesterol; infinitely many solutions

Chapter 4 Concept Review Questions, page 274

1. Maximized; nonnegative; less than; equal to

2. Equations; slack variables; $-c_1x_1 - c_2x_2 - \cdots - c_nx_n + P = 0$; below; augmented

3. Minimized; nonnegative; greater than; equal to

4. Dual; objective; optimal value

Chapter 4 Review Exercises, page 274

1. $x = 3, y = 4, u = 0, v = 0$, and $P = 25$

2. $x = 3, y = 6, u = 4, v = 0, w = 0$, and $P = 36$

3. $x = \frac{56}{5}, y = \frac{2}{5}, z = 0, u = 0, v = 0$, and $P = 23\frac{3}{5}$

4. $x = 0, y = \frac{11}{3}, z = \frac{25}{6}, u = \frac{37}{6}, v = 0, w = 0$ and $P = \frac{119}{6}$

5. $x = \frac{3}{2}, y = 1, u = \frac{1}{4}, v = \frac{5}{4}$, and $C = \frac{13}{2}$

6. $x = \frac{32}{11}, y = \frac{36}{11}, u = \frac{2}{11}, v = 0$, and $C = \frac{104}{11}$

7. $x = \frac{3}{4}, y = 0, z = \frac{7}{4}, u = 6, v = 6$, and $C = 60$

8. $x = 0, y = 2, z = 0, u = 1, v = 0, w = 0$, and $C = 4$

9. $x = 45, y = 0, u = 0, v = 35$, and $P = 135$

10. $x = 4, y = 4, u = 2, v = 0, w = 0$, and $C = 20$

11. $x = 5, y = 2, u = 0, v = 0$, and $P = 16$

12. $x = 20, y = 25, z = 0, u = 0, v = 30, w = 10$, and $C = -160$

13. Saddle Mine: 4 days; Horseshoe Mine: 6 days; $152,000

14. $70,000 in blue-chip stocks; $0 in growth stocks; $30,000 in speculative stocks; maximum return: $13,000

15. 0 unit product A, 30 units product B, 0 unit product C; $P = \$180$

16. $50,000 in stocks, $100,000 in bonds, $50,000 in money market funds; $P = \$21,500$

Chapter 4 Before Moving On, page 275

1.

x	y	z	u	v	w	P	Constant
2	①	-1	1	0	0	0	3
1	-2	3	0	1	0	0	1
3	2	4	0	0	1	0	17
-1	-2	3	0	0	0	1	0

2. $x = 2; y = 0; z = 11; u = 2; v = 0; w = 0; P = 28$

3. Max: $x = 6, y = 2; P = 34$ 4. Min: $x = 3, y = 0; C = 3$

5. Max: $x = 10, y = 0; P = 20$

CHAPTER 5

Exercises 5.1, page 289

1. $80; $580   3. $836   5. $1000   7. 146 days

9. 10%/yr   11. $1718.19   13. $4974.47

15. $27,566.93   17. $261,751.04   19. $214,986.69

21. $10\frac{1}{4}$%/yr   23. 8.3%/yr   25. $29,277.61

27. $30,255.95   29. $6885.64   31. 6.08%/yr

33. 2.06 yr   35. 24%/yr   37. $123,600

39. 5%/yr   41. $705.28   43. $255,256

45. $2.58 million   47. $22,163.75   49. $26,267.49

51. a. $34,626.88   b. $33,886.16   c. $33,506.76

53. Acme Mutual Fund   55. $23,329.48   57. 4.2%

59. 5.75%   61. 5.12%/yr   63. $115.3 billion

65. Investment A   67. $33,885.14; $33,565.38

69. $80,000e^{(\sqrt{t}/2 - 0.09t)}$; $151,718   73. True   75. True

Using Technology Exercises 5.1, page 296

1. $5872.78   3. $475.49   5. 8.95%/yr

7. 10.20%/yr   9. $29,743.30   11. $53,303.25

Exercises 5.2, page 304

1. $15,937.42   3. $54,759.35   5. $37,965.57

7. $137,209.97   9. $28,733.19   11. $15,558.61

13. $15,011.29   15. $109,658.91   17. $455.70

19. $44,526.45   21. Karen   23. $9850.12   25. $608.54

27. Between $383,242 and $469,053

29. Between $347,014 and $413,768　　31. $17,887.62

33. False

Using Technology Exercises 5.2, page 308

1. $59,622.15　　3. $8453.59　　5. $35,607.23

7. $13,828.60

Exercises 5.3, page 315

1. $14,902.95　　3. $444.24　　5. $622.13

7. $731.79　　9. $1491.19　　11. $516.76

13. $172.95　　15. $1957.36　　17. $3450.87

19. $16,274.54　　21. **a.** $212.27　　**b.** $1316.36; $438.79

23. **a.** $387.21; $304.35　　**b.** $1939.56; $2608.80

25. $1761.03; $41,833; $59,461; $124,853　　27. $60,982.31

29. $3135.48　　31. $242.23　　33. $199.07

35. $2090.41; $4280.21　　37. $33,835.20

39. $167,341.33　　41. $152,018.20

43. **a.** $1681.71　　**b.** $194,282.67　　**c.** $1260.11　　**d.** $421.60

45. $71,799

Using Technology Exercises 5.3, page 321

1. $628.02　　3. $1685.47　　5. $2234.40

7. $1436.41　　9. $18,288.92

Exercises 5.4, page 328

1. 30　　3. −4.5　　5. −3, 8, 19, 30, 41　　7. $x + 6y$

9. 795　　11. 792　　13. 550

15. **a.** 275　　**b.** −280　　17. 37 wk　　19. $7.90

21. **b.** $800　　23. GP; 256; 508　　25. Not a GP

27. GP; 1/3; $364\frac{1}{3}$　　29. 3; 0　　31. 293,866

33. $50,150.20　　35. Annual raise of 8%/yr

37. **a.** $20,113.57　　**b.** $87,537.38　　39. $25,165.82

41. $39,321.60; $110,678.40　　43. True

Chapter 5 Concept Review Questions, page 331

1. **a.** Original; $P(1 + rt)$　　**b.** Interest; $P(1 + i)^n$; $A(1 + i)^{-n}$

2. Simple; one; nominal; m; $\left(1 + \dfrac{r}{m}\right)^m - 1$

3. Annuity; ordinary annuity; simple annuity.

4. **a.** $R\left[\dfrac{(1 + i)^n - 1}{i}\right]$　　**b.** $R\left[\dfrac{1 - (1 + i)^{-n}}{i}\right]$

5. $\dfrac{Pi}{1 - (1 + i)^{-n}}$　　6. Future; $\dfrac{iS}{(1 + i)^n - 1}$

7. Constant d; $a + (n - 1)d$; $\dfrac{n}{2}[2a + (n - 1)d]$

8. Constant r; ar^{n-1}; $\dfrac{a(1 - r^n)}{1 - r}$

Chapter 5 Review Exercises, page 332

1. **a.** $7320.50　　**b.** $7387.28　　**c.** $7422.53
　　d. $7446.77

2. **a.** $19,859.95　　**b.** $20,018.07　　**c.** $20,100.14
　　d. $20,156.03

3. **a.** 12%　　**b.** 12.36%　　**c.** 12.5509%　　**d.** 12.6825%

4. **a.** 11.5%　　**b.** 11.8306%　　**c.** 12.0055%　　**d.** 12.1259%

5. $30,000.29　　6. $39,999.95　　7. $5557.68

8. $23,221.71　　9. $7861.70　　10. $173,804.43

11. $694.49　　12. $318.93　　13. $332.73　　14. $208.44

15. 7.179%　　16. 9.563%　　17. 12%/yr　　18. $80,000

19. $2,592,702; $8,612,002　　20. $5,491,922　　21. $2982.73

22. $15,000　　23. $5000　　24. 7.6%　　25. $218.64

26. $73,178.41　　27. $13,026.89　　28. $2000

29. **a.** $965.55　　**b.** $227,598　　**c.** $42,684

30. **a.** $1217.12　　**b.** $99,081.60　　**c.** $91,367

31. $19,573.56　　32. $4727.67　　33. $205.09; 20.27%/yr

34. $2203.83

Chapter 5 Before Moving On, page 334

1. $2540.47　　2. 6.2%/yr　　3. $569,565.47　　4. $1213.28

5. $35.13　　6. **a.** 210　　**b.** 127.5

CHAPTER 6

Exercises 6.1, page 342

1. $\{x \mid x$ is a gold medalist in the 2006 Winter Olympic Games$\}$

3. $\{x \mid x$ is an integer greater than 2 and less than 8$\}$

5. $\{2, 3, 4, 5, 6\}$　　7. $\{-2\}$

9. **a.** True　　**b.** False　　11. **a.** False　　**b.** False

13. True　　15. **a.** True　　**b.** False

17. **a.** and **b.**

19. **a.** $\varnothing, \{1\}, \{2\}, \{1, 2\}$
　　b. $\varnothing, \{1\}, \{2\}, \{3\}, \{1, 2\}, \{1, 3\}, \{2, 3\}, \{1, 2, 3\}$
　　c. $\varnothing, \{1\}, \{2\}, \{3\}, \{4\}, \{1, 2\}, \{1, 3\}, \{1, 4\}, \{2, 3\}, \{2, 4\}, \{3, 4\},$
　　　$\{1, 2, 3\}, \{1, 2, 4\}, \{1, 3, 4\}, \{2, 3, 4\}, \{1, 2, 3, 4\}$

21. $\{1, 2, 3, 4, 6, 8, 10\}$

23. $\{$Jill, John, Jack, Susan, Sharon$\}$

25. a.

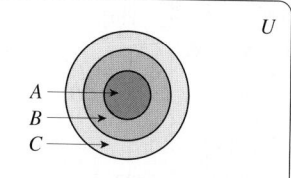

b.

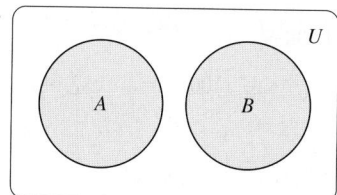

c.

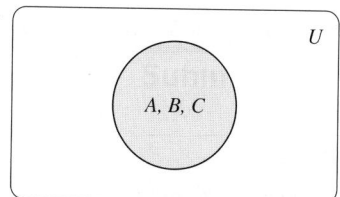

27. a.

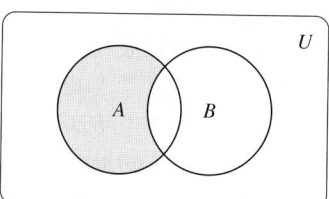

b.

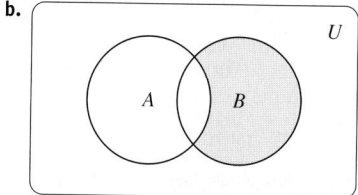

29. a.

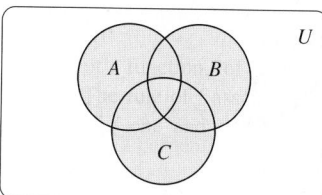

b.

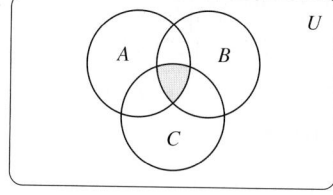

31. a.

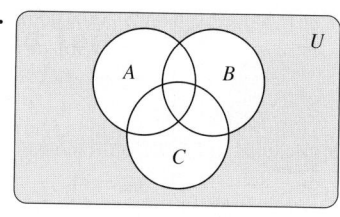

b.

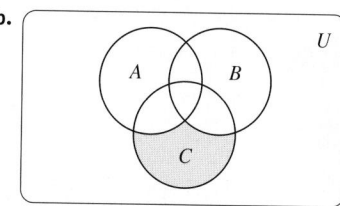

33. a. $\{2, 4, 6, 8, 10\}$ **b.** $\{1, 2, 4, 5, 6, 8, 9, 10\}$
c. U

35. a. $C = \{1, 2, 4, 5, 8, 9\}$ **b.** $\varnothing$ **c.** U

37. a. Not disjoint   **b.** Disjoint

39. a. The set of all employees at Universal Life Insurance who do not drink tea
b. The set of all employees at Universal Life Insurance who do not drink coffee

41. a. The set of all employees at Universal Life Insurance who drink tea but not coffee
b. The set of all employees at Universal Life Insurance who drink coffee but not tea

43. a. The set of all employees in a hospital who are not doctors
b. The set of all employees in a hospital who are not nurses

45. a. The set of all employees in a hospital who are female doctors
b. The set of all employees in a hospital who are both doctors and administrators

47. a. $D \cap F$ **b.** $R \cap F^c \cap L^c$

49. a. B^c **b.** $A \cap B$ **c.** $A \cap B \cap C^c$

51. a. $A \cap B \cap C$; the set of tourists who have taken the underground, a cab, and a bus over a 1-wk period in London
b. $A \cap C$; the set of tourists who have taken the underground and a bus over a 1-wk period in London
c. B^c; the set of tourists who have not taken a cab over a 1-wk period in London

53. a.

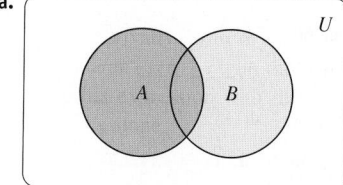

b.

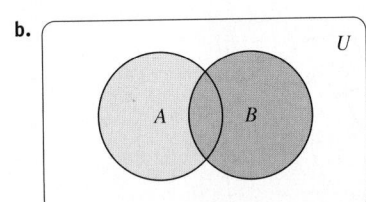

55.

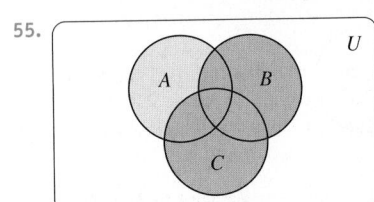

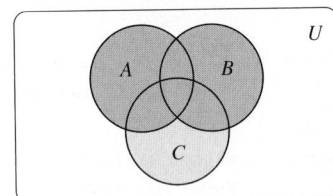

57.

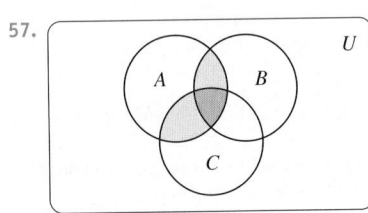

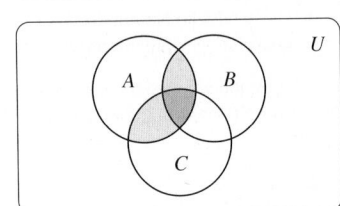

61. **a.** x, y, v, r, w, u **b.** v, r

63. **a.** s, t, y **b.** t, z, w, x, s 65. $A \subset C$

67. False   69. True   71. True

Exercises 6.2, page 349

3. **a.** 4   **b.** 5   **c.** 7   **d.** 2

7. 20

9. **a.** 140   **b.** 100   **c.** 60

11. 13   13. 0   15. 13   17. 61

19. **a.** 106   **b.** 64   **c.** 38   **d.** 14

21. **a.** 182   **b.** 118   **c.** 56   **d.** 18   23. 30

25. **a.** 64   **b.** 10   27. **a.** 36   **b.** 36   29. 5

31. **a.** 62   **b.** 33   **c.** 25   **d.** 38

33. **a.** 108   **b.** 15   **c.** 45   **d.** 12

35. **a.** 22   **b.** 80

37. True   39. True

Exercises 6.3, page 357

1. 12   3. 64   5. 24

7. 24   9. 60   11. 1 billion   13. 5^{50}

15. 30   17. 9990

19. **a.** 17,576,000   **b.** 17,576,000

21. 1024; 59,049   23. 2730

25. 217   27. True

Exercises 6.4, page 369

1. 360   3. 10   5. 120

7. 20   9. n 11. 1

13. 35   15. 1   17. 84

19. $\dfrac{n(n-1)}{2}$ 21. $\dfrac{n!}{2}$

23. Permutation   25. Combination

27. Permutation   29. Combination

31. $P(4, 4) = 24$ 33. $P(4, 4) = 24$

35. $P(9, 9) = 362,880$ 37. $C(12, 3) = 220$

39. 151,200   41. $C(12, 3) = 220$

43. $C(100, 3) = 161,700$ 45. $P(6, 6) = 720$

47. $P(12, 6) = 665,280$

49. **a.** $P(10, 10) = 3,628,800$
    b. $P(3, 3)P(4, 4)P(3, 3)P(3, 3) = 5184$

51. **a.** $P(20, 20) = 20!$
    b. $P(5, 5)P[(4, 4)]^5 = 5!(4!)^5 = 955,514,880$

53. $P(2, 1)P(3, 1) = 6$

55. $C(3,3)[C(8, 6) + C(8, 7) + C(8, 8)] = 37$

57. **a.** $C(12, 3) = 220$ **b.** $C(11, 2)] = 55$
    c. $C(5, 1)C(7, 2) + C(5, 2)C(7, 1) + C(5, 3) = 185$

59. $P(7, 3) + C(7, 2)P(3, 2) = 336$

61. $[C(5, 1)C(3, 1)C(6, 2)][C(4, 1) + C(3, 1)] = 1575$

63. $C(10, 8) + C(10, 9) + C(10, 10) = 56$

65. $10C(4, 1) = 40$

67. $4C(13, 5) - 40 = 5108$

69. $13C(4, 3)12C(4, 2) = 3744$

71. $C(6, 2) = 15$

73. $C(12, 6) + C(12, 7) + C(12, 8) + C(12, 9) +$
$C(12, 10) + C(12, 11) + C(12, 12) = 2510$

75. $4! = 24$ **79.** True   **81.** True

Using Technology Exercises 6.4, page 374

1. $1.307674368 \times 10^{12}$ **3.** $2.56094948229 \times 10^{16}$

5. 674,274,182,400   **7.** 133,784,560

9. 4,656,960

11. 658,337,004,000

Chapter 6 Concept Review Questions, page 375

1. Set; elements; set   **2.** Equal   **3.** Subset

4. a. No   **b.** All   **5.** Union; intersection

6. Complement   **7.** $A^C \cap B^C \cap C^C$

8. Permutation; combination

Chapter 6 Review Exercises, page 375

1. {3}   **2.** {A, E, H, L, S, T}

3. {4, 6, 8, 10}   **4.** {−4}   **5.** Yes

6. Yes   **7.** Yes   **8.** No

9.

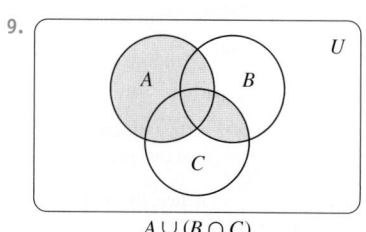

$A \cup (B \cap C)$

10.

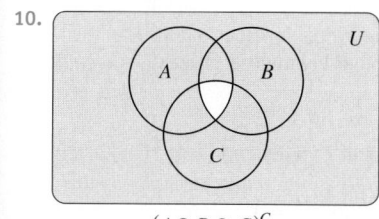

$(A \cap B \cap C)^C$

11.

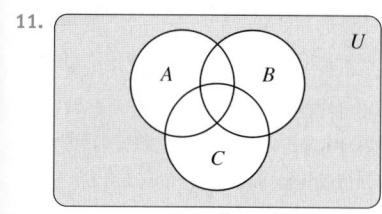

$A^C \cap B^C \cap C^C$

12.

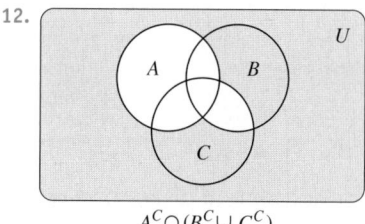

$A^C \cap (B^C \cup C^C)$

17. The set of all participants in a consumer-behavior survey who both avoided buying a product because it is not recyclable and boycotted a company's products because of its record on the environment.

18. The set of all participants in a consumer-behavior survey who avoided buying a product because it is not recyclable and/or voluntarily recycled their garbage.

19. The set of all participants in a consumer-behavior survey who both did not use cloth diapers rather than disposable diapers and voluntarily recycled their garbage.

20. The set of all participants in a consumer-behavior survey who did not boycott a company's products because of the company's record on the environment and/or who do not voluntarily recycle their garbage.

21. 150   **22.** 230   **23.** 270   **24.** 30   **25.** 70   **26.** 200

27. 190   **28.** 181,440   **29.** 120   **30.** 8400   **31.** None

32. a. 446   **b.** 377   **c.** 34   **33.** 720   **34.** 20

35. a. 50,400   **b.** 5040   **36. a.** 60   **b.** 125

37. 80   **38. a.** 1287   **b.** 288

39. 720   **40.** 1050   **41. a.** 5040   **b.** 3600

42. a. 487,635   **b.** 550   **c.** 341,055

43. a. $C(15, 4) = 1365$ **b.** $C(15, 4) - C(10, 4) = 1155$

Chapter 6 Before Moving On, page 377

1. a. {d, f, g}   **b.** {b, c, d, e, f, g}   **c.** {b, c, e}

2. 3   **3.** 360   **4.** 264   **5.** 200

CHAPTER 7

Exercises 7.1, page 385

1. {a, b, d, f}; {a}   **3.** {b, c, e}; {a}   **5.** No   **7.** S

9. ∅   **11.** Yes   **13.** Yes   **15.** $E \cup F$ **17.** G^c

19. $(E \cup F \cup G)^c$

21. a. {(2, 1), (3, 1), (4, 1), (5, 1), (6, 1), (3, 2), (4, 2), (5, 2), (6, 2), (4, 3), (5, 3), (6, 3), (5, 4), (6, 4), (6, 5)}
b. {(1, 2), (2, 4), (3, 6)}

23. ∅, {a}, {b}, {c}, {a, b}, {a, c}, {b, c}, S

25. a. $S = \{B, R\}$ **b.** ∅, {B}, {R}, S

27. **a.** $S = \{(H, 1), (H, 2), (H, 3), (H, 4), (H, 5), (H, 6), (T, 1),$
    $(T, 2), (T, 3), (T, 4), (T, 5), (T, 6)\}$
    b. $\{(H, 2), (H, 4), (H, 6)\}$

29. **a.** No   **b.** No

31. $S = \{ddd, ddn, dnd, ndd, dnn, ndn, nnd, nnn\}$

33. **a.** $\{ABC, ABD, ABE, ACD, ACE, ADE, BCD, BCE, BDE, CDE\}$
    b. 6   **c.** 3   **d.** 6

35. **a.** E^c **b.** $E^c \cap F^c$ **c.** $E \cup F$
    d. $(E \cap F^c) \cup (E^c \cap F)$

37. **a.** $\{t \mid t > 0\}$ **b.** $\{t \mid 0 < t \le 2\}$ **c.** $\{t \mid t > 2\}$

39. **a.** $S = \{0, 1, 2, 3, \ldots, 10\}$ **b.** $E = \{0, 1, 2, 3\}$
    c. $F = \{5, 6, 7, 8, 9, 10\}$

41. **a.** $S = \{0, 1, 2, \ldots, 20\}$
    b. $E = \{0, 1, 2, \ldots, 9\}$ **c.** $F = \{20\}$

47. False

Exercises 7.2, page 393

1. $\{(H, H)\}, \{(H, T)\}, \{(T, H)\}, \{(T, T)\}$

3. $\{(D, m)\}, \{(D, f)\}, \{(R, m)\}, \{(R, f)\}, \{(I, m)\}, \{(I, f)\}$

5. $\{(1, i)\}, \{(1, d)\}, \{(1, s)\}, \{(2, i)\}, \{(2, d)\}, \{(2, s)\}, \ldots,$
   $\{(5, i)\}, \{(5, d)\}, \{(5, s)\}$

7. $\{(A, Rh^+)\}, \{(A, Rh^-)\}, \{(B, Rh^+)\}, \{(B, Rh^-)\},$
   $\{(AB, Rh^+)\}, \{(AB, Rh^-)\}, \{(O, Rh^+)\}, \{(O, Rh^-)\}$

9.

Grade	A	B	C	D	F
Probability	.10	.25	.45	.15	.05

11. **a.** $S = \{(0 < x \le 200), (200 < x \le 400),$
    $(400 < x \le 600), (600 < x \le 800),$
    $(800 < x \le 1000), (x > 1000)\}$

b.

Cars, x	Probability
$0 < x \le 200$	.075
$200 < x \le 400$	.1
$400 < x \le 600$	.175
$600 < x \le 800$	.35
$800 < x \le 1000$	.225
$x > 1000$	.075

13.

Opinion	Favor	Oppose	Don't know
Probability	.47	.46	.07

15.

Event	A	B	C	D	E
Probability of an Event	.026	.199	.570	.193	.012

17.

Figures Produced (in dozens)	30	31	32
Probability	.125	0	.1875

Figures Produced (in dozens)	33	34	35	36
Probability	.25	.1875	.125	.125

19. .469

21.

Income, \$	0–24,999	25,000–49,999	50,000–74,999	75,000–99,999
Probability	.287	.293	.195	.102

Income, \$	100,000–124,999	125,000–149,999	150,000–199,999	200,000 or more
Probability	.052	.025	.022	.024

23. **a.** .856   **b.** .144   25. .46

27. **a.** $\frac{1}{4}$ **b.** $\frac{1}{2}$ **c.** $\frac{1}{13}$ 29. $\frac{3}{8}$

31. .95   33. **a.** .633   **b.** .276

35. There are two ways of obtaining a sum of 7.

37. No   39. No   41. **a.** $\frac{3}{8}$ **b.** $\frac{1}{2}$ **c.** $\frac{1}{4}$

43. .783   45. False

Exercises 7.3, page 403

1. $\frac{1}{2}$ 3. $\frac{1}{36}$ 5. $\frac{1}{9}$ 7. $\frac{1}{52}$

9. $\frac{3}{13}$ 11. $\frac{12}{13}$ 13. .002; .998

15. $P(a) + P(b) + P(c) \ne 1$

17. Since the five events are not mutually exclusive, Property (3) cannot be used; that is, he could win more than one purse.

19. The two events are not mutually exclusive; hence, the probability of the given event is $\frac{1}{6} + \frac{1}{6} - \frac{1}{36} = \frac{11}{36}$.

21. $E^C \cap F^C = \{e\} \ne \varnothing$

23. $P(G \cup C)^C \ne 1 - P(G) - P(C)$; he has not considered the case in which a customer buys both glasses and contact lenses.

25. **a.** 0   **b.** .7   **c.** .8   **d.** .3

27. **a.** $\frac{1}{2}, \frac{3}{8}$ **b.** $\frac{1}{2}, \frac{5}{8}$ **c.** $\frac{1}{8}$ **d.** $\frac{3}{4}$ 29. .33

31. **a.** .16   **b.** .38   **c.** .22   33. **a.** .38   **b.** .58

35. **a.** .2   **b.** .34   37. **a.** .68   **b.** .87

39. **a.** .90   **b.** .40   **c.** .40

41. **a.** .6   **b.** .332   **c.** .232   **d.** .6

45. True   47. False

Exercises 7.4, page 412

1. $\frac{1}{32}$ 3. $\frac{31}{32}$

5. $P(A) = 13C(4, 2)/C(52, 2) \approx .0588$

7. $C(26, 2)/C(52, 2) = .245$

9. $[C(3, 2)C(5, 2)]/C(8, 4) = 3/7$

11. $[C(5, 3)C(3, 1)]/C(8, 4) = 3/7$ **13.** $[C(3, 2)C(1, 1)]/8 = 3/8$

15. 1/8   **17.** $C(10, 6)/2^{10} \approx .205$

19. a. $C(4, 2)/C(24, 2) \approx .022$
b. $1 - C(20, 2)/C(24, 2) \approx .312$

21. a. $C(6, 2)/C(80, 2) \approx .005$
b. $1 - C(74, 2)/C(80, 2) \approx .145$

23. a. .12; $C(98, 10)/C(100, 12) \approx .013$
b. .15; .015

25. $[C(12, 8)C(8, 2) + C(12, 9)C(8, 1) + C(12, 10)]/C(20, 10) \approx .085$

27. a. $\frac{3}{5}$ **b.** $C(3, 1)/C(5, 3) = .3$ **c.** $1 - C(3, 3)/C(5, 3) = .9$

29. $\frac{1}{729}$ **31.** .0001   **33.** .10   **35.** $40/C(52, 5) \approx .0000154$

37. $[4C(13, 5) - 40]/C(52, 5) \approx .00197$

39. $[13C(4, 3)12C(4, 2)]/C(52, 5) \approx .00144$

41. a. .618   **b.** .059   **43.** .03

Exercises 7.5, page 425

1. a. .4   **b.** .33   **3.** .3   **5.** Independent

7. Independent   **9. a.** .24   **b.** .76

11. a. .5   **b.** .4   **c.** .2   **d.** .35   **e.** No   **f.** No

13. a. .4   **b.** .3   **c.** .12   **d.** .30   **e.** Yes   **f.** Yes

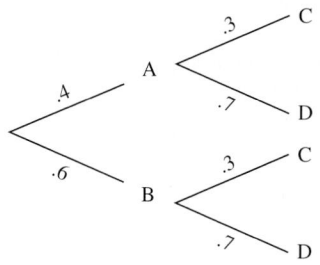

15. a. $\frac{1}{12}$ **b.** $\frac{1}{36}$ **c.** $\frac{1}{6}$ **d.** $\frac{1}{6}$ **e.** No

17. $\frac{4}{11}$ **19.** Independent   **21.** Not independent   **23.** .1875

25. a. $\frac{4}{9}$ **b.** $\frac{4}{9}$ **27. a.** $\frac{1}{21}$ **b.** $\frac{1}{3}$ **29.** .25   **31.** $\frac{1}{7}$

33. a.

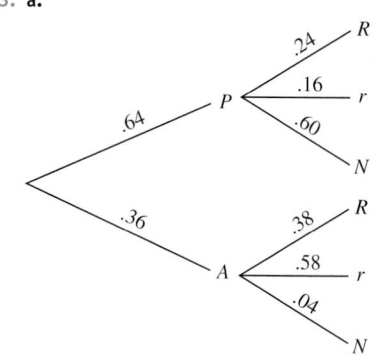

P = Professional
A = Amateur
R = Recovered within 48 hr
r = Recovered after 48 hr
N = Never recovered

b. .24   **c.** .40

35. a. .16   **b.** .424   **c.** .1696

37. a. .092   **b.** $\approx$.008

39. a. .280; .390; .180; .643; .292   **b.** Not independent

41. Not independent   **43.** .0000068   **45. a.** $\frac{7}{10}$ **b.** $\frac{1}{5}$

47. 3   **51.** 1   **53.** True   **55.** True

Exercises 7.6, page 433

1.

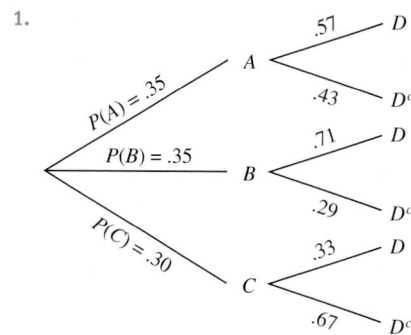

3. a. .45   **b.** .22   **5. a.** .48   **b.** .33

7. a. .08   **b.** .15   **c.** .35

9. a. $\frac{1}{12}$ **b.** $\frac{1}{4}$ **c.** $\frac{1}{18}$ **d.** $\frac{3}{14}$

11. $\frac{4}{17}$ **13.** .0784

15.

(tree diagram with $\frac{2}{5}$, $\frac{4}{9}$ W, $\frac{5}{9}$ B, $\frac{3}{5}$, $\frac{1}{3}$ W, $\frac{2}{3}$ B)

17. .53   **19.** .422   **21. a.** $\frac{3}{4}$ **b.** $\frac{2}{9}$ **23.** .856

25. a. .03   **b.** .29

27. .35   **29. a.** .30   **b.** .10

31. a. .513   **b.** .390   **33.** .65   **35.** .93

37. .3010   **39.** .3758   **41.** .2056

Chapter 7 Concept Review Questions, page 439

1. Experiment; sample; space; event   **2.** $\varnothing$ **3.** Uniform; $\frac{1}{n}$

4. Conditional   **5.** Independent   **6.** A posteriori probability

Chapter 7 Review Exercises, page 439

1. a. 0   **b.** .6   **c.** .6   **d.** .4   **e.** 1

2. a. .35   **b.** .65   **c.** .05

3. a. .49   **b.** .39   **c.** .48

4. $\frac{2}{7}$ **5. a.** .019   **b.** .981   **6.** .364

7. .18   **8.** .25   **9.** .06   **10.** .367   **11.** .49

12. **a.** $\frac{7}{8}$ **b.** $\frac{7}{8}$ **c.** No   13. **a.** .284   **b.** .984

14. .150   15. $\frac{2}{15}$ 16. $\frac{1}{24}$ 17. $\frac{1}{52}$ 18. .00018

19. .00995   20. .2451   21. .510   22. .2451

23. **a.** .17   **b.** .16   24. .457   25. .368

26. **a.** .68   **b.** .053   27. .32   28. .619

Chapter 7 Before Moving On, page 441

1. $\frac{5}{12}$ 2. $\frac{4}{13}$ 3. **a.** .9   **b.** .3   4. .72   5. .3077

CHAPTER 8

Exercises 8.1, page 449

1. **a.** See part (b)

b.

Outcome	GGG	GGR	GRG	RGG
Value	3	2	2	2

Outcome	GRR	RGR	RRG	RRR
Value	1	1	1	0

c. {GGG}

3. Any positive integer   5. $\frac{1}{6}$

7. Any positive integer; infinite discrete

9. $0 \le x < \infty$; continuous

11. Any positive integer; infinite discrete

13. **a.** .20   **b.** .60   **c.** .30   **d.** 1

15.

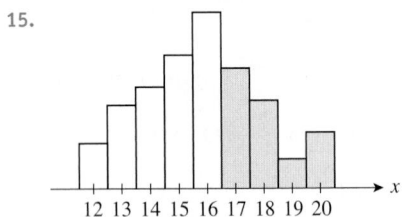

17. **a.**

x	1	2	3	4	5	6
$P(X = x)$	$\frac{1}{6}$	$\frac{1}{6}$	$\frac{1}{6}$	$\frac{1}{6}$	$\frac{1}{6}$	$\frac{1}{6}$

y	1	2	3	4	5	6
$P(Y = y)$	$\frac{1}{6}$	$\frac{1}{6}$	$\frac{1}{6}$	$\frac{1}{6}$	$\frac{1}{6}$	$\frac{1}{6}$

b.

$x + y$	2	3	4	5	6	7
$P(X + Y = x + y)$	$\frac{1}{36}$	$\frac{2}{36}$	$\frac{3}{36}$	$\frac{4}{36}$	$\frac{5}{36}$	$\frac{6}{36}$

$x + y$	8	9	10	11	12
$P(X + Y = x + y)$	$\frac{5}{36}$	$\frac{4}{36}$	$\frac{3}{36}$	$\frac{2}{36}$	$\frac{1}{36}$

19. **a.**

x	0	1	2	3	4
$P(X = x)$	.017	.067	.033	.117	.233

x	5	6	7	8	9	10
$P(X = x)$	.133	.167	.1	.05	.067	.017

b.

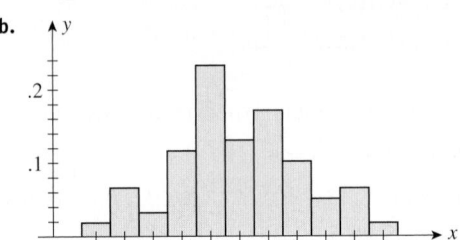

21.

x	1	2	3	4	5
$P(X = x)$	.007	.029	.021	.079	.164

x	6	7	8	9	10
$P(X = x)$	.15	.20	.207	.114	.029

23. True

Using Technology Exercises 8.1, page 453

Graphing Utility

1.

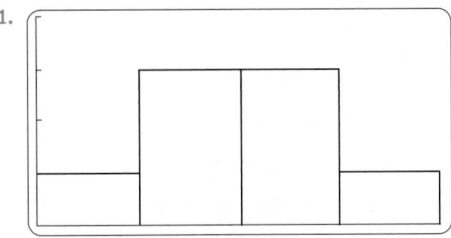

3.

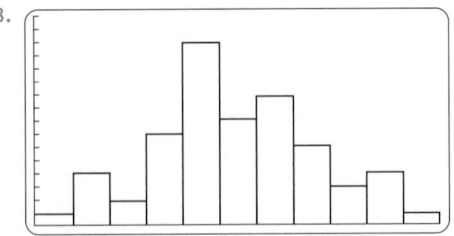

Excel

1.

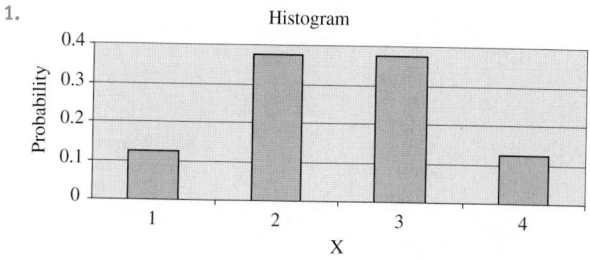

Histogram

3.

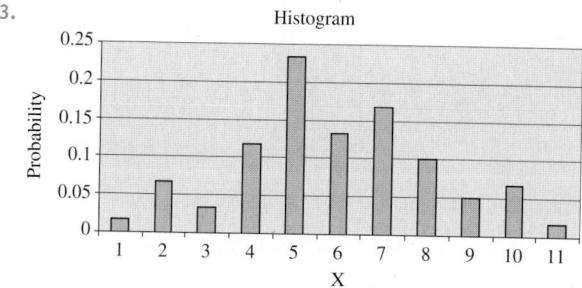

Histogram

Exercises 8.2, page 463

1. a. 2.6

b.

x	0	1	2	3	4	
$P(X = x)$	0	.1	.4	.3	.2	; 2.6

3. 0.86   **5.** $78.50   **7.** 0.91

9. 0.12   **11.** 1.73   **13.** 5.16%

15. −39¢   **17.** $50   **19.** $12,000

21. City B   **23.** Company B   **25.** 2.86%

27. −5.3¢   **29.** −2.7¢   **31.** 2 to 3; 3 to 2

33. 0.4   **35.** .5833   **37.** ≈.3571

39. a. Mean: 74; mode: 85; median: 80   **b.** Mode

41. 3; close   **43.** 16; 16; 16   **45.** True

Exercises 8.3, page 473

1. $\mu = 2$, Var$(X) = 1$, $\sigma = 1$

3. $\mu = 0$, Var$(X) = 1$, $\sigma = 1$

5. $\mu = 518$, Var$(X) = 1891$, $\sigma \approx 43.5$

7. Figure (a)   **9.** 1.56

11. $\mu = 4.5$, Var$(X) = 5.25$

13. a. Let X = the annual birthrate during the years 1991–2000

b.

x	14.5	14.6	14.7	14.8	15.2	15.5	15.9	16.3
$P(X = x)$	.2	.1	.2	.1	.1	.1	.1	.1

c. $\mu \approx 15.07$, Var$(X) \approx 0.3621$, $\sigma \approx 0.6017$

15. a. Mutual fund A: $\mu = \$620$, Var$(X) = \$267,600$;
Mutual fund B: $\mu = \$520$, Var$(X) = \$137,600$
b. Mutual fund A
c. Mutual fund B

17. 1

19. $\mu = \$339,600$; Var$(X) = \$1,443,840,000$; $\sigma \approx \$37,998$

21. $\mu \approx 77.17$; $\sigma \approx 10.62$

23. $\mu = 1607.33$; $\sigma \approx 182.29$

25. $\mu = 5.452$; $\sigma \approx 0.1713$

27. 16.88 million/mo; 0.6841 million

29. a. At least .75
b. At least .96

31. $c = 7$ **33.** At least 7/16

35. .9375   **37.** True

Using Technology Exercises 8.3, page 479

1. a.

b. $\mu = 4$,
$\sigma = 1.40$

3. a.

b. $\mu = 17.34$,
$\sigma = 1.11$

5. a. Let X denote the random variable that gives the weight of a carton of sugar.

b.

x	4.96	4.97	4.98	4.99	5.00	5.01
$P(X = x)$	$\frac{3}{30}$	$\frac{4}{30}$	$\frac{4}{30}$	$\frac{1}{30}$	$\frac{1}{30}$	$\frac{5}{30}$

x	5.02	5.03	5.04	5.05	5.06
$P(X = x)$	$\frac{3}{30}$	$\frac{3}{30}$	$\frac{4}{30}$	$\frac{1}{30}$	$\frac{1}{30}$

c. $\mu \approx 5.00$; $\sigma \approx 0.03$

7. a.

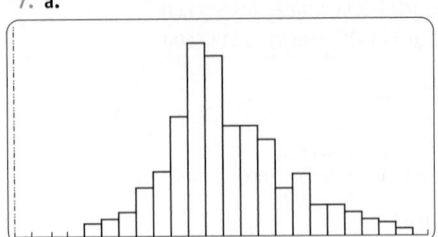

b. 65.875; 1.73

Exercises 8.4, page 487

1. Yes

3. No. There are more than two outcomes to the experiment.

5. No. The probability of an accident on a clear day is not the same as the probability of an accident on a rainy day.

7. .296   **9.** .0512   **11.** .132

13. $\frac{21}{32}$ **15.** .0041   **17.** .116

19. a. $P(X = 0) \approx .08$; $P(X = 1) \approx .26$;
$P(X = 2) \approx .35$; $P(X = 3) \approx .23$;
$P(X = 4) \approx .08$; $P(X = 5) \approx .01$

b.

x	0	1	2	3	4	5
$P(X = x)$	.08	.26	.35	.23	.08	.01

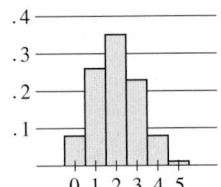

c. $\mu = 2$; $\sigma \approx 1.1$

21. No. The probability that at most 1 is defective is $P(X = 0) + P(X = 1) = .74$.

23. .165   **25.** $\approx$.0002

27. a. $\approx$.633   **b.** $\approx$.367   **29. a.** $\approx$.0329   **b.** $\approx$.3292

31. a. $\approx$.273   **b.** $\approx$.650   **33. a.** $\approx$.817   **b.** $\approx$.999

35. a. $\approx$.075   **b.** $\approx$.012   **37. a.** $\approx$.1216   **b.** $\approx$.3585

39. $\approx$.3487   **41.** $\mu = 375$; $\sigma \approx 9.68$

43. False   **45.** False

Exercises 8.5, page 497

1. .9265   **3.** .0401   **5.** .8657

7. a.

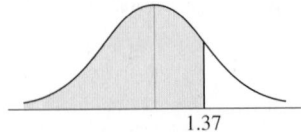

b. .9147

9. a.

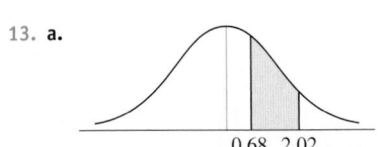

b. .2578

11. a.

b. .8944

13. a.

b. .2266

15. a. 1.23   **b.** −0.81   **17. a.** 1.9   **b.** −1.9

19. a. .9772   **b.** .9192   **c.** .7333

Exercises 8.6, page 505

1. a. .2206   **b.** .2206   **c.** .2960

3. a. .0228   **b.** .0228   **c.** .4772   **d.** .7258

5. a. .0038   **b.** .0918   **c.** .4082   **d.** .2514

7. .6247   **9.** 0.62%   **11.** A: 80; B: 77; C: 73; D: 62; F: 54

13. a. .4207   **b.** .4254   **c.** .0125

15. a. .2877   **b.** .0008   **c.** .7287

17. .9265   **19.** .8686

21. a. .0037   **b.** The drug is very effective.   **23.** 2142

Chapter 8 Concept Review Questions, page 508

1. Random   **2.** Finite; infinite; continuous   **3.** Sum; .75

4. a. $\dfrac{P(E)}{P(E^C)}$ **b.** $\dfrac{a}{a + b}$

5. $p_1(x_1 - \mu)^2 + p_2(x_2 - \mu)^2 + \cdots + p_n(x_n - \mu)^2$; $\sqrt{\text{Var}(X)}$

6. Fixed; two; same; independent

7. Continuous; probability density function; set

8. Normal; large; 0; 1

Chapter 8 Review Exercises, page 508

1. a. {WWW, BWW, WBW, WWB, BBW, BWB, WBB, BBB}

b.

Outcome	WWW	BWW	WBW	WWB
Value of X	0	1	1	1

Outcome	BBW	BWB	WBB	BBB
Value of X	2	2	2	3

c.

x	0	1	2	3
$P(X = x)$	$\frac{1}{35}$	$\frac{12}{35}$	$\frac{18}{35}$	$\frac{4}{35}$

d.

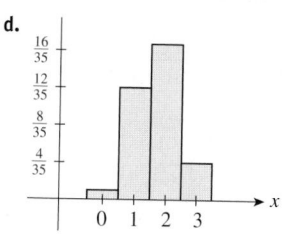

2. $100

3. a. .8　**b.** $\mu = 2.7$; $\sigma \approx 1.42$

4. a.

x	0	1	2	3	4
$P(X = x)$	.1296	.3456	.3456	.1536	.0256

b. $\mu = 1.6$; $V(x) = 0.96$; $\sigma \approx 0.9798$

5. .6915

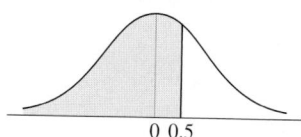

6. .2266

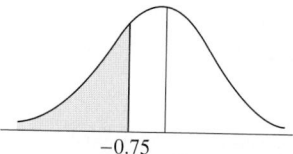

7. .4649

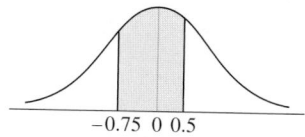

8. .4082

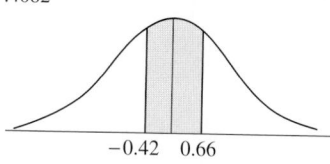

9. 2.42　**10.** −1.05　**11.** −2.03　**12.** 1.42　**13.** .6915

14. .8413　**15.** .2417　**16.** .7333　**17.** .2646; .9163

18. 15.87%　**19.** At least .75　**20.** $\mu = 27.87$; $\sigma = 6.41$

21. .677　**22.** $\mu = 120$; $\sigma \approx 10.1$　**23.** 0.6%

24. .9738　**25.** .9997

Chapter 8 Before Moving On, page 509

1.

x	−3	−2	0	1	2	3
$P(X = x)$	.05	.1	.25	.3	.2	.1

2. a. .8　**b.** .92　**3.** 0.44; 4.0064; 2

4. a. .2401; .4116; .2646; .0756; .0081　**b.** .12; .917

5. a. .9772　**b.** .9772　**c.** .9544

6. a. .0228　**b.** .5086　**c.** .0228

CHAPTER 9

Exercises 9.1, page 519

1. Yes　**3.** Yes　**5.** No　**7.** Yes　**9.** No

11. a. Given that the outcome state 1 has occurred, the conditional probability that the outcome state 1 will occur is .3.

b. .7　**c.** $\begin{bmatrix} .48 \\ .52 \end{bmatrix}$

13. $TX_0 = \begin{bmatrix} .4 \\ .6 \end{bmatrix}$;

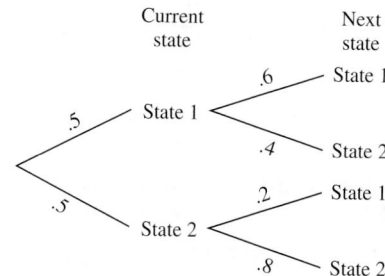

15. $X_2 = \begin{bmatrix} .576 \\ .424 \end{bmatrix}$　**17.** $X_2 = \begin{bmatrix} \frac{5}{16} \\ \frac{27}{64} \\ \frac{17}{64} \end{bmatrix}$

19. a.

	Current state		Next state
	.5	State 1	.8 State 1
			.2 State 2
	.5	State 2	.9 State 1
			.1 State 2

b. $T = \begin{matrix} \text{L} \\ \text{R} \end{matrix} \begin{bmatrix} .8 & .9 \\ .2 & .1 \end{bmatrix}$ (L R)　**c.** $X_0 = \begin{matrix} \text{L} \\ \text{R} \end{matrix} \begin{bmatrix} .5 \\ .5 \end{bmatrix}$　**d.** .85

21. a. Vote is evenly split.　**b.** Democrat

23. 50% in zone I, 30% in zone II, 20% in zone III

25. University: 37%, Campus: 35%, Book Mart: 28%; University: 34.5%, Campus: 31.35%, Book Mart: 34.15%

27. Business: 36%, Humanities: 23.8%, Education: 15%, Natural Sciences and others: 25.1%

29. False

Using Technology Exercises 9.1, page 523

1. $X_5 = \begin{bmatrix} .204489 \\ .131869 \\ .261028 \\ .186814 \\ .2158 \end{bmatrix}$

3. Manufacturer A will have 23.95% of the market share, manufacturer B will have 49.71% of the market share, and manufacturer C will have 26.34% of the market share.

Exercises 9.2, page 530

1. Regular **3.** Not regular **5.** Regular

7. Not regular **9.** $\begin{bmatrix} \frac{3}{11} \\ \frac{8}{11} \end{bmatrix}$ **11.** $\begin{bmatrix} \frac{2}{7} \\ \frac{5}{7} \end{bmatrix}$

13. $\begin{bmatrix} \frac{3}{13} \\ \frac{8}{13} \\ \frac{2}{13} \end{bmatrix}$ **15.** $\begin{bmatrix} \frac{3}{19} \\ \frac{8}{19} \\ \frac{8}{19} \end{bmatrix}$ **17.** 81.8%

19. 40.8% one wage earner and 59.2% two wage earners; 30% one wage earner and 70% two wage earners

21. 72.5% in single-family homes and 27.5% in condominiums; 70% in single-family homes and 30% in condominiums

23. a. 31.7% ABC, 37.35% CBS, 30.95% NBC
b. $33\frac{1}{3}$% ABC, $33\frac{1}{3}$% CBS, $33\frac{1}{3}$% NBC

25. 25% red, 50% pink, 25% white **27.** False

Using Technology Exercises 9.2, page 534

1. $X_5 = \begin{bmatrix} .2045 \\ .1319 \\ .2610 \\ .1868 \\ .2158 \end{bmatrix}$

Exercises 9.3, page 540

1. Yes **3.** Yes **5.** Yes **7.** Yes

9. $\begin{bmatrix} 1 & .4 \\ 0 & .6 \end{bmatrix}$, $R = [.6]$, and $S = [.4]$

11. $\begin{bmatrix} 1 & .4 & .5 \\ 0 & .4 & .5 \\ 0 & .2 & 0 \end{bmatrix}$, $R = \begin{bmatrix} .4 & .5 \\ .2 & 0 \end{bmatrix}$, and $S = [.4 \quad .5]$, or

$\begin{bmatrix} 1 & .5 & .4 \\ 0 & 0 & .2 \\ 0 & .5 & .4 \end{bmatrix}$, $R = \begin{bmatrix} 0 & .2 \\ .5 & .4 \end{bmatrix}$, and $S = [.5 \quad .4]$

13. $\begin{bmatrix} 1 & 0 & .2 & .4 \\ 0 & 1 & .3 & 0 \\ 0 & 0 & .3 & .2 \\ 0 & 0 & .2 & .4 \end{bmatrix}$,

$R = \begin{bmatrix} .3 & .2 \\ .2 & .4 \end{bmatrix}$ and $S = \begin{bmatrix} .2 & .4 \\ .3 & 0 \end{bmatrix}$, or

$\begin{bmatrix} 1 & 0 & .4 & .2 \\ 0 & 1 & 0 & .3 \\ 0 & 0 & .4 & .2 \\ 0 & 0 & .2 & .3 \end{bmatrix}$,

$R = \begin{bmatrix} .4 & .2 \\ .2 & .3 \end{bmatrix}$, $S = \begin{bmatrix} .4 & .2 \\ 0 & .3 \end{bmatrix}$, and so forth

15. $\begin{bmatrix} 1 & 1 \\ 0 & 0 \end{bmatrix}$ **17.** $\begin{bmatrix} 1 & 1 & 1 \\ 0 & 0 & 0 \\ 0 & 0 & 0 \end{bmatrix}$

19. $\begin{bmatrix} 1 & 0 & 1 & 1 \\ 0 & 1 & 0 & 0 \\ 0 & 0 & 0 & 0 \\ 0 & 0 & 0 & 0 \end{bmatrix}$ **21.** $\begin{bmatrix} 1 & 0 & \frac{1}{2} & \frac{1}{2} \\ 0 & 1 & \frac{1}{2} & \frac{1}{2} \\ 0 & 0 & 0 & 0 \\ 0 & 0 & 0 & 0 \end{bmatrix}$

23. $\begin{bmatrix} 1 & 0 & 0 & \frac{7}{22} & \frac{3}{11} \\ 0 & 1 & 0 & \frac{5}{22} & \frac{9}{22} \\ 0 & 0 & 1 & \frac{5}{11} & \frac{7}{22} \\ 0 & 0 & 0 & 0 & 0 \\ 0 & 0 & 0 & 0 & 0 \end{bmatrix}$

25. a. $\begin{array}{cc} & \text{UL} \quad \text{L} \end{array}$ $\begin{array}{c} \text{UL} \\ \text{L} \end{array}\begin{bmatrix} 1 & .2 \\ 0 & .8 \end{bmatrix}$, $R = [.8]$, and $S = [.2]$

b. $\begin{bmatrix} 1 & 1 \\ 0 & 0 \end{bmatrix}$; eventually, only unleaded fuel will be used.

27. .25; .50; .75

29. a. $\begin{array}{cccc} & \text{D} & \text{G} & 1 & 2 \end{array}$ $\begin{array}{c} \text{D} \\ \text{G} \\ 1 \\ 2 \end{array}\begin{bmatrix} 1 & 0 & .25 & .1 \\ 0 & 1 & 0 & .9 \\ 0 & 0 & 0 & 0 \\ 0 & 0 & .75 & 0 \end{bmatrix}$

b. $\begin{bmatrix} 1 & 0 & .325 & .1 \\ 0 & 1 & .675 & .9 \\ 0 & 0 & 0 & 0 \\ 0 & 0 & 0 & 0 \end{bmatrix}$

c. .675

31. False

Chapter 9 Concept Review Questions, page 543

1. Probabilities; preceding **2.** State; state **3.** Transition

4. $n \times n$; nonnegative; 1 **5.** Distribution; steady-state

6. Regular; rows; equal; positive; $TX = X$; elements; 1

7. Absorbing; leave; stages

Chapter 9 Review Exercises, page 544

1. Not regular   2. Regular   3. Regular   4. Not regular

5. $\begin{bmatrix} .3675 \\ .36 \\ .2725 \end{bmatrix}$ 6. $\begin{bmatrix} .1915 \\ .4215 \\ .387 \end{bmatrix}$

7. Yes   8. No   9. No   10. Yes

11. $\begin{bmatrix} \frac{3}{7} & \frac{3}{7} \\ \frac{4}{7} & \frac{4}{7} \end{bmatrix}$ 12. $\begin{bmatrix} \frac{4}{9} & \frac{4}{9} \\ \frac{5}{9} & \frac{5}{9} \end{bmatrix}$

13. $\begin{bmatrix} .457 & .457 & .457 \\ .200 & .200 & .200 \\ .343 & .343 & .343 \end{bmatrix}$ 14. $\begin{bmatrix} .323 & .323 & .323 \\ .290 & .290 & .290 \\ .387 & .387 & .387 \end{bmatrix}$

15. **a.**

$\begin{array}{c} \\ A \\ U \\ N \end{array} \begin{array}{ccc} A & U & N \\ \end{array}$

$\begin{array}{c} A \\ U \\ N \end{array} \begin{bmatrix} .85 & 0 & .10 \\ .10 & .95 & .05 \\ .05 & .05 & .85 \end{bmatrix} \begin{array}{l} A = \text{Agriculture} \\ U = \text{Urban} \\ N = \text{Nonagricultural} \end{array}$

b. $\begin{array}{c} A \\ U \\ N \end{array} \begin{bmatrix} .50 \\ .15 \\ .35 \end{bmatrix}$ **c.** $\begin{array}{c} A \\ U \\ N \end{array} \begin{bmatrix} .424 \\ .262 \\ .314 \end{bmatrix}$

16. 12.5% large cars, 30.36% intermediate-sized cars, 57.14% small cars

Chapter 9 Before Moving On, page 545

1. $\begin{bmatrix} .366 \\ .634 \end{bmatrix}$ 2. $\begin{bmatrix} \frac{3}{11} \\ \frac{8}{11} \end{bmatrix}$ 3. $\begin{bmatrix} 1 & 1 & 1 \\ 0 & 0 & 0 \\ 0 & 0 & 0 \end{bmatrix}$

CHAPTER 10

Exercises 10.1, page 551

1. 9   3. 1   5. 4

7. 7   9. $\frac{1}{5}$ 11. 2

13. 2   15. 1   17. True

19. False   21. False   23. False

25. False   27. $\frac{1}{(xy)^2}$ 29. $\frac{1}{x^{5/6}}$

31. $\frac{1}{(s + t)^3}$ 33. $x^{13/3}$ 35. $\frac{1}{x^3}$

37. x 39. $\frac{9}{x^2 y^4}$ 41. $\frac{y^8}{x^{10}}$

43. $2x^{11/6}$ 45. $-2xy^2$ 47. $2x^{4/3} y^{1/2}$

49. 2.828   51. 5.196   53. 31.62

55. 316.2   57. $\frac{3\sqrt{x}}{2x}$ 59. $\frac{2\sqrt{3y}}{3}$

61. $\frac{\sqrt[3]{x^2}}{x}$ 63. $\frac{2x}{3\sqrt{x}}$ 65. $\frac{2y}{\sqrt{2xy}}$

67. $\frac{xz}{y\sqrt[3]{xz^2}}$

Exercises 10.2, page 559

1. $9x^2 + 3x + 1$ 3. $4y^2 + y + 8$

5. $-x - 1$ 7. $\frac{2}{3} + e - e^{-1}$

9. $6\sqrt{2} + 8 + \frac{1}{2}\sqrt{x} - \frac{11}{4}\sqrt{y}$ 11. $x^2 + 6x - 16$

13. $a^2 + 10a + 25$ 15. $x^2 + 4xy + 4y^2$

17. $4x^2 - y^2$ 19. $-2x$

21. $2t(2\sqrt{t} + 1)$ 23. $2x^3(2x^2 - 6x - 3)$

25. $7a^2(a^2 + 7ab - 6b^2)$ 27. $e^{-x}(1 - x)$

29. $\frac{1}{2}x^{-5/2}(4 - 3x)$ 31. $(2a + b)(3c - 2d)$

33. $(2a + b)(2a - b)$ 35. $-2(3x + 5)(2x - 1)$

37. $3(x - 4)(x + 2)$ 39. $2(3x - 5)(2x + 3)$

41. $(3x - 4y)(3x + 4y)$ 43. $(x^2 + 5)(x^4 - 5x^2 + 25)$

45. $x^3 - xy^2$ 47. $4(x - 1)(3x - 1)(2x + 2)^3$

49. $4(x - 1)(3x - 1)(2x + 2)^3$ 51. $2x(x^2 + 2)^2(5x^4 + 20x^2 + 17)$

53. -4 and 3 55. -1 and $\frac{1}{2}$

57. 2 and 2   59. -2 and $\frac{3}{4}$

61. $\frac{1}{2} + \frac{1}{4}\sqrt{10}$ and $\frac{1}{2} - \frac{1}{4}\sqrt{10}$ 63. $-1 + \frac{1}{2}\sqrt{10}$ and $-1 - \frac{1}{2}\sqrt{10}$

Exercises 10.3, page 566

1. $\frac{x - 1}{x - 2}$ 3. $\frac{3(2t + 1)}{2t - 1}$ 5. $-\frac{7}{(4x - 1)^2}$ 7. $\frac{1}{(2x + 3)^2}$

9. $\frac{e^x(1 - 2e^x)}{(e^x + 1)^3}$ 11. -8 13. $\frac{3x - 1}{2}$ 15. $\frac{t + 20}{3t + 2}$

17. $-\frac{x(2x - 13)}{(2x - 1)(2x + 5)}$ 19. $\frac{x^4 - 1}{x}$ 21. $-\frac{x + 27}{(x - 3)^2(x + 3)}$

23. $-\frac{3a + 10}{(a + 2)(a - 2)}$ 25. $\frac{x + 1}{x - 1}$ 27. $\frac{y - x}{x^2 y^2}$

29. $\frac{4x^2 + 7}{\sqrt{2x^2 + 7}}$ 31. $\frac{x - 1}{x^2\sqrt{x + 1}}$ 33. $\frac{x - 1}{(2x + 1)^{3/2}}$

39. $\frac{\sqrt{3} + 1}{2}$ 41. $\frac{\sqrt{x} + \sqrt{y}}{x - y}$ 43. $\frac{(\sqrt{a} + \sqrt{b})^2}{a - b}$

45. $\frac{x}{3\sqrt{x}}$ 47. $-\frac{2}{3(1 + \sqrt{3})}$ 49. $-\frac{x + 1}{\sqrt{x + 2}(1 - \sqrt{x + 2})}$

Exercises 10.4, page 572

1. False   3. False

5.

7.

9.

11. $(-\infty, 2)$ 13. $(-\infty, -5]$ 15. $(-4, 6)$

17. $(-\infty, -3) \cup (3, \infty)$ 19. $(-2, 3)$ 21. $[-3, 5]$

23. $(-\infty, 1] \cup [\frac{3}{2}, \infty)$ 25. $(-\infty, -3] \cup (2, \infty)$

27. $(-\infty, 0] \cup (1, \infty)$ 29. 4

31. 2   33. $5\sqrt{3}$ 35. $\pi + 1$ 37. 2

39. False   41. False   43. True   45. False

47. True   49. False   51. $[362, 488.7]$

53. $12,300   55. $64,000   57. $|x - 0.5| < 0.01$

59. Between 1000 and 4000 units

Chapter 10 Review Exercises, page 574

1. $\frac{27}{8}$ 2. 25   3. $\frac{1}{144}$ 4. -32 5. $\frac{1}{4}$ 6. $3\sqrt[3]{3}$

7. $4(x^2 + y)^2$ 8. $\frac{a^{15}}{b^{11}}$ 9. $\frac{2x}{3z}$ 10. $-x^{1/2}$ 11. $6xy^7$

12. $\frac{9}{2}a^5b^8$ 13. $9x^2y^4$ 14. $\frac{x}{y^{1/2}}$ 15. $-2\pi r^2(\pi r - 50)$

16. $2vw(v^2 + w^2 + u^2)$ 17. $(4 - x)(4 + x)$

18. $6t(2t - 3)(t + 1)$ 19. $-2(x - 1)(x + 3)$

20. $4(3x - 5)(x - 6)$ 21. $\frac{180}{(t + 6)^2}$ 22. $\frac{15x^2 + 24x + 2}{4(3x^2 + 2)(x + 2)}$

23. $\frac{78x^2 - 8x - 27}{3(2x^2 - 1)(3x - 1)}$ 24. $\frac{2(x + 2)}{\sqrt{x + 1}}$ 25. $\frac{1}{2}; -\frac{3}{4}$

26. $-2; \frac{1}{3}$ 27. 0; 1; -3 28. $\pm\sqrt{2}/2$ 29. $[-2, \infty)$

30. $[-1, 2]$ 31. $(-\infty, -4) \cup (5, \infty)$ 32. $(-\infty, -5) \cup (5, \infty)$

33. 4   34. 1   35. $\pi - 6$

36. $8 - 3\sqrt{3}$ 37. $[-2, \frac{1}{2}]$

38. $-2 < x < -\frac{3}{2}$ 39. $(-1, 4)$

40. $\frac{3}{2}; \frac{2}{3}$ 41. $\frac{1}{\sqrt{x} + 1}$ 42. $\frac{x - \sqrt{x}}{2x}$

43. $1 + \sqrt{6}, 1 - \sqrt{6}$ 44. $-2 \pm \frac{1}{2}\sqrt{2}$

45. $100   46. $400

Chapter 10 Before Moving On, page 575

1. **a.** $\sqrt{3} + \sqrt{2} - \pi$ **b.** -3

2. **a.** $12x^5y$ **b.** $\frac{b^5}{a^3}$

3. **a.** $\frac{2x\sqrt{y}}{3y}$ **b.** $\frac{x(\sqrt{x} + 4)}{x - 16}$

4. **a.** $\frac{1 - 3x^2}{2\sqrt{x}(x^2 + 1)^2}$ **b.** $\frac{6\sqrt{x} + 2}{x + 2}$

5. $\frac{x - y}{(\sqrt{x} - \sqrt{y})^2}$ 6. **a.** $2x(3x + 2)(2x - 3)$ **b.** $(2b + 3c)(x - y)$

7. **a.** $x = -\frac{1}{4}$, or 1   **b.** $\frac{5 \pm \sqrt{13}}{6}$ 8. $\left[-\frac{2}{3}, \frac{3}{2}\right]$

CHAPTER 11

Exercises 11.1, page 586

1. $21, -9, 5a + 6, -5a + 6, 5a + 21$

3. $-3, 6, 3a^2 - 6a - 3, 3a^2 + 6a - 3, 3x^2 - 6$

5. $2a + 2h + 5, -2a + 5, 2a^2 + 5, 2a - 4h + 5, 4a - 2h + 5$

7. $\frac{8}{15}, 0, \frac{2a}{a^2 - 1}, \frac{2(2 + a)}{a^2 + 4a + 3}, \frac{2(t + 1)}{t(t + 2)}$

9. $8, \frac{2a^2}{\sqrt{a - 1}}, \frac{2(x + 1)^2}{\sqrt{x}}, \frac{2(x - 1)^2}{\sqrt{x - 2}}$

11. 5, 1, 1   13. $\frac{5}{2}, 3, 3, 9$

15. **a.** -2 **b.** (i) $x = 2$; (ii) $x = 1$ **c.** $[0, 6]$ **d.** $[-2, 6]$

17. Yes   19. Yes

21. $(-\infty, \infty)$ 23. $(-\infty, 0) \cup (0, \infty)$ 25. $(-\infty, \infty)$

27. $(-\infty, 5]$ 29. $(-\infty, -1) \cup (-1, 1) \cup (1, \infty)$

31. $[-3, \infty)$ 33. $(-\infty, -2) \cup (-2, 1]$

35. **a.** $(-\infty, \infty)$
   b. $6, 0, -4, -6, -\frac{25}{4}, -6, -4, 0$
   c.

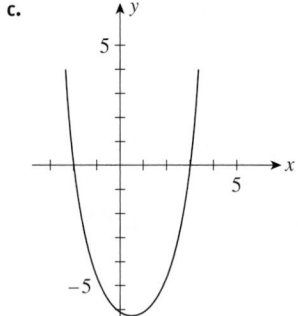

37.

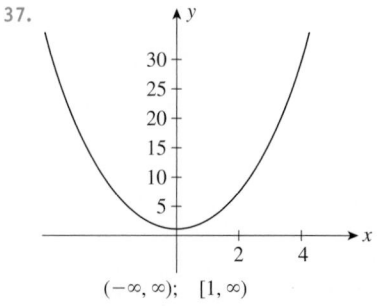

$(-\infty, \infty); [1, \infty)$

39.

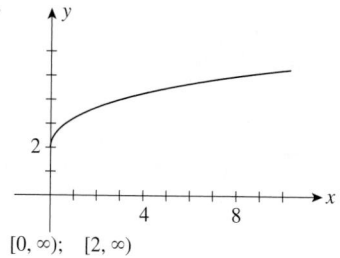

$[0, \infty); \quad [2, \infty)$

41.

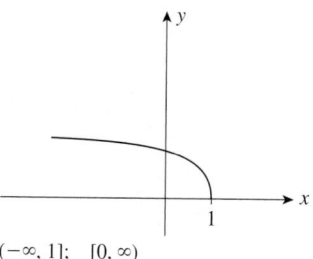

$(-\infty, 1]; \quad [0, \infty)$

43.

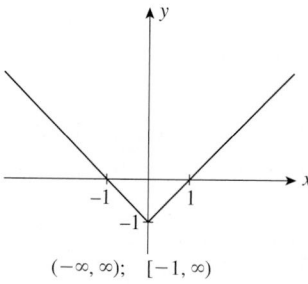

$(-\infty, \infty); \quad [-1, \infty)$

45.

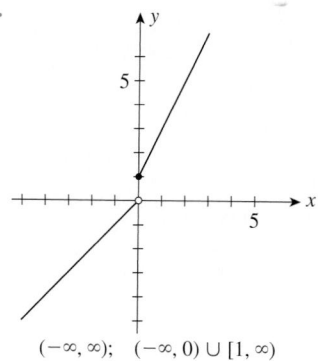

$(-\infty, \infty); \quad (-\infty, 0) \cup [1, \infty)$

47.

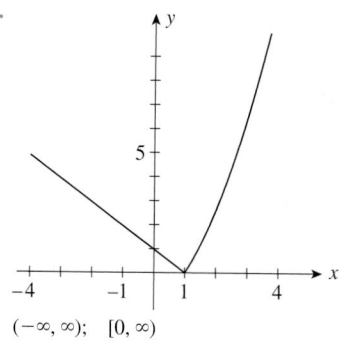

$(-\infty, \infty); \quad [0, \infty)$

49. Yes   **51.** No   **53.** Yes   **55.** Yes

57. 10π in.   **59.** 8   **61. a.** From 1985 to 1990
   b. From 1990 on
   c. 1990; $3.5 billion

63. a. $f(t) = \begin{cases} 0.0185t + 0.58 & \text{if } 0 \le t \le 20 \\ 0.015t + 0.65 & \text{if } 20 < t \le 30 \end{cases}$
   b. 0.0185/yr from 1960 through 1980; 0.015/yr from 1980 through 1990
   c. 1983

65. a. $0.06x$ **b.** $12.00; $0.34   **67.** 160 mg

69. a. $f(t) = 7.5t + 20$ **b.** 65 million

71. a. $V = -12{,}000n + 120{,}000$
   b.

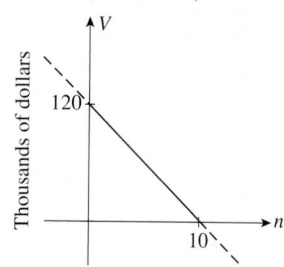

   c. $48,000   **d.** $12,000/yr

73. $(0, \infty)$

75. a. 3.6 million; 9.5 million   **b.** 11.2 million

77. 0.77   **79. a.** 4.1 million   **b.** 5.47 million

81. a. 130 tons/day; 100 tons/day; 40 tons/day
   b.

83. True   **85.** False

Using Technology Exercises 11.1, page 595

1.

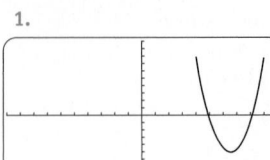

3.

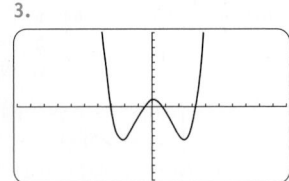

5. a.

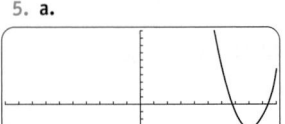

b.

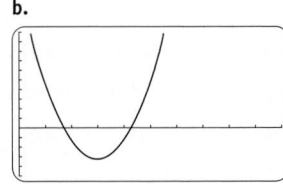

7. a.

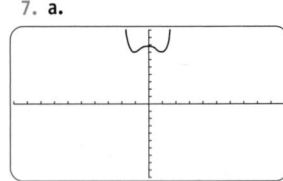

b.

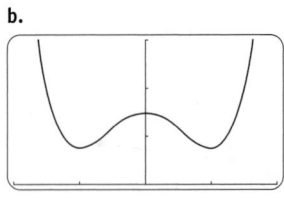

9. a.

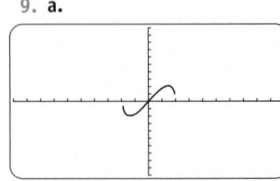

b.

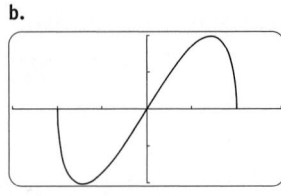

11.

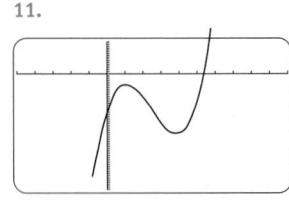

13.

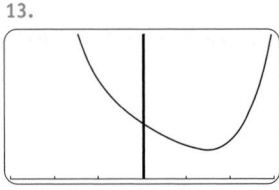

15.

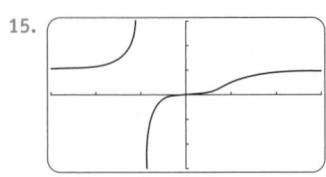

17. 18　　**19.** 2　　**21.** 18.5505　　**23.** 4.1616

25. a.

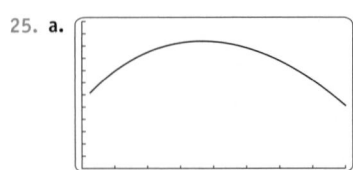

b. 9.41%; 8.71%

27. a.

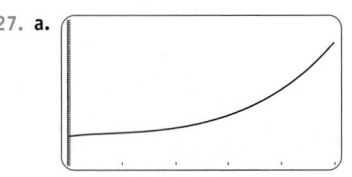

b. 44.7; 52.7; 129.2

Exercises 11.2, page 601

1. $f(x) + g(x) = x^3 + x^2 + 3$

3. $f(x)g(x) = x^5 - 2x^3 + 5x^2 - 10$

5. $\dfrac{f(x)}{g(x)} = \dfrac{x^3 + 5}{x^2 - 2}$

7. $\dfrac{f(x)g(x)}{h(x)} = \dfrac{x^5 - 2x^3 + 5x^2 - 10}{2x + 4}$

9. $f(x) + g(x) = x - 1 + \sqrt{x + 1}$

11. $f(x)g(x) = (x - 1)\sqrt{x + 1}$

13. $\dfrac{g(x)}{h(x)} = \dfrac{\sqrt{x + 1}}{2x^3 - 1}$

15. $\dfrac{f(x)g(x)}{h(x)} = \dfrac{(x - 1)\sqrt{x + 1}}{2x^3 - 1}$

17. $\dfrac{f(x) - h(x)}{g(x)} = \dfrac{x - 2x^3}{\sqrt{x + 1}}$

19. $f(x) + g(x) = x^2 + \sqrt{x} + 3$;
$f(x) - g(x) = x^2 - \sqrt{x} + 7$;
$f(x)g(x) = (x^2 + 5)(\sqrt{x} - 2)$; $\dfrac{f(x)}{g(x)} = \dfrac{x^2 + 5}{\sqrt{x} - 2}$

21. $f(x) + g(x) = \dfrac{(x - 1)\sqrt{x + 3} + 1}{x - 1}$;
$f(x) - g(x) = \dfrac{(x - 1)\sqrt{x + 3} - 1}{x - 1}$;
$f(x)g(x) = \dfrac{\sqrt{x + 3}}{x - 1}$; $\dfrac{f(x)}{g(x)} = (x - 1)\sqrt{x + 3}$

23. $f(x) + g(x) = \dfrac{2(x^2 - 2)}{(x - 1)(x - 2)}$;
$f(x) - g(x) = \dfrac{-2x}{(x - 1)(x - 2)}$;
$f(x)g(x) = \dfrac{(x + 1)(x + 2)}{(x - 1)(x - 2)}$; $\dfrac{f(x)}{g(x)} = \dfrac{(x + 1)(x - 2)}{(x - 1)(x + 2)}$

25. $f(g(x)) = x^4 + x^2 + 1$; $g(f(x)) = (x^2 + x + 1)^2$

27. $f(g(x)) = \sqrt{x^2 - 1} + 1$; $g(f(x)) = x + 2\sqrt{x}$

29. $f(g(x)) = \dfrac{x}{x^2 + 1}$; $g(f(x)) = \dfrac{x^2 + 1}{x}$

31. 49　　**33.** $\dfrac{\sqrt{5}}{5}$

35. $f(x) = 2x^3 + x^2 + 1$ and $g(x) = x^5$

37. $f(x) = x^2 - 1$ and $g(x) = \sqrt{x}$

39. $f(x) = x^2 - 1$ and $g(x) = \dfrac{1}{x}$

41. $f(x) = 3x^2 + 2$ and $g(x) = \dfrac{1}{x^{3/2}}$

43. $3h$ **45.** $-h(2a + h)$ **47.** $2a + h$

49. $3a^2 + 3ah + h^2 - 1$ **51.** $-\dfrac{1}{a(a+h)}$

53. The total revenue in dollars from both restaurants at time t

55. The value in dollars of Nancy's shares of IBM at time t

57. The carbon monoxide pollution in parts per million at time t

59. $C(x) = 0.6x + 12{,}100$

61. a. $f(t) = 267$; $g(t) = 2t^2 + 46t + 733$
   b. $f(t) + g(t) = 2t^2 + 46t + 1000$
   c. 1936 tons

63. a. $P(x) = -0.000003x^3 - 0.07x^2 + 300x - 100{,}000$
   b. \$182,375

65. a. $3.5t^2 + 2.4t + 71.2$
   b. 71,200; 109,900

67. a. 55%; 98.2%
   b. \$444,700; \$1,167,600

69. True   **71.** False

Exercises 11.3, page 611

1. Polynomial function; degree 6

3. Polynomial function; degree 6

5. Some other function   **7.** \$43,200   **9.** \$128,000

11. 123,780,000 kWh; 175,820,000 kWh

13. 0.7; 7.95   **15. a.** 320,000   **b.** 3,923,200

17. 582,650; 1,605,590

19. a.

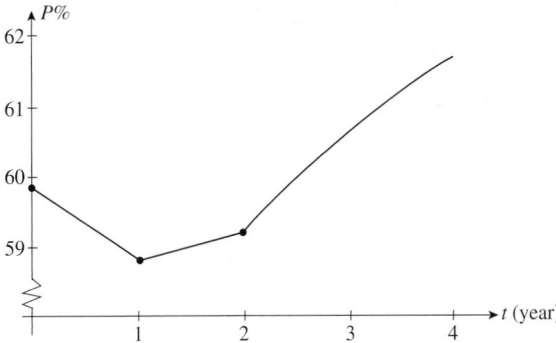

b. 599.1 million

21. a. $R(x) = \dfrac{100x}{40 + x}$ **b.** 60%

23. \$4770; \$6400; \$7560

25. a. \$16.4 billion; \$17.6 billion; \$18.3 billion; \$18.8 billion; \$19.3 billion
   b.

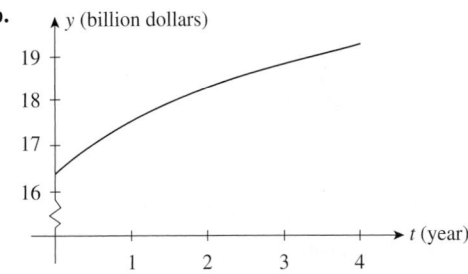

27. a.

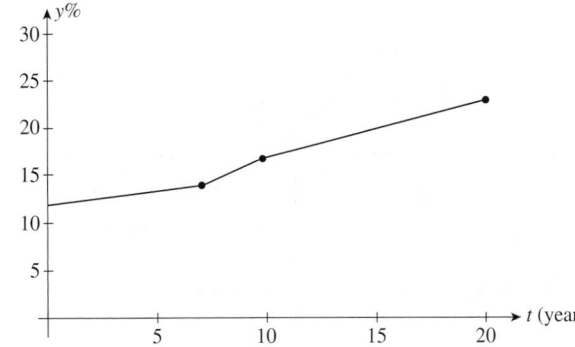

   b. 13.43%; 20%

29. $\dfrac{110}{\frac{1}{2}t + 1} - 26\left(\dfrac{1}{4}t^2 - 1\right)^2 - 52$; \$32, \$6.71, \$3; the gap was closing.

31. a. 59.8%; 58.9%; 59.2%; 60.7%, 61.7%
   b.

   c. 60.66%

33. a.

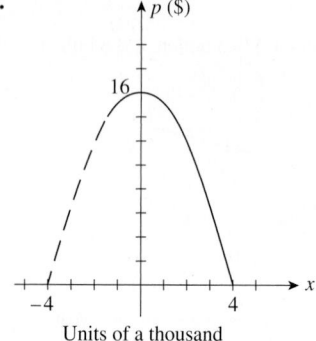

Units of a thousand

b. 3000 units

35. a.

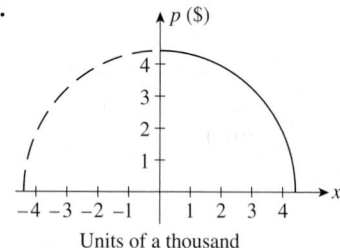

Units of a thousand

b. 3000

37. a.

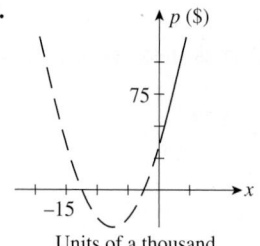

Units of a thousand

b. $76

39. a.

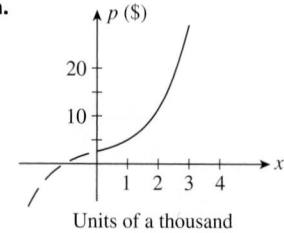

Units of a thousand

b. $15

41. a.

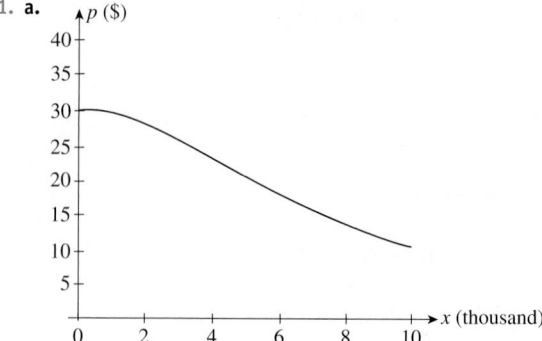

b. $10

43. a.   **b.** $20

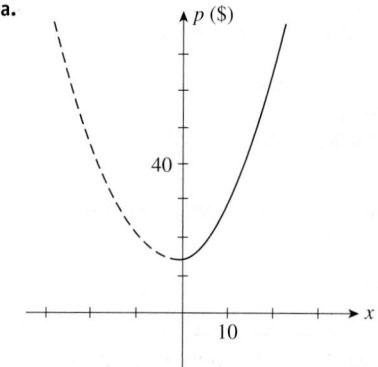

Units of a thousand

45. 5000; $65   **47.** 2000; $52

49. 500; $32.50   **51.** $40x - x^2$; [0, 40]

53. $(15 - 2x)(8 - 2x)x$ **55.** $x\left(28 - \dfrac{\pi}{2}x - 2x\right)$

57. $-2x + 52 - \dfrac{50}{x}; \left[1, \frac{25}{2}\right]$

59. a. $-4x^2 + 520x + 12{,}000$ **b.** $26,400   **c.** $28,800

61. True   **63.** False

Using Technology Exercises 11.3, page 619

1. (−3.0414, 0.1503); (3.0414, 7.4497)

3. (−2.3371, 2.4117); (6.0514, −2.5015)

5. (−1.0219, −6.3461); (1.2414, −1.5931); (5.7805, 7.9391)

7. a.

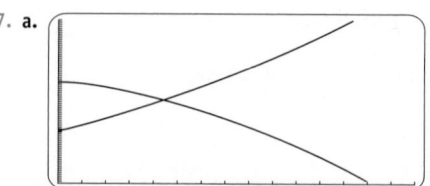

b. 438 wall clocks; $40.92

9. a. $y = 0.1554t^2 + 0.5861t + 3.1607$

b.

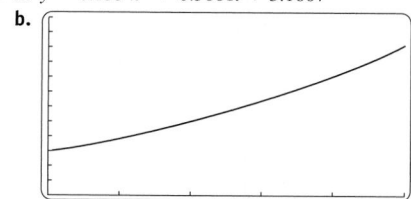

c. 3.16; 3.9; 4.95; 6.32; 7.99; 9.98

11. a. $y = -0.02028t^3 + 0.31393t^2 + 0.40873t + 0.66024$

b.

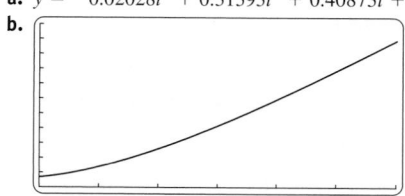

c. 0.66; 1.36; 2.57; 4.16; 6.02; 8.02; 10.03

13. a. $0.001532t^3 - 0.0588t^2 + 0.5208t + 2.55$

b.

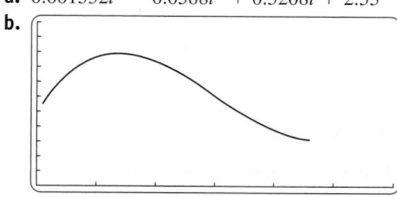

c. $2.55 trillion; $3.84 trillion; $3.42 trillion; $1.702 trillion

15. a. $y = 0.05833t^3 - 0.325t^2 + 1.8881t + 5.07143$

b.

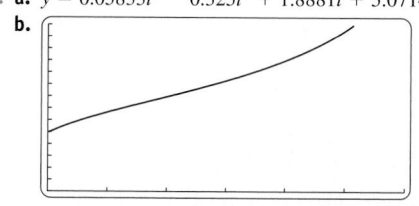

c. 6.7; 8.0; 9.4; 11.2; 13.7

17. a. $y = 0.0125t^4 - 0.01389t^3 + 0.55417t^2 + 0.53294t + 4.95238$
$(0 \le t \le 5)$

b.

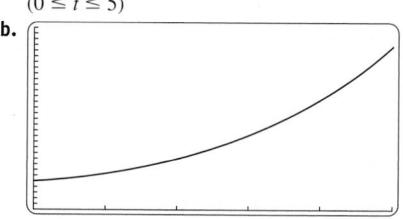

c. 5.0; 6.0; 8.3; 12.2; 18.3; 27.5

Exercises 11.4, page 635

1. $\lim\limits_{x \to -2} f(x) = 3$ **3.** $\lim\limits_{x \to 3} f(x) = 3$

5. $\lim\limits_{x \to -2} f(x) = 3$

7. The limit does not exist.

9.

x	1.9	1.99	1.999
$f(x)$	4.61	4.9601	4.9960

x	2.001	2.01	2.1
$f(x)$	5.004	5.0401	5.41

$\lim\limits_{x \to 2} (x^2 + 1) = 5$

11.

x	-0.1	-0.01	-0.001
$f(x)$	-1	-1	-1

x	0.001	0.01	0.1
$f(x)$	1	1	1

The limit does not exist.

13.

x	0.9	0.99	0.999
$f(x)$	100	10,000	1,000,000

x	1.001	1.01	1.1
$f(x)$	1,000,000	10,000	100

The limit does not exist.

15.

x	0.9	0.99	0.999	1.001	1.01	1.1
$f(x)$	2.9	2.99	2.999	3.001	3.01	3.1

$\lim\limits_{x \to 1} \dfrac{x^2 + x - 2}{x - 1} = 3$

17.

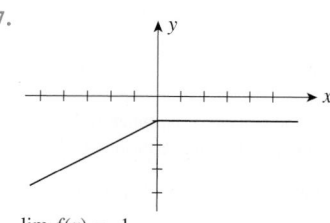

$\lim\limits_{x \to 0} f(x) = -1$

19.

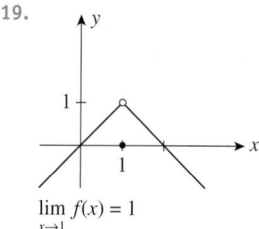

$\lim\limits_{x \to 1} f(x) = 1$

21.

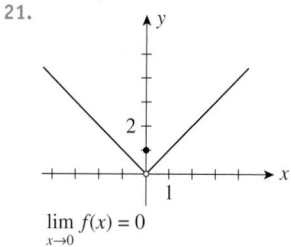

$\lim\limits_{x \to 0} f(x) = 0$

23. 3   25. 3   27. −1   29. 2   31. −4   33. $\frac{5}{4}$

35. 2   37. $\sqrt{171} = 3\sqrt{19}$ 39. $\frac{3}{2}$ 41. −1

43. −6   45. 2   47. $\frac{1}{6}$ 49. 2   51. −1

53. −10   55. The limit does not exist.   57. $\frac{5}{3}$

59. $\frac{1}{2}$ 61. $\frac{1}{3}$ 63. $\lim\limits_{x\to\infty} f(x) = \infty$; $\lim\limits_{x\to-\infty} f(x) = \infty$

65. 0; 0   67. $\lim\limits_{x\to\infty} f(x) = -\infty$; $\lim\limits_{x\to-\infty} f(x) = -\infty$

69.

x	1	10	100	1000
$f(x)$	0.5	0.009901	0.0001	0.000001

x	−1	−10	−100	−1000
$f(x)$	0.5	0.009901	0.0001	0.000001

$\lim\limits_{x\to\infty} f(x) = 0$ and $\lim\limits_{x\to-\infty} f(x) = 0$

71.

x	1	5	10	100
$f(x)$	12	360	2910	2.99×10^6

x	1000	−1	−5
$f(x)$	2.999×10^9	6	−390

x	−10	−100	−1000
$f(x)$	−3090	-3.01×10^6	-3.0×10^9

$\lim\limits_{x\to\infty} f(x) = \infty$ and $\lim\limits_{x\to-\infty} f(x) = -\infty$

73. 3   75. 3   77. $\lim\limits_{x\to-\infty} f(x) = -\infty$ 79. 0

81. **a.** $0.5 million; $0.75 million; $1,166,667; $2 million; $4.5 million; $9.5 million
    b. The limit does not exist; as the percent of pollutant to be removed approaches 100, the cost becomes astronomical.

83. $2.20; the average cost of producing x DVDs will approach $2.20/disc in the long run.

85. **a.** $24 million; $60 million; $83.1 million
    b. $120 million

87. **a.** 76.1¢/mi; 30.5¢/mi; 23¢/mi; 20.6¢/mi; 19.5¢/mi
    b.

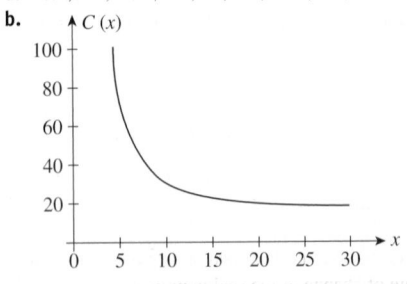

    c. It approaches 17.8¢/mi.

89. False   91. True   93. True

95. a moles/liter/second   97. No

Using Technology Exercises 11.4, page 642

1. 5   3. 3   5. $\frac{2}{3}$ 7. $\frac{1}{2}$ 9. e^2

13. **a.**

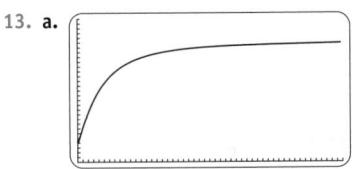

b. 25,000

Exercises 11.5, page 651

1. 3; 2; the limit does not exist.

3. The limit does not exist; 2; the limit does not exist.

5. 0; 2; the limit does not exist.

7. −2; 2; the limit does not exist.

9. True   11. True   13. False   15. True

17. False   19. True   21. 6   23. $-\frac{1}{4}$

25. The limit does not exist.   27. −1   29. 0

31. −4   33. The limit does not exist.   35. 4

37. 0; 0   39. $x = 0$; conditions 2 and 3

41. Continuous everywhere   43. $x = 0$; condition 3

45. $(-\infty, \infty)$ 47. $(-\infty, \infty)$

49. $\left(-\infty, \frac{1}{2}\right) \cup \left(\frac{1}{2}, \infty\right)$

51. $(-\infty, -2) \cup (-2, 1) \cup (1, \infty)$

53. $(-\infty, \infty)$ 55. $(-\infty, \infty)$

57. −1 and 1   59. 1 and 2

61. f is discontinuous at $x = 1, 2, \ldots, 11$.

63. Michael makes progress toward solving the problem until $x = x_1$. Between $x = x_1$ and $x = x_2$, he makes no further progress. But at $x = x_2$ he suddenly achieves a breakthrough, and at $x = x_3$ he proceeds to complete the problem.

65. Conditions 2 and 3 are not satisfied at each of these points.

67.

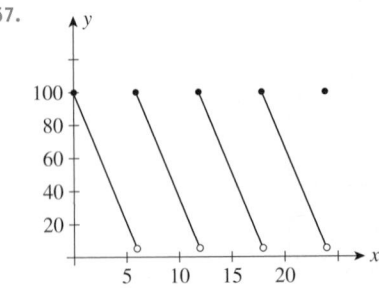

69.

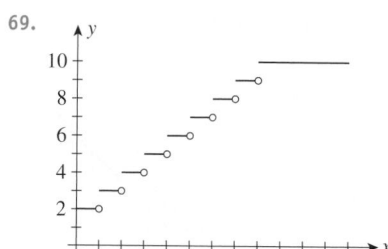

f is discontinuous at $x = \frac{1}{2}, 1, 1\frac{1}{2}, \ldots, 4$.

71. a. ∞; As the time taken to excite the tissue is made smaller and smaller, the strength of the electric current gets stronger and stronger.
 b. b; As the time taken to excite the tissue is made larger and larger, the strength of the electric current gets smaller and smaller and approaches b.

73. 3   **75. a.** Yes   **b.** No

77. a. f is a polynomial of degree 2.
 b. $f(1) = 3$ and $f(3) = -1$

79. a. f is a polynomial of degree 3.
 b. $f(-1) = -4$ and $f(1) = 4$

81. $x \approx 0.59$ **83.** ≈ 1.34

85. c. $\frac{1}{2}; \frac{7}{2}$; Joan sees the ball on its way up $\frac{1}{2}$ sec after it was thrown and again $3\frac{1}{2}$ sec later.

87. False   **89.** False   **91.** False

93. False   **95. c.** $\pm\dfrac{\sqrt{2}}{2}$

Using Technology Exercises 11.5, page 658

1. $x = 0, 1$ **3.** $x = 2$ **5.** $x = 0, \frac{1}{2}$

7. $x = -\frac{1}{2}, 2$ **9.** $x = -2, 1$

11.

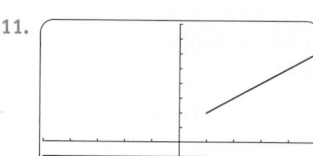

13.

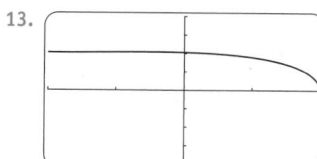

15.

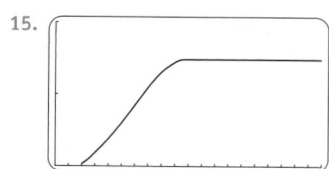

Exercises 11.6, page 671

1. 1.5 lb/mo; 0.58 lb/mo; 1.25 lb/mo

3. 3.1%/hr; -21.2%/hr

5. a. Car A   **b.** They are traveling at the same speed.
 c. Car B   **d.** Both cars covered the same distance.

7. a. P_2 **b.** P_1 **c.** Bactericide B; bactericide A

9. 0   **11.** 2   **13.** $6x$ **15.** $-2x + 3$

17. 2; $y = 2x + 7$ **19.** 6; $y = 6x - 3$

21. $\frac{1}{9}$; $y = \frac{1}{9}x - \frac{2}{3}$

23. a. $4x$ **b.** $y = 4x - 1$
 c.

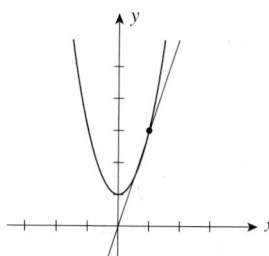

25. a. $2x - 2$ **b.** $(1, 0)$
 c.

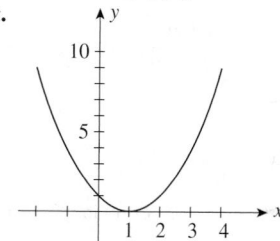

 d. 0

27. a. 6; 5.5; 5.1   **b.** 5
 c. The computations in part (a) show that as h approaches zero, the average velocity approaches the instantaneous velocity.

29. a. 130 ft/sec; 128.2 ft/sec; 128.02 ft/sec   **b.** 128 ft/sec
 c. The computations in part (a) show that as the time intervals over which the average velocity are computed become smaller and smaller, the average velocity approaches the instantaneous velocity of the car at $t = 20$.

31. a. 5 sec   **b.** 80 ft/sec   **c.** 160 ft/sec

33. a. $-\frac{1}{6}$ liter/atmosphere   **b.** $-\frac{1}{4}$ liter/atmosphere

35. a. $-\frac{2}{3}x + 7$ **b.** 333/quarter; $-$13,000$/quarter

37. $6 billion/yr; $10 billion/yr

39. Average rate of change of the seal population over $[a, a + h]$; instantaneous rate of change of the seal population at $x = a$

41. Average rate of change of the country's industrial production over $[a, a + h]$; instantaneous rate of change of the country's industrial production at $x = a$

43. Average rate of change of atmospheric pressure over $[a, a + h]$; instantaneous rate of change of atmospheric pressure at $x = a$

45. a. Yes **b.** No **c.** No

47. a. Yes **b.** Yes **c.** No

49. a. No **b.** No **c.** No

51. 32.1, 30.939, 30.814, 30.8014, 30.8001, 30.8000; 30.8 ft/sec

53. False

55.

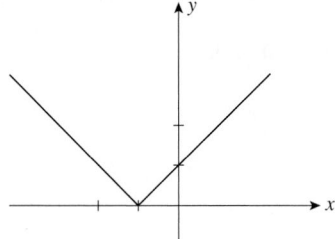

57. $a = 2, b = -1$

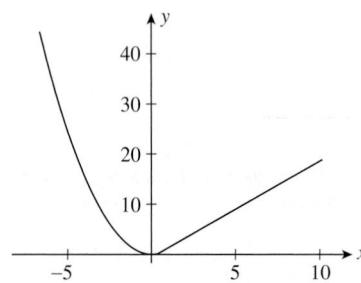

Using Technology Exercises 11.6, page 678

1. a. $y = 4x - 3$
b.
3. a. $y = 9x - 11$
b.

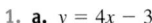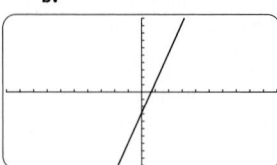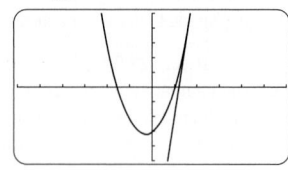

5. a. $y = \frac{1}{4}x + 1$
b.
7. a. 4
b. $y = 4x - 1$
c.

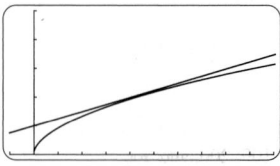

9. a. 0.75
b. $y = 0.75x - 1$
c.

11. a. 4.02
b. $y = 4.02x - 3.57$
c.

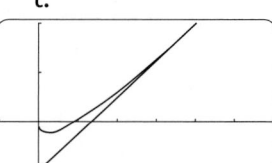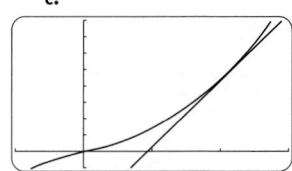

13. a.

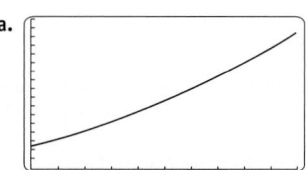

b. 41.22¢/mi **c.** 1.22¢/mi/yr

Chapter 11 Concept Review, page 679

1. Domain; range; B **2.** Domain, $f(x)$; vertical, point

3. $f(x) \pm g(x)$; $f(x)g(x)$; $\dfrac{f(x)}{g(x)}$; $A \cap B$; $A \cap B$; 0

4. $g[f(x)]$; f; $f(x)$; g

5. a. $P(x) = a_0x^n + a_1x^{n-1} + \cdots + a_{n-1}x + a_n$
  $(a_0 \neq 0, n,$ a positive integer)
b. Linear; quadratic; cubic
c. Quotient; polynomials
d. x^r (r, a real number)

6. L; $f(x)$; L; a

7. a. L^r **b.** $L \pm M$ **c.** LM

  d. $\dfrac{L}{M}$; $M \neq 0$

8. a. L; x **b.** M; negative; absolute

9. a. Right **b.** Left **c.** L; L

10. a. Continuous **b.** Discontinuous **c.** Every

11. a. a; a; $g(a)$; **b.** Everywhere **c.** $Q(x)$

12. a. $[a, b]$; $f(c) = M$ **b.** $f(x) = 0$; (a, b)

13. a. $f'(a)$ **b.** $y = f(a) + m(x - a)$

14. a. $\dfrac{f(a + h) - f(a)}{h}$ **b.** $\lim\limits_{h \to 0} \dfrac{f(a + h) - f(a)}{h}$

Chapter 11 Review Exercises, page 680

1. a. $(-\infty, 9]$ **b.** $(-\infty, -1) \cup \left(-1, \frac{3}{2}\right) \cup \left(\frac{3}{2}, \infty\right)$

2. a. 0 **b.** $3a^2 + 17a + 20$
  c. $12a^2 + 10a - 2$
  d. $3a^2 + 6ah + 3h^2 + 5a + 5h - 2$

3. a.

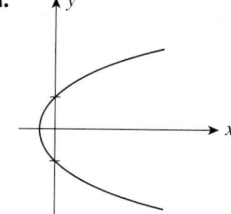

b. No   **c.** Yes

4.

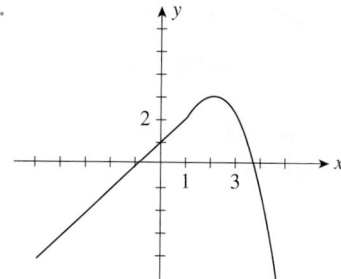

5. a. $\dfrac{2x + 3}{x}$ **b.** $\dfrac{1}{x(2x + 3)}$ **c.** $\dfrac{1}{2x + 3}$ **d.** $\dfrac{2}{x} + 3$

6. −3   **7.** 2   **8.** −21   **9.** 0

10. −1   **11.** The limit does not exist.

12. 7   **13.** $\frac{9}{2}$ **14.** 1   **15.** $\frac{1}{2}$

16. 1   **17.** 1   **18.** $\frac{3}{2}$

19. The limit does not exist.

20.

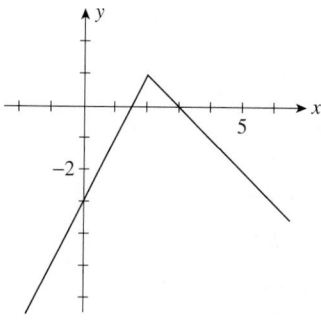

1; 1; 1

21.

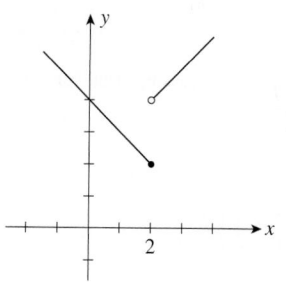

4; 2; the limit does not exist.

22. $x = 2$ **23.** $x = -\frac{1}{2}, 1$ **24.** $x = -1$ **25.** $x = 0$

26. a. 3; 2.5; 2.1   **b.** 2   **27.** 3

28. $\dfrac{1}{x^2}$ **29.** $\dfrac{3}{2}; y = \dfrac{3}{2}x + 5$

30. −4; $y = -4x + 4$ **31. a.** Yes   **b.** No   **32.** 54,000

33. a. $S(t) = t + 2.4$ **b.** $5.4 million

34. a. $C(x) = 6x + 30,000$ **b.** $R(x) = 10x$
   c. $P(x) = 4x - 30,000$ **d.** ($6000); $2000; $18,000

35. $\left(6, \dfrac{21}{2}\right)$ **36.** $P(x) = 8x - 20,000$ **37.** 6000; $22

38. 117 mg   **39.** $400,000   **40.** $45,000

41. 400; 800   **42.** 990; 2240

43.

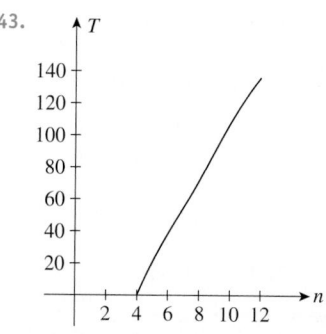

As the length of the list increases, the time taken to learn the list increases by a very large amount.

44. After $5\frac{1}{2}$ years

45. 5000; $20

46. a. πr^2 **b.** $2t$ **c.** $4\pi t^2$ **d.** 3600π ft^2

47. $C(x) = \begin{cases} 5 & \text{if } \quad 1 \le x \le 100 \\ 9 & \text{if } 100 < x \le 200 \\ 12.50 & \text{if } 200 < x \le 300 \\ 15.00 & \text{if } 300 < x \le 400 \\ 7 + 0.02x & \text{if } \qquad x > 400 \end{cases}$

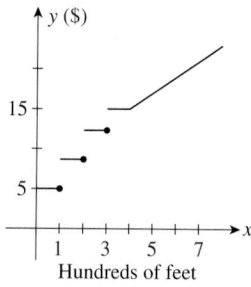

The function is discontinuous at $x = 100, 200,$ and 300.

48. 20

Chapter 11 Before Moving On, page 682

1. **a.** 3   **b.** 2   **c.** $\frac{17}{4}$

2. **a.** $\frac{1}{x+1} + x^2 + 1$ **b.** $\frac{x^2+1}{x+1}$ **c.** $\frac{1}{x^2+2}$ **d.** $\frac{1}{(x+1)^2} + 1$

3. $108x^2 - 4x^3$ 4. 2   5. **a.** 0   **b.** 1; no

6. $-1; y = -x$

CHAPTER 12

Exercises 12.1, page 691

1. 0   3. $5x^4$ 5. $2.1x^{1.1}$ 7. $6x$ 9. $2\pi r$

11. $\frac{3}{x^{2/3}}$ 13. $\frac{3}{2\sqrt{x}}$ 15. $-84x^{-13}$ 17. $10x - 3$

19. $-3x^2 + 4x$ 21. $0.06x - 0.4$ 23. $2x - 4 - \frac{3}{x^2}$

25. $16x^3 - 7.5x^{3/2}$ 27. $-\frac{3}{x^2} - \frac{8}{x^3}$ 29. $-\frac{16}{t^5} + \frac{9}{t^4} - \frac{2}{t^2}$

31. $2 - \frac{5}{2\sqrt{x}}$ 33. $-\frac{4}{x^3} + \frac{1}{x^{4/3}}$

35. **a.** 20   **b.** -4 **c.** 20   37. 3   39. 11

41. $m = 5; y = 5x - 4$ 43. $m = -2; y = -2x + 2$

45. **a.** $(0, 0)$
   b.
   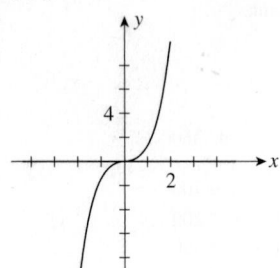

47. **a.** $(-2, -7), (2, 9)$
   b. $y = 12x + 17$ and $y = 12x - 15$
   c.
   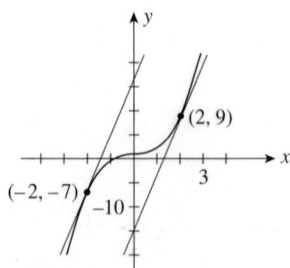

49. **a.** $(0, 0); (1, -\frac{13}{12})$
   b. $(0, 0); (2, -\frac{8}{3}); (-1, -\frac{5}{12})$
   c. $(0, 0); (4, \frac{80}{3}); (-3, \frac{81}{4})$

51. **a.** $\frac{16\pi}{9}$ cm³/cm   **b.** $\frac{25\pi}{4}$ cm³/cm

53. **a.** 16.3 million   **b.** 14.3 million/yr
   c. 66.8 million   **d.** 11.7 million/yr

55. **a.** 49.6%; 41.13%; 36.87%; 34.11%
   b. -5.55%/yr; -3.32%/yr

57. **a.** 51.5%   **b.** 1.95%/yr

59. **a.** $120 - 30t$ **b.** 120 ft/sec   **c.** 240 ft

61. **a.** 5%; 11.3%; 15.5%   **b.** 0.63%/yr; 0.525%/yr

63. **a.** -0.9 thousand metric tons/yr; 20.3 thousand metric tons/yr
   b. Yes

65. **a.** 15 pts/yr; 12.6 pts/yr; 0 pts/yr   **b.** 10 pts/yr

67. **a.** $(0.0001)(\frac{5}{4})x^{1/4}$ **b.** \$0.00125/radio

69. **a.** $20\left(1 - \frac{1}{\sqrt{t}}\right)$
   b. 50 mph; 30 mph; 33.43 mph
   c. -8.28; 0; 5.86; at 6:30 a.m., the average velocity is decreasing at the rate of 8.28 mph/hr; at 7 a.m., it is unchanged; and at 8 a.m., it is increasing at the rate of 5.86 mph.

71. 32 turtles/yr; 428 turtles/yr; 3260 turtles

73. **a.** 12%; 23.9%   **b.** 0.8%/yr; 1.1%/yr

75. True

Using Technology Exercises 12.1, page 696

1. 1   3. 0.4226   5. 0.1613

7. **a.**
   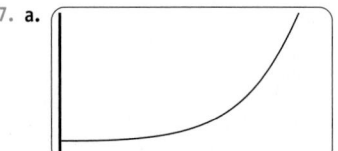
   b. 3.4295 ppm; 105.4332 ppm

9. **a.**
   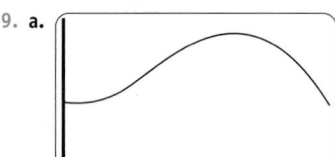
   b. Decreasing at the rate of 9 days/yr; increasing at the rate of 13 days/yr

11. **a.**
   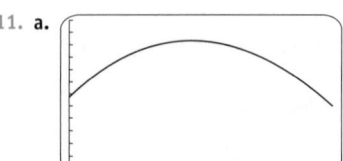
   b. Increasing at the rate of 1.1557%/yr; decreasing at the rate of 0.2116%/yr

Exercises 12.2, page 704

1. $2x(2x) + (x^2 + 1)(2)$, or $6x^2 + 2$

3. $(t - 1)(2) + (2t + 1)(1)$, or $4t - 1$

5. $(3x + 1)(2x) + (x^2 - 2)(3)$, or $9x^2 + 2x - 6$

7. $(x^3 - 1)(1) + (x + 1)(3x^2)$, or $4x^3 + 3x^2 - 1$

9. $(w^3 - w^2 + w - 1)(2w) + (w^2 + 2)(3w^2 - 2w + 1)$, or
$5w^4 - 4w^3 + 9w^2 - 6w + 2$

11. $(5x^2 + 1)(x^{-1/2}) + (2x^{1/2} - 1)(10x)$, or $\dfrac{25x^2 - 10x\sqrt{x} + 1}{\sqrt{x}}$

13. $\dfrac{(x^2 - 5x + 2)(x^2 + 2)}{x^2} + \dfrac{(x^2 - 2)(2x - 5)}{x}$, or $\dfrac{3x^4 - 10x^3 + 4}{x^2}$

15. $\dfrac{-1}{(x - 2)^2}$ 17. $\dfrac{2x + 1 - (x - 1)(2)}{(2x + 1)^2}$, or $\dfrac{3}{(2x + 1)^2}$

19. $-\dfrac{2x}{(x^2 + 1)^2}$ 21. $\dfrac{s^2 + 2s + 4}{(s + 1)^2}$

23. $\dfrac{(\frac{1}{2}x^{-1/2})[(x^2 + 1) - 4x^2]}{(x^2 + 1)^2}$, or $\dfrac{1 - 3x^2}{2\sqrt{x}(x^2 + 1)^2}$

25. $\dfrac{2x^3 + 2x^2 + 2x - 2x^3 - x^2 - 4x - 2}{(x^2 + x + 1)^2}$, or $\dfrac{x^2 - 2x - 2}{(x^2 + x + 1)^2}$

27. $\dfrac{(x - 2)(3x^2 + 2x + 1) - (x^3 + x^2 + x + 1)}{(x - 2)^2}$, or
$\dfrac{2x^3 - 5x^2 - 4x - 3}{(x - 2)^2}$

29. $\dfrac{(x^2 - 4)(x^2 + 4)(2x + 8) - (x^2 + 8x - 4)(4x^3)}{(x^2 - 4)^2(x^2 + 4)^2}$, or
$\dfrac{-2x^5 - 24x^4 + 16x^3 - 32x - 128}{(x^2 - 4)^2(x^2 + 4)^2}$

31. 8   33. -9 35. $2(3x^2 - x + 3)$; 10

37. $\dfrac{-3x^4 + 2x^2 - 1}{(x^4 - 2x^2 - 1)^2}$; $-\dfrac{1}{2}$

39. 60; $y = 60x - 102$

41. $-\dfrac{1}{2}$; $y = -\dfrac{1}{2}x + \dfrac{3}{2}$

43. $y = 7x - 5$ 45. $(\frac{1}{3}, \frac{50}{27})$; $(1, 2)$

47. $(\frac{4}{3}, -\frac{770}{27})$; $(2, -30)$

49. $y = -\dfrac{1}{2}x + 1$; $y = 2x - \dfrac{3}{2}$

51. 0.125, 0.5, 2, 50; the cost of removing all of the pollutant is prohibitively high.

53. -5000/min; -1600/min; 7000; 4000

55. **a.** $\dfrac{180}{(t + 6)^2}$ **b.** 3.7; 2.2; 1.8; 1.1

c.

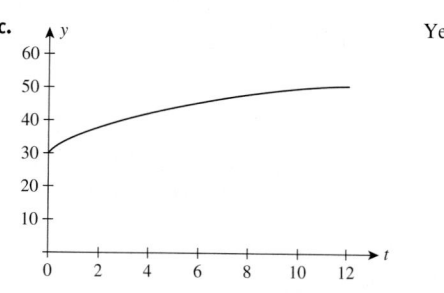

Yes

d. 50 words/min

57. Dropping at the rate of 0.0375 ppm/yr; dropping at the rate of 0.006 ppm/yr

59. False   61. False

Using Technology Exercises 12.2, page 708

1. 0.8750   3. 0.0774   5. -0.5000

7. 87,322/yr

Exercises 12.3, page 716

1. $8(2x - 1)^3$ 3. $10x(x^2 + 2)^4$

5. $3(2x - x^2)^2(2 - 2x)$, or $6x^2(1 - x)(2 - x)^2$ 7. $\dfrac{-4}{(2x + 1)^3}$

9. $3x\sqrt{x^2 - 4}$ 11. $\dfrac{3}{2\sqrt{3x - 2}}$ 13. $\dfrac{-2x}{3(1 - x^2)^{2/3}}$

15. $-\dfrac{6}{(2x + 3)^4}$ 17. $\dfrac{-1}{(2t - 3)^{3/2}}$ 19. $-\dfrac{3(16x^3 + 1)}{2(4x^4 + x)^{5/2}}$

21. $-2(3x^2 + 2x + 1)^{-3}(6x + 2) = -4(3x + 1)(3x^2 + 2x + 1)^{-3}$

23. $3(x^2 + 1)^2(2x) - 2(x^3 + 1)(3x^2)$, or $6x(2x^2 - x + 1)$

25. $3(t^{-1} - t^{-2})^2(-t^{-2} + 2t^{-3})$ 27. $\dfrac{1}{2\sqrt{x - 1}} + \dfrac{1}{2\sqrt{x + 1}}$

29. $2x^2(4)(3 - 4x)^3(-4) + (3 - 4x)^4(4x)$, or $(-12x)(4x - 1)(3 - 4x)^3$

31. $8(x - 1)^2(2x + 1)^3 + 2(x - 1)(2x + 1)^4$, or
$6(x - 1)(2x - 1)(2x + 1)^3$

33. $3\left(\dfrac{x + 3}{x - 2}\right)^2\left[\dfrac{(x - 2)(1) - (x + 3)(1)}{(x - 2)^2}\right]$, or $-\dfrac{15(x + 3)^2}{(x - 2)^4}$

35. $\dfrac{3}{2}\left(\dfrac{t}{2t + 1}\right)^{1/2}\left[\dfrac{(2t + 1)(1) - t(2)}{(2t + 1)^2}\right]$, or $\dfrac{3t^{1/2}}{2(2t + 1)^{5/2}}$

37. $\dfrac{1}{2}\left(\dfrac{u + 1}{3u + 2}\right)^{-1/2}\left[\dfrac{(3u + 2)(1) - (u + 1)(3)}{(3u + 2)^2}\right]$, or
$-\dfrac{1}{2\sqrt{u + 1}(3u + 2)^{3/2}}$

39. $\dfrac{(x^2 - 1)^4(2x) - x^2(4)(x^2 - 1)^3(2x)}{(x^2 - 1)^8}$, or $\dfrac{(-2x)(3x^2 + 1)}{(x^2 - 1)^5}$

41. $\dfrac{2x(x^2-1)^3(3x^2+1)^2[9(x^2-1)-4(3x^2+1)]}{(x^2-1)^8}$, or

$-\dfrac{2x(3x^2+13)(3x^2+1)^2}{(x^2-1)^5}$

43. $\dfrac{(2x+1)^{-1/2}[(x^2-1)-(2x+1)(2x)]}{(x^2-1)^2}$, or

$-\dfrac{3x^2+2x+1}{\sqrt{2x+1}(x^2-1)^2}$

45. $\dfrac{(t^2+1)^{1/2}(\frac{1}{2})(t+1)^{-1/2}(1)-(t+1)^{1/2}(\frac{1}{2})(t^2+1)^{-1/2}(2t)}{t^2+1}$, or

$-\dfrac{t^2+2t-1}{2\sqrt{t+1}(t^2+1)^{3/2}}$

47. $4(3x+1)^3(3)(x^2-x+1)^3+(3x+1)^4(3)(x^2-x+1)^2(2x-1)$, or $3(3x+1)^3(x^2-x+1)^2(10x^2-5x+3)$

49. $\frac{4}{3}u^{1/3}$; $6x$; $8x(3x^2-1)^{1/3}$

51. $-\dfrac{2}{3u^{5/3}}$; $6x^2-1$; $-\dfrac{2(6x^2-1)}{3(2x^3-x+1)^{5/3}}$

53. $\frac{1}{2}u^{-1/2}-\frac{1}{2}u^{-3/2}$; $3x^2-1$; $\dfrac{(3x^2-1)(x^3-x-1)}{2(x^3-x)^{3/2}}$

55. -12 **57.** 6 **59.** No

61. $y=-33x+57$ **63.** $y=\frac{43}{5}x-\frac{54}{5}$

65. 0.333 million/wk; 0.305 million/wk; 16 million; 22.7 million

67. $\dfrac{6.87775}{(5+t)^{0.795}}$; 0.53%/yr; 64.9%

69. a. $8.7 billion/yr   **b.** $92.3 billion

71. a. $0.027(0.2t^2+4t+64)^{-1/3}(0.1t+1)$
b. 0.0091 ppm

73. a. $0.03[3t^2(t-7)^4+t^3(4)(t-7)^3]$, or $0.21t^2(t-3)(t-7)^3$
b. 90.72; 0; −90.72; at 8 a.m. the level of nitrogen dioxide is increasing; at 10 a.m. the level stops increasing; at 11 a.m. the level is decreasing.

75.
$300\left[\dfrac{(t+25)\frac{1}{2}(\frac{1}{2}t^2+2t+25)^{-1/2}(t+2)-(\frac{1}{2}t^2+2t+25)^{1/2}(1)}{(t+25)^2}\right]$,

or $\dfrac{3450t}{(t+25)^2\sqrt{\frac{1}{2}t^2+2t+25}}$; 2.9 beats/min², 0.7 beats/min²,
0.2 beats/min², 179 beats/min

77. 160π ft²/sec   **79.** −27 mph/decade; 19 mph

81.
$(1.42)\left[\dfrac{(3t^2+80t+550)(14t+140)-(7t^2+140t+700)(6t+80)}{(3t^2+80t+550)^2}\right]$,

or $\dfrac{1.42(140t^2+3500t+21,000)}{(3t^2+80t+550)^2}$; 31,312 jobs/yr

83. −400 wristwatches/(dollar price increase)

85. True   **87.** True

Using Technology Exercises 12.3, page 721

1. 0.5774   **3.** 0.9390   **5.** −4.9498

7. 10,146,200/decade; 7,810,520/decade

Exercises 12.4, page 732

1. a. $C(x)$ is always increasing because as the number of units x produced increases, the amount of money that must be spent on production also increases.
b. 4000

3. a. $1.80; $1.60
b. $1.80; $1.60

5. a. $100+\dfrac{200,000}{x}$
b. $-\dfrac{200,000}{x^2}$
c. $\overline{C}(x)$ approaches $100 if the production level is very high.

7. $\dfrac{2000}{x}+2-0.0001x$; $-\dfrac{2000}{x^2}-0.0001$

9. a. $8000-200x$
b. 200, 0, −200   **c.** $40

11. a. $-0.04x^2+600x-300,000$
b. $-0.08x+600$ **c.** 200; −40
d. The profit increases as production increases, peaking at 7500 units; beyond this level, profit falls.

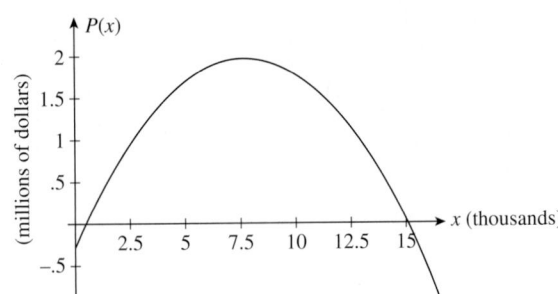

13. a. $600x-0.05x^2$; $-0.000002x^3-0.02x^2+200x-80,000$
b. $0.000006x^2-0.06x+400$; $600-0.1x$; $-0.000006x^2-0.04x+200$
c. 304; 400; 96

d.

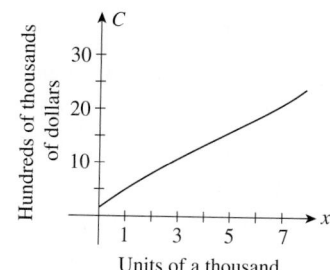

15. $0.000002x^2 - 0.03x + 400 + \dfrac{80,000}{x}$

a. $0.000004x - 0.03x - \dfrac{80,000}{x^2}$

b. -0.0132; 0.0092; the marginal average cost is negative (average cost is decreasing) when 5000 units are produced and positive (average cost is increasing) when 10,000 units are produced.

c.

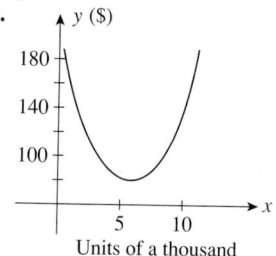

17. a. $\dfrac{50x}{0.01x^2 + 1}$ **b.** $\dfrac{50 - 0.5x^2}{(0.01x^2 + 1)^2}$

c. $44,380$; when the level of production is 2000 units, the revenue increases at the rate of \$44,380 per additional 1000 units produced.

19. \$1.21 billion/billion dollars

21. \$0.288 billion/billion dollars

23. $\frac{5}{3}$; elastic   **25.** 1; unitary   **27.** 0.104; inelastic

29. a. Inelastic; elastic   **b.** When $p = 8.66$
     c. Increase   **d.** Increase

31. a. Inelastic   **b.** Increase

33. $\dfrac{2p^2}{9 - p^2}$; for $p < \sqrt{3}$, demand is inelastic; for $p = \sqrt{3}$, demand is unitary; and for $p > \sqrt{3}$, demand is elastic.

35. True

Exercises 12.5, page 740

1. $8x - 2$; 8   **3.** $6x^2 - 6x$; $6(2x - 1)$

5. $4t^3 - 6t^2 + 12t - 3$; $12(t^2 - t + 1)$

7. $10x(x^2 + 2)^4$; $10(x^2 + 2)^3(9x^2 + 2)$

9. $6t(2t^2 - 1)(6t^2 - 1)$; $6(60t^4 - 24t^2 + 1)$

11. $14x(2x^2 + 2)^{5/2}$; $28(2x^2 + 2)^{3/2}(6x^2 + 1)$

13. $(x^2 + 1)(5x^2 + 1)$; $4x(5x^2 + 3)$

15. $\dfrac{1}{(2x + 1)^2}$; $-\dfrac{4}{(2x + 1)^3}$ **17.** $\dfrac{2}{(s + 1)^2}$; $-\dfrac{4}{(s + 1)^3}$

19. $-\dfrac{3}{2(4 - 3u)^{1/2}}$; $-\dfrac{9}{4(4 - 3u)^{3/2}}$

21. $72x - 24$ **23.** $-\dfrac{6}{x^4}$

25. $\frac{81}{8}(3s - 2)^{-5/2}$ **27.** $192(2x - 3)$

29. 128 ft/sec; 32 ft/sec^2

31. a. and b.

t	0	1	2	3	4	5	6	7
$N'(t)$	0	2.7	4.8	6.3	7.2	7.5	7.2	6.3
$N''(t)$				0.6	0	−0.6	−1.2

33. 8.1 million; 0.204 million/yr; −0.03 million/yr^2. At the beginning of 1998, there were 8.1 million people receiving disability benefits; the number was increasing at the rate of 0.2 million/yr; the rate of the rate of change of the number of people was decreasing at the rate of 0.03 million people/yr^2.

35. a. $\frac{1}{4}t^3 - 3t^2 + 8t$
     b. 0 ft/sec; 0 ft/sec; 0 ft/sec
     c. $\frac{3}{4}t^2 - 6t + 8$
     d. 8 ft/sec^2; −4 ft/sec^2; 8 ft/sec^2
     e. 0 ft; 16 ft; 0 ft

37. −0.01%/yr^2. The rate of the rate of change of the percent of Americans aged 55 and over decreases at the rate of 0.01%/yr^2.

39. False   **41.** True   **43.** True   **45.** $f(x) = x^{n+1/2}$

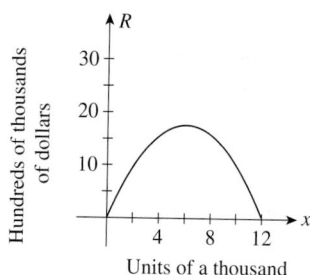

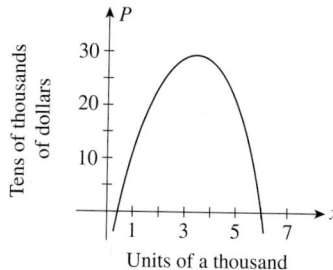

Using Technology Exercises 12.5, page 744

1. -18 3. 15.2762 5. -0.6255 7. 0.1973

9. -68.46214; at the beginning of 1988, the rate of the rate of the rate at which banks were failing was 68 banks/yr/yr/yr.

Exercises 12.6, page 753

1. a. $-\frac{1}{2}$ b. $-\frac{1}{2}$ 3. a. $-\frac{1}{x^2}$ b. $-\frac{y}{x}$

5. a. $2x - 1 + \frac{4}{x^2}$ b. $3x - 2 - \frac{y}{x}$

7. a. $\frac{1-x^2}{(1+x^2)^2}$ b. $-2y^2 + \frac{y}{x}$

9. $-\frac{x}{y}$ 11. $\frac{x}{2y}$ 13. $1 - \frac{y}{x}$ 15. $-\frac{y}{x}$

17. $-\frac{\sqrt{y}}{\sqrt{x}}$ 19. $2\sqrt{x+y} - 1$ 21. $-\frac{y^3}{x^3}$

23. $\frac{2\sqrt{xy} - y}{x - 2\sqrt{xy}}$ 25. $\frac{6x - 3y - 1}{3x + 1}$ 27. $\frac{2(2x - y^{3/2})}{3x\sqrt{y} - 4y}$

29. $-\frac{2x^2 + 2xy + y^2}{x^2 + 2xy + 2y^2}$ 31. $y = 2$ 33. $y = -\frac{3}{2}x + \frac{5}{2}$

35. $\frac{2y}{x^2}$ 37. $\frac{2y(y-x)}{(2y-x)^3}$

39. a. $\frac{dV}{dt} = \pi r\left(r\frac{dh}{dt} + 2h\frac{dr}{dt}\right)$ b. 3.6π cu in./sec

41. Dropping at the rate of 111 tires/wk

43. Increasing at the rate of 44 ten packs/wk

45. Dropping at the rate of 3.7¢/carton/wk

47. 0.37; inelastic 49. 160π ft²/sec 51. 17 ft/sec

53. 7.69 ft/sec 57. 196.8 ft/sec 59. 9 ft/sec

61. 19.2 ft/sec 63. True

Exercises 12.7, page 762

1. $4x\,dx$ 3. $(3x^2 - 1)\,dx$ 5. $\frac{dx}{2\sqrt{x+1}}$

7. $\frac{6x + 1}{2\sqrt{x}}\,dx$ 9. $\frac{x^2 - 2}{x^2}\,dx$

11. $\frac{-x^2 + 2x + 1}{(x^2 + 1)^2}\,dx$ 13. $\frac{6x - 1}{2\sqrt{3x^2 - x}}\,dx$

15. a. $2x\,dx$ b. 0.04 c. 0.0404

17. a. $-\frac{dx}{x^2}$ b. -0.05 c. -0.05263

19. 3.167 21. 7.0357 23. 1.983 25. 0.298

27. 2.50375 29. ± 8.64 cm³ 31. 18.85 ft³

33. It will drop by 40%. 35. 274 sec 37. 111,595

39. Decrease of $1.33 41. $\pm$$64,800

43. Decrease of 11 crimes/yr

45. a. $100,000\left(1 + \frac{r}{12}\right)^{119} dr$ b. $220.49; $440.99; $661.48

47. True

Using Technology Exercises 12.7, page 766

1. 7.5787 3. 0.031220185778 5. -0.0198761598

7. $26.60/mo; $35.47/mo; $44.34/mo 9. 625

Chapter 12 Concept Review, page 767

1. a. 0 b. nx^{n-1} c. $cf'(x)$ d. $f'(x) \pm g'(x)$

2. a. $f(x)g'(x) + g(x)f'(x)$ b. $\frac{g(x)f'(x) - f(x)g'(x)}{[g(x)]^2}$

3. a. $g'[f(x)]f'(x)$ b. $n[f(x)]^{n-1}f'(x)$

4. Marginal cost; marginal revenue; marginal profit; marginal average cost

5. a. $-p\frac{f'(p)}{f(p)}$ b. Elastic; unitary; inelastic

6. Both sides; $\frac{dy}{dx}$ 7. y; $\frac{dy}{dt}$; a 8. $-\frac{x}{y}$; $-\frac{y}{x}$

9. a. $x_2 - x_1$ b. $f(x + \Delta x) - f(x)$

10. Δx; Δx; x; $f'(x)\,dx$

Chapter 12 Review Exercises, page 767

1. $15x^4 - 8x^3 + 6x - 2$ 2. $24x^5 + 8x^3 + 6x$

3. $\frac{6}{x^4} - \frac{3}{x^2}$ 4. $4t - 9t^2 + \frac{1}{2}t^{-3/2}$ 5. $-\frac{1}{t^{3/2}} - \frac{6}{t^{5/2}}$

6. $2x - \frac{2}{x^2}$ 7. $1 - \frac{2}{t^2} - \frac{6}{t^3}$ 8. $4s + \frac{4}{s^2} - \frac{1}{s^{3/2}}$

9. $2x + \frac{3}{x^{5/2}}$ 10. $\frac{(2x-1)(1) - (x+1)(2)}{(2x-1)^2}$, or $-\frac{3}{(2x-1)^2}$

11. $\frac{(2t^2+1)(2t) - t^2(4t)}{(2t^2+1)^2}$, or $\frac{2t}{(2t^2+1)^2}$

12. $\frac{(t^{1/2}+1)\frac{1}{2}t^{-1/2} - t^{1/2}(\frac{1}{2}t^{-1/2})}{(t^{1/2}+1)^2}$, or $\frac{1}{2\sqrt{t}(\sqrt{t}+1)^2}$

13. $\frac{(x^{1/2}+1)(\frac{1}{2}x^{-1/2}) - (x^{1/2}-1)(\frac{1}{2}x^{-1/2})}{(x^{1/2}+1)^2}$, or $\frac{1}{\sqrt{x}(\sqrt{x}+1)^2}$

14. $\frac{(2t^2+1)(1) - t(4t)}{(2t^2+1)^2}$, or $\frac{1 - 2t^2}{(2t^2+1)^2}$

15. $\frac{(x^2-1)(4x^3+2x) - (x^4+x^2)(2x)}{(x^2-1)^2}$, or $\frac{2x(x^4-2x^2-1)}{(x^2-1)^2}$

16. $3(4x+1)(2x^2+x)^2$ 17. $8(3x^3-2)^7(9x^2)$, or $72x^2(3x^3-2)^7$

18. $5(x^{1/2}+2)^4 \cdot \frac{1}{2}x^{-1/2}$, or $\frac{5(\sqrt{x}+2)^4}{2\sqrt{x}}$

19. $\frac{1}{2}(2t^2+1)^{-1/2}(4t)$, or $\frac{2t}{\sqrt{2t^2+1}}$

20. $\frac{1}{3}(1 - 2t^3)^{-2/3}(-6t^2)$, or $-2t^2(1 - 2t^3)^{-2/3}$

21. $-4(3t^2 - 2t + 5)^{-3}(3t - 1)$, or $-\dfrac{4(3t - 1)}{(3t^2 - 2t + 5)^3}$

22. $-\frac{3}{2}(2x^3 - 3x^2 + 1)^{-5/2}(6x^2 - 6x)$, or $-9x(x - 1)(2x^3 - 3x^2 + 1)^{-5/2}$

23. $2\left(x + \dfrac{1}{x}\right)\left(1 - \dfrac{1}{x^2}\right)$, or $\dfrac{2(x^2 + 1)(x^2 - 1)}{x^3}$

24. $\dfrac{(2x^2 + 1)^2(1) - (1 + x)2(2x^2 + 1)(4x)}{(2x^2 + 1)^4}$, or $-\dfrac{6x^2 + 8x - 1}{(2x^2 + 1)^3}$

25. $(t^2 + t)^4(4t) + 2t^2 \cdot 4(t^2 + t)^3(2t + 1)$, or $4t^2(5t + 3)(t^2 + t)^3$

26. $(2x + 1)^3 \cdot 2(x^2 + x)(2x + 1) + (x^2 + x)^2 3(2x + 1)^2(2)$, or $2(2x + 1)^2(x^2 + x)(7x^2 + 7x + 1)$

27. $x^{1/2} \cdot 3(x^2 - 1)^2(2x) + (x^2 - 1)^3 \cdot \frac{1}{2}x^{-1/2}$, or $\dfrac{(13x^2 - 1)(x^2 - 1)^2}{2\sqrt{x}}$

28. $\dfrac{(x^3 + 2)^{1/2}(1) - x \cdot \frac{1}{2}(x^3 + 2)^{-1/2} \cdot 3x^2}{x^3 + 2}$, or $\dfrac{4 - x^3}{2(x^3 + 2)^{3/2}}$

29. $\dfrac{(4x - 3)\frac{1}{2}(3x + 2)^{-1/2}(3) - (3x + 2)^{1/2}(4)}{(4x - 3)^2}$, or $-\dfrac{12x + 25}{2\sqrt{3x + 2}(4x - 3)^2}$

30. $\dfrac{(t + 1)^3\frac{1}{2}(2t + 1)^{-1/2}(2) - (2t + 1)^{1/2} \cdot 3(t + 1)^2(1)}{(t + 1)^6}$, or $-\dfrac{5t + 2}{\sqrt{2t + 1}(t + 1)^4}$

31. $2(12x^2 - 9x + 2)$ 32. $-\dfrac{1}{4x^{3/2}} + \dfrac{3}{4x^{5/2}}$

33. $\dfrac{(t^2 + 4)^2(-2t) - (4 - t^2)2(t^2 + 4)(2t)}{(t^2 + 4)^4}$, or $\dfrac{2t(t^2 - 12)}{(t^2 + 4)^3}$

34. $2(15x^4 + 12x^2 + 6x + 1)$

35. $2(2x^2 + 1)^{-1/2} + 2x\left(-\dfrac{1}{2}\right)(2x^2 + 1)^{-3/2}(4x)$, or $\dfrac{2}{(2x^2 + 1)^{3/2}}$

36. $(t^2 + 1)(14t) + (7t^2 + 1)(2)(t^2 + 1)(2t)$, or $6t(t^2 + 1)(7t^2 + 3)$

37. $\dfrac{2x}{y}$ 38. $\dfrac{2x^2 - y}{x}$ 39. $-\dfrac{2x}{y^2 - 1}$ 40. $-\dfrac{x(1 + 2y^2)}{y(2x^2 + 1)}$

41. $\dfrac{x - 2y}{2x + y}$ 42. $\dfrac{4y - 6xy - 1}{3x^2 - 4x - 2}$ 43. $\dfrac{2(x^4 - 1)}{x^3}\,dx$

44. $-\dfrac{3x^2}{2(x^3 + 1)^{3/2}}\,dx$

45. **a.** $\dfrac{2x}{\sqrt{2(x^2 + 2)}}\,dx$ **b.** 0.1333   **c.** 0.1335; differ by 0.0002

46. 2.9926   47. **a.** $(2, -25)$ and $(-1, 14)$
    b. $y = -4x - 17; y = -4x + 10$

48. **a.** $\left(-2, \frac{25}{3}\right)$ and $\left(1, -\frac{13}{6}\right)$
    b. $y = -2x + \frac{13}{3}; y = -2x - \frac{1}{6}$

49. $y = -\dfrac{\sqrt{3}}{3}x + \dfrac{4}{3}\sqrt{3}$ 50. $y = 112x - 80$

51. $-\dfrac{48}{(2x - 1)^4}; (-\infty; \frac{1}{2}) \cup (\frac{1}{2}, \infty)$

52. **a.** $\frac{1}{3}$; inelastic   **b.** 1; unitary   **c.** 3; elastic

53. $\dfrac{25}{2(25 - \sqrt{p})}$; for $p > 156.25$, demand is elastic; for $p = 156.25$, demand is unitary; and for $p < 156.25$, demand is inelastic.

54. **a.** Inelastic   **b.** Increase   55. **a.** Elastic   **b.** Decrease

56. **a.** 15%; 31.99%   **b.** 0.51%/yr; 1.04%/yr

57. 200 subscribers/wk

58. $-14.346(1 + t)^{-1.45}$; $-2.92¢$/min; $19.45¢$/min

59. $\approx$75 yr; 0.07 yr/yr

60. **a.** \$2.20; \$2.20   **b.** $\dfrac{2500}{x} + 2.2$; $-\dfrac{2500}{x^2}$
    c. $\lim\limits_{x\to\infty}\left(\dfrac{2500}{x} + 2.2\right) = 2.2$

61. **a.** $-0.02x^2 + 600x$ **b.** $-0.04x + 600$
    c. 200; the sale of the 10,001st phone will bring a revenue of \$200.

62. **a.** $2000x - 0.04x^2$; $-0.000002x^3 - 0.02x^2 + 1000x - 120,000$; $0.000002x^2 - 0.02x + 1000 + \dfrac{120,000}{x}$
    b. $0.000006x^2 - 0.04x + 1000$; $2000 - 0.08x$; $-0.000006x^2 - 0.04x + 1000$; $0.000004x - 0.02 - \dfrac{120,000}{x^2}$
    c. 934; 1760; 826
    d. -0.0048; 0.010125; at a production level of 5000, the average cost is decreasing by 0.48¢/unit; at a production level of 8000, the average cost is increasing by 1.0125¢/unit.

Chapter 12 Before Moving On, page 769

1. $6x^2 - \dfrac{1}{x^{2/3}} - \dfrac{10}{3x^{5/3}}$ 2. $\dfrac{4x^2 - 1}{\sqrt{2x^2 - 1}}$

3. $-\dfrac{2x^2 + 2x - 1}{(x^2 + x + 1)^2}$

4. $-\dfrac{1}{2(x + 1)^{3/2}}; \dfrac{3}{4(x + 1)^{5/2}}; -\dfrac{15}{8(x + 1)^{7/2}}$

5. $\dfrac{-y^2 + 2xy - 3x^2}{2xy - x^2}$

6. **a.** $\dfrac{2x^2 + 5}{\sqrt{x^2 + 5}}\,dx$ **b.** 0.0433

CHAPTER 13

Exercises 13.1, page 782

1. Decreasing on $(-\infty, 0)$ and increasing on $(0, \infty)$

3. Increasing on $(-\infty, -1) \cup (1, \infty)$ and decreasing on $(-1, 1)$

5. Decreasing on $(-\infty, 0) \cup (2, \infty)$ and increasing on $(0, 2)$

7. Decreasing on $(-\infty, -1) \cup (1, \infty)$ and increasing on $(-1, 1)$

9. Increasing on $(20.2, 20.6) \cup (21.7, 21.8)$, constant on $(19.6, 20.2) \cup (20.6, 21.1)$, and decreasing on $(21.1, 21.7) \cup (21.8, 22.7)$

11. Increasing on $(-\infty, \infty)$

13. Decreasing on $(-\infty, \frac{3}{2})$ and increasing on $(\frac{3}{2}, \infty)$

15. Decreasing on $(-\infty, -\sqrt{3}/3) \cup (\sqrt{3}/3, \infty)$ and increasing on $(-\sqrt{3}/3, \sqrt{3}/3)$

17. Increasing on $(-\infty, -2) \cup (0, \infty)$ and decreasing on $(-2, 0)$

19. Increasing on $(-\infty, 3) \cup (3, \infty)$

21. Decreasing on $(-\infty, 0) \cup (0, 3)$ and increasing on $(3, \infty)$

23. Decreasing on $(-\infty, 2) \cup (2, \infty)$

25. Decreasing on $(-\infty, 1) \cup (1, \infty)$

27. Increasing on $(-\infty, 0) \cup (0, \infty)$

29. Increasing on $(-1, \infty)$

31. Increasing on $(-4, 0)$; decreasing on $(0, 4)$

33. Increasing on $(-\infty, 0) \cup (0, \infty)$

35. Relative maximum: $f(0) = 1$; relative minima: $f(-1) = 0$ and $f(1) = 0$

37. Relative maximum: $f(-1) = 2$; relative minimum: $f(1) = -2$

39. Relative maximum: $f(1) = 3$; relative minimum: $f(2) = 2$

41. Relative minimum: $f(0) = 2$ 43. a   45. d

47. Relative minimum: $f(2) = -4$

49. Relative maximum: $f(3) = 15$

51. None

53. Relative maximum: $g(0) = 4$; relative minimum: $g(2) = 0$

55. Relative maximum: $f(0) = 0$; relative minimum: $f(-1) = -\frac{1}{2}$ and $f(1) = -\frac{1}{2}$

57. Relative minimum: $F(3) = -5$; relative maximum: $F(-1) = \frac{17}{3}$

59. Relative minimum: $g(3) = -19$ 61. None

63. Relative maximum: $f(-3) = -4$; relative minimum: $f(3) = 8$

65. Relative maximum: $f(1) = \frac{1}{2}$; relative minimum: $f(-1) = -\frac{1}{2}$

67. Relative minimum: $f(1) = 0$

69.

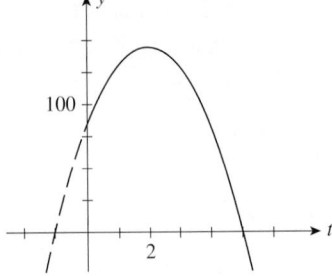

Rising in the time interval $(0, 2)$; falling in the time interval $(2, 5)$; when $t = 5$ sec

71. The percent of the U.S. population age 65 and over afflicted by the disease increases with age.

73. Rising on $(0, 33)$ and descending on $(33, T)$ for some positive number T.

75. f is decreasing on $(0, 1)$ and increasing on $(1, 4)$. The average speed decreases from 6 a.m. to 7 a.m. and then picks up from 7 a.m. to 10 a.m.

77. **a.** Increasing on $(0, 6)$ **b.** Sales will be increasing.

79. **a.** Decreasing on $(0, 21.4)$; increasing on $(21.4, 30)$
    b. The percent of men 65 years and older in the workforce was decreasing from 1970 until mid-1991 and increasing from mid-1991 through 2000.

83. Spending was increasing from 2001 to 2006.

85. Increasing on $(0, 1)$ and decreasing on $(1, 4)$

87. Increasing on $(0, 4.5)$ and decreasing on $(4.5, 11)$; the pollution is increasing from 7 a.m. to 11:30 a.m. and decreasing from 11:30 a.m. to 6 p.m.

89. **a.** $0.0021t^2 - 0.0061t + 0.1$
    b. Decreasing on $(0, 1.5)$ and increasing on $(1.5, 15)$. The gap (shortage of nurses) was decreasing from 2000 to mid-2001 and is expected to be increasing from mid-2001 to 2015.
    c. $(1.5, 0.096)$. The gap was smallest ($\approx 96{,}000$) in mid-2001.

91. True   93. True   95. False

99. **a.** $-2x$ if $x \neq 0$ **b.** No

Using Technology Exercises 13.1, page 790

1. **a.** f is decreasing on $(-\infty, -0.2934)$ and increasing on $(-0.2934, \infty)$.
   b. Relative minimum: $f(-0.2934) = -2.5435$

3. **a.** f is increasing on $(-\infty, -1.6144) \cup (0.2390, \infty)$ and decreasing on $(-1.6144, 0.2390)$.
   b. Relative maximum: $f(-1.6144) = 26.7991$; relative minimum: $f(0.2390) = 1.6733$

5. **a.** f is decreasing on $(-\infty, -1) \cup (0.33, \infty)$ and increasing on $(-1, 0.33)$.
   b. Relative maximum: $f(0.33) = 1.11$; relative minimum: $f(-1) = -0.63$

7. a. f is decreasing on $(-1, -0.71)$ and increasing on $(-0.71, 1)$.
   b. Relative minimum: $f(-0.71) = -1.41$

9. a.

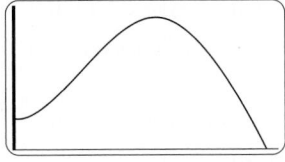

   b. f is decreasing on $(0, 0.2398) \cup (6.8758, 12)$ and increasing on $(0.2398, 6.8758)$.
   c. $(6.8758, 200.14)$

11. a.

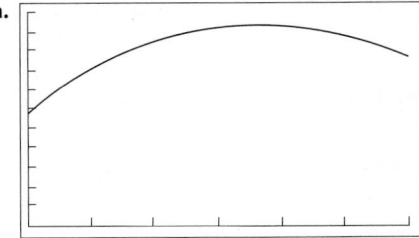

   b. Increasing on $(0, 3.6676)$ and decreasing on $(3.6676, 6)$

13. f is decreasing on $(0, 1)$ and increasing on $(1, 4)$; relative minimum: 32. The speed of the traffic flow drops from 6 a.m. to 7 a.m., reaching a low of 32 mph. Thereafter it increases till 10 a.m.

Exercises 13.2, page 800

1. Concave downward on $(-\infty, 0)$ and concave upward on $(0, \infty)$; inflection point: $(0, 0)$

3. Concave downward on $(-\infty, 0) \cup (0, \infty)$

5. Concave upward on $(-\infty, 0) \cup (1, \infty)$ and concave downward on $(0, 1)$; inflection points: $(0, 0)$ and $(1, -1)$

7. Concave downward on $(-\infty, -2) \cup (-2, 2) \cup (2, \infty)$

9. a   **11.** b

13. a. $D_1'(t) > 0$, $D_2'(t) > 0$, $D_1''(t) > 0$, and $D_2''(t) < 0$ on $(0, 12)$
   b. With or without the proposed promotional campaign, the deposits will increase; with the promotion, the deposits will increase at an increasing rate; without the promotion, the deposits will increase at a decreasing rate.

15. At the time t_0, corresponding to its t-coordinate, the restoration process is working at its peak.

21. Concave upward on $(-\infty, \infty)$

23. Concave downward on $(-\infty, 0)$; concave upward on $(0, \infty)$

25. Concave upward on $(-\infty, 0) \cup (3, \infty)$; concave downward on $(0, 3)$

27. Concave downward on $(-\infty, 0) \cup (0, \infty)$

29. Concave downward on $(-\infty, 4)$

31. Concave downward on $(-\infty, 2)$; concave upward on $(2, \infty)$

33. Concave upward on $(-\infty, -\sqrt{6}/3) \cup (\sqrt{6}/3, \infty)$; concave downward on $(-\sqrt{6}/3, \sqrt{6}/3)$

35. Concave downward on $(-\infty, 1)$; concave upward on $(1, \infty)$

37. Concave upward on $(-\infty, 0) \cup (0, \infty)$

39. Concave upward on $(-\infty, 2)$; concave downward on $(2, \infty)$

41. $(0, -2)$ **43.** $(1, -15)$ **45.** $(0, 1)$ and $\left(\frac{2}{3}, \frac{11}{27}\right)$

47. $(0, 0)$ **49.** $(1, 2)$ **51.** $(-\sqrt{3}/3, 3/2)$ and $(\sqrt{3}/3, 3/2)$

53. Relative maximum: $f(1) = 5$

55. None

57. Relative maximum: $f(-1) = -\frac{22}{3}$, relative minimum: $f(5) = -\frac{130}{3}$

59. Relative maximum: $g(-3) = -6$; relative minimum: $g(3) = 6$

61. None

63. Relative minimum: $f(-2) = 12$

65. Relative maximum: $g(1) = \frac{1}{2}$; relative minimum: $g(-1) = -\frac{1}{2}$

67. Relative maximum: $f(0) = 0$; relative minimum: $f\left(\frac{4}{3}\right) = \frac{256}{27}$

69.

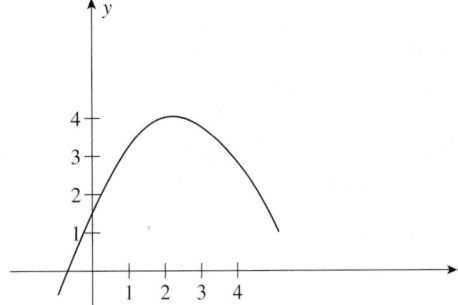

71.

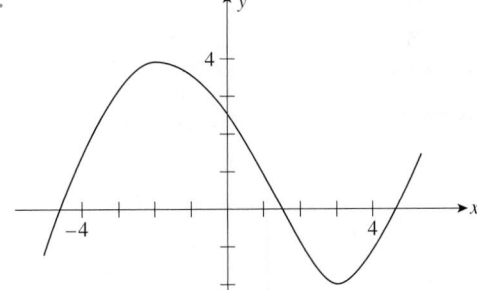

73.

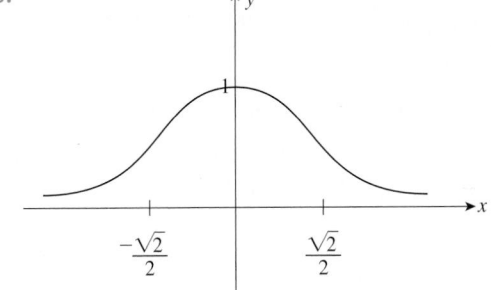

75. **a.** N is increasing on $(0, 12)$.
  b. $N''(t) < 0$ on $(0, 6)$ and $N''(t) > 0$ on $(6, 12)$
  c. The rate of growth of the number of help-wanted advertisements was decreasing over the first 6 mo of the year and increasing over the last 6 mo.

77. $f(t)$ increases at an increasing rate until the water level reaches the middle of the vase at which time (corresponding to the inflection point) $f(t)$ is increasing at the fastest rate. After that, $f(t)$ increases at a decreasing rate until the vase is filled.

79. **b.** The rate was increasing.

81. 10 a.m.

83. The rate of business spending on technology was increasing from 2000 through 2005.

85. **a.** Concave upward on $(0, 150)$; concave downward on $(150, 400)$; $(150, 28,550)$
  b. \$140,000

89. $(1.9, 784.9)$; the rate of annual pharmacy spending slowed down near the end of 2000.

91. **a.** $74.925t^2 - 99.62t + 41.25$; $149.85t - 99.62$
  c. $(0.66, 12.91)$; the rate was increasing least rapidly around August 1999.

97. True   99. True

Using Technology Exercises 13.2, page 808

1. **a.** f is concave upward on $(-\infty, 0) \cup (1.1667, \infty)$ and concave downward on $(0, 1.1667)$.
  b. $(1.1667, 1.1153)$; $(0, 2)$

3. **a.** f is concave downward on $(-\infty, 0)$ and concave upward on $(0, \infty)$.
  b. $(0, 2)$

5. **a.** f is concave downward on $(-\infty, 0)$ and concave upward on $(0, \infty)$.
  b. $(0, 0)$

7. **a.** f is concave downward on $(-\infty, -2.4495) \cup (0, 2.4495)$ and concave upward on $(-2.4495, 0) \cup (2.4495, \infty)$.
  b. $(2.4495, 0.3402)$; $(-2.4495, -0.3402)$

9. **a.**
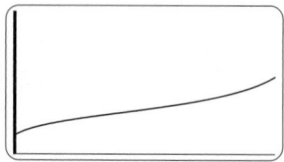
11. **a.**
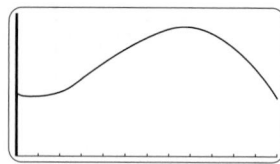

  b. $(5.5318, 35.9483)$ **b.** $(3.9024, 77.0919)$
  c. $t = 5.5318$

13. **a.**
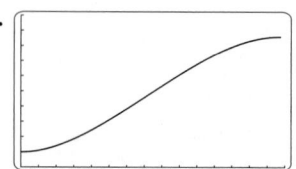

  b. April 1993 $(t = 7.36)$

Exercises 13.3, page 817

1. Horizontal asymptote: $y = 0$

3. Horizontal asymptote: $y = 0$; vertical asymptote: $x = 0$

5. Horizontal asymptote: $y = 0$; vertical asymptotes: $x = -1$ and $x = 1$

7. Horizontal asymptote: $y = 3$; vertical asymptote: $x = 0$

9. Horizontal asymptotes: $y = 1$ and $y = -1$

11. Horizontal asymptote: $y = 0$; vertical asymptote: $x = 0$

13. Horizontal asymptote: $y = 0$; vertical asymptote: $x = 0$

15. Horizontal asymptote: $y = 1$; vertical asymptote: $x = -1$

17. None

19. Horizontal asymptote: $y = 1$; vertical asymptotes: $t = -3$ and $t = 3$

21. Horizontal asymptote: $y = 0$; vertical asymptotes: $x = -2$ and $x = 3$

23. Horizontal asymptote: $y = 2$; vertical asymptote: $t = 2$

25. Horizontal asymptote: $y = 1$; vertical asymptotes: $x = -2$ and $x = 2$

27. None

29. f is the derivative function of the function g.

31.
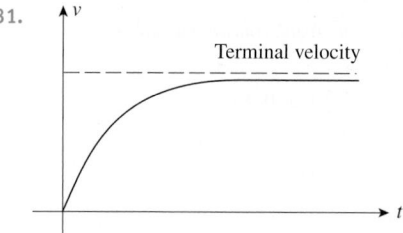

33.
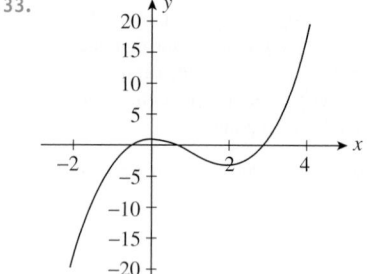

35.

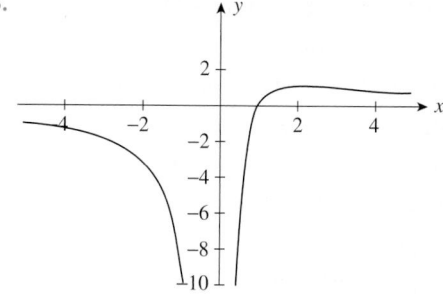

37.

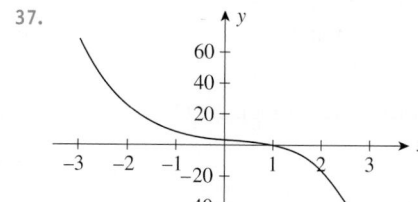

39.

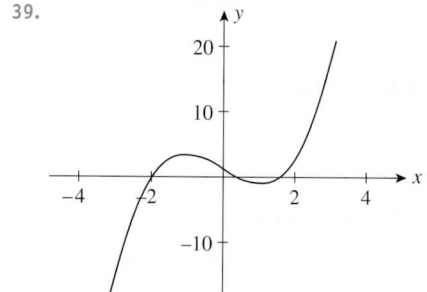

41.

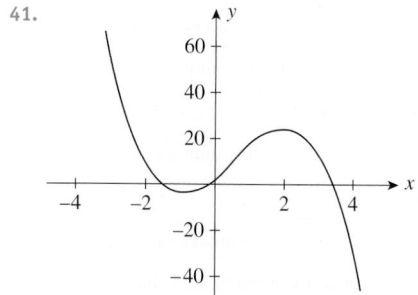

43.

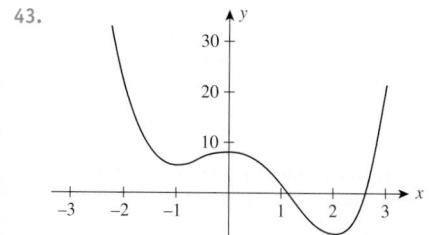

45.

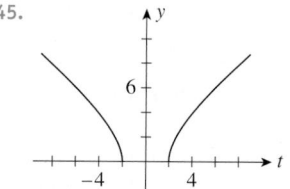

47.

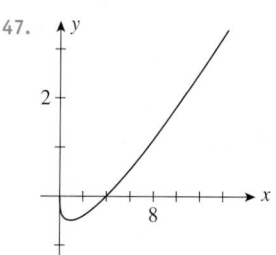

49.

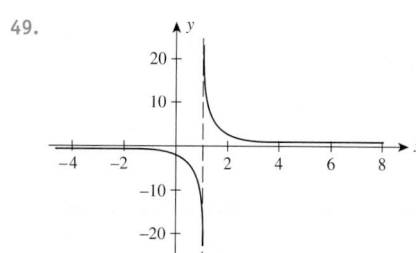

51.

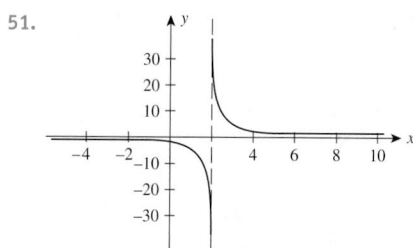

53.

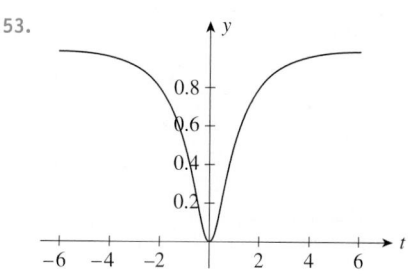

55.

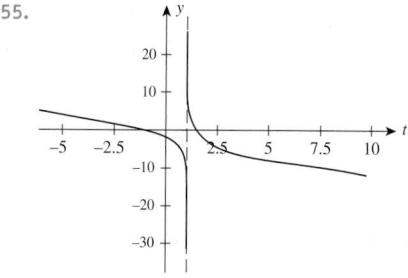

57.

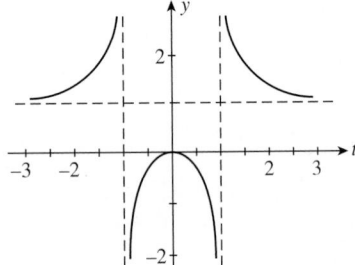

59.

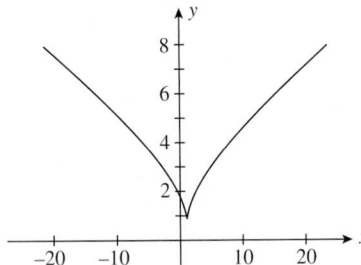

61. a. $x = 100$ **b.** No

63. a. $y = 0$

b. As time passes, the concentration of the drug decreases and approaches zero.

65.

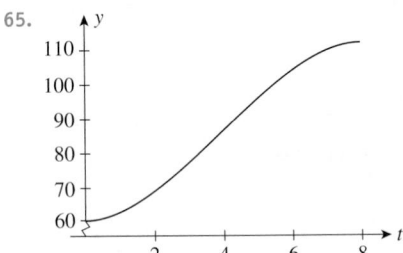

67.

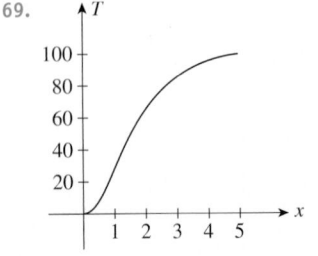

69.

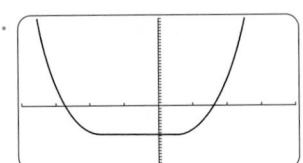

Using Technology Exercises 13.3, page 823

1.

3.

5. -0.9733; 2.3165, 4.6569 **7.** -1.1301; 2.9267

9. 1.5142

Exercises 13.4, page 832

1. None

3. Absolute minimum value: 0

5. Absolute maximum value: 3; absolute minimum value: -2

7. Absolute maximum value: 3; absolute minimum value: $-\frac{27}{16}$

9. Absolute minimum value: $-\frac{41}{8}$

11. No absolute extrema

13. Absolute maximum value: 1

15. Absolute maximum value: 5; absolute minimum value: -4

17. Absolute maximum value: 10; absolute minimum value: 1

19. Absolute maximum value: 19; absolute minimum value: -1

21. Absolute maximum value: 16; absolute minimum value: -1

23. Absolute maximum value: 3; absolute minimum value: $\frac{5}{3}$

25. Absolute maximum value: $\frac{37}{3}$; absolute minimum value: 5

27. Absolute maximum value ≈ 1.04; absolute minimum value: -1.5

29. No absolute extrema

31. Absolute maximum value: 1; absolute minimum value: 0

33. Absolute maximum value: 0; absolute minimum value: -3

35. Absolute maximum value: $\sqrt{2}/4 \approx 0.35$; absolute minimum value: $-\frac{1}{3}$

37. Absolute maximum value: $\sqrt{2}/2$; absolute minimum value: $-\sqrt{2}/2$

39. 144 ft   **41.** 17.72%

43. $f(6) = 3.60$, $f(0.5) = 1.13$; the number of nonfarm, full-time, self-employed women over the time interval from 1963 to 1993 reached its highest level, 3.6 million, in 1993.

45. $3600   **47.** 6000   **49.** 3333

51. a. $0.0025x + 80 + \dfrac{10,000}{x}$ **b.** 2000

c. 2000   **d.** Same

53. 533

55. a. 2 days after the organic waste was dumped into the pond

b. 3.5 days after the organic waste was dumped into the pond

63. $52.79/sq ft

65. a. 2000; $105.8 billion   **b.** 1995; $7.6 billion

69. False   **71.** False

75. c.

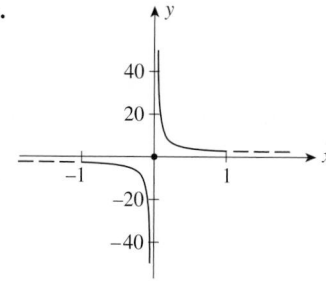

Using Technology Exercises 13.4, page 838

1. Absolute maximum value: 145.8985; absolute minimum value: -4.3834

3. Absolute maximum value: 16; absolute minimum value: -0.1257

5. Absolute maximum value: 2.8889; absolute minimum value: 0

7. a.

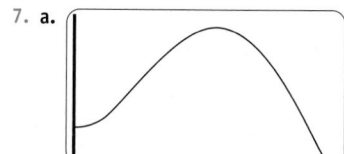

b. 200.1410 banks/yr

9. a.

b. 21.51%

11. b. 1145

13. a.

b. 2029

Exercises 13.5, page 846

1. 750 yd $\times$ 1500 yd; 1,125,000 yd^2

3. $10\sqrt{2}$ ft $\times$ $40\sqrt{2}$ ft

5. $\frac{16}{3}$ in. $\times$ $\frac{16}{3}$ in. $\times$ $\frac{4}{3}$ in.

7. 5.04 in. $\times$ 5.04 in. $\times$ 5.04 in.

9. 18 in. $\times$ 18 in. $\times$ 36 in.; 11,664 in.^{3}

11. $r = \dfrac{36}{\pi}$ in.; $l = 36$ in.; $\dfrac{46,656}{\pi}$ in.^{3}

13. $\frac{2}{3}\sqrt[3]{9}$ ft $\times$ $\sqrt[3]{9}$ ft $\times$ $\frac{2}{5}\sqrt[3]{9}$ ft   **15.** 250; $62,500; $250

17. 85; $28,900; $340

19. $w \approx 13.86$ in.; $h \approx 19.60$ in.

21. $x = 2250$ ft   **23.** $x \approx 2.68$

25. 440 ft; 140 ft; 184,874 sq ft   **27.** 45, 44,445

Chapter 13 Concept Review, page 851

1. a. $f(x_1) < f(x_2)$ **b.** $f(x_1) > f(x_2)$

2. a. Increasing   **b.** $f'(x) < 0$ **c.** Constant

3. a. $f(x) \leq f(c)$ **b.** $f(x) \geq f(c)$

4. a. Domain; $= 0$; exist   **b.** Critical number

c. Relative extremum

5. a. $f'(x)$ **b.** > 0 **c.** Concavity

d. Relative maximum; relative extremum

6. $\pm\infty$; $\pm\infty$ **7.** 0; 0   **8.** b; b

9. a. $f(x) \leq f(c)$; absolute maximum value

b. $f(x) \geq f(c)$; open interval

10. Continuous; absolute; absolute

Chapter 13 Review Exercises, page 852

1. a. f is increasing on $(-\infty, 1) \cup (1, \infty)$

b. No relative extrema

c. Concave down on $(-\infty, 1)$; concave up on $(1, \infty)$

d. $(1, -\frac{17}{3})$

2. a. f is increasing on $(-\infty, 2) \cup (2, \infty)$

b. No relative extrema

c. Concave down on $(-\infty, 2)$; concave up on $(2, \infty)$

d. $(2, 0)$

3. a. f is increasing on $(-1, 0) \cup (1, \infty)$ and decreasing on $(-\infty, -1) \cup (0, 1)$

b. Relative maximum value: 0; relative minimum value: -1

c. Concave up on $\left(-\infty, -\dfrac{\sqrt{3}}{3}\right) \cup \left(\dfrac{\sqrt{3}}{3}, \infty\right)$; concave down on $\left(-\dfrac{\sqrt{3}}{3}, \dfrac{\sqrt{3}}{3}\right)$

d. $\left(-\dfrac{\sqrt{3}}{3}, -\dfrac{5}{9}\right)$; $\left(\dfrac{\sqrt{3}}{3}, -\dfrac{5}{9}\right)$

4. **a.** f is increasing on $(-\infty, -2) \cup (2, \infty)$ and decreasing on $(-2, 0)$ $\cup (0, 2)$
   b. Relative maximum value: -4; relative minimum value: 4
   c. Concave down on $(-\infty, 0)$; concave up on $(0, \infty)$
   d. None

5. **a.** f is increasing on $(-\infty, 0) \cup (2, \infty)$; decreasing on $(0, 1) \cup (1, 2)$
   b. Relative maximum value: 0; relative minimum value: 4
   c. Concave up on $(1, \infty)$; concave down on $(-\infty, 1)$
   d. None

6. **a.** f is increasing on $(1, \infty)$ **b.** No relative extrema
   c. Concave down on $(1, \infty)$ **d.** None

7. **a.** f is decreasing on $(-\infty, 1) \cup (1, \infty)$
   b. No relative extrema
   c. Concave down on $(-\infty, 1)$; concave up on $(1, \infty)$
   d. $(1, 0)$

8. **a.** f is increasing on $(1, \infty)$
   b. No relative extrema
   c. Concave down on $(1, \frac{4}{3})$; concave up on $(\frac{4}{3}, \infty)$
   d. $\left(\dfrac{4}{3}, \dfrac{4\sqrt{3}}{9}\right)$

9. **a.** f is increasing on $(-\infty, -1) \cup (-1, \infty)$
   b. No relative extrema
   c. Concave down on $(-1, \infty)$; concave up on $(-\infty, -1)$
   d. None

10. **a.** f is decreasing on $(-\infty, 0)$ and increasing on $(0, \infty)$
    b. Relative minimum value: -1
    c. Concave down on $\left(-\infty, -\dfrac{1}{\sqrt{3}}\right) \cup \left(\dfrac{1}{\sqrt{3}}, \infty\right)$; concave up on $\left(-\dfrac{1}{\sqrt{3}}, \dfrac{1}{\sqrt{3}}\right)$
    d. $\left(-\dfrac{1}{\sqrt{3}}, -\dfrac{3}{4}\right); \left(\dfrac{1}{\sqrt{3}}, -\dfrac{3}{4}\right)$

11.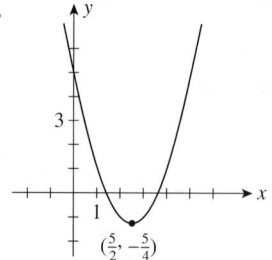

12.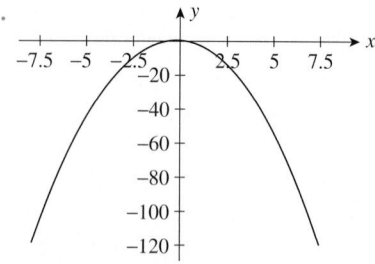

13.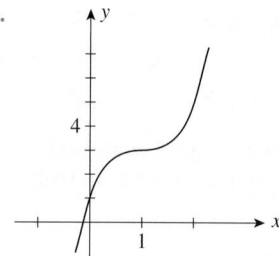

14.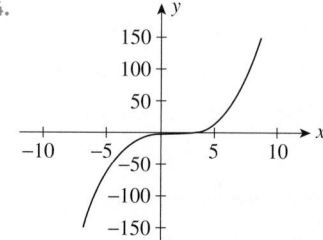

15.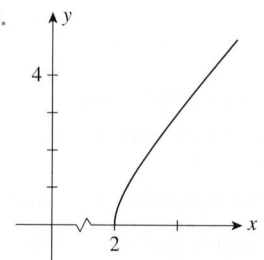

16.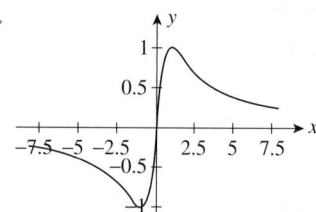

17.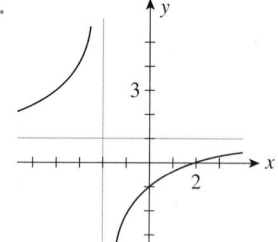

18.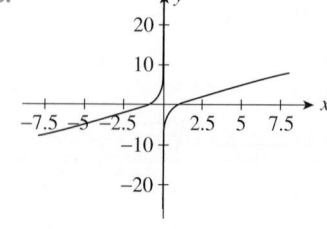

19. Vertical asymptote: $x = -\frac{3}{2}$; horizontal asymptote: $y = 0$

20. Horizontal asymptote: $y = 2$; vertical asymptote: $x = -1$

21. Vertical asymptotes: $x = -2$, $x = 4$, horizontal asymptote: $y = 0$

22. Horizontal asymptote: $y = 1$; vertical asymptote: $x = 1$

23. Absolute minimum value: $-\frac{25}{8}$

24. Absolute minimum value: 0

25. Absolute maximum value: 5; absolute minimum value: 0

26. Absolute maximum value: $\frac{5}{3}$; absolute minimum value: 1

27. Absolute maximum value: -16; absolute minimum value: -32

28. Absolute maximum value: $\frac{1}{2}$; absolute minimum value: 0

29. Absolute maximum value: $\frac{8}{3}$; absolute minimum value: 0

30. Absolute maximum value: $\frac{215}{9}$; absolute minimum value: 7

31. Absolute maximum value: $\frac{1}{2}$; absolute minimum value: $-\frac{1}{2}$

32. No absolute extrema

33. $4000

34. **c.** Online travel spending is expected to increase at an increasing rate over that period of time.

35. **a.** $16.25t + 24.625$; sales were increasing.
    b. 16.25; the rate of sales was increasing from 2002 to 2005.

36. (100, 4600); sales increase rapidly until $100,000 is spent on advertising; after that, any additional expenditure results in increased sales but at a slower rate of increase.

37. (266.67, 11,874.08); the rate of increase is lowest when 267 calculators are produced.

38. **a.** $I'(t) = -\dfrac{200t}{(t^2 + 10)^2}$

    b. $I''(t) = \dfrac{-200(10 - 3t^2)}{(t^2 + 10)^3}$; concave up on $(\sqrt{10/3}, \infty)$; concave down on $(0, \sqrt{10/3})$

    c.

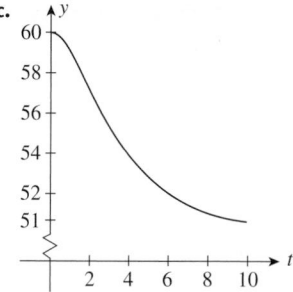

    d. The rate of decline in the environmental quality of the wildlife was increasing the first 1.8 yr. After that time the rate of decline decreased.

39. 168   40. 3000

41. **a.** $0.001x + 100 + \dfrac{4000}{x}$ **b.** 2000

42. 10 a.m.

43. **a.** Decreasing on (0, 12.7); increasing on (12.7, 30)
    b. (12.7, 7.9)
    c. The percent of women 65 years and older in the workforce was decreasing from 1970 to Sept. 1982 and increasing from Sept. 1982 to 2000. It reached a minimum value of 7.9% in Sept. 1982.

45. 74.07 in.^{3} 46. Radius: 2 ft; height: 8 ft

47. 1 ft $\times$ 2 ft $\times$ 2 ft   48. 20,000 cases

49. If $a > 0$, f is decreasing on $\left(-\infty, -\frac{b}{2a}\right)$ and increasing on $\left(-\frac{b}{2a}, \infty\right)$; if $a < 0$, f is increasing on $\left(-\infty, -\frac{b}{2a}\right)$ and decreasing on $\left(-\frac{b}{2a}, \infty\right)$.

50. **a.** $f'(x) = 3x^2$ if $x \neq 0$ **b.** No

Chapter 13 Before Moving On, page 854

1. Decreasing on $(-\infty, 0) \cup (2, \infty)$; increasing on $(0, 1) \cup (1, 2)$

2. Rel. min: $(1, -10)$

3. Concave downward on $\left(-\infty, \frac{1}{4}\right)$; concave upward on $\left(\frac{1}{4}, \infty\right)$; $\left(\frac{1}{4}, \frac{83}{96}\right)$

4.

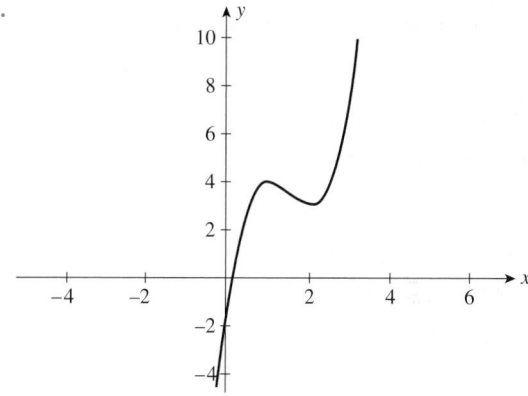

5. Abs. min. value: -5; abs. max value: 80

6. $r = h = \dfrac{1}{\sqrt[3]{\pi}}$ (ft)

CHAPTER 14

Exercises 14.1, page 859

1. **a.** 16   **b.** 27   3. **a.** 3   **b.** $\sqrt{5}$

5. **a.** -3 **b.** 8   7. **a.** 25   **b.** $4^{1.8}$

9. **a.** $4x^3$ **b.** $5xy^2\sqrt{x}$ 11. **a.** $\dfrac{2}{a^2}$ **b.** $\frac{1}{3}b^2$

13. **a.** $8x^9y^6$ **b.** $16x^4y^4z^6$

15. **a.** $\dfrac{64x^6}{y^4}$ **b.** $(x - y)(x + y)$

17. 2   **19.** 3   **21.** 3   **23.** $\frac{5}{4}$ **25.** 1 or 2

27.

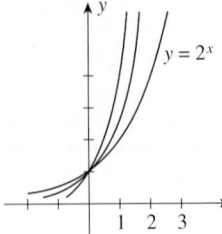

29.

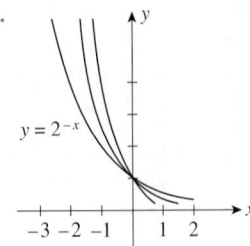

31.

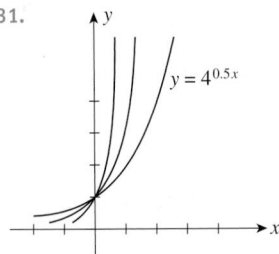

33.

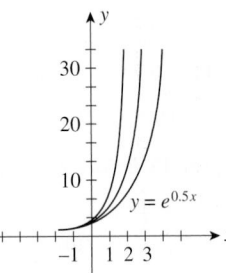

35.

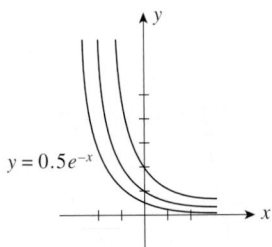

37. a. 26.3%; 24.67%; 21.71%; 19.72%
b.

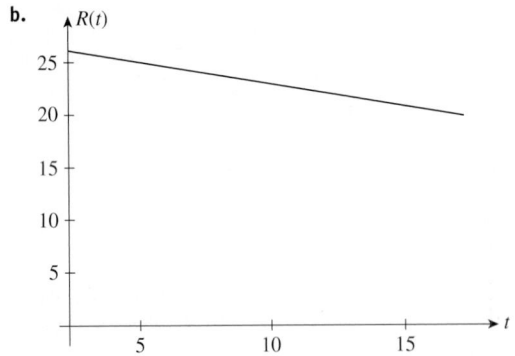

39. a.

Year	0	1	2	3	4	5
Web Addresses (billions)	0.45	0.80	1.41	2.49	4.39	7.76

b.

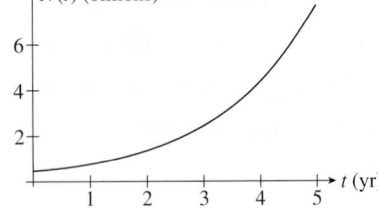

41. 34,210,000

43. a. 0.08 g/cm³   **b.** 0.12 g/cm³   **c.** 0.2 g/cm³
d.

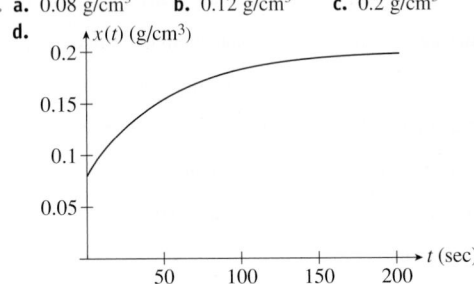

45. False   **47.** True

Using Technology Exercises 14.1, page 862

1.

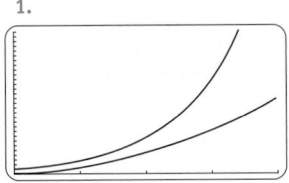

3.

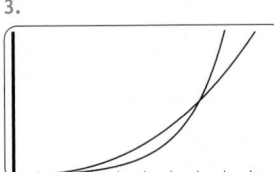

5.

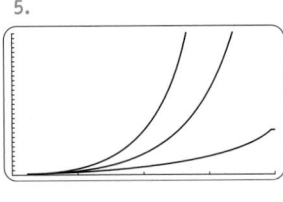

7.

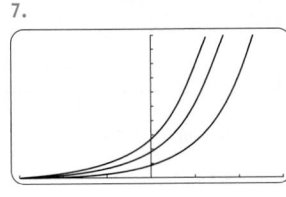

9.

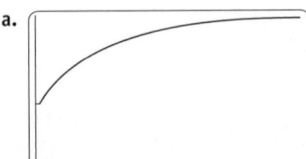

11. a.

b. 0.08 g/cm³   **c.** 0.12 g/cm³   **d.** 0.2 g/cm³

13. **a.**

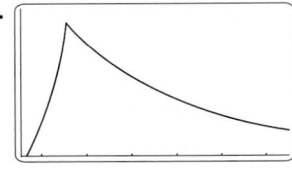

    b. 20 sec   **c.** 35.1 sec

Exercises 14.2, page 870

1. $\log_2 64 = 6$ 3. $\log_3 \frac{1}{9} = -2$ 5. $\log_{1/3} \frac{1}{3} = 1$

7. $\log_{32} 8 = \frac{3}{5}$ 9. $\log_{10} 0.001 = -3$ 11. 1.0792

13. 1.2042   15. 1.6813   17. $\ln a^2 b^3$ 19. $\ln \dfrac{3\sqrt{x}\, y}{\sqrt[3]{z}}$

21. $\log x + 4 \log(x + 1)$

23. $\frac{1}{2} \log(x + 1) - \log(x^2 + 1)$ 25. $\ln x - x^2$

27. $-\frac{3}{2} \ln x - \frac{1}{2} \ln(1 + x^2)$

29.

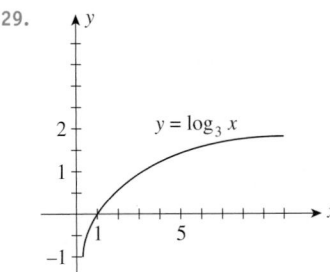

31.

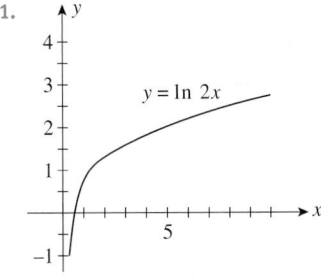

33.

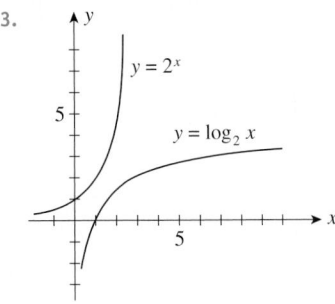

35. 5.1986   37. -0.0912 39. -8.0472

41. -4.9041 43. $-2 \ln\left(\dfrac{A}{B}\right)$ 45. 13.59%/yr

47. 12.1%/yr   49. 14.87%/yr   51. 2.2 yr

53. 7.7 yr   55. 105.7 mm

57. **a.** $10^3 I_0$
    b. 100,000 times greater
    c. 10,000,000 times greater

59. 27.40 yr   61. 6.44 yr

63. **a.** 9.12 sec   **b.** 20.27 sec

65. False   67. True   69. **a.** ln 2

Exercises 14.3, page 879

1. $3e^{3x}$ 3. $-e^{-t}$ 5. $e^x + 1$ 7. $x^2 e^x (x + 3)$

9. $\dfrac{2e^x(x - 1)}{x^2}$ 11. $3(e^x - e^{-x})$ 13. $-\dfrac{1}{e^w}$

15. $6e^{3x-1}$ 17. $-2xe^{-x^2}$ 19. $\dfrac{3e^{-1/x}}{x^2}$

21. $25e^x(e^x + 1)^{24}$ 23. $\dfrac{e^{\sqrt{x}}}{2\sqrt{x}}$ 25. $e^{3x+2}(3x - 2)$

27. $\dfrac{2e^x}{(e^x + 1)^2}$ 29. $2(8e^{-4x} + 9e^{3x})$ 31. $6e^{3x}(3x + 2)$

33. $y = 2x - 2$

35. f is increasing on $(-\infty, 0)$ and decreasing on $(0, \infty)$.

37. Concave downward on $(-\infty, 0)$; concave upward on $(0, \infty)$

39. $(1, e^{-2})$

41. $y = e^{-1/2}(-\sqrt{2}x + 2)$; $y = e^{-1/2}(\sqrt{2}x + 2)$

43. Absolute maximum value: 1; absolute minimum value: e^{-1}

45. Absolute minimum value: -1; absolute maximum value: $2e^{-3/2}$

47.   49.

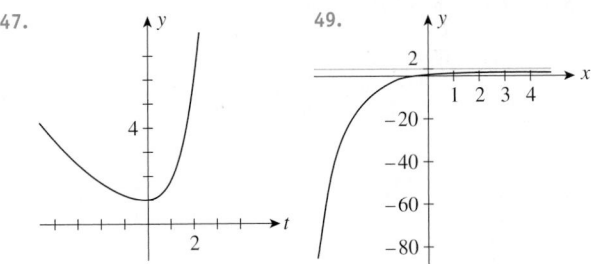

51. -0.1694, -0.1549, -0.1415; the percentage of the total population relocating was decreasing at the rate of 0.17%/yr in 1970, 0.15%/yr in 1980, and 0.14%/yr in 1990.

53. **a.** 70,000; 353,716   **b.** 37,800/decade; 191,000/decade

55. **a.** $4.6 trillion; $3.62 trillion
    b. $184 billion/yr; $145 billion/yr

57. **a.** $-$6065/day; $-$3679/day; $-$2231/day; $-$1353/day
    b. 2 days

59. **b.** 4505/yr; 273 cases/yr   61. 10,000; $367,879

63. **a.** -1.68¢/case   **b.** $40.36/case

65. a. 45.6 kg   **b.** 4.2 kg   **67.** 7.72 yr; $160,207.69

69. 1.8; −0.11; −0.23; −0.13; the rate of change of the amount of oil used is 1.8 barrels/$1000 of output/decade in 1965; it is decreasing at the rate of 0.11 barrel/$1000 of output/decade in 1966; and so on.

71. a. $12/unit
  b. Decreasing at the rate of $7/wk
  c. $8/unit

75. a. 0.05 g/cm³/sec   **b.** −0.01 g/cm³/sec
  c. 20 sec   **d.** 0.90 g/cm³

77. False   **79.** False

Using Technology Exercises 14.3, page 884

1. 5.4366   **3.** 12.3929   **5.** 0.1861

7. a. 50   **c.**

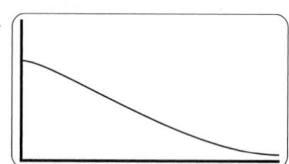

9. a.

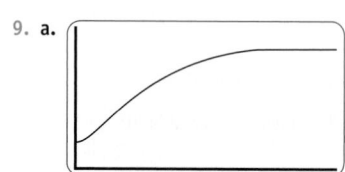

  b. 4.2720 billion/half century

11. a. 153,024; 235,181
  b. 634; 18,401

13. a. 69.63%   **b.** 5.094%/decade

Exercises 14.4, page 890

1. $\dfrac{5}{x}$ **3.** $\dfrac{1}{x+1}$ **5.** $\dfrac{8}{x}$ **7.** $\dfrac{1}{2x}$ **9.** $\dfrac{-2}{x}$

11. $\dfrac{2(4x-3)}{4x^2-6x+3}$ **13.** $\dfrac{1}{x(x+1)}$ **15.** $x(1+2\ln x)$

17. $\dfrac{2(1-\ln x)}{x^2}$ **19.** $\dfrac{3}{u-2}$ **21.** $\dfrac{1}{2x\sqrt{\ln x}}$

23. $\dfrac{3(\ln x)^2}{x}$ **25.** $\dfrac{3x^2}{x^3+1}$ **27.** $\dfrac{(x\ln x+1)e^x}{x}$

29. $\dfrac{e^{2t}[2(t+1)\ln(t+1)+1]}{t+1}$ **31.** $\dfrac{1-\ln x}{x^2}$ **33.** $-\dfrac{1}{x^2}$

35. $\dfrac{2(2-x^2)}{(x^2+2)^2}$ **37.** $(x+1)(5x+7)(x+2)^2$

39. $(x-1)(x+1)^2(x+3)^3(9x^2+14x-7)$

41. $\dfrac{(2x^2-1)^4(38x^2+40x+1)}{2(x+1)^{3/2}}$ **43.** $3^x\ln 3$

45. $(x^2+1)^{x-1}[2x^2+(x^2+1)\ln(x^2+1)]$

47. $y=x-1$

49. f is decreasing on $(-\infty, 0)$ and increasing on $(0, \infty)$.

51. Concave up: $(-\infty, -1) \cup (1, \infty)$; concave down: $(-1, 0) \cup (0, 1)$

53. $(-1, \ln 2)$ and $(1, \ln 2)$ **55.** $y=4x-3$

57. Absolute minimum value: 1; absolute maximum value: $3 - \ln 3$

59. 0.0580%/kg; 0.0133%/kg

61. a. 78.82 million   **b.** 3.95 million/yr

63. b.

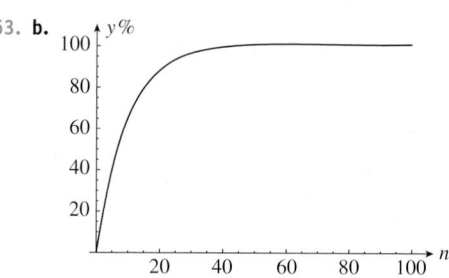

  c. 100

65. a. 6   **b.** 276,310 I_0

67.

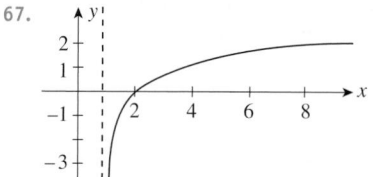

69. False

Exercises 14.5, page 900

1. a. 0.05
  b. 400
  c.

t	0	10	20	100	1000
Q	400	660	1087	59,365	2.07×10^{24}

3. a. $Q(t)=100e^{0.035t}$
  b. 266 min
  c. $Q(t)=1000e^{0.035t}$

5. a. 54.9 yr   **b.** 14.25 billion   **7.** 8.7 lb/in.²; 0.0004 lb/in.²/ft

9. $Q(t)=100e^{-0.049t}$; 70.7 g; 3.451 g/day   **11.** 13,412 yr ago

13.

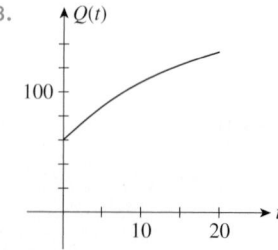

a. 60 words/min
b. 107 words/min
c. 136 words/min

15.

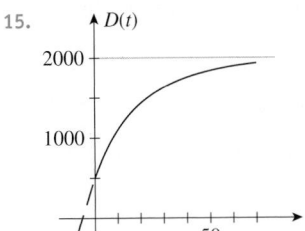

a. 573 computers; 1177 computers; 1548 computers; 1925 computers
b. 2000 computers
c. 45 computers/mo

17. **a.** 122.3 cm   **b.** 14 cm/yr   **c.** 200 cm

19. **a.** 86.1%   **b.** 10.44%/yr   **c.** 1970

21. 76.4 million   23. 1080; 280   25. 15 lb

27. **a.** $\dfrac{\ln\frac{b}{a}}{b-a}$ min   **b.** $\dfrac{k}{b-a}$ g/cm^3 29. **b.** 5599 yr

Using Technology Exercises 14.5, page 904

1. **a.**

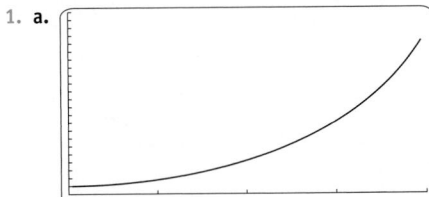

b. 12.146%/yr   **c.** 9.474%/yr

3. **a.**

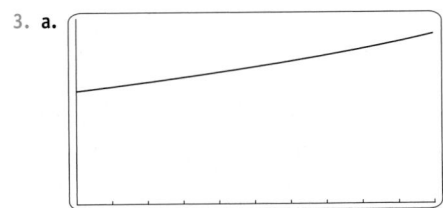

b. 666 million, 818.8 million   **c.** 33.8 million/yr

5. **a.**

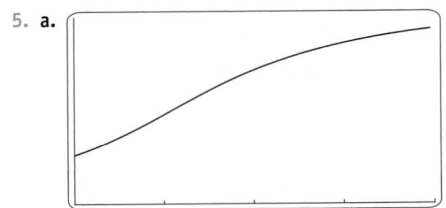

b. 86.12%/yr   **c.** 10.44%/yr   **d.** 1970

7. **a.**

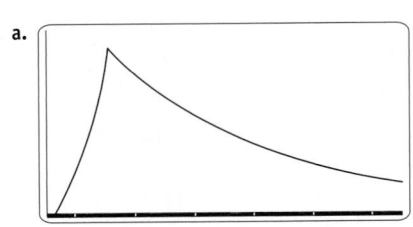

b. 325 million   **c.** 76.84 million/decade

9. **a.**

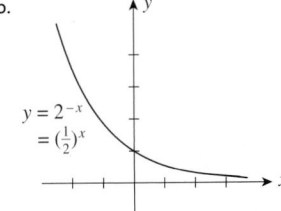

b. 0   **c.** 0.237 g/cm^3 **d.** 0.760 g/cm^3 **e.** 0

Chapter 14 Concept Review, page 906

1. Power; 0; 1; exponential

2. **a.** $(-\infty, \infty)$; $(0, \infty)$ **b.** $(0, 1)$; $(-\infty, \infty)$

3. **a.** $(0, \infty)$; $(-\infty, \infty)$; $(1, 0)$ **b.** $<1; >1$

4. **a.** x **b.** x

5. **a.** $e^{f(x)}f'(x)$ **b.** $\dfrac{f'(x)}{f(x)}$

6. **a.** Initially; growth   **b.** Decay   **c.** Time; one half

7. **a.** Horizontal asymptote: C
b. Horizontal asymptote; A, carrying capacity

Chapter 14 Review Exercises, page 907

1. a. and b.

$y = 2^{-x} = (\frac{1}{2})^x$

2. $\log_{2/3}(\frac{27}{8}) = -3$ 3. $\log_{16} 0.125 = -\frac{3}{4}$ 4. $\frac{15}{2}$ 5. $x = 2$

6. $x + y + z$ 7. $x + 2y - z$ 8. $y + 2z$

9.

$y = \log_2(x+3)$

10.

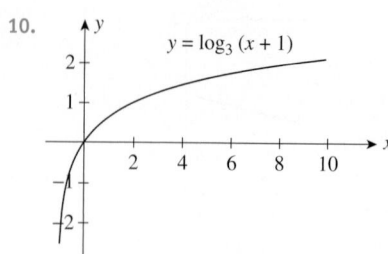

$y = \log_3 (x + 1)$

11. $(2x + 1)e^{2x}$ **12.** $\dfrac{e^t}{2\sqrt{t}} + \sqrt{t}e^t + 1$ **13.** $\dfrac{1 - 4t}{2\sqrt{t}e^{2t}}$

14. $\dfrac{e^x(x^2 + x + 1)}{\sqrt{1 + x^2}}$ **15.** $\dfrac{2(e^{2x} + 2)}{(1 + e^{-2x})^2}$ **16.** $4xe^{2x^2-1}$

17. $(1 - 2x^2)e^{-x^2}$ **18.** $3e^{2x}(1 + e^{2x})^{1/2}$ **19.** $(x + 1)^2e^x$

20. $\ln t + 1$ **21.** $\dfrac{2xe^{x^2}}{e^{x^2} + 1}$ **22.** $\dfrac{\ln x - 1}{(\ln x)^2}$

23. $\dfrac{x - x \ln x + 1}{x(x + 1)^2}$ **24.** $(x + 2)e^x$ **25.** $\dfrac{4e^{4x}}{e^{4x} + 3}$

26. $\dfrac{(r^3 - r^2 + r + 1)e^r}{(1 + r^2)^2}$ **27.** $\dfrac{1 + e^x(1 - x \ln x)}{x(1 + e^x)^2}$

28. $\dfrac{(2x^2 + 2x^2 \cdot \ln x - 1)e^{x^2}}{x(1 + \ln x)^2}$ **29.** $-\dfrac{9}{(3x + 1)^2}$

30. $\dfrac{1}{x}$ **31.** 0 **32.** -2 **33.** $6x(x^2 + 2)^2(3x^3 + 2x + 1)$

34. $\dfrac{(4x^3 - 5x^2 + 2)(x^2 - 2)}{(x - 1)^2}$ **35.** $y = -(2x - 3)e^{-2}$

36. $y = \dfrac{1}{e}$

37.

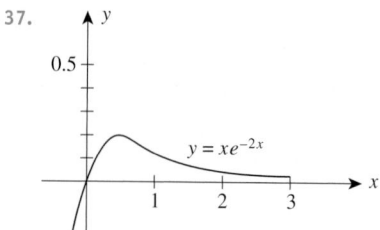

$y = xe^{-2x}$

38.

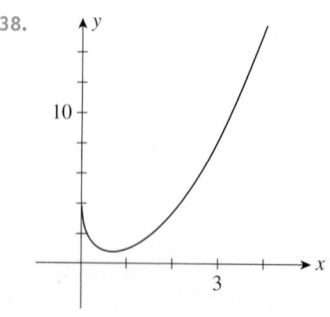

39. Absolute maximum value: $\dfrac{1}{e}$

40. Absolute maximum value: $\dfrac{\ln 2}{2}$; absolute minimum value: 0

41. 9.58 yr

42. a. $Q(t) = 2000e^{0.01831t}$ **b.** 161,992   **43.** 0.0004332

44.

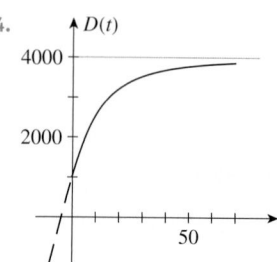

a. 1175, 2540, 3289
b. 4000

45. 970

46. a. 12.5/1000 live births; 9.3/1000 live births; 6.9/1000 live births
b.

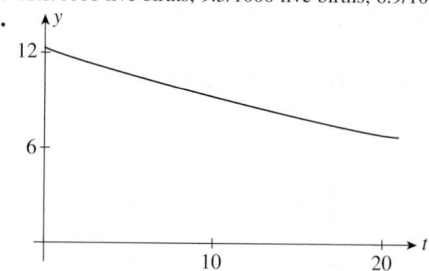

47. a. 0 g/cm³   **b.** 0.0361 g/cm³   **c.** 0.08 g/cm³
d.

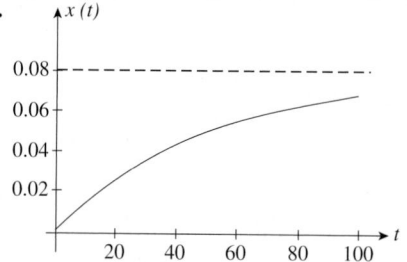

Chapter 14 Before Moving On, page 908

1. -0.9589 **2.** $\dfrac{e^{\sqrt{x}}}{2\sqrt{x}}$ **3.** $1 + \ln 2$

4. $e^{2x}\left(\dfrac{4x^2\ln 3x + 4x - 1}{x^2}\right)$ **5.** 8.7 min

CHAPTER 15

Exercises 15.1, page 918

5. **b.** $y = 2x + C$

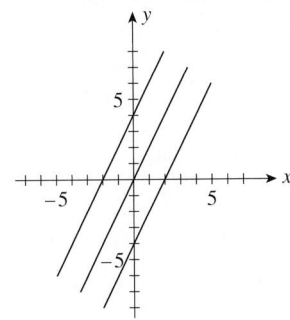

7. **b.** $y = \frac{1}{3}x^3 + C$

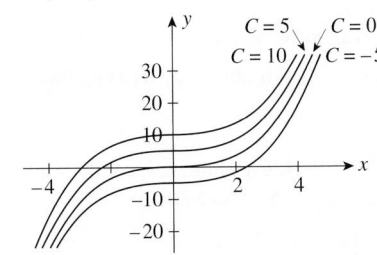

9. $6x + C$ 11. $\frac{1}{4}x^4 + C$ 13. $-\frac{1}{3x^3} + C$

15. $\frac{3}{5}x^{5/3} + C$ 17. $-\frac{4}{x^{1/4}} + C$ 19. $-\frac{2}{x} + C$

21. $\frac{2}{3}\pi t^{3/2} + C$ 23. $3x - x^2 + C$

25. $\frac{1}{3}x^3 + \frac{1}{2}x^2 - \frac{1}{2x^2} + C$ 27. $4e^x + C$

29. $x + \frac{1}{2}x^2 + e^x + C$ 31. $x^4 + \frac{2}{x} - x + C$

33. $\frac{2}{7}x^{7/2} + \frac{4}{5}x^{5/2} - \frac{1}{2}x^2 + C$ 35. $\frac{2}{3}x^{3/2} + 6\sqrt{x} + C$

37. $\frac{1}{9}u^3 + \frac{1}{3}u^2 - \frac{1}{3}u + C$ 39. $\frac{2}{3}t^3 - \frac{3}{2}t^2 - 2t + C$

41. $\frac{1}{3}x^3 - 2x - \frac{1}{x} + C$ 43. $\frac{1}{3}s^3 + s^2 + s + C$

45. $e^t + \frac{t^{e+1}}{e+1} + C$ 47. $\frac{1}{2}x^2 + x - \ln|x| - \frac{1}{x} + C$

49. $\ln|x| + \frac{4}{\sqrt{x}} - \frac{1}{x} + C$ 51. $x^2 + x + 1$

53. $x^3 + 2x^2 - x - 5$ 55. $x - \frac{1}{x} + 2$ 57. $x + \ln|x|$

59. $\sqrt{x}$ 61. $e^x + \frac{1}{2}x^2 + 2$ 63. Branch A

65. $s(t) = \frac{4}{3}t^{3/2}$ 67. $\$3370$ 69. 5000 units; $\$34,000$

71. **a.** $-t^3 + 6t^2 + 45t$ **b.** 212

73. **a.** $-0.125t^3 + 1.05t^2 + 2.45t + 1.5$ **b.** 24.375 million

75. **a.** $-1.493t^3 + 34.9t^2 + 279.5t + 2917$ **b.** $\$9168$

77. **a.** $3.133t^3 - 6.7t^2 + 14.07t + 36.7$ **b.** 103,201

79. **a.** $y = 4.096t^3 - 75.2797t^2 + 695.23t + 3142$ **b.** $\$4264.11$

81. 21,960

83. $1.0974t^3 - 0.0915t^4$ 85. $-t^3 + 96t^2 + 120t$; 63,000 ft

87. **a.** $9.3e^{-0.02t}$ **b.** 7030   **c.** 6619

89. $\frac{1}{2}k(R^2 - r^2)$ 91. $9\frac{7}{9}$ ft/sec^2; 396 ft

93. 0.924 ft/sec^2 95. **a.** $\frac{2t}{t+4}$ **b.** $\frac{2}{5}$ in.; $\frac{2}{3}$ in.

97. True   99. True

Exercises 15.2, page 931

1. $\frac{1}{5}(4x + 3)^5 + C$ 3. $\frac{1}{3}(x^3 - 2x)^3 + C$

5. $-\frac{1}{2(2x^2 + 3)^2} + C$ 7. $\frac{2}{3}(t^3 + 2)^{3/2} + C$

9. $\frac{1}{20}(x^2 - 1)^{10} + C$ 11. $-\frac{1}{5}\ln|1 - x^5| + C$

13. $\ln(x - 2)^2 + C$ 15. $\frac{1}{2}\ln(0.3x^2 - 0.4x + 2) + C$

17. $\frac{1}{6}\ln|3x^2 - 1| + C$ 19. $-\frac{1}{2}e^{-2x} + C$

21. $-e^{2-x} + C$ 23. $-\frac{1}{2}e^{-x^2} + C$ 25. $e^x + e^{-x} + C$

27. $\ln(1 + e^x) + C$ 29. $2e^{\sqrt{x}} + C$

31. $-\frac{1}{6(e^{3x} + x^3)^2} + C$ 33. $\frac{1}{8}(e^{2x} + 1)^4 + C$

35. $\frac{1}{2}(\ln 5x)^2 + C$ 37. $\ln|\ln x| + C$

39. $\frac{2}{3}(\ln x)^{3/2} + C$ 41. $\frac{1}{2}e^{x^2} - \frac{1}{2}\ln(x^2 + 2) + C$

43. $\frac{2}{3}(\sqrt{x} - 1)^3 + 3(\sqrt{x} - 1)^2 + 8(\sqrt{x} - 1) + 4\ln|\sqrt{x} - 1| + C$

45. $\frac{(6x + 1)(x - 1)^6}{42} + C$

47. $5 + 4\sqrt{x} - x - 4\ln(1 + \sqrt{x}) + C$

49. $-\frac{1}{252}(1 - v)^7(28v^2 + 7v + 1) + C$

51. $\frac{1}{2}[(2x - 1)^5 + 5]$ 53. $e^{-x^2+1} - 1$ 55. 17,341,000

57. $21,000 - \frac{20,000}{\sqrt{1 + 0.2t}}$; 6858   59. $\frac{250}{\sqrt{16 + x^2}}$

61. $30(\sqrt{2t + 4} - 2)$; $14,400\pi$ ft^2

63. $\frac{65.8794}{1 + 2.449e^{-0.3277t}} + 0.3$; 56.22 in.   65. $\frac{r}{a}(1 - e^{-at})$

Exercises 15.3, page 941

1. 4.27 sq units

3. **a.** 6 sq units

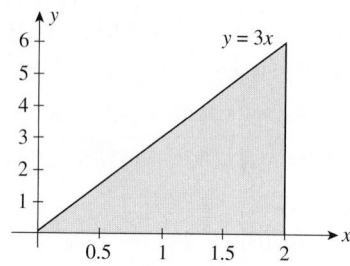
$y = 3x$

b. 4.5 sq units   **c.** 5.25 sq units   **d.** Yes

5. **a.** 4 sq units

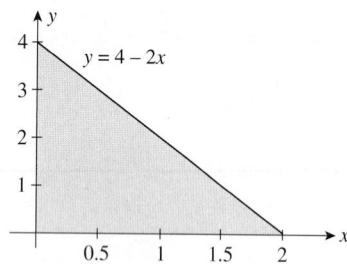
$y = 4 - 2x$

b. 4.8 sq units   **c.** 4.4 sq units   **d.** Yes

7. **a.** 18.5 sq units   **b.** 18.64 sq units
   c. 18.66 sq units   **d.** $\approx$18.7 sq units

9. **a.** 25 sq units   **b.** 21.12 sq units
   c. 19.88 sq units   **d.** $\approx$19.9 sq units

11. **a.** 0.0625 sq unit   **b.** 0.16 sq unit
    c. 0.2025 sq unit   **d.** $\approx$0.2 sq unit

13. 4.64 sq units   15. 0.95 sq unit   17. 9400 sq ft

Exercises 15.4, page 951

1. 6 sq units   3. 8 sq units   5. 12 sq units

7. 9 sq units   9. ln 2 sq units   11. $17\frac{1}{3}$ sq units

13. $18\frac{1}{4}$ sq units   15. $(e^2 - 1)$ sq units   17. 6

19. 14   21. $18\frac{2}{3}$ 23. $\frac{4}{3}$ 25. 45   27. $\frac{7}{12}$

29. ln 2   31. 56   33. $\frac{256}{15}$ 35. $\frac{2}{3}$ 37. $2\frac{2}{3}$

39. $19\frac{1}{2}$ 41. **a.** $4100   **b.** $900

43. **a.** $2800   **b.** $219.20   45. $10,133\frac{1}{3}$ ft

47. **a.** $0.2833t^3 - 1.936t^2 + 5t + 5.6$ **b.** 12.8%   **c.** 5.2%

49. 49.7 million   51. $\frac{23}{15}$ sq units   53. False   55. False

Using Technology Exercises 15.4, page 954

1. 6.1787   3. 0.7873   5. -0.5888 7. 2.7044

9. 3.9973   11. 46%; 24%   13. 333,209   15. 903,213

Exercises 15.5, page 961

1. 10   3. $\frac{19}{15}$ 5. $32\frac{4}{15}$ 7. $\sqrt{3} - 1$ 9. $24\frac{1}{5}$

11. $\frac{32}{15}$ 13. $18\frac{2}{15}$ 15. $\frac{1}{2}(e^4 - 1)$ 17. $\frac{1}{2}e^2 + \frac{5}{6}$

19. 0   21. 2 ln 4   23. $\frac{1}{3}(\ln 19 - \ln 3)$

25. $2e^4 - 2e^2 - \ln 2$ 27. $\frac{1}{2}(e^{-4} - e^{-8} - 1)$ 29. 5

31. $\frac{17}{3}$ 33. -1 35. $\frac{13}{6}$ 37. $\frac{1}{4}(e^4 - 1)$

39. 120.3 billion metric tons   41. $\approx$$2.24 million

43. $40,339.50   45. 49.69 ft/sec   47. 1367

49. $6\frac{1}{3}$ million   51. 16 ft/sec   53. $14.78   55. 80.7%

63. Property 5   65. 0   67. **a.** -1 **b.** 5   **c.** -13

69. True   71. False   73. True

Using Technology Exercises 15.5, page 965

1. 7.71667   3. 17.56487   5. 10,140   7. 60.45 mg/day

Exercises 15.6, page 972

1. 108 sq units   3. $\frac{2}{3}$ sq units   5. $2\frac{2}{3}$ sq units

7. $1\frac{1}{2}$ sq units   9. 3   11. $3\frac{1}{3}$ 13. 27

15. $2(e^2 - e^{-1})$ 17. $12\frac{2}{3}$ 19. $3\frac{1}{3}$ 21. $4\frac{3}{4}$

23. $12 - \ln 4$ 25. $e^2 - e - \ln 2$ 27. $2\frac{1}{2}$ 29. $7\frac{1}{3}$

31. $\frac{3}{2}$ 33. $e^3 - 4 + \frac{1}{e}$ 35. $20\frac{5}{6}$

37. $\frac{1}{12}$ 39. $\frac{71}{6}$ 41. 18

43. S is the additional revenue that Odyssey Travel could realize by switching to the new agency; $S = \int_0^b [g(x) - f(x)]\,dx$

45. Shortfall $= \int_{2010}^{2050} [f(t) - g(t)]\,dt$

47. **a.** $A_2 - A_1$
    b. The distance car 2 is ahead of car 1 after t sec

49. 30 ft/sec faster   51. 21,850   53. True

Using Technology Exercises 15.6, page 977

1. **a.**

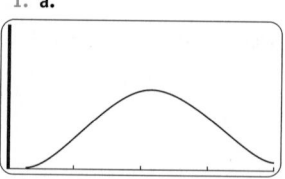

b. 1074.2857

3. **a.**

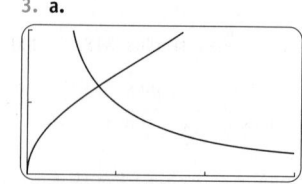

b. 0.9961

5. a.
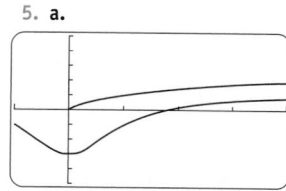
b. 5.4603

7. a.
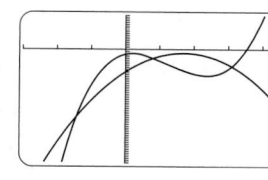
b. 25.8549

9. a.
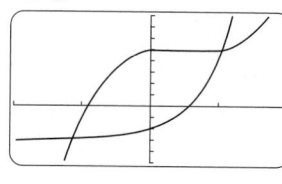
b. 10.5144

11. a.
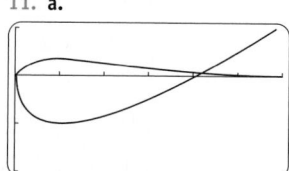
b. 3.5799

13. 207.43

Exercises 15.7, page 988

1. $11,667   **3.** $6667   **5.** $11,667

7. Consumers' surplus: $13,333; producers' surplus: $11,667

9. $824,200   **11.** $148,239   **13.** $52,203

15. $76,615   **17.** $111,869   **19.** $20,964

21. a.
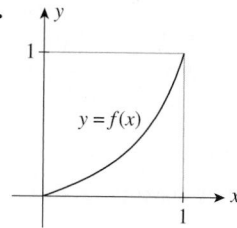
$y = f(x)$
b. 0.175; 0.816

23. a.
$y = f(x)$
b. 0.104; 0.504

Using Technology Exercises 15.7, page 990

1. Consumers' surplus: $18,000,000; producers' surplus: $11,700,000

3. Consumers' surplus: $33,120; producers' surplus: $2880

5. Investment A

Chapter 15 Concept Review, page 993

1. a. $F'(x) = f(x)$ **b.** $F(x) + C$

2. a. $c \int f(x)\, dx$ **b.** $\int f(x)\, dx \pm \int g(x)\, dx$

3. a. Unknown   **b.** Function

4. $g'(x)\, dx;\ \int f(u)\, du$ **5. a.** $\int_a^b f(x)\, dx$ **b.** Minus

6. a. $F(b) - F(a)$; antiderivative   **b.** $\int_a^b f'(x)\, dx$

7. a. $\dfrac{1}{b-a} \int_a^b f(x)\, dx$ **b.** Area; area

8. $\int_a^b [f(x) - g(x)]\, dx$

9. a. $\int_0^{\bar x} D(x)\, dx - \bar p\, \bar x$ **b.** $\bar p\, \bar x - \int_0^{\bar x} S(x)\, dx$

10. a. $e^{rT} \int_0^T R(t)e^{-rt}\, dt$ **b.** $\int_0^T R(t)e^{-rt}\, dt$

11. $A = \dfrac{mP}{r}(e^{rT} - 1)$ **12.** $L = 2 \int_0^1 [x - f(x)]\, dx$

Chapter 15 Review Exercises, page 993

1. $\frac{1}{4}x^4 + \frac{2}{3}x^3 - \frac{1}{2}x^2 + C$ **2.** $\frac{1}{12}x^4 - \frac{2}{3}x^3 + 8x + C$

3. $\frac{1}{5}x^5 - \frac{1}{2}x^4 - \frac{1}{x} + C$ **4.** $\frac{3}{4}x^{4/3} - \frac{2}{3}x^{3/2} + 4x + C$

5. $\frac{1}{2}x^4 + \frac{2}{5}x^{5/2} + C$ **6.** $\frac{2}{7}x^{7/2} - \frac{1}{3}x^3 + \frac{2}{3}x^{3/2} - x + C$

7. $\frac{1}{3}x^3 - \frac{1}{2}x^2 + 2\ln|x| + 5x + C$

8. $\frac{1}{3}(2x + 1)^{3/2} + C$ **9.** $\frac{3}{8}(3x^2 - 2x + 1)^{4/3} + C$

10. $\dfrac{(x^3 + 2)^{11}}{33} + C$ **11.** $\frac{1}{2}\ln(x^2 - 2x + 5) + C$

12. $-e^{-2x} + C$ **13.** $\frac{1}{2}e^{x^2+x+1} + C$ **14.** $\dfrac{1}{e^{-x} + x} + C$

15. $\frac{1}{6}(\ln x)^6 + C$ **16.** $(\ln x)^2 + C$ **17.** $\dfrac{(11x^2 - 1)(x^2 + 1)^{11}}{264} + C$

18. $\frac{2}{15}(3x - 2)(x + 1)^{3/2} + C$ **19.** $\frac{2}{3}(x + 4)\sqrt{x - 2} + C$

20. $2(x - 2)\sqrt{x + 1} + C$ **21.** $\frac{1}{2}$ **22.** -6 **23.** $5\frac{2}{3}$

24. 242   **25.** -80 **26.** $\frac{132}{5}$ **27.** $\frac{1}{2}\ln 5$ **28.** $\frac{1}{15}$ **29.** 4

30. $1 - \dfrac{1}{e^2}$ **31.** $\dfrac{e - 1}{2(1 + e)}$ **32.** $\frac{1}{2}$

33. $f(x) = x^3 - 2x^2 + x + 1$ **34.** $f(x) = \sqrt{x^2 + 1}$

35. $f(x) = x + e^{-x} + 1$ **36.** $f(x) = \frac{1}{2}(\ln x)^2 - 2$

37. -4.28 **38.** $6740

39. a. $-0.015x^2 + 60x$;   **b.** $p = -0.015x + 60$

40. $V(t) = 1900(t - 10)^2 + 10,000$; $40,400

41. $3000t - 50,000(1 - e^{-0.04t})$; 16,939

42. $N(t) = 15,000\sqrt{1 + 0.4t} + 85,000$; 112,659

43. 26,027   **44.** $3100   **45.** 37.7 million

46. 19.4 billion metric tons   47. 15 sq units

48. $\frac{1}{2}(e^4 - 1)$ sq units   49. $\frac{2}{3}$ sq units   50. $\frac{9}{2}$ sq units

51. $(e^2 - 3)$ sq units   52. $\frac{3}{10}$ sq units   53. $\frac{1}{2}$ sq units

54. 234,500 barrels   55. $\frac{1}{3}$ sq units   56. 26°F   57. $270,000

58. Consumers' surplus: $2083; producers' surplus: $3333

59. $197,652   60. $98,973   61. $505,696

62. **a.**

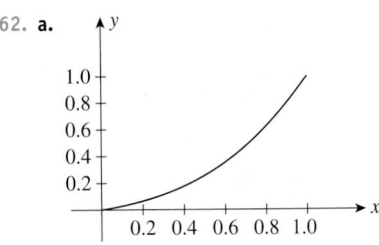

b. 0.1017; 0.3733   **c.** 0.315

63. 70,784

Chapter 15 Before Moving On, page 996

1. $\frac{1}{2}x^4 + \frac{2}{3}x^{3/2} + 2 \ln |x| - 4\sqrt{x} + C$ 2. $e^x + \frac{1}{2}x^2 + 1$

3. $\sqrt{x^2 + 1} + C$ 4. $\frac{1}{3}(2\sqrt{2} - 1)$ 5. $\frac{9}{2}$ sq units

CHAPTER 16

Exercises 16.1, page 1003

1. $\frac{1}{4}e^{2x}(2x - 1) + C$ 3. $4(x - 4)e^{x/4} + C$

5. $\frac{1}{2}e^{2x} - 2(x - 1)e^x + \frac{1}{3}x^3 + C$ 7. $xe^x + C$

9. $\dfrac{2(x + 2)}{\sqrt{x + 1}} + C$ 11. $\frac{2}{3}x(x - 5)^{3/2} - \frac{4}{15}(x - 5)^{5/2} + C$

13. $\dfrac{x^2}{4}(2 \ln 2x - 1) + C$ 15. $\dfrac{x^4}{16}(4 \ln x - 1) + C$

17. $\frac{2}{9}x^{3/2}(3 \ln \sqrt{x} - 1) + C$ 19. $-\dfrac{1}{x}(\ln x + 1) + C$

21. $x(\ln x - 1) + C$ 23. $-(x^2 + 2x + 2)e^{-x} + C$

25. $\frac{1}{4}x^2[2(\ln x)^2 - 2 \ln x + 1] + C$ 27. $2 \ln 2 - 1$

29. $4 \ln 4 - 3$ 31. $\frac{1}{4}(3e^4 + 1)$

33. $-\frac{1}{2}xe^{-2x} - \frac{1}{4}e^{-2x} + \frac{13}{4}$ 35. $5 \ln 5 - 4$

37. 1485 ft   39. 2.04 mg/mL

41. $-20e^{-0.1t}(t + 10) + 200$ 43. $131,324   45. 101,606

47. $(c_2 - c_1)\left[\dfrac{r_1}{r_2 - r_1} + \dfrac{1}{\ln r_1 - \ln r_2}\right] + c_2$ moles/L   49. True

Exercises 16.2, page 1011

1. $\frac{2}{9}[2 + 3x - 2 \ln|2 + 3x|] + C$

3. $\frac{3}{32}[(1 + 2x)^2 - 4(1 + 2x) + 2 \ln|1 + 2x|] + C$

5. $2\left[\dfrac{x}{8}\left(\dfrac{9}{4} + 2x^2\right)\sqrt{\dfrac{9}{4} + x^2} - \dfrac{81}{128}\ln\left(x + \sqrt{\dfrac{9}{4} + x^2}\right)\right] + C$

7. $\ln\left(\dfrac{\sqrt{1 + 4x} - 1}{\sqrt{1 + 4x} + 1}\right) + C$ 9. $\frac{1}{2}\ln 3$

11. $\dfrac{x}{9\sqrt{9 - x^2}} + C$

13. $\dfrac{x}{8}(2x^2 - 4)\sqrt{x^2 - 4} - 2 \ln|x + \sqrt{x^2 - 4}| + C$

15. $\sqrt{4 - x^2} - 2 \ln\left|\dfrac{2 + \sqrt{4 - x^2}}{x}\right| + C$

17. $\frac{1}{4}(2x - 1)e^{2x} + C$ 19. $\ln|\ln(1 + x)| + C$

21. $\dfrac{1}{9}\left[\dfrac{1}{1 + 3e^x} + \ln(1 + 3e^x)\right] + C$

23. $6[e^{(1/2)x} - \ln(1 + e^{(1/2)x})] + C$

25. $\frac{1}{9}(2 + 3 \ln x - 2 \ln|2 + 3 \ln x|) + C$

27. $e - 2$ 29. $\dfrac{x^3}{9}(3 \ln x - 1) + C$

31. $x[(\ln x)^3 - 3(\ln x)^2 + 6 \ln x - 6] + C$

33. $\approx$$2329   35. 27,136   37. 44; 49   39. 26,157

41. $418,444

Exercises 16.3, page 1024

1. 2.7037; 2.6667; $2\frac{2}{3}$ 3. 0.2656; 0.2500; $\frac{1}{4}$

5. 0.6970; 0.6933; $\approx$0.6931   7. 0.5090; 0.5004; $\frac{1}{2}$

9. 5.2650; 5.3046; $\frac{16}{3}$ 11. 0.6336; 0.6321; $\approx$0.6321

13. 0.3837; 0.3863; $\approx$0.3863   15. 1.1170; 1.1114

17. 1.3973; 1.4052   19. 0.8806; 0.8818

21. 3.7757; 3.7625   23. **a.** 3.6   **b.** 0.0324

25. **a.** 0.013   **b.** 0.00043

27. **a.** 0.0078125   **b.** 0.0002848

29. 52.84 mi   31. 21.65 mpg

33. 17.1 million barrels   35. 1922.4 ft³/sec

37. **a.** $51,558   **b.** $51,708

39. 474.77 million barrels   41. $\approx$50%   43. 6.42 L/min

45. False   47. True

Exercises 16.4, page 1034

1. $\frac{2}{3}$ sq unit   3. 1 sq unit   5. 2 sq units

7. $\frac{2}{3}$ sq unit   9. $\frac{1}{2}e^4$ sq units   11. 1 sq unit

13. **a.** $\frac{2}{3}b^{3/2}$ 15. 1   17. 2   19. Divergent

21. $-\frac{1}{8}$ 23. 1   25. 1   27. $\frac{1}{2}$ 29. Divergent

31. -1 **33.** Divergent   **35.** 0   **37.** 0

39. Divergent   **41.** Convergent   **43.** $18,750

45. $\dfrac{10{,}000r + 4000}{r^2}$ dollars   **47.** True

49. False   **51. b.** $83,333

Exercises 16.5, page 1044

11. a. $k = \frac{1}{8}$ **b.** $\frac{1}{2}$ **13. a.** $k = \frac{\ln 2}{4}$ **b.** .2676

15. a. $f(x) = \frac{1}{15}e^{-x/15}$ **b.** .06   **c.** .37   **17.** $3\frac{1}{3}$ min

19. 2500 lb   **21.** $\frac{5}{16}$ **23.** 3 yr   **25.** 0.37

27. 0.18   **29.** False   **31.** False

Chapter 16 Concept Review, page 1048

1. Product: $uv - \int v\, du$; u; easy to integrate   **2.** $x^2 + 1$; $2x\, dx$; (27)

3. $\dfrac{\Delta x}{2}\left[f(x_0) + 2f(x_1) + \cdots + 2f(x_{n-1}) + f(x_n)\right]$; $\dfrac{M(b-a)^3}{12n^2}$

4. $\dfrac{\Delta x}{3}\left[f(x_0) + 4f(x_1) + 2f(x_2) + 4f(x_3) + 2f(x_4)\right.$
$\left. + \cdots + 4f(x_{n-1}) + f(x_n)\right]$; even; $\dfrac{M(b-a)^5}{180n^4}$

5. $\displaystyle\lim_{a\to -\infty}\int_a^b f(x)\, dx$; $\displaystyle\lim_{b\to\infty}\int_a^b f(x)\, dx$; $\displaystyle\int_{-\infty}^{c} f(x)dx + \int_c^{\infty} f(x)\, dx$

Chapter 16 Review Exercises, page 1048

1. $-2(1 + x)e^{-x} + C$ **2.** $\frac{1}{16}(4x - 1)e^{4x} + C$

3. $x(\ln 5x - 1) + C$ **4.** $4 \ln 8 - \ln 2 - 3$ **5.** $\frac{1}{4}(1 - 3e^{-2})$

6. $\frac{1}{4}(1 + 3e^4)$ **7.** $2\sqrt{x}(\ln x - 2) + 2$

8. $-\frac{1}{3}xe^{-3x} - \frac{1}{9}e^{-3x} + \frac{1}{9}$

9. $\dfrac{1}{8}\left[3 + 2x - \dfrac{9}{3 + 2x} - 6\ln|3 + 2x|\right] + C$

10. $\frac{2}{3}(x - 3)\sqrt{2x + 3} + C$ **11.** $\frac{1}{32}e^{4x}(8x^2 - 4x + 1) + C$

12. $-\dfrac{x}{25\sqrt{x^2 - 25}} + C$ **13.** $\dfrac{1}{4}\dfrac{\sqrt{x^2 - 4}}{x} + C$

14. $\frac{1}{2}x^4(4\ln 2x - 1) + C$ **15.** $\frac{1}{2}$ **16.** $\frac{1}{3}$

17. Divergent   **18.** 1   **19.** $\frac{1}{10}$ **20.** 3

21. 0.8421; 0.8404   **22.** 1.491; 1.464   **23.** 2.2379; 2.1791

24. 8.1310; 8.041   **25. a.** 0.002604   **b.** 0.000033

28. a. $k = \frac{3}{8}$ **b.** 0.6495

29. a. $\frac{1}{2}$ **b.** 0.3178   **30. a.** $k = \frac{4}{27}$ **b.** 0.4815

31. a. 0.22   **b.** 0.39   **c.** 4 days   **32.** $1,157,641

33. 48,761   **34.** $41,100   **35.** 274,000 ft²; 278,667 ft²

36. 7850 sq ft   **37.** $111,111

Chapter 16 Before Moving On, page 1050

1. $\frac{1}{3}x^3 \ln x - \frac{1}{9}x^3 + C$ **2.** $-\dfrac{\sqrt{8 + 2x^2}}{8x} + C$ **3.** 6.3367

4. 3.00358   **5.** $\dfrac{1}{2e^2}$ **6. b.** $\frac{31}{32}$

CHAPTER 17

Exercises 17.1, page 1058

1. $f(0, 0) = -4$; $f(1, 0) = -2$; $f(0, 1) = -1$; $f(1, 2) = 4$;
$f(2, -1) = -3$

3. $f(1, 2) = 7$; $f(2, 1) = 9$; $f(-1, 2) = 1$; $f(2, -1) = 1$

5. $g(1, 2) = 4 + 3\sqrt{2}$; $g(2, 1) = 8 + \sqrt{2}$; $g(0, 4) = 2$;
$g(4, 9) = 56$

7. $h(1, e) = 1$; $h(e, 1) = -1$; $h(e, e) = 0$

9. $g(1, 1, 1) = e$; $g(1, 0, 1) = 1$; $g(-1, -1, -1) = -e$

11. All real values of x and y

13. All real values of u and v except those satisfying the equation $u = v$

15. All real values of r and s satisfying $rs \geq 0$

17. All real values of x and y satisfying $x + y > 5$

19.   **21.**

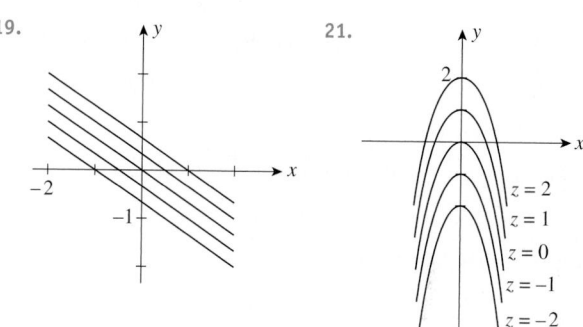

23.

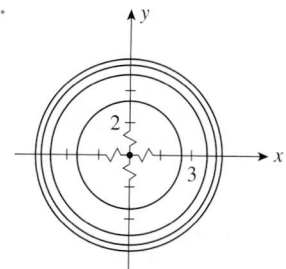

25. 9π ft³   **27. a.** 24.69   **b.** 81 kg

29. a. $-\frac{1}{5}x^2 - \frac{1}{4}y^2 - \frac{1}{5}xy + 200x + 160y$

b. The set of all points (x, y) satisfying $200 - \frac{1}{5}x - \frac{1}{10}y \geq 0$,
$160 - \frac{1}{10}x - \frac{1}{4}y \geq 0$

31. **a.** $-0.005x^2 - 0.003y^2 - 0.002xy + 20x + 15y$
    b. The set of all ordered pairs (x, y) for which
    $$20 - 0.005x - 0.001y \geq 0$$
    $$15 - 0.001x - 0.003y \geq 0$$

33. **a.** The set of all ordered pairs (P, T), where P and T are positive numbers   **b.** 11.10 L

35. $7200 billion   37. 103

39. **a.** $733.76; $877.57   **b.** $836.44

41. 40.28 times gravity   43. False   45. False

Exercises 17.2, page 1071

1. 2; 3   3. $4x$; 4   5. $-\dfrac{4y}{x^3}; \dfrac{2}{x^2}$

7. $\dfrac{2v}{(u+v)^2}; -\dfrac{2u}{(u+v)^2}$

9. $3(2s-t)(s^2-st+t^2)^2$; $3(2t-s)(s^2-st+t^2)^2$

11. $\dfrac{4x}{3(x^2+y^2)^{1/3}}; \dfrac{4y}{3(x^2+y^2)^{1/3}}$ 13. $ye^{xy+1}; xe^{xy+1}$

15. $\ln y + \dfrac{y}{x}; \dfrac{x}{y} + \ln x$ 17. $e^u \ln v; \dfrac{e^u}{v}$

19. $yz + y^2 + 2xz$; $xz + 2xy + z^2$; $xy + 2yz + x^2$

21. ste^{rst}; rte^{rst}; rse^{rst} 23. $f_x(1, 2) = 8; f_y(1, 2) = 5$

25. $f_x(2, 1) = 1; f_y(2, 1) = 3$ 27. $f_x(1, 2) = \frac{1}{2}; f_y(1, 2) = -\frac{1}{4}$

29. $f_x(1, 1) = e; f_y(1, 1) = e$

31. $f_x(1, 0, 2) = 0; f_y(1, 0, 2) = 8; f_z(1, 0, 2) = 0$

33. $f_{xx} = 2y; f_{xy} = 2x + 3y^2 = f_{yx}; f_{yy} = 6xy$

35. $f_{xx} = 2; f_{xy} = f_{yx} = -2; f_{yy} = 4$

37. $f_{xx} = \dfrac{y^2}{(x^2+y^2)^{3/2}}; f_{xy} = f_{yx} = -\dfrac{xy}{(x^2+y^2)^{3/2}};$
    $f_{yy} = \dfrac{x^2}{(x^2+y^2)^{3/2}}$

39. $f_{xx} = \dfrac{1}{y^2}e^{-x/y}; f_{xy} = \dfrac{y-x}{y^3}e^{-x/y} = f_{yx};$
    $f_{yy} = \dfrac{x}{y^3}\left(\dfrac{x}{y} - 2\right)e^{-x/y}$

41. **a.** $f_x = 7.5; f_y = 40$ **b.** Yes

43. $p_x = 10$—at $(0, 1)$, the price of land is changing at the rate of $10/ft²/mile change to the right; $p_y = 0$—at $(0, 1)$, the price of land is constant/mile change upward.

45. Complementary commodities

47. $30/unit change in finished desks; $-$25/unit change in unfinished desks. The weekly revenue increases by $30/unit for each additional finished desk produced (beyond 300) when the level of production of unfinished desks remains fixed at 250; the revenue decreases by $25/unit when each additional unfinished desk (beyond 250) is produced and the level of production of finished desks remains fixed at 300.

49. **a.** $\approx 20°F$ **b.** $\approx -0.3°F$

51. 0.039 L/degree; -0.015 L/mm of mercury. The volume increases by 0.039 L when the temperature increases by 1 degree (beyond 300 K) and the pressure is fixed at 800 mm of mercury. The volume decreases by 0.015 L when the pressure increases by 1 mm of mercury (beyond 800 mm) and the temperature is fixed at 300 K.

55. False   57. True

Using Technology Exercises 17.2, page 1074

1. 1.3124; 0.4038   3. $-1.8889; 0.7778$

5. $-0.3863; -0.8497$

Exercises 17.3, page 1083

1. $(0, 0)$; relative maximum value: $f(0, 0) = 1$

3. $(1, 2)$; saddle point: $f(1, 2) = 4$

5. $(8, -6)$; relative minimum value: $f(8, -6) = -41$

7. $(1, 2)$ and $(2, 2)$; saddle point: $f(1, 2) = -1$; relative minimum value: $f(2, 2) = -2$

9. $\left(-\frac{1}{3}, \frac{11}{3}\right)$ and $(1, 5)$; saddle point: $f\left(-\frac{1}{3}, \frac{11}{3}\right) = -\frac{319}{27}$; relative minimum value: $f(1, 5) = -13$

11. $(0, 0)$ and $(1, 1)$; saddle point: $f(0, 0) = -2$; relative minimum value: $f(1, 1) = -3$

13. $(2, 1)$; relative minimum value: $f(2, 1) = 6$

15. $(0, 0)$; saddle point: $f(0, 0) = -1$

17. $(0, 0)$; relative minimum value: $f(0, 0) = 1$

19. $(0, 0)$; relative minimum value: $f(0, 0) = 0$

21. 200 finished units and 100 unfinished units; $10,500

23. Price of land ($200/ft²) is highest at $\left(\frac{1}{2}, 1\right)$.

25. $(0, 1)$ gives desired location.

27. $10'' \times 10'' \times 5''$; 500 in.³

29. $30'' \times 40'' \times 10''$; $7200   31. False

Exercises 17.4, page 1095

1. Min. of $\frac{3}{4}$ at $\left(\frac{3}{4}, \frac{1}{4}\right)$ 3. Max. of $-\frac{7}{4}$ at $\left(2, \frac{7}{2}\right)$

5. Min. of 4 at $\left(\sqrt{2}, \sqrt{2}/2\right)$ and $\left(-\sqrt{2}, -\sqrt{2}/2\right)$

7. Max. of $-\frac{3}{4}$ at $\left(\frac{3}{2}, 1\right)$

9. Max. of $2\sqrt{3}$ at $\left(\sqrt{3}/3, -\sqrt{6}\right)$ and $\left(\sqrt{3}/3, \sqrt{6}\right)$

11. Max. of 8 at $(2\sqrt{2}, 2\sqrt{2})$ and $(-2\sqrt{2}, -2\sqrt{2})$; min. of -8 at $(2\sqrt{2}, -2\sqrt{2})$ and $(-2\sqrt{2}, 2\sqrt{2})$

13. Max.: $\dfrac{2\sqrt{3}}{9}$; min.: $-\dfrac{2\sqrt{3}}{9}$

15. Min. of $\frac{18}{7}$ at $\left(\frac{9}{7}, \frac{6}{7}, \frac{3}{7}\right)$

17. 140 finished and 60 unfinished units

19. $10\sqrt{2}$ ft $\times$ $40\sqrt{2}$ ft

21. $r = \frac{4}{3}\sqrt[3]{\frac{18}{\pi}}$ in.; $h = 2\sqrt[3]{\frac{18}{\pi}}$ in.

23. $\frac{2}{3}\sqrt[3]{9} \times \frac{2}{3}\sqrt[3]{9} \times \sqrt[3]{9}$

25. 1500 units on labor and 250 units of capital

27. False

Exercises 17.5, page 1107

1. $\frac{7}{2}$ 3. 0   5. $4\frac{1}{2}$ 7. $(e^2 - 1)(1 - e^{-2})$ 9. 1

11. $\frac{2}{3}$ 13. $\frac{188}{3}$ 15. $\frac{84}{5}$ 17. $2\frac{2}{3}$ 19. 1   21. $\frac{1}{2}(3 - e)$

23. $\frac{1}{4}(e^4 - 1)$ 25. $\frac{2}{3}(e - 1)$ 27. 48

29. 6   31. $\frac{19}{6}$ 33. $5(1 - 2e^{-2} + e^{-4})$ 35. 16

37. $\frac{1}{2}(e - 1)^2$ 39. $\frac{1}{2}\ln 17$ 41. $\frac{64}{3}$ 43. $\frac{4}{3}$

45. $1 - \frac{1}{e}$ 47. $\frac{1}{8}(9 \ln 3 - 4)$ 49. ≈ 2166 people/mi^2

51. \$10,460/wk   53. True   55. True

Chapter 17 Concept Review, page 1111

1. xy; ordered pair; real number; $f(x, y)$

2. Independent; dependent; value

3. $z = f(x, y)$; f; surface

4. $f(x, y) = c$; level curve; level curves; c

5. Fixed number; x 6. Slope; $(a, b, f(a, b))$; x; b

7. $\leq$; (a, b); $\leq$; domain

8. Domain; $f_x(a, b) = 0$ and $f_y(a, b) = 0$; exist; candidate

9. $g(x, y) = 0$; $f(x, y) + \lambda g(x, y)$; $F_x = 0$; $F_y = 0$; $F_\lambda = 0$; extrema

10. Volume; solid   11. Iterated; $\int_3^5 \int_0^1 (2x + y^2) \, dx \, dy$

Chapter 17 Review Exercises, page 1112

1. $0, 0, \frac{1}{2}$; no   2. $e, \dfrac{e^2}{1 + \ln 2}, \dfrac{2e}{1 + \ln 2}$; no

3. $2, -(e + 1), -(e + 1)$

4. The set of all ordered pairs (u, v) such that $u \neq v$

5. The set of all ordered pairs (x, y) such that $y \neq -x$

6. The set of all ordered pairs (x, y) such that $x \leq 1$ and $y \geq 0$

7. The set of all triplets (x, y, z) such that $z \geq 0$ and $x \neq 1$, $y \neq 1$, and $z \neq 1$

8. $2x + 3y = z$

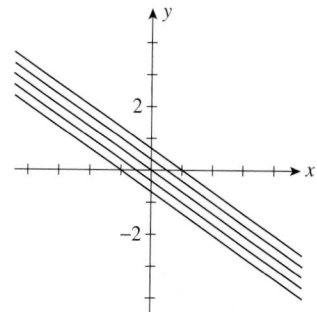

9. $z = y - x^2$

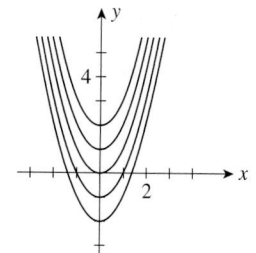

10. $z = \sqrt{x^2 + y^2}$

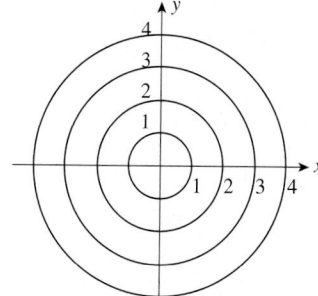

11. $z = e^{xy}$

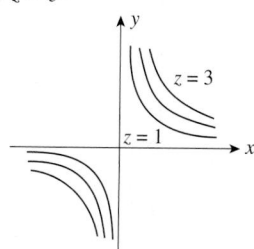

12. $f_x = 2xy^3 + 3y^2 + \dfrac{1}{y}$; $f_y = 3x^2y^2 + 6xy - \dfrac{x}{y^2}$

13. $f_x = \sqrt{y} + \dfrac{y}{2\sqrt{x}}$; $f_y = \dfrac{x}{2\sqrt{y}} + \sqrt{x}$

14. $f_u = \dfrac{v^2 - 2}{2\sqrt{uv^2 - 2u}}$; $f_v = \dfrac{uv}{\sqrt{uv^2 - 2u}}$

15. $f_x = \dfrac{3y}{(y + 2x)^2}$; $f_y = -\dfrac{3x}{(y + 2x)^2}$

16. $g_x = \dfrac{y(y^2 - x^2)}{(x^2 + y^2)^2}$; $g_y = \dfrac{x(x^2 - y^2)}{(x^2 + y^2)^2}$

17. $h_x = 10y(2xy + 3y^2)^4$; $h_y = 10(x + 3y)(2xy + 3y^2)^4$

18. $f_x = \dfrac{e^y}{2(xe^y + 1)^{1/2}}$; $f_y = \dfrac{xe^y}{2(xe^y + 1)^{1/2}}$

19. $f_x = 2x(1 + x^2 + y^2)e^{x^2+y^2}$; $f_y = 2y(1 + x^2 + y^2)e^{x^2+y^2}$

20. $f_x = \dfrac{4x}{1 + 2x^2 + 4y^4}$; $f_y = \dfrac{16y^3}{1 + 2x^2 + 4y^4}$

21. $f_x = \dfrac{2x}{x^2 + y^2}$; $f_y = -\dfrac{2x^2}{y(x^2 + y^2)}$

22. $f_{xx} = 6x - 4y$; $f_{xy} = -4x = f_{yx}$; $f_{yy} = 2$

23. $f_{xx} = 12x^2 + 4y^2$; $f_{xy} = 8xy = f_{yx}$; $f_{yy} = 4x^2 - 12y^2$

24. $f_{xx} = 12(2x^2 + 3y^2)(10x^2 + 3y^2)$;
$f_{xy} = 144xy(2x^2 + 3y^2) = f_{yx}$;
$f_{yy} = 18(2x^2 + 3y^2)(2x^2 + 15y^2)$

25. $g_{xx} = \dfrac{-2y^2}{(x + y^2)^3}$; $g_{xy} = \dfrac{2y(x - y^2)}{(x + y^2)^3} = g_{yx}$;
$g_{yy} = \dfrac{2x(3y^2 - x)}{(x + y^2)^3}$

26. $g_{xx} = 2(1 + 2x^2)e^{x^2+y^2}$; $g_{xy} = 4xye^{x^2+y^2} = g_{yx}$;
$g_{yy} = 2(1 + 2y^2)e^{x^2+y^2}$

27. $h_{ss} = -\dfrac{1}{s^2}$; $h_{st} = h_{ts} = 0$; $h_{tt} = \dfrac{1}{t^2}$

28. $f_x(1, 1, 0) = 3$; $f_y(1, 1, 0) = 3$; $f_z(1, 1, 0) = -2$

29. $(2, 3)$; relative minimum value: $f(2, 3) = -13$

30. $(8, -2)$; saddle point at $f(8, -2) = -8$

31. $(0, 0)$ and $\left(\frac{3}{2}, \frac{9}{4}\right)$; saddle point at $f(0, 0) = 0$; relative minimum value: $f\left(\frac{3}{2}, \frac{9}{4}\right) = -\frac{27}{16}$

32. $\left(-\frac{1}{3}, \frac{13}{3}\right)$, $(3, 11)$; saddle point at $f\left(-\frac{1}{3}, \frac{13}{3}\right) = \left(-\frac{445}{27}\right)$; relative minimum value: $f(3, 11) = -35$

33. $(0, 0)$; relative minimum value: $f(0, 0) = 1$

34. $(1, 1)$; relative minimum value: $f(1, 1) = \ln 2$

35. $f\left(\frac{12}{11}, \frac{20}{11}\right) = -\frac{32}{11}$

36. $f\left(\frac{1}{22}, \frac{21}{22}\right) = \frac{179}{44}$

37. Relative maximum value: $f(5, -5) = 26$; relative minimum value: $f(-5, 5) = -24$

38. Relative maximum value: $f\left(\dfrac{\sqrt{2}}{2}, -\dfrac{\sqrt{2}}{2}\right) = e^{\sqrt{2}}$; relative minimum value: $f\left(-\dfrac{\sqrt{2}}{2}, \dfrac{\sqrt{2}}{2}\right) = e^{-\sqrt{2}}$

39. 48   40. $\frac{1}{2}(e^{-2} - 1)^2$ 41. $\frac{2}{63}$ 42. $\frac{1}{4}(3 - 2 \ln 2)$

43. $\frac{34}{3}$ 44. $10\frac{4}{5}$ cu units   45. 3

46. **a.** $R(x, y) = -0.02x^2 - 0.2xy - 0.05y^2 + 80x + 60y$
b. The set of all points satisfying $0.02x + 0.1y \le 80$, $0.1x + 0.05y \le 60$, $x \ge 0$, $y \ge 0$
c. 15,300; the revenue realized from the sale of 100 16-speed and 300 10-speed electric blenders is \$15,300.

47. Complementary

48. The company should spend \$11,000 on advertising and employ 14 agents in order to maximize its revenue.

49. 337.5 yd $\times$ 900 yd   50. 75 units on labor; 25 units on capital

Chapter 17 Before Moving On, page 1114

1. All real values of x and y satisfying $x \ge 0$, $x \ne 1$, $y \ge 0$, $y \ne 2$

2. $f_x = 2xy + ye^{xy}$; $f_{xx} = 2y + y^2e^{xy}$; $f_{xy} = 2x + (xy + 1)e^{xy} = f_{yx}$; $f_y = x^2 + xe^{xy}$; $f_{yy} = x^2e^{xy}$

3. Rel. min.: $(1, 1, -7)$

4. $f\left(\frac{1}{2}, \frac{1}{2}\right) = \frac{5}{2}$ 5. $\frac{1}{8}$

Index

FORMULAS

Equation of a Straight Line

a. point-slope form $y - y_1 = m(x - x_1)$
b. slope-intercept form $y = mx + b$
c. general form $Ax + By + C = 0$

Equation of the Least-Squares Line

$$y = mx + b$$

where m and b satisfy the **normal equations**

$$nb + (x_1 + x_2 + \cdots + x_n)m = y_1 + y_2 + \cdots + y_n$$
$$(x_1 + x_2 + \cdots + x_n)b + (x_1^2 + x_2^2 + \cdots + x_n^2)m = x_1y_1 + x_2y_2 + \cdots + x_ny_n$$

Compound Interest

$$A = P(1 + i)^n \qquad (i = r/m, \, n = mt)$$

where A is the accumulated amount at the end of n conversion periods, P is the principal, r is the interest rate per year, m is the number of conversion periods per year, and t is the number of years.

Effective Rate of Interest

$$r_{\text{eff}} = \left(1 + \frac{r}{m}\right)^m - 1$$

where r_{eff} is the effective rate of interest, r is the nominal interest rate per year, and m is the number of conversion periods per year.

Future Value of an Annuity

$$S = R\left[\frac{(1 + i)^n - 1}{i}\right]$$

Present Value of an Annuity

$$P = R\left[\frac{1 - (1 + i)^{-n}}{i}\right]$$

Amortization Formula

$$R = \frac{Pi}{1 - (1 + i)^{-n}}$$

Sinking Fund Payment

$$R = \frac{iS}{(1 + i)^n - 1}$$